# PHYSICAL, FUNDAMENTAL, AND MATHEMA[T]

| quantity | symbol | customary U.S. | |
|---|---|---|---|

**Charge**

| | | | |
|---|---|---|---|
| electron | $e$ | | |
| proton | $p$ | | $+1.60218 \times 10$ |

**Density**

| | | | |
|---|---|---|---|
| air [STP] [$32°$F, ($0°$C)] | | $0.0805$ lbm/ft$^3$ | $1.29$ kg/m$^3$ |
| air [$70°$F, ($20°$C), 1 atm] | | $0.0749$ lbm/ft$^3$ | $1.20$ kg/m$^3$ |
| earth [mean] | | $345$ lbm/ft$^3$ | $5520$ kg/m$^3$ |
| mercury | | $849$ lbm/ft$^3$ | $1.360 \times 10^4$ kg/m$^3$ |
| seawater | | $64.0$ lbm/ft$^3$ | $1025$ kg/m$^3$ |
| water [mean] | | $62.4$ lbm/ft$^3$ | $1000$ kg/m$^3$ |

**Distance** [mean]

| | | | |
|---|---|---|---|
| earth radius | | $2.09 \times 10^7$ ft | $6.371 \times 10^6$ m |
| earth-moon separation | | $1.26 \times 10^9$ ft | $3.844 \times 10^8$ m |
| earth-sun separation | | $4.89 \times 10^{11}$ ft | $1.49 \times 10^{11}$ m |
| moon radius | | $5.71 \times 10^6$ ft | $1.737 \times 10^6$ m |
| sun radius | | $2.28 \times 10^9$ ft | $6.96 \times 10^8$ m |
| first Bohr radius | $a_0$ | $1.736 \times 10^{-10}$ ft | $5.292 \times 10^{-11}$ m |

**Gravitational Acceleration**

| | | | |
|---|---|---|---|
| earth [mean] | $g$ | $32.174$ ft/sec$^2$ | $9.8067$ m/s$^2$ |
| moon [mean] | | $5.321$ ft/sec$^2$ | $1.622$ m/s$^2$ |

**Mass**

| | | | |
|---|---|---|---|
| atomic mass unit | $u$ | $3.66 \times 10^{-27}$ lbm | $1.6605 \times 10^{-27}$ kg |
| earth | | $4.11 \times 10^{23}$ slugs | $5.97 \times 10^{24}$ kg |
| earth [customary U.S.] | | $1.32 \times 10^{25}$ lbm | n.a. |
| electron [rest] | $m_e$ | $2.008 \times 10^{-30}$ lbm | $9.1094 \times 10^{-31}$ kg |
| moon | | $1.623 \times 10^{23}$ lbm | $7.347 \times 10^{22}$ kg |
| neutron [rest] | $m_n$ | $3.693 \times 10^{-27}$ lbm | $1.6749 \times 10^{-27}$ kg |
| proton [rest] | $m_p$ | $3.688 \times 10^{-27}$ lbm | $1.6726 \times 10^{-27}$ kg |
| sun | | $4.387 \times 10^{30}$ lbm | $1.989 \times 10^{30}$ kg |

**Pressure,** atmospheric

| | | | |
|---|---|---|---|
| | | $14.696$ lbf/in$^2$ | $1.01323 \times 10^5$ Pa |

**Temperature,** standard

| | | | |
|---|---|---|---|
| | | $32°$F ($492°$R) | $0°$C (273K) |

**Velocity**

| | | | |
|---|---|---|---|
| earth escape | | $3.67 \times 10^4$ ft/sec | $1.12 \times 10^4$ m/s |
| light [vacuum] | $c$ | $9.84 \times 10^8$ ft/sec | $2.9979 \times 10^8$ m/s |
| sound [air, STP] | $a$ | $1090$ ft/sec | $331$ m/s |
| [air, $70°$F ($20°$C)] | | $1130$ ft/sec | $344$ m/s |

**Volume,** molal ideal gas [STP]

| | | | |
|---|---|---|---|
| | | $359$ ft$^3$/lbmol | $22.41$ m$^3$/kmol |

**Fundamental Constants**

| | | | |
|---|---|---|---|
| Avogadro's number | $N_A$ | | $6.0221 \times 10^{23}$ mol$^{-1}$ |
| Bohr magneton | $\mu_B$ | | $9.2740 \times 10^{-24}$ J/T |
| Boltzmann constant | $k$ | $5.65 \times 10^{-24}$ ft-lbf/$°$R | $1.38065 \times 10^{-23}$ J/K |
| Faraday constant | $F$ | | $96\,485$ C/mol |
| gravitational constant | $g_c$ | $32.174$ lbm-ft/lbf-sec$^2$ | |
| gravitational constant | $G$ | $3.44 \times 10^{-8}$ ft$^4$/lbf-sec$^4$ | $6.674 \times 10^{-11}$ N·m$^2$/kg$^2$ |
| nuclear magneton | $\mu_N$ | | $5.050 \times 10^{-27}$ J/T |
| permeability of a vacuum | $\mu_0$ | | $1.2566 \times 10^{-6}$ N/A$^2$ (H/m) |
| permittivity of a vacuum | $\epsilon_0$ | | $8.8542 \times 10^{-12}$ C$^2$/N·m$^2$ (F/m) |
| Planck's constant | $h$ | | $6.6260 \times 10^{-34}$ J·s |
| Rydberg constant | $R_\infty$ | | $1.0974 \times 10^7$ m$^{-1}$ |
| specific gas constant, air | $R$ | $53.3$ ft-lbf/lbm-$°$R | $287.03$ J/kg·K |
| Stefan-Boltzmann constant | | $1.71 \times 10^{-9}$ Btu/ft$^2$-hr-$°$R$^4$ | $5.670 \times 10^{-8}$ W/m$^2$·K$^4$ |
| triple point, water | | $32.02°$F, $0.0888$ psia | $0.01109°$C, $0.61173$ kPa |
| universal gas constant | $R^*$ | $1545$ ft-lbf/lbmol-$°$R | $8314$ J/kmol·K |
| | $R^*$ | $1.986$ Btu/lbmol-$°$R | $0.08206$ atm·L/mol·K |

**Mathematical Constants**

| | | | |
|---|---|---|---|
| Archimedes' number (pi) | $\pi$ | | $3.14159\,26536$ |
| base of natural logs | $e$ | | $2.71828\,18285$ |
| Euler constant | $\gamma$ | | $0.57721\,56649$ |

# CORE ENGINEERING CONCEPTS
## FOR STUDENTS AND PROFESSIONALS

**MICHAEL R. LINDEBURG, PE**

The Power to Pass™
www.ppi2pass.com

Professional Publications, Inc. • Belmont, California

## Benefit by Registering This Book with PPI

- Get book updates and corrections
- Receive special discounts

Register your book at **www.ppi2pass.com/register**.

## Report Errors and View Corrections for This Book

PPI is grateful to every reader who notifies us of a possible error. Your feedback allows us to improve the quality and accuracy of our products. You can report errata and view corrections at **www.ppi2pass.com/errata**.

**CORE ENGINEERING CONCEPTS FOR STUDENTS AND PROFESSIONALS**

Printing History

| edition number | printing number | update |
|---|---|---|
| 1 | 1 | New book. |

Printed in the United States of America.

PPI
1250 Fifth Avenue, Belmont, CA 94002
(650) 593-9119
www.ppi2pass.com

ISBN: 978-1-59126-190-2

Library of Congress Control Number: 2010920727

# Topics

Background and Support

Material Science

Engineering Management

Mathematics

Mechanics of Materials

Engineering Licensure

Fluids

Dynamics

Thermodynamics

Circuits

Chemistry

Physics

Biology

Systems Analysis

Heat Transfer

Computer Programming

Statics

Atomic Theory

# Table of Contents

# Appendices
# Table of Contents

# Preface

*Core Engineering Concepts for Students and Professionals* is intended to provide a comprehensive review of engineering concepts. It contains more than 80 chapters covering thousands of engineering topics, and there are more than 520 example problems to illustrate key concepts. It emphasizes the subject matter typically encountered in four- and five-year engineering degree programs, as well as situations typically encountered in engineering practice. It is basic enough for you to learn new subjects and complete enough for you to learn the nuances of those subjects.

## WHY I WROTE THIS BOOK

The audience for this book includes engineering majors needing a general, all-inclusive engineering reference book, as well as professionals desiring a convenient summary of familiar, but faintly remembered engineering subjects. This book is intended to be useful from your first engineering course through your consulting or teaching careers. I hope you will hold on to it, storing it initially in your dorm room and later in your company office or home den, and pull it out whenever you need more information. I want it to be the first place you go when looking for more information and the one book you would never sell back to the bookstore, auction on eBay®, or advertise on Craigslist.

I can remember people saying (when I graduated) that engineering knowledge, particularly electrical engineering knowledge, was doubling every five years. What those people *didn't* say was that I would also likely forget half of my engineering knowledge every five years. Fortunately, I became involved in engineering education, and my career path has helped maintain my engineering cranial contents. If you are like most engineers, however, you are not so fortunate.

This book grew from 30 years of working with engineers preparing for the Fundamentals of Engineering (FE) exam. This exam used to be called the "Engineer-in-Training exam," but regardless of its name, there has always been a similar refrain sung about it: "I should have taken the exam sooner, while I still remembered what I learned in school."

I wrote this book to help you remember what most engineers forget: Their engineering education.

## WHY I WROTE THIS BOOK NOW

Publishers have always been able to produce elegant technical books from authors' manuscripts. In the early 1980s, Donald Knuth's generous gifts to technical composition of TeX® (one of the first open-source programs I ever heard about), Metafont, and the glorious computer modern typeface removed all excuses for ugly Smith-Corona and IBM Selectric typed doctoral theses, mathematical treatises, and engineering textbooks. But, even TeX wasn't enough to produce a book such as this one, especially in this modern era of electronic books, online access, web presentation, PDA and cell phone applications, and custom books. Therefore, while I have been thinking about this book for nearly three decades, I've had to wait until a particular technology developed in order to produce it.

The innovative technology this book needed (and to which I am referring) affects the way this book is stored electronically by PPI, the interface available to authors, and the ease with which changes can be made. I won't bore you with PPI's publishing process. But because a book such as this one changes with advances, discoveries, and methods of analysis, design, and production, as well as (to a limited extent) with code and standard changes, just believe me when I say PPI has the technology to keep this book current. A book that purports to capture the essence of your engineering degree has to be flexible enough to change when engineering degree programs change. So, when the technology to do this became available, I wrote this book.

## WHAT'S NEW IN THIS EDITION

This is where I would normally explain how this edition differs from the previous edition. However, since there wasn't a previous edition of this title, I am at a disadvantage. However, I can tell you that this book uses a tried-and-true format, style guide, and presentation that over many years have proven themselves to maximize usefulness.

While I have gathered material for this book for many years, *Core Engineering Concepts* is not a thrown-together hodgepodge of words and figures. It represents thousands of conscious decisions (and probably trillions of unconscious synaptic decisions) about style, wording, order and sequence, suitability of material, and method of presentation. In that sense, all of the decisions made for this book are new.

## FOR WHOM THIS BOOK IS WRITTEN

I had you in mind when I wrote *Core Engineering Concepts*. Yes, you. Having been "there," I had a clear vision of this book's purpose, who you are, and how you would use it. The book's purpose is simple: It is a compendium of everything you will learn or have learned as an engineering major. "Everything you learned" is probably too high a bar to reach, since everyone's "everything" is different, particularly when most engineers probably don't even take some of the courses covered by chapters in this book. So, perhaps I was thinking about everything you "should have learned." Or, perhaps I was thinking about the things that most people forget after they graduate. Anyway, I'm sure you get the idea because you wouldn't be reading this preface, otherwise. This book contains everything that I don't want to forget, and by extension, everything that I don't want you to forget.

By the way, I think you'll find the chapters easy to follow. They're logically arranged and properly supported by tables, footnotes, and appendices, and (in my own opinion) just plain fascinating to read. I wrote this book hoping that, even if you never took a course in thermodynamics, mechanics of materials, operations research, or electricity, you will be amazed at how simple some of the concepts truly are.

As for who "you" are, I tried to simultaneously (1) keep in mind what I needed when I was an engineering major in college, and (2) remember all those times that I had to look up something as a practicing engineer or educator. I remember struggling to recall a formula that was hiding at the periphery of my consciousness, to understand a concept or term that I had heard muttered but did not really understand, or to implement a procedure that I had never properly learned in the first place.

## HOW THIS BOOK GOT ITS TITLE

This book could have been titled a dozen different ways. During the planning and writing process, I referred to it interchangeably as *The Engineering Major's Bible, Your Engineering Degree in a Box, Engineering Degree in a Can, Everything Engineers Forget from College,* and even, *Everything You Never Learned.* In my heart, I knew these titles would never be chosen, but my sense of engineering humor kept them alive, just as I enjoyed referring to my first environmental reference manual as "Envaroom" during the writing process. All along, I believed that *Engineering Fundamentals Desktop Reference, Engineering Major's First Handbook,* or something similar would be selected as the title. Eventually, *Core Engineering Concepts for Students and Professionals* was chosen for its blend of political correctness, accuracy, and marketing appeal. And, I think it's a great title.

## THIS BOOK'S HERITAGE

In truth, this book shares some chapters with other reference manuals that I have written. In particular, the *Engineer-in-Training Reference Manual* went through eight editions and sold hundreds of thousands of copies before parts of it were incorporated into this book. I hope you won't complain, but I really appreciated not having to write a new chapter on Linear Algebra when I already had one in that book. However, lest you think that this book is nothing more than a mismatched assemblage of unrelated, thrown-together chapters, I can tell you that (1) everything in this book is as up to date as I could make it, (2) everything flows and is consistent, and (3) dozens of people have been working for years editing, typesetting, illustrating, indexing, and proofreading this book in order to bring you something truly useful. This book's heritage is that of a purebred, not of a mongrel.

## LOOKING FORWARD TO THE FUTURE

Future editions of this book will be very much shaped by what you and others want to see in it. I am braced for the influx of comments and suggestions from readers who (1) want more topics, (2) want more detail in existing topics, and (3) think that continuing to include some topics is just plain lame. Computer hardware and programming change quickly, as does the grand unified (atomic) theory of everything. In particular, the chapters on analog logic and Fortran are certain to generate some kind of debate.

I realize that even though I have tried to walk the fine line between being timeless and useful, but not being so specific that the material becomes obsolete overnight, it is inevitable that I'll offend someone who is closer to "the truth" now than when I was writing. I expect that readers from other cultures and countries will make significant contributions. In that regard, I need to hear from you. I already have a file of changes that I want to make for the next edition. For an author, that's pretty common fare. This book was planned with future changes in mind, so I hope you won't be bashful with your comments and suggestions. This book capitalizes on the wealth of information and suggestions provided to me by engineers for over 30 years, and I will continue to welcome your suggestions.

First printings of first editions result in humbling experiences. Despite the utmost care having been taken in writing and preparing this book, more than 30 years of publishing experience has taught me you will find something as obvious as the nose mistakenly connected to the shinbone, or something as illogical as $2 + 2 = 5$. Like most authors of professional books, I cringe, cry, and take responsibility for every mistake in *Core Engineering Concepts*. And I apologize.

PPI has a cool way for you to tell me about my blunders and make suggestions for future editions. Just go to **www.ppi2pass.com/errata**. You'll be able to report what you've found and get name recognition for your contributions. PPI maintains a web listing of errata for all of its books, so any mistakes you report will also be posted for others to benefit from.

Until I hear from you, best wishes in your career.

Michael R. Lindeburg, PE

# Acknowledgments

The number of people required to produce a good technical book is astonishing. If I had known how difficult it was to write a good book, I probably would have become a dentist. *Core Engineering Concepts for Students and Professionals* required almost two years to plan, prepare for, and produce—and even that time was preceded by several years of intense work by PPI's information technology department. Along the way, many caring and supportive people touched this book, and in so doing, touched your life and mine.

Here are just a few of the people we owe much to. To start with, there are many people at PPI (aka, Professional Publications, Inc.) who worked hard for us. Mitch Bakos, Director of Information Technology (IT), did much more than just manage PPI's IT department. He was the prime architect in developing PPI's book development software. He was also the first person to see many of this book's chapters, whipping them into shape by tweaking applications and templates. Along the way, he's probably seen the chapters more times than I have. Also in the IT department, I'm grateful for the contributions of Syed Hussain and Henry Mok who coded special output routines that streamlined the production process, allowing for blazingly fast retrieval of chapters for review and modification.

Cathy Schrott, in her capacity as Director of Production, oversaw all aspects of production, including scheduling, typesetting and illustrating, coordinating with vendors and freelancers, and setting and enforcing quality standards. She also worked tirelessly, long into the evenings and on weekends for the better part of a year to assemble and massage all of the book's many pieces in a production environment that was still in development. Within PPI's Production Department itself, Kate Hayes and Tom Bergstrom did what used to be called "typesetting," and then "composition," but is now called "coding" and is not much different than building a website. Tom also rendered many hundreds of illustrations, along with Amy Schwertman who designed the wonderful cover. Nicole Collins did production coordination on the final pieces.

Sarah Hubbard, Director of New Product Development, oversaw the entire editorial process, contributing significantly to the new procedures and infrastructure, as well as the quality and content. Sarah enforced the style guide and ensured permissions were obtained from other publishers for figures and tables of data borrowed from their books. Jenny Lindeburg King, Editorial Project Supervisor, slogged through chapter after chapter laboriously checking that everything was dotted, crossed, and indexed as the author intended. She worked with a dedicated, heavy-lifting project team that included Meaghan Banks, Editorial Coordinator; Scott Marley, Proofreader; Courtnee Crystal, Proofreader; Sesa Pabalan, Proofreader; and Megan Synnestvedt, Proofreader. The members of this project team put in many extra evening and weekend hours editing, proofreading, and indexing to bring out a quality product. They saw the chapters so many times and in so many different forms that they are probably sorry they ever joined the company.

Greg Monteforte, Director of Marketing, and Laryssa Polika, Engineering Marketing Manager, contributed their expertise to the titling, appearance, and commercial success of this book. They tried to get into your head to anticipate what you need and want, making *Core Engineering Concepts for Students and Professionals* truly a market-driven book. And, they've been singing the book's virtues since it was first announced.

Last but not least at PPI, Patty Steinhardt, Director of Operations, has been with this project from the beginning, witnessing all of its fits and starts, and banging the coxswain's drum incessantly. Having her at the project helm has allowed me to devote my time to authorship. We oarsmen of the galley all rowed to her beat. Without her confidence in its importance, this book would have been seriously delayed.

External to PPI, I am pleased to acknowledge the contributions of Robin Corcoran at GPSL for his extensive help in formatting this book, John Camara (whose *Electrical Engineering Reference Manual's* "Digital Logic" chapter formed the basis of a chapter in this book), Gregg Wagener (who wrote the "Pulse Circuits: Waveform Shaping and Logic" chapter), "Sis" Margaret Hyde (who came out of retirement to proofread many of this book's chapters), and John Ruark (who reminded me that, as an author, pain is inevitable, but suffering is optional). Finally, looking to the future, I mention Linda Monahan, Barb and Bill Emanuel, Chris Miller, and Larry Struck—all friends who waited patiently for this book to be completed before we could go out and start having some fun.

Although they are acknowledged in earlier books that I have written, I reluctantly fail to mention several hundred people (in industry, in academia, and at PPI) who helped me in the past. Their suggestions are just as valuable today as they were when they were first contributed one, two, or three decades ago. I hope that they know how much they have helped me. Their assistance and contributions are not forgotten.

Most importantly of all, my wife and children—Elizabeth, Jenny, and Katie—have survived another book. The time I spent working on this book can never be recovered. I could have been a more attentive husband and a more involved father, but such behavior is not in the stars. They must sometimes wonder what living with a normal husband or father would have been like, but they'll never know. The best I can offer is to love them unconditionally.

My sincerest thanks to you all!

Michael R. Lindeburg, PE

# Introduction

## HOW TO USE THIS BOOK

*Core Engineering Concepts for Students and Professionals* is meant to be used as a reference tool to supplement your engineering study and professional practice. You can either use it as your go-to reference when you need to refresh your memory of a specific subject, or you can read and study it from the beginning, in chapter order, for a thorough review of engineering. Unless you are taking a course or preparing for an exam, you will probably use it the first way. Much like you use a dictionary to look up words you're unfamiliar with, you should use this book to look up the engineering subjects you need to know more about. You can use *Core Engineering Concepts*, and its detailed index, to find answers fast.

Using this book isn't going to be that much different from using any other well-conceived reference. You will typically search for a topic, go to and read the appropriate material, use the nomenclature at the start of the chapter to clarify the meaning of variables and the choice of units, and then refer to relevant data in the accompanying tables and appendices. Some of those steps may require some explanation, which I've provided in the following paragraphs.

## ORGANIZATION

Regardless of how you use *Core Engineering Concepts*, you should start by familiarizing yourself with the Table of Contents. To help you work through various topics, chapters have been combined into related groups. While it may not seem obvious, the groups have been logically arranged, and even though the book starts out with mathematics, the groups certainly do not increase in difficulty. They're all about the same. The printed subject tabs on each page will help you keep track of where you are and where you need to go. The appendices are accumulated at the end of the book. I've used the endpapers (inside front and back covers of the hardcover book) to duplicate text material that I think you'll need to reference frequently.

## PRECEDENCE

As much as is practical, each chapter develops linearly and consecutively. By this, I mean that you usually won't need concepts presented later in a chapter to understand the earlier concepts. Also, the chapters pretty much develop linearly within their groups of chapters. Even the groups are sequenced to build on one another. However, I say "as much as practical," "usually," and "pretty much" because this isn't always possible. For example, the chapter on Electrostatics and Electromagnetics refers to an Ohm's law analogy, even though Ohm's law is presented in the subsequent chapter. In some cases, data to support an example are drawn from a subsequent chapter's appendix. In such cases, in order to maintain the book's linearity, I have provided forward-references (e.g., "See Sec. X.X") when referring to something that will be presented later in the chapter.

Of course, if you use the index and jump into the middle of a chapter, all bets are off. You may have to backtrack in order to move forward.

## NOMENCLATURE

*Core Engineering Concepts* uses industry-standard (also known as "normal and customary") units, symbols, and terminology. Almost every chapter starts with lists of nomenclature, symbols, and subscripts that define the variables and units used in text, formulas, illustrations, and tables. The nomenclature is as industry-standard as possible; symbols are consistent between chapters, particularly between related chapters. The equations are consistent with the units defined in the nomenclature. You should never have to play the "What's this variable?" game or wonder what units work in an equation.

## UNITS

When it comes to units, your environment will determine what is normal and customary. As a student, most of your coursework will undoubtedly be in SI units, because that is what the engineering education system uses in the United States. As a practicing engineer, you will probably work in customary U.S. ("English" or "British") units. That's just the way it is. For a multi-purpose book such as this, that disparity means a little duplication here and there.

Equations in this book are given for both customary U.S. and SI units. Examples, figures, tables, and appendices are similarly supported with dual values, conversions, and dual dimensioning. No matter what set of units you prefer, this book has you covered.

Exact conversions are not used in examples. Rather, an emphasis is placed on making the problems parallel in terms of complexity. For example, a 1 kg mass might be dual-dimensioned as a 2 lbm mass, even though 2.2 lbm is a more accurate conversion. For dual-dimensioned examples, the SI solution is presented before the customary U.S. solution.

## SUPPORTING DATA

In keeping with this book's goal to be a complete reference, hundreds of tables and more than a hundred appendices are included to support your real-world needs for data. The appendices are consolidated after the last chapter, and they are numbered (labeled) with the chapter number they support. It is my intention that the tables of data and appendices will save you from having to do more research once you've refreshed your memory regarding the topic.

## INDEXING

Some subjects appear in more than one chapter, and some tidbits are buried within unrelated sections. You can use the index liberally to locate all content related to a particular subject. This book has been extensively indexed, backward and forward, and with the names and logical alternatives of all key concepts. For example, "Torricelli's speed of efflux," "Efflux, speed of," "Speed of efflux," "Speed of a jet," "Velocity, jet," and "Discharge velocity" all refer to the same place, even though those exact words might not appear anywhere on the page. So when you are looking for something, the odds are, no matter what your search phrase, you'll find your subject.

## RELEVANCE AND CODE DEPENDENCY

If you are researching a subject that has legal or statutory implications, or is enforceable or regulated (e.g., the maximum sound power level that employees can be exposed to for 8 hours), you can start with this book, but always follow up by referencing current codes, standards, and regulations. For the most part, though, the subjects in this book are not code-dependent or subject to change. Regardless of when and how you use this book—every day during your engineering degree program or on a case-by-case basis in your professional career—the information in this book will be appropriate and applicable.

## IF YOU ARE NOT FAMILIAR WITH CUSTOMARY U.S. UNITS

If you are making a transition from SI units, Ch. 2 ("Systems of Units") will be particularly helpful in understanding the peculiarities of the various "English" systems of units. The list of abbreviations in "Engineering Abbreviations, Acronyms, and Units" will help you with conventions (e.g., pcf, gpm, and psia) that U.S. engineers take for granted.

You will find that many of the customary U.S. equations in this book contain the ratio $g/g_c$. For calculations at standard gravity, the numerical value of this ratio is 1.00. Therefore, it is necessary to incorporate this quantity only when you are being meticulous with units and when you are working problems with nonstandard gravities.

## IF YOU ARE NOT FAMILIAR WITH SI UNITS

If you are transitioning from customary U.S. to SI units, Ch. 2 will be helpful, but probably the most useful part this book will be the table of "Equivalent Units of Derived and Common SI Units" located on the inside back cover ("endpaper").

## IF YOU ARE A STUDENT

If you are a student, you can use *Core Engineering Concepts* for supplemental lectures or as an alternative to visiting your professor or teaching assistant during office hours. I don't suggest that you use it as an alternative to attending class, of course. It is inevitable that the day you cut class will be the day of a pop quiz, the day the professor tells a really good joke, or the day you would have been invited on a great double date.

## IF YOU ARE A PRACTICING ENGINEER

As I described in the Preface, if you are a practicing engineer, I expect you will store this book until you have a need for it. However, until you are promoted into management or your job deteriorates into rote boredom, it is quite likely that your need for this book will be frequent. Unless you are part of the 1% of engineering graduates who have kept all their textbooks, I hope that this becomes the most referenced book of your career.

## IF YOU ARE PREPARING FOR THE FE EXAM

It is inevitable that you or others will want to know whether *Core Engineering Concepts* can be used to prepare for the Fundamentals of Engineering (FE) exam, also known as the Engineer-in-Training exam. Although this book was not written specifically for the exam, it was written with the exam in mind, as evidenced by Ch. 82 ("The FE Exam"). So, with only a few caveats, the answer is, "Of course you can use this book to prepare for the FE exam!"

In fact, for someone who has been out of school for a long time, or for someone who managed to graduate without taking a course in thermodynamics or any one of a number of other dreaded subjects, this book represents one of the easiest ways to become familiar with new topics. One of the caveats is that you'll end up being over-prepared, because this book goes wider and deeper into engineering than does the FE exam. Another caveat is that, despite having hundreds of solved example problems, this book does not have a corresponding collection of multiple-choice, exam-like practice problems. However, PPI (**www.ppi2pass.com**) has a number of problem-oriented books to supplement your study.

## IF YOU ARE PREPARING FOR THE PE EXAM

While adequate in breadth and depth for an FE exam review, this book is not adequate for PE exam review. Visit the PPI website at **www.ppi2pass.com** to see all of the reference manuals, practice problem books, sample exams, and other products for the PE exam.

## IF YOU ARE AN INSTRUCTOR OF AN ENGINEERING SURVEY COURSE

If you are an instructor of an Introduction to Engineering, Survey of Engineering, or a Senior Seminar (capstone) course, *Core Engineering Concepts* will support just about any lecture on engineering principles that you can come up with. As someone who has stood at the front of a classroom filled with budding engineers, I am confident that this book will be well-received by your students. Not only will they have a great exam review tool, but they'll also have a textbook that will remain useful throughout their entire careers.

There are various ways of organizing such courses. Some courses are instructed by a single instructor; others use a tag-team approach. Some are exam-specific, limiting what is covered to the bare essentials; others are general reviews of engineering. Some courses deal with what is happening in industry; others are theoretical. Some courses start off solving problems, while others start with theory and concepts. Some courses depend almost entirely on the assigned textbook, while others are based on handouts prepared by the instructor. There is no "best" way. But some ways are easier than others.

There are some topics that will require more time in class than you have to cover them, and some students will need more time in a topic than other students. With over 80 chapters of important engineering topics, *Core Engineering Concepts* is a reference tool your students can use to augment both their understanding of topics you don't have time to cover in class, as well as to review those topics in which they need more practice.

## IF YOU ARE AN INSTRUCTOR OF AN FE REVIEW COURSE

Long ago, I prepared my first set of handouts for an engineering fundamentals course I taught in California. Those handouts referenced all the long formulas, illustrations, and tables of data that I did not have the inclination to put on the chalkboard. In subsequent administrations of courses, I reorganized the handouts into chapters that closely paralleled the organization, contents, and emphasis of my lectures. Those chapters eventually became a mature book—the *Engineer-in-Training Reference Manual* (affectionately known as the "Big Yellow Book"), inarguably the most popular FE exam preparation tool ever published. The text in some chapters so closely paralleled my verbal presentation that I was occasionally accused of giving lectures by reading from my own book. Oh well.

The advantages that *Core Engineering Concepts* gives you are as follows. (1) You don't have to prepare any handout notes. (2) Formulas, terminology, methods, and units are largely consistent with industry standards and the FE exam. (3) The book is organized progressively, with subsequent chapters building on previous chapters. (4) The nomenclature and other authoring conventions are consistent throughout. Once you get used to the presentation in one chapter, you and your students won't be surprised by subsequent chapters. (5) This book's foundation includes the best of the "Big Yellow Book" whose material has been refined and augmented over decades by legions of examinees.

The difficulties you will face in using this book in an FE review course are as follows. (1) You will have to limit your instructional coverage to the scope of the FE exam. (2) You will need to obtain exam-like sample problems for your lecture, homework problems, and a sample exam (if you give one) from another source. Both of these are quite satisfactorily addressed in a number of ways. First and foremost, if you are teaching a review course for the FE exam without the benefit of recent, first-hand experience, you can download a free FE instructor's kit at **www.ppi2pass.com/FERC**. This kit contains suggestions for organizing a course, lectures, example problems, and homework assignments, as well as overheads (slides, transparencies, etc.) in PowerPoint format. The lectures in the instructor's kit will keep you from straying outside the exam's scope. And the kit's homework problem sets will provide the exposure to exam-like problems that your students need.

You'll find that lecture coverage of some exam subjects is necessarily brief. For one thing, time is not on your side in a review course. For another, the benefit to covering certain subjects is only minimal. For example, how many exam questions do you really think will be on eigenvectors? Unless you have two quarters in which to teach your FE review course, your students' time can be better spent covering common exam fare.

A second option addressing students' seemingly insatiable need for ever more example and practice problems and sample exams is by assigning problems from PPI's online Exam Cafe or from other PPI books. This second option will be limited by economics and the manner by which your course is administered.

Depending on the available time, budget, and intended audience, there are many ways to organize an FE review course. However, all good course formats have the same result: the students struggle with the work load during the course . . . and then they breeze through the examination.

# Engineering Abbreviations, Acronyms, and Units

This list of abbreviations, acronyms, and nonstandard units is intended to help you navigate through this book, as well as through other U.S. engineering publications. There is considerable inconsistency among engineers in capitalization and use of periods, dashes, and other punctuation. Many of the conventional and traditional units listed do not follow the standard SI conventions (which are presented in parentheses). Only the most obscure SI units are listed.

| | |
|---|---|
| a | year (SI abbreviation) |
| AA | arithmetic average |
| AAES | American Association of Engineering Societies |
| ABET | Accreditation Board for Engineering and Technology |
| abs | absolute |
| ABS | acrylonitrile-butadiene-styrene |
| ac, a-c, AC, or A-C | alternating current |
| ACEC | American Consulting Engineers Council |
| ACI | American Concrete Institute |
| ACRS | accelerated cost recovery system |
| adj | adjoint |
| ADP | apparatus dew point |
| AEA | American Electronics Association |
| AFR | air-fuel ratio |
| AGC | American General Contractors of America |
| AI | artificial intelligence |
| AIA | American Institute of Architects |
| AIChE | American Institute of Chemical Engineers |
| AISC | American Institute of Steel Construction |
| AISI | American Iron and Steel Institute |
| ALU | arithmetic logical unit |
| am, a.m., or AM | morning |
| AM | amplitude modulation |
| amp | ampere |
| amu | atomic mass unit |
| ANSI | American National Standards Institute |
| API | American Petroleum Institute |
| App. | appendix |
| ASCE | American Society of Civil Engineers |
| ASCII | American Standard Code for Information Interchange |
| ASHRAE | American Society of Heating, Refrigerating and Air-Conditioning Engineers |
| ASIC | application-specific integrated circuit |
| ASME | American Society of Mechanical Engineers |
| ASTM | American Society for Testing and Materials |
| atm | atmosphere or atmospheres |
| ave | average |
| avg | average |
| avoir | avoirdupois |
| AW or A.W. | atomic weight |
| AWG | American Wire Gage |
| bbl or BBL | barrel |
| B/C | benefit/cost |
| BCC | body-centered cubic |
| BCD | binary-coded decimal |
| BCT | body-centered tetragonal |
| BDC | bottom dead center |
| Bé | Baumé |
| BES | British Engineering System (English units) |
| BeV | billion electron volts (GeV) |
| BHN | Brinell hardness number |
| bhp or BHP | brake horsepower |
| BIOS | basic input-output system |
| bps | bits per second |
| BJT | bipolar junction transistor |
| BkW | brake kilowatt |
| BMEP | brake mean effective pressure |
| BOD | biochemical oxygen demand |
| bp | boiling point |
| bps | bits per second |
| BSFC | brake specific fuel consumption |
| BSI | British Standard Institute |
| BTU or Btu | British thermal unit |
| BUE | built-up edge |
| BV | book value |
| BW | bandwidth |
| BWR | boiling water reactor |
| CAD | computer-aided design |
| caf | compound amount factor |
| CAI | computer-aided instruction |
| cal | calorie |
| CAM | computer-aided manufacturing |
| CASE | computer-aided software engineering |
| CB or C.B. | center of buoyancy |
| CBC | California Building Code |
| cc | cubic centimeter |
| CC | capitalized cost |
| CCD | charge-coupled device |
| CCF | hundred cubic feet |

| | | | |
|---|---|---|---|
| CCITT | International Telegraph and Telephone Consultative Committee | d | day |
| CCT | controlled cooling transformation | da or d.a. | dry air |
| ccw or CCW | counterclockwise | DA or D.A. | double-acting |
| CD | compact disc | DAT | digital audio tape |
| C-D | converging-diverging | db or dB | decibel (dB) or dry bulb |
| CDF | continuous distribution function | DB | declining balance or database |
| CE | civil engineer | dBW | decibels above one watt |
| C.E. | consulting engineer | dc, d-c, DC, or D-C | direct current |
| CERN | European Laboratory for Particle Physics | DCTL | direct-coupled transistor logic |
| cfd | cubic feet per day | DDB | double declining balance |
| cfm | cubic feet per minute | DF or D.F. | degree of freedom |
| cfs | cubic feet per second | dia | diameter |
| cg, c.g., CG, or C.G. | center of gravity | dif. eq. | differential equation |
| cgs | centimeter-gram-second | dim | dimension(s) |
| Chap. | chapter | DIN | Deutsche Industrie Normen (German Standards Organization) |
| CHU | caloric heat unit | DIP | dual in-line package |
| CI | cast iron or coefficient of the instrument | dis | disintegration |
| CIM | computer-integrated manufacturing | DMA | direct memory access |
| cir | circular | DMS | drill manufacturers' size |
| CIS | copper indium selenide | dns or DNS | do not scale |
| CISC | complex instruction-set computing | DO | dissolved oxygen |
| CL | centerline, confidence limit, or clearance | DOD | Department of Defense (U.S.) |
| CLA | centerline average | DOT | Department of Transportation (U.S.) |
| CLK | clock | dp or DP | dew point, degree of polymerization, or deep |
| cm or c.m. | circular mil | | |
| CM or C.M. | center of mass | dps | disintegrations per second |
| cmil | circular mil | DR | depreciation recovery |
| CML | current mode logic | DTL | diode-transistor logic |
| CMOS | complementary metallic oxide logic | DTR | data terminal ready |
| COD | chemical oxygen demand | EAA | equivalent annual amount |
| COGS | cost of goods sold | EBCDIC | extended binary coded decimal interchange code |
| COMFET | conductivity-modulated field-effect transistor | EC | European Community (Common Market) or electrical conductor |
| COP | coefficient of performance | | |
| cp or CP | candlepower | ECG | electrochemical grinding |
| cP | centipoise | ECL | emitter-coupled logic |
| CP | center of pressure | ECM | electrochemical machining |
| CPH | close-packed hexagonal | ECPD | Engineers' Council for Professional Development (obsolete; see ABET) |
| cpm | cycles per minute | | |
| CPM | critical path method | EDIF | electronic-design interchange format |
| cps | cycles per second (Hz) | EDM | electrical discharge machining |
| CPU | central processing unit | EDP | electronic data processing |
| CRC | cycling redundancy checking | EER | energy-efficiency ratio |
| CRF | capital recovery factor | EF | earliest finish |
| crit | critical | eff | efficiency or effective |
| CRT | cathode ray tube | e.g. | for example |
| cS | centistoke | EGL | energy grade line |
| CS | Canadian Standard or cutting speed | ehp or EHP | electrical horsepower |
| CSI | Construction Specifications Institute (U.S.) | EIA | Electronic Industries Association (U.S.) |
| | | EIRP | equivalent isotropically radiated power |
| ctc | center-to-center | EIS | executive information system |
| CTL | complementary transistor logic | EIT or E-I-T | engineer-in-training |
| CTS | clear-to-send or copper tube size | EL | energy line |
| cu. ft. | cubic feet | EMF | electromotive force |
| cu. in. | cubic inch | EMI | electromagnetic interference |
| CVD | chemical vapor deposition | emu | electromagnetic unit |
| CW | clockwise or carrier wave | EOF | end-of-file |
| c.w.g. | carbureted water gas | EOP | end of period |
| cwt | hundredweight | EOQ | economic order quantity |

| | | | |
|---|---|---|---|
| EPROM | erasable-programmable read-only memory | gps | gallons per second |
| Eq. | equation | gr | grain |
| erf | error function | GTO | gate turn-off thyristor |
| ERP | effective radiated power | HBL | hydrodynamic boundary layer |
| ES | earliest start | HCP | hexagonal close-packed |
| esu | electrostatic unit | hdbk | handbook |
| ETSI | European Telecommunications Standards Institute | HDTV | high-definition television |
| | | HERF | high-energy rate forming |
| EUAC | equivalent uniform annual cost | hex | hexadecimal |
| EUT | equipment under test | HF | high frequency |
| EW or E.W. | equivalent weight | Hg | mercury |
| Ex. | example | HGL | hydraulic grade line |
| exp | exponent of $e$ | hhp or HHP | hydraulic horsepower |
| FAR | fuel-air ratio | HHV | higher heating value |
| FAT | file allocation table | HI or H.I. | height of the instrument |
| FATT | fraction appearance transition temperature | hp, h.p., or HP | horsepower |
| | | h-p or H-P | high pressure |
| FB | feedback | HPBT | high-power bipolar transistor |
| FBD | free-body diagram | hp-hr | horsepower hour |
| fc or FC | foot-candle | hr | hour |
| FCC | face-centered cubic or Federal Communications Commission (U.S.) | HS | high-strength |
| | | HSS | high-speed steel |
| | | HTAH | high-temperature air heater |
| FDM | frequency division multiplexing | HTGR | high-temperature gas-cooled reactor |
| FDMA | frequency division multiple access | HTL | high-threshold logic |
| FE or F.E. | fundamentals of engineering | HV | heating value (gross) or high-voltage |
| FEM | fixed-end moment | HWR | heavy water reactor |
| FET | field effect transistor | IACS | International Annealed Copper Standard |
| FFT | fast Fourier transform | IBC | International Building Code |
| fhp or FHP | fluid horsepower or friction horsepower | IBG | interblock gap |
| FIFO | first in, first out | IC or I.C. | integrated circuit or instantaneous center |
| Fig. | figure | ICFH | inlet cubic feet per hour |
| fl, fL or f.l. | footlambert | ICFM | inlet cubic feet per minute |
| FLOPS | floating point operations per second | ICFS | inlet cubic feet per second |
| FM | frequency modulation | id, i.d., ID, or I.D. | inside diameter |
| fp or FP | freezing point | i.e. | that is |
| fpm | feet per minute | I.E. | intern engineer or industrial engineer |
| fps | feet per second | IEEE | Institute of Electrical and Electronic Engineers |
| FS | factor of safety | | |
| ft | feet or foot | | |
| FTP | fracture transition plastic | IGFET | insulated gate field effect transistor |
| ftn | footnote | ihp or IHP | indicated horsepower |
| FW | future worth or formula weight | IME | indirect manufacturing expense |
| g | gravity | IMEP | indicated mean effective pressure |
| gee | gravity | Imp | impulse |
| G | specific gravity | in | inch |
| GaAs | gallium arsenide | int'l | international |
| gal | gallon | inv | inverse |
| GAO | Government Accounting Office (U.S.) | I/O | input/output |
| GATT | General Agreement on Trade and Tariffs | ips | inches per second |
| GB | gigabyte | IPS | iron pipe size |
| GCR | gas cooled reactor | IR | instruction register or infrared |
| GeV | giga-electron volts | ISA | International Standard Association or Instrument Society of America |
| gew or GEW | gram-equivalent weight | | |
| GFE | general flow equation | ISDN | integrated services digital network |
| GIGO | garbage in, garbage out | ISFC | indicated specific fuel consumption |
| GMAW | gas-metal arc welding | ISO | International Standards Organization |
| gmole | gram-mole (mol) | IST | international steam tables |
| gpd | gallons per day | ITC | investment tax credit |
| gph | gallons per hour | IUPAC | International Union of Pure and Applied Chemistry |
| gpm | gallons per minute | | |

| | |
|---|---|
| iwg or i.w.g. | inches water gage (inches of water) |
| JCL | job control language |
| JFET | junction field-effect transistor |
| k | 1000 or 1000 pounds (kip) |
| kB | kilobyte |
| kc | kilocycle or kilocycles per second (kHz) |
| KCL | Kirchhoff's current law |
| kcmil | 1000 circular mills |
| kcs | kilocycles per second (kHz) |
| KE or K.E. | kinetic energy |
| kip | kilopound (1000 pounds) |
| kph or KPH | kilometers per hour |
| ksi | kilopounds per square inch |
| kv or kV | kilovolt (kV) |
| kva or kVA | kilovolt-ampere (kVA) |
| kvar or kVAR | kilovolt-ampere reactive (kVAr) |
| KVL | Kirchhoff's voltage law |
| kw, k.w., or kW | kilowatt or kiloword |
| kWh or kWhr | kilowatt-hour (kW·h) |
| LAN | local area network |
| lat | latitude |
| lb | pound |
| lbf | pound of force |
| lbm | pound of mass |
| lbmole | pound-mole |
| LCD | liquid crystal display |
| LCL | lower confidence limit |
| LCR | inductor-capacitor-resistor |
| LED | light-emitting diode |
| LEL | lower explosive limit |
| LF | load factor |
| lh or LH | left-hand |
| lhs or LHS | left-hand side |
| LHV | lower heating value |
| LILO | last in, last out |
| lim | limit |
| LMFBR | liquid metal fast breeder reactor |
| LMR | liquid metal reactor |
| LMTD | logarithmic mean temperature difference |
| ln | natural logarithm |
| log | base-10 logarithm |
| l-p, LP, or L-P | low pressure |
| LP | liquefied petroleum |
| LPG | liquefied petroleum gas |
| LSB | least-significant bit |
| LSD | least-significant digit |
| LSI | large scale integration |
| LV | low-voltage |
| LW | lost work |
| MACRS | modified accelerated cost recovery system |
| MARR | minimum attractive rate of return |
| max | maximum |
| MB | megabyte |
| Mbps | millions of bits per second |
| MC | megacycle, megacycles per second (MHz), or moisture content |
| Mcf or MCF | thousand cubic feet |
| MCM | 1000 circular mills |
| MDF | multiple degree of freedom |
| MDOF | multiple degree of freedom |

| | |
|---|---|
| MDR | minimum daily requirements |
| mep or MEP | mean effective pressure |
| meq | milligram equivalent weight |
| MeV | millions of electron volts |
| mf | millifarad (mF) |
| MGD | millions of gallons per day |
| mh | millihenry (mH) |
| MHD | magnetohydrodynamics |
| MHN | Meyer hardness number |
| MHz | megahertz |
| mi | mile |
| MICR | magnetic ink character recognition |
| MIG | metal-inert gas |
| min | minimum or minute |
| MIPS | millions of instructions per second |
| MIS | management information system |
| MIT | Massachusetts Institute of Technology (U.S.) |
| MJ | metric bolt with radiused root |
| mks | meter-kilogram-second (metric system) |
| MM | million ($10^6$) |
| MMCF | million cubic feet |
| MMF | magnetomotive force |
| MMIC | microwave monolithic integrated circuit |
| MMSCFD | million standard cubic feet per day |
| mo | month or method of operation |
| mol | mole (gram-mole) |
| mol. wt. | molecular weight |
| mom | momentum |
| MOS | metallic oxide semiconductor |
| MOSFET | metallic-oxide semiconductor field effect transistor |
| MPD | maximum permissible dose |
| mph | miles per hour |
| MS | margin of safety (FS − 1) |
| MSB | most-significant bit |
| MSD | most-significant digit |
| MSI | medium scale integration |
| MSW | municipal solid waste |
| MTBF | mean time before failure |
| MTTF | mean time to failure |
| mux | multiplexer |
| MW or M.W. | molecular weight |
| n.a. or NA | neutral axis |
| NBS | National Bureau of Standards (U.S., obsolete, see NIST) |
| NC | numerically-controlled |
| NCEE | National Council of Engineering Examiners (U.S., obsolete, see NCEES) |
| NCEES | National Council of Examiners for Engineering and Surveying (U.S.) |
| NDT | nondestructive testing |
| NEC | National Electric Code (U.S.) |
| NEMA | National Electrical Manufacturers Association (U.S.) |
| NFPA | National Fire Protection Association (U.S.) |
| NIPA | net inlet pressure available |
| NIPR | net inlet pressure required |

| | |
|---|---|
| NIST | National Institute of Standards and Technology (U.S.) |
| NLD | nonlinear device |
| NMOS | n-channel metallic oxide semiconductor |
| nom | nominal |
| NPSH | net positive suction head |
| NPSHA | net positive suction head available |
| NPSHR | net positive suction head required |
| NRTL | nationally recognized testing lab (U.S.) |
| NRZI | non-return-to-zero inverted |
| NS | nominal size |
| NSPE | National Society of Professional Engineers (U.S.) |
| NTS | not to scale |
| o.c. or OC | on center |
| OCR | optical character recognition or organic liquid cooled reactor |
| od, o.d., OD, or O.D. | outside diameter |
| O&M | operating and maintenance |
| op amp | operational amplifier |
| OS | operating system |
| OSHA | Occupational Safety and Health Administration (U.S.) |
| oz or OZ | ounce |
| p or P | poise |
| PC | point of curvature, personal computer, or program counter |
| PCB | printed circuit board |
| pcf | pounds (mass) per cubic foot |
| PDF | probability density function |
| PE or P.E. | potential energy, pressure energy, or professional engineer |
| P/E | price/earnings |
| PERT | program evaluation and review technique |
| pf, p.f., or PF | power factor |
| pH | hydrogen ion concentration |
| P.I. | point of intersection |
| PIEV | perception, identification, emotion, and volition |
| PIV | peak inverse voltage |
| P&L | profit and loss |
| pm, p.m., PM | afternoon |
| PM | preventive maintenance |
| pmole | pound-mole |
| PMOS | p-channel metallic oxide semiconductor |
| pOH | hydroxide ion concentration |
| POS | point-of-sale |
| P&P | principles and practice (exam) |
| ppb | parts per billion ($10^9$) |
| ppm | parts per million ($10^6$) |
| PROM | programmable read-only memory |
| PRV | peak reverse voltage |
| psych | psychrometric |
| psf | pounds (force) per square foot |
| psi | pounds (force) per square inch |
| psia | pounds (force) per square inch absolute |
| psig | pounds (force) per square inch gage |
| pt | point |
| PT or P.T. | point of tangency |
| PTO or P.T.O. | power takeoff |

| | |
|---|---|
| p-V | pressure-volume |
| PV | photovoltaic or present value |
| PVC | polyvinyl chloride |
| PW | present worth |
| PWR | pressurized water reactor |
| QA | quality assurance |
| QC | quality control |
| QIC | quarter-inch cartridge |
| rad | radian |
| RAD | radiation absorbed dose |
| RAM | random access memory |
| RBE | relative biological effectiveness |
| RCTL | resistor-capacitor-transistor logic |
| RE or R.E. | registered engineer |
| redox | reduction-oxidation |
| ref | reference |
| REM | radiation effective man |
| rev | revolution or reverse |
| RF | radio frequency |
| RFI | radio frequency interference |
| rh or RH | right-hand or relative humidity |
| rhs or RHS | right-hand side |
| RISC | reduced instruction-set computing |
| rms | root-mean-square |
| R/O | read-only |
| ROI | return on investment |
| ROM | read-only memory |
| ROR | rate of return |
| rph | revolutions per hour |
| rpm | revolutions per minute |
| rps | revolutions per second |
| RTL | resistor-transistor logic |
| R/W | read/write |
| S | stoke |
| S.A. | single-acting |
| SAE | Society of Automotive Engineers (U.S.) |
| sat | saturation or saturated |
| SAW | surface acoustic wave or submerged arc welding |
| SC | standard conditions or simple cubic |
| scf or SCF | standard cubic feet |
| SCFH | standard cubic feet per hour |
| SCFM | standard cubic feet per minute |
| SCR | silicon-controlled rectifier |
| SDF | single degree of freedom |
| SDOF | single degree of freedom |
| sec | second |
| Sec. | section |
| SF | safety factor or sinking fund |
| SFC | specific fuel consumption |
| SFEE | steady flow energy equation |
| SFS | Saybolt Furol seconds |
| SG or S.G. | specific gravity |
| SHM | simple harmonic motion |
| shp or SHP | shaft horsepower |
| SHR | sensible heat ratio |
| SI | Système International d'Unités |
| SL | straight line |
| SLAC | Stanford Linear Accelerator Center (U.S.) |
| SMAW | shielded metal arc welding |

| | | | |
|---|---|---|---|
| SME | Society of Manufacturing Engineers (U.S.) | UNC | Unified Coarse |
| sop or SOP | standard operating procedure or seat-of-the-pants | UNEF | Unified Extra Fine |
| | | UNF | Unified Fine |
| SOYD | sum-of-the-years' digits | uno | unless noted otherwise |
| sp. gr. | specific gravity | UNR | Unified bolt with radiused root |
| sp. ht. | specific heat | UNS | Unified Special |
| sq, SQ, or □ | square | uos | unless otherwise specified |
| SQC | statistical quality control | UPC | Uniform Plumbing Code (U.S.) |
| sr | steradian | USM | ultrasonic machining |
| SR | slenderness ratio or speed regulation | uso | unless specified otherwise |
| SSC | Superconducting Super Collider | UTS | ultimate tensile strength |
| SSD | saturated surface dry | UV | ultraviolet |
| SSF | seconds Saybolt Furol | VAR | volt-amperes reactive |
| SSU | seconds Saybolt Universal | VDU | visual display unit |
| sta or Sta | station | VHN | Vickers hardness number |
| std | standard | VHSIC | very-high-speed integrated circuit |
| STP | standard temperature and pressure | VLSI | very-large-scale integration |
| SUV | Saybolt Universal viscosity | VM | virtual machine |
| SW | shortwave or standing wave | VMOSFET | vertical MOSFET |
| SWR | standing wave ratio | vp or VP | vapor pressure |
| TC | tax credit | VPH | Vickers penetration hardness |
| TDC | top dead center | VR | voltage regulation |
| TDD | time division duplex | VS | virtual storage |
| TDM | time division multiplexing | VSAT | very-small-aperture terminals |
| TDMA | time division multiple access | VSWR | voltage standing wave ratio |
| TH'D | thread or threaded | VTL | variable threshold logic |
| THM | trihalomethane | VTOL | vertical takeoff and landing |
| TIG | tungsten-inert gas | wb | wet bulb |
| tpd | tons per day | w.c. | water column |
| tr | transpose | WF | wide flange |
| tsf | tons per square foot | w.g. | water gage or water gas |
| TTD | terminal temperature difference | whp or WHP | water horsepower |
| TTL | transistor-transistor logic | WIP | work in process |
| TTT | time-temperature-transformation | wk | week |
| TTY | teletype | w/o | without |
| Tw | Twaddell degrees | WORM | write once, read many |
| TWT | traveling-wave tube | wrt | with respect to |
| TWX | telex | wt | weight |
| typ | typical | YAG | yttrium-aluminum-garnet |
| UBC | Uniform Building Code (U.S., obsolete, see IBC) | yd | yard |
| | | yp or YP | yield point |
| UCL | upper confidence limit | yr | year |
| UL or U.L. | Underwriters Labs (U.S.) | YS | yield strength |
| UN | Unified | | |

# Topic I: Background and Support

Chapter

# 1 Engineering Drawing Practice

**Figure 1.1** *Intersecting and Non-Intersecting Lines*

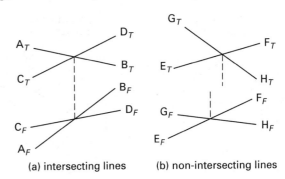

(a) intersecting lines     (b) non-intersecting lines

## 1. NORMAL VIEWS OF LINES AND PLANES[1]

A *normal view* of a line is a perpendicular projection of the line onto a viewing plane parallel to the line. In the normal view, all points of the line are equidistant from the observer. Therefore, the true length of a line is viewed and can be measured.

Generally, however, a line will be viewed from an oblique position and will appear shorter than it actually is. The normal view can be constructed by drawing an auxiliary view (see Sec. 1.5) from the orthographic view.[2]

Similarly, a normal view of a plane figure is a perpendicular projection of the figure onto a viewing plane parallel to the plane of the figure. All points of the plane are equidistant from the observer. Therefore, the true size and shape of any figure in the plane can be determined.

## 2. INTERSECTING AND PERPENDICULAR LINES

A single orthographic view is not sufficient to determine whether two lines intersect. However, if two or more views show the lines as having the same common point (i.e., crossing at the same position in space), then the lines intersect. In Fig. 1.1, the subscripts $F$ and $T$ refer to front and top views, respectively.

According to the *perpendicular line principle*, two perpendicular lines appear perpendicular only in a normal view of either one or both of the lines. Conversely, if two lines appear perpendicular in any view, the lines are perpendicular only if the view is a normal view of one or both of the lines.

## 3. TYPES OF VIEWS

Objects can be illustrated in several different ways depending on the number of views, the angle of observation, and the degree of artistic latitude taken for the purpose of simplifying the drawing process.[3] Table 1.1 categorizes the types of views.

**Table 1.1** *Types of Views of Objects*

orthographic views
    principal views
    auxiliary views
    oblique views
        cavalier projection
        cabinet projection
        clinographic projection
        axonometric views
            isometric
            dimetric
            trimetric
perspective views
    parallel perspective
    angular perspective

---

[1]This chapter is not meant to show "how to do it" as much as it is to present the conventions and symbols of engineering drawing.
[2]The technique for constructing a normal view is covered in engineering drafting texts.

[3]The omission of perspective from a drawing is an example of a step taken to simplify the drawing process.

The different types of views are easily distinguished by their *projectors* (i.e., projections of parallel lines on the object). For a cube, there are three sets of projectors corresponding to the three perpendicular axes. In an *orthographic (orthogonal) view*, the projectors are parallel. In a *perspective (central) view*, some or all of the projectors converge to a point. Figure 1.2 illustrates the orthographic and perspective views of a block.

**Figure 1.2** *Orthographic and Perspective Views of a Block*

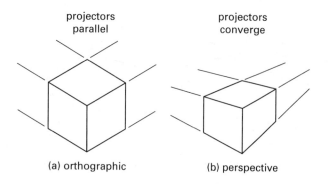

(a) orthographic          (b) perspective

## 4. PRINCIPAL (ORTHOGRAPHIC) VIEWS

In a *principal view* (also known as a *planar view*), one of the sets of projectors is normal to the view. That is, one of the planes of the object is seen in a normal view. The other two sets of projectors are orthogonal and are usually oriented horizontally and vertically on the paper. Because background details of an object may not be visible in a principal view, it is necessary to have at least three principal views to completely illustrate a symmetrical object. At most, six principal views will be needed to illustrate complex objects.

The relative positions of the six views have been standardized and are shown in Fig. 1.3, which also defines the *width* (also known as *depth*), *height*, and *length* of the object. The views that are not needed to illustrate features or provide dimensions (i.e., *redundant views*) can be omitted. The usual combination selected consists of the top, front, and right side views.

**Figure 1.3** *Positions of Standard Orthographic Views*

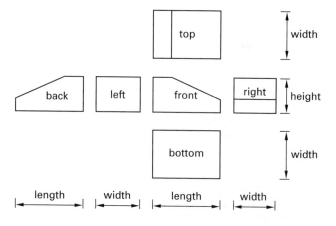

It is common to refer to the front, side, and back views as *elevations* and to the top and bottom views as *plan views*. These terms are not absolute since any plane can be selected as the front.

## 5. AUXILIARY (ORTHOGRAPHIC) VIEWS

An *auxiliary view* is needed when an object has an inclined plane or curved feature or when there are more details than can be shown in the six principal views. As with the other orthographic views, the auxiliary view is a normal (face-on) view of the inclined plane. Figure 1.4 illustrates an auxiliary view.

**Figure 1.4** *Auxiliary View*

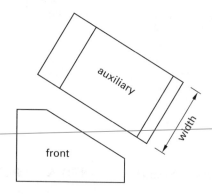

The projectors in an auxiliary view are perpendicular to only one of the directions in which a principal view is observed. Accordingly, only one of the three dimensions of width, height, and depth can be measured (scaled). In a *profile auxiliary view*, the object's width can be measured. In a *horizontal auxiliary view (auxiliary elevation)*, the object's height can be measured. In a *frontal auxiliary view*, the depth of the object can be measured.

## 6. OBLIQUE (ORTHOGRAPHIC) VIEWS

If the object is turned so that three principal planes are visible, it can be completely illustrated by a single *oblique view*.[4] In an oblique view, the direction from which the object is observed is not (necessarily) parallel to any of the directions from which principal and auxiliary views are observed.

In two common methods of oblique illustration, one of the view planes coincides with an orthographic view plane. Two of the drawing axes are at right angles to each other; one of these is vertical, and the other (the *oblique axis*) is oriented at 30° or 45° (originally chosen to coincide with standard drawing triangles). The ratio of scales used for the horizontal, vertical, and oblique axes can be 1:1:1 or 1:1:$\frac{1}{2}$. The latter ratio helps to overcome the visual distortion due to the absence of perspective in the oblique direction.

---

[4]Oblique views are not unique in this capability—perspective drawings share it. Oblique and perspective drawings are known as *pictorial drawings* because they give depth to the object by illustrating it in three dimensions.

*Cavalier* (45° oblique axis and 1:1:1 scale ratio) and *cabinet* (45° oblique axis and 1:1:$^1/_2$ scale ratio) *projections* are the two common types of oblique views that incorporate one of the orthographic views. If an angle of 9.5° is used (as in illustrating crystalline lattice structures), the technique is known as *clinographic projection.* Figure 1.5 illustrates cavalier and cabinet oblique drawings.

**Figure 1.5** *Cavalier and Cabinet Oblique Drawings*

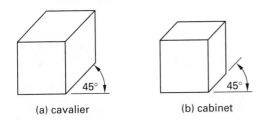

(a) cavalier　　　　(b) cabinet

## 7. AXONOMETRIC (ORTHOGRAPHIC OBLIQUE) VIEWS

In axonometric views, the view plane is not parallel to any of the principal orthographic planes. Figure 1.6 illustrates types of axonometric views. Axonometric views and axonometric drawings are not the same. In a *view (projection)*, one or more of the face lengths is foreshortened. In a *drawing*, the lengths are drawn full length, resulting in a distorted illustration. Table 1.2 lists the proper ratios that should be observed.

**Table 1.2** *Axonometric Foreshortening*

| view | projector intersection angles | proper ratio of sides |
|------|-------------------------------|-----------------------|
| isometric | 120°, 120°, 120° | 0.82:0.82:0.82 |
| dimetric | 131°25′, 131°25′, 97°10′ | 1:1:$^1/_2$ |
|  | 103°38′, 103°38′, 152°44′ | $^3/_4$:$^3/_4$:1 |
| trimetric | 102°28′, 144°16′, 113°16′ | 1:$^2/_3$:$^7/_8$ |
|  | 138°14′, 114°46′, 107° | 1:$^3/_4$:$^7/_8$ |

In an *isometric view*, the three projectors intersect at equal angles (120°) with the plane. This simplifies construction with standard 30° drawing triangles. All of the faces are foreshortened an equal amount, to $\sqrt{2/3}$, or approximately 81.6% of the true length. In a *dimetric view*, two of the projectors intersect at equal angles, and only two of the faces are equally reduced in length. In a *trimetric view*, all three intersection angles are different, and all three faces are reduced different amounts.

## 8. PERSPECTIVE VIEWS

In a *perspective view*, one or more sets of projectors converge to a fixed point known as the *center of vision.* In the *parallel perspective*, all vertical lines remain vertical in the picture; all horizontal frontal lines remain

**Figure 1.6** *Types of Axonometric Views*

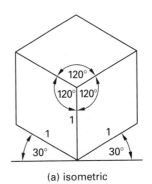

(a) isometric

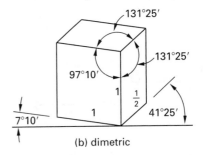

(b) dimetric

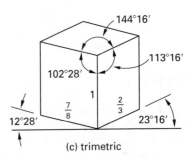

(c) trimetric

horizontal. Therefore, one face is parallel to the observer and only one set of projectors converges. In the *angular perspective*, two sets of projectors converge. In the little-used *oblique perspective*, all three sets of projectors converge. Figure 1.7 illustrates types of perspective views.

**Figure 1.7** *Types of Perspective Views*

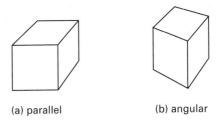

(a) parallel　　　　(b) angular

## 9. SECTIONS

A *section* is an imaginary cut taken through an object to reveal the shape or interior construction.[5] Figure 1.8 illustrates the standard symbol for a *sectioning cut* and the resulting sectional view. Section arrows are perpendicular to the cutting plane and indicate the viewing direction.

**Figure 1.8** *Sectioning Cut Symbol and Sectional View*

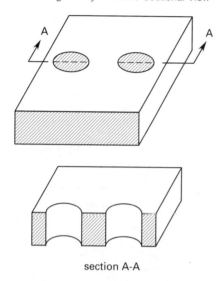

section A-A

## 10. TOLERANCES

The *tolerance* for a dimension is the total permissible variation or difference between the acceptable limits. The tolerance for a dimension can be specified in two ways: either as a general rule in the title block (e.g., ± 0.001 in unless otherwise specified) or as specific limits that are given with each dimension (e.g., 2.575 in ± 0.005 in).

## 11. SURFACE FINISH

ANSI B46.1 specifies surface finish by a combination of parameters.[6] The basic symbol for designating these factors is shown in Fig. 1.9. In the symbol, $A$ is the maximum *roughness height index*, $B$ is the optional minimum *roughness height*, $C$ is the peak-to-valley *waviness height*, $D$ is the optional peak-to-valley *waviness*

**Figure 1.9** *Surface Finish Designations*

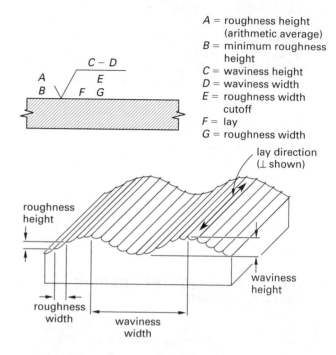

A = roughness height (arithmetic average)
B = minimum roughness height
C = waviness height
D = waviness width
E = roughness width cutoff
F = lay
G = roughness width

*spacing (width)* rating, $E$ is the optional *roughness width cutoff (roughness sampling length)*, $F$ is the *lay*, and $G$ is the *roughness width*. Unless minimums are specified, all parameters are maximum allowable values, and all lesser values are permitted.

Since the roughness varies, the waviness height is an arithmetic average within a sampled square, and the designation $A$ is known as the *roughness weight*, $R_a$.[7] Values are normally given in microns ($\mu$m) or microinches ($\mu$in) in SI or customary U.S. units, respectively. A value for the roughness width cutoff of 0.80 mm (0.03 in) is assumed when $E$ is not specified. Other standard values in common use are 0.25 mm (0.010 in) and 0.08 mm (0.003 in). The lay symbol, $F$, can be = (parallel to indicated surface), $\perp$ (perpendicular), C (circular), M (multidirectional), P (pitted), R (radial), or X (crosshatch).

If a small circle is placed at the $A$ position, no machining is allowed and only cast, forged, die-cast, injection-molded, and other unfinished surfaces are acceptable.

---

[5]The term *section* is also used to mean a *cross section*—a slice of finite but negligible thickness that is taken from an object to show the cross section or interior construction at the plane of the slice.
[6]Specification does not indicate appearance (i.e., color, luster) or performance (i.e., hardness, corrosion resistance, microstructure).

[7]The symbol $R_a$ is the same as the AA (arithmetic average) and CLA (centerline average) terms used in other (and earlier) standards.

Background and Support

# 2 Systems of Units

## 1. INTRODUCTION

The purpose of this chapter is to eliminate some of the confusion regarding the many units available for each engineering variable. In particular, an effort has been made to clarify the use of the so-called English systems, which for years have used the *pound* unit both for force and mass—a practice that has resulted in confusion even for those familiar with it.

## 2. COMMON UNITS OF MASS

The choice of a mass unit is the major factor in determining which system of units will be used in solving a problem. It is obvious that one will not easily end up with a force in pounds if the rest of the problem is stated in meters and kilograms. Actually, the choice of a mass unit determines more than whether a conversion factor will be necessary to convert from one system to another (e.g., between SI and English units). An inappropriate choice of a mass unit may actually require a conversion factor *within* the system of units.

The common units of mass are the gram, pound, kilogram, and slug.[1] There is nothing mysterious about these units. All represent different quantities of matter,

[1]Normally, one does not distinguish between a unit and a multiple of that unit, as is done here with the gram and the kilogram. However, these two units actually are bases for different consistent systems.

as Fig. 2.1 illustrates. In particular, note that the pound and slug do not represent the same quantity of matter.[2]

**Figure 2.1** *Common Units of Mass*

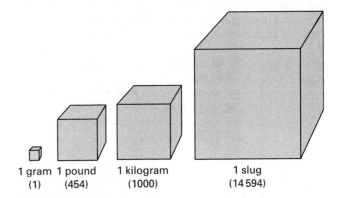

1 gram 1 pound    1 kilogram        1 slug
(1)    (454)      (1000)         (14 594)

## 3. MASS AND WEIGHT

In SI, *kilograms* are used for mass and *newtons* for weight (force). The units are different, and there is no confusion between the variables. However, for years the term *pound* has been used for both mass and weight. This usage has obscured the distinction between the two: mass is a constant property of an object; weight varies with the gravitational field. Even the conventional use of the abbreviations *lbm* and *lbf* (to distinguish between pounds-mass and pounds-force) has not helped eliminate the confusion.

It is true that an object with a mass of one pound will have an earthly weight of one pound, but this is true only on the earth. The weight of the same object will be much less on the moon. Therefore, care must be taken when working with mass and force in the same problem.

The relationship that converts mass to weight is familiar to every engineering student.

$$W = mg \qquad\qquad 2.1$$

Equation 2.1 illustrates that an object's weight will depend on the local acceleration of gravity as well as the object's mass. The mass will be constant, but gravity will depend on location. Mass and weight are not the same.

[2]A slug is approximately equal to 32.1740 pounds-mass.

## 4. ACCELERATION OF GRAVITY

Gravitational acceleration on the earth's surface is usually taken as 32.2 ft/sec$^2$ or 9.81 m/s$^2$. These values are rounded from the more exact standard values of 32.1740 ft/sec$^2$ and 9.8066 m/s$^2$. However, the need for greater accuracy must be evaluated on a problem-by-problem basis. Usually, three significant digits are adequate, since gravitational acceleration is not constant anyway but is affected by location (primarily latitude and altitude) and major geographical features.

The term *standard gravity*, $g_0$, is derived from the acceleration at essentially any point at sea level and approximately 45° N latitude. If additional accuracy is needed, the gravitational acceleration can be calculated from Eq. 2.2. This equation neglects the effects of large land and water masses. $\phi$ is the latitude in degrees.

$$g_{\text{surface}} = g'(1 + (5.305 \times 10^{-3})\sin^2\phi$$
$$- (5.9 \times 10^{-6})\sin^2 2\phi)$$
$$g' = 32.0881 \text{ ft/sec}^2$$
$$= 9.78045 \text{ m/s}^2 \qquad 2.2$$

If the effects of the earth's rotation are neglected, the gravitational acceleration at an altitude $h$ above the earth's surface is given by Eq. 2.3. $R_e$ is the earth's radius.

$$g_h = g_{\text{surface}}\left(\frac{R_e}{R_e + h}\right)^2$$
$$R_e = 3960 \text{ mi}$$
$$= 6.37 \times 10^6 \text{ m} \qquad 2.3$$

## 5. CONSISTENT SYSTEMS OF UNITS

A set of units used in a calculation is said to be *consistent* if no conversion factors are needed.[3] For example, a moment is calculated as the product of a force and a lever arm length.

$$M = Fd \qquad 2.4$$

A calculation using Eq. 2.4 would be consistent if $M$ was in newton-meters, $F$ was in newtons, and $d$ was in meters. The calculation would be inconsistent if $M$ was in ft-kips, $F$ was in kips, and $d$ was in inches (because a conversion factor of 1/12 would be required).

The concept of a consistent calculation can be extended to a system of units. A *consistent system of units* is one in which no conversion factors are needed for any calculation. For example, Newton's second law of motion can be written without conversion factors. Newton's second law simply states that the force required to accelerate an object is proportional to the acceleration of the object. The constant of proportionality is the object's mass.

$$F = ma \qquad 2.5$$

Notice that Eq. 2.5 is $F = ma$, not $F = Wa/g$ or $F = ma/g_c$. Equation 2.5 is consistent: it requires no conversion factors. This means that in a consistent system where conversion factors are not used, once the units of $m$ and $a$ have been selected, the units of $F$ are fixed. This has the effect of establishing units of work and energy, power, fluid properties, and so on.

It should be mentioned that the decision to work with a consistent set of units is desirable but unnecessary, depending on tradition and environment. Problems in fluid flow and thermodynamics are routinely solved in the United States with inconsistent units. This causes no more of a problem than working with inches and feet when calculating a moment. It is necessary only to use the proper conversion factors.

## 6. THE ENGLISH ENGINEERING SYSTEM

Through common and widespread use, pounds-mass (lbm) and pounds-force (lbf) have become the standard units for mass and force in the *English Engineering System*. (The English Engineering System is used in this book.)

There are subjects in the United States where the practice of using pounds for mass is firmly entrenched. For example, most thermodynamics, fluid flow, and heat transfer problems have traditionally been solved using the units of lbm/ft$^3$ for density, Btu/lbm for enthalpy, and Btu/lbm-°F for specific heat. Unfortunately, some equations contain both lbm-related and lbf-related variables, as does the steady flow conservation of energy equation, which combines enthalpy in Btu/lbm with pressure in lbf/ft$^2$.

The units of pounds-mass and pounds-force are as different as the units of gallons and feet, and they cannot be canceled. A mass conversion factor, $g_c$, is needed to make the equations containing lbf and lbm dimensionally consistent. This factor is known as the *gravitational constant* and has a value of 32.1740 ft-lbm/lbf-sec$^2$. The numerical value is the same as the standard acceleration of gravity, but $g_c$ is not the local gravitational acceleration, $g$.[4] $g_c$ is a conversion constant, just as 12.0 is the conversion factor between feet and inches.

The English Engineering System is an inconsistent system as defined according to Newton's second law.

---

[3]The terms *homogeneous* and *coherent* are also used to describe a consistent set of units.

[4]It is acceptable (and recommended) that $g_c$ be rounded to the same number of significant digits as $g$. Therefore, a value of 32.2 for $g_c$ would typically be used.

$F = ma$ cannot be written if lbf, lbm, and ft/sec$^2$ are the units used. The $g_c$ term must be included.

$$F \text{ in lbf} = \frac{(m \text{ in lbm})\left(a \text{ in } \frac{\text{ft}}{\text{sec}^2}\right)}{g_c \text{ in } \frac{\text{ft-lbm}}{\text{lbf-sec}^2}} \qquad 2.6$$

It is important to note in Eq. 2.6 that $g_c$ does more than "fix the units." Since $g_c$ has a numerical value of 32.174, it actually changes the calculation numerically. A force of 1.0 pound will not accelerate a 1.0-pound mass at the rate of 1.0 ft/sec$^2$.

In the English Engineering System, work and energy are typically measured in ft-lbf (mechanical systems) or in British thermal units, Btu (thermal and fluid systems). One Btu is equal to 778.17 ft-lbf.

### Example 2.1

Calculate the weight in lbf of a 1.00 lbm object in a gravitational field of 27.5 ft/sec$^2$.

*Solution*

From Eq. 2.6,

$$F = \frac{ma}{g_c} = \frac{(1.00 \text{ lbm})\left(27.5 \frac{\text{ft}}{\text{sec}^2}\right)}{32.2 \frac{\text{ft-lbm}}{\text{lbf-sec}^2}}$$

$$= 0.854 \text{ lbf}$$

## 7. OTHER FORMULAS AFFECTED BY INCONSISTENCY

It is not a significant burden to include $g_c$ in a calculation, but it may be difficult to remember when $g_c$ should be used. Knowing when to include the gravitational constant can be learned through repeated exposure to the formulas in which it is needed, but it is safer to carry the units along in every calculation.

The following is a representative (but not exhaustive) listing of formulas that require the $g_c$ term. In all cases, it is assumed that the standard English Engineering System units will be used.

- kinetic energy

$$E = \frac{m\text{v}^2}{2g_c} \quad (\text{in ft-lbf}) \qquad 2.7$$

- potential energy

$$E = \frac{mgz}{g_c} \quad (\text{in ft-lbf}) \qquad 2.8$$

- pressure at a depth

$$p = \frac{\rho gh}{g_c} \quad (\text{in lbf/ft}^2) \qquad 2.9$$

### Example 2.2

A rocket that has a mass of 4000 lbm travels at 27,000 ft/sec. What is its kinetic energy in ft-lbf?

*Solution*

From Eq. 2.7,

$$E_k = \frac{m\text{v}^2}{2g_c} = \frac{(4000 \text{ lbm})\left(27,000 \frac{\text{ft}}{\text{sec}}\right)^2}{(2)\left(32.2 \frac{\text{ft-lbm}}{\text{lbf-sec}^2}\right)}$$

$$= 4.53 \times 10^{10} \text{ ft-lbf}$$

## 8. WEIGHT AND WEIGHT DENSITY

Weight is a force exerted on an object due to its placement in a gravitational field. If a consistent set of units is used, Eq. 2.1 can be used to calculate the weight of a mass. In the English Engineering System, however, Eq. 2.10 must be used.

$$W = \frac{mg}{g_c} \qquad 2.10$$

Both sides of Eq. 2.10 can be divided by the volume of an object to derive the *weight density*, $\gamma$, of the object. Equation 2.11 illustrates that the weight density (in lbf/ft$^3$) can also be calculated by multiplying the mass density (in lbm/ft$^3$) by $g/g_c$. Since $g$ and $g_c$ usually have the same numerical values, the only effect of Eq. 2.12 is to change the units of density.

$$\frac{W}{V} = \left(\frac{m}{V}\right)\left(\frac{g}{g_c}\right) \qquad 2.11$$

$$\gamma = \frac{W}{V} = \left(\frac{m}{V}\right)\left(\frac{g}{g_c}\right) = \frac{\rho g}{g_c} \qquad 2.12$$

Weight does not occupy volume. Only mass has volume. The concept of weight density has evolved to simplify certain calculations, particularly fluid calculations. For example, pressure at a depth is calculated from Eq. 2.13. (Compare this with Eq. 2.9.)

$$p = \gamma h \qquad 2.13$$

## 9. THE ENGLISH GRAVITATIONAL SYSTEM

Not all English systems are inconsistent. Pounds can still be used as the unit of force as long as pounds are not used as the unit of mass. Such is the case with the consistent *English Gravitational System*.

If acceleration is given in ft/sec$^2$, the units of mass for a consistent system of units can be determined from Newton's second law. The combination of units in Eq. 2.14 is known as a *slug*. $g_c$ is not needed at all since this system is consistent. It would be needed only to convert slugs to another mass unit.

$$\text{units of } m = \frac{\text{units of } F}{\text{units of } a}$$

$$= \frac{\text{lbf}}{\frac{\text{ft}}{\text{sec}^2}} = \frac{\text{lbf-sec}^2}{\text{ft}} \qquad 2.14$$

Slugs and pounds-mass are not the same, as Fig. 2.1 illustrates. However, both are units for the same quantity: mass. Equation 2.15 will convert between slugs and pounds-mass.

$$\text{no. of slugs} = \frac{\text{no. of lbm}}{g_c} \qquad 2.15$$

It is important to recognize that the number of slugs is not derived by dividing the number of pounds-mass by the local gravity. $g_c$ is used regardless of the local gravity. The conversion between feet and inches is not dependent on local gravity; neither is the conversion between slugs and pounds-mass.

Since the English Gravitational System is consistent, Eq. 2.16 can be used to calculate weight. Notice that the local gravitational acceleration is used.

$$W \text{ in lbf} = (m \text{ in slugs})\left(g \text{ in } \frac{\text{ft}}{\text{sec}^2}\right) \qquad 2.16$$

## 10. THE ABSOLUTE ENGLISH SYSTEM

The obscure *Absolute English System* takes the approach that mass must have units of pounds-mass (lbm) and the units of force can be derived from Newton's second law. The units for $F$ cannot be simplified any more than they are in Eq. 2.17. This particular combination of units is known as a *poundal*.[5] A poundal is not the same as a pound.

$$\text{units of } F = (\text{units of } m)(\text{units of } a)$$

$$= (\text{lbm})\left(\frac{\text{ft}}{\text{sec}^2}\right)$$

$$= \frac{\text{lbm-ft}}{\text{sec}^2} \qquad 2.17$$

Poundals have not seen widespread use in the United States. The English Gravitational System (using slugs for mass) has greatly eclipsed the Absolute English System in popularity. Both are consistent systems, but there seems to be little need for poundals in modern engineering. Figure 2.2 shows the poundal in comparison to other common units of force.

---

[5]A poundal is equal to 0.03108 pounds-force.

**Figure 2.2** *Common Force Units*

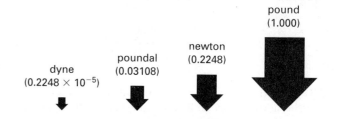

## 11. METRIC SYSTEMS OF UNITS

Strictly speaking, a *metric system* is any system of units that is based on meters or parts of meters. This broad definition includes *mks systems* (based on meters, kilograms, and seconds) as well as *cgs systems* (based on centimeters, grams, and seconds).

Metric systems avoid the pounds-mass versus pounds-force ambiguity in two ways. First, a unit of weight is not established at all. All quantities of matter are specified as mass. Second, force and mass units do not share a common name.

The term *metric system* is not explicit enough to define which units are to be used for any given variable. For example, within the cgs system there is variation in how certain electrical and magnetic quantities are represented (resulting in the ESU and EMU systems). Also, within the mks system, it is common practice in some industries to use kilocalories as the unit of thermal energy, while the SI unit for thermal energy is the joule. Thus, there is a lack of uniformity even within the metricated engineering community.[6]

The "metric" parts of this book use SI, which is the most developed and codified of the so-called metric systems.[7] There will be occasional variances with local engineering custom, but it is difficult to anticipate such variances within a book that must itself be consistent.[8]

## 12. THE cgs SYSTEM

The *cgs system* is used widely by chemists and physicists. It is named for the three primary units used to construct its derived variables: the centimeter, the gram, and the second.

---

[6]In the "field test" of the metric system conducted over the past 200 years, other conventions are to use kilograms-force (kgf) instead of newtons and kgf/cm² for pressure (instead of pascals).
[7]SI units are an outgrowth of the *General Conference of Weights and Measures*, an international treaty organization that established the *Système International d'Unités (International System of Units)* in 1960. The United States subscribed to this treaty in 1975.
[8]Conversion to pure SI units is essentially complete in Australia, Canada, New Zealand, and South Africa. The use of nonstandard metric units is more common among European civil engineers, who have had little need to deal with the inertial properties of mass. However, even the American Society of Civil Engineers declared its support for SI units in 1985.

---

When Newton's second law is written in the cgs system, the following combination of units results.

$$\text{units of force} = (m \text{ in g})\left(a \text{ in } \frac{\text{cm}}{\text{s}^2}\right)$$
$$= \text{g·cm/s}^2 \qquad 2.18$$

This combination of units for force is known as a *dyne*. Energy variables in the cgs system have units of dyne·cm or, equivalently, $\text{g·cm}^2/\text{s}^2$. This combination is known as an *erg*. There is no uniformly accepted unit of power in the cgs system, although calories per second is frequently used.

The fundamental volume unit in the cgs system is the cubic centimeter (cc). Since this is the same volume as one thousandth of a liter, units of milliliters (mL) are also used.

## 13. SI UNITS (THE mks SYSTEM)

*SI units* comprise an *mks system* (so named because it uses the meter, kilogram, and second as base units). All other units are derived from the base units, which are completely listed in Table 2.1. This system is fully consistent, and there is only one recognized unit for each physical quantity (variable).

**Table 2.1** SI Base Units

| quantity | name | symbol |
|---|---|---|
| length | meter | m |
| mass | kilogram | kg |
| time | second | s |
| electric current | ampere | A |
| temperature | kelvin | K |
| amount of substance | mole | mol |
| luminous intensity | candela | cd |

Two types of units are used: base units and derived units. The *base units* are dependent only on accepted standards or reproducible phenomena. The *derived units* (Table 2.2 and Table 2.3) are made up of combinations of base units. The old *supplementary units* (Table 2.4) were classified as derived units in 1995.

In addition, there is a set of non-SI units that may be used. This concession is primarily due to the significance and widespread acceptance of these units. Use of the non-SI units listed in Table 2.5 will usually create an inconsistent expression requiring conversion factors.

The SI unit of force can be derived from Newton's second law. This combination of units for force is known as a *newton*.

$$\text{units of force} = (m \text{ in kg})\left(a \text{ in } \frac{\text{m}}{\text{s}^2}\right)$$
$$= \text{kg·m/s}^2 \qquad 2.19$$

**Table 2.2** Some SI Derived Units with Special Names

| quantity | name | symbol | expressed in terms of other units |
|---|---|---|---|
| frequency | hertz | Hz | 1/s |
| force | newton | N | kg·m/s$^2$ |
| pressure, stress | pascal | Pa | N/m$^2$ |
| energy, work, quantity of heat | joule | J | N·m |
| power, radiant flux | watt | W | J/s |
| quantity of electricity, electric charge | coulomb | C | |
| electric potential, potential difference, electromotive force | volt | V | W/A |
| electric capacitance | farad | F | C/V |
| electric resistance | ohm | Ω | V/A |
| electric conductance | siemens | S | A/V |
| magnetic flux | weber | Wb | V·s |
| magnetic flux density | tesla | T | Wb/m$^2$ |
| inductance | henry | H | Wb/A |
| luminous flux | lumen | lm | |
| illuminance | lux | lx | lm/m$^2$ |

Energy variables in SI units have units of N·m or, equivalently, $\text{kg·m}^2/\text{s}^2$. Both of these combinations are known as a *joule*. The units of power are joules per second, equivalent to a *watt*.

### Example 2.3

A 10 kg block hangs from a cable. What is the tension in the cable? (Standard gravity equals 9.81 m/s$^2$.)

*Solution*

$$F = mg$$
$$= (10 \text{ kg})\left(9.81 \frac{\text{m}}{\text{s}^2}\right)$$
$$= 98.1 \text{ kg·m/s}^2 \quad (98.1 \text{ N})$$

### Example 2.4

A 10 kg block is raised vertically 3 m. What is the change in potential energy?

*Solution*

$$\Delta E_p = mg\Delta h$$
$$= (10 \text{ kg})\left(9.81 \frac{\text{m}}{\text{s}^2}\right)(3 \text{ m})$$
$$= 294 \text{ kg·m}^2/\text{s}^2 \quad (294 \text{ J})$$

**Table 2.3** Some SI Derived Units

| quantity | description | symbol |
|---|---|---|
| area | square meter | $m^2$ |
| volume | cubic meter | $m^3$ |
| speed—linear | meter per second | m/s |
| speed—angular | radian per second | rad/s |
| acceleration—linear | meter per second squared | $m/s^2$ |
| acceleration—angular | radian per second squared | $rad/s^2$ |
| density, mass density | kilogram per cubic meter | $kg/m^3$ |
| concentration (of amount of substance) | mole per cubic meter | $mol/m^3$ |
| specific volume | cubic meter per kilogram | $m^3/kg$ |
| luminance | candela per square meter | $cd/m^2$ |
| absolute viscosity | pascal second | Pa·s |
| kinematic viscosity | square meters per second | $m^2/s$ |
| moment of force | newton meter | N·m |
| surface tension | newton per meter | N/m |
| heat flux density, irradiance | watt per square meter | $W/m^2$ |
| heat capacity, entropy | joule per kelvin | J/K |
| specific heat capacity, specific entropy | joule per kilogram kelvin | J/kg·K |
| specific energy | joule per kilogram | J/kg |
| thermal conductivity | watt per meter kelvin | W/m·K |
| energy density | joule per cubic meter | $J/m^3$ |
| electric field strength | volt per meter | V/m |
| electric charge density | coulomb per cubic meter | $C/m^3$ |
| surface density of charge, flux density | coulomb per square meter | $C/m^2$ |
| permittivity | farad per meter | F/m |
| current density | ampere per square meter | $A/m^2$ |
| magnetic field strength | ampere per meter | A/m |
| permeability | henry per meter | H/m |
| molar energy | joule per mole | J/mol |
| molar entropy, molar heat capacity | joule per mole kelvin | J/mol·K |
| radiant intensity | watt per steradian | W/sr |

**Table 2.4** SI Supplementary Units[*]

| quantity | name | symbol |
|---|---|---|
| plane angle | radian | rad |
| solid angle | steradian | sr |

[*]classified as derived units in 1995.

**Table 2.5** Acceptable Non-SI Units

| quantity | unit name | symbol or abbreviation | relationship to SI unit |
|---|---|---|---|
| area | hectare | ha | $1\ ha = 10\,000\ m^2$ |
| energy | kilowatt-hour | kW·h | $1\ kW\cdot h = 3.6\ MJ$ |
| mass | metric ton[a] | t | $1\ t = 1000\ kg$ |
| plane angle | degree (of arc) | ° | $1° = 0.017\,453\ rad$ |
| speed of rotation | revolution per minute | r/min | $1\ r/min = 2\pi/60\ rad/s$ |
| temperature interval | degree Celsius | °C | $1°C = 1K$ |
| time | minute | min | $1\ min = 60\ s$ |
| | hour | h | $1\ h = 3600\ s$ |
| | day (mean solar) | d | $1\ d = 86\,400\ s$ |
| | year (calendar) | a | $1\ a = 31\,536\,000\ s$ |
| velocity | kilometer per hour | km/h | $1\ km/h = 0.278\ m/s$ |
| volume | liter[b] | L | $1\ L = 0.001\ m^3$ |

[a]The international name for metric ton is *tonne*. The metric ton is equal to the *megagram* (Mg).
[b]The international symbol for liter is the lowercase l, which can be easily confused with the numeral 1. Several English-speaking countries have adopted the script ℓ or uppercase L (as does this book) as a symbol for liter in order to avoid any misinterpretation.

## 14. RULES FOR USING SI UNITS

In addition to having standardized units, the set of SI units also has rigid syntax rules for writing the units and combinations of units. Each unit is abbreviated with a specific symbol. The following rules for writing and combining these symbols should be adhered to.

- The expressions for derived units in symbolic form are obtained by using the mathematical signs of multiplication and division. For example, units of velocity are m/s. Units of torque are N·m (not N-m or Nm).

- Scaling of most units is done in multiples of 1000.

- The symbols are always printed in roman type, regardless of the type used in the rest of the text. The only exception to this is in the use of the symbol for liter, where the use of the lower case "el" (l) may be confused with the numeral one (1). In this case, "liter" should be written out in full, or the script ℓ or L used. (L is used in this book.)

- Symbols are not pluralized: 1 kg, 45 kg (not 45 kgs).

- A period after a symbol is not used, except when the symbol occurs at the end of a sentence.

- When symbols consist of letters, there must always be a full space between the quantity and the symbols: 45 kg (not 45kg). However, for planar angle designations, no space is left: 32°C (not 32° C or 32 °C); or 42°12′45″ (not 42 ° 12 ′ 45 ″).

- All symbols are written in lowercase, except when the unit is derived from a proper name: m for meter; s for second; A for ampere, Wb for weber, N for newton, W for watt.

- Prefixes are printed without spacing between the prefix and the unit symbol (e.g., km is the symbol for kilometer). Table 2.6 lists common prefixes, their symbols, and their values.

**Table 2.6** SI Prefixes[*]

| prefix | symbol | value |
|--------|--------|-------|
| exa | E | $10^{18}$ |
| peta | P | $10^{15}$ |
| tera | T | $10^{12}$ |
| giga | G | $10^{9}$ |
| mega | M | $10^{6}$ |
| kilo | k | $10^{3}$ |
| hecto | h | $10^{2}$ |
| deka (or "deca") | da | $10^{1}$ |
| deci | d | $10^{-1}$ |
| centi | c | $10^{-2}$ |
| milli | m | $10^{-3}$ |
| micro | $\mu$ | $10^{-6}$ |
| nano | n | $10^{-9}$ |
| pico | p | $10^{-12}$ |
| femto | f | $10^{-15}$ |
| atto | a | $10^{-18}$ |

[*]There is no "B" (billion) prefix. In fact, the word "billion" means $10^{9}$ in the United States but $10^{12}$ in most other countries. This unfortunate ambiguity is handled by avoiding the use of the term billion.

- In text, when no number is involved, the unit should be spelled out. Example: Carpet is sold by the square meter, not by the $m^2$.

- Where a decimal fraction of a unit is used, a zero should always be placed before the decimal marker: 0.45 kg (not .45 kg). This practice draws attention to the decimal marker and helps avoid errors of scale.

- A practice in some countries is to use a comma as a decimal marker, while the practice in North America, the United Kingdom, and some other countries is to use a period (or dot) as the decimal marker. Furthermore, in some countries that use the decimal comma, a dot is frequently used to divide long numbers into groups of three. Because of these differing practices, spaces must be used instead of commas to separate long lines of digits into easily readable blocks of three digits with respect to the decimal marker: 32 453.246 072 5. A space (half-space preferred) is optional with a four-digit number: 1 234 or 1234.

- Some confusion may arise with the word "tonne" (1000 kg). When this word occurs in French text of Canadian origin, the meaning may be a ton of 2000 pounds.

## 15. PRIMARY DIMENSIONS

Regardless of the system of units chosen, each variable representing a physical quantity will have the same *primary dimensions*. For example, velocity may be expressed in miles per hour (mph) or meters per second (m/s), but both units have dimensions of length per unit time. Length and time are two of the primary dimensions, as neither can be broken down into more basic dimensions. The concept of primary dimensions is useful when converting little-used variables between different systems of units, as well as in correlating experimental results (i.e., dimensional analysis).

There are three different sets of primary dimensions in use.[9] In the $ML\theta T$ system, the primary dimensions are mass ($M$), length ($L$), time ($\theta$), and temperature ($T$). Notice that all symbols are uppercase. In order to avoid confusion between time and temperature, the Greek letter theta is used for time.[10]

All other physical quantities can be derived from these primary dimensions.[11] For example, work in SI units has units of N·m. Since a newton is a kg·m/s², the primary dimensions of work are $ML^2/\theta^2$. The primary dimensions for many important engineering variables are shown in Table 2.7. If it is more convenient to stay with traditional English units, it may be more desirable to work in the $FML\theta TQ$ system (sometimes called the *engineering dimensional system*). This system adds the primary dimensions of force ($F$) and heat ($Q$). Thus, work (ft-lbf in the English system) has the primary dimensions of $FL$. (Compare this with the primary dimensions for work in the $ML\theta T$ system.) Thermodynamic variables are similarly simplified.

Dimensional analysis will be more conveniently carried out when one of the four-dimension systems ($ML\theta T$ or $FL\theta T$) is used. Whether the $ML\theta T$, $FL\theta T$, or $FML\theta TQ$ system is used depends on what is being derived and who will be using it, and whether or not a consistent set of variables is desired. Conversion constants such as $g_c$ and $J$ will almost certainly be required if the $ML\theta T$ system is used to generate variables for use in the English systems. It is also much more convenient to use the $FML\theta TQ$ system when working in the fields of thermodynamics, fluid flow, heat transfer, and so on.

---

[9]One of these, the $FL\theta T$ system, is not discussed here.

[10]This is the most common usage. There is a lack of consistency in the engineering world about the symbols for the primary dimensions in dimensional analysis. Some writers use $t$ for time instead of $\theta$. Some use $H$ for heat instead of $Q$. And, in the worst mix-up of all, some have reversed the use of $T$ and $\theta$.

[11]A *primary dimension* is the same as a *base unit* in the SI set of units. The SI units add several other base units, as shown in Table 2.1, to deal with variables that are difficult to derive in terms of the four primary base units.

*Table 2.7* Dimensions of Common Variables

| variable | dimensional system | | |
|---|---|---|---|
| | $ML\theta T$ | $FL\theta T$ | $FMLT\theta Q$ |
| mass ($m$) | $M$ | $F\theta^2/L$ | $M$ |
| force ($F$) | $ML/\theta^2$ | $F$ | $F$ |
| length ($L$) | $L$ | $L$ | $L$ |
| time ($\theta$) | $\theta$ | $\theta$ | $\theta$ |
| temperature ($T$) | $T$ | $T$ | $T$ |
| work ($W$) | $ML^2/\theta^2$ | $FL$ | $FL$ |
| heat ($Q$) | $ML^2/\theta^2$ | $FL$ | $Q$ |
| acceleration ($a$) | $L/\theta^2$ | $L/\theta^2$ | $L/\theta^2$ |
| frequency ($N$) | $1/\theta$ | $1/\theta$ | $1/\theta$ |
| area ($A$) | $L^2$ | $L^2$ | $L^2$ |
| coefficient of thermal expansion ($\beta$) | $1/T$ | $1/T$ | $1/T$ |
| density ($\rho$) | $M/L^3$ | $F\theta^2/L^4$ | $M/L^3$ |
| dimensional constant ($g_c$) | 1.0 | 1.0 | $ML/\theta^2 F$ |
| specific heat at constant pressure ($c_p$); at constant volume ($c_v$) | $L^2/\theta^2 T$ | $L^2/\theta^2 T$ | $Q/MT$ |
| heat transfer coefficient ($h$); overall ($U$) | $M/\theta^3 T$ | $F/\theta LT$ | $Q/\theta L^2 T$ |
| power ($P$) | $ML^2/\theta^3$ | $FL/\theta$ | $FL/\theta$ |
| heat flow rate ($\dot{Q}$) | $ML^2/\theta^3$ | $FL/\theta$ | $Q/\theta$ |
| kinematic viscosity ($\nu$) | $L^2/\theta$ | $L^2/\theta$ | $L^2/\theta$ |
| mass flow rate ($\dot{m}$) | $M/\theta$ | $F\theta/L$ | $M/\theta$ |
| mechanical equivalent of heat ($J$) | – | – | $FL/Q$ |
| pressure ($p$) | $M/L\theta^2$ | $F/L^2$ | $F/L^2$ |
| surface tension ($\sigma$) | $M/\theta^2$ | $F/L$ | $F/L$ |
| angular velocity ($\omega$) | $1/\theta$ | $1/\theta$ | $1/\theta$ |
| volumetric flow rate ($\dot{m}/\rho = \dot{V}$) | $L^3/\theta$ | $L^3/\theta$ | $L^3/\theta$ |
| conductivity ($k$) | $ML/\theta^3 T$ | $F/\theta T$ | $Q/L\theta T$ |
| thermal diffusivity ($\alpha$) | $L^2/\theta$ | $L^2/\theta$ | $L^2/\theta$ |
| velocity (v) | $L/\theta$ | $L/\theta$ | $L/\theta$ |
| viscosity, absolute ($\mu$) | $M/L\theta$ | $F\theta/L^2$ | $F\theta/L^2$ |
| volume ($V$) | $L^3$ | $L^3$ | $L^3$ |

## 16. LINEAL AND BOARD FOOT MEASUREMENTS

The term *lineal* is often mistaken as a typographical error for *linear*. Although "lineal" has its own specific meaning slightly different from "linear," the two are often used interchangeably by engineers.[12] The adjective *lineal* is often encountered in the building trade (e.g., 12 lineal feet of lumber), where the term is used to distinguish it from board feet measurement.

A *board foot* (abbreviated bd-ft) is not a measure of length. Rather, it is a measure of volume used with lumber. Specifically, a board foot is equal to 144 in$^3$ ($2.36 \times 10^{-3}$ m$^3$). The name is derived from the volume of a board 1 foot square and 1 inch thick. In that sense, it is parallel in concept to the acre-foot. Since lumber cost is directly related to lumber weight and volume, the board foot unit is used in determining the overall lumber cost.

## 17. DIMENSIONAL ANALYSIS

*Dimensional analysis* is a means of obtaining an equation that describes some phenomenon without understanding the mechanism of the phenomenon. The most serious limitation is the need to know beforehand which variables influence the phenomenon. Once these are known or assumed, dimensional analysis can be applied by a routine procedure.

The first step is to select a system of primary dimensions. (See Sec. 2.15.) Usually the $ML\theta T$ system is used, although this choice may require the use of $g_c$ and $J$ in the final results.

The second step is to write a functional relationship between the dependent variable and the independent variable, $x_i$.

$$y = f(x_1, x_2, ..., x_m) \qquad 2.20$$

This function can be expressed as an exponentiated series. The $C_1$, $a_i$, $b_i$, ..., $z_i$ in Eq. 2.21 are unknown constants.

$$y = C_1 x_1^{a_1} x_2^{b_1} x_3^{c_1} ... x_m^{z_1} + C_2 x_1^{a_2} x_2^{b_2} x_3^{c_2} ... x_m^{z_2} + ... \qquad 2.21$$

The key to solving Eq. 2.21 is that each term on the right-hand side must have the same dimensions as $y$. Simultaneous equations are used to determine some of the $a_i$, $b_i$, $c_i$, and $z_i$. Experimental data are required to determine the $C_i$ and remaining exponents. In most analyses, it is assumed that the $C_i = 0$ for $i \geq 2$.

Since this method requires working with $m$ different variables and $n$ different independent dimensional quantities (such as $M$, $L$, $\theta$, and $T$), an easier method is desirable. One simplification is to combine the $m$ variables into dimensionless groups called *pi-groups*. (See Table 2.7.)

If these dimensionless groups are represented by $\pi_1$, $\pi_2$, $\pi_3$, ..., $\pi_k$, the equation expressing the relationship between the variables is given by the *Buckingham $\pi$-theorem*.

$$f(\pi_1, \pi_2, \pi_3, ..., \pi_k) = 0 \qquad 2.22$$

$$k = m - n \qquad 2.23$$

The dimensionless pi-groups are usually found from the $m$ variables according to an intuitive process.

---

[12]*Lineal* is best used when discussing a line of succession (e.g., a lineal descendant of a particular person). *Linear* is best used when discussing length (e.g., a linear dimension of a room).

## Example 2.5

A solid sphere rolls down a submerged incline. Find an equation for the velocity, v.

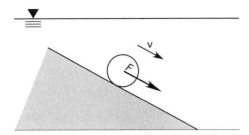

*Solution*

Assume that the velocity depends on the force, $F$, due to gravity, the diameter of the sphere, $D$, the density of the fluid, $\rho$, and the viscosity of the fluid, $\mu$.

$$\text{v} = f(F, D, \rho, \mu) = CF^a D^b \rho^c \mu^d$$

This equation can be written in terms of the primary dimensions of the variables.

$$\frac{L}{\theta} = C\left(\frac{ML}{\theta^2}\right)^a L^b \left(\frac{M}{L^3}\right)^c \left(\frac{M}{L\theta}\right)^d$$

Since $L$ on the left-hand side has an implied exponent of one, a necessary condition is

$$1 = a + b - 3c - d \quad (L)$$

Similarly, the other necessary conditions are

$$-1 = -2a - d \quad (\theta)$$
$$0 = a + c + d \quad (M)$$

Solving simultaneously yields

$$b = -1$$
$$c = a - 1$$
$$d = 1 - 2a$$
$$\text{v} = CF^a D^{-1} \rho^{a-1} \mu^{1-2a}$$
$$= C\left(\frac{\mu}{D\rho}\right)\left(\frac{F\rho}{\mu^2}\right)^a$$

$C$ and $a$ must be determined experimentally.

# 3 Transport Phenomena

## Nomenclature

| | | | |
|---|---|---|---|
| $a$ | acceleration | ft/sec$^2$ | m/s$^2$ |
| $A$ | area | ft$^2$ | m$^2$ |
| $c_m$ | concentration | mol/ft$^3$ | mol/m$^3$ |
| $c_p$ | specific heat | Btu/lbm-°R | kJ/kg·K |
| $D$ | diameter | ft | m |
| $D_m$ | diffusion coefficient | ft$^2$/sec | m$^2$/s |
| $f$ | Moody friction factor | – | – |
| $F$ | force | lbf | N |
| $G$ | mass velocity | lbm/ft$^2$-sec | kg/m$^2$·s |
| $h$ | film coefficient | Btu/hr-ft$^2$-°R | W/m$^2$·K |
| $h$ | head loss | ft | m |
| $j_H$ | heat $j$-number | – | – |
| $j_M$ | mass $j$-number | – | – |
| $k$ | thermal conductivity | Btu/hr-ft$^2$-°R | W/m$^2$·K |
| $L$ | length | ft | m |
| $m$ | mass | lbm | kg |
| $p$ | pressure | lbf/ft$^2$ | Pa |
| Pr | Prandtl number | – | – |
| $\dot{Q}$ | heat transfer | Btu/hr | W |
| Re | Reynolds number | – | – |
| Sc | Schmidt number | – | – |
| Sh | Sherwood number | – | – |
| St | Stanton number | – | – |
| $t$ | time | sec | s |
| $T$ | temperature | °R | K |
| v | velocity | ft/sec | m/s |
| $y$ | distance | ft | m |
| $y$ | radial distance from inner wall to tube centerline | ft | m |

## Symbols

| | | | |
|---|---|---|---|
| $\rho$ | density | lbm/ft$^3$ | kg/m$^3$ |
| $\mu$ | absolute viscosity | lbf-sec/ft$^2$ | N·s/m$^2$ |
| $\tau$ | shear stress | lbf/ft$^2$ | N/m$^2$ |

## Subscripts

| | |
|---|---|
| $f$ | friction |
| $H$ | heat |
| $m$ | mass |
| $M$ | mass |
| $w$ | wall |

## 1. INTRODUCTION

Transport phenomena traditionally include the processes of energy transfer, momentum transfer, and mass transfer. Not all engineering students have studied all of these subjects; for instance, mass transfer is typically taught only to chemical engineers. Some students have studied a subject without recognizing it was a subset of transport phenomena; for example, heat transfer is an energy transport phenomenon.

*Energy transfer* is concerned with the transfer of thermal energy from one location to another. *Mass transfer* is concerned with the transfer of mass from one phase to another. *Momentum transfer* is concerned with the transfer of momentum from one location to another.

All of the transport phenomena have basic concepts in common. The forms of the equations and the methods of solutions are similar. Only the names of the constants and coefficients vary. An analogy can be drawn between the phenomena.

## 2. DIMENSIONLESS GROUPS

A *dimensionless group* is derived as a ratio of two forces or other quantities. Considerable use of dimensionless groups is made in certain subjects, notably fluid mechanics and heat transfer. For example, the Reynolds number, Mach number, and Froude number are used to distinguish between distinctly different flow regimes in pipe flow, compressible flow, and open channel flow, respectively.

Table 3.1 contains information about the most common dimensionless groups used in fluid mechanics and heat transfer.

## 3. UNIT OPERATIONS

There are many applications of transport phenomena. Regardless of the industry or product, transport phenomena can be divided into separate and distinct steps called *unit operations*. The most important unit operations are:

- fluid flow

- heat transfer

- evaporation

- drying

- distillation
- absorption
- membrane separation
- liquid-liquid extraction
- liquid-solid leaching
- crystallization
- mechanical separation processes

## 4. MOMENTUM TRANSFER

*Momentum transfer* is generally studied and referred to as "fluid mechanics." The concepts of flow regime (laminar or turbulent) and viscosity are basic. The flow regime is determined by calculating the *Reynolds number*, Re, which is the ratio of inertia force to viscous force.

$$\text{Re} = \frac{Dv\rho}{\mu} \qquad 3.1$$

When a turbulent fluid flows in a circular tube, the fluid shear stress at the wall is a function of the velocity gradient, $dv/dy$. The constant of proportionality, $\mu$, is the *absolute viscosity*.

$$\frac{F}{A} = \tau = -\mu\frac{dv}{dy} \qquad 3.2$$

Equation 3.2 is taught in every fluid dynamics course. The linear relationship is intrinsic to the definition of a *Newtonian fluid*. However, the connection with momentum transfer may not be obvious. For a constant mass flow rate, *Newton's second law* can be written as

$$\begin{aligned} F = ma &= m\frac{dv}{dt} = \frac{d(mv)}{dt} \\ &= \frac{d(\text{momentum})}{dt} \end{aligned} \qquad 3.3$$

Therefore, the force, $F$, can be interpreted as the flux of fluid momentum in the radial direction (i.e., away from the wall). The *shear stress*, $F/A$, can be interpreted as the flux of fluid momentum in the radial direction (i.e., away from the wall) per unit area. The negative sign in Eq. 3.2 indicates that momentum transfer is high when the velocity is low (i.e., at the wall), and vice versa.

The *Fanning friction factor* is defined as the shear stress divided by the kinetic energy per unit volume. The *Moody friction factor*, $f$, is four times the Fanning friction factor, hence the factor "4" in Eq. 3.4.

$$f = \frac{4\tau}{\frac{\frac{1}{2}mv^2}{V}} = \frac{8\tau}{\rho v^2} \qquad 3.4$$

The Moody friction factor is used to calculate friction loss. The traditional *Darcy equation* for friction head loss along a length, $L$, of pipe is

$$h_f = \frac{\Delta p}{\rho g} = \frac{fLv^2}{2Dg} \qquad 3.5$$

Since $h_f = \Delta p/\rho g$, the Moody friction factor can also be written as

$$f = \frac{2D\Delta p}{\rho L v^2} \qquad 3.6$$

Combining Eq. 3.2, Eq. 3.4, and Eq. 3.6, the momentum transfer in the radial direction per unit area is

$$\begin{aligned} \frac{F}{A} = \tau &= -\mu\frac{dv}{dy} \\ &= \frac{-f\rho v^2}{8} = \frac{-D\Delta p}{4L} \end{aligned} \qquad 3.7$$

## 5. ENERGY TRANSFER

Energy transfer is generally referred to as "heat transfer." The concepts of transfer regime (laminar or turbulent) and thermal conductivity are basic. An important dimensionless number is the *Prandtl number*, Pr. The Prandtl number is the ratio of the shear component of momentum diffusivity to the heat diffusivity (i.e., the ratio of diffusion of momentum to the diffusion of heat).

$$\text{Pr} = \frac{c_p\mu}{k} \qquad 3.8$$

The *heat flux* depends on the *thermal gradient* (i.e., a *temperature gradient*), $dT/dy$, in a material. The constant of proportionality, $k$, is the *thermal conductivity* of the substance.

$$\frac{\dot{Q}}{A} = -k\frac{dT}{dy} \qquad 3.9$$

While Eq. 3.9 is valid specifically for conduction in any substance, the relationship is somewhat different for heat transfer to or from a fluid moving past a solid surface (i.e., a turbulent fluid moving through a tube). The *heat transfer film coefficient*, $h$, usually just referred to as the *film coefficient*, is defined as the rate of heat transfer per degree temperature of difference. Unlike with Eq. 3.9, it is not necessary to know the thickness of the film in order to use the film coefficient. In Eq. 3.10 and Eq. 3.11, $\Delta T$ is the difference in tube wall and bulk fluid temperatures.

$$h = \frac{\dfrac{\dot{Q}}{A}}{\Delta T} \qquad 3.10$$

$$\frac{\dot{Q}}{A} = h\Delta T \qquad 3.11$$

## 6. MASS TRANSFER

Mass transfer is generally studied by chemical engineers as background for designing and analyzing such processes as distillation, absorption, drying, and extraction. Mass transfer deals with the migration of a single substance from one location or phase to another.

Quantities of mass transfer are usually calculated per unit area. Although the units of molecules per unit volume (per unit time) could be used, it is more common to work with units of moles per unit volume (per unit time). For very large quantities, units of mass per unit area (per unit time) can be used. The symbol $G$ and the names *mass velocity* or *mass flow rate per unit area* are generally used to represent mass transfer in units of lbm/ft²-sec or kg/m²·s.

The flow regime for the *diffusion* of a small amount of substance 1 through substance 2 (i.e., a dilute "mixture" or *dilute solution*) is determined by calculating the *Schmidt number*, Sc. This is the dimensionless ratio of the *molecular momentum diffusivity* ($\mu/\rho$) to the *molecular mass diffusivity*, quantified by the *diffusion coefficient* for substance 1 moving through substance 2, $D_m$. Typical values for gases range from 0.5 to 2.0. For liquids, the range is about 100 to more than 10,000 for viscous liquids. The Schmidt number is essentially independent of temperature and pressure within "normal" operating conditions.

$$\text{Sc} = \frac{\mu}{\rho D_m} \qquad 3.12$$

*Fick's law* describes molecular diffusion of mass for dilute solutions. The number of molecules of substance 1 moving through substance 2 (i.e., the *molecular diffusion*) per unit area in the $y$-direction is given by Eq. 3.13. $c_m$ is the concentration of substance 1 at any particular point. $dc_m/dy$ is the *concentration gradient* of substance 1 in the $y$-direction. When moles are the units used, $N/A$ is referred to as the *molar flux*.

$$\frac{N}{A} = -D_m \frac{dc_m}{dy} \qquad 3.13$$

For a fluid flowing past a solid surface (as a turbulent fluid flowing in a pipe), the relationship for the inward mass transfer flux, $N/A$, at the wall is very similar to Eq. 3.11. $k_m$ is the *mass transfer coefficient* for the process, and $\Delta c_m$ is the difference in the concentrations between the wall and the bulk fluid.

$$\frac{N}{A} = k_m \Delta c_m \qquad 3.14$$

For convenience, two other dimensionless numbers are used with mass transfer. The dimensionless *Sherwood number*, Sh, is the ratio of *mass diffusivity* to molecular diffusivity and is given by Eq. 3.15. For turbulent flow through a pipe, $D$ is the pipe inside diameter.

$$\text{Sh} = \frac{k_m D}{D_m} \qquad 3.15$$

The *Stanton number*, St, is the ratio of heat transfer at the wall to the energy transported by the mass flow.

$$\text{St} = \frac{h}{c_p G} \qquad 3.16$$

## 7. TRANSPORT PHENOMENA ANALOGIES

It is clear that the structures of the equations that describe the three transport processes are similar. In that sense, momentum, energy, and mass transport processes are *analogous* processes. However, the phrase "transport phenomena analogy" actually has a slightly different meaning.

In some fluid processes, two or three transport processes occur simultaneously. For example, a turbulent fluid flowing through a cooled pipe will experience both frictional and thermal energy losses. It is convenient to use known data from one of the processes to predict the performance of the other process(es). For example, since fluid friction factors (a momentum transport property) are easily calculated, it is desirable to use them to predict the (more difficult) heat and mass transfer rates. Such correlations between the performance data are known as "analogies."

In the popular Chilton and Colburn analogy, several dimensionless numbers are correlated with dimensionless "*j*-factors" and friction factors.

$$j_H = (\text{St})(\text{Pr})^{2/3} = \frac{f}{8} \qquad 3.17$$

$$j_M = (\text{Sh})(\text{Sc})^{2/3} = \frac{f}{8} \qquad 3.18$$

Equation 3.17 and Eq. 3.18 are supported by theoretical derivations for laminar flow over flat plates as well as experimental data. For turbulent flow, the analogies are supported by experimental data for liquids and gases.

**Table 3.1** *Common Dimensionless Groups*

| name | symbol | formula | interpretation |
|------|--------|---------|----------------|
| Biot number | Bi | $\dfrac{hL}{k_s}$ | $\dfrac{\text{surface conductance}}{\text{internal conduction of solid}}$ |
| Cauchy number | Ca | $\dfrac{\text{v}^2}{\dfrac{B_s}{\rho}} = \dfrac{\text{v}^2}{a^2}$ | $\dfrac{\text{inertia force}}{\text{compressive force}} = \text{Mach number}^2$ |
| Eckert number | Ec | $\dfrac{\text{v}^2}{2c_p\Delta T}$ | $\dfrac{\text{temperature rise due to energy conversion}}{\text{temperature difference}}$ |
| Eötvös number | Eo | $\dfrac{\rho g L^2}{\sigma}$ | $\dfrac{\text{buoyancy}}{\text{surface tension}}$ |
| Euler number | Eu | $\dfrac{\Delta p}{\rho \text{v}^2}$ | $\dfrac{\text{pressure force}}{\text{inertia force}}$ |
| Fourier number | Fo | $\dfrac{kt}{\rho c_p L^2} = \dfrac{\alpha t}{L^2}$ | $\dfrac{\text{rate of conduction of heat}}{\text{rate of storage of energy}}$ |
| Froude number[*] | Fr | $\dfrac{\text{v}^2}{gL}$ | $\dfrac{\text{inertia force}}{\text{gravity force}}$ |
| Graetz number[*] | Gz | $\left(\dfrac{D}{L}\right)\left(\dfrac{\text{v}\rho c_p D}{k}\right)$ | $\dfrac{(\text{Re})(\text{Pr})}{L/D}$<br>$\dfrac{\text{heat transfer by convection in entrance region}}{\text{heat transfer by conduction}}$ |
| Grashof number[*] | Gr | $\dfrac{g\beta\Delta T L^3}{\nu^2}$ | $\dfrac{\text{buoyancy force}}{\text{viscous force}}$ |
| Knudsen number | Kn | $\dfrac{\lambda}{L}$ | $\dfrac{\text{mean free path of molecules}}{\text{characteristic length of object}}$ |
| Lewis number[*] | Le | $\dfrac{\alpha}{D_c}$ | $\dfrac{\text{thermal diffusivity}}{\text{molecular diffusivity}}$ |
| Mach number | M | $\dfrac{\text{v}}{a}$ | $\dfrac{\text{macroscopic velocity}}{\text{speed of sound}}$ |
| Nusselt number | Nu | $\dfrac{hL}{k}$ | $\dfrac{\text{temperature gradient at wall}}{\text{overall temperature difference}}$ |
| Péclet number | Pé | $\dfrac{\text{v}\rho c_p D}{k}$ | $\dfrac{(\text{Re})(\text{Pr})}{\dfrac{\text{heat transfer by convection}}{\text{heat transfer by conduction}}}$ |
| Prandtl number | Pr | $\dfrac{\mu c_p}{k} = \dfrac{\nu}{\alpha}$ | $\dfrac{\text{diffusion of momentum}}{\text{diffusion of heat}}$ |
| Reynolds number | Re | $\dfrac{\rho \text{v} L}{\mu} = \dfrac{\text{v} L}{\nu}$ | $\dfrac{\text{inertia force}}{\text{viscous force}}$ |
| Schmidt number | Sc | $\dfrac{\mu}{\rho D_c} = \dfrac{\nu}{D_c}$ | $\dfrac{\text{diffusion of momentum}}{\text{diffusion of mass}}$ |
| Sherwood number[*] | Sh | $\dfrac{k_D L}{D_c}$ | $\dfrac{\text{mass diffusivity}}{\text{molecular diffusivity}}$ |
| Stanton number | St | $\dfrac{h}{\text{v}\rho c_p} = \dfrac{h}{c_p G}$ | $\dfrac{\text{heat transfer at wall}}{\text{energy transported by stream}}$ |
| Stokes number | Sk | $\dfrac{\Delta p L}{\mu \text{v}}$ | $\dfrac{\text{pressure force}}{\text{viscous force}}$ |
| Strouhal number[*] | Sl | $\dfrac{L}{t\text{v}} = \dfrac{L\omega}{\text{v}}$ | $\dfrac{\text{frequency of vibration}}{\text{characteristic frequency}}$ |
| Weber number | We | $\dfrac{\rho \text{v}^2 L}{\sigma}$ | $\dfrac{\text{inertia force}}{\text{surface tension force}}$ |

[*]Multiple definitions exist.

# 4 Energy, Work, and Power

## Nomenclature

| | | | |
|---|---|---|---|
| $c$ | specific heat | Btu/lbm-°F | J/kg·°C |
| $C$ | molar specific heat | Btu/lbmol-°F | J/kmol·°C |
| $E$ | energy | ft-lbf | J |
| $F$ | force | lbf | N |
| $g$ | gravitational acceleration | ft/sec$^2$ | m/s$^2$ |
| $g_c$ | gravitational constant | ft-lbm/lbf-sec$^2$ | n.a. |
| $h$ | height | ft | m |
| $I$ | mass moment of inertia | lbm-ft$^2$ | kg·m$^2$ |
| $J$ | Joule's constant | ft-lbf/Btu | n.a. |
| $k$ | spring constant | lbf/ft | N/m |
| $m$ | mass | lbm | kg |
| MW | molecular weight | lbm/lbmol | kg/kmol |
| $p$ | pressure | lbf/ft$^2$ | Pa |
| $P$ | power | ft-lbf/sec | W |
| $q$ | heat | Btu/lbm | J/kg |
| $Q$ | heat | Btu | J |
| $r$ | radius | ft | m |
| $s$ | distance | ft | m |
| $t$ | time | sec | s |
| $T$ | temperature | °F | °C |
| $T$ | torque | ft-lbf | N·m |
| $u$ | specific energy | ft-lbf/lbm | J/kg |
| $U$ | internal energy | Btu | J |
| $v$ | velocity | ft/sec | m/s |
| $W$ | work | ft-lbf | J |

## Symbols

| | | | |
|---|---|---|---|
| $\delta$ | deflection | ft | m |
| $\eta$ | efficiency | – | – |
| $\theta$ | angular position | rad | rad |
| $\rho$ | mass density | lbm/ft$^3$ | kg/m$^3$ |
| $\upsilon$ | specific volume | ft$^3$/lbm | m$^3$/kg |
| $\phi$ | angle | deg | deg |
| $\omega$ | angular velocity | rad/sec | rad/s |

## Subscripts

| | |
|---|---|
| $f$ | frictional |
| $p$ | constant pressure |
| $v$ | constant volume |

## 1. ENERGY OF A MASS

The *energy* of a mass represents the capacity of the mass to do work. Such energy can be stored and released. There are many forms that it can take, including mechanical, thermal, electrical, and magnetic energies. Energy is a positive, scalar quantity (although the change in energy can be either positive or negative).

The total energy of a body can be calculated from its mass, $m$, and its *specific energy*, $u$ (i.e., the energy per unit mass).[1]

$$E = mu \qquad 4.1$$

Typical units of mechanical energy are foot-pounds and joules. (A joule is equivalent to the units of N·m and kg·m$^2$/s$^2$.) In traditional English-unit countries, the *British thermal unit* (Btu) is used for thermal energy, whereas the kilocalorie (kcal) is still used in some applications in SI countries. *Joule's constant* or the *Joule equivalent* (778.26 ft-lbf/Btu, usually shortened to 778, three significant digits) is used to convert between English mechanical and thermal energy units.

$$\text{energy in Btu} = \frac{\text{energy in ft-lbf}}{J} \qquad 4.2$$

Two other units of large amounts of energy are the therm and the quad. A *therm* is $10^5$ Btu ($1.055 \times 10^8$ J). A *quad* is equal to a quadrillion ($10^{15}$) Btu. This is $1.055 \times 10^{18}$ J, or roughly the energy contained in 200 million barrels of oil.

## 2. LAW OF CONSERVATION OF ENERGY

The *law of conservation of energy* says that energy cannot be created or destroyed. However, energy can be converted into different forms. Therefore, the sum of all energy forms is constant.

$$\sum E = \text{constant} \qquad 4.3$$

---

[1]The use of symbols $E$ and $U$ for energy is not consistent in the engineering field.

## 3. WORK

*Work*, $W$, is the act of changing the energy of a particle, body, or system. For a mechanical system, *external work* is work done by an external force, whereas *internal work* is done by an internal force. Work is a signed, scalar quantity. Typical units are inch-pounds, foot-pounds, and joules. Mechanical work is seldom expressed in British thermal units or kilocalories.

For a mechanical system, work is positive when a force acts in the direction of motion and helps a body move from one location to another. Work is negative when a force acts to oppose motion. (Friction, for example, always opposes the direction of motion and can do only negative work.) The work done on a body by more than one force can be found by superposition.

From a thermodynamic standpoint, work is positive if a particle or body does work on its surroundings. Work is negative if the surroundings do work on the object. (Thus, blowing up a balloon represents negative work to the balloon.) Although this may be a difficult concept, it is consistent with the conservation of energy, since the sum of negative work and the positive energy increase is zero (i.e., no net energy change in the system).[2]

The work performed by a variable force or torque is calculated from the dot products of Eq. 4.4 and Eq. 4.5,

$$W_{\text{variable force}} = \int \mathbf{F} \cdot d\mathbf{s} \quad \text{[linear systems]} \qquad 4.4$$

$$W_{\text{variable torque}} = \int \mathbf{T} \cdot d\theta \quad \text{[rotational systems]} \qquad 4.5$$

The work done by a force or torque of constant magnitude is

$$W_{\text{constant force}} = \mathbf{F} \cdot \mathbf{s} = Fs\cos\phi \quad \text{[linear systems]} \qquad 4.6$$

$$W_{\text{constant torque}} = \mathbf{T} \cdot \theta$$
$$= Fr\theta\cos\phi \quad \text{[rotational systems]} \qquad 4.7$$

The nonvector forms, Eq. 4.6 and Eq. 4.7, illustrate that only the component of force or torque in the direction of motion contributes to work. (See Fig. 4.1.)

Common applications of the work done by a constant force are frictional work and gravitational work. The

**Figure 4.1** *Work of a Constant Force*

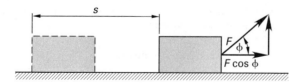

work to move an object a distance $s$ against a frictional force of $F_f$ is

$$W_{\text{friction}} = F_f s \qquad 4.8$$

The work done against gravity when a mass $m$ changes in elevation from $h_1$ to $h_2$ is

$$W_{\text{gravity}} = mg(h_2 - h_1) \qquad \text{[SI]} \quad 4.9(a)$$

$$W_{\text{gravity}} = \left(\frac{mg}{g_c}\right)(h_2 - h_1) \qquad \text{[U.S.]} \quad 4.9(b)$$

The work done by or on a *linear spring* whose length or deflection changes from $\delta_1$ to $\delta_2$ is given by Eq. 4.10.[3] It does not make any difference whether the spring is a compression spring or an extension spring.

$$W_{\text{spring}} = \tfrac{1}{2}k(\delta_2^2 - \delta_1^2) \qquad 4.10$$

### Example 4.1

A lawn mower engine is started by pulling a cord wrapped around a sheave. The sheave radius is 8.0 cm. The cord is wrapped around the sheave two times. If a constant tension of 90 N is maintained in the cord during starting, what work is done?

*Solution*

The starting torque on the engine is

$$T = Fr = (90 \text{ N})\left(\frac{8 \text{ cm}}{100 \frac{\text{cm}}{\text{m}}}\right)$$
$$= 7.2 \text{ N·m}$$

The cord wraps around the sheave $(2)(2\pi) = 12.6$ rad. From Eq. 4.7, the work done by a constant torque is

$$W = T\theta = (7.2 \text{ N·m})(12 \text{ rad})$$
$$= 90.7 \text{ J}$$

### Example 4.2

A 200 lbm crate is pushed 25 ft at constant velocity across a warehouse floor. There is a frictional force of 60 lbf between the crate and floor. What work is done by the frictional force on the crate?

*Solution*

From Eq. 4.8,

$$W_{\text{friction}} = F_f s = (60 \text{ lbf})(25 \text{ ft})$$
$$= 1500 \text{ ft-lbf}$$

---

[2]This is just a partial statement of the *first law of thermodynamics*.

[3]A *linear spring* is one for which the linear relationship $F = kx$ is valid.

## 4. POTENTIAL ENERGY OF A MASS

*Potential energy* (*gravitational energy*) is a form of mechanical energy possessed by a body due to its relative position in a gravitational field. Potential energy is lost when the elevation of a body decreases. The lost potential energy usually is converted to kinetic energy or heat.

$$E_{\text{potential}} = mgh \qquad \text{[SI]} \quad \textit{4.11(a)}$$

$$E_{\text{potential}} = \frac{mgh}{g_c} \qquad \text{[U.S.]} \quad \textit{4.11(b)}$$

In the absence of friction and other nonconservative forces, the change in potential energy of a body is equal to the work required to change the elevation of the body.

$$W = \Delta E_{\text{potential}} \qquad \textit{4.12}$$

## 5. KINETIC ENERGY OF A MASS

*Kinetic energy* is a form of mechanical energy associated with a moving or rotating body. The kinetic energy of a body moving with instantaneous linear velocity v is

$$E_{\text{kinetic}} = \tfrac{1}{2}mv^2 \qquad \text{[SI]} \quad \textit{4.13(a)}$$

$$E_{\text{kinetic}} = \frac{mv^2}{2g_c} \qquad \text{[U.S.]} \quad \textit{4.13(b)}$$

A body can also have rotational kinetic energy.

$$E_{\text{rotational}} = \tfrac{1}{2}I\omega^2 \qquad \text{[SI]} \quad \textit{4.14(a)}$$

$$E_{\text{rotational}} = \frac{I\omega^2}{2g_c} \qquad \text{[U.S.]} \quad \textit{4.14(b)}$$

According to the *work-energy principle* (see Sec. 4.9), the kinetic energy is equal to the work necessary to initially accelerate a stationary body or to bring a moving body to rest.

$$W = \Delta E_{\text{kinetic}} \qquad \textit{4.15}$$

### Example 4.3

A solid disk flywheel ($I = 200 \text{ kg·m}^2$) is rotating with a speed of 900 rpm. What is its rotational kinetic energy?

*Solution*

The angular rotational velocity is

$$\omega = \frac{(900 \text{ rpm})\left(2\pi \, \dfrac{\text{rad}}{\text{rev}}\right)}{60 \, \dfrac{\text{s}}{\text{min}}} = 94.25 \text{ rad/s}$$

From Eq. 4.14, the rotational kinetic energy is

$$E = \tfrac{1}{2}I\omega^2 = (\tfrac{1}{2})(200 \text{ kg·m}^2)\left(94.25 \, \frac{\text{rad}}{\text{s}}\right)^2$$

$$= 888 \times 10^3 \text{ J} \quad (888 \text{ kJ})$$

## 6. SPRING ENERGY

A spring is an energy storage device, since the spring has the ability to perform work. In a perfect spring, the amount of energy stored is equal to the work required to compress the spring initially. The stored spring energy does not depend on the mass of the spring. Given a spring with spring constant (stiffness) $k$, the *spring energy* is

$$E_{\text{spring}} = \tfrac{1}{2}k\delta^2 \qquad \textit{4.16}$$

### Example 4.4

A body of mass $m$ falls from height $h$ onto a massless, simply supported beam. The mass adheres to the beam. If the beam has a lateral stiffness $k$, what will be the deflection, $\delta$, of the beam?

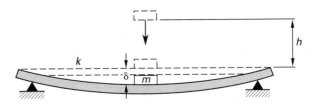

*Solution*

The initial energy of the system consists of only the potential energy of the body. Using consistent units, the change in potential energy is

$$E = mg(h + \delta) \quad \text{[consistent units]}$$

All of this energy is stored in the spring. Therefore,

$$\tfrac{1}{2}k\delta^2 = mg(h + \delta)$$

Solving for the deflection,

$$\delta = \frac{mg \pm \sqrt{mg(2hk + mg)}}{k}$$

## 7. PRESSURE ENERGY OF A MASS

Since work is done in increasing the pressure of a system (e.g., it takes work to blow up a balloon), mechanical energy can be stored in pressure form. This is known as *pressure energy, static energy, flow energy, flow work,* and *p-V work (energy)*. For a system of pressurized mass $m$, the pressure energy is

$$E_{\text{flow}} = \frac{mp}{\rho} = mpv \quad [v = \text{specific volume}] \qquad 4.17$$

## 8. INTERNAL ENERGY OF A MASS

The total internal energy, usually given the symbol $U$, of a body increases when the body's temperature increases.[4] In the absence of any work done on or by the body, the change in internal energy is equal to the heat flow, $Q$, into the body. $Q$ is positive if the heat flow is into the body and negative otherwise.

$$U_2 - U_1 = Q \qquad 4.18$$

The property of internal energy is encountered primarily in thermodynamics problems. Typical units are British thermal units, joules, and kilocalories.

An increase in internal energy is needed to cause a rise in temperature. Different substances differ in the quantity of heat needed to produce a given temperature increase. The ratio of heat, $Q$, required to change the temperature of a mass, $m$, by an amount $\Delta T$ is called the *specific heat (heat capacity)* of the substance, $c$.

Because specific heats of solids and liquids are slightly temperature dependent, the mean specific heats are used when evaluating processes covering a large temperature range.

$$Q = mc\Delta T \qquad 4.19$$

$$c = \frac{Q}{m\Delta T} \qquad 4.20$$

The lowercase $c$ implies that the units are Btu/lbm-°F or J/kg·°C. Typical values of specific heat are given in Table 4.1. The *molar specific heat*, designated by the symbol $C$, has units of Btu/lbmol-°F or J/kmol·°C.

$$C = (\text{MW}) \times c \qquad 4.21$$

For gases, the specific heat depends on the type of process during which the heat exchange occurs. Specific heats for constant-volume and constant-pressure processes are designated by $c_v$ and $c_p$, respectively.

$$Q = mc_v\Delta T \quad [\text{constant-volume process}] \qquad 4.22$$

$$Q = mc_p\Delta T \quad [\text{constant-pressure process}] \qquad 4.23$$

Approximate values of $c_p$ and $c_v$ for solids and liquids are essentially the same. However, the designation $c_p$ is often encountered for solids and liquids. Table 4.1 gives values of $c_p$ for selected liquids and solids.

---

[4]The *thermal energy*, represented by the body's enthalpy, is the sum of internal and pressure energies.

**Table 4.1** *Approximate Specific Heats of Selected Liquids and Solids**

| substance | $c_p$ Btu/lbm-°F | $c_p$ kJ/kg·°C |
|---|---|---|
| aluminum, pure | 0.23 | 0.96 |
| aluminum, 2024-T4 | 0.2 | 0.84 |
| ammonia | 1.16 | 4.86 |
| asbestos | 0.20 | 0.84 |
| benzene | 0.41 | 1.72 |
| brass, red | 0.093 | 0.39 |
| bronze | 0.082 | 0.34 |
| concrete | 0.21 | 0.88 |
| copper, pure | 0.094 | 0.39 |
| Freon-12 | 0.24 | 1.00 |
| gasoline | 0.53 | 2.20 |
| glass | 0.18 | 0.75 |
| gold, pure | 0.031 | 0.13 |
| ice | 0.49 | 2.05 |
| iron, pure | 0.11 | 0.46 |
| iron, cast (4% C) | 0.10 | 0.42 |
| lead, pure | 0.031 | 0.13 |
| magnesium, pure | 0.24 | 1.00 |
| mercury | 0.033 | 0.14 |
| oil, light hydrocarbon | 0.5 | 2.09 |
| silver, pure | 0.06 | 0.25 |
| steel, 1010 | 0.10 | 0.42 |
| steel, stainless 301 | 0.11 | 0.46 |
| tin, pure | 0.055 | 0.23 |
| titanium, pure | 0.13 | 0.54 |
| tungsten, pure | 0.032 | 0.13 |
| water | 1.0 | 4.19 |
| wood (typical) | 0.6 | 2.50 |
| zinc, pure | 0.088 | 0.37 |

(Multiply Btu/lbm-°F by 4.1868 to obtain kJ/kg·°C.)

*Values in cal/g·°C are the same as Btu/lbm-°F. Values in kJ/kg·°C are the same as kJ/kg·K.

## 9. WORK-ENERGY PRINCIPLE

Since energy can neither be created nor destroyed, external work performed on a conservative system goes into changing the system's total energy. This is known as the *work-energy principle* (or *principle of work and energy*).

$$W = \Delta E = E_2 - E_1 \qquad 4.24$$

Generally, the term *work-energy principle* is limited to use with mechanical energy problems (i.e., conversion of work into kinetic or potential energies). When energy is limited to kinetic energy, the work-energy principle is a direct consequence of Newton's second law but is valid for only inertial reference systems.

By directly relating forces, displacements, and velocities, the work-energy principle introduces some simplifications into many mechanical problems.

- It is not necessary to calculate or know the acceleration of a body to calculate the work performed on it.

- Forces that do not contribute to work (e.g., are normal to the direction of motion) are irrelevant.

- Only scalar quantities are involved.

- It is not necessary to individually analyze the particles or component parts in a complex system.

### Example 4.5

A 4000 kg elevator starts from rest, accelerates uniformly to a constant speed of 2.0 m/s, and then decelerates uniformly to a stop 20 m above its initial position. Neglecting friction and other losses, what work was done on the elevator?

*Solution*

By the work-energy principle, the work done on the elevator is equal to the change in the elevator's energy. Since the initial and final kinetic energies are zero, the only mechanical energy change is the potential energy change.

Taking the initial elevation of the elevator as the reference (i.e., $h_1 = 0$),

$$W = E_{2,\text{potential}} - E_{1,\text{potential}} = mg(h_2 - h_1)$$
$$= (4000 \text{ kg})\left(9.81 \ \frac{\text{m}}{\text{s}^2}\right)(20 \text{ m})$$
$$= 785 \times 10^3 \text{ J} \quad (785 \text{ kJ})$$

### 10. CONVERSION BETWEEN ENERGY FORMS

Conversion of one form of energy into another does not violate the conservation of energy law. However, most problems involving conversion of energy are really just special cases of the work-energy principle. For example, consider a falling body that is acted upon by a gravitational force. The conversion of potential energy into kinetic energy can be interpreted as equating the work done by the constant gravitational force to the change in kinetic energy.

In general terms, *Joule's law* states that one energy form can be converted without loss into another. There are two specific formulations of Joule's law. As related to electricity, $P = I^2R = V^2/R$ is the common formulation of Joule's law. As related to thermodynamics and ideal gases, Joule's law states that "the change in internal energy of an ideal gas is a function of the temperature change, not of the volume." This latter form can also be stated more formally as "at constant temperature, the internal energy of a gas approaches a finite value that is independent of the volume as the pressure goes to zero."

### Example 4.6

A 2.0 lbm projectile is launched straight up with an initial velocity of 700 ft/sec. Neglecting air friction, calculate the (a) kinetic energy immediately after launch, (b) kinetic energy at maximum height, (c) potential energy at maximum height, (d) total energy at an elevation where the velocity has dropped to 300 ft/sec, and (e) maximum height attained.

*Solution*

(a) From Eq. 4.13, the kinetic energy is

$$e_{\text{kinetic}} = \frac{mv^2}{2g_c} = \frac{(2 \text{ lbm})\left(700 \ \frac{\text{ft}}{\text{sec}}\right)^2}{(2)\left(32.2 \ \frac{\text{ft-lbm}}{\text{lbf-sec}^2}\right)}$$
$$= 15{,}217 \text{ ft-lbf}$$

(b) The velocity is zero at the maximum height. Therefore, the kinetic energy is zero.

(c) At the maximum height, all of the kinetic energy has been converted into potential energy. Therefore, the potential energy is 15,217 ft-lbf.

(d) Although some of the kinetic energy has been transformed into potential energy, the total energy is still 15,217 ft-lbf.

(e) Since all of the kinetic energy has been converted into potential energy, the maximum height can be found from Eq. 4.11.

$$E_{\text{potential}} = \frac{mgh}{g_c}$$
$$15{,}217 \text{ ft-lbf} = \frac{(2 \text{ lbm})\left(32.2 \ \frac{\text{ft}}{\text{sec}^2}\right)h}{32.2 \ \frac{\text{ft-lbm}}{\text{lbf-sec}^2}}$$
$$h = 7609 \text{ ft}$$

### Example 4.7

A 4500 kg ore car rolls down an incline and passes point A traveling at 1.2 m/s. The ore car is stopped by a spring bumper that compresses 0.6 m. A constant friction force of 220 N acts on the ore car at all times. What spring constant is required?

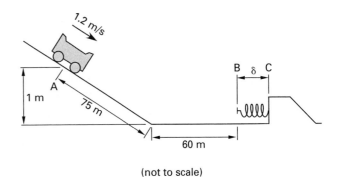

(not to scale)

*Solution*

The car's total energy at point A is the sum of the kinetic and potential energies.

$$E_{\text{total,A}} = E_{\text{kinetic}} + E_{\text{potential}}$$
$$= \tfrac{1}{2}mv^2 + mgh$$
$$= \left(\tfrac{1}{2}\right)(4500 \text{ kg})\left(1.2 \ \frac{\text{m}}{\text{s}}\right)^2$$
$$+ (4500 \text{ kg})\left(9.81 \ \frac{\text{m}}{\text{s}^2}\right)(1 \text{ m})$$
$$= 47\,385 \text{ J}$$

At point B, the potential energy has been converted into additional kinetic energy. However, except for friction, the total energy is the same as at point A. Since the frictional force does negative work, the total energy remaining at point B is

$$E_{\text{total,B}} = E_{\text{total,A}} - W_{\text{friction}}$$
$$= 47\,385 \text{ J} - (220 \text{ N})(75 \text{ m} + 60 \text{ m})$$
$$= 17\,685 \text{ J}$$

At point C, the maximum compression point, the remaining energy has gone into compressing the spring a distance $\delta = 0.6$ m and performing a small amount of frictional work.

$$E_{\text{total,B}} = E_{\text{total,C}} = W_{\text{spring}} + W_{\text{friction}}$$
$$= \tfrac{1}{2}k\delta^2 + F_f\delta$$

$$17\,685 \text{ J} = \left(\tfrac{1}{2}\right)(k)(0.6 \text{ m})^2 + (220 \text{ N})(0.6 \text{ m})$$

The spring constant can be determined directly.

$$k = 97\,520 \text{ N/m} \quad (97.5 \text{ kN/m})$$

## 11. POWER

*Power* is the amount of work done per unit time. It is a scalar quantity. (Although power is calculated from two vectors, the vector dot-product operation is seldom needed.)

$$P = \frac{W}{\Delta t} \qquad \text{4.25}$$

For a body acted upon by a force or torque, the instantaneous power can be calculated from the velocity.

$$P = Fv \quad \text{[linear systems]} \qquad \text{4.26}$$
$$P = T\omega \quad \text{[rotational systems]} \qquad \text{4.27}$$

For a fluid flowing at a rate of $\dot{m}$, the unit of time is already incorporated into the flow rate (e.g., lbm/sec). If the fluid experiences a specific energy change of $\Delta u$, the power generated or dissipated will be

$$P = \dot{m}\Delta u \qquad \text{4.28}$$

Typical basic units of power are ft-lbf/sec and watts (J/s), although *horsepower* is widely used. Table 4.2 can be used to convert units of power.

**Table 4.2** *Useful Power Conversion Formulas*

| | |
|---|---|
| 1 hp | = 550 ft-lbf/sec |
| | = 33,000 ft-lbf/min |
| | = 0.7457 kW |
| | = 0.7068 Btu/sec |
| 1 kW | = 737.6 ft-lbf/sec |
| | = 44,250 ft-lbf/min |
| | = 1.341 hp |
| | = 0.9483 Btu/sec |
| 1 Btu/sec | = 778.26 ft-lbf/sec |
| | = 46,680 ft-lbf/min |
| | = 1.415 hp |

## Example 4.8

When traveling at 100 km/h, a car supplies a constant horizontal force of 50 N to the hitch of a trailer. What tractive power (in horsepower) is required for the trailer alone?

*Solution*

From Eq. 4.26, the power being generated is

$$P = Fv = \frac{(50 \text{ N})\left(100 \ \frac{\text{km}}{\text{h}}\right)\left(1000 \ \frac{\text{m}}{\text{km}}\right)}{\left(60 \ \frac{\text{s}}{\text{min}}\right)\left(60 \ \frac{\text{min}}{\text{h}}\right)\left(1000 \ \frac{\text{W}}{\text{kW}}\right)}$$
$$= 1.389 \text{ kW}$$

Using a conversion from Table 4.2, the horsepower is

$$P = \left(1.341 \ \frac{\text{hp}}{\text{kW}}\right)(1.389 \text{ kW})$$
$$= 1.86 \text{ hp}$$

## 12. EFFICIENCY

For energy-using systems (such as cars, electrical motors, elevators, etc.), the *energy-use efficiency*, $\eta$, of a system is the ratio of an ideal property to an actual property. The property used is commonly work, power, or, for thermodynamics problems, heat. When the rate of work is constant, either work or power can be used to calculate the efficiency. Otherwise, power should be used. Except in rare instances, the numerator and denominator of the ratio must have the same units.[5]

$$\eta = \frac{P_{\text{ideal}}}{P_{\text{actual}}} \quad [P_{\text{actual}} \geq P_{\text{ideal}}] \qquad 4.29$$

For energy-producing systems (such as electrical generators, prime movers, and hydroelectric plants), the *energy-production efficiency* is

$$\eta = \frac{P_{\text{actual}}}{P_{\text{ideal}}} \quad [P_{\text{ideal}} \geq P_{\text{actual}}] \qquad 4.30$$

The efficiency of an *ideal machine* is 1.0 (100%). However, all *real machines* have efficiencies of less than 1.0.

Background and Support

---

[5]The *energy-efficiency ratio* used to evaluate refrigerators, air conditioners, and heat pumps, for example, has units of Btu per watt-hour (Btu/W-hr).

# Topic II: Mathematics

Chapter

# 5 Algebra

## 1. INTRODUCTION

Engineers working in design and analysis encounter mathematical problems on a daily basis. Although algebra and simple trigonometry are often sufficient for routine calculations, there are many instances when certain advanced subjects are needed. This chapter and the following, in addition to supporting the calculations used in other chapters, consolidate the mathematical concepts most often needed by engineers.

## 2. SYMBOLS USED IN THIS BOOK

Many symbols, letters, and Greek characters are used to represent variables in the formulas used throughout this book. These symbols and characters are defined in the nomenclature section of each chapter. However, some of the other symbols in this book are listed in Table 5.2.

## 3. GREEK ALPHABET

Table 5.1 lists the Greek Alphabet.

***Table 5.1*** *The Greek Alphabet*

| | | | | | |
|---|---|---|---|---|---|
| $A$ | $\alpha$ | alpha | $N$ | $\nu$ | nu |
| $B$ | $\beta$ | beta | $\Xi$ | $\xi$ | xi |
| $\Gamma$ | $\gamma$ | gamma | $O$ | $o$ | omicron |
| $\Delta$ | $\delta$ | delta | $\Pi$ | $\pi$ | pi |
| $E$ | $\epsilon$ | epsilon | $P$ | $\rho$ | rho |
| $Z$ | $\zeta$ | zeta | $\Sigma$ | $\sigma$ | sigma |
| $H$ | $\eta$ | eta | $T$ | $\tau$ | tau |
| $\Theta$ | $\theta$ | theta | $\Upsilon$ | $\upsilon$ | upsilon |
| $I$ | $\iota$ | iota | $\Phi$ | $\phi$ | phi |
| $K$ | $\kappa$ | kappa | $X$ | $\chi$ | chi |
| $\Lambda$ | $\lambda$ | lambda | $\Psi$ | $\psi$ | psi |
| $M$ | $\mu$ | mu | $\Omega$ | $\omega$ | omega |

## 4. TYPES OF NUMBERS

The *numbering system* consists of three types of numbers: real, imaginary, and complex. *Real numbers*, in turn, consist of rational numbers and irrational numbers. *Rational real numbers* are numbers that can be written as the ratio of two integers (e.g., 4, $^2/_5$, and $^1/_3$).[1] *Irrational real numbers* are nonterminating, nonrepeating numbers that cannot be expressed as the ratio of two integers (e.g., $\pi$ and $\sqrt{2}$). Real numbers can be positive or negative.

*Imaginary numbers* are square roots of negative numbers. The symbols $i$ and $j$ are both used to represent the square root of $-1$.[2] For example, $\sqrt{-5} = \sqrt{5}\sqrt{-1} = \sqrt{5}i$. *Complex numbers* consist of combinations of real and imaginary numbers (e.g., $3 - 7i$).

## 5. SIGNIFICANT DIGITS

The significant digits in a number include the leftmost, nonzero digits to the rightmost digit written. Final answers from computations should be rounded off to the number of decimal places justified by the data. The answer can be no more accurate than the least accurate number in the data. Of course, rounding should

---

[1]Notice that 0.3333333 is a nonterminating number, but as it can be expressed as a ratio of two integers (i.e., 1/3), it is a rational number.
[2]The symbol $j$ is used to represent the square root of $-1$ in electrical calculations to avoid confusion with the current variable, $i$.

Mathematics

**Table 5.2** Symbols Used in This Book

| symbol | name | use | example |
|--------|------|-----|---------|
| $\Sigma$ | sigma | series summation | $\sum\limits_{i=1}^{3} x_i = x_1 + x_2 + x_3$ |
| $\pi$ | pi | 3.1415927... | $p = \pi D$ |
| $e$ | base of natural logs | 2.71828... | |
| $\Pi$ | pi | series multiplication | $\prod\limits_{i=1}^{3} x_i = x_1 x_2 x_3$ |
| $\Delta$ | delta | change in quantity | $\Delta h = h_2 - h_1$ |
| $-$ | over bar | average value | $\bar{x}$ |
| $\cdot$ | over dot | per unit time | $\dot{m}$ = mass flowing per second |
| ! | factorial[a] | | $x! = x(x-1)(x-2)...(2)(1)$ |
| \| \| | absolute value[b] | | $\lvert -3 \rvert = +3$ |
| $\sim$ | similarity | | $\Delta ABC \sim \Delta DEF$ |
| $\approx$ | approximately equal to | | $x \approx 1.5$ |
| $\cong$ | congruency | | $ST \cong UV$ |
| $\propto$ | proportional to | | $x \propto y$ |
| $\equiv$ | equivalent to | | $a + bi \equiv r e^{i\theta}$ |
| $\infty$ | infinity | | $x \to \infty$ |
| log | base 10 logarithm | | $\log(5.74)$ |
| ln | natural logarithm | | $\ln(5.74)$ |
| exp | exponential power | | $\exp(x) = e^x$ |
| rms | root-mean-square | $\sqrt{\dfrac{1}{n}\sum\limits_{i=1}^{n} x_i^2}$ | $V_{\text{rms}}$ |
| $\angle$ | phasor or angle | | $\angle 53°$ |

[a]*Zero factorial* (0!) is frequently encountered in the form of $(n-n)!$ when calculating permutations and combinations. Zero factorial is defined as 1.
[b]The notation abs$(x)$ is also used to indicate the absolute value.

**Table 5.3** Examples of Significant Digits

| number as written | number of significant digits | implied range |
|-------------------|------------------------------|---------------|
| 341 | 3 | 340.5 to 341.5 |
| 34.1 | 3 | 34.05 to 34.15 |
| 0.00341 | 3 | 0.003405 to 0.003415 |
| $341 \times 10^7$ | 3 | $340.5 \times 10^7$ to $341.5 \times 10^7$ |
| $3.41 \times 10^{-2}$ | 3 | $3.405 \times 10^{-2}$ to $3.415 \times 10^{-2}$ |
| 3410 | 3 | 3405 to 3415 |
| 3410[*] | 4 | 3409.5 to 3410.5 |
| 341.0 | 4 | 340.95 to 341.05 |

[*]It is permitted to write "3410." to distinguish the number from its 3-significant digit form, although this is rarely done.

be done on final calculation results only. It should not be done on interim results.

There are two ways that significant digits can affect calculations. For the operations of multiplication and division, the final answer is rounded to the number of significant digits in the least significant multiplicand, divisor, or dividend. So, $2.0 \times 13.2 = 26$ since the first multiplicand (2.0) has two significant digits only.

For the operations of addition and subtraction, the final answer is rounded to the position of the least significant digit in the addenda, minuend, or subtrahend. So, $2.0 + 13.2 = 15.2$ because both addenda are significant to the tenth's position; but $2 + 13.4 = 15$ since the 2 is significant only in the ones' position.

The multiplication rule should not be used for addition or subtraction, as this can result in strange answers. For example, it would be incorrect to round $1700 + 0.1$ to 2000 simply because 0.1 has only one significant digit. Table 5.3 gives examples of significant digits.

## 6. EQUATIONS

An *equation* is a mathematical statement of equality, such as $5 = 3 + 2$. *Algebraic equations* are written in terms of *variables*. In the equation $y = x^2 + 3$, the value of variable $y$ depends on the value of variable $x$. Therefore, $y$ is the *dependent variable* and $x$ is the *independent variable*. The dependency of $y$ on $x$ is clearer when the equation is written in *functional form*: $y = f(x)$.

A *parametric equation* uses one or more independent variables (*parameters*) to describe a function.[3] For example, the parameter $\theta$ can be used to write the parametric equations of a unit circle.

$$x = \cos\theta \qquad \qquad 5.1$$

$$y = \sin\theta \qquad \qquad 5.2$$

---

[3]As used in this section, there is no difference between a parameter and an independent variable. However, the term *parameter* is also used as a descriptive measurement that determines or characterizes the form, size, or content of a function. For example, the radius is a parameter of a circle, and mean and variance are parameters of a probability distribution. Once these parameters are specified, the function is completely defined.

A unit circle can also be described by a *nonparametric equation*.[4]

$$x^2 + y^2 = 1 \qquad 5.3$$

# 7. FUNDAMENTAL ALGEBRAIC LAWS

Algebra provides the rules that allow complex mathematical relationships to be expanded or condensed. Algebraic laws may be applied to complex numbers, variables, and real numbers. The general rules for changing the form of a mathematical relationship are given as follows.

- commutative law for addition:

$$A + B = B + A \qquad 5.4$$

- commutative law for multiplication:

$$AB = BA \qquad 5.5$$

- associative law for addition:

$$A + (B + C) = (A + B) + C \qquad 5.6$$

- associative law for multiplication:

$$A(BC) = (AB)C \qquad 5.7$$

- distributive law:

$$A(B + C) = AB + AC \qquad 5.8$$

# 8. POLYNOMIALS

A *polynomial* is a rational expression—usually the sum of several variable terms known as *monomials*—that does not involve division. The *degree of the polynomial* is the highest power to which a variable in the expression is raised. The following *standard polynomial forms* are useful when trying to find the roots of an equation.

$$(a + b)(a - b) = a^2 - b^2 \qquad 5.9$$

$$(a \pm b)^2 = a^2 \pm 2ab + b^2 \qquad 5.10$$

$$(a \pm b)^3 = a^3 \pm 3a^2b + 3ab^2 \pm b^3 \qquad 5.11$$

$$(a^3 \pm b^3) = (a \pm b)(a^2 \mp ab + b^2) \qquad 5.12$$

$$(a^n - b^n) = (a - b)\left( \begin{array}{c} a^{n-1} + a^{n-2}b + a^{n-3}b^2 \\ + \cdots + b^{n-1} \end{array} \right)$$

[$n$ is any positive integer] $\qquad 5.13$

$$(a^n + b^n) = (a + b)\left( \begin{array}{c} a^{n-1} - a^{n-2}b + a^{n-3}b^2 \\ - \cdots + b^{n-1} \end{array} \right)$$

[$n$ is any positive odd integer] $\qquad 5.14$

The *binomial theorem* defines a polynomial of the form $(a + b)^n$.

$$(a + b)^n = \underset{[i=0]}{a^n} + \underset{[i=1]}{na^{n-1}b} + \underset{[i=2]}{C_2 a^{n-2}b^2} + \cdots$$
$$+ C_i a^{n-i}b^i + \cdots + nab^{n-1} + b^n \qquad 5.15$$

$$C_i = \frac{n!}{i!(n-1)!} \qquad [i = 0,\ 1,\ 2, ..., n] \qquad 5.16$$

The coefficients of the expansion can be determined quickly from *Pascal's triangle*. (See Fig. 5.1.) Notice that each entry is the sum of the two entries directly above it.

**Figure 5.1** *Pascal's Triangle*

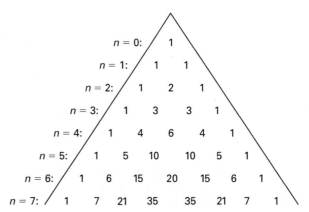

The values $r_1$, $r_2$, ..., $r_n$ of the independent variable $x$ that satisfy a polynomial equation $f(x) = 0$ are known as *roots* or *zeros* of the polynomial. A polynomial of degree $n$ with real coefficients will have at most $n$ real roots, although they need not all be distinctly different.

# 9. ROOTS OF QUADRATIC EQUATIONS

A *quadratic equation* is an equation of the general form $ax^2 + bx + c = 0$ [$a \neq 0$]. The *roots*, $x_1$ and $x_2$, of the equation are the two values of $x$ that satisfy it.

$$x_1, x_2 = \frac{-b \pm \sqrt{b^2 - 4ac}}{2a} \qquad 5.17$$

$$x_1 + x_2 = -\frac{b}{a} \qquad 5.18$$

$$x_1 x_2 = \frac{c}{a} \qquad 5.19$$

---

[4]Since only the coordinate variables are used, this equation is also said to be in *Cartesian equation form*.

Mathematics

The types of roots of the equation can be determined from the *discriminant* (i.e., the quantity under the radical in Eq. 5.17).

- If $(b^2 - 4ac) > 0$, the roots are real and unequal.

- If $(b^2 - 4ac) = 0$, the roots are real and equal. This is known as a *double root*.

- If $(b^2 - 4ac) < 0$, the roots are complex and unequal.

## 10. ROOTS OF GENERAL POLYNOMIALS

It is more difficult to find roots of cubic and higher-degree polynomials because few general techniques exist.

- *inspection*: Finding roots by inspection is equivalent to making reasonable guesses about the roots and substituting into the polynomial.

- *graphing*: If the value of a polynomial $f(x)$ is calculated and plotted for different values of $x$, an approximate value of a root can be determined as the value of $x$ at which the plot crosses the $x$-axis.

- *numerical methods*: If an approximate value of a root is known, numerical methods (bisection method, Newton's method, etc.) can be used to refine the value. The more efficient techniques are too complex to be performed by hand.

- *factoring*: If at least one root (say, $x = r$) of a polynomial $f(x)$ is known, the quantity $(x - r)$ can be factored out of $f(x)$ by long division. The resulting quotient will be lower by one degree, and the remaining roots may be easier to determine. This method is particularly applicable if the polynomial is in one of the standard forms presented in Sec. 5.8.

- *special cases*: Certain polynomial forms can be simplified by substitution or solved by standard formulas if they are recognized as being special cases. (The standard solution to the quadratic equation is such a special case.) For example, $ax^4 + bx^2 + c = 0$ can be reduced to a polynomial of degree 2 if the substitution $u = x^2$ is made.

## 11. EXTRANEOUS ROOTS

With simple equalities, it may appear possible to derive roots by basic algebraic manipulations.[5] However, multiplying each side of an equality by a power of a variable may introduce *extraneous roots*. Such roots do not satisfy the original equation even though they are derived according to the rules of algebra. Checking a calculated root is always a good idea, but is particularly necessary if the equation has been multiplied by one of its own variables.

---

[5]In this sentence, *equality* means a combination of two expressions containing an equal sign. Any two expressions can be linked in this manner, even those that are not actually equal. For example, the expressions for two non-intersecting ellipses can be equated even though there is no intersection point. Finding extraneous roots is more likely when the underlying equality is false to begin with.

## Example 5.1

Use algebraic operations to determine a value that satisfies the following equation. Determine if the value is a valid or extraneous root.

$$\sqrt{x - 2} = \sqrt{x} + 2$$

*Solution*

Square both sides.

$$x - 2 = x + 4\sqrt{x} + 4$$

Subtract $x$ from each side, and combine the constants.

$$4\sqrt{x} = -6$$

Solve for $x$.

$$x = \left(\frac{-6}{4}\right)^2 = \frac{9}{4}$$

Substitute $x = 9/4$ into the original equation.

$$\sqrt{\frac{9}{4} - 2} = \sqrt{\frac{9}{4}} + 2$$

$$\frac{1}{2} = \frac{7}{2}$$

Since the equality is not established, $x = 9/4$ is an extraneous root.

## 12. DESCARTES' RULE OF SIGNS

*Descartes' rule of signs* determines the maximum number of positive (and negative) real roots that a polynomial will have by counting the number of sign reversals (i.e., changes in sign from one term to the next) in the polynomial. The polynomial $f(x) = 0$ must have real coefficients and must be arranged in terms of descending powers of $x$.

- The number of positive roots of the polynomial equation $f(x) = 0$ will not exceed the number of sign reversals.

- The difference between the number of sign reversals and the number of positive roots is an even number.

- The number of negative roots of the polynomial equation $f(x) = 0$ will not exceed the number of sign reversals in the polynomial $f(-x)$.

- The difference between the number of sign reversals in $f(-x)$ and the number of negative roots is an even number.

## Example 5.2

Determine the possible numbers of positive and negative roots that satisfy the following polynomial equation.

$$4x^5 - 5x^4 + 3x^3 - 8x^2 - 2x + 3 = 0$$

*Solution*

There are four sign reversals, so up to four positive roots exist. To keep the difference between the number of positive roots and the number of sign reversals an even number, the number of positive real roots is limited to zero, two, and four.

Substituting $-x$ for $x$ in the polynomial results in

$$-4x^5 - 5x^4 - 3x^3 - 8x^2 + 2x + 3 = 0$$

There is only one sign reversal, so the number of negative roots cannot exceed one. There must be exactly one negative real root in order to keep the difference to an even number (zero in this case).

## 13. RULES FOR EXPONENTS AND RADICALS

In the expression $b^n = a$, $b$ is known as the *base* and $n$ is the *exponent* or *power*. In Eq. 5.20 through Eq. 5.33, $a$, $b$, $m$, and $n$ are any real numbers with limitations listed.

$$b^0 = 1 \quad [b \neq 0] \tag{5.20}$$

$$b^1 = b \tag{5.21}$$

$$b^{-n} = \frac{1}{b^n} = \left(\frac{1}{b}\right)^n \quad [b \neq 0] \tag{5.22}$$

$$\left(\frac{a}{b}\right)^n = \frac{a^n}{b^n} \quad [b \neq 0] \tag{5.23}$$

$$(ab)^n = a^n b^n \tag{5.24}$$

$$b^{m/n} = \sqrt[n]{b^m} = \left(\sqrt[n]{b}\right)^m \tag{5.25}$$

$$(b^n)^m = b^{nm} \tag{5.26}$$

$$b^m b^n = b^{m+n} \tag{5.27}$$

$$\frac{b^m}{b^n} = b^{m-n} \quad [b \neq 0] \tag{5.28}$$

$$\sqrt[n]{b} = b^{1/n} \tag{5.29}$$

$$\left(\sqrt[n]{b}\right)^n = \left(b^{1/n}\right)^n = b \tag{5.30}$$

$$\sqrt[n]{ab} = \sqrt[n]{a}\sqrt[n]{b} = a^{1/n} b^{1/n}$$
$$= (ab)^{1/n} \tag{5.31}$$

$$\sqrt[n]{\frac{a}{b}} = \frac{\sqrt[n]{a}}{\sqrt[n]{b}} = \left(\frac{a}{b}\right)^{1/n} \quad [b \neq 0] \tag{5.32}$$

$$\sqrt[m]{\sqrt[n]{b}} = \sqrt[mn]{b} = b^{1/mn} \tag{5.33}$$

## 14. LOGARITHMS

Logarithms can be considered to be exponents. For example, the exponent $n$ in the expression $b^n = a$ is the logarithm of $a$ to the base $b$. Therefore, the two expressions $\log_b a = n$ and $b^n = a$ are equivalent.

The base for *common logs* is 10. Usually, "log" will be written when common logs are desired, although "$\log_{10}$" appears occasionally. The base for *natural (Napierian) logs* is 2.71828..., a number which is given the symbol $e$. When natural logs are desired, usually "ln" will be written, although "$\log_e$" is also used.

Most logarithms will contain an integer part (the *characteristic*) and a decimal part (the *mantissa*). The common and natural logarithms of any number less than one are negative. If the number is greater than one, its common and natural logarithms are positive. Although the logarithm may be negative, the mantissa is always positive. For negative logarithms, the characteristic is found by expressing the logarithm as the sum of a negative characteristic and a positive mantissa.

For common logarithms of numbers greater than one, the characteristics will be positive and equal to one less than the number of digits in front of the decimal. If the number is less than one, the characteristic will be negative and equal to one more than the number of zeros immediately following the decimal point.

If a negative logarithm is to be used in a calculation, it must first be converted to *operational form* by adding the characteristic and mantissa. The operational form should be used in all calculations and is the form displayed by scientific calculators.

The logarithm of a negative number is a complex number.

## Example 5.3

Use logarithm tables to determine the operational form of $\log_{10}(0.05)$.

*Solution*

Since the number is less than one and there is one leading zero, the characteristic is found by observation to be $-2$. From a book of logarithm tables, the mantissa of 5.0 is 0.699. Two ways of combining the mantissa and characteristic are possible.

*method 1:* $\bar{2}.699$

*method 2:* $8.699 - 10$

The operational form of this logarithm is $-2 + 0.699 = -1.301$.

## 15. LOGARITHM IDENTITIES

Prior to the widespread availability of calculating devices, logarithm identities were used to solve complex calculations by reducing the solution method to table

Mathematics

look-up, addition, and subtraction. Logarithm identities are still useful in simplifying expressions containing exponentials and other logarithms. In Eq. 5.34 through Eq. 5.45, $a \neq 1$, $b \neq 1$, $x > 0$, and $y > 0$.

$$\log_b(b) = 1 \qquad \qquad 5.34$$

$$\log_b(1) = 0 \qquad \qquad 5.35$$

$$\log_b(b^n) = n \qquad \qquad 5.36$$

$$\log(x^a) = a \log(x) \qquad \qquad 5.37$$

$$\log(\sqrt[n]{x}) = \log(x^{1/n}) = \frac{\log(x)}{n} \qquad 5.38$$

$$b^{n \log_b(x)} = x^n = \text{antilog}\left(n \log_b(x)\right) \qquad 5.39$$

$$b^{\log_b(x)/n} = x^{1/n} \qquad \qquad 5.40$$

$$\log(xy) = \log(x) + \log(y) \qquad \qquad 5.41$$

$$\log\left(\frac{x}{y}\right) = \log(x) - \log(y) \qquad 5.42$$

$$\log_a(x) = \log_b(x)\log_a(b) \qquad \qquad 5.43$$

$$\ln(x) = \log_{10}(x)\ln(10) \approx 2.3026 \log_{10}(x) \qquad 5.44$$

$$\log_{10}(x) = \ln(x)\log_{10}(e) \approx 0.4343 \ln(x) \qquad 5.45$$

### Example 5.4

The surviving fraction, $x$, of a radioactive isotope is given by $x = e^{-0.005t}$. For what value of $t$ will the surviving percentage be 7%?

*Solution*

$$x = 0.07 = e^{-0.005t}$$

Take the natural log of both sides.

$$\ln(0.07) = \ln(e^{-0.005t})$$

From Eq. 5.36, $\ln e^x = x$. Therefore,

$$-2.66 = -0.005t$$

$$t = 532$$

### 16. PARTIAL FRACTIONS

The method of *partial fractions* is used to transform a proper polynomial fraction of two polynomials into a sum of simpler expressions, a procedure known as *resolution* .[6,7] The technique can be considered to be the act of "unadding" a sum to obtain all of the addends.

---

[6]To be a *proper polynomial fraction*, the degree of the numerator must be less than the degree of the denominator. If the polynomial fraction is improper, the denominator can be divided into the numerator to obtain whole and fractional polynomials. The method of partial fractions can then be used to reduce the fractional polynomial.
[7]This technique is particularly useful for calculating integrals and inverse Laplace transforms in subsequent chapters.

Suppose $H(x)$ is a proper polynomial fraction of the form $P(x)/Q(x)$. The object of the resolution is to determine the partial fractions $u_1/v_1$, $u_2/v_2$, etc., such that

$$H(x) = \frac{P(x)}{Q(x)} = \frac{u_1}{v_1} + \frac{u_2}{v_2} + \frac{u_3}{v_3} + \cdots \qquad 5.46$$

The form of the denominator polynomial $Q(x)$ will be the main factor in determining the form of the partial fractions. The task of finding the $u_i$ and $v_i$ is simplified by categorizing the possible forms of $Q(x)$.

*case 1:* $Q(x)$ factors into $n$ different linear terms.

$$Q(x) = (x - a_1)(x - a_2) \cdots (x - a_n) \qquad 5.47$$

Then,

$$H(x) = \sum_{i=1}^{n} \frac{A_i}{x - a_i} \qquad \qquad 5.48$$

*case 2:* $Q(x)$ factors into $n$ identical linear terms.

$$Q(x) = (x - a)(x - a) \cdots (x - a) \qquad 5.49$$

Then,

$$H(x) = \sum_{i=1}^{n} \frac{A_i}{(x - a)^i} \qquad \qquad 5.50$$

*case 3:* $Q(x)$ factors into $n$ different quadratic terms, $x^2 + p_i x + q_i$. Then,

$$H(x) = \sum_{i=1}^{n} \frac{A_i x + B_i}{x^2 + p_i x + q_i} \qquad 5.51$$

*case 4:* $Q(x)$ factors into $n$ identical quadratic terms, $x^2 + px + q$. Then,

$$H(x) = \sum_{i=1}^{n} \frac{A_i x + B_i}{(x^2 + px + q)^i} \qquad 5.52$$

Once the general forms of the partial fractions have been determined from inspection, the *method of undetermined coefficients* is used. The partial fractions are all cross-multiplied to obtain $Q(x)$ as the denominator, and the coefficients are found by equating $P(x)$ and the cross-multiplied numerator.

### Example 5.5

Resolve $H(x)$ into partial fractions.

$$H(x) = \frac{x^2 + 2x + 3}{x^4 + x^3 + 2x^2}$$

*Solution*

Here, $Q(x) = x^4 + x^3 + 2x^2$ factors into $x^2(x^2 + x + 2)$. This is a combination of cases 2 and 3.

$$H(x) = \frac{A_1}{x} + \frac{A_2}{x^2} + \frac{A_3 + A_4 x}{x^2 + x + 2}$$

Cross-multiplying to obtain a common denominator yields

$$\frac{(A_1 + A_4)x^3 + (A_1 + A_2 + A_3)x^2 + (2A_1 + A_2)x + 2A_2}{x^4 + x^3 + 2x^2}$$

Since the original numerator is known, the following simultaneous equations result.

$$A_1 + A_4 = 0$$
$$A_1 + A_2 + A_3 = 1$$
$$2A_1 + A_2 = 2$$
$$2A_2 = 3$$

The solutions are $A_1 = 0.25$; $A_2 = 1.50$; $A_3 = -0.75$; and $A_4 = -0.25$.

$$H(x) = \frac{1}{4x} + \frac{3}{2x^2} - \frac{x+3}{4(x^2 + x + 2)}$$

## 17. SIMULTANEOUS LINEAR EQUATIONS

A *linear equation* with $n$ variables is a polynomial of degree 1 describing a geometric shape in $n$-space. A *homogeneous linear equation* is one that has no constant term, and a *nonhomogeneous linear equation* has a constant term.

A solution to a set of simultaneous linear equations represents the intersection point of the geometric shapes in $n$-space. For example, if the equations are limited to two variables (e.g., $y = 4x - 5$), they describe straight lines. The solution to two simultaneous linear equations in 2-space is the point where the two lines intersect. The set of the two equations is said to be a *consistent system* when there is such an intersection.[8]

Simultaneous equations do not always have unique solutions, and some have none at all. In addition to crossing in 2-space, lines can be parallel or they can be the same line expressed in a different equation format (i.e., dependent equations). In some cases, parallelism and dependency can be determined by inspection. In most cases, however, matrix and other advanced methods must be used to determine whether a solution exists. A set of linear equations with no simultaneous solution is known as an *inconsistent system*.

[8]A homogeneous system always has at least one solution: the *trivial solution*, in which all variables have a value of zero.

Several methods exist for solving linear equations simultaneously by hand.[9]

- *graphing:* The equations are plotted and the intersection point is read from the graph. This method is possible only with two-dimensional problems.

- *substitution:* An equation is rearranged so that one variable is expressed as a combination of the other variables. The expression is then substituted into the remaining equations wherever the selected variable appears.

- *reduction:* All terms in the equations are multiplied by constants chosen to eliminate one or more variables when the equations are added or subtracted. The remaining sum can then be solved for the other variables. This method is also known as *eliminating the unknowns.*

- *Cramer's rule:* This is a procedure in linear algebra that calculates determinants of the original coefficient matrix **A** and of the $n$ matrices resulting from the systematic replacement of column **A** by the constant matrix **B**.

### Example 5.6

Solve the following set of linear equations by (a) substitution and (b) reduction.

$$2x + 3y = 12 \quad [\text{Eq. 1}]$$

$$3x + 4y = 8 \quad [\text{Eq. 2}]$$

*Solution*

(a) From Eq. 1, solve for variable $x$.

$$x = 6 - 1.5y \quad [\text{Eq. 3}]$$

Substitute $6 - 1.5y$ into Eq. 2 wherever $x$ appears.

$$(3)(6 - 1.5y) + 4y = 8$$
$$18 - 4.5y + 4y = 8$$
$$y = 20$$

Substitute 20 for $y$ in Eq. 3.

$$x = 6 - (1.5)(20) = -24$$

The solution $(-24, 20)$ should be checked to verify that it satisfies both original equations.

(b) Eliminate variable $x$ by multiplying Eq. 1 by 3 and Eq. 2 by 2.

$$3 \times \text{Eq. 1: } 6x + 9y = 36 \quad [\text{Eq. 1}']$$
$$2 \times \text{Eq. 2: } 6x + 8y = 16 \quad [\text{Eq. 2}']$$

[9]Other methods exist, but they require a computer.

Subtract Eq. 2' from Eq. 1'.

$$y = 20 \quad [\text{Eq. } 1' - \text{Eq. } 2']$$

Substitute $y = 20$ into Eq. 1'.

$$6x + (9)(20) = 36$$

$$x = -24$$

The solution $(-24, 20)$ should be checked to verify that it satisfies both original equations.

## 18. COMPLEX NUMBERS

A *complex number*, **Z**, is a combination of real and imaginary numbers. When expressed as a sum (e.g., $a + bi$), the complex number is said to be in *rectangular* or *trigonometric form*. The complex number can be plotted on the real-imaginary coordinate system known as the *complex plane*, as illustrated in Fig. 5.2.

**Figure 5.2** *A Complex Number in the Complex Plane*

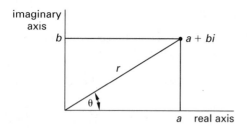

The complex number $\mathbf{Z} = a + bi$ can also be expressed in *exponential form*.[10] The quantity $r$ is known as the *modulus* of **Z**; $\theta$ is the *argument*.

$$a + bi \equiv re^{i\theta} \qquad 5.53$$

$$r = \text{mod}(\mathbf{Z}) = \sqrt{a^2 + b^2} \qquad 5.54$$

$$\theta = \arg(\mathbf{Z}) = \arctan\left(\frac{b}{a}\right) \qquad 5.55$$

Similarly, the *phasor form* (also known as the *polar form*) is

$$\mathbf{Z} = r\angle\theta \qquad 5.56$$

The *rectangular form* can be determined from $r$ and $\theta$.

$$a = r\cos\theta \qquad 5.57$$

$$b = r\sin\theta \qquad 5.58$$

$$\mathbf{Z} = a + bi = r\cos\theta + ir\sin\theta$$

$$= r(\cos\theta + i\sin\theta) \qquad 5.59$$

---

[10]The terms *polar form*, *phasor form*, and *exponential form* are all used somewhat interchangeably.

The *cis form* is a shorthand method of writing a complex number in rectangular (trigonometric) form.

$$a + bi = r(\cos\theta + i\sin\theta) = r\,\text{cis}\,\theta \qquad 5.60$$

*Euler's equation* (see Eq. 5.61) expresses the equality of complex numbers in exponential and trigonometric form.

$$e^{i\theta} = \cos\theta + i\sin\theta \qquad 5.61$$

Related expressions are

$$e^{-i\theta} = \cos\theta - i\sin\theta \qquad 5.62$$

$$\cos\theta = \frac{e^{i\theta} + e^{-i\theta}}{2} \qquad 5.63$$

$$\sin\theta = \frac{e^{i\theta} - e^{-i\theta}}{2i} \qquad 5.64$$

### Example 5.7

What is the exponential form of the complex number $\mathbf{Z} = 3 + 4i$?

*Solution*

$$r = \sqrt{a^2 + b^2} = \sqrt{3^2 + 4^2} = \sqrt{25} = 5$$

$$\theta = \arctan\left(\frac{b}{a}\right) = \arctan\left(\frac{4}{3}\right) = 0.927 \text{ rad}$$

$$\mathbf{Z} = re^{i\theta} = 5e^{i(0.927)}$$

## 19. OPERATIONS ON COMPLEX NUMBERS

Most algebraic operations (addition, multiplication, exponentiation, etc.) work with complex numbers, but notable exceptions are the inequality operators. The concept of one complex number being less than or greater than another complex number is meaningless.

When adding two complex numbers, real parts are added to real parts, and imaginary parts are added to imaginary parts.

$$(a_1 + ib_1) + (a_2 + ib_2) = (a_1 + a_2) + i(b_1 + b_2) \qquad 5.65$$

$$(a_1 + ib_1) - (a_2 + ib_2) = (a_1 - a_2) + i(b_1 - b_2) \qquad 5.66$$

Multiplication of two complex numbers in rectangular form is accomplished by the use of the algebraic distributive law, remembering that $i^2 = -1$.

Division of complex numbers in rectangular form requires use of the *complex conjugate*. The complex conjugate of the complex number $(a + bi)$ is $(a - bi)$. By multiplying the numerator and the denominator by the complex conjugate, the denominator will be converted to the real number $a^2 + b^2$. This technique is known as *rationalizing* the denominator and is illustrated in Ex. 5.8(c).

Multiplication and division are often more convenient when the complex numbers are in exponential or phasor forms, as Eq. 5.67 and Eq. 5.68 show.

$$(r_1 e^{i\theta_1})(r_2 e^{i\theta_2}) = r_1 r_2 e^{i(\theta_1 + \theta_2)} \qquad 5.67$$

$$\frac{r_1 e^{i\theta_1}}{r_2 e^{i\theta_2}} = \left(\frac{r_1}{r_2}\right) e^{i(\theta_1 - \theta_2)} \qquad 5.68$$

Taking powers and roots of complex numbers requires *de Moivre's theorem*, Eq. 5.69 and Eq. 5.70.

$$\mathbf{Z}^n = (re^{i\theta})^n = r^n e^{in\theta} \qquad 5.69$$

$$\sqrt[n]{\mathbf{Z}} = (re^{i\theta})^{1/n} = \sqrt[n]{r} e^{i(\theta + k(360°)/n)}$$

$$[k = 0, 1, 2, ..., n-1] \qquad 5.70$$

### Example 5.8

Perform the following complex arithmetic.

(a) $(3 + 4i) + (2 + i)$

(b) $(7 + 2i)(5 - 3i)$

(c) $\dfrac{2 + 3i}{4 - 5i}$

*Solution*

(a)
$$(3 + 4i) + (2 + i) = (3 + 2) + (4 + 1)i$$
$$= 5 + 5i$$

(b)
$$(7 + 2i)(5 - 3i) = (7)(5) - (7)(3i) + (2i)(5)$$
$$- (2i)(3i)$$
$$= 35 - 21i + 10i - 6i^2$$
$$= 35 - 21i + 10i - (6)(-1)$$
$$= 41 - 11i$$

(c) Multiply the numerator and denominator by the complex conjugate of the denominator.

$$\frac{2 + 3i}{4 - 5i} = \frac{(2 + 3i)(4 + 5i)}{(4 - 5i)(4 + 5i)}$$

$$= \frac{-7 + 22i}{(4)^2 + (5)^2}$$

$$= \frac{-7}{41} + i\frac{22}{41}$$

## 20. LIMITS

A *limit* (*limiting value*) is the value a function approaches when an independent variable approaches a target value. For example, suppose the value of $y = x^2$ is desired as $x$ approaches 5. This could be written as

$$\lim_{x \to 5} x^2 \qquad 5.71$$

The power of limit theory is wasted on simple calculations such as this but is appreciated when the function is undefined at the target value. The object of limit theory is to determine the limit without having to evaluate the function at the target. The general case of a limit evaluated as $x$ approaches the target value $a$ is written as

$$\lim_{x \to a} f(x) \qquad 5.72$$

It is not necessary for the actual value $f(a)$ to exist for the limit to be calculated. The function $f(x)$ may be undefined at point $a$. However, it is necessary that $f(x)$ be defined on both sides of point $a$ for the limit to exist. If $f(x)$ is undefined on one side, or if $f(x)$ is discontinuous at $x = a$ (as in Fig. 5.3(c) and Fig. 5.3(d)), the limit does not exist.

**Figure 5.3** *Existence of Limits*

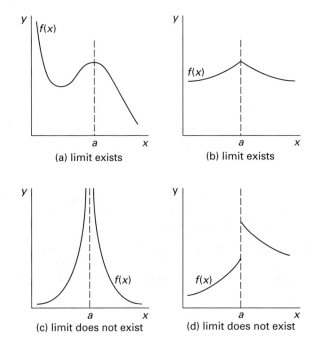

(a) limit exists

(b) limit exists

(c) limit does not exist

(d) limit does not exist

The following theorems can be used to simplify expressions when calculating limits.

$$\lim_{x \to a} x = a \qquad 5.73$$

$$\lim_{x \to a} (mx + b) = ma + b \qquad 5.74$$

$$\lim_{x \to a} b = b \qquad 5.75$$

$$\lim_{x \to a} (kF(x)) = k \lim_{x \to a} F(x) \qquad 5.76$$

Mathematics

$$\lim_{x \to a} \left( F_1(x) \begin{Bmatrix} + \\ - \\ \times \\ \div \end{Bmatrix} F_2(x) \right)$$

$$= \lim_{x \to a} (F_1(x)) \begin{Bmatrix} + \\ - \\ \times \\ \div \end{Bmatrix} \lim_{x \to a} (F_2(x)) \qquad 5.77$$

The following identities can be used to simplify limits of trigonometric expressions.

$$\lim_{x \to 0} \sin x = 0 \qquad 5.78$$

$$\lim_{x \to 0} \left( \frac{\sin x}{x} \right) = 1 \qquad 5.79$$

$$\lim_{x \to 0} \cos x = 1 \qquad 5.80$$

The following standard methods (tricks) can be used to determine limits.

- If the limit is taken to infinity, all terms can be divided by the largest power of $x$ in the expression. This will leave at least one constant. Any quantity divided by a power of $x$ vanishes as $x$ approaches infinity.

- If the expression is a quotient of two expressions, any common factors should be eliminated from the numerator and denominator.

- *L'Hôpital's rule*, Eq. 5.81, should be used when the numerator and denominator of the expression both approach zero or both approach infinity.[11] $P^k(x)$ and $Q^k(x)$ are the $k$th derivatives of the functions $P(x)$ and $Q(x)$, respectively. (L'Hôpital's rule can be applied repeatedly as required.)

$$\lim_{x \to a} \left( \frac{P(x)}{Q(x)} \right) = \lim_{x \to a} \left( \frac{P^k(x)}{Q^k(x)} \right) \qquad 5.81$$

**Example 5.9**

Evaluate the following limits.

(a) $\lim\limits_{x \to 3} \left( \dfrac{x^3 - 27}{x^2 - 9} \right)$

(b) $\lim\limits_{x \to \infty} \left( \dfrac{3x - 2}{4x + 3} \right)$

(c) $\lim\limits_{x \to 2} \left( \dfrac{x^2 + x - 6}{x^2 - 3x + 2} \right)$

---

[11]L'Hôpital's rule should not be used when only the denominator approaches zero. In that case, the limit approaches infinity regardless of the numerator.

*Solution*

(a) Factor the numerator and denominator. (L'Hôpital's rule can also be used.)

$$\lim_{x \to 3} \left( \frac{x^3 - 27}{x^2 - 9} \right) = \lim_{x \to 3} \left( \frac{(x - 3)(x^2 + 3x + 9)}{(x - 3)(x + 3)} \right)$$

$$= \lim_{x \to 3} \left( \frac{x^2 + 3x + 9}{x + 3} \right) = \frac{(3)^2 + (3)(3) + 9}{3 + 3}$$

$$= 9/2$$

(b) Divide through by the largest power of $x$. (L'Hôpital's rule can also be used.)

$$\lim_{x \to \infty} \left( \frac{3x - 2}{4x + 3} \right) = \lim_{x \to \infty} \left( \frac{3 - \dfrac{2}{x}}{4 + \dfrac{3}{x}} \right)$$

$$= \frac{3 - \dfrac{2}{\infty}}{4 + \dfrac{3}{\infty}}$$

$$= \frac{3 - 0}{4 + 0} = \frac{3}{4}$$

(c) Use L'Hôpital's rule. (Factoring can also be used.) Take the first derivative of the numerator and denominator.

$$\lim_{x \to 2} \left( \frac{x^2 + x - 6}{x^2 - 3x + 2} \right) = \lim_{x \to 2} \left( \frac{2x + 1}{2x - 3} \right)$$

$$= \frac{(2)(2) + 1}{(2)(2) - 3}$$

$$= \frac{5}{1} = 5$$

## 21. SEQUENCES AND PROGRESSIONS

A *sequence* is an ordered *progression* of numbers, $a_i$, such as 1, 4, 9, 16, 25, ... The *terms* in a sequence can be all positive, all negative, or of alternating signs. $a_n$ is known as the *general term* of the sequence.

$$\{A\} = \{a_1, a_2, a_3, ..., a_n\} \qquad 5.82$$

A sequence is said to *diverge* (i.e., be *divergent*) if the terms approach infinity or if the terms fail to approach any finite value, and it is said to *converge* (i.e., be *convergent*) if the terms approach any finite value (including zero). That is, the sequence converges if the limit defined by Eq. 5.83 exists.

$$\lim_{n\to\infty} a_n \begin{cases} \text{converges if } L \text{ is finite} \\ \text{diverges if } L \text{ is infinite} \\ \text{or does not exist} \end{cases} \quad 5.83$$

The main task associated with a sequence is determining the next (or the general) term. If several terms of a sequence are known, the next (unknown) term must usually be found by intuitively determining the pattern of the sequence. In some cases, though, the method of *Rth-order differences* can be used to determine the next term. This method consists of subtracting each term from the following term to obtain a set of differences. If the differences are not all equal, the next order of differences can be calculated.

### Example 5.10

What is the general term of the sequence $A$?

$$\{A\} = \left\{ 3, \frac{9}{2}, \frac{27}{6}, \frac{81}{24}, \ldots \right\}$$

*Solution*

The solution is purely intuitive. The numerator is recognized as a power series based on the number 3. The denominator is recognized as the factorial sequence. The general term is

$$a_n = \frac{3^n}{n!}$$

### Example 5.11

Find the sixth term in the sequence $\{7, 16, 29, 46, 67, a_6\}$.

*Solution*

The sixth term is not intuitively obvious, so the method of $R$th-order differences is tried. The pattern is not obvious from the first order differences, but the second order differences are all 4.

$$\delta_5 - 21 = 4$$
$$\delta_5 = 25$$
$$a_6 - 67 = \delta_5 = 25$$
$$a_6 = 92$$

### Example 5.12

Does the sequence with general term $e^n/n$ converge or diverge?

*Solution*

See if the limit exists.

$$\lim_{n\to\infty}\left(\frac{e^n}{n}\right) = \frac{\infty}{\infty}$$

Since $\infty/\infty$ is inconclusive, apply L'Hôpital's rule. Take the derivative of both the numerator and the denominator with respect to $n$.

$$\lim_{n\to\infty}\left(\frac{e^n}{1}\right) = \frac{\infty}{1}$$
$$= \infty$$

The sequence diverges.

## 22. STANDARD SEQUENCES

There are four standard sequences: the geometric, arithmetic, harmonic, and $p$-sequence.

- *geometric sequence:* The geometric sequence converges for $-1 < r \le 1$ and diverges otherwise. $a$ is known as the *first term*; $r$ is known as the *common ratio*.

$$a_n = ar^{n-1} \quad \begin{bmatrix} a \text{ is a constant} \\ n = 1, 2, 3, \ldots, \infty \end{bmatrix} \quad 5.84$$

  *example:* $\{1, 2, 4, 8, 16, 32\} \quad (a = 1, \quad r = 2)$

- *arithmetic sequence:* The arithmetic sequence always diverges.

$$a_n = a + (n-1)d \quad \begin{bmatrix} a \text{ and } d \text{ are constants} \\ n = 1, 2, 3, \ldots, \infty \end{bmatrix} \quad 5.85$$

  *example:* $\{2, 7, 12, 17, 22, 27\} \quad (a = 2, \quad d = 5)$

- *harmonic sequence:* The harmonic sequence always converges.

$$a_n = \frac{1}{n} \quad [n = 1, 2, 3, \ldots, \infty] \quad 5.86$$

  *example:* $\{1, 1/2, 1/3, 1/4, 1/5, 1/6\}$

- *p-sequence:* The $p$-sequence converges if $p \ge 0$ and diverges if $p < 0$. (Notice that this is different than the $p$-series.)

$$a_n = \frac{1}{n^p} \quad [n = 1, 2, 3, \ldots, \infty] \quad 5.87$$

  *example:* $\{1, 1/4, 1/9, 1/16, 1/25, 1/36\} \quad (p = 2)$

## 23. SERIES

A *series* is the sum of terms in a sequence. There are two types of series. A *finite series* has a finite number of terms, and an *infinite series* has an infinite number of terms.[12] The main tasks associated with series are determining the sum of the terms and whether the series converges. A series is said to *converge* (be *convergent*) if the sum, $S_n$, of its term exists.[13] A finite series is always convergent.

The performance of a series based on standard sequences (defined in Sec. 5.22) is well known.

- *geometric series:*

$$S_n = \sum_{i=1}^{n} ar^{i-1} = \frac{a(1 - r^n)}{r} \quad \text{[finite]} \qquad 5.88$$

$$S_n = \sum_{i=1}^{\infty} ar^{i-1} = \frac{a}{1 - r} \quad \begin{bmatrix} \text{infinite} \\ -1 < r < 1 \end{bmatrix} \qquad 5.89$$

- *arithmetic series:* The infinite series diverges unless $a = d = 0$.

$$S_n = \sum_{i=1}^{n} (a + (i - 1)d) = \frac{n(2a + (n - 1)d)}{2} \quad \text{[finite]}$$

$$5.90$$

- *harmonic series:* The infinite series diverges.

- *p-series:* The infinite series diverges if $p \le 1$. The infinite series converges if $p > 1$. (Notice that this is different than the *p*-sequence.)

## 24. TESTS FOR SERIES CONVERGENCE

It is obvious that all *finite series* (i.e., series having a finite number of terms) converge. That is, the sum, $S_n$, defined by Eq. 5.91 exists.

$$S_n = \sum_{i=1}^{n} a_i \qquad 5.91$$

Convergence of an infinite series can be determined by taking the limit of the sum. If the limit exists, the series converges; otherwise, it diverges.

$$\lim_{n \to \infty} S_n = \lim_{n \to \infty} \sum_{i=1}^{n} a_i \qquad 5.92$$

In most cases, the expression for the general term $a_n$ will be known, but there will be no simple expression for the sum $S_n$. Therefore, Eq. 5.92 cannot be used to determine convergence. It is helpful, but not conclusive, to look at the limit of the general term. If $L$, as defined in Eq. 5.93, is nonzero, the series diverges. If $L$ equals zero, the series

[12]The term *infinite series* does not imply the sum is infinite.
[13]This is different from the definition of convergence for a sequence where only the last term was evaluated.

may either converge or diverge. Additional testing is needed in that case.

$$\lim_{n \to \infty} a_n \begin{cases} = 0 & \text{inconclusive} \\ \ne 0 & \text{diverges} \end{cases} \qquad 5.93$$

Two tests can be used independently or after Eq. 5.93 has proven inconclusive: the ratio and comparison tests. The *ratio test* calculates the limit of the ratio of two consecutive terms.

$$\lim_{n \to \infty} \frac{a_{n+1}}{a_n} \begin{cases} < 1 & \text{converges} \\ = 1 & \text{inconclusive} \\ > 1 & \text{diverges} \end{cases} \qquad 5.94$$

The *comparison test* is an indirect method of determining convergence of an unknown series. It compares a standard series (geometric and *p*-series are commonly used) against the unknown series. If all terms in a positive standard series are smaller than the terms in the unknown series and the standard series diverges, the unknown series must also diverge. Similarly, if all terms in the standard series are larger than the terms in the unknown series and the standard series converges, then the unknown series also converges.

In mathematical terms, if $A$ and $B$ are both series of positive terms such that $a_n < b_n$ for all values of $n$, then (a) $B$ diverges if $A$ diverges, and (b) $A$ converges if $B$ converges.

### Example 5.13

Does the infinite series $A$ converge or diverge?

$$A = 3 + \frac{9}{2} + \frac{27}{6} + \frac{81}{24} + \cdots$$

*Solution*

The general term was found in Ex. 5.10 to be

$$a_n = \frac{3^n}{n!}$$

Since limits of factorials are not easily determined, use the ratio test.

$$\lim_{n \to \infty} \left( \frac{a_{n+1}}{a_n} \right) = \lim_{n \to \infty} \left( \frac{\dfrac{3^{n+1}}{(n+1)!}}{\dfrac{3^n}{n!}} \right) = \lim_{n \to \infty} \left( \frac{3}{n+1} \right)$$

$$= \frac{3}{\infty} = 0$$

Since the limit is less than 1, the infinite series converges.

## Example 5.14

Does the infinite series $A$ converge or diverge?

$$A = 2 + \frac{3}{4} + \frac{4}{9} + \frac{5}{16} + \cdots$$

*Solution*

By observation, the general term is

$$a_n = \frac{1+n}{n^2}$$

The general term can be expanded by partial fractions to

$$a_n = \frac{1}{n} + \frac{1}{n^2}$$

However, $1/n$ is the harmonic series. Since the harmonic series is divergent and this series is larger than the harmonic series (by the term $1/n^2$), this series also diverges.

## 25. SERIES OF ALTERNATING SIGNS[14]

Some series contain both positive and negative terms. The ratio and comparison tests can both be used to determine if a series with alternating signs converges. If a series containing all positive terms converges, then the same series with some negative terms also converges. Therefore, the all-positive series should be tested for convergence. If the all-positive series converges, the original series is said to be *absolutely convergent*. (If the all-positive series diverges and the original series converges, the original series is said to be *conditionally convergent*.)

Alternatively, the ratio test can be used with the absolute value of the ratio. The same criteria apply.

$$\lim_{n \to \infty} \left| \frac{a_{n+1}}{a_n} \right| \begin{cases} < 1 & \text{converges} \\ = 1 & \text{inconclusive} \\ > 1 & \text{diverges} \end{cases} \qquad 5.95$$

---

[14]This terminology is commonly used even though it is not necessary that the signs strictly alternate.

# 6 Linear Algebra

## 1. MATRICES

A *matrix* is an ordered set of *entries* (*elements*) arranged rectangularly and set off by brackets.[1] The entries can be variables or numbers. A matrix by itself has no particular value—it is merely a convenient method of representing a set of numbers.

The size of a matrix is given by the number of rows and columns, and the nomenclature $m \times n$ is used for a matrix with $m$ rows and $n$ columns. For a *square matrix*, the number of rows and columns will be the same, a quantity known as the *order* of the matrix.

Bold uppercase letters are used to represent matrices, while lowercase letters represent the entries. For example, $a_{23}$ would be the entry in the second row and third column of matrix $\mathbf{A}$.

$$\mathbf{A} = \begin{bmatrix} a_{11} & a_{12} & a_{13} \\ a_{21} & a_{22} & a_{23} \\ a_{31} & a_{32} & a_{33} \end{bmatrix}$$

A *submatrix* is the matrix that remains when selected rows or columns are removed from the original matrix.[2] For example, for matrix $\mathbf{A}$, the submatrix remaining after the second row and second column have been removed is

$$\begin{bmatrix} a_{11} & a_{13} \\ a_{31} & a_{33} \end{bmatrix}$$

An *augmented matrix* results when the original matrix is extended by repeating one or more of its rows or columns or by adding rows and columns from another matrix. For example, for the matrix $\mathbf{A}$, the augmented matrix created by repeating the first and second columns is

$$\left[ \begin{array}{ccc|cc} a_{11} & a_{12} & a_{13} & a_{11} & a_{12} \\ a_{21} & a_{22} & a_{23} & a_{21} & a_{22} \\ a_{31} & a_{32} & a_{33} & a_{31} & a_{32} \end{array} \right]$$

## 2. SPECIAL TYPES OF MATRICES

Certain types of matrices are given special designations.

- *cofactor matrix:* the matrix formed when every entry is replaced by the cofactor (see Sec. 6.4) of that entry

- *column matrix:* a matrix with only one column

- *complex matrix:* a matrix with complex number entries

- *diagonal matrix:* a square matrix with all zero entries except for the $a_{ij}$ for which $i = j$

- *echelon matrix:* a matrix in which the number of zeros preceding the first nonzero entry of a row increases row by row until only zero rows remain. A *row-reduced echelon matrix* is an echelon matrix in which the first nonzero entry in each row is a 1 and all other entries in the columns are zero.

- *identity matrix:* a diagonal (square) matrix with all nonzero entries equal to 1, usually designated as $\mathbf{I}$, having the property that $\mathbf{AI} = \mathbf{IA} = \mathbf{A}$

- *null matrix:* the same as a zero matrix

- *row matrix:* a matrix with only one row

- *scalar matrix:*[3] a diagonal (square) matrix with all diagonal entries equal to some scalar $k$

---

[1]The term *array* is synonymous with *matrix*, although the former is more likely to be used in computer applications.
[2]By definition, a matrix is a submatrix of itself.

[3]Although the term *complex matrix* means a matrix with complex entries, the term *scalar matrix* means more than a matrix with scalar entries.

- *singular matrix:* a matrix whose determinant is zero (see Sec. 6.10)

- *skew symmetric matrix:* a square matrix whose transpose (see Sec. 6.9) is equal to the negative of itself (i.e., $\mathbf{A} = -\mathbf{A}^t$)

- *square matrix:* a matrix with the same number of rows and columns (i.e., $m = n$)

- *symmetric(al) matrix:* a square matrix whose transpose is equal to itself (i.e., $\mathbf{A}^t = \mathbf{A}$), which occurs only when $a_{ij} = a_{ji}$

- *triangular matrix:* a square matrix with zeros in all positions above or below the diagonal

- *unit matrix:* the same as the identity matrix

- *zero matrix:* a matrix with all zero entries

Figure 6.1 shows examples of special matrices.

**Figure 6.1** *Examples of Special Matrices*

$$
\begin{bmatrix} 9 & 0 & 0 & 0 \\ 0 & -6 & 0 & 0 \\ 0 & 0 & 1 & 0 \\ 0 & 0 & 0 & 5 \end{bmatrix}
\quad
\begin{bmatrix} 2 & 18 & 2 & 18 \\ 0 & 0 & 1 & 9 \\ 0 & 0 & 0 & 9 \\ 0 & 0 & 0 & 0 \end{bmatrix}
\quad
\begin{bmatrix} 1 & 9 & 0 & 0 \\ 0 & 0 & 1 & 0 \\ 0 & 0 & 0 & 1 \\ 0 & 0 & 0 & 0 \end{bmatrix}
$$

(a) diagonal        (b) echelon        (c) row-reduced echelon

$$
\begin{bmatrix} 1 & 0 & 0 & 0 \\ 0 & 1 & 0 & 0 \\ 0 & 0 & 1 & 0 \\ 0 & 0 & 0 & 1 \end{bmatrix}
\quad
\begin{bmatrix} 3 & 0 & 0 & 0 \\ 0 & 3 & 0 & 0 \\ 0 & 0 & 3 & 0 \\ 0 & 0 & 0 & 3 \end{bmatrix}
\quad
\begin{bmatrix} 2 & 0 & 0 & 0 \\ 7 & 6 & 0 & 0 \\ 9 & 1 & 1 & 0 \\ 8 & 0 & 4 & 5 \end{bmatrix}
$$

(d) identity        (e) scalar        (f) triangular

## 3. ROW EQUIVALENT MATRICES

A matrix $\mathbf{B}$ is said to be *row equivalent* to a matrix $\mathbf{A}$ if it is obtained by a finite sequence of *elementary row operations* on $\mathbf{A}$:

- interchanging the $i$th and $j$th rows

- multiplying the $i$th row by a nonzero scalar

- replacing the $i$th row by the sum of the original $i$th row and $k$ times the $j$th row

However, two matrices that are row equivalent as defined do not necessarily have the same determinants. (See Sec. 6.5.)

*Gauss-Jordan elimination* is the process of using these elementary row operations to row-reduce a matrix to echelon or row-reduced echelon forms, as illustrated in Ex. 6.8. When a matrix has been converted to a row-reduced echelon matrix, it is said to be in *row canonical form.* Thus, the terms *row-reduced echelon form* and *row canonical form* are synonymous.

## 4. MINORS AND COFACTORS

Minors and cofactors are determinants of submatrices associated with particular entries in the original square matrix. The *minor* of entry $a_{ij}$ is the determinant of a submatrix resulting from the elimination of the single row $i$ and the single column $j$. For example, the minor corresponding to entry $a_{12}$ in a $3 \times 3$ matrix $\mathbf{A}$ is the determinant of the matrix created by eliminating row 1 and column 2.

$$
\text{minor of } a_{12} = \begin{vmatrix} a_{21} & a_{23} \\ a_{31} & a_{33} \end{vmatrix}
\qquad 6.1
$$

The *cofactor* of entry $a_{ij}$ is the minor of $a_{ij}$ multiplied by either $+1$ or $-1$, depending on the position of the entry. (That is, the cofactor either exactly equals the minor or it differs only in sign.) The sign is determined according to the following positional matrix.[4]

$$
\begin{bmatrix} +1 & -1 & +1 & \cdots \\ -1 & +1 & -1 & \cdots \\ +1 & -1 & +1 & \cdots \\ \vdots & \vdots & \vdots & \end{bmatrix}
$$

For example, the cofactor of entry $a_{12}$ in matrix $\mathbf{A}$ (described in Sec. 6.4) is

$$
\text{cofactor of } a_{12} = -\begin{vmatrix} a_{21} & a_{23} \\ a_{31} & a_{33} \end{vmatrix}
\qquad 6.2
$$

**Example 6.1**

What is the cofactor corresponding to the $-3$ entry in the following matrix?

$$
\mathbf{A} = \begin{bmatrix} 2 & 9 & 1 \\ -3 & 4 & 0 \\ 7 & 5 & 9 \end{bmatrix}
$$

*Solution*

The minor's submatrix is created by eliminating the row and column of the $-3$ entry.

$$
\mathbf{M} = \begin{bmatrix} 9 & 1 \\ 5 & 9 \end{bmatrix}
$$

The minor is the determinant of $\mathbf{M}$.

$$
|\mathbf{M}| = (9)(9) - (5)(1) = 76
$$

The sign corresponding to the $-3$ position is negative. Therefore, the cofactor is $-76$.

---

[4]The sign of the cofactor $a_{ij}$ is positive if $(i + j)$ is even and is negative if $(i + j)$ is odd.

## 5. DETERMINANTS

A *determinant* is a scalar calculated from a square matrix. The determinant of matrix $\mathbf{A}$ can be represented as $D\{\mathbf{A}\}$, Det $(\mathbf{A})$, $\Delta\mathbf{A}$, or $|\mathbf{A}|$.[5] The following rules can be used to simplify the calculation of determinants.

- If $\mathbf{A}$ has a row or column of zeros, the determinant is zero.

- If $\mathbf{A}$ has two identical rows or columns, the determinant is zero.

- If $\mathbf{B}$ is obtained from $\mathbf{A}$ by adding a multiple of a row (column) to another row (column) in $\mathbf{A}$, then $|\mathbf{B}| = |\mathbf{A}|$.

- If $\mathbf{A}$ is triangular, the determinant is equal to the product of the diagonal entries.

- If $\mathbf{B}$ is obtained from $\mathbf{A}$ by multiplying one row or column in $\mathbf{A}$ by a scalar $k$, then $|\mathbf{B}| = k|\mathbf{A}|$.

- If $\mathbf{B}$ is obtained from the $n \times n$ matrix $\mathbf{A}$ by multiplying by the scalar matrix $k$, then $|k\mathbf{A}| = k^n|\mathbf{A}|$.

- If $\mathbf{B}$ is obtained from $\mathbf{A}$ by switching two rows or columns in $\mathbf{A}$, then $|\mathbf{B}| = -|\mathbf{A}|$.

Calculation of determinants is laborious for all but the smallest or simplest of matrices. For a $2 \times 2$ matrix, the formula used to calculate the determinant is easy to remember.

$$\mathbf{A} = \begin{bmatrix} a & b \\ c & d \end{bmatrix}$$

$$|\mathbf{A}| = \begin{vmatrix} a & b \\ c & d \end{vmatrix} = ad - bc \qquad 6.3$$

Two methods are commonly used for calculating the determinant of $3 \times 3$ matrices by hand. The first uses an augmented matrix constructed from the original matrix and the first two columns (as presented in Sec. 6.1).[6] The determinant is calculated as the sum of the products in the left-to-right downward diagonals less the sum of the products in the left-to-right upward diagonals.

$$\mathbf{A} = \begin{bmatrix} a & b & c \\ d & e & f \\ g & h & i \end{bmatrix}$$

$$|\mathbf{A}| = aei + bfg + cdh - gec - hfa - idb \qquad 6.5$$

The second method of calculating the determinant is somewhat slower than the first for a $3 \times 3$ matrix but illustrates the method that must be used to calculate determinants of $4 \times 4$ and larger matrices. This method is known as *expansion by cofactors*. One row (column) is selected as the base row (column). The selection is arbitrary, but the number of calculations required to obtain the determinant can be minimized by choosing the row (column) with the most zeros. The determinant is equal to the sum of the products of the entries in the base row (column) and their corresponding cofactors.

$$\mathbf{A} = \begin{bmatrix} a & b & c \\ d & e & f \\ g & h & i \end{bmatrix}$$

$$|\mathbf{A}| = a\begin{vmatrix} e & f \\ h & i \end{vmatrix} - d\begin{vmatrix} b & c \\ h & i \end{vmatrix} + g\begin{vmatrix} b & c \\ e & f \end{vmatrix} \qquad 6.6$$

### Example 6.2

Calculate the determinant of matrix $\mathbf{A}$ (a) by cofactor expansion, and (b) by the augmented matrix method.

$$\mathbf{A} = \begin{bmatrix} 2 & 3 & -4 \\ 3 & -1 & -2 \\ 4 & -7 & -6 \end{bmatrix}$$

*Solution*

(a) Since there are no zero entries, it does not matter which row or column is chosen as the base. Choose the first column as the base.

$$\begin{aligned} |\mathbf{A}| &= 2\begin{vmatrix} -1 & -2 \\ -7 & -6 \end{vmatrix} - 3\begin{vmatrix} 3 & -4 \\ -7 & -6 \end{vmatrix} + 4\begin{vmatrix} 3 & -4 \\ -1 & -2 \end{vmatrix} \\ &= (2)(6 - 14) - (3)(-18 - 28) + (4)(-6 - 4) \\ &= 82 \end{aligned}$$

---

[5]The vertical bars should not be confused with the square brackets used to set off a matrix, nor with absolute value.
[6]It is not actually necessary to construct the augmented matrix, but doing so helps avoid errors.

(b)

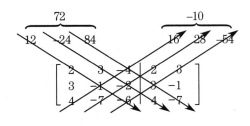

$$|\mathbf{A}| = 72 - (-10) = 82$$

## 6. MATRIX ALGEBRA[7]

Matrix algebra differs somewhat from standard algebra.

- *equality:* Two matrices, **A** and **B**, are equal only if they have the same numbers of rows and columns *and* if all corresponding entries are equal.

- *inequality:* The > and < operators are not used in matrix algebra.

- *commutative law of addition:*

$$\mathbf{A} + \mathbf{B} = \mathbf{B} + \mathbf{A} \qquad 6.7$$

- *associative law of addition:*

$$\mathbf{A} + (\mathbf{B} + \mathbf{C}) = (\mathbf{A} + \mathbf{B}) + \mathbf{C} \qquad 6.8$$

- *associative law of multiplication:*

$$(\mathbf{AB})\mathbf{C} = \mathbf{A}(\mathbf{BC}) \qquad 6.9$$

- *left distributive law:*

$$\mathbf{A}(\mathbf{B} + \mathbf{C}) = \mathbf{AB} + \mathbf{AC} \qquad 6.10$$

- *right distributive law:*

$$(\mathbf{B} + \mathbf{C})\mathbf{A} = \mathbf{BA} + \mathbf{CA} \qquad 6.11$$

- *scalar multiplication:*

$$k(\mathbf{AB}) = (k\mathbf{A})\mathbf{B} = \mathbf{A}(k\mathbf{B}) \qquad 6.12$$

It is important to recognize that, except for trivial and special cases, matrix multiplication is not commutative. That is,

$$\mathbf{AB} \neq \mathbf{BA}$$

## 7. MATRIX ADDITION AND SUBTRACTION

Addition and subtraction of two matrices is possible only if both matrices have the same numbers of rows and columns (i.e., order). They are accomplished by adding or subtracting the corresponding entries of the two matrices.

---

[7]Since matrices are used to simplify the presentation and solution of sets of linear equations, matrix algebra is also known as *linear algebra*.

## 8. MATRIX MULTIPLICATION

A matrix can be multiplied by a scalar, an operation known as *scalar multiplication*, in which case all entries of the matrix are multiplied by that scalar. For example, for the $2 \times 2$ matrix **A**,

$$k\mathbf{A} = \begin{bmatrix} ka_{11} & ka_{12} \\ ka_{21} & ka_{22} \end{bmatrix}$$

A matrix can be multiplied by another matrix, but only if the left-hand matrix has the same number of columns as the right-hand matrix has rows. *Matrix multiplication* occurs by multiplying the elements in each left-hand matrix row by the entries in each right-hand matrix column, adding the products, and placing the sum at the intersection point of the participating row and column.

*Matrix division* can only be accomplished by multiplying by the inverse of the denominator matrix. There is no specific division operation in matrix algebra.

**Example 6.3**

Determine the product matrix **C**.

$$\mathbf{C} = \begin{bmatrix} 1 & 4 & 3 \\ 5 & 2 & 6 \end{bmatrix} \begin{bmatrix} 7 & 12 \\ 11 & 8 \\ 9 & 10 \end{bmatrix}$$

*Solution*

The left-hand matrix has three columns, and the right-hand matrix has three rows. Therefore, the two matrices can be multiplied.

The first row of the left-hand matrix and the first column of the right-hand matrix are worked with first. The corresponding entries are multiplied, and the products are summed.

$$c_{11} = (1)(7) + (4)(11) + (3)(9) = 78$$

The intersection of the top row and left column is the entry in the upper left-hand corner of the matrix **C**.

The remaining entries are calculated similarly.

$$c_{12} = (1)(12) + (4)(8) + (3)(10) = 74$$
$$c_{21} = (5)(7) + (2)(11) + (6)(9) = 111$$
$$c_{22} = (5)(12) + (2)(8) + (6)(10) = 136$$

The product matrix is

$$\mathbf{C} = \begin{bmatrix} 78 & 74 \\ 111 & 136 \end{bmatrix}$$

## 9. TRANSPOSE

The *transpose*, $\mathbf{A}^t$, of an $m \times n$ matrix $\mathbf{A}$ is an $n \times m$ matrix constructed by taking the $i$th row and making it the $i$th column. The diagonal is unchanged. For example,

$$\mathbf{A} = \begin{bmatrix} 1 & 6 & 9 \\ 2 & 3 & 4 \\ 7 & 1 & 5 \end{bmatrix}$$

$$\mathbf{A}^t = \begin{bmatrix} 1 & 2 & 7 \\ 6 & 3 & 1 \\ 9 & 4 & 5 \end{bmatrix}$$

Transpose operations have the following characteristics.

$$(\mathbf{A}^t)^t = \mathbf{A} \qquad 6.13$$

$$(k\mathbf{A})^t = k(\mathbf{A}^t) \qquad 6.14$$

$$\mathbf{I}^t = \mathbf{I} \qquad 6.15$$

$$(\mathbf{A}\mathbf{B})^t = \mathbf{B}^t\mathbf{A}^t \qquad 6.16$$

$$(\mathbf{A} + \mathbf{B})^t = \mathbf{A}^t + \mathbf{B}^t \qquad 6.17$$

$$|\mathbf{A}^t| = |\mathbf{A}| \qquad 6.18$$

## 10. SINGULARITY AND RANK

A *singular matrix* is one whose determinant is zero. Similarly, a *nonsingular* matrix is one whose determinant is nonzero.

The *rank* of a matrix is the maximum number of linearly independent row or column vectors.[8] A matrix has rank $r$ if it has at least one nonsingular square submatrix of order $r$ but has no nonsingular square submatrix of order more than $r$. While the submatrix must be square (in order to calculate the determinant), the original matrix need not be.

The rank of an $m \times n$ matrix will be, at most, the smaller of $m$ and $n$. The rank of a null matrix is zero. The ranks of a matrix and its transpose are the same. If a matrix is in echelon form, the rank will be equal to the number of rows containing at least one nonzero entry. For a $3 \times 3$ matrix, the rank can either be 3 (if it is nonsingular), 2 (if any one of its $2 \times 2$ submatrices is nonsingular), 1 (if it and all $2 \times 2$ submatrices are singular), or 0 (if it is null).

The determination of rank is laborious if done by hand. Either the matrix is reduced to echelon form by using elementary row operations, or exhaustive enumeration is used to create the submatrices and many determinants are calculated. If a matrix has more rows than columns and row-reduction is used, the work required to

put the matrix in echelon form can be reduced by working with the transpose of the original matrix.

### Example 6.4

What is the rank of matrix $\mathbf{A}$?

$$\mathbf{A} = \begin{bmatrix} 1 & -2 & -1 \\ -3 & 3 & 0 \\ 2 & 2 & 4 \end{bmatrix}$$

*Solution*

Matrix $\mathbf{A}$ is singular because $|\mathbf{A}| = 0$. However, there is at least one $2 \times 2$ nonsingular submatrix:

$$\begin{vmatrix} 1 & -2 \\ -3 & 3 \end{vmatrix} = (1)(3) - (-3)(-2) = -3$$

Therefore, the rank is 2.

### Example 6.5

Determine the rank of matrix $\mathbf{A}$ by reducing it to echelon form.

$$\mathbf{A} = \begin{bmatrix} 7 & 4 & 9 & 1 \\ 0 & 2 & -5 & 3 \\ 0 & 4 & -10 & 6 \end{bmatrix}$$

*Solution*

By inspection, the matrix can be row-reduced by subtracting two times the second row from the third row. The matrix cannot be further reduced. Since there are two nonzero rows, the rank is 2.

$$\begin{bmatrix} 7 & 4 & 9 & 1 \\ 0 & 2 & -5 & 3 \\ 0 & 0 & 0 & 0 \end{bmatrix}$$

## 11. CLASSICAL ADJOINT

The *classical adjoint* is the transpose of the cofactor matrix. The resulting matrix can be designated as $\mathbf{A}_{\text{adj}}$, $\text{adj}\{\mathbf{A}\}$, or $\mathbf{A}^{\text{adj}}$.

### Example 6.6

What is the classical adjoint of matrix $\mathbf{A}$?

$$\mathbf{A} = \begin{bmatrix} 2 & 3 & -4 \\ 0 & -4 & 2 \\ 1 & -1 & 5 \end{bmatrix}$$

---

[8]The *row rank* and *column rank* are the same.

*Solution*

The matrix of cofactors is

$$\begin{bmatrix} -18 & 2 & 4 \\ -11 & 14 & 5 \\ -10 & -4 & -8 \end{bmatrix}$$

The transpose of the matrix of cofactors is

$$\mathbf{A}_{adj} = \begin{bmatrix} -18 & -11 & -10 \\ 2 & 14 & -4 \\ 4 & 5 & -8 \end{bmatrix}$$

## 12. INVERSE

The product of a matrix $\mathbf{A}$ and its inverse, $\mathbf{A}^{-1}$, is the identity matrix, $\mathbf{I}$. Only square matrices have inverses, but not all square matrices are invertible. A matrix has an inverse if and only if it is nonsingular (i.e., its determinant is nonzero).

$$\mathbf{A}\mathbf{A}^{-1} = \mathbf{A}^{-1}\mathbf{A} = \mathbf{I} \qquad 6.19$$

$$(\mathbf{A}\mathbf{B})^{-1} = \mathbf{B}^{-1}\mathbf{A}^{-1} \qquad 6.20$$

The inverse of a $2 \times 2$ matrix is easily determined by formula.

$$\mathbf{A} = \begin{bmatrix} a & b \\ c & d \end{bmatrix}$$

$$\mathbf{A}^{-1} = \frac{\begin{bmatrix} d & -b \\ -c & a \end{bmatrix}}{|\mathbf{A}|} \qquad 6.21$$

For a $3 \times 3$ or larger matrix, the inverse is determined by dividing every entry in the classical adjoint by the determinant of the original matrix.

$$\mathbf{A}^{-1} = \frac{\mathbf{A}_{adj}}{|\mathbf{A}|} \qquad 6.22$$

**Example 6.7**

What is the inverse of matrix $\mathbf{A}$?

$$\mathbf{A} = \begin{bmatrix} 4 & 5 \\ 2 & 3 \end{bmatrix}$$

*Solution*

The determinant is calculated as

$$|\mathbf{A}| = (4)(3) - (2)(5) = 2$$

Using Eq. 6.22, the inverse is

$$\mathbf{A}^{-1} = \frac{\begin{bmatrix} 3 & -5 \\ -2 & 4 \end{bmatrix}}{2} = \begin{bmatrix} \frac{3}{2} & -\frac{5}{2} \\ -1 & 2 \end{bmatrix}$$

Check.

$$\mathbf{A}\mathbf{A}^{-1} = \begin{bmatrix} 4 & 5 \\ 2 & 3 \end{bmatrix} \begin{bmatrix} \frac{3}{2} & -\frac{5}{2} \\ -1 & 2 \end{bmatrix} = \begin{bmatrix} 6-5 & -10+10 \\ 3-3 & -5+6 \end{bmatrix}$$

$$= \begin{bmatrix} 1 & 0 \\ 0 & 1 \end{bmatrix} = \mathbf{I} \qquad [\text{OK}]$$

## 13. WRITING SIMULTANEOUS LINEAR EQUATIONS IN MATRIX FORM

Matrices are used to simplify the presentation and solution of sets of simultaneous linear equations. For example, the following three methods of presenting simultaneous linear equations are equivalent:

$$a_{11}x_1 + a_{12}x_2 = b_1$$

$$a_{21}x_1 + a_{22}x_2 = b_2$$

$$\begin{bmatrix} a_{11} & a_{12} \\ a_{21} & a_{22} \end{bmatrix} \begin{bmatrix} x_1 \\ x_2 \end{bmatrix} = \begin{bmatrix} b_1 \\ b_2 \end{bmatrix}$$

$$\mathbf{A}\mathbf{X} = \mathbf{B}$$

In the second and third representations, $\mathbf{A}$ is known as the *coefficient matrix*, $\mathbf{X}$ as the *variable matrix*, and $\mathbf{B}$ as the *constant matrix*.

Not all systems of simultaneous equations have solutions, and those that do may not have unique solutions. The existence of a solution can be determined by calculating the determinant of the coefficient matrix. These rules are summarized in Table 6.1.

- If the system of linear equations is homogeneous (i.e., $\mathbf{B}$ is a zero matrix) and $|\mathbf{A}|$ is zero, there are an infinite number of solutions.

- If the system is homogeneous and $|\mathbf{A}|$ is nonzero, only the trivial solution exists.

- If the system of linear equations is nonhomogeneous (i.e., $\mathbf{B}$ is not a zero matrix) and $|\mathbf{A}|$ is nonzero, there is a unique solution to the set of simultaneous equations.

- If $|\mathbf{A}|$ is zero, a nonhomogeneous system of simultaneous equations may still have a solution. The requirement is that the determinants of all substitutional matrices (see Sec. 6.14) are zero, in which case there will be an infinite number of solutions. Otherwise, no solution exists.

**Table 6.1** *Solution Existence Rules for Simultaneous Equations*

|  | $\mathbf{B} = 0$ | $\mathbf{B} \neq 0$ |
|---|---|---|
| $|\mathbf{A}| = 0$ | infinite number of solutions (linearly dependent equations) | either an infinite number of solutions or no solution at all |
| $|\mathbf{A}| \neq 0$ | trivial solution only ($x_i = 0$) | unique nonzero solution |

## 14. SOLVING SIMULTANEOUS LINEAR EQUATIONS

Gauss-Jordan elimination can be used to obtain the solution to a set of simultaneous linear equations. The coefficient matrix is augmented by the constant matrix. Then, elementary row operations are used to reduce the coefficient matrix to canonical form. All of the operations performed on the coefficient matrix are performed on the constant matrix. The variable values that satisfy the simultaneous equations will be the entries in the constant matrix when the coefficient matrix is in canonical form.

Determinants are used to calculate the solution to linear simultaneous equations through a procedure known as *Cramer's rule*.

The procedure is to calculate determinants of the original coefficient matrix $\mathbf{A}$ and of the $n$ matrices resulting from the systematic replacement of a column in $\mathbf{A}$ by the constant matrix $\mathbf{B}$. For a system of three equations in three unknowns, there are three substitutional matrices, $\mathbf{A}_1$, $\mathbf{A}_2$, and $\mathbf{A}_3$, as well as the original coefficient matrix, for a total of four matrices whose determinants must be calculated.

The values of the unknowns that simultaneously satisfy all of the linear equations are

$$x_1 = \frac{|\mathbf{A}_1|}{|\mathbf{A}|} \qquad \qquad 6.23$$

$$x_2 = \frac{|\mathbf{A}_2|}{|\mathbf{A}|} \qquad \qquad 6.24$$

$$x_3 = \frac{|\mathbf{A}_3|}{|\mathbf{A}|} \qquad \qquad 6.25$$

### Example 6.8

Use Gauss-Jordan elimination to solve the following system of simultaneous equations.

$$2x + 3y - 4z = 1$$
$$3x - y - 2z = 4$$
$$4x - 7y - 6z = -7$$

*Solution*

The augmented matrix is created by appending the constant matrix to the coefficient matrix.

$$\begin{bmatrix} 2 & 3 & -4 & | & 1 \\ 3 & -1 & -2 & | & 4 \\ 4 & -7 & -6 & | & -7 \end{bmatrix}$$

Elementary row operations are used to reduce the coefficient matrix to canonical form. For example, two times the first row is subtracted from the third row. This step obtains the 0 needed in the $a_{31}$ position.

$$\begin{bmatrix} 2 & 3 & -4 & | & 1 \\ 3 & -1 & -2 & | & 4 \\ 0 & -13 & 2 & | & -9 \end{bmatrix}$$

This process continues until the following form is obtained.

$$\begin{bmatrix} 1 & 0 & 0 & | & 3 \\ 0 & 1 & 0 & | & 1 \\ 0 & 0 & 1 & | & 2 \end{bmatrix}$$

$x = 3$, $y = 1$, and $z = 2$ satisfy this system of equations.

### Example 6.9

Use Cramer's rule to solve the following system of simultaneous equations.

$$2x + 3y - 4z = 1$$
$$3x - y - 2z = 4$$
$$4x - 7y - 6z = -7$$

*Solution*

The determinant of the coefficient matrix is

$$|\mathbf{A}| = \begin{vmatrix} 2 & 3 & -4 \\ 3 & -1 & -2 \\ 4 & -7 & -6 \end{vmatrix} = 82$$

The determinants of the substitutional matrices are

$$|\mathbf{A}_1| = \begin{vmatrix} 1 & 3 & -4 \\ 4 & -1 & -2 \\ -7 & -7 & -6 \end{vmatrix} = 246$$

$$|\mathbf{A}_2| = \begin{vmatrix} 2 & 1 & -4 \\ 3 & 4 & -2 \\ 4 & -7 & -6 \end{vmatrix} = 82$$

$$|\mathbf{A}_3| = \begin{vmatrix} 2 & 3 & 1 \\ 3 & -1 & 4 \\ 4 & -7 & -7 \end{vmatrix} = 164$$

The values of $x$, $y$, and $z$ that will satisfy the linear equations are

$$x = \frac{246}{82} = 3$$

$$y = \frac{82}{82} = 1$$

$$z = \frac{164}{82} = 2$$

## 15. EIGENVALUES AND EIGENVECTORS

Eigenvalues and eigenvectors (also known as *characteristic values* and *characteristic vectors*) of a square matrix $\mathbf{A}$ are the scalars $k$ and matrices $\mathbf{X}$ such that

$$\mathbf{AX} = k\mathbf{X} \qquad 6.26$$

The scalar $k$ is an eigenvalue of $\mathbf{A}$ if and only if the matrix $(k\mathbf{I} - \mathbf{A})$ is singular; that is, if $|k\mathbf{I} - \mathbf{A}| = 0$. This equation is called the *characteristic equation* of the matrix $\mathbf{A}$. When expanded, the determinant is called the *characteristic polynomial*. The method of using the characteristic polynomial to find eigenvalues and eigenvectors is illustrated in Ex. 6.10.

If all of the eigenvalues are unique (i.e., nonrepeating), then Eq. 6.27 is valid.

$$[k\mathbf{I} - \mathbf{A}]\mathbf{X} = 0 \qquad 6.27$$

**Example 6.10**

Find the eigenvalues and nonzero eigenvectors of the matrix $\mathbf{A}$.

$$\mathbf{A} = \begin{bmatrix} 2 & 4 \\ 6 & 4 \end{bmatrix}$$

*Solution*

$$k\mathbf{I} - \mathbf{A} = \begin{bmatrix} k & 0 \\ 0 & k \end{bmatrix} - \begin{bmatrix} 2 & 4 \\ 6 & 4 \end{bmatrix} = \begin{bmatrix} k-2 & -4 \\ -6 & k-4 \end{bmatrix}$$

The characteristic polynomial is found by setting the determinant $|k\mathbf{I} - \mathbf{A}|$ equal to zero.

$$(k-2)(k-4) - (-6)(-4) = 0$$
$$k^2 - 6k - 16 = (k-8)(k+2) = 0$$

The roots of the characteristic polynomial are $k = +8$ and $k = -2$. These are the eigenvalues of $\mathbf{A}$.

Substituting $k = 8$,

$$k\mathbf{I} - \mathbf{A} = \begin{bmatrix} 8-2 & -4 \\ -6 & 8-4 \end{bmatrix} = \begin{bmatrix} 6 & -4 \\ -6 & 4 \end{bmatrix}$$

The resulting system can be interpreted as the linear equation $6x_1 - 4x_2 = 0$. The values of $x$ that satisfy this equation define the eigenvector. An eigenvector $\mathbf{X}$ associated with the eigenvalue $+8$ is

$$\mathbf{X} = \begin{bmatrix} x_1 \\ x_2 \end{bmatrix} = \begin{bmatrix} 4 \\ 6 \end{bmatrix}$$

All other eigenvectors for this eigenvalue are multiples of $\mathbf{X}$. Normally $\mathbf{X}$ is reduced to smallest integers.

$$\mathbf{X} = \begin{bmatrix} 2 \\ 3 \end{bmatrix}$$

Similarly, the eigenvector associated with the eigenvalue $-2$ is

$$\mathbf{X} = \begin{bmatrix} x_1 \\ x_2 \end{bmatrix} = \begin{bmatrix} +4 \\ -4 \end{bmatrix}$$

Reducing this to smallest integers gives

$$\mathbf{X} = \begin{bmatrix} +1 \\ -1 \end{bmatrix}$$

# 7 Vectors

## 1. INTRODUCTION

A physical property or quantity can be described by a scalar, vector, or tensor. A *scalar* has only magnitude. Knowing its value is sufficient to define a scalar. Mass, enthalpy, density, and speed are examples of scalars.

Force, momentum, displacement, and velocity are examples of vectors. A *vector* is a directed straight line with a specific magnitude. Thus, a vector is specified completely by its direction (consisting of the vector's *angular orientation* and its *sense*) and magnitude. A vector's *point of application* (*terminal point*) is not needed to define the vector.[1] Two vectors with the same direction and magnitude are said to be *equal vectors* even though their *lines of action* may be different.[2]

A vector can be designated by a boldface variable (as in this book) or as a combination of the variable and some other symbol. For example, the notations $\mathbf{V}$, $\overline{V}$, $\hat{V}$, $\vec{V}$, and $\underline{V}$ are used by different authorities to represent vectors. In this book, the magnitude of a vector can be designated by either $|\mathbf{V}|$ or $V$ (italic but not bold).

Stress, dielectric constant, and magnetic susceptibility are examples of tensors. A *tensor* has magnitude in a specific direction but the direction is not unique. Tensors are frequently associated with *anisotropic materials* that have different properties in different directions. A tensor in three-dimensional space is defined by nine components, compared with the three that are required

to define vectors. These components are written in matrix form. Stress, $\sigma$, at a point, for example, would be defined by the following tensor matrix.

$$\sigma \equiv \begin{pmatrix} \sigma_{xx} & \sigma_{xy} & \sigma_{xz} \\ \sigma_{yx} & \sigma_{yy} & \sigma_{yz} \\ \sigma_{zx} & \sigma_{zy} & \sigma_{zz} \end{pmatrix}$$

## 2. VECTORS IN $n$-SPACE

In some cases, a vector, $\mathbf{V}$, will be designated by its two endpoints in $n$-dimensional vector space. The usual vector space is three-dimensional force-space. Usually, one of the points will be the origin, in which case the vector is said to be "based at the origin," "origin-based," or "zero-based."[3] If one of the endpoints is the origin, specifying a terminal point P would represent a force directed from the origin to point P.

If a coordinate system is superimposed on the vector space, a vector can be specified in terms of the $n$ coordinates of its two endpoints. The magnitude of the vector $\mathbf{V}$ is the distance in vector space between the two points, as given by Eq. 7.1. Similarly, the direction is defined by the angle the vector makes with one of the axes. Figure 7.1 illustrates a vector in two dimensions.

$$|\mathbf{V}| = \sqrt{(x_2 - x_1)^2 + (y_2 - y_1)^2} \qquad 7.1$$

$$\phi = \arctan\left(\frac{y_2 - y_1}{x_2 - x_1}\right) \qquad 7.2$$

**Figure 7.1** *Vector in Two-Dimensional Space*

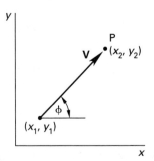

---

[1]A vector that is constrained to act at or through a certain point is a *bound vector (fixed vector)*. A *sliding vector (transmissible vector)* can be applied anywhere along its line of action. A *free vector* is not constrained and can be applied at any point in space.

[2]A distinction is sometimes made between equal vectors and equivalent vectors. *Equivalent vectors* produce the same effect but are not necessarily equal.

[3]Any vector directed from $P_1$ to $P_2$ can be transformed into a zero-based vector by subtracting the coordinates of point $P_1$ from the coordinates of terminal point $P_2$. The transformed vector will be equivalent to the original vector.

The *components* of a vector are the projections of the vector on the coordinate axes. (For a zero-based vector, the components and the coordinates of the endpoint are the same.) Simple trigonometric principles are used to resolve a vector into its components. A vector reconstructed from its components is known as a *resultant vector*. Figure 7.2 shows the location of direction angles.

$$V_x = |\mathbf{V}|\cos\phi_x \qquad 7.3$$

$$V_y = |\mathbf{V}|\cos\phi_y \qquad 7.4$$

$$V_z = |\mathbf{V}|\cos\phi_z \qquad 7.5$$

$$|\mathbf{V}| = \sqrt{V_x^2 + V_y^2 + V_z^2} \qquad 7.6$$

**Figure 7.2** *Direction Angles of a Vector*

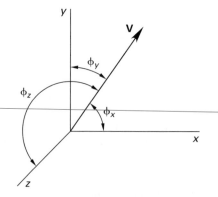

In Eq. 7.3 through Eq. 7.5, $\phi_x$, $\phi_y$, and $\phi_z$ are the *direction angles*—the angles between the vector and the $x$-, $y$-, and $z$-axes, respectively. The cosines of these angles are known as *direction cosines*. The sum of the squares of the direction cosines is equal to 1.

$$\cos^2\phi_x + \cos^2\phi_y + \cos^2\phi_z = 1 \qquad 7.7$$

## 3. UNIT VECTORS

*Unit vectors* are vectors with unit magnitudes (i.e., magnitudes of 1). They are represented in the same notation as other vectors. (Unit vectors in this book are written in boldface type.) Although they can have any direction, the standard unit vectors (the *Cartesian unit vectors* $\mathbf{i}$, $\mathbf{j}$, and $\mathbf{k}$) have the directions of the $x$-, $y$-, and $z$-coordinate axes and constitute the *Cartesian triad*, as illustrated in Fig. 7.3.

A vector $\mathbf{V}$ can be written in terms of unit vectors and its components.

$$\mathbf{V} = |\mathbf{V}|\mathbf{a} = V_x\mathbf{i} + V_y\mathbf{j} + V_z\mathbf{k} \qquad 7.8$$

**Figure 7.3** *Cartesian Unit Vectors*

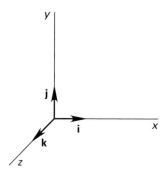

The unit vector, $\mathbf{a}$, has the same direction as the vector $\mathbf{V}$ but has a length of 1. This unit vector is calculated by dividing the original vector, $\mathbf{V}$, by its magnitude, $|\mathbf{V}|$.

$$\mathbf{a} = \frac{\mathbf{V}}{|\mathbf{V}|} = \frac{V_x\mathbf{i} + V_y\mathbf{j} + V_z\mathbf{k}}{\sqrt{V_x^2 + V_y^2 + V_z^2}} \qquad 7.9$$

## 4. VECTOR REPRESENTATION

The most common method of representing a vector is by writing it in *rectangular form*—a vector sum of its orthogonal components. In rectangular form, each of the orthogonal components has the same units as the resultant vector.

$$\mathbf{A} \equiv A_x\mathbf{i} + A_y\mathbf{j} + A_z\mathbf{k} \quad \text{[three dimensions]}$$

However, the vector is also completely defined by its magnitude and associated angle. These two quantities can be written together in *phasor form*, sometimes referred to as *polar form*.

$$\mathbf{A} \equiv |\mathbf{A}|\angle\phi = A\angle\phi$$

## 5. CONVERSION BETWEEN SYSTEMS

The choice of the $\mathbf{ijk}$ triad may be convenient but is arbitrary. A vector can be expressed in terms of any other set of unit vectors, $\mathbf{uvw}$.

$$\mathbf{V} = V_x\mathbf{i} + V_y\mathbf{j} + V_z\mathbf{k} = V_x'\mathbf{u} + V_y'\mathbf{v} + V_z'\mathbf{w} \qquad 7.10$$

The two representations are related.

$$V_x' = \mathbf{V}\cdot\mathbf{u} = (\mathbf{i}\cdot\mathbf{u})V_x + (\mathbf{j}\cdot\mathbf{u})V_y + (\mathbf{k}\cdot\mathbf{u})V_z$$
$$7.11$$

$$V_y' = \mathbf{V}\cdot\mathbf{v} = (\mathbf{i}\cdot\mathbf{v})V_x + (\mathbf{j}\cdot\mathbf{v})V_y + (\mathbf{k}\cdot\mathbf{v})V_z$$
$$7.12$$

$$V_z' = \mathbf{V}\cdot\mathbf{w} = (\mathbf{i}\cdot\mathbf{w})V_x + (\mathbf{j}\cdot\mathbf{w})V_y + (\mathbf{k}\cdot\mathbf{w})V_z$$
$$7.13$$

Equation 7.11 through Eq. 7.13 can be expressed in matrix form. The dot products are known as the *coefficients of transformation*, and the matrix containing them is the *transformation matrix*.

$$\begin{pmatrix} V'_x \\ V'_y \\ V'_z \end{pmatrix} = \begin{pmatrix} \mathbf{i} \cdot \mathbf{u} & \mathbf{j} \cdot \mathbf{u} & \mathbf{k} \cdot \mathbf{u} \\ \mathbf{i} \cdot \mathbf{v} & \mathbf{j} \cdot \mathbf{v} & \mathbf{k} \cdot \mathbf{v} \\ \mathbf{i} \cdot \mathbf{w} & \mathbf{j} \cdot \mathbf{w} & \mathbf{k} \cdot \mathbf{w} \end{pmatrix} \begin{pmatrix} V_x \\ V_y \\ V_z \end{pmatrix} \qquad 7.14$$

## 6. VECTOR ADDITION

Addition of two vectors by the *polygon method* is accomplished by placing the tail of the second vector at the head (tip) of the first. The sum (i.e., the *resultant vector*) is a vector extending from the tail of the first vector to the head of the second, as shown in Fig. 7.4. Alternatively, the two vectors can be considered as the two sides of a parallelogram, while the sum represents the diagonal. This is known as addition by the *parallelogram method*.

**Figure 7.4** Addition of Two Vectors

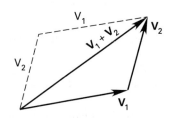

The components of the resultant vector are the sums of the components of the added vectors (that is, $V_{1x} + V_{2x}$, $V_{1y} + V_{2y}$, $V_{1z} + V_{2z}$).

Vector addition is both commutative and associative.

$$\mathbf{V}_1 + \mathbf{V}_2 = \mathbf{V}_2 + \mathbf{V}_1 \qquad 7.15$$
$$\mathbf{V}_1 + (\mathbf{V}_2 + \mathbf{V}_3) = (\mathbf{V}_1 + \mathbf{V}_2) + \mathbf{V}_3 \qquad 7.16$$

## 7. MULTIPLICATION BY A SCALAR

A vector, $\mathbf{V}$, can be multiplied by a scalar, $c$. If the original vector is represented by its components, each of the components is multiplied by $c$.

$$c\mathbf{V} = c|\mathbf{V}|\mathbf{a} = cV_x\mathbf{i} + cV_y\mathbf{j} + cV_z\mathbf{k} \qquad 7.17$$

Scalar multiplication is distributive.

$$c(\mathbf{V}_1 + \mathbf{V}_2) = c\mathbf{V}_1 + c\mathbf{V}_2 \qquad 7.18$$

## 8. VECTOR DOT PRODUCT

The *dot product (scalar product)*, $\mathbf{V}_1 \cdot \mathbf{V}_2$, of two vectors is a scalar that is proportional to the length of the projection of the first vector onto the second vector, as illustrated in Fig. 7.5.[4]

**Figure 7.5** Vector Dot Product

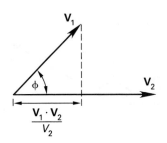

The dot product is commutative and distributive.

$$\mathbf{V}_1 \cdot \mathbf{V}_2 = \mathbf{V}_2 \cdot \mathbf{V}_1 \qquad 7.19$$
$$\mathbf{V}_1 \cdot (\mathbf{V}_2 + \mathbf{V}_3) = \mathbf{V}_1 \cdot \mathbf{V}_2 + \mathbf{V}_1 \cdot \mathbf{V}_3 \qquad 7.20$$

The dot product can be calculated in two ways, as Eq. 7.21 indicates. $\phi$ is limited to 180° and is the angle between the two vectors.

$$\mathbf{V}_1 \cdot \mathbf{V}_2 = |\mathbf{V}_1|\,|\mathbf{V}_2| \cos\phi$$
$$= V_{1x}V_{2x} + V_{1y}V_{2y} + V_{1z}V_{2z} \qquad 7.21$$

When Eq. 7.21 is solved for the angle between the two vectors, $\phi$, it is known as the *Cauchy-Schwartz theorem*.

$$\cos\phi = \frac{V_{1x}V_{2x} + V_{1y}V_{2y} + V_{1z}V_{2z}}{|\mathbf{V}_1|\,|\mathbf{V}_2|} \qquad 7.22$$

The dot product can be used to determine whether a vector is a unit vector and to show that two vectors are orthogonal (perpendicular). For any unit vector, $\mathbf{u}$,

$$\mathbf{u} \cdot \mathbf{u} = 1 \qquad 7.23$$

For two non-null orthogonal vectors,

$$\mathbf{V}_1 \cdot \mathbf{V}_2 = 0 \qquad 7.24$$

Equation 7.23 and Eq. 7.24 can be extended to the Cartesian unit vectors.

$$\mathbf{i} \cdot \mathbf{i} = 1 \qquad 7.25$$
$$\mathbf{j} \cdot \mathbf{j} = 1 \qquad 7.26$$
$$\mathbf{k} \cdot \mathbf{k} = 1 \qquad 7.27$$
$$\mathbf{i} \cdot \mathbf{j} = 0 \qquad 7.28$$

---

[4]The dot product is also written in parentheses without a dot, that is, $(\mathbf{V}_1\mathbf{V}_2)$.

$$\mathbf{i} \cdot \mathbf{k} = 0 \qquad \textit{7.29}$$

$$\mathbf{j} \cdot \mathbf{k} = 0 \qquad \textit{7.30}$$

## Example 7.1

What is the angle between the zero-based vectors $\mathbf{V}_1 = (-\sqrt{3}, 1)$ and $\mathbf{V}_2 = (2\sqrt{3}, 2)$?

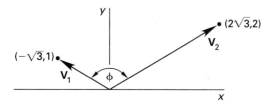

*Solution*

From Eq. 7.22,

$$\cos\phi = \frac{V_{1x}V_{2x} + V_{1y}V_{2y}}{|\mathbf{V}_1|\,|\mathbf{V}_2|} = \frac{V_{1x}V_{2x} + V_{1y}V_{2y}}{\sqrt{V_{1x}^2 + V_{1y}^2}\sqrt{V_{2x}^2 + V_{2y}^2}}$$

$$= \frac{(-\sqrt{3})(2\sqrt{3}) + (1)(2)}{\sqrt{(-\sqrt{3})^2 + (1)^2}\sqrt{(2\sqrt{3})^2 + (2)^2}}$$

$$= \frac{-4}{8} = -\frac{1}{2}$$

$$\phi = \arccos\left(-\frac{1}{2}\right) = 120°$$

## 9. VECTOR CROSS PRODUCT

The *cross product (vector product)*, $\mathbf{V}_1 \times \mathbf{V}_2$, of two vectors is a vector that is orthogonal (perpendicular) to the plane of the two vectors.[5] The unit vector representation of the cross product can be calculated as a third-order determinant. Figure 7.6 illustrates the vector cross product.

$$\mathbf{V}_1 \times \mathbf{V}_2 = \begin{vmatrix} \mathbf{i} & V_{1x} & V_{2x} \\ \mathbf{j} & V_{1y} & V_{2y} \\ \mathbf{k} & V_{1z} & V_{2z} \end{vmatrix} \qquad \textit{7.31}$$

The direction of the cross-product vector corresponds to the direction a right-hand screw would progress if vectors $\mathbf{V}_1$ and $\mathbf{V}_2$ are placed tail-to-tail in the plane they define and $\mathbf{V}_1$ is rotated into $\mathbf{V}_2$. The direction can also be found from the *right-hand rule*.

The magnitude of the cross product can be determined from Eq. 7.32, in which $\phi$ is the angle between the two vectors and is limited to 180°. The magnitude

---

[5]The cross product is also written in square brackets without a cross, that is, $[\mathbf{V}_1\mathbf{V}_2]$.

**Figure 7.6** *Vector Cross Product*

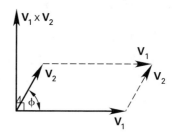

corresponds to the area of a parallelogram that has $\mathbf{V}_1$ and $\mathbf{V}_2$ as two of its sides.

$$|\mathbf{V}_1 \times \mathbf{V}_2| = |\mathbf{V}_1||\mathbf{V}_2|\sin\phi \qquad \textit{7.32}$$

Vector cross multiplication is distributive but not commutative.

$$\mathbf{V}_1 \times \mathbf{V}_2 = -(\mathbf{V}_2 \times \mathbf{V}_1) \qquad \textit{7.33}$$

$$c(\mathbf{V}_1 \times \mathbf{V}_2) = (c\mathbf{V}_1) \times \mathbf{V}_2 = \mathbf{V}_1 \times (c\mathbf{V}_2) \qquad \textit{7.34}$$

$$\mathbf{V}_1 \times (\mathbf{V}_2 + \mathbf{V}_3) = \mathbf{V}_1 \times \mathbf{V}_2 + \mathbf{V}_1 \times \mathbf{V}_3 \qquad \textit{7.35}$$

If the two vectors are parallel, their cross product will be zero.

$$\mathbf{i} \times \mathbf{i} = \mathbf{j} \times \mathbf{j} = \mathbf{k} \times \mathbf{k} = 0 \qquad \textit{7.36}$$

Equation 7.31 and Eq. 7.33 can be extended to the unit vectors.

$$\mathbf{i} \times \mathbf{j} = -\mathbf{j} \times \mathbf{i} = \mathbf{k} \qquad \textit{7.37}$$

$$\mathbf{j} \times \mathbf{k} = -\mathbf{k} \times \mathbf{j} = \mathbf{i} \qquad \textit{7.38}$$

$$\mathbf{k} \times \mathbf{i} = -\mathbf{i} \times \mathbf{k} = \mathbf{j} \qquad \textit{7.39}$$

## Example 7.2

Find a unit vector orthogonal to $\mathbf{V}_1 = \mathbf{i} - \mathbf{j} + 2\mathbf{k}$ and $\mathbf{V}_2 = 3\mathbf{j} - \mathbf{k}$.

*Solution*

The cross product is a vector orthogonal to $\mathbf{V}_1$ and $\mathbf{V}_2$.

$$\mathbf{V}_1 \times \mathbf{V}_2 = \begin{vmatrix} \mathbf{i} & 1 & 0 \\ \mathbf{j} & -1 & 3 \\ \mathbf{k} & 2 & -1 \end{vmatrix}$$

$$= -5\mathbf{i} + \mathbf{j} + 3\mathbf{k}$$

Check to see whether this is a unit vector.

$$|\mathbf{V}_1 \times \mathbf{V}_2| = \sqrt{(-5)^2 + (1)^2 + (3)^2} = \sqrt{35}$$

Since its length is $\sqrt{35}$, the vector must be divided by $\sqrt{35}$ to obtain a unit vector.

$$\mathbf{a} = \frac{-5\mathbf{i} + \mathbf{j} + 3\mathbf{k}}{\sqrt{35}}$$

## 10. MIXED TRIPLE PRODUCT

The *mixed triple product* (*triple scalar product* or just *triple product*) of three vectors is a scalar quantity representing the volume of a parallelepiped with the three vectors making up the sides. It is calculated as a determinant. Since Eq. 7.40 can be negative, the absolute value must be used to obtain the volume in that case. Figure 7.7 shows a mixed triple product.

$$\mathbf{V}_1 \cdot (\mathbf{V}_2 \times \mathbf{V}_3) = \begin{vmatrix} V_{1x} & V_{1y} & V_{1z} \\ V_{2x} & V_{2y} & V_{2z} \\ V_{3x} & V_{3y} & V_{3z} \end{vmatrix} \qquad 7.40$$

**Figure 7.7** *Vector Mixed Triple Product*

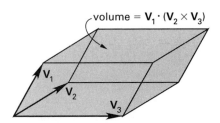

The mixed triple product has the property of *circular permutation*, as defined by Eq. 7.41.

$$\mathbf{V}_1 \cdot (\mathbf{V}_2 \times \mathbf{V}_3) = (\mathbf{V}_1 \times \mathbf{V}_2) \cdot \mathbf{V}_3 \qquad 7.41$$

## 11. VECTOR TRIPLE PRODUCT

The *vector triple product* is a vector defined by Eq. 7.42. The quantities in parentheses on the right-hand side are scalars.

$$\mathbf{V}_1 \times (\mathbf{V}_2 \times \mathbf{V}_3) = (\mathbf{V}_1 \cdot \mathbf{V}_3)\mathbf{V}_2 - (\mathbf{V}_1 \cdot \mathbf{V}_2)\mathbf{V}_3 \qquad 7.42$$

## 12. VECTOR FUNCTIONS

A vector can be a function of another parameter. For example, a vector $\mathbf{V}$ is a function of variable $t$ when its $V_x$, $V_y$, and $V_z$ are functions of $t$.

$$\mathbf{V}(t) = (2t - 3)\mathbf{i} + (t^2 + 1)\mathbf{j} + (-7t + 5)\mathbf{k} \qquad 7.43$$

When the functions of $t$ are differentiated (or integrated) with respect to $t$, the vector itself is differentiated (integrated).[6]

$$\frac{d\mathbf{V}(t)}{dt} = \left(\frac{dV_x}{dt}\right)\mathbf{i} + \left(\frac{dV_y}{dt}\right)\mathbf{j} + \left(\frac{dV_z}{dt}\right)\mathbf{k} \qquad 7.44$$

Similarly, the integral of the vector is

$$\int \mathbf{V}(t)\, dt = \mathbf{i} \int V_x\, dt + \mathbf{j} \int V_y\, dt + \mathbf{k} \int V_z\, dt \qquad 7.45$$

---

[6]This is particularly valuable when converting among position, velocity, and acceleration vectors.

# 8 Trigonometry

**Figure 8.1** *Radians and Area of Unit Circle*

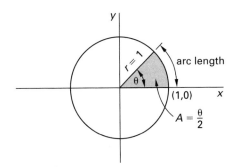

**Figure 8.2** *Angle*

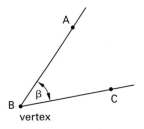

## 1. DEGREES AND RADIANS

*Degrees* and *radians* are two units for measuring angles. One complete circle is divided into 360 degrees (written $360°$) or $2\pi$ radians (abbreviated *rad*).[1] The conversions between degrees and radians are

| multiply | by | to obtain |
|----------|-----|-----------|
| radians | $\dfrac{180}{\pi}$ | degrees |
| degrees | $\dfrac{\pi}{180}$ | radians |

The number of radians in an angle $\theta$ corresponds to two times the area within a circular sector with arc length $\theta$ and a radius of one, as shown in Fig. 8.1. Alternatively, the area of a sector with central angle $\theta$ radians is $\theta/2$ for a *unit circle* (i.e., a circle with a radius of one unit).

## 2. PLANE ANGLES

A *plane angle* (usually referred to as just an *angle*) consists of two intersecting lines and an intersection point known as the *vertex*. The angle can be referred to by a capital letter representing the vertex (e.g., B in Fig. 8.2), a letter representing the angular measure (e.g., B or $\beta$), or by three capital letters, where the middle letter is the vertex and the other two letters are two points on different lines, and either the symbol $\angle$ or $\sphericalangle$ (e.g., $\sphericalangle$ ABC).

The angle between two intersecting lines generally is understood to be the smaller angle created.[2] Angles have been classified as follows.

- *acute angle:* an angle less than $90°$ ($\pi/2$ rad)

- *obtuse angle:* an angle more than $90°$ ($\pi/2$ rad) but less than $180°$ ($\pi$ rad)

- *reflex angle:* an angle more than $180°$ ($\pi$ rad) but less than $360°$ ($2\pi$ rad)

- *related angle:* an angle that differs from another by some multiple of $90°$ ($\pi/2$ rad)

- *right angle:* an angle equal to $90°$ ($\pi/2$ rad)

- *straight angle:* an angle equal to $180°$ ($\pi$ rad), that is, a straight line

*Complementary angles* are two angles whose sum is $90°$ ($\pi/2$ rad). *Supplementary angles* are two angles whose sum is $180°$ ($\pi$ rad). *Adjacent angles* share a common vertex and one (the interior) side. Adjacent angles are supplementary only if their exterior sides form a straight line.

---

[1]The abbreviation *rad* is also used to represent *radiation absorbed dose*, a measure of radiation exposure.

[2]In books on geometry, the term *ray* is used instead of *line*.

*Vertical angles* are the two angles with a common vertex and with sides made up by two intersecting straight lines, as shown in Fig. 8.3. Vertical angles are equal.

**Figure 8.3** *Vertical Angles*

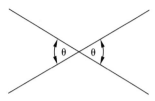

*Angle of elevation* and *angle of depression* are surveying terms referring to the angle above and below the horizontal plane of the observer, respectively.

## 3. TRIANGLES

A *triangle* is a three-sided closed polygon with three angles whose sum is 180° (π rad). Triangles are identified by their vertices and the symbol Δ (e.g., ΔABC in Fig. 8.4). A side is designated by its two endpoints (e.g., AB in Fig. 8.4) or by a lowercase letter corresponding to the capital letter of the opposite vertex (e.g., *c*).

In *similar triangles*, the corresponding angles are equal and the corresponding sides are in proportion. (Since there are only two independent angles in a triangle, showing that two angles of one triangle are equal to two angles of the other triangle is sufficient to show similarity.) The symbol for similarity is ~. In Fig. 8.4, ΔABC ~ ΔDEF (i.e., ΔABC is similar to ΔDEF).

**Figure 8.4** *Similar Triangles*

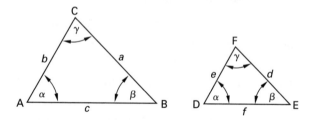

## 4. RIGHT TRIANGLES

A *right triangle* is a triangle in which one of the angles is 90° (π/2 rad). The remaining two angles are complementary. If one of the acute angles is chosen as the reference, the sides forming the right angle are known as the *adjacent side*, $x$, and the *opposite side*, $y$. The longest side is known as the *hypotenuse*, $r$. The *Pythagorean theorem* relates the lengths of these sides.

$$x^2 + y^2 = r^2 \qquad 8.1$$

In certain cases, the lengths of unknown sides of right triangles can be determined by inspection.[3] This occurs when the lengths of the sides are in the ratios of 3:4:5, 1:1:$\sqrt{2}$, 1:$\sqrt{3}$:2, and 5:12:13. Figure 8.5 illustrates a 3:4:5 triangle.

**Figure 8.5** *3:4:5 Right Triangle*

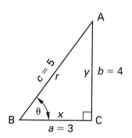

## 5. CIRCULAR TRANSCENDENTAL FUNCTIONS

The *circular transcendental functions* (usually referred to as the *transcendental functions*, *trigonometric functions*, or *functions of an angle*) are calculated from the sides of a right triangle. Equation 8.2 through Eq. 8.7 refer to Fig. 8.5.

$$\text{sine: } \sin\theta = \frac{y}{r} = \frac{\text{opposite}}{\text{hypotenuse}} \qquad 8.2$$

$$\text{cosine: } \cos\theta = \frac{x}{r} = \frac{\text{adjacent}}{\text{hypotenuse}} \qquad 8.3$$

$$\text{tangent: } \tan\theta = \frac{y}{x} = \frac{\text{opposite}}{\text{adjacent}} \qquad 8.4$$

$$\text{cotangent: } \cot\theta = \frac{x}{y} = \frac{\text{adjacent}}{\text{opposite}} \qquad 8.5$$

$$\text{secant: } \sec\theta = \frac{r}{x} = \frac{\text{hypotenuse}}{\text{adjacent}} \qquad 8.6$$

$$\text{cosecant: } \csc\theta = \frac{r}{y} = \frac{\text{hypotenuse}}{\text{opposite}} \qquad 8.7$$

Three of the transcendental functions are reciprocals of the others. Notice that while the tangent and cotangent functions are reciprocals of each other, the sine and cosine functions are not.

$$\cot\theta = \frac{1}{\tan\theta} \qquad 8.8$$

$$\sec\theta = \frac{1}{\cos\theta} \qquad 8.9$$

$$\csc\theta = \frac{1}{\sin\theta} \qquad 8.10$$

---

[3]These cases are almost always contrived examples. There is nothing intrinsic in nature to cause the formation of triangles with these proportions.

The trigonometric functions correspond to the lengths of various line segments in a right triangle with a unit hypotenuse. Figure 8.6 shows such a triangle inscribed in a unit circle.

**Figure 8.6** *Trigonometric Functions in a Unit Circle*

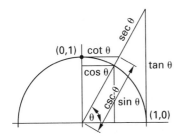

## 6. SMALL ANGLE APPROXIMATIONS

When an angle is very small, the hypotenuse and adjacent sides are essentially equal in length, and certain approximations can be made. (The angle $\theta$ must be expressed in radians in Eq. 8.11 and Eq. 8.12.)

$$\sin\theta \approx \tan\theta \approx \theta\big|_{\theta < 10° \ (0.175 \ \mathrm{rad})} \qquad 8.11$$

$$\cos\theta \approx 1\big|_{\theta < 5° \ (0.0873 \ \mathrm{rad})} \qquad 8.12$$

## 7. GRAPHS OF THE FUNCTIONS

Figure 8.7 illustrates the periodicity of the sine, cosine, and tangent functions.[4]

**Figure 8.7** *Graphs of Sine, Cosine, and Tangent Functions*

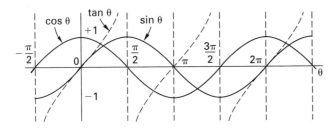

## 8. SIGNS OF THE FUNCTIONS

Table 8.1 shows how the sine, cosine, and tangent functions vary in sign with different values of $\theta$. All three functions are positive for angles $0° \leq \theta \leq 90°$ ($0 \leq \theta \leq \pi/2$ rad), but only the sine is positive for angles $90° < \theta \leq 180°$ ($\pi/2$ rad $< \theta \leq \pi$ rad). The concept of quadrants is used to summarize the signs of the functions: angles up to $90°$ ($\pi/2$ rad) are in quadrant I, between $90°$ and $180°$ ($\pi/2$ and $\pi$ rad) are in quadrant II, and so on.

**Table 8.1** *Signs of the Functions by Quadrant*

| function | quadrant | | | |
|---|---|---|---|---|
| | I | II | III | IV |
| sine | + | + | − | − |
| cosine | + | − | − | + |
| tangent | + | − | + | − |

## 9. FUNCTIONS OF RELATED ANGLES

Figure 8.7 shows that the sine, cosine, and tangent curves are symmetrical with respect to the horizontal axis. Furthermore, portions of the curves are symmetrical with respect to a vertical axis. The values of the sine and cosine functions repeat every $360°$ ($2\pi$ rad), and the absolute values repeat every $180°$ ($\pi$ rad). This can be written as

$$\sin(\theta + 180°) = -\sin\theta \qquad 8.13$$

Similarly, the tangent function repeats every $180°$ ($\pi$ rad), and its absolute value repeats every $90°$ ($\pi/2$ rad).

Table 8.2 summarizes the functions of the related angles.

**Table 8.2** *Functions of Related Angles*

| $f(\theta)$ | $-\theta$ | $90°-\theta$ | $90°+\theta$ | $180°-\theta$ | $180°+\theta$ |
|---|---|---|---|---|---|
| sin | $-\sin\theta$ | $\cos\theta$ | $\cos\theta$ | $\sin\theta$ | $-\sin\theta$ |
| cos | $\cos\theta$ | $\sin\theta$ | $-\sin\theta$ | $-\cos\theta$ | $-\cos\theta$ |
| tan | $-\tan\theta$ | $\cot\theta$ | $-\cot\theta$ | $-\tan\theta$ | $\tan\theta$ |

## 10. TRIGONOMETRIC IDENTITIES

There are many relationships between trigonometric functions. For example, Eq. 8.14 through Eq. 8.16 are well known.

$$\sin^2\theta + \cos^2\theta = 1 \qquad 8.14$$

$$1 + \tan^2\theta = \sec^2\theta \qquad 8.15$$

$$1 + \cot^2\theta = \csc^2\theta \qquad 8.16$$

Other relatively common identities are listed as follows.[5]

- *double-angle formulas*

$$\sin 2\theta = 2\sin\theta\cos\theta = \frac{2\tan\theta}{1 + \tan^2\theta} \qquad 8.17$$

$$\cos 2\theta = \cos^2\theta - \sin^2\theta = 1 - 2\sin^2\theta$$
$$= 2\cos^2\theta - 1 = \frac{1 - \tan^2\theta}{1 + \tan^2\theta} \qquad 8.18$$

$$\tan 2\theta = \frac{2\tan\theta}{1 - \tan^2\theta} \qquad 8.19$$

$$\cot 2\theta = \frac{\cot^2\theta - 1}{2\cot\theta} \qquad 8.20$$

---

[4]The remaining functions, being reciprocals of these three functions, are also periodic.

[5]It is an idiosyncrasy of the trade that these formulas are conventionally referred to as *formulas*, not *identities*.

- *two-angle formulas*

$$\sin(\theta \pm \phi) = \sin\theta\cos\phi \pm \cos\theta\sin\phi \qquad 8.21$$

$$\cos(\theta \pm \phi) = \cos\theta\cos\phi \mp \sin\theta\sin\phi \qquad 8.22$$

$$\tan(\theta \pm \phi) = \frac{\tan\theta \pm \tan\phi}{1 \mp \tan\theta\tan\phi} \qquad 8.23$$

$$\cot(\theta \pm \phi) = \frac{\cot\phi\cot\theta \mp 1}{\cot\phi \pm \cot\theta} \qquad 8.24$$

- *half-angle formulas* $(\theta < 180°)$

$$\sin\frac{\theta}{2} = \sqrt{\frac{1 - \cos\theta}{2}} \qquad 8.25$$

$$\cos\frac{\theta}{2} = \sqrt{\frac{1 + \cos\theta}{2}} \qquad 8.26$$

$$\tan\frac{\theta}{2} = \sqrt{\frac{1 - \cos\theta}{1 + \cos\theta}} = \frac{\sin\theta}{1 + \cos\theta} = \frac{1 - \cos\theta}{\sin\theta} \qquad 8.27$$

- *miscellaneous formulas* $(\theta < 90°)$

$$\sin\theta = 2\sin\frac{\theta}{2}\cos\frac{\theta}{2} \qquad 8.28$$

$$\sin\theta = \sqrt{\frac{1 - \cos 2\theta}{2}} \qquad 8.29$$

$$\cos\theta = \cos^2\frac{\theta}{2} - \sin^2\frac{\theta}{2} \qquad 8.30$$

$$\cos\theta = \sqrt{\frac{1 + \cos 2\theta}{2}} \qquad 8.31$$

$$\tan\theta = \frac{2\tan\frac{\theta}{2}}{1 - \tan^2\frac{\theta}{2}}$$

$$= \frac{2\sin\frac{\theta}{2}\cos\frac{\theta}{2}}{\cos^2\frac{\theta}{2} - \sin^2\frac{\theta}{2}} \qquad 8.32$$

$$\tan\theta = \sqrt{\frac{1 - \cos 2\theta}{1 + \cos 2\theta}}$$

$$= \frac{\sin 2\theta}{1 + \cos 2\theta} = \frac{1 - \cos 2\theta}{\sin 2\theta} \qquad 8.33$$

$$\cot\theta = \frac{\cot^2\frac{\theta}{2} - 1}{2\cot\frac{\theta}{2}}$$

$$= \frac{\cos^2\frac{\theta}{2} - \sin^2\frac{\theta}{2}}{2\sin\frac{\theta}{2}\cos\frac{\theta}{2}} \qquad 8.34$$

$$\cot\theta = \sqrt{\frac{1 + \cos 2\theta}{1 - \cos 2\theta}}$$

$$= \frac{1 + \cos 2\theta}{\sin 2\theta} = \frac{\sin 2\theta}{1 - \cos 2\theta} \qquad 8.35$$

## 11. INVERSE TRIGONOMETRIC FUNCTIONS

Finding an angle from a known trigonometric function is a common operation known as an *inverse trigonometric operation*. The inverse function can be designated by adding "inverse," "arc-," or the superscript −1 to the name of the function. For example,

$$\text{inverse }\sin(0.5) = \arcsin(0.5) = \sin^{-1}(0.5) = 30°$$

## 12. HYPERBOLIC TRANSCENDENTAL FUNCTIONS

*Hyperbolic transcendental functions* (normally referred to as *hyperbolic functions*) are specific equations containing combinations of the terms $e^\theta$ and $e^{-\theta}$. These combinations appear regularly in certain types of problems (e.g., analysis of cables and heat transfer from fins) and are given specific names and symbols to simplify presentation.[6]

hyperbolic sine: $\sinh\theta = \dfrac{e^\theta - e^{-\theta}}{2}$ $\qquad 8.36$

hyperbolic cosine: $\cosh\theta = \dfrac{e^\theta + e^{-\theta}}{2}$ $\qquad 8.37$

hyperbolic tangent: $\tanh\theta = \dfrac{e^\theta - e^{-\theta}}{e^\theta + e^{-\theta}} = \dfrac{\sinh\theta}{\cosh\theta}$ $\qquad 8.38$

hyperbolic cotangent: $\coth\theta = \dfrac{e^\theta + e^{-\theta}}{e^\theta - e^{-\theta}} = \dfrac{\cosh\theta}{\sinh\theta}$ $\qquad 8.39$

hyperbolic secant: $\operatorname{sech}\theta = \dfrac{2}{e^\theta + e^{-\theta}} = \dfrac{1}{\cosh\theta}$ $\qquad 8.40$

hyperbolic cosecant: $\operatorname{csch}\theta = \dfrac{2}{e^\theta - e^{-\theta}} = \dfrac{1}{\sinh\theta}$ $\qquad 8.41$

Hyperbolic functions cannot be related to a right triangle, but they are related to a rectangular (equilateral) hyperbola, as shown in Fig. 8.8. The shaded area has a value of $\theta/2$ and is sometimes given the units of *hyperbolic radians*.

$$\sinh\theta = \frac{y}{a} \qquad 8.42$$

$$\cosh\theta = \frac{x}{a} \qquad 8.43$$

$$\tanh\theta = \frac{y}{x} \qquad 8.44$$

$$\coth\theta = \frac{x}{y} \qquad 8.45$$

---

[6]The hyperbolic sine and cosine functions are pronounced (by some) as "sinch" and "cosh," respectively.

Mathematics

$$\operatorname{sech} \theta = \frac{a}{x} \qquad \textit{8.46}$$

$$\operatorname{csch} \theta = \frac{a}{y} \qquad \textit{8.47}$$

**Figure 8.8** *Equilateral Hyperbola and Hyperbolic Functions*

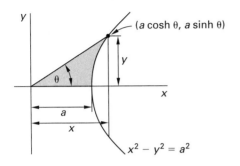

## 13. HYPERBOLIC IDENTITIES

The hyperbolic identities are different from the standard trigonometric identities. Some of the most important identities are presented as follows.

$$\cosh^2 \theta - \sinh^2 \theta = 1 \qquad \textit{8.48}$$

$$1 - \tanh^2 \theta = \operatorname{sech}^2 \theta \qquad \textit{8.49}$$

$$1 - \coth^2 \theta = -\operatorname{csch}^2 \theta \qquad \textit{8.50}$$

$$\cosh \theta + \sinh \theta = e^{\theta} \qquad \textit{8.51}$$

$$\cosh \theta - \sinh \theta = e^{-\theta} \qquad \textit{8.52}$$

$$\sinh(\theta \pm \phi) = \sinh \theta \cosh \phi \pm \cosh \theta \sinh \phi \qquad \textit{8.53}$$

$$\cosh(\theta \pm \phi) = \cosh \theta \cosh \phi \pm \sinh \theta \sinh \phi \qquad \textit{8.54}$$

$$\tanh(\theta \pm \phi) = \frac{\tanh \theta \pm \tanh \phi}{1 \pm \tanh \theta \tanh \phi} \qquad \textit{8.55}$$

## 14. GENERAL TRIANGLES

A *general triangle* (also known as an *oblique triangle*) is one that is not specifically a right triangle, as shown in Fig. 8.9. Equation 8.56 calculates the area of a general triangle.

$$\text{area} = \tfrac{1}{2}ab \sin C = \tfrac{1}{2}bc \sin A = \tfrac{1}{2}ca \sin B \qquad \textit{8.56}$$

**Figure 8.9** *General Triangle*

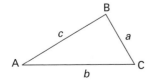

The *law of sines* (see Eq. 8.57) relates the sides and the sines of the angles.

$$\frac{\sin A}{a} = \frac{\sin B}{b} = \frac{\sin C}{c} \qquad \textit{8.57}$$

The *law of cosines* relates the cosine of an angle to an opposite side. (Equation 8.58 can be extended to the two remaining sides.)

$$a^2 = b^2 + c^2 - 2bc \cos A \qquad \textit{8.58}$$

The *law of tangents* relates the sum and difference of two sides. (Equation 8.59 can be extended to the two remaining sides.)

$$\frac{a - b}{a + b} = \frac{\tan\left(\dfrac{A - B}{2}\right)}{\tan\left(\dfrac{A + B}{2}\right)} \qquad \textit{8.59}$$

## 15. SPHERICAL TRIGONOMETRY

A *spherical triangle* is a triangle that has been drawn on the surface of a sphere, as shown in Fig. 8.10. The *trihedral angle* O–ABC is formed when the vertices A, B, and C are joined to the center of the sphere. The *face angles* (BOC, COA, and AOB in Fig. 8.10) are used to measure the sides ($a$, $b$, and $c$ in Fig. 8.10). The *vertex angles* are A, B, and C. Thus, angles are used to measure both vertex angles and sides.

**Figure 8.10** *Spherical Triangle ABC*

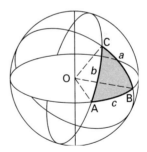

The following rules are valid for spherical triangles for which each side and angle is less than 180°.

- The sum of the three vertex angles is greater than 180° and less than 540°.

$$180° < A + B + C < 540° \qquad \textit{8.60}$$

- The sum of any two sides is greater than the third side.

- The sum of the three sides is less than 360°.

$$0° < a + b + c < 360° \qquad \textit{8.61}$$

- If the two sides are equal, the corresponding angles opposite are equal, and the converse is also true.

- If two sides are unequal, the corresponding angles opposite are unequal. The greater angle is opposite the greater side.

The *spherical excess*, $\epsilon$, is the amount by which the sum of the vertex angles exceeds 180°. The *spherical defect*, $d$, is the amount by which the sum of the sides differs from 360°.

$$\epsilon = A + B + C - 180° \qquad 8.62$$

$$d = 360° - (a + b + c) \qquad 8.63$$

There are many trigonometric identities that define the relationships between angles in a spherical triangle. Some of the more common identities are presented as follows.

- *law of sines*

$$\frac{\sin A}{\sin a} = \frac{\sin B}{\sin b} = \frac{\sin C}{\sin c} \qquad 8.64$$

- *first law of cosines*

$$\cos a = \cos b \cos c + \sin b \sin c \cos A \qquad 8.65$$

- *second law of cosines*

$$\cos A = -\cos B \cos C + \sin B \sin C \cos a \qquad 8.66$$

## 16. SOLID ANGLES

A *solid angle*, $\omega$, is a measure of the angle subtended at the vertex of a cone, as shown in Fig. 8.11. The solid angle has units of *steradians* (abbreviated *sr*). A steradian is the solid angle subtended at the center of a unit sphere (i.e., a sphere with a radius of one) by a unit area on its surface. Since the surface area of a sphere of radius $r$ is $r^2$ times the surface area of a unit sphere, the solid angle is equal to the area cut out by the cone divided by $r^2$.

$$\omega = \frac{\text{surface area}}{r^2} \qquad 8.67$$

*Figure 8.11* Solid Angle

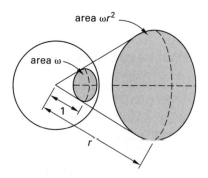

# Analytic Geometry

*Solution*

Points O, B, and C constitute a circular segment and are used to find the central angle of the circular segment.

$$\tfrac{1}{2}\angle\text{BOC} = \arccos\tfrac{1}{3} = 70.53°$$

$$\phi = \angle\text{BOC} = \frac{(2)(70.53°)(2\pi)}{360°} = 2.462 \text{ rad}$$

From App. 9.A, the area in flow and arc length are

$$A = \tfrac{1}{2}r^2(\phi - \sin\phi)$$
$$= \left(\tfrac{1}{2}\right)(3 \text{ in})^2\big(2.462 \text{ rad} - \sin(2.462 \text{ rad})\big)$$
$$= 8.251 \text{ in}^2$$
$$s = r\phi$$
$$= (3 \text{ in})(2.462 \text{ rad}) = 7.386 \text{ in}$$

The hydraulic radius is

$$r_h = \frac{A}{s}$$
$$= \frac{8.251 \text{ in}^2}{7.386 \text{ in}} = 1.12 \text{ in}$$

## 1. MENSURATION OF REGULAR SHAPES

The dimensions, perimeter, area, and other geometric properties constitute the *mensuration* (i.e., the measurements) of a geometric shape. Appendix 9.A and App. 9.B contain formulas and tables used to calculate these properties.

### Example 9.1

In the study of open channel fluid flow, the hydraulic radius is defined as the ratio of flow area to wetted perimeter. What is the hydraulic radius of a 6 in inside diameter pipe filled to a depth of 2 in?

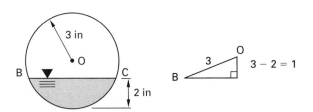

## 2. AREAS WITH IRREGULAR BOUNDARIES

Areas of sections with irregular boundaries (such as creek banks) cannot be determined precisely, and approximation methods must be used. If the irregular side can be divided into a series of cells of equal width, either the trapezoidal rule or Simpson's rule can be used. Figure 9.1 shows an example of an irregular area.

**Figure 9.1** *Irregular Areas*

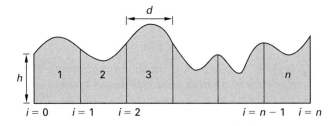

Mathematics

If the irregular side of each cell is fairly straight, the *trapezoidal rule* is appropriate.

$$A = \frac{d}{2}\left(h_0 + h_n + 2\sum_{i=1}^{n-1} h_i\right) \qquad 9.1$$

If the irregular side of each cell is curved (parabolic), *Simpson's rule* should be used. ($n$ must be even to use Simpson's rule.)

$$A = \frac{d}{3}\left(h_0 + h_n + 4\sum_{\substack{i \text{ odd} \\ i=1}}^{n-1} h_i + 2\sum_{\substack{i \text{ even} \\ i=2}}^{n-2} h_i\right) \qquad 9.2$$

## 3. GEOMETRIC DEFINITIONS

The following terms are used in this book to describe the relationship or orientation of one geometric figure to another. Figure 9.2 illustrates some of the following geometric definitions.

- *abscissa:* the horizontal coordinate, typically designated as $x$ in a rectangular coordinate system

- *asymptote:* a straight line that is approached but not intersected by a curved line

- *asymptotic:* approaching the slope of another line; attaining the slope of another line in the limit

- *center:* a point equidistant from all other points

- *collinear:* falling on the same line

- *concave:* curved inward (in the direction indicated)[1]

- *convex:* curved outward (in the direction indicated)

- *convex hull:* a closed figure whose surface is convex everywhere

- *coplanar:* falling on the same plane

- *inflection point:* a point where the second derivative changes sign or the curve changes from concave to convex; also known as a *point of contraflexure*

- *locus of points:* a set or collection of points having some common property and being so infinitely close together as to be indistinguishable from a line

- *node:* a point on a line from which other lines enter or leave

- *normal:* rotated 90°; being at right angles

- *ordinate:* the vertical coordinate, typically designated as $y$ in a rectangular coordinate system

- *orthogonal:* rotated 90°; being at right angles

- *saddle point:* a point in three-dimensional space where all adjacent points are higher in one direction

---

[1]This is easily remembered since one must go inside to explore a cave.

(the direction of the saddle) and lower in an orthogonal direction (the direction of the sides)

- *tangent point:* having equal slopes at a common point

**Figure 9.2** *Geometric Definitions*

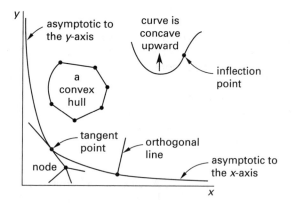

## 4. CONCAVE CURVES

*Concavity* is a term that is applied to curved lines. A *concave up curve* is one whose function's first derivative increases continuously from negative to positive values. Straight lines drawn tangent to concave up curves are all below the curve. The graph of such a function may be thought of as being able to "hold water."

The first derivative of a *concave down curve* decreases continuously from positive to negative. A graph of a concave down function may be thought of as "spilling water." (See Fig. 9.3.)

**Figure 9.3** *Concave Curves*

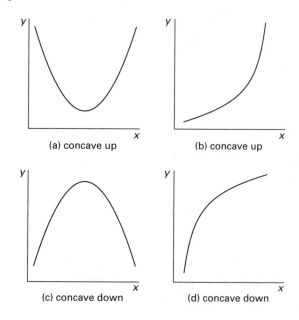

## 5. CONVEX REGIONS

*Convexity* is a term that is applied to sets and regions.[2] It plays an important role in many mathematics subjects. A set or multidimensional region is *convex* if it contains the line segment joining any two of its points; that is, if a straight line is drawn connecting any two points in a convex region, that line will lie entirely within the region. For example, the interior of a parabola is a convex region, as is a solid sphere. The *void* or *null region* (i.e., an empty set of points), single points, and straight lines are convex sets. A convex region bounded by separate, connected line segments is known as a *convex hull*. (See Fig. 9.4.)

Within a convex region, a local maximum is also the global maximum. Similarly, a local minimum is also the global minimum. The intersection of two convex regions is also convex.

**Figure 9.4** *Convexity*

| convex region | convex hull | nonconvex region |

## 6. CONGRUENCY

*Congruence* in geometric figures is analogous to *equality* in algebraic expressions. Congruent line segments are segments that have the same length. Congruent angles have the same angular measure. Congruent triangles have the same vertex angles and side lengths.

In general, *congruency*, indicated by the symbol $\cong$, means that there is one-to-one correspondence between all points on two objects. This correspondence is defined by the *mapping function* or *isometry*, which can be a translation, rotation, or reflection. Since the identity function is a valid mapping function, every geometric shape is congruent to itself.

Two congruent objects can be in different spaces. For example, a triangular area in three-dimensional space can be mapped into a triangle in two-dimensional space.

## 7. COORDINATE SYSTEMS

The manner in which a geometric figure is described depends on the coordinate system that is used. The three-dimensional system (also known as the *rectangular coordinate system* and *Cartesian coordinate system*) with its $x$-, $y$-, and $z$-coordinates is the most commonly

---

[2]It is tempting to define regions that fail the convexity test as being "concave." However, it is more proper to define such regions as "nonconvex." In any case, it is important to recognize that convexity depends on the reference point: An observer within a sphere will see the spherical boundary as convex; an observer outside the sphere may see the boundary as nonconvex.

used in engineering. Table 9.1 summarizes the components needed to specify a point in the various coordinate systems. Figure 9.5 illustrates the use of and conversion between the coordinate systems.

**Table 9.1** *Components of Coordinate Systems*

| name | dimensions | components |
|------|------------|------------|
| rectangular | 2 | $x, y$ |
| rectangular | 3 | $x, y, z$ |
| polar | 2 | $r, \theta$ |
| cylindrical | 3 | $r, \theta, z$ |
| spherical | 3 | $r, \theta, \phi$ |

**Figure 9.5** *Different Coordinate Systems*

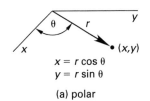

$$x = r \cos \theta$$
$$y = r \sin \theta$$

(a) polar

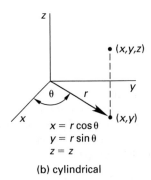

$$x = r \cos \theta$$
$$y = r \sin \theta$$
$$z = z$$

(b) cylindrical

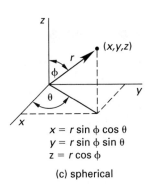

$$x = r \sin \phi \cos \theta$$
$$y = r \sin \phi \sin \theta$$
$$z = r \cos \phi$$

(c) spherical

## 8. CURVES

A *curve* (commonly called a *line*) is a function over a finite or infinite range of the independent variable. When a curve is drawn in two- or three-dimensional space, it is known as a *graph of the curve*. It may or may not be possible to describe the curve mathematically. The *degree of a curve* is the highest exponent in the function. For example, Eq. 9.3 is a fourth-degree curve.

$$f(x) = 2x^4 + 7x^3 + 6x^2 + 3x + 9 = 0 \qquad 9.3$$

An *ordinary cycloid* ("wheel line") is a curve traced out by a point on the rim of a wheel that rolls without slipping. (See Fig. 9.6.) Cycloids that start with the tracing point down (i.e., on the $x$-axis) are described in parametric form by Eq. 9.4 and Eq. 9.5 and in rectangular form by Eq. 9.6. In Eq. 9.4 through Eq. 9.6, using the minus sign results in a bottom (downward) cusp at the origin; using the plus sign results in a vertex (trough) at the origin and a top (upward) cusp.

$$x = r(\theta \pm \sin\theta) \qquad 9.4$$

$$y = r(1 \pm \cos\theta) \qquad 9.5$$

$$x = r\arccos\left(\frac{r-y}{r}\right) \pm \sqrt{2ry - y^2} \qquad 9.6$$

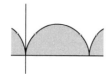

**Figure 9.6** *Cycloid (cusp at origin shown)*

An *epicycloid* is a curve generated by a point on the rim of a wheel that rolls on the outside of a circle. A *hypocycloid* is a curve generated by a point on the rim of a wheel that rolls on the inside of a circle. The equation of a hypocycloid of four cusps is

$$x^{2/3} + y^{2/3} = r^{2/3} \quad [\text{4 cusps}] \qquad 9.7$$

## 9. SYMMETRY OF CURVES

Two points, P and Q, are symmetrical with respect to a line if the line is a perpendicular bisector of the line segment PQ. If the graph of a curve is unchanged when $y$ is replaced with $-y$, the curve is symmetrical with respect to the $x$-axis. If the curve is unchanged when $x$ is replaced with $-x$, the curve is symmetrical with respect to the $y$-axis.

Repeating waveforms can be symmetrical with respect to the $y$-axis. A curve $f(x)$ is said to have *even symmetry* if $f(x) = f(-x)$. (Alternatively, $f(x)$ is said to be a *symmetrical function*.) With even symmetry, the function to the left of $x = 0$ is a reflection of the function to the right of $x = 0$. (In effect, the $y$-axis is a mirror.) The cosine curve is an example of a curve with even symmetry.

A curve is said to have *odd symmetry* if $f(x) = -f(-x)$. (Alternatively, $f(x)$ is said to be an *asymmetrical function*.[3]) The sine curve is an example of a curve with odd symmetry.

A curve is said to have *rotational symmetry (half-wave symmetry)* if $f(x) = -f(x + \pi)$.[4] Curves of this type are identical except for a sign reversal on alternate half-cycles. (See Fig. 9.7.)

**Figure 9.7** *Waveform Symmetry*

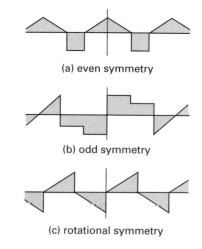

(a) even symmetry

(b) odd symmetry

(c) rotational symmetry

Table 9.2 describes the type of function resulting from the combination of two functions.

**Table 9.2** *Combinations of Functions*

| | operation | | | |
|---|---|---|---|---|
| | $+$ | $-$ | $\times$ | $\div$ |
| $f_1(x)$ even, $f_2(x)$ even | even | even | even | even |
| $f_1(x)$ odd, $f_2(x)$ odd | odd | odd | even | even |
| $f_1(x)$ even, $f_2(x)$ odd | neither | neither | odd | odd |

## 10. STRAIGHT LINES

Figure 9.8 illustrates a straight line in two-dimensional space. The *slope* of the line is $m$, the *y-intercept* is $b$, and the *x-intercept* is $a$. The equation of the line can be represented in several forms. The procedure for finding the equation depends on the form chosen to represent the line. In general, the procedure involves substituting one or more known points on the line into the equation in order to determine the coefficients.

---

[3]Although they have the same meaning, the semantics of "even symmetry" and "asymmetrical function" are contradictory.

[4]The symbol $\pi$ represents half of a full cycle of the waveform, not the value 3.141 . . . .

**Figure 9.8** Straight Line

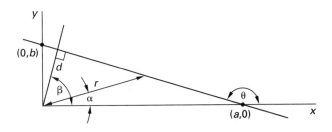

- *general form*

$$Ax + By + C = 0 \qquad 9.8$$
$$A = -mB \qquad 9.9$$
$$B = \frac{-C}{b} \qquad 9.10$$
$$C = -aA = -bB \qquad 9.11$$

- *slope-intercept form*

$$y = mx + b \qquad 9.12$$
$$m = \frac{-A}{B} = \tan\theta = \frac{y_2 - y_1}{x_2 - x_1} \qquad 9.13$$
$$b = \frac{-C}{B} \qquad 9.14$$
$$a = \frac{-C}{A} \qquad 9.15$$

- *point-slope form*

$$y - y_1 = m(x - x_1) \qquad 9.16$$

- *intercept form*

$$\frac{x}{a} + \frac{y}{b} = 1 \qquad 9.17$$

- *two-point form*

$$\frac{y - y_1}{x - x_1} = \frac{y_2 - y_1}{x_2 - x_1} \qquad 9.18$$

- *normal form*

$$x\cos\beta + y\sin\beta - d = 0 \qquad 9.19$$

($d$ and $\beta$ are constants; $x$ and $y$ are variables.)

- *polar form*

$$r = \frac{d}{\cos(\beta - \alpha)} \qquad 9.20$$

($d$ and $\beta$ are constants; $r$ and $\alpha$ are variables.)

## 11. DIRECTION NUMBERS, ANGLES, AND COSINES

Given a directed line from $(x_1, y_1, z_1)$ to $(x_2, y_2, z_2)$, the *direction numbers* are

$$L = x_2 - x_1 \qquad 9.21$$

$$M = y_2 - y_1 \qquad 9.22$$
$$N = z_2 - z_1 \qquad 9.23$$

The distance between two points is

$$d = \sqrt{L^2 + M^2 + N^2} \qquad 9.24$$

The *direction cosines* are

$$\cos\alpha = \frac{L}{d} \qquad 9.25$$
$$\cos\beta = \frac{M}{d} \qquad 9.26$$
$$\cos\gamma = \frac{N}{d} \qquad 9.27$$

Note that

$$\cos^2\alpha + \cos^2\beta + \cos^2\gamma = 1 \qquad 9.28$$

The *direction angles* are the angles between the axes and the lines. They are found from the inverse functions of the direction cosines.

$$\alpha = \arccos\left(\frac{L}{d}\right) \qquad 9.29$$
$$\beta = \arccos\left(\frac{M}{d}\right) \qquad 9.30$$
$$\gamma = \arccos\left(\frac{N}{d}\right) \qquad 9.31$$

The direction cosines can be used to write the equation of the straight line in terms of the unit vectors. The line **R** would be defined as

$$\mathbf{R} = d(\mathbf{i}\cos\alpha + \mathbf{j}\cos\beta + \mathbf{k}\cos\gamma) \qquad 9.32$$

Similarly, the line may be written in terms of its direction numbers.

$$\mathbf{R} = L\mathbf{i} + M\mathbf{j} + N\mathbf{k} \qquad 9.33$$

**Example 9.2**

A line passes through the points $(4, 7, 9)$ and $(0, 1, 6)$. Write the equation of the line in terms of its (a) direction numbers and (b) direction cosines.

*Solution*

(a) The direction numbers are

$$L = 4 - 0 = 4$$
$$M = 7 - 1 = 6$$
$$N = 9 - 6 = 3$$

Using Eq. 9.33,

$$\mathbf{R} = 4\mathbf{i} + 6\mathbf{j} + 3\mathbf{k}$$

(b) The distance between the two points is

$$d = \sqrt{(4)^2 + (6)^2 + (3)^2} = 7.81$$

The line in terms of its direction cosines is

$$\mathbf{R} = \frac{4\mathbf{i} + 6\mathbf{j} + 3\mathbf{k}}{7.81}$$

$$= 0.512\mathbf{i} + 0.768\mathbf{j} + 0.384\mathbf{k}$$

## 12. INTERSECTION OF TWO LINES

The intersection of two lines is a point. The location of the intersection point can be determined by setting the two equations equal and solving them in terms of a common variable. Alternatively, Eq. 9.34 and Eq. 9.35 can be used to calculate the coordinates of the intersection point.

$$x = \frac{B_2 C_1 - B_1 C_2}{A_2 B_1 - A_1 B_2} \qquad \text{9.34}$$

$$y = \frac{A_1 C_2 - A_2 C_1}{A_2 B_1 - A_1 B_2} \qquad \text{9.35}$$

## 13. PLANES

A *plane* in three-dimensional space (see Fig. 9.9) is completely determined by one of the following:

• three noncollinear points

• two nonparallel vectors $\mathbf{V}_1$ and $\mathbf{V}_2$ and their intersection point $P_0$

• a point $P_0$ and a vector, $\mathbf{N}$, normal to the plane (i.e., the *normal vector*)

**Figure 9.9** *Plane in Three-Dimensional Space*

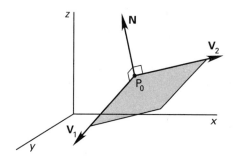

The plane can be specified mathematically in one of two ways: in rectangular form or as a parametric equation. The general form is

$$A(x - x_0) + B(y - y_0) + C(z - z_0) = 0 \qquad \text{9.36}$$

$x_0$, $y_0$, and $z_0$ are the coordinates of the intersection point of any two vectors in the plane. The coefficients

$A$, $B$, and $C$ are the same as the coefficients of the normal vector, $\mathbf{N}$.

$$\mathbf{N} = \mathbf{V}_1 \times \mathbf{V}_2 = A\mathbf{i} + B\mathbf{j} + C\mathbf{k} \qquad \text{9.37}$$

Equation 9.36 can be simplified as follows.

$$Ax + By + Cz + D = 0 \qquad \text{9.38}$$

$$D = -(Ax_0 + By_0 + Cz_0) \qquad \text{9.39}$$

The following procedure can be used to determine the equation of a plane from three noncollinear points, $P_1$, $P_2$, and $P_3$, or from a normal vector and a single point.

*step 1:* (If the normal vector is known, go to step 3.) Determine the equations of the vectors $\mathbf{V}_1$ and $\mathbf{V}_2$ from two pairs of the points. For example, determine $\mathbf{V}_1$ from points $P_1$ and $P_2$, and determine $\mathbf{V}_2$ from $P_1$ and $P_3$. Express the vectors in the form $A\mathbf{i} + B\mathbf{j} + C\mathbf{k}$.

$$\mathbf{V}_1 = (x_2 - x_1)\mathbf{i} + (y_2 - y_1)\mathbf{j}$$
$$+ (z_2 - z_1)\mathbf{k} \qquad \text{9.40}$$
$$\mathbf{V}_2 - (x_3 - x_1)\mathbf{i} + (y_3 - y_1)\mathbf{j}$$
$$+ (z_3 - z_1)\mathbf{k} \qquad \text{9.41}$$

*step 2:* Find the normal vector, $N$, as the cross product of the two vectors.

$$\mathbf{N} = \mathbf{V}_1 \times \mathbf{V}_2$$
$$= \begin{vmatrix} \mathbf{i} & (x_2 - x_1) & (x_3 - x_1) \\ \mathbf{j} & (y_2 - y_1) & (y_3 - y_1) \\ \mathbf{k} & (z_2 - z_1) & (z_3 - z_1) \end{vmatrix} \qquad \text{9.42}$$

*step 3:* Write the general equation of the plane in rectangular form (see Eq. 9.36) using the coefficients $A$, $B$, and $C$ from the normal vector and any one of the three points as $P_0$.

The parametric equations of a plane also can be written as a linear combination of the components of two vectors in the plane. Referring to Fig. 9.9, the two known vectors are

$$\mathbf{V}_1 = V_{1x}\mathbf{i} + V_{1y}\mathbf{j} + V_{1z}\mathbf{k} \qquad \text{9.43}$$

$$\mathbf{V}_2 = V_{2x}\mathbf{i} + V_{2y}\mathbf{j} + V_{2z}\mathbf{k} \qquad \text{9.44}$$

If $s$ and $t$ are scalars, the coordinates of each point in the plane can be written as Eq. 9.45 through Eq. 9.47. These are the parametric equations of the plane.

$$x = x_0 + sV_{1x} + tV_{2x} \qquad \text{9.45}$$

$$y = y_0 + sV_{1y} + tV_{2y} \qquad \text{9.46}$$

$$z = z_0 + sV_{1z} + tV_{2z} \qquad \text{9.47}$$

Mathematics

## Example 9.3

The following points are coplanar.

$$P_1 = (2, 1, -4)$$
$$P_2 = (4, -2, -3)$$
$$P_3 = (2, 3, -8)$$

Determine the equation of the plane in (a) general form and (b) parametric form.

*Solution*

(a) Use the first two points to find a vector, $\mathbf{V}_1$.

$$\mathbf{V}_1 = (x_2 - x_1)\mathbf{i} + (y_2 - y_1)\mathbf{j} + (z_2 - z_1)\mathbf{k}$$
$$= (4 - 2)\mathbf{i} + (-2 - 1)\mathbf{j} + \left(-3 - (-4)\right)\mathbf{k}$$
$$= 2\mathbf{i} - 3\mathbf{j} + 1\mathbf{k}$$

Similarly, use the first and third points to find $\mathbf{V}_2$.

$$\mathbf{V}_2 = (x_3 - x_1)\mathbf{i} + (y_3 - y_1)\mathbf{j} + (z_3 - z_1)\mathbf{k}$$
$$= (2 - 2)\mathbf{i} + (3 - 1)\mathbf{j} + \left(-8 - (-4)\right)\mathbf{k}$$
$$= 0\mathbf{i} + 2\mathbf{j} - 4\mathbf{k}$$

From Eq. 9.42, determine the normal vector as a determinant.

$$\mathbf{N} = \begin{vmatrix} \mathbf{i} & 2 & 0 \\ \mathbf{j} & -3 & 2 \\ \mathbf{k} & 1 & -4 \end{vmatrix}$$

Expand the determinant across the top row.

$$\mathbf{N} = \mathbf{i}(12 - 2) - 2(-4\mathbf{j} - 2\mathbf{k})$$
$$= 10\mathbf{i} + 8\mathbf{j} + 4\mathbf{k}$$

The rectangular form of the equation of the plane uses the same constants as in the normal vector. Use the first point and write the equation of the plane in the form of Eq. 9.36.

$$(10)(x - 2) + (8)(y - 1) + (4)(z + 4) = 0$$

The three constant terms can be combined by using Eq. 9.39.

$$D = -\left((10)(2) + (8)(1) + (4)(-4)\right) = -12$$

The equation of the plane is

$$10x + 8y + 4z - 12 = 0$$

(b) The parametric equations based on the first point and for any values of $s$ and $t$ are

$$x = 2 + 2s + 0t$$
$$y = 1 - 3s + 2t$$
$$z = -4 + 1s - 4t$$

The scalars $s$ and $t$ are not unique. Two of the three coordinates can also be chosen as the parameters. Dividing the rectangular form of the plane's equation by 4 to isolate $z$ results in an alternate set of parametric equations.

$$x = x$$
$$y = y$$
$$z = 3 - 2.5x - 2y$$

## 14. DISTANCES BETWEEN GEOMETRIC FIGURES

The smallest distance, $d$, between various geometric figures is given by the following equations.

- between two points in $(x, y, z)$ format:

$$d = \sqrt{(x_2 - x_1)^2 + (y_2 - y_1)^2 + (z_2 - z_1)^2} \quad \text{9.48}$$

- between a point $(x_0, y_0)$ and a line $Ax + By + C = 0$:

$$d = \frac{|Ax_0 + By_0 + C|}{\sqrt{A^2 + B^2}} \quad \text{9.49}$$

- between a point $(x_0, y_0, z_0)$ and a plane $Ax + By + Cz + D = 0$:

$$d = \frac{|Ax_0 + By_0 + Cz_0 + D|}{\sqrt{A^2 + B^2 + C^2}} \quad \text{9.50}$$

- between two parallel lines $Ax + By + C = 0$:

$$d = \left| \frac{|C_2|}{\sqrt{A_2^2 + B_2^2}} - \frac{|C_1|}{\sqrt{A_1^2 + B_1^2}} \right| \quad \text{9.51}$$

**Example 9.4**

What is the minimum distance between the line $y = 2x + 3$ and the origin $(0, 0)$?

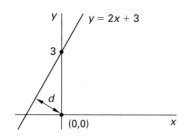

*Solution*

Put the equation in general form.

$$Ax + By + C = 2x - y + 3 = 0$$

Use Eq. 9.49 with $(x, y) = (0, 0)$.

$$d = \frac{|Ax + By + C|}{\sqrt{A^2 + B^2}} = \frac{(2)(0) + (-1)(0) + 3}{\sqrt{(2)^2 + (-1)^2}}$$

$$= \frac{3}{\sqrt{5}}$$

## 15. ANGLES BETWEEN GEOMETRIC FIGURES

The angle, $\phi$, between various geometric figures is given by the following equations.

- between two lines in $Ax + By + C = 0$, $y = mx + b$, or direction angle formats:

$$\phi = \arctan\left(\frac{A_1 B_2 - A_2 B_1}{A_1 A_2 + B_1 B_2}\right) \qquad 9.52$$

$$\phi = \arctan\left(\frac{m_2 - m_1}{1 + m_1 m_2}\right) \qquad 9.53$$

$$\phi = |\arctan(m_1) - \arctan(m_2)| \qquad 9.54$$

$$\phi = \arccos\left(\frac{L_1 L_2 + M_1 M_2 + N_1 N_2}{d_1 d_2}\right) \qquad 9.55$$

$$\phi = \arccos\left(\begin{array}{c} \cos\alpha_1 \cos\alpha_2 + \cos\beta_1 \cos\beta_2 \\ + \cos\gamma_1 \cos\gamma_2 \end{array}\right) \qquad 9.56$$

If the lines are parallel, then $\phi = 0$.

$$\frac{A_1}{A_2} = \frac{B_1}{B_2} \qquad 9.57$$

$$m_1 = m_2 \qquad 9.58$$

$$\alpha_1 = \alpha_2;\ \beta_1 = \beta_2;\ \gamma_1 = \gamma_2 \qquad 9.59$$

If the lines are perpendicular, then $\phi = 90°$.

$$A_1 A_2 = -B_1 B_2 \qquad 9.60$$

$$m_1 = \frac{-1}{m_2} \qquad 9.61$$

$$\alpha_1 + \alpha_2 = \beta_1 + \beta_2 = \gamma_1 + \gamma_2 = 90° \qquad 9.62$$

- between two planes in $A\mathbf{i} + B\mathbf{j} + C\mathbf{k} = 0$ format, the coefficients $A$, $B$, and $C$ are the same as the coefficients for the normal vector. (See Eq. 9.37.) $\phi$ is equal to the angle between the two normal vectors.

$$\cos\phi = \frac{|A_1 A_2 + B_1 B_2 + C_1 C_2|}{\sqrt{A_1^2 + B_1^2 + C_1^2}\sqrt{A_2^2 + B_2^2 + C_2^2}} \qquad 9.63$$

**Example 9.5**

Use Eq. 9.52, Eq. 9.53, and Eq. 9.54 to find the angle between the lines.

$$y = -0.577x + 2$$

$$y = +0.577x - 5$$

*Solution*

Write both equations in general form.

$$-0.577x - y + 2 = 0$$

$$0.577x - y - 5 = 0$$

(a) From Eq. 9.52,

$$\phi = \arctan\left(\frac{A_1 B_2 - A_2 B_1}{A_1 A_2 + B_1 B_2}\right)$$

$$= \arctan\left(\frac{(-0.577)(-1) - (0.577)(-1)}{(-0.577)(0.577) + (-1)(-1)}\right) = 60°$$

(b) Use Eq. 9.53.

$$\phi = \arctan\left(\frac{m_2 - m_1}{1 + m_1 m_2}\right)$$

$$= \arctan\left(\frac{0.577 - (-0.577)}{1 + (0.577)(-0.577)}\right) = 60°$$

(c) Use Eq. 9.54.

$$\phi = |\arctan(m_1) - \arctan(m_2)|$$

$$= |\arctan(-0.577) - \arctan(0.577)|$$

$$= |-30° - 30°| = 60°$$

## 16. CONIC SECTIONS

A *conic section* is any one of several curves produced by passing a plane through a cone as shown in Fig. 9.10. If $\alpha$ is the angle between the vertical axis and the cutting plane and $\beta$ is the cone generating angle, Eq. 9.64 gives the *eccentricity*, $\epsilon$, of the conic section. Values of the eccentricity are given in Fig. 9.10.

$$\epsilon = \frac{\cos \alpha}{\cos \beta} \qquad 9.64$$

**Figure 9.10** *Conic Sections Produced by Cutting Planes*

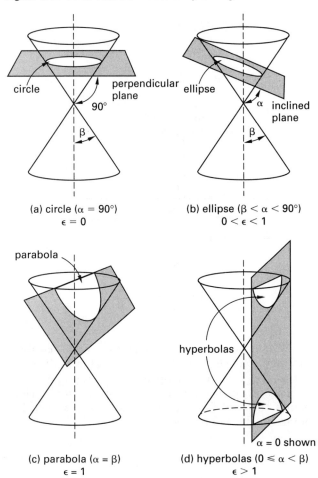

(a) circle ($\alpha = 90°$)
$\epsilon = 0$

(b) ellipse ($\beta < \alpha < 90°$)
$0 < \epsilon < 1$

(c) parabola ($\alpha = \beta$)
$\epsilon = 1$

(d) hyperbolas ($0 \le \alpha < \beta$)
$\epsilon > 1$

All conic sections are described by second-degree polynomials (i.e., are *quadratic equations*) of the following form.[5]

$$Ax^2 + Bxy + Cy^2 + Dx + Ey + F = 0 \qquad 9.65$$

This is the *general form*, which allows the figure axes to be at any angle relative to the coordinate axes. The *standard forms* presented in the following sections pertain to figures whose axes coincide with the coordinate axes, thereby eliminating certain terms of the general equation.

Figure 9.11 can be used to determine which conic section is described by the quadratic function. The quantity $B^2 - 4AC$ is known as the *discriminant*. Figure 9.11

[5]One or more straight lines are produced when the cutting plane passes through the cone's vertex. Straight lines can be considered to be quadratic functions without second-degree terms.

determines only the type of conic section; it does not determine whether the conic section is degenerate (e.g., a circle with a negative radius).

**Figure 9.11** *Determining Conic Sections from Quadratic Equations*

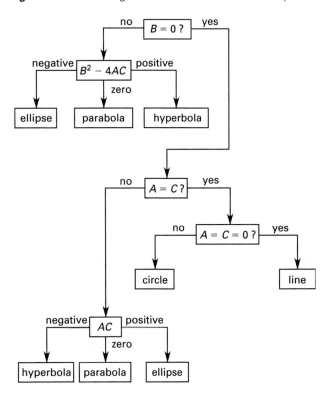

### Example 9.6

What geometric figures are described by the following equations?

(a) $4y^2 - 12y + 16x + 41 = 0$

(b) $x^2 - 10xy + y^2 + x + y + 1 = 0$

(c) $x^2 + 4y^2 + 2x - 8y + 1 = 0$

(d) $x^2 + y^2 - 6x + 8y + 20 = 0$

*Solution*

(a) Referring to Fig. 9.11, $B = 0$ since there is no $xy$ term, $A = 0$ since there is no $x^2$ term, and $AC = (0)(4) = 0$. This is a parabola.

(b) $B \ne 0$; $B^2 - 4AC = (-10)^2 - (4)(1)(1) = +96$. This is a hyperbola.

(c) $B = 0$; $A \ne C$; $AC = (1)(4) = +4$. This is an ellipse.

(d) $B = 0$; $A = C$; $A = C = 1 (\ne 0)$. This is a circle.

## 17. CIRCLE

The general form of the equation of a circle (see Fig. 9.12) is

$$Ax^2 + Ay^2 + Dx + Ey + F = 0 \qquad 9.66$$

**Figure 9.12** Circle

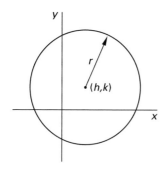

The *center-radius form* of the equation of a circle with radius $r$ and center at $(h, k)$ is

$$(x - h)^2 + (y - k)^2 = r^2 \qquad 9.67$$

The two forms can be converted by use of Eq. 9.68 through Eq. 9.70.

$$h = \frac{-D}{2A} \qquad 9.68$$

$$k = \frac{-E}{2A} \qquad 9.69$$

$$r^2 = \frac{D^2 + E^2 - 4AF}{4A^2} \qquad 9.70$$

If the right-hand side of Eq. 9.70 is positive, the figure is a circle. If it is zero, the circle shrinks to a point. If the right-hand side is negative, the figure is imaginary. A *degenerate circle* is one in which the right-hand side is less than or equal to zero.

## 18. PARABOLA

A *parabola* is the locus of points equidistant from the *focus* (point F in Fig. 9.13) and a line called the *directrix*. A parabola is symmetric with respect to its *parabolic axis*. The line normal to the parabolic axis and passing through the focus is known as the *latus rectum*. The eccentricity of a parabola is 1.

There are two common types of parabolas in the Cartesian plane—those that open right and left, and those that open up and down. Equation 9.65 is the general form of the equation of a parabola. With Eq. 9.71, the parabola points horizontally to the right if $CD > 0$ and

to the left if $CD < 0$. With Eq. 9.72, the parabola points vertically up if $AE > 0$ and down if $AE < 0$.

$$Cy^2 + Dx + Ey + F = 0 \Big|_{\substack{C, D \neq 0 \\ \text{opens horizontally}}} \qquad 9.71$$

$$Ax^2 + Dx + Ey + F = 0 \Big|_{\substack{A, E \neq 0 \\ \text{opens vertically}}} \qquad 9.72$$

**Figure 9.13** Parabola

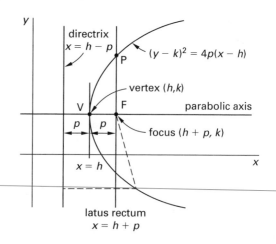

The *standard form* of the equation of a parabola with vertex at $(h, k)$, focus at $(h + p, k)$, and directrix at $x = h - p$, and that opens to the right or left is given by Eq. 9.73. The parabola opens to the right (points to the left) if $p > 0$ and opens to the left (points to the right) if $p < 0$.

$$(y - k)^2 = 4p(x - h) \Big|_{\text{opens horizontally}} \qquad 9.73$$

$$y^2 = 4px \Big|_{\substack{\text{vertex at origin} \\ h = k = 0}} \qquad 9.74$$

The *standard form* of the equation of a parabola with vertex at $(h, k)$, focus at $(h, k + p)$, and directrix at $y = k - p$, and that opens up or down is given by Eq. 9.75. The parabola opens up (points down) if $p > 0$ and opens down (points up) if $p < 0$.

$$(x - h)^2 = 4p(y - k) \Big|_{\text{opens vertically}} \qquad 9.75$$

$$x^2 = 4py \Big|_{\text{vertex at origin}} \qquad 9.76$$

The general and vertex forms of the equations can be reconciled with Eq. 9.77 through Eq. 9.79. Whether the first or second forms of these equations are used depends on whether the parabola opens horizontally or vertically (i.e., whether $A = 0$ or $C = 0$), respectively.

$$h = \begin{cases} \dfrac{E^2 - 4CF}{4CD} & \text{[opens horizontally]} \\[2ex] \dfrac{-D}{2A} & \text{[opens vertically]} \end{cases} \qquad 9.77$$

$$k = \begin{cases} \dfrac{-E}{2C} & \text{[opens horizontally]} \\[2ex] \dfrac{D^2 - 4AF}{4AE} & \text{[opens vertically]} \end{cases} \qquad 9.78$$

$$p = \begin{cases} \dfrac{-D}{4C} & \text{[opens horizontally]} \\[2ex] \dfrac{-E}{4A} & \text{[opens vertically]} \end{cases} \qquad 9.79$$

## 19. ELLIPSE

An *ellipse* has two foci separated along the *major axis* by a distance $2c$. The line perpendicular to the major axis passing through the center of the ellipse is the *minor axis*. The lines perpendicular to the major axis passing through the foci are the *latus recta*. The distance between the two vertices is $2a$. The ellipse is the locus of points such that the sum of the distances from the two foci is $2a$. Referring to Fig. 9.14,

$$\mathrm{F_1P} + \mathrm{PF_2} = 2a \qquad 9.80$$

**Figure 9.14** *Ellipse*

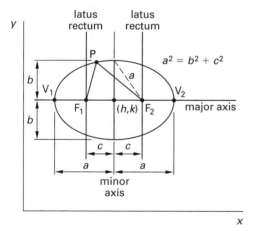

Equation 9.81 is the standard equation for an ellipse with axes parallel to the coordinate axes. (Equation 9.65 is the general form.) $F$ is not independent of $A$, $C$, $D$, and $E$ for the ellipse.

$$Ax^2 + Cy^2 + Dx + Ey + F = 0 \Big|_{\substack{AC > 0 \\ A \neq C}} \qquad 9.81$$

Equation 9.82 gives the standard form of the equation of an ellipse centered at $(h, k)$. Distances $a$ and $b$ are known as the *semimajor distance* and *semiminor distance*, respectively.

$$\frac{(x - h)^2}{a^2} + \frac{(y - k)^2}{b^2} = 1 \qquad 9.82$$

The distance between the two foci is $2c$.

$$2c = 2\sqrt{a^2 - b^2} \qquad 9.83$$

The *aspect ratio* of the ellipse is

$$\text{aspect ratio} = \frac{a}{b} \qquad 9.84$$

The *eccentricity*, $\epsilon$, of the ellipse is always less than 1. If the eccentricity is zero, the figure is a circle (another form of a *degenerative ellipse*).

$$\epsilon = \frac{\sqrt{a^2 - b^2}}{a} < 1 \qquad 9.85$$

The standard and center forms of the equations of an ellipse can be reconciled by using Eq. 9.86 through Eq. 9.89.

$$h = \frac{-D}{2A} \qquad 9.86$$

$$k = \frac{-E}{2C} \qquad 9.87$$

$$a = \sqrt{C} \qquad 9.88$$

$$b = \sqrt{A} \qquad 9.89$$

## 20. HYPERBOLA

A *hyperbola* has two foci separated along the *transverse axis* by a distance $2c$. Lines perpendicular to the transverse axis passing through the foci are the *conjugate axes*. The distance between the two vertices is $2a$, and the distance along a conjugate axis passing through each vertex between two points on the asymptotes is $2b$. The hyperbola is the locus of points such that the difference in distances from the two foci is $2a$. Referring to Fig. 9.15,

$$\mathrm{F_2P} - \mathrm{PF_1} = 2a \qquad 9.90$$

Equation 9.91 is the standard equation of a hyperbola. Coefficients $A$ and $C$ have opposite signs.

$$Ax^2 + Cy^2 + Dx + Ey + F = 0 \big|_{AC < 0} \qquad 9.91$$

**Figure 9.15** *Hyperbola*

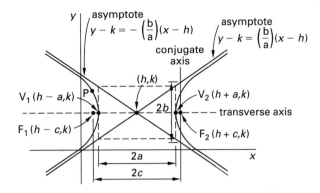

Equation 9.92 gives the standard form of the equation of a hyperbola centered at $(h, k)$ and opening to the left and right.

$$\frac{(x-h)^2}{a^2} - \frac{(y-k)^2}{b^2} = 1 \bigg|_{\text{opens horizontally}} \qquad 9.92$$

Equation 9.93 gives the standard form of the equation of a hyperbola centered at $(h, k)$ and opening up and down.

$$\frac{(y-k)^2}{a^2} - \frac{(y-k)^2}{b^2} = 1 \bigg|_{\text{opens vertically}} \qquad 9.93$$

The distance between the two foci is $2c$.

$$2c = 2\sqrt{a^2 + b^2} \qquad 9.94$$

The *eccentricity*, $\epsilon$, of the hyperbola is calculated from Eq. 9.95 and is always greater than 1.

$$\epsilon = \frac{c}{a} = \frac{\sqrt{a^2 + b^2}}{a} > 1 \qquad 9.95$$

The hyperbola is asymptotic to the lines given by Eq. 9.96 and Eq. 9.97.

$$y = \pm \frac{b}{a}(x-h) + k \bigg|_{\text{opens horizontally}} \qquad 9.96$$

$$y = \pm \frac{a}{b}(x-h) + k \bigg|_{\text{opens vertically}} \qquad 9.97$$

For a *rectangular (equilateral) hyperbola*, the asymptotes are perpendicular, $a = b$, $c = \sqrt{2}a$, and the eccentricity is $\epsilon = \sqrt{2}$. If the hyperbola is centered at the origin (i.e., $h = k = 0$), then the equations are $x^2 - y^2 = a^2$ (opens horizontally) and $y^2 - x^2 = a^2$ (opens vertically).

If the asymptotes are the $x$- and $y$-axes, the equation of the hyperbola is simply

$$xy = \pm \frac{a^2}{2} \qquad 9.98$$

The general and center forms of the equations of a hyperbola can be reconciled by using Eq. 9.99 through Eq. 9.103. Whether the hyperbola opens left and right or up and down depends on whether $M/A$ or $M/C$ is positive, respectively, where $M$ is defined by Eq. 9.99.

$$M = \frac{D^2}{4A} + \frac{E^2}{4C} - F \qquad 9.99$$

$$h = \frac{-D}{2A} \qquad 9.100$$

$$k = \frac{-E}{2C} \qquad 9.101$$

$$a = \begin{cases} \sqrt{-C} & \text{[opens horizontally]} \\ \sqrt{-A} & \text{[opens vertically]} \end{cases} \qquad 9.102$$

$$b = \begin{cases} \sqrt{A} & \text{[opens horizontally]} \\ \sqrt{C} & \text{[opens vertically]} \end{cases} \qquad 9.103$$

## 21. SPHERE

Equation 9.104 is the general equation of a sphere. The coefficient $A$ cannot be zero.

$$Ax^2 + Ay^2 + Az^2 + Bx + Cy + Dz + E = 0 \qquad 9.104$$

Equation 9.105 gives the standard form of the equation of a sphere centered at $(h, k, l)$ with radius $r$.

$$(x-h)^2 + (y-k)^2 + (z-l)^2 = r^2 \qquad 9.105$$

The general and center forms of the equations of a sphere can be reconciled by using Eq. 9.106 through Eq. 9.109.

$$h = \frac{-B}{2A} \qquad 9.106$$

$$k = \frac{-C}{2A} \qquad 9.107$$

$$l = \frac{-D}{2A} \qquad 9.108$$

$$r = \sqrt{\frac{B^2 + C^2 + D^2}{4A^2} - \frac{E}{A}} \qquad 9.109$$

## 22. HELIX

A *helix* is a curve generated by a point moving on, around, and along a cylinder such that the distance the point moves parallel to the cylindrical axis is proportional to the angle of rotation about that axis. (See Fig. 9.16.) For a cylinder of radius $r$, Eq. 9.110 through Eq. 9.112 define the three-dimensional positions of points along the helix. The quantity $2\pi k$ is the *pitch* of the helix.

$$x = r\cos\theta \qquad 9.110$$

$$y = r\sin\theta \qquad 9.111$$

$$z = k\theta \qquad 9.112$$

**Figure 9.16** *Helix*

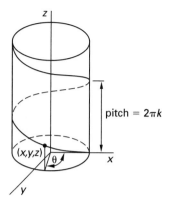

**Mathematics**

# 10 Differential Calculus

is undefined is called a *singular point*, as Fig. 10.1 illustrates.

**Figure 10.1** *Derivatives and Singular Points*

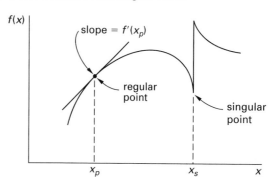

## 1. DERIVATIVE OF A FUNCTION

In most cases, it is possible to transform a continuous function, $f(x_1, x_2, x_3, ...)$, of one or more independent variables into a derivative function.[1] In simple cases, the *derivative* can be interpreted as the slope (tangent or rate of change) of the curve described by the original function. Since the slope of the curve depends on $x$, the derivative function will also depend on $x$. The derivative, $f'(x)$, of a function $f(x)$ is defined mathematically by Eq. 10.1. However, limit theory is seldom needed to actually calculate derivatives.

$$f'(x) = \lim_{\Delta x \to 0} \frac{\Delta f(x)}{\Delta x} \qquad \textit{10.1}$$

The derivative of a function $f(x)$, also known as the *first derivative*, is written in various ways, including

$$f'(x), \frac{df(x)}{dx}, \frac{df}{dx}, \mathbf{D}f(x), \mathbf{D}_x f(x), \dot{f}(x), sf(s)$$

A *second derivative* may exist if the derivative operation is performed on the first derivative—that is, a derivative is taken of a derivative function. This is written as

$$f''(x), \frac{d^2 f(x)}{dx^2}, \frac{d^2 f}{dx^2}, \mathbf{D}^2 f(x), \mathbf{D}_x^2 f(x), \ddot{f}(x), s^2 f(s)$$

A *regular* (*analytic* or *holomorphic*) *function* possesses a derivative. A point at which a function's derivative

---

[1] A function, $f(x)$, of one independent variable, $x$, is used in this section to simplify the discussion. Although the derivative is taken with respect to $x$, the independent variable can be anything.

## 2. ELEMENTARY DERIVATIVE OPERATIONS

Equation 10.2 through Eq. 10.5 summarize the elementary derivative operations on polynomials and exponentials. Equation 10.2 and Eq. 10.3 are particularly useful. ($a$, $n$, and $k$ represent constants. $f(x)$ and $g(x)$ are functions of $x$.)

$$\mathbf{D}k = 0 \qquad \textit{10.2}$$

$$\mathbf{D}x^n = nx^{n-1} \qquad \textit{10.3}$$

$$\mathbf{D}\ln x = \frac{1}{x} \qquad \textit{10.4}$$

$$\mathbf{D}e^{ax} = ae^{ax} \qquad \textit{10.5}$$

Equation 10.6 through Eq. 10.17 summarize the elementary derivative operations on transcendental (trigonometric) functions.

$$\mathbf{D}\sin x = \cos x \qquad \textit{10.6}$$

$$\mathbf{D}\cos x = -\sin x \qquad \textit{10.7}$$

$$\mathbf{D}\tan x = \sec^2 x \qquad \textit{10.8}$$

$$\mathbf{D}\cot x = -\csc^2 x \qquad \textit{10.9}$$

$$\mathbf{D}\sec x = \sec x \tan x \qquad \textit{10.10}$$

$$\mathbf{D}\csc x = -\csc x \cot x \qquad \textit{10.11}$$

$$\mathbf{D}\arcsin x = \frac{1}{\sqrt{1 - x^2}} \qquad \textit{10.12}$$

$$\mathbf{D}\arccos x = -\mathbf{D}\arcsin x \qquad \textit{10.13}$$

Mathematics

$$\mathbf{D}\arctan x = \frac{1}{1+x^2} \qquad \text{10.14}$$

$$\mathbf{D}\operatorname{arccot} x = -\mathbf{D}\arctan x \qquad \text{10.15}$$

$$\mathbf{D}\operatorname{arcsec} x = \frac{1}{x\sqrt{x^2-1}} \qquad \text{10.16}$$

$$\mathbf{D}\operatorname{arccsc} x = -\mathbf{D}\operatorname{arcsec} x \qquad \text{10.17}$$

Equation 10.18 through Eq. 10.23 summarize the elementary derivative operations on hyperbolic transcendental functions. Derivatives of hyperbolic functions are not completely analogous to those of the regular transcendental functions.

$$\mathbf{D}\sinh x = \cosh x \qquad \text{10.18}$$

$$\mathbf{D}\cosh x = \sinh x \qquad \text{10.19}$$

$$\mathbf{D}\tanh x = \operatorname{sech}^2 x \qquad \text{10.20}$$

$$\mathbf{D}\coth x = -\operatorname{csch}^2 x \qquad \text{10.21}$$

$$\mathbf{D}\operatorname{sech} x = -\operatorname{sech} x \tanh x \qquad \text{10.22}$$

$$\mathbf{D}\operatorname{csch} x = -\operatorname{csch} x \coth x \qquad \text{10.23}$$

Equation 10.24 through Eq. 10.29 summarize the elementary derivative operations on functions and combinations of functions.

$$\mathbf{D}kf(x) = k\mathbf{D}f(x) \qquad \text{10.24}$$

$$\mathbf{D}(f(x) \pm g(x)) = \mathbf{D}f(x) \pm \mathbf{D}g(x) \qquad \text{10.25}$$

$$\mathbf{D}(f(x){\cdot}g(x)) = f(x)\mathbf{D}g(x) + g(x)\mathbf{D}f(x) \qquad \text{10.26}$$

$$\mathbf{D}\left(\frac{f(x)}{g(x)}\right) = \frac{g(x)\mathbf{D}f(x) - f(x)\mathbf{D}g(x)}{(g(x))^2} \qquad \text{10.27}$$

$$\mathbf{D}(f(x))^n = n(f(x))^{n-1}\mathbf{D}f(x) \qquad \text{10.28}$$

$$\mathbf{D}f(g(x)) = \mathbf{D}_g f(g)\mathbf{D}_x g(x) \qquad \text{10.29}$$

### Example 10.1

What is the slope at $x = 3$ of the curve $f(x) = x^3 - 2x$?

*Solution*

The derivative function found from Eq. 10.3 determines the slope.

$$f'(x) = 3x^2 - 2$$

The slope at $x = 3$ is

$$f'(3) = (3)(3)^2 - 2 = 25$$

### Example 10.2

What are the derivatives of the following functions?

(a) $f(x) = 5\sqrt[3]{x^5}$

(b) $f(x) = \sin x \cos^2 x$

(c) $f(x) = \ln(\cos e^x)$

*Solution*

(a) Using Eq. 10.3 and Eq. 10.24,

$$\begin{aligned}
f'(x) &= 5\mathbf{D}\sqrt[3]{x^5} = 5\mathbf{D}\left(x^5\right)^{1/3} \\
&= (5)\left(\tfrac{1}{3}\right)\left(x^5\right)^{-2/3}\mathbf{D}x^5 \\
&= (5)\left(\tfrac{1}{3}\right)\left(x^5\right)^{-2/3}(5)\left(x^4\right) \\
&= \frac{25x^{2/3}}{3}
\end{aligned}$$

(b) Using Eq. 10.26,

$$\begin{aligned}
f'(x) &= \sin x\mathbf{D}\cos^2 x + \cos^2 x\mathbf{D}\sin x \\
&= (\sin x)(2\cos x)(\mathbf{D}\cos x) + \cos^2 x \cos x \\
&= (\sin x)(2\cos x)(-\sin x) + \cos^2 x \cos x \\
&= -2\sin^2 x \cos x + \cos^3 x
\end{aligned}$$

(c) Using Eq. 10.29,

$$\begin{aligned}
f'(x) &= \left(\frac{1}{\cos e^x}\right)\mathbf{D}\cos e^x \\
&= \left(\frac{1}{\cos e^x}\right)(-\sin e^x)\mathbf{D}e^x \\
&= \left(\frac{-\sin e^x}{\cos e^x}\right)e^x \\
&= -e^x \tan e^x
\end{aligned}$$

## 3. CRITICAL POINTS

Derivatives are used to locate the local *critical points* of functions of one variable—that is, *extreme points* (also known as *maximum* and *minimum* points) as well as the *inflection points* (*points of contraflexure*). The plurals *extrema*, *maxima*, and *minima* are used without the word "points." These points are illustrated in Fig. 10.2. There is usually an inflection point between two adjacent local extrema.

**Figure 10.2** *Extreme and Inflection Points*

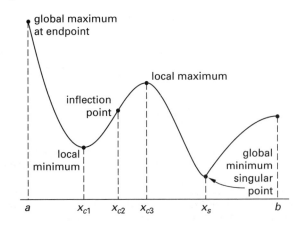

The first derivative is calculated to determine the locations of the critical points. The second derivative is calculated to determine whether a critical point is a local maximum, minimum, or inflection point, according to the following conditions. With this method, no distinction is made between local and global extrema. Therefore, the extrema should be compared with the function values at the endpoints of the interval, as illustrated in Ex. 10.3.[2] Note that $f'(x) \neq 0$ at an inflection point.

$$f'(x_c) = 0 \text{ at any extreme point, } x_c \qquad 10.30$$

$$f''(x_c) < 0 \text{ at a maximum point} \qquad 10.31$$

$$f''(x_c) > 0 \text{ at a minimum point} \qquad 10.32$$

$$f''(x_c) = 0 \text{ at an inflection point} \qquad 10.33$$

**Example 10.3**

Find the global extrema of the function $f(x)$ on the interval $[-2, +2]$.

$$f(x) = x^3 + x^2 - x + 1$$

*Solution*

The first derivative is

$$f'(x) = 3x^2 + 2x - 1$$

Since the first derivative is zero at extreme points, set $f'(x)$ equal to zero and solve for the roots of the quadratic equation.

$$3x^2 + 2x - 1 = (3x - 1)(x + 1) = 0$$

The roots are $x_1 = \frac{1}{3}$, $x_2 = -1$. These are the locations of the two extrema.

The second derivative is

$$f''(x) = 6x + 2$$

Substituting $x_1$ and $x_2$ into $f''(x)$,

$$f''(x_1) = (6)\left(\tfrac{1}{3}\right) + 2 = 4$$

$$f''(x_2) = (6)(-1) + 2 = -4$$

Therefore, $x_1$ is a local minimum point (because $f''(x_1)$ is positive), and $x_2$ is a local maximum point (because $f''(x_2)$ is negative). The inflection point between these two extrema is found by setting $f''(x)$ equal to zero.

$$f''(x) = 6x + 2 = 0 \text{ or } x = -\tfrac{1}{3}$$

Since the question asked for the global extreme points, it is necessary to compare the values of $f(x)$ at the local extrema with the values at the endpoints.

$$f(-2) = -1$$

$$f(-1) = 2$$

$$f\left(\tfrac{1}{3}\right) = 22/27$$

$$f(2) = +11$$

Therefore, the actual global extrema are the endpoints.

## 4. DERIVATIVES OF PARAMETRIC EQUATIONS

The derivative of a function $f(x_1, x_2, ..., x_n)$ can be calculated from the derivatives of the parametric equations $f_1(s), f_2(s), ..., f_n(s)$. The derivative will be expressed in terms of the parameter, $s$, unless the derivatives of the parametric equations can be expressed explicitly in terms of the independent variables.

**Example 10.4**

A circle is expressed parametrically by the equations

$$x = 5\cos\theta$$

$$y = 5\sin\theta$$

Express the derivative $dy/dx$ (a) as a function of the parameter $\theta$ and (b) as a function of $x$ and $y$.

*Solution*

(a) Taking the derivative of each parametric equation with respect to $\theta$,

$$\frac{dx}{d\theta} = -5\sin\theta$$

$$\frac{dy}{d\theta} = 5\cos\theta$$

Then,

$$\frac{dy}{dx} = \frac{\frac{dy}{d\theta}}{\frac{dx}{d\theta}} = \frac{5\cos\theta}{-5\sin\theta} = -\cot\theta$$

(b) The derivatives of the parametric equations are closely related to the original parametric equations.

$$\frac{dx}{d\theta} = -5\sin\theta = -y$$

$$\frac{dy}{d\theta} = 5\cos\theta = x$$

---

[2]It is also necessary to check the values of the function at singular points (i.e., points where the derivative does not exist).

$$\frac{dy}{dx} = \frac{\dfrac{dy}{d\theta}}{\dfrac{dx}{d\theta}} = \frac{-x}{y}$$

## 5. PARTIAL DIFFERENTIATION

Derivatives can be taken with respect to only one independent variable at a time. For example, $f'(x)$ is the derivative of $f(x)$ and is taken with respect to the independent variable $x$. If a function, $f(x_1, x_2, x_3, ...)$, has more than one independent variable, a *partial derivative* can be found, but only with respect to one of the independent variables. All other variables are treated as constants. Symbols for a partial derivative of $f$ taken with respect to variable $x$ are $\partial f/\partial x$ and $f_x(x,y)$.

The geometric interpretation of a partial derivative $\partial f/\partial x$ is the slope of a line tangent to the surface (a sphere, ellipsoid, etc.) described by the function when all variables except $x$ are held constant. In three-dimensional space with a function described by $z = f(x,y)$, the partial derivative $\partial f/\partial x$ (equivalent to $\partial f/\partial y$) is the slope of the line tangent to the surface in a plane of constant $y$. Similarly, the partial derivative $\partial f/\partial y$ (equivalent to $\partial z/\partial y$) is the slope of the line tangent to the surface in a plane of constant $x$.

### Example 10.5

What is the partial derivative $\partial z/\partial x$ of the following function?

$$z = 3x^2 - 6y^2 + xy + 5y - 9$$

*Solution*

The partial derivative with respect to $x$ is found by considering all variables other than $x$ to be constants.

$$\frac{\partial z}{\partial x} = 6x - 0 + y + 0 - 0 = 6x + y$$

### Example 10.6

A surface has the equation $x^2 + y^2 + z^2 - 9 = 0$. What is the slope of a line that lies in a plane of constant $y$ and is tangent to the surface at $(x, y, z) = (1, 2, 2)$?[3]

*Solution*

Solve for the dependent variable. Then, consider variable $y$ to be a constant.

$$z = \sqrt{9 - x^2 - y^2}$$

---

[3]Although only implied, it is required that the point actually be on the surface (i.e., it must satisfy the equation $f(x, y, z) = 0$).

$$\frac{\partial z}{\partial x} = \frac{\partial (9 - x^2 - y^2)^{1/2}}{\partial x}$$

$$= \left(\tfrac{1}{2}\right)(9 - x^2 - y^2)^{-1/2}\left(\frac{\partial (9 - x^2 - y^2)}{\partial x}\right)$$

$$= \left(\tfrac{1}{2}\right)(9 - x^2 - y^2)^{-1/2}(-2x)$$

$$= \frac{-x}{\sqrt{9 - x^2 - y^2}}$$

At the point $(1, 2, 2)$, $x = 1$ and $y = 2$.

$$\left.\frac{\partial z}{\partial x}\right|_{(1,2,2)} = \frac{-1}{\sqrt{9 - (1)^2 - (2)^2}} = -\frac{1}{2}$$

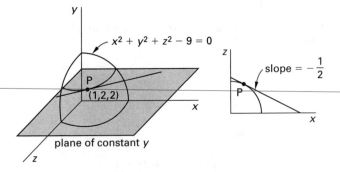

## 6. IMPLICIT DIFFERENTIATION

When a relationship between $n$ variables cannot be manipulated to yield an explicit function of $n - 1$ independent variables, that relationship implicitly defines the $n$th variable. Finding the derivative of the implicit variable with respect to any other independent variable is known as *implicit differentiation*.

An implicit derivative is the quotient of two partial derivatives. The two partial derivatives are chosen so that dividing one by the other eliminates a common differential. For example, if $z$ cannot be explicitly extracted from $f(x, y, z) = 0$, the partial derivatives $\partial z/\partial x$ and $\partial z/\partial y$ can still be found as follows.

$$\frac{\partial z}{\partial x} = \frac{-\dfrac{\partial f}{\partial x}}{\dfrac{\partial f}{\partial z}} \qquad 10.34$$

$$\frac{\partial z}{\partial y} = \frac{-\dfrac{\partial f}{\partial y}}{\dfrac{\partial f}{\partial z}} \qquad 10.35$$

### Example 10.7

Find the derivative $dy/dx$ of

$$f(x, y) = x^2 + xy + y^3$$

*Solution*

Implicit differentiation is required because $x$ cannot be extracted from $f(x, y)$.

$$\frac{\partial f}{\partial x} = 2x + y$$

$$\frac{\partial f}{\partial y} = x + 3y^2$$

$$\frac{dy}{dx} = \frac{-\dfrac{\partial f}{\partial x}}{\dfrac{\partial f}{\partial y}} = \frac{-(2x + y)}{x + 3y^2}$$

**Example 10.8**

Solve Ex. 10.6 using implicit differentiation.

*Solution*

$$f(x, y, z) = x^2 + y^2 + z^2 - 9 = 0$$

$$\frac{\partial f}{\partial x} = 2x$$

$$\frac{\partial f}{\partial z} = 2z$$

$$\frac{\partial z}{\partial x} = \frac{-\dfrac{\partial f}{\partial x}}{\dfrac{\partial f}{\partial z}} = \frac{-2x}{2z} = -\frac{x}{z}$$

At the point $(1, 2, 2)$,

$$\frac{\partial z}{\partial x} = -\frac{1}{2}$$

## 7. TANGENT PLANE FUNCTION

Partial derivatives can be used to find the equation of a plane tangent to a three-dimensional surface defined by $f(x, y, z) = 0$ at some point, $P_0$.

$$T(x_0, y_0, z_0) = (x - x_0)\frac{\partial f(x, y, z)}{\partial x}\Big|_{P_0}$$

$$+ (y - y_0)\frac{\partial f(x, y, z)}{\partial y}\Big|_{P_0}$$

$$+ (z - z_0)\frac{\partial f(x, y, z)}{\partial z}\Big|_{P_0}$$

$$= 0 \qquad\qquad 10.36$$

The coefficients of $x$, $y$, and $z$ are the same as the coefficients of $\mathbf{i}$, $\mathbf{j}$, and $\mathbf{k}$ of the normal vector at point $P_0$. (See Sec. 10.10.)

**Example 10.9**

What is the equation of the plane that is tangent to the surface defined by $f(x, y, z) = 4x^2 + y^2 - 16z = 0$ at the point $(2, 4, 2)$?

*Solution*

First, calculate the partial derivatives and substitute the coordinates of the point.

$$\frac{\partial f(x, y, z)}{\partial x}\Big|_{P_0} = 8x|_{(2, 4, 2)} = (8)(2) = 16$$

$$\frac{\partial f(x, y, z)}{\partial y}\Big|_{P_0} = 2y|_{(2, 4, 2)} = (2)(4) = 8$$

$$\frac{\partial f(x, y, z)}{\partial z}\Big|_{P_0} = -16|_{(2, 4, 2)} = -16$$

$$T(2, 4, 2) = (16)(x - 2) + (8)(y - 4) - (16)(z - 2)$$

$$= 2x + y - 2z - 4$$

Next, substitute into Eq. 10.36.

$$2x + y - 2z - 4 = 0$$

## 8. GRADIENT VECTOR

The slope of a function is the change in one variable with respect to a distance in a chosen direction. Usually, the direction is parallel to a coordinate axis. However, the maximum slope at a point on a surface may not be in a direction parallel to one of the coordinate axes.

The *gradient vector function* $\nabla f(x, y, z)$ (pronounced "del f") gives the maximum rate of change of the function $f(x, y, z)$.

$$\nabla f(x, y, z) = \left(\frac{\partial f(x, y, z)}{\partial x}\right)\mathbf{i} + \left(\frac{\partial f(x, y, z)}{\partial y}\right)\mathbf{j}$$

$$+ \left(\frac{\partial f(x, y, z)}{\partial z}\right)\mathbf{k} \qquad 10.37$$

**Example 10.10**

A two-dimensional function is defined as

$$f(x, y) = 2x^2 - y^2 + 3x - y$$

(a) What is the gradient vector for this function? (b) What is the direction of the line passing through the point $(1, -2)$ that has a maximum slope? (c) What is the maximum slope at the point $(1, -2)$?

*Solution*

(a) It is necessary to calculate two partial derivatives in order to use Eq. 10.37.

$$\frac{\partial f(x, y)}{\partial x} = 4x + 3$$

$$\frac{\partial f(x, y)}{\partial y} = -2y - 1$$

$$\nabla f(x, y) = (4x + 3)\mathbf{i} + (-2y - 1)\mathbf{j}$$

(b) Find the direction of the line passing through $(1, -2)$ with maximum slope by inserting $x = 1$ and $y = -2$ into the gradient vector function.

$$\mathbf{V} = \left((4)(1) + 3\right)\mathbf{i} + \left((-2)(-2) - 1\right)\mathbf{j}$$

$$= 7\mathbf{i} + 3\mathbf{j}$$

(c) The magnitude of the slope is

$$|\mathbf{V}| = \sqrt{(7)^2 + (3)^2} = 7.62$$

## 9. DIRECTIONAL DERIVATIVE

Unlike the gradient vector (covered in Sec. 10.8), which calculates the maximum rate of change of a function, the *directional derivative*, indicated by $\nabla_u f(x, y, z)$, $D_u f(x, y, z)$, or $f'_u(x, y, z)$, gives the rate of change in the direction of a given vector, $\mathbf{u}$ or $\mathbf{U}$. The subscript $u$ implies that the direction vector is a unit vector, but it does not need to be, as only the direction cosines are calculated from it.

$$\nabla_u f(x, y, z) = \left(\frac{\partial f(x, y, z)}{\partial x}\right)\cos \alpha$$

$$+ \left(\frac{\partial f(x, y, z)}{\partial y}\right)\cos \beta$$

$$+ \left(\frac{\partial f(x, y, z)}{\partial z}\right)\cos \gamma \qquad 10.38$$

$$\mathbf{U} = U_x\mathbf{i} + U_y\mathbf{j} + U_z\mathbf{k} \qquad 10.39$$

$$\cos \alpha = \frac{U_x}{|\mathbf{U}|} = \frac{U_x}{\sqrt{U_x^2 + U_y^2 + U_z^2}} \qquad 10.40$$

$$\cos \beta = \frac{U_y}{|\mathbf{U}|} \qquad 10.41$$

$$\cos \gamma = \frac{U_z}{|\mathbf{U}|} \qquad 10.42$$

**Example 10.11**

What is the rate of change of $f(x, y) = 3x^2 + xy - 2y^2$ at the point $(1, -2)$ in the direction $4\mathbf{i} + 3\mathbf{j}$?

*Solution*

The direction cosines are given by Eq. 10.40 and Eq. 10.41.

$$\cos \alpha = \frac{U_x}{|\mathbf{U}|} = \frac{4}{\sqrt{(4)^2 + (3)^2}} = \frac{4}{5}$$

$$\cos \beta = \frac{U_y}{|\mathbf{U}|} = \frac{3}{5}$$

The partial derivatives are

$$\frac{\partial f(x, y)}{\partial x} = 6x + y$$

$$\frac{\partial f(x, y)}{\partial y} = x - 4y$$

The directional derivative is given by Eq. 10.38.

$$\nabla_u f(x, y) = \left(\tfrac{4}{5}\right)(6x + y) + \left(\tfrac{3}{5}\right)(x - 4y)$$

Substituting the given values of $x = 1$ and $y = -2$,

$$\nabla_u f(1, -2) = \left(\tfrac{4}{5}\right)\left((6)(1) - 2\right) + \left(\tfrac{3}{5}\right)\left(1 - (4)(-2)\right)$$

$$= \frac{43}{5} = 8.6$$

## 10. NORMAL LINE VECTOR

Partial derivatives can be used to find the vector normal to a three-dimensional surface defined by $f(x, y, z) = 0$ at some point $P_0$. Notice that the coefficients of $\mathbf{i}$, $\mathbf{j}$, and $\mathbf{k}$ are the same as the coefficients of $x$, $y$, and $z$ calculated for the equation of the tangent plane at point $P_0$. (See Sec. 10.7.)

$$\mathbf{N} = \left.\frac{\partial f(x, y, z)}{\partial x}\right|_{P_0}\mathbf{i} + \left.\frac{\partial f(x, y, z)}{\partial y}\right|_{P_0}\mathbf{j}$$

$$+ \left.\frac{\partial f(x, y, z)}{\partial z}\right|_{P_0}\mathbf{k} \qquad 10.43$$

**Example 10.12**

What is the vector normal to the surface of $f(x, y, z) = 4x^2 + y^2 - 16z = 0$ at the point $(2, 4, 2)$?

*Solution*

The equation of the tangent plane at this point was calculated in Ex. 10.9 to be

$$T(2, 4, 2) = 2x + y - 2z - 4 = 0$$

A vector normal to the tangent plane through this point is

$$\mathbf{N} = 2\mathbf{i} + \mathbf{j} - 2\mathbf{k}$$

Mathematics

## 11. DIVERGENCE OF A VECTOR FIELD

The *divergence*, div $\mathbf{F}$, of a vector field $\mathbf{F}(x, y, z)$ is a scalar function defined by Eq. 10.44 through Eq. 10.46.[4] The divergence of $\mathbf{F}$ can be interpreted as the *accumulation* of flux (i.e., a flowing substance) in a small region (i.e., at a point). One of the uses of the divergence is to determine whether flow (represented in direction and magnitude by $\mathbf{F}$) is compressible. Flow is incompressible if div $\mathbf{F} = 0$, since the substance is not accumulating.

$$\mathbf{F} = P(x, y, z)\mathbf{i} + Q(x, y, z)\mathbf{j} + R(x, y, z)\mathbf{k} \qquad 10.44$$

$$\text{div } \mathbf{F} = \frac{\partial P}{\partial x} + \frac{\partial Q}{\partial y} + \frac{\partial R}{\partial z} \qquad 10.45$$

It may be easier to calculate the divergence from Eq. 10.46.

$$\text{div } \mathbf{F} = \nabla \cdot \mathbf{F} \qquad 10.46$$

The vector del operator, $\nabla$ is defined as

$$\nabla = \frac{\partial}{\partial x}\mathbf{i} + \frac{\partial}{\partial y}\mathbf{j} + \frac{\partial}{\partial z}\mathbf{k} \qquad 10.47$$

If there is no divergence, then the dot product calculated in Eq. 10.46 is zero.

### Example 10.13

Calculate the divergence of the following vector function.

$$\mathbf{F}(x, y, z) = xz\mathbf{i} + e^x y\mathbf{j} + 7x^3 y\mathbf{k}$$

*Solution*

From Eq. 10.45,

$$\text{div } \mathbf{F} = \frac{\partial}{\partial x}(xz) + \frac{\partial}{\partial y}(e^x y) + \frac{\partial}{\partial z}(7x^3 y)$$

$$= z + e^x + 0 = z + e^x$$

## 12. CURL OF A VECTOR FIELD

The *curl*, curl $\mathbf{F}$, of a vector field $\mathbf{F}(x, y, z)$ is a vector field defined by Eq. 10.49 and Eq. 10.50. The curl $\mathbf{F}$ can be interpreted as the *vorticity* per unit area of flux (i.e., a flowing substance) in a small region (i.e., at a point). One of the uses of the curl is to determine whether flow (represented in direction and magnitude by $\mathbf{F}$) is rotational. Flow is irrotational if curl $\mathbf{F} = 0$.

$$\mathbf{F} = P(x, y, z)\mathbf{i} + Q(x, y, z)\mathbf{j} + R(x, y, z)\mathbf{k} \qquad 10.48$$

---

[4]Notice that a bold letter, $\mathbf{F}$, is used to indicate that the vector is a function of $x$, $y$, and $z$.

$$\text{curl } \mathbf{F} = \left(\frac{\partial R}{\partial y} - \frac{\partial Q}{\partial z}\right)\mathbf{i} + \left(\frac{\partial P}{\partial z} - \frac{\partial R}{\partial x}\right)\mathbf{j}$$

$$+ \left(\frac{\partial Q}{\partial x} - \frac{\partial P}{\partial y}\right)\mathbf{k} \qquad 10.49$$

It may be easier to calculate the curl from Eq. 10.49. (The vector del operator, $\nabla$, was defined in Eq. 10.47.)

$$\text{curl } \mathbf{F} = \nabla \times \mathbf{F}$$

$$= \begin{vmatrix} \mathbf{i} & \mathbf{j} & \mathbf{k} \\ \dfrac{\partial}{\partial x} & \dfrac{\partial}{\partial y} & \dfrac{\partial}{\partial z} \\ P(x, y, z) & Q(x, y, z) & R(x, y, z) \end{vmatrix} \qquad 10.50$$

If the velocity vector is $\mathbf{V}$, then the vorticity is

$$\omega = \nabla \times \mathbf{V} = \omega_x\mathbf{i} + \omega_y\mathbf{j} + \omega_z\mathbf{k} \qquad 10.51$$

The circulation is the line integral of the velocity $\mathbf{V}$ along a closed curve.

$$\Gamma = \oint V \cdot ds = \oint \omega \cdot dA \qquad 10.52$$

### Example 10.14

Calculate the curl of the following vector function.

$$\mathbf{F}(x, y, z) = 3x^2\mathbf{i} + 7e^x y\mathbf{j}$$

*Solution*

Using Eq. 10.50,

$$\text{curl } \mathbf{F} = \begin{vmatrix} \mathbf{i} & \mathbf{j} & \mathbf{k} \\ \dfrac{\partial}{\partial x} & \dfrac{\partial}{\partial y} & \dfrac{\partial}{\partial z} \\ 3x^2 & 7e^x y & 0 \end{vmatrix}$$

Expand the determinant across the top row.

$$\mathbf{i}\left(\frac{\partial}{\partial y}(0) - \frac{\partial}{\partial z}(7e^x y)\right) - \mathbf{j}\left(\frac{\partial}{\partial x}(0) - \frac{\partial}{\partial z}(3x^2)\right)$$

$$+ \mathbf{k}\left(\frac{\partial}{\partial x}(7e^x y) - \frac{\partial}{\partial y}(3x^2)\right)$$

$$= \mathbf{i}(0 - 0) - \mathbf{j}(0 - 0) + \mathbf{k}(7e^x y - 0) = 7e^x y\mathbf{k}$$

## 13. TAYLOR'S FORMULA

*Taylor's formula* (*series*) can be used to expand a function around a point (i.e., approximate the function at one point based on the function's value at another point). The approximation consists of a series, each term composed of a derivative of the original function and a polynomial. Using Taylor's formula requires that the original function be continuous in the interval $[a,b]$ and have the required number of derivatives. To expand a function, $f(x)$, around a point, $a$, in order to obtain $f(b)$, Taylor's formula is

$$f(b) = f(a) + \frac{f'(a)}{1!}(b-a) + \frac{f''(a)}{2!}(b-a)^2$$
$$+ \cdots + \frac{f^n(a)}{n!}(b-a)^n + R_n(b) \qquad 10.53$$

In Eq. 10.53, the expression $f^n$ designates the $n$th derivative of the function $f(x)$. To be a useful approximation, point $a$ must satisfy two requirements: It must be relatively close to point $b$, and the function and its derivatives must be known or easy to calculate. The last term, $R_n(b)$, is the uncalculated remainder after $n$ derivatives. It is the difference between the exact and approximate values. By using enough terms, the remainder can be made arbitrarily small. That is, $R_n(b)$ approaches zero as $n$ approaches infinity.

It can be shown that the remainder term can be calculated from Eq. 10.54, where $c$ is some number in the interval $[a,b]$. With certain functions, the constant $c$ can be completely determined. In most cases, however, it is possible only to calculate an upper bound on the remainder from Eq. 10.55. $M_n$ is the maximum (positive) value of $f^{n+1}(x)$ on the interval $[a,b]$.

$$R_n(b) = \frac{f^{n+1}(c)}{(n+1)!}(b-a)^{n+1} \qquad 10.54$$

$$|R_n(b)| \le M_n \frac{|(b-a)^{n+1}|}{(n+1)!} \qquad 10.55$$

## 14. MACLAURIN POWER APPROXIMATIONS

If $a = 0$ in the Taylor series, Eq. 10.53 is known as the *Maclaurin series*. The Maclaurin series can be used to approximate functions at some value of $x$ between 0 and 1. The following common approximations may be referred to as Maclaurin series, Taylor series, or power series approximations.

$$\sin x \approx x - \frac{x^3}{3!} + \frac{x^5}{5!} - \frac{x^7}{7!} + \cdots$$
$$+ (-1)^n \frac{x^{2n+1}}{(2n+1)!} \qquad 10.56$$

$$\cos x \approx 1 - \frac{x^2}{2!} + \frac{x^4}{4!} - \frac{x^6}{6!} + \cdots + (-1)^n \frac{x^{2n}}{(2n)!} \qquad 10.57$$

$$\sinh x \approx x + \frac{x^3}{3!} + \frac{x^5}{5!} + \frac{x^7}{7!} + \cdots + \frac{x^{2n+1}}{(2n+1)!} \qquad 10.58$$

$$\cosh x \approx 1 + \frac{x^2}{2!} + \frac{x^4}{4!} + \frac{x^6}{6!} + \cdots + \frac{x^{2n}}{(2n)!} \qquad 10.59$$

$$e^x \approx 1 + x + \frac{x^2}{2!} + \frac{x^3}{3!} + \cdots + \frac{x^n}{n!} \qquad 10.60$$

$$\ln(1+x) \approx x - \frac{x^2}{2} + \frac{x^3}{3} - \frac{x^4}{4} + \cdots + (-1)^{n+1}\frac{x^n}{n} \qquad 10.61$$

$$\frac{1}{1-x} \approx 1 + x + x^2 + x^3 + \cdots + x^n \qquad 10.62$$

# 11 Integral Calculus

## 2. ELEMENTARY OPERATIONS

Equation 11.2 through Eq. 11.8 summarize the elementary integration operations on polynomials and exponentials.[2]

Equation 11.2 and Eq. 11.3 are particularly useful. ($C$ and $k$ represent constants. $f(x)$ and $g(x)$ are functions of $x$.)

$$\int k \, dx = kx + C \qquad \text{11.2}$$

$$\int x^m \, dx = \frac{x^{m+1}}{m+1} + C \quad [m \neq -1] \qquad \text{11.3}$$

$$\int \frac{1}{x} \, dx = \ln |x| + C \qquad \text{11.4}$$

$$\int e^{kx} \, dx = \frac{e^{kx}}{k} + C \qquad \text{11.5}$$

$$\int x e^{kx} \, dx = \frac{e^{kx}(kx - 1)}{k^2} + C \qquad \text{11.6}$$

$$\int k^{ax} \, dx = \frac{k^{ax}}{a \ln k} + C \qquad \text{11.7}$$

$$\int \ln x \, dx = x \ln x - x + C \qquad \text{11.8}$$

Equation 11.9 through Eq. 11.20 summarize the elementary integration operations on transcendental functions.

$$\int \sin x \, dx = -\cos x + C \qquad \text{11.9}$$

$$\int \cos x \, dx = \sin x + C \qquad \text{11.10}$$

$$\int \tan x \, dx = \ln |\sec x| + C$$
$$= -\ln |\cos x| + C \qquad \text{11.11}$$

$$\int \cot x \, dx = \ln |\sin x| + C \qquad \text{11.12}$$

$$\int \sec x \, dx = \ln |(\sec x + \tan x)| + C$$
$$= \ln \left| \tan \left( \frac{x}{2} + \frac{\pi}{4} \right) \right| + C \qquad \text{11.13}$$

## 1. INTEGRATION

*Integration* is the inverse operation of differentiation. For that reason, *indefinite integrals* are sometimes referred to as *antiderivatives*.[1] Although expressions can be functions of several variables, integrals can only be taken with respect to one variable at a time. The *differential term* ($dx$ in Eq. 11.1) indicates that variable. In Eq. 11.1, the function $f'(x)$ is the *integrand*, and $x$ is the variable of integration.

$$\int f'(x) \, dx = f(x) + C \qquad \text{11.1}$$

While most of a function, $f(x)$, can be "recovered" through integration of its derivative, $f'(x)$, a constant term will be lost. This is because the derivative of a constant term vanishes (i.e., is zero), leaving nothing to recover from. A *constant of integration*, $C$, is added to the integral to recognize the possibility of such a constant term.

---

[1]The difference between an indefinite and definite integral (covered in Sec. 11.7) is simple: An *indefinite integral* is a function, while a *definite integral* is a number.

[2]More extensive listings, known as *tables of integrals*, are widely available. (See App. 11.A.)

$$\int \csc x \, dx = \ln |(\csc x - \cot x)| + C$$

$$= \ln \left| \tan \frac{x}{2} \right| + C \qquad \textit{11.14}$$

$$\int \frac{dx}{k^2 + x^2} = \frac{1}{k} \arctan \frac{x}{k} + C \qquad \textit{11.15}$$

$$\int \frac{dx}{\sqrt{k^2 - x^2}} = \arcsin \frac{x}{k} + C \quad [k^2 > x^2] \qquad \textit{11.16}$$

$$\int \frac{dx}{x\sqrt{x^2 - k^2}} = \frac{1}{k} \text{arcsec} \frac{x}{k} + C \quad [x^2 > k^2] \qquad \textit{11.17}$$

$$\int \sin^2 x \, dx = \frac{1}{2}x - \frac{1}{4}\sin 2x + C \qquad \textit{11.18}$$

$$\int \cos^2 x \, dx = \frac{1}{2}x + \frac{1}{4}\sin 2x + C \qquad \textit{11.19}$$

$$\int \tan^2 x \, dx = \tan x - x + C \qquad \textit{11.20}$$

Equation 11.21 through Eq. 11.26 summarize the elementary integration operations on hyperbolic transcendental functions. Integrals of hyperbolic functions are not completely analogous to those of the regular transcendental functions.

$$\int \sinh x \, dx = \cosh x + C \qquad \textit{11.21}$$

$$\int \cosh x \, dx = \sinh x + C \qquad \textit{11.22}$$

$$\int \tanh x \, dx = \ln |\cosh x| + C \qquad \textit{11.23}$$

$$\int \coth x \, dx = \ln |\sinh x| + C \qquad \textit{11.24}$$

$$\int \text{sech} \, x \, dx = \arctan(\sinh x) + C \qquad \textit{11.25}$$

$$\int \text{csch} \, x \, dx = \ln \left| \tanh \left( \frac{x}{2} \right) \right| + C \qquad \textit{11.26}$$

Equation 11.27 through Eq. 11.31 summarize the elementary integration operations on functions and combinations of functions.

$$\int kf(x) \, dx = k \int f(x) \, dx \qquad \textit{11.27}$$

$$\int (f(x) + g(x)) \, dx = \int f(x) \, dx + \int g(x) \, dx \qquad \textit{11.28}$$

$$\int \frac{f'(x)}{f(x)} \, dx = \ln |f(x)| + C \qquad \textit{11.29}$$

$$\int f(x) \, dg(x) = f(x) \int dg(x) - \int g(x) \, df(x) + C$$

$$= f(x)g(x) - \int g(x) \, df(x) + C \qquad \textit{11.30}$$

### Example 11.1

Find the integral with respect to $x$ of

$$3x^2 + \tfrac{1}{3}x - 7 = 0$$

*Solution*

This is a polynomial function, and Eq. 11.3 can be applied to each of the three terms.

$$\int \left(3x^2 + \tfrac{1}{3}x - 7\right) dx = x^3 + \tfrac{1}{6}x^2 - 7x + C$$

## 3. INTEGRATION BY PARTS

Equation 11.30, repeated here, is known as *integration by parts*. $f(x)$ and $g(x)$ are functions. The use of this method is illustrated by Ex. 11.2.

$$\int f(x) \, dg(x) = f(x)g(x) - \int g(x) \, df(x) + C \qquad \textit{11.31}$$

### Example 11.2

Find the following integral.

$$\int x^2 e^x \, dx$$

*Solution*

$x^2 e^x$ is factored into two parts so that integration by parts can be used.

$$f(x) = x^2$$
$$dg(x) = e^x \, dx$$
$$df(x) = 2x \, dx$$
$$g(x) = \int dg(x) = \int e^x \, dx = e^x$$

From Eq. 11.31, disregarding the constant of integration (which cannot be evaluated),

$$\int f(x) \, dg(x) = f(x)g(x) - \int g(x) \, df(x)$$

$$\int x^2 e^x \, dx = x^2 e^x - \int e^x (2x) \, dx$$

The second term is also factored into two parts, and integration by parts is used again. This time,

$$f(x) = x$$
$$dg(x) = e^x \, dx$$
$$df(x) = dx$$
$$g(x) = \int dg(x) = \int e^x \, dx = e^x$$

From Eq. 11.31,

$$\int 2xe^x \, dx = 2 \int xe^x \, dx$$

$$= 2\left( xe^x - \int e^x \, dx \right)$$

$$= 2(xe^x - e^x)$$

Then, the complete integral is

$$\int x^2 e^x \, dx = x^2 e^x - 2(xe^x - e^x) + C$$

$$= e^x(x^2 - 2x + 2) + C$$

## 4. SEPARATION OF TERMS

Equation 11.28 shows that the integral of a sum of terms is equal to a sum of integrals. This technique is known as *separation of terms*. In many cases, terms are easily separated. In other cases, the technique of *partial fractions* can be used to obtain individual terms. These techniques are illustrated by Ex. 11.3 and Ex. 11.4.

### Example 11.3

Find the following integral.

$$\int \frac{(2x^2 + 3)^2}{x} \, dx$$

*Solution*

$$\int \frac{(2x^2 + 3)^2}{x} \, dx = \int \frac{4x^4 + 12x^2 + 9}{x} \, dx$$

$$= \int \left( 4x^3 + 12x + \frac{9}{x} \right) dx$$

$$= x^4 + 6x^2 + 9\ln|x| + C$$

### Example 11.4

Find the following integral.

$$\int \frac{3x + 2}{3x - 2} \, dx$$

*Solution*

The integrand is larger than 1, so use long division to simplify it.

$$
\begin{array}{r}
1 \text{ rem } 4 \quad \left( 1 + \dfrac{4}{3x - 2} \right) \\
3x - 2 \,\overline{\big)\, 3x + 2} \\
\underline{3x - 2} \\
4 \text{ remainder}
\end{array}
$$

$$\int \frac{3x + 2}{3x - 2} \, dx = \int \left( 1 + \frac{4}{3x - 2} \right) dx$$

$$= \int dx + \int \frac{4}{3x - 2} \, dx$$

$$= x + \tfrac{4}{3}\ln|(3x - 2)| + C$$

## 5. DOUBLE AND HIGHER-ORDER INTEGRALS

A function can be successively integrated. (This is analogous to successive differentiation.) A function that is integrated twice is known as a *double integral*; if integrated three times, it is a *triple integral*; and so on. Double and triple integrals are used to calculate areas and volumes, respectively.

The successive integrations do not need to be with respect to the same variable. Variables not included in the integration are treated as constants.

There are several notations used for a multiple integral, particularly when the product of length differentials represents a differential area or volume. A double integral (i.e., two successive integrations) can be represented by one of the following notations.

$$\iint f(x, y) \, dx \, dy, \quad \int_{R^2} f(x, y) \, dx \, dy,$$

$$\text{or } \iint_{R^2} f(x, y) \, dA$$

A triple integral can be represented by one of the following notations.

$$\iiint f(x, y, z) \, dx \, dy \, dz,$$

$$\int_{R^3} f(x, y, z) \, dx \, dy \, dz,$$

$$\text{or } \iiint_{R^3} f(x, y, z) \, dV$$

### Example 11.5

Find the following double integral.

$$\iint (x^2 + y^3 x) \, dx \, dy$$

*Solution*

$$\int (x^2 + y^3 x) \, dx = \tfrac{1}{3}x^3 + \tfrac{1}{2}y^3 x^2 + C_1$$

$$\int \left( \tfrac{1}{3}x^3 + \tfrac{1}{2}y^3 x^2 + C_1 \right) dy = \tfrac{1}{3}yx^3 + \tfrac{1}{8}y^4 x^2 + C_1 y + C_2$$

So,

$$\iint (x^2 + y^3 x)\,dx\,dy = \tfrac{1}{3}yx^3 + \tfrac{1}{8}y^4 x^2 + C_1 y + C_2$$

## 6. INITIAL VALUES

The constant of integration, $C$, can be found only if the value of the function $f(x)$ is known for some value of $x_0$. The value $f(x_0)$ is known as an *initial value* or *initial condition*. To completely define a function, as many initial values, $f(x_0)$, $f'(x_0)$, $f''(x_0)$, and so on, as there are integrations are needed.

### Example 11.6

It is known that $f(x) = 4$ when $x = 2$ (i.e., the initial value is $f(2) = 4$). Find the original function.

$$\int (3x^3 - 7x)\,dx$$

*Solution*

The function is

$$f(x) = \int (3x^3 - 7x)\,dx = \tfrac{3}{4}x^4 - \tfrac{7}{2}x^2 + C$$

Substituting the initial value determines $C$.

$$4 = \left(\tfrac{3}{4}\right)(2)^4 - \left(\tfrac{7}{2}\right)(2)^2 + C$$
$$4 = 12 - 14 + C$$
$$C = 6$$

The function is

$$f(x) = \tfrac{3}{4}x^4 - \tfrac{7}{2}x^2 + 6$$

## 7. DEFINITE INTEGRALS

A *definite integral* is restricted to a specific range of the independent variable. (Unrestricted integrals of the types shown in all preceding examples are known as *indefinite integrals*.) A definite integral restricted to the region bounded by *lower* and *upper limits* (also known as *bounds*), $x_1$ and $x_2$, is written as

$$\int_{x_1}^{x_2} f(x)\,dx$$

Equation 11.32 indicates how definite integrals are evaluated. It is known as the *fundamental theorem of calculus*.

$$\int_{x_1}^{x_2} f'(x)\,dx = f(x)\Big|_{x_1}^{x_2} = f(x_2) - f(x_1) \qquad 11.32$$

A common use of a definite integral is the calculation of work performed by a force, $F$, that moves an object from position $x_1$ to $x_2$.

$$W = \int_{x_1}^{x_2} F\,dx \qquad 11.33$$

### Example 11.7

Evaluate the following definite integral.

$$\int_{\pi/4}^{\pi/3} \sin x\,dx$$

*Solution*

From Eq. 11.32,

$$\int_{\pi/4}^{\pi/3} \sin x\,dx = -\cos x\Big|_{\pi/4}^{\pi/3}$$
$$= -\cos \tfrac{\pi}{3} - \left(-\cos \tfrac{\pi}{4}\right)$$
$$= -0.5 - (-0.707) = 0.207$$

## 8. AVERAGE VALUE

The average value of a function $f(x)$ that is integrable over the interval $[a, b]$ is

$$\text{average value} = \frac{1}{b-a}\int_a^b f(x)\,dx \qquad 11.34$$

## 9. AREA

Equation 11.35 calculates the area, $A$, bounded by $x = a$, $x = b$, $f_1(x)$ above and $f_2(x)$ below. ($f_2(x) = 0$ if the area is bounded by the $x$-axis.) This is illustrated in Fig. 11.1.

$$A = \int_a^b (f_1(x) - f_2(x))\,dx \qquad 11.35$$

**Figure 11.1** *Area Between Two Curves*

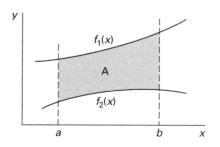

## Example 11.8

Find the area between the $x$-axis and the parabola $y = x^2$ in the interval $[0, 4]$.

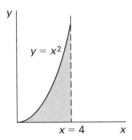

*Solution*

Referring to Eq. 11.35,

$$f_1(x) = x^2$$
$$f_2(x) = 0$$

$$A = \int_a^b (f_1(x) - f_2(x))\, dx = \int_0^4 x^2\, dx$$
$$= \frac{x^3}{3}\bigg|_0^4 = 64/3$$

## 10. ARC LENGTH

Equation 11.36 gives the length of a curve defined by $f(x)$ whose derivative exists in the interval $[a, b]$.

$$\text{length} = \int_a^b \sqrt{1 + \left(f'(x)\right)^2}\, dx \qquad \text{11.36}$$

## 11. PAPPUS' THEOREMS[3]

The first and second theorems of Pappus are:[4]

- *First Theorem:* Given a curve, $C$, that does not intersect the $y$-axis, the area of the *surface of revolution* generated by revolving $C$ around the $y$-axis is equal to the product of the length of the curve and the circumference of the circle traced by the centroid of curve $C$.

$$A = \text{length} \times \text{circumference}$$
$$= \text{length} \times 2\pi \times \text{radius} \qquad \text{11.37}$$

- *Second Theorem:* Given a plane region, $R$, that does not intersect the $y$-axis, the *volume of revolution* generated by revolving $R$ around the $y$-axis is equal

---

[3]This section is an introduction to surfaces and volumes of revolution but does not involve integration.
[4]Some authorities call the first theorem the second and vice versa.

---

to the product of the area and the circumference of the circle traced by the centroid of area $R$.

$$V = \text{area} \times \text{circumference}$$
$$= \text{area} \times 2\pi \times \text{radius} \qquad \text{11.38}$$

## 12. SURFACE OF REVOLUTION

The surface area obtained by rotating $f(x)$ about the $x$-axis is

$$A = 2\pi \int_{x=a}^{x=b} f(x)\sqrt{1 + \left(f'(x)\right)^2}\, dx \qquad \text{11.39}$$

The surface area obtained by rotating $f(y)$ about the $y$-axis is

$$A = 2\pi \int_{y=c}^{y=d} f(y)\sqrt{1 + \left(f'(y)\right)^2}\, dy \qquad \text{11.40}$$

## Example 11.9

The curve $f(x) = \frac{1}{2}x$ over the region $x = [0, 4]$ is rotated about the $x$-axis. What is the surface of revolution?

*Solution*

The surface of revolution is

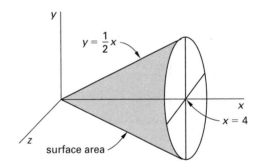

Since $f(x) = \frac{1}{2}x$, $f'(x) = \frac{1}{2}$. From Eq. 11.39, the area is

$$A = 2\pi \int_{x=a}^{x=b} f(x)\sqrt{1 + \left(f'(x)\right)^2}\, dx$$
$$= 2\pi \int_0^4 \frac{1}{2}x\sqrt{1 + \left(\frac{1}{2}\right)^2}\, dx$$
$$= \frac{\sqrt{5}}{2}\pi \int_0^4 x\, dx$$
$$= \frac{\sqrt{5}}{2}\pi \frac{x^2}{2}\bigg|_0^4$$
$$= \frac{\sqrt{5}}{2}\pi \left(\frac{(4)^2 - (0)^2}{2}\right)$$
$$= 4\sqrt{5}\pi$$

Mathematics

## 13. VOLUME OF REVOLUTION

The volume obtained by rotating $f(x)$ about the $x$-axis is given by Eq. 11.41. $f^2(x)$ is the square of the function, not the second derivative. Equation 11.41 is known as the *method of discs*.

$$V = \pi \int_{x=a}^{x=b} f^2(x)\,dx \qquad 11.41$$

The volume obtained by rotating $f(x)$ about the $y$-axis can be found from Eq. 11.41 (i.e., using the method of discs) by rewriting the limits and equation in terms of $y$, or alternatively, the *method of shells* can be used, resulting in the second form of Eq. 11.42.

$$V = \pi \int_{y=c}^{y=d} f^2(y)\,dy$$
$$= 2\pi \int_{x=a}^{x=b} xf(x)\,dx \qquad 11.42$$

### Example 11.10

The curve $f(x) = x^2$ over the region $x = [0, 4]$ is rotated about the $x$-axis. What is the volume of revolution?

*Solution*

The volume of revolution is

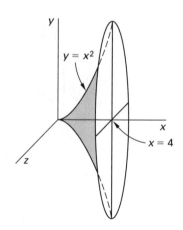

$$V = \pi \int_a^b f^2(x)\,dx = \pi \int_0^4 (x^2)^2\,dx$$
$$= \pi \frac{x^5}{5}\Big|_0^4 = \pi\left(\frac{1024}{5} - 0\right) = 204.8\pi$$

## 14. MOMENTS OF A FUNCTION

The *first moment of a function* is a concept used in finding centroids and centers of gravity. Equation 11.43 and Eq. 11.44 are for one- and two-dimensional

problems, respectively. It is the exponent of $x$ (1 in this case) that gives the moment its name.

$$\text{first moment} = \int xf(x)\,dx \qquad 11.43$$

$$\text{first moment} = \iint xf(x, y)\,dx\,dy \qquad 11.44$$

The *second moment of a function* is a concept used in finding moments of inertia with respect to an axis. Equation 11.45 and Eq. 11.46 are for two- and three-dimensional problems, respectively. Second moments with respect to other axes are analogous.

$$(\text{second moment})_x = \iint y^2 f(x, y)\,dy\,dx \qquad 11.45$$

$$(\text{second moment})_x = \iiint (y^2 + z^2) f(x, y, z)\,dy\,dz\,dx$$
$$11.46$$

## 15. FOURIER SERIES

Any periodic waveform can be written as the sum of an infinite number of sinusoidal terms, known as *harmonic terms* (i.e., an infinite series). Such a sum of terms is known as a *Fourier series*, and the process of finding the terms is *Fourier analysis*. (Extracting the original waveform from the series is known as *Fourier inversion*.) Since most series converge rapidly, it is possible to obtain a good approximation to the original waveform with a limited number of sinusoidal terms.

*Fourier's theorem* is Eq. 11.47.[5] The object of a Fourier analysis is to determine the coefficients $a_n$ and $b_n$. The constant $a_0$ can often be determined by inspection since it is the average value of the waveform.

$$f(t) = a_0 + a_1 \cos\omega t + a_2 \cos 2\omega t + \cdots$$
$$+ b_1 \sin\omega t + b_2 \sin 2\omega t + \cdots \qquad 11.47$$

$\omega$ is the *natural (fundamental) frequency* of the waveform. It depends on the actual waveform period, $T$.

$$\omega = \frac{2\pi}{T} \qquad 11.48$$

To simplify the analysis, the time domain can be normalized to the radian scale. The normalized scale is obtained by dividing all frequencies by $\omega$. Then the Fourier series becomes

$$f(t) = a_0 + a_1 \cos t + a_2 \cos 2t + \cdots$$
$$+ b_1 \sin t + b_2 \sin 2t + \cdots \qquad 11.49$$

---

[5]The independent variable used in this section is $t$, since Fourier analysis is most frequently used in the time domain.

The coefficients $a_n$ and $b_n$ are found from the following relationships.

$$a_0 = \frac{1}{2\pi}\int_0^{2\pi} f(t)\,dt$$

$$= \frac{1}{T}\int_0^{T} f(t)\,dt \qquad 11.50$$

$$a_n = \frac{1}{\pi}\int_0^{2\pi} f(t)\cos nt\,dt$$

$$= \frac{2}{T}\int_0^{T} f(t)\cos nt\,dt \quad [n\geq 1] \qquad 11.51$$

$$b_n = \frac{1}{\pi}\int_0^{2\pi} f(t)\sin nt\,dt$$

$$= \frac{2}{T}\int_0^{T} f(t)\sin nt\,dt \quad [n\geq 1] \qquad 11.52$$

While Eq. 11.51 and Eq. 11.52 are always valid, the work of integrating and finding $a_n$ and $b_n$ can be greatly simplified if the waveform is recognized as being symmetrical. Table 11.1 summarizes the simplifications.

### Example 11.11

Find the first four terms of a Fourier series that approximates the repetitive step function illustrated.

$$f(t) = \left\{ \begin{array}{ll} 1 & 0 < t < \pi \\ 0 & \pi < t < 2\pi \end{array} \right\}$$

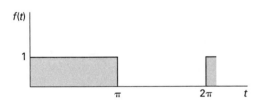

*Solution*

From Eq. 11.50,

$$a_0 = \frac{1}{2\pi}\int_0^{\pi} (1)\,dt + \frac{1}{2\pi}\int_{\pi}^{2\pi} (0)\,dt = \tfrac{1}{2}$$

This value of $^1/_2$ corresponds to the average value of $f(t)$. It could have been found by observation.

$$a_1 = \frac{1}{\pi}\int_0^{\pi} (1)\cos t\,dt + \frac{1}{\pi}\int_{\pi}^{2\pi} (0)\cos t\,dt$$

$$= \frac{1}{\pi}\sin t\Big|_0^{\pi} + 0 = 0$$

**Table 11.1** *Fourier Analysis Simplifications for Symmetrical Waveforms*

| | even symmetry $f(-t) = f(t)$ | odd symmetry $f(-t) = -f(t)$ |
|---|---|---|
| **full-wave symmetry*** $f(t + 2\pi) = f(t)$ $\lvert A_2\rvert = \lvert A_1\rvert$ $\lvert A_{\text{total}}\rvert = \lvert A_1\rvert$ *any repeating wave form | $b_n = 0$ [all $n$] $a_n = \frac{1}{\pi}\int_0^{2\pi} f(t)\cos nt\,dt$ [all $n$] | $a_0 = 0$ $a_n = 0$ [all $n$] $b_n = \frac{1}{\pi}\int_0^{2\pi} f(t)\sin nt\,dt$ [all $n$] |
| **half-wave symmetry*** $f(t + \pi) = -f(t)$ $\lvert A_2\rvert = \lvert A_1\rvert$ $\lvert A_{\text{total}}\rvert = 2\lvert A_1\rvert$ *same as rotational symmetry | $a_n = 0$ [even $n$] $b_n = 0$ [all $n$] $a_n = \frac{2}{\pi}\int_0^{\pi} f(t)\cos nt\,dt$ [odd $n$] | $a_0 = 0$ $a_n = 0$ [all $n$] $b_n = 0$ [even $n$] $b_n = \frac{2}{\pi}\int_0^{\pi} f(t)\sin nt\,dt$ [odd $n$] |
| **quarter-wave symmetry** $f(t + \pi) = -f(t)$ $\lvert A_2\rvert = \lvert A_1\rvert$ $\lvert A_{\text{total}}\rvert = 4\lvert A_1\rvert$ | $a_0 = 0$ $a_n = 0$ [even $n$] $b_n = 0$ [all $n$] $a_n = \frac{4}{\pi}\int_0^{\frac{\pi}{2}} f(t)\cos nt\,dt$ [odd $n$] | $a_0 = 0$ $a_n = 0$ [all $n$] $b_n = 0$ [even $n$] $b_n = \frac{4}{\pi}\int_0^{\frac{\pi}{2}} f(t)\sin nt\,dt$ [odd $n$] |

In general,

$$a_n = \frac{1}{\pi}\frac{\sin nt}{n}\bigg|_0^{\pi} = 0$$

$$b_1 = \frac{1}{\pi}\int_0^{\pi} (1)\sin t\,dt + \frac{1}{\pi}\int_{\pi}^{2\pi} (0)\sin t\,dt$$

$$= \frac{1}{\pi} - \cos t\Big|_0^{\pi} = \frac{2}{\pi}$$

In general,

$$b_n = \frac{1}{\pi}\frac{-\cos nt}{n}\bigg|_0^{\pi} = \left\{ \begin{array}{ll} 0 & \text{for } n \text{ even} \\ \dfrac{2}{\pi n} & \text{for } n \text{ odd} \end{array} \right.$$

Mathematics

The series is

$$f(t) = \tfrac{1}{2} + \tfrac{2}{\pi}(\sin t + \tfrac{1}{3}\sin 3t + \tfrac{1}{5}\sin 5t + \cdots)$$

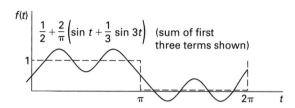

## 16. FAST FOURIER TRANSFORMS

Many mathematical operations are needed to implement a true Fourier transform. While the terms of a Fourier series might be slowly derived by integration, a faster method is needed to analyze real-time data. The *fast Fourier transform* (FFT) is a computer algorithm implemented in *spectrum analyzers (signal analyzers or FFT analyzers)* and replaces integration and multiplication operations with table look-ups and additions.[6]

Since the complexity of the transform is reduced, the transformation occurs more quickly, enabling efficient analysis of waveforms with little or no periodicity.[7]

Using a spectrum analyzer requires choosing the frequency band (e.g., 0 kHz to 20 kHz) to be monitored. (This step automatically selects the sampling period. The lower the frequencies sampled, the longer the sampling period.) If they are not fixed by the analyzer, the numbers of time-dependent input variable samples (e.g., 1024) and frequency-dependent output variable values (e.g., 400) are chosen.[8] There are half as many frequency lines as data points because each line contains two pieces of information—real (amplitude) and imaginary (phase). The *resolution* of the resulting frequency analysis is

$$\text{resolution} = \frac{\text{frequency bandwidth}}{\text{no. of output variable values}} \qquad \textit{11.53}$$

## 17. INTEGRAL FUNCTIONS

Integrals that cannot be evaluated as finite combinations of elementary functions are called *integral functions*. These functions are evaluated by series expansion. Some of the more common functions are listed as follows.[9,10]

- *integral sine function*

$$\text{Si}(x) = \int_0^x \frac{\sin x}{x}\,dx$$

$$= x - \frac{x^3}{3\cdot 3!} + \frac{x^5}{5\cdot 5!} - \frac{x^7}{7\cdot 7!} + \cdots \qquad \textit{11.54}$$

- *integral cosine function*

$$\text{Ci}(x) = \int_{-\infty}^x \frac{\cos x}{x}\,dx = -\int_x^\infty \frac{\cos x}{x}\,dx$$

$$= C_E + \ln x - \frac{x^2}{2\cdot 2!} + \frac{x^4}{4\cdot 4!} - \cdots \qquad \textit{11.55}$$

- *integral exponential function*

$$\text{Fi}(x) = \int_{-\infty}^x \frac{e^x}{x}\,dx = -\int_{-x}^\infty \frac{e^{-x}}{x}\,dx$$

$$= C_E + \ln x + x + \frac{x^2}{2\cdot 2!} + \frac{x^3}{3\cdot 3!} + \cdots \qquad \textit{11.56}$$

- *error function*

$$\text{erf}(x) = \frac{2}{\sqrt{\pi}} \int_0^x e^{-x^2}\,dx$$

$$= \left(\frac{2}{\sqrt{\pi}}\right)\left(\frac{x}{1\cdot 0!} - \frac{x^3}{3\cdot 1!} + \frac{x^5}{5\cdot 2!} - \frac{x^7}{7\cdot 3!} + \cdots\right)$$

$$\textit{11.57}$$

---

[6]*Spectrum analysis*, also known as *frequency analysis, signature analysis*, and *time-series analysis*, develops a relationship (usually graphical) between some property (e.g., amplitude or phase shift) versus frequency.

[7]Hours and days of manual computations are compressed into milliseconds.

[8]Two samples per time-dependent cycle (at the maximum frequency) is the lower theoretical limit for sampling, but the practical minimum rate is approximately 2.5 samples per cycle. This will ensure that *alias components* (i.e., low-level frequency signals) do not show up in the frequency band of interest.

---

[9]Other integral functions include the Fresnel integral, gamma function, and elliptic integral.

[10]$C_E$ in Eq. 11.55 and Eq. 11.56 is *Euler's constant.*

$$C_E = \int_{+\infty}^0 e^{-x} \ln x\,dx$$

$$= \lim_{m\to\infty}\left(1 + \tfrac{1}{2} + \tfrac{1}{3} + \cdots + \tfrac{1}{m} - \ln m\right)$$

$$= 0.577215665$$

# 12 Differential Equations

## 1. TYPES OF DIFFERENTIAL EQUATIONS

A *differential equation* is a mathematical expression combining a function (e.g., $y = f(x)$) and one or more of its derivatives. The *order* of a differential equation is the highest derivative in it. *First-order differential equations* contain only first derivatives of the function, *second-order differential equations* contain second derivatives (and may contain first derivatives as well), and so on.

A *linear differential equation* can be written as a sum of products of multipliers of the function and its derivatives. If the multipliers are scalars, the differential equation is said to have *constant coefficients*. If the function or one of its derivatives is raised to some power (other than one) or is embedded in another function (e.g., $y$ embedded in $\sin y$ or $e^y$), the equation is said to be *nonlinear*.

Each term of a *homogeneous differential equation* contains either the function ($y$) or one of its derivatives;

that is, the sum of derivative terms is equal to zero. In a *nonhomogeneous differential equation*, the sum of derivative terms is equal to a nonzero *forcing function* of the independent variable (e.g., $g(x)$). In order to solve a nonhomogeneous equation, it is often necessary to solve the homogeneous equation first. The homogeneous equation corresponding to a nonhomogeneous equation is known as a *reduced equation* or *complementary equation*.

The following examples illustrate the types of differential equations.

$y' - 7y = 0$       homogeneous, first-order linear, with constant coefficients

$y'' - 2y' + 8y = \sin 2x$       nonhomogeneous, second-order linear, with constant coefficients

$y'' - (x^2 - 1)y^2 = \sin 4x$       nonhomogeneous, second-order, nonlinear

An *auxiliary equation* (also called the *characteristic equation*) can be written for a homogeneous linear differential equation with constant coefficients, regardless of order. This auxiliary equation is simply the polynomial formed by replacing all derivatives with variables raised to the power of their respective derivatives.

The purpose of solving a differential equation is to derive an expression for the function in terms of the independent variable. The expression does not need to be explicit in the function, but there can be no derivatives in the expression. Since, in the simplest cases, solving a differential equation is equivalent to finding an indefinite integral, it is not surprising that *constants of integration* must be evaluated from knowledge of how the system behaves. Additional data are known as *initial values*, and any problem that includes them is known as an *initial value problem*.[1]

Most differential equations require lengthy solutions and are not efficiently solved by hand. However, several types are fairly simple and are presented in this chapter.

---

[1]The term *initial* implies that time is the independent variable. While this may explain the origin of the term, initial value problems are not limited to the time domain. A *boundary value problem* is similar, except that the data come from different points. For example, additional data in the form $y(x_0)$ and $y'(x_0)$ or $y(x_0)$ and $y'(x_1)$ that need to be simultaneously satisfied constitute an initial value problem. Data of the form $y(x_0)$ and $y(x_1)$ constitute a boundary value problem. Until solved, it is difficult to know whether a boundary value problem has zero, one, or more than one solution.

### Example 12.1

Write the complementary differential equation for the following nonhomogeneous differential equation.

$$y'' + 6y' + 9y = e^{-14x} \sin 5x$$

*Solution*

The complementary equation is found by eliminating the forcing function, $e^{-14x} \sin 5x$.

$$y'' + 6y' + 9y = 0$$

### Example 12.2

Write the auxiliary equation to the following differential equation.

$$y'' + 4y' + y = 0$$

*Solution*

Replacing each derivative with a polynomial term whose degree equals the original order, the auxiliary equation is

$$r^2 + 4r + 1 = 0$$

## 2. HOMOGENEOUS, FIRST-ORDER LINEAR DIFFERENTIAL EQUATIONS WITH CONSTANT COEFFICIENTS

A homogeneous, first-order linear differential equation with constant coefficients will have the general form of Eq. 12.1.

$$y' + ky = 0 \qquad\qquad 12.1$$

The auxiliary equation is $r + k = 0$ and it has a root of $r = -k$. Equation 12.2 is the solution.

$$y = Ae^{rx} = Ae^{-kx} \qquad\qquad 12.2$$

If the initial condition is known to be $y(0) = y_0$, the solution is

$$y = y_0 e^{-kx} \qquad\qquad 12.3$$

## 3. FIRST-ORDER LINEAR DIFFERENTIAL EQUATIONS

A first-order linear differential equation has the general form of Eq. 12.4. $p(x)$ and $g(x)$ can be constants or any function of $x$ (but not of $y$). However, if $p(x)$ is a constant and $g(x)$ is zero, it is easier to solve the equation as shown in Sec. 12.2.

$$y' + p(x)y = g(x) \qquad\qquad 12.4$$

The *integrating factor* (which is usually a function) to this differential equation is

$$u(x) = \exp\left(\int p(x)\,dx\right) \qquad\qquad 12.5$$

The closed-form solution to Eq. 12.4 is

$$y = \frac{1}{u(x)}\left(\int u(x)g(x)\,dx + C\right) \qquad\qquad 12.6$$

For the special case where $p(x)$ and $g(x)$ are both constants, Eq. 12.4 becomes

$$y' + ay = b \qquad\qquad 12.7$$

If the initial condition is $y(0) = y_0$, then the solution to Eq. 12.7 is

$$y = \left(\frac{b}{a}\right)\left(1 - e^{-ax}\right) + y_0 e^{-ax} \qquad\qquad 12.8$$

### Example 12.3

Find a solution to the following differential equation.

$$y' - y = 2xe^{2x} \qquad y(0) = 1$$

*Solution*

This is a first-order linear equation with $p(x) = -1$ and $g(x) = 2xe^{2x}$. The integrating factor is

$$u(x) = \exp\left(\int p(x)\,dx\right) = \exp\left(\int -1\,dx\right) = e^{-x}$$

The solution is given by Eq. 12.6.

$$\begin{aligned}
y &= \frac{1}{u(x)}\left(\int u(x)g(x)\,dx + C\right)\\
&= \frac{1}{e^{-x}}\left(\int e^{-x}2xe^{2x}\,dx + C\right)\\
&= e^x\left(2\int xe^x\,dx + C\right)\\
&= e^x(2xe^x - 2e^x + C)\\
&= e^x(2e^x(x - 1) + C)
\end{aligned}$$

From the initial condition,

$$y(0) = 1$$
$$e^0\big((2)(e^0)(0 - 1) + C\big) = 1$$
$$1\big((2)(1)(-1) + C\big) = 1$$

Therefore, $C = 3$. The complete solution is

$$y = e^x\big(2e^x(x - 1) + 3\big)$$

## 4. FIRST-ORDER SEPARABLE DIFFERENTIAL EQUATIONS

*First-order separable differential equations* can be placed in the form of Eq. 12.9. For clarity and convenience, $y'$ is written as $dy/dx$.

$$m(x) + n(y)\frac{dy}{dx} = 0 \qquad 12.9$$

Equation 12.9 can be placed in the form of Eq. 12.10, both sides of which are easily integrated. An initial value will establish the constant of integration.

$$m(x)\,dx = -n(y)\,dy \qquad 12.10$$

## 5. FIRST-ORDER EXACT DIFFERENTIAL EQUATIONS

A *first-order exact differential equation* has the form

$$f_x(x, y) + f_y(x, y)y' = 0 \qquad 12.11$$

Notice that $f_x(x, y)$ is the exact derivative of $f(x, y)$ with respect to $x$, and $f_y(x, y)$ is the exact derivative of $f(x, y)$ with respect to $y$. The solution is

$$f(x, y) - C = 0 \qquad 12.12$$

## 6. HOMOGENEOUS, SECOND-ORDER LINEAR DIFFERENTIAL EQUATIONS WITH CONSTANT COEFFICIENTS

*Homogeneous second-order linear differential equations with constant coefficients* have the form of Eq. 12.13. They are most easily solved by finding the two roots of the auxiliary equation. (See Eq. 12.14.)

$$y'' + k_1 y' + k_2 y = 0 \qquad 12.13$$

$$r^2 + k_1 r + k_2 = 0 \qquad 12.14$$

There are three cases. If the two roots of Eq. 12.14 are real and different, the solution is

$$y = A_1 e^{r_1 x} + A_2 e^{r_2 x} \qquad 12.15$$

If the two roots are real and the same, the solution is

$$y = A_1 e^{rx} + A_2 x e^{rx} \qquad 12.16$$

$$r = \frac{-k_1}{2} \qquad 12.17$$

If the two roots are imaginary, they will be of the form $(\alpha + i\omega)$ and $(\alpha - i\omega)$, and the solution is

$$y = A_1 e^{\alpha x} \cos \omega x + A_2 e^{\alpha x} \sin \omega x \qquad 12.18$$

In all three cases, $A_1$ and $A_2$ must be found from the two initial conditions.

### Example 12.4

Solve the following differential equation.

$$y'' + 6y' + 9y = 0$$

$$y(0) = 0 \qquad y'(0) = 1$$

*Solution*

The auxiliary equation is

$$r^2 + 6r + 9 = 0$$

$$(r + 3)(r + 3) = 0$$

The roots to the auxiliary equation are $r_1 = r_2 = -3$. Therefore, the solution has the form of Eq. 12.16.

$$y = A_1 e^{-3x} + A_2 x e^{-3x}$$

The first initial condition is

$$y(0) = 0$$

$$A_1 e^0 + A_2(0) e^0 = 0$$

$$A_1 + 0 = 0$$

$$A_1 = 0$$

To use the second initial condition, the derivative of the equation is needed. Making use of the known fact that $A_1 = 0$,

$$y' = \frac{d}{dx}(A_2 x e^{-3x}) = -3A_2 x e^{-3x} + A_2 e^{-3x}$$

Using the second initial condition,

$$y'(0) = 1$$

$$-3A_2(0) e^0 + A_2 e^0 = 1$$

$$0 + A_2 = 1$$

$$A_2 = 1$$

The solution is

$$y = x e^{-3x}$$

## 7. NONHOMOGENEOUS DIFFERENTIAL EQUATIONS

A nonhomogeneous equation has the form of Eq. 12.19. $f(x)$ is known as the *forcing function*.

$$y'' + p(x)y' + q(x)y = f(x) \qquad 12.19$$

The solution to Eq. 12.19 is the sum of two equations. The *complementary solution*, $y_c$, solves the complementary (i.e., homogeneous) problem. The *particular solution*, $y_p$, is any specific solution to the nonhomogeneous Eq. 12.19 that is known or can be found. Initial values are used to evaluate any unknown coefficients in the complementary solution *after* $y_c$ and $y_p$ have been combined. (The particular solution will not have any unknown coefficients.)

$$y = y_c + y_p \qquad 12.20$$

Two methods are available for finding a particular solution. The *method of undetermined coefficients*, as presented here, can be used only when $p(x)$ and $q(x)$ are constant coefficients and $f(x)$ takes on one of the forms in Table 12.1.

The particular solution can be read from Table 12.1 if the forcing function is of one of the forms given. Of course, the coefficients $A_i$ and $B_i$ are not known—these are the *undetermined coefficients*. The exponent $s$ is the smallest nonnegative number (and will be 0, 1, or 2), which ensures that no term in the particular solution, $y_p$, is also a solution to the complementary equation, $y_c$. $s$ must be determined prior to proceeding with the solution procedure.

**Table 12.1** *Particular Solutions*[*]

| form of $f(x)$ | form of $y_p$ |
|---|---|
| $P_n(x) = a_0 x^n + a_1 x^{n-1}$ $+ \cdots + a_n$ | $x^s \left( \dfrac{A_0 x^n + A_1 x^{n-1} + \ldots}{+A_n} \right)$ |
| $P_n(x) e^{\alpha x}$ | $x^s \left( \dfrac{A_0 x^n + A_1 x^{n-1} + \cdots}{+A_n} \right) e^{\alpha x}$ |
| $P_n(x) e^{\alpha x} \begin{Bmatrix} \sin \omega x \\ \cos \omega x \end{Bmatrix}$ | $x^s \Big[ (A_0 x^n + A_1 x^{n-1} + \cdots \\ + A_n) e^{\alpha x} \cos \omega x \\ + (B_0 x^n + B_1 x^{n-1} + \cdots \\ + B_n) e^{\alpha x} \sin \omega x \Big]$ |

[*]$P_n(x)$ is a polynomial of degree $n$.

Once $y_p$ (including $s$) is known, it is differentiated to obtain $y_p'$ and $y_p''$, and all three functions are substituted into the original nonhomogeneous equation. The resulting equation is rearranged to match the forcing function, $f(x)$, and the unknown coefficients are determined, usually by solving simultaneous equations.

If the forcing function, $f(x)$, is more complex than the forms shown in Table 12.1, or if either $p(x)$ or $q(x)$ is a function of $x$, the method of *variation of parameters* should be used. This complex and time-consuming method is not covered in this book.

### Example 12.5

Solve the following nonhomogeneous differential equation.

$$y'' + 2y' + y = e^x \cos x$$

*Solution*

*step 1:* Find the solution to the complementary (homogeneous) differential equation.

$$y'' + 2y' + y = 0$$

Since this is a differential equation with constant coefficients, write the auxiliary equation.

$$r^2 + 2r + 1 = 0$$

The auxiliary equation factors in $(r + 1)^2 = 0$ with two identical roots at $r = -1$. Therefore, the solution to the homogeneous differential equation is

$$y_c(x) = C_1 e^{-x} + C_2 x e^{-x}$$

*step 2:* Use Table 12.1 to determine the form of a particular solution. Since the forcing function has the form $P_n(x) e^{\alpha x} \cos \omega x$ with $P_n(x) = 1$ (equivalent to $n = 0$), $\alpha = 1$, and $\omega = 1$, the particular solution has the form

$$y_p(x) = x^s (A e^x \cos x + B e^x \sin x)$$

*step 3:* Determine the value of $s$. Check to see if any of the terms in $y_p(x)$ will themselves solve the homogeneous equation. Try $A e^x \cos x$ first.

$$\frac{d}{dx} (A e^x \cos x) = A e^x \cos x - A e^x \sin x$$

$$\frac{d^2}{dx^2} (A e^x \cos x) = -2 A e^x \sin x$$

Substitute these quantities into the homogeneous equation.

$$y'' + 2y' + y = 0$$
$$-2 A e^x \sin x + 2 A e^x \cos x$$
$$- 2 A e^x \sin x + A e^x \cos x = 0$$
$$3 A e^x \cos x - 4 A e^x \sin x = 0$$

Disregarding the trivial (i.e., $A = 0$) solution, $A e^x \cos x$ does not solve the homogeneous equation.

Next, try $B e^x \sin x$.

$$\frac{d}{dx} (B e^x \sin x) = B e^x \cos x + B e^x \sin x$$

$$\frac{d^2}{dx^2} (B e^x \sin x) = 2 B e^x \cos x$$

Substitute these quantities into the homogeneous equation.

$$y'' + 2y' + y = 0$$

$$2Be^x \cos x + 2Be^x \cos x$$

$$+ 2Be^x \sin x + Be^x \sin x = 0$$

$$3Be^x \sin x + 4Be^x \cos x = 0$$

Disregarding the trivial $(B = 0)$ case, $Be^x \sin x$ does not solve the homogeneous equation.

Since none of the terms in $y_p(x)$ solve the homogeneous equation, $s = 0$, and a particular solution has the form

$$y_p(x) = Ae^x \cos x + Be^x \sin x$$

*step 4:* Use the method of unknown coefficients to determine $A$ and $B$ in the particular solution. Drawing on the previous steps, substitute the quantities derived from the particular solution into the nonhomogeneous equation.

$$y'' + 2y' + y = e^x \cos x$$

$$-2Ae^x \sin x + 2Be^x \cos x$$

$$+ 2Ae^x \cos x - 2Ae^x \sin x$$

$$+ 2Be^x \cos x + 2Be^x \sin x$$

$$+ Ae^x \cos x + Be^x \sin x = e^x \cos x$$

Combining terms,

$$(-4A + 3B)e^x \sin x + (3A + 4B)e^x \cos x$$

$$= e^x \cos x$$

Equating the coefficients of like terms on either side of the equal sign results in the following simultaneous equations.

$$-4A + 3B = 0$$

$$3A + 4B = 1$$

The solution to these equations is

$$A = \tfrac{3}{25}$$

$$B = \tfrac{4}{25}$$

A particular solution is

$$y_p(x) = \left(\tfrac{3}{25}\right)\left(e^x \cos x\right) + \left(\tfrac{4}{25}\right)\left(e^x \sin x\right)$$

*step 5:* Write the general solution.

$$y(x) = y_c(x) + y_p(x)$$

$$= C_1 e^{-x} + C_2 x e^{-x} + \left(\tfrac{3}{25}\right)\left(e^x \cos x\right)$$

$$+ \left(\tfrac{4}{25}\right)\left(e^x \sin x\right)$$

The values of $C_1$ and $C_2$ would be determined at this time if initial conditions were known.

## 8. NAMED DIFFERENTIAL EQUATIONS

Some differential equations with specific forms are named after the individuals who developed solution techniques for them.

- *Bessel equation of order $\nu$*

$$x^2 y'' + xy' + (x^2 - \nu^2)y = 0 \qquad \text{12.21}$$

- *Cauchy equation*

$$a_0 x^n \frac{d^n y}{dx^n} + a_1 x^{n-1} \frac{d^{n-1}y}{dx^{n-1}} + \cdots$$

$$+ a_{n-1} x \frac{dy}{dx} + a_n y = f(x) \qquad \text{12.22}$$

- *Euler equation*

$$x^2 y'' + \alpha xy' + \beta y = 0 \qquad \text{12.23}$$

- *Gauss' hypergeometric equation*

$$x(1-x)y'' + (c - (a+b+1)x)y' - aby = 0 \qquad \text{12.24}$$

- *Legendre equation of order $\lambda$*

$$(1 - x^2)y'' - 2xy' + \lambda(\lambda + 1)y = 0 \quad [-1 < x < 1]$$

$$\text{12.25}$$

## 9. LAPLACE TRANSFORMS

Traditional methods of solving nonhomogeneous differential equations by hand are usually difficult and/or time consuming. *Laplace transforms* can be used to reduce many solution procedures to simple algebra.

Every mathematical function, $f(t)$, for which Eq. 12.26 exists has a Laplace transform, written as $\mathcal{L}(f)$ or $F(s)$. The transform is written in the *s*-domain, regardless of the independent variable in the original function.[2] (The variable $s$ is equivalent to a derivative operator, although it may be handled in the equations as a simple

---

[2]It is traditional to write the original function as a function of the independent variable $t$ rather than $x$. However, Laplace transforms are not limited to functions of time.

variable.) Equation 12.26 converts a function into a Laplace transform.

$$\mathcal{L}(f(t)) = F(s) = \int_0^\infty e^{-st} f(t) dt \qquad \textit{12.26}$$

Equation 12.26 is not often needed because tables of transforms are readily available. (Appendix 12.A contains some of the most common transforms.)

Extracting a function from its transform is the *inverse Laplace transform* operation. Although other methods exist, this operation is almost always done by finding the transform in a set of tables.[3]

$$f(t) = \mathcal{L}^{-1}(F(s)) \qquad \textit{12.27}$$

### Example 12.6

Find the Laplace transform of the following function.

$$f(t) = e^{at} \qquad [s > a]$$

*Solution*

Applying Eq. 12.26,

$$\mathcal{L}(e^{at}) = \int_0^\infty e^{-st} e^{at} dt = \int_0^\infty e^{-(s-a)t} dt$$

$$= -\frac{e^{-(s-a)t}}{s-a} \bigg|_0^\infty = \frac{1}{s-a} \qquad [s > a]$$

## 10. STEP AND IMPULSE FUNCTIONS

Many forcing functions are sinusoidal or exponential in nature; others, however, can only be represented by a step or impulse function. A *unit step function*, $u_t$, is a function describing the disturbance of magnitude 1 that is not present before time $t$ but is suddenly there after time $t$. A step of magnitude 5 at time $t = 3$ would be represented as $5u_3$. (The notation $5u(t-3)$ is used in some books.)

The *unit impulse function*, $\delta_t$, is a function describing a disturbance of magnitude 1 that is applied and removed so quickly as to be instantaneous. An impulse of magnitude 5 at time 3 would be represented by $5\delta_3$. (The notation $5\delta(t-3)$ is used in some books.)

### Example 12.7

What is the notation for a forcing function of magnitude 6 that is applied at $t = 2$ and that is completely removed at $t = 7$?

*Solution*

The notation is $f(t) = 6(u_2 - u_7)$.

---

[3]Other methods include integration in the complex plane, convolution, and simplification by partial fractions.

### Example 12.8

Find the Laplace transform of $u_0$, a unit step at $t = 0$.

$$f(t) = 0 \text{ for } t < 0$$

$$f(t) = 1 \text{ for } t \geq 0$$

*Solution*

Since the Laplace transform is an integral that starts at $t = 0$, the value of $f(t)$ prior to $t = 0$ is irrelevant.

$$\mathcal{L}(u_0) = \int_0^\infty e^{-st}(1) dt = -\frac{e^{-st}}{s} \bigg|_0^\infty$$

$$= 0 - \left(\frac{-1}{s}\right) = \frac{1}{s}$$

## 11. ALGEBRA OF LAPLACE TRANSFORMS

Equations containing Laplace transforms can be simplified by applying the following principles.

- *linearity theorem* ($c$ is a constant.)

$$\mathcal{L}(cf(t)) = c\mathcal{L}(f(t)) = cF(s) \qquad \textit{12.28}$$

- *superposition theorem* ($f(t)$ and $g(t)$ are different functions.)

$$\mathcal{L}(f(t)) \pm g(t)) = \mathcal{L}(f(t)) \pm \mathcal{L}(g(t))$$

$$= F(s) \pm G(s) \qquad \textit{12.29}$$

- *time-shifting theorem (delay theorem)*

$$\mathcal{L}(f(t-b)u_b) = e^{-bs}F(s) \qquad \textit{12.30}$$

- *Laplace transform of a derivative*

$$\mathcal{L}(f^n(t)) = -f^{n-1}(0) - sf^{n-2}(0) - \cdots$$

$$- s^{n-1}f(0) + s^n F(s) \qquad \textit{12.31}$$

- *other properties*

$$\mathcal{L}\left(\int_0^t f(u) du\right) = \left(\frac{1}{s}\right) F(s) \qquad \textit{12.32}$$

$$\mathcal{L}(tf(t)) = -\frac{dF}{ds} \qquad \textit{12.33}$$

$$\mathcal{L}\left(\frac{1}{t}f(t)\right) = \int_0^\infty F(u) du \qquad \textit{12.34}$$

## 12. CONVOLUTION INTEGRAL

A complex Laplace transform, $F(s)$, will often be recognized as the product of two other transforms, $F_1(s)$ and $F_2(s)$, whose corresponding functions $f_1(t)$ and $f_2(t)$ are known. Unfortunately, Laplace transforms cannot be computed with ordinary multiplication. That is, $f(t) \neq f_1(t)f_2(t)$ even though $F(s) = F_1(s)F_2(s)$.

However, it is possible to extract $f(t)$ from its *convolution*, $h(t)$, as calculated from either of the *convolution integrals* in Eq. 12.35. This process is demonstrated in Ex. 12.9. $\chi$ is a dummy variable.

$$
\begin{aligned}
f(t) &= \mathcal{L}^{-1}(F_1(s)F_2(s)) \\
&= \int_0^t f_1(t-\chi)f_2(\chi)\,d\chi \\
&= \int_0^t f_1(\chi)f_2(t-\chi)\,d\chi \qquad \text{12.35}
\end{aligned}
$$

### Example 12.9

Use the convolution integral to find the inverse transform of

$$
F(s) = \frac{3}{s^2(s^2+9)}
$$

*Solution*

$F(s)$ can be factored as

$$
F_1(s)F_2(s) = \left(\frac{1}{s^2}\right)\left(\frac{3}{s^2+9}\right)
$$

As the inverse transforms of $F_1(s)$ and $F_2(s)$ are $f_1(t) = t$ and $f_2(t) = \sin 3t$, respectively, the convolution integral from Eq. 12.35 is

$$
\begin{aligned}
f(t) &= \int_0^t (t-\chi)\sin 3\chi\,d\chi \\
&= \int_0^t (t\sin 3\chi - \chi \sin 3\chi)\,d\chi \\
&= t\int_0^t \sin 3\chi\,d\chi - \int_0^t \chi \sin 3\chi\,d\chi
\end{aligned}
$$

Expand using integration by parts.

$$
\begin{aligned}
f(t) &= -\tfrac{1}{3}t\cos 3\chi + \tfrac{1}{3}\chi\cos 3\chi - \tfrac{1}{9}\sin 3\chi\Big|_0^t \\
&= \frac{3t - \sin 3t}{9}
\end{aligned}
$$

## 13. USING LAPLACE TRANSFORMS

Any nonhomogeneous linear differential equation with constant coefficients can be solved with the following procedure, which reduces the solution to simple algebra. A complete table of transforms simplifies or eliminates step 5.

*step 1:* Put the differential equation in standard form (i.e., isolate the $y''$ term).

$$
y'' + k_1 y' + k_2 y = f(t) \qquad \text{12.36}
$$

*step 2:* Take the Laplace transform of both sides. Use the linearity and superposition theorems. (See Eq. 12.28 and Eq. 12.29.)

$$
\mathcal{L}(y'') + k_1\mathcal{L}(y') + k_2\mathcal{L}(y) = \mathcal{L}(f(t)) \qquad \text{12.37}
$$

*step 3:* Use Eq. 12.38 and Eq. 12.39 to expand the equation. (These are specific forms of Eq. 12.31.) Use a table to evaluate the transform of the forcing function.

$$
\begin{aligned}
\mathcal{L}(y'') &= s^2\mathcal{L}(y) - sy(0) - y'(0) \qquad \text{12.38} \\
\mathcal{L}(y') &= s\mathcal{L}(y) - y(0) \qquad \text{12.39}
\end{aligned}
$$

*step 4:* Use algebra to solve for $\mathcal{L}(y)$.

*step 5:* If needed, use partial fractions to simplify the expression for $\mathcal{L}(y)$.

*step 6:* Take the inverse transform to find $y(t)$.

$$
y(t) = \mathcal{L}^{-1}(\mathcal{L}(y)) \qquad \text{12.40}
$$

### Example 12.10

Find $y(t)$ for the following differential equation.

$$
\begin{aligned}
y'' + 2y' + 2y &= \cos t \\
y(0) = 1 \qquad y'(0) &= 0
\end{aligned}
$$

*Solution*

*step 1:* The equation is already in standard form.

*step 2:* $\mathcal{L}(y'') + 2\mathcal{L}(y') + 2\mathcal{L}(y) = \mathcal{L}(\cos t)$

*step 3:* Use Eq. 12.38 and Eq. 12.39. Use App. 12.A to find the transform of $\cos t$.

$$
\begin{aligned}
&s^2\mathcal{L}(y) - sy(0) - y'(0) + 2s\mathcal{L}(y) - 2y(0) + 2\mathcal{L}(y) \\
&= \frac{s}{s^2+1}
\end{aligned}
$$

But, $y(0) = 1$ and $y'(0) = 0$.

$$
s^2\mathcal{L}(y) - s + 2s\mathcal{L}(y) - 2 + 2\mathcal{L}(y) = \frac{s}{s^2+1}
$$

*step 4:* Combine terms and solve for $\mathcal{L}(y)$.

$$
\mathcal{L}(y)(s^2 + 2s + 2) - s - 2 = \frac{s}{s^2+1}
$$

$$
\begin{aligned}
\mathcal{L}(y) &= \frac{\dfrac{s}{s^2+1} + s + 2}{s^2 + 2s + 2} \\
&= \frac{s^3 + 2s^2 + 2s + 2}{(s^2+1)(s^2+2s+2)}
\end{aligned}
$$

**Mathematics**

*step 5:* Expand the expression for $\mathcal{L}(y)$ by partial fractions.

$$\mathcal{L}(y) = \frac{s^3 + 2s^2 + 2s + 2}{(s^2 + 1)(s^2 + 2s + 2)}$$

$$= \frac{A_1 s + B_1}{s^2 + 1} + \frac{A_2 s + B_2}{s^2 + 2s + 2}$$

$$= \frac{\begin{array}{c} s^3(A_1 + A_2) + s^2(2A_1 + B_1 + B_2) \\ + s(2A_1 + 2B_1 + A_2) + (2B_1 + B_2) \end{array}}{(s^2 + 1)(s^2 + 2s + 2)}$$

The following simultaneous equations result.

$$
\begin{array}{rcrcrcrcl}
A_1 & + & A_2 & & & & & = & 1 \\
2A_1 & & & + & B_1 & + & B_2 & = & 2 \\
2A_1 & + & A_2 & + & 2B_1 & & & = & 2 \\
& & & & 2B_1 & + & B_2 & = & 2
\end{array}
$$

These equations have the solutions $A_1 = \frac{1}{5}$, $A_2 = \frac{4}{5}$, $B_1 = \frac{2}{5}$, and $B_2 = \frac{6}{5}$.

*step 6:* Refer to App. 12.A and take the inverse transforms. (The numerator of the second term is rewritten from $(4s + 6)$ to $((4s + 4) + 2)$.)

$$y = \mathcal{L}^{-1}\Big(\mathcal{L}(y)\Big)$$

$$= \mathcal{L}^{-1}\left( \frac{\left(\frac{1}{5}\right)(s+2)}{s^2 + 1} + \frac{\left(\frac{1}{5}\right)(4s+6)}{s^2 + 2s + 2} \right)$$

$$= \left(\tfrac{1}{5}\right)\left( \begin{array}{c} \mathcal{L}^{-1}\left(\dfrac{s}{s^2+1}\right) + 2\mathcal{L}^{-1}\left(\dfrac{1}{s^2+1}\right) \\[2mm] + 4\mathcal{L}^{-1}\left(\dfrac{s-(-1)}{\left(s-(-1)\right)^2 + 1}\right) \\[2mm] + 2\mathcal{L}^{-1}\left(\dfrac{1}{\left(s-(-1)\right)^2 + 1}\right) \end{array} \right)$$

$$= \left(\tfrac{1}{5}\right)(\cos t + 2\sin t + 4e^{-t}\cos t + 2e^{-t}\sin t)$$

## 14. THIRD- AND HIGHER-ORDER LINEAR DIFFERENTIAL EQUATIONS WITH CONSTANT COEFFICIENTS

The solutions of third- and higher-order linear differential equations with constant coefficients are extensions of the solutions for second-order equations of this type. Specifically, if an equation is homogeneous, the auxiliary equation is written and its roots are found. If the equation is nonhomogeneous, Laplace transforms can be used to simplify the solution.

Consider the following homogeneous differential equation with constant coefficients.

$$y^n + k_1 y^{n-1} + \cdots + k_{n-1} y' + k_n y = 0 \qquad 12.41$$

The auxiliary equation to Eq. 12.41 is

$$r^n + k_1 r^{n-1} + \cdots + k_{n-1} r + k_n = 0 \qquad 12.42$$

For each real and distinct root $r$, the solution contains the term

$$y = A e^{rx} \qquad 12.43$$

For each real root $r$ that repeats $m$ times, the solution contains the term

$$y = (A_1 + A_2 x + A_3 x^2 + \cdots + A_m x^{m-1}) e^{rx} \qquad 12.44$$

For each pair of complex roots of the form $r = \alpha \pm i\omega$ the solution contains the terms

$$y = e^{\alpha x}(A_1 \sin \omega x + A_2 \cos \omega x) \qquad 12.45$$

## 15. APPLICATION: ENGINEERING SYSTEMS

There is a wide variety of engineering systems (mechanical, electrical, fluid flow, heat transfer, and so on) whose behavior is described by linear differential equations with constant coefficients.

## 16. APPLICATION: MIXING

A typical mixing problem involves a liquid-filled tank. The liquid may initially be pure or contain some solute. Liquid (either pure or as a solution) enters the tank at a known rate. A drain may be present to remove thoroughly mixed liquid. The concentration of the solution (or, equivalently, the amount of solute in the tank) at some given time is generally unknown. (See Fig. 12.1.)

If $m(t)$ is the mass of solute in the tank at time $t$, the rate of solute change will be $m'(t)$. If the solute is being added at the rate of $a(t)$ and being removed at the rate of $r(t)$, the rate of change is

$$m'(t) = \text{rate of addition} - \text{rate of removal}$$

$$= a(t) - r(t) \qquad 12.46$$

The rate of solute addition $a(t)$ must be known and, in fact, may be constant or zero. However, $r(t)$ depends on the concentration, $c(t)$, of the mixture and volumetric flow rates at time $t$. If $o(t)$ is the volumetric flow rate out of the tank, then

$$r(t) = c(t)o(t) \qquad 12.47$$

However, the concentration depends on the mass of solute in the tank at time $t$. Recognizing that the

**Figure 12.1** *Fluid Mixture Problem*

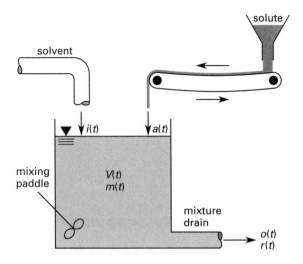

volume, $V(t)$, of the liquid in the tank may be changing with time,

$$c(t) = \frac{m(t)}{V(t)} \qquad 12.48$$

The differential equation describing this problem is

$$m'(t) = a(t) - \frac{m(t)o(t)}{V(t)} \qquad 12.49$$

## Example 12.11

A tank contains 100 gal of pure water at the beginning of an experiment. Pure water flows into the tank at a rate of 1 gal/min. Brine containing $1/4$ lbm of salt per gallon enters the tank from a second source at a rate of 1 gal/min. A perfectly mixed solution drains from the tank at a rate of 2 gal/min. How much salt is in the tank 8 min after the experiment begins?

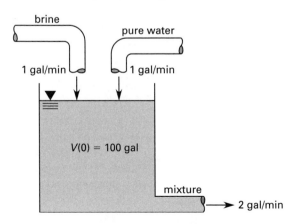

*Solution*

Let $m(t)$ represent the mass of salt in the tank at time $t$. 0.25 lbm of salt enters the tank per minute (that is,

$a(t) = 0.25$ lbm/min). The salt removal rate depends on the concentration in the tank. That is,

$$r(t) = o(t)c(t)$$
$$= \left(2 \ \frac{\text{gal}}{\text{min}}\right)\left(\frac{m(t)}{100 \ \text{gal}}\right) = \left(0.02 \ \frac{1}{\text{min}}\right)m(t)$$

From Eq. 12.46, the rate of change of salt in the tank is

$$m'(t) = a(t) - r(t)$$
$$= 0.25 \ \frac{\text{lbm}}{\text{min}} - \left(0.02 \ \frac{1}{\text{min}}\right)m(t)$$

$$m'(t) + \left(0.02 \ \frac{1}{\text{min}}\right)m(t) = 0.25 \ \text{lbm/min}$$

This is a first-order linear differential equation of the form of Eq. 12.7. Since the initial condition is $m(0) = 0$, the solution is

$$m(t) = \left(\frac{0.25 \ \dfrac{\text{lbm}}{\text{min}}}{0.02 \ \dfrac{1}{\text{min}}}\right)\left(1 - e^{-\left(0.02 \frac{1}{\text{min}}\right)t}\right)$$
$$= (12.5 \ \text{lbm})\left(1 - e^{-\left(0.01 \frac{1}{\text{min}}\right)t}\right)$$

At $t = 8$,

$$m(t) = (12.5 \ \text{lbm})\left(1 - e^{-\left(0.02 \frac{1}{\text{min}}\right)(8 \ \text{min})}\right)$$
$$= (12.5 \ \text{lbm})(1 - 0.852)$$
$$= 1.85 \ \text{lbm}$$

## 17. APPLICATION: EXPONENTIAL GROWTH AND DECAY

Equation 12.50 describes the behavior of a substance whose quantity, $m(t)$, changes at a rate proportional to the quantity present. The constant of proportionality, $k$, will be negative for decay (e.g., radioactive decay) and positive for growth (e.g., compound interest).

$$m'(t) = km(t) \qquad 12.50$$
$$m'(t) - km(t) = 0 \qquad 12.51$$

If the initial quantity of substance is $m(0) = m_0$, then Eq. 12.51 has the solution

$$m(t) = m_0 e^{kt} \qquad 12.52$$

If $m(t)$ is known for some time $t$, the constant of proportionality is

$$k = \left(\frac{1}{t}\right)\ln\left(\frac{m(t)}{m_0}\right) \qquad 12.53$$

For the case of a decay, the *half-life*, $t_{1/2}$, is the time at which only half of the substance remains. The relationship between $k$ and $t_{1/2}$ is

$$kt_{1/2} = \ln\left(\tfrac{1}{2}\right) = -0.693 \qquad 12.54$$

## 18. APPLICATION: EPIDEMICS

During an epidemic in a population of $n$ people, the density of sick (contaminated, contagious, affected, etc.) individuals is $\rho_s(t) = s(t)/n$, where $s(t)$ is the number of sick individuals at time $t$. Similarly, the density of well (uncontaminated, unaffected, susceptible, etc.) individuals is $\rho_w(t) = w(t)/n$, where $w(t)$ is the number of well individuals. Assuming there is no quarantine, the population size is constant, individuals move about freely, and sickness does not limit the activities of individuals, the rate of contagion, $\rho_s'(t)$, will be $k\rho_s(t)\rho_w(t)$, where $k$ is a proportionality constant.

$$\rho_s'(t) = k\rho_s(t)\rho_w(t) = k\rho_s(t)(1 - \rho_s(t)) \qquad 12.55$$

This is a separable differential equation that has the solution

$$\rho_s(t) = \frac{\rho_s(0)}{\rho_s(0) + (1 - \rho_s(0))e^{-kt}} \qquad 12.56$$

## 19. APPLICATION: SURFACE TEMPERATURE

*Newton's law of cooling* states that the surface temperature, $T$, of a cooling object changes at a rate proportional to the difference between the surface and ambient temperatures. The constant $k$ is a positive number.

$$T'(t) = -k(T(t) - T_{\text{ambient}}) \quad [k>0] \qquad 12.57$$

$$T'(t) + kT(t) - kT_{\text{ambient}} = 0 \quad [k>0] \qquad 12.58$$

This first-order linear differential equation with constant coefficients has the following solution (from Eq. 12.8).

$$T(t) = T_{\text{ambient}} + (T(0) - T_{\text{ambient}})e^{-kt} \qquad 12.59$$

If the temperature is known at some time $t$, the constant $k$ can be found from Eq. 12.60.

$$k = \left(\frac{-1}{t}\right)\ln\left(\frac{T(t) - T_{\text{ambient}}}{T(0) - T_{\text{ambient}}}\right) \qquad 12.60$$

## 20. APPLICATION: EVAPORATION

The mass of liquid evaporated from a liquid surface is proportional to the exposed surface area. Since quantity, mass, and remaining volume are all proportional, the differential equation is

$$\frac{dV}{dt} = -kA \qquad 12.61$$

For a spherical drop of radius $r$, Eq. 12.61 reduces to

$$\frac{dr}{dt} = -k \qquad 12.62$$

$$r(t) = r(0) - kt \qquad 12.63$$

For a cube with sides of length $s$, Eq. 12.61 reduces to

$$\frac{ds}{dt} = -2k \qquad 12.64$$

$$s(t) = s(0) - 2kt \qquad 12.65$$

# 13 Probability and Statistical Analysis of Data

$A$ or $B$ or both. The *intersection of two sets*, denoted by $A \cap B$ and shown in Fig. 13.1(c), is the set of all elements that belong to both $A$ and $B$. If $A \cap B = \emptyset$, $A$ and $B$ are said to be *disjoint sets*.

**Figure 13.1** *Venn Diagrams*

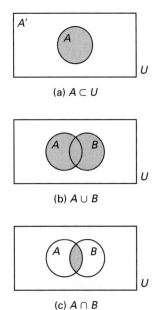

(a) $A \subset U$

(b) $A \cup B$

(c) $A \cap B$

## 1. SET THEORY

A *set* (usually designated by a capital letter) is a population or collection of individual items known as *elements* or *members*. The *null set*, $\emptyset$, is empty (i.e., contains no members). If $A$ and $B$ are two sets, $A$ is a *subset* of $B$ if every member in $A$ is also in $B$. $A$ is a *proper subset* of $B$ if $B$ consists of more than the elements in $A$. These relationships are denoted as follows.

$$A \subseteq B \quad \text{[subset]}$$

$$A \subset B \quad \text{[proper subset]}$$

The *universal set*, $U$, is one from which other sets draw their members. If $A$ is a subset of $U$, then $A'$ (also designated as $A^{-1}$, $\tilde{A}$, $-A$, and $\overline{A}$) is the *complement* of $A$ and consists of all elements in $U$ that are not in $A$. This is illustrated in a *Venn diagram* in Fig. 13.1(a).

The *union of two sets*, denoted by $A \cup B$ and shown in Fig. 13.1(b), is the set of all elements that are either in

If $A$, $B$, and $C$ are subsets of the universal set, the following laws apply.

- *identity laws*

$$A \cup \emptyset = A \qquad \text{13.1}$$

$$A \cup U = U \qquad \text{13.2}$$

$$A \cap \emptyset = \emptyset \qquad \text{13.3}$$

$$A \cap U = A \qquad \text{13.4}$$

- *idempotent laws*

$$A \cup A = A \qquad \text{13.5}$$

$$A \cap A = A \qquad \text{13.6}$$

- *complement laws*

$$A \cup A' = U \qquad \text{13.7}$$
$$(A')' = A \qquad \text{13.8}$$
$$A \cap A' = \emptyset \qquad \text{13.9}$$
$$U' = \emptyset \qquad \text{13.10}$$

- *commutative laws*

$$A \cup B = B \cup A \qquad \text{13.11}$$
$$A \cap B = B \cap A \qquad \text{13.12}$$

- *associative laws*

$$(A \cup B) \cup C = A \cup (B \cup C) \qquad \text{13.13}$$
$$(A \cap B) \cap C = A \cap (B \cap C) \qquad \text{13.14}$$

- *distributive laws*

$$A \cup (B \cap C) = (A \cup B) \cap (A \cup C) \qquad \text{13.15}$$
$$A \cap (B \cup C) = (A \cap B) \cup (A \cap C) \qquad \text{13.16}$$

- *de Morgan's laws*

$$(A \cup B)' - A' \cap B' \qquad \text{13.17}$$
$$(A \cap B)' = A' \cup B' \qquad \text{13.18}$$

## 2. COMBINATIONS OF ELEMENTS

There are a finite number of ways in which $n$ elements can be combined into distinctly different groups of $r$ items. For example, suppose a farmer has a hen, a rooster, a duck, and a cage that holds only two birds. The possible *combinations* of three birds taken two at a time are (hen, rooster), (hen, duck), and (rooster, duck). The birds in the cage will not remain stationary, and the combination (rooster, hen) is not distinctly different from (hen, rooster). That is, the groups are not *order conscious*.

The number of combinations of $n$ items taken $r$ at a time is written $C(n, r)$, $C_r^n$, $_nC_r$, or $\binom{n}{r}$ (pronounced "$n$ choose $r$") and given by Eq. 13.19. It is sometimes referred to as the *binomial coefficient*.

$$\binom{n}{r} = C(n, r) = \frac{n!}{(n-r)!\,r!} \qquad [\text{for } r \leq n] \qquad \text{13.19}$$

### Example 13.1

Six people are on a sinking yacht. There are four life jackets. How many combinations of survivors are there?

*Solution*

The groups are not order-conscious. From Eq. 13.19,

$$C(6, 4) = \frac{6!}{(6-4)!\,4!} = \frac{6 \cdot 5 \cdot 4 \cdot 3 \cdot 2 \cdot 1}{(2 \cdot 1)(4 \cdot 3 \cdot 2 \cdot 1)}$$
$$= 15$$

## 3. PERMUTATIONS

An order-conscious subset of $r$ items taken from a set of $n$ items is the *permutation* $P(n, r)$, also written $P_r^n$ and $_nP_r$. The permutation is order conscious because the arrangement of two items (say $a_i$ and $b_i$) as $a_ib_i$ is different from the arrangement $b_ia_i$. The number of permutations is

$$P(n, r) = \frac{n!}{(n-r)!} \qquad [\text{for } r \leq n] \qquad \text{13.20}$$

If groups of the entire set of $n$ items are being enumerated, the number of permutations of $n$ items taken $n$ at a time is

$$P(n, n) = \frac{n!}{(n-n)!} = \frac{n!}{0!} = n! \qquad \text{13.21}$$

A *ring permutation* is a special case of $n$ items taken $n$ at a time. There is no identifiable beginning or end, and the number of permutations is divided by $n$.

$$P_{\text{ring}}(n, n) = \frac{P(n, n)}{n} = (n-1)! \qquad \text{13.22}$$

### Example 13.2

A pianist knows four pieces but will have enough stage time to play only three of them. Pieces played in a different order constitute a different program. How many different programs can be arranged?

*Solution*

The groups are order conscious. From Eq. 13.20,

$$P(4, 3) = \frac{4!}{(4-3)!} = \frac{4 \cdot 3 \cdot 2 \cdot 1}{1} = 24$$

### Example 13.3

Seven diplomats from different countries enter a circular room. The only furnishings are seven chairs arranged around a circular table. How many ways are there of arranging the diplomats?

*Solution*

All seven diplomats must be seated, so the groups are permutations of seven objects taken seven at a time. Since there is no head chair, the groups are ring permutations. From Eq. 13.22,

$$P_{\text{ring}}(7, 7) = (7-1)! = 6 \cdot 5 \cdot 4 \cdot 3 \cdot 2 \cdot 1 = 720$$

## 4. PROBABILITY THEORY

The act of conducting an experiment (trial) or taking a measurement is known as *sampling*. *Probability theory* determines the relative likelihood that a particular event will occur. An *event*, $e$, is one of the possible outcomes of the *trial*. Taken together, all of the possible events constitute a finite *sample space*, $E = [e_1, e_2, ..., e_n]$. The trial is drawn from the *population* or *universe*. Populations can be finite or infinite in size.

Events can be numerical or nonnumerical, discrete or continuous, and dependent or independent. An example of a nonnumerical event is getting tails on a coin toss. The number from a roll of a die is a discrete numerical event. The measured diameter of a bolt produced from an automatic screw machine is a numerical event. Since the diameter can (within reasonable limits) take on any value, its measured value is a continuous numerical event.

An event is *independent* if its outcome is unaffected by previous outcomes (i.e., previous runs of the experiment) and *dependent* otherwise. Whether or not an event is independent depends on the population size and how the sampling is conducted. Sampling (a trial) from an infinite population is implicitly independent. When the population is finite, *sampling with replacement* produces independent events, while *sampling without replacement* changes the population and produces dependent events.

The terms *success* and *failure* are loosely used in probability theory to designate obtaining and not obtaining, respectively, the tested-for condition. "Failure" is not the same as a *null event* (i.e., one that has a zero probability of occurrence).

The *probability* of event $e_1$ occurring is designated as $p\{e_1\}$ and is calculated as the ratio of the total number of ways the event can occur to the total number of outcomes in the sample space.

**Example 13.4**

There are 380 students in a rural school—200 girls and 180 boys. One student is chosen at random and is checked for gender and height. (a) Define and categorize the population. (b) Define and categorize the sample space. (c) Define the trials. (d) Define and categorize the events. (e) In determining the probability that the student chosen is a boy, define success and failure. (f) What is the probability that the student is a boy?

*Solution*

(a) The population consists of 380 students and is finite.

(b) In determining the gender of the student, the sample space consists of the two outcomes $E = $ [girl, boy]. This sample space is nonnumerical and discrete. In determining the height, the sample space consists of a range of values and is numerical and continuous.

(c) The trial is the actual sampling (i.e., the determination of gender and height).

(d) The events are the outcomes of the trials (i.e., the gender and height of the student). These events are independent if each student returns to the population

prior to the random selection of the next student; otherwise, the events are dependent.

(e) The event is a success if the student is a boy and is a failure otherwise.

(f) From the definition of probability,

$$p\{\text{boy}\} = \frac{\text{no. of boys}}{\text{no. of students}} = \frac{180}{380} = \frac{9}{19} = 0.47$$

## 5. JOINT PROBABILITY

*Joint probability* rules specify the probability of a combination of events. If $n$ mutually exclusive events from the set $E$ have probabilities $p\{e_i\}$, the probability of any one of these events occurring in a given trial is the sum of the individual probabilities. Notice that the events in Eq. 13.23 come from a single sample space and are linked by the word *or*.

$$p\{e_1 \text{ or } e_2 \text{ or } \cdots \text{ or } e_k\} = p\{e_1\} + p\{e_2\} + \cdots + p\{e_k\}$$
$$\textit{13.23}$$

Given two independent sets of events, $E$ and $G$, Eq. 13.24 gives the probability that either event $e_i$ or $g_i$, or both, will occur. Notice that the events in Eq. 13.24 come from two different sample spaces and are linked by the word *or*.

$$p\{e_i \text{ or } g_i\} = p\{e_i\} + p\{g_i\} - p\{e_i\}p\{g_i\} \qquad \textit{13.24}$$

Given two independent sets of events, $E$ and $G$, Eq. 13.25 gives the probability that events $e_i$ and $g_i$ will both occur. Notice that the events in Eq. 13.25 come from two different sample spaces and are linked by the word *and*.

$$p\{e_i \text{ and } g_i\} = p\{e_i\}p\{g_i\} \qquad \textit{13.25}$$

**Example 13.5**

A bowl contains five white balls, two red balls, and three green balls. What is the probability of getting either a white ball or a red ball in one draw from the bowl?

*Solution*

The two possible events are mutually exclusive and come from the same sample space, so Eq. 13.23 can be used.

$$p\{\text{white or red}\} = p\{\text{white}\} + p\{\text{red}\} = \frac{5}{10} + \frac{2}{10} = \frac{7}{10}$$

**Example 13.6**

One bowl contains five white balls, two red balls, and three green balls. Another bowl contains three yellow balls and seven black balls. What is the probability of getting a red ball from the first bowl and a yellow ball from the second bowl in one draw from each bowl?

*Solution*

The two trials are independent, so Eq. 13.25 can be used.

$$p\{\text{red and yellow}\} = p\{\text{red}\}p\{\text{yellow}\}$$
$$= \left(\frac{2}{10}\right)\left(\frac{3}{10}\right) = \frac{6}{100}$$

## 6. COMPLEMENTARY PROBABILITIES

The probability of an event occurring is equal to one minus the probability of the event not occurring. This is known as *complementary probability.*

$$p\{e_i\} = 1 - p\{\text{not } e_i\} \qquad \textbf{13.26}$$

Equation 13.26 can be used to simplify some probability calculations. Specifically, calculation of the probability of numerical events being "greater than" or "less than" or quantities being "at least" a certain number can often be simplified by calculating the probability of complementary event.

### Example 13.7

A fair coin is tossed five times.[1] What is the probability of getting at least one tail?

*Solution*

The probability of getting at least one tail in five tosses could be calculated as

$$p\{\text{at least 1 tail}\} = p\{1 \text{ tail}\} + p\{2 \text{ tails}\}$$
$$+ p\{3 \text{ tails}\} + p\{4 \text{ tails}\}$$
$$+ p\{5 \text{ tails}\}$$

However, it is easier to calculate the complementary probability of getting no tails (i.e., getting all heads).

$$p\{\text{at least 1 tail}\} = 1 - p\{0 \text{ tails}\}$$
$$= 1 - (0.5)^5 = 0.96875$$

## 7. CONDITIONAL PROBABILITY

Given two dependent sets of events, $E$ and $G$, the probability that event $e_k$ will occur given the fact that the dependent event $g$ has already occurred is written as $p\{e_k|g\}$ and given by *Bayes' theorem*, Eq. 13.27.

$$p\{e_k|g\} = \frac{p\{e_k \text{ and } g\}}{p\{g\}} = \frac{p\{g|e_k\}p\{e_k\}}{\sum_{i=1}^{n} p\{g|e_i\}p\{e_i\}} \qquad \textbf{13.27}$$

---

[1]It makes no difference whether one coin is tossed five times or five coins are each tossed once.

## 8. PROBABILITY DENSITY FUNCTIONS

A *density function* is a nonnegative function whose integral taken over the entire range of the independent variable is unity. A *probability density function* (PDF) is a mathematical formula that gives the probability of a discrete numerical event. A *numerical event* is an occurrence that can be described (usually) by an integer. For example, 27 cars passing through a bridge toll booth in an hour is a discrete numerical event. Figure 13.2 shows a graph of a typical probability density function.

*Figure 13.2 Probability Density Function*

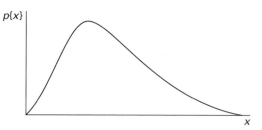

A probability density function, $f(x)$, gives the probability that discrete event $x$ will occur. That is, $p\{x\} = f(x)$. Important discrete probability density functions are the binomial, hypergeometric, and Poisson distributions.

## 9. BINOMIAL DISTRIBUTION

The *binomial probability density function* is used when all outcomes can be categorized as either successes or failures. The probability of success in a single trial is $\hat{p}$, and the probability of failure is the complement, $\hat{q} = 1 - \hat{p}$. The population is assumed to be infinite in size so that sampling does not change the values of $\hat{p}$ and $\hat{q}$. (The binomial distribution can also be used with finite populations when sampling with replacement.)

Equation 13.28 gives the probability of $x$ successes in $n$ independent *successive trials*. The quantity $\binom{n}{x}$ is the binomial coefficient, identical to the number of combinations of $n$ items taken $x$ at a time.

$$p\{x\} = f(x) = \binom{n}{x}\hat{p}^x\hat{q}^{n-x} \qquad \textbf{13.28}$$

$$\binom{n}{x} = \frac{n!}{(n-x)!x!} \qquad \textbf{13.29}$$

Equation 13.28 is a true (discrete) distribution, taking on values for each integer value up to $n$. The mean, $\mu$, and variance, $\sigma^2$ (see Sec. 13.22 and Sec. 13.23), of this distribution are

$$\mu = n\hat{p} \qquad \textbf{13.30}$$

$$\sigma^2 = n\hat{p}\hat{q} \qquad \textbf{13.31}$$

## Example 13.8

Five percent of a large batch of high-strength steel bolts purchased for bridge construction are defective. (a) If seven bolts are randomly sampled, what is the probability that exactly three will be defective? (b) What is the probability that two or more bolts will be defective?

*Solution*

(a) The bolts are either defective or not, so the binomial distribution can be applied.

$$\hat{p} = 0.05 \quad [\text{success} = \text{defective}]$$

$$\hat{q} = 1 - 0.05 = 0.95 \quad [\text{failure} = \text{not defective}]$$

From Eq. 13.28,

$$p\{3\} = f(3) = \binom{7}{3}\hat{p}^3\hat{q}^{7-3}$$

$$= \left(\frac{7 \cdot 6 \cdot 5 \cdot 4 \cdot 3 \cdot 2 \cdot 1}{4 \cdot 3 \cdot 2 \cdot 1 \cdot 3 \cdot 2 \cdot 1}\right)(0.05)^3(0.95)^4$$

$$= 0.00356$$

(b) The probability that two or more bolts will be defective could be calculated as

$$p\{x \geq 2\} = p\{2\} + p\{3\} + p\{4\} + p\{5\} + p\{6\} + p\{7\}$$

This method would require six probability calculations. It is easier to use the complement of the desired probability.

$$p\{x \geq 2\} = 1 - p\{x \leq 1\} = 1 - (p\{0\} + p\{1\})$$

$$p\{0\} = \binom{7}{0}(0.05)^0(0.95)^7 = (0.95)^7$$

$$p\{1\} = \binom{7}{1}(0.05)^1(0.95)^6 = (7)(0.05)(0.95)^6$$

$$p\{x \geq 2\} = 1 - \left((0.95)^7 + (7)(0.05)(0.95)^6\right)$$

$$= 1 - (0.6983 + 0.2573)$$

$$= 0.0444$$

## 10. HYPERGEOMETRIC DISTRIBUTION

Probabilities associated with sampling from a finite population without replacement are calculated from the *hypergeometric distribution*. If a population of finite size $M$ contains $K$ items with a given characteristic (e.g., red color, defective construction), then the probability

of finding $x$ items with that characteristic in a sample of $n$ items is

$$p\{x\} = f(x) = \frac{\binom{K}{x}\binom{M-K}{n-x}}{\binom{M}{n}} \quad [\text{for } x \leq n] \quad \textit{13.32}$$

## 11. MULTIPLE HYPERGEOMETRIC DISTRIBUTION

Sampling without replacement from finite populations containing several different types of items is handled by the *multiple hypergeometric distribution*. If a population of finite size $M$ contains $K_i$ items of type $i$ (such that $\Sigma K_i = M$), the probability of finding $x_1$ items of type 1, $x_2$ items of type 2, and so on, in a sample size $n$ (such that $\Sigma x_i = n$) is

$$p\{x_1, x_2, x_3, ...\} = \frac{\binom{K_1}{x_1}\binom{K_2}{x_2}\binom{K_3}{x_3}\cdots}{\binom{M}{n}} \quad \textit{13.33}$$

## 12. POISSON DISTRIBUTION

Certain events occur relatively infrequently but at a relatively regular rate. The probability of such an event occurring is given by the *Poisson distribution*. Suppose an event occurs, on the average, $\lambda$ times per period. The probability that the event will occur $x$ times per period is

$$p\{x\} = f(x) = \frac{e^{-\lambda}\lambda^x}{x!} \quad [\lambda > 0] \quad \textit{13.34}$$

$\lambda$ is both the mean and the variance of the Poisson distribution.

## Example 13.9

The number of customers arriving at a hamburger stand in the next period is a Poisson distribution having a mean of eight. What is the probability that exactly six customers will arrive in the next period?

*Solution*

$\lambda = 8$ and $x = 6$. From Eq. 13.34,

$$p\{6\} = \frac{e^{-\lambda}\lambda^x}{x!} = \frac{e^{-8}(8)^6}{6!} = 0.122$$

## 13. CONTINUOUS DISTRIBUTION FUNCTIONS

Most numerical events are *continuously distributed* and are not constrained to discrete or integer values. For example, the resistance of a 10% 1 $\Omega$ resistor may be any value between 0.9 and 1.1 $\Omega$. The probability of an exact numerical event is zero for continuously distributed variables. That is, there is no chance that a numerical event will be *exactly* $x$.[2] It is possible to determine only the probability that a numerical event will be less than $x$, greater than $x$, or between the values of $x_1$ and $x_2$, but not exactly equal to $x$.

While an expression, $f(x)$, for a probability density function can be written, it is used to derive the *continuous distribution function* (CDF), $F(x_0)$, which gives the probability of numerical event $x_0$ or less occurring, as illustrated in Fig. 13.3.

$$p\{X < x_0\} = F(x_0) = \int_0^{x_0} f(x)\,dx \qquad 13.35$$

$$f(x) = \frac{dF(x)}{dx} \qquad 13.36$$

**Figure 13.3** *Continuous Distribution Function*

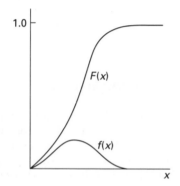

## 14. EXPONENTIAL DISTRIBUTION

The continuous *exponential distribution* is given by its probability density and continuous distribution functions.

$$f(x) = \lambda e^{-\lambda x} \qquad 13.37$$

$$p\{X < x\} = F(x) = 1 - e^{-\lambda x} \qquad 13.38$$

The mean and variance of the exponential distribution are

$$\mu = \frac{1}{\lambda} \qquad 13.39$$

$$\sigma^2 = \frac{1}{\lambda^2} \qquad 13.40$$

---

[2]It is important to understand the rationale behind this statement. Since the variable can take on any value and has an infinite number of significant digits, we can infinitely continue to increase the precision of the value. For example, the probability is zero that a resistance will be exactly 1 $\Omega$ because the resistance is really 1.03 or 1.0260008 or 1.02600080005, and so on.

## 15. NORMAL DISTRIBUTION

The *normal distribution (Gaussian distribution)* is a symmetrical distribution commonly referred to as the *bell-shaped curve*, which represents the distribution of outcomes of many experiments, processes, and phenomena. (See Fig. 13.4.) The probability density and continuous distribution functions for the normal distribution with mean $\mu$ and variance $\sigma^2$ are

$$f(x) = \frac{e^{-\frac{1}{2}\left(\frac{x-\mu}{\sigma}\right)^2}}{\sigma\sqrt{2\pi}} \qquad [-\infty < x < +\infty] \qquad 13.41$$

$$p\{\mu < X < x_0\} = F(x_0)$$

$$= \frac{1}{\sigma\sqrt{2\pi}} \int_0^{x_0} e^{-\frac{1}{2}\left(\frac{x-\mu}{\sigma}\right)^2} dx \qquad 13.42$$

**Figure 13.4** *Normal Distribution*

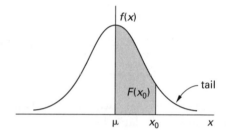

Since $f(x)$ is difficult to integrate, Eq. 13.41 is seldom used directly, and a *standard normal table* (see App. 13.A) is used instead. The standard normal table is based on a normal distribution with a mean of zero and a standard deviation of 1. Since the range of values from an experiment or phenomenon will not generally correspond to the standard normal table, a value, $x_0$, must be converted to a *standard normal value*, $z$. In Eq. 13.43, $\mu$ and $\sigma$ are the mean and standard deviation, respectively, of the distribution from which $x_0$ comes. For all practical purposes, all normal distributions are completely bounded by $\mu \pm 3\sigma$.

$$z = \frac{x_0 - \mu}{\sigma} \qquad 13.43$$

Numbers in the standard normal table given by App. 13.A are the probabilities of the normalized $x$ being between zero and $z$ and represent the areas under the curve up to point $z$. When $x$ is less than $\mu$, $z$ will be negative. However, the curve is symmetrical, so the table value corresponding to positive $z$ can be used. The probability of $x$ being greater than $z$ is the complement of the table value. The curve area past point $z$ is known as the *tail of the curve*.

The *error function*, $\text{erf}(x)$, and its complement, $\text{erfc}(x)$, are defined by Eq. 13.44 and Eq. 13.45. The error function is used to determine the probable error of a measurement.

$$\operatorname{erf}(x_0) = \frac{2}{\sqrt{\pi}} \int_0^{x_0} e^{-x^2} \, dx \qquad \textit{13.44}$$

$$\operatorname{erfc}(x_0) = 1 - \operatorname{erf}(x_0) \qquad \textit{13.45}$$

### Example 13.10

The mass, $m$, of a particular hand-laid fiberglass (Fibreglas™) part is normally distributed with a mean of 66 kg and a standard deviation of 5 kg. (a) What percent of the parts will have a mass less than 72 kg? (b) What percent of the parts will have a mass in excess of 72 kg? (c) What percent of the parts will have a mass between 61 kg and 72 kg?

*Solution*

(a) The value of 72 kg must be normalized by using Eq. 13.43. The standard normal variable is

$$z = \frac{x - \mu}{\sigma} = \frac{72\,\text{kg} - 66\,\text{kg}}{5\,\text{kg}} = 1.2$$

Reading from App. 13.A, the area under the normal curve is 0.3849. This represents the probability of the mass, $m$, being between 66 kg and 72 kg (i.e., $z$ being between 0 and 1.2). However, the probability of the mass being less than 66 kg is also needed. Since the curve is symmetrical, this probability is 0.5. Therefore,

$$p\{m < 72\,\text{kg}\} = p\{z < 1.2\} = 0.5 + 0.3849 = 0.8849$$

(b) The probability of the mass exceeding 72 kg is the area under the tail past point $z$.

$$p\{m > 72\,\text{kg}\} = p\{z > 1.2\} = 0.5 - 0.3849 = 0.1151$$

(c) The standard normal variable corresponding to $m = 61$ kg is

$$z = \frac{x - \mu}{\sigma} = \frac{61\,\text{kg} - 66\,\text{kg}}{5\,\text{kg}} = -1$$

Since the two masses are on opposite sides of the mean, the probability will have to be determined in two parts.

$$p\{61 < m < 72\} = p\{61 < m < 66\} + p\{66 < m < 72\}$$
$$= p\{-1 < z < 0\} + p\{0 < z < 1.2\}$$
$$= 0.3413 + 0.3849 = 0.7262$$

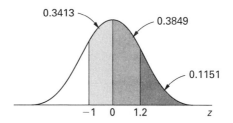

## 16. APPLICATION: RELIABILITY

### Introduction

*Reliability*, $R\{t\}$, is the probability that an item will continue to operate satisfactorily up to time $t$. The *bathtub distribution*, Fig. 13.5, is often used to model the probability of failure of an item (or, the number of failures from a large population of items) as a function of time. Items initially fail at a high rate, a phenomenon known as *infant mortality*. For the majority of the operating time, known as the *steady-state operation*, the failure rate is constant (i.e., is due to random causes). After a long period of time, the items begin to deteriorate and the failure rate increases. (No mathematical distribution describes all three of these phases simultaneously.)

***Figure 13.5*** *Bathtub Reliability Curve*

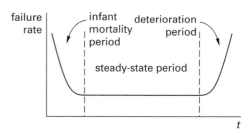

The *hazard function*, $z\{t\}$, represents the *conditional probability of failure*—the probability of failure in the next time interval, given that no failure has occurred thus far.[3]

$$z\{t\} = \frac{f(t)}{R(t)} = \frac{\dfrac{dF(t)}{dt}}{1 - F(t)} \qquad \textit{13.46}$$

### Exponential Reliability

Steady-state reliability is often described by the *negative exponential distribution*. This assumption is appropriate whenever an item fails only by random causes and does not experience deterioration during its life. The

---

[3]The symbol $z\{t\}$ is traditionally used for the hazard function and is not related to the standard normal variable.

parameter $\lambda$ is related to the *mean time to failure* (MTTF) of the item.[4]

$$R\{t\} = e^{-\lambda t} = e^{-t/\text{MTTF}} \qquad 13.47$$

$$\lambda = \frac{1}{\text{MTTF}} \qquad 13.48$$

Equation 13.47 and the exponential continuous distribution function, Eq. 13.38, are complementary.

$$R\{t\} = 1 - F(t) = 1 - \left(1 - e^{-\lambda t}\right) = e^{-\lambda t} \qquad 13.49$$

The hazard function for the negative exponential distribution is

$$z\{t\} = \lambda \qquad 13.50$$

Thus, the hazard function for exponential reliability is constant and does not depend on $t$ (i.e., on the age of the item). In other words, the expected future life of an item is independent of the previous history (length of operation). This lack of memory is consistent with the assumption that only random causes contribute to failure during steady-state operations. And since random causes are unlikely discrete events, their probability of occurrence can be represented by a Poisson distribution with mean $\lambda$. That is, the probability of having $x$ failures in any given period is

$$p\{x\} = \frac{e^{-\lambda}\lambda^x}{x!} \qquad 13.51$$

### Serial System Reliability

In the analysis of system reliability, the binary variable $X_i$ is defined as one if item $i$ operates satisfactorily and zero if otherwise. Similarly, the binary variable $\Phi$ is one only if the entire system operates satisfactorily. Thus, $\Phi$ will depend on a *performance function* containing the $X_i$.

A *serial system* is one for which all items must operate correctly for the system to operate. Each item has its own reliability, $R_i$. For a serial system of $n$ items, the performance function is

$$\Phi = X_1 X_2 X_3 \cdots X_n = \min(X_i) \qquad 13.52$$

The probability of a serial system operating correctly is

$$p\{\Phi = 1\} = R_{\text{serial system}} = R_1 R_2 R_3 \cdots R_n \qquad 13.53$$

### Parallel System Reliability

A *parallel system* with $n$ items will fail only if all $n$ items fail. Such a system is said to be *redundant* to the $n$th degree. Using redundancy, a highly reliable system can be produced from components with relatively low individual reliabilities.

---

[4]The term "mean time *between* failures" is improper. However, the term *mean time before failure* (MTBF) is acceptable.

The performance function of a redundant system is

$$\Phi = 1 - (1 - X_1)(1 - X_2)(1 - X_3)\cdots(1 - X_n)$$
$$= \max(X_i) \qquad 13.54$$

The reliability of the parallel system is

$$R = p\{\Phi = 1\}$$
$$= 1 - (1 - R_1)(1 - R_2)(1 - R_3)\cdots(1 - R_n) \qquad 13.55$$

### Example 13.11

The reliability of an item is exponentially distributed with mean time to failure (MTTF) of 1000 hr. What is the probability that the item will not have failed after 1200 hr of operation?

*Solution*

The probability of not having failed before time $t$ is the reliability. From Eq. 13.48 and Eq. 13.49,

$$\lambda - \frac{1}{\text{MTTF}} = \frac{1}{1000 \text{ hr}} = 0.001 \text{ hr}^{-1}$$

$$R\{1200\} = e^{-\lambda t} = e^{(-0.001 \text{ hr}^{-1})(1200 \text{ hr})} = 0.3$$

### Example 13.12

What are the reliabilities of the following systems?

(a)

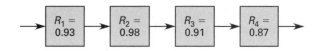

(b)

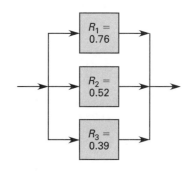

*Solution*

(a) This is a serial system. From Eq. 13.53,

$$R = R_1 R_2 R_3 R_4 = (0.93)(0.98)(0.91)(0.87)$$
$$= 0.72$$

(b) This is a parallel system. From Eq. 13.55,

$$R = 1 - (1 - R_1)(1 - R_2)(1 - R_3)$$
$$= 1 - (1 - 0.76)(1 - 0.52)(1 - 0.39)$$
$$= 0.93$$

## 17. ANALYSIS OF EXPERIMENTAL DATA

Experiments can take on many forms. An experiment might consist of measuring the mass of one cubic foot of concrete or measuring the speed of a car on a roadway. Generally, such experiments are performed more than once to increase the precision and accuracy of the results.

Both systematic and random variations in the process being measured will cause the observations to vary, and the experiment would not be expected to yield the same result each time it was performed. Eventually, a collection of experimental outcomes (observations) will be available for analysis.

The *frequency distribution* is a systematic method for ordering the observations from small to large, according to some convenient numerical characteristic. The *step interval* should be chosen so that the data are presented in a meaningful manner. If there are too many intervals, many of them will have zero frequencies; if there are too few intervals, the frequency distribution will have little value. Generally, 10 to 15 intervals are used.

Once the frequency distribution is complete, it can be represented graphically as a *histogram*. The procedure in drawing a histogram is to mark off the interval limits (also known as *class limits*) on a number line and then draw contiguous bars with lengths that are proportional to the frequencies in the intervals and that are centered on the midpoints of their respective intervals. The continuous nature of the data can be depicted by a *frequency polygon*. The number or percentage of observations that occur up to and including some value can be shown in a *cumulative frequency table*.

### Example 13.13

The number of cars that travel through an intersection between 12 noon and 1 p.m. is measured for 30 consecutive working days. The results of the 30 observations are

79, 66, 72, 70, 68, 66, 68, 76, 73, 71, 74, 70, 71, 69, 67, 74, 70, 68, 69, 64, 75, 70, 68, 69, 64, 69, 62, 63, 63, 61

(a) What are the frequency and cumulative distributions? (Use a distribution interval of two cars per hour.) (b) Draw the histogram. (Use a cell size of two cars per hour.) (c) Draw the frequency polygon. (d) Graph the cumulative frequency distribution.

*Solution*

(a)

| cars per hour | frequency | cumulative frequency | cumulative percent |
|---|---|---|---|
| 60–61 | 1 | 1 | 3 |
| 62–63 | 3 | 4 | 13 |
| 64–65 | 2 | 6 | 20 |
| 66–67 | 3 | 9 | 30 |
| 68–69 | 8 | 17 | 57 |
| 70–71 | 6 | 23 | 77 |
| 72–73 | 2 | 25 | 83 |
| 74–75 | 3 | 28 | 93 |
| 76–77 | 1 | 29 | 97 |
| 78–79 | 1 | 30 | 100 |

(b)

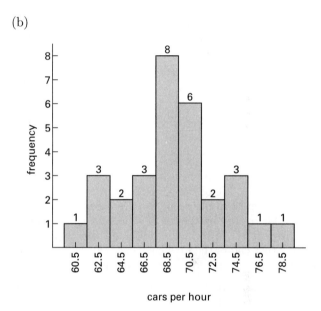

(c)

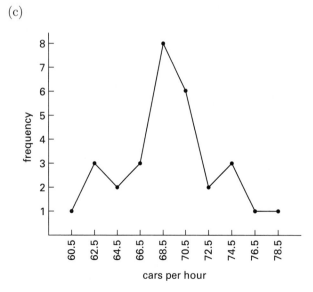

(d)

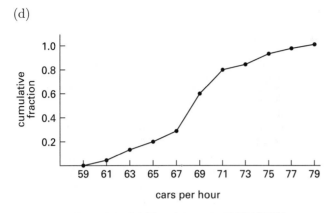

## 18. MEASURES OF CENTRAL TENDENCY

It is often unnecessary to present the experimental data in their entirety, either in tabular or graphical form. In such cases, the data and distribution can be represented by various parameters. One type of parameter is a measure of *central tendency*. Mode, median, and mean are measures of central tendency.

The *mode* is the observed value that occurs most frequently. The mode may vary greatly between series of observations. Therefore, its main use is as a quick measure of the central value since little or no computation is required to find it. Beyond this, the usefulness of the mode is limited.

The *median* is the point in the distribution that partitions the total set of observations into two parts containing equal numbers of observations. It is not influenced by the extremity of scores on either side of the distribution. The median is found by counting up (from either end of the frequency distribution) until half of the observations have been accounted for.

For even numbers of observations, the median is estimated as some value (i.e., the average) between the two center observations.

Similar in concept to the median are *percentiles* (*percentile ranks*), *quartiles*, and *deciles*. The median could also have been called the *50th percentile* observation. Similarly, the 80th percentile would be the observed value (e.g., the number of cars per hour) for which the cumulative frequency was 80%. The quartile and decile points on the distribution divide the observations or distribution into segments of 25% and 10%, respectively.

The *arithmetic mean* is the arithmetic average of the observations. The sample mean, $\bar{x}$, can be used as an unbiased estimator of the population mean, $\mu$. The *mean* may be found without ordering the data (as was necessary to find the mode and median). The mean can be found from the following formula.

$$\bar{x} = \left(\frac{1}{n}\right)(x_1 + x_2 + \cdots + x_n) = \frac{\sum x_i}{n} \qquad 13.56$$

The *geometric mean* is used occasionally when it is necessary to average ratios. The geometric mean is calculated as

$$\text{geometric mean} = \sqrt[n]{x_1 x_2 x_3 \cdots x_n} \quad [x_i > 0] \qquad 13.57$$

The *harmonic mean* is defined as

$$\text{harmonic mean} = \frac{n}{\dfrac{1}{x_1} + \dfrac{1}{x_2} + \cdots + \dfrac{1}{x_n}} \qquad 13.58$$

The *root-mean-squared* (rms) *value* of a series of observations is defined as

$$x_{\text{rms}} = \sqrt{\frac{\sum x_i^2}{n}} \qquad 13.59$$

### Example 13.14

Find the mode, median, and arithmetic mean of the distribution represented by the data given in Ex. 13.13.

*Solution*

First, resequence the observations in increasing order.

61, 62, 63, 63, 64, 64, 66, 66, 67, 68, 68, 68, 68, 69, 69, 69, 69, 70, 70, 70, 70, 71, 71, 72, 73, 74, 74, 75, 76, 79

The mode is the interval 68–69, since this interval has the highest frequency. If 68.5 is taken as the interval center, then 68.5 would be the mode.

The 15th and 16th observations are both 69, so the median is

$$\frac{69 + 69}{2} = 69$$

The mean can be found from the raw data or from the grouped data using the interval center as the assumed observation value. Using the raw data,

$$\bar{x} = \frac{\sum x}{n} = \frac{2069}{30} = 68.97$$

## 19. MEASURES OF DISPERSION

The simplest statistical parameter that describes the variability in observed data is the *range*. The range is found by subtracting the smallest value from the largest. Since the range is influenced by extreme (low probability) observations, its use as a measure of variability is limited.

The *standard deviation* is a better estimate of variability because it considers every observation. That is, $N$ in Eq. 13.60 is the total population size, not the sample size, $n$.

$$\sigma = \sqrt{\frac{\sum (x_i - \mu)^2}{N}} = \sqrt{\frac{\sum x_i^2}{N} - \mu^2} \qquad 13.60$$

The standard deviation of a sample (particularly a small sample) is a biased (i.e., is not a good) estimator of the population standard deviation. An *unbiased estimator* of the population standard deviation is the *sample standard deviation, s.*[5]

$$s = \sqrt{\frac{\sum (x_i - \bar{x})^2}{n-1}} = \sqrt{\frac{\sum x_i^2 - \frac{(\sum x_i)^2}{n}}{n-1}} \qquad \text{13.61}$$

If the sample standard deviation, $s$, is known, the standard deviation of the sample, $\sigma_{\text{sample}}$, can be calculated.

$$\sigma_{\text{sample}} = s\sqrt{\frac{n-1}{n}} \qquad \text{13.62}$$

The *variance* is the square of the standard deviation. Since there are two standard deviations, there are two variances. The *variance of the sample* is $\sigma^2$, and the *sample variance* is $s^2$.

The *relative dispersion* is defined as a measure of dispersion divided by a measure of central tendency. The *coefficient of variation* is a relative dispersion calculated from the sample standard deviation and the mean.

$$\text{coefficient of variation} = \frac{s}{\bar{x}} \qquad \text{13.63}$$

*Skewness* is a measure of a frequency distribution's lack of symmetry.

$$\text{skewness} = \frac{\bar{x} - \text{mode}}{s}$$

$$\approx \frac{3(\bar{x} - \text{median})}{s} \qquad \text{13.64}$$

## Example 13.15

For the data given in Ex. 13.13, calculate (a) the sample range, (b) the standard deviation of the sample, (c) an unbiased estimator of the population standard deviation, (d) the variance of the sample, and (e) the sample variance.

*Solution*

$$\sum x_i = 2069$$

$$\left(\sum x_i\right)^2 = (2069)^2 = 4{,}280{,}761$$

---

[5]There is a subtle yet significant difference between *standard deviation of the sample*, $\sigma$ (obtained from Eq. 13.60 for a finite sample drawn from a larger population) and the *sample standard deviation*, $s$ (obtained from Eq. 13.61). While $\sigma$ can be calculated, it has no significance or use as an estimator. It is true that the difference between $\sigma$ and $s$ approaches zero when the sample size, $n$, is large, but this convergence does nothing to legitimize the use of $\sigma$ as an estimator of the true standard deviation. (Some people say "large" is 30, others say 50 or 100.)

$$\sum x_i^2 = 143{,}225$$

$$n = 30$$

$$\bar{x} = \frac{2069}{30} = 68.967$$

(a) $$R = x_{\max} - x_{\min} = 79 - 61 = 18$$

(b) $$\sigma = \sqrt{\frac{\sum x_i^2}{n} - (\bar{x})^2} = \sqrt{\frac{143{,}225}{30} - \left(\frac{2069}{30}\right)^2}$$
$$= \sqrt{17.766} = 4.215$$

(c) $$s = \sqrt{\frac{\sum x_i^2 - \frac{(\sum x_i)^2}{n}}{n-1}}$$
$$= \sqrt{\frac{143{,}225 - \frac{4{,}280{,}761}{30}}{29}}$$
$$= \sqrt{18.378} = 4.287$$

(d) $$\sigma^2 = 17.77$$

(e) $$s^2 = 18.38$$

## 20. CENTRAL LIMIT THEOREM

Measuring a sample of $n$ items from a population with mean $\mu$ and standard deviation $\sigma$ is the general concept of an experiment. The sample mean, $\bar{x}$, is one of the parameters that can be derived from the experiment. This experiment can be repeated $k$ times, yielding a set of averages $(\bar{x}_1, \bar{x}_2, ..., \bar{x}_k)$. The $k$ numbers in the set themselves represent samples from distributions of averages. The average of averages, $\bar{\bar{x}}$, and sample standard deviation of averages, $s_{\bar{x}}$ (known as the *standard error of the mean*), can be calculated.

The *central limit theorem* characterizes the distribution of the sample averages. The theorem can be stated in several ways, but the essential elements are the following points.

1. The averages, $\bar{x}_i$, are normally distributed variables, even if the original data from which they are calculated are not normally distributed.

2. The grand average, $\bar{\bar{x}}$ (i.e., the average of the averages), approaches and is an unbiased estimator of $\mu$.

$$\mu \approx \bar{\bar{x}} \qquad \text{13.65}$$

The standard deviation of the original distribution, $\sigma$, is much larger than the standard error of the mean.

$$\sigma \approx \sqrt{n}s_{\overline{x}} \qquad 13.66$$

## 21. CONFIDENCE LEVEL

The results of experiments are seldom correct 100% of the time. Recognizing this, researchers accept a certain probability of being wrong. In order to minimize this probability, the experiment is repeated several times. The number of repetitions depends on the desired level of confidence in the results.

If the results have a 5% probability of being wrong, the *confidence level*, $C$, is 95% that the results are correct, in which case the results are said to be *significant*. If the results have only a 1% probability of being wrong, the confidence level is 99%, and the results are said to be *highly significant*. Other confidence levels (90%, 99.5%, etc.) are used as appropriate.

## 22. APPLICATION: CONFIDENCE LIMITS

As a consequence of the central limit theorem, sample means of $n$ items taken from a normal distribution with mean $\mu$ and standard deviation $\sigma$ will be normally distributed with mean $\mu$ and variance $\sigma^2/n$. Thus, the probability that any given average, $\overline{x}$, exceeds some value, $L$, is

$$p\{\overline{x} > L\} = p\left\{ z > \left| \frac{L - \mu}{\frac{\sigma}{\sqrt{n}}} \right| \right\} \qquad 13.67$$

$L$ is the *confidence limit* for the confidence level $1 - p\{\overline{x} > L\}$ (expressed as a percent). Values of $z$ are read directly from the standard normal table. As an example, $z = 1.645$ for a 95% confidence level since only 5% of the curve is above that $z$ in the upper tail. Similar values are given in Table 13.1. This is known as a *one-tail confidence limit* because all of the probability is given to one side of the variation.

**Table 13.1** *Values of z for Various Confidence Levels*

| confidence level, $C$ | one-tail limit, $z$ | two-tail limit, $z$ |
|---|---|---|
| 90% | 1.28 | 1.645 |
| 95% | 1.645 | 1.96 |
| 97.5% | 1.96 | 2.17 |
| 99% | 2.33 | 2.575 |
| 99.5% | 2.575 | 2.81 |
| 99.75% | 2.81 | 3.00 |

With *two-tail confidence limits*, the probability is split between the two sides of variation. There will be upper and lower confidence limits, UCL and LCL, respectively.

$$p\{\text{LCL} < \overline{x} < \text{UCL}\} = p\left\{ \frac{\text{LCL} - \mu}{\frac{\sigma}{\sqrt{n}}} < z < \frac{\text{UCL} - \mu}{\frac{\sigma}{\sqrt{n}}} \right\}$$

$$13.68$$

## 23. APPLICATION: BASIC HYPOTHESIS TESTING

A *hypothesis test* is a procedure that answers the question, "Did these data come from [a particular type of] distribution?" There are many types of tests, depending on the distribution and parameter being evaluated. The simplest hypothesis test determines whether an average value obtained from $n$ repetitions of an experiment could have come from a population with known mean, $\mu$, and standard deviation, $\sigma$. A practical application of this question is whether a manufacturing process has changed from what it used to be or should be. Of course, the answer (i.e., "yes" or "no") cannot be given with absolute certainty—there will be a confidence level associated with the answer.

The following procedure is used to determine whether the average of $n$ measurements can be assumed (with a given confidence level) to have come from a known population.

*step 1:* Assume random sampling from a normal population.

*step 2:* Choose the desired confidence level, $C$.

*step 3:* Decide on a one-tail or two-tail test. If the hypothesis being tested is that the average has or has not *increased* or *decreased*, choose a one-tail test. If the hypothesis being tested is that the average has or has not *changed*, choose a two-tail test.

*step 4:* Use Table 13.1 or the standard normal table to determine the $z$-value corresponding to the confidence level and number of tails.

*step 5:* Calculate the actual standard normal variable, $z'$.

$$z' = \frac{\overline{x} - \mu}{\frac{\sigma}{\sqrt{n}}} \qquad 13.69$$

*step 6:* If $z' \geq z$, the average can be assumed (with confidence level $C$) to have come from a different distribution.

## Example 13.16

When it is operating properly, a cement plant has a daily production rate that is normally distributed with a mean of 880 tons/day and a standard deviation of 21 tons/day. During an analysis period, the output is measured on 50 consecutive days, and the mean output is found to be 871 tons/day. With a 95% confidence level, determine whether the plant is operating properly.

*Solution*

*step 1:* Given.

*step 2:* $C = 0.95$ is given.

*step 3:* Since a specific direction in the variation is not given (i.e., the example does not ask whether the average has decreased), use a two-tail hypothesis test.

*step 4:* From Table 13.1, $z = 1.96$.

*step 5:* From Eq. 13.69,

$$z' = \left| \frac{\overline{x} - \mu}{\frac{\sigma}{\sqrt{n}}} \right| = \left| \frac{871 - 880}{\frac{21}{\sqrt{50}}} \right| = 3.03$$

Since $3.03 > 1.96$, the distributions are not the same. There is at least a 95% probability that the plant is not operating correctly.

## 24. APPLICATION: STATISTICAL PROCESS CONTROL

All manufacturing processes contain variation due to random and nonrandom causes. Random variation cannot be eliminated. *Statistical process control* (SPC) is the act of monitoring and adjusting the performance of a process to detect and eliminate nonrandom variation.

Statistical process control is based on taking regular (hourly, daily, etc.) samples of $n$ items and calculating the mean, $\overline{x}$, and range, $R$, of the sample. To simplify the calculations, the range is used as a measure of the dispersion. These two parameters are graphed on their respective *x-bar* and *R-control charts* (see Fig. 13.6).[6] Confidence limits are drawn at $\pm 3\sigma/\sqrt{n}$. From a statistical standpoint, the control chart tests a hypothesis each time a point is plotted. When a point falls outside these limits, there is a 99.75% probability that the process is out of control. Until a point exceeds the control limits, no action is taken.[7]

---

[6]Other charts (e.g., the *sigma chart*, *p-chart*, and *c-chart*) are less common but are used as required.

[7]Other indications that a correction may be required are seven measurements on one side of the average and seven consecutively increasing measurements. Rules such as these detect shifts and trends.

**Figure 13.6** *Typical Statistical Process Control Charts*

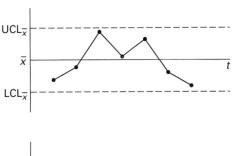

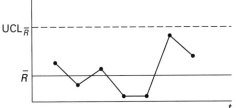

## 25. MEASURES OF EXPERIMENTAL ADEQUACY

An experiment is said to be *accurate* if it is unaffected by experimental error. In this case, *error* is not synonymous with *mistake*, but rather includes all variations not within the experimenter's control.

For example, suppose a gun is aimed at a point on a target and five shots are fired. The mean distance from the point of impact to the sight in point is a measure of the alignment accuracy between the barrel and sights. The difference between the actual value and the experimental value is known as *bias*.

*Precision* is not synonymous with accuracy. Precision is concerned with the repeatability of the experimental results. If an experiment is repeated with identical results, the experiment is said to be precise.

The average distance of each impact from the centroid of the impact group is a measure of the precision of the experiment. Thus, it is possible to have a highly precise experiment with a large bias.

Most of the techniques applied to experiments in order to improve the accuracy (i.e., reduce bias) of the experimental results (e.g., repeating the experiment, refining the experimental methods, or reducing variability) actually increase the precision.

Sometimes the word *reliability* is used with regard to the precision of an experiment. Thus, a "reliable estimate" is used in the same sense as a "precise estimate."

*Stability* and *insensitivity* are synonymous terms. A stable experiment will be insensitive to minor changes in the experimental parameters. For example, suppose the centroid of a bullet group is 2.1 in from the target point at 65°F and 2.3 in away at 80°F. The sensitivity of the experiment to temperature change would be

$$\text{sensitivity} = \frac{\Delta x}{\Delta T} = \frac{2.3\,\text{in} - 2.1\,\text{in}}{80°\text{F} - 65°\text{F}} = 0.0133\,\text{in}/°\text{F}$$

## 26. LINEAR REGRESSION

If it is necessary to draw a straight line $(y = mx + b)$ through $n$ data points $(x_1, y_1), (x_2, y_2), ..., (x_n, y_n)$, the following method based on the *method of least squares* can be used.

*step 1:* Calculate the following nine quantities.

$$\sum x_i \quad \sum x_i^2 \quad \left(\sum x_i\right)^2 \quad \overline{x} = \frac{\sum x_i}{n} \quad \sum x_i y_i$$

$$\sum y_i \quad \sum y_i^2 \quad \left(\sum y_i\right)^2 \quad \overline{y} = \frac{\sum y_i}{n}$$

*step 2:* Calculate the slope, $m$, of the line.

$$m = \frac{n\sum(x_i y_i) - \left(\sum x_i\right)\left(\sum y_i\right)}{n\sum x_i^2 - \left(\sum x_i\right)^2} \qquad 13.70$$

*step 3:* Calculate the $y$-intercept, $b$.

$$b = \overline{y} - m\overline{x} \qquad 13.71$$

*step 4:* To determine the goodness of fit, calculate the correlation coefficient, $r$.

$$r = \frac{n\sum(x_i y_i) - \left(\sum x_i\right)\left(\sum y_i\right)}{\sqrt{\left(n\sum x_i^2 - \left(\sum x_i\right)^2\right)\left(n\sum y_i^2 - \left(\sum y_i\right)^2\right)}}$$

$$13.72$$

If $m$ is positive, $r$ will be positive; if $m$ is negative, $r$ will be negative. As a general rule, if the absolute value of $r$ exceeds 0.85, the fit is good; otherwise, the fit is poor. $r$ equals 1.0 if the fit is a perfect straight line.

A low value of $r$ does not eliminate the possibility of a nonlinear relationship existing between $x$ and $y$. It is possible that the data describe a parabolic, logarithmic, or other nonlinear relationship. (Usually this will be apparent if the data are graphed.) It may be necessary to convert one or both variables to new variables by taking squares, square roots, cubes, or logarithms, to name a few of the possibilities, in order to obtain a linear relationship. The apparent shape of the line through the data will give a clue to the type of variable transformation that is required. The curves in Fig. 13.7 may be used as guides to some of the simpler variable transformations.

Figure 13.8 illustrates several common problems encountered in trying to fit and evaluate curves from

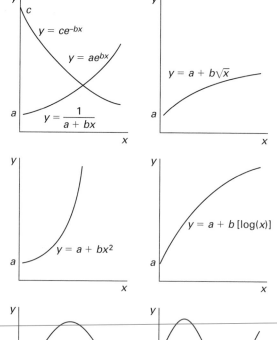

**Figure 13.7** *Nonlinear Data Curves*

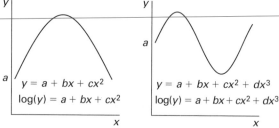

**Figure 13.8** *Common Regression Difficulties*

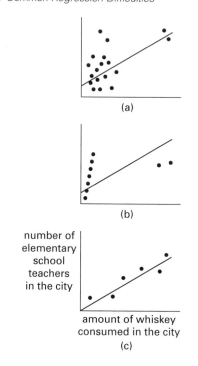

experimental data. Figure 13.8(a) shows a graph of clustered data with several extreme points. There will be moderate correlation due to the weighting of the extreme points, although there is little actual correlation at low values of the variables. The extreme data should be excluded, or the range should be extended by obtaining more data.

Figure 13.8(b) shows that good correlation exists in general, but extreme points are missed, and the overall correlation is moderate. If the results within the small linear range can be used, the extreme points should be excluded. Otherwise, additional data points are needed, and curvilinear relationships should be investigated.

Figure 13.8(c) illustrates the problem of drawing conclusions of cause and effect. There may be a predictable relationship between variables, but that does not imply a cause and effect relationship. In the case shown, both variables are functions of a third variable, the city population. But there is no direct relationship between the plotted variables.

## Example 13.17

An experiment is performed in which the dependent variable $y$ is measured against the independent variable $x$. The results are as follows.

| $x$ | $y$ |
|-----|-----|
| 1.2 | 0.602 |
| 4.7 | 5.107 |
| 8.3 | 6.984 |
| 20.9 | 10.031 |

(a) What is the least squares straight line equation that represents this data? (b) What is the correlation coefficient?

*Solution*

(a)
$$\sum x_i = 35.1$$
$$\sum y_i = 22.72$$
$$\sum x_i^2 = 529.23$$
$$\sum y_i^2 = 175.84$$
$$\left(\sum x_i\right)^2 = 1232.01$$
$$\left(\sum y_i\right)^2 = 516.38$$
$$\bar{x} = 8.775$$
$$\bar{y} = 5.681$$
$$\sum x_i y_i = 292.34$$
$$n = 4$$

From Eq. 13.70, the slope is

$$m = \frac{(4)(292.34) - (35.1)(22.72)}{(4)(529.23) - (35.1)^2} = 0.42$$

From Eq. 13.71, the $y$-intercept is

$$b = 5.681 - (0.42)(8.775) = 2.0$$

The equation of the line is

$$y = 0.42x + 2.0$$

(b) From Eq. 13.72, the correlation coefficient is

$$r = \frac{(4)(292.34) - (35.1)(22.72)}{\sqrt{\begin{array}{c}((4)(529.23) - 1232.01)\\ \times ((4)(175.84) - 516.38)\end{array}}} = 0.914$$

## Example 13.18

Repeat Ex. 13.17 assuming the relationship between the variables is nonlinear.

*Solution*

The first step is to graph the data. Since the graph has the appearance of the fourth case in Fig. 13.7, it can be assumed that the relationship between the variables has the form of $y = a + b[\log(x)]$. Therefore, the variable change $z = \log(x)$ is made, resulting in the following set of data.

| $z$ | $y$ |
|-----|-----|
| 0.0792 | 0.602 |
| 0.672 | 5.107 |
| 0.919 | 6.984 |
| 1.32 | 10.031 |

If the regression analysis is performed on this set of data, the resulting equation and correlation coefficient are

$$y = 7.599z + 0.000247$$

$$r = 0.999$$

This is a very good fit. The relationship between the variable $x$ and $y$ is approximately

$$y = 7.599 \log(x) + 0.000247$$

# 14 Numbering Systems

## 1. POSITIONAL NUMBERING SYSTEMS

A *base-b number*, $N_b$, is made up of individual *digits*. In a *positional numbering system*, the position of a digit in the number determines that digit's contribution to the total value of the number. Specifically, the position of the digit determines the power to which the *base* (also known as the *radix*), $b$, is raised. For *decimal numbers*, the radix is 10, hence the description *base-10 numbers*.

$$(a_n a_{n-1} \cdots a_2 a_1 a_0)_b = a_n b^n + a_{n-1} b^{n-1} + \cdots$$
$$+ a_2 b^2 + a_1 b + a_0 \qquad 14.1$$

The leftmost digit, $a_n$, contributes the greatest to the number's magnitude and is known as the *most significant digit* (MSD). The rightmost digit, $a_0$, contributes the least and is known as the *least significant digit* (LSD).

## 2. CONVERTING BASE-$b$ NUMBERS TO BASE-10

Equation 14.1 converts base-$b$ numbers to base-10 numbers. The calculation of the right-hand side of Eq. 14.1 is performed in the base-10 arithmetic and is known as the *expansion method*.[1]

---

[1]Equation 14.1 works with any base number. The *double-dabble (double and add) method* is a specialized method of converting from base-2 to base-10 numbers.

Converting base-$b$ numbers (i.e., decimals) to base-10 is similar to converting whole numbers and is accomplished by Eq. 14.2.

$$(0.a_1 a_2 \cdots a_m)_b = a_1 b^{-1} + a_2 b^{-2} + \cdots + a_m b^{-m}$$
$$14.2$$

## 3. CONVERTING BASE-10 NUMBERS TO BASE-$b$

The *remainder method* is used to convert base-10 numbers to base-$b$ numbers. This method consists of successive divisions by the base, $b$, until the quotient is zero. The base-$b$ number is found by taking the remainders in the reverse order from which they were found. This method is illustrated in Ex. 14.1 and Ex. 14.3.

Converting a base-10 fraction to base-$b$ requires multiplication of the base-10 fraction and subsequent fractional parts by the base. The base-$b$ fraction is formed from the integer parts of the products taken in the same order in which they were determined. This is illustrated in Ex. 14.2(d).

## 4. BINARY NUMBER SYSTEM

There are only two *binary digits (bits)* in the *binary number system:* zero and one.[2] Thus, all binary numbers consist of strings of bits (i.e., zeros and ones). The leftmost bit is known as the *most significant bit* (MSB), and the rightmost bit is the *least significant bit* (LSB).

As with digits from other numbering systems, bits can be added, subtracted, multiplied, and divided, although only digits 0 and 1 are allowed in the results. The rules of bit addition are

$$0 + 0 = 0$$
$$0 + 1 = 1$$
$$1 + 0 = 1$$
$$1 + 1 = 0 \text{ carry } 1$$

### Example 14.1

(a) Convert $(1011)_2$ to base-10. (b) Convert $(75)_{10}$ to base-2.

---

[2]Alternatively, the binary states may be called *true* and *false, on* and *off, high* and *low,* or *positive* and *negative.*

*Solution*

(a) Using Eq. 14.1 with $b = 2$,

$$(1)(2)^3 + (0)(2)^2 + (1)(2)^1 + 1 = 11$$

(b) Use the remainder method. (See Sec. 14.3.)

$$75 \div 2 = 37 \text{ remainder } 1$$
$$37 \div 2 = 18 \text{ remainder } 1$$
$$18 \div 2 = 9 \text{ remainder } 0$$
$$9 \div 2 = 4 \text{ remainder } 1$$
$$4 \div 2 = 2 \text{ remainder } 0$$
$$2 \div 2 = 1 \text{ remainder } 0$$
$$1 \div 2 = 0 \text{ remainder } 1$$

The binary representation of $(75)_{10}$ is $(1001011)_2$.

## 5. OCTAL NUMBER SYSTEM

The *octal (base-8) system* is one of the alternatives to working with long binary numbers. Only the digits 0 through 7 are used. The rules for addition in the octal system are the same as for the decimal system except that the digits 8 and 9 do not exist. For example,

$$7 + 1 = 6 + 2 = 5 + 3 = (10)_8$$
$$7 + 2 = 6 + 3 = 5 + 4 = (11)_8$$
$$7 + 3 = 6 + 4 = 5 + 5 = (12)_8$$

**Example 14.2**

Perform the following operations.

(a) $(2)_8 + (5)_8$

(b) $(7)_8 + (6)_8$

(c) Convert $(75)_{10}$ to base-8.

(d) Convert $(0.14)_{10}$ to base-8.

(e) Convert $(13)_8$ to base-10.

(f) Convert $(27.52)_8$ to base-10.

*Solution*

(a) The sum of 2 and 5 in base-10 is 7, which is less than 8 and, therefore, is a valid number in the octal system. The answer is $(7)_8$.

(b) The sum of 7 and 6 in base-10 is 13, which is greater than 8 (and, therefore, needs to be converted). Using the remainder method (see Sec. 14.3),

$$13 \div 8 = 1 \text{ remainder } 5$$
$$1 \div 8 = 0 \text{ remainder } 1$$

The answer is $(15)_8$.

(c) Use the remainder method (see Sec. 14.3).

$$75 \div 8 = 9 \text{ remainder } 3$$
$$9 \div 8 = 1 \text{ remainder } 1$$
$$1 \div 8 = 0 \text{ remainder } 1$$

The answer is $(113)_8$.

(d) Refer to Sec. 14.3.

$$0.14 \times 8 = 1.12$$
$$0.12 \times 8 = 0.96$$
$$0.96 \times 8 = 7.68$$
$$0.68 \times 8 = 5.44$$
$$0.44 \times 8 = \text{etc.}$$

The answer, $(0.1075\cdots)_8$, is constructed from the integer parts of the products.

(e) Use Eq. 14.1.

$$(1)(8) + 3 = (11)_{10}$$

(f) Use Eq. 14.1 and Eq. 14.2.

$$(2)(8)^1 + (7)(8)^0$$
$$+ (5)(8)^{-1} + (2)(8)^{-2} = 16 + 7 + \frac{5}{8} + \frac{2}{64}$$
$$= (23.656)_{10}$$

## 6. HEXADECIMAL NUMBER SYSTEM

The *hexadecimal (base-16) system* is a shorthand method of representing the value of four binary digits at a time.[3] Since 16 distinctly different characters are needed, the capital letters A through F are used to represent the decimal numbers 10 through 15. The progression of hexadecimal numbers is illustrated in Table 14.1.

**Example 14.3**

(a) Convert $(4D3)_{16}$ to base-10. (b) Convert $(1475)_{10}$ to base-16. (c) Convert $(0.8)_{10}$ to base-16.

*Solution*

(a) The hexadecimal number D is 13 in base-10. Using Eq. 14.1,

$$(4)(16)^2 + (13)(16)^1 + 3 = (1235)_{10}$$

---

[3]The term *hex number* is often heard.

(b) Use the remainder method. (See Sec. 14.3.)

$$1475 \div 16 = 92 \text{ remainder } 3$$
$$92 \div 16 = 5 \text{ remainder } 12$$
$$5 \div 16 = 0 \text{ remainder } 5$$

Since $(12)_{10}$ is $(C)_{16}$, (or hex C), the answer is $(5C3)_{16}$.

(c) Refer to Sec. 14.3.

$$0.8 \times 16 = 12.8$$
$$0.8 \times 16 = 12.8$$
$$0.8 \times 16 = \text{etc.}$$

Since $(12)_{10} = (C)_{16}$, the answer is $(0.CCCCC\cdots)_{16}$.

**Table 14.1** *Binary, Octal, Decimal, and Hexadecimal Equivalents*

| binary | octal | decimal | hexadecimal |
|--------|-------|---------|-------------|
| 0 | 0 | 0 | 0 |
| 1 | 1 | 1 | 1 |
| 10 | 2 | 2 | 2 |
| 11 | 3 | 3 | 3 |
| 100 | 4 | 4 | 4 |
| 101 | 5 | 5 | 5 |
| 110 | 6 | 6 | 6 |
| 111 | 7 | 7 | 7 |
| 1000 | 10 | 8 | 8 |
| 1001 | 11 | 9 | 9 |
| 1010 | 12 | 10 | A |
| 1011 | 13 | 11 | B |
| 1100 | 14 | 12 | C |
| 1101 | 15 | 13 | D |
| 1110 | 16 | 14 | E |
| 1111 | 17 | 15 | F |
| 10000 | 20 | 16 | 10 |

## 7. CONVERSIONS AMONG BINARY, OCTAL, AND HEXADECIMAL NUMBERS

The octal system is closely related to the binary system since $(2)^3 = 8$. Conversion from a binary to an octal number is accomplished directly by starting at the LSB (right-hand bit) and grouping the bits in threes. Each group of three bits corresponds to an octal digit. Similarly, each digit in an octal number generates three bits in the equivalent binary number.

Conversion from a binary to a hexadecimal number starts by grouping the bits (starting at the LSB) into fours. Each group of four bits corresponds to a hexadecimal digit. Similarly, each digit in a hexadecimal number generates four bits in the equivalent binary number.

Conversion between octal and hexadecimal numbers is easiest when the number is first converted to a binary number.

**Example 14.4**

(a) Convert $(5431)_8$ to base-2. (b) Convert $(1001011)_2$ to base-8. (c) Convert $(1011111101111001)_2$ to base-16.

*Solution*

(a) Convert each octal digit to binary digits.

$$(5)_8 = (101)_2$$
$$(4)_8 = (100)_2$$
$$(3)_8 = (011)_2$$
$$(1)_8 = (001)_2$$

The answer is $(101100011001)_2$.

(b) Group the bits into threes starting at the LSB.

$$1 \quad 001 \quad 011$$

Convert these groups into their octal equivalents.

$$(1)_2 = (1)_8$$
$$(001)_2 = (1)_8$$
$$(011)_2 = (3)_8$$

The answer is $(113)_8$.

(c) Group the bits into fours starting at the LSB.

$$1011 \quad 1111 \quad 0111 \quad 1001$$

Convert these groups into their hexadecimal equivalents.

$$(1011)_2 = (B)_{16}$$
$$(1111)_2 = (F)_{16}$$
$$(0111)_2 = (7)_{16}$$
$$(1001)_2 = (9)_{16}$$

The answer is $(BF79)_{16}$.

## 8. COMPLEMENT OF A NUMBER

The *complement*, $N^*$, of a number, $N$, depends on the machine (computer, calculator, etc.) being used. Assuming that the machine has a maximum number, $n$, of digits per integer number stored, the $b$'s and $(b-1)$'s complements are

$$N_b^* = b^n - N \qquad 14.3$$
$$N_{b-1}^* = N_b^* - 1 \qquad 14.4$$

For a machine that works in base-10 arithmetic, the *tens* and *nines complements* are

$$N_{10}^* = 10^n - N \qquad 14.5$$
$$N_9^* = N_{10}^* - 1 \qquad 14.6$$

For a machine that works in base-2 arithmetic, the *twos* and *ones complements* are

$$N_2^* = 2^n - N \qquad \textit{14.7}$$

$$N_1^* = N_2^* - 1 \qquad \textit{14.8}$$

## 9. APPLICATION OF COMPLEMENTS TO COMPUTER ARITHMETIC

Equation 14.9 and Eq. 14.10 are the practical applications of complements to computer arithmetic.

$$(N^*)^* = N \qquad \textit{14.9}$$

$$M - N = M + N^* \qquad \textit{14.10}$$

The binary ones complement is easily found by switching all of the ones and zeros to zeros and ones, respectively. It can be combined with a technique known as *end-around carry* to perform subtraction. End-around carry is the addition of the *overflow bit* to the sum of $N$ and its ones complement.

### Example 14.5

(a) Simulate the operation of a base-10 machine with a capacity of four digits per number and calculate the difference $(18)_{10} - (6)_{10}$ with tens complements.

(b) Simulate the operation of a base-2 machine with a capacity of five digits per number and calculate the difference $(01101)_2 - (01010)_2$ with twos complements.

(c) Solve part (b) with a ones complement and end-around carry.

*Solution*

(a) The tens complement of 6 is

$$(6)_{10}^* = (10)^4 - 6 = 10{,}000 - 6 = 9994$$

Using Eq. 14.10,

$$18 - 6 = 18 + (6)_{10}^* = 18 + 9994 = 10{,}012$$

However, the machine has a maximum capacity of four digits. Therefore, the leading 1 is dropped, leaving 0012 as the answer.

(b) The twos complement of $(01010)_2$ is

$$N_2^* = (2)^5 - N = (32)_{10} - N$$
$$= (100000)_2 - (01010)_2$$
$$= (10110)_2$$

From Eq. 14.10,

$$(01101)_2 - (01010)_2 = (01101)_2 + (10110)_2$$
$$= (100011)_2$$

Since the machine has a capacity of only five bits, the leftmost bit is dropped, leaving $(00011)_2$ as the difference.

(c) The ones complement is found by reversing all the digits.

$$(01010)_1^* = (10101)_2$$

Adding the ones complement,

$$(01101)_2 + (10101)_2 = (100010)_2$$

The leading bit is the overflow bit, which is removed and added to give the difference.

$$(00010)_2 + (1)_2 = (00011)_2$$

## 10. COMPUTER REPRESENTATION OF NEGATIVE NUMBERS

On paper, a minus sign indicates a negative number. This representation is not possible in a machine. Hence, one of the $n$ digits, usually the MSB, is reserved for sign representation. (This reduces the machine's capacity to represent numbers to $n - 1$ bits per number.) It is arbitrary whether the sign bit is 1 or 0 for negative numbers as long as the MSB is different for positive and negative numbers.

The ones complement is ideal for forming a negative number since it automatically reverses the MSB. For example, $(00011)_2$ is a five-bit representation of decimal 3. The ones complement is $(11100)_2$, which is recognized as a negative number because the MSB is 1.

### Example 14.6

Simulate the operation of a six-digit binary machine that uses ones complements for negative numbers.

(a) What is the machine representation of $(-27)_{10}$? (b) What is the decimal equivalent of the twos complement of $(-27)_{10}$? (c) What is the decimal equivalent of the ones complement of $(-27)_{10}$?

*Solution*

(a) $(27)_{10} = (011011)_2$. The negative of this number is the same as the ones complement: $(100100)_2$.

(b) The twos complement is one more than the ones complement. (See Eq. 14.7 and Eq. 14.8.) Therefore, the twos complement is

$$(100100)_2 + 1 = (100101)_2$$

This represents $(-26)_{10}$.

(c) From Eq. 14.9, the complement of a complement of a number is the original number. Therefore, the decimal equivalent is $-27$.

# 15 Numerical Analysis

## 1. NUMERICAL METHODS

Although the roots of second-degree polynomials are easily found by a variety of methods (by factoring, completing the square, or using the quadratic equation), easy methods of solving cubic and higher-order equations exist only for specialized cases. However, cubic and higher-order equations occur frequently in engineering, and they are difficult to factor. Trial and error solutions, including graphing, are usually satisfactory for finding only the general region in which the root occurs.

*Numerical analysis* is a general subject that covers, among other things, iterative methods for evaluating roots to equations. The most efficient numerical methods are too complex to present and, in any case, work by hand. However, some of the simpler methods are presented here. Except in critical problems that must be solved in real time, a few extra calculator or computer iterations will make no difference.[1]

## 2. FINDING ROOTS: BISECTION METHOD

The *bisection method* is an iterative method that "brackets" (also known as "straddles") an interval containing the *root* or *zero* of a particular equation.[2] The size of the interval is halved after each iteration. As the method's name suggests, the best estimate of the root after any iteration is the midpoint of the interval. The maximum error is half the interval length. The procedure continues until the size of the maximum error is "acceptable."[3]

---

[1]Most advanced hand-held calculators have "root finder" functions that use numerical methods to iteratively solve equations.

[2]The equation does not have to be a pure polynomial. The bisection method requires only that the equation be defined and determinable at all points in the interval.

[3]The bisection method is not a closed method. Unless the root actually falls on the midpoint of one iteration's interval, the method continues indefinitely. Eventually, the magnitude of the maximum error is small enough not to matter.

The disadvantages of the bisection method are (a) the slowness in converging to the root, (b) the need to know the interval containing the root before starting, and (c) the inability to determine the existence of or find other real roots in the starting interval.

The bisection method starts with two values of the independent variable, $x = L_0$ and $x = R_0$, which straddle a root. Since the function passes through zero at a root, $f(L_0)$ and $f(R_0)$ will have opposite signs. The following algorithm describes the remainder of the bisection method.

Let $n$ be the iteration number. Then, for $n = 0, 1, 2, ...,$ perform the following steps until sufficient accuracy is attained.

*step 1:* Set $m = \frac{1}{2}(L_n + R_n)$.

*step 2:* Calculate $f(m)$.

*step 3:* If $f(L_n)f(m) \le 0$, set $L_{n+1} = L_n$ and $R_{n+1} = m$. Otherwise, set $L_{n+1} = m$ and $R_{n+1} = R_n$.

*step 4:* $f(x)$ has at least one root in the interval $[L_{n+1}, R_{n+1}]$. The estimated value of that root, $x^*$, is

$$x^* \approx \tfrac{1}{2}(L_{n+1} + R_{n+1})$$

The maximum error is $\frac{1}{2}(R_{n+1} - L_{n+1})$.

### Example 15.1

Use two iterations of the bisection method to find a root of

$$f(x) = x^3 - 2x - 7$$

*Solution*

The first step is to find $L_0$ and $R_0$, which are the values of $x$ that straddle a root and have opposite signs. A table can be made and values of $f(x)$ calculated for random values of $x$.

| $x$ | $-2$ | $-1$ | $0$ | $+1$ | $+2$ | $+3$ |
|------|------|------|-----|------|------|------|
| $f(x)$ | $-11$ | $-6$ | $-7$ | $-8$ | $-3$ | $+14$ |

Since $f(x)$ changes sign between $x = 2$ and $x = 3$, $L_0 = 2$ and $R_0 = 3$.

First iteration, $n = 0$:

$$m = \left(\tfrac{1}{2}\right)(2 + 3) = 2.5$$
$$f(2.5) = (2.5)^3 - (2)(2.5) - 7 = 3.625$$

Since $f(2.5)$ is positive, a root must exist in the interval $[2, 2.5]$. Therefore, $L_1 = 2$ and $R_1 = 2.5$. At this point, the best estimate of the root is

$$x^* \approx \left(\tfrac{1}{2}\right)(2 + 2.5) = 2.25$$

The maximum error is $\left(\tfrac{1}{2}\right)(2.5 - 2) = 0.25$.

Second iteration, $n = 1$:

$$m = \left(\tfrac{1}{2}\right)(2 + 2.5) = 2.25$$

$$f(2.25) = (2.25)^3 - (2)(2.25) - 7 = -0.1094$$

Since $f(2.25)$ is negative, a root must exist in the interval $[2.25, 2.5]$. Therefore, $L_2 = 2.25$ and $R_2 = 2.5$. The best estimate of the root is

$$x^* \approx \left(\tfrac{1}{2}\right)(2.25 + 2.5) = 2.375$$

The maximum error is $\left(\tfrac{1}{2}\right)(2.5 - 2.25) = 0.125$.

## 3. FINDING ROOTS: NEWTON'S METHOD

Many other methods have been developed to overcome one or more of the disadvantages of the bisection method. These methods have their own disadvantages.[4]

*Newton's method* is a particular form of *fixed-point iteration*. In this sense, "fixed point" is often used as a synonym for "root" or "zero." However, fixed-point iterations get their name from functions with the characteristic property $x = g(x)$ such that the limit of $g(x)$ is the fixed point (i.e., is the root).

All fixed-point techniques require a starting point. Preferably, the starting point will be close to the actual root.[5] And, while Newton's method converges quickly, it requires the function to be continuously differentiable.

Newton's method algorithm is simple. At each iteration ($n = 0, 1, 2$, etc.), Eq. 15.1 estimates the root. The maximum error is determined by looking at how much the estimate changes after each iteration. If the change between the previous and current estimates (representing the magnitude of error in the estimate) is too large, the current estimate is used as the independent variable for the subsequent iteration.[6]

$$x_{n+1} = g(x_n) = x_n - \frac{f(x_n)}{f'(x_n)} \qquad 15.1$$

---

[4]The *regula falsi (false position) method* converges faster than the bisection method but is unable to specify a small interval containing the root. The *secant method* is prone to round-off errors and gives no indication of the remaining distance to the root.

[5]Theoretically, the only penalty for choosing a starting point too far away from the root will be a slower convergence to the root.

[6]Actually, the equation defining the maximum error is more definite than this. For example, for a large enough value of $n$, the error decreases approximately linearly. Therefore, the consecutive values of $x_n$ converge linearly to the root as well.

### Example 15.2

Solve Ex. 15.1 using two iterations of Newton's method. Use $x_0 = 2$.

*Solution*

The function and its first derivative are

$$f(x) = x^3 - 2x - 7$$

$$f'(x) = 3x^2 - 2$$

First iteration, $n = 0$:

$$x_0 = 2$$

$$f(x_0) = f(2) = (2)^3 - (2)(2) - 7 = -3$$

$$f'(x_0) = f'(2) = (3)(2)^2 - 2 = 10$$

$$x_1 = x_0 - \frac{f(x_0)}{f'(x_0)} = 2 - \frac{-3}{10} = 2.3$$

Second iteration, $n = 1$:

$$x_1 = 2.3$$

$$f(x_1) = (2.3)^3 - (2)(2.3) - 7 = 0.567$$

$$f'(x_1) = (3)(2.3)^2 - 2 = 13.87$$

$$x_2 = x_1 - \frac{f(x_1)}{f'(x_1)} = 2.3 - \frac{0.567}{13.87} = 2.259$$

## 4. NONLINEAR INTERPOLATION: LAGRANGIAN INTERPOLATING POLYNOMIAL

Interpolating between two points of known data is common in engineering. Primarily due to its simplicity and speed, straight-line interpolation is used most often. Even if more than two points on the curve are explicitly known, they are not used. Since straight-line interpolation ignores all but two of the points on the curve, it ignores the effects of curvature.

A more powerful technique that accounts for the curvature is the *Lagrangian interpolating polynomial*. This method uses an $n$th degree parabola (polynomial) as the interpolating curve.[7] This method requires that $f(x)$ be continuous and real-valued on the interval

---

[7]The Lagrangian interpolating polynomial reduces to straight-line interpolation if only two points are used.

$[x_0, x_n]$ and that $n + 1$ values of $f(x)$ are known corresponding to $x_0, x_1, x_2, ..., x_n$.

The procedure for calculating $f(x)$ at some intermediate point $x^*$ starts by calculating the Lagrangian interpolating polynomial for each known point.

$$L_k(x^*) = \prod_{\substack{i=0 \\ i \neq k}}^{n} \frac{x^* - x_i}{x_k - x_i} \qquad 15.2$$

The value of $f(x)$ at $x^*$ is calculated from Eq. 15.3.

$$f(x^*) = \sum_{k=0}^{n} f(x_k) L_k(x^*) \qquad 15.3$$

The Lagrangian interpolating polynomial has two primary disadvantages. The first is that a large number of additions and multiplications are needed.[8] The second is that the method does not indicate how many interpolating points should be (or should have been) used. Other interpolating methods have been developed that overcome these disadvantages.[9]

### Example 15.3

A real-valued function has the following values.

$$f(1) = 3.5709$$

$$f(4) = 3.5727$$

$$f(6) = 3.5751$$

Use the Lagrangian interpolating polynomial to determine the value of the function at 3.5.

*Solution*

The procedure for applying Eq. 15.2, the Lagrangian interpolating polynomial, is illustrated in tabular form. Notice that the term corresponding to $i = k$ is omitted from the product.

$k = 0$:       $i = 0$   $i = 1$   $i = 2$

$$L_0(3.5) = \left(\frac{3.5 - 1}{1 - 1}\right)\left(\frac{3.5 - 4}{1 - 4}\right)\left(\frac{3.5 - 6}{1 - 6}\right)$$
$$= 0.08333$$

$k = 1$:

$$L_1(3.5) = \left(\frac{3.5 - 1}{4 - 1}\right)\left(\frac{3.5 - 4}{4 - 4}\right)\left(\frac{3.5 - 6}{4 - 6}\right)$$
$$= 1.04167$$

$k = 2$:

$$L_2(3.5) = \left(\frac{3.5 - 1}{6 - 1}\right)\left(\frac{3.5 - 4}{6 - 4}\right)\left(\frac{3.5 - 6}{6 - 6}\right)$$
$$= -0.12500$$

Equation 15.3 is used to calculate the estimate.

$$f(3.5) = (3.5709)(0.08333) + (3.5727)(1.04167)$$
$$+ (3.5751)(-0.12500)$$
$$= 3.57225$$

### 5. NONLINEAR INTERPOLATION: NEWTON'S INTERPOLATING POLYNOMIAL

Newton's form of the interpolating polynomial is more efficient than the Lagrangian method of interpolating between known points.[10] Given $n + 1$ known points for $f(x)$, the *Newton form of the interpolating polynomial* is

$$f(x^*) = \sum_{i=0}^{n}\left(f[x_0, x_1, ..., x_i]\prod_{j=0}^{i-1}(x^* - x_j)\right) \qquad 15.4$$

$f[x_0, x_1, ..., x_i]$ is known as the *i*th *divided difference*.

$$f[x_0, x_1, ..., x_i] = \sum_{k=0}^{i}\left(\frac{f(x_k)}{(x_k - x_0)\cdots(x_k - x_{k-1})} \times (x_k - x_{k+1})\cdots(x_k - x_i)\right)$$
$$15.5$$

It is necessary to define the following two terms.

$$f[x_0] = f(x_0) \qquad 15.6$$
$$\prod(x^* - x_j) = 1 \quad [i = 0] \qquad 15.7$$

### Example 15.4

Repeat Ex. 15.3 using Newton's form of the interpolating polynomial.

---

[8]As with the numerical methods for finding roots previously discussed, the number of calculations probably will not be an issue if the work is performed by a calculator or computer.
[9]Other common methods for performing interpolation include the *Newton form* and *divided difference table*.

[10]In this case, "efficiency" relates to the ease in adding new known points without having to repeat all previous calculations.

*Solution*

Since there are $n + 1 = 3$ data points, $n = 2$. Evaluate the terms for $i = 0$ to 2.

$i = 0$:

$$f[x_0]\prod_{j=0}^{-1}(x^* - x_j) = f(x_0)(1) = f(x_0)$$

$i = 1$:

$$f[x_0, x_1]\prod_{j=0}^{0}(x^* - x_j) = f[x_0, x_1](x^* - x_0)$$

$$f[x_0, x_1] = \frac{f(x_0)}{x_0 - x_1} + \frac{f(x_1)}{x_1 - x_0}$$

$i = 2$:

$$f[x_0, x_1, x_2]\prod_{j=0}^{1}(x^* - x_j) = f[x_0, x_1, x_2](x^* - x_0)$$

$$\times (x^* - x_1)$$

$$f[x_0, x_1, x_2] = \frac{f(x_0)}{(x_0 - x_1)(x_0 - x_2)}$$

$$+ \frac{f(x_1)}{(x_1 - x_0)(x_1 - x_2)}$$

$$+ \frac{f(x_2)}{(x_2 - x_0)(x_2 - x_1)}$$

Substitute known values.

$$f(3.5) = 3.5709 + \left(\frac{3.5709}{1 - 4} + \frac{3.5727}{4 - 1}\right)(3.5 - 1)$$

$$+ \left(\frac{3.5709}{(1 - 4)(1 - 6)} + \frac{3.5727}{(4 - 1)(4 - 6)}\right.$$

$$\left. + \frac{3.5751}{(6 - 1)(6 - 4)}\right)$$

$$\times (3.5 - 1)(3.5 - 4)$$

$$= 3.57225$$

This answer is the same as that determined in Ex. 15.3.

# Topic III: Fluids

**Fluids**

# 16 Fluid Properties

## Nomenclature

| | | | |
|---|---|---|---|
| $a$ | speed of sound | ft/sec | m/s |
| $A$ | area | ft$^2$ | m$^2$ |
| $d$ | diameter | ft | m |
| $E$ | bulk modulus | lbf/ft$^2$ | Pa |
| $F$ | force | lbf | N |
| $g$ | gravitational acceleration | ft/sec$^2$ | m/s$^2$ |
| $g_c$ | gravitational conversion constant | ft-lbm/lbf-sec$^2$ | n.a. |
| $h$ | height | ft | m |
| $k$ | ratio of specific heats | – | – |
| $L$ | length | ft | m |
| M | Mach number | – | – |
| $M$ | molar concentration | lbmol/ft$^3$ | kmol/m$^3$ |
| MW | molecular weight | lbm/lbmol | kg/kmol |
| $n$ | number of moles | – | – |
| $p$ | pressure | lbf/ft$^2$ | Pa |
| $r$ | radius | ft | m |
| $R$ | specific gas constant | ft-lbf/lbm-°R | J/kg·K |
| $R^*$ | universal gas constant | ft-lbf/lbm-°R | J/kmol·K |
| SG | specific gravity | – | – |
| $T$ | absolute temperature | °R | K |
| v | velocity | ft/sec | m/s |
| $V$ | volume | ft$^3$ | m$^3$ |
| VI | viscosity index | various | various |
| $x$ | mole fraction | – | – |
| $y$ | distance | ft | m |
| $Z$ | compressibility factor | – | – |

## Symbols

| | | | |
|---|---|---|---|
| $\beta$ | compressibility | ft$^2$/lbf | Pa$^{-1}$ |
| $\beta$ | contact angle | deg | deg |
| $\gamma$ | specific weight | lbf/ft$^3$ | n.a. |
| $\mu$ | absolute viscosity | lbf-sec/ft$^2$ | Pa·s |
| $\nu$ | kinematic viscosity | ft$^2$/sec | m$^2$/s |
| $\pi$ | osmotic pressure | lbf/ft$^2$ | Pa |
| $\rho$ | density | lbm/ft$^3$ | kg/m$^3$ |
| $\sigma$ | surface tension | lbf/ft | N/m |
| $\tau$ | shear stress | lbf/ft$^2$ | Pa |
| $v$ | specific volume | ft$^3$/lbm | m$^3$/kg |

## Subscripts

| | |
|---|---|
| $c$ | critical |
| $o$ | original |
| $p$ | constant pressure |
| $T$ | constant temperature |

## 1. CHARACTERISTICS OF A FLUID

Liquids and gases can both be categorized as fluids, although this chapter is primarily concerned with incompressible liquids. There are certain characteristics shared by all fluids, and these characteristics can be used, if necessary, to distinguish between liquids and gases.[1]

- *Compressibility:* Liquids are only slightly compressible and are assumed to be incompressible for most purposes. Gases are highly compressible.

- *Shear resistance:* Liquids and gases cannot support shear, and they deform continuously to minimize applied shear forces.

- *Shape and volume:* As a consequence of their inability to support shear forces, liquids and gases take on the shapes of their containers. Only liquids have free surfaces. Liquids have fixed volumes, regardless of their container volumes, and these volumes are not significantly affected by temperature and pressure. Unlike liquids, gases take on the volumes of their containers. If allowed to do so, gas densities will change as temperature and pressure are varied.

- *Resistance to motion:* Due to viscosity, liquids resist instantaneous changes in velocity, but the resistance stops when liquid motion stops. Gases have very low viscosities.

---

[1]The differences between liquids and gases become smaller as temperature and pressure are increased. Gas and liquid properties become the same at the critical temperature and pressure.

Fluids

- *Molecular spacing:* Molecules in liquids are relatively close together and are held together with strong forces of attraction. Liquid molecules have low kinetic energy. The distance each liquid molecule travels between collisions is small. In gases, the molecules are relatively far apart and the attractive forces are weak. Kinetic energy of the molecules is high. Gas molecules travel larger distances between collisions.

- *Pressure:* The pressure at a point in a fluid is the same in all directions. Pressure exerted by a fluid on a solid surface (e.g., container wall) is always normal to that surface.

## 2. TYPES OF FLUIDS

For computational convenience, fluids are generally divided into two categories: ideal fluids and real fluids. (See Fig. 16.1.) *Ideal fluids* are assumed to have no viscosity (and hence, no resistance to shear), be incompressible, and have uniform velocity distributions when flowing. In an ideal fluid, there is no friction between moving layers of fluid, and there are no eddy currents or turbulence.

*Real fluids* exhibit finite viscosities and non-uniform velocity distributions, are compressible, and experience friction and turbulence in flow. Real fluids are further divided into *Newtonian fluids* and *non-Newtonian fluids*, depending on their viscous behavior. The differences between Newtonian and the various types of non-Newtonian fluids are described in Sec. 16.9.

For convenience, most fluid problems assume real fluids with Newtonian characteristics. This is an appropriate assumption for water, air, gases, steam, and other simple fluids (alcohol, gasoline, acid solutions, etc.). However, slurries, pastes, gels, suspensions, and polymer/electrolyte solutions may not behave according to simple fluid relationships.

**Figure 16.1** *Types of Fluids*

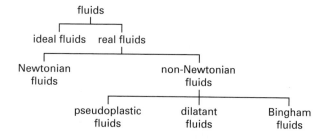

## 3. FLUID PRESSURE AND VACUUM

In the English system, fluid pressure is measured in pounds per square inch (lbf/in$^2$ or psi) and pounds per square foot (lbf/ft$^2$ or psf), although tons (2000 pounds) per square foot (tsf) are occasionally used. In SI units, pressure is measured in pascals (Pa). Because a pascal is

very small, kilopascals (kPa) are usually used. Other units of pressure are bars, millibars, atmospheres, inches and feet of water, millimeters, centimeters, and inches of mercury, and torrs. (See Fig. 16.2.)

**Figure 16.2** *Relative Sizes of Pressure Units*

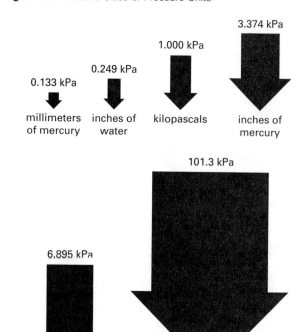

Fluid pressures are measured with respect to two pressure references: zero pressure and atmospheric pressure. Pressures measured with respect to a true zero pressure reference are known as *absolute pressures*. Pressures measured with respect to atmospheric pressure are known as *gage pressures*.[2] Most pressure gauges read the excess of the test pressure over atmospheric pressure (i.e., the gage pressure). To distinguish between these two pressure measurements, the letters "a" and "g" are traditionally added to the the unit symbols in the English unit system (e.g., 14.7 psia and 4015 psfg). For SI units, the actual words "gage" and "absolute" can be added to the measurement (e.g., 25.1 kPa absolute). Alternatively, the pressure is assumed to be absolute unless the "g" is used (e.g., 15 Pag).

Absolute and gage pressures are related by Eq. 16.1. It should be mentioned that $p_{\text{atmospheric}}$ in Eq. 16.1 is the actual atmospheric pressure existing when the gage measurement is taken. It is not standard atmospheric pressure, unless that pressure is implicitly or explicitly

---

[2]The spelling *gage* persists even though pressures are measured with *gauges*. In some countries, the term *meter pressure* is used instead of gage pressure.

applicable. Also, since a barometer measures atmospheric pressure, *barometric pressure* is synonymous with atmospheric pressure. Table 16.1 lists standard atmospheric pressure in various units.

$$p_{\text{absolute}} = p_{\text{gage}} + p_{\text{atmospheric}} \qquad 16.1$$

**Table 16.1** *Standard Atmospheric Pressure*

| | |
|---|---|
| 1.000 atm | (atmosphere) |
| 14.696 psia | (pounds per square inch absolute) |
| 2116.2 psfa | (pounds per square foot absolute) |
| 407.1 in wg | (inches of water, inches water gage) |
| 33.93 ft wg | (feet of water, feet water gage) |
| 29.921 in Hg | (inches of mercury) |
| 760.0 mm Hg | (millimeters of mercury) |
| 760.0 torr | |
| 1.013 bars | |
| 1013 millibars | |
| $1.013 \times 10^5$ Pa | (pascals) |
| 101.3 kPa | (kilopascals) |

A *vacuum* measurement is implicitly a pressure below atmospheric (i.e., a negative gage pressure). It must be assumed that any measured quantity given as a vacuum is a quantity to be subtracted from the atmospheric pressure. Thus, when a condenser is operating with a vacuum of 4.0 in Hg (4 in of mercury), the absolute pressure is 29.92 in Hg − 4.0 in Hg = 25.92 in Hg. Vacuums are generally stated as positive numbers.

$$p_{\text{absolute}} = p_{\text{atmospheric}} - p_{\text{vacuum}} \qquad 16.2$$

A difference in two pressures may be reported with units of *psid* (i.e., a *differential* in psi).

## 4. DENSITY

The *density*, $\rho$, of a fluid is its mass per unit volume.[3] In SI units, density is measured in $kg/m^3$. In a consistent English system, density is measured in $slugs/ft^3$, even though fluid density is traditionally reported in $lbm/ft^3$.

$$\rho = \text{fluid density} \qquad 16.3$$

The density of a fluid in a liquid form is usually given, known in advance, or easily obtained from tables in any one of a number of sources. (See Table 16.2.) Most English fluid data are reported on a per pound basis, and the data included in this book follow that tradition. To make the conversion from pounds to slugs, divide by

---

[3]Mass is an absolute property of a substance. Weight is not absolute, since it depends on the local gravity. Some fluids books continue to use $\gamma$ as the symbol for weight density. The equations using $\gamma$ that result (such as Bernoulli's equation) cannot be used with SI data, since the equations are not consistent. Thus, engineers end up with two different equations for the same thing.

$g_c$ as an implied step whenever using pound-basis fluid data.

$$\rho_{\text{slugs}} = \frac{\rho_{\text{lbm}}}{g_c} \qquad 16.4$$

The density of an ideal gas can be found from the specific gas constant and the ideal gas law.

$$\rho = \frac{p}{RT} \qquad 16.5$$

**Table 16.2** *Approximate Room-Temperature Densities of Common Fluids*

| fluid | $lbm/ft^3$ | $kg/m^3$ |
|---|---|---|
| air (STP) | 0.0807 | 1.29 |
| air (70°F, 1 atm) | 0.075 | 1.20 |
| alcohol | 49.3 | 790 |
| ammonia | 38 | 602 |
| gasoline | 44.9 | 720 |
| glycerin | 78.8 | 1260 |
| mercury | 848 | 13 600 |
| water | 62.4 | 1000 |

(Multiply $lbm/ft^3$ by 16.01 to obtain $kg/m^3$.)

### Example 16.1

The density of water is typically taken as 62.4 $lbm/ft^3$ for engineering problems where greater accuracy is not required. What is the value in (a) $slugs/ft^3$ and (b) $kg/m^3$?

*Solution*

(a) Equation 16.4 can be used to calculate the slug-density of water.

$$\rho = \frac{\rho_{\text{lbm}}}{g_c} = \frac{62.4 \dfrac{\text{lbm}}{\text{ft}^3}}{32.2 \dfrac{\text{ft-lbm}}{\text{lbf-sec}^2}} = 1.94 \text{ lbf-sec}^2/\text{ft-ft}^3$$

$$= 1.94 \text{ slugs/ft}^3$$

(b) The conversion between $lbm/ft^3$ and $kg/m^3$ is approximately 16.0, derived as follows.

$$\rho = \left( 62.4 \ \frac{\text{lbm}}{\text{ft}^3} \right) \left( \frac{35.31 \dfrac{\text{ft}^3}{\text{m}^3}}{2.205 \dfrac{\text{lbm}}{\text{kg}}} \right)$$

$$= \left( 62.4 \ \frac{\text{lbm}}{\text{ft}^3} \right) \left( 16.01 \ \frac{\text{kg-ft}^3}{\text{m}^3\text{-lbm}} \right) = 999 \text{ kg/m}^3$$

In SI problems, it is common to take the density of water as 1000 $kg/m^3$.

## 5. SPECIFIC VOLUME

*Specific volume*, $v$, is the volume occupied by a unit mass of fluid.[4] Since specific volume is the reciprocal of density, typical units will be ft³/lbm, ft³/lbmole, or m³/kg.[5]

$$v = \frac{1}{\rho} \qquad 16.6$$

## 6. SPECIFIC GRAVITY

*Specific gravity* (SG) is a dimensionless ratio of a fluid's density to some standard reference density.[6] For liquids and solids, the reference is the density of pure water. There is some variation in this reference density, however, since the temperature at which the water density is evaluated is not standardized. Temperatures of 39.2°F (4°C), 60°F (16.5°C), and 70°F (21.1°C) have been reported.[7]

Fortunately, the density of water is the same to three significant digits over the normal ambient temperature range: 62.4 lbm/ft³ or 1000 kg/m³. However, to be precise, the temperature of both the fluid and water should be specified (e.g., "... the specific gravity of the 20°C fluid is 1.05 referred to 4°C water ...").

$$SG_{liquid} = \frac{\rho_{liquid}}{\rho_{water}} \qquad 16.7$$

Since the SI density of water is very nearly 1.000 g/cm³ (1000 kg/m³), the numerical values of density in g/cm³ and specific gravity are the same. Such is not the case with English units.

The standard reference used to calculate the specific gravity of gases is the density of air. Since the density of a gas depends on temperature and pressure, both must be specified for the gas and air (i.e., two temperatures and two pressures must be specified). While STP (standard temperature and pressure) conditions are commonly specified, they are not universal.[8] Table 16.3 lists several common sets of standard conditions.

$$SG_{gas} = \frac{\rho_{gas}}{\rho_{air}} \qquad 16.8$$

If it is known or implied that the temperature and pressure of the air and gas are the same, the specific

**Table 16.3** *Commonly Quoted Values of Standard Temperature and Pressure*

| system | temperature | pressure |
|---|---|---|
| SI | 273.15K | 101.325 kPa |
| scientific | 0.0°C | 760 mm Hg |
| U.S. engineering | 32°F | 14.696 psia |
| natural gas industry (U.S.) | 60°F | 14.65, 14.73, or 15.025 psia |
| natural gas industry (Canada) | 60°F | 14.696 psia |

gravity of the gas will be equal to the ratio of molecular weights and the inverse ratio of specific gas constants. The density of air evaluated at STP is listed in Table 16.2. At 70°F (21.1°C) and 1.0 atm, the density is approximately 0.075 lbm/ft³ (1.20 kg/m³).

$$SG_{gas} = \frac{MW_{gas}}{MW_{air}} = \frac{MW_{gas}}{29.0}$$

$$= \frac{R_{air}}{R_{gas}} = \frac{53.3 \frac{ft\text{-}lbf}{lbm\text{-}°R}}{R_{gas}} \qquad 16.9$$

Specific gravities of petroleum liquids and aqueous solutions (of acid, antifreeze, salts, etc.) can be determined by use of a *hydrometer*. (See Fig. 16.3.) In its simplest form, a hydrometer is constructed as a graduated scale weighted at one end so it will float vertically. The height at which the hydrometer floats depends on the density of the fluid, and the graduated scale can be calibrated directly in specific gravity.[9]

**Figure 16.3** *Hydrometer*

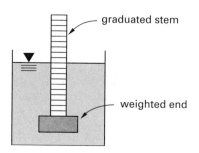

graduated stem

weighted end

There are two standardized hydrometer scales (i.e., methods for calibrating the hydrometer stem).[10] Both state specific gravity in degrees, although temperature is not being measured. The *American Petroleum Institute* (API) scale (°API) may be used with all liquids, not only with oils or other hydrocarbons. For the specific

---

[4]Care must be taken to distinguish between the symbol upsilon, $v$, used for specific volume, and italic roman "vee," $v$, used for velocity in many engineering textbooks.
[5]Units of ft³/slug are also possible, but this combination of units is almost never encountered.
[6]The symbols S.G., sp.gr., S, and G are also used. In fact, petroleum engineers in the United States use $\gamma$, a symbol that civil engineers use for specific weight. There is no standard engineering symbol for specific gravity.
[7]Density of liquids is sufficiently independent of pressure to make consideration of pressure in specific gravity calculations unnecessary.
[8]The abbreviation "SC" (standard conditions) is interchangeable with "STP."

[9]This is a direct result of the buoyancy principle of Archimedes.
[10]In addition to °Be and °API mentioned in this chapter, the *Twaddell scale* (°Tw) is used in chemical processing, the *Brix* and *Balling's scales* are used in the sugar industry, and the *Salometer scale* is used to measure salt (NaCl and CaCl₂) solutions.

gravity value, a standard reference temperature of $60°F$ $(15.6°C)$ is implied for both the liquid and the water.

$$°API = \frac{141.5}{SG} - 131.5 \qquad 16.10$$

$$SG = \frac{141.5}{°API + 131.5} \qquad 16.11$$

The *Baumé scale* (°Be) is used in the wine, honey, and acid industries. It is somewhat confusing because there are actually two Baumé scales—one for liquids heavier than water and another for liquids lighter than water. (There is also a discontinuity in the scales at $SG = 1.00$.) As with the API scale, the specific gravity value assumes $60°F$ $(15.6°C)$ is the standard temperature for both scales.

$$SG = \frac{140.0}{130.0 + °Be} \quad [SG \le 1.00] \qquad 16.12$$

$$SG = \frac{145.0}{145.0 - °Be} \quad [SG \ge 1.00] \qquad 16.13$$

### Example 16.2

Determine the specific gravity of carbon dioxide gas (molecular weight = 44) at $66°C$ $(150°F)$ and 138 kPa (20 psia) using STP air as a reference.

*SI Solution*

$$R^* = 8314 \text{ J/kmol·K}$$

$$R = \frac{R^*}{MW} = \frac{8314 \dfrac{\text{J}}{\text{kmol·K}}}{44 \dfrac{\text{kg}}{\text{kmol}}} = 189.0 \text{ J/kg·K}$$

$$\rho = \frac{p}{RT} = \frac{1.38 \times 10^5 \text{ Pa}}{\left(189.0 \dfrac{\text{J}}{\text{kg·K}}\right)(66°C + 273°)}$$

$$= 2.15 \text{ kg/m}^3$$

$$SG = \frac{2.15 \dfrac{\text{kg}}{\text{m}^3}}{1.29 \dfrac{\text{kg}}{\text{m}^3}} = 1.67$$

*Customary U.S. Solution*

Since the conditions of the carbon dioxide and air are different, Eq. 16.9 cannot be used. Therefore, it is necessary to calculate the density of the carbon dioxide from

Eq. 16.5. The specific gas constant of carbon dioxide is 35.1 ft-lbf/lbm-°R. The density is

$$\rho = \frac{p}{RT} = \frac{\left(20 \dfrac{\text{lbf}}{\text{in}^2}\right)\left(144 \dfrac{\text{in}^2}{\text{ft}^2}\right)}{\left(35.1 \dfrac{\text{ft-lbf}}{\text{lbm-°R}}\right)(150°F + 460°)}$$

$$= 0.135 \text{ lbm/ft}^3$$

From Table 16.2, the density of STP air is 0.0807 lbm/ft$^3$. From Eq. 16.8, the specific gravity of carbon dioxide at the conditions given is

$$SG = \frac{\rho_{\text{gas}}}{\rho_{\text{air}}} = \frac{0.135 \dfrac{\text{lbm}}{\text{ft}^3}}{0.0807 \dfrac{\text{lbm}}{\text{ft}^3}} = 1.67$$

### 7. SPECIFIC WEIGHT

*Specific weight*, $\gamma$, is the weight of fluid per unit volume. The use of specific weight is most often encountered in civil engineering projects in the United States, where it is commonly called "density." Mechanical and chemical engineers seldom encounter the term. The usual units of specific weight are lbf/ft$^3$.[11] Specific weight is not an absolute property of a fluid, since it depends not only on the fluid but on the local gravitational field as well.

$$\gamma = g\rho \qquad \text{[SI]} \quad 16.14(a)$$

$$\gamma = \rho \times \frac{g}{g_c} \qquad \text{[U.S.]} \quad 16.14(b)$$

If the gravitational acceleration is $32.2$ ft/sec$^2$, as it is almost everywhere on the earth, the specific weight in lbf/ft$^3$ will be numerically equal to the density in lbm/ft$^3$. This is illustrated in Ex. 16.3.

### Example 16.3

What is the sea level ($g = 32.2$ ft/sec$^2$) specific weight (in lbf/ft$^3$) of liquids with densities of (a) 1.95 slug/ft$^3$, and (b) 58.3 lbm/ft$^3$?

*Solution*

(a) Equation 16.14(a) can be used with any consistent set of units, including densities involving slugs.

$$\gamma = g\rho = \left(32.2 \frac{\text{ft}}{\text{sec}^2}\right)\left(1.95 \frac{\text{slug}}{\text{ft}^3}\right)$$

$$= \left(32.2 \frac{\text{ft}}{\text{sec}^2}\right)\left(1.95 \frac{\text{lbf-sec}^2}{\text{ft-ft}^3}\right) = 62.8 \text{ lbf/ft}^3$$

---

[11]Notice that the units are lbf/ft$^3$, not lbm/ft$^3$. Pound-mass (lbm) is a mass unit, not a weight (force) unit.

(b) From Eq. 16.14(b),

$$\gamma = \rho \times \frac{g}{g_c}$$

$$= \left(58.3\ \frac{\text{lbm}}{\text{ft}^3}\right) \left(\frac{32.2\ \dfrac{\text{ft}}{\text{sec}^2}}{32.2\ \dfrac{\text{ft-lbm}}{\text{lbf-sec}^2}}\right) = 58.3\ \text{lbf/ft}^3$$

## 8. MOLE FRACTION

*Mole fraction* is an important parameter in many practical engineering problems, particularly in chemistry and chemical engineering. The composition of a fluid consisting of two or more distinctly different substances, A, B, C, and so on, can be described by the mole fractions, $x_A$, $x_B$, $x_C$, and so on, of each substance. (There are also other methods of specifying the composition.) The mole fraction of component A is the number of moles of that component, $n_A$, divided by the total number of moles in the combined fluid mixture.

$$x_A = \frac{n_A}{n_A + n_B + n_C + \cdots} \qquad \textit{16.15}$$

*Mole fraction* is a number between 0 and 1.000. *Mole percent* is the mole fraction multiplied by 100, expressed in percent.

## 9. VISCOSITY

The *viscosity* of a fluid is a measure of that fluid's resistance to flow when acted upon by an external force such as a pressure differential or gravity. Some fluids, such as heavy oils, jellies, and syrups, are very viscous. Other fluids, such as water, lighter hydrocarbons, and gases, are not as viscous.

Most viscous liquids will flow more easily when their temperatures are raised. However, the behavior of a fluid when temperature, pressure, or stress is varied will depend on the type of fluid. The different types of fluids can be determined with a *sliding plate viscometer test.*[12]

The more viscous the fluid, the more time will be required for the fluid to leak out of a container. *Saybolt Seconds Universal* (SSU) and *Saybolt Seconds Furol* (SSF) are scales of such viscosity measurement based on the smaller and larger orifices, respectively. Seconds can be converted (empirically) to viscosity in other units. The following relations are approxmiate conversions between SSU, stokes, and poise.

---

[12]This test is conceptually simple but is not always practical, since the liquid leaks out between the plates. In research work with liquids, it is common to determine viscosity with a *concentric cylinder viscometer*, also known as a *cup-and-bob viscometer*. Viscosities of perfect gases can be predicted by the kinetic theory of gases. Viscosity can also be measured by a *Saybolt viscometer*, which is essentially a container that allows a given quantity of fluid to leak out through one of two different-sized orifices.

- For SSU < 100 sec,

$$\nu_{\text{stokes}} = (0.00226)(\text{SSU}) - \frac{1.95}{\text{SSU}}$$

$$\mu_{\text{poise}} = (\text{SG})(\nu_{\text{stokes}})$$

- For SSU > 100 sec,

$$\nu_{\text{stokes}} = (0.00220)(\text{SSU}) - \frac{1.35}{\text{SSU}}$$

$$\mu_{\text{poise}} = (\text{SG})(\nu_{\text{stokes}})$$

Consider two plates of area A separated by a fluid with thickness $y_0$, as shown in Fig. 16.4. The bottom plate is fixed, and the top plate is kept in motion at a constant velocity, $v_0$, by a force, F.

**Figure 16.4** *Sliding Plate Viscometer*

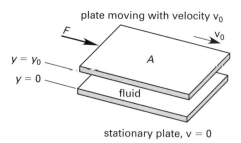

Experiments with Newtonian fluids have shown that the force, F, required to maintain the velocity, $v_0$, is proportional to the velocity and the area and is inversely proportional to the separation of the plates. That is,

$$\frac{F}{A} \propto \frac{d\text{v}}{dy} \qquad \textit{16.16}$$

The constant of proportionality needed to make Eq. 16.16 an equality is the *absolute viscosity*, $\mu$, also known as the *coefficient of viscosity*.[13] The reciprocal of absolute viscosity, $1/\mu$, is known as the *fluidity*.

$$\frac{F}{A} = \mu \frac{d\text{v}}{dy} \qquad \textit{16.17}$$

$F/A$ is the *fluid shear stress*, $\tau$. The quantity $d\text{v}/dy$ ($\text{v}_0/y_0$) is known by various names, including *rate of strain, shear rate, velocity gradient*, and *rate of shear formation*. Equation 16.17 is known as *Newton's law of viscosity*, from which Newtonian fluids get their name. Sometimes Eq. 16.18 is written with a minus sign to compare viscous behavior with other behavior.

---

[13]Another name for absolute viscosity is *dynamic viscosity*. The name *absolute viscosity* is preferred, if for no other reason than to avoid confusion with *kinematic viscosity*.

However, the direction of positive shear stress is arbitrary. Equation 16.18 is simply the equation of a straight line.

$$\tau = \mu \frac{dv}{dy} \qquad \qquad 16.18$$

Not all fluids are Newtonian (although most common fluids are), and Eq. 16.18 is not universally applicable. Figure 16.5 (known as the *rheogram*) illustrates how differences in fluid shear stress behavior (at constant temperature and pressure) can be used to define Bingham, pseudoplastic, and dilatant fluids, as well as Newtonian fluids.

Gases, water, alcohol, and benzene are examples of *Newtonian fluids*. In fact, all liquids with a simple chemical formula are Newtonian. Also, most solutions of simple compounds, such as sugar and salt, are Newtonian. For a highly viscous fluid, the straight line (see Fig. 16.5) will be closer to the $\tau$ axis (i.e., the slope will be higher). For low-viscosity fluids, the straight line will be closer to the $dv/dy$ axis (i.e., the slope will be lower).

**Figure 16.5** *Shear Stress Behavior for Different Types of Fluids*

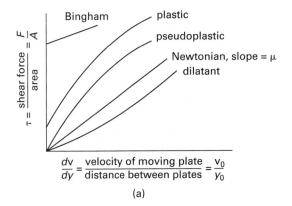

(a)

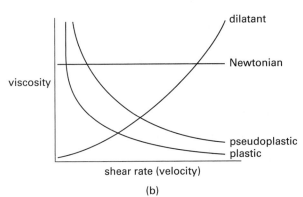

(b)

*Pseudoplastic fluids* (muds, motor oils, polymer solutions, natural gums, and most slurries) exhibit viscosities that decrease with an increasing velocity gradient. Such fluids present no serious pumping problems.

*Plastic materials*, such as tomato catsup, behave similarly to pseudoplastic fluids once movement begins; that

is, their viscosities decrease with agitation. However, a finite force must be applied before any fluid movement occurs.

*Bingham fluids* (Bingham plastics), typified by toothpaste, jellies, bread dough, and some slurries, are capable of indefinitely resisting a small shear stress but move easily when the stress becomes large—that is, Bingham fluids become pseudoplastic when the stress increases.

*Dilatant fluids* are rare but include clay slurries, various starches, some paints, milk chocolate with nuts, and other candy compounds. They exhibit viscosities that increase with increasing agitation (i.e., with increasing velocity gradients), but they return rapidly to their normal viscosity after the agitation ceases. Pump selection is critical for dilatant fluids because these fluids can become almost solid if the shear rate is high enough.

Viscosity can also change with time (all other conditions being constant). If viscosity decreases with time, the fluid is said to be a *thixotropic fluid*. If viscosity increases (usually up to a finite value) with time, the fluid is a *rheopectic fluid*. Viscosity does not change in time-independent fluids. *Colloidal materials*, such as gelatinous compounds, lotions, shampoos, and low-temperature solutions of soaps in water and oil, behave like *thixotropic liquids*—their viscosities decrease as the rate of shear is increased. However, viscosity does not return to its original state after the agitation ceases.

Molecular cohesion is the dominating cause of viscosity in liquids. As the temperature of a liquid increases, these cohesive forces decrease, resulting in a decrease in viscosity.

In gases, the dominant cause of viscosity is random collisions between gas molecules. This molecular agitation increases with increases in temperature. Therefore, viscosity in gases increases with temperature.

Although viscosity of liquids increases slightly with pressure, the increase is insignificant over moderate pressure ranges. Therefore, the absolute viscosity of both gases and liquids is usually considered to be essentially independent of pressure.[14]

The units of absolute viscosity, as derived from Eq. 16.18, are lbf-sec/ft$^2$. Such units are actually used in the English engineering system.[15] Another common unit used throughout the world is the *poise* (abbreviated P), equal to a dyne·s/cm$^2$. These dimensions are the same primary dimensions as in the English system, $F\theta/L^2$ or $M/L\theta$, and are functionally the same as a g/cm·s. Since the poise is a large unit, the *centipoise* (abbreviated cP) scale is generally used. By coincidence, the viscosity of pure water at room temperature is approximately 1 cP.

---

[14]This is not true for kinematic viscosity, however.
[15]Units of lbm/ft-sec are also used for absolute viscosity in the English system, although it is difficult to see how such units are derived from the sliding-plate test. These units are obtained by dividing lbf-sec/ft$^2$ units by $g_c$.

Absolute viscosity is measured in pascal-seconds (Pa·s) in SI units.

### Example 16.4

A liquid ($\mu = 5.2 \times 10^{-5}$ lbf-sec/ft$^2$) is flowing in a rectangular duct. The equation of the symmetrical velocity (in ft/sec) is approximately v $= 3y^{0.7}$ ft/sec, where $y$ is in inches. (a) What is the velocity gradient at $y = 3.0$ in from the duct wall? (b) What is the shear stress in the fluid at that point?

*Solution*

(a) The velocity is not a linear function of $y$, so $dv/dy$ must be calculated as a derivative.

$$\frac{dv}{dy} = \frac{d}{dy}(3y^{0.7})$$
$$= (3)(0.7y^{-0.3}) = 2.1y^{-0.3}$$

At $y = 3$ in,

$$\frac{dv}{dy} = (2.1)(3)^{-0.3} = 1.51 \text{ ft/sec-in}$$

(b) From Eq. 16.18, the shear stress is

$$\tau = \mu \frac{dv}{dy}$$

$$= \left(5.2 \times 10^{-5} \frac{\text{lbf-sec}}{\text{ft}^2}\right)\left(1.51 \frac{\text{ft}}{\text{sec-in}}\right)\left(12 \frac{\text{in}}{\text{ft}}\right)$$

$$= 9.42 \times 10^{-4} \text{ lbf/ft}^2$$

### 10. KINEMATIC VISCOSITY

Another quantity with the name *viscosity* is the ratio of absolute viscosity to mass density. This combination of variables, known as *kinematic viscosity*, $\nu$, appears sufficiently often in fluids and other problems as to warrant its own symbol and name. Thus, kinematic viscosity is the name given to a frequently occurring combination of variables.

$$\nu = \frac{\mu}{\rho} \qquad \text{[SI]} \quad 16.19(a)$$

$$\nu = \frac{\mu g_c}{\rho} \qquad \text{[U.S.]} \quad 16.19(b)$$

The primary dimensions of kinematic viscosity are $L^2/\theta$. Typical units are ft$^2$/sec and cm$^2$/s (the *stoke*, St). It is also common to give kinematic viscosity in *centistokes*, cSt. The SI units of kinematic viscosity are m$^2$/s.

It is essential that consistent units be used with Eq. 16.19(a). The following sets of units are consistent:

$$\text{ft}^2/\text{sec} = \frac{\text{lbf-sec/ft}^2}{\text{slugs/ft}^3}$$

$$\text{m}^2/\text{s} = \frac{\text{Pa·s}}{\text{kg/m}^3}$$

$$\text{St (stoke)} = \frac{\text{P (poise)}}{\text{g/cm}^3}$$

$$\text{cSt (centistokes)} = \frac{\text{cP (centipoise)}}{\text{g/cm}^3}$$

Unlike absolute viscosity, kinematic viscosity is greatly dependent on both temperature and pressure, since these variables affect the density of the fluid. Referring to Eq. 16.19, even if absolute viscosity is independent of temperature or pressure, the change in density will change the kinematic viscosity.

### 11. VISCOSITY CONVERSIONS

The most common units of absolute and kinematic viscosity are listed in Table 16.4.

*Table 16.4* Common Viscosity Units

|  | absolute ($\mu$) | kinematic ($\nu$) |
|---|---|---|
| English | lbf-sec/ft$^2$ (slug/ft-sec) | ft$^2$/sec |
| conventional metric | dyne·s/cm$^2$ (poise) | cm$^2$/sec (stoke) |
| SI | Pa·s (N·s/m$^2$) | m$^2$/s |

Table 16.5 contains conversions between the various viscosity units.

### Example 16.5

Water at 60°F has a specific gravity of 0.999 and a kinematic viscosity of 1.12 cSt. What is the absolute viscosity in lbf-sec/ft$^2$?

*Solution*

The density of a liquid expressed in g/cm$^3$ is numerically equal to its specific gravity.

$$\rho = 0.999 \text{ g/cm}^3$$

**Table 16.5** Viscosity Conversions[*]

| multiply | by | to obtain |
|---|---|---|
| *absolute viscosity, $\mu$* | | |
| dyne·s/cm$^2$ | 0.10 | Pa·s |
| lbf-sec/ft$^2$ | 478.8 | P |
| lbf-sec/ft$^2$ | 47,880 | cP |
| lbf-sec/ft$^2$ | 47.88 | Pa·s |
| slug/ft-sec | 47.88 | Pa·s |
| lbm/ft-sec | 1.488 | Pa·s |
| cP | $1.0197 \times 10^{-4}$ | kgf·s/m$^2$ |
| cP | $2.0885 \times 10^{-5}$ | lbf-sec/ft$^2$ |
| cP | 0.001 | Pa·s |
| Pa·s | 0.020885 | lbf-sec/ft$^2$ |
| Pa·s | 1000 | cP |
| *kinematic viscosity, $\nu$* | | |
| ft$^2$/sec | 92,903 | cSt |
| ft$^2$/sec | 0.092903 | m$^2$/s |
| m$^2$/s | 10.7639 | ft$^2$/sec |
| m$^2$/s | $1 \times 10^6$ | cSt |
| cSt | $1 \times 10^{-6}$ | m$^2$/s |
| cSt | $1.0764 \times 10^{-5}$ | ft$^2$/sec |
| *absolute viscosity to kinematic viscosity* | | |
| cP | $1/\rho$ in g/cm$^3$ | cSt |
| cP | $6.7195 \times 10^{-4}/\rho$ in lbm/ft$^3$ | ft$^2$/sec |
| lbf-sec/ft$^2$ | $32.174/\rho$ in lbm/ft$^3$ | ft$^2$/sec |
| kgf·s/m$^2$ | $9.807/\rho$ in kg/m$^3$ | m$^2$/s |
| Pa·s | $1000/\rho$ in g/cm$^3$ | cSt |
| *kinematic viscosity to absolute viscosity* | | |
| cSt | $\rho$ in g/cm$^3$ | cP |
| cSt | $0.001 \times \rho$ in g/cm$^3$ | Pa·s |
| m$^2$/s | $0.10197 \times \rho$ in kg/m$^3$ | kgf·s/m$^2$ |
| m$^2$/s | $1000 \times \rho$ in g/cm$^3$ | Pa·s |
| ft$^2$/sec | $0.031081 \times \rho$ in lbm/ft$^3$ | lbf-sec/ft$^2$ |
| ft$^2$/sec | $1488.2 \times \rho$ in lbm/ft$^3$ | cP |

[*]cP: centipoise; cSt: centistoke; P: poise

The centistoke (cSt) is a measure of kinematic viscosity. Kinematic viscosity is converted first to the absolute viscosity units of centipoise. From Table 16.5,

$$\mu_{cP} = \nu_{cSt}\rho_{g/cm^3}$$

$$= (1.12 \text{ cSt})\left(0.99 \ \frac{g}{cm^3}\right) = 1.119 \text{ cP}$$

Next, centipoise is converted to lbf-sec/ft$^2$.

$$\mu_{lbf\text{-}sec/ft^2} = \mu_{cP}(2.0885 \times 10^{-5})$$

$$= (1.119 \text{ cP})(2.0885 \times 10^{-5})$$

$$= 2.34 \times 10^{-5} \text{ lbf-sec/ft}^2$$

## 12. VISCOSITY INDEX

*Viscosity index* (VI) is a measure of a fluid's sensitivity to changes in viscosity with changes in temperature. It has traditionally been applied to crude and refined oils through use of a 100-point scale.[16] The viscosity is measured at two temperatures: 100°F and 210°F (38°C and 99°C). These viscosities are converted into a viscosity index in accordance with standard ASTM D2270.

## 13. VAPOR PRESSURE

Molecular activity in a liquid will allow some of the molecules to escape the liquid surface. Strictly speaking, a small portion of the liquid vaporizes. Molecules of the vapor also condense back into the liquid. The vaporization and condensation at constant temperature are equilibrium processes. The equilibrium pressure exerted by these free molecules is known as the *vapor pressure* or *saturation pressure*. (Vapor pressure does not include the pressure of other substances in the mixture.)

Some liquids, such as propane, butane, ammonia, and Freon, have significant vapor pressures at normal temperatures. Liquids near their boiling points or that vaporize easily are said to be *volatile liquids*.[17] Other liquids, such as mercury, have insignificant vapor pressures at the same temperature. Liquids with low vapor pressures are used in accurate barometers.

The tendency toward vaporization is dependent on the temperature of the liquid. *Boiling* occurs when the liquid temperature is increased to the point that the vapor pressure is equal to the local ambient pressure. Thus, a liquid's boiling temperature depends on the local ambient pressure as well as on the liquid's tendency to vaporize.

Vapor pressure is usually considered to be a nonlinear function of temperature only. It is possible to derive correlations between vapor pressure and temperature, and such correlations usually involve a logarithmic transformation of vapor pressure.[18] Vapor pressure can also be graphed against temperature in a (logarithmic) *Cox chart* when values are needed over larger temperature extremes. Although there is also some variation with external pressure, the external pressure effect is negligible under normal conditions.

Typical values of vapor pressure are given in Table 16.6.

---

[16]Use of the *viscosity index* has been adopted by other parts of the chemical process industry (CPI), including in the manufacture of solvents, polymers, and other synthetics. The 100-point scale may be exceeded (on both ends) for these uses. Refer to standard ASTM D2270 for calculating extreme values of the viscosity index.

[17]Because a liquid that vaporizes easily has an aroma, the term *aromatic liquid* is also occasionally used.

[18]The *Clausius-Clapeyron equation* and *Antoine equation* are two such logarithmic correlations of vapor pressure with temperature.

**Table 16.6** *Typical Vapor Pressures*

| fluid | lbf/ft², 68°F | kPa, 20°C |
|---|---|---|
| mercury | 0.00362 | 0.000173 |
| turpentine | 1.115 | 0.0534 |
| water | 48.9 | 2.34 |
| ethyl alcohol | 122.4 | 5.86 |
| ether | 1231 | 58.9 |
| butane | 4550 | 218 |
| Freon-12 | 12,200 | 584 |
| propane | 17,900 | 855 |
| ammonia | 18,550 | 888 |

## 14. OSMOTIC PRESSURE

*Osmosis* is a special case of diffusion in which molecules of the *solvent* move under pressure from one fluid to another (i.e., from the *solvent* to the *solution*) in one direction only, usually through a *semipermeable membrane*.[19] Osmosis continues until sufficient solvent has passed through the membrane to make the activity (or solvent pressure) of the solution equal to that of the solvent.[20] The pressure at equilibrium is known as the *osmotic pressure*, $\pi$.

Figure 16.6 illustrates an *osmotic pressure apparatus*. The fluid column can be interpreted as the result of an osmotic pressure that has developed through diffusion into the solution. The fluid column will continue to increase in height until equilibrium is reached. Alternatively, the fluid column can be adjusted so that the solution pressure just equals the osmotic pressure that would develop otherwise, in order to prevent the flow of solvent. For the arrangement in Fig. 16.6, the osmotic pressure can be calculated from the difference in fluid level heights, $h$.

$$\pi = \rho g h \qquad \text{[SI]} \qquad 16.20(a)$$

$$\pi = \frac{\rho g h}{g_c} \qquad \text{[U.S.]} \qquad 16.20(b)$$

**Figure 16.6** *Osmotic Pressure Apparatus*

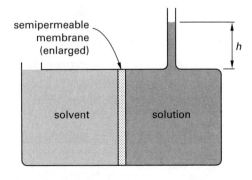

In dilute solutions, osmotic pressure obeys the ideal gas law. The solute acts like a gas in exerting pressure against the membrane. The solvent exerts no pressure since it can pass through. In Eq. 16.21, $M$ is the molarity (concentration). Consistent units must be used.

$$\pi = MR^*T \qquad 16.21$$

### Example 16.6

An aqueous solution is in isopiestic equilibrium with a 0.1 molarity sucrose solution at 22°C. What is the osmotic pressure?

*Solution*

Referring to Eq. 16.21,

$$M = 0.1 \text{ mol/L of solution}$$

$$R^* = 0.0821 \text{ atm·L/mol·K}$$

$$T = 22°C + 273° = 295\text{K}$$

$$\pi = MR^*T = \left(0.1 \frac{\text{mol}}{\text{L}}\right)\left(0.0821 \frac{\text{atm·L}}{\text{mol·K}}\right)(295\text{K})$$

$$= 2.42 \text{ atm}$$

## 15. SURFACE TENSION

The membrane or "skin" that seems to form on the free surface of a fluid is due to the intermolecular cohesive forces and is known as *surface tension*, $\sigma$. Surface tension is the reason that insects are able to sit on water and a needle is able to float on it. Surface tension also causes bubbles and droplets to take on a spherical shape, since any other shape would have more surface area per unit volume.

Data on the surface tension of liquids is important in determining the performance of heat-, mass-, and momentum-transfer equipment, including heat transfer devices.[21] Surface tension data is needed to calculate the nucleate boiling point (i.e., the initiation of boiling) of liquids in a pool (using the *Rohsenow equation*) and the maximum heat flux of boiling liquids in a pool (using the *Zuber equation*).

Surface tension can be interpreted as the tension between two points a unit distance apart on the surface or as the amount of work required to form a new unit of surface area in an apparatus similar to that shown in Fig. 16.7. Typical units of surface tension are lbf/ft (ft-lbf/ft²) dyne/cm, and N/m.

---

[19]A semipermeable membrane will be impermeable to the solute but permeable for the solvent.

[20]Two solutions in equilibrium (i.e., whose activities are equal) are said to be in *isopiestic equilibrium*.

[21]Surface tension plays a role in processes involving dispersion, emulsion, flocculation, foaming, and solubilization. It is not surprising that surface tension data are particularly important in determining the performance of equipment in the chemical process industry (CPI), such as distillation columns, packed towers, wetted-wall columns, strippers, and phase-separation equipment.

The apparatus shown in Fig. 16.7 consists of a wire frame with a sliding side that has been dipped in a liquid to form a film. Surface tension is determined by measuring the force necessary to keep the sliding side stationary against the surface tension pull of the film.[22] (The film does not act like a spring, since the force, $F$, does not increase as the film is stretched.) Since the film has two surfaces (i.e., two surface tensions), the surface tension is

$$\sigma = \frac{F}{2L} \qquad 16.22$$

**Figure 16.7** *Wire Frame for Stretching a Film*

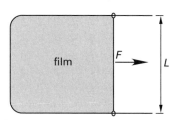

Alternatively, surface tension can also be determined by measuring the force required to pull a wire ring out of the liquid, as shown in Fig. 16.8.[23] Since the ring's inner and outer sides are in contact with the liquid, the wetted perimeter is twice the circumference. The surface tension is

$$\sigma = \frac{F}{4\pi r} \qquad 16.23$$

**Figure 16.8** *Du Nouy Ring Surface Tension Apparatus*

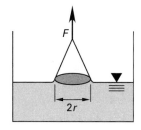

Surface tension depends slightly on the gas in contact with the free surface. Surface tension values are usually quoted for air contact. Typical values of surface tension are listed in Table 16.7.

At temperatures below freezing, the substance will be a solid, so surface tension is a moot point. As the temperature of a liquid is raised, the surface tension

**Table 16.7** *Approximate Values of Surface Tension (air contact)*

| fluid | lbf/ft, 68°F | N/m, 20°C |
|---|---|---|
| *n*-octane | 0.00149 | 0.0217 |
| ethyl alcohol | 0.00156 | 0.0227 |
| acetone | 0.00162 | 0.0236 |
| kerosene | 0.00178 | 0.0260 |
| carbon tetrachloride | 0.00185 | 0.0270 |
| turpentine | 0.00186 | 0.0271 |
| toluene | 0.00195 | 0.0285 |
| benzene | 0.00198 | 0.0289 |
| olive oil | 0.0023 | 0.034 |
| glycerin | 0.00432 | 0.0631 |
| water | 0.00499 | 0.0728 |
| mercury | 0.0356 | 0.519 |

(Multiply lbf/ft by 14.59 to obtain N/m.)
(Multiply dyne/cm by 0.001 to obtain N/m.)

decreases because the cohesive forces decrease. Surface tension is zero at a substance's critical temperature. If a substance's critical temperature is known, the *Othmer correlation*, Eq. 16.24, can be used to determine the surface tension at one temperature from the surface tension at another temperature.

$$\sigma_2 = \sigma_1 \left( \frac{T_c - T_2}{T_c - T_1} \right)^{11/9} \qquad 16.24$$

Surface tension is the reason that the pressure on the inside of bubbles and droplets is greater than on the outside. Equation 16.25 gives the relationship between the surface tension in a hollow bubble surrounded by a gas and the difference between the inside and outside pressures. For a spherical droplet or a bubble in a liquid, where in both cases there is only one surface in tension, the surface tension is twice as large. ($r$ is the radius of the bubble or droplet.)

$$\sigma_{\text{bubble}} = \frac{r(p_{\text{inside}} - p_{\text{outside}})}{4} \qquad 16.25$$

$$\sigma_{\text{droplet}} = \frac{r(p_{\text{inside}} - p_{\text{outside}})}{2} \qquad 16.26$$

## 16. CAPILLARY ACTION

*Capillary action (capillarity)* is the name given to the behavior of a liquid in a thin-bore tube. Capillary action is caused by surface tension between the liquid and a vertical solid surface.[24] In the case of liquid water in a glass tube, the adhesive forces between the liquid molecules and the surface are greater than (i.e., dominate) the cohesive forces between the water molecules themselves.[25] The adhesive forces cause the water to attach

---

[22]The force includes the weight of the sliding side wire if the frame is oriented vertically, with gravity acting on the sliding side wire to stretch the film.
[23]This apparatus is known as a *Du Nouy torsion balance*. The ring is made of platinum with a diameter of 4.00 cm.

[24]In fact, observing the rise of liquid in a capillary tube is another method of determining the surface tension of a liquid.
[25]*Adhesion* is the attractive force between molecules of different substances. *Cohesion* is the attractive force between molecules of the same substance.

itself to and climb a solid vertical surface. It can be said that the water "reaches up and tries to wet as much of the interior surface as it can." In so doing, the water rises above the general water surface level. This is illustrated in Fig. 16.9.

**Figure 16.9** *Capillarity of Liquids*

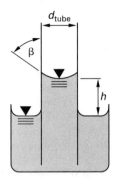 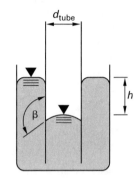

(a) adhesive force dominates     (b) cohesive force dominates

Figure 16.9 also illustrates that the same surface tension forces that keep a droplet spherical are at work on the surface of the liquid in the tube. The curved liquid surface, known as the *meniscus*, can be considered to be an incomplete droplet. If the inside diameter of the tube is less than approximately 0.1 in (2.5 mm), the meniscus is essentially hemispherical, and $r_{\text{meniscus}} = r_{\text{tube}}$.

For a few other liquids, such as mercury, the molecules have a strong affinity for each other (i.e., the cohesive forces dominate). The liquid avoids contact with the tube surface. In such liquids, the meniscus in the tube will be below the general surface level.

The *angle of contact*, $\beta$, is an indication of whether adhesive or cohesive forces dominate. For contact angles less than 90°, adhesive forces dominate. For contact angles greater than 90°, cohesive forces dominate.

Equation 16.27 can be used to predict the capillary rise in a small-bore tube. Surface tension and contact angles can be obtained from Table 16.7 and Table 16.8, respectively.

$$h = \frac{4\sigma \cos \beta}{\rho d_{\text{tube}} g} \qquad \text{[SI]} \quad 16.27(a)$$

$$h = \left(\frac{4\sigma \cos \beta}{\rho d_{\text{tube}}}\right) \times \left(\frac{g_c}{g}\right) \qquad \text{[U.S.]} \quad 16.27(b)$$

$$\sigma = \frac{h \rho d_{\text{tube}} g}{4 \cos \beta} \qquad \text{[SI]} \quad 16.28(a)$$

$$\sigma = \left(\frac{h \rho d_{\text{tube}}}{4 \cos \beta}\right) \times \left(\frac{g}{g_c}\right) \qquad \text{[U.S.]} \quad 16.28(b)$$

$$r_{\text{meniscus}} = \frac{d_{\text{tube}}}{2 \cos \beta} \qquad 16.29$$

**Table 16.8** *Contact Angles, $\beta$*

| materials | angle |
|---|---|
| mercury–glass | 140° |
| water–paraffin | 107° |
| water–silver | 90° |
| kerosene–glass | 26° |
| glycerin–glass | 19° |
| water–glass | 0° |
| ethyl alcohol–glass | 0° |

If it is assumed that the meniscus is hemispherical, then $r_{\text{meniscus}} = r_{\text{tube}}$, $\beta = 0°$, and $\cos \beta = 1.0$, and the above equations can be simplified. (Such an assumption can only be made when the diameter of the capillary tube is less than 0.1 in.)

### Example 16.7

To what height will 68°F (20°C) ethyl alcohol rise in a 0.005 in (0.127 mm) internal diameter glass capillary tube? The density of the alcohol is 49 lbm/ft³ (790 kg/m³).

*SI Solution*

$$\sigma = 0.0227 \text{ N/m}$$

$$\beta = 0°$$

The acceleration due to gravity is 9.81 m/s². From Eq. 16.27, the height is

$$h = \frac{4\sigma \cos \beta}{\rho d_{\text{tube}} g}$$

$$= \frac{(4)\left(0.0227 \frac{\text{N}}{\text{m}}\right)(1.0)\left(1000 \frac{\text{mm}}{\text{m}}\right)}{\left(790 \frac{\text{kg}}{\text{m}^3}\right)(0.127 \text{ mm})\left(9.81 \frac{\text{m}}{\text{sec}^2}\right)}$$

$$= 0.0923 \text{ m}$$

*Customary U.S. Solution*

From Table 16.7 and Table 16.8, respectively, the surface tension and contact angle are

$$\sigma = 0.00156 \text{ lbf/ft}$$

$$\beta = 0°$$

The acceleration due to gravity is 32.2 ft/sec$^2$. From Eq. 16.27, the height is

$$h = \frac{4\sigma \cos\beta g_c}{\rho d_{\text{tube}} g}$$

$$= \frac{(4)\left(0.00156 \frac{\text{lbf}}{\text{ft}}\right)(1.0)\left(32.2 \frac{\text{ft-lbm}}{\text{lbf-sec}^2}\right)\left(12 \frac{\text{in}}{\text{ft}}\right)}{\left(49 \frac{\text{lbm}}{\text{ft}^3}\right)(0.005 \text{ in})\left(32.2 \frac{\text{ft}}{\text{sec}^2}\right)}$$

$$= 0.306 \text{ ft}$$

## 17. COMPRESSIBILITY[26]

*Compressibility* (also known as the *coefficient of compressibility*), $\beta$, is the fractional change in the volume of a fluid per unit change in pressure in a constant-temperature process.[27] Typical units are 1/psi, 1/psf, 1/atm, and 1/kPa. (See Table 16.9.) It is the reciprocal of the bulk modulus, a quantity that is more commonly tabulated than compressibility. Equation 16.30 is written with a negative sign to show that volume decreases as pressure increases.

$$\beta = \frac{-\frac{\Delta V}{V_0}}{\Delta p} = \frac{1}{E} \qquad 16.30$$

Compressibility can also be written in terms of partial derivatives.

$$\beta = \left(\frac{-1}{V_0}\right)\left(\frac{\partial V}{\partial p}\right)_T = \left(\frac{1}{\rho_0}\right)\left(\frac{\partial \rho}{\partial p}\right)_T \qquad 16.31$$

Compressibility changes only slightly with temperature. The small compressibility of liquids is typically considered to be insignificant, giving rise to the common understanding that liquids are incompressible.

The density of a compressible fluid depends on the fluid's pressure. For small changes in pressure, the density at one pressure can be calculated from the density at another pressure from Eq. 16.32.

$$p_2 \approx p_1\left(1 + \beta(p_2 - p_1)\right) \qquad 16.32$$

Gases, of course, are easily compressed. The compressibility of an ideal gas depends on its pressure, $p$, its ratio of specific heats, $k$, and the nature of the process.[28]

---

[26]Compressibility should not be confused with the *thermal coefficient of expansion*, $(1/V_0)(\partial V/\partial T)_p$, which is the fractional change in volume per unit temperature change in a constant-pressure process (with units of 1/°F or 1/°C), or the dimensionless *compressibility factor*, $Z$, used with the ideal gas law.
[27]Other symbols used for compressibility are $c$, $C$, and $K$.
[28]For air, $k = 1.4$.

**Table 16.9** *Approximate Compressibility of Common Liquids at 1 atm*

| liquid | temperature | $\beta$, 1/psi | $\beta$, 1/atm |
|---|---|---|---|
| mercury | 32°F | $0.027 \times 10^{-5}$ | $0.39 \times 10^{-5}$ |
| glycerin | 60°F | $0.16 \times 10^{-5}$ | $2.4 \times 10^{-5}$ |
| water | 60°F | $0.33 \times 10^{-5}$ | $4.9 \times 10^{-5}$ |
| ethyl alcohol | 32°F | $0.68 \times 10^{-5}$ | $10 \times 10^{-5}$ |
| chloroform | 32°F | $0.68 \times 10^{-5}$ | $10 \times 10^{-5}$ |
| gasoline | 60°F | $1.0 \times 10^{-5}$ | $15 \times 10^{-5}$ |
| hydrogen | 20K | $11 \times 10^{-5}$ | $160 \times 10^{-5}$ |
| helium | 2.1K | $48 \times 10^{-5}$ | $700 \times 10^{-5}$ |

(Multiply 1/psi by 0.14504 to obtain 1/kPa.)
(Multiply 1/psi by 14.696 to obtain 1/atm.)

Depending on the process, the compressibility may be known as *isothermal compressibility* or *(adiabatic) isentropic compressibility*. Of course, compressibility is zero for constant-volume processes and is infinite (or undefined) for constant-pressure processes.

$$\beta_T = \frac{1}{p} \quad \text{[isothermal ideal gas processes]} \qquad 16.33$$

$$\beta_s = \frac{1}{kp} \quad \text{[adiabatic ideal gas processes]} \qquad 16.34$$

### Example 16.8

Water at 68°F (20°C) and 1 atm has a density of 62.3 lbm/ft$^3$ (997 kg/m$^3$). What is the new density if the pressure is isothermally increased from 14.7 psi (100 kPa) to 400 psi (2760 kPa)? Assume that the bulk modulus has a constant value of 320,000 psi ($2.2 \times 10^6$ kPa).

*SI Solution*

Compressibility is the reciprocal of the bulk modulus.

$$\beta = \frac{1}{E} = \frac{1}{2.2 \times 10^6 \text{ kPa}} = 4.55 \times 10^{-7} \text{ 1/kPa}$$

All other information needed to use Eq. 16.32 is provided.

$$\rho_2 = \rho_1\left(1 + \beta(p_2 - p_1)\right)$$

$$= \left(997 \frac{\text{kg}}{\text{m}^3}\right)\left(1 + \left(4.55 \times 10^{-7}\frac{1}{\text{kPa}}\right) \times (2760 \text{ kPa} - 100 \text{ kPa})\right)$$

$$= 998.2 \text{ kg/m}^3$$

*Customary U.S. Solution*

Compressibility is the reciprocal of the bulk modulus.

$$\beta = \frac{1}{E} = \frac{1}{320{,}000 \frac{\text{lbf}}{\text{in}^2}} = 0.3125 \times 10^{-5} \text{ in}^2/\text{lbf}$$

All other information needed to use Eq. 16.32 is provided.

$$\rho_2 = \rho_1 \left(1 + \beta(p_2 - p_1)\right)$$
$$= \left(62.3 \frac{\text{lbm}}{\text{ft}^3}\right) \begin{pmatrix} 1 + \left(0.3125 \times 10^{-5} \frac{\text{in}^2}{\text{lbf}}\right) \\ \times \left(400 \frac{\text{lbf}}{\text{in}^2} - 14.7 \frac{\text{lbf}}{\text{in}^2}\right) \end{pmatrix}$$
$$= 62.38 \text{ lbm/ft}^3$$

## 18. BULK MODULUS

The *bulk modulus, E,* of a fluid is analogous to the modulus of elasticity of a solid.[29] Typical units are psi, atm, and kPa. The term $\Delta p$ in Eq. 16.35 represents an increase in stress. The term $\Delta V / V_0$ is a *volumetric strain.* Analogous to Hooke's law describing elastic formation, the *bulk modulus* of a fluid (liquid or gas) is given by Eq. 16.35.

$$E = \frac{\text{stress}}{\text{strain}} = \frac{-\Delta p}{\frac{\Delta V}{V_0}} \qquad 16.35(a)$$

$$E = -V_0 \left(\frac{\partial p}{\partial V}\right)_T \qquad 16.35(b)$$

The term *secant bulk modulus* is associated with Eq. 16.35(a) (using finite differences), while the terms *tangent bulk modulus* and *point bulk modulus* are associated with Eq. 16.35(b) (using partial derivatives).

**Table 16.10** *Approximate Bulk Modulus of Water*

| pressure (psi) | 32°F | 68°F | 120°F | 200°F | 300°F |
|---|---|---|---|---|---|
| | (thousands of psi) | | | | |
| 15 | 292 | 320 | 332 | 308 | – |
| 1500 | 300 | 330 | 340 | 319 | 218 |
| 4500 | 317 | 348 | 362 | 338 | 271 |
| 15,000 | 380 | 410 | 420 | 405 | 350 |

(Multiply psi by 6.8948 to obtain kPa.)
Reprinted with permission from Victor L. Streeter, *Handbook of Fluid Dynamics,* © 1961, by McGraw-Hill Book Company.

The bulk modulus is the reciprocal of compressibility.

$$E = \frac{1}{\beta} \qquad 16.36$$

The bulk modulus changes only slightly with temperature. Water's bulk modulus is usually taken as 300,000 psi ($2.1 \times 10^6$ kPa) unless greater accuracy is required, in which case Table 16.10 or App. 16.A can be used.

## 19. SPEED OF SOUND

The *speed of sound (acoustical velocity* or *sonic velocity), a,* in a fluid is a function of its bulk modulus (or, equivalently, of its compressibility).[30] Equation 16.37 gives the speed of sound through a liquid.

$$a = \sqrt{\frac{E}{\rho}} = \sqrt{\frac{1}{\beta\rho}} \qquad \text{[SI]} \quad 16.37(a)$$

$$a = \sqrt{\frac{Eg_c}{\rho}} = \sqrt{\frac{g_c}{\beta\rho}} \qquad \text{[U.S.]} \quad 16.37(b)$$

Equation 16.38 gives the speed of sound in an ideal gas. The temperature, $T$, must be in degrees absolute (i.e., °R or K). For air, the ratio of specific heats is $k = 1.4$, the molecular weight is 28.967. The universal gas constant is $R^* = 1545.4$ ft-lbf/lbmol-°R (8314.57 J/kmol·K).

$$a = \sqrt{\frac{E}{\rho}} = \sqrt{\frac{kp}{\rho}}$$
$$= \sqrt{kRT} = \sqrt{\frac{kR^*T}{\text{MW}}} \qquad \text{[SI]} \quad 16.38(a)$$

$$a = \sqrt{\frac{Eg_c}{\rho}} = \sqrt{\frac{kg_c p}{\rho}}$$
$$= \sqrt{kg_c RT} = \sqrt{\frac{kg_c R^*T}{\text{MW}}} \qquad \text{[U.S.]} \quad 16.38(b)$$

Since $k$ and $R$ are constant for an ideal gas, the speed of sound is a function of temperature only. Equation 16.39 can be used to calculate the new speed of sound when temperature is varied.

$$\frac{a_1}{a_2} = \sqrt{\frac{T_1}{T_2}} \qquad 16.39$$

---

[29]To distinguish it from the modulus of elasticity, the bulk modulus is represented by the symbol $B$ when dealing with solids.

[30]The symbol $c$ is also used for the sonic velocity.

The *Mach number* of an object is the ratio of the object's speed to the speed of sound in the medium through which it is traveling. (See Table 16.11.)

$$M = \frac{v}{a} \qquad 16.40$$

The term *subsonic travel* implies $M < 1$.[31] Similarly, *supersonic travel* implies $M > 1$, but usually $M < 5$. Travel above $M = 5$ is known as *hypersonic travel*. Travel in the transition region between subsonic and supersonic (i.e., $0.8 < M < 1.2$) is known as *transonic travel*. A *sonic boom* (a shock-wave phenomenon) occurs when an object travels at supersonic speed.

**Table 16.11** Approximate Speeds of Sound (at one atmospheric pressure)

| material | speed of sound | |
|---|---|---|
| | m/s | ft/sec |
| air | 330 at 0°C | 1130 at 70°F |
| aluminum | 4990 | 16,400 |
| carbon dioxide | 260 at 0°C | 870 at 70°F |
| hydrogen | 1260 at 0°C | 3310 at 70°F |
| steel | 5150 | 16,900 |
| water | 1490 at 20°C | 4880 at 70°F |

### Example 16.9

What is the speed of sound in 150°F (66°C) water? The density is 61.2 lbm/ft$^3$ (980 kg/m$^3$), and the bulk modulus is 328,000 psi ($2.26 \times 10^6$ kPa).

*SI Solution*

$$a = \sqrt{\frac{E}{\rho}}$$

$$= \sqrt{\frac{(2.26 \times 10^6 \text{ kPa})\left(1000 \frac{\text{Pa}}{\text{kPa}}\right)}{980 \frac{\text{kg}}{\text{m}^3}}}$$

$$= 1519 \text{ m/s}$$

*Customary U.S. Solution*

From Eq. 16.37,

$$a = \sqrt{\frac{E g_c}{\rho}}$$

$$= \sqrt{\frac{\left(328,000 \frac{\text{lbf}}{\text{in}^2}\right)\left(144 \frac{\frac{\text{lbf}}{\text{ft}^2}}{\frac{\text{lbf}}{\text{in}^2}}\right)\left(32.2 \frac{\text{ft-lbm}}{\text{lbf-sec}^2}\right)}{61.2 \frac{\text{lbm}}{\text{ft}^3}}}$$

$$= 4985 \text{ ft/sec}$$

---

[31]In the language of compressible fluid flow, this is known as the *subsonic flow regime*.

### Example 16.10

What is the speed of sound in 150°F (66°C) air at standard atmospheric pressure?

*SI Solution*

$$R = \frac{8314.57 \frac{\text{J}}{\text{kmol·K}}}{28.967 \frac{\text{kg}}{\text{kmol}}} = 287.03 \text{ J/kg·K}$$

$$T = 66°C + 273° = 339\text{K}$$

$$a = \sqrt{kRT} = \sqrt{(1.4)\left(287.03 \frac{\text{J}}{\text{kg·K}}\right)(339\text{K})}$$

$$= 369 \text{ m/s}$$

*Customary U.S. Solution*

The specific gas constant, $R$, for air is

$$R = \frac{R^*}{MW} = \frac{1545.4 \frac{\text{ft-lbf}}{\text{lbmol-°R}}}{28.967 \frac{\text{lbm}}{\text{lbmol}}}$$

$$= 53.35 \text{ ft-lbf/lbm-°R}$$

The absolute temperature is

$$T = 150°F + 460° = 610°R$$

From Eq. 16.38(b),

$$a = \sqrt{k g_c R T}$$

$$= \sqrt{(1.4)\left(32.2 \frac{\text{ft-lbm}}{\text{lbf-sec}^2}\right)\left(53.35 \frac{\text{ft-lbf}}{\text{lbm-°R}}\right)(610°R)}$$

$$= 1211 \text{ ft/sec}$$

## 20. PROPERTIES OF MIXTURES OF NONREACTING FLUIDS

There are few convenient ways of predicting the properties of nonreacting, nonvolatile organic and aqueous solutions (acids, brines, alcohol mixtures, coolants, etc.) and mixtures from the individual properties of the components.

Volumes of two combining organic liquids (e.g., acetone and chloroform) are essentially additive. The volume change upon mixing will seldom be more than a few tenths of a percent. The volume change in aqueous solutions is often slightly greater, but is still limited to a few percent (e.g., 3% for some solutions of methanol and water).

Thus, the specific gravity (density, specific weight, etc.) can be considered to be a volumetric weighting of the individual specific gravities. Most times, however, the specific gravity of a known solution must be calculated from known data regarding one of the various density scales or determined through research.

Most other important fluid properties of aqueous solutions, such as viscosity, compressibility, surface tension, and vapor pressure, have been measured and are usually determined through research.[32] It is important to be aware of the operating conditions of the solution. Data for one concentration or condition should not be used for another concentration or condition.

---

[32]There is no substitute for a complete fluid properties data book.

# 17 Fluid Statics

## Nomenclature

| $a$ | acceleration | ft/sec$^2$ | m/s$^2$ |
|---|---|---|---|
| $A$ | area | ft$^2$ | m$^2$ |
| $b$ | base length | ft | m |
| $d$ | diameter | ft | m |
| $e$ | eccentricity | ft | m |
| $F$ | force | lbf | N |
| FS | factor of safety | – | – |
| $g$ | gravitational acceleration | ft/sec$^2$ | m/s$^2$ |
| $g_c$ | gravitational conversion constant (32.2) | ft-lbm/lbf-sec$^2$ | n.a. |
| $h$ | height | ft | m |
| $I$ | moment of inertia | ft$^4$ | m$^4$ |
| $J$ | polar moment of inertia | ft$^4$ | m$^4$ |
| $k$ | radius of gyration | ft | m |
| $k$ | ratio of specific heats | – | – |
| $L$ | length | ft | m |
| $m$ | mass | lbm | kg |
| $M$ | mechanical advantage | – | – |
| $M$ | moment | ft-lbf | N·m |
| $n$ | polytropic exponent | – | – |
| $N$ | normal force | lbf | N |
| $p$ | pressure | lbf/ft$^2$ | Pa |
| $r$ | radius | ft | m |
| $R$ | resultant force | lbf | N |
| $R$ | specific gas constant | ft-lbf/lbm-°R | J/kg·K |
| SG | specific gravity | – | – |
| $t$ | wall thickness | ft | m |
| $T$ | temperature | °R | K |
| v | velocity | ft/sec | m/s |
| $V$ | volume | ft$^3$ | m$^3$ |
| $W$ | weight | lbf | N |
| $x$ | distance | ft | m |
| $x$ | fraction | – | – |
| $y$ | distance | ft | m |

### Symbols

| $\gamma$ | specific weight | lbm/ft$^3$ | n.a. |
|---|---|---|---|
| $\eta$ | efficiency | – | – |
| $\theta$ | angle | deg | deg |
| $\mu$ | coefficient of friction | – | – |
| $\rho$ | density | lbm/ft$^3$ | kg/m$^3$ |
| $\sigma$ | stress | lbm/ft$^2$ | Pa |
| $\nu$ | specific volume | ft$^3$/lbm | m$^3$/kg |
| $\omega$ | angular velocity | rad/sec | rad/s |

### Subscripts

| $a$ | atmospheric |
|---|---|
| $b$ | buoyant |
| bg | between CB and CG |
| $c$ | centroidal |
| $f$ | frictional |
| $F$ | force |
| $h$ | hoop |
| $l$ | lever or longitudinal |
| $m$ | manometer fluid, mercury, or metacentric |
| $p$ | plunger |
| $r$ | ram |
| $R$ | resultant |
| $t$ | tank |
| v | vapor or vertical |
| $w$ | water |

## 1. PRESSURE-MEASURING DEVICES

There are many devices for measuring and indicating fluid pressure. Some devices measure gage pressure; others measure absolute pressure. The effects of non-standard atmospheric pressure and nonstandard gravitational acceleration must be determined, particularly for devices relying on columns of liquid to indicate pressure. Table 17.1 lists the common types of devices and the ranges of pressure appropriate for each.

Fluids

**Table 17.1** Common Pressure-Measuring Devices

| device | approximate range (in atm) |
|---|---|
| water manometer | 0–0.1 |
| mercury barometer | 0–1 |
| mercury manometer | 0.001–1 |
| metallic diaphragm | 0.01–200 |
| transducer | 0.001–700 |
| Bourdon pressure gauge | 1–3000 |
| Bourdon vacuum gauge | 0.1–1 |

The *Bourdon pressure gauge* (see Fig. 17.1) is the most common pressure-indicating device. This mechanical device consists of a coiled hollow tube that tends to straighten out (i.e., unwind) when the tube is subjected to an internal pressure. The degree to which the coiled tube unwinds depends on the difference between the internal and external pressures. A Bourdon gauge directly indicates *gage pressure*. Extreme accuracy is generally not a characteristic of Bourdon gauges.

**Figure 17.1** C-Bourdon Pressure Gauge

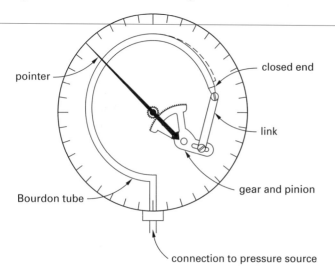

In non-SI installations, gauges are always calibrated in psi, unless the dial is marked "altitude" (measuring in feet of water) or "vacuum" (measuring in inches of mercury) on its face. To avoid confusion, the gauge dial will be clearly marked if other units are indicated.

The *barometer* is a common device for measuring the absolute pressure of the atmosphere.[1] It is constructed by filling a long tube open at one end with mercury (or alcohol, or some other liquid) and inverting the tube so that the open end is below the level of a mercury-filled container. If the vapor pressure of the mercury in the tube is neglected, the fluid column will be supported

only by the atmospheric pressure transmitted through the container fluid at the lower, open end.

*Strain gauges*, *diaphragm gauges*, *quartz-crystal transducers*, and other devices using the *piezoelectric effect* are also used to measure stress and pressure, particularly when pressure fluctuates quickly (e.g., as in a rocket combustion chamber). With these devices, calibration is required to interpret pressure from voltage generation or changes in resistance, capacitance, or inductance. These devices are generally unaffected by atmospheric pressure or gravitational acceleration.

*Manometers (U-tube manometers)* can also be used to indicate small pressure differences, and for this purpose they provide great accuracy. (Manometers are not suitable for measuring pressures much larger than 10 psi (70 kPa), however.) A difference in manometer fluid surface heights is converted into a pressure difference. If one end of a manometer is open to the atmosphere, the manometer indicates gage pressure. It is theoretically possible, but impractical, to have a manometer indicate absolute pressure, since one end of the manometer would have to be exposed to a perfect vacuum.

A *static pressure tube (piezometer tube)* is a variation of the manometer. (See Fig. 17.2.) It is a simple method of determining the static pressure in a pipe or other vessel, regardless of fluid motion in the pipe. A vertical transparent tube is connected to a hole in the pipe wall.[2] (None of the tube projects into the pipe.) The static pressure will force the contents of the pipe up into the tube. The height of the contents will be an indication of gage pressure in the pipe.

**Figure 17.2** Static Pressure Tube

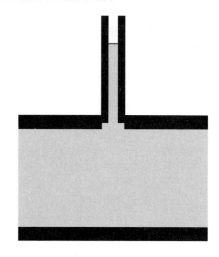

The device used to measure the pressure should not be confused with the method used to obtain exposure to the pressure. For example, a static pressure *tap* in a pipe is merely a hole in the pipe wall. A Bourdon gauge,

---

[1]A barometer can be used to measure the pressure inside any vessel. However, the barometer must be completely enclosed in the vessel, which may not be possible. Also, it is difficult to read a barometer enclosed within a tank.

[2]Where greater accuracy is required, multiple holes may be drilled around the circumference of the pipe and connected through a manifold (*piezometer ring*) to the pressure-measuring device.

manometer, or transducer can then be used with the tap to indicate pressure.

Tap holes are generally $^1/_8$ in to $^1/_4$ in in diameter, drilled at right angles to the wall, and smooth and flush with the pipe wall. No part of the gauge or connection projects into the pipe. The tap holes should be at least 5 to 10 pipe diameters downstream from any source of turbulence (e.g., a bend, fitting, or valve).

## 2. MANOMETERS

Figure 17.3 illustrates a simple U-tube manometer used to measure the difference in pressure between two vessels. When both ends of the manometer are connected to pressure sources, the name *differential manometer* is used. If one end of the manometer is open to the atmosphere, the name *open manometer* is used.[3] The open manometer implicitly measures gage pressures.

**Figure 17.3** Simple U-Tube Manometer

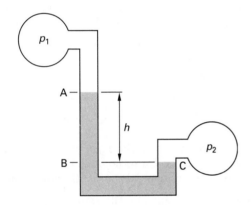

Since the pressure at point B in Fig. 17.3 is the same as at point C, the pressure differential produces the vertical fluid column of height $h$. In Eq. 17.2, $A$ is the area of the tube. Equation 17.2 and Eq. 17.3 assume consistent density units. In the absence of any capillary action, the inside diameters of the manometer tubes are irrelevant.

$$F_{\text{net}} = \text{weight of fluid column} \qquad 17.1$$

$$(p_2 - p_1) \times A = \rho_m g h \times A \qquad 17.2$$

$$p_2 - p_1 = \rho_m g h \qquad \text{[SI]} \quad 17.3$$

In countries that do not use SI units, densities are commonly quoted in pounds per cubic foot. In that case, Eq. 17.3 can be written as

$$p_2 - p_1 = \rho_m \times \frac{g}{g_c} \times h$$
$$= \gamma_m h \qquad \text{[U.S.]} \quad 17.4$$

The quantity $g/g_c$ has a value of 1.0 lbf/lbm in almost all cases, and thus $\gamma_m$ is numerically equal to $\rho_m$, with units of lbf/ft$^3$.

Equation 17.3 and Eq. 17.4 assume that the manometer fluid height is small, or that only low-density gases fill the tubes above the manometer fluid. If a high-density fluid (such as water) is present above the measuring fluid, or if the columns $h_1$ or $h_2$ are very long, corrections will be necessary. (See Fig. 17.4.)

**Figure 17.4** Manometer Requiring Corrections

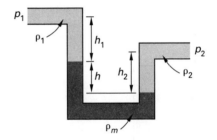

Fluid column $h_2$ "sits on top" of the manometer fluid, forcing the manometer fluid to the left. This increase must be subtracted out. Similarly, the column $h_1$ restricts the movement of the manometer fluid. The observed measurement must be increased to correct for this restriction.

$$p_2 - p_1 = g(\rho_m h + \rho_1 h_1 - \rho_2 h_2) \qquad \text{[SI]} \quad 17.5(a)$$

$$p_2 - p_1 = \frac{g}{g_c} \times (\rho_m h + \rho_1 h_1 - \rho_2 h_2)$$
$$= \gamma_m h + \gamma_1 h_1 - \gamma_2 h_2 \qquad \text{[U.S.]} \quad 17.5(b)$$

When a manometer is used to measure the pressure difference across an orifice or other fitting where the same liquid exists in both manometer sides (as in Fig. 17.5), it is not necessary to correct the manometer reading for all of the liquid present above the manometer fluid. This is because parts of the correction for both sides of the manometer are the same. Thus, the distance $y$ in Fig. 17.5 is an irrelevant distance.

Manometer tubes are generally large enough in diameter to avoid significant capillary effects. Corrections for capillarity are seldom necessary.

---

[3]If one of the manometer legs is inclined, the term *inclined manometer* or *draft gauge* is used. Although only the vertical distance between the manometer fluid surfaces should be used to calculate the pressure difference, with small pressure differences it may be more accurate to read the inclined distance (which is larger than the vertical distance) and compute the vertical distance from the angle of inclination.

**Figure 17.5** *Irrelevant Distance*

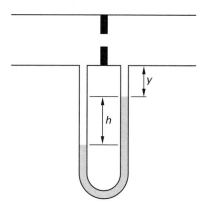

### Example 17.1

The pressure at the bottom of a water tank (62.4 lbm/ft$^3$; $\rho = 998$ kg/m$^3$) is measured with a mercury manometer. (The density of mercury is 848 lbm/ft$^3$; 13 575 kg/m$^3$.) What is the gage pressure at the bottom of the water tank?

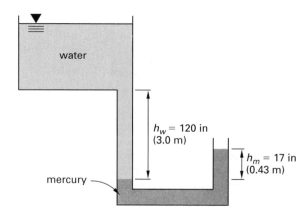

*SI Solution*

From Eq. 17.5(a),

$$\Delta p = g(\rho_m h_m - \rho_w h_w)$$

$$= \left(9.81 \ \frac{\text{m}}{\text{s}^2}\right) \left( \begin{array}{l} \left(13\,575 \ \frac{\text{kg}}{\text{m}^3}\right)(0.43 \ \text{m}) \\ - \left(998 \ \frac{\text{kg}}{\text{m}^3}\right)(3.0 \ \text{m}) \end{array} \right)$$

$$= 27\,892 \ \text{Pa} \ (27.9 \ \text{kPa gage})$$

*Customary U.S. Solution*

From Eq. 17.5(b),

$$\Delta p = \gamma_m h_m - \gamma_w h_w$$

$$= \frac{\left(848 \ \frac{\text{lbf}}{\text{ft}^3}\right)(17 \ \text{in}) - \left(62.4 \ \frac{\text{lbf}}{\text{ft}^3}\right)(120 \ \text{in})}{\left(12 \ \frac{\text{in}}{\text{ft}}\right)^3}$$

$$= 4.01 \ \text{lbf/in}^2 \ \text{(psig)}$$

## 3. HYDROSTATIC PRESSURE

*Hydrostatic pressure* is the pressure a fluid exerts on an immersed object or container walls.[4] Pressure is equal to the force per unit area of surface.

$$p = \frac{F}{A} \qquad \qquad 17.6$$

Hydrostatic pressure in a stationary, incompressible fluid behaves according to the following characteristics.

- Pressure is a function of vertical depth (and density) only. The pressure will be the same at two points with identical depths.

- Pressure varies linearly with (vertical) depth.

- Pressure is independent of an object's area and size and the weight (mass) of water above the object. Figure 17.6 illustrates the *hydrostatic paradox*. The pressures at depth $h$ are the same in all four columns because pressure depends on depth, not volume.

- Pressure at a point has the same magnitude in all directions (*Pascal's law*). Thus, pressure is a scalar quantity.

- Pressure is always normal to a surface, regardless of the surface's shape or orientation. (This is a result of the fluid's inability to support shear stress.)

- The resultant of the pressure distribution acts through the *center of pressure*.

**Figure 17.6** *Hydrostatic Paradox*

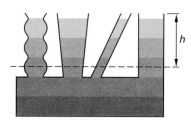

---

[4]The term *hydrostatic* is used with all fluids, not only with water.

## 4. FLUID HEIGHT EQUIVALENT TO PRESSURE

Pressure varies linearly with depth. The relationship between pressure and depth (i.e., the *hydrostatic head*) for an incompressible fluid is given by Eq. 17.7.

$$p = \rho g h \qquad \text{[SI]} \quad 17.7(a)$$

$$p = \frac{\rho g h}{g_c} = \gamma h \qquad \text{[U.S.]} \quad 17.7(b)$$

Since $\rho$ and $g$ are constants, Eq. 17.7 shows that $p$ and $h$ are linearly related. Knowing one determines the other.[5] For example, the height of a fluid column needed to produce a pressure is

$$h = \frac{p}{\rho g} \qquad \text{[SI]} \quad 17.8(a)$$

$$h = \frac{p g_c}{\rho g} = \frac{p}{\gamma} \qquad \text{[U.S.]} \quad 17.8(b)$$

Table 17.2 lists six important fluid height equivalents that many engineers commit to memory.[6]

*Table 17.2* Approximate Fluid Height Equivalents at 68°F (20°C)

| liquid | height equivalents | |
|---|---|---|
| water | 0.0361 psi/in | 27.70 in/psi |
| water | 62.4 psf/ft | 0.01603 ft/psf |
| water | 9.81 kPa/m | 0.1019 m/kPa |
| water | 0.4329 psi/ft | 2.31 ft/psi |
| mercury | 0.491 psi/in | 2.036 in/psi |
| mercury | 133.3 kPa/m | 0.00750 m/kPa |

A barometer is an example of the measurement of pressure by the height of a fluid column. If the vapor pressure of the barometer liquid is neglected, the atmospheric pressure will be given by Eq. 17.9.

$$p_a = \rho g h \qquad \text{[SI]} \quad 17.9(a)$$

$$p_a = \frac{\rho g h}{g_c} = \gamma h \qquad \text{[U.S.]} \quad 17.9(b)$$

If the vapor pressure of the barometer liquid is significant (as it would be with alcohol or water), the vapor pressure effectively reduces the height of the fluid column, as Eq. 17.10 illustrates.

$$p_a - p_v = \rho g h \qquad \text{[SI]} \quad 17.10(a)$$

$$p_a - p_v = \frac{\rho g h}{g_c} = \gamma h \qquad \text{[U.S.]} \quad 17.10(b)$$

---

[5]In fact, pressure and height of a fluid column can be used interchangeably. The height of a fluid column is known as *head*. For example: "The fan developed a static head of 3 inches of water," or "The pressure head at the base of the water tank was 8 meters." When the term "head" is used, it is essential to specify the fluid.

[6]Of course, these values are recognized to be the approximate specific weights of the liquids.

## Example 17.2

A vacuum pump is used to drain a flooded mine shaft of 68°F (20°C) water.[7] The vapor pressure of water at this temperature is 0.34 psi (2.34 kPa). The pump is incapable of lifting the water higher than 400 in (10.16 m). What is the atmospheric pressure?

*SI Solution*

From Eq. 17.10,

$$p_a = p_v + \rho g h$$

$$= 2.34 \text{ kPa} + \frac{\left(998 \frac{\text{kg}}{\text{m}^3}\right)\left(9.81 \frac{\text{m}}{\text{s}^2}\right)(10.16 \text{ m})}{1000 \frac{\text{Pa}}{\text{kPa}}}$$

$$= 101.8 \text{ kPa}$$

*(alternate SI solution, using Table 17.2)*

$$p_a = p_v + \rho g h$$

$$= 2.34 \text{ kPa} + \left(9.81 \frac{\text{kPa}}{\text{m}}\right)(10.16 \text{ m})$$

$$= 102 \text{ kPa}$$

*Customary U.S. Solution*

From Table 17.2, the height equivalent of water is approximately 0.0361 psi/in. Notice that psi/in is the same as $\text{lbf/in}^3$, the units of $\gamma$. From Eq. 17.10, the atmospheric pressure is

$$p_a = p_v + \rho g h = p_v + \gamma h$$

$$= 0.34 \frac{\text{lbf}}{\text{in}^2} + \left(0.0361 \frac{\text{lbf}}{\text{in}^3}\right)(400 \text{ in})$$

$$= 14.78 \text{ lbf/in}^2 \quad \text{(psia)}$$

## 5. MULTIFLUID BAROMETERS

It is theoretically possible to fill a barometer tube with several different immiscible fluids.[8] Upon inversion, the fluids will separate, leaving the most dense fluid at the bottom and the least dense fluid at the top. All of the fluids will contribute, by superposition, to the balance between the external atmospheric pressure and the weight of the fluid column.

$$p_a - p_v = g \sum \rho_i h_i \qquad \text{[SI]} \quad 17.11(a)$$

$$p_a - p_v = \frac{g}{g_c} \sum \rho_i h_i = \sum \gamma_i h_i \qquad \text{[U.S.]} \quad 17.11(b)$$

---

[7]A reciprocating or other direct-displacement pump would be a better choice to drain a mine.

[8]In practice, barometers are never constructed this way. This theory is more applicable to a category of problems dealing with up-ended containers, as illustrated in Ex. 17.3.

The pressure at any intermediate point within the fluid column is found by starting at a location where the pressure is known, and then adding or subtracting $\rho gh$ to get to the point where the pressure is needed. Usually, the known pressure will be the atmospheric pressure located in the barometer barrel at the level (elevation) of the fluid outside of the barometer.

### Example 17.3

Neglecting vapor pressure, what is the pressure of the air (at point E) in the container shown? The external pressure is 1.0 atm.

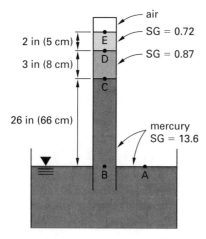

*SI Solution*

$$p_E = p_{atm} - g\rho_{water}\sum(SG_i)h_i$$

$$= 101\,300 \text{ Pa} - \left(9.81 \frac{m}{s^2}\right)\left(1000 \frac{kg}{m^3}\right)$$

$$\times \left(\begin{array}{c}(0.66 \text{ m})(13.6) + (0.08 \text{ m})(0.87) \\ + (0.05 \text{ m})(0.72)\end{array}\right)$$

$$= 12\,210 \text{ Pa} \quad (12.2 \text{ kPa})$$

*Customary U.S. Solution*

The pressure at point B is the same as the pressure at point A—1.0 atm. The density of mercury is $13.6 \times 0.0361$ lbm/in³ = 0.491 lbm/in³. The pressure at point C is

$$p_C = 14.7 \text{ psia} - (26 \text{ in})\left(0.491 \frac{\frac{\text{lbf}}{\text{in}^2}}{\text{in}}\right)$$

$$= 1.93 \text{ lbf/in}^2 \quad (\text{psia})$$

Similarly, the pressure at point E (and anywhere within the captive air space) is

$$p_E = 14.7 \frac{\text{lbf}}{\text{in}^2} - (26 \text{ in})\left(0.491 \frac{\text{lbf}}{\text{in}^3}\right)$$

$$- (3 \text{ in})(0.87)\left(0.0361 \frac{\text{lbf}}{\text{in}^3}\right)$$

$$- (2 \text{ in})(0.72)\left(0.0361 \frac{\text{lbf}}{\text{in}^3}\right)$$

$$= 1.79 \text{ lbf/in}^2 \quad (\text{psia})$$

## 6. PRESSURE ON A HORIZONTAL PLANE SURFACE

The pressure on a horizontal plane surface is uniform over the surface because the depth of the fluid is uniform. (See Fig. 17.7.) The resultant of the pressure distribution acts through the center of pressure of the surface, which corresponds to the centroid of the surface.

*Figure 17.7* Hydrostatic Pressure on a Horizontal Plane Surface

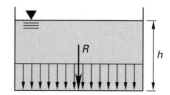

The uniform pressure at depth $h$ is given by Eq. 17.12.[9]

$$p = \rho gh \qquad\qquad \text{[SI]} \quad 17.12(a)$$

$$p = \frac{\rho gh}{g_c} = \gamma h \qquad \text{[U.S.]} \quad 17.12(b)$$

The total vertical force on the horizontal plane of area $A$ is given by Eq. 17.13.

$$R = pA \qquad\qquad 17.13$$

It is tempting, but not always correct, to calculate the vertical force on a submerged surface as the weight of the fluid above it. Such an approach works only when there is no change in the cross-sectional area of the fluid above the surface. This is a direct result of the *hydrostatic paradox*. (See Sec. 17.3.) Figure 17.8 illustrates two containers with the same pressure distribution (force) on their bottom surfaces.

---

[9]The phrase *pressure at a depth* is universally understood to mean the *gage pressure*, as given by Eq. 17.12.

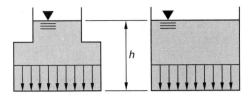

*Figure 17.8* Two Containers with the Same Pressure Distribution

## 7. PRESSURE ON A RECTANGULAR VERTICAL PLANE SURFACE

The pressure on a vertical rectangular plane surface increases linearly with depth. The pressure distribution will be triangular, as in Fig. 17.9(a), if the plane surface extends to the surface; otherwise, the distribution will be trapezoidal, as in Fig. 17.9(b).

*Figure 17.9* Hydrostatic Pressure on a Vertical Plane Surface

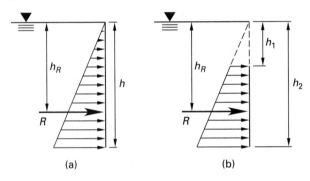

(a) (b)

The resultant force is calculated from the *average pressure.*

$$\overline{p} = \tfrac{1}{2}(p_1 + p_2) \qquad 17.14$$

$$\overline{p} = \tfrac{1}{2}\rho g(h_1 + h_2) \qquad [\text{SI}] \quad 17.15(a)$$

$$\overline{p} = \frac{\tfrac{1}{2}\rho g(h_1 + h_2)}{g_c} = \tfrac{1}{2}\gamma(h_1 + h_2) \quad [\text{U.S.}] \quad 17.15(b)$$

$$R = \overline{p}A \qquad 17.16$$

Although the resultant is calculated from the average depth, it does not act at the average depth. The resultant of the pressure distribution passes through the centroid of the pressure distribution. For the triangular distribution of Fig. 17.9(a), the resultant is located at a depth of $h_R = {}^2\!/_3 h$. For the more general case of Fig. 17.9(b), the resultant is located from Eq. 17.17.

$$h_R = \tfrac{2}{3}\left(h_1 + h_2 - \frac{h_1 h_2}{h_1 + h_2}\right) \qquad 17.17$$

## 8. PRESSURE ON A RECTANGULAR INCLINED PLANE SURFACE

The average pressure and resultant force on an inclined rectangular plane surface (see Fig. 17.10) are calculated in much the same fashion as for the vertical plane surface. The pressure varies linearly with depth. The resultant is calculated from the average pressure, which, in turn, depends on the average depth.

*Figure 17.10* Hydrostatic Pressure on an Inclined Rectangular Plane Surface

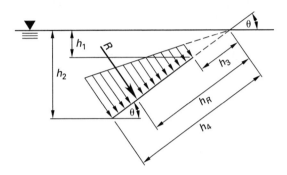

The average pressure and resultant are given by Eq. 17.18 through Eq. 17.20.

$$\overline{p} = \tfrac{1}{2}(p_1 + p_2) \qquad 17.18$$

$$\overline{p} = \tfrac{1}{2}\rho g(h_1 + h_2) = \tfrac{1}{2}\rho g(h_3 + h_4)\sin\theta \qquad [\text{SI}] \quad 17.19(a)$$

$$\overline{p} = \frac{\tfrac{1}{2}\rho g(h_1 + h_2)}{g_c} = \frac{\tfrac{1}{2}\rho g(h_3 + h_4)\sin\theta}{g_c}$$
$$= \tfrac{1}{2}\gamma(h_1 + h_2) = \tfrac{1}{2}\gamma(h_3 + h_4)\sin\theta \quad [\text{U.S.}] \quad 17.19(b)$$

$$R = \overline{p}A \qquad 17.20$$

As with the vertical plane surface, the resultant acts at the centroid of the pressure distribution, not at the average depth. Equation 17.17 is rewritten in terms of inclined depths.[10]

$$h_R = \left(\frac{\tfrac{2}{3}}{\sin\theta}\right)\left(h_1 + h_2 - \frac{h_1 h_2}{h_1 + h_2}\right)$$
$$= \tfrac{2}{3}\left(h_3 + h_4 - \frac{h_3 h_4}{h_3 + h_4}\right) \qquad 17.21$$

### Example 17.4

The tank shown is filled with water ($\rho = 62.4$ lbm/ft³; $\rho = 1000$ kg/m³). (a) What is the total force on a 1 ft

[10]Notice that $h_R$ is an inclined distance. If a vertical distance is wanted, it must usually be calculated from $h_R$ and $\sin\theta$. Equation 17.21 can be derived simply by dividing Eq. 17.17 by $\sin\theta$.

(1 m) width of the inclined portion of the wall?[11] (b) At what depth (vertical distance) is the resultant force located?

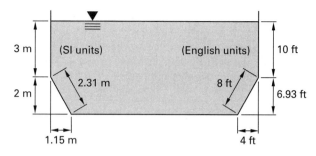

*SI Solution*

(a) The depth of the tank bottom is

$$h_2 = 3 \text{ m} + 2 \text{ m} = 5 \text{ m}$$

From Eq. 17.21, the average gage pressure on the inclined section is

$$\bar{p} = \left(\tfrac{1}{2}\right)\left(1000 \ \frac{\text{kg}}{\text{m}^3}\right)\left(9.81 \ \frac{\text{m}}{\text{s}^2}\right)(3 \text{ m} + 5 \text{ m})$$

$$= 39\,240 \text{ Pa} \quad \text{(gage)}$$

The total force on a 1 m section of wall is

$$R = \bar{p}A = (39\,240 \text{ Pa})(2.31 \text{ m})(1 \text{ m})$$

$$= 90\,644 \text{ N} \quad (90.6 \text{ kN})$$

(b) $\theta$ must be known to determine $h_R$.

$$\theta = \arctan\left(\frac{2 \text{ m}}{1.15 \text{ m}}\right) = 60°$$

The location of the resultant can be calculated from Eq. 17.21 once $h_3$ and $h_4$ are known.

$$h_3 = \frac{3 \text{ m}}{\sin 60°} = 3.464 \text{ m}$$

$$h_4 = \frac{5 \text{ m}}{\sin 60°} = 5.774 \text{ m}$$

$$h_R = \left(\tfrac{2}{3}\right)\left(3.464 \text{ m} + 5.774 \text{ m} - \frac{(3.464 \text{ m})(5.774 \text{ m})}{3.464 \text{ m} + 5.774 \text{ m}}\right)$$

$$= 4.715 \text{ m} \quad \text{[inclined]}$$

The vertical depth at which the resultant acts is

$$h = h_R \sin \theta = (4.715 \text{ m})(\sin 60°)$$

$$= 4.08 \text{ m} \quad \text{[vertical]}$$

---

[11]Since the width of the tank (the distance into and out of the illustration) is unknown, it is common to calculate the pressure or force per unit width of tank wall. This is the same as calculating the pressure on a 1 ft (1 m) wide section of wall.

*Customary U.S. Solution*

(a) The water density is given in traditional U.S. mass units. The specific weight, $\gamma$, is

$$\gamma = \frac{\rho g}{g_c} = \frac{\left(62.4 \ \frac{\text{lbm}}{\text{ft}^3}\right)\left(32.2 \ \frac{\text{ft}}{\text{sec}^2}\right)}{32.2 \ \frac{\text{lbm-ft}}{\text{lbf-sec}^2}}$$

$$= 62.4 \text{ lbf/ft}^3$$

The depth of the tank bottom is

$$h_2 = 10 \text{ ft} + 6.93 \text{ ft} = 16.93 \text{ ft}$$

From Eq. 17.21, the average gage pressure on the inclined section is

$$\bar{p} = \left(\tfrac{1}{2}\right)\left(62.4 \ \frac{\text{lbf}}{\text{ft}^3}\right)(10 \text{ ft} + 16.93 \text{ ft})$$

$$= 840.2 \text{ lbf/ft}^2 \quad \text{(gage)}$$

The total force on a 1 ft section of wall is

$$R = \bar{p}A = \left(840.2 \ \frac{\text{lbf}}{\text{ft}^2}\right)(8 \text{ ft})(1 \text{ ft})$$

$$= 6722 \text{ lbf}$$

(b) $\theta$ must be known to determine $h_R$.

$$\theta = \arctan\left(\frac{6.93 \text{ ft}}{4 \text{ ft}}\right) = 60°$$

The location of the resultant can be calculated from Eq. 17.21 once $h_3$ and $h_4$ are known.

$$h_3 = \frac{10 \text{ ft}}{\sin 60°} = 11.55 \text{ ft}$$

$$h_4 = \frac{16.93 \text{ ft}}{\sin 60°} = 19.55 \text{ ft}$$

From Eq. 17.21,

$$h_R = \left(\tfrac{2}{3}\right)\left(11.55 \text{ ft} + 19.55 \text{ ft} - \frac{(11.55 \text{ ft})(19.55 \text{ ft})}{11.55 \text{ ft} + 19.55 \text{ ft}}\right)$$

$$= 15.89 \text{ ft} \quad \text{[inclined]}$$

The vertical depth at which the resultant acts is

$$h = h_R \sin \theta = (15.89 \text{ ft})(\sin 60°)$$

$$= 13.76 \text{ ft} \quad \text{[vertical]}$$

## 9. PRESSURE ON A GENERAL PLANE SURFACE

Figure 17.11 illustrates a nonrectangular plane surface that may or may not extend to the liquid surface and that may or may not be inclined, as shown in Fig. 17.10.

**Figure 17.11** General Plane Surface

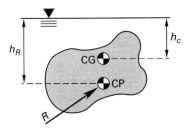

As with other regular surfaces, the resultant force depends on the average pressure and acts through the *center of pressure* (CP). The average pressure is calculated from the depth of the surface's centroid (center of gravity, CG).

$$\overline{p} = \rho g h_c \sin\theta \qquad \text{[SI]} \quad 17.22(a)$$

$$\overline{p} = \frac{\rho g h_c \sin\theta}{g_c} = \gamma h_c \sin\theta \qquad \text{[U.S.]} \quad 17.22(b)$$

$$R = \overline{p}A \qquad 17.23$$

The resultant force acts at depth $h_R$ normal to the plane surface. $I_c$ in Eq. 17.24 is the centroidal area moment of inertia, with dimensions of $L^4$ (length$^4$) about an axis parallel to the surface. Both $h_c$ and $h_R$ are measured parallel to the plane surface. That is, if the plane surface is inclined, $h_c$ and $h_R$ are inclined distances.

$$h_R = h_c + \frac{I_c}{A h_c} \qquad 17.24$$

**Example 17.5**

The top edge of a vertical circular observation window in a submarine is located 4.0 ft (1.25 m) below the surface of the water. The window is 1.0 ft (0.3 m) in diameter. The water's density is 62.4 lbm/ft$^3$ (1000 kg/m$^3$). Neglect the salinity of the water. (a) What is the resultant force on the window? (b) At what depth does the resultant force act?

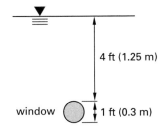

*SI Solution*

(a) The radius of the window is

$$r = \frac{0.3 \text{ m}}{2} = 0.15 \text{ m}$$

$$h_c = 1.25 \text{ m} + 0.15 \text{ m} = 1.4 \text{ m}$$

$$A = \pi(0.15 \text{ m})^2 = 0.0707 \text{ m}^2$$

$$\overline{p} = \rho g h_c = \left(1000 \ \frac{\text{kg}}{\text{m}^3}\right)\left(9.81 \ \frac{\text{m}}{\text{s}^2}\right)(1.4 \text{ m})$$
$$= 13\,734 \text{ Pa} \quad \text{(gage)}$$

$$R = \overline{p}A = (13\,734 \text{ Pa})(0.0707 \text{ m}^2) = 971 \text{ N}$$

(b)

$$I_c = \frac{\pi}{4}r^4 = \left(\frac{\pi}{4}\right)(0.15 \text{ m})^4 = 3.976 \times 10^{-4} \text{ m}^4$$

From Eq. 17.24,

$$h_R = 1.4 \text{ m} + \frac{3.976 \times 10^{-4} \text{ m}^4}{(0.0707 \text{ m}^2)(1.4 \text{ m})}$$
$$= 1.404 \text{ m}$$

*Customary U.S. Solution*

(a) The radius of the window is

$$r = \frac{1 \text{ ft}}{2} = 0.5 \text{ ft}$$

The depth at which the centroid of the circular window is located is

$$h_c = 4.0 \text{ ft} + 0.5 \text{ ft} = 4.5 \text{ ft}$$

The area of the circular window is

$$A = \pi r^2 = \pi(0.5 \text{ ft})^2$$
$$= 0.7854 \text{ ft}^2$$

As in Ex. 17.4, the resultant force will be calculated using $\gamma$ as the weight density.

The average pressure is

$$\overline{p} = \gamma h_c = \left(62.4 \ \frac{\text{lbf}}{\text{ft}^3}\right)(4.5 \text{ ft})$$
$$= 280.8 \text{ lbf/ft}^2 \quad \text{(psfg)}$$

The resultant is calculated from Eq. 17.20.

$$R = \overline{p}A = \left(280.8 \ \frac{\text{lbf}}{\text{ft}^2}\right)(0.7854 \text{ ft}^2)$$
$$= 220.5 \text{ lbf}$$

(b) The centroidal area moment of inertia of a circle is

$$I_c = \frac{\pi}{4}r^4 = \left(\frac{\pi}{4}\right)(0.5 \text{ ft})^4$$
$$= 0.049 \text{ ft}^4$$

From Eq. 17.24, the depth at which the resultant force acts is

$$h_R = 4.5 \text{ ft} + \frac{0.049 \text{ ft}^4}{(0.7854 \text{ ft}^2)(4.5 \text{ ft})}$$

$$= 4.514 \text{ ft}$$

## 10. SPECIAL CASES: VERTICAL SURFACES

Several simple wall shapes and configurations recur frequently. Figure 17.12 indicates the location of their hydrostatic pressure resultants (*centers of pressure*). In all cases, the surfaces are vertical and extend to the liquid's surface.

*Figure 17.12 Centers of Pressure for Common Configurations*

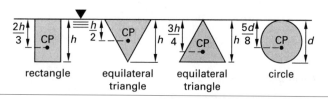

rectangle   equilateral triangle   equilateral triangle   circle

## 11. FORCES ON CURVED AND COMPOUND SURFACES

Figure 17.13 illustrates a curved surface, BA. The resultant force acting on such a curved surface is not difficult to determine, although the $x$- and $y$-components of the resultant usually must be calculated first. The magnitude and direction of the resultant are found by conventional methods.

$$R = \sqrt{R_x^2 + R_y^2} \qquad 17.25$$

$$\theta = \arctan\left(\frac{R_y}{R_x}\right) \qquad 17.26$$

*Figure 17.13 Pressure Distributions on a Curved Surface*

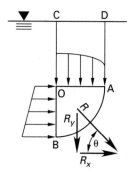

The horizontal component of the resultant hydrostatic force is found in the same manner as for a vertical plane surface.

The fact that the surface is curved does not affect the calculation of the horizontal force. In Fig. 17.13, the horizontal pressure distribution on curved surface BA is the same as the horizontal pressure distribution on imaginary projected surface BO.

The vertical component of force on the curved surface is most easily calculated as the weight of the liquid above it.[12] In Fig. 17.13, the vertical component of force on the curved surface BA is the weight of liquid within the area ABCD, with a vertical line of action passing through the centroid of the area ABCD.

Figure 17.14 illustrates a curved surface with no liquid above it. However, it is not difficult to show that the resultant force acting upward on the curved surface HG is equal in magnitude (and opposite in direction) to the force that would be acting downward due to the missing area EFGH. Such an imaginary area used to calculate hydrostatic pressure is known as an *equivalent area*.

*Figure 17.14 Equivalent Area*

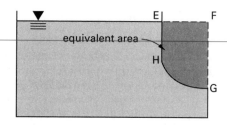

### Example 17.6

What is the total force on a 1 ft section of the wall in Example 17.4?

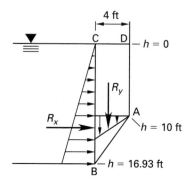

*Solution*

The average depth is

$$\overline{h} = \left(\tfrac{1}{2}\right)(0 + 16.93 \text{ ft}) = 8.465 \text{ ft}$$

---

[12]Calculating the vertical force component is not in conflict with the hydrostatic paradox as long as the cross-sectional area of liquid above the curved surface does not decrease between the curved surface and the liquid's free surface. If there is a change in the cross-sectional area, the vertical component of force is equal to the weight of fluid in an unchanged cross-sectional area (i.e., the equivalent area).

The average pressure and horizontal component of the resultant on a 1 ft section of wall are

$$\overline{p} = \gamma \overline{h} = \left( 62.4 \ \frac{\text{lbf}}{\text{ft}^3} \right)(8.465 \ \text{ft})$$

$$= 528.2 \ \text{lbf/ft}^2 \ (\text{psfg})$$

$$R_x = \overline{p} A = \left( 528.2 \ \frac{\text{lbf}}{\text{ft}^2} \right)(16.93 \ \text{ft})(1 \ \text{ft})$$

$$= 8942 \ \text{lbf}$$

The volume of a 1 ft section of area ABCD is

$$V_{\text{ABCD}} = (1 \ \text{ft})\left( (4 \ \text{ft})(10 \ \text{ft}) + \left(\tfrac{1}{2}\right)(4 \ \text{ft})(6.93 \ \text{ft}) \right)$$

$$= 53.86 \ \text{ft}^3$$

The vertical component is

$$R_y = \gamma V = \left( 62.4 \ \frac{\text{lbf}}{\text{ft}^3} \right)(53.86 \ \text{ft}^3)$$

$$= 3361 \ \text{lbf}$$

The total resultant force is

$$R = \sqrt{(8942 \ \text{lbf})^2 + (3361 \ \text{lbf})^2}$$

$$= 9553 \ \text{lbf}$$

## 12. TORQUE ON A GATE

When an openable gate or door is submerged in such a manner as to have unequal depths of liquid on either of its sides, or when there is no liquid present on one side of a gate or door, the hydrostatic pressure will act to either open or close the door. If the gate does not open, this pressure is resisted, usually by a latching mechanism on the gate itself.[13] The magnitude of the resisting latch force can be determined from the *hydrostatic torque (hydrostatic moment)* acting on the gate. (See Fig. 17.15.) The moment is almost always taken with respect to the gate hinges.

The applied moment is calculated as the product of the resultant force on the gate and the distance from the hinge to the resultant on the gate. This applied moment is balanced by the resisting moment, calculated as the latch force times the separation of the latch and hinge.

$$M_{\text{applied}} = M_{\text{resisting}} \qquad \textit{17.27}$$

$$R \times y_R = F_{\text{latch}} \times y_F \qquad \textit{17.28}$$

---

[13]Any contribution to resisting force from stiff hinges or other sources of friction is typically neglected.

**Figure 17.15** Torque on a Hinge (Gate)

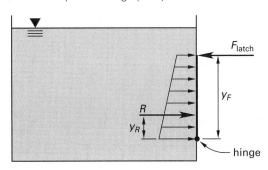

## 13. HYDROSTATIC FORCES ON A DAM

The concepts presented in the preceding sections are applicable to dams. That is, the horizontal force on the dam face can be found as in Ex. 17.6, regardless of inclination or curvature of the dam face. The vertical force on the dam face is calculated as the weight of the water above the dam face. Of course, the vertical force is zero if the dam face is vertical.

Figure 17.16 illustrates a typical dam, defining its *heel*, *toe*, and *crest*. $x_{\text{CG}}$, the distance to the dam's center of gravity, is not shown.

However, there are other considerations for gravity dams.[14] Most notably, the dam must not tip over or slide away due to the hydrostatic pressure. Furthermore, the pressure distribution within the soil under the dam is not uniform, and soil loading must not be excessive.

The *overturning moment* is a measure of the horizontal pressure's tendency to tip the dam over, pivoting it about the toe of the dam (point B in Fig. 17.16). (Usually, moments are calculated with respect to the pivot point.) The overturning moment is calculated as the product of the horizontal component of hydrostatic pressure (i.e., the $x$-component of the resultant) and the vertical distance between the toe and the line of action of the force.

$$M_{\text{overturning}} = R_x \times y_{R_x} \qquad \textit{17.29}$$

In most configurations, the overturning is resisted jointly by moments from the dam's own weight, $W$, and the vertical component of the resultant, $R_y$ (i.e., the weight of area EAD in Fig. 17.16).[15]

$$M_{\text{resisting}} = (R_y \times x_{R_y}) + (W \times x_{\text{CG}}) \qquad \textit{17.30}$$

---

[14]A *gravity dam* is one that is held in place and orientation by its own mass (weight) and the friction between its base and the ground.
[15]The density of concrete or masonry with steel reinforcing is usually taken to be approximately 150 lbm/ft$^3$ (2400 kg/m$^3$).

**Figure 17.16** Dam

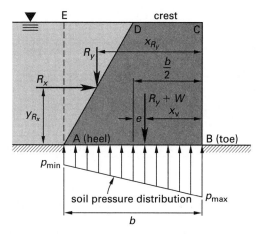

The *factor of safety against overturning* is

$$(\text{FS})_{\text{overturning}} = \frac{M_{\text{resisting}}}{M_{\text{overturning}}} \qquad 17.31$$

In addition to causing the dam to tip over, the horizontal component of hydrostatic force will also cause the dam to tend to slide along the ground. This tendency is resisted by the frictional force between the dam bottom and soil. The frictional force, $F_f$, is calculated as the product of the *normal force*, $N$, and the *coefficient of static friction*, $\mu$.

$$F_f = \mu_{\text{static}} N = \mu_{\text{static}}(W + R_y) \qquad 17.32$$

The *factor of safety against sliding* is

$$(\text{FS})_{\text{sliding}} = \frac{F_f}{R_x} \qquad 17.33$$

The soil pressure distribution beneath the dam is usually assumed to vary linearly from a minimum to a maximum value. (The minimum value must be greater than zero, since soil cannot be in a state of tension. The maximum pressure should not exceed the allowable soil pressure.) Equation 17.34 predicts the minimum and maximum soil pressures.

$$p_{\text{max}}, p_{\text{min}} = \left(\frac{R_y + W}{b}\right)\left(1 \pm \frac{6e}{b}\right) \quad \text{[per unit width]}$$
$$17.34$$

The *eccentricity*, $e$, in Eq. 17.34 is the distance between the mid-length of the dam and the line of action of the total vertical force, $W + R_y$. (The eccentricity must be less than $b/6$ for the entire base to be in compression.) Notice that distances $x_v$ and $x_{\text{CG}}$ are different.

$$e = \frac{b}{2} - x_v \qquad 17.35$$

$$x_v = \frac{M_{\text{resisting}}}{R_y + W} \qquad 17.36$$

## 14. PRESSURE DUE TO SEVERAL IMMISCIBLE LIQUIDS

Figure 17.17 illustrates the non-uniform pressure distribution due to two immiscible liquids (e.g., oil on top and water below).

**Figure 17.17** Pressure Distribution from Two Immiscible Liquids

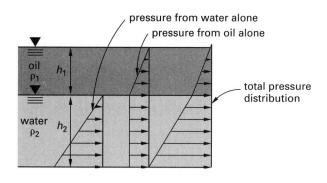

The pressure due to the upper liquid (oil), once calculated, serves as a *surcharge* to the liquid below (water). The pressure at the tank bottom is given by Eq. 17.37. (The principle can be extended to three or more immiscible liquids as well.)

$$p_{\text{bottom}} = \rho_1 g h_1 + \rho_2 g h_2 \qquad \text{[SI]} \quad 17.37(a)$$

$$p_{\text{bottom}} = \frac{\rho_1 g h_1}{g_c} + \frac{\rho_2 g h_2}{g_c}$$
$$= \gamma_1 h_1 + \gamma_2 h_2 \qquad \text{[U.S.]} \quad 17.37(b)$$

## 15. PRESSURE FROM COMPRESSIBLE FLUIDS

Fluid density, thus far, has been assumed to be independent of pressure. In reality, even "incompressible" liquids are slightly compressible. Sometimes, the effect of this compressibility cannot be neglected.

The familiar $p = \rho g h$ equation is a special case of Eq. 17.38. (It is assumed that $h_2 > h_1$. The minus sign in Eq. 17.38 indicates that pressure decreases as elevation (height) increases.)

$$\int_{p_1}^{p_2} \frac{dp}{\rho g} = -(h_2 - h_1) \qquad \text{[SI]} \quad 17.38(a)$$

$$\int_{p_1}^{p_2} \frac{g_c \, dp}{\rho g} = -(h_2 - h_1) \qquad \text{[U.S.]} \quad 17.38(b)$$

If the fluid is a perfect gas, and if compression is an isothermal (i.e., constant temperature) process, then the relationship between pressure and density is given by Eq. 17.39. The isothermal assumption is appropriate, for example, for the earth's *stratosphere* (i.e., above 35,000 ft or 11 000 m), where the temperature is assumed to be constant at approximately $-67°F$ $(-55°C)$.

$$pv = \frac{p}{\rho} = RT = \text{constant} \qquad 17.39$$

In the isothermal case, Eq. 17.39 can be rewritten as Eq. 17.40. (For air, $R = 53.35$ ft-lbf/lbm-°R; $R = 287.03$ J/kg·K.) Of course, the temperature, $T$, must be in degrees absolute (i.e., in °R or K). Equation 17.40 is known as the *barometric height relationship*[16] because knowledge of atmospheric temperature and the pressures at two points is sufficient to determine the height difference between the two points.

$$h_2 - h_1 = \left(\frac{RT}{g}\right)\ln\left(\frac{p_1}{p_2}\right) \qquad \text{[SI]} \qquad 17.40(a)$$

$$h_2 - h_1 = \left(\frac{g_c RT}{g}\right)\ln\left(\frac{p_1}{p_2}\right) \qquad \text{[U.S.]} \qquad 17.40(b)$$

The pressure at an elevation (height) $h_2$ in a layer of perfect gas that has been isothermally compressed is given by Eq. 17.41.

$$p_2 = p_1 e^{g(h_1 - h_2)/RT} \qquad \text{[SI]} \qquad 17.41(a)$$

$$p_2 = p_1 e^{g(h_1 - h_2)/g_c RT} \qquad \text{[U.S.]} \qquad 17.41(b)$$

If the fluid is a perfect gas, and if compression is an *adiabatic process*, the relationship between pressure and density is given by Eq. 17.42[17] where $k$ is the *ratio of specific heats*, a property of the gas. ($k = 1.4$ for air, hydrogen, oxygen, and carbon monoxide, among others.)

$$pv^k = p\left(\frac{1}{\rho}\right)^k = \text{constant} \qquad 17.42$$

The following three equations apply to adiabatic compression of an ideal gas.

$$h_2 - h_1 = \left(\frac{k}{k-1}\right)\left(\frac{RT_1}{g}\right)\left(1 - \left(\frac{p_2}{p_1}\right)^{k-1/k}\right)$$
$$\text{[SI]} \qquad 17.43(a)$$

$$h_2 - h_1 = \left(\frac{k}{k-1}\right)\left(\frac{g_c}{g}\right)RT_1\left(1 - \left(\frac{p_2}{p_1}\right)^{k-1/k}\right)$$
$$\text{[U.S.]} \qquad 17.43(b)$$

$$p_2 = p_1\left(1 - \left(\frac{k-1}{k}\right)\left(\frac{g}{RT_1}\right)(h_2 - h_1)\right)^{k/k-1}$$
$$\text{[SI]} \qquad 17.44(a)$$

$$p_2 = p_1\left(1 - \left(\frac{k-1}{k}\right)\left(\frac{g}{g_c}\right)\left(\frac{h_2 - h_1}{RT_1}\right)\right)^{k/k-1}$$
$$\text{[U.S.]} \qquad 17.44(b)$$

$$T_2 = T_1\left(1 - \left(\frac{k-1}{k}\right)\left(\frac{g}{RT_1}\right)(h_2 - h_1)\right) \qquad \text{[SI]} \qquad 17.45(a)$$

$$T_2 = T_1\left(1 - \left(\frac{k-1}{k}\right)\left(\frac{g}{g_c}\right)\left(\frac{h_2 - h_1}{RT_1}\right)\right) \qquad \text{[U.S.]} \qquad 17.45(b)$$

The three adiabatic compression equations can be used for the more general *polytropic compression* case simply by substituting the *polytropic exponent*, $n$, for $k$.[18] Unlike the ratio of specific heats, the polytropic exponent is a function of the process, not of the gas. The polytropic compression assumption is appropriate for the earth's *troposphere*.[19] Assuming a linear decrease in temperature along with an altitude of $-0.00356°F/ft$ $(-0.00649°C/m)$, a polytropic exponent of $n = 1.235$ can be derived.

### Example 17.7

The air pressure and temperature at sea level are 1.0 standard atmosphere and 68°F (20°C), respectively. Assume polytropic compression with $n = 1.235$. What is the pressure at an altitude of 5000 ft (1525 m)?

*SI Solution*

The absolute temperature of the air is $20°C + 273° = 293$K. From Eq. 17.44 (substituting $k = n = 1.235$ for polytropic compression), the pressure at 1525 m altitude is

$$p_2 = (1.0 \text{ atm})$$
$$\times\left(1 - \left(\frac{1.235 - 1}{1.235}\right)\left(9.81 \frac{\text{m}}{\text{s}^2}\right) \times \left(\frac{1525 \text{ m}}{\left(287.03 \frac{\text{J}}{\text{kg·K}}\right)(293\text{K})}\right)\right)^{1.235/(1.235-1)}$$

$$= 0.834 \text{ atm}$$

---

[16]You may recognize this as being equivalent to the work done in an isothermal compression process. The elevation (height) difference, $h_2 - h_1$ (with units of feet), can be interpreted as the work done per unit mass during compression (with units of ft-lbf/lbm).

[17]There is no heat or energy transfer to or from the ideal gas in an adiabatic process. However, this is not the same as an isothermal process.

[18]Actually, polytropic compression is the general process. Isothermal compression is a special case ($n = 1$) of the polytropic process, as is adiabatic compression ($n = k$).

[19]The *troposphere* is the part of the earth's atmosphere we live in and where most atmospheric disturbances occur. The *stratosphere*, starting at approximately 35,000 ft (11 000 m), is cold, clear, dry, and still. Between the troposphere and the stratosphere is the *tropopause*, a transition layer that contains most of the atmosphere's dust and moisture. Temperature actually increases with altitude in the stratosphere and decreases with altitude in the troposphere, but is constant in the tropopause.

*Customary U.S. Solution*

The absolute temperature of the air is $68°F + 460° = 528°R$. From Eq. 17.44 (substituting $k = n = 1.235$ for polytropic compression), the pressure at 5000 ft altitude is

$$p_2 = (1.0 \text{ atm})$$

$$\times \left( 1 - \left(\frac{1.235-1}{1.235}\right)\left(\frac{32.2 \frac{\text{ft}}{\text{sec}^2}}{32.2 \frac{\text{ft-lbm}}{\text{lbf-sec}^2}}\right) \right.$$
$$\left. \times \left(\frac{5000 \text{ ft}}{\left(53.35 \frac{\text{ft-lbf}}{\text{lbm-°R}}\right)(528°R)}\right) \right)^{1.235/(1.235-1)}$$

$$= 0.835 \text{ atm}$$

## 16. EXTERNALLY PRESSURIZED LIQUIDS

If the gas above a liquid in a closed tank is pressurized to a gage pressure of $p_t$, this pressure will add to the hydrostatic pressure anywhere in the fluid. The pressure at the tank bottom illustrated in Fig. 17.18 is given by Eq. 17.46.

$$p_{\text{bottom}} = p_t + \rho g h \qquad \text{[SI]} \quad 17.46(a)$$

$$p_{\text{bottom}} = p_t + \frac{\rho g h}{g_c} = p_t + \gamma h \qquad \text{[U.S.]} \quad 17.46(b)$$

**Figure 17.18** *Externally Pressurized Liquid*

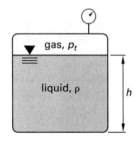

## 17. HYDRAULIC RAM

A *hydraulic ram* (*hydraulic jack, hydraulic press, fluid press*, etc.) is illustrated in Fig. 17.19. This is a force-multiplying device. A force, $F_p$, is applied to the *plunger*, and a useful force, $F_r$, appears at the *ram*. Even though the pressure in the hydraulic fluid is the same on the ram and plunger, the forces at the two cylinders will be proportional to their respective cross-sectional areas.[20]

**Figure 17.19** *Hydraulic Ram*

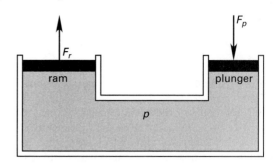

Since the pressure is the same everywhere, Eq. 17.47 can be solved for $p$ for the ram and plunger.

$$F = pA = p\pi r^2 \qquad 17.47$$

$$p_p = p_r \qquad 17.48$$

$$\frac{F_p}{A_p} = \frac{F_r}{A_r} \qquad 17.49$$

$$\frac{F_p}{d_p^2} = \frac{F_r}{d_r^2} \qquad 17.50$$

Small, manually actuated hydraulic rams usually have a lever handle to increase the mechanical advantage of the ram from 1.0 to $M$, as illustrated in Fig. 17.20. In most cases, the pivot and connection mechanism will not be frictionless, and some of the applied force will be used to overcome the friction. This friction loss is accounted for by a *lever efficiency* or *lever effectiveness*, $\eta$.

$$M = \frac{L_1}{L_2} \qquad 17.51$$

$$F_p = \eta M F_l \qquad 17.52$$

$$\frac{\eta M F_l}{A_p} = \frac{F_r}{A_r} \qquad 17.53$$

**Figure 17.20** *Hydraulic Ram with Mechanical Advantage*

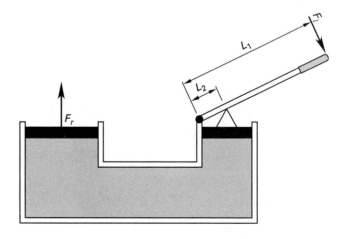

---

[20] *Pascal's law:* Pressure in a liquid is the same in all directions.

## 18. BUOYANCY

Buoyant force is an upward force that acts on all objects that are partially or completely submerged in a fluid. The fluid can be a liquid, as in the case of a ship floating at sea, or the fluid can be a gas, as in a balloon floating in the atmosphere.

There is a buoyant force on all submerged objects, not just those that are stationary or ascending. There will be, for example, a buoyant force on a rock sitting at the bottom of a pond. There will also be a buoyant force on a rock sitting exposed on the ground, since the rock is "submerged" in air. A buoyant force due to displaced air also exists, although it may be insignificant, in the case of partially exposed floating objects such as icebergs.

Buoyant force always acts to counteract an object's weight (i.e., buoyancy acts against gravity). The magnitude of the buoyant force is predicted from *Archimedes' principle (the buoyancy theorem)*: The buoyant force on a submerged object is equal to the weight of the displaced fluid.[21] An equivalent statement of Archimedes' principle is: A floating object displaces liquid equal in weight to its own weight.

$$F_{\text{buoyant}} = \rho g V_{\text{displaced}} \quad \text{[SI]} \quad \textit{17.54(a)}$$

$$F_{\text{buoyant}} = \frac{\rho g V_{\text{displaced}}}{g_c} = \gamma V_{\text{displaced}} \quad \text{[U.S.]} \quad \textit{17.54(b)}$$

In the case of stationary (i.e., not moving vertically) floating or submerged objects, the buoyant force and object weight are in equilibrium. If the forces are not in equilibrium, the object will rise or fall until equilibrium is reached. That is, the object will sink until its remaining weight is supported by the bottom, or it will rise until the weight of displaced liquid is reduced by breaking the surface.[22]

The specific gravity (SG) of an object submerged in water can be determined from its dry and submerged weights. Neglecting the buoyancy of any surrounding gases,

$$SG = \frac{W_{\text{dry}}}{W_{\text{dry}} - W_{\text{submerged}}} \quad \textit{17.55}$$

Figure 17.21 illustrates an object floating partially exposed in a liquid. Neglecting the insignificant buoyant force from the displaced air (or other gas), the fractions, $x$, of volume exposed and submerged are easily determined.

[21]The volume term in Eq. 17.54 is the total volume of the object only in the case of complete submergence.
[22]An object can also stop rising or falling due to a change in the fluid's density. The buoyant force will increase with increasing depth in the ocean due to an increase in density at great depths. The buoyant force will decrease with increasing altitude in the atmosphere due to a decrease in density at great heights.

*Figure 17.21* Partially Submerged Object

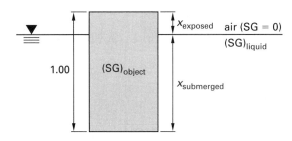

$$x_{\text{submerged}} = \frac{\rho_{\text{object}}}{\rho_{\text{liquid}}} = \frac{(\text{SG})_{\text{object}}}{(\text{SG})_{\text{liquid}}} \quad \textit{17.56}$$

$$x_{\text{exposed}} = 1 - x_{\text{submerged}} \quad \textit{17.57}$$

Figure 17.22 illustrates a somewhat more complicated situation—that of an object floating at the interface between two liquids of different densities. The fractions of immersion in each liquid are given by the following equations.

$$x_1 = \frac{(\text{SG})_2 - (\text{SG})_{\text{object}}}{(\text{SG})_2 - (\text{SG})_1} \quad \textit{17.58}$$

$$x_2 = 1 - x_1 = \frac{(\text{SG})_{\text{object}} - (\text{SG})_1}{(\text{SG})_2 - (\text{SG})_1} \quad \textit{17.59}$$

*Figure 17.22* Object Floating in Two Liquids

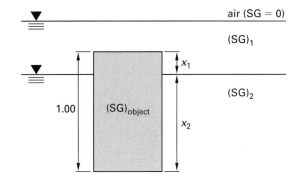

A more general case of a floating object is shown in Fig. 17.23. Situations of this type are easily evaluated by equating the object's weight with the sum of the buoyant forces.

In the case of Fig. 17.23 (with two liquids), the following relationships apply, where $x_0$ is the fraction, if any, extending into the air above.

$$(\text{SG})_{\text{object}} = x_1(\text{SG})_1 + x_2(\text{SG})_2 \quad \textit{17.60}$$

$$(\text{SG})_{\text{object}} = (1 - x_0 - x_2)(\text{SG})_1 + x_2(\text{SG})_2 \quad \textit{17.61}$$

$$(\text{SG})_{\text{object}} = x_1(\text{SG})_1 + (1 - x_0 - x_1)(\text{SG})_2 \quad \textit{17.62}$$

**Figure 17.23** *General Two-Liquid Buoyancy Problem*

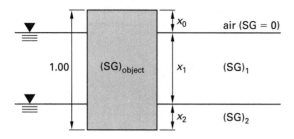

**Fluids**

### Example 17.8

An empty polyethylene telemetry balloon and payload have a mass of 500 lbm (225 kg). The balloon is filled with helium when the atmospheric conditions are 60°F (15.6°C) and 14.8 psia (102 kPa). The specific gas constant of helium is 2079 J/kg·K (386.3 ft-lbf/lbm-°R). What volume of helium is required for lift-off from a sea-level platform?

*SI Solution*

$$\rho_{\text{air}} = \frac{p}{RT} = \frac{1.02 \times 10^5 \text{ Pa}}{\left(287.03 \dfrac{\text{J}}{\text{kg·K}}\right)(15.6°\text{C} + 273°)}$$

$$= 1.231 \text{ kg/m}^3$$

$$\rho_{\text{helium}} = \frac{1.02 \times 10^5 \text{ Pa}}{\left(2079 \dfrac{\text{J}}{\text{kg·K}}\right)(288.6\text{K})} = 0.17 \text{ kg/m}^3$$

$$m = 225 \text{ kg} + \left(0.17 \dfrac{\text{kg}}{\text{m}^3}\right) V_{\text{He}}$$

$$m_b = \left(1.231 \dfrac{\text{kg}}{\text{m}^3}\right) V_{\text{He}}$$

$$225 \text{ kg} + \left(0.17 \dfrac{\text{kg}}{\text{m}^3}\right) V_{\text{He}} = \left(1.231 \dfrac{\text{kg}}{\text{m}^3}\right) V_{\text{He}}$$

$$V_{\text{He}} = 212.1 \text{ m}^3$$

*Customary U.S. Solution*

The gas densities are

$$\rho_{\text{air}} = \frac{p}{RT} = \frac{\left(14.8 \dfrac{\text{lbf}}{\text{in}^2}\right)\left(12 \dfrac{\text{in}}{\text{ft}}\right)^2}{\left(53.35 \dfrac{\text{ft-lbf}}{\text{lbm-°R}}\right)(60°\text{F} + 460°)}$$

$$= 0.07682 \text{ lbm/ft}^3$$

$$\gamma_{\text{air}} = \rho \times \frac{g}{g_c} = 0.07682 \text{ lbf/ft}^3$$

$$\rho_{\text{helium}} = \frac{\left(14.8 \dfrac{\text{lbf}}{\text{in}^2}\right)\left(12 \dfrac{\text{in}}{\text{ft}}\right)^2}{\left(386.3 \dfrac{\text{ft-lbf}}{\text{lbm-°R}}\right)(520°\text{R})}$$

$$= 0.01061 \text{ lbm/ft}^3$$

$$\gamma_{\text{helium}} = 0.01061 \text{ lbf/ft}^3$$

The total weight of the balloon, payload, and helium is

$$W = 500 \text{ lbf} + \left(0.01061 \dfrac{\text{lbf}}{\text{ft}^3}\right) V_{\text{He}}$$

The buoyant force is the weight of the displaced air. Neglecting the payload volume, the displaced air volume is the same as the helium volume.

$$F_b = \left(0.07682 \dfrac{\text{lbf}}{\text{ft}^3}\right) V_{\text{He}}$$

At lift-off, the weight of the balloon is just equal to the buoyant force.

$$W = F_b$$

$$500 \text{ lbf} + \left(0.01061 \dfrac{\text{lbf}}{\text{ft}^3}\right) V_{\text{He}} = \left(0.07682 \dfrac{\text{lbf}}{\text{ft}^3}\right) V_{\text{He}}$$

$$V_{\text{He}} = 7552 \text{ ft}^3$$

## 19. BUOYANCY OF SUBMERGED PIPELINES

Whenever possible, submerged pipelines for river crossings should be completely buried at a level below river scour. This will reduce or eliminate loads and movement due to flutter, scour and fill, drag, collisions, and buoyancy. Submerged pipelines should cross at right angles to the river. For maximum flexibility, ductility, and weighting, pipelines should be made of thick-walled mild steel.

Submerged pipelines should be weighted to achieve a minimum of 20% negative buoyancy (i.e., an average density of 1.2 times the environment, approximately 72 lbm/ft³ or 1200 kg/m³). Metal or concrete clamps can be used for this purpose, as well as concrete coatings. Thick steel clamps have the advantage of a smaller lateral exposed area (resulting in less drag from river flow), while brittle concrete coatings are sensitive to pipeline flutter and temperature fluctuations.

Due to the critical nature of many pipelines and the difficulty in accessing submerged pipelines for repair, it is common to provide a parallel auxiliary line. The auxiliary and main lines are provided with crossover and mainline valves, respectively, on high ground at both sides of the river to permit either or both lines to be used.

## 20. INTACT STABILITY: STABILITY OF FLOATING OBJECTS

A stationary object is said to be in *static equilibrium*. However, an object in static equilibrium is not necessarily stable. For example, a coin balanced on edge is in static equilibrium, but it will not return to the balanced position if it is disturbed. An object is said to be *stable* (i.e., in *stable equilibrium*) if it tends to return to the equilibrium position when slightly displaced.

Stability of floating and submerged objects is known as *intact stability*.[23] There are two forces acting on a stationary floating object: the buoyant force and the object's weight. The buoyant force acts upward through the centroid of the displaced volume (not the object's volume). This centroid is known as the *center of buoyancy*. The gravitational force on the object (i.e., the object's weight) acts downward through the entire object's center of gravity.

For a totally submerged object (as in the balloon and submarine shown in Fig. 17.24) to be stable, the center of buoyancy must be above the center of gravity. The object will be stable because a righting moment will be created if the object tips over, since the center of buoyancy will move outward from the center of gravity.

**Figure 17.24** *Stability of a Submerged Object*

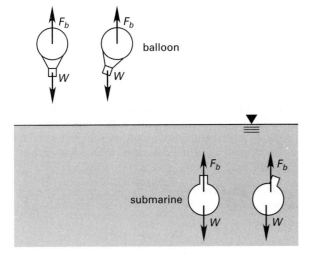

The stability criterion is different for partially submerged objects (e.g., surface ships). For partially submerged objects to be stable, the *metacenter* must be above the center of gravity. If the vessel shown in Fig. 17.25 heels (i.e., lists or rolls), the location of the center of gravity of the object does not change.[24]

---

[23]The subject of intact stability, being a part of naval architecture curriculum, is not covered extensively in most fluids books. However, it is covered extensively in basic ship design and naval architecture books.

[24]The verbs *roll*, *list*, and *heel* are synonymous.

**Figure 17.25** *Stability of a Partially Submerged Floating Object*

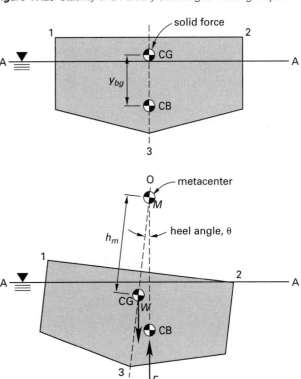

However, the center of buoyancy shifts to the centroid of the new submerged section 123. The centers of buoyancy and gravity are no longer in line. The righting couple resists further overturning.

This righting couple exists when the extension of the buoyant force, $F_b$, intersects line O-O above the center of gravity at $M$, the *metacenter*. If $M$ lies below the center of gravity, an overturning couple will exist. The distance between the center of gravity and the metacenter is called the *metacentric height*, and it is reasonably constant for heel angles less than $10°$. Also, for angles less than $10°$, the center of buoyancy follows a locus for which the metacenter is the instantaneous center.

The metacentric height is one of the most important and basic parameters in ship design. It determines the ship's ability to remain upright as well as the ship's roll and pitch characteristics.

"Acceptable" minimum values of the metacentric height have been established from experience, and these depend on the ship type and class. For example, many submarines are required to have a metacentric height of 1 ft (0.3 m) when surfaced. This will increase to approximately 3.5 ft (1.2 m) for some of the largest surface ships. If an acceptable metacentric height is not achieved initially, the center of gravity must be lowered or the keel depth increased. The beam width can also be increased slightly to increase the waterplane moment of inertia.

Fluids

For a surface vessel rolling through an angle less than approximately 10°, the distance between the vertical center of gravity and the metacenter can be found from Eq. 17.63. Variable $I$ is the centroidal area moment of inertia of the original waterline (free surface) cross section about a longitudinal (fore and aft) waterline axis; $V$ is the displaced volume.

If the distance, $y_{bg}$, separating the centers of buoyancy and gravity is known, Eq. 17.63 can be solved for the metacentric height. $y_{bg}$ is positive when the center of gravity is above the center of buoyancy. This is the normal case. Otherwise, $y_{bg}$ is negative.

$$y_{bg} + h_m = \frac{I}{V} \qquad \text{17.63}$$

The *righting moment* (also known as the *restoring moment*) is the stabilizing moment exerted when the ship rolls. Values of the righting moment are typically specified with units of foot-tons (MN·m).

$$M_{\text{righting}} = h_m \gamma_w V_{\text{displaced}} \sin \theta \qquad \text{17.64}$$

The transverse (roll) and longitudinal (pitch) *periods* also depend on the metacentric height. The roll characteristics are found from the differential equation formed by equating the righting moment to the product of the ship's transverse mass moment of inertia and the angular acceleration. Larger metacentric heights result in lower roll periods. If $k$ is the radius of gyration about the roll axis, the roll period is

$$T_{\text{roll}} = \frac{2\pi k}{\sqrt{gh_m}} \qquad \text{17.65}$$

The roll and pitch periods must be adjusted for the appropriate level of crew and passenger comfort. A "beamy" ice-breaking ship will have a metacentric height much larger than normally required for intact stability, resulting in a long, nauseating, roll period. The designer of a passenger ship, however, would have to decrease the intact stability (i.e., decrease the metacentric height) in order to achieve an acceptable ride characteristic. This requires a metacentric height that is less than approximately 6% of the beam length.

### Example 17.9

A 600,000 lbm (280 000 kg) rectangular barge has external dimensions of 24 ft width, 98 ft length, and 12 ft height (7 m × 30 m × 3.6 m). It floats in seawater ($\gamma_w = 64.0$ lbf/ft$^3$; $\rho_w = 1024$ kg/m$^3$). The center of gravity is 7.8 ft (2.4 m) from the top of the barge as loaded. Find (a) the location of the center of buoyancy when the barge is floating on an even keel, and (b) the approximate location of the metacenter when the barge experiences a 5° heel.

*SI Solution*

(a) Refer to the following diagram. Let dimension $y$ represent the depth of the submerged barge.

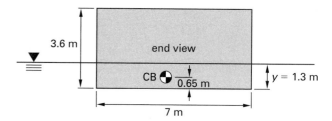

From Archimedes' principle, the buoyant force equals the weight of the barge. This, in turn, equals the weight of the displaced seawater.

$$F_b = W = V\gamma_w = V\rho_w g$$

$$(280\,000 \text{ kg})\left(9.81 \frac{\text{m}}{\text{s}^2}\right) = y(7 \text{ m})(30 \text{ m})\left(1024 \frac{\text{kg}}{\text{m}^3}\right)$$
$$\times \left(9.81 \frac{\text{m}}{\text{s}^2}\right)$$

$$y = 1.30 \text{ m}$$

The center of buoyancy is located at the centroid of the submerged cross section. When floating on an even keel, the submerged cross section is rectangular with a height of 1.30 m. The height of the center of buoyancy above the keel is

$$\frac{1.30 \text{ m}}{2} = 0.65 \text{ m}$$

(b) While the location of the new center of buoyancy can be determined, the location of the metacenter does not change significantly for small angles of heel. Therefore, for approximate calculations, the angle of heel is not significant.

The area moment of inertia of the longitudinal waterline cross section is

$$I = \frac{Lw^3}{12} = \frac{(30 \text{ m})(7 \text{ m})^3}{12}$$
$$= 858 \text{ m}^4$$

The submerged volume is

$$V = (1.3 \text{ m})(7 \text{ m})(30 \text{ m}) = 273 \text{ m}^3$$

The distance between the center of gravity and the center of buoyancy is

$$y_{bg} = 3.6 \text{ m} - 2.4 \text{ m} - 0.65 \text{ m} = 0.55 \text{ m}$$

The metacentric height measured above the center of gravity is

$$h_m = \frac{I}{V} - y_{bg}$$

$$= \frac{858 \text{ m}^4}{273 \text{ m}^3} - 0.55 \text{ m}$$

$$= 2.6 \text{ m}$$

*Customary U.S. Solution*

(a) Refer to the following diagram. Let dimension $y$ represent the depth of the submerged barge.

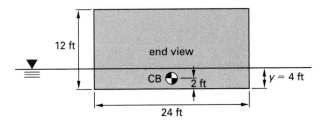

From Archimedes' principle, the buoyant force equals the weight of the barge. This, in turn, equals the weight of the displaced seawater.

$$F_b = W = V\gamma_w$$

$$600{,}000 \text{ lbf} = y(24 \text{ ft})(98 \text{ ft})\left(64 \frac{\text{lbf}}{\text{ft}^3}\right)$$

$$y = 4.00 \text{ ft}$$

The center of buoyancy is located at the centroid of the submerged cross section. When floating on an even keel, the submerged cross section is rectangular with a height of 4.00 ft. The height of the center of buoyancy above the keel is

$$\frac{4.00 \text{ ft}}{2} = 2.00 \text{ ft}$$

(b) While the location of the new center of buoyancy can be determined, the location of the metacenter does not change significantly for small angles of heel. Therefore, for approximate calculations, the angle of heel is not significant.

The area moment of inertia of the longitudinal waterline cross section is

$$I = \frac{Lw^3}{12} = \frac{(98 \text{ ft})(24 \text{ ft})^3}{12}$$

$$= 112{,}900 \text{ ft}^4$$

The submerged volume is

$$V = (4 \text{ ft})(24 \text{ ft})(98 \text{ ft}) = 9408 \text{ ft}^3$$

The distance between the center of gravity and the center of buoyancy is

$$y_{bg} = 12 \text{ ft} - 7.8 \text{ ft} - 2.0 \text{ ft} = 2.2 \text{ ft}$$

The metacentric height measured above the center of gravity is

$$h_m = \frac{I}{V} - y_{bg}$$

$$= \frac{112{,}900 \text{ ft}^4}{9408 \text{ ft}^3} - 2.2 \text{ ft}$$

$$= 9.8 \text{ ft}$$

## 21. FLUID MASSES UNDER EXTERNAL ACCELERATION

Up to this point, fluid masses have been stationary, and gravity has been the only force acting on them. If a fluid mass is subjected to an external acceleration (moved sideways, rotated, etc.), an additional force will be introduced. This force will change the equilibrium position of the fluid surface as well as the hydrostatic pressure distribution.

Figure 17.26 illustrates a liquid mass subjected to constant accelerations in the vertical and/or horizontal directions. ($a_y$ is negative if the acceleration is downward.) The surface is inclined at the angle predicted by Eq. 17.66. The planes of equal hydrostatic pressure beneath the surface are also inclined at the same angle.[25]

$$\theta = \arctan\left(\frac{a_x}{a_y + g}\right) \qquad \textit{17.66}$$

$$p = \rho g h\left(1 + \frac{a_y}{g}\right) \qquad \text{[SI]} \quad \textit{17.67(a)}$$

$$p = \left(\frac{\rho g h}{g_c}\right)\left(1 + \frac{a_y}{g}\right) = \gamma h\left(1 + \frac{a_y}{g}\right) \quad \text{[U.S.]} \quad \textit{17.67(b)}$$

**Figure 17.26** *Fluid Mass Under Constant Linear Acceleration*

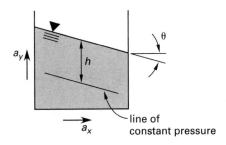

---

[25]Once the orientation of the surface is known, the pressure distribution can be determined without considering the acceleration. The hydrostatic pressure at a point depends only on the height of the liquid above that point. The acceleration affects that height but does not change the $p = \rho g h$ relationship.

Figure 17.27 illustrates a fluid mass rotating about a vertical axis at constant angular velocity, $\omega$ in rad/sec.[26] The resulting surface is parabolic in shape. The elevation of the fluid surface at point A at distance $r$ from the axis of rotation is given by Eq. 17.69. The distance $h$ in Fig. 17.27 is measured from the lowest fluid elevation during rotation. $h$ is not measured from the original elevation of the stationary fluid.

$$\theta = \arctan\left(\frac{\omega^2 r}{g}\right) \qquad \text{17.68}$$

$$h = \frac{(\omega r)^2}{2g} = \frac{v^2}{2g} \qquad \text{17.69}$$

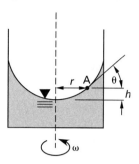

**Figure 17.27** Rotating Fluid Mass

---

[26]Even though the rotational speed is not increasing, the fluid mass experiences a constant *centripetal acceleration* radially outward from the axis of rotation.

# 18 Fluid Flow Parameters

## Nomenclature

| | | | |
|---|---|---|---|
| $A$ | area | ft$^2$ | m$^2$ |
| $C$ | correction factor | – | – |
| $d$ | depth | ft | m |
| $D$ | diameter | ft | m |
| $E$ | specific energy | ft-lbf/lbm | J/kg |
| $g$ | gravitational acceleration | ft/sec$^2$ | m/s$^2$ |
| $g_c$ | gravitational conversion constant | ft-lbm/lbf-sec$^2$ | n.a. |
| $G$ | mass flow rate per unit area | lbm/ft$^2$-sec | kg/m$^2$·s |
| $h$ | height or head | ft | m |
| $L$ | length | ft | m |
| $p$ | pressure | lbf/ft$^2$ | Pa |
| $r$ | radius | ft | m |
| Re | Reynolds number | – | – |
| $s$ | wetted perimeter | ft | m |
| v | velocity | ft/sec | m/s |
| $\dot{V}$ | volumetric flow rate | ft$^3$/sec | m$^3$/s |
| $y$ | distance | ft | m |
| $z$ | elevation | ft | m |

## Symbols

| | | | |
|---|---|---|---|
| $\alpha$ | angle | rad | rad |
| $\theta$ | time | sec | s |
| $\mu$ | absolute viscosity | lbf-sec/ft$^2$ | Pa·s |
| $\nu$ | kinematic viscosity | ft$^2$/sec | m$^2$/s |
| $\rho$ | density | lbm/ft$^3$ | kg/m$^3$ |
| $\phi$ | angle | rad | rad |

## Subscripts

| | |
|---|---|
| $e$ | equivalent |
| $h$ | hydraulic |
| $i$ | impact or inner |
| $o$ | outer |
| $p$ | pressure |
| $r$ | radius |
| $s$ | static |
| $t$ | total |
| v | velocity |
| $z$ | potential |

## 1. INTRODUCTION TO FLUID ENERGY UNITS

Several important fluids and thermodynamics equations, such as Bernoulli's equation and the steady-flow energy equation, are special applications of the *conservation of energy* concept. However, it is not always obvious how some formulations of these equations can be termed "energy." For example, elevation, $z$, with units of feet, is often called *gravitational energy*.[1]

Since every fluids problem is different, energy must be expressed per unit mass (i.e., *specific energy*). With SI formulations, the choice of units is unambiguous: J/kg is the only choice.

$$\frac{\text{J}}{\text{kg}} = \frac{\text{N·m}}{\text{kg}} = \frac{\text{m}^2}{\text{s}^2} \qquad 18.1$$

With problems formulated in English units, specific energy units will be ft-lbf/lbm.[2] (If the consistent set of units ft-lbf/slug is chosen, the same equations can be used with SI and consistent English units, since the primary dimensions are the same, that is, $L^2/\theta^2$.)

$$\frac{\text{ft-lbf}}{\text{slug}} = \frac{\text{ft}^2}{\text{sec}^2} \qquad 18.2$$

The gravitational conversion constant, $g_c$, must be used if the units ft-lbf/lbm are used for specific fluid energy.

In many cases, the ratio $g/g_c$ appears in equations. Since $g$ and $g_c$ have the same numerical value in most

---

[1]Foot and ft-lbf/lbm may be thought of as one and the same. Certainly, the set of units ft-lbf/lbm represents energy per unit mass. Unfortunately, lbf and lbm do not really cancel out to yield ft.

[2]Btu could be used for energy, as is common in thermodynamics problems, instead of ft-lbf. However, this is almost never done in fluids problems.

instances, this ratio affects the units without affecting the calculation. Because of this, some engineers omit the term $g/g_c$ entirely. While it is easy to justify such a practice, the resulting equations are not dimensionally consistent.[3]

## 2. KINETIC ENERGY

Energy is required to accelerate a stationary body. Thus, a moving mass of fluid possesses more energy than an identical, stationary mass. The energy difference is the *kinetic energy* of the fluid.[4] If the kinetic energy is evaluated per unit mass, the term *specific kinetic energy* is used. Equation 18.3 gives the specific kinetic energy corresponding to a fluid flow with uniform (i.e., turbulent) velocity, v.

$$E_v = \frac{v^2}{2} \qquad [SI] \qquad 18.3(a)$$

$$E_v = \frac{v^2}{2g_c} \qquad [U.S.] \qquad 18.3(b)$$

The units of specific kinetic energy in consistent units are clearly $m^2/s^2$ or $ft^2/sec^2$, which coincide with Eq. 18.1 and Eq. 18.2 as representing energy per consistent mass unit. The units in traditional English units (see Eq. 18.3(b)) are

$$\frac{\left(\dfrac{ft}{sec}\right)^2}{\dfrac{ft\text{-}lbm}{lbf\text{-}sec^2}} = \frac{ft\text{-}lbf}{lbm} \qquad 18.4$$

## 3. POTENTIAL ENERGY

Work is performed in elevating a body. Thus, a mass of fluid at a high elevation will have more energy than an identical mass of fluid at a lower elevation. The energy difference is the *potential energy* of the fluid.[5] Like kinetic energy, potential energy is usually expressed per unit mass. Equation 18.5 gives the potential energy of fluid at an elevation $z$.[6]

$$E_z = zg \qquad [SI] \qquad 18.5(a)$$

$$E_z = \frac{zg}{g_c} \qquad [U.S.] \qquad 18.5(b)$$

The units of potential energy in a consistent system are again $m^2/s^2$ or $ft^2/sec^2$. The units in a traditional English system are ft-lbf/lbm.

$z$ is the elevation of the fluid. The reference point (i.e., zero elevation point) is entirely arbitrary and can be chosen for convenience. This is because potential energy always appears in a difference equation (i.e., $\Delta E_z$), and the reference point cancels out.

## 4. PRESSURE ENERGY

Work is performed and energy is added when a substance is compressed. Thus, a mass of fluid at a high pressure will have more energy than an identical mass of fluid at a lower pressure. The energy difference is the *pressure energy* of the fluid.[7] Pressure energy is usually found in equations along with kinetic and potential energies and is expressed as energy per unit mass. Equation 18.6 gives the pressure energy of fluid at pressure $p$.

$$E_p = \frac{p}{\rho} \qquad 18.6$$

The consistent SI units of pressure energy are

$$\frac{Pa \cdot m^3}{kg} = \frac{N \cdot m}{kg} = \frac{J}{kg} = \frac{m^2}{s^2} \qquad 18.7$$

The units of pressure energy in the consistent English system are

$$\frac{lbf\text{-}ft^3}{ft^2\text{-}slug} = \frac{ft\text{-}lbf}{slug} = \frac{ft^2}{sec^2} \qquad 18.8$$

The units of pressure energy in the traditional English system are

$$\frac{lbf\text{-}ft^3}{ft^2\text{-}lbm} = \frac{ft\text{-}lbf}{lbm} \qquad 18.9$$

## 5. BERNOULLI EQUATION

The *Bernoulli equation* is an energy conservation equation based on several reasonable assumptions. The equation assumes the following.

- The fluid is incompressible.
- There is no fluid friction.
- Changes in thermal energy are negligible.[8]

The Bernoulli equation states that the *total energy* of a fluid flowing without friction losses in a pipe is constant.[9] The total energy possessed by the fluid is the sum of its pressure, kinetic, and potential energies.

---

[3]More than being dimensionally inconsistent, the resulting equations will be numerically incorrect in any nonstandard (other than one gravity) gravitational field.
[4]The terms *velocity energy* and *dynamic energy* are used less often.
[5]The term *gravitational energy* is also used.
[6]Since $g = g_c$ (numerically), it is tempting to write $E_z = z$. In fact, many engineers in the United States do just that.

[7]The terms *static energy* and *flow energy* are also used. The name *flow energy* results from the need to push (pressurize) a fluid to get it to flow through a pipe. However, flow energy and kinetic energy are not the same.
[8]In thermodynamics, the fluid flow is said to be *adiabatic*.
[9]Strictly speaking, this is the *total specific energy*, since the energy is per unit mass. However, the word "specific," being understood, is seldom used. Of course, "the total energy of the system" means something else and requires knowing the fluid mass in the system.

Drawing on Eq. 18.3, Eq. 18.5, and Eq. 18.6, the Bernoulli equation is written as

$$E_t = E_p + E_v + E_z \qquad \textit{18.10}$$

$$E_t = \frac{p}{\rho} + \frac{v^2}{2} + zg \qquad \text{[SI]} \quad \textit{18.11(a)}$$

$$E_t = \frac{p}{\rho} + \frac{v^2}{2g_c} + \frac{zg}{g_c} \qquad \text{[U.S.]} \quad \textit{18.11(b)}$$

Equation 18.11 is valid for both laminar and turbulent flows. It can also be used for gases and vapors if the incompressibility assumption is valid.[10]

The quantities known as *total head*, $h_t$, and *total pressure*, $p_t$, can be calculated from total energy.

$$h_t = \frac{E_t}{g} \qquad \text{[SI]} \quad \textit{18.12(a)}$$

$$h_t = E_t \times \frac{g_c}{g} \qquad \text{[U.S.]} \quad \textit{18.12(b)}$$

$$p_t = \rho g h_t \qquad \text{[SI]} \quad \textit{18.13(a)}$$

$$p_t = \rho h_t \times \frac{g}{g_c} \qquad \text{[U.S.]} \quad \textit{18.13(b)}$$

## Example 18.1

A pipe draws water from the bottom of a reservoir and discharges it freely at point C, 100 ft (30 m) below the surface. The flow is frictionless. (a) What is the total specific energy at an elevation 50 ft (15 m) below the water surface (i.e., point B)? (b) What is the velocity at point C?

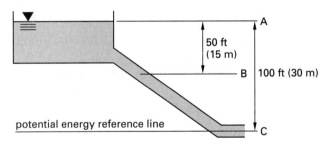

*SI Solution*

(a) At point A, the velocity and gage pressure are both zero. Therefore, the total energy consists only of potential energy. Point C is chosen as the reference $(z = 0)$ elevation.

$$E_A = z_A g = (30 \text{ m})\left(9.81 \ \frac{\text{m}}{\text{s}^2}\right)$$

$$= 294.3 \text{ m}^2/\text{s}^2 \quad (\text{J/kg})$$

---

[10]A gas or vapor can be considered to be incompressible as long as its pressure does not change by more than 10% between the entrance and exit, and its velocity is less than Mach 0.3 everywhere.

At point B, the fluid is moving and possesses kinetic energy. The fluid is also under hydrostatic pressure and possesses pressure energy. These energy forms have come at the expense of potential energy. (This is a direct result of the Bernoulli equation.) Also, the flow is frictionless. Thus, there is no net change in the total energy between points A and B.

$$E_B = E_A = 294.3 \text{ m}^2/\text{s}^2 \quad (\text{J/kg})$$

(b) At point C, the gage pressure and pressure energy are again zero, since the discharge is at atmospheric pressure. The potential energy is zero, since $z = 0$. The total energy of the system has been converted to kinetic energy. From Eq. 18.11,

$$E_t = 294.3 \ \frac{\text{m}^2}{\text{s}^2} = 0 + \frac{v^2}{2} + 0$$

$$v = 24.3 \text{ m/s}$$

*Customary U.S. Solution*

(a)
$$E_A = \frac{z_A g}{g_c} = \frac{(100 \text{ ft})\left(32.2 \ \frac{\text{ft}}{\text{sec}^2}\right)}{32.2 \ \frac{\text{ft-lbm}}{\text{lbf-sec}^2}}$$

$$= 100 \text{ ft-lbf/lbm}$$

$$E_B = E_A = 100 \text{ ft-lbf/lbm}$$

(b)
$$E_t = 100 \ \frac{\text{ft-lbf}}{\text{lbm}} = 0 + \frac{v^2}{2g_c} + 0$$

$$v^2 = (2)\left(32.2 \ \frac{\text{ft-lbm}}{\text{lbf-sec}^2}\right)\left(100 \ \frac{\text{ft-lbf}}{\text{lbm}}\right)$$

$$= 6440 \text{ ft}^2/\text{sec}^2$$

$$v = 80.2 \text{ ft/sec}$$

## Example 18.2

Water (62.4 lbm/ft³; 1000 kg/m³) is pumped up a hillside into a reservoir. The pump discharges water at the rate of 6 ft/sec (2 m/s) and with a pressure of 150 psig (1000 kPa). Disregarding friction, what is the maximum elevation (above the centerline of the pump's discharge) of the reservoir's water surface?

*SI Solution*

At the centerline of the pump's discharge, the potential energy is zero. The pressure and velocity energies are

$$E_p = \frac{p}{\rho} = \frac{(1000 \text{ kPa})\left(1000 \ \frac{\text{Pa}}{\text{kPa}}\right)}{1000 \ \frac{\text{kg}}{\text{m}^3}} = 1000 \text{ J/kg}$$

$$E_v = \frac{v^2}{2} = \frac{\left(2 \ \frac{\text{m}}{\text{s}}\right)^2}{2} = 2 \text{ J/kg}$$

The total energy at the pump's discharge is

$$E_{t,1} = E_p + E_v = 1000 \ \frac{\text{J}}{\text{kg}} + 2 \ \frac{\text{J}}{\text{kg}}$$

$$= 1002 \ \text{J/kg}$$

Since the flow is frictionless, the same energy is possessed by the water at the reservoir's surface. Since the velocity and gage pressure at the surface are zero, all of the available energy has been converted to potential energy.

$$E_{t,2} = E_{t,1}$$

$$z_2 g = 1002 \ \text{J/kg}$$

$$z_2 = \frac{E_{t,2}}{g} = \frac{1002 \ \frac{\text{J}}{\text{kg}}}{9.81 \ \frac{\text{m}}{\text{s}^2}} = 102.1 \ \text{m}$$

Notice that the volumetric flow rate of the water is not relevant since the water velocity was known. Similarly, the pipe size is not needed.

*Customary U.S. Solution*

$$E_p = \frac{p}{\rho} = \frac{\left(150 \ \frac{\text{lbf}}{\text{in}^2}\right)\left(12 \ \frac{\text{in}}{\text{ft}}\right)^2}{62.4 \ \frac{\text{lbm}}{\text{ft}^3}}$$

$$= 346.15 \ \text{ft-lbf/lbm}$$

$$E_v = \frac{v^2}{2g_c} = \frac{\left(6 \ \frac{\text{ft}}{\text{sec}}\right)^2}{(2)\left(32.2 \ \frac{\text{ft-lbm}}{\text{lbf-sec}^2}\right)}$$

$$= 0.56 \ \text{ft-lbf/lbm}$$

$$E_{t,1} = E_p + E_v = 346.15 \ \frac{\text{ft-lbf}}{\text{lbm}} + 0.56 \ \frac{\text{ft-lbf}}{\text{lbm}}$$

$$= 346.71 \ \text{ft-lbf/lbm}$$

$$E_{t,2} = E_{t,1}$$

$$\frac{z_2 g}{g_c} = 346.71 \ \text{ft-lbf/lbm}$$

$$z_2 = \frac{E_{t,2} g_c}{g} = \frac{\left(346.71 \ \frac{\text{ft-lbf}}{\text{lbm}}\right)\left(32.2 \ \frac{\text{ft-lbm}}{\text{lbf-sec}^2}\right)}{32.2 \ \frac{\text{ft}}{\text{sec}^2}}$$

$$= 346.71 \ \text{ft}$$

## 6. PITOT TUBE

A *pitot tube* (also known as an *impact tube* or *stagnation tube*) is simply a hollow tube that is placed longitudinally in the direction of fluid flow, allowing the flow to enter one end at the fluid's *velocity of approach*. (See Fig. 18.1.) It is used to measure velocity of flow and finds uses in both subsonic and supersonic applications.

**Figure 18.1** *Pitot Tube*

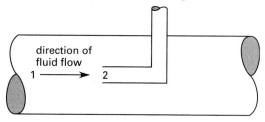

to a pressure measuring device

direction of fluid flow

When the fluid enters the pitot tube, it is forced to come to a stop (at the *stagnation point*), and the velocity energy is transformed into pressure energy. If the fluid is a low-velocity gas, the stagnation is assumed to occur without compression heating of the gas. If there is no friction (the common assumption), the process is said to be adiabatic.

Bernoulli's equation can be used to predict the static pressure at the stagnation point. Since the velocity of the fluid within the pitot tube is zero, the upstream velocity can be calculated if the static and stagnation pressures are known.

$$\frac{p_1}{\rho} + \frac{v_1^2}{2} = \frac{p_2}{\rho} \qquad \qquad 18.14$$

$$v_1 = \sqrt{\frac{2(p_2 - p_1)}{\rho}} \qquad \text{[SI]} \quad 18.15(a)$$

$$v_1 = \sqrt{\frac{2g_c(p_2 - p_1)}{\rho}} \qquad \text{[U.S.]} \quad 18.15(b)$$

In reality, both friction and heating occur, and the fluid may be compressible. These errors are taken care of by a correction factor known as the *impact factor*, $C_i$, which is applied to the derived velocity. $C_i$ is usually very close to 1.00 (e.g., 0.99 or 0.995).

$$v_{\text{actual}} = C_i v_{\text{indicated}}$$

Since accurate measurements of fluid velocity are dependent on one-dimensional fluid flow, it is essential that any obstructions or pipe bends be more than ten pipe diameters upstream from the pitot tube.

## 7. IMPACT ENERGY

*Impact energy, $E_i$,* (also known as *stagnation energy* and *total energy*), is the sum of the kinetic and pressure energy terms.[11] Equation 18.17 is applicable to liquids and gases flowing with velocities less than approximately Mach 0.3.

$$E_i = E_p + E_v \qquad \qquad 18.16$$

$$E_i = \frac{p}{\rho} + \frac{v^2}{2} \qquad \text{[SI]} \quad 18.17(a)$$

$$E_i = \frac{p}{\rho} + \frac{v^2}{2g_c} \qquad \text{[U.S.]} \quad 18.17(b)$$

*Impact head, $h_i$,* is calculated from the impact energy in a manner analogous to Eq. 18.12. Impact head represents the height the liquid will rise in a piezometer-pitot tube when the liquid has been brought to rest (i.e., stagnated) in an adiabatic manner. Such a case is illustrated in Fig. 18.2. If a gas or high-velocity, high-pressure liquid is flowing, it will be necessary to use a mercury manometer or pressure gauge to measure stagnation head.

**Figure 18.2** *Pitot Tube-Piezometer Apparatus*

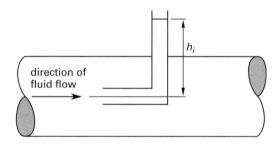

**Example 18.3**

The static pressure of air ($0.075$ lbm/ft$^3$; $1.2$ kg/m$^3$) flowing in a pipe is measured by a precision gauge to be $10.00$ psig ($68.95$ kPa). A pitot tube-manometer indicates $20.6$ in ($0.52$ m) of mercury. The density of mercury is $0.491$ lbm/in$^3$ ($13\,600$ kg/m$^3$). Losses are insignificant. What is the velocity of the air in the pipe?

---

[11]It is confusing to label Eq. 18.16 *total* when the gravitational energy term has been omitted. However, the reference point for gravitational energy is arbitrary, and in this application the reference coincides with the centerline of the fluid flow. In truth, the effective pressure developed in a fluid which has been brought to rest adiabatically does not depend on the elevation or altitude of the fluid. This situation is seldom ambiguous. The application will determine which definition of total head or total energy is intended.

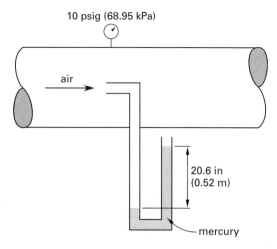

*SI Solution*

The density of mercury is $13\,600$ kg/m$^3$. The impact pressure is

$$p_i = \rho g h$$

$$= \left(13\,600\ \frac{\text{kg}}{\text{m}^3}\right)\left(9.81\ \frac{\text{m}}{\text{s}^2}\right)\left(\frac{0.52\ \text{m}}{1000\ \frac{\text{Pa}}{\text{kPa}}}\right)$$

$$= 69.38\ \text{kPa}$$

From Eq. 18.15, the velocity is

$$v = \sqrt{\frac{2(p_i - p_s)}{\rho}}$$

$$= \sqrt{\frac{(2)(69.38\ \text{kPa} - 68.95\ \text{kPa})\left(1000\ \frac{\text{Pa}}{\text{kPa}}\right)}{1.2\ \frac{\text{kg}}{\text{m}^3}}}$$

$$= 26.8\ \text{m/s}$$

*Customary U.S. Solution*

The pitot tube measures impact (stagnation) pressure. The impact pressure could be calculated from $p = \gamma h$. Alternatively the fluid height equivalent (specific weight) of mercury is $0.491$ psi/in. Therefore, the impact pressure is

$$p_i = (20.6\ \text{in})\left(0.491\ \frac{\text{psi}}{\text{in}}\right) = 10.11\ \text{psig}$$

Since impact pressure is the sum of the static and kinetic (velocity) pressures, the kinetic pressure is

$$p_v = p_i - p_s$$

$$= 10.11\ \text{psig} - 10.00\ \text{psig} = 0.11\ \text{psi}$$

From Eq. 18.15, the velocity is

$$v = \sqrt{\frac{2g_c(p_i - p_s)}{\rho}}$$

$$= \sqrt{\frac{(2)\left(32.2 \frac{\text{ft-lbm}}{\text{lbf-sec}^2}\right)\left(0.11 \frac{\text{lbf}}{\text{in}^2}\right)\left(12 \frac{\text{in}}{\text{ft}}\right)^2}{0.075 \frac{\text{lbm}}{\text{ft}^3}}}$$

$$= 117 \text{ ft/sec}$$

## 8. HYDRAULIC RADIUS

The *hydraulic radius* is defined as the area in flow divided by the *wetted perimeter*.[12] (The hydraulic radius is not the same as the radius of a pipe.) The area in flow is the cross-sectional area of the fluid flowing. When a fluid is flowing under pressure in a pipe (i.e., *pressure flow* in a *pressure conduit*), the area in flow will be the internal area of the pipe. However, the fluid may not completely fill the pipe and may flow simply because of a sloped surface (i.e., *gravity flow* or *open channel flow*).

The wetted perimeter is the length of the line representing the interface between the fluid and the pipe or channel. It does not include the *free surface* length (i.e., the interface between fluid and atmosphere).

$$r_h = \frac{\text{area in flow}}{\text{wetted perimeter}} = \frac{A}{s} \qquad \textit{18.18}$$

Consider a circular pipe flowing completely full. The area in flow is $\pi r^2$. The wetted perimeter is the entire circumference, $2\pi r$. The hydraulic radius is

$$r_{h,\text{pipe}} = \frac{\pi r^2}{2\pi r} = \frac{r}{2} = \frac{D}{4} \qquad \textit{18.19}$$

The hydraulic radius of a pipe flowing half full is also $r/2$, since the flow area and wetted perimeter are both halved. However, it is time-consuming to calculate the hydraulic radius for pipe flow at any intermediate depth, due to the difficulty in evaluating the flow area and wetted perimeter. Appendix 18.A greatly simplifies such calculations.

### Example 18.4

A pipe (internal diameter = 6 units) carries water with a depth of 2 units flowing under the influence of gravity. (a) Calculate the hydraulic radius analytically. (b) Verify the result by using App. 18.A.

---

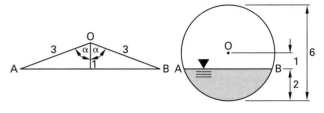

*Solution*

(a) The equations for a circular segment must be used. The radius is $6/2 = 3$.

Points A, O, and B are used to find the central angle of the circular segment.

$$\phi = 2\alpha = 2\left(\arccos\tfrac{1}{3}\right)$$
$$= (2)(70.53°) = 141.06°$$

$\phi$ must be expressed in radians.

$$\phi = 2\pi\left(\frac{141.06°}{360°}\right) = 2.46 \text{ rad}$$

The area of the circular segment (i.e., the area in flow) is

$$A = \tfrac{1}{2}r^2(\phi - \sin\phi) \quad [\phi \text{ in radians}]$$
$$= (0.5)(3)^2(2.46 \text{ rad} - \sin(2.46 \text{ rad}))$$
$$= 8.235 \text{ units}^2$$

The arc length (i.e., the wetted perimeter) is

$$s = r\phi = (3)(2.46 \text{ rad}) = 7.38 \text{ units}$$

The hydraulic radius is

$$r_h = \frac{A}{s} = \frac{8.235 \text{ units}^2}{7.38 \text{ units}} = 1.12 \text{ units}$$

(b) The ratio $d/D$ is needed to use App. 18.A.

$$\frac{d}{D} = \frac{2 \text{ units}}{6 \text{ units}} = 0.333$$

From App. 18.A,

$$\frac{r_h}{D} \approx 0.186$$

$$r_h = (0.186)(6 \text{ units}) = 1.12 \text{ units}$$

## 9. HYDRAULIC DIAMETER

Many fluid, thermodynamic, and heat transfer processes are dependent on the physical length of an object. The general name for this controlling variable is *characteristic dimension*. The characteristic dimension in evaluating fluid flow is the *hydraulic diameter* (also known as

---

[12]The hydraulic radius can also be calculated as one-fourth of the hydraulic diameter of the pipe or channel, as will be subsequently shown. That is, $r_h = \frac{1}{4}D_h$.

the *equivalent hydraulic diameter*).[13] The hydraulic diameter for a full-flowing pipe is simply its inside diameter. The hydraulic diameters of other cross sections in flow are given in Table 18.1. If the hydraulic radius is known, it can be used to calculate the hydraulic diameter.

$$D_h = 4r_h \qquad 18.20$$

**Table 18.1** *Hydraulic Diameters for Common Conduit Shapes*

| conduit cross section | $D_h$ |
|---|---|
| *flowing full* | |
| circle | $D$ |
| annulus (outer diameter $D_o$, inner diameter $D_i$) | $D_o - D_i$ |
| square (side $L$) | $L$ |
| rectangle (sides $L_1$) and $L_2$ | $\dfrac{2L_1 L_2}{L_1 + L_2}$ |
| *flowing partially full* | |
| half-filled circle (diameter $D$) | $D$ |
| rectangle ($h$ deep, $L$ wide) | $\dfrac{4hL}{L + 2h}$ |
| wide, shallow stream ($h$ deep) | $4h$ |
| triangle ($h$ deep, $L$ broad, $s$ side) | $\dfrac{hL}{s}$ |
| trapezoid ($h$ deep, $a$ wide at top, $b$ wide at bottom, $s$ side) | $\dfrac{2h(a+b)}{b+2s}$ |

**Example 18.5**

Determine the hydraulic diameter and hydraulic radius for the open trapezoidal channel shown.

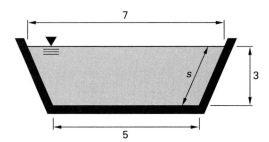

*Solution*

The batter of the inclined walls is $(7 - 5)/2$ walls $= 1$.

$$s = \sqrt{(3)^2 + (1)^2} = 3.16$$

Using Table 18.1,

$$D_h = \frac{2h(a+b)}{b+2s} = \frac{(2)(3)(7+5)}{5+(2)(3.16)} = 6.36$$

From Eq. 18.20,

$$r_h = \frac{D_h}{4} = \frac{6.36}{4} = 1.59$$

## 10. REYNOLDS NUMBER

The *Reynolds number*, Re, is a dimensionless number interpreted as the ratio of inertial forces to viscous forces in the fluid.[14]

$$\text{Re} = \frac{\text{inertial forces}}{\text{viscous forces}} \qquad 18.21$$

The inertial forces are proportional to the flow diameter, velocity, and fluid density. (Increasing these variables will increase the momentum of the fluid in flow.) The viscous force is represented by the fluid's absolute viscosity, $\mu$. Thus, the Reynolds number is calculated as

$$\text{Re} = \frac{D_h \text{v} \rho}{\mu} \qquad \text{[SI]} \quad 18.22(a)$$

$$\text{Re} = \frac{D_h \text{v} \rho}{g_c \mu} \qquad \text{[U.S.]} \quad 18.22(b)$$

Since $\mu/\rho$ is defined as the *kinematic viscosity*, $\nu$, Eq. 18.22 can be simplified.[15]

$$\text{Re} = \frac{D_h \text{v}}{\nu} \qquad 18.23$$

Occasionally, the *mass flow rate per unit area*, $G = \rho\text{v}$, will be known. This variable expresses the quantity of fluid flowing in kg/m²·s or lbm/ft²-sec.

$$\text{Re} = \frac{D_h G}{\mu} \qquad \text{[SI]} \quad 18.24(a)$$

$$\text{Re} = \frac{D_h G}{g_c \mu} \qquad \text{[U.S.]} \quad 18.24(b)$$

---

[13]The engineering community is very inconsistent, but the three terms —hydraulic depth, hydraulic diameter, and equivalent diameter—do not have the same meanings. Hydraulic depth (flow area divided by exposed surface width) is a characteristic length used in Froude number and other open channel flow calculations. Hydraulic diameter (four times the area in flow divided by the wetted surface) is a characteristic length used in Reynolds number and friction loss calculations. Equivalent diameter $(1.3(ab)^{0.625}/(a+b)^{0.25})$ is the diameter of a round duct or pipe that will have the same friction loss per unit length as a noncircular duct. Unfortunately, these terms are often used interchangeably.

[14]Engineering authors are not in agreement about the symbol for the Reynolds number. In addition to Re (used in this book), engineers commonly use **Re**, R, $\Re$, $N_{Re}$, and $N_R$.

[15]This simplification implies a caveat as well. If the viscosity is known or is given in a problem, the units must be used to determine if this viscosity is $\mu$ or $\nu$.

Fluids

## 11. LAMINAR FLOW

*Laminar flow* gets its name from the word *laminae* (layers). If all of the fluid particles move in paths parallel to the overall flow direction (i.e., in layers), the flow is said to be *laminar*. (The terms *viscous flow* and *streamline flow* are also used.) This occurs in pipeline flow when the Reynolds number is less than (approximately) 2100. Laminar flow is typical when the flow channel is small, the velocity is low, and the fluid is viscous. Viscous forces are dominant in laminar flow.

In laminar flow, a stream of dye inserted in the flow will continue from the source in a continuous, unbroken line with very little mixing of the dye and surrounding liquid. The fluid particle paths coincide with imaginary *streamlines*. (Streamlines and velocity vectors are always tangent to each other.) A "bundle" of these streamlines (i.e., a *streamtube*) constitutes a complete fluid flow.

## 12. TURBULENT FLOW

A fluid is said to be in *turbulent flow* if the Reynolds number is greater than (approximately) 4000. (This is the most common situation.) Turbulent flow is characterized by a three-dimensional movement of the fluid particles superimposed on the overall direction of motion. A stream of dye injected into a turbulent flow will quickly disperse and uniformly mix with the surrounding flow. Inertial forces dominate in turbulent flow. At very high Reynolds numbers, the flow is said to be *fully turbulent*.

## 13. CRITICAL FLOW

The flow is said to be in a *critical zone* or *transition region* when the Reynolds number is between 2100 and 4000. These numbers are known as the lower and upper *critical Reynolds numbers* for fluid flow, respectively. (Critical Reynolds numbers for other processes are different.) It is difficult to design for the transition region, since fluid behavior is not consistent and few processes operate in the critical zone. In the event a critical zone design is required, the conservative assumption of turbulent flow will result in the greatest value of friction loss.

## 14. FLUID VELOCITY DISTRIBUTION IN PIPES

With laminar flow, the viscous effects make some fluid particles adhere to the pipe wall. The closer a particle is to the pipe wall, the greater the tendency will be for the fluid to adhere to the pipe wall. The following statements characterize laminar flow.

- The velocity distribution is parabolic.

- The velocity is zero at the pipe wall.

- The velocity is maximum at the center and equal to twice the average velocity.

$$v_{ave} = \frac{\dot{V}}{A} = \frac{v_{max}}{2} \quad \text{[laminar]} \qquad 18.25$$

With turbulent flow, there is generally no distinction made between the velocities of particles near the pipe wall and particles at the pipe centerline.[16] All of the fluid particles are assumed to have the same velocity. This velocity is known as the *average* or *bulk velocity*. It can be calculated from the volume flowing.

$$v_{ave} = \frac{\dot{V}}{A} \quad \text{[turbulent]} \qquad 18.26$$

Laminar and turbulent velocity distributions are shown in Fig. 18.3. In actuality, no flow is completely turbulent, and there is a difference between the *centerline velocity* and the average velocity. The error decreases as the Reynolds number increases. The ratio $v_{ave}/v_{max}$ starts at approximately 0.75 for $Re = 4000$ and increases to approximately 0.86 at $Re = 10^6$. Most problems ignore the difference between $v_{ave}$ and $v_{max}$, but care should be taken when a centerline measurement (as from a pitot tube) is used to evaluate the average velocity.

**Figure 18.3** *Laminar and Turbulent Velocity Distributions*

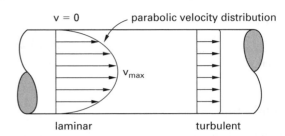

For turbulent flow ($Re \approx 10^5$) in a smooth, circular pipe of radius $r_o$, the velocity at a radial distance $r$ from the centerline is given by the $1/7$-*power law*.

$$v_r = v_{max} \left( \frac{r_o - r}{r_o} \right)^{1/7} \quad \text{[turbulent flow]} \qquad 18.27$$

The fluid's *velocity profile*, given by Eq. 18.27, is valid in smooth pipes up to a Reynolds number of approximately 100,000. Above that, up to a Reynolds number of approximately 400,000, an exponent of $1/8$ fits experimental data better. For rough pipes, the exponent is larger (e.g., $1/5$).

The ratio of the average velocity to maximum velocity is known as the *pipe coefficient* or *pipe factor*. Considering all the other coefficients used in pipe flow, these names are somewhat vague and ambiguous. Therefore, they are not in widespread use.

---

[16]This disregards the *boundary layer*, a thin layer near the pipe wall, where the velocity goes from zero to $v_{ave}$.

Equation 18.27 can be integrated to determine the average velocity.

$$\mathrm{v_{ave}} = \left(\frac{49}{60}\right)\mathrm{v_{max}} = 0.817\mathrm{v_{max}} \quad \text{[turbulent flow]} \qquad \textit{18.28}$$

When the flow is laminar, the velocity profile within a pipe will be parabolic and of the form of Eq. 18.29. (Equation 18.27 is for turbulent flow and does not describe a parabolic velocity profile.) The velocity at a radial distance $r$ from the centerline in a pipe of radius $r_o$ is

$$\mathrm{v}_r = \mathrm{v_{max}}\left(\frac{r_o^2 - r^2}{r_o^2}\right) \quad \text{[laminar flow]} \qquad \textit{18.29}$$

When the velocity profile is parabolic, the flow rate and pressure drop can easily be determined. The average velocity is half of the maximum velocity given in the velocity profile equation.

$$\mathrm{v_{ave}} = \tfrac{1}{2}\mathrm{v_{max}} \quad \text{[laminar flow]} \qquad \textit{18.30}$$

The average velocity is used to determine the flow quantity and friction loss. The friction loss is determined by traditional means.

$$\dot{V} = A\mathrm{v_{ave}} \qquad \textit{18.31}$$

The kinetic energy of laminar flow can be found by integrating the velocity profile equation, resulting in Eq. 18.32.

$$E_\mathrm{v} = \mathrm{v_{ave}^2} \quad \text{[laminar flow]} \qquad \text{[SI]} \quad \textit{18.32(a)}$$

$$E_\mathrm{v} = \frac{\mathrm{v_{ave}^2}}{g_c} \quad \text{[laminar flow]} \qquad \text{[U.S.]} \quad \textit{18.32(b)}$$

## 15. ENERGY GRADE LINE

The *energy grade line* (EGL) is a graph of the total energy (total specific energy) along a length of pipe.[17] In a frictionless pipe without pumps or turbines, the total specific energy is constant, and the EGL will be horizontal. (This is a restatement of the Bernoulli equation.)

$$\text{elevation of EGL} = h_p + h_\mathrm{v} + h_z \qquad \textit{18.33}$$

The *hydraulic grade line* (HGL) is the graph of the sum of the pressure and gravitational heads, plotted as a position along the pipeline. Since the pressure head can increase at the expense of the velocity head, the HGL can increase in elevation if the flow area is increased.

$$\text{elevation of HGL} = h_p + h_z \qquad \textit{18.34}$$

---

[17]The term *energy line* (EL) is also used.

The difference between the EGL and the HGL is the velocity head, $h_\mathrm{v}$, of the fluid.

$$h_\mathrm{v} = \text{elevation of EGL} - \text{elevation of HGL} \qquad \textit{18.35}$$

The following rules apply to these grade lines in a frictionless environment, in a pipe flowing full (i.e., under pressure), without pumps or turbines. (See Fig. 18.4.)

- The EGL is always horizontal.
- The HGL is always equal to or below the EGL.
- For still ($\mathrm{v} = 0$) fluid at a free surface, EGL = HGL (i.e., the EGL coincides with the fluid surface in a reservoir).
- If flow velocity is constant (i.e., flow in a constant-area pipe), the HGL will be horizontal and parallel to the EGL, regardless of pipe orientation or elevation.
- When the flow area decreases, the HGL decreases.
- When the flow area increases, the HGL increases.
- In a free jet (i.e., a stream of water from a hose), the HGL coincides with the jet elevation, following a parabolic path.

**Figure 18.4** *Energy and Hydraulic Grade Lines Without Friction*

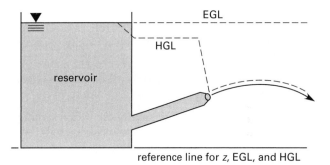

reference line for *z*, EGL, and HGL

## 16. SPECIFIC ENERGY

*Specific energy* is a term that is used primarily with open channel flow. It is the total energy with respect to the channel bottom, consisting of pressure and velocity energy terms only.

$$E_\text{specific} = E_p + E_\mathrm{v} \qquad \textit{18.36}$$

Since the channel bottom is chosen as the reference elevation ($z = 0$) for gravitational energy, there is no contribution by gravitational energy to specific energy.

$$E_\text{specific} = \frac{p}{\rho} + \frac{\mathrm{v}^2}{2} \qquad \text{[SI]} \quad \textit{18.37(a)}$$

$$E_\text{specific} = \frac{p}{\rho} + \frac{\mathrm{v}^2}{2g_c} \qquad \text{[U.S.]} \quad \textit{18.37(b)}$$

However, since $p$ is the hydrostatic pressure at the channel bottom due to a fluid depth, $d$, $p/\rho$ can be interpreted as the depth of the fluid.

$$E_{\text{specific}} = d + \frac{\text{v}^2}{2g} \quad \text{[open channel]} \qquad 18.38$$

Specific energy is constant when the flow depth and width are constant (i.e., *uniform flow*). A change in channel width will cause a change in flow depth, and since width is not part of the equation for specific energy, there will be a corresponding change in specific energy. There are other ways that specific energy can decrease, also.[18]

## 17. PIPE MATERIALS AND SIZES

Many materials are used for pipes. The material used depends on the application. Water supply distribution, wastewater collection, and air conditioning refrigerant lines all place different demands on pipe material performance. Pipe materials are chosen on the basis of strength to withstand internal pressures, strength to withstand external loads from backfill and traffic, smoothness, corrosion resistance, chemical inertness, cost, and other factors.

The following are characteristics of the major types of commercial pipe materials that are in use.

- *asbestos cement:* immune to electrolysis and corrosion, light in weight but weak structurally; environmentally limited

- *concrete:* durable, water-tight, low maintenance, smooth interior

- *copper and brass:* used primarily for water, condensate, and refrigerant lines; in some cases, easily bent by hand, good thermal conductivity

- *ductile cast iron:* long-lived, strong, impervious, heavy, scour-resistant, but costly

- *plastic* (PVC and ABS):[19] chemically inert, resistant to corrosion, very smooth, lightweight, low cost

- *steel:* high strength, ductile, resistant to shock, very smooth interior, but susceptible to corrosion

- *vitrified clay:* resistant to corrosion, acids (e.g., hydrogen sulfide from septic sewage), scour, and erosion

Table 18.2 lists recommendations for pipe materials in several common applications.

The required wall thickness of a pipe is proportional to the pressure the pipe must carry. However, not all pipes operate at high pressures. Therefore, pipes and tubing may be available in different wall thicknesses (*schedules,*

---

[18]Specific energy changes dramatically in a *hydraulic jump* or *hydraulic drop.*
[19]PVC: polyvinyl chloride; ABS: acrylonitrile-butadiene-styrene.

*Table 18.2* Recommended Pipe Materials by Application

| service | pipe material |
|---|---|
| most refrigerants (suction, liquid, and hot gas lines) | hard copper tubing (type L);[a] standard wall steel pipe, lap-welded or seamless |
| chilled water | hard copper tubing; plain (black) or galvanized steel pipe[b] |
| condenser or make-up water | hard copper tubing; plain or galvanized steel pipe[b] |
| steam or condensate | hard copper tubing; steel pipe[b] |
| hot water | hard copper tubing; steel pipe[b] |

[a]Soft copper may be used for $^1/_4$ in (6.3 mm) and $^3/_8$ in (9.5 mm) (outside diameter) with wall thicknesses of 0.30 in (7.6 mm) and 0.32 in (8.1 mm), respectively. Soft copper refrigeration lines are commonly used up to $1^3/_8$ in (35 mm) (outside diameter). Mechanical joints should not be used with soft copper tubing larger than $^7/_8$ in (22 mm).
[b]Standard wall steel pipe or type-M hard copper tubing are usually satisfactory for air conditioning applications. However, the pressure rating of the pipe material should be checked at the design temperature.

*series,* or *types*). Steel pipe, for example, is available in schedules 40, 80, and others.[20]

For initial estimates, the approximate schedule of steel pipe can be calculated from Eq. 18.39. $p$ is the operating pressure in psig; $S$ is the allowable stress in the pipe material; and $E$ is the *joint efficiency*, also known as the *joint quality factor* (typically 1.00 for seamless pipe, 0.85 for electric resistance-welded pipe, 0.80 for electric fusion-welded pipe, and 0.60 for furnace butt-welded pipe). For seamless carbon steel (A53) pipe used below 650°F (340°C), the allowable stress is approximately 12,000 psi to 15,000 psi. So, with butt-welded joints, a value of 6500 psi is often used for the product $SE$.

$$\text{schedule} \approx \frac{1000p}{SE} \qquad 18.39$$

Steel pipe is available in black (i.e., plain "black pipe") and galvanized (inside, outside, or both) varieties. Steel pipe is manufactured in plain-carbon and stainless varieties. AISI 316 stainless is particularly corrosion resistant.

The actual dimensions of some pipes (concrete, clay, some cast iron, etc.) coincide with their *nominal dimensions*. For example, a 12 in concrete pipe has an inside diameter of 12 in, and no further refinement is needed. However, some pipes and tubing (e.g., steel pipe, copper and brass tubing, and some cast iron) are called out by a nominal diameter that has nothing to do with the internal diameter of the pipe. For example, a 16 in schedule-40 steel pipe has an actual inside diameter of 15 in. In

---

[20]Other schedules of steel pipe, such as 30, 60, 120, etc., also exist, but in limited sizes, as Table 18.3 indicates. Schedule-40 pipe roughly corresponds to the standard weight (S) designation used in the past. Schedule-80 roughly corresponds to the extra-strong (X) designation. There is no uniform replacement designation for double-extra-strong (XX) pipe.

**Table 18.3** Dimensions of Commercial Steel Pipe* (English Units)

| nominal diameter | outside diameter | schedule 10 | 20 | 30 | 40 | 60 | 80 | 100 | 120 | 140 | 160 |
|---|---|---|---|---|---|---|---|---|---|---|---|
| | | wall thickness (in) | | | | | | | | | |
| $\frac{1}{2}$ | 0.840 | . . . . | . . . . | . . . . | 0.109 | . . . . | 0.147 | . . . . | . . . . | . . . . | 0.187 |
| $\frac{3}{4}$ | 1.05 | . . . . | . . . . | . . . . | 0.113 | . . . . | 0.154 | . . . . | . . . . | . . . . | 0.218 |
| 1 | 1.315 | . . . . | . . . . | . . . . | 0.133 | . . . . | 0.179 | . . . . | . . . . | . . . . | 0.250 |
| $1\frac{1}{4}$ | 1.660 | . . . . | . . . . | . . . . | 0.140 | . . . . | 0.191 | . . . . | . . . . | . . . . | 0.250 |
| $1\frac{1}{2}$ | 1.900 | . . . . | . . . . | . . . . | 0.145 | . . . . | 0.200 | . . . . | . . . . | . . . . | 0.281 |
| 2 | 2.375 | . . . . | . . . . | . . . . | 0.154 | . . . . | 0.218 | . . . . | . . . . | . . . . | 0.343 |
| $2\frac{1}{2}$ | 2.875 | . . . . | . . . . | . . . . | 0.203 | . . . . | 0.276 | . . . . | . . . . | . . . . | 0.375 |
| 3 | 3.500 | . . . . | . . . . | . . . . | 0.216 | . . . . | 0.300 | . . . . | . . . . | . . . . | 0.437 |
| $3\frac{1}{2}$ | 4.000 | . . . . | . . . . | . . . . | 0.226 | . . . . | 0.318 | . . . . | . . . . | . . . . | . . . . |
| 4 | 4.500 | . . . . | . . . . | . . . . | 0.237 | . . . . | 0.337 | . . . . | 0.437 | . . . . | 0.531 |
| 5 | 5.563 | . . . . | . . . . | . . . . | 0.258 | . . . . | 0.375 | . . . . | 0.500 | . . . . | 0.625 |
| 6 | 6.625 | . . . . | . . . . | . . . . | 0.280 | . . . . | 0.432 | . . . . | 0.562 | . . . . | 0.718 |
| 8 | 8.625 | . . . . | 0.250 | 0.277 | 0.322 | 0.406 | 0.500 | 0.593 | 0.718 | 0.812 | 0.906 |
| 10 | 10.75 | . . . . | 0.250 | 0.307 | 0.365 | 0.500 | 0.593 | 0.718 | 0.843 | 1.000 | 1.125 |
| 12 | 12.75 | . . . . | 0.250 | 0.330 | 0.406 | 0.562 | 0.687 | 0.843 | 1.000 | 1.125 | 1.312 |
| 14 | 14.00 | 0.250 | 0.312 | 0.375 | 0.437 | 0.593 | 0.750 | 0.937 | 1.062 | 1.250 | 1.406 |
| 16 | 16.00 | 0.250 | 0.312 | 0.375 | 0.500 | 0.656 | 0.843 | 1.031 | 1.218 | 1.437 | 1.562 |
| 18 | 18.00 | 0.250 | 0.312 | 0.437 | 0.562 | 0.718 | 0.937 | 1.156 | 1.343 | 1.562 | 1.750 |
| 20 | 20.00 | 0.250 | 0.375 | 0.500 | 0.593 | 0.812 | 1.031 | 1.250 | 1.500 | 1.750 | 1.937 |
| 24 | 24.00 | 0.250 | 0.375 | 0.562 | 0.687 | 0.937 | 1.218 | 1.500 | 1.750 | 2.062 | 2.312 |

*Also, see App. 18.B and App. 18.C.

some cases, the nominal size does not coincide with the external diameter, either.

PVC (polyvinyl chloride) pipe is used extensively as water and sewer pipe due to its combination of strength, ductility, and corrosion resistance. Manufactured lengths are approximately 10 ft to 13 ft (3 m to 3.9 m) for sewer pipe and 20 ft (6 m) for water pipe, with integral gasketed joints or solvent-weld bells. Infiltration is very low (less than 50 gal/in-mile-day), even in the wettest environments. The low Manning's roughness constant (0.009 typical) allows PVC sewer pipe to be used with flatter grades or smaller diameters. PVC pipe is resistant to corrosive soils and sewerage gases and is generally resistant to abrasion from pipe-cleaning tools.

It is essential that tables of pipe sizes, such as App. 18.B, be used when working problems involving steel and copper pipes since there is no other way to obtain the inside diameters of such pipes.[21]

## 18. MANUFACTURED PIPE STANDARDS

There are many different standards governing pipe diameters and wall thicknesses. A pipe's nominal outside diameter is rarely sufficient to determine the internal dimensions of the pipe. A manufacturing specification and class or category are usually needed to completely specify pipe dimensions.

Cast iron pipe was formerly produced to ANSI/AWWA C106/A21.6 but is now obsolete. Ductile iron (CI/DI) pipes are produced to ANSI/AWWA C150/A21.50 and C151/A21.51 standards. Gasketed PVC sewer pipe up to 15 in inside diameter is produced to ASTM D3034 standards. Gasketed sewer PVC pipe from 18 in to 48 in is produced to ASTM F679 standards. PVC pressure pipe for water distribution is manufactured to ANSI/ AWWA C900 standards. ABS and PVC truss pipe for unpressurized use is manufactured to ASTM D2680 standards. Reinforced concrete pipe (RCP) for culvert, storm drain, and sewer applications is manufactured to ASTM/AASHTO C76/MI70 standards.

## 19. CORRUGATED METAL PIPE

Corrugated metal pipe (CMP, also known as corrugated steel pipe) is frequently used for culverts. Pipe is made from corrugated sheets of galvanized steel that are rolled and riveted together along a longitudinal seam. Aluminized steel may also be used in certain ranges of soil pH. Standard round pipe diameters range from 8 in to 96 in (200 mm to 2450 mm). Metric dimensions of standard diameters are usually rounded to the nearest 25 mm or 50 mm (e.g., a 42 in culvert would be specified as a 1050 mm culvert, not 1067 mm).

---

[21]It is a characteristic of standard steel pipes that the schedule number does not affect the outside diameter of the pipe. An 8 in schedule-40 pipe has the same exterior dimensions as an 8 in schedule-80 pipe. However, the interior flow area will be less for the schedule-80 pipe.

Larger and noncircular culverts can be created out of curved steel plate. Standard section lengths are 10 ft to 20 ft (3 m to 6 m). Though most corrugations are transverse (i.e., annular), helical corrugations are also used. Metal gages of 8, 10, 12, 14, and 16 are commonly used, depending on the depth of burial.

The most common corrugated steel pipe has transverse corrugations that are $^1/_2$ in (13 mm) deep and $2^2/_3$ in (68 mm) from crest to crest. These are referred to as "$2^1/_2$ inch" or "$68 \times 13$" corrugations. For larger culverts, corrugations with a 2 in (25 mm) depth and 3 in, 5 in, and 6 in (76 mm, 125 mm, or 152 mm) pitches are used. Plate-based products using 6 in by 2 in (152 mm by 51 mm) corrugations are known as *structural plate corrugated steel pipe* (SPCSP) and *multiplate* after the trade-named product "Multi-Plate™."

The flow area for circular culverts is based on the nominal culvert diameter, regardless of the gage of the plate metal used to construct the pipe. Flow area is calculated to (at most) three significant digits.

A Hazen-Williams coefficient, $C$, of 60 is typically used with all sizes of corrugated pipe. Values of $C$ and Manning's constant, $n$, for corrugated pipe are generally not affected by age. *Design Charts for Open Channel Flow* (U.S. Department of Transportation, 1979) recommends a Manning constant of $n = 0.024$ for all cases. The U.S. Department of the Interior recommends the following values. For standard ($2^2/_3$ in by $^1/_2$ in or 68 mm by 13 mm) corrugated pipe with the diameters given: 12 in (457 mm), 0.027; 24 in (610 mm), 0.025; 36–48 in (914–1219 mm), 0.024; 60–84 in (1524–2134 mm), 0.023; 96 in (2438 mm), 0.022. For (6 in by 2 in or 152 mm by 51 mm) multiplate construction with the diameters given: 5–6 ft (1.5–1.8 m), 0.034; 7–8 ft (2.1–2.4 m), 0.033; 9–11 ft (2.7–3.3 m), 0.032; 12–13 ft (3.6–3.9 m), 0.031; 14–15 ft (4.2–4.5 m), 0.030; 16–18 ft (4.8–5.4 m), 0.029; 19–20 ft (5.8–6.0 m), 0.028; 21–22 ft (6.3–6.6 m), 0.027.

If the inside of the corrugated pipe has been asphalted completely smooth 360° circumferentially, Manning's $n$ ranges from 0.009 to 0.011. For culverts with 40% asphalted inverts, $n = 0.019$. For other percentages of paved invert, the resulting value is proportional to the percentage and the values normally corresponding to that diameter pipe. For field-bolted corrugated metal pipe arches, $n = 0.025$.

It is also possible to calculate the Darcy friction loss if the corrugation depth, 0.5 in (13 mm) for standard corrugations and 2.0 in (51 mm) for multiplate, is taken as the specific roughness.

## 20. PAINTS, COATINGS, AND LININGS

Various materials are used to protect steel and ductile iron pipes against rust and other forms of corrosion. *Red primer* is a shop-applied, rust-inhibiting primer applied to prevent short-term rust prior to shipment and the application of subsequent coatings. *Asphaltic coating* ("tar" coating) is applied to the exterior of underground pipes. *Bituminous coating* refers to a similar coating made from tar pitch. Both asphaltic and bituminous coatings should be completely removed or sealed with a synthetic resin prior to the pipe being finish-coated, since their oils may bleed through otherwise.

Though bituminous materials (i.e., asphaltic materials) continue to be cost effective, epoxy-based products are now extensively used. Epoxy products are delivered as a two-part formulation (a polyamide resin and liquid chemical hardener) that is mixed together prior to application. *Coal tar epoxy*, also referred to as *epoxy coal tar*, a generic name, sees frequent use in pipes exposed to high humidity, seawater, other salt solutions, and crude oil. Though suitable for coating steel penstocks of hydroelectric installations, coal tar epoxy is generally not suitable for potable water delivery systems. Though it is self-priming, appropriate surface preparation is required for adequate adhesion. Coal tar epoxy has a density of 1.9 lbm/gal to 2.3 lbm/gal (230 g/L to 280 g/L).

## 21. TYPES OF VALVES

Valves used for *shutoff service* (e.g., gate, plug, ball, and butterfly valves) are used fully open or fully closed. *Gate valves* offer minimum resistance to flow. They are used in clean fluid and slurry services when valve operation is infrequent. Many turns of the handwheels are required to raise or lower their gates. *Plug valves* provide for tight shutoff. A 90° turn of their handles is sufficient to rotate the plugs fully open or closed. *Eccentric plug valves*, in which the plug rotates out of the fluid path when open, are among the most common wastewater valves. *Plug cock valves* have a hollow passageway in their plugs through which fluid can flow. Both eccentric plug valves and plug cock valves are referred to as "plug valves." *Ball valves* offer an unobstructed flow path and tight shutoff. They are often used with slurries and viscous fluids, as well as with cryogenic fluids. A 90° turn of their handles rotates the balls fully open or closed. *Butterfly valves* (when specially designed with appropriate seats) can be used for shutoff operation. They are particularly applicable to large flows of low-pressure (vacuum up to 200 psig (1.4 MPa)) gases or liquids, although high performance butterfly valves can operate as high as 600 psig (4.1 MPa). Their straight-through, open-disk design results in minimal solids build-up and low pressure drops.

Other valve types (e.g., globe, needle, Y-, angle, and butterfly valves) are more suitable for *throttling service*. *Globe valves* provide positive shutoff and precise metering on clean fluids. However, since the seat is parallel to the direction of flow and the fluid makes two right-angle turns, there is substantial resistance and pressure drop through them, as well as relatively fast erosion of the seat. Globe valves are intended for frequent operation. *Needle valves* are similar to globe valves, except that the plug is a tapered, needle-like cone. Needle valves provide accurate metering of small flows of clean fluids. Needle valves are applicable to cryogenic fluids. *Y-valves* are similar to globe valves in operation, but their seats are inclined to the direction of flow, offering more of a straight-through passage and unobstructed flow than the globe valve. *Angle valves* are essentially globe valves where the fluid makes a 90° turn. They can be used for throttling and shut-off of clean or viscous fluids and slurries. *Butterfly valves* are often used for throttling

services with the same limitations and benefits as those listed for shutoff use.

Other valves are of the *check (nonreverse-flow, antireversal)* variety. These react automatically to changes in pressure to prevent reversals of flow. Special check valves can also prevent excess flow. Figure 18.5 illustrates *swing*, *lift*, and *angle lift check valves*, and Table 18.4 gives typical characteristics of common valve types.

**Figure 18.5** *Types of Valves*

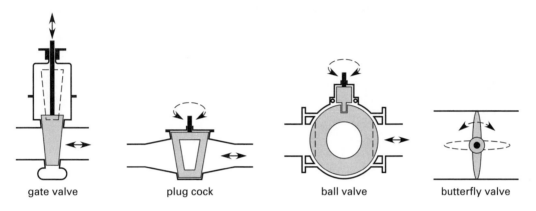

(a) valves for shut-off service

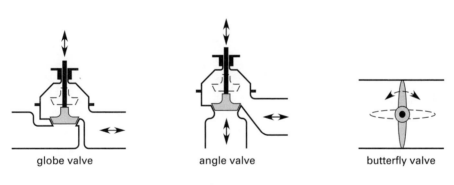

(b) valves for throttle service

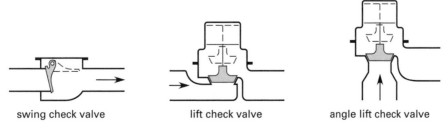

(c) valves for antireversal service

**Table 18.4** *Typical Characteristics of Common Valve Types*

| valve type | fluid condition | switching frequency | pressure drop (fully open) | typical control response | typical maximum pressure, atm | typical maximum temperature, °C |
|---|---|---|---|---|---|---|
| ball | clean | low | low | very poor | 160 | 300 |
| butterfly | clean | low | low | poor | 200 | 400 |
| diaphragm* (not shown) | clean to slurried | very high | low to medium | very good | 16 | 150 |
| gate | clean | low | low | very poor | 50 | 400 |
| globe | clean | high | medium to high | very good | 80 | 300 |
| plug | clean | low | low | very poor | 160 | 300 |

*Diaphragm valves use a flexible diaphragm to block the flow path. The diaphragm may be manually or pneumatically actuated. Such valves are suitable for both throttling and shut-off service.

# 19 Fluid Dynamics

## Nomenclature

| | | | |
|---|---|---|---|
| $a$ | length | ft | m |
| $a$ | speed of sound | ft/sec | m/s |
| $A$ | area | ft$^2$ | m$^2$ |
| $C$ | coefficient | – | – |
| $C$ | Hazen-Williams coefficient | – | – |
| $d$ | diameter | in | cm |
| $D$ | diameter | ft | m |
| $E$ | bulk modulus | lbf/ft$^2$ | Pa |
| $E$ | specific energy | ft-lbf/lbm | J/kg |
| $f$ | Darcy friction factor | – | – |
| $f$ | fraction split | – | – |
| $F$ | force | lbf | N |
| Fr | Froude number | – | – |
| $g$ | gravitational acceleration | ft/sec$^2$ | m/s$^2$ |
| $g_c$ | gravitational conversion constant | ft-lbm/lbf-sec$^2$ | n.a. |
| $G$ | mass flow rate per unit area | lbm/ft$^2$-sec | kg/m$^2$·s |
| $h$ | height or head | ft | m |
| $I$ | impulse | lbf-sec | N·s |
| $k$ | ratio of specific heats | – | – |
| $K$ | minor loss coefficient | – | – |
| $l$ | length | ft | m |
| $L$ | length | ft | m |
| $m$ | mass | lbm | kg |
| $\dot{m}$ | mass flow rate | lbm/sec | kg/s |
| MW | molecular weight | lbm/lbmol | kg/kmol |
| $n$ | Manning roughness constant | – | – |
| $n$ | flow rate exponent | – | – |
| $p$ | pressure | lbf/ft$^2$ | Pa |

Fluids

| $P$ | momentum | lbm-ft/sec | kg·m/s |
| $P$ | power | ft-lbf/sec | W |
| $Q$ | flow rate | gal/min | n.a. |
| $r$ | radius | ft | m |
| rpm | rotational speed | rev/min | rev/min |
| $R$ | resultant force | lbf | N |
| $R^*$ | universal gas constant | ft-lbf/lbmol-°R | J/kmol·K |
| Re | Reynolds number | – | – |
| SG | specific gravity | – | – |
| $t$ | thickness | ft | m |
| $t$ | time | sec | s |
| $T$ | absolute temperature | °R | K |
| $u$ | $x$-component of velocity | ft/sec | m/s |
| $v$ | $y$-component of velocity | ft/sec | m/s |
| v | velocity | ft/sec | m/s |
| $V$ | volume | ft$^3$ | m$^3$ |
| $\dot{V}$ | volumetric flow rate | ft$^3$/sec | m$^3$/s |
| $W$ | work | ft-lbf | J |
| We | Weber number | – | – |
| WHP | water horsepower | hp | n.a. |
| $x$ | $x$-coordinate of position | ft | m |
| $y$ | $y$-coordinate of position | ft | m |
| $Y$ | expansion factor | – | – |
| $z$ | elevation | ft | m |

**Symbols**

| $\beta$ | diameter ratio | – | – |
| $\gamma$ | specific weight | lbf/ft$^3$ | N/m$^3$ |
| $\Gamma$ | circulation | ft$^2$/sec | m$^2$/s |
| $\epsilon$ | specific roughness | ft | m |
| $\eta$ | efficiency | – | – |
| $\eta$ | non-Newtonian viscosity | lbf-sec/ft$^2$ | Pa·s |
| $\theta$ | angle | deg | deg |
| $\mu$ | absolute viscosity | lbf-sec/ft$^2$ | Pa·s |
| $\nu$ | kinematic viscosity | ft$^2$/sec | m$^2$/s |
| $\rho$ | density | lbm/ft$^3$ | kg/m$^3$ |
| $\sigma$ | surface tension | lbf/ft | N/m |
| $\tau$ | shear stress | lbf/ft$^2$ | Pa |
| $\upsilon$ | specific volume | ft$^3$/lbm | m$^3$/kg |
| $\phi$ | angle | deg | deg |
| $\Phi$ | stream potential | – | – |
| $\psi$ | sphericity | – | – |
| $\Psi$ | stream function | – | – |
| $\omega$ | angular velocity | rad/sec | rad/s |

**Subscripts**

| $A$ | added (by pump) |
| $b$ | blade or buoyant |
| $c$ | contraction |
| $d$ | discharge |
| $D$ | drag |
| $e$ | equivalent |
| $E$ | extracted (by turbine) |
| $f$ | friction or flow |
| $i$ | inside |
| $I$ | instrument |
| $L$ | lift |
| $m$ | minor, model, or manometer fluid |
| $o$ | orifice or outside |
| $p$ | pressure or prototype |

| $r$ | ratio |
| $s$ | static |
| $t$ | total, tank, or theoretical |
| v | velocity |
| va | velocity of approach |
| $z$ | potential |

## 1. HYDRAULICS AND HYDRODYNAMICS

This chapter covers fluid moving through pipes, measurements with venturis and orifices, and other motion-related topics such as model theory, lift and drag, and pumps. In a strict interpretation, any fluid-related phenomenon that is not hydro*statics* should be hydro*dynamics*. However, tradition has separated the study of moving fluids into the fields of hydraulics and hydrodynamics.

In a general sense, *hydraulics* is the study of the practical laws of fluid flow and resistance in pipes and open channels. Hydraulic formulas are often developed from experimentation, empirical factors, and curve fitting, without an attempt to justify why the fluid behaves the way it does.

On the other hand, *hydrodynamics* is the study of fluid behavior based on theoretical considerations. Hydrodynamicists start with Newton's laws of motion and try to develop models of fluid behavior. Models developed in this manner are complicated greatly by the inclusion of viscous friction and compressibility. Therefore, hydrodynamic models assume a perfect fluid with constant density and zero viscosity. The conclusions reached by hydrodynamicists can differ greatly from those reached by hydraulicians.[1]

## 2. CONSERVATION OF MASS

Fluid mass is always conserved in fluid systems, regardless of the pipeline complexity, orientation of the flow, or which fluid is flowing. This single concept is often sufficient to solve simple fluid problems.

$$\dot{m}_1 = \dot{m}_2 \qquad \textbf{19.1}$$

When applied to fluid flow, the conservation of mass law is known as the *continuity equation*.

$$\rho_1 A_1 v_1 = \rho_2 A_2 v_2 \qquad \textbf{19.2}$$

If the fluid is incompressible, then $\rho_1 = \rho_2$.

$$A_1 v_1 = A_2 v_2 \qquad \textbf{19.3}$$

$$\dot{V}_1 = \dot{V}_2 \qquad \textbf{19.4}$$

Various units and symbols are used for *volumetric flow rate*. (Though this book uses $\dot{V}$, the symbol $Q$ is often used when the flow rate is expressed in gallons.) MGD (millions of gallons per day) and MGPCD

---

[1]Perhaps the most disparate conclusion is *D'Alembert's paradox*. In 1744, D'Alembert derived theoretical results "proving" that there is no resistance to bodies moving through an ideal (non-viscous) fluid.

(millions of gallons per capita day) are units commonly used in municipal water works problems. MMSCFD (millions of standard cubic feet per day) may be used to express gas flows.

Calculation of flow rates is often complicated by the interdependence between flow rate and friction loss. Each affects the other. Hence, many pipe flow problems must be solved iteratively. Usually, a reasonable friction factor is assumed and is used to calculate an initial flow rate. The flow rate establishes the flow velocity, from which a revised friction factor can be determined.

## 3. TYPICAL VELOCITIES IN PIPES

Fluid friction in pipes is kept at acceptable levels by maintaining reasonable fluid velocities. Table 19.1 lists typical maximum fluid velocities. Higher velocities may be observed in practice, but only with a corresponding excessive increase in friction and pumping power.

*Table 19.1* Typical Fluid Velocities

| fluid and application | ft/sec | m/s |
|---|---|---|
| water: city service | 2–7 | 0.6–2.1 |
| water: boiler feed | 8–15 | 2.4–4.5 |
| air: compressor suction | 75–200 | 23–60 |
| air: compressor discharge | 100–250 | 30–75 |
| refrigerant: suction | 15–35 | 4.5–11 |
| refrigerant: discharge | 35–60 | 11–18 |
| steam, saturated: heating | 65–100 | 20–30 |
| steam, saturated: miscellaneous | 100–200 | 30–60 |
| steam, superheated: turbine feed | 160–250 | 50–75 |

## 4. STREAM POTENTIAL AND STREAM FUNCTION

An application of hydrodynamic theory is the derivation of the stream function from stream potential. The *stream potential function (velocity potential function)*, $\Phi$, is the algebraic sum of the component velocity potential functions.[2]

$$\Phi = \Phi_x(x, y) + \Phi_y(x, y) \qquad \textit{19.5}$$

The velocity component of the resultant in the $x$-direction is

$$u = \frac{\partial \Phi}{\partial x} \qquad \textit{19.6}$$

The velocity component of the resultant in the $y$-direction is

$$v = \frac{\partial \Phi}{\partial y} \qquad \textit{19.7}$$

[2]The two-dimensional derivation of the stream function can be extended to three dimensions, if necessary. The stream function can also be expressed in the cylindrical coordinate system.

The total derivative of the stream potential function is

$$d\Phi = \frac{\partial \Phi}{\partial x}\, dx + \frac{\partial \Phi}{\partial y}\, dy$$
$$= u\, dx + v\, dy \qquad \textit{19.8}$$

An *equipotential line* is a line along which the function $\Phi$ is constant (i.e., $d\Phi = 0$). The slope of the equipotential line is derived from Eq. 19.8.

$$\left.\frac{dy}{dx}\right|_{\text{equipotential}} = -\frac{u}{v} \qquad \textit{19.9}$$

For flow through a porous, permeable medium, pressure will be constant along equipotential lines (i.e., along lines of constant $\Phi$). (See Fig. 19.1.) However, for an ideal, nonviscous fluid flowing in a frictionless environment, $\Phi$ has no physical significance. Even though $\Phi$ has a theoretical basis, it does not coincide with any measurable physical quantity.

*Figure 19.1* Equipotential Lines and Streamlines

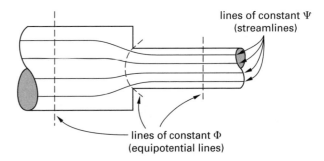

lines of constant $\Psi$ (streamlines)

lines of constant $\Phi$ (equipotential lines)

The *stream function (Lagrange stream function)*, $\Psi(x, y)$, defines the direction of flow at a point.

$$u = \frac{\partial \Psi}{\partial y} \qquad \textit{19.10}$$

$$v = -\frac{\partial \Psi}{\partial x} \qquad \textit{19.11}$$

The stream function can also be written in total derivative form.

$$d\Psi = \frac{\partial \Psi}{\partial x}\, dx + \frac{\partial \Psi}{\partial y}\, dy$$
$$= -v\, dx + u\, dy \qquad \textit{19.12}$$

The stream function, $\Psi(x, y)$, satisfies Eq. 19.12. For a given streamline, $d\Psi = 0$, and each streamline is a line representing a constant value of $\Psi$. A streamline is perpendicular to an equipotential line.

$$\left.\frac{dy}{dx}\right|_{\text{streamline}} = -\frac{v}{u} \qquad \textit{19.13}$$

**Example 19.1**

The stream potential function for water flowing through a particular valve is $\Phi = 3xy - 2y$. What is the stream function, $\Psi$?

*Solution*

First, work with $\Phi$ to obtain $u$ and $v$.

$$u = \frac{\partial \Phi}{\partial x} = \frac{\partial (3xy - 2y)}{\partial x} = 3y$$

$$v = \frac{\partial \Phi}{\partial y} = 3x - 2$$

$u$ and $v$ are also related to the stream function, $\Psi$. From Eq. 19.10,

$$u = \frac{\partial \Psi}{\partial y}$$

$$\partial \Psi = u\, \partial y$$

$$\Psi = \int 3y\, dy - \tfrac{3}{2}y^2 + \text{some function of } x + C_1$$

Similarly, from Eq. 19.11,

$$v = -\frac{\partial \Psi}{\partial x}$$

$$\partial \Psi = -v\, \partial x$$

$$\Psi = -\int (3x - 2)\, dx$$

$$= 2x - \tfrac{3}{2}x^2 + \text{some function of } y + C_2$$

$\Psi$ is found by superposition of these two results.

$$\Psi = \tfrac{3}{2}y^2 + 2x - \tfrac{3}{2}x^2 + C$$

## 5. HEAD LOSS DUE TO FRICTION

The original Bernoulli equation was based on an assumption of frictionless flow. In actual practice, friction occurs during fluid flow. This friction "robs" the fluid of energy, so that the fluid at the end of a pipe section has less energy than it does at the beginning.[3]

$$E_1 > E_2 \qquad\qquad 19.14$$

Most formulas for calculating friction loss use the symbol $h_f$ to represent the *head loss due to friction*.[4] This

---

[3]The friction generates minute amounts of heat. The heat is lost to the surroundings.
[4]Other names and symbols for this friction loss are *friction head loss* ($h_L$), *lost work* (LW), *friction heating* ($\mathcal{F}$), *skin friction loss* ($F_f$), and *pressure drop due to friction* ($\Delta p_f$). All terms and symbols essentially mean the same thing, although the units may be different.

loss is added into the original Bernoulli equation to restore the equality. Of course, the units of $h_f$ must be the same as the units for the other terms in the Bernoulli equation. (See Eq. 19.23.) If the Bernoulli equation is written in terms of energy, the units will be ft-lbf/lbm or J/kg.

$$E_1 = E_2 + E_f \qquad\qquad 19.15$$

Consider the constant-diameter, horizontal pipe in Fig. 19.2. An incompressible fluid is flowing at a steady rate. Since the elevation of the pipe does not change, the potential energy is constant. Since the pipe has a constant area, the kinetic energy (velocity) is constant. Therefore, the friction energy loss must show up as a decrease in pressure energy. Since the fluid is incompressible, this can only occur if the pressure decreases in the direction of flow.

*Figure 19.2 Pressure Drop in a Pipe*

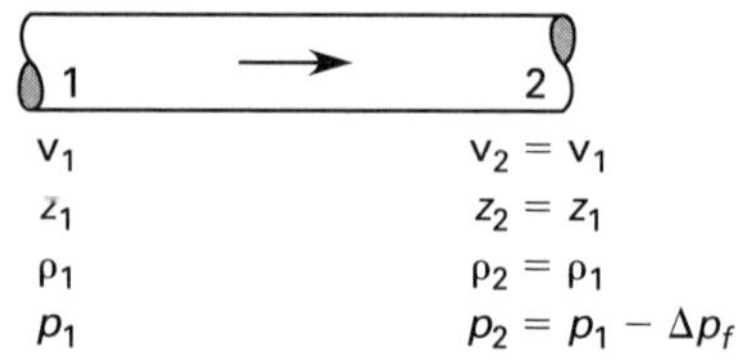

| 1 | → | 2 |
|---|---|---|
| $v_1$ | | $v_2 = v_1$ |
| $z_1$ | | $z_2 = z_1$ |
| $\rho_1$ | | $\rho_2 = \rho_1$ |
| $p_1$ | | $p_2 = p_1 - \Delta p_f$ |

## 6. RELATIVE ROUGHNESS

It is intuitive that pipes with rough inside surfaces will experience greater friction losses than smooth pipes.[5] *Specific roughness*, $\epsilon$, is a parameter that measures the average size of imperfections inside the pipe. Table 19.2 lists values of $\epsilon$ for common pipe materials. (Also, see App. 19.A.)

*Table 19.2 Values of Specific Roughness for Common Pipe Materials*

| material | $\epsilon$ ft | $\epsilon$ m |
|---|---|---|
| plastic (PVC, ABS) | 0.000005 | $1.5 \times 10^{-6}$ |
| copper and brass | 0.000005 | $1.5 \times 10^{-6}$ |
| steel | 0.0002 | $6.0 \times 10^{-5}$ |
| plain cast iron | 0.0008 | $2.4 \times 10^{-4}$ |
| concrete | 0.004 | $1.2 \times 10^{-3}$ |

However, an imperfection the size of a sand grain will have much more effect in a small-diameter hydraulic line than in a large-diameter sewer. Therefore, the *relative roughness*, $\epsilon/D$, is a better indicator of pipe roughness. Both $\epsilon$ and $D$ have units of length (e.g., feet or meters), and the relative roughness is dimensionless.

---

[5]Surprisingly, this intuitive statement is valid only for turbulent flow. The roughness does not (ideally) affect the friction loss for laminar flow.

## 7. FRICTION FACTOR

The *Darcy friction factor*, *f*, is one of the parameters used to calculate friction loss.[6] The friction factor is not constant but decreases as the Reynolds number (fluid velocity) increases, up to a certain point known as *fully turbulent flow* (or *rough-pipe flow*). Once the flow is fully turbulent, the friction factor remains constant and depends only on the relative roughness and not on the Reynolds number. (See Fig. 19.3.) For very smooth pipes, fully turbulent flow is achieved only at very high Reynolds numbers.

*Figure 19.3* Friction Factor as a Function of Reynolds Number

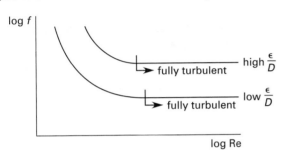

The friction factor is not dependent on the material of the pipe but is affected by the roughness. For example, for a given Reynolds number, the friction factor will be the same for any smooth pipe material (glass, plastic, smooth brass and copper, etc.).

The friction factor is determined from the relative roughness, $\epsilon/D$, and the Reynolds number, Re, by various methods. These methods include explicit and implicit equations, the Moody diagram, and tables. The values obtained are based on experimentation, primarily the work of J. Nikuradse in the early 1930s.

When a moving fluid initially encounters a parallel surface (as when a moving gas encounters a flat plate or when a fluid first enters the mouth of a pipe), the flow will generally not be turbulent, even for very rough surfaces. The flow will be laminar for a certain *critical distance* before becoming turbulent.

### Friction Factors for Laminar Flow

The easiest method of obtaining the friction factor for laminar flow (Re < 2100) is to calculate it. Equation 19.16 illustrates that roughness is not a factor in determining the frictional loss in ideal laminar flow.

$$f = \frac{64}{\text{Re}} \qquad \text{19.16}$$

### Friction Factors for Turbulent Flow: by Formula

One of the earliest attempts to predict the friction factor for turbulent flow in smooth pipes resulted in the *Blasius equation* (claimed "valid" for 3000 < Re < 100,000).

$$f = \frac{0.316}{\text{Re}^{0.25}} \qquad \text{19.17}$$

The *Nikuradse equation* can also be used to determine the friction factor for smooth pipes (i.e., when $\epsilon/D = 0$). Unfortunately, this equation is implicit in *f* and must be solved iteratively.

$$\frac{1}{\sqrt{f}} = 2.0 \log_{10}(\text{Re}\sqrt{f}) - 0.80 \qquad \text{19.18}$$

The *Karman-Nikuradse equation* predicts the fully turbulent friction factor (i.e., when Re is very large).

$$\frac{1}{\sqrt{f}} = 1.74 - 2\log_{10}\left(\frac{2\epsilon}{D}\right) \qquad \text{19.19}$$

The most widely known method of calculating the friction factor for any pipe roughness and Reynolds number is another implicit formula, the *Colebrook equation*. Most other equations are variations of this equation. (Notice that the relative roughness, $\epsilon/D$, is used to calculate *f*.)

$$\frac{1}{\sqrt{f}} = -2\log_{10}\left(\frac{\frac{\epsilon}{D}}{3.7} + \frac{2.51}{\text{Re}\sqrt{f}}\right) \qquad \text{19.20}$$

A suitable approximation would appear to be the *Swamee-Jain equation*, which claims to have less than 1% error (as measured against the Colebrook equation) for relative roughnesses between 0.000001 and 0.01, and for Reynolds numbers between 5000 and 100,000,000.[7] Even with a 1% error, this equation produces more accurate results than can be read from the Moody friction factor chart.

$$f = \frac{0.25}{\left(\log_{10}\left(\frac{\frac{\epsilon}{D}}{3.7} + \frac{5.74}{\text{Re}^{0.9}}\right)\right)^2} \qquad \text{19.21}$$

### Friction Factors for Turbulent Flow: by Moody Chart

The *Moody friction factor chart*, Fig. 19.4, presents the friction factor graphically as a function of Reynolds number and relative roughness. There are different lines for selected discrete values of relative roughness. Due to the complexity of this graph, it is easy to mislocate the Reynolds number or use the wrong curve. Nevertheless,

---

[6]There are actually two friction factors: the Darcy friction factor and the *Fanning friction factor*, $f_{\text{Fanning}}$, also known as the *skin friction coefficient* and *wall shear stress factor*. Both factors are in widespread use, sharing the same symbol, *f*. Civil and (most) mechanical engineers use the Darcy friction factor. The Fanning friction factor is encountered more often in the chemical industry. One can be derived from the other: $f_{\text{Darcy}} = 4f_{\text{Fanning}}$.

[7]*ASCE Hydraulic Division Journal*, Vol. 102, May 1976, p. 657. This is not the only explicit approximation to the Colebrook equation in existence.

the Moody chart remains the most common method of obtaining the friction factor.

## Friction Factors for Turbulent Flow: by Table

Appendix 19.B (based on the Colebrook equation), or a similar table, will usually be the most convenient method of obtaining friction factors for turbulent flow.

### Example 19.2

Determine the friction factor for a Reynolds number of Re = 400,000 and a relative roughness of $\epsilon/D = 0.004$ using (a) the Moody diagram, (b) Appendix 19.B, and (c) the Swamee-Jain approximation. (d) Check the table value of $f$ with the Colebrook equation.

*Solution*

(a) From Fig. 19.4, the friction factor is approximately 0.028.

(b) Appendix 19.B lists the friction factor as 0.0287.

(c) From Eq. 19.21,

$$f = \frac{0.25}{\left(\log_{10}\left(\dfrac{0.004}{3.7} + \dfrac{5.74}{(400{,}000)^{0.9}}\right)\right)^2}$$
$$= 0.0288$$

(d) From Eq. 19.20,

$$\frac{1}{\sqrt{0.0287}} = -2\log_{10}\left(\frac{0.004}{3.7} + \frac{2.51}{400{,}000\sqrt{0.0287}}\right)$$
$$5.903 = 5.903$$

## 8. ENERGY LOSS DUE TO FRICTION: LAMINAR FLOW

Two methods are available for calculating the frictional energy loss for fluids experiencing laminar flow. The most common is the *Darcy equation* (which is also known as the *Weisbach equation* or the *Darcy-Weisbach*

**Figure 19.4** *Moody Friction Factor Chart*

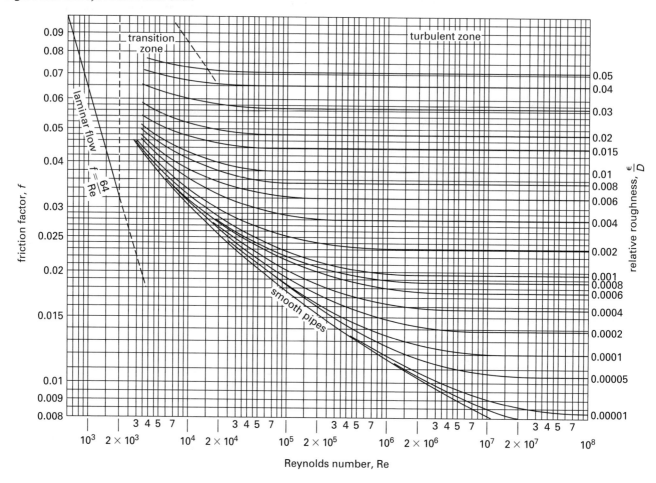

equation), which can be used for both laminar and turbulent flow.[8] One of the advantages of using the Darcy equation is that the assumption of laminar flow does not need to be confirmed if $f$ is known.

$$h_f = \frac{fLv^2}{2Dg} \qquad 19.22$$

$$E_f = h_f g = \frac{fLv^2}{2D} \qquad [\text{SI}] \quad 19.23(a)$$

$$E_f = h_f \times \left(\frac{g}{g_c}\right) = \frac{fLv^2}{2Dg_c} \qquad [\text{U.S.}] \quad 19.23(b)$$

If the flow is truly laminar and the fluid is flowing in a circular pipe, then the *Hagen-Poiseuille equation* can be used.

$$E_f = \frac{32\mu v L}{D^2 \rho} \qquad [\text{SI}] \quad 19.24(a)$$

$$E_f = \frac{32\mu v L g_c}{D^2 \rho} \qquad [\text{U.S.}] \quad 19.24(b)$$

An alternate form of the Hagen-Poiseuille equation substitutes $\dot{V}/A$ for v.

$$E_f = \frac{128\mu \dot{V} L}{\pi D^4 \rho} \qquad [\text{SI}] \quad 19.25(a)$$

$$E_f = \frac{128\mu \dot{V} L g_c}{\pi D^4 \rho} \qquad [\text{U.S.}] \quad 19.25(b)$$

If necessary, $h_f$ can be converted to an actual pressure drop in psi or Pa by multiplying by the fluid density.

$$\Delta p = h_f \times \rho g \qquad [\text{SI}] \quad 19.26(a)$$

$$\Delta p = h_f \times \rho \left(\frac{g}{g_c}\right) \qquad [\text{U.S.}] \quad 19.26(b)$$

Values of the Darcy friction factor, $f$, are often quoted for new, clean pipe. The friction head losses and pumping power requirements calculated from these values are minimal values. Depending on the nature of the service, scale and impurity buildup within pipes may decrease the pipe diameters over time. Since the frictional loss is proportional to the fifth power of the diameter, such diameter decreases can produce dramatic increases in the friction loss.

$$\frac{h_{f,\text{scaled}}}{h_{f,\text{new}}} = \left(\frac{D_{\text{new}}}{D_{\text{scaled}}}\right)^5 \qquad 19.27$$

Equation 19.27 accounts only for the decrease in diameter. Any increase in roughness (i.e., friction factor) will produce a proportional increase in friction loss.

---

[8]The difference is that the friction factor can be derived by hydrodynamics: $f = 64/\text{Re}$. For turbulent flow, $f$ is empirical.

Because the "new, clean" condition is transitory in most applications, an uprating factor of 10% to 30% is often applied to either the friction factor, $f$, or the head loss, $h_f$. Of course, even larger increases should be considered when extreme fouling is expected.

Another approach eliminates the need to estimate the scaled pipe diameter. This simplistic approach multiplies the initial friction loss by a factor based on the age of the pipe. For example, for schedule-40 pipe 4 in to 10 in (10 cm to 25 cm) in diameter, the multipliers of 1.4, 2.2, and 5.0 have been proposed for pipe ages of 5, 10, and 20 years, respectively. For larger pipes, the corresponding multipliers are 1.3, 1.6, and 2.0. Obviously, use of these values should be based on a clear understanding of the method's limitations.

## 9. ENERGY LOSS DUE TO FRICTION: TURBULENT FLOW

The *Darcy equation* is used almost exclusively to calculate the head loss due to friction for turbulent flow.

$$h_f = \frac{fLv^2}{2Dg} \qquad 19.28$$

The head loss can be converted to pressure drop.

$$\Delta p = h_f \times \rho g \qquad [\text{SI}] \quad 19.29(a)$$

$$\Delta p = h_f \times \rho \left(\frac{g}{g_c}\right) \qquad [\text{U.S.}] \quad 19.29(b)$$

In problems where the pipe size is unknown, it will be impossible to obtain an accurate initial value of the friction factor, $f$ (since $f$ depends on velocity). In such problems, an iterative solution will be necessary.

It is not uncommon for civil engineers to use the *Hazen-Williams equation* to calculate head loss. This method requires knowledge of the Hazen-Williams *roughness coefficient*, $C$, values of which are widely tabulated.[9] (See App. 19.A.) The advantage of using this equation is that $C$ does not depend on the Reynolds number.

$$h_{f,\text{feet}} = \frac{3.022 v_{\text{ft/sec}}^{1.85} L_{\text{ft}}}{C^{1.85} D_{\text{ft}}^{1.17}} \qquad [\text{U.S.}] \quad 19.30$$

Or, in terms of other units,

$$h_{f,\text{feet}} = \frac{10.44 L_{\text{ft}} Q_{\text{gpm}}^{1.85}}{C^{1.185} d_{\text{inches}}^{4.87}} \qquad [\text{U.S.}] \quad 19.31$$

---

[9]An approximate value of $C = 140$ is often chosen for initial calculations for new water pipe. $C = 100$ is more appropriate for water pipe that has been in service for some time. For sludge, $C$ values are 20% to 40% lower than the equivalent water pipe values.

The Hazen-Williams equation is empirical and is not dimensionally homogeneous. It is taken as a matter of faith that the units of $h_f$ are feet.

The Hazen-Williams equation should be used only for turbulent flow. It gives good results for liquids that have kinematic viscosities around $1.2 \times 10^{-5}$ ft$^2$/sec ($1.1 \times 10^{-6}$ m$^2$/s), which corresponds to the viscosity of 60°F (16°C) water. At extremely high and low temperatures, the Hazen-Williams equation can be 20% or more in error for water.

### Example 19.3

50°F water is pumped through 1000 ft of 4 in, schedule-40 welded steel pipe at the rate of 300 gpm. What friction loss (in ft-lbf/lbm) is predicted by the Darcy equation?

*Solution*

First, it is necessary to collect data on the pipe and water. The fluid viscosity, pipe dimensions, and other parameters can be found from the appendices.

$$\nu = 1.41 \times 10^{-5} \text{ ft}^2/\text{sec} \quad \text{[Appendix 16.A]}$$

$$\epsilon = 0.0002 \text{ ft} \quad \text{[Appendix 19.A]}$$

$$D = 0.3355 \text{ ft} \quad \text{[Appendix 18.B]}$$

$$A = 0.0884 \text{ ft}^2 \quad \text{[Appendix 18.B]}$$

The flow quantity is converted from gallons per minute to cubic feet per second.

$$\dot{V} = (300 \text{ gpm})\left(0.002228 \frac{\frac{\text{ft}^3}{\text{sec}}}{\text{gpm}}\right) = 0.6684 \text{ ft}^3/\text{sec}$$

The velocity is

$$\text{v} = \frac{\dot{V}}{A} = \frac{0.6684 \frac{\text{ft}^3}{\text{sec}}}{0.0884 \text{ ft}^2} = 7.56 \text{ ft/sec}$$

The Reynolds number is

$$\text{Re} = \frac{D\text{v}}{\nu} = \frac{(0.3355 \text{ ft})\left(7.56 \frac{\text{ft}}{\text{sec}}\right)}{1.41 \times 10^{-5} \frac{\text{ft}^2}{\text{sec}}}$$

$$= 1.8 \times 10^5$$

The relative roughness is

$$\frac{\epsilon}{D} = \frac{0.0002}{0.3355} = 0.0006$$

From the friction factor table (or the Moody friction factor chart), $f = 0.0195$. Equation 19.23(b) is used to calculate the friction loss.

$$E_f = h_f \times \left(\frac{g}{g_c}\right) = \frac{fL\text{v}^2}{2Dg_c}$$

$$= \frac{(0.0195)(1000 \text{ ft})\left(7.56 \frac{\text{ft}}{\text{sec}}\right)^2}{(2)(0.3355 \text{ ft})\left(32.2 \frac{\text{ft-lbm}}{\text{lbf-sec}^2}\right)}$$

$$= 51.6 \text{ ft-lbf/lbm}$$

### Example 19.4

Calculate the head loss due to friction for the pipe in Ex. 19.3 using the Hazen-Williams formula. Assume $C = 100$.

*Solution*

Substituting the parameters derived in Ex. 19.3 into Eq. 19.30,

$$h_f = \frac{(3.022)\left(7.56 \frac{\text{ft}}{\text{sec}}\right)^{1.85}(1000 \text{ ft})}{(100)^{1.85}(0.3355 \text{ ft})^{1.17}} = 91.3 \text{ ft}$$

Alternatively, the given data can be substituted directly into Eq. 19.31.

$$h_f = \frac{(10.44)(1000 \text{ ft})(300 \text{ gpm})^{1.85}}{(100)^{1.85}(4.026 \text{ in})^{4.87}} = 90.3 \text{ ft}$$

## 10. FRICTION LOSS FOR WATER FLOW IN STEEL PIPES

Friction loss and velocity for water flowing through steel pipe (as well as for other liquids and other pipe materials) in table and chart form are widely available. (Appendix 19.C is an example of such a table.) These tables and charts are unable to compensate for the effects of fluid temperature and different pipe roughness. Unfortunately, the assumptions made in developing the tables and charts are seldom listed. Another disadvantage is that the values can be read to only a few significant figures.

Since water's specific volume is essentially constant within the normal temperature range, tables and charts can be used to determine water velocity. Friction loss data, however, should be considered accurate to only ± 20%. Alternatively, a 20% safety margin should be established in choosing pumps and motors.

Tables and charts almost always give the friction loss per 100 ft or 10 m of pipe. The pressure drop is proportional to the length, so the value read can be scaled for other pipe lengths. Flow velocity is independent of pipe length.

## 11. FRICTION LOSS IN NONCIRCULAR DUCTS

The frictional energy loss by a fluid flowing in a rectangular, annular, or other noncircular duct can be calculated from the Darcy equation by using the *hydraulic diameter*, $D_h$, in place of the diameter variable, $D$.[10] The friction factor, $f$, is determined in any of the conventional manners.

## 12. FRICTION LOSS FOR STEAM AND GASES

The Darcy equation can be used to calculate the frictional energy loss for all incompressible liquids, not just for water. Alcohol, gasoline, fuel oil, and refrigerants, for example, are all handled well, since the effect of viscosity is considered in determining the friction factor, $f$.[11]

In fact, the Darcy equation is commonly used with noncondensing vapors and compressed gases, such as air, nitrogen, and steam.[12] In such cases, reasonable accuracy will be achieved as long as the fluid is not moving too fast (i.e., less than Mach 0.3) and is incompressible. The fluid is assumed to be incompressible if the pressure (or density) change along the section of interest is less than 10% of the starting pressure.

If possible, it is preferred to base all calculations on the average properties of the fluid.[13] Specifically, the fluid velocity would normally be calculated as

$$\text{v} = \frac{\dot{m}}{\rho_{\text{ave}} A} \qquad 19.32$$

However, the average density of a gas depends on the average pressure, which is unknown at the start of a problem. The solution is to write the Reynolds number and Darcy equation in terms of the constant mass flow rate per unit area, $G$, instead of velocity, v, which varies.

$$G = \text{v}_{\text{ave}} \rho_{\text{ave}} \qquad 19.33$$

$$\text{Re} = \frac{DG}{\mu} \qquad \text{[SI]} \qquad 19.34(a)$$

$$\text{Re} = \frac{DG}{g_c \mu} \qquad \text{[U.S.]} \qquad 19.34(b)$$

$$\Delta p_f = p_1 - p_2 = \rho_{\text{ave}} h_f g = \frac{fLG^2}{2D\rho_{\text{ave}}} \qquad \text{[SI]} \qquad 19.35(a)$$

$$\Delta p_f = p_1 - p_2 = \gamma_{\text{ave}} h_f = \rho_{\text{ave}} h_f \frac{g}{g_c}$$

$$= \frac{fLG^2}{2D\rho_{\text{ave}} g_c} \qquad \text{[U.S.]} \qquad 19.35(b)$$

Assuming a perfect gas with a molecular weight of MW, the ideal gas law can be used to calculate $\rho_{\text{ave}}$ from the absolute temperature, $T$, and $p_{\text{ave}} = (p_1 + p_2)/2$.

$$p_1^2 - p_2^2 = \frac{fLG^2 R^* T}{D(\text{MW})} \qquad \text{[SI]} \qquad 19.36(a)$$

$$p_1^2 - p_2^2 = \frac{fLG^2 R^* T}{Dg_c(\text{MW})} \qquad \text{[U.S.]} \qquad 19.36(b)$$

To summarize, use the following guidelines when working with compressible gases or vapors flowing in a pipe or duct. (a) If the pressure drop, based on the entrance pressure, is less than 10%, the fluid can be assumed to be incompressible, and the gas properties can be evaluated at any point known along the pipe. (b) If the pressure drop is between 10% and 40%, use of the midpoint properties will yield reasonably accurate friction losses. (c) If the pressure drop is greater than 40%, the pipe can be divided into shorter sections and the losses calculated for each section, or exact calculations based on compressible flow theory must be made.

Calculating a friction loss for steam flow can be frustrating if steam viscosity data are unavailable. Generally, the steam viscosities listed in compilations of heat transfer data are sufficiently accurate. Various empirical methods are also in use. For example, the *Babcock formula* (see Eq. 19.37) for pressure drop when steam with a specific volume of $v$ flows in a pipe of diameter $d$ is

$$\Delta p_{\text{psi}} = 0.470 \left( \frac{d_{\text{in}} + 3.6}{d_{\text{in}}^6} \right) (\dot{m}_{\text{lbm/sec}})^2 L_{\text{ft}} v_{\text{ft}^3/\text{lbm}} \qquad 19.37$$

Use of empirical formulas is not limited to steam. Theoretical formulas (e.g., the *complete isothermal flow equation*) and specialized empirical formulas (e.g., the *Weymouth*, *Panhandle*, and *Spitzglass formulas*) have been developed, particularly by the gas pipeline industry. Each of these provides reasonable accuracy within their operating limits. However, none should be used without knowing the assumptions and operational limitations that were used in their derivations.

### Example 19.5

0.0011 kg/s of 25°C nitrogen gas flows isothermally through a 175 m section of smooth tubing (inside diameter = 0.012 m). The viscosity of the nitrogen is $1.8 \times 10^{-5}$ Pa·s. The pressure of the nitrogen is 200 kPa originally. At what pressure is the nitrogen delivered?

---

[10]Although it is used for both, this approach is better suited for turbulent flow than for laminar flow. Also, the accuracy of this method decreases as the flow area becomes more noncircular. The friction drop in long, narrow slit passageways is poorly predicted, for example. However, there is no other convenient method of predicting friction drop. Experimentation should be used with a particular flow geometry if extreme accuracy is required.

[11]Since viscosity is not an explicit factor in the formula, it should be obvious that the Hazen-Williams equation is primarily used for water.

[12]Use of the Darcy equation is limited only by the availability of the viscosity data needed to calculate the Reynolds number.

[13]Of course, the entrance (or exit) conditions can be used if great accuracy is not needed.

*SI Solution*

The flow area of the pipe is

$$A = \frac{\pi}{4}D^2 = \frac{\pi(0.012 \text{ m})^2}{4} = 1.131 \times 10^{-4} \text{ m}^2$$

The mass flow rate per unit area is

$$G = \frac{\dot{m}}{A} = \frac{0.0011 \dfrac{\text{kg}}{\text{s}}}{1.131 \times 10^{-4} \text{ m}^2} = 9.73 \text{ kg/m}^2\text{·s}$$

The Reynolds number is

$$\text{Re} = \frac{DG}{\mu} = \frac{(0.012 \text{ m})\left(9.73 \dfrac{\text{kg}}{\text{m}^2\text{·s}}\right)}{1.8 \times 10^{-5} \text{ Pa·s}} = 6487$$

The flow is turbulent, and the pipe is said to be smooth. Therefore, the friction factor is interpolated (from App. 19.B) as 0.0347.

Since two atoms of nitrogen form a molecule of nitrogen gas, the molecular weight of nitrogen is twice the atomic weight, or 28.0 kg/kgmol. The temperature must be in degrees absolute: $T = 25°\text{C} + 273° = 298\text{K}$. The universal gas constant is 8314.3 J/kmol·K.

From Eq. 19.36, the final pressure is

$$p_2^2 = p_1^2 - \frac{fLG^2 R^* T}{D(\text{MW})}$$

$$= (200\,000 \text{ Pa})^2$$

$$- \frac{(0.0347)(175 \text{ m})\left(9.73 \dfrac{\text{kg}}{\text{m}^2\text{·s}}\right)^2 \times \left(8314.3 \dfrac{\text{J}}{\text{kmol·K}}\right)(298\text{K})}{(0.012 \text{ m})\left(28 \dfrac{\text{kg}}{\text{kmol}}\right)}$$

$$= 4 \times 10^{10} \text{ Pa}^2 - 4.24 \times 10^9 \text{ Pa}^2 = 3.576 \times 10^{10} \text{ Pa}^2$$

$$p_2 = \sqrt{3.576 \times 10^{10} \text{ Pa}^2} = 1.89 \times 10^5 \text{ Pa} \quad (189 \text{ kPa})$$

The percentage drop in pressure should not be more than 10%.

$$\frac{200 \text{ kPa} - 189 \text{ kPa}}{200 \text{ kPa}} = 0.055 \ (5.5\%) \quad [\text{OK}]$$

### Example 19.6

Superheated steam at 140 psi and 500°F enters a 200 ft long steel pipe with an internal diameter of 3.826 in. The pipe is insulated so that there is no heat loss. (a) Use the Babcock formula to determine the maximum velocity and mass flow rate such that the steam does not experience more than a 10% drop in pressure. (b) Verify the velocity by calculating the pressure drop with the Darcy equation.

*Solution*

(a) From superheated steam tables, the specific volume of the steam is 3.954 ft³/lbm. The maximum pressure drop is 10% of 140 psi or 14 psi.

From Eq. 19.37,

$$\Delta p_{\text{psi}} = 0.470\left(\frac{d_{\text{in}} + 3.6}{d_{\text{in}}^6}\right)(\dot{m}_{\text{lbm/sec}})^2 L_{\text{ft}} v$$

$$14 \text{ psi} = (0.470)\left(\frac{3.826 \text{ in} + 3.6}{(3.826 \text{ in})^6}\right)\dot{m}^2$$

$$\times (200 \text{ ft})\left(3.954 \frac{\text{ft}^3}{\text{lbm}}\right)$$

$$\dot{m} = 3.99 \text{ lbm/sec} \quad (4 \text{ lbm/sec})$$

$$v = \frac{Q}{A} = \frac{\dot{m}}{\rho A} = \frac{\dot{m}v}{A}$$

$$= \frac{\left(4 \dfrac{\text{lbm}}{\text{sec}}\right)\left(3.954 \dfrac{\text{ft}^3}{\text{lbm}}\right)}{\left(\dfrac{\pi}{4}\right)\left(\dfrac{3.826 \text{ in}}{12 \dfrac{\text{in}}{\text{ft}}}\right)^2}$$

$$= 198 \text{ ft/sec}$$

(b) Assume a Darcy friction factor of 0.02 (typical for turbulent flow in steel pipe). The steam flow velocity is

$$h_f = \frac{fLv^2}{2Dg}$$

$$= \frac{(0.02)(200 \text{ ft})\left(198 \dfrac{\text{ft}}{\text{sec}}\right)^2}{(2)\left(\dfrac{3.826 \text{ in}}{12 \dfrac{\text{in}}{\text{ft}}}\right)\left(32.2 \dfrac{\text{ft}}{\text{sec}^2}\right)}$$

$$= 7637 \text{ ft of steam}$$

$$\Delta p = \rho h_f \times \left(\frac{g}{g_c}\right) = \left(\frac{h_f}{v}\right) \times \left(\frac{g}{g_c}\right)$$

$$= \left(\frac{7637 \text{ ft}}{\left(3.954 \dfrac{\text{ft}^3}{\text{lbm}}\right)\left(12 \dfrac{\text{in}}{\text{ft}}\right)^2}\right)$$

$$\times \left(\frac{32.2 \dfrac{\text{ft}}{\text{sec}^2}}{32.2 \dfrac{\text{ft-lbm}}{\text{lbf-sec}^2}}\right)$$

$$= 13.4 \text{ lbf/in}^2 \quad (\text{psi})$$

## 13. EFFECT OF VISCOSITY ON HEAD LOSS

Friction loss in a pipe is affected by the fluid viscosity. For both laminar and turbulent flow, viscosity is considered when the Reynolds number is calculated. When viscosities substantially increase without a corresponding decrease in flow rate, two things usually happen: (a) the friction loss greatly increases, and (b) the flow becomes laminar.

It is sometimes necessary to estimate head loss for a new fluid viscosity based on head loss at an old fluid viscosity. The estimation procedure used depends on the flow regimes for the new and old fluids.

For laminar flow, the friction factor is directly proportional to the viscosity. If the flow is laminar for both fluids, the ratio of new-to-old head losses will be equal to the ratio of new-to-old viscosities. Thus, if a flow is already known to be laminar at one viscosity and the fluid viscosity increases, a simple ratio will define the new friction loss.

If both flows are fully turbulent, the friction factor will not change. If flow is fully turbulent and the viscosity decreases, the Reynolds number will increase. Theoretically, this will have no effect on the friction loss.

There are no analytical ways of estimating the change in friction loss when the flow regime changes between laminar and turbulent or between semiturbulent and fully turbulent. Various graphical methods are used, particularly by the pump industry, for calculating power requirements.

## 14. FRICTION LOSS WITH SLURRIES AND NON-NEWTONIAN FLUIDS

A *slurry* is a mixture of a liquid (usually water) and a solid (e.g., coal, paper pulp, foodstuffs). The liquid is generally used as the transport mechanism (i.e., the *carrier*) for the solid.

Friction loss calculations for slurries vary in sophistication depending on what information is available. In many cases, only the slurry's specific gravity is known. In that case, use is made of the fact that friction loss can be reasonably predicted by multiplying the friction loss based on the pure carrier (e.g., water) by the specific gravity of the slurry.

Another approach is possible if the density and viscosity in the operating range are known. The traditional Darcy equation (see Eq. 19.28) and Reynolds number can be used for thin slurries as long as the flow velocity is high enough to keep solids from settling. (Settling is more of a concern for laminar flow. With turbulent flow, the direction of velocity components fluctuates, assisting the solids to remain in suspension.)

The most analytical approach to slurries or other non-Newtonian fluids requires laboratory-derived rheological data. Non-Newtonian viscosity ($\eta$, in Pa·s) is fitted to data of the shear rate ($dv/dy$, in s$^{-1}$) according to two common models: the power-law model and the Bingham-plastic model. These two models are applicable to both laminar and turbulent flow, although each has its advantages and disadvantages.

The *power-law model* has two empirical constants, $m$ and $n$, that must be determined.

$$\eta = m\left(\frac{dv}{dy}\right)^{n-1} \qquad 19.38$$

The *Bingham-plastic model* also requires finding two empirical constants: the yield stress $\tau_0$ (in Pa) and the Bingham-plastic limiting viscosity $\mu_\infty$ (in Pa·s).

$$\eta = \frac{\tau_0}{\dfrac{dv}{dy}} + \mu_\infty \qquad 19.39$$

Once $m$ and $n$ (or $\tau_0$ and $\mu_\infty$) have been determined, the friction factor is determined from one of various models (e.g., Buckingham-Reiner, Dodge-Metzner, Metzner-Reed, Hanks-Ricks, Darby, or Hanks-Dadia). Specialized texts and articles cover these models in greater detail. The friction loss is calculated from the traditional Moody equation.

## 15. MINOR LOSSES

In addition to the frictional energy lost due to viscous effects, friction losses also result from fittings in the line, changes in direction, and changes in flow area. These losses are known as *minor losses* or *local losses*, since they are usually much smaller in magnitude than the pipe wall frictional loss.[14] Two methods are used to calculate minor losses: equivalent lengths and loss coefficients.

With the *method of equivalent lengths*, each fitting or other flow variation is assumed to produce friction equal to the pipe wall friction from an *equivalent length* of pipe. For example, a 2 in globe valve may produce the same amount of friction as 54 ft (its equivalent length) of 2 in pipe. The equivalent lengths for all minor losses are added to the pipe length term, $L$, in the Darcy equation. This method can be used with all liquids, but it is generally limited to turbulent flow.

$$L_t = L + \sum L_e \qquad 19.40$$

Equivalent lengths are simple to use, but the method depends on having a table of equivalent length values. The actual value for a fitting will depend on the fitting manufacturer, as well as the fitting material (e.g., brass, cast iron, or steel) and the method of attachment (e.g., weld, thread, or flange).[15] Because of these many variations, it may be necessary to use a "generic table" of

---

[14]Example and practice problems often include the instruction to "Ignore minor losses." In some industries, valves are considered to be "components," not fittings. In such cases, instructions to "Ignore minor losses in fittings" would be ambiguous, since minor losses in valves would be included in the calculations. However, this interpretation is rare in examples and practice problems.

[15]In the language of pipe fittings, a *threaded fitting* is known as a *screwed fitting*, even though no screws are used.

equivalent lengths during the initial design stages. (See Table 19.3 and App. 19.D.)

**Table 19.3** *Typical Equivalent Lengths (schedule-40, screwed steel fittings)*

| | pipe size | | |
|---|---|---|---|
| | 1 in | 2 in | 4 in |
| fitting type | equivalent length, ft | | |
| angle valve | 17.0 | 18.0 | 18.0 |
| coupling or union | 0.29 | 0.45 | 0.65 |
| gate valve | 0.84 | 1.5 | 2.5 |
| globe valve | 29.0 | 54.0 | 110.0 |
| long radius 90° elbow | 2.7 | 3.6 | 4.6 |
| regular 45° elbow | 1.3 | 2.7 | 5.5 |
| regular 90° elbow | 5.2 | 8.5 | 13.0 |
| swing check valve | 11.0 | 19.0 | 38.0 |
| tee, flow through line (run) | 3.2 | 7.7 | 17.0 |
| tee, flow through stem | 6.6 | 12.0 | 21.0 |
| 180° return bend | 5.2 | 8.5 | 13.0 |

An alternative method of calculating the minor loss for a fitting is to use the *method of loss coefficients*. Each fitting has a *loss coefficient*, $K$, associated with it, which, when multiplied by the kinetic energy, gives the loss. (See Table 19.4.) Thus, a loss coefficient is the minor loss expressed in fractions (or multiples) of the velocity head.

$$h_m = K h_v \qquad 19.41$$

The loss coefficient for any minor loss can be calculated if the equivalent length is known. However, there is no advantage to using one method over the other, except for consistency in calculations.

$$K = \frac{f L_e}{D} \qquad 19.42$$

Exact friction loss coefficients for bends, fittings, and valves are unique to each manufacturer. Furthermore, except for contractions, enlargements, exits, and entrances, the coefficients decrease fairly significantly (according to the fourth power of the diameter ratio) with increases in valve size. Therefore, a single $K$ value is seldom applicable to an entire family of valves. Nevertheless, generic tables and charts have been developed. These compilations can be used for initial estimates as long as the general nature of the data is recognized.

Loss coefficients for specific fittings and valves must be known in order to be used. They cannot be derived theoretically. However, the loss coefficients for certain changes in flow area can be calculated from the following equations.[16]

**Table 19.4** *Typical Loss Coefficients*[a]

| device | $K$ |
|---|---|
| angle valve | 5 |
| bend, close return | 2.2 |
| butterfly valve,[b] 2 to 8 in | $45f_t$ |
| butterfly valve, 10 to 14 in | $35f_t$ |
| butterfly valve, 16 to 24 in | $25f_t$ |
| check valve, swing, fully open | 2.3 |
| corrugated bends | 1.3 to 1.6 times value for smooth bend |
| standard 90° elbow | 0.9 |
| long radius 90° elbow | 0.6 |
| 45° elbow | 0.42 |
| gate valve, fully open | 0.19 |
| gate valve, $1/4$ closed | 1.15 |
| gate valve, $1/2$ closed | 5.6 |
| gate valve, $3/4$ closed | 24 |
| globe valve | 10 |
| meter disk or wobble | 3.4 to 10 |
| meter, rotary (star or cog-wheel piston) | 10 |
| meter, reciprocating piston | 15 |
| meter, turbine wheel (double flow) | 5 to 7.5 |
| tee, standard | 1.8 |

[a]The actual loss coefficient will usually depend on the size of the valve. Average values are given.
[b]Loss coefficients for butterfly valves are calculated from the friction factors for the pipes with complete turbulent flow.

- *sudden enlargements:* ($D_1$ is the smaller of the two diameters)

$$K = \left(1 - \left(\frac{D_1}{D_2}\right)^2\right)^2 \qquad 19.43$$

- *sudden contractions:* ($D_1$ is the smaller of the two diameters)

$$K = \frac{1}{2}\left(1 - \left(\frac{D_1}{D_2}\right)^2\right) \qquad 19.44$$

- *pipe exit:* (projecting exit, sharp-edged or rounded)

$$K = 1.0 \qquad 19.45$$

- *pipe entrance:*

reentrant: $K = 0.78$
sharp-edged: $K = 0.50$
rounded:

| bend radius $\dfrac{}{D}$ | $K$ |
|---|---|
| 0.02 | 0.28 |
| 0.04 | 0.24 |
| 0.06 | 0.15 |
| 0.10 | 0.09 |
| 0.15 | 0.04 |

---

[16]No attempt is made to imply great accuracy with these equations. Correlation between actual and theoretical losses is fair.

- *tapered diameter changes:*

$$\beta = \frac{\text{small diameter}}{\text{large diameter}} = \frac{D_1}{D_2}$$

$$\phi = \text{wall-to-horizontal angle}$$

enlargement, $\phi \leq 22°$:

$$K = 2.6 \sin \phi (1 - \beta^2)^2 \qquad \textit{19.46}$$

enlargement, $\phi > 22°$:

$$K = (1 - \beta^2)^2 \qquad \textit{19.47}$$

contraction, $\phi \leq 22°$:

$$K = 0.8 \sin \phi \, (1 - \beta^2) \qquad \textit{19.48}$$

contraction, $\phi > 22°$:

$$K = 0.5 \, \sqrt{\sin \phi} \, (1 - \beta^2) \qquad \textit{19.49}$$

### Example 19.7

Determine the total equivalent length of the piping system shown. The pipeline contains one gate valve, five regular 90° elbows, one tee (flow through the run), and 228 ft of straight pipe. All fittings are 1 in screwed steel pipe. Disregard entrance and exit losses.

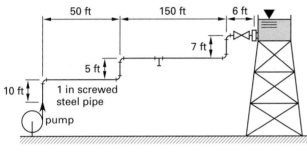

(not to scale)

*Solution*

From Table 19.3, the individual and total equivalent lengths are

| | | | | |
|---|---|---|---|---|
| 1 | gate valve | $1 \times 0.84$ ft | = | 0.84 ft |
| 5 | regular elbows | $5 \times 5.2$ ft | = | 26.00 ft |
| 1 | tee run | $1 \times 3.2$ ft | = | 3.20 ft |
| | straight pipe | | = | 228.00 ft |
| | total $L_t$ | | = | 258.04 ft |

## 16. VALVE FLOW COEFFICIENTS

Valve flow capacities depend on the geometry of the inside of the valve. The *flow coefficient*, $C_v$, for a valve (particularly a control valve) relates the flow quantity (in gallons per minute) of a fluid with specific gravity to the pressure drop (in pounds per square inch). (The flow coefficient for a valve is not the same as the coefficient of flow for an orifice or venturi meter.) As Eq. 19.50 shows,

the flow coefficient is not dimensionally homogeneous and is specifically limited to English units.

$$Q_{\text{gpm}} = C_v \sqrt{\frac{\Delta p_{\text{psi}}}{\text{SG}}} \qquad \textit{19.50}$$

When selecting a control valve for a particular application, the value of $C_v$ is first calculated. Depending on the application and installation, $C_v$ may be further modified by dividing by *piping geometry* and *Reynolds number factors*. (These additional procedures are often specified by the valve manufacturer.) Then, a valve with the required value of $C_v$ is selected.

Although the flow coefficient concept is generally limited to control valves, its use can be extended to all fittings and valves. The relationship between $C_v$ and the loss coefficient, $K$, is

$$C_v = \frac{29.9 \, d_{\text{in}}^2}{\sqrt{K}} \qquad \textit{19.51}$$

## 17. SHEAR STRESS IN CIRCULAR PIPES

Shear stress in fluid always acts to oppose the motion of the fluid. (That is the reason the term *frictional force* is occasionally used.) Shear stress for a fluid in laminar flow can always be calculated from the basic definition of absolute viscosity.

$$\tau = \mu \frac{d\text{v}}{dy} \qquad \textit{19.52}$$

In the case of the flow in a circular pipe, $dr$ can be substituted for $dy$ in the expression for *shear rate (velocity gradient)*, $d\text{v}/dr$.

$$\tau = \mu \frac{d\text{v}}{dr} \qquad \textit{19.53}$$

Equation 19.54 calculates the shear stress between fluid layers a distance $r$ from the pipe centerline from the pressure drop across a length $L$ of the pipe.[17] Equation 19.54 is valid for both laminar and turbulent flows.

$$\tau = \frac{(p_1 - p_2)r}{2L} \qquad \left[r \leq \frac{D}{2}\right] \qquad \textit{19.54}$$

The quantity $(p_1 - p_2)$ can be calculated from the Darcy equation. (See Eq. 19.28.) If v is the average flow velocity, the shear stress at the wall (where $r = D/2$) is

$$\tau_{\text{wall}} = \frac{f \rho \text{v}^2}{8} \qquad \text{[SI]} \qquad \textit{19.55(a)}$$

---

[17]In highly turbulent flow, shear stress is not caused by viscous effects but rather by momentum effects. Equation 19.54 is derived from a shell momentum balance. Such an analysis requires the concept of *momentum flux*. In a circular pipe with laminar flow, momentum flux is maximum at the pipe wall, zero at the flow centerline, and varies linearly in between.

$$\tau_{\mathrm{wall}} = \frac{f\rho \mathrm{v}^2}{8g_c} \qquad \text{[U.S.]} \quad 19.55(b)$$

Equation 19.54 can be rearranged somewhat to give the relationship between the pressure gradient along the flow path and the shear stress at the wall.

$$\frac{dp}{dL} = \frac{4\tau_{\mathrm{wall}}}{D} \qquad\qquad 19.56$$

Equation 19.55 can be combined with the Hagen-Poiseuille equation (see Eq. 19.24) if the flow is laminar. (v in Eq. 19.57 is the average velocity of fluid flow.)

$$\tau = \frac{16\mu \mathrm{v} r}{D^2} \quad \left[ r \le \frac{D}{2} \right] \qquad 19.57$$

At the pipe wall, $r = D/2$ and the shear stress is maximum. (See Fig. 19.5.) Therefore,

$$\tau_{\mathrm{wall}} = \frac{8\mu \mathrm{v}}{D} \qquad\qquad 19.58$$

**Figure 19.5** *Shear Stress Distribution in a Circular Pipe*

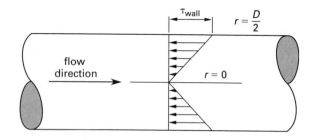

## 18. INTRODUCTION TO PUMPS AND TURBINES[18]

A *pump* adds energy to the fluid flowing through it. (See Fig. 19.6.) The amount of energy that a pump puts into the fluid stream can be determined by the difference between the total energy on either side of the pump. In most situations, a pump will add primarily pressure energy. The specific energy added (a positive number) on a per-unit mass basis (i.e., ft-lbf/lbm or J/kg) is given by Eq. 19.59.

$$E_A = E_{t,2} - E_{t,1} \qquad\qquad 19.59$$

The *head added* by a pump is

$$h_A = \frac{E_A}{g} \qquad \text{[SI]} \quad 19.60(a)$$

$$h_A = \frac{E_A g_c}{g} \qquad \text{[U.S.]} \quad 19.60(b)$$

The specific energy added by a pump can also be calculated from the input power if the mass flow rate is

---

[18]Greater detail on pumps and turbines can be found in books on hydraulic machines. This section is only an introduction.

**Figure 19.6** *Pump and Turbine Representation*

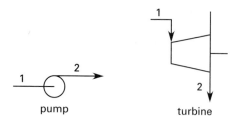

known. The input power to the pump will be the output power of the electric motor or engine driving the pump.

$$E_A = \frac{\left(1000 \; \dfrac{\mathrm{W}}{\mathrm{kW}}\right) P_{\mathrm{kW,input}} \eta_{\mathrm{pump}}}{\dot{m}} \qquad \text{[SI]} \quad 19.61(a)$$

$$E_A = \frac{\left(550 \; \dfrac{\mathrm{ft\text{-}lbf}}{\mathrm{sec\text{-}hp}}\right) P_{\mathrm{hp,input}} \eta_{\mathrm{pump}}}{\dot{m}} \qquad \text{[U.S.]} \quad 19.61(b)$$

The *water horsepower* (WHP, also known as the *hydraulic horsepower* and *theoretical horsepower*) is the amount of power actually entering the fluid.

$$\mathrm{WHP} = (P_{\mathrm{hp,input}})\eta_{\mathrm{pump}} \qquad\qquad 19.62$$

A *turbine* extracts energy from the fluid flowing through it. As with a pump, the energy extraction can be obtained by evaluating the Bernoulli equation on both sides of the turbine and taking the difference. The energy extracted (a positive number) on a per-unit mass basis is given by Eq. 19.63.

$$E_E = E_{t,1} - E_{t,2} \qquad\qquad 19.63$$

## 19. EXTENDED BERNOULLI EQUATION

The original Bernoulli equation assumes frictionless flow and does not consider the effects of pumps and turbines. When friction is present and when there are minor losses such as fittings and other energy-related devices in a pipeline, the energy balance is affected. The *extended Bernoulli equation* takes these additional factors into account.

$$(E_p + E_{\mathrm{v}} + E_z)_1 + E_A$$
$$= (E_p + E_{\mathrm{v}} + E_z)_2 + E_E + E_f + E_m \qquad 19.64$$

$$\frac{p_1}{\rho} + \frac{\mathrm{v}_1^2}{2} + z_1 g + E_A$$
$$= \frac{p_2}{\rho} + \frac{\mathrm{v}_2^2}{2} + z_2 g + E_E + E_f + E_m \qquad \text{[SI]} \quad 19.65(a)$$

$$\frac{p_1}{\rho} + \frac{\mathrm{v}_1^2}{2g_c} + \frac{z_1 g}{g_c} + E_A$$

$$= \frac{p_2}{\rho} + \frac{\mathrm{v}_2^2}{2g_c} + \frac{z_2 g}{g_c} + E_E + E_f + E_m$$

[U.S.] *19.65(b)*

As defined, $E_A$, $E_E$, and $E_f$ are all positive terms. None of the terms in Eq. 19.64 are negative.

The concepts of sources and sinks can be used to decide whether the friction, pump, and turbine terms appear on the left or right side of the Bernoulli equation. An *energy source* puts energy into the system. The incoming fluid and a pump contribute energy to the system. An *energy sink* removes energy from the system. The leaving fluid, friction, and a turbine remove energy from the system. In an energy balance, all energy must be accounted for, and the energy sources just equal the energy sinks.

$$\sum E_{\text{sources}} = \sum E_{\text{sinks}} \qquad 19.66$$

Therefore, the energy added by a pump always appears on the entrance side of the Bernoulli equation. Similarly, the frictional energy loss always appears on the discharge side.

## 20. ENERGY AND HYDRAULIC GRADE LINES WITH FRICTION[19]

The *energy grade line* (EGL, also known as *total energy line*) is a graph of the total energy versus position in a pipeline. Since a pitot tube measures total (stagnation) energy, EGL will always coincide with the elevation of a pitot-piezometer fluid column. When friction is present, the EGL will always slope down, in the direction of flow. Figure 19.7 illustrates the EGL for a complex pipe network. The difference between $\text{EGL}_{\text{frictionless}}$ and $\text{EGL}_{\text{with friction}}$ is the energy loss due to friction.

Notice that the EGL line in Fig. 19.7 is discontinuous at point 2, since the friction in pipe section B-C cannot be portrayed without disturbing the spatial correlation of points in the figure. Since the friction loss is proportional to $\mathrm{v}^2$, the slope is steeper when the fluid velocity increases (i.e., when the pipe decreases in flow area), as it does in section D-E. Disregarding air friction, the EGL becomes horizontal at point 6 when the fluid becomes a free jet.

The *hydraulic grade line* (HGL) is a graph of the sum of pressure and potential energies versus position in the pipeline. (That is, the EGL and HGL differ by the kinetic energy.) The HGL will always coincide with the height of the fluid column in a static piezometer tube. The reference point for elevation is arbitrary, and the pressure energy is usually referenced to atmospheric

---

[19]This section covers only the energy and hydraulic grade lines with friction, not the frictionless case.

*Figure 19.7 Energy and Hydraulic Grade Lines*

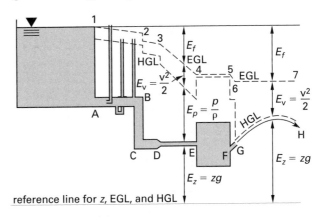

reference line for $z$, EGL, and HGL

pressure. Thus, the pressure energy, $E_p$, for a free jet will be zero, and the HGL will consist only of the potential energy, as shown in section G-H.

The easiest way to draw the energy and hydraulic grade lines is to start with the EGL. The EGL can be drawn simply by recognizing that the rate of divergence from the horizontal $\text{EGL}_{\text{frictionless}}$ line is proportional to $\mathrm{v}^2$. Then, since EGL and HGL differ by the velocity head, the HGL can be drawn parallel to the EGL when the pipe diameter is constant. The larger the pipe diameter, the closer the two lines will be.

The EGL for a pump will increase in elevation by $E_A$ across the pump. (The actual energy "path" taken by the fluid is unknown, and a dotted line is used to indicate a lack of knowledge about what really happens in the pump.) The placement of the HGL for a pump will depend on whether the pump increases the fluid velocity and elevation, as well as the fluid pressure. In most cases, only the pressure will be increased. Figure 19.8 illustrates the HGL for the case of a pressure increase only.

*Figure 19.8 EGL and HGL for a Pump*

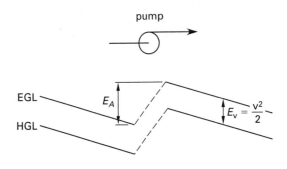

The EGL and HGL for minor losses (fittings, contractions, expansions, etc.) are shown in Fig. 19.9.

## 21. DISCHARGE FROM TANKS

The velocity of a jet issuing from an orifice in a tank can be determined by comparing the total energies at the

**Figure 19.9** *EGL and HGL for Minor Losses*

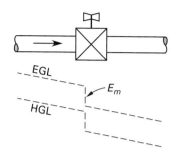

(a) valve, fitting, or obstruction

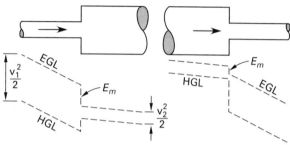

(b) sudden enlargement    (c) sudden contraction

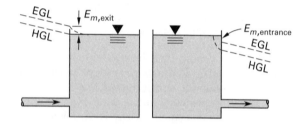

(d) transition to reservoir    (e) transition to pipeline

free fluid surface and the jet itself. (See Fig. 19.10.) At the fluid surface, $p_1 = 0$ (atmospheric) and $v_1 = 0$. ($v_1$ is known as the *velocity of approach*.) The only energy the fluid has is potential energy. At the jet, $p_2 = 0$. All of the potential energy difference ($z_1 - z_2$) has been converted to kinetic energy. The theoretical velocity of the jet can be derived from the Bernoulli equation. Equation 19.67 is known as the equation for *Torricelli's speed of efflux*.

$$v_t = \sqrt{2gh} \qquad 19.67$$

$$h = z_1 - z_2 \qquad 19.68$$

The actual jet velocity is affected by the orifice geometry. The *coefficient of velocity*, $C_v$, is an empirical factor that accounts for the friction and turbulence at the orifice. Typical values of $C_v$ are given in Table 19.5.

$$v_o = C_v \sqrt{2gh} \qquad 19.69$$

$$C_v = \frac{\text{actual velocity}}{\text{theoretical velocity}} = \frac{v_o}{v_t} \qquad 19.70$$

**Figure 19.10** *Discharge from a Tank*

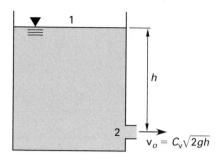

The specific energy loss due to turbulence and friction at the orifice is calculated as a multiple of the jet's kinetic energy.

$$E_f = \left(\frac{1}{C_v^2} - 1\right)\left(\frac{v_o^2}{2}\right) = (1 - C_v^2)gh \qquad \text{[SI]} \quad 19.71(a)$$

$$E_f = \left(\frac{1}{C_v^2} - 1\right)\left(\frac{v_o^2}{2g_c}\right) = (1 - C_v^2)h \times \left(\frac{g}{g_c}\right) \qquad \text{[U.S.]} \quad 19.71(b)$$

The total head producing discharge (*effective head*) is the difference in elevations that would produce the same velocity from a frictionless orifice.

$$h_{\text{effective}} = C_v^2 h \qquad 19.72$$

The orifice guides quiescent water from the tank into the jet geometry. Unless the orifice is very smooth and the transition is gradual, momentum effects will continue to cause the jet to contract after it has passed through. The velocity calculated from Eq. 19.69 is usually assumed to be the velocity at the *vena contracta*, the section of smallest cross-sectional area. (See Fig. 19.11.)

**Figure 19.11** *Vena Contracta of a Fluid Jet*

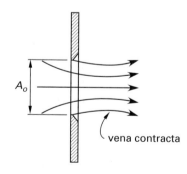

For a thin plate or sharp-edged orifice, the vena contracta is often assumed to be located approximately one half an orifice diameter past the orifice, although the actual distance can vary from $0.3D_o$ to $0.8D_o$. The area of the vena contracta can be calculated from the orifice area and the *coefficient of contraction*, $C_c$. For water flowing with a high Reynolds number through a small

**Table 19.5** Approximate Orifice Coefficients for Turbulent Water

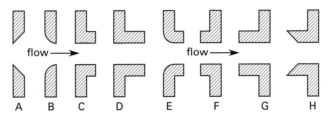

| illustration | description | $C_d$ | $C_c$ | $C_v$ |
|---|---|---|---|---|
| A | sharp-edged | 0.62 | 0.63 | 0.98 |
| B | round-edged | 0.98 | 1.00 | 0.98 |
| C | short tube[*] (fluid separates from walls) | 0.61 | 1.00 | 0.61 |
| D | sharp tube (no separation) | 0.82 | 1.00 | 0.82 |
| E | short tube with rounded entrance | 0.97 | 0.99 | 0.98 |
| F | reentrant tube, length less than one-half of pipe diameter | 0.54 | 0.55 | 0.99 |
| G | reentrant tube, length 2 to 3 pipe diameters | 0.72 | 1.00 | 0.72 |
| H | Borda | 0.51 | 0.52 | 0.98 |
| (none) | smooth, well-tapered nozzle | 0.98 | 0.99 | 0.99 |

[*]A short tube has a length less than 2 to 3 diameters.

sharp-edged orifice, the contracted area is approximately 61% to 63% of the orifice area.

$$A_{\text{vena contracta}} = C_c A_o \qquad 19.73$$

$$C_c = \frac{\text{area of vena contracta}}{\text{orifice area}} \qquad 19.74$$

The theoretical discharge rate from a tank is $\dot{V} = A_o\sqrt{2gh}$. However, this relationship needs to be corrected for friction and contraction by multiplying by $C_v$ and $C_c$. The *coefficient of discharge*, $C_d$, is the product of the coefficients of velocity and contraction.

$$\dot{V} = C_c v_o A_o = C_d v_t A_o = C_d A_o \sqrt{2gh} \qquad 19.75$$

$$\begin{aligned} C_d &= C_v C_c \\ &= \frac{\text{actual discharge}}{\text{theoretical discharge}} \end{aligned} \qquad 19.76$$

## 22. DISCHARGE FROM PRESSURIZED TANKS

If the gas or vapor above the liquid in a tank is at gage pressure $p$, and the discharge is to atmospheric pressure, the head causing discharge will be

$$h = z_1 - z_2 + \frac{p}{\rho g} \qquad \text{[SI]} \quad 19.77(a)$$

$$h = z_1 - z_2 + \left(\frac{p}{\rho}\right) \times \left(\frac{g_c}{g}\right) \qquad \text{[U.S.]} \quad 19.77(b)$$

The discharge velocity can be calculated from Eq. 19.69 using the increased discharge head. (See Fig. 19.12.)

**Figure 19.12** Discharge from a Pressurized Tank

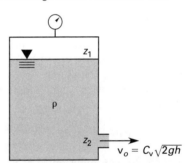

## 23. COORDINATES OF A FLUID STREAM

Fluid discharged from an orifice in a tank gets its initial velocity from the conversion of potential energy. After discharge, no additional energy conversion occurs, and all subsequent velocity changes are due to external forces. (See Fig. 19.13.)

**Figure 19.13** Coordinates of a Fluid Stream

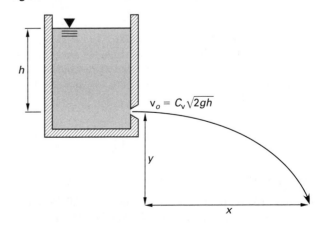

In the absence of air friction (drag), there are no retarding or accelerating forces in the $x$-direction on the fluid stream. The $x$-component of velocity is constant. Projectile motion equations can be used to predict the path of the fluid stream.

$$v_x = v_o \quad \text{[horizontal discharge]} \qquad 19.78$$

$$x = v_o t = v_o \sqrt{\frac{2y}{g}} = 2C_v \sqrt{hy} \qquad 19.79$$

After discharge, the fluid stream is acted upon by a constant gravitational acceleration. The $y$-component of velocity is zero at discharge but increases linearly with time.

$$v_y = gt \qquad 19.80$$

$$y = \frac{gt^2}{2} = \frac{gx^2}{2v_o^2} = \frac{x^2}{4hC_v^2} \qquad 19.81$$

## 24. TIME TO EMPTY A TANK

If the fluid in an open or vented tank is not replenished at the rate of discharge, the static head forcing discharge through the orifice will decrease with time. If the tank has a varying cross section, $A_t$, Eq. 19.82 specifies the basic relationship between the change in elevation and elapsed time. (The negative sign indicates that $z$ decreases as $t$ increases.)

$$\dot{V} dt = -A_t dz \qquad 19.82$$

If $A_t$ can be expressed as a function of $h$, Eq. 19.83 can be used to determine the time to lower the fluid elevation from $z_1$ to $z_2$.

$$t = \int_{z_1}^{z_2} \frac{-A_t dz}{C_d A_o \sqrt{2gz}} \qquad 19.83$$

For a tank with a constant cross-sectional area, $A_t$, the time required to lower the fluid elevation is

$$t = \frac{2A_t(\sqrt{z_1} - \sqrt{z_2})}{C_d A_o \sqrt{2g}} \qquad 19.84$$

If a tank is replenished at a rate of $\dot{V}_{in}$, Eq. 19.85 can be used to calculate the discharge time. If the tank is replenished at a rate greater than the discharge rate, $t$ in Eq. 19.85 will represent the time to raise the fluid level from $z_1$ to $z_2$.

$$t = \int_{z_1}^{z_2} \frac{A_t dz}{(C_d A_o \sqrt{2gz}) - \dot{V}_{in}} \qquad 19.85$$

**Example 19.8**

A tank 15 ft in diameter discharges 150°F water ($\rho = 61.20$ lbm/ft$^3$) through a sharp-edged 1.0 in diameter orifice ($C_d = 0.62$) in the bottom. The original

water depth is 12 ft. The tank is continually pressurized to 50 psig. What is the time to empty the tank?

*Solution*

The area of the orifice is

$$A_o = \frac{\pi D^2}{4} = \frac{\pi (1 \text{ in})^2}{(4)\left(12 \frac{\text{in}}{\text{ft}}\right)^2} = 0.00545 \text{ ft}^2$$

The tank area constant with respect to $z$ is

$$A_t = \frac{\pi D^2}{4} = \frac{\pi (15 \text{ ft})^2}{4} = 176.7 \text{ ft}^2$$

The total initial head includes the effect of the pressurization. Use Eq. 19.77.

$$h_1 = 12 \text{ ft} + \left(\frac{\left(50 \frac{\text{lbf}}{\text{in}^2}\right)\left(12 \frac{\text{in}}{\text{ft}}\right)^2}{61.2 \frac{\text{lbm}}{\text{ft}^3}}\right)\left(\frac{32.2 \frac{\text{ft-lbm}}{\text{lbf-sec}^2}}{32.2 \frac{\text{ft}}{\text{sec}^2}}\right)$$

$$= 12 \text{ ft} + 117.6 \text{ ft} = 129.6 \text{ ft}$$

When the fluid has reached the level of the orifice, the fluid potential head will be zero, but the pressurization will remain.

$$h_2 = 117.6 \text{ ft}$$

The time to empty the tank is given by Eq. 19.84.

$$t = \frac{(2)(176.7 \text{ ft}^2)(\sqrt{129.6 \text{ ft}} - \sqrt{117.6 \text{ ft}})}{(0.62)(0.00545 \text{ ft}^2)\sqrt{(2)\left(32.2 \frac{\text{ft}}{\text{sec}^2}\right)}}$$

$$= 7036 \text{ sec}$$

## 25. DISCHARGE FROM LARGE ORIFICES

When an orifice diameter is large compared with the discharge head, the jet velocity at the top edge of the orifice will be less than the velocity at the bottom edge. Since the velocity is related to the square root of the head, the distance used to calculate the effective jet velocity should be measured from the fluid surface to a point above the centerline of the orifice.

This correction is generally neglected, however, since it is small for heads of more than twice the orifice diameter. Furthermore, if an orifice is intended to work regularly with small heads, the orifice should be calibrated in place. The discrepancy can then be absorbed into the discharge coefficient, $C_d$.

## 26. CULVERTS

A *culvert* is a water path (usually a large diameter pipe) used to channel water around or through an obstructing feature. (See Fig. 19.14.) In most instances, a culvert is used to restore a natural water path obstructed by a manufactured feature. For example, when a road is built across (perpendicular to) a natural ravine or arroyo, a culvert can be used to channel water under the road.

**Figure 19.14** Simple Pipe Culvert

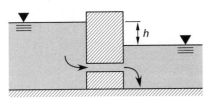

Because culverts usually operate only partially full and with low heads, Torricelli's equation does not apply. Therefore, most culvert designs are empirical. However, if the entrance and exit of a culvert are both submerged, the culvert will flow full, and the discharge will be independent of the barrel slope. Equation 19.86 can be used to calculate the discharge.

$$\dot{V} = C_d A \sqrt{2gh} \qquad 19.86$$

If the culvert is long (more than 60 ft or 20 m), or if the entrance is not gradual, the available energy will be divided between friction and velocity heads. The effective head used in Eq. 19.86 should be

$$h_{\text{effective}} = h - h_{f,\text{barrel}} - h_{m,\text{entrance}} \qquad 19.87$$

The friction loss in the barrel can be found in the usual manner, from either the Darcy equation or the Hazen-Williams equation. The entrance loss is calculated using the standard method of loss coefficients. Representative values of the loss coefficient, $K$, are given in Table 19.6. Since the fluid velocity is not initially known but is needed to find the friction factor, a trial-and-error solution will be necessary.

**Table 19.6** Representative Loss Coefficients for Culvert Entrances

| entrance | $K$ |
|---|---|
| smooth and gradual transition | 0.08 |
| flush vee or bell shape | 0.10 |
| projecting vee or bell shape | 0.15 |
| flush, square-edged | 0.50 |
| projecting, square-edged | 0.90 |

## 27. SIPHONS

A *siphon* is a bent or curved tube that carries fluid from a container at a high elevation to another container at a lower elevation. Normally, it would not seem difficult to have a fluid flow to a lower elevation. However, the fluid seems to flow "uphill" in a portion of a siphon. Figure 19.15 illustrates a siphon.

Starting a siphon requires the tube to be completely filled with liquid. Then, since the fluid weight is greater in the longer arm than in the shorter arm, the fluid in

**Figure 19.15** Siphon

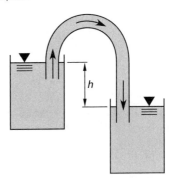

the longer arm "falls" out of the siphon, "pulling" more liquid into the shorter arm and over the bend.

Operation of a siphon is essentially independent of atmospheric pressure. The theoretical discharge is the same as predicted by the Torricelli equation. A correction for discharge is necessary, but little data is available on typical values of $C_d$. Therefore, siphons should be tested and calibrated in place.

$$\dot{V} = C_d A \text{v} = C_d A \sqrt{2gh} \qquad 19.88$$

## 28. SERIES PIPE SYSTEMS

A system of pipes in series consists of two or more lengths of different-diameter pipes connected together. In the case of the series pipe from a reservoir discharging to the atmosphere shown in Fig. 19.16, the available head will be split between the velocity head and the friction loss.

**Figure 19.16** Series Pipe System

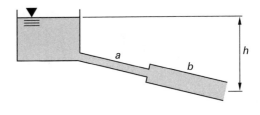

$$h = h_{\text{v}} + h_f \qquad 19.89$$

If the flow rate or velocity in any part of the system is known, the friction loss can easily be found as the sum of the friction losses in the individual sections. The solution is somewhat more simple than first appears, since the velocity of all sections can be written in terms of only one velocity.

$$h_{f,t} = h_{f,a} + h_{f,b} \qquad 19.90$$

$$A_a \text{v}_a = A_b \text{v}_b \qquad 19.91$$

If neither the velocity nor the flow quantity is known, a trial-and-error solution will be required, since a friction factor must be known to calculate $h_f$. A good starting point is to assume fully turbulent flow.

When velocity and flow rate are both unknown, the following procedure using the Darcy friction factor can be used.[20]

*step 1:* Calculate the relative roughness, $\epsilon/D$, for each section. Use the Moody diagram to determine $f_a$ and $f_b$ for fully turbulent flow (i.e., the horizontal portion of the curve).

*step 2:* Write all of the velocities in terms of one unknown velocity.

$$\dot{V}_a = \dot{V}_b \qquad 19.92$$

$$\mathrm{v}_b = \left(\frac{A_a}{A_b}\right)\mathrm{v}_a \qquad 19.93$$

*step 3:* Write the total friction loss in terms of the unknown velocity.

$$h_{f,t} = \frac{f_a L_a \mathrm{v}_a^2}{2 D_a g} + \left(\frac{f_b L_b}{2 D_b g}\right)\left(\frac{A_a}{A_b}\right)^2 \mathrm{v}_a^2$$

$$= \left(\frac{\mathrm{v}_a^2}{2g}\right)\left(\left(\frac{f_a L_a}{D_a}\right) + \left(\frac{f_b L_b}{D_b}\right)\left(\frac{A_a}{A_b}\right)^2\right) \qquad 19.94$$

*step 4:* Solve for the unknown velocity using the Bernoulli equation between the free reservoir surface ($p = 0$, $\mathrm{v} = 0$, $z = h$) and the discharge point ($p = 0$, if free discharge; $z = 0$). Include pipe friction, but disregard minor losses for convenience.

$$h = \frac{\mathrm{v}_b^2}{2g} + h_{f,t}$$

$$= \left(\frac{\mathrm{v}_a^2}{2g}\right)\left(\left(\frac{A_a}{A_b}\right)^2\left(1 + \frac{f_b L_b}{D_b}\right) + \frac{f_a L_a}{D_a}\right) \qquad 19.95$$

*step 5:* Using the value of $\mathrm{v}_a$, calculate $\mathrm{v}_b$. Calculate the Reynolds number and check the values of $f_a$ and $f_b$ from step 4. Repeat steps 3 and 4 if necessary.

## 29. PARALLEL PIPE SYSTEMS

A *pipe loop* is a set of two pipes placed in parallel, both originating and terminating at the same junction. (See Fig. 19.17.) Adding a second pipe in parallel with a first is a standard method of increasing the capacity of a line.

---

[20]If Hazen-Williams constants are given for the pipe sections, the procedure for finding the unknown velocities is similar, although considerably more difficult since $\mathrm{v}^2$ and $\mathrm{v}^{1.85}$ cannot be combined. A first approximation, however, can be obtained by replacing $\mathrm{v}^{1.85}$ in the Hazen-Williams equation for friction loss. A trial and error method can then be used to find velocity.

*Figure 19.17* Parallel Pipe System

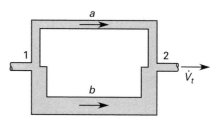

There are three principles that govern the distribution of flow between the two branches.

- The flow divides in such a manner as to make the head loss in each branch the same.

$$h_{f,a} = h_{f,b} \qquad 19.96$$

- The head loss between the junctions 1 and 2 is the same as the head loss in branches $a$ and $b$.

$$h_{f,1-2} = h_{f,a} = h_{f,b} \qquad 19.97$$

- The total flow rate is the sum of the flow rates in the two branches.

$$\dot{V}_t = \dot{V}_a + \dot{V}_b \qquad 19.98$$

If the pipe diameters are known, Eq. 19.96 and Eq. 19.98 can be solved simultaneously for the branch velocities. In such problems, it is common to neglect minor losses, the velocity head, and the variation in the friction factor, $f$, with velocity.

If the parallel system has only two branches, the unknown branch flows can be determined by solving Eq. 19.99 and Eq. 19.101 simultaneously.

$$\frac{f_a L_a \mathrm{v}_a^2}{2 D_a g} = \frac{f_b L_b \mathrm{v}_b^2}{2 D_b g} \qquad 19.99$$

$$\dot{V}_a + \dot{V}_b = \dot{V}_t \qquad 19.100$$

$$\frac{\pi}{4}(D_a^2 \mathrm{v}_a + D_b^2 \mathrm{v}_b) = \dot{V}_t \qquad 19.101$$

However, if the parallel system has three or more branches, it is easier to use the following iterative procedure. This procedure can be used for problems (a) where the flow rate is unknown but the pressure drop between the two junctions is known, or (b) where the total flow rate is known but the pressure drop and velocity are both unknown. In both cases, the solution iteratively determines the friction coefficients ($f$).

*step 1:* Solve the friction head loss $(h_f)$ expression (either Darcy or Hazen-Williams) for velocity in each branch. If the pressure drop is known, first convert it to friction head loss.

$$\text{v} = \sqrt{\frac{2Dgh_f}{fL}} \quad \text{[Darcy]} \quad\quad\quad 19.102$$

$$\text{v} = \frac{0.355\,CD^{0.63}h_j^{0.54}}{L^{0.54}} \quad \text{[Hazen-Williams; SI]} \quad\quad\quad 19.103(a)$$

$$\text{v} = \frac{0.550\,CD^{0.63}h_f^{0.54}}{L^{0.54}} \quad \text{[Hazen-Williams; U.S.]} \quad\quad\quad 19.103(b)$$

*step 2:* Solve for the flow rate in each branch. If they are unknown, friction factors, $f$, must be assumed for each branch. The fully turbulent assumption provides a good initial estimate. (The value of $k'$ will be different for each branch.)

$$\dot{V} = A\text{v} = A\sqrt{\frac{2Dgh_f}{fL}}$$

$$= k'\sqrt{h_f} \quad \text{[Darcy]} \quad\quad\quad 19.104$$

*step 3:* Write the expression for the conservation of flow. Calculate the friction head loss from the total flow rate. For example, for a three-branch system,

$$\dot{V}_t = \dot{V}_1 + \dot{V}_2 + \dot{V}_3$$

$$= (k'_1 + k'_2 + k'_3)\sqrt{h_f} \quad\quad\quad 19.105$$

*step 4:* Check the assumed values of the friction factor. Repeat as necessary.

### Example 19.9

3.0 ft³/sec of water enter the parallel pipe network shown. All pipes are schedule-40 steel with the nominal sizes shown. Minor losses are insignificant. What is the total friction head loss between junctions A and B?

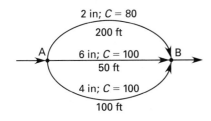

2 in; $C = 80$
200 ft

A    6 in; $C = 100$    B
50 ft

4 in; $C = 100$
100 ft

*Solution*

*step 1:* Collect the pipe dimensions.

|  | 2 in | 4 in | 6 in |
|---|---|---|---|
| flow area | 0.0233 ft² | 0.0884 ft² | 0.2006 ft² |
| diameter | 0.1723 ft | 0.3355 ft | 0.5054 ft |

Follow the procedure given in Sec. 19.29. Since the Hazen-Williams loss coefficients are given for each branch, the Hazen-Williams friction loss equation must be used.

$$h_f = \frac{3.022\,\text{v}^{1.85}L}{C^{1.85}D^{1.165}}$$

$$\text{v} = \frac{0.550\,CD^{0.63}h_f^{0.54}}{L^{0.54}}$$

The velocity (expressed in ft/sec) in the 2 in pipe branch is

$$\text{v}_{2\,\text{in}} = \frac{(0.550)(80)(0.1723\ \text{ft})^{0.63}h_f^{0.54}}{(200\ \text{ft})^{0.54}}$$

$$= 0.831h_f^{0.54}$$

The velocities in the other two branches are

$$\text{v}_{6\,\text{in}} = 4.327h_f^{0.54}$$

$$\text{v}_{4\,\text{in}} = 2.299h_f^{0.54}$$

*step 2:* The flow rates are

$$\dot{V} = A\text{v}$$

$$\dot{V}_{2\,\text{in}} = (0.0233\ \text{ft}^2)(0.831)h_f^{0.54}$$

$$= 0.0194h_f^{0.54}$$

$$\dot{V}_{6\,\text{in}} = (0.2006\ \text{ft}^2)(4.327)h_f^{0.54}$$

$$= 0.8680h_f^{0.54}$$

$$\dot{V}_{4\,\text{in}} = (0.0884\ \text{ft}^2)(2.299)h_f^{0.54}$$

$$= 0.2032h_f^{0.54}$$

*step 3:*

$$\dot{V}_t = \dot{V}_{2\,\text{in}} + \dot{V}_{6\,\text{in}} + \dot{V}_{4\,\text{in}}$$

$$3\ \text{ft}^3/\text{sec} = 0.0194h_f^{0.54} + 0.8680h_f^{0.54} + 0.2032h_f^{0.54}$$

The friction head loss is the same in all parallel branches.

$$3\ \text{ft}^3/\text{sec} = (0.0194 + 0.8680 + 0.2032)h_f^{0.54}$$

$$h_f = 6.5\ \text{ft}$$

## 30. MULTIPLE RESERVOIR SYSTEMS

In the *three-reservoir problem*, there are many possible choices for the unknown quantity (pipe length, diameter, head, flow rate, etc.). In all but the simplest cases, the solution technique is by trial and error based on conservation of mass and energy. (See Fig. 19.18.)

**Figure 19.18** *Three-Reservoir System*

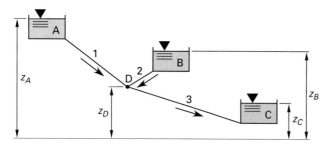

For simplification, velocity heads and minor losses are usually insignificant and can be neglected. However, the presence of a pump in any of the lines must be included in the solution procedure. This is most easily done by adding the pump head to the elevation of the reservoir feeding the pump. If the pump head is not known or depends on the flow rate, it must be determined iteratively as part of the solution procedure.

**Case 1:** Given all lengths, diameters, and elevations, find all flow rates.

Although an analytical solution method is possible, this type of problem is easily solved iteratively. The following procedure makes an initial estimate of a flow rate and uses it to calculate $p_D$. Since this method may not converge if the initial estimate of $\dot{V}_1$ is significantly in error, it is helpful to use other information (e.g., normal pipe velocities; see Sec. 19.3) to obtain the initial estimate. An alternate procedure is simply to make several estimates of $p_D$ and calculate the corresponding values of flow rate.

*step 1:* Assume a reasonable value for $\dot{V}_1$. Calculate the corresponding friction loss, $h_{f,1}$. Use the Bernoulli equation to find the corresponding value of $p_D$. Disregard minor losses and velocity head.

$$\mathrm{v}_1 = \frac{\dot{V}_1}{A_1} \qquad 19.106$$

$$z_A = z_D + \frac{p_D}{\gamma} + h_{f,1} \qquad 19.107$$

*step 2:* Use the value of $p_D$ to calculate $h_{f,2}$. Use the friction loss to determine $\mathrm{v}_2$. Use $\mathrm{v}_2$ to determine $\dot{V}_2$.

$$z_B = z_D + \frac{p_D}{\gamma} \pm h_{f,2} \qquad 19.108$$

$$\dot{V}_2 = \mathrm{v}_2 A_2 \qquad 19.109$$

If flow is out of reservoir $B$, $h_{f,2}$ should be added. If $z_D + (p_D/\gamma) > z_B$, flow will be into reservoir $B$. In this case, $h_{f,2}$ should be subtracted.

*step 3:* Similarly, use the value of $p_D$ to calculate $h_{f,3}$. Use the friction loss to determine $\mathrm{v}_3$. Use $\mathrm{v}_3$ to determine $\dot{V}_3$.

$$z_C = z_D + \frac{p_D}{\gamma} - h_{f,3} \qquad 19.110$$

$$\dot{V}_3 = \mathrm{v}_3 A_3 \qquad 19.111$$

*step 4:* Check that $\dot{V}_1 \pm \dot{V}_2 = \dot{V}_3$. If it does not, repeat steps 1 through 4. After the second iteration, plot $\dot{V}_1 \pm \dot{V}_2 - \dot{V}_3$ versus $\dot{V}_1$. Interpolate or extrapolate the value of $\dot{V}_1$ that makes the difference zero.

**Case 2:** Given $\dot{V}_1$ and all lengths, diameters, and elevations except $z_C$, find $z_C$.

*step 1:* Calculate $\mathrm{v}_1$.

$$\mathrm{v}_1 = \frac{\dot{V}_1}{A_1} \qquad 19.112$$

*step 2:* Calculate the corresponding friction loss, $h_{f,1}$. Use the Bernoulli equation to find the corresponding value of $p_D$. Disregard minor losses and velocity head.

$$z_A = z_D + \frac{p_D}{\gamma} + h_{f,1} \qquad 19.113$$

*step 3:* Use the value $p_D$ to calculate $h_{f,2}$. Use the friction loss to determine $\mathrm{v}_2$. Use $\mathrm{v}_2$ to determine $\dot{V}_2 A_2$.

$$z_B = z_D + \frac{p_D}{\gamma} \pm h_{f,2} \qquad 19.114$$

$$\dot{V}_2 = \mathrm{v}_2 A_2 \qquad 19.115$$

If flow is out of reservoir $B$, $h_{f,2}$ should be added. If $z_D + (p_D/\gamma) > z_B$, flow will be into reservoir $B$. In this case, $h_{f,2}$ should be subtracted.

*step 4:*

$$\dot{V}_3 = \dot{V}_1 \pm \dot{V}_2 \qquad 19.116$$

*step 5:*

$$\mathrm{v}_3 = \frac{\dot{V}_3}{A_3} \qquad 19.117$$

*step 6:* Calculate $h_{f,3}$.

*step 7:*

$$z_C = z_D + \frac{p_D}{\gamma} - h_{f,3} \qquad 19.118$$

**Case 3:** Given $\dot{V}_1$, all lengths, all elevations, and all diameters except $D_3$, find $D_3$.

*step 1:* Repeat step 1 from case 2.

*step 2:* Repeat step 2 from case 2.

*step 3:* Repeat step 3 from case 2.

*step 4:* Repeat step 4 from case 2.

*step 5:* Calculate $h_{f,3}$ from

$$z_C = z_D + \frac{p_D}{\gamma} - h_{f,3} \qquad 19.119$$

*step 6:* Calculate $D_3$ from $h_{f,3}$.

**Case 4:** Given all lengths, diameters, and elevations except $z_D$, find all flow rates.

*step 1:* Calculate the head loss between each reservoir and junction $D$. Combine as many terms as possible into constant $k'$.

$$\dot{V} = A\mathrm{v} = A\sqrt{\frac{2Dgh_f}{f_L}}$$

$$= k'\sqrt{h_f} \quad \text{[Darcy]} \qquad 19.120$$

$$\dot{V} = A\mathrm{v} = \frac{A(0.550)CD^{0.63}h_f^{0.54}}{L^{0.54}}$$

$$= k'h_f^{0.54} \quad \text{[Hazen-Williams; U.S.]} \qquad 19.121$$

*step 2:* Assume that the flow direction in all three pipes is toward junction $D$. Write the conservation equation for junction $D$.

$$\dot{V}_{D,t} = \dot{V}_1 + \dot{V}_2 + \dot{V}_3 = 0 \qquad 19.122$$

$$k'_1\sqrt{h_{f,1}} + k'_2\sqrt{h_{f,2}}$$
$$+ k'_3\sqrt{h_{f,3}} = 0 \quad \text{[Darcy]} \qquad 19.123$$

$$k'_1 h_{f,1}^{0.54} + k'_2 h_{f,2}^{0.54}$$
$$+ k'_3 h_{f,3}^{0.54} = 0 \quad \text{[Hazen-Williams]} \qquad 19.124$$

*step 3:* Write the Bernoulli equation between each reservoir and junction $D$. Since $p_A = p_B = p_C = 0$, and $\mathrm{v}_A = \mathrm{v}_B = \mathrm{v}_C = 0$, the friction loss in branch 1 is

$$h_{f,1} = z_A - z_D - \frac{p_D}{\gamma} \qquad 19.125$$

However, $z_D$ and $p_D$ can be combined since they are related constants in any particular situation. Define $R_D$ as

$$R_D = z_D + \frac{p_D}{\gamma} \qquad 19.126$$

Then, the friction head losses in the branches are

$$h_{f,1} = z_A - R_D \qquad 19.127$$

$$h_{f,2} = z_B - R_D \qquad 19.128$$

$$h_{f,3} = z_C - R_D \qquad 19.129$$

*step 4:* Assume a value for $R_D$. Calculate the corresponding $h_f$ values. Use Eq. 19.120 to find $\dot{V}_1$, $\dot{V}_2$, and $\dot{V}_3$. Calculate the corresponding $\dot{V}_t$ value. Repeat until $\dot{V}_t$ converges to zero. It

is not necessary to calculate $p_D$ or $z_D$ once all of the flow rates are known.

## 31. PIPE NETWORKS

Network flows in a *multiloop system* cannot be determined by any closed-form equation. (See Fig. 19.19.) Most real-world problems involving multiloop systems are analyzed iteratively on a computer. Computer programs are based on the *Hardy Cross method*, which can also be performed manually when there are only a few loops. In this method, flows in all of the branches are first assumed, and adjustments are made in consecutive iterations to the assumed flow.

**Figure 19.19** *Multiloop System*

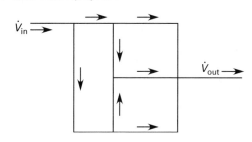

The Hardy Cross method is based on the following principles.

*Principle 1:* Conservation—The flows entering a junction equal the flows leaving the junction.

*Principle 2:* The algebraic sum of head losses around any closed loop is zero.

The friction head loss has the form $h_f = K'\dot{V}^n$, with $h_f$ having units of feet. (Note that $k'$ used in Eq. 19.104 is equal to $\sqrt{1/K'}$.) The Darcy friction factor, $f$, is usually assumed to be the same in all parts of the network. For a Darcy head loss, the exponent is $n = 2$. For a Hazen-Williams loss, $n = 1.85$.

- For $\dot{V}$ in ft³/sec, $L$ in feet, and $D$ in feet, the friction coefficient is

$$K' = \frac{0.02517fL}{D^5} \quad \text{[Darcy]} \qquad 19.130$$

$$K' = \frac{4.727L}{D^{4.8655}C^{1.85}} \quad \text{[Hazen-Williams]} \qquad 19.131$$

- For $\dot{V}$ in gal/min, $L$ in feet, and $d$ in inches, the friction coefficient is

$$K' = \frac{0.03109fL}{d^5} \quad \text{[Darcy]} \qquad 19.132$$

$$K' = \frac{10.44L}{d^{4.8655}C^{1.85}} \quad \text{[Hazen-Williams]} \qquad 19.133$$

- For $\dot{V}$ in gal/min, $L$ in feet, and $D$ in feet, the friction coefficient is

$$K' = \frac{1.251 \times 10^{-7} fL}{D^5} \quad \text{[Darcy]} \qquad 19.134$$

$$K' = \frac{5.862 \times 10^{-5} L}{D^{4.8655} C^{1.85}} \quad \text{[Hazen-Williams]} \qquad 19.135$$

- For $\dot{V}$ in MGD (millions of gallons per day), $L$ in feet, and $D$ in feet, the friction coefficient is

$$K' = \frac{0.06026 fL}{D^5} \quad \text{[Darcy]} \qquad 19.136$$

$$K' = \frac{10.59 L}{D^{4.8655} C^{1.85}} \quad \text{[Hazen-Williams]} \qquad 19.137$$

If $\dot{V}_a$ is the assumed flow in a pipe, the true value, $\dot{V}$, can be calculated from the difference (correction), $\delta$.

$$\dot{V} = \dot{V}_a + \delta \qquad 19.138$$

The friction loss term for the assumed value and its correction can be expanded as a series. Since the correction is small, higher order terms can be omitted.

$$h_f = K'(\dot{V}_a + \delta)^n$$

$$\approx K' \dot{V}_a^n + nK'\delta \dot{V}_a^{n-1} \qquad 19.139$$

From Principle 2 (see Sec. 19.31), the sum of the friction drops is zero around a loop. The correction, $\delta$, is the same for all pipes in the loop and can be taken out of the summation. Since the loop closes on itself, all elevations can be omitted.

$$\sum h_f = \sum K' \dot{V}_a^n + n\delta \sum K' \dot{V}_a^{n-1} = 0 \qquad 19.140$$

This equation can be solved for $\delta$.

$$\delta = \frac{-\sum K' \dot{V}_a^n}{n \sum \left| K' \dot{V}_a^{n-1} \right|}$$

$$= -\frac{\sum h_f}{n \sum \left| \dfrac{h_f}{\dot{V}_a} \right|} \qquad 19.141$$

The Hardy Cross procedure is as follows.

step 1: Determine the value of $n$. For a Darcy head loss, the exponent is $n = 2$. For a Hazen-Williams loss, $n = 1.85$.

step 2: Arbitrarily select a positive direction (e.g., clockwise).

step 3: Label all branches and junctions in the network.

step 4: Separate the network into independent loops such that each branch is included in at least one loop.

step 5: Calculate $K'$ for each branch in the network.

step 6: Assume consistent and reasonable flow rates and directions for each branch in the network.

step 7: Calculate the correction, $\delta$, for each independent loop. (The numerator is the sum of head losses around the loop, taking signs into consideration.) It is not necessary for the loop to be at the same elevation everywhere. Disregard elevations. Since the loop closes on itself, all elevations can be omitted.

step 8: Apply the correction, $\delta$, to each branch in the loop. The correction must be applied in the same sense to each branch in the loop. If clockwise has been taken as the positive direction, then $\delta$ is added to clockwise flows and subtracted from counterclockwise flows.

step 9: Repeat steps 7 and 8 until the correction is sufficiently small.

### Example 19.10

A two-loop pipe network is shown. All junctions are at the same elevation. The Darcy friction factor is 0.02 for all pipes in the network. For convenience, use the nominal pipe sizes shown in the figure. Determine the flows in all branches.

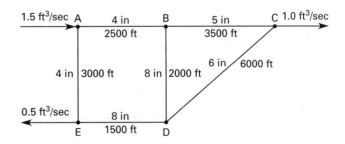

*Solution*

step 1: The Darcy friction factor is given, so $n = 2$.

step 2: Select clockwise as the positive direction.

step 3: Use the junction letters in the illustration.

step 4: Two independent loops are needed. Work with loops ABDE and BCD. (Loop ABCDE could also be used but would be more complex than loop BCD.)

step 5: Work with branch AB.

$$D = \frac{4 \text{ in}}{12 \ \dfrac{\text{in}}{\text{ft}}} = 0.3333 \text{ ft}$$

Use Eq. 19.130.

$$K'_{AB} = \frac{0.0252 fL}{D^5}$$

$$= \frac{(0.0252)(0.02)(2500 \text{ ft})}{(0.3333 \text{ ft})^5}$$

$$= 306.2$$

Similarly,

$$K'_{BC} = 140.5$$
$$K'_{DC} = 96.8$$
$$K'_{BD} = 7.7$$
$$K'_{ED} = 5.7$$
$$K'_{AE} = 367.4$$

*step 6:* Assume the direction and flow rates shown.

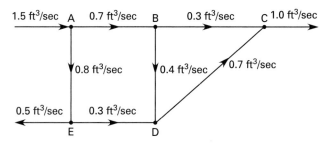

*step 7:* Use Eq. 19.141.

$$\delta = \frac{-\sum K' \dot{V}_a^n}{n \sum \left| K' \dot{V}_a^{n-1} \right|}$$

$$\delta_{ABDE} = \frac{-\left( (306.2)(0.7)^2 + (7.7)(0.4)^2 \\ - (5.7)(0.3)^2 - (367.4)(0.8)^2 \right)}{(2)\left( (306.2)(0.7) + (7.7)(0.4) \\ + (5.7)(0.3) + (367.4)(0.8) \right)}$$

$$= 0.08 \text{ ft}^3/\text{sec}$$

$$\delta_{BCD} = \frac{-\left( (140.5)(0.3)^2 - (96.8)(0.7)^2 \\ - (7.7)(0.4)^2 \right)}{(2)\left( (140.5)(0.3) + (96.8)(0.7) \\ + (7.7)(0.4) \right)}$$

$$= 0.16 \text{ ft}^3/\text{sec}$$

*step 8:* The corrected flows are

$$\dot{V}_{AB} = 0.7 \frac{\text{ft}^3}{\text{sec}} + 0.08 \frac{\text{ft}^3}{\text{sec}} = 0.78 \text{ ft}^3/\text{sec}$$

$$\dot{V}_{BC} = 0.3 \frac{\text{ft}^3}{\text{sec}} + 0.16 \frac{\text{ft}^3}{\text{sec}} = 0.46 \text{ ft}^3/\text{sec}$$

$$\dot{V}_{DC} = 0.7 \frac{\text{ft}^3}{\text{sec}} - 0.16 \frac{\text{ft}^3}{\text{sec}} = 0.54 \text{ ft}^3/\text{sec}$$

$$\dot{V}_{BD} = 0.4 \frac{\text{ft}^3}{\text{sec}} + 0.08 \frac{\text{ft}^3}{\text{sec}} - 0.16 \frac{\text{ft}^3}{\text{sec}}$$
$$= 0.32 \text{ ft}^3/\text{sec}$$

$$\dot{V}_{ED} = 0.3 \frac{\text{ft}^3}{\text{sec}} - 0.08 \frac{\text{ft}^3}{\text{sec}} = 0.22 \text{ ft}^3/\text{sec}$$

$$\dot{V}_{AE} = 0.8 \frac{\text{ft}^3}{\text{sec}} - 0.08 \frac{\text{ft}^3}{\text{sec}} = 0.72 \text{ ft}^3/\text{sec}$$

## 32. FLOW MEASURING DEVICES

A device that measures flow can be calibrated to indicate either velocity or volumetric flow rate. There are many methods available to obtain the flow rate. Some are indirect, requiring the use of transducers and solid-state electronics, and others can be evaluated using the Bernoulli equation. Some are more appropriate for one variety of fluid than others, and some are limited to specific ranges of temperature and pressure.

Table 19.7 categorizes a few common flow measurement methods. Many other methods and variations thereof exist, particularly for specialized industries. Some of the methods listed are so basic that only a passing mention will be made of them. Others, particularly those that can be analyzed with energy and mass conservation laws, will be covered in greater detail in subsequent sections.

The utility meters used to measure gas and water usage are examples of *displacement meters*. Such devices are cyclical, fixed-volume devices with counters to record the numbers of cycles. Displacement devices are generally unpowered, drawing on only the pressure energy to overcome mechanical friction. Most configurations for positive-displacement pumps (e.g., reciprocating piston, helical screw, and nutating disk) have also been converted to measurement devices.

The venturi nozzle, orifice plate, and flow nozzle are examples of *obstruction meters*. These devices rely on a decrease in static pressure to measure the flow velocity. One disadvantage of these devices is that the pressure drop is proportional to the square of the velocity, limiting the range over which any particular device can be used.

**Table 19.7** *Flow Measuring Devices*

I   direct (primary) measurements
     positive-displacement meters
     volume tanks
     weight and mass scales
II  indirect (secondary) measurements
     obstruction meters
       – flow nozzles
       – orifice plate meters
       – variable-area meters
       – venturi meters
     velocity probes
       – direction sensing probes
       – pitot-static meters
       – pitot tubes
       – static pressure probes
     miscellaneous methods
       – hot-wire meters
       – magnetic flow meters
       – mass flow meters
       – sonic flow meters
       – turbine and propeller meters

An obstruction meter that somewhat overcomes the velocity range limitation is the *variable-area meter*, also known as a *rotameter*, illustrated in Fig. 19.20.[21] This device consists of a float (which is actually more dense than the fluid) and a transparent sight tube. With proper design, the effects of fluid density and viscosity can be minimized. The sight glass can be directly calibrated in volumetric flow rate, or the height of the float above the zero position can be used in a volumetric calculation.

It is necessary to be able to measure static pressures in order to use obstruction meters and pitot-static tubes. In some cases, a *static pressure probe* is used. Figure 19.21 illustrates a simplified static pressure probe. In practice, such probes are sensitive to burrs and irregularities in the tap openings, orientation to the flow (i.e., *yaw*), and interaction with the pipe walls and other probes. A *direction-sensing probe* overcomes some of these problems.

A weather station *anemometer* used to measure wind velocity is an example of a simple *turbine meter*. Similar devices are used to measure the speed of a stream or river, in which case the name *current meter* may be used. Turbine meters are further divided into cup-type meters and propeller-type meters, depending on the orientation of the turbine axis relative to the flow direction. (The turbine axis and flow direction are parallel for propeller-type meters; they are perpendicular for cup-type meters.) Since the wheel motion is proportional to the flow velocity, the velocity is determined by counting the number of revolutions made by the wheel per unit time.

---

[21]The rotameter has its own disadvantages, however. It must be installed vertically; the fluid cannot be opaque; and it is more difficult to manufacture for use with high-temperature, high-pressure fluids.

**Figure 19.20** *Variable-Area Rotameter*

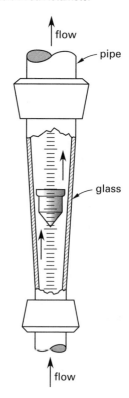

**Figure 19.21** *Simple Static Pressure Probe*

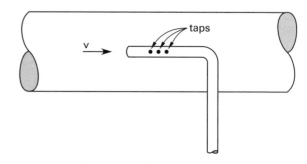

A more sophisticated turbine flowmeter uses a reluctance-type pickup coil to detect wheel motion. The permeability of a magnetic circuit changes each time a wheel blade passes the pole of a permanent magnet in the meter body. This change is detected to indicate velocity or flow rate.

A *hot-wire anemometer* measures velocity by determining the cooling effect of fluid (usually a gas) flowing over an electrically heated tungsten, platinum, or nickel wire. Cooling is primarily by convection; radiation and conduction are neglected. Circuitry can be used either to keep the current constant (in which case, the changing resistance is measured) or to keep the temperature constant (in which case, the changing current is measured). Additional circuitry can be used to compensate for thermal lag if the velocity changes rapidly.

A voltage proportional to the velocity will be generated when a conductor passes through a magnetic field.[22] This characteristic can be used to measure flow velocity if the fluid is electrically conductive. *Magnetic flowmeters* are ideal for measuring the flow of liquid metals, but variations of the device can also be used when the fluid is only slightly conductive. In some cases, precise quantities of conductive ions can be added to the fluid to permit measurement by this method.

In an *ultrasonic flowmeter*, two electric or magnetic transducers are placed a short distance apart on the outside of the pipe. One transducer serves as a transmitter of ultrasonic waves; the other transducer is a receiver. As an ultrasonic wave travels from the transmitter to the receiver, its velocity will be increased (or decreased) by the relative motion of the fluid. The phase shift between the fluid-carried waves and the waves passing through a stationary medium can be measured and converted to fluid velocity.

## 33. PITOT-STATIC GAUGE

Measurements from pitot tubes are used to determine total (stagnation) energy. Piezometer tubes and wall taps are used to measure static pressure energy. The difference between the total and static energies is the kinetic energy of the flow. Figure 19.22 illustrates a comparative method of directly measuring the velocity head for an incompressible fluid.

**Figure 19.22** *Comparative Velocity Head Measurement*

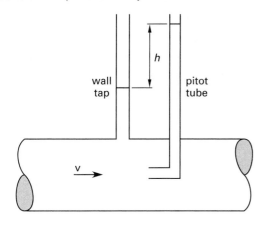

$$\frac{v^2}{2} = \frac{p_t - p_s}{\rho} = hg \qquad \text{[SI]} \qquad 19.142(a)$$

$$\frac{v^2}{2g_c} = \frac{p_t - p_s}{\rho} = h \times \left(\frac{g}{g_c}\right) \qquad \text{[U.S.]} \qquad 19.142(b)$$

$$v = \sqrt{2gh} \qquad\qquad 19.143$$

The pitot tube and static pressure tap shown in Fig. 19.22 can be combined into a *pitot-static gauge*. (See

---
[22]The magnitude of this induced voltage is predicted by *Faraday's law*.

Fig. 19.23.) In a pitot-static gauge, one end of the manometer is acted upon by the static pressure (also referred to as the *transverse pressure*). The other end of the manometer experiences the total pressure. The difference in elevations of the manometer fluid columns is the velocity head. This distance must be corrected if the density of the flowing fluid is significant.

**Figure 19.23** *Pitot-Static Gauge*

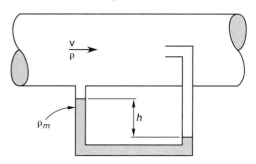

$$\frac{v^2}{2} = \frac{p_t - p_s}{\rho} = \frac{h(\rho_m - \rho)g}{\rho} \qquad \text{[SI]} \qquad 19.144(a)$$

$$\frac{v^2}{2g_c} = \frac{p_t - p_s}{\rho} = \frac{h(\rho_m - \rho)}{\rho} \times \left(\frac{g}{g_c}\right) \qquad \text{[U.S.]} \qquad 19.144(b)$$

$$v = \sqrt{\frac{2gh(\rho_m - \rho)}{\rho}} \qquad\qquad 19.145$$

Another correction, which is seldom made, is to multiply the velocity calculated from Eq. 19.145 by $C_I$, the *coefficient of the instrument*. Since the flow past the pitot-static tube is slightly faster than the free-fluid velocity, the static pressure measured will be slightly lower than the true value. This makes the indicated velocity slightly higher than the true value. $C_I$, a number close to but less than 1.0, corrects for this.

A pitot-static tube indicates the velocity at only one point in a pipe. If the flow is laminar, and if the pitot-static tube is in the center of the pipe, $v_{max}$ will be determined. The average velocity, however, will be only half the maximum value.

Pitot tube measurements are sensitive to the condition of the opening and errors in installation alignment. The *yaw angle* (i.e., the acute angle between the pitot tube axis and the flow streamline) should be zero.

### Example 19.11

Water (62.4 lbm/ft³; $\rho = 1000$ kg/m³) is flowing through a pipe. A pitot-static gauge registers 3.0 in (0.076 m) of mercury. What is the velocity of the water in the pipe?

*SI Solution*

The density of mercury is $\rho = 13\,580$ kg/m$^3$. The velocity can be calculated directly from Eq. 19.145.

$$v = \sqrt{\frac{(2)\left(9.81\,\frac{\text{m}}{\text{s}^2}\right)(0.076\,\text{m}) \times \left(13\,580\,\frac{\text{kg}}{\text{m}^3} - 1000\,\frac{\text{kg}}{\text{m}^3}\right)}{1000\,\frac{\text{kg}}{\text{m}^3}}}$$

$$= 4.33\ \text{m/s}$$

*Customary U.S. Solution*

The density of mercury is $\rho = 848.6$ lbm/ft$^3$.

From Eq. 19.145,

$$v = \sqrt{\frac{(2)\left(32.2\,\frac{\text{ft}}{\text{sec}^2}\right)\left((3\,\text{in})\left(\frac{1\,\text{ft}}{12\,\text{in}}\right)\right) \times \left(848.6\,\frac{\text{lbm}}{\text{ft}^3} - 62.4\,\frac{\text{lbm}}{\text{ft}^3}\right)}{62.4\,\frac{\text{lbm}}{\text{ft}^3}}}$$

$$= 14.24\ \text{ft/sec}$$

## 34. VENTURI METER

Figure 19.24 illustrates a simple *venturi*. (Sometimes the venturi is called a *converging-diverging nozzle*.) This flow measuring device can be inserted directly into a pipeline. Since the diameter changes are gradual, there is very little friction loss.[23] Static pressure measurements are taken at the throat and upstream of the diameter change. These measurements are traditionally made by manometer.

**Figure 19.24** *Venturi Meter*

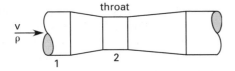

The analysis of *venturi meter* performance is relatively simple. The traditional derivation of upstream velocity starts by assuming a horizontal orientation and frictionless, incompressible, and turbulent flow. Then the Bernoulli equation is written for points 1 and 2. Equation

19.146 shows that the static pressure decreases as the velocity increases. This is known as the *venturi effect.*

$$\frac{v_1^2}{2} + \frac{p_1}{\rho} = \frac{v_2^2}{2} + \frac{p_2}{\rho} \qquad \text{[SI]} \quad 19.146(a)$$

$$\frac{v_1^2}{2g_c} + \frac{p_1}{\rho} = \frac{v_2^2}{2g_c} + \frac{p_2}{\rho} \qquad \text{[U.S.]} \quad 19.146(b)$$

The two velocities are related by the continuity equation.

$$A_1 v_1 = A_2 v_2 \qquad 19.147$$

Combining Eq. 19.146 and Eq. 19.147 and eliminating the unknown $v_1$ produces an expression for the throat velocity. Also, a *coefficient of velocity* is used to account for the small effect of friction. ($C_v$ is very close to 1.0, usually 0.98 or 0.99.)

$$v_2 = C_v v_{2,\text{ideal}}$$

$$= \left(\frac{C_v}{\sqrt{1 - \left(\frac{A_2}{A_1}\right)^2}}\right)\sqrt{\frac{2(p_1 - p_2)}{\rho}}$$

$$\text{[SI]} \quad 19.148(a)$$

$$v_2 = \left(\frac{C_v}{\sqrt{1 - \left(\frac{A_2}{A_1}\right)^2}}\right)\sqrt{\frac{2g_c(p_1 - p_2)}{\rho}} \quad \text{[U.S.]} \quad 19.148(b)$$

The *velocity of approach factor*, $F_{va}$, also known as the *meter constant*, is the reciprocal of the denominator of the first term of Eq. 19.148. The *beta ratio* can be incorporated into the formula for $F_{va}$.

$$\beta = \frac{D_2}{D_1} \qquad 19.149$$

$$F_{va} = \frac{1}{\sqrt{1 - \left(\frac{A_2}{A_1}\right)^2}} = \frac{1}{\sqrt{1 - \beta^4}} \qquad 19.150$$

If a manometer is used to measure the pressure difference directly, Eq. 19.148 can be rewritten in terms of the manometer fluid reading. (See Fig. 19.25.)

$$v_2 = \left(\frac{C_v}{\sqrt{1 - \beta^4}}\right)\sqrt{\frac{2g(\rho_m - \rho)h}{\rho}}$$

$$= C_v F_{va}\sqrt{\frac{2g(\rho_m - \rho)h}{\rho}} \qquad 19.151$$

The flow rate through a venturi meter can be calculated from the throat area. There is an insignificant

---

[23]The actual friction loss is approximately 10% of the pressure difference $p_1 - p_2$.

**Figure 19.25** *Venturi Meter with Manometer*

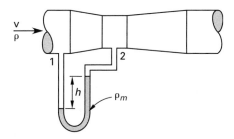

**Figure 19.27** *Orifice Meter with Differential Manometer*

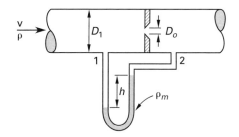

amount of contraction of the flow as it passes through the throat, and the *coefficient of contraction* is seldom encountered in venturi meter work. The *coefficient of discharge* ($C_d = C_c C_v$) is quoted, nevertheless. Values of $C_d$ range from slightly less than 0.90 to over 0.99, depending on the Reynolds number. $C_d$ is seldom less than 0.95 for turbulent flow. (See Fig. 19.26.)

**Figure 19.26** *Typical Venturi Meter Discharge Coefficients (Long Radius Venturi Meter)*

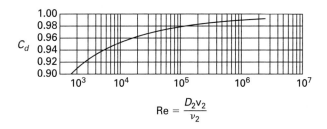

$$\dot{V} = C_d A_2 v_{2,\text{ideal}} \qquad 19.152$$

The product $C_d F_{va}$ is known as the *coefficient of flow* or *flow coefficient*, not to be confused with the coefficient of discharge.[24] This factor is used for convenience, since it combines the losses with the meter constant.

$$C_f = C_d F_{va} = \frac{C_d}{\sqrt{1 - \beta^4}} \qquad 19.153$$

$$\dot{V} = C_f A_2 \sqrt{\frac{2g(\rho_m - \rho)h}{\rho}} \qquad 19.154$$

## 35. ORIFICE METER

The *orifice meter* (or *orifice plate*) is used more frequently than the venturi meter to measure flow rates in small pipes. It consists of a thin or sharp-edged plate with a central, round hole through which the fluid flows. Such a plate is easily clamped between two flanges in an existing pipeline. (See Fig. 19.27.)

While (for small pipes) the orifice meter may consist of a thin plate without significant thickness, various types of

---

[24]Some writers use the symbol $K$ for the flow coefficient.

bevels and rounded edges are also used with thicker plates. There is no significant difference in the analysis procedure between "flat plate," "sharp-edged," or "square-edged" orifice meters. Any effect that the orifice edges have is accounted for in the discharge and flow coefficient correlations. Similarly, the direction of the bevel will affect the coefficients but not the analysis method.

As with the venturi meter, pressure taps are used to obtain the static pressure upstream of the orifice plate and at the *vena contracta* (i.e., at the point of minimum pressure).[25] A differential manometer connected to the two taps conveniently indicates the difference in static pressures.

The derivation of the governing equations for an orifice meter is similar to that of the venturi meter. (The obvious falsity of assuming frictionless flow through the orifice is corrected by the coefficient of discharge.) The major difference is that the coefficient of contraction is taken into consideration in writing the mass continuity equation, since the pressure is measured at the vena contracta, not the orifice.

$$A_2 = C_c A_o \qquad 19.155$$

$$v_o = \left( \frac{C_v}{\sqrt{1 - \left(\frac{C_c A_o}{A_1}\right)^2}} \right) \sqrt{\frac{2(p_1 - p_2)}{\rho}} \qquad \text{[SI]} \quad 19.156(a)$$

---

[25]Calibration of the orifice meter is sensitive to tap placement. Upstream taps are placed between one-half and two pipe diameters upstream from the orifice. (An upstream distance of one pipe diameter is often quoted and used.) There are three tap-placement options: flange, vena contracta, and standardized. Flange taps are used with prefabricated orifice meters that are inserted (by flange bolting) in pipes. If the location of the vena contracta is known, a tap can be placed there. However, the location of the vena contracta depends on the diameter ratio $\beta = D_o/D$ and varies from approximately 0.4 to 0.7 pipe diameters downstream. Due to the difficulty of locating the vena contracta, the standardized $1D\text{-}^1/_2D$ configuration is often used. The upstream tap is one diameter before the orifice; the downstream tap is one-half diameter after the orifice. Since approaching flow should be stable and uniform, care must be taken not to install the orifice meter less than approximately five diameters after a bend or elbow.

$$v_o = \left( \frac{C_v}{\sqrt{1 - \left( \frac{C_c A_o}{A_1} \right)^2}} \right) \sqrt{\frac{2g_c(p_1 - p_2)}{\rho}}$$

$$\text{[U.S.]} \quad 19.156(b)$$

If a manometer is used to indicate the differential pressure $p_1 - p_2$, the velocity at the vena contracta can be calculated from Eq. 19.157.

$$v_o = \left( \frac{C_v}{\sqrt{1 - \left( \frac{C_c A_o}{A_1} \right)^2}} \right) \sqrt{\frac{2g(\rho_m - \rho)h}{\rho}} \qquad 19.157$$

Although the orifice meter is simpler and less expensive than a venturi meter, its discharge coefficient is much less than that of a venturi meter. $C_d$ usually ranges from 0.55 to 0.75, with values of 0.60 and 0.61 often being quoted. (The coefficient of contraction has a large effect, since $C_d = C_v C_c$.) Also, its pressure recovery is poor (i.e., there is a permanent pressure reduction), and it is susceptible to inaccuracies from wear and abrasion.[26]

The *velocity of approach factor*, $F_{va}$, for an orifice meter is defined differently than for a venturi meter, since it takes into consideration the contraction of the flow. However, the velocity of approach factor is still combined with the coefficient of discharge into the flow coefficient, $C_f$. Figure 19.28 illustrates how the flow coefficient varies with the area ratio and the Reynolds number.

$$F_{va} = \frac{1}{\sqrt{1 - \left( \frac{C_c A_o}{A_1} \right)^2}} \qquad 19.158$$

$$C_f = C_d F_{va} \qquad 19.159$$

The flow rate through an orifice meter is given by Eq. 19.160.

$$\dot{V} = C_f A_o \sqrt{\frac{2g(\rho_m - \rho)h}{\rho}} = C_f A_o \sqrt{\frac{2(p_1 - p_2)}{\rho}}$$

$$\text{[SI]} \quad 19.160(a)$$

[26]The actual loss varies from 40% to 90% of the differential pressure. The loss depends on the diameter ratio $\beta = D_o/D_1$, and is not particularly sensitive to the Reynolds number for turbulent flow. For $\beta = 0.5$, the loss is 73% of the measured pressure difference, $p_1 - p_2$. This decreases to approximately 56% of the pressure difference when $\beta = 0.65$ and to 38%, when $\beta = 0.8$. For any diameter ratio, the pressure drop coefficient, $K$, in multiples of the orifice velocity head is $K = (1 - \beta^2)/C_f^2$.

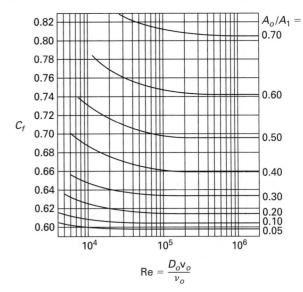

**Figure 19.28** *Typical Flow Coefficients for Orifice Plates*

$$\dot{V} = C_f A_o \sqrt{\frac{2g(\rho_m - \rho)h}{\rho}} = C_f A_o \sqrt{\frac{2g_c(p_1 - p_2)}{\rho}}$$

$$\text{[U.S.]} \quad 19.160(b)$$

### Example 19.12

150°F water ($\rho = 61.2$ lbm/ft³) flows in an 8 in schedule-40 steel pipe at the rate of 2.23 ft³/sec. A sharp-edged orifice with a 7 in diameter hole is placed in the line. A mercury differential manometer is used to record the pressure difference. If the orifice has a flow coefficient, $C_f$, of 0.62, what deflection in inches of mercury is observed? (Mercury has a density of 848.6 lbm/ft³.)

*Solution*

The orifice area is

$$A_o = \frac{\pi}{4} D_o^2 = \left( \frac{\pi}{4} \right) \left( \frac{7 \text{ in}}{12 \frac{\text{in}}{\text{ft}}} \right)^2$$

$$= 0.2673 \text{ ft}^2$$

Equation 19.160 is solved for $h$.

$$h = \frac{\dot{V}^2 \rho}{2g C_f^2 A_o^2 (\rho_m - \rho)}$$

$$= \frac{\left( 2.23 \frac{\text{ft}^3}{\text{sec}} \right)^2 \left( 61.2 \frac{\text{lbm}}{\text{ft}^3} \right) \left( 12 \frac{\text{in}}{\text{ft}} \right)}{(2) \left( 32.2 \frac{\text{ft}}{\text{sec}^2} \right) (0.62)^2} \times (0.2673 \text{ ft}^2)^2 \left( 848.6 \frac{\text{lbm}}{\text{ft}^3} - 61.2 \frac{\text{lbm}}{\text{ft}^3} \right)$$

$$= 2.62 \text{ in}$$

## 36. FLOW NOZZLE

A typical flow nozzle is illustrated in Fig. 19.29. This device consists only of a converging section. It is somewhat between an orifice plate and a venturi meter in performance, possessing some of the advantages and disadvantages of each. The venturi performance equations can be used for the flow nozzle.

**Figure 19.29** *Flow Nozzle*

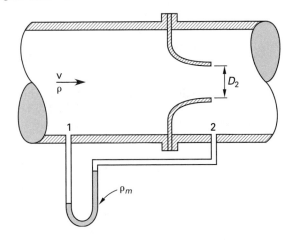

The geometry of the nozzle entrance is chosen to prevent separation of the fluid from the wall. The converging portion and the subsequent parallel section keep the coefficients of velocity and contraction close to 1.0. However, the absence of a diffuser section disrupts the orderly return of fluid to its original condition. The permanent pressure drop is more similar to that of the orifice meter than the venturi meter.

Since the nozzle geometry greatly affects the performance, values of $C_d$ and $C_f$ have been established for only a limited number of specific proprietary nozzles.[27]

## 37. FLOW MEASUREMENTS OF COMPRESSIBLE FLUIDS

Volume measurements of compressible fluids (i.e., gases) are not very meaningful. The volume of a gas will depend on its temperature and pressure. For that reason, flow quantities of gases discharged should be stated as mass flow rates.

$$\dot{m} = \rho_2 A_2 \mathrm{v}_2 \qquad 19.161$$

Equation 19.161 requires that the velocity and area be measured at the same point. More important, the density must be measured at that point, as well. However, it is common practice in flow measurement work to use the density of the upstream fluid at position 1. (Note that this is not the stagnation density.)

The significant error introduced by this simplification is corrected by the use of an *expansion factor*, $Y$. For venturi meters and flow nozzles, values of the expansion factor are generally calculated theoretical values. Values of $Y$ are determined experimentally for orifice plates. (See Fig. 19.30.)

$$\dot{m} = Y\rho_1 A_2 \mathrm{v}_2 \qquad 19.162$$

Derivation of the theoretical formula for the expansion factor for venturi meters and flow nozzles is based on thermodynamic principles and an assumption of adiabatic flow.

$$Y = \sqrt{\frac{(1-\beta^4)\left(\left(\dfrac{p_2}{p_1}\right)^{2/k} - \left(\dfrac{p_2}{p_1}\right)^{(k+1)/k}\right)}{\left(\dfrac{k-1}{k}\right)\left(1-\dfrac{p_2}{p_1}\right)\left(1-\beta^4\left(\dfrac{p_2}{p_1}\right)^{2/k}\right)}} \qquad 19.163$$

**Figure 19.30** *Approximate Expansion Factors*

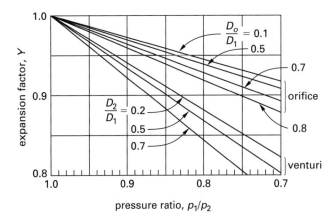

*Transport Processes and Unit Operations*, 3rd ed., © 1993. Reprinted by permission of Pearson Education, Inc. Upper Saddle River, NJ.

Once the expansion factor is known, it can be used with the standard flow rate ($\dot{V}$) equations for venturi meters, flow nozzles, and orifice meters. For example, for a venturi meter, the mass flow rate would be calculated from Eq. 19.164.

$$\dot{m} = Y\dot{m}_{\text{ideal}} = \left(\frac{YC_dA_2}{\sqrt{1-\beta^4}}\right)\sqrt{2\rho_1(p_1-p_2)}$$

$$[\text{SI}] \quad 19.164(a)$$

$$\dot{m} = Y\dot{m}_{\text{ideal}} = \left(\frac{YC_dA_2}{\sqrt{1-\beta^4}}\right)\sqrt{2g_c\rho_1(p_1-p_2)}$$

$$[\text{U.S.}] \quad 19.164(b)$$

---

[27]Some of the proprietary nozzles for which detailed performance data exist are the ASME long-radius nozzle (low-$\beta$ and high-$\beta$ series) and the International Standards Association (ISA) nozzle (German standard nozzle).

## 38. IMPULSE-MOMENTUM PRINCIPLE

(The convention of this section is to make $F$ and $x$ positive when they are directed toward the right. $F$ and $y$ are positive when directed upward. Also, the fluid is assumed to flow horizontally from left to right, and it has no initial $y$-component of velocity.)

The *momentum*, $\mathbf{P}$ (also known as *linear momentum* to distinguish it from *angular momentum*, which is not considered here), of a moving object is a vector quantity defined as the product of the object's mass and velocity.[28]

$$\mathbf{P} = m\mathbf{v} \qquad \text{[SI]} \qquad 19.165(a)$$

$$\mathbf{P} = \frac{m\mathbf{v}}{g_c} \qquad \text{[U.S.]} \qquad 19.165(b)$$

The *impulse*, $I$, of a constant force is calculated as the product of the force's magnitude and the length of time the force is applied.

$$\mathbf{I} = \mathbf{F}\Delta t \qquad\qquad 19.166$$

The *impulse-momentum principle* states that the impulse applied to a body is equal to the change in momentum. (This is also known as the *law of conservation of momentum*, even though fluid momentum is not always conserved.) Equation 19.167 is one way of stating Newton's second law.

$$\mathbf{I} = \Delta \mathbf{P} \qquad\qquad 19.167$$

$$F\Delta t = m\Delta\mathrm{v} = m(\mathrm{v}_2 - \mathrm{v}_1) \qquad \text{[SI]} \qquad 19.168(a)$$

$$F\Delta t = \frac{m\Delta\mathrm{v}}{g_c} = \frac{m(\mathrm{v}_2 - \mathrm{v}_1)}{g_c} \qquad \text{[U.S.]} \qquad 19.168(b)$$

For fluid flow, there is a mass flow rate, $\dot{m}$, but no mass per se. Since $\dot{m} = m/\Delta t$, the impulse-momentum equation can be rewritten as follows.

$$F = \dot{m}\Delta\mathrm{v} \qquad \text{[SI]} \qquad 19.169(a)$$

$$F = \frac{\dot{m}\Delta\mathrm{v}}{g_c} \qquad \text{[U.S.]} \qquad 19.169(b)$$

Equation 19.169 calculates the constant force required to accelerate or retard a fluid stream. This would occur when fluid enters a reduced or enlarged flow area. If the flow area decreases, for example, the fluid will be accelerated by a wall force up to the new velocity. Ultimately, this force must be resisted by the pipe supports.

As Eq. 19.169 illustrates, fluid momentum is not always conserved, since it is generated by the external force, $F$. Examples of external forces are gravity (considered zero for horizontal pipes), gage pressure, friction, and turning

---

[28]The symbol $B$ is also used for momentum. In many texts, however, momentum is given no symbol at all.

forces from walls and vanes. Only if these external forces are absent is fluid momentum conserved.

Since force is a vector, it can be resolved into its $x$- and $y$-components of force.

$$F_x = \dot{m}\Delta\mathrm{v}_x \qquad \text{[SI]} \qquad 19.170(a)$$

$$F_x = \frac{\dot{m}\Delta\mathrm{v}_x}{g_c} \qquad \text{[U.S.]} \qquad 19.170(b)$$

$$F_y = \dot{m}\Delta\mathrm{v}_y \qquad \text{[SI]} \qquad 19.171(a)$$

$$F_y = \frac{\dot{m}\Delta\mathrm{v}_y}{g_c} \qquad \text{[U.S.]} \qquad 19.171(b)$$

If the flow is initially at velocity v but is directed through an angle $\theta$ with respect to the original direction, the $x$- and $y$-components of velocity can be calculated from Eq. 19.172 and Eq. 19.173.

$$\Delta\mathrm{v}_x = \mathrm{v}(\cos\theta - 1) \qquad\qquad 19.172$$

$$\Delta\mathrm{v}_y = \mathrm{v}\sin\theta \qquad\qquad 19.173$$

Since $F$ and v are vector quantities and $\Delta t$ and $m$ are scalars, $F$ must have the same direction as $\mathrm{v}_2 - \mathrm{v}_1$. (See Fig. 19.31.) This provides an intuitive method of determining the direction in which the force acts. Essentially, one needs to ask, "In which direction must the force act in order to push the fluid stream into its new direction?" (The force, $F$, is the force on the fluid. The force on the pipe walls or pipe supports has the same magnitude but is opposite in direction.)

**Figure 19.31** *Force on a Confined Fluid Stream*

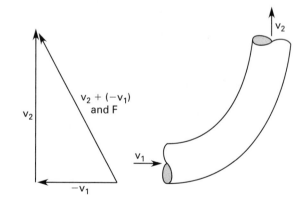

If a jet is completely stopped by a flat plate placed perpendicular to its flow, then $\theta = 90°$ and $\Delta\mathrm{v}_x = -\mathrm{v}$. If a jet is turned around so that it ends up returning to where it originated, then $\theta = 180°$ and $\Delta\mathrm{v}_x = -2\mathrm{v}$. Notice that a positive $\Delta\mathrm{v}$ indicates an increase in velocity. A negative $\Delta\mathrm{v}$ indicates a decrease in velocity.

## 39. JET PROPULSION

A basic application of the impulse-momentum principle is the analysis of jet propulsion. Air enters a jet engine and is mixed with a small amount of jet fuel. The air and fuel mixture is compressed and ignited, and the exhaust products leave the engine at a greater velocity than was possessed by the original air. The change in momentum of the air produces a force on the engine. (See Fig. 19.32.)

**Figure 19.32** *Jet Engine*

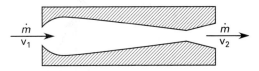

The governing equation for a jet engine is Eq. 19.174. In the special case of VTOL (vertical takeoff and landing) aircraft, there will also be a $y$-component of force. The mass of the jet fuel is small compared with the air mass, and the fuel mass is commonly disregarded.

$$F_x = \dot{m}(v_2 - v_1) \qquad\qquad 19.174$$

$$F_x = \dot{V}_2\rho_2 v_2 - \dot{V}_1\rho_1 v_1 \qquad \text{[SI]} \quad 19.175(a)$$

$$F_x = \frac{\dot{V}_2\rho_2 v_2 - \dot{V}_1\rho_1 v_1}{g_c} \qquad \text{[U.S.]} \quad 19.175(b)$$

## 40. OPEN JET ON A VERTICAL FLAT PLATE

Figure 19.33 illustrates an open jet on a vertical flat plate. The fluid approaches the plate with no vertical component of velocity; it leaves the plate with no horizontal component of velocity. (This is another way of saying there is no splash-back.) Thus, all of the velocity in the $x$-direction is canceled. (The minus sign in Eq. 19.176 indicates that the force is opposite the initial velocity direction.)

**Figure 19.33** *Jet on a Vertical Plate*

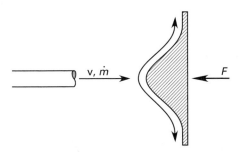

$$\Delta v = -v \qquad\qquad 19.176$$

$$F_x = -\dot{m}v \qquad \text{[SI]} \quad 19.177(a)$$

$$F_x = \frac{-\dot{m}v}{g_c} \qquad \text{[U.S.]} \quad 19.177(b)$$

Since the flow is divided, half going up and half going down, the net velocity change in the $y$-direction is zero. There is no force in the $y$-direction on the fluid.

## 41. OPEN JET ON A HORIZONTAL FLAT PLATE

If a jet of fluid is directed upward, its velocity will decrease due to the effect of gravity. The force exerted on the fluid by the plate will depend on the fluid velocity at the plate surface, $v_y$, not the original jet velocity, $v_o$. All of this velocity is canceled. Since the flow divides evenly in both horizontal directions ($\Delta v_x = 0$), there is no force component in the $x$-direction. (See Fig. 19.34.)

$$v_y = \sqrt{v_o^2 - 2gh} \qquad\qquad 19.178$$

$$\Delta v_y = -\sqrt{v_o^2 - 2gh} \qquad\qquad 19.179$$

$$F_y = -\dot{m}\sqrt{v_o^2 - 2gh} \qquad \text{[SI]} \quad 19.180(a)$$

$$F_y = \frac{-\dot{m}\sqrt{v_o^2 - 2gh}}{g_c} \qquad \text{[U.S.]} \quad 19.180(b)$$

**Figure 19.34** *Open Jet on a Horizontal Plate*

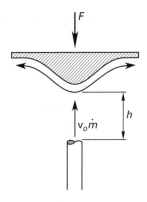

## 42. OPEN JET ON AN INCLINED PLATE

An open jet will be diverted both up and down (but not laterally) a stationary, inclined plate, as shown in Fig. 19.35. In the absence of friction, the velocity in each diverted flow will be $v$, the same as in the approaching jet. The fractions $f_1$ and $f_2$ of the jet that

**Figure 19.35** *Open Jet on an Inclined Plate*

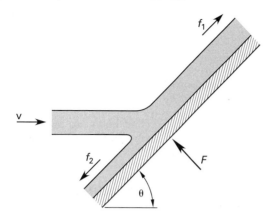

are diverted up and down can be found from Eq. 19.181 through Eq. 19.184.

$$f_1 = \frac{1 + \cos\theta}{2} \qquad 19.181$$

$$f_2 = \frac{1 - \cos\theta}{2} \qquad 19.182$$

$$f_1 - f_2 = \cos\theta \qquad 19.183$$

$$f_1 + f_2 = 1.0 \qquad 19.184$$

If the flow along the plate is frictionless, there will be no force component parallel to the plate. The force perpendicular to the plate is given by Eq. 19.185.

$$F = \dot{m}\mathrm{v}\sin\theta \qquad \text{[SI]} \quad 19.185(a)$$

$$F = \frac{\dot{m}\mathrm{v}\sin\theta}{g_c} \qquad \text{[U.S.]} \quad 19.185(b)$$

### 43. OPEN JET ON A SINGLE STATIONARY BLADE

Figure 19.36 illustrates a fluid jet being turned through an angle $\theta$ by a stationary blade (also called a *vane*). It is common to assume that $|\mathrm{v}_2| = |\mathrm{v}_1|$, although this will not be strictly true if friction between the blade and fluid is considered. Since the fluid is both retarded (in the $x$-direction) and accelerated (in the $y$-direction), there will be two components of force on the fluid.

$$\Delta\mathrm{v}_x = \mathrm{v}_2\cos\theta - \mathrm{v}_1 \qquad 19.186$$

$$\Delta\mathrm{v}_y = \mathrm{v}_2\sin\theta \qquad 19.187$$

$$F_x = \dot{m}(\mathrm{v}_2\cos\theta - \mathrm{v}_1) \qquad \text{[SI]} \quad 19.188(a)$$

$$F_x = \frac{\dot{m}(\mathrm{v}_2\cos\theta - \mathrm{v}_1)}{g_c} \qquad \text{[U.S.]} \quad 19.188(b)$$

$$F_y = \dot{m}\mathrm{v}_2\sin\theta \qquad \text{[SI]} \quad 19.189(a)$$

$$F_y = \frac{\dot{m}\mathrm{v}_2\sin\theta}{g_c} \qquad \text{[U.S.]} \quad 19.189(b)$$

$$F = \sqrt{F_x^2 + F_y^2} \qquad 19.190$$

**Figure 19.36** *Open Jet on a Stationary Blade*

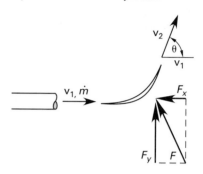

### 44. OPEN JET ON A SINGLE MOVING BLADE

If a blade is moving away at velocity $\mathrm{v}_b$ from the source of the fluid jet, only the *relative velocity difference* between the jet and blade produces a momentum change. Furthermore, not all of the fluid jet overtakes the moving blade. The equations used for the single stationary blade can be used by substituting $(\mathrm{v} - \mathrm{v}_b)$ for v and by using the effective mass flow rate, $\dot{m}_{\mathrm{eff}}$. (See Fig. 19.37.)

$$\Delta\mathrm{v}_x = (\mathrm{v} - \mathrm{v}_b)(\cos\theta - 1) \qquad 19.191$$

$$\Delta\mathrm{v}_y = (\mathrm{v} - \mathrm{v}_b)\sin\theta \qquad 19.192$$

$$\dot{m}_{\mathrm{eff}} = \left(\frac{\mathrm{v} - \mathrm{v}_b}{\mathrm{v}}\right)\dot{m} \qquad 19.193$$

$$F_x = \dot{m}_{\mathrm{eff}}(\mathrm{v} - \mathrm{v}_b)(\cos\theta - 1) \qquad \text{[SI]} \quad 19.194(a)$$

$$F_x = \frac{\dot{m}_{\mathrm{eff}}(\mathrm{v} - \mathrm{v}_b)(\cos\theta - 1)}{g_c} \qquad \text{[U.S.]} \quad 19.194(b)$$

$$F_y = \dot{m}_{\mathrm{eff}}(\mathrm{v} - \mathrm{v}_b)\sin\theta \qquad \text{[SI]} \quad 19.195(a)$$

$$F_y = \frac{\dot{m}_{\mathrm{eff}}(\mathrm{v} - \mathrm{v}_b)\sin\theta}{g_c} \qquad \text{[U.S.]} \quad 19.195(b)$$

**Figure 19.37** *Open Jet on a Moving Blade*

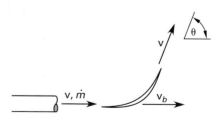

## 45. OPEN JET ON A MULTIPLE-BLADED WHEEL

An *impulse turbine* consists of a series of blades (buckets or vanes) mounted around a wheel. (See Fig. 19.38.) The tangential velocity of the blades is approximately parallel to the jet. The effective mass flow rate, $\dot{m}_{eff}$, used in calculating the reaction force is the full discharge rate, since when one blade moves away from the jet, other blades will have moved into position. Thus, all of the fluid discharged is captured by the blades. Equation 19.194 and Eq. 19.195 are applicable if the total flow rate is used. The tangential blade velocity is

$$v_b = \frac{\text{rpm} \times 2\pi r}{60} = \omega r \qquad 19.196$$

*Figure 19.38* Impulse Turbine

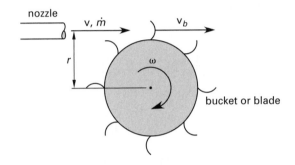

## 46. IMPULSE TURBINE POWER

The total power potential of a fluid jet can be calculated from the kinetic energy of the jet and the mass flow rate.[29] (This neglects the pressure energy, which is small by comparison.)

$$P_{jet} = \frac{\dot{m}v^2}{2} \qquad \text{[SI]} \qquad 19.197(a)$$

$$P_{jet} = \frac{\dot{m}v^2}{2g_c} \qquad \text{[U.S.]} \qquad 19.197(b)$$

The power transferred from a fluid jet to the blades of a turbine is calculated from the $x$-component of force on the blades. The $y$-component of force does no work.

$$P = F_x v_b \qquad 19.198$$

$$P = \dot{m}v_b(v - v_b)(1 - \cos\theta) \qquad \text{[SI]} \qquad 19.199(a)$$

$$P = \frac{\dot{m}v_b(v - v_b)(1 - \cos\theta)}{g_c} \qquad \text{[U.S.]} \qquad 19.199(b)$$

The maximum theoretical blade velocity is the velocity of the jet: $v_b = v$. This is known as the *runaway speed* and can only occur when the turbine is unloaded. If Eq. 19.199 is maximized with respect to $v_b$, the maximum power will be found to occur when the blade is traveling at half of the jet velocity: $v_b = v/2$. The power (force) is also affected by the deflection angle of the blade. Power is maximized when $\theta = 180°$. Figure 19.39 illustrates the relationship between power and the variables $\theta$ and $v_b$.

*Figure 19.39* Turbine Power

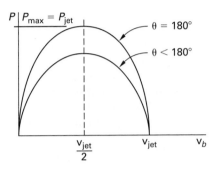

Putting $\theta = 180°$ and $v_b = v/2$ into Eq. 19.199 results in $P_{max} = \dot{m}v^2/2$, which is the same as $P_{jet}$ in Eq. 19.197. If the machine is 100% efficient, 100% of the jet power can be transferred to the machine.

## 47. CONFINED STREAMS IN PIPE BENDS

As presented in Sec. 19.38, momentum can also be changed by pressure forces. Such is the case when fluid enters a pipe fitting or bend. (See Fig. 19.40.) Since the fluid is confined, the forces due to static pressure must be included in the analysis. (The effects of gravity and friction are neglected.)

$$F_x = p_2 A_2 \cos\theta - p_1 A_1 + \dot{m}(v_2 \cos\theta - v_1)$$
$$\text{[SI]} \qquad 19.200(a)$$

$$F_x = p_2 A_2 \cos\theta - p_1 A_1 + \frac{\dot{m}(v_2 \cos\theta - v_1)}{g_c}$$
$$\text{[U.S.]} \qquad 19.200(b)$$

$$F_y = (p_2 A_2 + \dot{m}v_2)\sin\theta \qquad \text{[SI]} \qquad 19.201(a)$$

$$F_y = \left(p_2 A_2 + \frac{\dot{m}v_2}{g_c}\right)\sin\theta \qquad \text{[U.S.]} \qquad 19.201(b)$$

---

[29]The full jet discharge is used in this section. If only a single blade is involved, the effective mass flow rate, $\dot{m}_{eff}$, must be used.

**Figure 19.40** *Pipe Bend*

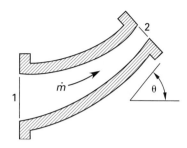

## Example 19.13

60°F water ($\rho = 62.4$ lbm/ft$^3$) at 40 psig enters a 12 in × 8 in reducing elbow at 8 ft/sec and is turned through an angle of 30°. Water leaves 26 in higher in elevation. (a) What is the resultant force exerted on the water by the elbow? (b) What other forces should be considered in the design of supports for the fitting?

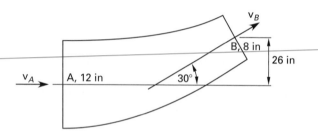

*Solution*

(a) The velocity and pressure at point B are both needed. The velocity is easily calculated from the continuity equation.

$$A_A = \frac{\pi}{4} D_A^2 = \left(\frac{\pi}{4}\right)\left(\frac{12 \text{ in}}{12 \frac{\text{in}}{\text{ft}}}\right)^2 = 0.7854 \text{ ft}^2$$

$$A_B = \left(\frac{\pi}{4}\right)\left(\frac{8}{12}\right)^2 = 0.3491 \text{ ft}^2$$

$$v_B = \frac{v_A A_A}{A_B}$$

$$= \left(8 \frac{\text{ft}}{\text{sec}}\right)\left(\frac{0.7854 \text{ ft}^2}{0.3491 \text{ ft}^2}\right) = 18 \text{ ft/sec}$$

$$p_A = \left(40 \frac{\text{lbf}}{\text{in}^2}\right)\left(12 \frac{\text{in}}{\text{ft}}\right)^2 = 5760 \text{ lbf/ft}^2$$

The Bernoulli equation is used to calculate $p_B$. (Notice that gage pressures are used. Absolute pressures could

also be used, but the addition of $p_{\text{atm}}/\rho$ to both sides of the Bernoulli equation would not affect $p_B$.)

$$\frac{5760 \frac{\text{lbf}}{\text{ft}^2}}{62.4 \frac{\text{lbm}}{\text{ft}^3}} + \frac{\left(8 \frac{\text{ft}}{\text{sec}}\right)^2}{(2)\left(32.2 \frac{\text{ft-lbm}}{\text{lbf-sec}^2}\right)}$$

$$= \frac{p_B}{62.4 \frac{\text{lbm}}{\text{ft}^3}} + \frac{\left(18 \frac{\text{ft}}{\text{sec}}\right)^2}{(2)\left(32.2 \frac{\text{ft-lbm}}{\text{lbf-sec}^2}\right)}$$

$$+ \left(\frac{26 \text{ in}}{12 \frac{\text{in}}{\text{ft}}}\right) \times \left(\frac{g}{g_c}\right)$$

$$p_B = 5373 \text{ lbf/ft}^2$$

The mass flow rate is

$$\dot{m} = \dot{V}\rho = vA\rho$$

$$= \left(8 \frac{\text{ft}}{\text{sec}}\right)(0.7854 \text{ ft}^2)\left(62.4 \frac{\text{lbm}}{\text{ft}^3}\right)$$

$$= 392.1 \text{ lbm/sec}$$

From Eq. 19.200,

$$F_x = \left(5373 \frac{\text{lbf}}{\text{ft}^2}\right)(0.3492 \text{ ft}^2)(\cos 30°)$$

$$- \left(5760 \frac{\text{lbf}}{\text{ft}^2}\right)(0.7854 \text{ ft}^2)$$

$$+ \frac{\left(392.1 \frac{\text{lbm}}{\text{sec}}\right)\left(\left(18 \frac{\text{ft}}{\text{sec}}\right)(\cos 30°) - 8 \frac{\text{ft}}{\text{sec}}\right)}{32.2 \frac{\text{ft-lbm}}{\text{lbf-sec}^2}}$$

$$= -2807 \text{ lbf}$$

From Eq. 19.201,

$$F_y = \left(\begin{array}{c} \left(5373 \frac{\text{lbf}}{\text{ft}^2}\right)(0.3491 \text{ ft}^2) \\[1em] + \frac{\left(392.1 \frac{\text{lbm}}{\text{sec}}\right)\left(18 \frac{\text{ft}}{\text{sec}}\right)}{32.2 \frac{\text{ft-lbm}}{\text{lbf-sec}^2}} \end{array}\right)$$

$$\times (\sin 30°)$$

$$= 1047 \text{ lbf}$$

The resultant force on the water is

$$R = \sqrt{F_x^2 + F_y^2} = \sqrt{(-2807 \text{ lbf})^2 + (1047 \text{ lbf})^2}$$

$$= 2996 \text{ lbf}$$

(b) In addition to counteracting the resultant force, $R$, the support should be designed to carry the weight of the elbow and the water in it. Also, the support must carry a part of the pipe and water weight tributary to the elbow.

## 48. WATER HAMMER

*Water hammer* in a long pipe is an increase in fluid pressure caused by a sudden velocity decrease. (See Fig. 19.41.) The sudden velocity decrease will usually be caused by a valve closing. Analysis of the water hammer phenomenon can take two approaches, depending on whether or not the pipe material is assumed to be elastic.

If the pipe material is assumed to be inelastic (i.e., rigid pipe), the time required for the water hammer shock wave to travel from the suddenly closed valve to a point of interest depends only on the velocity of sound in the fluid ($a$) and the distance ($L$) between the two points. This is also the time required to bring all of the fluid in the pipe to rest.

$$t = \frac{L}{a} \qquad \textit{19.202}$$

When the water hammer shock wave reaches the original source of water, the pressure wave will dissipate. A rarefaction wave (at the pressure of the water source) will return at velocity $a$ to the valve. The time for the compression shock wave to travel to the source and the rarefaction wave to return to the valve is given by Eq. 19.203. This is also the length of time that the pressure is constant at the valve.

$$t = \frac{2L}{a} \qquad \textit{19.203}$$

The fluid pressure increase resulting from the shock wave is calculated by equating the kinetic energy change of the fluid with the average pressure during the compression process. The pressure increase is independent of the length of pipe. If the velocity is decreased by an amount $\Delta v$ instantaneously, the increase in pressure will be

$$\Delta p = \rho a \Delta v \qquad \text{[SI]} \qquad \textit{19.204(a)}$$

$$\Delta p = \frac{\rho a \Delta v}{g_c} \qquad \text{[U.S.]} \qquad \textit{19.204(b)}$$

It is interesting that the pressure increase at the valve depends on $\Delta v$ but not on the actual length of time it takes to close the valve, as long as the valve is closed when the wave returns to it. Therefore, there is no difference in pressure buildups at the valve for an "instantaneous closure," "rapid closure," or "sudden closure."[30] It is only necessary for the closure to occur rapidly.

Having a very long pipe is equivalent to assuming an instantaneous closure. When the pipe is long, the time for the shock wave to travel round-trip is much longer than the time to close the valve. Thus, the valve will be closed when the rarefaction wave returns to the valve.

If the pipe is short, it will be difficult to close the valve before the rarefaction wave returns to the valve. With a short pipe, the pressure buildup will be less than is predicted by Eq. 19.204. (Having a short pipe is equivalent to the case of "slow closure.") The actual pressure history is complex, and no simple method exists for calculating the pressure buildup in short pipes.

Installing a *surge tank*, *accumulator*, *slow-closing valve* (e.g., a gate valve), or *pressure-relief valve* in the line will protect against water hammer damage. (See Fig. 19.42.) The surge tank (or *surge chamber*) is an open tank or reservoir. Since the water is unconfined, large pressure buildups do not occur. An accumulator is a closed tank that is partially filled with air. Since the air is much more compressible than the water, it will be compressed by the water hammer shock wave. The energy of the shock wave is dissipated when the air is compressed.

*Figure 19.41* Water Hammer

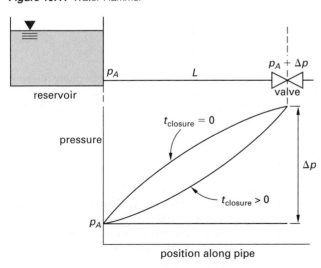

*Figure 19.42* Water Hammer Protective Devices

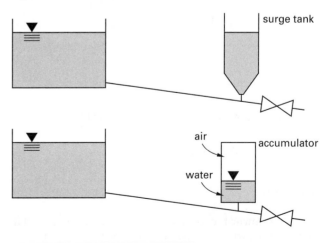

---

[30]The pressure elsewhere along the pipe, however, will be lower for slow closures than for instantaneous closures.

If the pipe material is elastic (the typical assumption for steel and plastic), the previous analysis of water hammer effects ($t$ and $\Delta p$) is still valid. However, the calculation of the speed of sound in water must account for the elasticity of the pipe material. This is accomplished by using Eq. 19.205 for the modulus of elasticity (bulk modulus) when calculating the speed of sound, $a$. (In Eq. 19.205, $t$ is the pipe wall thickness and $D$ is the pipe inside diameter.) At room temperature, the modulus of elasticity of ductile steel is approximately $2.9 \times 10^7$ lbf/in$^2$ (200 GPa); for ductile cast iron, it is $2.2$–$2.5 \times 10^7$ lbf/in$^2$ (150–170 GPa); for PVC, it is $3.5$–$4.1 \times 10^5$ lbf/in$^2$ (2.4–2.8 GPa); for ABS it is $3.2$–$3.5 \times 10^5$ lbf/in$^2$ (2.2–2.4 GPa).

$$E = \frac{E_{\text{water}} t_{\text{pipe}} E_{\text{pipe}}}{t_{\text{pipe}} E_{\text{pipe}} + D_{\text{pipe}} E_{\text{water}}} \qquad 19.205$$

Equation 19.205 indicates that $E$ (and, hence, the effect of water hammer) can be reduced by using a larger diameter pipe. The size of the valve does not affect the wave velocity.

### Example 19.14

Water ($\rho = 1000$ kg/m$^3$, $E = 2 \times 10^9$ Pa), is flowing at 4 m/s through a long length of 4 in schedule-40 steel pipe ($D_i = 0.102$ m, $t = 0.00602$ m, $E = 2 \times 10^9$ Pa) when a valve suddenly closes completely. What is the theoretical increase in pressure?

*Solution*

From Eq. 19.205, the modulus of elasticity to be used in calculating the speed of sound is

$$E = \frac{(2 \times 10^9 \text{ Pa})(0.00602 \text{ m})(2 \times 10^{11} \text{ Pa})}{(0.00602 \text{ m})(2 \times 10^{11} \text{ Pa})}$$
$$+ (0.102 \text{ m})(2 \times 10^9 \text{ Pa})$$
$$= 1.71 \times 10^9 \text{ Pa}$$

The speed of sound in the pipe is

$$a = \sqrt{\frac{E}{\rho}} = \sqrt{\frac{1.71 \times 10^9 \text{ Pa}}{1000 \ \frac{\text{kg}}{\text{m}^3}}}$$
$$= 1308 \text{ m/s}$$

From Eq. 19.204, the pressure increase is

$$\Delta p = \rho a \Delta v = \left(1000 \ \frac{\text{kg}}{\text{m}^3}\right)\left(1308 \ \frac{\text{m}}{\text{s}}\right)\left(4 \ \frac{\text{m}}{\text{s}}\right)$$
$$= 5.23 \times 10^6 \text{ Pa}$$

## 49. LIFT

*Lift* is an upward force that is exerted on an object (flat plate, airfoil, rotating cylinder, etc.) as the object passes through a fluid. Lift combines with drag to form the resultant force on the object, as shown in Fig. 19.43.

**Figure 19.43** *Lift and Drag on an Airfoil*

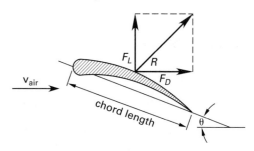

The generation of lift from air flowing over an airfoil is predicted by Bernoulli's equation. Air molecules must travel a longer distance over the top surface of the airfoil than over the lower surface, and, therefore, they travel faster over the top surface. Since the total energy of the air is constant, the increase in kinetic energy comes at the expense of pressure energy. The static pressure on the top of the airfoil is reduced, and a net upward force is produced.

Within practical limits, the lift produced can be increased at lower speeds by increasing the curvature of the wing. This increased curvature is achieved by the use of *flaps*. (See Fig. 19.44.) When a plane is traveling slowly (e.g., during take-off or landing), its flaps are extended to create the lift needed.

**Figure 19.44** *Use of Flaps in an Airfoil*

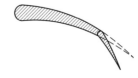

The lift produced can be calculated from Eq. 19.206, whose use is not limited to airfoils.

$$F_L = \frac{C_L A \rho v^2}{2} \qquad \text{[SI]} \quad 19.206(a)$$

$$F_L = \frac{C_L A \rho v^2}{2 g_c} \qquad \text{[U.S.]} \quad 19.206(b)$$

The dimensions of an airfoil or wing are frequently given in terms of chord length and aspect ratio. The *chord length* is the front-to-back dimension of the airfoil. The *aspect ratio* is the ratio of the *span* (wing length) to chord length. The area, $A$, in Eq. 19.206 is the airfoil's area projected onto the plane of the chord. Thus, for a rectangular airfoil, $A = \text{chord} \times \text{span}$.

The dimensionless *coefficient of lift*, $C_L$, is used to measure the effectiveness of the airfoil. The coefficient of lift depends on the shape of the airfoil and the Reynolds number. No simple relationship can be given for calculating the coefficient of lift for airfoils, but the theoretical coefficient of lift for a thin plate in two-dimensional flow at a low angle of attack, $\theta$, is given by Eq. 19.207. Actual airfoils are able to achieve only 80% to 90% of this theoretical value.

$$C_L = 2\pi \sin \theta \qquad 19.207$$

The coefficient of lift for an airfoil cannot be increased without limit merely by increasing $\theta$. Eventually, the *stall angle* is reached, at which point the coefficient of lift decreases dramatically. (See Fig. 19.45.)

*Figure 19.45* Typical Plot of Lift Coefficient

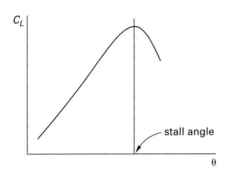

## 50. CIRCULATION

A theoretical concept for calculating the lift generated by an object (an airfoil, propeller, turbine blade, etc.) is *circulation*. Circulation, $\Gamma$, is defined by Eq. 19.208.[31] Its units are length$^2$/time.

$$\Gamma = \oint v \cos \theta \, dl \qquad 19.208$$

Figure 19.46 illustrates an arbitrary closed curve drawn around a point (or body) in steady flow. The tangential components of velocity, $v$, at all points around the curve are $v \cos \theta$. It is a fundamental theorem that circulation has the same value for every closed curve that can be drawn around a body.

Lift on a body traveling with relative velocity $v$ through a fluid of density $\rho$ can be calculated from the circulation by using Eq. 19.209.[32]

$$F_L = \rho v \Gamma \times \text{chord length} \qquad 19.209$$

There is no actual circulation of air "around" an airfoil, but this mathematical concept can be used, nevertheless. However, since the flow of air "around" an airfoil is not symmetrical in path or velocity, experimental determination of $C_L$ is favored over theoretical calculations of circulation.

---

[31]Equation 19.208 is analogous to the calculation of work being done by a constant force moving around a curve. If the force makes an angle of $\theta$ with the direction of motion, the work done as the force moves a distance $dl$ around a curve is $W = \oint F \cos \theta \, dl$. In calculating circulation, velocity takes the place of force.

[32]$U$ is the traditional symbol of velocity in circulation studies.

*Figure 19.46* Circulation Around a Point

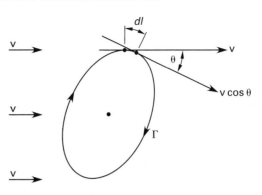

## 51. LIFT FROM ROTATING CYLINDERS

When a cylinder is placed transversely to an airflow traveling at velocity $v_\infty$, the velocity at a point on the surface of the cylinder is $2v_\infty \sin \theta$. Since the flow is symmetrical, however, no lift is produced. (See Fig. 19.47.)

*Figure 19.47* Flow Over a Cylinder

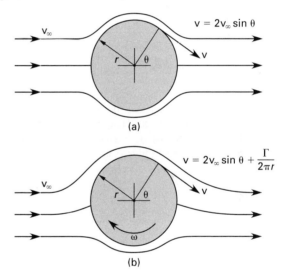

If the cylinder with radius $r$ is rotating at $\omega$ rad/sec while it moves with a relative velocity $v_\infty$ through the air, the *Kutta-Joukowsky result* (theorem) can be used to calculate the lift per unit length of cylinder.[33] This is known as the *Magnus effect*.

$$F_L(\text{per unit length}) = \rho v_\infty \Gamma \qquad \text{[SI]} \quad 19.210(a)$$

$$F_L = \frac{\rho v_\infty \Gamma}{g_c} \qquad \text{[U.S.]} \quad 19.210(b)$$

$$\Gamma = 2\pi r^2 \omega \qquad 19.211$$

---

[33]A similar analysis can be used to explain why a pitched baseball curves. The rotation of the ball produces a force that changes the path of the ball as it travels.

Equation 19.210 assumes that there is no slip (i.e., that the air drawn around the cylinder by rotation moves at $\omega$), and in that ideal case, the maximum coefficient of lift is $4\pi$. Practical rotating devices, however, seldom achieve a coefficient of lift in excess of 9 or 10, and even then, the power expenditure is excessive.

## 52. DRAG

*Drag* is a frictional force that acts parallel but opposite to the direction of motion. It combines with the lift (acting perpendicular to the direction of motion) to produce a resultant force on the object. The total drag force is made up of *skin friction* and *pressure drag* (also known as *form drag*). These components, in turn, can be subdivided and categorized into *wake drag*, *induced drag*, and *profile drag*. (See Fig. 19.48.)

**Figure 19.48** *Components of Total Drag*

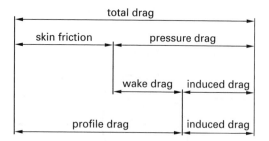

Most aeronautical engineering books contain descriptions of these drag terms. However, the difference between the situations where either skin friction drag or pressure drag predominates is illustrated in Fig. 19.49.

**Figure 19.49** *Extreme Cases of Pressure Drag and Skin Friction*

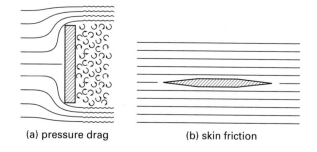

(a) pressure drag          (b) skin friction

Total drag is most easily calculated from the dimensionless *drag coefficient*, $C_D$. It can be shown by dimensional analysis that the drag coefficient depends only on the Reynolds number.

$$F_D = \frac{C_D A \rho v^2}{2} \qquad \text{[SI]} \qquad 19.212(a)$$

$$F_D = \frac{C_D A \rho v^2}{2 g_c} \qquad \text{[U.S.]} \qquad 19.212(b)$$

In most cases, the area, $A$, in Eq. 19.212 is the projected area (i.e., the *frontal area*) normal to the stream. This is appropriate for spheres, cylinders, and automobiles. In a few cases (e.g., for airfoils and flat plates), the area is a projection of the object onto a plane parallel to the stream.

Typical drag coefficients for production cars vary from approximately 0.25 to approximately 0.70, with most modern cars being nearer the lower end. By comparison, other low-speed drag coefficients are approximately 0.05 (aircraft wing), 0.10 (sphere in turbulent flow), and 1.2 (flat plate).

*Aero horsepower* is a term used by automobile manufacturers to designate the power required to move a car horizontally at 50 mi/hr (80.5 km/h) against the drag force. Aero horsepower varies from approximately 7 hp (5.2 kW) for a streamlined subcompact car to approximately 100 hp (75 kW) for a box-shaped truck.

## 53. DRAG ON SPHERES AND DISKS

The drag coefficient varies linearly with the Reynolds number for laminar flow around a sphere or disk. In this region, the drag is almost entirely due to skin friction. For Reynolds numbers below approximately 0.4, experiments have shown that the drag coefficient can be calculated from Eq. 19.213.[34] In calculating the Reynolds number, the sphere or disk diameter should be used as the characteristic dimension.

$$C_D = \frac{24}{\text{Re}} \qquad 19.213$$

Substituting this value of $C_D$ into Eq. 19.212 results in *Stokes' law*, which is applicable to slow motion (ascent or descent) of spherical particles and bubbles traveling at velocity v through a fluid. Stokes' law is based on the assumptions that (a) flow is laminar, (b) Newton's law of viscosity is valid, and (c) all higher-order velocity terms ($v^2$, etc.) are negligible.

$$F_D = 3\pi\mu v D \qquad 19.214$$

The drag coefficients for disks and spheres operating outside the region covered by Stokes' law have been determined experimentally. In the turbulent region, pressure drag is predominant. Figure 19.50 can be used to obtain approximate values for $C_D$.

Figure 19.50 shows that there is a dramatic drop in the drag coefficient around $\text{Re} = 10^5$. The explanation for this is that the point of separation of the boundary layer shifts, decreasing the width of the wake. (See Fig. 19.51.) Since the drag force is primarily pressure drag at higher Reynolds numbers, a reduction in the wake reduces the pressure drag. Thus, anything that can be done to a sphere (scuffing or wetting a baseball, dimpling a golf ball, etc.) to induce a smaller wake will

---

[34]Some sources report that the region in which Stokes' law applies extends to $\text{Re} = 1.0$.

*Figure 19.50 Drag Coefficients for Spheres and Circular Flat Disks*

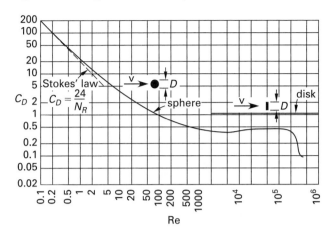

Used with permission from *Fluid Mechanics*, by Richard C. Binder, PhD., Pearson Education, Upper Saddle River, NJ, © 1962.

*Figure 19.51 Turbulent Flow Around a Sphere at Various Reynolds Numbers*

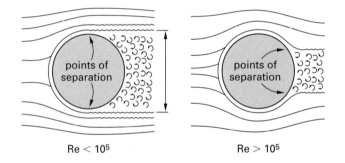

reduce the drag. There can be no shift in the boundary layer separation point for a thin disk, since the disk has no depth in the direction of flow. Therefore, the drag coefficient remains the same at all turbulent Reynolds numbers.

## 54. TERMINAL VELOCITY

The velocity of an object falling through a fluid will continue to increase until the drag force equals the net downward force (i.e., the weight less the buoyant force). The maximum velocity attained is known as the *terminal velocity*. At terminal velocity,

$$F_D = mg - F_b \qquad \text{[SI]} \qquad 19.215(a)$$

$$F_D = m \times \left(\frac{g}{g_c}\right) - F_b \qquad \text{[U.S.]} \qquad 19.215(b)$$

If the drag coefficient is known, the terminal velocity can be calculated from Eq. 19.216. For small, heavy objects falling in air, the buoyant force can be neglected.

$$v = \sqrt{\frac{2(mg - F_b)}{C_D A \rho_{\text{fluid}}}} = \sqrt{\frac{2Vg(\rho_{\text{object}} - \rho_{\text{fluid}})}{C_D A \rho_{\text{fluid}}}} \qquad 19.216$$

For a sphere of diameter $D$, the terminal velocity is

$$v = \sqrt{\frac{4Dg(\rho_{\text{sphere}} - \rho_{\text{fluid}})}{3C_D \rho_{\text{fluid}}}} \qquad 19.217$$

If the spherical particle is very small, Stokes' law may apply. In that case, the terminal velocity can be calculated from Eq. 19.218.

$$v = \frac{D^2 g(\rho_{\text{sphere}} - \rho_{\text{fluid}})}{18\mu} \qquad \text{[SI]} \qquad 19.218(a)$$

$$v = \frac{D^2(\rho_{\text{sphere}} - \rho_{\text{fluid}})}{18\mu} \times \left(\frac{g}{g_c}\right) \qquad \text{[U.S.]} \qquad 19.218(b)$$

## 55. NONSPHERICAL PARTICLES

Only the most simple bodies can be modeled as spheres. (See Fig. 19.51.) One method of overcoming the complexity of dealing with the flow of real particles is to correlate performance with *sphericity*. For a particle and a sphere with the same volume, the sphericity is defined by Eq. 19.219. Sphericity will always be less than or equal to 1.0.

$$\psi = \text{sphericity} = \frac{A_{\text{sphere}}}{A_{\text{particle}}} \qquad 19.219$$

*Figure 19.52 Drag Coefficients for Nonspherical Particles*

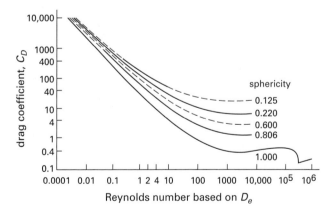

Used with permission from *Unit Operations* by George Granger Brown, et al., John Wiley & Sons, Inc., © 1950.

Another parameter that can be used to describe the deviation from ideal spherical behavior is the ratio of equivalent to average diameters, $D_e/D_{\text{ave}}$. The *average diameter* can be determined by screening a sample of the particles and evaluating the size distribution. The *equivalent diameter* is the diameter of a sphere having the same volume as the particle. (See Table 19.8.)

**Table 19.8** *Sphericity and $D_e/D_{ave}$ Ratios for Nonspherical Particles*

| shape | sphericity | $D_e/D_{ave}$ |
|---|---|---|
| sphere | 1.00 | 1.00 |
| octahedron | 0.817 | 0.965 |
| cube | 0.806 | 1.24 |
| prisms | | |
| $\quad a \times a \times 2a$ | 0.767 | 1.564 |
| $\quad a \times 2a \times 2a$ | 0.761 | 0.985 |
| $\quad a \times 2a \times 3a$ | 0.725 | 1.127 |
| cylinders | | |
| $\quad h = 2r$ | 0.874 | 1.135 |
| $\quad h = 3r$ | 0.860 | 1.31 |
| $\quad h = 10r$ | 0.691 | 1.96 |
| $\quad h = 20r$ | 0.580 | 2.592 |
| disks | | |
| $\quad h = 1.33r$ | 0.858 | 1.00 |
| $\quad h = r$ | 0.827 | 0.909 |
| $\quad h = r/3$ | 0.594 | 0.630 |
| $\quad h = r/10$ | 0.323 | 0.422 |
| $\quad h = r/15$ | 0.254 | 0.368 |

Used with permission from *Unit Operations* by George Granger Brown, et al., John Wiley & Sons, Inc., © 1950.

## 56. FLOW AROUND A CYLINDER

The characteristic drag coefficient plot for cylinders placed normal to the fluid flow is similar to the plot for spheres. The plot shown in Fig. 19.53 is for infinitely long cylinders, since there is additional wake drag at the cylinder ends. In calculating the Reynolds number, the cylinder diameter should be used as the characteristic dimension.

**Figure 19.53** *Drag Coefficient for a Cylinder*

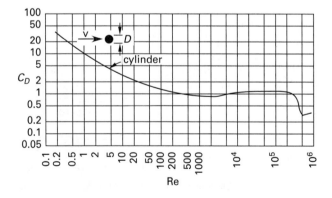

**Example 19.15**

A 50 ft (15 m) high flagpole is constructed of a uniformly smooth cylinder 10 in (25 cm) in diameter. The surrounding air is at 40°F (0°C) and 14.6 psia (100 kPa). What is the total drag force on the flagpole in a 30 mi/hr (50 km/h) gust? (Neglect variations in the wind speed with height above the ground.)

*SI Solution*

At 0°C, the absolute viscosity of air is

$$\mu_{p,0°C} = 1.709 \times 10^{-5} \text{ Pa·s}$$

The density of the air is

$$\rho_p = \frac{p}{RT} = \frac{(100 \text{ kPa})\left(1000 \dfrac{\text{Pa}}{\text{kPa}}\right)}{\left(287 \dfrac{\text{J}}{\text{kg·K}}\right)(0°C + 273°)}$$

$$= 1.276 \text{ kg/m}^3$$

The kinematic viscosity is

$$\nu_p = \frac{\mu}{\rho} = \frac{1.709 \times 10^{-5} \text{ Pa·s}}{1.276 \dfrac{\text{kg}}{\text{m}^3}}$$

$$= 1.339 \times 10^{-5} \text{ m}^2/\text{s}$$

The wind speed is

$$v = \frac{\left(50 \dfrac{\text{km}}{\text{h}}\right)\left(1000 \dfrac{\text{m}}{\text{km}}\right)}{3600 \dfrac{\text{s}}{\text{h}}} = 13.89 \text{ m/s}$$

The characteristic dimension of the flagpole is its diameter. The Reynolds number is

$$Re = \frac{Lv}{\nu} = \frac{Dv}{\nu} = \frac{\left(\dfrac{25 \text{ cm}}{100 \dfrac{\text{cm}}{\text{m}}}\right)\left(13.89 \dfrac{\text{m}}{\text{s}}\right)}{1.339 \times 10^{-5} \dfrac{\text{m}^2}{\text{s}}}$$

$$= 2.59 \times 10^5$$

From Fig. 19.53 for this Reynolds number, the drag coefficient is approximately 1.2.

The frontal area of the flagpole is

$$A = DL = \frac{(25 \text{ cm})(15 \text{ m})}{100 \dfrac{\text{cm}}{\text{m}}}$$

$$= 3.75 \text{ m}^2$$

From Eq. 19.212(a), the drag on the flagpole is

$$F_D = \frac{C_D A \rho v^2}{2}$$

$$= \frac{(1.2)(3.75 \text{ m}^2)\left(1.276 \dfrac{\text{kg}}{\text{m}^3}\right)\left(13.89 \dfrac{\text{m}}{\text{s}}\right)^2}{2}$$

$$= 554 \text{ N}$$

*Customary U.S. Solution*

At 40°F, the absolute viscosity of air is

$$\mu_{p,40°F} = 3.62 \times 10^{-7} \text{ lbf-sec/ft}^2$$

The density of the air is

$$\rho_p = \frac{p}{RT} = \frac{\left(14.6 \frac{\text{lbf}}{\text{in}^2}\right)\left(12 \frac{\text{in}}{\text{ft}}\right)^2}{\left(53.3 \frac{\text{ft-lbf}}{\text{lbm-°R}}\right)(40°F + 460°)}$$

$$= 0.07889 \text{ lbm/ft}^3$$

The kinematic viscosity is

$$\nu_p = \frac{\mu g_c}{\rho} = \frac{\left(3.62 \times 10^{-7} \frac{\text{lbf-sec}}{\text{ft}^2}\right)\left(32.2 \frac{\text{ft-lbm}}{\text{lbf-sec}^2}\right)}{0.07889 \frac{\text{lbm}}{\text{ft}^3}}$$

$$= 1.478 \times 10^{-4} \text{ ft}^2/\text{sec}$$

The wind speed is

$$v = \frac{\left(30 \frac{\text{mi}}{\text{hr}}\right)\left(5280 \frac{\text{ft}}{\text{mi}}\right)}{3600 \frac{\text{sec}}{\text{hr}}} = 44 \text{ ft/sec}$$

The characteristic dimension of the flagpole is its diameter. The Reynolds number is

$$\text{Re} = \frac{Lv}{\nu} = \frac{Dv}{\nu} = \frac{\left(\frac{10 \text{ in}}{12 \frac{\text{in}}{\text{ft}}}\right)\left(44 \frac{\text{ft}}{\text{sec}}\right)}{1.478 \times 10^{-4} \frac{\text{ft}^2}{\text{sec}}}$$

$$= 2.48 \times 10^5$$

From Fig. 19.53 for this Reynolds number, the drag coefficient is approximately 1.2.

The frontal area of the flagpole is

$$A = DL = \frac{(10 \text{ in})(50 \text{ ft})}{12 \frac{\text{in}}{\text{ft}}} = 41.67 \text{ ft}^2$$

From Eq. 19.212(b), the drag on the flagpole is

$$F_D = \frac{C_D A \rho v^2}{2g_c}$$

$$= \frac{(1.2)(41.67 \text{ ft}^2)\left(0.07889 \frac{\text{lbm}}{\text{ft}^3}\right)\left(44 \frac{\text{ft}}{\text{sec}}\right)^2}{(2)\left(32.2 \frac{\text{ft-lbm}}{\text{lbf-sec}^2}\right)}$$

$$= 118.6 \text{ lbf}$$

## 57. FLOW OVER A PARALLEL FLAT PLATE

The drag experienced by a flat plate oriented parallel to the direction of flow is almost totally skin friction drag. (See Fig. 19.54.) Prandtl's *boundary layer theory* can be used to evaluate the frictional effects. Such an analysis shows that the shape of the boundary layer is a function of the Reynolds number. The characteristic dimension used in calculating the Reynolds number is the chord length, $L$ (i.e., the dimension of the plate parallel to the flow).

*Figure 19.54* Flow Over a Parallel Flat Plate (one side)

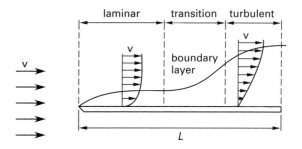

When skin friction predominates, it is common to use the symbol $C_f$ (i.e., the *skin friction coefficient*) for the drag coefficient. For laminar flow over a smooth, flat plate, the drag coefficient based on the boundary layer theory is given by Eq. 19.220, which is known as the *Blasius solution*.[35] The critical Reynolds number for laminar flow is often reported to be 530,000. However, the transition region between laminar flow and turbulent flow actually occupies the Reynolds number range of 100,000 to 1,000,000.

$$C_f = \frac{1.328}{\sqrt{\text{Re}}} \qquad \textit{19.220}$$

Prandtl reported that the skin friction coefficient for turbulent flow is

$$C_f = \frac{0.455}{(\log \text{Re})^{2.58}} \qquad \textit{19.221}$$

The drag force is calculated from Eq. 19.222. The factor 2 appears because there is friction on two sides of the flat plate.

$$F_D = 2\left(\frac{C_f A \rho v^2}{2}\right) = C_f A \rho v^2 \qquad \text{[SI]} \quad \textit{19.222(a)}$$

$$F_D = 2\left(\frac{C_f A \rho v^2}{2g_c}\right) = \frac{C_f A \rho v^2}{g_c} \qquad \text{[U.S.]} \quad \textit{19.222(b)}$$

---

[35]Other correlations substitute the coefficient 1.44, the original value calculated by Prandtl, for 1.328 in Eq. 19.220. The Blasius solution is considered to be more accurate.

## 58. SIMILARITY

Similarity considerations between a *model* (subscript *m*) and a full-sized object (subscript *p*, for *prototype*) imply that the model can be used to predict the performance of the prototype. Such a model is said to be *mechanically similar* to the prototype.

Complete *mechanical similarity* requires both geometric and dynamic similarity.[36] *Geometric similarity* means that the model is true to scale in length, area, and volume. The *model scale (length ratio)* is defined as

$$L_r = \frac{\text{size of model}}{\text{size of prototype}} \qquad 19.223$$

The area and volume ratios are based on the model scale.

$$\frac{A_m}{A_p} = (L_r)^2 \qquad 19.224$$

$$\frac{V_m}{V_p} = (L_r)^3 \qquad 19.225$$

*Dynamic similarity* means that the ratios of all types of forces are equal for the model and the prototype. These forces result from inertia, gravity, viscosity, elasticity (i.e., fluid compressibility), surface tension, and pressure.

The number of possible ratios of forces is large. For example, the ratios of viscosity/inertia, inertia/gravity, and inertia/surface tension are only three of the ratios of forces that must match for every corresponding point on the model and prototype. Fortunately, some force ratios can be neglected because the forces are negligible or are self-canceling.

In some cases, the geometric scale may be deliberately distorted. For example, with scale models of harbors and rivers, the water depth might be only a fraction of an inch if the scale is followed loyally. Not only will the surface tension be excessive, but the shifting of the harbor or river bed may not be properly observed. Therefore, the vertical scale is chosen to be different from the horizontal scale. Such models are known as *distorted models*. Experience is needed to interpret observations made of distorted models.

The following sections deal with the most common similarity problems, but not all. For example, similarity of steady laminar flow in a horizontal pipe requires the Stokes number, similarity of high-speed (near-sonic) aircraft requires the Mach number, and similarity of capillary rise in a tube requires the *Eötvös number* $(g\rho L^2/\sigma)$.

---

[36]Complete mechanical similarity also requires kinematic and thermal similarity, which are not discussed in this book.

## 59. VISCOUS AND INERTIAL FORCES DOMINATE

Consider the testing of a completely submerged object, such as the items listed in Table 19.9. Surface tension will be negligible. The fluid can be assumed to be incompressible for low velocity. Gravity does not change the path of the fluid particles significantly during the passage of the object.

**Table 19.9** *Applications of Reynolds Number Similarity*

> aircraft (subsonic)
> airfoils (subsonic)
> closed-pipe flow (turbulent)
> drainage through tank orifices
> fans
> flow meters
> open channel flow (without wave action)
> pumps
> submarines
> torpedoes
> turbines

Only inertial, viscous, and pressure forces are significant. Because they are the only forces that are acting, these three forces are in equilibrium. Since they are in equilibrium, knowing any two forces will define the third force. This third force is dependent and can be omitted from the similarity analysis. For submerged objects, pressure is traditionally chosen as the dependent force.

The dimensionless ratio of the inertial forces to the viscous forces is the Reynolds number. Equating the model's and prototype's Reynolds numbers will ensure similarity.[37]

$$\text{Re}_m = \text{Re}_p \qquad 19.226$$

$$\frac{L_m \text{v}_m}{\nu_m} = \frac{L_p \text{v}_p}{\nu_p} \qquad 19.227$$

If the model is tested in the same fluid and at the same temperature in which the prototype is expected to operate, setting the Reynolds numbers equal is equivalent to setting $L_m \text{v}_m = L_p \text{v}_p$.

### Example 19.16

A $^1/_{30}$ size scale model of a helicopter fuselage is tested in a wind tunnel at 120 mi/hr (190 km/h). The conditions in the wind tunnel are 50 psia and 100°F (350 kPa and 50°C). What is the corresponding speed of a prototype traveling in 14.0 psia and 40°F still air (100 kPa, 0°C)?

---

[37]An implied assumption is that the drag coefficients are the same for the model and the prototype. In the case of pipe flow, it is assumed that flow will be in the turbulent region with the same relative roughness.

*SI Solution*

The absolute viscosity of air at atmospheric pressure is

$$\mu_{p,0^\circ C} = 1.709 \times 10^{-5} \ \text{Pa·s}$$

$$\mu_{m,50^\circ C} = 1.951 \times 10^{-5} \ \text{Pa·s}$$

The densities of air at the two conditions are

$$\rho_p = \frac{p}{RT} = \frac{(100 \ \text{kPa})\left(1000 \ \frac{\text{Pa}}{\text{kPa}}\right)}{\left(287 \ \frac{\text{J}}{\text{kg·K}}\right)(0^\circ C + 273^\circ)}$$

$$= 1.276 \ \text{kg/m}^3$$

$$\rho_m = \frac{p}{RT} = \frac{(350 \ \text{kPa})\left(1000 \ \frac{\text{Pa}}{\text{kPa}}\right)}{\left(287 \ \frac{\text{J}}{\text{kg·K}}\right)(50^\circ C + 273^\circ)}$$

$$= 3.776 \ \text{kg/m}^3$$

The kinematic viscosities are

$$\nu_p = \frac{\mu}{\rho} = \frac{1.709 \times 10^{-5} \ \text{Pa·s}}{1.276 \ \frac{\text{kg}}{\text{m}^3}}$$

$$= 1.339 \times 10^{-5} \ \text{m}^2/\text{s}$$

$$\nu_m = \frac{\mu}{\rho} = \frac{1.1951 \times 10^{-5} \ \text{Pa·s}}{3.776 \ \frac{\text{kg}}{\text{m}^3}}$$

$$= 5.167 \times 10^{-6} \ \text{m}^2/\text{s}$$

From Eq. 19.227,

$$v_p = v_m \left(\frac{L_m}{L_p}\right)\left(\frac{\nu_p}{\nu_m}\right)$$

$$= \frac{\left(190 \ \frac{\text{km}}{\text{h}}\right)\left(\frac{1}{30}\right)\left(1.339 \times 10^{-5} \ \frac{\text{m}^2}{\text{s}}\right)}{5.167 \times 10^{-6} \ \frac{\text{m}^2}{\text{s}}}$$

$$= 16.4 \ \text{km/h}$$

*Customary U.S. Solution*

Since surface tension and gravitational forces on the air particles are insignificant and since the flow velocities are low, viscous and inertial forces dominate. The Reynolds numbers of the model and prototype are equated.

The kinematic viscosity of air must be evaluated at the respective temperatures and pressures. As kinematic viscosity tables for air are not readily available, the viscosity must be calculated. Although kinematic

viscosity depends on the temperature and pressure, absolute viscosity is essentially independent of pressure.

$$\mu_{p,40^\circ F} = 3.62 \times 10^{-7} \ \text{lbf-sec/ft}^2$$

$$\mu_{m,100^\circ F} = 3.96 \times 10^{-7} \ \text{lbf-sec/ft}^2$$

The densities of air at the two conditions are

$$\rho_p = \frac{p}{RT} = \frac{\left(14.0 \ \frac{\text{lbf}}{\text{in}^2}\right)\left(12 \ \frac{\text{in}}{\text{ft}}\right)^2}{\left(53.3 \ \frac{\text{ft-lbf}}{\text{lbm-}^\circ R}\right)(40^\circ F + 460^\circ)}$$

$$= 0.0756 \ \text{lbm/ft}^3$$

$$\rho_m = \frac{p}{RT} = \frac{\left(50.0 \ \frac{\text{lbf}}{\text{in}^2}\right)\left(12 \ \frac{\text{in}}{\text{ft}}\right)^2}{\left(53.3 \ \frac{\text{ft-lbf}}{\text{lbm-}^\circ R}\right)(100^\circ F + 460^\circ)}$$

$$= 0.2412 \ \text{lbm/ft}^2$$

The kinematic viscosities are

$$\nu_p = \frac{\mu g_c}{\rho} = \frac{\left(3.62 \times 10^{-7} \ \frac{\text{lbf-sec}}{\text{ft}^2}\right)g_c}{0.0756 \ \frac{\text{lbm}}{\text{ft}^3}}$$

$$= 4.79 \times 10^{-6} \ \text{ft}^2/\text{sec} \times g_c$$

$$\nu_m = \frac{\mu g_c}{\rho} = \frac{\left(3.96 \times 10^{-7} \ \frac{\text{lbf-sec}}{\text{ft}^3}\right)g_c}{0.2412 \ \frac{\text{lbm}}{\text{ft}^3}}$$

$$= 1.64 \times 10^{-6} \ \text{ft}^2/\text{sec} \times g_c$$

From Eq. 19.227,

$$v_p = v_m \left(\frac{L_m}{L_p}\right)\left(\frac{\nu_p}{\nu_m}\right)$$

$$= \frac{\left(120 \ \frac{\text{mi}}{\text{hr}}\right)\left(\frac{1}{30}\right)\left(4.79 \times 10^{-6} \ \frac{\text{ft}^2}{\text{sec}} \times g_c\right)}{\left(1.64 \times 10^{-6} \ \frac{\text{ft}^2}{\text{sec}} \times g_c\right)}$$

$$= 11.7 \ \text{mi/hr}$$

## 60. INERTIAL AND GRAVITATIONAL FORCES DOMINATE

Table 19.10 lists the cases when elasticity and surface tension forces can be neglected but gravitational forces cannot. Omitting these two forces from the similarity calculations leaves pressure, inertia, viscosity, and gravity forces, which are in equilibrium. Pressure is chosen as the dependent force and is omitted from the analysis.

There are only two possible combinations of the remaining three forces. The ratio of inertial to viscous forces is

**Fluids**

**Table 19.10** *Applications of Froude Number Similarity*

bow waves from ships
flow over spillways
flow over weirs
motion of a fluid jet
open channel flow with varying surface levels
oscillatory wave action
seaplane hulls
surface ships
surface wave action
surge and flood waves

the Reynolds number. The ratio of the inertial forces to the gravitational forces is the *Froude number*, Fr.[38] The Froude number is used when gravitational forces are significant, such as in wave motion produced by a ship or seaplane hull.

$$\text{Fr} = \frac{\text{v}^2}{Lg} \qquad \qquad 19.228$$

Thus, similarity is ensured when Eq. 19.229 and Eq. 19.230 are satisfied.

$$\text{Re}_m = \text{Re}_p \qquad \qquad 19.229$$

$$\text{Fr}_m = \text{Fr}_p \qquad \qquad 19.230$$

As an alternative, Eq. 19.229 and Eq. 19.230 can be solved simultaneously. This results in the following requirement for similarity, which indicates that it is necessary to test the model in a manufactured liquid with a specific viscosity.

$$\frac{\nu_m}{\nu_p} = \left(\frac{L_m}{L_p}\right)^{3/2} = (L_r)^{3/2} \qquad \qquad 19.231$$

Sometimes it is not possible to satisfy Eq. 19.229 and Eq. 19.230. This occurs when a model fluid viscosity is called for that is not available. If only one of the equations is satisfied, the model is said to be *partially similar*. In such a case, corrections based on other factors are used.

Another problem with trying to achieve similarity in open channel flow problems is the need to scale surface drag. It can be shown that the ratio of Manning's roughness constants is given by Eq. 19.232. In some cases, it may not be possible to create a surface smooth enough to satisfy this requirement.

$$n_r = (L_r)^{1/6} \qquad \qquad 19.232$$

## 61. SURFACE TENSION FORCE DOMINATES

Table 19.11 lists some of the cases where surface tension is the predominant force. Such cases can be handled by equating the Weber numbers, We, of the model and prototype.[39] (The *Weber number* is the ratio of inertial force to surface tension.)

$$\text{We} = \frac{\text{v}^2 L \rho}{\sigma} \qquad \qquad 19.233$$

$$\text{We}_m = \text{We}_p \qquad \qquad 19.234$$

**Table 19.11** *Applications of Weber Number Similarity*

air entrainment
bubbles
droplets
waves

---

[38]There are two definitions of the Froude number. Dimensional analysis determines the Froude number as Eq. 19.228 ($\text{v}^2/Lg$), a form that is used in model similitude. However, in open channel flow studies performed by civil engineers, the Froude number is taken as the square root of Eq. 19.228. Whether the derived form or its square root is used can sometimes be determined from the application. If the Froude number is squared (e.g., as in $dE/dx = 1 - \text{Fr}^2$), the square root form is probably needed. In similarity problems, it doesn't make any difference which definition is used.

[39]There are two definitions of the Weber number. The alternate definition is the square root of Eq. 19.233. In similarity problems, it does not make any difference which definition of the Weber number is used.

# 20

# Hydraulic Machines

## Nomenclature

| | | | |
|---|---|---|---|
| $C$ | coefficient | – | – |
| $D$ | diameter | ft | m |
| $E$ | specific energy | ft-lbf/lbm | J/kg |
| $f$ | Darcy friction factor | – | – |
| $f$ | frequency | Hz | Hz |
| $g$ | gravitational acceleration | ft/sec$^2$ | m/s$^2$ |
| $g_c$ | gravitational constant | ft-lbm/lbf-sec$^2$ | n.a. |
| $h$ | height or head | ft | m |
| $h_{\mathrm{ac}}$ | acceleration head | ft | m |
| $K$ | dimensionless factor | – | – |
| $L$ | length | ft | m |
| $m$ | mass | lbm | kg |
| $\dot{m}$ | mass flow rate | lbm/sec | kg/s |
| $n$ | dimensionless exponent | – | – |
| $n$ | rotational speed | rpm | rpm |

| | | | |
|---|---|---|---|
| NPSHA | net positive suction head available | ft | m |
| NPSHR | net positive suction head required | ft | m |
| $p$ | pressure | lbf/ft$^2$ | Pa |
| $P$ | power | ft-lbf/sec | W |
| $Q$ | volumetric flow rate | gal/min | L/s |
| $r$ | radius | ft | m |
| SA | suction specific speed available | rpm | rpm |
| SG | specific gravity | – | – |
| $t$ | time | sec | s |
| $T$ | transmitted torque | ft-lbf | N·m |
| v | velocity | ft/sec | m/s |
| $\dot{V}$ | volumetric flow rate | ft$^3$/sec | m$^3$/s |
| $W$ | work | ft-lbf | kW·h |
| WHP | water horsepower | hp | n.a. |
| WkW | water kilowatts | n.a. | kW |
| $z$ | elevation | ft | m |

## Symbols

| | | | |
|---|---|---|---|
| $\gamma$ | specific weight | lbf/ft$^3$ | n.a. |
| $\eta$ | efficiency | – | – |
| $\theta$ | angle | deg | deg |
| $\nu$ | kinematic viscosity | ft$^2$/sec | m$^2$/s |
| $\rho$ | density | lbm/ft$^3$ | kg/m$^3$ |
| $\sigma$ | cavitation number | – | – |
| $\omega$ | angular velocity | rad/sec | rad/s |

## Subscripts

| | |
|---|---|
| atm | atmospheric |
| $A$ | added (by pump) |
| $b$ | blade |
| cr | critical |
| $d$ | discharge |
| $f$ | friction |
| $i$ | inlet |
| $j$ | jet |
| $m$ | motor |
| $n$ | nozzle |
| $o$ | outlet |
| $p$ | pressure or pump |
| $s$ | suction or specific |
| ss | suction specific |
| $t$ | total or tangential |
| th | theoretical |
| v | velocity or volumetric |
| vp | vapor pressure |
| $z$ | potential |

**Fluids**

## 1. HYDRAULIC MACHINES

Pumps and turbines are the two basic types of hydraulic machines discussed in this chapter. Pumps convert mechanical energy into fluid energy, increasing the energy possessed by the fluid. Turbines convert fluid energy into mechanical energy, extracting energy from the fluid.

## 2. TYPES OF PUMPS

Pumps can be classified according to the method by which pumping energy is transferred to the fluid. This classification separates pumps into positive displacement pumps and kinetic pumps.

The most common types of *positive displacement pumps* are *reciprocating action pumps* (which use pistons, plungers, diaphragms, or bellows) and *rotary action pumps* (using vanes, screws, lobes, or progressing cavities). Such pumps discharge a fixed volume for each stroke or revolution. Energy is added intermittently to the fluid.

*Kinetic pumps* transform fluid kinetic energy into fluid static pressure energy. The pump imparts the kinetic energy; the pump mechanism or housing is constructed in a manner that causes the transformation. *Jet pumps* and *ejector pumps* fall into the kinetic pump category, but centrifugal pumps are the primary examples.

In the operation of a *centrifugal pump*, liquid flowing into the *suction side* (the *inlet*) is captured by the *impeller* and thrown to the outside of the pump casing. Within the casing, the velocity imparted to the fluid by the impeller is converted into pressure energy. The fluid leaves the pump through the *discharge line* (the *exit*). It is a characteristic of most centrifugal pumps that the fluid is turned approximately 90° from the original flow direction. (See Table 20.1.)

**Table 20.1** *Generalized Characteristics of Positive Displacement and Kinetic Pumps*

| characteristic | positive displacement pumps | kinetic pumps |
|---|---|---|
| flow rate | low | high |
| pressure rise per stage | high | low |
| constant quantity over operating range | flow rate | pressure rise |
| self-priming | yes | no |
| discharge stream | pulsing | steady |
| works with high viscosity fluids | yes | no |

## 3. POSITIVE DISPLACEMENT RECIPROCATING PUMPS

*Reciprocating positive displacement* (PD) pumps can be used with all fluids, and are useful with viscous fluids and slurries (up to about 8000 SSU), when the fluid is sensitive to shear, and when a high discharge pressure is required.[1] By entrapping a volume of liquid in the cylinder, reciprocating pumps provide a fixed-displacement volume per cycle. They are self-priming and inherently leak-free. Within the pressure limits of the line and pressure relief valve and the current capacity of the motor circuit, reciprocating pumps can provide an infinite discharge pressure.[2]

There are three main types of reciprocating pumps: power, direct-acting, and diaphragm. A *power pump* is a *cylinder-operated pump*. It can be single-acting or double-acting. A *single-acting pump* discharges liquid (or takes suction) only on one side of the piston, and there is only one transfer operation per crankshaft revolution. A *double-acting pump* discharges from both sides, and there are two transfers per revolution of the crank.

Traditional reciprocating pumps with pistons and rods can be either single-acting or double-acting and are suitable up to approximately 2000 psi (14 MPa). *Plunger pumps* are only single-acting and are suitable up to approximately 10,000 psi (70 MPa).

*Simplex pumps* have one cylinder, *duplex pumps* have two cylinders, *triplex pumps* have three cylinders, and so forth. *Direct-acting pumps* (sometimes referred to as *steam pumps*) are always double-acting. They use steam, unburned fuel gas, or compressed air as a motive fluid.

PD pumps are limited by both their NPSHR characteristics, acceleration head, and (for rotary pumps) slip.[3] Because the flow is unsteady, a certain amount of energy, the *acceleration head* ($h_{ac}$) is required to accelerate the fluid flow each stroke or cycle. If the acceleration head is too large, the NPSHR requirements may not be attainable. Acceleration head can be reduced by increasing the pipe diameter, shortening the suction piping, decreasing the pump speed, or placing a *pulsation damper (stabilizer)* in the suction line.[4]

Generally, friction losses with pulsating flow are calculated based on the maximum velocity attained by the fluid. Since this is difficult to determine, the maximum velocity can be approximated by multiplying the

---

[1]For viscosities of Saybolt seconds universal (SSU) greater than 240, multiply SSU viscosity by 0.216 to get viscosity in centistokes.
[2]For this reason, a relief valve should be included in every installation of positive displacement pumps. Rotary pumps typically have integral relief valves, but external relief valves are often installed to provide easier adjusting, cleaning, and checking.
[3]Manufacturers of PD pumps prefer the term *net positive inlet pressure* (NPIP) to NPSH. NPIPA corresponds to NPSHA; NPIPR corresponds to NPSHR. Pressure and head are related by $p = \gamma h$.
[4]Pulsation dampers are not needed with rotary-action PD pumps, as the discharge is essentially constant.

Fluids

average velocity (calculated from the rated capacity) by the factors in Table 20.2.

**Table 20.2** Typical $v_{max}/v_{ave}$ Velocity Ratios[a,b]

| pump type | single-acting | double-acting |
|---|---|---|
| simplex | 3.2 | 2.0 |
| duplex | 1.6 | 1.3 |
| triplex | 1.1 | 1.1 |
| quadriplex | 1.1 | 1.1 |
| quintuplex and up | 1.05 | 1.05 |

[a]Without stabilization. With properly sized stabilizers, use 1.05 to 1.1 for all cases.
[b]Multiply the values by 1.3 for metering pumps where lost fluid motion is relied on for capacity control.

When the suction line is "short," the acceleration head can be calculated from the length of the suction line, the average velocity in the line, and the rotational speed.[5]

In Eq. 20.1, $C$ and $K$ are dimensionless factors. $K$ represents the relative compressibility of the liquid. (Typical values are 1.4 for hot water; 1.5 for amine, glycol, and cold water; and 2.5 for hot oil.) Values of $C$ are given in Table 20.3.

$$h_{ac} = \left(\frac{C}{K}\right)\left(\frac{L_{suction} v_{ave} n_{rpm}}{g}\right) \qquad 20.1$$

**Table 20.3** Typical Acceleration Head C-Values[*]

| pump type | single-acting | double-acting |
|---|---|---|
| simplex | 0.4 | 0.2 |
| duplex | 0.2 | 0.115 |
| triplex | 0.066 | 0.066 |
| quadriplex | 0.040 | 0.040 |
| quintuplex and up | 0.028 | 0.028 |

[*]Typical values for common connecting rod lengths and crank radii.

## 4. ROTARY PUMPS

Rotary pumps are *positive displacement* (PD) pumps that move fluid by means of screws, progressing cavities, gears, lobes, or vanes turning within a fixed casing (the *stator*). Rotary pumps are useful for high viscosities (up to $4 \times 10^6$ SSU for screw pumps). The rotation creates a cavity of fixed volume near the pump input; atmospheric or external pressure forces the liquid into that cavity. Near the outlet, the cavity is collapsed, forcing the liquid out. Figure 20.1 illustrates the external circumferential piston rotary pump.

Discharge from rotary pumps is relatively smooth. Acceleration head is negligible. Pulsation dampers and suction stabilizers are not required.

*Slip* in rotary pumps is the amount (sometimes expressed as a percentage) of each rotational fluid volume that "leaks" back to the suction line on each

[5]With a properly designed pulsation damper, the effective length of the suction line is reduced to approximately 10 pipe diameters.

**Figure 20.1** External Circumferential Piston Rotary Pump

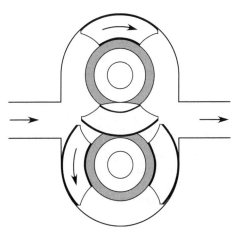

revolution. Slip reduces pump capacity. It is a function of clearance, differential pressure, and viscosity. Slip is proportional to the third power of the clearance between the rotating element and the casing. Slip decreases with increases in viscosity; it increases linearly with increases in differential pressure. Slip is not affected by rotational speed. The *volumetric efficiency* is defined by Eq. 20.2. Figure 20.2 illustrates the relationship between flow rate, speed, slip, and differential pressure.

$$\eta_v = \frac{Q_{actual}}{Q_{ideal}} = \frac{Q_{ideal} - Q_{slip}}{Q_{ideal}} \qquad 20.2$$

**Figure 20.2** Slip in Rotary Pumps

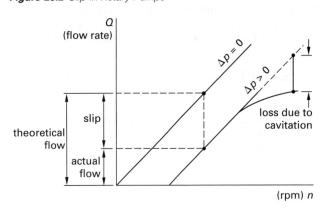

Except for screw pumps, rotary pumps are generally not used for handling abrasive fluids or materials with suspended solids. Few rotary pumps are suitable when variable flow is required.

## 5. DIAPHRAGM PUMPS

Hydraulically operated *diaphragm pumps* have a diaphragm that completely separates the pumped fluid from the rest of the pump. A reciprocating plunger pressurizes and moves a hydraulic fluid that, in turn, flexes the diaphragm. Single-ball check valves in the suction and discharge lines determine the direction of flow during both phases of the diaphragm action.

Metering is a common application of diaphragm pumps. They have no packing and are essentially leak-proof. This makes them ideal when fugitive emissions are undesirable. Diaphragm pumps are suitable for pumping a wide range of materials, from liquefied gases to coal slurries, though the upper viscosity limit is approximately 3500 SSU. Within the limits of their reactivities, hazardous and reactive materials can also be handled.

Diaphragm pumps are limited by capacity, suction pressure, and discharge pressure and temperature. Because of their construction and size, most diaphragm pumps are limited to discharge pressures of 5000 psi (35 MPa) or less, and most high-capacity pumps are limited to 2000 psi (14 MPa). Suction pressures are similarly limited to 5000 psi (35 MPa). A minimum pressure of 3 psi to 9 psi (20 kPa to 60 kPa) is often quoted as the minimum liquid-side pressure for metering applications.

The discharge is inherently pulsating, and the dampers or stabilizers are often used. (The acceleration head term is required when calculating NPSHR.) The discharge can be smoothed out somewhat by using two or three (i.e., duplex or triplex) plungers.

Diaphragms are commonly manufactured from stainless steel (type 316) and polytetrafluorethylene (PTFE) or other elastomers. PTFE diaphragms are suitable in the range of −50°F to 300°F (−45°C to 150°C) while metal diaphragms (and some ketone resin diaphragms) are used up to approximately 400°F (200°C) with life expectancy being reduced at higher temperatures. Although most diaphragm pumps usually operate below 200 spm (strokes per minute), diaphragm life will be improved by limiting the maximum speed to 100 spm.

## 6. CENTRIFUGAL PUMPS

Centrifugal pumps can be classified according to the way their impellers impart energy to the fluid. Each category of pump is suitable for a different application and (specific) speed range. Figure 20.3 illustrates a typical centrifugal pump and its schematic symbol.

**Figure 20.3** Centrifugal Pump and Symbol

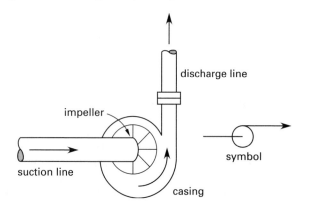

*Radial-flow impellers* impart energy primarily by centrifugal force. Liquid enters the impeller at the hub and flows radially to the outside of the casing. Radial-flow pumps are suitable for adding high pressure at low fluid flow rates. *Axial-flow impellers* impart energy to the fluid by acting as compressors. Fluid enters and exits along the axis of rotation. Axial-flow pumps are suitable for adding low pressures at high fluid flow rates.[6]

Radial-flow pumps can be designed for either single- or double-suction operation. In a *single-suction pump*, fluid enters from only one side of the impeller. In a *double-suction pump*, fluid enters from both sides of the impeller.[7] (That is, the impeller is two-sided.) Operation is similar to having two single-suction pumps in parallel. (See Fig. 20.4.)

**Figure 20.4** Radial- and Axial-Flow Impellers

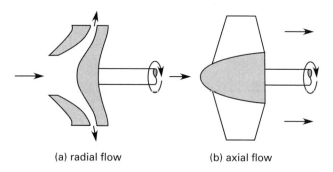

(a) radial flow      (b) axial flow

A *multiple-stage pump* consists of two or more impellers within a single casing. The discharge of one stage feeds the input of the next stage, and operation is similar to having several pumps in series. In this manner, higher heads are achieved than would be possible with a single impeller.

The *circular blade pitch* is the impeller's circumference divided by the number of impeller vanes. The impeller *tip speed* (not to be confused with the specific and suction specific speeds) is easily calculated from the impeller diameter and rotational speed. The impeller "tip speed" is actually the tangential velocity at the periphery. Tip speed is typically somewhat less than 1000 ft/sec (300 m/s).

$$v_{tip} = \frac{\pi D n}{60 \frac{\text{sec}}{\text{min}}} = \frac{D\omega}{2} \qquad 20.3$$

---

[6]There is a third category of centrifugal pumps known as *mixed flow pumps*. Mixed flow pumps have operational characteristics between those of radial flow and axial flow pumps.
[7]The double-suction pump can handle a greater fluid flow rate than a single-suction pump with the same specific speed. Also, the double-suction pump will have a lower NPSHR.

## 7. SEWAGE PUMPS

The primary consideration in choosing a pump for sewage and large solids is resistance to clogging. Centrifugal pumps should always be the single-suction type with nonclog, open impellers. (Double-suction pumps are prone to clogging because rags catch and wrap around the shaft extending through the impeller eye.) Clogging can be further minimized by limiting the number of impeller blades to two or three, providing for large passageways, and using a bar screen ahead of the pump.

Though made of heavy construction, nonclog pumps are constructed for ease of cleaning and repair. Horizontal pumps usually have a split casing, half of which can be removed for maintenance. A hand-sized cleanout opening may also be built into the casing. Although designed for long life, a sewage pump should normally be used with a grit chamber for prolonged bearing life.

The solids-handling capacity of a pump may be specified in terms of the largest sphere that can pass through it without clogging, usually about 80% of the inlet diameter. For example, a wastewater pump with a 6 in (150 mm) inlet should be able to pass a 4 in (100 mm) sphere. The pump must be capable of handling spheres with diameters slightly larger than the bar screen spacing.

Figure 20.5 shows a simplified wastewater pump installation. Not shown are instrumentation and water level measurement devices, baffles, lighting, drains for the dry well, electrical power, pump lubrication equipment, and access ports. (Totally submerged pumps do not require dry wells. However, such pumps without dry wells are more difficult to access, service, and repair.)

**Figure 20.5** Typical Wastewater Pump Installation (greatly simplified)

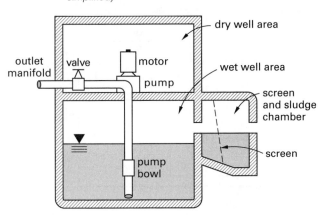

The multiplicity and redundancy of pumping equipment is not apparent from Fig. 20.5. The number of pumps used in a wastewater installation largely depends on the expected demand, pump capacity, and design criteria for backup operation. Although there may be state and federal regulations affecting the design, it is considered good practice to install pumps in sets of two, with a third backup pump being available for each set of pumps that performs the same function. The number of pumps and their capacities should be able to handle the peak flow when one pump in the set is out of service.

## 8. SLUDGE PUMPS AND GRAVITY FLOW

Centrifugal and reciprocating pumps are extensively used for pumping sludge. Progressive cavity screw impeller pumps are also used.

As described in Sec. 20.28, the pumping power is proportional to the specific gravity. Accordingly, pumping power for dilute and well-digested sludges is typically only 10% to 25% higher than for water. However, most sludges are non-Newtonian fluids, often flow in a laminar mode, and have characteristics that may change with the season. Also, sludge characteristics change greatly during the pumping cycle. Therefore, engineering judgment and rules of thumb are often important in choosing sludge pumps. For example, a general rule is to choose sludge pumps capable of developing at least 50% to 100% excess head.

One method of determining the required pumping power is to multiply the power required for pumping pure water by a numerical factor. Empirical data is the best method of selecting this factor. Choice of initial values is a matter of judgment. Guidelines are listed in Table 20.4.

**Table 20.4** Pumping Power Multiplicative Factors

| solids concentration | digested sludge | untreated, primary, and concentrated sludge |
|---|---|---|
| 0% | 1.0 | 1.0 |
| 2% | 1.2 | 1.4 |
| 4% | 1.3 | 2.5 |
| 6% | 1.7 | 4.1 |
| 8% | 2.2 | 7.0 |
| 10% | 3.0 | 10.0 |

Derived from *Wastewater Engineering: Treatment, Disposal, Reuse*, 3rd ed., by Metcalf & Eddy, et al., © 1991, with permission from The McGraw-Hill Companies.

Generally, sludge will thin out during a pumping cycle. The most dense sludge components will be pumped first, with more watery sludge appearing at the end of the pumping cycle. With a constant power input, the reduction in load at the end of pumping cycles may cause centrifugal pumps to operate far from the desired operating point and experience overload failures. The operating point should be evaluated with high-, medium-, and low-density sludges.

To avoid cavitation, sludge pumps should always be under a positive suction head of at least 4 ft (1.2 m), and suction lifts should be avoided. The minimum diameters of suction and discharge lines for pumped sludge are typically 6 in (150 mm) and 4 in (100 mm), respectively.

Not all sludge is moved by pump action. Some installations rely on gravity flow to move sludge. The minimum diameter of sludge gravity transfer lines is typically 8 in (200 mm), and the recommended minimum slope is 3%.

To avoid clogging due to settling, flow should be above the transition from laminar to turbulent flow, known as the *critical velocity*. The critical velocity for most sludges is approximately 3.5 ft/sec (1.1 m/s). Velocities of 5 to 8 ft/sec (1.5 to 2.4 m/s) are typical and adequate.

## 9. TERMINOLOGY OF HYDRAULIC MACHINES

A pump will always have an inlet (designated the *suction*) and an outlet (designated the *discharge*). The subscripts $s$ and $d$ refer to the inlet and outlet of the pump, not of the pipeline.

All of the terms that are discussed in this section are *head* terms and, as such, have units of length. When working with hydraulic machines, it is common to hear such phrases as "a pressure head of 50 feet" and "a static discharge head of 15 meters." The term *head* is often substituted for pressure or pressure drop. Any head term (*pressure head, atmospheric head, vapor pressure head,* etc.) can be calculated from pressure by using Equation 20.4.[8]

$$h = \frac{p}{\gamma} \qquad 20.4$$

$$h = \frac{p}{g\rho} \qquad \text{[SI]} \quad 20.5(a)$$

$$h = \frac{p}{\rho} \times \frac{g_c}{g} \qquad \text{[U.S.]} \quad 20.5(b)$$

Some of the terms used in the description of pipelines appear to be similar and may be initially confusing (e.g., suction head and total suction head). The following general rules will help to clarify the meanings.

*rule 1:* The word *suction* or *discharge* limits the quantity to the suction line or discharge line, respectively. The absence of either word implies that both the suction and discharge lines are included. Example: discharge head.

*rule 2:* The word *static* means that static head only is included (not velocity head, friction head, etc.). Example: static suction head.

*rule 3:* The word *total* means that static head, velocity head, and friction head are all included. (Note that total does not mean the combination of suction and discharge.) Example: total suction head.

---

[8]Equation 20.5 can be used to define *pressure head, atmospheric head,* and *vapor pressure head,* whose meanings and derivations should be obvious.

The following terms are commonly encountered.

- *friction head* ($h_f$): The head required to overcome resistance to flow in the pipes, fittings, valves, entrances, and exits.

$$h_f = \frac{fL\text{v}^2}{2Dg} \qquad 20.6$$

- *velocity head* ($h_\text{v}$): The specific kinetic energy of the fluid. Also known as *dynamic head.* [9]

$$h_\text{v} = \frac{\text{v}^2}{2g} \qquad 20.7$$

- *static suction head* ($h_{z(s)}$): The vertical distance above the centerline of the pump inlet to the free level of the fluid source. If the free level of the fluid is below the pump inlet, $h_z$ will be negative and is known as *static suction lift.* (See Fig. 20.6.)

**Figure 20.6** *Static Suction Lift*

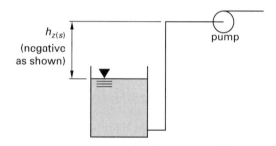

- *static discharge head* ($h_{z(d)}$): The vertical distance above the centerline of the pump inlet to the point of free discharge or surface level of the discharge tank. (See Fig. 20.7.)

**Figure 20.7** *Static Discharge Head*

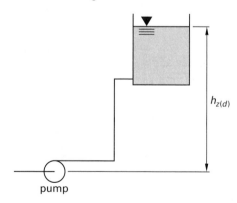

The ambiguous term *effective head* is not commonly used when discussing hydraulic machines, but when used, the term most closely means *net head* (i.e., starting head less losses). Consider a hydroelectric turbine that is fed by water with a static head of $H$.

---

[9]The term *dynamic* is not as consistently applied as are the terms described in the rules. In particular, it is not clear whether friction head is to be included with the dynamic head.

After frictional and other losses, the net head acting on the turbine will be less than $H$. The turbine output will coincide with an ideal turbine being acted upon by the net or effective head. Similarly, the actual increase in pressure across a pump will be the effective head added (i.e., the head net of internal losses and geometric effects).

### Example 20.1

Write the symbolic equations for the following terms: (a) the total suction head, (b) the total discharge head, and (c) the total head.

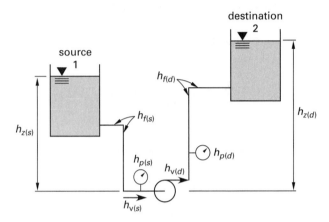

*Solution*

(a) The *total suction head* at the pump inlet is the sum of static (pressure) head and velocity head at the pump suction.

$$h_{t(s)} = h_{p(s)} + h_{v(s)}$$

Total suction head can also be calculated from the conditions existing at the source (1), in which case suction line friction would also be considered. (With an open reservoir, $h_{p(1)}$ will be zero if gage pressures are used, and $h_{v(1)}$ will be zero if the source is large.)

$$h_{t(s)} = h_{p(1)} + h_{z(s)} + h_{v(1)} - h_{f(s)}$$

(b) The *total discharge head* at the pump outlet is the sum of the static (pressure) and velocity heads at the pump outlet. Friction head is zero since the fluid has not yet traveled through any length of pipe when it is discharged.

$$h_{t(d)} = h_{p(d)} + h_{v(d)}$$

The total discharge head can also be evaluated at the destination (2) if the friction head, $h_{f(d)}$, between the discharge and the destination is known. (With an open reservoir, $h_{p(2)}$ will be zero if gage pressures are used, and $h_{v(2)}$ will be zero if the destination is large.)

$$h_{t(d)} = h_{p(2)} + h_{z(d)} + h_{v(2)} + h_{f(d)}$$

(c) The *total head added* by the pump is the total discharge head less the total suction head. Assuming

suction from and discharge to reservoirs exposed to the atmosphere and assuming negligible reservoir velocities,

$$h_t = h_A = h_{t(d)} - h_{t(s)} \approx h_{z(d)} - h_{z(s)} + h_{f(d)} + h_{f(s)}$$

## 10. PUMPING POWER

The energy (head) added by a pump can be determined from the difference in total energy on either side of the pump. Writing the Bernoulli equation for the discharge and suction conditions produces Eq. 20.8, an equation for the *total dynamic head*, often abbreviated TDH.

$$h_A = h_t = h_{t(d)} - h_{t(s)} \qquad \text{20.8}$$

$$h_A = \frac{p_d - p_s}{\rho g} + \frac{v_d^2 - v_s^2}{2g} + z_d - z_s \qquad \text{[SI]} \quad \text{20.9(a)}$$

$$h_A = \frac{(p_d - p_s)g_c}{\rho g} + \frac{v_d^2 - v_s^2}{2g} + z_d - z_s \qquad \text{[U.S.]} \quad \text{20.9(b)}$$

In most applications, the change in velocity and potential heads is either zero or small in comparison to the increase in pressure head. Equation 20.9 then reduces to Eq. 20.10.

$$h_A = \frac{p_d - p_s}{\rho g} \qquad \text{[SI]} \quad \text{20.10(a)}$$

$$h_A = \frac{p_d - p_s}{\rho} \times \frac{g_c}{g} \qquad \text{[U.S.]} \quad \text{20.10(b)}$$

It is important to recognize that the variables in Eq. 20.9 and Eq. 20.10 refer to the conditions at the pump's immediate inlet and discharge, not to the distant ends of the suction and discharge lines. However, the total dynamic head added by a pump can be calculated in another way. For example, for a pump raising water from one open reservoir to another, the total dynamic head would consider the total elevation rise, the velocity head (often negligible), and the friction losses in the suction and discharge lines.

The head added by the pump can also be calculated from the impeller and fluid speeds. Equation 20.11 is useful for radial- and mixed-flow pumps for which the incoming fluid has little or no rotational velocity component (i.e., up to a specific speed of approximately 2000 U.S. or 40 SI). In Eq. 20.11, $v_{impeller}$ is the tangential impeller velocity at the radius being considered, and $v_{fluid}$ is the average tangential velocity imparted to the fluid by the impeller. The impeller efficiency, $\eta_{impeller}$, is typically 0.85 to 0.95. This is much higher than the total pump efficiency (see Sec. 20.11) because it does not include mechanical and fluid friction losses.

$$h_A = \frac{\eta_{impeller} v_{impeller} v_{fluid}}{g} \qquad \text{20.11}$$

The pumping power depends on the head added, $h_A$, and the mass flow rate. For example, the product $\dot{m}h_A$

has the units of foot-pounds per second (in customary U.S. units), which can be easily converted to horsepower. Pump output power is known as *hydraulic power* or *water power*. Hydraulic power is the net power actually transferred to the fluid.

Horsepower is the unit of power in the United States and other non-SI countries, which gives rise to the terms *hydraulic horsepower* and *water horsepower*, WHP. Various relationships for finding the hydraulic horsepower are given in Table 20.5.

The unit of power in SI units is the watt (kilowatt). Table 20.6 can be used to determine *hydraulic kilowatts*, WkW.

**Table 20.5** Hydraulic Horsepower Equations[a]

| | $Q$ (gal/min) | $\dot{m}$ (lbm/sec) | $\dot{V}$ (ft²/sec) |
|---|---|---|---|
| $h_A$ in feet | $\dfrac{h_A Q(\text{SG})}{3956}$ | $\dfrac{h_A \dot{m}}{550} \times \dfrac{g}{g_c}$ | $\dfrac{h_A \dot{V}(\text{SG})}{8.814}$ |
| $\Delta p$ in psi[b] | $\dfrac{\Delta p Q}{1714}$ | $\dfrac{\Delta p \dot{m}}{(238.3)(\text{SG})} \times \dfrac{g}{g_c}$ | $\dfrac{\Delta p \dot{V}}{3.819}$ |
| $\Delta p$ in psf[b] | $\dfrac{\Delta p Q}{2.468 \times 10^5}$ | $\dfrac{\Delta p \dot{m}}{(34{,}320)(\text{SG})} \times \dfrac{g}{g_c}$ | $\dfrac{\Delta p \dot{V}}{550}$ |
| $W$ in $\dfrac{\text{ft-lbf}}{\text{lbm}}$ | $\dfrac{W Q(\text{SG})}{3956}$ | $\dfrac{W \dot{m}}{550}$ | $\dfrac{W \dot{V}(\text{SG})}{8.814}$ |

(Multiply horsepower by 0.7457 to obtain kilowatts.)
[a]based on $\rho_{\text{water}} = 62.4$ lbm/ft³ and $g = 32.2$ ft/sec²
[b]Velocity head changes must be included in $\Delta p$.

**Table 20.6** Hydraulic Kilowatt Equations[a]

| | $Q$ (L/s) | $\dot{m}$ (kg/s) | $\dot{V}$ (m³/s) |
|---|---|---|---|
| $h_A$ in meters | $\dfrac{(9.81)h_A Q(\text{SG})}{1000}$ | $\dfrac{(9.81)h_A \dot{m}}{1000}$ | $(9.81)h_A \dot{V}(\text{SG})$ |
| $\Delta p$ in kPa[b] | $\dfrac{\Delta p Q}{1000}$ | $\dfrac{\Delta p \dot{m}}{1000(\text{SG})}$ | $\Delta p \dot{V}$ |
| $W$ in $\dfrac{\text{J}}{\text{kg}}$[b] | $\dfrac{W Q(\text{SG})}{1000}$ | $\dfrac{W \dot{m}}{1000}$ | $W \dot{V}(\text{SG})$ |

(Multiply kilowatts by 1.341 to obtain horsepower.)
[a]based on $\rho_{\text{water}} = 1000$ kg/m³ and $g = 9.81$ m/s²
[b]Velocity head changes must be included in $\Delta p$.

### Example 20.2

A pump adds 550 ft of pressure head to 100 lbm/sec of water. (a) Complete the following table of performance data. (b) What is the hydraulic power in horsepower and kilowatts? (Assume $\rho = 62.4$ lbm/ft³ or 1000 kg/m³, and $g = 9.81$ m/s².)

| item | customary U.S. | SI |
|---|---|---|
| $\dot{m}$ | 100 lbm/sec | ___ kg/s |
| $h$ | 550 ft | ___ m |
| $\Delta p$ | ___ lbf/ft² | ___ kPa |
| $\dot{V}$ | ___ ft³/sec | ___ m³/s |
| $W$ | ___ ft-lbf/lbm | ___ J/kg |
| $P$ | ___ hp | ___ kW |

*Solution*

(a) Work initially with the customary U.S. data.

$$\Delta p = \rho h \times \frac{g}{g_c} = \left(62.4 \ \frac{\text{lbm}}{\text{ft}^3}\right)(550 \text{ ft}) \times \frac{g}{g_c}$$
$$= 34{,}320 \text{ lbf/ft}^2$$

$$\dot{V} = \frac{\dot{m}}{\rho} = \frac{100 \ \dfrac{\text{lbm}}{\text{sec}}}{62.4 \ \dfrac{\text{lbm}}{\text{ft}^3}} = 1.603 \text{ ft}^3/\text{sec}$$

$$W = h \times \frac{g}{g_c} = 550 \text{ ft-lbf/lbm}$$

Now, convert to SI units.

$$\dot{m} = \frac{100 \ \dfrac{\text{lbm}}{\text{sec}}}{2.201 \ \dfrac{\text{lbm}}{\text{kg}}} = 45.43 \text{ kg/s}$$

$$h = \frac{550 \text{ ft}}{3.281 \ \dfrac{\text{ft}}{\text{m}}} = 167.6 \text{ m}$$

$$\Delta p = \left(34{,}320 \ \frac{\text{lbf}}{\text{ft}^2}\right)\left(\frac{1}{\left(12 \ \dfrac{\text{in}}{\text{ft}}\right)^2}\right)\left(6.895 \ \frac{\text{kPa}}{\dfrac{\text{lbf}}{\text{in}^2}}\right)$$
$$= 1643 \text{ kPa}$$

$$\dot{V} = \left(1.603 \ \frac{\text{ft}^3}{\text{sec}}\right)\left(0.0283 \ \frac{\text{m}^3}{\text{ft}^3}\right) = 0.0454 \text{ m}^3/\text{s}$$

$$W = \left(550 \ \frac{\text{ft-lbf}}{\text{lbm}}\right)\left(1.356 \ \frac{\text{J}}{\text{ft-lbf}}\right)\left(2.201 \ \frac{\text{lbm}}{\text{kg}}\right)$$
$$= 1642 \text{ J/kg}$$

(b) From Table 20.5, the hydraulic horsepower is

$$\text{WHP} = \frac{h_A \dot{m}}{550} \times \frac{g}{g_c}$$
$$= \frac{(550 \text{ ft})\left(100 \ \dfrac{\text{lbm}}{\text{sec}}\right)}{550 \ \dfrac{\text{ft-lbf}}{\text{hp-sec}}} \times \frac{g}{g_c}$$
$$= 100 \text{ hp}$$

From Table 20.6, the power is

$$\text{WkW} = \frac{\Delta p \dot{m}}{(1000)(\text{SG})} = \frac{(1643 \text{ kPa})\left(45.43 \ \frac{\text{kg}}{\text{s}}\right)}{\left(1000 \ \frac{\text{W}}{\text{kW}}\right)(1.0)}$$

$$= 74.6 \text{ kW}$$

## 11. PUMPING EFFICIENCY

Hydraulic power is the net energy actually transferred to the fluid per unit time. The input power delivered by the motor to the pump is known as the *brake pump power*. (For example, the term *brake horsepower*, bhp, is commonly used.) Due to frictional losses between the fluid and the pump and mechanical losses in the pump itself, the brake pump power will be greater than the hydraulic power.

The ratio of hydraulic power to brake pump power is the pump efficiency, $\eta_p$. Figure 20.8 gives typical pump efficiencies as a function of the pump's specific speed. The difference between the brake and hydraulic powers is known as the *friction power* (or *friction horsepower*).

**Figure 20.8** *Average Pump Efficiency Versus Specific Speed*

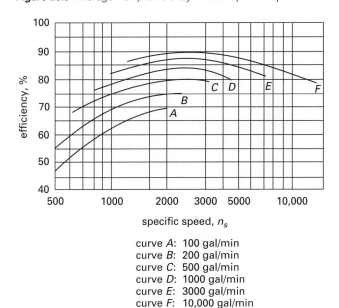

curve A: 100 gal/min
curve B: 200 gal/min
curve C: 500 gal/min
curve D: 1000 gal/min
curve E: 3000 gal/min
curve F: 10,000 gal/min

$$\text{brake pump power} = \frac{\text{hydraulic power}}{\eta_p} \qquad 20.12$$

friction power

$$= \text{brake pump power} - \text{hydraulic power} \qquad 20.13$$

Pumping efficiency is not constant for any specific pump; rather, it depends on the operating point (see Sec. 20.23) and the speed of the pump. A specific

pump's characteristic efficiency curves will be published by its manufacturer.

With pump characteristic curves given by the manufacturer, the efficiency is not determined from the intersection of the system curve (see Sec. 20.22) and the efficiency curve. Rather, the efficiency is a function of only the flow rate. Therefore, the operating efficiency is read from the efficiency curve directly above or below the operating point. (See Sec. 20.23 and Fig. 20.9.)

**Figure 20.9** *Typical Centrifugal Pump Efficiency Curves*

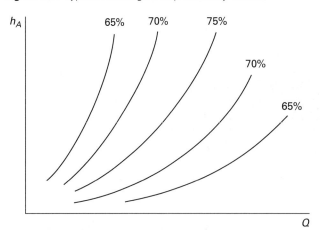

Unfortunately, efficiency curves published by a manufacturer may not be representative of the actual installed efficiency. The manufacturer's efficiency may not include such losses in the suction elbow, discharge diffuser, couplings, bearing frame, seals, or pillow blocks. Up to 15% of the motor horsepower may be lost to these factors. Therefore, the manufacturer should be requested to provide the pump's installed *wire-to-water efficiency* (i.e., the fraction of the electrical power drawn that is converted to hydraulic power).

The pump must be driven by an engine or motor.[10] The power delivered to the motor is greater than the power delivered to the pump, as accounted for by the motor efficiency, $\eta_m$. If the pump motor is electrical, its input power requirements will be stated in kilowatts.[11]

$$\text{motor input power}$$
$$= \frac{\text{brake pump power}}{\eta_m} \qquad 20.14$$

$$(\text{motor input power})_\text{kW}$$
$$= \frac{0.7457 \times (\text{brake pump power})_\text{hp}}{\eta_m} \qquad 20.15$$

---

[10]The source of power is sometimes called the *prime mover*.
[11]A *watt* is a joule per second.

The *overall efficiency* of the pump installation is the product of the pump and motor efficiencies.

$$\eta = \eta_p \eta_m = \frac{\text{hydraulic horsepower}}{\text{motor horsepower}} \qquad \textit{20.16}$$

## 12. COST OF ELECTRICITY

The power utilization of pump motors is usually measured in kilowatts. The kilowatt usage represents the rate that energy is transferred by the pump motor. The total amount of work, $W$, done by the pump motor is found by multiplying the rate of energy usage by the length of time the pump is in operation.

$$W = Pt \qquad \textit{20.17}$$

Although the units horsepower-hours are occasionally encountered, it is more common to measure electrical work in *kilowatt-hours* (kW-hr). Accordingly, the cost of electrical energy is stated per kW-hr (e.g., $0.10 per kW-hr).

$$\text{cost} = \frac{W_{\text{kW-hr}} \times \text{cost per kW-hr}}{\eta_m} \qquad \textit{20.18}$$

## 13. STANDARD MOTOR SIZES AND SPEEDS

An effort should be made to specify standard motor sizes when selecting the source of pumping power. Table 20.7 lists NEMA (National Electrical Manufacturers Association) standard motor sizes by horsepower *nameplate rating*.[12] The rated horsepower is the maximum power the motor can provide without incurring damage. Other motor sizes may also be available by special order.

*Table 20.7* NEMA Standard Motor Sizes (brake horsepower)

| $\frac{1}{8}$, | $\frac{1}{6}$, | $\frac{1}{4}$, | $\frac{1}{3}$ | | | | |
|------|------|------|------|------|------|------|------|
| 0.5, | 0.75, | 1, | 1.5, | 2, | 3, | 5, | 7.5 |
| 10, | 15, | 20, | 25, | 30, | 40, | 50, | 60 |
| 75, | 100, | 125, | 150, | 200, | 250, | 300, | 350 |
| 400, | 450, | 500, | 600, | 700, | 800, | 900, | 1000 |
| 1250, | 1500, | 1750, | 2000, | 2250, | 2500, | 2750, | 3000 |
| 3500, | 4000, | 4500, | 5000, | 6000, | 7000, | 8000 | |

Larger horsepower motors are usually three-phase induction motors. The *synchronous speed*, $n$ in rpm, of such motors is the speed of the rotating field, which depends on the number of poles per stator phase and the frequency, $f$. The number of poles must be an even number. The frequency is typically 60 Hz, as in the

United States, or 50 Hz, as in European countries. Table 20.8 lists common synchronous speeds.

$$n = \frac{120 \times f}{\text{no. of poles}} \qquad \textit{20.19}$$

*Table 20.8* Common Synchronous Speeds

| number of poles | $n$ (rpm) 60 Hz | 50 Hz |
|:---:|:---:|:---:|
| 2 | 3600 | 3000 |
| 4 | 1800 | 1500 |
| 6 | 1200 | 1000 |
| 8 | 900 | 750 |
| 10 | 720 | 600 |
| 12 | 600 | 500 |
| 14 | 514 | 428 |
| 18 | 400 | 333 |
| 24 | 300 | 250 |
| 48 | 150 | 125 |

Induction motors do not run at their synchronous speeds when loaded. Rather, they run at slightly less than synchronous speed. The deviation is known as the *slip*. Slip is typically around 4% and is seldom greater than 10% at full load for motors in the 1 hp to 75 hp range.

slip (in rpm)

$$= \text{synchronous speed} - \text{actual speed} \qquad \textit{20.20}$$

slip (in percent)

$$= 100\% \times \frac{\text{synchronous speed} - \text{actual speed}}{\text{synchronous speed}} \qquad \textit{20.21}$$

Induction motors may also be specified in terms of their kVA (kilovolt-amp) ratings. The kVA rating is not the same as the power in kilowatts, although one can be derived from the other if the motor's power factor is known. Such power factors typically range from 0.8 to 0.9, depending on the installation and motor size.

$$\text{kVA rating} = \frac{\text{motor power in kW}}{\text{power factor}} \qquad \textit{20.22}$$

## Example 20.3

A pump driven by an electrical motor moves 25 gal/min of water from reservoir A to reservoir B, lifting the water a total of 245 ft. The efficiencies of the pump and motor are 64% and 84%, respectively. Electricity costs $0.08/kW-hr. Neglect velocity head, friction, and minor losses. (a) What size motor is required? (b) How much does it cost to operate the pump for 6 hr?

---

[12]The nameplate rating gets its name from the information stamped on the motor's identification plate. Besides the horsepower rating, other nameplate data used to classify the motor are the service class, voltage, full-load current, speed, number of phases, frequency of the current, and maximum ambient temperature (or, for older motors, the motor temperature rise).

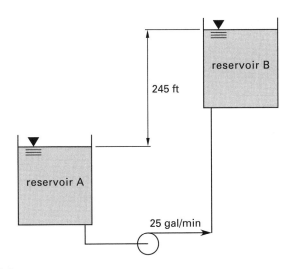

*Solution*

(a) The head added is 245 ft.

Using Table 20.5 and incorporating the pump efficacy, the motor power required is

$$P = \frac{h_A Q (\text{SG})}{3956\eta_p} = \frac{(245 \text{ ft})\left(25 \dfrac{\text{gal}}{\text{min}}\right)(1.0)}{(3956)(0.64)}$$

$$= 2.42 \text{ hp}$$

From Table 20.7, select a 3 hp motor.

(b) From Eq. 20.17 and Eq. 20.18,

$$\text{cost} = \text{cost per kW-hr} \times \frac{Pt}{\eta_m}$$

$$= \frac{\left(0.08 \dfrac{\$}{\text{kW-hr}}\right)\left(0.7457 \dfrac{\text{kW}}{\text{hp}}\right)(2.42 \text{ hp})(6 \text{ hr})}{0.84}$$

$$= \$1.03$$

Notice that the developed power, not the motor's rated power, is used.

## 14. PUMP SHAFT LOADING

The torque on a pump or motor shaft, brake power, and speed are all related. The power used in Eq. 20.23 through Eq. 20.25 can be either the brake (shaft) power or the hydraulic power developed.

$$T_{\text{in-lbf}} = \frac{63{,}025 \times P_{\text{hp}}}{n} \qquad 20.23$$

$$T_{\text{ft-lbf}} = \frac{5252 \times P_{\text{hp}}}{n} \qquad 20.24$$

$$T_{\text{N·m}} = \frac{9549 \times P_{\text{kW}}}{n} \qquad 20.25$$

The actual (developed) torque can be calculated from the change in momentum of the fluid flow. For radial

impellers, the fluid enters through the eye and is turned 90°. The direction change is related to the increase in momentum and the shaft torque. When fluid is introduced axially through the eye of the impeller, the tangential velocity at the inlet (eye), $v_{t(i)}$, is zero.[13]

$$T_{\text{actual}} = \frac{\dot{m}}{g_c}\left(v_{t(d)} r_{\text{impeller}} - v_{t(i)} r_{\text{eye}}\right) \qquad 20.26$$

Centrifugal pumps may be driven directly from a motor, or a speed changer may be used. Rotary pumps generally require a speed reduction. *Gear motors* have integral speed reducers. V-belt drives are widely used because of their initial low cost, although timing belts and chains can be used in some applications.

When a belt or chain is used, the pump's and motor's maximum overhung loads must be checked. This is particularly important for high-power, low-speed applications (such as rotary pumps). *Overhung load* is the side load (force) put on shafts and bearings. The overhung load is calculated from Eq. 20.27. The empirical factor $K$ is 1.0 for chain drives, 1.25 for timing belts, and 1.5 for V-belts.

$$\text{overhung load} = \frac{2KT}{D_{\text{sheave}}} \qquad 20.27$$

If a direct drive cannot be used and the overhung load is excessive, the installation can incorporate a jack shaft or outboard bearing.

### Example 20.4

A centrifugal pump delivers 275 lbm/sec (125 kg/s) of water while turning at 850 rpm. The impeller has straight radial vanes and an outside diameter of 10 in (25.4 cm). Water enters the impeller through the eye. The driving motor delivers 30 hp (22 kW). What are the (a) theoretical torque, (b) pump efficiency, and (c) total developed dynamic head?

*SI Solution*

(a) From Eq. 20.3, the impeller's tangential velocity is

$$v_t = \frac{\pi D n}{60 \dfrac{\text{s}}{\text{min}}} = \frac{\pi (25.4 \text{ cm})(850 \text{ rpm})}{\left(60 \dfrac{\text{s}}{\text{min}}\right)\left(100 \dfrac{\text{cm}}{\text{m}}\right)}$$

$$= 11.3 \text{ m/s}$$

---

[13]The tangential component of fluid velocity is sometimes referred to as the *velocity of whirl*.

Since water enters axially, the incoming water has no tangential component. From Eq. 20.26, the developed torque is

$$T = \dot{m}\mathrm{v}_{t(d)}r_{\text{impeller}}$$

$$= \frac{\left(125 \; \dfrac{\text{kg}}{\text{s}}\right)\left(11.3 \; \dfrac{\text{m}}{\text{s}}\right)\left(\dfrac{25.4 \; \text{cm}}{2}\right)}{100 \; \dfrac{\text{cm}}{\text{m}}}$$

$$= 179.4 \; \text{N·m}$$

(b) From Eq. 20.25, the developed power is

$$P_{\text{kW}} = \frac{nT_{\text{N·m}}}{9549} = \frac{(850 \; \text{rpm})(179.4 \; \text{N·m})}{9549}$$

$$= 15.97 \; \text{kW}$$

The pump efficiency is

$$\eta_p = \frac{P_{\text{developed}}}{P_{\text{input}}}$$

$$= \frac{15.97 \; \text{kW}}{22 \; \text{kW}} = 0.726 \quad (72.6\%)$$

(c) From Table 20.6, the total dynamic head is

$$h_A = \frac{P_{\text{kW}}\left(1000 \; \dfrac{\text{W}}{\text{kW}}\right)}{\dot{m}\left(9.81 \; \dfrac{\text{m}}{\text{s}^2}\right)}$$

$$= \frac{(15.97 \; \text{kW})\left(1000 \; \dfrac{\text{W}}{\text{kW}}\right)}{\left(125 \; \dfrac{\text{kg}}{\text{s}}\right)\left(9.81 \; \dfrac{\text{m}}{\text{s}^2}\right)}$$

$$= 13.0 \; \text{m}$$

*Customary U.S. Solution*

(a) From Eq. 20.3, the impeller's tangential velocity is

$$\mathrm{v}_t = \frac{\pi D n}{60 \; \dfrac{\text{sec}}{\text{min}}} = \frac{\pi(10 \; \text{in})(850 \; \text{rpm})}{\left(60 \; \dfrac{\text{sec}}{\text{min}}\right)\left(12 \; \dfrac{\text{in}}{\text{ft}}\right)}$$

$$= 37.08 \; \text{ft/sec}$$

Since water enters axially, the incoming water has no tangential component. From Eq. 20.26, the developed torque is

$$T = \frac{\dot{m}\mathrm{v}_{t(d)}r_{\text{impeller}}}{g_c}$$

$$= \frac{\left(275 \; \dfrac{\text{lbm}}{\text{sec}}\right)\left(37.08 \; \dfrac{\text{ft}}{\text{sec}}\right)\left(\dfrac{10 \; \text{in}}{2}\right)}{\left(32.2 \; \dfrac{\text{ft-lbm}}{\text{lbf-sec}^2}\right)\left(12 \; \dfrac{\text{in}}{\text{ft}}\right)}$$

$$= 131.9 \; \text{ft-lbf}$$

(b) From Eq. 20.24, the developed power is

$$P_{\text{hp}} = \frac{T_{\text{ft-lbf}}\, n}{5252}$$

$$= \frac{(131.9 \; \text{ft-lbf})(850 \; \text{rpm})}{5252} = 21.35 \; \text{hp}$$

The pump efficiency is

$$\eta_p = \frac{P_{\text{developed}}}{P_{\text{input}}}$$

$$= \frac{21.35 \; \text{hp}}{30 \; \text{hp}} = 0.712 \quad (71.2\%)$$

(c) From Table 20.5, the total dynamic head is

$$h_A = \frac{\left(550 \; \dfrac{\text{ft-lbf}}{\text{hp-sec}}\right)P_{\text{hp}}}{\dot{m}} \times \frac{g_c}{g}$$

$$= \left(\frac{\left(550 \; \dfrac{\text{ft-lbf}}{\text{hp-sec}}\right)(21.35 \; \text{hp})}{275 \; \dfrac{\text{lbm}}{\text{sec}}}\right)$$

$$\times \left(\frac{32.2 \; \dfrac{\text{ft-lbm}}{\text{lbf-sec}^2}}{32.2 \; \dfrac{\text{ft}}{\text{sec}^2}}\right)$$

$$= 42.7 \; \text{ft}$$

## 15. SPECIFIC SPEED

The capacity and efficiency of a centrifugal pump are partially governed by the impeller design. For a desired flow rate and added head, there will be one optimum impeller design. The quantitative index used to optimize the impeller design is known as *specific speed*, $n_s$, also known as *impeller specific speed*. Table 20.9 lists the impeller designs that are appropriate for different specific speeds.[14]

*Table 20.9* Specific Speed versus Impeller Design

| | approximate range of specific speed (rpm) | |
| impeller type | customary U.S. units | SI units |
|---|---|---|
| radial vane | 500 to 1000 | 10 to 20 |
| Francis (mixed) vane | 2000 to 3000 | 40 to 60 |
| mixed flow | 4000 to 7000 | 80 to 140 |
| axial flow | 9000 and above | 180 and above |

[14]Specific speed is useful for more than just selecting an impeller type. Maximum suction lift, pump efficiency, and net positive suction head required (NPSHR) can be correlated with specific speed.

Fluids

Highest heads per stage are developed at low specific speeds. However, for best efficiency, specific speed should be greater than 650 (13 in SI units). If the specific speed for a given set of conditions drops below 650 (13), a multiple-stage pump should be selected.[15]

Specific speed is a function of a pump's capacity, head, and rotational speed at peak efficiency, as shown in Equation 20.28. For a given pump and impeller configuration, the specific speed remains essentially constant over a range of flow rates and heads.

While specific speed is not dimensionless, the units are meaningless. Specific speed may be assigned units of rpm, but most often it is expressed simply as a pure number. ($Q$ or $\dot{V}$ in Eq. 20.28 is half of the full flow rate for double-suction pumps.)

$$n_s = \frac{n\sqrt{\dot{V}}}{h_A^{0.75}} \qquad \text{[SI]} \quad 20.28(a)$$

$$n_s = \frac{n\sqrt{Q}}{h_A^{0.75}} \qquad \text{[U.S.]} \quad 20.28(b)$$

The numerical range of acceptable performance for each impeller type is redefined when SI units are used. The SI specific speed is obtained by dividing the customary U.S. specific speed by 51.66.

A common definition of specific speed is the speed (in rpm) at which a *homologous pump* would have to turn in order to deliver one gallon per minute at one foot total added head.[16] This definition is implicit to Equation 20.28 but is not very useful otherwise.

Specific speed can be used to determine the type of impeller needed. Once a pump is selected, its specific speed and Eq. 20.28 can be used to determine other operational parameters (e.g., maximum rotational speed). Specific speed can be used with Fig. 20.8 to obtain an approximate pump efficiency.

### Example 20.5

A centrifugal pump powered by a direct-drive induction motor is needed to discharge 150 gal/min against a 300 ft total head when turning at the fully loaded speed of 3500 rpm. What type of pump should be selected?

*Solution*

From Eq. 20.28, the specific speed is

$$n_s = \frac{n\sqrt{Q}}{h_A^{0.75}} = \frac{(3500\text{ rpm})\sqrt{150\ \dfrac{\text{gal}}{\text{min}}}}{(300\text{ ft})^{0.75}} = 595$$

[15]*Partial emission, forced vortex centrifugal pumps* allow operation down to specific speeds of 150 (3 in SI). Such pumps have been used for low-flow, high-head applications, such as high-pressure petrochemical cracking processes.

[16]*Homologous pumps* are geometrically similar. This means that each pump is a scaled up or down version of the others. Such pumps are said to belong to a *homologous family*.

From Table 20.9, the pump should be a radial vane type. However, pumps achieve their highest efficiencies when specific speed exceeds 650. (See Fig. 20.8.) To increase the specific speed, the rotational speed can be increased, or the total added head can be decreased. Since the pump is direct-driven and 3600 rpm is the maximum speed for induction motors (see Table 20.8), the total added head should be divided evenly between two stages, or two pumps should be used in series.

In a two-stage system, the specific speed would be

$$n_s = \frac{(3500\text{ rpm})\sqrt{150\ \dfrac{\text{gal}}{\text{min}}}}{(150\text{ ft})^{0.75}} = 1000$$

This is satisfactory for a radial vane pump.

### Example 20.6

An induction motor turning at 1200 rpm is to be selected to drive a single-stage, single-suction centrifugal water pump through a direct drive. The total dynamic head added by the pump is 26 ft. The flow rate is 900 gal/min. What size motor should be selected?

*Solution*

The specific speed is

$$n_s = \frac{n\sqrt{Q}}{h_A^{0.75}} = \frac{(1200\text{ rpm})\sqrt{900\ \dfrac{\text{gal}}{\text{min}}}}{(26\text{ ft})^{0.75}}$$
$$= 3127$$

From Fig. 20.8, the pump efficiency will be approximately 82%.

From Table 20.5, the minimum motor horsepower is

$$P_{\text{hp}} = \frac{h_A\,Q(\text{SG})}{\left(3956\ \dfrac{\text{ft-gal}}{\text{hp-min}}\right)\eta_p}$$
$$= \frac{(26\text{ ft})\left(900\ \dfrac{\text{gal}}{\text{min}}\right)(1.0)}{\left(3956\ \dfrac{\text{ft-gal}}{\text{hp-min}}\right)(0.82)}$$
$$= 7.2\text{ hp}$$

From Table 20.7, select a 7.5 hp or larger motor.

### Example 20.7

A single-stage pump driven by a 3600 rpm motor is currently delivering 150 gal/min. The total dynamic head is 430 ft. What would be the approximate increase in efficiency per stage if the single-stage pump is replaced by a double-stage pump?

*Solution*

The specific speed is

$$n_s = \frac{n\sqrt{Q}}{h_A^{0.75}} = \frac{(3600 \text{ rpm})\sqrt{150 \frac{\text{gal}}{\text{min}}}}{(430 \text{ ft})^{0.75}}$$
$$= 467$$

From Fig. 20.8, the approximate efficiency is 45%.

In a two-stage pump, each stage adds half of the head. The specific speed per stage would be

$$n_s = \frac{n\sqrt{Q}}{h_A^{0.75}} = \frac{(3600 \text{ rpm})\sqrt{150 \frac{\text{gal}}{\text{min}}}}{\left(\frac{430 \text{ ft}}{2}\right)^{0.75}}$$
$$= 785$$

From Fig. 20.8, the efficiency for this configuration is approximately 60%.

The increase in stage efficiency is 60% − 45% = 15%. Whether or not the cost of multistaging is worthwhile in this low-volume application would have to be determined. The overall efficiency of the two-stage pump may actually be lower than the efficiency of the one-stage pump.

## 16. CAVITATION

*Cavitation* is the spontaneous vaporization of the fluid, resulting in a degradation of pump performance. If the fluid pressure is less than the vapor pressure, small pockets of vapor will form. These pockets usually form only within the pump itself, although cavitation slightly upstream within the suction line is also possible. As the vapor pockets reach the surface of the impeller, the local high fluid pressure collapses them. Noise, vibration, impeller pitting, and structural damage to the pump casing are manifestations of cavitation.

Cavitation can be caused by any of the following conditions.

- discharge head far below the pump head at peak efficiency

- high suction lift or low suction head

- excessive pump speed

- high liquid temperature (i.e., high vapor pressure)

## 17. NET POSITIVE SUCTION HEAD

The occurrence of cavitation is predictable. Cavitation will occur when the net pressure in the fluid drops below the vapor pressure. This criterion is commonly stated in terms of head: Cavitation occurs when the available head is less than the required head for satisfactory operation.

$$\text{available head} < \text{required head} \quad \begin{bmatrix} \text{criterion for} \\ \text{cavitation} \end{bmatrix} \quad \text{20.29}$$

The minimum fluid energy required at the pump inlet for satisfactory operation (i.e., the required head) is known as the *net positive suction head required*, NPSHR.[17] NPSHR is a function of the pump and will be given by the pump manufacturer as part of the pump performance data.[18] NPSHR is dependent on the flow rate. However, if NPSHR is known for one flow rate, it can be determined for another flow rate from Eq. 20.30.

$$\frac{\text{NPSHR}_2}{\text{NPSHR}_1} = \left(\frac{Q_2}{Q_1}\right)^2 \quad \text{20.30}$$

*Net positive suction head available*, NPSHA, is the actual total fluid energy at the inlet. There are two different methods for calculating NPSHA, both of which are correct and will yield identical answers. Equation 20.31(a) is based on the conditions at the fluid surface at the top of an open fluid source (e.g., tank or reservoir). There is a potential energy term but no kinetic energy term. Equation 20.31(b) is based on the conditions at the immediate entrance (suction, subscript *s*) to the pump. At that point, some of the potential head has been converted to velocity head. Frictional losses are implicitly part of the reduced pressure head, as is the atmospheric pressure head. Since the pressure head, $h_{p(s)}$, is absolute, it includes the atmospheric pressure head, and the effect of higher altitudes is explicit in Eq. 20.31(a) and implicit in Eq. 20.31(b). If the source was pressurized instead of being open to the atmosphere, the pressure head would replace $h_{\text{atm}}$ in Eq. 20.31(a) but would be implicit in $h_{p(s)}$ in Eq. 20.31(b).

$$\text{NPSHA} = h_{\text{atm}} + h_{z(s)} - h_{f(s)} - h_{\text{vp}} \quad \text{20.31(a)}$$

$$\text{NPSHA} = h_{p(s)} + h_{v(s)} - h_{\text{vp}} \quad \text{20.31(b)}$$

The net positive suction head available (NPSHA) for most positive displacement pumps includes a term for acceleration head.[19]

$$\text{NPSHA} = h_{\text{atm}} + h_{z(s)} - h_{f(s)} - h_{\text{vp}} - h_{\text{ac}} \quad \text{20.32}$$

$$\text{NPSHA} = h_{p(s)} + h_{v(s)} - h_{\text{vp}} - h_{\text{ac}} \quad \text{20.33}$$

---

[17]If NPSHR (a head term) is multiplied by the fluid specific weight, it is known as the *net inlet pressure required*, NIPR. Similarly, NPSHA can be converted to NIPA.

[18]It is also possible to calculate NPSHR from other information, such as suction specific speed. However, this still depends on information provided by the manufacturer.

[19]The friction loss and the acceleration are both maximum values, but they do not occur in phase. Combining them is conservative.

If NPSHA is less than NPSHR, the fluid will cavitate. The criterion for cavitation is given by Eq. 20.34. (In practice, it is desirable to have a safety margin.)

$$\text{NPSHA} < \text{NPSHR} \quad \begin{bmatrix} \text{criterion for} \\ \text{cavitation} \end{bmatrix} \qquad 20.34$$

**Example 20.8**

2.0 ft$^3$/sec (56 L/s) of 60°F (16°C) water are pumped from an elevated feed tank to an open reservoir through 6 in (15.2 cm), schedule-40 steel pipe, as shown. The friction loss for the piping and fittings in the suction line is 2.6 ft (0.9 m). The friction loss for the piping and fittings in the discharge line is 13 ft (4.3 m). The atmospheric pressure is 14.7 psia (101 kPa). What is the NPSHA?

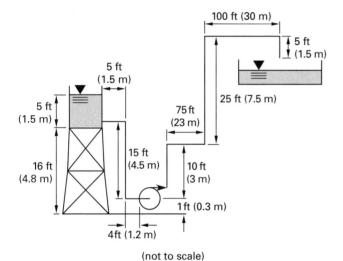

(not to scale)

*SI Solution*

The density of water is approximately 1000 kg/m$^3$. The atmospheric head is

$$h_{\text{atm}} = \frac{p}{\rho g} = \frac{(101 \text{ kPa})\left(1000 \dfrac{\text{Pa}}{\text{kPa}}\right)}{\left(1000 \dfrac{\text{kg}}{\text{m}^3}\right)\left(9.81 \dfrac{\text{m}}{\text{s}^2}\right)}$$

$$= 10.3 \text{ m}$$

For 16°C water, the vapor pressure is approximately 0.01818 bars. The vapor pressure head is

$$h_{\text{vp}} = \frac{p}{\rho g} = \frac{(0.01818 \text{ bar})\left(1 \times 10^5 \dfrac{\text{Pa}}{\text{bar}}\right)}{\left(1000 \dfrac{\text{kg}}{\text{m}^3}\right)\left(9.81 \dfrac{\text{m}}{\text{s}^2}\right)}$$

$$= 0.2 \text{ m}$$

From Eq. 20.31(a), the NPSHA is

$$\text{NPSHA} = h_{\text{atm}} + h_{z(s)} - h_{f(s)} - h_{\text{vp}}$$

$$= 10.3 \text{ m} + 1.5 \text{ m} + 4.8 \text{ m} - 0.3 \text{ m}$$

$$- 0.9 \text{ m} - 0.2 \text{ m}$$

$$= 15.2 \text{ m}$$

*Customary U.S. Solution*

The specific weight of water is approximately 62.4 lbf/ft$^3$. The atmospheric head is

$$h_{\text{atm}} = \frac{p}{\gamma} = \frac{\left(14.7 \dfrac{\text{lbf}}{\text{in}^2}\right)\left(12 \dfrac{\text{in}}{\text{ft}}\right)^2}{62.4 \dfrac{\text{lbf}}{\text{ft}^3}}$$

$$= 33.9 \text{ ft}$$

For 60°F water, the vapor pressure head is 0.59 ft. Use 0.6 ft.

From Eq. 20.31(a), the NPSHA is

$$\text{NPSHA} = h_{\text{atm}} + h_{z(s)} - h_{f(s)} - h_{\text{vp}}$$

$$= 33.9 \text{ ft} + 5 \text{ ft} + 16 \text{ ft} - 1 \text{ ft} - 2.6 \text{ ft} - 0.6 \text{ ft}$$

$$= 50.7 \text{ ft}$$

## 18. PREVENTING CAVITATION

Cavitation is eliminated by increasing NPSHA or decreasing NPSHR. NPSHA can be increased by:

- increasing the height of the fluid source
- lowering the pump
- reducing friction and minor losses by shortening the suction line or using a larger pipe size
- reducing the temperature of the fluid at the pump entrance
- pressurizing the fluid supply tank
- reducing the flow rate or velocity (i.e., reducing the pump speed)

NPSHR can be reduced by:

- placing a throttling valve or restriction in the discharge line[20]
- using an oversized pump
- using a double-suction pump
- using an impeller with a larger eye
- using an inducer

---

[20]This will increase the total head ($h_A$) added by the pump, thereby reducing the pump's output and driving the pump's operating point into a region of lower NPSHR.

High NPSHR applications, such as boiler feed pumps needing 150 ft to 250 ft (50 m to 80 m), should use one or more booster pumps in front of each high-NPSHR pump. Such booster pumps are typically single-stage, double-suction pumps running at low speed. Their NPSHR can be 25 ft (8 m) or less.

Throttling the input line to a pump and venting or evacuating the receiving tank both increase cavitation. Throttling the input line increases the friction head and decreases NPSHA. Evacuating the receiving tank increases the flow rate, increasing NPSHR while simultaneously increasing the friction head and reducing NPSHA.

### 19. CAVITATION COEFFICIENT

The *cavitation coefficient* (or *cavitation number*), $\sigma$, is a dimensionless number that can be used in modeling and extrapolating experimental results. The actual cavitation coefficient is compared with the critical cavitation number obtained experimentally. If the actual cavitation number is less than the critical cavitation number, cavitation will occur. Absolute pressure must be used in calculating $\sigma$.

$$\sigma = \frac{2(p - p_{\mathrm{vp}})}{\rho \mathrm{v}^2} = \frac{\text{NPSHA}}{h_A} \qquad \text{[SI]} \quad 20.35(a)$$

$$\sigma = \frac{2g_c(p - p_{\mathrm{vp}})}{\rho \mathrm{v}^2} = \frac{\text{NPSHA}}{h_A} \qquad \text{[U.S.]} \quad 20.35(b)$$

$$\sigma < \sigma_{\mathrm{cr}} \quad \text{[criterion for cavitation]} \qquad 20.36$$

The two forms of Eq. 20.35 yield slightly different results. The first form is essentially the ratio of the net pressure available for collapsing a vapor bubble to the velocity pressure creating the vapor. It is useful in model experiments. The second form is applicable to tests of production model pumps.

### 20. SUCTION SPECIFIC SPEED

The formula for *suction specific speed*, $n_{\mathrm{ss}}$, can be derived by substituting NPSHR for total head in the expression for specific speed. $Q$ and $\dot{V}$ are halved for double-suction pumps.

$$n_{\mathrm{ss}} = \frac{n\sqrt{\dot{V}}}{(\text{NPSHR in m})^{0.75}} \qquad \text{[SI]} \quad 20.37(a)$$

$$n_{\mathrm{ss}} = \frac{n\sqrt{Q}}{(\text{NPSHR in ft})^{0.75}} \qquad \text{[U.S.]} \quad 20.37(b)$$

Suction specific speed is an index of the suction characteristics of the impeller. Ideally, it should be approximately 8500 (165 in SI) for both single- and double-suction pumps. This assumes the pump is operating at or near its point of optimum efficiency.

Suction specific speed can be used to determine the maximum recommended operating speed by substituting 8500 (165 in SI) for $n_{\mathrm{ss}}$ in Eq. 20.37 and solving for $n$.

If the suction specific speed is known, it can be used to determine the NPSHR. If the pump is known to be operating at or near its optimum efficiency, an approximate NPSHR value can be found by substituting 8500 (165 in SI) for $n_{\mathrm{ss}}$ in Eq. 20.37 and solving for NPSHR.

*Suction specific speed available*, SA, is obtained when NPSHA is substituted for total head in the expression for specific speed. The suction specific speed available must be less than the suction specific speed required to prevent cavitation.[21]

### 21. PUMP PERFORMANCE CURVES

For a given impeller diameter and constant speed, the head added will decrease as the flow rate increases. This can be shown graphically on the *pump performance curve (pump curve)* supplied by the pump manufacturer. Other operating characteristics (e.g., power requirement, NPSHR, and efficiency) also vary with flow rate, and these are usually plotted on a common graph, as shown in Fig. 20.10.[22] Manufacturers' pump curves show performance over a limited number of calibration speeds. If an operating point is outside the range of published curves, the affinity laws can be used to estimate the speed at which the pump gives the required performance.

*Figure 20.10* Pump Performance Curves

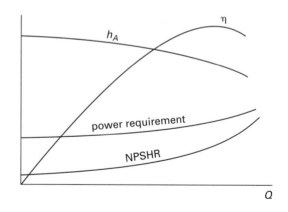

On the pump curve, the *shutoff point* (also known as *churn*) corresponds to a closed discharge valve (i.e., zero flow); the *rated point* is where the pump operates with rated 100% of capacity and head; the *overload point* corresponds to 65% of the rated head.

---

[21]Since speed and flow rate are constants, this is another way of saying NPSHA must equal or exceed NPSHR.

[22]The term *pump curve* is commonly used to designate the $h_A$ versus $Q$ characteristics, whereas *pump characteristics curve* implies all of the pump data.

Figure 20.10 is for a pump with a fixed impeller diameter and rotational speed. The characteristics of a pump operated over a range of speeds or for different impeller diameters are illustrated in Fig. 20.11.

**Figure 20.11** *Centrifugal Pump Characteristics Curves*

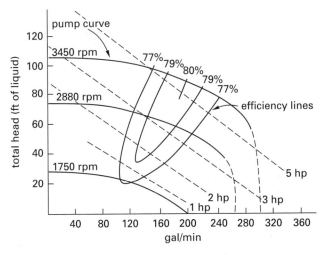

(a) variable speed

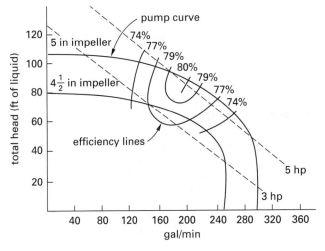

(b) variable impeller diameter

## 22. SYSTEM CURVES

A *system curve* (or *system performance curve*) is a plot of the static and friction energy losses experienced by the fluid for different flow rates. Unlike the pump curve, which depends only on the pump, the system curve depends only on the configuration of the suction and discharge lines. (The following equations assume equal pressures at the fluid source and destination surfaces, which is the case for pumping from one atmospheric reservoir to another. The velocity head is insignificant and is disregarded.)

$$h_A = h_z + h_f \qquad 20.38$$

$$h_z = h_{z(d)} - h_{z(s)} \qquad 20.39$$

$$h_f = h_{f(s)} + h_{f(d)} \qquad 20.40$$

If the fluid reservoirs are large, or if the fluid reservoir levels are continually replenished, the net static suction head $(h_{z(1)} - h_{z(2)})$ will be constant for all flow rates. The friction loss, $h_f$, varies with $v^2$ (and hence with $Q^2$) in the Darcy friction formula. This makes it easy to find friction losses for other flow rates (subscript 2) once one friction loss (subscript 1) is known.[23]

$$\frac{h_{f,1}}{h_{f,2}} = \left(\frac{Q_1}{Q_2}\right)^2 \qquad 20.41$$

Figure 20.12 illustrates a system curve with a negative suction head (i.e., a fluid source below the fluid destination). The system curve is shifted upward, intercepting the vertical axis at some positive value of $h_A$.

**Figure 20.12** *System Curve*

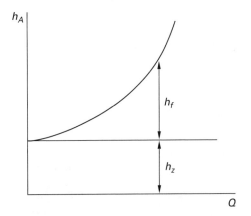

## 23. OPERATING POINT

The intersection of the pump curve and the system curve determines the *operating point*, as shown in Fig. 20.13. The operating point defines the system head and system flow rate.

When selecting a pump, the system curve is plotted on manufacturers' pump curves for different speeds and/or impeller diameters (i.e., Fig. 20.11). There will be several possible operating points corresponding to the various pump curves shown. Generally, the design operating point should be close to the highest pump efficiency. This, in turn, will determine speed and impeller diameter.

In many systems, the static head will vary as the source reservoir is drained or as the destination reservoir fills. The system head is then defined by a pair of matching

---

[23]Equation 20.41 implicitly assumes that the friction factor, $f$, is constant. This may be true over a limited range of flow rates, but it is not true over large ranges unless the Hazen-Williams friction loss equation is being used. Nevertheless, Eq. 20.41 is often used to quickly construct preliminary versions of the system curve.

**Figure 20.13** *Extreme Operating Points*

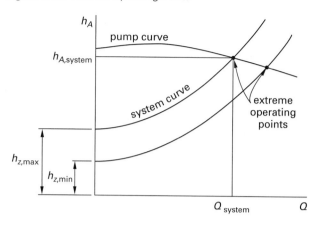

system friction curves intersecting the pump curve. The two intersection points are the *extreme operating points* —the maximum and minimum capacity requirements.

After a pump is installed, it may be desired to change the operating point. This can be done without replacing the pump by placing a throttling valve in the discharge line. The operating point can then be moved along the pump curve by partially opening or closing the valve, as illustrated in Fig. 20.14. (A throttling valve should never be placed in the suction line since that would reduce NPSHA.)

**Figure 20.14** *Throttling the Discharge*

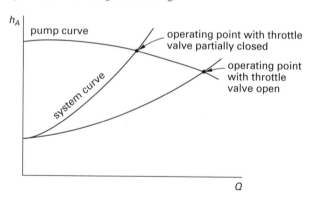

## 24. PUMPS IN PARALLEL

Parallel operation is obtained by having two pumps discharging into a common header. This type of connection is advantageous when the system demand varies greatly or when high reliability is required. A single pump providing total flow would have to operate far from its optimum efficiency at one point or another. With two pumps in parallel, one can be shut down during low demand. This allows the remaining pump to operate close to its optimum efficiency point.

Figure 20.15 illustrates that parallel operation increases the capacity of the system while maintaining the same total head.

**Figure 20.15** *Pumps Operating in Parallel*

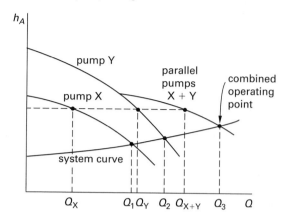

The performance curve for a set of pumps in parallel can be plotted by adding the capacities of the two pumps at various heads. Capacity does not increase at heads above the maximum head of the smaller pump.

Furthermore, a second pump will operate only when its discharge head is greater than the discharge head of the pump already running.

When the parallel performance curve is plotted with the system head curve, the operating point is the intersection of the system curve with the X + Y curve. With pump X operating alone, the capacity is given by $Q_1$. When pump Y is added, the capacity increases to $Q_3$ with a slight increase in total head.

## 25. PUMPS IN SERIES

Series operation is achieved by having one pump discharge into the suction of the next. This arrangement is used primarily to increase the discharge head, although a small increase in capacity also results. (See Fig. 20.16.)

The performance curve for a set of pumps in series can be plotted by adding the heads of the two pumps at various capacities.

**Figure 20.16** *Pumps Operating in Series*

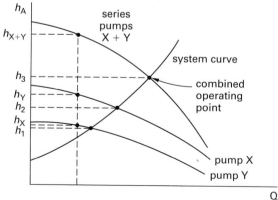

## 26. AFFINITY LAWS

Most parameters (impeller diameter, speed, and flow rate) determining a specific pump's performance can vary. If the impeller diameter is held constant and the speed is varied, the following ratios are maintained with no change in efficiency.

$$\frac{Q_2}{Q_1} = \frac{n_2}{n_1} \qquad \textit{20.42}$$

$$\frac{h_2}{h_1} = \left(\frac{n_2}{n_1}\right)^2 = \left(\frac{Q_2}{Q_1}\right)^2 \qquad \textit{20.43}$$

$$\frac{P_2}{P_1} = \left(\frac{n_2}{n_1}\right)^3 = \left(\frac{Q_2}{Q_1}\right)^3 \qquad \textit{20.44}$$

If the speed is held constant and the impeller size is varied,

$$\frac{Q_2}{Q_1} = \frac{D_2}{D_1} \qquad \textit{20.45}$$

$$\frac{h_2}{h_1} = \left(\frac{D_2}{D_1}\right)^2 \qquad \textit{20.46}$$

$$\frac{P_2}{P_1} = \left(\frac{D_2}{D_1}\right)^3 \qquad \textit{20.47}$$

The affinity laws are based on the assumption that the efficiencies of the two pumps are the same. In reality, larger pumps are somewhat more efficient than smaller pumps. Therefore, extrapolations to greatly different sized pumps should be avoided. Equation 20.48 can be used to estimate the efficiency of a different sized pump. The dimensionless exponent, $n$, varies from 0 to approximately 0.26, with 0.2 being a typical value.

$$\frac{1 - \eta_{\text{smaller}}}{1 - \eta_{\text{larger}}} = \left(\frac{D_{\text{larger}}}{D_{\text{smaller}}}\right)^n \qquad \textit{20.48}$$

### Example 20.9

A pump operating at 1770 rpm delivers 500 gal/min against a total head of 200 ft. Changes in the piping system have increased the total head to 375 ft. At what speed should this pump be operated to achieve this new head at the same efficiency?

*Solution*

From Eq. 20.43,

$$n_2 = n_1 \sqrt{\frac{h_2}{h_1}} = (1770 \text{ rpm}) \sqrt{\frac{375 \text{ ft}}{200 \text{ ft}}}$$

$$= 2424 \text{ rpm}$$

### Example 20.10

A pump is required to pump 500 gal/min while providing a total dynamic head of 425 ft. The hydraulic system has no static head change. Only the 1750 rpm performance curve is known for the pump. At what speed must the pump be turned to achieve the desired performance with no change in efficiency or impeller size?

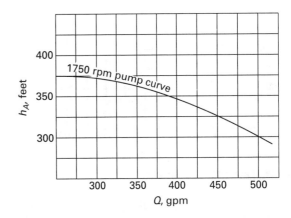

*Solution*

A flow of 500 gal/min with a head of 425 ft does not correspond to any point on the 1750 rpm curve.

From Eq. 20.43, the quantity $h/Q^2$ is constant.

$$\frac{h}{Q^2} = \frac{425 \text{ ft}}{\left(500 \ \dfrac{\text{gal}}{\text{min}}\right)^2} = 1.7 \times 10^{-3} \quad \text{[mixed units]}$$

In order to use the affinity laws, the operating point on the 1750 rpm curve must be determined. Random values of $Q$ are chosen and the corresponding values of $h$ are determined such that the ratio $h/Q^2$ is unchanged.

| $Q$ | $h$ |
|-----|-----|
| 475 | 383 |
| 450 | 344 |
| 425 | 307 |
| 400 | 272 |

These points are plotted and connected to draw the system curve. The intersection of the system and 1750 rpm pump curve at 440 gal/min defines the operating point at that speed. From Eq. 20.42,

$$n_2 = \frac{n_1 Q_2}{Q_1} = \frac{(1750 \text{ rpm})\left(500 \ \dfrac{\text{gal}}{\text{min}}\right)}{440 \ \dfrac{\text{gal}}{\text{min}}}$$

$$= 1989 \text{ rpm}$$

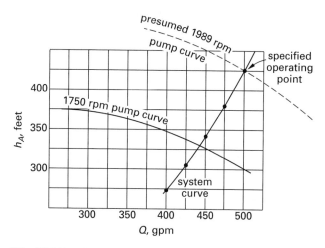

## 27. PUMP SIMILARITY

The performance of one pump can be used to predict the performance of a *dynamically similar (homologous) pump*. This can be done by using Eq. 20.49 through Eq. 20.54.

$$\frac{n_1 D_1}{\sqrt{h_1}} = \frac{n_2 D_2}{\sqrt{h_2}} \qquad 20.49$$

$$\frac{Q_1}{D_1^2 \sqrt{h_1}} = \frac{Q_2}{D_2^2 \sqrt{h_2}} \qquad 20.50$$

$$\frac{P_1}{\rho_1 D_1^2 h_1^{1.5}} = \frac{P_2}{\rho_2 D_2^2 h_2^{1.5}} \qquad 20.51$$

$$\frac{Q_1}{n_1 D_1^3} = \frac{Q_2}{n_2 D_2^3} \qquad 20.52$$

$$\frac{P_1}{\rho_1 n_1^3 D_1^5} = \frac{P_2}{\rho_2 n_2^3 D_2^5} \qquad 20.53$$

$$\frac{n_1 \sqrt{Q_1}}{h_1^{0.75}} = \frac{n_2 \sqrt{Q_2}}{h_2^{0.75}} \qquad 20.54$$

These *similarity laws* assume that both pumps:

- operate in the turbulent region
- have the same pump efficiency
- operate at the same percentage of wide-open flow

Similar pumps also will have the same specific speed and cavitation number.

As with the affinity laws, these relationships assume that the efficiencies of the larger and smaller pumps are the same. In reality, larger pumps will be more efficient than smaller pumps. Therefore, extrapolations to much larger or much smaller sizes should be avoided.

### Example 20.11

A 6 in pump operating at 1770 rpm discharges 1500 gal/min of cold water (SG = 1.0) against an 80 ft head at 85% efficiency. A homologous 8 in pump operating at 1170 rpm is being considered as a replacement.

(a) What total head and capacity can be expected from the new pump? (b) What would be the new horsepower requirement?

*Solution*

(a) From Eq. 20.49,

$$h_2 = \left(\frac{D_2 n_2}{D_1 n_1}\right)^2 h_1 = \left(\frac{(8 \text{ in})(1170 \text{ rpm})}{(6 \text{ in})(1770 \text{ rpm})}\right)^2 (80 \text{ ft})$$
$$= 62.14 \text{ ft}$$

From Eq. 20.52,

$$Q_2 = \left(\frac{n_2 D_2^3}{n_1 D_1^3}\right) Q_1$$
$$= \left(\frac{(1170 \text{ rpm})(8 \text{ in})^3}{(1770 \text{ rpm})(6 \text{ in})^3}\right) \left(1500 \frac{\text{gal}}{\text{min}}\right)$$
$$= 2350.3 \text{ gal/min}$$

(b) From Table 20.5, the hydraulic horsepower is

$$\text{WHP}_2 = \frac{h_2 Q_2 (\text{SG})}{3956} = \frac{(62.14 \text{ ft})\left(2350.3 \frac{\text{gal}}{\text{min}}\right)(1.0)}{3956 \frac{\text{ft-gal}}{\text{hp-min}}}$$
$$= 36.92 \text{ hp}$$

From Eq. 20.48,

$$\eta_{\text{larger}} = 1 - \frac{1 - \eta_{\text{smaller}}}{\left(\frac{D_{\text{larger}}}{D_{\text{smaller}}}\right)^{0.2}}$$
$$= 1 - \frac{1 - 0.85}{\left(\frac{8 \text{ in}}{6 \text{ in}}\right)^{0.2}} = 0.858$$

$$\text{BHP}_2 = \frac{\text{WHP}_2}{\eta_p} = \frac{36.92 \text{ hp}}{0.858} = 43.0 \text{ hp}$$

## 28. PUMPING LIQUIDS OTHER THAN COLD WATER

Many pump parameters are determined from tests with cold, clear water at 85°F (29°C). The following guidelines can be used when pumping water at other temperatures or when pumping other fluids.

- Head developed is independent of the liquid's specific gravity. Pump performance curves from tests with water can be used with other Newtonian fluids (e.g., gasoline, alcohol, and aqueous solutions) having similar viscosities.

- Head, flow rate, and efficiency are all reduced when pumping highly viscous non-Newtonian fluids. No exact method exists for determining the reduction factors, other than actual tests of an installation using both fluids. Some sources have published charts of correction factors based on tests over limited viscosity and size ranges.[24]

- The hydraulic horsepower depends on the specific gravity of the fluid. If the pump characteristic curve is used to find the operating point, multiply the horsepower reading by the specific gravity. Table 20.5 and Table 20.6 incorporate the specific gravity term in the calculation of hydraulic power where required.

- Efficiency is not affected by changes in temperature that cause only the specific gravity to change.

- Efficiency is nominally affected by changes in temperature that cause the viscosity to change. Equation 20.55 is an approximate relationship suggested by the Hydraulics Institute when extrapolating the efficiency (in decimal form) from cold water to hot water. $n$ is an experimental exponent established by the pump manufacturer, generally in the range of 0.05 to 0.1.

$$\eta_{\mathrm{hot}} = 1 - (1 - \eta_{\mathrm{cold}}) \left( \frac{\nu_{\mathrm{hot}}}{\nu_{\mathrm{cold}}} \right)^{n} \qquad 20.55$$

- NPSHR is not significantly affected by minor variations in the water temperature.

- When hydrocarbons are pumped, the NPSHR determined from cold water can usually be reduced. This reduction is apparently due to the slow vapor release of complex organic liquids. If the hydrocarbon's vapor pressure at the pumping temperature is known, Fig. 20.17 will give the percentage of the cold-water NPSHR.

- Pumping many fluids requires expertise that goes far beyond simply extrapolating parameters in proportion to the fluid's specific gravity. Such special cases include pumping liquids containing abrasives, liquids that solidify, highly corrosive liquids, liquids with vapor or gas, highly viscous fluids, paper stock, and hazardous fluids.

### Example 20.12

A centrifugal pump has an NIPR (NPSHR) of 12 psi based on cold water. 10°F isobutane has a specific gravity of 0.60 and a vapor pressure of 15 psia. What NPSHR should be used with 10°F liquid isobutane?

*Solution*

From Fig. 20.17, the intersection of a specific gravity of 0.60 and 15 psia is above the horizontal 100% line. The full NIPR of 12 psi should be used.

---

[24]A chart published by the Hydraulics Institute is widely distributed.

*Figure 20.17* Hydrocarbon NPSHR Correction Factor

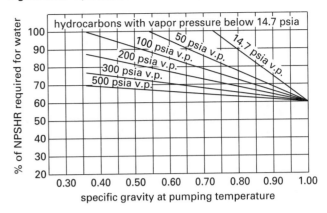

### 29. HYDROELECTRIC GENERATING PLANTS

In a typical hydroelectric generating plant using reaction turbines (see Sec. 20.32), the turbine is generally housed in a *powerhouse*, with water conducted to the turbine through the *penstock* piping. Water originates in a reservoir, dam, or *forebay* (in the instance where the reservoir is a long distance from the turbine). (See Fig. 20.18.)

*Figure 20.18* Typical Hydroelectric Plant

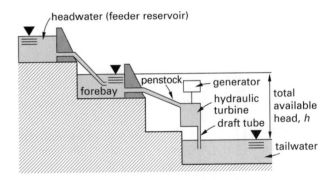

After the water passes through the turbine, it is discharged through the draft tube to the receiving reservoir, known as the *tailwater*. The *draft tube* is used to keep the turbine up to 15 ft (5 m) above the tailwater surface, while still being able to extract the total available head. If a draft tube is not employed, water may be returned to the tailwater by way of a channel known as the *tail race*. The turbine, draft tube, and all related parts comprise what is known as the *setting*.

When a forebay is not part of the generating plant's design, it will be desirable to provide a *surge chamber* in order to relieve the effects of rapid changes in flow rate. In the case of a sudden power demand, the surge chamber would provide an immediate source of water, without waiting for a contribution from the feeder reservoir.

Similarly, in the case of a sudden decrease in discharge through the turbine, the excess water would surge back into the surge chamber.

## 30. TURBINE SPECIFIC SPEED

Like centrifugal pumps, turbines are classified according to the manner in which the impeller extracts energy from the fluid flow. This is measured by the turbine-specific speed equation, which is different from the equation used to calculate specific speed for pumps.

$$n_s = \frac{n\sqrt{P \text{ in kW}}}{h_t^{1.25}} \qquad \text{[SI]} \quad 20.56(a)$$

$$n_s = \frac{n\sqrt{P \text{ in hp}}}{h_t^{1.25}} \qquad \text{[U.S.]} \quad 20.56(b)$$

## 31. IMPULSE TURBINES

An *impulse turbine* consists of a rotating shaft (called a *turbine runner*) on which buckets or blades are mounted. (This is commonly called a *Pelton wheel*.)[25] A jet of water (or other fluid) hits the buckets and causes the turbine to rotate. The kinetic energy of the jet is converted into rotational kinetic energy. The jet is essentially at atmospheric pressure. (See Fig. 20.19.)

**Figure 20.19** *Impulse Turbine Installation*

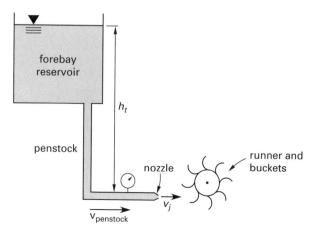

Impulse turbines are generally employed where the available head is very high, above 800 ft to 1600 ft (250 m to 500 m). (There is no exact value for the limiting head, hence the range. What is important is that impulse turbines are *high-head turbines*.)

The total available head in this installation is $h_t$, but not all of this energy can be extracted. Some of the energy is lost to friction in the penstock. Minor losses also occur, but these small losses are usually disregarded. In the penstock, immediately before entering the nozzle, the

remaining head is divided between the pressure head and the velocity head.[26]

$$h' = h_t - h_f = (h_p + h_v)_{\text{penstock}} \qquad 20.57$$

Another loss, $h_n$, occurs in the nozzle itself. The head remaining to turn the turbine is

$$\begin{aligned} h'' &= h' - h_n \\ &= h_t - h_f - h_n \end{aligned} \qquad 20.58$$

$h_f$ is calculated from either the Darcy or the Hazen-Williams equation. The nozzle loss is calculated from the *nozzle coefficient*, $C_v$.

$$h_n = h'(1 - C_v^2) \qquad 20.59$$

The total head at the nozzle exit is converted into velocity head according to the Torricelli equation.

$$v_j = \sqrt{2gh''} = C_v\sqrt{2gh'} \qquad 20.60$$

In order to maximize power output, buckets are usually designed to partially reverse the direction of the water jet flow. The forces on the turbine buckets can be found from the impulse-momentum equations. If the water is turned through an angle $\theta$ and the wheel's *tangential velocity* is $v_b$, the energy transmitted by each unit mass of water to the turbine runner is[27]

$$E = v_b(v_j - v_b)(1 - \cos\theta) \qquad \text{[SI]} \quad 20.61(a)$$

$$E = \left(\frac{v_b(v_j - v_b)}{g_c}\right)(1 - \cos\theta) \qquad \text{[U.S.]} \quad 20.61(b)$$

$$v_b = \frac{2\pi n r}{60} = \omega r \qquad 20.62$$

The theoretical *turbine power* is found by multiplying Eq. 20.61 by the mass flow rate. The actual power will be less than the theoretical output. Efficiencies are in the range of 80% to 90%, with the higher efficiencies being associated with turbines having two or more jets per runner. For a mass flow rate in kg/s, the theoretical power (in kilowatts) will be as shown in Eq. 20.63(a). For a mass flow rate in lbm/sec, the theoretical horsepower will be as shown in Eq. 20.63(b).

$$P_{\text{th}} = \frac{\dot{m}E}{1000} \qquad \text{[SI]} \quad 20.63(a)$$

$$P_{\text{th}} = \frac{\dot{m}E}{550} \qquad \text{[U.S.]} \quad 20.63(b)$$

---

[25]In a Pelton wheel turbine, the spoon-shaped buckets are divided into two halves, with a ridge between the halves. Half of the water is thrown to each side of the bucket. A Pelton wheel is known as a *tangential turbine (tangential wheel)* because the centerline of the jet is directed at the centers of the buckets.

[26]Care must be taken to distinguish between the conditions existing in the penstock, the nozzle throat, and the jet itself. The velocity in the nozzle throat and jet will be the same, but this is different from the penstock velocity. Similarly, the pressure in the jet is zero, although it is nonzero in the penstock.

[27]$\theta = 180°$ would be ideal. However, the actual angle is limited to approximately 165° to keep the deflected jet out of the way of the incoming jet.

## Example 20.13

A Pelton wheel impulse turbine develops 100 hp (brake) while turning at 500 rpm. The water is supplied from a penstock with an internal area of 0.3474 ft$^2$. The water subsequently enters a nozzle with a reduced flow area. The total head is 200 ft before nozzle loss. The turbine efficiency is 80%, and the nozzle coefficient, $C_v$, is 0.95. Disregard penstock friction losses. What are the (a) flow rate (in ft$^3$/sec), (b) area of the jet, and (c) pressure head in the penstock just before the nozzle?

*Solution*

(a) From Eq. 20.60 with $h' = h_t = 200$ ft, the jet velocity is

$$v_j = C_v\sqrt{2gh'} = 0.95\sqrt{(2)\left(32.2 \frac{\text{ft}}{\text{sec}^2}\right)(200 \text{ ft})}$$

$$= 107.8 \text{ ft/sec}$$

From Eq. 20.59, the nozzle loss is

$$h_n = h'(1 - C_v^2) = (200 \text{ ft})(1 - (0.95)^2)$$

$$= 19.5 \text{ ft}$$

From Table 20.5, the flow rate is

$$\dot{V} = \frac{8.814P}{h_A(\text{SG})\eta} = \frac{(8.814)(100 \text{ hp})}{(200 \text{ ft} - 19.5 \text{ ft})(1)(0.8)}$$

$$= 6.104 \text{ ft}^3/\text{sec}$$

(b) The jet area is

$$A_j = \frac{\dot{V}}{v_j} = \frac{6.104 \frac{\text{ft}^3}{\text{sec}}}{107.8 \frac{\text{ft}}{\text{sec}}} = 0.0566 \text{ ft}^2$$

(c) The velocity in the penstock is

$$v_{\text{penstock}} = \frac{\dot{V}}{A} = \frac{6.104 \frac{\text{ft}^3}{\text{sec}}}{0.3474 \text{ ft}^2}$$

$$= 17.57 \text{ ft/sec}$$

The pressure head in the penstock is

$$h_p = h' - h_v = h' - \frac{v^2}{2g}$$

$$= 200 \text{ ft} - \frac{\left(17.57 \frac{\text{ft}}{\text{sec}}\right)^2}{(2)\left(32.2 \frac{\text{ft}}{\text{sec}^2}\right)}$$

$$= 195.2 \text{ ft}$$

## 32. REACTION TURBINES

*Reaction turbines* (also known as *Francis turbines* or *radial-flow turbines*) are essentially centrifugal pumps operating in reverse. (See Fig. 20.20.) They are used when the total available head is small, typically below 600 ft to 800 ft (183 m to 244 m). However, their energy conversion efficiency is higher than that of impulse turbines, typically in the 85% to 95% range.

**Figure 20.20** *Reaction Turbine*

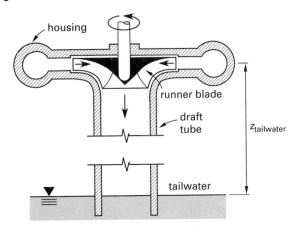

In a reaction turbine, water enters the turbine housing with a pressure greater than atmospheric pressure. The water completely surrounds the turbine runner (impeller) and continues through the draft tube. There is no vacuum or air pocket between the turbine and the tailwater.

All of the power, affinity, and similarity relationships used with centrifugal pumps can be used with reaction turbines.

## Example 20.14

A reaction turbine with a draft tube develops 500 hp (brake) when 50 ft$^3$/sec water flow through it. Water enters the turbine at 20 ft/sec with a 100 ft pressure head. The elevation of the turbine above the tailwater level is 10 ft. Disregarding friction, what are the (a) total available head and (b) turbine efficiency?

*Solution*

(a) The available head is the difference between the forebay and tailwater elevations. The tailwater depression is known, but the height of the forebay above the turbine is not known. At the turbine entrance, this unknown potential energy has been converted to pressure and velocity head. Therefore, the total available head (exclusive of friction) is

$$h_t = z_{\text{forebay}} - z_{\text{tailwater}}$$

$$= h_p + h_{\text{v}} - z_{\text{tailwater}}$$

$$= 100 \text{ ft} + \frac{\left(20 \; \dfrac{\text{ft}}{\text{sec}}\right)^2}{(2)\left(32.2 \; \dfrac{\text{ft}}{\text{sec}^2}\right)} - (-10 \text{ ft})$$

$$= 116.2 \text{ ft}$$

(b) From Table 20.5, the theoretical hydraulic horsepower is

$$P_{\text{th}} = \frac{h_A \, \dot{V}(\text{SG})}{8.814} = \frac{(116.2 \text{ ft})\left(50 \; \dfrac{\text{ft}^3}{\text{sec}}\right)(1.0)}{8.814}$$

$$= 659.2 \text{ hp}$$

The efficiency of the turbine is

$$\eta = \frac{P_{\text{brake}}}{P_{\text{th}}} = \frac{500 \text{ hp}}{659.2 \text{ hp}} = 0.758 \quad (75.8\%)$$

## 33. TYPES OF REACTION TURBINES

Each of the three types of turbines is associated with a range of specific speeds.

- *Axial-flow reaction turbines* (also known as *propeller turbines*) are used for low heads, high rotational speeds, and large flow rates. (See Fig. 20.21.) These propeller turbines operate with specific speeds in the

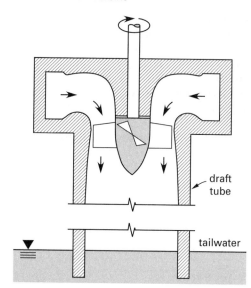

**Figure 20.21** *Axial-Flow Turbine*

70 to 260 range (266 to 988 in SI). Their best efficiencies, however, are produced with specific speeds between 120 and 160 (460 and 610 in SI).

- For *mixed-flow reaction turbines*, the specific speed varies from 10 to 90 (38 and 342 in SI). Best efficiencies are found in the 40 to 60 (150 and 230 in SI) range with heads below 600 to 800 ft (180 to 240 m).

- *Radial-flow reaction turbines* have the lowest flow rates and specific speeds but are used when heads are high. These turbines have specific speeds between 1 and 20 (3.8 and 76 in SI).

# 21 Open Channel Flow

## Nomenclature

| | | | |
|---|---|---|---|
| $A$ | area | $ft^2$ | $m^2$ |
| $b$ | weir or channel width | ft | m |
| $C$ | coefficient | $ft^{1/2}/sec$ | $m^{1/2}/s$ |
| $d$ | depth of flow | ft | m |
| $d$ | diameter | in | m |
| $D$ | diameter | ft | m |
| $E$ | specific energy | ft | m |

| | | | |
|---|---|---|---|
| Fr | Froude number | – | – |
| $g$ | acceleration due to gravity | $ft/sec^2$ | $m/s^2$ |
| $g_c$ | gravitational constant | $ft\text{-}lbm/lbf\text{-}sec^2$ | n.a. |
| $h$ | head | ft | m |
| $H$ | total hydraulic head | ft | m |
| $k$ | minor loss coefficient | – | – |
| $K$ | conveyance | $ft^3/sec$ | $m^3/s$ |
| $K'$ | modified conveyance | – | – |
| $L$ | channel length | ft | m |
| $m$ | cotangent of side slope angle | – | – |
| $n$ | Manning roughness coefficient | – | – |
| $N$ | number of end contractions | – | – |
| $p$ | pressure | $lbf/ft^2$ | Pa |
| $P$ | wetted perimeter | ft | m |
| $Q$ | flow quantity | $ft^3/sec$ | $m^3/s$ |
| $R$ | hydraulic radius | ft | m |
| $S$ | slope of energy line (energy gradient) | – | – |
| $S_0$ | channel slope | – | – |
| $T$ | width of surface | ft | m |
| v | velocity | ft/sec | m/s |
| $w$ | channel width | ft | m |
| $x$ | distance | ft | m |
| $Y$ | weir height | ft | m |
| $z$ | height above datum | ft | m |

## Symbols

| | | | |
|---|---|---|---|
| $\alpha$ | velocity-head coefficient | – | – |
| $\gamma$ | specific weight | $lbf/ft^3$ | n.a. |
| $\rho$ | density | $lbm/ft^3$ | $kg/m^3$ |
| $\theta$ | angle | deg | deg |

## Subscripts

| | |
|---|---|
| $b$ | brink |
| $c$ | critical or composite |
| $d$ | discharge |
| $e$ | entrance |
| $f$ | friction |
| $h$ | hydraulic |
| $n$ | normal |
| $o$ | channel or culvert barrel |
| $s$ | spillway |
| $t$ | total |
| $w$ | weir |

## 1. INTRODUCTION

An *open channel* is a fluid passageway that allows part of the fluid to be exposed to the atmosphere. This type of channel includes natural waterways, canals, culverts, flumes, and pipes flowing under the influence of gravity (as opposed to pressure conduits, which always flow full). A *reach* is a straight section of open channel with uniform shape, depth, slope, and flow quantity.

There are difficulties in evaluating open channel flow. The unlimited geometric cross sections and variations in roughness have contributed to a relatively small number of scientific observations upon which to estimate the required coefficients and exponents. Therefore, the analysis of open channel flow is more empirical and less exact than that of pressure conduit flow. This lack of precision, however, is more than offset by the percentage error in runoff calculations that generally precede the channel calculations.

Flow can be categorized on the basis of the channel material, for example, concrete or metal pipe or earth material. Except for a short discussion of erodible canals in Sec. 21.36, this chapter assumes the channel is non-erodible.

## 2. TYPES OF FLOW

Flow in open channels is almost always turbulent; laminar flow will occur only in very shallow channels or at very low fluid velocities. However, within the turbulent category are many somewhat confusing categories of flow. Flow can be a function of time and location. If the flow quantity (volume per unit of time across an area in flow) is invariant, it is said to be *steady flow*. (Flow that varies with time, such as stream flow during a storm, known as *varied flow*, is not covered in this chapter.) If the flow cross section does not depend on the location along the channel, it is said to be *uniform flow*. Steady flow can also be *nonuniform flow*, as in the case of a river with a varying cross section or on a steep slope. Furthermore, uniform channel construction does not ensure uniform flow, as will be seen in the case of hydraulic jumps.

Table 21.1 summarizes some of the more common categories and names of steady open channel flow. All of the subcategories are based on variations in depth and flow area with respect to location along the channel.

**Table 21.1** *Categories of Steady Open Channel Flow*

   *subcritical flow (tranquil flow)*
      uniform flow
         normal flow
      nonuniform flow
         accelerating flow
         decelerating flow (retarded flow)
   *critical flow*
   *supercritical flow (rapid flow, shooting flow)*
      uniform flow
         normal flow
      nonuniform flow
         accelerating flow
         decelerating flow

## 3. MINIMUM VELOCITIES

The minimum permissible velocity in a sewer or other nonerodible channel is the lowest that prevents sedimentation and plant growth. Velocities of 2 ft/sec to 3 ft/sec (0.6 m/s to 0.9 m/s) keep all but the heaviest silts in suspension. 2.5 ft/sec (0.75 m/s) is considered the minimum to prevent plant growth.

## 4. VELOCITY DISTRIBUTION

Due to the adhesion between the wetted surface of the channel and the water, the velocity will not be uniform across the area in flow. The velocity term used in this chapter is the *mean velocity*. The mean velocity, when multiplied by the flow area, gives the flow quantity.

$$Q = A\mathrm{v} \qquad\qquad 21.1$$

The location of the mean velocity depends on the distribution of velocities in the waterway, which is generally quite complex. The procedure for measuring the velocity of a channel (called *stream gauging*) involves measuring the average channel velocity at multiple locations across the channel width. These subaverage velocities are averaged to give a grand average (mean) flow velocity. (See Fig. 21.1.)

**Figure 21.1** *Velocity Distribution in an Open Channel*

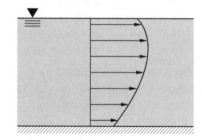

## 5. PARAMETERS USED IN OPEN CHANNEL FLOW

The *hydraulic radius* is the ratio of the area in flow to the wetted perimeter.[1]

$$R = \frac{A}{P} \qquad\qquad 21.2$$

For a circular channel flowing either full or half-full, the hydraulic radius is one-fourth of the *hydraulic diameter*, $D_h/4$. The hydraulic radii of other channel shapes are easily calculated from the basic definition. Table 21.2 summarizes parameters for the basic shapes. For very wide channels such as rivers, the hydraulic radius is approximately equal to the depth.

---

[1]The hydraulic radius is also referred to as the *hydraulic mean depth*. However, this name is easily confused with "mean depth" and "hydraulic depth," both of which have different meanings. Therefore, the term "hydraulic mean depth" is not used in this chapter.

**Table 21.2** *Hydraulic Parameters of Basic Channel Sections*

| section | area, $A$ | wetted perimeter, $P$ | hydraulic radius, $R$ |
|---|---|---|---|
| rectangle | $dw$ | $2d + w$ | $\dfrac{dw}{w + 2d}$ |
| trapezoid | $\left(b + \dfrac{d}{\tan\theta}\right)d$ | $b + 2\left(\dfrac{d}{\sin\theta}\right)$ | $\dfrac{bd\sin\theta + d^2\cos\theta}{b\sin\theta + 2d}$ |
| triangle | $\dfrac{d^2}{\tan\theta}$ | $\dfrac{2d}{\sin\theta}$ | $\dfrac{d\cos\theta}{2}$ |
| circle | $\frac{1}{8}(\theta - \sin\theta)D^2$ [$\theta$ in radians] | $\frac{1}{2}\theta D$ [$\theta$ in radians] | $\frac{1}{4}\left(1 - \dfrac{\sin\theta}{\theta}\right)D$ [$\theta$ in radians] |

The *hydraulic depth* is the ratio of the area in flow to the width of the channel at the fluid surface.[2]

$$D_h = \frac{A}{w} \qquad 21.3$$

The uniform flow *section factor* represents a frequently occurring variable group. The section factor is often evaluated against depth of flow when working with discharge from irregular cross sections.

$$\text{section factor} = AR^{2/3} \quad \text{[general uniform flow]} \qquad 21.4$$

$$\text{section factor} = A\sqrt{D_h} \quad \text{[critical flow only]} \qquad 21.5$$

The *slope*, $S$, is the gradient of the energy line. In general, the slope can be calculated from the Bernoulli equation as the energy loss per unit length of channel. For small slopes typical of almost all natural waterways,

the channel length and horizontal run are essentially identical.

$$S = \frac{dE}{dL} \qquad 21.6$$

If the flow is uniform, the slope of the energy line will parallel the water surface and channel bottom, and the *energy gradient* will equal the *geometric slope*, $S_0$.

$$S_0 = \frac{\Delta z}{L} = S \quad \text{[uniform flow]} \qquad 21.7$$

Any open channel performance equation can be written using the geometric slope, $S_0$, instead of the hydraulic slope, $S$, but only under the condition of uniform flow.

In most problems, the slope is a function of the terrain and is known. However, it may be necessary to calculate the slope that results in some other specific parameter. The slope that produces flow at some normal depth, $d$, is called the *normal slope*. The slope that produces flow at some critical depth, $d_c$, is called the *critical slope*. Both are determined by solving the Manning equation for slope.

---

[2]For a rectangular channel, $D_h = d$.

## 6. GOVERNING EQUATIONS FOR UNIFORM FLOW

Since water is incompressible, the continuity equation is

$$A_1 v_1 = A_2 v_2 \qquad 21.8$$

The most common equation used to calculate the flow velocity in open channels is the 1768 *Chezy equation*.[3]

$$v = C\sqrt{RS} \qquad 21.9$$

Various methods for evaluating the *Chezy coefficient, C,* or "Chezy's C," have been proposed.[4] If the channel is small and very smooth, Chezy's own formula can be used. The friction factor, *f*, is dependent on the Reynolds number and can be found in the usual manner from the Moody diagram.

$$C = \sqrt{\frac{8g}{f}} \qquad 21.10$$

If the channel is large and the flow is fully turbulent, the friction loss will not depend so much on the Reynolds number as on the channel roughness. The 1888 Manning formula is frequently used to evaluate the constant $C$.[5] The value of $C$ depends only on the channel roughness and geometry. (The conversion constant 1.49 in Eq. 21.11(b) is reported as 1.486 by some authorities. 1.486 is the correct SI-to-English conversion, but it is doubtful whether this equation warrants four significant digits.)

$$C = \left(\frac{1.00}{n}\right) R^{1/6} \qquad \text{[SI]} \quad 21.11(a)$$

$$C = \left(\frac{1.49}{n}\right) R^{1/6} \qquad \text{[U.S.]} \quad 21.11(b)$$

$n$ is the *Manning roughness coefficient (Manning constant)*. Typical values of Manning's $n$ are given in App. 21.A. Judgment is needed in selecting values since tabulated values often differ by as much as 30%. More important to recognize for sewer work is the layer of slime that often coats the sewer walls. Since the slime characteristics can change with location in the sewer, there can be variations in Manning's roughness coefficient along the sewer length.

---

[3]Pronounced "Shay'-zee." This equation does not appear to be dimensionally consistent. However, the coefficient $C$ is not a pure number. Rather, it has units of $(\text{length})^{1/2}/\text{time}$ (i.e., $(\text{acceleration})^{1/2}$).

[4]Other methods of evaluating $C$ include the *Kutter equation* (also known as the *G.K. formula*) and the *Bazin formula*. These methods are interesting from a historical viewpoint, but both have been replaced by the Manning equation.

[5]This equation was originally proposed in 1868 by Gaukler and again in 1881 by Hagen, both working independently. For some reason, the Frenchman Flamant attributed the equation to an Irishman, R. Manning. In Europe and many other places, the Manning equation may be known as the *Strickler equation*.

Combining Eq. 21.9 and Eq. 21.11 produces the *Manning equation*, also known as the *Chezy-Manning equation*.

$$v = \left(\frac{1.00}{n}\right) R^{2/3}\sqrt{S} \qquad \text{[SI]} \quad 21.12(a)$$

$$v = \left(\frac{1.49}{n}\right) R^{2/3}\sqrt{S} \qquad \text{[U.S.]} \quad 21.12(b)$$

All of the coefficients and constants in the Manning equation may be combined into the *conveyance, K*.

$$Q = vA = \left(\frac{1.00}{n}\right) A R^{2/3}\sqrt{S}$$
$$= K\sqrt{S} \qquad \text{[SI]} \quad 21.13(a)$$

$$Q = vA = \left(\frac{1.49}{n}\right) A R^{2/3}\sqrt{S}$$
$$= K\sqrt{S} \qquad \text{[U.S.]} \quad 21.13(b)$$

### Example 21.1

A rectangular channel on a 0.002 slope is contructed of finished concrete. The channel is 8 ft (2.4 m) wide. Water flows at a depth of 5 ft (1.5 m). What is the flow rate?

*SI Solution*

The hydraulic radius is

$$R = \frac{A}{P} = \frac{(2.4 \text{ m})(1.5 \text{ m})}{1.5 \text{ m} + 2.4 \text{ m} + 1.5 \text{ m}}$$
$$= 0.67 \text{ m}$$

From App. 21.A, the roughness coefficient for finished concrete is 0.012. The Manning coefficient is determined by Eq. 21.11(a).

$$C = \left(\frac{1.00}{n}\right) R^{1/6} = \left(\frac{1.00}{0.012}\right)(0.67 \text{ m})^{1/6}$$
$$= 77.9$$

The discharge is

$$Q = vA = C\sqrt{RS}\,A$$
$$= \left(77.9 \, \frac{\sqrt{\text{m}}}{\text{s}}\right)\left(\sqrt{(0.67 \text{ m})(0.002)}\right)(1.5 \text{ m})(2.4 \text{ m})$$
$$= 10.3 \text{ m}^3/\text{s}$$

*Customary U.S. Solution*

The hydraulic radius is

$$R = \frac{A}{P} = \frac{(8 \text{ ft})(5 \text{ ft})}{5 \text{ ft} + 8 \text{ ft} + 5 \text{ ft}}$$
$$= 2.22 \text{ ft}$$

From App. 21.A, the roughness coefficient for finished concrete is 0.012. The Manning coefficient is determined by Eq. 21.11(b).

$$C = \left(\frac{1.49}{n}\right) R^{1/6}$$
$$= \left(\frac{1.49}{0.012}\right)(2.22 \text{ ft})^{1/6}$$
$$= 141.8$$

The discharge is

$$Q = vA = C\sqrt{RS}A$$
$$= \left(141.8 \ \frac{\sqrt{\text{ft}}}{\text{sec}}\right)\left(\sqrt{(2.22 \text{ ft})(0.002)}\right)(8 \text{ ft})(5 \text{ ft})$$
$$= 377.9 \text{ ft}^3/\text{sec}$$

## 7. VARIATIONS IN THE MANNING CONSTANT

The value of $n$ also depends on the depth of flow, leading to a value ($n_{\text{full}}$) specifically intended for use with full flow. (It is seldom clear from tabulations, such as App. 21.A, whether the values are for full flow or general use.) The variation in $n$ can be taken into consideration using *Camp's correction*, shown in App. 21.C. However, this degree of sophistication cannot be incorporated into an analysis problem unless a specific value of $n$ is known for a specific depth of flow.

For most calculations, however, $n$ is assumed to be constant. The accuracy of other parameters used in open-flow calculations often does not warrant considering the variation of $n$ with depth, and the choice to use a constant or varying $n$-value is left to the individual designer.

If it is desired to acknowledge variations in $n$ with respect to depth, it is expedient to use tables or graphs of hydraulic elements prepared for that purpose. Table 21.3 lists such hydraulic elements under the assumption that $n$ varies. (Appendix 21.C can be used for both varying and constant $n$.)

**Table 21.3** Circular Channel Ratios (varying n)

| $\dfrac{d}{D}$ | $\dfrac{Q}{Q_{\text{full}}}$ | $\dfrac{v}{v_{\text{full}}}$ |
|---|---|---|
| 0.1 | 0.02 | 0.31 |
| 0.2 | 0.07 | 0.48 |
| 0.3 | 0.14 | 0.61 |
| 0.4 | 0.26 | 0.71 |
| 0.5 | 0.41 | 0.80 |
| 0.6 | 0.56 | 0.88 |
| 0.7 | 0.72 | 0.95 |
| 0.8 | 0.87 | 1.01 |
| 0.9 | 0.99 | 1.04 |
| 0.95 | 1.02 | 1.03 |
| 1.00 | 1.00 | 1.00 |

### Example 21.2

2.5 ft³/sec (0.07 m³/s) of water flow in a 20 in (0.5 m) sewer line ($n = 0.015$, $S = 0.001$). The Manning coefficient, $n$, varies with depth. Flow is uniform and steady. What are the velocity and depth?

*SI Solution*

The hydraulic radius is

$$R = \frac{D}{4} = \frac{0.5 \text{ m}}{4} = 0.125 \text{ m}$$

From Eq. 21.12(a),

$$v_{\text{full}} = \left(\frac{1.00}{n}\right) R^{2/3}\sqrt{S}$$
$$= \left(\frac{1.00}{0.015}\right)(0.125 \text{ m})^{2/3}\sqrt{0.001}$$
$$= 0.53 \text{ m/s}$$

If the pipe were flowing full, it would carry $Q_{\text{full}}$.

$$Q_{\text{full}} = v_{\text{full}} A$$
$$= \left(0.53 \ \frac{\text{m}}{\text{s}}\right)\left(\frac{\pi}{4}\right)(0.5 \text{ m})^2$$
$$= 0.10 \text{ m}^3/\text{s}$$

$$\frac{Q}{Q_{\text{full}}} = \frac{0.07 \ \frac{\text{m}^3}{\text{s}}}{0.10 \ \frac{\text{m}^3}{\text{s}}} = 0.7$$

From App. 21.C, $d/D = 0.68$ and $v/v_{\text{full}} = 0.94$.

$$v = (0.94)\left(0.53 \ \frac{\text{m}}{\text{s}}\right) = 0.50 \text{ m/s}$$
$$d = (0.68)(0.5 \text{ m}) = 0.34 \text{ m}$$

*Customary U.S. Solution*

The hydraulic radius is

$$R = \frac{D}{4} = \frac{\dfrac{20 \text{ in}}{12 \ \frac{\text{in}}{\text{ft}}}}{4} = 0.417 \text{ ft}$$

From Eq. 21.12(b),

$$v_{\text{full}} = \left(\frac{1.49}{n}\right) R^{2/3}\sqrt{S}$$
$$= \left(\frac{1.49}{0.015}\right)(0.417 \text{ ft})^{2/3}\sqrt{0.001}$$
$$= 1.75 \text{ ft/sec}$$

Fluids

If the pipe were flowing full, it would carry $Q_{full}$.

$$Q_{full} = v_{full}A$$

$$= \left(1.75 \ \frac{ft}{sec}\right)\left(\frac{\pi}{4}\right)\left(\frac{20 \ in}{12 \ \frac{in}{ft}}\right)^2$$

$$= 3.83 \ ft^3/sec$$

$$\frac{Q}{Q_{full}} = \frac{2.5 \ \frac{ft^3}{sec}}{3.83 \ \frac{ft^3}{sec}} = 0.65$$

From App. 21.C, $d/D = 0.66$ and $v/v_{full} = 0.92$.

$$v = (0.92)\left(1.75 \ \frac{ft}{sec}\right) = 1.61 \ ft/sec$$

$$d = (0.66)(20 \ in) = 13.2 \ in$$

## 8. HAZEN-WILLIAMS VELOCITY

The empirical Hazen-Williams open channel velocity equation was developed in the early 1920s. It is still occasionally used in the United States for sizing gravity sewers. It is applicable to water flows at reasonably high Reynolds numbers and is based on sound dimensional analysis. However, the constants and exponents were developed experimentally.

The equation uses the Hazen-Williams constant, $C$, to characterize the roughness of the channel. Since the equation is used only for water within "normal" ambient conditions, the effects of temperature, pressure, and viscosity are disregarded. The primary advantage of this approach is that the constant, $C$, depends only on the roughness, not on the fluid characteristics. This is also the method's main disadvantage, since professional judgment is required in choosing the value of $C$.

$$v = 0.85CR^{0.63}S_0^{0.54} \qquad \text{[SI]} \qquad \textit{21.14(a)}$$

$$v = 1.318CR^{0.63}S_0^{0.54} \qquad \text{[U.S.]} \qquad \textit{21.14(b)}$$

## 9. NORMAL DEPTH

When the depth of flow is constant along the length of the channel (i.e., the depth is neither increasing nor decreasing), the flow is said to be *uniform*. The depth of flow in that case is known as the *normal depth, $d_n$*. If the normal depth is known, it can be compared with the actual depth of flow to determine if the flow is uniform.[6]

The difficulty with which the normal depth is calculated depends on the cross section of the channel. If the width is very large compared to the depth, the flow cross section will essentially be rectangular and the Manning equation can be used. (Equation 21.15 assumes that the hydraulic radius equals the normal depth.)

$$d_n = \left(\frac{nQ}{w\sqrt{S}}\right)^{3/5} \qquad [w \gg d_n] \qquad \text{[SI]} \qquad \textit{21.15(a)}$$

$$d_n = 0.788\left(\frac{nQ}{w\sqrt{S}}\right)^{3/5} \qquad [w \gg d_n] \qquad \text{[U.S.]} \qquad \textit{21.15(b)}$$

Normal depth in circular channels can be calculated directly only under limited conditions. If the circular channel is flowing full, the normal depth is the inside pipe diameter.

$$D = d_n = 1.548\left(\frac{nQ}{\sqrt{S}}\right)^{3/8} \qquad \text{[full]} \qquad \text{[SI]} \qquad \textit{21.16(a)}$$

$$D = d_n = 1.335\left(\frac{nQ}{\sqrt{S}}\right)^{3/8} \qquad \text{[full]} \qquad \text{[U.S.]} \qquad \textit{21.16(b)}$$

If a circular channel is flowing half full, the normal depth is half of the inside pipe diameter.

$$D = 2d_n = 2.008\left(\frac{nQ}{\sqrt{S}}\right)^{3/8} \qquad \text{[half full]} \qquad \text{[SI]} \qquad \textit{21.17(a)}$$

$$D = 2d_n = 1.731\left(\frac{nQ}{\sqrt{S}}\right)^{3/8} \qquad \text{[half full]} \qquad \text{[U.S.]} \qquad \textit{21.17(b)}$$

For other cases of uniform flow (trapezoidal, triangular, etc.), it is more difficult to determine normal depth. Various researchers have prepared tables and figures to assist in the calculations. For example, Table 21.3 is derived from App. 21.C and can be used for circular channels flowing other than full or half full.

In the absence of tables or figures, trial-and-error solutions are required. The appropriate expressions for the flow area and hydraulic radius are used in the Manning equation. Trial values are used in conjunction with graphical techniques, linear interpolation, or extrapolation to determine the normal depth. The Manning equation is solved for flow rate with various assumed values of $d_n$. The calculated value is compared to the actual known flow quantity, and the normal depth is approached iteratively.

For a rectangular channel whose width is small compared to the depth, the hydraulic radius and area in flow are

$$R = \frac{wd_n}{w + 2d_n} \qquad \textit{21.18}$$

$$A = wd_n \qquad \textit{21.19}$$

$$Q = \left(\frac{1.00}{n}\right)(wd_n)\left(\frac{wd_n}{w + 2d_n}\right)^{2/3}\sqrt{S} \qquad \text{[rectangular]}$$

$$\text{[SI]} \qquad \textit{21.20(a)}$$

---

[6]Normal depth is a term that applies only to uniform flow. The two alternate depths that can occur in nonuniform flow are not normal depths.

$$Q = \left(\frac{1.49}{n}\right)(wd_n)\left(\frac{wd_n}{w + 2d_n}\right)^{2/3}\sqrt{S} \quad \text{[rectangular]}$$

[U.S.]   *21.20(b)*

For a trapezoidal channel with exposed surface width $w$, base width $b$, side length $s$, and normal depth of flow $d_n$, the hydraulic radius and area in flow are

$$R = \frac{d_n(b + w)}{2(b + 2s)} \quad \text{[trapezoidal]}$$   *21.21*

$$A = \frac{d_n(w + b)}{2} \quad \text{[trapezoidal]}$$   *21.22*

For a symmetrical triangular channel with exposed surface width $w$, side slope 1:$z$ (vertical:horizontal), and normal depth of flow $d_n$, the hydraulic radius and area in flow are

$$R = \frac{zd_n}{2\sqrt{1 + z^2}}$$   *21.23*

$$A = zd_n^2$$   *21.24*

## 10. ENERGY AND FRICTION RELATIONSHIPS

Bernoulli's equation is an expression for the conservation of energy along a fluid streamline. The Bernoulli equation can also be written for two points along the bottom of an open channel.

$$\frac{p_1}{\rho g} + \frac{v_1^2}{2g} + z_1 = \frac{p_2}{\rho g} + \frac{v_2^2}{2g} + z_2 + h_f \quad \text{[SI]}$$   *21.25(a)*

$$\frac{p_1}{\gamma} + \frac{v_1^2}{2g} + z_1 = \frac{p_2}{\gamma} + \frac{v_2^2}{2g} + z_2 + h_f \quad \text{[U.S.]}$$   *21.25(b)*

However, $p/\rho g = d$.

$$d_1 + \frac{v_1^2}{2g} + z_1 = d_2 + \frac{v_2^2}{2g} + z_2 + h_f$$   *21.26*

And since $d_1 = d_2$ and $v_1 = v_2$ for uniform flow at the bottom of a channel,

$$h_f = z_1 - z_2$$   *21.27*

$$S_0 = \frac{z_1 - z_2}{L}$$   *21.28*

The channel slope, $S_0$, and the hydraulic energy gradient, $S$, are numerically the same for uniform flow. Therefore, the total friction loss along a channel is

$$h_f = LS$$   *21.29*

Combining Eq. 21.29 with the Manning equation (see Eq. 21.12) results in a method for calculating friction loss.

$$h_f = \frac{Ln^2v^2}{R^{4/3}} \quad \text{[SI]}$$   *21.30(a)*

$$h_f = \frac{Ln^2v^2}{2.208R^{4/3}} \quad \text{[U.S.]}$$   *21.30(b)*

### Example 21.3

The velocities upstream and downstream, $v_1$ and $v_2$, of a 12 ft (4.0 m) wide sluice gate are both unknown. The upstream and downstream depths are 6 ft (2.0 m) and 2 ft (0.6 m), respectively. Flow is uniform and steady. What is the downstream velocity, $v_2$?

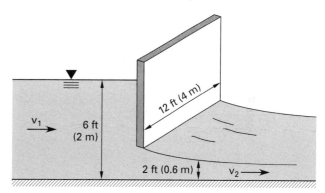

*SI Solution*

Since the channel bottom is essentially level on either side of the gate, $z_1 = z_2$. Bernoulli's equation reduces to

$$d_1 + \frac{v_1^2}{2g} = d_2 + \frac{v_2^2}{2g}$$

$$2 \text{ m} + \frac{v_1^2}{2g} = 0.6 \text{ m} + \frac{v_2^2}{2g}$$

$v_1$ and $v_2$ are related by continuity.

$$Q_1 = Q_2$$

$$A_1v_1 = A_2v_2$$

$$(2 \text{ m})(4 \text{ m})v_1 = (0.6 \text{ m})(4 \text{ m})v_2$$

$$v_1 = 0.3v_2$$

Substituting the expression for $v_1$ into the Bernoulli equation gives

$$2 \text{ m} + \frac{(0.3v_2)^2}{(2)\left(9.81 \frac{m}{s^2}\right)} = 0.6 \text{ m} + \frac{v_2^2}{(2)\left(9.81 \frac{m}{s^2}\right)}$$

$$2 \text{ m} + 0.004587v_2^2 = 0.6 \text{ m} + 0.050968v_2^2$$

$$v_2 = 5.5 \text{ m/s}$$

*Customary U.S. Solution*

Since the channel bottom is essentially level on either side of the gate, $z_1 = z_2$. Bernoulli's equation reduces to

$$d_1 + \frac{v_1^2}{2g} = d_2 + \frac{v_2^2}{2g}$$

$$6 \text{ ft} + \frac{v_1^2}{2g} = 2 \text{ ft} + \frac{v_2^2}{2g}$$

$v_1$ and $v_2$ are related by continuity.

$$Q_1 = Q_2$$
$$A_1 v_1 = A_2 v_2$$
$$(6 \text{ ft})(12 \text{ ft})v_1 = (2 \text{ ft})(12 \text{ ft})v_2$$
$$v_1 = \frac{v_2}{3}$$

Substituting the expression for $v_1$ into the Bernoulli equation gives

$$6 \text{ ft} + \frac{v_2^2}{(3)^2(2)\left(32.2 \frac{\text{ft}}{\text{sec}^2}\right)} = 2 \text{ ft} + \frac{v_2^2}{(2)\left(32.2 \frac{\text{ft}}{\text{sec}^2}\right)}$$

$$6 \text{ ft} + 0.00173 v_2^2 = 2 \text{ ft} + 0.0155 v_2^2$$

$$v_2 = 17.0 \text{ ft/sec}$$

**Example 21.4**

In Ex. 21.1, the open channel experiencing normal flow had the following characteristics: $S = 0.002$, $n = 0.012$, $v = 9.447$ ft/sec (2.9 m/s), and $R = 2.22$ ft (0.68 m). What is the energy loss per 1000 ft (100 m)?

*SI Solution*

There are two methods for finding the energy loss. From Eq. 21.29,

$$h_f = LS = (100 \text{ m})(0.002)$$
$$= 0.2 \text{ m}$$

From Eq. 21.30(a),

$$h_f = \frac{Ln^2 v^2}{R^{4/3}}$$

$$= \frac{(100 \text{ m})(0.012)^2 \left(2.9 \frac{\text{m}}{\text{s}}\right)^2}{(0.68 \text{ m})^{4/3}}$$

$$= 0.2 \text{ m}$$

*Customary U.S. Solution*

There are two methods for finding the energy loss. From Eq. 21.29,

$$h_f = LS = (1000 \text{ ft})(0.002)$$
$$= 2 \text{ ft}$$

From Eq. 21.30(b),

$$h_f = \frac{Ln^2 v^2}{2.208 R^{4/3}}$$

$$= \frac{(1000 \text{ ft})(0.012)^2 \left(9.447 \frac{\text{ft}}{\text{sec}}\right)^2}{(2.208)(2.22 \text{ ft})^{4/3}}$$

$$= 2 \text{ ft}$$

## 11. SIZING TRAPEZOIDAL AND RECTANGULAR CHANNELS

Trapezoidal and rectangular cross sections are commonly used for artificial surface channels. The flow through a trapezoidal channel is easily determined from the Manning equation when the cross section is known. However, when the cross section or uniform depth is unknown, a trial-and-error solution is required. (See Fig. 21.2.)

*Figure 21.2 Trapezoidal Cross Section*

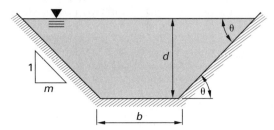

For such problems involving rectangular and trapezoidal channels, it is common to calculate and plot the *conveyance, K* (or alternatively, the product $Kn$), against depth. For trapezoidal sections, it is particularly convenient to write the uniform flow, $Q$, in terms of a modified conveyance, $K'$. $b$ is the base width of the channel, $d$ is the depth of flow, and $m$ is the cotangent of the side slope angle. $m$ and the ratio $d/b$ are treated as independent variables. Values of $K'$ are tabulated in App. 21.F.

$$Q = \frac{K' b^{8/3} \sqrt{S_0}}{n} \qquad 21.31$$

$$K' = \left( \frac{\left(1 + m\left(\frac{d}{b}\right)\right)^{5/3}}{\left(1 + 2\left(\frac{d}{b}\right)\sqrt{1+m^2}\right)^{2/3}} \right) \left(\frac{d}{b}\right)^{5/3} \qquad \text{[SI]} \quad 21.32(a)$$

$$K' = \left( \frac{1.49\left(1 + m\left(\frac{d}{b}\right)\right)^{5/3}}{\left(1 + 2\left(\frac{d}{b}\right)\sqrt{1 + m^2}\right)^{2/3}} \right) \left(\frac{d}{b}\right)^{5/3} \quad \text{[U.S.]} \quad 21.32(b)$$

$$m = \cot\ \theta \qquad 21.33$$

For any fixed value of $m$, enough values of $K'$ are calculated over a reasonable range of the $d/b$ ratio $(0.05 < d/b < 0.5)$ to define a curve. Given specific values of $Q$, $n$, $S_0$, and $b$, the value of $K'$ can be calculated from the expression for $Q$. The graph is used to determine the ratio $d/b$, giving the depth of uniform flow, $d$, since $b$ is known.

When the ratio of $d/b$ is very small (less than 0.02), it is satisfactory to consider the trapezoidal channel as a wide rectangular channel with area $A = bd$.

## 12. MOST EFFICIENT CROSS SECTION

The most efficient open channel cross section will maximize the flow for a given Manning coefficient, slope, and flow area. Accordingly, the Manning equation requires that the hydraulic radius be maximum. For a given flow area, the wetted perimeter will be minimum.

Semicircular cross sections have the smallest wetted perimeter; therefore, the cross section with the highest efficiency is the semicircle. Although such a shape can be constructed with concrete, it cannot be used with earth channels.

The most efficient cross section is also generally assumed to minimize construction cost. This is true only in the most simplified cases, however, since the labor and material costs of excavation and formwork must be considered. Rectangular and trapezoidal channels are much easier to form than semicircular channels. So in this sense the "least efficient" (i.e, most expensive) cross section (i.e., semicircular) is also the "most efficient." (See Fig. 21.3.)

**Figure 21.3** *Circles Inscribed in Efficient Channels*

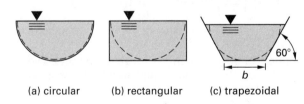

| (a) circular | (b) rectangular | (c) trapezoidal |
| --- | --- | --- |

The most efficient rectangle is one having depth equal to one-half of the width (i.e., is one-half of a square).

$$d = \frac{w}{2} \quad \text{[most efficient rectangle]} \qquad 21.34$$

$$A = dw = \frac{w^2}{2} = 2d^2 \qquad 21.35$$

$$P = d + w + d = 2w = 4d \qquad 21.36$$

$$R = \frac{w}{4} = \frac{d}{2} \qquad 21.37$$

The most efficient trapezoidal channel is always one in which the flow depth is twice the hydraulic radius. If the side slope is adjustable, the sides of the most efficient trapezoid should be inclined at 60° from the horizontal. Since the surface width will be equal to twice the sloping side length, the most efficient trapezoidal channel will be half of a regular hexagon (i.e., three adjacent equilateral triangles of side length $2d/\sqrt{3}$). If the side slope is any other angle, only the $d = 2R$ criterion is applicable.

$$d = 2R \quad \text{[most efficient trapezoid]} \qquad 21.38$$

$$b = \frac{2d}{\sqrt{3}} \qquad 21.39$$

$$A = \sqrt{3}d^2 \qquad 21.40$$

$$P = 3b = 2\sqrt{3}d \quad \text{[most efficient trapezoid]} \qquad 21.41$$

$$R = \frac{d}{2} \qquad 21.42$$

A semicircle with its center at the middle of the water surface can always be inscribed in a cross section with maximum efficiency.

### Example 21.5

A rubble masonry open channel is being designed to carry 500 ft³/sec (14 m³/s) of water on a 0.0001 slope. Using $n = 0.017$, find the most efficient dimensions for a rectangular channel.

*SI Solution*

Let the depth and width be $d$ and $w$, respectively. For an efficient rectangle, $d = w/2$.

$$A = dw = \left(\frac{w}{2}\right)w = \frac{w^2}{2}$$

$$P = d + w + d = \frac{w}{2} + w + \frac{w}{2} = 2w$$

$$R = \frac{A}{P} = \frac{\frac{w^2}{2}}{2w} = \frac{w}{4}$$

Using Eq. 21.13(a),

$$Q = \left(\frac{1.00}{n}\right)AR^{2/3}\sqrt{S}$$

$$14 \text{ m}^3/\text{s} = \left(\frac{1.00}{0.017}\right)\left(\frac{w^2}{2}\right)\left(\frac{w}{4}\right)^{2/3}\sqrt{0.0001}$$

$$14 \text{ m}^3/\text{s} = 0.1167w^{8/3}$$

$$w = 6.02 \text{ m}$$

$$d = \frac{w}{2} = \frac{6.02 \text{ m}}{2} = 3.01 \text{ m}$$

*Customary U.S. Solution*

Let the depth and width be $d$ and $w$, respectively. For an efficient rectangle, $d = w/2$.

$$A = dw = \left(\frac{w}{2}\right)w = \frac{w^2}{2}$$

$$P = d + w + d = \frac{w}{2} + w + \frac{w}{2} = 2w$$

$$R = \frac{A}{P} = \frac{\dfrac{w^2}{2}}{2w} = \frac{w}{4}$$

Using Eq. 21.13(b),

$$Q = \left(\frac{1.49}{n}\right)AR^{2/3}\sqrt{S}$$

$$500 \text{ ft}^3/\text{sec} = \left(\frac{1.49}{0.017}\right)\left(\frac{w^2}{2}\right)\left(\frac{w}{4}\right)^{2/3}\sqrt{0.0001}$$

$$500 \text{ ft}^3/\text{sec} = 0.1739 w^{8/3}$$

$$w = 19.82 \text{ ft}$$

$$d = \frac{w}{2} = \frac{19.82 \text{ ft}}{2} = 9.91 \text{ ft}$$

## 13. ANALYSIS OF NATURAL WATERCOURSES

Natural watercourses do not have uniform paths or cross sections. This complicates their analysis considerably. Frequently, analyzing the flow from a river is a matter of making the most logical assumptions. Many evaluations can be solved with a reasonable amount of error.

As was seen in Eq. 21.30, the friction loss (and hence the hydraulic gradient) depends on the square of the roughness coefficient. Therefore, an attempt must be made to evaluate the roughness constant as accurately as possible. If the channel consists of a river with flood plains (see Fig. 21.4), it should be treated as parallel channels. The flow from each subdivision can be calculated independently and the separate values added to obtain the total flow. (The common interface between adjacent subdivisions is not included in the wetted perimeter.) Alternatively, a composite value of the roughness coefficient, $n_c$, can be approximated from the individual values of $n$ and the corresponding wetted perimeters.

$$n_c = \left(\frac{\sum P_i(n_i)^{3/2}}{\sum P_i}\right)^{2/3} \qquad 21.43$$

If the channel is divided (see Fig. 21.5) by an island into two channels, some combination of flows will usually be known. For example, if the total flow, $Q$, is known, $Q_1$ and $Q_2$ may be unknown. If the slope is known, $Q_1$ and $Q_2$ may be known. Iterative trial-and-error solutions are often required.

**Figure 21.4** *River with Flood Plain*

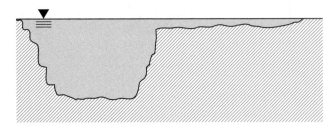

Since the elevation drop $(z_A - z_B)$ between points A and B is the same regardless of flow path,

$$S_1 = \frac{z_A - z_B}{L_1} \qquad 21.44$$

$$S_2 = \frac{z_A - z_B}{L_2} \qquad 21.45$$

Once the slopes are known, initial estimates $Q_1$ and $Q_2$ can be calculated from Eq. 21.13. The sum of $Q_1$ and $Q_2$ will probably not be the same as the given flow quantity, $Q$. In that case, $Q$ should be prorated according to the ratios of $Q_1$ and $Q_2$ to $Q_1 + Q_2$.

**Figure 21.5** *Divided Channel*

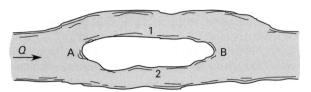

If the lengths $L_1$ and $L_2$ are the same or almost so, the Manning equation may be solved for the slope by writing Eq. 21.46.

$$Q = Q_1 + Q_2$$
$$= \left(\left(\frac{A_1}{n_1}\right)(R_1)^{2/3} + \left(\frac{A_2}{n_2}\right)(R_2)^{2/3}\right)\sqrt{S} \quad \text{[SI]} \quad 21.46(a)$$

$$Q = Q_1 + Q_2$$
$$= 1.49\left(\left(\frac{A_1}{n_1}\right)(R_1)^{2/3} + \left(\frac{A_2}{n_2}\right)(R_2)^{2/3}\right)\sqrt{S}$$

$$\text{[U.S.]} \quad 21.46(b)$$

Equation 21.46 yields only a rough estimate of the flow quantity, as the geometry and roughness of a natural channel changes considerably along its course.

## 14. FLOW MEASUREMENT WITH WEIRS

A *weir* is an obstruction in an open channel over which flow occurs. Although a dam spillway is a specific type of weir, most weirs are intended specifically for flow measurement.

Measurement weirs consist of a vertical flat plate with sharp edges. Because of their construction, they are called *sharp-crested weirs*. Sharp-crested weirs are most

Fluids

frequently rectangular, consisting of a straight, horizontal crest. However, weirs may also have trapezoidal and triangular openings.

For any given width of weir opening (referred to as the *weir length*), the discharge will be a function of the head over the weir. The head (or sometimes surface elevation) can be determined by a standard *staff gauge* mounted adjacent to the weir.

The full channel flow usually goes over the weir. However, it is also possible to divert a small portion of the total flow through a measurement channel. The full channel flow rate can be extrapolated from a knowledge of the split fractions.

If a rectangular weir is constructed with an opening width less than the channel width, the falling liquid sheet (called the *nappe*) decreases in width as it falls. Because of this *contraction* of the nappe, these weirs are known as *contracted weirs*, although it is the nappe that is actually contracted, not the weir. If the opening of the weir extends the full channel width, the weir is known as a *suppressed weir*, since the contractions are suppressed. (See Fig. 21.6.)

The derivation of an expression for the quantity flowing over a weir is dependent on many simplifying assumptions. The basic weir equation (see Eq. 21.47 or Eq. 21.48) is, therefore, an approximate result requiring correction by experimental coefficients.

If it is assumed that the contractions are suppressed, upstream velocity is uniform, flow is laminar over the crest, nappe pressure is zero, the nappe is fully ventilated, and viscosity, turbulence, and surface tension effects are negligible, then the following equation may be derived from the Bernoulli equation.

$$Q = \tfrac{2}{3}b\sqrt{2g}\left(\left(H + \frac{v_1^2}{2g}\right)^{3/2} - \left(\frac{v_1^2}{2g}\right)^{3/2}\right) \qquad 21.47$$

If the velocity of approach, $v_1$, is negligible, then

$$Q = \tfrac{2}{3}b\sqrt{2g}H^{3/2} \qquad 21.48$$

Equation 21.48 must be corrected for all of the assumptions made, primarily for a nonuniform velocity distribution. This is done by introducing an empirical discharge coefficient, $C_1$. Equation 21.49 is known as the *Francis weir equation*.

$$Q = \tfrac{2}{3}C_1 b\sqrt{2g}H^{3/2} \qquad 21.49$$

**Figure 21.6** *Contracted and Suppressed Weirs*

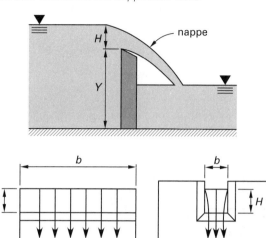

Many investigations have been done to evaluate $C_1$ analytically. Perhaps the most widely known is the coefficient formula developed by *Rehbock*.[7]

$$C_1 = \left(0.6035 + 0.0813\left(\frac{H}{Y}\right) + \frac{0.000295}{Y}\right)$$
$$\times \left(1 + \frac{0.00361}{H}\right)^{3/2} \quad \text{[U.S. only]}$$
$$\approx 0.602 + 0.083\left(\frac{H}{Y}\right) \quad \text{[U.S. and SI]} \qquad 21.50$$

When $H/Y < 0.2$, $C_1$ approaches 0.61 to 0.62. In most cases, a value in this range is adequate. Other constants (i.e., $^2/_3$ and $\sqrt{2g}$) can be taken out of Eq. 21.49. In that case,

$$Q \approx 1.84bh^{3/2} \qquad \text{[SI]} \qquad 21.51(a)$$

$$Q \approx 3.33bh^{3/2} \qquad \text{[U.S.]} \qquad 21.51(b)$$

If the contractions are not suppressed (i.e., one or both sides do not extend to the channel sides) then the actual width, $b$, should be replaced with the *effective width*. In Eq. 21.52, $N$ is 1 if one side is contracted and $N$ is 2 if there are two end contractions.

$$b_{\text{effective}} = b_{\text{actual}} - 0.1NH \qquad 21.52$$

A *submerged rectangular weir* requires a more complex analysis because of the difficulty in measuring $H$ and because the discharge depends on both the upstream

---

[7]There is much variation in how different investigators calculate the discharge coefficient, $C_1$. For ratios of $H/b$ less than 5, $C_1 = 0.622$ gives a reasonable value. With the questionable accuracy of some of the other variables used in open channel flow problems, the pursuit of greater accuracy is of dubious value.

and downstream depths. (See Fig. 21.7.) The following equation, however, may be used with little difficulty.

$$Q_{\text{submerged}} = Q_{\text{free flow}} \left(1 - \left(\frac{H_{\text{downstream}}}{H_{\text{upstream}}}\right)^{3/2}\right)^{0.385}$$

$$21.53$$

Equation 21.53 is used by first finding the flow rate, $Q$, from Eq. 21.49 and then correcting it with the bracketed quantity.

**Figure 21.7** *Submerged Weir*

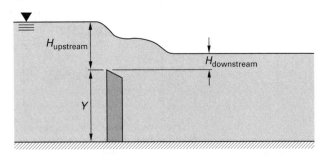

**Example 21.6**

The crest of a sharp-crested, rectangular weir with two contractions is 2.5 ft (1.0 m) high above the channel bottom. The crest is 4 ft (1.6 m) long. A 4 in (100 mm) head exists over the weir. What is the velocity of approach?

*SI Solution*

$$H = \frac{100 \text{ mm}}{1000 \frac{\text{mm}}{\text{m}}} = 0.1 \text{ m}$$

The number of contractions, $N$, is 2. From Eq. 21.52, the effective width is

$$b_{\text{effective}} = b_{\text{actual}} - 0.1NH = 1.6 \text{ m} - (0.1)(2)(0.1 \text{ m})$$

$$= 1.58 \text{ m}$$

$$C_1 \approx 0.602 + 0.083\left(\frac{H}{Y}\right)$$

$$= 0.602 + (0.083)\left(\frac{0.1}{1}\right) = 0.61$$

From Eq. 21.49, the flow is

$$Q = \tfrac{2}{3}C_1 b\sqrt{2g}H^{3/2}$$

$$= \left(\tfrac{2}{3}\right)(0.61)(1.58 \text{ m})\sqrt{(2)\left(9.81 \frac{\text{m}}{\text{s}^2}\right)}(0.10 \text{ m})^{3/2}$$

$$= 0.090 \text{ m}^3/\text{s}$$

$$v = \frac{Q}{A}$$

$$= \frac{0.090 \frac{\text{m}^3}{\text{s}}}{(1.6 \text{ m})(1.0 \text{ m} + 0.1 \text{ m})}$$

$$= 0.05 \text{ m/s}$$

*Customary U.S. Solution*

$$H = \frac{4 \text{ in}}{12 \frac{\text{in}}{\text{ft}}} = 0.333 \text{ ft}$$

The number of contractions, $N$, is 2. From Eq. 21.52, the effective width is

$$b_{\text{effective}} = b_{\text{actual}} - 0.1NH = 4 \text{ ft} - (0.1)(2)(0.333 \text{ ft})$$

$$= 3.93 \text{ ft}$$

The Rehbock coefficient (see Eq. 21.50) is

$$C_1 = \left(0.6035 + 0.0813\left(\frac{H}{Y}\right) + \frac{0.000295}{Y}\right)$$

$$\times \left(1 + \frac{0.00361}{H}\right)^{3/2}$$

$$= \left(0.6035 + (0.0813)\left(\frac{0.333 \text{ ft}}{2.5 \text{ ft}}\right) + \frac{0.000295}{2.5 \text{ ft}}\right)$$

$$\times \left(1 + \frac{0.00361}{0.333 \text{ ft}}\right)^{3/2}$$

$$= 0.624$$

From Eq. 21.49, the flow is

$$Q = \tfrac{2}{3}C_1 b\sqrt{2g}H^{3/2}$$

$$= \left(\tfrac{2}{3}\right)(0.624)(3.93 \text{ ft})\sqrt{(2)\left(32.2 \frac{\text{ft}}{\text{sec}^2}\right)}(0.333 \text{ ft})^{3/2}$$

$$= 2.52 \text{ ft}^3/\text{sec}$$

$$v = \frac{Q}{A}$$

$$= \frac{2.52 \frac{\text{ft}^3}{\text{sec}}}{(4 \text{ ft})(2.5 \text{ ft} + 0.333 \text{ ft})}$$

$$= 0.222 \text{ ft/sec}$$

## 15. TRIANGULAR WEIRS

*Triangular weirs* (*V-notch weirs*) should be used when small flow rates are to be measured. The flow coefficient over a triangular weir depends on the notch angle, $\theta$, but generally varies from 0.58 to 0.61. For a 90° weir, $C_2 \approx 0.593$. (See Fig. 21.8.)

$$Q = C_2 \left(\frac{8}{15}\right) \tan\left(\frac{\theta}{2}\right) \sqrt{2g} H^{5/2} \qquad 21.54$$

$$Q \approx 1.4 H^{2.5} \quad [\text{90° weir}] \qquad [\text{SI}] \quad 21.55(a)$$

$$Q \approx 2.5 H^{2.5} \quad [\text{90° weir}] \qquad [\text{U.S.}] \quad 21.55(b)$$

**Figure 21.8** Triangular Weir

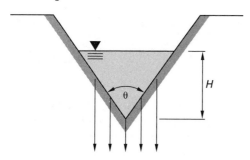

## 16. TRAPEZOIDAL WEIRS

A *trapezoidal weir* is essentially a rectangular weir with a triangular weir on either side. (See Fig. 21.9.) If the angle of the sides from the vertical is approximately 14° (i.e., 4 vertical and 1 horizontal), the weir is known as a *Cipoletti weir*. The discharge from the triangular ends of a Cipoletti weir approximately make up for the contractions that would reduce the flow over a rectangular weir. Therefore, no correction is theoretically necessary. This is not completely accurate, and for this reason, Cipoletti weirs are not used where great accuracy is required. The discharge is

$$Q = \tfrac{2}{3} C_d b \sqrt{2g} H^{3/2} \qquad 21.56$$

**Figure 21.9** Trapezoidal Weir

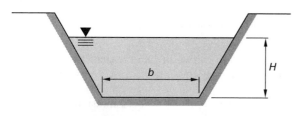

The average value of the discharge coefficient is 0.63. The discharge from a Cipoletti weir is given by Eq. 21.57.

$$Q = 1.86 b H^{3/2} \qquad [\text{SI}] \quad 21.57(a)$$

$$Q = 3.367 b H^{3/2} \qquad [\text{U.S.}] \quad 21.57(b)$$

## 17. BROAD-CRESTED WEIRS AND SPILLWAYS

Most weirs used for flow measurement are sharp-crested. However, the flow over spillways, broad-crested weirs, and similar features can be calculated from Eq. 21.49 even though flow measurement is not the primary function of the feature. (A weir is broad-crested if the weir thickness is greater than half of the head, $H$.)

A *dam's spillway (overflow spillway)* is designed for a capacity based on the dam's inflow hydrograph, turbine capacity, and storage capacity. Spillways frequently have a cross section known as an *ogee*, which closely approximates the underside of a nappe from a sharp-crested weir. This cross section minimizes the cavitation that is likely to occur if the water surface breaks contact with the spillway due to upstream heads that are higher than designed for.[8]

Discharge from an overflow spillway is derived in the same manner as for a weir. Equation 21.58 can be used for broad-crested weirs ($C_1 = 0.5$ to 0.57) and ogee spillways ($C_1 = 0.60$ to 0.75).

$$Q = \tfrac{2}{3} C_1 b \sqrt{2g} H^{3/2} \qquad 21.58$$

The *Horton equation* (see Eq. 21.59) for broad-crested weirs combines all of the coefficients into a spillway (weir) coefficient and adds the velocity of approach to the upstream head. The *Horton coefficient*, $C_s$, is specific to the Horton equation. (Notice that $C_s$ and $C_1$ differ by a factor of about 5 and cannot easily be mistaken for each other.)

$$Q = C_s b \left(H + \frac{v^2}{2g}\right)^{3/2} \qquad 21.59$$

If the velocity of approach is insignificant, the discharge is

$$Q = C_s b H^{3/2} \qquad 21.60$$

$C_s$ is a *spillway coefficient*, which varies from about 3.3–3.98 ft$^{0.5}$/sec (1.8–2.2 m$^{0.5}$/s) for ogee spillways. 3.97 ft$^{0.5}$/sec (2.2 m$^{0.5}$/s) is frequently used for first approximations. For broad-crested weirs, $C_s$ varies between 2.63–3.33 ft$^{0.5}$/sec (1.45–1.84 m$^{0.5}$/s). (Use 3.33 ft$^{0.5}$/sec (1.84 m$^{0.5}$/sec) for initial estimates.) $C_s$ increases as

---

[8]Cavitation and separation will not normally occur as long as the actual head, $H$, is less than twice the design value. The shape of the ogee spillway will be a function of the design head.

the upstream design head above the spillway top, $H$, increases, and the larger values apply to the higher heads.

Broad-crested weirs and spillways can be calibrated to obtain greater accuracy in predicting flow rates.

*Scour protection* is usually needed at the toe of a spillway to protect the area exposed to a hydraulic jump. This protection usually takes the form of an extended horizontal or sloping apron. Other measures, however, are needed if the tailwater exhibits large variations in depth.

## 18. PROPORTIONAL WEIRS

The *proportional weir (Sutro weir)* is used in water level control because it demonstrates a linear relationship between $Q$ and $H$. Figure 21.10 illustrates a proportional weir whose sides are hyperbolic in shape.

$$Q = C_d K \left(\frac{\pi}{2}\right) \sqrt{2gH} \qquad 21.61$$

$$K = 2x\sqrt{y} \qquad 21.62$$

**Figure 21.10** *Proportional Weir*

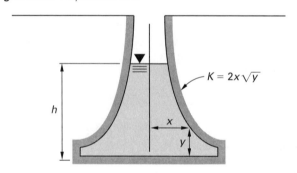

## 19. FLOW MEASUREMENT WITH PARSHALL FLUMES

The Parshall flume is widely used for measuring open channel wastewater flows. It performs well when head losses must be kept to a minimum and when there are high amounts of suspended solids. (See Fig. 21.11.)

The Parshall flume is constructed with a converging upstream section, a throat, and a diverging downstream section. The walls of the flume are vertical, but the floor of the throat section drops. The length, width, and height of the flume are essentially predefined by the anticipated flow rate.[9]

The throat geometry in a Parshall flume has been designed to force the occurrence of critical flow (see Sec. 21.25) at that point. Following the critical section is a short length of supercritical flow followed by a

---

[9]This chapter does not attempt to design the Parshall flume, only to predict flow rates through its use.

**Figure 21.11** *Parshall Flume*

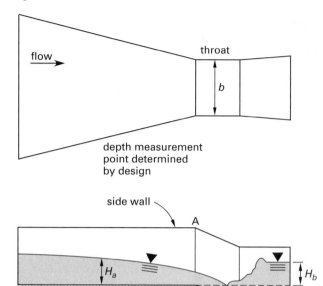

hydraulic jump. (See Sec. 21.33.) This design eliminates any dead water region where debris and silt can accumulate (as are common with flat-topped weirs).

The discharge relationship for a Parshall flume is given by for submergence ratios of $H_b/H_a$ up to 0.7. Above 0.7, the true discharge is less than predicted by Eq. 21.63. Values of $K$ are given in Table 21.4, although using a value of 4.0 is accurate for most purposes.

$$Q = KbH_a^n \qquad 21.63$$

$$n = 1.522b^{0.026} \qquad 21.64$$

**Table 21.4** *Parshall Flume K-Values*

| $b$, ft (m) | $K$ |
|---|---|
| 0.25 (0.075) | 3.97 |
| 0.50 (0.15) | 4.12 |
| 0.75 (0.225) | 4.09 |
| 1.0 (0.3) | 4.00 |
| 1.5 (0.45) | 4.00 |
| 2.0 (0.6) | 4.00 |
| 3.0 (0.9) | 4.00 |
| 4.0 (1.2) | 4.00 |

(Multiply ft by 0.3 to obtain m.)

Above a certain tailwater height, the Parshall flume no longer operates in the *free-flow mode*. Rather, it operates in a *submerged mode*. A very high tailwater reduces the flow rate through the flume. Equation 21.63 predicts the flow rate with reasonable accuracy, however, even for 50% to 80% submergence (calculated as $H_b/H_a$). For large submergence, the tailwater height must be known and a different analysis method must be used.

## 20. UNIFORM AND NONUNIFORM STEADY FLOW

Steady flow is constant-volume flow. However, the flow may be uniform or nonuniform (varied) in depth. There may be significant variations over long and short distances without any change in the flow rate.

Figure 21.12 illustrates the three definitions of "slope" existing for open channel flow. These three slopes are the slope of the channel bottom, the slope of the water surface, and the slope of the energy gradient line.

**Figure 21.12** *Slopes Used in Open Channel Flow*

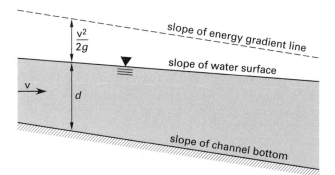

Under conditions of uniform flow, all of these three slopes are equal since the flow quantity and flow depth are constant along the length of flow.[10] With nonuniform flow, however, the flow velocity and depth vary along the length of channel and the three slopes are not necessarily equal.

If water is introduced down a path with a steep slope (as after flowing over a spillway), the effect of gravity will cause the velocity to increase. As the velocity increases, the depth decreases in accordance with the continuity of flow equation. The downward velocity is opposed by friction. Because the gravitational force is constant but friction varies with the square of velocity, these two forces eventually become equal. When equal, the velocity stops increasing, the depth stops decreasing, and the flow becomes uniform. Until they become equal, however, the flow is nonuniform (varied).

## 21. SPECIFIC ENERGY

The total head possessed by a fluid is given by the Bernoulli equation.

$$E = \frac{p}{\rho g} + \frac{v^2}{2g} + z \qquad \text{[SI]} \qquad 21.65(a)$$

$$E = \frac{p}{\gamma} + \frac{v^2}{2g} + z \qquad \text{[U.S.]} \qquad 21.65(b)$$

---

[10]As a simplification, this chapter deals only with channels of constant width. If the width is varied, changes in flow depth may not coincide with changes in flow quantity.

*Specific energy*, $E$, is defined as the total head with respect to the channel bottom. In this case, $z = 0$ and $p/\gamma = d$.

$$E = d + \frac{v^2}{2g} \qquad 21.66$$

Equation 21.66 is not meant to imply that the potential energy is an unimportant factor in open channel flow problems. The concept of specific energy is used for convenience only, and it should be clear that the Bernoulli equation is still the valid energy conservation equation.

In uniform flow, total head also decreases due to the frictional effects, but specific energy is constant. In non-uniform flow, total head also decreases, but specific energy may increase or decrease.

Since $v = Q/A$, Eq. 21.66 can be written as

$$E = d + \frac{Q^2}{2gA^2} \qquad \text{[general case]} \qquad 21.67$$

For a rectangular channel, the velocity can be written in terms of the width and flow depth.

$$v = \frac{Q}{A} = \frac{Q}{wd} \qquad 21.68$$

The specific energy equation for a rectangular channel is given by Eq. 21.69 and shown in Fig. 21.13.

$$E = d + \frac{Q^2}{2g(wd)^2} \qquad \text{[rectangular]} \qquad 21.69$$

Specific energy can be used to differentiate between flow regimes. Figure 21.13 illustrates how specific energy is affected by depth, and accordingly, how specific energy relates to critical depth (see Sec. 21.25) and the Froude number (see Sec. 21.27).

**Figure 21.13** *Specific Energy Diagram*

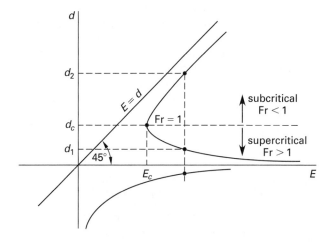

## 22. SPECIFIC FORCE

The *specific force* of a general channel section is the total force per unit weight acting on the water. Equivalently, specific force is the total force that a submerged object would experience. In Eq. 21.70, $\overline{d}$ is the distance from the free surface to the centroid of the flowing area cross section, $A$.

$$\frac{F}{g\rho} = \frac{Q^2}{gA} + \overline{d}A \qquad \text{[SI]} \quad 21.70(a)$$

$$\frac{F}{\gamma} = \frac{Q^2}{gA} + \overline{d}A \qquad \text{[U.S.]} \quad 21.70(b)$$

The first term in Eq. 21.70 represents the momentum flow through the channel per unit time and per unit mass of water. The second term is the pressure force per unit mass of water. Graphs of specific force and specific energy are similar in appearance and predict equivalent results for the critical and alternate depths. (See Sec. 21.24.)

## 23. CHANNEL TRANSITIONS

Sudden changes in channel width or bottom elevation are known as *channel transitions*. (Contractions in width are not covered in this chapter.) For sudden vertical steps in channel bottom, the Bernoulli equation, written in terms of the specific energy, is used to predict the flow behavior (i.e., the depth).

$$E_1 + z_1 = E_2 + z_2 \qquad 21.71$$

$$E_1 - E_2 = z_2 - z_1 \qquad 21.72$$

The maximum possible change in bottom elevation without affecting the energy equality occurs when the depth of flow over the step is equal to the critical depth ($d_2$ equals $d_c$). (See Sec. 21.25.)

## 24. ALTERNATE DEPTHS

Since the area depends on the depth, fixing the channel shape and slope and assuming a depth will determine the flow rate, $Q$, as well as the specific energy. Since Eq. 21.69 is a cubic equation, there are three values of depth of flow, $d$, that will satisfy it. One of them is negative, as Fig. 21.13 shows. Since depth cannot be negative, that value can be discarded. The two remaining values are known as *alternate depths*.

For a given flow rate, the two alternate depths have the same energy. One represents a high velocity with low depth; the other represents a low velocity with high depth. The former is called *supercritical (rapid) flow*; the latter is called *subcritical (tranquil) flow*.

The Bernoulli equation cannot predict which of the two alternate depths will occur for any given flow quantity. The concept of *accessibility* is required to evaluate the two depths. Specifically, the upper and lower limbs of the energy curve are not accessible from each other unless there is a local restriction in the flow.

Energy curves can be drawn for different flow quantities, as shown in Fig. 21.14 for flow quantities $Q_A$ and $Q_B$. Suppose that flow is initially at point 1. Since the flow is on the upper limb, the flow is initially subcritical. If there is a step up in the channel bottom, Eq. 21.72 predicts that the specific energy will decrease.

**Figure 21.14** *Specific Energy Curve Families*

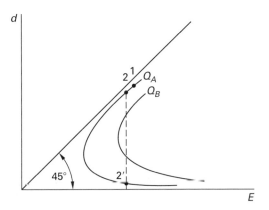

However, the flow cannot arrive at point $2'$ without the flow quantity changing (i.e., going through a specific energy curve for a different flow quantity). Therefore, point $2'$ is not accessible from point 1 without going through point 2 first.[11]

If the flow is well up on the top limb of the specific energy curve (as it is in Ex. 21.7), the water level will drop only slightly. Since the upper limb is asymptotic to a 45° diagonal line, any change in specific energy will result in almost the same change in depth.[12] Therefore, the surface level will remain almost the same.

$$\Delta d \approx \Delta E \qquad \text{[fully subcritical]} \qquad 21.73$$

However, if the initial point on the limb is close to the critical point (i.e., the nose of the curve), then a small change in the specific energy (such as might be caused by a small variation in the channel floor) will cause a large change in depth. That is why severe turbulence commonly occurs near points of critical flow.

### Example 21.7

4 ft/sec (1.2 m/s) of water flows in a 7 ft (2.1 m) wide, 6 ft (1.8 m) deep open channel. The flow encounters a 1.0 ft (0.3 m) step in the channel bottom. What is the depth of flow above the step?

---

[11]Actually, specific energy curves are typically plotted for flow per unit width, $q = Q/w$. If that is the case, a jump from one limb to the other could take place if the width were allowed to change as well as depth.

[12]A rise in the channel bottom does not always produce a drop in the water surface. Only if the flow is initially subcritical will the water surface drop upon encountering a step. The water surface will rise if the flow is initially supercritical.

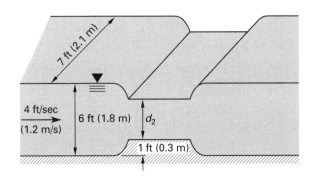

*SI Solution*

The initial specific energy is found from Eq. 21.66.

$$E_1 = d + \frac{\text{v}^2}{2g}$$

$$= 1.8 \text{ m} + \frac{\left(1.2 \ \dfrac{\text{m}}{\text{s}}\right)^2}{(2)\left(9.81 \ \dfrac{\text{m}}{\text{s}^2}\right)}$$

$$= 1.87 \text{ m}$$

From Eq. 21.72, the specific energy above the step is

$$E_2 = E_1 + z_1 - z_2 = 1.87 \text{ m} + 0 - 0.3 \text{ m}$$

$$= 1.57 \text{ m}$$

The quantity flowing is

$$Q = A\text{v} = (2.1 \text{ m})(1.8 \text{ m})\left(1.2 \ \frac{\text{m}}{\text{s}}\right)$$

$$= 4.54 \text{ m}^3/\text{s}$$

Substituting $Q$ into Eq. 21.69 gives

$$E = d + \frac{Q^2}{2g(wd)^2}$$

$$1.57 \text{ m} = d_2 + \frac{\left(4.54 \ \dfrac{\text{m}^3}{\text{s}}\right)^2}{(2)\left(9.81 \ \dfrac{\text{m}}{\text{s}^2}\right)(2.1 \text{ m})^2(d_2)^2}$$

By trial and error or a calculator's equation solver, the alternate depths are $d_2 = 0.46$ m, 1.46 m.

Since the 0.46 m depth is not accessible from the initial depth of 1.8 m, the depth over the step is 1.5 m. The drop in the water level is

$$1.8 \text{ m} - (1.46 \text{ m} + 0.3 \text{ m}) = 0.04 \text{ m}$$

*Customary U.S. Solution*

The initial specific energy is found from Eq. 21.66.

$$E_1 = d + \frac{\text{v}^2}{2g}$$

$$= 6 \text{ ft} + \frac{\left(4 \ \dfrac{\text{ft}}{\text{sec}}\right)^2}{(2)\left(32.2 \ \dfrac{\text{ft}}{\text{sec}^2}\right)}$$

$$= 6.25 \text{ ft}$$

From Eq. 21.72, the specific energy over the step is

$$E_2 = E_1 + z_1 - z_2 = 6.25 \text{ ft} + 0 - 1 \text{ ft}$$

$$= 5.25 \text{ ft}$$

The quantity flowing is

$$Q = A\text{v} = (7 \text{ ft})(6 \text{ ft})\left(4 \ \frac{\text{ft}}{\text{sec}}\right)$$

$$= 168 \text{ ft}^3/\text{sec}$$

Substituting $Q$ into Eq. 21.69 gives

$$E = d + \frac{Q^2}{2g(wd)^2}$$

$$5.25 \text{ ft} = d_2 + \frac{\left(168 \ \dfrac{\text{ft}^3}{\text{sec}}\right)^2}{(2)\left(32.2 \ \dfrac{\text{ft}}{\text{sec}^2}\right)(7 \text{ ft})^2(d_2)^2}$$

By trial and error or a calculator's equation solver, the alternate depths are $d_2 = 1.6$ ft, 4.9 ft.

Since the 1.6 ft depth is not accessible from the initial depth of 6 ft, the depth over the step is 4.9 ft. The drop in the water level is

$$6 \text{ ft} - (4.9 \text{ ft} + 1 \text{ ft}) = 0.1 \text{ ft}$$

## 25. CRITICAL FLOW AND CRITICAL DEPTH IN RECTANGULAR CHANNELS

There is one depth, known as the *critical depth*, that minimizes the energy of flow. (The depth is not minimized, however.) The critical depth for a given flow depends on the shape of the channel.

For a rectangular channel, if Eq. 21.69 is differentiated with respect to depth in order to minimize the specific energy, Eq. 21.74 results.

$$d_c^3 = \frac{Q^2}{gw^2} \quad \text{[rectangular]} \qquad 21.74$$

Geometrical and analytical methods can be used to correlate the critical depth and the minimum specific energy.

$$d_c = \tfrac{2}{3} E_c \qquad\qquad 21.75$$

For a rectangular channel, $Q = d_c w v_c$. Substituting this into Eq. 21.74 produces an equation for the *critical velocity*.

$$v_c = \sqrt{g d_c} \qquad\qquad 21.76$$

The expression for critical velocity also coincides with the expression for the velocity of a low-amplitude *surface wave (surge wave)* moving in a liquid of depth $d_c$. Since surface disturbances are transmitted as ripples upstream (and downstream) at velocity $v_c$, it is apparent that a surge wave will be stationary in a channel moving at the critical velocity. Such motionless waves are known as *standing waves*.

If the flow velocity is less than the surge wave velocity (for the actual depth), then a ripple can make its way upstream. If the flow velocity exceeds the surge wave velocity, the ripple will be swept downstream.

**Example 21.8**

500 ft$^3$/sec (14 m$^3$/s) of water flow in a 20 ft (6 m) wide rectangular channel. What are the (a) critical depth and (b) critical velocity?

*SI Solution*

(a) From Eq. 21.74, the critical depth is

$$d_c^3 = \frac{Q^2}{gw^2}$$

$$= \frac{\left(14 \; \frac{\text{m}^3}{\text{s}}\right)^2}{\left(9.81 \; \frac{\text{m}}{\text{s}^2}\right)(6 \text{ m})^2}$$

$$d_c = 0.822 \text{ m}$$

(b) From Eq. 21.76, the critical velocity is

$$v_c = \sqrt{g d_c} = \sqrt{\left(9.81 \; \frac{\text{m}}{\text{s}^2}\right)(0.822 \text{ m})}$$

$$= 2.84 \text{ m/s}$$

*Customary U.S. Solution*

(a) From Eq. 21.74, the critical depth is

$$d_c^3 = \frac{Q^2}{gw^2}$$

$$= \frac{\left(500 \; \frac{\text{ft}^3}{\text{sec}}\right)^2}{\left(32.2 \; \frac{\text{ft}}{\text{sec}^2}\right)(20 \text{ ft})^2}$$

$$d_c = 2.687 \text{ ft}$$

(b) From Eq. 21.76, the critical velocity is

$$v_c = \sqrt{g d_c} = \sqrt{\left(32.2 \; \frac{\text{ft}}{\text{sec}^2}\right)(2.687 \text{ ft})}$$

$$= 9.30 \text{ ft/sec}$$

## 26. CRITICAL FLOW AND CRITICAL DEPTH IN NONRECTANGULAR CHANNELS

For nonrectangular shapes (including trapezoidal channels), the critical depth can be found by trial and error from the following equation in which $T$ is the surface width. To use Eq. 21.77, assume trial values of the critical depth, use them to calculate dependent quantities in the equation, and then verify the equality.

$$\frac{Q^2}{g} = \frac{A^3}{T} \quad \text{[nonrectangular]} \qquad 21.77$$

Equation 21.77 is particularly difficult to use with circular channels. Appendix 21.D is a convenient method of determining critical depth in circular channels.

## 27. FROUDE NUMBER

The dimensionless *Froude number*, Fr, is a convenient index of the flow regime. It can be used to determine whether the flow is subcritical or supercritical. $L$ is the *characteristic length*, also referred to as the *characteristic (length) scale*, hydraulic depth, mean hydraulic depth, and others, depending on the channel configuration. $d$ is the depth corresponding to velocity v. For circular channels flowing half full, $L = \pi D/8$. For a rectangular channel, $L = d$. For trapezoidal and semicircular channels, and in general, $L$ is the area in flow divided by the top width, $T$.

$$\text{Fr} = \frac{v}{\sqrt{gL}} \qquad\qquad 21.78$$

When the Froude number is less than one, the flow is subcritical (i.e., the depth of flow is greater than the critical depth) and the velocity is less than the critical velocity.

For convenience, the Froude number can be written in terms of the flow rate per average unit width.

$$\text{Fr} = \frac{\dfrac{Q}{b}}{\sqrt{gd^3}} \quad \text{[rectangular]} \qquad 21.79$$

$$\text{Fr} = \frac{\dfrac{Q}{b_{\text{ave}}}}{\sqrt{g\left(\dfrac{A}{b_{\text{ave}}}\right)^3}} \quad \text{[nonrectangular]} \qquad 21.80$$

When the Froude number is greater than one, the flow is supercritical. The depth is less than critical depth, and the flow velocity is greater than the critical velocity.

When the Froude number is equal to one, the flow is critical.[13]

The Froude number has another form. Dimensional analysis determines it to be $v^2/gL$, a form that is also used in analyzing similarity of models. Whether the derived form or the square root form is used can sometimes be determined by observing the form of the intended application. If the Froude number is squared (as it is in Eq. 21.81), then the square root form is probably intended. For open channel flow, the Froude number is always the square root of the derived form.

## 28. PREDICTING OPEN CHANNEL FLOW BEHAVIOR

Upon encountering a variation in the channel bottom, the behavior of an open channel flow is dependent on whether the flow is initially subcritical or supercritical. Open channel flow is governed by Eq. 21.81 in which the Froude number is the primary independent variable.

$$\frac{dd}{dx}(1 - \text{Fr}^2) + \frac{dz}{dx} = 0 \qquad 21.81$$

The quantity $dd/dx$ is the slope of the surface (i.e., it is the derivative of the depth with respect to the channel length). The quantity $dz/dx$ is the slope of the channel bottom.

For an upward step, $dz/dx > 0$. If the flow is initially subcritical (i.e., $\text{Fr} < 1$), then Eq. 21.81 requires that $dd/dx < 0$, a drop in depth.

This logic can be repeated for other combinations of the terms. Table 21.5 lists the various behaviors of open channel flow surface levels based on Eq. 21.81.

If $dz/dx = 0$ (i.e., a horizontal slope), then either the depth must be constant or the Froude number must be unity. The former case is obvious. The latter case predicts critical flow. Such critical flow actually occurs where the slope is horizontal over broad-crested weirs

---

[13]The similarity of the Froude number to the Mach number used to classify gas flows is more than coincidental. Both bodies of knowledge employ parallel concepts.

**Table 21.5** *Surface Level Change Behavior*

| initial flow | step up | step down |
|---|---|---|
| subcritical | surface drops | surface rises |
| supercritical | surface rises | surface drops |

(see Fig. 21.17) and at the top of a rounded spillway. Since broad-crested weirs and spillways produce critical flow, they represent a class of controls on flow.

## 29. OCCURRENCES OF CRITICAL FLOW

The critical depth not only minimizes the energy of flow, but also maximizes the quantity flowing for a given cross section and slope. Critical flow is generally quite turbulent because of the large changes in energy that occur with small changes in elevation and depth. Critical depth flow is often characterized by successive water surface undulations over a very short stretch of channel. (See Fig. 21.15.)

**Figure 21.15** *Occurrence of Critical Depth*

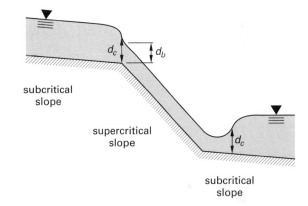

For any given discharge and cross section, there is a unique slope that will produce and maintain flow at critical depth. Once $d_c$ is known, this critical slope can be found from the Manning equation. In all of the instances of critical depth, Eq. 21.76 can be used to calculate the actual velocity.

Critical depth occurs at free outfall from a channel of mild slope. The occurrence is at the point of curvature inversion, just upstream from the brink. (See Fig. 21.16.) For mild slopes, the *brink depth* is approximately

$$d_b = 0.715 d_c \qquad 21.82$$

Critical flow can occur across a broad-crested weir, as shown in Fig. 21.17.[14] With no obstruction to hold the water, it falls from the normal depth to the critical depth, but it can fall no more than that because there is no source to increase the specific energy (to increase the velocity). This is not a contradiction of the previous

---

[14]Figure 21.17 is an example of a *hydraulic drop*, the opposite of a hydraulic jump. A hydraulic drop can be recognized by the sudden decrease in depth over a short length of channel.

**Figure 21.16** *Free Outfall*

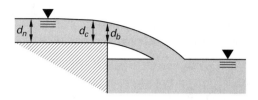

free outfall case where the brink depth is less than the critical depth. The flow curvatures in free outfall are a result of the constant gravitational acceleration.

**Figure 21.17** *Broad-Crested Weir*

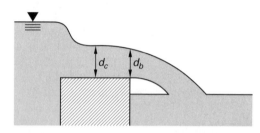

Critical depth can also occur when a channel bottom has been raised sufficiently to choke the flow. A raised channel bottom is essentially a broad-crested weir. (See Fig. 21.18.)

**Figure 21.18** *Raised Channel Bottom with Choked Flow*

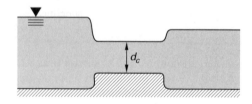

**Example 21.9**

At a particular point in an open rectangular channel ($n = 0.013$, $S = 0.002$, and $w = 10$ ft (3 m)), the flow is 250 ft$^3$/sec (7 m$^3$/s) and the depth is 4.2 ft (1.3 m).

(a) Is the flow tranquil, critical, or rapid?

(b) What is the normal depth?

(c) If the channel ends in a free outfall, what is the brink depth?

*SI Solution*

(a) From Eq. 21.74, the critical depth is

$$d_c = \left(\frac{Q^2}{gw^2}\right)^{1/3}$$

$$= \left(\frac{\left(7\ \frac{\text{m}^3}{\text{s}}\right)^2}{(3\ \text{m})^2\left(9.81\ \frac{\text{m}}{\text{s}^2}\right)}\right)^{1/3}$$

$$= 0.82\ \text{m}$$

Since the actual depth exceeds the critical depth, the flow is tranquil.

(b) From Eq. 21.13,

$$Q = \left(\frac{1.00}{n}\right)AR^{2/3}\sqrt{S}$$

$$R = \frac{A}{P} = \frac{d_n\,(3\ \text{m})}{2d_n + 3\ \text{m}}$$

Substitute the expression for $R$ into Eq. 21.13 and solve for $d_n$.

$$7\ \frac{\text{m}^3}{\text{s}} = \left(\frac{1.00}{0.013}\right)d_n(3\ \text{m})\left(\frac{d_n(3\ \text{m})}{2d_n + 3\ \text{m}}\right)^{2/3}\sqrt{0.002}$$

By trial and error or a calculator's equation solver, $d_n = 0.97$ m. Since the actual and normal depths are different, the flow is nonuniform.

(c) From Eq. 21.82, the brink depth is

$$d_b = 0.715d_c = (0.715)(0.82\ \text{m})$$

$$= 0.59\ \text{m}$$

*Customary U.S. Solution*

(a) From Eq. 21.74, the critical depth is

$$d_c = \left(\frac{Q^2}{gw^2}\right)^{1/3}$$

$$= \left(\frac{\left(250\ \frac{\text{ft}^3}{\text{sec}}\right)^2}{\left(32.2\ \frac{\text{ft}}{\text{sec}^2}\right)(10\ \text{ft})^2}\right)^{1/3}$$

$$= 2.69\ \text{ft}$$

Since the actual depth exceeds the critical depth, the flow is tranquil.

(b) From Eq. 21.13,

$$Q = \left(\frac{1.49}{n}\right) A R^{2/3} \sqrt{S}$$

$$R = \frac{A}{P} = \frac{d_n(10 \text{ ft})}{2d_n + 10 \text{ ft}}$$

Substitute the expression for $R$ into Eq. 21.13 and solve for $d_n$.

$$250 \; \frac{\text{ft}^3}{\text{sec}} = \left(\frac{1.49}{0.013}\right) d_n(10 \text{ ft}) \left(\frac{d_n(10 \text{ ft})}{2d_n + 10 \text{ ft}}\right)^{2/3} \sqrt{0.002}$$

By trial and error or a calculator's equation solver, $d_n = 3.1$ ft. Since the actual and normal depths are different, the flow is nonuniform.

(c) From Eq. 21.82, the brink depth is

$$d_b = 0.715 d_c$$
$$= (0.715)(2.69 \text{ ft})$$
$$= 1.92 \text{ ft}$$

## 30. CONTROLS ON FLOW

In general, any feature that affects depth and discharge rates is known as a *control on flow*. Controls may consist of constructed control structures (weirs, gates, sluices, etc.), forced flow through critical depth (as in a free outfall), sudden changes of slope (which forces a hydraulic jump or hydraulic drop to the new normal depth), or free flow between reservoirs of different surface elevations. A downstream control may also be an upstream control, as Fig. 21.19 shows.

**Figure 21.19** *Control on Flow*

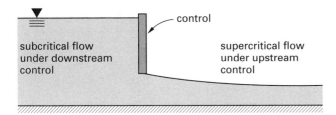

If flow is subcritical, then a disturbance downstream will be able to affect the upstream conditions. Since the flow velocity is less than the critical velocity, a ripple will be able to propagate upstream to signal a change in the downstream conditions. Any object downstream that affects the flow rate, velocity, or depth upstream is known as a *downstream control*.

If a flow is supercritical, then a downstream obstruction will have no effect upstream, since disturbances cannot propagate upstream faster than the flow velocity. The only effect on supercritical flow is from an upstream obstruction. Such an obstruction is said to be an *upstream control*.

## 31. FLOW CHOKING

A channel feature that causes critical flow to occur is known as a *choke*, and the corresponding flow past the feature and downstream is known as *choked flow*.

In the case of vertical transitions (i.e., upward or downward steps in the channel bottom), choked flow will occur when the step size is equal to the difference between the upstream specific energy and the critical flow energy.

$$\Delta z = E_1 - E_c \quad \text{[choked flow]} \qquad 21.83$$

In the case of a rectangular channel, the maximum variation in channel bottom will be

$$\Delta z = E_1 - \left(d_c + \frac{\text{v}_c^2}{2g}\right)$$
$$= E_1 - \tfrac{3}{2}d_c \qquad 21.84$$

The flow downstream from a choke point can be subcritical or supercritical, depending on the downstream conditions. If there is a downstream control, such as a sluice gate, the flow downstream will be subcritical. If there is additional gravitational acceleration (as with flow down the side of a dam spillway), then the flow will be supercritical.

## 32. VARIED FLOW

*Accelerated flow* occurs in any channel where the actual slope exceeds the friction loss per foot.

$$S_0 > \frac{h_f}{L} \qquad 21.85$$

*Retarded flow* occurs when the actual slope is less than the unit friction loss.

$$S_0 < \frac{h_f}{L} \qquad 21.86$$

In sections AB and CD of Fig. 21.20, the slopes are less than the energy gradient, so the flows are retarded. In section BC, the slope is greater than the energy gradient, so the velocity increases (i.e., the flow is accelerated). If section BC were long enough, the friction loss would eventually become equal to the accelerating energy and the flow would become uniform.

The distance between points 1 and 2 with two known depths in accelerated or retarded flow can be determined from the average velocity. Equation 21.87 and

**Figure 21.20** Varied Flow

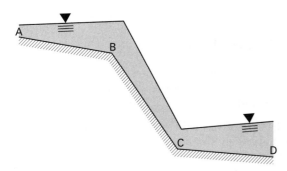

Eq. 21.88 assume that the friction losses are the same for varied flow as for uniform flow.

$$S_{\text{ave}} = \left(\frac{n v_{\text{ave}}}{R_{\text{ave}}^{2/3}}\right)^2 \qquad \text{[SI]} \quad 21.87(a)$$

$$S_{\text{ave}} = \left(\frac{n v_{\text{ave}}}{1.49 R_{\text{ave}}^{2/3}}\right)^2 \qquad \text{[U.S.]} \quad 21.87(b)$$

$$v_{\text{ave}} = \tfrac{1}{2}(v_1 + v_2) \qquad 21.88$$

$S$ is the slope of the energy gradient from Eq. 21.87, not the channel slope $S_0$. The usual method of finding the *depth profile* is to start at a point in the channel where $d_2$ and $v_2$ are known. Then, assume a depth $d_1$, find $v_1$ and $S$, and solve for $L$. Repeat as needed.

$$L = \frac{\left(d_1 + \dfrac{v_1^2}{2g}\right) - \left(d_2 + \dfrac{v_2^2}{2g}\right)}{S - S_0}$$

$$= \frac{E_1 - E_2}{S - S_0} \qquad 21.89$$

In Eq. 21.88 and Eq. 21.89, $d_1$ is always the smaller of the two depths.

### Example 21.10

How far from the point described in Ex. 21.9 will the depth be 4 ft (1.2 m)?

*SI Solution*

The difference between 1.3 m and 1.2 m is small, so a one-step calculation will probably be sufficient.

$$d_1 = 1.2 \text{ m}$$

$$v_1 = \frac{Q}{A} = \frac{7 \, \dfrac{\text{m}^3}{\text{s}}}{(1.2 \text{ m})(3 \text{ m})}$$

$$= 1.94 \text{ m/s}$$

$$E_1 = d_1 + \frac{v_1^2}{2g} = 1.2 \text{ m} + \frac{\left(1.94 \, \dfrac{\text{m}}{\text{s}}\right)^2}{(2)\left(9.81 \, \dfrac{\text{m}}{\text{s}^2}\right)}$$

$$= 1.39 \text{ m}$$

$$R_1 = \frac{A_1}{P_1} = \frac{(1.2 \text{ m})(3 \text{ m})}{1.2 \text{ m} + 3 \text{ m} + 1.2 \text{ m}}$$

$$= 0.67 \text{ m}$$

$$d_2 = 1.3 \text{ m}$$

$$v_2 = \frac{7 \, \dfrac{\text{m}^3}{\text{s}}}{(1.3 \text{ m})(3 \text{ m})}$$

$$= 1.79 \text{ m/s}$$

$$E_2 = d_2 + \frac{v_2^2}{2g} = 1.3 \text{ m} + \frac{\left(1.79 \, \dfrac{\text{m}}{\text{s}}\right)^2}{(2)\left(9.81 \, \dfrac{\text{m}}{\text{s}^2}\right)}$$

$$= 1.46 \text{ m}$$

$$R_2 = \frac{A_2}{P_2} = \frac{(1.3 \text{ m})(3 \text{ m})}{1.3 \text{ m} + 3 \text{ m} + 1.3 \text{ m}}$$

$$= 0.70 \text{ m}$$

$$v_{\text{ave}} = \tfrac{1}{2}(v_1 + v_2) = \left(\tfrac{1}{2}\right)\left(1.94 \, \dfrac{\text{m}}{\text{s}} + 1.79 \, \dfrac{\text{m}}{\text{s}}\right)$$

$$= 1.865 \text{ m/s}$$

$$R_{\text{ave}} = \tfrac{1}{2}(R_1 + R_2) = \left(\tfrac{1}{2}\right)(0.70 \text{ m} + 0.67 \text{ m})$$

$$= 0.685 \text{ m}$$

From Eq. 21.87,

$$S = \left(\frac{n v_{\text{ave}}}{R_{\text{ave}}^{2/3}}\right)^2$$

$$= \left(\frac{(0.013)\left(1.865 \, \dfrac{\text{m}}{\text{s}}\right)}{(1.00)(0.685 \text{ m})^{2/3}}\right)^2$$

$$= 0.000973$$

From Eq. 21.89,

$$L = \frac{E_1 - E_2}{S - S_0} = \frac{1.39 \text{ m} - 1.46 \text{ m}}{0.000973 - 0.002}$$

$$= 68.2 \text{ m}$$

*Customary U.S. Solution*

The difference between 4 ft and 4.2 ft is small, so a one-step calculation will probably be sufficient.

$$d_1 = 4 \text{ ft}$$

$$v_1 = \frac{Q}{A} = \frac{250 \, \dfrac{\text{ft}^3}{\text{sec}}}{(4 \text{ ft})(10 \text{ ft})}$$

$$= 6.25 \text{ ft/sec}$$

$$E_1 = d_1 + \frac{\text{v}_1^2}{2g}$$

$$= 4 \text{ ft} + \frac{\left(6.25 \; \dfrac{\text{ft}}{\text{sec}}\right)^2}{(2)\left(32.2 \; \dfrac{\text{ft}}{\text{sec}^2}\right)}$$

$$= 4.607 \text{ ft}$$

$$R_1 = \frac{A_1}{P_1} = \frac{(4 \text{ ft})(10 \text{ ft})}{4 \text{ ft} + 10 \text{ ft} + 4 \text{ ft}}$$

$$= 2.22 \text{ ft}$$

$$d_2 = 4.2 \text{ ft}$$

$$\text{v}_2 = \frac{250 \; \dfrac{\text{ft}^3}{\text{sec}}}{(4.2 \text{ ft})(10 \text{ ft})}$$

$$= 5.95 \text{ ft/sec}$$

$$E_2 = d_2 + \frac{\text{v}_2^2}{2g} = 4.2 \text{ ft} + \frac{\left(5.95 \; \dfrac{\text{ft}}{\text{sec}}\right)^2}{(2)\left(32.2 \; \dfrac{\text{ft}}{\text{sec}^2}\right)}$$

$$= 4.75 \text{ ft}$$

$$R_2 = \frac{A_2}{P_2} = \frac{(4.2 \text{ ft})(10 \text{ ft})}{4.2 \text{ ft} + 10 \text{ ft} + 4.2 \text{ ft}}$$

$$= 2.28 \text{ ft}$$

$$\text{v}_{\text{ave}} = \tfrac{1}{2}(\text{v}_1 + \text{v}_2)$$

$$= \left(\tfrac{1}{2}\right)\left(6.25 \; \frac{\text{ft}}{\text{sec}} + 5.95 \; \frac{\text{ft}}{\text{sec}}\right)$$

$$= 6.1 \text{ ft/sec}$$

$$R_{\text{ave}} = \tfrac{1}{2}(R_1 + R_2)$$

$$= \left(\tfrac{1}{2}\right)(2.22 \text{ ft} + 2.28 \text{ ft})$$

$$= 2.25 \text{ ft}$$

From Eq. 21.87,

$$S = \left(\frac{n\text{v}_{\text{ave}}}{1.49 R_{\text{ave}}^{2/3}}\right)^2$$

$$= \left(\frac{(0.013)\left(6.1 \; \dfrac{\text{ft}}{\text{sec}}\right)}{(1.49)(2.25 \text{ ft})^{2/3}}\right)^2$$

$$= 0.000965$$

From Eq. 21.89,

$$L = \frac{E_1 - E_2}{S - S_0}$$

$$= \frac{4.607 \text{ ft} - 4.75 \text{ ft}}{0.000965 - 0.002} = 138 \text{ ft}$$

## 33. HYDRAULIC JUMP

If water is introduced at high (supercritical) velocity to a section of slow-moving (subcritical) flow (as in Fig. 21.21), the velocity will be reduced rapidly over a short length of channel. The abrupt rise in the water surface is known as a *hydraulic jump*. The increase in depth is always from below the critical depth to above the critical depth.[15] The depths on either side of the hydraulic jump are known as *conjugate depths*. The conjugate depths and the relationship between them are as follows.

$$d_1 = -\tfrac{1}{2}d_2 + \sqrt{\frac{2\text{v}_2^2 d_2}{g} + \frac{d_2^2}{4}} \quad \begin{bmatrix} \text{rectangular} \\ \text{channels} \end{bmatrix} \qquad 21.90$$

$$d_2 = -\tfrac{1}{2}d_1 + \sqrt{\frac{2\text{v}_1^2 d_1}{g} + \frac{d_1^2}{4}} \quad \begin{bmatrix} \text{rectangular} \\ \text{channels} \end{bmatrix} \qquad 21.91$$

$$\frac{d_2}{d_1} = \tfrac{1}{2}\left(\sqrt{1 + 8(\text{Fr}_1)^2} - 1\right) \quad \begin{bmatrix} \text{rectangular} \\ \text{channels} \end{bmatrix} \qquad 21.92(a)$$

$$\frac{d_1}{d_2} = \tfrac{1}{2}\left(\sqrt{1 + 8(\text{Fr}_2)^2} - 1\right) \quad \begin{bmatrix} \text{rectangular} \\ \text{channels} \end{bmatrix} \qquad 21.92(b)$$

**Figure 21.21** *Conjugate Depths*

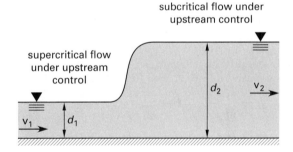

If the depths $d_1$ and $d_2$ are known, then the upstream velocity can be found from Eq. 21.93.

$$\text{v}_1^2 = \left(\frac{gd_2}{2d_1}\right)(d_1 + d_2) \quad \begin{bmatrix} \text{rectangular} \\ \text{channels} \end{bmatrix} \qquad 21.93$$

Conjugate depths are not the same as alternate depths. Alternate depths are derived from the conservation of energy equation (i.e., a variation of the Bernoulli equation). Conjugate depths (calculated in Eq. 21.90 through Eq. 21.92) are derived from a conservation of momentum equation. Conjugate depths are calculated only when there has been an abrupt energy loss such as occurs in a hydraulic jump or drop.

---

[15]This provides a way of determining if a hydraulic jump can occur in a channel. If the original depth is above the critical depth, the flow is already subcritical. Therefore, a hydraulic jump cannot form. Only a hydraulic drop could occur.

Hydraulic jumps have practical applications in the design of stilling basins. Stilling basins are designed to intentionally reduce energy of flow through hydraulic jumps. In the case of a concrete apron at the bottom of a dam spillway, the apron friction is usually low, and the water velocity will decrease only gradually. However, supercritical velocities can be reduced to much slower velocities by having the flow cross a series of baffles on the channel bottom.

The specific energy lost in the jump is the energy lost per pound of water flowing.

$$\Delta E = \left(d_1 + \frac{v_1^2}{2g}\right) - \left(d_2 + \frac{v_2^2}{2g}\right) \approx \frac{(d_2 - d_1)^3}{4d_1 d_2} \qquad 21.94$$

Evaluation of hydraulic jumps in stilling basins starts by determining the depth at the toe. The depth of flow at the toe of a spillway is found from an energy balance. Neglecting friction, the total energy at the toe equals the total upstream energy before the spillway. The total upstream energy before the spillway is

$$E_{\text{upstream}} = E_{\text{toe}} \qquad 21.95$$

$$y_{\text{crest}} + H + \frac{v^2}{2g} = d_{\text{toe}} + \frac{v_{\text{toe}}^2}{2g} \qquad 21.96$$

The upstream velocity, v, is the velocity before the spill-way (which is essentially zero), not the velocity over the brink. If the brink depth is known, the velocity over the brink can be used with the continuity equation to calculate the upstream velocity, but the velocity over the brink should not be used with $H$ to determine total energy since $v_{\text{brink}} \neq v$. (See Fig. 21.22.)

**Figure 21.22** *Total Energy Upstream of a Spillway*

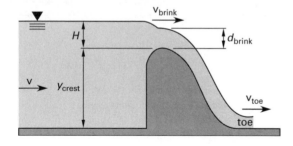

If water is drained quickly from the apron so that the tailwater depth is small or zero, no hydraulic jump will form. This is because the tailwater depth is already less than the critical depth.

A hydraulic jump will form along the apron at the bottom of the spillway when the actual tailwater depth equals the conjugate depth $d_2$ corresponding to the depth at the toe. That is, the jump is located at the toe when $d_2 = d_{\text{tailwater}}$, where $d_2$ and $d_1 = d_{\text{toe}}$ are conjugate depths. The tailwater and toe depths are implicitly the conjugate depths. This is shown in

Fig. 21.23(a) and is the proper condition for energy dissipation in a stilling basin.

When the actual tailwater depth is less than the conjugate depth $d_2$ corresponding to $d_{\text{toe}}$, but still greater than the critical depth, flow will continue along the apron until the depth increases to conjugate depth $d_1$ corresponding to the actual tailwater depth. This is shown in Fig. 21.23(b). A hydraulic jump will form at that point to increase the depth to the tailwater depth. (Another way of saying this is that the hydraulic jump moves downstream from the toe.) This is an undesirable condition, since the location of the jump is often inadequately protected from scour.

If the tailwater depth is greater than the conjugate depth corresponding to the depth at the toe, as in Fig. 21.23(c), the hydraulic jump may occur up on the spill-way, or it may be completely submerged (i.e., it will not occur at all).

**Figure 21.23** *Hydraulic Jump to Reach Tailwater Level*

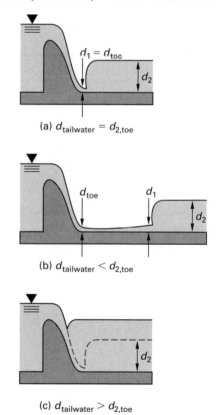

(a) $d_{\text{tailwater}} = d_{2,\text{toe}}$

(b) $d_{\text{tailwater}} < d_{2,\text{toe}}$

(c) $d_{\text{tailwater}} > d_{2,\text{toe}}$

### Example 21.11

A hydraulic jump is produced at a point in a 10 ft (3 m) wide channel where the depth is 1 ft (0.3 m). The flow rate is 200 ft³/sec (5.7 m³/s). (a) What is the depth after the jump? (b) What is the total power dissipated?

*SI Solution*

(a) From Eq. 21.68,

$$v_1 = \frac{Q}{A} = \frac{5.7 \ \frac{m^3}{s}}{(3 \ m)(0.3 \ m)}$$
$$= 6.33 \ m/s$$

From Eq. 21.91,

$$d_2 = -\tfrac{1}{2}d_1 + \sqrt{\frac{2v_1^2 d_1}{g} + \frac{d_1^2}{4}}$$

$$= -\left(\tfrac{1}{2}\right)(0.3 \ m)$$

$$+ \sqrt{\frac{(2)\left(6.33 \ \frac{m}{s}\right)^2 (0.3 \ m)}{9.81 \ \frac{m}{s^2}} + \frac{(0.3 \ m)^2}{4}}$$

$$= 1.42 \ m$$

(b) The mass flow rate is

$$\dot{m} = \left(5.7 \ \frac{m^3}{s}\right)\left(1000 \ \frac{kg}{m^3}\right)$$
$$= 5700 \ kg/s$$

The velocity after the jump is

$$v_2 = \frac{Q}{A_2} = \frac{5.7 \ \frac{m^3}{s}}{(3 \ m)(1.42 \ m)}$$
$$= 1.33 \ m/s$$

From Eq. 21.94, the change in specific energy is

$$\Delta E = \left(d_1 + \frac{v_1^2}{2g}\right) - \left(d_2 + \frac{v_2^2}{2g}\right)$$

$$= \left(0.3 \ m + \frac{\left(6.33 \ \frac{m}{s}\right)^2}{(2)\left(9.81 \ \frac{m}{s^2}\right)}\right)$$

$$- \left(1.423 \ m + \frac{\left(1.33 \ \frac{m}{s}\right)^2}{(2)\left(9.81 \ \frac{m}{s^2}\right)}\right)$$

$$= 0.83 \ m$$

The total power dissipated is

$$P = \dot{m}g\Delta E$$

$$= \left(5700 \ \frac{kg}{s}\right)\left(9.81 \ \frac{m}{s^2}\right)\left(\frac{0.83 \ m}{1000 \ \frac{W}{kW}}\right)$$

$$= 46.4 \ kW$$

*Customary U.S. Solution*

(a) From Eq. 21.68,

$$v_1 = \frac{Q}{A} = \frac{200 \ \frac{ft^3}{sec}}{(10 \ ft)(1 \ ft)}$$
$$= 20 \ ft/sec$$

From Eq. 21.91,

$$d_2 = -\tfrac{1}{2}d_1 + \sqrt{\frac{2v_1^2 d_1}{g} + \frac{d_1^2}{4}}$$

$$= -\left(\tfrac{1}{2}\right)(1 \ ft) + \sqrt{\frac{(2)\left(20 \ \frac{ft}{sec}\right)^2 (1 \ ft)}{32.2 \ \frac{ft}{sec^2}} + \frac{(1 \ ft)^2}{4}}$$

$$= 4.51 \ ft$$

(b) The mass flow rate is

$$\dot{m} = \left(200 \ \frac{ft^3}{sec}\right)\left(62.4 \ \frac{lbm}{ft^3}\right)$$
$$= 12{,}480 \ lbm/sec$$

The velocity after the jump is

$$v_2 = \frac{Q}{A_2} = \frac{200 \ \frac{ft^3}{sec}}{(10 \ ft)(4.51 \ ft)}$$
$$= 4.43 \ ft/sec$$

From Eq. 21.94, the change in specific energy is

$$\Delta E = \left(d_1 + \frac{v_1^2}{2g}\right) - \left(d_2 + \frac{v_2^2}{2g}\right)$$

$$= \left(1 \ ft + \frac{\left(20 \ \frac{ft}{sec}\right)^2}{(2)\left(32.2 \ \frac{ft}{sec^2}\right)}\right)$$

$$- \left(4.51 \ ft + \frac{\left(4.43 \ \frac{ft}{sec}\right)^2}{(2)\left(32.2 \ \frac{ft}{sec^2}\right)}\right)$$

$$= 2.4 \ ft$$

The total power dissipated is

$$P = \frac{\dot{m}g\Delta E}{g_c}$$

$$= \frac{\left(12{,}480 \ \frac{\text{lbm}}{\text{sec}}\right)\left(32.2 \ \frac{\text{ft}}{\text{sec}^2}\right)(2.4 \ \text{ft})}{\left(32.2 \ \frac{\text{ft-lbm}}{\text{lbf-sec}^2}\right)\left(550 \ \frac{\text{ft-lbf}}{\text{hp-sec}}\right)}$$

$$= 54.5 \ \text{hp}$$

## 34. LENGTH OF HYDRAULIC JUMP

For practical stilling basin design, it is helpful to have an estimate of the length of the hydraulic jump. Lengths of hydraulic jumps are difficult to measure because of the difficulty in defining the endpoints of the jumps. However, the length of the jump, $L$, varies within the limits of $5 < L/d_2 < 6.5$, in which $d_2$ is the conjugate depth after the jump. Where greater accuracy is warranted, Table 21.6 can be used. This table correlates the length of the jump to the upstream Froude number.

**Table 21.6** Approximate Lengths of Hydraulic Jumps

| $\text{Fr}_1$ | $L/d_2$ |
|---|---|
| 3 | 5.25 |
| 4 | 5.8 |
| 5 | 6.0 |
| 6 | 6.1 |
| 7 | 6.15 |
| 8 | 6.15 |

## 35. HYDRAULIC DROP

A *hydraulic drop* is the reverse of a hydraulic jump. If water is introduced at low (subcritical) velocity to a section of fast-moving (supercritical) flow, the velocity will be increased rapidly over a short length of channel. The abrupt drop in the water surface is known as a hydraulic drop. The decrease in depth is always from above the critical depth to below the critical depth.

Water flowing over a spillway and down a long, steep chute typically experiences a hydraulic drop, with critical depth occurring just before the brink. This is illustrated in Fig. 21.17 and Fig. 21.22.

The depths on either side of the hydraulic drop are the *conjugate depths*, which are determined from Eq. 21.90 and Eq. 21.91. The equations for calculating specific energy and power changes are the same for hydraulic jumps and drops.

## 36. ERODIBLE CHANNELS

Given an appropriate value of the Manning coefficient, the analysis of channels constructed of erodible materials is similar to that for concrete or pipe channels.

However, for design problems, maximum velocities and permissible side slopes must also be considered. The present state of knowledge is not sufficiently sophisticated to allow for precise designs. The usual uniform flow equations are insufficient because the stability of erodible channels is dependent on the properties of the channel material rather than on the hydraulics of flow. Two methods of design exist: (a) the tractive force method and (b) the simpler maximum permissible velocity method. Maximum velocities that should be used with erodible channels are given in Table 21.7.

**Table 21.7** Suggested Maximum Velocities

| soil type or lining (earth; no vegetation) | maximum permissible velocities (ft/sec) | | |
|---|---|---|---|
| | clear water | water carrying fine silts | water carrying sand and gravel |
| fine sand (noncolloidal) | 1.5 | 2.5 | 1.5 |
| sandy loam (noncolloidal) | 1.7 | 2.5 | 2.0 |
| silt loam (noncolloidal) | 2.0 | 3.0 | 2.0 |
| ordinary firm loam | 2.5 | 3.5 | 2.2 |
| volcanic ash | 2.5 | 3.5 | 2.0 |
| fine gravel | 2.5 | 5.0 | 3.7 |
| stiff clay (very colloidal) | 3.7 | 5.0 | 3.0 |
| graded, loam to cobbles (noncolloidal) | 3.7 | 5.0 | 5.0 |
| graded, silt to cobbles (colloidal) | 4.0 | 5.5 | 5.0 |
| alluvial silts (noncolloidal) | 2.0 | 3.5 | 2.0 |
| alluvial silts (colloidal) | 3.7 | 5.0 | 3.0 |
| coarse gravel (noncolloidal) | 4.0 | 6.0 | 6.5 |
| cobbles and shingles | 5.0 | 5.5 | 6.5 |
| shales and hard pans | 6.0 | 6.0 | 5.0 |

(Multiply ft/sec by 0.3 to obtain m/s.)

Source: Special Committee on Irrigation Research, ASCE, 1926.

The sides of the channel should not have a slope exceeding the natural angle of repose for the material used. Although there are other factors that determine the maximum permissible side slope, Table 21.8 lists some guidelines.

## 37. CULVERTS[16]

A *culvert* is a pipe that carries water under or through some feature (usually a road or highway) that would otherwise block the flow of water. For example,

---

[16]The methods of culvert flow analysis in this chapter are based on *Measurement of Peak Discharge at Culverts by Indirect Methods*, U.S. Department of the Interior (1968).

*Table 21.8* Recommended Side Slopes

| type of channel | side slope (horizontal:vertical) |
|---|---|
| firm rock | vertical to $\frac{1}{4}$:1 |
| concrete-lined stiff clay | $\frac{1}{2}$:1 |
| fissured rock | $\frac{1}{2}$:1 |
| firm earth with stone lining | 1:1 |
| firm earth, large channels | 1:1 |
| firm earth, small channels | $1\frac{1}{2}$:1 |
| loose, sandy earth | 2:1 |
| sandy, porous loam | 3:1 |

highways are often built at right angles to ravines draining hillsides and other watersheds. Culverts under the highway keep the construction fill from blocking the natural runoff.

Culverts are classified according to which of their ends controls the discharge capacity: inlet control or outlet control. If water can flow through and out of the culvert faster than it can enter, the culvert is under *inlet control*. If water can flow into the culvert faster than it can flow through and out, the culvert is under *outlet control*. Culverts under inlet control will always flow partially full. Culverts under outlet control can flow either partially full or full.

The culvert length is one of the most important factors in determining whether the culvert flows full. A culvert may be known as "hydraulically long" if it runs full and "hydraulically short" if it does not.[17]

All culvert design theory is closely dependent on energy conservation. However, due to the numerous variables involved, no single formula or procedure can be used to design a culvert. Culvert design is often an empirical, trial-and-error process. Figure 21.24 illustrates some of the important variables that affect culvert performance.

*Figure 21.24* Flow Profiles in Culvert Design

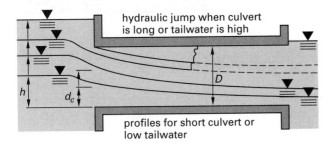

hydraulic jump when culvert is long or tailwater is high

profiles for short culvert or low tailwater

A culvert can operate with its entrance partially or totally submerged. Similarly, the exit can be partially or totally submerged, or it can have free outfall. The upstream head, $h$, is the water surface level above the lowest part of the culvert barrel, known as the *invert*.[18]

---

[17]Proper design of culvert entrances can reduce the importance of length on culvert filling.
[18]The highest part of the culvert barrel is known as the *soffit* or *crown*.

In Fig. 21.24, the three lowermost surface level profiles are of the type that would be produced with inlet control. Such a situation can occur if the culvert is short and the slope is steep. Flow at the entrance is critical as the water falls over the brink. Since critical flow occurs, the flow is choked and the inlet controls the flow rate. Downstream variations cannot be transmitted past the critical section.

If the tailwater covers the culvert exit completely (i.e., a submerged exit), the culvert will be full at that point, even though the inlet control forces the culvert to be only partially full at the inlet. The transition from partially full to totally full occurs in a hydraulic jump, the location of which depends on the flow resistance and water levels. If the flow resistance is very high, or if the headwater and tailwater levels are high enough, the jump will occur close to or at the entrance.

If the flow in a culvert is full for its entire length, then the flow is under outlet control. The discharge will be a function of the differences in tailwater and headwater levels, as well as the flow resistance along the barrel length.

## 38. DETERMINING TYPE OF CULVERT FLOW

For convenience, culvert flow is classified into six different types on the basis of the type of control, the steepness of the barrel, the relative tailwater and headwater heights, and in some cases, the relationship between critical depth and culvert size. These parameters are quantified through the use of the ratios in Table 21.9.[19] The six types are illustrated in Fig. 21.25. Identification of the type of flow beyond the guidelines in Table 21.9 requires a trial-and-error procedure.

In the following cases, several variables appear repeatedly. $C_d$ is the discharge coefficient, a function of the barrel inlet geometry. Orifice data can be used to approximate the discharge coefficient when specific information is unavailable. $v_1$ is the average velocity of the water approaching the culvert entrance and is often insignificant. The velocity-head coefficient, $\alpha$, also called the *Coriolis coefficient*, accounts for a nonuniform distribution of velocities over the channel section. However, it represents only a second-order correction and is normally neglected (i.e., assumed equal to 1.0). $d_c$ is the critical depth, which may not correspond to the actual depth of flow. (It must be calculated from the flow conditions.) $h_f$ is the friction loss in the identified section. For culverts flowing full, the friction loss can be found in the usual manner developed for pipe flow: from the Darcy formula and the Moody friction factor chart. For partial flow, the Manning equation and its variations (e.g., Eq. 21.30) can also be used. The Manning equation is particularly useful since it eliminates the

---

[19]The six cases presented here do not exhaust the various possibilities for entrance and exit control. Culvert design is complicated by this multiplicity of possible flows. Since only the easiest problems can be immediately categorized as one of the six cases, each situation needs to be carefully evaluated.

**Table 21.9** Culvert Flow Classification Parameters

| flow type | $\dfrac{h_1 - z}{D}$ | $\dfrac{h_4}{h_c}$ | $\dfrac{h_4}{D}$ | culvert slope | barrel flow | location of control | kind of control |
|---|---|---|---|---|---|---|---|
| 1 | < 1.5 | < 1.0 | ≤ 1.0 | steep | partial | inlet | critical depth |
| 2 | < 1.5 | < 1.0 | ≤ 1.0 | mild | partial | outlet | critical depth |
| 3 | < 1.5 | > 1.0 | ≤ 1.0 | mild | partial | outlet | backwater |
| 4 | > 1.0 | | ≥ 1.0 | any | full | outlet | backwater |
| 5 | ≥ 1.5 | | ≤ 1.0 | any | partial | inlet | entrance geometry |
| 6 | ≥ 1.5 | | ≤ 1.0 | any | full | outlet | entrance and barrel geometry |

**Figure 21.25** Culvert Flow Classifications

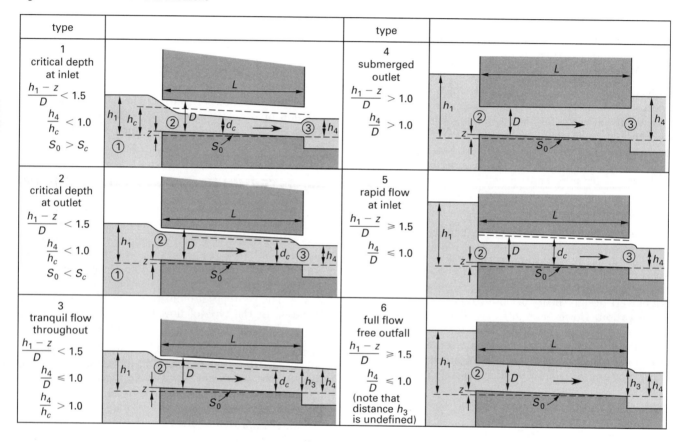

need for trial-and-error solutions. The friction head loss between sections 1 and 2, for example, can be calculated from Eq. 21.97.

$$h_{f,1-2} = \frac{LQ^2}{K_1 K_2} \qquad 21.97$$

$$K = \left(\frac{1.00}{n}\right) R^{2/3} A \qquad \text{[SI]} \quad 21.98(a)$$

$$K = \left(\frac{1.49}{n}\right) R^{2/3} A \qquad \text{[U.S.]} \quad 21.98(b)$$

The total hydraulic head available, $H$, is divided between the velocity head in the culvert, the entrance loss from Table 21.10 (if considered), and the friction.

$$H = \frac{\mathrm{v}^2}{2g} + k_e\left(\frac{\mathrm{v}^2}{2g}\right) + \frac{\mathrm{v}^2 n^2 L}{R^{4/3}} \qquad \text{[SI]} \quad 21.99(a)$$

$$H = \frac{\mathrm{v}^2}{2g} + k_e\left(\frac{\mathrm{v}^2}{2g}\right) + \frac{\mathrm{v}^2 n^2 L}{2.21 R^{4/3}} \qquad \text{[U.S.]} \quad 21.99(b)$$

Equation 21.99 can be solved directly for the velocity. Equation 21.100 is valid for culverts of any shape.

$$\mathrm{v} = \sqrt{\frac{H}{\dfrac{1 + k_e}{2g} + \dfrac{n^2 L}{R^{4/3}}}} \qquad \text{[SI]} \quad 21.100(a)$$

$$\mathrm{v} = \sqrt{\frac{H}{\dfrac{1 + k_e}{2g} + \dfrac{n^2 L}{2.21 R^{4/3}}}} \qquad \text{[U.S.]} \quad 21.100(b)$$

**Table 21.10** Minor Entrance Loss Coefficients

| $k_e$ | condition of entrance |
|---|---|
| 0.08 | smooth, tapered |
| 0.10 | flush concrete groove |
| 0.10 | flush concrete bell |
| 0.15 | projecting concrete groove |
| 0.15 | projecting concrete bell |
| 0.50 | flush, square-edged |
| 0.90 | projecting, square-edged |

## A. Type-1 Flow

Water passes through the critical depth near the culvert entrance, and the culvert flows partially full. The slope of the culvert barrel is greater than the critical slope, and the tailwater elevation is less than the elevation of the water surface at the control section.

The discharge is

$$Q = C_d A_c \sqrt{2g\left(h_1 - z + \frac{\alpha v_1^2}{2g} - d_c - h_{f,1-2}\right)} \quad \text{21.101}$$

The area, $A$, used in the discharge equation is not the culvert area since the culvert does not flow full. $A_c$ is the area in flow at the critical section.

## B. Type-2 Flow

As in type-1 flow, flow passes through the critical depth at the culvert outlet, and the barrel flows partially full. The slope of the culvert is less than critical, and the tailwater elevation does not exceed the elevation of the water surface at the control section.

$$Q = C_d A_c \sqrt{2g\left(h_1 + \frac{\alpha v_1^2}{2g} - d_c - h_{f,1-2} - h_{f,2-3}\right)} \quad \text{21.102}$$

The area, $A$, used in the discharge equation is not the culvert area since the culvert does not flow full. $A_c$ is the area in flow at the critical section.

## C. Type-3 Flow

When backwater is the controlling factor in culvert flow, the critical depth cannot occur. The upstream water-surface elevation for a given discharge is a function of the height of the tailwater. For type-3 flow, flow is subcritical for the entire length of the culvert, with the flow being partial. The outlet is not submerged, but the tailwater elevation does exceed the elevation of critical depth at the terminal section.

$$Q = C_d A_3 \sqrt{2g\left(h_1 + \frac{\alpha v_1^2}{2g} - h_3 - h_{f,1-2} - h_{f,2-3}\right)} \quad \text{21.103}$$

The area, $A$, used in the discharge equation is not the culvert area since the culvert does not flow full. $A_3$ is the area in flow at numbered section 3 (i.e., the exit).

## D. Type-4 Flow

As in type-3 flow, the backwater elevation is the controlling factor in this case. Critical depth cannot occur, and the upstream water surface elevation for a given discharge is a function of the tailwater elevation. Discharge is independent of barrel slope. The culvert is submerged at both the headwater and the tailwater. No differentiation between low head and high head is made for this case. If the velocity head at section 1 (the entrance), the entrance friction loss, and the exit friction loss are neglected, the discharge can be calculated. $A_o$ is the culvert area.

$$Q = C_d A_o \sqrt{2g\left(\frac{h_1 - h_4}{1 + \frac{29C_d^2 n^2 L}{R^{4/3}}}\right)} \quad \text{21.104}$$

The complicated term in the denominator corrects for friction. For rough estimates and for culverts less than 50 ft (15 m) long, the friction loss can be ignored.

$$Q = C_d A_o \sqrt{2g(h_1 - h_4)} \quad \text{21.105}$$

## E. Type-5 Flow

Partially full flow under a high head is classified as type-5 flow. The flow pattern is similar to the flow downstream from a sluice gate, with rapid flow near the entrance. Usually, type-5 flow requires a relatively square entrance that causes contraction of the flow area to less than the culvert area. In addition, the barrel length, roughness, and bed slope must be sufficient to keep the velocity high throughout the culvert.

It is difficult to distinguish in advance between type-5 and type-6 flow. Within a range of the important parameters, either flow can occur.[20] $A_o$ is the culvert area.

$$Q = C_d A_o \sqrt{2g(h_1 - z)} \quad \text{21.106}$$

## F. Type-6 Flow

Type-6 flow, like type-5 flow, is considered a high-head flow. The culvert is full under pressure with free outfall. The discharge is

$$Q = C_d A_o \sqrt{2g(h_1 - h_3 - h_{f,2-3})} \quad \text{21.107}$$

Equation 21.107 is inconvenient because $h_3$ (the true piezometric head at the outfall) is difficult to evaluate

---

[20]If the water surface ever touches the top of the culvert, the passage of air in the culvert will be prevented and the culvert will flow full everywhere. This is type-6 flow.

without special graphical aids. The actual hydraulic head driving the culvert flow is a function of the Froude number. For conservative first approximations, $h_3$ can be taken as the barrel diameter. This will give the minimum hydraulic head. In reality, $h_3$ varies from somewhat less than half the barrel diameter to the full diameter.

If $h_3$ is taken as the barrel diameter, the total hydraulic head $(H = h_1 - h_3)$ will be split between the velocity head and friction. In that case, Eq. 21.100 can be used to calculate the velocity. The discharge is easily calculated from Eq. 21.108.[21] $A_o$ is the culvert area.

$$Q = A_o v \qquad 21.108$$

### Example 21.12

Size a square culvert with an entrance fluid level 5 ft above the barrel top and a free exit to operate with the following characteristics.

$$\text{slope} = 0.01$$
$$\text{length} = 250 \text{ ft}$$
$$\text{capacity} = 45 \text{ ft}^3/\text{sec}$$
$$n = 0.013$$

*Solution*

Since the $h_1$ dimension is measured from the culvert invert, it is difficult to classify the type of flow at this point. However, either type 5 or type 6 is likely since the head is high.

step 1: Assume a trial culvert size. Select a square opening with 1.0 ft sides.

step 2: Calculate the flow assuming case 5 (entrance control). The entrance will act like an orifice.

$$A_o = (1 \text{ ft})(1 \text{ ft}) = 1 \text{ ft}^2$$
$$H = h_1 - z$$
$$= \left(5 \text{ ft} + 1 \text{ ft} + (0.01)(250 \text{ ft})\right)$$
$$\quad - (0.01)(250 \text{ ft})$$
$$= 6 \text{ ft}$$

$C_d$ is approximately 0.62 for square-edged openings with separation from the wall. From Eq. 21.106,

$$Q = (0.62)(1 \text{ ft}^2)\sqrt{(2)\left(32.2 \frac{\text{ft}}{\text{sec}^2}\right)(6 \text{ ft})}$$
$$= 12.2 \text{ ft}^3/\text{sec}$$

Since this size has insufficient capacity, try a larger culvert. Choose a square opening with 2.0 ft sides.

$$A_o = (2 \text{ ft})(2 \text{ ft}) = 4 \text{ ft}^2$$
$$H = h_1 - z = 5 \text{ ft} + 2 \text{ ft} = 7 \text{ ft}$$
$$Q = C_d A_o \sqrt{2g(h_1 - z)}$$
$$= (0.62)(4 \text{ ft}^2)\sqrt{(2)\left(32.2 \frac{\text{ft}}{\text{sec}^2}\right)(7 \text{ ft})}$$
$$= 52.7 \text{ ft}^3/\text{sec}$$

step 3: Begin checking the entrance control assumption by calculating the maximum hydraulic radius. The upper surface of the culvert is not wetted because the flow is entrance controlled. The hydraulic radius is maximum at the entrance.

$$R = \frac{A_o}{P}$$
$$= \frac{4 \text{ ft}^2}{2 \text{ ft} + 2 \text{ ft} + 2 \text{ ft}}$$
$$= 0.667 \text{ ft}$$

step 4: Calculate the velocity using the Manning equation for open channel flow. Since the hydraulic radius is maximum, the velocity will also be maximum.

$$v = \frac{1.49}{n} R^{2/3}\sqrt{S}$$
$$= \left(\frac{1.49}{0.013}\right)(0.667 \text{ ft})^{2/3}\sqrt{0.01}$$
$$= 8.75 \text{ ft}/\text{sec}$$

step 5: Calculate the normal depth, $d_n$.

$$d_n = \frac{Q}{vw} = \frac{45 \frac{\text{ft}^3}{\text{sec}}}{\left(8.75 \frac{\text{ft}}{\text{sec}}\right)(2 \text{ ft})}$$
$$= 2.57 \text{ ft}$$

Since the normal depth is greater than the culvert size, the culvert will flow full under pressure. (It was not necessary to calculate the critical depth since the flow is implicitly subcritical.) The entrance control assumption was, therefore, not valid for this size culvert.[22] At this point, two things can be done: A larger culvert can be chosen if entrance control is

---

[21]Equation 21.108 does not include the discharge coefficient. Velocity, v, when calculated from Eq. 21.100, is implicitly the velocity in the barrel.

[22]If the normal depth had been less than the barrel diameter, it would still be necessary to determine the critical depth of flow. If the normal depth was less than the critical depth, the entrance control assumption would have been valid.

desired, or the solution can continue by checking to see if the culvert has the required capacity as a pressure conduit.

*step 6:* Check the capacity as a pressure conduit. $H$ is the total available head.

$$H = h_1 - h_3$$
$$= \left(5 \text{ ft} + 2 \text{ ft} + (0.01)(250 \text{ ft})\right) - 2 \text{ ft}$$
$$= 7.5 \text{ ft}$$

*step 7:* Since the pipe is flowing full, the hydraulic radius is

$$R = \frac{A}{P} = \frac{4 \text{ ft}^2}{8 \text{ ft}} = 0.5 \text{ ft}$$

*step 8:* Equation 21.100 can be used to calculate the flow velocity. Since the culvert has a square-edged entrance, a loss coefficient of $k_e = 0.5$ is used. However, this does not greatly affect the velocity.

$$\text{v} = \sqrt{\frac{H}{\dfrac{1 + k_e}{2g} + \dfrac{n^2 L}{2.21 R^{4/3}}}}$$
$$= \sqrt{\frac{7.5 \text{ ft}}{\dfrac{1 + 0.5}{(2)\left(32.2 \ \dfrac{\text{ft}}{\text{sec}}\right)} + \dfrac{(0.013)^2 (250 \text{ ft})}{(2.21)(0.5 \text{ ft})^{4/3}}}}$$
$$= 10.24 \text{ ft/sec}$$

*step 9:* Check the capacity.

$$Q = \text{v} A_o$$
$$= \left(10.24 \ \frac{\text{ft}}{\text{sec}}\right)(4 \text{ ft}^2)$$
$$= 40.96 \text{ ft}^3/\text{sec}$$

The culvert size is not acceptable since its discharge under the maximum head does not have a capacity of 45 ft$^3$/sec.

*step 10:* Repeat from step 2, trying a larger-size culvert. With a 2.5 ft side, the following values are obtained.

$$A_o = (2.5 \text{ ft})(2.5 \text{ ft}) = 6.25 \text{ ft}^2$$
$$H = 5 \text{ ft} + 2.5 \text{ ft} = 7.5 \text{ ft}$$
$$Q = (0.62)(6.25 \text{ ft}^2)$$
$$\times \sqrt{(2)\left(32.2 \ \frac{\text{ft}}{\text{sec}^2}\right)(7.5 \text{ ft})}$$
$$= 85.2 \text{ ft}^3/\text{sec}$$
$$R = \frac{6.25 \text{ ft}^2}{7.5 \text{ ft}} = 0.833 \text{ ft}$$
$$\text{v} = \left(\frac{1.49}{0.013}\right)(0.833 \text{ ft})^{2/3}\sqrt{0.01}$$
$$= 10.12 \text{ ft/sec}$$
$$d_n = \frac{\left(45 \ \dfrac{\text{ft}^3}{\text{sec}}\right)}{\left(10.12 \ \dfrac{\text{ft}}{\text{sec}}\right)(2.5 \text{ ft})}$$
$$= 1.78 \text{ ft}$$

*step 11:* Calculate the critical depth. For rectangular channels, Eq. 21.74 can be used.

$$d_c = \left(\frac{Q^2}{gw^2}\right)^{1/3}$$
$$= \left(\frac{\left(45 \ \dfrac{\text{ft}^3}{\text{sec}}\right)^2}{\left(32.2 \ \dfrac{\text{ft}}{\text{sec}^2}\right)(2.5 \text{ ft})^2}\right)^{1/3}$$
$$= 2.16 \text{ ft}$$

Since the normal depth is less than the critical depth, the flow is supercritical. The entrance control assumption was correct for the culvert. The culvert has sufficient capacity to carry 45 ft$^3$/sec.

## 39. CULVERT DESIGN

Designing a culvert is somewhat easier than culvert analysis because of common restrictions placed on designers and the flexibility to change almost everything else. For example, culverts may be required to (a) never be more than 50% full (deep), (b) always be under inlet control, or (c) always operate with some minimum head (above the centerline or crown). In the absence of any specific guidelines, a culvert may be designed using the following procedure.

*step 1:* Determine the required flow rate.

*step 2:* Determine all water surface elevations, lengths, and other geometric characteristics.

*step 3:* Determine the material to be used for the culvert and its roughness.

*step 4:* Assume type 1 flow (inlet control).

*step 5:* Select a trial diameter.

*step 6:* Assume a reasonable slope.

*step 7:* Position the culvert entrance such that the ratio of headwater depth (inlet to water surface) to culvert diameter is 1:2 to 1:2.5.

*step 8:* Calculate the flow. Repeat steps 5 through 7 until the capacity is adequate.

*step 9:* Determine the location of the outlet. Check for outlet control. If the culvert is outlet controlled, repeat steps 5 through 7 using a different flow model.

*step 10:* Calculate the discharge velocity. Specify rip-rap, concrete, or other protection to prevent erosion at the outlet.

# Topic IV: Thermodynamics

Chapter

Thermodynamics

# 22 Thermodynamic Properties of Substances

## Nomenclature

| $a$ | van der Waals factor | atm-ft$^6$/lbmol | Pa·m$^6$/kmol |
| AW | atomic weight | lbm/lbmol | kg/kmol |
| $b$ | van der Waals factor | ft$^3$/lbmol | m$^3$/kmol |
| $B$ | volumetric fraction | – | – |
| $c$ | specific heat | Btu/lbm-°F | kJ/kg·°C |
| $C$ | molar specific heat | Btu/lbmol-°F | kJ/kmol·°C |
| $E$ | energy | Btu | kJ |
| $G$ | gravimetric fraction | – | – |
| $h$ | enthalpy | Btu/lbm | kJ/kg |
| $H$ | molar enthalpy | Btu/lbmol | kJ/kmol |
| $J$ | Joule's constant (778.17) | ft-lbf/Btu | n.a. |
| $k$ | ratio of specific heats | – | – |
| $\kappa$ | Boltzmann constant | n.a. | kJ/molecule·K |
| $m$ | mass | lbm | kg |
| MW | molecular weight | lbm/lbmol | kg/kmol |
| $n$ | number of moles | – | – |
| $N$ | number of molecules | – | – |
| $N_A$ | Avogadro's number | molecules/ lbmol | molecules/ kmol |
| $p$ | absolute pressure | lbf/ft$^2$ | Pa |
| $q$ | heat | Btu/lbm | kJ/kg |
| $Q$ | molar heat | Btu/lbmol | kJ/kmol |
| $Q$ | total heat | Btu | kJ |
| $R$ | specific gas constant | ft-lbf/lbm-°R | kJ/kg·K |
| $R^*$ | universal gas constant | ft-lbf/lbmol-°R | kJ/kmol·K |
| $s$ | entropy | Btu/lbm-°R | kJ/kg·K |
| $S$ | molar entropy | Btu/lbmol-°R | kJ/kmol·K |
| $T$ | temperature | °F | °C |
| $T$ | absolute temperature | °R | K |
| $u$ | internal energy | Btu/lbm | kJ/kg |
| $U$ | molar internal energy | Btu/lbmol | kJ/kmol |
| v | velocity | ft/sec | m/s |
| $V$ | molar specific volume | ft$^3$/lbmol | m$^3$/kmol |
| $V$ | volume | ft$^3$ | m$^3$ |
| $x$ | mole fraction | – | – |
| $x$ | quality | – | – |
| $Z$ | compressibility factor | – | – |

### Symbols

| $\alpha$ | isentropic compressibility | 1/°R | 1/K |
| $\beta$ | isobaric compressibility | 1/°R | 1/K |
| $\kappa$ | isothermal compressibility | 1/°R | 1/K |
| $\rho$ | density | lbm/ft$^3$ | kg/m$^3$ |
| $v$ | specific volume | ft$^3$/lbm | m$^3$/kg |
| $\phi$ | entropy function | Btu/lbm-°R | kJ/kg·K |

### Subscripts

| $A$ | Avogadro |
| $c$ | critical |
| $f$ | fluid (liquid) |
| $fg$ | liquid-to-gas (vaporization) |

Thermodynamics

$g$      gas (vapor)
$i$      ice
$k$      kinetic
$l$      latent
$m$      mean
$o$      outside (environment)
$p$      constant pressure or most probable
$r$      ratio or reduced
rms    root-mean-square
$s$      sensible or solid
sat     saturated
$t$      total
$v$      constant volume

## 1. PHASES OF A PURE SUBSTANCE

*Thermodynamics* is the study of a substance's energy-related properties. The properties of a substance and the procedures used to determine those properties depend on the state and the phase of the substance. The *thermodynamic state* of a substance is defined by two or more independent thermodynamic properties. For example, the temperature and pressure of a substance are two properties commonly used to define the state of a superheated vapor.

The common *phases* of a substance are solid, liquid, and gas. However, because substances behave according to different rules, it is convenient to categorize them into more than these three phases.[1]

*solid:* A solid does not take on the shape or volume of its container.

*subcooled liquid:* If a liquid is not saturated (i.e., the liquid is not at its boiling point), it is said to be subcooled. Water at 1 atm and room temperature is subcooled, as the addition of a small amount of heat will not cause vaporization.

*saturated liquid:* A saturated liquid has absorbed as much heat energy as it can without vaporizing. Liquid water at standard atmospheric pressure and 212°F (100°C) is an example of a saturated liquid.

*liquid-vapor mixture:* A liquid and vapor of the same substance can coexist at the same temperature and pressure. This is called a two-phase, liquid-vapor mixture.

*saturated vapor:* A vapor (e.g., steam at standard atmospheric pressure and 212°F (100°C)) that is on the verge of condensing is said to be saturated.

*superheated vapor:* A superheated vapor is one that has absorbed more heat than is needed merely to vaporize it. A superheated vapor will not condense when small amounts of heat are removed.

---

[1]Plasma, *cryogenic fluids (cryogens)* that boil at temperatures less than approximately 200°R (110K), and solids near absolute zero are not discussed in this chapter.

*ideal gas:* A gas is a highly superheated vapor. If the gas behaves according to the ideal gas law, $pV = R^*T$, it is called an ideal gas.

*real gas:* A real gas does not behave according to the ideal gas laws.

*gas mixtures:* Most gases mix together freely. Two or more pure gases together constitute a gas mixture.

*vapor/gas mixtures:* Atmospheric air is an example of a mixture of several gases and water vapor.

These phases and subphases can be illustrated with a pure substance in the piston/cylinder arrangement shown in Fig. 22.1. The pressure in this system is constant and is determined by the weight of the piston, which moves freely to permit volume changes.

**Figure 22.1** *Phase Changes at Constant Pressure*

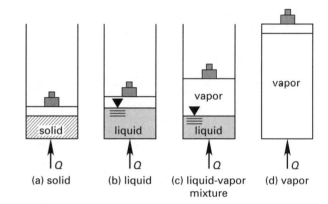

In Fig. 22.1(a) the volume is minimum. This is usually the solid phase. (Water is an exception. Solid ice has a lower density than liquid water.) The temperature will rise as heat, $Q$, is added to the solid. This increase in temperature is accompanied by a small increase in volume. The temperature increases until the melting point is reached.

The solid will begin to melt as heat is added to it at the melting point. The temperature will not increase until all of the solid has been turned into liquid. The liquid phase, with its small increase in volume, is illustrated by Fig. 22.1(b).

If the subcooled liquid continues to receive heat, its temperature will rise. This temperature increase continues until evaporation is imminent. The liquid at this point is said to be saturated. Any increase in heat energy will cause a portion of the liquid to vaporize. This is shown in Fig. 22.1(c), in which a liquid-vapor mixture exists.

As with melting, evaporation occurs at constant temperature and pressure but with a very large increase in volume. The temperature cannot increase until the last drop of liquid has been evaporated, at which point the vapor is said to be saturated. This is shown in Figure 22.1(d).

Additional heat will result in high-temperature *superheated vapor*. This vapor may or may not behave according to the ideal gas laws.

## 2. DETERMINING PHASE

It is possible to develop a three-dimensional surface that predicts the substance's phase based on the properties of pressure, temperature, and specific volume. Such an *equilibrium solid* is illustrated in Fig. 22.2.[2]

**Figure 22.2** *Equilibrium Solid*

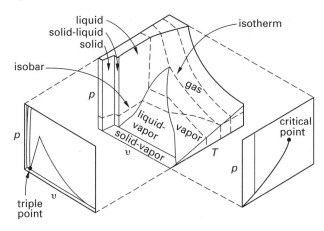

If one property is held constant during a process, a two-dimensional projection of the equilibrium solid can be used. This projection is known as an *equilibrium diagram* or a *phase diagram*, of which Fig. 22.3 is an example.

**Figure 22.3** *Constant Temperature Phase Diagram*

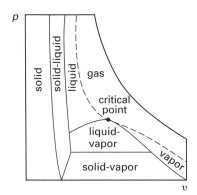

---

[2]Figure 22.2 is applicable for most substances, excluding water. In Fig. 22.2, the specific volume "steps in" (i.e., decreases) when the liquid turns to a solid as the temperature drops below freezing. However, water is unique in that it expands upon freezing. Thus, the $pVT$ diagram for water shows a "step out" instead of a "step in" upon freezing. The remainder of the $pVT$ diagram for water is the same as Fig. 22.2.

The most important part of a phase diagram is limited to the liquid-vapor region. A general phase diagram showing this region and the bell-shaped dividing line (known as the *vapor dome*) is shown in Fig. 22.4.

**Figure 22.4** *Vapor Dome with Isobars*

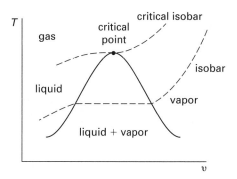

The vapor dome region can be drawn with many variables for the axes. For example, either temperature or pressure can be used for the vertical axis. Energy, specific volume, or entropy can be chosen for the horizontal axis. The principles presented here apply to all combinations.

The left-hand part of the vapor dome separates the liquid phase from the liquid-vapor phase. This part of the line is known as the *saturated liquid line*. Similarly, the right-hand part of the line separates the liquid-vapor phase from the vapor phase. This line is called the *saturated vapor line*.

Lines of constant pressure (*isobars*) can be superimposed on the vapor dome. Each isobar is horizontal as it passes through the two-phase region, indicating that both temperature and pressure remain unchanged as a liquid vaporizes.

Notice that there is no real dividing line between liquid and vapor at the top of the vapor dome. Far above the vapor dome, there is no distinction between liquids and gases, as their properties are identical. The phase is assumed to be a gas.

The implied dividing line between liquid and gas is the isobar that passes through the topmost part of the vapor dome. This is known as the *critical isobar*. The highest point of the vapor dome is known as the *critical point*. This critical isobar also provides a way to distinguish between a vapor and a gas. A substance below the critical isobar (but to the right of the vapor dome) is a vapor. Above the critical isobar, it is a gas.

Figure 22.5 illustrates a vapor dome for which pressure, $p$, has been chosen as the vertical axis, and enthalpy, $h$, has been chosen as the horizontal axis. The shape of the dome is essentially the same, but the lines of constant temperature (*isotherms*) have a different direction than isobars.

Figure 22.5 also illustrates the subscripting convention used to identify points on the saturation line. The subscript $f$ (fluid) is used to indicate a saturated liquid. The subscript $g$ (gas) is used to indicate a saturated vapor.[3] The subscript $fg$ is used to indicate the difference in saturation properties.

**Figure 22.5** Vapor Dome with Isotherms

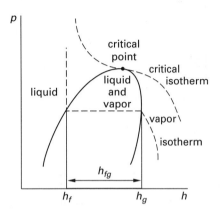

The vapor dome is a good tool for illustration, but it cannot be used to determine a substance's phase. Such a determination must be made based on the substance's pressure and temperature.

For example, consider water at a pressure of 1 atm. Its boiling temperature (the *saturation temperature*) at this pressure is 212°F (100°C). If the water has a lower temperature, for example, 85°F (29°C), the water will be liquid (i.e., will be subcooled). On the other hand, if the water's temperature (at 1 atm) is 270°F (132°C), the water must be in vapor form (i.e., must be superheated).

This example is valid only for water at atmospheric pressure. Water at other pressures will have other boiling temperatures. (The lower the pressure, the lower the boiling temperature.) However, the rules given here follow directly from the previous example and will become more meaningful as this chapter progresses.

*Rule 1:* A substance is a subcooled liquid if its temperature is less than the saturation temperature corresponding to its pressure.

*Rule 2:* A substance is in the liquid-vapor region if its temperature is equal to the saturation temperature corresponding to its pressure.

*Rule 3:* A substance is a superheated vapor if its temperature is greater than the saturation temperature corresponding to its pressure.

---

[3]Although this book makes it a rule never to call a vapor a gas, this convention is not adhered to in the field of thermodynamics. The subscript $g$ is standard for a saturated vapor.

The rules that follow can be stated using pressure as the determining variable.

*Rule 4:* A substance is a subcooled liquid if its pressure is greater than the saturation pressure corresponding to its temperature.

*Rule 5:* A substance is in the liquid-vapor region if its pressure is equal to the saturation pressure corresponding to its temperature.

*Rule 6:* A substance is a superheated vapor if its pressure is less than the saturation pressure corresponding to its temperature.

## 3. PROPERTIES OF A SUBSTANCE

The thermodynamic *state* or condition of a substance is determined by its properties. *Intensive properties* are independent of the amount of substance present. Temperature, pressure, and stress are examples of intensive properties. *Extensive properties* are dependent on the amount of substance present. Examples of extensive properties are volume, strain, charge, and mass.

In this chapter, and in most books on thermodynamics, both lowercase and uppercase forms of the same characters are used to represent property variables. The two forms are used to distinguish between the units of mass. For example, lowercase $h$ represents *specific enthalpy* (usually just called "enthalpy") in units of Btu/lbm or kJ/kg. Uppercase $H$ is used to represent the *molar enthalpy* in units of Btu/lbmol or kJ/kmol.

## 4. MASS: *m*

The mass of a substance is a measure of its quantity. Mass is independent of location and gravitational field strength. In thermodynamics, the customary U.S. and SI units of mass ($m$) are pound-mass (lbm) and kilogram (kg), respectively.

## 5. TEMPERATURE: *T*

Temperature is a thermodynamic property of a substance that depends on energy content. Heat energy entering a substance will increase the temperature of that substance. Normally, heat energy will flow only from a hot object to a cold object. If two objects are in *thermal equilibrium* (are at the same temperature), no heat will flow between them.

If two systems are in thermal equilibrium, they must be at the same temperature. If both systems are in equilibrium with a third, then all three are at the same temperature. This concept is known as the *Zeroth Law of Thermodynamics.*

The scales most commonly used for measuring temperature are the Fahrenheit and Celsius scales.[4] The relationship between these two scales is

$$T_{°F} = 32° + \frac{9}{5}T_{°C} \qquad 22.1$$

The *absolute temperature scale* defines temperature independently of the properties of any particular substance. This is unlike the Celsius and Fahrenheit scales, which are based on the freezing point of water. The absolute temperature scale should be used for all calculations.

In the customary U.S. system, the absolute scale is the *Rankine scale.*[5]

$$T_{°R} = T_{°F} + 459.67° \qquad 22.2$$

$$\Delta T_{°R} = \Delta T_{°F} \qquad 22.3$$

The absolute temperature scale in the SI system is the *Kelvin scale.*[6]

$$T_K = T_{°C} + 273.15° \qquad 22.4$$

$$\Delta T_K = \Delta T_{°C} \qquad 22.5$$

The relationships between temperature differences in the customary U.S. and SI systems are independent of the freezing point of water. (See Table 22.1.)

$$\Delta T_{°C} = \frac{5}{9}\Delta T_{°F} \qquad 22.6$$

$$\Delta T_K = \frac{5}{9}\Delta T_{°R} \qquad 22.7$$

**Table 22.1** *Temperature Scales*

|  | Kelvin | Celsius | Rankine | Fahrenheit |
|---|---|---|---|---|
| normal boiling point of water | 373.15K | 100.00°C | 671.67°R | 212.00°F |
| triple point of water (see Sec. 14) | 273.16K | 0.01°C | 491.69°R | 32.02°F |
|  | 273.15K | 0.00°C | 491.67°R | 32.00°F ice point |
| absolute zero | 0K | −273.15°C | 0°R | −459.67°F |

## 6. PRESSURE: *p*

Customary U.S. pressure units are pounds per square inch (psi). Standard SI pressure units are kPa or MPa, although bars are also used in tabulations of thermodynamic data.

---

[4]The term *centigrade* was replaced by the term *Celsius* in 1948.
[5]Normally, three significant temperature digits (i.e., 460°) are sufficient.
[6]Normally, three significant temperature digits (i.e., 273°) are sufficient.

## 7. DENSITY: *ρ*

Customary U.S. density units in tabulations of thermodynamic data are pounds per cubic foot (lbm/ft³). Standard SI density units are kilograms per cubic meter (kg/m³). Density is the reciprocal of specific volume.

$$\rho = \frac{1}{v} \qquad 22.8$$

## 8. SPECIFIC VOLUME: *v* AND *V*

Specific volume, $v$, is the volume occupied by one unit mass of a substance. Customary U.S. units in tabulations of thermodynamic data are cubic feet per pound (ft³/lbm). Standard SI specific volume units are cubic meters per kilogram (m³/kg). *Molar specific volume, V,* has units of ft³/lbmol (m³/kmol) and is seldom encountered. Specific volume is the reciprocal of density.

$$v = \frac{1}{\rho} \qquad 22.9$$

$$V = MW \times v \qquad 22.10$$

## 9. INTERNAL ENERGY: *u* AND *U*

*Internal energy* includes all of the potential and kinetic energies of the atoms or molecules in a substance. Energies in the translational, rotational, and vibrational modes are included. Since this movement increases as the temperature increases, internal energy is a function of temperature. It does not depend on the process or path taken to reach a particular temperature.

In the United States, the *British thermal unit,* Btu, is used to measure all forms of thermodynamic energy. (One Btu is approximately the energy given off by burning one wooden match.) Standard units of *specific internal energy, u,* are Btu/lbm and kJ/kg. The units of *molar internal energy, U,* are Btu/lbmol and kJ/kmol. Equation 22.11 gives the relationship between the specific and molar quantities.

$$U = MW \times u \qquad 22.11$$

## 10. ENTHALPY: *h* AND *H*

*Enthalpy* (also known at various times in history as *total heat* and *heat content*) represents the total useful energy of a substance. Useful energy consists of two parts—the internal energy, $u$, and the *flow energy* (also known as *flow work* and *p-V work*), $pV$. Therefore, enthalpy has the same units as internal energy.

$$h = u + pv \qquad 22.12$$

$$H = U + pV \qquad 22.13$$

$$H = MW \times h \qquad 22.14$$

Enthalpy is defined as useful energy because, ideally, all of it can be used to perform useful tasks. It takes energy to increase the temperature of a substance. If that internal energy is recovered, it can be used to heat something else (e.g., to vaporize water in a boiler). Also, it takes energy to increase pressure and volume (as in blowing up a balloon). If pressure and volume are decreased, useful energy is given up.

The customary U.S. units of Eq. 22.12 and Eq. 22.13 are not consistent, since flow work (as written) has units of ft-lbf/lbm, not Btu/lbm. (There is also a consistency problem if pressure is defined in lbf/ft$^2$ and given in lbf/in$^2$.) Strictly, Eq. 22.12 should be written as

$$h = u + \frac{pv}{J} \qquad \text{[U.S.]} \quad 22.15$$

The conversion factor, $J$, in Eq. 22.15 is known as *Joule's constant*. It has a value of 778.17 ft-lbf/Btu. (In SI units, Joule's constant has a value of 1.0 N·m/J and is unnecessary.) As in Eq. 22.12 and Eq. 22.13, Joule's constant is often omitted from the statement of generic thermodynamic equations, but it is always needed with customary U.S. units for dimensional consistency.

## 11. ENTROPY: $s$ AND $S$

*Absolute entropy* is a measure of the energy that is no longer available to perform useful work within the current environment. Other definitions used (the "disorder of the system," the "randomness of the system," etc.) are frequently quoted. Although these alternate definitions cannot be used in calculations, they are consistent with the *third law of thermodynamics* (also known as the *Nernst theorem*). This law states that the absolute entropy of a perfect crystalline solid in thermodynamic equilibrium is (approaches) zero when the temperature is (approaches) absolute zero.[7] Equation 22.16 expresses the third law mathematically.

$$\lim_{T \to 0K} s = 0 \qquad 22.16$$

An increase in entropy is known as *entropy production*. The total absolute entropy in a system is equal to the summation of all absolute entropy productions that have occurred over the life of the system.

$$s = \sum \Delta s \qquad 22.17$$

For an isothermal process taking place at a constant temperature $T_o$, the entropy production depends on the amount of energy transfer.

$$\Delta s = \frac{q}{T_o} \qquad 22.18$$

$$\Delta S = \frac{Q}{T_o} \qquad 22.19$$

$$\Delta S = \text{MW} \times \Delta s \qquad 22.20$$

For processes that occur over a varying temperature, the entropy production must be found by integration.

$$\Delta s = \int ds = \int \frac{dq}{T} \qquad 22.21$$

From Eq. 22.18 and Eq. 22.21, it is apparent that the units of specific absolute entropy are Btu/lbm-°R and kJ/kg·K. For molar absolute entropy, the units are Btu/lbmol-°R and kJ/kmol·K.

Unlike absolute entropy, *standardized entropy*, usually just called "entropy," is not referenced to absolute zero conditions but is measured with respect to some other convenient thermodynamic state. For water, the reference condition is the liquid phase at the triple point. (See Sec. 22.14.)

### Example 22.1

Three planets of identical size and mass are oriented in space such that radiant energy transfers can occur. The average temperatures of planets A, B, and C are 530°R, 520°R, and 510°R (294K, 289K, and 283K), respectively. All three planets are massive enough that small energy losses or gains can be considered to be isothermal processes (i.e., they will not change the average temperature).

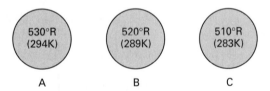

(a) Can a radiation energy transfer occur spontaneously from planet B to planet C? (b) What are the entropy productions for planets B and C if an energy transfer of 1000 Btu/lbm (2330 kJ/kg) occurs by radiation? (c) What is the overall entropy change as a result of the energy transfer in (b)? (d) Does entropy always increase? (e) Can planet B ever be returned to its original condition?

*Solution*

(a) A radiation transfer can occur spontaneously because planet B is hotter than planet C. Energy will flow spontaneously from a hot object to a cold object.

---

[7]A molecule with zero entropy exists in only one quantum state. The energy state is known precisely, without *uncertainty*.

Thermodynamics

(b) In SI units, the entropy productions are

$$\Delta s_B = \frac{q}{T_o} = \frac{-2330\ \frac{kJ}{kg}}{289K} = -8.062\ kJ/kg\cdot K$$

$$\Delta s_C = \frac{2330\ \frac{kJ}{kg}}{283K} = 8.233\ kJ/kg\cdot K$$

In customary U.S. units, the entropy productions are

$$\Delta s_B = \frac{q}{T_o} = \frac{-1000\ \frac{Btu}{lbm}}{520°R} = -1.923\ Btu/lbm\text{-}°R$$

$$\Delta s_C = \frac{1000\ \frac{Btu}{lbm}}{510°R} = 1.961\ Btu/lbm\text{-}°R$$

(c) The entropy change is not the same for the two planets. Entropy is not conserved in an energy transfer process. The overall entropy production is

$$\Delta s = \Delta s_B + \Delta s_C = -8.062\ \frac{kJ}{kg\cdot K} + 8.233\ \frac{kJ}{kg\cdot K}$$
$$= 0.171\ kJ/kg\cdot K$$

In customary U.S. units,

$$\Delta s = \Delta s_B + \Delta s_C$$
$$= -1.923\ \frac{Btu}{lbm\text{-}°R} + 1.961\ \frac{Btu}{lbm\text{-}°R}$$
$$= 0.038\ Btu/lbm\text{-}°R$$

(d) Local entropy can decrease, as shown by planet B's negative entropy production. However, overall entropy always increases when the total universe is considered, as shown in part (c).

(e) Planet B can be brought back to its original condition if 1000 Btu/lbm (2330 kJ/kg) of energy is transferred from planet A to planet B. (Heat will not flow spontaneously from planet C to planet B, as heat will not flow spontaneously from a cold object to a hot object.)[8]

## 12. SPECIFIC HEAT: c AND C

An increase in internal energy is needed to cause a rise in temperature. Different substances differ in the quantity of heat needed to produce a given temperature increase. The ratio of heat, $Q$, required to change the temperature of a mass, $m$, by an amount $\Delta T$ is called the *specific heat (heat capacity)* of the substance, $c$.

Because specific heats of solids and liquids are slightly temperature dependent, the mean specific heats are used

---
[8]This is one way of stating the *Second Law of Thermodynamics*.

when evaluating processes covering a large temperature range.

$$Q = mc\Delta T \qquad 22.22$$

$$c = \frac{Q}{m\Delta T} \qquad 22.23$$

The lowercase $c$ implies that the units are Btu/lbm-°F or J/kg·°C. Typical values of specific heat are given in Table 22.2. The *molar specific heat*, designated by the symbol $C$, has units of Btu/lbmol-°F or J/kmol·°C.

$$C = MW \times c \qquad 22.24$$

For gases, the specific heat depends on the type of process during which the heat exchange occurs. Specific heats for constant-volume and constant-pressure processes are designated by $c_v$ and $c_p$, respectively.

$$Q = mc_v\Delta T \quad \left[\begin{matrix}\text{perfect gas} \\ \text{constant-volume process}\end{matrix}\right] \qquad 22.25$$

$$Q = mc_p\Delta T \quad \left[\begin{matrix}\text{perfect gas} \\ \text{constant-pressure process}\end{matrix}\right] \qquad 22.26$$

Approximate values of $c_p$ and $c_v$ for common gases are given in Table 22.7. $c_v$ and $c_p$ for solids and liquids are essentially the same. However, the designation $c_p$ is often encountered for solids and liquids.

The law of *Dulong and Petit* predicts the approximate molar specific heat (in cal/mol·°C) at high temperatures from the atomic weight.[9] This law is valid for solid elements having atomic weights greater than 40 and for most metallic elements. It is not valid at room temperature for carbon, silicon, phosphorus, and sulfur. 6.3 cal/mol·°C is known as the *Dulong and Petit value*.

$$c \times AW \approx 6.3 \pm 0.1 \qquad 22.27$$

### Example 22.2

Compare the value of specific heat of pure iron calculated from Dulong and Petit's law with the value from Table 22.2.

*Solution*

The atomic weight of iron is 55.8. From Eq. 22.27,

$$c = \frac{6.3}{AW} = \frac{6.3\ \frac{cal}{mol\cdot°C}}{55.8\ \frac{g}{mol}} = 0.11\ cal/g\cdot°C$$

This is the same value as is given in Table 22.2.

---
[9]Dulong and Petit's law becomes valid at different temperatures for different substances, and a more specific definition of "high temperature" is impossible. For lead, the law is valid at 200K. For copper, it is not valid until above 400K.

**Table 22.2** *Approximate Specific Heats of Selected Liquids and Solids*[*]

| substance | $c_p$ Btu/lbm-°F | kJ/kg·°C |
|---|---|---|
| aluminum, pure | 0.23 | 0.96 |
| aluminum, 2024-T4 | 0.2 | 0.84 |
| ammonia | 1.16 | 4.86 |
| asbestos | 0.20 | 0.84 |
| benzene | 0.41 | 1.72 |
| brass, red | 0.093 | 0.39 |
| bronze | 0.082 | 0.34 |
| concrete | 0.21 | 0.88 |
| copper, pure | 0.094 | 0.39 |
| Freon-12 | 0.24 | 1.00 |
| gasoline | 0.53 | 2.20 |
| glass | 0.18 | 0.75 |
| gold, pure | 0.031 | 0.13 |
| ice | 0.49 | 2.05 |
| iron, pure | 0.11 | 0.46 |
| iron, cast (4% C) | 0.10 | 0.42 |
| lead, pure | 0.031 | 0.13 |
| magnesium, pure | 0.24 | 1.00 |
| mercury | 0.033 | 0.14 |
| oil, light hydrocarbon | 0.5 | 2.09 |
| silver, pure | 0.06 | 0.25 |
| steel, 1010 | 0.10 | 0.42 |
| steel, stainless 301 | 0.11 | 0.46 |
| tin, pure | 0.055 | 0.23 |
| titanium, pure | 0.13 | 0.54 |
| tungsten, pure | 0.032 | 0.13 |
| water | 1.0 | 4.19 |
| wood (typical) | 0.6 | 2.50 |
| zinc, pure | 0.088 | 0.37 |

(Multiply Btu/lbm-°F by 4.1868 to obtain kJ/kg·°C.)
[*]Values in cal/g·°C are the same as Btu/lbm-°F.

## 13. RATIO OF SPECIFIC HEATS: *k*

For gases, the *ratio of specific heats*, *k*, is defined by Eq. 22.28. Typical values are given in Table 22.7.

$$k = \frac{c_p}{c_v} \qquad 22.28$$

## 14. TRIPLE POINT PROPERTIES

The *triple point* of a substance is a unique state at which solid, liquid, and gaseous phases can coexist. Table 22.3 lists values of the triple point for several common substances.

## 15. CRITICAL PROPERTIES

If the temperature and pressure of a liquid are increased, a state will eventually be reached at which the liquid and gas phases are indistinguishable. This state is known as the *critical point*, and the properties

**Table 22.3** *Approximate Triple Points*

| substance | pressure atm | temperature °R | K |
|---|---|---|---|
| ammonia | 0.060 | 352 | 196 |
| argon | 0.676 | 151 | 84 |
| carbon dioxide | 5.10 | 390 | 217 |
| helium | 0.0508 | 4 | 2 |
| hydrogen | 0.0676 | 26 | 14 |
| nitrogen | 0.127 | 14 | 7.8 |
| oxygen | 0.00265 | 99 | 55 |
| water | 0.00592 | 492.02 | 273.34 |

(generally temperature, pressure, and specific volume) at that point are known as the *critical properties*. At the critical point, the heat of vaporization, $h_{fg}$, becomes zero. Above the critical temperature, the substance will be a gas no matter how high the pressure. Critical properties are listed in Table 22.4.

**Table 22.4** *Approximate Critical Properties*

| substance | pressure atm | temperature °R | K |
|---|---|---|---|
| air | 37.2 | 235.8 | 131.0 |
| ammonia | 111.5 | 730.1 | 405.6 |
| argon | 48.0 | 272.2 | 151.2 |
| carbon dioxide | 72.9 | 547.8 | 304.3 |
| carbon monoxide | 34.6 | 242.2 | 134.6 |
| chlorine | 75.9 | 751.0 | 417.2 |
| ethane | 48.8 | 549.8 | 305.4 |
| ethylene | 50.7 | 509.5 | 283.1 |
| helium | 2.3 | 10.0 | 5.56 |
| hydrogen | 12.8 | 60.5 | 33.6 |
| mercury | 180.0 | 2109.0 | 1171.7 |
| methane | 45.8 | 343.9 | 191.1 |
| neon | 25.7 | 79.0 | 43.9 |
| nitrogen | 33.5 | 227.2 | 126.2 |
| oxygen | 49.7 | 278.1 | 154.5 |
| propane | 42.0 | 666.3 | 370.2 |
| sulfur dioxide | 77.6 | 775.0 | 430.6 |
| water vapor | 218.2 | 1165.4 | 647.4 |
| xenon | 58.2 | 521.9 | 289.9 |

## 16. LATENT HEATS

The total energy (*total heat*, $Q_t$) entering a substance is the sum of the energy that changes the phase of the substance (*latent heat*, $Q_l$) and energy that changes the temperature of the substance (*sensible heat*, $Q_s$). During a phase change (solid to liquid, liquid to vapor, etc.), energy will be transferred to or from the substance without a change in temperature.[10]

$$Q_t = Q_s + Q_l \qquad 22.29$$

Examples of latent energies are the *latent heat of fusion* (i.e., change from solid to liquid), $h_{sl}$, *latent heat of*

[10]Changes in crystalline form are also latent changes.

vaporization, $h_{fg}$, and *latent heat of sublimation* (i.e., direct change from solid to vapor without becoming liquid), $h_{ig}$.[11,12] The energy required for these latent changes to occur in water is given in Table 22.5.

**Table 22.5** Latent Heats for Water at One Atmosphere

| effect | Btu/lbm | kJ/kg | cal/g |
|---|---|---|---|
| fusion | 143.4 | 333.5 | 79.7 |
| vaporization | 970.1 | 2256.5 | 539.0 |
| sublimation | 1220 | 2838 | 677.8 |

## Example 22.3

How much energy is required to just vaporize 1.0 lbm (0.45 kg) of water that is originally at 75°F (24°C) and 1 atm?

*SI Solution*

$$Q_s = mc(T_2 - T_1)$$
$$= (0.45 \text{ kg})\left(4.190 \, \frac{\text{kJ}}{\text{kg·°C}}\right)(100°C - 24°C)$$
$$= 143.3 \text{ kJ}$$
$$Q_l = mh_{fg} = (0.45 \text{ kg})\left(2256.5 \, \frac{\text{kJ}}{\text{kg}}\right) = 1015.4 \text{ kJ}$$
$$Q_t = Q_s + Q_l = 143.3 \text{ kJ} + 1015.4 \text{ kJ}$$
$$= 1158.7 \text{ kJ}$$

*Customary U.S. Solution*

The sensible heat required to raise the temperature of the water from 75°F to 212°F is given by Eq. 22.22.

$$Q_s = mc(T_2 - T_1)$$
$$= (1 \text{ lbm})\left(1.0 \, \frac{\text{Btu}}{\text{lbm-°F}}\right)(212°F - 75°F)$$
$$= 137.0 \text{ Btu}$$

From Table 22.5, the latent heat required to vaporize the water is

$$Q_l = mh_{fg} = (1 \text{ lbm})\left(970.1 \, \frac{\text{Btu}}{\text{lbm}}\right) = 970.1 \text{ Btu}$$

The total heat required is

$$Q_t = Q_s + Q_l = 137.0 \text{ Btu} + 970.1 \text{ Btu}$$
$$= 1107.1 \text{ Btu}$$

---

[11]The subscript $s$ (for "solid") is sometimes used in place of $i$ (for "ice").
[12]*Sublimation* can only occur below the triple point, where it is too cold for the liquid phase to exist at all.

## 17. QUALITY: *x*

Within the vapor dome, water is at its saturation pressure and temperature. There are an infinite number of thermodynamic states in which the water can simultaneously exist in liquid and vapor phases. The *quality* is the fraction by weight of the total mass that is vapor.

$$x = \frac{m_{\text{vapor}}}{m_{\text{vapor}} + m_{\text{liquid}}} \qquad 22.30$$

## 18. GIBBS FUNCTION: *g* AND *G*

The *Gibbs function* is defined for a pure substance by Eq. 22.31 through Eq. 22.33.

$$g = h - Ts = u + pv - Ts \qquad 22.31$$
$$G = H - TS = U + pV - TS \qquad 22.32$$
$$G = \text{MW} \times g \qquad 22.33$$

The Gibbs function is used in investigating latent changes and chemical reactions. For a constant-temperature, constant-pressure nonflow process approaching equilibrium, the Gibbs function approaches its minimum value. That is,

$$(dG)_{T,p} < 0 \qquad 22.34$$

Once the minimum value is obtained, the process will stop, and the Gibbs function will be constant. That is,

$$(dG)_{T,p} = 0\big|_{\text{equilibrium}} \qquad 22.35$$

Like enthalpy of formation, the Gibbs function, $G^0$, has been tabulated at the standard reference conditions of 25°C (77°F) and one atmosphere. A chemical reaction can occur spontaneously only if the change in Gibbs function is negative (i.e., the Gibbs function for the products is less than the Gibbs function for the reactants).

$$\sum_{\text{products}} nG^0 < \sum_{\text{reactants}} nG^0 \qquad 22.36$$

## 19. HELMHOLTZ FUNCTION: *a* AND *A*

The *Helmholtz function* is defined for a pure substance by Eq. 22.37 through Eq. 22.39.[13]

$$a = u - Ts = h - pv - Ts \qquad 22.37$$
$$A = U - TS = H - pV - TS \qquad 22.38$$
$$A = \text{MW} \times a \qquad 22.39$$

Like the Gibbs function, the Helmholtz function is used in investigating equilibrium conditions. For a

---

[13]In older references, the symbol $F$ was commonly used for the Helmholtz function.

Thermodynamics

constant-temperature, constant-volume nonflow process approaching equilibrium, the Helmholtz function approaches its minimum value. That is,

$$(dA)_{T,V} < 0 \qquad 22.40$$

Once the minimum value is obtained, the process will stop, and the Helmholtz function will be constant. That is,

$$(dA)_{T,V} = 0\big|_{\text{equilibrium}} \qquad 22.41$$

## 20. FREE ENERGY

The Helmholtz function has in the past also been known as the *free energy* of the system because its change in a reversible isothermal process equals the energy that can be "freed" and converted to mechanical work. Unfortunately, the same term has also been used for the Gibbs function under analogous conditions. For example, the difference in standard Gibbs functions of reactants and products has often been called the "free energy difference."

Since there is a great possibility for confusion, it is better to refer to the Gibbs and Helmholtz functions by their actual names.

## 21. FUGACITY AND ACTIVITY: *f*

The *fugacity*, *f*, of a substance is a modified pressure that accounts for non-ideal behavior. It has the same units as pressure. It is commonly used in conjunction with a generalized fugacity chart. Fugacity is defined by the *fugacity function* (see Eq. 22.42) and is related to the Gibbs function.

$$\lim_{p \to 0} f = p \qquad 22.42$$

$$(dG)_T = ZR^*T\,d(\ln p)_T = ZR^*T\,d(\ln f)_T \qquad 22.43$$

The ratio of fugacity at actual conditions to the fugacity at some reference state is known as the *activity*.

## 22. COMPRESSIBILITY: $\beta$, $\kappa$, AND $\alpha$

Three different compressibilities are distinguished. The *isobaric compressibility*, $\beta$, is defined as

$$\beta = \left(\frac{1}{v}\right)\left(\frac{\partial v}{\partial T}\right)_p \qquad 22.44$$

The *isothermal compressibility*, $\kappa$, is

$$\kappa = \left(-\frac{1}{v}\right)\left(\frac{\partial v}{\partial p}\right)_T \qquad 22.45$$

The *isentropic compressibility*, $\alpha$, is

$$\alpha = \left(-\frac{1}{v}\right)\left(\frac{\partial v}{\partial p}\right)_s \qquad 22.46$$

## 23. USING THE MOLLIER DIAGRAM

The *Mollier diagram (enthalpy-entropy diagram)* is a graph of enthalpy versus entropy for steam. (See App. 22.E.) It is particularly suitable for determining property changes between the superheated vapor and the liquid-vapor regions. For this reason, the Mollier diagram covers only a limited region.

The Mollier diagram plots the enthalpy for a unit mass of steam as the ordinate and plots the entropy as the abscissa. Lines of constant pressure (isobars) slope upward from left to right. Below the saturation line, curves of *constant moisture content* (the complement of quality) slope down from left to right. Above the saturation line are lines of constant temperature and lines of constant superheat.

### Example 22.4

Find the following properties using the Mollier diagram: (a) enthalpy and entropy of steam at 700 psia and 1000°F, (b) enthalpy of steam at 1 psia and 80% quality, and (c) final temperature of steam throttled from 700 psia and 1000°F to 450 psia.

*Solution*

(a) Reading directly from the Mollier diagram (see App. 22.E), $h = 1510$ Btu/lbm, and $s = 1.7$ Btu/lbm-°R.

(b) 80% quality is the same as 20% moisture. Reading at the intersection of 20% moisture and 1 psia, $h = 900$ Btu/lbm.

(c) By definition, a throttling process does not change the enthalpy. This process is represented by a horizontal line to the right on the Mollier diagram. Starting at the intersection of 700 psia and 1000°F and moving horizontally to the right until 450 psia is reached defines the endpoint of the process. The final temperature is interpolated as approximately 990°F.

## 24. USING SATURATION TABLES

The information presented graphically on an enthalpy-entropy diagram can be obtained with greater accuracy from *saturation tables*, also known as *property tables* and (in the case of water) *steam tables*. These tables represent extensive tabulations of data for liquid and vapor phases of a substance.

Saturation tables contain values of enthalpy, *h*, entropy, *s*, internal energy, *u*, and specific volume, *v*. Within the vapor dome, these properties are functions of temperature. Appendix 22.A and App. 22.N (for water) is organized in this manner.

However, as shown in Fig. 22.4, there is a unique pressure associated with each temperature (i.e., there is only one horizontal isobar for each temperature). Since the pressure does not vary in even increments when temperature is changed, the second column of App. 22.A and App. 22.N varies irregularly. A second property table, App. 22.B and App. 22.O, is set up so that the pressure increments are uniform.

In App. 22.A and App. 22.N, the first column after temperature gives the corresponding saturation pressure. The next two columns give specific volume. The first of these gives the specific volume of the saturated liquid, $v_f$; the second column gives the specific volume of the saturated vapor, $v_g$.

The relationship between $v_f$, $v_{fg}$, and $v_g$ is given by Eq. 22.47.

$$v_g = v_f + v_{fg} \qquad 22.47$$

The subsequent columns list the same data for enthalpy and entropy.

$$h_g = h_f + h_{fg} \qquad 22.48$$

$$s_g = s_f + s_{fg} \qquad 22.49$$

In App. 22.B and App. 22.O, the first column after the pressure gives the corresponding saturation temperature. The next two columns give specific volume in a manner similar to that of App. 22.A and App. 22.N.

## 25. USING SUPERHEAT TABLES

In the superheated region, pressure and temperature are independent properties. Therefore, for each pressure, a large number of temperatures is possible. Appendix 22.C and App. 22.P is a superheated steam table that gives the properties of specific volume, enthalpy, and entropy for various combinations of temperature and pressure. The *degrees of superheat* (e.g., "100°F of superheat") represents the difference between actual and saturation temperatures.

## 26. USING COMPRESSED LIQUID TABLES

A liquid whose pressure is greater than the pressure corresponding to its saturation pressure is known as a *compressed liquid* or *subcooled liquid*. (See Sec. 22.30.) Liquids are only slightly compressible. For most thermodynamic problems, changes in properties for a liquid are negligible. In problems where the exact values are needed, tables such as App. 22.D and App. 22.Q of compressed liquid properties are available. Such tables may present the properties directly or may give the properties at the saturation state and the corrections to those properties for various pressures.

### Example 22.5

A device compresses 1.0 lbm of 300°F (150°C) saturated water to 1000 psia (7.5 MPa). What is the final specific volume?

*SI Solution*

$$p = \frac{(7.5 \text{ MPa})\left(10^6 \ \dfrac{\text{Pa}}{\text{MPa}}\right)}{10^5 \ \dfrac{\text{Pa}}{\text{bar}}} = 75 \text{ bar}$$

From App. 22.Q, the specific volume of the water at 75 bars and 150°C is read directly as 1085.8 cm³/kg.

*Customary U.S. Solution*

From App. 22.D, the specific volume of the water at 1000 psia and 300°F is read directly as 0.01738 ft³/lbm.

## 27. USING GAS TABLES

Gas tables are essentially superheat tables for which an assumption has been made about the pressure. For example, App. 22.F and App. 22.S is a gas table for air at low pressures. "Low pressure" means less than several hundred psi pressure. However, reasonably good results can be expected even if pressures are higher.

Gas tables are indexed by temperature. That is, implicit in their use is the assumption that properties are functions of temperature only. Gas tables are not arranged in the same way as other property tables.

The *volume ratio*, $v_r$, and the *pressure ratio*, $p_r$, columns can be used when gases take part in an isentropic process. These terms should not be confused with the reduced variables introduced in Sec. 22.42. These two columns are not the specific volume or pressure. They are ratios with arbitrary references that make analysis of isentropic processes easier. Their use is illustrated in Ex. 22.7 and is based on Eq. 22.50 and Eq. 22.51.

$$\frac{v_{r,1}}{v_{r,2}} = \frac{V_1}{V_2} \quad [\Delta s = 0] \qquad 22.50$$

$$\frac{p_{r,1}}{p_{r,2}} = \frac{p_1}{p_2} \quad [\Delta s = 0] \qquad 22.51$$

Entropy is not listed at all in App. 22.F and App. 22.S. Instead, a column of *entropy function*, $\phi$, with the same units as entropy is given. The entropy function is not the same as specific entropy, but it can be used to calculate the change in entropy as the gas goes through a process. This entropy change is calculated with Eq. 22.52. Units of the gas constant must match those of $s$ and $\phi$.

$$s_2 - s_1 = \phi_2 - \phi_1 - R \ln\left(\frac{p_2}{p_1}\right) \qquad 22.52$$

### Example 22.6

What is the enthalpy of air at 100°F (310K) and 50 psia (345 kPa)?

*SI Solution*

Since the pressure is low (less than 20 atm), App. 22.S can be used. From the $T = 310$K line, the enthalpy is read directly as 310.24 kJ/kg.

*Customary U.S. Solution*

Since the pressure is low (less than 300 psia), App. 22.F can be used. 100°F is the same as 560°R. From the $T = 560$°R line, the enthalpy is read directly as 133.86 Btu/lbm.

### Example 22.7

Air is originally at 60°F and 14.7 psia. It is compressed isentropically to 86.5 psia. What are its new temperature and enthalpy?

*Solution*

From App. 22.F for $T = 60°F + 460° = 520°R$, $p_{r,1} = 1.2147$. Using Eq. 22.51,

$$p_{r,2} = \frac{p_{r,1}p_2}{p_1} = \frac{(1.2147)(86.5 \text{ psia})}{14.7 \text{ psia}}$$
$$= 7.148$$

Searching the $p_r$ column of App. 22.F results in $T = 860°R$ and $h = 206.46$ Btu/lbm.

### 28. USING PRESSURE-ENTHALPY CHARTS

For convenience (and by tradition), properties of refrigerants are typically shown graphically in a *pressure-enthalpy (p-h) diagram*. Figure 22.6 shows a skeleton *p-h* diagram. The vapor dome and saturation lines from Fig. 22.5 are recognizable. Lines of constant specific volume, constant entropy, and constant temperature are added.

**Figure 22.6** *Pressure-Enthalpy Diagram*

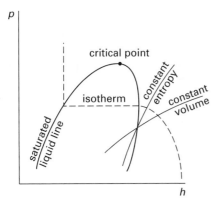

Isotherms are horizontal within the vapor dome, corresponding to the constant saturation pressure. In the subcooled-liquid region to the left of the vapor dome, isotherms are essentially vertical, since the temperature (not the pressure) of a subcooled liquid determines enthalpy. In the superheated region, the isotherms gradually approach vertical as the gaseous refrigerant becomes more like an ideal gas.

The enthalpy of superheated refrigerant is easily determined from the pressure and temperature. Isentropic compression follows a line of constant entropy. Throttling follows a line of constant enthalpy (i.e., a vertical line).

Enthalpies (plotted on the horizontal axis) may be scaled to an arbitrary "zero point." Comparing charts in traditional units and SI units can be perplexing. For most calculations, including refrigeration calculations, only the difference between two values is important. The absolute value at any particular point is seldom a useful value.

### 29. PROPERTIES OF SOLIDS

There are few mathematical relationships that predict the thermodynamic properties of solids. Properties such as temperature, specific heat, and density, usually are known. If the properties are not known, they must be found from tables.

The reference point for properties of solids is usually absolute zero temperature. That is, properties such as enthalpy and entropy are defined to be zero at 0°R (0K). This is an arbitrary convention. The choice of reference point does not affect the *change* in properties between two temperatures.

### 30. PROPERTIES OF SUBCOOLED LIQUIDS

A subcooled liquid is at a temperature less than the saturation temperature corresponding to its pressure. Unless the pressure of the liquid is very high, the various thermodynamic properties can be considered to be functions of only the liquid's temperature. In addition to compressed liquid tables (see Sec. 22.26), saturation tables (see Sec. 22.24) may be used to determine properties.

### Example 22.8

What is the enthalpy of water at 30 psia (2.0 bars, 0.2 MPa) and 240°F (110°C, 383K)?

*SI Solution*

The logic is the same as presented in the customary U.S. solution.

$$p = 0.2 \text{ MPa} \quad (2 \text{ bars})$$
$$T_{\text{sat},0.2\,\text{MPa}} = 120.2°\text{C} \quad [\text{App. 22.O}]$$
$$T_{\text{actual}} < T_{\text{sat}} \quad [\text{liquid phase}]$$
$$h_{f,110°\text{C}} = 461.42 \text{ kJ/kg} \quad [\text{App. 22.N}]$$

*Customary U.S. Solution*

Although the substance is water, its phase is unknown. It could be liquid, vapor, or a combination of the two. From App. 22.B , the saturation (boiling) temperature for 30 psia water is approximately 250°F. Since the actual water temperature is less than the saturation temperature, the water is liquid.

As a liquid, the properties are essentially functions of temperature only. From App. 22.A for 240°F, $h = 208.5$ Btu/lbm.

## 31. PROPERTIES OF SATURATED LIQUIDS

Either the temperature or the pressure of a saturated liquid must be known in order to identify its thermodynamic state. One defines the other, since there is a one-to-one relationship between saturation pressure and saturation temperature. The first two columns of App. 22.A and App. 22.B (for SI, App. 22.N and App. 22.O) can be used to determine saturation temperatures and pressures.

Since the saturation tables are set up specifically for saturated substances, enthalpy, entropy, internal energy, and specific volume can be read directly from the $h_f$, $s_f$, $u_f$, and $v_f$ columns, respectively. Density can be calculated as the reciprocal of the specific volume. The liquid's vapor pressure is the same as the saturation pressure listed in the table.

## 32. PROPERTIES OF LIQUID-VAPOR MIXTURES

When the thermodynamic state of a substance is within the vapor dome, there is a one-to-one correspondence between the saturation temperature and saturation pressure. One determines the other. The thermodynamic state is uniquely defined by any two independent properties (temperature and quality, pressure and enthalpy, entropy and quality, etc.).

If the quality of a liquid-vapor mixture is known, it can be used to calculate all of the primary thermodynamic properties. If a thermodynamic property has a value between the saturated liquid and saturated vapor values (i.e., $h$ is between $h_f$ and $h_g$), any of Eq. 22.53 through Eq. 22.56 can be solved for the quality.

$$h = h_f + x h_{fg} \quad \quad 22.53$$

$$s = s_f + x s_{fg} \quad \quad 22.54$$

$$u = u_f + x u_{fg} \quad \quad 22.55$$

$$v = v_f + x v_{fg} \quad \quad 22.56$$

### Example 22.9

What is the final enthalpy of superheated steam that is expanded isentropically (i.e., with no change in entropy) from 100 psia (700 kPa) and 500°F (250°C) to 3 psia (20 kPa)?

*SI Solution*

From App. 22.P, the entropy, $s_1$, of the superheated steam at 700 kPa and 250°C is 7.1070 kJ/kg·K. Since the expansion is isentropic, this is also the final entropy, $s_2$.

From App. 22.O, for 20 kPa (0.20 bars) vapor, the entropy of a saturated liquid, $s_f$, is 0.8320 kJ/kg·K; and the entropy of a saturated vapor, $s_g$, is 7.9072 kJ/kg·K. Since $s_2 < s_g$, the expanded steam is in the liquid-vapor region. The quality of the mixture is given by Eq. 22.54.

$$
\begin{aligned}
x &= \frac{s - s_f}{s_{fg}} = \frac{s - s_f}{s_g - s_f} \\
&= \frac{7.1070 \, \dfrac{\text{kJ}}{\text{kg·K}} - 0.8320 \, \dfrac{\text{kJ}}{\text{kg·K}}}{7.9072 \, \dfrac{\text{kJ}}{\text{kg·K}} - 0.8320 \, \dfrac{\text{kJ}}{\text{kg·K}}} \\
&= 0.8869
\end{aligned}
$$

From App. 22.O, the enthalpy of saturated liquid, $h_f$, at 20 kPa is 251.42 kJ/kg. The heat of vaporization, $h_{fg}$, is 2357.5 kJ/kg. The final enthalpy is given by Eq. 22.53.

$$
\begin{aligned}
h &= h_f + x h_{fg} \\
&= 251.42 \, \frac{\text{kJ}}{\text{kg}} + (0.8869)\left(2357.5 \, \frac{\text{kJ}}{\text{kg}}\right) \\
&= 2342.3 \text{ kJ/kg}
\end{aligned}
$$

*Customary U.S. Solution*

From App. 22.C, the entropy, $s_1$, of the superheated steam at 100 psia and 500°F is 1.7089 Btu/lbm-°R. Since the expansion is isentropic, this is also the final entropy, $s_2$.

From App. 22.B, for 3 psia vapor, the entropy of a saturated liquid, $s_f$, is 0.2009 Btu/lbm-°R; the entropy of vaporization, $s_{fg}$, is 1.6849 Btu/lbm-°R; and the entropy of a saturated vapor, $s_g$, is 1.8858 Btu/lbm-°R. Since

**Thermodynamics**

$s_2 < s_g$, the expanded steam is in the liquid-vapor region. The quality of the mixture is given by Eq. 22.54.

$$x = \frac{s - s_f}{s_{fg}}$$

$$= \frac{1.7089 \; \frac{\text{Btu}}{\text{lbm-}^\circ\text{R}} - 0.2009 \; \frac{\text{Btu}}{\text{lbm-}^\circ\text{R}}}{1.6849 \; \frac{\text{Btu}}{\text{lbm-}^\circ\text{R}}}$$

$$= 0.8950$$

From App. 22.B, the enthalpy of saturated liquid, $h_f$, at 3 psia is 109.4 Btu/lbm. The heat of vaporization, $h_{fg}$, is 1012.8 Btu/lbm. The final enthalpy is given by Eq. 22.53.

$$h = h_f + xh_{fg}$$

$$= 109.4 \; \frac{\text{Btu}}{\text{lbm}} + (0.8950)\left(1012.8 \; \frac{\text{Btu}}{\text{lbm}}\right)$$

$$= 1015.9 \; \text{Btu/lbm}$$

### Example 22.10

What is the enthalpy of 200°F (90°C, 363K) steam with a quality of 90%?

*SI Solution*

Using App. 22.N and Eq. 22.53,

$$h = h_f + xh_{fg}$$

$$= 377.04 \; \frac{\text{kJ}}{\text{kg}} + (0.9)\left(2282.5 \; \frac{\text{kJ}}{\text{kg}}\right) = 2431.3 \; \text{kJ/kg}$$

*Customary U.S. Solution*

Using App. 22.A and Eq. 22.53,

$$h = h_f + xh_{fg}$$

$$= 168.13 \; \frac{\text{Btu}}{\text{lbm}} + (0.9)\left(977.6 \; \frac{\text{Btu}}{\text{lbm}}\right)$$

$$= 1048.0 \; \text{Btu/lbm}$$

## 33. PROPERTIES OF SATURATED VAPORS

Properties of saturated vapors can be read directly from the saturation tables. The vapor's pressure or temperature can be used to define its thermodynamic state. Enthalpy, entropy, internal energy, and specific volume can be read directly as $h_g$, $s_g$, $u_g$, and $v_g$, respectively.

## 34. PROPERTIES OF SUPERHEATED VAPORS

Unless a vapor is highly superheated, its properties should be found from a superheat table, such as App. 22.C and App. 22.P (for water vapor). Since the temperature and pressure are independent for a superheated vapor, both must be known in order to define the thermodynamic state.

If the vapor's temperature and pressure do not correspond to the superheat table entries, single or double interpolation will be required. Such interpolation can be avoided by using more complete tables, but where required, linear interpolation is standard practice.

### Example 22.11

What is the enthalpy of water at 300 psia (20 bars, 2.0 MPa) and 900°F (500°C, 773K)?

*SI Solution*

The enthalpy can be read directly from App. 22.P as 3468.2 kJ/kg.

*Customary U.S. Solution*

It may not be obvious what phase the water is in. From App. 22.B , the saturation (boiling) temperature for 300 psia water is 417.35°F. Since the actual water temperature is higher than the saturation temperature, the water exists as a vapor.

From App. 22.C , the enthalpy can be read directly as 1473.9 Btu/lbm.

## 35. EQUATION OF STATE FOR IDEAL GASES

An *equation of state* is a relationship that predicts the state (a property such as pressure, temperature, volume, etc.) from a set of two other independent properties.

*Avogadro's law* states that equal volumes of different gases at the same temperature and pressure contain equal numbers of molecules. For one mole of any gas, Avogadro's law can be stated as the *equation of state for ideal gases*. (Temperature, $T$, in Eq. 22.57 must be absolute.)

$$\frac{pV}{T} = R^* \qquad \qquad 22.57$$

In Eq. 22.57, $R^*$ is known as the *universal gas constant*. It is "universal" (within a consistent system of units) because the same value can be used with any gas. Its value depends on the units used for pressure, temperature, and volume, as well as on the units of mass. (See Table 22.6.)

The ideal gas equation of state can be modified for more than one mole of gas. If there are $n$ moles,

$$pV = nR^*T \qquad \qquad 22.58$$

The number of moles can be calculated from the substance's mass and molecular weight.

$$n = \frac{m}{\text{MW}} \qquad \qquad 22.59$$

Thermodynamics

**Table 22.6** *Values of the Universal Gas Constant, R\**

*units in SI and other metric systems*
8.3143 kJ/kmol·K
8314.3 J/kmol·K
0.08206 atm·L/mol·K
1.986 cal/mol·K
8.314 J/mol·K
82.06 atm·cm$^3$/mol·K
0.08206 atm·m$^3$/kmol·K
8314.3 kg·m$^2$/s$^2$ kmol·K
8314.3 m$^3$·Pa/kmol·K
$8.314 \times 10^7$ erg/mol·K

*units in English systems*
1545.33 ft-lbf/lbmol-°R
1.986 Btu/lbmol-°R
0.7302 atm-ft$^3$/lbmol-°R
10.73 ft$^3$-lbf/in$^2$-lbmol-°R

Equation 22.58 and Eq. 22.59 can be combined. $R$ is the *specific gas constant*. It is specific because it is valid only for a gas with a molecular weight of MW.

$$pV = \frac{mR^*T}{\text{MW}} = m\left(\frac{R^*}{\text{MW}}\right)T = mRT \qquad 22.60$$

$$R = \frac{R^*}{\text{MW}} \qquad 22.61$$

Approximate values of the specific gas constant and the molecular weights of several common gases are given in Table 22.7.

### Example 22.12

What mass of nitrogen is contained in a 2000 ft$^3$ (57 m$^3$) tank if the pressure and temperature are 1 atm and 70°F (21°C), respectively?

*SI Solution*

First, convert to absolute temperature.

$$T = 21°C + 273° = 294K$$

From Table 22.7, $R = 297$ J/kg·K. From Eq. 22.60,

$$m = \frac{pV}{RT}$$

$$= \frac{(1\text{ atm})\left(1.013 \times 10^5 \frac{\text{Pa}}{\text{atm}}\right)(57\text{ m}^3)}{\left(297 \frac{\text{J}}{\text{kg·K}}\right)(294K)}$$

$$= 66.1\text{ kg}$$

*Customary U.S. Solution*

First, convert to absolute temperature.

$$T = 70°F + 460° = 530°R$$

From Table 22.7, $R = 55.16$ ft-lbf/lbm-°R. From Eq. 22.60,

$$m = \frac{pV}{RT}$$

$$= \frac{(1\text{ atm})\left(14.7 \frac{\text{lbf}}{\text{in}^2\text{-atm}}\right)\left(12 \frac{\text{in}}{\text{ft}}\right)^2(2000\text{ ft}^3)}{\left(55.16 \frac{\text{ft-lbf}}{\text{lbm-°R}}\right)(530°R)}$$

$$= 144.8\text{ lbm}$$

### Example 22.13

A 25 ft$^3$ (0.71 m$^3$) tank contains 10 lbm (4.5 kg) of an ideal gas. The gas has a molecular weight of 44 and is at 70°F (21°C). What is the pressure of the gas?

*SI Solution*

From Eq. 22.61, the specific gas constant is

$$R = \frac{R^*}{\text{MW}} = \frac{8314.3 \frac{\text{J}}{\text{kmol·K}}}{44 \frac{\text{kg}}{\text{kmol}}}$$

$$= 189\text{ J/kg·K}$$

The absolute temperature is $T = 21°C + 273° = 294K$. From Eq. 22.60, the pressure is

$$p = \frac{mRT}{V} = \frac{(4.5\text{ kg})\left(189 \frac{\text{J}}{\text{kg·K}}\right)(294K)}{(0.71\text{ m}^3)\left(1000 \frac{\text{Pa}}{\text{kPa}}\right)}$$

$$= 352.2\text{ kPa}$$

*Customary U.S. Solution*

$$R = \frac{R^*}{\text{MW}} = \frac{1545.33 \frac{\text{ft-lbf}}{\text{lbmol-°R}}}{44 \frac{\text{lbm}}{\text{lbmol}}}$$

$$= 35.12\text{ ft-lbf/lbm-°R}$$

$$T = 70°F + 460° = 530°R$$

$$p = \frac{mRT}{V} = \frac{(10\text{ lbm})\left(35.12 \frac{\text{ft-lbf}}{\text{lbm-°R}}\right)(530°R)}{25\text{ ft}^3}$$

$$= 7445\text{ lbf/ft}^2$$

### 36. PROPERTIES OF IDEAL GASES

A gas can be considered to behave ideally if its pressure is very low and the temperature is much higher than its critical temperature. (Otherwise, the substance is in vapor form.) Under these conditions, the molecule size is insignificant compared with the distance between

**Table 22.7** *Approximate Properties of Selected Gases*

| gas | symbol | temperature °F | MW | customary U.S. units $R$ ft-lbf/lbm-°R | $c_p$ Btu/lbm-°R | $c_v$ Btu/lbm-°R | SI units $R$ J/kg·K | $c_p$ J/kg·K | $c_v$ J/kg·K | $k$ |
|-----|--------|------|-----|------|------|------|------|------|------|------|
| acetylene | $C_2H_2$ | 68 | 26.038 | 59.35 | 0.350 | 0.274 | 319.32 | 1465 | 1146 | 1.279 |
| air | | 100 | 28.967 | 53.35 | 0.240 | 0.171 | 287.03 | 1005 | 718 | 1.400 |
| ammonia | $NH_3$ | 68 | 17.032 | 90.73 | 0.523 | 0.406 | 488.16 | 2190 | 1702 | 1.287 |
| argon | Ar | 68 | 39.944 | 38.69 | 0.124 | 0.074 | 208.15 | 519 | 311 | 1.669 |
| butane (-$n$) | $C_4H_{10}$ | 68 | 58.124 | 26.59 | 0.395 | 0.361 | 143.04 | 1654 | 1511 | 1.095 |
| carbon dioxide | $CO_2$ | 100 | 44.011 | 35.11 | 0.207 | 0.162 | 188.92 | 867 | 678 | 1.279 |
| carbon monoxide | CO | 100 | 28.011 | 55.17 | 0.249 | 0.178 | 296.82 | 1043 | 746 | 1.398 |
| chlorine | $Cl_2$ | 100 | 70.910 | 21.79 | 0.115 | 0.087 | 117.25 | 481 | 364 | 1.322 |
| ethane | $C_2H_6$ | 68 | 30.070 | 51.39 | 0.386 | 0.320 | 276.50 | 1616 | 1340 | 1.206 |
| ethylene | $C_2H_4$ | 68 | 28.054 | 55.08 | 0.400 | 0.329 | 296.37 | 1675 | 1378 | 1.215 |
| Freon (R-12)[*] | $CCl_2F_2$ | 200 | 120.925 | 12.78 | 0.159 | 0.143 | 68.76 | 666 | 597 | 1.115 |
| helium | He | 100 | 4.003 | 386.04 | 1.240 | 0.744 | 2077.03 | 5192 | 3115 | 1.667 |
| hydrogen | $H_2$ | 100 | 2.016 | 766.53 | 3.420 | 2.435 | 4124.18 | 14 319 | 10 195 | 1.405 |
| hydrogen sulfide | $H_2S$ | 68 | 34.082 | 45.34 | 0.243 | 0.185 | 243.95 | 1017 | 773 | 1.315 |
| krypton | Kr | | 83.800 | 18.44 | 0.059 | 0.035 | 99.22 | 247 | 148 | 1.671 |
| methane | $CH_4$ | 68 | 16.043 | 96.32 | 0.593 | 0.469 | 518.25 | 2483 | 1965 | 1.264 |
| neon | Ne | 68 | 20.183 | 76.57 | 0.248 | 0.150 | 411.94 | 1038 | 626 | 1.658 |
| nitrogen | $N_2$ | 100 | 28.016 | 55.16 | 0.249 | 0.178 | 296.77 | 1043 | 746 | 1.398 |
| nitric oxide | NO | 68 | 30.008 | 51.50 | 0.231 | 0.165 | 277.07 | 967 | 690 | 1.402 |
| nitrous oxide | $NO_2$ | 68 | 44.01 | 35.11 | 0.221 | 0.176 | 188.92 | 925 | 736 | 1.257 |
| octane vapor | $C_8H_{18}$ | | 114.232 | 13.53 | 0.407 | 0.390 | 72.78 | 1704 | 1631 | 1.045 |
| oxygen | $O_2$ | 100 | 32.000 | 48.29 | 0.220 | 0.158 | 259.82 | 921 | 661 | 1.393 |
| propane | $C_3H_8$ | 68 | 44.097 | 35.04 | 0.393 | 0.348 | 188.55 | 1645 | 1457 | 1.129 |
| sulfur dioxide | $SO_2$ | 100 | 64.066 | 24.12 | 0.149 | 0.118 | 129.78 | 624 | 494 | 1.263 |
| water vapor[*] | $H_2O$ | 212 | 18.016 | 85.78 | 0.445 | 0.335 | 461.50 | 1863 | 1402 | 1.329 |
| xenon | Xe | | 131.300 | 11.77 | 0.038 | 0.023 | 63.32 | 159 | 96 | 1.661 |

(Multiply Btu/lbm-°F by 4186.8 to obtain J/kg·K.) (Multiply ft-lbf/lbm-°R by 5.3803 to obtain J/kg·K.)
[*]Values for steam and Freon are approximate and should be used only for low pressures and high temperatures.

molecules, and molecules do not come into contact. By definition, an *ideal gas* behaves according to the various ideal gas laws.

Values of $h$, $u$, and $v$ for gases usually are read from gas tables. Since it contains a $pv$ term, enthalpy can be related to the equation of state. Depending on the units chosen, a conversion factor may be needed.

$$h = u + pv = u + RT \qquad 22.62$$

Furthermore, density is the reciprocal of specific volume, $v$. Therefore, the density of an ideal gas can be derived from the equation of state by setting $m = 1$ and solving for the reciprocal of volume.

$$\rho = \frac{1}{v} = \frac{p}{RT} \qquad 22.63$$

## 37. SPECIFIC HEATS OF IDEAL GASES

The specific heats of an ideal gas can be calculated from its specific gas constant. Depending on the units chosen, a conversion factor may or may not be needed in

Equation 22.64 through Eq. 22.67. By definition, a *perfect gas* is an ideal gas whose specific heats are constant.

$$c_p - c_v = R \qquad 22.64$$

$$C_p - C_v = R^* \qquad 22.65$$

$$c_p = \frac{Rk}{k-1} \qquad 22.66$$

$$C_p = \frac{R^* k}{k-1} \qquad 22.67$$

## 38. KINETIC GAS THEORY

The *kinetic gas theory* predicts the velocity distribution of gas molecules as a function of temperature. This theory makes the following assumptions.

- Gas molecules do not attract one another.

- The volume of the gas molecules is negligible compared with the volume of the gas.

- The molecules behave like hard spheres.

- The container volume is large enough that interactions with the wall are not a predominant activity of the molecules.

Equation 22.68 is the *Maxwell-Boltzmann distribution* (see Fig. 22.7) of gas molecule velocities. The variable $\kappa$ is the Boltzmann constant, which has a value of $1.3807 \times 10^{-23}$ J/molecule·K. (These units are the same as kg·m$^2$/s$^2$·molecule·K.) Kinetic gas theory calculations are traditionally performed in SI units. English units can be used, however, if all constants and units are consistent.

$$\frac{dN}{d\mathrm{v}} = \left(\frac{4N}{\sqrt{\pi}}\right) \mathrm{v}^2 \left(\frac{m}{2\kappa T}\right)^{3/2} e^{-m\mathrm{v}^2/2\kappa T} \qquad 22.68$$

$$\kappa = \frac{R^*}{N_A} \qquad 22.69$$

**Figure 22.7** *Maxwell-Boltzmann Velocity Distribution*

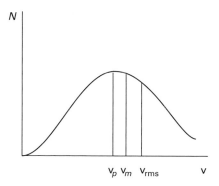

The *most probable speed* of a molecule with mass $m$ is

$$\mathrm{v}_p = \sqrt{\frac{2\kappa T}{m}} \qquad 22.70$$

The *mean speed* of a molecule with mass $m$ is

$$\mathrm{v}_m = 2\sqrt{\frac{2\kappa T}{\pi m}} \qquad 22.71$$

The *root-mean-square speed* of a molecule with mass $m$ is

$$\mathrm{v}_{\mathrm{rms}} = \sqrt{\frac{3\kappa T}{m}} \qquad 22.72$$

The three velocities are illustrated in Fig. 22.7 and are related.

$$\frac{\mathrm{v}_m}{\mathrm{v}_p} = 1.128 \qquad 22.73$$

$$\frac{\mathrm{v}_{\mathrm{rms}}}{\mathrm{v}_p} = 1.225 \qquad 22.74$$

*Temperature* has a molecular interpretation derived from the kinetic gas theory. It can be shown that the root-mean-square velocity is related to the absolute temperature. That is, absolute temperature is proportional to the square of the rms velocity.

$$T = \left(\frac{m}{3\kappa}\right) \mathrm{v}_{\mathrm{rms}}^2 \qquad 22.75$$

Since $\frac{1}{2}m\mathrm{v}^2$ is the definition of kinetic energy, the mean translational kinetic energy of the molecule is proportional to the mean absolute temperature.

$$E_k = \tfrac{1}{2}m\mathrm{v}_{\mathrm{rms}}^2 = \tfrac{3}{2}\kappa T \qquad 22.76$$

The pressure of a gas can be calculated as the total change in momentum of all the gas molecules bouncing off the walls of the container.

$$p = \frac{\rho \mathrm{v}_{\mathrm{rms}}^2}{3} = \frac{Nm\mathrm{v}_{\mathrm{rms}}^2}{3V} \qquad 22.77$$

A small-enough particle suspended in a fluid will exhibit small random movements due to the statistical (i.e., random) collisions of fluid molecules on the particle's surface. Such motion is known as *Brownian movement.*

### Example 22.14

What are the kinetic energy and rms velocity of 275K argon molecules (MW = 39.9)?

*Solution*

From Eq. 22.76,

$$E_k = \tfrac{3}{2}\kappa T$$
$$= (1.5)\left(1.3807 \times 10^{-23}\ \frac{\mathrm{J}}{\mathrm{molecule \cdot K}}\right)(275\mathrm{K})$$
$$= 5.70 \times 10^{-21}\ \mathrm{J/molecule}$$

The molecular mass of argon is its mass per mole divided by the number of molecules in a mole.

$$m = \frac{\mathrm{MW}}{N_A} = \frac{39.9\ \dfrac{\mathrm{kg}}{\mathrm{kmol}}}{\left(6.023 \times 10^{23}\ \dfrac{\mathrm{molecules}}{\mathrm{mol}}\right)\left(1000\ \dfrac{\mathrm{mol}}{\mathrm{kmol}}\right)}$$
$$= 6.62 \times 10^{-26}\ \mathrm{kg/molecule}$$

From Eq. 22.72, the rms velocity is

$$\mathrm{v}_{\mathrm{rms}} = \sqrt{\frac{3\kappa T}{m}}$$
$$= \sqrt{\frac{(3)\left(1.3807 \times 10^{-23}\ \dfrac{\mathrm{J}}{\mathrm{molecule \cdot K}}\right)(275\mathrm{K})}{6.62 \times 10^{-26}\ \dfrac{\mathrm{kg}}{\mathrm{molecule}}}}$$
$$= 414.8\ \mathrm{m/s}$$

**Thermodynamics**

## 39. GRAVIMETRIC, VOLUMETRIC, AND MOLE FRACTIONS

The *gravimetric fraction*, $G_A$ (also known as the *mass fraction*), of a component $A$ in a mixture of components $A$, $B$, $C$, and so on, is the ratio of the component's mass to the total mixture mass.

$$G_A = \frac{m_A}{m} = \frac{m_A}{m_A + m_B + m_C} \qquad 22.78$$

The *volumetric fraction*, $B_A$, of a component $A$ is the ratio of the component's partial volume to the overall mixture volume.

$$B_A = \frac{V_A}{V} = \frac{V_A}{V_A + V_B + V_C} \qquad 22.79$$

It is possible to convert between gravimetric and volumetric fractions.

$$G_A = \frac{B_A(\text{MW})_A}{B_A(\text{MW})_A + B_B(\text{MW})_B + B_C(\text{MW})_C} \qquad 22.80$$

$$B_A = \frac{\dfrac{G_A}{(\text{MW})_A}}{\dfrac{G_A}{(\text{MW})_A} + \dfrac{G_B}{(\text{MW})_B} + \dfrac{G_C}{(\text{MW})_C}} \qquad 22.81$$

The *mole fraction*, $x_A$, of a component $A$ is the ratio of the number of moles of substance $A$ to the total number of moles of all substances.

$$x_A = \frac{n_A}{n} = \frac{n_A}{n_A + n_B + n_C} \qquad 22.82$$

For nonreacting mixtures of ideal gases, the mole fraction and volumetric fraction (and partial pressure ratio) are the same.

$$x_A = B_A \Big|_{\text{ideal gases}} \qquad 22.83$$

## 40. PARTIAL PRESSURE AND PARTIAL VOLUME OF GAS MIXTURES

A *gas mixture* consists of an aggregation of molecules of each gas component, the molecules of any single component being distributed uniformly and moving as if they alone occupied the space. The *partial volume*, $V_A$, of a gas $A$ in a mixture of nonreacting gases $A$, $B$, $C$, and so on, is the volume that gas $A$ alone would occupy at the temperature and pressure of the mixture. (See Fig. 22.8.)

The partial volume can be calculated from the volumetric fraction and total volume.

$$V_A = B_A V \qquad 22.84$$

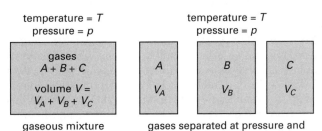

**Figure 22.8** *Mixture of Ideal Gases*

temperature = $T$
pressure = $p$

gases
$A + B + C$

volume $V$ =
$V_A + V_B + V_C$

gaseous mixture

temperature = $T$
pressure = $p$

$A$ $V_A$    $B$ $V_B$    $C$ $V_C$

gases separated at pressure and temperature of mixture

*Amagat's law* (also known as *Amagat-Leduc's rule*) states that the total volume of a mixture of nonreacting gases is equal to the sum of the partial volumes.

$$V = V_A + V_B + V_C \qquad 22.85$$

The *partial pressure*, $p_A$, of gas $A$ in a mixture of nonreacting gases $A$, $B$, $C$, and so on, is the pressure gas $A$ alone would exert in the total volume at the temperature of the mixture.

$$p_A = \frac{m_A R_A T}{V} = \frac{n_A R^* T}{V} \qquad 22.86$$

The partial pressure can also be calculated from the mole fraction and the total pressure. However, for ideal gases, the partial pressure ratio, mole fraction, and volumetric fraction are the same.

$$\frac{p_A}{p} = x_A = B_A \qquad 22.87$$

If the average specific gas constant, $\overline{R}$, for the gas mixture is known, it can be used with the gravimetric fraction to calculate the partial pressure.

$$p_A = G_A \frac{\overline{R}_A p}{\overline{R}} \qquad 22.88$$

According to *Dalton's law of partial pressures*, the *total pressure* of a gas mixture is the sum of the partial pressures.

$$p = p_A + p_B + p_C \qquad 22.89$$

## 41. PROPERTIES OF NONREACTING IDEAL GAS MIXTURES

A mixture's average molecular weight and density are volumetrically weighted averages of its components' values.

$$\text{MW} = \sum B_i (\text{MW})_i \qquad 22.90$$

$$\rho = \sum B_i \rho_i \qquad 22.91$$

On the other hand, a mixture's internal energy, enthalpy, specific heats, and specific gas constant are equal to the sum of the values of its individual components (i.e., the mixture average is gravimetrically weighted).

$$u = \frac{\sum m_i u_i}{m} = \sum_i G_i u_i \qquad 22.92$$

$$h = \frac{\sum m_i h_i}{m} = \sum_i G_i h_i \qquad 22.93$$

$$c_v = \frac{\sum m_i c_{v,i}}{m} = \sum_i G_i c_{v,i} \qquad 22.94$$

$$c_p = \frac{\sum m_i c_{p,i}}{m} = \sum_i G_i c_{p,i} \qquad 22.95$$

$$R = \frac{\sum m_i R_i}{m} = \sum_i G_i R_i \qquad 22.96$$

If the mixing is reversible and adiabatic, the entropy will also be equal to the sum of the individual entropies. However, each individual entropy, $s_i$, must be evaluated at the temperature and volume of the mixture and at the individual partial pressure, $p_i$.

$$s = \frac{\sum m_i s_i}{m} = \sum_i G_i s_i \qquad 22.97$$

Equation 22.92, Eq. 22.93, and Eq. 22.97 are mathematical formulations of *Gibbs theorem* (also known as *Gibbs rule*). This theorem states that the total property (e.g., *U*, *H*, or *S*) of a mixture of nonreacting ideal gases is the sum of the properties that the individual gases would have if each occupied the total mixture volume alone at the same temperature.

Table 22.8 summarizes the composite gas properties. Notice that the ratio of specific heats, *k*, is not a composite gas property. It is most expedient to find the composite ratio of specific heats from the gravimetrically weighted specific heats.

**Table 22.8** Summary of Composite Ideal Gas Properties

| gravimetrically (mass) weighted | volumetrically (mole fraction) weighted |
|---|---|
| $u$ | $U$ |
| $h$ | $H$ |
| $c_p$ | $C_p$ |
| $c_v$ | $C_v$ |
| $R$ | MW |
| $s$ | $S$ |
| | $\rho$ |

**Example 22.15**

0.14 lbm (0.064 kg) of octane vapor (MW = 114) is mixed with 2.0 lbm (0.91 kg) of air (MW = 29.0) in the manifold of an engine. The total pressure in the manifold is 12.5 psia (86.1 kPa), and the temperature is 520°R (290K). Assume octane behaves ideally. (a) What is the total volume of this mixture? (b) What is the partial pressure of the air in the mixture?

*SI Solution*

(a) The number of moles of octane and air are

$$n_{\text{octane}} = \frac{m}{\text{MW}} = \frac{0.064 \text{ kg}}{114 \, \frac{\text{kg}}{\text{kmol}}} = 5.61 \times 10^{-4} \text{ kmol}$$

$$n_{\text{air}} = \frac{0.91 \text{ kg}}{29 \, \frac{\text{kg}}{\text{kmol}}} = 0.0314 \text{ kmol}$$

From Eq. 22.58, the total volume of any gas is

$$V = \frac{n R^* T}{p}$$

$$= (5.61 \times 10^{-4} \text{ kmol} + 0.0314 \text{ kmol})$$

$$\times \left( \frac{\left( 8314.3 \, \frac{\text{J}}{\text{kmol·K}} \right)(290\text{K})}{86.1 \times 10^3 \text{ Pa}} \right)$$

$$= 0.895 \text{ m}^3$$

(b) The mole fraction of the air is

$$x_{\text{air}} = \frac{n_{\text{air}}}{n_{\text{total}}} = \frac{0.0314 \text{ kmol}}{5.61 \times 10^{-4} \text{ kmol} + 0.0314 \text{ kmol}}$$

$$= 0.982$$

The partial pressure of air is

$$p_{\text{air}} = x_{\text{air}} p = (0.982)(86.1 \text{ kPa})$$

$$= 84.6 \text{ kPa}$$

*Customary U.S. Solution*

(a) The number of moles of octane and air are

$$n_{\text{octane}} = \frac{m}{\text{MW}} = \frac{0.14 \text{ lbm}}{114 \, \frac{\text{lbm}}{\text{lbmol}}} = 0.001228 \text{ lbmol}$$

$$n_{\text{air}} = \frac{2.0 \text{ lbm}}{29.0 \, \frac{\text{lbm}}{\text{lbmol}}} = 0.068966 \text{ lbmol}$$

From Eq. 22.58, the total volume of any gas is

$$V = \frac{nR^*T}{p}$$

$$= (0.001228 \text{ lbmol} + 0.068966 \text{ lbmol})$$

$$\times \left( \frac{\left( 1545.33 \ \frac{\text{ft-lbf}}{\text{lbmol-}^\circ\text{R}} \right)(520^\circ\text{R})}{\left( 12.5 \ \frac{\text{lbf}}{\text{in}^2} \right)\left( 12 \ \frac{\text{in}}{\text{ft}} \right)^2} \right)$$

$$= 31.34 \text{ ft}^3$$

(b) The mole fraction of the air is

$$x_{\text{air}} = \frac{n_{\text{air}}}{n} = \frac{0.068966 \text{ lbmol}}{0.001228 \text{ lbmol} + 0.068966 \text{ lbmol}}$$

$$= 0.9825$$

The partial air pressure is

$$p_{\text{air}} = x_{\text{air}}p = (0.9825)\left( 12.5 \ \frac{\text{lbf}}{\text{in}^2} \right)$$

$$= 12.3 \text{ psia}$$

## 42. EQUATION OF STATE FOR REAL GASES

Real gases do not meet the basic assumptions defining an ideal gas. Specifically, the molecules of a real gas occupy a volume that is not negligible in comparison with the total volume of the gas. (This is especially true for gases at low temperatures.) Furthermore, real gases are subject to *van der Waals' forces*, which are attractive forces between gas molecules.

There are two methods of accounting for real gas behavior. The first method is to modify the ideal gas equation of state with various empirical correction factors. Since the modifications are empirical, the resulting equations of state are known as *correlations*. One well-known correlation is *van der Waals' equation of state*.

$$\left( p + \frac{a}{V^2} \right)(V - b) = nR^*T \qquad 22.98$$

The van der Waals corrections usually need to be made only when a gas is below its critical temperature. For an ideal gas, the $a$ and $b$ terms are zero. When the spacing between molecules is close, as it would be at low temperatures, the molecules attract each other and reduce the pressure exerted by the gas. The pressure is then corrected by the $a/V^2$ term. $b$ is a constant that accounts for the molecular volume in a dense state.

Other correlations of this type include the *Clausius, Bertholet, Dieterici*, and *Beattie-Bridgeman equations of state*. However, the most accurate empirical correlation is the *virial equation of state*, which has the form

shown in Eq. 22.99 and Eq. 22.100. The constants $B$, $C$, $D$, and so on, are called the *virial coefficients*.

$$pV = nR^*T \left( 1 + \frac{B}{V} + \frac{C}{V^2} + \frac{D}{V^3} + \cdots \right) \qquad 22.99$$

$$pV = nR^*T(1 + B'p + C'p^2 + D'p^3 + \cdots) \qquad 22.100$$

The *principle (law) of corresponding states* provides a second method of correcting for real gas behavior. This law states that the *reduced properties* of all gases are identical (i.e., all gases behave similarly when their reduced variables are used). Specifically, there is one property, the *compressibility factor*, $Z$, that has the same value for all gases when evaluated at the same values of the *reduced variables*.[14] (The reduced variables are not the same as the ratios defined in Sec. 22.27.)

$$Z = f(T_r, p_r, v_r) \qquad 22.101$$

$$T_r = \frac{T}{T_c} \qquad 22.102$$

$$p_r = \frac{p}{p_c} \qquad 22.103$$

$$v_r = \frac{v}{v_c} \qquad 22.104$$

Compressibility factors are almost always read from *generalized compressibility charts* such as App. 22.Z. The compressibility factor can then be used to correct the ideal gas equation of state.

$$pv = ZRT \qquad 22.105$$

$$pV = mZRT \qquad 22.106$$

### Example 22.16

What is the specific volume of carbon dioxide at 2680 psia (182 atm) and 300°F (150°C)?

*SI Solution*

The absolute temperature is

$$T = 150^\circ\text{C} + 273^\circ = 423\text{K}$$

From Table 22.4, the critical temperature and pressure of carbon dioxide are 304.3K and 72.9 atm, respectively. The reduced variables are

$$T_r = \frac{T}{T_c} = \frac{423\text{K}}{304.3\text{K}} = 1.39$$

$$p_r = \frac{p}{p_c} = \frac{182 \text{ atm}}{72.9 \text{ atm}} = 2.5$$

---

[14]Some engineering disciplines (e.g., petroleum engineering) use the symbol $z$ for compressibility factor.

From App. 22.Z, $Z$ is read as 0.75. $R$ is read from Table 22.7. Solving Eq. 22.105 for $v$,

$$v = \frac{ZRT}{p} = \frac{(0.75)\left(189 \; \frac{J}{kg \cdot K}\right)(423K)}{(182 \; atm)\left(101.3 \; \times \; 10^3 \; \frac{Pa}{atm}\right)}$$

$$= 3.25 \times 10^{-3} \; m^3/kg$$

*Customary U.S. Solution*

The absolute temperature is

$$T = 300°F + 460° = 760°R$$

From Table 22.4, the critical temperature and pressure of carbon dioxide are 547.8°R and 72.9 atm, respectively. The reduced variables are

$$T_r = \frac{T}{T_c} = \frac{760°R}{547.8°R} = 1.39$$

$$p_r = \frac{p}{p_c} = \frac{2680 \; psia}{(72.9 \; atm)\left(14.7 \; \frac{psia}{atm}\right)} = 2.5$$

From App. 22.Z, $Z$ is read as 0.75. $R$ is read from Table 22.7. Solving Eq. 22.105 for $v$,

$$v = \frac{ZRT}{p} = \frac{(0.75)\left(35.11 \; \frac{ft\text{-}lbf}{lbm\text{-}°R}\right)(760°R)}{\left(2680 \; \frac{lbf}{in^2}\right)\left(12 \; \frac{in}{ft}\right)^2}$$

$$= 0.0519 \; ft^3/lbm$$

## 43. SPECIFIC HEATS OF REAL GASES

By definition, the specific heat of an ideal gas is independent of temperature. However, the specific heat of a real gas varies with temperature and (slightly) with pressure. There are several ways to find the specific heat of a gas at different temperatures: tables, graphs, and correlations.[15]

Equation 22.107 is a typical temperature correlation. The variation with pressure, being small at low pressures, is disregarded. The coefficients $A$, $B$, $C$, and $D$ will depend on the units, the temperature range over which the correlation is to be used, and the desired accuracy. Typical coefficients are given in Table 22.9.

$$c_p = A + BT + CT^2 + \frac{D}{\sqrt{T}} \qquad \textit{22.107}$$

$$c_v = c_p - R \qquad \text{[SI]} \quad \textit{22.108(a)}$$

$$c_v = c_p - \frac{R}{J} \qquad \text{[U.S.]} \quad \textit{22.108(b)}$$

[15]It is also possible to calculate the specific heat over a small temperature range if the enthalpies are known (e.g., from an air table) from $c_p = \Delta h/\Delta T$. However, if the enthalpies are known, it is unlikely that the specific heat will be needed.

**Table 22.9** *Correlation Coefficients for Calculating Specific Heat (Btu/lbm-°R)*

| gas | temperature range | | $A$ | $B$ | $C$ | $D$ |
| --- | --- | --- | --- | --- | --- | --- |
| | °R | K | | | | |
| air | 400 to 1200 | 220 to 670 | 0.2405 | $-1.186 \times 10^{-5}$ | $20.1 \times 10^{-9}$ | 0 |
| | 1200 to 4000 | 670 to 2220 | 0.2459 | $3.22 \times 10^{-5}$ | $-3.74 \times 10^{-9}$ | $-0.833$ |
| $CH_4$ | 400 to 1000 | 220 to 560 | 0.453 | $0.62 \times 10^{-5}$ | $268.8 \times 10^{-9}$ | 0 |
| | 1000 to 4000 | 560 to 2220 | 1.152 | $32.58 \times 10^{-5}$ | $-41.29 \times 10^{-9}$ | $-22.42$ |
| CO | 400 to 1200 | 220 to 670 | 0.2534 | $-2.35 \times 10^{-5}$ | $26.88 \times 10^{-9}$ | 0 |
| | 1200 to 4000 | 670 to 2220 | 0.2763 | $3.04 \times 10^{-5}$ | $-3.89 \times 10^{-9}$ | $-1.5$ |
| $CO_2$ | 400 to 4000 | 220 to 2220 | 0.328 | $3.2 \times 10^{-5}$ | $-4.4 \times 10^{-9}$ | $-3.33$ |
| $H_2$ | 400 to 1000 | 220 to 560 | 2.853 | $145 \times 10^{-5}$ | $-883 \times 10^{-9}$ | 0 |
| | 1000 to 2500 | 560 to 1390 | 3.447 | $-4.7 \times 10^{-5}$ | $70.3 \times 10^{-9}$ | 0 |
| | 2500 to 4000 | 1390 to 2220 | 2.841 | $45 \times 10^{-5}$ | $-31.2 \times 10^{-9}$ | 0 |
| $H_2O$ | 400 to 1800 | 220 to 1000 | 0.4267 | $2.425 \times 10^{-5}$ | $23.85 \times 10^{-9}$ | 0 |
| | 1800 to 4000 | 1000 to 2220 | 0.3275 | $14.67 \times 10^{-5}$ | $-13.59 \times 10^{-9}$ | 0 |
| $N_2$ | 400 to 1200 | 220 to 670 | 0.2510 | $-1.63 \times 10^{-5}$ | $20.4 \times 10^{-9}$ | 0 |
| | 1200 to 4000 | 670 to 2220 | 0.2192 | $4.38 \times 10^{-5}$ | $-5.14 \times 10^{-9}$ | $-0.124$ |
| $O_2$ | 400 to 1200 | 220 to 670 | 0.213 | $0.188 \times 10^{-5}$ | $20.3 \times 10^{-9}$ | 0 |
| | 1200 to 4000 | 670 to 2220 | 0.340 | $-0.36 \times 10^{-5}$ | $0.616 \times 10^{-9}$ | $-3.19$ |

(Multiply values in this table by 4.187 to obtain coefficients for specific heats in kJ/kg·K.)

Adapted from a table published in *Engineering Thermodynamics*, C. O. Mackay, W. N. Barnard, and F. O. Ellenwood (John Wiley, New York, 1957).

# 23 Changes in Thermodynamic Properties

## Nomenclature

| | | | |
|---|---|---|---|
| $c$ | specific heat | Btu/lbm-°F | kJ/kg·K |
| $C$ | molar specific heat | Btu/lbmol-°F | kJ/kmol·K |
| $E$ | energy | Btu | kJ |
| $g$ | acceleration of gravity | ft/sec$^2$ | m/s$^2$ |
| $g_c$ | gravitational constant | ft-lbm/lbf-sec$^2$ | n.a. |
| $h$ | enthalpy | Btu/lbm | kJ/kg |
| $H$ | molar enthalpy | Btu/lbmol | kJ/kmol |
| $J$ | Joule's constant (778.17) | ft-lbf/Btu | n.a. |
| $k$ | ratio of specific heats | – | – |
| $m$ | mass | lbm | kg |
| $n$ | number of moles | – | – |
| $n$ | polytropic exponent | – | – |
| $p$ | pressure | lbf/ft$^2$ | kPa |
| $P$ | power | Btu/sec | kW |
| $q$ | heat per unit mass | Btu/lbm | kJ/kg |
| $Q$ | molar heat | Btu/lbmol | kJ/kmol |
| $Q$ | total heat | Btu | kJ |
| $R$ | specific gas constant | ft-lbf/lbm-°R | kJ/kg·K |
| $R^*$ | universal gas constant | ft-lbf/lbmol-°R | kJ/kmol·K |
| $s$ | entropy | Btu/lbm-°F | kJ/kg·K |
| $S$ | molar entropy | Btu/lbmol-°F | kJ/kmol·K |
| $T$ | temperature | °F | °C |
| $T$ | absolute temperature | °R | K |
| $u$ | specific internal energy | Btu/lbm | kJ/kg |
| $U$ | molar internal energy | Btu/lbmol | kJ/kmol |
| $v$ | velocity | ft/sec | m/s |
| $V$ | molar specific volume | ft$^3$/lbmol | m$^3$/kmol |
| $V$ | volume | ft$^3$ | m$^3$ |
| $w$ | specific work | ft-lbf/lbm | kJ/kg |
| $W$ | total work | ft-lbf | kJ |
| $z$ | elevation | ft | m |

## Symbols

| | | | |
|---|---|---|---|
| $\beta$ | isobaric compressibility | 1/°R | 1/K |
| $\eta$ | efficiency | – | – |
| $\mu_J$ | Joule-Thomson coefficient | °F-ft$^2$/lbf | K/Pa |
| $\upsilon$ | specific volume | ft$^3$/lbm | m$^3$/kg |
| $\Phi$ | availability function | Btu/lbm | kJ/kg |

## Subscripts

| | |
|---|---|
| al | aluminum |
| $H$ | hot (high-temperature) |
| $k$ | kinetic |
| $L$ | cold (low-temperature) |
| $n$ | polytropic |
| $o$ | outside (environment) |
| $p$ | constant pressure or potential |
| st | steel |
| $t$ | total |
| th | thermal |
| $v$ | constant volume |
| $w$ | water |

## 1. SYSTEMS

A *thermodynamic system* is defined as the matter enclosed within an arbitrary but precisely defined *control volume*. Everything external to the system is defined as the *surroundings*, *environment*, or *universe*. The environment and system are separated by the *system boundaries*. The surface of the control volume is known as the *control surface*. The control surface can be real (e.g., piston and cylinder walls) or imaginary.

If mass flows through the system across system boundaries, the system is an *open system*. Pumps, heat exchangers, and jet engines are examples of open systems. An important type of open system is the *steady-flow open system* in which matter enters and exits at the same rate. Pumps, turbines, heat exchangers, and boilers are all steady-flow open systems.

If no mass crosses the system boundaries, the system is said to be a *closed system*. The matter in a closed system may be referred to as a *control mass*. Closed systems can have variable volumes. The gas compressed by a piston in a cylinder is an example of a closed system with a variable control volume.

In most cases, energy in the form of heat, work, or electrical energy can enter or exit any open or closed system. Systems closed to both matter and energy transfer are known as *isolated systems*.

## 2. TYPES OF PROCESSES

Changes in thermodynamic properties of a system often depend on the type of process experienced. This is particularly true of gaseous systems. Several common types of processes are listed as follows, along with their heat, energy, and work relationships.[1]

- *adiabatic process*—a process in which no heat or other energy crosses the system boundary.[2] Adiabatic processes include isentropic and throttling processes.

$$Q = 0 \qquad\qquad 23.1$$

$$\Delta U = -W \qquad\qquad 23.2$$

- *constant pressure process*—also known as an *isobaric process*

$$\Delta p = 0 \qquad\qquad 23.3$$

$$Q = \Delta H \qquad\qquad 23.4$$

- *constant temperature process*—also known as an *isothermal process*

$$\Delta T = 0 \qquad\qquad 23.5$$

$$Q = W \qquad\qquad 23.6$$

- *constant volume process*—also known as an *isochoric* or *isometric process*

$$\Delta V = 0 \qquad\qquad 23.7$$

$$Q = \Delta U \qquad\qquad 23.8$$

$$W = 0 \qquad\qquad 23.9$$

- *isentropic process*—an adiabatic process in which there is no change in system entropy (i.e., is reversible)

$$\Delta S = 0 \qquad\qquad 23.10$$

$$Q = 0 \qquad\qquad 23.11$$

- *throttling process*—an adiabatic process in which there is no change in system enthalpy, but for which there is a significant pressure drop.[3]

$$\Delta H = 0 \qquad\qquad 23.12$$

$$p_2 < p_1 \qquad\qquad 23.13$$

A system that is in equilibrium at the start and finish of a process may or may not be in equilibrium during the process. A *quasistatic process (quasiequilibrium process)* is one that can be divided into a series of infinitesimal deviations (steps) from equilibrium. During each step, the property changes are small, and all intensive properties are uniform throughout the system. The interim equilibrium at each step is known as a *quasi-equilibrium*.

Transport processes are generally more complex than can be analyzed by the simple thermodynamic relationships in this chapter. In a *transport process*, there is a transfer of some quantity. Drying, distillation, and evaporation are examples of heat transfer processes. Fluid flow, mixing, and sedimentation are examples of momentum transfer. Distillation, absorption, extraction, and leaching are examples of mass transfer processes.

## 3. POLYTROPIC PROCESSES

A *polytropic process* is one that obeys Eq. 23.14, the *polytropic equation of state*. Gases always constitute the system in polytropic processes.

$$p_1 V_1^n = p_2 V_2^n \qquad\qquad 23.14$$

$n$ is the *polytropic exponent*, a property of the equipment, not of the gas. For efficient air compressors, $n$ is typically between 1.25 and 1.30.

Depending on the value of the polytropic exponent, the polytropic equation of state can also be used for other processes.

$$
\begin{array}{ll}
n = 0 & \text{[constant pressure process]} \\
n = 1 & \text{[constant temperature process]} \\
n = k & \text{[isentropic process]} \\
n = \infty & \text{[constant volume process]}
\end{array}
$$

The *polytropic specific heat*, $c_n$, is defined as

$$c_n = \frac{n-k}{n-1} \times c_v \qquad\qquad 23.15$$

$$Q = mc_n\Delta T \qquad\qquad 23.16$$

## 4. THROTTLING PROCESSES

For ideal gases, throttling processes are constant temperature processes. Some real gases at normal temperatures, however, decrease in temperature when

---

[1]The heat, energy, and work relationships are easily derived from the *first law of thermodynamics*.

[2]An adiabatic process is not the same as a constant temperature process, however.

[3]The classic throttling process is the expansion of a gas in a pipe through a porous plug that offers significant resistance to flow.

throttled.[4] Others increase in temperature. Whether or not an increase or a decrease in temperature occurs depends on the temperature and pressure at which the throttling occurs.

For any given set of initial conditions, there is one temperature at which no temperature change occurs when a real gas is throttled. This is called the *inversion temperature* or *inversion point*. The inversion temperature is dependent on the initial gas conditions. (See Table 23.1.)

**Table 23.1** *Approximate Maximum Experimental Inversion Temperatures*

| substance | °R | K |
|---|---|---|
| air | 1085 | 603 |
| argon | 1301 | 723 |
| carbon dioxide | ≈ 2700 | ≈ 1500 |
| helium | ≈ 72 | ≈ 40 |
| hydrogen | 364 | 202 |
| nitrogen | 1118 | 621 |

The *Joule-Thomson coefficient*, $\mu_J$, (also known as the *Joule-Kelvin coefficient*), is defined as the ratio of the change in temperature to the change in pressure when a real gas is throttled. The Joule-Thomson coefficient is zero for an ideal gas.

$$\mu_J = \left(\frac{\partial T}{\partial p}\right)_h \qquad \textit{23.17}$$

The Joule-Thomson coefficient can be calculated from the specific heat, $c_p$, and other properties.

$$\mu_J = \left(\frac{1}{c_p}\right)\left(T\left(\frac{\partial V}{\partial T}\right)_p - v\right) = \left(\frac{v}{c_p}\right)(T\beta - 1)$$

$$= \left(\frac{-1}{c_p}\right)\left(\frac{\partial h}{\partial p}\right)_T \qquad \textit{23.18}$$

It is convenient to plot lines of constant enthalpy (*isenthalpic curves*) on a *T-p* diagram. The various curves in Fig. 23.1 represent different sets of starting conditions ($p_1$ and $T_1$). The slope of a curve represents the Joule-Thomson coefficient, which can be zero, positive, or negative at that point. The points where the slope is zero are known as *inversion points*. The locus of inversion points is known as the *inversion curve*.

States to the right of an inversion point will have negative Joule-Thomson coefficients. States to the left of an inversion point will have positive Joule-Thomson coefficients. Throttling totally within the region bounded by the temperature axis and the inversion curve will always result in a decrease in temperature. Similarly, throttling totally outside of the inversion curve will always produce an increase in temperature. Throttling across the

inversion curve may produce a higher or lower final temperature, depending on initial and final conditions.

**Figure 23.1** *Inversion Curve*

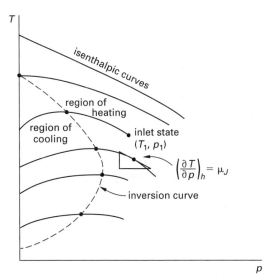

## 5. REVERSIBLE PROCESSES

Processes can be categorized as reversible and irreversible. A *reversible process* is one that is performed in such a way that at the conclusion of the process, *both* the system and the local surroundings can be restored to their initial states. A process that does not meet these requirements is an *irreversible process*. A reversible adiabatic process is implicitly isentropic.

Although all real-world processes result in an overall increase in entropy, it is possible to conceptualize processes that have zero entropy change. For a reversible process,

$$\Delta s = 0 \Big|_{\text{reversible}} \qquad \textit{23.19}$$

Processes that contain friction are never reversible. Other processes that are irreversible are listed as follows.

- stirring a viscous fluid
- slowing down a moving fluid
- unrestrained expansion of gas
- throttling
- changes of phase (freezing, condensation, etc.)
- chemical reaction
- diffusion
- current flow through electrical resistance
- electrical polarization
- magnetization with hysteresis

---

[4]This tendency can be used to liquefy gases and vapors (e.g., refrigerants) by passing them through an expansion valve.

- releasing a stretched spring

- inelastic deformation

- heat conduction

A system that has experienced an irreversible process can still be returned to its original state. An example of this is the water in a closed boiler-turbine installation. The entropy increases when the water is vaporized in the boiler (an irreversible process). The entropy decreases when the steam is condensed in the condenser. When the water returns to the boiler, its entropy is increased to its original value. However, the environment cannot be returned to its original condition; hence the cycle overall is irreversible.

## 6. FINDING WORK AND HEAT GRAPHICALLY

A process between two thermodynamic states can be shown graphically. The line representing the locus of quasiequilibrium states between the initial and final states is known as the *path* of the process.

It is sometimes convenient to see what happens to the pressure and volume of a system by plotting the path on a p-V diagram. In addition, the work done by or on the system can be determined from the graph. This is possible because the integral calculating p-V *work* represents area under a curve in the p-V plane.

$$W = \int_{V_1}^{V_2} p \, dV \qquad 23.20$$

Similarly, the amount of heat absorbed or released from a system can be determined as the area under the path on the T-s diagram.

$$Q = \int_{s_1}^{s_2} T \, ds \qquad 23.21$$

The variables $p$, $V$, $T$, and $s$ are *point functions* because their values are independent of the path taken to arrive at the thermodynamic state. Work and heat ($W$ and $Q$), however, are *path functions* because they depend on the path taken. (See Fig. 23.2.)

## 7. SIGN CONVENTION

A definite sign convention is used in calculating work, heat, and other property changes in systems.[5] This sign convention takes the system (not the environment) as the reference. (For example, a net heat gain would mean the system gained energy and the environment lost energy.)

- Heat, $Q$, is positive if heat flows into the system.

- Work, $W$, is positive if the system does work on the surroundings.

- Changes in enthalpy, entropy, and internal energy ($\Delta H$, $\Delta S$, and $\Delta U$) are positive if these properties increase within the system.

*Figure 23.2* Process Work and Heat

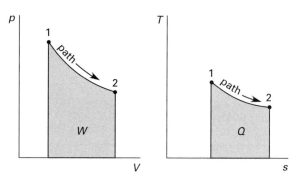

## 8. FIRST LAW OF THERMODYNAMICS FOR CLOSED SYSTEMS

There is a basic principle that underlies all property changes as a system experiences a process: all energy must be accounted for. Energy that enters a system must either leave the system or be stored in some manner. Energy cannot be created or destroyed. These statements are the primary manifestations of the *first law of thermodynamics*: the work done in an adiabatic process depends solely on the system's endpoint conditions, not on the nature of the process.

For closed systems, the first law can be written in differential form.[6] $Q$, $U$, and $W$ must all have the same units. This is less of a problem with SI units than with English units. For example, if $Q$ and $U$ are in Btu/lbm and $W$ is in ft-lbf/lbm, Joule's constant must be incorporated into the first law: $Q = \Delta U + W/J$ [U.S.].

Since $Q$ and $W$ are not properties, infinitesimal changes are designated as $\delta$.

$$\delta Q = dU + \delta W \qquad 23.22$$

Most simple thermodynamic problems can be solved without resorting to differential calculus. The first law can then be written in finite terms.

$$Q = \Delta U + W \qquad 23.23$$

Equation 23.23 says that the heat, $Q$, entering a closed system can either increase the temperature (i.e., increase $U$) or be used to perform work (increase $W$) on the surroundings. In a non-adiabatic closed system, heat energy entering the system also can leak to the surroundings. However, in Eq. 23.23, the $Q$ term is understood to be the net heat entering the system, exclusive of the loss.

---

[5]This sign convention is automatically followed if the formulas in this chapter are used.

[6]This formulation of the conservation of energy implicitly assumes that kinetic and potential energies are negligible.

In accordance with the standard sign convention given in Sec. 23.7, $Q$ will be negative if the net heat exchange to the system is a loss. $\Delta U$ will be negative if the internal energy of the system decreases. $W$ will be negative if the surroundings do work on the system (e.g., a piston compressing gas in a cylinder).

## 9. THERMAL EQUILIBRIUM

A simple application of the first law is the calculation of a thermal equilibrium point for a nonreacting system. *Thermal equilibrium* is reached when all parts of the system are at the same temperature.

The drive toward thermal equilibrium occurs spontaneously, without the addition of external work, whenever masses (two liquids, a solid and a liquid, etc.) with different temperatures are combined. Therefore, the work term, $W$, in the first law is zero.

The energy comes from the heat given off by the cooling mass (i.e., mass $A$). Energy is stored by increasing the temperature of the warmed mass (mass $B$). These changes are equal, but opposite, since the net change is zero.

$$Q_{\text{net}} = Q_{\text{in},A} - Q_{\text{out},B} = 0 \qquad \textit{23.24}$$

The form of the equation for $Q$ depends on the phases of the substances and the nature of the system. If the substances are either solid or liquid, Eq. 23.25 can be used. (For use with open systems, the $m$ in Eq. 23.25 should be replaced with $\dot{m}$.)

$$m_A c_{p,A}(T_{1,A} - T_{2,A}) = m_B c_{p,B}(T_{2,B} - T_{1,B}) \qquad \textit{23.25}$$

### Example 23.1

A block of steel is removed from a furnace and quenched in an insulated aluminum tank filled with water. The water and aluminum tank are initially in equilibrium at $75°F$ ($24°C$), and the final equilibrium temperature after quenching is $100°F$ ($38°C$). What is the initial temperature of the steel?

steel block mass:

| $m_{\text{st}}$ | 2.0 lbm | 0.9 kg |

steel specific heat:

| $c_{p,\text{st}}$ | 0.11 Btu/lbm-°F | 0.460 kJ/kg·K |

aluminum tank mass:

| $m_{\text{al}}$ | 5.0 lbm | 2.25 kg |

aluminum specific heat:

| $c_{p,\text{al}}$ | 0.21 Btu/lbm-°F | 0.880 kJ/kg·K |

water mass:

| $m_w$ | 12.0 lbm | 5.4 kg |

water specific heat:

| $c_{p,w}$ | 1.0 Btu/lbm-°F | 4.190 kJ/kg·K |

*SI Solution*

The heat lost by the steel is equal to the heat gained by the tank and water.

$$m_{\text{st}} c_{p,\text{st}}(T_{1,\text{st}} - T_{\text{eq}})$$
$$= (m_{\text{al}} c_{p,\text{al}} + m_w c_{p,w})(T_{\text{eq}} - T_{1,w})$$

$$(0.9\ \text{kg})\left(0.460\ \frac{\text{kJ}}{\text{kg·K}}\right)(T_{1,\text{st}} - 38°\text{C})$$
$$= \left(\begin{array}{l}(2.25\ \text{kg})\left(0.880\ \dfrac{\text{kJ}}{\text{kg·K}}\right)\\[2mm] + (5.4\ \text{kg})\left(4.190\ \dfrac{\text{kJ}}{\text{kg·K}}\right)\end{array}\right)(38°\text{C} - 24°\text{C})$$

$$T_{1,\text{st}} = 870°\text{C}$$

*Customary U.S. Solution*

Proceeding as in the SI solution,

$$(2.0\ \text{lbm})\left(0.11\ \frac{\text{Btu}}{\text{lbm-°F}}\right)(T_{1,\text{st}} - 100°\text{F})$$
$$= \left(\begin{array}{l}(5.0\ \text{lbm})\left(0.21\ \dfrac{\text{Btu}}{\text{lbm-°F}}\right)\\[2mm] + (12.0\ \text{lbm})\left(1\ \dfrac{\text{Btu}}{\text{lbm-°F}}\right)\end{array}\right)(100°\text{F} - 75°\text{F})$$

$$T_{1,\text{st}} = 1583°\text{F}$$

## 10. FIRST LAW OF THERMODYNAMICS FOR OPEN SYSTEMS

The first law of thermodynamics can also be written for open systems, but more terms are required to account for the many energy forms. The first law formulation is essentially the Bernoulli energy conservation equation extended to non-adiabatic processes.

If the mass flow rate is constant, the system is a *steady-flow system*, and the first law is known as the *steady-flow energy equation*, SFEE, Eq. 23.26. It is customary to express the SFEE in terms of energy per unit mass or per mole.

$$Q = \Delta U + \Delta E_p + \Delta E_k + W_{\text{flow}} + W_{\text{shaft}} \qquad \textit{23.26}$$

Some of the terms in Eq. 23.26 are illustrated in Fig. 23.3. $Q$ is the net heat flow into or out of the system, inclusive of any losses. It can be supplied from furnace flame, electrical heating, nuclear reaction, or other sources. If the system is adiabatic, $Q$ is zero.

$\Delta E_p$ and $\Delta E_k$ are the fluid's potential and kinetic energy changes. Generally, these terms are insignificant compared to the thermal energy transfers.

$W_{\text{shaft}}$ is the *shaft work*—work that the steady-flow device does on the surroundings. Its name is derived from the output shaft that serves to transmit energy

**Thermodynamics**

**Figure 23.3** Steady Flow Device

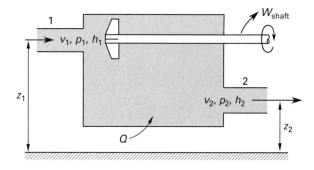

out of the system. For example, turbines and internal combustion engines have output shafts. $W_{\text{shaft}}$ can be negative, as in the case of a pump or compressor.

$W_{\text{flow}}$ is the *p-V work* (*flow energy, flow work*, etc.). There is a pressure, $p_2$, at the exit of the steady-flow device in Fig. 23.3. This exit pressure opposes the entrance of the fluid. Therefore, the flow work term represents the work required to cause the flow into the system against the exit pressure. The flow work can be calculated from Eq. 23.27. (Consistent units must be used.)

$$w_{\text{flow}} = p_2 v_2 - p_1 v_1 \qquad 23.27$$

$\Delta U$ is the change in the system's internal energy. Since the combination of internal energy and flow work constitutes enthalpy, it will seldom be necessary to work with either internal energy or flow work.

$$h = u + pv \qquad 23.28$$

$$\Delta h = \Delta u + w_{\text{flow}} \qquad 23.29$$

Equation 23.30 is a useful formulation of the SFEE. Units of power can be obtained by multiplying both sides by $\dot{m}$.

$$q = h_2 - h_1 + \frac{v_2^2 - v_1^2}{2} + (z_2 - z_1)g + w_{\text{shaft}} \qquad \text{[SI]}$$
$$23.30(a)$$

$$q = h_2 - h_1 + \frac{v_2^2 - v_1^2}{2g_c J} + \frac{(z_2 - z_1)g}{g_c J} + \frac{w_{\text{shaft}}}{J}$$
$$\text{[U.S.]} \quad 23.30(b)$$

**Example 23.2**

4 lbm/sec (1.8 kg/s) of steam enter a turbine with a velocity of 65 ft/sec (20 m/s) and an enthalpy of 1350 Btu/lbm (3140 kJ/kg). The steam enters the condenser after being expanded to 1075 Btu/lbm (2500 kJ/kg) at 125 ft/sec (38 m/s). There is a total heat loss from the turbine casing of 50 Btu/sec (53 kJ/s). Potential energy changes are insignificant. What power is generated at the turbine shaft?

*SI Solution*

Equation 23.30 can be solved for the shaft work.

$$P_{\text{shaft}} = \dot{m}w_{\text{shaft}} = \dot{Q}_t + \dot{m}\left(h_1 - h_2 + \frac{v_1^2 - v_2^2}{2}\right)$$

$$= -53\,000 \ \frac{\text{J}}{\text{s}}$$

$$+ \left(1.8 \ \frac{\text{kg}}{\text{s}}\right) \left( \begin{array}{c} 3140 \times 10^3 \ \dfrac{\text{J}}{\text{kg}} - 2500 \times 10^3 \ \dfrac{\text{J}}{\text{kg}} \\ + \dfrac{\left(20 \ \dfrac{\text{m}}{\text{s}}\right)^2 - \left(38 \ \dfrac{\text{m}}{\text{s}}\right)^2}{2} \end{array} \right)$$

$$= 1.098 \times 10^6 \ \text{W} \quad (1.1 \ \text{MW})$$

*Customary U.S. Solution*

Proceeding as in the SI solution,

$$P_{\text{shaft}} = \dot{m}w_{\text{shaft}} = \dot{Q}_t + \dot{m}\left(h_1 - h_2 + \frac{v_1^2 - v_2^2}{2g_c J}\right)$$

$$= -50 \ \frac{\text{Btu}}{\text{sec}}$$

$$+ \left(4.0 \ \frac{\text{lbm}}{\text{sec}}\right) \left( \begin{array}{c} 1350 \ \dfrac{\text{Btu}}{\text{lbm}} - 1075 \ \dfrac{\text{Btu}}{\text{lbm}} \\ + \dfrac{\left(65 \ \dfrac{\text{ft}}{\text{sec}}\right)^2 - \left(125 \ \dfrac{\text{ft}}{\text{sec}}\right)^2}{(2)\left(32.2 \ \dfrac{\text{ft-lbm}}{\text{lbf-sec}^2}\right)\left(778 \ \dfrac{\text{ft-lbf}}{\text{Btu}}\right)} \end{array} \right)$$

$$= 1049 \ \text{Btu/sec}$$

## 11. BASIC THERMODYNAMIC RELATIONS

The following relations for a simple compressible substance (though not necessarily an ideal gas) can be derived from the first law of thermodynamics and other basic definitions.[7]

$$du = T\,ds - p\,dv \qquad 23.31$$

$$dh = T\,ds + v\,dp \qquad 23.32$$

## 12. MAXWELL RELATIONS

Equation 23.35 through Eq. 23.38 are exact differentials with the same form.

$$dz = M\,dx + N\,dy \qquad 23.33$$

---

[7]It is impractical to list every relationship for property changes. Most formulas can be written in several forms. For example, all equations can be written either for a unit mass (i.e., per lbm or kg) or for a mole (i.e., per lbmol or kmol). For example, Eq. 23.31 on a molar basis would be written as $dU = T\,dS - p\,dV$.

For exact differentials of this form, it can be shown that the following relationship is valid.

$$\left(\frac{\partial M}{\partial y}\right)_x = \left(\frac{\partial N}{\partial x}\right)_y \qquad 23.34$$

It is possible to relate many of the thermodynamic properties of simple compressible substances. Four of these relationships are collectively known as *Maxwell relations*.

$$\left(\frac{\partial T}{\partial v}\right)_s = -\left(\frac{\partial p}{\partial s}\right)_v \qquad 23.35$$

$$\left(\frac{\partial T}{\partial p}\right)_s = \left(\frac{\partial v}{\partial s}\right)_p \qquad 23.36$$

$$\left(\frac{\partial s}{\partial v}\right)_T = \left(\frac{\partial p}{\partial T}\right)_v \qquad 23.37$$

$$\left(\frac{\partial v}{\partial T}\right)_p = -\left(\frac{\partial s}{\partial p}\right)_T \qquad 23.38$$

The Maxwell relations are useful in defining macroscopic properties (e.g., temperature) in terms of microscopic considerations and in designing experiments to measure thermodynamic properties. In particular, the relations show that pressure, temperature, and specific volume can be determined experimentally, but entropy must be derived from the other properties.

$$v = \left(\frac{\partial g}{\partial p}\right)_T = \left(\frac{\partial h}{\partial p}\right)_s \qquad 23.39$$

$$p = -\left(\frac{\partial a}{\partial v}\right)_T = -\left(\frac{\partial u}{\partial v}\right)_s \qquad 23.40$$

$$T = \left(\frac{\partial h}{\partial s}\right)_p = \left(\frac{\partial u}{\partial s}\right)_v \qquad 23.41$$

$$s = -\left(\frac{\partial a}{\partial T}\right)_v = -\left(\frac{\partial g}{\partial T}\right)_p \qquad 23.42$$

## 13. PROPERTY CHANGES IN IDEAL GASES

For real solids, liquids, and vapors, changes in most properties can be determined only by subtracting the initial property value from the final property value. For simple compressible substances (i.e., ideal gases), however, many changes in properties can be found directly, without knowing the initial and final property values.

Some of the relations for determining property changes do not depend on the type of process. For example, a general relationship that applies to any ideal gas experiencing any process is easily derived from the equation of state.

$$\frac{p_1 V_1}{T_1} = \frac{p_2 V_2}{T_2} \qquad 23.43$$

When temperature is held constant, Eq. 23.43 reduces to *Boyle's law*.

$$p_1 V_1 = p_2 V_2 \qquad 23.44$$

When pressure is held constant, Eq. 23.43 reduces to *Charles' law*.

$$\frac{V_1}{T_1} = \frac{V_2}{T_2} \qquad 23.45$$

Similarly, the changes in enthalpy, internal energy, and entropy are independent of the process. For perfect gases (i.e., ideal gases with constant specific heats),

$$\Delta h = c_p \Delta T \quad \text{[perfect gas]} \qquad 23.46$$

$$\Delta u = c_v \Delta T \quad \text{[perfect gas]} \qquad 23.47$$

$$\begin{aligned}\Delta s &= c_p \ln\left(\frac{T_2}{T_1}\right) - R\ln\left(\frac{p_2}{p_1}\right) \\ &= c_v \ln\left(\frac{T_2}{T_1}\right) + R\ln\left(\frac{v_2}{v_1}\right) \quad \text{[perfect gas]}\end{aligned} \qquad 23.48$$

Enthalpy and internal energy are *point functions*, so Eq. 23.46 and Eq. 23.47 are valid for all processes, even though the specific heats at constant pressure and volume are part of the calculation. These equations should not be confused with the heat transfer relations, which have a similar form. Heat, $Q$, is a path function and depends on the path taken.

$$q = c_p \Delta T \Big|_p \quad \text{[perfect gas]} \qquad 23.49$$

$$q = c_v \Delta T \Big|_v \quad \text{[perfect gas]} \qquad 23.50$$

Relationships between the properties follow for ideal gases experiencing specific processes. (The standard thermodynamic sign convention is automatically followed.) All relations can be written in slightly different forms, including per-unit mass and per-mole bases. For compactness, only the per-unit mass equations are listed. However, all equations can be converted to a molar basis by substituting $H$ for $h$, $V$ for $v$, $R^*$ for $R$, $C_p$ for $c_p$, and so on. Other forms can be derived by substituting the equation of state where appropriate.[8]

### Constant Pressure, Closed Systems

$$p_2 = p_1 \qquad 23.51$$

$$T_2 = T_1\left(\frac{v_2}{v_1}\right) \qquad 23.52$$

$$v_2 = v_1\left(\frac{T_2}{T_1}\right) \qquad 23.53$$

---

[8]For example, $RT$ can be substituted anywhere $pv$ appears.

$$q = h_2 - h_1 \qquad 23.54$$

$$= c_p(T_2 - T_1) \qquad 23.55$$

$$= c_v(T_2 - T_1) + p(v_2 - v_1) \qquad 23.56$$

$$u_2 - u_1 = c_v(T_2 - T_1) \qquad 23.57$$

$$= \frac{c_v p(v_2 - v_1)}{R} \qquad 23.58$$

$$= \frac{p(v_2 - v_1)}{k - 1} \qquad 23.59$$

$$w = p(v_2 - v_1) \qquad 23.60$$

$$= R(T_2 - T_1) \qquad 23.61$$

$$s_2 - s_1 = c_p \ln\left(\frac{T_2}{T_1}\right) \qquad 23.62$$

$$= c_p \ln\left(\frac{v_2}{v_1}\right) \qquad 23.63$$

$$h_2 - h_1 = q \qquad 23.64$$

$$= c_p(T_2 - T_1) \qquad 23.65$$

$$= \frac{kp(v_2 - v_1)}{k - 1} \qquad 23.66$$

## Constant Volume, Closed Systems

$$p_2 = p_1\left(\frac{T_2}{T_1}\right) \qquad 23.67$$

$$T_2 = T_1\left(\frac{p_2}{p_1}\right) \qquad 23.68$$

$$v_2 = v_1 \qquad 23.69$$

$$q = u_2 - u_1 \qquad 23.70$$

$$= c_v(T_2 - T_1) \qquad 23.71$$

$$u_2 - u_1 = q \qquad 23.72$$

$$= c_v(T_2 - T_1) \qquad 23.73$$

$$= \frac{c_v v(p_2 - p_1)}{R} \qquad 23.74$$

$$= \frac{v(p_2 - p_1)}{k - 1} \qquad 23.75$$

$$w = 0 \qquad 23.76$$

$$s_2 - s_1 = c_v \ln\left(\frac{T_2}{T_1}\right) \qquad 23.77$$

$$= c_v \ln\left(\frac{p_2}{p_1}\right) \qquad 23.78$$

$$h_2 - h_1 = c_p(T_2 - T_1) \qquad 23.79$$

$$= \frac{kv(p_2 - p_1)}{k - 1} \qquad 23.80$$

## Constant Temperature, Closed Systems

$$p_2 = p_1\left(\frac{v_1}{v_2}\right) \qquad 23.81$$

$$T_2 = T_1 \qquad 23.82$$

$$v_2 = v_1\left(\frac{p_1}{p_2}\right) \qquad 23.83$$

$$q = w \qquad 23.84$$

$$= T(s_2 - s_1) \qquad 23.85$$

$$= p_1 v_1 \ln\left(\frac{v_2}{v_1}\right) \qquad 23.86$$

$$= p_1 v_1 \ln\left(\frac{p_1}{p_2}\right) \qquad 23.87$$

$$= RT \ln\left(\frac{v_2}{v_1}\right) \qquad 23.88$$

$$= RT \ln\left(\frac{p_1}{p_2}\right) \qquad 23.89$$

$$u_2 - u_1 = 0 \qquad 23.90$$

$$w = q \qquad 23.91$$

$$= p_1 v_1 \ln\left(\frac{v_2}{v_1}\right) \qquad 23.92$$

$$= p_1 v_1 \ln\left(\frac{p_1}{p_2}\right) \qquad 23.93$$

$$= RT \ln\left(\frac{v_2}{v_1}\right) \qquad 23.94$$

$$= RT \ln\left(\frac{p_1}{p_2}\right) \qquad 23.95$$

$$s_2 - s_1 = \frac{q}{T} \qquad 23.96$$

$$= R \ln\left(\frac{v_2}{v_1}\right) \qquad 23.97$$

$$= R \ln\left(\frac{p_1}{p_2}\right) \qquad 23.98$$

$$h_2 - h_1 = 0 \qquad 23.99$$

## Isentropic, Closed Systems (Reversible Adiabatic)

$$p_2 = p_1\left(\frac{v_1}{v_2}\right)^k \qquad 23.100$$

$$= p_1\left(\frac{T_2}{T_1}\right)^{\frac{k}{k-1}} \qquad 23.101$$

Thermodynamics

$$T_2 = T_1 \left( \frac{v_1}{v_2} \right)^{k-1} \qquad \textit{23.102}$$

$$= T_1 \left( \frac{p_2}{p_1} \right)^{\frac{k-1}{k}} \qquad \textit{23.103}$$

$$v_2 = v_1 \left( \frac{p_1}{p_2} \right)^{\frac{1}{k}} \qquad \textit{23.104}$$

$$= v_1 \left( \frac{T_1}{T_2} \right)^{\frac{1}{k-1}} \qquad \textit{23.105}$$

$$q = 0 \qquad \textit{23.106}$$

$$u_2 - u_1 = -w \qquad \textit{23.107}$$

$$= c_v(T_2 - T_1) \qquad \textit{23.108}$$

$$= \frac{c_v(p_2 v_2 - p_1 v_1)}{R} \qquad \textit{23.109}$$

$$= \frac{p_2 v_2 - p_1 v_1}{k - 1} \qquad \textit{23.110}$$

$$w = u_1 - u_2 \qquad \textit{23.111}$$

$$= c_v(T_1 - T_2) \qquad \textit{23.112}$$

$$= \frac{p_1 v_1 - p_2 v_2}{k - 1} \qquad \textit{23.113}$$

$$= \frac{p_1 v_1}{k - 1} \left( 1 - \left( \frac{p_2}{p_1} \right)^{\frac{k-1}{k}} \right) \qquad \textit{23.114}$$

$$s_2 - s_1 = 0 \qquad \textit{23.115}$$

$$h_2 - h_1 = c_p(T_2 - T_1) \qquad \textit{23.116}$$

$$= \frac{k(p_2 v_2 - p_1 v_1)}{k - 1} \qquad \textit{23.117}$$

## Polytropic, Closed Systems

For $n = 1$, use constant temperature equations.

$$p_2 = p_1 \left( \frac{v_1}{v_2} \right)^n \qquad \textit{23.118}$$

$$= p_1 \left( \frac{T_2}{T_1} \right)^{\frac{n}{n-1}} \qquad \textit{23.119}$$

$$T_2 = T_1 \left( \frac{v_1}{v_2} \right)^{n-1} \qquad \textit{23.120}$$

$$= T_1 \left( \frac{p_2}{p_1} \right)^{\frac{n-1}{n}} \qquad \textit{23.121}$$

$$v_2 = v_1 \left( \frac{p_1}{p_2} \right)^{\frac{1}{n}} \qquad \textit{23.122}$$

$$= v_1 \left( \frac{T_1}{T_2} \right)^{\frac{1}{n-1}} \qquad \textit{23.123}$$

$$q = \frac{c_v(n - k)(T_2 - T_1)}{n - 1} \qquad \textit{23.124}$$

$$u_2 - u_1 = c_v(T_2 - T_1) \qquad \textit{23.125}$$

$$= q - w \qquad \textit{23.126}$$

$$w = \frac{R(T_1 - T_2)}{n - 1} \qquad \textit{23.127}$$

$$= \frac{p_1 v_1 - p_2 v_2}{n - 1} \qquad \textit{23.128}$$

$$= \frac{p_1 v_1}{n - 1} \left( 1 - \left( \frac{p_2}{p_1} \right)^{\frac{n-1}{n}} \right) \qquad \textit{23.129}$$

$$s_2 - s_1 = \frac{c_v(n - k)}{n - 1} \ln\left( \frac{T_2}{T_1} \right) \qquad \textit{23.130}$$

$$h_2 - h_1 = c_p(T_2 - T_1) \qquad \textit{23.131}$$

$$= \frac{n(p_2 v_2 - p_1 v_1)}{n - 1} \qquad \textit{23.132}$$

## Isentropic, Steady-Flow Systems

$p_2$, $v_2$, and $T_2$ are the same as for isentropic, closed systems.

$$q = 0 \qquad \textit{23.133}$$

$$w = h_1 - h_2 \qquad \textit{23.134}$$

$$= \frac{k p_1 v_1}{k - 1} \left( 1 - \left( \frac{p_2}{p_1} \right)^{\frac{k-1}{k}} \right)$$

$$= c_p T_1 \left( 1 - \left( \frac{p_2}{p_1} \right)^{\frac{k-1}{k}} \right) \qquad \textit{23.135}$$

$$u_2 - u_1 = c_v(T_2 - T_1) \qquad \textit{23.136}$$

$$h_2 - h_1 = -w \qquad \textit{23.137}$$

$$= c_p(T_2 - T_1) \qquad \textit{23.138}$$

$$= \frac{k(p_2 v_2 - p_1 v_1)}{k - 1} \qquad \textit{23.139}$$

$$s_2 - s_1 = 0 \qquad \textit{23.140}$$

## Polytropic, Steady-Flow Systems

$p_2$, $v_2$ and $T_2$ are the same as for polytropic, closed systems.

$$q = \frac{c_v(n - k)(T_2 - T_1)}{n - 1} \qquad \textit{23.141}$$

$$w = h_1 - h_2 \qquad \textit{23.142}$$

$$= \frac{nR(T_1 - T_2)}{n - 1} \qquad \textit{23.143}$$

$$u_2 - u_1 = c_v(T_2 - T_1) \qquad \textit{23.144}$$

**Thermodynamics**

$$h_2 - h_1 = -w \qquad\qquad 23.145$$

$$= c_p(T_2 - T_1) \qquad\qquad 23.146$$

$$= \frac{n(p_2 v_2 - p_1 v_1)}{n - 1} \qquad 23.147$$

$$s_2 - s_1 = \frac{c_v(n - k)}{n - 1} \ln\left(\frac{T_2}{T_1}\right) \qquad 23.148$$

## Throttling, Steady-Flow Systems

$$p_1 v_1 = p_2 v_2 \qquad\qquad 23.149$$

$$p_2 < p_1 \qquad\qquad 23.150$$

$$v_2 > v_1 \qquad\qquad 23.151$$

$$T_2 = T_1 \qquad\qquad 23.152$$

$$q = 0 \qquad\qquad 23.153$$

$$w = 0 \qquad\qquad 23.154$$

$$u_2 - u_1 = 0 \qquad\qquad 23.155$$

$$h_2 - h_1 = 0 \qquad\qquad 23.156$$

$$s_2 - s_1 = R \ln\left(\frac{p_1}{p_2}\right) \qquad 23.157$$

$$= R \ln\left(\frac{v_2}{v_1}\right) \qquad 23.158$$

### Example 23.3

2.0 lbm (0.9 kg) of hydrogen are cooled from 760°F to 660°F (400°C to 350°C) in a constant volume process. The specific heat at constant volume, $c_v$, is 2.435 Btu/lbm-°F (10.2 kJ/kg·K). How much heat is removed?

*SI Solution*

Use Eq. 23.71 on a per unit mass basis. The total heat transfer for $m$ kilograms is

$$Q = m c_v(T_2 - T_1)$$

$$= (0.9 \text{ kg})\left(10.2 \frac{\text{kJ}}{\text{kg·K}}\right)(350°C - 400°C)$$

$$= -459 \text{ kJ}$$

The minus sign is consistent with the convention that a heat loss is negative.

*Customary U.S. Solution*

The total heat transfer for $m$ lbm is

$$Q = m c_v(T_2 - T_1)$$

$$= (2.0 \text{ lbm})\left(2.435 \frac{\text{Btu}}{\text{lbm-°F}}\right)(660°F - 760°F)$$

$$= -487 \text{ Btu}$$

### Example 23.4

4 lbmol (4 kmol) of air initially at 1 atm and 530°R (295K) are compressed isothermally to 8 atm. How much total heat is removed during the compression?

*SI Solution*

Equation 23.95 applies to constant temperature compression. $R^*$ is used in place of $R$ to convert the calculations to a molar basis.

$$Q = n R^* T \ln\left(\frac{p_1}{p_2}\right)$$

$$= (4.0 \text{ kmol})\left(8.314 \frac{\text{kJ}}{\text{kmol·K}}\right)(295K) \ln\left(\frac{1 \text{ atm}}{8 \text{ atm}}\right)$$

$$= -20\,400 \text{ kJ}$$

*Customary U.S. Solution*

Writing Eq. 23.95 on a molar basis,

$$Q = n R^* T \ln\left(\frac{p_1}{p_2}\right)$$

$$= (4.0 \text{ lbmol})\left(1545 \frac{\text{ft-lbf}}{\text{lbmol-°R}}\right)(530°R) \ln\left(\frac{1 \text{ atm}}{8 \text{ atm}}\right)$$

$$= -6.81 \times 10^6 \text{ ft-lbf}$$

## 14. PROPERTY CHANGES IN INCOMPRESSIBLE FLUIDS AND SOLIDS

In order to simplify the solution to practical problems, enthalpy, entropy, internal energy, and specific volume in liquids and solids are often considered to be functions of temperature only. The effect of pressure is disregarded.

There are times, however, when it is necessary to evaluate the property changes in a solid or liquid system, no matter how small they may be. Changes for liquids can be evaluated by using compressed liquid tables. If certain assumptions are made, some property changes can also be calculated without knowing the initial and final values. The main assumptions are incompressibility and constant specific heats.

$$c_p = c_v = c \qquad\qquad 23.159$$

$$dv = 0 \qquad\qquad 23.160$$

$$du = c\,dT \qquad\qquad 23.161$$

$$ds = \frac{du}{T} \qquad\qquad 23.162$$

$$dh = c\,dT + v\,dp \qquad 23.163$$

If the specific heat is constant, Eq. 23.164 through Eq. 23.167 are valid.

$$v_2 - v_1 = 0 \qquad\qquad 23.164$$

$$u_2 - u_1 = c(T_2 - T_1) \qquad 23.165$$

$$s_2 - s_1 = c \ln\left(\frac{T_2}{T_1}\right) \qquad 23.166$$

$$h_2 - h_1 = c(T_2 - T_1) + v(p_2 - p_1) \qquad 23.167$$

## 15. HEAT RESERVOIRS

It is convenient to show a source of energy as an infinite constant-temperature *reservoir*. Figure 23.4 illustrates a source of energy (known as a *high-temperature reservoir* or *source reservoir*). By convention, the reservoir temperature is designated $T_H$, and the heat transfer from it is $Q_H$. The energy derived from such a theoretical source might actually be supplied by combustion, electrical heating, or nuclear reaction.

*Figure 23.4* Energy Reservoirs

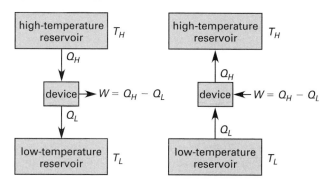

(a) power generation          (b) refrigeration effect

Similarly, energy is released to a *low-temperature reservoir* known as a *sink reservoir* or *energy sink*. The most common practical sink is the local environment. $T_L$ and $Q_L$ are used to represent the reservoir temperature and energy absorbed. It is common to refer to $Q_L$ as the rejected energy or energy released to the environment.

## 16. CYCLES

Although heat can be extracted and work performed in a single process, a cycle is necessary to obtain work in a useful quantity and duration. A *cycle* is a series of processes that eventually brings the system back to its original condition. Most cycles are continually repeated.

A cycle is completely defined by the working substance, the high- and low-temperature reservoirs, the means of doing work on the system, and the means of removing energy from the system.[9] (See Fig. 23.5.)

A cycle will appear as a closed curve when plotted on $p$-$V$ and $T$-$s$ diagrams. The area within the $p$-$V$ or $T$-$s$ curve represents the net work or net heat, respectively.

*Figure 23.5* Net Work and Net Heat

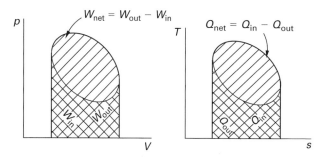

## 17. THERMAL EFFICIENCY

The *thermal efficiency* of a *power cycle* is defined as the ratio of useful work output to the supplied input energy.[10] $W$ in Eq. 23.168 is the *net work*, since some of the gross output work may be used to run certain parts of the cycle. For example, a small amount of turbine output power may run boiler feed pumps.

$$\eta_{\text{th}} = \frac{\text{net work output}}{\text{energy input}} = \frac{W_{\text{net}}}{Q_{\text{in}}}$$
$$= \frac{W_{\text{out}} - W_{\text{in}}}{Q_{\text{in}}} \qquad 23.168$$

The first law can be written as

$$Q_{\text{in}} = Q_{\text{out}} + W_{\text{net}} \qquad 23.169$$

Equation 23.168 and Eq. 23.169 can be combined to define the thermal efficiency in terms of heat variables alone.

$$\eta_{\text{th}} = \frac{Q_{\text{in}} - Q_{\text{out}}}{Q_{\text{in}}} \qquad 23.170$$

Equation 23.170 shows that obtaining the maximum efficiency requires minimizing the $Q_{\text{out}}$ term. The most efficient power cycle possible is the *Carnot cycle*.

## 18. SECOND LAW OF THERMODYNAMICS

The *second law of thermodynamics* can be stated in several ways. Equation 23.171 is the mathematical relation defining the second law. The equality holds for

---

[9]The *Carnot cycle* depends only on the source and sink temperatures, not on the working fluid. However, most practical cycles depend on the working fluid.

[10]The effectiveness of refrigeration and compression cycles is measured by other parameters.

reversible processes; the inequality holds for irreversible processes.

$$\Delta s \geq \int_{T_1}^{T_2} \frac{dq}{T} \qquad \textit{23.171}$$

Equation 23.171 effectively states that net entropy must always increase in practical (irreversible) cyclical processes.

> A natural process that starts in one equilibrium state and ends in another will go in the direction that causes the entropy of the system and the environment to increase.

The *Kelvin-Planck statement* of the second law effectively says that it is impossible to build a cyclical engine that will have a thermal efficiency of 100%.

> It is impossible to operate an engine operating in a cycle that will have no other effect than to extract heat from a reservoir and turn it into an equivalent amount of work.

This formulation is not a contradiction of the first law of thermodynamics. The first law does not preclude the possibility of converting heat entirely into work—it only denies the possibility of creating or destroying energy. The second law says that if some heat is converted entirely into work, some other energy must be rejected to a low-temperature sink (i.e., lost to the surroundings).

Figure 23.6 illustrates violations of the second law.

**Figure 23.6** Second Law Violations

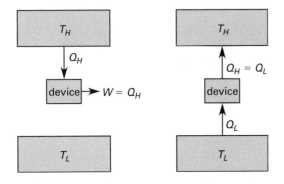

## 19. AVAILABILITY

The maximum possible work that can be obtained from a cycle is known as the *availability*. Availability is independent of the device but is dependent on the temperature of the local environment. Both the first and second law must be applied to determine availability.

In Fig. 23.7, the *heat engine* is surrounded by an environment at absolute temperature $T_L$. The net heat available for conversion to useful work is $Q = Q_H - Q_L$. Work, $W$, is performed at a constant rate on the environment by the engine. This is the traditional development of the availability concept. Actually, the symbols

$\dot{Q}$, $\dot{m}$, and $P$ (power) should be used to represent the time rate of heat, mass, and work. Assuming steady flow, and neglecting kinetic and potential energies, the first law can be written as

$$mh_1 + Q = mh_2 + W \qquad \textit{23.172}$$

**Figure 23.7** Typical Heat Engine

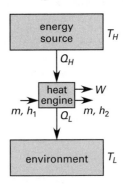

Entropy is not a part of the first law. The leaving entropy, however, can be calculated from the entering entropy and the entropy production.

$$ms_2 = ms_1 + \frac{Q}{T_L} \qquad \textit{23.173}$$

Since Eq. 23.172 and Eq. 23.173 both contain $Q$, they can be combined.

$$W = m(h_1 - T_L s_1 - h_2 + T_L s_2) \qquad \textit{23.174}$$

This equation can be simplified by introducing the *steady-flow availability function*, $\Phi$. The maximum work output (availability) is

$$W_{max} = \Phi_1 - \Phi_2 \qquad \textit{23.175}$$

$$\Phi = h - T_L s \qquad \textit{23.176}$$

If the equality in Eq. 23.175 holds, both the process within the control volume and the energy transfers between the system and the environment must be reversible. Maximum work output, therefore, will be obtained in a reversible process. The difference between the maximum and the actual work output is known as the *process irreversibility*.

### Example 23.5

What is the maximum useful work that can be produced per pound (per kilogram) of steam that enters a steady flow system at 800 psia (5.0 MPa) saturated and leaves in equilibrium with the environment at 70°F and 14.7 psia (20°C and 100 kPa)?

*SI Solution*

From the saturated steam table, $h_1 = 2794.2$ kJ/kg and $s_1 = 5.9737$ kJ/kg·K.

The final properties are obtained from the saturated steam table for water at 20°C. $h_2 = 83.91$ kJ/kg and $s_2 = 0.2965$ kJ/kg·K.

The availability is calculated from Eq. 23.174 using $T_L = 20°C + 273° = 293K$.

$$W_{max} = m\left(h_1 - h_2 - T_L(s_1 - s_2)\right)$$

$$= (1\ kg)\left(\begin{array}{c} 2794.2\ \dfrac{kJ}{kg} - 83.91\ \dfrac{kJ}{kg} \\[2mm] -(293K)\left(\begin{array}{c} 5.9737\ \dfrac{kJ}{kg\cdot K} \\[2mm] -\ 0.2965\ \dfrac{kJ}{kg\cdot K} \end{array}\right) \end{array}\right)$$

$$= 1047\ kJ$$

*Customary U.S. Solution*

From the saturated steam table, $h_1 = 1199.3$ Btu/lbm and $s_1 = 1.4162$ Btu/lbm-°R.

The final properties are obtained from the saturated steam table for water at 70°F. $h_2 = 38.08$ Btu/lbm and $s_2 = 0.007459$ Btu/lbm-°R.

The availability is calculated from Eq. 23.174 using $T_L = 70°F + 460° = 530°R$.

$$W_{max} = m\left(h_1 - h_2 - T_L(s_1 - s_2)\right)$$

$$= (1\ lbm)\left(\begin{array}{c} 1199.3\ \dfrac{Btu}{lbm} - 38.08\ \dfrac{Btu}{lbm} \\[2mm] -(530°R)\left(\begin{array}{c} 1.4162\ \dfrac{Btu}{lbm\text{-}°R} \\[2mm] -\ 0.07459\ \dfrac{Btu}{lbm\text{-}°R} \end{array}\right) \end{array}\right)$$

$$= 450.2\ Btu$$

# 24 Psychrometrics

## Nomenclature

| | | | |
|---|---|---|---|
| ADP | apparatus dew point | °F | °C |
| $B$ | volumetric fraction | – | – |
| BF | bypass factor | – | – |
| $C$ | concentration | ppm | mg/L |
| CF | contact factor | – | – |
| $c_p$ | specific heat | Btu/lbm-°F | kJ/kg·°C |
| $G$ | gravimetric fraction | – | – |
| $h$ | enthalpy | Btu/lbm | kJ/kg |
| $m$ | mass | lbm | kg |
| $n$ | number of moles | – | – |
| $p$ | pressure | lbf/ft$^2$ | kPa |
| PF | performance factor | – | – |
| $q$ | heat | Btu/lbm | J/kg |
| $Q$ | volumetric flow rate | gal/min | L/s |
| $R$ | specific gas constant | ft-lbf/lbm-°R | kJ/kg·K |
| RF | rating factor | – | – |
| SHR | sensible heat ratio | – | – |
| $T$ | temperature | °F | °C |
| TDS | total dissolved solids | ppm | mg/L |
| TU | tower units | – | – |
| $V$ | volume | ft$^3$ | m$^3$ |
| $x$ | mole fraction | – | – |

## Symbols

| | | | |
|---|---|---|---|
| $\eta$ | efficiency | – | – |
| $\mu$ | degree of saturation | – | – |
| $\rho$ | mass density | lbm/ft$^3$ | kg/m$^3$ |
| $v$ | specific volume | ft$^3$/lbm | m$^3$/kg |
| $\phi$ | relative humidity | – | – |
| $\omega$ | humidity ratio | lbm/lbm | kg/kg |

## Subscripts

| | |
|---|---|
| $a$ | dry air |
| db | dry-bulb |
| dp | dew-point |
| $fg$ | vaporization |
| $l$ | latent |
| $s$ | sensible |
| sat | saturation |
| $t$ | total |
| unsat | unsaturated |
| $v$ | vapor |
| $w$ | water |
| wb | wet-bulb |

## 1. INTRODUCTION TO PSYCHROMETRICS

Atmospheric air contains small amounts of moisture and can be considered to be a mixture of two ideal gases—dry air and water vapor. All of the thermodynamic rules relating to the behavior of nonreacting gas mixtures apply to atmospheric air. From Dalton's law, for example, the total atmospheric pressure is the sum of the dry air partial pressure and the water vapor pressure.[1]

$$p = p_a + p_w \qquad 24.1$$

The study of the properties and behavior of atmospheric air is known as *psychrometrics*. Properties of atmospheric air are seldom evaluated, however, from theoretical thermodynamic principles. Rather, specialized techniques and charts have been developed for that purpose.

## 2. PROPERTIES OF ATMOSPHERIC AIR

At first, psychrometrics seems complicated by three different definitions of temperature. These three terms are not interchangeable.

---

[1]Equation 24.1 points out a problem in semantics. The term *air* means *dry air*. The term *atmosphere* refers to the combination of dry air and water vapor. It is common to refer to the atmosphere as *moist air*.

- *dry-bulb temperature*, $T_{db}$: This is the equilibrium temperature that a regular thermometer measures if exposed to atmospheric air.

- *wet-bulb temperature*, $T_{wb}$: This is the temperature of air that has gone through an adiabatic saturation process. (See Sec. 24.13.)

- *dew-point temperature*, $T_{dp}$: This is the dry-bulb temperature at which water starts to condense out when moist air is cooled in a constant pressure process.

For every temperature, there is a unique vapor pressure, $p_{sat}$, which represents the maximum pressure the water vapor can exert. The actual vapor pressure, $p_w$, can be less than or equal to, but not greater than, the saturation value. The saturation pressure is found from steam tables as the pressure corresponding to the dry-bulb temperature of the atmospheric air.

$$p_w \leq p_{sat} \qquad 24.2$$

If the vapor pressure equals the saturation pressure, the air is said to be saturated.[2] *Saturated air* is a mixture of dry air and saturated water vapor. When the air is saturated, all three temperatures are equal.

$$T_{db} = T_{wb} = T_{dp} \Big|_{sat} \qquad 24.3$$

*Unsaturated air* is a mixture of dry air and superheated water vapor.[3] When the air is unsaturated, the dew-point temperature will be less than the wet-bulb temperature. The *wet-bulb depression* is the difference between the dry-bulb and wet-bulb temperatures.

$$T_{dp} < T_{wb} < T_{db} \Big|_{unsat} \qquad 24.4$$

The amount of water vapor in atmospheric air is specified by three different parameters. The *humidity ratio*, $\omega$ (also known as the *specific humidity*), is the mass ratio of water vapor to dry air. If both masses are expressed in pounds (kilograms), the units of humidity ratio are lbm/lbm (kg/kg). However, since there is so little water vapor, the water vapor mass is often reported in *grains* of water. (There are 7000 grains per pound.) Accordingly, the humidity ratio will have the units of grains per pound.

$$\omega = \frac{m_w}{m_a} \qquad 24.5$$

Since $m = \rho V$, and since $V_w = V_a$, the humidity ratio can be written as

$$\omega = \frac{\rho_w}{\rho_a} \qquad 24.6$$

From the equation of state for an ideal gas, $m = pV/RT$. Since $V_w = V_a$ and $T_w = T_a$, the humidity ratio can be written in one additional form.

$$\omega = \frac{R_a p_w}{R_w p_a} = \frac{53.35 p_w}{85.78 p_a} = 0.622 \left( \frac{p_w}{p_a} \right) \qquad 24.7$$

The *degree of saturation*, $\mu$ (also known as the *saturation ratio* and the *percentage humidity*), is the ratio of the actual humidity ratio to the saturated humidity ratio at the same temperature and pressure.

$$\mu = \frac{\omega}{\omega_{sat}} \qquad 24.8$$

A third index of moisture content is the *relative humidity*—the partial pressure of the water vapor divided by the saturation pressure.

$$\phi = \frac{p_w}{p_{sat}} \qquad 24.9$$

From the equation of state for an ideal gas, $\rho = p/RT$, so the relative humidity can be written as

$$\phi = \frac{\rho_w}{\rho_{sat}} \qquad 24.10$$

Combining the definitions of specific and relative humidities,

$$\phi = 1.608 \omega \left( \frac{p_a}{p_{sat}} \right) \qquad 24.11$$

## 3. VAPOR PRESSURE

There are at least six ways of determining the partial pressure, $p_w$, of the water vapor in the air. The first method, derived from Eq. 24.9, is to multiply the relative humidity, $\phi$, by the water's saturation pressure. The saturation pressure, in turn, is obtained from steam tables as the pressure corresponding to the air's dry-bulb temperature.

$$p_w = \phi p_{sat,db} \qquad 24.12$$

A more direct method is to read the saturation pressure (from the steam tables) corresponding to the air's dew-point temperature.

$$p_w = p_{sat,dp} \qquad 24.13$$

The third method can be used if water's mole (volumetric) fraction is known.

$$p_w = x_w p_t = B_w p_t \qquad 24.14$$

---

[2]Actually, the water vapor is saturated, not the air. However, this particular inconsistency in terms is characteristic of psychrometrics.
[3]As strange as it sounds, atmospheric water vapor is almost always superheated. This can be shown by drawing an isotherm passing through the vapor dome on a $p$-$V$ diagram. The only place where the water vapor pressure is less than the saturation pressure is in the superheated region.

The fourth method is to calculate the actual vapor pressure from the empirical *Carrier equation*, valid for customary U.S. units only.[4]

$$p_w = p_{sat,wb} - \frac{(p_t - p_{sat,wb})(T_{db} - T_{wb})}{2830 - 1.44\, T_{wb}} \qquad \text{[U.S.]} \quad \textit{24.15}$$

The fifth method is based on the humidity ratio.

$$p_w = \frac{p_t \omega}{0.622 + \omega} \qquad \textit{24.16}$$

The sixth (and easiest) method is to read the water vapor pressure from a psychrometric chart. Some, but not all psychrometric charts, have water vapor scales.

### Example 24.1

Use the methods described in the previous section to determine the partial pressure of water vapor in standard atmospheric air at 60°F (16°C) dry-bulb and 50% relative humidity.

*SI Solution*

*method 1:* From the steam tables, the saturation pressure corresponding to 16°C is 0.01819 bars. The partial pressure of the vapor is

$$p_w = \phi p_{sat} = (0.50)(0.01819 \text{ bars})\left(100\, \frac{\text{kPa}}{\text{bar}}\right)$$
$$= 0.910 \text{ kPa}$$

*method 2:* The dew-point temperature (reading straight across on the psychrometric chart) is approximately 5°C. The saturation pressure from the steam table corresponding to 5°C is approximately 0.0087 bars (0.87 kPa).

*method 3:* The humidity ratio is 0.0056 kg/kg. From Eq. 24.16,

$$p_w = \frac{p_t \omega}{0.622 + \omega}$$
$$= \frac{(101.3 \text{ kPa})\left(0.0056\, \frac{\text{kg}}{\text{kg}}\right)}{0.622 + 0.0056\, \frac{\text{kg}}{\text{kg}}}$$
$$= 0.904 \text{ kPa}$$

*Customary U.S. Solution*

*method 1:* From the steam tables, the saturation pressure corresponding to 60°F is 0.2564 lbf/in². The partial pressure of the vapor is

$$p_w = \phi p_{sat} = (0.50)\left(0.2564\, \frac{\text{lbf}}{\text{in}^2}\right)$$
$$= 0.128 \text{ lbf/in}^2$$

*method 2:* The dew-point temperature (reading straight across the psychrometric chart) is approximately 41°F. The saturation pressure from the steam table corresponding to 41°F is approximately 0.127 lbf/in².

*method 3:* Use the Carrier equation. The wet-bulb temperature of the air is approximately 50°F. From the steam tables, the saturation pressure corresponding to that temperature is 0.1780 lbf/in².

$$p_w = p_{sat,wb} - \frac{(p_t - p_{sat,wb})(T_{db} - T_{wb})}{2830 - 1.44\, T_{wb}}$$

$$= 0.1780\, \frac{\text{lbf}}{\text{in}^2} - \frac{\left(14.7\, \frac{\text{lbf}}{\text{in}^2} - 0.1780\, \frac{\text{lbf}}{\text{in}^2}\right)}{2830 - (1.44)(50°F)} \times (60°F - 50°F)$$

$$= 0.125 \text{ lbf/in}^2$$

## 4. ENERGY CONTENT OF AIR

Since moist air is a mixture of dry air and water vapor, its total enthalpy, $h$ (i.e., energy content), takes both components into consideration. Total enthalpy is conveniently shown on the diagonal scales of the psychrometric chart, but it can also be calculated. As Eq. 24.18 indicates, the reference temperature (i.e., the temperature that corresponds to a zero enthalpy) for the enthalpy of dry air is 0°F (0°C). Steam properties correspond to a low-pressure superheated vapor at room temperature.

$$h_t = h_a + \omega h_w \qquad \textit{24.17}$$

$$h_a = c_{p,air} T \approx \left(1.005\, \frac{\text{kJ}}{\text{kg-°C}}\right) T_{°C} \qquad \text{[SI]} \quad \textit{24.18(a)}$$

$$h_a = c_{p,air} T \approx \left(0.240\, \frac{\text{Btu}}{\text{lbm-°F}}\right) T_{°F} \qquad \text{[U.S.]} \quad \textit{24.18(b)}$$

$$h_w = c_{p,water\,vapor} T + h_{fg}$$
$$\approx \left(1.805\, \frac{\text{kJ}}{\text{kg-°C}}\right) T_{°C} + 2501\, \frac{\text{kJ}}{\text{kg}} \qquad \text{[SI]} \quad \textit{24.19(a)}$$

$$h_w = c_{p,water\,vapor} T + h_{fg}$$
$$\approx \left(0.444\, \frac{\text{Btu}}{\text{lbm-°F}}\right) T_{°F} + 1061\, \frac{\text{Btu}}{\text{lbm}} \qquad \text{[U.S.]} \quad \textit{24.19(b)}$$

[4]Equation 24.15 uses updated constants and is more accurate than the equation originally published by Carrier.

## 5. THE PSYCHROMETRIC CHART

It is possible to develop mathematical relationships for enthalpy and specific volume (the two most useful thermodynamic properties) for atmospheric air. However, these relationships are almost never used. Rather, psychrometric properties can be read directly from *psychrometric charts* ("psych charts," as they are usually referred to), as illustrated in App. 24.A and App. 24.B. There are different psychrometric charts for low, medium, and high temperature ranges, as well as charts for different atmospheric pressures (i.e., elevations).

The usage of several scales varies somewhat from chart to chart. In particular, the use of the enthalpy scale depends on the chart used. Furthermore, not all psychrometric charts contain all scales.

A psychrometric chart is easy to use, despite the multiplicity of scales. The thermodynamic state (i.e., the position on the chart) is defined by specifying the values of any two parameters on intersecting scales (e.g., dry-bulb and wet-bulb temperature, or dry-bulb temperature and relative humidity). Once the state point has been located on the chart, all other properties can be read directly.

## 6. ENTHALPY CORRECTIONS

Some psychrometric charts have separate lines or scales for wet-bulb temperature and enthalpy. However, the deviation between lines of constant wet-bulb temperature and lines of constant enthalpy is small. Therefore, other psychrometric charts use only one set of diagonal lines for both scales. The error introduced is small—seldom greater than 0.1–0.2 Btu/lbm (0.23–0.46 kJ/kg). When extreme precision is needed, correction factors from the psychrometric chart can be used.

### Example 24.2

Air at 50°F (10°C) dry bulb has a humidity ratio of 0.006 lbm/lbm (0.006 kg/kg). (a) Use the psychrometric chart to determine the enthalpy of the air. (b) Calculate the enthalpy of the air directly. (c) How much heat is needed to heat one unit mass of the air from 50°F to 140°F (10°C to 60°C) without changing the moisture content?

*SI Solution*

(a) Use the moisture content and dry-bulb temperature scales to locate the point corresponding to the original conditions. From the psychrometric chart, the enthalpy is approximately 25 kJ/kg.

(b) Use Eq. 24.17 and Eq. 24.18.

$$h_a = c_{p,\text{air}} T$$
$$\approx \left(1.005 \ \frac{\text{kJ}}{\text{kg·°C}}\right) T_{\text{°C}}$$
$$= \left(1.005 \ \frac{\text{kJ}}{\text{kg·°C}}\right)(10°\text{C}) = 10 \ \text{kJ/kg}$$

$$h_w = c_{p,\text{water vapor}} T + h_{fg}$$
$$\approx \left(1.805 \ \frac{\text{kJ}}{\text{kg·°C}}\right) T_{\text{°C}} + 2501 \ \frac{\text{kJ}}{\text{kg}}$$
$$= \left(1.805 \ \frac{\text{kJ}}{\text{kg·°C}}\right)(10°\text{C}) + 2501 \ \frac{\text{kJ}}{\text{kg}}$$
$$= 2519 \ \text{kJ/kg}$$

$$h_t = h_a + \omega h_w$$
$$= 10 \ \frac{\text{kJ}}{\text{kg}} + \left(0.006 \ \frac{\text{kg}}{\text{kg}}\right)\left(2519 \ \frac{\text{kJ}}{\text{kg}}\right)$$
$$= 25.1 \ \text{kJ/kg}$$

(c) The psychrometric chart does not go up to 60°C. Therefore, the energy difference must be calculated mathematically. Although the initial enthalpy could be subtracted from the calculated final enthalpy, it is equivalent merely to calculate the difference based on the variable terms.

$$q = h_{t,2} - h_{t,1}$$
$$= (c_{p,\text{air}} + \omega c_{p,\text{water vapor}})(T_2 - T_1)$$
$$= \left(\left(1.005 \ \frac{\text{kJ}}{\text{kg·°C}}\right) + \left(0.006 \ \frac{\text{kg}}{\text{kg}}\right)\left(1.805 \ \frac{\text{kJ}}{\text{kg·°C}}\right)\right)$$
$$\times (60°\text{C} - 10°\text{C})$$
$$= 50.8 \ \text{kJ/kg}$$

*Customary U.S. Solution*

(a) Use the moisture content and dry-bulb temperature scales to locate the point corresponding to the original conditions. From the psychrometric chart, the enthalpy is approximately 18.5 Btu/lbm.

(b) Use Eq. 24.17 and Eq. 24.18.

$$h_a = c_{p,\text{air}} T$$
$$\approx \left(0.240 \ \frac{\text{Btu}}{\text{lbm-°F}}\right) T_{\text{°F}}$$
$$= \left(0.240 \ \frac{\text{Btu}}{\text{lbm-°F}}\right)(50°\text{F}) = 12 \ \text{Btu/lbm}$$

$$h_w = c_{p,\text{water vapor}} T + h_{fg}$$

$$\approx \left(0.444 \; \frac{\text{Btu}}{\text{lbm-}^\circ\text{F}}\right) T_{^\circ\text{F}} + 1061 \; \frac{\text{Btu}}{\text{lbm}}$$

$$= \left(0.444 \; \frac{\text{Btu}}{\text{lbm-}^\circ\text{F}}\right)(50^\circ\text{F}) + 1061 \; \frac{\text{Btu}}{\text{lbm}}$$

$$= 1083.2 \; \text{Btu/lbm}$$

$$h_t = h_a + \omega h_w$$

$$= 12 \; \frac{\text{Btu}}{\text{lbm}} + \left(0.006 \; \frac{\text{lbm}}{\text{lbm}}\right)\left(1083.2 \; \frac{\text{Btu}}{\text{lbm}}\right)$$

$$= 18.5 \; \text{Btu/lbm}$$

(c) Psychrometric charts for room temperature do not go up to $140^\circ\text{F}$. (Appendix 24.D could be used.) Therefore, the energy difference must be calculated mathematically. Although the initial enthalpy could be subtracted from the calculated final enthalpy, it is equivalent merely to calculate the difference based on the variable terms.

$$q = h_{t,2} - h_{t,1}$$

$$= (c_{p,\text{air}} + \omega c_{p,\text{water vapor}})(T_2 - T_1)$$

$$= \left( \begin{array}{c} 0.240 \; \dfrac{\text{Btu}}{\text{lbm-}^\circ\text{F}} + \left(0.006 \; \dfrac{\text{lbm}}{\text{lbm}}\right) \\ \times \left(0.444 \; \dfrac{\text{Btu}}{\text{lbm-}^\circ\text{F}}\right) \end{array} \right)$$
$$\times (140^\circ\text{F} - 50^\circ\text{F})$$

$$= 21.84 \; \text{Btu/lbm}$$

## 7. BASIS OF PROPERTIES

Several of the properties read from the psychrometric chart (specific volume, enthalpy, etc.) are given "per pound of dry air." This basis does not mean that the water vapor's contribution is absent. For example, if the enthalpy of atmospheric air is 28.0 Btu per pound of dry air, the energy content of the water vapor has been included. However, to get the energy of a mass of moist air, the enthalpy of 28 Btu/lbm would be multiplied by the mass of the dry air $(m_a)$ only, not by the combined air and water masses.

$$h_t = m_a h_{\text{chart}} \qquad \qquad \textit{24.20}$$

### Example 24.3

During the summer, air in a room reaches $75^\circ\text{F}$ and 50% relative humidity. Find the air's (a) wet-bulb temperature, (b) humidity ratio, (c) enthalpy, (d) specific volume, (e) dew-point temperature, (f) actual vapor pressure, and (g) degree of saturation.

*Solution*

Locate the point where the $75^\circ\text{F}$ vertical line intersects the curved 50% humidity line. Read all other values directly from the chart.

(a) Follow the diagonal line up to the left until it intersects the wet-bulb temperature scale. Read $T_{\text{wb}} = 62.6^\circ\text{F}$.

(b) Follow the horizontal line to the right until it intersects the humidity ratio scale. Read $\omega = 64.8$ gr (0.0093 lbm) of moisture per pound of dry air.

(c) Finding the enthalpy is different on different charts. Some charts use the same diagonal lines for wet-bulb temperature and humidity. Corrections are required in such cases. Other charts employ two alignment scales to use in conjunction with a straightedge. Read 28.1 Btu per pound of dry air.

(d) Interpolate between diagonal specific volume lines. Read $v = 13.68$ cubic feet per pound of dry air.

(e) Follow the horizontal line to the left until it intersects the dew-point scale. Read $T_{\text{dp}} = 55.1^\circ\text{F}$.

(f) From the steam tables, the saturation pressure corresponding to a dry-bulb temperature of $75^\circ\text{F}$ is approximately 0.43 psia. From Eq. 24.9, the water vapor pressure is

$$p_w = \phi p_{\text{sat}} = (0.50)(0.43 \; \text{psia}) = 0.215 \; \text{psia}$$

(g) The humidity ratio at $75^\circ\text{F}$ saturated is 131.5 gr (0.0188 lbm) per pound. From Eq. 24.8, the degree of saturation is

$$\mu = \frac{\omega}{\omega_{\text{sat}}} = \frac{64.8 \; \dfrac{\text{gr}}{\text{lbm}}}{131.5 \; \dfrac{\text{gr}}{\text{lbm}}} = 0.49$$

## 8. LEVER RULE

With few exceptions (e.g., relative humidity and enthalpy correction), the scales on a psychrometric chart are linear. Because they are linear, any one property can be used as the basis for interpolation or extrapolation for another property on an intersecting linear scale. This applies regardless of orientation of the scales. The scales do not have to be orthogonal.

Furthermore, since psychrometric properties are extensive properties (i.e., they depend on the quantity of air present), the mass of air can be used as the basis for interpolation or extrapolation. This principle, known as the *lever rule* or *inverse lever rule*, is used when determining the properties of a mixture. (See Sec. 24.9.)

## 9. ADIABATIC MIXING OF TWO AIR STREAMS

Figure 24.1 shows the mixing of two moist air streams. The state of the mixture can be determined if the flow rates and psychrometric properties of the two component streams are known.

**Figure 24.1** *Mixing of Two Air Streams*

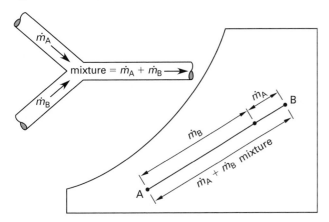

The two input states are located on the psychrometric chart and a straight line is drawn between them. The state of the mixture air will be on the straight line. The lever rule based on air masses is used to locate the mixture point. (Since the water vapor adds little to the mixture mass, the ratio of moist air masses can be approximated by the ratio of dry air masses, which, in turn, can be approximated by the ratio of air flow volumes.)

The lever rule can be used to find the mixture properties algebraically. Density changes can generally be disregarded, allowing volumetric flow rates to be used in place of mass flow rates. For the dry-bulb temperature (or any other property with a linear scale), the mixture temperature is

$$T_{\text{mixture}} = T_A + \left(\frac{\dot{m}_B}{\dot{m}_A + \dot{m}_B}\right)(T_B - T_A)$$

$$\approx T_A + \left(\frac{\dot{V}_B}{\dot{V}_A + \dot{V}_B}\right)(T_B - T_A) \qquad 24.21$$

### Example 24.4

5000 ft³/min (2.36 m³/s) of air at 40°F (4°C) dry-bulb and 35°F (2°C) wet-bulb are mixed with 15,000 ft³/min (7.08 m³/s) of air at 75°F (24°C) dry-bulb and 50% relative humidity. Find the mixture dry-bulb temperature.

*SI Solution*

An approximate mixture temperature can be found by disregarding the change in density and taking a volumetrically weighted average. (The psychrometric chart can

be used to determine a more precise value mixture temperature. The more precise approach is used in the customary U.S. solution.)

$$T_{\text{mixture}} \approx \frac{\dot{V}_A T_A + \dot{V}_B T_B}{\dot{V}_A + \dot{V}_B}$$

$$= \frac{\left(2.36 \, \frac{\text{m}^3}{\text{s}}\right)(4°C) + \left(7.08 \, \frac{\text{m}^3}{\text{s}}\right)(24°C)}{2.36 \, \frac{\text{m}^3}{\text{s}} + 7.08 \, \frac{\text{m}^3}{\text{s}}}$$

$$= 19°C$$

*Customary U.S. Solution*

Locate the two points on the psychrometric chart, and draw a line between them. Estimate the specific volumes.

$$v_A = 12.65 \, \text{ft}^3/\text{lbm}$$

$$v_B = 13.68 \, \text{ft}^3/\text{lbm}$$

Calculate the dry air masses.

$$\dot{m}_A = \frac{\dot{V}_A}{v_A} = \frac{5000 \, \frac{\text{ft}^3}{\text{min}}}{12.65 \, \frac{\text{ft}^3}{\text{lbm}}}$$

$$= 395 \, \text{lbm/min}$$

$$\dot{m}_B = \frac{\dot{V}_B}{v_B} = \frac{15,000 \, \frac{\text{ft}^3}{\text{min}}}{13.68 \, \frac{\text{ft}^3}{\text{lbm}}}$$

$$= 1096 \, \text{lbm/min}$$

Use Eq. 24.21.

$$T_{\text{mixture}} = T_A + \left(\frac{\dot{m}_B}{\dot{m}_A + \dot{m}_B}\right)(T_B - T_A)$$

$$= 40°F + \left(\frac{1096 \, \frac{\text{lbm}}{\text{min}}}{1096 \, \frac{\text{lbm}}{\text{min}} + 395 \, \frac{\text{lbm}}{\text{min}}}\right)$$

$$\times (75°F - 40°F)$$

$$= 65.7°F$$

## 10. AIR CONDITIONING PROCESSES

The psychrometric chart is particularly useful in analyzing air conditioning processes because the paths of many processes are straight lines. Sensible heating and cooling processes, for example, follow horizontal straight lines. Adiabatic saturation processes follow lines of constant enthalpy (essentially parallel to lines of constant wet-bulb temperature). The paths of pure humidification and dehumidification follow vertical paths. Figure 24.2 summarizes the directions of these paths.

**Figure 24.2** *Air Conditioning Processes*

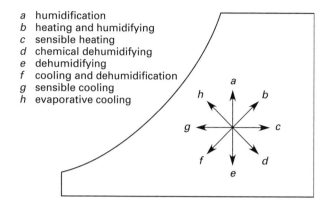

*a* humidification
*b* heating and humidifying
*c* sensible heating
*d* chemical dehumidifying
*e* dehumidifying
*f* cooling and dehumidification
*g* sensible cooling
*h* evaporative cooling

## 11. SENSIBLE HEAT RATIO

In general, the slope of any process line on the psychrometric chart is determined from the *sensible heat ratio*, also known as the *sensible heat factor*, SHF, and *sensible-total ratio*, S/T, scale on the chart. In an air conditioning process, the sensible heat ratio, SHR, is the ratio of sensible heat added (or removed) to total heat added (or removed). (The use of such scales varies from chart to chart. The process slope is determined from the sensible heat factor protractor and then translated (i.e., moved) to the appropriate point on the chart.)

$$\text{SHR} = \frac{q_s}{q_t} = \frac{q_s}{q_s + q_l} \qquad \textit{24.22}$$

The sensible heat ratio is always the slope of the line representing the change from the beginning point to the ending point on the psychrometric chart. Different designations are given to the sensible heat ratio, however, depending on where the changes occur.

If the sensible and latent energies change as the air passes through an occupied room, the term *room sensible heat ratio*, RSHR, is used. If the changes occur as the air passes through an air conditioning coil (apparatus), the term *coil* (or *apparatus*) *sensible heat ratio* is used, CSHR. Since the air conditioning apparatus usually removes heat and moisture from both the conditioned room and from outside makeup air, the term *grand sensible heat ratio*, GSHR, can be used in place of the coil sensible heat ratio. The *effective sensible heat ratio*, ESHR, is the slope of the line between the apparatus dew point on the saturation line and the design conditions of the conditioned space.

The sensible heat ratio is a psychrometric slope; it is not a geometric slope.

### Example 24.5

During the summer, air from a conditioner enters an occupied space at 55°F (13°C) dry-bulb and 30% relative humidity. The ratio of sensible to total loads in the space is 0.45:1. The humidity ratio of the air leaving the room is 60 gr/lbm (8.6 g/kg). What is the dry-bulb temperature of the leaving air?

*SI Solution*

The sensible heat ratio is 0.45. Use the psychrometric chart (see App. 24.B) to determine the slope corresponding to this ratio. Draw a temporary line from the center of the protractor to the 0.45 mark on the sensible heat factor (inside) scale.

Locate 13°C dry-bulb and 30% relative humidity on the psychrometric chart. Draw a line through this point parallel to the temporary line, which is drawn with a slope of 0.45. The intersection of this line and the horizontal line corresponding to 8.6 g/kg determines the condition of the leaving air. The dry-bulb temperature is approximately 25.2°C.

*Customary U.S. Solution*

The sensible heat ratio is 0.45. Use the psychrometric chart (see App. 24.A) to determine the slope corresponding to this ratio. Draw a temporary line from the center of the protractor to 0.45 on the sensible heat factor (inside) scale.

Locate 55°F dry-bulb and 30% relative humidity on the psychrometric chart. Draw a line through this point parallel to the temporary line, which is drawn with a slope of 0.45. The intersection of this line and the horizontal line corresponding to 60 gr/lbm determines the condition of the leaving air. The dry-bulb temperature is approximately 76°F.

## 12. STRAIGHT HUMIDIFICATION

Straight (pure) *humidification* increases the water content of the air without changing the dry-bulb temperature. This is represented by a vertical condition line on the psychrometric chart. The *humidification load* is the mass of water added to the air per unit time (usually per hour).

## 13. BYPASS FACTOR AND COIL EFFICIENCY

Conditioning of air is accomplished by passing it through cooling or heating coils. Ideally, all of the air will come into contact with the coil for a long enough time and will leave at the coil temperature. In reality, this does not occur, and the air does not reach the coil temperature. The *bypass factor* can be thought of as the percentage of the air that is not cooled (or heated) by the coil. Under this interpretation, the remaining air (which is cooled or heated by the coil) is assumed to reach the coil temperature. The bypass factor expressed in decimal form is

$$\text{BF} = \frac{T_{\text{db,out}} - T_{\text{coil}}}{T_{\text{db,in}} - T_{\text{coil}}} \qquad \textit{24.23}$$

Bypass factors depend largely on the type of coil used. Bypass factors for large commercial units (such as those used in department stores) are small—around 10%. For small residential units, they are approximately 35%.

The *coil efficiency* is the complement of the bypass factor.

$$\eta_{\text{coil}} = 1.0 - \text{BF} \qquad \textit{24.24}$$

## 14. SENSIBLE COOLING AND HEATING

There is no change in the dew point or moisture content of the air with *sensible heating* and *cooling*. Since the moisture content is constant, these processes are represented by horizontal condition lines on the psychrometric chart (moving right for heating and left for cooling). (See Fig. 24.3.)

The energy change during the process can be calculated from enthalpies read directly from the psychrometric chart or approximated from the dry-bulb temperatures. In Eq. 24.25, $c_{p,\text{air}}$ is usually taken as 0.240 Btu/lbm-°F (1.005 kJ/kg·°C), and $c_{p,\text{moisture}}$ is taken as approximately 0.444 Btu/lbm-°F (1.805 kJ/kg·°C).

$$q = m_a(h_2 - h_1)$$
$$= m_a(c_{p,\text{air}} + \omega c_{p,\text{moisture}})(T_2 - T_1) \qquad \textit{24.25}$$

**Figure 24.3** *Sensible Cooling*

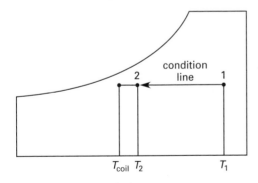

## 15. COOLING WITH COIL DEHUMIDIFICATION

If the cooling coil's temperature is below the air's dew point (as is usually the case), moisture will condense on the coil. The effective coil temperature in this instance is referred to as the *apparatus dew point*, ADP, and is determined from the intersection of the condition line (i.e., *coil load line*) and the curved saturation line on the psychrometric chart. The apparatus dew point is the

temperature to which the air would be cooled if 100% of it contacted the coil.

The mass of condensing water will be

$$m_w = m_a(\omega_1 - \omega_2) \qquad \textit{24.26}$$

The total energy removed from the air includes both sensible and latent components. The latent heat is calculated from the heat of vaporization evaluated at the pressure of the water vapor.

$$q_t = q_s + q_l = m_a(h_1 - h_2) \qquad \textit{24.27}$$
$$q_l = m_a(\omega_1 - \omega_2)h_{fg} \qquad \textit{24.28}$$

Referring to Fig. 24.4, it is convenient to think of air experiencing sensible cooling from point 1 to point 3, after which the air follows the saturation line down from point 3 to point 4 (the apparatus dew point). Water condenses out between points 3 and 4. For convenience, the condition line is drawn as a straight line between points 1 and 4. The slope of the ADP-2-1 line corresponds to the sensible heat ratio. (See Sec. 24.11.) Since some of the air does not contact the coil at all, the final condition of the air will actually be at point 2 on the condition line.

**Figure 24.4** *Cooling and Dehumidification*

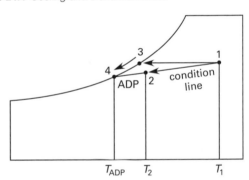

In practice, point 1 is usually known and either point 2 or point 4 are unknown. If point 2 is known, point 4 (the apparatus dew point) can be found graphically by extending the condition line over to the saturation line. (In some cases, the sensible heat ratio must be used to locate the apparatus dew point.) If point 4 is known, point 2 can be found from the bypass factor. The *contact factor*, CF, is essentially a dehumidification efficiency, calculated as the complement of the bypass factor.

$$\text{CF} = 1 - \text{BF} = 1 - \frac{T_{2,\text{db}} - \text{ADP}}{T_{1,\text{db}} - \text{ADP}} \qquad \textit{24.29}$$

Water condenses out over the entire temperature range from point 3 to point 4. The temperature of the water being removed is assumed to be the dew-point temperature at point 2.

## Example 24.6

A coil has a bypass factor of 20% and an apparatus dew point of 55°F (13°C). Air enters the coil at 85°F (29°C) dry-bulb and 69°F (21°C) wet-bulb. What are the (a) latent heat loss, (b) sensible heat loss, and (c) sensible heat ratio?

*SI Solution*

(a) Locate the point corresponding to the entering air on the psychrometric chart. The enthalpy and humidity ratio are approximately

$$h_1 = 60.4 \text{ kJ/kg}$$

$$\omega_1 = 0.0123 \text{ kg/kg}$$

Use Eq. 24.29 to calculate the dry-bulb temperature of the air leaving the coil.

$$
\begin{aligned}
T_{2,\text{db}} &= \text{ADP} + \text{BF}(T_{1,\text{db}} - \text{ADP}) \\
&= 13°\text{C} + (0.20)(29°\text{C} - 13°\text{C}) \\
&= 16.2°\text{C}
\end{aligned}
$$

Draw a condition line between the entering air and the apparatus dew point on the psychrometric chart. Locate the point corresponding to 16.2°C dry-bulb on the condition line. The leaving enthalpy and humidity ratio are approximately

$$h_2 = 40.8 \text{ kJ/kg}$$

$$\omega_2 = 0.0100 \text{ kg/kg}$$

The total energy loss per kilogram is

$$
\begin{aligned}
q_t = h_1 - h_2 &= 60.4 \, \frac{\text{kJ}}{\text{kg}} - 40.8 \, \frac{\text{kJ}}{\text{kg}} \\
&= 19.6 \text{ kJ/kg of dry air}
\end{aligned}
$$

Since the partial pressure of the water vapor is unknown, estimate $h_{fg} \approx 2501$ kJ/kg.

From Eq. 24.28, on a kilogram basis,

$$
\begin{aligned}
\frac{q_l}{m_a} &= (\omega_1 - \omega_2)h_{fg} \\
&= \left(0.0123 \, \frac{\text{kg}}{\text{kg}} - 0.0100 \, \frac{\text{kg}}{\text{kg}}\right)\left(2501 \, \frac{\text{kJ}}{\text{kg}}\right) \\
&= 5.75 \text{ kJ/kg of dry air}
\end{aligned}
$$

(b) The sensible heat loss is

$$
\begin{aligned}
q_s &= q_t - q_l \\
&= 19.6 \, \frac{\text{kJ}}{\text{kg}} - 5.75 \, \frac{\text{kJ}}{\text{kg}} = 13.9 \text{ kJ/kg of dry air}
\end{aligned}
$$

(c) The sensible heat ratio is

$$
\begin{aligned}
\text{SHR} = \frac{q_s}{q_t} &= \frac{13.9 \, \dfrac{\text{kJ}}{\text{kg}}}{19.6 \, \dfrac{\text{kJ}}{\text{kg}}} \\
&= 0.71
\end{aligned}
$$

*Customary U.S. Solution*

(a) Locate the point corresponding to the entering air on the psychrometric chart. The enthalpy and humidity ratio are approximately

$$h_1 = 33.1 \text{ Btu/lbm}$$

$$\omega_1 = 0.0116 \text{ lbm/lbm}$$

Use Eq. 24.29 to calculate the dry-bulb temperature of the air leaving the coil.

$$
\begin{aligned}
T_{2,\text{db}} &= \text{ADP} + \text{BF}(T_{1,\text{db}} - \text{ADP}) \\
&= 55°\text{F} + (0.20)(85°\text{F} - 55°\text{F}) \\
&= 61°\text{F}
\end{aligned}
$$

Draw a condition line between the entering air and apparatus dew point on the psychrometric chart. Locate the point corresponding to 61°F dry-bulb on the condition line. The leaving enthalpy and humidity ratio are approximately

$$h_2 = 25.1 \text{ Btu/lbm}$$

$$\omega_2 = 0.0097 \text{ lbm/lbm}$$

The total energy loss per pound is

$$
\begin{aligned}
q_t = h_1 - h_2 &= 33.1 \, \frac{\text{Btu}}{\text{lbm}} - 25.1 \, \frac{\text{Btu}}{\text{lbm}} \\
&= 8.0 \text{ Btu/lbm of dry air}
\end{aligned}
$$

Since the partial pressure of the water vapor is not known, estimate $h_{fg} \approx 1060$ Btu/lbm.

From Eq. 24.28, on a pound basis,

$$
\begin{aligned}
\frac{q_l}{m_a} &= (\omega_1 - \omega_2)h_{fg} \\
&= \left(0.0116 \, \frac{\text{lbm}}{\text{lbm}} - 0.0097 \, \frac{\text{lbm}}{\text{lbm}}\right)\left(1060 \, \frac{\text{Btu}}{\text{lbm}}\right) \\
&= 2.01 \text{ Btu/lbm of dry air}
\end{aligned}
$$

(b) The sensible heat loss is

$$
\begin{aligned}
q_s &= q_t - q_l \\
&= 8.0 \, \frac{\text{Btu}}{\text{lbm}} - 2.0 \, \frac{\text{Btu}}{\text{lbm}} = 6.0 \text{ Btu/lbm of dry air}
\end{aligned}
$$

**Thermodynamics**

(c) The sensible heat ratio is

$$\mathrm{SHR} = \frac{q_s}{q_t} = \frac{6.0 \; \dfrac{\mathrm{Btu}}{\mathrm{lbm}}}{8.0 \; \dfrac{\mathrm{Btu}}{\mathrm{lbm}}}$$

$$= 0.75$$

## 16. ADIABATIC SATURATION PROCESSES

To measure the wet-bulb temperature, air must experience an *adiabatic saturation process*, also known as *evaporative cooling*. Adiabatic saturation processes occur in cooling towers, air washers, and evaporative coolers ("swamp coolers"). To become saturated, the air must pick up the maximum amount of moisture it can hold at that temperature. This moisture comes from the vaporization of liquid water. For the process to be adiabatic, there can be no external source of energy to vaporize the liquid water needed to saturate the air.

At first analysis, the terms adiabatic and saturation seem contradictory. Adiabatic saturation is possible, however, if the latent heat of vaporization comes from the air itself. If the air gives up sensible heat, that energy can be used to vaporize liquid water. Of course, the air temperature decreases when sensible heat is given up. That is the reason that the wet-bulb temperature is generally less than the dry-bulb temperature. Only when the air is saturated will the two temperatures be equal.

An adiabatic saturation process can be produced with a *sling psychrometer*, which is essentially a regular thermometer with its bulb wrapped in wet cotton or gauze. Rapidly twirling the thermometer through the air at the end of a cord will cause the water in the gauze to evaporate. The latent heat needed to vaporize the water will come from the sensible heat of the air, and the thermometer will measure the wet-bulb temperature.

Since the increase in the water vapor's latent heat content equals the decrease in the air's sensible heat, the total enthalpies before and after adiabatic saturation are the same. Therefore, an adiabatic saturation process follows a line of constant enthalpy on the psychrometric chart. These lines are, for approximation purposes, parallel to lines of constant wet-bulb temperature.

The bypass factor concept is not used with adiabatic saturation processes. Instead, the *saturation efficiency (humidification efficiency)* is used. The saturation efficiency of large commercial air washers is typically 90% to 95%.

$$\eta_{\mathrm{sat}} = \frac{T_{\mathrm{db,air,in}} - T_{\mathrm{db,air,out}}}{T_{\mathrm{db,air,in}} - T_w} \qquad 24.30$$

## 17. AIR WASHERS

An *air washer* is a device that passes air through a dense spray of recirculating water. The water is used to change the properties of the air. Air washers are used in air purifying and cleaning processes (i.e., removal of solids, liquids, gases, vapors, and odors), as well as for evaporative cooling and dew-point control.[5]

The difference between a spray humidifier and spray dehumidifier is the temperature of the spray water. In an *adiabatic air washer*, the spray water is recirculated without being heated or cooled. After equilibrium is reached, the water temperature will be equal to the air's entering wet-bulb temperature. The air will be cooled and humidified, leaving partially or completely saturated at its entering wet-bulb temperature. However, if the spray water is chilled, the air will be cooled and dehumidified. And, if the spray water is heated, the air will be humidified and (possibly) heated.

An air washer's *saturation efficiency*, typically 90% to 95%, is measured by the drop in dry-bulb temperature relative to the entering wet-bulb depression.

$$\eta_{\mathrm{sat}} = \frac{T_{\mathrm{in,db}} - T_{\mathrm{out,db}}}{T_{\mathrm{in,db}} - T_{\mathrm{in,wb}}} \qquad 24.31$$

Air velocity through washers is approximately 500 ft/min (2.6 m/s). Velocities outside the range of 300 ft/min to 750 ft/min (1.5 m/s to 3.8 m/s) are probably faulty. The water pressure is typically 20 psig to 40 psig (140 kPa to 280 kPa). The spray quantity per bank of nozzles is in the range of 1.5 gal/min to 5 gal/min per 1000 ft$^3$ (3.3 L/s to 11 L/s per 1000 m$^3$) of air. Screens, louvers, and mist eliminator plates will generate a static pressure drop of approximately 0.2 in wg to 0.5 in wg (50 kPa to 125 kPa) at 500 ft/min (2.6 m/s). Other operating parameters used to describe air washer performance include air mass flow rate per unit area (lbm/hr-ft$^2$ or kg/m$^2$·s), air and liquid heat transfer coefficients per volume of chamber (Btu/hr-°F-ft$^3$ or kW/°C·m$^3$), and the *spray ratio* (the mass of water sprayed to the mass of air passing through the washer per unit time).

## 18. COOLING WITH HUMIDIFICATION

When air passes through a water spray (as in an *air washer*), an *adiabatic saturation process* known as *evaporative cooling* occurs.[6] (See Sec. 24.16.) The air leaves with a lower temperature and a higher moisture content. This is represented on the psychrometric chart by

---

[5]Air washers are generally not used for removing carbonaceous or greasy particles.

[6]An *air washer* is basically a *spray chamber* through which air passes. When supplied with chilled water from a refrigeration source, the air washer can cool, dehumidify, or humidify the air. Air washers can be used without refrigeration to cool and humidify the air through an evaporative cooling process.

a condition line parallel to the lines of constant enthalpy (essentially constant wet-bulb temperature).

Adiabatic saturation is a constant-enthalpy process, since any evaporation of the water requires heat to be drawn from the air. Since the removed heat goes into the remaining water, the water temperature increases. When the water spray is continuously recirculated, the water temperature gradually increases to the wet-bulb temperature of the incoming air. The minimum leaving air temperature will be the water temperature (i.e., the wet-bulb temperature of the incoming air).

During steady-state operation, the temperature of the water spray will normally be stable at the air's wet-bulb temperature. However, the water temperature can also be artificially maintained by refrigeration at less than the wet-bulb temperature (but more than the dew-point temperature). Line 1–3 in Fig. 24.5 illustrates such a process.

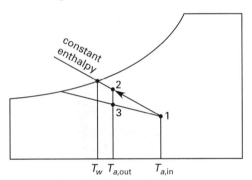

**Figure 24.5** *Cooling with Humidification (adiabatic saturation)*

To prevent ice buildup, the cooled air temperature should be kept from dropping below the freezing point of water. The entering wet-bulb temperature should be kept above 35°F (1.7°C).

### Example 24.7

Air at 90°F (32°C) dry-bulb and 65°F (18°C) wet-bulb enters an evaporative cooler. The air leaves at 90% relative humidity. The continuously recirculated spray water is stable at 65°F (18°C). What is the dry-bulb temperature of the leaving air?

*Solution*

Since the spray water is the same temperature as the wet-bulb temperature of the entering air, the cooler has reached its steady-state operating conditions. Locate the entering point on the psychrometric chart and draw a line of constant enthalpy (or constant 65°F (18°C) wet-bulb temperature) up to the 90% relative humidity curve. Read the dry-bulb temperature as approximately 67°F (19°C).

## 19. COOLING WITH SPRAY DEHUMIDIFICATION

If air passes through a water spray whose temperature is less than the entering air's wet-bulb temperature, both the dry-bulb and wet-bulb temperatures will decrease.[7] If the leaving water temperature is below the entering air's dew point, dehumidification will occur. As with any evaporative cooling, the air will give up thermal energy to the water. The final water temperature will depend on the thermal energy pickup and water flow rate. All air temperatures decrease, and some moisture condenses. The *performance factor* is defined as

$$\mathrm{PF} = 1 - \frac{T_{\mathrm{air,wb,out}} - T_{w,\mathrm{out}}}{T_{\mathrm{air,wb,in}} - T_{w,\mathrm{in}}} \qquad 24.32$$

## 20. HEATING WITH HUMIDIFICATION

If air is humidified by injecting steam (*steam humidification*) or by passing the air through a hot water spray, the dry-bulb temperature and enthalpy of the air will increase.[8] The final air enthalpy and/or the required steam enthalpy can be determined from a conservation of energy equation. In Eq. 24.33, the mass of the air used is the dry air mass, which does not change. $h_a$, though expressed per pound of dry air, includes the energy of all vaporized water.

$$m_a h_{a,\mathrm{in}} + m_w h_w = m_a h_{a,\mathrm{out}} \qquad 24.33$$

From a conservation of mass for the water,

$$m_a \omega_{\mathrm{in}} + m_w = m_a \omega_{\mathrm{out}} \qquad 24.34$$

Figure 24.6 illustrates that the condition line will be above the line of constant enthalpy that radiates from the point corresponding to the incoming air. However, even though heat is added to the water, the air temperature can either decrease (as in the 1–2 process shown) or increase (as in the 1–3 process shown).

**Figure 24.6** *Heating with Steam Humidification*

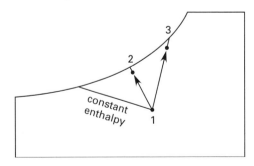

---

[7]This can unintentionally occur during the start-up of an air washer used for humidification, or the water can be kept intentionally chilled.
[8]When a spray of hot water is used, the water must be continually heated. Unlike a cold water spray, a natural equilibrium water temperature is not achieved.

## 21. HEATING AND DEHUMIDIFICATION

Air passing through a solid or liquid *adsorbent bed*, such as silica gel or activated alumina, will decrease in humidity. This is sometimes referred to as *chemical dehumidification, chemical dehydration,* or *"adsorbent" dehumidification.*[9] If only latent heat was involved, this process would be the reverse of an adiabatic saturation process. However, as moisture is removed, exothermic chemical energy is generated in addition to the heat of vaporization liberated. Since thermal energy is generated, this is not an adiabatic process. (See Fig. 24.7.)

**Figure 24.7** *Heating and Dehumidification*

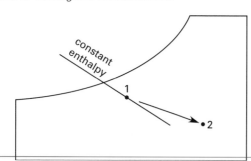

## 22. WET COOLING TOWERS

Conventional *wet cooling towers* cool warm water by exposing it to colder air.[10] They are usually used to provide cold water to power plant and large refrigeration condensers. The air is used to change the properties of the water, which leaves cooler. As it leaves, the saturated (or nearly saturated) warmed air takes sensible and latent heat from the water.

Cooling towers are generally counterflow, crossflow, or a combination. Though natural-draft and atmospheric towers exist, limited space usually requires that cooling towers operate with mechanical draft. Fans are located at the base of *forced draft towers* and blow air into the water cascading down. With *induced mechanical draft*, fans are located at the top of the tower, drawing air upward. Some portion of the exhaust air might reenter the cooling tower. This is known as *recycle air (recirculation air)*. Recycle air decreases the efficiency of the tower.

During countercurrent operation, warm water is introduced at the top of the tower and is distributed by troughs or spray nozzles. The water passes over

staggered slats or interior *fill* (also known as *packing*).[11] Air flows upward, contacting the water on its downward path. A portion of the water evaporates, cooling the remainder of the water. The water temperature cannot decrease below the wet-bulb temperature. The actual final water temperature depends on a number of factors, including the state of the incoming air, the heat load, and the design (and efficiency) of the cooling tower. (See Fig. 24.8.)

**Figure 24.8** *Counterflow Wet Cooling Tower*

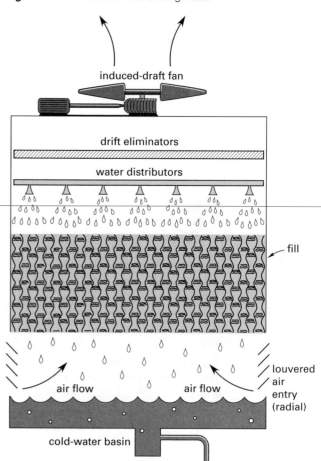

There are several environmental issues associated with wet cooling towers. Makeup water, though relatively little is needed, may be difficult to obtain. Moist plume discharges cause shadowing of adjacent areas and fogging and icing on nearby highways. Disposal of blowdown wastewater is also problematic.

Equation 24.35 is a per-unit energy balance that can be used to evaluate cooling tower performance. Each term,

---

[9]The correct term for a substance that collects water on its surface is an *adsorbent*. By virtue of their great porosities, adsorbent particles have large surface areas. The attractive forces on the surfaces of these solids cause a thin layer of condensed water to form. Adsorbents are reactivated by heating.

[10]Though larger in size, a cooling tower is similar in operation to an air washer. In fact, an air washer can be used to cool water. Since air washer operation is not countercurrent, however, larger air flows are required to obtain the same cooling effect.

[11]Modern *filled towers* use corrugated *cellular fill* to maximize the air-water contact area. Standard polyvinyl chloride (PVC) fill is useful up to about 125°F (52°C). From 125°F to 140°F (52°C to 60°C), chlorinated PVC fill is recommended. Polypropylene fill should be used above 140°F (60°C). "Fill-less" towers, where the sprayed water merely falls through oncoming air, are used in some industries (food, steel, and paper processes) where a high-product carryover can lead to coating or buildup on the fill material.

*Thermodynamics*

including the circulating water flow rate, is per unit mass (e.g., pound or kilogram) of dry air. Since the energy contribution of the makeup water is small, that term can be omitted for a first approximation.

$$m_{w,\text{in}}h_{w,\text{in}} + h_{a,\text{in}} + (\omega_{a,\text{out}} - \omega_{a,\text{in}})h_{\text{makeup}}$$
$$= m_{w,\text{out}}h_{w,\text{out}} + h_{a,\text{out}}$$
$$= (m_{w,\text{in}} + \omega_{a,\text{in}} - \omega_{a,\text{out}})h_{w,\text{out}} + h_{a,\text{out}} \quad 24.35$$

If operation is at standard pressure, a psychrometric chart can be used to obtain the air enthalpies. For operation at different altitudes (i.e., different atmospheric pressures), the mathematical psychrometric relationships in Sec. 24.4 can be used to calculate the enthalpy. From Eq. 24.16, the humidity ratio is

$$\omega = \frac{0.622 p_{\text{water vapor}}}{p_{\text{total}} - p_{\text{water vapor}}} \quad 24.36$$

When a cooling tower is used to provide cold water for the condenser of a refrigeration system, the water circulation will be approximately 3 gal/min per ton (0.19 L/s) of refrigeration. Approximately 2 gal/min to 4 gal/min of water are distributed per square foot (1.4 L/s to 2.7 L/s per square meter) of tower, and the air velocity should be approximately 700 ft/min (3.6 m/s) through the net free area. Coolants for condensers in reciprocating refrigeration systems usually call for an 85°F to 90°F (29°C to 32°C) water temperature. (This corresponds to a condensing temperature of approximately 100°F to 110°F (38°C to 43°C).) Various valves, mixing, louvers, and dampers are used to maintain a constant output water temperature.

The lowest temperature to which water can be cooled by purely evaporative means is the wet-bulb temperature of the entering air. The *cooling efficiency*, $\eta_w$, is based on the water temperature. The *water range (cooling range* or *range)* is defined as the actual difference between the entering and leaving water temperatures. (For water-cooled refrigeration condensers, this is equal to the water's temperature increase in the condenser.) The *approach* is defined as the difference between the leaving water temperature and the entering air wet-bulb temperatures.[12] Cooling efficiency is typically 50% to 70%.[13] Natural draft towers can cool the water to within 10°F to 12°F (5.5°C to 6.7°C) of the wet-bulb temperature. Forced draft towers can cool the water to within 5°F to 6°F (2.8°C to 3.3°C).

$$\eta_w = \frac{\text{range}}{\text{approach} + \text{range}}$$
$$= \frac{T_{w,\text{in}} - T_{w,\text{out}}}{T_{w,\text{in}} - T_{\text{air,wb,in}}} \quad 24.37$$

As Eq. 24.37 indicates, the actual wet-bulb temperature of the cooling air is particularly important in determining cooling tower performance. The higher the wet-bulb temperature, the lower the efficiency. (This is because when the denominator in Eq. 24.37 decreases, the numerator decreases even more.) In rating their cooling towers, most manufacturers have adopted the practice of using wet-bulb temperatures that will be exceeded only 2.5% of the time or less.

Performance of a cooling tower also depends on the relative humidity of the air. High relative humidities decrease the water evaporation rate, decreasing the efficiency.

The *heat load (tower load* or *cooling duty)* is calculated from the range and the water mass flow rate.

$$q = m_w c_p (T_{w,\text{in}} - T_{w,\text{out}})$$
$$= m_w (h_{w,\text{in}} - h_{w,\text{out}}) \quad 24.38$$

Cooling towers are sometimes rated in tower units, which are essentially proportional to the tower cost. The number of *tower units,* TU, is equal to a rating factor multiplied by the flow rate. *Rating factors* define the relative difficulty in cooling, essentially the relative amount of contact area or fill volume required. Manufacturers provide charts showing the relationship between rating factor, approach, range, and wet-bulb temperature.

$$\text{TU} = \text{RF} \times Q_{\text{gpm}} \quad 24.39$$

## 23. COOLING TOWER BLOWDOWN

Water losses occur from evaporation, windage, and blowdown. *Evaporation loss* can be calculated from the humidity ratio increase and is approximately 0.1% per °F (0.18% per °C) decrease in water temperature.[14] *Windage loss*, also known as *drift*, is water lost in small droplets and carried away by the air flow. Windage loss is typically in the 0.1% to 0.3% range for mechanical draft towers. Since windage droplets are a mechanical mixture (not a thermodynamic solution of two gases), they are not adequately accounted for by the humidity ratio.

Makeup water must be provided to replace all water losses. As more and more water enters the system, *total dissolved solids*, TDS (e.g., chlorides), will build up over time. Water can be treated to prevent deposit, and a portion of the water can be periodically or continuously bled off. *Cycles of concentration (ratio of concentration)*, C, is the ratio of total dissolved solids in the

---

[12]Thus, approach for a cooling tower is analogous to the terminal temperature difference in the surface condenser.
[13]The term "thermal efficiency" is sometimes used here inappropriately.

[14]This value is approximate and is reported in various ways. Some authorities state "0.1% per degree Fahrenheit"; others say "1% per 10°F"; and yet others, "1% per 10°F to 13°F."

recirculating water to the total dissolved solids in the makeup water.[15]

$$C = \frac{(\text{TDS})_{\text{recirculating}}}{(\text{TDS})_{\text{makeup}}}$$

$$= \frac{m_{\text{evaporation}} + m_{\text{blowdown}} + m_{\text{windage}}}{m_{\text{blowdown}} + m_{\text{windage}}} \quad 24.40$$

Though windage removes some of the solids, most must be removed by bleeding some of the water off. This is known as *blowdown* or *bleed-off*. If the maximum cycles of concentration are known, the blowdown is

$$m_{\text{blowdown}} = \frac{m_{\text{evaporation}} + (1 - C_{\text{max}})m_{\text{windage}}}{C_{\text{max}} - 1} \quad 24.41$$

Additives should be used to prevent specific problems encountered, such as scale buildup, corrosion, biological growth, foaming, and discoloration.

## 24. DRY COOLING TOWERS

Dry cooling is used when environmental protection and water conservation are issues. It is used primarily by nonutility generators (e.g., waste-to-energy and cogeneration plants).

There are two types of dry cooling towers. Both use finned-tube heat exchangers. In a *direct-condensing tower*, steam travels through large-diameter "trunks" to a crossflow heat exchanger where it is condensed and cooled by the cooler air.[16] In an *indirect-condensing dry cooling tower*, steam is condensed by cold water jets (surface or jet condenser) and is subsequently cooled by air. The hot condensate is then pumped to crossflow heat exchangers where it is sensibly cooled

(no condensation) by the air. Air flow may be mechanical or natural draft. Most U.S. installations are direct-condensing. Worldwide, natural-draft indirect systems are more predominant, particularly for power plants with capacities in excess of 100 MW. (See Fig. 24.9.)

**Figure 24.9** *Dry Cooling Towers*

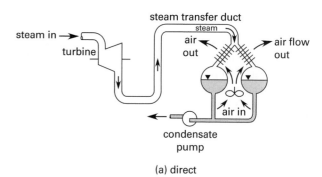

(a) direct

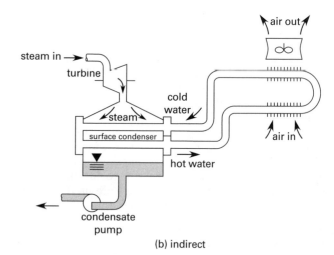

(b) indirect

---

[15]Multiply grains/gallon (gr/gal) by 17.1 to obtain parts per million (ppm) or milligrams per liter (mg/L).

[16]The term "direct contact" does not mean that the air and steam are combined in a single vessel.

# 25 Compressible Fluid Dynamics

## Nomenclature

| | | | |
|---|---|---|---|
| $a$ | speed of sound | ft/sec | m/s |
| $A$ | area | ft$^2$ | m$^2$ |
| $c$ | specific heat | Btu/lbm-°F | kJ/kg·K |
| $C$ | coefficient | – | – |
| $D$ | diameter | ft | m |
| $f$ | Fanning friction factor | – | – |
| $F$ | thrust force | lbf | N |
| $g$ | acceleration of gravity | ft/sec$^2$ | m/s$^2$ |
| $g_c$ | gravitational constant | ft-lbm/lbf-sec$^2$ | n.a. |
| $h$ | enthalpy | Btu/lbm | kJ/kg |
| $h$ | head | ft | m |
| $I$ | impulse | lbf-sec | N·s |
| $I_{sp}$ | specific impulse | sec | s |
| $J$ | Joule's constant (778.17) | ft-lbf/Btu | n.a. |
| $k$ | ratio of specific heats | – | – |
| $L$ | distance | ft | m |
| $L$ | length | ft | m |
| $m$ | mass | lbm | kg |
| M | Mach number | – | – |
| MW | molecular weight | lbm/lbmol | kg/kmol |
| $p$ | pressure | lbf/ft$^2$ | Pa |
| $q$ | heat per unit mass | Btu/lbm | kJ/kg |
| $R$ | specific gas constant | ft-lbf/lbm-°R | kJ/kg·K |
| $R^*$ | universal gas constant | ft-lbf/lbmol-°R | kJ/kmol·K |
| $R_{cp}$ | critical pressure ratio | – | – |
| $s$ | entropy | Btu/lbm-°F | kJ/kg·K |
| $t$ | time | sec | s |
| $T$ | absolute temperature | °R | K |
| v | velocity | ft/sec | m/s |
| $V$ | volume | ft$^3$ | m$^3$ |
| $z$ | elevation | ft | m |

## Symbols

| | | | |
|---|---|---|---|
| $\gamma$ | specific weight | lbf/ft$^3$ | n.a. |
| $\delta$ | semivertex angle | deg | deg |
| $\eta$ | efficiency | – | – |
| $\theta$ | shock semivertex angle | deg | deg |
| $\rho$ | mass density | lbf/ft$^3$ | kg/m$^3$ |
| $v$ | specific volume | ft$^3$/lbm | m$^3$/kg |

## Subscripts

| | |
|---|---|
| 0 | total (stagnation) |
| $a$ | ambient |
| cp | critical pressure |
| $d$ | discharge |
| $e$ | exit |
| eff | effective |
| $f$ | friction |
| $F$ | thrust |
| $h$ | hydraulic |
| $i$ | inlet |
| $p$ | constant pressure |
| sp | specific |
| $t$ | throat |
| v | constant volume, velocity |
| $x$ | before the shock wave |
| $y$ | after the shock wave |

## 1. INTRODUCTION

A *high-velocity gas* is defined as a gas moving with a velocity in excess of approximately 300 ft/sec (100 m/s). A high gas velocity is often achieved at the expense of internal energy. A drop in internal energy, $u$, is seen as a drop in enthalpy, $h$, since $h = u + pv$. Since the Bernoulli equation does not account for this conversion, it cannot be used to predict the thermodynamic properties of the gas. Furthermore, density changes and shock waves complicate the use of traditional evaluation tools such as energy and momentum conservation equations.

## 2. STEADY-FLOW ENERGY EQUATION

Consider a gas or vapor flowing from a high-pressure reservoir through a duct, as shown in Fig. 25.1. The properties in the source reservoir, designated by the

Thermodynamics

subscript zero (0), are known alternatively as the *total properties*, *stagnation properties*, or *chamber properties*.

**Figure 25.1** *High-Velocity Flow Locations*

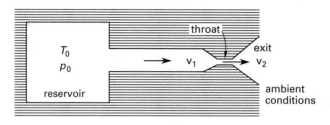

The flow can be assumed to be one dimensional as long as the flow cross section varies slowly along the path length. The gas possesses energy in static, kinetic, and internal (thermal) forms. Since the flow is fast, there is no time for significant heat transfer to occur, so an assumption of adiabatic flow is appropriate. If the duct run is short, there will be little or no friction. Equation 25.1 is an energy balance of an adiabatic open-flow system for one unit mass of the gas. This is the *steady-flow energy equation*, SFEE, for an adiabatic system.[1]

$$h_1 + \frac{v_1^2}{2} + z_1 g = h_2 + \frac{v_2^2}{2} + z_2 g \quad \text{[SI]} \quad \textbf{25.1(a)}$$

$$J h_1 + \frac{v_1^2}{2g_c} + z_1 \left(\frac{g}{g_c}\right)$$
$$= J h_2 + \frac{v_2^2}{2g_c} + z_2 \left(\frac{g}{g_c}\right) \quad \text{[U.S.]} \quad \textbf{25.1(b)}$$

It is often necessary to write the velocity at a point in terms of the total properties. The velocity is zero in the source reservoir. Since the gas density is small, the potential energy terms are always disregarded. Equation 25.2 can be used to calculate the discharge velocity for high-pressure steam and other substances for which the ideal gas assumption would be inappropriate. The *theoretical maximum velocity* is achieved when all internal and pressure energies are converted to kinetic energy (i.e., when $h_2 = 0$). This is never achieved in practice, however.

$$v_2 = \sqrt{2(h_0 - h_2)} \quad \text{[SI]} \quad \textbf{25.2(a)}$$

$$v_2 = \sqrt{2 g_c J (h_0 - h_2)} = 223.8 \sqrt{h_0 - h_2} \quad \text{[U.S.]} \quad \textbf{25.2(b)}$$

If it is appropriate to assume constant values of specific heat or specific gas constant, ideal gas relationships can be substituted into the SFEE.

---

[1]The *steady-flow energy equation*, SFEE, is also known as the *general flow equation*, GFE.

$$v_2 = \sqrt{2 c_p (T_0 - T_2)} = \sqrt{\frac{2 R k (T_0 - T_2)}{k - 1}}$$
$$= \sqrt{\left(\frac{2k}{k-1}\right)\left(\frac{p_0}{\rho_0} - \frac{p_2}{\rho_2}\right)} \quad \text{[SI]} \quad \textbf{25.3(a)}$$

$$v_2 = \sqrt{2 g_c J c_p (T_0 - T_2)}$$
$$= \sqrt{\frac{2 g_c R k (T_0 - T_2)}{k - 1}}$$
$$= \sqrt{\left(\frac{2 g_c k}{k-1}\right)\left(\frac{p_0}{\rho_0} - \frac{p_2}{\rho_2}\right)} \quad \text{[U.S.]} \quad \textbf{25.3(b)}$$

## Example 25.1

Steam at 200 psia (1.4 MPa) and 500°F (250°C) enters an insulated nozzle with negligible velocity and is expanded to 20 psia (150 kPa) and 98% quality. Find the steam's exit velocity.

*SI Solution*

The initial enthalpy is found from the superheat tables or a Mollier diagram. The final enthalpy is found from property tables or from a Mollier diagram.

$$h_0 = 2927.7 \text{ kJ/kg}$$
$$h_2 = 2648.6 \text{ kJ/kg}$$

From Eq. 25.2, the exit velocity is

$$v_2 = \sqrt{2(h_0 - h_2)}$$
$$= \sqrt{(2)\left(2927.7 \frac{\text{kJ}}{\text{kg}} - 2648.6 \frac{\text{kJ}}{\text{kg}}\right)\left(1000 \frac{\text{J}}{\text{kJ}}\right)}$$
$$= 747 \text{ m/s}$$

*Customary U.S. Solution*

The initial enthalpy is found from the superheat tables or a Mollier diagram. The final enthalpy is found from property tables or from a Mollier diagram.

$$h_0 = 1269.0 \text{ Btu/lbm}$$
$$h_2 = 1137.0 \text{ Btu/lbm}$$

From Eq. 25.2, the exit velocity is

$$v_2 = \sqrt{2 g_c J (h_0 - h_2)}$$
$$= \sqrt{\begin{array}{c} (2)\left(32.2 \frac{\text{ft-lbm}}{\text{lbf-sec}^2}\right)\left(778 \frac{\text{ft-lbf}}{\text{Btu}}\right) \\ \times \left(1269.0 \frac{\text{Btu}}{\text{lbm}} - 1137.0 \frac{\text{Btu}}{\text{lbm}}\right) \end{array}}$$
$$= 2572 \text{ ft/sec}$$

## 3. ISENTROPIC FLOW

If the gas flow is adiabatic and frictionless (that is, reversible), the entropy change is zero and the flow is known as *isentropic flow*. As a practical matter, completely isentropic flow does not exist. However, some high-velocity, steady-state flow processes proceed with little increase in entropy and are considered to be isentropic. The irreversible effects are accounted for by various correction factors, such as nozzle and discharge coefficients.

## 4. ISENTROPIC FLOW FACTORS

In isentropic flow, total pressure, total temperature, and total density remain constant, regardless of the flow area and velocity. The instantaneous properties, known as *static properties*, do change along the flow path, however.[2] Equation 25.5 through Eq. 25.7 predict these static properties as functions of the Mach number, M, for ideal gas flow.[3]

$$M = \frac{v}{a} = \frac{v}{\sqrt{kRT}} = \frac{v}{\sqrt{\dfrac{kR^*T}{MW}}} \qquad \text{[SI]} \quad 25.4(a)$$

$$M = \frac{v}{a} = \frac{v}{\sqrt{kg_cRT}} = \frac{v}{\sqrt{\dfrac{kg_cR^*T}{MW}}} \qquad \text{[U.S.]} \quad 25.4(b)$$

$$\left[\frac{T_0}{T}\right] = \tfrac{1}{2}(k-1)M^2 + 1 \qquad 25.5$$

$$\left[\frac{p_0}{p}\right] = \left(\tfrac{1}{2}(k-1)M^2 + 1\right)^{k/(k-1)} = \left[\frac{T_0}{T}\right]^{k/(k-1)} \qquad 25.6$$

$$\left[\frac{\rho_0}{\rho}\right] = \left(\tfrac{1}{2}(k-1)M^2 + 1\right)^{1/(k-1)} = \left[\frac{T_0}{T}\right]^{1/(k-1)} \qquad 25.7$$

The isentropic flow ratios given by Eq. 25.5 through Eq. 25.7 are functions only of the Mach number, M, and ratio of specific heats, $k$. Therefore, the ratios can be easily tabulated, as has been done in App. 25.A. The numbers in such tables are known as *isentropic flow factors*.

### Example 25.2

Air ($k = 1.4$, MW = 29.0) flows isentropically from a large tank at 530°R (294K) through a convergent-divergent nozzle and is expanded to supersonic velocities. At a point where the Mach number is 2.5, what are the (a) gas temperature and (b) actual velocity?

*SI Solution*

(a) The temperature isentropic flow factor, $T_0/T$, can be calculated from Eq. 25.5 or its inverse read directly from the M = 2.5 line in App. 25.B.

$$\left[\frac{T_0}{T}\right] = \tfrac{1}{2}(k-1)M^2 + 1$$

$$= \left(\tfrac{1}{2}\right)(1.4 - 1)(2.5)^2 + 1 = 2.25$$

The temperature is

$$T = \frac{T_0}{\left[\dfrac{T_0}{T}\right]} = \frac{294\text{K}}{2.25} = 130.7\text{K}$$

(b) Since the Mach number is 2.5, the gas velocity is 2.5 times the speed of sound. This speed of sound is calculated from the static temperature, not from the total temperature. From Eq. 25.4,

$$v = Ma = M\sqrt{\frac{kR^*T}{MW}}$$

$$= 2.5\sqrt{\frac{(1.4)\left(8314\ \dfrac{\text{J}}{\text{kmol·K}}\right)(130.7\text{K})}{29.0\ \dfrac{\text{kg}}{\text{kmol}}}} = 573\text{ m/s}$$

*Customary U.S. Solution*

(a) The temperature isentropic flow factor, $T_0/T$, can be read from the M = 2.5 line in App. 25.B.

$$\left[\frac{T}{T_0}\right] = 0.4444$$

$$T = T_0\left[\frac{T}{T_0}\right] = (530°\text{R})(0.4444) = 235.5°\text{R}$$

(b) Since the Mach number is 2.5, the gas velocity is

$$v = M\sqrt{\frac{kg_cR^*T}{MW}}$$

$$= 2.5\sqrt{\frac{(1.4)\left(32.2\ \dfrac{\text{ft-lbm}}{\text{lbf-sec}^2}\right)}{29.0\ \dfrac{\text{lbm}}{\text{lbmol}}} \times \left(1545\ \dfrac{\text{ft-lbf}}{\text{lbmol-°R}}\right)(235.5°\text{R})}$$

$$= 1880\text{ ft/sec}$$

---

[2]The *static properties* are not the same as the *stagnation properties*.
[3]For example, Eq. 25.5 can be derived from Eq. 25.1 by dividing both sides by the speed of sound.

## 5. RATIO OF SPECIFIC HEATS

The ratio of specific heats, $k = c_p/c_v$ is remarkably similar for gases with similar structures. For monatomic gases (He, Ar, Ne, Kr, etc.), it is approximately 1.67. For diatomic gases ($N_2$, $O_2$, $H_2$, CO, NO, and air), it is approximately 1.4. For triatomic gases ($H_2O$ and $CO_2$), it is approximately 1.3. The value of $k$ is less than 1.3 for more complex gases.

In most problems, the ratio of specific heats is assumed to be constant over all temperatures and pressures encountered. However, when the gas experiences a large temperature change, the ratio of specific heats should be evaluated at the average temperature.

Rather than using mathematical relationships to calculate exact solutions, it is common to use factors from tables appropriate for that ratio of specific heat. Since tables are typically available only for $k = 1.0, 1.1, 1.2, 1.3, 1.4$, and $1.67$, a solution will not be exact when the ratio of specific heats is some intermediate value.

## 6. CRITICAL CONSTANTS

The location where sonic velocity has been achieved (i.e., M = 1) is known as a *critical point*. Sonic properties are designated by an asterisk, *. The insentropic flow factors at that point are known as *critical ratios* or *critical constants*. For example, the *critical pressure ratio* can be calculated from Eq. 25.8.

$$R_{cp} = \left[\frac{p^*}{p_0}\right] = \left(\frac{2}{k+1}\right)^{k/(k-1)} \qquad 25.8$$

The *critical temperature ratio* and *critical density ratio* are also easily derived.

$$\left[\frac{T^*}{T_0}\right] = \frac{2}{k+1} \qquad 25.9$$

$$\left[\frac{\rho^*}{\rho_0}\right] = \left(\frac{2}{k+1}\right)^{1/(k-1)} \qquad 25.10$$

**Example 25.3**

What is the critical pressure ratio for air ($k = 1.4$)?

*Solution*

From Eq. 25.8,

$$R_{cp} = \left(\frac{2}{k+1}\right)^{k/(k-1)} = \left(\frac{2}{1.4+1}\right)^{1.4/(1.4-1)} = 0.5283$$

## 7. CHOKED FLOW

Equation 25.6 seems to indicate that the gas velocity (and hence, the mass flow rate) in a discharge duct can be increased by increasing the ratio of source reservoir to ambient pressures (or by decreasing the ratio of ambient

to source reservoir pressures). This is a logical conclusion, but it is valid only up to a certain gas velocity in the duct—the *sonic velocity* (i.e., speed of sound).

Changes in ambient pressure required to change the gas velocity in the duct travel upstream at sonic velocity. Therefore, if the gas in the duct is already traveling at sonic velocity, the fact that the ambient pressure has been changed cannot be transmitted back to the source reservoir. Sonic velocity will occur when the ratio of ambient to source pressures drops to the *critical pressure ratio* (0.5283 for air). Once sonic velocity has been achieved in a duct or nozzle throat, the mass flow rate will be at its maximum and will remain constant for all subsequent decreases in ambient pressure.[4] This condition is called *choked flow*.

## 8. CHANGING DUCT AREA

Equation 25.11 defines the relationship between the change in flow area and the initial gas velocity for isentropic flow.

$$\frac{dA}{A} = \frac{dp}{\rho v^2}(1 - M^2) \qquad 25.11$$

Table 25.1 is based on Eq. 25.11 and summarizes the effects of changing flow area on the various thermodynamic properties for a gas in isentropic flow.

**Table 25.1** *Effect of Changing Duct Area on Isentropic Flow*

| | initial Mach no. | |
|---|---|---|
| | M < 1 | M > 1 |
| *area decreasing* | | |
| total properties constant | total properties constant | |
| velocity increases | velocity decreases | |
| Mach no. increases | Mach no. decreases | |
| pressure decreases | pressure increases | |
| density decreases | density increases | |
| temperature decreases | temperature increases | |
| enthalpy decreases | enthalpy increases | |
| internal energy decreases | internal energy increases | |
| entropy is constant | entropy is constant | |
| *area increasing* | | |
| total properties constant | total properties constant | |
| velocity decreases | velocity increases | |
| Mach no. decreases | Mach no. increases | |
| pressure increases | pressure decreases | |
| density increases | density decreases | |
| temperature increases | temperature decreases | |
| enthalpy increases | enthalpy decreases | |
| internal energy increases | internal energy decreases | |
| entropy is constant | entropy is constant | |

---

[4]When the flow is already choked and the back pressure is lowered, neither the throat velocity nor the mass flow rate will increase. When the flow is already choked and the source pressure is increased, the throat velocity will remain constant. However, the increased density in the source will cause the mass flow rate to increase.

Thermodynamics

## 9. CONVERGENT-DIVERGENT NOZZLES

Sonic velocity from a reservoir can be achieved with almost any configuration of reservoir orifice. All that is necessary to achieve sonic velocity in an orifice is to lower the ambient pressure until the pressure ratio equals the critical pressure ratio. Achieving supersonic flow, however, will specifically require a *convergent-divergent nozzle* (also known as a *C-D nozzle* or *venturi nozzle*) in addition to the proper pressure conditions.

The properties at the point of flow constriction (i.e., the *throat*, subscript *t*) in a C-D nozzle are known as the *throat properties*. If sonic velocity is achieved in the throat, the properties are known as *critical properties*. Similarly, the *exit properties* (subscript *e*) represent the gas properties at the discharge point.

The rules for property changes given in Table 25.1, as well as the following statements, apply to a C-D nozzle attached to a source of (subsonic) pressurized gas or vapor.

- If sonic velocity occurs anywhere in the nozzle, it occurs in the throat.

- If sonic velocity occurs in the throat, the velocity may or may not be supersonic in the diverging section.

- If supersonic velocity occurs anywhere in the nozzle, it occurs in the diverging section.[5]

- If supersonic velocity occurs in the diverging section of the nozzle, the velocity is sonic in the throat.

From these four statements, it is clear that having sonic velocity in the throat is a necessary but not a sufficient condition for achieving supersonic velocity in the diverging section. In order to ensure supersonic velocity existing everywhere in the diverging section, there is an additional necessary condition. Specifically, the nozzle must be designed to expand the gases to the *design pressure ratio*, equal to $p_{\text{ambient}}/p_0$.

## 10. DESIGN OF SUPERSONIC NOZZLES

In order to design a nozzle capable of expanding a gas to some given velocity or Mach number, it is sufficient to have an expression for the flow area versus Mach number. Since the reservoir cross-sectional area is an unrelated variable, it is not possible to use the stagnation area as a reference area and to develop the ratio $[A_0/A]$ as was done for temperature, pressure, and density. The usual choice for a reference area is the *critical area*—the

area at which the gas velocity is (or could be) sonic. This area is designated as $A^*$.

$$\frac{A}{A^*} = \frac{1}{\text{M}} \left( \frac{\frac{1}{2}(k-1)\text{M}^2 + 1}{\frac{1}{2}(k-1) + 1} \right)^{(k+1)/2(k-1)} \qquad 25.12$$

Figure 25.2 is a plot of Eq. 25.12 versus Mach number. As long as the Mach number is less than 1.0, the area must decrease in order for the velocity to increase. However, if the Mach number is greater than 1.0, the area must increase in order for the velocity to increase. This is consistent with Table 25.1.

**Figure 25.2** A/A* versus M

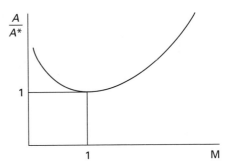

It is not possible to change the Mach number to any desired value over any arbitrary distance along the axis of a converging-diverging nozzle. If the rate of change, $dA/dx$, is too great, the assumptions of one-dimensional flow become invalid. Usually, the converging section has a steeper angle (known as the *convergent angle*) than the diverging section. If the diverging angle is too great, a normal shock wave may form in that part of the nozzle.

Equation 25.13 gives the ratio of mass flow rate to the critical area. This equation is useful when the flow rate is known and the critical area is needed.

$$\frac{\dot{m}}{A^*} = \frac{\rho_0 \sqrt{kRT_0}}{\left(\frac{1}{2}(k-1) + 1\right)^{(k+1)/2(k-1)}} \qquad \text{[SI]} \quad 25.13(a)$$

$$\frac{\dot{m}}{A^*} = \frac{\rho_0 \sqrt{kg_cRT_0}}{\left(\frac{1}{2}(k-1) + 1\right)^{(k+1)/2(k-1)}} \qquad \text{[U.S.]} \quad 25.13(b)$$

Equation 25.14 gives the ratio of $v/a^*$—the ratio of gas velocity at some point to the sonic velocity in the throat. This formula is useful if the total properties and velocity at some point are known and the Mach number is needed. It is possible to solve for the static temperature

---

[5]This is a valid statement only if the flow in the converging section is subsonic. If a C-D nozzle is placed in an existing supersonic flow, there will be supersonic flow in the converging section of the nozzle.

and speed of sound, but this is more tedious than is first apparent.[6]

$$\frac{\mathrm{v}}{a^*} = \sqrt{\frac{\frac{1}{2}(k+1)\mathrm{M}^2}{\frac{1}{2}(k-1)\mathrm{M}^2+1}} \qquad 25.14$$

By now, it should be apparent that the algebraic burden of solving gas dynamics problems is tremendous. The easiest way to solve such problems is with isentropic flow tables, such as App. 25.A and App. 25.B. The notation and symbols in these appendices, or in any other table that might be available, should be noted carefully.

### Example 25.4

The Mach number is 0.8 at a point in a nozzle where the area is 1.5 square units. Air ($k = 1.4$) is flowing. (a) What would the throat area have to be in order to achieve sonic velocity there? (b) What is the area at a point where the Mach number is 0.4?

*Solution*

(a) From the isentropic flow table for M = 0.8, $[A/A^*]$ = 1.0382. Therefore, the throat area would have to be

$$A_t = \frac{A_{\mathrm{M}=0.8}}{\left[\frac{A}{A^*}\right]_{\mathrm{M}=0.8}} = \frac{1.5 \text{ units}^2}{1.0382} = 1.445 \text{ units}^2$$

(b) In this case, the critical point will be used as a reference point, even though sonic velocity may not actually be achieved. For M = 0.4, $[A/A^*]$ = 1.5901.

$$A_{\mathrm{M}=0.4} = \left[\frac{A}{A^*}\right]_{\mathrm{M}=0.4} \times A_{\mathrm{M}=1}$$

$$= \frac{A_{\mathrm{M}=0.8} \times \left[\frac{A}{A^*}\right]_{\mathrm{M}=0.4}}{\left[\frac{A}{A^*}\right]_{\mathrm{M}=0.8}}$$

$$= \frac{(1.5 \text{ units}^2)(1.5901)}{1.0382} = 2.30 \text{ units}^2$$

### Example 25.5

A frictionless, adiabatic nozzle receives 10 lbm/sec (5.0 kg/s) of air from a reservoir whose stagnation properties are 200°F and 30 psia (100°C and 200 kPa). At a particular point in the nozzle, a supersonic velocity of 1400 ft/sec (450 m/s) is attained. The atmospheric pressure is 14.7 psia (101.3 kPa). What are the (a) Mach number, (b) pressure, and (c) cross-sectional area at that point?

---

[6]The speed of sound in a sonic throat, $a^*$, depends on the throat temperature, not on the total temperature.

*SI Solution*

The absolute total temperature is

$$T_0 = 100°\mathrm{C} + 273° = 373\mathrm{K}$$

The total density in the reservoir is

$$\rho_0 = \frac{p_0}{RT_0}$$

$$= \frac{(200 \text{ kPa})\left(1000 \, \frac{\mathrm{Pa}}{\mathrm{kPa}}\right)}{\left(287 \, \frac{\mathrm{J}}{\mathrm{kg \cdot K}}\right)(373\mathrm{K})}$$

$$= 1.868 \text{ kg/m}^3$$

Read the property ratios for Mach 1 from the isentropic flow table: $[T/T_0] = 0.8333$, and $[\rho/\rho_0] = 0.6339$.

The sonic properties at the throat are

$$T^* = \left[\frac{T}{T_0}\right]T_0 = (0.8333)(373\mathrm{K}) = 310.8\mathrm{K}$$

$$\rho^* = \left[\frac{\rho}{\rho_0}\right]\rho_0$$

$$= (0.6339)\left(1.868 \, \frac{\mathrm{kg}}{\mathrm{m}^3}\right) = 1.184 \text{ kg/m}^3$$

The sonic velocity at the throat is

$$a^* = \sqrt{kRT^*}$$

$$= \sqrt{(1.4)\left(287 \, \frac{\mathrm{J}}{\mathrm{kg \cdot K}}\right)(310.8\mathrm{K})}$$

$$= 353.4 \text{ m/s}$$

The throat area is

$$A^* = \frac{\dot{m}}{\rho^* a^*}$$

$$= \frac{5.0 \, \frac{\mathrm{kg}}{\mathrm{s}}}{\left(1.184 \, \frac{\mathrm{kg}}{\mathrm{m}^3}\right)\left(353.4 \, \frac{\mathrm{m}}{\mathrm{s}}\right)}$$

$$= 0.01195 \text{ m}^2$$

The ratio of actual-to-sonic throat velocities is

$$\left[\frac{\mathrm{v}}{a^*}\right] = \frac{450 \, \frac{\mathrm{m}}{\mathrm{s}}}{353.4 \, \frac{\mathrm{m}}{\mathrm{s}}} = 1.273$$

Searching the $[\mathrm{v}/a^*]$ column of the isentropic flow tables for this value gives M = 1.36. The corresponding ratios are $[p/p_0] = 0.3323$ and $[A/A^*] = 1.094$. At the point where the velocity is 450 m/s, the static pressure and area are

$$p_{450} = \left[\frac{p}{p_0}\right] p_0 = (0.3323)(200 \text{ kPa})$$

$$= 66.46 \text{ kPa}$$

$$A_{450} = \left[\frac{A}{A^*}\right] A^* = (1.094)(0.01195 \text{ m}^2)$$

$$= 0.01307 \text{ m}^2$$

*Customary U.S. Solution*

The absolute total temperature is

$$T_0 = 200°\text{F} + 460° = 660°\text{R}$$

The total density in the reservoir is

$$\rho_0 = \frac{p_0}{RT_0}$$

$$= \frac{\left(30 \frac{\text{lbf}}{\text{in}^2}\right)\left(12 \frac{\text{in}}{\text{ft}}\right)^2}{\left(53.3 \frac{\text{ft-lbf}}{\text{lbm-°R}}\right)(660°\text{R})}$$

$$= 0.1228 \text{ lbm/ft}^3$$

Read the property ratios for Mach 1 from the isentropic flow table: $[T/T_0] = 0.8333$, and $[\rho/\rho_0] = 0.6339$.

The sonic properties at the throat are

$$T^* = \left[\frac{T}{T_0}\right] T_0 = (0.8333)(660°\text{R}) = 550°\text{R}$$

$$\rho^* = \left[\frac{\rho}{\rho_0}\right]\rho_0$$

$$= (0.6339)\left(0.1228 \frac{\text{lbm}}{\text{ft}^3}\right) = 0.07784 \text{ lbm/ft}^3$$

The sonic velocity at the throat is

$$a^* = \sqrt{kg_c RT^*}$$

$$= \sqrt{(1.4)\left(32.2 \frac{\text{ft-lbm}}{\text{lbf-sec}^2}\right)\left(53.3 \frac{\text{ft-lbf}}{\text{lbm-°R}}\right)(550°\text{R})}$$

$$= 1149.6 \text{ ft/sec}$$

The throat area is

$$A^* = \frac{\dot{m}}{\rho^* a^*}$$

$$= \frac{10 \frac{\text{lbm}}{\text{sec}}}{\left(0.07784 \frac{\text{lbm}}{\text{ft}^3}\right)\left(1149.6 \frac{\text{ft}}{\text{sec}}\right)}$$

$$= 0.1118 \text{ ft}^2$$

The ratio of actual-to-sonic throat velocities is

$$\left[\frac{\text{v}}{a^*}\right] = \frac{1400 \frac{\text{ft}}{\text{sec}}}{1149.6 \frac{\text{ft}}{\text{sec}}} = 1.218$$

Searching the $[\text{v}/a^*]$ column of the isentropic flow tables for this value gives M = 1.28. The corresponding ratios are $[p/p_0] = 0.3708$ and $[A/A^*] = 1.058$. At the point where the velocity is 1400 ft/sec, the static pressure and area are

$$p_{1400} = \left[\frac{p}{p_0}\right] p_0 = (0.3708)\left(30 \frac{\text{lbf}}{\text{in}^2}\right)$$

$$= 11.12 \text{ lbf/in}^2$$

$$A_{1400} = \left[\frac{A}{A^*}\right] A^* = (1.058)(0.1118 \text{ ft}^2)$$

$$= 0.1183 \text{ ft}^2$$

**Example 25.6**

An attitude-adjustment jet in a satellite uses high-pressure gas with a ratio of specific heats of 1.4 and a molecular weight of 21.0 lbm/mol (kg/kmol). The chamber conditions are 450 psia and 4700°R (3.2 MPa and 2600K). The gas is expanded supersonically to 2.97 psia (20.5 kPa) at the nozzle exit. What are the (a) sonic velocity in the throat, (b) required exit-to-throat area ratio, and (c) exit velocity?

*SI Solution*

(a) The specific gas constant is

$$R = \frac{R^*}{\text{MW}} = \frac{8314.3 \frac{\text{J}}{\text{kmol·K}}}{21 \frac{\text{kg}}{\text{kmol}}}$$

$$= 395.9 \text{ J/kg·K}$$

Use the chamber properties as the total properties. Since the exit velocity is supersonic, sonic flow is achieved in the throat. From the isentropic flow tables at Mach 1, $[T/T_0] = 0.8333$. The temperature at the throat is

$$T_t = \left[\frac{T}{T_0}\right] T_0 = (0.8333)(2600\text{K})$$
$$= 2167\text{K}$$

The velocity is sonic at the throat.

$$a^* = \sqrt{kRT_t}$$
$$= \sqrt{(1.4)\left(395.9 \ \frac{\text{J}}{\text{kg·K}}\right)(2167\text{K})}$$
$$= 1096 \ \text{m/s}$$

(b) The static pressure ratio at the exit is

$$\frac{p}{p_0} = \frac{20.5 \ \text{kPa}}{(3.2 \ \text{MPa})\left(1000 \ \frac{\text{kPa}}{\text{MPa}}\right)}$$
$$= 0.006406$$

Locate this pressure ratio in the supersonic region of the isentropic flow table. The corresponding Mach number is approximately 4.0. Read $[A/A^*] = 10.7187$.

(c) At Mach 4, $[T/T_0] = 0.2381$. The exit temperature is

$$T_e = \left[\frac{T}{T_0}\right] T_0 = (0.2381)(2600\text{K})$$
$$= 619.1\text{K}$$

The exit velocity is

$$v_e = Ma_e = M\sqrt{kRT_e}$$
$$= 4\sqrt{(1.4)\left(395.9 \ \frac{\text{J}}{\text{kg·K}}\right)(619.1\text{K})}$$
$$= 2343 \ \text{m/s}$$

*Customary U.S. Solution*

(a) The specific gas constant is

$$R = \frac{R^*}{MW} = \frac{1545 \ \frac{\text{ft-lbf}}{\text{lbmol-°R}}}{21 \ \frac{\text{lbm}}{\text{lbmol}}}$$
$$= 73.57 \ \text{ft-lbf/lbm-°R}$$

Use the chamber properties as the total properties. Since the exit velocity is supersonic, sonic flow is achieved in

the throat. From the isentropic flow tables at Mach 1, $[T/T_0] = 0.8333$. The temperature at the throat is

$$T_t = \left[\frac{T}{T_0}\right] T_0 = (0.8333)(4700°\text{R})$$
$$= 3917°\text{R}$$

The velocity is sonic at the throat.

$$a^* = \sqrt{kg_cRT}$$
$$= \sqrt{(1.4)\left(32.2 \ \frac{\text{ft-lbm}}{\text{lbf-sec}^2}\right)\left(73.57 \ \frac{\text{ft-lbf}}{\text{lbm-°R}}\right)(3917°\text{R})}$$
$$= 3604 \ \text{ft/sec}$$

(b) The static pressure ratio at the exit is

$$\frac{p}{p_0} = \frac{2.97 \ \frac{\text{lbf}}{\text{in}^2}}{450 \ \frac{\text{lbf}}{\text{in}^2}} = 0.0066$$

Locate this pressure ratio in the supersonic region of the isentropic flow table. The corresponding Mach number is approximately 4.0. Read $[A/A^*] = 10.7187$.

(c) At Mach 4, $[T/T_0] = 0.2381$. The exit temperature is

$$T_e = \left[\frac{T}{T_0}\right] T_0 = (0.2381)(4700°\text{R})$$
$$= 1119°\text{R}$$

The exit velocity is

$$v_e = Ma_e = M\sqrt{kg_cRT_e}$$
$$= 4\sqrt{(1.4)\left(32.2 \ \frac{\text{ft-lbm}}{\text{lbf-sec}^2}\right)\left(73.57 \ \frac{\text{ft-lbf}}{\text{lbm-°R}}\right)(1119°\text{R})}$$
$$= 7706 \ \text{ft/sec}$$

## 11. NOZZLE PERFORMANCE CHARACTERISTICS

The *nozzle velocity coefficient* is the ratio of actual to ideal (isentropic) exit velocities.

$$C_v = \frac{v_{actual}}{v_{ideal}} \qquad 25.15$$

The *nozzle discharge coefficient*, $C_d$, is the ratio of actual to ideal (isentropic) mass flow rates. $C_d$ is typically high for well-designed nozzles, in the 0.97 to 0.98 range.

$$C_d = \frac{\dot{m}_{\text{actual}}}{\dot{m}_{\text{ideal}}} \qquad 25.16$$

$$\dot{m}_{\text{actual}} = \rho_e \text{v}_e A_e \qquad 25.17$$

The *nozzle efficiency, η,* is defined as the ratio of the actual to ideal (isentropic) energies extracted from the flowing gas. From Eq. 25.18, this is equal to the ratio of the square of actual to ideal velocities.

$$\eta = \frac{\Delta h_{\text{actual}}}{\Delta h_{\text{ideal}}} = \left(\frac{\text{v}_{\text{actual}}}{\text{v}_{\text{ideal}}}\right)^2 = C_{\text{v}}^2 \qquad 25.18$$

The *effective exhaust velocity,* $\text{v}_{\text{eff}}$, is determined from the thrust, $F$, and the actual mass flow rate.[7]

$$\text{v}_{\text{eff}} = \frac{F}{\dot{m}_{\text{actual}}} \qquad \text{[SI]} \quad 25.19(a)$$

$$\text{v}_{\text{eff}} = \frac{F g_c}{\dot{m}_{\text{actual}}} \qquad \text{[U.S.]} \quad 25.19(b)$$

Equation 25.20 gives the *thrust, F,* developed by the nozzle. (If the nozzle is operating at its design pressure ratio, then the exit and ambient pressures will be equal.)

$$F = \dot{m}_{\text{actual}} \text{v}_e + A_e(p_e - p_a)$$
$$= \dot{m}_{\text{actual}} \text{v}_{\text{eff}} \qquad \text{[SI]} \quad 25.20(a)$$

$$F = \frac{\dot{m}_{\text{actual}} \text{v}_e}{g_c} + A_e(p_e - p_a)$$
$$= \frac{\dot{m}_{\text{actual}} \text{v}_{\text{eff}}}{g_c} \qquad \text{[U.S.]} \quad 25.20(b)$$

The *coefficient of thrust,* $C_F$, is defined as

$$C_F = \frac{F}{p_0 A_t} \qquad 25.21$$

The *specific impulse,* $I_{\text{sp}}$, has units of seconds and is the ratio of thrust to rate of fuel consumption.

$$I_{\text{sp}} = \frac{F}{\dot{m}_{\text{actual}} g} = \frac{\text{v}_{\text{eff}}}{g} \qquad \text{[SI]} \quad 25.22(a)$$

$$I_{\text{sp}} = \frac{F}{\dot{m}_{\text{actual}}} \times \frac{g_c}{g} = \frac{\text{v}_{\text{eff}}}{g} \qquad \text{[U.S.]} \quad 25.22(b)$$

For constant-thrust engines, the *total impulse,* $I_{\text{total}}$, is the product of thrust and effective length of time, $t_{\text{eff}}$, over which the thrust is produced. For variable-thrust

engines, the total impulse is the integral of thrust with respect to time.

$$I_{\text{total}} = \int F(t)\,dt$$
$$= F_{\text{ave}} t_{\text{eff}} = I_{\text{sp}} \times \text{total fuel weight} \qquad 25.23$$

The *effective time,* $t_{\text{eff}}$, is

$$t_{\text{eff}} = \frac{I_{\text{total}}}{F_{\text{ave}}} = \frac{I_{\text{sp}} \times \text{total fuel weight}}{F_{\text{ave}}} \qquad 25.24$$

The *characteristic exhaust velocity,* $\text{v}^*$, is defined as

$$\text{v}^* = \frac{\text{v}_{\text{eff}}}{C_F} = \frac{g I_{\text{sp}}}{C_F} = \frac{p_0 A_t}{\dot{m}_{\text{actual}}} \qquad \text{[SI]} \quad 25.25(a)$$

$$\text{v}^* = \frac{\text{v}_{\text{eff}}}{C_F} = \frac{g I_{\text{sp}}}{C_F} = \frac{g_c p_0 A_t}{\dot{m}_{\text{actual}}} \qquad \text{[U.S.]} \quad 25.25(b)$$

The *characteristic length,* $L^*$, of the combustion chamber is

$$L^* = \frac{V_{\text{chamber}}}{A_t} \qquad 25.26$$

### Example 25.7

What are the (a) nozzle efficiency and (b) coefficient of velocity for the steam nozzle in Ex. 25.1?

*SI Solution*

(a) If the expansion through the nozzle had been isentropic (that is, straight down on the Mollier diagram), the exit enthalpy would have been approximately 2510 kJ/kg. The nozzle efficiency is defined by Eq. 25.18.

$$\eta_{\text{nozzle}} = \frac{h_0 - h_2'}{h_0 - h_2} = \frac{2927 \frac{\text{kJ}}{\text{kg}} - 2649 \frac{\text{kJ}}{\text{kg}}}{2927 \frac{\text{kJ}}{\text{kg}} - 2510 \frac{\text{kJ}}{\text{kg}}} = 0.66667$$

(b) The coefficient of velocity is defined by Eq. 25.18.

$$C_{\text{v}} = \sqrt{\eta_{\text{nozzle}}} = \sqrt{0.66667} = 0.816$$

*Customary U.S. Solution*

(a) If the expansion through the nozzle had been isentropic (that is, straight down on the Mollier diagram), the exit enthalpy would have been approximately 1082.2 Btu/lbm. The nozzle efficiency is defined by Eq. 25.18.

$$\eta_{\text{nozzle}} = \frac{h_0 - h_2'}{h_0 - h_2} = \frac{1268.8 \frac{\text{Btu}}{\text{lbm}} - 1137.2 \frac{\text{Btu}}{\text{lbm}}}{1268.8 \frac{\text{Btu}}{\text{lbm}} - 1082.2 \frac{\text{Btu}}{\text{lbm}}} = 0.70525$$

---

[7]Symbols used in this section have been chosen to be consistent with other symbols in this chapter. However, it is traditional in the field of propulsion system design to use the following symbols: $c$, effective exhaust velocity; $c^*$, characteristic exhaust velocity; $I_t$, total impulse; and $p_o$, ambient (outside) pressure.

**Thermodynamics**

(b) The coefficient of velocity is defined by Eq. 25.18.

$$C_v = \sqrt{\eta_{\text{nozzle}}} = \sqrt{0.70525} = 0.840$$

## 12. SHOCK WAVES

In Sec. 25.9 it was stated that a nozzle must be designed to the design pressure ratio in order to keep the flow supersonic in the diverging section of the nozzle. It is possible, though, to have supersonic velocity only in part of the diverging section. Once the flow is supersonic in a part of the diverging section, however, it cannot become subsonic by an isentropic process.

Therefore, the gas experiences a *shock wave* as the velocity drops from supersonic to subsonic. Shock waves are very thin (several molecules thick) and separate areas of radically different thermodynamic properties. Since the shock wave forms normal to the flow direction, it is known as a *normal shock wave*. The strength of a shock wave is measured by the change in Mach number across it.

The velocity always changes from supersonic to subsonic across a shock wave. Since there is no loss of heat energy, a shock wave is an adiabatic process, and total temperature is constant. However, the process is not isentropic, and total pressure decreases. Momentum is also conserved. Table 25.2 lists the property changes across a shock wave.

**Table 25.2** *Property Changes Across a Normal Shock Wave*

| property | change |
|---|---|
| total temperature | is constant |
| total pressure | decreases |
| total density | decreases |
| velocity | decreases |
| Mach number | decreases |
| pressure | increases |
| density | increases |
| temperature | increases |
| entropy | increases |
| internal energy | increases |
| enthalpy | is constant |
| momentum | is constant |

The properties before a shock wave are given the subscript $x$. The subscript $y$ is used for the properties after a shock wave. These subscripts may be combined with the subscript 0 to represent a total (stagnation) property. (For example, $p_{0,x} < p_{0,y}$.)

Although it is possible to calculate the ratio of a property before and after a shock wave, the relationships are complex, and it is more convenient to use *normal shock factors* from a *normal shock table* such as App. 25.B. These factors are dependent only on the ratio of specific heats, and the Mach number, $M_x$, immediately before the shock wave.

## Example 25.8

A shock wave in air ($k = 1.4$) occurs at a point where the Mach number is 2.4. If the static pressure before the shock wave is 5.0 atm, what are the (a) Mach number after the shock, (b) static pressure after the shock, and (c) total pressure after the shock?

*Solution*

All of the factors are read from the normal shock table (see App. 25.B) for $M_x = 2.4$.

(a) The Mach number after the shock is read directly as $M_y = 0.5231$.

(b) The ratio of static pressures is read as $[p_y/p_x] = 6.553$. The static pressure after the shock is

$$p_y = \left[\frac{p_y}{p_x}\right]p_x = (6.553)(5.0) = 32.8 \text{ atm}$$

(c) The ratio of static pressure before the shock to total pressure after the shock is read as $[p_x/p_{0,y}] = 0.1266$. The total pressure after the shock is

$$p_{0,y} = \frac{p_x}{\left[\dfrac{p_x}{p_{0,y}}\right]} = \frac{5.0 \text{ atm}}{0.1266} = 39.5 \text{ atm}$$

## Example 25.9

The temperature and pressure in a desert prior to an explosive test are 70°F and 14.7 psia (21°C and 101.3 kPa), respectively. At the center of the blast, the pressure is raised to 2 atm. (a) What is the wave velocity near the explosion? What are the (b) pressure, (c) temperature, and (d) velocity behind the wave?

*SI Solution*

(a) Even though the expansion is spherical, the values from the normal shock tables can be used near the blast point, as long as $x$ and $y$ are located on a line projecting radially from the center. Let the $x$-subscript correspond to the still air into which the shock wave is expanding. Let the $y$-subscript correspond to the gas inside the expanding high-pressure region.

The absolute static temperature of the desert air is

$$T_x = 21°C + 273° = 294\text{K}$$

The speed of sound in the desert air is

$$a = \sqrt{kRT_x}$$

$$= \sqrt{(1.4)\left(287 \frac{J}{kg \cdot K}\right)(294K)}$$

$$= 343.7 \text{ m/s}$$

From the normal shock tables for $[p_y/p_x] = 2$, the Mach number is 1.36. The velocity of the expanding shock wave is

$$v = (1.36)\left(343.7 \frac{m}{s}\right) = 467.4 \text{ m/s}$$

(b) Behind the wave, the pressure is

$$(2)(101.3 \text{ kPa}) = 202.6 \text{ kPa}$$

(c) From the isentropic flow tables at M = 1.36, $[T_y/T_x] = 1.229$. The temperature behind the shock wave is

$$T = \left[\frac{T_y}{T_x}\right]T_x = (1.229)(294K)$$

$$= 361.3K$$

(d) The velocity of sound behind the shock wave is

$$a = \sqrt{kRT_y}$$

$$= \sqrt{(1.4)\left(287 \frac{J}{kg \cdot K}\right)(361.3K)}$$

$$= 381.0 \text{ m/s}$$

From the isentropic flow tables from the M = 1.36 row, $M_y = 0.75718$. The velocity of the air behind the shock wave is

$$v_y = M_y a = (0.75718)\left(381.0 \frac{m}{s}\right)$$

$$= 288.5 \text{ m/s}$$

*Customary U.S. Solution*

(a) Use the normal shock tables. Let the $x$-subscript correspond to the still air into which the shock wave is expanding. Let the $y$-subscript correspond to the gas inside the expanding high-pressure region.

The absolute static temperature of the desert air is

$$T_x = 70°F + 460° = 530°R$$

The speed of sound in the desert air is

$$a = \sqrt{kg_cRT_x}$$

$$= \sqrt{(1.4)\left(32.2 \frac{\text{ft-lbm}}{\text{lbf-sec}^2}\right)\left(53.3 \frac{\text{ft-lbf}}{\text{lbm-}°R}\right)(530°R)}$$

$$= 1128 \text{ ft/sec}$$

From the normal shock tables for $[p_y/p_x] = 2$, the Mach number is 1.36. The velocity of the expanding shock wave is

$$v = (1.36)\left(1128 \frac{\text{ft}}{\text{sec}}\right) = 1534 \text{ ft/sec}$$

(b) Behind the wave, the pressure is

$$(2)\left(14.7 \frac{\text{lbf}}{\text{in}^2}\right) = 29.4 \text{ lbf/in}^2$$

(c) From the isentropic flow tables at M = 1.36, $[T_y/T_x] = 1.229$. The temperature behind the shock wave is

$$T = \left[\frac{T_y}{T_x}\right]T_x = (1.229)(530°R)$$

$$= 651.4°R$$

(d) The velocity of sound behind the shock wave is

$$a = \sqrt{kg_cRT_y}$$

$$= \sqrt{(1.4)\left(32.2 \frac{\text{ft-lbm}}{\text{lbf-sec}^2}\right)\left(53.3 \frac{\text{ft-lbf}}{\text{lbm-}°R}\right)(651.4°R)}$$

$$= 1251 \text{ ft/sec}$$

From the isentropic flow tables from the M = 1.36 row, $M_y = 0.75718$. The velocity of the air behind the shock wave is

$$v_y = M_y a = (0.75718)\left(1251 \frac{\text{ft}}{\text{sec}}\right)$$

$$= 947.2 \text{ ft/sec}$$

## 13. OBLIQUE SHOCK WAVES

The shock waves described in Sec. 25.12 are called *normal shock waves* because they are normal to the direction of flow. Supersonic flow past wedges and cones creates *oblique shock waves* that are inclined from the direction of flow. For a wedge, the shock wave is in the form of two inclined planes. For a conical projectile, the shock wave is in the form of a conical envelope. The shock wave may be attached or may form in front of the object.

There is no convenient formula for determining the *shock angle*, $\theta$, although it can be obtained graphically from Fig. 25.3.

**Figure 25.3** *Shock Angle for Wedges (k = 1.4)*

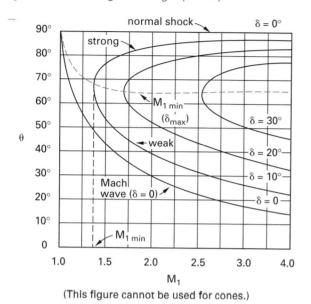

(This figure cannot be used for cones.)

Figure 25.3 illustrates that, for higher Mach numbers, there are two angles corresponding to each Mach number. Below the broken line on the chart, a *weak shock wave* (more like a Mach wave than a shock wave) exists.[8] Above the broken line, a strong oblique shock wave (more like a normal shock wave) exists. Though both can theoretically exist simultaneously, it is believed that only the weaker wave can exist attached to the body.

If the *semivertex angle* (also known as the *deviation angle*), $\delta$, is small enough, the shock wave will be attached. Attached shock waves are more likely to occur with wedges and narrow cones. Detached shock waves are more likely to form on thick wedges and objects with blunt leading edges.[9]

For the attached shock waves that form above the detachment Mach number on sharp-edged bodies, the conditions (static pressure, static temperature, etc.) after the shock wave can be found from the normal shock tables if $M_1 \sin\theta$ is used in place of $M_x$. If the

shock wave is inclined (i.e., not normal to the flow direction) but not attached, the shock tables cannot be used.

Unlike normal shock waves, the velocity after an oblique shock wave will not necessarily be subsonic. (See Fig. 25.4.) The Mach number for the normal component of velocity after an oblique shock wave will be less than 1, but the Mach number of the resultant velocity may exceed 1. The Mach number after the shock is

$$M_2 = \frac{M_y \text{ from shock table}}{\sin(\theta - \delta)} \qquad 25.27$$

**Figure 25.4** *Attached and Detached Oblique Shock Waves*

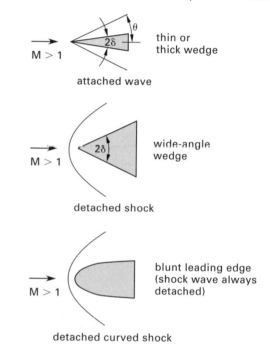

**Example 25.10**

A wedge with a semivertex angle of $10°$ is traveling through air at Mach 2. An attached oblique shock wave forms. What are the (a) shock angle, (b) Mach number after the shock wave, and (c) ratio of static pressures before and after the shock wave?

*Solution*

(a) From Fig. 25.3, the semivertex shock angle will be approximately $40°$.

(b) Use the normal shock table with $M_1 \sin\theta$ in place of $M_x$.

$$M_1 \sin\theta = 2\sin 40° = 1.29$$

---

[8]A *Mach wave* is an infinitesimal oblique shock wave whose normal component of the Mach number is 1.0. That is, $M_1 \sin\theta$ is 1.0, even though $M_1 > 1$. Changes in flow properties across a Mach wave are negligible. A Mach wave is not a discontinuity in the flow, and flow through it is isentropic. Therefore, Mach waves can exist anywhere in supersonic flow. The inclination of a Mach wave to the initial direction of flow is known as the *Mach angle* and is given by $\theta = \sin^{-1}(1/M_1)$.
[9]For wedges in air, the shock wave will be attached if the Mach number is greater than the minimum value indicated by Fig. 25.3 for the wedge semivertex angle. The shock wave will be detached if the Mach number is less than this value, or if the wedge semivertex angle is greater than $\sin^{-1}(1/k)$, which is $45.58°$ for air. For cones, the semivertex angle must be less than $57.5°$ for attached shock waves.

Thermodynamics

For $M_x = 1.29$, $M_y = 0.79108$. From Eq. 25.27,

$$M_2 = \frac{M_y}{\sin(\theta - \delta)}$$

$$= \frac{0.79108}{\sin(40° - 10°)}$$

$$= 1.58$$

(c) The ratio of static pressures before and after the shock wave is read from the normal shock table for Mach 1.29: $[p_y/p_x] = 1.7748$.

## 14. CRITICAL BACK-PRESSURE RATIO

Figure 25.5(a) illustrates a converging-diverging nozzle separating high- and low-pressure reservoirs. The total pressure in the high-pressure reservoir is constant. The pressure in the low-pressure reservoir (i.e., the *back pressure*) is variable. The nozzle geometry, particularly the throat and exit areas, $A_t$ and $A_e$, is assumed to be known. Figure 25.5(b) illustrates the ratio of static pressure at the corresponding point to the total pressure $(p/p_0)$ in the first reservoir. The following eight cases (*flow regimes*) represent decidedly different performances.

**Figure 25.5** *Flow Regimes in a C-D Nozzle*

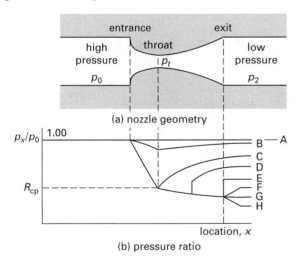

(a) nozzle geometry

(b) pressure ratio

*case A:* If $p_2 = p_0$, there will be no flow.

*case B:* When $p_2$ is lowered to just slightly less than $p_0$, flow will be initiated. Refer to Fig. 25.5. As long as the pressure ratio $p_t/p_0$ is greater than the first critical pressure ratio, $R_{cp}$ (0.5283 for air), flow in the divergent section will be subsonic and the exit pressure, $p_e$, will be equal to $p_2$. Flow will be isentropic, and the nozzle acts like a venturi. The conditions at any point in the nozzle where the area is known can be determined in the following manner.

*step 1:* Calculate the exit pressure ratio. (For any other point, substitute the known pressure for $p_2$.)

$$\frac{p_e}{p_0} = \frac{p_2}{p_0} \qquad 25.28$$

*step 2:* Locate the pressure ratio calculated in step 1 in the isentropic flow table. Read the exit Mach number, $M_e$, and the critical area ratio $[A/A^*]_e$. (Since sonic velocity is not actually achieved, the theoretical critical area $A^*$ does not exist in this nozzle.)

*step 3:* Calculate the ratio of actual throat area to the hypothetical critical throat area.

$$\frac{A_t}{A^*} = \frac{A_t\left[\dfrac{A}{A^*}\right]_e}{A_e} \qquad 25.29$$

*step 4:* Locate the area ratio calculated in step 3 in the isentropic flow table. Read the throat Mach number, $M_t$.

*case C:* If $p_2$ is lowered so that $p_t/p_0$ equals the *first critical (back) pressure ratio* (0.5283 for air), sonic velocity will be achieved in the throat. Flow will be choked and isentropic. The gas properties in the nozzle can be found from the procedure given in case B. The exit velocity will be subsonic, and the exit pressure will equal the back pressure ($p_e = p_2$).

*case D:* If $p_2$ is lowered slightly below that of case C (but not below the pressure ratio corresponding to the area ratio $A_e/A^*$), the mass flow rate will not increase above that of case C because the flow will be choked. However, since the back pressure $p_2$ is lower, the velocity will be higher. Supersonic velocity will be achieved in some parts of the nozzle. Flow will drop back to subsonic via a normal shock wave. The exit velocity will be subsonic, and the exit pressure will equal the back pressure ($p_e = p_2$). The location (i.e., the area) in the divergent section where the shock wave occurs can be found from the following trial-and-error procedure.

*step 1:* Assume a Mach number, $M_x$, at which the shock wave occurs.

*step 2:* For $M_x$, read the ratio of total pressures $[p_{0,y}/p_{0,x}]$ from the normal shock table.

*step 3:* Calculate the ratio of static exit pressure to total pressure after the shock wave. $p_{0,x}$ is the total pressure in the source reservoir.

$$\frac{p_e}{p_{0,y}} = \frac{p_e}{p_{0,x}\left[\dfrac{p_{0,y}}{p_{0,x}}\right]} \qquad 25.30$$

*step 4:* Determine the exit Mach number, $M_e$, by locating the ratio calculated in step 3 in the $[p/p_0]$ column of the normal shock tables. Read the area ratio $[A/A^*]$.

Thermodynamics

*step 5:* Calculate the area ratio.

$$\frac{A_e}{A_t}\left[\frac{p_{0,y}}{p_{0,x}}\right] \qquad 25.31$$

*step 6:* Compare the area ratio from steps 4 and 5. If the values differ, repeat from step 1. (The area ratio from step 5 decreases when the assumed Mach number increases.) If the values are equal, the shock wave occurs at the assumed Mach number, $M_x$.

*step 7:* Calculate the area at which the shock wave occurs.

$$A = A_t\left[\frac{A}{A^*}\right] \qquad 25.32$$

*case E:* If the back pressure is equal to the *second critical (back) pressure ratio* (i.e., the pressure ratio corresponding to the area ratio $(A_e/A^*)$, the shock wave will stand at the exit. The mass flow rate is the same as in case C.

*case F:* If the back pressure ratio is less than the second critical back pressure ratio but greater than the third, the condition is known as *over-expansion*. The exit area is too large, and gas will be discharged at a pressure less than the ambient conditions. The gas will "pop" back up to the ambient pressure through an *oblique compression shock wave* outside the nozzle, but this will not change the conditions inside the nozzle. The mass flow rate is the same as in case C.

*case G:* If the back pressure ratio is equal to the *third critical (back) pressure ratio*, there will be no shock wave. Pressure $p_2$ is known as the *design pressure* or the *isentropic pressure* for the nozzle. Flow will be isentropic. The mass flow rate is the same as in case C.

*case H:* If the pressure ratio is less than the third critical back pressure ratio, the condition is known as *under-expansion*. The exit area is too small for the back pressure. Expansion will be incomplete in the nozzle, and the gas will be discharged at a pressure above the local ambient pressure. Expansion continues outside the nozzle, and the pressure drops down to the ambient pressure by way of a *rarefaction shock wave* (also known as an *expansion wave* or an *oblique expansion shock wave*) outside the nozzle. The mass flow rate is the same as in case C.

Both over- and under-expansion reduce the efficiency of the nozzle. The effect of over-expansion is to reduce the gas exit velocity. In the case of rocket propulsion nozzles, the decreased exit velocity produces a proportional decrease in thrust (see Eq. 25.20). In the case of a fixed steam nozzle, the energy available to the steam turbine is reduced.

A given percentage of over-expansion (based on the ratio of actual-to-theoretical exit areas) can reduce the

available energy as much as ten times the reduction for the same percentage of under-expansion. For that reason, steam nozzles feeding turbines are sometimes designed 10–20% too small to ensure under-expansion under light or partial loads.

## Example 25.11

A supersonic wind tunnel is constructed so that the back pressure can be varied during the testing of a nozzle. The wind tunnel is fed from a reservoir with a total pressure of 1000 psia (7 MPa) and a total temperature of 1000°R (560K). A nozzle in the wind tunnel has a throat area of 1.5 in$^2$ (9.68 cm$^2$) and an exit area of 3.457 in$^2$ (22.26 cm$^2$). What is the highest back pressure for which the flow will remain supersonic throughout the entire length of the nozzle?

*SI Solution*

The desired operating conditions (i.e., that of no shock wave in the nozzle) can be satisfied by case E, where there is a shock wave at (or just outside of) the exit.[10] Since the velocity is supersonic in the diverging section, the velocity is sonic in the throat. The ratio of exit-to-throat areas, corresponding to $[A/A^*]$ since Mach 1 has been achieved, is

$$\frac{A_e}{A_t} = \frac{22.26 \text{ cm}^2}{9.68 \text{ cm}^2} = 2.3$$

Searching the isentropic flow tables for this value, the exit Mach number is found to be 2.35. For that point, $[p/p_0] = 0.07396$. The exit pressure before the shock wave is

$$p_{x,e} = \left[\frac{p}{p_0}\right]p_0 = (0.07396)(7.0 \text{ MPa})$$
$$= 0.5177 \text{ MPa}$$

This pressure is the design pressure for the nozzle and corresponds to case G. However, if there is a shock wave at the exit, the pressure after the shock wave will be higher. From the normal shock table for $M_x = 2.35$, $[p_y/p_x] = 6.276$. The maximum back pressure is

$$p_y = \left[\frac{p_y}{p_x}\right]p_{x,e} = (6.276)(0.5177 \text{ MPa})$$
$$= 3.249 \text{ MPa}$$

*Customary U.S. Solution*

The desired operating conditions (i.e., that of no shock wave in the nozzle) can be satisfied by case E, where there is a shock wave at (or just outside of) the exit.[11] Since the velocity is supersonic in the diverging section, the velocity will be sonic in the throat. The ratio of

---

[10]Therefore, this example is *not* correctly solved by assuming case G and finding the design pressure.
[11]See Ftn. 10.

exit-to-throat areas, corresponding to $[A/A^*]$ since Mach 1 has been achieved, is

$$\frac{A_e}{A_t} = \frac{3.457 \text{ in}^2}{1.5 \text{ in}^2} = 2.3$$

Searching the isentropic flow tables for this value, the exit Mach number is found to be 2.35. For that point, $[p/p_0] = 0.07396$. The exit pressure before the shock wave is

$$p_{x,e} = \left[\frac{p}{p_0}\right] p_0 = (0.07396)\left(1000 \ \frac{\text{lbf}}{\text{in}^2}\right)$$
$$= 73.96 \text{ lbf/in}^2$$

This pressure is the design pressure for the nozzle and corresponds to case G. However, if there is a shock wave at the exit, the pressure after the shock wave will be higher. From the normal shock table for $M_x = 2.35$, $[p_y/p_x] = 6.276$. The maximum back pressure is

$$p_y = \left[\frac{p_y}{p_x}\right] p_{x,e} = (6.276)\left(73.96 \ \frac{\text{lbf}}{\text{in}^2}\right)$$
$$= 464.2 \text{ lbf/in}^2$$

### Example 25.12

A nozzle is fed from a large reservoir containing 100 psia (0.7 MPa) air. The back pressure is adjustable. A shock wave stands at the exit of the nozzle. The Mach number just before the shock wave is 3.0. For which values of the back pressure will the shock wave at the exit disappear completely?

*SI Solution*

There are two ways the shock wave can disappear. If the back pressure is decreased to the design pressure (case G), the shock wave will move out of the nozzle and dissipate. At Mach 3, $[p/p_0] = 0.02722$. Therefore, the design pressure for this nozzle is

$$p_{\text{design}} = \left[\frac{p}{p_0}\right] p_0 = (0.02722)(0.7 \text{ MPa})\left(1000 \ \frac{\text{kPa}}{\text{MPa}}\right)$$
$$= 19.05 \text{ kPa}$$

If the back pressure is increased, the shock wave will move upstream and vanish at the throat (case C). From the supersonic portion of the isentropic flow tables at Mach 3, $[A/A^*] = 4.235$. The same value occurs in the subsonic part of the isentropic flow table. The Mach number corresponding to this value in the subsonic part of the table is 0.138. At this Mach number, the shock wave will vanish and the flow will be subsonic

everywhere in the nozzle. At Mach 0.138, $[p/p_0] = 0.987$. The back pressure is

$$p_{\text{back}} = \left[\frac{p}{p_0}\right] p_0 = (0.987)(0.7 \text{ MPa})$$
$$= 0.6909 \text{ MPa}$$

*Customary U.S. Solution*

Shock wave at the exit (case G): At Mach 3, $[p/p_0] = 0.02722$. Therefore, the design pressure is

$$p_{\text{design}} = \left[\frac{p}{p_0}\right] p_0 = (0.02722)\left(100 \ \frac{\text{lbf}}{\text{in}^2}\right)$$
$$= 2.722 \text{ lbf/in}^2$$

Shock wave at the throat (case C): At Mach 3, $[A/A^*] = 4.235$. The Mach number corresponding to this value in the subsonic part of the table is 0.138. At Mach 0.138, $[p/p_0] = 0.987$. The back pressure is

$$p_{\text{back}} = \left[\frac{p}{p_0}\right] p_0 = (0.987)\left(100 \ \frac{\text{lbf}}{\text{in}^2}\right)$$
$$= 98.7 \text{ lbf/in}^2$$

## 15. COMPRESSIBLE FLOW THROUGH ORIFICES

Compressible flow through orifices (simple holes, perforations, re-entrant tubes, etc.) is similar to flow through nozzles, cases A, B, and C. If the ratio of ambient-to-source pressures is less than the first critical pressure ratio, the velocity will be sonic in the orifice. However, since there is no diverging section, the flow can never become supersonic. Due to the high turbulence at the orifice, the coefficient of velocity and nozzle efficiency may be relatively low. This will be manifested in a lower-than-ideal mass flow rate.

## 16. ADIABATIC FLOW WITH FRICTION IN CONSTANT-AREA DUCTS

Although flow through nozzles is assumed to be frictionless (because nozzles are short), friction cannot be disregarded in long constant-area ducts. If the velocity is high or if the duct is insulated, flow may still be adiabatic, however. Adiabatic flow with friction is known as *Fanno flow*. In Fanno flow, total pressure always decreases in the direction of flow. Total temperature, however, is constant, regardless of the velocity of the entering gas.

The *Fanno line* is a plot of temperature (or enthalpy) versus entropy for a specific flow rate. The line illustrates how the velocity tends toward Mach 1, regardless of the entering velocity. (See Fig. 25.6.)

If the velocity of the gas entering the duct is subsonic, the pressure and temperature will drop in the direction

**Figure 25.6** Fanno Line

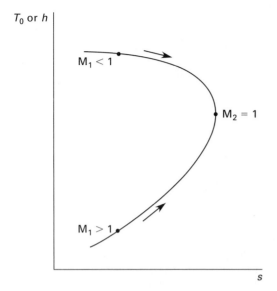

of flow. Since the density also drops, the velocity will increase to maintain the same mass flow rate. Eventually, the velocity in the duct becomes sonic, the flow is choked, and no further increase in mass flow rate can occur.

If the velocity of the gas entering the duct is supersonic, friction causes the pressure and temperature to increase in the direction of flow. Since the density also increases, the velocity will decrease to maintain the same mass flow rate. Eventually, the velocity in the duct becomes sonic.

When the flow is Mach 1 at the entrance, friction causes the pressure, density, and temperature to increase and the velocity to decrease. The Mach number is 1 at the exit.

Since the flow is adiabatic, the total temperature is constant. Equation 25.5 can be used to find the temperature at any point along the duct. The ratio of static pressures for two points along the duct is given by Eq. 25.33.

$$\frac{p_2}{p_1} = \left(\frac{\rho_1}{\rho_2}\right)\left(1 + \left(\tfrac{1}{2}(k-1)M_1^2\right)\left(1 - \left(\frac{\rho_1}{\rho_2}\right)^2\right)\right) \quad \textit{25.33}$$

The duct length required for the flow to reach Mach 1 is found by solving Eq. 25.34 for $L_{max}$. For round ducts, the *hydraulic diameter*, $D_h$, is equal to the inside diameter. $f_{Fanning}$ is the *Fanning friction factor*, also known as the *coefficient of friction*, $C_f$.[12]

---

[12]Multiplying the Fanning friction factor by 4, as is done in Eq. 25.34, converts it to the Darcy friction factor.

$$\frac{4f_{Fanning}L_{max}}{D_h} = \left(\frac{1-M^2}{kM^2}\right)$$
$$+ \left(\frac{1+k}{2k}\right)\ln\left(\frac{(1+k)M^2}{2\left(1+\tfrac{1}{2}(k-1)M^2\right)}\right)$$
$$\textit{25.34}$$

The length of duct between two points with known Mach numbers can be found from Eq. 25.35. The quantity $4f_{Fanning}L_{max}/D_h$ is found from Eq. 25.34, tables, or graphs.

$$x = \left(\frac{D_h}{4f_{Fanning}}\right)\left(\left(\frac{4f_{Fanning}x_{max}}{D_h}\right)_1 - \left(\frac{4f_{Fanning}x_{max}}{D_h}\right)_2\right)$$
$$\textit{25.35}$$

The static gas properties at any point can be related to the sonic properties farther downstream.

$$\frac{p}{p^*} = \frac{1}{M}\sqrt{\frac{1+k}{2\left(1+\tfrac{1}{2}(k-1)M^2\right)}} \quad \textit{25.36}$$

$$\frac{T}{T^*} = \frac{1+k}{2\left(1+\tfrac{1}{2}(k-1)M^2\right)} \quad \textit{25.37}$$

$$\frac{p_0}{p_0^*} = \left(\frac{1}{M}\right)\left(\left(\frac{2}{1+k}\right)\left(1+\tfrac{1}{2}(k-1)M^2\right)\right)^{(1+k)/2(k-1)}$$
$$\textit{25.38}$$

**Example 25.13**

Air enters an adiabatic round duct with an inside diameter of 0.3 ft (0.09 m). The Fanning friction factor is 0.003. The entrance air pressure is 12 psia (80 kPa), and the entrance Mach number is 0.3. What are the (a) Mach number and (b) pressure 50 ft (15 m) downstream?

*SI Solution*

(a) From App. 25.C for M = 0.3,

$$\frac{4f_{Fanning}x_{max}}{D_h} = 5.299$$

The value 15 m farther into the duct is

$$\frac{4f_{Fanning}x}{D_h} = \frac{(4)(0.003)(15\text{ m})}{0.09\text{ m}}$$
$$= 2.000$$

From Eq. 25.35,

$$\left(\frac{4f_{Fanning}x_{max}}{D_h}\right)_2 = 5.299 - 2.000$$
$$= 3.299$$

Interpolating in App. 25.C, $M_2$ is approximately 0.356.

(b) At the entrance, for Mach 0.3, the ratio $[p/p^*]$ is 3.619. Therefore, the sonic pressure is

$$p^* = \frac{p}{\left[\dfrac{p}{p^*}\right]} = \frac{80 \text{ kPa}}{3.619} = 22.11 \text{ kPa}$$

15 m downstream, at Mach 0.35, the ratio $[p/p^*]$ is 3.092. $p^*$ is unchanged.

$$p_2 = \left[\frac{p}{p^*}\right] p^* = (3.092)(22.11 \text{ kPa})$$
$$= 68.36 \text{ kPa}$$

*Customary U.S. Solution*

(a) From App. 25.C for M = 0.3,

$$\frac{4f_{\text{Fanning}} x_{\max}}{D_h} = 5.299$$

The value 50 ft farther into the duct is

$$\frac{4f_{\text{Fanning}} x}{D_h} = \frac{(4)(0.003)(50 \text{ ft})}{0.3 \text{ ft}}$$
$$= 2.000$$

From Eq. 25.35,

$$\left(\frac{4f_{\text{Fanning}} x_{\max}}{D_h}\right)_2 = 5.299 - 2.000$$
$$= 3.299$$

Interpolating in App. 25.C, $M_2$ is approximately 0.356.

(b) At the entrance, for Mach 0.3, the ratio $[p/p^*]$ is 3.619. Therefore, the sonic pressure is

$$p^* = \frac{p}{\left[\dfrac{p}{p^*}\right]} = \frac{12 \text{ psia}}{3.619} = 3.316 \text{ psia}$$

50 ft downstream, at Mach 0.35, the ratio $[p/p^*]$ is 3.092. $p^*$ is unchanged.

$$p_2 = \left[\frac{p}{p^*}\right] p^* = (3.092)(3.316 \text{ psia})$$
$$= 10.25 \text{ psia}$$

## Example 25.14

Air enters a round duct at Mach 2. The inside diameter of the duct is 2.0 in (5 cm), and the Fanning friction factor is 0.005. The total pressure and total temperature at the duct entrance are 78.25 psia and 1080°R (540 kPa and 600K), respectively. What are the (a) temperature, (b) pressure at the entrance, (c) static pressure at a point where the Mach number is 1.75, and (d) distance

between the points where the Mach numbers are 2.0 and 1.75?

*SI Solution*

(a) From the isentropic flow table at Mach 2, $[T/T_0] = 0.5556$.

$$T = \left[\frac{T}{T_0}\right] T_0 = (0.5556)(600\text{K})$$
$$= 333.4\text{K}$$

(b) From the isentropic flow table at Mach 2, the pressure ratio is $[p/p_0] = 0.1278$.

$$p = \left[\frac{p}{p_0}\right] p_0 = (0.1278)(540 \text{ kPa})$$
$$= 69.0 \text{ kPa}$$

(c) It is not known if the duct is long enough for choked flow (M = 1) to occur. However, the conditions at the hypothetical choke point can be used as a reference for other conditions. The following values are read from a Fanno flow table.

From App. 25.C, at Mach 2.0, $[p/p^*] = 0.408$. At Mach 1.75, $[p/p^*] = 0.493$.

$$p_{\text{M}=1.75} = \frac{\left[\dfrac{p}{p^*}\right]_{\text{M}=1.75} p_{\text{M}=2.0}}{\left[\dfrac{p}{p^*}\right]_{\text{M}=2.0}}$$
$$= \frac{(0.493)(69.0 \text{ kPa})}{0.408}$$
$$= 83.4 \text{ kPa}$$

(d) At Mach 2.0, $[4fL/D] = 0.305$. At Mach 1.75, $[4fL/D] = 0.225$. At Mach 2, the distance to reach Mach 1 is

$$L_{\text{M}=2.0} = \frac{\left[\dfrac{4f_{\text{Fanning}} L}{D}\right] D}{4f} = \frac{(0.305)(5 \text{ cm})}{(4)(0.005)}$$
$$= 76.25 \text{ cm}$$

At Mach 1.75, the distance to reach Mach 1 is

$$L_{\text{M}=1.75} = \frac{\left[\dfrac{4f_{\text{Fanning}} L}{D}\right] D}{4f} = \frac{(0.225)(5 \text{ cm})}{(4)(0.005)}$$
$$= 56.25 \text{ cm}$$

**Thermodynamics**

The distance between the two points is

$$L = L_{M=2.0} - L_{M=1.75}$$
$$= 76.25 \text{ cm} - 56.25 \text{ cm} = 20 \text{ cm}$$

*Customary U.S. Solution*

(a) From the isentropic flow table at Mach 2, $[T/T_0] = 0.5556$.

$$T = \left[\frac{T}{T_0}\right] T_0 = (0.5556)(1080°\text{R})$$
$$= 600°\text{R}$$

(b) From the isentropic flow table at Mach 2, the pressure ratio is $[p/p_0] = 0.1278$.

$$p = \left[\frac{p}{p_0}\right] p_0 = (0.1278)\left(78.25 \ \frac{\text{lbf}}{\text{in}^2}\right)$$
$$= 10 \ \text{lbf/in}^2$$

(c) From App. 25.C, at Mach 2.0, $[p/p^*] = 0.408$. At Mach 1.75, $[p/p^*] = 0.493$.

$$p_{M=1.75} = \frac{\left[\dfrac{p}{p^*}\right]_{M=1.75} p_{M=2.0}}{\left[\dfrac{p}{p^*}\right]_{M=2.0}}$$
$$= \frac{(0.493)\left(10 \ \dfrac{\text{lbf}}{\text{in}^2}\right)}{0.408}$$
$$= 12.1 \ \text{lbf/in}^2$$

(d) At Mach 2.0, $[4fL/D] = 0.305$. At Mach 1.75, $[4fL/D] = 0.225$. At Mach 2, the distance to reach Mach 1 is

$$L_{M=2.0} = \frac{\left[\dfrac{4f_{\text{Fanning}}L}{D}\right]D}{4f} = \frac{(0.305)(2 \text{ in})}{(4)(0.005)}$$
$$= 30.5 \text{ in}$$

At Mach 1.75, the distance to reach Mach 1 is

$$L_{M=1.75} = \frac{\left[\dfrac{4f_{\text{Fanning}}L}{D}\right]D}{4f} = \frac{(0.225)(2 \text{ in})}{(4)(0.005)}$$
$$= 22.5 \text{ in}$$

The distance between the two points is

$$L = L_{M=2.0} - L_{M=1.75}$$
$$= 30.5 \text{ in} - 22.5 \text{ in} = 8 \text{ in}$$

## 17. ISOTHERMAL FLOW WITH FRICTION

As the Mach number approaches 1.0, an infinite heat transfer rate would be required to keep the flow isothermal. For that reason, gas flow in most systems is more of an adiabatic process than an isothermal process. A notable exception is encountered with long-distance gas pipelines.[13] The earth acts like a large heat reservoir and supplies the energy needed to keep the flow isothermal.

For flow that is initially subsonic with a Mach number less than $1/\sqrt{k}$, pressure, density, and total pressure decrease in the direction of flow. The velocity, Mach number, and total temperature all increase. The Mach number approaches $1/\sqrt{k}$.

For flow that is initially subsonic with a Mach number greater than $1/\sqrt{k}$, and for flow that is initially supersonic, pressure and density increase, while velocity, Mach number, and total temperature decrease. The total pressure increases for Mach numbers less than $\sqrt{2}/(1+k)$ and decreases otherwise. The Mach number approaches $1/\sqrt{k}$.

The Darcy equation can be used to calculate the friction loss, and the thermodynamic relationships for isothermal processes are all applicable. For pipelines with diameters greater than 24 in (60 cm) experiencing turbulent flow, for any given pressure drop between two points a distance $L$ apart, the mass flow rate is given by the *Weymouth equation*.

$$\dot{m} = \sqrt{\frac{(p_1^2 - p_2^2)D^5 g_c \left(\dfrac{\pi}{4}\right)^2}{4f_{\text{Fanning}}LRT}} \qquad \text{25.39}$$

The *Fanning friction factor*, $f_{\text{Fanning}}$, in the Weymouth equation is

$$f_{\text{Fanning}} = \frac{0.00349}{(D_{\text{ft}})^{0.333}} \qquad \text{25.40}$$

The *transmission factor* is defined as

$$\text{transmission factor} = \sqrt{\frac{1}{f_{\text{Fanning}}}} \qquad \text{25.41}$$

### Example 25.15

A 40 in (100 cm) inside diameter pipeline carries natural gas (essentially methane with a molecular weight of 16) between compressor stations 75 mi (120 km) apart. The gas leaves the upstream station at 650 psia (4.5 MPa). The pressure has decreased to 450 psia (3.1 MPa) by the time it reaches the next station. The gas temperature is

---

[13]The low velocities (typically less than 20 ft/sec (6 m/s)) typical of natural gas pipelines seem out of place in this chapter. However, the gas experiences a density change along the length of pipeline, qualifying it for coverage in this chapter.

40°F (4°C) along the entire pipeline. (a) Calculate the mass flow rate from the Weymouth equation. (b) Calculate the mass flow rate from the Bernoulli equation using the average gas properties along the pipeline.

*Solution*

The pipeline diameter is

$$D = \frac{40 \text{ in}}{12 \frac{\text{in}}{\text{ft}}} = 3.333 \text{ ft}$$

The flow area of the pipeline is

$$A = \frac{\pi D^2}{4} = \frac{\pi (3.333 \text{ ft})^2}{4}$$
$$= 8.725 \text{ ft}^2$$

The absolute temperature of the natural gas is

$$T = 40°\text{F} + 460° = 500°\text{R}$$

Convert the pressures.

$$p_1 = \left(650 \frac{\text{lbf}}{\text{in}^2}\right)\left(12 \frac{\text{in}}{\text{ft}}\right)^2$$
$$= 9.36 \times 10^4 \text{ lbf/ft}^2$$

$$p_2 = \left(450 \frac{\text{lbf}}{\text{in}^2}\right)\left(12 \frac{\text{in}}{\text{ft}}\right)^2$$
$$= 6.48 \times 10^4 \text{ lbf/ft}^2$$

The specific gas constant for methane is

$$R = \frac{R^*}{\text{MW}} = \frac{1545 \frac{\text{ft-lbf}}{\text{lbmol-°R}}}{16 \frac{\text{lbm}}{\text{lbmol}}}$$
$$= 96.56 \text{ ft-lbf/lbm-°R}$$

(a) The Fanning friction factor is given by Eq. 25.40.

$$f_{\text{Fanning}} = \frac{0.00349}{(D_{\text{ft}})^{0.333}}$$
$$= \frac{0.0349}{(3.333 \text{ ft})^{0.333}} = 0.00234$$

From Eq. 25.39, the mass flow rate is

$$\dot{m} = \sqrt{\frac{(p_1^2 - p_2^2)D^5 g_c \left(\frac{\pi}{4}\right)^2}{4 f_{\text{Fanning}} LRT}}$$

$$= \sqrt{\frac{\begin{array}{c}\left(\left(9.36 \times 10^4 \frac{\text{lbf}}{\text{ft}^2}\right)^2 - \left(6.48 \times 10^4 \frac{\text{lbf}}{\text{ft}^2}\right)^2\right) \\ \times (3.333 \text{ ft})^5 \left(32.2 \frac{\text{ft-lbm}}{\text{lbf-sec}^2}\right)\left(\frac{\pi}{4}\right)^2\end{array}}{\begin{array}{c}(4)(0.00234)(75 \text{ mi})\left(5280 \frac{\text{ft}}{\text{mi}}\right) \\ \times \left(96.56 \frac{\text{ft-lbf}}{\text{lbm-°R}}\right)(500°\text{R})\end{array}}}$$

$$= 456 \text{ lbm/sec}$$

(b) Since the pressure drops approximately 30% along the pipeline, use the average pressure to calculate the average density.

$$\rho_{\text{ave}} = \frac{p}{RT}$$
$$= \frac{9.36 \times 10^4 \frac{\text{lbf}}{\text{ft}^2} + 6.48 \times 10^4 \frac{\text{lbf}}{\text{ft}^2}}{(2)\left(96.56 \frac{\text{ft-lbf}}{\text{lbm-°R}}\right)(500°\text{R})}$$
$$= 1.640 \text{ lbm/ft}^3$$

The entire pressure drop is due to friction. Neglecting the low kinetic energy and unknown changes in potential energy, the Bernoulli equation can be written for the two ends of the pipeline. (The specific weight and density are numerically equal.)

$$\frac{p_1}{\gamma} = \frac{p_2}{\gamma} + h_f$$

$$\frac{9.36 \times 10^4 \frac{\text{lbf}}{\text{ft}^2}}{1.640 \frac{\text{lbm}}{\text{ft}^3}} = \frac{6.48 \times 10^4 \frac{\text{lbf}}{\text{ft}^2}}{1.640 \frac{\text{lbm}}{\text{ft}^3}} + h_f$$

$$h_f = 17{,}560 \text{ ft}$$

Use the same friction factor as in part (a).

$$f_{\text{Darcy}} = 4 f_{\text{Fanning}} = (4)(0.00234) = 0.00936$$

*Thermodynamics*

Solve for the average flow velocity from the Darcy equation.

$$h_f = \frac{fLv^2}{2Dg}$$

$$17{,}560 \text{ ft} = \frac{(0.00936)(75 \text{ mi})\left(5280 \frac{\text{ft}}{\text{mi}}\right)v^2}{(2)(3.333 \text{ ft})\left(32.2 \frac{\text{ft}}{\text{sec}^2}\right)}$$

$$v_{ave} = 31.89 \text{ ft/sec}$$

The mass flow rate is

$$\dot{m} = \rho_{ave} v_{ave} A$$

$$= \left(1.64 \frac{\text{lbm}}{\text{ft}^3}\right)\left(31.89 \frac{\text{ft}}{\text{sec}}\right)(8.725 \text{ ft}^2)$$

$$= 456 \text{ lbm/sec}$$

## 18. FRICTIONLESS FLOW WITH HEATING OR COOLING

If heat is added to or removed from a gas flowing through a frictionless constant-area duct, the flow is said to be *diabatic flow* (flow *with* heat transfer) or *Rayleigh flow*. Heating or cooling can be caused either by chemical reactions taking place in the gas, phase changes, electrical sources, or by heating or refrigeration acting through the duct walls. (Heating due to friction is not Rayleigh flow.) The heat transfer rate per unit mass is

$$q = c_p(T_{0,2} - T_{0,1}) \qquad 25.42$$

The *Rayleigh line* is a plot of temperature (or enthalpy) versus entropy for a specific flow rate. (See Fig. 25.7.) It illustrates how the velocity tends toward Mach 1 for both subsonic and supersonic entrances. Entropy is maximum at the "nose" corresponding to Mach 1. The maximum static temperature occurs at $M = 1/\sqrt{k}$.

Equation 25.43 through Eq. 25.46 give the ratios of static-to-sonic parameters. Since it is difficult to extract the Mach number from these expressions, graphs and tables are often used to solve Rayleigh flow problems.

$$\frac{T}{T^*} = \left(\frac{M(1+k)}{1+kM^2}\right)^2 \qquad 25.43$$

$$\frac{T_0}{T_0^*} = \frac{2(1+k)M^2\left(1 + \left(\frac{1}{2}\right)(k-1)M^2\right)}{(1+kM^2)^2} \qquad 25.44$$

$$\frac{p}{p^*} = \frac{1+k}{1+kM^2} \qquad 25.45$$

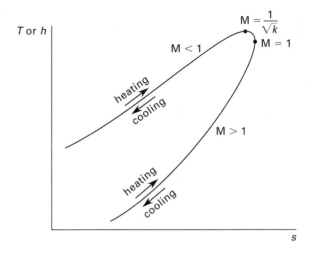

**Figure 25.7** *Rayleigh Line*

$$\frac{p_0}{p_0^*} = \left(\frac{2}{1+k}\right)^{k/(k-1)} \left(\frac{1+k}{1+kM^2}\right) \qquad 25.46$$
$$\times \left(1 + \tfrac{1}{2}(k-1)M^2\right)^{k/(k-1)}$$

## Example 25.16

Air with a specific heat of 0.24 Btu/lbm-°R (1 kJ/kg·K) enters a constant-area duct at 50°F (10°C). While passing through the duct, the air's enthalpy is increased by 150 Btu/lbm (350 kJ/kg). If the inlet Mach number is 0.3, what are the (a) exit temperature, (b) exit Mach number, and (c) exit stagnation temperature?

*SI Solution*

From the isentropic flow table (not a Rayleigh table) for Mach 0.3, $[T/T_0]$ is 0.9823. The total temperature at the inlet is

$$T_{0,i} = \frac{T}{\left[\frac{T}{T_0}\right]}$$

$$= \frac{10°\text{C} + 273°}{0.9823} = 288\text{K}$$

Sonic velocity may not be achieved if the duct is not long enough. The sonic temperature is determined as a reference point. From App. 25.D at Mach 0.3, the ratio of total temperatures at the inlet and sonic points is $[T_0/T_0^*]$ is 0.3468. The total temperature at the sonic point is

$$T_0^* = \frac{T_{0,i}}{\left[\frac{T_0}{T_0^*}\right]}$$

$$= \frac{288\text{K}}{0.3468} = 830.4\text{K}$$

From App. 25.D, at Mach 0.3, the ratio of static temperatures at the inlet and sonic points is $[T/T^*] = 0.409$. The static temperature at the sonic point is

$$T^* = \frac{T_i}{\left[\dfrac{T}{T^*}\right]}$$

$$= \frac{10°\text{C} + 273°}{0.409} = 691.9\text{K}$$

The total (stagnation) temperature at the exit is

$$T_{0,e} = T_{0,i} + \frac{q}{c_p}$$

$$= 288\text{K} + \frac{350 \; \dfrac{\text{kJ}}{\text{kg}}}{1 \; \dfrac{\text{kJ}}{\text{kg·K}}}$$

$$= 638\text{K}$$

The ratio of total temperature at the exit and sonic points is

$$\frac{T_{0,e}}{T_0^*} = \frac{638\text{K}}{830.3\text{K}} = 0.7684$$

Interpolating from App. 25.D, the Mach number corresponding to this ratio of total temperatures is read as $M_e = 0.557$ (use 0.56). At that Mach number, $[T/T^*] = 0.872$. The temperature at the exit is

$$T_e = \left[\frac{T}{T^*}\right] T^* = (0.872)(691.9\text{K})$$

$$= 603\text{K}$$

*Customary U.S. Solution*

From the isentropic flow table for Mach 0.3, $[T/T_0]$ is 0.9823. The total temperature at the inlet is

$$T_{0,i} = \frac{T}{\left[\dfrac{T}{T_0}\right]}$$

$$= \frac{50°\text{F} + 460°}{0.9823} = 519.2°\text{R}$$

From App. 25.D at Mach 0.3, the ratio of total temperatures at the inlet and sonic points is $[T/T_0^*]$ is 0.3468. The total temperature at the sonic point is

$$T_0^* = \frac{T_{0,i}}{\left[\dfrac{T_0}{T_0^*}\right]}$$

$$= \frac{519.2°\text{R}}{0.3468} = 1497°\text{R}$$

From App. 25.D, at Mach 0.3, the ratio of static temperatures at the inlet and sonic points is $[T/T^*] = 0.409$. The static temperature at the sonic point is

$$T^* = \frac{T_i}{\left[\dfrac{T}{T^*}\right]}$$

$$= \frac{50°\text{F} + 460°}{0.409} = 1247°\text{R}$$

The total (stagnation) temperature at the exit is

$$T_{0,e} = T_{0,i} + \frac{q}{c_p}$$

$$= 519.1°\text{R} + \frac{150 \; \dfrac{\text{Btu}}{\text{lbm}}}{0.24 \; \dfrac{\text{Btu}}{\text{lbm-°R}}}$$

$$= 1144°\text{R}$$

The ratio of total temperature at the exit and sonic points is

$$\frac{T_{0,e}}{T_0^*} = \frac{1144°\text{R}}{1497°\text{R}} = 0.7642$$

Interpolating from App. 25.D, the Mach number corresponding to this ratio of total temperatures is read as $M_e = 0.554$ (use 0.56). At that Mach number, $[T/T^*] = 0.872$. The temperature at the exit is

$$T_e = \left[\frac{T}{T^*}\right] T^* = (0.872)(1247°\text{R})$$

$$= 1087°\text{R}$$

## 19. STEAM FLOW THROUGH NOZZLES

If steam is assumed to be an ideal gas, then all of the relationships presented in this chapter can be used. For low temperatures, steam's ratio of specific heats is approximately 1.33. However, the value is lower at the temperatures normally encountered in steam nozzles.

The actual value of the ratio of specific heats can be readily found. Since $\Delta h = c_p \Delta T$, the specific heat, $c_p$, can be calculated from the superheat tables as $\Delta h/\Delta T$ for states in close proximity to the actual steam condition. Then, $c_v$ and $k$ can be calculated from the ideal gas relationships. For steam, the specific gas constant is approximately 85.78 ft-lbf/lbm-°R (0.4615 kJ/kg·K).

Equation 25.1 and Eq. 25.2 (if the entrance velocity is small) are always applicable to the flow of steam through nozzles. However, since steam expansion is generally not isentropic, the exit velocity will be less than calculated, and Eq. 25.15 through Eq. 25.18 should be used. Isentropic efficiencies of 85% to 95% are typical, as are coefficients of velocity, $C_v$, of 0.95 to 0.99.

The critical pressure ratio can be calculated from Eq. 25.8 using an average ratio of specific heats. However, for typical steam inlet pressures, the critical pressure ratio is often taken as 0.5457 (corresponding to the value $k = 1.3$) for highly superheated steam, 0.57 (corresponding to $k = 1.18$) for moderately superheated steam, and 0.577 to 0.58 (corresponding to $k = 1.13$) for saturated (i.e., "wet") steam.[14] Thus, sonic velocity at the throat can be achieved if the back pressure is 54% to 57% of the inlet pressure.

It is often necessary to calculate the throat area that will produce sonic velocity for a given mass flow rate. Treating steam as an ideal gas, the ratio of mass flow to throat area for sonic flow is

$$\frac{\dot{m}}{A^*} = \rho_t a = \rho_t \sqrt{kRT_t} \qquad \text{[SI]} \quad 25.47(a)$$

$$\frac{\dot{m}}{A^*} = \rho_t a = \rho_t \sqrt{kg_c RT_t} \qquad \text{[U.S.]} \quad 25.47(b)$$

Generally, only the inlet conditions are known, not the throat conditions. The ratios calculated from Eq. 25.4 and Eq. 25.5 (or from a table for the appropriate value of $k$, if available) are used to write the throat temperature and pressure in terms of the inlet temperature and pressure. Using 1.3 for the ratio of specific heats, 85.8 ft-lbf/lbm-°R (0.4615 kJ/kg·K) as the specific gas constant, 0.6276 as the ratio of throat-to-inlet densities, and 0.8696 as the ratio of throat-to-inlet temperatures, the approximate ratio of mass flow to throat area is given by Eq. 25.48.

$$\frac{\dot{m}}{A^*} \approx 14.34 \rho_i \sqrt{T_i} \qquad \text{[SI]} \quad 25.48(a)$$

$$\frac{\dot{m}}{A^*} \approx 35.1 \rho_i \sqrt{T_i} \qquad \text{[U.S.]} \quad 25.48(b)$$

The exit area can be found from the mass flow rate and the exit conditions. Modern turbines are multistage, and the condenser pressure is the back pressure for the last stage only. For other stages, the back pressure must be specified.

## 20. PITOT TUBES

When the initial velocity of a gas is subsonic, air flow is brought to rest in a pitot tube through an isentropic

process. Equation 25.49 and Eq. 25.50 can be used to calculate the velocity and Mach number, respectively.

$$v = \sqrt{\frac{2kRT}{k-1}\left(\left(\frac{p_0}{p}\right)^{\frac{k-1}{k}} - 1\right)} \qquad \text{[SI]} \quad 25.49(a)$$

$$v = \sqrt{\frac{2kg_c RT}{k-1}\left(\left(\frac{p_0}{p}\right)^{\frac{k-1}{k}} - 1\right)} \qquad \text{[U.S.]} \quad 25.49(b)$$

$$M = \sqrt{\frac{2\left(\left(\frac{p_0}{p}\right)^{\frac{k-1}{k}} - 1\right)}{k-1}} \qquad 25.50$$

When the initial velocity is supersonic, the gas cannot be brought to rest isentropically at the pitot tube inlet. (An isentropic stagnation process would require a converging-diverging nozzle at the pitot tube inlet.) Rather, a shock wave forms in front of the pitot tube. The pressure measured by the pitot tube is the downstream stagnation pressure, not the upstream stagnation pressure. Normal shock tables tabulate the value of $[p_x/p_{0,y}]$ so that the Mach number can be read directly.

### Example 25.17

A pitot tube measures an air pressure of 140 kPa. The corresponding static pressure is 30 kPa. (a) Is the flow subsonic or supersonic? (b) What is the Mach number?

*Solution*

(a) The ratio of static-to-total pressure is

$$\frac{p}{p_0} = \frac{30 \text{ kPa}}{140 \text{ kPa}}$$
$$= 0.214$$

Searching the isentropic flow table for this value, the Mach number appears to be approximately 1.66. However, since this is greater than 1.0, flow is supersonic, and there is a normal shock wave before the pitot tube.

(b) Searching the $[p_x/p_{0,y}]$ column in the normal shock tables, M = 1.80.

---

[14]Different values of $k$ should be used to design the converging and diverging sections if the steam is dry in the converging section and wet in the diverging section.

# 26 Vapor Power Equipment

## Nomenclature

| | | | |
|---|---|---|---|
| $c_p$ | specific heat | Btu/lbm-°R | kJ/kg·K |
| $C$ | concentration | ppm | ppm |
| $g_c$ | gravitational constant (32.2) | ft-lbm/lbf-°R | n.a. |
| $h$ | enthalpy | Btu/lbm | kJ/kg |
| HV | heating value | Btu/lbm | kJ/kg |
| HWD | hot well depression | °F | °C |
| $J$ | Joule's constant (778) | ft-lbf/Btu | n.a. |
| $m$ | mass | lbm | kg |
| $\dot{m}$ | mass flow rate | lbm/sec | kg/s |
| $p$ | pressure | lbf/ft² | kPa |
| $P$ | power | Btu/sec | kW |
| $Q$ | heat | Btu | kJ |
| $Q$ | heat per unit mass | Btu/lbm | kJ/kg |
| $\dot{Q}$ | heat flow rate | Btu/sec | kW |
| $t$ | time (duration) | sec | s |
| $T$ | temperature | °R | K |
| TTD | terminal temperature difference | °F | °C |
| v | velocity | ft/sec | m/s |
| $W$ | work | Btu | kJ |
| $W$ | work per unit mass | Btu/lbm | kJ/kg |
| $y$ | bleed fraction | – | – |

## Symbols

| | | | |
|---|---|---|---|
| $\eta$ | efficiency | – | – |
| $v$ | specific volume | ft³/lbm | m³/kg |

## Subscripts

| | |
|---|---|
| $f$ | saturated fluid |
| $fg$ | fluid-to-gas (vaporization) |
| $m$ | mechanical or intermediate |
| $p$ | at pressure $p$ or at constant pressure |
| $s$ | isentropic |
| sat | saturated |

## 1. TYPICAL SYSTEM INTEGRATION

This chapter takes a quick look at some of the equipment necessary to implement a typical utility power-generating plant. It is impractical to show all of the lines, valves, tanks, and redundant elements in even a simple power-generating system. There are numerous ways of combining the devices described in this chapter into a working power-generating system. However, Fig. 26.1(a) illustrates the main elements and interconnections for *unit operation* (i.e., in a typical drum boiler unit).

The integration of elements in high-pressure *once-through boilers* is somewhat different, as Fig. 26.1(b) illustrates. (The term "once-through" does not mean the steam is discarded after use. It means that the steam is heated along one long progressive transfer path.) Once-through boilers do not have steam drums. The steam flow rate is controlled by the boilerfeed pump. At start-up and at low load, the steam generator produces more steam than the turbine requires. At low loads, superheated steam is passed to a flash tank where it becomes available for feedwater heating, or it is passed directly to the condenser and returned to the boiler. This lets the flash tank act like a drum in a drum boiler unit. Valving and valve sequencing is more complex with once-through operation.

Power plant output may be qualified by the terms "thermal," "electrical," "gross," or "net." For example, a 1,000,000 Btu/hr *thermal* (i.e, "1 MBth" or "MBt") power plant indicates the energy transfer to the feedwater. A 1000 MW *electrical* (i.e., "1000 MWe") power plant refers to the generator output. Similarly, "MWt" refers to thermal power in megawatts.

Some of the steam and some of the generated electrical power are used to drive auxiliary devices (motors, fans, soot-blowers, pumps, etc.). The *gross electrical output* is

**Figure 26.1** *Typical Integration of Power-Generating Elements*

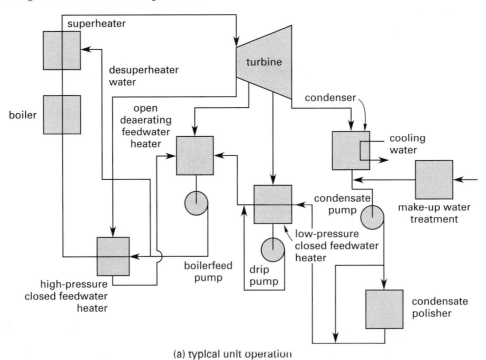

(a) typical unit operation

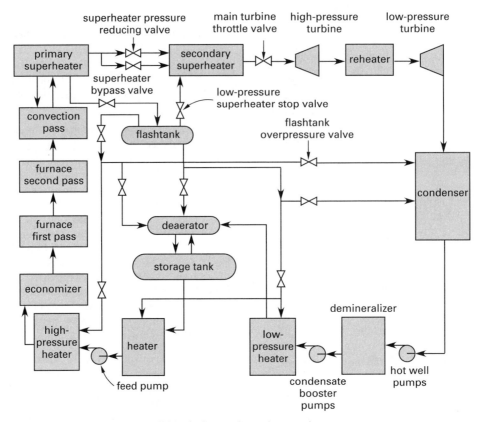

(b) typical once-through operation
(bleeds not shown)

the power before the auxiliary loads have been removed; the *net electrical output* is after.

## 2. FURNACES

Most modern power-generating plants burn coal; some burn natural gas, and a few burn oil. In the past, solid fuel furnaces have been dominated by "pile burning" on grates, and more recently, by "suspension burning" of pulverized coal. Current fluidized-bed combustion offers additional benefits (particularly environmental) over both traditional methods.

The amount of air introduced into the combustion chamber affects the completeness of combustion, pollutant production, and furnace efficiency. For efficient combustion, air may be introduced at more than one point along the combustion path. *Primary air* is introduced first, followed by *secondary air* and *tertiary air*.

A high turndown ratio is desirable for furnaces that are frequently started cold.[1] For gaseous fuels, the *turndown ratio* is the ratio of the maximum-to-minimum fuel-to-air ratios over which the burner will operate satisfactorily. The maximum turndown ratio is limited by the flame velocity. *Flame blow-off* results when a fuel-air mixture is injected too fast. *Flash-back* occurs when the flame velocity exceeds the mixture velocity.

In older lump-coal furnaces, coal was introduced to the furnace by a *stoker*. Stoking is the act of adding and distributing coal. In low-volume home furnaces and early fire-tube boilers, stoking was accomplished by hand shoveling. Stoking in lump-coal plants and in modern refuse-burning plants is performed mechanically and automatically.

Figure 26.2 illustrates several types of ram stokers for lump coal and refuse. Fuel can also be brought into the furnace by screws and sprinklers. With *overfeed stoking*, coal is placed above the air flow. With *underfeed stoking (retort stoking)*, coal is pushed into position under the air flow by plungers. Air enters the combustion chamber through ports in the furnace walls known as *tuyères*. The combustion area holding burning coal is known as the *retort*.

Coal in all modern power-generating plants is pulverized to some extent prior to use. In some cases, coal is ground into a fine or microfine powder.[2] The finely ground coal is suspended in a gaseous atmosphere while burning. There are two general ways of feeding *pulverized-coal furnaces*. The *bin system (central system)* stores dried

and pulverized coal for later use.[3] Air transport, pressure pulse, and screw conveyor systems can be used to transfer pulverized coal to the furnace as needed.[4] *Direct-firing systems (unit systems)* pulverize coal on demand. Whether the bin or direct-firing system is used depends on the type of coal and the reliability of the pulverizing and feed equipment.[5]

**Figure 26.2** Stokers

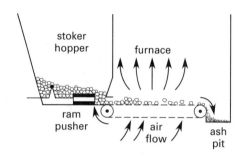

(a) overfeed traveling bed grate
with ram stoker

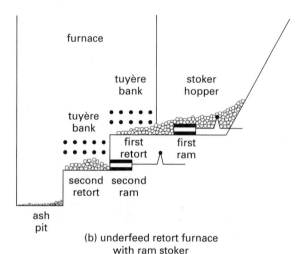

(b) underfeed retort furnace
with ram stoker

High-sulfur, high-ash coals, "scrap" fuels, and toxic materials can be burned in *fluidized-bed furnaces*. Solid fuel is turned into a turbulent, fluid-like mass by mixing it with a bed material and blowing air through the mixture at a controlled rate. The bed material can be any solid substance that is not consumed. Sand and limestone are the most common options. The mixture of fuel and bed material becomes fluidized and assumes free-flowing properties when the air flow is at the *fluidizing velocity*.

*Fluidized-bed combustion*, FBC, using limestone as the bed material considerably reduces some pollutants,

---

[1]The turndown ratio is not of much interest for continuously fired furnaces.

[2]Generally, approximately 65% to 70% of the pulverized coal will pass through a 200-mesh (74-micron) sieve, which has 200 openings per inch. The nominal aperture for 200 mesh is 0.0029 in (0.074 mm), which is like talcum powder. Anthracitic coals can be ground so that up to 90% passes through a 200 mesh sieve. *Micronized coal* is even finer: 80% to 90% will pass through a 325-mesh sieve (43 microns).

[3]An *eductor* is a special type of jet pump that uses air at 20 psig to 80 psig (140 kPa to 550 kPa) to withdraw pulverized coal from an unpressurized storage hopper, transport it in dense phase through a small-diameter feed line, and inject it into the burner. The coal flow rate is controlled by the air pressure. A control valve is not needed.

[4]Ground coal can also be carried for longer distances in slurry form.

[5]Reliability of pulverizers is not the problem it was when pulverizing was first introduced. Pulverizing on demand is now the norm.

reducing the need for expensive air-pollution control equipment. Approximately 90% of the sulfur is absorbed by the limestone, forming calcium sulfate. Combustion takes place rapidly at 1500°F to 1750°F (815°C to 950°C), about 400°F (220°C) lower than conventional boilers. This is below the point at which nitrogen oxides are formed.

## 3. STEAM GENERATORS

A *steam generator* is a combination of a furnace and a boiler. The *boiler* is constructed so that the combustion heat is transferred to feedwater. Modern boilers are *water-tube boilers* (i.e., water passes through tubes surrounded by combustion products).[6] Some of the tubes may be surface mounted or embedded in the walls of the furnace (known as the *setting* or "brickwork").

Other tubes may be placed across the path of the combustion gases. In the typical *cross-drum boiler*, the combustion gases flow perpendicular to the water tubes. Tubes may be straight or bent (i.e., *straight-tube boilers* or *bent-tube boilers*). Steam accumulates in one or more headers at the top of the boiler (known as the *steam drums*).[7] (See Fig. 26.3.)

***Figure 26.3*** *Cross-Drum, Straight-Tube, Water-Tube Steam Generator*

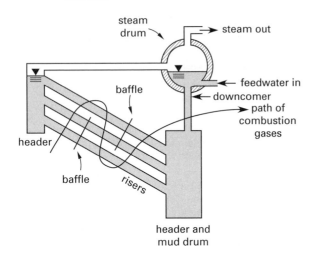

Circulation of water through the boiler can occur in one of four ways. For pressures below approximately 2800 psi (19.3 MPa), *natural-circulation boilers* can be used. Density is the force driving circulation up the *riser* to the steam drum and back through the *downcomer* to the water-supply header or mud drum. With *controlled-circulation boilers*, common between 2400 psi and 2800

psi (16.6 MPa and 19.3 MPa), a pump keeps the water circulating between the steam drum and water-supply header. In *once-through boilers* (with moisture separators but no steam drums) and *supercritical-pressure boilers* (with neither steam drums nor separators) that operate above 3200 psi (22.1 MPa) and 1000°F (540°C), water circulates by use of the boilerfeed pump and must be ultrapure to prevent solids build-up.[8] Supercritical boilers are also known as *Benson boilers*. Ultrasupercritical (USC) boilers routinely operate at 1112°F (600°C) and 4420 psig (30.5 MPa), while demonstration units have operated as high as 1400°F (760°C) and 6090 psig (42 MPa).

Capacities of steam generators are given in mass of steam per hour. This is not an exact determination of the thermodynamic output, of course, unless the conditions of the entering water and leaving steam are also specified. The net energy input to a steam generator, known as the *heat absorption*, is given by Eq. 26.1.[9]

$$Q_{\text{in,net}} = \dot{m}_{\text{steam}}(h_{\text{steam}} - h_{\text{feed}}) \qquad 26.1$$

Not all of the combustion energy released in a steam generator will be transferred to the water. The *boiler efficiency*, which ranges from 75% to 90%, is calculated from Eq. 26.2.

$$\eta_{\text{boiler}} = \frac{Q_{\text{in,net}}}{Q_{\text{in,gross}}}$$

$$= \frac{\dot{m}_{\text{steam}}(h_{\text{steam}} - h_{\text{feed}})}{\dot{m}_{\text{fuel}}\text{HV}} \qquad 26.2$$

The steam mass flow rate in Eq. 26.2 will be as high as it can be at *maximum capacity*. At *normal capacity*, the boiler efficiency will be as high as it can be.

There are several obsolete units that are occasionally encountered. These units compare energy and power on the basis of steam produced at one atmospheric pressure. One *boiler horsepower* is equal to 33,479 Btu/hr (9.81 kW). The *factor of evaporation* is defined by Eq. 26.3. $h_{fg,1\,\text{atm}}$ has a value of 970.3 Btu/lbm (2257 MJ/kg).

$$\text{factor of evaporation} = \frac{h_{\text{steam}} - h_{\text{feed}}}{h_{fg,1\,\text{atm}}} \qquad 26.3$$

---

[6]*Fire-tube boilers*, in which the combustion products pass through small-diameter tubes and transfer heat to a surrounding water jacket, are seldom used today because they are limited to low-pressure steam. The maximum operating pressure for fire-tube boilers is approximately 150 psi (1 MPa) for riveted construction and 400 psia (3 MPa) for welded construction.

[7]A drum at the bottom of the boiler for collecting sediment is called a *mud drum*.

[8]In the not-too-distant past, commercial operating temperatures have been restricted to about 1050°F (570°C) by metallurgical considerations. Similarly, commercial pressures were limited to about 3500 psig (24.1 MPa). These limitations no longer restrict supercritical and ultracritical boilers.

[9]In the United States, various confusing and ambiguous abbreviations for units of heat absorption have been used, including kB, kBtu, kBH, kB/hr, and MBH (or MBh) (1000 Btus per hour); and, mB, MB, mBtu, MBtu, MMBH, and MB/hr (1,000,000 Btus per hour). In some particularly unfortunate cases, "MB" is used to mean 1000 Btus (no hour rate), and "MMB" is used to mean 1,000,000 Btus (no hour rate). The actual meaning often has to be determined from the context.

The *equivalent evaporation* is

$$\text{equivalent evaporation} = \frac{\dot{m}_{\text{steam}}(h_{\text{steam}} - h_{\text{feed}})}{h_{fg,1 \text{ atm}}} \qquad 26.4$$

For an electrical generating system, the *heat rate* (*station rate*, *plant heat rate*, etc.) in Btu/kW-hr (kJ/kW·s) is defined as the total energy input to the steam generator divided by the electrical energy output. Typical full-throttle values in modern, large coal-fired plants are around 7500–8500 Btu/kW-hr (2.2–2.5 kJ/kW·s), with the lower values being more efficient. Values for partial throttle operation, and values for older plants, are higher. If the output is mechanical, the heat rate is the total energy input divided by the horsepower output.

Water passing through the boiler experiences a small pressure drop due to the friction and expansion. This pressure drop slightly increases the power to pump the water. Calculations are typically based on conditions at the boiler outlet. The error due to neglecting the pressure drop through the boiler is negligible.

## 4. BLOWDOWN

As water is evaporated in the boiler, dissolved and suspended solids are left behind. To prevent these solids from accumulating and causing fouling, constriction, and corrosion, some of the boiler water is bled off, a process known as *blowdown* (or *blowoff*). Blowdown may be intermittent or continuous. The blowdown rate is easily determined from the circulation rate (steam or water) and actual and permitted concentrations. With intermittent blowdown, the mass of water to be released after a period, $\Delta t$, of accumulation is given by Eq. 26.5. Concentrations of total solids, $C$, are typically given in ppm or mg/L.

$$m_{\text{blowdown}} = \left( \frac{C_{\text{feedwater}}}{C_{\text{maximum limit}} - C_{\text{feedwater}}} \right) \Delta t \dot{m}_{\text{steam}}$$
$$26.5$$

For continuous blowdown, the blowdown rate is

$$\dot{m}_{\text{blowdown}} = \left( \frac{C_{\text{feedwater}}}{C_{\text{maximum limit}} - C_{\text{feedwater}}} \right) \dot{m}_{\text{steam}}$$
$$26.6$$

Automatic (unattended) blowdown systems work in conjunction with total dissolved solids (TDS) monitors that continuously evaluate the ionic conductivity of the water. Sodium is completely soluble and is the most common ion in boiler water. Automatic blowdown systems operate by maintaining a preselected conductivity set-point in the range of 2400–2800 $\mu$S.

### Example 26.1

The total solids concentration in a boiler may not exceed 2500 mg/L. The concentration of total solids in the make-up water is 175 mg/L. The concentration of

total solids in the condensate return is 25 mg/L. The boiler produces steam at the rate of 20,000 lbm/hr (2.5 kg/s), of which 5000 lbm/hr (0.63 kg/s) are used for process heating elsewhere. What should be the rate of continuous blowdown?

*SI Solution*

Assume a blowdown rate of 0.06 kg/s. Then, the mass of the make-up water will be

$$\dot{m}_{\text{make-up water}} = 0.63 \,\frac{\text{kg}}{\text{s}} + 0.06 \,\frac{\text{kg}}{\text{s}} = 0.69 \text{ kg/s}$$

The rate of condensate return is

$$\dot{m}_{\text{condensate return}} = 2.5 \,\frac{\text{kg}}{\text{s}} - 0.63 \,\frac{\text{kg}}{\text{s}} = 1.87 \text{ kg/s}$$

The concentration of total solids in the boiler feedwater is a weighted average of the concentration of the make-up water and the condensate return.

$$C_{\text{feedwater}} = \frac{\begin{array}{c} \dot{m}_{\text{make-up water}} C_{\text{make-up water}} \\ + \dot{m}_{\text{condensate return}} C_{\text{condensate return}} \end{array}}{\dot{m}_{\text{make-up water}} + \dot{m}_{\text{condensate return}}}$$

$$= \frac{\left( 0.69 \,\frac{\text{kg}}{\text{s}} \right) \left( 175 \,\frac{\text{mg}}{\text{L}} \right) + \left( 1.87 \,\frac{\text{kg}}{\text{s}} \right) \left( 25 \,\frac{\text{mg}}{\text{L}} \right)}{0.69 \,\frac{\text{kg}}{\text{s}} + 1.87 \,\frac{\text{kg}}{\text{s}}}$$

$$= 65.4 \text{ mg/L}$$

From Eq. 26.6,

$$\dot{m}_{\text{blowdown}} = \left( \frac{C_{\text{feedwater}}}{C_{\text{maximum limit}} - C_{\text{feedwater}}} \right) \dot{m}_{\text{steam}}$$

$$= \left( \frac{65.4 \,\frac{\text{mg}}{\text{L}}}{2500 \,\frac{\text{mg}}{\text{L}} - 65.4 \,\frac{\text{mg}}{\text{L}}} \right) \left( 2.5 \,\frac{\text{kg}}{\text{s}} \right)$$

$$= 0.067 \text{ kg/s}$$

If necessary, perform a second iteration with this new blowdown rate to derive a more precise value.

*Customary U.S. Solution*

Assume a blowdown rate of 500 lbm/hr. Then, the mass of the make-up water will be

$$\dot{m}_{\text{make-up water}} = 5000 \,\frac{\text{lbm}}{\text{hr}} + 500 \,\frac{\text{lbm}}{\text{hr}} = 5500 \text{ lbm/hr}$$

The rate of condensate return is

$$\dot{m}_{\text{condensate return}} = 20{,}000 \,\frac{\text{lbm}}{\text{hr}} - 5000 \,\frac{\text{lbm}}{\text{hr}}$$

$$= 15{,}000 \text{ lbm/hr}$$

The concentration of total solids in the boiler feedwater is a weighted average of the concentration of the make-up water and the condensate return.

$$C_{\text{feedwater}} = \frac{\dot{m}_{\text{make-up water}} C_{\text{make-up water}} + \dot{m}_{\text{condensate return}} C_{\text{condensate return}}}{\dot{m}_{\text{make-up water}} + \dot{m}_{\text{condensate return}}}$$

$$= \frac{\left(5500 \frac{\text{lbm}}{\text{hr}}\right)\left(175 \frac{\text{mg}}{\text{L}}\right) + \left(15{,}000 \frac{\text{lbm}}{\text{hr}}\right)\left(25 \frac{\text{mg}}{\text{L}}\right)}{5500 \frac{\text{lbm}}{\text{hr}} + 15{,}000 \frac{\text{lbm}}{\text{hr}}}$$

$$= 65.2 \text{ mg/L}$$

From Eq. 26.6,

$$\dot{m}_{\text{blowdown}} = \left(\frac{C_{\text{feedwater}}}{C_{\text{maximum limit}} - C_{\text{feedwater}}}\right)\dot{m}_{\text{steam}}$$

$$= \left(\frac{65.2 \frac{\text{mg}}{\text{L}}}{2500 \frac{\text{mg}}{\text{L}} - 65.2 \frac{\text{mg}}{\text{L}}}\right)\left(20{,}000 \frac{\text{lbm}}{\text{hr}}\right)$$

$$= 536 \text{ lbm/hr}$$

If necessary, perform a second iteration with this new blowdown rate to derive a more precise value.

## 5. PUMPS

Centrifugal pumps are used in power-generation plants. Positive displacement pumps have become nearly obsolete for this application.[10]

The purpose of a pump is to increase the total energy content of the fluid flowing through it. Pumps can be considered adiabatic devices because the fluid gains (or loses) very little heat during the short time it passes through them. If the inlet and outlet are the same size and at the same elevation, the kinetic and potential energy terms can be neglected.[11] Then, the steady flow energy equation, SFEE, reduces to Eq. 26.7. (The second form of Eq. 26.7 is applicable to incompressible liquids only.)

$$W_{\text{pump}} = m(h_2 - h_1) = mv(p_2 - p_1) \qquad 26.7$$

$$P_{\text{pump}} = \dot{m}(h_2 - h_1) \qquad 26.8$$

Equation 26.7 assumes that the pump is capable of isentropic compression. However, due to inefficiencies, some of the input energy is converted to heat. This heat energy raises the temperature (and hence, the internal

[10]Obsolescence is due to the difficulty in controlling reciprocating pumps during start-up and partial-load. With modern variable-frequency drives, however, this may no longer be the case.
[11]Even if the pump inlet and outlet are different sizes and at different elevations, the kinetic and potential energy changes are small compared to the pressure energy increase.

energy portion of enthalpy) without increasing the pressure. The actual exit enthalpy, $h_2'$, takes the inefficiency into consideration.

$$h_2' = h_1 + \frac{h_2 - h_1}{\eta_s} \qquad 26.9$$

Combining Eq. 26.8 and Eq. 26.9, the actual input pump power is

$$P_{\text{pump}}' = \dot{m}(h_2' - h_1)$$
$$= \frac{\dot{m}(h_2 - h_1)}{\eta_s \eta_m} \qquad 26.10$$

The ideal exit enthalpy, $h_2$, can be calculated from the specific volume and the exit pressure, $p_2$. The calculation is simplified if the fluid is assumed to be incompressible. (Equation 26.11 cannot be used with compression of gases and vapors.)

$$h_2 = h_1 + v(p_2 - p_1)\Big|_{\substack{\text{isentropic}\\\text{incompressible}}} \quad \text{[SI]} \quad 26.11(a)$$

$$h_2 = h_1 + \frac{v(p_2 - p_1)}{J}\Big|_{\substack{\text{isentropic}\\\text{incompressible}}} \quad \text{[U.S.]} \quad 26.11(b)$$

## 6. PUMP EFFICIENCY

Line 1-to-2 in Fig. 26.4 illustrates the ideal condition line for a pump. Since the compression is isentropic, the line is directed vertically upward. The actual work, however, results in an increase in entropy. Line 1-to-2' is the actual condition line for non-isentropic compression.

**Figure 26.4** *Compression by a Pump*

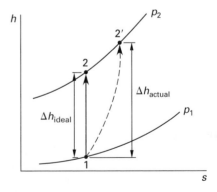

The ideal and actual work for a pump are

$$W_{\text{ideal}} = m(h_2 - h_1) \qquad 26.12$$

$$W_{\text{actual}} = m(h_2' - h_1) \qquad 26.13$$

The definition of *isentropic efficiency*, $\eta_s$, for a pump is the inverse of what it is for a turbine.

$$\eta_s = \frac{W_{\text{ideal}}}{W_{\text{actual}}}$$
$$= \frac{h_2 - h_1}{h_2' - h_1} \qquad 26.14$$

Equation 26.15 defines the pump efficiency. Since $W_{\text{ideal}}$ is small—less than 5 Btu/lbm (10 kJ/kg)—and does not affect the thermal efficiency greatly, the pump mechanical efficiency is often ignored, and $\eta_{\text{pump}}$ is taken as $\eta_s$.

$$\eta_{\text{pump}} = \eta_s \eta_m \qquad 26.15$$

## 7. TEMPERATURE INCREASE IN PRESSURIZED LIQUIDS

The temperature of water (or any other liquid) increases when it is pressurized (i.e., passes through a pump). However, the increase in temperature is very small—far less than what would occur with a gas or vapor.

The increase in temperature can be thought of as being caused by two processes: an isentropic compression and an irreversible compression (if the pump is not 100% efficient). The increase due to isentropic compression is so small as to be negligible. For example, the increase in temperature of 100°F (38°C) water compressed isentropically from saturation to 1000 psi (6.9 MPa) is approximately 0.3°F (0.17°C).

The irreversible portion of the process incorporates the additional energy that must be put into the water to overcome friction and turbulence. These factors produce a profound heating effect (as compared to the isentropic portion). It is proper to consider only this irreversible work when calculating the increase in water temperature. This is because the change in enthalpy is a combination of changes in internal energy and flow work (i.e., $h = v + pV$), but only the internal energy change manifests as a temperature change.

### Example 26.2

A boilerfeed pump increases the pressure of 90°F (30°C) water from 1 atm to 150 psia (1000 kPa). The pump's isentropic efficiency is 80%. The water's specific heat is 1.0 Btu/lbm-°F (4.187 kJ/kg·K). What is the temperature of the water leaving the pump?

*SI Solution*

Since the properties of a liquid are essentially independent of pressure, the properties of 30°C water can be read from a saturated steam table.

$$h_{30°C} = 125.73 \text{ kJ/kg}$$

$$v_{30°C} = 1.0044 \text{ cm}^3/\text{g} = 0.0010044 \text{ m}^3/\text{kg}$$

The ideal enthalpy of the feedwater entering the boiler (point 2 on Fig. 26.4) is equal to the enthalpy at point 1 plus the energy put into the water by the pump. Assuming the water is incompressible, the specific volumes at points 1 and 2 are the same. From Eq. 26.11,

$$h_2 = h_1 + v_1(p_2 - p_1)$$
$$= 125.73 \text{ kJ/kg}$$
$$+ \frac{\left(0.0010044 \ \frac{\text{m}^3}{\text{kg}}\right)(1000 \text{ kPa} - 101.3 \text{ kPa})}{1000 \ \frac{\text{J}}{\text{kJ}}}$$
$$\times \left(1000 \ \frac{\text{Pa}}{\text{kPa}}\right)$$
$$= 125.73 \ \frac{\text{kJ}}{\text{kg}} + 0.903 \ \frac{\text{kJ}}{\text{kg}} = 126.63 \text{ kJ/kg}$$

This calculation assumes that the pump is capable of isentropic compression. Because of the pump's inefficiency, not all of the 0.903 kJ/kg goes into raising the pressure. Some of the energy goes into raising the temperature. (Since $h = u + pv$, both energy contributions increase the enthalpy.) Therefore, to get to 1000 kPa, more than 0.903 kJ/kg must be added to the water. The actual enthalpy at point 2′ is

$$h_2' = h_1 + \frac{h_2 - h_1}{\eta_s}$$
$$= 125.73 \ \frac{\text{kJ}}{\text{kg}} + \frac{0.903 \ \frac{\text{kJ}}{\text{kg}}}{0.80}$$
$$= 125.73 \ \frac{\text{kJ}}{\text{kg}} + 1.13 \ \frac{\text{kJ}}{\text{kg}} = 126.86 \text{ kJ/kg}$$

The enthalpy was increased by 1.13 kJ/kg, but only the irreversible portion is considered when determining the temperature change.

$$h_{\text{irreversible}} = 1.13 \ \frac{\text{kJ}}{\text{kg}} - 0.903 \ \frac{\text{kJ}}{\text{kg}}$$
$$= 0.227 \text{ kJ/kg}$$

Since the specific heat is 4.187 kJ/kg·K, the final temperature is

$$T_2' = T_1 + \frac{\Delta h}{c_p} = 30°C + \frac{0.227 \ \frac{\text{kJ}}{\text{kg}}}{4.187 \ \frac{\text{kJ}}{\text{kg·K}}}$$
$$= 30.05°C$$

**Thermodynamics**

*Customary U.S. Solution*

Since the properties of a liquid are essentially independent of pressure, the properties of 90°F water can be read from a saturated steam table.

$$h_{90°F} = 58.05 \text{ Btu/lbm}$$

$$v_{90°F} = 0.01610 \text{ ft}^3/\text{lbm}$$

The ideal enthalpy of the feedwater entering the boiler (point 2 on Fig. 26.4) is equal to the enthalpy at point 1 plus the energy put into the water by the pump. Assuming the water is incompressible, the specific volumes at points 1 and 2 are the same. From Eq. 26.11,

$$h_2 = h_1 + \frac{v_1(p_2 - p_1)}{J}$$

$$= 58.05 \text{ Btu/lbm}$$

$$+ \frac{\left(0.01610 \frac{\text{ft}^3}{\text{lbm}}\right)\left(150 \frac{\text{lbf}}{\text{in}^2} - 14.7 \frac{\text{lbf}}{\text{in}^2}\right)\left(12 \frac{\text{in}}{\text{ft}}\right)^2}{778 \frac{\text{ft-lbf}}{\text{Btu}}}$$

$$= 58.05 \frac{\text{Btu}}{\text{lbm}} + 0.403 \frac{\text{Btu}}{\text{lbm}} = 58.45 \text{ Btu/lbm}$$

This calculation assumes that the pump is capable of isentropic compression. Because of the pump's inefficiency, not all of the 0.403 Btu/lbm goes into raising the pressure. Some of the energy goes into raising the temperature. (Since $h = u + pv$, both energy contributions increase the enthalpy.) Therefore, to get to 150 psia, more than 0.403 Btu/lbm must be added to the water. The actual enthalpy at point 2' is

$$h_2' = h_1 + \frac{h_2 - h_1}{\eta_s}$$

$$= 58.05 \frac{\text{Btu}}{\text{lbm}} + \frac{0.403 \frac{\text{Btu}}{\text{lbm}}}{0.80}$$

$$= 58.05 \frac{\text{Btu}}{\text{lbm}} + 0.5 \frac{\text{Btu}}{\text{lbm}} = 58.55 \text{ Btu/lbm}$$

The enthalpy was increased by 0.5 Btu/lbm, but only the irreversible portion is considered when determining the temperature change.

$$\Delta h_{\text{irreversible}} = 0.5 \frac{\text{Btu}}{\text{lbm}} - 0.403 \frac{\text{Btu}}{\text{lbm}}$$

$$= 0.907 \text{ Btu/lbm}$$

Since the specific heat is 1.0 Btu/lbm-°F, the corresponding temperature increase is 0.5°F. The final temperature is 90°F + 0.0907°F ≈ 90.1°F.

## 8. TURBINES

There are two general categories of steam turbine operation. A *reaction turbine* consists of a rotating drum with small nozzles (reaction jets) located around the drum's periphery. Steam is discharged from the nozzles, and the drum turns in reaction to the steam's action. An *impulse turbine* is characterized by stationary jets discharging against vanes mounted on the periphery of a wheel. Modern steam turbines use both principles to extract energy from the steam.

There are several designations given to turbines. A *back pressure turbine* (*topping turbine* or *superposed turbine*) exhausts either to a high-pressure industrial process or to a second turbine operating in a lower-pressure range.[12] In a two-turbine installation, the first is designated as the *high-pressure (HP) turbine*, and the second is known as the *low-pressure (LP) turbine*. (If there are three turbines in series, the middle one is known as an *intermediate-pressure (IP) turbine*.) Turbines that operate at supercritical conditions are known as *super-critical-pressure (SP) turbines*.

Some of the steam entering a turbine may be removed before it has expanded to the turbine exit pressure. This removal is known as a *bleed*, and the steam is known as *bleed steam*. The *extraction rate* or *bleed rate* is the rate, usually expressed in lbm/hr (kg/h) or in Btu/hr (kW), at which the partially expanded steam is bled off. A turbine with one or more bleeds is known as an *extraction turbine*.

In a *reheating turbine*, steam is removed from the turbine, returned to its original temperature in a reheater, and returned to a lower-pressure portion of the same turbine.

## 9. TURBINE WORK, POWER, AND EFFICIENCY

Turbines can generally be thought of as pumps operating in reverse. A turbine extracts energy from the fluid which, in turn, decreases in temperature and pressure. The process is essentially adiabatic because the fluid loses very little heat during the short time it passes through the turbine.

Figure 26.5 diagrams the expansion of steam in a turbine from a high pressure, $p_1$, to a low pressure, $p_2$. The broken line is the *condition line*—the locus of all states of steam during the expansion process. The ideal condition line is isentropic, which is represented by a vertical line downward on the Mollier ($h$ versus $s$) diagram.

Changes in potential and kinetic energies are small and can be neglected. The ideal steady flow energy equation, SFEE, reduces to Eq. 26.16 and predicts the work output of the turbine. The work is known as *shaft work*

---

[12]Topping turbines are added when older, low-pressure plants are "repowered." The furnace and steam generator are replaced so that high-pressure steam is produced, and a topping turbine is added. The topping turbine uses the high pressure steam. The original low-pressure turbine uses the exhaust steam from the topping turbine. Most of the remaining equipment in the cycle is unchanged.

**Figure 26.5** *Single-Stage Turbine Expansion*

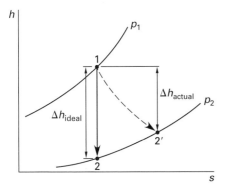

because it is transmitted from the turbine to a generator through a shaft.[13]

$$W_{\text{ideal}} = m(h_1 - h_2) \qquad 26.16$$

Equation 26.16 assumes that the turbine is capable of isentropic expansion. However, not all of the available fluid energy can be extracted. Friction within the turbine increases the entropy without decreasing the pressure. The actual exit enthalpy, $h_2'$, takes this inefficiency into consideration. If the expansion process is not isentropic, entropy will increase and the actual final enthalpy, $h_2'$, will be higher than the ideal final enthalpy, $h_2$.

The *isentropic efficiency (adiabatic efficiency)*, $\eta_s$, of a turbine is the ratio of actual to ideal energy extractions.[14] Actual isentropic efficiencies vary from approximately 65% for 1 MW unit to over 80% for 100 MW units.

$$\eta_s = \frac{W_{\text{actual}}}{W_{\text{ideal}}} = \frac{h_1 - h_2'}{h_1 - h_2} \qquad 26.17$$

The actual energy extracted is

$$W_{\text{actual}} = m(h_1 - h_2') = m\eta_s(h_1 - h_2) \qquad 26.18$$

$$h_2' = h_1 - \eta_s(h_1 - h_2) \qquad 26.19$$

In addition to thermodynamic friction losses, there are additional friction losses in bearings and other moving mechanical parts. The net turbine work is

$$W_{\text{turbine}} = W_{\text{actual}} - W_{\text{friction}}$$
$$= \eta_m W_{\text{actual}} \qquad 26.20$$

Most steam turbines have high mechanical efficiencies—in the order of 98%. Inasmuch as the friction losses are very small, the isentropic and turbine efficiencies are

essentially identical. However, the *overall turbine efficiency* incorporates the mechanical friction losses.

$$\eta_{\text{turbine}} = \eta_s \eta_m \qquad 26.21$$

The isentropic efficiency of a device affects both the flow rate and the thermodynamic properties of the substance flowing through the device. The mechanical efficiency affects the flow rate, but it does not affect the thermodynamic properties and should not be used to determine final enthalpies. Both the isentropic and mechanical efficiencies reduce the amount of useful energy that can be generated in a turbine, and they affect the amount of substance flowing through the turbine. The actual power generated in the turbine is

$$P_{\text{turbine}} = \dot{m}_{\text{actual}} \eta_{\text{turbine}}(h_1 - h_2) \qquad 26.22$$

$$\dot{m}_{\text{actual}} = \frac{\dot{m}_{\text{ideal}}}{\eta_{\text{turbine}}}$$
$$= \frac{\dot{m}_{\text{ideal}}}{\eta_s \eta_m} \qquad 26.23$$

## 10. TWO-STAGE EXPANSION

If expansion through the turbine is multiple stage, or if the turbine has a bleed at an intermediate pressure, $p_m$, the expansion process for each stage will be as illustrated by Fig. 26.6. In this figure, a fraction $y$ of the original steam mass is removed after expanding to pressure $p_m$. The remaining fraction $1 - y$ expands to pressure $p_2$.

**Figure 26.6** *Two-Stage Turbine Expansion*

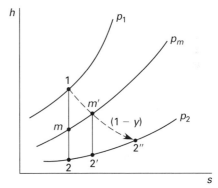

The ideal and actual work generated in the first stage are

$$W_{\text{ideal},1} = m(h_1 - h_m) \qquad 26.24$$

$$W_{\text{actual},1} = m(h_1 - h_m')$$
$$= \eta_{s,1} m(h_1 - h_m) \qquad 26.25$$

---

[13]Other names for *shaft work* are *brake work*, *useful work*, and *net work*.
[14]The terminology "adiabatic efficiency" is ambiguous because an "adiabatic system" does not require a 100% efficient turbine.

The ideal and actual work generated by the remaining fraction, $1 - y$, in the second stage are

$$W_{\text{ideal},2} = (1 - y)m(h'_m - h'_2) \qquad 26.26$$

$$W_{\text{actual},2} = (1 - y)m(h'_m - h''_2)$$
$$= (1 - y)m\eta_{s,2}(h'_m - h'_2) \qquad 26.27$$

The total work done per unit mass of the steam is

$$W_{\text{turbine}} = W_{\text{actual},1} + W_{\text{actual},2}$$
$$= m(h_1 - h'_m + (1 - y)(h'_m - h''_2))$$
$$= m(\eta_{s,1}(h_1 - h_m) + (1 - y)\eta_{s,2}(h'_m - h'_2))$$
$$26.28$$

## Example 26.3

Steam is expanded from 700°F (360°C) and 200 psia (1500 kPa) to 5 psia (50 kPa) in an 87% efficient turbine. What is the final enthalpy of the steam?

*SI Solution*

Refer to Fig. 26.5. From the superheated steam tables,

$$h_1 = 3169.8 \text{ kJ/kg}$$
$$s_1 = 7.1382 \text{ kJ/kg·K}$$

At this point, proceed as if the turbine is 100% efficient. From the saturated steam tables for 50 kPa,

$$s_f = 1.0912 \text{ kJ/kg·K}$$
$$s_g = 7.5939 \text{ kJ/kg·K}$$
$$h_f = 340.54 \text{ kJ/kg}$$
$$h_{fg} = 2304.7 \text{ kJ/kg}$$

Since at this point it is assumed that the expansion is isentropic (i.e., 100% efficient), $s_2 = s_1$. The quality at point 2 can be found as

$$x = \frac{s_2 - s_f}{s_g - s_f} = \frac{7.1382 \dfrac{\text{kJ}}{\text{kg·K}} - 1.0912 \dfrac{\text{kJ}}{\text{kg·K}}}{7.5939 \dfrac{\text{kJ}}{\text{kg·K}} - 1.0912 \dfrac{\text{kJ}}{\text{kg·K}}}$$
$$= 0.9299$$

The ideal final enthalpy can be found from the quality.

$$h_2 = h_f + x h_{fg}$$
$$= 340.54 \frac{\text{kJ}}{\text{kg}} + (0.9299)\left(2304.7 \frac{\text{kJ}}{\text{kg}}\right)$$
$$= 2483.7 \text{ kJ/kg}$$

However, this value of $h_2$ assumes the expansion through the turbine is isentropic. Since the turbine is capable of extracting only 87% of the ideal energy, the actual final enthalpy is

$$h'_2 = h_1 - \eta_s(h_1 - h_2)$$
$$= 3169.8 \frac{\text{kJ}}{\text{kg}} - (0.87)\left(3169.8 \frac{\text{kJ}}{\text{kg}} - 2483.7 \frac{\text{kJ}}{\text{kg}}\right)$$
$$= 2572.9 \text{ kJ/kg}$$

*Customary U.S. Solution*

Refer to Fig. 26.5. From the superheated steam tables,

$$h_1 = 1374.1 \text{ Btu/lbm}$$
$$s_1 = 1.7238 \text{ Btu/lbm-°F}$$

At this point, proceed as though the turbine is 100% efficient. From the saturated steam tables for 5 psia,

$$s_f = 0.2349 \text{ Btu/lbm-°F}$$
$$s_{fg} = 1.6089 \text{ Btu/lbm-°F}$$
$$h_f = 130.2 \text{ Btu/lbm}$$
$$h_{fg} = 1000.5 \text{ Btu/lbm}$$

Since it is assumed that the expansion is isentropic (i.e., 100% efficient), $s_2 = s_1$. The quality at point 2 can be found as

$$x = \frac{s_2 - s_f}{s_{fg}} = \frac{1.7238 \dfrac{\text{Btu}}{\text{lbm-°F}} - 0.2349 \dfrac{\text{Btu}}{\text{lbm-°F}}}{1.6089 \dfrac{\text{Btu}}{\text{lbm-°F}}}$$
$$= 0.9254$$

The ideal final enthalpy can be found from the quality.

$$h_2 = h_f + x h_{fg} = 130.2 \frac{\text{Btu}}{\text{lbm}} + (0.9254)\left(1000.5 \frac{\text{Btu}}{\text{lbm}}\right)$$
$$= 1056.1 \text{ Btu/lbm}$$

However, this value of $h_2$ assumes the expansion through the turbine is isentropic. Since the turbine is capable of extracting only 87% of the ideal energy, the actual final enthalpy is

$$h'_2 = h_1 - \eta_s(h_1 - h_2)$$
$$= 1374.1 \frac{\text{Btu}}{\text{lbm}} - (0.87)\left(1374.1 \frac{\text{Btu}}{\text{lbm}} - 1056.1 \frac{\text{Btu}}{\text{lbm}}\right)$$
$$= 1097.4 \text{ Btu/lbm}$$

## Example 26.4

Repeat Ex. 27.3 using the Mollier ($h$ versus $s$) diagram.

*SI Solution*

$h_1$ is read directly from the Mollier diagram at the intersection of 360°C and 1500 kPa. (Greater accuracy is possible with a large Mollier diagram.)

$$h_1 \approx 3170 \text{ kJ/kg}$$

The ideal final enthalpy, $h_2$, is found by dropping straight down to the 50 kPa line. $h_2$ is read as approximately 2485 kJ/kg. Since the turbine is capable of extracting only 87% of the ideal energy, the actual final enthalpy is calculated from Eq. 26.19.

$$
\begin{aligned}
h_2' &= h_1 - \eta_s(h_1 - h_2) \\
&= 3170 \; \frac{\text{kJ}}{\text{kg}} - (0.87)\left(3170 \; \frac{\text{kJ}}{\text{kg}} - 2485 \; \frac{\text{kJ}}{\text{kg}}\right) \\
&= 2574 \text{ kJ/kg}
\end{aligned}
$$

*Customary U.S. Solution*

$h_1$ is read directly from the Mollier diagram at the intersection of 700°F and 200 psia. (Greater accuracy is possible with a large Mollier diagram.)

$$h_1 \approx 1374 \text{ Btu/lbm}$$

The ideal final enthalpy, $h_2$, is found by dropping straight down to the 5 psia line. $h_2$ is read as approximately 1055 Btu/lbm. Since the turbine is capable of extracting only 87% of the ideal energy, the actual final enthalpy is calculated from Eq. 26.19.

$$
\begin{aligned}
h_2' &= h_1 - \eta_s(h_1 - h_2) \\
&= 1374 \; \frac{\text{Btu}}{\text{lbm}} - (0.87)\left(1374 \; \frac{\text{Btu}}{\text{lbm}} - 1055 \; \frac{\text{Btu}}{\text{lbm}}\right) \\
&= 1096 \text{ Btu/lbm}
\end{aligned}
$$

## 11. HEAT EXCHANGERS

A *heat exchanger* transfers energy from one fluid to another through a wall separating them. If the heat transfer is assumed to be adiabatic, the total energy of both input streams must be the same as the total energy of both output streams. No work is done within a heat exchanger, and the potential and kinetic energies of the fluids can be ignored.

$$\dot{m}_A h_{A,\text{in}} + \dot{m}_B h_{B,\text{in}} = \dot{m}_A h_{A,\text{out}} + \dot{m}_B h_{B,\text{out}} \qquad 26.29$$

## 12. CONDENSERS

Condensers are special-purpose heat exchangers that remove the heat of vaporization from steam.[15] Steam passes over tubes in which cooling water (or, occasionally, air) passes. Liquid water falls to the lower portion of the condenser known as the *hot well* or *condensate well*. Heat is transferred through the tubing walls to the cooling water and is then rejected to the environment. Since the condensation takes place on a cold surface, the term *surface condenser* is occasionally encountered.[16]

The heat flow, $\dot{Q}$, out of a condenser is referred to as the *heat load* and the *condenser duty*.

$$\dot{Q} = \dot{m}_{\text{steam}}(h_1 - h_2) \qquad 26.30$$

Equation 26.30 uses the steam properties to calculate the heat load. The heat load can also be calculated from the temperature increase in the cooling water. The *range* is the initial difference between the incoming cooling water temperature and the saturation temperature corresponding to the condenser pressure. The *terminal temperature difference* is the difference in temperatures of the cooling water and the condensed liquid leaving the condenser. It is seldom less than 5°F (2.7°C). A terminal temperature difference of 10°F (5.6°C) is usually assumed in initial performance estimates. The *rise* is the increase in the cooling water temperature as it passes through the condenser. The rise seldom exceeds 20°F (11°C).

$$\dot{Q} = \dot{m}_{\text{cooling water}} c_p (T_{\text{out}} - T_{\text{in}}) \qquad 26.31$$

The *hot well depression*, HWD, also known as the *condensate depression, condenser hot well subcooling*, and *degrees of freedom*, is the difference in temperature between the saturation temperature corresponding to the condenser pressure and the steam condensate in the condenser's hot well. It typically varies between 5°F and 10°F (2.7°C and 5.4°C). Large hot well depressions are undesirable because energy is wasted and because the solubility of gases in the condensate is higher at lower temperatures. To keep the temperature high, some steam is bypassed to the hot well.

$$\text{HWD} = T_{\text{sat},p} - T_{\text{condensate,out}} \qquad 26.32$$

The *cleanliness factor* is the ratio (expressed as a percentage) of the actual to the ideal (new, clean, etc.) heat transfer coefficient. For a given installation, it is the ratio of the actual to the ideal heat loads.

Condensation is assumed to occur at constant pressure. It is also assumed that the water will leave the

---

[15] *Dump condensers* are separate condensers used to condense steam when the turbine is not in service (i.e., are tripped).

[16] Besides surface condensers, *jet condensers* constitute the other major condenser category. While surface condensers both produce a vacuum and recover the condensate, jet condensers produce only a vacuum. The condensate escapes with the jet. Since the condensate is lost, jet condensers are not used in power-generating systems.

condenser as a saturated liquid (corresponding to the condenser pressure).[17] In real power-generating systems, subcooling represents wastage of energy, as nothing is gained by subcooling the water.

Due to the volume decrease that occurs during condensation, the condenser pressure can be low and is usually far below atmospheric. The *back pressure* is the absolute pressure in the condenser, usually expressed in inches of mercury (bars). When the back pressure is lowered, more energy can be extracted from the steam.

Utility condensers operate at a high vacuum. It is inevitable that some *noncondensing gases* (primarily oxygen and other air components) will enter the condenser. These gases, commonly referred to as "air" regardless of composition, enter dissolved in the steam and through air leaks.[18] At high temperatures, these gases are highly corrosive, particularly to the copper used in condenser tubing. In addition, the effect of these gases is to increase the back pressure. As Fig. 26.7 shows, increasing the back pressure reduces the energy available to the power cycle. Noncondensing gases are removed from condensers with steam-jet air ejectors. A general rule of thumb is that the air-removal rate should be less than 1 SCFM (0.5 L/s) per 100 MW thermal.

**Figure 26.7** *Effect of Noncondensing Gases on Power Cycle Performance*

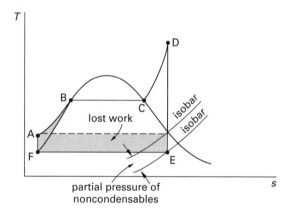

Turbine output cannot be increased indefinitely by decreasing the back pressure, however. As back pressure is lowered below what produces sonic velocity in the last turbine stage, condensate pressure is lowered without any increase in turbine output. This results in a loss similar to that of subcooling the condensate.

---

[17]The term "straight condensing" is sometimes used to mean converting saturated steam to saturated water at the same pressure and temperature. However, this may be confused with the *straight condensing cycle*. Therefore, "pure condensing" is probably a better choice. Besides pure condensing, other modes of condenser operation are desuperheating with condensing, condensing with subcooling, and desuperheating with condensing and subcooling.

[18]The most common *leak points* are cracks in expansion joints between the turbine and condenser, cracks in the condenser shell at condensate-return lines, low-pressure heater vents, and condensate pump seals.

## 13. CONDENSER FOULING

As Fig. 26.7 illustrates, an increase in condenser pressure will decrease the energy extracted from the steam. The pressure increase is caused by an accumulation of noncondensing gases, but it can also occur when the condenser is unable to remove the heat fast enough due to fouling. Within the vapor dome, reductions in enthalpy, temperature, and pressure go hand-in-hand. When the enthalpy is not reduced, the pressure is not reduced.

The most common causes of fouling on the water side of water-cooled, steam surface condensers are microbiological growth, scale formation, and tube pluggage by debris and aquatic organisms. These problems are more intense in once-through condensers that draw water from lakes, rivers, and oceans.

Since condensers are warm, they are prone to microbiological growth (slime and biofilm). This *microfouling* also provides a base for most *microbiologically influenced corrosion* (MIC). Microfouling is easily detected by a gradual increase in the terminal temperature difference. Growth of biological organisms is controlled along the coolant intake by continuous injection and *shock treatment* (sudden injection) of *biocides*.

*Scaling* is the formation of mineral compounds, most notably calcium carbonate and (to a lesser extent) calcium sulfate and phosphate, manganese compounds, and silicates, on the condenser surfaces. Sometimes adding sulfuric acid to the coolant will keep the pH low enough to prevent calcium carbonate deposits. Phosphates, phosphonates, and polymers can also be used. As with make-up water treatment, though, no generic treatment works for all installations. Once formed, scale must be mechanically removed or treated chemically. Calcium carbonate dissolves in acid, but other deposits require more exotic treatments.

Tube pluggage from leaves, vegetation, and anything else that can pass through intake screens will reduce coolant flow. This can be detected by weekly pressure-drop checks or sudden changes in terminal temperature difference. *Macrofouling* is a term referring to infestation by Asian clams, saltwater barnacles, marine blue mussels, zebra mussels, and similar marine life.

*Zebra mussels*, accidentally introduced in the United States in 1986, are particularly troublesome for several reasons. First, young mussels are microscopic and easily pass through intake screens. Second, they attach to anything, even other mussels, to produce thick colonies. Third, adult mussels quickly sense some biocides, most notably those that are halogen-based (including chlorine), and quickly close and remain closed for days or weeks. An ongoing biocide program aimed at pre-adult mussels, combined with slippery polymer-based surface coatings, is probably the best approach at prevention. Once a condenser is colonized, mechanical removal by scraping or water blasting is the only practical option.

Chlorine is a common treatment for once-through systems. However, chlorine reacts with organic precursors to produce trihalorganics, including trihalomethanes, which are carcinogenic compounds, and its use may be prohibited in the United States. Or, when chlorine is used, a *dehalogenation agent* (e.g., sodium bisulfite) may be needed downstream of the condenser. In addition to specialized *molluscicides*, other more expensive substitutes for chlorine include bromine, chlorine dioxide, ozone, nonoxidizing biocides, copper, cyanuric acid, hydrogen peroxide, polymers, potassium permanganate, and sodium hypochlorite.

If the condenser can be taken out of service for a few hours or more, heat treatment (*thermal backflushing*) is another alternative for zebra mussel prevention. Exposure to 95°F to 105°F (35°C to 41°C) water results in 100% mortality within approximately an hour. A temperature of 89°F (32°C) for six hours or more also appears to be effective. Water can be heated to these temperatures by recirculating condenser discharge, or for more distant points, by steam sparging.

## 14. NOZZLES, ORIFICES, AND VALVES

Since a flowing fluid is in contact with nozzle, orifice, and valve walls for only a very short period of time, flow through them can be considered adiabatic. No work is done on the fluid as it passes through. If the potential energy changes are neglected, the SFEE reduces to Eq. 26.33.

$$h_1 + \frac{v_1^2}{2} = h_2 + \frac{v_2^2}{2} \qquad \text{[SI]} \quad 26.33(a)$$

$$h_1 + \frac{v_1^2}{2 g_c J} = h_2 + \frac{v_2^2}{2 g_c J} \qquad \text{[U.S.]} \quad 26.33(b)$$

If the kinetic energy changes are neglected, $h_1 = h_2$. A constant enthalpy pressure drop is characteristic of a *throttling valve*. Pressure-reduction and superheater-bypass valves used to control turbine output in sliding-pressure operation are typically of the "tortuous-path" variety. Steam makes numerous abrupt direction changes, and pressure is reduced in a throttling process.

## 15. SUPERHEATERS

A *superheater* is essentially a heat exchanger used to increase the energy of the steam. A *radiant superheater* is exposed directly to combustion flame and receives energy by radiation. A *convection superheater* is typically exposed to the first pass of stack gases and is screened from direct view of the combustion flame by rows of boiler tubes. Alternatively, a superheater can be a separately fired, independent unit.

Since the overall heat transfer coefficient increases with increases in combustion product flow, the steam temperature increases with convection superheaters when the combustion rate increases. The opposite is true for radiant superheaters (since furnace temperature does not increase rapidly enough with increasing steam flow). In order to maintain a constant superheat, modern superheaters use both radiant and convection sections in series. Damper control and tiltable burners may also be used to maintain constant superheat.

A superheater does not increase steam pressure; it only increases temperature. In fact, steam pressure will decrease a slight amount (e.g., 2% to 5%) by virtue of the superheater's resistance to flow.

## 16. REHEATERS

The name *reheater (resuperheater)* is given to specific parts of a furnace or to separately fired superheaters that add energy to steam after it has already passed at least once through a turbine. This is done to reduce the moisture content of steam that, after partial expansion, may have already entered the vapor dome. Reheaters do not differ much from convection-type superheaters, except that the volume of steam handled is greater (since the pressure and density are less). In fact, reheaters are typically placed after the primary superheater tubing in the stack and before the secondary superheater tubing, if any. Radiant-type reheaters are also occasionally used.

The *reheat temperature* is usually the same as the original steam temperature (known as the *throttle temperature*), usually 1000°F to 1050°F (540°C to 570°C) in traditional drum boiler units and 1000°F to 1100°F (540°C to 590°C) or more in supercritical boilers. Optimum reheat pressure is approximately 25% of the throttle pressure. *Multiple reheat*, double or triple, is applicable only with supercritical throttle pressures (i.e., above 3206 psi (22.1 MPa)). Otherwise, multiple reheat will result in superheated exhaust.

## 17. DESUPERHEATERS

Some power plant auxiliary equipment is designed to use saturated (as opposed to superheated) steam. It is common to bleed steam from the superheater and run it through a *desuperheater (attemperator)* to decrease the steam enthalpy. This can be done by looping the steam back through cooler water in the steam drum or a feedwater heater, or more commonly, by injecting cooling water (i.e., a *fixed-orifice desuperheater*).[19] Since the water and steam leave at the same properties, the heat balance equation for water injection is given by Eq. 26.34. Kinetic energy is neglected, although it can be significant in some situations.

$$m_{\text{water}} h_{\text{water}} + m_{\text{steam}} h_{\text{steam}}$$
$$= (m_{\text{water}} + m_{\text{steam}}) h_{\text{desuperheated steam}} \qquad 26.34$$

---

[19]Other types of desuperheaters are surface-water absorption, steam atomizing, ejector-recycle, venturi, and variable orifice.

The *turndown ratio (capacity turndown)* is the ratio of maximum-to-minimum mass steam flow rates at which the temperature can be accurately controlled by the desuperheater. Minimum flow rates are influenced by the lowest velocity at which water droplets can be held in suspension.

Desuperheaters are usually needed to run auxiliary equipment (e.g., turbine-driven boilerfeed pumps); they should also be installed across high-pressure turbines. If a high-pressure turbine trips or is taken out of service, the high-pressure steam can be desuperheated and routed to the low-pressure turbine. Desuperheaters are also used to temper steam from intermediate- and low-pressure turbines on imbalanced or start-up conditions.

## 18. FEEDWATER HEATERS

A *feedwater heater* uses steam to increase the temperature of water entering the steam generator, thus improving the boiler efficiency. Each $10°F$ ($5.6°C$) increase in water temperature increases a corresponding improvement in boiler efficiency of approximately 1%. In a normal condensing cycle, the water being heated comes from the condenser. Steam for heating is bled off from an *extraction turbine*. Eight or more stages of feedwater heating, using both open and closed heaters, may be used in modern plants. The actual number of heaters is an economic decision, as the incremental savings from increased thermal efficiency eventually drops below the incremental installation cost.

*Open heaters* (also known as *direct-contact heaters* and *mixing heaters*) physically mix the steam and water, as shown in Fig. 26.8. Because they are essentially large boxes under internal steam pressure, open heaters are usually closest to the condenser (i.e., on the suction side of the boilerfeed pumps), where the pressure is low. Each open heater requires its own feed pump. Because open heaters are not highly pressurized, the temperature of the leaving water is seldom over $220°F$ ($104°C$) and is usually near $212°F$ ($100°C$).

**Figure 26.8** *Open Feedwater Heater*

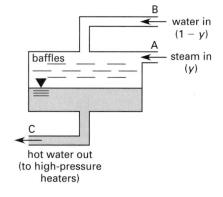

hot water out
(to high-pressure
heaters)

Open heaters may be simply vented to the atmosphere, or dissolved gases that come out of solution may be extracted by a vent condenser or a vacuum pump.

When saturated or nearly saturated water is reduced in pressure, some of it may "flash" into steam. The reduction in pressure can occur in condensate return lines due to fluid friction or from pressure losses through valves and traps. When flashing occurs, the steam will impede liquid flow, a condition known as "binding."[20] To prevent boilerfeed pumps handling water near the boiling temperature from racing due to lack of feed liquid, open feedwater heaters are mounted quite high (i.e., 12 ft to 20 ft (3.6 m to 6 m) up or higher). This provides the necessary NPSHR for the pumps below.

The adiabatic energy balance around the open feedwater heater shown in Fig. 26.8 is

$$(1 - y)h_{\mathrm{B}} + yh_{\mathrm{A}} = h_{\mathrm{C}} \qquad 26.35$$

If the three enthalpies are known, the *bleed fraction, y*, can be determined.

$$y = \frac{h_{\mathrm{C}} - h_{\mathrm{B}}}{h_{\mathrm{A}} - h_{\mathrm{B}}} \qquad 26.36$$

A *closed feedwater heater* is a heat exchanger that can operate at either high or low pressures. There is no mixing of the water and steam in the feedwater heater. The cooled steam (known as the *drips*) leaves the feedwater heater in liquid form.

There are various methods of disposal of the drips. The condensate may be combined with the feedwater (as in Fig. 26.9) after passing through a *drip pump* to raise its pressure. (Since the drip pump work is small, it can be omitted for first approximations.) Alternatively, the drips can be returned to the condenser *hot well*. For the closed feedwater shown in Fig. 26.9 receiving bleed fraction $y$ of steam, the adiabatic heat balance is

$$h_{\mathrm{C}} = yh_{\mathrm{A}} + (1 - y)h_{\mathrm{D}} + W_{\mathrm{drip\ pump\ per\ pound}} \qquad 26.37$$

For a feedwater heater, the *terminal temperature difference*, TTD (also known as the *approach*), is defined as the difference in saturation temperature corresponding to the steam pressure and the temperature of the leaving water. It typically varies between $5°F$ and $20°F$ ($2.8°C$ and $11°C$).

$$\mathrm{TTD} = T_{\mathrm{sat},p,\mathrm{A}} - T_{\mathrm{C}} = T_{\mathrm{B}} - T_{\mathrm{C}} \qquad 26.38$$

---

[20]Flashing is intentional in a *flash evaporator* (*flash tank, flash chamber*, etc.). Since the temperature of the flash steam produced depends on the pressure maintained in the flash tank, precise temperature control can be maintained where needed. The enthalpy of the flash steam is the same as live steam at that pressure and temperature. The fraction of condensate that will flash when dropped from pressure $p_1$ to $p_2$ is

$$\text{flash fraction} = \frac{h_{f,p_1} - h_{f,p_2}}{h_{fg,p_2}}$$

**Figure 26.9** *Closed Feedwater Heater*

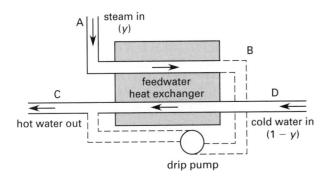

In the United States, *tubesheet feedwater heaters* have been used almost exclusively. Since tubesheet heaters are limited to temperature change rates of about 10°F/min (5.6°C/min), they are susceptible to thermal stresses when used with cycling plants. With *header-type feedwater heaters*, all the tubes are welded to headers. This type of design tolerates temperature change rates up to 30°F/min (17°C/min) and is more suitable for cycling loads.

## 19. EVAPORATORS

An *evaporator* is a closed shell-and-tube heat exchanger that is used to produce distilled boiler feedwater. Steam bled off from other points is the source of heating. Steam passes through the tubes. As the water is evaporated, the distilled water steam is routed to a holding tank, an open feedwater heater, or a special condenser.

*Single-effect (single-stage) evaporators* produce distilled water from the steam of a single evaporator. *Multiple-effect evaporators* are more common and use the steam produced in one evaporator as the heat source for the next evaporator. (See Fig. 26.10.) In that way, the heat of evaporization is used several times. Three and four effects in series are typical for power-generating plants.

**Figure 26.10** *Double-Effect Evaporator*

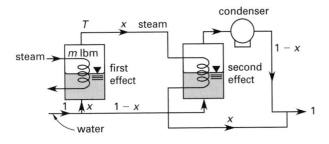

## 20. DEAERATORS

*Deaerators (deaerating heaters)* are a special category of open feedwater heaters. (Although the pressure is higher than in normal open feedwater heaters, deaerators are still direct-contact heaters.) They remove dissolved gases (e.g., oxygen and carbon dioxide) from feedwater.

This is done in a baffled chamber in which the water is broken into droplets and heated by exposure to high-temperature steam. Gas solubility is essentially zero at the saturation temperature.[21] Gases are removed by a vent condenser.

## 21. ECONOMIZERS

An *economizer* is a water-tube heat exchanger consisting of tubes heated by the last pass of the combustion gases. It increases the temperature of boiler feedwater entering the steam generator and reduces the required combustion energy input by utilizing energy that would otherwise be wasted. The heat transfer takes place in the downstream of the boiler. Because an economizer can add significant resistance to the flow of stack gases, a forced draft fan is usually required.

The overall coefficient of heat transfer is in the range of approximately 2–3 Btu/hr-ft²-°F (11–17 W/m²·°C) for flue gases passing at 2000 lbm/hr-ft² (2.7 kg/s·m²) to approximately 5–7 Btu/hr-ft²-°F (28–40 W/m²·°C) for flue gases passing at 6000 lbm/hr-ft² (8.1 kg/s·m²).

External corrosion of the economizer heat transfer surface is avoided by keeping the temperature high enough (i.e., above the dew point of the stack gases) to prevent acid formation. This is generally 275°F to 350°F (135°C to 175°C), depending on the amount of sulfur dioxide in the stack gases. Internal corrosion is avoided by maintaining proper feedwater chemistry.[22]

## 22. AIR HEATERS

Increasing the temperature of combustion air increases combustion efficiency. While the temperature of incoming air for stoker furnaces is limited to approximately 250°F to 350°F (120°C to 175°C) by mechanical cooling considerations, at least some portion of combustion air used with pulverized coal must be heated to approximately 600°F (315°C) or more.

An *air heater (air preheater)* recovers energy from the stack gases and transfers it to incoming combustion air. *Convection preheaters* (also known as *recuperative heaters*) are conventional heat exchangers that transfer energy through tubes or flat plates. *Regenerative preheaters* use a slowly rotating drum with honeycomb-like passageways that are progressively exposed to incoming air and outgoing stack gases. (See Fig. 26.11.) The overall heat transfer coefficient for both types is generally low, varying from approximately 1.5 Btu/hr-ft²-°F (8.5 W/m²·°C) for air passing at 2000 lbm/hr-ft² (2.7 kg/s·m²) to approximately 3.0 Btu/hr-ft²-°F

[21]After heating, the residual oxygen concentration will be approximately 0.005 mg/L. Vacuum deaerators using steam-jet extractors or vacuum pumps without heating leave water with residual concentrations of approximately 0.2 mg/L of oxygen and 2 mg/L to 10 mg/L of carbon dioxide.
[22]Originally, economizer tubes were cast iron to protect them from corrosion. Now, with better furnace oxygen control, cast-iron tubes have been essentially replaced by steel tubes.

(17 W/m$^2$·°C) for air passing at 6000 lbm/hr-ft$^2$ (8.1 kg/s·m$^2$).

**Figure 26.11** *Regenerative Air Preheater*

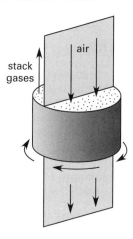

## 23. ELECTRICAL GENERATORS

*Electrical generators* convert energy extracted from the steam into electrical energy. Generator efficiencies are high—around 95% for 1 MW unit, 96% for 10 MW units, to 98% for 100 MW units and above. The *steam rate (water rate)* is defined as the steam mass flow rate divided by the generator output in kilowatts.[23] Typical units for steam rate are lbm/kW-hr (kg/kW·h). Values depend greatly on the throttle steam conditions and condenser pressure. For 1000°F (540°C) steam and condenser pressures of 2 in to 4 in of mercury, the steam rate is approximately 5.6 lbm/kW-hr to 5.9 lbm/kW-hr (7.1 g/kW·s to 7.4 g/kW·s), with the lower values corresponding to the lower condenser pressures.

## 24. LOAD AND CAPACITY

Electrical demand varies with time of day, month, and season. The *load curve* is a curve of fluctuating instantaneous load (electrical demand in MW or steam requirements) versus time of day for an average day in a particular season. The *load duration curve* is a curve of load (in MW) versus the number of hours (in a year) the load was experienced. Large power demand variations are met by taking entire power-generating plants on the electrical grid online or offline. Smaller variations are met at the plant level by taking individual boilers and turbines online or offline. Capacity at the plant and regional levels is categorized into *firm capacity* (always available, even in emergencies), *spinning reserve* (floating capacity on the electrical bus), *hot reserve* (in operation but not in service), and *cold reserve* (not in operation but operational).

The *load factor* is the ratio of the peak-to-average loads. The *capacity factor (plant factor)* of a unit is the ratio of the average load to rated capacity. The *demand factor* is the ratio of the maximum demand to the total maximum possible connected load (assuming everybody turned everything on at once). The *output factor (use factor)* accounts for downtime and is the ratio of actual energy output during a particular period to the output that would have occurred during that period if the plant had been operating at full load rating. The *operation factor* is a ratio of operational time to total time (including downtime).

A *base-load unit* is a unit or plant intended to satisfy a constant fixed demand by running continuously. A base-load unit will usually have a capacity factor of 50% or more. A *peaking unit* is intended to satisfy only peak loads.

## 25. CONTROL AND REGULATION OF TURBINES

Traditionally, turbines in the United States have been designed to operate at *constant throttle* pressure. Base-load units were not intended to drop below the minimum load they were designed for, typically 25% to 30% of capacity. However, daily cycling of base-load units (down to as low as 10%) has now become common.

*Sliding load units* are turbines that can infinitely adjust their output, perhaps down to as low as 10% of the rated capacities. With the commonly used *sliding pressure operation*, also known as *variable throttle pressure operation*, steam generator pressure is varied to change the turbine output. Since temperature and pressure are related, this subjects the turbine and other equipment to thermal-stress damage due to rapid and repeated temperature swings.

An alternative to using sliding pressure operation is using control valves to redirect the steam flow through different components.[24] The amount of steam entering the turbine is controlled by *boiler throttle valves* or *turbine (steam admission) throttle valves*. For example, at low loads, steam may be routed around the second superheater; at intermediate loads, superheated steam may pass through a pressure reduction valve; and at high loads, the turbine may receive fully superheated steam at full pressure. The process of implementing this control method in older plants is known as *cycle redesign*, *backfitting*, and *retrofitting*.

Different methods of governing can be used to maintain the turbine operating point. With *throttle control* (also known as *cut-off governing* and *full-arc admission*), the pressure and temperature of the steam admitted to the turbine are constant and all turbine nozzles are operational, but valves control the steam flow rate through

---

[23]For turbines acting as prime movers, the steam flow rate is divided by the horsepower output, with units of lbm/hp-hr.

[24]Actually, most modern plants use a combination of sliding pressure and backfitting, a system known as *hybrid throttle pressure operation*.

each nozzle.[25] With *nozzle control*, also known as *partial-arc admission*, the steam flow rate to the turbine is controlled by "cutting out" some of the steam nozzles. With *bypass governing*, steam pressure and temperature are constant, but the stage at which steam is introduced to the turbine(s) is changed. With *blast governing*, the steam is not admitted to the turbine continuously. Rather, the steam flow is repeatedly turned on and off in a ratio that achieves the desired duty cycle.

At the steam level, variations in output are achieved by varying the firing (fueling) rate, by varying the input air and/or draft, and by using separately fired superheaters, desuperheaters, movable burner heads, and other techniques.

At the individual turbine, steam accumulators can supply short-term peak loads.[26] This method is commonly used in once-through boilers, but may also be used with sliding pressure operation to supply all low-load steam.

---

[25]The terminology is confusing. A *throttle valve* controls the mass flow rate, but a *throttling valve* changes the pressure. Actually, partly closing any valve will result in reductions in both the mass flow rate and the pressure. Phrases such as "...at 85% throttle..." usually refer to the capacity ratio, not the mass flow rate.

[26]A volume of saturated water at the system pressure is kept heated in the accumulator by boiler steam. When the load peaks and the system pressure drops (below the saturation pressure), some of the accumulator water flashes into steam, satisfying the majority of the increase in demand.

# 27 Vapor Power Cycles

## Nomenclature

| | | | |
|---|---|---|---|
| $h$ | enthalpy | Btu/lbm | kJ/kg |
| HR | heat rate | Btu/kW-hr | MJ/kW·h |
| $p$ | pressure | lbf/ft$^2$ | Pa |
| $Q$ | heat | Btu | kJ |
| $Q$ | heat per unit mass | Btu/lbm | kJ/kg |
| RF | reheat factor | – | – |
| $s$ | entropy | Btu/lbm-°R | kJ/kg·K |
| $T$ | temperature | °R | K |
| $W$ | work | Btu | kJ |
| $W$ | work per unit mass | Btu/lbm | kJ/kg |
| $x$ | quality | – | – |
| $y$ | bleed fraction | – | – |

## Symbols

| | | | |
|---|---|---|---|
| $\eta$ | efficiency | – | – |
| $\upsilon$ | specific volume | ft$^3$/lbm | m$^3$/kg |

## Subscripts

| | |
|---|---|
| $f$ | fluid (liquid) |
| $fg$ | fluid-to-gas (vaporization) |
| $s$ | isentropic |
| th | thermal |

## 1. GENERAL VAPOR POWER CYCLES

The most general form of a vapor (almost always steam) power cycle incorporates a vapor generator, turbine, condenser, and feed pump as illustrated in Fig. 27.1.

The following thermodynamic processes take place in a typical steam vapor power cycle.

**Figure 27.1** *Simplified Vapor Power Cycle*

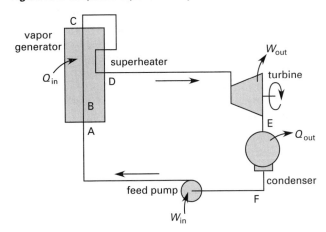

A to B:  Subcooled water is heated to the saturation temperature corresponding to the pressure in the steam generator.

B to C:  Saturated water is vaporized in the steam generator, producing saturated vapor.

C to D:  An optional superheating process increases the steam temperature and enthalpy.

D to E:  Steam expands in the turbine and does work as it decreases in temperature, pressure, and quality.

E to F:  Vapor is returned to a liquid phase in the condenser.

F to A:  The pressure of the liquid water is brought up to the steam generator pressure by the boilerfeed pump.

After making a full cycle, the steam is brought back to the exact same thermodynamic conditions it started the cycle with. While the entropy of the steam returns to its original value after each full cycle, the entropy of the environment increases due to the heat transfers (losses, etc.) from the steam generator and condenser. Thus, even if the pump compression and turbine expansion are isentropic processes, the cycle does irreversible thermal "damage" to the environment.

## 2. SIGN CONVENTION

The sign convention normally adhered to in textbooks for changes in thermodynamic properties is not followed for pump work. Since the pump does work on the water, the pump work would be negative according to the traditional sign convention. However, since the power industry customarily refers to pump work as a positive quantity, that convention is followed in this chapter.[1]

## 3. THERMAL EFFICIENCIES

The *thermal efficiency*, $\eta_{th}$, of a power cycle is the ratio of net work out (or net heat in) divided by the heat input.[2] The upper limit of thermal efficiency is the thermal efficiency of the theoretical Carnot cycle. (See Eq. 27.8.) Typical actual values of thermal efficiency depend to a great extent on the operating temperature and pressure and on the degree of cycle sophistication. Cycles with high-pressure superheat, reheat, and multiple stages of regeneration will be far more efficient in converting thermal energy to electricity than the simple low-pressure cycles that prevailed before 1960. However, the best single-cycle plants seldom have thermal efficiencies in excess of 40%.

$$\eta_{th} = \frac{W_{out} - W_{in}}{Q_{in}}$$
$$= \frac{Q_{in} - Q_{out}}{Q_{in}} \qquad 27.1$$

The *heat rate*, HR (or *station rate*), is the ratio of the total energy input divided by the net electrical output. This is not merely the reciprocal of the thermal efficiency because of the units for heat rate. Heat rate has units of Btu/kW-hr (MJ/kW·h). It can be calculated from the thermal efficiency and the appropriate conversion factor.[3]

$$\text{HR} = \frac{Q_{in}}{W_{out} - W_{in}} = \frac{3.6 \, \dfrac{\text{MJ}}{\text{kW·h}}}{\eta_{th}} \qquad \text{[SI]} \quad 27.2(a)$$

$$\text{HR} = \frac{Q_{in}}{W_{out} - W_{in}} = \frac{3413 \, \dfrac{\text{Btu}}{\text{kW-hr}}}{\eta_{th}} \qquad \text{[U.S.]} \quad 27.2(b)$$

For heat rates expressed in Btu/hp-hr, the relationship is

$$\text{HR} = \frac{2543 \, \dfrac{\text{Btu}}{\text{hp-hr}}}{\eta_{th}} \qquad 27.3$$

## 4. CYCLE DESIGNATIONS

There are numerous descriptive terms used to describe categories of cycles. If a condenser is not part of the power-generating process, the steam cannot expand

below atmospheric pressure. A cycle without a condenser is known as a *noncondensing cycle*. The steam may be used for process heating or feedwater heating, or may simply be discharged to the atmosphere. Steam may condense elsewhere (in a feedwater heater, for example). It is the absence of a condenser that identifies a noncondensing cycle, not the absence of condensation.

If there is a condenser, the cycle is known as a *condensing cycle*. If there are no bleeds (i.e., no extraction of steam prior to steam entering the condenser), the cycle is known as *straight condensing*. In an *extraction cycle*, steam is withdrawn from the turbine, usually for feedwater heating, although the bleed steam can also be used for other process heating.

## 5. CARNOT CYCLE

The Carnot cycle is an ideal power cycle that is impractical to implement. However, its theoretical work output sets the maximum attainable from any heat engine, as evidenced by the isentropic (reversible) processes between states (D and A) and (B and C) in Fig. 27.2. The working fluid in a Carnot cycle is irrelevant.

The processes involved are:

A to B:  isothermal expansion of saturated liquid to saturated vapor
B to C:  isentropic expansion of vapor
C to D:  isothermal compression of vapor
D to A:  isentropic compression

The properties at the various states can be found from the following solution methods. The letters $f$ and $g$ stand for saturated fluid and saturated gas, respectively. They do not correspond to states F and G on any cycle diagram. The Carnot cycle can be evaluated by working around, finding $T$, $p$, $x$, $h$, and $s$ at each node.

At A:  From the property table for $T_{high}$, read $p_A$, $h_A$, and $s_A$ for a saturated fluid.
At B:  $T_B = T_A$; $p_B = p_A$; $x = 1$; $h_B$ is read from the table as $h_g$; $s_B$ is read as $s_g$.
At C:  Either $p_C$ or $T_C$ must be known. Read $p_C$ from the $T_C$ line on the property table or vice versa; $x_C = (s_B - s_f)/s_{fg}$; $h_C = h_f + x_C h_{fg}$; $s_C = s_B$.
At D:  $T_D = T_C$; $p_D = p_C$; $x_D = (s_A - s_f)/s_{fg}$; $h_D = h_f + x_D h_{fg}$; $s_D = s_A$.

The turbine and pump work terms per unit mass are

$$W_{turbine} = h_B - h_C \qquad 27.4$$
$$W_{pump} = h_A - h_D \qquad 27.5$$

The heat flows per unit mass into and out of the system are

$$Q_{in} = T_{high}(s_B - s_A) = h_B - h_A \qquad 27.6$$

---

[1]Nobody says "The pump work is negative four Btus per pound."
[2]Only a *cycle* (i.e., an integrated system of devices) can have a *thermal* efficiency. A single device (e.g., a pump) can have mechanical and isentropic efficiencies, but it cannot have a thermal efficiency.
[3]The conversion between kW and Btu/hr is given in some sources as 3412. The conversion between hp and Btu/hr is given in some sources as 2544.

*Figure 27.2* Carnot Cycle

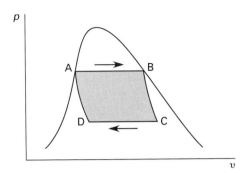

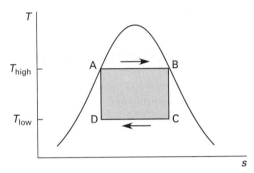

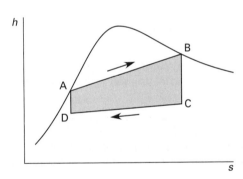

$$Q_{\text{out}} = T_{\text{low}}(s_C - s_D) = T_{\text{low}}(s_B - s_A)$$
$$= h_C - h_D \qquad \textit{27.7}$$

The thermal efficiency of the entire cycle is

$$\eta_{\text{th}} = \frac{Q_{\text{in}} - Q_{\text{out}}}{Q_{\text{in}}} = \frac{W_{s,\text{turbine}} - W_{s,\text{pump}}}{Q_{\text{in}}}$$
$$= \frac{(h_B - h_C) - (h_A - h_D)}{h_B - h_A} = \frac{T_{\text{high}} - T_{\text{low}}}{T_{\text{high}}} \qquad \textit{27.8}$$

If isentropic efficiencies for the pump and turbine are known, proceed as follows. Calculate all properties, assuming that the efficiencies are 100%. Then, modify $h_C$ and $h_A$ as given in Eq. 27.9 and Eq. 27.10. Use the new values to find the actual thermal efficiency of the cycle.

$$h'_C = h_B - \eta_{s,\text{turbine}}(h_B - h_C) \qquad \textit{27.9}$$

$$h'_A = h_D + \frac{h_A - h_D}{\eta_{s,\text{pump}}} \qquad \textit{27.10}$$

$$W'_{\text{turbine}} = h_B - h'_C \qquad \textit{27.11}$$

$$W'_{\text{pump}} = h'_A - h_D \qquad \textit{27.12}$$

## 6. BASIC RANKINE CYCLE

The basic Rankine cycle (see Fig. 27.3 and Fig. 27.4) is similar to the Carnot cycle except that the compression process occurs in the liquid region. The Rankine cycle is closely approximated in steam turbine plants. The efficiency of the Rankine cycle is lower than that of a Carnot cycle operating between the same temperature limits because the mean temperature at which heat is added to the system is lower than $T_{\text{high}}$.

*Figure 27.3* Basic Rankine Cycle

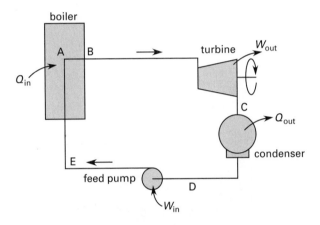

The processes used in the basic Rankine cycle are:

A to B:  vaporization in the boiler
B to C:  adiabatic expansion in the turbine
C to D:  condensation
D to E:  adiabatic compression to boiler pressure
E to A:  heating liquid to saturation temperature

The properties at each point can be found from the following procedure. The letters $f$ and $g$ refer to saturated fluid and saturated gas, respectively. They do not correspond to locations F and G on any diagram. Usually $T_{\text{high}}$ and $T_{\text{low}}$ are known. The procedure is to work around the cycle, finding $T$, $p$, $x$, $h$, and $s$ at each node.

At A:  From the property table for $T_{\text{high}}$, read $p_A$, $h_A$, and $s_A$ for a saturated liquid.
At B:  $T_B = T_A$; $p_B = p_A$; $x = 1$; $h_B$ is read from the table as $h_g$; $s_B$ is read as $s_g$.
At C:  Either $p_C$ or $T_C$ must be known. Read $p_C$ from the $T_C$ line on the property table or vice versa; $x_C = (s_B - s_f)/s_{fg}$; $h_C = h_f + x_C h_{fg}$; $s_C = s_B$.

Thermodynamics

At D: $T_D = T_C$; $p_D = p_C$; $x_D = 0$; $h_D$ is read as $h_f$; $s_D$ is read as $s_f$; $v_D$ is read as $v_f$.

At E: $p_E = p_A$; $s_E = s_D$; $h_E = h_D + W_{pump} = h_D + v_D(p_E - p_D)$ (consistent units); $T_E$ is found as the saturation temperature for a liquid with enthalpy equal to $h_E$.

**Figure 27.4** Basic Rankine Cycle

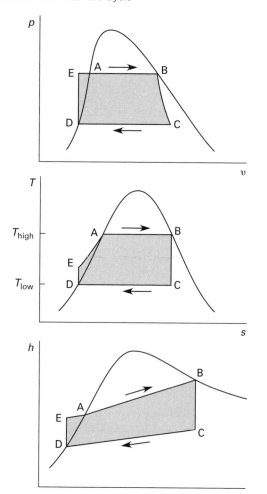

The per unit mass work and heat flow terms are

$$W_{turbine} = h_B - h_C \qquad 27.13$$

$$W_{pump} = h_E - h_D \approx v_D(p_E - p_D) \quad \text{[consistent units]}$$
$$\qquad 27.14$$

$$Q_{in} = h_B - h_E \qquad 27.15$$

$$Q_{out} = h_C - h_D \qquad 27.16$$

The thermal efficiency of the entire cycle is

$$\eta_{th} = \frac{Q_{in} - Q_{out}}{Q_{in}} = \frac{W_{turbine} - W_{pump}}{Q_{in}}$$
$$= \frac{(h_B - h_C) - (h_E - h_D)}{h_B - h_E} \qquad 27.17$$

If isentropic efficiencies for the pump and the turbine are known, calculate all properties as if these efficiencies were 100%. Then use the following relationships to modify $h_C$ and $h_E$. Use the new values to recalculate the thermal efficiency.

$$h'_C = h_B - \eta_{s,turbine}(h_B - h_C) \qquad 27.18$$

$$h'_E = h_D + \frac{h_E - h_D}{\eta_{s,pump}} \qquad 27.19$$

$$W'_{turbine} = h_B - h'_C \qquad 27.20$$

$$W'_{pump} = h'_E - h_D \qquad 27.21$$

$$Q'_{in} = h_B - h'_E \qquad 27.22$$

## 7. RANKINE CYCLE WITH SUPERHEAT

Superheating occurs when heat in excess of that required to produce saturated vapor is added to the water. Superheat is used to raise the vapor above the critical temperature, to raise the mean effective temperature at which heat is added, and to keep the expansion primarily in the vapor region to reduce wear on the turbine blades. A maximum practical metallurgical limit on superheat is approximately 1150°F (625°C), although some ultrasupercritical boilers operate above this temperature.

The processes in the Rankine cycle with superheat are similar to the basic Rankine cycle. (See Fig. 27.5 and Fig. 27.6.)

A to B: heating water to the saturation temperature in the boiler
B to C: vaporization of water in the boiler
C to D: superheating steam in the superheater region of the boiler
D to E: adiabatic expansion in the turbine
E to F: condensation
F to A: adiabatic compression to boiler pressure

**Figure 27.5** Rankine Cycle with Superheat

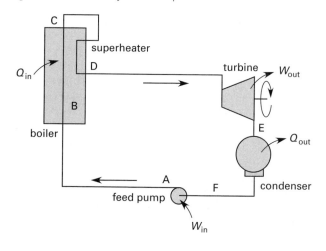

**Figure 27.6** *Rankine Cycle with Superheat*

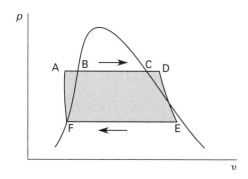

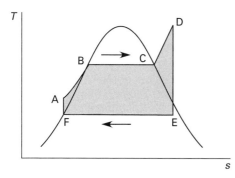

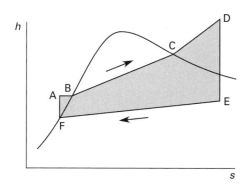

At F: $T_F = T_E$; $p_F = p_E$; $x = 0$; $h_F$, $s_F$, and $v_F$ are read as $h_f$, $s_f$, and $v_f$ from the saturated table.

At A: $p_A = p_B$; $h_A = h_F + v_F(p_A - p_F)$ (consistent units); $s_A = s_F$. $T_A$ is equal to the saturation temperature for a liquid with enthalpy equal to $h_A$.

The per unit mass work and heat flow terms are

$$W_{\text{turbine}} = h_D - h_E \qquad 27.23$$

$$W_{\text{pump}} = h_A - h_F = v_F(p_A - p_F) \qquad [\text{consistent units}]$$
$$27.24$$

$$Q_{\text{in}} = h_D - h_A \qquad 27.25$$

$$Q_{\text{out}} = h_E - h_F \qquad 27.26$$

The thermal efficiency of the entire cycle is

$$\eta_{\text{th}} = \frac{Q_{\text{in}} - Q_{\text{out}}}{Q_{\text{in}}} = \frac{W_{\text{turbine}} - W_{\text{pump}}}{Q_{\text{in}}}$$

$$= \frac{(h_D - h_A) - (h_E - h_F)}{h_D - h_A} \qquad 27.27$$

If pump and turbine isentropic efficiencies are known, calculate all quantities as if those efficiencies were 100%. Then modify $h_E$ and $h_A$ prior to recalculating the thermal efficiency.

$$h'_E = h_D - \eta_{\text{s,turbine}}(h_D - h_E) \qquad 27.28$$

$$h'_A = h_F + \frac{h_A - h_F}{\eta_{\text{s,pump}}} \qquad 27.29$$

$$W'_{\text{turbine}} = h_D - h'_E \qquad 27.30$$

$$W'_{\text{pump}} = h'_A - h_F \qquad 27.31$$

$$Q'_{\text{in}} = h_D - h'_A \qquad 27.32$$

## 8. RANKINE CYCLE WITH SUPERHEAT AND REHEAT

*Reheat* is used to increase the mean effective temperature at which heat is added without producing significant expansion in the liquid-vapor region. The analysis given assumes that $T_D = T_F$, as is usually the case. It is possible, however, that the two temperatures will be different. (See Fig. 27.7 and Fig. 27.8.)

The properties at each point can be found from the following procedure.

At A: Point A is covered below. Start the analysis at point B.

At B: From the vapor table for $T_B$ or $p_B$, read $h_B$ and $s_B$ for a saturated fluid.

At C: $T_C = T_B$; $p_C = p_B$; $x_C = 1$; $h_C$ is read as $h_g$; $s_C$ is read as $s_g$.

At D: $T_D$ usually is known; $p_D = p_C$; $h_D$ is read from superheat tables with $T_D$ and $p_D$ known. Same for $s_D$ and $v_D$.

The properties at each point can be found from the following procedure. The subscripts $f$ and $g$ refer to saturated fluid and saturated gas, respectively. They do not correspond to any points on the property plot diagram.

At A: Point A is covered below. Start the analysis at point B.

At B: From the property table for $T_B$ or $p_B$, read $h_B$ and $s_B$ for a saturated fluid.

At C: $T_C = T_B$; $p_C = p_B$; $x = 1$; $h_C$ is read as $h_g$; $s_C$ is read as $s_g$.

At D: $T_D$ usually is known; $p_D = p_C$; $h_D$ is read from superheat tables with $T_D$ and $p_D$ known. Same for $s_D$ and $v_D$.

At E: $T_E$ is usually known; $p_E$ is read from the saturated table for $T_E$; $x_E = (s_E - s_f)/s_{fg}$; $h_E = h_f + x_E h_{fg}$; $s_E = s_D$. If point E is in the superheated region, the Mollier diagram should be used to find the properties at point E.

Thermodynamics

*Figure 27.7* Reheat Cycle

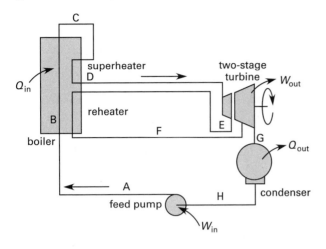

*Figure 27.8* Reheat Cycle

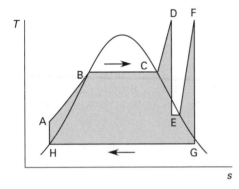

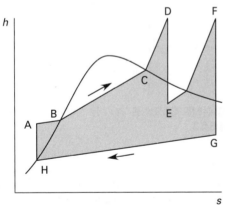

At E:  $p_E$ is usually known; $T_E$ is read from the property table for $p_E$; $s_E = s_D$; $x_E = (s_E - s_f)/s_{fg}$; $h_E = h_f + x_E h_{fg}$. (Use the Mollier diagram if superheated.)

At F:  $T_F$ is usually known; $p_F = p_E$; $h_F$ is read from superheat tables with $T_F$ and $p_F$ known. Same for $s_F$ and $v_F$.

At G:  $T_G$ is usually known; $p_G$ is read from property tables for $T_G$; $s_G = s_F$. $x_G = (s_G - s_f)/s_{fg}$; $h_G = h_f + x_G h_{fg}$. (Use the Mollier diagram if superheated.)

At H:  $T_H = T_G$; $p_H = p_G$; $x_H = 0$; $h_H$, $s_H$, and $v_H$ are read as $h_f$, $s_f$, and $v_f$.

At A:  $p_A = p_B$; $h_A = h_H + v_H(p_A - p_H)$ (consistent units); $s_A = s_H$. $T_A$ is equal to the saturation temperature for a liquid with enthalpy equal to $h_A$.

The per unit mass work and heat flow terms are

$$W_{\text{turbine}} = (h_D - h_E) + (h_F - h_G) \qquad 27.33$$

$$W_{\text{pump}} = (h_A - h_H) = v_H(p_A - p_H) \qquad 27.34$$

$$Q_{\text{in}} = (h_D - h_A) + (h_F - h_E) \qquad 27.35$$

$$Q_{\text{out}} = h_G - h_H \qquad 27.36$$

The thermal efficiency of the entire cycle is

$$\eta_{\text{th}} = \frac{W_{\text{turbine}} - W_{\text{pump}}}{Q_{\text{in}}} = \frac{Q_{\text{in}} - Q_{\text{out}}}{Q_{\text{in}}}$$

$$= \frac{(h_D - h_A) + (h_F - h_E) - (h_G - h_H)}{(h_D - h_A) + (h_F - h_E)} \qquad 27.37$$

If pump and turbine isentropic efficiencies are known, calculate all quantities as if those efficiencies were 100%. Then modify $h_E$ and $h_A$ prior to recalculating the thermal efficiency. Then modify $h_E$, $h_G$, and $h_A$ before recalculating the thermal efficiency.

$$h_E' = h_D - \eta_{s,\text{turbine}}(h_D - h_E) \qquad 27.38$$

$$h_G' = h_F - \eta_{s,\text{turbine}}(h_F - h_G) \qquad 27.39$$

$$h_A' = h_H + \frac{h_A - h_H}{\eta_{s,\text{pump}}} \qquad 27.40$$

$$W_{\text{turbine}}' = (h_D - h_E') + (h_F - h_G') \qquad 27.41$$

$$W_{\text{pump}}' = h_A' - h_H \qquad 27.42$$

$$Q_{\text{in}}' = (h_D - h_A') + (h_F - h_E') \qquad 27.43$$

The *reheat factor*, RF, is the ratio of the actual turbine work (with all of the reheat stages) in a multistage expansion to the ideal turbine work assuming a one-stage, isentropic expansion from the same entering conditions to the same condenser pressure. The fractional improvement due to the reheat is RF − 1. The reheat factor depends on the steam properties and complexity of the cycle. Values are typically in the range of 1.05 to 1.10.

## 9. RANKINE CYCLE WITH REGENERATION (REGENERATIVE CYCLE)

If the mean effective temperature at which heat is added can be increased, the overall thermal efficiency of the cycle will be improved. This can be accomplished by raising the temperature at which the condensed fluid enters the boiler.

In the regenerative cycle, portions of the steam in the turbine are withdrawn at various points. Heat is transferred from this bleed stream to the feedwater coming from the condenser. Although only two bleeds are used in the following analysis, seven or more exchange locations can be used in a large installation. The regenerative cycle always involves superheating, although conceptually it does not need to. (See Fig. 27.9.)

*Figure 27.9* Regenerative Cycle with Two Feedwater Heaters

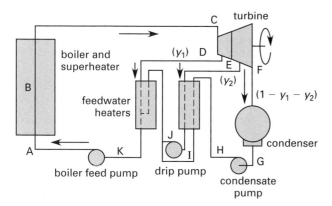

In the following analysis, $y_1$ is the first bleed fraction, and analysis, $y_2$ is the second bleed fraction. (See Fig. 27.10.)

*Figure 27.10* Regenerative Cycle Property Plot

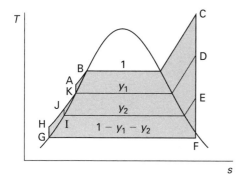

The ideal per unit mass work and heat flow terms for the regenerative cycle are

$$W_{\text{turbine}} = (h_C - h_D) + (1 - y_1)(h_D - h_E)$$
$$+ (1 - y_1 - y_2)(h_E - h_F) \qquad 27.44$$

$$W_{\text{pumps}} = (h_A - h_K) + y_2(h_J - h_I)$$
$$+ (1 - y_1 - y_2)(h_H - h_G) \qquad 27.45$$

$$Q_{\text{in}} = h_C - h_A \qquad 27.46$$

$$Q_{\text{out}} = (1 - y_1 - y_2)(h_F - h_G) \qquad 27.47$$

The thermal efficiency of the entire cycle is

$$\eta_{\text{th}} = \frac{Q_{\text{in}} - Q_{\text{out}}}{Q_{\text{in}}}$$
$$= \frac{(h_C - h_A) - (1 - y_1 - y_2)(h_F - h_G)}{h_C - h_A} \qquad 27.48$$

## 10. SUPERCRITICAL AND ULTRASUPERCRITICAL CYCLES

The thermal efficiency is increased by raising the average temperature at which heat is added. This is evident from Eq. 27.8, which shows that the ideal thermal efficiency depends on the highest temperature achieved in the cycle. Therefore, some modern plants are designed to achieve supercritical temperatures. The operating temperatures are limited only by metallurgical considerations.

Figure 27.11 shows that no heat is added at constant temperature. Reheating (resuperheating) is used to keep the steam quality high when it expands from point B.

*Figure 27.11* Supercritical Rankine Cycle Property Plot

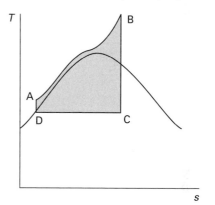

## 11. BINARY CYCLE

The binary cycle utilizes two different fluids, such as mercury and water, achieving conditions unobtainable with a single working fluid.[4] The binary cycle is essentially two Rankine cycles, and Rankine procedures should be used to evaluate it. (See Fig. 27.12 and Fig. 27.13.)

A to B: Mercury is heated to its saturation temperature.
B to C: Mercury is vaporized in the boiler.
C to D: Mercury expands adiabatically in the turbine.

---

[4]The mercury binary cycle is of academic interest. However, there are no longer any commercial installations using this design. The last binary mercury/steam power plant to be built was the Schiller power station at Portsmouth, New Hampshire, 1950, which was retired in 1968.

D to E:   Mercury condenses in the combination condenser-boiler.

E to A:   Mercury is compressed adiabatically.

F to G:   Water is heated to its saturation temperature.

G to H:   Water vaporizes in a boiler.

H to I:   Steam is superheated.

I to J:   Steam expands adiabatically in the turbine.

J to K:   Water condenses.

K to F:   Water is compressed adiabatically.

**Figure 27.12** *Binary Cycle*

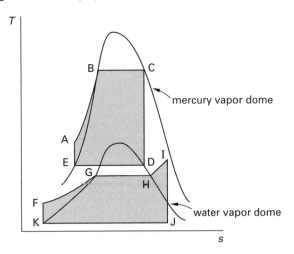

**Figure 27.13** *Binary Cycle*

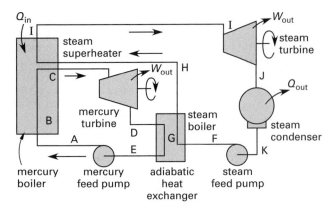

If $m$ mass units of mercury flow for every mass unit of steam and the heat transfer between the mercury and steam is adiabatic, the thermal efficiency of the binary cycle is

$$\eta_{\text{th}} = \frac{W_{\text{turbines}} - W_{\text{pumps}}}{Q_{\text{in}}}$$

$$= \frac{m(h_{\text{C}} - h_{\text{D}}) + (h_{\text{I}} - h_{\text{J}}) - m(h_{\text{A}} - h_{\text{E}}) - (h_{\text{F}} - h_{\text{K}})}{m(h_{\text{C}} - h_{\text{A}}) + (h_{\text{I}} - h_{\text{H}})}$$

$$27.49$$

# 28 Combustion Power Cycles

## Nomenclature

| | | | |
|---|---|---|---|
| $A$ | bore area | in$^2$ | m$^2$ |
| AFR | air-fuel ratio | lbm/lbm | kg/kg |
| BHP | brake horsepower | hp | n.a. |
| BkW | brake kilowatts | n.a. | kW |
| BSFC | brake specific fuel consumption | lbm/hp-hr | kg/kW·h |
| BWR | back-work ratio | – | – |
| $c$ | clearance | – | – |
| $c$ | specific heat | Btu/lbm-°F | kJ/kg·K |
| $D$ | diameter | ft | m |
| $F$ | force | lbf | N |
| FAR | fuel-air ratio | lbm/lbm | kg/kg |
| FHP | friction (horse) power | hp | kW |
| FU | fuel utilization | – | – |
| $h$ | enthalpy | Btu/lbm | kJ/kg |
| HV | heating value | Btu/lbm | kJ/kg |
| IHP | indicated (horse) power | hp | kW |
| $k$ | ratio of specific heats | – | – |
| $L$ | stroke length | ft | m |
| $\dot{m}$ | mass flow rate | lbm/sec | kg/s |
| MEP | mean effective pressure | lbf/ft$^2$ | Pa |
| $n$ | engine speed (rpm) | rev/min | r/min |
| $N$ | number of power strokes per minute | min$^{-1}$ | min$^{-1}$ |
| $p$ | mean effective pressure | psig | Pa |
| $p$ | pressure | lbf/ft$^2$ | Pa |
| $P$ | power | hp | kW |
| PTR | power-to-heat ratio | – | – |
| $q$ | heat | Btu/lbm | kJ/kg |
| $Q$ | heat | Btu | kJ |
| $r$ | moment arm or radius | ft | m |
| $r$ | ratio | – | – |
| $R$ | specific gas constant | ft-lbf/lbm-°R | kJ/kg·K |
| $s$ | entropy | Btu/lbm-°F | kJ/kg·K |
| $T$ | temperature | °R | K |
| $T$ | torque | ft-lbf | N·m |
| $u$ | internal energy | Btu/lbm | kJ/kg |
| $U$ | overall coefficient of heat transfer | Btu/ft$^2$-hr-°F | W/m$^2$·K |
| $V$ | volume | ft$^3$ | m$^3$ |
| $\dot{V}$ | volumetric flow rate | ft$^3$/sec | m$^3$/s |
| $W$ | work | Btu | kJ |
| $W$ | work per unit mass | Btu/lbm | kJ/kg |

## Symbols

| | | | |
|---|---|---|---|
| $\eta$ | efficiency | – | – |
| $\rho$ | mass density | lbm/ft$^3$ | kg/m$^3$ |
| $v$ | specific volume | ft$^3$/lbm | m$^3$/kg |

## Subscripts

| | |
|---|---|
| $a$ | air |
| $c$ | cut-off |
| $f$ | fuel |
| $m$ | mechanical |
| $p$ | pressure or constant pressure |
| $r$ | ratio or relative |
| $s$ | isentropic |
| th | thermal |
| $v$ | volume, volumetric compression, or constant volume |

## 1. INTRODUCTION TO AIR-STANDARD CYCLES

Combustion power cycles differ from vapor power cycles in that the combustion products cannot be returned to their initial conditions for reuse. Combustion power cycles are often analyzed as air-standard cycles, due to the computational difficulties of working with mixtures of fuel vapor, combustion products, and air.

An *air-standard cycle* is a closed system using a fixed amount of ideal air as the working fluid. In contrast to a combustion process, the heat of combustion is included in the calculations without consideration of the heat source or delivery mechanism. (That is, the combustion process is replaced by a process of instantaneous heat transfer from high-temperature surroundings.) Similarly, the cycle ends with an instantaneous transfer of waste heat to the surroundings. All processes are considered to be internally reversible. Because the air is ideal, it has a constant specific heat.[1]

Actual engine efficiencies for internal combustion engine cycles may be up to 50% lower than the efficiencies calculated from air-standard analyses. Empirical corrections must be applied to theoretical calculations based on the characteristics of the engine. The large amount of excess air used in turbine combustion cycles results in better agreement between actual and ideal performance than for reciprocating engines.

## 2. ENGINE TERMINOLOGY

Internal combustion (IC) engines can be categorized into *spark ignition* (SI) and *compression ignition* (CI) categories. SI engines (i.e., typical gasoline engines) use a spark to ignite the air-fuel mixture, while the heat of compression ignites the air-fuel mixture in CI engines (i.e., typical diesel engines).

The diameter of the circular cylinder is the *bore.* The maximum distance traveled by the piston is the *stroke.* An engine with a bore of diameter $D$ and stroke of length $L$ is sometimes referred to as a $D \times L$ engine. The product of the cylinder area and stroke is the *swept volume.* The fractional *clearance* is the ratio of *clearance (clear) volume* to the *swept volume,* usually expressed as a percentage. At *top-dead-center* (TDC), the piston is at maximum reach from the crankshaft. At *bottom-dead-center* (BDC), the piston is closest to the crankshaft.

The expanding combustion gases act on one end of the piston in *single-acting engines* and on both ends of the piston in *double-acting engines.* Double-acting internal combustion engines are essentially nonexistent due to difficulties in sealing and power transmission. However, some double-acting reciprocating steam engines are used for stationary applications, and double-acting Stirling machines are common.[2] The total power generated in a double-acting engine is the sum of the power generated by the *head end* and *crank end.*

## 3. FOUR- AND TWO-STROKE ENGINES

Engines are categorized as either four-stroke or two-stroke. In a *four-stroke cycle,* four separate piston movements (*strokes*) and two complete crankshaft revolutions are required to accomplish all of the processes: *intake, compression, power,* and *exhaust strokes.*[3]

*Two-stroke cycles* are used in small gasoline engines (such as in lawn mowers and other garden equipment, outboards, and motorcycles) as well as in some diesel engines. In a two-stroke cycle, only two piston movements and one crankshaft revolution occur per power stroke. The air-fuel mixture is drawn in through the intake port, and exhaust gases expand out through the exhaust port, near the end of the power stroke. The intake, exhaust, and power strokes overlap. The term *scavenging* is used to describe the act of blowing the exhaust products out with the air-fuel mixture.

Because power strokes occur twice as often, a two-stroke engine produces more power (for a given engine weight and displacement) than a four-stroke engine. However, the increase in power is only approximately 70% to 90% (not 100%) due to various inefficiencies, including incomplete mixing and scavenging.

In the typical commercial two-stroke engine, some of the exhaust gases dilute the air-fuel mixture. Also, oil is mixed with the fuel to provide engine lubrication. Both of these practices lead to poorer fuel economy and higher emissions.

With the proper modifications (e.g., supercharging, indirect fuel injection, scavenging with pure air, lean-burning *stratified charge combustion,* and *exhaust gas dilution*), however, two-stroke gasoline engines may someday be an attractive alternative to traditional automobile engines. The primary benefits (as perceived by automobile engine manufacturers) of two-stroke engines are reduced engine vibration, more uniform torque and power, and the ability to achieve reduced air pollution emissions, particularly nitrogen oxides.[4] This theoretically eliminates the need for a nitrogen-reducing catalyst.[5]

## 4. CARBURETION AND FUEL INJECTION

In a normally aspirated engine, fuel is mixed with air outside of the cylinder in the *carburetor.* Mixing continues in the manifold. A butterfly valve in the

---

[1]The term *cold air-standard cycle* is sometimes used to describe a cycle where the specific heats are considered to be constant at their room-temperature values.

[2]Stirling engines are not very common in the first place. But among Stirling engines, double action is common.

[3]The intake and exhaust strokes do not affect the thermodynamics of the cycle. These strokes do not appear on the cycle diagrams. The four strokes do not correspond to the four processes in the Otto and diesel cycles.

[4]Another benefit, that of fewer parts and similar design, is sacrificed in commercial vehicle designs because of the need for strict emissions control.

[5]*Two-way catalytic converters* (using platinum and palladium) are still required for hydrocarbon and carbon monoxide reduction. However, these catalysts are less expensive than the *three-way converters* that also use rhodium for nitrogen control.

carburetor throttles the flow of air. The air-fuel mixture enters through ports closed by flat-headed valves.[6]

In *fuel-injected* (FI) *engines*, pressurized fuel is injected directly into the cylinder at just the right moment in the cycle. The timing may be controlled electronically or mechanically (by distributor disk, camshaft lobes, etc.). Pressurization may be supplied by a fuel pump or by injector pistons (either spring-loaded or cam-actuated). With *direct injection*, fuel is injected directly into the combustion chamber, typically into the crown at the top of the piston. With *indirect injection*, fuel is injected into a *precombustion chamber (prechamber)* where it is mixed with air prior to moving into the combustion chamber. Diesel engines using indirect injection are quieter and less polluting, though they are slightly less fuel-efficient.

## 5. OPERATING CHARACTERISTICS OF COMBUSTION ENGINES

The *specific fuel consumption* (SFC) is the fuel usage rate divided by the power generated. Typical units are lbm/hp-hr and kg/kW·h.

The *air-fuel ratio* (AFR) is the ratio of the air mass that enters the engine to each mass of fuel burned. The *fuel-air ratio* (FAR) is the reciprocal of the air-fuel ratio.

$$\text{AFR} = \frac{\dot{m}_{\text{air}}}{\dot{m}_{\text{fuel}}} = \frac{1}{\text{FAR}} \qquad 28.1$$

For Otto and diesel cycles, the *compression ratio*, $r_v$, is the ratio of two volumes.[7]

$$r_v = \frac{V_{\text{max}}}{V_{\text{min}}} \qquad 28.2$$

The *thermal efficiency*, $\eta_{\text{th}}$, of a combustion engine cycle is the ratio of net output power to input energy, both expressed in the same units.

$$\eta_{\text{th}} = \frac{W_{\text{out}} - W_{\text{in}}}{Q_{\text{in}}} = \frac{Q_{\text{in}} - Q_{\text{out}}}{Q_{\text{in}}} \qquad 28.3$$

The *mechanical efficiency*, $\eta_m$, of a combustion engine is

$$\eta_m = \frac{\text{BHP}}{\text{IHP}} = \frac{\text{actual power developed}}{\text{frictionless power developed}} \qquad 28.4$$

In reciprocating engines, the *volumetric efficiency*, $\eta_v$, is the ratio of the actual to ideal volumes of entering gases.

$$\eta_v = \frac{\dot{V}_{\text{actual}}}{\dot{V}_{\text{ideal}}} \qquad 28.5$$

The *relative efficiency*, $\eta_r$, is the ratio of the actual and ideal thermal efficiencies. The *ideal efficiency*, $\eta_i$, can usually be calculated from other cycle properties (e.g., the compression ratio).

$$\eta_r = \frac{\eta_{\text{th,actual}}}{\eta_{\text{th,ideal}}} \qquad 28.6$$

## 6. BRAKE AND INDICATED PROPERTIES

The performance characteristics (e.g., horsepower) of internal combustion engines can be reported with or without the effect of power-reducing friction and other losses. A value of a property that includes the effect of friction is known as a *brake value*. If the effect of friction is removed, the property is known as an *indicated value*.[8]

Common brake properties are *brake horsepower* (BHP), *brake specific fuel consumption* (BSFC), and *brake mean effective pressure* (BMEP). Common indicated properties are *indicated horsepower* (IHP), *indicated specific fuel consumption* (ISFC), and *indicated mean effective pressure* (IMEP).

The brake and indicated horsepowers differ by the *friction horsepower* (FHP).

$$\text{FHP} = \text{IHP} - \text{BHP} \qquad 28.7$$

Except for Eq. 28.7, indicated and brake properties are usually not combined in calculations since they refer to two different operating conditions. For example, the calculation of the actual fuel mass flow rate requires two brake parameters.

$$\dot{m}_f = (\text{BSFC})(\text{BHP}) \qquad 28.8$$

## 7. ENGINE POWER AND TORQUE

The power and torque curves are graphs of maximum power and maximum torque that the engine can develop over its speed range. For any given speed, the maximum power that the engine can develop can be determined by loading the engine with progressively higher loads until stalling occurs. This is done on a *dynamometer* or with a *prony brake*, which applies a frictional resistance to a rotating drum attached to the engine's power takeoff, as shown in Fig. 28.1. Since the prony brake measures net available power (inclusive of frictional losses), the operating characteristics derived are brake values (i.e., brake torque and brake horsepower).

The *brake torque* developed is

$$T = rF \qquad 28.9$$

---

[6]The term *poppet valve* (i.e., a valve that rises and returns to its seat) is almost never heard anymore.

[7]For compressors and compression cycles, the term *compression ratio* is the ratio of pressures.

[8]It may be helpful to think of the *i* in "indicated" as meaning "ideal."

**Figure 28.1** *Idealized Prony Brake*

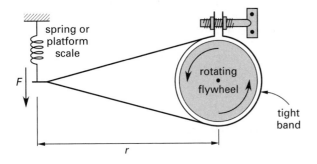

The brake torque can also be calculated from the brake horsepower (brake kilowatts) and rotational speed, $n$, in rpm.

$$T_{\text{N·m}} = \frac{(60\,000)(\text{BkW})}{2\pi n_{\text{rpm}}} = \frac{(9549)(\text{BkW})}{n_{\text{rpm}}} \quad \text{[SI]} \quad \textbf{\textit{28.10(a)}}$$

$$T_{\text{ft-lbf}} = \frac{(33,000)(\text{BHP})}{2\pi n_{\text{rpm}}} = \frac{(5252)(\text{BHP})}{n_{\text{rpm}}} \quad \text{[U.S.]} \quad \textbf{\textit{28.10(b)}}$$

The *brake horsepower (brake kilowatts)* can be calculated from the brake torque or directly from the results of the prony brake test. Notice that $rF$ is the brake torque.

$$\text{BkW} = \frac{2\pi r F n}{60\,000} = \frac{rFn}{9549} \quad \text{[SI]} \quad \textbf{\textit{28.11(a)}}$$

$$\text{BHP} = \frac{2\pi r F n}{33,000} = \frac{rFn}{5252} \quad \text{[U.S.]} \quad \textbf{\textit{28.11(b)}}$$

Power ratings listed for most commercial engines correspond to the "standard conditions" of 500 ft (150 m) altitude, a dry barometric pressure of 29.00 in of mercury, water vapor pressure of 0.38 in of mercury, and temperature of 85°F (29°C).[9,10]

The *continuous duty rating* is the rated power that the manufacturer claims the engine is able to provide on a continuous (governed or steady rpm) basis without incurring damage. The *intermittent rating* represents the peak power that can be produced on an occasional basis.[11]

## 8. AIR-STANDARD CARNOT CYCLE

Theoretically, the *air-standard Carnot combustion cycle* can be implemented either as a reciprocating or steady-flow device. However, like the Carnot vapor cycle, the air-standard Carnot combustion cycle is not a practical

---

[9]This "yet another" definition of standard conditions is specified in SAE Standard J816b.
[10]Some manufacturers rate their engines at sea level and 60°F (15.6°C). This increases the reported power by approximately 4%.
[11]Although more appropriate in jet aircraft flying without using after-burners, the terms "military power" and "full military power" refer to an intermittent maximum power setting above the normally continuous duty setting.

---

engine cycle.[12] The value of the cycle is in establishing a maximum thermal efficiency against which all other cycles can be compared. The Carnot cycle consists of the following processes.

    A to B:  isentropic compression
    B to C:  isothermal expansion power stroke
    C to D:  isentropic expansion
    D to A:  isothermal compression

The *T-s* diagram shown in Fig. 28.2 is the same for any Carnot cycle. However, the isothermal expansion and compression do not occur within a vapor dome as they do in a vapor power cycle and, therefore, do not occur at constant pressure. Thus, the *p-v* diagram is different than for a vapor cycle.

**Figure 28.2** *Air-Standard Carnot Cycle*

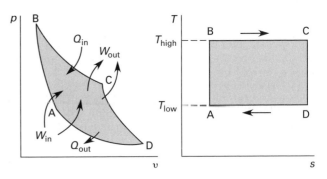

The *isentropic pressure ratio* is a ratio of pressures.

$$r_{p,s} = \frac{p_{\text{B}}}{p_{\text{A}}} = \frac{p_{\text{C}}}{p_{\text{D}}} = \left(\frac{T_{\text{D}}}{T_{\text{C}}}\right)^{k/(1-k)} \quad \textbf{\textit{28.12}}$$

The *isentropic compression ratio* for reciprocating equipment is a ratio of volumes.

$$r_{v,s} = \frac{v_{\text{A}}}{v_{\text{B}}} = \frac{v_{\text{D}}}{v_{\text{C}}} = \left(\frac{T_{\text{D}}}{T_{\text{C}}}\right)^{1/(1-k)} \quad \textbf{\textit{28.13}}$$

Pressure, temperature, and volume can be calculated from the isentropic and isothermal relationships for ideal gases.

At A:

$$p_{\text{A}} = p_{\text{D}}\left(\frac{v_{\text{D}}}{v_{\text{A}}}\right) = p_{\text{B}}\left(\frac{v_{\text{B}}}{v_{\text{A}}}\right)^{k} = p_{\text{B}}\left(\frac{T_{\text{A}}}{T_{\text{B}}}\right)^{k/(k-1)} \quad \textbf{\textit{28.14}}$$

$$T_{\text{A}} = T_{\text{D}} = T_{\text{B}}\left(\frac{v_{\text{B}}}{v_{\text{A}}}\right)^{k-1} = T_{\text{B}}\left(\frac{p_{\text{A}}}{p_{\text{B}}}\right)^{(k-1)/k} \quad \textbf{\textit{28.15}}$$

---

[12]In particular, it is not possible to design equipment that will transfer heat to a working fluid at a constant temperature in a reversible process over a reasonably finite time.

$$v_A = v_D \left(\frac{p_D}{p_A}\right) = v_B \left(\frac{p_B}{p_A}\right)^{1/k} = v_B \left(\frac{T_B}{T_A}\right)^{1/(k-1)}$$
$$\text{28.16}$$

At B:

$$p_B = p_A \left(\frac{v_A}{v_B}\right)^k = p_A \left(\frac{T_B}{T_A}\right)^{k/(k-1)} = p_C \left(\frac{v_C}{v_B}\right) \quad \text{28.17}$$

$$T_B = T_A \left(\frac{v_A}{v_B}\right)^{k-1} = T_A \left(\frac{p_B}{p_A}\right)^{(k-1)/k} = T_C \quad \text{28.18}$$

$$v_B = v_A \left(\frac{p_A}{p_B}\right)^{1/k} = v_A \left(\frac{T_A}{T_B}\right)^{1/(k-1)} = v_C \left(\frac{p_C}{p_B}\right) \quad \text{28.19}$$

At C:

$$p_C = p_B \left(\frac{v_B}{v_C}\right) = p_D \left(\frac{v_D}{v_C}\right)^k = p_D \left(\frac{T_C}{T_D}\right)^{k/(k-1)} \quad \text{28.20}$$

$$T_C = T_B = T_D \left(\frac{v_D}{v_C}\right)^{k-1} = T_D \left(\frac{p_C}{p_D}\right)^{(k-1)/k} \quad \text{28.21}$$

$$v_C = v_B \left(\frac{p_B}{p_C}\right) = v_D \left(\frac{p_D}{p_C}\right)^{1/k} = v_D \left(\frac{T_D}{T_C}\right)^{1/(k-1)}$$
$$\text{28.22}$$

At D:

$$p_D = p_C \left(\frac{v_C}{v_D}\right)^k = p_C \left(\frac{T_D}{T_C}\right)^{k/(k-1)} = p_A \left(\frac{v_A}{v_D}\right) \quad \text{28.23}$$

$$T_D = T_C \left(\frac{v_C}{v_D}\right)^{k-1} = T_C \left(\frac{p_D}{p_C}\right)^{(k-1)/k} = T_A \quad \text{28.24}$$

$$v_D = v_C \left(\frac{p_C}{p_D}\right)^{1/k} = v_C \left(\frac{T_C}{T_D}\right)^{1/(k-1)} = v_A \left(\frac{p_A}{p_D}\right) \quad \text{28.25}$$

The work and heat flow terms per unit mass of working fluid are

$$W_{out} = c_v (T_C - T_D) + T_B (s_C - s_B) \quad \text{28.26}$$

$$W_{in} = |c_v (T_A - T_B)| + T_D (s_D - s_A) \quad \text{28.27}$$

$$q_{out,D-A} = |T_{low}(s_A - s_D)| = \left| p_D v_D \ln \frac{v_A}{v_D} \right| \quad \text{28.28}$$

$$q_{in,B-C} = T_{high}(s_C - s_B) = p_B v_B \ln \frac{v_C}{v_B} \quad \text{28.29}$$

The thermal efficiency of the air-standard Carnot cycle can be calculated from the heat and work terms, but it is easier to work with the two temperature extremes.

The temperature forms of the thermal efficiency equation are easy to derive since $q = T\Delta s$.

$$\eta_{th} = \frac{W_{out} - W_{in}}{Q_{in}} = \frac{Q_{in} - Q_{out}}{Q_{in}}$$
$$= \frac{T_B - T_A}{T_B} = \frac{T_C - T_D}{T_C} \quad \text{28.30}$$

The ideal thermal efficiency can also be calculated from the compression ratio.

$$\eta_{th} = 1 - r_{v,s}^{1-k} = 1 - r_{p,s}^{(1-k)/k} \quad \text{28.31}$$

## 9. AIR-STANDARD OTTO CYCLE

The *air-standard Otto cycle* consists of the following processes and is illustrated in Fig. 28.3. The Otto cycle is a *four-stroke cycle*.

A to B: isentropic compression
B to C: constant volume heat addition
C to D: isentropic expansion
D to A: constant volume heat rejection

**Figure 28.3** Air-Standard Otto Cycle

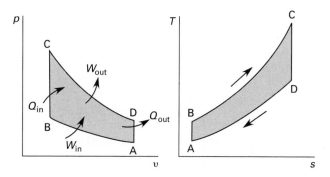

Although it is possible to define a pressure ratio for the Otto cycle, usually only the volumetric *compression ratio*, $r_v$, is used.

$$r_v = \frac{V_A}{V_B} = \frac{V_D}{V_C} \quad \text{28.32}$$

Depending on the octane, compression ratios are typically in the range of 5:1 to 10.5:1. Most modern automobile engines designed to run on unleaded gasoline have compression ratios of approximately 8:1 to 8.5:1.

The ratios of absolute temperatures are equal because the piston travels a fixed distance, regardless of which stroke the cycle is on.

$$\frac{T_D}{T_C} = \frac{T_A}{T_B} \quad \text{28.33}$$

Pressure, volume, and temperature relationships for the isentropic and constant volume ideal gas processes can be evaluated using air tables or the ideal gas equations. Using the ideal gas equations makes the implicit assumption that the specific heats are constant during the four processes, while the specific heats actually vary greatly over the wide temperature extremes encountered. By using the air tables this source of error is eliminated, but because the working fluid is not actually pure air, the additional refinement is probably unwarranted.

Assuming an ideal gas, the work and heat flow terms are

$$q_{in,B-C} = c_v(T_C - T_B) \qquad \textit{28.34}$$

$$q_{out,D-A} = |c_v(T_A - T_D)| \qquad \textit{28.35}$$

$$W_{in,A-B} = c_v(T_B - T_A) = \frac{p_B v_B - p_A v_A}{k-1} \qquad \textit{28.36}$$

$$W_{out,C-D} = |c_v(T_D - T_C)| = \left|\frac{p_D v_D - p_C v_C}{k-1}\right| \qquad \textit{28.37}$$

The thermal efficiency for the Otto cycle can be calculated in a number of ways.

$$\eta_{th} = \frac{Q_{in} - Q_{out}}{Q_{in}} = \frac{W_{out} - W_{in}}{Q_{in}}$$
$$= \frac{T_C - T_D}{T_C} = \frac{T_B - T_A}{T_B} \qquad \textit{28.38}$$

The ideal thermal efficiency can be calculated from the compression ratio.

$$\eta_{th} = 1 - r_v^{1-k} \qquad \textit{28.39}$$

From Eq. 28.39, it can be inferred that the ideal thermal efficiency of an Otto cycle can be increased by increasing either the ratio of specific heats or the compression ratio or both. Increasing the ratio of specific heats can only be accomplished by substituting another gas (e.g., helium) for the atmospheric nitrogen. This is not a practical alternative. Increases in the compression ratio are limited to approximately 10:1 by fuel detonation.[13] (Cylinder sealing is a problem only for compression ratios above approximately 16:1.)

The air-standard Otto cycle is less efficient than the Carnot cycle operating between the same temperature limits. However, Eq. 28.31 and Eq. 28.39 show that equal efficiencies are obtained if the compression ratios are made equal.

## Example 28.1

An internal combustion engine is to be evaluated on the basis of an air-standard Otto cycle. The pressure and temperature at the intake are 14.7 psia and 600°R

(101.3 kPa and 330K), respectively. The maximum pressure and temperature in the cycle are 340 psia and 2610°R (2.3 MPa and 1450K), respectively. Consider air to be an ideal gas. What are the (a) pressure at the end of the compression stroke, (b) temperature at the end of the compression stroke, and (c) cycle efficiency?

*SI Solution*

Refer to Fig. 28.3. The pressure at point B can be related to the temperatures at points A and B. $p_B$ and $p_C$ must have the same units.

$$p_B = p_A \left(\frac{T_B}{T_A}\right)^{k/(k-1)}$$
$$= (101.3 \text{ kPa})\left(\frac{T_B}{330K}\right)^{1.4/(1.4-1)}$$
$$= (0.1013 \text{ MPa})\left(\frac{T_B}{330K}\right)^{3.5}$$

The temperature at point B can be related to the pressures at points B and C.

$$T_B = \frac{T_C p_B}{p_C} = \frac{(1450K)(p_B)}{2.3 \text{ MPa}}$$
$$= 630.4 p_B$$

Solve these two equations simultaneously.

$$p_B = 1.01 \text{ MPa}$$
$$T_B = 637K$$

Use Eq. 28.38.

$$\eta_{th} = \frac{T_B - T_A}{T_B} = \frac{637K - 330K}{637K}$$
$$= 0.482 \quad (48.2\%)$$

*Customary U.S. Solution*

Refer to Fig. 28.3. The pressure at point B can be related to the temperatures at points A and B. $p_B$ and $p_C$ must have the same units.

$$p_B = p_A \left(\frac{T_B}{T_A}\right)^{k/(k-1)}$$
$$= (14.7 \text{ psia})\left(\frac{T_B}{600°R}\right)^{1.4/(1.4-1)}$$
$$= (14.7 \text{ psia})\left(\frac{T_B}{600°R}\right)^{3.5}$$

---

[13]*Detonation* is a premature autoignition of the fuel. After the spark ignites one portion of the air-fuel mixture, the advancing flame front compresses the remainder of the mixture. If the compression ratio is too high, the compressed remainder may detonate.

The temperature at point B can be related to the pressures at points B and C.

$$T_B = \frac{T_C p_B}{p_C} = \frac{(2610°R)(p_B)}{340 \text{ psia}}$$
$$= 7.676 p_B$$

Solve these two equations simultaneously.

$$p_B = 152.5 \text{ psia}$$
$$T_B = 1171°R$$

Use Eq. 28.38.

$$\eta_{th} = \frac{T_B - T_A}{T_B} = \frac{1171°R - 600°R}{1171°R}$$
$$= 0.488 \quad (48.8\%)$$

**Example 28.2**

An Otto engine has 15% clearance. Air enters the engine at 14.0 psia and 580°R (96.6 kPa and 320K). The air-fuel ratio is 16.67. The heating value of gasoline is 19,500 Btu/lbm (45.5 MJ/kg), and the combustion efficiency is 76.9%. What are the temperatures and pressures at all points in the cycle?

*SI Solution*

(This solution takes a cold air-standard approach. The customary U.S. solution uses an air table.)

Refer to Fig. 28.3.

At A:

$$V_A = V_{\text{swept}} + V_{\text{clearance}} = (1 + c) V_{\text{swept}}$$
$$= (1 + 0.15) V_{\text{swept}} = 1.15 V_{\text{swept}}$$
$$p_A = 96.6 \text{ kPa} \quad \text{[given]}$$
$$T_A = 320K \quad \text{[given]}$$

At B:

$$V_B = 0.15 V_{\text{swept}}$$

The compression ratio is

$$r_v = \frac{V_A}{V_B} = \frac{1.15 V_{\text{swept}}}{0.15 V_{\text{swept}}}$$
$$= 7.67$$

The compression from A to B is isentropic.

$$T_B = T_A \left(\frac{V_A}{V_B}\right)^{k-1} = (320K)(7.67)^{1.4-1}$$
$$= 722.9K$$
$$p_B = p_A \left(\frac{V_A}{V_B}\right)^{k} = (96.6 \text{ kPa})(7.67)^{1.4}$$
$$= 1674 \text{ kPa} \ (1.674 \text{ MPa})$$

At C: The heat added by the fuel per kg of air is

$$q = \left(\frac{HV}{AFR}\right)\eta = \left(\frac{45.5 \ \frac{MJ}{\text{kg fuel}}}{16.67 \ \frac{\text{kg air}}{\text{kg fuel}}}\right)(0.769)$$
$$= 2.099 \text{ MJ/kg air}$$

The combustion heat increases the internal energy of the air. For any process, $\Delta u = c_v \Delta T$. The cold specific heat (0.718 kJ/kg·K) of air is used.

$$T_C = T_B + \frac{\Delta u}{c_v} = T_B + \frac{q}{c_v}$$
$$= 722.9K + \frac{\left(2.099 \ \frac{MJ}{\text{kg}}\right)\left(1000 \ \frac{kJ}{MJ}\right)}{0.718 \ \frac{kJ}{\text{kg·K}}}$$
$$= 3646K$$
$$p_C = \frac{p_B T_C}{T_B} = \frac{(1.674 \text{ MPa})(3646K)}{722.9K}$$
$$= 8.44 \text{ MPa}$$

At D: The process from C to D is isentropic. Since the piston travels the same distance going from A to B as it does from C to D, the C-D expansion ratio is the same as the A-B compression ratio.

$$T_D = T_C \left(\frac{V_C}{V_D}\right)^{k-1} = (3646K)\left(\frac{1}{7.67}\right)^{1.4-1}$$
$$= 1614K$$
$$p_D = p_C \left(\frac{V_C}{V_D}\right)^{k} = (8.44 \text{ MPa})\left(\frac{1}{7.67}\right)^{1.4}$$
$$= 0.487 \text{ MPa}$$

*Customary U.S. Solution*

(This solution uses an air table. The SI solution takes a cold air-standard cycle approach.)

Refer to Fig. 28.3.

At A:

$$V_A = V_{swept} + V_{clearance} = (1 + c)V_{swept}$$
$$= (1 + 0.15)V_{swept} = 1.15\,V_{swept}$$

$$p_A = 14.0 \text{ psia} \quad \text{[given]}$$

$$T_A = 580°\text{R} \quad \text{[given]}$$

From the air table at $580°\text{R}$, $v_{r,A} = 120.7$ and $p_{r,A} = 1.78$.

At B:

$$V_B = 0.15\,V_{swept}$$

The compression ratio is

$$r_v = \frac{V_A}{V_B} = \frac{1.15\,V_{swept}}{0.15\,V_{swept}}$$
$$= 7.67$$

Consider the compression from A to B to be isentropic. Then,

$$v_{r,B} = \frac{v_{r,A}}{r_v}$$
$$= \frac{120.7}{7.67} = 15.74$$

The temperature corresponding to this volume ratio in the air table is $1273°\text{R}$.

$$T_B = 1273°\text{R}$$

$$u_B = 222.72 \text{ Btu/lbm}$$

$$p_{r,B} = 29.93$$

$$p_B = \frac{p_A p_{r,B}}{p_{r,A}} = \frac{(14.0 \text{ psia})(29.93)}{1.78}$$
$$= 235.4 \text{ psia}$$

At C: The heat added by the fuel per pound of air is

$$q = \left(\frac{\text{HV}}{\text{AFR}}\right)\eta = \left(\frac{19,500 \frac{\text{Btu}}{\text{lbm fuel}}}{16.67 \frac{\text{lbm air}}{\text{lbm fuel}}}\right)(0.769)$$
$$= 899.55 \text{ Btu/lbm air}$$

Enthalpy is the sum of internal energy and $pV$ energy. Even though the enthalpy at point B could be found from the air table, the pressure at point C is not yet known. Therefore, this heat addition cannot merely be added to the enthalpy at point B. However, internal energy does not include the $pV$ term. The internal energy at the end of the heat addition is

$$u_C = u_B + q = 222.72 \frac{\text{Btu}}{\text{lbm}} + 899.55 \frac{\text{Btu}}{\text{lbm}}$$
$$= 1122.27 \text{ Btu/lbm}$$

Locate this value of internal energy in the air table.

$$T_C = 5300°\text{R}$$

$$v_{r,C} = 0.1710$$

$$p_{r,C} = 11,481$$

Use the ideal gas law to find the pressure at C.

$$p_C = \frac{p_B T_C}{T_B} = \frac{(235.4 \text{ psia})(5300°\text{R})}{1273°\text{R}}$$
$$= 980.1 \text{ psia}$$

At D: The process from C to D is isentropic.

$$v_{r,D} = v_{r,C}r_v$$
$$= (0.1710)(7.67) = 1.3116$$

Locate this volume ratio in the air table.

$$T_D \approx 2900°\text{R}$$

$$p_{r,D} = 814.8$$

$$p_D = \frac{p_C p_{r,D}}{p_{r,C}} = \frac{(980.1 \text{ psia})(814.8)}{11,481}$$
$$= 69.6 \text{ psia}$$

## 10. PLAN FORMULA

The performance of an internal combustion engine operating on the Otto cycle can be predicted from the PLAN formula, Eq. 28.40.[14] It is necessary to know the engine bore area for one cylinder ($A$ in square inches), stroke ($L$ in feet), number of engine power strokes per minute ($N$), and mean effective pressure ($p$ in psig). The product $LA$ is the *swept volume* of one cylinder. If the *engine displacement*, $V$, is known, it should be divided by the number of cylinders before being substituted for $LA$. It is essential to use the units as defined.[15] The *brake horsepower* (BHP) will be calculated if the *brake mean effective pressure* (BMEP) is used. The *indicated horsepower* (IHP) will be calculated if the *indicated mean effective pressure* (IMEP) is used.

$$P_{kW} = \frac{pLAN}{60} \qquad \text{[SI]} \quad 28.40(a)$$

---

[14]The PLAN formula is also applicable to reciprocating steam engines.
[15]The PLAN formula can be used with SI units if consistent units are used.

$$P_{\text{hp}} = \frac{pLAN}{33{,}000} \qquad \text{[U.S.]} \quad 28.40(b)$$

Equation 28.41 gives the number of engine power strokes per minute for two- and four-stroke engines.

$$N = \frac{(2n)(\text{no. cylinders})}{\text{no. strokes per cycle}} \qquad 28.41$$

The indicated *mean effective pressure*, $p$, is determined graphically from an actual $p$-$v$ plot (known as an *indicator drawing*, *indicator diagram*, or *indicator card*) of the actual cycle. This is done by taking the overall area of the indicator drawing and dividing it by the total width of the drawing. The resulting number is then multiplied by the indicator spring constant (scale) to obtain the net work per cycle, $W_{\text{net}}$.

The mean effective pressure is a theoretical average pressure that would produce the same amount of net work during one stroke as the varying pressure does in the entire cycle. The mean effective pressure can be calculated if the cycle performance is known. The mean effective pressure in nonsupercharged engines is typically limited to approximately 100 psi (700 kPa).[16]

$$\text{MEP} = \frac{W_{\text{net}}}{V_A - V_B} \qquad 28.42$$

In some cases, the power generated by an engine is correlated with $CND^2$, where $C$ is a constant. This is the same as Eq. 28.40, with the coefficient $C$ incorporating all of the constant terms.

Figure 28.4 illustrates how the actual Otto cycle deviates from the ideal cycle (shown by the broken line) and the causes of the deviations. Due to heat transfers during the A-to-B and C-to-D processes, the lines are not true adiabats. Rather, these two processes are polytropic, with a polytropic exponent of approximately 1.3. The net effect of all of these deviations is to reduce the actual efficiency to approximately half of the ideal efficiency.

**Example 28.3**

A single-cylinder, four-stroke engine has a 10 in (25.4 cm) bore and an 18 in (45.7 cm) stroke. The engine is run at 200 rpm while being tested on a prony brake. The gross force exerted by the brake is 140 lbf (620 N), the tare is 25 lbf (110 N), and the arm length is 66 in (1.7 m). The indicator card shows an area of 1.20 in$^2$ (7.7 cm$^2$) with an overall length of 3 in (7.6 cm). The spring scale used to draw the indicator card is 200 psi/in (540 kPa/cm). What are the (a) indicated main effective pressure, (b) indicated horsepower, (c) brake horsepower, and (d) mechanical efficiency?

---

[16]With the advent of supercharging, *peak firing pressure*, also known as *peak cylinder pressure* (PCP), has become a benchmark for engine performance.

**Figure 28.4** *Otto Cycle Indicator Drawing*

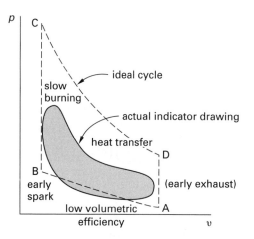

*SI Solution*

(a) The indicated mean effective pressure is

$$p = \frac{(\text{diagram area})(\text{pressure scale factor})}{\text{diagram length}}$$

$$= \frac{(7.7 \text{ cm}^2)\left(540 \ \dfrac{\text{kPa}}{\text{cm}}\right)}{7.6 \text{ cm}}$$

$$= 547 \text{ kPa}$$

(b) The stroke is

$$L = \frac{45.7 \text{ cm}}{100 \ \dfrac{\text{cm}}{\text{m}}} = 0.457 \text{ m}$$

The cylinder area is

$$A = \frac{\pi D^2}{4} = \frac{\pi(25.4 \text{ cm})^2}{4}$$

$$= 506.7 \text{ cm}^2 \ (0.05067 \text{ m}^2)$$

The number of power strokes per minute is

$$N = \frac{(2n)(\text{no. cylinders})}{\text{no. strokes per cycle}}$$

$$= \frac{\left(2 \ \dfrac{\text{strokes}}{\text{rev}}\right)\left(200 \ \dfrac{\text{rev}}{\text{min}}\right)(1 \text{ cylinder})}{4 \text{ strokes per power stroke}}$$

$$= 100 \text{ power strokes/min}$$

Equation 28.40 gives the indicated power. The conversion to horsepower is not needed.

indicated power

$$= pLAN$$

$$= \frac{(547 \text{ kPa})\left(1000 \frac{\text{Pa}}{\text{kPa}}\right)(0.457 \text{ m})}{\left(60 \frac{\text{s}}{\text{min}}\right)\left(1000 \frac{\text{W}}{\text{kW}}\right)}$$

$$= 21.1 \text{ kW}$$

(c) The brake horsepower is found from the brake information. The brake arm length is

$$r = 1.7 \text{ m}$$

The force on the brake caused by the engine excludes the tare weight.

$$F = 620 \text{ N} - 110 \text{ N} = 510 \text{ N}$$

The brake power is

brake power $= 2\pi rFn$

$$= \frac{(2\pi)(1.7 \text{ m})(510 \text{ N})\left(200 \frac{\text{rev}}{\text{min}}\right)}{\left(60 \frac{\text{s}}{\text{min}}\right)\left(1000 \frac{\text{W}}{\text{kW}}\right)}$$

$$= 18.16 \text{ kW}$$

(d) The mechanical efficiency is

$$\eta_m = \frac{\text{brake power}}{\text{indicated power}} = \frac{18.16 \text{ kW}}{21.1 \text{ kW}}$$

$$= 0.861 \ (86.1\%)$$

*Customary U.S. Solution*

(a) The indicated mean effective pressure is

$$p = \frac{(\text{diagram area})(\text{pressure scale factor})}{\text{diagram length}}$$

$$= \frac{(1.2 \text{ in}^2)\left(200 \frac{\text{lbf}}{\text{in}^3}\right)}{3 \text{ in}}$$

$$= 80 \text{ lbf/in}^2$$

(b) The stroke is

$$L = \frac{18 \text{ in}}{12 \frac{\text{in}}{\text{ft}}} = 1.5 \text{ ft}$$

The cylinder area is

$$A = \frac{\pi D^2}{4} = \frac{\pi(10 \text{ in})^2}{4}$$

$$= 78.54 \text{ in}^2$$

The number of power strokes per minute is

$$N = \frac{(2n)(\text{no. cylinders})}{\text{no. strokes per cycle}}$$

$$= \frac{\left(2 \frac{\text{strokes}}{\text{rev}}\right)\left(200 \frac{\text{rev}}{\text{min}}\right)(1 \text{ cylinder})}{4 \text{ strokes per power stroke}}$$

$$= 100 \text{ power strokes/min}$$

From Eq. 28.40, the indicated horsepower is

$$\text{IHP} = \frac{pLAN}{33,000}$$

$$= \frac{\left(80 \frac{\text{lbf}}{\text{in}^2}\right)(1.5 \text{ ft})(78.54 \text{ in}^2)}{33,000 \frac{\text{ft-lbf}}{\text{hp-min}}}$$

$$= 28.56 \text{ hp}$$

(c) The brake horsepower is found from the brake information. The brake arm length is

$$r = \frac{66 \text{ in}}{12 \frac{\text{in}}{\text{ft}}} = 5.5 \text{ ft}$$

The force on the brake caused by the engine excludes the tare weight.

$$F = 140 \text{ lbf} - 25 \text{ lbf} = 115 \text{ lbf}$$

From Eq. 28.11, the brake horsepower is

$$\text{BHP} = \frac{2\pi rFn}{33,000}$$

$$= \frac{(2\pi)(5.5 \text{ ft})(115 \text{ lbf})\left(200 \frac{\text{rev}}{\text{min}}\right)}{33,000 \frac{\text{ft-lbf}}{\text{hp-min}}}$$

$$= 24.09 \text{ hp}$$

(d) The mechanical efficiency is

$$\eta_m = \frac{\text{BHP}}{\text{IHP}} = \frac{24.09 \text{ hp}}{28.56 \text{ hp}}$$

$$= 0.843 \ (84.3\%)$$

## 11. INTRODUCTION TO DIESEL ENGINES

A diesel engine is a compression-ignition, internal combustion engine. Spark plugs are not used for ignition.[17] Rather, high compression ratios produce auto-ignition of the air-fuel mixture. The higher the compression ratio, the higher the thermal efficiency. Compression ratios for diesel engines are ratios of volumes. The compression ratio varies from about 13.5:1 to 17.5:1, depending on whether or not the engine is turbocharged. Diesel engines can burn a variety of fuels including No. 6 heavy fuel oil, natural gas, and light distillate fuel oils.

Four-stroke diesels offer many significant advantages over two-stroke diesels. (1) With two-stroke engines, a significant amount of energy is expended by a blower to supply air for scavenging. When the load on the engine drops, the blower continues to run at its peak load, reducing engine efficiency even more. (2) Two-cycle engines often require premium fuels, compared to No. 2 oil that most four-stroke engines can use. (3) Valve, piston, and ring burning are more common in two-stroke engines since there is less time for cooling during the cycle. Four-stroke engines can use aluminum pistons, while heavy cast-iron pistons are needed for heat dissipation in two-stroke engines. (4) In a two-stroke engine, piston rings pass over the intake ports on every stroke, leading to increased ring wear. (5) Since the valves and injectors operate every cycle with two-stroke engines, overhauls are required more frequently.

*Supercharging* compresses and increases the amount of air that enters the cylinder per stroke. (This increases the volumetric efficiency above 100%.) *Turbocharging* is a form of supercharging in which the exhaust gases drive the supercharger. With more air, more fuel can be burned, increasing the power per stroke. Supercharging also helps deliver air for combustion at higher altitudes. This results in better fuel economy than normally aspirated diesels. Supercharging also reduces smoke, particularly during lugging.

The temperature of compressed air from a turbocharger can be decreased by passing it through an aftercooler. An *aftercooler* is a closed heat exchanger that transfers heat from compressed air to cooler air. In vehicles, cool air flow is found in the manifold.

Emissions from diesel engines include nitrogen oxides, unburned hydrocarbons (*polycyclic aromatic hydrocarbons*—PAH), and carbon monoxide, much the same as from gasoline engines. Particulate emission (smoke or soot) and noise are also characteristic diesel pollutants.

## 12. STATIONARY AND MARINE GAS DIESELS

Large stationary engines operating on the diesel cycle can be built to run on natural gas.[18] There are two types of gas-burning diesels: gas-diesel and lean-burn engines. Natural gas under pressure has poor ignition and combustion characteristics. Therefore, in the *gas-diesel engine*, a small amount (about 3%) of oil is burned in addition to the natural gas. The oil is injected into the regular chamber or a precombustion chamber. Compression ignites the oil, and the resulting flame "torch" ignites the natural gas.

Since most gas-diesel engines produce less than 10 MW of power, they can be used for small electrical generation plants, gas pipeline compression, refrigeration chillers, and so on.[19]

In the *lean-burn engine*, a spark ignites a very lean mixture in a precombustion chamber. The resulting flame jet issuing from the precombustion chamber ignites the main fuel mixture. Most commercial lean-burn engines produce power in the range of 400 kW to 3000 kW.

## 13. AIR-STANDARD DIESEL CYCLE

The processes in an air-standard diesel cycle are illustrated in Fig. 28.5.

  A to B:  isentropic compression
  B to C:  constant pressure heating
  C to D:  isentropic expansion
  D to A:  constant volume cooling

**Figure 28.5** *Air-Standard Diesel Cycle*

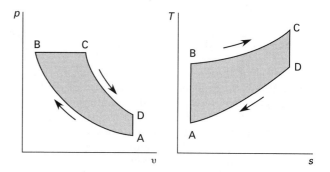

Diesel engines can be either four-stroke or two-stroke devices. The four-stroke and two-stroke engines are analyzed the same way, since the four processes on the $p$-$v$ and $T$-$s$ diagrams do not correspond to the four strokes.

---

[17]A *glow plug* may be used to improve cold-weather starting.

[18]Large diesels are those producing up to approximately 10 MW. This is large for diesels but small for gas turbines. In a *combined diesel or gas turbine system* (CODOG) for marine propulsion, diesels are used for low power and cruise operation; the turbine takes over when high speeds are needed.
[19]This is an awkwardly small size for gas turbines. Power turbines are not very practical in this size range.

The main difference is the number of power strokes per revolution—one for the two-stroke engine and one-half for the four-stroke engine.

The diesel cycle *compression ratio*, $r_v$, is

$$r_v = \frac{V_A}{V_B} \qquad \text{28.43}$$

The *cut-off ratio*, $r_{\text{cut-off}}$, is defined by Eq. 28.44. The volume $V_C$ is known as the *cut-off volume*.

$$r_{\text{cut-off}} = \frac{V_C}{V_B} = \frac{T_C}{T_B} \qquad \text{28.44}$$

Pressure, volume, and temperature at each point in the cycle can be evaluated with ideal gas equations. The work and heat flow terms are evaluated with the following equations. (Equation 28.45 uses $c_p$ because B-to-C is a constant pressure process.)

$$q_{\text{in}} = c_p(T_C - T_B) \qquad \text{28.45}$$

$$q_{\text{out}} = |c_v(T_A - T_D)| \qquad \text{28.46}$$

$$W_{\text{in}} = c_v(T_B - T_A) \qquad \text{28.47}$$

$$W_{\text{out}} = |c_v(T_D - T_C) + (c_p - c_v)(T_C - T_B)| \qquad \text{28.48}$$

The air-standard diesel cycle is always less efficient than the air-standard Otto cycle for equal compression ratios. However, for a specific maximum cylinder pressure, the diesel cycle is more efficient than the Otto cycle. The thermal efficiency can be calculated in a number of ways.

$$\eta_{\text{th}} = \frac{Q_{\text{in}} - Q_{\text{out}}}{Q_{\text{in}}} = \frac{W_{\text{out}} - W_{\text{in}}}{Q_{\text{in}}}$$

$$= 1 - \frac{T_D - T_A}{k(T_C - T_B)} \qquad \text{28.49}$$

In traditional diesels (i.e., those constructed through about 1980), thermal efficiency was not much higher than approximately 35%. Most manufacturers of stationary diesels are now reporting full-load thermal efficiencies of approximately 45%, and some as high as 50% for experimental test bed engines. Production diesel engines for commercial trucks routinely achieve efficiencies of 41–42%. These are the highest thermal efficiencies of any type of prime mover (including turbines). Significant technological advances in equipment, materials, and controls are required to achieve the highest efficiencies.

## 14. AIR-STANDARD DUAL CYCLE

The *air-standard dual cycle* shown in Fig. 28.6 is a combination of the diesel and Otto cycles. It more accurately predicts the performance of a spark-ignited internal combustion engine, since the combustion energy is added partly at constant volume and partly at constant pressure.

**Figure 28.6** *Air-Standard Dual Cycle*

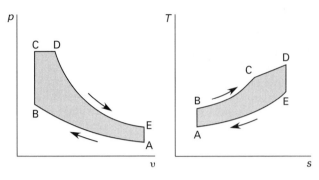

The processes in the dual cycle are

    A to B: isentropic compression
    B to C: constant volume heating
    C to D: constant pressure heating
    D to E: isentropic expansion
    E to A: constant volume cooling

As with the Otto and diesel cycles, the *compression ratio*, $r_v$, is defined as a ratio of volumes.

$$r_v = \frac{V_A}{V_B} \qquad \text{28.50}$$

The *pressure ratio*, $r_p$, is

$$r_p = \frac{p_C}{p_B} \qquad \text{28.51}$$

The *cut-off ratio*, $r_{\text{cut-off}}$, is

$$r_{\text{cut-off}} = \frac{V_D}{V_C} \qquad \text{28.52}$$

Pressure, temperature, and volume at various points in the cycle can be evaluated with ideal gas relationships. The work and heat terms are defined as follows.

$$q_{\text{in}} = c_v(T_C - T_B) + c_p(T_D - T_C) \qquad \text{28.53}$$

$$q_{\text{out}} = |c_v(T_A - T_E)| \qquad \text{28.54}$$

$$W_{\text{in}} = c_v(T_B - T_A) \qquad \text{28.55}$$

$$W_{\text{out}} = |p_C(v_D - v_C) + c_v(T_D - T_E)|$$

$$= |(c_p - c_v)(T_D - T_C) + c_v(T_D - T_E)| \qquad \text{28.56}$$

The efficiency of a dual cycle is between those of the Otto and diesel cycles.

$$\eta_{\text{th}} = \frac{Q_{\text{in}} - Q_{\text{out}}}{Q_{\text{in}}} = \frac{W_{\text{out}} - W_{\text{in}}}{Q_{\text{in}}}$$

$$= 1 - \frac{T_E - T_A}{(T_C - T_B) + k(T_D - T_C)} \qquad \text{28.57}$$

## 15. STIRLING ENGINES

During a typical Stirling cycle, the gaseous working fluid (e.g., helium or hydrogen) is shuttled through a heat exchanger circuit consisting of a heat acceptor, a regenerator, and a heat rejector. A piston compresses the working fluid in a compression space near the *heat rejector*. A *displacer* then shuttles the gas through a heat exchanger circuit to an expansion space. Along the way, the gas absorbs heat stored in the *regenerator*.[20] The piston is then moved so that the gas is expanded. During the expansion, heat is absorbed through the *acceptor walls*. The displacer then shuttles the gas back through the heat exchanger circuit at a constant volume, heating the regenerator along the way. This returns the Stirling machine to its original condition.

Heating of the working fluid occurs externally. The heat source can be from a combustion, nuclear (typically radioisotope), or solar process. Cooling is usually by a water-glycol mixture. When combustion provides the energy, Stirling machines draw power from combustion in a separate *combustor*. Since the combustion process is external, Stirling engines are able to use a variety of fuels and burn them at lower temperatures, resulting in lower emissions.

There are two methods of configuring Stirling machines. In a *kinematic Stirling engine*, double-acting pistons function separately as the compressor and displacer. Sealing of kinematic engines is a major challenge. *Free-piston engines* avoid leakage problems by eliminating the mechanical connection with the piston. Magnetic coupling (magnetic flux linkage) between the piston (and anything moving with it) and the surroundings limit the output of free-piston machines to electrical power generation.

Though Stirling engines are unlikely to be widely used in commercial vehicles, they are used to limited extents in small cryogenic applications. Since solar energy can be used as the heating source, Stirling engines are attractive for generating small amounts of electricity (i.e., 20 kW or less per high-efficiency engine/generator).

## 16. STIRLING CYCLE

The *Stirling cycle* has a thermal efficiency that can equal that of the Carnot cycle. The processes are shown in Fig. 28.7.

A to B:  constant volume heating
B to C:  isothermal heating and expansion
C to D:  constant volume cooling
D to A:  isothermal cooling and compression

---

[20]A *regenerator* is a device that captures heat energy and transfers it to the working fluid. In the case of the Stirling cycle, the heated gas passes through and heats a wire or ceramic mesh. When cool gas is passed back over the mesh, the heat transfer is reversed, and the gas is heated.

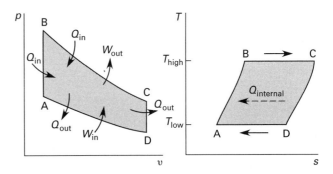

***Figure 28.7*** *Stirling Cycle*

Pressure, temperature, and volume for the various points on the cycle can be evaluated from ideal gas relationships.

The work and heat terms are given by Eq. 28.58 through Eq. 28.61. There is also another heat transfer to the working fluid in the A-to-B process. However, this heat transfer takes place in a reversible regenerator, and the heat transferred is from the working fluid's C-to-D process. Since the heat remains within the system boundary, it is not included in the $Q_{in}$ or $Q_{out}$ terms.

$$q_{in,B-C} = T_{high}(s_C - s_B) = RT_{high} \ln \frac{v_C}{v_B}$$

$$= RT_{high} \ln \frac{p_B}{p_C} \qquad \textit{28.58}$$

$$q_{out,D-A} = \left| T_{low}(s_A - s_D) \right| = \left| RT_{low} \ln \frac{v_A}{v_D} \right|$$

$$= \left| RT_{low} \ln \frac{p_D}{p_A} \right| \qquad \textit{28.59}$$

$$W_{in,D-A} = q_{out,D-A} \qquad \textit{28.60}$$

$$W_{out,B-C} = q_{in,B-C} \qquad \textit{28.61}$$

The thermal efficiency of the Stirling cycle is equal to the Carnot cycle efficiency if the regenerator used to transfer heat from the C-to-D process to the A-to-B process is reversible. With a reversible regenerator, the thermal efficiency is

$$\eta_{th} = \frac{Q_{in} - Q_{out}}{Q_{in}} = \frac{W_{out} - W_{in}}{Q_{in}}$$

$$= \frac{T_{high} - T_{low}}{T_{high}} \qquad \textit{28.62}$$

## 17. ERICSSON CYCLE

The processes of the Ericsson cycle are shown in Fig. 28.8. This cycle offers the best chance of achieving a thermal efficiency approaching that of the Carnot cycle.[21]

A to B: isothermal compression
B to C: constant pressure heating
C to D: isothermal expansion
D to A: constant pressure cooling

**Figure 28.8** *Ericsson Cycle*

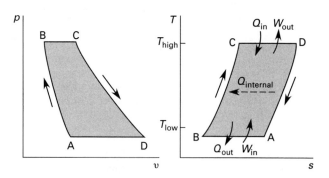

Pressure, temperature, and volume for the various points on the cycle can be evaluated from ideal gas relationships.

The work and heat terms are given by Eq. 28.63 through Eq. 28.66. There is also a heat transfer to the working fluid in the B-to-C process. However, if a reversible regenerator is used to capture heat from the D-to-A process, no external heat will be required for this.

$$q_{in,C-D} = T_{high}(s_D - s_C) = RT_{high}\ln\frac{v_D}{v_C}$$

$$= RT_{high}\ln\frac{p_C}{p_D} \qquad 28.63$$

$$q_{out,A-B} = \left|T_{low}(s_B - s_A)\right| = \left|RT_{low}\ln\frac{v_B}{v_A}\right|$$

$$= \left|RT_{low}\ln\frac{p_A}{p_B}\right| \qquad 28.64$$

$$W_{in,A-B} = q_{out,A-B} \qquad 28.65$$

$$W_{out,C-D} = q_{in,C-D} \qquad 28.66$$

With a reversible regenerator, the thermal efficiency of the Ericsson cycle is equal to that of the Carnot cycle.

$$\eta_{th} = \frac{Q_{in} - Q_{out}}{Q_{in}} = \frac{W_{out} - W_{in}}{Q_{in}}$$

$$= \frac{T_{high} - T_{low}}{T_{high}} \qquad 28.67$$

---

[21]The Ericsson cycle is approximated if the Brayton gas turbine cycle (covered in Sec. 28.23) is modified to include regenerative heat exchange, intercooling, and reheating.

## 18. EFFECT OF ALTITUDE ON OUTPUT POWER

Since a lower atmospheric pressure decreases atmospheric density, the oxygen per intake stroke available to engines operating on the Otto, diesel, and dual cycles decreases with altitude. The following steps constitute a procedure for determining the variation in power when the altitude is changed. It is assumed that the engine speed is constant and that ideal gas behavior applies.

*step 1:* Let 1 and 2 be the lower and higher altitudes, respectively.

*step 2:* Calculate the frictionless power.

$$IHP_1 = \frac{BHP_1}{\eta_{m1}} \qquad 28.68$$

*step 3:* Calculate the friction power, which is assumed to be constant at constant speed.

$$FHP = IHP_1 - BHP_1 \qquad 28.69$$

*step 4:* Calculate the air densities $\rho_{a1}$ and $\rho_{a2}$ from App. 20.A.

*step 5:* Calculate the new frictionless power.

$$IHP_2 = IHP_1\left(\frac{\rho_{a2}}{\rho_{a1}}\right) \qquad 28.70$$

*step 6:* Calculate the new net power.

$$BHP_2 = IHP_2 - FHP \qquad 28.71$$

*step 7:* Calculate the new mechanical efficiency.

$$\eta_{m2} = \frac{BHP_2}{IHP_2} \qquad 28.72$$

*step 8:* The volumetric air flow rates are the same.

$$\dot{V}_{a2} = \dot{V}_{a1} \qquad 28.73$$

*step 9:* The original air and fuel rates are

$$\dot{m}_{f1} = (BSFC_1)(BHP_1) \qquad 28.74$$
$$\dot{m}_{a1} = (AFR)(\dot{m}_{f1}) \qquad 28.75$$
$$\dot{V}_{a1} = \frac{\dot{m}_{a1}}{\rho_{a1}} \qquad 28.76$$

*step 10:* The new air mass flow rate is

$$\dot{m}_{a2} = \dot{V}_{a2}\rho_{a2} \quad [\text{see step 8}] \qquad 28.77$$

*step 11:* For engines with metered injection and without air/fuel ratio (i.e., "wide range") sensors, $\dot{m}_{f2} = \dot{m}_{f1}$. For engines with carburetors, $\dot{m}_{f2} \approx \dot{m}_{f1}$. For engines with air/fuel ratio sensors,

$$\dot{m}_{f2} = \frac{\dot{m}_{a2}}{AFR} \qquad 28.78$$

*step 12:* The new fuel consumption is

$$\text{BSFC}_2 = \frac{\dot{m}_{f2}}{\text{BHP}_2} \qquad 28.79$$

## 19. GAS TURBINES

*Combustion turbines* (CTs), or "gas" turbines (GTs), are the preferred combustion engines in applications much above 10 MW. Large units regularly operate in the 100 MW to 200 MW range (up to approximately 340 MW).[22] Some smaller CTs—typically less than 40,000 hp (30 MW)—for such applications as marine propulsion and pipeline compression are rated in standard horsepower. Turbine size for the purpose of specifying environmental regulations is categorized by "MMBTU/hr," millions of Btus per hour.

There are two general CT categories. The traditional heavy-duty, industrial CT, and the smaller, lighter aeroderivative CT. *Heavy-duty turbines* typically have a single shaft. The rotor is supported on two bearings, and the thrust bearing is on the compressor end. The generator is direct-driven from the compressor (i.e, *cold-end drive*). Small individual combustors surround the hot end radially. The *axial exhaust* duct goes directly into the steam generator in cogeneration and combined cycle plants.

The *aeroderivative combustion turbine* is basically a jet engine that exhausts into a turbine generator. Output is less than 50 MW per unit, and most aeroderivative CTs produce less than 40 MW. Split shafts are common in this range. The power turbine and generator are usually mounted at the "hot end" of the gas generator. The generator is run from a gearbox, allowing the turbine to run at higher, more efficient speeds.

The compression ratio (based on pressures) in the compression stage is typically 11:1 to 16:1, with most heavy-duty turbines in the 14:1 to 15:1 range. Aeroderivative turbines have higher compression ratios—typically 19:1 to 21:1, even as high as 30:1. Most heavy-duty combustion turbines have 16 to 18 compression stages. However, the Mach number at the tip of the of the first-stage rotor has risen from approximately 0.9 to approximately 1.4, reducing the number of compression stages to approximately 12 to 15 in modern transsonic turbines.

The temperature of the gas entering the expander section is typically 2200°F to 2350°F (1200°C to 1290°C).[23] The exhaust temperature is typically 1000°F to 1100°F (540°C to 590°C), which makes the exhaust an ideal heat source for combined cycles. Most combustion turbines have three to four expander stages. The exhaust

flow rate in modern heavy-duty turbines per 100 MW is approximately 525 lbm/sec to 550 lbm/sec (240 kg/s to 250 kg/s).

*Combustors* vary widely in design. Large, single chambers (i.e., "cans") have large residence times and allow heavy fuels to be burned completely. Smaller, multiple chambers and annular burners perform better for gaseous and distillate fuels. Film cooling is no longer adequate for wall cooling. Intensive convection cooling and thermal barrier coatings, including ceramic tiles, are needed.

## 20. STEAM INJECTION

Steam and water can be injected into the combustor to lower the exhaust temperature, inhibiting the formation of nitrogen oxides. An injection rate on the order of one-half pound of water per pound of fuel is sufficient to keep NOx emissions below older limits in the 75 ppm to 150 ppm range. More water can be used to reduce the emissions somewhat below those values. However, to reach the strictest limits (e.g., less than 10 ppm), selective catalytic reduction (SCR) is needed. NOx emissions can also be controlled to intermediate values (to approximately 25 ppm) without steam in *"dry combustors"* *(low-NOx burners)* based on *staged lean combustion,* also known as *sequential combustion.*

Steam is also injected into the expansion section (sometimes referred to as the *expander*) of CTs, a process known as *power-boosting.* Power-boost steam injection is particularly applicable with aeroderivative turbines where steam pressures are consistent with higher (20:1 or more) compression ratios. Steam injection is popular in cogeneration plants where process steam use is variable, and excess steam can be routed back to the turbine. NOx reduction occurs, but this is not the primary purpose. Steam, which absorbs heat better than air, is also widely used for cooling gas turbines.

## 21. TURBINE FUELS

Modern turbines can burn a wide range of gaseous and distillate fuels and can switch from one fuel to another over the entire load range. Natural gas is the most economical and, therefore, the most common fuel in combustion turbines used for electrical power generation. Propane, No. 2 oil, and kerosene are used as backup fuels.

Turbines can also be partially fueled by *synthetic gas* ("syngas") generated from coal. The *gasifier* may be a fixed-bed, fluid-bed, or entrained flow type. Both air-blown and oxygen-blown gasification processes can be used, although the oxygen-blown process requires a separate oxygen plant. The heating value of syngas is low, approximately 240 Btu/scf (8.9 MJ/m³). So, combustion turbines cannot achieve full load operation on syngas alone.

---

[22]In the past, gas turbines were plagued by poor reliability, availability, and maintainability (RAM). The RAM record of modern turbines, particularly aeroderivative types, is excellent.

[23]New materials, coatings, and other devices in *advanced turbine systems* (ATS) have pushed the upper limit of temperatures in commercially available turbines to the 2600°F (1425°C) mark, while simultaneously dropping NOx emissions below 10 ppm.

*Thermodynamics*

In the *gasification process*, coal-water slurry is pumped into the gasifier. Oxygen or air is added, forming a hot, partially burned gas consisting of carbon monoxide, hydrogen sulfide, and carbonyl sulfide. Most of the non-carbon material in the coal melts and flows out of the gasifier as *slag*. Hot-gas clean-up equipment removes particulates, sulfur, and other impurities from the gas. Gasification is relatively insensitive to coal feedstock.

In future advances, finely ground coal may be used in *direct coal-fueled turbines* (DCFTs). In the past, attempts to inject coal directly into the turbine have been plagued by severe erosion of turbine blades by tiny (3–10 $\mu$m) ash particles, pluggage of the gas flow passages by ash deposits, and corrosion of high-temperature metallic surfaces by alkali compounds. This complication may be addressed in the future by coal-cleaning technologies, such as the thermal extraction/solid-liquid separation hyper-coal process. Other considerations are increased emissions and cost.

Traditional gas turbines burn fuel within the envelope of the engine. However, the in-line combustor does not provide adequate residence time to burn coal. Therefore, an "external combustor" must be used with DCFTs.[24] The shortcomings associated with burning ground coal may be overcome by some form of staged combustion, sometimes referred to as "rich-quench-lean" (RQL) firing.

With RQL external combustion, finely ground coal in powder or slurry form (known as *coal-water mixture* or CWM) is burned in a fuel-rich, oxygen-starved first-stage combustor at about 3000°F (1650°C). The low oxygen level inhibits NOx formation from fuel-bound nitrogen. The fuel-rich gas is quenched with water to approximately 2000°F (1100°C), inhibiting thermal NOx formation, and solidifying coal ash so that it can be removed in a cyclone separator, inertial slag separator, or ceramic filter.[25] Clean gas is then sent to the fuel-lean second-stage combustor where additional air is injected, and the temperature increases to approximately 2800°F (1540°C) as the carbon monoxide and hydrogen components burn. Sulfur emissions are controlled by injection of calcium-based sorbents in either the first- or second-stage combustors.

## 22. INDIRECT COAL-FIRED TURBINES

With *indirect coal-fired turbines* (ICFTs), the coal combustion products never enter the turbine expander. Rather, heat from combustion is transferred through a closed heat exchanger to the compressed air. Only clean air flows through the expander. Since the heat transfer occurs at over 2000°F (1100°C), well over the 1600°F (870°C) limit of traditional metallic heat exchangers, a special ceramic heat exchanger must be used. When

---

[24]The external combuster is located "off base" on a separate combustion "island."
[25]There is also promise in *pressurized slagging combustors* where slag is removed in liquid form.

used in a combined cycle, the term *externally fired combined cycle* (EFCC) is used to describe this process. (See Fig. 28.9.)

Theoretically, the thermal efficiency of the cycle can be increased by closing the process (i.e., making it a *closed turbine cycle*) where the exhaust from the expander is returned to the compressor for reuse.

**Figure 28.9** *Direct and Indirect Coal-Fired Turbines*

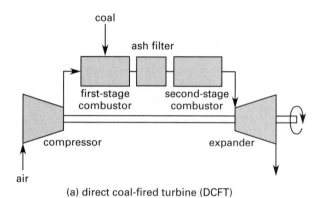

(a) direct coal-fired turbine (DCFT)

(b) indirect coal-fired turbine (ICFT)

## 23. BRAYTON GAS TURBINE CYCLE

Strictly speaking, the *Brayton gas turbine cycle* (also known as the *Joule cycle*) is an internal combustion cycle. It differs from the previous cycles in that each process is carried out in a different location, airflow and fuel injection are steady, and air-standard calculations are realistic since a large air-fuel ratio is used to keep combustion temperatures below metallurgical limits.

Figure 28.10 illustrates the physical arrangement of components used to achieve the Brayton cycle. Almost all installations drive the compressor from the turbine. (Approximately 50% to 75% of the turbine power is required to drive the high-efficiency compressor.) The actual arrangement differs from the air-standard property plot shown in Fig. 28.11 in that the exhaust products exiting at point D are not cooled and returned to the compressor at point A. The processes in the air-standard Brayton cycle follow.

**Figure 28.10** *Gas Turbine*

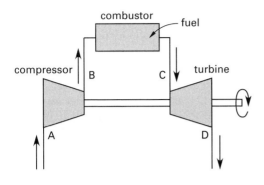

**Figure 28.11** *Brayton Gas Turbine Cycle*

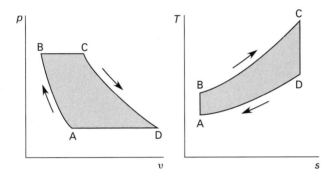

A to B: isentropic compression in the compressor
B to C: constant pressure heat addition in the combustor
C to D: isentropic expansion in the turbine
D to A: constant pressure cooling to original conditions

Combustors are said to be *open* if the incoming air and combustion products flow to the turbine. A high percentage, 30% to 60%, of excess air is needed to cool the gases. Depending on the turbine construction details, the temperature of the air entering the turbine will be between 1200°F and 1800°F (650°C and 1000°C). Turbojet and turboprop engines typically use *open combustors*.

A *closed combustor* is a heat exchanger. Air flowing to the turbine is not combined with the combustion gases. This allows any type of fuel to be used, but the bulk of the heat exchanger limits closed combustors to stationary power plants and pipeline pumping stations.

Ideal gas relationships for steady-flow systems can be used to evaluate the $p$, $V$, and $T$ properties at points A, B, C, and D. Constant values of $k$, $c_p$, and $c_v$ can be assumed if air is considered ideal.[26] An air table also can be used if the isentropic efficiencies are known. Usually, it is assumed that $p_B$ and $p_C$ are equal.

---

[26]Another typical assumption, justified on the basis of large amounts of excess air, is that the gas is all nitrogen.

At A:

$$p_A = p_D \text{ [usually atmospheric]} = p_B \left(\frac{v_B}{v_A}\right)^k$$

$$= p_B \left(\frac{T_A}{T_B}\right)^{k/(k-1)} \qquad 28.80$$

$$T_A = T_D \left(\frac{v_A}{v_D}\right) = T_B \left(\frac{v_B}{v_A}\right)^{k-1}$$

$$= T_B \left(\frac{p_A}{p_B}\right)^{(k-1)/k} \qquad 28.81$$

$$v_A = v_D \left(\frac{T_A}{T_D}\right) = v_B \left(\frac{p_B}{p_A}\right)^{1/k}$$

$$= v_B \left(\frac{T_B}{T_A}\right)^{1/(k-1)} \qquad 28.82$$

At B:

$$p_B = p_C = p_A \left(\frac{v_A}{v_B}\right)^k = p_A \left(\frac{T_B}{T_A}\right)^{k/(k-1)} \qquad 28.83$$

$$T_B = T_A \left(\frac{v_A}{v_B}\right)^{k-1} = T_A \left(\frac{p_B}{p_A}\right)^{(k-1)/k}$$

$$= T_C \left(\frac{v_B}{v_C}\right) \qquad 28.84$$

$$v_B = v_A \left(\frac{p_A}{p_B}\right)^{1/k} = v_A \left(\frac{T_A}{T_B}\right)^{1/(k-1)}$$

$$= v_C \left(\frac{T_B}{T_C}\right) \qquad 28.85$$

At C:

$$p_C = p_B = p_D \left(\frac{v_D}{v_C}\right)^k = p_D \left(\frac{T_C}{T_D}\right)^{k/(k-1)} \qquad 28.86$$

$$T_C = T_B \left(\frac{v_C}{v_B}\right) = T_D \left(\frac{v_D}{v_C}\right)^{k-1}$$

$$= T_D \left(\frac{p_C}{p_D}\right)^{(k-1)/k} \qquad 28.87$$

$$v_C = v_B \left(\frac{T_C}{T_B}\right) = v_D \left(\frac{p_D}{p_C}\right)^{1/k}$$

$$= v_D \left(\frac{T_D}{T_C}\right)^{1/(k-1)} \qquad 28.88$$

At D:

$$p_D = p_C \left(\frac{v_C}{v_D}\right)^k = p_C \left(\frac{T_D}{T_C}\right)^{k/(k-1)} \qquad 28.89$$

$$T_D = T_C \left(\frac{v_C}{v_D}\right)^{k-1} = T_C \left(\frac{p_D}{p_C}\right)^{(k-1)/k}$$

$$= T_A \left(\frac{v_D}{v_A}\right) \qquad 28.90$$

$$v_D = v_C \left(\frac{p_C}{p_D}\right)^{1/k} = v_C \left(\frac{T_C}{T_D}\right)^{1/(k-1)}$$

$$= v_A \left(\frac{T_D}{T_A}\right) \qquad 28.91$$

The work and heat flow terms are

$$q_{in} = c_p(T_C - T_B) = h_C - h_B \qquad 28.92$$

$$q_{out} = \left| c_p(T_A - T_D) \right| = \left| h_A - h_D \right| \qquad 28.93$$

$$W_{turbine} = \left| c_p(T_D - T_C) \right| = \left| h_D - h_C \right| \qquad 28.94$$

$$W_{compressor} = c_p(T_B - T_A) = h_B - h_A \qquad 28.95$$

$$\eta_{th} = \frac{Q_{in} - Q_{out}}{Q_{in}} = \frac{W_{turbine} - W_{compressor}}{Q_{in}}$$

$$= \frac{(h_C - h_B) - (h_D - h_A)}{h_C - h_B} \qquad 28.96$$

If the gas is ideal so that $c_p$ is constant, then

$$\eta_{th} = \frac{(T_C - T_B) - (T_D - T_A)}{T_C - T_B} \qquad 28.97$$

If the isentropic efficiency is less than 100% for either or both the compressor and turbine, the actual enthalpies or temperatures must be used to calculate the work, heat, and thermal efficiency.

$$h'_B = h_A + \frac{h_B - h_A}{\eta_{s,compressor}} \qquad 28.98$$

$$T'_B = T_A + \frac{T_B - T_A}{\eta_{s,compressor}} \qquad 28.99$$

$$h'_D = h_C - \eta_{s,turbine}(h_C - h_D) \qquad 28.100$$

$$T'_D = T_C - \eta_{s,turbine}(T_C - T_D) \qquad 28.101$$

*Simple cycle* (SC) operation means that the turbine is not part of a cogeneration or combined cycle system.[27] The full-load thermal efficiency of existing heavy-duty combustion turbines in simple cycles is approximately 34% to 36%, while new turbines on the cutting edge of technology (i.e., *advanced turbine systems*, ATS) are able to achieve 38% to 38.5%, with 40% being a reasonable future goal. Aeroderivative turbines commonly achieve efficiencies up to 41%. The heat rate can be found from the cycle efficiency.

---

[27]When integrated into combined cycle systems, overall thermal efficiencies are higher; however, these "combined efficiencies" are not the turbine thermal efficiencies.

The *back work ratio* (BWR), which is approximately 50% to 75%, is the ratio of the compressor work to the turbine expansion work.

**Example 28.4**

Air enters the compressor of a gas turbine at 14.7 psia and 540°R (101.3 kPa and 300K). The pressure ratio is 4.5:1. The conditions at the turbine inlet are 64 psia and 2200°R (440 kPa and 1220K). The turbine's expansion pressure ratio is 1:4, and exhaust is to the atmosphere. The isentropic efficiency of the turbine is 85%. Compression is isentropic. What is the thermal efficiency of the cycle?

*SI Solution*

Assume an ideal gas. (The customary U.S. solution solves this example using air tables.)

Refer to Fig. 28.11.

At A:

$$T_A = 300K \quad \text{[given]}$$

$$p_A = 101.3 \text{ kPa} \quad \text{[given]}$$

At B:

$$T_B = T_A \left(\frac{p_B}{p_A}\right)^{(k-1)/k}$$

$$= (300K)(4.5)^{(1.4-1)/1.4} = 461K$$

$$p_B = (4.5)(101.3 \text{ kPa}) = 456 \text{ kPa}$$

At C:

$$T_C = 1220K \quad \text{[given]}$$

$$p_C = 440 \text{ kPa} \quad \text{[given]}$$

At D:

$$p_D = \frac{440 \text{ kPa}}{4} = 110 \text{ kPa}$$

If the expansion had been isentropic, the temperature would have been

$$T_D = T_C \left(\frac{p_D}{p_C}\right)^{(k-1)/k}$$

$$= (1220K)\left(\frac{1}{4}\right)^{(1.4-1)/1.4}$$

$$= 821K$$

For ideal gases, the specific heats are constant. Therefore, the change in internal energy (and enthalpy, approximately) is proportional to the change in temperature. The actual temperature is

$$T'_D = T_C - \eta_{s,\text{turbine}}(T_C - T_D)$$
$$= 1220\text{K} - (0.85)(1220\text{K} - 821\text{K})$$
$$= 881\text{K}$$

The thermal efficiency is given by Eq. 28.97.

$$\eta_{\text{th}} = \frac{(T_C - T_B) - (T'_D - T_A)}{T_C - T_B}$$
$$= \frac{(1220\text{K} - 461\text{K}) - (881\text{K} - 300\text{K})}{1220\text{K} - 461\text{K}}$$
$$= 0.235 \ (23.5\%)$$

*Customary U.S. Solution*

(Use an air table. The SI solution assumes an ideal gas.)

At A:

$$T_A = 540°\text{R} \quad [\text{given}]$$
$$p_A = 14.7 \text{ psia} \quad [\text{given}]$$
$$h_A = 129.06 \text{ Btu/lbm}$$
$$p_{r,A} = 1.3860$$

At B: The A-to-B process is isentropic.

$$p_{r,B} = p_{r,A}\left(\frac{p_B}{p_A}\right) = (1.3860)(4.5)$$
$$= 6.237$$

Locate this pressure ratio in the air table.

$$T_B = 827.6°\text{R}$$
$$h_B = 198.5 \text{ Btu/lbm}$$
$$p_B = (4.5)(14.7 \text{ psia}) = 66.15 \text{ psia}$$

At C:

$$T_C = 2200°\text{R} \quad [\text{given}]$$
$$p_C = 64 \text{ psia} \quad [\text{given}]$$
$$h_C = 560.59 \text{ Btu/lbm}$$
$$p_{r,C} = 256.6$$

At D:

$$p_{r,D} = p_{r,C}\left(\frac{p_D}{p_C}\right)$$
$$= \frac{256.6}{4} = 64.15$$

Locate this pressure ratio in the air table.

$$T_D = 1554.6°\text{R}$$
$$h_D = 383.66 \text{ Btu/lbm}$$

Since the efficiency of the expansion process is 85%,

$$h'_D = h_C - \eta_{s,\text{turbine}}(h_C - h_D)$$
$$= 560.59 \frac{\text{Btu}}{\text{lbm}} - (0.85)\left(560.59 \frac{\text{Btu}}{\text{lbm}} - 383.66 \frac{\text{Btu}}{\text{lbm}}\right)$$
$$= 410.2 \text{ Btu/lbm}$$

The thermal efficiency is

$$\eta_{\text{th}} = \frac{h_C - h_B - (h'_D - h_A)}{h_C - h_B}$$
$$= \frac{\left(560.59 \frac{\text{Btu}}{\text{lbm}} - 198.5 \frac{\text{Btu}}{\text{lbm}}\right) - \left(410.2 \frac{\text{Btu}}{\text{lbm}} - 129.06 \frac{\text{Btu}}{\text{lbm}}\right)}{560.59 \frac{\text{Btu}}{\text{lbm}} - 198.5 \frac{\text{Btu}}{\text{lbm}}}$$
$$= 0.224 \ (22.4\%)$$

## 24. AIR-STANDARD BRAYTON CYCLE WITH REGENERATION

Regeneration is used to improve the efficiency of the Brayton cycle. Regeneration involves transferring some of the heat from the exhaust products to the air in the compressor. The transfer occurs in a *regenerator*, which is a crossflow heat exchanger. There is no effect on turbine work, compressor work, or net output. However, the cycle is more efficient since less heat is added. Of course, $T_B$ cannot be greater than $T_F$. Similarly, $T_C$ cannot be greater than $T_E$. (See Fig. 28.12 and Fig. 28.13.)

**Figure 28.12** Gas Turbine with Regeneration

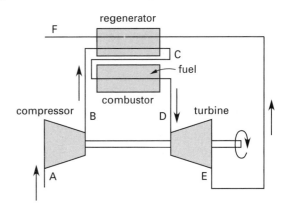

**Figure 28.13** *Brayton Cycle with 100% Efficient Regeneration*

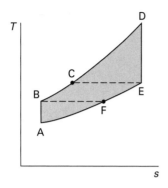

**Figure 28.14** *Augmented Gas Turbine*

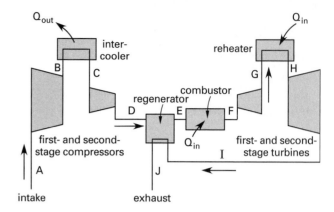

The processes are:

A to B: isentropic compression
B to C: constant pressure heat addition in regenerator
C to D: constant pressure heat addition in combustor
D to E: isentropic expansion
E to F: constant pressure heat removal in the regenerator
F to A: constant pressure heat removal in the sink

If $T_C = T_E$, the regenerator is said to be 100% efficient. Otherwise, the *regenerator efficiency*, also known as *regeneration effectiveness*, is calculated from Eq. 28.102. Actual regeneration efficiency rarely exceeds 75%.

$$\eta_{\text{regenerator}} = \frac{h_C - h_B}{h_E - h_B} \qquad 28.102$$

The thermal efficiency of the air-standard Brayton cycle with regeneration is

$$\eta_{\text{th}} = \frac{W_{\text{out}} - W_{\text{in}}}{Q_{\text{in}}} = \frac{(h_D - h_E) - (h_B - h_A)}{h_D - h_C} \qquad 28.103$$

If air is considered to be an ideal gas, temperatures can be substituted for enthalpies in Eq. 28.102 and Eq. 28.103.

## 25. BRAYTON CYCLE WITH REGENERATION, INTERCOOLING, AND REHEATING

Multiple stages of compression and expansion can be used to improve the efficiency of the Brayton cycle. Physical limitations usually preclude more than two stages of intercooling and reheat. (This section assumes only one stage of each.) The physical arrangement is shown in Fig. 28.14.

The processes are:

A to B: isentropic compression
B to C: cooling at constant pressure (usually back to $T_A$)
C to D: isentropic compression

D to F: constant pressure heat addition
F to G: isentropic expansion
G to H: reheating at constant pressure in combustor or reheater (usually back to $T_F$)
H to I: isentropic expansion
I to A: constant pressure heat rejection

Calculation of the work and heat flow terms and of thermal efficiency is similar to that in the previous cycles, except that there are two $W_{\text{turbine}}$, $W_{\text{compressor}}$, and $Q_{\text{in}}$ terms. If efficiencies for the compressor, turbine, and regenerator are given, the following relationships are required.

$$p'_B = p_B \qquad 28.104$$

$$h'_B = h_A + \frac{h_B - h_A}{\eta_{s,\text{compressor}}} \qquad 28.105$$

$$p'_D = p_D \qquad 28.106$$

$$h'_D = h_C + \frac{h_D - h_C}{\eta_{s,\text{compressor}}} \qquad 28.107$$

$$p'_E = p_E \qquad 28.108$$

$$h'_E = h_D + \eta_{\text{regenerator}}(h_I - h_D) \qquad 28.109$$

$$p'_G = p_G \qquad 28.110$$

$$h'_G = h_F - \eta_{s,\text{turbine}}(h_F - h_G) \qquad 28.111$$

$$p'_I = p_I \qquad 28.112$$

$$h'_I = h_H - \eta_{s,\text{turbine}}(h_H - h_I) \qquad 28.113$$

Optimum performance improvement with multiple staging is achieved when the pressure ratio across each turbine stage (either compression stages or expansion stages) is the same. Referring to Fig. 28.14 and Fig. 28.15, for the compressors, $p_B = p_C = \sqrt{p_A p_D}$. For the turbine, $p_G = p_H = \sqrt{p_F p_I}$.

**Figure 28.15** *Augmented Brayton Cycle*

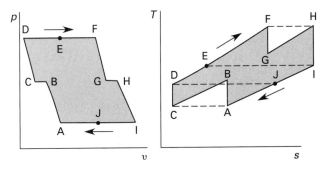

## 26. HEAT RECOVERY STEAM GENERATORS

All simple power cycles end up discarding most of the incoming heat energy. For example, a thermal efficiency of 40% means that 60% of the heat energy is lost. Combustion cycles (gas turbine cycles in particular) produce high-temperature, 800°F to 1100°F (430°C to 590°C) exhaust streams.[28] Some of this energy can be recovered.

All cogeneration and combined cycles make use of some type of *heat-recovery steam generator* (HRSG), also known as a *waste-heat boiler*. Although fire-tube designs are applicable to low gas flow rates—less than 100,000 lbm/hr (12.6 kg/s)—due to the large mass flow rates of exhaust gases typically encountered and the high-pressure 1000 psi to 2800 psi (7 MPa to 19 MPa) steam generated, HRSGs are typically of the water-tube design. When turbines are fueled by natural gas, the exhaust is clean and does not pose a corrosion problem for HRSGs.[29] (See Fig. 28.16.)

The exhaust gases pass sequentially through superheater, high-pressure evaporator, economizer, and low-pressure evaporator sections. (A selective catalytic conversion (SCR) section may also be used, though no heat transfer occurs in it.) To extract more heat energy, extended surfaces (i.e., fins) may be used on the tubes. The maximum fin temperature occurs at the tip, and fin material limits the operating temperature of the HRSG. Since the tube-side heat transfer coefficient is high, on the order of 1000 Btu/hr-ft²-°F to 3000 Btu/hr-ft²-°F (1.7 kW/m·K to 5.2 kW/m·K) in economizers and LP evaporators, the *fin density* is also high—on the order of 2 to 5 fins per inch. The fin density is lower in the superheater section.

---

[28]The exhaust from municipal waste incinerators is approximately 1800°F (980°C), and some chemical waste incinerators produce 2000°F to 2400°F (1090°C to 1320°C) combustion gases.
[29]Combustion gas from municipal incinerators contains particulates that can cause slagging, and such gas is corrosive at both high and low temperatures. Above 800°F (430°C), hydrogen chloride (HCl) formed from the combustion of waste plastics, is very corrosive. Chlorine, formed when incinerating chlorinated wastes, is even more corrosive than HCl on carbon steel above 400°F to 450°F (200°C to 230°C). Therefore, HRSGs associated with the burning of such wastes must operate in a narrow temperature band below 400°F (200°C) and above the *acid-vapor dew point*.

As in traditional steam generators, steam drums in HRSGs can be either natural or forced circulation in design. The *circulation ratio* is the ratio of the mass of circulating steam-water (in the risers and downcomers) to the mass of generated steam. It varies from 10 to 40 for natural circulation and from 3 to 10 for forced circulation.

The temperature difference between the gas side and steam sides varies along the run of the HRSG and with variations in gas flow, inlet temperature, and extent of supplemental firing. The *pinch point* is the minimum temperature difference. The *approach point* is the difference between the saturation temperature and the temperature of the leaving water (in the economizer). Trial-and-error solutions are usually required to determine the entering and leaving temperatures at each section. For trial-and-error solutions, the pinch and approach points may initially be assumed to be typical values, 20°F and 15°F (11°C and 8°C), respectively.

HRSGs may be unfired, supplementary fired, or furnace fired. (Fired HRSGs are more common with combined cycles than with cogeneration.) *Unfired HRSGs* are usually one- or two-pass designs. Since the turbine exhaust contains approximately 16% oxygen by volume, additional fuel can be sprayed into the exhaust stream, usually after the superheater portion of the HRSG. This option is known as a *supplementary-fired HRSG*. Temperatures are limited to approximately 1700°F (930°C) by the liner material. Other than that, supplementary-fired units are similar to unfired versions.

Temperatures in excess of 1700°F (930°C) can be achieved with *furnace-fired HRSGs*. Temperatures up to about 2300°F (1260°C) can be achieved with a duct burner and water-cooled membrane walls; higher temperatures require special register burners with their own air chambers.

Referring to Fig. 28.17, the heat balance in the superheater and evaporator sections is

$$\dot{m}_{gas} c_{p,gas} (T_{gas,1} - T_{gas,3})$$
$$= \dot{m}_{steam} (h_{steam,out} - h_{water,2}) \qquad 28.114$$

The heat balance for the complete HRSG is

$$\dot{m}_{gas} c_{p,gas} (T_{gas,1} - T_{gas,4})$$
$$= \dot{m}_{steam} (h_{steam,out} - h_{water,1}) \qquad 28.115$$

The *transferred duty* (or just "duty") of the HRSG is the total heat transfer rate for all components (superheater, evaporator, economizer, etc.).

HRSGs often operate far from their design point. Equation 28.116 can be used to predict the performance of

**Figure 28.16** *Typical Heat Recovery Steam Generator*

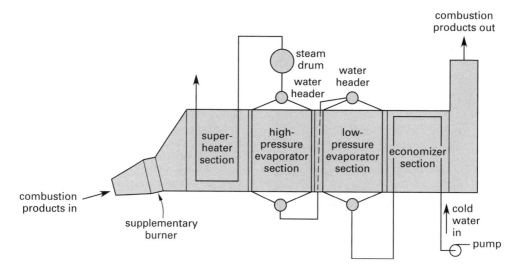

**Figure 28.17** *Pinch and Approach Points*

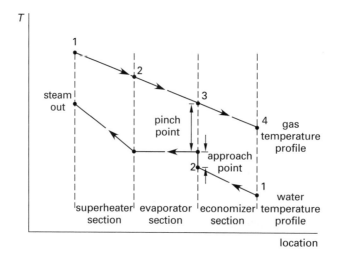

each section of the HRSG at one mass flow rate based on known performance at another mass flow rate.

$$\frac{U_1 A_1}{U_2 A_2} = \frac{\dfrac{\dot{Q}_1}{T_1}}{\dfrac{\dot{Q}_2}{T_2}} \approx \left(\frac{\dot{m}_1}{\dot{m}_2}\right)^{0.65} \qquad 28.116$$

## 27. COGENERATION CYCLES

In *cogeneration* plants, HRSGs convert waste heat (generally from turbine exhaust) into low-pressure high-quality or saturated process steam with pressures of 10 psig to 300 psig (70 kPa to 210 kPa).[30] *Process steam* is steam that is used for some process other than electrical power generation,[31] such as *space heating* (also known as *district heating*). (See Fig. 28.18.)

Cogeneration systems can be configured as topping or bottoming cycles. In a *topping cycle*, the primary fuel produces electricity first. The turbine exhaust (either steam or gas) is captured to make steam. In a *bottoming cycle*, the waste heat from an industrial process is captured to produce steam first for the process and then to power a turbine. Bottoming cycles, being less efficient and requiring supplementary firing, are less common.

---

[30]Not all combined cycles utilize exhaust from gas turbines. High-temperature diesel engine exhaust is rich in oxygen—ideal for use as preheated combustion air in smaller plants.

[31]Theoretically, the steam could be used for cooling in an absorption system. This probably is never done in practice, however.

**Figure 28.18** *Simple Cogeneration Process*

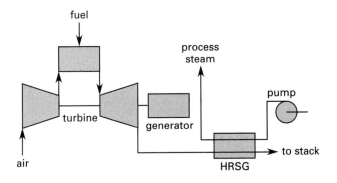

The *power-to-heat ratio* (PTR) is

$$\text{PTR} = \frac{P_{\text{turbine}}}{\dot{Q}_{\text{recovered}}} \qquad 28.117$$

The *fuel utilization* (FU) is not the same as thermal efficiency. Fuel utilizations for different units should be compared only for equal PTRs. Highly efficient cycles have fuel utilizations above 80%.

$$\text{FU} = \frac{P_{\text{turbine}} + \dot{Q}_{\text{recovered}}}{\dot{Q}_{\text{in}}} \qquad 28.118$$

## 28. COMBINED CYCLES

In a *combined cycle*, the heat recovered in an HRSG is used to vaporize water for use in another Rankine steam cycle. Steam from the HRSG is high pressure, high temperature, exceeding 750 psi and 700°F (5 MPa and 370°C). Supplementary firing in the HRSG may be used. Combined cycle efficiencies are much higher than simple cycles, typically 45% to 55%, with plants providing only electrical power output achieving the highest efficiencies. Some state-of-the-art prototype ("proof of concept") installations are able to achieve combined efficiencies up to 59%. A combined efficiency of 60% appears to be achievable.[32] (See Fig. 28.19.)

A plethora of confusing acronyms identify combined cycle (CC) variations, including GFCC (gas-fired turbine combined cycle), GTCC (gas turbine combined cycles), GCC (gasification combined cycle), CGCC (coal gasification combined cycle), DCCC (diesel coal combined cycle), EFCC (externally fired combined cycle), and IGCC (integrated gasification combined cycle). Some of these terms are synonymous.

Figure 28.20 illustrates an IGCC cycle. "Integrated" refers to the fact that the gasifier uses part of the compressor discharge rather than operating as an

independent source. IGCC is an inherently low-emissions process because ash is removed in the gasification process and sulfur is removed from the fuel gas, not the flue gas. Overall efficiencies with IGCC are approximately 42%—lower than some combined cycles, but still much higher than the traditional coal-fired plant with emission controls.

## 29. HIGH-PERFORMANCE POWER SYSTEMS

It may be possible to combine all of the advanced technologies into a combined cycle and make coal combustion as efficient as natural gas.[33] In the prototypical *high-performance power system* (HIPPS), a *high-temperature advanced furnace* (HITAF, which integrates the combustion of coal, heat transfer, and emissions control into a single unit) provides most of the heat for a Brayton topping cycle. Supplementary firing with clean fuel (such as natural gas) boosts the air temperature to achieve optimum gas turbine performance. Energy in the HITAF can also be used for superheating and reheating in the conventional Rankine bottoming cycle. Gas turbine exhaust is routed to the HITAF as combustion air.

## 30. REPOWERING AND LIFE EXTENSION

The average useful life of most coal-fired utility power plants is 30 to 40 years. However, because of increasing demand, utilities are unlikely to retire any significant portion of their old capacity. Rather, the aging plants are improved. *Repowering* is a popular term used by utilities to mean upgrading an existing plant. In a sense, repowering is a variation of *life extension*.[34] However, life extension also includes programs to increase reliability, availability, and maintainability (RAM) without changing equipment. For example, many programs concentrate on weak areas, such as boiler tube failure (BTF), the leading cause of forced outages.

There are still smaller electrical generating plants in the United States that do not even have reheat boilers. Repowering concentrates on replacing the furnace and boiler with clean-burning, more efficient coal combustion (as in fluidized-bed combustion) or gasification while retaining the remainder of the existing plant (i.e., the coal handling, steam cycle, and generating equipment).

In the face of massive boiler repairs or replacement, a utility may choose to repower by replacing boilers with gas turbines and HRSGs, using the existing steam turbine in a topping cycle. This is a good option if the existing steam cycle is below that of the HRSG output,

---

[32]The practical upper limit for combined cycle efficiencies is considered to be approximately 70%. The Carnot efficiency sets the ideal upper limit. This is 81% for cycles operating between 70°F (21°C) and 2350°F (1290°C).

[33]However, it may not be economical. Time will tell.

[34]Utilities in the United States are averse to using the term *life extension*. The reason lies in the 1970 Clean Air Act. Power plants built prior to 1971 were exempted from emissions control requirements. These "grandfathered" plants lose their exemption if they are significantly modified. "Refurbishment" and "repair," maybe; "life extension," no.

**Figure 28.19** *Combined Cycle Process*

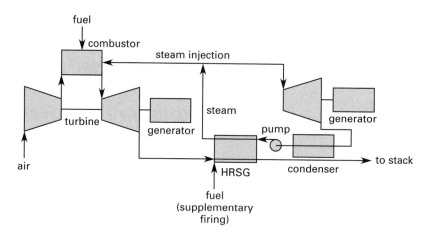

**Figure 28.20** *Integrated Gasification Combined Cycle*

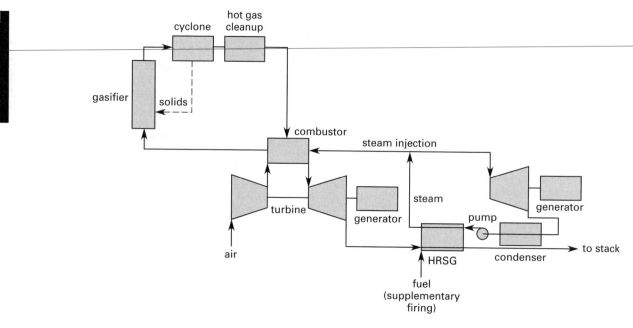

generally, less than 1450 psig (10 MPa). There are three general ways the gas turbine can be integrated into the existing plant. (1) In the *hot windbox system,* some or all of the turbine exhaust is routed to the furnace as combustion air. (2) Turbine exhaust can be used for feedwater heating. (3) Turbine exhaust can be used as the heat source in a *supplementary boiler* or separate superheater.

Though often overlooked, the fuel flexibility and efficiency of large diesel engines make diesel combined cycles (DCC) ideal candidates in repowering options in smaller power plants.

# 29 Nuclear Power Cycles

## 1. INTRODUCTION

A *nuclear power cycle* is a Rankine vapor power cycle in which a nuclear reactor core replaces the traditional furnace. Because there is no combustion, nuclear plants do not emit carbon dioxide, sulfur dioxide, or nitrogen oxides. However, they continue to be plagued by technical problems, poor economics, licensing difficulties, and poor public acceptance.

In the United States, approximately 19% of all electricity is generated by nuclear plants.[1] Commercial nuclear reactors produce 600 MW to 1500 MW of power, though most operate in the 800 MW to 1000 MW range. Limitations associated with the method of heat extraction have kept pressures and temperatures lower than what is achievable in simple Rankine cycles. This limits the maximum attainable thermal efficiency to (approximately) 30%.

## 2. REACTOR TYPES[2,3]

Nuclear reactor systems are characterized by their neutron moderators and methods used to transfer heat from the reactor to the steam. Although there are a number of different reactor designs, only *light-water (moderated)*

reactors (LWRs), which includes pressured water reactors and boiling water reactors, have been commercialized in any significant numbers.

## 3. BOILING WATER REACTORS

In a *boiling water reactor* (BWR), steam is generated directly within the nuclear core. It is passed through a series of separators and driers (integral to the reactor vessel) before entering the turbine. (See Fig. 29.1.) The pressure in a typical BWR is approximately 1000 psia (6.9 MPa). With this configuration, the turbine and all the equipment processing water and steam eventually become radioactive. Radioactivity in steam is primarily N-16, an isotope with a half-life of only 7 sec. Therefore, radioactivity in steam exists only during power generation.

**Figure 29.1** *Boiling Water Reactor*

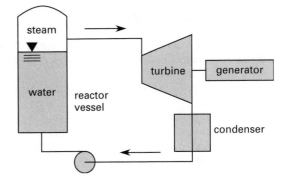

## 4. PRESSURIZED WATER REACTORS

In a *pressurized water reactor* (PWR), heated liquid water in the *primary loop* travels from the reactor through a heat exchanger, which serves as the *steam generator*. (See Fig. 29.2.) Steam flowing in the *secondary loop* absorbs heat from the primary loop water. Since water in the primary loop cannot be permitted to vaporize, the primary loop is limited to approximately 2000–2250 psia (13.8–15.5 MPa), which corresponds to 640–650°F (340–345°C). Water at approximately 1850 psi (12.7 MPa) and 530–550°F (275–290°C) enters the steam generator and leaves as 590–625°F (310–330°C) steam. State-of-the-art systems are generating steam at approximately 1040 psig (7.2 MPa).

Steam produced in the secondary loop is passed through a *moisture separator* before entering the turbine. These

---

[1]This figure is low compared to many other countries that don't have the large coal reserves that the United States has. In France, for example, the percentage is approximately 78%.

[2]Successful, but noncommercial, designs include heavy water reactors (HWR), gas-cooled reactors (GCR) and high-temperature gas-cooled reactors (HTGR), organic liquid-cooled reactors (OCR), and liquid metal-cooled reactors (LMR). The only commercial gas-cooled reactor to operate in the United States—the 330 MW, helium-cooled Ft. St. Vrain unit run by the Colorado Public Service Company—was permanently closed in 1989. The U.S. Department of Energy (DOE) *liquid metal fast breeder reactor* (LMFBR) program was canceled in 1983.

[3]Slightly more than half of the nuclear power plants in the world are pressurized water reactors, while most of the remainder are boiling water reactors. A handful of breeder reactors are in various stages of design, development, and commercial use throughout the world, though none in the United States, and most as test, proof-of-concept facilities.

**Figure 29.2** *Pressurized Water Reactor*

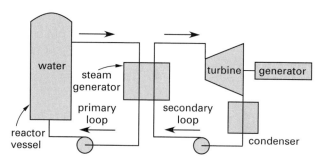

are typically "passive swirl" (i.e., centrifugal cyclone) in nature and produce steam with a quality of 99.5% or higher.

If the steam were to expand directly to the condenser pressure, the final steam quality would be approximately 75% to 80%. To avoid turbine problems associated with a high moisture content, steam is treated between the HP and LP turbine elements, when the pressure drops to approximately 10% to 25% of the throttle pressure.[4] The steam passes through an external *moisture separator reheater*. Reheating to approximately 25°F (14°C) below the saturation temperature occurs in a shell-and-tube exchanger. Steam from the throttle (*single-stage reheat*) or extraction steam is the source of reheat.

## 5. STANDARDIZED REACTORS

In order to lower cost, shorten the licensing period, shorten construction time, and increase safety, standardized reactors are being designed, built, and evaluated.[5] Manufacturers have coined the names *advanced light water reactor* (ALWR), *advanced boiling water reactor* (ABWR), *advanced water-cooled reactor* (AWCR), and *safe integral reactor* (SIR). These designs are also collectively referred to as *simplified boiling water reactor* (SBWR) designs.

SBWRs use prefabricated construction and standardized components. The numbers of pumps, valves, transducers, and instruments are reduced. Control rooms are highly computerized. Multiplexed cabling or fiber optics provides communication between devices. Related equipment is located in a single building. Safety systems are inherent and redundant.

SBWRs can be categorized as evolutionary or passive.[6] *Evolutionary designs* take the best, proven design features from existing technology. They rely on pumps for cooling during normal operation. *Passive designs* rely on natural forces (e.g., gravity and natural convection) for core and reactor cooling during normal operation. These are significant departures from traditional reactor designs.

Standardized reactors have been designed for intrinsic safety. In the event of an accident, the safety systems operate without the need for pumps, traditional heat exchangers, electricity, or even operator intervention—typically for up to three days. For example, in a *loss-of-coolant accident* (LOCA), cooling water from tanks above the reactor would be automatically released, flowing downward by gravity. Cooling air would flow upward by natural convection. Flooding by borated water would ensure reactor shutdown.

## 6. CORROSION AND EMBRITTLEMENT

The two major causes of reactor outages are refueling and repair or replacement of the steam generator. Most nuclear plants eventually suffer from various corrosion-related problems in their primary loops. This includes *intergranular stress corrosion cracking* (IGSCC) and *neutron embrittlement*.

Neutron irradiation of reactor vessels reduces steel's toughness. The problem is most severe at the reactor's *beltline* where the neutron flux is highest. Beltline welds are considered to be the limiting features. Ongoing embrittlement can be reduced by *fuel management* (i.e., loading new fuel rods in the center of the core and moving spent fuel rods to the outside) and by the addition of internal shielding. Existing embrittlement can be eliminated by in situ thermal annealing of the reactor vessel.[7]

## 7. EXTERNALITIES

Some states have mandated that the effects of externalities be included when new power plants are being economically evaluated. *Externalities* (also known as

---

[4]The moisture separation-reheating process using steam-to-steam results in a loss (or, at best, a very small gain) in thermal efficiency. Its function is primarily moisture separation, not efficiency improvement.

[5]Although standardization has been used for years in other countries (e.g., France), the first advanced standardized nuclear plant were the 1315–1356 MW ABWR, named KK-6 and KK-7 for the cities of Kashiwazaki and Kariway on the coast of the Sea of Japan, where a series of seven nuclear power plants reside. KK-6 and KK-7 became operational in 1996 and 1997, respectively. The reactors were built for the Tokyo Electric Power Company jointly by a consortium that included General Electric (GE), Hitachi, and Toshiba. They and their sister plants were shut down for inspection and repairs after experiencing the 6.6–6.8 magnitude ($M_w$) Chuetsu earthquake in July 2007, which generated forces that exceeded the reactors' design specifications.

---

[6]The term "natural forces" has been proposed as a replacement for "passive," which connotes a "we don't care" attitude to the general public.

[7]Thermal annealing is an emerging technology. The embrittled area is heated to approximately 850°F (450°C) and is held at that temperature for as long as is necessary to anneal the metal.

*Pace values*) are difficult-to-evaluate economic factors.[8] In engineering economics problems, these might be known as "nonquantifiables." A typical externality for a fossil-fueled plant is the effect of carbon dioxide on global warming.

When the value of externalities associated with fossil-fuel plants is included, nuclear power becomes economically viable. Of course, nuclear power has its own set of unevaluated externalities. These include terrorism, low-level waste disposal, radon emissions, poor public acceptance, and severe accidents.

---

[8]Pace values are named after the Pace University Center for Environmental Studies.

# 30 Advanced and Alternative Power Generating Systems

## Nomenclature

| | | | |
|---|---|---|---|
| $a$ | interference factor | – | – |
| $A$ | area | ft$^2$ | m$^2$ |
| $B$ | magnetic flux density | – | T |
| $c_p$ | specific heat | Btu/lbm-°F | kJ/kg·K |
| $C_P$ | power factor | – | – |
| $E$ | electric field | – | V/m |
| $F_L$ | loss factor | – | – |
| $F_R$ | removal factor | – | – |
| $F_s$ | shading factor | – | – |
| $g$ | acceleration of gravity | ft/sec$^2$ | m/s$^2$ |
| $I$ | incident energy | Btu/ft$^2$-hr | W/m$^2$ |
| $k$ | ratio of specific heats | – | – |
| $k$ | thermal conductivity | Btu-ft/hr-ft$^2$-°F | W/m·K |
| $K$ | loading factor | – | – |
| KE | kinetic energy | ft-lbf/lbm | J/kg |
| $\dot{m}$ | mass flow rate | lbm/sec | kg/s |
| $p$ | pressure | lbf/ft$^2$ | Pa |
| $P$ | power | ft-lbf/sec | W |
| $q$ | heat transfer | Btu/hr | W |
| $\dot{Q}$ | volumetric flow rate | ft$^3$/sec | m$^3$/s |
| $r$ | radius | ft | m |
| $S$ | Seebeck coefficient | – | V/K |
| $T$ | temperature | °F | K |
| $U$ | overall coefficient of heat transfer | Btu/hr-ft$^2$-°F | W/m$^2$·K |
| v | velocity | ft/sec | m/s |
| $Z$ | figure of merit | 1/°R | 1/K |

## Symbols

| | | | |
|---|---|---|---|
| $\alpha$ | absorptance | – | – |
| $\eta$ | efficiency | – | – |
| $\rho$ | density | lbm/ft$^3$ | kg/m$^3$ |
| $\rho$ | electrical resistivity | Ω-ft | Ω·m |
| $\tau$ | transmittance | – | – |

## Subscripts

$s$    isentropic

## 1. INTRODUCTION

Many advanced energy technologies are mature but uneconomical, are limited to specific applications, or are still in various stages of development. The various *renewables* (e.g., *renewable energy sources* such as wind and solar energy) are in the first category. Radioisotope sources (used primarily for space probes) are in the second. Emerging energy systems (e.g., fusion) are in the third category. Magnetohydrodynamics is in all three categories.[1]

The 1973 energy crisis and oil embargo illuminated the need for alternative energy sources. However, with the large fossil fuel reserves in the United States, wide geographic distribution of these reserves, and the high efficiencies being achieved in combined cycle plants, it is likely that coal and natural gas will continue to be the most economic source of baseload electricity well into the future. Of all the other renewable energy sources, wind energy comes closest in price. Even so, and even with tax credit incentives (i.e., "production credits"), wind power is still 25% to 100% more expensive than coal/gas power.

Renewables are limited by localization and capacity factor. (An installation's *capacity factor* is the actual power output over some period of time divided by the theoretical maximum output. A wind turbine's capacity factor, for example, is affected by the percentages of time the wind does not blow.) Not only are most renewables confined to specific locations, but even then, their capacity factors are often below 30%. (By comparison, new coal plants have capacity factors in excess of 85%.)

## 2. SOLAR THERMAL ENERGY

Solar thermal energy arrives at the outside of the earth's atmosphere at an average rate of 429.2 Btu/ft$^2$-hr (1.354 kW/m$^2$), a value known as the *solar constant*. 40% to 70% of this energy survives absorption in and reflection from the atmosphere and reaches the earth's surface. The actual incident energy, $I$, sometimes referred to as *insolation*, depends on many factors, including geographic location, tilt and orientation of

---

[1]Land-based hydroelectric power is a big exception. This mature renewable energy source is limited by site availability, not by economics.

the receiving surface, calendar day, time of day, and weather conditions. Average values for clear days are given in maps and tabulations.[2] Some of this energy can be captured in *active solar systems* and used for space heating, domestic hot water (DHW) generation, and cooling (using heat pumps and absorption chillers).[3]

Solar thermal energy is captured in *solar collectors*. The sun's energy enters the collector and, because of the *greenhouse effect*, is trapped inside.[4] Heat is absorbed by *heat transfer fluid* pumped through tubes mounted on the *absorber plate*.[5] The tubes and absorber plate can be left uncoated, painted black, or treated with a *selective surface* coating that absorbs more energy than it reradiates.[6] Water at 100°F to 150°F (38°C to 66°C) can easily be generated in this manner. Thermal energy in the heat transfer fluid can be used directly (as in swimming pool heaters), or it can be transferred to and stored in a tank of water or rock pebble beds.

Most collectors in simple heating systems are *flat-plate collectors*. These are essentially wide, flat boxes with clear plastic or glass coverings known as the *glazing*. *Concentrating (focusing) collectors* use mirrors and/or lenses to focus the sun's energy on a small absorber area. Except for some parabolic mirror designs, focusing collectors use a controller and motor to track the sun across the sky. *Evacuated-tube collectors* are more complex, but their efficiencies are higher. A U-shaped tube carries the transfer fluid through an air-filled transparent cylinder, which itself is enclosed in a transparent vacuum cylinder. Evacuated collectors are useful when extremely hot transfer fluid is needed and are generally limited to commercial projects.

Ideally, collectors should point due south and be tilted (from the horizontal) an angle equal to the latitude. For winter use, the tilt should be somewhat higher (so that the maximum energy is received on the coldest days).

---

[2]Some insolation maps use units of langleys/day. A *langley* is equal to 1 cal/cm², 3.69 Btu/ft², and 41 840 J/m².

[3]*Passive solar systems*, which include strategically oriented buildings, walls, and thermal collectors, and which rely on natural convection and conduction for storage and heat transfer, are not included in this chapter.

[4]As received, solar radiation has wavelengths of 0.2 μm to 3.0 μm. Thermal radiation reradiated from the collector plate has a wavelength of approximately 3 μm. Good covering materials have high transmittance (85% to 95%) of received radiation and significantly lower transmittance (less than 2%) of reradiated radiation. White crystal glass, low-iron tempered and sheet glasses, and tempered float glass satisfy these requirements. Polycarbonates, acrylics, and fiberglass also perform well but suffer from weathering and durability problems.

[5]Water can be used as the heat transfer medium, but it is subject to freezing, boiling, and chemical breakdown; and the system is subject to corrosion. To counteract these problems, ethylene glycol-water and glycerine-water mixtures are often used. Ethylene glycol, however, is toxic, and a heat exchanger must be used to keep the heat transfer fluid separate from domestic water. "State-of-the-art" fluids include silicones, hydrocarbon (aromatic and paraffinic) oils, and change-of-phase refrigerants (see Table 30.1).

[6]Common selective surface coatings with low-to-moderate costs include copper oxide, black nickel, black chrome, lead oxide, and aluminum conversion.

The variations from south-pointing and latitude tilt may be ± 10° to 15° without significant degradation in performance.

The heat absorbed by the solar collector can be calculated from the incident energy, the *absorptance*, $\alpha$, of the absorber, and the *transmittance*, $\tau$, of the cover plate. The *shading factor*, $F_s$, in Eq. 30.1 has a value of approximately 0.95 to 0.97 and accounts for dirt on the cover plates and shading from the glazing supports. It is neglected in most initial studies.

$$q_{\text{absorbed}} = F_s A \alpha \tau I \qquad 30.1$$

The heat absorbed by the transfer fluid is

$$q_{\text{fluid}} = \dot{m} c_p (T_{\text{out}} - T_{\text{in}}) \qquad 30.2$$

The difference between the heat absorbed by the collector and transfer fluid constitutes the conduction and convection losses.

$$q_{\text{loss}} = q_{\text{absorbed}} - q_{\text{fluid}}$$
$$= U_L A (T_{\text{plate}} - T_{\text{air}}) \qquad 30.3$$

The average plate temperature, $T_{\text{plate}}$, in Eq. 30.3 is seldom known. However, the incoming fluid temperature is usually known and is used by convention in place of the plate temperature. The collector *heat removal efficiency factor*, $F_R$, is used to correct for the substitution in variables. For liquid collectors, $F_R$ ranges from 0.8 to 0.95 and is usually specified by the collector manufacturer.

$$q_{\text{fluid}} = q_{\text{absorbed}} - q_{\text{loss}}$$
$$= F_R F_s A \alpha \tau I - F_R U_L A (T_{\text{in}} - T_{\text{air}}) \qquad 30.4$$

The *collector efficiency* is the ratio of energy absorbed by the transfer fluid to the original incident energy striking the collector.

$$\eta = \frac{q_{\text{fluid}}}{I A} \qquad 30.5$$

Since convective and radiation losses increase at higher collector temperatures, the collector efficiency decreases as the difference between ambient air and average plate (or inlet) temperatures increases. When the efficiency is plotted against the ratio of inlet-ambient temperature difference to incident energy, $(T_{\text{in}} - T_{\text{ambient}})/I$, the falloff rate is approximately linear with a slope of $-F_R U_L$. (See Fig. 30.1.) Typical values of $F_R U_L$ for flat-plate collectors range from 0.6 Btu/hr-ft²-°F to 1.1 Btu/hr-ft²-°F (3.4 W/m²·K to 6.2 W/m²·K). When the temperature difference is zero, $q_{\text{loss}}$ will also be zero, and the $y$-intercept will be the theoretical maximum efficiency, $F_R \tau \alpha$. Typical values of $F_R \tau \alpha$ range from 0.50 to 0.75.

**Table 30.1** *Properties of Transfer Fluids*

| fluid | specific gravity | viscosity[a] cP | specific heat Btu/lbm-°F | specific heat kJ/kg·K | freezing point °F | freezing point °C |
|---|---|---|---|---|---|---|
| water | 1.00 | 0.5–0.9 | 1.00 | 4.19 | +32 | 0 |
| 50% water-50% ethylene glycol | 1.05 | 1.2–4.4 | 0.83 | 3.5 | −33 | −36 |
| 50% water-50% propylene glycol | 1.02 | 1.4–7.0 | 0.85 | 3.6 | −28 | −33 |
| paraffinic oil | 0.82 | 12–30 | 0.51 | 2.1 | +15 | −9 |
| aromatic oil[b] | 0.85 | 0.6–0.8 | 0.45 | 1.9 | −100 | −73 |
| silicone oil[b] | 0.94 | 10–20 | 0.38 | 1.6 | −120 | −84 |

[a]Viscosity in the range of 80°F to 140°F (27°C to 60°C).
[b]Typical values given.

**Figure 30.1** *Solar Collector Efficiency Curve*

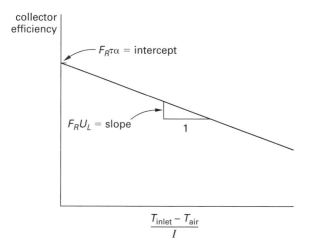

## 3. SOLAR POWER CYCLES

*Solar energy generating systems* (SEGS) use solar energy to generate electricity in traditional power cycles. Three main approaches are taken: *distributed collector systems* (DCS), also known as "trough-electric systems"; *central receiver systems* (CRS), also known as "power-tower systems"; and *dish/Stirling* (D/S) systems.[7] In a *trough-electric system*, parabolic tracking trough concentrators focus sunlight on evacuated glass tubes that run along the collectors' focal lines. Synthetic oil in tubes in the glass cylinders is heated to 575°F to 750°F (300°C to 400°C). Heat is transferred in a heat exchanger to steam used in a traditional Rankine cycle. Trough-electric technology is relatively mature, but due to the low temperatures, average annual thermal efficiencies are only 10% to 15%.

In a *power-tower system*, a field of *heliostats* (tracking mirrors) concentrates solar energy onto a receiver on a central tower. Since directly generated steam would be subject to variations in solar flux from passing clouds, heat generation and use are decoupled. Heat can be stored in molten eutectic salts such as sodium nitrate, which has a melting point of approximately 430°F (220°C), or mixtures of sodium and potassium nitrate. In one scenario, molten salt would be heated to approximately 1050°F (560°C) and cooled to approximately 550°F (290°C) in the heat exchanger. With these higher temperatures, typical thermal efficiencies of 15% to 20% are possible.

In a *dish-engine system*, the heat engine and electrical generator are located at the focus of a parabolic dish. Most installations use Stirling engines (kinematic and free-piston designs) since they are externally heated. Heat can be transferred to the Stirling cycle through direct irradiation, heat pipes, or pool boiling.[8] Due to size and wind loading, 25 kW to 50 kW is about the highest expected output per dish. However, since the units are modular, they can be grouped in a field to produce any desired output. With operating temperatures up to 1400°F (800°C), thermal efficiencies of 24% to 28% are already being realized, and over 30% has been achieved in some prototypes.

In addition to technical and economic problems, SEGS require large areas. Modern trough systems would require approximately 4500 ac (19 km$^2$) to generate 1000 MWe. Even with this space, practical and economic issues limit trough-electric systems to about 200 MWe and tower systems to approximately 100 MWe to 300 MWe.[9]

---

[7]This discussion excludes the ultra-high concentrations and temperatures that are available with high-tech optical, two-stage collectors. These systems produce extremely high temperatures on extremely small targets. The discussion also omits the *solar chimney* concept in which collectors establish a 54°F to 72°F (30°C to 40°C) thermal gradient in the air within tall towers. The stack effect produces air movement, which drives fan-driven generators.

[8]One dish-engine design uses a *reflux pool-boiling receiver*. The receiver uses a pool of molten liquid metal (e.g., sodium, potassium, or a mixture thereof) to transfer heat from the exposed face of the receiver to the helium-filled heater tubes of the Stirling engine. As the liquid metal boils, it vaporizes. It gives up heat when it condenses on the heater tubes. Condensed liquid drips back to the pool by gravity (hence the term "reflux").
[9]The SEGS in southern California's Mojave desert has a total combined capacity of about 350 MW.

**Thermodynamics**

## 4. PHOTOVOLTAIC ENERGY CONVERSION

A *photovoltaic cell (PV cell* or *solar cell)* generates a voltage from incident light, usually light in the visible region. Traditional PV cells are sliced from monolithic silicon crystals and (when cost is unimportant) from gallium arsenide (GaAs) in crystalline form. Thin vacuum-deposited semiconductor films (known as *amorphous thin films*), including silicon, copper indium diselenide (CIS), and cadmium telluride, are less costly than monolithic crystal slices but have much lower efficiencies.

Each silicon cell produces the same current and a voltage of approximately $1/2$ V. To obtain larger voltages or currents, cells are connected into *modules*. Large collections of modules are known as *arrays*.

Efficiency is measured by electrical energy output as a fraction of solar energy input. Although maximum efficiencies of single-crystal and thin-film cells are approximately 33% and 15%, respectively, module efficiencies are much less than single-cell efficiencies due to electrical and optical losses. Most commercial single-crystal and thin-film cells have efficiencies of approximately 20% and 10%, respectively.

Several approaches are taken to increase module efficiencies: (1) optical concentrating of solar energy onto existing technology cells, and (2) production of more efficient solar cells. Module efficiencies as high as 40% have been achieved with *triple junction solar cells* (cells with multiple layers, each responding to a different wavelength) and 40% to 43% using optical concentration. Commercial terrestrial concentration solar cells are available with 38% ratings. Since PV cells are wavelength sensitive, higher efficiencies can be achieved in layered *triple junction cells*.[10]

Existing utility PV "plants" (most of which are less than 1 MW) have capacity factors of approximately 18% to 22%. (Desert sites and two-axis tracking might increase capacity factors to 35%.) Availability is high, 90% to 97%. PV cell output is direct current. Output must be inverted (i.e., converted to alternating current) when connecting to the electric grid.

While PV power cannot compete economically with baseload power generation, it might someday be competitive with peak power alternatives. It is significant that peak PV output coincides with some peak power uses (e.g., air conditioning peaks on hot days). The main problems facing commercialization of PV technologies are high cost, limited lifetimes due to deterioration in performance, and storage of generated energy.

[10]For example, blue and green light could be captured in amorphous silicon layers, while a silicon-germanium layer could intercept infrared light.

## 5. WIND POWER

*Wind energy conversion systems* (WECS), also known as "wind turbines," usually consist of a conventional induction generator driven through a drivetrain by a large rotor. Most rotors consist of a hub and two or three blades. The main shaft, gearbox, and generator are protected from the elements by an enclosure known as a *nacelle*. The WECS is mounted on a tower. Most WECS feeding the electrical grid turn at constant speed.

WECS are classified by the orientation of their rotational axes.[11] With *upwind machines (head-on, horizontal-axis rotors* or *wind-axis rotors)*, the axis of rotation is parallel to the windstream and the blades face the direction from which the wind is coming. With *downwind machines*, the blades face away from the source of the wind. With *vertical-axis rotors (Darrieus rotors* or "eggbeater windmills"), the axis is perpendicular to both the surface of the earth and the windstream. (See Fig. 30.2.)

**Figure 30.2** *Types of Rotors*

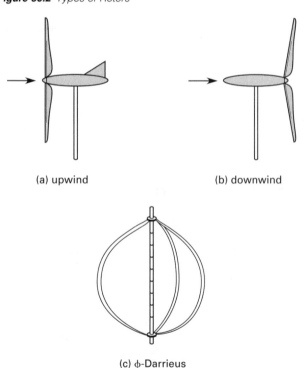

(a) upwind                (b) downwind

(c) φ-Darrieus

Most WECS are upwind machines, even though they require yaw drives to keep them facing into the wind. However, the propeller turbulence (*wash*) can put a strain on the tower. Downwind machines suffer from *tower shadow*. The tower eclipses some of the windstream, and the forces on the up- and down-blades are

[11]*Cross-wind horizontal-axis rotors*, including *Davonius rotor* machines, where the axis is horizontal and perpendicular to the windstream, are very uncommon.

different.[12] The resulting vibration can be severe. However, downwind machines can yaw freely without a yaw drive, and blade stresses are less.

Vertical-axis rotors do not have to be turned to face the wind and require no support tower. They have low starting torques, but may not be self-starting. Their high tip/wind speed ratio produces relatively high power outputs. Since the rotor shape is the shape that a flexible cable would take if spun, the bending stresses are low, no matter how fast the rotor turns. The drive-train and generator are located on the ground, providing for easy maintenance. However, most often they are not installed high enough above the ground to catch the highest velocity wind.

Because of frictional and inertial effects, WECS will not operate much below 5 mph (8 kph), and generally wind speeds of 10 mph (16 kph) or more are needed. 15 mph (24 kph) is generally considered an ideal average speed. Since power increases with the cube of the velocity, WECS are particularly efficient at high velocities. However, forces on WECS become unacceptably high much above 30 mph (48 kph). Various methods are used to limit speed. Most commercial designs are either *stall-limited* (stall-controlled) or *pitch-limited* so that the blades are automatically feathered at higher speeds. Other options include use of blade ailerons, yawing the rotor out of the wind, mechanical braking of the shaft, blade tip brakes, and electrical dynamic braking.

Most sites have a positive *wind shear*—the wind velocity increases exponentially (for a distance) with altitude. Because of this, WECS manufacturers use tower heights of 200 ft to 400 ft (60 m to 120 m) to generate the increased power available at higher altitudes.

Graphs of the number of hours per year that the wind reaches each hourly mean velocity are known as *annual average velocity duration (AAVD) curves*. Curves showing the distribution of annual average wind power per unit subtended area as a function of wind speed are called *annual average power density distribution (AAPD) curves*. The annual average wind energy density distribution is equal to the annual average power density distribution multiplied by the number of hours per year the corresponding wind speeds occur.

The total ideal power available in a windstream is found by multiplying the kinetic energy per unit mass by the flow rate.

$$P_{\text{ideal}} = \dot{m}(\text{KE}) = \dot{Q}\rho(\text{KE})$$

$$= Av\rho\left(\frac{v^2}{2}\right) = \frac{A\rho v^3}{2}$$

$$= \frac{\pi(r_{\text{rotor}})^2 \rho v^3}{2} \qquad \text{[SI]} \quad 30.6(a)$$

$$P_{\text{ideal}} = \dot{m}(\text{KE}) = \dot{Q}\rho(\text{KE})$$

$$= Av\rho\left(\frac{v^2}{2g_c}\right) = \frac{A\rho v^3}{2g_c}$$

$$= \frac{\pi(r_{\text{rotor}})^2 \rho v^3}{2g_c} \qquad \text{[U.S.]} \quad 30.6(b)$$

WECS are unable to extract the ideal power from the airstream. To do so would require decelerating the air flow to zero velocity (i.e., removing all of the kinetic energy). The actual power can be determined if the actual pressures and velocities before and after the WECS are known.

$$P_{\text{actual}} = Av\rho\left(\frac{p_2 - p_1}{\rho} + \frac{v_2^2 - v_1^2}{2}\right) \qquad \text{[SI]} \quad 30.7(a)$$

$$P_{\text{actual}} = Av\rho\left(\frac{p_2 - p_1}{\rho} + \frac{v_2^2 - v_1^2}{2g_c}\right) \qquad \text{[U.S.]} \quad 30.7(b)$$

The actual power generated will depend on the *power coefficient*, $C_P$. The power coefficient of an ideal wind machine rotor with a propeller rotor varies with the ratio of blade tip speed to free-flow windstream speed and approaches the maximum value of 0.593:1 (known as the *Betz coefficient*) when this ratio reaches a value of 5:1 or 6:1. The maximum power coefficients for the best two-blade rotors is approximately 0.47, and for Darrieus vertical-axis rotors, the maximum value is approximately 0.35. Figure 30.3 illustrates the variation in power coefficient for rotors.

$$P_{\text{actual}} = C_P P_{\text{ideal}} \qquad\qquad 30.8$$

**Figure 30.3** Power Coefficients

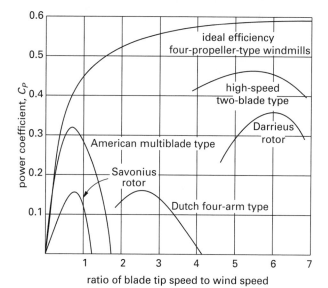

(National Technical Information Service)

---

[12]This happens to a certain extent in all horizontal-axis machines because the wind speed is lower closer to the ground.

The velocity in the immediate wake of the rotor depends on the *interference factor, a,* and is given by $v_2 = v_1(1 - a)$. If some simplifying assumptions are made, the power coefficient can be calculated from the interference factor.[13]

$$C_P = 4a(1 - a)^2 \qquad 30.9$$

Commercial WECS come in all sizes. Units for residential use are seldom larger than 2.5 kW. Modern units connected to utility grids are generally in the 1 MW to 2 MW range (as limited by economics and mechanical stresses), though some mammoth 7 MW to 8 MW units have been built over the years. Individual units can be combined into large "wind farms" to produce any amount of power needed.

Like solar energy, WECS produce little or no environmental impact.[14] However, they require vast amounts of space. Structural corrosion from air pollutants and fatigue failures from normal operation and random gust loads are prevalent. During normal operation, WECS are plagued by blade pitting and soiling by insects and air pollutants. Many first-generation units built in the late 1970s and early 1980s (the heyday of U.S. construction) have failed because of underestimated operating conditions and forces.

*Teetering (hinged or pivoted) two-blade machines* produce lower stresses. Blades are also being redesigned to eliminate areas of laminar flow at all speeds. Only laminar flow is sensitive to blade roughness due to pitting and soiling.

## 6. OCEAN THERMAL ENERGY CONVERSION

During the summer, the temperature at the surface of tropical oceans is almost constant at 77°F (25°C). 2000 ft to 3000 ft (600 m to 900 m) below, however, the temperature can be as low as 38°F to 40°F (4°C to 5°C). An *ocean thermal energy conversion* (OTEC) system uses the difference in temperatures to drive a vapor cycle.

In a *closed-cycle plant,* warm water from the surface supplies heat to a boiler; cold water from the depths provides cooling in the condenser. A traditional low-pressure vapor cycle operating with ammonia, propylene, or other refrigerant drives a generator.

In an *open-cycle plant,* the warm seawater is injected into a near vacuum. The water vaporizes and the resulting steam expands through a low-pressure turbogenerator before being condensed with the cold seawater.[15]

The net-to-gross power ratio, typically around 0.20, is primarily a function of the power needed to pump water from the depths. In addition to low efficiencies due to the small temperature differential, other problems associated with OTEC systems include high component and material costs, corrosion, biofouling, and long-distance power transmission. Production of electrically intensive products (e.g., hydrogen from electrolysis of seawater or even desalinated fresh water) has been proposed as a solution to the latter problem. The major environmental effect would be a slight cooling at the surface of the ocean and a warming of the depths.

## 7. TIDAL AND WAVE POWER PLANTS

Electric power can be generated by capturing water brought into a lagoon or bay by tidal action, and then releasing the water through traditional hydroelectric generators when the tide recedes. The technology is neither new nor complex. However, only about a hundred sites in the world have large enough rises and falls for tidal power to be practical.

With *wave power plants,* advancing and receding ocean waves repeatedly compress and force air through air turbines. (The turbines spin in the same direction regardless of the wave direction.) The turbines drive electrical generators. Wave machines do not suffer from the power transmission problem of OTEC units, since they can be located close to shore.

## 8. GEOTHERMAL ENERGY

There are three general types of geothermal energy sources: vapor-dominated, liquid-dominated, and hot rock sources.[16] The 750 MW "Geysers" in northern California is an example of a *vapor-dominated reservoir* driving a *direct steam cycle.* Multiple wells provide steam at 100 psi to 120 psi (690 kPa to 820 kPa) and 400°F (205°C) which is collected, separated to remove liquid, filtered to remove abrasive particles, and passed through turbines. Evaporative air cooling towers are used to cool condensing water. Water lost in evaporation is replaced by condensed water, and approximately 80% of the steam is evaporated in this manner. The remaining 20%, containing minerals, is reinjected into the ground through deep injection wells. Substantial noncondensible gases (primarily hydrogen sulfide and ammonia, but also including carbon monoxide, hydrogen, methane, and nitrogen) constitute up to 5% of the steam volume and are removed by steam ejectors, vacuum pumps, or compressors and released through the cooling tower.[17] Thermal efficiency is approximately

---

[13]The radial pressure gradient and the kinetic energy of the swirl velocity component in the rotor's wake are neglected.

[14]Many birds are killed by spinning rotors.

[15]The closed process was originally proposed in 1881 by French engineer Jacques-Arsene D'Arsonval. The open process is sometimes referred to as the *Claude cycle,* named after Georges Claude, a student of D'Arsonval, who demonstrated the practicality (at a net power loss) of the open process off the coast of Cuba in 1929.

[16]*Hydrothermal sources* (steam and water), *hot dry rock sources* (steam only), and *geopressured sources* (high-pressure liquid) are three slightly different categorizations of geothermal sources.

[17]Because of the high noncondensible component of steam at the Geysers, removal of noncondensibles is the largest single use of steam —up to 6%.

22%, although 16% is more typical of similar plants worldwide.

Most geothermal sites are *liquid-dominated reservoirs* and produce mixtures of hot water (i.e., brine) and hot steam. In *hot water systems*, the hot water is discarded and only the steam is used. If the hot water temperature is approximately 330°F (165°C) or higher, a *flash steam cycle* can be used. Hot water and steam are pumped to an evacuated *flash tank*. The resulting high-pressure flash steam drives the turbine. In the *dual-flash cycle*, low-pressure hot water remaining in the first flash tank is routed to a second, low-pressure flash tank. Low-pressure steam is used in the LP turbine section. Liquid-dominated reservoirs typically do not produce large quantities of noncondensible gases.

If the temperature of the hot water is between approximately 250°F and 330°F (120°C and 165°C), a *binary cycle* using a separate heat transfer fluid, such as pentane or isobutane, is required. Hot water passes through a closed heat exchanger, vaporizing the heat transfer fluid. Cooled water is discarded in injection wells. The vapor drives a binary turbine generator in a traditional Rankine cycle. An air-cooled condenser maintains the turbine back pressure.

Fewer than ten naturally occurring vapor-dominated reservoirs are known to exist in the world, and only three or four of them are commercially exploited. With drilling, however, hot geological rock can be found anywhere. The temperature of the earth's crust increases by 30°F (16.7°C) for every kilometer of depth. Anomalous temperatures of 570°F to 1300°F (300°C to 700°C) can be found within 20,000 ft (6000 m) of the surface at some locations. In *hot rock systems*, water is injected through injection wells into artificially made fractured rock beds 0.6 mi to 3.6 mi (1 km to 6 km) below the surface. Pressurized hot water at approximately 400°F (200°C) would be removed from an adjacent production well. Steam flashing from the hot water drives turbines located at the surface.

Since there are very few dry steam (i.e., vapor-dominated) sites, but many locations where very hot water is available, current geothermal energy plants exploit either binary cycle equipment, low-pressure flash operation, or both.

Difficulties associated with natural geothermal energy include the scarcity of natural sites, method of cooling, corrosion and fouling by mineral-laden water, and high moisture contents during steam expansion. Environmental problems include disposal of high-mineral waste water, pollution of the air by noncondensing noxious gases, drift, waste heat release, noise from steam venting, and ground subsidence.[18] Hot rock systems, still in their infancy, have many technical difficulties to

overcome, including drilling and controlled fracturing techniques and loss of water.[19]

## 9. FUEL CELLS

A *fuel cell* converts chemical energy directly into electrical energy. With two electrodes separated by an electrolyte, a fuel cell is similar to a continuously fueled battery. (See Fig. 30.4.) Unlike a battery, however, the chemical process continues indefinitely as the spent reactants are replaced. Fuel (e.g., hydrogen or methane) is supplied to the anode (the negative terminal), and oxidizer (e.g., oxygen) is supplied to the cathode (the positive terminal). The anode and cathode are both porous to allow diffusion of the fuel and oxygen through them. Fuel reacts with the electrolyte at the cathode; oxygen reacts at the anode. Liquid water is produced, along with heat and electricity, in a process that is essentially the reverse of electrolysis. (Combustion does not occur in the fuel cell.) The voltage produced by a hydrogen-oxygen fuel cell is approximately 0.7 V.

**Figure 30.4** *Fuel Cell*

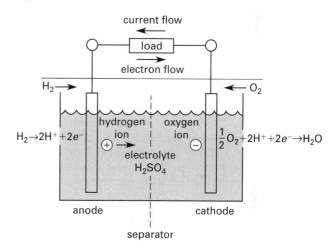

When free hydrogen gas is not available, hydrogen-rich gas can be generated from fossil fuels in a *reformer (reformer reactor)*. Reforming ("front-end reforming") can take place within the fuel cell if the temperature is high enough. This is referred to as the *direct fuel cycle* or *internal fuel reforming*. If the cell temperature is too low or if the fuel is not gaseous, a separate reformer is needed. Steam reforming of natural gas, adiabatic reforming of heavy distillates, and coal gasification are all applicable processes.

The major factor limiting the amount of power generated is the speed of the diffusion processes through the electrolyte between the electrodes. Because of this, fuel cells are named for the electrolyte used.

---

[18]The salt content of geothermal waters varies from 2% to about 20%. By comparison, the salt content of seawater is approximately 3.3%. Hydrogen sulfide is soluble in water and escapes by evaporation during the cooling process.

[19]Nuclear blasts are no longer considered a viable method of producing a cavity deep within the earth.

Commercial units combine individual cells, each of which is about 0.5 cm thick and generates approximately 700 mV, into *stacks*. Stacks containing hundreds of cells are used to generate up to 200 kW to 250 kW of power.[20] Multiple stacks can be used to produce any amount of power desired.

Since fuel cells are not heat engines, they are not limited by Carnot efficiencies. Theoretical fuel cell efficiencies are high, ideally on the order of 80% for hydrogen, 90% for methane, and up to 98% for more complex hydrocarbons.[21] However, not all of the heat generated can be recovered. This and other factors reduce the practical stack efficiency to 35% to 45% based on higher heating values (HHV).

Small hydrogen-oxygen fuel cells have been used in spacecraft for years. First-generation commercial utility-sized *phosphoric acid fuel cells* (PAFCs) operating at 400°F (200°C) are a mature technology. Efficiencies are typically around 40%.

Second-generation *molten carbonate fuel cells* (MCFCs), operating at 1200°F (650°C), offer higher efficiencies. By reforming coal, MCFCs could theoretically replace gas turbines in *integrated coal gasification combined cycle* (ICGCC) plants. HHV efficiencies of carbonate fuel cell stacks are projected to be approximately 50% to 55%.

Third-generation monolithic *solid oxide fuel cells* (MSOFC or SOFC) operating at 1800°F (1100°C) are being developed. A separate reformer reactor is not needed; the high temperature with or without a catalyst can reform the fuel gas (natural gas or gasified coal) internally. The high temperatures also make SOFCs ideal candidates for cogeneration and combined cycle plants, where overall efficiencies may eventually reach 50% to 70%.

Ongoing research continues with *alkaline fuel cells* (which power the U.S. space shuttle) and *proton exchange membrane* (PEM) cells (originally called *solid polymer fuel cells*).

Fuel cells are attractive in the transportation sector because they produce clean energy from readily available reactants. Fuel can be supplied by hydrogen-rich liquids such as methanol, and oxygen is supplied by the air. Temperatures are low enough to keep NOx from being a problem, carbon dioxide is one third of that from hydrocarbon combustion, and all other polluting emissions are substantially reduced. However, because of size, mass, poor start-up and transient responses, and expensive catalysts, fuel cells have been incorporated into only a few experimental and/or demonstration vehicles.

## 10. MAGNETOHYDRODYNAMICS

In a *magnetohydrodynamic* (MHD) *generator*, a high-temperature ionized plasma flows through a supersonic nozzle.[22] The high-velocity plasma in the expansion channel passes through a magnetic field generated by toroidal coil or plates. A constant electric field is generated perpendicularly to the magnetic field (in the direction of the moving gas) by the moving ions.[23] Direct current is generated, so inverters are needed to produce grid-quality power.[24]

High magnetic fields (5 T) and high temperatures—4600°F (2800K) at the generator entrance and 4400°F (2700K) at the exit—are required to create the plasma. The high temperatures are achieved by combustion of gasified coal (with or without oxygen enrichment) in a high-temperature air heater (HTAH) prior to the additional combustion of fossil fuels (coal or natural gas) in a subsequent combustor.

The combustion products are doped (seeded) with salts of easily ionized elements, such as potassium or cesium, to obtain the necessary (1%) ionization. When supplied at approximately 150% of the stoichiometric rate, these alkali metal ions also combine with virtually all sulfur in the fuel. The resulting compounds (potassium sulfate) can be recovered in electrostatic precipitators for reuse as seed. With all sulfur removed, it may be possible to allow the stack gas to cool to as low as 395°F (200°C).

NOx generation was once thought to be problematic. Large amounts (10,000 ppm to 12,000 ppm) of NOx are produced at the high plasma temperatures. However, various modern control methods, including staged fuel-rich combustion, and decomposition in the radiant furnace apparently can reduce NOx to acceptable levels.

MHD generators are attractive because there are no highly stressed rotating parts and because the higher temperatures are compatible with bottoming cycles. Most of the heat remaining in the combustion gases can be recovered in a high-temperature air heater for use in cogeneration or combined bottoming cycles.

From a thermodynamic standpoint, the MHD cycle is the same as the Brayton gas turbine cycle, except that the work output during the expansion is electrical, not mechanical. Since the movement of ionized particles through a magnetic field results in a conversion of mechanical energy to electrical energy, the efficiency of that process is not limited to the Carnot maximum. The mechanical-electrical conversion efficiency, known as the *loading factor, K*, of this process is high: 0.80 to

---

[20]For example, a stack of 470 1 m² cells will generate approximately 750 kW.
[21]All fuel cell efficiencies are temperature-dependent.

[22]MHD machines are not limited to plasmas. In MHD marine (submarine) propulsion systems, the cycle is reversed. Seawater flowing through the duct is acted upon by magnetic and electrical fields. The fields force the water through the duct, propelling the vessel forward.
[23]Generation of this electric field is known as the *Hall effect*.
[24]Momentum in MHD research has decreased significantly since the 1980s due to low efficiencies and pollution concerns.

0.90 (i.e., 80% to 90%). For an isentropic process, $K = 1$. For an isothermal process, $K = 0$.

$$K = \frac{E}{vB} \qquad 30.10$$

The temperature change along the expansion channel is

$$\frac{T_2}{T_1} = \left(\frac{p_2}{p_1}\right)^{K(k-1)/k} \qquad 30.11$$

In a constant-velocity generator, the electrical energy removed decreases the enthalpy between the entrance and exit of the generator. Assuming a constant velocity generator and an ideal gas, the isentropic efficiency of the expansion process is

$$\eta_s = \frac{\Delta h_{\text{actual}}}{\Delta h_{\text{ideal}}} = \frac{\Delta T_{\text{actual}}}{\Delta T_{\text{ideal}}} \qquad 30.12$$

While the mechanical-electrical conversion efficiency is not limited to the Carnot efficiency, the thermal-electrical conversion efficiency is. Experimental installations have achieved a maximum 22% thermal efficiency, although 17% is more typical of coal-fired units. Since Rankine steam cycles can achieve 40% efficiency, MHD channels by themselves are not thermally attractive. When MHD generation is integrated with cogeneration and combined cycles, thermal efficiencies of up to 60% seem reasonable.

## 11. THERMOELECTRIC AND THERMIONIC GENERATORS

A *thermocouple* generates electric potential directly from heat and is an example of a *thermoelectric generator*. Depending on the temperature, certain semiconductors also exhibit thermoelectric effects: bismuth telluride and selenide at room temperature, lead telluride alloys at 390°F to 930°F (200°C to 500°C), and silicon germanium alloys at 750°F to 1830°F (400°C to 1000°C).

The *figure of merit*, $Z$, for a thermoelectric material is an indicator of the effectiveness of the thermoelectric conversion process. Typical values are $3 \times 10^{-3}$ K$^{-1}$ for bismuth telluride alloys to $1 \times 10^{-3}$ K$^{-1}$. In Eq. 30.13, $S$ is the *Seebeck coefficient*, and $k$ is the thermal conductivity.

$$Z = \frac{S^2}{\rho k} \qquad 30.13$$

Current *radioisotope thermoelectric generators* (RTGs) used in space probes contain radiatively coupled *unicouples* using plutonium dioxide as the heat source.[25] Typical output per unicouple is approximately 2.5 W at 3.5 V. Theoretically, any number of unicouples can be combined into a *multicouple*, and multicouples can be

stacked to obtain power in any amount. Most RTG stacks for space missions produce less than 1 kWe.

*Thermionic generators*, also known as *thermionic energy converters* (TECs), consist of two closely spaced tungsten plates (an emitter cathode and a collector anode) known as *shoes*. The hot shoe is maintained at approximately 2600°F to 3150°F (1700K to 2000K), and the cold shoe is maintained at approximately 1350°F (1000K). Current is generated when electrons boil off the hot shoe by thermionic emission and flow to the cold shoe. Since the plates are separated by a low-pressure ionized gas (usually cesium vapor), the device is sometimes referred to as a *plasma diode*.

Thermoelectric and thermionic power generators are power cycles that convert heat into work. Therefore, their efficiencies are limited to the Carnot efficiency. Most RTGs have efficiencies of 10%, and TECs have efficiencies around 15%. For space missions, these low efficiencies are offset by other desirable operating characteristics.

## 12. METHODS OF ENERGY STORAGE

Thermal energy is stored in heated beds. In simple residential solar installations, pebble and rock beds or tanks filled with water and/or other high-density materials can be used. In power utility applications, large tanks or caverns containing oil and rock have been proposed. In high-tech applications, molten salt and liquid metals can be used. Molten salt is particularly attractive in two-stage solar electric plants.

Electrical energy can be stored electrically, mechanically, or chemically. Large energy storage capacitor banks and (theoretically) superconducting coils can store electrical energy in electric and magnetic field forms. Mechanical storage involves converting electrical energy into potential, gravitational, or kinetic energy. Tanks ("accumulators") or caverns of compressed air, gas, or liquid store energy in pressure.[26] Elevated water ("pumped hydro" storage) stores gravitational energy. Flywheels store rotational kinetic energy. Electrical energy can be used to charge batteries, or it can be used to produce hydrogen through electrolysis for later use in fuel cells.

## 13. BATTERIES

Batteries store and produce electrical energy electrochemically. The common flashlight cell and mercury-oxide "button" cells do not have enough energy or power for *electric vehicles* (EVs), however. In addition to traditional lead-acid batteries and the lithium-ion batteries widely used in commercial EVs, other technologies include nickel-cadmium, nickel-metal hydride, zinc-air, and molten sodium-sulfur. (See Table 30.2.) All other

---

[25]Some of the Pioneer and Voyager space probes used RTGs.

[26]CAES is the acronym used in the electrical power industry for *compressed-air energy storage*.

*Table 30.2* Battery Technology Comparison

| battery system | negative electrode | positive electrode | electrolyte | nominal voltage (V) | theoretical specific energy (W·h/kg) | practical specific energy (W·h/kg) | practical energy density (W·h/L) | major issues |
|---|---|---|---|---|---|---|---|---|
| lead-acid | Pb | $PbO_2$ | $H_2SO_4$ | 2.0 | 252 | 35 | 70 | heavy, low cycle life, toxic materials |
| nickel iron | Fe | NiOOH | KOH | 1.2 | 313 | 45 | 60 | heavy, high maintenance |
| nickel cadmium | Cd | NiOOH | KOH | 1.2 | 244 | 50 | 75 | toxic materials, maintenance, cost |
| nickel hydrogen | $H_2$ | NiOOH | KOH | 1.2 | 434 | 55 | 60 | cost, high pressure hydrogen, bulky |
| nickel metal hydride | H (as MH) | NiOOH | KOH | 1.2 | 278–800 (depends on MH) | 70 | 170 | cost |
| nickel zinc | Zn | NiOOH | KOH | 1.6 | 372 | 60 | 120 | low cycle life |
| silver zinc | Zn | AgO | KOH | 1.9 | 524 | 100 | 180 | very expensive, limited life |
| zinc air | Zn | $O_2$ | KOH | 1.1 | 1320 | 110 | 80 | low power, limited cycle life, bulky |
| zinc bromine | Zn | bromine complex | $ZnBr_2$ | 1.6 | 450 | 70 | 60 | low power, hazardous components, bulky |
| lithium ion | Li | $Li_xCoO_2$ | PC or DMC w/$LiPF_6$ | 4.0 | 766 | 120 | 200 | safety issues, calendar life, cost |
| sodium sulfur | Na | S | beta alumina | 2.0 | 792 | 100 | >150 | high temperature battery, safety, low power electrolyte |
| sodium nickel chloride | Na | $NiCl_2$ | beta alumina | 2.5 | 787 | 90 | >150 | high temperature operation, low power |

exotic batteries appear to be too expensive, toxic, or limited for commercial use.

Indicators of battery performance are their *specific power* (in W/kg); *specific energy*, also known as *energy density* (in W·h/kg); and *specific capacity* (in A·h/kg). Generally, specific energy decreases as specific power increases. Thus, batteries can have either high specific power or high specific energy. The best commercial EV batteries, NiMH and lithium-ion, operate with typical specific powers of 150–200 W/kg and 250–300 W/kg (and higher in the lab), respectively, and practical specific energies up to 70 W·h/kg and 120 W·h/kg, respectively.[27]

---

[27]For EVs, goals of 400 W/kg and 200 W·h/kg have been established.

The specific capacity of lead-acid batteries, the most commercially mature technology, is theoretically around 120 A·h/kg; the typical energy density is around 30 W·h/kg. Drawbacks to using lead-acid batteries in EVs are many: high mass, limited recharging cycles, low range, long recharge time, and life reduction if fully discharged.

Lithium-ion and nickel-metal hydride batteries appear to be the most advanced EV-capable batteries. They have long life spans and high power and energy densities. They recharge fully in approximately an hour.

Fiber-nickel-cadmium (FNC) batteries are already in limited use, particularly in military and civilian aircraft. FNC batteries can be recharged several thousand times, and they operate over a wide temperature range. Cadmium constitutes a disposal problem.

Sodium-sulfur batteries are attractive because they have three to four times the storage capacity of lead-acid batteries. Their major drawback is that they must be operated at elevated temperatures of 662°F to 716°F (350°C to 380°C) in order to keep the sulfur electrode molten. They also need a liquid cooling system.

Polymer-electrolyte technology, such as that used in flat cells for powering instant film packs and smart credit cards, hold promise for the future.

# 31 Gas Compression Processes

## Nomenclature

| | | | |
|---|---|---|---|
| $c$ | clearance | percent | percent |
| $c_p$ | specific heat at constant pressure | Btu/lbm-°F | kJ/kg·K |
| $h$ | enthalpy | Btu/lbm | kJ/kg |
| $H$ | head | ft | m |
| $k$ | ratio of specific heats | – | – |
| $m$ | mass | lbm | kg |
| $\dot{m}$ | mass flow rate | lbm/sec | kg/s |
| MW | molecular weight | lbm/lbmol | kg/kmol |
| $n$ | polytropic exponent | – | – |
| $p$ | pressure | lbf/ft² | Pa |
| $P$ | power | Btu/lbm-sec | W/kg |
| $r$ | ratio | – | – |
| $R$ | specific gas constant | ft-lbf/lbm-°R | kJ/kg·K |
| $R^*$ | universal gas constant | ft-lbf/lbmol-°R | kJ/kmol·K |
| $s$ | entropy | Btu/lbm-°F | kJ/kg·K |
| $T$ | temperature | °R | K |
| $V$ | volume | ft³ | m³ |
| $W$ | work | Btu/lbm | kJ/kg |
| $Z$ | compressibility factor | – | – |

## Symbols

| | | | |
|---|---|---|---|
| $\eta$ | efficiency | – | – |
| $v$ | specific volume | ft³/lbm | m³/kg |

## Subscripts

| | |
|---|---|
| $c$ | compression |
| $m$ | mechanical |
| $n$ | polytropic |
| $p$ | pressure |
| $s$ | isentropic |
| $v$ | volumetric |

## 1. TYPES OF COMPRESSORS

There are two main types of gas compressors: reciprocating and centrifugal.[1] *Reciprocating compressors* are appropriate for high-pressure, low-volume applications, such as air conditioning systems. *Rotating compressors* (also known as *dynamic compressors* and *blowers*) are used in low- to moderate-pressure, high-volume applications such as gas turbines and turbojets.[2,3]

Since any compressor will be limited to operating at a finite *pressure ratio*, higher pressures are achieved by routing the output of one compression stage to the input of a subsequent compression stage. Compressors are categorized by the number of *compression stages* (e.g., a two-stage compressor). Subsequent compressions usually occur in different parts of a multistage compressor and are seldom carried out in separate compressors.

Like pumps and fans, dynamic compressors can be radial or axial in design.[4] In radial compressors (i.e., centrifugal compressors), air enters through the eye, is accelerated by the impeller blades, and leaves 90° from the original inlet direction. By mounting two or more impellers on a single shaft, centrifugal compressors can easily be constructed as multistage machines. In *axial-flow compressors*, air flows essentially straight through as it is compressed by blades in the rotor. Axial-flow compressors are explicitly multistage machines, since each row of blades represents a single stage.

In comparison, radial (centrifugal) machines compress small volumes through large pressure ratios, and axial machines compress large volumes through small pressure ratios. Combined *axial-centrifugal compressors* are used when large volumes need to be compressed through large pressure ratios. The axial stages first reduce the volume; the subsequent radial stages (on the same shaft) complete the pressure increase.

---

[1]Other types of *rotary positive-displacement compressors*, such as screw, globe, water-ring, and sliding vane designs, are not covered in this chapter.

[2]There is much overlap in the operating characteristics of all types of compressors. The decision to use one type in favor of another will also be influenced by cost, speed, and available driving mechanisms.

[3]The terms *turbocompressor* and *turboblower* are also used to describe these dynamic machines. The "turbo" prefix does not require that the input power comes from engine exhaust or combustion products (i.e., a turbocompressor is not a *turbocharger*). Input power can be from any shaft drive.

[4]Traditional radial (centrifugal) compressors are used for compression of air and other gases (nitrogen, oxygen, carbon dioxide, etc.). Axial compressors are traditionally found in the steel industry to produce blast furnace air and in plants with large air-separation processes.

## 2. BOOSTERS, INTENSIFIERS, AND AMPLIFIERS

Most compressed air equipment (pneumatic tools, painting, instrumentation, etc.) operates in the range of 90 psig to 125 psig (620 kPa to 860 kPa). Some higher-pressure processes (e.g., plastic molding, metal-working, and pressure testing machines) may require up to 500 psig (3.5 MPa). This ultra-high pressure plant air can be obtained by dedicated multistage compressors, but a power savings can be achieved by starting with the lower-pressure plant air.

A *pressure booster* is essentially another electrically driven single-stage compressor (usually air-cooled) drawing on compressed plant air. Pressure ratios are typically 2:1 to 7:1. Flow rates up to 500 cfm (240 L/s) are typical.

A *pressure amplifier (pressure intensifier)* is one method of producing high-pressure air from a lower-pressure source. A large piston is driven by plant air at low pressures. This piston, in turn, drives a smaller piston, producing a smaller quantity of higher-pressure air. Output is seldom higher than 25 cfm (12 L/s). The output pressure is determined by the input pressure and the ratio of piston areas. Amplifiers do not require any input electrical power. However, they exhaust substantial amounts of compressed plant air, and they require complex valving.

## 3. COMPRESSOR CONTROL

Reciprocating compressors used to supply plant air can either run continuously or intermittently. There is essentially no "control" when running continuously, although pressure will be limited to the relief valve setting, usually just above the highest working pressure required. For reciprocating compressors operating intermittently, the maximum pressure will be the cutoff pressure, usually 15 psi to 20 psi (100 kPa to 140 kPa) higher than the highest pressure needed.

There are three general ways of controlling dynamic compressors. The most effective method is *speed control*, which is applicable to drivers (i.e., prime movers) such as steam or combustion turbines whose speeds can be varied. Figure 31.5 illustrates the variation in performance with changes in rotational speed.

When the driving speed is fixed, as it is with electrical motors, guide vane control and suction throttling can be used. With *guide vane control*, vanes in the compressor impart rotation to the inlet stream. By changing the guide vane settings, inlet velocity, flow rate, and pressure are changed. By imparting either a rotation or counter-rotation to the inlet stream, guide-vane control can either decrease or increase the flow and pressure, respectively. With the inefficient *suction throttling* control method, a butterfly valve in the suction line is partially closed. This decreases the entering suction pressure and flow rate.

## 4. STORAGE TANKS

Compressed-air storage tanks (*receiver tanks*) are used to reduce the duty cycle of plant compressors. The compressor starts when the tank pressure is reduced to the *cut-in pressure* and runs until the tank pressure reaches the *cut-out pressure*. Larger tanks result in longer off-periods for the compressor and fewer compressor on-cycles. However, they increase the running time per cycle.

## 5. COMPRESSOR CAPACITY

In the United States, the capacity of a compressor can be expressed in any of several ways.

- mass flow rate: $\dot{m}$

- volumetric flow rate referred to standard conditions (usually 14.7 psia and 60°F): SCFM (standard cubic feet per minute); SCFH (standard cubic feet per hour); MMSCFD (million standard cubic feet per 24-hour day)[5]

- volumetric flow rate referred to inlet conditions: ICFS (inlet cubic feet per second); ICFM (inlet cubic feet per minute); ICFH (inlet cubic feet per hour)

## 6. COMPRESSION PROCESSES

The thermodynamic behavior of a gas in a *reciprocating compressor* is controlled by the movement of the piston and heat transfer to compressor surfaces. Such compression and expansion are polytropic processes. The work done per cycle depends on the polytropic exponent, $n$, which is a function of the compressor. Efficient air compressors have polytropic exponents between 1.25 and 1.30, but values up to 1.35 are not uncommon. If the compression is known to be isentropic, the gas' ratio of specific heats, $k$, should be used in place of the polytropic exponent, $n$.

Ideal compression in an uncooled *dynamic compressor* is usually considered to be adiabatic (isentropic), with deviations from ideal performance accounted for by the *adiabatic (isentropic) compression efficiency*.[6] With intercooling (see Sec. 31.11), compression in dynamic compressors can also approach an isothermal process. Deviations from ideal isothermal performance are accounted for by the *isothermal compression efficiency*.

$$\eta_{\text{adiabatic}} = \frac{W_{\text{adiabatic}}}{W_{\text{actual}}} \qquad 31.1$$

---

[5]The ASME Power Test Code defines "standard air" as air at a temperature of 68°F (20°C), a total pressure of 14.7 psia (101.3 kPa), and a relative humidity of 36%. The term *free air* is also used by manufacturers to describe the air demands of their equipment. Free air is air at atmospheric pressure and ambient temperature.

[6]Equation 31.1 is the definition of isentropic efficiency. For some reason, in the study of gas compression, the term *adiabatic* is used rather than the terms *reversible adiabatic* or *isentropic*.

$$\eta_{\text{isothermal}} = \frac{W_{\text{isothermal}}}{W_{\text{actual}}} \qquad 31.2$$

Figure 31.1 shows that the work of an isothermal compression is less than the work of an adiabatic compression.[7] The work of a polytropic compression will be somewhere between the two.

**Figure 31.1** *Comparison of Isothermal and Adiabatic Work (zero clearance)*

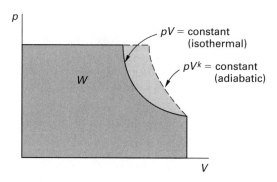

## 7. CLEARANCE AND CLEARANCE VOLUME

Reciprocating compressors are characterized by their *clearance volume*.[8] (See Fig. 31.2.) This is the volume (volume $V_D$ in Fig. 31.3) at the head of the cylinder when the piston is at its most extended position in its stroke. (This position is known as *top-dead-center* or TDC.)[9] The gases remaining in the clearance volume after the discharge valve closes at top-dead-center are known as the *residual gases*.

**Figure 31.2** *Clearance Volume*

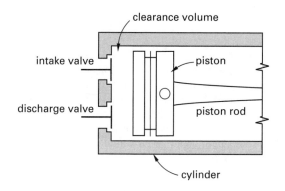

----

[7]Note this confusing point: The work done *in* an adiabatic (isentropic) *process* is less than the work done in an isothermal *process*, but (for the same pressure limits) the work done *by* an isentropic *compressor* is more than the work done by an isothermal *compressor*.
[8]Although there is clearance between the casing and blades of a dynamic compressor, the terms *clearance volume* and *clearance percent* always refer to reciprocating compressors.
[9]The piston is said to be at *bottom-dead-center* (BDC) when it is at its most retracted position in the stroke.

The *percent clearance* (also known as *clearance*), $c$, is calculated from the *swept volume (piston displacement)*, $V_B - V_D$ in Eq. 31.3. Swept volume is not the total volume of the free gas.

$$c = 100\% \times \frac{\text{clearance volume}}{\text{swept volume}} = 100\% \times \frac{V_D}{V_B - V_D} \qquad 31.3$$

The residual gases expand along with the next intake of gas during the intake stroke to reduce the volumetric capacity per stroke. However, clearance affects only the volumetric efficiency; it does not affect the required input power. Two compressors with the same gas flow rate but with different clearances will require the same input power because the expanding residual gases give back their work of compression when they expand. The power requirement depends on only the mass of the gas passing through the compressor.

During steady-state compressor operation, the mass of gas entering the cylinder equals the mass of gas discharged, and clearance is not considered.

## 8. RECIPROCATING, SINGLE-STAGE COMPRESSION

The following processes describe a single stage of compression in a reciprocating compressor.

A to B:  constant pressure suction (intake valve open, discharge valve closed)
B to C:  polytropic compression (both valves closed)
C to D:  constant pressure delivery (exhaust valve open, intake valve closed)
D to A:  polytropic expansion (both valves closed)

The processes describing the compression do not constitute a true cycle because the gas mass in the B-to-C process is not the same as the gas mass in the C-to-D process. For that reason, the broken lines used in Fig. 31.3 show the cylinder volume, not the gas conditions.

The main parameters affecting performance are the polytropic exponent, $n$, compression ratio, $r_p$, and volumetric efficiency, $\eta_v$. For convenience, the polytropic exponents for the expansion and compression processes are assumed to be the same. (However, it is not difficult to consider different values.)

The *compression ratio* for reciprocating and rotating compressors, defined by Eq. 31.4, is the ratio of pressures (not the ratio of volumes as it is for internal combustion engines).[10]

$$r_p = \frac{p_D}{p_A} = \frac{p_C}{p_B} \qquad 31.4$$

The *volumetric efficiency*, $\eta_v$, is the ratio of the actual mass of gas compressed to the mass of gas in the swept

----

[10]The term *ratio of compression*, $r_c$, is often used to distinguish the ratio of pressures from the *ratio of volumes*.

Thermodynamics

**Figure 31.3** *Single-Stage Compression*

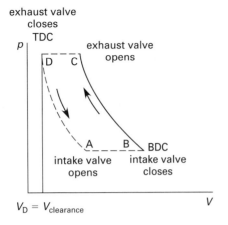

volume. Volumetric efficiency and capacity are evaluated at the inlet conditions unless otherwise noted.

$$\eta_v = \frac{\text{actual mass of gas compressed}}{\text{mass of gas in swept volume}}\bigg|_{\text{inlet conditions}}$$
31.5

$$\eta_v = \frac{V_B - V_A}{V_B - V_D} = 1 - (r_p^{1/n} - 1)\left(\frac{c}{100\%}\right)$$
31.6

The gas mass flow rate through the compressor can be determined from the mass per stroke and the number of strokes per minute. Assuming an ideal gas and evaluating the gas at the compressor's inlet conditions, the ideal and actual gas masses per stroke are

$$m_{\text{stroke,ideal}} = \frac{p_A(V_B - V_D)}{RT_A}$$
31.7

$$m_{\text{stroke,actual}} = \eta_v m_{\text{stroke,ideal}}$$
31.8

The mass flow rate is

$$\dot{m} = m_{\text{stroke,actual}} \times \text{rotational speed}$$
31.9

The relationships between the pressures and volumes are predicted by the ideal gas laws for polytropic processes.

$$\frac{V_A}{V_D} = \left(\frac{p_D}{p_A}\right)^{1/n} = r_p^{1/n}$$
31.10

$$\frac{V_B}{V_C} = \left(\frac{p_C}{p_B}\right)^{1/n} = r_p^{1/n}$$
31.11

Equation 31.12 through Eq. 31.14 give the ideal work of compression for a polytropic *steady-flow process*. (From a thermodynamic standpoint, work is negative if the surroundings compress the system. The work of

compression in these equations is expressed as a positive number.) The specific heat is constant for an ideal gas.

$$W_{BC} = h_C - h_B = c_p(T_C - T_B) = c_p(T_C - T_A)$$
31.12

$$W_{BC} = \left(\frac{n}{1-n}\right)(p_C v_C - p_B v_B)$$
31.13

$$W_{BC} = \left(\frac{n p_B v_B}{1-n}\right)\left(\left(\frac{p_C}{p_B}\right)^{(n-1)/n} - 1\right)$$
31.14

During the D-to-A recovery stroke, the residual gases expand and do work on the piston. However, the gas mass has been greatly reduced, and the expansion work is disregarded. Therefore, the net work per cycle is

$$W_{\text{net}} \approx W_{BC}$$
31.15

"Work done by compressor" almost always means the change in energy imparted to the air. It does not mean the work supplied to the compressor (i.e., it does not include the compression efficiencies).

In addition to the work of compression, the compressor motor supplies power to overcome friction, discharge the gas against the receiver pressure, and move the piston during noncompression parts of the cycle. Equation 31.16 gives the *brake work*, which is expressed in brake horsepower (BHP) and brake kilowatts (BkW).

$$\text{brake work} = \frac{W_{\text{net}}}{\eta_m \eta_s}$$
31.16

Since brake work is expressed per unit mass, the power requirement is

$$P = \dot{m} \times \text{brake work}$$
31.17

Typical single-stage compressors for common plant processes produce approximately 4 cfm to 5 cfm of 90 psig to 125 psig (2 L/s to 2.5 L/s of 620 kPa to 860 kPa) air per brake horsepower delivered. Single-stage pressure boosters starting with compressed plant air are highly efficient and can produce as much as 13 cfm (6.1 L/s) per brake horsepower.

### Example 31.1

Air enters a reciprocating compressor at 14.7 psia and 70°F (101.3 kPa and 21°C). 100 ft³/min (50 L/s) of free air is compressed isentropically to 55 psia (379 kPa). The clearance is 6%. What is the volumetric efficiency?

*SI Solution*

Solve by using Eq. 31.6. (The customary U.S. solution takes a different approach.) Since the compression is isentropic, the polytropic exponent, $n$, is equal to the ratio of specific heats, $k$.

$$\eta_v = 1 - (r_p^{1/n} - 1)\left(\frac{c}{100\%}\right)$$

$$= 1 - \left(\left(\frac{379 \text{ kPa}}{101.3 \text{ kPa}}\right)^{1/1.4} - 1\right)\left(\frac{6\%}{100\%}\right)$$

$$= 0.906 \ (90.6\%)$$

*Customary U.S. Solution*

(The SI solution takes a different approach.) Refer to Fig. 31.3. Calculate the volumes per minute. Since the compression is isentropic, the polytropic exponent, $n$, is equal to the ratio of specific heats, $k$.

$$\frac{V_A}{V_D} = \left(\frac{p_D}{p_A}\right)^{1/k}$$

$$= \left(\frac{55 \ \frac{\text{lbf}}{\text{in}^2}}{14.7 \ \frac{\text{lbf}}{\text{in}^2}}\right)^{1/1.4} = 2.566$$

$$V_A = 2.566 \, V_D$$

Since $p_C = p_D$ and $p_B = p_A$, $V_B/V_C = 2.566$ as well.

The actual volumetric flow rate entering the compressor each minute is 100 ft³ of free air.

$$V_B - V_A = 100 \text{ ft}^3$$

$$V_B = 100 \text{ ft}^3 + V_A$$

$$= 100 \text{ ft}^3 + 2.566 \, V_D$$

The swept volume (piston displacement) is

$$\text{swept volume} = V_B - V_D$$

$V_D$ is the clearance volume, which is 6% of the swept volume.

$$V_D = \frac{c}{100\%}(\text{swept volume}) = \left(\frac{6\%}{100\%}\right)(V_B - V_D)$$

Solving for $V_B$,

$$V_B = 17.667 \, V_D$$

Combine the two equations for $V_B$.

$$17.667 \, V_D = 100 \text{ ft}^3 + 2.566 \, V_D$$

$$V_D = 6.622 \text{ ft}^3$$

$$V_B = 17.667 \, V_D = (17.667)(6.622 \text{ ft}^3)$$

$$= 117.0 \text{ ft}^3$$

The volumetric efficiency (calculated from volumes per minute) is

$$\eta_v = \frac{V_B - V_A}{V_B - V_D}$$

$$= \frac{100 \text{ ft}^3}{117.0 \text{ ft}^3 - 6.622 \text{ ft}^3}$$

$$= 0.906 \ (90.6\%)$$

## 9. NET WORK AND POLYTROPIC HEAD

The net work can be calculated per unit mass from Eq. 31.18. $Z_{\text{ave}}$ is the compressibility factor averaged over the inlet and discharge conditions.

$$W_{\text{net}} = \left(\frac{n}{n-1}\right) Z_{\text{ave}} R T_{\text{inlet}} (r_p^{(n-1)/n} - 1) \qquad 31.18$$

When the net work calculated in Eq. 31.15 or Eq. 31.18 is expressed in feet (numerically equivalent to ft-lbf/lbm) or meters, it may be referred to as the *polytropic head* or (if the process is adiabatic) as the *adiabatic head*. While the discharge pressure depends on the gas, the head does not.

## 10. POLYTROPIC EFFICIENCY

The *polytropic efficiency*, $\eta_n$, is the ratio of ideal polytropic compression work to actual compression work. (For isentropic processes, the polytropic efficiency is 100%.)

$$\eta_n = \frac{W_{\text{ideal}}}{W_{\text{actual}}} = \frac{\dot{m}H}{P_{\text{actual}}} \qquad 31.19$$

The polytropic efficiency can also be calculated from the polytropic exponent, $n$.

$$\frac{n}{n-1} = \frac{\eta_n k}{k-1} \qquad 31.20$$

**Example 31.2**

A dynamic compressor receives 50,000 cfm (23.5 kL/s) of a gas at 15 psia (104 kPa) and 70°F (21°C). The pressure is increased by the compressor to 85 psia (590 kPa). The gas has a molecular weight of 31 and a ratio of specific heats of 1.17. The compressibility factor at the inlet is 0.99, and the average compressibility factor over the compression process is 0.98. The actual shaft power is 9225 hp (6880 kW). 183 hp (136 kW) are required to overcome friction and windage losses. What are the (a) theoretical compression power, (b) polytropic head, and (c) polytropic efficiency?

*SI Solution*

The absolute inlet temperature is

$$T_{\text{inlet}} = 21°\text{C} + 273° = 294\text{K}$$

The compression ratio is

$$r_p = \frac{p_{discharge}}{p_{inlet}}$$

$$= \frac{590 \text{ kPa}}{104 \text{ kPa}} = 5.673$$

The specific gas constant is

$$R = \frac{R^*}{MW} = \frac{8314 \frac{J}{kmol \cdot K}}{31 \frac{kg}{kmol}}$$

$$= 268.2 \frac{J}{kg \cdot K}$$

The inlet density is

$$\rho_{inlet} = \frac{p_{inlet}}{ZRT_{inlet}}$$

$$= \frac{(104 \text{ kPa})\left(1000 \frac{Pa}{kPa}\right)}{(0.99)\left(268.2 \frac{J}{kg \cdot K}\right)(294K)}$$

$$= 1.332 \text{ kg/m}^3$$

The mass flow rate is

$$\dot{m} = \dot{V}\rho$$

$$= \left(23.5 \frac{kL}{s}\right)\left(1 \frac{kL}{m^3}\right)\left(1.332 \frac{kg}{m^3}\right)$$

$$= 31.3 \text{ kg/s}$$

This process is not known to be adiabatic or isentropic. The polytropic exponent is not known. An iterative process is required. Assume the polytropic compression efficiency, $\eta_n$, is 80%. Use Eq. 31.18 and Eq. 31.20 to determine the theoretical polytropic head for this iteration using the assumed efficiency.

$$\frac{n}{n-1} = \frac{\eta_n k}{k-1}$$

$$= \frac{(0.80)(1.17)}{1.17-1} = 5.506$$

$$W_{net} = \left(\frac{n}{n-1}\right)Z_{ave}RT_{inlet}\left(r_p^{(n-1)/n} - 1\right)$$

$$= (5.506)(0.98)\left(268.2 \frac{J}{kg \cdot K}\right)(294K)$$

$$\times \left((5.673)^{1/5.506} - 1\right)$$

$$= 1.577 \times 10^5 \text{ J/kg}$$

$$H = \frac{W_{net}}{g} = \frac{1.577 \times 10^5 \frac{J}{mg}}{9.81 \frac{m}{s^2}} = 16\,075 \text{ m}$$

The theoretical power for this iteration is

$$P_{theoretical} = \frac{P_{ideal}}{\eta_n}$$

$$= \frac{\dot{m}W_{net}}{\eta_n}$$

$$= \frac{\left(31.3 \frac{kg}{s}\right)\left(1.577 \times 10^5 \frac{J}{kg}\right)}{(0.80)\left(1000 \frac{W}{kW}\right)}$$

$$= 6170 \text{ kW}$$

The actual compression power is known to be

$$P_{actual} = P_{shaft} - P_{friction}$$

$$= 6880 \text{ kW} - 136 \text{ kW} = 6744 \text{ kW}$$

Since the actual power is higher, the assumed efficiency is too high.

For the next iteration, use an assumed efficiency of

$$\eta_{n,2} = \eta_{n,1}\left(\frac{P_{theoretical}}{P_{actual}}\right)$$

$$= (0.80)\left(\frac{6170 \text{ kW}}{6744 \text{ kW}}\right) = 0.732$$

*Customary U.S. Solution*

The absolute inlet temperature is

$$T_{inlet} = 70°F + 460° = 530°R$$

The compression ratio is

$$r_p = \frac{p_{discharge}}{p_{inlet}}$$

$$= \frac{85 \text{ psia}}{15 \text{ psia}} = 5.667$$

The specific gas constant is

$$R = \frac{R^*}{MW} = \frac{1545 \frac{ft\text{-}lbf}{lbmol\text{-}°R}}{31 \frac{lbm}{lbmol}}$$

$$= 49.84 \text{ ft-lbf/lbm-}°R$$

The inlet density is

$$\rho_{\text{inlet}} = \frac{p_{\text{inlet}}}{ZRT_{\text{inlet}}}$$

$$= \frac{\left(15 \; \frac{\text{lbf}}{\text{in}^2}\right)\left(12 \; \frac{\text{in}}{\text{ft}}\right)^2}{(0.99)\left(49.84 \; \frac{\text{ft-lbf}}{\text{lbm-}^\circ\text{R}}\right)(530^\circ\text{R})}$$

$$= 0.0826 \; \text{lbm/ft}^3$$

The mass flow rate is

$$\dot{m} = \dot{V}\rho$$

$$= \left(50,000 \; \frac{\text{ft}^3}{\text{min}}\right)\left(0.0826 \; \frac{\text{lbm}}{\text{ft}^3}\right)$$

$$= 4130 \; \text{lbm/min}$$

This process is not known to be adiabatic or isentropic. The polytropic exponent is not known. An iterative process can be used. Assume the polytropic compression efficiency, $\eta_n$, is 80%. Use Eq. 31.18 and Eq. 31.20 to determine the theoretical polytropic head for this iteration using the assumed efficiency.

$$\frac{n}{n-1} = \frac{\eta_n k}{k-1}$$

$$= \frac{(0.80)(1.17)}{1.17 - 1} = 5.506$$

$$W_{\text{net}} = \left(\frac{n}{n-1}\right) Z_{\text{ave}} R T_{\text{inlet}} \left(r_p^{(n-1)/n} - 1\right)$$

$$= (5.506)(0.98)\left(49.84 \; \frac{\text{ft-lbf}}{\text{lbm-}^\circ\text{R}}\right)(530^\circ\text{R})$$

$$\times \left((5.667)^{1/5.506} - 1\right)$$

$$= 52,784 \; \text{ft-lbf/lbm}$$

$$H = 52,784 \; \text{ft}$$

The theoretical horsepower for this iteration is

$$P_{\text{theoretical}} = \frac{P_{\text{ideal}}}{\eta_n}$$

$$= \frac{\dot{m}W_{\text{net}}}{\eta_n}$$

$$= \frac{\left(4130 \; \frac{\text{lbm}}{\text{min}}\right)\left(52,784 \; \frac{\text{ft-lbf}}{\text{lbm}}\right)}{(0.80)\left(33,000 \; \frac{\text{ft-lbf}}{\text{hp-min}}\right)}$$

$$= 8257 \; \text{hp}$$

The actual compression power is known to be

$$P_{\text{actual}} = P_{\text{shaft}} - P_{\text{friction}}$$

$$= 9225 \; \text{hp} - 183 \; \text{hp} = 9042 \; \text{hp}$$

Since the actual power is higher, the assumed efficiency is too high.

For the next iteration, use an assumed efficiency of

$$\eta_{n,2} = \eta_{n,1}\left(\frac{P_{\text{theoretical}}}{P_{\text{actual}}}\right)$$

$$= (0.80)\left(\frac{8257 \; \text{hp}}{9042 \; \text{hp}}\right) = 0.731$$

After two additional iterations, the values are essentially stable.

$$P_{\text{theoretical}} = 9000 \; \text{hp} \quad [9042 \; \text{hp ideally}]$$

$$H = 53,460 \; \text{ft}$$

$$\eta_n = 0.741$$

## 11. COOLING AND INTERCOOLING

For a given compression ratio, the work for isothermal compression is less than the work for adiabatic compression.[11] Furthermore, standard hydrocarbon lubricants used in air compressors are limited to approximately 365°F (185°C). Therefore, cooling is often used in compressors whose compression ratio exceeds 4.0. (A third benefit from cooling is a reduction in ring and valve loading.) For reciprocating compressors, the cooling is accomplished by surrounding the cylinder with a water jacket or by incorporating finned heat radiators in the jacket design.

Where a partially compressed gas is withdrawn, cooled, and compressed further, the term *intercooling* is used. Intercoolers are typically used with centrifugal compressors. The term *perfect intercooling* refers to the case where the gas is cooled to the original inlet temperature (i.e., $T_B = T_D$ in Fig. 31.4).[12]

If necessary, an *aftercooler* removes heat (and reduces the pressure) from the compressed gas after the compression process is complete.

Compressed air for plant use usually leaves the aftercooler at 25°F to 75°F (15°C to 45°C) above the ambient temperature.

## 12. MULTISTAGE COMPRESSION

Figure 31.4 shows the path of a gas experiencing two-stage compression and intercooling. For a reciprocating compressor, the curved solid lines represent polytropic compressions, and the broken lines represent the cylinder volume. For a centrifugal compressor, the curved solid lines represent isentropic (or isothermal) compressions and the broken lines have no meaning.

[11]In practice, the additional cost of the intercooler apparatus must be compared with the power savings.

[12]The discharge from an intercooler will usually be approximately 20°F (10°C) higher than the jacket water temperature.

**Figure 31.4** *Two-Stage Compression with Intercooling*

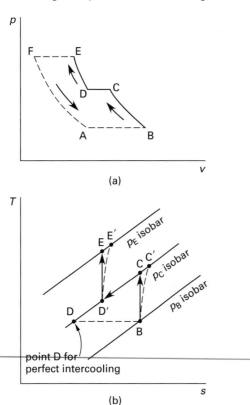

(a)

(b)

Analysis of a multistage compressor is similar to that of a single-stage compressor. The minimum work occurs when the compression ratios of all stages are the same. This is known as *optimum staging*. In that case,

$$r_p = \frac{p_C}{p_B} = \frac{p_E}{p_D} \qquad \text{31.21}$$

Since $p_C = p_D$,

$$p_C^2 = p_E p_B \qquad \text{31.22}$$

Typical water-cooled two-stage compressors for common plant processes produce approximately 2 to 3 cfm of 200 to 500 psig (1 to 1.5 L/s of 1.4 to 3.5 MPa) air per brake horsepower delivered.

## 13. DYNAMIC COMPRESSORS

The ideal gas relationships for adiabatic and isothermal processes can be used to analyze the performance of dynamic gas compressors. In fact, most of the equations in Sec. 31.8 can be used for isentropic processes if the ratio of specific heats, $k$, is used in place of the polytropic exponent, $n$.

## 14. COMPRESSOR CHARACTERISTIC CURVES

Performance of compressors, like that of pumps and fans, is described by characteristic curves. Compressor manufacturers usually provide curves of discharge pressure versus inlet volumetric flow rate and brake power versus flow rate. Alternatively, head versus flow rate or pressure ratio versus flow rate may be substituted for the discharge pressure curve.

Characteristic curves are different for reciprocating and centrifugal compressors. Reciprocating compressors are essentially constant-volume, variable-pressure devices. Since their capacities are essentially fixed, the characteristic curve is quite steep. This is shown in Fig. 31.5.

Figure 31.5 illustrates the characteristic curve for a dynamic compressor. It is similar to curves for pumps and fans. The *surge limit* is the leftmost area on the curve. In surge, flow becomes unstable and the compressor pressure intermittently drops below the system pressure, resulting in backflow.[13] Figure 31.5 also illustrates that for any given speed, even with decreasing discharge pressure, the flow rate does not increase indefinitely. Flow approaches a maximum flow rate at the *choke* or *stone wall point.*

**Figure 31.5** *Dynamic Compressor Head-Capacity Curves*

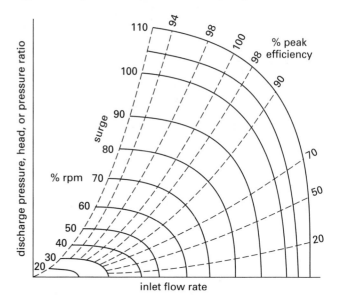

The system curve (not shown) will intersect the characteristic curve at the operating point, as it does with pumps and fans. Ideally, the *operating point* should coincide with the maximum efficiency line. Since each stage of compression will have its own characteristic curve, there will be as many operating points as there are stages. In a properly matched multistage

---

[13]Dynamic compressors should have antisurge controls to open a recirculation valve. These increase the flow sufficiently to keep operation out of surge.

compressor, all of the stages will be operating at high efficiencies. If the stages are not well matched or if the flow rate or inlet pressure change, one or more of the stages may operate "off" its design point. In the worst cases, a stage may be in a surge or choke condition.

## 15. OPERATION AT CHANGED INLET CONDITIONS

Discharge pressure and power curves can only be used for the gas and inlet conditions (typically, "standard" inlet conditions) intended. Those curves cannot be used if the gas or any of the inlet conditions are changed. In most cases, curves applicable for the new conditions are not available. In those cases, a curve for head versus inlet flow rate can be laboriously generated for the new conditions.[14]

Ideally, energy transfer (head) per unit mass is a function of only the velocity of the impeller. Head is the same for all gases, regardless of density. Therefore, while the gas properties affect the discharge pressure, they do not affect the head. Moving a compressor to another altitude will affect the discharge pressure, mass flow rate, and power requirement. It will not affect the ideal compression ratio or the volumetric flow rate.

In reality, compressors (particularly piston compressors) do not deliver as much flow at higher altitudes as they do at sea level. As the air viscosity decreases, leakage through pump clearances increases. The effect is known as *droop*. (See Fig. 31.6.) The disparity between sea-level and high-altitude performance becomes more pronounced at higher discharge pressures.

*Figure 31.6* *Reciprocating Compressor Droop*

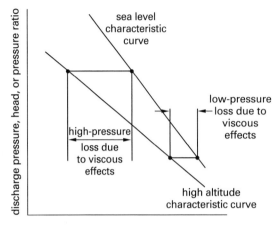

---

[14]Generating the head-flow rate curve is not as easy as simply applying Eq. 31.18. The polytropic exponent, $n$, must simultaneously satisfy Eq. 31.18, Eq. 31.19, and Eq. 31.20. The actual compression power is taken from the existing power-flow rate curve provided by the manufacturer. Therefore, calculating even a single point on the head-flow rate curve is an iterative process. This is illustrated in Ex. 31.2.

# 32 Refrigeration Cycles

## Nomenclature

| | | | |
|---|---|---|---|
| $A$ | area | $\text{ft}^2$ | $\text{m}^2$ |
| $c_p$ | specific heat | Btu/lbm-°F | kJ/kg·K |
| COP | coefficient of performance | – | – |
| $D$ | diameter | ft | m |
| EER | energy efficiency ratio | Btu/W-hr | n.a. |
| $h$ | enthalpy | Btu/lbm | kJ/kg |
| $k$ | ratio of specific heats | – | – |
| $L$ | stroke | ft | m |
| $\dot{m}$ | mass flow rate | lbm/sec | kg/s |
| $n$ | rotational speed | rpm | rpm |
| $p$ | pressure | $\text{lbf/ft}^2$ | kPa |
| $q$ | heat | Btu/lbm | kJ/kg |
| $Q$ | heat | Btu | kJ |
| $r$ | ratio | – | – |
| $s$ | entropy | Btu/lbm-°F | kJ/kg·K |
| $T$ | temperature | °R | K |
| $\dot{V}$ | volumetric flow rate | $\text{ft}^3/\text{min}$ | L/s |
| $W$ | work | Btu/lbm | kJ/kg |
| $x$ | quality | – | – |

## Symbols

| | | | |
|---|---|---|---|
| $\eta$ | efficiency | – | – |
| $\upsilon$ | specific volume | $\text{ft}^3/\text{lbm}$ | $\text{m}^3/\text{kg}$ |

## Subscripts

| | |
|---|---|
| $f$ | saturated liquid |
| $g$ | saturated vapor |
| $p$ | pressure or constant pressure |
| $s$ | isentropic |
| $v$ | volumetric |

## 1. INTRODUCTION TO REFRIGERATION

*Refrigeration* is the process of transferring heat from a low-temperature area to a high-temperature area. Since heat flows spontaneously only from high- to low-temperature areas (according to the second law of thermodynamics), refrigeration needs an external energy source to force the heat transfer to occur. This energy source is a pump or compressor that does work in compressing the refrigerant. (See Fig. 32.1.) It is necessary to perform this work on the refrigerant in order to get it to discharge energy to the high-temperature area.

**Figure 32.1** *Device Operating on a Refrigeration Cycle*

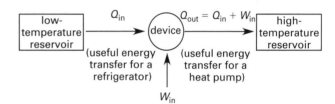

In a power cycle, heat from combustion is the input and work is the desired effect. Refrigeration cycles, though, are power cycles in reverse; work is the input and cooling is the desired effect. (For every power cycle, there is a corresponding refrigeration cycle.) In a refrigerator, the heat is absorbed from a low-temperature area and is rejected to a high-temperature area.[1] The pump work is also rejected to the high-temperature area.

General refrigeration devices consist of a coil (the *evaporator*) that absorbs heat, a *condenser* that rejects heat, a compressor, and a pressure-reduction device (the *expansion valve* or *throttling valve*).[2] (See Fig. 32.2.)

In operation, liquid refrigerant passes through the evaporator where it picks up heat from the low-temperature area and vaporizes, becoming slightly superheated. The vaporized refrigerant, that is, the "suction gas," is compressed by the compressor and, in so doing, increases

---

[1] It is common thermodynamic jargon to refer to the low- and high-temperature areas as *environments* or *thermal reservoirs*. In particular, a low-temperature area is called a *source*, and a high-temperature area is referred to as a *sink*.

[2] In home refrigerators, the *expansion valve* takes the form of a long capillary tube. Other components include a motor to drive the compressor, a fan for the evaporator, a fan for the condenser, and the refrigerant itself.

*Figure 32.2* *Refrigeration Device*

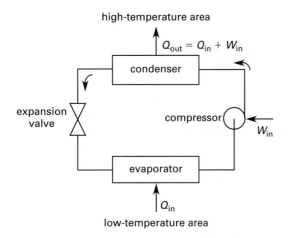

even more in temperature. The high-pressure, high-temperature refrigerant passes through the condenser coils, and, being hotter than the high-temperature environment, loses energy. Finally, the pressure is reduced in the expansion valve, where some of the liquid refrigerant also flashes into a vapor.

If the low-temperature area from which the heat is being removed is air from occupied space (that is, air is being cooled), the device is known as an *air conditioner*. If the heat is being removed from water, the device is known as a *chiller*. An air conditioner produces cold air; a chiller produces cold water.

In small refrigeration systems, such as those used for in-window home air conditioning, the compressor, condenser, evaporator (cooling) coil, and fan are combined into a single housing. These are referred to as *unit* (or *unitary*) *air conditioners*. In large commercial systems, the compressor is separate from the evaporation coil.

## 2. REFRIGERANTS

There are more than three hundred commercial refrigerants commonly available. The refrigerant used depends on the temperatures of the low- and high-temperature areas, as well as the cooling load and the compressor power.

Prior to the 1987 Montreal Protocol, chlorofluorocarbon (CFC) refrigerants R-11 and R-12, and hydrochlorofluorocarbon (HCFC) R-22, in pure form or in blends, were the economical and efficient choices for residential, automotive, appliance, and commercial applications.[3] All these formulations are now subject to production and use limitations or outright bans, with replacements coming from the hydrofluorocarbon (HFC) families, most notably HFC-134a, HFC-407C, and HFC-410A. (HFC-141b is used as a blowing agent, not a refrigerant.) Production of all HCFCs will cease in 2030.

Originally concerned only with ozone depletion potential (ODP), production and use limitations for CFCs and HCFCs have been extended even to HFCs due to an awareness of *global warming potential* (GWP). For example, due to its modest GWP, the sale of R-134a for refrigeration use has been banned in some states and countries. However, these restrictions are not specifically related to the Montreal Protocol. There are currently no restrictions on equipment or use of R-134a, R-407C, R410A, and R-417A.

R-11 was used in large (plant) centrifugal compressors. Worldwide production stopped in 1995.[4] Although there is no direct substitute (without incurring a performance loss), R-11 has been replaced by HFC-123 and, to a lesser extent, HFC-134a.

R-12 was used in small- to medium-sized centrifugal compressors up to about 100 tons. It was commonly used in automotive air conditioning. Worldwide production stopped in 1995. It has been replaced for automotive use with R-134a. R-423A and R-437A are other replacements.

R-22 was the refrigerant "of choice" in unitary air conditioners (i.e., residential window units), chillers, freezers, residential air conditioning and heat pump units, and commercial and transport refrigeration and cooling. Under the Clean Air Act, R-22 may still be produced in the United States until 2020, but only for the purpose of servicing existing equipment; after the first day of 2020, it may not be produced or imported. R-410A is a replacement, as are R-407C, R-417A, R-422A, R-422D, R-500, and R-502.

R-114 was the most commonly used refrigerant for high-temperature coolers and heat pumps. As with R-11, because of the diversity of applications, there is no single replacement for R-114.

Ammonia was used in early residential refrigerators, but because it is toxic, it was replaced in the 1930s by Dupont's Freon R-12. However, ammonia is still used when extremely low temperatures are required and toxicity concerns are minimal, such as in industrial freezing and storage equipment, marine cargo ships, and ice rinks. It has potential, also, as a replacement for R-22.

The actual schedule for implementing production and use restrictions on these refrigerants is complicated by several factors. Developed countries have earlier compliance dates than developing countries. All HCFCs are grouped together in the phase-out schedule. Limitations are based on mass production, so a country can phase out one compound (e.g., a blowing agent) more quickly in order to continue producing another compound (e.g., a refrigerant). Compounds remain available indefinitely for servicing older units, as long as stocks of refrigerant come from reclaimed equipment.

---

[3]Refrigerants often have a number of designations and trade names. R-22, for example, may be referred to as HCFC-22, Freon-22, and SUVA-22.

[4]R-11 and R-22 are still available for servicing equipment already in use, with existing inventory being reclaimed refrigerant recovered from "scrapped" equipment.

Depending on the operating conditions, backfilling older equipment with replacement refrigerants (e.g., substituting R-134a for R-12) will generally result in a 5% to 30% decrease in cooling capacities. If this reduction is unacceptable, performance can be restored by (1) substituting a more efficient impeller, (2) increasing the compressor's rotational speed, and (3) substituting a more efficient heat exchanger and tubing.

## 3. HEAT PUMPS

*Heat pumps* also operate on refrigeration cycles. Like standard refrigerators, they transfer heat from low-temperature areas to high-temperature areas. The device shown in Fig. 32.1 could represent either a heat pump or a refrigerator. There is no significant difference in the mechanisms or construction of heat pumps and refrigerators. The only difference is their purpose.

A refrigerator's main function is to cool the low-temperature area. The *useful energy transfer* for a refrigerator is the heat removed from the cold area. A heat pump's main function is to warm the high-temperature area. The useful energy transfer is the heat rejected to the high-temperature area. A heat pump is almost always used when *space heating* of an occupied area is needed. The attraction of *heat pump devices* is that they can provide both heating in the winter and cooling in the summer. During the summer, however, they run refrigeration cycles and are technically refrigerators. Strictly speaking, *heat pump cycles* are used only for space heating.

## 4. COEFFICIENT OF PERFORMANCE

The concept of thermal efficiency is not used with devices operating on refrigeration cycles. Rather, the *coefficient of performance* (COP) is defined as the ratio of useful energy transfer (as defined in Sec. 32.3) to the work input. The higher the coefficient of performance, the greater will be the effect for a given work input. Since the useful energy transfer is different for refrigerators and heat pumps, the coefficients of performance will also be different.

In calculating coefficients of performance, the refrigerant is considered to be the system. Therefore, $Q_{in}$ is the energy that enters the refrigerant. ($Q_{in}$ is not the energy that enters the high-temperature area because that area is not the system.)

$$\text{COP}_{\text{refrigerator}} = \frac{Q_{in}}{W_{in}} = \frac{Q_{in}}{Q_{out} - Q_{in}}$$
$$= \text{COP}_{\text{heat pump}} - 1 \qquad 32.1$$

The coefficient of performance for a heat pump includes the desired heating effect of the pump work input.

$$\text{COP}_{\text{heat pump}} = \frac{Q_{in} + W_{in}}{W_{in}} = \frac{Q_{in} + W_{in}}{Q_{out} - Q_{in}}$$
$$= \text{COP}_{\text{refrigerator}} + 1 \qquad 32.2$$

## 5. ENERGY EFFICIENCY RATIO

The *energy efficiency ratio* (EER) is defined as the useful energy transfer in Btu/hr divided by input power in watts. This is just the coefficient of performance expressed in mixed units.

$$\text{EER} = 3.41 \times \text{COP} \qquad 32.3$$

As with the coefficient of performance, the definition of useful energy transfer depends on whether the device is being used as a refrigerator or as a heat pump.

$$\text{EER}_{\text{refrigerator}} = \frac{\dot{Q}_{in,\text{Btu/hr}}}{\dot{W}_{in,\text{watts}}}$$
$$= \text{EER}_{\text{heat pump}} - 1 \qquad 32.4$$

$$\text{EER}_{\text{heat pump}} = \frac{(\dot{Q}_{in} + \dot{W}_{in})_{\text{Btu/hr}}}{\dot{W}_{in,\text{watts}}}$$
$$= \text{EER}_{\text{refrigerator}} + 1 \qquad 32.5$$

## 6. REFRIGERATION CAPACITY

The rate of energy removal from the low-temperature area is known as the *refrigeration capacity* or *refrigeration effect*. While the kilowatt is the appropriate SI unit, in the United States capacity is measured in *refrigeration tons*, where one ton is equal to 200 Btu/min or 12,000 Btu/hr (3.517 kW) of heat removal. The ton is derived from the heat flow required to melt one ton of ice in 24 hours.

Since refrigeration capacity has the units of power (e.g., Btu/hr), it can be combined with the coefficient of performance to calculate the required pump horsepower.

$$\text{pump horsepower} = \frac{4.715\dot{Q}_{in,\text{tons}}}{\text{COP}} = \frac{4.715\dot{Q}_{in,\text{Btu/hr}}}{(12,000)(\text{COP})} \qquad 32.6$$

The rate of energy removal can be used to calculate the refrigerant mass flow rate.

$$\dot{m} = \frac{\dot{Q}_{in}}{\Delta h_{\text{evaporator}}} \qquad 32.7$$

**Thermodynamics**

## 7. CARNOT REFRIGERATION CYCLE

The *Carnot refrigeration cycle* is a Carnot power cycle running in reverse. Because it is reversible, the Carnot refrigeration cycle has the highest coefficient of performance for any given temperature limits of all the refrigeration cycles. As shown in Fig. 32.3, the processes all occur within the vapor dome.

A to B: isentropic expansion
B to C: isothermal heating (vaporization)
C to D: isentropic compression
D to A: isothermal cooling (condensation)

**Figure 32.3** *Carnot Refrigeration Cycle*

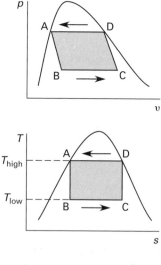

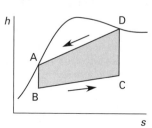

The analysis procedure for the Carnot refrigeration cycle is reversed from, but otherwise identical to, the procedure for analyzing the Carnot power cycle. The coefficients of performance are

$$\text{COP}_{\text{refrigerator}} = \frac{Q_{\text{in}}}{W_{\text{in}}} = \frac{Q_{\text{in}}}{Q_{\text{out}} - Q_{\text{in}}}$$

$$= \frac{T_{\text{low}}}{T_{\text{high}} - T_{\text{low}}} = \text{COP}_{\text{heat pump}} - 1$$

$$32.8$$

$$\text{COP}_{\text{heat pump}} = \frac{Q_{\text{in}} + W_{\text{in}}}{W_{\text{in}}} = \frac{Q_{\text{out}}}{Q_{\text{out}} - Q_{\text{in}}}$$

$$= \frac{T_{\text{high}}}{T_{\text{high}} - T_{\text{low}}} = \text{COP}_{\text{refrigerator}} + 1$$

$$32.9$$

## 8. VAPOR COMPRESSION CYCLE

The *vapor compression cycle* is essentially a reversed Rankine vapor cycle.[5] It is the most common type of refrigeration cycle, finding application in household refrigerators, air conditioners for cars and houses, chillers, and so on. The processes are illustrated in Fig. 32.4.

A to B: isenthalpic expansion
B to C: constant pressure heating (vaporization)
C to D: isentropic compression
D to A: constant pressure cooling (condensation)

**Figure 32.4** *Wet Vapor Compression Cycle*

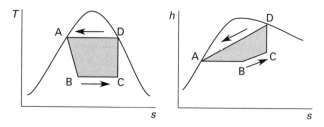

If the refrigerant leaves the evaporator (see Fig. 32.4, point C) with a quality of less than 1.0, the cycle is known as a *wet vapor compression cycle*. *Wet compression* is undesirable because of compressor wear and performance problems. For that reason, refrigerators are designed so that the refrigerant leaves the evaporator either saturated or slightly superheated, as shown by point C in Fig. 32.5. Compression of saturated or superheated vapor is said to be *dry compression*, and the cycle is known as a *dry vapor compression cycle*.

**Figure 32.5** *Dry Vapor Compression Cycle*

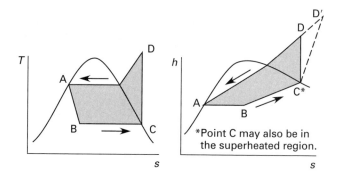

*Point C may also be in the superheated region.

---

[5]An irreversible expansion through a throttling valve takes the place of the boiler.

If the refrigerant is saturated when it leaves the evaporator, the following solution method can be used. (The subscripts $f$ and $g$ refer to saturated fluid and vapor properties read from a refrigerant table. They do not correspond to any point on the figures shown in this chapter.)

At A: $x = 0$ (saturated liquid); either $p_A$ or $T_A$ must be known, or they can be found as the saturation temperature and pressure having an enthalpy of $h_A$. $h_A = h_B$; $p_A = p_D$.

At B: $T_B = T_C$; $h_B = h_A$ (because A-to-B is isenthalpic).

At C: $x = 1$ (saturated vapor); $T_C = T_B$; $p_C = p_{sat}$ for temperature $T_C$; $h_C$ and $s_C$ are read as $h_g$ and $s_g$ from the refrigerant table.

At D: $p_D = p_A$; if $s_D$ and either $p_D$ or $T_D$ are known, $h_D$ can be found by searching the refrigerant superheat table. $s_D = s_C$ (because C-to-D is isentropic).

The refrigerant mass flow rate is

$$\dot{m} = \frac{\dot{Q}_{in}}{h_C - h_B} \qquad 32.10$$

The compressor power is

$$\dot{W}_{in} = \dot{m}(h_D - h_C) \qquad 32.11$$

The refrigeration effect is

$$\dot{Q}_{in} = \dot{m}(h_C - h_B) \qquad 32.12$$

The coefficient of performance as a refrigerator is

$$COP_{refrigerator} = \frac{\dot{Q}_{in}}{\dot{W}_{in}} = \frac{h_C - h_B}{h_D - h_C} \qquad 32.13$$

If the compression is not isentropic (i.e., an isentropic compressor efficiency is known), $h_D$ will be affected. This, in turn, will determine $T_D$. ($p_D$ is not affected.)

$$h'_D = h_C + \frac{h_D - h_C}{\eta_{s,compressor}} \qquad 32.14$$

### Example 32.1

A refrigerator using HFC-134a has a cooling effect of 10,000 Btu/hr (2.9 kW). Refrigerant leaves the evaporator saturated at 0°F (−18°C). The pressure of the refrigerant entering the condenser is 120 psia (830 kPa; 8.3 bars). The refrigerant leaves the condenser as a saturated liquid. The compressor's isentropic efficiency is 80%. Find (a) the temperature of the refrigerant immediately after compression, (b) the refrigerant flow rate, (c) the compressor input shaft power, and (d) the refrigerator's coefficient of performance.

*SI Solution*

Refer to Fig. 32.5 and a pressure-enthalpy diagram for refrigerant HFC-134a (see App. 22.X). Values will vary somewhat with the precision to which the pressure-enthalpy chart can be read.

At point C: Locate the intersection of the horizontal −18°C line and saturated vapor line.

$$T_C = -18°C \quad \text{[given]}$$
$$p_C = 1.6 \text{ bars}$$
$$h_C = 385 \text{ kJ/kg}$$

At point D: Follow a line of constant entropy upward and to the right to the horizontal 8.3 bars line.

$$T_D = 40°C$$
$$p_D = 8.3 \text{ bars} \quad \text{[given]}$$
$$h_D = 425 \text{ kJ/kg}$$
$$h'_D = h_C + \frac{h_D - h_C}{\eta_{s,compressor}}$$
$$= 385 \ \frac{kJ}{kg} + \frac{425 \ \frac{kJ}{kg} - 385 \ \frac{kJ}{kg}}{0.80}$$
$$= 435 \ \text{kJ/kg}$$

At point A: Move horizontally to the saturated liquid line.

$$p_A = p_D = 8.3 \text{ bars}$$
$$T_A = 32°C$$
$$h_A = 244 \text{ kJ/kg}$$

At point B: Throttling processes are constant-enthalpy processes. Follow a vertical line downward to the horizontal 1.6 bar line.

$$p_B = p_C = 1.6 \text{ bar}$$
$$h_B = h_A = 244 \text{ kJ/kg} \quad \text{[throttling process]}$$

(a) Use the chart to find the temperature corresponding to 8.3 bars and 435 kJ/kg. The temperature immediately after the compression process is $T'_D \approx 50°C$.

(b) The required refrigerant flow is found from the total heat load and the enthalpy change per pound.

$$\dot{m} = \frac{Q_{in}}{h_C - h_B}$$
$$= \frac{(2.9 \text{ kW})\left(1.0 \ \frac{kJ}{kW \cdot s}\right)}{385 \ \frac{kJ}{kg} - 244 \ \frac{kJ}{kg}}$$
$$= 0.0206 \text{ kg/s}$$

(c) The compressor power is

$$\dot{W}_{\mathrm{in}} = \dot{m}(h'_{\mathrm{D}} - h_{\mathrm{C}})$$

$$= \left(0.0206 \ \frac{\mathrm{kg}}{\mathrm{s}}\right)\left(435 \ \frac{\mathrm{kJ}}{\mathrm{kg}} - 385 \ \frac{\mathrm{kJ}}{\mathrm{kg}}\right)$$

$$= 1.03 \ \mathrm{kW}$$

(d) The coefficient of performance of the refrigerator is

$$\mathrm{COP} = \frac{\dot{Q}_{\mathrm{in}}}{\dot{W}_{\mathrm{in}}} = \frac{2.9 \ \mathrm{kW}}{1.03 \ \mathrm{kW}}$$

$$= 2.8$$

*Customary U.S. Solution*

Refer to Fig. 32.5 and a pressure-enthalpy diagram for refrigerant HFC-134a (see App. 22.M). Values will vary somewhat with the precision to which the pressure-enthalpy chart can be read.

At point C: Locate the intersection of the horizontal 0°F line and saturated vapor line.

$$T_{\mathrm{C}} = 0°\mathrm{F} \quad [\text{given}]$$

$$p_{\mathrm{C}} = 21 \ \mathrm{psia}$$

$$h_{\mathrm{C}} = 102 \ \mathrm{Btu/lbm}$$

At point D: Follow a line of constant entropy upward and to the right to the horizontal 120 psia line.

$$T_{\mathrm{D}} = 103°\mathrm{F}$$

$$p_{\mathrm{D}} = 120 \ \mathrm{psia} \quad [\text{given}]$$

$$h_{\mathrm{D}} = 119 \ \mathrm{Btu/lbm}$$

$$h'_{\mathrm{D}} = h_{\mathrm{C}} + \frac{h_{\mathrm{D}} - h_{\mathrm{C}}}{\eta_{s,\mathrm{compressor}}}$$

$$= 102 \ \frac{\mathrm{Btu}}{\mathrm{lbm}} + \frac{119 \ \frac{\mathrm{Btu}}{\mathrm{lbm}} - 102 \ \frac{\mathrm{Btu}}{\mathrm{lbm}}}{0.80}$$

$$= 123 \ \mathrm{Btu/lbm}$$

($T'_{\mathrm{D}}$ could be found if needed.)

At point A: Move horizontally to the saturated liquid line.

$$p_{\mathrm{A}} = p_{\mathrm{D}} = 120 \ \mathrm{psia}$$

$$T_{\mathrm{A}} = 91°\mathrm{F}$$

$$h_{\mathrm{A}} = 41 \ \mathrm{Btu/lbm}$$

At point B: Throttling processes are constant-enthalpy processes. Follow a vertical line downward to the horizontal 21 psia line.

$$p_{\mathrm{B}} = p_{\mathrm{C}} = 21 \ \mathrm{psia}$$

$$h_{\mathrm{B}} = h_{\mathrm{A}} = 41 \ \mathrm{Btu/lbm} \quad [\text{throttling process}]$$

(a) Use the chart to find the temperature corresponding to 120 psia and 123 Btu/lbm. The temperature immediately after the compression process is $T'_{\mathrm{D}} \approx 117°\mathrm{F}$.

(b) The required refrigerant flow is found from the total heat load and the enthalpy change per pound.

$$\dot{m} = \frac{Q_{\mathrm{in}}}{h_{\mathrm{C}} - h_{\mathrm{B}}}$$

$$= \frac{10,000 \ \frac{\mathrm{Btu}}{\mathrm{hr}}}{102 \ \frac{\mathrm{Btu}}{\mathrm{lbm}} - 41 \ \frac{\mathrm{Btu}}{\mathrm{lbm}}}$$

$$= 164 \ \mathrm{lbm/hr}$$

(c) The compressor power is

$$\dot{W}_{\mathrm{in}} = \dot{m}(h'_{\mathrm{D}} - h_{\mathrm{C}})$$

$$= \left(164 \ \frac{\mathrm{lbm}}{\mathrm{hr}}\right)\left(123 \ \frac{\mathrm{Btu}}{\mathrm{lbm}} - 102 \ \frac{\mathrm{Btu}}{\mathrm{lbm}}\right)$$

$$= 3444 \ \mathrm{Btu/hr}$$

(d) The coefficient of performance of the refrigerator is

$$\mathrm{COP} = \frac{\dot{Q}_{\mathrm{in}}}{\dot{W}_{\mathrm{in}}} = \frac{10,000 \ \frac{\mathrm{Btu}}{\mathrm{hr}}}{3444 \ \frac{\mathrm{Btu}}{\mathrm{hr}}}$$

$$= 2.9$$

## 9. AIR REFRIGERATION CYCLE

Any gas will cool when expanded. This is the principle behind the air refrigeration cycle. The *air refrigeration cycle* (also known as a *Brayton cooling cycle*) is essentially a reversed Brayton turbine cycle. It is not common because of its high power consumption. However, air is nonflammable, readily available, and nontoxic. Therefore, the air refrigeration cycle is often used in aircraft air conditioning and gas liquefaction applications where a refrigerant with such characteristics is needed.

Compressed-air air conditioners, now essentially non-existent, were common prior to the development of Freon and other chlorofluorocarbon refrigerants. These units consisted of a compressor, heat exchanger, and expansion turbine. The heat increase due to compression was dissipated by the heat exchanger. The

Thermodynamics

expanded, cooler air was discharged directly into the room. (See Fig. 32.6.)

A to B: isentropic expansion
B to C: constant pressure heating
C to D: isentropic compression
D to A: constant pressure cooling

**Figure 32.6** *Air Refrigeration Cycle*

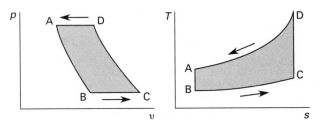

If air is considered to be an ideal gas, the ideal gas equations can be used to find the properties at each point.

$$\frac{T_A}{T_B} = \frac{T_D}{T_C} = \left(\frac{p_{\text{high}}}{p_{\text{low}}}\right)^{(k-1)/k} \qquad 32.15$$

Since $\Delta h = c_p \Delta T$ for an ideal gas, the coefficient of performance can be calculated from Eq. 32.16.

$$\text{COP}_{\text{refrigerator}} = \frac{T_C - T_B}{(T_D - T_A) - (T_C - T_B)} \qquad 32.16$$

If air is not considered to be an ideal gas, an air table must be used to find the coefficient of performance.

$$\text{COP}_{\text{refrigerator}} = \frac{h_C - h_B}{(h_D - h_A) - (h_C - h_B)} \qquad 32.17$$

The coefficient of performance can also be calculated from the pressure ratio, $r_p$. (This formula cannot be easily corrected for nonisentropic expansion or compression.)

$$\text{COP}_{\text{refrigerator}} = \frac{1}{r_p^{(k-1)/k} - 1} \qquad 32.18$$

$$r_p = \frac{p_{\text{high}}}{p_{\text{low}}} \qquad 32.19$$

## 10. HEAT-DRIVEN REFRIGERATION CYCLES

A *heat-driven refrigeration cycle*, also known as a *heat-activated refrigeration cycle*, is practical when large quantities of waste or inexpensive heat energy are available. In addition to using waste heat, combustion of natural gas and LPG, solar collectors, geothermal sources, and waste steam can provide the energy to drive a refrigeration cycle.

One method of obtaining refrigeration is to use the available heat to drive a Stirling heat engine. The work from the engine drives the refrigerator's compressor. Thus, the apparatus consists of combined power-generating and refrigeration units, as shown in Fig. 32.7.

**Figure 32.7** *Heat-Driven Cooling Cycle*

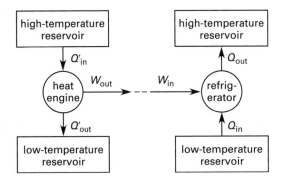

## 11. ABSORPTION CYCLE

The *absorption cycle* is similar to the vapor compression cycle, with one major change—there is no compressor, and no external work is used to compress the refrigerant.[6] Rather, a generator-absorber-recuperator apparatus (see Fig. 32.8) produces a solution of refrigerant in another liquid, and external heat superheats the refrigerant.

**Figure 32.8** *Absorption Cycle Apparatus*

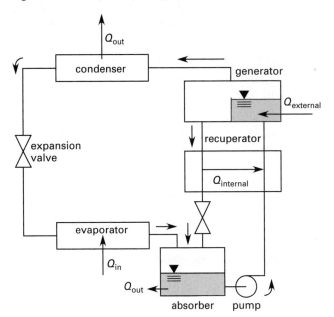

---

[6]A very small amount of work may be used by the liquid return pump between the absorber and generator shown in Fig. 32.8. However, even this pump can be replaced by a *thermosiphon* (i.e., using an inert gas that circulates the liquid by expanding and contracting in response to heat).

Two working fluids are required in the absorption cycle.[7] As in the vapor-compression cycle, the cycle starts with the refrigerant passing through the evaporator, removing heat from the low-temperature area. The refrigerant (in saturated vapor condition) then enters an *absorber* where it is absorbed by the liquid *absorbent*.

The mixture of liquid refrigerant and absorbent is next pumped into the higher pressure of the *generator*. (This operation requires very little work because the fluid is liquid.) Heat from an external source (e.g., solar collectors, geothermal generators, or combustion of natural gas) drives off the refrigerant in a superheated (though usually subatmospheric) condition. The absorbent left behind is returned to the absorber.

An optional heat exchanger (known as the *recuperator*) may be used to transfer heat from the returning absorbent to the mixture entering the generator. This heat transfer helps improve the system efficiency since the absorbent can carry more refrigerant at a lower temperature. If a recuperator is not used, it may be necessary to cool the liquid in the absorber by some other means (e.g., passing cooling water from another source through the absorber).

An optional *rectifier* (not shown) between the generator and condenser is needed in ammonia-water systems to remove any remaining traces of absorbent from the refrigerant.

The absorption cycle is not very efficient (e.g., coefficients of performance less than 1.0) and the initial equipment costs are higher than for the vapor compression cycle, but when cheap or free energy is available, the cycle can be economical. The cycle is also bulky and involves toxic fluids; hence it is unsuitable for home and auto cooling.

In Fig. 32.8, the coefficient of performance is

$$\text{COP}_{\text{refrigerator}} = \frac{Q_{\text{in}}}{Q_{\text{external}}} = \frac{Q_{\text{in}}}{Q_{\text{out}} - Q_{\text{in}}} \qquad 32.20$$

The *generator-absorber-heat exchanger* (GAX) *cycle* is an advanced variation of the absorption cycle. With an ammonia-water absorption cycle, the lower range of the high-temperature absorption process overlaps the higher range of the low-temperature heat input process. With the proper equipment, a portion of the rejected absorber heat can be used to supply heat to the generator.

## 12. HIGH-SIDE EQUIPMENT

*High-side equipment*, primarily the compressor and condenser, is the equipment that operates at the high pressure in the refrigerator. Reciprocating piston compressors are used for small- and medium-sized refrigerators. For the smallest units, for example, those used

in kitchen refrigerators and other unitary machines, hermetically sealed single-stage motor and compressor packages have traditionally been used. Sealed units eliminate the leakage through mechanical seals that is common in large commercial units. Compressor operators, power requirements, volumetric efficiency, and other thermodynamic considerations are similar to those for other compressors. Centrifugal compressors are used in the largest machines (usually 200 tons or larger).

*Scroll compressors*, also referred to as *scroll pumps* and *spiral compressors*, products of computer-aided manufacturing and precision machining, were introduced commercially in the late 1980s as replacements for reciprocating compressors in small residential air conditioners. They are approximately 10% more efficient than piston compressors, have higher reliabilities, and are suitable for use with modern refrigerants such as R-134a, R-407C, and R-410A.

Condensers used in small- and medium-sized (i.e., up to approximately 100 tons) refrigerators are typically aircooled heat exchangers. For efficient operation, the condensing temperature should not be lower than 10°F (5°C) and not more than 30°F (17°C) above the initial air temperature. For larger capacities, water-cooled condensers are used.

### Example 32.2

Refrigerant HFC-134a is used in a single-acting, two-cylinder compressor running at 300 rpm. The bore diameter is 6 in (152 mm), and the stroke length is 7 in (178 mm). Saturated vapor enters the compressor and is compressed isentropically to 150 psia (1.0 MPa; 10 bars). Saturated liquid enters the expansion valve. Evaporation occurs at 50 psia (350 kPa; 3.5 bars). The cooling effect is 17 tons (60.4 kW). What is the volumetric efficiency?

*SI Solution*

The swept volume per stroke is

$$V = AL = \frac{\pi D^2 L}{4}$$

$$= \frac{\pi (152 \text{ mm})^2 \left(178 \, \dfrac{\text{mm}}{\text{stroke-cylinder}}\right)}{(4)\left(1000 \, \dfrac{\text{mm}}{\text{m}}\right)^3}$$

$$= 0.00323 \text{ m}^3/\text{stroke-cylinder}$$

---

[7]Common pairs are ammonia (the refrigerant) and water (the absorbent), and water (the refrigerant) and lithium bromide (the absorbent).

Since the compressor is single-acting, each cylinder contributes one compression stroke each revolution. The ideal volumetric flow rate is

$$\dot{V}_{\text{ideal}} = Vn$$

$$= \frac{\left(0.00323 \; \dfrac{\text{m}^3}{\text{stroke-cylinder}}\right)}{60 \; \dfrac{\text{s}}{\text{min}}} \times (300 \text{ rpm})(2 \text{ cylinders})$$

$$= 0.0323 \text{ m}^3/\text{s}$$

See Fig. 32.5. At point C, the pressure is 3.5 bars. From a pressure-enthalpy diagram for refrigerant HFC-134a (see App. 22.M), the enthalpy of saturated 3.5 bar vapor is approximately 400 kJ/kg. The specific volume is approximately 0.060 m$^3$/kg. At point A, the pressure is 10 bars. The enthalpy of saturated 10 bar liquid is approximately 255 kJ/kg. This is also the enthalpy at point B.

The actual volumetric flow rate is

$$\dot{V}_{\text{actual}} = \dot{m}_{\text{C}} v_{\text{C}} = \frac{\dot{Q} v_{\text{C}}}{h_{\text{C}} - h_{\text{B}}}$$

$$= \frac{(60.4 \text{ kW})\left(0.060 \; \dfrac{\text{m}^3}{\text{kg}}\right)}{400 \; \dfrac{\text{kJ}}{\text{kg}} - 255 \; \dfrac{\text{kJ}}{\text{kg}}}$$

$$= 0.0250 \text{ m}^3/\text{s}$$

The volumetric efficiency is

$$\eta_v = \frac{\dot{V}_{\text{actual}}}{\dot{V}_{\text{ideal}}}$$

$$= \frac{0.0250 \; \dfrac{\text{m}^3}{\text{s}}}{0.0323 \; \dfrac{\text{m}^3}{\text{s}}}$$

$$= 0.774 \quad (77.4\%)$$

*Customary U.S. Solution*

The swept volume per stroke is

$$V = AL = \frac{\pi D^2 L}{4}$$

$$= \frac{\pi (6 \text{ in})^2 \left(7 \; \dfrac{\text{in}}{\text{stroke-cylinder}}\right)}{(4)\left(12 \; \dfrac{\text{in}}{\text{ft}}\right)^3}$$

$$= 0.1145 \text{ ft}^3/\text{stroke-cylinder}$$

Since the compressor is single-acting, each cylinder contributes one compression stroke each revolution. The ideal volumetric flow rate is

$$\dot{V}_{\text{ideal}} = Vn$$

$$= \left(0.1145 \; \dfrac{\text{ft}^3}{\text{stroke-cylinder}}\right)(300 \text{ rpm})$$

$$\times (2 \text{ cylinders})$$

$$= 68.7 \text{ ft}^3/\text{min}$$

See Fig. 32.5. At point C, the pressure is 50 psia. From a pressure-enthalpy diagram for refrigerant HFC-134a (see App. 22.X), the enthalpy of saturated 50 psia vapor is approximately 107 Btu/lbm. The specific volume is approximately 0.94 ft$^3$/lbm. At point A, the pressure is 150 psia. The enthalpy of saturated 150 psia liquid is approximately 46 Btu/lbm. This is also the enthalpy at point B.

The actual volumetric flow rate is

$$\dot{V}_{\text{actual}} = \dot{m}_{\text{C}} v_{\text{C}} = \frac{\dot{Q} v_{\text{C}}}{h_{\text{C}} - h_{\text{B}}}$$

$$= \frac{(17 \text{ tons})\left(200 \; \dfrac{\text{Btu}}{\text{min-ton}}\right)\left(0.94 \; \dfrac{\text{ft}^3}{\text{lbm}}\right)}{107 \; \dfrac{\text{Btu}}{\text{lbm}} - 46 \; \dfrac{\text{Btu}}{\text{lbm}}}$$

$$= 52.4 \text{ ft}^3/\text{min}$$

The volumetric efficiency is

$$\eta_v = \frac{\dot{V}_{\text{actual}}}{\dot{V}_{\text{ideal}}}$$

$$= \frac{52.4 \; \dfrac{\text{ft}^3}{\text{min}}}{68.7 \; \dfrac{\text{ft}^3}{\text{min}}}$$

$$= 0.763 \quad (76.3\%)$$

## 13. LOW-SIDE EQUIPMENT

*Low-side equipment*, primarily the evaporator, is the equipment that operates at the low pressure in the refrigerator. *Evaporators* for cooling air are usually externally finned and are known as *cooling coils*. Evaporators for cooling water and other liquids are mostly shell-and-coil and shell-and-tube heat exchangers known as *coolers* and *chillers*.[8] Tubes are generally smooth inside.[9]

Coolers and chillers for water generally operate with an average temperature difference of 6°F to 20°F (3°C to

---

[8]In industrial settings, the shell-and-coil design is less desirable because the coil cannot be mechanically cleaned.

[9]*Microfins* inside the evaporator tube increase the heat transfer efficiency.

11°C), and optimally with a 10°F to 14°F (5°C to 8°C) difference. To avoid freezing problems, the temperature of the entering refrigerant should be above 28°F (−2°C).

## 14. REFRIGERANT LINE SIZING

The size of the refrigerant return line (from the condenser) is not a critical design parameter. The pressure drop in the throttling valve is far greater than the pressure drop from the fluid flow.

A small suction line, however, can greatly decrease compressor capacity. The suction line between the evaporator and the compressor is typically sized from tables based on the allowing pressure drop, refrigeration capacity, and length. Tables are presented in different formats depending on the source, and different tables are needed for each different refrigerant. Maximum allowable pressure drops of 3 psi (20 kPa) on high-temperature units and 1.5 psi (10 kPa) on low-temperature units are the established rules of thumb, and tables are given for this range of pressure drops.

Suction lines should not be sized too large, as a reasonable velocity is needed to carry oil from the evaporator back to the compressor. For horizontal suction lines, 750 ft/min (3.8 m/s) is a recommended minimum velocity. For vertical suction lines, the velocity should be 1200 ft/min to 1400 ft/min (6.1 m/s to 7.1 m/s).

# Topic V: Chemistry

**Chemistry**

# 33

# Inorganic Chemistry

Chemistry

## Nomenclature

| | | | |
|---|---|---|---|
| $A$ | Arrhenius constant | n.a. | various |
| $A$ | atomic weight | lbm/lbmol | kg/kmol |
| $C$ | concentration | n.a. | mol/L |
| $\mathcal{E}$ | cell potential | V | V |
| $E$ | energy | ft-lbf | J |
| EW | equivalent weight | lbm/lbmol | kg/kmol |
| $F$ | force | lbf | N |
| $F$ | formality | n.a. | FW/L |
| FW | formula weight | lbm/lbmol | kg/kmol |
| GEW | gram equivalent weight | – | – |
| $H$ | enthalpy | Btu/lbmol | kcal/mol |
| $H$ | Henry's law constant | 1/atm | 1/atm |
| $i$ | van't Hoff factor | – | – |
| $I$ | current | A | A |
| $k$ | reaction rate constant | various | various |
| $K$ | constant | various | various |
| $K$ | equilibrium constant[1] | – | – |
| $m$ | mass | lbm | kg |
| $m$ | molality | n.a. | mol/1000 g |
| $M$ | molarity | n.a. | mol/L |
| MW | molecular weight | lbm/lbmol | kg/kmol |
| $n$ | number of moles | – | – |
| $n$ | principal quantum number | – | – |
| $N$ | normality | n.a. | GEW/L |
| $N$ | number | – | – |
| $N_A$ | Avogadro's number | 1/mol | 1/mol |
| $p$ | pressure | lbf/ft$^2$ | Pa |
| $R^*$ | universal gas constant | n.a. | Pa·L/mol·K |
| $t$ | time | sec | s |
| $T$ | absolute temperature | °R | K |
| v | rate of reaction | n.a. | mol/L·s |
| $V$ | volume | ft$^3$ | m$^3$ |
| $V$ | volume (traditional cgs) | n.a. | L |
| $x$ | distance or position | ft | m |
| $x$ | gravimetric fraction | – | – |
| $x$ | mole fraction | – | – |
| $x$ | relative abundance | – | – |
| $X$ | fraction ionized | – | – |
| $Z$ | atomic number | – | – |

## Symbols

| | | | |
|---|---|---|---|
| $\pi$ | osmotic pressure | lbf/ft$^2$ | Pa |

## Subscripts

| | |
|---|---|
| $a$ | activation or acid |
| $A$ | Avogadro |
| $b$ | boiling or base |
| eq | equilibrium |
| $f$ | formation or freezing |
| $p$ | pressure |
| $r$ | reaction |
| sp | solubility product |
| $t$ | total |
| $x$ | compound $x$ |

## Superscripts

| | |
|---|---|
| 0 | pure vapor |

[1]The equilibrium constant may actually have units of concentration raised to some positive or negative integer value. However, it is always reported without units, reflecting its definition as the ratio of two reaction rates.

## 1. ATOMIC STRUCTURE

An *element* is a substance that cannot be decomposed into simpler substances during ordinary chemical reactions.[2] An *atom* is the smallest subdivision of an element that can take part in a chemical reaction. A *molecule* is the smallest subdivision of an element or compound that can exist in a natural state.

The atomic nucleus consists of neutrons and protons, known as *nucleons*. The masses of neutrons and protons are essentially the same—one *atomic mass unit*, amu. One amu is exactly $^1/_{12}$ of the mass of an atom of carbon-12, approximately equal to $1.66 \times 10^{-27}$ kg.[3] The *relative atomic weight* or *atomic weight*, $A$, of an atom is approximately equal to the number of protons and neutrons in the nucleus.[4] The *atomic number*, $Z$, of an atom is equal to the number of protons in the nucleus.

The atomic number and atomic weight of an element E are written in symbolic form as $_ZE^A$, $E_Z^A$, or $_Z^AE$. For example, carbon is the sixth element; radioactive carbon has an atomic mass of 14. Therefore, the symbol for carbon-14 is $C_6^{14}$. Since the atomic number is superfluous if the chemical symbol is given, the atomic number can be omitted (e.g., $C^{14}$).

## 2. ISOTOPES

Although an element can have only a single atomic number, atoms of that element can have different atomic weights. Many elements possess *isotopes*. The nuclei of isotopes differ from one another only in the number of neutrons. Isotopes behave the same way chemically.[5] Therefore, isotope separation must be done physically (e.g., by centrifugation or gaseous diffusion) rather than chemically.

Hydrogen has three isotopes. $H_1^1$ is *normal hydrogen* with a single proton nucleus. $H_1^2$ is known as *deuterium* (*heavy hydrogen*), with a nucleus of a proton and neutron. (This nucleus is known as a *deuteron*.) Finally, $H_1^3$ (*tritium*) has two neutrons in the nucleus. While normal hydrogen and deuterium are stable, tritium is radioactive. Many elements have more than one stable isotope. Tin, for example, has ten.

The *relative abundance*, $x_i$, of an isotope $i$ is equal to the fraction of that isotope in a naturally occurring sample

[2]Atoms of an element can be decomposed into subatomic particles in nuclear reactions.
[3]Until 1961, the atomic mass unit was defined as $^1/_{16}$ of the mass of one atom of oxygen-16. The carbon-12 reference has now been universally adopted.
[4]The term *weight* is used even though all chemical calculations involve mass. The atomic weight of an atom includes the mass of the electrons. Published *chemical atomic weights* of elements are averages of all the atomic weights of stable isotopes, taking into consideration the relative abundances of the isotopes.
[5]There are slight differences, known as *isotope effects*, in the chemical behavior of isotopes. These effects usually influence only the rate of reaction, not the kind of reaction.

of the element. The *chemical atomic weight* is the weighted average of the isotope weights.

$$A_{\text{average}} = x_1 A_1 + x_2 A_2 + \cdots \qquad 33.1$$

## 3. INERT ELEMENTS

It is now known that electrons travel randomly in the vicinity of the nucleus. Because of the nature of the electron motion around the nucleus, the word "orbit" is not used; rather, it is said that an electron occupies an *orbital*. The hydrogen orbital, named the $1s$ orbital, has a maximum capacity of two electrons. The "1" in the $1s$ classification is the *principal quantum number*, $n$. The $s$ orbital is the most stable state an electron can occupy.

The electrons in the outermost orbitals (i.e., the outer *shell*) are known as the *valence electrons*. These valence electrons are important in bond formation as well as in electrical and thermal conduction.

An atom with two electrons in its $s$ orbital will be in a lower energy state than an atom with only one $s$ orbital electron. In fact, the $1s^2$ element (helium) is so stable that it will not combine with any other element.

For the principal quantum number $n = 2$, there are two subshells: the $s$ and $p$ subshells. Since each orbital holds at most two electrons, and there are three $p$ orbitals, a total of eight electrons can occupy the $s$ and $p$ orbitals. The element that has the $s$ and $p$ subshells completely filled is neon, which also will not combine with other elements.

It can be generalized that very stable atomic structures result when all of the orbitals corresponding to a quantum number are filled. The elements that possess a filled structure (e.g., helium, neon, argon, krypton, xenon, and radon) are known as the *inert gases (noble gases)* because they do not normally combine with other elements to form compounds.[6]

## 4. PERIODIC TABLE

The *periodic table*, as shown in Table 33.1, is organized around the *periodic law*: The properties of the elements depend on the atomic structure and vary with the atomic number in a systematic way. Elements are arranged in order of increasing atomic numbers from left to right. Adjacent elements in horizontal rows differ decidedly in both physical and chemical properties. However, elements in the same column have similar properties. Graduations in properties, both physical and chemical, are most pronounced in the *periods* (i.e., the horizontal rows).

The electron-attracting power of an atom is called its *electronegativity*, which is measured on an arbitrary scale of 0 to 4. Metals have electronegativities less

than 2. Group VIIA elements (fluorine, chlorine, etc.) are most strongly electronegative. The alkali metals (Group IA) are the most weakly electronegative. Generally, the most *electronegative elements* are those at the right ends of the periods. Elements with low electronegativities are found at the beginning (i.e., left end) of the periods. Electronegativity decreases as you go down a group.

The vertical columns are known as *groups*, numbered in Roman numerals. Elements in a group are called *cogeners*. Each vertical group except 0 and VIII has A and B subgroups (*families*). The elements of a family resemble each other more than they resemble elements in the other family of the same group. Graduations in properties are definite but less pronounced in vertical families. The trend in any family is toward more *metallic properties* as the atomic weight increases.

*Metals* (elements at the left end of the periodic chart) have low electron affinities and electronegativities, are reducing agents, form positive ions, and have positive oxidation numbers. They have high electrical conductivities, luster, generally high melting points, ductility, and malleability.

*Nonmetals* (elements at the right end of the periodic chart) have high electron affinities and electronegativities, are oxidizing agents, form negative ions, and have negative oxidation numbers. They are poor electrical conductors, have little or no luster, and form brittle solids. Of the common nonmetals, fluorine has the highest electronic affinity and electronegativity, with oxygen having the next highest values.

The *metalloids* (e.g., boron, silicon, germanium, arsenic, antimony, tellurium, and polonium) have characteristics of both metals and nonmetals. Electrically, they are semiconductors.

Elements in the periodic table are often categorized into the following groups.

- *actinides:* same as actinons
- *actinons:* elements 90 to 103[7]
- *alkali metals:* group IA
- *alkaline earth metals:* group IIA
- *halogens:* group VIIA
- *heavy metals:* metals near the center of the chart
- *inner transition elements:* same as transition metals
- *lanthanides:* same as lanthanons
- *lanthanons:* elements 58 to 71[8]
- *light metals:* elements in the first two groups
- *metals:* everything except the nonmetals

---

[6]Some compounds of inert gases, primarily fluorides of xenon, have been synthetically produced.

[7]The *actinons* resemble element 89, *actinium*. Therefore, element 89 is sometimes included as an actinon.

[8]The *lanthanons* resemble element 57, *lanthanum*. Therefore element 57 is sometimes included as a lanthanon.

**Table 33.1** *Periodic Table of the Elements (referred to Carbon-12)*

The Periodic Table of Elements (Long Form)

The number of electrons in filled shells is shown in the column at the extreme left; the remaining electrons for each element are shown immediately below the symbol for each element. Atomic numbers are enclosed in brackets. Atomic weights (rounded, based on carbon-12) are shown above the symbols. Atomic weight values in parentheses are those of the isotopes of longest half-life for certain radioactive elements whose atomic weights cannot be precisely quoted without knowledge of origin of the element.

metals — transition metals — nonmetals

| periods | I A | II A | III B | IV B | V B | VI B | VII B | VIII B | | | I B | II B | III A | IV A | V A | VI A | VII A | 0 |
|---|---|---|---|---|---|---|---|---|---|---|---|---|---|---|---|---|---|---|
| 1 / 0 | 1.0079 H[1] 1 | | | | | | | | | | | | | | | | | 4.0026 He[2] 2 |
| 2 / 2 | 6.939 Li[3] 1 | 9.0122 Be[4] 2 | | | | | | | | | | | 10.81 B[5] 3 | 12.0115 C[6] 4 | 14.0067 N[7] 5 | 15.9994 O[8] 6 | 18.994 F[9] 7 | 20.183 Ne[10] 8 |
| 3 / 2,8 | 22.9898 Na[11] 1 | 24.312 Mg[12] 2 | | | | | | | | | | | 26.9815 Al[13] 3 | 28.086 Si[14] 4 | 30.9738 P[15] 5 | 32.064 S[16] 6 | 35.453 Cl[17] 7 | 39.948 Ar[18] 8 |
| 4 / 2,8 | 39.098 K[19] 8,1 | 40.08 Ca[20] 8,2 | 44.956 Sc[21] 9,2 | 47.90 Ti[22] 10,2 | 50.942 V[23] 11,2 | 51.996 Cr[24] 13,1 | 54.938 Mn[25] 13,2 | 55.847 Fe[26] 14,2 | 58.933 Co[27] 15,2 | 58.71 Ni[28] 16,2 | 63.546 Cu[29] 18,1 | 65.38 Zn[30] 18,2 | 69.72 Ga[31] 18,3 | 72.59 Ge[32] 18,4 | 74.922 As[33] 18,5 | 78.96 Se[34] 18,6 | 79.904 Br[35] 18,7 | 83.80 Kr[36] 18,8 |
| 5 / 2,8,18 | 85.47 Rb[37] 8,1 | 87.62 Sr[38] 8,2 | 88.905 Y[39] 9,2 | 91.22 Zr[40] 10,2 | 92.906 Nb[41] 12,1 | 95.94 Mo[42] 13,1 | (98) Tc[43] 14,1 | 101.07 Ru[44] 15,1 | 102.905 Rh[45] 16,1 | 106.4 Pd[46] 18 | 107.868 Ag[47] 18,1 | 112.40 Cd[48] 18,2 | 114.82 In[49] 18,3 | 118.69 Sn[50] 18,4 | 121.75 Sb[51] 18,5 | 127.60 Te[52] 18,6 | 126.904 I[53] 18,7 | 131.30 Xe[54] 18,8 |
| 6 / 2,8,18 | 132.905 Cs[55] 18,8,1 | 137.34 Ba[56] 18,8,2 | * (57-71) | 178.49 Hf[72] 32,10,2 | 180.948 Ta[73] 32,11,2 | 183.85 W[74] 32,12,2 | 186.2 Re[75] 32,13,2 | 190.2 Os[76] 32,14,2 | 192.2 Ir[77] 32,15,2 | 195.09 Pt[78] 32,17,1 | 196.967 Au[79] 32,18,1 | 200.59 Hg[80] 32,18,2 | 204.37 Tl[81] 32,18,3 | 207.19 Pb[82] 32,18,4 | 208.980 Bi[83] 32,18,5 | (210) Po[84] 32,18,6 | (210) At[85] 32,18,7 | (222) Rn[86] 32,18,8 |
| 7 / 2,8,18,32 | (223) Fr[87] 18,8,1 | 226.025 Ra[88] 18,8,2 | † (89-103) | Rf[104] 32,10,2 | Ha[105] 32,11,2 | [106] 32,12,2 | [107] | [108] | | | | | | | | | | |

*lathanide series

| 138.91 La[57] 18,9,2 | 140.12 Ce[58] 20,8,2 | 140.907 Pr[59] 21,8,2 | 144.24 Nd[60] 22,8,2 | (147) Pm[61] 23,8,2 | 150.35 Sm[62] 24,8,2 | 151.96 Eu[63] 25,8,2 | 157.25 Gd[64] 25,9,2 | 158.924 Tb[65] 27,8,2 | 162.50 Dy[66] 28,8,2 | 164.930 Ho[67] 29,8,2 | 167.26 Er[68] 30,8,2 | 168.934 Tm[69] 31,8,2 | 173.04 Yb[70] 32,8,2 | 174.97 Lu[71] 32,9,2 |
|---|---|---|---|---|---|---|---|---|---|---|---|---|---|---|

†actinide series

| (227) Ac[89] 18,9,2 | 232.038 Th[90] 18,10,2 | 231.036 Pa[91] 20,9,2 | 238.03 U[92] 21,9,2 | 237.048 Np[93] 23,8,2 | (242) Pu[94] 24,8,2 | (243) Am[95] 25,8,2 | (247) Cm[96] 25,9,2 | (247) Bk[97] 26,9,2 | (249) Cf[98] 28,8,2 | (254) Es[99] 29,8,2 | (253) Fm[100] 30,8,2 | (256) Md[101] 31,8,2 | (254) No[102] 31,8,2 | (257) Lr[103] 32,9,2 |
|---|---|---|---|---|---|---|---|---|---|---|---|---|---|---|

**Chemistry**

- *metalloids:* elements along the dark line in the chart separating metals and nonmetals

- *noble gases:* group 0

- *nonmetals:* elements 2, 5–10, 14–18, 33–36, 52–54, 85, and 86

- *rare earths:* same as lanthanons

- *transition elements:* same as transition metals

- *transition metals:* all B families and group VIII B[9]

## 5. SUMMARY OF TRENDS IN THE PERIODIC TABLE

As a general rule, as one moves from left to right in the periodic table, (1) the atomic radius decreases, (2) the ionization energy (see Sec. 33.16) increases, and (3) the electronegativity (see Sec. 33.18) increases. As one moves from top to bottom, (1) the atomic radius increases, (2) the ionization energy decreases, and (3) the electronegativity decreases.

## 6. OXIDATION NUMBER

The *oxidation number* (*oxidation state*) is an electrical charge assigned by a set of prescribed rules. It is actually the charge assuming all bonding is ionic. The sum of the oxidation numbers equals the net charge. For monoatomic ions, the oxidation number is equal to the charge. The oxidation numbers of some common ions and radicals are given in Table 33.2.

In covalent compounds, all of the bonding electrons are assigned to the ion with the greater electronegativity. For example, non-metals are more electronegative than metals. Carbon is more electronegative than hydrogen.

For atoms in a free-state molecule, the oxidation number is zero. Hydrogen gas is a diatomic molecule, $H_2$. Thus, the oxidation number of the hydrogen molecule, $H_2$, is zero. The same is true for the atoms in $O_2$, $N_2$, $Cl_2$, and so on. Also, the sum of all the oxidation numbers of atoms in a neutral molecule is zero.

Fluorine is the most electronegative element, and it has an oxidation number of −1. Oxygen is second only to fluorine in electronegativity. Usually, the oxidation number of oxygen is −2, except in peroxides, where it is −1, and when combined with fluorine, where it is +2. Hydrogen is usually +1, except in hydrides, where it is −1.

For a charged *radical* (a group of atoms that combine as a single unit), the net oxidation number is equal to the charge on the radical.

**Table 33.2** *Oxidation Numbers of Selected Atoms and Charge Numbers of Radicals*

| name | symbol | oxidation or charge number |
|---|---|---|
| acetate | $C_2H_3O_2$ | −1 |
| aluminum | Al | +3 |
| ammonium | $NH_4$ | +1 |
| barium | Ba | +2 |
| borate | $BO_3$ | −3 |
| boron | B | +3 |
| bromine | Br | −1 |
| calcium | Ca | +2 |
| carbon | C | +4, −4 |
| carbonate | $CO_3$ | −2 |
| chlorate | $ClO_3$ | −1 |
| chlorine | Cl | −1 |
| chlorite | $ClO_2$ | −1 |
| chromate | $CrO_4$ | −2 |
| chromium | Cr | +2, +3, +6 |
| copper | Cu | +1, +2 |
| cyanide | CN | −1 |
| dichromate | $Cr_2O_7$ | −2 |
| fluorine | F | −1 |
| gold | Au | +1, +3 |
| hydrogen | H | +1 |
| hydroxide | OH | −1 |
| hypochlorite | ClO | −1 |
| iron | Fe | +2, +3 |
| lead | Pb | +2, +4 |
| lithium | Li | +1 |
| magnesium | Mg | +2 |
| mercury | Hg | +1, +2 |
| nickel | Ni | +2, +3 |
| nitrate | $NO_3$ | −1 |
| nitrite | $NO_2$ | −1 |
| nitrogen | N | −3, +1, +2, +3, +4, +5 |
| oxygen | O | −2 (−1 in peroxides) |
| perchlorate | $ClO_4$ | −1 |
| permanganate | $MnO_4$ | −1 |
| phosphate | $PO_4$ | −3 |
| phosphorus | P | −3, +3, +5 |
| potassium | K | +1 |
| silicon | Si | +4, −4 |
| silver | Ag | +1 |
| sodium | Na | +1 |
| sulfate | $SO_4$ | −2 |
| sulfite | $SO_3$ | −2 |
| sulfur | S | −2, +4, +6 |
| tin | Sn | +2, +4 |
| zinc | Zn | +2 |

**Example 33.1**

What are the oxidation numbers of all the elements in the chlorate ($ClO_3^{-1}$) and permanganate ($MnO_4^{-1}$) ions?

*Solution*

For the chlorate ion, the oxygen is more electronegative than the chlorine. (Only fluorine is more electronegative than oxygen.) Therefore, the oxidation number of

---

[9]The *transition metals* are elements whose electrons occupy the d sublevel. They can have various oxidation numbers, including +2, +3, +4, +6, and +7.

oxygen is $-2$. In order for the net oxidation number to be $-1$, the chlorine must have an oxidation number of $+5$.

For the permanganate ion, the oxygen is more electronegative than the manganese. Therefore, the oxidation number of oxygen is $-2$. For the net oxidation number to be $-1$, the manganese must have an oxidation number of $+7$.

## 7. COMPOUNDS

Combinations of elements are known as *compounds*. *Binary compounds* contain two elements; *ternary (tertiary) compounds* contain three elements. A *chemical formula* is a representation of the relative numbers of each element in the compound. For example, the formula $CaCl_2$ shows that there are one calcium atom and two chlorine atoms in one molecule of calcium chloride.

Generally, the numbers of atoms are reduced to their lowest terms. However, there are exceptions. For example, acetylene is $C_2H_2$ and hydrogen peroxide is $H_2O_2$.

For binary compounds with a metallic element, the positive metallic element is listed first. The chemical name ends in the suffix "-ide." For example, $NaCl$ is sodium chloride. If the metal has two oxidation states, the suffix "-ous" is used for the lower state, and "-ic" is used for the higher state. Alternatively, the element name can be used with the oxidation number written in Roman numerals. For example,

$FeCl_2$: ferrous chloride, or iron (II) chloride
$FeCl_3$: ferric chloride, or iron (III) chloride

For binary compounds formed between two nonmetals, the more positive element is listed first. The number of atoms of each element is specified by the prefixes "di-" (2), "tri-" (3), "tetra-" (4), and "penta-" (5), etc. For example,

$N_2O_5$: dinitrogen pentoxide

*Binary acids* start with the prefix "hydro-," list the name of the nonmetallic element, and end with the suffix "-ic." For example,

$HCl$: hydrochloric acid

Ternary compounds generally consist of an element and a radical. The positive part is listed first in the formula. *Ternary acids* (also known as *oxyacids*) usually contain hydrogen, a nonmetal, and oxygen, and can be grouped into families with different numbers of oxygen atoms. The most common acid in a family (i.e., the root acid) has the name of the nonmetal and the suffix "-ic." The acid with one more oxygen atom than the root is given the prefix "per-" and the suffix "-ic." The acid containing one less oxygen atom than the root is given the ending

"-ous." The acid containing two less oxygen atoms than the root is given the prefix "hypo-" and the suffix "-ous."

For example,

$HClO$:   hypochlorous acid
$HClO_2$: chlorous acid
$HClO_3$: chloric acid (the root)
$HClO_4$: perchloric acid

## 8. FORMATION OF COMPOUNDS

Compounds form according to the *law of definite (constant) proportions:* A pure compound is always composed of the same elements combined in a definite proportion by mass. For example, common table salt is always $NaCl$. It is not sometimes $NaCl$ and other times $Na_2Cl$ or $NaCl_3$ (which do not exist, in any case).

Furthermore, compounds form according to the *law of (simple) multiple proportions:* When two elements combine to form more than one compound, the masses of one element that combine with the same mass of the other are in the ratios of small integers.

In order to evaluate whether a compound formula is valid, it is necessary to know the *oxidation numbers* of the interacting atoms. Although some atoms have more than one possible oxidation number, most do not.

The sum of the oxidation numbers must be zero if a neutral compound is to form. For example, $H_2O$ is a valid compound since the two hydrogen atoms have a total positive oxidation number of $2 \times 1 = +2$. The oxygen ion has an oxidation number of $-2$. These oxidation numbers sum to zero.

On the other hand, $NaCO_3$ is not a valid compound formula. The sodium ($Na$) ion has an oxidation number of $+1$. However, the carbonate radical has a *charge number* of $-2$. The correct sodium carbonate molecule is $Na_2CO_3$.

## 9. MOLES AND AVOGADRO'S LAW

The *mole* is a measure of the quantity of an element or compound. Specifically, a mole of an element will have a mass equal to the element's atomic (or molecular) weight. The three main types of moles are based on mass being measured in grams, kilograms, and pounds.[10] Obviously, a gram-based mole of carbon (12.0 grams) is not the same quantity as a pound-based mole of carbon (12.0 pounds). Although "mol" is understood in SI countries to mean a gram-mole, the term *mole* is ambiguous, and the units mol (gmol), kmol (kgmol), or lbmol must be specified, or the type of mole must be spelled out.[11]

---

[10]Theoretically, a slug-mole could be defined, but it is not used.
[11]There are also variations on the presentation of these units, such as g mol, gmole, g-mole, kmole, kg-mol, lb-mole, pound-mole, and p-mole. In most cases, the intent is clear.

One gram-mole of any substance has a number of particles (atoms, molecules, ions, electrons, etc.) equal to $6.022 \times 10^{23}$, *Avogadro's number*, $N_a$. A pound-mole contains approximately 454 times the number of particles in a gram-mole.

*Avogadro's law (hypothesis)* holds that equal volumes of all gases at the same temperature and pressure contain equal numbers of gas molecules. Specifically, at standard scientific conditions (1.0 atm and 0°C), one gram-mole of any gas contains $6.022 \times 10^{23}$ molecules and occupies 22.4 L. Of course, a pound-mole occupies 454 times that volume, 359 ft$^3$.

"Molar" is used as an adjective when describing properties of a mole. For example, a *molar volume* is the volume of a mole.

### Example 33.2

How many electrons are in 0.01 g of gold? ($A = 196.97$; $Z = 79$.)

*Solution*

The number of gram-moles of gold present is

$$n = \frac{0.01 \text{ g}}{196.97 \ \frac{\text{g}}{\text{mol}}} = 5.077 \times 10^{-5} \text{ mol}$$

The number of gold nuclei is

$$N = nN_A = \left(5.077 \times 10^{-5} \text{ mol}\right)\left(6.022 \times 10^{23} \ \frac{\text{nuclei}}{\text{mol}}\right)$$

$$= 3.057 \times 10^{19} \text{ nuclei}$$

Since the atomic number is 79, there are 79 protons and 79 electrons in each gold atom. The number of electrons is

$$N_{\text{electrons}} = (3.057 \times 10^{19})(79) = 2.42 \times 10^{21}$$

## 10. FORMULA AND MOLECULAR WEIGHTS

The *formula weight*, FW, of a molecule (compound) is the sum of the atomic weights of all elements in the molecule. The *molecular weight*, MW, is generally the same as the formula weight. The units of molecular weight are actually g/mol, kg/kmol, or lbm/lbmol. However, units are sometimes omitted because weights are relative. For example,

$$\text{CaCO}_3: \text{FW} = \text{MW} = 40.1 + 12 + 3 \times 16 = 100.1$$

An *ultimate analysis* (which determines how much of each element is present in a compound) will not determine the molecular formula. It will determine only the relative proportions of each element. Therefore, except for hydrated molecules and other linked structures, the molecular weight will be an integer multiple of the formula weight.

For example, an ultimate analysis of hydrogen peroxide ($H_2O_2$) will show that the compound has one oxygen atom for each hydrogen atom. In this case, the formula would be assumed to be HO and the formula weight would be approximately 17, although the actual molecular weight is 34.

For *hydrated molecules* (e.g., $FeSO_4 \cdot 7H_2O$), the mass of the *water of hydration* (also known as the *water of crystallization*) is included in the formula and in the molecular weight.

## 11. EQUIVALENT WEIGHT

The *equivalent weight*, EW (i.e., an *equivalent*), is the amount of substance (in grams) that supplies one gram-mole (i.e., $6.022 \times 10^{23}$) of reacting units. For acid-base reactions, an acid equivalent supplies one gram-mole of $H^+$ ions. A base equivalent supplies one gram-mole of $OH^-$ ions. In oxidation-reduction reactions, an equivalent of a substance gains or loses a gram-mole of electrons. Similarly, in electrolysis reactions an equivalent weight is the weight of substance that either receives or donates one gram-mole of electrons at an electrode.

The equivalent weight can be calculated as the molecular weight divided by the change in oxidation number experienced in a chemical reaction. A substance can have several equivalent weights.

$$\text{EW} = \frac{\text{MW}}{\Delta \text{ oxidation number}} \qquad 33.2$$

### Example 33.3

What are the equivalent weights of the following compounds?

(a) Al in the reaction

$$\text{Al}^{+++} + 3e^- \rightarrow \text{Al}$$

(b) $H_2SO_4$ in the reaction

$$\text{H}_2\text{SO}_4 + \text{H}_2\text{O} \rightarrow 2\text{H}^+ + \text{SO}_4^{-2} + \text{H}_2\text{O}$$

(c) NaOH in the reaction

$$\text{NaOH} + \text{H}_2\text{O} \rightarrow \text{Na}^+ + \text{OH}^- + \text{H}_2\text{O}$$

*Solution*

(a) The atomic weight of aluminum is approximately 27. Since the change in the oxidation number is 3, the equivalent weight is $27/3 = 9$.

(b) The molecular weight of sulfuric acid is approximately 98. Since the acid changes from a neutral molecule to ions with two charges each, the equivalent weight is $98/2 = 49$.

(c) Sodium hydroxide has a molecular weight of approximately 40. The originally neutral molecule goes to a singly charged state. Therefore, the equivalent weight is $40/1 = 40$.

## 12. DENSITY

The density, $\rho$, of substance is its mass per unit volume.

$$\rho = \frac{m}{V} \qquad 33.3$$

Density can be derived from the atomic or molecular weight and Avogadro's number, $N_A$.

$$\begin{pmatrix} \text{molecular} \\ \text{weight} \end{pmatrix} = \begin{pmatrix} \text{mass per} \\ \text{molecule} \end{pmatrix} \begin{pmatrix} \text{number of} \\ \text{molecules} \\ \text{in a mole} \end{pmatrix} \qquad 33.4$$

$$\text{MW} = m_{\text{molecule}} N_A \qquad 33.5$$

$$\text{MW} = \rho V_{\text{molecule}} N_A \qquad 33.6$$

$$\text{MW} = \rho V_{\text{mole}} \qquad 33.7$$

### Example 33.4

17 grams (0.017 kg) of magnesium occupy 9.77 cm$^3$ ($9.77 \times 10^{-6}$ m$^3$) of space. What is the average volume of a magnesium atom?

*Solution*

From Eq. 33.3, the density of magnesium is

$$\rho = \frac{m}{V} = \frac{0.017 \text{ kg}}{9.77 \times 10^{-6} \text{ m}^3}$$
$$= 1740 \text{ kg/m}^3$$

The atomic weight of magnesium is 24.3. From Eq. 33.6, the volume of an atom is

$$\text{volume} = \frac{\text{MW}}{\rho N_A}$$

$$= \frac{\left( 24.3 \, \frac{\text{kg}}{\text{mol}} \right) \left( 0.001 \, \frac{\text{kmol}}{\text{mol}} \right)}{\left( 1740 \, \frac{\text{kg}}{\text{m}^3} \right) \left( 6.022 \times 10^{23} \, \frac{\text{atoms}}{\text{mol}} \right)}$$

$$= 2.32 \times 10^{-29} \text{ m}^3/\text{atom}$$

## 13. GRAVIMETRIC FRACTION

The *gravimetric fraction*, $x_i$, of an element $i$ in a compound is the fraction by weight of that element in the compound. The gravimetric fraction is found from an *ultimate analysis* (also known as a *gravimetric analysis*) of the compound.

$$x_i = \frac{m_i}{m_1 + m_2 + \cdots + m_i + \cdots + m_n} = \frac{m_i}{m_t} \qquad 33.8$$

The *percentage composition* is the gravimetric fraction converted to percentage.

$$\% \text{ composition} = x_i \times 100\% \qquad 33.9$$

If the gravimetric fractions are known for all elements in a compound, the *combining weights* of each element can be calculated. (The term *weight* is used even though mass is the traditional unit of measurement.)

$$m_i = x_i m_t \qquad 33.10$$

## 14. EMPIRICAL FORMULA DEVELOPMENT

It is relatively simple to determine the *empirical formula* of a compound from the atomic and combining weights of elements in the compound. The empirical formula gives the relative number of atoms (i.e., the formula weight is calculated from the empirical formula).

*step 1:* Divide the gravimetric fractions (or percentage compositions) by the atomic weight of each respective element.

*step 2:* Determine the smallest ratio from step 1.

*step 3:* Divide all of the ratios from step 1 by the smallest ratio.

*step 4:* Write the chemical formula using the results from step 3 as the numbers of atoms. Multiply through as required to obtain all integer numbers of atoms.

### Example 33.5

A clear liquid is analyzed, and the following gravimetric percentage compositions are recorded: carbon, 37.5%; hydrogen, 12.5%; oxygen, 50%. What is the chemical formula for the liquid?

*Solution*

*step 1:* Divide the percentage compositions by the atomic weight.

$$\text{C: } \frac{37.5}{12} = 3.125$$

$$\text{H: } \frac{12.5}{1} = 12.5$$

$$\text{O: } \frac{50}{16} = 3.125$$

*step 2:* The smallest ratio is 3.125.

*step 3:* Divide all ratios by 3.125.

$$\text{C: } \frac{3.125}{3.125} = 1$$

$$\text{H: } \frac{12.5}{3.125} = 4$$

$$\text{O: } \frac{3.125}{3.125} = 1$$

*step 4:* The empirical formula is $CH_4O$.

If it had been known that the liquid behaved chemically as though it had a hydroxyl (OH) radical, the formula would have been written as $CH_3OH$. This is recognized as methyl alcohol.

## 15. ISOMERS

*Isomers* are different arrangements of the same atoms. For example, ethyl alcohol ($C_2H_5OH$) and dimethyl ether (($CH_3)_2O$) are isomers because they both have two carbon atoms, six hydrogen atoms, and one oxygen atom. Isomers are actually different substances, having different physical properties (e.g., density, melting point, etc.) and different chemical properties (e.g., reactivity).

There are two primary types of isomers: structural and geometric. *Structural isomers* differ in bonding arrangements and are characteristic of compounds with single carbon-carbon bonds. Within any molecule, there is a practically free rotation of the periphery atoms and groups about any C-C bond. Figure 33.1 illustrates structural isomers.

**Figure 33.1** *Structural Isomers*

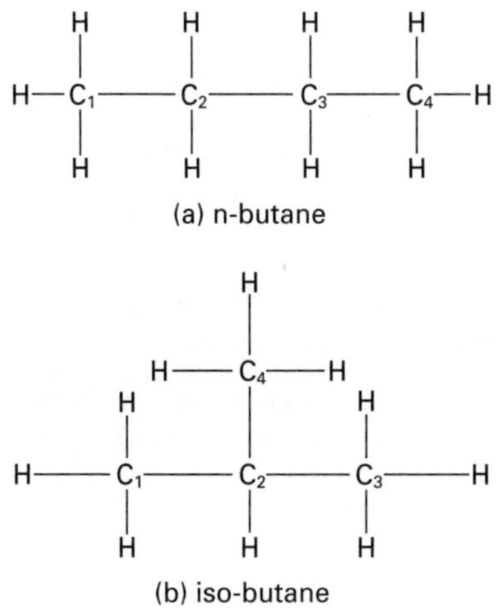

(a) n-butane

(b) iso-butane

The second form of isomerism is *stereoisomerism*, which implies that the isomers are mirror images of each other. It is manifested in two ways: geometric and optical isomerism.

*Geometric isomers* are characteristic of compounds with strong double carbon-carbon bonds. The double bond gives such molecules a more stable, flat geometry. These isomers have identical central bonds, but the location of each functional group differs. However, the differences cannot be visualized as having occurred through rotation of the periphery atoms around the double bond. Figure 33.2 illustrates geometric isomers.

*Optical isomerism* is found in organic and transition-metal compounds that lack planes or points of symmetry. They are described as "optical" because one isomer

rotates a plane of polarized light to the left while the other isomer rotates the light to the right.

Isomers are distinguished by using prefixes with the names. For example, the three isomers of phthalic acids (used in the manufacture of glyptal resins for the paint and varnish industry) are named ortho-, para-, and meta-phthalic acid. (These prefixes may be abbreviated *o-*, *p-*, and *m-*, respectively.) Other isomers may use the prefixes "normal-," abbreviated "*n-*" as in "*n*-butane," and "iso-" as in "iso-butane." The prefixes "cis" and "trans" are also used to distinguish between isomers of the same compound (e.g., cis- and trans-1,2-dichloroethylene shown in Fig. 33.2).

**Figure 33.2** *Geometric Isomers*

## 16. IONS AND ELECTRON AFFINITY

The atomic number, $Z$, of chlorine is 17, which means there are 17 protons in the nucleus of a chlorine atom. There are also 17 electrons in the various shells surrounding the nucleus. The atomic structure is written as $1s^2 2s^2 2p^6 3s^2 3p^5$.

Notice that there are only seven electrons in the outer shell corresponding to principal quantum number $n = 3$. A stable shell requires eight electrons. In order to achieve this stable configuration, chlorine atoms tend to attract electrons from other atoms, a tendency known as *electron affinity*. The energy required to remove an electron from a neighboring atom is known as the *ionization energy*. The electrons attracted by chlorine atoms come from neighboring atoms with low ionization energies.

Chlorine, prior to taking a neighboring atom's electron, is electrically neutral. The fact that it needs one electron to complete its outer shell does not mean that chlorine needs an electron to become neutral. On the contrary, the chlorine atom becomes negatively charged when it takes the electron. An atomic nucleus with a charge is known as an *ion*.

Negatively charged ions are known as *anions*. Anions lose electrons at the *anode* during electro-chemical reactions. Anions must lose electrons to become neutral. The loss of electrons is known as *oxidation*.

The charge on an anion is equal to the number of electrons taken from a neighboring atom. In the past, this charge was known as the *valence*. Valence is equal to the number of electrons that must be gained for charge neutrality. For a chlorine ion, the valence is $-1$. (The term *charge* can usually be substituted for *valence*.)

**Chemistry**

Sodium has an electron configuration of $1s^2 2s^2 2p^6 3s^1$. The orbitals corresponding to $n = 1$ and $n = 2$ are completely filled. The last electron, which occupies the $3s$ orbital, has a low ionization energy and is very easily removed.

If its outer electron is removed, sodium becomes positively charged. (For a sodium ion, the charge is +1.) Positively charged ions are known as *cations*. Cations gain electrons at the *cathode* in electro-chemical reactions. The gaining of electrons is known as *reduction*. Cations must gain electrons to become neutral.

## 17. ELECTRONEGATIVITY

The measure of attraction of an atom for the electrons already involved in a chemical bond is known as the *electronegativity*. The greater the attraction, the higher the electronegativity value will be. The electronegativity is related to ionization energies, because atoms with low ionization energies do not strongly hold onto the outer electrons. As a general rule, atoms with larger radii will have lower ionization energies (and, hence, lower electronegativities) because the electrostatic attraction of an electron drops off with the square of the distance from the nucleus. Values of electronegativity (see Table 33.3) are not absolute, but relative, and they have no units. Two tables may have different values of electronegativities for a list of the same elements, but the order will be the same, as should be the approximate ratios of values in the tables.

## 18. IONIC BONDS

If a chlorine atom becomes an anion by attracting an electron from a sodium atom (which becomes a cation), the two ions will be attracted to each other by electrostatic force. The electrostatic attraction of the positive sodium to the negative chlorine effectively bonds the two ions together. This type of bonding, in which electrostatic attraction is predominant, is known as *ionic bonding* or *electrovalent bonding*. In an ionic bond, one or more electrons are transferred from the valence shell of one atom to the valence shell of another. There is no sharing of electrons between atoms.

Ionic bonding is characteristic of compounds of atoms with high electron affinities and atoms with low ionization energies. Specifically, the difference in electronegativities must be approximately 1.7 or greater for the bond to be classified as ionic.

## 19. IONIC COMPOUNDS

*Ionic solids* typically do not have distinct molecules—they consist of groups of charged ions in crystalline lattices. The formulas (e.g., NaCl) are indications of the relative numbers of atoms in the compound.

Ionic compounds commonly have the following characteristics.

- They are usually hard, brittle crystalline solids.
- They have large lattice energies due to the strong electrostatic attractive force.
- They have high melting points.
- They have high boiling points.
- They are nonvolatile and have low vapor pressures.
- They are poor electrical conductors in the solid phase.
- They become electrically conductive when dissociated into the component ions.
- Those that are soluble in water form electrolytic solutions that conduct electricity (i.e., they are *electrolytes*).

**Table 33.3** *Electronegativities of Selected Elements*

|     |     |         |     |     |     |       |     |     |     |     |     |     |     |     | H 2.1 |
|-----|-----|---------|-----|-----|-----|-------|-----|-----|-----|-----|-----|-----|-----|-----|-----|

| Li 1.0 | Be 1.5 | B 2.0 |     |     |     |       |     |     |     |     |     |     | C 2.5 | N 3.0 | O 3.5 | F 4.0 |
|--------|--------|-------|-----|-----|-----|-------|-----|-----|-----|-----|-----|-----|-------|-------|-------|-------|
| Na 0.9 | Mg 1.2 | Al 1.5 |     |     |     |       |     |     |     |     |     |     | Si 1.8 | P 2.1 | S 2.5 | Cl 3.0 |
| K 0.8 | Ca 1.0 | Sc 1.3 | Ti 1.5 | V 1.6 | Cr 1.6 | Mn 1.5 | Fe 1.8 | Co 1.8 | Ni 1.8 | Cu 1.9 | Zn 1.6 | Ga 1.6 | Ge 1.8 | As 2.0 | Se 2.4 | Br 2.8 |
| Rb 0.8 | Sr 1.0 | Y 1.2 | Zr 1.4 | Nb 1.6 | Mo 1.8 | Tc 1.9 | Ru 2.2 | Rh 2.2 | Pd 2.2 | Ag 1.9 | Cd 1.7 | In 1.7 | Sn 1.8 | Sb 1.9 | Te 2.1 | I 2.5 |
| Cs 0.7 | Ba 0.9 | La–Lu 1.0–1.2 | Hf 1.3 | Ta 1.5 | W 1.7 | Re 1.9 | Os 2.2 | Ir 2.2 | Pt 2.2 | Au 2.4 | Hg 1.9 | Tl 1.8 | Pb 1.9 | Bi 1.9 | Po 2.0 | At 2.2 |
| Fr 0.7 | Ra 0.9 | Ac 1.1 | Th 1.3 | Pa 1.5 | U 1.7 | Np–No 1.5–1.3 |     |     |     |     |     |     |     |     |     |     |

Used with permission from *College Chemistry*, by Linus Pauling, published by W. H. Freeman and Company, 1964.

## 20. BONDING ENERGY

In an ionic bond, the anion and cation are drawn together until the Coulomb attraction is exactly balanced by the repulsion of the electrons in the outer shells. The final separation distance is known as the *equilibrium distance*, $x_{eq}$. The *bonding energy* is the sum of these forces integrated from infinity to the equilibrium distance (i.e., the work done in separating the ions). At the equilibrium distance, the bonding energy is minimum.

$$\text{bonding energy} = \int_{\infty}^{x_{eq}} (F_{Coulomb} + F_{repulsive})\,dx$$

*33.11*

The equilibrium distance is dependent on several factors.

- *temperature:* As temperature increases, the equilibrium distance increases.

- *ionic charge:* Since the electron repulsive forces diminish as an ion loses electrons, the equilibrium distance will decrease as an ion becomes more positive. (See Fig. 33.3.)

- *coordination number:* The equilibrium spacing increases as the number of adjacent ions in the molecule increases. This is due to the increase in repulsive electronic forces.

- *bond type:* The stronger the bond, the smaller will be the equilibrium spacing. The double covalent bond between the two carbon atoms in ethylene gas ($C_2H_4$) is smaller than the single covalent bond between the two carbon atoms in ethane gas ($C_2H_6$).

*Figure 33.3 Equilibrium Spacing in Ionic Solids*

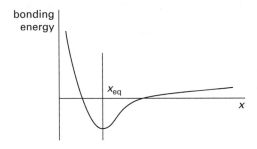

## 21. COVALENT BONDS

Several common gases in their free states exist as *diatomic molecules*. Examples are hydrogen ($H_2$), oxygen ($O_2$), nitrogen ($N_2$), and chlorine ($Cl_2$).[12] Since two atoms of the same element will have the same

electronegativity and ionization energy, it is unlikely that one atom will take electrons from the other. Therefore, the bond formed is not ionic.

The electrons in these diatomic molecules are shared equally in order to fill the outer shells. This type of bonding, in which sharing of electrons is the predominant characteristic, is known as *covalent bonding*. Covalent bonds are typical of bonds formed in organic compounds. Specifically, the difference in electronegativities must be less than approximately 1.7 for the bond to be classified as covalent.

The oxidation number of an atom that forms a covalent bond is equal to the number of *shared electron pairs*. For example, each hydrogen atom has one electron. There are two electrons (i.e., a single shared electron pair) in each carbon-hydrogen bond in methane ($CH_4$). Therefore, the oxidation number of hydrogen is 1.

If both atoms forming a covalent bond are the same element, the electrons will be shared equally. This is known as a *non-polar covalent bond*. If the atoms are not both the same element, the electrons will not be shared equally, resulting in a *polar covalent bond*. For example, the bond between hydrogen and chlorine in HCl is partially covalent and partially ionic in nature. Thus, there is no sharp dividing line between ionic and covalent bonds for most compounds.

## 22. DIPOLES

When electrons are not shared equally, the electrons tend to spend more time around the more electronegative atom. Thus, one atom will generally be negative, while the other will generally be positive. This creates a *dipole* structure, sometimes referred to as a *Debye molecule*. The more electronegative element in the bond is the negative end of the dipole. The unit of molecular dipole moment is the *debye* (symbol D), equal to $3.3356 \times 10^{-30}$ C·m. Water, ammonia, ethanol, acetone, and sulfur dioxide are dipolar in nature.

## 23. SYMBOLS FOR ELECTRONS AND COVALENT BONDS

It is convenient to represent the valence electrons as dots surrounding the chemical symbol. These primitive representations are known as *electron dot structures*. For example, oxygen has 6 valence electrons. Its symbol would be

$$:\ddot{O}:$$

Similarly, the electron dot structure for the compound HF shows that only two electrons are active in the bond. The other six electrons are "nonbonding."

$$H\ :\ddot{F}:$$

---

[12]Oxygen also forms an $O_3$ molecule known as *ozone*. Different molecular forms of the same element are known as *allotropes*. Allotropes are not the same as isotopes, which have different atomic weights. In the case of ozone, all of the oxygen atoms have the same atomic weight as the oxygen atoms in the $O_2$ molecule.

**Chemistry**

A shared electron pair (i.e., a covalent bond) is shown as a straight line. For example, hydrogen gas can be represented in one of two ways.

$$H:H \quad \text{or} \quad H\!-\!H$$

## 24. DOUBLE AND TRIPLE BONDS

Sharing of two pairs of electrons produces a *double bond.* For example, carbon dioxide ($CO_2$) can be represented two ways.

$$\ddot{O}::C::\ddot{O} \quad \text{or} \quad O\!=\!C\!=\!O$$

Sharing of three pairs of electrons produces a *triple bond.* For example, acetylene ($C_2H_2$) can be represented two ways.

$$H:C:::C:H \quad \text{or} \quad H\!-\!C\!\equiv\!C\!-\!H$$

It takes more energy to break a double bond than a single bond, and it takes more energy to break a triple bond than a double bond. Also, the carbon-carbon spacing decreases as the bond degree increases.

### Example 33.6

Write the chemical reaction for the formation of a nitrogen gas molecule from nitrogen atoms. Use the electron dot structure.

*Solution*

Nitrogen has an atomic number of seven and an electronic configuration of $1s^2 2s^2 2p^3$. There are five electrons in the $n = 2$ shell. In order to form a full set of eight electrons, three more electrons would have to be obtained from another nitrogen atom. These three electrons are shared.

$$:\dot{N}\cdot + :\dot{N}\cdot \rightarrow :N\!\equiv\!N:$$

## 25. OTHER TYPES OF BONDING

In addition to ionic and covalent bonding, there are other types of bonding. *Metallic bonding* occurs in metals when metal atoms lose electrons and the metallic ions are attracted to a "sea" of delocalized electrons. Such bonding is non-directional, since electrons are free to move from atom to atom.

There are also intermolecular bonds (bonds between molecules) that are much weaker than the intramolecular forces. *Van der Waals* bonding is a weak bonding that occurs in inert gases and other elements with full shells, primarily due to attraction between dipole structures (i.e., molecules with unsymmetrical structures such that the nucleus of one atom can attract the electrons of a neighboring molecule). *Hydrogen bonding* occurs when a hydrogen atom is strongly attracted to two different atoms. The hydrogen atom acts as the bond (i.e., the "glue" or "bridge") between the two atoms. A hydrogen bond is a strong dipole-dipole interaction.

## 26. RESONANCE HYBRIDS

Some molecules implicitly seem to change back and forth between two or three structures. Such changing is known as *resonance*, and the configurations that the molecules assume are known as *resonance hybrids*. The electronic structure of these molecules does not correspond to any of the resonant hybrids. Rather, the molecules behave as if their structure was an "average" (i.e., hybrid) structure.

Experiments with ozone have shown that the two end oxygen atoms are equidistant from the central atom. If a single theoretical electronic structure was valid, the end oxygen atom with a single bond should be farther away. The conclusion is that ozone resonates between the two different structures shown in Fig. 33.4. Many other compounds and ions are resonant, including $SO_2$, $CO_3^{-2}$, $H_3^{+1}$, $NO_3^{-1}$, and $C_6H_6$.

*Figure 33.4* *Resonance Hybrids of Ozone*

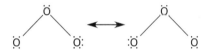

## 27. CHEMICAL REACTIONS

During chemical reactions, bonds between atoms are broken and new bonds are usually formed. The starting substances are known as *reactants*; the ending substances are known as *products*. In a chemical reaction, reactants are either converted to simpler products or synthesized into more complex compounds. There are four common types of reactions.

- *direct combination* (or *synthesis*): This is the simplest type of reaction where two elements or compounds combine directly to form a compound.

$$2H_2 + O_2 \rightarrow 2H_2O$$
$$SO_2 + H_2O \rightarrow H_2SO_3$$

- *decomposition* (or *analysis*): Bonds within a compound are disrupted by heat or other energy to produce simpler compounds or elements.

$$2HgO \rightarrow 2Hg + O_2$$
$$H_2CO_3 \rightarrow H_2O + CO_2$$

- *single displacement* (or *replacement*[13]): This type of reaction has one element and one compound as reactants.

$$2Na + 2H_2O \rightarrow 2NaOH + H_2$$

$$2KI + Cl_2 \rightarrow 2KCl + I_2$$

- *double displacement* (or *replacement*): These are reactions with two compounds as reactants and two compounds as products.

$$AgNO_3 + NaCl \rightarrow AgCl + NaNO_3$$

$$H_2SO_4 + ZnS \rightarrow H_2S + ZnSO_4$$

## 28. BALANCING CHEMICAL EQUATIONS

The coefficients in front of element and compound symbols in chemical reaction equations are the numbers of molecules or moles taking part in the reaction. (For gaseous reactants and products, the coefficients also represent the numbers of volumes. This is a direct result of Avogadro's hypothesis that equal numbers of molecules in the gas phase occupy equal volumes under the same conditions.)[14]

Since atoms cannot be changed in a normal chemical reaction (i.e., mass is conserved), the numbers of each element must match on both sides of the equation. When the numbers of each element match, the equation is said to be "balanced." The total atomic weights on both sides of the equation will be equal when the equation is balanced.

Balancing simple chemical equations is largely a matter of deductive trial and error. More complex reactions require use of oxidation numbers.[15]

### Example 33.7

Balance the following reaction equation.

$$Al + H_2SO_4 \rightarrow Al_2(SO_4)_3 + H_2$$

*Solution*

As written, the reaction is not balanced. For example, there is one aluminum on the left, but there are two on the right. The starting element in the balancing procedure is chosen somewhat arbitrarily.

*step 1:* Since there are two aluminums on the right, multiply Al by 2.

$$2Al + H_2SO_4 \rightarrow Al_2(SO_4)_3 + H_2$$

*step 2:* Since there are three sulfate radicals ($SO_4$) on the right, multiply $H_2SO_4$ by 3.

$$2Al + 3H_2SO_4 \rightarrow Al_2(SO_4)_3 + H_2$$

*step 3:* Now there are six hydrogens on the left, so multiply $H_2$ by 3 to balance the equation.

$$2Al + 3H_2SO_4 \rightarrow Al_2(SO_4)_3 + 3H_2$$

## 29. OXIDATION-REDUCTION REACTIONS

*Oxidation-reduction reactions* (also known as *redox reactions*) involve the transfer of electrons from one element or compound to another. Specifically, one reactant is oxidized, and the other reactant is reduced.

In *oxidation*, a substance's oxidation state increases, the substance loses electrons, and the substance becomes less negative. Oxidation occurs at the anode (positive terminal) in electrolytic reactions.

In *reduction*, a substance's oxidation state decreases, the substance gains electrons, and the substance becomes more negative. Reduction occurs at the cathode (negative terminal) in electrolytic reactions.

Whenever oxidation occurs in a chemical reaction, reduction must also occur. For example, consider the formation of sodium chloride from sodium and chlorine. This reaction is a combination of oxidation of sodium and reduction of chlorine. Notice that the electron released during oxidation is used up in the reduction reaction.

$$2Na + Cl_2 \rightarrow 2NaCl$$

$$Na \rightarrow Na^+ + e^-$$

$$Cl + e^- \rightarrow Cl^-$$

The substance (chlorine in the example) that causes oxidation to occur is called the *oxidizing agent* and is itself reduced (i.e., becomes more negative) in the process. The substance (sodium in the example) that causes reduction to occur is called the *reducing agent* and is itself oxidized (i.e., becomes less negative) in the process.

### Example 33.8

The balanced equation for a reaction between nitric acid and hydrogen sulfide is

$$2HNO_3 + 3H_2S \rightarrow 2NO + 4H_2O + 3S$$

(a) What is oxidized? (b) What is reduced? (c) What is the oxidizing agent? (d) What is the reducing agent?

---

[13]Another name for replacement is *metathesis*.

[14]When water is part of the reaction, the interpretation that the coefficients are volumes is valid only if the reaction takes place at a high enough temperature to vaporize the water.

[15]See Sec. 33.30, "Balancing Oxidation-Reduction Reactions," in this chapter.

Chemistry

*Solution*

First, recognize that the nitric acid exists as dissociated positive and negative ions, not as molecules of $HNO_3$. (Otherwise, the entire molecule might be chosen as one of the agents.)

$$2H^+ + 2NO_3^- + 3H_2S \rightarrow 2NO + 4H_2O + 3S$$

It is clear that the nitrogen is split out of the nitrate ion. Since the nitrate radical ($NO_3^-$) has a net charge number of $-1$, and the oxidation number of oxygen is (usually) $-2$, the nitrogen must have an initial oxidation number of $+5$.

In the NO molecule, the oxygen is still at $-2$, but the nitrogen has changed to $+2$.

In the $H_2S$ molecule, the oxidation number of hydrogen is $+1$, so the sulfur has an oxidation number of $-2$. As a product in the free state, sulfur has an oxidation number of 0.

(a) The sulfur becomes less negative. Sulfur is oxidized.

(b) The nitrogen becomes more negative (i.e., less positive). Nitrogen is reduced.

(c) The oxidizing agent is the nitrate radical, $NO_3^-$, since it contains nitrogen which decreases in oxidation state. (Notice that $HNO_3$ is not the oxidizing agent, because the molecule has dissociated.)

(d) The reducing agent is $H_2S$ since it contains sulfur, which increases in oxidation number. ($H_2S$ essentially does not dissociate. Therefore, the entire molecule is chosen as the reducing agent, rather than just $S^{-2}$.)

## 30. BALANCING OXIDATION-REDUCTION REACTIONS

The total number of electrons lost during oxidation must equal the total number of electrons gained during reduction. This is the main principle used in balancing redox reactions. Although there are several formal methods of applying this principle, balancing an oxidation-reduction equation remains somewhat intuitive and iterative.[16]

The *oxidation number change method* of balancing consists of the following steps.

*step 1:* Write an unbalanced equation that includes all reactants and products.

*step 2:* Assign oxidation numbers to each atom in the unbalanced equation.

*step 3:* Note which atoms change oxidation numbers, and calculate the amount of change for each atom. (When more than one atom of an element that changes oxidation number is present in a formula, calculate the change in oxidation number for that atom per formula unit.)

---

[16]The *ion-electron method* of balancing is not presented in this book.

*step 4:* Balance the equation so that the number of electrons gained equals the number lost.

*step 5:* Balance (by inspection) the remainder of the chemical equation as required.

### Example 33.9

How many $AgNO_3$ molecules are formed per NO molecule in the reaction of silver and nitric acid?

*Solution*

The unbalanced reaction is written with constants $c_i$ to represent the unknown coefficients.

$$c_1Ag + c_2HNO_3 \rightarrow c_3AgNO_3 + c_4NO + c_5H_2O$$

The oxidation number of Ag as a reactant is zero. The oxidation number of Ag in $AgNO_3$ is $+1$. Therefore, silver has become less negative (more positive) and has been oxidized through the loss of one electron.

$$Ag \rightarrow Ag^+ + e^-$$

The N in $HNO_3$ has an oxidation number of $+5$. The N in NO has an oxidation number of $+2$. The nitrogen has become more negative (less positive) and has been reduced through the gain of three electrons.

$$N^{+5} + 3e^- \rightarrow N^{+2}$$

Since three electrons are required for every NO molecule formed, it is necessary that $c_3 = 3c_4$.

## 31. STOICHIOMETRIC REACTIONS

*Stoichiometry* is the study of the proportions in which elements and compounds react and are formed. A *stoichiometric reaction* (also known as a *perfect reaction* or an *ideal reaction*) is one in which just the right amounts of reactants are present. After the reaction stops, there are no unused reactants.

Stoichiometric problems are known as *weight and proportion problems* because their solutions use simple ratios to determine the masses of reactants required to produce given masses of products, or vice versa. The procedure for solving these problems is essentially the same regardless of the reaction.

*step 1:* Write and balance the chemical equation.

*step 2:* Determine the atomic (molecular) weight of each element (compound) in the equation.

*step 3:* Multiply the atomic (molecular) weights by their respective coefficients and write the products under the formulas.

*step 4:* Write the given mass data under the weights determined in step 3.

*step 5:* Fill in the missing information by calculating simple ratios.

## Example 33.10

Caustic soda (NaOH) is made from sodium carbonate ($Na_2CO_3$) and slaked lime ($Ca(OH)_2$) according to the given reaction. How many kilograms of caustic soda can be made from 2000 kg of sodium carbonate?

*Solution*

$$Na_2CO_3 + Ca(OH)_2 \rightarrow 2NaOH + CaCO_3$$

| | | | | |
|---|---|---|---|---|
| molecular weights | 106 | 74 | $2 \times 40$ | 100 |
| given data | 2000 kg | | $X$ kg | |

The simple ratio used is

$$\frac{NaOH}{Na_2CO_3} = \frac{80}{106} = \frac{X}{2000}$$

Solving for the unknown mass, $X = 1509$ kg.

## 32. NONSTOICHIOMETRIC REACTIONS

In many cases, it is not realistic to assume a stoichiometric reaction because an excess of one or more reactants is necessary to assure that all of the remaining reactants take part in the reaction. Combustion is an example where the stoichiometric assumption is, more often than not, invalid. Excess air is generally needed to ensure that all of the fuel is burned.

With nonstoichiometric reactions, the reactant that is used up first is called the *limiting reactant*. The amount of product will be dependent on (limited by) the limiting reactant.

The *theoretical yield* or *ideal yield* of a product is the maximum amount of product per unit amount of limiting reactant that can be obtained from a given reaction if the reaction goes to completion. The *percentage yield* is a measure of the efficiency of the actual reaction.

$$\text{percentage yield} = \frac{\text{actual yield} \times 100\%}{\text{theoretical yield}} \qquad 33.12$$

## 33. SOLUTIONS OF GASES IN LIQUIDS

*Henry's law* states that the amount (i.e., mole fraction) of a slightly soluble gas dissolved in a liquid is proportional to the partial pressure of the gas. This law applies separately to each gas to which the liquid is exposed, as if each gas were present alone. The algebraic form of Henry's law is given by Eq. 33.13, in which $H$ is the *Henry's law constant* in mole fractions/atmosphere.

$$p_i = Hx_i \qquad 33.13$$

Generally, the solubility of gases in liquids decreases with increasing temperature.

The volume of gas absorbed at a partial pressure of 1 atm and 0°C is known as the *absorption coefficient*.

Typical absorption coefficients for solutions in water are: $H_2$, 0.017 L/L; He, 0.009 L/L; $N_2$, 0.015 L/L; $O_2$, 0.028 L/L; CO, 0.025 L/L; and $CO_2$, 0.88 L/L.

The amount of gas dissolved in a liquid varies with the temperature of the liquid and the concentration of dissolved salts in the liquid. Appendix 33.C lists the saturation values of dissolved oxygen in water at various temperatures and for various amounts of chloride ion (also referred to as *salinity*).

## Example 33.11

At 20°C and 1 atm, 1 L of water will absorb 0.043 g of oxygen or 0.019 g of nitrogen. Atmospheric air is 20.9% oxygen by volume, and the remainder is assumed to be nitrogen. What masses of oxygen and nitrogen will be absorbed by 1 L of water exposed to 20°C air at 1 atm?

*Solution*

Since partial pressure is volumetrically weighted,

$$m_{\text{oxygen}} = (0.209)\left(0.043 \ \frac{g}{L}\right) = 0.009 \ g/L$$

$$m_{\text{nitrogen}} = (1.000 - 0.209)\left(0.019 \ \frac{g}{L}\right) = 0.015 \ g/L$$

## Example 33.12

At an elevation of 4000 ft, the barometric pressure is 660 mm Hg. What is the dissolved oxygen concentration of 18°C water with a 800 mg/L chloride concentration at that elevation?

*Solution*

From App. 33.C, oxygen's saturation concentration for 18°C water corrected for a 800 mg/L chloride concentration is

$$C_s = 9.5 \ \frac{mg}{L} - (8)\left(0.009 \ \frac{mg}{L}\right) = 9.4 \ mg/L$$

Use the appendix footnote to correct for the barometric pressure.

$$C_s' = \left(9.4 \ \frac{mg}{L}\right)\left(\frac{660 \ mm - 16 \ mm}{760 \ mm - 16 \ mm}\right)$$
$$= 8.1 \ mg/L$$

## 34. PROPERTIES OF MIXTURES OF NONREACTING LIQUIDS

There are few convenient ways of predicting the properties of nonreacting, nonvolatile organic and aqueous solutions (acids, brines, alcohol mixtures, coolants, etc.) and mixtures from the individual properties of the components.

Volumes of two combining organic liquids (e.g., acetone and chloroform) are essentially additive. The volume

Chemistry

change upon mixing will seldom be more than a few tenths of a percent. The volume change in aqueous solutions is often slightly greater, but is still limited to a few percent (e.g., 3% for some solutions of methanol and water).

Thus, the specific gravity (density, specific weight, etc.) can be considered to be a volumetric weighting of the individual specific gravities. Most times, however, the specific gravity of a known solution must be calculated from known data regarding one of the various density scales or determined through research.

Most other important fluid properties of aqueous solutions, such as viscosity, compressibility, surface tension, and vapor pressure, have been measured and are usually determined through research.[17] It is important to be aware of the operating conditions of the solution. Data for one concentration or condition should not be used for another concentration or condition.

## 35. DISTILLATION OF MIXTURES OF LIQUIDS

*Distillation* is a method of separating a mixture of two or more non-reacting liquids into its components. The mixture is heated in a *distillation apparatus*, and the component with the lowest boiling point evaporates first, leaving all other components behind. The vapor is captured, cooled, and condensed by the distillation apparatus. The liquid temperature is then increased to the next-lowest component's boiling point, and the process repeated.

An *azeotrope (azeotropic mixture)* is a combination of two or more liquids, the composition of which does not change upon attempted distillation. When the liquid combination is boiled, the vapor has the same composition as the liquid.

## 36. SOLUTIONS OF SOLIDS IN LIQUIDS

When a solid is added to a liquid, the solid is known as the *solute* and the liquid is known as the *solvent*.[18] If the dispersion of the solute throughout the solvent is at the molecular level, the mixture is known as a *solution*. If the solute particles are larger than molecules, the mixture is known as a *suspension*.[19]

In some solutions, the solvent and solute molecules bond loosely together. This loose bonding is known as *solvation*. If water is the solvent, the bonding process is also known as *aquation* or *hydration*.

---

[17]There is no substitute for a complete fluid properties data book.

[18]The term *solvent* is often associated with volatile liquids, but the term is more general than that. (A *volatile liquid* evaporates rapidly and readily at normal temperatures.) Water is the solvent in aqueous solutions.

[19]An *emulsion* is not mixture of a solid in a liquid. It is a mixture of two immiscible liquids.

The solubility of most solids in liquid solvents usually increases with increasing temperature. Pressure has very little effect on the solubility of solids in liquids.

When the solvent has dissolved as much solute as it can, it is a *saturated solution*.[20] Adding more solute to an already saturated solution will cause the excess solute to settle to the bottom of the container, a process known as *precipitation*. Other changes (in temperature, concentration, etc.) can be made to cause precipitation from saturated and unsaturated solutions. Precipitation in a chemical reaction is indicated by a downward arrow (i.e., "↓"). For example, the precipitation of silver chloride from an aqueous solution of silver nitrate ($AgNO_3$) and potassium chloride (KCl) would be written:

$$AgNO_3(aq) + KCl(aq) \rightarrow AgCl(s){\downarrow} + KNO_3(aq)$$

## 37. UNITS OF CONCENTRATION

There are many units of concentration to express solution strengths.

F— *formality:* The number of gram formula weights (i.e., molecular weights in grams) per liter of solution.

m— *molality:* The number of gram-moles of solute per 1000 grams of solvent. A "molal" solution contains 1 gram-mole per 1000 grams of solvent.

M— *molarity:* The number of gram-moles of solute per liter of solution. A "molar" (i.e., 1 M) solution contains 1 gram-mole per liter of solution. Molarity is related to normality: $N = M \times \Delta$ oxidation number.

N— *normality:* The number of gram equivalent weights of solute per liter of solution. A solution is "normal" (i.e., 1 N) if there is exactly one gram equivalent weight per liter of solution.

x— *mole fraction:* The number of moles of solute divided by the number of moles of solvent and all solutes.

meq/L— *milligram equivalent weights of solute per liter of solution:* calculated by multiplying normality by 1000 or dividing concentration in mg/L by equivalent weight.

mg/L— *milligrams per liter:* The number of milligrams of solute per liter of solution. Same as ppm for solutions of water.

ppm— *parts per million:* The number of pounds (or grams) of solute per million pounds (or grams) of solution. Same as mg/L for solutions of water.

For compounds whose molecules do not dissociate in solution (e.g., table sugar), there is no difference between molarity and formality. There is a difference, however, for compounds that dissociate into ions (e.g., table salt). Consider a solution derived from 1 gmol of magnesium nitrate $Mg(NO_3)_2$ in enough water to bring

---

[20]Under certain circumstances, a *supersaturated solution* can exist for a limited amount of time.

the volume to 1 L. The formality is 1.0 (i.e., the solution is 1.0 formal). However, 3 moles of ions will be produced —1 mole of $Mg^{++}$ ions and 2 moles of $NO_3^-$ ions. Therefore, molarity is 1.0 for the magnesium ion and 2.0 for the nitrate ion.

The use of formality avoids the ambiguity in specifying concentrations for ionic solutions. Also, the use of formality avoids the problem of determining a molecular weight when there are no discernible molecules (e.g., as in a crystalline solid such as NaCl). However, in their quest for uniformity in nomenclature, most modern chemists do not make the distinction between molarity and formality, and molarity is used as if it were formality.

### Example 33.13

A solution is made by dissolving 0.353 g of $Al_2(SO_4)_3$ in 730 g of water. Assuming 100% ionization, what is the concentration expressed as normality, molarity, and mg/L?

*Solution*

The molecular weight of $Al_2(SO_4)_3$ is

$$
\begin{aligned}
MW &= (2)\left(26.98 \ \frac{g}{mol}\right) + (3)\left(32.06 \ \frac{g}{mol}\right) \\
&+ (4)\left(16 \ \frac{g}{mol}\right) = 342.14 \ g/mol
\end{aligned}
$$

The equivalent weight is

$$
EW = \frac{342.14 \ \frac{g}{mol}}{6} = 57.02 \ g/mol
$$

The number of gram equivalent weights used is

$$
\frac{0.353 \ g}{57.02 \ \frac{g}{mol}} = 6.19 \times 10^{-3} \ GEW
$$

The number of liters of solution (same as the solvent volume if the small amount of solute is neglected) is 0.73.

The normality is

$$
N = \frac{6.19 \times 10^{-3} \ GEW}{0.73 \ L} = 8.48 \times 10^{-3}
$$

The number of moles of solute used is

$$
\frac{0.353 \ g}{342.14 \ \frac{g}{mol}} = 1.03 \times 10^{-3} \ mol
$$

The molarity is

$$
M = \frac{1.03 \times 10^{-3} \ mol}{0.73 \ L} = 1.41 \times 10^{-3}
$$

The mass is

$$
m = \frac{0.353 \ g}{0.001 \ \frac{g}{mg}} = 353 \ mg
$$

The concentration is

$$
C = \frac{m}{V} = \frac{353 \ mg}{0.73 \ L} = 483.6 \ mg/L
$$

## 38. HEAT OF SOLUTION

The *heat of solution*, $\Delta H$, is an amount of energy that is absorbed or released when a substance enters a solution. It can be calculated from the enthalpies of formation of the solution components.[21] For example, the heat of solution associated with the formation of dilute hydrochloric acid from HCl gas and large amounts of water can be calculated as follows.

$$
HCl(g) \xrightarrow{H_2O} HCl(aq) + \Delta H
$$

$$
\Delta H = -17.21 \ kcal/mol
$$

If a heat of solution is negative (as it is for all aqueous solutions of gases), heat is given off when the solute dissolves in the solvent. This is an *exothermic reaction*. If the heat of solution is positive, heat is absorbed when the solute dissolves in the solvent. This is an *endothermic reaction*.

If the heat of solution is very large, the forces between the molecules in solution will be strong. If the heat of solution is very small, the forces between the molecules in solution will be weak. *Ideal solutions* are those for which the interaction forces between all solution components (solute-solute, solute-solvent, and solvent-solvent) are the same.

For dilute solutions, the value of the heat of solution is constant; for strong solutions, it depends on the concentration. Therefore, it is a general practice to state the number of moles of solvent in which one mole of solute is dissolved.

$$
HCl(g) + 50H_2O(l) \rightarrow HCl(50H_2O) + \Delta H
$$

To be more specific, the *integral heat of solution* is the increase in energy per mole of solute $x$ when forming a solution of a particular concentration. This is the secant slope $(\Delta H_x / n_x)$ of the curve in Fig. 33.5 at a particular point. The *differential heat of solution* is the increase in energy when one mole of solute is dissolved in such a large volume of solvent that there is no appreciable change in the concentration. This is the slope $(d\Delta H_x / dn_x)$ of the curve.

---

[21]The heat of solution is a special case of the heat of reaction. See Sec. 33.62 for more information on heats of reactions.

Chemistry

**Figure 33.5** *Heats of Solution*

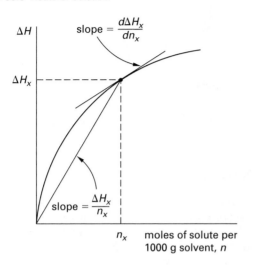

It is common to experience an exothermic reaction when moisture is absorbed by a desiccant. A *desiccant* is a substance that is hygroscopic (i.e., has a high affinity for the moisture in air). Desiccants are used as drying agents. Common desiccants are silica gel (produced from silicic acid), activated alumina, and anhydrous calcium sulfate. Calcium chloride can be used but it is not a powerful desiccant. Almost all perchlorates ($ClO_4^-$ ion) are hygroscopic. However, because perchlorates are also powerful oxidizing agents and create explosion hazards, they are not used as desiccants.

The crystals of a desiccant retain their shape when they absorb moisture. A *deliquescent substance* melts (dissolves) when it absorbs moisture from the air. (An *efflorescent substance* crumbles to a powder when it loses its water of crystallization.)

## 39. SOLVENT VAPOR PRESSURE

Vapor pressure is the pressure exerted by the solvent's vapor molecules when they are in equilibrium with the liquid. The symbol for the vapor pressure of a pure vapor over a pure solvent is $p^0$. Vapor pressure increases with increasing temperature.

*Raoult's law* gives the drop in a solvent's vapor pressure as the solute is added and the mole fraction, $x$, is increased. In the case of non-volatile, non-electrolytic liquids, Raoult's law is

$$\Delta p_{\text{vapor}} = p^0 - p_{\text{vapor,solution}}$$
$$= x_{\text{solute}} p^0 \qquad \qquad \textit{33.14}$$

$$p_{\text{vapor,solution}} = x_{\text{solvent}} p^0 \qquad \qquad \textit{33.15}$$

## 40. BOILING AND FREEZING POINTS

A liquid boils when its vapor pressure is equal to the surrounding pressure. Since the addition of a solute to a solvent decreases the vapor pressure (Raoult's law), the temperature of the solution must be increased to maintain the same vapor pressure. Thus, the *boiling point* (temperature), $T_b$, of a solution is higher than the boiling point of the pure solvent at the same pressure. The *boiling point elevation* is given by Eq. 33.16, in which $m$ is the molality. $K_b$ is the *molal boiling point constant*, a property of the solvent only.[22] The molal boiling point constant for water is $0.512°C/m$.

$$\Delta T_b = mK_b = \frac{m_{\text{solute,in g}} K_b}{(\text{MW}) m_{\text{solvent,in kg}}} \quad \text{[increase]} \qquad \textit{33.16}$$

Similarly, the *freezing (melting) point*, $T_f$, will be lower for the solution than for the pure solvent. The *freezing point depression* depends on the *molal freezing point constant*, $K_f$, a property of the solvent only. Table 33.4 lists approximate boiling and freezing point constants.

$$\Delta T_f = -mK_f = \frac{-m_{\text{solute,in g}} K_f}{(\text{MW}) m_{\text{solvent,in kg}}} \quad \text{[decrease]} \qquad \textit{33.17}$$

**Table 33.4** *Approximate Boiling and Freezing Point Constants*

| solvent | normal boiling point, °C | $K_b$ °C/m | normal freezing point, °C | $K_f$ °C/m |
|---|---|---|---|---|
| water | 100.0 | 0.512 | 0.0 | 1.86 |
| acetic acid | 118.5 | 3.07 | 16.7 | 3.9 |
| benzene | 80.2 | 2.53 | 5.5 | 5.12 |
| camphor | 208.3 | 5.95 | 178.4 | 40.0 |
| chloroform | 60.2 | 3.63 | −63.5 | 4.68 |
| ethyl alcohol (ethanol) | 78.3 | 1.22 | | |
| ethyl ether | 34.4 | 2.02 | | |
| methyl alcohol (methanol) | 64.7 | 0.83 | | |
| naphthalene | 218.0 | 5.65 | 80.2 | 6.9 |
| nitrobenzene | 210.9 | 5.24 | 5.7 | 8.1 |
| phenol | 181.2 | 3.56 | 42.0 | 7.27 |

Equation 33.16 and Eq. 33.17 are for dilute, non-electrolytic solutions and nonvolatile solutes. The ratio of the actual change in freezing (boiling) point to the ideal change is known as the *van't Hoff factor*, $i$. Typically, sucrose is used as the standard non-ionizing solute in determining the ideal change.

$$i = \frac{\Delta T_{\text{actual}}}{\Delta T_{\text{ideal}}} \qquad \qquad \textit{33.18}$$

The actual change in temperature will be one to two times larger than ideal changes (i.e., $i = 2$ to $3$), depending on the numbers of ions produced upon dissociation.

---

[22]Vapor pressure, boiling point, and freezing point of a solution differ from those of a pure solvent by amounts that are directly proportional to the molal concentration of the solute. Such properties are known as *colligative properties*.

A mole of sodium chloride, for example, produces 2 moles of ions (1 mole each of $Na^+$ and $Cl^-$). Therefore, the van't Hoff factor will be 2.0.

### Example 33.14

4.2 g of a non-ionizing solute was dissolved in 112 g of acetone ($K_b = 1.71°C/m$, $T_b = 55.95°C$). The boiling point of the solution increased to 56.7°C. What is the approximate molecular weight of the solute?

*Solution*

From Eq. 33.16,

$$MW = \frac{m_{solute,in\,g}\,K_b}{m_{solvent,in\,kg}\,\Delta T_b}$$

$$= \frac{(4.2\ g)(1.71°C)}{(0.112\ kg)(56.7°C - 55.95°C)}$$

$$= 85.5\ g/mol$$

## 41. OSMOTIC PRESSURE

*Osmosis* is the diffusion of a solvent into a stronger solution in an attempt to equalize the two concentrations. *Osmotic pressure*, $\pi$, is the pressure exerted by the pure solvent or dilute solution, causing the diffusion. Alternatively, the osmotic pressure can be defined as the minimum pressure that must be applied to the stronger solution to prevent the diffusion of solvent. Applying a greater pressure causes pure solvent to leave the solution. This is known as *reverse osmosis*.

Osmotic pressure can be measured by allowing a solvent to diffuse through a *semipermeable membrane*. The osmotic pressure will depend on the concentration, $C$, the ideal gas constant, $R^*$, and the absolute temperature. The units of $C$ depend on the units of $R^*$ and can be expressed as molality or molarity.

$$\pi = CR^*T \qquad 33.19$$

Solutions that have the same osmotic pressure are called *isotonic solutions*.

## 42. ACIDS AND BASES

An *acid* is any compound that dissociates in water into $H^+$ ions. (The combination of $H^+$ and water, $H_3O^+$, is known as the *hydronium ion*.) This is known as the *Arrhenius theory* of acids.[23] Acids with 1, 2, and 3 ionizable hydrogen atoms are called *monoprotic, diprotic*, and *triprotic acids*, respectively.

The properties of acids are as follows.

- Acids conduct electricity in aqueous solutions.

- Acids have a sour taste.

- Acids turn blue litmus paper red.

- Acids have a pH between 0 and 7.

- Acids neutralize bases.

- Acids react with active metals to form hydrogen.

$$2H^+ + Zn \rightarrow Zn^{++} + H_2$$

- Acids react with oxides and hydroxides of metals to form salts and water.

$$2H^+ + 2Cl^- + FeO \rightarrow Fe^{++} + 2Cl^- + H_2O$$

- Acids react with salts of either weaker or more volatile acids (such as carbonates and sulfides) to give a new salt and a new acid.

$$2H^+ + 2Cl^- + CaCO_3 \rightarrow H_2CO_3 + Ca^{++} + 2Cl^-$$

The relative strengths of acids depend on the number of oxygen atoms bound in the hydroxy radical. For example, $H_2SO_4$ with four oxygens is a stronger acid than $H_2SO_3$ (with three) or HCl (with none). Common inorganic and organic acids are listed in Table 33.5.

*Table 33.5* Common Inorganic and Organic Acids

| | |
|---|---|
| acetic | $CH_3COOH$ |
| acrylic | $C_2H_3COOH$ |
| benzene sulfonic | $C_6H_5SO_3H$ |
| benzoic | $C_6H_5COOH$ |
| butyric | $C_3H_7COOH$ |
| carbolic | $C_6H_5OH$ |
| carbonic | $H_2CO_3$ |
| chloric | $HClO_3$ |
| formic | $HCOOH$ |
| hydrobromic | $HBr$ |
| hydrochloric | $HCl$ |
| hydrosulfuric | $H_2S$ |
| nitric | $HNO_3$ |
| oleic | $C_{17}H_{33}COOH$ |
| oxalic | $H_2C_2O_4$ |
| perchloric | $HClO_4$ |
| phenol | $C_6H_5OH$ |
| phosphoric | $H_3PO_4$ |
| propionic | $C_2H_5COOH$ |
| stearic | $C_{17}H_{35}COOH$ |
| sulfuric | $H_2SO_4$ |
| sulfurous | $H_2SO_3$ |
| valeric | $C_4H_9COOH$ |

A *base* is any compound that dissociates in water into $OH^-$ ions. This is known as the *Arrhenius theory of bases*.[24] Bases with one, two, and three replaceable hydroxide ions are called *monohydroxic, dihydroxic*, and *trihydroxic* bases, respectively.

---

[23]The *Bronsted-Lowry theory* defines acids as *proton donors*. The *Lewis theory* defines acids as *electron-pair acceptors*.

[24]The *Bronsted-Lowry theory* defines bases as *proton acceptors*. The *Lewis theory* defines bases as *electron-pair donors*.

The properties of bases are as follows.

- Bases conduct electricity in aqueous solutions.
- Bases have a bitter taste.
- Bases turn red litmus paper blue.
- Bases have a pH between 7 and 14.
- Bases neutralize acids, forming salts and water.

## 43. CONJUGATE ACIDS AND BASES

According to the *Bronsted-Lowry theory*, the acid that results when a base accepts a proton is called the *conjugate acid* of the base. The base that results when an acid donates its proton is called the *conjugate base* of the acid. These definitions are sufficiently broad to encompass any positively and negatively charged ions in the products.

If the acid is weak, the conjugate base tends to be strong; if the acid is strong, the conjugate base tends to be weak. For example, acetic acid, $HC_2H_3O_2$, is a moderately weak acid, and the acetate ion, $C_2H_3O_2^-$, is a moderately strong base.

## 44. pH AND pOH

A measure of the strength of an acid or base is the number of hydrogen or hydroxide ions in a liter of solution. Since these are very small numbers, a logarithmic scale is used.

$$pH = -\log_{10}[H^+] = \log_{10}\left(\frac{1}{[H^+]}\right) \qquad 33.20$$

$$pOH = -\log_{10}[OH^-] = \log_{10}\left(\frac{1}{[OH^-]}\right) \qquad 33.21$$

The quantities $[H^+]$ and $[OH^-]$ in square brackets are the *ionic concentrations* in moles of ions per liter. The number of moles can be calculated from Avogadro's law by dividing the actual number of ions per liter by $6.022 \times 10^{23}$. Alternatively, for a partially ionized compound in a solution of known molarity, $M$, the ionic concentration is

$$[ion] = XM \qquad 33.22$$

A *neutral solution* has a pH of 7.[25] Solutions with a pH below 7 are acidic; the smaller the pH, the more acidic the solution. Solutions with a pH above 7 are basic.

The relationship between pH and pOH is

$$pH + pOH = 14 \qquad 33.23$$

---

[25]The pH of a neutral solution depends on the temperature. For example, at 25°C, the pH is 7. When the temperature is higher (lower) than 25°C, the pH will be less (greater) than 7.

### Example 33.15

A 4.2% ionized 0.01M ammonia solution is prepared from ammonium hydroxide ($NH^4OH$). Calculate the pH, pOH, and concentrations of $[H^+]$ and $[OH^-]$.

*Solution*

From Eq. 33.22,

$$[OH^-] = XM = (0.042)(0.01)$$
$$= 4.2 \times 10^{-4} \text{ mol/L}$$

From Eq. 33.21,

$$pOH = -\log[OH^-] = -\log(4.2 \times 10^{-4})$$
$$= 3.38$$

From Eq. 33.23,

$$pH = 14 - pOH = 14 - 3.38$$
$$= 10.62$$

The $[H^+]$ ionic concentration can be extracted from the definition of pH.

$$[H^+] = 10^{-pH} = 10^{-10.62}$$
$$= 2.4 \times 10^{-11} \text{ mol/L}$$

## 45. BUFFERS

A *buffer solution* resists changes in acidity and maintains a relatively constant pH when a small amount of an acid or base is added to it. Buffers are usually combinations of weak acids and their salts. A buffer is most effective when the acid and salt concentrations are equal.

## 46. NEUTRALIZATION

Acids and bases neutralize each other to form water.

$$H^+ + OH^- \rightarrow H_2O$$

Assuming 100% ionization of the solute, the volumes, $V$, required for complete neutralization can be calculated from the normalities, $N$, or the molarities, $M$.

$$V_{base}N_{base} = V_{acid}N_{acid} \qquad 33.24$$

$$V_{base}M_{base}\Delta_{base\ charge} = V_{acid}M_{acid}\Delta_{acid\ charge} \qquad 33.25$$

## 47. TITRATION AND INDICATOR SOLUTIONS

A solution of known concentration is called a *standard solution*. The *neutralizing solution* is the solution that is added in order to neutralize the standard solution.

*Titration* is the process of determining how much neutralizing solution is required.

An *equivalent* of a substance is defined as the mass (in grams for a gram-equivalent weight) of that substance that releases one mole of either $H^+$ or $OH^-$ ions. When equal numbers of equivalents have reacted, the solution will have reached the *equivalence point* and will be neutral (i.e., have a pH of 7).

Figure 33.6 is a *titration curve* of pH versus the volume of a base added to an acidic standard solution. *Indicators* (also known as *indicator solutions* and *colorimetric indicators*) are used to signal having reached the equivalence point.[26] The observed color change is called the *endpoint*. An indicator is usually a weak organic acid or base that changes color over a narrow pH range (i.e., approximately two pH units). Table 33.6 lists common indicator solutions.

**Figure 33.6** *Typical Titration Curve*

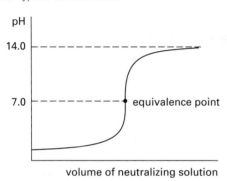

**Table 33.6** *Common Indicator Solutions*

| common name | pH visual transition interval | color acidic | color basic |
|---|---|---|---|
| cresol red | 0.2–1.8 | red | yellow |
| thymol blue | 1.2–2.8 | red | yellow |
| methyl yellow | 2.4–4.0 | red | yellow |
| bromophenol blue | 3.0–4.6 | yellow | blue |
| methyl orange | 3.2–4.4 | red | yellow-orange |
| methyl orange + xylene cyanole FF, 40:56 | $(3.8–4.1)^a$ | violet | green |
| bromocresol green | 3.9–5.4 | yellow | blue |
| methyl red | 4.2–6.2 | pink | yellow |
| methyl red + methylene blue, 1:1 | $(\approx 5.3)^a$ | red-violet | green |
| bromocresol purple | 5.2–6.8 | yellow | purple |
| bromothymol blue | 6.0–7.6 | yellow | blue |
| cresol red | 7.2–8.8 | yellow | red |
| phenol red | 6.8–8.2 | yellow | red |
| thymol blue | 8.0–9.2 | yellow | blue |
| phenolphthalein | $(8.0–9.8)^b$ | colorless | red-violet |
| phenolphthalein + methylene green, 1:2 | $(8.8)^c$ | green | violet |
| thymolphthalein | $(9.0–10.5)^b$ | colorless | blue |
| eriochrome black T | 7–10 | blue | wine-red |
| alizarin yellow | 10.1–12 | yellow | red |

$^a$screened indicator, neutral gray at stated pH
$^b$based on addition of 1 or 2 drops of a 0.1% indicator solution to 10 ml of aqueous solution
$^c$screened indicator, pale blue at stated pH

## 48. SALTS

Salts are ionic compounds formed during the complete or partial neutralization of acids. (However, a salt can also be formed by direct union of elements, reactions of acids and salts, and reactions between different salts.) During neutralization, the $H^+$ acid ions are replaced with metal or other electropositive ions. For example, sodium chloride (NaCl) is formed during the neutralization of hydrochloric acid (HCl) and sodium hydroxide (NaOH).

$$NaOH + HCl \rightarrow NaCl + H_2O$$

Salts formed during complete neutralization, such as sodium sulfate ($Na_2SO_4$), are known as *normal salts*. Salts formed during incomplete neutralization, such as sodium hydrogen sulfate, also known as sodium acid sulfate ($NaHSO_4$), still contain some of the original $H^+$ ions from the acid and are known as *acid salts*.

## 49. HYDROLYSIS

The term *hydrolysis* refers to salts of weak acids (bases) dissolving in water to form basic (acidic) solutions. Hydrolysis can also be considered to be the incomplete neutralization of acids and bases. Typically, sulfides ($S^-$), carbonates ($CO_3^{-2}$), phosphates ($PO_4^{-3}$), and salts of the transition elements hydrolyze.

## 50. REVERSIBLE REACTIONS

*Reversible reactions* are capable of going in either direction and do so to varying degrees (depending on the concentrations and temperature) simultaneously. These reactions are characterized by the simultaneous presence of all reactants and all products. For example, the chemical equation for the exothermic formation of ammonia from nitrogen and hydrogen is

$$N_2 + 3H_2 \rightleftharpoons 2NH_3 + \Delta H = -24.5 \text{ kcal}$$

At *chemical equilibrium*, reactants and products are both present. However, the concentrations of the reactants and products do not change after equilibrium is reached.

---

[26]When a signal of complete neutralization is needed, the indicator should be active in the pH range of 6 to 8. However, various indicators can be used when other solutions with different pHs are needed.

## 51. LE CHATELIER'S PRINCIPLE

*Le Châtelier's principle* predicts the direction in which a reversible reaction at equilibrium will go when some condition (e.g., temperature, pressure, concentration) is "stressed" (i.e., changed). The principle says that when an equilibrium state is stressed by a change, a new equilibrium is formed that reduces that stress.

Consider the formation of ammonia from nitrogen and hydrogen. When the reaction proceeds in the forward direction, energy in the form of heat is released and the temperature increases. If the reaction proceeds in the reverse direction, heat is absorbed and the temperature decreases. If the system is stressed by increasing the temperature, the reaction will proceed in the reverse direction because that direction absorbs heat and reduces the temperature.

For reactions that involve gases, the reaction equation coefficients can be interpreted as volumes. In the nitrogen-hydrogen reaction (see Sec. 33.50), four volumes combine to form two volumes. If the equilibrium system is stressed by increasing the pressure, then the forward reaction will occur because this direction reduces the volume and pressure.[27]

If the concentration of any substance is increased, the reaction proceeds in a direction away from the substance with the increase in concentration. (For example, an increase in the concentration of the reactants shifts the equilibrium to the right, increasing the amount of products formed.)

The *common ion effect* is a special case of Le Châtelier's principle. If a salt containing a common ion is added to a solution of a weak acid, almost all of the salt will dissociate, adding large quantities of the common ion to the solution. Ionization of the acid will be greatly suppressed, a consequence of the need to have an unchanged equilibrium constant.

## 52. IRREVERSIBLE REACTION KINETICS

The rate at which a compound is formed or used up in an irreversible (one-way) reaction is known as the *rate of reaction*, also known as the *speed of reaction*, *reaction velocity*, and so on. The rate, v, is the change in concentration per unit time, usually measured in mol/L·s.

$$\text{v} = \text{change in concentration/time} \qquad 33.26$$

According to the *law of mass action*, the rate of reaction varies with the concentrations of the reactants and products. Specifically, the rate is proportional to the molar concentrations (i.e., the molarities). The rate of the formation or a conversion of substance A is represented in various forms, such as $r_A$, $dA/dt$, and $d[A]/dt$, where the variable A or [A] can represent either the

mass or the concentration of substance A. Substance A can be either a pure element or a compound.

The rate of reaction is generally not affected by pressure, but does depend on five other factors.

- *type of substances in the reaction:* Some substances are more reactive than others.

- *exposed surface area:* The rate of reaction is proportional to the amount of contact between the reactants.

- *concentrations:* The rate of reaction increases with increases in concentration.

- *temperature:* The rate of reaction approximately doubles with every 10°C increase in temperature.

- *catalysts:* If a catalyst is present, the rate of reaction increases. However, the equilibrium point is not changed. (A catalyst is a substance that increases the reaction rate without being consumed in the reaction.)

## 53. ORDER OF THE REACTION

The *order of the reaction* is the total number of reacting molecules in or before the slowest step in the mechanism.[28] The order must be determined experimentally. However, for an irreversible elementary reaction, the order is usually assumed from the stoichiometric reaction equation as the sum of the combining coefficients for the reactants.[29,30] For example, for the reaction $mA + nB \rightarrow pC$, the overall order of the forward reaction is assumed to be $m + n$.

Many reactions (e.g., dissolving metals in acid or the evaporation of condensed materials) have *zero-order reaction rates*. These reactions do not depend on the concentrations or temperature at all, but rather, are affected by other factors such as the availability of reactive surfaces or the absorption of radiation. The formation (conversion) rate of a compound in a zero-order reaction is constant. That is, $dA/dt = -k_0$. $k_0$ is known as the *reaction rate constant*. (The subscript "0" refers to the zero-order.) Since the concentration (amount) of the substance decreases with time, $dA/dt$ is negative. Since the negative sign is explicit in rate equations, the reaction rate constant is always considered to be a positive number.

---

[27]The exception to this rule is the addition of an inert gas to a gaseous equilibrium system. Although there is an increase in total pressure, the position of the equilibrium is not affected.

[28]This definition is valid for elementary reactions. For complex reactions, the order is an empirical number that need not be an integer.
[29]The overall order of the reaction is the sum of the orders with respect to the individual reactants. For example, in the reaction $2NO + O_2 \rightarrow 2NO_2$, the reaction is second order with respect to NO, first order with respect to $O_2$, and third order overall.
[30]In practice, the order of the reaction must be known, given, or determined experimentally. It is not always equal to the sum of the combining coefficients for the reactants. For example, in the reaction $H_2 + I_2 \rightarrow 2HI$, the overall order of the reaction is indeed 2, as expected. However, in the reaction $H_2 + Br_2 \rightarrow 2HBr$, the overall order is found experimentally to be 3/2, even though the two reactions have the same stoichiometry, and despite the similarities of iodine and bromine.

**Table 33.7** *Reaction Rates and Half-Life Equations*

| reaction | order | rate equation | integrated forms |
|---|---|---|---|
| $A \to B$ | zero | $\dfrac{d[A]}{dt} = -k_0$ | $[A] = [A]_0 - k_0 t$ <br> $t_{1/2} = \dfrac{[A]_0}{2k_0}$ |
| $A \to B$ | first | $\dfrac{d[A]}{dt} = -k_1[A]$ | $\ln \dfrac{[A]}{[A]_0} = k_1 t$ <br> $t_{1/2} = \dfrac{1}{k_1} \ln 2$ |
| $A + A \to P$ | second, type I | $\dfrac{d[A]}{dt} = -k_2[A]^2$ | $\dfrac{1}{[A]} - \dfrac{1}{[A]_0} = k_2 t$ <br> $t_{1/2} = \dfrac{1}{k_2[A]_0}$ |
| $aA + bB \to P$ | second, type II | $\dfrac{d[A]}{dt} = -k_2[A][B]$ | $\ln \dfrac{[A]_0 - [B]}{[B]_0 - \left(\frac{b}{a}\right)[X]} = \ln \dfrac{[A]}{[B]}$ <br> $= \left(\dfrac{b[A]_0 - a[B]_0}{a}\right) k_2 t + \ln \dfrac{[A]_0}{[B]_0}$ <br> $t_{1/2} = \left(\dfrac{a}{k_2(b[A]_0 - a[B]_0)}\right) \ln \left(\dfrac{a[B]_0}{2a[B]_0 - b[A]_0}\right)$ |

Table 33.7 contains reaction rate and half-life equations for various types of low-order reactions. Once a reaction rate equation is known, it can be integrated to obtain an expression for the concentration (mass) of the substance at various times. The time for half of the substance to be formed (or converted) is the *half-life*, $t_{1/2}$.

### Example 33.16

Nitrogen pentoxide decomposes according to the following first-order reaction.

$$N_2O_5 \to 2NO_2 + \tfrac{1}{2}O_2$$

At a particular temperature, the decomposition of nitrogen pentoxide is 85% complete at the end of 11 min. The reaction rate constant is to be determined.

*Solution*

The reaction is given as first order. Use the integrated reaction rate equation from Table 33.7. Since the decomposition reaction is 85% complete, the surviving fraction is 15% (0.15).

$$\ln \frac{[A]}{[A]_0} = k_1 t$$

$$\ln(0.15) = k(11 \text{ min})$$

$$k = -0.172 \text{ 1/min} \quad (0.172 \text{ 1/min})$$

(The rate constant is considered to be a positive number.)

## 54. ACTIVATION ENERGY

The *activation energy*, $E_a$, is the energy required to cause the reaction to occur. A catalyst lowers the activation energy in both directions (for a reversible reaction). Figure 33.7 shows that the activation energy is different in the forward and reverse directions and that the difference is the *heat of reaction*, $\Delta H_r$.

**Figure 33.7** *Activation Energy*

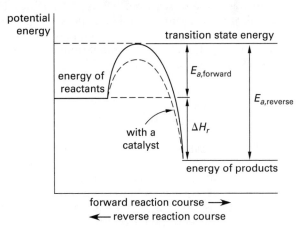

## 55. ARRHENIUS EQUATION

The *Arrhenius equation* defines the relationship between the rate constant, $k$, and temperature.

$$k = A e^{-E_a / R^* T} \qquad \text{33.27}$$

The constant $A$ is the *Arrhenius constant*, representing the fraction of effective collisions between reacting molecules.

Taking natural logs of both sides and rearranging,

$$\ln k = \ln A - \frac{E_a}{R^* T} \qquad 33.28$$

$$\ln \frac{k_2}{k_1} = \frac{-E_a}{R^*}\left(\frac{1}{T_2} - \frac{1}{T_1}\right) \qquad 33.29$$

## 56. REVERSIBLE REACTION KINETICS

Consider the following reversible reaction.

$$a\mathrm{A} + b\mathrm{B} \rightleftharpoons c\mathrm{C} + d\mathrm{D} \qquad 33.30$$

In Eq. 33.31 and Eq. 33.32, the *reaction rate constants* are $k_{\mathrm{forward}}$ and $k_{\mathrm{reverse}}$. The order of the forward reaction is $a + b$; the order of the reverse reaction is $c + d$.

$$\mathrm{v}_{\mathrm{forward}} = k_{\mathrm{forward}}[\mathrm{A}]^a[\mathrm{B}]^b \qquad 33.31$$

$$\mathrm{v}_{\mathrm{reverse}} = k_{\mathrm{reverse}}[\mathrm{C}]^c[\mathrm{D}]^d \qquad 33.32$$

At equilibrium, the forward and reverse speeds of reaction are equal.

$$\mathrm{v}_{\mathrm{forward}} = \mathrm{v}_{\mathrm{reverse}}|_{\mathrm{equilibrium}} \qquad 33.33$$

## 57. EQUILIBRIUM CONSTANT

For reversible reactions, the *equilibrium constant*, $K$, is proportional to the ratio of the reverse rate of reaction to the forward rate of reaction.[31] Except for catalysis, the equilibrium constant depends on the same factors affecting the reaction rate. For the complex reversible reaction given by Eq. 33.30, the equilibrium constant is given by the *law of mass action*.

$$K = \frac{[\mathrm{C}]^c[\mathrm{D}]^d}{[\mathrm{A}]^a[\mathrm{B}]^b} = \frac{k_{\mathrm{forward}}}{k_{\mathrm{reverse}}} \qquad 33.34$$

If any of the reactants or products are in pure solid or pure liquid phases, their concentrations are omitted from the calculation of the equilibrium constant. For example, in weak aqueous solutions, the concentration of water, $H_2O$, is very large and essentially constant; therefore, that concentration is omitted.

For gaseous reactants and products, the concentrations (i.e., the numbers of atoms) will be proportional to the partial pressures. Therefore, an equilibrium constant can be calculated directly from the partial pressures and is given the symbol $K_p$. For example, for the

formation of ammonia gas from nitrogen and hydrogen, the equilibrium constant is

$$K_p = \frac{[p_{\mathrm{NH_3}}]^2}{[p_{\mathrm{N_2}}][p_{\mathrm{H_2}}]^3} \qquad 33.35$$

$K$ and $K_p$ are not numerically the same, but they are related by Eq. 33.36. $\Delta n$ is the number of moles of products minus the number of moles of reactants.

$$K_p = K(R^* T)^{\Delta n} \qquad 33.36$$

**Example 33.17**

A particularly weak solution of acetic acid ($HC_2H_3O_2$) in water has the ionic concentrations (in mol/L) given. What is the equilibrium constant?

$$\mathrm{HC_2H_3O_2} + \mathrm{H_2O} \rightleftharpoons \mathrm{H_3O^+} + \mathrm{C_2H_3O_2^-}$$
$$[\mathrm{HC_2H_3O_2}] = 0.09866$$
$$[\mathrm{H_2O}] = 55.5555$$
$$[\mathrm{H_3O^+}] = 0.00134$$
$$[\mathrm{C_2H_3O_2^-}] = 0.00134$$

*Solution*

The concentration of the water molecules is not included in the calculation of the equilibrium or ionization constant. Therefore, the equilibrium constant is

$$K = K_a = \frac{[\mathrm{H_3O^+}][\mathrm{C_2H_3O_2^-}]}{[\mathrm{HC_2H_3O_2}]}$$
$$= \frac{(0.00134)(0.00134)}{0.09866} = 1.82 \times 10^{-5}$$

## 58. IONIZATION CONSTANT

The equilibrium constant for a weak solution is essentially constant and is known as the *ionization constant* (also known as a *dissociation constant*). (See Table 33.8.) For weak acids, the symbol $K_a$ and name *acid constant* are used. For weak bases, the symbol $K_b$ and the name *base constant* are used. For example, for the ionization of hydrocyanic acid,

$$\mathrm{HCN} \rightleftharpoons \mathrm{H^+} + \mathrm{CN^-}$$
$$K_a = \frac{[\mathrm{H^+}][\mathrm{CN^-}]}{[\mathrm{HCN}]}$$

Pure water is itself a very weak electrolyte and ionizes only slightly.

$$2\mathrm{H_2O} \rightleftharpoons \mathrm{H_3O^+} + \mathrm{OH^-} \qquad 33.37$$

---

[31]The symbols $K_c$ (in molarity units) and $K_{\mathrm{eq}}$ are occasionally used for the equilibrium constant.

**Table 33.8** Approximate Ionization Constants

| substance | 0°C | 5°C | 10°C | 15°C | 20°C | 25°C |
|---|---|---|---|---|---|---|
| Ca(OH)$_2$ | | | | | | $3.74 \times 10^{-3}$ |
| HClO | $2.0 \times 10^{-8}$ | $2.3 \times 10^{-8}$ | $2.6 \times 10^{-8}$ | $3.0 \times 10^{-8}$ | $3.3 \times 10^{-8}$ | $3.7 \times 10^{-8}$ |
| HC$_2$H$_3$O$_2$ | $1.67 \times 10^{-5}$ | $1.70 \times 10^{-5}$ | $1.73 \times 10^{-5}$ | $1.75 \times 10^{-5}$ | $1.75 \times 10^{-5}$ | $1.75 \times 10^{-5}$ |
| HBrO | | | | | $\approx 2 \times 10^{-9}$ | |
| H$_2$CO$_3$ ($K_1$) | $2.6 \times 10^{-7}$ | $3.04 \times 10^{-7}$ | $3.44 \times 10^{-7}$ | $3.81 \times 10^{-7}$ | $4.16 \times 10^{-7}$ | $4.45 \times 10^{-7}$ |
| HClO$_2$ | | | | | $\approx 1.1 \times 10^{-2}$ | |
| NH$_3$ | $1.37 \times 10^{-5}$ | $1.48 \times 10^{-5}$ | $1.57 \times 10^{-5}$ | $1.65 \times 10^{-5}$ | $1.71 \times 10^{-5}$ | $1.77 \times 10^{-5}$ |
| NH$_4$OH | | | | | | $1.79 \times 10^{-5}$ |
| water* | 14.9435 | 14.7338 | 14.5346 | 14.3463 | 14.1669 | 13.9965 |

*$-\log_{10} K$ given

At equilibrium, the ionic concentrations are equal.

$$[H_3O^+] = 10^{-7}$$

$$[OH^-] = 10^{-7}$$

From Eq. 33.34, the ionization constant (*ion product*) for pure water is

$$K_w = K_{a,\text{water}} = [H_3O^+][OH^-]$$
$$= (10^{-7})(10^{-7})$$
$$= 10^{-14} \qquad \textit{33.38}$$

If the molarity, $M$, and *fraction ionization*, $X$, are known, the ionization constant can be calculated from Eq. 33.39.

$$K_{\text{ionization}} = \frac{MX^2}{1-X} \quad [K_a \text{ or } K_b] \qquad \textit{33.39}$$

The reciprocal of the ionization constant is the *stability constant (overall stability constant)*, also known as the *formation constant*. Stability constants are used to describe complex ions that dissociate readily.

**Example 33.18**

A 0.1 molar (0.1M) acetic acid solution is 1.34% ionized. Find the (a) hydrogen ion concentration, (b) acetate ion concentration, (c) un-ionized acid concentration, and (d) ionization constant.

*Solution*

(a) From Eq. 33.22, the hydrogen ion concentration is

$$[H_3O^+] = XM$$
$$= (0.0134)(0.1) = 0.00134 \text{ mol/L}$$

(b) Since every hydronium ion has a corresponding acetate ion, the acetate and hydronium ion concentrations are the same.

$$[C_2H_3O_2^-] = [H_3O^+] = 0.00134 \text{ mol/L}$$

(c) The concentration of un-ionized acid can be derived from Eq. 33.22.

$$[HC_2H_3O_2] = 1 - XM$$
$$= (1 - 0.0134)(0.1) = 0.09866 \text{ mol/L}$$

(d) The ionization constant is calculated from Eq. 33.39.

$$K_a = \frac{MX^2}{1-X}$$
$$= \frac{(0.1)(0.0134)^2}{1 - 0.0134} = 1.82 \times 10^{-5}$$

**Example 33.19**

The ionization constant for acetic acid is $1.82 \times 10^{-5}$. What is the hydrogen ion concentration for a 0.2M solution?

*Solution*

From Eq. 33.39,

$$K_a = \frac{MX^2}{1-X}$$
$$1.82 \times 10^{-5} = \frac{0.2X^2}{1-X}$$

Since acetic acid is a weak acid, $X$ is known to be small. Therefore, the computational effort can be reduced by assuming that $1 - X \approx 1$.

$$1.82 \times 10^{-5} = 0.2X^2$$
$$X = 9.49 \times 10^{-3}$$

From Eq. 33.22, the concentration of the hydrogen ion is

$$[H_3O^+] = XM = (9.49 \times 10^{-3})(0.2)$$
$$= 1.9 \times 10^{-3} \text{ mol/L}$$

Chemistry

**Example 33.20**

The ionization constant for acetic acid ($HC_2H_3O_2$) is $1.82 \times 10^{-5}$. What is the hydrogen ion concentration of a solution with 0.1 mol of 80% ionized ammonium acetate ($NH_4C_2H_3O_2$) in one liter of 0.1M acetic acid?

*Solution*

The acetate ion ($C_2H_3O_2^-$) is a common ion, since it is supplied by both the acetic acid and the ammonium acetate. Both sources contribute to the ionic concentration. However, the ammonium acetate's contribution dominates. Since the acid dissociates into an equal number of hydrogen and acetate ions,

$$
\begin{aligned}
[C_2H_3O_2^-]_{total} &= [C_2H_3O_2^-]_{acid} \\
&\quad + [C_2H_3O_2^-]_{ammonium\ acetate} \\
&= [H_3O^+] + (0.8)(0.1) \\
&\approx (0.8)(0.1) = 0.08
\end{aligned}
$$

As a result of the common ion effect and Le Châtelier's law, the acid's dissociation is essentially suppressed by the addition of the ammonium acetate. The concentration of un-ionized acid is

$$[HC_2H_3O_2] = 0.1 - [H_3O^+]$$
$$\approx 0.1$$

The ionization constant is unaffected by the number of sources of the acetate ion.

$$K_a = \frac{[H_3O^+][C_2H_3O_2^-]}{[HC_2H_3O_2]}$$

$$1.82 \times 10^{-5} = \frac{[H_3O^+](0.08)}{0.1}$$

$$[H_3O^+] = 2.3 \times 10^{-5}\ \text{mol/L}$$

## 59. IONIZATION CONSTANTS FOR POLYPROTIC ACIDS

A *polyprotic acid* has as many ionization constants as it has acidic hydrogen atoms. For oxyacids (see Sec. 33.7), each successive ionization constant is approximately $10^5$ times smaller than the preceding one. For example, phosphoric acid ($H_3PO_4$) has three ionization constants.

$$K_1 = 7.1 \times 10^{-3} \quad (H_3PO_4)$$
$$K_2 = 6.3 \times 10^{-8} \quad (H_2PO_4^-)$$
$$K_3 = 4.4 \times 10^{-13} \quad (HPO_4^{-2})$$

## 60. SOLUBILITY PRODUCT

When an ionic solid is dissolved in a solvent, it dissociates. For example, consider the ionization of silver chloride in water.

$$AgCl(s) \rightleftharpoons Ag^+(aq) + Cl^-(aq)$$

If the equilibrium constant is calculated, the terms for pure solids and liquids (in this case $[AgCl]$ and $[H_2O]$) are omitted. Thus, the *solubility product*, $K_{sp}$, consists only of the ionic concentrations. As with the general case of ionization constants, the solubility product for slightly soluble solutes is essentially constant at a standard value.

$$K_{sp} = [Ag^+][Cl^-] \qquad \textit{33.40}$$

When the product of terms exceeds the standard value of the solubility product, solute will precipitate out until the product of the remaining ion concentrations attains the standard value. If the product is less than the standard value, the solution is not saturated.

The solubility products of nonhydrolyzing compounds are relatively easy to calculate. (Example 33.20 illustrates a method.) Such is the case for chromates ($CrO_4^{-2}$), halides ($F^-, Cl^-, Br^-, I^-$), sulfates ($SO_4^{-2}$), and iodates ($IO_3^-$). However, compounds that hydrolyze (i.e., combine with water molecules) must be treated differently. The method used in Ex. 33.19 cannot be used for hydrolyzing compounds.

**Example 33.21**

At a particular temperature, it takes 0.038 grams of lead sulfate ($PbSO_4$, molecular weight = 303.25) per liter of water to prepare a saturated solution. What is the solubility product of lead sulfate if all of the lead sulfate ionizes?

*Solution*

Sulfates are not one of the hydrolyzing ions. Therefore, the solubility product can be calculated from the concentrations.

Since one liter of water has a mass of 1 kg, the number of moles of lead sulfate dissolved per saturated liter of solution is

$$n = \frac{m}{MW} = \frac{0.038\ \text{g}}{303.25\ \frac{\text{g}}{\text{mol}}}$$
$$= 1.25 \times 10^{-4}\ \text{mol}$$

Lead sulfate ionizes according to the following reaction.

$$PbSO_4(s) \rightleftharpoons Pb^{+2}(aq) + SO_4^-(aq) \quad [\text{in water}]$$

Since all of the lead sulfate ionizes, the number of moles of each ion is the same as the number of moles of lead sulfate. Therefore,

$$K_{sp} = [Pb^{+2}][SO_4^{-2}] = (1.25 \times 10^{-4})(1.25 \times 10^{-4})$$
$$= 1.56 \times 10^{-8}$$

## 61. ENTHALPY OF FORMATION

*Enthalpy, H,* is the potential energy that a substance possesses by virtue of its temperature, pressure, and phase.[32] The *enthalpy of formation (heat of formation),* $\Delta H_f$, of a compound is the energy absorbed during the formation of 1 gmol of the compound from the elements.[33] The enthalpy of formation is assigned a value of zero for elements in their free states at 25°C and 1 atm. This is the so-called *standard state* for enthalpies of formation.

Table 33.9 contains enthalpies of formation for some common elements and compounds. The enthalpy of formation depends on the temperature and phase of the compound. A standard temperature of 25°C is used in most tables of enthalpies of formation.[34] Compounds are solid (*s*) unless indicated to be gaseous (*g*) or liquid (*l*). Some aqueous (*aq*) values are also encountered.

## 62. ENTHALPY OF REACTION

The *enthalpy of reaction (heat of reaction),* $\Delta H_r$, is the energy absorbed during a chemical reaction under constant volume conditions. It is found by summing the enthalpies of formation of all products and subtracting the sum of enthalpies of formation of all reactants. This is essentially a restatement of the energy conservation principle and is known as *Hess' law of energy summation.*

$$\Delta H_r = \sum \Delta H_{f,\text{products}} - \sum \Delta H_{f,\text{reactants}} \qquad 33.41$$

Reactions that give off energy (i.e., have negative enthalpies of reaction) are known as *exothermic reactions.* Many (but not all) exothermic reactions begin spontaneously. On the other hand, *endothermic reactions* absorb energy and require heat or electrical energy to begin.

### Example 33.22

Using enthalpies of formation, calculate the heat of stoichiometric combustion (standardized to 25°C) of gaseous methane ($CH_4$) and oxygen.

---

[32]The older term *heat* is rarely encountered today.
[33]The symbol $H$ is used to denote molar enthalpies. The symbol $h$ is used for specific enthalpies (i.e., energy per kilogram or per pound).
[34]It is possible to correct the enthalpies of formation to account for other reaction temperatures.

**Table 33.9** *Standard Enthalpies of Formation (kcal/mol at 25°C)*

| element/compound | $\Delta H_f$ |
|---|---|
| Al (*s*) | 0.00 |
| $Al_2O_3$ (*s*) | −399.09 |
| C (graphite) | 0.00 |
| C (diamond) | 0.45 |
| C (*g*) | 171.70 |
| CO (*g*) | −26.42 |
| $CO_2$ (*g*) | −94.05 |
| $CH_4$ (*g*) | −17.90 |
| $C_2H_2$ (*g*) | 54.19 |
| $C_2H_4$ (*g*) | 12.50 |
| $C_2H_6$ (*g*) | −20.24 |
| $CCl_4$ (*g*) | −25.5 |
| $CHCl_4$ (*g*) | −24 |
| $CH_2Cl_2$ (*g*) | −21 |
| $CH_3Cl$ (*g*) | −19.6 |
| $CS_2$ (*g*) | 27.55 |
| COS (*g*) | −32.80 |
| $(CH_3)_2S$ (*g*) | −8.98 |
| $CH_3OH$ (*g*) | −48.08 |
| $C_2H_5OH$ (*g*) | −56.63 |
| $(CH_3)_2O$ (*g*) | −44.3 |
| $C_3H_6$ (*g*) | 9.0 |
| $C_6H_{12}$ (*g*) | −29.98 |
| $C_6H_{10}$ (*g*) | −1.39 |
| $C_6H_6$ (*g*) | 19.82 |
| Fe (*s*) | 0.00 |
| Fe (*g*) | 99.5 |
| $Fe_2O_3$ (*s*) | −196.8 |
| $Fe_3O_4$ (*s*) | −267.8 |
| $H_2$ (*g*) | 0.00 |
| $H_2O$ (*g*) | −57.80 |
| $H_2O$ (*l*) | −68.32 |
| $H_2O_2$ (*g*) | −31.83 |
| $H_2S$ (*g*) | −4.82 |
| $N_2$ (*g*) | 0.00 |
| NO (*g*) | 21.60 |
| $NO_2$ (*g*) | 8.09 |
| $NO_3$ (*g*) | 13 |
| $NH_3$ (*g*) | −11.04 |
| $O_2$ (*g*) | 0.00 |
| $O_3$ (*g*) | 34.0 |
| S (*g*) | 0.00 |
| $SO_2$ (*g*) | −70.96 |
| $SO_3$ (*g*) | −94.45 |

(Multiply kcal/mol by 4.184 to obtain kJ/mol.)
(Multiply kcal/mol by 1800/MW to obtain Btu/lbm.)

*Solution*

The balanced chemical equation for the stoichiometric combustion of methane is

$$CH_4 + 2O_2 \rightarrow 2H_2O + CO_2$$

The enthalpy of formation of oxygen gas (its free-state configuration) is zero. Using enthalpies of formation from Table 33.9 in Eq. 33.41, the enthalpy of reaction per mole of methane is

$$\Delta H_r = 2\Delta H_{f,\text{H}_2\text{O}} + \Delta H_{f,\text{CO}_2} - \Delta H_{f,\text{CH}_4} - 2\Delta H_{f,\text{O}_2}$$

$$= (2)\left(-57.80\ \frac{\text{kcal}}{\text{mol}}\right) + \left(-94.05\ \frac{\text{kcal}}{\text{mol}}\right)$$
$$- \left(-17.90\ \frac{\text{kcal}}{\text{mol}}\right) - (2)(0)$$

$$= -191.75\ \text{kcal/mol CH}_4 \quad [\text{exothermic}]$$

Using the footnote to Table 33.9, this value can be converted to Btu/lbm. The molecular weight of methane is

$$\text{MW}_{\text{CH}_4} = 12 + (4)(1) = 16$$

$$\text{higher heating value} = \frac{\left(191.75\ \frac{\text{kcal}}{\text{mol}}\right)\left(1800\ \frac{\text{Btu-mol}}{\text{lbm-kcal}}\right)}{16}$$
$$= 21{,}570\ \text{Btu/lbm}$$

## 63. ELECTROLYSIS

An *electrolyte* is a substance that dissociates in solution to produce positive and negative ions. It can be an aqueous solution of a soluble salt, or it can be an ionic substance in molten form.

*Electrolysis* is the passage of an electric current through an electrolyte caused by an external voltage source. Electrolysis occurs when the positive terminal (the *anode*) and negative terminal (the *cathode*) of a voltage source are placed in an electrolyte. (These definitions are the opposite of galvanic cell terminals.) Negative ions (anions) will be attracted to the anode, where they are oxidized. Positive ions (cations) will be attracted to the cathode, where they will be reduced. The passage of the ions constitutes the current.

Some reactions that do not proceed spontaneously can be forced to proceed by supplying electrical energy. Such reactions are called *electrolytic (electrochemical) reactions*.

The combination of an anode and cathode (i.e., the *electrodes*), an electrolyte, and a voltage source is known as an *electrolytic cell*. (See Fig. 33.8.) Electrical energy is converted into chemical energy in an electrolytic cell. The minimum voltage that must be applied to cause the reaction to proceed is the *cell potential*.

## 64. FARADAY'S LAWS OF ELECTROLYSIS

*Faraday's laws of electrolysis* can be used to predict the duration and magnitude of a direct current needed to complete an electrolytic reaction.

*Figure 33.8* An Electrolytic Cell

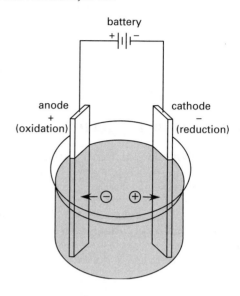

*law 1:* The mass of a substance generated by electrolysis is proportional to the amount of electricity used.

*law 2:* For any constant amount of electricity, the mass of substance generated is proportional to its equivalent weight.

*law 3:* One *faraday* of electricity (96 485 C or 96 485 A·s) will produce one gram equivalent weight.[35]

The mass in grams of a substance produced at an electrode in an electrolytic reaction can be found from Equation 33.42 and Eq. 33.43.

$$m_{\text{grams}} = \frac{It(\text{MW})}{(96\,485)(\text{change in oxidation state})} \qquad 33.42$$

$$m_{\text{grams}} = (\text{no. of faradays})(\text{GEW}) \qquad 33.43$$

The number of gram moles produced is

$$n = \frac{m}{\text{MW}} = \frac{\text{no. of faradays}}{\text{change in oxidation state}}$$
$$= \frac{It}{(96\,485)(\text{change in oxidation state})} \qquad 33.44$$

### Example 33.23

What current is required to produce two grams of metallic copper (atomic weight = 63.6 g/mol) from a copper sulfate solution in 1.5 hours?

---

[35]An electron has $1.6022 \times 10^{-19}$ C of charge. A *faraday* is the charge associated with one mole of electrons. Therefore, a faraday is

$$\text{faraday} = \left(1.6022 \times 10^{-19}\ \frac{\text{C}}{\text{electron}}\right)(6.022 \times 10^{23}\ \text{electrons})$$
$$= 96\,485\ \text{C}$$

*Solution*

The electrolysis reaction is

$$Cu^{+2} + 2e^- \rightarrow Cu$$

Since the change in charge on the copper is 2, the equivalent weight of copper is

$$EW_{Cu} = \dfrac{63.6 \; \dfrac{g}{mol}}{2} = 31.8 \; g/EW$$

From Eq. 33.42,

$$m_{grams} = \dfrac{It(MW)}{(96\,485)(\text{change in oxidation state})}$$

$$2 \; g = \dfrac{I(1.5 \; h)\left(3600 \; \dfrac{s}{h}\right)\left(31.8 \; \dfrac{g}{EW}\right)}{96\,485 \; \dfrac{A \cdot s}{EW}}$$

$$I = 1.12 \; A$$

## 65. GALVANIC CELLS

A *galvanic cell (voltaic cell)* is a device that produces electrical current by way of an oxidation-reduction reaction. That is, chemical energy is converted into electrical energy. Galvanic cells typically have the following characteristics.

- The oxidizing agent is separate from the reducing agent.

- Each agent has its own electrolyte and metallic electrode, and the combination is known as a *half-cell*.

- Each agent can be in solid (i.e., paste), liquid, or gaseous form, or can consist simply of the electrode.

- The ions can pass between the electrolytes of the two half-cells. The connection can be through a porous substance, salt bridge, another electrolyte, or other method.

Figure 33.9 illustrates how a *zinc-copper galvanic cell*, known as the *Daniell cell*, could be constructed. The complete reaction is

$$Zn + CuSO_4(aq) \rightarrow ZnSO_4 + Cu \qquad 33.45$$

The cell shown uses an aqueous electrolyte (e.g., KCl or $NH_4NO_3$) as a *salt bridge*. (To keep the salt bridge in its tube, the ends of the tube can be plugged with metallic caps. Alternatively, the electrolyte can be in gelatin form.)

In the zinc-copper cell, the left beaker contains a zinc electrode and a solution of zinc sulfate ($ZnSO_4$). The right beaker contains a copper electrode and a solution of copper sulfate ($CuSO_4$). The zinc electrode dissolves, and the $Zn^{+2}$ concentration increases. Electrons are

**Figure 33.9** *A Zinc-Copper Galvanic Cell*

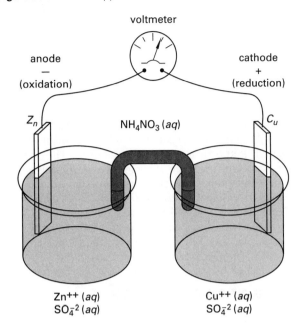

given up as the zinc electrode dissolves (i.e., the zinc electrode oxidizes).

$$Zn \rightarrow Zn^{+2} + 2e^-$$

Copper is deposited on the copper electrode, and the $Cu^{+2}$ concentration decreases. Electrons are received as copper is deposited (i.e., the copper ion is reduced).

$$Cu^{+2} + 2e^- \rightarrow Cu$$

The amount of current generated by a half-cell depends on the electrode material and the oxidation-reduction reaction taking place in the cell. (The temperature and chemical concentrations also affect the reaction.) The current-producing ability is known as the *oxidation potential, reduction potential,*[36] or *half-cell potential,* $\mathcal{E}^0$. *Standard oxidation potentials* have a zero reference voltage corresponding to the potential of a *standard hydrogen electrode*.

$$H_2 \leftrightarrow 2H^+ + 2e^- \; (0.000 \; V)$$

Since the copper ion will normally be depleted before the solid zinc electrode, the zinc-copper battery goes dead when the concentration of $Cu^{+2}$ diminishes.

In a galvanic cell, the anode is negative and the cathode is positive, the opposite of electrolytic cells, where the reaction must be forced.

---

[36]Reduction potentials are equal in magnitude but opposite in sign to oxidation potentials. It is a common practice to assign the positive sign for the reaction that proceeds spontaneously. This is the traditional approach in the United States, and one that seems logical for physical chemists. Such values are referred to as *standard oxidation potentials*. The SI convention is to report *standard reduction potentials*. The numerical values are the same, but the signs are reversed. The symbol $\nu^0$ is used in most European countries.

The voltage developed by a galvanic cell at its terminals, the *cell potential* or *electromotive force*, $\mathcal{E}$, is the difference of the larger and smaller half-cell potentials. For a zinc-copper cell, the cell potential is

$$\mathcal{E}_{cell} = \mathcal{E}_{Zn}^0 - \mathcal{E}_{Cu}^0$$
$$= 0.763 \text{ V} - (-0.337 \text{ V}) = 1.100 \text{ V} \qquad 33.46$$

In the zinc-copper cell, the zinc electrode gradually dissolves. An electrode that is gradually consumed or deposited is known as an *active electrode*. However, it is not necessary that one of the electrodes dissolve. It is only necessary that oxidation and reduction occur at the electrodes. An *inert electrode (sensing electrode)* is left unchanged by a half-cell reaction.[37] Platinum and carbon are the two most common inert electrode materials. For example, the $Fe^{+3}$ ferric ion can be reduced to the ferrous $Fe^{+2}$ ion at a platinum electrode.

## 66. THE NERNST EQUATION

For reaction $aA + bB \rightarrow cC + dD$, the Nernst equation calculates the standard (25°C) cell voltage as

$$\mathcal{E}_{cell} = \mathcal{E}^0 - \frac{0.059}{n} \log \frac{[C]^c [D]^d}{[A]^a [B]^b}$$
$$= \mathcal{E}^0 - \frac{0.059}{n} \log K \qquad 33.47$$

For the half-cell reaction, the Nernst equation determines the standard half-cell potential from the ionic concentration of the metal ion. In Eq. 33.48, $C$ is the molar ionic concentration (i.e., the molarity), $K$ is an equilibrium constant, and $n$ is the oxidation number change experienced by the ion (i.e., the number of electrons transferred in the reaction). The standard potentials listed in Table 33.10 were derived by using $C = 1$. For other than unit concentrations, the Nernst equation is given by Eq. 33.48. $n$ is the change in charge experienced by the ion.

$$\mathcal{E}_{cell,C} = \frac{0.059}{n} \log \frac{C}{K} = \mathcal{E}^0 + \frac{0.059}{n} \log C \qquad 33.48$$

The Nernst equation indicates that a cell can be constructed from two half-cells using the same material for both electrodes, as long as the ionic concentrations are different for both half-cells. Such a cell is known as a *concentration cell*.

Equation 33.48 can be combined with Eq. 33.47 to predict the terminal voltage of a cell. Alternatively, the ratio of concentrations (molarities) at which the battery runs down and dies can be determined by setting $\mathcal{E}_{cell} = 0$. Since this represents a state of equilibrium, this procedure also evaluates the equilibrium constant.

---

[37]Another type of inert electrode not covered in this book is a *gas electrode*.

**Table 33.10** *Representative Standard Half-Cell Oxidation Potentials at 25°C (anodic to cathodic)*

| half-reaction | $\mathcal{E}^0$ (volts) |
|---|---|
| $Li \rightarrow Li + e^-$ | +3.045 |
| $Na \rightarrow Na^+ + e^-$ | +2.714 |
| $Mg \rightarrow Mg^{++} + 2e^-$ | +2.37 |
| $Al \rightarrow Al^{+3} + 3e^-$ | +1.66 |
| $Mn \rightarrow Mn^{++} + 2e^-$ | +1.18 |
| $Zn \rightarrow Zn^{++} + 2e^-$ | +0.763 |
| $Cr \rightarrow Cr^{+3} + 3e^-$ | +0.74 |
| $Fe \rightarrow Fe^{++} + 2e^-$ | +0.440 |
| $Cd \rightarrow Cd^{++} + 2e^-$ | +0.403 |
| $Co \rightarrow Co^{++} + 2e^-$ | +0.277 |
| $Ni \rightarrow Ni^{++} + 2e^-$ | +0.250 |
| $Sn \rightarrow Sn^{++} + 2e^-$ | +0.136 |
| $Pb \rightarrow Pb^{++} + 2e^-$ | +0.126 |
| $H_2 \rightarrow 2H^+ + 2e^-$ | 0 (definition) |
| $Cu \rightarrow Cu^{++} + 2e^-$ | -0.337 |
| $2I^- \rightarrow I_2 + 2e^-$ | -0.536 |
| $H_2O_2 \rightarrow O_2 + 2H^+ + 2e^-$ | -0.682 |
| $Fe^{++} \rightarrow Fe^{+3} + e^-$ | -0.771 |
| $2Hg \rightarrow Hg_2^{++} + 2e^-$ | -0.789 |
| $Ag \rightarrow Ag^+ + e^-$ | -0.799 |
| $Hg_2^{++} \rightarrow 2Hg^{++} + 2e^-$ | -0.920 |
| $2Cl^- \rightarrow Cl_2 + 2e^-$ | -1.36 |
| $Cc^{+3} \rightarrow Ce^{+4} + e^-$ | -1.61 |
| $O_2 + H_2O \rightarrow O_3 + 2H^+ + 2e^-$ | -2.07 |
| $2F^- \rightarrow F_2 + 2e^-$ | -2.87 |

### Example 33.24

At what ionic concentration ratio does the zinc-copper cell become dead?

*Solution*

From Eq. 33.47 and Eq. 33.48,

$$\mathcal{E}_{cell} = \mathcal{E}_{Zn} - \mathcal{E}_{Cu}$$
$$= \left( \mathcal{E}_{Zn}^0 + \left( \frac{0.059}{n} \right) \log [Zn^{+2}] \right)$$
$$- \left( \mathcal{E}_{Cu}^0 + \left( \frac{0.059}{n} \right) \log [Cu^{+2}] \right)$$

When the battery is dead, $\mathcal{E}_{cell} = 0$. Also, the charge change, $n$, for both reactions is 2.

$$0 = \left( -0.763 + \left( \frac{0.059}{2} \right) \log [Zn^{+2}] \right)$$
$$- \left( 0.337 + \left( \frac{0.059}{2} \right) \log [Cu^{+2}] \right)$$

$$1.100 = 0.0295 \log \frac{[Zn^{+2}]}{[Cu^{+2}]}$$

$$37.3 = \log \frac{[Zn^{+2}]}{[Cu^{+2}]}$$

$$\frac{[Zn^{+2}]}{[Cu^{+2}]} = 2 \times 10^{37} = K$$

Chemistry

## 67. BATTERIES

A common type of battery is the *lead-acid battery* known as a *storage cell*, typically used in automobiles. This battery is recharged when the normal flow of electrons is reversed. The lead-acid battery consists of a negative electrode of lead-alloy and a positive electrode of lead dioxide, both immersed in sulfuric acid. Depending on the concentration of the acid, the cell voltage varies from 1.8 V to 2.3 V. Multiple cells are connected in series to achieve the voltage (i.e., 6 V, 12 V, or 24 V) required by the automobile.

The two half-cell reactions are

$$Pb \rightarrow Pb^{+2} + 2e^-$$

$$PbO_2 + H_2SO_4 \rightarrow H_2O + PbO$$

Reaction with the acid produces lead sulfate at both plates.

$$Pb^{+2} + H_2SO_4 \rightarrow PbSO_4 + 2H^+$$

Figure 33.10 illustrates the construction of a common flashlight "dry cell." This battery consists of a carbon rod cathode and a zinc anode separated by a moist paste of $MnO_2$ and $NH_4Cl$ (plus carbon and water). The anode reaction is

$$Zn(s) \rightarrow Zn^{+2}(aq) + 2e^-$$

The cathode reaction is

$$MnO_2(s) + 4NH_4^+(aq) + e^-$$
$$\rightarrow Mn^{+3}(aq) + 4NH_3(aq) + 2H_2O$$

**Figure 33.10** *A Common Dry Cell*

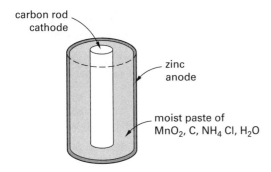

Nickel-cadmium batteries are rechargeable batteries for portable electronic devices. During the recharge process, hydrated CdO becomes pure Cd, and NiO is oxidized to NiOOH.

*Lithium-ion ("Li-ion") batteries* are popularly used in consumer electronics because of their high energy/weight ratio, long charge retention, and absence of a charging memory effect. In these batteries, a lithium ion moves from the anode to the cathode during

discharge and from the cathode to the anode when charging.

## 68. CELL POTENTIAL AND EQUILIBRIUM CONSTANT

The Nernst equation can be solved for the equilibrium constant.

$$K = \frac{10^{n\mathcal{E}^0}}{0.059} \qquad \textit{33.49}$$

Equation 33.49 shows that the equilibrium constant is related to the standard oxidation potential. A large value of the equilibrium constant means that there is a strong tendency for the reaction to proceed. Since a large $K$ is produced when $\mathcal{E}^0$ is positive, these two criteria are equivalent. Conversely, a small value of $K$ and a negative $\mathcal{E}^0$ imply that the reaction will not occur spontaneously.[38]

## 69. CORROSION

*Corrosion* is an undesirable degradation of a material resulting from a chemical or physical reaction with the environment. Conditions within the crystalline structure can accentuate or retard corrosion. The main types of corrosion are listed in subsequent sections. Corrosion rates are reported in units of mils per year (mpy) and micrometers per year ($\mu$m/y).

## 70. UNIFORM ATTACK CORROSION

Uniform rusting of steel and oxidation of aluminum over entire exposed surfaces are examples of *uniform attack corrosion*. Uniform attack is usually prevented by the use of paint, plating, and other protective coatings.

## 71. INTERGRANULAR CORROSION

Some metals are particularly sensitive to *intergranular corrosion*, IGC—selective or localized attack at metal-grain boundaries. For example, the $Cr_2O_3$ oxide film on stainless steel contains numerous imperfections at grain boundaries, and these boundaries can be attacked and enlarged by chlorides.

Intergranular corrosion may occur after a metal has been heated, in which case it may be known as *weld decay*. In the case of type 304 austenitic stainless steels, heating to 930°F to 1300°F (500°C to 700°C) in a welding process causes chromium carbides to precipitate out, reducing the corrosion resistance.[39] Reheating to 1830°F

---

[38]This is only partially true. The reaction might occur, but the products formed will be in concentrations less than the standard values.

[39]*Austenitic stainless steels* are the 300 series. They consist of chromium nickel alloys with up to 8% nickel. They are not hardenable by heat treatment, are nonmagnetic, and offer the greatest resistance to corrosion. *Martensitic stainless steels* are hardenable and magnetic. *Ferritic stainless steels* are magnetic and not hardenable.

to 2010°F (1000°C to 1100°C) followed by rapid cooling will redissolve the chromium carbides and restore corrosion resistance.

## 72. PITTING

*Pitting* is a localized perforation on the surface. It can occur even where there is little or no other visible damage. Chlorides and other halogens (e.g., HF and HCl) in the presence of water foster pitting in passive alloys, especially in stainless steels and aluminum alloys.

## 73. CONCENTRATION-CELL CORROSION

*Concentration-cell corrosion* (also known as *crevice corrosion* and *intergranular attack*, IGA) occurs when a metal is in contact with different electrolyte concentrations. It usually occurs in crevices, between two assembled parts, under riveted joints, or where there are scale and surface deposits that create stagnant areas in a corrosive medium.

## 74. SELECTIVE LEACHING

*Selective leaching* is the dealloying process in which one of the alloy ingredients is removed from the solid solution. This occurs because the lost ingredient has a lower corrosion resistance than the remaining ingredient. *Dezincification* is the classic case where zinc is selectively destroyed in brass. Other examples are the dealloying of nickel from copper-nickel alloys, iron from steel, and aluminum from copper-aluminum alloys.

## 75. HYDROGEN EMBRITTLEMENT

*Hydrogen damage* occurs when hydrogen gas diffuses through and decarburizes steel (i.e., reacts with carbon to form methane). *Hydrogen embrittlement* (also known as *caustic embrittlement*) is hydrogen damage from hydrogen produced by caustic corrosion.

## 76. GALVANIC ACTION

*Galvanic action* (*galvanic corrosion* or *two-metal corrosion*) results from the difference in oxidation potentials of metallic ions. The greater the difference in oxidation potentials, the greater will be the galvanic corrosion. If two metals with different oxidation potentials are placed in an electrolytic medium (e.g., seawater), a galvanic cell will be created. The metal with the higher potential (i.e., the more "active" metal) will act as an anode and will corrode. The metal with the lower potential (the more "noble" metal), being the cathode, will be unchanged. In one extreme type of intergranular corrosion known as *exfoliation*, open endgrains separate into layers.

Metals are often classified according to their positions in the *galvanic series* listed in Table 33.11. As would be

*Table 33.11* Galvanic Series in Seawater (top to bottom anodic (sacrificial, active) to cathodic (noble, passive))

magnesium
zinc
Alclad 3S
cadmium
2024 aluminum alloy
low-carbon steel
cast iron
stainless steels (active)
    no. 410
    no. 430
    no. 404
    no. 316
Hastelloy A
lead
lead-tin alloys
tin
nickel
brass (copper-zinc)
copper
bronze (copper-tin)
90/10 copper-nickel
70/30 copper-nickel
Inconel
silver solder
silver
stainless steels (passive)
Monel metal
Hastelloy C
titanium
graphite
gold

expected, the metals in this series are in approximately the same order as their half-cell potentials. However, alloys and proprietary metals are also included in the series.

Precautionary measures can be taken to inhibit or eliminate galvanic action when use of dissimilar metals is unavoidable.

- Use dissimilar metals that are close neighbors in the galvanic series.

- Use *sacrificial anodes*. In marine saltwater applications, sacrificial zinc plates can be used.

- Use protective coatings, oxides, platings, or inert spacers to reduce or eliminate the access of corrosive environments to the metals.[40]

## 77. STRESS CORROSION

When subjected to sustained surface tensile stresses (including low residual stresses from manufacturing) in corrosive environments, certain metals exhibit

---

[40]While cadmium, nickel, chromium, and zinc are often used as protective deposits on steel, porosities in the surfaces can act as small galvanic cells, resulting in invisible subsurface corrosion.

catastrophic *stress corrosion* cracking, SCC. When the stresses are cyclic, this type of corrosion is called *corrosion fatigue*, which can lead to fatigue failures well below normal yield stresses.

Stress corrosion occurs because the more highly stressed grains (at the crack tip) are slightly more anodic than neighboring grains with lower stresses. Although intergranular cracking (at grain boundaries) is more common, corrosion cracking may be *intergranular* (between the grains), *transgranular* (through the grains), or a combination of the two, depending on the alloy. Cracks propagate, often with extensive branching, until failure occurs.

The precautionary measures that can be taken to inhibit or eliminate stress corrosion are as follows.

- Avoid using metals that are susceptible to stress corrosion. These include austenitic stainless steels without heat treatment in seawater; certain tempers of the aluminum alloys 2124, 2219, 7049, and 7075 in seawater; and copper alloys exposed to ammonia.

- Protect open-grain surfaces from the environment. For example, press-fitted parts in drilled holes can be assembled with wet zinc chromate paste. Also, weldable aluminum can be "buttered" with pure aluminum rod.

- Stress-relieve by annealing heat treatment after welding or cold working.

## 78. EROSION CORROSION

*Erosion corrosion* is the deterioration of metals buffeted by the entrained solids in a corrosive medium.

## 79. FRETTING CORROSION

*Fretting corrosion* occurs when two highly loaded members have a common surface at which rubbing and sliding take place. The phenomenon is a combination of wear and chemical corrosion. Metals that depend on a film of surface oxide for protection, such as aluminum and stainless steel, are especially susceptible.

Fretting corrosion can be reduced by the following methods.

- Lubricate the rubbing surfaces.
- Seal the surfaces.
- Reduce vibration and movement.

## 80. CAVITATION CORROSION

*Cavitation* is the formation and collapse of minute bubbles of vapor in liquids. It is caused by a combination of reduced pressure and increased velocity in the fluid. In effect, very small amounts of the fluid vaporize (i.e.,

boil) and almost immediately condense. The repeated collapse of the bubbles hammers and work-hardens the surface.

When the surface work-hardens, it becomes brittle. Small amounts of the surface flake away, and the surface becomes pitted. This is known as *cavitation corrosion*. Eventually, the entire piece may work-harden and become brittle, leading to structural failure.

## 81. WATER SUPPLY CHEMISTRY

Most water supply composition data is not given in units of molarity, normality, molality, and so on. Rather, the most common measure of solution strength is the *$CaCO_3$ equivalent* measurement. With this method, substances are reported in milligrams per liter (mg/L, same as parts per million, ppm) "as $CaCO_3$," even when $CaCO_3$ is unrelated to the substance or reaction that produced the substance.

Actual gravimetric amounts of a substance can be converted to amounts as $CaCO_3$ by use of the conversion factors in App. 33.B. These factors are easily derived from stoichiometric principles.

The reason for converting all substance quantities to amounts as $CaCO_3$ is that equal $CaCO_3$ amounts constitute stoichiometric reaction quantities. For example, 100 mg/L as $CaCO_3$ of sodium ion ($Na^+$) will react with 100 mg/L as $CaCO_3$ of chloride ion ($Cl^-$) to produce 100 mg/L as $CaCO_3$ of salt (NaCl), even though the gravimetric quantities differ and $CaCO_3$ is not part of the reaction.

### Example 33.25

Lime is added to water to remove carbon dioxide gas.

$$CO_2 + Ca(OH)_2 \rightarrow CaCO_3\downarrow + H_2O$$

If water contains 5 mg/L of $CO_2$, how much lime is required for its removal?

*Solution*

From App. 33.B, the factor that converts $CO_2$ as substance to $CO_2$ as $CaCO_3$ is 2.27.

$$CO_2 \text{ as } CaCO_3 \text{ equivalent} = (2.27)\left(5 \; \frac{mg}{L}\right)$$
$$= 11.35 \text{ mg/L as } CaCO_3$$

Therefore, the $CaCO_3$ equivalent of lime required will also be 11.35 mg/L.

From App. 33.B again, the factor that converts lime as $CaCO_3$ to lime as substance is (1/1.35).

$$Ca(OH)_2 \text{ substance} = \frac{11.35 \; \frac{mg}{L}}{1.35}$$
$$= 8.41 \text{ mg/L as substance}$$

This problem could also have been solved stoichiometrically.

## 82. ACIDITY AND ALKALINITY IN WATER SUPPLIES

*Acidity* is a measure of acids in solutions. Acidity in surface water (e.g., lakes and streams) is caused by formation of *carbonic acid* ($H_2CO_3$) from carbon dioxide in the air.[41] Acidity in water is typically given in terms of the $CaCO_3$ equivalent that would neutralize the acid.

$$CO_2 + H_2O \rightarrow H_2CO_3 \qquad 33.50$$

$$H_2CO_3 + H_2O \rightarrow HCO_3^- + H_3O^+ \quad [pH > 4.5] \qquad 33.51$$

$$HCO_3^- + H_2O \rightarrow CO_3^{--} + H_3O^+ \quad [pH > 8.3] \qquad 33.52$$

*Alkalinity* is a measure of the amount of negative (basic) ions in the water. Specifically, $OH^-$, $CO_3^{--}$, and $HCO_3^-$ all contribute to alkalinity.[42] The measure of alkalinity is the sum of concentrations of each of the substances measured as $CaCO_3$.

Alkalinity and acidity of a titrated sample is determined from color changes in indicators added to the titrant.

### Example 33.26

Water from a city well is analyzed and is found to contain 20 mg/L as substance of $HCO_3^-$ and 40 mg/L as substance of $CO_3^{--}$. What is the alkalinity of this water as $CaCO_3$?

*Solution*

From App. 33.B, the factors converting $HCO_3^-$ and $CO_3^{--}$ ions to $CaCO_3$ equivalents are 0.82 and 1.67, respectively.

$$\text{alkalinity} = (0.82)\left(20 \ \frac{mg}{L}\right) + (1.67)\left(40 \ \frac{mg}{L}\right)$$
$$= 83.2 \ \text{mg/L as } CaCO_3$$

## 83. WATER HARDNESS

Water hardness is caused by multivalent (doubly charged, triply charged, etc., but not singly charged) positive metallic ions such as calcium, magnesium, iron, and manganese. (Iron and manganese are not as common, however.) Hardness reacts with soap to reduce its

cleansing effectiveness and to form scum on the water surface and a ring around the bathtub.

Water containing bicarbonate ($HCO_3^-$) ions can be heated to precipitate a carbonate molecule.[43] This hardness is known as *temporary hardness* or *carbonate hardness*.[44]

$$Ca^{++} + 2HCO_3^- + \text{heat} \rightarrow CaCO_3\downarrow + CO_2 + H_2O$$
$$33.53$$

$$Mg^{++} + 2HCO_3^- + \text{heat} \rightarrow MgCO_3\downarrow + CO_2 + H_2O$$
$$33.54$$

Remaining hardness due to sulfates, chlorides, and nitrates is known as *permanent hardness* or *noncarbonate hardness* because it cannot be removed by heating. The amount of permanent hardness can be determined numerically by causing precipitation, drying, and then weighing the precipitate.

$$Ca^{++} + SO_4^{--} + Na_2CO_3 \rightarrow 2Na^+ + SO_4^{--} + CaCO_3\downarrow$$
$$33.55$$

$$Mg^{++} + 2Cl^- + 2NaOH \rightarrow 2Na^+ + 2Cl^- + Mg(OH)_2\downarrow$$
$$33.56$$

*Total hardness* is the sum of temporary and permanent hardnesses, both expressed in mg/L as $CaCO_3$.

## 84. COMPARISON OF ALKALINITY AND HARDNESS

Hardness measures the presence of positive, multivalent ions in the water supply. Alkalinity measures the presence of negative (basic) ions such as hydrates, carbonates, and bicarbonates. Since positive and negative ions coexist, an alkaline water can also be hard.

If certain assumptions are made, then it is possible to draw conclusions about the water composition from the hardness and alkalinity. For example, if the effects of $Fe^{+2}$ and $OH^-$ are neglected, the following rules apply. (All concentrations are measured as $CaCO_3$.)

- *hardness = alkalinity:* There is no noncarbonate hardness. There are no $SO_4^-$, $Cl^-$ or $NO_3^-$ ions present.

- *hardness > alkalinity:* Noncarbonate hardness is present.

- *hardness < alkalinity:* All hardness is carbonate hardness. The extra $HCO_3^-$ comes from other sources (e.g., $NaHCO_3$).

Titration with indicator solutions is used to determine the alkalinity. The *phenolphthalein alkalinity* (or

---

[41]Carbonic acid is very aggressive and must be neutralized to eliminate the cause of water pipe corrosion. If the pH of water is greater than 4.5, carbonic acid ionizes to form bicarbonate. (See Eq. 33.51.) If the pH is greater than 8.3, carbonate ions form that cause water hardness by combining with calcium. (See Eq. 33.52.)

[42]Other ions, such as $NO_3^-$, also contribute to alkalinity, but their presence is rare. If detected, they should be included in the calculation of alkalinity.

[43]Hard water forms scale when heated. This scale, if it forms in pipes, eventually restricts water flow. Even in small quantities, the scale insulates boiler tubes. Therefore, water used in steam-producing equipment must be essentially hardness-free.

[44]The hardness is known as *carbonate* hardness even though it is caused by *bicarbonate* ions, not carbonate ions.

"P reading" in mg/L as $CaCO_3$) measures hydrate alkalinity and half of the carbonate alkalinity. The *methyl orange alkalinity* (or "M reading" in mg/L as $CaCO_3$) measures the total alkalinity (including the phenolphthalein alkalinity). Table 33.12 can be used to interpret these tests.[45]

*Table 33.12* Interpretation of Alkalinity Tests

| case | hydrate as $CaCO_3$ | carbonate as $CaCO_3$ | bicarbonate as $CaCO_3$ |
|---|---|---|---|
| $P = 0$ | 0 | 0 | M |
| $0 < P < \dfrac{M}{2}$ | 0 | 2P | M − 2P |
| $P = \dfrac{M}{2}$ | 0 | 2P | 0 |
| $\dfrac{M}{2} < P < M$ | 2P − M | 2(M − P) | 0 |
| $P = M$ | M | 0 | 0 |

## 85. WATER SOFTENING WITH LIME

Water softening can be accomplished with lime and soda ash to precipitate calcium and magnesium ions from the solution. Lime treatment has the added benefits of disinfection, iron removal, and clarification. Practical limits of *precipitation softening* are 30 mg/L of $CaCO_3$ and 10 mg/L of $Mg(OH)_2$ (as $CaCO_3$) because of intrinsic solubilities. Water treated by this method usually leaves the softening apparatus with a hardness of between 50 and 80 mg/L as $CaCO_3$.

*Lime* (CaO) is available as granular *quicklime* (90% CaO, 10% MgO) or *hydrated lime* (68% CaO, 32% water). Both forms are *slaked* prior to use, which means that water is added to form a lime slurry in an exothermic reaction.

$$CaO + H_2O \rightarrow Ca(OH)_2 + heat \qquad 33.57$$

*Soda ash* is usually available as 98% pure sodium carbonate ($Na_2CO_3$).

In the *first stage treatment*, lime added to water reacts with free carbon dioxide to form calcium carbonate precipitate.[46]

$$CO_2 + Ca(OH)_2 \rightarrow CaCO_3\downarrow + H_2O \qquad 33.58$$

Next, the lime reacts with the calcium bicarbonate.

$$Ca(HCO_3)_2 + Ca(OH)_2 \rightarrow 2CaCO_3\downarrow + 2H_2O \qquad 33.59$$

[45]The titration may be affected by the presence of silica and phosphates, which also contribute to alkalinity. The effect is small, but the titration may not be a completely accurate measure of carbonates and bicarbonates.
[46]First stage softening removes the carbonate hardness. A second softening process is necessary if there is *non-carbonate hardness* resulting from sulfates and chlorides.

Magnesium hardness is next removed.

$$Mg(HCO_3)_2 + Ca(OH)_2 \rightarrow CaCO_3\downarrow + 2H_2O + MgCO_3 \qquad 33.60$$

To remove the soluble $MgCO_3$, the pH must be above 10.8. This is accomplished by adding an excess of CaO or $Ca(OH)_2$.

$$MgCO_3 + Ca(OH)_2 \rightarrow CaCO_3\downarrow + Mg(OH)_2\downarrow \qquad 33.61$$

## 86. WATER SOFTENING BY ION EXCHANGE

In the *ion exchange process* (also known as *zeolite process* or *base exchange method*), water is passed through a filter bed of exchange material. This exchange material is known as *zeolite*. Ions in the insoluble exchange material are displaced by ions in the water.

The processed water will have a zero hardness. However, if there is no need for water with zero hardness (as in municipal water supply systems), some water can be bypassed around the unit.

There are three types of ion exchange materials. *Greensand (glauconite)* is a natural substance that is mined and treated with manganese dioxide. *Siliceous-gel zeolite* is an artificial solid used in small volume deionizer columns. *Polystyrene resins* are also synthetic and dominate the softening field.

The earliest synthetic zeolites were gelular *ion exchange resins* using a three-dimensional copolymer (e.g., styrene-divinyl benzene). Porosity through the continuous-phase gel was near zero, and dry contact surface areas of 500 $ft^2/lbm$ (0.1 $m^2/g$) or less were common.

*Macroreticular synthetic resins* are discontinuous, three-dimensional copolymer beads in a rigid-sponge type formation. Each bead is made up of thousands of micro-spheres of the gel resin. Porosity is increased, and dry contact surface areas are approximately 270,000 to 320,000 $ft^2/lbm$ (55 to 65 $m^2/g$).

During operation, the calcium and magnesium ions are removed according to the following reaction in which R is the zeolite anion.

$$\begin{Bmatrix} Ca \\ Mg \end{Bmatrix} \begin{Bmatrix} (HCO_3)_2 \\ SO_4 \\ Cl_2 \end{Bmatrix} + Na_2R$$

$$\rightarrow Na_2 \begin{Bmatrix} (HCO_3)_2 \\ SO_4 \\ Cl_2 \end{Bmatrix} + \begin{Bmatrix} Ca \\ Mg \end{Bmatrix} R \qquad 33.62$$

The resulting sodium compounds are soluble.

Typical saturation capacities of synthetic resins are 1.0 to 1.5 meq/mL for anion exchange resins and 1.7 to 1.9 meq/mL for cation exchange resins. However, working capacities are more realistic measures. Working capacities are approximately 10 to 15 kilograins/ft$^3$ (23 to 35 kg/m$^3$) before regeneration.

Flow rates through the bed are typically 1 to 6 gpm/ft$^3$ (2 to 13 L/s·m$^3$) of resin volume. The flow rate in terms of gpm/ft$^2$ (L/s·m$^2$) across the exposed surface will depend on the geometry of the bed, but values of 3 to 15 gpm/ft$^2$ (2 to 10 L/s·m$^2$) are typical.[47]

### Example 33.27

A municipal plant processes water with a total initial hardness of 200 mg/L. The designed discharge hardness is 50 mg/L. If an ion exchange unit is used, what is the bypass factor?

*Solution*

The water passing through the ion exchange unit is reduced to zero hardness. If $x$ is the water fraction bypassed around the zeolite bed,

$$(1 - x)\left(0 \; \frac{\text{mg}}{\text{L}}\right) + x\left(200 \; \frac{\text{mg}}{\text{L}}\right) = 50 \; \text{mg/L}$$
$$x = 0.25$$

### 87. REGENERATION OF ION EXCHANGE RESINS

Ion exchange material has a finite capacity for ion removal. When the zeolite is saturated or has reached some other prescribed limit, it must be regenerated (rejuvenated).

Standard ion exchange units are regenerated when the alkalinity of their effluent increases to the *set point*. Most condensate polishing units that also collect crud are operated to a *pressure-drop endpoint*. The pressure drop through the ion exchange unit is primarily dependent on the amount of crud collected. When the pressure drop reaches a set point, the resin is regenerated.

Regeneration of synthetic ion exchange resins is accomplished by passing a *regenerating solution* over/through the resin. Although regeneration can occur in the ion exchange unit itself, external regeneration is becoming common. This involves removing the bed contents hydraulically, backwashing to separate the components (for mixed beds), regenerating the bed components separately, washing, then recombining and transferring the bed components back into service.

Common regeneration compounds are NaCl (for water hardness removal units), $H_2SO_4$ (for cation exchange resins), and NaOH (for anion exchange resins). The amount of regeneration solution depends on the resin's degree of saturation. A rule of thumb is to expect to use 6 to 10 lbm of regeneration compound per cubic foot of

resin (100 to 160 kg per cubic meter). Alternatively, dosage of the regeneration compound may be specified in terms of hardness removed (e.g., 0.4 lbm of salt per 1000 grains of hardness removed). These rates are applicable to deionization plants for boiler make-up water. For condensate polishing, saturation levels of 10 to 25 lbm/ft$^3$ (160 to 400 kg/m$^3$) are used.

### 88. OTHER WATER SUPPLY PROPERTIES

In addition to alkalinity and hardness, the quality of drinking water is also affected by the presence of other compounds and ions. Tests for iron, manganese, chlorides, phosphorus, nitrogen, and fluorides can be made. Other tests for color, suspended solids, *turbidity* (i.e., lack of visual clarity), viruses, and bacteria can be made.

### 89. TRIHALOMETHANES

Only four trihalomethane (THM) compounds are normally found in chlorinated waters.

| | |
|---|---|
| $CHCl_3$ | trichloromethane (chloroform) |
| $CHBrCl_2$ | bromodichloromethane |
| $CHBr_2Cl$ | dibromochloromethane |
| $CHBr_3$ | tribromomethane (bromoform) |

Trihalomethanes are regulated by the EPA under the National Primary Drinking Water Standards. Standards for total *haloacetic acids* (referred to as "HAA5" in consideration of the five compounds identified) have also been established.

### 90. DISINFECTION OF DRINKING WATER

Chlorine is added to the drinking water distribution system for disinfection. It can be added as a gas or a solid (i.e., from sodium and calcium hypochlorite).

Alternatives to chlorination (or subsequent dechlorination) have become popular since THMs have been traced to the *organic compounds* that react during the chlorination process. The alternatives include chlorine dioxide gas, ozone, and low-volume exotics (e.g., silver oxide and gamma and ultraviolet radiation). All are more expensive than chlorine gas. Bromine and iodine are also alternative disinfectants, but they do not eliminate the THM problem.

### 91. AVOIDANCE OF DISINFECTION BYPRODUCTS IN DRINKING WATER

Reduction of DBPs can best be achieved by avoiding their production in the first place. The best strategy dictates using source water with few or no organic precursors.

Often source water choices are limited, necessitating tailoring treatment processes to produce the desired result. This entails removing the precursors prior to the application of chlorine, applying chlorine at certain

---

[47]Much higher values, up to 15 to 20 gpm/ft$^3$ or 40 to 50 gpm/ft$^2$ (33 to 45 L/s·m$^2$ or 27 to 34 L/s·m$^2$), may occur in certain types of units and at certain times (e.g., start-up and leak conditions).

points in the treatment process that minimize production of DBPs, using disinfectants that do not produce significant DBPs, or a combination of these techniques.

Removal of precursors is achieved by preventing growth of vegetative material (algae, plankton, etc.) in the source water and by collecting source water at various depths to avoid concentrations of precursors. Oxidizers such as potassium permanganate and chlorine dioxide can often reduce the concentration of the precursors without forming the DBPs. Under some instances, application of powdered activated carbon or a pH-adjustment process can reduce the impact of chlorination.

Chlorine application should be delayed if possible until after the flocculation, coagulation, settling, and filtration processes have been completed. In this manner, turbidity and common precursors will be reduced. If it is necessary to chlorinate early to facilitate treatment processes, chlorination can be followed by dechlorination to reduce contact time. Granular activated carbon has been used to some extent to remove DBPs after they form, but the carbon needs frequent regeneration.

Alternative disinfectants include ozone, chloramines, chlorine dioxide, potassium permanganate, and ultraviolet radiation. Ozone and chloramines, singularly or together, are often used for control of THMs. However, ozone creates other DBPs, including aldehydes, hydrogen peroxide, carboxylic acids, ketones, and phenols. When ozone is used as the primary disinfectant, a secondary disinfectant such as chlorine or chloramine must be used to provide an active residual that can be measured within the distribution system.

## 92. CHARACTERISTICS OF BOILER FEEDWATER

Water that circulates continuously and is converted to steam in a boiler is known as *feedwater*. Water that has been added to replace actual losses is known as *make-up water*.

The purity of the feedwater returned to the boiler (after condensing) depends on the purity of the make-up water, since impurities continually build up. *Blowdown* is the intentional periodic release of some of the feedwater in order to remove chemicals whose concentrations have built up over time.

Water impurities cause *scaling* (which reduces fluid flow and heat transfer rates) and corrosion (which reduces strength). Deposits from calcium, magnesium, and silica compounds are particularly troublesome.[48] As feedwater impurity increases, deposits from copper, iron, and nickel oxides (corrosion products from pipe and equipment) become more problematic.

Water can also contain dissolved oxygen, nitrogen, and carbon dioxide. The nitrogen is inert and does not need to be considered. However, both the oxygen and carbon dioxide need to be removed. High-temperature dissolved oxygen readily attacks pipe and boiler metal. Most of the oxygen in make-up water is removed by heating the water. Although the solubility of oxygen in water decreases with temperature, water at high pressures can hold large amounts of oxygen. Hydrazine ($N_2H_4$) and sodium sulfite ($Na_2SO_3$) are used for *oxygen scavenging*.[49] Because sodium sulfite forms sodium sulfate at high pressures, only hydrazine should be used above 1500 to 1800 psi (10.3 to 12.4 MPa).

Carbon dioxide combines with water to form *carbonic acid*. In power plants, this is more likely to occur in condenser return lines than elsewhere. Acidity of the condensate is reduced by *neutralizing amines* such as morpholine, cyclohexylamine, ethanolamine (the tetra-sodium salt of ethylenadiamine tetra-acetic acid, also known as $Na_4EDTA$), and diethylaminoethanol.

Most types of corrosion are greatly reduced when the pH is within the range of 9 to 10. Below this range, corrosion and deposits of sulfates, carbonates, and silicates become a major problem. For this reason, boilers are often shut down when the pH drops below 8.

*Filming amines (polar amines)* such as octadecylamine do not neutralize acidity or raise pH. They form a non-wettable layer which protects surfaces from corrosive compounds. Filming and neutralizing amines are often used in conjunction, though they are injected at different locations in the system.

Other purity requirements depend on the boiler equipment, and in particular, the steam pressure. Also, specific locations within the system can tolerate different concentrations.[50] The following guidelines apply to boilers operating in the 900 to 2500 psig (6.2 to 17.3 MPa) range.[51]

- All water feedwater entering the boiler should be free from dissolved oxygen, carbon dioxide, suspended solids, and hardness.

- pH of water entering the boiler should be in the range of 8.5 to 9.0. pH of water in the boiler should be in the range of 10.8 to 11.5.

- Silica in boiler water should be limited to 5 ppm (as $CaCO_3$) at 900 psig (6.2 MPa), with a gradual reduction to 1 ppm (as $CaCO_3$) at 2500 psig (17.3 MPa).

---

[48]Silica is most troubling because silicate deposits cannot be removed by chemical means. They must be removed mechanically.

[49]8 ppm of sodium nitrate react with 1 ppm of oxygen. To achieve a complete reaction, an excess of 2 to 3 ppm of sodium nitrate is required. Hydrazine reacts on a one-to-one basis and is commonly available as a 35% solution. Therefore, approximately 3 ppm of solution is required per ppm of oxygen.
[50]For example, silica in the boiler feedwater may be limited to 1 to 5 ppm, but the silica content in steam should be limited to 0.01 to 0.03 ppm.
[51]Tolerable concentrations at cold start-up can be as much as 5 to 100 times higher.

## 93. PRODUCTION AND REGENERATION OF BOILER FEEDWATER

The processes used to treat raw water for use in power-generating plants depend on the incoming water quality. Depending on need, filters may be used to remove solids, softeners or exchange units may remove permanent (sulfate) hardness, and activated carbon may remove organics.

Bicarbonate hardness is converted to sludge when the water is boiled, is removed during blow-down, and does not need to be chemically removed. However, a strong *dispersant (sludge conditioner)* must be added to prevent scale formation.[52] Alternatively, the calcium salts can be converted to sludges of phosphate salts by adding phosphates. Prior to the 1970s, *caustic phosphate treatments* using sodium phosphate, SP, in the form of $Na_3PO_4$, $NaHPO_4$, and $NaH_2PO_4$, was the most common method used to treat low-pressure boiler feedwater with high solids. Sodium phosphate converts impurities and buffers pH.

However, magnesium phosphate is a very sticky sludge, and using phosphates may cause more problems than not using them. Also, with higher pressures and temperatures, *wasting* becomes problematic.[53] Therefore, many modern plants now use an *all-volatile treatment*, AVT, also known as a *zero-solids treatment*. Early AVTs used ammonia ($NH_3$) for pH control, but ammonia causes *denting* in systems that operate without blowdown.[54]

*Chelant AVTs* do not add any solid chemicals to the boiler. They work by keeping calcium and magnesium in solution. The compounds are removed through continuous blowdown.[55] Chelant treatments can be used in boilers up to approximately 1500 psi (10.3 MPa) but require high-purity (low-solids) feedwater.[56]

Subsequent advances in corrosion protection have been based on use of stainless steel or titanium-tubed condensers, condensate polishing, and even better AVT formulations. Supplementing ammonia-AVTs are volatile amines, chelants, and boric acid treatments. Boric acid effectively combats denting, intergranular attack, and intergranular stress corrosion cracking.

Water that is processed through an ion-exchange unit is known as *deionized water*. Deionized water is "hungry" for minerals and picks up contamination as it passes through the power plant. When the operating pressure is 1500 psig (10.3 MPa) or above, condensed steam cannot be reused without additional demineralization.[57] *Demineralization* is not generally needed for lower-pressure units, but some *condensate polishing* is still necessary.

In low-pressure plants, condensate polishing may consist of passing the water through a filter (of sand, anthracite, or pre-coat cellulose filters), a strong-acid cation unit, and a mixed-bed unit. The filter removes gross particles (referred to as *crud*), the strong-acid unit removes the dissolved iron, and the mixed-bed unit polishes the condensate.[58]

In modern high-pressure plants, the filtration step is often omitted, and crud removal occurs in the strong-acid unit. Furthermore, if the mixed-bed unit is made large enough or the cation component is increased, the strong-acid unit can also be omitted. A mixed-bed unit operating by itself is known as a *naked mixed-bed unit*. Demineralization of condensed steam is commonly accomplished with naked mixed beds.

## 94. MONITORING OF BOILER FEEDWATER QUALITY

Water chemistry is monitored through continuous *inline sampling* and periodic *grab sampling*. Water impurities are expressed in milligrams per liter (mg/L), parts per million (ppm), parts per billion (ppb), unit equivalents per million (epm), and milliequivalents per milliliter (meq/mL).[59,60] ppm and ppb can be "as $CaCO_3$" equivalents or "as substance." Electrical conductivity (a measure of total dissolved solids) is measured in microsiemens[61] ($\mu$S) or ppm.

Water quality is maintained by monitoring pH, electrical resistivity silica, and hardness (calcium, magnesium, and bicarbonates), the primary corrosion species (sulfates, chlorides, and sodium), dissolved gases (primarily oxygen), and the concentrations of buffering chemicals (e.g., phosphate, hydrazine, or AVT).

---

[52]Typical sludge conditioner dispersants are natural organics (lignins, tannins, and starches) and synthetics (sodium polyacrylate, sodium polymethacrylate, sulfonated polystyrene, and maleic anhydride).

[53]*Wasting* is a term used in the power-generating industry to describe the process of general pipe wall thinning.

[54]*Denting* is the constricting of the intersection between tubes and support plates in boilers due to a buildup of corrosion.

[55]AVTs work in a completely different way than phosphates. It is not necessary to use phosphates and AVTs simultaneously.

[56]If the feedwater becomes contaminated, a supplementary phosphate treatment will be required.

[57]The term *deionization* refers to the process that produces makeup water. *Demineralization* refers to the process used to prepare condensed steam for reuse.

[58]*Crud* is primarily iron-corrosion products ranging from dissolved to particulate matter.

[59]Milligrams per liter (mg/L) is essentially the same as parts per million (ppm). The older units of grains per gallon are still occasionally encountered. 1 grain equals l/7000th of a pound. Multiply grains per gallon (gpg) by 17.1 to get ppm. Unit equivalents per million are derived by dividing the concentration in ppm by the equivalent weight. Equivalent weight is the molecular weight divided by the valence or oxidation number.

[60]Use the following relationship to convert meq/mL to ppm.

$$ppm = (1000) \left( \frac{meq}{mL} \right) \text{(equivalent weight)}$$

$$= \frac{(1000) \left( \frac{meq}{mL} \right) \text{(formula weight)}}{valence}$$

[61]Microsiemens were previously known as "micromhos," where "ohms" is spelled backward to indicate its inverse.

Continuous monitoring of ionic concentrations is complicated by a process known as *hideout,* in which a chemical species (e.g., *crevice salts*) disappears by precipitation or adsorption during low-flow and high heat transfer. Upon cooling (as during a shut-down), the hideout chemicals reappear.

## 95. WASTEWATER QUALITY CHARACTERISTICS

The primary parameters used to classify wastewater quality and determine treatment steps are as follows.

- dissolved oxygen
- biochemical oxygen demand
- chemical oxygen demand
- chlorine demand

Aquatic life requires oxygen. The biological decomposition of organic solids is also dependent on oxygen. If the *dissolved oxygen* content of water is less than saturated, it is likely that the water is organically polluted.

Biological organisms remove oxygen from water when they oxidize organic waste matter. Therefore, the oxygen depletion is an indication of the organic waste content. The more oxygen that is removed over a given period, the greater will be the amount of organic food present in the water.

*Biochemical oxygen demand* (BOD) is determined by adding a measured amount of wastewater (which supplies the organic material) to a measured amount of dilution water (which reduces toxicity and supplies dissolved oxygen). An oxygen use curve similar to Fig. 33.11 will result. The standard BOD test typically calls for a five-day incubation period at $20°C$.

**Figure 33.11** *BOD Time Curve*

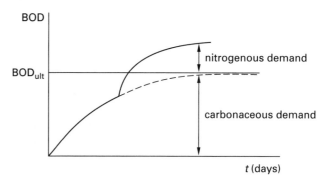

Unlike BOD, which is a measure of oxygen removed by biological organisms, *chemical oxygen demand* (COD) is a measure of the total oxidizable substances. COD is a good measure of effluent strength when chemical contamination is present.

Chlorination destroys bacteria, hydrogen sulfide, and other noxious substances by oxidation. For example, hydrogen sulfide is oxidized according to

$$H_2S + 4H_2O + 4Cl_2 \rightarrow H_2SO_4 + 8HCl \qquad 33.63$$

The *chlorine demand* is the amount of chlorine required to give a 2 mg/L residual after fifteen minutes of contact time. Fifteen minutes is the recommended contact and mixing time prior to discharge since this period will kill nearly all pathogenic bacteria in the water.

The chlorine dose applied to wastewater is frequently determined by the water's breakpoint. *Breakpoint chlorination* implies that chlorine is added to the water until free chlorine residuals begin to appear.

**Chemistry**

# 34  Organic Chemistry

## 1. INTRODUCTION TO ORGANIC CHEMISTRY

Organic chemistry deals with the formation and reaction of compounds of carbon, many of which are produced by living organisms. Organic compounds typically have one or more of the following characteristics.

- They are insoluble in water.[1]

- They are soluble in concentrated acids.

- They are relatively non-ionizing.

- They are unstable at high temperatures.

The method of naming organic compounds was standardized in 1930 at the International Union Chemistry meeting in Belgium. Names conforming to the established guidelines are known as *IUC names* or *IUPAC names*.[2]

## 2. FUNCTIONAL GROUPS

Certain combinations (groups) of atoms occur repeatedly in organic compounds and remain intact during reactions. Such combinations are called *functional groups* or *moieties*.[3] For example, the radical $OH^-$ is known as a *hydroxyl group*. Table 34.1 contains some common functional groups. In this table and others similar to it, the symbols R and R′ usually denote an attached hydrogen atom or other hydrocarbon chain of any length. They may also denote some other group of atoms.

---

[1]This is especially true for hydrocarbons. However, many organic compounds containing oxygen are water soluble. The sugar family is an example of water-soluble compounds.

[2]IUPAC stands for *International Union of Pure and Applied Chemistry*.

[3]Although "moiety" is often used synonymously with "functional group," there is a subtle difference. When a functional group combines into a compound, its moiety may gain/lose an atom from/to its combinant. Thus, moieties differ from their original functional groups by one or more atoms.

**Table 34.1** *Selected Functional Groups*

| name | standard symbol | formula | number of single bonding sites |
|---|---|---|---|
| aldehyde | | CHO | 1 |
| alkyl | [R] | $C_nH_{2n+1}$ | 1 |
| alkoxy | [RO] | $C_nH_{2n+1}O$ | 1 |
| amine (amino, n = 2) | | $NH_n$ | $3-n$ $[n = 0,1,2]$ |
| aryl (benzene ring) | [Ar] | $C_6H_5$ | 1 |
| carbinol | | COH | 3 |
| carbonyl (keto) | [CO] | CO | 2 |
| carboxyl | | COOH | 1 |
| ester | | COO | 1 |
| ether | | O | 2 |
| halogen (halide) | [X] | Cl, Br, I, or F | 1 |
| hydroxyl | | OH | 1 |
| nitrile | | CN | 1 |
| nitro | | $NO_2$ | 1 |

## 3. FAMILIES OF ORGANIC COMPOUNDS

For convenience, organic compounds are categorized into families, or *chemical classes*. "R" is an abbreviation for the word "radical," though it is unrelated to ionic radicals. Compounds within each family have similar structures, being based on similar combinations of groups. For example, all compounds in the alcohol family have the structure [R]-OH, where [R] is any alkyl group and -OH is the hydroxyl group.

Families of compounds can be further subdivided into subfamilies. For example, the hydrocarbons are classified into alkanes (single carbon-carbon bond), alkenes (double carbon-carbon bond), and alkynes (triple carbon-carbon bond).[4]

Table 34.2 contains some common organic families, and Table 34.3 gives synthesis routes for various classes of organic compounds.

## 4. SYMBOLIC REPRESENTATION

The nature and structure of organic groups and families cannot be explained fully without showing the types of bonds between the elements. Figure 34.1 illustrates the symbolic representation of some of the functional groups

---

[4]Hydrocarbons with two double carbon-carbon bonds are known as *dienes*.

Chemistry

**Table 34.2** *Families (Chemical Classes) of Organic Compounds*

| family (chemical classes) | structure[a] | example |
|---|---|---|
| organic acids | | |
|    carboxylic acids | [R]-COOH | acetic acid $((CH_3)COOH)$ |
|    fatty acids | [Ar]-COOH | benzoic acid $(C_6H_5COOH)$ |
| alcohols | | |
|    aliphatic | [R]-OH | methanol $(CH_3OH)$ |
|    aromatic | [Ar]-[R]-OH | benzyl alcohol $(C_6H_5CH_2OH)$ |
| aldehydes | [R]-CHO | formaldehyde $(HCHO)$ |
| alkyl halides | | |
|    (haloalkanes) | [R]-[X] | chloromethane $(CH_3Cl)$ |
| amides | $[R]\text{-}CO\text{-}NH_n$ | $\beta$-methylbutyramide $(C_4H_9CONH_2)$ |
| amines | $[R]_{3-n}\text{-}NH_n$ | methylamine $(CH_3NH_2)$ |
| | $[Ar]_{3-n}\text{-}NH_n$ | aniline $(C_6H_5NH_2)$ |
|      primary amines | $n = 2$ | |
|      secondary amines | $n = 1$ | |
|      tertiary amines | $n = 0$ | |
| amino acids | $CH\text{-}[R]\text{-}(NH_2)COOH$ | glycine $(CH_2(NH_2)COOH)$ |
| anhydrides | [R]-CO-O-CO-[R'] | acetic anhydride $(CH_3CO)_2O$ |
| arene (aromatics) | $ArH = C_nH_{2n-6}$ | benzene $(C_6H_6)$ |
| aryl halides | [AR]-[X] | fluorobenzene $(C_6H_5F)$ |
| carbohydrates | $C_x(H_2O)_y$ | dextrose $(C_6H_{12}O_6)$ |
|    sugars | | |
|    polysaccharides | | |
| esters | [R]-COO-[R'] | methyl acetate $(CH_3COOCH_3)$ |
| ethers | [R]-O-[R] | diethyl ether $(C_2H_5OC_2H_5)$ |
| | [Ar]-O-[R] | methyl phenyl ether $(CH_3OC_6H_5)$ |
| | [Ar]-O-[Ar] | diphenyl ether $(C_6H_5OC_6H_5)$ |
| glycols | $C_nH_{2n}(OH)_2$ | ethylene glycol $(C_2H_4(OH)_2)$ |
| hydrocarbons | | |
|    alkanes (single bonds)[b] | $RH = C_nH_{2n+2}$ | octane $(C_8H_{18})$ |
|      saturated hydrocarbons | | |
|      cycloalkanes (cycloparaffins) | | |
| | $C_nH_{2n}$ | cyclohexane $(C_6H_{12})$ |
|    alkenes (double bonds between two | $C_nH_{2n}$ | ethylene $(C_2H_4)$ |
|     carbons)[c] | | |
|      unsaturated hydrocarbons | | |
|      cycloalkenes | $C_nH_{2n-2}$ | cyclohexene $(C_6H_{10})$ |
|    alkynes (triple bonds between two | $C_nH_{2n-2}$ | acetylene $(C_2H_2)$ |
|     carbons) | | |
|      unsaturated hydrocarbons | | |
| ketones | [R]-[CO]-[R] | acetone $((CH_3)_2CO)$ |
| nitriles | [R]-CN | acetonitrile $(CH_3CN)$ |
| phenols | [Ar]-OH | phenol $(C_6H_5OH)$ |

[a]See Table 34.1 for definitions of [R], [Ar], [X], and [CO].
[b]Alkanes are also known as the *paraffin series* and *methane series*.
[c]Alkenes are also known as the *olefin series*.

and families, and Fig. 34.2 illustrates reactions between organic compounds.

## 5. FORMATION OF ORGANIC COMPOUNDS

There are usually many ways of producing an organic compound. The types of reactions contained in this section deal only with the interactions between the organic families. The following processes are referred to.

- *oxidation*: replacement of a hydrogen atom with a hydroxyl group

- *reduction*: replacement of a hydroxyl group with a hydrogen atom

- *hydrolysis*: addition of one or more water molecules

- *dehydration*: removal of one or more water molecules

**Figure 34.1** *Representation of Functional Groups and Families*

| group | group representation | family | family representation |
|---|---|---|---|
| aldehyde | — CH or — C(H)=O | aldehyde | R — C(H)=O |
| amino | — NH₂ | amine | R — NH₂ |
| aryl (benzene ring) | ⬡ | aryl halide | ⬡—X |
| carbonyl (keto) | — C — or — C(R)=O | ketone | R — C(R)=O |
| carboxyl | — C — OH or — C(O–H)=O | organic acid | R — C(O–H)=O |
| ester | — C — OR or — C(O–R)=O | ester | R — C(O–R)=O |
| hydroxyl | — OH | alcohol | R — OH |

**Figure 34.2** *Reactions Between Organic Compounds*

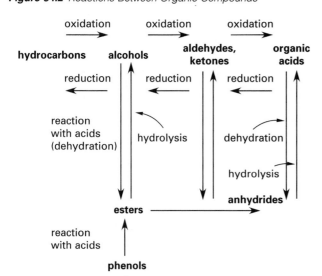

**Table 34.3** *Synthesis Routes for Various Classes of Organic Compounds*

*organic acids*
  oxidation of primary alcohols
  oxidation of ketones
  oxidation of aldehydes
  hydrolysis of esters
*alcohols*
  oxidation of hydrocarbons
  reduction of aldehydes
  reduction of organic acids
  hydrolysis of esters
  hydrolysis of alkyl halides
  hydrolysis of alkenes (aromatic hydrocarbons)
*aldehydes*
  oxidation of primary and tertiary alcohols
  oxidation of esters
  reduction of organic acids
*amides*
  replacement of hydroxyl group in an acid with an amino group
*anhydrides*
  dehydration of organic acids (withdrawal of one water molecule from two acid molecules)
*carbohydrates*
  oxidation of alcohols
*esters*
  reaction of acids with alcohols (*ester alcohols*)*
  reaction of acids with phenols (*ester phenols*)
  dehydration of alcohols
  dehydration of organic acids
*ethers*
  dehydration of alcohol
*hydrocarbons*
  alkanes:  reduction of alcohols and organic acids
       hydrogenation of alkenes
  alkenes:  dehydration of alcohols
       dehydrogenation of alkanes
*ketones*
  oxidation of secondary and tertiary alcohols
  reduction of organic acids
*phenols*
  hydrolysis of aryl halides

*The reaction of an organic acid with an alcohol is called *esterification*.

## 6. AMINO ACIDS AND PROTEINS

*Amino acids* are difunctional organic compounds. *Peptides* are amides formed from amino and carboxyl groups. Two amino acids produce a *dipeptide*; three acids produce a *tripeptide*. A *polypeptide* is produced from numerous linked amino acids. *Proteins* are polypeptides of great size and molecular weight (in excess of 10,000).

# 35 Fuels and Combustion

## Nomenclature

| | | | |
|---|---|---|---|
| $A$ | area | ft$^2$ | m$^2$ |
| $B$ | volumetric fraction | – | – |
| $c$ | specific heat | Btu/lbm-°F | kJ/kg·°C |
| $d$ | diameter | in | cm |
| $D$ | diameter | ft | m |
| $D$ | draft | in wg | kPa |
| $g$ | acceleration of gravity | ft/sec$^2$ | m/s$^2$ |
| $g_c$ | gravitational constant | ft-lbm/lbf-sec$^2$ | n.a. |
| $G$ | gravimetric fraction | – | – |
| $h$ | enthalpy | Btu/lbm | kJ/kg |
| $h$ | head | ft | m |
| $H$ | height | ft | m |
| HHV | higher heating value | Btu/lbm | kJ/kg |
| HV | heating value | Btu/lbm | kJ/kg |
| $J$ | gravimetric air-fuel ratio | – | – |
| $K$ | volumetric air-fuel ratio | – | – |
| $L$ | length | ft | m |
| LHV | lower heating value | Btu/lbm | kJ/kg |
| $m$ | mass | lbm | kg |
| $M$ | moisture fraction | – | – |
| ON | octane number | – | – |
| $p$ | pressure | lbf/ft$^2$ | Pa |
| $P$ | power | Btu/sec | kW |
| PN | performance number | – | – |
| $q$ | heat loss | Btu/lbm | kJ/kg |
| $Q$ | flow rate | ft$^3$/sec | m$^3$/s |
| $R$ | ratio | lbm/lbm | kg/kg |
| $R$ | specific gas constant | ft-lbf/lbm-°R | kJ/kg·K |
| $T$ | temperature | °F | °C |
| v | velocity | ft/sec | m/s |

## Symbols

| | | | |
|---|---|---|---|
| $\gamma$ | specific weight | lbf/ft$^3$ | n.a. |
| $\eta$ | efficiency | – | – |
| $\omega$ | humidity ratio | – | – |

## Subscripts

| | |
|---|---|
| $a/f$ | air/fuel |
| $fg$ | vaporization |
| $g$ | gas |
| $i$ | initial |
| $p$ | constant pressure |
| SE | stack effect |

## 1. HYDROCARBONS

With the exception of sulfur and related compounds, most fuels are hydrocarbons. Hydrocarbons are further categorized into subfamilies such as *alkynes* ($C_nH_{2n-2}$, such as acetylene $C_2H_2$), *alkenes* ($C_nH_{2n}$, such as ethylene $C_2H_4$), and *alkanes* ($C_nH_{2n+2}$, such as octane $C_8H_{18}$). The alkynes and alkenes are referred to as *unsaturated hydrocarbons*, while the alkanes are referred to as *saturated hydrocarbons*. The alkanes are also known as the *paraffin series* and *methane series*. The alkenes are subdivided into the chain-structured *olefin*

Chemistry

**Table 35.1** Approximate Specific Heats (at Constant Pressure) of Gases ($c_p$ in Btu/lbm-°R; at 1 atm)

| gas | temperature (°R) | | | | | | | |
|---|---|---|---|---|---|---|---|---|
| | 500 | 1000 | 1500 | 2000 | 2500 | 3000 | 4000 | 5000 |
| air | 0.240 | 0.249 | 0.264 | 0.277 | 0.286 | 0.294 | 0.302 | – |
| carbon dioxide | 0.196 | 0.251 | 0.282 | 0.302 | 0.314 | 0.322 | 0.332 | 0.339 |
| carbon monoxide | 0.248 | 0.257 | 0.274 | 0.288 | 0.298 | 0.304 | 0.312 | 0.316 |
| hydrogen | 3.39 | 3.47 | 3.52 | 3.63 | 3.77 | 3.91 | 4.14 | 4.30 |
| nitrogen | 0.248 | 0.255 | 0.270 | 0.284 | 0.294 | 0.301 | 0.310 | 0.315 |
| oxygen | 0.218 | 0.236 | 0.253 | 0.264 | 0.271 | 0.276 | 0.286 | 0.294 |
| sulfur dioxide | 0.15 | 0.16 | 0.18 | 0.19 | 0.20 | 0.21 | 0.23 | – |
| water vapor | 0.444 | 0.475 | 0.519 | 0.566 | 0.609 | 0.645 | 0.696 | 0.729 |

(Multiply Btu/lbm-°R by 4.187 to obtain kJ/kg·K.)

*series* and the ring-structured *naphthalene series*. *Aromatic hydrocarbons* ($C_nH_{2n-6}$, such as benzene $C_6H_6$) constitute another subfamily. Names for common hydrocarbon compounds are listed in App. 35.A.

## 2. CRACKING OF HYDROCARBONS

*Cracking* is the process of splitting hydrocarbon molecules into smaller molecules. For example, alkane molecules crack into a smaller member of the alkane subfamily and a member of the alkene subfamily. Cracking is used to obtain lighter hydrocarbons (such as those in gasoline) from heavy hydrocarbons (e.g., crude oil).

Cracking can proceed under the influence of high temperatures (*thermal cracking*) or catalysts (*catalytic cracking* or "cat cracking"). Since (from Le Châtelier's principle) cracking at high pressure favors recombination, catalytic cracking is performed at pressures near atmospheric. Catalytic cracking produces gasolines with better antiknock properties than does thermal cracking.

## 3. FUEL ANALYSIS

Fuel analyses are reported as either percentages by weight (for liquid and solid fuels) or percentages by volume (for gaseous fuels). Percentages by weight are known as *gravimetric analyses*, while percentages by volume are known as *volumetric analyses*. An *ultimate analysis* is a type of gravimetric analysis in which the constituents are reported by atomic species rather than by compound. In an ultimate analysis, combined hydrogen from moisture in the fuel is added to hydrogen from the combustive compounds. (See Sec. 35.6.)

A *proximate analysis* (not "approximate") gives the gravimetric fraction of moisture, volatile matter, fixed carbon, and ash. Sulfur may be combined with the ash or may be specified separately.

A *combustible analysis* considers only the combustible components, disregarding moisture and ash.

Gas analyses are typically specified as volumetric fractions. For a gas in a mixture, its *volumetric fraction* (i.e., its *volumetric percentage*) is the same as its *mole fraction* and *partial pressure fraction*.

A volumetric fraction can be converted to a gravimetric fraction by multiplying by the molecular weight and then dividing by the sum of the products of all the volumetric fractions and molecular weights.

## 4. WEIGHTING OF THERMODYNAMIC PROPERTIES

Many gaseous fuels (and all gaseous combustion products) are mixtures of different compounds. Some thermodynamic properties of mixtures are gravimetrically weighted, while others are volumetrically weighted. Specific heat, specific gas constant, enthalpy, internal energy, and entropy are gravimetrically weighted. For gases, molecular weight, density, and all molar properties are volumetrically weighted.[1]

When a compound experiences a large temperature change, the thermodynamic properties should be evaluated at the average temperature. Table 35.1 can be used to find the specific heat of gases at various temperatures.

## 5. STANDARD CONDITIONS

Though "standard conditions" usually means 70°F (21°C) and 1 atm pressure, *standard temperature and pressure*, STP, for manufactured fuel gases is 60°F (16°C) and 1 atm pressure. Since this convention is not well standardized, the actual temperature and pressure should be stated.[2]

Some combustion equipment (e.g., particulate collectors) operates within a narrow range of temperatures and pressures. These conditions are referred to as *normal temperature and pressure*, NTP.

---

[1]For gases, molar properties include molar specific heats, enthalpy per mole, and internal energy per mole.

[2]Both of these are different from the standard temperature and pressure used for scientific work. The standard temperature in that case is 32°F (0°C).

Chemistry

## 6. MOISTURE

If an ultimate analysis of a solid or liquid fuel is given, all of the oxygen is assumed to be in the form of free water.[3] The amount of hydrogen combined as free water is assumed to be one-eighth of the oxygen weight.[4] All remaining hydrogen, known as the *available hydrogen*, is assumed to be combustible.

$$G_{H,combined} = \frac{G_O}{8} \qquad \textit{35.1}$$

$$G_{H,available} = G_{H,total} - \frac{G_O}{8} \qquad \textit{35.2}$$

For coal, the *"bed" moisture level* refers to the moisture level when the coal is mined. The terms *dry* and *as fired* are often used in commercial coal specifications. The "as fired" condition corresponds to a specific moisture content when placed in the furnace. The "as fired" heating value should be used, since the moisture actually decreases the combustion heat. The approximate relationship between the two heating values is displayed in Eq. 35.3, where $M$ is the moisture content from a proximate analysis.[5]

$$HV_{as\ fired} = HV_{dry}(1 - M) \qquad \textit{35.3}$$

Moisture in fuel is undesirable because it increases fuel weight (transportation costs) and decreases available combustion heat.[6]

## 7. ASH AND MINERAL MATTER

*Mineral matter* is the noncombustible material in a fuel. *Ash* is the residue remaining after combustion. Ash may contain some combustible carbon as well as the original mineral matter. The two terms ("mineral matter" and "ash") are often used interchangeably when reporting fuel analyses.

Ash may also be categorized according to where it is recovered. Dry and wet *bottom ashes* are recovered from *ash pits*. However, as little as 10% of the total ash content may be recovered in the ash pit. *Flyash* is carried out of the boiler by the flue gas. Flyash can be deposited on walls and heat transfer surfaces. It will be discharged from the stack if not captured. *Economizer ash* and *air heater ash* are recovered from the devices the ash is named after.

The finely powdered ash that covers combustion grates protects them from high temperatures.[7] If the ash has a low (i.e., below 2200°F; 1200°C) fusion temperature (melting point), it may form *clinkers* in the furnace and/or *slag* in other high-temperature areas. In extreme cases, it can adhere to the surfaces. Ashes with high melting (fusion) temperatures (i.e., above 2600°F; 1430°C) are known as *refractory ashes*. The $T_{250}$ *temperature* is used as an index of slagging tendencies of an ash. This is the temperature at which the slag becomes molten with a viscosity of 250 poise. Slagging will be experienced when the $T_{250}$ temperature is exceeded.

The actual melting point depends on the ash composition. Ash is primarily a mixture of silica ($SiO_2$), alumina ($Al_2O_3$), and ferric oxide ($Fe_2O_3$).[8] The relative proportions of each will determine the melting point, with lower melting points resulting from high amounts of ferric oxide and calcium oxide. The melting points of pure alumina and pure silica are in the 2700°F to 2800°F (1480°C to 1540°C) range.

*Coal ash* is either of a bituminous type or lignite type. Bituminous-type ash (from midwestern and eastern coals) contains more ferric oxide than lime and magnesia. Lignite-type ash (from western coals) contains more lime and magnesia than ferric oxide.

## 8. SULFUR

Several forms of sulfur are present in coal and fuel oils. *Pyritic sulfur* ($FeS_2$) is the primary form. *Organic sulfur* is combined with hydrogen and carbon in other compounds. *Sulfate sulfur* is iron sulfate and gypsum ($CaSO_4 \cdot 2H_2O$). Sulfur in elemental, organic, and pyritic forms oxidizes to sulfur dioxide. *Sulfur trioxide* can be formed under certain conditions. Sulfur trioxide combines with water to form sulfuric acid and is a major source of boiler/stack corrosion and acid rain.

$$SO_3 + H_2O \rightarrow H_2SO_4$$

## 9. WOOD

Wood is not an industrial fuel, though it may be used in small quantities in developing countries. Most woods have heating values around 8300 Btu/lbm (19 MJ/kg), with specific values depending on the species and moisture content. Variations in wood properties are so great that generalized properties are meaningless.

---

[3]This assumes that none of the oxygen is in the form of carbonates.
[4]The value of 1/8 follows directly from the combustion reaction of hydrogen and oxygen.
[5]Equation 35.3 corrects for the portion of the as-fired coal that isn't combustible, but it neglects other moisture-related losses. Section 35.35 and Sec. 35.37 contain more rigorous discussions.
[6]A moisture content up to 5% is reported to be beneficial in some mechanically fired boilers. The moisture content contributes to lower temperatures, protecting grates from slag formation and sintering.

[7]Some boiler manufacturers rely on the thermal protection the ash provides. For example, coal burned in cyclone boilers should have a minimum ash content of 7% to cover and protect the cyclone barrel tubes. Boiler wear and ash carryover will increase with lower ash contents.
[8]Calcium oxide (CaO), magnesium oxide ("magnesia," MgO), titanium oxide ("titania," $TiO_2$), ferrous oxide (FeO), and alkalies ($Na_2O$ and $K_2O$) may be present in smaller amounts.

## 10. WASTE FUELS

*Waste fuels* are increasingly being burned or incinerated in industrial boilers and furnaces. Such fuels include digester and landfill gases, waste process gases, flammable waste liquids, and volatile organic compounds (VOCs) such as benzene, toluene, xylene, ethanol, and methane. Other waste fuels include oil shale, tar sands, green wood, seed and rice hulls, biomass refuse, peat, tire shreddings, and shingle/roofing waste.

The term *refuse-derived fuels*, RDF, is used to describe fuel produced from municipal waste. After separation (removal of glass, plastics, metals, corrugated cardboard, etc.), the waste is merely pushed into the combustion chamber. If the waste is to be burned elsewhere, it is compressed and baled.

The heating value of RDF depends on the moisture content and fraction of combustible material. For RDFs derived from typical municipal wastes, the heating value will range from 3000 Btu/lbm to 6000 Btu/lbm (7 MJ/kg to 14 MJ/kg). Higher ranges (7500 Btu/lbm to 8500 Btu/lbm (17.5 MJ/kg to 19.8 MJ/kg)) can be obtained by careful selection of ingredients. Pelletized RDF (containing some coal and a limestone binder) with heating values around 8000 Btu/lbm (18.6 MJ/kg) can be used as a supplemental fuel in coal-fired units.

Scrap tires are an attractive fuel source due to their high heating values—12,000 Btu/lbm to 16,000 Btu/lbm (28 MJ/Kg to 37 MJ/kg). To be compatible with existing coal-loading equipment, tires are chipped or shredded to 1 in (25 mm) size. Tires in this form are known as *tire-derived fuel*, TDF. Metal (from tire reinforcement) may or may not be present.

TDF has been shown capable of supplying up to 90% of a steam-generating plant's total Btu input without any deterioration in particulate emissions, pollutants, and stack opacity. In fact, compared with some low-quality coals (e.g., lignite), TDF is far superior: about 2.5 times the heating value and about 2.5 times less sulfur per Btu.

## 11. INCINERATION

Many toxic wastes are incinerated rather than "burned." Incineration and combustion are not the same. *Incineration* is the term used to describe a disposal process that uses combustion to render wastes ineffective (nonharmful, nontoxic, etc.). Wastes and combustible fuel are combined in a furnace, and the heat of combustion destroys the waste.[9] Wastes may themselves be combustible, though they may not be self-sustaining if the moisture content is too high.

Incinerated wastes are categorized into seven types. Type 0 is *trash* (highly combustible paper and wood, with 10% or less moisture); type 1 is *rubbish* (combustible waste with up to 25% moisture); type 2 is *refuse* (a mixture of rubbish and garbage, with up to 50% moisture); type 3 is *garbage* (residential waste with up to 70% moisture); type 4 is animal solids and pathological wastes (85% moisture); type 5 is industrial process wastes in gaseous, liquid, and semiliquid form; and type 6 is industrial process wastes in solid and semisolid form requiring incineration in hearth, retort, or grate burning equipment.

## 12. COAL

Coal consists of volatile matter, fixed carbon, moisture, noncombustible mineral matter ("ash"), and sulfur. *Volatile matter* is driven off as a vapor when the coal is heated, and it is directly responsible for flame size. *Fixed carbon* is the combustible portion of the solid remaining after the volatile matter is driven off. Moisture is present in the coal as free water and (for some mineral compounds) as water of hydration. Sulfur, an undesirable component, contributes to heat content.

Coals are categorized into anthracitic, bituminous, and lignitic types. *Anthracite coal* is clean, dense, and hard. It is comparatively difficult to ignite but burns uniformly and smokelessly with a short flame. *Bituminous coal* varies in composition, but generally has a higher volatile content than anthracite, starts easily, and burns freely with a long flame. Smoke and soot are possible if bituminous coal is improperly fired. *Lignite coal* is a coal of woody structure, very high in moisture and with a low heating value. It normally ignites slowly due to its moisture, breaks apart when burning, and burns with little smoke or soot.

Coal is burned efficiently in a particular furnace only if it is uniform in size. Screen sizes are used to grade coal, but descriptive terms can also be used.[10] *Run-of-mine coal*, ROM, is coal as mined. *Lump coal* is in the 1 in to 6 in (25 mm to 150 mm) range. *Nut coal* is smaller, followed by even smaller *pea coal screenings*, and *fines* (dust).

## 13. LOW-SULFUR COAL

Switching to low-sulfur coal is a way to meet strict sulfur emission standards. Western and eastern low-sulfur coals have different properties.[11,12] Eastern low-sulfur coals are generally low-impact coals (that is, few changes need to be made to the power plant when

---

[9]Rotary kilns can accept waste in many forms. They are "workhorse" incinerators.

[10]The problem with descriptive terms is that one company's "pea coal" may be as small as 1/4 in (6 mm), while another's may start at 1/2 in (13 mm).

[11]In the United States, low-sulfur coals predominantly come from the western United States ("western subbituminous"), although some come from the east ("eastern bituminous").

[12]Some parameters dependent on coal type are coal preparation, firing rate, ash volume and handling, slagging, corrosion rates, dust collection and suppression, and fire and explosion prevention.

switching to them). Western coals are generally high-impact coals. Properties of typical high- and low-sulfur fuels are shown in Table 35.2.

The lower sulfur content results in less boiler corrosion. However, of all the coal variables, the different ash characteristics are the most significant with regard to the steam generator components. The slagging and fouling tendencies are prime concerns.

**Table 35.2** Typical Properties of High- and Low-Sulfur Coals[*]

| | | low-sulfur | |
| property | high-sulfur | eastern | western |
| --- | --- | --- | --- |
| higher heating value, | | | |
|   Btu/lbm | 10,500 | 13,400 | 8000 |
|   (MJ/kg) | (24.4) | (31.2) | (18.6) |
| moisture content, % | 11.7 | 6.9 | 30.4 |
| ash content, % | 11.8 | 4.5 | 6.4 |
| sulfur content, % | 3.2 | 0.7 | 0.5 |
| slag melting | | | |
|   temperature, °F | 2400 | 2900 | 2900 |
|   (°C) | (1320) | (1590) | (1590) |

(Multiply Btu/lbm by 2.326 to obtain kJ/kg.)

[*]All properties are "as received."

## 14. CLEAN COAL TECHNOLOGIES

A lot of effort has been put into developing technologies that will reduce acid rain, pollution and air toxics (NOx and $SO_2$), and global warming. These technologies are loosely labeled as *clean coal technologies*, CCTs. Whether or not these technologies can be retrofitted into an existing plant or designed into a new plant depends on the economics of the process.

With *coal cleaning*, coal is ground to ultrafine sizes to remove sulfur and ash-bearing minerals.[13] However, finely ground coal creates problems in handling, storage, and dust production. The risk of fire and explosion increases. Different approaches to reducing the problems associated with transporting and storing finely ground coal include the use of dust suppression chemicals, pelletizing, transportation of coal in liquid slurry form, and pelletizing followed by reslurrying. Some of these technologies may not be suitable for retrofit into existing installations.

With *coal upgrading*, moisture is thermally removed from low-rank coal (e.g., lignite or subbituminous coal). With some technologies, sulfur and ash are also removed when the coal is upgraded.

Reduction in sulfur dioxide emissions is the goal of *$SO_2$ control* technologies. These technologies include conventional use of lime and limestone in *flue gas desulfurization* (FGD) systems, *furnace sorbent-injection* (FSI) and *duct sorbent-injection*. *Advanced scrubbing* is included in FGD technologies.

Redesigned burners and injectors and adjustment of the flame zone are typical types of *NOx control*. Use of secondary air, injection of ammonia or urea, and selective catalytic reduction (SCR) are also effective in NOx reduction.

*Fluidized-bed combustion* (FBC) reduces NOx emissions by reducing combustion temperatures to around 1500°F (815°C). FBC is also effective in removing up to 90% of the $SO_2$. *Atmospheric FBC* operates at atmospheric pressure, but higher thermal efficiencies are achieved in *pressurized FBC* units operating at pressures up to 10 atm.

*Integrated gasification/combined cycle* (IGCC) processes are able to remove 99% of all sulfur while reducing $NO_x$ to well below current emission standards. *Synthetic gas (syngas)* is derived from coal. Syngas has a lower heating value than natural gas, but it can be used to drive gas turbines in combined cycles or as a reactant in the production of other liquid fuels.

## 15. COKE

*Coke*, typically used in blast furnaces, is produced by heating coal in the absence of oxygen. The heavy hydrocarbons crack (i.e., the hydrogen is driven off), leaving only a carbonaceous residue containing ash and sulfur. Coke burns smokelessly. *Breeze* is coke smaller than 5/8 in (16 mm). It is not suitable for use in blast furnaces, but steam boilers can be adapted to use it. *Char* is produced from coal in a 900°F (500°C) carbonization process. The volatile matter is removed, but there is little cracking. The process is used to solidify tars, bitumens, and some gases.

## 16. LIQUID FUELS

Liquid fuels are lighter hydrocarbon products refined from crude petroleum oil. They include liquefied petroleum gas (LPG), gasoline, kerosene, jet fuel, diesel fuels, and heating oils. Important characteristics of a liquid fuel are its composition, ignition temperature, flash point,[14] viscosity, and heating value.

## 17. FUEL OILS

In the United States, fuel oils are categorized into grades 1 through 6 according to their viscosities.[15] Viscosity is the major factor in determining firing rate and the need for preheating for pumping or atomizing prior to burning. Grades 1 and 2 can be easily pumped at ambient

---

[13]80% or more of *micronized coal* is 44 microns or less in size.

[14]This is different from the *flash point* that is the temperature at which fuel oils generate enough vapor to sustain ignition in the presence of spark or flame.

[15]Grade 3 became obsolete in 1948. Grade 5 is also subdivided into light and heavy categories.

**Table 35.3** *Typical Properties of Common Commercial Fuels*

| | butane | no. 1 diesel | no. 2 diesel | ethanol | gasoline | JP-4 | methanol | propane |
|---|---|---|---|---|---|---|---|---|
| chemical formula | $C_4H_{10}$ | – | – | $C_2H_5OH$ | – | – | $CH_3OH$ | $C_3H_8$ |
| molecular weight | 58.12 | $\approx 170$ | $\approx 184$ | 46.07 | $\approx 126$ | | | 44.09 |
| heating value | | | | | | | | |
|    higher Btu/lbm | 21,240 | 19,240 | 19,110 | 12,800 | 20,260 | | | 21,646 |
|    lower Btu/lbm | 19,620 | 18,250 | 18,000 | 11,500 | 18,900 | 18,400 | 9078 | 19,916 |
|    lower Btu/gal | 102,400 | 133,332 | 138,110 | 76,152 | 116,485 | 123,400 | 60,050 | 81,855 |
| latent heat of | | | | | | | | |
|   vaporization | | 115 | 105 | 361 | 142 | | | 147 |
|   Btu/lbm | | | | | | | | |
| specific gravity[*] | 2.01 | 0.876 | 0.920 | 0.794 | 0.68-0.74 | 0.8017 | 0.793 | 1.55 |

(Multiply Btu/lbm by 2.326 to obtain kJ/kg.)
(Multiply Btu/gal by 0.2786 to obtain $MJ/m^3$.)

[*]Specific gravities of propane and butane are with respect to air.

temperatures. In the United States, the heaviest fuel oil used is grade 6, also known as *Bunker C oil*.[16,17]

Fuel oils are also classified according to their viscosities as *distillate oils* (lighter) and *residual fuel oils* (heavier).

Like coal, fuel oils contain sulfur and ash that may cause pollution, slagging on the hot end of the boiler, and corrosion in the cold end. Table 35.3 lists typical properties of common commercial fuels, while Table 35.4 lists typical properties of fuel oils.

**Table 35.4** *Typical Properties of Fuel Oils[a]*

| | | heating value | |
|---|---|---|---|
| grade | specific gravity | MBtu/gal[b] | $GJ/m^3$ |
| 1 | 0.805 | 134 | 37.3 |
| 2 | 0.850 | 139 | 38.6 |
| 4 | 0.903 | 145 | 40.4 |
| 5 | 0.933 | 148 | 41.2 |
| 6 | 0.965 | 151 | 41.9 |

(Multiply MBtu/gal by 0.2786 to obtain $GJ/m^3$.)

[a]Actual values will vary depending on composition.
[b]One MBtu equals one thousand Btus.

## 18. GASOLINE

*Gasoline* is not a pure compound. It is a mixture of various hydrocarbons blended to give a desired flammability, volatility, heating value, and octane rating. There is an infinite number of blends that can be used to produce gasoline.

Gasoline's heating value depends only slightly on composition. Within a variation of $1^1/_2\%$, the heating value can be taken as 20,200 Btu/lbm (47.0 MJ/kg) for regular gasoline and as 20,300 Btu/lbm (47.2 MJ/kg) for high-octane aviation fuel.

Since gasoline is a mixture of hydrocarbons, different fractions will evaporate at different temperatures. The *volatility* is the percentage of the fuel that evaporates by a given temperature. Typical volatility specifications call for 10% at 167°F (75°C), 50% at 221°F (105°C), and 90% at 275°F (135°C). Low volatility causes difficulty in starting and poor engine performance at low temperatures.

The *octane number*, ON, is a measure of knock resistance. It is based on comparison, performed in a standardized one-cylinder engine, with the burning of isooctane and *n*-heptane. *n*-heptane, $C_7H_{16}$, is rated zero and produces violent knocking. Isooctane, $C_8H_{18}$, is rated 100 and produces relatively knock-free operation. The percentage blend by volume of these fuels that matches the performance of the gasoline is the octane rating. The *research octane number* (RON) is a measure of the fuel's antiknock characteristics while idling; the *motor octane number* (MON) applies to high-speed, high-acceleration operations. The octane rating reported for commercial gasoline is an average of the two.

The *performance number* (PN) of gasoline containing antiknock compounds (e.g., tetraethyl lead, TEL) is related to the octane number.

$$\text{ON} = 100 + \frac{\text{PN} - 100}{3} \qquad 35.4$$

---

[16]120°F (48°C) is the optimum temperature for pumping no. 6 fuel oil. At that temperature, no. 6 oil has a viscosity of approximately 3000 SSU. Further heating is necessary to lower the viscosity to 150 SSU to 350 SSU for atomizing.
[17]To avoid *coking* of oil, heating coils in contact with oil should not be hotter than 240°F (116°C).

**Table 35.5** *Typical Properties of Common Oxygenates*

| | Ethanol | MTBE[a] | TAME | ETBE | TAEE |
|---|---|---|---|---|---|
| specific gravity | 0.794 | 0.744 | 0.740 | 0.770 | 0.791 |
| octane | 115 | 110 | 112 | 105 | 100 |
| heating value (MBtu/gal)[b] | 76.2 | 93.6 | | | |
| Reid vapor pressure (psig)[c] | 18 | 8 | 15–4 | 3–4 | 2 |
| percent oxygen by weight | 34.73 | 18.15 | 15.66 | 15.66 | 13.8 |
| volumetric percent needed to achieve gasoline | | | | | |
|    2.7% oxygen by weight | | 15.1 | 17.2 | 17.2 | 19.4 |
|    2.0% oxygen by weight | | 11.0 | 12.4 | 12.7 | 13.0 |

(Multiply MBtu/gal by 0.2786 to obtain MJ/m$^3$.)

[a]MTBE is water soluble and does not degrade. As a suspected carcinogen, it is a serious threat to water supplies. It has been legislatively banned in some states, including California and Washington.
[b]One MBtu equals one thousand Btus.
[c]The Reid vapor pressure is the vapor pressure when heated to 100°F (38°C). This may also be referred to as the "blending vapor pressure."

## 19. OXYGENATED GASOLINE

In parts of the United States, gasoline is "oxygenated" during the cold winter months. This has led to use of the term, "winterized gasoline." The addition of *oxygenates* raises the combustion temperature, reducing carbon-monoxide and unburned hydrocarbons.[18] Common oxygenates used in *reformulated gasoline* (RFG) include methyl tertiary-butyl ether (MTBE) and ethanol. Methanol, ethyl tertiary-butyl ether (ETBE) tertiary-amyl methyl ether (TAME) and tertiary-amyl ethyl ether (TAEE) may also be used. (See Table 35.5.) Oxygenates are added to bring the minimum oxygen level to 2% to 3% by weight.[19]

## 20. DIESEL FUEL

Properties and specifications for various grades of diesel fuel oil are similar to specifications for fuel oils. Grade 1-D ("D" for diesel) is a light distillate oil for high-speed engines in service requiring frequent speed and load changes. Grade 2-D is a distillate of lower volatility for engines in industrial and heavy mobile service. Grade 4-D is for use in medium speed engines under sustained loads.

Diesel oils are specified by a *cetane number*, which is a measure of the ignition quality (ignition delay) of a fuel. Like the octane number for gasoline, the cetane number is determined by comparison with standard fuels. Cetane, $C_{16}H_{34}$, has a cetane number of 100. *n*-methyl-naphthalene, $C_{11}H_{10}$, has a cetane number of zero.

A cetane number of approximately 30 is required for satisfactory operation of low-speed diesel engines. High-speed engines, such as those used in cars, require a cetane number of 45 or more. The cetane number can be increased by use of such additives as amyl nitrate, ethyl nitrate, and ether.

A diesel fuel's *pour point* number refers to its viscosity. A fuel with a pour point of 10°F (−12°C) will flow freely above that temperature. A fuel with a high pour point will thicken in cold temperatures.

The *cloud point* refers to the temperature at which wax crystals cloud the fuel at lower temperatures. The cloud point should be 20°F (−7°C) or higher. Below that temperature, the engine will not run well.

## 21. ALCOHOL

Both methanol and ethanol can be used in internal combustion engines. *Methanol (methyl alcohol)* is produced from natural gas and coal, although it can also be produced from wood and organic debris. *Ethanol (ethyl alcohol, grain alcohol)* is distilled from grain, sugar cane, potatoes, and other agricultural products containing various amounts of sugars, starches, and cellulose.

Although methanol generally works as well as ethanol, only ethanol can be produced in large quantities from inexpensive agricultural products and by-products.

Alcohol is water soluble. The concentration of alcohol is measured by its *proof*, where 200 proof is pure alcohol. (180 proof is 90% alcohol and 10% water.)

*Gasohol* is a mixture of approximately 90% gasoline and 10% alcohol (generally ethanol).[20] Alcohol's heating value is less than gasoline's, so fuel consumption (per distance traveled) is higher with gasohol. Also, since alcohol absorbs moisture more readily than gasoline, corrosion of fuel tanks becomes problematic. In some engines, significantly higher percentages of alcohol may require such modifications as including larger carburetor

---

[18]Oxygenation may not be successful in reducing carbon dioxide. Since the heating value of the oxygenates is lower, fuel consumption of oxygenated fuels is higher. On a per-gallon (per-liter) basis, oxygenation reduces carbon dioxide. On a per-mile (per-kilometer) basis, however, oxygenation appears to increase carbon dioxide. In any case, claims of $CO_2$ reduction are highly controversial, as the $CO_2$ footprint required to plant, harvest, dispose of decaying roots, stalks, and leaves (i.e., silage), and refine alcohol is generally ignored.
[19]Other restrictions on gasoline during the winter months intended to reduce pollution may include maximum percentages of benzene and total aromatics, and limits on Reid vapor pressure, as well as specifications covering volatile organic compounds, nitric oxide (NOx), and toxins.

[20]In fact, oxygenated gasoline may use more than 10% alcohol.

Chemistry

jets, timing advances, heaters for preheating fuel in cold weather, tank lining to prevent rusting, and alcohol-resistant gaskets.

Mixtures of gasoline and alcohol can be designated by the first letter and the fraction of the alcohol. E10 is a mixture of 10% ethanol and 90% gasoline. M85 is a blend of 85% methanol and 15% gasoline.

Alcohol is a poor substitute for diesel fuel because alcohol's cetane number is low—from −20 to +8. Straight injection of alcohol results in poor performance and heavy knocking.

## 22. GASEOUS FUELS

Various gaseous fuels are used as energy sources, but most applications are limited to natural gas and *liquefied petroleum gases*, LPGs (i.e., propane, butane, and mixtures of the two).[21,22] Natural gas is a mixture of methane (55% to 95%), higher hydrocarbons (primarily ethane), and other noncombustible gases. Typical heating values for natural gas range from 950 Btu/ft$^3$ to 1100 Btu/ft$^3$ (35 MJ/m$^3$ to 41 MJ/m$^3$).

The production of *synthetic gas (syngas)* through coal gasification may be applicable to large power generating plants. The cost of gasification, though justifiable to reduce sulfur and other pollutants, is too high for syngas to become a widespread substitute for natural gas.

## 23. IGNITION TEMPERATURE

The *ignition temperature (autoignition temperature)* is the minimum temperature at which combustion can be sustained. It is the temperature at which more heat is generated by the combustion reaction than is lost to the surroundings, after which combustion becomes self-sustaining. For coal, the minimum ignition temperature varies from around 800°F (425°C) for bituminous varieties to 900°F to 1100°F (480°C to 590°C) for anthracite. For sulfur and charcoal, the ignition temperatures are approximately 470°F (240°C) and 650°F (340°C), respectively.

For gaseous fuels, the ignition temperature depends on the air/fuel ratio, temperature, pressure, and length of time the source of heat is applied. Ignition can be instantaneous or with a lag, depending on the temperature. Generalizations can be made for any gas, but the generalized temperatures will be meaningless without specifying all of these factors.

## 24. ATMOSPHERIC AIR

It is important to make a distinction between "air" and "oxygen." Atmospheric air is a mixture of oxygen, nitrogen, and small amounts of carbon dioxide, water vapor, argon, and other inert ("rare") gases. For the purpose of combustion calculations, all constituents except oxygen are grouped with nitrogen. (See Table 35.6.) It is necessary to supply 4.32 (i.e., 1/0.2315) masses of air to obtain one mass of oxygen. Similarly, it is necessary to supply 4.773 volumes of air to obtain one volume of oxygen. The average molecular weight of air is 28.97, and the specific gas constant is 53.35 ft-lbf/lbm-°R (287.03 J/kg·K).

*Table 35.6* Composition of Dry Air[a]

| component | percent by weight | percent by volume |
|---|---|---|
| oxygen | 23.15 | 20.95 |
| nitrogen/inerts | 76.85 | 79.05 |
| ratio of nitrogen to oxygen | 3.320 | 3.773[b] |
| ratio of air to oxygen | 4.320 | 4.773 |

[a]Inert gases and $CO_2$ included as $N_2$.
[b]The value is also reported by various sources as 3.76, 3.78, and 3.784.

## 25. COMBUSTION REACTIONS

A limited number of elements appear in combustion reactions. Carbon, hydrogen, sulfur, hydrocarbons, and oxygen are the reactants. Carbon dioxide and water vapor are the main products, with carbon monoxide, sulfur dioxide, and sulfur trioxide occurring in lesser amounts. Nitrogen and excess oxygen emerge hotter but unchanged from the stack.

Combustion reactions occur according to the normal chemical reaction principles. Balancing combustion reactions is usually easiest if carbon is balanced first, followed by hydrogen and then by oxygen. When a gaseous fuel has several combustible gases, the volumetric fuel composition can be used as coefficients in the chemical equation.

Table 35.7 lists ideal combustion reactions. These reactions do not include any nitrogen or water vapor that are present in the combustion air.

### Example 35.1

A gaseous fuel is 20% hydrogen and 80% methane by volume. What volume of oxygen is required to burn 120 volumes of fuel at the same conditions?

*Solution*

Write the unbalanced combustion reaction.

$$H_2 + CH_4 + O_2 \rightarrow CO_2 + H_2O$$

---

[21]A number of *manufactured gases* are of practical (and historical) interest in specific industries, including *coke-oven gas, blast-furnace gas, water gas, producer gas,* and *town gas.* However, these gases are not now in widespread use.
[22]At atmospheric pressure, propane boils at −44°F (−42°C), while butane boils at 31°F (−0.5°C).

Use the volumetric analysis as coefficients of the fuel.

$$0.2H_2 + 0.8CH_4 + O_2 \rightarrow CO_2 + H_2O$$

Balance the carbons.

$$0.2H_2 + 0.8CH_4 + O_2 \rightarrow 0.8CO_2 + H_2O$$

Balance the hydrogens.

$$0.2H_2 + 0.8CH_4 + O_2 \rightarrow 0.8CO_2 + 1.8H_2O$$

Balance the oxygens.

$$0.2H_2 + 0.8CH_4 + 1.7O_2 \rightarrow 0.8CO_2 + 1.8H_2O$$

For gaseous components, the coefficients correspond to the volumes. Since one $(0.2 + 0.8)$ volume of fuel requires 1.7 volumes of oxygen, the required oxygen is

$$(1.7)(120 \text{ volumes of fuel}) = 204 \text{ volumes of oxygen}$$

**Table 35.7** *Ideal Combustion Reactions*

| fuel | formula | reaction equation (excluding nitrogen) |
|------|---------|----------------------------------------|
| carbon (to CO) | C | $2C + O_2 \rightarrow 2CO$ |
| carbon (to $CO_2$) | C | $C + O_2 \rightarrow CO_2$ |
| sulfur (to $SO_2$) | S | $S + O_2 \rightarrow SO_2$ |
| sulfur (to $SO_3$) | S | $2S + 3O_2 \rightarrow 2SO_3$ |
| carbon monoxide | CO | $2CO + O_2 \rightarrow 2CO_2$ |
| methane | $CH_4$ | $CH_4 + 2O_2 \rightarrow CO_2 + 2H_2O$ |
| acetylene | $C_2H_2$ | $2C_2H_2 + 5O_2 \rightarrow 4CO_2 + 2H_2O$ |
| ethylene | $C_2H_4$ | $C_2H_4 + 3O_2 \rightarrow 2CO_2 + 2H_2O$ |
| ethane | $C_2H_6$ | $2C_2H_6 + 7O_2 \rightarrow 4CO_2 + 6H_2O$ |
| hydrogen | $H_2$ | $2H_2 + O_2 \rightarrow 2H_2O$ |
| hydrogen sulfide | $H_2S$ | $2H_2S + 3O_2 \rightarrow 2H_2O + 2SO_2$ |
| propane | $C_3H_8$ | $C_3H_8 + 5O_2 \rightarrow 3CO_2 + 4H_2O$ |
| $n$-butane | $C_4H_{10}$ | $2C_4H_{10} + 13O_2 \rightarrow 8CO_2 + 10H_2O$ |
| octane | $C_8H_{18}$ | $2C_8H_{18} + 25O_2 \rightarrow 16CO_2 + 18H_2O$ |
| olefin series | $C_nH_{2n}$ | $2C_nH_{2n} + 3nO_2 \rightarrow 2nCO_2 + 2nH_2O$ |
| paraffin series | $C_nH_{2n+2}$ | $2C_nH_{2n+2} + (3n+1)O_2 \rightarrow 2nCO_2 + (2n+2)H_2O$ |

(Multiply oxygen volumes by 3.773 to get nitrogen volumes.)

## 26. STOICHIOMETRIC REACTIONS

*Stoichiometric quantities (ideal quantities)* are the exact quantities of reactants that are needed to complete a combustion reaction without any reactants left over. Table 35.7 contains some of the more common chemical reactions. Stoichiometric volumes and masses can always be determined from the balanced chemical reaction equation. Table 35.8 can be used to quickly determine stoichiometric amounts for some fuels.

## 27. STOICHIOMETRIC AIR

*Stoichiometric air (ideal air)* is the air necessary to provide the exact amount of oxygen for complete combustion of a fuel. Stoichiometric air includes atmospheric nitrogen. For each volume of oxygen, 3.773 volumes of nitrogen pass unchanged through the reaction.[23]

Stoichiometric air can be stated in units of mass (pounds or kilograms of air) for solid and liquid fuels, and in units of volume (cubic feet or cubic meters of air) for gaseous fuels. When stated in terms of mass, the stoichiometric ratio of air to fuel masses is known as the ideal *air/fuel ratio*, $R_{a/f}$.

$$R_{a/f,\text{ideal}} = \frac{m_{\text{air,ideal}}}{m_{\text{fuel}}} \qquad 35.5$$

The ideal air/fuel ratio can be determined from the combustion reaction equation. It can also be determined by adding the oxygen and nitrogen amounts listed in Table 35.8.

For fuels whose ultimate analysis is known, the approximate stoichiometric air (oxygen and nitrogen) requirement in pounds of air per pound of fuel (kilograms of air per kilogram of fuel) can be quickly calculated by using Eq. 35.6.[24] All oxygen in the fuel is assumed to be free moisture. All of the reported oxygen is assumed to be locked up in the form of water. Any free oxygen (i.e., oxygen dissolved in liquid fuels) is subtracted from the oxygen requirements.

$$R_{a/f,\text{ideal}} = (34.5)\left(\frac{G_C}{3} + G_H - \frac{G_O}{8} + \frac{G_S}{8}\right) \qquad 35.6$$

[solid and liquid fuels]

For fuels consisting of a mixture of gases, Eq. 35.7 and the constants $J_i$ from Table 35.9 can be used to quickly determine the stoichiometric air requirements.

$$R_{a/f,\text{ideal}} = \sum J_i G_i \quad \text{[gaseous fuels]} \qquad 35.7$$

[23]The only major change in the nitrogen gas is its increase in temperature. Dissociation of nitrogen and formation of nitrogen compounds can occur but are essentially insignificant.

[24]This is a "compromise" equation. Variations in the atomic weights will affect the coefficients slightly. The coefficient 34.5 is reported as 34.43 in some older books. 34.5 is the exact value needed for carbon and hydrogen, which constitute the bulk of the fuel. 34.43 is the correct value for sulfur, but the error is small and is disregarded in this equation.

*Chemistry*

*Table 35.8* Consolidated Combustion Data[a,b,c,d]

| fuel | units of fuel | for 1 mole of fuel — air O₂ | air N₂ | CO₂ | H₂O | SO₂ | for 1 ft³ of fuel[e] — air O₂ | air N₂ | CO₂ | H₂O | for 1 lbm of fuel — air O₂ | air N₂ | CO₂ | H₂O | SO₂ |
|---|---|---|---|---|---|---|---|---|---|---|---|---|---|---|---|
| C carbon | moles | 1.0 | 3.773 | 1.0 | | | | | | | 0.0833 | 0.3143 | 0.0833 | | |
| | ft³ | 379.5 | 1432 | 379.5 | | | | | | | 31.63 | 119.3 | 31.63 | | |
| | lbm | 32.0 | 106 | 44.0 | | | | | | | 2.667 | 8.883 | 3.667 | | |
| H₂ hydrogen | moles | 0.5 | 1.887 | | 1.0 | | 0.001317 | 0.004969 | | 0.002635 | 0.248 | 0.9357 | | 0.496 | |
| | ft³ | 189.8 | 716.1 | | 379.5 | | 0.5 | 1.887 | | 1.0 | 94.12 | 355.1 | | 188.25 | |
| | lbm | 16.0 | 53.0 | | 18.0 | | 0.04216 | 0.1397 | | 0.04747 | 7.936 | 26.29 | | 8.936 | |
| S sulfur | moles | 1.0 | 3.773 | | | 1.0 | | | | | 0.03119 | 0.1177 | | | 0.03119 |
| | ft³ | 379.5 | 1432 | | | 379.5 | | | | | 11.84 | 44.67 | | | 11.84 |
| | lbm | 32.0 | 106.0 | | | 64.06 | | | | | 0.998 | 3.306 | | | 1.998 |
| CO carbon monoxide | moles | 0.5 | 1.887 | 1.0 | | | 0.001317 | 0.004969 | 0.002635 | | 0.01785 | 0.06735 | 0.03570 | | |
| | ft³ | 189.8 | 716.1 | 379.5 | | | 0.5 | 1.887 | 1.0 | | 6.774 | 25.56 | 13.55 | | |
| | lbm | 16.0 | 53.0 | 44.01 | | | 0.04216 | 0.1397 | 0.1160 | | 0.5712 | 1.892 | 1.572 | | |
| CH₄ methane | moles | 2.0 | 7.546 | 1.0 | 2.0 | | 0.00527 | 0.01988 | 0.002635 | 0.00527 | 0.1247 | 0.4705 | 0.06233 | 0.1247 | |
| | ft³ | 759 | 2864 | 379.5 | 758 | | 2.0 | 7.546 | 1.0 | 2.0 | 47.31 | 178.5 | 23.66 | 47.31 | |
| | lbm | 64.0 | 212.0 | 44.01 | 36.03 | | 0.1686 | 0.5586 | 0.1160 | 0.0949 | 3.989 | 13.21 | 2.743 | 2.246 | |
| C₂H₂ acetylene | moles | 2.5 | 9.433 | 2.0 | 1.0 | | 0.006588 | 0.02486 | 0.00527 | 0.002635 | 0.09601 | 0.3622 | 0.07681 | 0.03841 | |
| | ft³ | 948.8 | 3580 | 758 | 379.5 | | 2.5 | 9.443 | 2.0 | 1.0 | 36.44 | 137.5 | 29.15 | 14.57 | |
| | lbm | 80.0 | 265.0 | 88.02 | 18.02 | | 0.2108 | 0.6983 | 0.2319 | 0.04747 | 3.072 | 10.18 | 3.380 | 0.6919 | |
| C₂H₄ ethylene | moles | 3.0 | 11.32 | 2.0 | 2.0 | | 0.007905 | 0.02983 | 0.00527 | 0.00527 | 0.1069 | 0.4033 | 0.07129 | 0.07129 | |
| | ft³ | 1139 | 4297 | 758 | 758 | | 3.0 | 11.32 | 2.0 | 2.0 | 40.58 | 153.1 | 27.05 | 27.05 | |
| | lbm | 96.0 | 318.0 | 88.02 | 36.03 | | 0.2530 | 0.8380 | 0.2319 | 0.0949 | 3.422 | 11.34 | 3.137 | 1.284 | |
| C₂H₆ ethane | moles | 3.5 | 13.21 | 2.0 | 3.0 | | 0.009223 | 0.03480 | 0.00527 | 0.007905 | 0.1164 | 0.4392 | 0.06651 | 0.09977 | |
| | ft³ | 1328 | 5010 | 758 | 1139 | | 3.5 | 13.21 | 2.0 | 3.0 | 44.17 | 166.7 | 25.24 | 37.86 | |
| | lbm | 112.0 | 371.0 | 88.02 | 54.05 | | 0.2951 | 0.9776 | 0.2319 | 0.1424 | 3.724 | 12.34 | 2.927 | 1.797 | |

(Multiply lbm/ft³ by 0.06243 to obtain kg/m³.)

[a] Rounding of molecular weights and air composition may introduce slight inconsistencies in the table values. This table is based on atomic weights with at least four significant digits, a ratio of 3.773 volumes of nitrogen per volume of oxygen, and 379.5 ft³ per mole at 1 atm and 60°F.

[b] Volumes per unit mass are at 1 atm and 60°F (16°C). To obtain volumes at other temperatures, multiply by $(T_F + 460°)/520°$ or $(T_C + 273°)/289°$.

[c] The volume of water applies only when the combustion products are at such high temperatures that all of the water is in vapor form.

[d] This table can be used to directly determine some SI ratios. For kg/kg ratios, the values are the same as lbm/lbm. For mixed units (e.g., ft³/lbm), conversions are required. For L/L or m³/m³, use ft³/ft³. For mol/mol, use mole/mole.

[e] Sulfur is not used in gaseous form.

**Chemistry**

**Table 35.9** Approximate Air/Fuel Ratio Coefficients for Components of Natural Gas[*]

| fuel component | $J$ (gravimetric) | $K$ (volumetric) |
|---|---|---|
| acetylene, $C_2H_2$ | 13.25 | 11.945 |
| butane, $C_4H_{10}$ | 15.43 | 31.06 |
| carbon monoxide, CO | 2.463 | 2.389 |
| ethane, $C_2H_6$ | 16.06 | 16.723 |
| ethylene, $C_2H_4$ | 14.76 | 14.33 |
| hydrogen, $H_2$ | 34.23 | 2.389 |
| hydrogen sulfide, $H_2S$ | 6.074 | 7.167 |
| methane, $CH_4$ | 17.20 | 9.556 |
| oxygen, $O_2$ | −4.320 | −4.773 |
| propane, $C_3H_8$ | 15.65 | 23.89 |

[*]Rounding of molecular weights and air composition may introduce slight inconsistencies in the table values. This table is based on atomic weights with at least four significant digits and a ratio of 3.773 volumes of nitrogen per volume of oxygen.

For fuels consisting of a mixture of gases, the air/fuel ratio can also be expressed in volumes of air per volume of fuel.

$$\text{volumetric air/fuel ratio} = \sum K_i B_i \quad \text{[gaseous fuels]}$$

*35.8*

**Example 35.2**

Use Table 35.8 to determine the theoretical volume of 90°F (32°C) air required to burn 1 volume of 60°F (16°C) carbon monoxide to carbon dioxide.

*Solution*

From Table 35.8, 0.5 volumes of oxygen are required to burn 1 volume of carbon monoxide to carbon dioxide. 1.887 volumes of nitrogen accompany the oxygen. The total amount of air at the temperature of the fuel is $0.5 + 1.887 = 2.387$ volumes.

This volume will expand at the higher temperature. The volume at the higher temperature is

$$V_2 = \frac{T_2 V_1}{T_1}$$
$$= \frac{(90°F + 460°)(2.387 \text{ volumes})}{60°F + 460°}$$
$$= 2.53 \text{ volumes}$$

**Example 35.3**

How much air is required for the ideal combustion of (a) coal with an ultimate analysis of 93.5% carbon, 2.6% hydrogen, 2.3% oxygen, 0.9% nitrogen, and 0.7% sulfur, (b) fuel oil with a gravimetric analysis of 84% carbon, 15.3% hydrogen, 0.4% nitrogen, and 0.3% sulfur, and (c) natural gas with a volumetric analysis of 86.92% methane, 7.95% ethane, 2.81% nitrogen, 2.16% propane, and 0.16% butane?

*Solution*

(a) Use Eq. 35.6.

$$R_{a/f,\text{ideal}} = (34.5)\left(\frac{G_C}{3} + G_H - \frac{G_O}{8} + \frac{G_S}{8}\right)$$
$$= (34.5)\left(\frac{0.935}{3} + 0.026 - \frac{0.023}{8} + \frac{0.007}{8}\right)$$
$$= 11.58 \text{ lbm/lbm (kg/kg)}$$

(b) Use Eq. 35.6.

$$R_{a/f,\text{ideal}} = (34.5)\left(\frac{G_C}{3} + G_H + \frac{G_S}{8}\right)$$
$$= (34.5)\left(\frac{0.84}{3} + 0.153 + \frac{0.003}{8}\right)$$
$$= 14.95 \text{ lbm/lbm (kg/kg)}$$

(c) Use Eq. 35.8 and the coefficients from Table 35.9.

$$\frac{\text{volumetric}}{\text{air/fuel ratio}} = \sum K_i B_i$$
$$= (0.8692)(9.556) + (0.0795)(16.723)$$
$$\quad + (0.0216)(23.89) + (0.0016)(31.06)$$
$$= 10.20 \text{ ft}^3/\text{ft}^3 \text{ (m}^3/\text{m}^3)$$

## 28. INCOMPLETE COMBUSTION

*Incomplete combustion* occurs when there is insufficient oxygen to burn all of the hydrogen, carbon, and sulfur in the fuel. Without enough available oxygen, carbon burns to carbon monoxide.[25] Carbon monoxide in the flue gas indicates incomplete and inefficient combustion. Incomplete combustion is caused by cold furnaces, low combustion temperatures, poor air supply, smothering from improperly vented stacks, and insufficient mixing of air and fuel.

## 29. SMOKE

The amount of smoke can be used as an indicator of combustion cleanliness. Smoky combustion may indicate improper air/fuel ratio, insufficient draft, leaks, insufficient preheat, or misadjustment of the fuel system.

Smoke measurements are made in a variety of ways, with the standards depending on the equipment used. Photoelectric sensors in the stack are used to continuously monitor smoke. The *smoke spot number* (SSN) and ASTM smoke scale are used with continuous stack monitors. For coal-fired furnaces, the maximum desirable smoke number is SSN 4. For grade 2 fuel oil, the SSN should be less than 1; for grade 4, SSN 4; for grades

[25]Toxic alcohols, ketones, and aldehydes may also be formed during incomplete combustion.

Chemistry

5L, 5H, and low-sulfur residual fuels, SSN 3; for grade 6, SSN 4.

The *Ringelmann scale* is a subjective method in which the smoke density is visually compared to five standardized white-black grids. Ringelmann chart no. 0 is solid white; chart no. 5 is solid black. Ringelmann chart no. 1, which is 20% black, is the preferred (and required) operating point for most power plants.

## 30. FLUE GAS ANALYSIS

Combustion products that pass through a furnace's exhaust system are known as *flue gases (stack gases)*. Flue gases are almost all nitrogen.[26] (Nitrogen oxides are not present in large enough amounts to be included separately in combustion reactions.)

The actual composition of flue gases can be obtained in a number of ways, including by modern electronic detectors, less expensive "length-of-stain" detectors, and direct sampling with an Orsat apparatus.

The antiquated *Orsat apparatus* determines the volumetric percentages of $CO_2$, CO, $O_2$, and $N_2$ in a flue gas. The sampled flue gas passes through a series of chemical compounds. The first compound absorbs only $CO_2$, the next only $O_2$, and the third only CO. The unabsorbed gas is assumed to be $N_2$ and is found by subtracting the volumetric percentages of all other components from 100%. An Orsat analysis is a dry analysis; the percentage of water vapor is not usually determined. A wet volumetric analysis (needed to compute the dew-point temperature) can be derived if the volume of water vapor is added to the Orsat volumes.

The Orsat procedure is now rarely used, although the term "Orsat" may be generally used to refer to any flue gas analyzer. Modern electronic analyzers can determine free oxygen (and other gases) independently of the other gases. Because the relationship between oxygen and excess air is relatively insensitive to fuel composition, oxygen measurements are replacing standard carbon dioxide measurements in determining combustion efficiency.

## 31. ACTUAL AND EXCESS AIR

*Complete combustion* occurs when all of the fuel is burned. Usually, *excess air* is required to achieve complete combustion. Excess air is expressed as a percentage of the theoretical air requirements. Different fuel types burn more efficiently with different amounts of excess air. Coal-fired boilers need approximately 30% to 35% excess air, oil-based units need about 15%, and natural gas burners need about 10%.

The actual air/fuel ratio for dry, solid fuels with no unburned carbon can be estimated from the volumetric flue gas analysis and the gravimetric fractions of carbon and sulfur in the fuel.

$$R_{a/f,\text{actual}} = \frac{m_{\text{air,actual}}}{m_{\text{fuel}}}$$

$$= \frac{3.04 B_{N_2}\left(G_C + \dfrac{G_S}{1.833}\right)}{B_{CO_2} + B_{CO}} \qquad 35.9$$

Too much free oxygen or too little carbon dioxide in the flue gas is indicative of excess air. The relationship between excess air and oxygen in the flue gases is not highly affected by fuel composition.[27] The relationship between excess air and the volumetric fraction of oxygen in the flue gas is given in Table 35.10.

**Table 35.10** *Approximate Volumetric Percentage of Oxygen in Stack Gas*

| fuel[*] | excess air | | | | | | | |
|---|---|---|---|---|---|---|---|---|
| | 0% | 1% | 5% | 10% | 20% | 50% | 100% | 200% |
| fuel oils, | | | | | | | | |
| no. 2–6 | 0 | 0.22 | 1.06 | 2.02 | 3.69 | 7.29 | 10.8 | 14.2 |
| natural gas | 0 | 0.25 | 1.18 | 2.23 | 4.04 | 7.83 | 11.4 | 14.7 |
| propane | 0 | 0.23 | 1.08 | 2.06 | 3.75 | 7.38 | 10.9 | 14.3 |

[*]Values for coal are only marginally lower than the values for fuel oils.

Reducing the air/fuel ratio will have several outcomes. (a) The furnace temperature will increase due to a reduction in cooling air. (b) The flue gas will decrease in quantity. (c) The heat loss will decrease. (d) The furnace efficiency will increase. (e) Pollutants will (usually) decrease.

With a properly adjusted furnace and good mixing, the flue gas will contain no carbon monoxide, and the amount of carbon dioxide will be maximized. The stoichiometric amount of carbon dioxide in the flue gas is known as the *ultimate $CO_2$*. The air/fuel mixture should be adjusted until the maximum level of carbon dioxide is attained.

### Example 35.4

Propane ($C_3H_8$) is burned completely with 20% excess air. What is the volumetric fraction of carbon dioxide in the flue gas?

*Solution*

The balanced chemical reaction equation is

$$C_3H_8 + 5O_2 \rightarrow 3CO_2 + 4H_2O$$

With 20% excess air, the oxygen volume is $(1.2)(5) = 6$.

$$C_3H_8 + 6O_2 \rightarrow 3CO_2 + 4H_2O + O_2$$

---

[26]This assumption is helpful in making quick determinations of the thermodynamic properties of flue gases.

[27]The relationship between excess air and $CO_2$ is much more dependent on fuel type and composition.

From Table 35.6, there are 3.773 volumes of nitrogen for every volume of oxygen.

$$(3.773)(6) = 22.6$$

$$C_3H_8 + 6O_2 + 22.6N_2 \rightarrow 3CO_2 + 4H_2O + O_2 + 22.6N_2$$

For gases, the coefficients can be interpreted as volumes. The volumetric fraction of carbon dioxide is

$$B_{CO_2} = \frac{3}{3 + 4 + 1 + 22.6}$$
$$= 0.0980 \quad (9.8\%)$$

## 32. CALCULATIONS BASED ON FLUE GAS ANALYSIS

Equation 35.10 gives the approximate percentage (by volume) of actual excess air.

$$\frac{\text{actual excess air}}{\% \text{ by volume}} = \frac{(100\%)(B_{O_2} - 0.5B_{CO})}{0.264B_{N_2} - B_{O_2} + 0.5B_{CO}} \quad \textbf{35.10}$$

The ultimate $CO_2$ (i.e., the maximum theoretical carbon dioxide) can be determined from Eq. 35.11.

$$\frac{\text{ultimate } CO_2,}{\% \text{ by volume}} = \frac{(100\%)B_{CO_2,\text{actual}}}{1 - 4.773B_{O_2,\text{actual}}} \quad \textbf{35.11}$$

The mass ratio of dry flue gases to solid fuel is given by Eq. 35.12.

$$\frac{\text{mass of flue gas}}{\text{mass of solid fuel}} = \frac{\left(11B_{CO_2} + 8B_{O_2} + 7(B_{CO} + B_{N_2})\right) \times \left(G_C + \dfrac{G_S}{1.833}\right)}{3(B_{CO_2} + B_{CO})} \quad \textbf{35.12}$$

### Example 35.5

A sulfur-free coal has a proximate analysis of 75% carbon. The volumetric analysis of the flue gas is 80.2% nitrogen, 12.6% carbon dioxide, 6.2% oxygen, and 1.0% carbon monoxide. Calculate the (a) actual air/fuel ratio, (b) percentage excess air, (c) ultimate carbon dioxide, and (d) mass of flue gas per mass of fuel.

*Solution*

(a) Use Eq. 35.9.

$$R_{a/f,\text{actual}} = \frac{3.04B_{N_2}G_C}{B_{CO_2} + B_{CO}}$$
$$= \frac{(3.04)(0.802)(0.75)}{0.126 + 0.01}$$
$$= 13.4 \text{ lbm air/lbm fuel} \quad (\text{kg air/kg fuel})$$

(b) Use Eq. 35.10.

$$\frac{\text{actual}}{\text{excess air}} = \frac{(100\%)(B_{O_2} - 0.5B_{CO})}{0.264B_{N_2} - B_{O_2} + 0.5B_{CO}}$$
$$= \frac{(100\%)\big(0.062 - (0.5)(0.01)\big)}{(0.264)(0.802) - 0.062 + (0.5)(0.01)}$$
$$= 36.8\% \text{ by volume}$$

(c) Use Eq. 35.11.

$$\text{ultimate } CO_2 = \frac{(100\%)B_{CO_2,\text{actual}}}{1 - 4.773B_{O_2,\text{actual}}}$$
$$= \frac{(100\%)(0.126)}{1 - (4.773)(0.062)}$$
$$= 17.9\% \text{ by volume}$$

(d) Use Eq. 35.12.

$$\frac{\text{mass of flue gas}}{\text{mass of solid fuel}} = \frac{\left(11B_{CO_2} + 8B_{O_2} + 7(B_{CO} + B_{N_2})\right) \times \left(G_C + \dfrac{G_S}{1.833}\right)}{3(B_{CO_2} + B_{CO})}$$
$$= \frac{\big((11)(0.126) + (8)(0.062) + (7)(0.01 + 0.802)\big)(0.75)}{(3)(0.126 + 0.01)}$$
$$= 13.9 \text{ lbm flue gas/lbm fuel}$$
$$\quad (\text{kg flue gas/kg fuel})$$

## 33. TEMPERATURE OF FLUE GAS

The temperature of the gas at the furnace outlet—before the gas reaches any other equipment—should be approximately 550°F (300°C). Overly low temperatures mean there is too much excess air. Overly high temperatures—above 750°F (400°C)—mean that heat is being wasted to the atmosphere and indicate other problems (ineffective heat transfer surfaces, overfiring, defective combustion chamber, etc.).

The *net stack temperature* is the difference between the stack and local environment temperatures. The net stack temperature should be as low as possible without causing corrosion of the low end.

## 34. DEW POINT OF FLUE GAS MOISTURE

The *dew point* is the temperature at which the water vapor in the flue gas begins to condense in a constant pressure process. To avoid condensation and corrosion in the stack, the temperature of the flue gases must be above the dew point.

**Chemistry**

*Dalton's law* predicts the dew point of moisture in the flue gas. The partial pressure of the water vapor depends on the mole fraction (i.e., the volumetric fraction) of water vapor. The higher the water vapor pressure, the higher the dew point. Air entering a furnace can also contribute to moisture in the flue gas. This moisture should be added to the water vapor from combustion when calculating the mole fraction.

$$\text{partial pressure} = (\text{water vapor mole fraction})$$
$$\times (\text{flue gas pressure})$$

$$35.13$$

Once the water vapor's partial pressure is known, the dew point can be found from steam tables as the saturation temperature corresponding to the partial pressure.

When there is no sulfur in the fuel, the dew point is typically around $100°F$ ($40°C$). The presence of sulfur in virtually any quantity increases the actual dew point to approximately $300°F$ ($150°C$).[28]

## 35. HEAT OF COMBUSTION

The *heating value* of a fuel can be determined experimentally in a *bomb calorimeter*, or it can be estimated from the fuel's chemical analysis. The *higher heating value*, HHV (or *gross heating value*), of a fuel includes the heat of vaporization (condensation) of the water vapor formed from the combustion of hydrogen in the fuel. The *lower heating value*, LHV (or *net heating value*), assumes that all the products of combustion remain gaseous. The LHV is generally the value to use in calculations of thermal energy generated, since the heat of vaporization is not recovered within the furnace.

Traditionally, heating values have been reported on an HHV basis for coal-fired systems but on an LHV basis for natural gas-fired combustion turbines. There is an 11% difference between HHV and LHV thermal efficiencies for gas-fired systems and a 4% difference for coal-fired systems, approximately.

The HHV can be calculated from the LHV if the enthalpy of vaporization, $h_{fg}$, is known at the pressure of the water vapor.[29] In Eq. 35.14, $m_{\text{water}}$ is the mass of water produced per unit (lbm, mole, $m^3$, etc.) of fuel.

$$\text{HHV} = \text{LHV} + m_{\text{water}} h_{fg} \qquad 35.14$$

Only the hydrogen that is not locked up with oxygen in the form of water is combustible. This is known as the *available hydrogen*. The correct percentage of combustible hydrogen, $G_{\text{H,available}}$, is calculated from the hydrogen and oxygen fraction. Equation 35.15 assumes that all of the oxygen is present in the form of water.

$$G_{\text{H,available}} = G_{\text{H,total}} - \frac{G_O}{8} \qquad 35.15$$

*Dulong's formula* calculates the higher heating value of coals and coke with a 2% to 3% accuracy for moisture contents below approximately 10%.[30] The gravimetric or volumetric analysis percentages for each combustible element (including sulfur) are multiplied by the heating value per unit (mass or volume) from App. 35.A and summed.

$$\text{HHV}_{\text{MJ/kg}} = 32.78 G_C + 141.8 \left( G_H - \frac{G_O}{8} \right)$$
$$+ 9.264 G_S$$

$$[\text{SI}] \quad 35.16(a)$$

$$\text{HHV}_{\text{Btu/lbm}} = 14{,}093 G_C + 60{,}958 \left( G_H - \frac{G_O}{8} \right)$$
$$+ 3983 G_S$$

$$[\text{U.S.}] \quad 35.16(b)$$

The higher heating value of gasoline can be approximated from the Baumé specific gravity.

$$\text{HHV}_{\text{gasoline,MJ/kg}} = 42.61 + 0.093(°\text{Baumé} - 10)$$

$$[\text{SI}] \quad 35.17(a)$$

$$\text{HHV}_{\text{gasoline,Btu/lbm}} = 18{,}320 + 40(°\text{Baumé} - 10)$$

$$[\text{U.S.}] \quad 35.17(b)$$

The heating value of petroleum oils (including diesel fuel) can also be approximately determined from the oil's specific gravity. The values derived by using Eq. 35.18 may not exactly agree with values for specific oils because the equation does not account for refining methods and sulfur content. Equation 35.18 was originally intended for combustion at constant volume, as in a gasoline engine. However, variations in heating values for different oils are very small, and Eq. 35.18 is widely used as an approximation for all types of combustion, including constant pressure combustion in industrial boilers.

$$\text{HHV}_{\text{fuel oil,MJ/kg}} = 51.92 - 8.792(\text{SG})^2 \quad [\text{SI}] \quad 35.18(a)$$

$$\text{HHV}_{\text{fuel oil,Btu/lbm}} = 22{,}320 - 3780(\text{SG})^2 \ [\text{U.S.}] \quad 35.18(b)$$

---

[28]The theoretical dew point is even higher—up to $350°F$ to $400°F$ ($175°C$ to $200°C$). For complex reasons, the theoretical value is not attained.

[29]For the purpose of initial studies, the heat of vaporization is usually assumed to be 1040 Btu/lbm (2.42 kJ/kg). This corresponds to a partial pressure of approximately 1 psia (7 kPa) and a dew point of $100°F$ ($38°C$).

[30]The coefficients in Eq. 35.16 are slightly different from the coefficients originally proposed by Dulong. Equation 35.16 reflects currently accepted heating values that were unavailable when Dulong developed his formula. Equation 35.16 makes these assumptions: (1) None of the oxygen is in carbonate form. (2) There is no free oxygen. (3) The hydrogen and carbon are not combined as hydrocarbons. (4) Carbon is amorphous, not graphitic. (5) Sulfur is not in sulfate form. (6) Sulfur burns to sulfur dioxide.

## Example 35.6

A coal has an ultimate analysis of 93.9% carbon, 2.1% hydrogen, 2.3% oxygen, 0.3% nitrogen, and 1.4% ash. What are its (a) higher and (b) lower heating values in Btu/lbm?

*Solution*

(a) The noncombustible ash, oxygen, and nitrogen do not contribute to heating value. Some of the hydrogen is in the form of water. From Eq. 35.2, the available hydrogen fraction is

$$G_{H,available} = G_{H,total} - \frac{G_O}{8}$$

$$= 2.1\% - \frac{2.3\%}{8} = 1.8\%$$

From App. 35.A, the higher heating values of carbon and hydrogen are 14,093 Btu/lbm and 60,958 Btu/lbm, respectively. From Eq. 35.16, the total heating value per pound of coal is

$$HHV = (0.939)\left(14,093 \; \frac{Btu}{lbm}\right)$$

$$+ (0.018)\left(60,958 \; \frac{Btu}{lbm}\right)$$

$$= 14,331 \; Btu/lbm$$

(b) All of the combustible hydrogen forms water vapor. The mass of water produced is equal to the hydrogen mass plus eight times as much oxygen.

$$m_{water} = m_{H,available} + m_{oxygen}$$

$$= 0.018 + (8)(0.018)$$

$$= 0.162 \; lbm \; water/lbm \; coal$$

Assume that the partial pressure of the water vapor is approximately 1 psia (7 kPa). Then, from steam tables, the heat of condensation will be 1040 Btu/lbm. From Eq. 35.14, the lower heating value is approximately

$$LHV = HHV - m_{water}h_{fg}$$

$$= 14,331 \; \frac{Btu}{lbm} - \left(0.162 \; \frac{lbm}{lbm}\right)\left(1040 \; \frac{Btu}{lbm}\right)$$

$$= 14,163 \; Btu/lbm$$

Alternatively, the lower heating value of the coal can be calculated from by substituting the lower (net) hydrogen heating value from App. 35.A, 51,623 Btu/lbm, for the gross heating value of 60,598 Btu/lbm. This yields 14,160 Btu/lbm.

## 36. MAXIMUM THEORETICAL COMBUSTION (FLAME) TEMPERATURE

It can be assumed that the maximum theoretical increase in flue gas temperature will occur if all of the combustion energy is absorbed adiabatically by the smallest possible quantity of combustion products. This provides a method of estimating the *maximum theoretical combustion temperature*, also sometimes called the *maximum flame temperature* or *adiabatic flame temperature*.[31]

In Eq. 35.19, the mass of the products is the sum of the fuel, oxygen, and nitrogen masses for stoichiometric combustion. The mean specific heat is a gravimetrically weighted average of the values of $c_p$ for all combustion gases. (Since nitrogen comprises the majority of the combustion gases, the mixture's specific heat will be approximately that of nitrogen.) The heat of combustion can be found either from the lower heating value, LHV, or from a difference in air enthalpies across the furnace.

$$T_{max} = T_i + \frac{c_{p,mean}(\text{lower heat of combustion})}{m_{products}}$$

$$35.19$$

Due to thermal losses, incomplete combustion, and excess air, actual flame temperatures are always lower than the theoretical temperature. Most fuels produce flame temperatures in the range of 3350°F to 3800°F (1850°C to 2100°C).

## 37. COMBUSTION LOSSES

A portion of the combustion energy is lost in heating the dry flue gases, dfg.[32] This is known as *dry flue gas loss*. In Eq. 35.20, $m_{flue\;gas}$ is the mass of dry flue gas per unit mass of fuel. It can be estimated from Eq. 35.12. Although the full temperature difference is used, the specific heat should be evaluated at the average temperature of the flue gas. For quick estimates, the dry flue gas can be assumed to be pure nitrogen.

$$q_1 = m_{flue\;gas}c_p(T_{flue\;gas} - T_{incoming\;air}) \qquad 35.20$$

Heat is lost in the vapor formed during the combustion of hydrogen. In Eq. 35.21, $m_{vapor}$ is the mass of vapor per pound of fuel. $G_H$ is the gravimetric fraction of hydrogen in the fuel. The coefficient 8.94 is essentially $8 + 1 = 9$ and converts the gravimetric mass of hydrogen to gravimetric mass of water formed. $h_g$ is the enthalpy of superheated steam at the flue gas temperature and the partial pressure of the water vapor. $h_f$ is the enthalpy of saturated liquid at the air's entrance temperature.

$$q_2 = m_{vapor}(h_g - h_f) = 8.94G_H(h_g - h_f) \qquad 35.21$$

---

[31]Flame temperature is limited by the dissociation of common reaction products ($CO_2$, $N_2$, etc.). At high enough temperatures (3400°F to 3800°F; 1880°C to 2090°C), the endothermic dissociation process reabsorbs combustion heat and the temperature stops increasing. The temperature at which this occurs is known as the *dissociation temperature* (*maximum flame temperature*). This definition of flame temperature is not a function of heating values and flow rates.

[32]The abbreviation "dfg" for dry flue gas is peculiar to the combustion industry. It may not be recognized outside of that field.

**Chemistry**

Heat is lost when it is absorbed by moisture originally in the combustion air (and by free moisture in the fuel, if any). In Eq. 35.22, $m_{\text{combustion air}}$ is the mass of combustion air per pound of fuel. $\omega$ is the humidity ratio. $h'_g$ is the enthalpy of superheated steam at the air's entrance temperature and partial pressure of the water vapor.

$$q_3 = m_{\text{atmospheric water vapor}}(h_g - h'_g)$$
$$= \omega m_{\text{combustion air}}(h_g - h'_g) \qquad 35.22$$

When carbon monoxide appears in the flue gas, potential energy is lost in incomplete combustion. The difference in the two heating values in Eq. 35.23 is 9746 Btu/lbm (24.67 MJ/kg).[33]

$$q_4 = \frac{(\text{HHV}_C - \text{HHV}_{CO})G_C B_{CO}}{B_{CO_2} + B_{CO}} \qquad 35.23$$

For solid fuels, potential energy is lost in unburned carbon in the ash. (Some carbon may be carried away in the flue gas, as well.) This is known as *combustible loss* or *unburned fuel loss*. In Eq. 35.24, $m_{\text{ash}}$ is the mass of ash produced per pound of fuel consumed. $G_{C,\text{ash}}$ is the gravimetric fraction of carbon in the ash. The heating value of carbon is 14,093 Btu/lbm (32.8 MJ/kg).

$$q_5 = \text{HHV}_C m_{\text{ash}} G_{C,\text{ash}} \qquad 35.24$$

Energy is also lost through radiation from the exterior boiler surfaces. This can be calculated if enough information is known. The *radiation loss* is fairly insensitive to different firing rates, and once calculated it can be considered constant for different conditions.

Other conditions where energy can be lost include air leaks, poor pulverizer operation, excessive blowdown, steam leaks, missing or loose insulation, and excessive soot-blower operation. Losses due to these sources must be evaluated on a case-by-case basis.

## 38. COMBUSTION EFFICIENCY

The *combustion efficiency* (also referred to as *boiler efficiency*, *furnace efficiency*, and *thermal efficiency*) is the overall thermal efficiency of the combustion reaction. Furnace/boilers for all fuels (coal, oil, gas) with air heaters and economizers have 75% to 85% efficiencies, with all modern installation trending to the higher end of the range.

In Eq. 35.25, $m_{\text{steam}}$ is the mass of steam produced per pound of fuel burned. The useful heat may also be determined from the boiler rating. One *boiler*

*horsepower* is equal to approximately 33,475 Btu/hr (9.808 kW).[34]

$$\eta = \frac{\text{useful heat extracted}}{\text{heating value}}$$
$$= \frac{m_{\text{steam}}(h_{\text{steam}} - h_{\text{feedwater}})}{\text{HHV}} \qquad 35.25$$

Calculating the efficiency by subtracting all known losses is known as the *loss method*. Minor sources of thermal energy such as the entering air and feedwater are essentially disregarded.

$$\eta = \frac{\text{HHV} - q_1 - q_2 - q_3 - q_4 - q_5 - \text{radiation}}{\text{HHV}}$$
$$= \frac{\text{LHV} - q_1 - q_4 - q_5 - \text{radiation}}{\text{HHV}} \qquad 35.26$$

Combustion efficiency can be improved by decreasing either the temperature or the volume of the flue gas or both. Since the latent heat of moisture is a loss, and since the amount of moisture generated corresponds to the hydrogen content of the fuel, a minimum efficiency loss due to moisture formation cannot be eliminated. This minimum loss is approximately 13% for natural gas, 8% for oil, and 6% for coal.

## 39. DRAFT

The amount of air that flows through a furnace is determined by the furnace draft. *Draft* is the difference in static pressures that causes the flue gases to flow.[35] It is usually stated in inches of water (kPa). *Natural draft* (ND) furnaces rely on the *stack effect (chimney draft)* to draw off combustion gases. Air flows through the furnace and out the stack due to the pressure differential caused by reduced densities in the stack.[36]

*Forced draft* (FD) fans located before the furnace are used to supply air for burning. Combustion occurs under pressure, hence the descriptive term "pressure-fired unit" for such fans. FD fans are run at relatively high speeds (1200 rpm to 1800 rpm) with direct-drive motors. Two or more fans are used in parallel to provide for efficient operation at low furnace demand.

FD fans create a positive pressure (e.g., 2 in wg to 10 in wg; 0.5 kPa to 2.5 kPa). This pressure is reduced to a very small negative pressure after passing through the air heater, ducts, and windbox system. The negative pressure in the furnace serves to keep combustion gases from leaking out into the furnace and boiler areas. The pressure continues to drop as it passes through the

---

[33]Obsolete values such as 10,150 Btu/lbm (23.61 MJ/kg) and 10,190 Btu/lbm (23.70 MJ/kg) are still encountered for the difference in heating values between carbon and carbon monoxide.

[34]Boiler horsepower is sometimes equated with a gross heating rate of 44,633 Btu/hr (13.08 kW). However, this is the total incoming heating value assuming a standardized 75% combustion efficiency.
[35]The British term "draught" is synonymous with "draft."
[36]Chimneys that rely on natural draft are sometimes referred to as *gravity chimneys*.

Chemistry

boiler, economizer, air heater, and pollution-control equipment.

Whereas FD fans force air through the system, *induced draft* (ID) fans are used to draw combustion products through the furnace bed, stack, and pollution control system by injecting air into the stack after combustion.

The term "suction units" is used with ID fans. ID fans are located after dust collectors and precipitators (often at the base of the stack) in order to reduce the abrasive effects of fly ash. They are run at slower speeds than forced draft fans in order to reduce the abrasive effects even further. Unlike FD fans, ID fans are usually very large and powerful because they have to handle all of the combustion gases, not just combustion air.

Pure FD and ID systems are rarely used. Modern stack systems operate in a condition of *balanced draft.* Balanced draft is the term used when the static pressure is equal to atmospheric pressure. This requires the use of both ID and FD fans. In order to keep combustion products inside the combustion chamber and stack system, balanced draft systems may actually operate with a slight negative pressure.

Modern welded stacks are essentially "air-tight" and can operate at pressures above atmospheric, often eliminating the need for ID fans.

The *draft loss* is the static pressure drop due to friction through the boiler and stack. The *available draft* is the difference between the theoretical draft (stack effect) and the draft loss. The available draft is zero in a balanced system.

$$D_{\text{available}} = D_{\text{theoretical}} - D_{\text{friction}} \qquad 35.27$$

The *net rating* or *fan boost* of the fan is the total pressure (the difference of draft losses and draft gains) supplied by the fan at maximum operating conditions. The fan power for ID and FD fans is calculated, as with any fan application, from the net rating and the flow rate. It is customary in sizing ID and FD draft fans to include increases of approximately 15% for flow rate, 30% for pressure, and 20°C (10°C) for temperature.

## 40. STACK EFFECT

*Stack effect (chimney action* or *natural draft)* is a pressure difference caused by the difference in atmospheric air and flue gas densities. It can be relied on to draw combustion products at least partially up through the stack. The stack effect is determined with no flow of flue gas. With flow, some of the stack effect is converted to velocity head, and the remainder is used to overcome friction.

Generally, the higher the chimney, the greater the stack effect. However, the stack effect in modern plants is greatly reduced by the friction and cooling effects of economizers, air heaters, and precipitators. Fans supply the needed pressure differential. Therefore, the primary

function of modern stacks is to carry the combustion products a sufficient distance upward to dilute the combustion products, not to generate draft. Modern stacks for coal-fired power plants are seldom built shorter than 200 ft (60 m) in order to meet their dispersion requirements, and most are higher than 500 ft (150 m).

The theoretical stack effect, $D_{\text{theoretical}}$, is calculated from the densities of the flue gas and atmospheric air. In Eq. 35.28, $H_{\text{stack}}$ is the total height of the stack. The average temperature is used to determine the average density in Eq. 35.28.[37]

$$D_{\text{theoretical}} = H_{\text{stack}}(\gamma_{\text{air}} - \gamma_{\text{flue gas,ave}}) \qquad 35.28$$

If a few assumptions are made, the stack effect can be calculated from the average temperature of the flue gas (i.e., the temperature at the average stack elevation).[38] The average flue gas temperature is the temperature halfway up the stack. $H_{\text{stack}}$ is the total height of the stack.

$$D_{\text{theoretical}} = \left(\frac{p_{\text{air}}H_{\text{stack}}g}{R_{\text{air}}}\right)\left(\frac{1}{T_{\text{air}}} - \frac{1}{T_{\text{flue gas,ave}}}\right)$$
$$[\text{SI}] \quad 35.29(a)$$

$$D_{\text{theoretical}} = \left(\frac{p_{\text{air}}H_{\text{stack}}g}{R_{\text{air}}g_c}\right)\left(\frac{1}{T_{\text{air}}} - \frac{1}{T_{\text{flue gas,ave}}}\right)$$
$$[\text{U.S.}] \quad 35.29(b)$$

The *stack effect head,* $h_{\text{SE}}$, is used to calculate the stack velocity. The coefficient of velocity, $C_v$, is approximately 0.30 to 0.50.

$$h_{\text{SE}} = \frac{D_{\text{theoretical}}}{\gamma_{\text{flue gas,ave}}} \qquad 35.30$$

$$v = C_v v_{\text{ideal}} = C_v \sqrt{2gh_{\text{SE}}} \qquad 35.31$$

The required stack area is

$$A_{\text{stack}} = \frac{Q_{\text{flue gas}}}{v} \qquad 35.32$$

### Example 35.7

The air surrounding an 80 ft (24 m) tall vertical stack is at 70°F (21°C) and one standard atmospheric pressure. The temperature of the flue gas at the average stack elevation is 400°F (200°C). What is the theoretical natural draft?

---

[37]The actual stack effect will be less than the value calculated with Eq. 35.28. For realistic problems, the achievable stack effect probably should be considered to be 80% of the ideal.

[38]One assumption is that the pressure in the stack is equal to atmospheric pressure. Then, the density difference will be due to only the temperature difference. Also, the flue gas is assumed to be air. This makes it unnecessary to know the flue gas composition, molecular weight, and so on.

**Chemistry**

*SI Solution*

The absolute temperatures are

$$T_{\text{air}} = 21°C + 273° = 294K$$

$$T_{\text{flue gas}} = 200°C + 273° = 473K$$

Calculate the stack effect from Eq. 35.29.

$$D_{\text{theoretical}} = \left(\frac{p_{\text{air}}H_{\text{stack}}g}{R_{\text{air}}}\right)\left(\frac{1}{T_{\text{air}}} - \frac{1}{T_{\text{flue gas}}}\right)$$

$$= \left(\frac{(101.3 \text{ kPa})(24 \text{ m})\left(9.81 \frac{\text{m}}{\text{s}^2}\right)}{287 \frac{\text{J}}{\text{kg·K}}}\right)$$

$$\times \left(\frac{1}{294K} - \frac{1}{473K}\right)$$

$$= 0.107 \text{ kPa}$$

*Customary U.S Solution*

The absolute temperatures are

$$T_{\text{air}} = 70°F + 460° = 530°R$$

$$T_{\text{flue gas}} = 400°F + 460° = 860°R$$

Calculate the stack effect from Eq. 35.29.

$$D_{\text{theoretical}} = \left(\frac{p_{\text{air}}H_{\text{stack}}g}{R_{\text{air}}g_c}\right)\left(\frac{1}{T_{\text{air}}} - \frac{1}{T_{\text{flue gas}}}\right)$$

$$= \left(\frac{\left(14.7 \frac{\text{lbf}}{\text{in}^2}\right)\left(12 \frac{\text{in}}{\text{ft}}\right)^2(80 \text{ ft})}{53.3 \frac{\text{ft-lbf}}{\text{lbm-°R}}}\right)$$

$$\times \left(\frac{1}{530°R} - \frac{1}{860°R}\right)\left(\frac{32.2 \frac{\text{ft}}{\text{sec}^2}}{32.2 \frac{\text{ft-lbm}}{\text{lbf-sec}^2}}\right)$$

$$= 2.3 \text{ lbf/ft}^2$$

Convert this to inches of water.

$$D_{\text{theoretical}} = \frac{p}{\gamma} = \frac{\left(2.3 \frac{\text{lbf}}{\text{ft}^2}\right)\left(12 \frac{\text{in}}{\text{ft}}\right)}{62.4 \frac{\text{lbf}}{\text{ft}^3}}$$

$$= 0.44 \text{ in wg}$$

## 41. STACK FRICTION

Flue gas velocities of 15 ft/sec to 40 ft/sec (4.5 m/s to 12 m/s) are typical. As with most turbulent fluids, the friction pressure drop, $D$, through stack "piping" is proportional to the velocity head.[39]

$$D_{\text{friction}} = \frac{Kv^2}{2g} \qquad 35.33$$

Values of the friction coefficient, $K$, for chimney fittings are similar to those used in ventilating ductwork. Values for specific pieces of equipment (burners, pressure regulators, vents, etc.) must be provided by the manufacturers.

The approximate friction loss coefficient for a circular section of ductwork or chimney of length $L$ and diameter $d$ is given by Eq. 35.34. Equation 35.34 assumes a Darcy friction factor of 0.0233. For other friction factors, the constant terms can be scaled up or down proportionately. For noncircular ducts, the hydraulic diameter, $d$, can be replaced by four times the hydraulic radius (i.e., four times the duct area divided by the duct perimeter).

$$D_{\text{friction,kPa}} = \frac{0.119 L_{\text{m}}\rho_{\text{kg/m}^3}Q_{\text{L/s}}^2}{d_{\text{cm}}^5} \qquad \text{[SI]} \quad 35.34(a)$$

$$D_{\text{friction,in wg}} = \frac{28 L_{\text{ft}}\gamma_{\text{lbf/ft}^3}Q_{\text{cfs}}^2}{d_{\text{in}}^5} \qquad \text{[U.S.]} \quad 35.34(b)$$

# Topic VI: Biology

Biology

# 36 Cellular Biology

## Nomenclature

| | | |
|---|---|---|
| BOD | biochemical oxygen demand | mg/L |
| DO | dissolved oxygen | mg/L |
| $F$ | fraction of influent | – |
| $k$ | logistic growth rate constant | $d^{-1}$ |
| $K$ | biodegradation rate constant | $d^{-1}$ |
| $K$ | reaction rate constant | $d^{-1}$ |
| $K_a$ | partition coefficient | – |
| $K_{eq}$ | equilibrium constant | – |
| $L$ | ultimate BOD | mg/L |
| MW | molecular weight | g/mol |
| $pK_a$ | ionization constant | – |
| $P$ | fraction of digester suspended solids | – |
| $Q$ | volumetric flow rate | $m^3/s$ or $m^3/d$ |
| $t$ | time | d |
| TWA | time-weighted average | ppm |
| $x$ | cell or organism number or concentration | – |
| $x_0$ | initial concentration | mg/L |
| $x_\infty$ | carrying capacity | mg/L |
| $X$ | suspended solids | – |
| $Y$ | yield coefficient | – |

## Symbols

| | | |
|---|---|---|
| $\gamma$ | degree of reduction | – |
| $\mu$ | specific growth rate | $d^{-1}$ |
| $k$ | logistic growth constant | d |

## Subscripts

| | |
|---|---|
| 5 | 5 days |
| $b$ | biomass |
| $d$ | digester or diluted |
| $i$ | influent |
| $p$ | product |
| $s$ | substrate |
| $v$ | volatile |

## 1. CELL STRUCTURE

A cell is the fundamental unit of living organisms. Organisms are classified as either prokaryotes or eukaryotes. A *prokaryote* is a cellular organism that does not have a distinct nucleus. Examples of prokaryotes are bacteria and blue-green algae. Figure 36.1 shows the features of a typical prokaryotic cell.

**Figure 36.1** *Prokaryotic Cell Features*

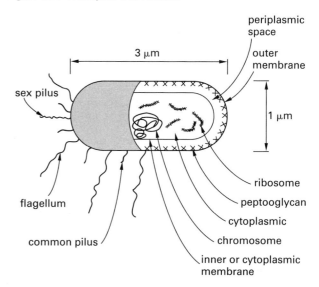

A *eukaryote* is an organism composed of one or more cells containing visibly evident nuclei and *organelles* (structures with specialized functions). Eukaryotic cells are found in protozoa, fungi, plants, and animals. For a long time, it was thought that eukaryotic cells were composed of an outer membrane, an inner nucleus, and a large mass of cytoplasm within the cell. Better experimental techniques revealed that eukaryotic cells also contain many organelles. Figure 36.2 shows the features of a typical animal and a typical plant cell. Note the size scale for the cells in Fig. 36.1 and Fig. 36.2.

A cell membrane consists of a double phospholipid layer in which the polar ends of the molecules point to the outer and inner surfaces of the membrane, and the nonpolar ends point to the center. A cell membrane also contains proteins, cholesterol, and glycoproteins. Some organisms, such as plant cells, have a cell wall made of carbohydrates that surrounds the cell membrane. Many single-celled organisms have specialized structures called *flagella* that extend outside the cell and help them to move.

**Biology**

**Figure 36.2** *Eukaryotic Cell Features*

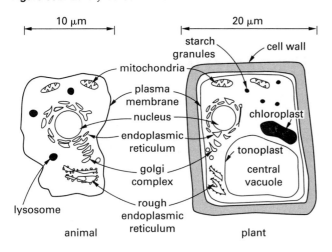

Some cell membranes can be stained crystal violet (i.e., red) for ease of research in the gram stain process. Many pathogenic cells, however, do not readily accept such staining and are referred to as *gram-negative cells.*

Cell size is limited by the transport of materials through the cell membrane. The volume of a cell is proportional to the cubic power of its average linear dimension; the surface area of a cell is proportional to the square of the average linear dimension. The material (nutrients, etc.) transported through a cell's membrane is proportional to its surface area, while the material composing the cell is proportional to the volume of the cell.

The *cytoplasm* includes all the contents of the cell other than the nucleus. *Cytosol* is the watery solution of the cytoplasm. It contains dissolved glucose and other nutrients, salts, enzymes, carbon dioxide, and oxygen. In eukaryotic cells, the cytoplasm also surrounds the organelles.

The *endoplasmic reticulum* is one example of an organelle that is found in all eukaryotic cells. A rough endoplasmic reticulum has sub-microscopic organelles called ribosomes attached to it. Amino acids are bound together at the surface of ribosomes to form proteins.

The *Golgi apparatus*, or Golgi complex, is an organelle found in most eukaryotic cells. It looks like a stack of flattened sacks. The Golgi apparatus is responsible for accepting materials (mostly proteins), making modifications, and packaging the materials for transport to specific areas of the cell.

The *mitochondria* are specialized organelles that are the major site of energy production via aerobic respiration in eukaryotic cells. They convert organic materials into energy.

*Lysosomes* are other organelles located within eukaryotic cells that serve the specialized function of digestion within the cell. They break lipids, carbohydrates, and proteins into smaller particles that can be used by the rest of the cell.

*Chloroplasts* are organelles located within plant cells and algae that conduct photosynthesis. Chlorophyll is the green pigment that absorbs energy from sunlight during photosynthesis. During photosynthesis, the energy from light is captured and eventually stored as sugar.

*Vacuoles* are found in some eukaryotic cells. They serve many functions including storage, separation of harmful materials from the remainder of the cell, and maintaining fluid balance or cell size. Vacuoles are separated from the cytoplasm by a single membrane called the tonoplast. Most mature plant cells contain a large central vacuole that occupies the largest volume of any single structure within the cell.

Two main types of nucleic acids are found in cells—*deoxyribonucleic acid* (DNA) and *ribonucleic acid*, (RNA). DNA is found within the nucleus of eukaryotic cells and within the cytoplasm in prokaryotic cells. DNA contains the genetic sequence that is passed on during reproduction. This sequence governs the functions of cells by determining the sequence of amino acids that are combined to form proteins. Each species manufactures its own unique proteins.

RNA is found both within the nucleus and in the cytoplasm of prokaryotic and eukaryotic cells. Most RNA molecules are involved in protein synthesis. Messenger RNA (mRNA) serves the function of carrying the DNA sequence from the nucleus to the rest of the cell. Ribosomal RNA (rRNA) makes up part of the ribosomes, where amino acids are bound together to form proteins. Transfer RNA (tRNA) carries amino acids to the ribosomes for protein synthesis.

## 2. CELL TRANSPORT

The transfer of materials across membrane barriers occurs by means of several mechanisms.

- *Passive diffusion* in cells is similar to the transfer that occurs in non-living systems. Material moves spontaneously from a region of high concentration to a region of low concentration. The rate of transfer obeys Fick's Law, the principle governing passive diffusion in dilute solutions. The rate is proportional to the concentration gradient across the membrane. In the case of living beings, passive diffusion is affected by lipid solubility (high solubility increases the rate of transport), the size of the molecules (the rate of transport increases with decreasing size of molecules), and the degree of ionization (the rate of transport increases with decreasing ionization). A *partition coefficient*, $K_a$, or the relative solubility of the solute in lipid to its solubility in water, can be used to describe the effect of lipid solubility on transport.

$$K_a = \frac{\text{concentration in lipid}}{\text{concentration in water}} \qquad 36.1$$

It is more common to report $\log K_a$ than $K_a$ because the relationship between $pK_a$ and permeability is fairly linear.

$$pK_a = \log K_a = \log_{10}\left(\frac{\text{concentration in lipid}}{\text{concentration in water}}\right) \quad \textit{36.2}$$

The permeability of a molecule across a membrane of thickness $x$ is

$$\text{permeability}_{\text{cm/s}} = \frac{pK_a\left(\begin{array}{c}\text{diffusion}\\ \text{coefficient}\\ \text{in water}\end{array}\right)_{\text{cm}^2/\text{s}}}{x_{\text{cm}}} \quad \textit{36.3}$$

The pH of a solution will have an effect on the partition coefficient. Although the relationship is complex, for weakly acidic and basic solutions $pK_a$ and pH are approximately related by the *Henderson-Hasselbach equation*. In Eq. 36.4 and Eq. 36.5, [X] designates the molar concentration of component X. Equation 36.4 is based on the dissociation of a weak acid according to

$$HA + H_2O \rightleftharpoons H_3O^+ + A^-$$

Or,

$$HA \rightleftharpoons H^+ + A^-$$

Equation 36.5 is based on the dissociation of the salt of a weak base according to

$$B + H_3O^+ \rightleftharpoons H_2O + BH^+$$

Or,

$$B + H^+ \rightleftharpoons BH^+$$

$$pK_a - pH = \log_{10}\left(\frac{\text{nonionized form}}{\text{ionized form}}\right)$$
$$= \log_{10}\frac{[HA]}{[A^-]} \quad \text{[weakly acidic]} \quad \textit{36.4}$$

$$pK_a - pH = \log_{10}\left(\frac{\text{ionized form}}{\text{nonionized form}}\right)$$
$$= \log_{10}\frac{[HB^+]}{[B]} \quad \text{[weakly basic]} \quad \textit{36.5}$$

- *Facilitated diffusion* is transfer during which a permease or membrane enzyme carries the substance across the membrane.

- Active diffusion, of which there are three categories, is transfer forced by a pressure gradient or "piggyback" function.

*Membrane pumping*, where cell membrane proteins called *permeases* transport the substance in a direction opposite the direction of passive diffusion.

*Endocytosis*, a process where cells absorb a material by surrounding (i.e., engulfing) it with their membrane.

*Exocytosis*, during which a secretory vesicle expels material that was within the cell to an area *outside the cell*.

- Other specialized mechanisms for specific organs.

In the case of many gram-negative bacteria with dual (i.e., inner and outer) wall membranes, including *Escherichia coli (E. coli)*, sugars and amino acids are transported across the inner plasma membrane by water-soluble proteins located in the periplasmic space between the two membranes.

### Example 36.1

$HC_2H_3O_2$ has an acid dissociation constant of $1.8 \times 10^{-5}$. Calculate the pH of a buffer solution made from 0.25M $HC_2H_3O_2$ and 0.050M $C_2H_3O_2^-$.

*Solution*

$HC_2H_3O_2$ is acetic acid, a weak acid. Use the Henderson-Hasselbach equation with Eq. 36.2.

$$pH = pK_a + \log\frac{[A^-]}{[HA]}$$
$$= pK_a + \log\frac{[C_2H_3O_2^-]}{[HC_2H_3O_2]}$$
$$= -\log\left(1.8 \times 10^{-5}\right) + \log\frac{0.50M}{0.25M}$$
$$= 4.7 + 0.30 = 5.0$$

### 3. ORGANISMAL GROWTH IN A BATCH CULTURE

Bacterial organisms can be grown (cultured) in a nutritive medium. The rate of growth follows the phases depicted in Fig. 36.3.

**Figure 36.3** *Organismal Growth in Batch Culture*

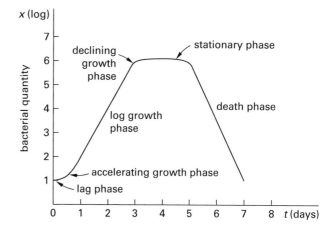

As Fig. 36.3 shows, the *lag phase* begins immediately after inoculation of the microbes into the nutrient medium. In this period, the microbial cells adapt to their new environment. The microbes might have to produce new enzymes to take advantage of new nutrients they are being exposed to, or they may need to adapt to different concentrations of solutes or to different temperatures than they are accustomed to. The length of the lag phase depends on the differences between the conditions the microbes experience before and after inoculation. The cells will start to divide when the requirements for growth are satisfied.

The *exponential*, or *logarithmic, growth phase* follows the lag phase. During this phase the number of cells increases according to the following equation.

$$\mu = \frac{1}{x}\,\frac{dx}{dt} \qquad\qquad 36.6$$

In this equation, $x$ is the cell or organism number or concentration, $t$ is time, and $\mu$ is the specific growth rate during the exponential phase.

The *declining growth phase* follows the log phase. During this phase, one or more essential nutrients are depleted and/or waste products are accumulated at levels that slow cell growth.

The *stationary phase* begins after the decelerating growth phase. During the stationary phase, the net growth rate of the cells is zero. New cells are created at the same rate at which cells are dying.

The *death phase* follows the stationary phase. During this phase, the death rate exceeds the growth rate.

The *logistic equation* is used to represent the population quantity up to (but excluding) the death phase. The curve takes on a *sigmoidal shape*, also known as an *S-curve* or *bounded exponential growth curve*. In the logistic formulation, the specific growth rate is related to the *carrying capacity*, $x_\infty$, which is the maximum population the environment can support. Carrying capacity depends on the specific culture, medium, and conditions. The equation for exponential growth rate is

$$\mu = k\left(1 - \frac{x}{x_\infty}\right) \qquad\qquad 36.7$$

In this equation, $k$ is the *logistic growth rate constant*, and $x$ is the number of organisms at time $t$. Therefore, Eq. 36.7 can be written for growth including the initial stationary phase as

$$\frac{dx}{dt} = kx\left(1 - \frac{x}{x_\infty}\right) \qquad\qquad 36.8$$

Integration of Eq. 36.8 gives the equation for the number of cells or organisms as a function of time.

$$x = \frac{x_0}{x_\infty}\left(1 - e^{kt}\right) \qquad\qquad 36.9$$

## 4. MICROORGANISMS

Microorganisms include viruses, bacteria, fungi, algae, protozoa, worms, rotifers, and crustaceans. Microorganisms are organized into three broad groups based on their structural and functional differences. The groups are called *kingdoms*. The three kingdoms are animals (rotifers and crustaceans), plants (mosses and ferns), and *Protista* (bacteria, algae, fungi, and protozoa). Bacteria and protozoa of the kingdom Protista make up the major groups of microorganisms in the biological system that is used in secondary treatment of wastewater.

### Pathogens

Organisms causing infectious diseases are categorized as *pathogens*. Pathogens are found in fecal wastes that are transmitted by exposure to wastewater. Pathogens will proliferate in areas where sanitary disposal of feces is not adequately practiced and where contamination of water supply from infected individuals is not properly controlled. The wastes may also be improperly discharged into surface waters, making the water *nonpotable* (unfit for drinking). Certain shellfish can become toxic when they concentrate pathogenic organisms in their tissues, increasing the toxic levels much higher than the levels in the surrounding waters.

Organisms that are considered to be pathogens include bacteria, protozoa, viruses, and helminths (worms). Table 36.1 lists potential waterborne diseases, the causative organisms, and the typical infection sources.

Not all microorganisms are considered pathogens. Some microorganisms are exploited for their usefulness in wastewater processing. Most wastewater engineering (and an increasing portion of environmental engineering) involves designing processes and facilities that use microorganisms to destroy organic and inorganic substances.

### Microbe Categorization

Carbon is the basic building block for cell synthesis, and it is prevalent in large quantities in wastewater. Wastewater treatment mixes carbon with microorganisms that are subsequently removed from the water by settling. Therefore, the growth of organisms that use organic material as energy is encouraged.

If a microorganism uses organic material as its carbon supply, it is *heterotrophic*. *Autotrophs* require only carbon to supply their energy needs. Organisms that rely only on the sun for energy are called *phototrophs*. *Chemotrophs* extract energy from organic or inorganic oxidation/reduction (redox) reactions. *Organotrophs* use organic materials, while *lithotrophs* oxidize inorganic compounds. Figure 36.4 may be used to categorize microbes.

Most microorganisms in wastewater treatment processes are bacteria. Conditions in the treatment plant are readjusted so that chemoheterotrophs predominate.

**Table 36.1** *Potential Pathogens*

| name of organism | major disease | source |
|---|---|---|
| **Bacteria** | | |
| *Salmonella typhi* | typhoid fever | human feces |
| *Salmonella paratyphi* | paratyphoid fever | human feces |
| other *Salmonella* | salmonellosis | human/animal feces |
| *Shigella* | bacillary dysentery | human feces |
| *Vibriocholerae* | cholera | human feces |
| *Enteropathogeniccoli* | gastroenteritis | human feces |
| *Yersiniaenterocolitica* | gastroenteritis | human/animal feces |
| *Campylobacterjejuni* | gastroenteritis | human/animal feces |
| *Legionella pneumophila* | acute respiratory illness | thermally enriched waters |
| *Mycobacterium* | tuberculosis | human respiratory exudates |
| other *Mycobacteria* | pulmonary illness | soil and water |
| Opportunistic bacteria | variable | natural waters |
| **Enteric Viruses/Enteroviruses** | | |
| *Polioviruses* | poliomyelitis | human feces |
| *Coxsackieviruses A* | aseptic meningitis | human feces |
| *Coxsackieviruses B* | aseptic meningitis | human feces |
| *Echoviruses* | aseptic meningitis | human feces |
| other *Enteroviruses* | encephalitis | human feces |
| *Reoviruses* | upper respiratory and gastrointestinal illness | human/animal feces |
| *Rotaviruses* | gastroenteritis | human feces |
| *Adenoviruses* | upper respiratory and gastrointestinal illness | human feces |
| *Hepatitis A* virus | infectious hepatitis | human feces |
| *Norwalk* and related gastrointestinal viruses | gastroenteritis | human feces |
| **Fungi** | | |
| *Aspergillus* | ear, sinus, lung, and skin infections | airborne spores |
| *Candida* | yeast infections | various |
| **Protozoa** | | |
| *Acanthamoeba castellani* | amoebic meningoencephalitis | soil and water |
| *Balantidium coli* | balantidosis (dysentery) | human feces |
| *Cryptosporidium*[*] | cryptosporidiosis | human/animal feces |
| *Entamoeba histolytica* | amoebic dysentery | human feces |
| *Giardia lamblia* | giardiasis (gastroenteritis) | human/animal feces |
| *Naegleria fowleri* | amoebic meningoencephalitis | soil and water |
| **Algae (blue-green)** | | |
| *Anabaena flos-aquae* | gastroenteritis (possible) | natural waters |
| *Microcystis aeruginosa* | gastroenteritis (possible) | natural waters |
| *Alphanizomenon flos-aquae* | gastroenteritis (possible) | natural waters |
| *Schizothrix calciola* | gastroenteritis (possible) | natural waters |
| **Helminths (intestinal parasites/worms)** | | |
| *Ascaris lumbricoides* (roundworm) | digestive disturbances | ingested worm eggs |
| *E. vericularis* (pinworm) | any part of the body | ingested worm eggs |
| Hookworm | pneumonia, anemia | ingested worm eggs |
| Threadworm | abdominal pain, nausea, weight loss | ingested worm eggs |
| *T. trichiuro* (whipworm) | trichinosis | ingested worm eggs |
| Tapeworm | digestive disturbances | ingested worm eggs |

[*]Disinfectants have little effect on *Cryptosporidia*. Most large systems now use filtration, the most effective treatment to date against *Cryptosporidia*.

**Biology**

**Figure 36.4** *Microbe Categorization Decision Tree*

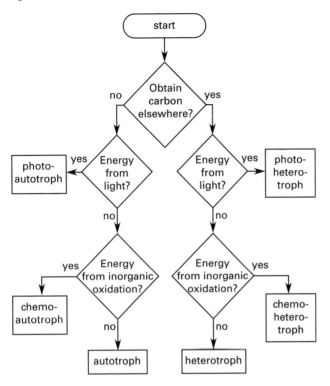

Each species of bacteria reproduces most efficiently within a limited range of temperatures. Table 36.2 shows these types and their most viable temperature ranges.

**Table 36.2** *Best Temperatures for Bacterial Growth*

| bacteria type | best temperature range for growth |
|---|---|
| psychrophiles | below 68°F (20°C) |
| mesophiles | 68°F (20°C) to 113°F (45°C) |
| thermophiles | 113°F (45°C) to 140°F (60°C) |
| stenothermophiles | above 140°F (60°C) |

Because most reactions proceed slowly at some temperatures, cells use *enzymes* to speed up the reactions and control the rate of growth. Enzymes are proteins, ranging from simple structures to complex conjugates, and are specialized for the reactions they catalyze.

The temperature ranges in Table 36.2 are qualitative and somewhat subjective. The growth range of facultative thermophiles extends from the thermophilic range into the mesophilic range. Bacteria will grow and survive in a very large range of temperatures. *E. coli,* for example, is classified as a mesophile. It grows best at temperatures between 68°F (20°C) and 122°F (50°C) but can continue to reproduce at temperatures down to 32°F (0°C).

*Nonphotosynthetic bacteria* are classified into two groups, heterotrophic and autotrophic, by their sources of nutrients and energy. *Heterotrophs* use organic matter as both an energy source and a carbon source for synthesis.

Heterotrophs are further subdivided into groups depending on their behavior toward free oxygen: aerobes, anaerobes, and facultative bacteria. *Obligate aerobes* require free dissolved oxygen while they decompose organic matter to gain energy for growth and reproduction. *Obligate anaerobes* oxidize organics in the complete absence of dissolved oxygen by using the oxygen bound in other compounds, such as nitrate and sulfate. *Facultative bacteria* comprise a group that uses free dissolved oxygen when available but that can also behave anaerobically in the absence of free dissolved oxygen (also known as *anoxic conditions*). Under anoxic conditions, a group of facultative anaerobes, called *denitrifiers,* uses nitrites and nitrates instead of oxygen. Nitrate nitrogen is converted to nitrogen gas in the absence of oxygen. This process is called *anoxic denitrification.*

*Autotrophic bacteria (autotrophs)* oxidize inorganic compounds for energy, use free oxygen, and use carbon dioxide as a carbon source. Significant members of this group are the *Leptothrix* and *Crenothrix* families of *iron bacteria.* These have the ability to oxidize soluble ferrous iron into insoluble ferric iron. Because soluble iron is often found in well waters and iron pipe, these bacteria deserve some attention. They thrive in water pipes where dissolved iron is available as an energy source and bicarbonates are available as a carbon source. As the colonies die and decompose, they release foul tastes and odors and have the potential to cause staining of porcelain or fabrics.

Table 36.3 lists selected microbial cells and describes some of their characteristics.

## Viruses

*Viruses* are parasitic organisms that can only be seen with an electron microscope and grow and reproduce only inside living cells, although they can survive outside the host. They are not cells, but particles composed of a protein sheath surrounding a nucleic-acid core. Most viruses of interest in supply water range in size from 10 nm to 25 nm. They pass through filters that retain bacteria.

Viruses invade living cells, and the viral genetic material redirects cell activities toward production of new viral particles. A large number of viruses are released when the infected cell dies. Viruses are host-specific, attacking only one type of organism.

There are more than 100 types of human enteric viruses. Those of interest in drinking water are *Hepatitis A, Norwalk*-type viruses, *Rotaviruses, Adenoviruses, Enteroviruses,* and *Reoviruses.*

## Bacteria

*Bacteria* are microscopic plants having round, rodlike, spiral, or filamentous single-celled or noncellular bodies. They are often aggregated into colonies. Bacteria use soluble food and reproduce through binary fission. Most

Biology

**Table 36.3** Characteristics of Selected Microbial Cells

| organism genus or type | type | metabolism[a] | gram reaction[b] | morphological characteristics[c] |
|---|---|---|---|---|
| aspergillus | mold | chemoorganotroph-aerobic and facultative | − | filamentous fan-like or cylindrical conidia and various spores |
| bacillus | bacteria | chemoorganotroph-aerobic | positive | rod—usually motile; spore; can be significant extracellular material |
| candida | yeast | chemoorganotroph-aerobic and facultative | − | usually oval, but can form elongated cells, mycelia and various spores |
| chromatium | bacteria | photoautotroph-anaerobic | n/a | rods—motile; some extracellular material |
| clostridium | bacteria | chemoorganotroph-anaerobic | positive | rods—usually motile; spore; some extracellular slime |
| enterobacter | bacteria | chemoorganotroph-facultative | negative | rod—motile; significant extracellular material |
| escherichia | bacteria | chemoorganotroph-facultative | negative | rod—may or may not be motile, variable extracellular material |
| lactobacillus | bacteria | chemoogranotroph-facultative | variable | rod—chains—usually nonmotile; little extracellular material |
| methanobacterium | bacteria | chemoautotroph-anaerobic | unknown | rods or cocci—motility unknown; some extracellular slime |
| nitrobacter | bacteria | chemoautotroph-aerobic; can use nitrite as electron donor | negative | short rod—usually nonmotile; little extracellular material |
| pseudomonas | bacteria | chemoorganotroph-aerobic and some chemolithotroph facultative (using $NO_3$ as electron acceptor) | negative | rods—motile; little extracellular slime |
| rhizobium | bacteria | chemoorganotroph-aerobic; nitrogen fixing | negative | rods—motile; copius extracellular slime |
| saccharomyces | yeast | chemoorganotroph-facultative | − | spherical or ellipsoidal; reproduced by budding; can form various spores |
| spirogyra | algae | photoautotroph-aerobic | n/a | rod/filaments; little extracellular material |
| staphylococcus | bacteria | chemoogranotroph-facultative | positive | cocci—nonmotile; moderate extracellular material |
| thiobacillus | bacteria | chemoautotroph-facultative | negative | rods—motile; little extracellular slime |

[a]aerobic—requires or can use oxygen as an electron receptor facultative—can vary the electron receptor from oxygen to organic materials anaerobic—organic or inorganics other than oxygen serve as electron acceptor
chemoorganotrophs—derive energy and carbon from organic materials
chemoautotrophs—derive energy from organic carbons and carbon from carbon dioxide Some species can also derive energy from inorganic sources.
photolithotrophs—derive energy from light and carbon from $CO_2$. May be aerobic or anaerobic.
[b]Gram negative indicates a complex cell wall with a lipopolychaccharide outer layer; gram positive indicates a less complicated cell wall with a peptide-based outer layer.
[c]Extracellular material production usually increases with reduced oxygen levels (e.g., facultative). Carbon source also affects production; extracellular material may be polysaccharides and/or proteins; statements are to be understood as general in nature.

bacteria are not pathogenic to humans, but they do play a significant role in the decomposition of organic material and can have an impact on the aesthetic quality of water.

## Fungi

*Fungi* are aerobic, multicellular, nonphotosynthetic, heterotrophic, eukaryotic protists. Most fungi are saprophytes that degrade dead organic matter. Fungi grow in low-moisture areas, and they are tolerant of low-pH environments. Fungi release carbon dioxide and nitrogen during the breakdown of organic material.

Fungi are obligate aerobes that reproduce by a variety of methods including fission, budding, and spore formation. They form normal cell material with one-half the nitrogen required by bacteria. In nitrogen-deficient

wastewater, they may replace bacteria as the dominant species.

## Algae

*Algae* are autotrophic, photosynthetic organisms (*photoautotrophs*) and may be either unicellular or multicellular. They take on the color of the pigment that is the catalyst for photosynthesis. In addition to chlorophyll (green), different algae have different pigments, such as carotenes (orange), phycocyanin (blue), phycoerythrin (red), fucoxanthin (brown), and xanthophylls (yellow).

Algae derive carbon from carbon dioxide and bicarbonates in water. The energy required for cell synthesis is obtained through photosynthesis. Algae and bacteria have a symbiotic relationship in aquatic systems,

Biology

with the algae producing oxygen used by the bacterial population.

In the presence of sunlight, the photosynthetic production of oxygen is greater than the amount used in respiration. At night algae use up oxygen in respiration. If the daylight hours exceed the night hours by a reasonable amount, there is a net production of oxygen.

Excessive algal growth (*algal blooms*) can result in supersaturated oxygen conditions in the daytime and anaerobic conditions at night.

Some algae create tastes and odors in natural water. While they are not generally considered pathogenic to humans, algae do cause turbidity, and turbidity favors microorganisms that are pathogenic.

## Protozoa

*Protozoa* are single-celled animals that reproduce by *binary fission* (dividing in two). Most are aerobic chemoheterotrophs (*facultative heterotrophs*). Protozoa have complex digestive systems and use solid organic matter, including algae and bacteria, as food. Therefore, they are desirable in wastewater effluent because they act as polishers by consuming remaining bacteria.

Protozoa are categorized into *flagellates*, *amoeboids*, *sporozoans*, and *ciliates* according to their means of locomotion.

*Flagellated protozoa* are the smallest protozoans. Their *flagella* (long hairlike strands) provide mobility through a whiplike action. *Amoeba* move and take in food through the action of a mobile protoplasm. Free-swimming protozoa have *cilia* (small hairlike features) used for propulsion and gathering in organic matter. Sporozoans do not locomote under their own power at all.

## Worms and Rotifers

A number of worms and rotifers are of importance to water quality. *Rotifers* are aerobic, multicellular chemoheterotrophs. The rotifer derives its name from the apparent rotating motion of two sets of cilia on its head. The cilia provide mobility and a mechanism for catching food. Rotifers consume bacteria and small particles of organic matter.

Many worms are aquatic parasites. *Flatworms* of the class *Trematoda* are known as *flukes*, and the *Cestoda* are tapeworms. *Nematodes* of public health concern are *Trichinella*, which causes trichinosis; *Necator*, which causes pneumonia; *Ascaris*, which is the common roundworm; and *Filaria*, which causes filariasis.

## Mollusks

*Mollusks*, such as mussels and clams, are characterized by a shell structure. They are aerobic chemoheterotrophs that feed on bacteria and algae. They are a source of food for fish and are not found in wastewater treatment systems to any extent, except in underloaded lagoons. Their presence is indicative of a high level of dissolved oxygen and a very low level of organic matter.

*Macrofouling* is a term referring to infestation of water inlets and outlets by clams and mussels. For example, *zebra mussels* were accidentally introduced into the United States in 1986 and are particularly troublesome for several reasons. First, young zebra mussels are microscopic and can easily pass through intake screens. Second, they attach to anything, even other mussels, which produces thick mussel colonies. Third, adult zebra mussels quickly sense biocides, most notably those that are halogen-based, like chlorine. They quickly close and remain closed for days or weeks.

The use of biocides to control the growth of zebra mussels is controversial. Chlorination treatment is recommended with some caution since it results in increased toxicity, affecting other species, and THM (trihalomethane) production. An ongoing biocide program aimed at pre-adult mussels, combined with slippery polymer-based surface coatings, is most likely the best approach to prevention. Once a pipe is colonized, mechanical removal by scraping or water blasting is the only practical option.

## Indicator Organisms

The techniques for comprehensive bacteriological examination for pathogens are complex and time consuming. Isolating and identifying specific pathogenic microorganisms is a difficult and lengthy task. Many of these organisms require sophisticated tests that take several days to produce results. Because of these difficulties, and also because the number of pathogens relative to other microorganisms in water can be very small, *indicator organisms* are used as a measure of the quality of the water. The primary function of an indicator organism is to provide evidence of recent fecal contamination from warm-blooded animals.

Characteristics of a good indicator organism are:

(a) The indicator is always present when the pathogenic organism of concern is present. It is absent in clean, uncontaminated water.

(b) The indicator is present in fecal material in large numbers.

(c) The indicator responds to natural environmental conditions and to treatment processes in a manner similar to the pathogens of interest.

(d) The indicator is easy to isolate, identify, and enumerate.

(e) The ratio of indicator to pathogen should be high.

(f) The indicator and pathogen should come from the same source, such as gastrointestinal tract.

While there are several microorganisms that meet these criteria, *total coliform* and *fecal coliform* are the indicators generally used. *Total coliform* refers to the group of aerobic and facultatively anaerobic, gram-negative, non-spore-forming, rod-shaped bacteria that ferment lactose with gas formation within 48 hr at 95°F (35°C). This encompasses a variety of organisms, mostly of intestinal origin, including *E. coli*, which is the most numerous facultative bacterium in the feces of warm-blooded animals. Unfortunately, this group also includes *Enterobacter, Klebsiella,* and *Citrobacter,* which are present in wastewater but can be derived from other environmental sources such as soil and plant materials.

*Fecal coliforms* are a subgroup of the total coliforms that come from the intestines of warm-blooded animals. They are measured by running the standard total coliform fermentation test at an elevated temperature of 112°F (44.5°C), providing a means to distinguish false positives in the total coliform test.

Results of fermentation tests are reported as a *most probable number index* (MPN). This is an index of the number of coliform bacteria that, more than any other number, would give the results shown by the laboratory examination. MPN is not an actual enumeration.

## Metabolism/Metabolic Processes

*Metabolism* is a term given to describe all chemical activities performed by a cell. The cell uses *adenosine triphosphate* (ATP) as the principal energy currency in all processes. Those processes that allow the bacterium to synthesize new cells from the energy stored within its body are called *anabolic*. All biochemical processes in which cells convert substrate into useful energy and waste products are called *catabolic*.

## Decomposition of Waste

Decomposition of waste involves oxidation/reduction reactions and is classified as aerobic or anaerobic. The type of electron acceptor available for catabolism determines the type of decomposition used by a mixed culture of microorganisms. Each type of decomposition has peculiar characteristics that affect its use in waste treatment.

## Aerobic Decomposition

Molecular oxygen, $O_2$, must be present in order for decomposition to proceed by aerobic oxidation. The chemical end products of decomposition are primarily carbon dioxide, water, and new cell material as shown in Table 36.4. Odoriferous, gaseous end-products are kept to a minimum. In healthy natural water systems, aerobic decomposition is the principal means of self-purification.

A wide spectrum of organic material can be oxidized by aerobic decomposition. Aerobic oxidation releases large amounts of energy, meaning most aerobic organisms are capable of high growth rates. Consequently, there is a relatively large production of new cells in comparison with the other oxidation systems. This means that more biological sludge is generated in aerobic oxidation than in the other oxidation systems.

A laboratory analysis of organic matter in water often includes a biochemical oxygen demand (BOD) test.

Water quality laboratories have a ready supply of water that is saturated with oxygen, obtained by sparging air overnight through the water. To measure BOD, a sample of wastewater is diluted with oxygen-saturated

**Biology**

*Table 36.4 Waste Decomposition End Products*

| | representative end products | | |
| --- | --- | --- | --- |
| substrates | aerobic decomposition | anoxic decomposition | anaerobic decomposition |
| proteins and other organic nitrogen compounds | amino acids<br>ammonia → nitrites → nitrates<br>alcohols<br>organic acids $\Big\} \to CO_2 + H_2O$ | amino acids<br>nitrates → nitrites → $N_2$<br>alcohols<br>organic acids $\Big\} \to CO_2 + H_2O$ | amino acids<br>ammonia<br>hydrogen sulfide<br>methane<br>carbon dioxide<br>alcohols<br>organic acids |
| carbohydrates | alcohols<br>fatty acids $\Big\} \to CO_2 + H_2O$ | alcohols<br>fatty acids $\Big\} \to CO_2 + H_2O$ | carbon dioxide<br>alcohols<br>fatty acids |
| fats and related substances | fatty acids + glycerol<br>alcohols<br>lower fatty acids $\Big\} \to CO_2 + H_2O$ | fatty acids + glycerol<br>alcohols<br>lower fatty acids $\Big\} \to CO_2 + H_2O$ | fatty acids + glycerol<br>carbon dioxide<br>alcohols<br>lower fatty acids |

water, and a small amount of bacteria is added to the sample. Oxygen concentrations are measured at the beginning of the test and also on a daily basis. The difference between the oxygen concentration at the initial time and at time $t$ reported in mg/L gives a measure of the concentration of organic compounds in the water and is called the *BOD exerted* at time $t$ (typically 5 days). In this way, BOD provides a measure of the concentration of organic compounds in water without the complexity of analyzing the different compounds.

Aerobic decomposition is the preferred method for large quantities of dilute ($BOD_5 < 500$ mg/L) wastewater because decomposition is rapid and efficient and has a low odor potential. For concentrated wastewater ($BOD_5 > 1000$ mg/L), aerobic decomposition is not suitable because of the difficulty in supplying enough oxygen and because of the large amount of biological sludge that is produced.

### Anoxic Decomposition

Some microorganisms can use nitrates in the absence of oxygen to oxidize carbon. In wastewater treatment for the removal of nitrogen compounds, this is known as *denitrification*. The end products from denitrification are nitrogen gas, carbon dioxide, water, and new cell material. The amount of energy made available to the cell during denitrification is about the same as that made during aerobic decomposition. The production of cells, though not as high as in aerobic decomposition, is relatively high.

Denitrification is especially important in wastewater treatment when nitrogen must be removed. In such cases, a separate treatment process is used. An important consideration regarding anoxic decomposition relates to the final clarification of the treated wastewater. If the final clarifier becomes anoxic, the formation of nitrogen gas will cause large masses of sludge to float to the surface and escape from the treatment plant into the receiving water. Thus, it is necessary to ensure that anoxic conditions do not develop in the final clarifier.

### Anaerobic Decomposition

In order to achieve anaerobic decomposition, molecular oxygen and nitrate must not be present. Sulfate, carbon dioxide, and organic compounds that can be reduced serve as terminal electron acceptors. The reduction of sulfate results in the production of hydrogen sulfide, $H_2S$, and a group of equally odoriferous organic sulfur compounds called *mercaptans*.

The anaerobic decomposition of organic matter, also known as *fermentation*, is generally considered to be a two-step process. In the first step, complex organic compounds are fermented to low molecular weight *fatty acids (volatile acids)*. In the second step, the organic acids are converted to methane. Carbon dioxide serves as the electron acceptor.

Anaerobic decomposition produces carbon dioxide, methane, and water as the major end products. Additional end products include ammonia, hydrogen sulfide, and mercaptans. As a consequence of these last three compounds, anaerobic decomposition is characterized by a malodorous stench.

Because only small amounts of energy are released during anaerobic oxidation, the amount of cell production is low. Thus, sludge production is correspondingly low. Wastewater treatment based on anaerobic decomposition is used to stabilize sludge produced during aerobic and anoxic decomposition.

Direct anaerobic decomposition of wastewater generally is not feasible for dilute waste. The optimum growth temperature for the anaerobic bacteria is at the upper end of the mesophilic range. Therefore, to get reasonable biodegradation, the temperature of the culture must first be elevated. For dilute wastewater, this is not practical. Anaerobic digestion is quite appropriate for concentrated wastes ($BOD_5 > 1000$ mg/L).

## 5. FACTORS AFFECTING DISEASE TRANSMISSION

Waterborne disease transmission is influenced by the latency, persistence, and quantity (dose) of the pathogens. *Latency* is the period of time between excretion of a pathogen and its becoming infectious to a new host. *Persistence* is the length of time that a pathogen remains viable in the environment outside a human host. The *infective dose* is the number of organisms that must be ingested to result in disease.

### AIDS

*Acquired immunodeficiency syndrome* (AIDS) is caused by the *human immunodeficiency virus* (HIV). HIV is present in virtually all body excretions of infected persons, and therefore, is present in wastewater. However, the risk of contracting AIDS through contact with wastewater or working at a wastewater plant is very small, based on the following facts.

- HIV is relatively weak and does not remain viable for long in harsh environments such as wastewater.

- HIV is quickly inactivated by alcohol, chlorine, and exposure to air. The chlorine concentration present in many toilets, for example, is enough to inactivate HIV.

- HIV that survives disinfection has been found to be too dilute to be infectious.

- HIV replicates in white blood cells, not in the human intestinal tract. It has not been found to reproduce in wastewater.

- There is no evidence that HIV can be transmitted through water, air, food, or casual contact. HIV must enter the bloodstream directly, through a wound. It cannot enter through unbroken skin or through respiration.

- HIV is less infectious than the hepatitis virus.

- There are no reported AIDS cases linked to occupational exposure in wastewater collection and treatment.

## 6. STOICHIOMETRY OF SELECTED BIOLOGICAL SYSTEMS

This section shows four classes of biological reactions involving microorganisms, each with simplified stoichiometric equations.

Stoichiometric problems are known as *weight and proportion problems* because their solutions use simple ratios to determine the masses of reactants required to produce given masses of products, or vice versa. The procedure for solving these problems is essentially the same regardless of the reaction.

*step 1:* Write and balance the chemical equation. (For convenience in using the degree of reduction equation, reduce the reactant's chemical formula to a single carbon atom.)

*step 2:* Determine the atomic (molecular) weight of each element (compound) in the equation.

*step 3:* Multiply the atomic (molecular) weights by their respective coefficients and write the products under the formulas.

*step 4:* Write the given mass data under the weights determined in step 3.

*step 5:* Fill in the missing information by calculating simple ratios.

The first biological reaction is the production of biomass with a single extracellular product. Water and carbon dioxide are also produced as shown in the reaction equation.

$$\underset{[\text{substrate}]}{CH_mO_n} + aO_2 + bNH_3$$

$$\rightarrow c\underset{[\text{biomass}]}{CH_\alpha O_\beta N_\delta} + d\underset{[\text{product}]}{CH_x O_y N_z} + eH_2O + fCO_2$$

*36.10*

Equation 36.11 through Eq. 36.13 are used to calculate degrees of reduction (available electrons per unit of carbon) for substrate $(s)$, biomass $(b)$, and product $(p)$.

$$\gamma_s = 4 + m - 2n \qquad 36.11$$

$$\gamma_b = 4 + \alpha - 2\beta - 3\delta \qquad 36.12$$

$$\gamma_p = 4 + x - 2y - 3z \qquad 36.13$$

Table 36.5 shows typical degrees of reduction, $\gamma$. A high *degree of reduction* denotes a low degree of oxidation. Solving for the coefficients in Eq. 36.10 requires satisfying the carbon, nitrogen and electron balances, plus knowing the respiratory coefficient and a yield coefficient. The key biomass production and reduction factors involved in determining carbon, nitrogen, electron, and energy balances are shown in Eq. 36.14 through Eq. 36.17.

**Table 36.5** *Composition Data for Biomass and Selected Organic Compounds*

| compound | molecular formula | degree of reduction, $\gamma$ | molecular weight (MW) |
|---|---|---|---|
| generic biomass[*] | $CH_{1.64}N_{0.16}O_{0.52}$ | 4.17 ($NH_3$) | 24.5 |
| | $P_{0.0054}S_{0.005}$ | 4.65 ($N_3$) | |
| | | 5.45 ($HNO_3$) | |
| methane | $CH_4$ | 8 | 16.0 |
| $n$-alkane | $C_4H_{32}$ | 6.13 | 14.1 |
| methanol | $CH_4O$ | 6.0 | 32.0 |
| ethanol | $C_2H_6O$ | 6.0 | 23.0 |
| glycerol | $C_2H_6O_3$ | 4.67 | 30.7 |
| mannitol | $C_6H_{14}O_6$ | 4.33 | 30.3 |
| acetic acid | $C_2H_4O_2$ | 4.0 | 30.0 |
| lactic acid | $C_3H_6O_3$ | 4.0 | 30.0 |
| glucose | $C_6H_{12}O_6$ | 4.0 | 30.0 |
| formaldehyde | $CH_2O$ | 4.0 | 30.0 |
| gluconic acid | $C_6H_{12}O_7$ | 3.67 | 32.7 |
| succinic acid | $C_4H_6O_4$ | 3.50 | 29.5 |
| citric acid | $C_6H_8O_7$ | 3.0 | 32.0 |
| malic acid | $C_4H_6O_5$ | 3.0 | 33.5 |
| formic acid | $CH_2O_2$ | 2.0 | 46.0 |
| oxalic acid | $C_2H_2O_4$ | 1.0 | 45.0 |

[*]Sulfur is present in proteins.

Adapted from *Biochemical Engineering and Biotechnology Handbook* by B. Atkinson and F. Mavitona, Macmillan, Inc., 1983.

**Biology**

$$c + d + f = 1 \quad \text{[carbon]} \qquad 36.14$$

$$c\delta + dz = b \quad \text{[nitrogen]} \qquad 36.15$$

$$c\gamma_b + d\gamma_p = \gamma_s - 4a \quad \text{[electron]} \qquad 36.16$$

$$Q_o c\gamma_b + Q_o d\gamma_p = Q_o \gamma_s - Q_o 4a \quad \text{[energy]} \qquad 36.17$$

$Q_o$ is the heat evolved per equivalent (gram mole) of available electrons, approximately 26.95 kcal/mole of electrons.

The *respiratory quotient* (RQ) is the $CO_2$ produced per unit of $O_2$.

$$RQ = \frac{f}{a} \qquad 36.18$$

The coefficients $c$ and $d$ in Eq. 36.10 are referred to as maximum theoretical *yield coefficients* when expressed per gram of substrate. The yield coefficient can be given either as grams of cells or grams of product per gram of substrate.

$$Y_{\text{ideal},b} = \frac{m_b}{m_s} \qquad 36.19$$

$$Y_{\text{ideal},p} = \frac{m_p}{m_s} \qquad 36.20$$

The ideal yield coefficients are related to the actual yield coefficients by the *yield factor*.

$$\text{yield factor} = \frac{Y_{\text{actual}}}{Y_{\text{ideal}}} \qquad 36.21$$

The second reaction is the aerobic biodegradation of glucose in the presence of oxygen and ammonia. In this reaction, cells are formed, and carbon dioxide and water are the only products. The stoichiometric equation is

$$C_6H_{12}O_6 + aO_2 + bNH_3$$
$$\text{[substrate]}$$
$$\rightarrow cCH_{1.8}O_{0.5}N_{0.2} + dCO_2 + eH_2O \qquad 36.22$$
$$\text{[cells]}$$

For Eq. 36.19,

$$a = 1.94$$
$$b = 0.77$$
$$c = 3.88$$
$$d = 2.13$$
$$e = 3.68$$

The coefficient $c$ is the theoretical maximum yield coefficient, which may be reduced by a yield factor.

The third reaction is the anaerobic (no oxygen) biodegradation of organic wastes with incomplete stabilization (i.e., incomplete treatment). Methane, carbon dioxide, ammonia, and water as well as smaller organic waste

molecules are the products. The stoichiometric equation is

$$C_aH_bO_cN_d$$
$$\rightarrow nC_wH_xO_yN_z + mCH_4$$
$$+ sCO_2 + rH_2O + (d - nx)NH_3 \qquad 36.23$$

$$s = a - nw - m \qquad 36.24$$

$$r = c - ny - 2s \qquad 36.25$$

Knowledge of product composition, yield coefficient, and methane $CO_2$ ratio is needed.

The fourth reaction is the anaerobic biodegradation of organic wastes with complete stabilization. Besides organic waste, water is consumed in this reaction and the products are methane, carbon dioxide, and ammonia. The stoichiometric equation is

$$C_aH_bO_cN_d + rH_2O$$
$$\rightarrow mCH_4 + sCO_2 + dNH_3 \qquad 36.26$$

$$r = \frac{4a - b - 2c + 3d}{4} \qquad 36.27$$

$$s = \frac{4a - b + 2c + 3d}{8} \qquad 36.28$$

$$m = \frac{4a + b - 2c - 3d}{8} \qquad 36.29$$

Composition data for biomass and selected organic compounds is given in Table 36.5.

### Example 36.2

An organic compound has an empirical formula of $CH_{1.8}O_{0.5}N_{0.2}$. During combustion in oxygen gas, the compound dissociates and nitrogen is produced. What is the standard heat of combustion for one gram of this compound expressed in kJ/g?

*Solution*

Since this organic compound's chemical formula contains nitrogen, it matches the generic empirical formula for a "biomass" in Eq. 36.10. However, the compound is the reactant, not the product, and it should be considered as a substrate.

Write Eq. 36.10 for this compound with $c = \alpha = \beta = \delta = x = y = 0$.

$$CH_{1.8}O_{0.5}N_{0.2} + aO_2 \rightarrow dN_2 + eH_2O + fCO_2$$

Balance this reaction.

carbon, C: $f = 1$
nitrogen, N: $2d = 0.2$; $d = 0.1$
hydrogen, H: $2e = 1.8$; $e = 0.9$
oxygen, O: $2f + e = 0.5 + 2a$; $a = 1.2$

The balanced stoichiometric combustion reaction is

$$CH_{1.8}O_{0.5}N_{0.2} + 1.2O_2 \rightarrow 0.1N_2 + 0.9H_2O + CO_2$$

Use Eq. 36.11 (with $m = 1.8$ and $n = 0.5$) to determine the degree of reduction for a substrate with a single carbon atom.

$$\gamma_s = 4 + m - 2n$$
$$= 4 + 1.8 - (2)(0.5)$$
$$= 4.8$$

Now, use Eq. 36.17 to determine the molar heat of combustion.

molar heat of combustion

$$= -\gamma_s Q_o$$
$$= (4.8)\left(26.95 \frac{kcal}{mol}\right)\left(4.1868 \frac{kJ}{kcal}\right)$$
$$= 541.6 \text{ kJ/mol}$$

The approximate molecular weight of this compound is

$$MW = 12 + (1.8)(1) + (0.5)(16) + (0.2)(14) = 24.6$$

The ideal heat of combustion per unit mass is

$$\frac{541.6 \frac{kJ}{mol}}{24.6 \frac{g}{mol}} = 22.0 \text{ kJ/g}$$

## Example 36.3

Verify the heat of combustion from Ex. 36.2 using traditional combustion reaction methods.

*Solution*

The gravimetric fractions, $G$, of the elements in this fuel are

  carbon, C: $12/24.6 = 0.488$
  hydrogen, H: $1.8/24.6 = 0.073$
  oxygen, O: $8/24.6 = 0.325$
  nitrogen, N: $2.8/24.6 = 0.114$

The higher heat of combustion of a solid fuel in MJ/mg or kJ/g is approximately

$$HHV = 32.78 G_C + 141.8\left(G_H - \frac{G_O}{8}\right)$$
$$= (32.78)(0.488) + (141.8)\left(0.073 - \left(\frac{0.325}{8}\right)\right)$$
$$= 20.59 \text{ kJ/g}$$

**Biology**

# 37 Toxicology

## Nomenclature

| | | |
|---|---|---|
| AT | averaging time | yr |
| BCF | bioconcentration factor (mg/kg) in tissue/(mg/L) in water | L/kg |
| BW | body mass (weight) | kg |
| $C$ | concentration | ppm, mg/m$^3$ |
| CDI | chronic daily intake | mg/kg·d |
| CPF | carcinogen potency factor (same as slope factor) | (mg/kg·d)$^{-1}$ |
| CR | contact rate | day$^{-1}$ |
| $E$ | time weighted average exposure | ppm, mg/m$^3$ |
| ED | exposure duration | yr |
| EED | estimated exposure dose | mg/kg·d |
| EF | exposure factor | – |
| $E_m$ | equivalent exposure of mixture | – |
| EP | exposed population | – |
| $k$ | decay constant | d$^{-1}$ |
| $L$ | exposure limit of particular contaminants | mg/m$^3$ |
| LC$_{50}$ | lethal concentration | mg/m$^3$ |
| LD$_{50}$ | lethal dose | mg/kg |
| LOAEL | lowest observed adverse effect level | mg/kg·d |
| LOEL | lowest observed effect level | mg/kg·d |
| MF | modifying factor | – |
| NOAEL | no observed adverse effect level | mg/kg·d |
| NOEL | no observed effect level | mg/kg·d |
| $R$ | risk; probability of excess cancer | – |
| SF | slope factor (same as carcinogen potency factor) | (mg/kg·d)$^{-1}$ |
| $T$ | time of exposure | s |
| UF | uncertainty factor | – |

## Subscripts

| | |
|---|---|
| $m$ | mixture |
| $o$ | initial condition |
| org | organism |
| $t$ | time |
| $w$ | water (or other medium) |
| $x$ | toxicant |

## 1. INTRODUCTION

Toxicology is defined as the study of adverse effects of chemicals on living organisms.

This chapter examines the pathways of human exposure available to chemicals and other toxicants, the effects on workers of exposure to toxicants, dose-response relationships, methods determining safe human doses, and standards for worker protection.

## 2. EXPOSURE PATHWAYS

The human body has three primary exposure pathways: dermal absorption, inhalation, and ingestion. (See Fig. 37.1.) The eyes are an additional exposure pathway because they are particularly vulnerable to damage in the workplace.

### Dermal Absorption

The skin is composed of the epidermis, the dermis, and the subcutaneous layer. The *epidermis* is the upper layer, which is composed of several layers of flattened and scale-like cells. These cells do not contain blood vessels; they obtain their nutrients from the underlying dermis. The cells of the epidermis migrate to the surface, die, and leave behind a protein called *keratin*. Keratin is the most insoluble of all proteins, and together with the scale-like cells, provides extreme resistance to substances and environmental conditions. Beneath the epidermis lies the *dermis*, which contains blood vessels, connective tissue, hair follicles, sweat glands, and other glands. The dermis supplies the nutrients for itself and for the epidermis. The innermost layer is called the *subcutaneous fatty tissue*, which provides a cushion for the skin and connection to the underlying tissue.

The condition of the skin, and the chemical nature of any toxic substance it contacts, affect whether and at what rate the skin absorbs that substance. The epidermis is impermeable to many gases, water, and chemicals. However, if the epidermis is damaged by cuts and abrasions, or is broken down by repeated exposure to soaps, detergents, or organic solvents, toxic substances can readily penetrate and enter the bloodstream.

Chemical burns, such as those from acids, can also destroy the protection afforded by the epidermis, allowing toxicants to enter the bloodstream. Inorganic chemicals (and organic chemicals that are dissolved in water) are not readily absorbed through healthy skin.

**Biology**

**Figure 37.1** *Exposure Routes for Chemical Agents*

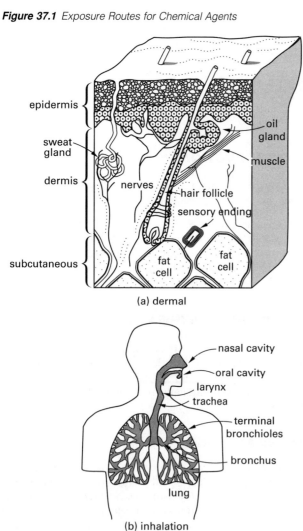

(a) dermal

(b) inhalation

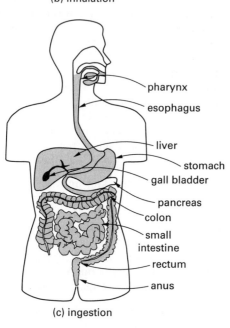

(c) ingestion

However, many organic solvents are lipid- (fat-) soluble and can easily penetrate skin cells and enter the body. After a toxicant has penetrated the skin and entered the bloodstream, the blood can transport it to target organs in the body.

## Inhalation

The respiratory tract consists of the nasal cavity, pharynx, larynx, trachea, primary bronchi, bronchioles, and alveoli. Molecular transfer of oxygen and carbon dioxide between the bloodstream and the lungs occurs in the millions of air sacs, known as *alveoli*. Toxicants that reach the alveoli can be transferred to the blood through the respiratory system.

Toxicants that reach the alveoli will not be transferred to the blood at the same rate. Various factors can increase or decrease the transfer rate of one toxicant relative to another. Both the respiration rate and the duration of exposure will affect the mass of toxicant transferred to the bloodstream over a given period.

## Ingestion

The small intestine is the principal site in the digestive tract where the beneficial nutrients from food, and toxics from contaminated substances, are absorbed. Within the small intestine, millions of *villi* (projections) provide a huge surface area to absorb substances into the bloodstream.

Toxic substances will be absorbed in the intestines at various rates depending on the specific toxicant, its molecular size, and its degree of lipid (fat) solubility. Small molecular size and high lipid solubility facilitate diffusion of toxicants in the digestive tract.

## The Eye

Transparent tissue in the front of the eye is known as the *cornea* and is the most likely eye tissue to come in contact with toxic substances.

## Systemic Effects

For systemic effects to occur, the rate of accumulation of toxicants must exceed the body's ability to excrete (eliminate) it or to biotransform it (transform it to less harmful substances). A toxicant can be eliminated from the body through the *kidneys*, which are the primary organs for eliminating toxicants from the body. The kidneys biotransform a toxicant into a water-soluble form and then eliminate it though the urine. The *liver* is also an important organ for eliminating toxicants from the body, first by biotransformation, then by excretion into the bile where it is eliminated through the small intestine as feces.

A toxicant may also be stored in tissues for long periods before an effect occurs. Toxic substances stored in tissue (primarily in fat but also in bones, the liver, and the kidneys) may exert no effect for many years, or at all within the affected person's life. DDT, a pesticide, for instance, can be stored in body fat for many years and not exert any adverse effect on the body.

When toxic substances are not eliminated fast enough to keep up with the exposure, the liver, kidneys, and central nervous system are the target organs and are commonly affected systemically.

## 3. EFFECTS OF EXPOSURE TO TOXICANTS

After toxicants are absorbed into the body through one or more of the pathways, a wide variety of effects on the human body are possible. When the toxic agents concentrate in target tissue or organs, the agents may interfere with the normal functioning of enzymes and cells or may cause genetic mutations.

### Pulmonary Toxicity

*Pulmonary toxicity* refers to adverse effects on the respiratory system from toxic agents. Examples are

- Damage to the nasal passages and nerve cells
- *Nasal cancer*
- *Bronchitis*, excessive mucus secretion
- *Pulmonary edema*, the excessive accumulation of fluid in the alveoli of the lungs
- *Fibrosis*, an increased amount of connective tissue
- *Silicosis*, the deposition of connective tissue around alveoli
- *Emphysema*, the inability of lungs to expand and contract

### Cardiotoxicity

*Cardiotoxicity* refers to the effects of toxic agents on the heart.

- The heart rate may be changed, and the strength of contractions may be diminished.
- Certain metals can affect the contractions of the heart and can interfere with cell metabolism.
- Carbon monoxide can result in a decrease in the oxygen supply, causing improper functioning of the nervous system controlling the heart rate.

### Hematoxicity

*Hematoxicity* refers to damage to the body's blood supply, which includes red blood cells, white blood cells, platelets, and plasma. The *red blood cells* transport oxygen to the body's cells and carbon dioxide to the lungs. *White blood cells* perform a variety of functions associated with the immune system.

- *Platelets* are important in blood clotting.
- *Plasma* is the noncellular portion of blood and contains proteins, nutrients, gases, and waste products.

- Benzene, lead, methylene chloride, nitrobenzene, naphthalene, and insecticides are capable of red blood cell destruction and can cause a decrease in the oxygen-carrying capacity of the blood. The resulting anemia can affect normal nerve cell functioning and control of the heart rate, and can cause shortness of breath, pale skin, and fatigue.
- Carbon tetrachloride, pesticides, benzene, and ionizing radiation can affect the ability of the bone marrow to produce red blood cells.
- Mercury, cadmium, and other toxicants can affect the ability of the kidneys to stimulate the bone marrow to produce more red blood cells when needed to counteract low oxygen levels in the blood.
- Some chemicals, including carbon monoxide, can interfere with the blood's capacity to carry oxygen, resulting in lowered blood pressure, dizziness, fainting, increased heart rate, muscular weakness, nausea, and after prolonged exposure, death.
- Hydrogen cyanide and hydrogen sulfide can stimulate cells in the aorta, causing increased heart and respiratory rate. At high concentrations, death can result from respiratory failure.
- Benzene, carbon tetrachloride, and trinitrotoluene can suppress stem cell production and the production of white blood cells. This can affect the clotting mechanism and the immune system.
- Benzene can cause high levels of white blood cells, a condition known as *leukemia*.

### Hepatoxicity

*Hepatoxicity* refers to adverse effects on the liver that impede its ability to function properly. The liver converts carbohydrates, fats, and proteins to maintain the proper levels of glucose in the blood and converts excess protein and carbohydrates to fat. It also converts excess amino acids to ammonia and urea, which are removed in the kidneys. The liver also provides storage of vitamins and beneficial metals, as well as carbohydrates, fats, and proteins. Red blood cells that have degenerated are removed by the liver. Substances needed for other metabolic processes are provided by the liver. Finally, the liver detoxifies metabolically produced substances and toxicants that enter the body.

- Hexavalent chromium and arsenic cause cell damage in the liver.
- Carbon tetrachloride and alcohol can cause damage and death of liver cells, a condition known as *cirrhosis of the liver.*
- Chemicals or viruses can cause inflammation of the liver, known as *hepatitis.* Cell death and enlargement of the liver can occur.

Biology

## Nephrotoxicity

*Nephrotoxicity* refers to adverse effects on the kidneys. The kidneys excrete ammonia as urea to rid the body of metabolic wastes. They maintain blood pH by exchanging hydrogen ions for sodium ions, and maintain the ion and water balance by excreting excess ions or water as needed. They also secrete hormones needed to regulate blood pressure. Like the liver, the kidneys function to detoxify substances.

- Heavy metals—primarily lead, mercury, and cadmium—cause impaired cell function and cell death. These metals can be stored in the kidneys, interfering with the functioning of enzymes in the kidneys.

- Chloroform and other organic substances can cause cell dysfunction, cell death, and cancer.

- Ethylene glycol can cause renal failure from obstruction of the normal flow of liquid through the kidneys.

## Neurotoxicity

*Neurotoxicity* refers to toxic effects on the nervous system, which consists of the *central nervous system* (CNS) and the *peripheral nervous system* (PNS). The central nervous system includes the brain and the spinal cord, while the peripheral nervous system includes the remaining nerves, which are distinguished as sensory and motor nerves.

Neurotoxic effects fall into two basic types: *destruction* of nerve cells and *interference* with neurotransmission.

## Immunotoxicity

*Immunotoxicity* refers to toxic effects on the immune system, which includes the lymph system, blood cells, and antibodies in the blood.

## Reproductive Toxicity

The effect of toxicants on the male or female reproductive system is known as *reproductive toxicity*. For the male reproductive system, toxicants primarily affect the division of sperm cells and the development of healthy sperm. For the female reproductive system, toxicants can affect the endocrine system, the brain, and the reproductive tract.

## Toxic Effects on the Eye

There are a wide variety of toxic substances that can cause damage to the eye through contact with the cornea.

## 4. DOSE-RESPONSE RELATIONSHIPS

Dose-response relationships can be used to relate the response of an organism to increasing dose levels of toxicants.

## Dose-Response Curves

An objective of toxicity tests is to establish the *dose-response curve*, as illustrated in Fig. 37.2.

**Figure 37.2** *Dose-Response Relationships*

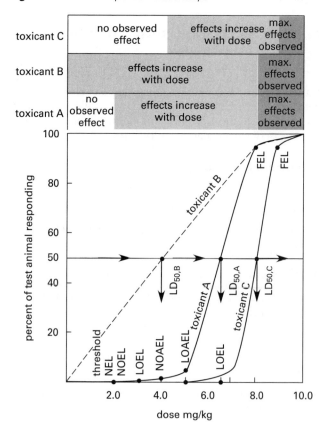

Several important features of a dose-response curve are illustrated and described for toxicant A of Fig. 37.2 as follows.

- *Response:* The ordinate.

- *Dose:* The abscissa.

- *No Observed Effect:* The range of the curve below which no effect is observed is the range of no observed effect. The upper end of the range is known as the *threshold*, the *no effect level* (NEL), or the *no observed effect level* (NOEL).

- *Lowest Observed Effect:* The dose where minor effects first can be measured, but the effects are not directly related to the response being measured, is known as the *lowest observed effect level* (LOEL).

- *No Observed Adverse Effect:* The dose where effects related to the response being measured first can be measured is known as the *no observed adverse effect level* (NOAEL). However, at this level, the effects observed at the higher doses are not observed.

- *Lowest Observed Adverse Effect:* The dose where effects related to the response being measured first can be measured, and are the same effects as the effects observed at the higher doses, is known as the *lowest observed adverse effect level* (LOAEL).

- *Frank Effect:* The *frank effect level* (FEL) dose marks the point where maximum effects are observed with little increase in effect for increasing dose.

For some toxicants, primarily those believed to be carcinogens, there is no apparent threshold. Any dose is considered to have an effect even though such effect may be unmeasurable at low doses. Such toxicants have no safe exposure level. This is illustrated as toxicant B in Fig. 37.2. Lead is an example of a toxicant with no threshold dose.

The *lethal dose* or *lethal concentration* is the concentration of toxicant at which a specified percentage of test animals die. The lethal dose is expressed as the mass of toxicant per unit mass of test animal. Thus, $LD_{50}$ means the dose in milligrams of toxicant per kilogram of body mass at which 50% of the test animals died.

For acute tests involving inhalation as the exposure pathway, the concentration, in parts per million, of the toxicant in air is used. If the toxicant is in particulate form, the concentration in milligrams of toxic particles per cubic meter of air is used. Thus, $LC_{50}$ means the concentration of the toxicant in air at which 50% of the test animals died.

## 5. SAFE HUMAN DOSE

### EPA Methods

Several approaches exist for selecting a safe human dose from the data obtained from toxicological and epidemiological studies. The approach most likely to be encountered by the environmental engineer is the one recommended by the U.S. Environmental Protection Agency (EPA).

*Carcinogens* generally do not exhibit thresholds of response at low doses of exposure and any exposure is assumed to have an associated risk. The EPA model used to evaluate carcinogenic risk assumes no threshold and a linear response to any amount of exposure. *Noncarcinogens* (systemic toxicants) are chemicals that do not produce tumors (or gene mutations) but instead adversely interfere with the functions of enzymes in the body, which thereby causes abnormal metabolic responses. Noncarcinogens have a dose threshold below which no adverse health response can be measured.

### Noncarcinogens

The threshold below which adverse health effects in humans are not measurable or observable is defined by the EPA as the *reference dose* (RfD). The reference dose is the safe daily intake that is believed not to cause adverse health effects. The reference dose relates to the ingestion and dermal contact pathways and is route

specific. For gases and vapors (exposure by the inhalation pathway), the threshold may be defined as the *reference concentration* (RfC). Sometimes the term "reference dose" is also used for exposure by the inhalation pathway.

After an RfD has been identified for one or more toxicants, the *hazard ratio* (HR) can be determined to assess whether exposures indicate an unacceptable hazard. The hazard ratio is the *estimated exposure dose* (EED) divided by the reference dose for each of the toxicants from all routes of exposure. If the sum of the ratios exceeds 1.0, the risk is unacceptable. Calculating the hazard ratio should be considered a preliminary assessment.

$$HR = \frac{EED}{RfD} \qquad 37.1$$

### Carcinogens

The distinguishing feature of cancer is the uncontrolled growth of cells into masses of tissue called *tumors*. Tumors may be *benign*, in which the mass of cells remains localized, or *malignant*, in which the tumors spread through the bloodstream to other sites within the body. This latter process is known as *metastasis* and determines whether the disease is characterized as cancer. The term *neoplasm* (new and abnormal tissue) is also used to describe tumors.

Cancer occurs in three stages: initiation, promotion, and progression. During the *initiation* stage, a cell mutates and the DNA is not repaired by the body's normal DNA repair mechanisms. During the *promotion* stage, the mutated cells increase in number and undergo differentiation to create new genes. During *progression*, the cancer cells invade adjacent tissue and move through the bloodstream to other sites in the body. It is believed that continued exposure to the agent that initiated genetic mutation is necessary for progression to continue. Many mutations are believed to be required for the progression of cancer cells to occur at remote sites in the body.

### Direct Human Exposure

The EPA's classification system for carcinogenicity is based on a consensus of expert opinion called *weight of evidence*.

The EPA maintains a database of toxicological information known as the Integrated Risk Information System (IRIS). The IRIS data include chemical names, chemical abstract service registry numbers (CASRN), reference doses for systemic toxicants, carcinogen potency factors (CPF) for carcinogens, and the carcinogenicity group classification, which is shown in Table 37.1.

The dose-response for carcinogens differs substantially from that of noncarcinogens. For carcinogens it is believed that any dose can cause a response (mutation of DNA). Since there are no levels (no thresholds) of

Biology

*Table 37.1* EPA Carcinogenicity Classification System

| group | description |
|-------|-------------|
| A | human carcinogen |
| B1 or B2 | probable human carcinogen |
| | B1 indicates that human data are available. |
| | B2 indicates sufficient evidence in animals and inadequate or no evidence in humans. |
| C | possible human carcinogen |
| D | not classifiable as to human carcinogenicity |
| E | evidence of noncarcinogenicity for humans |

carcinogens that could be considered safe for continued human exposure, a judgment must be made as to the acceptable level of exposure, which is typically chosen to be an excess lifetime cancer risk of $1 \times 10^{-6}$ (0.0001%). *Excess lifetime cancer risk* refers to the incidence of cancers developed in the exposed animals minus the incidence in the unexposed control animals. For whole populations exposed to carcinogens, the number of total excess cancers, EC, is the product of the probability of excess cancer, $R$, and the total exposed population, EP.

$$\text{EC} = \text{EP} \times R \qquad 37.2$$

Under the EPA approach, the *carcinogen potency factor* (CPF) is the slope of the dose-response curve at very low exposures. The CPF is also called the *potency factor* or *slope factor* and has units of $(\text{mg/kg·d})^{-1}$. The CPF is pathway (route) specific. The CPF is obtained by extrapolation from the high doses typically used in toxicological studies. (See Table 37.2.)

*Table 37.2* EPA Standard Values for Intake Calculations

| parameter | standard value |
|-----------|----------------|
| average body weight, adult | 70 kg |
| average body weight, child | 10 kg |
| daily water ingestion, adult | 2 L |
| daily water ingestion, child | 1 L |
| daily air breathed, adult | 20 m$^3$ |
| daily air breathed, child | 5 m$^3$ |
| daily fish consumed, adult | 6.5 g |
| lifetime exposure period | 70 yr |

The CPF is the probability of risk produced by lifetime exposure to 1.0 mg/kg·d of the known or potential human carcinogen. Thus, the slope factor can be multiplied by the long-term daily intake (*chronic daily intake*, CDI) to obtain the lifetime probability of risk, $R$, for daily doses other than 1.0 mg/kg·d. The CDI can be calculated from Eq. 37.3.

$$\text{CDI}_{a/w} = \frac{C(\text{CR})(\text{EF})(\text{ED})}{(\text{BW})(\text{AT})} \qquad 37.3$$

For less than lifetime exposure, the *exposure duration* must be used to calculate the total intake, which must be divided by the *averaging duration* of 70 years for carcinogens. For noncarcinogens, the averaging duration is the same as the exposure duration.

Once the CDI is known, the probable risk of additional cancers for adults and children is found using Eq. 37.4.

$$R = (\text{SF})(\text{CDI}) \qquad 37.4$$

### Bioconcentration Factors

Besides setting factors for direct human exposure to toxicants through water ingestion, inhalation, and skin contact, the EPA also has developed *bioconcentration factors* (BCF) (also referred to as *steady-state BCF*) so that the human intake from consumption of fish and other foods can be determined. Bioconcentration factors have been developed for many toxicants and provide a relationship between the toxicant concentration in the tissue of the organism and the concentration in the medium (e.g., water). The concentration in the organism equals the product of the BCF and the concentration in the medium. Not all chemicals or other substances will bioaccumulate, and the BCF pertains to a specific organism, such as fish.

$$C_{\text{org}} = \text{BCF} \times C_w \qquad 37.5$$

Selected bioconcentration factors (BCF) for selected chemicals in fish are given in Table 37.3. The substances are arranged in descending order of BCFs to illustrate the substances that have a high potential to bioaccumulate in fish. These substances are of great importance when the oral pathway is present in a particular situation.

The BCF factors can be applied to determine the total dose to humans who ingest fish from water contaminated with toxicants that bioaccumulate. This dose would be added to the dose received from drinking the contaminated water.

### ACGIH Methods

The American Conference of Governmental Industrial Hygienists (ACGIH) uses methods for determining the safe human dose that are somewhat different from the EPA methods previously described.

### Threshold Limit Values

The ACGIH method uses predetermined *threshold limit values* (TLV) for both noncarcinogens and carcinogens. The TLVs are the concentrations in air that workers could be repeatedly exposed to on a daily basis without adverse health effects. The term TLV-TWA means the maximum time-weighted average concentration that all workers may be exposed to during an 8 hour day and 40 hour week. The TLV-TWA is for the inhalation route of exposure.

Biology

**Table 37.3** *Typical Bioconcentration Factors for Fish*[*]

| substance | BCF (L/kg) |
|---|---|
| polychlorinated biphenyls | 100 000 |
| 4,4′ DDT | 54 000 |
| DDE | 51 000 |
| heptachlor | 15 700 |
| chlordane | 14 000 |
| toxaphene | 13 100 |
| mercury | 5500 |
| 2,3,7,8 tetrachlorodibenzo-p-dioxin (TCDD) | 5000 |
| dieldrin | 4760 |
| copper | 200 |
| cadmium | 81 |
| lead | 49 |
| zinc | 47 |
| arsenic | 44 |
| tetrachloroethylene | 31 |
| aldrin | 28 |
| carbon tetrachloride | 19 |
| chromium | 16 |
| chlorobenzene | 10 |
| benzene | 5.2 |
| chloroform | 3.75 |
| vinyl chloride | 1.17 |
| antimony | 1 |

[*]For illustrative purpose only. Subject to change without notice. Local regulations may be more restrictive than federal.

ACGIH also determines *short-term exposure limits* (TLV-STEL) for airborne toxicants, which are the recommended concentrations workers may be exposed to for short periods during the workday without suffering certain adverse health effects (e.g., irritation, chronic tissue damage, and narcosis). The TLV-STEL is the TWA concentration in air that should not be exceeded for more than 15 minutes of the workday. The TLV-STEL should not occur more than four times daily, and there should be at least 60 minutes between successive STEL exposures. In such cases, the excursions may exceed three times the TLV-TWA for no more than a total of 30 minutes during the workday, but shall not exceed five times the TLV-TWA under any circumstances. In all cases, the TLV-TWA may not be exceeded. Short-term exposure limits have not been established by ACGIH for some toxicants.

ACGIH also publishes *ceiling threshold limit values* (TLV-C) that should not be exceeded at any time during the workday. If instantaneous sampling is infeasible, the sampling period for the TLV-C can be up to 15 minutes in duration. Also, the TLV-TWA should not be exceeded.

For mixtures of substances, the *equivalent exposure* over 8 hours is the sum of the individual exposures.

$$E = \frac{1}{8} \sum_{i=1}^{n} C_i T_i \qquad \qquad 37.6$$

The *hazard ratio* is the concentration of the contaminant divided by the exposure limit of the contaminant. For mixtures of substances, the total hazard ratio is the sum of the individual hazard ratios and must not exceed unity. This is known as the *law of additive effects*. For this law to apply, the effects from the individual substances in the mixture must act on the same organ. If the effects do not act on the same organ, then each of the individual hazard ratios must not exceed unity. The equivalent exposure of a mixture of gases is

$$E_m = \sum_{i=1}^{n} \frac{C_i}{L_i} \qquad \qquad 37.7$$

## NIOSH Methods

The National Institute for Occupational Safety and Health (NIOSH) was established by the Occupational Safety and Health Act of 1970. NIOSH is part of the Centers for Disease Control and Prevention (CDC) and is the only federal institute responsible for conducting research and making recommendations for the prevention of work-related illnesses and injuries. The Institute's responsibilities include

- investigating hazardous working conditions as requested by employers or workers

- evaluating hazards ranging from chemicals to machinery

- creating and disseminating methods for preventing disease, injury, and disability

- conducting research and providing recommendations for protecting workers

- providing education and training to persons preparing for or actively working in the field of occupational safety and health

The NIOSH recommended exposure limits (RELs) are time-weighted average (TWA) concentrations for up to a 10 hour workday during a 40 hour workweek. A short-term exposure limit (STEL) is a 15 minute TWA exposure that should not be exceeded at any time during a workday. A ceiling REL should not be exceeded at any time. The "skin" designation means there is a potential for dermal absorption, so skin exposure should be prevented as necessary through the use of good work practices and gloves, coveralls, goggles, and other appropriate equipment.

**Biology**

## 6. LEGAL STANDARDS FOR WORKER PROTECTION

While the EPA provides exposure limitations and risk factors for environmental cleanup projects, ACGIH provides recommendations to industrial hygienists about workplace exposure, and NIOSH provides research and recommendations for workplace exposure limits (recommended exposure limits), the *Occupational Safety and Health Administration* (OSHA) sets the legally enforceable workplace exposure limits. OSHA standards are given in Title 29 of the Code of Federal Regulations (CFR).

The calculation procedure for use of the *permissible exposure limits* (PELs) is the same as for ACGIH–TLVs. The procedure for mixtures is also the same.

# 38

# Industrial Hygiene

## Nomenclature

| | | |
|---|---|---|
| $a$ | speed of sound | m/s |
| $A$ | activity metabolism | W |
| $A$ | radioactivity | Bq |
| AM | asymmetry multiplier for lifting | – |
| $B$ | basal metabolism | W |
| $C$ | concentration | ppm, mg/m$^3$ |
| $C$ | constant for calculating sound intensity | – |
| $C$ | time of noise exposure at specified level | s |
| CL | ceiling heat limit | °C |
| CM | coupling multiplier for lifting | – |
| DM | distance multiplier for lifting | – |
| $E$ | exposure | – |
| ECT | equivalent chill temperature | °C |
| $f$ | frequency | Hz |
| FM | frequency multiplier for lifting | – |
| HM | horizontal multiplier for lifting | – |
| $I$ | intensity | W/m$^2$ |
| IL | insertion loss | dB |
| $k$ | ratio of specific heats | – |
| $L$ | level | dB |
| LC | load constant for lifting | kg |
| $m$ | mass | kg |
| MW | molecular weight | g/mol |
| $n$ | number of moles | – |
| $p$ | pressure or partial pressure | Pa |
| $p$ | total number of observations | – |
| $P$ | posture metabolism | W |
| PEL | permissible exposure limit | mg/m$^3$ |
| $Q$ | heat flow | W |
| $r$ | distance | m |
| $R$ | specific gas constant | kJ/kg·K |
| $R^*$ | universal gas constant | kJ/kmol·K |
| RAL | recommended heat alert limit | °C |
| REL | recommended heat exposure limit | °C |
| RWL | recommended weight limit | kg |
| $t$ | rest time, percent of period | % |
| $t$ | time | s |
| $T$ | temperature | °C, K |
| TLV | threshhold limit value | – |
| $V$ | velocity metabolism | W |
| $V$ | volume | L, m$^3$ |
| VM | vertical multiplier for lifting | – |
| $W$ | power | W |
| WBGT | wet-bulb globe temperature | °C |
| $x$ | mole fraction | – |
| $x_{rms}$ | root mean square value of $n$ observations | – |

## Symbols

| | | |
|---|---|---|
| $\rho$ | density | kg/m$^3$ |
| $\lambda$ | decay constant | s$^{-1}$ |
| $\lambda$ | wavelength | m |

## Subscripts

| | | |
|---|---|---|
| 0 | initial condition or reference | |
| $C$ | convection | |
| db | dry bulb | |
| $E$ | evaporation | |
| $g$ | globe | |
| in | indoor | |
| $m$ | mixed | |
| max | maximum | |
| $M$ | metabolic | |
| M/V | mass per unit volume | |
| nwb | natural wet bulb | |
| $p$ | sound pressure | |
| $R$ | radiation | |
| rest | resting | |
| rms | root-mean-square | |
| rms-ref | reference rms | |
| $S$ | storage | |
| $t$ | time, time period | |
| $W$ | sound power | |

## 1. INDUSTRIAL HYGIENE

*Industrial hygiene* is the art and science of identifying, evaluating, and controlling environmental factors (including stress) that may cause sickness, health impairment, or discomfort among workers or citizens of the community. Industrial hygiene involves the recognition of health hazards associated with work operations and processes, evaluations, measurements of the magnitude of hazards, and determining applicable control methods. Occupational health seeks to reduce hazards leading to illness or impairment for which a worker may be compensated under a worker protection program.

Biology

The fundamental law governing worker protection in the U.S. is the 1970 federal *Occupational Safety and Health Act*. It requires employers to provide a workplace that is free from hazards by complying with specified safety and health standards. Employees must also comply with standards that apply to their own conduct. The federal regulatory agency responsible for administering the Occupational Safety and Health Act is OSHA, the Occupational Safety and Health Administration. OSHA sets standards, investigates violations of the standards, performs inspections of plants and other facilities, investigates complaints, and takes enforcement action against violators. OSHA also funds state programs, which are permitted if they are at least as stringent as the federal program.

The 1970 Act also established the *National Institute for Occupational Safety and Health* (NIOSH). NIOSH is responsible for safety and health research and makes recommendations for regulations. The recommendations are known as *Recommended Exposure Limits* (RELs). Among other activities, NIOSH also publishes health and safety criteria and notifications of health hazard alerts, and is responsible for testing and certifying respiratory protective equipment.

## 2. HAZARD IDENTIFICATION

### Overview of Hazards

There are four basic types of hazards with which industrial hygiene is concerned: chemical hazards, physical hazards, ergonomic hazards, and biological hazards. *Chemical hazards* result from chemicals such as gases, vapors, or particulates in harmful concentrations. Besides inhalation, chemical hazards may affect workers by absorption through the skin. *Physical hazards* include radiation, noise, vibration, and excessive heat or cold. *Ergonomic hazards* include work procedures and arrangements that require motions that result in biomechanical stress and injury. *Biological hazards* include exposure to biological organisms that may lead to illness.

In respect to chemical hazards, the terms "toxicity" and "hazard" are not synonymous. *Toxicity* is the capacity of the chemical to produce harm when it has reached a sufficient concentration at a particular site in the body. *Hazard* refers to the probability that this concentration will occur.

### Hazard Communication

Two important preventative measures required by OSHA are *Material Safety Data Sheets* (MSDS) and labeling of containers of hazardous materials. A third OSHA requirement is that all covered employers must provide the necessary information and training to affected workers. The OSHA Hazard Communication Standard is given in Title 29 of the *Code of Federal Regulations* (CFR) Part 1910.1200. Other OSHA requirements are also given in Title 29.

### MSDS

MSDS sheets provide key information about a chemical or substance so that users or emergency responders can determine safe use procedures and necessary emergency response actions. An MSDS provides information on the identification of the material and its manufacturer, identification of hazardous components and their characteristics, physical and chemical characteristics of the ingredients, fire and explosion hazard data, reactivity data, health hazard data, precautions for safe handling and use, and recommended control measures for use of the material.

### Container Labeling

Labels are required on hazardous material containers. Labels should provide essential information for the safe use and storage of hazardous materials. Failure to provide adequate labeling of hazardous material containers is a common violation of OSHA standards.

### Worker Information and Training

The OSHA standard requires that employers provide workers with information about the potential health hazards from exposure to hazardous chemicals that they use in the workplace. It also requires employers to provide adequate training to workers on how to safely handle and use hazardous materials.

## 3. EXPOSURE LIMITS

Two statutory limits quantify the concentration of a gas in air that a worker can be safely exposed to. These are *threshold limit value* (TLV) and *permissible exposure limit* (PEL). The TLV is a concentration in air that nearly all workers can be exposed to daily without any adverse effects. The PEL is a regulatory exposure limit for workers; OSHA publishes PELs as standards. Table 38.1 is a representative listing of toxic materials and their TLVs.

The American Conference of Governmental Industrial Hygienists (ACGIH) has established TLVs that should not be exceeded. OSHA establishes PELs for safe exposure levels in the workplace.

## 4. GASES, VAPORS, AND SOLVENTS

### Exposure Factors for Gases and Vapors

The most frequently encountered hazard in the workplace is exposure to gases and vapors from solvents and chemicals. Several factors define the exposure potential for gases and vapors. The most important are how a material is used and what engineering or personal protective controls exist. If the inhalation route of entry is controlled, dermal contact may still be a major route of exposure.

*Vapor pressure* of a substance is related to temperature. Vapor pressure affects the concentration of the substance in vapor form above the liquid and is dependent upon the temperature and the properties of the substance. Processes that operate at lower temperatures are inherently less hazardous than processes that operate at higher temperatures.

*Reactivity* affects the hazard potential because the products may be volatile or nonvolatile depending on the properties of the combining substances.

## Solvents

Solvents are widely used throughout industry for many purposes, and their safe use is an important industrial hygiene concern. It is essential that accurate MSDS information be provided to employees on the physical properties and the toxicological effects of exposure to solvents.

## Gases and Flammable or Combustible Liquids

Hazardous gases fall into four main types: cryogenic liquids, simple asphyxiants, chemical asphyxiants, and all other gases whose hazards depend on their properties.

*Cryogenic liquids* can vaporize rapidly, producing a cold gas that is more dense than air and displacing oxygen in confined spaces.

*Simple asphyxiants*, which include helium, neon, nitrogen, hydrogen, and methane, can dilute or displace oxygen. *Chemical asphyxiants*, which include carbon monoxide, hydrogen cyanide, and hydrogen sulfide, can pass into blood cells and tissue and interfere with blood-carrying oxygen.

The term *flammable* refers to the ability of an ignition source to propagate a flame throughout the vapor-air mixture and have a closed-cup flash point below 37.8°C (100°F) and a vapor pressure not exceeding 272 atm at 37.8°C (100°F). The phrase *closed-cup flash point* refers to a method of testing for flash points of liquids. The term *combustible* refers to liquids with flash points above 37.8°C (100°F).

For each airborne flammable substance there are minimum and maximum concentrations in air between which flame propagation will occur. The lower concentration in air is known as the *lower explosive limit* (LEL) or *lower flammable limit* (LFL). The upper limit is known as the *upper explosive limit* (UEL) or *upper flammable limit* (UFL). Below the LEL, there is not enough fuel to propagate a flame. Above the UEL, there is not enough air to propagate a flame. The lower the LEL, the greater the hazard from a flammable liquid. For many common liquids and gases, the LEL is a few percent and the UEL is 6% to 12%. Note that if a concentration in air is less than the PEL or the TLV, the concentration will be less than the LEL. The occupational safety requirements for handling and using flammable and combustible liquids are given in Subpart H of 29 CFR 1910.106.

See Table 38.1 for a representative listing of combustible materials and their LELs and UELs.

## Evaluation and Control of Hazards

The toxicological effects from aqueous solutions include dermatitis, throat irritation, and bronchitis.

### Vapor-Hazard Ratio

One indicator of hazards from vapors and gases from solvents is the *vapor-hazard ratio number*, which is the equilibrium vapor pressure in ppm at 25°C (77°F) divided by the TLV in ppm. The higher the ratio, the greater the hazard. The vapor-hazard ratio accounts for the volatility of a solvent as well as its toxicity. To assess the overall hazard, the vapor-hazard ratio should be evaluated in conjunction with the TLV, ignition temperature, flash point, toxicological information, and degree of exposure.

The best control method is not to use a solvent that is hazardous. Sometimes a process can be redesigned to eliminate the use of a solvent. The following evaluation steps are recommended.

- Use water or an aqueous solution when possible.
- Use a *safety solvent* if it is not possible to use water. Safety solvents have inhibitors and high flash points.
- Use a different process when possible to avoid use of a hazardous solvent.
- Provide a properly designed ventilation system if toxic solvents must be used.
- Never use highly toxic or highly flammable solvents (benzene, carbon tetrachloride, gasoline).

### Ventilation

The most effective way to prevent inhalation of vapors from solvents is to provide closed systems or adequate local exhaust ventilation. If limitations exist on the use of closed systems or local exhaust ventilation, then workers should be provided with personal protective equipment.

### Personal Protective Equipment

Respirators provide emergency and backup protection but are unreliable as a primary source of protection from hazardous vapors because they leak around the edges of the face mask, can become contaminated around the edges, reduce the efficiency of the worker, and increase the lack of oxygen in oxygen-deficient areas. Other drawbacks are the need to have the respirator properly fitted to the worker and the need for the worker to be trained in its proper use. Additionally, the worker may feel a false sense of security while wearing a respirator.

**Biology**

**Table 38.1** *Representative Hazardous Concentrations in Air*[*]

| | combustibles | | | | | | toxics | | | | |
|---|---|---|---|---|---|---|---|---|---|---|---|
| material | LEL (%/vol) | UEL (%/vol) | TLV/ TWA (ppm) | IDLH (ppm) | specific gravity (air=1.0) | material | TLV/ TWA (ppm) | IDLH (ppm) | LEL (ppm) | LEL (%/vol) | specific gravity (air=1.0) |
| Acetone | 2.5 | 12.8 | 750 | 2500 | 2.0 | Acetone | 750 | 2500 | 25 000 | 2.5 | 2.0 |
| Acetylene | 2.5 | 100.0 | -A- | -A- | 0.9 | Ammonia | 25 | 300 | 160 000 | 16.0 | 0.6 |
| Ammonia | 15.0 | 28.0 | 25 | 300 | 0.6 | Benzene | 1.0 | -C- | 12 000 | 1.2 | 2.6 |
| Benzene | 1.2 | 7.8 | 1.0 | 500 | 2.6 | Butane | 800 | -U- | 16 000 | 1.6 | 2.0 |
| Butane | 1.6 | 8.4 | 800 | -U- | 2.0 | n-Butyl Acetate | 150 | 1700 | 17 000 | 1.7 | 4.0 |
| n-Butyl Acetate | 1.7 | 7.6 | 150 | 1700 | 4.0 | Carbon Dioxide | 5000 | 40 000 | N/C | N/C | 1.5 |
| Diborane | 0.8 | 88.0 | 0.1 | 15 | 1.0 | Carbon Monoxide | 25 | 1200 | 125 000 | 12.5 | 1.0 |
| Ethane | 3.0 | 12.5 | -A- | -A- | 1.0 | Chlorine | 0.5 | 10 | N/C | N/C | 2.5 |
| Ethanol | 3.3 | 19.0 | 1000 | -U- | 1.6 | Ethylene Oxide | 1 | -C- | 30 000 | 3.0 | 1.5 |
| Ethyl Acetate | 2.0 | 11.5 | 400 | 2000 | 3.0 | Ethyl Ether | 400 | 19 000 | 19 000 | 1.9 | 2.6 |
| Ethyl Ether | 1.9 | 36.0 | 400 | 1900 | 2.6 | Gasoline | 300 | -U- | 14 000 | 1.4 | 3–4.0 |
| Ethylene Oxide | 3.0 | 100.0 | 1 | -C- | 1.5 | Heptane | 400 | 750 | 10 500 | 1.05 | 3.5 |
| Gasoline | 1.4 | 7.6 | 300 | -U- | 3–4.0 | Hexane | 50 | 1100 | 11 000 | 1.0 | 3.0 |
| Heptane | 1.05 | 6.7 | 400 | 750 | 3.5 | Hydrogen Cyanide | 10 | 50 | 56 000 | 5.6 | 0.9 |
| Hexane | 1.1 | 7.5 | 50 | 1100 | 3.0 | Hydrogen Sulfide | 10 | 100 | 40 000 | 4.0 | 1.2 |
| Hydrogen | 4.0 | 75.0 | -A- | -A- | 0.1 | Isopropyl Alcohol | 400 | 2000 | 20 000 | 2.0 | 2.1 |
| Isopropyl Alcohol | 2.0 | 12.0 | 400 | 2000 | 2.1 | Methyl Acetate | 200 | 3100 | 31 000 | 3.1 | 2.6 |
| Methane | 5.0 | 15.0 | -A- | -A- | 0.6 | Methanol | 200 | 6000 | 60 000 | 6.0 | 1.1 |
| Methanol | 6.0 | 36.0 | 200 | 6000 | 1.1 | Methyl Chloride | 50 | 2000 | 81 000 | 8.1 | 1.8 |
| Methyl Ethyl Ketone | 1.4 | 11.4 | 200 | 3000 | 2.5 | Methyl Ethyl Ketone | 200 | 3000 | 14 000 | 1.4 | 2.5 |
| Pentane | 1.5 | 7.8 | 600 | 15 000 | 2.5 | Methyl Methacrylate | 100 | 1000 | 17 000 | 1.7 | 3.5 |
| Propane | 2.1 | 9.5 | 1000 | 2100 | 1.6 | Nitric Oxide | 25 | 100 | N/C | N/C | 1.0 |
| Propylene Oxide | 2.3 | 36.0 | 20 | 400 | 2.0 | Nitrogen Dioxide | 3 | 20 | N/C | N/C | 1.6 |
| Styrene | 0.9 | 6.8 | 50 | 700 | 3.6 | Pentane | 600 | 15 000 | 15 000 | 1.5 | 2.5 |
| Toluene | 1.1 | 7.1 | 50 | 500 | 3.1 | n-Propyl Acetate | 200 | 1700 | 17 000 | 1.7 | 3.5 |
| Turpentine | 0.8 | -U- | 100 | 800 | 4.7 | Styrene | 50 | 700 | 9000 | 0.9 | 3.6 |
| Vinyl Acetate | 2.6 | 13.4 | 10 | -U- | 3.0 | Sulfur Dioxide | 2 | 100 | N/C | N/C | 2.2 |
| Vinyl Chloride | 3.6 | 33.0 | 1.0 | -C- | 2.2 | 1,1,1-Trichloroethane | 350 | 700 | 75 000 | 7.5 | 4.6 |
| Xylene | 0.9 | 6.7 | 100 | 900 | 3.7 | Toluene | 50 | 500 | 11 000 | 1.1 | 3.2 |
| | | | | | | Trichloroethylene | 50 | 1000 | 80 000 | 8.0 | 4.5 |
| | | | | | | Turpentine | 100 | 800 | 8000 | 0.8 | 4.7 |
| | | | | | | Vinyl Chloride | 1.0 | -C- | 36 000 | 3.6 | 2.2 |
| | | | | | | Xylene | 100 | 900 | 9000 | 0.9 | 3.7 |

Key: A, asphyxiant; C, carcinogen; U, data not available; N/C, noncombustible

[*]subject to change without notice

Besides inhalation, dermal contact is an important concern when working with hazardous solvents. Mechanical equipment should be provided to keep the worker isolated from contact with the solvent. However, since some contact may occur even with mechanical equipment in use, protective clothing should be provided.

Protective clothing includes aprons, face shields, goggles, and gloves. The manufacturer's recommendations should be followed for use of all protective clothing and equipment.

One common problem with protective clothing is incorrect selection or misuse of gloves. The time for particular solvents to penetrate gloves that are commonly thought of as "protective" is surprisingly short. Both the permeability and the abrasion resistance of gloves must be considered in their selection and use. For example, methyl chloride will permeate a neoprene glove in less than 15 minutes. The manufacturer should provide the *breakthrough time* and the *permeation rates* for the glove being evaluated. The breakthrough time and permeation rate are dependent on the specific chemical and the composition and thickness of the glove.

Protective eyewear should be provided where the risk of splashing of chemicals is present. Of course, mechanical equipment, barriers, guards, and other engineering measures should be provided as the first line of defense. For chemical splash protection, unvented chemical goggles, indirect-vented chemical goggles, or indirect-vented eyecup goggles should be used. A face shield may also be needed. Direct-vented goggles and normal eyeglasses should not be used, and contact lenses should not be worn.

## 5. PARTICULATES

*Particulates* include dusts, fumes, fibers, and mists. *Dusts* have a wide range of sizes and usually result from a mechanical process such as grinding. *Fumes* are extremely small particles, less than 1 $\mu$m in diameter, and result from combustion and other processes. *Fibers* are thin and long particulates, with asbestos being a prime example. *Mists* are suspended liquids that float in air, such as from the atomization of cutting oil. All of these types of particulate can pose an inhalation hazard if they reach the lungs.

With one known exception, particles larger than approximately 5 $\mu$m cannot reach the alveoli or inner recesses of the lungs before being trapped and expelled from the body through the digestive system or from the mouth and nose. Protection is afforded by the presence of mucus and cilia in the nasal passages, throat, larynx, trachea, and bronchi. The exception is asbestos fibers, which can reach the alveoli even though fibers may be larger. Particles smaller than 5 $\mu$m are considered respirable dusts and pose an exposure hazard when present in the breathing zone.

There are four factors that affect the health risk from exposure to particulates: the types of particulate, the length of exposure, the concentration of particulates in the breathing zone, and the size of particulates in the breathing zone.

The type of particulate can determine the type of health effect that may result from the exposure. Both organic and inorganic dusts can produce allergic effects, dermatitis, and systemic toxic effects. Particulates that contain free silica can produce pneumoconiosis from chronic exposure. *Pneumoconiosis* is lung disease caused by fibrosis from exposure to both organic and inorganic particulates. Other particulates can cause systemic toxicity to the kidneys, blood, and central nervous system. Asbestos fibers can cause lung scarring and cancer.

The critical duration of exposure varies with the type of particulate.

The concentration of particulates in the breathing zone is the primary factor in determining the health risk from particulates.

The fourth exposure factor is the size of the particulates. Particles larger than 5 $\mu$m will normally be filtered out through the upper respiratory system before reaching the alveoli of the lungs.

### Silica

Silica ($SiO_2$) has several associated health hazards. The crystalline form of free silica (quartz) deposited in the lungs causes the growth of fibrous tissue around the deposit. The fibrous tissue reduces the amount of normal lung tissue, thereby reducing the ability of the lungs to transfer oxygen. When the heart tries to pump more blood to compensate, heart strain and permanent damage or death may result. This condition is known as *silicosis*. Mycobacterial infection occurs in about 25% of silicosis cases. Smokers exposed to silica dust have a significantly increased chance of developing lung cancer.

### Asbestos

Asbestos is generically described as a naturally occurring, fibrous, hydrated mineral silicate. Asbestos mining, construction activities, and working in shipyards are possible exposure activities. Inhalation of short asbestos fibers can cause *asbestosis*, a kind of pneumoconiosis, as a nonmalignant scarring of the lungs. *Bronchogenic carcinoma* is a malignancy (cancer) of the lining of the lung's air passages. *Mesothelioma* is a diffuse malignancy of the lining of the chest cavity or the lining of the abdomen.

The onset of illness seems to be correlated with length and diameter of inhaled asbestos fibers. Fibers 2 $\mu$m in length cause asbestosis. Mesothelioma is associated with fibers 5 $\mu$m long. Fibers longer than 10 $\mu$m produce lung cancer. Fiber diameters greater than 3 $\mu$m are more likely to cause asbestosis or lung cancer, while fibers 3 $\mu$m or less in diameter are associated with mesothelioma.

The OSHA regulations for protection from exposure to asbestos are extensive. They require an employer to perform a negative exposure assessment in many cases. Monitoring must be performed by a competent person who is capable of identifying asbestos hazards and selecting control strategies and who has the authority to make corrective changes. The regulations also specify when medical surveillance is required, when personal protection must be provided, and the engineering controls and work practices that must be implemented.

**Biology**

## Lead

The body does not use lead for any metabolic purpose, so any exposure to lead is undesirable. Lead dust and fumes can pose a severe hazard. Acute large doses of lead can cause systemic poisoning or seizures. Chronic exposure can damage the blood-forming bone marrow and the urinary, reproductive, and nervous systems. Lead is probably a human carcinogen, although whether it is causative or facilitative is subject to research.

## Beryllium

Inhalation of metallic beryllium, beryllium oxide, or soluble beryllium compounds can lead to *chronic beryllium disease* (berylliosis). Ingestion and dermal contact do not pose a documented hazard, so maintaining beryllium dusts and fumes below the TLV in the breathing zone is a critical protection measure.

Chronic beryllium disease is characterized by granulomas on the lungs, skin, and other organs. The disease can result in lung and heart dysfunction and enlargement of certain organs. Beryllium has been classified as a suspected human carcinogen.

## Coal Dust

Coal dust can cause chronic bronchitis, silicosis, and *coal worker's pneumoconiosis*, also known as *black lung disease.*

## Welding Fumes

Exposure to welding fumes can cause a disease known as *metal fume fever.* This disease results from inhalation of extremely fine oxide particles that have been freshly formed as fume. Zinc oxide fume is the most common source, but magnesium oxide, copper oxide, and other metallic oxides can also cause metal fume fever. Metal fume fever is of short duration, with symptoms including fever and shaking chills appearing 4 to 12 hours after exposure.

## Radioactive Dusts

Radioactive dusts can cause toxicity in addition to the effects from ionizing radiation. Inhalation of radioactive dust can result in deposition of the radionuclide in the body, which may enter the bloodstream and affect individual organs.

Control measures should be instituted to prevent workers from inhaling radioactive dust, either by restricting access or by providing appropriate personal protection such as respirators. Engineering controls to capture radioactive dust are an absolute necessity to minimize worker exposure.

## Biological Particulates

A wide variety of biological organisms can be inhaled as particulates causing respiratory diseases and allergies.

Examples include dust that contains anthrax spores from the wool or bones of infected animals, and fungi spores from grain and other agricultural produce.

## Control of Particulates

### Ventilation

Ventilation is the most effective method for controlling particulates. Enclosed processes should be used wherever possible. Equipment can be enclosed so that only the feed and discharge openings are open. With adequate pressure, enclosed equipment can be nearly as effective as closed processes. Large automated equipment can sometimes be placed in separate enclosures. Workers would have to wear personal protection to enter the enclosures. Local exhaust ventilation with hooded enclosures can be very effective at controlling particulate emissions into general work areas. Where complete enclosure and local exhaust methods are not sufficient, *dilution ventilation* will be necessary to control particulates in the work area. In some instances, the work process can be changed from a dry to a wet process to reduce particulate generation.

### Personal Protection

Respirators are an effective means of controlling worker exposure to particulates that remain in the work area after engineering controls have been applied or when access to dusty areas is intermittent. Respirators may also be used to provide additional protection or comfort to workers in areas where local or general ventilation is effective. The NIOSH guidelines for selection of respirators should be followed to ensure that the respirator will be effective at removing the specific particulate to which the workers are exposed.

## 6. SOUND AND NOISE

### Characteristics of Sound

*Sound* is pressure variation in air, water, or some other medium that the human ear can detect. *Noise* is unwanted, unpleasant, or painful sound. The *frequency* of sound is the number of pressure variations per second, measured in cycles per second, or hertz (Hz). The frequency range of human audible sound is approximately 20 Hz to 20 000 Hz.

Sound passes through a medium at the *speed of sound, a,* equal to the product of the wavelength and the frequency.

$$a = f\lambda \hspace{3em} \textit{38.1}$$

The speed of sound is dependent upon the medium, as illustrated in Table 38.2.

*Table 38.2* Speed of Sound in Various Media

| medium | speed of sound (m/s) | condition |
|---|---|---|
| air | 330 | 1 atm, 0°C |
| water | 1490 | 1 atm, 20°C |
| aluminum | 4990 | 1 atm |
| steel | 5150 | 1 atm |

## Sound Pressure

*Sound pressure* measures the intensity of sound and is the variation in atmospheric pressure. (See Fig. 38.1.) The *root-mean-square (rms) pressure* is used.

*Figure 38.1* Characteristics of a Sound Wave

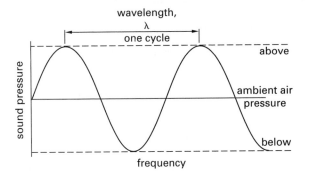

$$p_{\text{rms}} = \sqrt{\frac{\sum p_i^2}{n}} \qquad 38.2$$

*Sound pressure level* $(L_p)$ is measured in decibels relative to a reference level, $p_0$, of 20 $\mu$Pa, the threshold of hearing at a reference frequency of 1000 Hz, as follows.

$$L_p = 10 \log\left(\frac{p}{p_0}\right)^2$$
$$= 20 \log\left(\frac{p}{p_0}\right) \qquad 38.3$$

## Sound Power

*Sound power*, $W$, is the absorbed or transmitted sound energy per unit time (in watts). The sound power level, $L_W$, is measured in decibels relative to a reference level, $W_0$, of $10^{-12}$ W.

$$L_W = 10 \log\left(\frac{W}{W_0}\right) \qquad 38.4$$

## Sound Intensity

*Sound intensity* is an areal function of the sound power of a source.

$$I = \frac{W}{4\pi r^2} \qquad 38.5$$

## Combining Sound Sources

The combined sound pressure level or sound power level from two or more sound sources can be determined by

$$L = 10 \log \sum 10^{L_i/10} \qquad 38.6$$

## Loudness

*Loudness* in humans is primarily a function of sound pressure but is also affected by frequency because the human ear is more sensitive to high-frequency sounds than low-frequency sounds.

## Noise

Noise can cause psychological and physiological damage and can interfere with workers' communication, thereby affecting safety. Exposure to excessive noise for a sufficient time can result in hearing loss. The permissible OSHA noise exposures are given in Table 38.3.

*Table 38.3* Permissible Noise Exposures

| duration per day (h) | sound level (dBA slow response)[*] |
|---|---|
| 8 | 90 |
| 6 | 92 |
| 4 | 95 |
| 3 | 97 |
| 2 | 100 |
| 1.5 | 102 |
| 1 | 105 |
| 0.5 | 110 |
| 0.25 or less | 115 |

[*]"Slow response" refers to the metering circuit being set on slow mode to reduce rapid, hard-to-read needle excursions. Compliance with OSHA regulations requires measurement set on slow response mode.

When the daily noise exposure is composed of two or more periods of noise exposure at different levels, their combined effect should be considered, rather than the individual effect of each. If the sum of the fractions $C_1/t_1 + C_2/t_2 \ldots C_n/t_n$ exceeds unity, then the mixed exposure should be considered to exceed the limit value. $C_n$ indicates the total time of exposure at a specified noise level, and $t_n$ indicates the total time of exposure permitted at that level.

**Biology**

## Hearing Loss

Hearing loss can be caused by sudden intense noise over only a few exposures. This type of loss is known as *acoustic trauma*. Hearing loss can also be caused by exposure over a long duration (months or years) to hazardous noise levels. This type is known as *noise-induced hearing loss*. The permanence and nature of the injury depends on the type of hearing loss.

The main risk factors associated with hearing loss are the intensity of the noise (sound pressure level), the type of noise (frequency), daily exposure time (hours per day), and the total work duration (years of exposure). These are known as *noise exposure factors*. Generally, exposure to sound levels above 115 dBA is considered hazardous, and exposure to levels below 70–75 dBA is considered safe from risk of permanent hearing loss. Also, noise with predominant frequencies above 500 Hz is considered to have a greater potential to cause hearing loss than lower-frequency sounds.

## Classes of Noise Exposure

Noise exposure can be classified as continuous noise, intermittent noise, and impact noise. *Continuous noise* is broadband noise of a nearly constant sound pressure level and frequency to which a worker is exposed 8 hours daily and 40 hours weekly. *Intermittent noise* involves exposure to a specific broadband sound pressure level several times a day. *Impact noise* is a sharp burst of short duration sound.

OSHA has established permissible noise exposures for a specific duration each workday, known as *permissible exposure levels* (PELs). The PELs are based on continuous 8 hour exposure at a sound pressure level of 90 dBA, which is established as 100%. For other exposure durations, OSHA has established relationships between the sound level and the exposure time. Every 5 dBA increase in noise cuts the allowable exposure time in half. Sound pressure levels below 90 dBA are not considered hazardous and do not have to be determined.

When workers are exposed to different noise levels during the day, the mixed exposure, $E_m$, must be calculated. If $E_m$ equals or exceeds 1, the mixed exposure exceeds the OSHA standard.

$$E_m = \sum_{i=1}^{n} \frac{C_i}{t_i} \qquad 38.7$$

For intermittent noise, the time characteristics of the noise must also be determined. Both short-term and long-term exposure must be measured. A dosimeter is typically used for intermittent noises.

For impact noises, workers should not be exposed to peaks of more than 140 dBA under any circumstances. The threshold limit value for impulse noise should not exceed the values provided in Table 38.4.

*Table 38.4* Typical Threshold Limit Values for Impact Noise

| peak sound level (dB) | maximum number of daily impacts |
|---|---|
| 140 | 100 |
| 130 | 1000 |
| 120 | 10 000 |

## Noise Control

A *hearing conservation program* should include noise measurements, noise control measures, hearing protection, audiometric testing of workers, and information and training programs. Employees are required to properly use the protective equipment provided by employers.

After the noise exposure is compared with acceptable noise levels, the degree of noise reduction needed can be determined. Noise reduction measures can comprise the following three basic methods applied in order.

1. changing the process or equipment

2. limiting the exposure

3. using hearing protection

Administrative controls include changing the exposure of workers to high noise levels by modifying work schedules or locations so as to reduce workers' exposure times. Administrative controls include any administrative decision that limits a worker's exposure to noise.

Personal hearing protection is the final noise control measure, to be implemented only after engineering controls are implemented. Protective devices do not reduce the noise hazard and may not be totally effective. Therefore, engineering controls are preferred over hearing protection. Protective devices include helmets, earplugs, canal caps, and earmuffs. Earplugs may be used with helmets to increase the level of noise reduction.

An important characteristic of personal hearing protection is the *noise reduction rating* (NRR). The NRR is established by the Environmental Protection Agency (EPA) and must be printed on the package of a device. The NRR can be used to determine whether a device provides sufficient hearing protection.

## Audiometry

*Audiometry* is the measurement of hearing acuity. It is used to assess a worker's hearing ability by measuring the individual's threshold sound pressure level at various frequencies (250 Hz to 6000 Hz). The threshold audiogram can be used to create a baseline of hearing ability and to determine changes over time and identify changes resulting from noise control measures. Baseline and annual hearing tests are required where workers are exposed to more than a *time-weighted average* (TWA) over 8 hours of 85 dBA. The average change from the baseline is used to measure the degree of hearing impairment.

## 7. RADIATION

Radiation can be either nonionizing or ionizing. *Nonionizing radiation* includes electric fields, magnetic fields, electromagnetic radiation, radio frequency and microwave radiation, and optical radiation and lasers.

Dealing with *ionizing radiation* from nuclear sources requires special skills and knowledge. A specialist in health physics should be consulted whenever ionizing radiation is encountered.

### Nuclear Radiation

*Nuclear radiation* is a term that applies to all forms of radiation energy that originate in the nuclei of radioactive atoms. Nuclear radiation includes alpha particles, beta particles, neutrons, X-rays, and gamma rays. The common property of all nuclear radiation is an ability to be absorbed by and transfer energy to the absorbing body.

The preferred unit of ionizing radiation given in the National Council on Radiation Protection's (NCRP's) *Recommended Limitations for Exposure to Ionizing Radiation*, is the mSv. Sv is the symbol for sievert, which is the SI unit of absorbed dose times the *quality factor* of the radiation as compared to gamma radiation. The absorbed dose is measured in grays (Gy). The gray is equal to 1 J of absorbed energy per kilogram of matter. A summary of ionizing radiation units is given in Table 38.5.

*Table 38.5* Units for Measuring Ionizing Radiation

| property | SI |
|---|---|
| energy absorbed | gray (Gy)<br>1 J/kg<br>1 Gy = 100 rad (obsolete) |
| biological effect | sievert (Sv)<br>Gy × quality factor<br>1 Sv = 100 rem (obsolete) |

### Alpha Particles

*Alpha particles* consist of two protons and two neutrons, with an atomic mass of four. Alpha particles combine with electrons from the absorbed material and become helium atoms. Alpha particles have a positive charge of two units and react electrically with human tissue. Because of their large mass, they can travel only about 10 cm in air and are stopped by the outer layer of the skin. Alpha-emitters are considered to be only internal radiation hazards, which requires alpha particles to be ingested by eating or breathing. They affect the bones, kidney, liver, lungs, and spleen.

### Beta Particles

*Beta particles* are electrically charged particles ejected from the nuclei of radioactive atoms during disintegration. They have a negative charge of one unit and the mass of an electron. High-energy beta particles can penetrate in human tissue to a depth of 20 mm to 130 mm and travel up to 9 m in air. Skin burns can result from an extremely high dose of low-energy beta radiation, and some high-energy beta sources can penetrate deep into the body, but beta-emitters are primarily internal radiation hazards, which would require them to be ingested. Beta particles are more hazardous than alpha particles because they can penetrate deeper into tissue. High-energy beta radiation can produce a secondary radiation called *bremsstrahlung*. These are X-rays produced when electrons (i.e., beta particles) pass near the nuclei of other atoms. Bremsstrahlung radiation is proportional to the energy of the beta particle and the atomic number of the adjacent nucleus. Therefore, materials with low atomic numbers (e.g., plexiglass) are preferred shielding materials.

### Neutrons

*Neutron particles* have no electrical charge and are released upon disintegration of certain radioactive materials. Their range in air and in human tissue depends on their kinetic energy, but the average depth of penetration in human tissue is 60 mm. Neutrons lose velocity when they are absorbed or deflected by the nuclei with which they collide. However, the nuclei are left with higher energy that is later released as protons gamma rays, beta particles, or alpha particles. It is these secondary emissions from neutrons that produce damage in tissue.

### X-Rays

*X-rays* are produced by electron bombardment of target materials and are highly penetrating electromagnetic radiation. X-rays have a valuable scientific and commercial use in producing shadow pictures of objects. The energy of an X-ray is inversely proportional to its wavelength. X-rays of short wavelength are called *hard*, and they can penetrate several centimeters of steel. Long wavelength X-rays are called *soft*, and they are less penetrating. The power of X-rays and gamma rays to penetrate matter is called *quality*. *Intensity* is the energy flux density.

### Gamma Rays

*Gamma rays*, or gamma radiation, are a class of electromagnetic photons (radiation) emitted from the nuclei of radioactive atoms. They are highly penetrating and are an external radiation hazard. Gamma rays are emitted spontaneously from radioactive materials, and the energy emitted is specific to the radionuclide. Gamma rays present an internal exposure problem because of their deep penetrating ability.

### Radioactive Decay

Radioactive decay is measured in terms of *half-life*, the time to lose half of the activity of the original material.

Decay activity can be calculated from the *decay constant,* $\gamma$.

$$A = A_o e^{-\lambda t} \qquad 38.8$$

The half-life can be calculated from the decay constant.

$$t_{1/2} = 0.693/\lambda \qquad 38.9$$

## Radiation Effects on Humans

Ionizing radiation transfers energy to human tissue when it passes through the body. *Dose* refers to the amount of radiation that a body absorbs when exposed to ionizing radiation. The effects on the body from external radiation are quite different from the effects from internal radiation. Internal radiation is spread throughout the body to tissues and organs according to the chemical properties of the radiation. The effects of internal radiation depend on the energy and the residence time within the body. The principal effect of radiation on the body is destruction of or damage to cells. Damage may affect reproduction of cells or cause mutation of cells.

The effects of ionizing radiation on individuals include skin, lung, and other cancers; bone damage; cataracts; and a shortening of life. Effects on the population as a whole include possible damage to human reproductive elements, thereby affecting the genes of future generations.

## Safety Factors

An environmental engineer should be aware of the basic safety factors for limiting dose. These factors are time, distance, and shielding.

The dose received is directly related to the time exposed, so reducing the time of exposure will reduce the dose. An individual's time of exposure can also be limited by spreading the exposure time among more workers.

Distance is another safety factor that can be changed to reduce the dose. The intensity of external radiation decreases as the inverse of the square of the distance. By increasing the distance to a source from 2 m to 20 m, for example, the exposure would be reduced to 1% $(2 \text{ m}/20 \text{ m})^2$.

Shielding involves placing a mass of material between a source and workers. The objective is to use a high-density material that will act as a barrier to X-ray and gamma-ray radiation. Lead and concrete are often used, with lead being the more effective material because of its greater density. For neutrons, different material is needed than for X-rays and gamma rays because neutrons produce secondary radiation from collisions with nuclei. Neutron shielding requires a light nucleus material. Typically water or graphite is used.

The shielding properties of materials are often compared using the *half-value thickness,* which is the thickness of the material required to reduce the radiation to half of the incident value. The half-value properties vary with the radiation source.

## 8. HEAT AND COLD STRESS

### Thermal Stress

Heat and cold, or *thermal,* stress involves three zones of consideration relative to industrial hygiene. In the middle is the *comfort zone,* where workers feel comfortable in the work environment. On either side of the comfort zone is a *discomfort zone* where workers feel uncomfortable with the heat or cold, but a health risk is not present. Outside of each discomfort zone is a *health risk zone* where there is a significant risk of health disorders due to heat or cold. Industrial hygiene is primarily concerned with controlling worker exposure in the health risk zone.

The analysis of thermal stress involves taking a *heat balance* of the human body with the objective of determining whether the net heat storage is positive, negative, or zero. A simplified form of the heat balance is

$$Q_S = Q_M + Q_R + Q_C + Q_E \qquad 38.10$$

If the storage, $Q_S$, is zero, heat gain is balanced by heat loss, and the body is in equilibrium. If $Q_S$ is positive, the body is gaining heat; and, if $Q_S$ is negative, the body is losing heat.

The heat balance is affected by environmental and climatic conditions, work demands, and clothing. The metabolic rate, $Q_M$, is more significant for heat stress than for cold stress when compared with radiation and convection. The metabolic rate can affect heat gain by one to two orders of magnitude compared to radiation and convection, but it affects heat loss to about the same extent as radiation and convection.

Clothing affects the thermal balance through insulation, permeability, and ventilation. *Insulation* provides resistance to heat flow by radiation, convection, and conduction. *Permeability* affects the movement of water vapor and the amount of evaporative cooling. *Ventilation* influences evaporative and convective cooling.

### Heat Stress

*Heat stress* can increase body temperature, heart rate, and sweating, which together constitute *heat strain*.

The most serious heat disorder is *heatstroke,* because it involves a high risk of death or permanent damage. Fortunately, heatstroke is rare. Of lesser severity, *heat exhaustion* is the most commonly observed heat disorder for which treatment is sought. *Dehydration* is usually not noticed or reported, but without restoration of water loss, dehydration leads to heat exhaustion. The symptoms of these key heat stress disorders are as follows.

- heatstroke: chills, restlessness, irritability

- heat exhaustion: fatigue, weakness, blurred vision, dizziness, headache

- dehydration: no early symptoms, fatigue or weakness, headache, dry mouth

Appropriate first aid and medical attention should be sought when any heat stress disorder is recognized.

## Control of Heat Stress

Controls that are applicable to any heat stress situation are known as *general controls*. General controls include worker training, heat stress hygiene, and medical monitoring.

*Specific controls* are controls that are put in place for a particular job. They include engineering controls, administrative controls, and personal protection. *Engineering controls* include changing the physical work demands to reduce the metabolic heat gain, reducing external heat gain from the air or surfaces, and enhancing external heat loss by increasing sweat evaporation and decreasing air temperature. *Administrative controls* include scheduling the work to allow worker acclimatization to occur, leveling work activity to reduce peak metabolic activity, and sharing or scheduling work so the heat exposure of individual workers is reduced. *Personal protection* includes using systems to circulate air or water through tubes or channels around the body, wearing ice garments, and wearing reflective clothing.

## Cold Stress

The body reacts to cold stress by reducing blood circulation to the skin to insulate itself. The body also shivers to increase metabolism. These mechanisms are ineffective against long-term extreme cold stress, so humans react by increasing clothing for more insulation, increasing body activity to increase metabolic heat gain, and finding a warmer location.

There are two main hazards from cold stress: hypothermia and tissue damage. *Hypothermia* depresses the central nervous system, causing sluggishness and slurred speech, and progresses to disorientation and unconsciousness. To avoid hypothermia, the minimum core body temperature must be above 96.8°F (36°C) for prolonged exposure and above 95°F (35°C) for occasional exposure of short duration.

Worker training, cold stress hygiene, and medical surveillance can control cold stress. Engineering controls, administrative controls, and personal protection measures can also be used to control cold stress.

## 9. ERGONOMICS

*Ergonomics* is the study of human characteristics to determine how a work environment should be designed to make work activities safe and efficient. It includes both physiological and psychological effects on the worker, as well as health and safety and productivity aspects.

## Work-Rest Cycles

Excessively heavy work should be broken by frequent short rest periods to reduce cumulative fatigue. The percentage of time a worker should rest can be estimated by the following equation.

$$t_{\text{rest}} = \frac{Q_{M,\text{max}} - Q_M}{Q_{M,\text{rest}} - Q_M} \times 100\% \qquad 38.11$$

## Manual Handling of Loads

On many projects, loads must be handled manually. Improper handling is the most common cause of injury and of the most severe injuries in the workplace.

Heavy loads can strain the body, particularly the lower back. Even light or small objects can cause risk of injury to the body if they are handled in a way that requires strain-inducing stretching, reaching, or lifting.

## Cumulative Trauma Disorders

*Cumulative trauma disorders* (CTD) can occur in almost any work situation. CTDs result from repeated stresses that are not excessive individually but, over time, cause disorders, injuries, and the inability to perform a job. High repetitiveness, or continuous use of the same body part results in fatigue followed by cumulative muscle strain. These cumulative injuries are usually incurred by tendons, tendon sheaths, and soft tissue. Moreover, the cumulative injuries can result in damage to nerves and restricted blood flow. CTDs are common in the hand, wrist, forearm, shoulder, neck, and back. Bone and the spinal vertebrae may also be damaged.

The manifestations of CTDs on soft tissues include stretched and strained muscles, rough or torn tendons, inflammation of tendon sheaths, irritation and inflammation of bursa, and stretched (sprained) ligaments. Nerves can be affected by pressure from tendons or other soft tissue, resulting in loss of muscle control, numbness, tingling, or pain, and loss of response of nerves that control automatic functions such as body temperature and sweating. Blood vessels may be compressed, resulting in restricted blood flow and impaired control of tissues (muscles) dependent on that blood supply. Vibration, such as from operating vibrating tools, can cause the arteries in the fingers and hands to close down, resulting in numbness, tingling, and eventually loss of sensation and control.

Industrial hygienists have defined *high repetitiveness* as a cycle time of less than 30 s, or more than 50% of a cycle time spent performing the same fundamental motion. If the work activity requires the muscles to remain contracted at about 15% to 20% of their maximum capability, circulation can be restricted, which also contributes to CTDs. Also, severe deviation of the wrists, forearms, and other body parts can contribute to CTDs.

## 10. TYPES OF CTDS

### Carpal Tunnel Syndrome

Carpal tunnel syndrome (CTS) is the best known CTD. The American Industrial Hygiene Association has described CTS as an occupational illness of the hand and arm system. CTS results from rapid, repetitious finger and wrist movements.

The wrist has a "tunnel" created by the carpal bones on the outer side and ligaments, which are firmly attached to the bones, across the inner side. In the *carpal tunnel*, which is roughly oval in shape, are tendons and tendon sheaths of the fingers, several nerves, and arteries. If the wrist is bent up or down or flexed from side to side, the space in the carpal tunnel is reduced. Swelling of the tendons or tendon sheaths can place pressure on the nerves, blood vessels, and tendons. Activities that can lead to this disorder include grinding, sanding, hammering, keyboarding, and assembly work.

### Cubital Tunnel Syndrome

This disorder occurs from compression of the nerve in the forearm below the elbow and results in tingling, numbness, or pain in the fingers. Leaning over a workbench and resting the forearm on a hard surface or edge typically causes this disorder.

### Epicondylitis

This disorder is also known as "tennis elbow" and "golfer's elbow." It results from irritation of the tendons of the elbow. It is caused by forceful wrist extensions, repeated straightening and bending of the elbow, and impacting throwing motions.

### Ganglionitis

Ganglionitis is a swelling of a tendon sheath in the wrist. Activities that can lead to this disorder include grinding, sanding, sawing, cutting, and using pliers and screwdrivers.

### Neck Tension Syndrome

This disorder is characterized by an irritation of the muscles of the neck. It commonly occurs after repeated or sustained overhead work.

### Pronator Syndrome

This disorder compresses a nerve in the forearm. It results from rapid and forceful strenuous flexing of the elbow and wrist. Activities that can lead to this disorder include buffing, grinding, polishing, and sanding.

### Tendonitis

This is an inflammation of a tendon where its surface becomes thickened, bumpy, and irregular. Tendon fibers may become frayed or torn. This disorder can result from repetitious, forceful movements, contact with hard surfaces, and vibrations.

*Shoulder tendonitis* is irritation and swelling of the tendon or bursa of the shoulder. It is caused by continuous elevation of the arm.

### Tenosynovitis

This disorder is characterized by swelling of tendon sheaths and irritation of the tendon. It is known as *DeQuervain's syndrome* when it affects the thumb. Activities that can lead to this disorder include grinding, polishing, sanding, sawing, cutting, and using screwdrivers.

*Trigger finger* is a special case of tenosynovitis that results in the tendon of the trigger finger becoming nearly locked so that its forced movement is jerky. It comes from using hand tools with sharp edges pressing into the tissue of the finger or where the tip of the finger is flexed but the middle part is straight.

### Thoracic Outlet Syndrome

This disorder is characterized by reduced blood flow to and from the arm due to compression of nerves and blood vessels between the collarbone and the ribs. It results in a numbing of the arm and constrains muscular activities.

### Ulnar Artery Aneurysm

This disorder is characterized by a weakening of an artery in the wrist, causing an expansion that presses on the nerve. This often occurs from pounding or pushing with the heel of the hand, as in assembly work.

### Ulnar Nerve Entrapment

This disorder involves pressure on a nerve in the wrist. It occurs from prolonged flexing of the wrist and repeated pressure on the palm. Activities that can lead to this disorder include carpentry, brick laying, and using pliers and hammers.

### White Finger

This disorder is also known as "dead finger," *Raynaud's syndrome*, or *vibration syndrome*. In this disorder, the finger turns cold and numb, tingles, and loses sensation and control. The cause is insufficient blood supply, which causes the finger to turn white. It results from closure of the arteries due to vibrations. Gripping vibrating tools, especially in the cold, is a common cause.

## 11. BIOLOGICAL HAZARDS

### Biological Agents

Approximately 200 biological agents are known to produce infectious, allergenic, toxic, and carcinogenic reactions in workers. These agents and their reactions are as follows.

- Microorganisms (viruses, bacteria, fungi) and the toxins they produce cause infection and allergic reactions.

- Arthropod (crustaceans), arachnids (spiders, scorpions, mites, and ticks), and insect bites and stings cause skin inflammation, systemic intoxication, transmission of infectious agents, and allergic reactions.

- Allergens and toxins from plants cause dermatitis from skin contact, rhinitis (inflammation of the nasal mucus membranes), and asthma from inhalation.

- Protein allergens (urine, feces, hair, saliva, and dander) from vertebrate animals cause allergic reactions.

Also posing potential biohazards are lower plants other than fungi (e.g., lichens, liverworts, and ferns) and invertebrate animals other than arthropods (e.g., parasites, flatworms, and roundworms).

Microorganisms may be divided into prokaryotes and eukaryotes. *Prokaryotes* are organisms having DNA that is not physically separated from its cytoplasm (cell plasma that does not include the nucleus). They are small, simple, one-celled structures, less than 5 $\mu$m in diameter, with a primitive nuclear area consisting of one chromosome. Reproduction is normally by binary fission in which the parent cell divides into two daughter cells. All bacteria, both single-celled and multicellular, are prokaryotes, as are blue-green algae.

*Eukaryotes* are organisms having a nucleus that is separated from the cytoplasm by a membrane. Eukaryotes are larger cells (greater than 20 $\mu$m) than prokaryotes, with a more complex structure, and each cell contains a distinct membrane-bound nucleus with many chromosomes. They may be single-celled or multicellular, reproduction may be asexual or sexual, and complex life cycles may exist. This class of microorganisms includes fungi, algae (except blue-green), and protozoa.

Since prokaryotes and eukaryotes have all of the enzymes and biological elements to produce metabolic energy, they are considered organisms.

In contrast, a *virus* does not contain all of the elements needed to reproduce or sustain itself and must depend on its host for these functions. Viruses are nucleic acid molecules enclosed in a protein coat. A virus is inert outside of a host cell and must invade the host cell and use its enzymes and other elements for the virus's own reproduction. Viruses can infect very small organisms such as bacteria, as well as humans and animals. Viruses are 20 $\mu$m to 300 $\mu$m in diameter.

Smaller than the viruses by an order of magnitude are *prions*, small proteinaceous infectious particles. Prions have properties similar to viruses and cause degenerative diseases in humans and animals.

### Infection

The invasion of the body by pathogenic microorganisms and the reaction of the body to them and to the toxins they produce is called an *infection*. Infection may be *endogenous* where microorganisms that are normally present in the body (*indigenous*) at a particular site (such as *E. coli* in the intestinal tract) reach another site (such as the urinary tract), causing infection there.

Infections from microorganisms not normally found on the body are called *exogenous infections*.

The most common routes of exposure to infectious agents are through cuts, punctures and bites (insect and animal), abrasions of the skin, inhalation of aerosols generated by accidents or work practices, contact between mucous membranes or contaminated material, and ingestion. In laboratory and medical settings, transmission of blood-borne pathogens can occur through handling of blood products and human tissue.

### Biohazardous Workplaces and Activities

Although engineers have long been concerned with waterborne diseases and their prevention in the design and operation of water supply and wastewater systems, pathogens may also be encountered in the workplace through air or direct contact.

#### Microbiology and Public Health Laboratories

Workers in laboratories handling infectious agents experience a risk of infection.

#### Health Care Facilities

Health care facilities such as hospitals, medical offices, blood banks, and outpatient clinics, present numerous opportunities for exposure to a wide variety of hazardous and toxic substances, as well as to infectious agents.

#### Biotechnology Facilities

Biotechnology is one of the newest technologies and involves a much greater scope and complexity than the historical use of microorganisms in the chemical and pharmaceutical industries. This technology now deals with DNA manipulation and the development of products for medicine, industry, and agriculture. The microorganisms used by the biotechnology industry often are genetically engineered plant and animal cells. Allergies can be a major health issue.

#### Animal Facilities

Workers exposed to animals are at risk for animal-related allergies and infectious agents. Occupations

Biology

include agricultural workers, veterinarians, workers in zoos and museums, taxidermists, and workers in animal-product processing plants.

*Zoonotic diseases* (diseases that affect both humans and animals) are the most common diseases reported by laboratory workers. Work acquired infections from non-human primates are common.

Some of the diseases of concern in animal facilities include Q fever, hantavirus, Ebola, Marburg viruses, and simian immunodeficiency viruses.

### Agriculture

Agricultural workers are exposed to infectious microorganisms through inhalation of aerosols, contact with broken skin or mucus membranes, and inoculation from injuries. Farmers and horticultural workers may be exposed to fungal diseases. Food and grain handlers may be exposed to parasitic diseases. Workers who process animal products may acquire bacterial skin diseases such as anthrax from contaminated hides, tularemia from skinning infected animals, and erysipelas from contaminated fish, shellfish, meat, or poultry. Infected turkeys, geese, and ducks can expose poultry workers to *psittacosis*, a bacterial infection. Workers handling grain may be exposed to *mycotoxins* from fungi and *endotoxins* from bacteria.

### Utility Workers

Workers maintaining water systems may be exposed to Legionella pneumophila (Legionnaires' disease). Sewage collection and treatment workers may be exposed to enteric bacteria, hepatitis A virus, infectious bacteria, parasitic protozoa (giardia), and allergenic fungi. Solid waste handling and disposal facility workers may be exposed to blood-borne pathogens from infectious wastes.

### Wood-Processing Facilities

Wood-processing workers may be exposed to bacterial endotoxins and allergenic fungi.

### Mining

Miners may be exposed to zoonotic bacteria, mycobacteria, fungi, and untreated runoff water and wastewater.

### Forestry

Forestry workers may be exposed to zoonotic diseases (rabies virus, Russian spring fever virus, Rocky Mountain spotted fever, Lyme disease, and tularemia) transmitted by ticks and fungi.

### Blood-Borne Pathogens

The risk from hepatitis B and human immunodeficiency virus (HIV) in health care and laboratory situations led OSHA to publish standards for occupational exposure to blood-borne pathogens. Some blood-borne pathogens are summarized as follows.

### Human Immunodeficiency Virus (HIV)

HIV is the blood-borne virus that causes acquired immunodeficiency syndrome (AIDS). Contact with infected blood or other body fluids can transmit HIV. Transmission may occur from unprotected sexual intercourse, sharing of infected needles, accidental puncture wounds from contaminated needles or sharp objects, or transfusion with contaminated blood.

Symptoms of HIV include swelling of lymph nodes, pneumonia, intermittent fever, intestinal infections, weight loss, and tuberculosis. Death typically occurs from severe infection causing respiratory failure due to pneumonia.

### Hepatitis

The hepatitis virus affects the liver. Symptoms of infection include jaundice, cirrhosis and liver failure, and liver cancer.

Hepatitis A can be contracted through contaminated food or water or by direct contact with blood or body fluids such as blood or saliva. Hepatitis B, known as *serum hepatitis*, may be transmitted through contact with infected blood, body fluids, and through blood transfusions. Hepatitis B is the most significant occupational infector of health care and laboratory workers. Hepatitis C is similar to hepatitis B, but can also be transmitted by shared needles, accidental puncture wounds, through blood transfusions, and unprotected sex.

Hepatitis D occurs when one of the other hepatitis viruses replicates. Individuals with chronic hepatitis D often develop cirrhosis of the liver. Chronic hepatitis may be present in carriers.

### Syphilis

The bacterium responsible for the transmission of syphilis is called *treponema pallidum pallidum*. (Treponema pallidum has four subspecies, so the extra "pallidum" indicates the virus that specifically causes syphilis.) Syphilis is almost always transmitted through sexual contact, though it may be transmitted in utero through the placenta from mother to fetus. This is known as *congenital syphilis*.

### Toxoplasmosis

Toxoplasmosis is caused by a parasitic organism called *Toxoplasma gondii*, which may be transmitted by ingestion of contaminated meat, across the placenta, and through blood transfusions and organ transplants.

## Rocky Mountain Spotted Fever

Ticks infected with the pathogen *Rickettsia rickettsii*, pass this disease from pets and other animals to humans. Symptoms and effects include headache, rash, fever, chills, nausea, vomiting, cardiac arrhythmia, and kidney dysfunction. Death may occur from renal failure and shock.

## Bacteremia

*Bacteremia* is the presence of bacteria in the bloodstream, whether associated with active disease or not.

## Bacteria- and Virus-Derived Toxins

Some toxins are derived from bacteria and viruses. The effects of these toxins vary from mild illness to debilitating illness or death.

## Botulism

The organism *Clostridium botulinum* produces the toxin that is responsible for botulism. There are four types of botulism: food-borne, infant, adult enteric (intestinal), and wound. *Food botulism* is associated with poorly preserved foods and is the most widely recognized form. *Infant botulism* can occur in the second month after birth when the bacteria colonize the intestinal tract and produce the toxin. *Adult enteric botulism* is similar to infant botulism. *Wound botulism* occurs when the spores enter a wound through contaminated soil or needles. The toxin is absorbed in the bloodstream and blocks the release of a neurotransmitter. Severe cases can result in respiratory paralysis and death.

## Lyme Disease

Lyme disease is transmitted to humans through bites of ticks infected with *Borrelia burgdorferi*.

## Tetanus

Tetanus occurs from infection by the bacterium *Clostridium tetani*, which produces two exotoxins, tetanolysin and tetanospasmin. Routine immunizations prevent the disease.

## Toxic Shock Syndrome

Toxic shock syndrome (TSS) is caused by the bacterium *Staphylococcus aureus*, which produces a pyrogenic toxin.

## Ebola (African Hemorrhagic Fever)

The Ebola and Marburg viruses produce an acute hemorrhagic fever in humans. Symptoms include headache, progressive fever, sore throat, and diarrhea.

## Hantavirus

The hantavirus is found in rodents and shrews of the southwest and is spread by contact with their excreta.

## Tuberculosis

Tuberculosis (TB) is a bacterial disease from *Mycobacterium tuberculosis*. Humans are the primary source of infection. TB affects a third of the world's population outside the United States. A drug-resistant strain is a serious problem worldwide, including in the United States. The risk of contracting active TB is increased among HIV-infected individuals.

## Legionnaires' Disease

Legionnaires' disease (legionellosis) is a type of pneumonia caused by inhaling the bacteria *Legionella pneumophilia*. Symptoms include fever, cough, headache, muscle aches, and abdominal pain. People usually recover in a few weeks and suffer no long-term consequences. *Legionellae* are common in nature and are associated with heat-transfer systems, warm-temperature water, and stagnant water. Sources of exposure include sprays from cooling towers or evaporative condensers and fine mists from showers and humidifiers. Proper design and operation of ventilation, humidification, and water-cooled heat-transfer equipment and other water systems equipment can reduce the risk. Good system maintenance includes regular cleaning and disinfection.

**Biology**

# 39

# Bioprocessing

## Nomenclature

| | | |
|---|---|---|
| $A$ | surface area | $m^2$ |
| $A_{plan}$ | area of cross-section of a packed bed | $m^2$ |
| $F$ | fraction of influent BODs consisting of raw primary sewage | – |
| $k$ | rate constant | $d^{-1}$ |
| $k_d$ | microbial death ratio, kinetic constant | – |
| $K_d$ | digester reaction rate coefficient | – |
| $L$ | ultimate BOD (BOD remaining at time $t = \infty$) | mg/L |
| MLSS | mixed liquor suspended solids | mg/L |
| $n$ | media characteristic coefficient | – |
| $P$ | population | persons |
| $P_v$ | volatile fraction of suspended solids | – |
| $q$ | hydraulic loading | $m^3/m^2 \cdot min$ |
| $Q$ | volumetric flow rate | $m^3/s$ |
| $R$ | recycle ratio | – |
| $S$ | BOD | mg/L |
| $S$ | concentration | mg/L |
| $S_0$ | initial BOD ultimate in mixing zone | mg/L |
| SVI | sludge volume index | mL/L |
| $t$ | time | s |
| $t_{1/2}$ | half-life | d |
| $V$ | volume | $m^3$ |
| $V_1$ | raw sludge input | $m^3/d$ |
| $V_2$ | digested sludge accumulation | $m^3/d$ |
| $X$ | concentration | ppb or $\mu g/kg$ |
| $Y$ | yield coefficient | – |
| $y_t$ | amount of BOD exerted at time $t$ | mg/L |

## Symbols

| | | |
|---|---|---|
| $\theta$ | hydraulic residence time | d |
| $\theta_c$ | solids (cell) residence time or sludge age | d |
| $\mu$ | specific growth rate | 1/s |
| $\rho$ | density | $kg/m^3$ |

## Subscripts

| | |
|---|---|
| $A$ | aeration basin |
| $d$ | death or digester |
| $e$ | effective or effluent |
| $i$ | influent |
| max | maximum |
| $o$ | influent |
| $r$ | reaction or residence |
| $R$ | recycle |
| $s$ | solute, sludge, or storage |
| $t$ | thicken |
| $T$ | at time $T$ |
| $v$ | volatile |
| $w$ | aqueous phase, water, or waste sludge |

## 1. INTRODUCTION

The term *bioprocessing* refers to systems that use living organisms, mostly microorganisms, to obtain desired results. One large integrater of bioprocessing systems is the pharmaceutical industry. Another category of systems constitutes those that are used in wastewater processing by local and regional government agencies. The typical wastewater processing units are described in this chapter.

## 2. ENVIRONMENTAL MICROBIOLOGY

Microorganisms play an important role in the biological treatment of wastewater. They are useful for the removal of organic matter, colloids, and nitrogen and phosphorus compounds.

The important microorganisms in wastewater treatment are bacteria. *Bacteria* are single-celled organisms that survive within a narrow range of pH and temperature. Their size varies from 0.5 $\mu$m to 1 $\mu$m, and they generally reproduce by fission (division). (*Fungi* are aerobic organisms, multicellular in nature, that can survive in a low-pH environment. *Protozoa* are generally unicellular organisms, aerobic in nature, that consume bacteria.)

Bacteria placed in a vessel in a nutrient medium go through five growth phases.

### BOD Exertion

A laboratory analysis of organic matter in water often includes a *biochemical oxygen demand* (BOD) test. To measure BOD, a sample of wastewater is diluted with oxygen-saturated water obtained by sparging air overnight through the water. A small amount of bacteria is fed to the sample. Oxygen concentrations are measured at the beginning of the test and subsequently on a daily basis. The difference between the oxygen concentrations at the initial time and at time $t$ gives a measure of the concentration of organic compounds in the water and is

**Biology**

called the BOD exerted at time $t$. In this way, BOD provides a measure of the concentration of organic compounds in water without the complexity of analyzing the different compounds.

If the biological depletion of organic compounds in wastewater can be assumed to be an approximately first-order reaction, and $L$ corresponds to BOD remaining, then

$$-\frac{dL}{dt} = kL \qquad 39.1$$

Integrating between the limits of $(t = 0, L = L)$ to $(t = t, L = L_t)$ gives

$$\frac{L_t}{L} = e^{-kt} \qquad 39.2$$

The value $(L - L_t)$ is called $y_t$, the amount of BOD exerted at time $t$. Therefore,

$$y_t = L(1 - e^{-kt}) \qquad 39.3$$

## Monod Kinetics

Mathematically, in a batch reactor, the rate of growth of bacteria is given by

$$\frac{dX}{dt} = \mu X \qquad 39.4$$

The value of $\mu$ is given by

$$\mu = \mu_{max} \frac{S}{K_S + S} \qquad 39.5$$

## Half-Life of a Biologically Degraded Contaminant, Assuming a First-Order Rate Constant

The *half-life* of a reaction is the time taken to reduce the concentration (in this case, the organic compound) to half of its initial amount. The differential equation for a first-order reaction is shown as Eq. 39.1.

Integrating between the limits of $(t = 0, L = L)$ to $(t = t_{1/2}, L = 0.5L)$ gives

$$0.5 = e^{kt_{1/2}} \qquad 39.6$$

Solving for $k$ gives

$$k = \frac{\ln 2}{t_{1/2}} = \frac{0.693}{t_{1/2}} \qquad 39.7$$

## 3. ACTIVATED SLUDGE

The activated sludge process in its simplest form, as shown in Fig. 39.1, consists of a mixing/aeration tank followed by a settler/clarifier. Mathematical analysis of the entire process is done by means of material balances

on the different components and also by using the Monod kinetic model.

**Figure 39.1** *Activated Sludge Process*

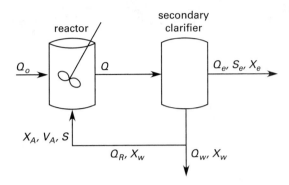

The results of the analysis for the reactor/aerator give a formula for the biomass concentration in the aerator, as well as the sludge flow rate and the solids residence time. The formulas for the organic loading rates are given. The biomass concentration in the aeration tank is

$$X_A = \frac{\theta_c Y(S_o - S_e)}{\theta(1 + k_d \theta_c)} \qquad 39.8$$

The *yield coefficient* is

$$Y = \frac{\text{mass of biomass}}{\text{mass of BOD consumed}} \qquad 39.9$$

The *hydraulic residence time* is

$$\theta = \frac{V}{Q} \qquad 39.10$$

The *solids residence time* is

$$\theta_c = \frac{V_A X_A}{Q_w X_w + Q_e X_e} \qquad 39.11$$

The *sludge volume* per day is

$$Q_s = \frac{M \times 100\%}{\rho_s(\% \text{ solids})} \qquad 39.12$$

$$\textit{solids loading rate} = \frac{QX}{A} \qquad 39.13$$

For an activated sludge secondary clarifier,

$$Q = Q_o + Q_R \qquad 39.14$$

$$\textit{organic loading rate (volumetric)} = \frac{Q_o S_o}{V} \qquad 39.15$$

$$\textit{organic loading rate (F:M)} = \frac{Q_o S_o}{V_A X_A} \qquad 39.16$$

$$\textit{organic loading rate (surface area)} = \frac{Q_o S_o}{A_M} \qquad 39.17$$

$$\text{SVI} = \frac{\left(\text{sludge volume after settling}_{(mL/L)}\right)\left(1000 \dfrac{\text{mg}}{\text{g}}\right)}{\text{MLSS}_{mg/L}}$$

$$39.18$$

The steady-state mass balance for a secondary clarifier is

$$(Q_o + Q_R)X_A = Q_e X_e + Q_R X_w + Q_w X_w \qquad 39.19$$

The *recycle ratio* is

$$R = \frac{Q_R}{Q_o} \qquad 39.20$$

The *recycle flow rate* is

$$Q_R = Q_o R \qquad 39.21$$

Design and operational parameters for activated-sludge treatment of municipal wastewater are given in Table 39.1.

## 4. FACULTATIVE POND

Facultative ponds, where the waste is kept in an open pond, are also called stabilization ponds. In these ponds, aerobic bacteria degrade the top layer, anaerobic bacteria degrade the bottom layer, and both aerobic and anaerobic bacteria degrade the middle layer.

### BOD Loading

The BOD loading is the $BOD_5$ application rate per unit area.

$$\text{BOD loading} = \frac{Q_{MGD}S_{mg/L}\left(8.345 \dfrac{\text{lbm-L}}{\text{MG-mg}}\right)}{A_{acres}} \qquad 39.22$$

A typical facultative pond system can process 35 lbm of $BOD_5$ per acre per day. A typical system contains a minimum of three ponds, which are 3 ft to 8 ft deep. The minimum residence time is 90–120 days.

## 5. BIOTOWER

Also called a *trickling filter*, a *biotower* operates by having the wastewater fall through a packed bed or tower filled with permeable packing. The packing has both aerobic and anaerobic microorganisms growing on it. Material balance equations can be formulated using the flux from both the connective and diffusive processes. The rate of reaction is formulated by using the Monod kinetic model. The biotowers can be operated either with or without recycling.

### Fixed-Film Equation without Recycle

$$\frac{S_e}{S_o} = e^{-kD/q^n} \qquad 39.23$$

### Fixed-Film Equation with Recycle

$$\frac{S_e}{S_a} = \frac{e^{-kD/q^n}}{(1+R) - R(e^{-kD/q^n})} \qquad 39.24$$

$$S_a = \frac{S_o + RS_e}{1+R} \qquad 39.25$$

Hydraulic loading with recycle is

$$q = \frac{Q_o + RQ_o}{A_{\text{plan}}} \qquad 39.26$$

The *treatability constant*, $k$, is given by Eq. 39.27, where the temperature, $T$, is in °C.

$$k_T = k_{20}(1.035)^{T-20°C} \qquad 39.27$$

As with activated sludge, the recycle ratio is

$$R = \frac{Q_R}{Q_o} \qquad 39.28$$

**Table 39.1** *Design and Operational Parameters for Activated-Sludge Treatment of Municipal Wastewater*

| type of process | mean-cell residence time, $\theta_c$ (d) | food-to-mass ratio (kg $BOD_5$/ kg MLSS) | volumetric loading, $V_L$ (kg $BOD_5$/m³) | hydraulic residence time in aeration basin, $\theta$ (h) | mixed liquor suspended solids, MLSS (mg/L) | recycle ratio ($Q_r/Q$) | flow regime | $BOD_5$ removal efficiency (%) | air supplied (m³/kg $BOD_5$) |
|---|---|---|---|---|---|---|---|---|---|
| tapered aeration | 5–15 | 0.2–0.4 | 0.3–0.6 | 4–8 | 1500–3000 | 0.25–0.5 | PF | 85–95 | 45–90 |
| conventional | 4–15 | 0.2–0.4 | 0.3–0.6 | 4–8 | 1500–3000 | 0.25–0.5 | PF | 85–95 | 45–90 |
| step aeration | 4–15 | 0.2–0.4 | 0.6–1.0 | 3–5 | 2000–3500 | 0.25–0.75 | PF | 85–95 | 45–90 |
| completely mixed | 4–15 | 0.2–0.4 | 0.8–2.0 | 3–5 | 3000–6000 | 0.25–1.0 | CM | 85–95 | 45–90 |
| contact stabilization | 4–15 | 0.2–0.6 | 1.0–1.2 | – | – | 0.25–1.0 | – | – | 45–90 |
|     contact basin | – | – | – | 0.5–1.0 | 1000–3000 | – | PF | 80–90 | – |
|     stabilization basin | – | – | – | 4–6 | 4000–10 000 | – | PF | – | – |
| high-rate aeration | 4–15 | 0.4–1.5 | 1.6–16 | 0.5–2.0 | 4000–10 000 | 1.0–5.0 | CM | 75–90 | 25–45 |
| pure oxygen | 8–20 | 0.2–1.0 | 1.6–4 | 1–3 | 6000–8000 | 0.25–0.5 | CM | 85–95 | – |
| extended aeration | 20–30 | 0.05–0.15 | 0.16–0.40 | 18–24 | 3000–6000 | 0.75–1.50 | CM | 75–90 | 90–125 |

Biology

## 6. ANAEROBIC DIGESTER

The sludge from the primary settlers and the biological treatment processes can be treated to obtain methane, carbon dioxide, and other products. In the standard-rate digester, the digester is unmixed and not externally heated; in the high-rate digester, there is external heating and stirring of the contents.

### Standard Rate

A standard rate digester must be sized to accommodate the raw sludge input, $V_1$, and the digested sludge accumulation, $V_2$, for the time for the sludge to digest and thicken (i.e., for the residence time, $t_r$) as well as to hold the accumulation for the period it is stored, $t_s$.

$$\text{reactor volume} = \left(\frac{V_1 + V_2}{2}\right)t_r + V_2 t_s \qquad 39.29$$

### High Rate

The first-stage reactor volume is selected to hold the raw sludge for as long as it takes for digestion to occur.

$$\text{reactor volume} = V_1 t_r \qquad 39.30$$

The second-stage reactor volume is selected based on the time it takes for thickening to occur, $t_t$.

$$\text{reactor volume} = \left(\frac{V_1 + V_2}{2}\right)t_t + V_2 t_s \qquad 39.31$$

## 7. AEROBIC DIGESTION

The aerobic digestion process is similar to the activated sludge process. Aeration is accomplished by means of diffusing equipment. The process can be either batch or continuous.

For a continuous process with residence time $\theta_c$, the tank volume is

$$V = \frac{Q_i(X_i + FS_i)}{X_d\left(K_d P_v + \dfrac{1}{\theta_c}\right)} \qquad 39.32$$

Biology

# Topic VII: Heat Transfer

Heat Transfer

# 40 Heat Transfer by Conduction

## Nomenclature

| | | | |
|---|---|---|---|
| $A$ | area | ft$^2$ | m$^2$ |
| $c_p$ | specific heat | Btu/lbm-°F | J/kg·K |
| $d$ | diameter | ft | m |
| $E$ | voltage | V | V |
| $h$ | film coefficient | Btu/hr-ft$^2$-°F | W/m$^2$·K |
| $I$ | current | A | A |
| $k$ | thermal conductivity | Btu-ft/ hr-ft$^2$-°F | W/m·K |
| $L$ | length | ft | m |
| $m$ | mass | lbm | kg |
| $q$ | heat transfer per unit area | Btu/hr-ft$^2$ | W/m$^2$ |
| $Q$ | heat transfer rate | Btu/hr | W |
| $r$ | radius | ft | m |
| $R$ | electrical resistance | Ω | Ω |
| $R$ | thermal resistance | hr-°F/Btu | K/W |
| $t$ | thickness | ft | m |
| $T$ | temperature | °F | K |
| $U$ | internal energy | Btu | J |
| $U$ | overall heat transfer coefficient | Btu/hr-ft$^2$-°F | W/m$^2$·K |
| $V$ | volume | ft$^3$ | m$^3$ |

## Symbols

| | | | |
|---|---|---|---|
| $\gamma$ | empirical constant | 1/°F | 1/K |
| $\rho$ | mass density | lbm/ft$^3$ | kg/m$^3$ |
| $\rho$ | electrical resistivity | Ω-in | Ω·cm |

## Subscripts

| | |
|---|---|
| 0 | initial |
| $c$ | characteristic |
| corr | corrected |
| $e$ | equivalent |
| $h$ | film |
| $i$ | inner or $i$th layer |
| $j$ | $j$th film |
| $L$ | per unit length |
| $m$ | logarithmic mean |
| $o$ | outer |
| $r$ | radiation, radius, or at radius $r$ |
| ref | reference |
| $s$ | surface |
| th | thermal |
| $T$ | at temperature $T$ |

## 1. HEAT TRANSFER MECHANISMS

*Heat* is thermal energy in motion. There are three distinct mechanisms by which thermal energy can move from one location to another. These mechanisms are distinguished by the media through which the energy moves.

If no medium (air, water, solid concrete, etc.) is required, the heat transfer occurs by *radiation*. If energy is transferred through a solid material by molecular vibration, the heat transfer mechanism is known as *conduction*. If energy is transferred from one point to another by a moving fluid, the mechanism is known as *convection*. *Natural convection* transfers heat by relying on density changes to cause fluid motion. *Forced convection* requires a pump, fan, or relative motion to move the fluid. (Change of phase—evaporation and condensation—is categorized as convection.)

Heat Transfer

In almost all problems, the heat transfer rate, $Q$, will initially vary with time.[1] This initial period is known as the *transient period*. Eventually, the rate of energy transfer becomes constant, and this period is known as the *steady-state* or *equilibrium rate*.

Whereas the total heat transfer depends on the area, the heat transfer per unit area $(q/A)$ has the units of power per unit area and is known as the *per unit power* of the source.

## 2. INTRODUCTION TO CONDUCTIVE HEAT TRANSFER

*Conduction* is the flow of heat through solids or stationary fluids. Thermal conductance in metallic solids is due to molecular vibrations within the metallic crystalline lattice and movement of free valence electrons through the lattice. Insulating solids, which have fewer free electrons, conduct heat primarily by the agitation of adjacent atoms vibrating about their equilibrium positions. This vibrational mode of heat transfer is several orders of magnitude less efficient than conduction by free electrons.

In stationary liquids, heat is transmitted by longitudinal vibrations, similar to sound waves. The *net transport theory* explains heat transfer through gases. Hot molecules move faster than cold molecules. Hot molecules travel to cold areas with greater frequency than cold molecules travel to hot areas.

## 3. SIMPLIFYING ASSUMPTIONS

Determining heat transfer by conduction can be an easy task if sufficient simplifying assumptions are made. Major discrepancies can arise, however, when the simplifying assumptions are not met. The following assumptions are commonly made in simple problems.

- The heat transfer is steady-state.
- The heat path is one-dimensional. (Objects are infinite in one or more directions and do not have any end effects.)
- The heat path has a constant area.
- The heat path consists of a homogeneous material with constant conductivity.
- The heat path consists of an isotropic material.[2]
- There is no internal heat generation.

Many real heat transfer cases violate one or more of these assumptions. Unfortunately, problems with closed-form solutions (suitable for working by hand) are in the minority. More complex problems must be solved by appropriate iterative, graphical, or numerical methods.[3]

## 4. CONDUCTIVITY

The *thermal conductivity* (also known as the *thermal conductance*), $k$, is a measure of the rate at which a substance transfers thermal energy through a unit thickness.[4] Units of thermal conductivity are Btu-ft/hr-ft$^2$-°F or Btu-in/hr-ft$^2$-°F (W/m·K or W·cm/m$^2$·K). The units of Btu-ft/hr-ft$^2$-°F are the same as Btu/hr-ft-°F.[5] The conductivity of a substance should not be confused with the *overall conductivity*, $U$, of an object. (See Sec. 40.17.) Appendix 40.A, App. 40.B, and App. 40.C list representative thermal conductivities for commonly encountered substances, as does Table 40.1.

**Table 40.1** Typical Thermal Conductivities at 32° F (0° C)

| | $k$ | |
| --- | --- | --- |
| substance | Btu/hr-ft-°F | W/m·K |
| silver | 242 | 419 |
| copper | 224 | 388 |
| aluminum | 117 | 202 |
| brass | 56 | 97 |
| steel (1% C) | 27 | 47 |
| lead | 20 | 35 |
| ice | 1.3 | 2.2 |
| glass | 0.63 | 1.1 |
| concrete | 0.50 | 0.87 |
| water | 0.32 | 0.55 |
| fiberglass | 0.030 | 0.052 |
| cork | 0.025 | 0.043 |
| air | 0.014 | 0.024 |

(Multiply Btu/hr-ft-°F by 1.731 to obtain W/m·K.)

## 5. VARYING CONDUCTIVITY

Solids exhibit conductivities that vary with temperature.[6] (See Table 40.2.) Over limited ranges, thermal conductivity in common solids is assumed to vary linearly with temperature, as indicated in Eq. 40.1. $k_{ref}$ is the conductivity at the reference temperature, usually 0°F ($-18$°C). Values of $\gamma$ and $\gamma'$ are not common, as graphs and tabulations of conductivity versus temperature are more readily available.

---

[1]The term "rate" is generally omitted and understood when "heat transfer" is used.

[2]Examples of *anisotropic materials*, materials whose heat transfer properties depend on the direction of heat flow, are crystals, plywood and other laminated sheets, and the core elements of some electrical transformers.

[3]Finite-difference methods are commonly used.

[4]Another (lesser-encountered) meaning for *conductivity* is the reciprocal of thermal resistance (i.e., conductivity = $kA/L$).

[5]Temperature units of °R can also be used in place of °F, since conductivity is always multiplied by $\Delta T$, and $\Delta T_{°F}$; = $\Delta T_{°R}$.

[6]Conductivity at absolute zero is zero because there is no atomic/molecular motion. For the first few degrees above absolute zero, conductivity increases with increases in temperature.

$$k_T = k_{\text{ref}}(1 + \gamma T)$$
$$= k_{\text{ref}} + \gamma' T \qquad \textit{40.1}$$

Conductivity decreases with temperature for pure metals and increases with temperature for most alloys. For insulating materials, it increases with temperature. Conductivity in water and aqueous solutions increases with increases in temperature up to approximately 250°F (120°C) and then gradually decreases. Conductivity decreases with increases in concentrations of aqueous solutions, as it does with most other liquids. Conductivity increases with increases in pressure. Of the nonmetallic liquids, water is the best thermal conductor. Conductivity in gases increases almost linearly with temperature but is fairly independent of pressure in common ranges.

Thermal conductivity is often assumed to be constant over the entire length of the transmission path. In most calculations, either the average thermal conductivity or the conductivity at the arithmetic mean temperature is used.[7] When $k$ varies linearly (or nearly so) with temperature, $k$ is evaluated at the average temperature $\frac{1}{2}(T_1 + T_2)$.

In rare cases where $k$ does not vary linearly with temperature, the solution must proceed iteratively. An initial estimate of the average conductivity yields a heat transfer which, in turn, is used to solve for the surface temperatures. These temperatures are used to determine a new average conductivity, and so on.

**Table 40.2** *Typical Ranges of Conductivity (See also App. 40.A, App. 40.B, and App. 40.C)*

| material | conductivity range | |
| --- | --- | --- |
| | Btu-ft/hr-ft$^2$-°F | W/m·K |
| gases (at 1 atm) | 0.004–0.10 | 0.007–0.17 |
| insulators | 0.02–0.12 | 0.03–0.21 |
| nonmetallic liquids | 0.05–0.40 | 0.09–0.70 |
| nonmetallic solids | 0.02–1.5 | 0.03–2.6 |
| liquid metals | 5.0–45 | 8.7–78 |
| metallic alloys | 8.0–70 | 14–120 |
| pure metals | 30–240 | 52–420 |

(Multiply Btu-ft/hr-ft$^2$-°F by 12 to get Btu-in/hr-ft$^2$-°F.)
(Multiply Btu-ft/hr-ft$^2$-°F by 1.73073 to get W/m·K.)
(Multiply Btu-ft/hr-ft$^2$-°F by 4.1365 × 10$^{-3}$ to get cal·cm/s·cm$^2$·°C.)

---

[7]The mean thermal conductivity is not the same as the thermal conductivity at the mean temperature. However, it is very nearly so, and this simplification is widely used.

## 6. THERMAL RESISTANCE

The *thermal resistance* is defined by Eq. 40.2. Typical units are °F-hr/Btu (K/W).

$$R_{\text{th}} = \frac{T_1 - T_2}{Q} \qquad \textit{40.2}$$

For a plane, the thermal resistance depends on the thickness (path length), $L$, and is

$$R_{\text{th}} = \frac{L}{kA} \qquad \textit{40.3}$$

For a film, the thermal resistance depends on the average film coefficient, $h$, and is

$$R_{\text{th,film}} = \frac{1}{hA} \qquad \textit{40.4}$$

For a curved layer (i.e., a layer of insulation on a pipe), the thermal resistance is

$$R_{\text{th}} = \frac{\ln\left(\dfrac{r_o}{r_i}\right)}{2\pi kL} \qquad \textit{40.5}$$

## 7. *R*-VALUE

The *R-value* of a substance is the thermal resistance on a unit area basis. Typical units are °F-ft$^2$-hr/Btu (K·m$^2$/W). The *R*-value concept is usually encountered in the construction industry as a means of comparing insulating materials. It is not the same as the thermal resistance.

$$R\text{-value} = \frac{T_1 - T_2}{\dfrac{Q}{A}} = R_{\text{th}}A \qquad \textit{40.6}$$

## 8. SPECIFIC HEAT

The *specific heat*, $c_p$, is the energy required to change the temperature of a unit mass of a body one degree. For solid bodies, $\Delta pV = 0$, so $Q = \Delta H = \Delta U$. Values of specific heat are given in App. 40.B and App. 40.C.

$$c_p = \frac{\Delta U}{m\Delta T} \qquad \textit{40.7}$$

## 9. CHARACTERISTIC DIMENSION

The *characteristic dimension* (*characteristic length*), $L_c$, of an object is the ratio of its volume to its surface area.

$$L_c = \frac{V}{A_s} \qquad \textit{40.8}$$

**Heat Transfer**

For a long cylinder of radius $r$, the characteristic dimension is

$$L_{c,\text{cylinder}} = \frac{V}{A_s} = \frac{\pi r^2 L}{2\pi r L}$$
$$= \frac{r}{2} \qquad 40.9$$

For a sphere, the characteristic dimension is one-third of the radius (i.e., $L_{c,\text{sphere}} = r/3$). For an infinite slab, the characteristic dimension is one-half of the slab thickness. For an infinite square rod, the characteristic dimension is a quarter of the rod thickness.

## 10. TEMPERATURE

The *ambient temperature* is the same as *environment temperature*, *far-field temperature*, *local temperature*, or (occasionally) *air temperature*. However, due to the thermal resistance of a film, it is not the same as the *surface temperature*. The *bulk temperature* is the temperature of a thoroughly mixed liquid or gas. The term "bulk temperature" is occasionally used to represent the ambient temperature.

In most heat transfer calculations, the temperature variable shows up as a change in temperature. It is convenient to recognize that $\Delta T_{\circ F} = \Delta T_{\circ R}$ and $\Delta T_{\circ C} = \Delta T_K$.

## 11. TEMPERATURE PROFILE

A *temperature profile* (see Fig. 40.1) is a graph of the temperature versus location within an object. A profile will often be drawn on a representation of the cross section of the object through which the energy transfer occurs. In uniform materials, the profile will consist of straight lines, with steeper lines representing materials with lower thermal conductivities. The temperature scale is omitted and replaced with values of temperature at the interface points.

## 12. HEAT TRANSFER

The term *heat transfer* is used in three different contexts. The intended interpretation must be determined from the needs of the investigation. The most common usage has units of energy per unit of time (e.g., Btu/hr).[8] Strictly speaking, this should be called a *heat transfer rate* and should be written $\dot{Q}$. However, common usage omits the rate symbol. A second usage, referred to in this book as *total heat transfer,* is the energy transfer taken over the entire body area or mass.

$$Q = qA \qquad 40.10$$

---

[8]In the United States, the units of Btu/hr are often written as Btuh. Similarly, MBh generally means millions of Btus per hour, but can mean thousands of Btus per hour in some industries (e.g., HVAC). The ambiguous unit of MMBtuh always means millions of Btus per hour.

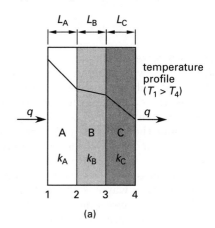

**Figure 40.1** *Typical Temperature Profile*

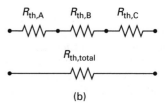

The term "heat transfer" is also used when meaning the total energy change (in Btu) over some period of time. In this book, the symbol $\Delta U$ (change in internal energy) is used for this purpose.

$$\Delta U = mc_p\Delta T \qquad 40.11$$

## 13. FOURIER'S LAW

If the assumptions listed in Sec. 40.3 are valid, *Fourier's law*, Eq. 40.12, is applicable. On its own, heat always flows from a higher temperature to a lower temperature. The heat transfer from high-temperature point 1 to lower-temperature point 2 through an infinite plane of thickness $L$ and homogeneous conductivity, $k$, is

$$q_{1-2} = \frac{-k(T_2 - T_1)}{L}$$
$$= \frac{k(T_2 - T_1)}{L} \qquad 40.12$$

$$Q_{1-2} = q_{1-2}A = \frac{kA(T_1 - T_2)}{L} \qquad 40.13$$

The temperature difference $T_2 - T_1$ is the *temperature gradient* or *thermal gradient*. Heat transfer is always positive. The minus sign in Eq. 40.12 indicates that the heat flow direction is opposite that of the thermal gradient. The direction of heat flow is obvious in most problems. Therefore, the minus sign is usually omitted.

## 14. ELECTRICAL ANALOGY

Ohm's law can be solved for current and written in terms of the electrical potential gradient, $E_1 - E_2$.

$$I = \frac{E_1 - E_2}{R} \qquad 40.14$$

Drawing on the concept of thermal resistance, Eq. 40.14 and Eq. 40.15 are analogous.

$$Q = \frac{T_1 - T_2}{R_{\text{th}}} \qquad 40.15$$

The electrical analogy is particularly valuable in analyzing heat transfer through multiple layers (e.g., a layered wall or cylinder with several layers of different insulation). Each layer is considered a series resistance. The sum of the individual thermal resistances is the total thermal resistance. (See Fig. 40.1(b).)

$$R_{\text{th,total}} = \sum R_{\text{th},i} \qquad 40.16$$

## 15. SANDWICHED PLANES

Figure 40.1(a) illustrates the cross section of a *sandwiched plane* with three homogeneous, but different, materials. This type of construction is sometimes known as a *composite wall*. The heat flow due to conduction through a series ($i = 1$ to $n$) of plane surfaces is

$$Q = \frac{T_1 - T_{n+1}}{R_{\text{th,total}}} = \frac{A(T_1 - T_{n+1})}{\displaystyle\sum_{i=1}^{n} \frac{L_i}{k_i}} \qquad 40.17$$

$$R_{\text{th,total}} = \sum_{i=1}^{n} \frac{L_i}{k_i A} \qquad 40.18$$

## 16. HEAT TRANSFER THROUGH A FILM

Heat conducted through solids is often removed by a physical transport process at an exposed fluid surface. For example, heat transmitted through a heat exchanger wall is removed by a moving coolant. Unless the fluid is extremely turbulent, the fluid molecules immediately adjacent to the exposed surface move much slower than molecules farther away. Molecules immediately adjacent to the wall may be stationary altogether. The fluid molecules that are affected by the exposed surface constitute a layer known as a film.[9] The film has a thermal resistance just like any other sandwiched plane.

Because the film thickness is not easily determined, the thermal resistance of a film is given by a *film coefficient* (*convective heat transfer coefficient* or *unit conductance*), $h$, with units of Btu/hr-ft$^2$-°F (W/m$^2$·K). If the

---

[9]The film coefficient concept can also be used to quantify the thermal resistance of imperfect bonding between two planes (i.e., *contact resistance*), a dust layer, or scale buildup.

---

film coefficient is not constant over the entire surface, the symbol for average film coefficient, $\bar{h}$ is used. Subscripts are used (e.g., $h_o$ or $h_i$) to indicate whether the film is located on the outside or inside of the object.

The heat flow through a film is

$$Q = hA(T_1 - T_2) \qquad 40.19$$

$$R_{\text{th}} = \frac{1}{hA} \qquad 40.20$$

If the fluid is extremely turbulent, the thermal resistance will be small (i.e., $h$ will be very large). In such cases, the wall temperature is essentially the same as the fluid temperature.

## 17. OVERALL COEFFICIENT OF HEAT TRANSFER

When conduction and convection (but not radiation) are the only modes of heat transfer, the *overall coefficient of heat transfer*, $U$, also known as the *overall conductivity*, *overall heat transmittance*, or simply *U-factor*, is defined by Eq. 40.21.

$$U = \frac{1}{A R_{\text{th,total}}}$$

$$= \frac{1}{\displaystyle\sum_i \frac{L_i}{k_i} + \sum_j \frac{1}{h_j}} \qquad 40.21$$

The overall coefficient is usually used in conjunction with the outside (exposed) surface area (because it is easier to measure), but not always. Therefore, it will generally be written with a subscript (e.g., $U_o$ or $U_i$) indicating which area it is based on. The heat transfer is

$$Q = U_o A_o (T_1 - T_2) = U_i A_i (T_1 - T_2) \qquad 40.22$$

When heat transfer occurs by two or three modes, the overall coefficient of heat transfer takes all active modes —conduction, convection, and radiation—into consideration. In that sense, the overall coefficient, $U$, is defined by Eq. 40.23, rather than being used in it.

$$U = \frac{Q}{A(T_1 - T_2)} \qquad 40.23$$

**Example 40.1**

A 100 ft$^2$ (9.3 m$^2$) wall consists of 4 in (10 cm) of red brick ($k = 0.38$ Btu-ft/hr-ft$^2$-°F; 0.66 W/m·K), 1 in (2.5 cm) of pine ($k = 0.06$ Btu-ft/hr-ft$^2$-°F; 0.10 W/m·K), and $1/2$ in (1.2 cm) of plasterboard ($k = 0.30$ Btu-ft/hr-ft$^2$-°F; 0.52 W/m·K). The internal and external film coefficients are 1.65 Btu/hr-ft$^2$-°F and 6.00 Btu/hr-ft$^2$-°F (9.38 W/m$^2$·K and 34.1 W/m$^2$·K), respectively. The inside and outside air temperatures are 72°F (22°C) and 30°F ($-1$°C), respectively. Determine the heat transfer.

*SI Solution*

Convert the thicknesses from centimeters to meters.

$$L_{\text{brick}} = \frac{10 \text{ cm}}{100 \frac{\text{cm}}{\text{m}}} = 0.10 \text{ m}$$

$$L_{\text{pine}} = \frac{2.5 \text{ cm}}{100 \frac{\text{cm}}{\text{m}}} = 0.025 \text{ m}$$

$$L_{\text{plasterboard}} = \frac{1.2 \text{ cm}}{100 \frac{\text{cm}}{\text{m}}} = 0.012 \text{ m}$$

From Eq. 40.21, the overall coefficient of heat transfer is

$$\frac{1}{U} = \frac{1}{\sum_i \frac{L_i}{k_i} + \sum_j \frac{1}{h_j}}$$

$$= \frac{0.10 \text{ m}}{0.66 \frac{\text{W}}{\text{m·K}}} + \frac{0.25 \text{ m}}{0.10 \frac{\text{W}}{\text{m·K}}} + \frac{0.012 \text{ m}}{0.52 \frac{\text{W}}{\text{m·K}}}$$

$$+ \frac{1}{9.38 \frac{\text{W}}{\text{m}^2\text{·K}}} + \frac{1}{34.1 \frac{\text{W}}{\text{m}^2\text{·K}}}$$

$$= 0.561 \frac{\text{m}^2\text{·K}}{\text{W}}$$

$$U = \frac{1}{0.561 \frac{\text{m}^2\text{·K}}{\text{W}}}$$

$$= 1.78 \text{ W/m}^2\text{·K}$$

From Eq. 40.22 (and recognizing that $\Delta T_{\circ\text{C}} = \Delta T_\text{K}$), the heat transfer is

$$Q = UA\Delta T = \left(1.78 \frac{\text{W}}{\text{m}^2\text{·K}}\right)(9.3 \text{ m}^2)$$

$$\times \left(22^\circ\text{C} - (-1^\circ\text{C})\right)$$

$$= 380.7 \text{ W}$$

*Customary U.S. Solution*

Convert the thicknesses from inches to feet.

$$L_{\text{brick}} = \frac{4 \text{ in}}{12 \frac{\text{in}}{\text{ft}}} = 0.333 \text{ ft}$$

$$L_{\text{pine}} = \frac{1 \text{ in}}{12 \frac{\text{in}}{\text{ft}}} = 0.083 \text{ ft}$$

$$L_{\text{plasterboard}} = \frac{0.5 \text{ in}}{12 \frac{\text{in}}{\text{ft}}} = 0.042 \text{ ft}$$

From Eq. 40.21, the overall coefficient of heat transfer is

$$\frac{1}{U} = \frac{1}{\sum_i \frac{L_i}{k_i} + \sum_j \frac{1}{h_j}}$$

$$= \frac{0.333 \text{ ft}}{0.38 \frac{\text{Btu-ft}}{\text{hr-ft}^2\text{-}^\circ\text{F}}} + \frac{0.083 \text{ ft}}{0.06 \frac{\text{Btu-ft}}{\text{hr-ft}^2\text{-}^\circ\text{F}}} + \frac{0.042 \text{ ft}}{0.30 \frac{\text{Btu-ft}}{\text{hr-ft}^2\text{-}^\circ\text{F}}}$$

$$+ \frac{1}{1.65 \frac{\text{Btu}}{\text{hr-ft}^2\text{-}^\circ\text{F}}} + \frac{1}{6.00 \frac{\text{Btu}}{\text{hr-ft}^2\text{-}^\circ\text{F}}}$$

$$= 3.17 \text{ hr-ft}^2\text{-}^\circ\text{F/Btu}$$

$$U = \frac{1}{3.17 \frac{\text{hr-ft}^2\text{-}^\circ\text{F}}{\text{Btu}}}$$

$$= 0.315 \text{ Btu/hr-ft}^2\text{-}^\circ\text{F}$$

From Eq. 40.22, the heat transfer is

$$Q = UA\Delta T$$

$$= \left(0.315 \frac{\text{Btu}}{\text{hr-ft}^2\text{-}^\circ\text{F}}\right)(100 \text{ ft}^2)(72^\circ\text{F} - 30^\circ\text{F})$$

$$= 1323 \text{ Btu/hr}$$

## 18. TEMPERATURE AT A POINT

The temperature at any point on or within a simple or complex wall can be found if the heat transfer, $q$ or $Q$, is known. The procedure is to calculate the thermal resistance up to the point of unknown temperature and then to solve Eq. 40.15 for the temperature difference. Since one of the temperatures is known, the unknown temperature is found from the temperature difference.

## 19. COMPLEX WALL

Figure 40.2 illustrates a complex wall. Heat transfer through such a wall is intrinsically a two-dimensional problem. Since the temperature profile (see Sec. 40.11) will not be the same for each heat transfer path, there will be adjacent areas with different temperatures. Heat will naturally flow from the hotter to the cooler areas.

There is no easy way to analytically obtain an exact value of heat transfer through a complex wall. A lower bound on the heat transfer can be obtained by considering the wall as having multiple one-dimensional transmission paths (i.e., ignoring the interaction of adjacent paths). Similarly, an upper bound can be obtained by assuming the entire transmission path consists of a homogeneous material whose conductivity is obtained by weighting each material's conductivity by its volume within the wall.

Thermal conductivities and resistances for common wall constructions are widely available in heating, ventilating, and air conditioning (HVAC) textbooks. Values in HVAC tabulations usually include the inside and

**Figure 40.2** *Complex Wall Construction*

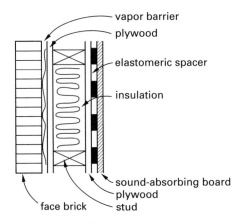

outside film coefficients and are based on some time of year and/or wind speed.

## 20. LOGARITHMIC MEAN AREA

Cylinders and spheres are examples of objects whose heat transfer paths increase in area from the inside to outside. In such instances, the *logarithmic mean area*, $A_m$, should be used.[10]

$$Q = UA_m(T_1 - T_2) \qquad 40.24$$

$$A_m = \frac{A_o - A_i}{\ln\left(\dfrac{A_o}{A_i}\right)} \qquad 40.25$$

## 21. RADIAL CONDUCTION THROUGH A SPHERICAL SHELL

The logarithmic mean area and heat transfer are

$$A_m = \sqrt{A_o A_i} \qquad 40.26$$

$$\begin{aligned}Q &= qA_m \\ &= k\sqrt{A_o A_i}\left(\frac{T_i - T_o}{r_o - r_i}\right) \\ &= \frac{4\pi k r_o r_i(T_i - T_o)}{r_o - r_i} \qquad 40.27\end{aligned}$$

The thermal resistance of each layer is

$$R_{\text{th}} = \left(\frac{1}{4\pi k}\right)\left(\frac{r_o - r_i}{r_o r_i}\right) \qquad 40.28$$

Spheres have the smallest surface area-to-volume ratio, so spherical tanks are used where heat transfer is to be minimized. Unlike for plane surfaces and cylindrical tanks, steady-state heat transfer from spheres cannot

[10]If $A_o/A_i \le 2$, using the arithmetic mean $\frac{1}{2}(A_o + A_i)$ in place of the logarithmic mean area will result in a maximum error of 4%, with the heat transfer being too high.

be reduced indefinitely by the addition of increasing amounts of insulation. For a sphere, steady-state conduction can never be less than the minimum value given by Eq. 40.29, no matter how thick the wall is. If $r_o$ is infinite, Eq. 40.27 becomes

$$Q_{\min} = 4\pi k r_i(T_i - T_o) \qquad 40.29$$

## 22. RADIAL CONDUCTION THROUGH A HOLLOW CYLINDER

The logarithmic mean area (excluding the ends) for a hollow cylinder of length $L$ and wall thickness $r_o - r_i$ is

$$A_m = \frac{2\pi L(r_o - r_i)}{\ln\left(\dfrac{r_o}{r_i}\right)} \qquad 40.30$$

The overall radial heat transfer through an uninsulated hollow cylinder without films is given by Eq. 40.31. This equation disregards heat transfer from the ends and assumes that the length is sufficiently large so that the heat transfer is radial at all locations.

$$\begin{aligned}Q = qA_m &= \frac{kA_m(T_1 - T_2)}{r_o - r_i} \\ &= \frac{2\pi kL(T_1 - T_2)}{\ln\dfrac{r_o}{r_i}} \qquad 40.31\end{aligned}$$

The temperature at a point within a layer a distance $r$ from the center is given by Eq. 40.32. $T_i$ is the temperature at the inside of the layer. For a pipe wall, the inside wall temperature is not necessarily the same as the temperature of the pipe's contents.

$$T_r = T_i - (T_i - T_o)\left(\frac{\ln\left(\dfrac{r}{r_i}\right)}{\ln\left(\dfrac{r_o}{r_i}\right)}\right) \qquad 40.32$$

## 23. RADIAL CONDUCTION THROUGH A COMPOSITE CYLINDER

A *composite cylinder* consists of two or more concentric, cylindrical layers. (See Fig. 40.3.) This configuration is typically encountered as an *insulated pipe*. The logarithmic mean area and heat transfer are

$$Q = qA_m = \frac{2\pi L(T_1 - T_2)}{\dfrac{\ln\left(\dfrac{r_b}{r_a}\right)}{k_{\text{pipe}}} + \dfrac{\ln\left(\dfrac{r_c}{r_b}\right)}{k_{\text{insulation}}}} \qquad 40.33$$

Heat Transfer

**Figure 40.3** *Composite Cylinder (Insulated Pipe)*

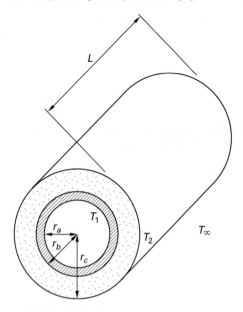

When films are present, Eq. 40.33 can be expanded. $h_b$ is a film coefficient representing contact or interface resistance, if any.

$$Q = \frac{2\pi L(T_1 - T_2)}{\dfrac{1}{r_a h_a} + \dfrac{\ln\left(\dfrac{r_b}{r_a}\right)}{k_{\text{pipe}}} + \dfrac{1}{r_b h_b} + \dfrac{\ln\left(\dfrac{r_c}{r_b}\right)}{k_{\text{insulation}}} + \dfrac{1}{r_c h_c}} \qquad 40.34$$

Compared to the thermal resistance of the insulation, the thermal resistance of metal pipe walls is negligible. Little accuracy will be lost if the pipe wall thickness term is omitted from the denominator.[11] An equivalent assumption is that there is no temperature change across the thickness of the pipe wall.

### Example 40.2

Liquid oxygen at $-290°F$ ($-180°C$) is stored in a cylindrical tank with a thermal conductance of 28.0 Btu-ft/hr-ft²-°F (48.0 W/m·K), 5 ft (1.5 m) inside diameter, and 20 ft (6.0 m) long. The wall thickness is $^3/_8$ in (1 cm). The tank is insulated with 1.0 ft (30 cm) of powdered diatomaceous silica with an average thermal conductivity of 0.022 Btu-ft/hr-ft²-°F (0.038 W/m·K). The surrounding air temperature is 70°F (21°C), and the outside film coefficient is 6.0 Btu/hr-ft²-°F (34 W/m²·K). The inside film coefficient is assumed to be infinite. Disregard heat transfer through the tank ends. Calculate the radial heat gain to the liquid oxygen.

---

*SI Solution*

The wall thickness is

$$t = \frac{10 \text{ cm}}{100 \frac{\text{cm}}{\text{m}}} = 0.01 \text{ m}$$

In Fig. 40.3, the corresponding radii are

$$r_a = \frac{1.5 \text{ m}}{2} = 0.75 \text{ m}$$
$$r_b = r_a + t = 0.75 \text{ m} + 0.01 \text{ m} = 0.76 \text{ m}$$
$$r_c = r_b + t_{\text{insulation}} = 0.76 \text{ m} + 0.30 \text{ m} = 1.06 \text{ m}$$

The heat transfer is calculated directly from Eq. 40.34.

$$Q = \frac{2\pi L(T_1 - T_2)}{\dfrac{1}{r_a h_a} + \dfrac{\ln\left(\dfrac{r_b}{r_a}\right)}{k_{\text{tank}}} + \dfrac{\ln\left(\dfrac{r_c}{r_b}\right)}{k_{\text{insulation}}} + \dfrac{1}{r_c h_c}}$$

$$= \frac{(2\pi)(6 \text{ m})\left(21°C - (-180°C)\right)}{\dfrac{1}{(0.75 \text{ m})(\infty)} + \dfrac{\ln\left(\dfrac{0.76 \text{ m}}{0.75 \text{ m}}\right)}{48.0 \dfrac{\text{W}}{\text{m·K}}} + \dfrac{\ln\left(\dfrac{1.06 \text{ m}}{0.76 \text{ m}}\right)}{0.038 \dfrac{\text{W}}{\text{m·K}}}}$$
$$\quad + \dfrac{1}{(1.06 \text{ m})\left(34 \dfrac{\text{W}}{\text{m}^2\text{·K}}\right)}$$

$$= 862.7 \text{ W}$$

*Customary U.S. Solution*

The wall thickness is

$$t = \frac{\frac{3}{8} \text{ in}}{12 \frac{\text{in}}{\text{ft}}}$$
$$= 0.031 \text{ ft}$$

In Fig. 40.3, the corresponding radii are

$$r_a = \frac{5 \text{ ft}}{2} = 2.5 \text{ ft}$$
$$r_b = r_a + t = 2.5 \text{ ft} + 0.031 \text{ ft} = 2.53 \text{ ft}$$
$$r_c = r_b + t_{\text{insulation}} = 2.53 \text{ ft} + 1 \text{ ft} = 3.53 \text{ ft}$$

---

[11]Since the thermal conductivity values may be in error by as much as 20%, it makes little sense to strive for perfection by including the thermal resistance of a thin-walled metal pipe.

The heat transfer is calculated directly from Eq. 40.34.

$$Q = \frac{2\pi L(T_1 - T_2)}{\dfrac{1}{r_a h_a} + \dfrac{\ln\left(\dfrac{r_b}{r_a}\right)}{k_{\text{tank}}} + \dfrac{\ln\left(\dfrac{r_c}{r_b}\right)}{k_{\text{insulation}}} + \dfrac{1}{r_c h_c}}$$

$$= \frac{(2\pi)(20\text{ ft})\left(70°F - (-290°F)\right)}{\dfrac{1}{(2.5\text{ ft})(\infty)} + \dfrac{\ln\left(\dfrac{2.53\text{ ft}}{2.5\text{ ft}}\right)}{28.0\ \dfrac{\text{Btu-ft}}{\text{hr-ft}^2\text{-}°F}}}$$

$$+ \dfrac{\ln\left(\dfrac{3.53\text{ ft}}{2.53\text{ ft}}\right)}{0.022\ \dfrac{\text{Btu-ft}}{\text{hr-ft}^2\text{-}°F}}$$

$$+ \dfrac{1}{(3.53\text{ ft})\left(6.0\ \dfrac{\text{Btu-ft}}{\text{hr-ft}^2\text{-}°F}\right)}$$

$$= 2979\text{ Btu/hr}$$

## 24. PIPE INSULATION

Commercial pipe insulation is usually fiberglass or calcium silicate.[12] Fiberglass has a high insulation value, is low in cost and low in weight, and is easy to install. However, conventional fiberglass is limited to uses below approximately 850°F (450°C).[13] Mineral wool (mineral-fiber), calcium silicate, and composite materials are used for high-temperature (i.e., above approximately 1000°F (540°C)) installations. Traditional calcium silicate, known as *cal sil*, can withstand more mechanical abuse than other insulating materials. However, it requires a saw for cutting, while fiberglass and mineral wool can easily be cut with a knife. An outer jacket (cover) of plastic (white Kraft all-service jacket, ASJ), aluminum, or stainless steel will protect insulation from dirt and moisture.[14,15]

Insulation products are available as pipe wrap, blankets, and boards. Integral protective jacketing of pipe wrap can be provided, with sealing tape closing the hinged wrap around the pipe. Insulating boards are designed for large flat surfaces, or, when manufactured with multiple strip hinges, for large curved surfaces such as cylindrical tanks. Compressible blankets are highly flexible and are used for irregular surfaces. (See Table 40.3.)

---

[12]Asbestos insulation, commonly used in the past, has fallen from favor. This includes *85% magnesia* insulation consisting of 85% magnesium carbonate and 15% asbestos fiber.

[13]High-temperature resin binders can increase the useful range of fiberglass insulation to approximately 1000°F (540°C).

[14]Metallic jacketing may represent a safety hazard for plant personnel since the surface is generally conductive and reflective. Jacket temperature should be less than 130°F (54°C) in areas where contact by personnel is possible.

[15]An exterior jacket should always be removed when adding a second layer of insulation. Retaining the original jacket may produce a condensation site or damage the jacket material or adhesive.

## 25. CRITICAL INSULATION THICKNESS

The addition of insulation to a bare pipe or wire increases the surface area. Adding insulation to a small-diameter pipe may actually increase the heat loss above bare-pipe levels. Adding insulation up to the *critical thickness* is dominated by the increase in surface area. Only adding insulation past the critical thickness will decrease heat loss.[16] The *critical radius* is usually very small (e.g., a few millimeters), and it is most relevant in the case of insulating thin wires. The critical radius, measured from the center of the pipe or wire, is

$$r_{\text{critical}} = \frac{k_{\text{insulation}}}{h} \qquad 40.35$$

## 26. ECONOMIC INSULATION THICKNESS

Optimizing an insulation installation usually requires choosing an insulating material and then selecting its thickness. Fiberglass and mineral wool insulations are commonly less expensive than traditional calcium silicate. However, other factors are included in the economic analysis, including cost of installation labor, thermal efficiency, current fuel cost, and useful life.[17]

Adding more insulation conserves more energy, but the costs of material and installation are also greater. Insulation thickness is optimized by balancing the heat losses against the cost of insulating. The *economic insulation thickness* is the thickness that minimizes the annual cost of ownership and operation. The economic thickness will vary with pipe diameter.

## 27. INSULATION THICKNESS TO PREVENT FREEZING OF WATER PIPE

Freezing of water flowing in a pipe will be prevented if the exit water temperature is kept from dropping below 32°F. The maximum allowable heat loss per unit mass from the water is

$$\frac{Q_{\max}}{\dot{m}} = c_p(T_{\text{entrance,water}} - 32°F) \qquad 40.36$$

If there is no water flow, a pipe exposed for long periods to subfreezing temperatures cannot be prevented from freezing.

---

[16]There is another, less commonly used, meaning for the term *critical thickness*: the thickest required insulation. In situations where the required insulation thickness is different for energy conservation, condensation control, personnel protection, and process temperature control, the "critical" thickness is the thickness that controls the design.

[17]The Thermal Insulation Manufacturer's Association (TIMA) can provide manual and computerized methods of evaluating the economics of different insulations.

Heat Transfer

**Table 40.3** Representative Conductivities of Pipe Insulation (Btu-ft/hr-ft$^2$-°F)

| | insulation temperature | | | | | |
|---|---|---|---|---|---|---|
| | 100°F (38°C) | 200°F (93°C) | 300°F (149°C) | 400°F (204°C) | 500°F (260°C) | 600°F (316°C) |
| calcium silicate | 0.033 | 0.037 | 0.041 | 0.046 | 0.057 | 0.060 |
| cellular glass | 0.039 | 0.047 | 0.055 | 0.064 | 0.074 | 0.085 |
| fiberglass | 0.026 | 0.030 | 0.034 | | | |
| magnesia, 85% | 0.034 | 0.037 | 0.041 | 0.044 | | |
| polyurethane | 0.016 | 0.016 | 0.016 | | | |

(Multiply Btu-ft/hr-ft$^2$-°F by 12 to get Btu in/hr-ft$^2$-°F.)
(Multiply Btu-ft/hr-ft$^2$-°F by 1.7307 to get W/m·K.)
(Multiply Btu-ft/hr-ft$^2$-°F by 4.1365 × 10$^{-3}$ to get cal·cm/s·cm$^2$·°C.)

## 28. INSULATION THICKNESS TO PREVENT SWEATING

*Sweating* is the condensation of moisture from the surrounding air on the surface of the pipe insulation. Sweating is prevented by keeping the surface temperature above the air's dew-point temperature.[18]

---

[18]The minimum thickness is found by equating the heat transfer through the insulation by conduction to the heat transfer from the surface by convection and radiation. As a first approximation, a total heat transfer coefficient, $h_{total}$, for convection and radiation of 0.65 Btu/hr-ft$^2$-°F (3.7 W/m$^2$·K) can be used.

# 41 Heat Transfer by Natural Convection

## Subscripts

| | |
|---|---|
| 0 | initial or zero gage pressure |
| $a$ | atmospheric |
| $b$ | boiling |
| $f$ | fluid |
| $h$ | film |
| $p$ | pressure |
| $s$ | surface |
| sat | saturated |
| $\infty$ | at infinity |

## Nomenclature

| | | | |
|---|---|---|---|
| $A$ | area | ft$^2$ | m$^2$ |
| $c_p$ | specific heat | Btu/lbm-°F | J/kg·K |
| $C$ | constant | – | – |
| $d$ | diameter | ft | m |
| $g$ | gravitational acceleration[1] | ft/hr$^2$ | m/s$^2$ |
| Gr | Grashof number | – | – |
| $h$ | film coefficient | Btu/hr-ft$^2$-°F | W/m$^2$·K |
| $k$ | thermal conductivity | Btu-ft/hr-ft$^2$-°F | W/m·K |
| $L$ | characteristic length | ft | m |
| $m$ | exponent | – | – |
| $n$ | exponent | – | – |
| $N$ | number of tube layers | – | – |
| Nu | Nusselt number | – | – |
| Pr | Prandtl number | – | – |
| $q$ | heat transfer per unit area | Btu/hr-ft$^2$ | W/m$^2$ |
| $Q$ | heat transfer rate | Btu/hr | W |
| $r$ | radius | ft | m |
| Ra | Rayleigh number | – | – |
| $T$ | temperature | °F | K |

## Symbols

| | | | |
|---|---|---|---|
| $\alpha$ | thermal diffusivity | ft$^2$/sec | m$^2$/s |
| $\beta$ | volumetric coefficient of expansion | 1/°R | 1/K |
| $\mu$ | viscosity[2,3,4] | lbm/hr-ft | kg/s·m |
| $\nu$ | kinematic viscosity | ft$^2$/sec | m$^2$/s |
| $\rho$ | mass density | lbm/ft$^3$ | kg/m$^3$ |

---

[1] $g$ has a value of $4.17 \times 10^8$ ft/hr$^2$ ($1.27 \times 10^8$ m/h$^2$).

[2] The use of mass units in viscosity values is typical in the subject of convective heat transfer.

[3] Most data compilations give fluid viscosity in units of seconds. In the United States, heat transfer is traditionally given on a per hour basis. Therefore, a conversion factor of 3600 is needed when calculating dimensionless numbers from table data.

[4] The combination of units kg/s·m is the same as a N·s/m$^2$ or Pa·s.

## 1. INTRODUCTION

*Natural convection* (also known as *free convection*) is the removal of heat from a surface by a fluid that moves vertically under the influence of a density gradient. As a fluid warms, it becomes lighter and rises from the heating surface. The fluid is acted upon by buoyant and gravitational forces. The fluid does not have a component of motion parallel to the surface.[5]

Natural convection is attractive from an engineering design standpoint because no motors, fans, pumps, or other equipment with moving parts are required. However, the transfer surface must be much larger than it would be with forced convection.[6]

## 2. HEAT TRANSFER BY NATURAL CONVECTION

Equation 41.1 is the basic equation used to calculate the steady-state heat transfer by natural convection in both heating and cooling configurations. The *film coefficient (heat transfer coefficient)*, $h$, is seldom known to great accuracy.[7] The average film coefficient, $\bar{h}$ is used where there are variations over the heat transfer surface.[8]

$$Q = qA = hA(T_s - T_\infty) \qquad 41.1$$

---

[5] Rotating spheres and cylinders and vertical plane walls are special categories of convective heat transfer where the fluid has a component of relative motion parallel to the heat transfer surface.

[6] Natural convection requires approximately 2 to 10 times more surface area than does forced convection.

[7] An error of up to 25% can be expected.

[8] Though $\bar{h}$ has traditionally been used in books on the subject of heat transfer, most modern books and this book use the symbol $h$. The fact that the film coefficient is an inaccurate, average value is implicit.

**Heat Transfer**

## 3. FILM COEFFICIENTS

Typical values of film coefficients for natural convection are listed in Table 41.1.

**Table 41.1** *Typical Film Coefficients for Natural Convection*[*]

|  | Btu/hr-ft²-°F | W/m²·K |
|---|---|---|
| no change in phase: |  |  |
| air, still | 0.8–4.4 | 5.0–25.0 |
| condensing: |  |  |
| steam |  |  |
| horizontal surface | 1700–4300 | 9600–24 400 |
| vertical | 700–2000 | 4000–11 300 |
| organic solvents | 150–500 | 850–2800 |
| ammonia | 500–1000 | 2800–5700 |
| evaporating: |  |  |
| water | 800–2000 | 4500–11 300 |
| organic solvents | 100–300 | 550–1700 |
| ammonia | 200–400 | 1100–2300 |

(Multiply Btu/hr-ft²-°F by 5.6783 to obtain W/m²·K.)
[*]Values outside of these ranges have been observed. However, these ranges are typical of those encountered in industrial processes.

## 4. NUSSELT NUMBER

The *Nusselt number,* Nu, is defined by Eq. 41.2. The Nusselt number is sometimes written with a subscript (e.g., $\text{Nu}_h$ or $\text{Nu}_f$) to indicate that the fluid properties are evaluated at the film temperature. (See Eq. 41.11.)

$$\text{Nu} = \frac{hd}{k} \qquad 41.2$$

## 5. PRANDTL NUMBER

The dimensionless *Prandtl number,* Pr, is defined by Eq. 41.3. It represents the ratio of momentum diffusion to thermal diffusion. The values used are for the fluid, not for the surface material. For gases, the values used in calculating the Prandtl number do not vary significantly with temperature, and hence neither does the Prandtl number itself.

$$\text{Pr} = \frac{c_p \mu}{k} = \frac{c_p \nu \rho}{k} = \frac{\nu}{\alpha} \qquad 41.3$$

## 6. GRASHOF NUMBER

The dimensionless *Grashof number,* Gr, is the ratio of buoyant to viscous forces. Dynamic similarity in free convection problems is assured by equating the Grashof numbers.

The *characteristic length, L,* is defined in Table 41.2 for various configurations.[9] The coefficient of volumetric

---

[9]The length of the side of a square, the mean length of a rectangle, and 90% of the diameter of a circle have historically been used as the *characteristic length.* However, the ratio of surface area to perimeter gives better agreement with experimental data.

expansion, $\beta$, for ideal gases is the reciprocal of the absolute film temperature. Gravity, $g$, and viscosity, $\mu$, must have the same unit of time in order to make Gr dimensionless. The quantity $g\beta\rho^2/\mu^2$ is tabulated in App. 41.A through App. 41.F, so the component values generally do not need to be evaluated individually.

$$\text{Gr} = \frac{L^3 g\beta\rho^2(T_s - T_\infty)}{\mu^2}$$
$$= \frac{L^3 g\beta(T_s - T_\infty)}{\nu^2} \qquad 41.4$$

The Grashof number may be written with a subscript indicating which dimension is to be used as the characteristic length. For example, the symbol $\text{Gr}_d$ or $\text{Gr}_D$ could be used to represent the Grashof number in which diameter is the characteristic length.

For air, the critical Grashof number for laminar flow is approximately $10^9$. Below $10^9$, the air flow will be laminar; above $10^9$, it will be turbulent.

## 7. RAYLEIGH NUMBER

The *Rayleigh number,* Ra, is the product of the Grashof and Prandtl numbers.

$$\text{Ra} = \text{GrPr} = \frac{L^3 g\beta\rho^2(T_s - T_\infty)c_p}{k\mu} \qquad 41.5$$

The quantity $g\beta\rho^2 c_p/k\mu$ is tabulated in some books and given the symbol $a$.

$$\text{Ra} = aL^3(T_s - T_\infty) \qquad 41.6$$

$$a = \frac{g\beta\rho^2 c_p}{k\mu} \qquad 41.7$$

## 8. REYNOLDS NUMBER

The free-stream velocity is always zero with natural convection, so the traditional Reynolds number is also always zero. The Grashof and Rayleigh numbers take the place of determining whether flow is laminar or turbulent.[10]

## 9. CORRELATIONS

Equation 41.1 is simple to use. The main difficulty is finding the film coefficient, $h$. Various theoretical, empirical, and semi-empirical correlations have been developed using dimensional analysis and experimentation. These correlations are of several forms.

Theoretical correlations are developed completely from dimensional analysis and theoretical considerations.

---

[10]Rising air nevertheless has a velocity. The critical Reynolds number for laminar flow of air is approximately 550 (corresponding to a Grashof number of $10^9$).

Empirical correlations are determined by fitting a curve through observed data points. The film coefficient has traditionally been correlated with the *heat flux*, $q = Q/A$ (in Btu/hr-ft$^2$ or kW/m$^2$), or with the difference in temperature between the heated surface and the fluid. For example, the convective film coefficient for boiling water can be predicted approximately by Eq. 41.8 and Eq. 41.9.[11]

$$\log(h_b) \approx -2.05 + 2.5\log(T_s - T_{\text{bulk}})$$
$$+ 0.014\,T_{\text{sat}} \quad \text{[U.S. only]} \qquad 41.8$$

$$h_b \approx 190 + 0.43q \quad \text{[U.S. only]} \qquad 41.9$$

Semi-empirical correlations, derived from dimension analysis with exponents and constants determined from experimentation, are the form of Eq. 41.10.[12] Exponents $m$ and $n$ are often sufficiently close so that a common value can be used.[13]

$$\text{Nu} = C \times (\text{Pr}^m\text{Gr}^n)$$
$$\approx C \times (\text{PrGr})^n = C \times (\text{Ra})^n \qquad 41.10$$

Each correlation can only be used in particular configurations (i.e., a correlation for horizontal cylinders cannot generally be used for vertical cylinders), and even then, the correlation will be valid only within a particular range of parameters (e.g., Prandtl or Grashof numbers).

The usefulness of a correlation depends on how well it predicts actual performance. Though correlations should always be accompanied by parameter ranges, the percentage accuracy of the correlation is generally not stated. Considering that correlations are often accurate to only $\pm 20\%$, a value derived from a correlation near the end of its applicable range should be considered "ballpark."

Often, two or more parameter ranges will be given for a particular configuration. (See the GrPr ranges in Table 41.2.) The lower range of parameters corresponds to laminar air flow, while the higher parameter ranges correspond to turbulent air flow. Correlations are less reliable near the transition region between the two regimes.

## 10. FILM TEMPERATURE

Film properties are evaluated at the average of the surface temperature, $T_s$, and the *bulk temperature*, $T_\infty$. When there is a variation of the surface temperature, as there could be along the length of a long tube used for heat transfer, the surface temperature is assumed to be the temperature at midlength along the tube.[14]

$$T_h = \tfrac{1}{2}(T_s + T_\infty) \qquad 41.11$$

## 11. NUSSELT EQUATION

The *Nusselt equation* and equations of its form are often used to find the film coefficient for natural convective heating and cooling. The thermal conductivity, $k$, in Eq. 41.12 is for the transfer fluid, not for the surface wall, and is evaluated at the film temperature, $T_h$.

$$\frac{hL}{k} = C(\text{GrPr})^n \qquad \text{[SI and U.S.]} \qquad 41.12$$

For laminar convection ($1000 < \text{GrPr} < 10^9$), $n$ has a value of approximately $1/4$. For turbulent convection ($\text{GrPr} > 10^9$), $n$ is approximately $1/3$. For sublaminar convection ($\text{GrPr} < 1000$), $n$ is less than $1/4$ (typically taken as $1/5$), and graphical solutions are commonly used.

The values of the dimensionless empirical constants $C$ and $n$ in Eq. 41.12 and given in Table 41.2 can be used with all fluids and any consistent systems of units. Table 41.2 is limited in application to single heat transfer surfaces (i.e., a single tube or a single plate).

Horizontal pipe diameters greater than approximately 8 in (20 cm) and plate heights greater than approximately 2 ft (0.6 m) have little effect on film coefficients.[15] Therefore, characteristic lengths for tall plates and large-diameter pipes should be limited to 2 ft (0.6 m) and 0.67 ft (0.2 m), respectively, when calculating the film coefficients for free convection.

---

[11]Equation 41.8 and Eq. 41.9 actually yield some "pretty good" initial estimates.
[12]Equation 41.10 in known as a *Nusselt-type correlation*.
[13]The implication of exponents $m$ and $n$ being identical is that the temperature differences and fluid velocities are small.

[14]Variations in the surface temperature with time, however, cannot be so easily handled.
[15]Some researchers report 3 ft (0.9 m) as the limiting value.

**Table 41.2** Parameters for the Nusselt Equation (any substance; isothermal surfaces, U.S. or SI units)

| configuration | $L$ | GrPr | $C$ | $n$ |
|---|---|---|---|---|
| vertical plate or vertical cylinder[a] | height[b] | $< 10^4$ | 1.36 | 0.20 |
| | | $10^4$ to $10^9$ | 0.59 | $1/4$ |
| | | $10^9$ to $10^{10}$ | 0.13 | $1/3$ |
| inclined plate ($\theta$ measured from horizontal) | Use vertical plate constants, substituting $\sin\theta\, \mathrm{Gr}$ for Gr. | | | |
| horizontal cylinder[c] | outside diameter | $10^3$ to $10^9$ | 0.53 | $1/4$ |
| | | $10^9$ to $10^{12}$ | 0.13 | $1/3$ |
| thin horizontal wire | diameter | $< 10^{-5}$ | 0.49 | 0 |
| | | $10^{-5}$ to $10^{-3}$ | 0.71 | 0.04 |
| | | $10^{-3}$ to 1 | 1.09 | 0.10 |
| | | 1 to $10^4$ | 1.09 | 0.20 |
| | | $10^4$ to $10^9$ | | |
| horizontal plate[d] hot surface facing up or cold surface facing down | $\frac{1}{2}(s_1 + s_2)$ or $0.9d$ | $10^5$ to $2 \times 10^7$ | 0.54 | $1/4$ |
| | | $2 \times 10^7$ to $3 \times 10^{10}$ | 0.14 | $1/3$ |
| horizontal plate[d] hot surface facing down or cold surface facing up | $\frac{1}{2}(s_1 + s_2)$ or $0.9d$ | $3 \times 10^5$ to $3 \times 10^{10}$ | 0.27 | $1/4$ |
| sphere[e] | radius | $10^3$ to $10^9$ | 0.53 | $1/4$ |
| | | $> 10^9$ | 0.15 | $1/3$ |

[a]A vertical cylinder can be considered a vertical plate as long as $d/L \geq 35/(\mathrm{Gr}_L)^{1/4}$.
[b]For short vertical plates, the characteristic length is approximately (height × width)/(height + width).
[c]The values for the laminar range can also be used for heat transfer to liquid metals.
[d]For a circular flat disc, the characteristic length is 90% of the disc diameter.
[e]Ranges and values reported by different researchers show significant variation. Some correlations use diameter as the characteristic length of the sphere. Values can also be used for short cylinders and blocks with a characteristic length of (height × width)/(height + width).

# 42 Heat Transfer by Forced Convection

## Nomenclature

| | | | |
|---|---|---|---|
| $A$ | area | $ft^2$ | $m^2$ |
| $c_p$ | specific heat | Btu/lbm-°F | J/kg·K |
| $C$ | a constant or coefficient | – | – |
| $d$ | diameter | ft | m |
| $F$ | factor | – | – |
| $G$ | mass flow rate per unit area | $lbm/hr\text{-}ft^2$ | $kg/s\cdot m^2$ |
| Gz | Graetz number | – | – |
| $h$ | film coefficient | $Btu/hr\text{-}ft^2\text{-}°F$ | $W/m^2\cdot K$ |
| $k$ | thermal conductivity | $Btu\text{-}ft/hr\text{-}ft^2\text{-}°F$ | $W/m\cdot K$ |
| $L$ | length | ft | m |
| $n$ | exponent | – | – |
| Nu | Nusselt number | – | – |
| Pe | Peclet number | – | – |
| Pr | Prandtl number | – | – |
| $q$ | heat transfer per unit area | $Btu/hr\text{-}ft^2$ | $W/m^2$ |
| $Q$ | heat transfer rate | Btu/hr | W |
| $r$ | radius | ft | m |
| Re | Reynolds number | – | – |
| St | Stanton number | – | – |
| $T$ | temperature[1,2] | °F | K |
| $U$ | overall coefficient of heat transfer | $Btu/hr\text{-}ft^2\text{-}°F$ | $W/m^2\cdot K$ |
| v | velocity | ft/hr | m/s |
| $x$ | distance $x$ | ft | m |

[1]The symbol θ is used for temperature in some books.
[2]It is common in heat exchanger literature to use lowercase $t$ as the cold side temperature. This eliminates the requirement for "cold" and "hot" designations.

## Symbols

| | | | |
|---|---|---|---|
| $\eta$ | efficiency | – | – |
| $\mu$ | viscosity[3,4,5] | lbm/hr-ft | kg/s·m |
| $\nu$ | kinematic viscosity | $ft^2$/sec | $m^2$/s |
| $\rho$ | mass density | $lbm/ft^3$ | $kg/m^3$ |

## Subscripts

| | |
|---|---|
| $A$ | at end $A$ |
| ave | average |
| $b$ | bulk |
| $B$ | at end $B$ |
| $c$ | correction |
| $d$ | based on diameter |
| $H$ | hydraulic |
| $i$ | inside |
| $L$ | over length $L$ |
| lm | log mean |
| $m$ | mean |
| max | maximum |
| min | minimum |
| $o$ | outside |
| $s$ | surface |
| $t$ | transverse or at time $t$ |
| $T$ | temperature |
| $V$ | constant volumetric flow rate |
| $x$ | at point $x$ |
| $\infty$ | free-stream (far field) or at time = $\infty$ |

## 1. INTRODUCTION

As with natural convection, *forced convection* depends on the movement of a fluid to remove heat from a surface. With forced convection, a fan, a pump, or relative motion causes the fluid motion. If the flow is over a flat surface, the fluid particles near the surface will flow more slowly due to friction with the surface. The *boundary layer* of slow-moving particles comprises the major thermal resistance. The thermal resistance of the tube and other heat exchanger components is often disregarded.

[3]The use of mass units in viscosity values is typical in the subject of convective heat transfer.
[4]Most data compilations give fluid viscosity in units of seconds. In the United States, heat transfer is traditionally stated on a per hour basis. Therefore, a conversion factor of 3600 sec/hr is needed when calculating dimensionless numbers from table data.
[5]The combination of units kg/s·m is the same as a Pa·s and N·s/$m^2$.

**Heat Transfer**

## 2. HEAT TRANSFER BY FORCED CONVECTION

*Newton's law of convection*, Eq. 42.1, gives the heat transfer for Newtonian fluids in forced convection over exterior surfaces.[6,7] The film coefficient, $h$, is also known as the *coefficient of forced convection*. $T_\infty$ is the *free-stream temperature*.

$$Q = qA = hA(T_s - T_\infty) \qquad 42.1$$

For flow within a tube, the more easily determined *bulk temperature* (see Sec. 42.5) is used in place of the free-stream temperature.

$$Q = qA = hA(T_s - T_b) \qquad 42.2$$

## 3. DIMENSIONLESS NUMBERS

The dimensionless Nusselt number, Nu, Prandtl number, Pr, and Reynolds number, Re, are

$$\text{Nu} = \frac{hd}{k} \qquad 42.3$$

$$\text{Pr} = \frac{c_p\mu}{k} = \frac{\nu}{\alpha} \qquad 42.4$$

$$\text{Re} = \frac{\text{v}d}{\nu} = \frac{dG}{\mu} \qquad 42.5$$

$$G = \text{v}_\infty\rho_\infty \qquad 42.6$$

The viscosity, $\mu$, used in Eq. 42.4 must have the same units of time as the conductivity, $k$.

The density and velocity used in Eq. 42.6 must correspond to the same point. In the most common case, both are free-stream values. It is incorrect to use the free-stream velocity with the density evaluated at the film temperature.

The *Peclet number*, Pe, is the product of the Reynolds and Prandtl numbers.

$$\text{Pe} = \text{Re}\,\text{Pr} = \frac{d\text{v}\rho c_p}{k} \qquad 42.7$$

The *Graetz number*, Gz, is used in the reporting of empirical data for laminar flow in tubes.

$$\text{Gz} = \text{Re}_d\,\text{Pr}\left(\frac{d}{L}\right) \qquad 42.8$$

The *Stanton number*, St, is encountered in correlations of fluid friction and heat transfer.

$$\text{St} = \frac{\text{Nu}}{\text{Re}\,\text{Pr}} = \frac{h}{\text{v}\rho c_p} \qquad 42.9$$

## 4. DIMENSIONLESS NUMBER RANGES

All heat transfer correlations have associated ranges of the Prandtl and Reynolds numbers, whether stated or not. The endpoints of these ranges are indistinct and depend on the fluid properties. For example, Eq. 42.16 can be used with a Reynolds number as low as 2100 as long as the Prandtl number is less than 10 (i.e., it cannot be used at that lower limit for fluids with viscosities more than twice that of water). Therefore, the lower limit is established as 10,000 instead of 2100, and the equation is deemed applicable to all fluids. This explains why researchers report different ranges for the same correlation.

## 5. BULK TEMPERATURE

The *bulk temperature*, $T_b$, also known as the *mixing temperature*, is the energy-average fluid temperature. The bulk temperature concept is usually encountered with tube flow where there is no free-stream temperature. The centerline temperature is a candidate for theoretical considerations, but it cannot be easily measured. Therefore, the bulk temperature is used to calculate the heat transfer for flow in tubes.

The bulk temperature used to calculate local properties is evaluated over the tube cross-sectional area at the point along the tube length being considered. The bulk temperature used in the calculation of an average film coefficient over the entire length of a tube or heat exchanger is evaluated as the average of the entrance and exit temperatures. For this reason, it is often referred to as the *mean bulk temperature*.

$$T_b = \tfrac{1}{2}(T_{\text{in}} + T_{\text{out}}) \qquad 42.10$$

The bulk temperature should be used to calculate the film coefficient of a fluid flowing in a tube or heat exchanger unless the properties change a lot (i.e., as they do with high-viscosity fluids).[8] It may be necessary to solve heat transfer equations iteratively in order to determine the bulk temperature, since the film coefficient (based on the bulk temperature) is needed in order to determine the outlet temperature.

The mass flow rate in a tube is constant everywhere. Where there are large variations in temperature along the length of a tube, the density, velocity, and temperature must all be consistent. It is incorrect to use a density evaluated at the midpoint temperature with an entrance velocity.

---

[6]Newton's law of convection is the same for natural and forced convection. Only the methods used to evaluate the film coefficient are different.

[7]The results of this chapter do not generally apply to non-Newtonian fluids.

[8]In that case, evaluate the film coefficient at the inlet and outlet and take the logarithmic mean $((h_2 - h_1)/(\ln(h_2/h_1)))$ of the two. This requires calculating the film coefficients twice.

## 6. FILM TEMPERATURE

As with natural convection, the free-field temperature, $T_\infty$, is used to calculate the film temperature in external flow configurations.

$$T_{\text{film}} = \tfrac{1}{2}(T_s + T_\infty) \qquad 42.11$$

Film properties inside tubes are evaluated at the average of the surface temperature, $T_s$, and the bulk temperature, $T_b$. When there is a variation of the surface temperature, as there could be along the length of a tube used for heat transfer, the surface temperature is assumed to be the temperature at midlength along the tube.

$$T_{\text{film}} = \tfrac{1}{2}(T_s + T_b) \qquad 42.12$$

## 7. FLOW OVER FLAT PLATES

The boundary layer of a fluid flowing over a flat plate is assumed to have a parabolic velocity distribution.[9] (See Fig. 42.1.) The layer has three distinct regions: laminar, transition, and turbulent. From the leading edge, the layer is laminar and the thickness increases gradually until the transition region where the thickness increases dramatically. Thereafter, the boundary layer is turbulent. The laminar region is always present, though its length decreases as velocity increases. Turbulent flow may not develop at all with short plates.

**Figure 42.1** *Flow Over a Flat Plate*

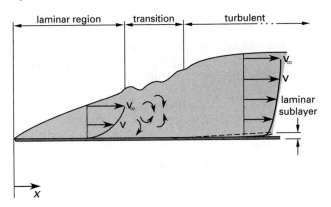

The Reynolds number is used to determine which of the three flow regimes is applicable. Laminar flow on smooth flat plates occurs for Reynolds numbers up to approximately $2 \times 10^5$; turbulent flow exists for Reynolds numbers greater than approximately $3 \times 10^6$.[10] Transition flow is in between. The distance from the leading edge at which turbulent flow is initially experienced is determined from the *critical Reynolds number*, commonly

taken as $\text{Re} = 5 \times 10^5$ for smooth flat plates, though the actual value is highly dependent on surface roughness. Distance, $x$, is measured from the leading edge.

$$\text{Re}_x = \frac{v_\infty x \rho}{\mu} = \frac{v_\infty x}{\nu} \qquad 42.13$$

The heat transfer from a flat plate is

$$Q = h_{\text{ave}} A (T_s - T_\infty) \qquad 42.14$$

## 8. TURBULENT FLOW INSIDE STRAIGHT TUBES

The theoretical *Nusselt equation* can be used to find the inside film coefficient for turbulent flow inside round, horizontal tubes.[11] All fluid properties except specific heat are evaluated at the film temperature. Specific heat is evaluated at the bulk temperature.

$$\text{Nu} = 0.0225\,\text{Re}^{0.8}\,\text{Pr}^{1/3}$$
$$\left[0.6 < \text{Pr} < 160; \text{Re} > 10^4; \frac{L}{d} > 60\right] \qquad 42.15$$

Equation 42.15[12] is difficult to use in design work because the film temperature is an inconvenient concept with tube flow. With tube flow, it is more common to base all fluid properties on the bulk temperature. The *Dittus-Boelter equation*, as it was modified by W. H. McAdams, evaluates all fluid properties at the bulk temperature.

$$\text{Nu} = 0.023\,\text{Re}^{0.8}\,\text{Pr}^n$$
$$\left[0.7 < \text{Pr} < 120; \text{Re} > 10^4; \frac{L}{d} > 60\right] \qquad 42.16$$

The exponent, $n$, in Eq. 42.16[13] has a value of 0.3 when the surface (wall) temperature is less than the bulk fluid temperature, and $n$ is 0.4 when the surface (wall) temperature is greater than the bulk fluid temperature.

Within the normal range of most gases, $\text{Pr}^n \approx 1.0$, resulting in Eq. 42.17.

$$\text{Nu} = 0.023\,\text{Re}^{0.8} \qquad 42.17$$

If there is a large change in viscosity during the heat transfer process, as there would be with oils and other viscous fluids heated in a long tube, Eq. 42.17 is modified into the *Sieder-Tate* (also known as *Seider-Tate*) *equation* for turbulent flow. All fluid properties in Eq. 42.18[14] are evaluated at the bulk temperature

---

[9]The velocity distribution does not have to be parabolic. In *Couette flow*, there are two closely spaced parallel surfaces, one which is stationary and the other moving with constant velocity. The velocity gradient is assumed to be linear between the plates.
[10]Turbulent flow can begin at Reynolds numbers less than $3 \times 10^5$ if the plate is rough. This discussion assumes the plate is smooth.

[11]Equation 42.15 can also be used to obtain conservative values for turbulent flow in vertical tubes.
[12]According to some authorities: $[0.5 < \text{Pr} < 100]$. According to some authorities: $[L/d > 10]$ when used with Eq. 42.19 and Eq. 42.20.
[13]According to some authorities: $[\text{Pr} < 100]$. According to some authorities: $[L/d > 10]$ when used with Eq. 42.19 and Eq. 42.20.
[14]Some authorities report the coefficient as 0.027 instead of 0.023. According to some authorities: $[\text{Pr} > 0.6]$. The upper Prandtl number limit is also reported as 700, 16,700, and 17,000. According to some authorities: $[L/d > 10]$ when used with Eq. 42.19 and Eq. 42.20.

except for $\mu_s$, which is evaluated at the surface temperature.

$$\text{Nu} = 0.023 \, \text{Re}^{0.8} \, \text{Pr}^{1/3} \left( \frac{\mu}{\mu_s} \right)^{0.14}$$

$$\left[ 0.7 < \text{Pr} < 160; \text{Re} > 10^4; \frac{L}{d} > 60 \right] \qquad 42.18$$

For $L/d$ ratios less than 60 occurring in pipes with sharp leading edges, the right-hand side of Eq. 42.18 can be multiplied by either Eq. 42.19 or Eq. 42.20.

$$1 + \left( \frac{d}{L} \right)^{0.7} \qquad \left[ 2 < \frac{L}{d} < 20 \right] \qquad 42.19$$

$$1 + \left( \frac{6d}{L} \right) \qquad \left[ 20 < \frac{L}{d} < 60 \right] \qquad 42.20$$

## 9. FLOW THROUGH NONCIRCULAR DUCTS

Dimensional analysis shows that a *characteristic length* is required in the Nusselt number, but it does not identify the length to be used. It has been common practice to correlate empirical pressure drop and heat transfer data with the *hydraulic diameter*, $d_H$, of noncircular (e.g., rectangular, square, elliptical, polygonal) ducts.[15,16] The Nusselt number for laminar and turbulent flow through noncircular ducts is given by Eq. 42.21.

$$\text{Nu} = \frac{h d_H}{k} \qquad 42.21$$

$$d_H = 4 \left( \frac{\text{area in flow}}{\text{wetted perimeter}} \right) \qquad 42.22$$

*Annular flow* is the flow of fluid through an annulus. Fluid flow is annular in simple tube-in-tube heat exchangers of the type shown in Fig. 42.4.[17] For an annulus, Eq. 42.23 gives the hydraulic diameter.[18]

$$d_H = d_{i,\text{shell}} - d_{o,\text{tube}} \qquad 42.23$$

---

[15]A *duct* is any closed channel through which a fluid flows. Tubes and pipes are examples of round ducts. "Ducts" are not limited to air conditioning ducts.

[16]The use of the hydraulic diameter as the characteristic length is convenient and logical. Though an approximation, empirical data support using the hydraulic diameter in most cases. Notable exceptions are flow through ducts with narrow angles (e.g., an equilateral triangle with a narrow angle) and flow *parallel* to banks of tubes.

[17]Flow is not annular through more complex shell-and-tube heat exchangers.

[18]Though the hydraulic diameter is widely used as the characteristic length in calculating the Nusselt number for annuli, it is not a universal choice. Some researchers recommend using an equivalent diameter defined as

$$D_{\text{equivalent}} = \frac{4(\text{area in flow})}{\text{heated perimeter}}$$

$$= \frac{d_{i,\text{shell}}^2 - d_{o,\text{tube}}^2}{d_{o,\text{tube}}}$$

The film coefficient for fully developed laminar flow through noncircular ducts can be calculated from the Nusselt numbers in Table 42.1.

**Table 42.1** *Nusselt Numbers for Fully Developed Laminar Flow Ducts*

| | Nusselt number | |
| configuration | constant wall temperature | constant heat flux |
| --- | --- | --- |
| circular | 3.658 | 4.364 |
| square | 2.98 | 3.63 |
| rectangular, aspect ratio | | |
| 1:$\sqrt{2}$ | – | 3.78 |
| 1:2 | 3.39 | 4.11 |
| 1:3 | – | 4.77 |
| 1:4 | 4.44 | 5.35 |
| 1:8 | 5.95 | 6.60 |
| parallel plates | 7.54 | 8.235 |
| triangular, isosceles | 2.35 | 3.00 |

## 10. CROSSFLOW OVER A SINGLE CYLINDER

Figure 42.2 illustrates a cylinder (e.g., a tube or wire) in crossflow. Equation 42.24 can be used with any fluid to calculate the film coefficient.[19] The fluid properties are evaluated at the film temperature. The entire surface area of the tube is used when calculating the heat transfer. Flow is laminar up to a Reynolds number of approximately $5 \times 10^5$. (See Fig. 42.3 and Table 42.2.)

$$\text{Nu} = C_1 (\text{Re}_d)^n \text{Pr}^{1/3} \qquad [\text{Pr} > 0.7] \qquad 42.24$$

Equation 42.24 can be simplified for air since $\text{Pr}^{1/3} \approx 1.00$. Equation 42.25 is sometimes referred to as the *Hilbert-Morgan equation*.

$$\text{Nu} = C_2 (\text{Re}_d)^n \qquad 42.25$$

$$C_1 \approx 1.1 C_2 \qquad 42.26$$

**Figure 42.2** *Single Cylinder in Crossflow*

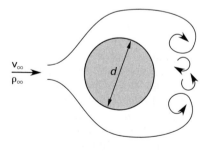

---

[19]There are more sophisticated correlations.

**Figure 42.3** *Average Nusselt Number for Cylinder in Crossflow (air)*

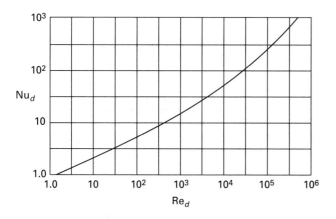

**Table 42.2** *Constants for Tubes in Crossflow (air and other gases)*

| $Re_d$ | $C_1$ | $C_2$ | $n$ |
|---|---|---|---|
| $0.4^a$–4 | 0.989 | 0.891 | 0.330 |
| 4–40 | 0.911 | 0.821 | 0.385 |
| 40–4000 | 0.683 | 0.615 | 0.466 |
| 4000–40,000 | 0.193 | 0.174 | 0.618 |
| 40,000–400,000$^b$ | 0.0266 | 0.0239 | 0.805 |

[a]Some sources give the lower limit as 1.0.
[b]Some sources give the upper limit as 250,000.

## 11. HEAT EXCHANGERS

Two fluids flow through or over a heat exchanger.[20] Heat from the hot fluid passes through the exchanger walls to the cold fluid.[21] The heat transfer mechanism is essentially completely forced convection.

Heat exchangers are categorized into simple *tube-in-tube heat exchangers* (also known as *jacketed pipe heat exchangers*), single-pass shell-and-tube heat exchangers, multiple-pass shell-and-tube heat exchangers, and crossflow heat exchangers.[22] *Shell-and-tube heat exchangers*, also known as *sathes* and *S & T heat exchangers*, consist of a large housing, the *shell*, with many smaller tubes running through it. The *tube fluid* passes through the tubes, while the *shell fluid* passes through the shell and around tubes.[23]

In a *single-pass heat exchanger*, each fluid is exposed to the other fluid only once (see Fig. 42.4). Operation is known as *parallel flow* (same as *cocurrent flow*) if both fluids flow in the same direction along the longitudinal axis of the exchanger and *counterflow* (same as *counter*

current flow*) if the fluids flow in opposite directions.[24,25] Counterflow is more efficient, and the heat transfer area required is less than that with parallel flow since the temperature gradient is more constant.

For increased efficiency, most exchangers are *multiple-pass heat exchangers*. The tubes pass through the shell more than once, and the shell fluid is routed around baffles.

**Figure 42.4** *Simple Heat Exchanger (single-pass, counterflow, tube-in-tube)*

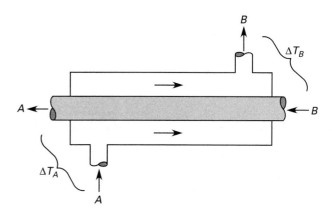

In a *crossflow heat exchanger*, one fluid flow is normal to the other.[26] Crossflow exchangers can operate with both fluids unmixed (typical when fluids are constrained to move through tubes and passageways), or one or both fluids may be mixed within the heat exchanger by forcing the fluids around tubes, baffles (see Fig. 42.5), or passages. If the fluid is mixed, its temperature is essentially uniform across the outlet. In TEMA X shells (see Sec. 42.12) experiencing pure crossflow and air-cooled exchangers, the fluids are generally unmixed.

One of the fluids in a crossflow heat exchanger can have multiple passes through the other fluid. Since the flow cannot be parallel in a crossflow heat exchanger, the designations used are *counter-crossflow* and *cocurrent-crossflow*. The distinction between mixed and unmixed fluids is further complicated by whether the fluids are mixed or unmixed between passes.

---

[20]These fluids do not have to be liquids. *Air-cooled exchangers* reduce water consumption in traditional cooling applications.
[21]A *recuperative heat exchanger*, typified by the traditional shell-and-tube exchanger, maintains separate flow channels for each of the fluids. A *regenerative heat exchanger* has only one flow path, to which the two fluids are exposed on an alternating basis.
[22]Fin coil heat exchangers are a special case of crossflow heat exchangers.
[23]Tubular heat exchangers are also known as *shell-and-tube heat exchangers*.

---

[24]Flow through shell-and-tube heat exchangers is neither purely parallel nor purely counterflow. Thus, these exchangers are sometimes designated as *parallel counterflow exchangers*.
[25]The designation *cocurrent* is not an abbreviation for *counter current*.
[26]An automobile radiator is an example of a crossflow exchanger.

*Figure 42.5* Types of Baffles

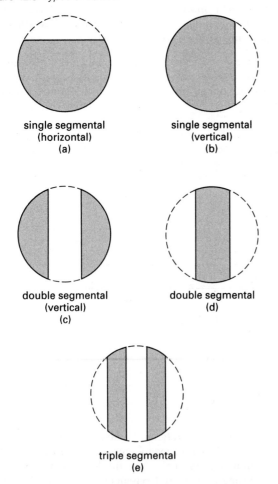

single segmental
(horizontal)
(a)

single segmental
(vertical)
(b)

double segmental
(vertical)
(c)

double segmental
(d)

triple segmental
(e)

## 12. HEAT EXCHANGER DESIGNATIONS

A heat exchanger with $X$ shell passes and $Y$ tube passes is designated as an *X-Y heat exchanger*. In addition, most manufacturers follow the TEMA standards for design, fabrication, and material selection.[27,28,29] Heat exchanger types can be described by a three-character TEMA designation. For example, a one-two TEMA E shell and tube heat exchanger would be a shell-and-tube heat exchanger with one shell pass and two tube passes.

---

[27]The Tubular Exchangers Manufacturers Association (TEMA) publishes the definitive standards for shell-and-tube heat exchanger construction and performance.

[28]Other applicable standards are published by the American Society of Mechanical Engineers (ASME) and the American Petroleum Institute (API).

[29]Similar to ASME's Pressure Vessel Code, TEMA standards B, C, and R are applicable to shell-and-tube heat exchangers with shell diameters not exceeding 60 in (152 cm), pressures not exceeding 3000 psi (20.67 MPa), and product of shell diameter and pressure not exceeding 60,000 lbf/in (10,500 N/mm).

## 13. TEMPERATURE DIFFERENCE TERMINOLOGY

There are several specialized temperature difference terms used in the analysis of heat transfer.

The difference in hot and cold fluid temperatures is seldom constant in a heat exchanger. The *approach* (*temperature approach* or *approach temperature*) is the smallest difference in temperature between the two fluids anywhere along the heat exchange path. For a double-pipe, single-pass counterflow heat exchanger, the temperature approach is defined by Eq. 42.27.

$$\Delta T_{\text{approach}} = T_{\text{hot,in}} - T_{\text{cold,out}} \qquad 42.27$$

Traditional cost-effective shell-and-tube heat exchangers seldom have temperature approaches less than 10°F (6°C). Exceptions are some refrigeration systems that work with temperature approaches of 5°F to 9°F (3°C to 5°C), and plate-and-frame heat exchangers that can work well with as little as a 2°F (1°C) temperature approach. In combustion air preheaters using flue gas as the heating source, the temperature approach should be approximately 36°F (20°C).

The *extreme temperature difference* for heat exchangers is defined as

$$\Delta T_{\text{extreme}} = T_{\text{hot,in}} - T_{\text{cold,in}} \qquad 42.28$$

The ratio of the cold fluid change to the extreme temperature difference, a form of "temperature efficiency," is[30]

$$\eta_T = \frac{T_{\text{cold,out}} - T_{\text{cold,in}}}{T_{\text{hot,in}} - T_{\text{cold,in}}} \qquad 42.29$$

## 14. TEMPERATURE CROSS

A *temperature cross* occurs when the exit temperature of the cold fluid is above the exit temperature of the hot fluid. This occurs predominantly with counterflow heat exchangers, although it can also occur with a shell-and-tube exchanger with one shell pass and multiple tube passes. A temperature cross indicates that there is a relatively small temperature difference between the two fluids. This requires either a large heat transfer area or a relative high fluid velocity (to increase the overall heat transfer coefficient).

## 15. LOGARITHMIC TEMPERATURE DIFFERENCE

The temperature difference between two fluids is not constant in a heat exchanger. When calculating the heat transfer for a tube whose temperature difference changes along its length, the *logarithmic mean temperature*

---

[30]The symbol $S$ is also used for this quantity by some authors.

Heat Transfer

*difference*, $\Delta T_{lm}$ or LMTD, is used.[31,32,33] (See Fig. 42.6.) In Eq. 42.30, $\Delta T_A$ and $\Delta T_B$ are the temperature differences at ends $A$ and $B$, respectively, regardless of whether the fluid flow is parallel or counterflow, as shown in Fig. 42.6.[34,35]

$$\Delta T_{lm} = \frac{\Delta T_A - \Delta T_B}{\ln \dfrac{\Delta T_A}{\Delta T_B}} \qquad 42.30$$

**Figure 42.6** *Temperature Profile for LMTD Calculation (single-pass, counterflow exchanger, no change of phase)*

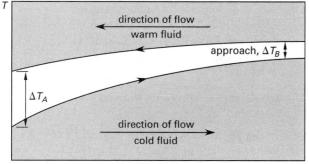

For multiple-pass and crossflow heat exchangers, a multiplicative correction factor, $F_c$, is required for $\Delta T_{lm}$. The correction factor is 1.00 for single-pass parallel and counterflow tube-in-tube heat exchangers. When one of the fluids does not change temperature, as in a feedwater heater or other condensation/evaporation environment, the correction factor is also 1.00 for parallel, counter-, and crossflow.[36] The procedure in all other cases is to calculate $\Delta T_{lm}$ as if the fluids were in counterflow. The correction factor, $F_c$, depends on the type of heat exchanger and is almost always given graphically.[37] (See App. 42.A and App. 42.B.)

## 16. HEAT TRANSFER IN HEAT EXCHANGERS

Equation 42.31 calculates the steady-state heat transfer (also known as the *heat duty* and *heat load*) in a heat exchanger or feedwater heater.[38,39,40] The *overall heat transfer coefficient*, $U$, also known as the *overall conductance* and the *overall coefficient of heat transfer*, can be specified for use with either the outside or inside tube areas.[41] The heat transfer is independent of whether the outside or inside area is used.

$$Q = U_o A_o F_c \Delta T_{lm} = U_i A_i F_c \Delta T_{lm} \qquad 42.31$$

## 17. OVERALL HEAT TRANSFER COEFFICIENT

The overall heat transfer coefficient, $U$, is calculated from the film coefficients and the tube material conductivities. (See Table 42.3.) The overall heat transfer coefficient, based on outside and inside areas and exclusive of a fouling factor, is

$$\frac{1}{U_o} = \frac{1}{h_o} + \left(\frac{r_o}{k_{tube}}\right)\ln\frac{r_o}{r_i} + \frac{r_o}{r_i h_i} \qquad 42.32$$

$$\frac{1}{U_i} = \frac{1}{h_i} + \left(\frac{r_i}{k_{tube}}\right)\ln\frac{r_o}{r_i} + \frac{r_i}{r_o h_o} \qquad 42.33$$

The second term in Eq. 42.32 and Eq. 42.33 is sometimes approximated by the term $t/k$ when the tube diameters are "large." However, the thermal resistance of the tube is very small and is often omitted entirely. If the tube resistance term is kept at all, it is not very difficult to use Eq. 42.32 and Eq. 42.33 as written.

---

[31]An exception occurs in HVAC calculations where $\Delta T$ at midlength has traditionally been used to calculate the heat transfer in air conditioning ducts. Considering the imprecise nature of HVAC calculations, the added sophistication of using the logarithmic mean temperature difference is probably unwarranted.

[32]The symbol $\Delta T_m$ is also widely used for the log-mean temperature difference. However, this can also be interpreted as the arithmetic mean temperature.

[33]The logarithmic temperature difference is used even with change of phase (e.g., boiling liquid or condensing vapor) and the temperature is constant in one tube.

[34]It doesn't make any difference which end is $A$ and which is $B$. If the numerator in Eq. 42.30 is negative, the denominator will also be negative.

[35]Using Eq. 42.30 presents many difficulties, particularly with computer-based heat transfer analysis. As $\Delta T_A$ and $\Delta T_B$ become equal, Eq. 42.30 becomes indeterminate, even though the correct relationship is $\Delta T_{lm} = \Delta T_A = \Delta T_B$. Also, the first derivative of Eq. 42.30, used in some calculations, is undefined when $\Delta T_A$ and $\Delta T_B$ are equal, even though the correct value is 0.5. A replacement expression (Underwood, 1933) that avoids these difficulties with (generally) less than a 0.3 % error is

$$\Delta T_{lm} \approx \left(\frac{\Delta T_A^{1/3} + \Delta T_B^{1/3}}{2}\right)^3$$

[36]A general rule for good designs of boiling/evaporative systems is that the logarithmic mean temperature difference should be kept below 110°F (60°C).

[37]Calculating the $F_c$ factor is preferred over reading it from a chart. The error of the $F_c$ factor read from a chart can be as great as 5%.

[38]*Closed feedwater heaters* are heat exchangers whose purpose is to heat water with condensing steam.

[39]There are three heat loads referred to in heat exchanger specifications: the *specific heat load*, which is the design heat transfer; the heat released by the hot fluid; and the heat absorbed by the cold fluid. All three would be the same if operation was adiabatic, but due to practical losses, they are not. If they differ by more than 10%, the cause of the discrepancy should be evaluated.

[40]Transient (i.e., start-up) performance of heat exchangers is poorly covered in most textbooks. An excellent article on this subject is Chester A. Plant's "Evaluate Heat-Exchanger Performance," *Chemical Engineering Magazine*, p. 104, July 1992.

[41]It is more common (and preferred) to use the outside tube area because the tube outside diameter is more easily measured.

**Heat Transfer**

**Table 42.3** *Typical Values of Overall Coefficient of Heat Transfer (U-values)* * (Btu-ft/hr-ft²-°F)

| heating applications | | clean surface | with normal fouling |
|---|---|---|---|
| *hot side* | *cold side* | | |
| steam | aqueous solution | 300–550 | 150–275 |
| steam | light oils | 110–140 | 60–110 |
| steam | medium lube oils | 110–130 | 50–100 |
| steam | Bunker C or no. 6 oil | 70–90 | 60–80 |
| steam | air or gases | 5–10 | 4–8 |
| hot water | aqueous solution | 200–250 | 110–160 |
| | | | |
| cooling applications | | | |
| *cold side* | *hot side* | | |
| water | aqueous solution | 200–250 | 105–155 |
| water | medium lube oil | 20–30 | 10–20 |
| water | air or gases | 5–10 | 4–8 |
| freon/ammonia | aqueous solution | 60–90 | 40–60 |
| calcium or sodium brine | aqueous solution | 175–200 | 80–125 |

(Multiply Btu/hr-ft²-°F by 5.6783 to obtain W/m²·K.)

*Overall heat transfer coefficient values are strongly dependent on the type of heat exchanger, as well as on the hot- and cold-side fluids.

If the tube thermal resistance term is omitted entirely and if the tube is thin-walled, Eq. 42.34 and Eq. 42.35 can be used as approximations.

$$\frac{1}{U_o} \approx \frac{1}{h_o} + \frac{r_o}{r_i h_i} \approx \frac{1}{h_o} + \frac{1}{h_i} \qquad \textbf{42.34}$$

$$\frac{1}{U_i} \approx \frac{1}{h_i} + \frac{r_i}{r_o h_o} \approx \frac{1}{h_i} + \frac{1}{h_o} \qquad \textbf{42.35}$$

In reality, it is very difficult to predict the heat transfer coefficient for most types of commercial heat exchangers. Values can be predicted by comparison with similar units, or "tried-and-true" rules of thumb can be used. One such rule of thumb for baffled shell-and-tube heat exchangers predicts the clean, average heat transfer coefficient as 60% of the value for the same arrangement of tubes in pure crossflow.

## 18. TUBE LENGTH REQUIRED

For a simple, tube-in-tube counterflow heat exchanger as shown in Fig. 42.4, the length of tube required for a fluid to change temperature from $T_{in}$ to $T_{out}$ is given by Eq. 42.36. The normal maximum length as limited by practical assembly and maintenance requirements for straight tubes is 20 ft (6 m). Multiple-pass heat exchangers are needed for longer lengths.

$$L = \frac{\rho v d_i^2 c_p (T_{out} - T_{in})}{4 U_o d_o \Delta T_{lm}} \qquad \textbf{42.36}$$

## Example 42.1

Water flows at 10 ft/sec (3 m/s) through the inside of a 2.00 in (51 mm) inside diameter, 2.125 in (54 mm) outside diameter tube. The tube wall temperature is 170°F (75°C) along its entire length. The water enters at 70°F (20°C) and is heated to 130°F (56°C). The inside film coefficient is 1757 Btu/hr-ft²-°F (9832 W/m²·K).

All thermal resistance other than the internal film for the tube can be disregarded. What tube length is required?

*SI Solution*

The two "end" temperature differences are

$$\Delta T_A = 75°C - 20°C = 55°C$$
$$\Delta T_B = 75°C - 56°C = 19°C$$

The logarithmic temperature difference is

$$\Delta T_{lm} = \frac{\Delta T_A - \Delta T_B}{\ln \dfrac{\Delta T_A}{\Delta T_B}} = \frac{55°C - 19°C}{\ln \dfrac{55°C}{19°C}}$$
$$= 33.9°C$$

Since $U_o A_o = U_i A_i$, $U_i = h_i$, and $A = \pi dL$,

$$U_o = \frac{h_i A_i}{A_o} = h_i \left(\frac{d_i}{d_o}\right)$$
$$= \frac{\left(9832 \dfrac{W}{m^2 \cdot K}\right)(51 \text{ mm})}{54 \text{ mm}}$$
$$= 9286 \text{ W/m}^2\text{·K}$$

The tube length required is

$$L = \frac{\rho v d_i^2 c_p (T_{out} - T_{in})}{4 U_o d_o \Delta T_{lm}}$$
$$= \left(994.7 \frac{\text{kg}}{\text{m}^3}\right)\left(3 \frac{\text{m}}{\text{s}}\right)(0.051 \text{ m})^2$$
$$\times \left(\frac{\left(4.183 \dfrac{\text{kJ}}{\text{kg·K}}\right)\left(1000 \dfrac{\text{J}}{\text{kJ}}\right)(56°C - 20°C)}{(4)\left(9286 \dfrac{\text{W}}{\text{m}^2\text{·K}}\right)(0.054 \text{ m})(33.9°C)}\right)$$
$$= 17.2 \text{ m}$$

*Customary U.S. Solution*

The two "end" temperature differences are

$$\Delta T_A = 170°F - 70°F = 100°F$$
$$\Delta T_B = 170°F - 130°F = 40°F$$

**Table 42.4** *Approximate Thermal Conductivity of Common Heat Exchanger Materials (Btu-ft/hr-ft²-°F)*

temperature, °F

| | 200 | 300 | 400 | 500 | 600 | 700 | 800 | 900 | 1000 | 1100 | 1200 | 1300 | 1400 | 1500 |
|---|---|---|---|---|---|---|---|---|---|---|---|---|---|---|
| **aluminum (annealed)** | | | | | | | | | | | | | | |
| type 1100-0 | 126 | 124 | 123 | 122 | 121 | 120 | 118 | | | | | | | |
| type 3003-0 | 111 | 111 | 111 | 111 | 111 | 111 | 111 | | | | | | | |
| type 3004-0 | 97 | 98 | 99 | 100 | 102 | 103 | 104 | | | | | | | |
| type 6061-0 | 102 | 103 | 104 | 105 | 106 | 106 | 106 | | | | | | | |
| **aluminum (tempered)** | | | | | | | | | | | | | | |
| type 1100 (all tempers) | 123 | 122 | 121 | 120 | 118 | 118 | 118 | | | | | | | |
| type 3003 (all tempers) | 96 | 97 | 98 | 99 | 100 | 102 | 104 | | | | | | | |
| type 3004 (all tempers) | 97 | 98 | 99 | 100 | 102 | 103 | 104 | | | | | | | |
| type 6061 (T4 & T6) | 95 | 96 | 97 | 98 | 99 | 100 | 102 | | | | | | | |
| type 6063 (T5 & T6) | 116 | 116 | 116 | 116 | 116 | 115 | 114 | | | | | | | |
| type 6063 (T42) | 111 | 111 | 111 | 111 | 111 | 111 | 111 | | | | | | | |
| **cast iron** | 31 | 31 | 30 | 29 | 28 | 27 | 26 | 25 | | | | | | |
| **carbon steel** | 29.2 | 28.4 | 27.6 | 26.6 | 25.6 | 24.6 | 23.5 | 22.5 | 21.4 | 20.2 | 19.0 | 17.6 | 16.2 | 15.6 |
| **carbon moly steel (½% C)** | 25.2 | 25.1 | 24.8 | 24.3 | 23.7 | 23.0 | 22.2 | 21.4 | 20.4 | 19.5 | 18.4 | 16.7 | 15.3 | 15.0 |
| **chrome moly steels** | | | | | | | | | | | | | | |
| 1% Cr, ½% Mo | 21.9 | 22.0 | 21.9 | 21.7 | 21.3 | 20.8 | 20.2 | 19.7 | 19.1 | 18.5 | 17.7 | 16.5 | 15.0 | 14.8 |
| 2¼% Cr, 1% Mo | 21.3 | 21.5 | 21.5 | 21.4 | 21.1 | 20.7 | 20.2 | 19.7 | 19.1 | 18.5 | 18.0 | 17.2 | 15.6 | 15.3 |
| 5% Cr, ½% Mo | 18.1 | 18.7 | 19.1 | 19.2 | 19.2 | 19.0 | 18.7 | 18.4 | 18.0 | 17.6 | 17.1 | 16.6 | 16.0 | 15.8 |
| 12% Cr | 14 | 15 | 15 | 15 | 16 | 16 | 16 | 16 | 17 | 17 | 17 | 18 | | |
| **austenitic stainless steels** | | | | | | | | | | | | | | |
| 18% Cr, 8% Ni | 9.3 | 9.8 | 10 | 11 | 11 | 12 | 12 | 13 | 13 | 14 | 14 | 14 | 15 | 15 |
| 25% Cr, 20% Ni | 7.8 | 8.4 | 8.9 | 9.5 | 10 | 11 | 11 | 12 | 12 | 13 | 14 | 14 | 15 | 15 |
| **admiralty brass** | 70 | 75 | 79 | 84 | 89 | | | | | | | | | |
| **naval brass** | 71 | 74 | 77 | 80 | 83 | | | | | | | | | |
| **copper** | 225 | 225 | 224 | 224 | 223 | 223 | | | | | | | | |
| **copper and nickel alloys** | | | | | | | | | | | | | | |
| 90% Cu, 10% Ni | 30 | 31 | 34 | 37 | 42 | 47 | 49 | 51 | 53 | | | | | |
| 80% Cu, 20% Ni | 22 | 23 | 25 | 27 | 29 | 31 | 34 | 37 | 40 | | | | | |
| 70% Cu, 30% Ni | 18 | 19 | 21 | 23 | 25 | 27 | 30 | 33 | 37 | | | | | |
| 30% Cu, 70% Ni Alloy 400 | 15 | 15 | 16 | 16 | 17 | 18 | 18 | 19 | 20 | 20 | | | | |

(Multiply Btu-ft/hr-ft²-°F by 1.731 to obtain W/m·K.)

Reprinted with permission from *Standards of the Tubular Exchanger Manufacturers Association*, 7th ed., © 1999, by Tubular Exchanger Manufacturers Association, Inc.

Heat Transfer

The logarithmic temperature difference is

$$\Delta T_{\rm lm} = \frac{\Delta T_A - \Delta T_B}{\ln \dfrac{\Delta T_A}{\Delta T_B}} = \frac{100°F - 40°F}{\ln \dfrac{100°F}{40°F}}$$

$$= 65.5°F$$

Since $U_o A_o = U_i A_i$, $U_i = h_i$, and $A = \pi d L$,

$$U_o = \frac{h_i A_i}{A_o} = h_i \left(\frac{d_i}{d_o}\right)$$

$$= \frac{\left(1757 \; \dfrac{\text{Btu}}{\text{hr-ft}^2\text{-}°F}\right)(2.00 \text{ in})}{2.125 \text{ in}}$$

$$= 1654 \text{ Btu/hr-ft}^2\text{-}°F$$

The tube length required is

$$L = \frac{\rho v d_i^2 c_p (T_{\rm out} - T_{\rm in})}{4 U_o d_o \Delta T_{\rm lm}}$$

$$= \left(62.0 \; \frac{\text{lbm}}{\text{ft}^3}\right)\left(10 \; \frac{\text{ft}}{\text{sec}}\right)\left(3600 \; \frac{\text{sec}}{\text{hr}}\right)\left(\frac{2.00 \text{ in}}{12 \; \dfrac{\text{in}}{\text{ft}}}\right)^2$$

$$\times \left(\frac{\left(0.998 \; \dfrac{\text{Btu}}{\text{lbm-}°F}\right)(130°F - 70°F)}{(4)\left(1654 \; \dfrac{\text{Btu}}{\text{hr-ft}^{2°}\text{-F}}\right)\left(\dfrac{2.125 \text{ in}}{12 \; \dfrac{\text{in}}{\text{ft}}}\right)(65.5°F)}\right)$$

$$= 48.4 \text{ ft}$$

**Heat Transfer**

# 43 Heat Transfer by Radiation

## Nomenclature

| | | | |
|---|---|---|---|
| $A$ | area | ft$^2$ | m$^2$ |
| $E$ | emissive power | Btu/hr-ft$^2$ | W/m$^2$ |
| $F$ | factor | – | – |
| $\Im$ | gray body shape factor | – | – |
| $G$ | geometric flux | ft$^2$ | m$^2$ |
| $h$ | film coefficient | Btu/hr-ft$^2$-°F | W/m$^2$·K |
| $q$ | unit heat transfer | Btu/hr-ft$^2$ | W/m$^2$ |
| $Q$ | heat transfer rate | Btu/hr | W |
| $T$ | temperature | °R | K |

## Symbols

| | | | |
|---|---|---|---|
| $\alpha$ | absorptivity | – | – |
| $\epsilon$ | emissivity | – | – |
| $\rho$ | reflectivity | – | – |
| $\sigma$ | Stefan-Boltzmann constant | Btu/hr-ft$^2$-°R$^4$ | W/m$^2$·K$^4$ |
| $\tau$ | transmissivity | – | – |

## Subscripts

| | |
|---|---|
| 12 | from body 1 to body 2 |
| 21 | from body 2 to body 1 |
| $a$ | arrangement |
| $e$ | emissivity |
| $i$ | inner |
| $o$ | outer |

## 1. THERMAL RADIATION

Thermal radiation is electromagnetic radiation with wavelengths in the 0.1 to 100 $\mu$m range. All bodies, even "cold" ones, radiate thermal radiation.

Thermal radiation incident to a body can be absorbed, reflected, or transmitted. The *radiation conservation law* is[1]

$$\alpha + \rho + \tau = 1 \qquad 43.1$$

## 2. BLACK, REAL, AND GRAY BODIES

The rate of thermal radiation emitted per unit area of a body is the *emissive power*, $E$.

$$E = \frac{Q_{\text{radiation}}}{A} \qquad 43.2$$

Since absorptivity, $\alpha$, cannot exceed 1.0, Kirchhoff's law places an upper limit on emissive power. Bodies that radiate at this upper limit (i.e., $\alpha = 1$) are known as *black bodies* or *ideal radiators*. A black body emits the maximum possible radiation for its temperature and absorbs all incident energy.[2]

*Real bodies* do not radiate at the ideal level. The ratio of actual to ideal emissive powers is the *emissivity*, $\epsilon$.

$$\epsilon = \frac{E_{\text{actual}}}{E_{\text{black}}} \qquad 43.3$$

Emissivity generally has the following characteristics.

- Emissivity varies widely with the surface condition of a material.

- Emissivity is low with highly polished metals.

- Emissivity is high with most nonmetals.

- Emissivity increases with increases in temperature.

The emissivity (and hence the emissive power) usually depends on the temperature of the body.[3] A body that emits at constant emissivity, regardless of wavelength, is known as a *gray body*.

*Kirchhoff's radiation law* states that for a body, the emissivity, $\epsilon$, and absorptivity, $\alpha$, are equal. At a given temperature, the ratios of emissive power to absorptivity

---

[1]Notice that emissivity, $\epsilon$, does not appear in the conservation law.
[2]Black body performance can be approximated but not achieved in practice.
[3]This is equivalent to saying the emissivity depends on the wavelength of the radiation.

**Heat Transfer**

for all bodies are equal. (Bodies at the same temperature are said to be in *thermal equilibrium*.)

$$\frac{\epsilon_1}{\alpha_1} = \frac{\epsilon_2}{\alpha_2} = \epsilon_{\text{black}}\Big|_T \qquad 43.4$$

For a black body, both emissivity and absorptivity are 1.0. However, emissivity also equals absorptivity for any body in thermal equilibrium.[4,5]

$$\epsilon = \alpha \text{ [thermal equilibrium]} \qquad 43.5$$

For a gray body, the reflectivity is constant and

$$\rho + \epsilon = 1 \qquad 43.6$$

## 3. RADIATION FROM A BODY

The *Stefan-Boltzmann law*, also known as the *fourth-power law*, gives the total emissive power, $E$, from a black body. The temperature, $T$, is expressed in degrees Rankine or in Kelvins. $\sigma$ is the *Stefan-Boltzmann constant*.

$$E_{\text{black}} = \sigma T^4 \qquad 43.7$$

$$\sigma = 0.1713 \times 10^{-8} \text{ Btu/hr-ft}^2\text{-}^\circ\text{R}^4 \qquad 43.8$$

$$\sigma = 5.67 \times 10^{-8} \text{ W/m}^2\text{·K}^4 \qquad 43.9$$

The radiation from a gray body follows directly from the definition of emissivity.

$$E_{\text{gray}} = \epsilon E_{\text{black}} = \epsilon\sigma T^4 \qquad 43.10$$

## 4. BLACK BODY SHAPE FACTOR

For two black bodies radiating to each other, a *shape factor*, $F_{12}$, accounts for the spatial arrangement of the two bodies. (Generally, the smaller body is designated as body 1.) The shape factor is the fraction of the total radiation leaving body 1 that will travel directly to body 2. The shape factor for two black bodies, therefore, is often referred to as the *arrangement factor, geometric factor, geometrical factor, geometric shape factor, angle factor, view factor, interaction factor,* and *configuration factor,* as well as the *black body shape factor.*

The shape factor is 1.0 for two infinite, parallel planes, two infinite coaxial cylinders, and two concentric spheres, since all emitted radiation is absorbed. It is more difficult to evaluate the black body shape factor for more complex arrangements of bodies and surfaces. However, many of the simpler cases have been solved, and their graphical solutions are available. (See Fig. 43.1 and Fig. 43.2.)

[4]"Steady-state operation" would be a better term here since the term "equilibrium" implies temperature equality with another body. In fact, the phrase "the bodies are in thermal equilibrium" means that the body temperatures are equal. In this case, the term "equilibrium" means that the body's temperature is constant.
[5]Equation 43.5 follows directly from Kirchhoff's law (see Eq. 43.4).

*Figure 43.1* Black Body Shape Factor Adjacent Perpendicular Rectangles

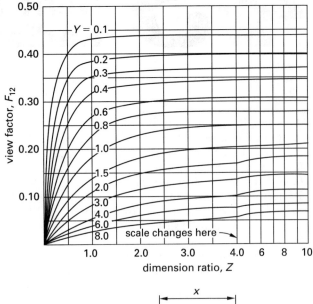

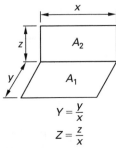

$$Y = \frac{y}{x}$$

$$Z = \frac{z}{x}$$

View factors for perpendicular rectangles with a common side. From H. C. Hottel, "Radiant Heat Transmission," *Mechanical Engineering* magazine, Volume 52 (1930). By permission of The American Society of Mechanical Engineers.

*Figure 43.2* Black Body Shape Factor (for directly opposed, parallel, finite surfaces)

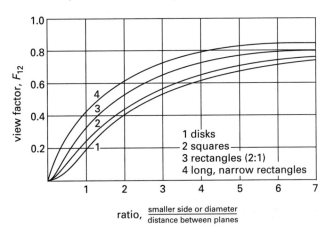

View factors for parallel squares, rectangles, and disks. From H. C. Hottel, "Radiant Heat Transmission," *Mechanical Engineering* magazine, Volume 52 (1930). By permission of The American Society of Mechanical Engineers.

**Table 43.1** *Arrangement and Emissivity Factors*

| arrangement | area | $F_a$ | $F_e$ |
|---|---|---|---|
| infinite parallel planes | $A_1$ or $A_2$ | 1 | $\dfrac{1}{\dfrac{1}{\epsilon_1}+\dfrac{1}{\epsilon_2}-1}$ |
| completely enclosed body; small compared with enclosure[a] | $A_1$ | 1 | $\epsilon_1$ |
| completely enclosed body; large compared with enclosure[a] | $A_1$ | 1 | $\dfrac{1}{\dfrac{1}{\epsilon_1}+\dfrac{1}{\epsilon_2}-1}$ |
| concentric spheres or infinite cylinders with diffuse radiation[a] | $A_1$ | 1 | $\dfrac{1}{\dfrac{1}{\epsilon_1}+\left(\dfrac{A_1}{A_2}\right)\left(\dfrac{1}{\epsilon_2}-1\right)}$ |
| concentric spheres or infinite cylinders with specular (mirror-like) radiation[a] | $A_1$ | 1 | $\dfrac{1}{\dfrac{1}{\epsilon_1}+\dfrac{1}{\epsilon_2}-1}$ |
| two perpendicular rectangles with a common edge | $A_1$ or $A_2$ | (See Fig. 43.1.)[b] | $\epsilon_1\epsilon_2$ |
| directly opposed, parallel disks, squares, or rectangles of equal size | $A_1$ or $A_2$ | (See Fig. 43.2.)[b] | $\epsilon_1\epsilon_2$ |
| directly opposed, parallel disks, squares, or rectangles of equal size, connected by nonconducting, reradiating walls | $A_1$ or $A_2$ | (See Fig. 43.3.)[b] | $\epsilon_1\epsilon_2$ |

[a]Object 1 is the smaller, enclosed body.
[b]Arrangement factors for these configurations are presented graphically in most heat transfer books.

# 5. GRAY BODY SHAPE FACTOR

Real bodies deviate from black body behavior. To account for the effect of less than ideal emissivities, the black body shape factor, $F_{12}$, is replaced by the gray body shape factor, $\Im_{12}$. The *gray body shape factor* accounts for the spatial arrangements of the bodies and their emissivities. The lower limit for the gray body shape factor is $\epsilon_1\epsilon_2$; the upper limit is 1.0.

The gray body shape factor can be written as the product of the black body shape factor (i.e., now referred to as the *arrangement factor*, $F_a$) and the *emissivity factor*, $F_e$. The emissivity factor accounts for the departure of the surface from black-body conditions (see Table 43.1).

$$\Im_{12} = F_{12}F_e = F_aF_e \qquad 43.11$$

For two gray bodies that radiate to each other (and to no others), the gray body shape factor can be calculated from the black body shape factor.[6] $A_1$ in Eq. 43.12 is the smaller area.

$$A_1\Im_{12} = \frac{1}{\dfrac{1-\epsilon_1}{\epsilon_1 A_1}+\dfrac{1}{A_1 F_{12}}+\dfrac{1-\epsilon_2}{\epsilon_2 A_2}} \qquad 43.12$$

A special case is two gray bodies with uniform thermal radiation and $F_{12}=1$. Examples of this case include infinite parallel gray plates, infinite length concentric gray cylinders, and concentric gray spheres. In such cases, Eq. 43.13 can be used. $A_1$ is the smaller area.

$$A_1\Im_{12} = \frac{1}{\dfrac{1-\epsilon_1}{\epsilon_1 A_1}+\dfrac{1}{A_1}+\dfrac{1-\epsilon_2}{\epsilon_2 A_2}} \qquad 43.13$$

With two infinite parallel plates, $A_1 = A_2$. Then, the gray body shape factor is

$$\Im_{12} = \frac{1}{\dfrac{1}{\epsilon_1}+\dfrac{1}{\epsilon_2}-1} \qquad 43.14$$

For a small gray body enclosed by a black body,

$$\Im_{12} = \epsilon_1 \qquad 43.15$$

# 6. NET RADIATION HEAT TRANSFER

The net heat transfer due to radiation between two gray bodies at different temperatures is given by Eq. 43.16. The area of body 1 must be used with $\Im_{12}$, and the area of body 2 must be used with $\Im_{21}$. Whether $\Im_{12}$ or $\Im_{21}$ is used depends on which is easier to evaluate.

$$E_{\text{net},12} = \sigma\Im_{12}\left(T_1^4 - T_2^4\right)$$
$$= \sigma F_a F_e\left(T_1^4 - T_2^4\right) \qquad 43.16$$

$$Q_{\text{net},12} = A_1 E_{\text{net},12} \qquad 43.17$$

[6]Although Eq. 43.12 is limited to two bodies that exchange heat with each other and with no other bodies, not all of each body's radiation has to reach the other body. This is evident in the presence of the black body shape factor, $F_{12}$.

Heat Transfer

## 7. RECIPROCITY THEOREM

Equation 43.19 is known as the *reciprocity theorem for radiation*.

$$Q_{\text{net}} = \sigma A_1 \Im_{12}\left(T_1^4 - T_2^4\right)$$
$$= \sigma A_2 \Im_{21}\left(T_1^4 - T_2^4\right) \qquad 43.18$$
$$A_1 \Im_{12} = A_2 \Im_{21} \qquad 43.19$$

The product of the area and the shape factor is known as the *geometric flux, G*.

$$G_{12} = A_1 \Im_{12} \qquad 43.20$$

The reciprocity theorem can be written in terms of the geometric flux.

$$G_{12} = G_{21} \qquad 43.21$$

## 8. RADIATION WITH REFLECTION/ RERADIATION

Surfaces that reradiate absorbed thermal radiation are known as *refractory materials* or *refractories*. (Furnace walls that reradiate almost all of the thermal energy they receive from combustion flames back to boiler tubes are examples of refractories.) The shape factor for cases with reradiation is traditionally given the symbol $\overline{F}_{12}$. Equation 43.22 (similar to Eq. 43.12) is used to calculate the gray body shape factor when there are two communicating refractory bodies (see Fig. 43.3).

$$A_1 \Im_{12} = \cfrac{1}{\cfrac{1-\epsilon_1}{\epsilon_1 A_1} + \cfrac{1}{A_1 \overline{F}_{12}} + \cfrac{1-\epsilon_2}{\epsilon_2 A_2}} \qquad 43.22$$

## 9. COMBINED HEAT TRANSFER

When heat is transferred by both radiation and convection, it is convenient to define the *radiant heat transfer coefficient*, $h_{\text{radiation}}$. $T_\infty$ is the free-stream (far-field) temperature for convective heat transfer.[7] $T_1$ and $T_2$ should both be expressed as absolute temperatures.

$$h_{\text{radiation}} = \frac{Q_{\text{net}}}{A_1(T_1 - T_\infty)} = \frac{E_{\text{net}}}{T_1 - T_\infty}$$
$$= \frac{\sigma F_a F_e\left(T_1^4 - T_2^4\right)}{T_1 - T_\infty} \qquad 43.23$$

[7]$T_\infty$ can be the same as either $T_1$ or $T_2$. In that case, the solutions to common types of problems will be greatly simplified since the heat transfer can be calculated separately. If the temperatures are different, trial and error will be necessary to determine the surface temperature.

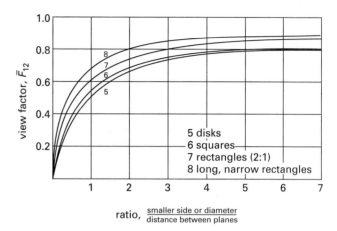

**Figure 43.3** *Black Body Shape Factor (for parallel finite squares, rectangles, and disks connected by nonconducting, reradiating wall)*

The *combined heat transfer coefficient* is[8]

$$h_{\text{total}} = h_{\text{radiation}} + h_{\text{convective}} \qquad 43.24$$

The combined radiation and convective heat transfer is

$$Q = Q_{\text{radiation}} + Q_{\text{convection}}$$
$$= h_{\text{total}} A_1 (T_1 - T_\infty) \qquad 43.25$$

### Example 43.1

A 1.0 ft (30 cm) diameter uninsulated horizontal duct carries hot air through a basement. The duct surface temperature is 200°F (95°C); the duct has an emissivity of 0.8. Air surrounding the duct in the basement is at 40°F (5°C); the basement walls are at 0°F (−20°C). The convective film coefficient on the exterior of the duct is 0.96 Btu/hr-ft²-°F (5.5 W/m²·K). What is the heat loss per unit area of duct?

*SI Solution*

Since the convective film coefficient for the duct is known, the heat losses from convection and radiation can be calculated independently.

The absolute temperatures are

$$T_{\text{duct}} = 95°C + 273° = 368\text{K}$$
$$T_\infty = 5°C + 273° = 278\text{K}$$
$$T_{\text{walls}} = -20°C + 273° = 253\text{K}$$

[8]It is understood that radiation and convection are the combination of heat transfer mechanisms. Conductive heat transfer does not use the film coefficient concept.

The convective heat loss is

$$q_{\text{convection}} = \frac{Q}{A} = h(T_{\text{duct}} - T_\infty)$$

$$= \left(5.5 \ \frac{\text{W}}{\text{m}^2 \cdot \text{K}}\right)(368\text{K} - 278\text{K})$$

$$= 495 \ \text{W/m}^2$$

Since the duct is entirely enclosed by the basement, the arrangement factor is $F_a = 1.0$. The emissivity factor is $F_e = \epsilon_{\text{duct}}$. The radiation heat transfer is

$$E_{\text{net}} = \sigma F_a F_e \left(T_{\text{duct}}^4 - T_{\text{walls}}^4\right)$$

$$= \left(5.67 \times 10^{-8} \ \frac{\text{W}}{\text{m}^2 \cdot \text{K}^4}\right)(0.8)(1.0)$$

$$\times \left((368\text{K})^4 - (253\text{K})^4\right)$$

$$= 646 \ \text{W/m}^2$$

The total heat loss is

$$q_{\text{total}} = q_{\text{convection}} + E_{\text{net}}$$

$$= 495 \ \frac{\text{W}}{\text{m}^2} + 646 \ \frac{\text{W}}{\text{m}^2}$$

$$= 1141 \ \text{W/m}^2$$

*Customary U.S. Solution*

Since the convective film coefficient for the duct is known, the heat losses from convection and radiation can be calculated independently.

The absolute temperatures are

$$T_{\text{duct}} = 200°\text{F} + 460° = 660°\text{R}$$

$$T_\infty = 40°\text{F} + 460° = 500°\text{R}$$

$$T_{\text{walls}} = 0°\text{F} + 460° = 460°\text{R}$$

The convective heat loss is

$$q_{\text{convection}} = \frac{Q}{A} = h(T_{\text{duct}} - T_\infty)$$

$$= \left(0.96 \ \frac{\text{Btu}}{\text{hr-ft}^2\text{-}°\text{F}}\right)(660°\text{R} - 500°\text{R})$$

$$= 153.6 \ \text{Btu/hr-ft}^2$$

Since the duct is entirely enclosed by the basement, the arrangement factor is $F_a = 1.0$. The emissivity factor is $F_e = \epsilon_{\text{duct}}$. The radiation heat transfer is

$$E_{\text{net}} = \sigma F_a F_e \left(T_{\text{duct}}^4 - T_{\text{walls}}^4\right)$$

$$= \left(0.1713 \times 10^{-8} \ \frac{\text{Btu}}{\text{hr-ft}^2\text{-}°\text{R}^4}\right)(0.8)(1.0)$$

$$\times \left((660°\text{R})^4 - (460°\text{R})^4\right)$$

$$= 198.7 \ \text{Btu/hr-ft}^2$$

The total heat loss is

$$q_{\text{total}} = q_{\text{convection}} + E_{\text{net}}$$

$$= 153.6 \ \frac{\text{Btu}}{\text{hr-ft}^2} + 198.7 \ \frac{\text{Btu}}{\text{hr-ft}^2}$$

$$= 352.3 \ \text{Btu/hr-ft}^2$$

### Example 43.2

The air temperature in the duct is increased, so that the duct in Ex. 43.1 loses heat at the rate of 500 Btu/ hr-ft$^2$ (1600 W/m$^2$). The duct temperature and film coefficient are unknown. All other values are the same. What is the duct surface temperature?

*SI Solution*

The unknown surface temperature cannot be extracted directly from the combined heat transfer equation. It is more convenient to solve this problem by trial and error.

Assume laminar convective heat transfer and a duct temperature of 393K (120°C). The convective film coefficient on the outside of the duct is approximately

$$h_{\text{convective}} = 1.32 \left(\frac{T_{\text{duct}} - T_\infty}{L}\right)^{0.25}$$

$$= (1.32)\left(\frac{393\text{K} - 278\text{K}}{0.3 \ \text{m}}\right)^{0.25}$$

$$= 5.84 \ \text{W/m}^2 \cdot \text{K}$$

The convective heat loss is

$$q_{\text{convection}} = \frac{Q}{A} = h(T_{\text{duct}} - T_\infty)$$

$$= \left(5.84 \ \frac{\text{W}}{\text{m}^2 \cdot \text{K}}\right)(393\text{K} - 278\text{K})$$

$$= 672 \ \text{W/m}^2$$

The radiation loss is

$$E_{\text{net}} = \sigma F_a F_e \left(T_{\text{duct}}^4 - T_{\text{walls}}^4\right)$$

$$= \left(5.67 \times 10^{-8} \ \frac{\text{W}}{\text{m}^2 \cdot \text{K}^4}\right)(0.8)(1.0)$$

$$\times \left((393\text{K})^4 - (253\text{K})^4\right)$$

$$= 896 \ \text{W/m}^2$$

The total heat loss for this iteration is

$$q_{\text{total}} = q_{\text{convection}} + E_{\text{net}}$$

$$= 672 \ \frac{\text{W}}{\text{m}^2} + 896 \ \frac{\text{W}}{\text{m}^2}$$

$$= 1568 \ \text{W/m}^2$$

**Heat Transfer**

Since the calculated heat loss agrees with the known heat loss, the assumed surface temperature is correct. (Usually, several trial and error iterations would be required to converge on the solution.)

*Customary U.S. Solution*

The unknown surface temperature cannot be extracted directly from the combined heat transfer equation. It is more convenient to solve this problem by trial and error.

Assume laminar convective heat transfer and a duct temperature of 710°R (250°F). The convective film coefficient on the outside of the duct is approximately

$$h_{\text{convective}} = 0.27\left(\frac{T_{\text{duct}} - T_\infty}{L}\right)^{0.25}$$

$$= (0.27)\left(\frac{710°\text{R} - 500°\text{R}}{1 \text{ ft}}\right)^{0.25}$$

$$= 1.03 \text{ Btu/hr-ft}^2\text{-}°\text{F}$$

The convective heat loss is

$$q_{\text{convection}} = \frac{Q}{A} = h(T_{\text{duct}} - T_\infty)$$

$$= \left(1.03 \frac{\text{Btu}}{\text{hr-ft}^2\text{-}°\text{F}}\right)(710°\text{R} - 500°\text{R})$$

$$= 216.3 \text{ Btu/hr-ft}^2$$

The radiation loss is

$$E_{\text{net}} = \sigma F_a F_e\left(T_{\text{duct}}^4 - T_{\text{walls}}^4\right)$$

$$= \left(0.1713 \times 10^{-8} \frac{\text{Btu}}{\text{hr-ft}^2\text{-}°\text{R}^4}\right)(0.8)(1.0)$$

$$\times \left((710°\text{R})^4 - (460°\text{R})^4\right)$$

$$= 286.9 \text{ Btu/hr-ft}^2$$

The total heat loss for this iteration is

$$q_{\text{total}} = q_{\text{convection}} + E_{\text{net}}$$

$$= 216.3 \frac{\text{Btu}}{\text{hr-ft}^2} + 286.9 \frac{\text{Btu}}{\text{hr-ft}^2}$$

$$= 503.2 \text{ Btu/hr-ft}^2$$

Since the calculated heat loss agrees with the known heat loss, the assumed surface temperature is correct. (Usually, several trial-and-error iterations would be required to converge on the solution.)

## 10. EQUILIBRIUM CONDITION WITH COMBINED HEAT TRANSFER

A single body that remains at a constant temperature is said to be in an equilibrium condition.[9] To be in equilibrium, the body must continually lose all of the energy gained. Depending on the situation, a body might

[9]See Ftn. 4.

radiate all of the energy gained by convection, or it may lose by convection all of the energy gained by radiation.

### Example 43.3

A small temperature probe with an emissivity of 0.8 measures the temperature of a gas flowing in a large pipe as 850°F (450°C). The pipe walls are 350°F (180°C). The convective film coefficient for the probe in the gas flow is 27 Btu/hr-ft²-°F (150 W/m²·K). The probe's surface temperature is constant. There is no heat transfer to the probe by conduction. What is the actual gas temperature?

*SI Solution*

The absolute temperatures are

$$T_{\text{probe}} = 450°\text{C} + 273° = 723\text{K}$$

$$T_{\text{pipe}} = 180°\text{C} + 273° = 453\text{K}$$

Since the probe is entirely enclosed by the large pipe, the arrangement factor is $F_a = 1$, and the emissivity factor is $F_e = 0.8$.

Since the probe's surface temperature is constant, the heat gained by the probe through convection from the gas stream is being lost through radiation to the cooler pipe walls.

$$q_{\text{gain,convection}} = E_{\text{loss}}$$

$$h(T_{\text{gas}} - T_{\text{probe}}) = \sigma F_a F_e\left(T_{\text{probe}}^4 - T_{\text{pipe}}^4\right)$$

$$\left(150 \frac{\text{W}}{\text{m}^2\text{·K}}\right)(T_{\text{gas}} - 723\text{K}) = \left(5.67 \times 10^{-8} \frac{\text{W}}{\text{m}^2\text{·K}}\right)$$

$$\times (0.8)(1.0)$$

$$\times \left((723\text{K})^4 - (453\text{K})^4\right)$$

Solving directly,

$$T_{\text{gas}} = 793\text{K}$$

*Customary U.S. Solution*

The absolute temperatures are

$$T_{\text{probe}} = 850°\text{F} + 460° = 1310°\text{R}$$

$$T_{\text{pipe}} = 350°\text{F} + 460° = 810°\text{R}$$

Since the probe is entirely enclosed by the large pipe, the arrangement factor is $F_a = 1$, and the emissivity factor is $F_e = 0.8$.

Since the probe's surface temperature is constant, the heat gained by the probe through convection from the gas stream is being lost through radiation to the cooler pipe walls.

$$q_{\text{gain,convection}} = E_{\text{loss}}$$

$$h(T_{\text{gas}} - T_{\text{probe}}) = \sigma F_a F \left( T_{\text{probe}}^4 - T_{\text{pipe}}^4 \right)$$

$$\left( 27\ \frac{\text{Btu}}{\text{hr-ft}^2\text{-}^\circ\text{F}} \right)(T_{\text{gas}} - 1310^\circ\text{R})$$
$$= \left( 0.1713 \times 10^{-8}\ \frac{\text{Btu}}{\text{hr-ft-}^\circ\text{R}^4} \right)(8.0)(1.0)$$
$$\times \left( (1310^\circ\text{R})^4 - (810^\circ\text{R})^4 \right)$$

Solving directly,

$$T_{\text{gas}} = 1438^\circ\text{R}$$

## 11. SOLAR RADIATION

The average solar energy hitting the outer edge of the earth's atmosphere is approximately 442 Btu/hr-ft$^2$ (1.41 kW/m$^2$) and is known as the *solar constant*. The actual instantaneous value reaching the surface depends on the altitude, latitude, time of year, time of day, sky conditions, and orientation angle of the receiving body.[10,11]

## 12. NOCTURNAL RADIATION

Measurements of the night sky show that the effective temperature of the sky for purposes of radiation is approximately 410°R (210K) on cold clear nights. Since this temperature is below the freezing point of water, when the air is calm it is possible for standing water and tree fruits to freeze even when the air temperature is above the freezing point of water.

### Example 43.4

On a cold, clear night, the surface of a pond has a convective film coefficient of 5 Btu/hr-ft$^2$-°R (28 W/m$^2$·K). The emissivity of the pond surface is 0.96. The air temperature is 60°F (15°C). The pond temperature remains the same all night long. Evaporative losses are negligible. What is the water temperature?

*SI Solution*

With an effective sky temperature of 210K, the pond will lose heat through radiation. Since the pond's temperature is constant, it gains heat from the air through

convection. Therefore, the pond temperature is less than 15°C.

The absolute temperatures, when the air is calm, are

$$T_{\text{sky}} = 210\text{K}$$
$$T_{\text{air}} = 15^\circ\text{C} + 273^\circ = 288\text{K}$$

The equilibrium equation is

$$q_{\text{gain,convection}} = E_{\text{loss}}$$

$$h(T_{\text{air}} - T_{\text{pond}}) = \sigma F_a F_e \left( T_{\text{pond}}^4 - T_{\text{sky}}^4 \right)$$
$$\times (1.0)\left( T_{\text{pond}}^4 - (210^\circ\text{K})^4 \right)$$

$$\left( 28\ \frac{\text{W}}{\text{m}^2\cdot\text{K}} \right)(288\text{K} - T_{\text{pond}}) = \left( 5.67 \times 10^{-8}\ \frac{\text{W}}{\text{m}^2\cdot\text{K}^4} \right)$$
$$\times (0.96)(1.0)$$
$$\times \left( T_{\text{pond}}^4 - (210\text{K})^4 \right)$$

Solving by trial and error,

$$T_{\text{pond}} \approx 280\text{K}\ (7^\circ\text{C})$$

*Customary U.S. Solution*

With an effective sky temperature of 410°R, the pond will lose heat through radiation. Since the pond's temperature is constant, it gains heat from the air through convection. Therefore, the pond temperature is less than 60°F.

The absolute temperatures when the air is calm are

$$T_{\text{sky}} = 410^\circ\text{R}$$
$$T_{\text{air}} = 60^\circ\text{F} + 460^\circ = 520^\circ\text{R}$$

The equilibrium equation is

$$q_{\text{gain,convection}} = E_{\text{loss}}$$

$$h(T_{\text{air}} - T_{\text{pond}}) = \sigma F_a F_e \left( T_{\text{pond}}^4 - T_{\text{sky}}^4 \right)$$
$$\times \left( T_{\text{pond}}^4 - (410^\circ\text{R})^4 \right)$$

$$\left( 5\ \frac{\text{Btu}}{\text{hr-ft}^2\text{-}^\circ\text{R}} \right)(520^\circ\text{R} - T_{\text{pond}})$$
$$= \left( 0.1713 \times 10^{-8}\ \frac{\text{Btu}}{\text{hr-ft}^2\text{-}^\circ\text{R}} \right)(0.96)(1.0)$$
$$\times \left( T_{\text{pond}}^4 - (410^\circ\text{R})^4 \right)$$

Solving by trial and error,

$$T_{\text{pond}} \approx 508^\circ\text{R}\ (48^\circ\text{F})$$

---

[10]The value of the solar constant given can be used with modification at altitudes higher than approximately 50,000 ft (15 000 m).
[11]Most heat transfer textbooks detail the method of calculating the exact instantaneous solar heat gain. Books on the subject of HVAC cover this subject particularly well.

**Heat Transfer**

# Topic VIII: Statics

Statics

# 44

# Determinate Statics

## Nomenclature

| | | | |
|---|---|---|---|
| $a$ | distance to lowest cable point | ft | m |
| $c$ | parameter of the catenary | ft | m |
| $d$ | distance or diameter | ft | m |
| $D$ | diameter | ft | m |
| $F$ | force | lbf | N |
| $H$ | horizontal cable force | lbf | N |
| $L$ | length | ft | m |
| $M$ | moment | ft-lbf | N·m |
| $n$ | number of sheaves | – | – |
| $r$ | position vector or radius | ft | m |
| $R$ | reaction force | lbf | N |
| $s$ | distance along cable | ft | m |
| $S$ | sag | ft | m |
| $T$ | tension | lbf | N |
| $w$ | load per unit length | lbf/ft | N/m |
| $W$ | weight | lbf | n.a. |
| $x$ | horizontal distance or position | ft | m |
| $y$ | vertical distance or position | ft | m |
| $z$ | distance or position along $z$-axis | ft | m |

## Symbols

| | | | |
|---|---|---|---|
| $\epsilon$ | pulley loss factor | – | – |
| $\eta$ | pulley efficiency | – | – |
| $\theta$ | angle | deg | deg |

## Subscripts

| | |
|---|---|
| O | origin |
| P | point P |
| R | resultant |

## 1. INTRODUCTION TO STATICS

*Statics* is a part of the subject known as *engineering mechanics*.[1] It is the study of rigid bodies that are stationary. To be stationary, a rigid body must be in static equilibrium. In the language of statics, a stationary rigid body has no *unbalanced forces* or moments acting on it.

## 2. INTERNAL AND EXTERNAL FORCES

An *external force* is a force on a rigid body caused by other bodies. The applied force can be due to physical contact (i.e., pushing) or close proximity (e.g., gravitational, magnetic, or electrostatic forces). If unbalanced, an external force will cause motion of the body.

An *internal force* is one that holds parts of the rigid body together. Internal forces are the tensile and

---

[1] Engineering mechanics also includes the subject of dynamics. Interestingly, the subject of mechanics of materials (i.e., strength of materials) is not part of engineering mechanics.

compressive forces within parts of the body as found from the product of stress and area. Although internal forces can cause deformation of a body, motion is never caused by internal forces.

## 3. UNIT VECTORS

A *unit vector* is a vector of unit length directed along a coordinate axis.[2] In the rectangular coordinate system, there are three unit vectors, **i**, **j**, and **k**, corresponding to the three coordinate axes, $x$, $y$, and $z$, respectively. (There are other methods of representing vectors, in addition to bold letters. For example, the unit vector **i** is represented as $\bar{i}$ or $\hat{i}$ in other sources.) Unit vectors are used in vector equations to indicate direction without affecting magnitude. For example, the vector representation of a 97 N force in the negative $x$-direction would be written as $\mathbf{F} = -97\mathbf{i}$.

## 4. CONCENTRATED FORCES

A *force* is a push or pull that one body exerts on another. A *concentrated force* also known as a *point force*, is a vector having magnitude, direction, and location (i.e., point of application) in three-dimensional space. (See Fig. 44.1.) In this chapter, the symbols **F** and $F$ will be used to represent the vector and its magnitude, respectively. (As with the unit vectors, the symbols **F**, $\overline{F}$, and $\hat{F}$ are used in other sources to represent the same vector.)

The vector representation of a three-dimensional force is given by Eq. 44.1. Of course, vector addition is required.

$$\mathbf{F} = F_x\mathbf{i} + F_y\mathbf{j} + F_z\mathbf{k} \qquad 44.1$$

**Figure 44.1** *Components and Direction Angles of a Force*

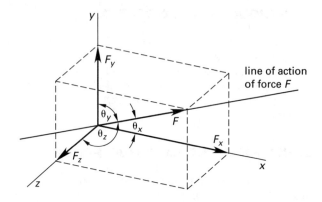

If **u** is a *unit vector* in the direction of the force, the force can be represented as

$$\mathbf{F} = F\mathbf{u} \qquad 44.2$$

---

[2]Although polar, cylindrical, and spherical coordinate systems can have unit vectors also, this chapter is concerned only with the rectangular coordinate system.

The components of the force can be found from the *direction cosines*, the cosines of the true angles made by the force vector with the $x$-, $y$-, and $z$-axes.

$$F_x = F\cos\theta_x \qquad 44.3$$

$$F_y = F\cos\theta_y \qquad 44.4$$

$$F_z = F\cos\theta_z \qquad 44.5$$

$$F = \sqrt{F_x^2 + F_y^2 + F_z^2} \qquad 44.6$$

The *line of action* of a force is the line in the direction of the force extended forward and backward. The force, **F**, and its unit vector, **u**, are along the line of action.

## 5. MOMENTS

*Moment* is the name given to the tendency of a force to rotate, turn, or twist a rigid body about an actual or assumed pivot point. (Another name for moment is *torque*, although torque is used mainly with shafts and other power-transmitting machines.) When acted upon by a moment, unrestrained bodies rotate. However, rotation is not required for the moment to exist. When a restrained body is acted upon by a moment, there is no rotation.

An object experiences a moment whenever a force is applied to it.[3] Only when the line of action of the force passes through the center of rotation (i.e., the actual or assumed pivot point) will the moment be zero.

Moments have primary dimensions of length × force. Typical units are foot-pounds, inch-pounds, and newton-meters.[4]

## 6. MOMENT OF A FORCE ABOUT A POINT

Moments are vectors. The moment vector, $\mathbf{M}_O$, for a force about point O is the *cross product* of the force, **F**, and the vector from point O to the point of application of the force, known as the *position vector*, **r**. The scalar product $|\mathbf{r}|\sin\phi$ is known as the *moment arm, d*.

$$\mathbf{M}_O = \mathbf{r} \times \mathbf{F} \qquad 44.7$$

$$M_O = |\mathbf{M}_O| = |\mathbf{r}||\mathbf{F}|\sin\theta = d|\mathbf{F}| \quad [\theta \le 180°] \qquad 44.8$$

The line of action of the moment vector is normal to the plane containing the force vector and the position vector. The sense (i.e., the direction) of the moment is determined from the *right-hand rule*.

---

[3]The moment may be zero, as when the moment arm length is zero, but there is a (trivial) moment nevertheless.

[4]Units of kilogram-force-meter have also been used in metricated countries. Foot-pounds and newton-meters are also the units of energy. To distinguish between moment and energy, some authors reverse the order of the units. Therefore, pound-feet and meter-newtons become the units of moment. This convention is not universal and is unnecessary since the context is adequate to distinguish between the two.

*Right-hand rule:* Place the position and force vectors tail to tail. Close your right hand and position it over the pivot point. Rotate the position vector into the force vector, and position your hand such that your fingers curl in the same direction as the position vector rotates. Your extended thumb will coincide with the direction of the moment.[5] (See Fig. 44.2.)

**Figure 44.2** *Right-Hand Rule*

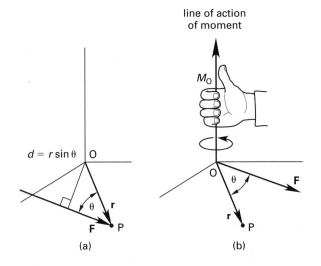

### 7. VARIGNON'S THEOREM

*Varignon's theorem* is a statement of how the total moment is derived from a number of forces acting simultaneously at a point.

*Varignon's theorem:* The sum of individual moments about a point caused by multiple concurrent forces is equal to the moment of the resultant force about the same point.

$$(\mathbf{r} \times \mathbf{F}_1) + (\mathbf{r} \times \mathbf{F}_2) + \cdots = \mathbf{r} \times (\mathbf{F}_1 + \mathbf{F}_2 + \cdots) \qquad 44.9$$

### 8. MOMENT OF A FORCE ABOUT A LINE

Most rotating machines (motors, pumps, flywheels, etc.) have a fixed rotational axis. That is, the machines turn around a line, not around a point. The moment of a force about the rotational axis is not the same as the moment of the force about a point. In particular, the moment about a line is a scalar.[6] (See Fig. 44.3.)

The moment $M_{\text{OL}}$ of a force $\mathbf{F}$ about a line OL is the projection OC of the moment $\mathbf{M}_{\text{O}}$ onto the line. Equation 44.10 gives the moment of a force about a line. $\mathbf{a}$ is the unit vector directed along the line, and $a_x$, $a_y$, and

---

[5]The direction of a moment also corresponds to the direction a right-hand screw would progress if it was turned in the direction that rotates $\mathbf{r}$ into $\mathbf{F}$.

[6]Some sources say that the moment of a force about a line can be interpreted as a moment directed along the line. However, this interpretation does not follow from vector operations.

**Figure 44.3** *Moment of a Force About a Line*

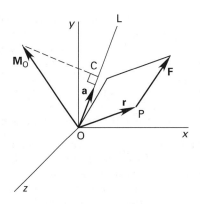

$a_z$ are the direction cosines of the axis OL. Notice that Eq. 44.10 is a dot product (i.e., a scalar).

$$M_{\text{OL}} = \mathbf{a} \cdot \mathbf{M}_{\text{O}} = \mathbf{a} \cdot (\mathbf{r} \times \mathbf{F})$$

$$= \begin{vmatrix} a_x & a_y & a_z \\ x_{\text{P}} - x_{\text{O}} & y_{\text{P}} - y_{\text{O}} & z_{\text{P}} - z_{\text{O}} \\ F_x & F_y & F_z \end{vmatrix} \qquad 44.10$$

If point O is the origin, Eq. 44.10 reduces to Eq. 44.11.

$$M_{\text{OL}} = \begin{vmatrix} a_x & a_y & a_z \\ x & y & z \\ F_x & F_y & F_z \end{vmatrix} \qquad 44.11$$

### 9. COMPONENTS OF A MOMENT

The direction cosines of a force (vector) can be used to determine the components of the moment about the coordinate axes.

$$M_x = M \cos \theta_x \qquad 44.12$$

$$M_y = M \cos \theta_y \qquad 44.13$$

$$M_z = M \cos \theta_z \qquad 44.14$$

Alternatively, the following three equations can be used to determine the components of the moment from a force applied at point $(x, y, z)$ referenced to an origin at $(0, 0, 0)$.

$$M_x = yF_z - zF_y \qquad 44.15$$

$$M_y = zF_x - xF_z \qquad 44.16$$

$$M_z = xF_y - yF_x \qquad 44.17$$

The resultant moment magnitude can be reconstituted from its components.

$$M = \sqrt{M_x^2 + M_y^2 + M_z^2} \qquad 44.18$$

**Statics**

## 10. COUPLES

Any pair of equal, opposite, and parallel forces constitutes a *couple*. A couple is equivalent to a single moment vector. Since the two forces are opposite in sign, the $x$-, $y$-, and $z$-components of the forces cancel out. Therefore, a body is induced to rotate without translation. A couple can be counteracted only by another couple. A couple can be moved to any location within the plane without affecting the equilibrium requirements.

In Fig. 44.4, the equal but opposite forces produce a moment vector $\mathbf{M}_O$ of magnitude $Fd$. The two forces can be replaced by this moment vector, which can be moved to any location on a body. (Such a moment is known as a *free moment, moment of a couple,* or *coupling moment.*)

$$M_O = 2rF\sin\theta = Fd \qquad 44.19$$

**Figure 44.4** *Couple*

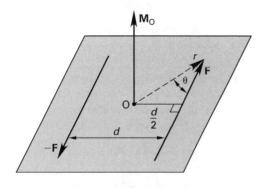

## 11. EQUIVALENCE OF FORCES AND FORCE-COUPLE SYSTEMS

If a force, $F$, is moved a distance $d$ from the original point of application, a couple, $M$, equal to $Fd$ must be added to counteract the induced couple. The combination of the moved force and the couple is known as a *force-couple system.* Alternatively, a force-couple system can be replaced by a single force located a distance $d = M/F$ away.

## 12. RESULTANT FORCE-COUPLE SYSTEMS

The equivalence described in the previous section can be extended to three dimensions and multiple forces. Any collection of forces and moments in three-dimensional space is statically equivalent to a single resultant force vector plus a single resultant moment vector. (Either or both of these resultants can be zero.)

The $x$-, $y$-, and $z$-components of the resultant force are the sums of the $x$-, $y$-, and $z$-components of the individual forces, respectively.

$$F_{R,x} = \sum_i (F\cos\theta_x)_i \qquad 44.20$$

$$F_{R,y} = \sum_i (F\cos\theta_y)_i \qquad 44.21$$

$$F_{R,z} = \sum_i (F\cos\theta_z)_i \qquad 44.22$$

The resultant moment vector is more complex. It includes the moments of all system forces around the reference axes plus the components of all system moments.

$$M_{R,x} = \sum_i (yF_z - zF_y)_i + \sum_i (M\cos\theta_x)_i \qquad 44.23$$

$$M_{R,y} = \sum_i (zF_x - xF_z)_i + \sum_i (M\cos\theta_y)_i \qquad 44.24$$

$$M_{R,z} = \sum_i (xF_y - yF_x)_i + \sum_i (M\cos\theta_z)_i \qquad 44.25$$

## 13. LINEAR FORCE SYSTEMS

A *linear force system* is one in which all forces are parallel and applied along a straight line. (See Fig. 44.5.) A straight beam loaded by several concentrated forces is an example of a linear force system.

**Figure 44.5** *Linear Force System*

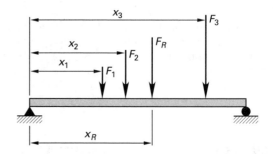

For the purposes of statics, all of the forces in a linear force system can be replaced by an *equivalent resultant force,* $F_R$, equal to the sum of the individual forces. The location of the equivalent force coincides with the location of the centroid of the force group.

$$F_R = \sum_i F_i \qquad 44.26$$

$$x_R = \frac{\sum_i F_i x_i}{\sum_i F_i} \qquad 44.27$$

Statics

## 14. DISTRIBUTED LOADS

If an object is continuously loaded over a portion of its length, it is subject to a *distributed load*. Distributed loads result from *dead load* (i.e., self-weight), hydrostatic pressure, and materials distributed over the object.

If the load per unit length at some point $x$ is $w(x)$, the statically equivalent concentrated load, $F_R$, can be found from Eq. 44.28. The equivalent load is the area under the loading curve. (See Fig. 44.6.)

$$F_R = \int_{x=0}^{x=L} w(x)\,dx \qquad 44.28$$

**Figure 44.6** *Distributed Loads on a Beam*

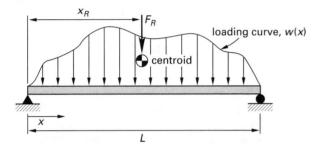

The location, $x_R$, of the equivalent load is calculated from Eq. 44.29. The location coincides with the centroid of the area under the loading curve and is referred to in some problems as the *center of pressure*.

$$x_R = \frac{\int_{x=0}^{x=L} x\,w(x)\,dx}{F_R} \qquad 44.29$$

For a straight beam of length $L$ under a uniform transverse loading of $w$ pounds per foot (newtons per meter),

$$F_R = wL \qquad 44.30$$

$$x_R = \frac{L}{2} \qquad 44.31$$

For a straight beam of length $L$ under a triangular distribution that increases from zero (at $x = 0$) to $w$ (at $x = L$) (see Fig. 44.7),

$$F_R = \frac{wL}{2} \qquad 44.32$$

$$x_R = \frac{2L}{3} \qquad 44.33$$

**Figure 44.7** *Special Cases of Distributed Loading*

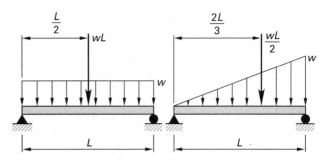

### Example 44.1

Find the magnitude and location of the two equivalent forces on the two spans of the beam.

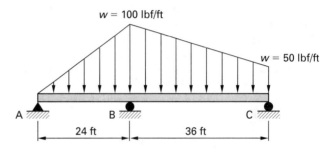

*Solution*

*Span A–B*

The area under the triangular loading curve is

$$A = \tfrac{1}{2}bh = \left(\tfrac{1}{2}\right)(24\text{ ft})\left(100\ \frac{\text{lbf}}{\text{ft}}\right)$$
$$= 1200\text{ lbf}$$

The centroid of the loading triangle is located at

$$x_{R,\text{A–B}} = \tfrac{2}{3}b = \left(\tfrac{2}{3}\right)(24\text{ ft})$$
$$= 16\text{ ft}$$

*Span B–C*

The area under the loading curve consists of a uniform load of 50 lbf/ft over the entire span B–C, plus a triangular load that starts at zero at point C and increases to 50 lbf/ft at point B. The area under the loading curve is

$$A = wL + \tfrac{1}{2}bh$$
$$= \left(50\ \frac{\text{lbf}}{\text{ft}}\right)(36\text{ ft}) + \left(\tfrac{1}{2}\right)(36\text{ ft})\left(50\ \frac{\text{lbf}}{\text{ft}}\right)$$
$$= 2700\text{ lbf}$$

**Statics**

The centroid of the trapezoidal loading curve is located at

$$x_{R,\mathrm{B-C}} = \left(\frac{h}{3}\right)\left(\frac{b+2t}{b+t}\right)$$
$$= \left(\frac{36 \text{ ft}}{3}\right)\left(\frac{50 \text{ ft} + (2)(100 \text{ ft})}{50 \text{ ft} + 100 \text{ ft}}\right) = 20 \text{ ft}$$

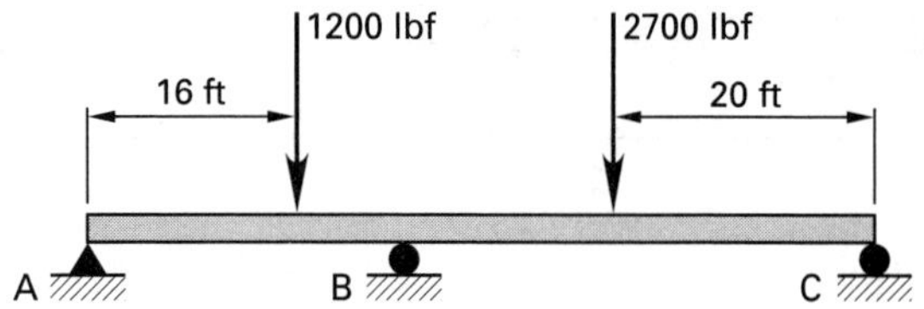


## 15. MOMENT FROM A DISTRIBUTED LOAD

The total force from a uniformly distributed load $w$ over a distance $x$ is $wx$. For the purposes of statics, the uniform load can be replaced by a concentrated force of $wx$ located at the centroid of the distributed load, that is, at the midpoint, $x/2$, of the load. Therefore, the moment taken about one end of the distributed load is

$$M_{\text{distributed load}} = \text{force} \times \text{distance}$$
$$= wx\left(\frac{x}{2}\right) = \tfrac{1}{2}wx^2 \qquad 44.34$$

In general, the moment of a distributed load, uniform or otherwise, is the product of the total force and the distance to the centroid of the distributed load.

## 16. TYPES OF FORCE SYSTEMS

The complexity of methods used to analyze a statics problem depends on the configuration and orientation of the forces. Force systems can be divided into the following categories.

- *concurrent force system:* All of the forces act at the same point.
- *collinear force system:* All of the forces share the same line of action.
- *parallel force system:* All of the forces are parallel (though not necessarily in the same direction).
- *coplanar force system:* All of the forces are in a plane.
- *general three-dimensional system:* This category includes all other combinations of nonconcurrent, nonparallel, and noncoplanar forces.

## 17. CONDITIONS OF EQUILIBRIUM

An object is static when it is stationary. To be stationary, all of the forces on the object must be in equilibrium.[7] For an object to be in equilibrium, the resultant force and moment vectors must both be zero.

$$\mathbf{F}_R = \sum \mathbf{F} = 0 \qquad 44.35$$

$$F_R = \sqrt{F_{R,x}^2 + F_{R,y}^2 + F_{R,z}^2} = 0 \qquad 44.36$$

$$\mathbf{M}_R = \sum \mathbf{M} = 0 \qquad 44.37$$

$$M_R = \sqrt{M_{R,x}^2 + M_{R,y}^2 + M_{R,z}^2} = 0 \qquad 44.38$$

Since the square of any nonzero quantity is positive, Eq. 44.39 through Eq. 44.44 follow directly from Eq. 44.36 and Eq. 44.38.

$$F_{R,x} = 0 \qquad 44.39$$

$$F_{R,y} = 0 \qquad 44.40$$

$$F_{R,z} = 0 \qquad 44.41$$

$$M_{R,x} = 0 \qquad 44.42$$

$$M_{R,y} = 0 \qquad 44.43$$

$$M_{R,z} = 0 \qquad 44.44$$

Equations Eq. 44.39 through Eq. 44.44 seem to imply that six simultaneous equations must be solved in order to determine whether a system is in equilibrium. While this is true for general three-dimensional systems, fewer equations are necessary with most problems. Table 44.1 can be used as a guide to determine which equations are most helpful in solving different categories of problems.

***Table 44.1*** *Number of Equilibrium Conditions Required to Solve Different Force Systems*

| type of force system | two-dimensional | three-dimensional |
|---|---|---|
| general | 3 | 6 |
| coplanar | 3 | 3 |
| concurrent | 2 | 3 |
| parallel | 2 | 3 |
| coplanar, parallel | 2 | 2 |
| coplanar, concurrent | 2 | 2 |
| collinear | 1 | 1 |

## 18. TWO- AND THREE-FORCE MEMBERS

Members limited to loading by two or three forces are special cases of equilibrium. A *two-force member* can be in equilibrium only if the two forces have the same line of action (i.e., are collinear) and are equal but opposite. In most cases, two-force members are loaded axially, and the line of action coincides with the member's longitudinal axis. By choosing the coordinate system so that one axis coincides with the line of action, only one equilibrium equation is needed.

A *three-force member* can be in equilibrium only if the three forces are concurrent or parallel. Stated another

[7]Thus, the term *static equilibrium*, though widely used, is redundant.

Statics

way, the force polygon of a three-force member in equilibrium must close on itself.

## 19. REACTIONS

The first step in solving most statics problems is to determine the reaction forces (i.e., the *reactions*) supporting the body. The manner in which a body is supported determines the type, location, and direction of the reactions. Conventional symbols are often used to define the type of support (such as pinned, roller, etc.). Examples of the symbols are shown in Table 44.2.

For beams, the two most common types of supports are the roller support and the pinned support. The *roller support*, shown as a cylinder supporting the beam, supports vertical forces only. Rather than support a horizontal force, a roller support simply rolls into a new equilibrium position. Only one equilibrium equation (i.e., the sum of vertical forces) is needed at a roller support. Generally, the terms *simple support* and *simply supported* refer to a roller support.

The *pinned support*, shown as a pin and clevis, supports both vertical and horizontal forces. Two equilibrium equations are needed.

Generally, there will be vertical and horizontal components of a reaction when one body touches another. However, when a body is in contact with a *frictionless surface*, there is no frictional force component parallel to the surface. Therefore, the reaction is normal to the contact surfaces. The assumption of frictionless contact is particularly useful when dealing with systems of spheres and cylinders in contact with rigid supports. Frictionless contact is also assumed for roller and rocker supports.[8]

## 20. DETERMINACY

When the equations of equilibrium are independent, a rigid body force system is said to be *statically determinate*. A statically determinate system can be solved for all unknowns, which are usually reactions supporting the body.

When the body has more supports than are necessary for equilibrium, the force system is said to be *statically indeterminate*. In a statically indeterminate system, one or more of the supports or members can be removed or reduced in restraint without affecting the equilibrium position.[9] Those supports and members are known as

Table 44.2 Types of Two-Dimensional Supports

| type of support | reactions and moments | number of unknowns[*] |
|---|---|---|
| simple, roller, rocker, ball, or frictionless surface | reaction normal to surface, no moment | 1 |
| cable in tension, or link | reaction in line with cable or link, no moment | 1 |
| frictionless guide or collar | reaction normal to rail, no moment | 1 |
| built-in, fixed support | two reaction components, one moment | 3 |
| frictionless hinge, pin connection, or rough surface | reaction in any direction, no moment | 2 |

[*]The number of unknowns is valid for two-dimensional problems only.

*redundant members.* The number of redundant members is known as the *degree of indeterminacy.* Figure 44.8 illustrates several common indeterminate structures.

A statically indeterminate body requires additional equations to supplement the equilibrium equations. The additional equations typically involve deflections and depend on mechanical properties of the body.

---

[8]Frictionless surface contact, which requires only one equilibrium equation, should not be confused with a frictionless pin connection, which requires two equilibrium equations. A pin connection with friction introduces a moment at the connection, increasing the number of required equilibrium equations to three.

[9]An example of a support reduced in restraint is a pinned joint replaced by a roller joint. The pinned joint restrains the body vertically and horizontally, requiring two equations of equilibrium. The roller joint restrains the body vertically only and requires one equilibrium equation.

Statics

**Figure 44.8** *Examples of Indeterminate Systems*

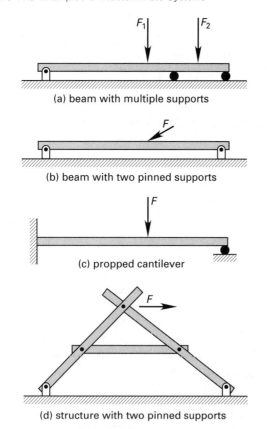

(a) beam with multiple supports

(b) beam with two pinned supports

(c) propped cantilever

(d) structure with two pinned supports

## 21. TYPES OF DETERMINATE BEAMS

Figure 44.9 illustrates the terms used to describe determinate beam types.

**Figure 44.9** *Types of Determinate Beams*

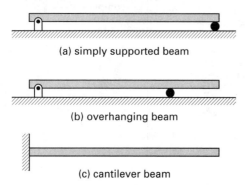

(a) simply supported beam

(b) overhanging beam

(c) cantilever beam

## 22. FREE-BODY DIAGRAMS

A *free-body diagram* is a representation of a body in equilibrium. It shows all applied forces, moments, and reactions. Free-body diagrams do not consider the internal structure or construction of the body, as Fig. 44.10 illustrates.

Since the body is in equilibrium, the resultants of all forces and moments on the free body are zero. In order

to maintain equilibrium, any portions of the body that are removed must be replaced by the forces and moments those portions impart to the body. Typically, the body is isolated from its physical supports in order to help evaluate the reaction forces. In other cases, the body may be sectioned (i.e., cut) in order to determine the forces at the section.

**Figure 44.10** *Bodies and Free Bodies*

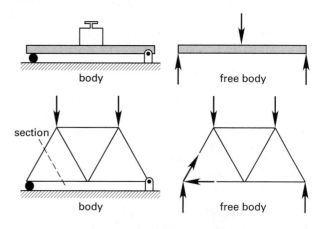

## 23. FINDING REACTIONS IN TWO DIMENSIONS

The procedure for finding determinate reactions in two-dimensional problems is straightforward. Determinate structures will have either a roller support and a pinned support or two roller supports.

*step 1:* Establish a convenient set of coordinate axes. (To simplify the analysis, one of the coordinate directions should coincide with the direction of the forces and reactions.)

*step 2:* Draw the free-body diagram.

*step 3:* Resolve the reaction at the pinned support (if any) into components normal and parallel to the coordinate axes.

*step 4:* Establish a positive direction of rotation (e.g., clockwise) for purposes of taking moments.

*step 5:* Write the equilibrium equation for moments about the pinned connection. (By choosing the pinned connection as the point about which to take moments, the pinned connection reactions do not enter into the equation.) This will usually determine the vertical reaction at the roller support.

*step 6:* Write the equilibrium equation for the forces in the vertical direction. Usually, this equation will have two unknown vertical reactions.

*step 7:* Substitute the known vertical reaction from step 5 into the equilibrium equation from step 6. This will determine the second vertical reaction.

Statics

*step 8:* Write the equilibrium equation for the forces in the horizontal direction. Since there is a maximum of one unknown reaction component in the horizontal direction, this step will determine that component.

*step 9:* If necessary, combine the vertical and horizontal force components at the pinned connection into a resultant reaction.

### Example 44.2

Determine the reactions, $R_1$ and $R_2$, on the following beam.

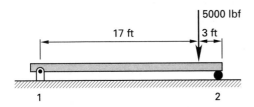

*Solution*

*step 1:* The $x$- and $y$-axes are established parallel and perpendicular to the beam.

*step 2:* The free-body diagram is

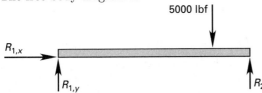

*step 3:* $R_1$ is a pinned support. Therefore, it has two components, $R_{1,x}$ and $R_{1,y}$.

*step 4:* Assume clockwise moments are positive.

*step 5:* Take moments about the left end and set them equal to zero. Use Eq. 44.37.

$$\sum M_{\text{left end}} = (5000 \text{ lbf})(17 \text{ ft}) - R_2(20 \text{ ft})$$
$$= 0$$
$$R_2 = 4250 \text{ lbf}$$

*step 6:* The equilibrium equation for the vertical direction is given by Eq. 44.43.

$$\sum F_y = R_{1,y} + R_2 - 5000 \text{ lbf}$$
$$= 0$$

*step 7:* Substituting $R_2$ into the vertical equilibrium equation,

$$R_{1,y} + 4250 \text{ lbf} - 5000 \text{ lbf} = 0$$
$$R_{1,y} = 750 \text{ lbf}$$

*step 8:* There are no applied forces in the horizontal direction. Therefore, the equilibrium equation is given by Eq. 44.39.

$$\sum F_x = R_{1,x} + 0$$
$$= 0$$
$$R_{1,x} = 0$$

### 24. COUPLES AND FREE MOMENTS

Once a couple on a body is known, the derivation and source of the couple are irrelevant. When the moment on a body is 80 N·m, it makes no difference whether the force is 40 N with a lever arm of 2 m, or 20 N with a lever arm of 4 m, and so on. Therefore, the point of application of a couple is disregarded when writing the moment equilibrium equation. For this reason, the term *free moment* is used synonymously with *couple*.

Figure 44.11 illustrates two diagrammatic methods of indicating the application of a free moment.

**Figure 44.11** *Free Moments*

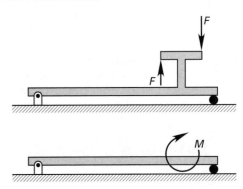

### Example 44.3

What is the reaction $R_2$ for the beam shown?

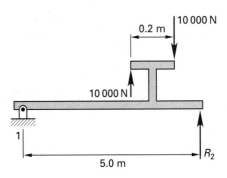

**Statics**

*Solution*

The two couple forces are equal and cancel each other as they come down the stem of the tee bracket. Therefore, there are no applied vertical forces.

The couple has a value given by Eq. 44.7.

$$M = (10\,000 \text{ N})(0.2 \text{ m}) = 2000 \text{ N·m} \quad \text{[clockwise]}$$

Choose clockwise as the direction for positive moments. Taking moments about the pinned connection and using Eq. 44.37,

$$\sum M = 2000 \text{ N·m} - R_2(5 \text{ m})$$
$$= 0$$
$$R_2 = 400 \text{ N}$$

## 25. INFLUENCE LINES FOR REACTIONS

An *influence line* (also known as an *influence graph* and *influence diagram*) is a graph of the magnitude of a reaction as a function of the load placement.[10] The $x$-axis of the graph corresponds to the location on the body (along the length of a beam). The $y$-axis corresponds to the magnitude of the reaction.

By convention (and to generalize the graph for use with any load), the load is taken as one force unit. Therefore, for an actual load of $F$ units, the actual reaction, $R$, is the product of the actual load and the influence line ordinate.

$$R = F \times \text{influence line ordinate} \qquad 44.45$$

### Example 44.4

Draw the influence line for the left reaction for the beam shown.

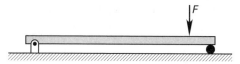

*Solution*

If the unit load is at the left end, the left reaction will be 1.0. If the unit load is at the right end, it will be supported entirely by the right reaction, so the left reaction will be zero. The influence line for the left reaction varies linearly for intermediate load placement.

---

[10]Influence diagrams can also be drawn for moments, shears, and deflections.

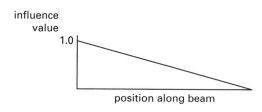

## 26. HINGES

*Hinges* are added to structures to prevent translation while permitting rotation. A frictionless hinge can support a force, but it cannot transmit a moment. Since the moment is zero at a hinge, a structure can be sectioned at the hinge and the remainder of the structure can be replaced by only a force.

### Example 44.5

Calculate the reaction $R_3$ and the hinge force on the two-span beam shown.

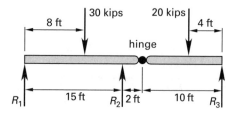

*Solution*

At first, this beam may appear to be statically indeterminate since it has three supports. However, the moment is known to be zero at the hinge. Therefore, the hinged portion of the span can be isolated.

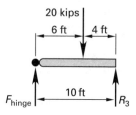

Reaction $R_3$ is found by taking moments about the hinge. Assume clockwise moments are positive.

$$\sum M_{\text{hinge}} = (20{,}000 \text{ lbf})(6 \text{ ft}) - R_3(10 \text{ ft}) = 0$$
$$R_3 = 12{,}000 \text{ lbf}$$

The hinge force is found by summing vertical forces on the isolated section.

$$\sum F_y = F_{\text{hinge}} + 12{,}000 \text{ lbf} - 20{,}000 \text{ lbf} = 0$$
$$F_{\text{hinge}} = 8000 \text{ lbf}$$

## 27. LEVERS

A *lever* is a simple mechanical machine able to increase an applied force. The ratio of the load-bearing force to applied force (i.e., the *effort*) is known as the *mechanical advantage* or *force amplification*. As Fig. 44.12 shows, the mechanical advantage is equal to the ratio of lever arms.

$$\begin{aligned} \text{mechanical} \atop \text{advantage} &= \frac{F_{\text{load}}}{F_{\text{applied}}} = \frac{\text{applied force lever arm}}{\text{load lever arm}} \\ &= \frac{\text{distance moved by applied force}}{\text{distance moved by load}} \end{aligned}$$

*44.46*

**Figure 44.12** Lever

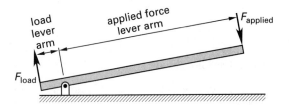

## 28. PULLEYS

A *pulley* (also known as a *sheave*) is used to change the direction of an applied tensile force. A series of pulleys working together (known as a *block and tackle*) can also provide *pulley advantage* (i.e., mechanical advantage). A *hoist* is any device used to raise or lower an object. A hoist may contain one or more pulleys.

If the pulley is attached by a bracket or cable to a fixed location, it is said to be a *fixed pulley*. If the pulley is attached to a load, or if the pulley is free to move, it is known as a *free pulley*.

Most simple problems disregard friction and assume that all ropes are parallel.[11] In such cases, the pulley advantage is equal to the number of ropes coming to and going from the load-carrying pulley. The diameters of the pulleys are not factors in calculating the pulley advantage. (See Table 44.3.)

In other cases, a *loss factor*, $\epsilon$, is used to account for rope rigidity. For most wire ropes and chains with 180° contact, the loss factor at low speeds varies between 1.03 and 1.06. The loss factor is the reciprocal of the *pulley efficiency*, $\eta$.

$$\epsilon = \frac{\text{applied force}}{\text{load}} = \frac{1}{\eta}$$

*44.47*

## 29. AXIAL MEMBERS

An *axial member* is capable of supporting axial forces only and is loaded only at its joints (i.e., ends). This type of performance can be achieved through the use of frictionless bearings or smooth pins at the ends. Since

[11]Although the term *rope* is used here, the principles apply equally well to wire rope, cables, chains, belts, etc.

**Table 44.3** Mechanical Advantages of Rope-Operated Machines

| | fixed sheave | free sheave | ordinary pulley block ($n$ sheaves) | differential pulley block |
|---|---|---|---|---|
| $F_{\text{ideal}}$ | $W$ | $\dfrac{W}{2}$ | $\dfrac{W}{n}$ | $\left(\dfrac{W}{2}\right)\left(1 - \dfrac{d}{D}\right)$ |
| $F$ to raise load | $\epsilon W$ | $\dfrac{\epsilon W}{1+\epsilon}$ | $\dfrac{\epsilon^{n}(\epsilon - 1)W}{\epsilon^{n} - 1}$ | $\dfrac{\left(\epsilon^{2} - \dfrac{d}{D}\right)W}{1+\epsilon}$ |
| $F$ to lower load | $\dfrac{W}{\epsilon}$ | $\dfrac{W}{1+\epsilon}$ | $\left(\dfrac{\frac{1}{\epsilon} - 1}{1 - \epsilon^{n}}\right)W$ | $\left(\dfrac{\epsilon W}{1+\epsilon}\right)\left(\dfrac{1}{\epsilon^{2}} - \dfrac{d}{D}\right)$ |
| ratio of distance of force to distance of load | $1$ | $2$ | $n$ | $\dfrac{2D}{D-d}$ |

the ends are assumed to be pinned (i.e., rotation-free), an axial member cannot support moments. The weight of the member is disregarded or is included in the joint loading.

An axial member can be in either tension or compression. It is common practice to label forces in axial members as (T) or (C) for tension or compression, respectively. Alternatively, tensile forces can be written as positive numbers, while compressive forces are written as negative numbers.

The members in simple trusses are assumed to be axial members. Each member is identified by its endpoints, and the force in a member is designated by the symbol for the two endpoints. For example, the axial force in a member connecting points C and D will be written as **CD**. Similarly, $\mathbf{EF}_y$ is the $y$-component of the force in the member connecting points E and F.

For equilibrium, the resultant forces at the two joints must be equal, opposite, and collinear. This applies to the total (resultant) force as well as to the $x$- and $y$-components at those joints.

## 30. FORCES IN AXIAL MEMBERS

The line of action of a force in an axial member coincides with the longitudinal axis of the member. Depending on the orientation of the coordinate axis system, the direction of the longitudinal axis will have both $x$- and

**Statics**

*y*-components. Therefore, the force in an axial member will generally have both *x*- and *y*-components.

The following four general principles are helpful in determining the force in an axial member.

- A horizontal member carries only horizontal loads. It cannot carry vertical loads.

- A vertical member carries only vertical loads. It cannot carry horizontal loads.

- The vertical component of an axial member's force is equal to the vertical component of the load applied to the member.

- The total and component forces in an inclined member are proportional to the sides of the triangle outlined by the member and the coordinate axes.[12]

### Example 44.6

Member BC is an inclined axial member pinned at both ends and oriented as shown. A vertical 1000 N force is applied to the top end. What are the *x*- and *y*-components of the force in member BC? What is the total force in the member?

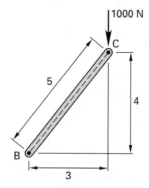

*Solution*

From the third principle,

$$\mathbf{BC}_y = 1000 \text{ N}$$

From the fourth principle,

$$\mathbf{BC}_x = \frac{3}{4}\mathbf{BC}_y$$

$$= \left(\frac{3}{4}\right)(1000 \text{ N})$$

$$= 750 \text{ N}$$

The resultant force in member BC can be calculated from the Pythagorean theorem. However, it is easier to use the fourth principle.

$$\mathbf{BC} = \frac{5}{4}\mathbf{BC}_y$$

$$= \left(\frac{5}{4}\right)(1000 \text{ N})$$

$$= 1250 \text{ N}$$

---

[12]This is an application of the principle of similar triangles.

### Example 44.7

The 12 ft long axial member FG supports an axial force of 180 lbf. What are the *x*- and *y*-components of the applied force?

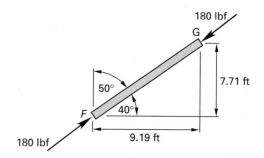

*Solution*

*method 1:* direction cosines

The *x*- and *y*-direction angles are 40° and 90° − 40° = 50°, respectively.

$$\mathbf{FG}_x = (180 \text{ lbf})\cos 40° = 137.9 \text{ lbf}$$

$$\mathbf{FG}_y = (180 \text{ lbf})\cos 50° = 115.7 \text{ lbf}$$

*method 2:* similar triangles

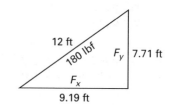

$$\mathbf{FG}_x = \left(\frac{9.19 \text{ ft}}{12 \text{ ft}}\right)(180 \text{ lbf}) = 137.9 \text{ lbf}$$

$$\mathbf{FG}_y = \left(\frac{7.71 \text{ ft}}{12 \text{ ft}}\right)(180 \text{ lbf}) = 115.7 \text{ lbf}$$

## 31. TRUSSES

A *truss* or *frame* is a set of *pin-connected axial members* (i.e., *two-force members*). The connection points are known as *joints*. Member weights are disregarded, and truss loads are applied only at joints. A *structural cell* consists of all members in a closed loop of members. For the truss to be stable (i.e., to be a *rigid truss*), all of the structural cells must be triangles. Figure 44.13 identifies *chords, end posts, panels*, and other elements of a typical *bridge truss*.

A *trestle* is a braced structure spanning a ravine, gorge, or other land depression in order to support a road or rail line. Trestles are usually indeterminate, have multiple earth contact points, have redundant members, and are more difficult to evaluate than simple trusses.

**Figure 44.13** *Parts of a Bridge Truss*

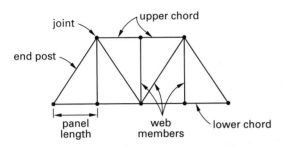

Several types of trusses have been given specific names. Some of the more common types of named trusses are shown in Fig. 44.14.

**Figure 44.14** *Special Types of Trusses*

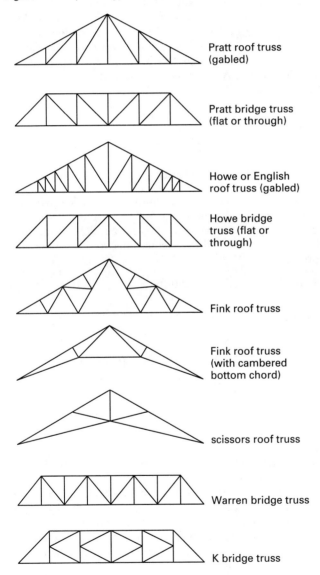

Pratt roof truss (gabled)

Pratt bridge truss (flat or through)

Howe or English roof truss (gabled)

Howe bridge truss (flat or through)

Fink roof truss

Fink roof truss (with cambered bottom chord)

scissors roof truss

Warren bridge truss

K bridge truss

Truss loads are considered to act only in the plane of a truss. Therefore, trusses are analyzed as two-dimensional structures. Forces in truss members hold the various truss parts together and are known as *internal forces*. The internal forces are found by applying equations of equilibrium to appropriate free-body diagrams.

Although free-body diagrams of truss members can be drawn, this is not usually done. Instead, free-body diagrams of the pins (i.e., the joints) are drawn. A pin in compression will be shown with force arrows pointing toward the pin, away from the member. (Similarly, a pin in tension will be shown with force arrows pointing away from the pin, toward the member.)[13]

With typical bridge trusses supported at the ends and loaded downward at the joints, the upper chords are almost always in compression, and the end panels and lower chords are almost always in tension.

## 32. DETERMINATE TRUSSES

A truss will be statically determinate if Eq. 44.48 holds.

$$\text{no. of members} = 2(\text{no. of joints}) - 3 \qquad 44.48$$

If the left-hand side is greater than the right-hand side (i.e., there are *redundant members*), the truss is statically indeterminate. If the left-hand side is less than the right-hand side, the truss is unstable and will collapse under certain types of loading.

Equation 44.48 is a special case of the following general criterion.

$$\begin{aligned} \text{no. of members} & \\ + \text{ no. of reactions} & \\ - 2(\text{no. of joints}) &= 0 \quad \text{[determinate]} \\ &> 0 \quad \text{[indeterminate]} \\ &< 0 \quad \text{[unstable]} \qquad 44.49 \end{aligned}$$

Furthermore, Eq. 44.48 is a necessary, but not sufficient, condition for truss stability. It is possible to arrange the members in such a manner as to not contribute to truss stability. This will seldom be the case in actual practice, however.

## 33. ZERO-FORCE MEMBERS

Forces in truss members can sometimes be determined by inspection. One of these cases is where there are *zero-force members*. A third member framing into a joint already connecting two collinear members carries no internal force unless there is a load applied at that joint. Similarly, both members forming an apex of the truss are zero-force members unless there is a load applied at the apex. (See Fig. 44.15.)

---

[13]The method of showing tension and compression on a truss drawing may appear incorrect. This is because the arrows show the forces on the pins, not on the members.

**Figure 44.15** Zero-Force Members

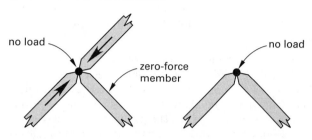

## 34. METHOD OF JOINTS

The *method of joints* is one of three methods that can be used to find the internal forces in each truss member. This method is useful when most or all of the truss member forces are to be calculated. Because this method advances from joint to adjacent joint, it is inconvenient when a single isolated member force is to be calculated.

The method of joints is a direct application of the equations of equilibrium in the $x$- and $y$-directions. Traditionally, the method starts by finding the reactions supporting the truss. Next, the joint at one of the reactions is evaluated, which determines all the member forces framing into the joint. Then, knowing one or more of the member forces from the previous step, an adjacent joint is analyzed. The process is repeated until all the unknown quantities are determined.

At a joint, there may be up to two unknown member forces, each of which can have dependent $x$- and $y$-components.[14] Since there are two equilibrium equations, the two unknown forces can be determined. Even though determinate, however, the sense of a force will often be unknown. If the sense cannot be determined by logic, an arbitrary decision can be made. If the incorrect direction is chosen, the calculated force will be negative.

### Example 44.8

Use the method of joints to calculate the force **BD** in the truss shown.

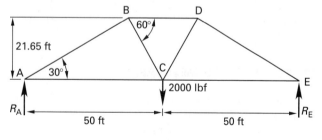

*Solution*

First, find the reactions. Assume clockwise is positive and take moments about point A.

$$\sum M_A = (2000 \text{ lbf})(50 \text{ ft}) - R_E(50 \text{ ft} + 50 \text{ ft}) = 0$$

---

[14]Occasionally, there will be three unknown member forces. In that case, an additional equation must be derived from an adjacent joint.

$$R_E = 1000 \text{ lbf}$$

Since the sum of forces in the $y$-direction is also zero,

$$\sum F_y = R_A + 1000 \text{ lbf} - 2000 \text{ lbf} = 0$$

$$R_A = 1000 \text{ lbf}$$

There are three unknowns at joint B (and also at D). Therefore, the analysis must start at joint A (or E) where there are only two unknowns (forces **AB** and **AC**).

The free-body diagram of pin A is shown. The direction of $R_A$ is known to be upward. The directions of forces **AB** and **AC** can be assumed, but logic can be used to determine them. Only the vertical component of **AB** can oppose $R_A$. Therefore, **AB** is directed downward. (This means that member AB is in compression.) Similarly, **AC** must oppose the horizontal component of **AB**. Therefore, **AC** is directed to the right. (This means that member AC is in tension.)

Resolve force **AB** into horizontal and vertical components using trigonometry, direction cosines, or similar triangles. ($R_A$ and **AC** are already parallel to an axis.) Then, use the equilibrium equations to determine the forces.

By inspection, $AB_y = 1000 \text{ lbf}$.

$$AB_y = AB \sin 30°$$
$$1000 \text{ lbf} = AB(0.5)$$
$$AB = 2000 \text{ lbf} \quad (C)$$
$$AB_x = AB \cos 30°$$
$$= (2000 \text{ lbf})(0.866)$$
$$= 1732 \text{ lbf}$$

Now, draw the free-body diagram of pin B. (Notice that the direction of force **AB** is toward the pin, just as it was for pin A.) Although the true directions of the forces are unknown, they can be determined logically. The direction of force **BC** is chosen to counteract the vertical component of force **AB**. The direction of force **BD** is chosen to counteract the horizontal components of forces **AB** and **BC**.

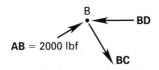

$AB_x$ and $AB_y$ are already known. Resolve the force **BC** into horizontal and vertical components.

$$BC_x = BC \sin 30° = BC(0.5)$$

$$BC_y = BC \cos 30° = BC(0.866)$$

Now, write the equations of equilibrium for point B.

$$\sum F_x = 1732 \text{ lbf} + 0.5BC - BD = 0$$

$$\sum F_y = 1000 \text{ lbf} - 0.866BC = 0$$

From the second equation, $BC = 1155$ lbf. Substituting this into the first equation,

$$1732 \text{ lbf} + (0.5)(1155 \text{ lbf}) - BD = 0$$

$$BD = 2310 \text{ lbf} \quad (C)$$

Since **BD** turned out to be positive, its direction was chosen correctly.

The direction of the arrow indicates that the member is compressing the pin. Consequently, the pin is compressing the member. Member BD is in compression.

If the process is continued, all forces can be determined. However, the truss is symmetrical, and it is not necessary to evaluate every joint to calculate all forces.

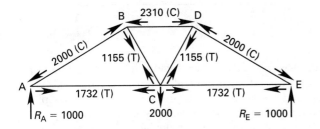

## 35. CUT-AND-SUM METHOD

The *cut-and-sum method* can be used to find forces in inclined members. This method is strictly an application of the vertical equilibrium condition ($\Sigma F_y = 0$).

The method starts by finding all of the support reactions on a truss. Then, a cut is made through the truss in such a way as to pass through one inclined or vertical member only. (At this point, it should be clear that the vertical component of the inclined member must balance all of the external vertical forces.) The equation for vertical equilibrium is written for the free body of the remaining truss portion.

## Example 44.9

Find the force in member BC for the truss in Ex. 44.8.

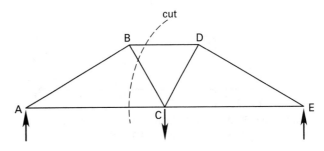

*Solution*

The reactions were determined in Ex. 44.8. The truss is cut in such a way as to pass through member BC but through no other inclined member. The free body of the remaining portion of the truss is

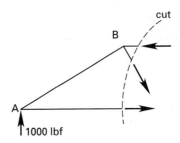

The vertical equilibrium equation is

$$\sum F_y = R_A - BC_y$$

$$= 0$$

$$= 1000 \text{ lbf} - 0.866BC$$

$$= 0$$

$$BC = 1155 \text{ lbf} \quad (T)$$

## 36. METHOD OF SECTIONS

The *method of sections* is a direct approach to finding forces in any truss member. This method is convenient when only a few truss member forces are unknown.

As with the previous two methods, the first step is to find the support reactions. Then, a cut is made through the truss, passing through the unknown member.[15] Finally, all three conditions of equilibrium are applied as needed to the remaining truss portion. (Since there are three equilibrium equations, the cut cannot pass through more than three members in which the forces are unknown.)

[15]Knowing where to cut the truss is the key part of this method. Such knowledge is developed only by practice.

## Example 44.10

Find the forces in members CD and CE. The support reactions have already been determined.

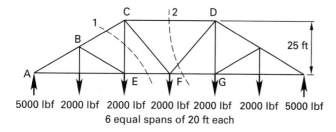

5000 lbf  2000 lbf  2000 lbf  2000 lbf  2000 lbf  2000 lbf  5000 lbf

6 equal spans of 20 ft each

*Solution*

To find the force **CE**, the truss is cut at section 1.

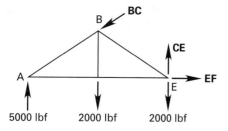

5000 lbf      2000 lbf      2000 lbf

Taking moments about point A will eliminate all of the unknown forces except **CE**. Assume clockwise moments are positive.

$$\sum M_A = (2000 \text{ lbf})(20 \text{ ft}) + (2000 \text{ lbf})(40 \text{ ft})$$
$$- (40 \text{ ft})\mathbf{CE} = 0$$

$$\mathbf{CE} = 3000 \text{ lbf} \quad (\text{T})$$

To find the force **CD**, the truss is cut at section 2.

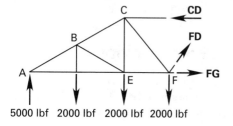

5000 lbf   2000 lbf   2000 lbf   2000 lbf

Taking moments about point F will eliminate all unknowns except **CD**. Assume clockwise moments are positive.

$$\sum M_F = (5000 \text{ lbf})(60 \text{ ft}) - (25)\mathbf{CD}$$
$$- (2000 \text{ lbf})(20 \text{ ft}) - (2000 \text{ lbf})(40 \text{ ft}) = 0$$

$$\mathbf{CD} = 7200 \text{ lbf} \quad (\text{C})$$

## 37. SUPERPOSITION OF LOADS

*Superposition* is a term used to describe the process of determining member forces by considering loads one at

a time. Suppose, for example, that the force in member FG is unknown and that the truss carries three loads. If the method of superposition is used, the force in member FG (call it **FG**$_1$) is determined with only the first load acting on the truss. **FG**$_2$ and **FG**$_3$ are similarly found. The true member force **FG** is found by adding **FG**$_1$, **FG**$_2$, and **FG**$_3$.

Superposition should be used with discretion since trusses can change shape under load. If a truss deflects such that the load application points are significantly different from those in the undeflected truss, superposition cannot be used for that truss.

In simple truss analysis, change of shape under load is neglected. Superposition, therefore, can be assumed to apply.

## 38. TRANSVERSE TRUSS MEMBER LOADS

Truss members are usually designed as axial members, not as beams. Trusses are traditionally considered to be loaded at joints only. Figure 44.16, however, illustrates cases of nontraditional *transverse loading* that can actually occur. For example, a truss member's own weight would contribute to a uniform load, as would a severe ice buildup.

***Figure 44.16*** *Transverse Truss Member Loads*

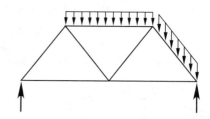

Transverse loads add two solution steps to a truss problem. First, the truss member must be individually considered as a beam simply supported at its pinned connections, and the reactions needed to support the transverse loading must be found. These reactions become additional loads applied to the truss joints, and the truss can then be evaluated in the normal manner.

The second step is to check the structural adequacy (deflection, bending stress, shear stress, buckling, etc.) of the truss member under transverse loading.

## 39. CABLES CARRYING CONCENTRATED LOADS

An *ideal* cable is assumed to be completely flexible, massless, and incapable of elongation. It, therefore, acts as an axial two-force tension member between points of concentrated loading. In fact, the term *tension* or *tensile force* is commonly used in place of "member force" when dealing with cables. (See Fig. 44.17.)

Statics

**Figure 44.17** *Cable with Concentrated Load*

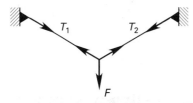

The methods of joints and sections used in truss analysis can be used to determine the tensions in cables carrying concentrated loads. After separating the reactions into $x$- and $y$-components, it is particularly useful to sum moments about one of the reaction points. All cables will be found to be in tension, and (with vertical loads only) the horizontal tension component will be the same in all cable segments. Unlike the case of a rope passing over a series of pulleys, however, the total tension in the cable will not be the same in every cable segment.

### Example 44.11

What are the tensions **AB**, **BC**, and **CD**?

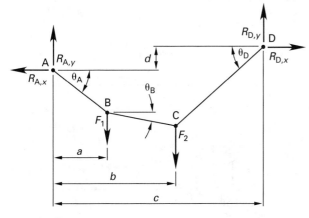

*Solution*

Separate the two reactions into $x$- and $y$-components. (The total reactions $R_A$ and $R_D$ are also the tensions **AB** and **CD**, respectively.)

$$R_{A,x} = -\mathbf{AB}\cos\theta_A$$
$$R_{A,y} = -\mathbf{AB}\sin\theta_A$$
$$R_{D,x} = -\mathbf{CD}\cos\theta_D$$
$$R_{D,y} = -\mathbf{CD}\sin\theta_D$$

Next, take moments about point A to find **CD**. Assume clockwise to be positive.

$$\sum M_A = aF_1 + bF_2 - d\mathbf{CD}_x + c\mathbf{CD}_y = 0$$

None of the applied loads are in the $x$-direction. Therefore, the only horizontal loads are the $x$-components of the reactions. To find tension **AB**, take the entire cable

as a free body. Then, sum the external forces in the $x$-direction.

$$\sum F_x = R_{D,x} - R_{A,x} = 0$$

$$\mathbf{CD}\cos\theta_D = \mathbf{AB}\cos\theta_A$$

The $x$-component of force is the same in all cable segments. To find **BC**, sum the $x$-direction forces at point B.

$$\sum F_x = \mathbf{BC}_x - \mathbf{AB}_x = 0$$

$$\mathbf{BC}\cos\theta_B = \mathbf{AB}\cos\theta_A$$

### 40. PARABOLIC CABLES

If the distributed load per unit length, $w$, on a cable is constant with respect to the horizontal axis (as is the load from a bridge floor), the cable will be parabolic in shape.[16] This is illustrated in Fig. 44.18.

**Figure 44.18** *Parabolic Cable*

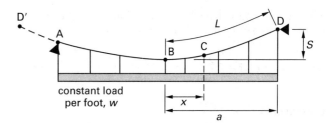

The maximum sag is designated as $S$. If the location of the maximum sag (i.e., the lowest cable point) is known, the horizontal component of tension, $H$, can be found by taking moments about a reaction point. If the cable is cut at the maximum sag point, B, the cable tension on the free body will be horizontal since there is no vertical component to the cable. Cutting the cable in Fig. 44.18 at point B and taking moments about point D will determine the minimum cable tension, $H$.

$$\sum M_D = wa\left(\frac{a}{2}\right) - HS = 0 \qquad 44.50$$

$$H = \frac{wa^2}{2S} \qquad 44.51$$

Since the load is vertical everywhere, the horizontal component of tension is constant everywhere in the cable. The tension, $T_C$, at any point C can be found

---

[16]The parabolic case can also be assumed with cables loaded only by their own weight (e.g., telephone and trolley wires), if both ends are at the same elevations and if the sag is no more than 10% of the distance between supports.

by applying the equilibrium conditions to the cable segment BC.

$$T_{C,x} = H = \frac{wa^2}{2S} \qquad \text{44.52}$$

$$T_{C,y} = wx \qquad \text{44.53}$$

$$T_C = \sqrt{T_{C,x}^2 + T_{C,y}^2}$$
$$= w\sqrt{\left(\frac{a^2}{2S}\right)^2 + x^2} \qquad \text{44.54}$$

The angle of the cable at any point is

$$\tan\theta = \frac{wx}{H} \qquad \text{44.55}$$

The tension and angle are maximum at the supports.

If the lowest sag point, point B, is used as the origin, the shape of the cable is

$$y(x) = \frac{wx^2}{2H} \qquad \text{44.56}$$

The approximate length of the cable from the lowest point to the support (i.e., length BD) is

$$L \approx a\left(1 + \left(\frac{2}{3}\right)\left(\frac{S}{a}\right)^2 - \left(\frac{2}{5}\right)\left(\frac{S}{a}\right)^4\right) \qquad \text{44.57}$$

### Example 44.12

A pedestrian foot bridge has two suspension cables and a flexible floor weighing 28 lbf/ft. The span of the bridge is 100 ft. When the bridge is empty, the tension at point C is 1500 lbf. Assuming a parabolic shape, what is the maximum cable sag, $S$?

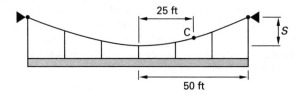

*Solution*

Since there are two cables, the floor weight per suspension cable is

$$w = \frac{28\ \dfrac{\text{lbf}}{\text{ft}}}{2}$$
$$= 14\ \text{lbf/ft}$$

From Eq. 44.54,

$$T_C = w\sqrt{\left(\frac{a^2}{2S}\right)^2 + x^2}$$

$$1500\ \text{lbf} = 14\ \frac{\text{lbf}}{\text{ft}}\sqrt{\left(\frac{(50\ \text{ft})^2}{2S}\right)^2 + (25\ \text{ft})^2}$$

$$S = 12\ \text{ft}$$

## 41. CABLES CARRYING DISTRIBUTED LOADS

An idealized tension cable with a distributed load is similar to a linkage made up of a very large number of axial members. The cable is an axial member in the sense that the internal tension acts tangentially to the cable everywhere.

Since the load is vertical everywhere, the horizontal component of cable tension is constant along the cable. The cable is horizontal at the point of lowest sag. There is no vertical tension component, and the cable tension is minimum. By similar reasoning, the cable tension is maximum at the supports.

Figure 44.19 illustrates a general cable with a distributed load. The shape of the cable will depend on the relative distribution of the load. A free-body diagram of segment BC is also shown.

*Figure 44.19* Cable with Distributed Load

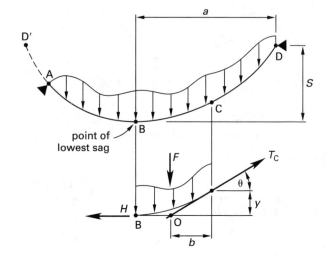

$F$ is the resultant of the distributed load on segment BC, $T$ is the cable tension at point C, and $H$ is the tension at the point of lowest sag (i.e., the point of minimum tension). Since segment BC taken as a free body is a three-force member, the three forces ($H$, $F$, and $T$) must be concurrent to be in equilibrium. The horizontal component of tension can be found by taking moments about point C.

$$\sum M_C = Fb - Hy = 0 \qquad \text{44.58}$$

Statics

$$H = \frac{Fb}{y} \qquad 44.59$$

Also, $\tan\theta = y/b$. Therefore,

$$H = \frac{F}{\tan\theta} \qquad 44.60$$

The basic equilibrium conditions can be applied to the free-body cable segment BC to determine the tension in the cable at point C.

$$\sum F_x = T_C\cos\theta - H = 0 \qquad 44.61$$

$$\sum F_y = T_C\sin\theta - F = 0 \qquad 44.62$$

The resultant tension at point C is

$$T_C = \sqrt{H^2 + F^2} \qquad 44.63$$

## 42. CATENARY CABLES

If the distributed load is constant along the length of the cable, as it is with a loose cable loaded by its own weight, the cable will have the shape of a *catenary*. A vertical axis catenary's shape is determined by Eq. 44.64, where $c$ is a constant and cosh is the *hyperbolic cosine*. The quantity $x/c$ is in radians.[17]

$$y(x) = c\cosh\left(\frac{x}{c}\right) \qquad 44.64$$

Referring to Fig. 44.20, the vertical distance, $y$, to any point C on the catenary is measured from a reference plane located a distance $c$ below the point of greatest sag, point B. The distance $c$ is known as the *parameter of the catenary*. Although the value of $c$ establishes the location of the $x$-axis, the value of $c$ does not correspond to any physical distance, nor is the reference plane the ground level.

In order to define the cable shape and determine cable tensions, it is necessary to have enough information to calculate $c$. For example, if $a$ and $S$ are known, Eq. 44.67 can be solved by trial and error for $c$.[18] Once $c$ is known, the cable geometry and forces are determined by the remaining equations.

For any point C, the equations most useful in determining the shape of the catenary are

$$y = \sqrt{s^2 + c^2} = c\left(\cosh\left(\frac{x}{c}\right)\right) \qquad 44.65$$

$$s = c\left(\sinh\left(\frac{x}{c}\right)\right) \qquad 44.66$$

---

[17]In order to use Eq. 44.64 through Eq. 44.67, you must reset your calculator from degrees to radians.
[18]Because obtaining the solution may require trial and error, it will be advantageous to assume a parabolic shape if the cable is taut. (See Ftn. 17.) The error will generally be small.

$$\text{sag} = S = y_D - c = c\left(\cosh\left(\frac{a}{c}\right) - 1\right) \qquad 44.67$$

$$\tan\theta = \frac{s}{c} \qquad 44.68$$

The equations most useful in determining the cable tensions are

$$H = wc \qquad 44.69$$

$$F = ws \qquad 44.70$$

$$T = wy \qquad 44.71$$

$$\tan\theta = \frac{ws}{H} \qquad 44.72$$

$$\cos\theta = \frac{H}{T} \qquad 44.73$$

**Figure 44.20** *Catenary Cable*

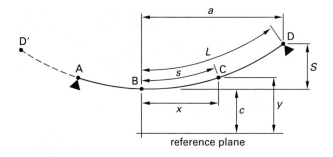

reference plane

## Example 44.13

A cable 100 m long is loaded by its own weight. The maximum sag is 25 m, and the supports are on the same level. What is the distance between the supports?

*Solution*

Since the two supports are on the same level, the cable length, $L$, between the point of maximum sag and support D is half of the total length.

$$L = \frac{100\text{ m}}{2} = 50\text{ m}$$

Writing Eq. 44.65 and Eq. 44.67 for point D (with $S = 25$ m),

$$y_D = c + S = \sqrt{L^2 + c^2}$$

$$c + 25\text{ m} = \sqrt{(50\text{ m})^2 + c^2}$$

$$c = 37.5\text{ m}$$

Substituting $a$ for $x$ and $L = 50$ for $s$ in Eq. 44.66,

$$s = c\left(\sinh\left(\frac{x}{c}\right)\right)$$

Statics

$$50 \text{ m} = (37.5 \text{ m})\left(\sinh\left(\frac{a}{37.5}\right)\right)$$

$$a = 41.2 \text{ m}$$

The distance between supports is

$$2a = (2)(41.2 \text{ m}) = 82.4 \text{ m}$$

## 43. CABLES WITH ENDS AT DIFFERENT ELEVATIONS

A cable will be asymmetrical if its ends are at different elevations. In some cases, as shown in Fig. 44.21, the cable segment will not include the lowest point B. However, if the location of the theoretical lowest point can be derived, the positions and elevations of the cable supports will not affect the analysis. The same procedure is used in proceeding from the theoretical point B to either support. In fact, once the theoretical shape of a cable has been determined, the supports can be relocated anywhere along the cable line without affecting the equilibrium of the supported segment.

**Figure 44.21** *Asymmetrical Segment of Symmetrical Cable*

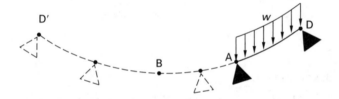

## 44. TWO-DIMENSIONAL MECHANISMS

A two-dimensional *mechanism (machine)* is a nonrigid structure. Although parts of the mechanism move, the relationships between forces in the mechanism can be determined by statics. In order to determine an unknown force, one or more of the mechanism components must be considered as a free body. All input forces and reactions must be included on this free body. In general, the resultant force on such a free body will not be in the direction of the member.

Several free bodies may be needed for complicated mechanisms. Sign conventions of acting and reacting forces must be strictly adhered to when determining the effect of one component on another.

### Example 44.14

A 70 N·m couple is applied to the mechanism shown. All connections are frictionless hinges. What are the $x$- and $y$-components of the reactions at B?

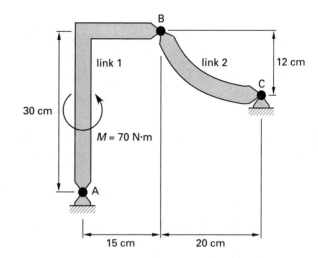

*Solution*

Isolate links 1 and 2 and draw their free bodies.

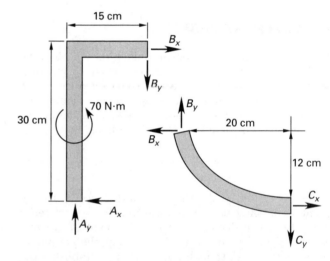

Assume clockwise moments are positive. Take moments about point A on link 1.

$$\sum M_{\text{A}} = B_x(0.3 \text{ m}) + B_y(0.15 \text{ m}) - 70 \text{ N·m} = 0$$

Assume clockwise moments are positive. Take moments about point C on link 2.

$$\sum M_{\text{C}} = B_y(0.20 \text{ m}) - B_x(0.12 \text{ m}) = 0$$

Solving these two equations simultaneously determines the force at joint B.

$$B_x = 179 \text{ N}$$

$$B_y = 108 \text{ N}$$

## 45. EQUILIBRIUM IN THREE DIMENSIONS

The basic equilibrium equations can be used with vector algebra to solve a three-dimensional statics problem. When a manual calculation is required, however, it is often more convenient to write the equilibrium equations for one orthogonal direction at a time, thereby avoiding the use of vector notation and reducing the problem to two dimensions. The following method can be used to analyze a three-dimensional structure.

*step 1:* Establish the $(0, 0, 0)$ origin for the structure.

*step 2:* Determine the $(x, y, z)$ coordinates of all load and reaction points.

*step 3:* Determine the $x$-, $y$-, and $z$-components of all loads and reactions. This is accomplished by using direction cosines calculated from the $(x, y, z)$ coordinates.

*step 4:* Draw a *coordinate free-body diagram* of the structure for each of the three coordinate axes. Include only forces, reactions, and moments that affect the coordinate freebody.

*step 5:* Apply the basic two-dimensional equilibrium equations.

### Example 44.15

Beam AC is supported at point A by a frictionless ball joint and at points B and C by cables. A 100 lbf load is applied vertically to point C, and a 180 lbf load is applied horizontally to point B. What are the cable tensions $T_1$ and $T_2$?

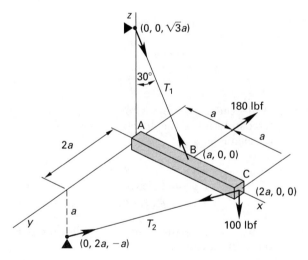

*Solution*

*step 1:* Point A has already been established as the origin. The locations of all support and load points are shown on the illustration.

*step 2:* See step 1.

*step 3:* By inspection, for the 180 lbf horizontal load at point B,

$$F_x = 0$$

$$F_y = -180 \text{ lbf}$$

$$F_z = 0$$

By inspection, for the 100 lbf vertical load at point C,

$$F_x = 0$$

$$F_y = 0$$

$$F_z = -100 \text{ lbf}$$

The length of cable 1 is

$$L_1 = \sqrt{(a-0)^2 + (0-0)^2 + (0-\sqrt{3}a)^2}$$
$$= 2a$$

The direction cosines of the force from cable 1 at point B are

$$\cos\theta_x = \frac{d_x}{L_1} = \frac{0-a}{2a} = -0.5$$

$$\cos\theta_y = \frac{d_y}{L_1} = \frac{0-0}{2a} = 0$$

$$\cos\theta_z = \frac{d_z}{L_1} = \frac{\sqrt{3}a-0}{2a} = 0.866$$

Therefore, the components of the tension in cable 1 are

$$T_{1,x} = -0.5\,T_1$$

$$T_{1,y} = 0$$

$$T_{1,z} = 0.866\,T_1$$

Similarly, for cable 2,

$$L_2 = \sqrt{(2a-0)^2 + (0-2a)^2 + (0-(-a))^2}$$
$$= 3a$$

The direction cosines for the force from cable 2 at point C are

$$\cos\theta_x = \frac{d_x}{L_2} = \frac{0-2a}{3a} = -0.667$$

$$\cos\theta_y = \frac{d_y}{L_2} = \frac{2a-0}{3a} = 0.667$$

$$\cos\theta_z = \frac{d_z}{L_2} = \frac{-a-0}{3a} = -0.333$$

Therefore, the components of the tension in cable 2 are

$$T_{2,x} = -0.667\,T_2$$

$$T_{2,y} = 0.667\,T_2$$

$$T_{2,z} = -0.333\,T_2$$

*step 4:* The three coordinate free-body diagrams are

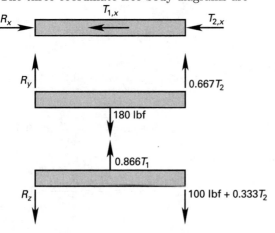

*step 5:* Tension $T_2$ can be found by taking moments about point A on the $y$-coordinate free body.

$$\sum M_A = (0.667\,T_2)(2a) - (180\text{ lbf})(a) = 0$$

$$T_2 = 135\text{ lbf}$$

Tension $T_1$ can be found by taking moments about point A on the $z$-coordinate free body.

$$\sum M_A = (0.866\ T_1)(a) - (0.333)(135\text{ lbf})(2a)$$
$$- (100\text{ lbf})(2a) = 0$$

$$T_1 = 335\text{ lbf}$$

## 46. TRIPODS

A *tripod* is a simple three-dimensional truss (frame) that consists of three axial members. (See Fig. 44.22.) One end of each member is connected at the *apex* of the tripod, while the other ends are attached to the supports. All connections are assumed to allow free rotation in all directions.

The general solution procedure given in the preceding section can be made more specific for tripods.

*step 1:* Establish the apex as the origin.

**Figure 44.22** *Tripod*

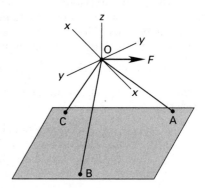

*step 2:* Determine the $x$-, $y$-, and $z$-components of the force applied to the apex.

*step 3:* Determine the $(x, y, z)$ coordinates of points A, B, and C—the three support points.

*step 4:* Determine the length of each tripod leg from the coordinates of the support points.

$$L = \sqrt{x^2 + y^2 + z^2} \qquad\qquad 44.74$$

*step 5:* Determine the direction cosines for the leg forces at the apex. For leg A, for example,

$$\cos\theta_{A,x} = \frac{x_A}{L} \qquad\qquad 44.75$$

$$\cos\theta_{A,y} = \frac{y_A}{L} \qquad\qquad 44.76$$

$$\cos\theta_{A,z} = \frac{z_A}{L} \qquad\qquad 44.77$$

*step 6:* Write the $x$-, $y$-, and $z$-components of each leg force in terms of the direction cosines. For leg A, for example,

$$F_{A,x} = F_A \cos\theta_{A,x} \qquad\qquad 44.78$$
$$F_{A,y} = F_A \cos\theta_{A,y} \qquad\qquad 44.79$$
$$F_{A,z} = F_A \cos\theta_{A,z} \qquad\qquad 44.80$$

*step 7:* Write the three sum-of-forces equilibrium equations for the apex.

$$F_{A,x} + F_{B,x} + F_{C,x} + F_x = 0 \qquad\qquad 44.81$$
$$F_{A,y} + F_{B,y} + F_{C,y} + F_y = 0 \qquad\qquad 44.82$$
$$F_{A,z} + F_{B,z} + F_{C,z} + F_z = 0 \qquad\qquad 44.83$$

Statics

**Example 44.16**

Determine the force in each leg of the tripod.

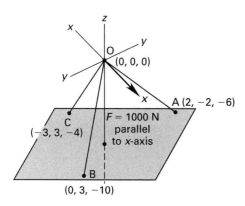

*Solution*

*step 1:* The origin is at the apex.

*step 2:* By inspection, $F_x = +1000$ N. All other components are zero.

*step 3:* The direction cosines for the tripod legs have been calculated and are presented in the following table.

| member | $x^2$ | $y^2$ | $z^2$ | $L^2$ | $L$ | $\cos\theta_x$ | $\cos\theta_y$ | $\cos\theta_z$ |
|--------|-------|-------|-------|-------|------|----------|----------|----------|
| OA | 4 | 4 | 36 | 44 | 6.63 | 0.3015 | −0.3015 | −0.9046 |
| OB | 0 | 9 | 100 | 109 | 10.44 | 0.0 | 0.2874 | −0.9579 |
| OC | 9 | 9 | 16 | 34 | 5.83 | −0.5146 | 0.5146 | −0.6861 |

*step 4:* See step 3.

*step 5:* The equilibrium equations are

$$0.3015F_A + \quad\;\; 0F_B - 0.5146\ F_C + 1000 = 0$$
$$-0.3015F_A + 0.2874F_B + 0.5146\ F_C \qquad\;\; = 0$$
$$-0.9046F_A - 0.9579F_B - 0.6861\ F_C \qquad\;\; = 0$$

The solution to these simultaneous equations is

$$F_A = +1531\ \text{N} \quad (\text{T})$$

$$F_B = -3480\ \text{N} \quad (\text{C})$$

$$F_C = +2841\ \text{N} \quad (\text{T})$$

*step 6:* See step 5.

Statics

# 45 Indeterminate Statics

## Nomenclature

| | | | |
|---|---|---|---|
| $A$ | area | ft$^2$ | m$^2$ |
| $d$ | distance | ft | m |
| $E$ | modulus of elasticity | lbf/ft$^2$ | Pa |
| $F$ | force | lbf | N |
| $I$ | area moment of inertia | ft$^4$ | m$^4$ |
| $L$ | length | ft | m |
| $M$ | moment | ft-lbf | N·m |
| $R$ | reaction | lbf | N |
| $S$ | force | lbf | N |
| $T$ | temperature | °F | °C |
| $u$ | force | lbf | N |
| $w$ | distributed load | lbf/ft | N/m |
| $y'$ | slope | ft/ft | m/m |

## Symbols

| | | | |
|---|---|---|---|
| $\alpha$ | coefficient of thermal expansion | 1/°F | 1/°C |
| $\delta$ | deformation | ft | m |
| $\theta$ | angle | deg | deg |

## Subscripts

| | |
|---|---|
| $c$ | concrete |
| $o$ | original |
| st | steel |

## 1. INTRODUCTION TO INDETERMINATE STATICS

A structure that is *statically indeterminate* is one for which the equations of statics are not sufficient to determine all reactions, moments, and internal forces. Additional formulas involving deflection are required to completely determine these variables.

Although there are many configurations of statically indeterminate structures, this chapter is primarily concerned with beams on more than two supports, trusses with more members than are required for rigidity, and miscellaneous composite structures.

## 2. DEGREE OF INDETERMINACY

The *degree of indeterminacy (degree of redundancy)* is equal to the number of reactions or members that would have to be removed in order to make the structure statically determinate. For example, a two-span beam on three simple supports is indeterminate (redundant) to the first degree. The degree of indeterminacy of a pin-connected truss is given by Eq. 45.1.

$$\begin{array}{c} \text{degree of} \\ \text{indeterminacy} \end{array} = 3 + \begin{array}{c} \text{no. of} \\ \text{members} \end{array} - \left( 2 \times \begin{array}{c} \text{no. of} \\ \text{joints} \end{array} \right) \quad \textbf{45.1}$$

## 3. INDETERMINATE BEAMS

Three common configurations of beams can easily be recognized as being statically indeterminate. These are the *continuous beam*, *propped cantilever beam*, and *fixed-end beam* illustrated in Fig. 45.1.

**Figure 45.1** *Types of Indeterminate Beams*

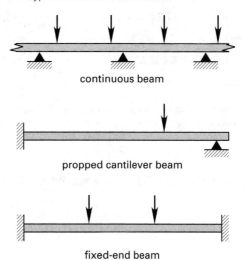

continuous beam

propped cantilever beam

fixed-end beam

Statics

## 4. REVIEW OF ELASTIC DEFORMATION

When an axial force, $F$, acts on an object with length $L$, cross-sectional area $A$, and modulus of elasticity $E$, the deformation[1] is

$$\delta = \frac{FL}{AE} \qquad\qquad 45.2$$

When an object with initial length $L_o$ and coefficient of thermal expansion $\alpha$ experiences a temperature change of $\Delta T$ degrees, the deformation is

$$\delta = \alpha L_o \Delta T \qquad\qquad 45.3$$

## 5. CONSISTENT DEFORMATION METHOD

The *consistent deformation method*, also known as the *compatibility method*, is one of the methods of solving indeterminate problems. This method is simple to learn and to apply. First, geometry is used to develop a relationship between the deflections of two different members (or for one member at two locations) in the structure. Then, the deflection equations for the two different members at a common point are written and equated, since the deformations must be the same at a common point. This method is illustrated by the following examples.

### Example 45.1

A pile carrying an axial compressive load is constructed of concrete with a steel jacket. The end caps are rigid, and the steel-concrete bond is perfect. What are the forces in the steel and concrete if a load $F$ is applied?

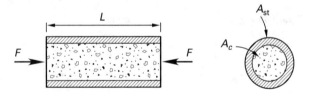

*Solution*

Let $F_c$ and $F_{st}$ be the loads carried by the concrete and steel, respectively. Then,

$$F_c + F_{st} = F$$

The deformation of the steel is given by Eq. 45.2.

$$\delta_{st} = \frac{F_{st}L}{A_{st}E_{st}}$$

Similarly, the deflection of the concrete is

$$\delta_c = \frac{F_c L}{A_c E_c}$$

---

[1]The terms *deformation* and *elongation* are often used interchangeably in this context.

But, $\delta_c = \delta_{st}$ since the bonding is perfect. Therefore,

$$\frac{F_c L}{A_c E_c} - \frac{F_{st}L}{A_{st}E_{st}} = 0$$

The first and last equations are solved simultaneously for $F_c$ and $F_{st}$.

$$F_c = \frac{F}{1 + \dfrac{A_{st}E_{st}}{A_c E_c}}$$

$$F_{st} = \frac{F}{1 + \dfrac{A_c E_c}{A_{st}E_{st}}}$$

### Example 45.2

A uniform bar is clamped at both ends and the axial load applied near one of the supports. What are the reactions?

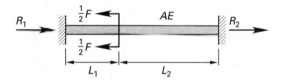

*Solution*

The first required equation is

$$R_1 + R_2 = F$$

The shortening of section 1 due to the reaction $R_1$ is

$$\delta_1 = \frac{-R_1 L_1}{AE}$$

The elongation of section 2 due to the reaction $R_2$ is

$$\delta_2 = \frac{R_2 L_2}{AE}$$

However, the bar is continuous, so $\delta_1 = -\delta_2$. Therefore,

$$R_1 L_1 = R_2 L_2$$

The first and last equations are solved simultaneously to find $R_1$ and $R_2$.

$$R_1 = \frac{F}{1 + \dfrac{L_1}{L_2}}$$

$$R_2 = \frac{F}{1 + \dfrac{L_2}{L_1}}$$

## Example 45.3

The non-uniform bar shown is clamped at both ends and constrained from changing length. What are the reactions if a temperature change of $\Delta T$ is experienced?

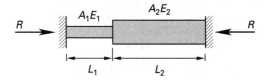

*Solution*

The thermal deformations of sections 1 and 2 can be calculated directly. Use Eq. 45.3.

$$\delta_1 = \alpha_1 L_1 \Delta T$$

$$\delta_1 = \alpha_1 L_2 \Delta T$$

The total deformation is $\delta = \delta_1 + \delta_2$. However, the deformation can also be calculated from the principles of mechanics.

$$\delta = \frac{R L_1}{A_1 E_1} + \frac{R L_2}{A_2 E_2}$$

These equations can be combined and solved directly for $R$.

$$R = \frac{(\alpha_1 L_1 + \alpha_2 L_2)\Delta T}{\dfrac{L_1}{A_1 E_1} + \dfrac{L_2}{A_2 E_2}}$$

## Example 45.4

The beam shown is supported by dissimilar members. The bar is rigid and remains horizontal.[2] The beam's mass is insignificant. What are the forces in the members?

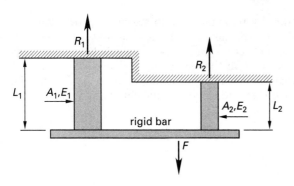

*Solution*

The required equilibrium condition is

$$R_1 + R_2 = F$$

The elongations of the two tension members are

$$\delta_1 = \frac{R_1 L_1}{A_1 E_1}$$

$$\delta_2 = \frac{R_2 L_2}{A_2 E_2}$$

Since the horizontal bar remains horizontal, $\delta_1 = \delta_2$.

$$\frac{R_1 L_1}{A_1 E_1} = \frac{R_2 L_2}{A_2 E_2}$$

The first and last equations are solved simultaneously to find $R_1$ and $R_2$.

$$R_1 = \frac{F}{1 + \dfrac{L_1 A_2 E_2}{L_2 A_1 E_1}}$$

$$R_2 = \frac{F}{1 + \dfrac{L_2 A_1 E_1}{L_1 A_2 E_2}}$$

## Example 45.5

The beam shown is supported by dissimilar members. The bar is rigid but is not constrained to remain horizontal. The beam's mass is insignificant. Develop the simultaneous equations needed to determine the reactions in the vertical members.

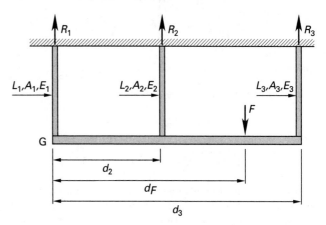

*Solution*

The forces in the supports are $R_1$, $R_2$, and $R_3$. Any of these may be tensile (positive) or compressive (negative).

$$R_1 + R_2 + R_3 = F$$

---

[2]This example is easily solved by summing moments about a point on the horizontal beam.

The changes in length are given by Eq. 45.2.

$$\delta_1 = \frac{R_1 L_1}{A_1 E_1}$$

$$\delta_2 = \frac{R_2 L_2}{A_2 E_2}$$

$$\delta_3 = \frac{R_3 L_3}{A_3 E_3}$$

Since the bar is rigid, the deflections will be proportional to the distance from point G.

$$\delta_2 = \delta_1 + \left(\frac{d_2}{d_3}\right)(\delta_3 - \delta_1)$$

Moments can be summed about point G to give a third equation.

$$M_{\mathrm{G}} = R_3 d_3 + R_2 d_2 - F d_F = 0$$

### Example 45.6

A load is supported by three tension members. Develop the simultaneous equations needed to find the forces in the three members.

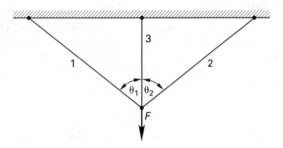

*Solution*

The equilibrium requirement is

$$F_{1y} + F_3 + F_{2y} = F$$

$$F_1 \cos\theta_1 + F_3 + F_2 \cos\theta_2 = F$$

Assuming the elongations are small compared to the member lengths, the angles $\theta_1$ and $\theta_2$ are unchanged. Then, the vertical deflections are the same for all three members.

$$\frac{F_1 L_1 \cos\theta_1}{A_1 E_1} = \frac{F_3 L_3}{A_3 E_3} = \frac{F_2 L_2 \cos\theta_2}{A_2 E_2}$$

These equations can be solved simultaneously to find $F_1$, $F_2$, and $F_3$. (It may be necessary to work with the $x$-components of the deflections in order to find a third equation.)

## 6. SUPERPOSITION METHOD

Two-span (three-support) beams and propped cantilevers are indeterminate to the first degree. Their reactions can be determined from a variation of the consistent deformation procedure known as the *superposition method*.[3] This method requires finding the deflection with one or more supports removed and then satisfying the known conditions.

*step 1:* Remove enough redundant supports to reduce the structure to a statically determinate condition.

*step 2:* Calculate the deflections at the locations of the previous redundant supports. Use consistent sign conventions.

*step 3:* Apply each redundant support as an isolated load, and find the deflections at the redundant support points as functions of the redundant support forces.

*step 4:* Use superposition to combine (i.e., add) the deflections due to the actual loads and the redundant support loads. The total deflections must agree with the known deflections (usually zero) at the redundant support points.

### Example 45.7

A propped cantilever is loaded by a concentrated force at midspan. Determine the reaction, $S$, at the prop.

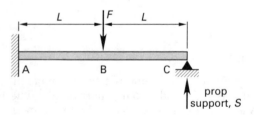

*Solution*

Start by removing the unknown prop reaction at point C. The cantilever beam is then statically determinate. The deflection and slope at point B can be found or derived from the elastic beam deflection equations. For a cantilever with end load (see App. 52.A, Case 1), the deflection and slope are calculated as follows.

$$\text{deflection:} \quad \delta_{\mathrm{B}} = \frac{-F L^3}{3EI}$$

$$\text{slope:} \quad y'_{\mathrm{B}} = \frac{-F L^2}{2EI}$$

---

[3]Superposition can also be used with higher-order indeterminate problems. However, the simultaneous equations that must be solved may make superposition unattractive for manual calculations.

Statics

The slope remains constant to the right of point B. Therefore, the deflection at point C due to the load at point B is

$$\delta_{\mathrm{C},F} = \delta_{\mathrm{B}} + y'_{\mathrm{B}}L$$
$$= \frac{-5FL^3}{6EI}$$

The upward deflection at the cantilever tip due to the prop support, $S$, alone is given by

$$\delta_{\mathrm{C},S} = \frac{S(2L)^3}{3EI} = \frac{8SL^3}{3EI}$$

Now, it is known that the actual deflection at point C is zero (the boundary condition). Therefore, the prop support, $S$, can be determined as a function of the applied load.

$$\delta_{\mathrm{C},S} + \delta_{\mathrm{C},F} = 0$$

$$\frac{8SL^3}{3EI} - \frac{5FL^3}{6EI} = 0$$

$$S = \frac{5F}{16}$$

## 7. THREE-MOMENT EQUATION

A *continuous beam* has two or more spans (i.e., three or more supports) and is statically indeterminate. (See Fig. 45.2.) The *three-moment equation* is a method of determining the reactions on continuous beams. It relates the moments at any three adjacent supports. The three-moment method can be used with a two-span beam to directly find all three reactions.

When a beam has more than two spans, the equation must be used with three adjacent supports at a time, starting with a support whose moment is known. (The moment is known to be zero at a simply supported end. For a cantilever end, the moment depends only on the loads on the cantilever portion.)

**Figure 45.2** *Portion of a Continuous Beam*

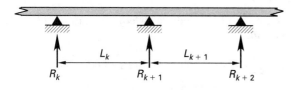

In its most general form, the three-moment equation is applicable to beams with non-uniform cross sections. In Eq. 45.4, $I_k$ is the moment of inertia of span $k$.

$$\frac{M_k L_k}{I_k} + (2M_{k+1})\left(\frac{L_k}{I_k} + \frac{L_{k+1}}{I_{k+1}}\right) + \frac{M_{k+2}L_{k+1}}{I_{k+1}}$$
$$= -6\left(\frac{A_k a}{I_k L_k} + \frac{A_{k+1}b}{I_{k+1}L_{k+1}}\right) \qquad 45.4$$

Equation 45.4 uses the following special nomenclature.

$a$  distance from the left support to the centroid of the moment diagram on the left span

$b$  distance from the right support to the centroid of the moment diagram on the right span

$I_k$  the moment of inertia of the open span between supports $k$ and $k + 1$

$L_k$  length of the span between supports $k$ and $k + 1$

$M_k$  bending moment at support $k$

$A_k$  area of moment diagram between supports $k$ and $k + 1$, assuming that the span is simply and independently supported

The products $Aa$ and $Ab$ are known as *first moments of the areas*. It is convenient to derive simplified expressions for $Aa$ and $Ab$ for commonly encountered configurations. Several are presented in Fig. 45.3.

For beams with uniform cross sections, the moment of inertia terms can be eliminated.

**Figure 45.3** *Simplified Three-Moment Equation Terms*

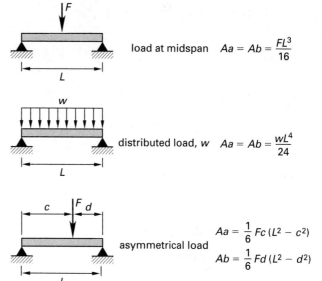

$$M_k L_k + (2M_{k+1})(L_k + L_{k+1}) + M_{k+2}L_{k+1}$$
$$= -6\left(\frac{A_k a}{L_k} + \frac{A_{k+1}b}{L_{k+1}}\right) \qquad 45.5$$

Statics

### Example 45.8

Find the four reactions supporting the beam. $EI$ is constant.

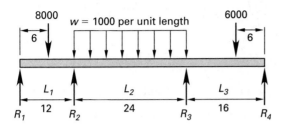

*Solution*

*Spans 1 and 2:*

Since the three-moment method can be applied to only two spans at a time, work first with the left and middle spans (spans 1 and 2).

From Fig. 45.3, the quantities $A_1a$ and $A_2b$ are

$$A_1 a = \frac{FL^3}{16} = \frac{(8000)(12)^3}{16} = 864{,}000$$

$$A_2 b = \frac{wL^4}{24} = \frac{(1000)(24)^4}{24} = 13{,}824{,}000$$

Since the left end of the beam is simply supported, $M_1$ is zero. Therefore, the three-moment equation (see Eq. 45.5) becomes

$$(2M_2)(L_1 + L_2) + M_3 L_2 = -6\left(\frac{A_1 a}{L_1} + \frac{A_2 b}{L_2}\right)$$

$$(2M_2)(12 + 24) + M_3(24) = (-6)\left(\frac{864{,}000}{12} + \frac{13{,}824{,}000}{24}\right)$$

After simplification,

$$3M_2 + M_3 = -162{,}000$$

*Spans 2 and 3:*

From the previous calculations,

$$A_2 a = A_2 b = 13{,}824{,}000$$

From Fig. 45.3 for the third span,

$$A_3 b = \tfrac{1}{6}Fd(L^2 - d^2)$$
$$= \left(\tfrac{1}{6}\right)(6000)(6)\left((16)^2 - (6)^2\right)$$
$$= 1{,}320{,}000$$

Since the right end is simply supported, $M_4 = 0$ and the three-moment equation is

$$M_2 L_2 + (2M_3)(L_2 + L_3) = -6\left(\frac{A_2 a}{L_2} + \frac{A_3 b}{L_3}\right)$$

$$M_2(24) + (2M_3)(24 + 16)$$
$$= (-6)\left(\frac{13{,}824{,}000}{24} + \frac{1{,}320{,}000}{16}\right)$$

After simplifying,

$$0.3M_2 + M_3 = -49{,}388$$

There are two equations in two unknowns ($M_2$ and $M_3$). A simultaneous solution yields

$$M_2 = -41{,}708$$

$$M_3 = -36{,}875$$

*Finding reactions:*

$M_2$ can be written in terms of the loads and reactions to the left of support 2. Assuming clockwise moments are positive,

$$M_2 = 12R_1 - (6)(8000) = -41{,}708$$

$$R_1 = 524.3$$

Now that $R_1$ is known, moments can be taken from support 3 to the left.

$$M_3 = (36)(524.3) + 24R_2 - (30)(8000) - (12)(24{,}000)$$
$$= -36{,}875$$

$$R_2 = 19{,}677.1$$

Similarly, $R_4$ and $R_3$ can be determined by working from the right end to the left. Assuming counterclockwise moments are positive,

$$M_3 = 16R_4 - (10)(6000) = -36{,}875$$

$$R_4 = 1445.3$$

$$M_2 = (40)(1445) + 24R_3 - (34)(6000) - (12)(24{,}000)$$
$$= -41{,}708$$

$$R_3 = 16{,}353.3$$

*Check:*

It is a good idea to check for equilibrium in the vertical direction.

$$\sum \text{loads} = 8000 + 24{,}000 + 6000$$
$$= 38{,}000$$
$$\sum \text{reactions} = 524.3 + 19{,}677.1 + 1445.3 + 16{,}353.3$$
$$= 38{,}000$$

## 8. FIXED-END MOMENTS

When the end of a beam is constrained against rotation, it is said to be a *fixed end* (also known as a *built-in end*). The ends of fixed-end beams are constrained to remain horizontal. Cantilever beams have a single fixed end. Some beams, as illustrated in Fig. 45.1, have two fixed ends and are known as *fixed-end beams*.[4]

Fixed-end beams are inherently indeterminate. To reduce the work required to find end moments and reactions, tables and books of fixed-end moments are often used.

## 9. INDETERMINATE TRUSSES

It is possible to manually calculate the forces in all members of an indeterminate truss. However, due to the time required, it is preferable to limit such manual calculations to trusses that are indeterminate to the first degree. The following *dummy unit load method* can be used to solve trusses with a single redundant member.

*step 1:* Draw the truss twice. Omit the redundant member on both trusses. (There may be a choice of redundant members.)

*step 2:* Load the first truss (which is now determinate) with the actual loads.

*step 3:* Calculate the force, $S$, in each of the members. Assign a positive sign to tensile forces.

*step 4:* Load the second truss with two unit forces acting collinearly toward each other along the line of the redundant member.

*step 5:* Calculate the force, $u$, in each of the members.

*step 6:* Calculate the force in the redundant member from Eq. 45.6.

$$S_{\text{redundant}} = \frac{-\sum \dfrac{SuL}{AE}}{\sum \dfrac{u^2 L}{AE}} \qquad 45.6$$

If $AE$ is the same for all members,

$$S_{\text{redundant}} = \frac{-\sum SuL}{\sum u^2 L} \qquad 45.7$$

The true force in member $j$ of the truss is

$$F_{j,\text{true}} = S_j + S_{\text{redundant}} u_j \qquad 45.8$$

### Example 45.9

Find the force in members BC and BD. $AE = 1$ for all members except for CB, which is 2, and AD, which is 1.5.

[4]The definition is loose. The term *fixed-end beam* can also be used to mean any indeterminate beam with at least one built-in end (e.g., a propped cantilever).

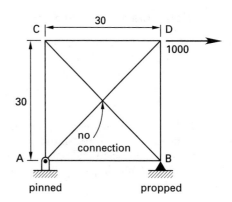

*Solution*

The two trusses are shown appropriately loaded.

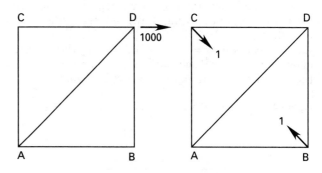

| member | $L$ | $AE$ | $S$ | $u$ | $\dfrac{SuL}{AE}$ | $\dfrac{u^2 L}{AE}$ |
|--------|-----|------|-----|-----|------|------|
| AB | 30 | 1 | 0 | −0.707 (C) | 0 | 15 |
| BD | 30 | 1 | −1000 (C) | −0.707 | 21,210 | 15 |
| DC | 30 | 1 | 0 | −0.707 | 0 | 15 |
| CA | 30 | 1 | 0 | −0.707 | 0 | 15 |
| CB | 42.43 | 2 | 0 | 1.0 | 0 | 21.22 |
| AD | 42.43 | 1.5 | 1414 (T) | 1.0 | 39,997 | 28.29 |
| | | | | | 61,207 | 109.51 |

From Eq. 45.6,

$$S_{\text{BC}} = \frac{-61{,}207}{109.51}$$
$$= -558.9 \ (C)$$

From Eq. 45.8,

$$F_{\text{BD,true}} = -1000 + (-558.9)(-0.707)$$
$$= -604.9 \ (C)$$

## 10. INFLUENCE DIAGRAMS

Shear, moment, and reaction influence diagrams (influence lines) can be drawn for any point on a beam or truss. This is a necessary first step in the evaluation of stresses induced by moving loads. It is important to realize, however, that the influence diagram applies only to one point on the beam or truss.

## Influence Diagrams for Beam Reactions

In a typical problem, the load is fixed in position and the reactions do not change. If a load is allowed to move across a beam, the reactions will vary. An influence diagram can be used to investigate the value of a chosen reaction as the load position varies.

To make the influence diagram as general in application as possible, a unit load is used. As an example, consider a 20 ft, simply supported beam and determine the effect on the left reaction of moving a 1 lbf load across the beam.

If the load is directly over the right reaction ($x = 0$), the left reaction will not carry any load. Therefore, the ordinate of the influence diagram is zero at that point. (Even though the right reaction supports 1 lbf, this influence diagram is being drawn for one point only—the left reaction.) Similarly, if the load is directly over the left reaction ($x = L$), the ordinate of the influence diagram will be 1. Basic statics can be used to complete the rest of the diagram, as shown in Fig. 45.4.

**Figure 45.4** *Influence Diagram for Reaction of Simple Beam*

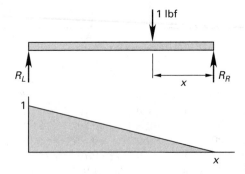

Use this rudimentary example of an influence diagram to calculate the left reaction for any placement of any load by multiplying the actual load by the ordinate of the influence diagram.

$$R_L = P \times \text{ordinate} \qquad 45.9$$

Even though the influence diagram was drawn for a point load, it can still be used when the beam carries a uniformly distributed load. In the case of a uniform load of $w$ distributed over the beam from $x_1$ to $x_2$, the left reaction can be calculated from Eq. 45.10.

$$R_L = \int_{x_1}^{x_2} (w \times \text{ordinate})\, dx$$
$$= w \times \text{area under curve} \qquad 45.10$$

## Example 45.10

A 500 lbf load is placed 15 ft from the right end of a 20 ft, simply supported beam. Use the influence diagram to determine the left reaction.

*Solution*

Since the influence line increases linearly from 0 to 1, the ordinate is the ratio of position to length. That is, the ordinate is $15/20 = 0.75$. The left reaction is

$$R_L = (0.75)(500 \text{ lbf}) = 375 \text{ lbf}$$

## Example 45.11

A uniform load of 15 lbf/ft is distributed between $x = 4$ ft and $x = 10$ ft along a 20 ft, simply supported beam. What is the left reaction?

*Solution*

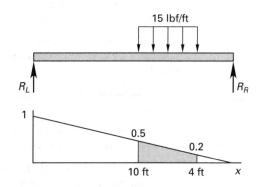

From Eq. 45.10, the left reaction can be calculated from the area under the influence diagram between the limits of loading.

$$\text{area} = \left(\tfrac{1}{2}\right)(10 \text{ ft})(0.5) - \left(\tfrac{1}{2}\right)(4 \text{ ft})(0.2) = 2.1 \text{ ft}$$

The left reaction is

$$R_L = \left(15 \, \frac{\text{lbf}}{\text{ft}}\right)(2.1 \text{ ft}) = 31.5 \text{ lbf}$$

## Finding Reaction Influence Diagrams Graphically

Since the reaction will always have a value of 1 when the unit load is directly over the reaction and since the reaction is always directly proportional to the distance $x$, the reaction influence diagram can be easily determined from the following steps.

*step 1:* Remove the support being investigated.

*step 2:* Displace (lift) the beam upward a distance of one unit at the support point. The resulting beam shape will be the shape of the reaction influence diagram.

## Example 45.12

What is the approximate shape of the reaction influence diagram for reaction 2?

*Solution*

Pushing up at reaction 2 such that the deflection is one unit results in the shown shape.

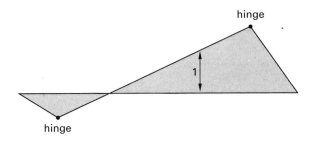

## Influence Diagrams for Beam Shears

A shear influence diagram (not the same as a shear diagram) illustrates the effect on the shear at a particular point in the beam of moving a load along the beam's length. As an illustration, consider point A along the simply supported beam of length 20.

In all cases, principles of statics can be used to calculate the shear at point A as the sum of loads and reactions on the beam from point A to the left end. (With the appropriate sign convention, summation to the right end could be used as well.) If the unit load is placed between the right end ($x = 0$) and point A, the shear at point A will consist only of the left reaction, since there are no other loads between point A and the left end. From the reaction influence diagram, the left reaction varies linearly. At $x = 12$ ft, the location of point A, the shear is $V = R_L = 12/20 = 0.6$.

When the unit load is between point A and the left end, the shear at point A is the sum of the left reaction (upward and positive) and the unit load itself (downward and negative). Therefore, $V = R_L - 1$. At $x = 12$ ft, the shear is $V = 0.6$ lbf $- 1$ lbf $= -0.4$ lbf.

Figure 45.5 is the shear influence diagram. In the diagram, notice that the shear goes through a reversal of 1. It is also helpful to note that the slopes of the two inclined sections are the same.

Shear influence diagrams are used in the same manner as reaction influence diagrams. The shear at point A for any position of the load can be calculated by multiplying the ordinate of the diagram by the actual load. Distributed loads are found by multiplying the uniform load by the area under the diagram between the limits of loading. If the loading extends over positive and negative parts of the curve, the sign of the area is considered when performing the final summation.

If it is necessary to determine the distribution of loading that will produce the maximum shear at a point whose influence diagram is available, the load should be positioned in order to maximize the area under the diagram.[5] This can be done by "covering" either all of the positive area or all of the negative area.[6]

**Figure 45.5** *Shear Influence Diagram for Simple Beam*

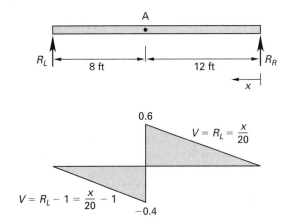

## Shear Influence Diagrams by Virtual Displacement

A difficulty in drawing shear influence diagrams for continuous beams on more than two supports is finding the reactions. The method of *virtual displacement* or *virtual work* can be used to find the influence diagram without going through that step.

*step 1:* Replace the point being investigated (i.e., point A) with an imaginary link with unit length. (It may be necessary to think of the link as having a length of 1 ft, but the link does not add to or subtract from any length of the beam.) If the point being investigated is a reaction, place a hinge at that point and lift the hinge upward a unit distance.

*step 2:* Push the two ends of the beam (with the link somewhere in between) toward each other a very small amount until the linkage is vertical. The distance between supports does not change, but the linkage allows the beam sections to assume a slope. The sections to the left and right of the linkage displace $\delta_1$ and $\delta_2$, respectively, from their equilibrium positions. The slope of both sections is the same. Points of support remain in contact with the beam.

---

[5]If the *minimum shear* is requested, the maximum negative shear is implied. The minimum shear is not zero in most cases.
[6]Usually, the dead load is assumed to extend over the entire length of the beam. The uniform live loads are distributed in any way that will cause the maximum shear.

*step 3:* Determine the ratio of $\delta_1$ and $\delta_2$. Since the slope on the two sections is the same, the longer section will have the larger deflection. If $L = a + b$ is the length of the beam, the relationships between the deflections can be determined from Eq. 45.11 through Eq. 45.13.

$$\delta_1 + \delta_2 = 1$$

$$\frac{\delta_1}{\delta_2} = \frac{a}{b} \qquad 45.11$$

$$\delta_1 = \left(\frac{a}{L}\right)\delta \qquad 45.12$$

$$\delta_2 = \left(\frac{b}{L}\right)\delta \qquad 45.13$$

Since $\delta = \delta_1 + \delta_2$ was chosen as 1, Eq. 45.12 and Eq. 45.13 really give the relative proportions of the unit link that extend below and above the reference line in Fig. 45.6.

**Figure 45.6** *Virtual Beam Displacements*

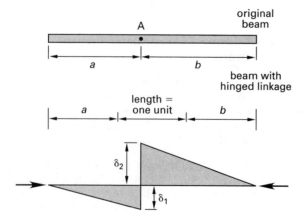

Knowing that the total shear reversal through point A is one unit and that the slopes are the same, the relative proportions of the reversal below and above the line will determine the shape of the displaced beam. The shape of the influence diagram is the shape taken on by the beam.

*step 4:* As required, use equations of straight lines to obtain the shear influence ordinate as a function of position along the beam.

## Example 45.13

For the simply supported beam shown, draw the shear influence diagram for a point 10 ft from the right end.

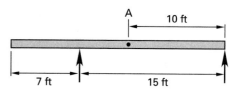

*Solution*

If a unit link is placed at point A and the beam ends are pushed together, the following shape will result. Notice that the beam must remain in contact with the points of support, and that the two slopes are the same.

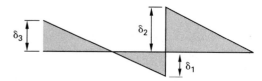

The overhanging 7 ft of beam do not change the shape of the shear influence diagram between the supports. The deflections can be evaluated assuming a 15 ft long beam.

$$\delta_1 = \frac{5 \text{ ft}}{15 \text{ ft}} = 0.33$$

$$\delta_2 = \frac{10 \text{ ft}}{15 \text{ ft}} = 0.67$$

The slope in both sections of the beam is the same. This slope can be used to calculate $\delta_3$.

$$m = \frac{\delta_1}{a} = \frac{0.33}{5 \text{ ft}} = 0.066 \text{ } 1/\text{ft}$$

$$\delta_3 = (7 \text{ ft})\left(0.066 \text{ } \frac{1}{\text{ft}}\right) = 0.46$$

## Example 45.14

Where should a uniformly distributed load be placed on the following beam to maximize the shear at section A?

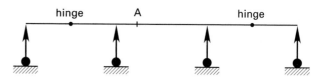

*Solution*

Using the principle of virtual displacement, the following shear influence diagram results by inspection. (It is not necessary to calculate the relative displacements to answer this question. It is only necessary to identify the positive and negative parts of the influence diagram.)

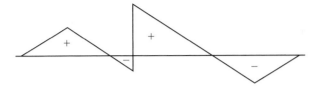

To maximize the shear, the uniform load should be distributed either over all positive or all negative sections of the influence diagram.

## Moment Influence Diagrams by Virtual Displacement

A moment influence diagram (not the same as a moment diagram) gives the moment at a particular point for any location of a unit load. The method of virtual displacement can be used in this situation to simplify finding the moment influence diagram. (See Fig. 45.7.)

**Figure 45.7** *Moment Influence Diagram by Virtual Displacement*

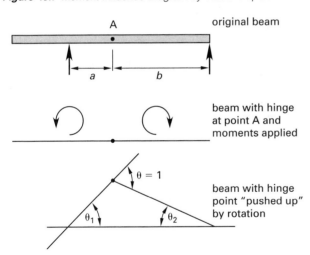

*step 1:* Replace the point being investigated (i.e., point A) with an imaginary hinge.

*step 2:* Rotate the beam one unit rotation by applying equal but opposite moments to each of the two beam sections. Except where the point being investigated is at a support, this unit rotation can be achieved simply by "pushing up" on the beam at the hinge point.

*step 3:* The angles made by the sections on either side of the hinge will be proportional to the lengths of the opposite sections. (Since the angle is small for a virtual displacement, the angle and its tangent, or slope, are the same.)

$$\theta_1 = \frac{b}{L} \qquad 45.14$$

$$\theta_2 = \frac{a}{L} \qquad 45.15$$

$$L = a + b \qquad 45.16$$

### Example 45.15

What are the approximate shapes of the moment influence diagrams for points A and B on the beam shown?

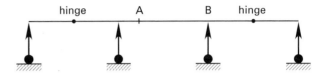

*Solution*

By placing an imaginary hinge at point A and rotating the two adjacent sections of the beam, the following shape results.

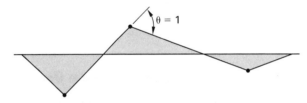

The moment influence diagram for point B is found by placing an imaginary hinge at point B and applying a rotating moment. Since the beam must remain in contact with all supports, and since there is no hinge between the two middle supports, the moment influence diagram must be horizontal in that region.

## Shear Influence Diagrams on Cross-Beam Decks

When girder-type construction is used to construct a road or bridge deck, the traffic loads will not be applied directly to the girder. Rather, the loads will be transmitted to the girder at panel points from cross beams (floor beams). Figure 45.8 shows a typical construction detail involving girders and cross beams.

**Statics**

**Figure 45.8** *Cross-Beam Decking*

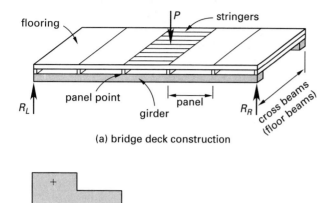

(a) bridge deck construction

(b) shear diagram for girder

A load applied to the deck stringers will be transmitted to the girder only at the panel points. Because the girder experiences a series of concentrated loads, the shear between panel points is horizontal. Since the shear is always constant between panel points, we speak of *panel shear* rather than shear at a point. Accordingly, shear influence diagrams are drawn for a panel, not for a point. Moment influence diagrams are similarly drawn for a panel.

### Influence Diagrams on Cross-Beam Decks

Shear and moment influence diagrams for girders with cross beams are identical to simple beams, except for the panel being investigated. Once the influence diagram has been drawn for the simple beam, the influence diagram ordinates at the ends of the panel being investigated are connected to obtain the influence diagram for the girder. This is illustrated in Fig. 45.9.

### Influence Diagrams for Truss Members

Since members in trusses are assumed to be axial members, they cannot carry shears or moments. Therefore, shear and moment influence diagrams do not exist for truss members. However, it is possible to obtain an influence diagram showing the variation in axial force in a given truss member as the load varies in position.

There are two general cases for finding forces in truss members. The force in a horizontal truss member is proportional to the moment across the member's panel. The force in an inclined truss member is proportional to the shear across that member's panel.

So, even though we may only want the axis load in a truss member, it is still necessary to construct the shear and moment influence diagrams for the entire truss in order to determine the applications of loading on the truss that produce the maximum shear and moment across the member's panel.

**Figure 45.9** *Comparison of Influence Diagrams for Simple Beams and Girders (Panel bc)*

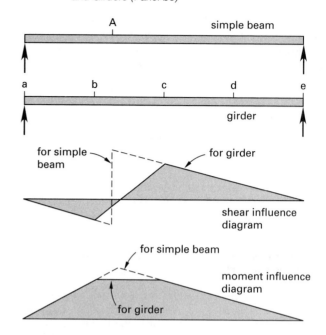

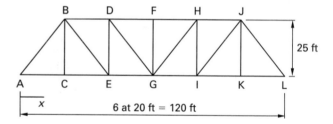

### Example 45.16

(a) Draw the influence diagram for vertical shear in panel DF of the through truss shown. (b) What is the maximum force in member DG if a 1000 lbf load moves across the truss?

*Solution*

(a) Allow a unit load to move from joint L to joint G along the lower chords. If the unit vertical load is at a distance $x$ from point L, the right reaction will be $+(1 - (x/120))$. The unit load itself has a value of $-1$, so the shear at distance $x$ is just $-x/120$.

Allow a unit load to move from joint A to joint E along the lower chords. If the unit load is a distance $x$ from point L, the left reaction will be $x/120$, and the shear at distance $x$ will be $(x/120) - 1$.

These two lines can be graphed.

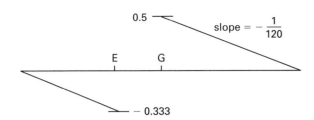

The influence line is completed by connecting the two lines as shown. Therefore, the maximum shear in panel DF will occur when a load is at point G on the truss.

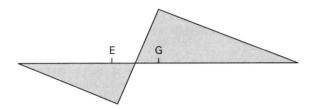

(b) If the 1000 lbf load is at point G, the two reactions at points A and L will each be 500 lbf. The cut-and-sum method can be used to calculate the force in member DG simply by evaluating the vertical forces on the free-body to the left of point G.

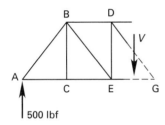

For equilibrium to occur, $V$ must be 500. This vertical shear is entirely carried by member DG. The length of member DG is

$$\sqrt{(20 \text{ ft})^2 + (25 \text{ ft})^2} = 32 \text{ ft}$$

The force in member DG is

$$\left(\frac{32 \text{ ft}}{25 \text{ ft}}\right)(500 \text{ lbf}) = 640 \text{ lbf}$$

### Example 45.17

(a) Draw the moment influence diagram for panel DF on the truss shown in Ex. 45.16. (b) What is the maximum force in member DF if a 1000 lbf load moves across the truss?

*Solution*

(a) The left reaction is $x/120$ where $x$ is the distance from the unit load to the right end. If the unit load is to the right of point G, the moment can be found by summing moments from point G to the left. The moment is $(x/120)(60) = 0.5x$.

If the unit load is to the left of point E, the moment will again be found by summing moments about point G. The distance between the unit load and point G is $x - 60$.

$$\left(\frac{x}{120}\right)(60) - (1)(x - 60) = 60 - 0.5x$$

These two lines can be graphed. The moment for a unit load between points E and G is obtained by connecting the two end points of the lines derived above. Therefore, the maximum moment in panel DF will occur when the load is at point G on the truss.

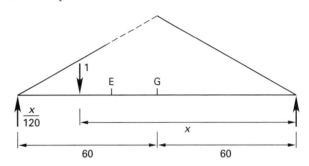

(b) If the 1000 lbf load is at point G, the two reactions at points A and L will each be 500 lbf. The method of sections can be used to calculate the force in member DF by taking moments about joint G.

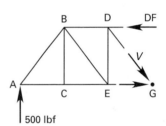

$$\sum M_{\text{G}}: (500 \text{ lbf})(60 \text{ ft}) - (\text{DF})(25 \text{ ft}) = 0$$
$$\text{DF} = 1200 \text{ lbf}$$

## 11. MOVING LOADS ON BEAMS

### Global Maximum Moment Anywhere on Beam

If a beam supports a single moving load, the maximum bending and shearing stresses at any point can be found by drawing the moment and shear influence diagrams for that point. Once the positions of maximum moment and maximum shear are known, the stress at the point in question can be found from $Mc/I$.

If a simply supported beam carries a set of moving loads (which remain equidistant as they travel across the beam), the following procedure can be used to find the *dominant load*. (The dominant load is the one that occurs directly over the point of maximum moment.)

*step 1:* Calculate and locate the resultant of the load group.

*step 2:* Assume that one of the loads is dominant. Place the group on the beam such that the distance from one support to the assumed dominant load is equal to the distance from the other support to the resultant of the load group.

*step 3:* Check to see that all loads are on the span and that the shear changes sign under the assumed dominant load. If the shear does not change sign under the assumed dominant load, the maximum moment may occur when only some of the load group is on the beam. If it does change sign, calculate the bending moment under the assumed dominant load.

*step 4:* Repeat steps 2 and 3, assuming that the other loads are dominant.

*step 5:* Find the maximum shear by placing the load group such that the resultant is a minimum distance from a support.

## Placement of Load Group to Maximize Local Moment

In the design of specific members or connections, it is necessary to place the load group in a position that will maximize the load on those members or connections. The procedure for finding these positions of local maximum loadings is different from the global maximum procedures.

The solution to the problem of local maximization is somewhat trial-and-error oriented. It is aided by use of the influence diagram. In general, the variable being evaluated (reaction, shear, or moment) is maximum when one of the wheels is at the location or section of interest.

When there are only two or three wheels in the load group, the various alternatives can be simply evaluated by using the influence diagram for the variable being evaluated. When there are many loads in the load group (e.g., a train loading), it may be advantageous to use heuristic rules for predicting the dominant wheel.

# Topic IX: Material Science

# 46 Engineering Materials

## Nomenclature

| | | | |
|---|---|---|---|
| $A$ | area | $in^2$ | $m^2$ |
| $c$ | distance from neutral axis to extreme fiber | in | m |
| $D$ | diameter | in | m |
| DP | degree of polymerization | – | – |
| $E$ | modulus of elasticity | $lbf/in^2$ | Pa |
| $f$ | fraction (moisture) | – | – |
| $f_c'$ | compressive strength | $lbf/in^2$ | Pa |
| $I$ | moment of inertia | $in^4$ | $m^4$ |
| $L$ | length | in | m |
| $m$ | mass | lbm | kg |
| $M$ | moment | in-lbf | N·m |
| MC | moisture content | % | % |
| MW | molecular weight | – | – |
| $p$ | pressure | $lbf/in^2$ | Pa |
| $P$ | force | lbf | N |
| $S_H$ | circumferential stress | $lbf/in^2$ | Pa |
| SG | specific gravity | – | – |
| $V$ | volume | $ft^3$ | $m^3$ |
| $w_c$ | specific weight of concrete (ACI 318 nomenclature) | $lbf/ft^3$ | $kg/m^3$ |
| $W$ | weight | lbf | n.a. |

## Symbols

| | | | |
|---|---|---|---|
| $\gamma$ | specific weight | $lbf/ft^3$ | n.a. |
| $\rho$ | density | $lbm/ft^3$ | $kg/m^3$ |

## Subscripts

| | |
|---|---|
| $c$ | concrete |
| $ct$ | cylinder tensile (splitting) |
| $r$ | rupture |
| SSD | saturated surface dry |

## 1. CHARACTERISTICS OF METALS

Metals are the most frequently used materials in engineering design. Steel is the most prevalent engineering metal because of the abundance of iron ore, simplicity of production, low cost, and predictable performance.

However, other metals play equally important parts in specific products.

Most metals are characterized by the properties in Table 46.1.

**Table 46.1** *Properties of Most Metals and Alloys*

high thermal conductivity (low thermal resistance)
high electrical conductivity (low electrical resistance)
high chemical reactivity[a]
high strength
high ductility[b]
high density
high radiation resistance
highly magnetic (ferrous alloys)
optically opaque
electromagnetically opaque

[a]Some alloys, such as stainless steel, are more resistant to chemical attack than pure metals.
[b]Brittle metals, such as some cast irons, are not ductile.

*Metallurgy* is the subject that encompasses the procurement and production of metals. *Extractive metallurgy* is the subject that covers the refinement of pure metals from their ores.

## 2. UNIFIED NUMBERING SYSTEM

The Unified Numbering System (UNS) was introduced in the mid-1970s to provide a consistent identification of metals and alloys for use throughout the world. The UNS designation consists of one of seventeen single uppercase letter prefixes followed by five digits. Many of the letters are suggestive of the family of metals, as Table 46.2 indicates.

**Table 46.2** *UNS Alloy Prefixes*

| | |
|---|---|
| A | aluminum |
| C | copper |
| E | rare-earth metals |
| F | cast irons |
| G | AISI and SAE carbon and alloy steels |
| H | AISI and SAE H-steels |
| J | cast steels (except tool steels) |
| K | miscellaneous steels and ferrous alloys |
| L | low-melting metals |
| M | miscellaneous nonferrous metals |
| N | nickel |
| P | precious metals |
| R | reactive and refractory metals |
| S | heat- and corrosion-resistant steels (stainless and valve steels and superalloys) |
| T | tool steels (wrought and cast) |
| W | welding filler metals |
| Z | zinc |

For each UNS designation, there is a specific percentage range of critical alloying elements. However, the UNS designation is a description, not a specification. Specifications are administered by the American Society of Testing and Materials (ASTM) and similar organizations. The UNS designation refers only to the major alloying and residual elements. It is not an exact specification. One manufacturer may produce an alloy in the middle of the UNS ranges, while another manufacturer may operate at the low or high end of the ranges.[1] The presence of small amounts of residual elements, directionality due to manufacturing processes (e.g., rolling), and heat treatments are also not part of the specification. Furthermore, the UNS designations are not always sufficiently unique to differentiate between two existing products. Additional specifications or trade names are still required in those instances.

A cross-reference index is generally needed to convert older designations to the UNS designation.[2] However, for many stainless steels, the first three digits are the same as the AISI numbering system. Straight AISI type 304 is written simply as UNS S30400. The last two digits may be used to designate some differentiating characteristic. For example, stainless 304L, with a maximum of 0.03% carbon, is designated as S30403.[3] The UNS designations for other families (e.g., aluminum, copper, and nickel) also incorporate some or all of the common designations in use prior to the UNS.

## 3. BLAST FURNACE IRON PRODUCTION

Iron (chemical symbol Fe) is obtained from its oxides $Fe_2O_3$ (*hematite*, 69.9% iron) and, to a lesser extent, $Fe_3O_4$ (*magnetite*, 72.4% iron).[4] Only about 50% of iron ore consists of iron oxides, the remainder being the gangue. *Gangue* is the earth and stone mixed with the iron oxides.

The process used to reduce iron oxides to pure iron takes place in a *blast furnace*. The furnace is charged with alternate layers of iron ore, coke, and limestone in the approximate ratio of 4:2:1, respectively.[5] The limestone serves as a flux for the gangue and the coke ash, enabling the molten gangue and impurities to be drawn off as *slag*.

---

[1]In addition to economically lowering the carbon content, *argon-oxygen-decarburization* (AOD) during refining has made it possible to control nitrogen and other alloying ingredients precisely. The percentage of expensive alloying ingredients (e.g., molybdenum in stainless steels) will generally be at the low ends of the allowable ranges.
[2]SAE and ASTM jointly publish *Metals and Alloys in the Unified Numbering System*, which includes a cross-reference index.
[3]Since straight type 304 has a maximum of 0.08% carbon, some engineers prefer to write S30408 for uniformity.
[4]Since an additional roasting process is needed to remove the sulfur, iron pyrite, $FeS_2$, is not used in iron production, despite its abundance. Other lower-grade ores, such as $FeCO_3$ (siderite, 48.3% iron) and $Fe_2O_3 \cdot n[H_2O]$ (limonite, 60 to 65% iron), are used only in the absence of better ores.
[5]*Coke* is coal that has been previously burned in an oxygen-poor environment. The remaining carbonaceous material has a high-combustion energy content. (Clean-air legislation has had a substantial impact on coke making processes.)

A traditional blast furnace is shown in Fig. 46.1. The top of the furnace is provided with a pair of conical bells for loading the charge (when open) and limiting the escape of gases (when closed). High-temperature air is injected through nozzles around the periphery of the lower portion of the furnace. These openings are known as *tuyères*.[6]

**Figure 46.1** *Blast Furnace*

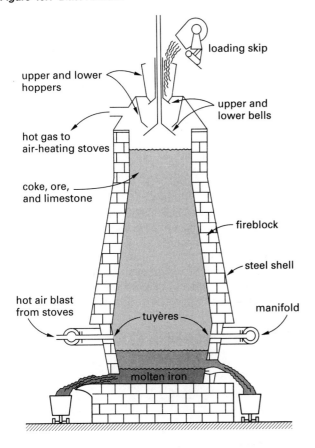

The hot combustion air is produced in preheaters (stoves) that adjoin the blast furnace. Generally, four stoves are provided for each furnace. Each stove is heated in rotation by burning the carbon monoxide-rich furnace gases. Cold air enters one stove while the remaining stoves are being heated. The air is heated to 1000°F to 1300°F (550°C to 700°C) before being injected into the blast furnace.

The injected air oxidizes the coke, producing heat and large amounts of carbon monoxide. The carbon monoxide rises to the top of the furnace and, at a temperature of approximately 600°F (300°C), reduces the iron oxide

to FeO. The following chemical reactions describe the production of FeO.

$$C + O_2 \rightarrow CO_2$$
$$CO_2 + C \rightarrow 2CO$$
$$2C + O_2 \rightarrow 2CO$$
$$3Fe_2O_3 + CO \rightarrow 2Fe_3O_4 + CO_2$$
$$Fe_3O_4 + CO \rightarrow 3FeO + CO_2$$

As the reduction process continues, the FeO temperature drops down to 1300°F to 1500°F (700°C to 800°C). The FeO is reduced to a spongy mass of pure iron by the carbon monoxide.

$$FeO + CO \rightarrow Fe + CO_2$$

The molten iron then drops into a region where the temperature is 1500°F to 2500°F (800°C to 1400°C). The iron becomes saturated with carbides and free carbon. The absorbed carbon lowers the melting point of the iron from approximately 2800°F (1550°C) to approximately 2100°F (1150°C) so that it runs as a liquid to the bottom of the furnace.

The slag melts at approximately the same temperature as the iron but, being less dense, floats on the liquid iron. This allows the slag and iron to be drawn off separately. The iron usually goes in a liquid state to a subsequent refinement process, but may be allowed to cool in molds (forming blocks of iron known as *pigs*). The slag is discarded in a *slag heap*.

Since liquid iron is an excellent solvent, pig iron contains all of the minerals that are not fluxed away by the liquid limestone. The approximate composition of pig iron is 3% to 4% carbon, 1% to 3% silicon, 0.1% to 2% phosphorus, 0.5% to 2% manganese, and 0.01% to 0.1% sulfur. The actual composition will depend on the gangue elements. Calcium, magnesium, and aluminum oxides are fluxed out by the molten limestone and appear in the slag.

The subsequent process that the pig iron undergoes depends on the desired end product (i.e., the desired carbon content). This is shown in Fig. 46.2. The Bessemer, oxygen, open hearth, and electric furnace processes are used to produce steel.

### 4. OXYGEN AND BESSEMER PROCESSES

Pig iron is saturated with carbon and contains other impurities. The *Bessemer* and *oxygen processes* (also known as the *dissolved oxygen process*, the L-D process, and the *Linz-Donawitz process*) are used to reduce the carbon content and purify the iron.[7] The conceptual differences between the oxygen and Bessemer processes shown in Table 46.3 are minor. However, the production

---

[6] *Tuyère* is pronounced twee-yer and too-ur.

[7] Linz and Donawitz are the two Austrian towns in which the oxygen process was perfected.

**Figure 46.2** *Methods of Refining Pig Iron*

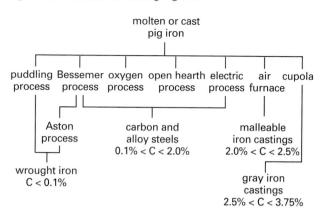

The impurities are completely oxidized in approximately 15 to 25 minutes (see Table 46.3), although loading and unloading extends the cycle time to approximately one hour. Since the refinement also eliminates beneficial elements, measured amounts of carbon, manganese, and other alloying ingredients are subsequently added at the end of the refinement process to obtain the desired steel grade.

**Table 46.3** *Comparison Between Oxygen and Bessemer Processes*

| characteristic | oxygen | Bessemer |
| --- | --- | --- |
| crucible lining | basic | acidic |
| oxidizer | pure oxygen | air |
| oxidizer supply | lance above | nozzles below |
| reaction time | 25 minutes | 10–15 minutes |
| typical charge size | 100 English tons | 25 English tons |
| temperature | higher | high |
| percent scrap steel in charge | up to 30% | 10–20% |

and economic advantages of the oxygen process are considerable.

The chemical refinement takes place in a pear-shaped steel crucible (a *converter*) lined with refractory material. The crucible is filled with a *bath* of molten pig iron at approximately 2200°F (1200°C), steel scrap, and lime. In the oxygen process, a water-cooled oxygen *lance* is lowered to within several feet (approximately a meter) of the bath surface. (See Fig. 46.3.)

Because a Bessemer reaction proceeds rapidly and because the process cannot be interrupted for analyses, the Bessemer process is considered to be crude. The progress of refinement is judged by the color and length of flame that issues from the mouth of the crucible. Also, phosphorus and sulfur are not affected by the Bessemer process, making further refinement often necessary. This process, when used at all, produces steel for less critical grades of sheet, wire, and pipe. (See Fig. 46.4.)

**Figure 46.3** *Oxygen Process Crucible*

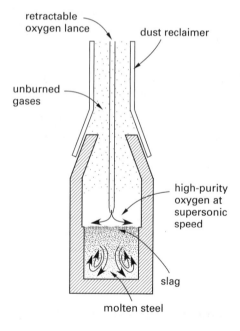

**Figure 46.4** *Bessemer Process Converter*

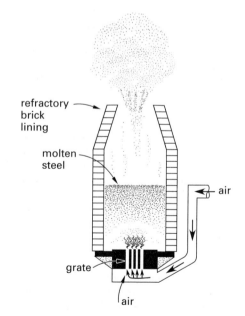

High-pressure oxygen flows through the lance at high velocity, pushing aside the molten slag and exposing the molten iron. (In the Bessemer process, air is injected from below.) The silicon and manganese impurities are oxidized first, causing a temperature rise to approximately 3500°F (1900°C). Carbon is oxidized at the higher temperatures. Since the reaction is violent, the bath churns and circulates naturally.

Bessemer steel that is subsequently refined in an open hearth process to remove sulfur, phosphorus, and iron oxide is known as *duplexed steel.*

Compared to the open hearth process, smaller batches of steel have traditionally been produced in the Bessemer process. However, larger vessels are now in use with the oxygen process. Also, since the refinement process proceeds faster than in an open hearth, the steel production rates are approximately the same.

## 5. OPEN HEARTH PROCESS

The major advantages of the open hearth process are control and flexibility in charge composition. A charge of 50 tons to 100 tons (45 Mg to 90 Mg) of scrap steel and limestone is placed in a fairly shallow furnace receptacle of large area.[8] Heat is provided by burning carbon monoxide (CO) above the charge. When the charge is molten, liquid pig iron from the blast furnace is added. Iron oxide (in the form of mill scale or iron ore) is added as an oxidizer. Gaseous oxygen from a lance is frequently used to accelerate the refinement.

The combustion air and fuel enter the hearth from one side, and a long flame plays over the charge. The combustion gases are withdrawn on the opposite side and pass through brick *checker chambers*, which absorb some of the residual heat. The incoming air is preheated to approximately 1800°F (1000°C) by passing through these chambers. When the chambers cool, the flow direction of air and fuel is reversed. (See Fig. 46.5.)

**Figure 46.5** *Open Hearth Furnace*

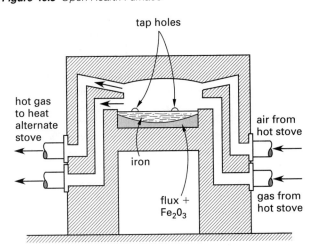

Since the combustion products contain very little free oxygen, oxidation of the impurities depends on the oxygen in the iron oxide. The carbon burns off as carbon monoxide and carbon dioxide. The molten limestone slag provides a protective covering over the melt (to prevent excessive oxidation and nitrogen absorption) and fluxes away the sulfur, phosphorus, and silicon in the steel.

Open hearth refinement takes eight to twelve hours, with melting and refining each taking approximately

half of the time. Continuous monitoring of the steel composition is possible, resulting in a high-grade steel. A typical final composition would be sulfur and phosphorus, less than 0.04% each; manganese, 0.05% to 0.35%; silicon, less than 0.01%; and carbon, as desired. The refined molten steel batch is known as a *heat of steel*.

When the steel is sufficiently pure, alloying elements are added to achieve the desired steel grade and properties. For example, manganese is added as necessary to combine with the sulfur remaining in the steel. Iron sulfide weakens the steel, whereas manganese sulfide does not.

Deoxidizers (ferromanganese and ferrosilicon) are added to compensate for the remaining high iron oxide content. If the steel is not deoxidized, the oxide reacts with carbon during solidification and produces large amounts of carbon monoxide gas. This gas is trapped in the steel, producing many voids and leading to the characteristic appearance of such *rimmed steel*. The gas voids usually do not constitute a defect since they are closed by welding during subsequent hot-working manufacturing processes.

Aluminum can be added to the steel to produce *killed steel*. No gas at all is evolved in killed steel during solidification. *Semi-killed steel* has some gas formation, but not as much as rimmed steel.

## 6. ELECTRIC ARC FURNACE

Electric furnaces common to "minimills" utilizing electric arc or induction heating are used to produce tool and special alloy steels.[9] It is possible to produce a high-quality steel because air and gaseous fuels are not required, and the impurities they introduce are eliminated.

To further increase the quality and to reduce the expensive refining time, the charge is usually select scrap or *direct-reduced iron* rather than molten pig iron. As with the open hearth process, iron oxide is added as an oxidizing agent, and the composition is modified following refinement.

In an *electric arc furnace*, heat is generated by electrical arcs from three electrodes (for three-phase current) extending through the furnace wall down into the charge space. Although the electrode voltage is low (approximately 40 V), the current is high (approximately 12,000 A). Coils surround an *induction furnace*, and the heating is created from eddy current flowing within the melt.

The two major problems associated with electric arc furnaces are (1) steel contamination from trace metals present in the charging scrap, and (2) the formation of ionized nitrogen in the arc, a cause of undesirable hardening.

---

[8]The charge can be as much as 500 English tons (450 Mg).

[9]The term "minimill" has become a misnomer. Electric arc furnaces can produce up to 130 tons/hr, not much less than the 150 tons/hr to 300 tons/hr capacity of a standard basic oxygen furnace.

## 7. ADVANCED STEEL-MAKING PROCESSES

Air-quality legislation has had a substantial impact on the processes used to produce coke. New steel-making technologies are being used to reduce or eliminate the need for coke entirely (as in *direct iron-making processes*). Existing blast furnaces can be retrofitted to use pulverized coal (*coal injection*). In *reduced-coke processes*, coal, iron pellets and fines, and limestone are added to an already molten iron bath. Carbon from the coal combines with oxygen from the ore to produce carbon monoxide and molten iron. Oxygen is injected to burn some of the gas before it leaves the vessel.

## 8. STEEL AND ALLOY STEEL GRADES

The properties of steel can be adjusted by the addition of alloying ingredients. Some steels are basically mixtures of iron and carbon. Other steels are produced with a variety of ingredients.

The simplest and most common grades of steel belong to the group of *carbon steels*. Carbon is the primary non-iron element, although sulfur, phosphorus, and manganese can also be present. Carbon steel can be subcategorized into *plain carbon steel (nonsulfurized carbon steel)*, *free-machining steel (resulfurized carbon steel)*, and *resulfurized and rephosphorized carbon steel*. Plain carbon steel is subcategorized into *low-carbon steel* (less than 0.30% carbon), *medium-carbon steel* (0.30% to 0.70% carbon), and *high-carbon steel* (0.70% to 1.40% carbon).

Low-carbon steels are used for wire, structural shapes, and screw machine parts. Medium-carbon steels are used for axles, gears, and similar parts requiring medium to high hardness and high strength. High-carbon steels are used for drills, cutting tools, and knives.

*Low-alloy steels* (containing less than 8.0% total alloying ingredients) include the majority of steel alloys but exclude the high-chromium content *corrosion-resistant (stainless) steels*. Generally, low-alloy steels will have higher strength (e.g., double the yield strength) of plain carbon steel. *Structural steel*, *high-strength steel*, and *ultrahigh-strength steel* are general types of low-alloy steel.[10]

*High-alloy steels* contain more than 8.0% total alloying ingredients.

Table 46.4 lists typical alloying ingredients and their effects on steel properties. The percentages represent typical values, not maximum solubilities.

---

[10]The *ultrahigh-strength steels*, also known as *maraging steels*, are very low-carbon (less than 0.03%) steels with 15% to 25% nickel and small amounts of cobalt, molybdenum, titanium, and aluminum. With precipitation hardening, ultimate tensile strengths up to 400,000 lbf/in$^2$ (2.8 GPa), yield strengths up to 250,000 lbf/in$^2$ (1.7 GPa), and elongations in excess of 10% are achieved. Maraging steels are used for rocket motor cases, aircraft and missile turbine housings, aircraft landing gear, and other applications requiring high strength, low weight, and toughness.

*Table 46.4* Steel Alloying Ingredients

| ingredient | range (%) | purpose |
|---|---|---|
| aluminum | – | deoxidation |
| boron | 0.001–0.003 | increase hardness |
| carbon | 0.1–4.0 | increase hardness and strength |
| chromium | 0.5–2 | increase hardness and strength |
| | 4–18 | increase corrosion resistance |
| copper | 0.1–0.4 | increase atmospheric corrosion resistance |
| iron sulfide | – | increase brittleness |
| manganese | 0.23–0.4 | reduce brittleness, combine with sulfur |
| | > 1.0 | increase hardness |
| manganese sulfide | 0.8–0.15 | increase machinability |
| molybdenum | 0.2–5 | increase dynamic and high-temperature strength and hardness |
| nickel | 2–5 | increase toughness, increase hardness |
| | 12–20 | increase corrosion resistance |
| | > 30 | reduce thermal expansion |
| phosphorus | 0.04–0.15 | increase hardness and corrosion resistance |
| silicon | 0.2–0.7 | increase strength |
| | 2 | increase spring steel strength |
| | 1–5 | improve magnetic properties |
| sulfur | – | (see *iron sulfide* and *manganese sulfide*) |
| titanium | – | fix carbon in inert particles; reduce martensitic hardness |
| tungsten | – | increase high-temperature hardness |
| vanadium | 0.15 | increase strength |

Since steel properties are dependent on composition, steels are designated by composition. Table 46.5 shows the AISI-SAE four-digit designations for typical steels and alloys.[11] The first two digits designate the type of steel; the last two digits designate the percentage of carbon in hundredths of a percent. (For example, AISI steel 1035 is a plain carbon steel with 0.35% carbon. This is also referred to as "35-point carbon" steel.)

---

[11]The abbreviations stand for the American Iron and Steel Institute and the Society of Automotive Engineers.

**Table 46.5** *AISI-SAE Steel Designations*

*carbon steels*
| | |
|---|---|
| 10XX | nonsulfurized carbon steel (plain-carbon) |
| 11XX | resulfurized carbon steel (free-machining) |
| 12XX | resulfurized and rephosphorized carbon steel |

*low-alloy steels*
| | |
|---|---|
| 13XX | manganese 1.75 |
| 23XX | nickel 3.50 |
| 25XX | nickel 1.25, chromium 0.65 |
| 31XX | nickel 3.50, chromium 1.55 |
| 33XX | nickel 3.50, chromium 1.55 |
| 40XX | molybdenum 0.25 |
| 41XX | chromium 0.50 or 0.95, molybdenum 0.12 or 0.20 |
| 43XX | nickel 1.80, chromium 0.50 or 0.80, molybdenum 0.25 |
| 46XX | nickel 1.55 or 1.80, molybdenum 0.20 or 0.25 |
| 47XX | nickel 1.05, chromium 0.45, molybdenum 0.20 |
| 48XX | nickel 3.50, molybdenum 0.25 |
| 50XX | chromium 0.38 or 0.40 |
| 51XX | chromium 0.80, 0.90, 0.95, 1.00, or 1.05 |
| 5XXXX | chromium 0.50, 1.00, or 1.45, carbon 1.00 |
| 61XX | chromium 0.60, vanadium 0.10–0.15; or chromium 0.95, vanadium 0.15 |
| 86XX | nickel 0.55, chromium 0.50 or 0.65, molybdenum 0.20 |
| 87XX | nickel 0.55, chromium 0.50, molybdenum 0.25 |
| 92XX | manganese 0.85, silicon 2.00 |
| 93XX | nickel 3.25, chromium 1.20, molybdenum 0.12 |
| 98XX | nickel 1.00, chromium 0.80, molybdenum 0.25 |

*heat-and corrosion-resistant steels*
| | |
|---|---|
| 2XX | chromium-nickel-manganese (nonhardenable, austenitic, nonmagnetic) |
| 3XX | chromium-nickel (nonhardenable, austenitic, nonmagnetic) |
| 4XX | chromium (hardenable, martensitic, magnetic) |
| 4XX | chromium (generally not hardenable, ferritic, magnetic) |
| 5XX | chromium (low-chromium, heat-resisting) |

A number following an alloying ingredient is the nominal percentage of that ingredient.

Also, an optional capital letter may be added as a prefix to designate the manufacturing process (A, acid Bessemer; B, basic Bessemer; C, basic open hearth; CB, either B or C at steel mill option; O, basic oxygen).[12]

# 9. TOOL STEEL

Each grade of tool steel is designed for a specific purpose, and as such, there are few generalizations that can be made about tool steel. Each tool steel exhibits its own blend of the three main performance criteria: toughness, wear resistance, and *hot hardness*.[13]

---

[12]The electric furnace process may be designated by a C, D, or E prefix.
[13]The ability of a steel to resist softening at high temperatures is known as *hot hardness* and *red hardness*.

Some of the few generalizations possible are listed as follows.

• An increase in carbon content increases wear resistance and reduces toughness.

• An increase in wear resistance reduces toughness.

• Hot hardness is independent of toughness.

• Hot hardness is independent of carbon content.

*Group A steels* are air-hardened, medium-alloy cold-work tool steels. Air-hardening allows the tool to develop a homogeneous hardness throughout, without distortion. This hardness is achieved by large amounts of alloying elements and comes at the expense of wear resistance.

*Group D steels* are high-carbon, high-chromium tool steels suitable for cold-working applications. These steels are high in abrasion resistance but low in machinability and ductility. Some steels in this group are air hardened, while others are oil quenched. Typical uses are blanking and cold-forming punches.

*Group H steels* are hot-work tool steels, capable of being used in the 1100°F to 2000°F (600°C to 1100°C) range. They possess good wear resistance, hot hardness, shock resistance, and resistance to surface cracking. Carbon content is low, between 0.35% and 0.65%. This group is subdivided according to the three primary alloying ingredients: chromium, tungsten, or molybdenum. For example, a particular steel might be designated as a "chromium hot-work tool steel."

*Group M steels* are molybdenum high-speed steels. Properties are very similar to the group T steels, but group M steels are less expensive since one part molybdenum can replace two parts tungsten. For that reason, most high-speed steel in common use is produced from the M group. Cobalt is added in large percentages (5% to 12%) to increase high-temperature cutting efficiency in heavy-cutting (high-pressure cutting) applications.

*Group O steels* are oil-hardened, cold-work tool steels. These high-carbon steels use alloying elements to permit oil quenching of large tools and are sometimes referred to as *nondeforming steels*. Chromium, tungsten, and silicon are typical alloying elements.

*Group S steels* are shock-resistant tool steels. Toughness (not hardness) is the main characteristic, and either water or oil may be used for quenching. Group S steels contain chromium and tungsten as alloying ingredients. Typical uses are hot header dies, shear blades, and chipping chisels.

*Group T steels* are tungsten high-speed tool steels that maintain a sharp hard cutting edge at temperatures in excess of 1000°F (550°C). The ubiquitous 18-4-1 grade T1 (named after the percentages of tungsten, chromium, and vanadium, respectively) is part of this group. Increases in hot hardness are achieved by simultaneous increases in carbon and vanadium (the key ingredient in these tool steels) and special, multiple-step heat treatments.[14]

*Group W steels* are water-hardened tool steels. These are plain high-carbon steels (modified with small amounts of vanadium or chromium, resulting in high surface hardness but low hardenability). The combination of high surface hardness and ductile core makes group W steels ideal for rock drills, pneumatic tools, and cold header dies. The limitation on this tool steel group is the loss of hardness that begins at temperatures above 300°F (150°C) and is complete at 600°F (300°C).

### Example 46.1

The composition of a group M tool steel is being formulated to replace the 18-4-1 group T steel. What are the percentages of alloying ingredients if two-thirds of the tungsten are to be replaced with molybdenum?

*Solution*

Since one part molybdenum replaces two parts tungsten, the alloy would be designated 6-6-4-1, representing 6% molybdenum, 6% tungsten, 4% chromium, and 1% vanadium.

### 10. STAINLESS STEEL

Adding chromium improves steel's corrosion resistance. Moderate corrosion resistance is obtained by adding 4% to 6% chromium to low-carbon steel. (Other elements, specifically less than 1% each of silicon and molybdenum, are also usually added.)

For superior corrosion resistance, larger amounts of chromium are needed. At a minimum level of 12% chromium, steel is *passivated* (i.e., an inert film of chromic oxide forms over the metal and inhibits further oxidation). The formation of this protective coating is the basis of the corrosion resistance of *stainless steel*.[15]

Passivity is enhanced by oxidizers and aeration but is reduced by abrasion that wears off the protective oxide coating. An increase in temperature may increase or decrease the passivity, depending on the abundance of oxygen.

Stainless steels are generally categorized into ferritic, martensitic (heat-treatable), austenitic, duplex, and high-alloy stainless steels.[16,17] Table 46.6 categorizes some of the more popular AISI grades of stainless steel.

**Table 46.6** *Characteristics of Common Stainless Steels*

| | AISI type | application |
|---|---|---|
| martensitic (hardenable by heat treatment) | 410 420 440C | general purpose |
| | | hardenable by heat treatment |
| ferritic (more corrosion resistant than martensitic; not hardenable by heat treatment) | 405 430 446 | |
| | | hardenable by cold working |
| austenitic (best corrosion resistance; hardenable only by cold working) | 201 202 301 302 302B 304 304L 310 316 321 | for elevated-temperature service |
| | | modified for welding |
| | | superior corrosion resistance |
| | 254SMo (UNS S31254) Alloy 904L (UNS N08904) AL-6XN (UNS N08367) | |
| duplex: austenitic-ferritic (tough, weldable, superior corrosion resistance) | Alloy 2205 (UNS S31803) Ferrallium 255 (UNS S32550) 329 (UNS S32900) 7-Mo Plus (UNS S32950) 44LN (UNS S31200) DP-3 (UNS S31260) 2304 (UNS S2304) SAF 2507 (UNS S32750) Code Plus Two® (UNS S32205) | |

*Ferritic stainless steels*, grouped with the AISI 400 series, contain more than 10% to 27% chromium. The body-centered cubic ferrite structure is stable (i.e., does not transform to austenite, a face-centered cubic structure) at all temperatures. For this reason, ferritic steels cannot be hardened significantly. Since ferritic stainless steels contain no nickel, they are less expensive than austenitic steels. Turbine blades are typical of the heat-resisting products manufactured from ferritic stainless steels.

The so-called *superferritics*, such as Alloy 2904C (S44735), Sea-Cure (UNS S44660), and Alloy 2903, are highly resistant to chloride pitting and crevice corrosion. Superferritics have been incorporated into marine

---

[14]For example, the 18-4-1 grade is heated to approximately 1050°F (550°C) for two hours, air cooled, and then heated again to the same temperature. The term *double-tempered steel* is used in reference to this process. Most heat treatments are more complex.

[15]Stainless steels are corrosion resistant in oxidizing environments. In reducing environments (such as with exposure to hydrochloric and other halide acids and salts), the steel will corrode.

[16]There is a fifth category, that of *precipitation-hardened stainless steels*, widely used in the aircraft industry. (Precipitation hardening is also known as *age hardening*.) These steels have been given the AISI designation 630. UNS designations for precipitation-hardened stainless steels include S13800, S15500, S17400, and S17700.

[17]The *sigma phase* structure that appears at very high chromium levels (e.g., 24% to 50%) is usually undesirable in stainless steels because it reduces corrosion resistance and impact strength. A notable exception is in the manufacture of automobile engine valves.

tubing and heat exchangers for power plant condensers. Like all ferritics, however, superferritics experience embrittlement above 885°F (475°C).

The *martensitic (heat-treatable) stainless steels* (also part of the AISI 400 series) contain no nickel and differ from ferritic stainless steels primarily in higher carbon contents. Cutlery and surgical instruments are typical applications requiring both corrosion resistance and hardness.

The *austenitic stainless steels* are commonly used for general corrosive applications. The stability of the austenite (a face-centered cubic structure) depends primarily on 4% to 22% nickel as an alloying ingredient. The basic composition is approximately 18% chromium and 8% nickel, hence the term "18 to 8 type." (See Table 46.7.)

**Table 46.7** *Typical Compositions of Stainless Steels*

| element | ferritic | martensitic | austenitic |
|---|---|---|---|
| carbon | 0.08–0.20% | 0.15–1.2% | 0.03–0.25% |
| manganese | 1–1.5% | 1% | 2% |
| silicon | 1% | 1% | 1–2% |
| chromium | 11–27% | 11.5–18% | 16–26% |
| nickel | – | – | 3.5–22% |
| phosphorus and sulfur | – | – | normal |
| molybdenum | – | – | some cases |
| titanium | – | – | some cases |

The so-called *superaustenitics*, such as the 317 and 316L series, achieve superior corrosion resistance by adding more molybdenum and nitrogen, respectively. AISI type 317 (UNS S31700) contains 3% to 4% molybdenum. Variants of type 317L (UNS S317XX) contain up to 5% molybdenum. Variants of type 316L achieve superior corrosion resistance by adding 10% to 14% nitrogen. "6-Mo" superaustenitics with approximately 6% molybdenum, 20% chromium, and 0.10% nitrogen are well established, particularly in the chemical process industry. "7-Mo" alloys probably represent the ultimate in corrosion resistance while still remaining commercially viable.

Austenitic stainless steels are the most weldable of the stainless steels but are nevertheless susceptible to sensitization. They are nonmagnetic and are hardenable only by cold working. They can be polished to a mirror finish, which makes them useful in food-industry applications. Because of the nickel, they are more expensive than ferritic stainless steels.

Welding stainless steels is possible when proper welding rod alloys are used, but is difficult for several reasons.

- Stainless steels, particularly austenitic types, possess relatively low thermal conductivities and high coefficients of thermal expansion. The maintenance of a high temperature gradient (because the heat is not readily dissipated) and a high expansion increases

the possibility of *weld bead cracking* (i.e., longitudinal cracking along the weld). Weld bead cracking can be minimized by welding at as low a temperature as possible.

- High temperature sensitizes the steel adjacent to welds, producing local chromium deficits. This phenomenon is known as *sensitization*, and the resulting corrosion is known as *weld decay*.

- Substantial grain growth occurs in ferritic steels, since there is no gamma-alpha transformation to keep grains small. Growth of grains is substantial at normal welding temperatures.

- When stainless steels cool from welding temperatures, martensite forms unless cooling is slowed down. The martensite makes the metal brittle and reduces its ductility.

Because of their nominal costs, austenitic AISI 304 and 316 are commonly used corrosion-resistant steels. However, their low strengths make them unsuitable for high-pressure applications. They are also susceptible to wear and galling, and they have limited resistance to localized corrosion and *stress corrosion cracking* (SCC), particularly *chloride stress corrosion cracking* (CSCC) above 130°F (54°C).[18] Alternatives include duplex and high-alloy austenitic stainless steels.

Second-generation *duplex stainless steels* are austenitic-ferritic stainless steels that have the toughness and weld-ability of austenitic stainless steels and yield strengths and corrosion/wear resistances greater than the 300 series.[19] Alloy 2205 is the most widely used. Types 2304 and 2507 (UNS S32304 and S32507, respectively) are third-generation duplex steels designed to reduce mill costs. With the highest percentages of nitrogen, molybdenum, and nickel of any duplex stainless steel, 2507 has been dubbed a "super-duplex" stainless steel.

The *high-alloy austenitic stainless steels* containing 22% to 28% chromium, 24% to 32% nickel, and 4% to 6% molybdenum provide superior corrosion resistance at lower cost than the nickel- and titanium-based alloys they replace.

## 11. CAST IRON

*Cast iron* is a general name given to a wide range of alloys containing iron, carbon, and silicon, and to a lesser extent, manganese, phosphorus, and sulfur. Generally, the carbon content will exceed 2%. The properties of cast iron depend on the amount of carbon

---

[18]Chloride stress corrosion cracking is an important issue in heat exchangers for use with ocean and inland water.

[19]The first generation duplex stainless steels developed in the 1930s, such as AISI type 329 (UNS S32900), lost much of their corrosion resistance after welding unless they were given a post-weld heat treatment. Second-generation stainless steels contain less carbon and 0.15% to 0.30% nitrogen and, when combined with proper welding technique, offer the same level of corrosion resistance as mill-annealed material.

present, as well as the form (i.e., graphite or carbide) of the carbon.

Carbon in the form of carbide is stable only at low temperatures. (The carbide is said to be a *metastable structure*.) At high temperatures, *graphitization* takes place according to the following reaction.

$$Fe_3C \rightarrow 3Fe + C \text{ (graphite)}$$

The most common type of cast iron is *gray cast iron*. The carbon in gray cast iron is in the form of graphite flakes. Graphite flakes are very soft and constitute points of weakness in the metal, which simultaneously improve machinability and decrease ductility. Gray cast iron is categorized into classes according to its tensile strength, as shown in Table 46.8. Compressive strength is three to five times the tensile strength.

**Table 46.8** Classes of Gray Cast Iron[*]

| class | minimum tensile strength | | tensile modulus of elasticity | |
|---|---|---|---|---|
| | $lbf/in^2$ | MPa | $lbf/in^2$ | GPa |
| 20 | 20,000 | 138 | $10\text{--}14 \times 10^6$ | 69–97 |
| 25 | 25,000 | 172 | $12\text{--}15 \times 10^6$ | 83–104 |
| 30 | 30,000 | 207 | $13\text{--}16.5 \times 10^6$ | 90–114 |
| 35 | 35,000 | 242 | $14.5\text{--}17 \times 10^6$ | 100–117 |
| 40 | 40,000 | 276 | $16\text{--}20 \times 10^6$ | 110–138 |
| 45 | 45,000 | 310 | – | – |
| 50 | 50,000 | 345 | $18.8\text{--}23 \times 10^6$ | 130–159 |
| 60 | 60,000 | 414 | $20.4\text{--}23.5 \times 10^6$ | 141–162 |

(Multiply $lbf/in^2$ by 0.006895 to obtain MPa.)
[*]ASTM specification A48

Magnesium and cerium can be added to improve the ductility of gray cast iron. The resulting *nodular cast iron* (also known as *ductile cast iron*) has the best tensile and yield strengths of all the cast irons. It also has good ductility (typically 5%) and machinability. Because of these properties, it is often used for automobile crankshafts. (See Table 46.9.)

*White cast iron* has been cooled quickly from a molten state. No graphite is produced from the cementite, and the carbon remains in the form of a carbide, $Fe_3C$.[20] The carbide is hard and is the reason that white cast iron is difficult to machine. White cast iron is used primarily in the production of malleable cast iron.

*Malleable cast iron* is produced by reheating white cast iron to between 1500°F and 1850°F (800°C and 1000°C) for several days, followed by slow cooling. During this treatment, the carbide is partially converted to nodules of graphitic carbon known as *temper carbon*. The tensile strength is increased to approximately

**Table 46.9** Common Grades of Ductile Iron[a]

| class/grade | minimum tensile strength | | minimum yield strength | | elongation |
|---|---|---|---|---|---|
| | ksi | MPa | ksi | MPa | % |
| 60-40-18[b] | 60 | 410 | 40 | 280 | 18 |
| 65-45-12[c] | 65 | 450 | 45 | 310 | 12 |
| 80-55-06[d] | 80 | 550 | 55 | 380 | 6 |
| 100-70-03[e] | 100 | 690 | 70 | 480 | 3 |
| 120-90-02[f] | 120 | 830 | 90 | 620 | 2 |

(Multiply ksi by 6.895 to obtain MPa.)
[a]ASTM A-536-70
[b]May be annealed after casting.
[c]An as-cast grade.
[d]An as-cast grade with higher manganese content.
[e]Usually obtained by a normalizing heat treatment.
[f]Oil quenched and tempered to specified hardness.

55,000 $lbf/in^2$ (380 MPa), and the elongation at fracture increases to approximately 18%.

*Mottled cast iron* contains both cementite and graphite and is between white and gray cast irons in composition and performance.

*Compacted graphitic iron* (CGI) is a unique form of cast iron with worm-shaped graphite particles. The shape of the graphite particles gives CGI the best properties of both gray and ductile cast iron: twice the strength of gray cast iron and half the cost of aluminum. The higher strength permits thinner sections. (Some engine blocks are 25% lighter than gray iron castings.) Using computer-controlled refining, volume production of CGI with the consistency needed for commercial applications is possible.

Silicon is the most important element affecting graphitization. The effects of various elements in cast iron are listed in Table 46.10. Most of the elements that increase hardness do so by promoting the formation of iron carbide.

**Table 46.10** Effects of Elements in Cast Iron

| element | effect |
|---|---|
| aluminum | deoxidizes molten cast iron |
| carbon | depending on form, affects machinability, ductility, and shrinkage |
| manganese | below 0.5%, reduces hardness by combining with sulfur; above 0.5%, increases hardness |
| phosphorus | increases fluidity and lowers melting temperature |
| silicon | below 3.25%, softens iron and increases ductility; above 3.25%, hardens iron; above 13%, increases acid and corrosion resistance |
| sulfur | increases hardness, sulfur is removed by addition of manganese |

---

[20]White and gray cast irons get their names from the coloration at a fracture.

## 12. WROUGHT IRON

*Wrought iron* is low-carbon (less than 0.1%) iron with small amounts (approximately 3%) of slag and gangue in the form of fibrous inclusions. It has good ductility and corrosion resistance. Prior to the use of steel, wrought iron was the most important structural metal.

In the ancient *puddling process*, wrought iron is produced in a *reverberatory furnace* similar to the open hearth furnace.[21] The molten iron floats on a layer of iron oxide that provides the oxygen for removal of almost all of the carbon, sulfur, and manganese. The limestone flux combines with silicon and phosphorus to form slag.

As the iron becomes purer, its melting temperature increases to above the furnace temperature. Spongy masses of congealing iron and slag are collected on the ends of rods inserted into the pool of molten metal. These masses are then removed and forged or hammered to squeeze out most of the slag. The remaining product consists of slag-coated iron particles welded together by the forging processes. The deformed slag particles contribute to the fatigue resistance of wrought iron.

In the modern *Aston process (Byers-Aston process)*, pig metal is melted in a cupola and is then highly purified in a Bessemer converter.[22] Simultaneously and separately, molten slag is prepared in an open hearth furnace and transferred to a mixing ladle. The molten iron is poured into the cooler slag in the mixing ladle (a process known as *shotting*), where the iron rapidly solidifies and releases dissolved gases. The gases fracture the solidifying iron, and molten slag enters the fissures. The excess molten slag is poured off, and the metal mass is pressed and rolled into blooms, billets, and slabs to remove most of the interior slag.[23] The slabs are hot-rolled together to form larger pieces of wrought iron.

## 13. PRODUCTION OF ALUMINUM

Aluminum is produced from *bauxite ore*, a mixture of hydroxides of aluminum ($Al_2O_3 \cdot n[H_2O]$) and oxides of iron, silicon, and titanium. Most of the nonrecycled aluminum produced today is produced in an electrochemical process known as the *Bayer process*.[24]

In the Bayer process, the ore is crushed and ground into a fine powder. It is then treated with a hot solution of sodium hydroxide, producing a solution of sodium aluminate. The solution is drawn off into a separate tank, leaving the remaining ore constituents (known as *red mud* because of the iron coloration) as a solid deposit to be discarded.

$$Al(OH)_3 + NaOH \rightarrow NaAl(OH)_4$$

As the solution cools, aluminum hydroxide precipitates, leaving a sodium hydroxide solution. The aluminum hydroxide is collected in solid, crystalline form and baked to form *alumina* (aluminum oxide, $Al_2O_3$). Because of its high melting temperature ($3720°F$, $2050°C$), alumina cannot be economically reduced in a furnace.

Final reduction (*smelting*) is accomplished through an electrolytic process using molten *cryolite* ($Na_3AlF_6$) as the electrolyte. Large carbon blocks act as anodes, and the carbon-lined steel tank acts as the cathode.

The aluminum oxide dissolves in the cryolite and is separated into molten aluminum and oxygen gas by the electric current. The aluminum collects in the bottom of the tank. Carbon dioxide is released at the anodes. The cryolite recomposes after decomposition and can be reused. (The following reactions disregard the cryolite.)

$$Al_2O_3 \rightarrow 2Al^{+++} + 3O^{--}$$
$$Al^{+++} + 3e^- \rightarrow Al$$
$$C + 2O^{--} \rightarrow CO_2 + 4e^-$$

## 14. PROPERTIES OF ALUMINUM

Aluminum satisfies applications requiring low weight, corrosion resistance, and good electrical and thermal conductivities. Its corrosion resistance derives from the oxide film that forms over the raw metal, inhibiting further oxidation. The primary disadvantages of aluminum are its cost and low strength.

In pure form, aluminum is soft, ductile, and not very strong. Copper, manganese, magnesium, and silicon can be added to increase its strength, at the expense of other properties, primarily corrosion resistance.[25] Aluminum is hardened by the *precipitation hardening (age hardening)* process.

The oxide coating that forms readily (particularly at high temperatures) on aluminum complicates welding. However, special processes that perform the welding under a blanket of inert gas (e.g., helium or argon) overcome this complication.[26]

---

[21]A *reverberatory furnace* is a cavernous, brick-lined chamber. The metal and ore are melted by flames that play over the top of the melt.
[22]A *cupola* is a tall, open-top vertical stack lined with furnace brick. An air blast is introduced at the base. Heat is produced from the combustion of coke mixed with the ore.
[23]In the rolling mill industry, a *bloom* is a long piece having a cross section greater than approximately 6 in (15 cm) square. A *billet* is smaller than a bloom, with a cross section greater than approximately 1.5 in (4 cm). A *slab* has a minimum thickness of 1.5 in (4 cm) and width between 10 and 15 in (25 and 40 cm). The width of a slab is always at least three times the thickness.
[24]Other metals that are refined in a similar electrochemical process are magnesium, copper, zinc, and (to lesser extents) gold and silver.

[25]One ingenious method of having both corrosion resistance and strength is to produce a composite material. *Alclad* is the name given to aluminum alloy that has a layer of pure aluminum bonded to the surface. The alloy provides the strength, and the pure aluminum provides the corrosion resistance.
[26]TIG (tungsten-inert gas) and MIG (metal-inert gas) processes are commonly used to weld aluminum.

## 15. ALUMINUM ALLOYS

Except for use in electrical work, most aluminum is alloyed with other elements, primarily copper, magnesium, and silicon.[27] Aluminum alloys are identified by a four-digit number and a letter suffix (e.g., 2014-T4). The number indicates the major alloying ingredient and chemical composition of the alloy, as determined from Table 46.11. The suffix indicates the condition of the alloy, as determined from Table 46.12 and Table 46.13.

**Table 46.11** *Aluminum Designations*

| designation | major alloying ingredient |
|---|---|
| 1XXX | commercially pure (99+%) |
| 2XXX | copper |
| 3XXX | manganese |
| 4XXX | silicon |
| 5XXX | magnesium |
| 6XXX | magnesium and silicon |
| 7XXX | zinc |
| 8XXX | other |

**Table 46.12** *Conditions of Aluminum Alloys*

| letter suffix | meaning |
|---|---|
| F | as fabricated |
| O | soft (after annealing) |
| H | strain hardened (cold worked) temper |
| T | heat treated |

**Table 46.13** *Aluminum Treatment Conditions*

| suffix | meaning |
|---|---|
| H1 | strain hardened by working to desired dimensions |
| H2 | strain hardened by cold working, followed by partial annealing |
| H3 | strain hardened and stabilized |
| T2 | annealed (castings only) |
| T3 | solution heat treated, followed by cold working (strain hardening) |
| T4 | solution heat treated, followed by natural aging at room temperature |
| T5 | artificial aging only |
| T6 | solution heat treated, followed by artificial aging |
| T7 | solution heat treated, followed by stabilizing by overaging heat treating |
| T8 | solution heat treated, followed by cold working and subsequent artificial aging |
| T9 | solution heat treated, followed by artificial aging and subsequent cold working |

Silicon occurs as a normal impurity in aluminum, and in natural amounts (less than 0.4%), it has little effect on properties. If moderate quantities (above 3%) of silicon are added, the molten aluminum will have high fluidity, making it ideal for castings. Above 12%, silicon improves the hardness and wear resistance of the alloy. When combined with copper and magnesium (as $Mg_2Si$ and AlCuMgSi) in the alloy, silicon improves age hardenability. Silicon has negligible effect on the corrosion resistance of aluminum.

Copper improves the age hardenability of aluminum, particularly in conjunction with silicon and magnesium. Thus, copper is a primary element in achieving high mechanical strength in aluminum alloys at elevated temperatures. Copper also increases the conductivity of aluminum, but decreases its corrosion resistance.

Magnesium is highly soluble in aluminum and is used to increase strength by improving age hardenability. Magnesium improves corrosion resistance and may be added when exposure to saltwater is anticipated.

Some aluminum alloys can be work-hardened (e.g., 1100, 3003, 5052). The ductility of these alloys decreases as strength is increased through working. Most aluminum alloys (e.g., 2014, 2017, 2024, 6061), however, must be precipitation-hardened.[28] The decrease in ductility with increased strength through heat treatment is small or nonexistent.

The letter suffixes H and T are followed by numbers that provide additional detail about the type of hardening process used to achieve the material properties. Table 46.13 lists the types of treatments associated with the H and T suffix letters.

## 16. PRODUCTION OF COPPER

Copper occurs in the free (metallic) state as well as in ores containing its oxides, sulfides, and carbonates. *Native copper* is recovered by the simple process of heating highly crushed ore. Molten copper flows to the bottom of the furnace.

Oxides and carbonates of copper are reduced in a blast or reverberatory furnace.

Sulfides of copper are heated in air, and the sulfur is replaced by oxygen. The product, which contains both copper and iron oxides, is further reduced in a reverberatory furnace. This step removes the oxygen but leaves some sulfur and iron. The final sulfur-removal process takes place in a furnace similar to a Bessemer converter in which air is injected into the molten copper. The iron oxide combines with a silica furnace liner.

Very low-grade ores are leached with sulfuric acid to recover the copper.

Regardless of the primary recovery method, *electrolysis (electrodeposition)* is generally required to remove the remaining impurities. Thick sheets of impure copper and thin sheets of pure copper are immersed together in an electrolyte of copper sulfate. The pure copper acts as the cathode. A direct current causes copper from the

---

[27]*EC (electrical-conductor) grade aluminum* consists of approximately 99.45% aluminum.

[28]These alloys are all known as *duralumin*.

impure sheets to migrate to the pure sheets. Impurities drop to the bottom of the tank as they are released.

## 17. ALLOYS OF COPPER

Zinc is the most common alloying ingredient in copper. It constitutes a significant part (up to 40% zinc) in brass.[29] (Brazing rod contains even more, approximately 45% to 50%, zinc.) Zinc increases copper's hardness and tensile strength. Up to approximately 30%, it increases the percent elongation at fracture. It decreases electrical conductivity considerably. *Dezincification*, a loss of zinc in the presence of certain corrosive media or at high temperatures, is a special problem that occurs in brasses containing more than 15% zinc.

Tin constitutes a major (up to 20%) component in most bronzes. Tin increases fluidity, which improves casting performance. In moderate amounts, corrosion resistance in saltwater is improved. (*Admiralty metal* has approximately 1%; *government bronze* and *phosphorus bronze* have approximately 10% tin.) In moderate amounts (less than 10%), tin increases the alloy's strength without sacrificing ductility. Above 15%, however, the alloy becomes brittle. For this reason, most bronzes contain less than 12% tin. Tin is more expensive than zinc as an alloying ingredient.

Lead is practically insoluble in solid copper. When present in small to moderate amounts, it forms minute soft particles that greatly improve machinability (2% to 3% lead) and wearing (bearing) properties (10% lead).

Silicon increases the mechanical properties of copper by a considerable amount. On a per unit basis, silicon is the most effective alloying ingredient in increasing hardness. *Silicon bronze* (96% copper, 3% silicon, 1% zinc) is used where high strength combined with corrosion resistance is needed (e.g., in boilers).

If aluminum is added in amounts of 9% to 10%, copper becomes extremely hard. Thus, *aluminum bronze* (as an example) trades an increase in brittleness for increased wearing qualities. Aluminum in solution with the copper makes it possible to precipitation harden the alloy.

Beryllium in small amounts (less than 2%) improves the strength and fatigue properties of copper. These properties make precipitation-hardened *copper-beryllium* (*beryllium-copper*, *beryllium bronze*, etc.) ideal for small springs. These alloys are also used for producing non-sparking tools.

## 18. NICKEL AND ITS ALLOYS

Like aluminum, nickel is largely hardened by precipitation hardening. Nickel is similar to iron in many of its properties, except that it has higher corrosion resistance

and a higher cost. Also, nickel alloys have special electrical and magnetic properties.

Copper and iron are completely miscible with nickel. Copper increases formability. Iron improves electrical and magnetic properties markedly.

Some of the better-known nickel alloys are *monel metal* (30% copper, used hot-rolled where saltwater corrosion resistance is needed), *K-monel metal* (29% copper, 3% aluminum, precipitation-hardened for use in valve stems), *inconel* (14% chromium, 6% iron, used hot-rolled in gas turbine parts), and *inconel-X* (15% chromium, 7% iron, 2.5% titanium, aged after hot rolling for springs and bolts subjected to corrosion). *Hastelloy* (22% chromium) is another well-known nickel alloy.[30]

*Nichrome* (15% to 20% chromium) has high electrical resistance, high corrosion resistance, and high strength at red heat temperatures, making it useful in resistance heating. *Constantan* (40% to 60% copper, the rest nickel) also has high electrical resistance and is used in thermocouples.

*Alnico* (14% nickel, 8% aluminum, 24% cobalt, 3% copper, the rest iron) and *cunife* (20% nickel, 60% copper, the rest iron) are two well-known nickel alloys with magnetic properties ideal for permanent magnets. Other magnetic nickel alloys are *permalloy* and *permivar*.

*Invar*, *Nilvar*, and *Elinvar* are nickel alloys with low or zero thermal expansion and are used in thermostats, instruments, and surveyors' measuring tapes.

For decades, C-family Alloy C-276 (Alloy 276) was the nickel-chromium-molybdenum workhorse for piping and reaction vessels in chemical process industries. Newer alloys from the C-family (e.g., Alloy C-22, also designated as Alloy 22, 622, and 5621 hMoW) with more chromium and less tungsten are extremely corrosion resistant, even in extremely aggressive, mixed-acid environments. Alloy 59 (UNS N06059), also referred to as Alloy 5923 hMo, maintains its corrosion resistance in strongly oxidizing environments, including the most severe corrosion conditions in modern pollution-control equipment.

## 19. REFRACTORY METALS

*Reactive* and *refractory metals* include alloys based on titanium, tantalum, zirconium, molybdenum, niobium (also known as columbium), and tungsten. These metals are used when superior properties (i.e., corrosion resistance) are needed. They are most often used where high-strength acids are used or manufactured.

## 20. NATURAL POLYMERS

A *polymer* is a large molecule in the form of a long chain of repeating units. The basic repeating unit is called a

---

[29]*Brass* is an alloy of copper and zinc. *Bronze* is an alloy of copper and tin. Unfortunately, brasses are often named for the color of the alloys, leading to some very misleading names. For example, *nickel silver*, *commercial bronze*, and *manganese bronze* are all brasses.

[30]K-monel is one of four special forms of monel metal. There are also H-monel, S-monel, and R-monel forms.

*monomer* or just *mer*. (A large molecule with two alternating mers is known as a *copolymer* or *interpolymer*. Vinyl chloride and vinyl acetate form one important family of copolymer plastics.)

Many of the natural organic materials (e.g., rubber and asphalt) are polymers. (Polymers with elastic properties similar to rubber are known as *elastomers*.) Natural rubber is a polymer of the *isoprene latex* mer (formula $[C_5H_8]_n$, repeating unit of $CH_2{=}CCH_3{-}CH{=}CH_2$, systematic name of 2-methyl-1,3-butadiene). The strength of natural polymers can be increased by causing the polymer chains to cross-link, restricting the motion of the polymers within the solid.

Cross-linking of natural rubber is known as *vulcanization*. Vulcanization is accomplished by heating raw rubber with small amounts of sulfur. The process raises the tensile strength of the material from approximately 300 lbf/in$^2$ (2.1 MPa) to approximately 3000 lbf/in$^2$ (21 MPa). The addition of carbon black as a reinforcing *filler* raises this value to approximately 4500 lbf/in$^2$ (31 MPa) and provides tear resistance and toughness.

The amount of cross-linking between the mers determines the properties of the solid. Figure 46.6 shows how sulfur joins two adjacent isoprene (natural rubber) mers in *complete cross-linking*.[31] If sulfur does not replace both of the double carbon bonds, *partial cross-linking* is said to have occurred.

**Figure 46.6** *Vulcanization of Natural Rubber*

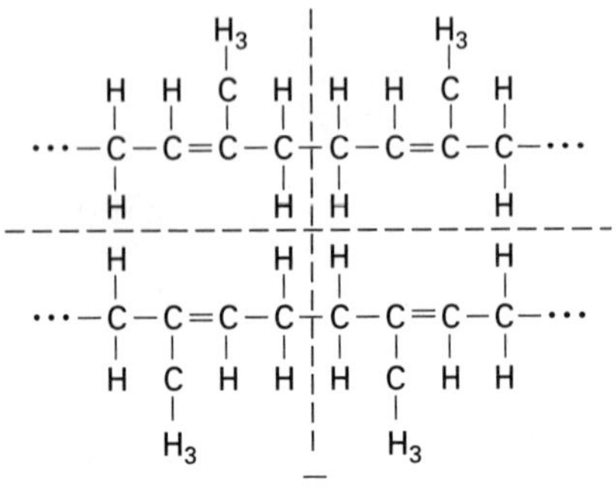

(natural) – 4 mers

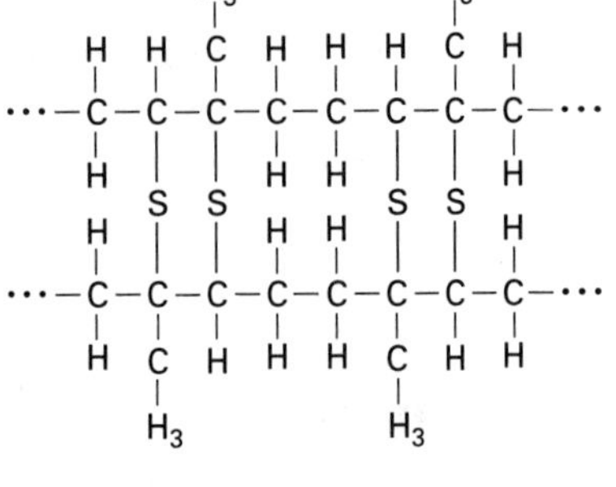

cross-linked

---

[31]A tire tread may contain 3% to 4% sulfur. Hard rubber products, which do not require flexibility, may contain as much as 40% to 50% sulfur.

**Example 46.2**

What is the approximate fraction (by mass) of sulfur in a completely cross-linked natural rubber polymer?

*Solution*

Figure 46.6 shows that, for complete cross-linking, each mer of the natural rubber mer requires one sulfur atom.

The atomic weight of sulfur is approximately 32 g/mol. The molecular weight of the rubber mer is

$$(5)\left(12\ \frac{g}{mol}\right) + (8)\left(1\ \frac{g}{mol}\right) = 68$$

The fraction of sulfur is

$$\frac{m_{sulfur}}{m_{sulfur}+m_{mer}} = \frac{32\ \frac{g}{mol}}{32\ \frac{g}{mol}+68\ \frac{g}{mol}}$$
$$= 0.32 \quad (32\%)$$

## 21. DEGREE OF POLYMERIZATION

The *degree of polymerization*, DP, is the average number of mers in the molecule, typically several hundred to several thousand.[32] (In general, compounds with degrees of less than ten are called *telenomers* or *oligomers*.) The degree of polymerization can be calculated from the mer and polymer molecular weights.

$$DP = \frac{MW_{polymer}}{MW_{mer}} \qquad 46.1$$

A polymer batch usually will contain molecules with different length chains. Therefore, the degree of polymerization will vary from molecule to molecule, and an average degree of polymerization is reported.

The stiffness and hardness of polymers vary with their degrees. Polymers with low degrees are liquids or oils. With increasing degree, they go through waxy to hard resin stages. High-degree polymers have hardness and strength qualities that make them useful for engineering applications. Tensile strength and melting (softening) point also increase with increasing degree of polymerization.

**Example 46.3**

A polyvinyl chloride molecule (mer molecular weight of 62.5) is found to contain 860 carbon atoms, 1290 hydrogen atoms, and 430 chlorine atoms. What is the degree of polymerization?

---

[32]Degrees of polymerization for commercial plastics are usually less than 1000.

*Solution*

The approximate atomic weights of carbon, hydrogen, and chlorine are 12 g/mol, 1 g/mol, and 35.5 g/mol, respectively. The molecular weight of the molecule is

$$MW_{\text{ploymer}} = (860)\left(12 \ \frac{g}{mol}\right) + (1290)\left(1 \ \frac{g}{mol}\right)$$
$$+ (430)\left(35.5 \ \frac{g}{mol}\right)$$
$$= 26{,}875 \ g/mol$$

The degree of polymerization is

$$DP = \frac{MW_{\text{polymer}}}{MW_{\text{mer}}} = \frac{26{,}875 \ \frac{g}{mol}}{62.5 \ \frac{g}{mol}} = 430$$

## Example 46.4

What is the degree of polymerization of a polyvinyl acetate (mer of $C_4H_6O_2$, molecular weight of 86 g/mol) sample having the following analysis of molecular weights?

| range of molecular weights (g/mol) | mole fraction |
|---|---|
| 5001–15,000 | 0.30 |
| 15,001–25,000 | 0.47 |
| 25,001–35,000 | 0.23 |

*Solution*

Using the midpoint of each range, the average polymer molecular weight is

$$MW_{\text{polymer}} = (0.30)\left(10{,}000 \ \frac{g}{mol}\right)$$
$$+ (0.47)\left(20{,}000 \ \frac{g}{mol}\right)$$
$$+ (0.23)\left(30{,}000 \ \frac{g}{mol}\right)$$
$$= 19{,}300 \ g/mol$$

The degree of polymerization is

$$DP = \frac{MW_{\text{polymer}}}{MW_{\text{mer}}} = \frac{19{,}300 \ \frac{g}{mol}}{86 \ \frac{g}{mol}} = 224$$

## 22. SYNTHETIC POLYMERS

Table 46.14 lists some of the common mers. Polymers are named by adding the prefix "poly" to the name of the basic mer. For example, $C_2H_4$ is the chemical formula for ethylene. Chains of $C_2H_4$ are called polyethylene.

***Table 46.14*** *Names of Common Mers*

| name | repeating unit | combined formula |
|---|---|---|
| ethylene | $CH_2CH_2$ | $C_2H_4$ |
| propylene | $CH_2(HCCH_3)$ | $C_3H_6$ |
| styrene | $CH_2CH(C_6H_5)$ | $C_8H_8$ |
| vinyl acetate | $CH_2CH(C_2H_3O_2)$ | $C_4H_6O_2$ |
| vinyl chloride | $CH_2CHCl$ | $C_2H_3Cl$ |
| isobutylene | $CH_2C(CH_3)_2$ | $C_4H_8$ |
| methyl methacrylate | $CH_2C(CH_3)(COOCH_3)$ | $C_5H_8O_2$ |
| acrylonitrile | $CH_2CHCN$ | $C_3H_3N$ |
| epoxide (ethoxylene) | $CH_2CH_2O$ | $C_2H_4O$ |
| amide (nylon) | $CONH_2$ or $CONH$ | $CONH_2$ or $CONH$ |

Polymers are able to form when double (covalent) bonds break and produce reaction sites. The number of bonds in the mer that can be broken open for attachment to other mers is known as the *functionality* of the mer. The ethylene mer in Fig. 46.7 is *bifunctional* since the C=C bond can be broken to form two reaction sites (e.g., –C–C–). Other mers are *trifunctional* or *tetrafunctional*. When combining into chains, bifunctional mers form *linear polymers*, whereas trifunctional and tetrafunctional mers form *network polymers*.

Two processes produce reaction sites: addition polymerization and condensation polymerization. With *addition polymerization*, mers simply combine sequentially into chains by breaking double (covalent) bonds. No other products are produced. For example, the formation of polyethylene is given by the following reaction and Fig. 46.7.

$$2CH_2=CH_2 \rightarrow -CH_2-CH_2-CH_2-CH_2-$$

***Figure 46.7*** *Steps in Forming Polyethylene*

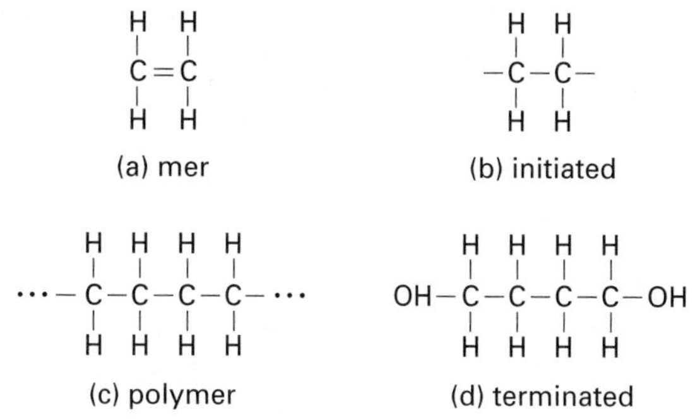

(a) mer     (b) initiated

(c) polymer     (d) terminated

Substances called *initiators* are used to start addition polymerization.[33] Initiators break down under heat, light, or other energy and provide free radicals. A free

---

[33]*Initiators* are usually *organic peroxides* such as *hydrogen peroxide* ($H_2O_2$) and *benzoyl peroxide* ($C_6H_5COO$).

radical disrupts the double bond in a mer and attaches to one side of the mer, opening up the mer's double bond and releasing enough energy to open another mer's double bond.

The process continues (*propagation*) until the initiator is used up, until the mers are used up, or until another free radical terminates the chain. The latter occurrence (i.e., when there is no possibility for additional joining) is known as *saturation*. Since the same radicals that initiated a polymer chain can terminate the chain, initiators are also known as *terminators*.

*Condensation polymerization (step reaction polymerization)* also requires opening bonds in two molecules to form a larger molecule. This formation is often accompanied by the release of small molecules such as $H_2O$, $CO_2$, or $N_2$. The repeating units derived from the condensation process are not the same as the monomers from which they are formed since portions of the original monomer form the small molecules that are lost.[34] The formation of *phenolic plastics* from formaldehyde and ammonia illustrates condensation polymerization.

$$6HCHO + 4NH_3 \rightarrow (CH_2)_6N_4 + 6H_2O$$

## 23. FLUOROPOLYMERS

Fluoropolymers (*fluoroplastics*) are a class of paraffinic, thermoplastic polymers in which some or all of the hydrogens have been replaced by fluorine.[35] There are seven major types of fluoropolymers, with overlapping characteristics and applications. They include the fully fluorinated fluorocarbon polymers of Teflon® PTFE (polytetrafluoroethylene), FEP (fluorinated ethylene propylene), and PFA (perfluoroalkoxy), as well as the partially fluorinated polymers of PCTFE (polychlorotrifluoroethylene), ETFE (ethylene tetrafluoroethylene), ECTFE (ethylene chlorotrifluoroethylene), and PVDF (polyvinylidene fluoride). (See Table 46.15.)

Fluoropolymers compete with metals, glass, and other polymers in providing corrosion resistance. Choosing the right fluoropolymer depends on the operating environment including temperature, chemical exposure, and mechanical stress.

PTFE, the first available fluoropolymer, is probably the most inert compound known. It has been used extensively for pipe and tank linings, fittings, gaskets, valves, and pump parts. It has the highest operating temperature—approximately 500°F (260°C). Unlike the other fluoropolymers, however, it is not a melt-processed polymer. Like a powdered metallurgy product, PTFE is processed by compression and isostatic molding, followed by sintering. PTFE is also the weakest of all the fluoropolymers.

---

[34]The formation of nylon and the famous *Bakelite material* from phenol ($C_6H_5OH$) are examples of condensation polymerization.
[35]*Fluoroelastomers* are uniquely different from fluoropolymers. They have their own areas of application.

**Table 46.15** *Characteristics of Fluoropolymers*

| fluoro-polymer | specific gravity | tensile strength (ksi) | flexural modulus (ksi) | upper service temperature[a] (°C) |
|---|---|---|---|---|
| PTFE[b] | 2.13–2.22 | 2.0–6.5 | 70–110 | 287 |
| FEP | 2.12–2.17 | 2.7–3.1 | 90 | 204 |
| PFA | 2.12–2.17 | 4.0–4.5 | 100 | 260 |
| PCTFE | 2.08–2.2 | 4.0–6.0 | 150–260 | 199 |
| ETFE | 1.70 | 6.5 | 200 | 149–182 |
| ECTFE | 1.68 | 6.6–7.8 | 240 | 149–179 |
| PVDF[c] | 1.76–1.78 | 3.5–6.2 | 70–320 | 150 |

(Multiply ksi by 6.895 to obtain MPa.)
[a]Upper service temperature is also known as *temperature of continuous heat resistance*.
[b]Properties of PTFE are highly variable, depending on type of resin and method of processing.
[c]Properties of PVDF include those of its copolymers, hence the range of values.

## 24. ELASTOMERIC COMPOUNDS

Flexible parts, such as gaskets and O-rings, are manufactured from *elastomeric compounds*. Common elastomerics include natural rubber, butyl rubber, buna-N (nitrile rubber), neoprene, ethylene-propylene-diene monomer (EPDM) rubber, chlorosulfonated polyethylene, and various fluoroelastomers. Chemical resistance, temperature, and pressure are the primary factors considered in choosing sealing compounds.

## 25. POLYMER CRYSTALLINITY

Under favorable conditions, polymer chains align in regular patterns, minimizing volume and increasing the attractive forces between atoms. This phenomenon is known as *polymer crystallinity*. Crystallinity is often only partial, however, because the chain structure interferes with complete alignment, and only weak van der Waals forces are available to drive the alignment.

Several factors affect the likelihood of crystallization.

- *Trans mers* have *unsaturated positions* (i.e., linkage points) on opposite sides of the mer, producing less tangling between chains. Trans mers are more likely to crystallize than *cis mers*, which have both unsaturated positions on the same side of the mer.

- When mers join with a high degree of regularity in chain structure and mer orientation, the result is an *isotactic polymer*. Isotactic polymers favor crystallization over the *syndiotactic polymers* (partial regularity) and *atactic polymers* (no regularity).

- Linear polymers are more likely to crystallize than network polymers.

Crystallization requires a physical rearrangement of polymer chains into regular layers. It should not be confused with *cross-linking* between adjacent polymers or with polymer *branching*. (See Fig. 46.8.)

**Figure 46.8** Polymer Crystallization, Cross-Linking, and Branching

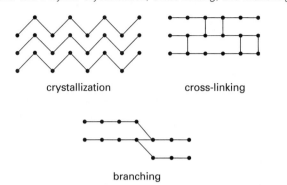

crystallization          cross-linking

branching

## 26. THERMOSETTING AND THERMOPLASTIC POLYMERS

Most polymers can be softened and formed by applying heat and pressure. These are known by various terms including *thermoplastics*, *thermoplastic resins*, and *thermoplastic polymers*. Polymers that are resistant to heat (and that actually harden or "kick over" through the formation of permanent cross-linking upon heating) are known as *thermosetting plastics*. Table 46.16 lists the common polymers in each category. Thermoplastic polymers retain their chain structures and do not experience any chemical change (i.e., bonding) upon repeated heating and subsequent cooling. Thermoplastics can be formed in a cavity mold, but the mold must be cooled before the product is removed. Thermoplastics are particularly suitable for injection molding. The mold is kept relatively cool, and the polymer solidifies almost instantly.

Thermosetting polymers form complex, three-dimensional networks. Thus, the complexity of the polymer increases dramatically, and a product manufactured from a thermosetting polymer may be thought of as one big molecule. Thermosetting plastics are rarely used with injection molding processes.

## 27. WOOD

Woods are classified broadly as softwoods or hardwoods, although it is difficult to define these terms exactly. *Softwoods* contain tube-like fibers (*tracheids*) oriented with the longitudinal axis (grain) and cemented together with *lignin*. *Hardwoods* contain more complex structures (e.g., storage cells) in addition to longitudinal fibers. Fibers in hardwoods are also much smaller and shorter than those in softwoods.

The mechanical properties of woods are influenced by moisture content and grain orientation. (Strengths of dry woods are approximately twice those of wet or green woods. Longitudinal strengths may be as much

**Table 46.16** Thermosetting and Thermoplastic Polymers

*thermosetting*
  epoxy
  melamine
  natural rubber (polyisoprene)
  phenolic (phenol formaldehyde, Bakelite®)
  polyester (DAP)
  silicone
  urea formaldehyde
*thermoplastic*
  acetal
  acrylic
  acrylonitrile-butadiene-styrene (ABS)
  cellulosics (e.g., cellophane)
  polyamide (nylon)
  polyarylate
  polycarbonate
  polyester (PBT and PET)
  polyethylene
  polymethyl-methacrylate (Plexiglas®, Lucite®)
  polypropylene
  polystyrene
  polytetrafluoroethylene (Teflon®)
  polyurethane
  polyvinyl chloride (PVC)
  synthetic rubber (Neoprene®)
  vinyl

as 40 times higher than cross-grain strengths.) *Moisture content*, MC, is defined by Eq. 46.2.

$$\text{MC} = \frac{W_{\text{wet}} - W_{\text{oven-dry}}}{W_{\text{oven-dry}}} \qquad 46.2$$

Wood is considered to be green if its moisture content is above 19%. Wood is considered to be dry when it has reached its *equilibrium moisture content*, generally between 12% and 15% moisture. Thus, moisture is not totally absent in dry wood.[36]

The approximate mechanical properties of several dry wood varieties are listed in Table 46.17.

## 28. GLASS

*Glass* is a term used to designate any material that has a volumetric expansion characteristic similar to Fig. 46.9. Glasses are sometimes considered to be *supercooled liquids* because their crystalline structures solidify in random orientation when cooled below their melting points. It is a direct result of the high liquid viscosities of oxides, silicates, borates, and phosphates that the molecules cannot move sufficiently to form large crystals with cleavage planes.

As a liquid glass is cooled, its atoms develop more efficient packing arrangements. This leads to a rapid decrease in volume (i.e., a steep slope on the temperature-volume curve). Since no crystallization occurs, the

---

[36]Oven-dry lumber is not used in construction.

**Table 46.17** Approximate Properties of Wood

| variety | flexural strength (lbf/in$^2$) | tensile strength (lbf/in$^2$)$^a$ | compressive strength (lbf/in$^2$)$^b$ | modulus of elasticity (lbf/in$^2$) |
|---------|---------|---------|---------|---------|
| cedar | 7700 | 220 | 5020 | $1.12 \times 10^6$ |
| douglas fir | 12,200 | 340 | 7430 | $1.95 \times 10^6$ |
| fir | 9800 | 300 | 5480 | $1.49 \times 10^6$ |
| mahogany | 11,460 | 740 | 6780 | $1.50 \times 10^6$ |
| oak (red) | 14,400 | 820 | 6920 | $1.81 \times 10^6$ |
| redwood | 10,000 | 240 | 6150 | $1.34 \times 10^6$ |
| spruce | 10,200 | 710 | 5610 | $1.57 \times 10^6$ |
| yellow pine | 12,800 | 470 | 7080 | $1.80 \times 10^6$ |

(Multiply lbf/in$^2$ by 6.894 to obtain kPa.)
$^a$perpendicular to grain
$^b$parallel to grain

**Figure 46.9** Behavior of a Glass

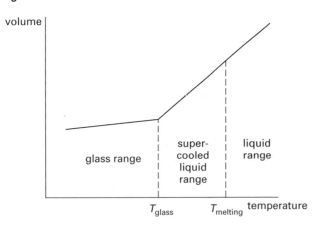

liquid glass simply solidifies without molecular change when cooled below the melting point. (This is known as *vitrification*.) The more efficient packing continues past the point of solidification.

At the *glass transition temperature* (*fictive temperature*), the glass viscosity increases suddenly by several orders of magnitude. Since the molecules are more restrained in movement, efficient atomic rearrangement is curtailed, and the volume-temperature curve changes slope. This temperature also divides the region into flexible and brittle regions. At the glass transition temperature, there is a 100-fold to 1000-fold increase in stiffness (modulus of elasticity).

Both organic and inorganic compounds may behave as glasses. *Common glasses* are mixtures of $SiO_2$, $B_2O_3$, and various other compounds to modify performance, as shown in Table 46.18.[37]

---

[37]This excludes lead-alkali glasses that contain 30% to 60% PbO.

## 29. CERAMICS

Ceramics are compounds of metallic and nonmetallic elements. Ceramics form crystalline structures but have no free valence electrons. All electrons are shared ionically or in covalent bonds. Common examples include brick, portland cement, refractories, and abrasives. (Glass is also considered a ceramic even though it does not crystallize.) Typical ceramic properties are listed in Table 46.19.

Although perfect ceramic crystals have extremely high tensile strengths (e.g., some glass fibers have ultimate strengths of 100,000 lbf/in$^2$ (700 MPa)), the multiplicity of cracks and other defects in natural crystals reduces their tensile strengths to near-zero levels.

Due to the absence of free electrons, ceramics are typically poor conductors of electrical current, although some (e.g., magnetite, $Fe_3O_4$) possess semiconductor properties. Other ceramics, such as $BaTiO_3$, $SiO_2$, and $PbZrO_3$, have *piezoelectric (ferroelectric) qualities* (i.e., generate a voltage when compressed).

Ceramics with similar structures behave similarly. Table 46.20 lists several structure designations (e.g., sodium chloride structure) as used in the study of ceramics.

*Polymorphs* are compounds that have the same chemical formula but have different physical structures. Some ceramics, of which *silica* ($SiO_2$) is a common example, exhibit *polymorphism*. At room temperature, silica is in the form of *quartz*. At 1607°F (875°C), the structure changes to *tridymite*. A change to a third structure, that of *cristobalite*, occurs at 2678°F (1470°C).

*Ferrimagnetic materials (ferrites, spinels, or ferrispinels)* are ceramics with valuable magnetic qualities. Advances in near-room-temperature superconductivity have been based on *lanthanum barium copper oxide* ($La_{2-x}Ba_xCuO_4$), a ceramic oxide, as well as compounds based on yttrium (Y-Ba-Cu-O), bismuth, thallium, and others.

Common ceramics are listed in Table 46.21.

## 30. ABRASIVES

An *abrasive* is a hard material that can cut other materials. Abrasives are typically ceramic compounds embedded in stiff binders. *Natural abrasives* include *emery* (50% to 60% $Al_2O_3$, rest iron oxide), corundum, quartz, garnets, and diamonds. *Artificial abrasives* include cemented carbides (e.g., SiC) and artificially made aluminum oxide ($Al_2O_3$).

Binders for rigid wheel abrasives are kiln-fired vitreous materials derived from clays or feldspars. Other grinding wheels use rubber or synthetic elastomers as the binders. Grains of abrasive are mixed thoroughly with a binder and molded into final form. Diamond wheels are often in the form of thin metal discs with the diamond grains bonded only to the periphery.

**Table 46.18** *Analyses and Properties of Representative Glasses*

| type of glass | analysis, percent by weight | | | | | softening temperature,[a] °C | coefficient of expansion, °C$^{-1}$ | characteristics | use |
|---|---|---|---|---|---|---|---|---|---|
| | SiO$_2$ | modifiers[a,b] | Al$_2$O$_3$ | B$_2$O$_3$ | PbO | | | | |
| fused silica | 99.9 | – | – | – | – | 1667 | $5.5 \times 10^{-7}$ | thermal shock resistant | laboratory equipment |
| 96% silica | 96.0 | – | – | 4.0 | – | 1500 | $8.0 \times 10^{-7}$ | thermal shock resistant | laboratory equipment |
| borosilicate (Pyrex®) | 80.5 | 4.2 | 2.2 | 12.9 | – | 820 | $32.0 \times 10^{-7}$ | thermal shock resistant, easy to form | cooking utensils |
| lead-alkali | 54.0 | 11.0 | – | – | 35 | 630 | $89.0 \times 10^{-7}$ | high index of refraction | cut glass |
| aluminasilicate | 57.7 | 9.5 | 25.3 | 7.4 | – | 915 | $42.0 \times 10^{-7}$ | thermal shock resistant | thermometers |
| soda-lime silica | 73.6 | 25.4 | 1.0 | – | – | 696 | $92.0 \times 10^{-7}$ | easy to form | plate, bulbs |
| lead-alkali | 35.0 | 7.0 | – | – | 58 | 580 | $91.0 \times 10^{-7}$ | dielectric | capacitors |

(Multiply °F$^{-1}$ by 9/5 to obtain °C$^{-1}$.)
[a]The temperature at which glass will sag appreciably under its own weight.
[b]Sum of Na$_2$O, K$_2$O, Ni$_2$O, CaO, MgO, and BaO.

Pyrex® is a trademark of Corning.
Data derived with permission from *Metals, Ceramics and Polymers*, by Oliver H. Wyatt and David Dew-Hughes, Cambridge University Press, © 1974. Reprinted with permission of Cambridge University Press.

**Table 46.19** *Properties of Typical Ceramics*

high melting point
high hardness
high compressive strength
high tensile strength (perfect crystals)
low ductility (brittleness)
high shear resistance (low slip)
low electrical conductivity
low thermal conductivity
high corrosion (acid) resistance
low coefficient of thermal expansion

**Table 46.20** *Structural Designations (A and B are metals; X is a nonmetal)*

| compound | formula | basic form | other examples |
|---|---|---|---|
| sodium chloride | NaCl | AX | FeO, MgO, CaO |
| cesium chloride | CsCl | AX | |
| calcium fluoride | CaF$_2$ | AX$_2$ | GeO$_2$, MgF$_2$,TO$_2$ |
| silica | SiO$_2$ | AX$_2$ | |
| corundum | Al$_2$O$_3$ | A$_2$X$_3$ | Cr$_2$O$_3$,Fe$_2$O$_3$ |
| spinels (ferrites) | MgAl$_2$O$_4$ | AB$_2$X$_4$ | |
| perovskite | CaTiO$_3$ | ABX$_3$ | BaTiO$_3$, PbZrO$_3$ |
| zircon | ZrSiO$_4$ | ABX$_4$ | |

**Table 46.21** *Common Ceramics*

| compound | mineral name | use[a] |
|---|---|---|
| Al$_2$O$_3$ | corundum, alumina | abrasives, firebrick |
| Al$_2$Si$_2$O$_5$(OH)$_4$ | kaolinite clay | porcelain paste |
| BaTiO$_3$ | barium titanate | piezoelectricity |
| BN | boron nitride | refractory |
| CaF$_2$ | fluorite | flux |
| CaO | | refractory |
| Fe$_2$O$_3$ | | refractory |
| Fe$_3$O$_4$ or FeFe$_2$O$_4$ | magnetite | thermistors |
| MgO | periclase | refractory |
| Mg$_2$SiO$_4$ | forsterite | refractory |
| MnFe$_2$O$_4$ | | ferrimagnetism |
| MgCr$_2$O$_4$ | magnesium chromate | piezoelectricity |
| MgFe$_2$O$_4$ | | antiferromagnetism |
| NiFe$_2$O$_4$ | | ferrimagnetism |
| NaCl | table salt | food, chemicals |
| PbZrO$_3$ | | piezoelectricity |
| SiC | silicon carbide | refractory |
| SiO$_2$ | quartz[b] | refractory |
| TiC | titanium carbide | refractory |
| TiO$_2$ | titanium dioxide | refractory |
| UO$_2$ | uranium dioxide | nuclear fuel |
| ZrN | zirconium nitride | refractory |
| ZnFe$_2$O$_4$ | | ferrimagnetism |

[a]The term *refractory* means the ceramic is used in firebrick, stoneware, and other containers intended for use at high temperatures.
[b]SiO$_2$ has several temperature-dependent polymorphs, including coesite, cristobalite, and tridymite.

*Carbides (cemented carbides, sintered carbides)* have extreme hardness, wear resistance, and thermal stability.[38] These properties make them useful for high-speed metal cutting. Silicon carbide (SiC, sold under the name of *carborundum*) is probably the best known. Carbides of tungsten, molybdenum, titanium, vanadium, tantalum, and zirconium are also widely used.

## 31. MODERN COMPOSITE MATERIALS

There are many types of modern composite material systems, including dispersion-strengthened, particle-strengthened, and fiber-strengthened materials. (Steel-reinforced concrete and steel-reinforced wood systems are also composite systems.)

In *dispersion-strengthened systems* (e.g., aluminum-aluminum oxide systems known as *SAP alloys, toughened ceramics, metal-metal systems* in which tungsten whiskers are blended in a copper alloy matrix, and $NiO_2ThO_2$ mixtures known as *TD-nickel*), the matrix is an actual load-bearing element.[39] The discrete particles (*dispersoids*) distributed throughout the matrix occupy less than 15% of the volume, and particle sizes are in the 0.01 to 0.1 $\mu$m range.

In *particle-strengthened systems* (e.g., tungsten carbide, known as WC, in a cobalt matrix, and *cermets* produced by sintering), the dispersoid particles are larger than 1.0 $\mu$m in size, and they occupy up to 25% of the material volume. The matrix is not the major load-carrying element, but it does contribute to strength.

In *fiber-strengthened systems* (e.g., glass-reinforced epoxies), reinforcing materials vary widely in size, and sizes may range up to several mils.[40] The matrix transmits the loads to the fibers and protects the fibers from chemical attack.

Initial attempts at fiber-reinforced polymers involved the impregnation of natural fibers (cotton, wood, etc.) with Bakelite and phenolic resins. The introduction of various types of glass (e.g., rovings, windings, and woven cloth) as reinforcement was the next development step.[41] Currently, higher-strength epoxy resins have been used as matrices with graphite, boron, beryllium, steel, titanium, aluminum, or magnesium fibers.

Many fiber-reinforced composites, particularly those involving cloth, are highly *anisotropic*. Properties vary with the orientation of lay-ups as well as with weave orientation in the reinforcing materials. In directions transverse to fiber orientation, tensile and compressive strengths are a function of the matrix material. Loads parallel to the fibers are carried by the reinforcement, while flexural strength is limited by the shear bond between the filaments and the matrix material.

Typical reinforcing materials for fiberglass include E- and S-glass.[42] *S-glass* is a silica-alumina-magnesia compound with improved tensile properties. It is used mainly in nonwoven, monodirectional, and wound configurations. *E-glass* is a lime-alumina-borosilicate ($CaOAl_2O_3$-$SiO_2$) compound used primarily in woven fabrics. (See Table 46.22.)

*Table 46.22* Typical Properties of E- and S-Glass Fibers at Room Temperature

| property | E-glass | S-glass |
|---|---|---|
| specific gravity | 2.54 | 2.48 |
| density (lbm/in$^3$) | 0.092 | 0.090 |
| ultimate tensile strength (lbf/in$^2$) | | |
| • monofilament | $5.0 \times 10^5$ | $6.6 \times 10^5$ |
| • 12-end roving | $3.7 \times 10^5$ | $5.5 \times 10^5$ |
| modulus of elasticity (lbf/in$^2$) | $10.5 \times 10^6$ | $12.5 \times 10^6$ |
| coefficient of thermal expansion (1/°F) | $2.8 \times 10^{-6}$ | $1.6$–$2.2 \times 10^{-6}$ |
| specific heat (Btu/lbm-°F) | 0.192 | 0.176 |

(Multiply lbm/in$^2$ by 27,680 to obtain kg/m$^3$.)
(Multiply lbf/in$^2$ by 0.006895 to obtain MPa.)

Graphite (i.e., carbon) fibers are used where high stiffnesses and low coefficients of thermal expansion are needed.[43] These advantages are balanced by the disadvantages of brittleness and high cost. The ultimate tensile strength for graphite varies inversely with modulus of elasticity. Graphite fibers range in strength from 180 ksi (1.2 GPa) for yarn configurations to 350 ksi (2.4 GPa) for tow, while the modulus of elasticity varies from $60 \times 10^6$ lbf/in$^2$ to $20 \times 10^6$ lbf/in$^2$ (410 GPa to 140 GPa).[44]

---

[38]*Sintering* is the process where a physical mixture of carbide and powdered metal is heated in order to solidify the powder into a single piece. When the metal melts, it acts as the binder for the carbide.

[39]Ceramics typically fail catastrophically, without warning. To counteract this tendency, *toughened ceramics* incorporate discrete solids such as SiC or TiC whiskers throughout the ceramic matrix. Cracks that start are arrested by the dispersoids. Dispersoids can increase ceramic toughness by as much as 40%.

[40]A *whisker* is a single crystal grown by vapor deposition. Although lengths of several millimeters are typical, diameters are only a few micrometers.

[41]A *roving* consists of a number of parallel strands of fiber. The strands are side by side (not interwoven or twisted together), forming a flat ribbon.

[42]*A-glass* is common soda-lime glass used for windows, bottles, and jars. Other types of glass used for reinforcement include *C-glass* (developed for greater chemical and corrosion resistance), *D-glass* (glass possessing a low dielectric constant), and *M-glass* (glass containing BeO to increase the elastic modulus).

[43]It is interesting that graphite in *fiber* or *whisker* form is used to provide strength, while graphite in *powder* form is used as a solid lubricant. The *laminar* structure of graphite permits particles to easily slide over one another. This laminar structure does not easily break down, making graphite particularly valuable as a lubricant at high temperatures and pressures—up to at least 3600°F (2000°C). In fact, graphite's coefficient of friction decreases with temperature. Another excellent solid lubricant, *molybdenum disulfide* ($MoS_2$), also has a laminar structure, but its friction coefficient increases sharply above 1600°F (900°C).

[44]*Tow* consists of loose, untwisted fibers.

There are three general categories of carbon fibers: standard, high-modulus, and high-strength. *Kevlar aramid* fibers provide strength and stiffness that are essentially in between that of glass and carbon fibers.

Resistance to chemical corrosion, rather than strength-to-weight ratio, is the primary factor in selecting *fiber-reinforced plastics* (also referred to as *fiber-reinforced polymers*, both abbreviated FRP) for process tanks, reaction vessels, and pipes.[45] The main resins used for this purpose are vinyl esters, epoxies, polyesters, furans, and phenolics. Vinyl esters, which can handle both acidic and basic fluids as well as strong oxidizers such as chlorine, are the most widely used. Epoxies have better thermal and mechanical properties, but epoxies cannot be used in highly acidic environments (i.e., pH less than 3). Polyesters, on the other hand, are acid resistant, but are not alkali resistant (i.e., pH more than 9). Furans and phenolics are relatively weak and are used in special applications.

A typical FRP tank or pipe consists of three layers: the veil, liner, and structural laminate (from the inside out). The *veil* consists of a thin layer (e.g., 0.25 mm) of glass or polyester fibers saturated with approximately 90% by weight of the resin. The 2.5 mm thick *liner* consists of glass in resin in a 3:1 ratio, respectively. The liner provides chemical resistance. The outermost fiberglass *structural laminate* bears all of the pressure, stresses, and mechanical forces imposed on the tank.[46] Although the structural laminate can be laid up by hand, filament-wound tanks are more popular because manufacturing costs are lower. (See Table 46.23.)

**Table 46.23** *Typical Properties of Laid-Up Composites (linear lay-up)*

| composite | fiber content (%) | specific gravity | modulus of elasticity (ksi) | | |
|---|---|---|---|---|---|
| | | | axial | transverse | shear |
| graphite-epoxy | | | | | |
| high strength | 65 | 1.58 | 20,000 | 1000 | 650 |
| high modulus | 65 | 1.61 | 29,000 | 1000 | 700 |
| ultrahigh modulus | 65 | 1.69 | 44,000 | 1000 | 950 |
| Kevlar™49-epoxy | 65 | 1.39 | 12,500 | 800 | 300 |
| E-glass-epoxy | 65 | 1.99 | 6000 | 1500 | 300 |
| chopped glass-polyester | 30 | 1.88 | 2500 | 2500 | 100 |
| sheet molding compound (SMC) | 65 | 1.99 | 3500 | 3500 | 150 |

(Multiply ksi by 6.895 to obtain MPa.)

[45]Other characteristics for which FRP may be used include cost, weight, and ease of manufacture of complex shapes.
[46]ASTM standards D3299 and D4097 both specify the common $pD/2S_H$ formula for wall thickness, where $S_H$ is the allowable circumferential (hoop) stress that produces a maximum strain of 0.0010.

## 32. CONCRETE

*Concrete* (*portland cement concrete*) is a mixture of cement, aggregates, water, and air. The cement paste consists of a mixture of portland cement and water. The paste binds the coarse and fine aggregates into a rock-like mass as the paste hardens during the chemical reaction (*hydration*). Table 46.24 lists the approximate volumetric percentage of each ingredient.

**Table 46.24** *Typical Volumetric Proportions of Concrete Ingredients*

| component | air-entrained | non-air-entrained |
|---|---|---|
| coarse aggregate | 31% | 31% |
| fine aggregate | 28% | 30% |
| water | 18% | 21% |
| cement | 15% | 15% |
| air | 8% | 3% |

## 33. CEMENT

*Portland cement* is produced by burning a mixture of lime and clay in a rotary kiln and grinding the resulting mass. Cement has a specific weight (density) of approximately 195 lbf/ft$^3$ (3120 kg/m$^3$) and is packaged in standard sacks ("bags") weighing 94 lbf (40 kg).

ASTM C-150 describes the five classifications of portland cement.

*Type I—Normal portland cement:* This is a general-purpose cement used whenever sulfate hazards are absent and when the heat of hydration will not produce a significant rise in the temperature of the cement. Typical uses are sidewalks, pavement, beams, columns, and culverts.

*Type II—Modified portland cement:* This cement has a moderate sulfate resistance, but is generally used in hot weather for the construction of large structures. Its heat rate and total heat generation are lower than those of normal portland cement.

*Type III—High-early strength portland cement:* This type develops its strength quickly. It is suitable for use when a structure must be put into early use or when long-term protection against cold temperatures is not feasible. Its shrinkage rate, however, is higher than those of types I and II, and extensive cracking may result.

*Type IV—Low-heat portland cement:* For massive concrete structures such as gravity dams, low-heat cement is required to maintain a low temperature during curing. The ultimate strength also develops more slowly than for the other types.

*Type V—Sulfate-resistant portland cement:* This type of cement is appropriate when exposure to sulfate concentration is expected. This typically occurs in regions having highly alkaline soils.

Types I, II, and III are available in two varieties: normal and air-entraining (designated by an "A" suffix).

The compositions of the three types of air-entraining portland cement (types IA, IIA, and IIIA) are similar to types I, II, and III, respectively, with the exception that an air-entraining admixture is added.

Many states use modified concrete mixes in critical locations in order to reduce *concrete-disintegration cracking* ("D-cracking") caused by the freeze-thaw cycle. Coarse aggregates are the primary cause of D-cracking, so the maximum coarse aggregate size is reduced. However, a higher cement paste content causes *shrinkage cracking* during setting, leading to increased water penetration and corrosion of reinforcing steel. The cracking can be reduced or eliminated by using *shrinkage-compensating cement*, known as "type-K cement" (named after ASTM C-846 type E-1(K)).

Type-K cement (often used in bridge decks) contains an aluminate that expands during setting, offsetting the shrinkage. The net volume change is near zero. The resulting concrete is referred to as *shrinkage-compensating concrete.*

Special cement formulations are needed to reduce *alkali-aggregate reactivity* (AAR)—the reaction of the alkalis in cement with compounds in the sand and gravel aggregate. AAR produces long-term distress in the forms of network cracking and spalling (popouts) in otherwise well-designed structures. AAR takes on two forms: the more common *alkali-silica reaction* (ASR) and the less-common *alkali-carbonate reaction* (ACR). ASR is countered by using low-alkali cement (ASTM C-150) with an equivalent alkali content of less than 0.60% (as sodium oxide), using lithium-based admixtures, or "sweetening" the mixture by replacing approximately 30% of the aggregate with crushed limestone. ACR is not effectively controlled by using low-alkali cements. Careful selection, blending, and sizing of the aggregate are needed to minimize ACR.

## 34. AGGREGATE

Because aggregate makes up 60–75% of the total concrete volume, its properties influence the behavior of freshly mixed concrete and the properties of hardened concrete. Aggregates should consist of particles with sufficient strength and resistance to exposure conditions such as freezing and thawing cycles. Also, they should not contain materials that will cause the concrete to deteriorate.

Most sand and rock aggregate has a specific weight of approximately 165 lbf/ft$^3$ (2640 kg/m$^3$) corresponding to a specific gravity of 2.64.

*Fine aggregate* consists of natural sand or crushed stone up to $^1/_4$ in (6 mm), with most particles being smaller than 0.2 in (5 mm). Aggregates, whether fine or coarse, must conform to certain standards to achieve the best engineering properties. They must be strong, clean, hard, and free of absorbed chemicals. Fine aggregates must meet the particle-size distribution (grading) requirements.

The seven standard ASTM C33 sieves for fine aggregates have openings ranging from 0.150 mm (no. 100 sieve) to $^3/_8$ in (9.5 mm). The fine aggregate should have not more than 45% passing any sieve and retained on the next consecutive sieve, and its fineness modulus should be not less than 2.3 or more than 3.1. The *fineness modulus* is an empirical factor obtained by adding the cumulative weight percentages retained on each of a specific series (usually no. 4, no. 8, no. 16, no. 30, no. 50, and no. 100 for the fine aggregate) of sieves and dividing the sum by 100. (The dust or pan percentage is not included in calculating the cumulative percentage retained.) The higher the fineness modulus, the coarser will be the gradation.

*Coarse aggregates* consist of natural gravel or crushed rock, with pieces large enough to be retained on a no. 4 sieve (openings of 0.2 in or 4.75 mm). In practice, coarse aggregate is generally between $^3/_8$ in and $1^1/_2$ in (9.5 to 38 mm) in size. Also, coarse aggregates should meet the gradation requirements of ASTM C33, which specifies 13 standard sieve sizes for coarse aggregate.

Coarse aggregate has three main functions in a concrete mix: (a) to act as relatively inexpensive filler, (b) to provide a mass of particles that are capable of resisting the applied loads, and (c) to reduce the volume changes that occur during the setting of the cement-water mixture.

## 35. WATER

Water in concrete has three functions: (a) Water reacts chemically with the cement. This chemical reaction is known as *hydration.* (b) Water wets the aggregate. (c) The water and cement mixture, which is known as *cement paste*, lubricates the concrete mixture and allows it to flow.

Water has a standard density of 62.4 lbf/ft$^3$ (1000 kg/m$^3$). 7.48 gallons occupy 1 ft$^3$ (1000 L occupy 1 m$^3$). One ton (2000 lbf) of water has a volume of 240 gal.

Any potable water that has no pronounced odor or taste can be used for producing concrete. (With some quality restrictions, the ACI code also allows nonpotable water to be used in concrete mixing.) Impurities in water may affect the setting time, strength, and corrosion resistance. Water used in mixing concrete should be clean and free from injurious amounts of oils, acids, alkalis, salt, organic materials, and other substances that could damage the concrete or reinforcing steel.

## 36. ADMIXTURES

*Admixtures* are routinely used to modify the performance of concrete. Advantages include higher strength, durability, chemical resistance, and workability; controlled rate of hydration; and reduced shrinkage and cracking. Accelerating and retarding admixtures fall into several different categories, as classified by ASTM C-494.

Type A: water-reducing
Type B: set-retarding
Type C: set-accelerating
Type D: water-reducing and set-retarding
Type E: water-reducing and set-accelerating
Type F: high-range water-reducing
Type G: high-range water-reducing and set-retarding

ASTM C-260 covers air-entraining admixtures, which enhance freeze-thaw durability. ASTM C-1017 deals exclusively with plasticizers to produce flowing concrete. ACI 212 recognizes additional categories, including corrosion inhibitors and damproofing. Finally, microsilica, fly ash, and synthetic fibers are routinely used in concrete.

*Water-reducing admixtures* disperse the cement particles throughout the plastic concrete, reducing water requirements by 5–10%. Water that would otherwise be trapped within the cement floc remains available to fluidize the concrete. Although water is necessary to produce concrete, using lesser amounts increases strength and durability and decreases permeability and shrinkage. The same slump can be obtained with less water.

*High-range water reducers*, also known as *superplasticizers*, function via the same mechanisms as regular water reducers. However, the possible water reduction is greater (e.g., 12–30%). Dramatic increases in slump, workability, and strength are achieved. *High-slump concrete* is suitable for use in sections that are heavily reinforced and in areas where consolidation cannot otherwise be attained. Also, concrete can be pumped at lower pump pressures, so the lift and pumping distance can be increased. Overall, superplasticizers reduce the cost of mixing, pumping, and finishing concrete.

*Set accelerators* increase the rate of cement hydration, shortening the setting time and increasing the rate of strength development. They are useful in cold weather (below 35–40°F or 2–4°C) or when urgent repairs are needed. While calcium chloride ($CaCl_2$) is a very effective accelerator, nonchloride, noncorrosive accelerators can provide comparable performance. (The ACI code does not allow chloride to be added to concrete used in prestressed construction, in concrete containing aluminum embedments, or in concrete cast against galvanized stay-in-place steel forms.)

*Set retarders* are used in hot environments and where the concrete must remain workable for an extended period of time, allowing extended haul and finishing times. A higher ultimate strength will also result. Most retarders also have water-reducing properties.

*Air entraining mixtures* create microscopic air bubbles in the concrete. This improves the durability of hardened concrete subject to freezing and thawing cycles. The wet workability is improved, while bleeding and segregation are reduced.

A waste product of coal-burning power-generation stations, *fly ash*, is the most common *pozzolanic additive*. As cement sets, calcium silicate hydrate and calcium hydroxide are formed. While the former is a binder that holds concrete together, calcium hydroxide does not contribute to binding. However, fly ash reacts with some of the calcium hydroxide to increase binding. Also, since fly ash acts as a microfiller between cement particles, strength and durability are increased while permeability is reduced. When used as a replacement for less than 45% of the portland cement, fly ash meeting ASTM C-618 enhances resistance to scaling from road deicing chemicals.

*Microsilica* (*silica fume*) is an extremely fine particulate material, approximately 1/100th the size of cement particles. It is a waste product of electric arc furnaces. It acts as a "super pozzolan." Adding 5–15% microsilica will increase the pozzolanic reaction as well as provide a microfiller to reduce permeability.

Microsilica reacts with calcium hydroxide in the same manner as fly ash. It is customarily used to achieve strengths in the 8000–9000 psi (55–62 MPa) range.

*Corrosion-resisting compounds* are intended to inhibit rusting of the reinforcing steel and prestressing strands. Calcium nitrate is commonly used to inhibit the corrosive action of chlorides. It acts by forming a passivating protective layer on the steel. Calcium nitrate has essentially no effect on the mechanical and plastic properties of concrete.

In the *DELVO*® admixture system, the cement particles are coated with a stabilizer, halting the hydration process indefinitely. Setting can be reinitiated at will hours or days later. The manufacturer claims that the treatment has no effect on the concrete when it hardens.

## 37. SLUMP

The four basic concrete components (cement, sand, coarse aggregate, and water) are mixed together to produce a homogeneous concrete mixture. The *consistency* and *workability* of the mixture affect the concrete's ability to be placed, consolidated, and finished without segregation or bleeding. The slump test is commonly used to determine consistency and workability.

The *slump test* consists of completely filling a slump cone mold in three layers of about one third of the mold volume. Each layer is rodded 25 times with a round, spherical-nosed steel rod of $^5/_8$ in (16 mm) diameter. When rodding the subsequent layers, the previous layers beneath are not penetrated by the rod. After rodding, the mold is removed by raising it carefully in the vertical direction. The slump is the difference in the mold height and the resulting concrete pile height. Typical values are 1–4 in (25–100 mm).

Concrete mixtures that do not slump appreciably are known as *stiff mixtures*. Stiff mixtures are inexpensive because of the large amounts of coarse aggregate.

However, placing time and workability are impaired. Mixtures with large slumps are known as *wet mixtures* (*watery mixtures*) and are needed for thin castings and structures with extensive reinforcing. Slumps for concrete that is machine-vibrated during placement can be approximately one third less than for concrete that is consolidated manually.

## 38. DENSITY

The density, also known as *weight density, unit weight*, and *specific weight*, of normal-weight concrete varies from about 140 lbf/ft$^3$ to about 160 lbf/ft$^3$ (2240 kg/m$^3$ to 2560 kg/m$^3$), depending on the specific gravities of the constituents. For most calculations involving normal-weight concrete, the density may be taken as 145 lbf/ft$^3$ to 150 lbf/ft$^3$ (2320 kg/m$^3$ to 2400 kg/m$^3$). Lightweight concrete can have a density as low as 90 lbf/ft$^3$ (1450 kg/m$^3$). Although steel has a density of more than three times that of concrete, due to the variability in concrete density values and the relatively small volume of steel, the density of steel-reinforced concrete is typically taken as 150 lbf/ft$^3$ (2400 kg/m$^3$) without any refinement for exact component contributions

## 39. COMPRESSIVE STRENGTH

The concrete's *compressive strength*, $f'_c$, is the maximum stress a concrete specimen can sustain in compressive axial loading. It is also the primary parameter used in ordering concrete. When one speaks of "6000 psi (41 MPa) concrete," the compressive strength is being referred to. Compressive strength is expressed in psi or MPa. SI compressive strength may be written as "Cxx" (e.g., "C20"), where xx is the compressive strength in MPa. (MPa is equivalent to N/mm$^2$, which is also commonly quoted.)

Typical compressive strengths range from 4000 psi to 6000 psi (27–41 MPa) for traditional structural concrete, though concrete for residential slabs-on-grade and foundations will be lower in strength (e.g., 3000 psi). 6000 psi (41 MPa) concrete is used in the manufacture of some concrete pipes, particularly those that are jacked in.

Cost is approximately proportional to concrete's compressive strength—a rule that applies to high-performance concrete as well as traditional concrete. For example, if 5000 psi (34 MPa) concrete costs $100 per cubic yard, then 14,000 psi concrete will cost approximately $280 per cubic yard.

Compressive strength is controlled by selective proportioning of the cement, coarse and fine aggregates, water, and various admixtures. However, the compressive strength of traditional concrete is primarily dependent on the mixture's water/cement ratio. (See Fig. 46.10.) Provided that the mix is of a workable consistency, strength varies directly with the cement/water ratio.

**Figure 46.10** *Typical Concrete Compressive Strength Characteristics*

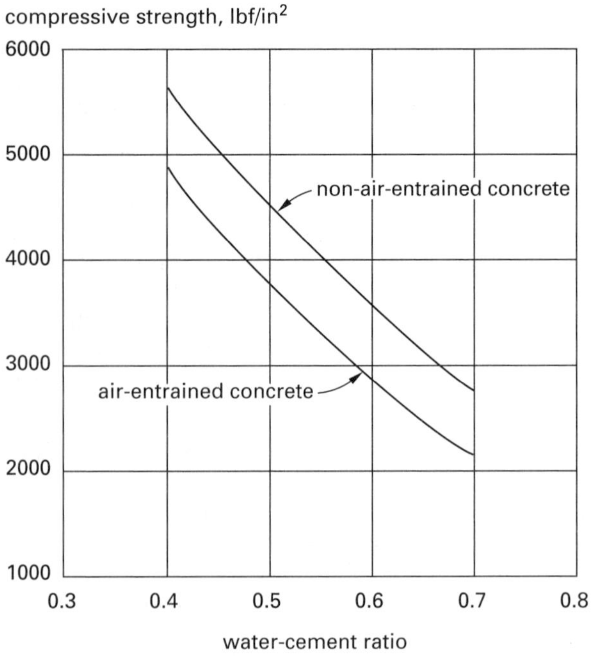

(This is *Abrams' strength law*, named after Dr. Duff Abrams, who formulated the law in 1918.)

The standard ASTM compressive test specimen mold is a cylinder with a 6 in (150 mm) diameter and a 12 in (300 mm) height. Steel molds are more expensive than plastic molds, but they provide greater rigidity. (Some experts say specimens from steel molds test 3 to 15% higher.) The concrete is cured for a specific amount of time (three days, a week, 28 days, or more) at a specific temperature. Plain or lime-saturated heated water baths, as well as dry "hot boxes" heated by incandescent lights, are used for this purpose at some testing labs. To ensure uniform loading, the ends are smoothed by grinding or are capped in sulfur. For testing of very high-strength concrete, the ends may need to be ground glass-smooth in lapidary machines.

Since the ultimate load for 15,000 psi (103 MPa) and higher concrete exceeds the capacity (typically 300,000 lbf (1.34 MN)) of most testing machines, testing firms are switching to smaller cylinders with diameters of 4 in (100 mm) and heights of 8 in (200 mm) rather than purchasing 400,000–600,000 lbf (1.8–2.7 MN) machines.

The specimen is axially loaded to failure at a specific rate. The compressive strength is calculated as the maximum axial load, $P$, divided by the cross-sectional area, $A$, of the cylinder. Since as little as 0.1 in difference in the diameter can affect the test results by 5%, the diameter must be measured precisely.

$$f'_c = \frac{P}{A} \qquad 46.3$$

Compressive strength is normally measured on the 28th day after the specimens are cast. Since the strength of concrete increases with time, all values of $f'_c$ must be stated with respect to a known age. If no age is given, a strength at a "standard" 28-day age is assumed.

The effect of the cement/water ratio on compressive strength (i.e., the more water the mix contains, the lower the compressive strength will be) is a different issue than the use of large amounts of surface water to cool the concrete during curing (i.e., *moist-curing*). The strength of newly poured concrete can be increased significantly (e.g., doubled) if the concrete is kept cool during part or all of curing. This is often accomplished by covering new concrete with wet burlap or by spraying with water. Although best results occur when the concrete is moist-cured for 28 days, it is seldom economical to do so. A substantial strength increase can be achieved if the concrete is kept moist for as little as 3 days. Externally applied curing retardants can also be used.

## 40. ABRASION RESISTANCE

Abrasion resistance is a function of the compressive strength. For traditionally manufactured concrete, generally, the higher the concrete's compressive strength, the lower its abrasion resistance will be. This is illustrated in Table 46.25. Replacing 7–10% by weight of cement with *silica fume* (also known as *microsilica*, a fine vitreous byproduct of the manufacture of silicon and ferrosilicon alloys) has been shown to raise the abrasion resistance. Abrasion resistance is not greatly affected by superplasticizers, curing compounds, or moderate replacement of cement (i.e., less than 30% by weight) by flyash. Results of tests on samples with larger amounts of flyash have not been consistent.

**Table 46.25** Typical Abrasion Resistance of Traditional Concrete

| compressive strength (lbf/in²) | approximate relative abrasion resistance |
|---|---|
| 1000 | 100% |
| 2000 | 62% |
| 3000 | 43% |
| 4000 | 30% |
| 5000 | 21% |
| 6000 | 15% |

## 41. STRESS-STRAIN RELATIONSHIP

The stress-strain relationship for concrete is dependent on its strength, age at testing, rate of loading, nature of the aggregates, cement properties, and type and size of specimens. Typical stress-strain curves for concrete specimens loaded in compression at 28 days of age under a normal rate of loading are shown in Fig. 46.11.

**Figure 46.11** Typical Concrete Stress-Strain Curves

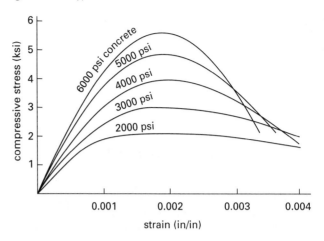

## 42. MODULUS OF ELASTICITY

The *modulus of elasticity* (also known as *Young's modulus*) is defined as the ratio of stress to strain in the elastic region. Unlike steel, the modulus of elasticity of concrete varies with compressive strength. Since the slope of the stress-strain curve varies with the applied stress, there are several ways of calculating the modulus of elasticity. Figure 46.12 shows a typical stress-strain curve for concrete with the *initial modulus*, the *tangent modulus*, and the *secant modulus* indicated.

**Figure 46.12** Concrete Moduli of Elasticity

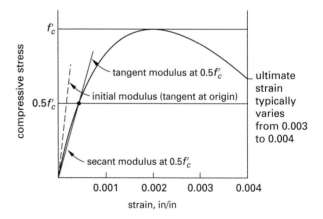

The *secant modulus of elasticity* is specified by the American Concrete Institute Code for use with specific weights that are between 90 and 155 lbf/ft³ (1440 and 2480 kg/m³). Equation 46.4 is used for both instantaneous and long-term deflection calculations. $w_c$ is in lbf/ft³ (kg/m³), and $E_c$ and $f'_c$ are in lbf/in² (MPa) [ACI 318 Sec. 8.5.1].

$$E_c = w_c^{1.5} 0.043 \sqrt{f'_c} \qquad \text{[SI]} \quad 46.4(a)$$

$$E_c = w_c^{1.5} 33 \sqrt{f'_c} \qquad \text{[U.S.]} \quad 46.4(b)$$

For normal-weight concrete, the ACI code suggests Eq. 46.5, corresponding to a specific weight of approximately 145 lbf/ft³ (2320 kg/m³) [ACI 318 Sec. 8.5.1].

$$E_c = 5000\sqrt{f'_c} \qquad \text{[SI]} \quad \textbf{46.5(a)}$$

$$E_c = 57{,}000\sqrt{f'_c} \qquad \text{[U.S.]} \quad \textbf{46.5(b)}$$

## 43. SPLITTING TENSILE STRENGTH

The extent and size of cracking in concrete structures are affected to a great extent by the tensile strength of the concrete. The ASTM C-496 *split cylinder testing procedure* is the standard test to determine the tensile strength of concrete. A 6 in × 12 in (150 mm × 300 mm) cast or drill-core cylinder is placed on its side as in Fig. 46.13, and the minimum load, $P$, that causes the cylinder to split in half is used to calculate the splitting tensile strength.

$$f_{ct} = \frac{2P}{\pi DL} \qquad \textbf{46.6}$$

ACI 318 Sec. 5.1.4 suggests that the splitting tensile strength $f_{ct}$ can be calculated from correlations with compressive strength, and ACI 318 Sec. 11.2 indirectly gives such correlations.

***Figure 46.13*** *Splitting Tensile Strength Test*

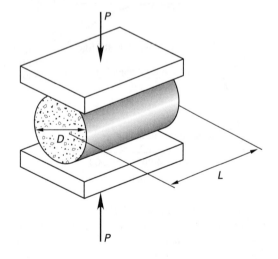

$$f_{ct,MPa} = 0.56\sqrt{f'_{c,MPa}} \quad \text{[normal weight]} \qquad \text{[SI]} \quad \textbf{46.7(a)}$$

$$f_{ct} = 6.7\sqrt{f'_c} \quad \text{[normal weight]} \qquad \text{[U.S.]} \quad \textbf{46.7(b)}$$

$$f_{ct,MPa} = 0.47\sqrt{f'_{c,MPa}} \quad \text{[sand lightweight]} \qquad \text{[SI]} \quad \textbf{46.8(a)}$$

$$f_{ct} = 5.7\sqrt{f'_c} \quad \text{[sand lightweight]} \qquad \text{[U.S.]} \quad \textbf{46.8(b)}$$

$$f_{ct,MPa} = 0.42\sqrt{f'_{c,MPa}} \quad \text{[all-lightweight]} \qquad \text{[SI]} \quad \textbf{46.9(a)}$$

$$f_{ct} = 5\sqrt{f'_c} \quad \text{[all-lightweight]} \qquad \text{[U.S.]} \quad \textbf{46.9(b)}$$

## 44. MODULUS OF RUPTURE

The tensile strength of concrete in flexure is known as the *modulus of rupture*, $f_r$, and is an important parameter for evaluating cracking and deflection in beams. The tensile strength of concrete is relatively low, about 10% to 15% (and occasionally up to 20%) of the compressive strength. ASTM C-78 gives the details of beam testing using *third point loading*. The modulus of rupture is calculated from Eq. 46.10.

$$f_r = \frac{Mc}{I} \quad \text{[tension]} \qquad \textbf{46.10}$$

Equation 46.10 gives higher values for tensile strength than the *splitting tensile strength test* because the stress distribution in concrete is not linear as is assumed in Eq. 46.6. For normal-weight concrete, the ACI code prescribes that Eq. 46.11 should be used for modulus of rupture calculations. For all-lightweight concrete, the modulus of rupture is taken as 75% of the calculated values. Other special rules for lightweight concrete may apply [ACI 318 Sec. 9.5.2.3].

$$f_r = 0.62\sqrt{f'_c} \qquad \text{[SI]} \quad \textbf{46.11(a)}$$

$$f_r = 7.5\sqrt{f'_c} \qquad \text{[U.S.]} \quad \textbf{46.11(b)}$$

## 45. SHEAR STRENGTH

Concrete's true *shear strength* is difficult to determine in the laboratory because shear failure is seldom pure and is typically affected by other stresses in addition to the shear stress. Reported values of shear strength vary greatly with the test method used, but they are a small percentage (e.g., 25% or less) of the ultimate compressive strength.

## 46. POISSON'S RATIO

*Poisson's ratio* is the ratio of the lateral strain to the axial strain. It varies in concrete from 0.11 to 0.23, with typical values being 0.17 to 0.21.

## 47. CONCRETE MIX DESIGN CONSIDERATIONS

Concrete can be designed for compressive strength or durability in its hardened state. In its wet state, concrete should have good workability. *Workability* relates to the effort required to transport, place, and finish wet concrete without segregation or bleeding. Workability is often closely correlated with slump. (See Sec. 46.37.) Table 46.26 gives typical slumps by application. All other design requirements being met, the most economical mix should be selected.

**Table 46.26** *Typical Slumps by Application*

| application | slump, in (mm) | |
| --- | --- | --- |
| | maximum | minimum |
| reinforced footings and foundations | 3 (76) | 1 (25) |
| plain footings and substructure walls | 3 (76) | 1 (25) |
| slabs, beams, and reinforced walls | 4 (102) | 1 (25) |
| reinforced columns | 4 (102) | 1 (25) |
| pavements and slabs | 3 (76) | 1 (25) |
| heavy mass construction | 2 (51) | 1 (25) |
| roller-compacted concrete | 0 | 0 |

(Multiply in by 25.4 to obtain mm.)

*Durability* is defined as the ability of concrete to resist environmental exposure or service loadings. One of the most destructive environmental factors is the freeze/thaw cycle. ASTM C-666, "Standard Test Method for Resistance of Concrete to Rapid Freezing and Thawing," is the standard laboratory procedure for determining the freeze-thaw durability of hardened concrete. This test determines a *durability factor*, the number of freeze-thaw cycles required to produce a certain amount of deterioration.

ACI 318 Sec. 4.2.2 places maximum limits on the water-cement ratio and minimum limits on the strength for concrete with special exposures, including concrete exposed to freeze-thaw cycles, deicing chemicals, and chloride, and installations requiring low permeability.

Specifying an air-entrained concrete will improve the durability of concrete subject to freeze-thaw cycles or deicing chemicals. The amount of entrained air needed will depend on the exposure conditions and the size of coarse aggregate, as prescribed in ACI 318 Sec. 4.2.1.

## 48. STRENGTH ACCEPTANCE TESTING

When extensive statistical data are not available, acceptance testing of laboratory-cured specimens can be evaluated per ACI 318 Sec. 5.6.3.3. This section states that the compressive strength of concrete is considered satisfactory if both of the following criteria are met: (a) No single test (average of two cylinders) falls below the specified compressive strength, $f'_c$, by more than 500 psi (3.45 MPa) when $f'_c$ is 5000 psi (34.5 MPa) or less, or by more than $0.10f'_c$ when $f'_c$ is more than 5000 psi

(34.5 MPa); (b) the average of any three consecutive test strengths equals or exceeds the specified compressive strength.

## 49. BATCHING

All concrete ingredients are weighed or volumetrically measured before being mixed, a process known as *batching*. Weighing is more common because it is simple and accurate. However, water and liquid admixtures can be added by either volume or weight. The following accuracies are commonly specified or assumed in concrete batching: cement, 1%; water, 1%; aggregates, 2%; and admixtures, 3%. ACI 318 Sec. 4.2.1 specifies the accuracy of air entrainment as $1^1/_2$%.

## 50. WATER-CEMENT RATIO AND CEMENT CONTENT

Concrete strength is inversely proportional to the *water-cement ratio* (ACI 318 uses the more general term *water-cementitious materials ratio*), the ratio of the amount of water to the amount of cement in a mixture, usually stated as a decimal by weight. Typical values are approximately 0.45–0.60 by weight. (Alternatively, the water-cement ratio may be stated as the number of gallons of water per 94 lbf sack of cement, in which case typical values are 5–7 gal.)

A mix is often described by the number of sacks of cement needed to produce 1 $yd^3$ of concrete. For example, a mix using 6 sacks of cement per cubic yard would be described as a "6-sack mix." Another method of describing the cement content is the *cement factor*, which is the number of cubic feet of cement per cubic yard of concrete.

## 51. PROPORTIONING MIXES

The oldest method of proportioning concrete is the *arbitrary proportions method* (*arbitrary volume method* or *arbitrary weight method*). Ingredients of average properties are assumed, and various proportions are logically selected. Tabular or historical knowledge of mixes and compressive strengths may be referred to. Proportions of cement, fine aggregate, and coarse aggregate are designated (in that sequence). For example, 1:2:3 means that one part of cement, two parts of fine aggregate, and three parts of coarse aggregate are combined. The proportions are generally in terms of weight; volumetric ratios are rarely used. (In the rare instances where the mix proportions are volumetric, the ratio values must be multiplied by the bulk densities to get the weights of the constituents. Then weight ratios may be calculated and the absolute volume method applied directly.)

For more critical applications, the actual ingredients can be tested in various logical proportions. This is known as the *trial batch method* or the *trial mix method*.

This method is more time-consuming initially, since cured specimens must be obtained for testing.

Although some "extra" concrete ends up being brought to the job site, once determined, mix quantities should be rounded up, not down. Otherwise, the concrete volume delivered may be short.

## 52. ABSOLUTE VOLUME METHOD

The *yield* is the volume of wet concrete produced in a batch. Typical units are cubic yards (referred to merely as "yards") but may also be cubic feet or cubic meters. The yield that results from mixing known quantities of ingredients can be found from the *absolute volume method*, also known as the *solid volume method* and *consolidated volume method*. This method uses the specific gravities or densities for all the ingredients to calculate the absolute volume each will occupy in a unit volume of concrete. The absolute volume (solid volume, consolidated volume, etc.) is

$$V_{\text{absolute}} = \frac{m}{(\text{SG})\rho_{\text{water}}} \qquad \text{[SI]} \quad \textbf{\textit{46.12(a)}}$$

$$V_{\text{absolute}} = \frac{W}{(\text{SG})\gamma_{\text{water}}} \qquad \text{[U.S.]} \quad \textbf{\textit{46.12(b)}}$$

The absolute volume method assumes that, for granular materials such as cement and aggregates, there will be no voids between particles. Therefore, the amount of concrete is the sum of the solid volumes of cement, sand, coarse aggregate, and water.

To use the absolute volume method, it is necessary to know the solid densities of the constituents. In the absence of other information, Table 46.27 can be used.

**Table 46.27** *Summary of Approximate Properties of Concrete Components*

cement
| | |
|---|---|
| specific weight | 195 lbf/ft$^3$ (3120 kg/m$^3$) |
| specific gravity | 3.13–3.15 |
| weight of one sack | 94 lbf (42 kg) |

fine aggregate
| | |
|---|---|
| specific weight | 165 lbf/ft$^3$ (2640 kg/m$^3$) |
| specific gravity | 2.64 |

coarse aggregate
| | |
|---|---|
| specific weight | 165 lbf/ft$^3$ (2640 kg/m$^3$) |
| specific gravity | 2.64 |

water
| | |
|---|---|
| specific weight | 62.4 lbf/ft$^3$ (1000 kg/m$^3$) |
| | 7.48 gal/ft$^3$ (1000 L/m$^3$) |
| | 8.34 lbf/gal (1 kg/L) |
| | 239.7 gal/ton (1 L/kg) |
| specific gravity | 1.00 |

(Multiply lbf/ft$^3$ by 16 to obtain kg/m$^3$.)
(Multiply lbf by 0.45 to obtain kg.)

## Example 46.5

A concrete mixture using 340 lbf of water per cubic yard has a water-cement ratio of 0.60 by weight. (a) Determine the ideal weight and volume of cement needed to produce 1.0 yd$^3$ of concrete. (b) How many sacks of cement are needed to produce a 4 in thick concrete slab 9 ft × 4 ft with this mix?

*Solution*

(a) The cement requirement is found from the water-cement ratio.

$$W_{\text{cement}} = \frac{W_{\text{water}}}{0.6} = \frac{340\,\frac{\text{lbf}}{\text{yd}^3}}{0.6}$$
$$= 567\ \text{lbf/yd}^3$$

From Table 46.27, the specific weight of cement is 195 lbf/ft$^3$. The volume of the cement is

$$V_{\text{cement}} = \frac{W_{\text{cement}}}{\gamma_{\text{cement}}} = \frac{567\,\frac{\text{lbf}}{\text{yd}^3}}{195\,\frac{\text{lbf}}{\text{ft}^3}}$$
$$= 2.91\ \text{ft}^3/\text{yd}^3$$

(b) The slab volume is

$$V_{\text{slab}} = \frac{(4\,\text{in})(9\,\text{ft})(4\,\text{ft})}{\left(12\,\frac{\text{in}}{\text{ft}}\right)\left(27\,\frac{\text{ft}^3}{\text{yd}^3}\right)}$$
$$= 0.444\ \text{yd}^3$$

The cement weight needed is

$$W_{\text{cement}} = (0.444\,\text{yd}^3)\left(567\,\frac{\text{lbf}}{\text{yd}^3}\right) = 252\ \text{lbf}$$

From Table 46.27, each sack weighs 94 lbf. The number of sacks is

$$\frac{252\,\text{lbf}}{94\,\frac{\text{lbf}}{\text{sack}}} = 2.7\ \text{sacks} \quad \text{[3 sacks]}$$

## 53. ADJUSTMENTS FOR WATER AND AIR

In most problems, the weights and volumes of the components must be adjusted for air entrainment and aggregate water content. This determines the *adjusted weights* of the components.

The *saturated, surface-dry* (SSD) condition occurs when the aggregate holds as much water as it can without trapping any free water between the aggregate particles. Calculations of yield should be based on the SSD

**Table 46.28** Dry and Wet Basis Calculations

| | dry basis | wet basis |
|---|---|---|
| fraction moisture, $f$ | $\dfrac{W_{\text{excess water}}}{W_{\text{SSD sand}}}$ | $\dfrac{W_{\text{excess water}}}{W_{\text{SSD sand}} + W_{\text{excess water}}}$ |
| weight of sand, $W_{\text{wet sand}}$ | $W_{\text{SSD sand}} + W_{\text{excess water}}$ | $W_{\text{SSD sand}} + W_{\text{excess water}}$ |
| | $(1 + f)\,W_{\text{SSD sand}}$ | $\left(\dfrac{1}{1-f}\right)W_{\text{SSD sand}}$ |
| weight of SSD sand, $W_{\text{SSD sand}}$ | $\dfrac{W_{\text{wet sand}}}{1+f}$ | $(1-f)\,W_{\text{wet sand}}$ |
| weight of excess water, $W_{\text{excess water}}$ | $fW_{\text{SSD sand}}$ | $fW_{\text{wet sand}}$ |
| | $\dfrac{fW_{\text{wet sand}}}{1+f}$ | $\dfrac{fW_{\text{SSD sand}}}{1-f}$ |

densities, and the water content should be adjusted (increased or decreased) to account for any deviation from the SSD condition. Any water in the aggregate above the SSD water content must be subtracted from the water requirements. Any moisture content deficit below the SSD water content must be added to the water requirements.

The phrase "5% excess water" (or similar) is ambiguous and can result in confusion when used to specify batch quantities. There are two methods in field use, differing in whether the percentage is calculated on a wet or dry basis. Table 46.28 summarizes the batching relationships for both bases. (Though Table 46.28 is written for sand, it can also be used for coarse aggregate.)

There is also a similar confusion about air entrainment. The volumetric basis can be calculated either with or without entrained air. If "5% air" means that the concrete volume is 5% air, then the solid volume is only 95% of the final volume. The solid volume should be divided by 0.95 to obtain the final volume. However, if "5% air" means that the concrete volume is increased by 5% when the air is added, the solid volume should be multiplied by 1.05 to obtain the final volume.

The bases of these percentages must be known. If they are not, then either definition could apply. Regardless, the final volume is not affected significantly by either interpretation.

**Example 46.6**

A mix is designed as 1:1.9:2.8 by weight. The water-cement ratio is 7 gal per sack. (a) What is the concrete yield in cubic feet? (b) How much sand, coarse aggregate, and water is needed to make 45 yd$^3$ of concrete?

*Solution*

(a) The solution can be tabulated. Refer to Table 46.27.

| material | ratio | weight per sack of cement (lbf) | solid density (lbf/ft$^3$) | absolute volume (ft$^3$/sack) |
|---|---|---|---|---|
| cement | 1.0 | $1 \times 94 = 94$ | 195 | $\dfrac{94}{195} = 0.48$ |
| sand | 1.9 | $1.9 \times 94 = 179$ | 165 | $\dfrac{179}{165} = 1.08$ |
| coarse | 2.8 | $2.8 \times 94 = 263$ | 165 | $\dfrac{263}{165} = 1.60$ |
| water | | | | $\dfrac{7}{7.48} = 0.98$ |
| | | | | 4.10 |

The solid yield is 4.10 ft$^3$ of concrete per sack of cement.

(b) The number of one-sack batches is

$$\frac{(45\text{ yd}^3)\left(27\,\dfrac{\text{ft}^3}{\text{yd}^3}\right)}{4.10\,\dfrac{\text{ft}^3}{\text{sack}}} = 296.3\text{ sacks}$$

Order a minimum of 297 sacks of cement. Calculate the remaining order quantities from the cement weight and the mix ratios.

$$W_{\text{sand}} = 1.9\,W_{\text{cement}}$$
$$= (1.9)\left(94\,\frac{\text{lbf}}{\text{sack}}\right)(297\text{ sacks})$$
$$= 53{,}044\text{ lbf}$$

$$W_{\text{coarse aggregate}} = (2.8)\left(94\,\frac{\text{lbf}}{\text{sack}}\right)(297\text{ sacks})$$
$$= 78{,}170\text{ lbf}$$

$$V_{\text{water}} = \left(7\,\frac{\text{gal}}{\text{sack}}\right)(297\text{ sacks}) = 2079\text{ gal}$$

**Example 46.7**

50 ft$^3$ of 1:2.5:4 (by weight) concrete are to be produced. The ingredients have the following properties.

| ingredient | SSD density (lbf/ft$^3$) | moisture (dry basis from SSD) |
|---|---|---|
| cement | 197 | – |
| fine aggregate | 164 | 5% excess (free moisture) |
| coarse aggregate | 168 | 2% deficit (absorption) |

5.5 gal of water are to be used per sack, and the mixture is to have 6% entrained air. What are the ideal order quantities expressed in tons?

*Solution*

Proceed as in Ex. 46.6.

| material | ratio | weight per sack of cement (lbf) | solid density (lbf/ft$^3$) | absolute volume (ft$^3$/sack) |
|---|---|---|---|---|
| cement | 1.0 | $1 \times 94 = 94$ | 197 | $\frac{94}{197} = 0.477$ |
| sand | 2.5 | $2.5 \times 94 = 235$ | 164 | $\frac{235}{164} = 1.433$ |
| coarse | 4.0 | $4.0 \times 94 = 376$ | 168 | $\frac{376}{168} = 2.238$ |
| water | | | | $\frac{5.5}{7.48} = 0.735$ |
| | | | | 4.883 |

The solid yield is 4.883 ft$^3$ of concrete per sack of cement. The yield with 6% air is

$$\frac{4.883 \ \dfrac{\text{ft}^3}{\text{sack}}}{1 - 0.06} = 5.19 \ \text{ft}^3/\text{sack}$$

The ideal number of one-sack batches required is

$$\frac{50 \ \text{ft}^3}{5.19 \ \dfrac{\text{ft}^3}{\text{sack}}} = 9.63 \ \text{sacks}$$

The required sand weight as ordered (not SSD) is

$$\frac{(9.63 \ \text{sacks})(1 + 0.05)\left(94 \ \dfrac{\text{lbf}}{\text{sack}}\right)(2.5)}{2000 \ \dfrac{\text{lbf}}{\text{ton}}} = 1.19 \ \text{tons}$$

The required coarse aggregate weight as ordered (not SSD) is

$$\frac{(9.63 \ \text{sacks})(1 - 0.02)\left(94 \ \dfrac{\text{lbf}}{\text{sack}}\right)(4.0)}{2000 \ \dfrac{\text{lbf}}{\text{ton}}} = 1.77 \ \text{tons}$$

From Table 46.27, the excess water contained in the sand is

$$(1.19 \ \text{tons})\left(\frac{0.05}{1 + 0.05}\right)\left(239.7 \ \frac{\text{gal}}{\text{ton}}\right) = 13.58 \ \text{gal}$$

The water needed to bring the coarse aggregate to SSD conditions is

$$(1.77 \ \text{tons})\left(\frac{0.02}{1 - 0.02}\right)\left(239.7 \ \frac{\text{gal}}{\text{ton}}\right) = 8.66 \ \text{gal}$$

The total water needed is

$$\left(5.5 \ \frac{\text{gal}}{\text{sack}}\right)(9.63 \ \text{sacks}) + 8.66 \ \text{gal} - 13.58 \ \text{gal} = 48.0 \ \text{gal}$$

## 54. REINFORCING STEEL

Steel is an alloy consisting almost entirely of iron. It also contains small quantities of carbon, silicon, manganese, sulfur, phosphorus, and other elements. Carbon has the greatest effect on the steel's properties. The carbon content is normally less than 0.5% by weight, with 0.2% to 0.3% being common percentages.

The density of steel is essentially unaffected by its composition, and a value of 0.283 lbf/in$^3$ (7820 kg/m$^3$) can be used.

Reinforcing steel for use in steel-reinforced concrete may be formed from billet steel, axle steel, or rail steel. Most modern reinforcing bars are made from new billet steel. (Special bars of titanium, stainless steel, corrosion-resistant alloys, and glass fiber composites may see extremely limited use in corrosion-sensitive applications.) The following ASTM designations are used for steel reinforcing bars. (See Table 46.29 for ASTM standards.)

> ASTM A-615: carbon steel, grades 40, 60, and 75 (symbol "S")
>
> ASTM A-996: rail steel, grades 50 and 60 (symbols "R" and "⊥"; only "R" is permitted to be used by ACI 318), and axle steel, grades 40 and 60 (symbol "A")
>
> ASTM A-706: low-alloy steel, grade 60 (symbol "W")

Reinforcing steel used for concrete structures comes in the form of bars (known as "rebar"), welded wire reinforcement, or wires. Reinforcing bars can be plain or deformed; however, most bars are manufactured deformed to increase the bond between concrete and steel. Figure 46.14 shows how the surface is deformed by rolling a pattern on the bar surface. The patterns used vary with the manufacturer.

Plain round reinforcing bars are designated by their nominal diameters in fractions of an inch or in millimeters (e.g., $1/2$ in, $5/8$ in, etc.). Deformed bars are also round, with sizes designed in numbers of eighths of an

**Table 46.29** ASTM Standards for Reinforcing Bars

| customary U.S. bar no. | soft metric bar no. | nominal diameter (in) | nominal diameter (mm) | nominal area (in²) | nominal weight (lbf/ft) |
|---|---|---|---|---|---|
| 3 | 10 | 0.375 | 9.5 | 0.11 | 0.376 |
| 4 | 13 | 0.500 | 12.7 | 0.20 | 0.668 |
| 5 | 16 | 0.625 | 15.9 | 0.31 | 1.043 |
| 6 | 19 | 0.750 | 19.1 | 0.44 | 1.502 |
| 7 | 22 | 0.875 | 22.2 | 0.60 | 2.044 |
| 8 | 25 | 1.000 | 25.4 | 0.79 | 2.670 |
| 9 | 29 | 1.128 | 28.7 | 1.00 | 3.400 |
| 10 | 32 | 1.270 | 32.3 | 1.27 | 4.303 |
| 11 | 36 | 1.410 | 35.8 | 1.56 | 5.313 |
| 14 | 43 | 1.693 | 43.0 | 2.25 | 7.650 |
| 18 | 57 | 2.257 | 57.3 | 4.00 | 13.600 |

**Figure 46.14** Deformed Bars

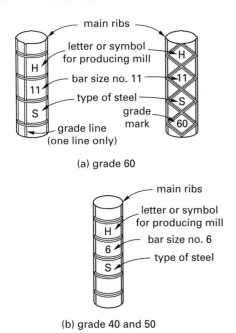

(a) grade 60

(b) grade 40 and 50

**Figure 46.15** Typical Stress-Strain Curve for Ductile Steel

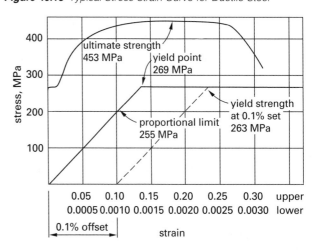

inch or in millimeters. Standard deformed bars are manufactured in sizes no. 3 to no. 11, with two special large sizes, no. 14 and no. 18, also available on special order. Metric (SI) bar designations are based on a "soft" conversion of the bar diameter to millimeters. For example, the traditional no. 3 bar has a nominal diameter of $^3/_8$ in, and this bar has a metric designation of no. 10 because it has a 9.5 mm diameter (approximately 10 mm). Hard metric conversions, where bars with slightly different diameters would be used in metric projects, were once considered but ultimately rejected. Hard conversions are used in Canadian bar sizes, which have different designations and sizes.

## 55. MECHANICAL PROPERTIES OF STEEL

A typical stress-strain curve for ductile structural steel is shown in Fig. 46.15. The curve consists of elastic, plastic, and strain-hardening regions.

The *elastic region* is the portion of the stress-strain curve where the steel will recover its size and shape upon release of load. Within the *plastic region*, the material shows substantial deformation without noticeable change of the stress. Plastic deformation, or *permanent set*, is any deformation that remains in the material after the load has been removed. In the *strain-hardening region*, additional stress is necessary to produce additional strain. This portion of the stress-strain diagram is not important from the design point of view, because so much strain occurs that the functionality of the material is affected.

The slope of the linear portion within the elastic range is the *modulus of elasticity*. The modulus of elasticity for all types of ductile reinforcing steels is taken to be 29 × 10⁶ psi (200 GPa) [ACI 318 Sec. 8.5.2]. (The modulus of elasticity is sometimes quoted as $30 \times 10^6$ psi (207 GPa). This value may apply to hard steels used for other purposes, but not to steel used for reinforcing concrete.)

The steel *grade* corresponds to its nominal yield tensile strength in ksi (thousands of pounds per square inch). Grade 60 steel with a yield strength of 60 ksi (413 MPa) is the most common, though grades 40 and 50 are also available upon request. SI values of steel strength are obtained using "soft" conversions. That is, steel properties are the same in customary U.S. and SI values. Only the designations are different.

## 56. CHLORIDE CORROSION

Chloride has been used as a concrete additive for a long time. In some fresh concrete, calcium chloride is deliberately added to the mix as a low-cost means of increasing early strength. In cold weather, chloride speeds up the initial set before the cement paste freezes.

However, corrosion from chlorides is a major problem for steel-reinforced concrete. On roads and bridges, deicing chemicals and environmental salt corrode from the outside in. In buildings, chloride from concrete accelerators works from the inside out. In both, chloride ions migrate through cracks to attack steel reinforcing bars.

Once corrosion has begun, the steel is transformed into expanding rust, putting pressure on surrounding concrete and causing the protective concrete layer to spall.

The ACI code limits chloride concentrations in four categories: prestressed concrete, reinforced concrete exposed to chloride, reinforced concrete protected from moisture in service, and other reinforced concrete construction. At the lowest end of these limits, the code permits 0.06% chloride in prestressed concrete. At the high end, the code permits 1.00% chloride in reinforced concrete that is kept dry. (These percentages are on a weight basis, as ion. A 1% chloride concentration is approximately equivalent to 2% calcium chloride, since $CaCl_2$ is approximately 63% chloride by weight.)

## 57. ELECTRICAL PROTECTION OF REBAR

There are two major methods of using electricity to prevent chloride corrosion. *Cathodic protection* lowers the active corrosion potential of the reinforcing steel to immune or passive levels. Current is supplied from an external direct-current source. The current flows from an anode embedded in the concrete or from a surface-mounted anode mesh covered with 1.5–2 in (38–51 mm) of concrete. From there, the current passes through the electrolyte (the water and salt in the pores of the concrete) to the steel. The steel acts as the cathode, which is protected.

Installation of cathodic protection is simple. Power consumption is very low. Maintenance is essentially zero. Large areas should be divided into "zones," each with its own power supply and monitoring equipment. This protection scheme has initial and ongoing expenses, as the system and power supply must remain in place throughout the life of the structure.

An alternative method, developed in Europe, also attaches steel mesh electrodes to the surface and uses a current. However, the treatment is maintained only for a limited amount of time (1 to 2 months). An electrolytic cellulose paste is sprayed on and the current is applied. Chloride ions migrate from the rebar and are replaced by alkali ions from the paste. This raises the pH and forms a passivating oxide layer around the rebar. The mesh and cellulose are removed and discarded when the treatment is complete. Protection lasts for years. This electrolytic treatment may not be effective with epoxy-coated bars, prestressed structures (where chemical reactions can cause embrittlement), and bridge decks (due to the duration of shutdown required for treatment).

## 58. COATED REBAR

Corrosion-resistant *epoxy-coated rebar* ("green bar" or "purple bar") has been in use in the United States since the 1970s, most of it in bridge decks and other structures open to traffic. After proper cleaning, standard rebar is given a 0.005–0.012 in (0.13–0.30 mm) electrostatic spray coating of epoxy. The epoxy is intended to be flexible enough to permit subsequent bending and cold working. Epoxy-coated bars must comply with ASTM standards A-775 or A-934.

Epoxy's effectiveness, once described as "maintenance-free in corrosive environments," is dependent on installation practices. Corrosion protection is less than expected when poor manufacturing quality and installation practices remove or damage portions of the coating. Corrosion of coated bars can be reduced by coating after bending and by using alternative "pipeline coatings," which are thicker but less flexible than epoxy.

Epoxy-coated rebar is only one of several fabrication methods needed to prevent or slow chloride corrosion in *bridge decks*, including cathodic protection, lateral and longitudinal prestressing, less-porous and low-slump concrete, thicker (3 in (75 mm) or more) topping layers, interlayer membranes and asphalt concrete, latex-modified or silica fume concrete overlays, corrosion-resistant additives, surface sealers (with or without overlays), galvanizing steel rebar, polymer impregnation, and polymer concrete.

# 47

# Crystallography and Atomic Bonding

## Nomenclature

| | | |
|---|---|---|
| $a$ | lattice constant | m |
| $A$ | area | $m^2$ |
| $A_e$ | effectivity product | $A \cdot Wb/m^3$ |
| AW | atomic weight | kg/kmol |
| $b$ | Burgers' vector | m |
| $B$ | magnetic flux density | T |
| $C$ | concentration | $1/m^3$ |
| CW | cold work index | – |
| $d$ | separation distance | m |
| $d_{hkl}$ | interplanar spacing | m |
| $D$ | diffusion constant | $m^2/s$ |
| $D_o$ | composite diffusion constant | $m^2/s$ |
| $E_D$ | molar activation energy | J/kmol |
| $E_g$ | energy gap | J |
| $F$ | normal force | N |
| $G$ | shear modulus | Pa |
| $H$ | magnetic field strength | A/m |
| $I$ | current | A |
| $J$ | defect flux | $1/m^2 \cdot s$ |
| $L$ | length | m |
| $m$ | mass | kg |
| $m$ | Schmid factor | – |
| $M$ | magnetization | A/m |
| $n$ | integer (order) | – |
| $N$ | number of atoms | – |
| $N_o$ | Avogadro's number ($6.022 \times 10^{23}$) | $mol^{-1}$ |
| $p_m$ | dipole moment | Bohr magnetons (equivalent to $A \cdot m^2$) |
| $Q$ | charge | C |
| $r$ | radius | m |
| $R^*$ | universal gas constant (8.314) | J/kmol·K |
| $t$ | time | s |
| $T$ | absolute temperature | K |
| $U$ | strain energy per unit length | J/m |
| $V$ | shearing force | N |
| $V$ | volume | $m^3$ |
| $x$ | distance | m |

## Symbols

| | | |
|---|---|---|
| $\theta$ | angle | deg |
| $\kappa$ | Boltzmann constant ($1.381 \times 10^{-23}$) | J/K |
| $\lambda$ | angle | deg |
| $\lambda$ | wavelength | m |
| $\mu$ | permeability | H/m |
| $\mu_o$ | permeability of free space | H/m |
| $\mu_r$ | permeability ratio | – |
| $\nu$ | Poisson's ratio | – |
| $\rho$ | density | $kg/m^3$ |
| $\rho$ | resistivity | ohm·m |
| $\tau$ | shear stress | Pa |
| $\phi$ | angle | deg |
| $\chi$ | susceptibility | – |
| $\omega$ | angle | deg |

## Subscripts

| | |
|---|---|
| $A$ | Avogadro |
| $c$ | coercive or critical |
| $D$ | diffusion |
| $g$ | gap |
| $i$ | intrinsic |
| $m$ | magnetic |
| $r$ | ratio or residual |

## 1. INTRODUCTION TO CRYSTALLINE LATTICES

The energy of a stable aggregation of atoms in a compound is lower than the energy of the individual atoms, and therefore, the aggregation is the more-stable configuration. The atoms configure themselves spontaneously in this stable configuration, without requiring external energy.

Common table salt, NaCl, is a stable crystalline solid. The formation of table salt from positive sodium and negative chlorine ions proceeds spontaneously upon mixture. The ions form a three-dimensional cubic lattice of alternating sodium and chlorine ions. Each ion of one sign has six neighbors of the opposite sign. The strong electrostatic attraction of these six neighbors provides the force causing the crystal to form. (See Fig. 47.1.)

*Figure 47.1* Basic Ionic Cubic Structure

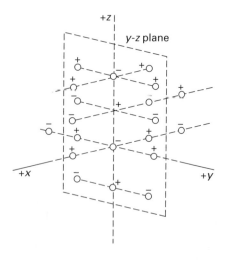

If the ions were merely positive and negative point charges, the arrangement of charges would collapse into itself. However, the inner electron shells of both positive and negative ions provide the repulsive force to keep the ions apart. At the *equilibrium position*, this repulsion just balances the ionic attraction.

## 2. TYPES OF CRYSTALLINE LATTICES

There are fourteen different three-dimensional crystalline structures, known as *Bravais lattices*. The smallest repeating unit of a Bravais lattice is known as a *cell* or *unit cell*, as illustrated in Fig. 47.2. There are seven different basic cell systems: the *cubic, tetragonal, orthorhombic, monoclinic, triclinic, hexagonal*, and *rhombohedral*. These systems are defined by Table 47.1.

Most metallic crystals form in one of three lattice structures, also known as *cells*: the *body-centered cubic* (BCC) cell, the *face-centered cubic* (FCC) cell, and

*Figure 47.2* Crystalline Lattice Structures

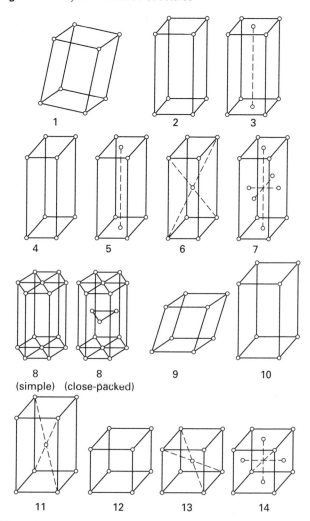

The 14 basic point-lattices are illustrated by a unit cell of each: (1) simple triclinic, (2) simple monoclinic, (3) base-centered monoclinic, (4) simple orthorhombic, (5) base-centered orthorhombic, (6) body-centered orthorhombic, (7) face-centered orthorhombic, (8) hexagonal, (9) rhombohedral, (10) simple tetragonal, (11) body-centered tetragonal, (12) simple cubic, (13) body-centered cubic, (14) face-centered cubic.

*Table 47.1* Characteristics of the Seven Different Crystal Systems

| system | axes | axial angles |
|---|---|---|
| cubic | $a_1 = a_2 = a_3$ | all angles = 90° |
| tetragonal | $a_1 = a_2 \neq c$ | all angles = 90° |
| orthorhombic | $a \neq b \neq c$ | all angles = 90° |
| monoclinic | $a \neq b \neq c$ | two angles = 90°; one angle $\neq$ 90° |
| triclinic | $a \neq b \neq c$ | all angles different; none equals 90° |
| hexagonal | $a_1 = a_2 = a_3 \neq c$ | angles = 90° and 120° |
| rhombohedral | $a_1 = a_2 = a_3$ | all angles equal, but not 90° |

the *hexagonal close-packed* (HCP) cell.[1] Also, some of the simpler ceramic compounds (e.g., MgO, TiC, and BaTiO$_3$) are cubic. Table 47.2 lists common materials and their crystalline forms.

All substances do not form crystalline lattices. *Amorphous substances* are noncrystalline and have neither definite form nor structure.

**Table 47.2** *Crystalline Structures of Common Materials*

   *body-centered cubic*
      chromium
      iron, alpha (below 1674°F; 912°C)
      iron, delta (above 2541°F; 1394°C)
      lithium
      molybdenum
      potassium
      sodium
      tantalum
      titanium, beta (above 1620°F; 882°C)
      tungsten, alpha
   *face-centered cubic*
      aluminum
      brass, alpha
      cobalt, beta
      copper
      gold
      iron, gamma (between 1674°F and 2541°F;
        912°C and 1394°C)
      lead
      nickel
      platinum
      salts: NaCl, KCl, AgCl
      silver
   *hexagonal close-packed*
      beryllium
      cadmium
      cobalt, alpha
      magnesium
      titanium, alpha (below 1620°F; 882°C)
      zinc

## 3. ALLOTROPES

*Allotropes* (also known as *polymorphs*) are different atomic arrangements of the same atoms. (This is different from *isotopes*, whose atoms have different atomic weights.) When an element changes from one structural form to another, usually at a significantly different temperature, it is said to have undergone an *allotropic (polymorphic) change* or *phase change*. Allotropic changes are accompanied by changes in volume and other properties.[2]

---

[1]Under certain circumstances, a *simple cubic (SC)* structure can form. Examples of this are alpha and beta manganese. However, the *packing factor* (defined in Sec. 47.6) is so low (only 52%) that the simple cubic structure rarely occurs.

[2]The property changes that occur in iron are significant and determine the nature of heat treatments used to obtain desired properties.

Iron is the best-known example of an element with allotropic changes, experiencing two allotropic changes as it is heated. *Alpha iron* is a BCC structure that exists up to 1674°F (912°C). Between 1674°F and 2541°F (912°C and 1394°C), *gamma iron* is an FCC structure. Above 2541°F (1394°C) and until it melts at 2800°F (1538°C), *delta iron* exists in a BCC structure.

Iron is not the only element with allotropes. Most metals and many nonmetals exhibit allotropic changes. Some elements may have no allotropes (e.g., magnesium, copper, and zinc), or (in rare cases) may have as many as six (e.g., plutonium). Most elements have only two or three allotropes.

## 4. ALLOTROPES OF CARBON— BUCKYBALLS

*Buckminsterfullerenes* (commonly referred to as *fullerenes* and *buckyballs*) are spherical molecules named after the architect best known for his geodesic dome designs. The appearance of these molecules is that of a soccer ball.

The most famous fullerene, discovered at Rice University in the mid-1980's, is a third allotrope of carbon, consisting of 60 (or more) hexagonal carbon atoms arranged as a sphere with pentagonal gaps between them. (Diamond and graphite are the other two allotropes of carbon.)

Carbon fullerenes are chemically inert because all of the carbon atom bonding positions are taken up by other carbon atoms in the molecule. The relatively large size of the molecule provides a cage-like structure in which other ions or atoms can be placed, although no efficient method of getting the "cargo" out has yet been devised.

Fullerenes deteriorate rapidly when exposed to oxygen. Mixed with metals, fullerenes become superconductors at temperatures as high as 59°R (33K). Fullerenes themselves are structurally very strong, but the forces binding them are very weak. This prevents fullerenes from being easily combined.

*Buckytubes*, also known as *carbon nanotubes*, are cylinders of carbon atoms arranged in rings.

## 5. DISTANCE BETWEEN ATOMS IN A CELL

When studying crystalline lattices, it is convenient to assume that the atoms have definite sizes and can be represented by hard spheres of radius $r$. The cell size then depends on the lattice type and the sizes of the touching spheres. On the basis of this representation, Fig. 47.3 illustrates three types of cubic structures and gives formulas for the center-to-center distances between the lattice atoms. The distance **a** is known as the *lattice constant*.

**Figure 47.3** *Cubic Lattice Dimensions (assuming hard touching spheres of radius r)*

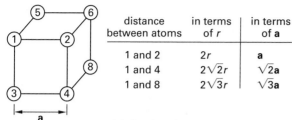

| distance between atoms | in terms of r | in terms of a |
|---|---|---|
| 1 and 2 | $2r$ | $a$ |
| 1 and 4 | $2\sqrt{2}r$ | $\sqrt{2}a$ |
| 1 and 8 | $2\sqrt{3}r$ | $\sqrt{3}a$ |

(a) simple cubic

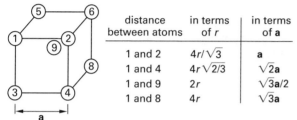

| distance between atoms | in terms of r | in terms of a |
|---|---|---|
| 1 and 2 | $4r/\sqrt{3}$ | $a$ |
| 1 and 4 | $4r\sqrt{2/3}$ | $\sqrt{2}a$ |
| 1 and 9 | $2r$ | $\sqrt{3}a/2$ |
| 1 and 8 | $4r$ | $\sqrt{3}a$ |

(b) body-centered cubic

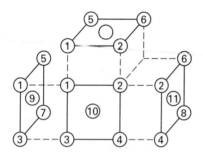

| distance between atoms | in terms of r | in terms of a |
|---|---|---|
| 1 and 2 | $2\sqrt{2}r$ | $a$ |
| 1 and 10 | $2r$ | $\sqrt{2}a/2$ |
| 1 and 4 | $4r$ | $\sqrt{2}a$ |
| 1 and 8 | $2\sqrt{6}r$ | $\sqrt{3}a$ |
| 1 and 11 | $2\sqrt{3}r$ | $\sqrt{3/2}a$ |
| 10 and 11 | $2r$ | $\sqrt{2}a/2$ |
| 9 and 11 | $2\sqrt{2}r$ | $a$ |

(c) face-centered cubic

## 6. CELL PACKING PARAMETERS

Some of the atoms in a unit cell are completely contained within the cell boundary (e.g., the center atom in a BCC structure). Other atoms are shared by adjacent cells (e.g., the corner atoms). Because of this sharing, the number of atoms attributable to a cell is not the number of whole atoms appearing in the lattice structures shown in Fig. 47.2.

For example, there are nine atoms shown for the BCC structure. Although the center atom is completely enclosed, each of the eight corner atoms is shared by

eight cells. Therefore, the number of atoms (also known as the number of *lattice points*) in a cell is $1 + \left(\frac{1}{8}\right)(8) = 2$.

The *coordination number* of an atom in an ionic compound is the number of closest (touching) atoms.[3] Since the atoms in an ionic solid are actually ions, another definition of coordination number is the number of anions surrounding each cation.

In reality, ions in a crystal are not all the same radius. The size of a cation that can fit in a site (known as *interstices* or *interstitial spaces*) between the anions is a function of the relative sizes of the ions. Figure 47.4 lists maximum coordination numbers as functions of ionic radius ratios.

**Figure 47.4** *Coordination Numbers versus Ionic Radius Ratios*

| ratio of cation radius to anion radius | disposition of ions about central ion | coordination number | |
|---|---|---|---|
| 1–0.732 | corners of cube | 8 | |
| 0.732–0.414 | corners of octahedron | 6 | |
| 0.414–0.225 | corners of tetrahedron | 4 | |
| 0.225–0.155 | corners of triangle | 3 | |

The *packing factor* is the volume of the atoms divided by the cell volume (i.e., $a^3$ for a cubic structure). These parameters are summarized in Table 47.3 for hard touching spheres of radius $r$.[4]

The low packing factor of the simple cubic and simple hexagonal structures indicates that these cells are

---

[3]It is also possible to use the coordination number for covalent structures. For example, the coordination number of carbon in a methane ($CH_4$) molecule is four. However, the coordination number of atoms in such compounds (e.g., gases, organic substances, and amorphous solids) has little to do with the density and other properties.

[4]The table is not valid unless all ions are the same size. For example, table salt, NaCl, forms in an FCC structure, but the packing factor is only 0.67 because the sodium ions are smaller than the chlorine ions.

**Table 47.3** *Cell Packing Parameters (assuming hard touching spheres of radius r)*

| type of cell | number of atoms in a cell | packing factor | coordination number |
|---|---|---|---|
| simple cubic | 1 | 0.52 | 6 |
| body-centered cubic | 2 | 0.68 | 8 |
| face-centered cubic | 4 | 0.74 | 12 |
| simple hexagonal | | | |
|    primitive cell | 1 | 0.52 | 8 |
|    total structure | 3 | 0.52 | 8 |
| hexagonal close-packed | | | |
|    primitive cell | 2 | 0.74 | 12 |
|    total structure | 6 | 0.74 | 12 |

wasteful of space. This is the primary reason simple cubic and simple hexagonal lattices seldom form naturally.

## 7. DENSITY OF IONIC SOLIDS

The density of an ionic solid can be calculated from the cell mass and volume. The cell mass is determined from the number of atoms per cell, the atomic weight, and Avogadro's number, $N_A$, ($6.022 \times 10^{23}$ atoms/mol). The cell volume depends on the lattice constant, **a**. Equation 47.1 is for cubic structures only.

$$\rho = \frac{m}{V} = \frac{\text{no. of atoms per cell} \times \text{AW}}{N_A \mathbf{a}^3 \times 1000 \text{ mol/kmol}} \qquad 47.1$$

### Example 47.1

The lattice constant of copper (AW = 63.54) has been determined from X-ray diffraction to be 0.361 nm. What is the theoretical density of copper in g/cm$^3$?

*Solution*

Copper forms an FCC structure (Table 47.2). The number of atoms in an FCC cell is 4 (Table 47.3). From Eq. 47.1, the density of copper is

$$\rho = \frac{\text{no. atoms per cell} \times \text{AW}}{N_A \mathbf{a}^3}$$

$$= \frac{\left(4 \, \frac{\text{atoms}}{\text{cell}}\right)\left(63.54 \, \frac{\text{kg}}{\text{kmol}}\right)\left(\frac{1 \text{ kmol}}{1000 \text{ mol}}\right)}{\left(6.022 \times 10^{23} \, \frac{\text{atoms}}{\text{mol}}\right)(0.361 \times 10^{-9}\text{m})^3}$$

$$= 8970 \text{ kg/m}^3 \quad (8.97 \text{ g/cm}^3)$$

## 8. LATTICE DIRECTIONS: [*uvw*] AND ⟨*uvw*⟩

Crystallography uses a system to designate directions of lines in a lattice cell. Lines are designated by numbers in square brackets, as illustrated in Fig. 47.5. The numbers are the relative intercepts of the line on the cell boundaries.

**Figure 47.5** *Sample Lattice Directions*

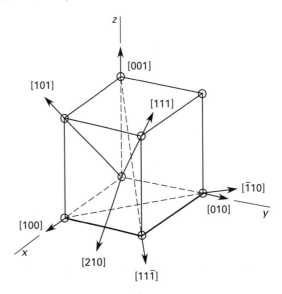

Consider an orthorhombic cell that has cell dimensions **a**, **b**, and **c** in the directions of the $x$-, $y$-, and $z$-axes, respectively. If a point in the cell is located at ($u$**a**, $v$**b**, $w$**c**) relative to the origin, where $u$, $v$, and $w$ are fractions or multiples of the unit cell dimensions **a**, **b**, and **c**, respectively, the direction taken by a line from the origin through that point is written as [$uvw$]. By convention, there are no fractional intercepts, so $u$, $v$, and $w$ are multiplied by the least common denominator to clear all fractions.

If a line does not pass through the origin of the cell, it is extended to pass through the origin of an adjacent cell. Alternatively, the line can be shifted parallel such that it does pass through an origin. The direction designation [$uvw$] is based on the intercepts of the cell where the line passes through the origin, not any other cell through which the line passes.

Shifting lines and using adjacent cells often results in negative intercepts. It is a crystallography convention that negative intercepts are written with overbars rather than with negative signs (e.g., [$u\bar{v}w$], not [$u -vw$]).

Since a cubic structure is symmetrical in all three directions, any corner can be used as the origin.[5] Thus, the lines [100] and [010] pass through the same sequence of atoms on their ways out of the crystal. These directions are said to be *crystallographically equivalent*. *Families of equivalent directions* are designated by angle brackets (e.g., [100] and [010] belong to the ⟨100⟩ family). Table 47.4 lists several families for cubic cells.

Sets of four numbers, [$uvwz$], are needed to designate directions in hexagonal cells. (See Fig. 47.6.) $u$, $v$, and $w$ are the respective intercepts on three axes **a₁**, **a₂**, and **a₃** 120° degrees apart on the base of the cell. $z$ corresponds

---

[5]Another way of saying this is that the cell can be rotated any multiple of 90° about any axis, and you will not be able to tell the difference.

Material Science

**Table 47.4** Cubic Cell Families

| family | equivalent directions | | |
|---|---|---|---|
| $\langle 100 \rangle$ | [100], | [010], | [001], |
| | [$\bar{1}$00], | [0$\bar{1}$0], | [00$\bar{1}$] |
| $\langle 110 \rangle$ | [110], | [101], | [011], |
| | [$\bar{1}$10], | [1$\bar{1}$0], | |
| | [$\bar{1}$01], | [10$\bar{1}$], | |
| | [0$\bar{1}$1], | [01$\bar{1}$] | |
| $\langle 111 \rangle$ | [111], | | |
| | [$\bar{1}$11], | [1$\bar{1}$1], | [11$\bar{1}$], |
| | [$\bar{1}\bar{1}$1], | [$\bar{1}$1$\bar{1}$], | [1$\bar{1}\bar{1}$], |
| | [$\bar{1}\bar{1}\bar{1}$] | | |

**Figure 47.6** Crystallographic Directions in Hexagonal Cells

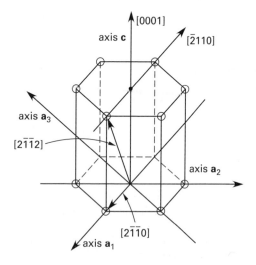

to **c**, the height of the hexagonal cell. With hexagonal systems, $w$ is always equal to $-(u + v)$.

## 9. LATTICE PLANES: (hkl) AND {hkl}

*Miller indices* are used to specify planes in crystalline lattices. Miller indices are calculated as the *reciprocals* of the plane intercepts on the unit cell axes, written in parentheses—for example, (011). If the plane does not intercept a cell axis, the intercept is infinity. As with the line directions, the numbers are cleared of all fractions, reduced to the smallest integers, and written without commas, with overbars used to designate negative numbers.

Due to the symmetry of cubic systems, there are many *crystallographically equivalent planes*. Families of equivalent planes are written as {hkl}. For example, the {100} family contains the planes (010) and (001). By substituting (hkl) and {hkl} for [uvw] and ⟨uvw⟩, Table 47.4 can be used to define equivalent cubic planes.

## Example 47.2

What are the Miller indices of the plane shown?

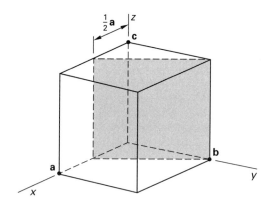

*Solution*

The intercepts along the $x$-, $y$-, and $z$- axes are **a**/2, **b**, and ∞. Dividing these intercepts by **a**, **b**, and **c**, respectively, produces quotients of 1/2, 1, and ∞. The reciprocals are $h = 2$, $k = 1$, and $l = 0$. The Miller indices are (210).

## 10. RELATIONSHIPS BETWEEN DIRECTIONS AND PLANES IN CUBIC CELLS

The edges of a cubic cell define a set of Cartesian axes, and it is not surprising that an [hkl] direction can be thought of as a type of vector notation. The following relationships can be derived directly from basic trigonometric and geometric relationships.

- A line [uvw] is normal to the plane (hkl) when $u = h$, $v = k$, and $w = l$. For example, [011] is normal to (011).

- [uvw] is parallel to the plane (i.e., lies in the plane) (hkl) if $hu + kv + lw = 0$. For example, [11$\bar{2}$] is parallel to (111).

- Two directions are normal if $u_1 u_2 + v_1 v_2 + w_1 w_2 = 0$. For example, [100] is normal to [001].

- Two planes are normal if $h_1 h_2 + k_1 k_2 + l_1 l_2 = 0$. For example, (010) is normal to (100).

- The angle, $\theta$, between two planes is given by Eq. 47.2. (This equation can also be used to calculate the angle between two lines by substituting $u$, $v$, and $w$ for $h$, $k$, and $l$.)

$$\cos\theta = \frac{h_1 h_2 + k_1 k_2 + l_1 l_2}{\sqrt{\left(h_1^2 + k_1^2 + l_1^2\right)\left(h_2^2 + k_2^2 + l_2^2\right)}} \qquad 47.2$$

## 11. X-RAY DIFFRACTION OF CRYSTALLINE PLANES

X-ray diffraction is used to determine the cell structure and dimensions of crystals. Diffraction of monochromatic X-ray radiation (wavelength of 0.01–1.0 nm) from surface planes is similar to reflection. However, radiation reflected from internal planes will emerge from the crystal out of phase with the incident radiation. Depending on the incident beam angle and crystal characteristics, the radiation may be reinforced or canceled.

Three layers of crystalline planes are shown in Fig. 47.7. The spacing between the layers (i.e., the *interplanar spacing*) is designated as $d_{hkl}$. For reinforcement to occur, the combined distance $AB + BC$ must be an integral number of wavelengths.[6]

$$n\lambda = AB + BC \qquad 47.3$$

**Figure 47.7** X-Ray Diffraction of a Crystal

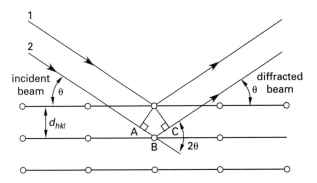

The distance $AB + BC$ can be calculated from trigonometry if the plane spacing is known.

$$AB + BC = 2d_{hkl} \sin \theta \qquad 47.4$$

Combining Eq. 47.3 and Eq. 47.4 produces *Bragg's law*, Eq. 47.5. The *order of diffraction* or *order of reflection*, $n$, is almost always 1 (i.e., *first-order reinforcement*).

$$n\lambda = 2d_{hkl} \sin \theta \qquad 47.5$$

The interplanar distance and the lattice constant, **a**, are not the same. For example, in a BCC crystal, there will be a plane of interior atoms as well as planes of the exterior (corner) atoms. Equation 47.6 relates the interplanar spacing to the lattice constant and the Miller indices.

$$d_{hkl} = \frac{\mathbf{a}}{\sqrt{h^2 + k^2 + l^2}} \qquad 47.6$$

If the wavelength of the X-rays is fixed, then the incident angle must be varied to obtain reinforcement. This is accomplished by using a random orientation of

crystals in powdered form. Reinforcement will then be observed at one or more emergent angles.[7] The minimum value of $\sin \theta$ corresponds to $n = 1$. The multiples of $\sin^2 \theta$ can be used to predict the type of lattice being studied. (See Table 47.5.)

**Table 47.5** Multiples of $\sin^2 \theta$ for Cubic Lattices

| lattice type | multiples of $\sin^2 \theta_{\min}$ |
| --- | --- |
| simple cubic | 1, 2, 3, 4, 5, 6, 8, 9, 10 |
| BCC | 2, 4, 6, 8, 10, 12, 14, 16, 18 |
| FCC | 3, 4, 8, 11, 12, 16, 19 |

## Example 47.3

The interplanar distance in a quartz crystal is 4.255 Å. What will be the first-order diffraction angle if 1.541 Å X-rays are used?

*Solution*

$n = 1$ for the first-order angle. From Eq. 47.5,

$$n\lambda = 2d_{hkl} \sin \theta$$
$$(1)(1.541 \text{ Å}) = (2)(4.255 \text{ Å})(\sin \theta)$$
$$\theta = \sin^{-1}(0.1811) = 10.43°$$

An angstrom (Å) is $10^{-10}$ m. However, it is not necessary to convert angstroms to any other unit, since angstroms appear on both sides of Bragg's law.

## 12. POINT DEFECTS

Real crystals possess a variety of imperfections and defects that affect *structure-sensitive properties*. Such properties include electrical conductivity, yield and ultimate strengths, creep strength, and semiconductor properties.[8] Most imperfections can be categorized into point, line, and planar (grain boundary) imperfections.[9]

The most important *point defects*, shown in Fig. 47.8, are listed here.

- *Vacancy:* A missing atom in the lattice. The number of vacant lattice sites increases with temperature and cold working.

- *Schottky defect:* A neutral defect of ionic lattices, consisting of two vacancies (one cation and one anion).

---

[6]*Reinforcement* is also known as *in-phase return, superposition, constructive interference*, and the presence of a *diffraction peak*.

[7]Figure 47.7 shows the reflected beam turning $2\theta$ away from the incident beam. In an X-ray diffraction study, the incident beam angle is assumed to be $\theta = 0°$. The angles of the observed reinforcements are divided by 2 to obtain $\theta$.

[8]The *structure-insensitive properties*, which are not affected by lattice defects, include melting point, density, specific heat, coefficient of thermal expansion, and shear and elastic moduli.

[9]There are many other types of defects, such as *stacking defects, sessile dislocations*, and *kinking*, which are not discussed here. Also, *bulk defects* (e.g., bubbles, cracks, or large quantities of foreign material), which are generally large enough to be visible, are not discussed.

- *Frenkel defect:* A vacant lattice site with the missing atom relocated to an interstitial point where no atom should be.

- *Substitutional atom:* The substitution of a foreign atom for an atom in the lattice.

- *Interstitial impurity atom:* A foreign atom occupying an interstitial site where no atom should exist.

- *Self-interstitial atom:* One of the crystalline atoms occupying an interstitial site where no atoms should exist. Such a defect can occur spontaneously or through diffusion.

**Figure 47.8** *Point Defects*

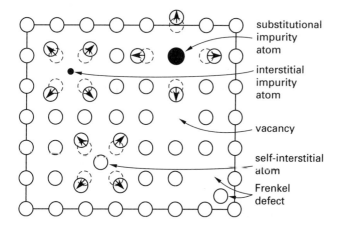

substitutional impurity atom

interstitial impurity atom

vacancy

self-interstitial atom

Frenkel defect

## 13. MOVEMENT OF POINT DEFECTS

All of the point defects shown in Fig. 47.8 can move individually and independently from one position to another through *diffusion.* The *activation energy* for such diffusion generally comes from heat and/or strain (i.e., bending or forming). In the absence of the activation energy, the defect will move very slowly, if at all.

Diffusion is governed by *Fick's laws.* The first law predicts the number of defects which will move across a unit surface area per unit time. This number is known as the *defect flux, J,* and is proportional to the defect concentration gradient $dC/dx$ in the direction of movement. The negative sign in Eq. 47.7 indicates that defects migrate to where the dislocation density is lower. *Fick's first law of diffusion* is

$$J = -D\left(\frac{dC}{dx}\right) \qquad 47.7$$

The *diffusion coefficient, D,* (also known as the *diffusivity)* is dependent on the material, activation energy, and temperature. (See Table 47.6.) It is calculated from the *composite constant, $D_o$,* and these other factors. Equation 47.8 is typical of the method used to determine the diffusion coefficient.

$$D = D_o e^{-\Delta E_D/R^*T} \qquad 47.8$$

As the defects migrate toward areas of lower defect concentration, the concentration changes. *Fick's second law of diffusion* describes the rate of change in concentration at a particular point.

$$\frac{dC}{dt} = D\left(\frac{d^2C}{dx^2}\right) \qquad 47.9$$

Equation 47.9 illustrates why equilibrium processes take a lot of time. As the composition becomes uniform, $d^2C/dx^2$ becomes small, and the concentration, $C$, changes slowly with time.

## 14. LINE DEFECTS

*Line defects (line dislocations)* are the most common types of crystal defects. In a line defect, the imperfection is repeated consistently in many adjacent cells and has extension in a particular direction. Line defects also move, but the movement requires all of the adjacent imperfections to move.

The most common line defects are the *screw dislocation* (see Fig. 47.9), the *edge dislocation* (see Fig. 47.10), and the *mixed dislocation (hybrid dislocation).*

**Figure 47.9** *Screw Dislocation*

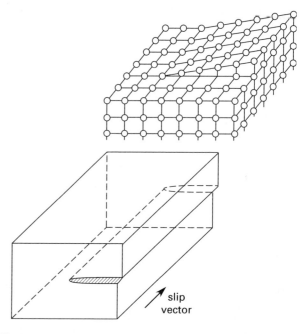

slip vector

**Figure 47.10** *A Positive Edge Dislocation*

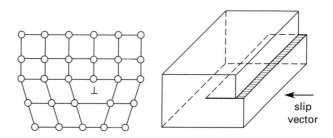

slip vector

An edge dislocation (also known as a *Taylor-Orowan dislocation*) is an incomplete plane of atoms within the lattice. If the incomplete plane is above the slip plane, it is known as a *positive edge dislocation* and is given the symbol ⊥. If the incomplete plane is below the slip plane, it is known as a *negative edge dislocation* and is given the symbol ⊤. The properties of an edge dislocation are:

- The Burgers' vector, **b** (see Sec. 47.15), is perpendicular to the dislocation line.

- The Burgers' vector, **b**, is parallel to the slip direction.

- The dislocation line movement is parallel to the slip direction.

- The edge dislocation leaves the slip plane by the process of climb.

In a screw dislocation, the lattice planes are nonplanar and, in fact, spiral around the dislocation line. The properties of a screw dislocation are:

- The Burgers' vector, **b**, is parallel to the dislocation line.

- The Burgers' vector, **b**, is parallel to the slip direction.

- The dislocation line movement is perpendicular to the slip direction.

- The edge dislocation leaves the slip plane by the process of cross-slip.

The *dislocation density* is the number of dislocations per unit area. For well-annealed crystals, it is typically on the order of $10^4$ to $10^6$ per square centimeter. This density can be increased to as much as $10^{12}$ dislocations per square centimeter through severe deformation and plastic strain. A high dislocation density represents a considerable increase in stored strain energy.

## 15. THE BURGERS' VECTOR

If a circuit is made around a dislocation in a crystal lattice, using an equal number of steps in the $x$- and $y$-directions, as in Fig. 47.11, the circuit will not close unless an additional *Burgers' vector*, **b**, is added. Thus, the Burgers' vector is a *closure vector*. The Burgers' vector is an important parameter in determining the type of dislocation, as well as in calculating the strain energy produced by the dislocations. The Burgers' vector is constant along the extent of the dislocation. In edge dislocations, Burgers' vectors are perpendicular to the dislocation line. In screw dislocations, the dislocation lines and their Burgers' vectors are parallel.

***Figure 47.11*** *The Burgers' Vector*

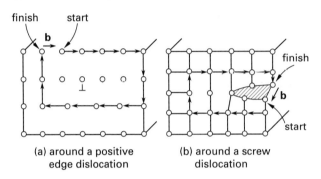

(a) around a positive edge dislocation

(b) around a screw dislocation

Equation 47.10 relates the Burgers' vector to the shearing force, $V$, per unit dislocation length and the shear stress, $\tau$.

$$\frac{\mathbf{V}}{L} = \mathbf{b}\tau \qquad 47.10$$

The energy per unit length of screw, edge, and mixed dislocations is proportional to the square of the Burgers' vector. (This is shown in Eq. 47.11 and Eq. 47.12.)

***Table 47.6*** *Typical Diffusivities*

| defect type | solute | solvent | $\Delta E_D$ (J/kmol) | $D_0$ (m²/s) | $D$ (m²/s) 500°C | 1000°C |
|---|---|---|---|---|---|---|
| interstitial | C | Fe (BCC) | $8.4 \times 10^7$ | $8 \times 10^{-7}$ | $1.8 \times 10^{-12}$ | $3 \times 10^{-10}$ |
| | N | Fe (BCC) | $7.5 \times 10^7$ | $7 \times 10^{-7}$ | $6 \times 10^{-12}$ | $5.7 \times 10^{-10}$ |
| | H | Fe (FCC) | $4.2 \times 10^6$ | $1 \times 10^{-6}$ | $1.5 \times 10^{-9}$ | $1.9 \times 10^{-8}$ |
| substitutional | Ni | Fe (FCC) | $2.8 \times 10^8$ | $5 \times 10^{-5}$ | $1 \times 10^{-23}$ | $2.5 \times 10^{-16}$ |
| | Co | Fe (BCC) | $2.3 \times 10^8$ | $2 \times 10^{-5}$ | $1.2 \times 10^{-20}$ | $9 \times 10^{-15}$ |
| | Si | Fe (BCC) | $2.0 \times 10^8$ | $4 \times 10^{-5}$ | $1.2 \times 10^{-18}$ | $2.2 \times 10^{-13}$ |
| | Al | Cu | $1.6 \times 10^8$ | $7 \times 10^{-6}$ | $5.6 \times 10^{-15}$ | $1.5 \times 10^{-12}$ |
| | S | GaAs | $3.9 \times 10^8$ | $0.79$ | $3.5 \times 10^{-27}$ | $1.6 \times 10^{-16}$ |
| | Zn | GaAs | $2.4 \times 10^8$ | $1.5 \times 10^{-12}$ | $1.3 \times 10^{-28}$ | $1.5 \times 10^{-22}$ |

(Multiply cm²/s by $10^{-4}$ to obtain m²/s.)
(Multiply kcal/mol by $4.187 \times 10^6$ to obtain J/kmol.)

Material Science

## 16. MOVEMENT OF LINE DEFECTS

Under the application of heat or shear stress, a dislocation can move (i.e., *slip* or *glide*). The *slip plane (glide plane)* coincides with a crystalline plane, and is the plane in which the displacement occurs. Thus, slip is the motion of an edge dislocation parallel to the slip plane. The magnitude and direction of each step of the slip coincides with the Burgers' vector. The slip is a plastic deformation (i.e., the dislocation does not return to its original position when the stress is removed).

In order to slip with the minimum amount of strain energy, crystals prefer to slip along closely packed planes in a particular direction. The combination of a preferred slip plane and slip direction is known as a *slip system*. Table 47.7 lists the most likely slip systems for common materials at low (room) temperatures. Other slip systems exist at higher temperatures.

**Table 47.7** *Common Room-Temperature Slip Systems*

| material | lattice | slip plane | slip direction |
|---|---|---|---|
| Al | FCC | (111) | $[10\bar{1}]$ |
| Cu | FCC | (111) | $[10\bar{1}]$ |
| Ag | FCC | (111) | $[10\bar{1}]$ |
| Au | FCC | (111) | $[10\bar{1}]$ |
| $\alpha$Fe | BCC | (101)(112)(123) | $[11\bar{1}]$ |
| Ta | BCC | (101)(112) | $[11\bar{1}]$ |
| W | BCC | (101)(112) | $[11\bar{1}]$ |
| Mg | HCP | (0001) | $[11\bar{2}0]$ |
| Zn | HCP | (0001) | $[11\bar{2}0]$ |
| Cd | HCP | (0001) | $[11\bar{2}0]$ |
| NaCl | FCC | (110) | $[1\bar{1}0]$ |
| LiF | FCC | (110) | $[1\bar{1}0]$ |
| MgO | FCC | (110) | $[1\bar{1}0]$ |

In FCC metals, the {111} family of planes and the ⟨110⟩ family of directions have the maximum density, and these define the FCC slip systems. In BCC structures, the slip always occurs in the ⟨111⟩ directions since this is the direction of closest packing.[10] However, there is no plane of maximum packing, so slip occurs along a number of planes.

The direction of the closest-packed plane rarely coincides with the direction of the applied stress, but the stress is always present to some extent in every direction.[11]

*Climb* is the motion of an edge dislocation perpendicular to the slip plane. (See Fig. 47.12.) It is most likely to take place in small angle tilt boundaries. Since climb requires extra vacancies (for the dislocation to climb up) or extra atoms (for the dislocation to climb down),

---

[10]Slip in iron in the ⟨111⟩ directions is known as *pencil glide*.

[11]As a direct result of the combined stress equation, shear stresses occur in solids even when the applied load is normal (i.e., tensile or compressive).

*Figure 47.12* Climb

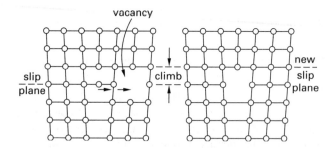

climb is controlled by diffusion. Except at high temperatures, climb is not as prevalent as slip.

Under the proper conditions, movement of dislocations can continue until movement is restricted by other dislocations. (See Fig. 47.13.) This movement concentrates dislocations and reduces their occurrence in the remainder of the crystal. Often, dislocations move all the way to a crystalline boundary, creating a band of dense dislocations known as a *pile-up*. The pile-up produces a *back stress* which opposes additional movement of dislocations along the slip plane (i.e., additional slippage is reduced). The intentional restriction of dislocation movement in metallic crystals is a prominent technique used to improve strength and ductility.

*Figure 47.13* *Movement of an Edge Dislocation by Glide and Climb*

## 17. GRAIN BOUNDARY DEFECTS

*Grain boundaries* (i.e., *planar defects* or *surface defects*) are the interfaces between two or more crystals. Grain boundaries occur naturally during solidification and, in the case of heat-treated metals, during recrystallization. During solidification, crystals form with random orientation. As the solidification continues, the growing crystals eventually reach the boundaries of one another.

The interface between two crystals is almost always a mismatch in crystalline structures. Also, the area between the crystals consists of a corridor of transitional structure or amorphous material (i.e., debris). In special cases called *twin boundaries*, however, the crystals are perfect up to where they meet.

The crystals on either side of the *twin plane* are mirror images of each other (at least for a few cell dimensions). Such twinning can occur when a crystal is stressed (i.e., *mechanical twins*) as shown in Fig. 47.14, or from an annealing operation (i.e., *annealing twins*). FCC metals do not normally twin, but BCC and HCP metals form

**Figure 47.14** Movement Mechanical Twinning

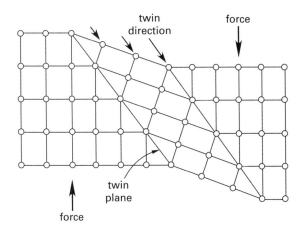

mechanical twins under shock loading and decreased temperature.[12] In some cases, the twinning click is audible (as in *tin cry*).

Another special case is a boundary consisting of a series of edge dislocations, as shown in Fig. 47.15. This is called a *tilt boundary, narrow-angle boundary,* or *small-angle boundary.*

**Figure 47.15** Tilt Boundary

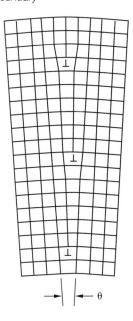

## 18. ENERGY OF DISLOCATIONS

Crystalline structures are stable because they permit the atoms to exist in lower energy states than would be experienced in the absence of the crystal formation. Imperfections introduce disorder into the lattice, and a higher energy state results. This energy is known as

---

[12]The major FCC twin-forming exceptions are gold-silver alloys and copper.

*strain energy.* Thus, the defects increase the energy storage of the lattice.

The energy contained per unit length of edge dislocation is given by Eq. 47.11, where $|\mathbf{b}|$ is the magnitude of the Burgers' vector.

$$U_{\mathrm{edge}} \approx \frac{G|\mathbf{b}|^2}{2(1-\nu)} \qquad \text{47.11}$$

The strain energy contained in per-unit length of screw dislocation is

$$U_{\mathrm{screw}} \approx \frac{G|\mathbf{b}|^2}{2} \qquad \text{47.12}$$

## 19. PLASTIC DEFORMATION

*Plastic deformation (plastic flow, inelastic flow, permanent set, plastic shear,* etc.) of a crystal is permanent deformation. If a high stress (i.e., higher than the *yield stress*) is applied to a crystal, the crystal will change shape, and the shape remains changed even after the stress is removed.[13] Plastic deformation in metals (which are crystalline in nature) occurs primarily through the mechanisms of *slip* (along planes of dense packing) and *twinning,* although *rotation* can also occur. Figure 47.16 illustrates severe slip in a normally loaded crystal.

**Figure 47.16** Severe Slip in a Crystal

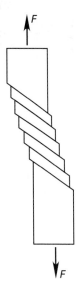

---

[13]The opposite of plastic deformation is *elastic deformation.* Elastic deformation is analogous to lightly stretching a spring or rubber band. Once the stretching force is removed, the spring will return to its original shape.

To slip, the shear stress in the slip system must exceed the *critical shear stress* (analogous to the yield stress in a common stress-strain test). In a normal stress test, the maximum shear stress will be $F/2A$ on a plane 45° from the specimen's longitudinal axis. However, the slip plane will not normally coincide with the plane of maximum shear stress.

Slip will occur on the slip plane. However, the planes can slide relative to each other in many directions. The shear stress in the slip system depends on the orientation of the slip plane and the slip direction, as illustrated in Fig. 47.17. Equation 47.13 (*Schmid's law*) determines the magnitude of the *resolved shear stress*, $\tau$, on the slip plane. The resolved shear stress is primarily a factor of the *glide strain (glide shear strain)* angle, $\lambda$.[14] The *Schmid factor*, $m$, is the ratio of resolved shear stress to the axial stress.

$$\tau = \frac{F}{A} \cos \lambda \cos \phi \qquad 47.13$$

$$m = \frac{\tau}{F/A} = \cos \lambda \cos \phi \qquad 47.14$$

**Figure 47.17** *Schmid Law Slip Angles*

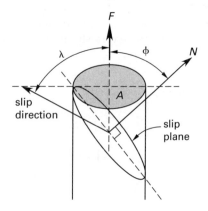

**Example 47.4**

Failure in a crystalline material is observed when a normal tensile stress of 16 MPa is applied in the $[1\bar{1}0]$ direction. The failure occurs on the $(1\bar{1}\bar{1})$ plane and in the $[0\bar{1}1]$ direction. What is the critical resolved shear stress?

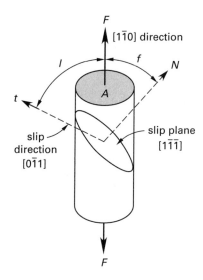

*Solution*

The cosine of the angle between the tensile axis $[1\bar{1}0]$ and the line normal to the $(1\bar{1}\bar{1})$ plane is

$$\cos \phi = \frac{(1)(1) + (-1)(-1) + (0)(-1)}{\sqrt{(1)^2 + (-1)^2 + (0)^2}\sqrt{(1)^2 + (-1)^2 + (-1)^2}}$$

$$= \frac{2}{\sqrt{2}\sqrt{3}} = 0.816$$

The cosine of the angle between the tensile axis $[1\bar{1}0]$ and the slip direction $[0\bar{1}1]$ is

$$\cos \lambda = \frac{(1)(0) + (-1)(-1) + (0)(1)}{\sqrt{(1)^2 + (-1)^2 + (0)^2}\sqrt{(0)^2 + (-1)^2 + (1)^2}}$$

$$= \frac{2}{\sqrt{2}\sqrt{2}} = 0.50$$

From Eq. 47.13, the critical resolved shear stress is

$$\tau = \frac{F}{A} \cos \lambda \cos \phi = (16 \text{ MPa})(0.50)(0.816)$$

$$= 6.53 \text{ MPa}$$

## 20. COLD AND HOT WORKING

Cold and hot working are both forming processes (rolling, bending, forging, extrusion, etc.). The term *hot working* implies that the forming process occurs above the *recrystallization temperature*. (The actual temperature depends on the rate of forming strain and the cooling period, if any.) *Cold working* (also known as *work hardening* and *strain hardening*) occurs below the recrystallization temperature.

Above the recrystallization temperature, almost all of the internal defects and imperfections caused by hot working are eliminated. In effect, hot working is a "self-healing" operation. Thus, a hot-worked part remains softer and has a greater ductility than a cold-worked

---

[14]Just as it is common to plot stress versus strain from a tensile test, resolved shear stress, $\tau$, is plotted against glide strain, $\lambda$.

part. Large strains are possible without strain hardening. Hot working is preferred when the part must go through a series of forming operations (passes or steps) or when large changes in size and shape are needed.

Since hot working occurs at an elevated temperature, surface oxidation will occur. In steel, such oxidation is known as *mill scale*. Rolled-in mill scale is one of the reasons that hot-worked parts may not have good surface finishes. In other cases, such as with titanium, *oxygen embrittlement* must be prevented entirely by performing the forming process in an inert atmosphere. Another disadvantage of hot working is the need for larger dimensional tolerances to account for dimension changes between the hot and cold conditions.

The hardness and toughness of a cold-worked part will be higher than a hot-worked part. Because the part temperature during cold working is uniform, the final microstructure will also be uniform. There are many times when these characteristics are desirable, and thus, hot working is not always the preferred forming method. In many cases, cold working will be the final operation after several steps of hot working.

Cold working is a true plastic deformation process. The *cold work index* (also known as the *reduction in area* and *percent cold work*) is the percentage change in the cross-sectional area of a part.

$$\text{CW} = \frac{A_{\text{initial}} - A_{\text{final}}}{A_{\text{initial}}} \times 100\% \qquad 47.15$$

## 21. CONDUCTIVITY AND RESISTIVITY

The electrical and magnetic characteristics of materials are influenced greatly by the *valence* (i.e., *outermost*) *electrons*. While it is convenient to view these valence electrons as belonging to a single nucleus, in reality, the electron clouds from adjoining atoms interact. The electrons that can move around are known as *free electrons*. The freedom of electrons to move from nucleus to nucleus is essential for electrical and thermal conductivity.

On the other hand, the electrons in *insulators* are not free to move. Therefore, electrical and thermal conduction is minimal in insulators.

Free electrons can be directed to move in a single direction by an electric field. The movement of electrons is known as *current*, I, with units of amperes.[15] An ampere is defined as the movement of one coulomb of charge, Q, per second.

$$I = \frac{dQ}{dt} \qquad 47.16$$

The mobility of free electrons is limited by interactions with other electrons and atoms. The *mean free path* of an electron in a lattice is a measure of this mobility. However, the common measures of electron mobility in a material are *conductivity* and its reciprocal, *resistivity*, $\rho$.

In metals, higher temperatures increase the thermal agitation of the lattice atoms. This subsequently reduces the mean free path of the electrons and reduces the conductivity. Thus, $d\rho/dT$ is positive for metals. For nonmetals at high temperatures, the added thermal energy breaks away additional electrons, increasing conductivity. Thus, $d\rho/dT$ is negative for nonmetals.

## 22. ELECTRON BAND MODEL

An electron in an atom possesses energy in several forms (translational, rotational, etc.). Energy is not available in arbitrary amounts, however, but is added to or removed from an electron in discrete amounts known as *quanta*. Thus, a *quantum of energy* is the smallest amount of energy that an electron can receive or lose. The characteristics of atomic valence electrons (as opposed to core electrons and free electrons) with quantized energies can be listed as follows.

- Electrons can exist at specific energy levels only. Electrons cannot have intermediate energies.

- Electrons fill the lowest energy levels first.

- Up to two, but no more, electrons can have the same energy level. If two electrons have the same energy, they must have opposite spins.

- Electrons can jump to higher energy levels if one or more quanta of energy are provided.

In a crystalline lattice, there is considerable interaction between the valence (outermost) electrons, and it is not possible to distinguish between one atom's and another's electrons. In effect, there is a very large population of free electrons. However, the requirement that two electrons have the same energy is still valid. Therefore, there are half as many filled energy levels as there are free electrons. These energy levels must differ by discrete amounts. Collectively, these energy levels are known as an *energy band*. Each electron exists within the energy band. (See Fig. 47.18.)

**Figure 47.18** *Energy Band Concept (core electrons not shown)*

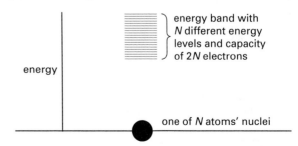

The number of total possible levels within the energy band is equal to the number of atoms, $N$, in the crystal. However, all of these levels need not be filled. For example, sodium and other alkali metals have only one valence electron. Therefore, the number of filled energy levels in a sodium energy band will be $N/2$.

Magnesium and other alkaline earth metals have two valence electrons, and the energy band is filled. In the case of silicon (with four valence electrons), two energy bands of $N$ levels each are needed to hold all of the electrons. Additional energy bands (with $N$ energy levels) are also needed to hold electrons that have been given more than their lowest (basement) state energy.

Current coincides with the movement of electrons. For an electron to move into the vicinity of another electron, it must have a different energy level. That is, it must move into a different position in the energy band. An applied electrical field is usually the source of this additional energy, although heat and light can also supply the energy.[16] In metals with only one valence electron, very little energy is needed since half of the energy band is empty, and the difference between energy levels within the band is very small (i.e., one quantum).

If a band is full, as it would be with any atom with an even number of valence electrons, movement of an electron requires the electron to move into a different energy band. (See Fig. 47.19.) This energy band is then known as a *conduction band*. In some cases (e.g., the alkaline earth metals), there is some *energy overlap* in the bands. The lowest part of the second band is actually lower than the highest part of the first band. Some of the first band's electrons fill the lower part of the second band in order to achieve the lowest energy levels. It is easy for an electron to move into higher levels, since empty levels exist in both energy bands, each level differing only by a quantum of energy.

*Figure 47.19* Multiple Energy Bands

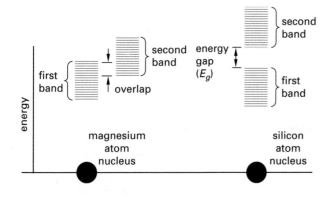

In other cases, a large amount of energy (i.e., more than a single quantum of energy) is needed to get an electron into the next energy band. In this case, there is an

energy gap, $E_g$, between the bands. Generally, a very strong electrical field is needed to supply the energy to jump over this energy gap into a higher band. For this reason, insulators, such as silicon, have high resistivities.

The existence of an energy gap does not always mean the material is an insulator. The material can be a good conductor as long as there are empty energy levels in the highest band. For example, there are energy gaps between the first, second, and third energy bands for aluminum. However, aluminum has three valence electrons, which means that the second band is only half full. Aluminum is a good conductor because the electrons can change locations by moving into remaining higher energy levels in the second band.

The size of the energy gap varies with the element. In the case of semiconductors, the gap at room temperatures is small enough that thermal energy can provide enough energy to allow the electrons to jump into the conduction band. This is the principle of *intrinsic conduction* by *thermal carrier generation*. The higher the temperature, the greater will be the number of electrons jumping into the conduction band. This principle is quantified by Eq. 47.17, which is a general form for predicting the number of intrinsic charge carriers. ($C$ is a constant over small temperature ranges.)

$$N_i = Cc^{-E_g/2\kappa T} \qquad 47.17$$

## 23. SEMICONDUCTORS AND INTENTIONAL IMPURITIES

Silicon and germanium, which form the basis of most semiconductors, are slightly more conductive than insulators, but they are not nearly as conductive as metals. Silicon and germanium each possess four valence electrons, leaving four unfilled electron positions. Their crystalline lattice structures depend on covalent bonding to fill the outer shell of eight electrons (the *octet*).

However, the crystalline lattice is not perfect. Occasional free electrons or missing electrons (known as *holes*) exist. Electron movement and hole movement are both *carriers* (i.e., contribute significantly to current when a voltage is applied). The formation of natural holes and free electrons increases as temperature increases. This formation is known as *thermal carrier generation*. The current due to natural defects is known as *intrinsic conduction*.

The addition of minute quantities (e.g., 1 in $10^8$) of impurities with either three or five valence electrons increases the defect density and increases conductivity. Intentional impurities are called *dopes* or *dopants*, and their addition is known as *doping*. Current carried by intentional impurities is known as *extrinsic conduction*.

When a five-valence dope, such as arsenic, antimony, or phosphorus, is used with germanium, an electron becomes available for conduction. This electron requires only a small amount of ionization energy and is easily

---

[16]*Thermal carrier generation* is important in semiconductors. Light is important in *photoconductors*.

removed. Thus, the dope contributes a negative carrier and is known as a *donor*. The resultant semiconductor is called an *n-type semiconductor*, since negative electrons are the *majority carriers*. Some holes will still be present, of course, and these are known as the *minority carriers*.

Aluminum, boron, indium, and gallium have only three valence electrons. A *vacancy* is created when one of these elements is used as a dope with germanium. The vacancy is easily shifted from one atom to another, contributing to current flow. Such dopes are called *acceptors*. Semiconductors manufactured with acceptor dopes are called *p-type semiconductors* since positive holes are the majority carriers. Electrons are the minority carriers in p-type semiconductors.

When gallium arsenide is the semiconductor substrate, donor dopants include sulfur, selenium, tellurium, and silicon. Acceptor dopants include magnesium, zinc, cadmium, and silicon.

## 24. SUPERCONDUCTIVITY

There are two cases when the resistivity of a substance can approach or reach zero. The first is when the temperature is absolute zero. Quantum theory predicts that the conductivity of a pure crystal (i.e., without lattice imperfections) is infinite at that temperature.

The second situation is when a real metal (with imperfections) exhibits zero resistivity at a temperature a few degrees above absolute zero. Once a current is started by the application of a voltage source, the current continues as long as the temperature is maintained, even after the voltage source is removed. This case is known as *superconductivity*. Impurities in the metal have no effect on superconductivity.

Superconductors are categorized as Type 1 and Type 2. *Type 1* (also known as *soft*) *superconductors* are pure metals below their *critical temperature* (also known as the *transition temperature*). They repel any external magnetic flux. A local external magnetic field will induce a current in the superconductor, creating an opposing magnetic field, known as the *Meissner-Oschsenfeld effect*. Demonstrations of magnetic levitation from this effect using floating supercooled magnets are common. Type 1 superconductivity disappears suddenly at the metal's critical temperature.[17] The low temperatures listed in Table 47.8 have limited the development of practical applications of superconductivity in pure metals, since those temperatures can only be maintained through use of expensive liquid helium (with a boiling point of 4.22K).

Interestingly, the more rigid metallic lattices (e.g., copper, silver, and gold) that contribute to high

**Table 47.8** Type 1 Superconducting Critical Temperatures

| element | critical temperature (K) | element | critical temperature (K) |
|---------|--------------------------|---------|--------------------------|
| Be | 0 | Al | 1.2 |
| Rh | 0 | Pa | 1.4 |
| W | 0.015 | Th | 1.4 |
| Ir | 0.1 | Re | 1.4 |
| Lu | 0.1 | Tl | 2.39 |
| Hf | 0.1 | In | 3.408 |
| U | 0.2 | Sn | 3.722 |
| Ti | 0.39 | Hg | 4.153 |
| Ru | 0.5 | Ta | 4.47 |
| Zr | 0.546 | V | 5.38 |
| Cd | 0.56 | La | 6.00 |
| Os | 0.7 | Pb | 7.193 |
| Zn | 0.85 | Tc | 7.77 |
| Mo | 0.92 | Nb | 9.46 |
| Ga | 1.083 | | |

conductivities at normal temperatures do not develop superconductivity. Less rigid lattices (e.g., tin and lead) are more superconductive.

Although the mechanism of superconductivity in metals is not fully understood, it appears that electrons initially set in motion by a voltage source form coupled pairs with equal but opposite momentum (known as Bose-coupled *Cooper pairs*). This highly quantum-dependent theory, initially proposed in the 1950s by John Bardeen, Leon Cooper, and Robert Schrieffer, is known as the *BCS theory*. The electron pairs move in unison through the lattice, avoiding collisions and wasteful electron scattering. From de Broglie's quantum momentum equation $\lambda p = h$, with a zero momentum sum, the wavelength becomes infinite. The BCS theory explains the effect of lattice rigidity on superconductivity.

In 1986, a new class of "Type 2" superconductors was discovered. (See Table 47.9.) These materials, which are primarily alloys and ceramic oxides, have two critical temperatures. Below their lower critical temperatures, the materials behave like Type 1 superconductors, repelling external magnetic flux. Above their upper critical temperatures, the materials behave like a normal alloy. In between, they exhibit *high temperature superconductivity* (i.e., superconductivity above 23K (41°R)). High temperature superconductivity has been observed at temperatures approaching 200K (360°R). Superconductivity can now be maintained with inexpensive liquid nitrogen, which boils at 78K (140°R). Additional research has reported that superconductivity can exist, at least tenuously and temporarily, at temperatures as high as 35°C (95°F).

Unlike Type 1 superconductors, Type 2 superconductors exhibit moderate permeability to magnetic flux. It is believed that the material exists in a *vortex state* of mixed regions (known as *cores*) of Type 1 and normal behavior, with the normal material permitting magnetic

---

[17]Critical temperatures exhibit an *isotope effect*, in that superconductivity appears and disappears at different temperatures depending on the element's atomic weight (isotope). Thus, the values in Table 47.8 depend slightly on isotope.

**Table 47.9** *Type 2 Superconducting Properties*

| alloy | critical temperature (K) | critical magnetic flux density (T) |
|---|---|---|
| CuS | 1.6 | |
| MoC | 8.0 | |
| NbTi | 10 | 15 |
| PbMoS | 14.4 | 6.0 |
| $V_3Ga$ | 14.8 | 2.1 |
| NbN | 15.7 | 1.5 |
| NbN | 16.0 | |
| $V_3Si$ | 16.9 | 2.35 |
| $V_3Si$ | 17.1 | |
| $PuCoGa_5$ | $\approx 18$ | |
| $Nb_3Sn$ | 18.0 | 24.5 |
| $Nb_3Al$ | 18.7 | 32.4 |
| $Nb_3AlGeO$ | 20.7 | 44 |
| $Nb_3Ge$ | 23.2 | 38 |
| LaBaCuO | 30 | |
| $MgBr_2$ | 39 | |
| YBaCuO (YBCO) | 92–98 | |
| BiSrCaCuO (BSCCO) | 95–107 | |

flux. Above the critical magnetic field intensity, however, superconductivity disappears.

The first family of Type 2 superconducting ceramic oxides is based on lanthanum-barium-copper-oxide compounds in the form $La_{2-x}Ba_xCuO_4$, with a transition temperature of approximately 30K (54°R).[18] The second family is based on yttrium-barium-copper-oxide (YBCO) compounds. The transition temperature for this class is approximately 98K (177°R). The third family is based on bismuth-strontium-calcium-copper-oxide (BSCCO) compounds. This class not only has the highest transition temperature (reported as 114K to 127K), but appears to be stronger, more stable, more resistant to corrosion, easier to manufacture in wire form, less expensive (because bismuth costs less than yttrium), and more capable of carrying higher currents than the other two classes.

The production of these classes of superconducting materials requires only mechanical mixing of the materials in the proper proportion, compression into shape, and sintering at 955°C (1750°F) for approximately one hour. The superconducting oxides are not easily commercialized, however. They are brittle and difficult to form once sintered. Therefore, shaping and coiling into wire form must be done when the compound is in powder form, prior to sintering. Also, the current density (ampacity) of the ceramics is very low—on the order of 200 A/cm$^2$ to 1000 A/cm$^2$. To be viable commercially, a current density of 10,000 A/cm$^2$ is needed, as are methods of joining two superconductor wires end to end without introducing resistance.

Commercial superconducting products (wires, coils, magnets, etc.) are most commonly manufactured from NbSn, NbTi, $MgB_2$, BSCCO, and YBCO compounds. Each has its own material cost, operating cost, current density, and magnetic field density characteristics and limitations. For example, YBCO compounds support current densities an order of magnitude greater than BSCCO compounds, albeit at a higher materials cost. Most high-energy nuclear particle accelerators now use helium-cooled, superconducting NbSn or NbTi magnets. Medical magnetic resonance imaging machines also routinely use superconducting magnets.

Commercial "flexible" superconductors (usually BSCCO) are manufactured using the *powder-in-tube* process (i.e., extruding a superconducting oxide powder within a small-bore silver (sheath) tube). Longer lengths and high current capacities are being achieved with multistranded wires with 20 to 100 fine filaments embedded in a copper matrix. Current capacities of up to 30,000 A/cm$^2$ have been achieved in this manner.

## 25. ATOMIC BASIS OF MAGNETISM

In a complete electron orbital, one electron spins in one direction and the other spins in the opposite direction. Thus, there will be no magnetic moment if a shell (containing an even number of electrons) is completely filled.[19] However, if a shell is incomplete, there will be an unpaired electron. This imbalance produces both magnetic and electric fields.

In a large mass of material, small groups of atoms may be aligned in the same direction. These groups are known as *magnetic domains* and are analogous to crystals in a large specimen. (See Fig. 47.20.) Despite the alignment of electron spins within magnetic domains, the orientation of domains is generally random. Thus, the magnetic fields from individual atoms tend to cancel each other out.

**Figure 47.20** *Magnetic Domains*

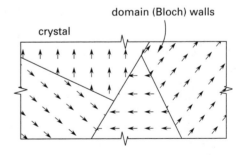

If the various domains are in alignment, the magnetic moments (i.e., fields) of the atoms will be reinforced, and the material will become magnetic. (This is the principle of *ferromagnetism*.) The alignment of atoms

---

[18]The development of this class of superconducting ceramics is generally credited to two Swiss IBM scientists, Johannes Georg Bednorz and Karl Alex Müller, in Zurich in 1986.

[19]It is not always the outer shell that is unbalanced. In the case of iron and cobalt, the 4s orbital contains a pair of electrons. However, the 3d orbitals have (four for iron or three for cobalt) unpaired electrons.

in all domains generally requires only the application of an external magnetic or electrical field. The alignment of atoms in all domains can be thought of as domain wall movement (*Bloch wall movement*).

## 26. PRINCIPLES OF MAGNETISM

A magnetic field can exist only with two opposite, equal poles, called the *north pole* and *south pole*. (This is unlike an electric field, which can be produced by a single charged object.) The combination of north and south poles linked together with a fixed distance between them is known as a *dipole*. The magnetic *dipole moment*, $p_m$, depends on the pole strength, $m$, and the distance, $d$, between the poles.

$$p_m = md \qquad 47.18$$

Magnetic moments have units of *Bohr magnetons*, also known as *Bohr-Procopiu magnetons*. A Bohr magneton in a ferromagnetic material is the moment produced by one unpaired electron (which is equivalent to $9.274 \times 10^{-24}$ J/T).

Figure 47.21 shows two common magnetic field configurations. It also illustrates the convention that the lines of *magnetic flux* are directed from the north pole (i.e., the magnetic source) to the south pole (i.e., the *magnetic sink*).

**Figure 47.21** *Typical Magnetic Fields*

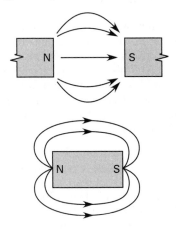

The *magnetic flux density*, $B$, (also known as *intensity of magnetization* and *dipole moment per unit volume*) for a bar magnet of pole strength $m$, length $L$, and cross-sectional area $A$ is easily calculated.

$$B = \frac{p_m}{V} = \frac{mL}{LA} = \frac{m}{A} \qquad 47.19$$

When material is placed in a magnetic field of strength $H_{applied}$, the magnetic flux density in the material is dependent on the *permeability* of the material.

$$B = \mu H_{applied} \qquad 47.20$$

$$\mu = \mu_r \mu_o \qquad 47.21$$

$\mu_o$ is the *permeability of a vacuum* (and air, for all practical purposes). $\mu_r$ is the *relative permeability* of the material, also known as the *permeability ratio*. $\mu_r = 1$ for air and vacuum. The relative permeability in general, however, is dependent on the flux density, $B$. Table 47.10 lists approximate relative permeabilities for common magnetic materials.

When a material (magnetic or otherwise) is placed in an external magnetic field of strength $H$, the strength of the *induced magnetic flux density (magnetic induction)*, $B$, will be the sum of two effects: the applied field and the material's *magnetization*, $M$.

$$B = \mu H = \mu_o(H + M) \qquad 47.22$$

For all practical purposes in ferromagnetic materials, $H \ll M$, so

$$B \approx \mu_o M \qquad 47.23$$

The ratio of the magnetization, $M$, to the external applied magnetic field, $H$, is known as the *susceptibility*, $\chi$. For diamagnetic and paramagnetic materials, the susceptibility is essentially constant. For ferromagnetic materials, the susceptibility is nonlinear with respect to the applied field, as is shown in Fig. 47.22.

$$\chi = \frac{M}{H} \qquad 47.24$$

Therefore, Eq. 47.22 can be written as

$$B = \mu_o H(1 + \chi) \qquad 47.25$$

## 27. HYSTERESIS

When a magnetic material is exposed to an external magnetic field, the material's magnetic domains align with the external field. A *permanent magnet* is so named because the material's domains remain aligned even after the external magnetic field is removed. Permanent magnets (e.g., alpha-iron, cobalt, gadolinium, alnico, and hard ferrites) are known as *magnetically hard materials*.

In some materials (e.g., silicon-iron alloys, permalloy, and soft ferrites), the domains do not remain aligned when the external field is removed. These materials can be remagnetized in any polarity from a subsequent exposure to a magnetic field. Such materials are said to be *magnetically soft materials*.

The graph of a material's magnetic response (i.e., internal magnetic flux density, $B$) to an applied magnetic field strength, $H$, is known as a *hysteresis loop*, or *B-H curve*, as illustrated in Fig. 47.23.

If a material without any internal magnetism is exposed to an external magnetic field, the material will become more or less magnetic. The internal magnetic flux, $B$,

**Table 47.10** *Typical Relative Permeabilities ($\mu_r$)*

| material | flux density, $B$ (T) | | | | | | | |
|---|---|---|---|---|---|---|---|---|
| | 0.2 | 0.4 | 0.6 | 0.8 | 1.0 | 1.2 | 1.4 | 1.6 |
| cast iron | 400 | 300 | 200 | 120 | 90 | 60 | 50 | 50 |
| cast steel | 750 | 1050 | 1200 | 1225 | 1175 | 950 | | |
| silicon sheet steel | 3000 | 5300 | 6000 | 5300 | 4000 | 2400 | | |

**Figure 47.22** *Magnetization Curves*

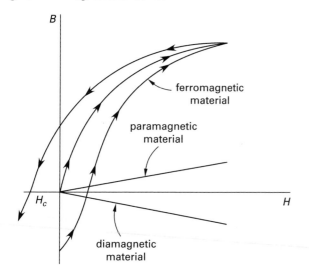

**Figure 47.23** *Typical Hysteresis Loops*

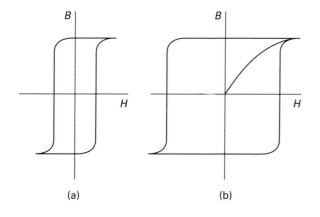

(a)                    (b)

will increase from zero to a maximum value (called the *saturation value*, $B_{max}$). The internal magnetism (i.e., flux) cannot be increased above the saturation value, no matter how high the external field.

If the external magnetic field is removed from a magnetically hard material, some of the internal magnetism will remain. This remaining magnetism is known as the *residual magnetism (residual magnetic flux or remnant induction)*, $B_r$. In order to totally remove the residual magnetism, a reverse external field with a strength equal to the *coercive field strength*, $H_c$, must be applied. The product of the residual magnetic flux and the coercive

field strength is known as the *effectivity product* or *effectivity area*.

$$A_e = B_r H_c \qquad 47.26$$

The area within the total loop represents the *hysteresis loss* (i.e., the energy lost to heat during each cycle). The hysteresis loss and effectivity product are very low for the magnetically soft materials used in transformers, solenoids, and motors. This is a desirable characteristic, because (with alternating currents) the field is reversed with a frequency of 60 Hz (50 Hz in Europe and other areas).

The hysteresis loss is high and the effectivity product is very large for permanent magnets.

The lifting power of a magnet is related to the area of the largest rectangle that can be drawn in the second quadrant of the hysteresis loop. This is quantified by the *BH product* and is known as the *power of the magnet*.

## 28. TYPES OF MAGNETISM

When some materials are exposed to a magnetic field, there is a spontaneous (and permanent) alignment of electron spins, creating a strongly magnetic material. (See Table 47.11.) This is seen primarily in alpha-iron, cobalt, nickel, and gadolinium, and is known as *ferromagnetism* (named after iron). The relative permeability of ferromagnetic materials is much greater than unity. (See Table 47.12.)

Ferromagnetic spin alignment is destroyed by random molecular vibrations when a ferromagnetic material is heated (i.e., annealed) above its *Curie temperature*, which is approximately 1414°F (770°C) for pure alpha-iron because the increased thermal activity promotes random alignment of the domain atoms.[20] Iron is paramagnetic above its Curie temperature.

An *antiferromagnetic material* will be only weakly attracted to a magnet. Such behavior is typical of salts of the transition elements and the elements of aluminum, copper, gold, and lead.

The strong magnetism that occurs in certain ceramic (crystalline) compounds such as *ferrites* (e.g., $MnFe_2O_4$ and $ZnFe_2O_4$), *spinels* ($MgAl_2O_4$), and garnet is known as *ferrimagnetism*. Although some ferrites are soft magnetically, the magnetically hard *square-loop ferrites* are best described by the shape of their hysteresis loops.

---

[20]The carbon content affects the Curie temperature considerably.

The weak attraction experienced by most alkali and transition metals when exposed to an external magnetic field is known as *paramagnetism*. Paramagnetism results solely from the magnetic moments created by the spinning electron dipoles and the atoms themselves, but not significant atomic alignment. Thus, paramagnetic materials are much less magnetic than ferromagnetic materials. Paramagnetic materials have positive susceptibilities and relative permeabilities slightly greater than unity. Magnetic susceptibilities of most paramagnetic materials are inversely proportional to their absolute temperatures, a fact known as *Curie's law*. Aluminum, beryllium, $MnSO_4$, and $NiCl_2$ are important paramagnetic materials.

The weakly repulsive effect experienced in most nonmetals and organic materials (e.g., bismuth, paraffin, silver, and wood) when exposed to a magnetic field is known as *diamagnetism. Diamagnetic materials* have negative susceptibilities. Specifically, the stronger the external magnetic field, the lower the internal magnetic field. The applied field interacts with the magnetic moment of the electron orbital motion, increasing the angular momentum of the electron and, in turn, increasing the orbital magnetic moment in a direction counter to the applied field. The relative permeabilities of these materials are slightly less than unity.

Table 47.12 summarizes the characteristics of magnetic materials.

*Table 47.11* Consistent Magnetic Units

| quantity | symbol | SI units | cgs units |
|---|---|---|---|
| pole strength (flux) | $m$ | Wb | lines[a] |
| flux density | $B$ | T[b] | gauss[c] |
| magnetization | $M$ | A/m[d] | oersted |
| field strength (intensity) | $H$ | A/m[e] | oersted |
| permeability | $\mu$ | H/m[f] | gauss/oersted |

[a]equivalent to $10^8$ Wb
[b]equivalent to $Wb/m^2$ and $V \cdot s/m^2$
[c]equivalent to $lines/cm^2$
[d]equivalent to N/Wb
[e]equivalent to N/Wb
[f]equivalent to $\Omega \cdot s/m$ and $WB/A \cdot m$

*Table 47.12* Characteristics of Magnetic Materials

| type of magnetic behavior | characteristics of magnetic susceptibility | | |
|---|---|---|---|
| | sign | magnitude | variability |
| diamagnetism | negative | small | constant |
| paramagnetism | positive | small | constant |
| ferromagnetism | positive | large | $f(H)$ |
| antiferromagnetism | positive | small | constant |
| ferrimagnetism | positive | large | $f(H)$ |

# *48* Material Testing

## Nomenclature

| | | | |
|---|---|---|---|
| $A$ | area | in$^2$ | m$^2$ |
| $b$ | width | in | m |
| $B$ | bulk modulus | lbf/in$^2$ | MPa |
| $C$ | constant | – | – |
| $C_V$ | impact energy | in-lbf | J |
| $d$ | diameter of impression | in | mm |
| $D$ | diameter | in | m |
| $e$ | engineering strain | in/in | m/m |
| $E$ | modulus of elasticity | lbf/in$^2$ | MPa |
| $F$ | force | lbf | N |
| $G$ | shear modulus | lbf/in$^2$ | MPa |
| $J$ | polar moment of inertia | in$^4$ | m$^4$ |
| $k$ | exponent | 1/hr | 1/h |
| $k$ | strength derating factor | – | – |
| $K$ | stress concentration factor | – | – |
| $K$ | strength coefficient | lbf/in$^2$ | MPa |
| $L$ | length | in | m |
| LYS | lower yield strength | lbf/in$^2$ | MPa |
| $n$ | exponent | – | – |
| $N$ | number of cycles | – | – |
| $P$ | force of impression | lbf | N |
| $q$ | fatigue notch sensitivity factor | – | – |
| $q$ | reduction in area | – | – |
| $r$ | radius | in | m |
| $s$ | stress | lbf/in$^2$ | MPa |
| $S$ | strength | lbf/in$^2$ | MPa |
| $t$ | depth (thickness) | in | mm |
| $t$ | time | hr | h |
| $T$ | torque | in-lbf | N·m |
| $U_R$ | modulus of resilience | lbf/in$^2$ | MPa |
| $U_T$ | modulus of toughness | lbf/in$^2$ | MPa |
| UYS | upper yield strength | lbf/in$^2$ | MPa |

## Symbols

| | | | |
|---|---|---|---|
| $\beta$ | Andrade's beta | hrs$^{-1/3}$ | h$^{-1/3}$ |
| $\gamma$ | angle of twist | rad | rad |
| $\delta$ | elongation | in | m |
| $\epsilon$ | true strain or creep | in/in | m/m |
| $\theta$ | angle of rupture | deg | deg |
| $\theta$ | shear strain | rad | rad |
| $\nu$ | Poisson's ratio | – | – |
| $\sigma$ | true stress | lbf/in$^2$ | MPa |
| $\tau$ | shear stress | lbf/in$^2$ | MPa |
| $\phi$ | angle of internal friction | deg | deg |

## Subscripts

| | |
|---|---|
| $c$ | compressive |
| $e$ | endurance |
| $f$ | fatigue or fracture |
| $o$ | original |
| $p$ | particular |
| $s$ | shear |
| $t$ | tensile |
| $u$ | ultimate |
| $y$ | yield |

## 1. CLASSIFICATION OF MATERIALS

When used to describe engineering materials, the terms "strong" and "tough" are not synonymous. Similarly, "weak," "soft," and "brittle" have different engineering meanings. A *strong material* has a high ultimate strength, whereas a *weak material* has a low ultimate strength. A *tough material* will yield greatly before breaking, whereas a *brittle material* will not. (A brittle material is one whose strain at fracture is less than approximately 0.5%.) A *hard material* has a high modulus of elasticity, whereas a *soft material* does not. Figure 48.1 illustrates some of the possible combinations of these classifications.

**Figure 48.1** *Types of Engineering Materials*

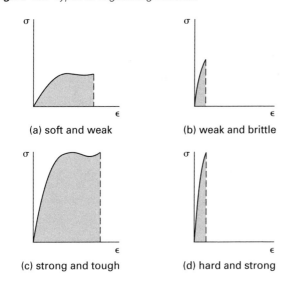

(a) soft and weak          (b) weak and brittle

(c) strong and tough       (d) hard and strong

## 2. TENSILE TEST

Many useful material properties are derived from the results of a standard *tensile test*. In this test, a prepared material sample (i.e., a *specimen*) is axially loaded in tension, and the resulting elongation, $\delta$, is measured as the load, $F$, increases. A *load-elongation curve* of tensile test data for a ductile ferrous material (e.g., low-carbon steel or other BCC transition metal) is shown in Fig. 48.2.

**Figure 48.2** *Typical Tensile Test Results for a Ductile Material*

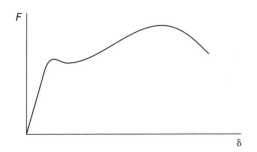

When elongation is plotted against the applied load, the graph is applicable only to an object with the same length and area as the test specimen. To generalize the test results, the data are converted to stresses and strains by the use of Eq. 48.1 and Eq. 48.2.[1]

*Engineering stress*, $s$ (usually called *stress*), is the load per unit original area. Typical engineering stress units are $lbf/in^2$ and MPa. *Engineering strain*, $e$ (usually called *strain*), is the elongation of the test specimen expressed as a percentage or decimal fraction of the

---

[1]The most common *test specimen* in the United States has a length of 2.00 in and a diameter of 0.505 in. Since the cross-sectional area of this *0.505 bar* is 0.2 in², the stress in lbf/in² is calculated by multiplying the force in pounds by five.

original length. The units in/in and m/m are also used for strain.

$$s = \frac{F}{A_o} \qquad 48.1$$

$$e = \frac{\delta}{L_o} \qquad 48.2$$

If the stress-strain data are plotted, the shape of the resulting line will be essentially the same as the force-elongation curve, although the scales will differ.

Segment O-A in Fig. 48.3 is a straight line. The relationship between the stress and strain in this linear region is given by *Hooke's law*, Eq. 48.3. The slope of line segment O-A is the *modulus of elasticity*, $E$, also known as *Young's modulus*. Table 48.1 lists approximate values of the modulus of elasticity for materials at room temperature. The modulus of elasticity will be lower at higher temperatures. For steel at higher temperatures, the modulus of elasticity is reduced approximately as shown in Table 48.2.

$$s = Ee \qquad 48.3$$

**Table 48.1** *Approximate Modulus of Elasticity of Representative Materials at Room Temperature*

| material | lbf/in² | GPa |
|---|---|---|
| aluminum alloys | $10–11 \times 10^6$ | 70–80 |
| brass | $15–16 \times 10^6$ | 100–110 |
| cast iron | $15–22 \times 10^6$ | 100–150 |
| cast iron, ductile | $22–25 \times 10^6$ | 150–170 |
| cast iron, malleable | $26–27 \times 10^6$ | 180–190 |
| copper alloys | $17–18 \times 10^6$ | 110–112 |
| glass | $7–12 \times 10^6$ | 50–80 |
| magnesium alloys | $6.5 \times 10^6$ | 45 |
| molybdenum | $47 \times 10^6$ | 320 |
| nickel alloys | $26–30 \times 10^6$ | 180–210 |
| steel, hard* | $30 \times 10^6$ | 210 |
| steel, soft* | $29 \times 10^6$ | 200 |
| steel, stainless | $28–30 \times 10^6$ | 190–210 |
| titanium | $15–17 \times 10^6$ | 100–110 |

(Multiply lbf/in² by $6.89 \times 10^{-6}$ to obtain GPa.)
*Common values are given.

**Table 48.2** *Approximate Reduction of Steel's Modulus of Elasticity at Higher Temperatures*

| temperature | | % of original value |
|---|---|---|
| °F | °C | |
| 70 | 20 | 100% |
| 400 | 200 | 90% |
| 800 | 425 | 75% |
| 1000 | 540 | 65% |
| 1200 | 650 | 60% |

The stress at point A in Fig. 48.3 is known as the *proportionality limit* (i.e, the maximum stress for which the linear relationship is valid). Strain in the *proportional region* is called *proportional strain*.

**Figure 48.3** *Typical Stress-Strain Curve for Steel*

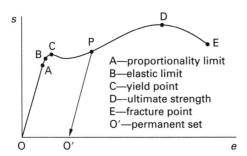

A—proportionality limit
B—elastic limit
C—yield point
D—ultimate strength
E—fracture point
O′—permanent set

**Table 48.3** *Approximate Yield Strengths of Representative Materials*

| material | yield strength | |
| --- | --- | --- |
| | lbf/in$^2$ | MPa |
| iron and steel | | |
| 1020 | 43,000 | 300 |
| A36 | 36,000 | 250 |
| stainless (304) | 43,000 | 300 |
| pure | 24,000 | 160 |
| copper | | |
| beryllium | 130,000 | 900 |
| brass | 11,000 | 75 |
| pure | 10,000 | 70 |
| aluminum | | |
| 2024 | 50,000 | 345 |
| 6061 | 21,000 | 145 |
| pure | 5000 | 35 |
| titanium | | |
| alloy 6% Al, 4% V | 160,000 | 1100 |
| pure | 20,000 | 140 |
| nickel | | |
| hastelloy | 55,000 | 380 |
| inconel | 40,000 | 280 |
| monel | 35,000 | 240 |
| pure | 20,000 | 140 |

(Multiply lbf/in$^2$ by $6.89 \times 10^{-3}$ to obtain MPa.)

The *elastic limit*, point B in Fig. 48.3, is slightly higher than the proportionality limit. As long as the stress is kept below the elastic limit, there will be no *permanent set* (permanent deformation) when the stress is removed. Strain that disappears when the stress is removed is known as *elastic strain*, and the stress is said to be in the *elastic region*. When the applied stress is removed, the *recovery* is 100%, and the material follows the original curve back to the origin.

If the applied stress exceeds the elastic limit, the recovery will be along a line parallel to the straight line portion of the curve, as shown in line segment P-O′. The strain that results (line O-O′) is *permanent set* (i.e., a permanent deformation). The terms *plastic strain* and *inelastic strain* are used to distinguish this behavior from elastic strain.

For steel, the *yield point*, point C, is very close to the elastic limit. For all practical purposes, the *yield strength* or *yield stress*, $S_y$ (or $S_{yt}$ to indicate yield in tension), can be taken as the stress that accompanies the beginning of plastic strain. Yield strengths are reported in lbf/in$^2$, ksi, and MPa.[2] (See Table 48.3.)

Figure 48.3 does not show the full complexity of the stress-strain curve near the yield point. Rather than being smooth, the curve is ragged near the yield point. At the upper yield strength, there is a pronounced drop (i.e., "drop of beam") in load-carrying ability to a plateau yield strength after the initial yielding occurs. The plateau value is known as the *lower yield strength* and is commonly reported as the yield strength. Figure 48.4 shows upper and lower yield strengths.

The *ultimate strength* or *tensile strength*, $S_u$ (or $S_{ut}$ to indicate an ultimate tensile strength), point D in Fig. 48.3, is the maximum stress the material can support without failure.

The *breaking strength* or *fracture strength*, $S_f$, is the stress at which the material actually fails (point E in Fig. 48.4). For ductile materials, the breaking strength is less than the ultimate strength due to the necking down in the cross-sectional area that accompanies high plastic strains.

**Figure 48.4** *Upper and Lower Yield Strengths*

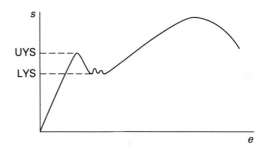

### 3. STRESS-STRAIN CHARACTERISTICS OF NONFERROUS METALS

Most nonferrous materials, such as aluminum, magnesium, copper, and other FCC and HCP metals, do not have well-defined yield points. The stress-strain curve starts to bend at low stresses, as illustrated by Fig. 48.5. In such cases, the yield strength is commonly defined as the stress that will cause a 0.2% *parallel offset* (i.e., a plastic strain of 0.002) for metals and 2% parallel offset for plastics.[3] However, the yield strength can also be defined by other offset values (e.g., 0.1% for metals and 1.0% for plastics).

The yield strength is found by extending a line from the offset strain value parallel to the linear portion of the curve until it intersects the curve.

---

[2]A *kip* is a thousand pounds. *ksi* is the abbreviation for kips per square inch (thousands of lbf/in$^2$).

[3]The 0.2% parallel offset strength is also known as the *proof stress*.

**Figure 48.5** *Typical Stress-Strain Curve for a Nonferrous Metal*

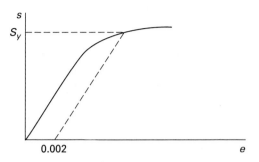

With nonferrous metals, the difference between parallel offset and total strain characteristics is important. Sometimes, the yield point will be defined as the stress accompanying *0.5% total strain* (i.e., a strain of 0.005) determined by extending a line from the stress-strain curve vertically downward to the strain axis.

## 4. STRESS-STRAIN CHARACTERISTICS OF BRITTLE MATERIALS

Brittle materials, such as glass, cast iron, and ceramics, can support only small strains before they fail catastrophically (i.e., without warning). As the stress is increased, the elongation is linear and Hooke's law (see Eq. 48.3) can be used to predict the strain. Failure occurs within the linear region, and there is very little, if any, necking down. Since the failure occurs at a low strain, brittle materials are not ductile. Figure 48.6 is typical of the stress-strain curve of a brittle material.

**Figure 48.6** *Stress-Strain Curve of a Brittle Material*

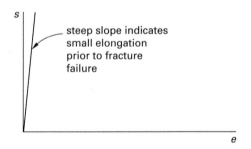

## 5. SECANT MODULUS

The modulus of elasticity, $E$, is usually determined from the steepest portion of the stress-strain curve. (This avoids the difficulty of locating the starting part of the curve.) For materials with variable modulus of elasticity, or for linear materials operating in the nonlinear region, the *secant modulus* gives the average ratio of stress to strain. The secant modulus is the slope of the straight line connecting the origin and the point of operation. Some designs using elastomers, concrete,

and prestressing wire may be based on the secant modulus. (See Fig. 48.7.)

**Figure 48.7** *Secant Modulus*

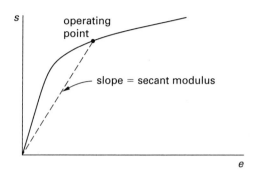

## 6. POISSON'S RATIO

As a specimen elongates axially during a tensile test, it will also decrease slightly in diameter or breadth. For any specific material, the percentage decrease in diameter, known as the *lateral strain*, will be a fraction of the *axial strain*. The ratio of the lateral strain to the axial strain is known as *Poisson's ratio*, $\nu$, which is approximately 0.3 for most metals. (See Table 48.4.)

**Table 48.4** *Approximate Values of Poisson's Ratio*

| material | Poisson's ratio |
|---|---|
| liquids | $0.50^a$ |
| rubber | 0.49 |
| thermosetting plastics | 0.40–0.45 |
| aluminum | $0.32–0.34 \ (0.33)^b$ |
| magnesium | 0.35 |
| copper | $0.33–0.36 \ (0.33)^b$ |
| titanium | 0.34 |
| brass | 0.33–0.36 |
| stainless steel | 0.30 |
| steel | $0.26–0.30 \ (0.30)^b$ |
| nickel | 0.30 |
| beryllium | 0.27 |
| cast iron | $0.21–0.33 \ (0.27)^b$ |
| glass ($SiO_2$) | $0.21–0.27 \ (0.23)^b$ |
| diamond | 0.20 |

[a]limiting value
[b]commonly used for design

$$\nu = \frac{e_{\text{lateral}}}{e_{\text{axial}}} = \frac{\dfrac{\Delta D}{D_o}}{\dfrac{\delta}{L_o}} \qquad 48.4$$

Poisson's ratio applies only to elastic strain. When the stress is removed, the lateral strain disappears along with the axial strain.

## 7. STRAIN HARDENING AND NECKING DOWN

When the applied stress exceeds the yield strength, the specimen will experience plastic deformation and will strain harden. (Plastic deformation is primarily due to the shear stress-induced movement of dislocations.) Since the specimen volume is constant (i.e., $A_oL_o = AL$), the cross-sectional area decreases. Initially, the strain hardening more than compensates for the decrease in area, so the material's strength increases and the engineering stress increases with larger strains.

Eventually, a point is reached when the available strain hardening and increase in strength cannot keep up with the decrease in cross-sectional area. The specimen then begins to neck down (at some local weak point), and all subsequent plastic deformation is concentrated at the neck. The cross-sectional area decreases even more rapidly thereafter since only a small portion of the specimen volume is strain hardening. The engineering stress decreases to failure.

Figure 48.8 shows necking down in two different specimens tested to failure. Very ductile materials pull out to a point, while most moderately ductile materials exhibit a cup-and-cone failure. (Failed brittle materials, not shown, do not exhibit any significant reduction in area.)

**Figure 48.8** *Types of Tensile Ductile Failure*

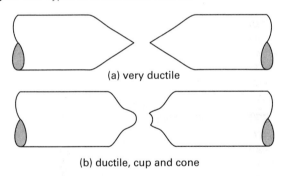

(a) very ductile

(b) ductile, cup and cone

## 8. TRUE STRESS AND STRAIN

Engineering stress, given by Eq. 48.1, is calculated for all stress levels from the original cross-sectional area. However, during a tensile test, the area of a specimen decreases as the stress increases. The decrease is only slight in the elastic region but is much more significant after plastic deformation begins.

If the stress is calculated from the instantaneous area, it is known as *true stress* or *physical stress*, $\sigma$. Equation 48.5 assumes a homogeneous strain distribution along the gage length and that there is no change in total volume with strain, which are valid up to the point of necking. ($q = (A_o - A)/A_o$, the fractional *reduction in area*, used in Eq. 48.5, is expressed as a number between 0 and 1.)

$$\sigma = \frac{F}{A} = \frac{F}{\left(1 - \dfrac{A_o - A}{A_o}\right)A_o} = \frac{F}{(1 - q)A_o}$$
$$= s(1 + e) \quad \text{[prior to necking, circular specimen]}$$
*48.5*

For circular or square specimens prior to necking, Eq. 48.5 may be written as Eq. 48.6

$$\sigma = \frac{s}{(1 - \nu e)^2} \qquad 48.6$$

Engineering strain, given by Eq. 48.2, is calculated from the original length, although the actual length increases during the tensile test. The *true strain* or *physical strain*, or *log strain*, $\epsilon$, is found from Eq. 48.7.

$$\epsilon = \int_{L_0}^{L} \frac{dL}{L} = \ln\left(\frac{L}{L_o}\right)$$
$$= \ln(1 + e) \quad \text{[prior to necking]}$$
*48.7*

Since the plastic deformation occurs through a shearing process, there is essentially no volume decrease during elongation.

$$A_oL_o = AL \qquad 48.8$$

Therefore, true strain can be calculated from the cross-sectional areas and, for a circular specimen, from diameters. If necking down has occurred, true strain must be calculated from the areas or diameters, not the lengths.

$$\epsilon = \ln\left(\frac{A_o}{A}\right) = \ln\left(\frac{D_o}{D}\right)^2 = 2\ln\left(\frac{D_o}{D}\right) \qquad 48.9$$

A graph of true stress and true strain is known as a *flow curve*. Log $\sigma$ can also be plotted against log $\epsilon$, resulting in a straight-line relationship.

The flow curve of many metals in the plastic region can be expressed by Eq. 48.10, known as a *power curve*. $K$ is known as the *strength coefficient*, and $n$ is the *strain-hardening exponent*. Values of both vary greatly with material, composition, and heat treatment. $n$ can vary from 0 (for a perfectly inelastic solid) to 1.0 (for an elastic solid). Typical values are between 0.1 and 0.5. For annealed steel with 0.05% carbon, for example, $K \approx 77{,}000$ psi and $n = 0.26$.

$$\sigma = K\epsilon^n \qquad 48.10$$

Figure 48.9 compares engineering and true stresses and strains for a ferrous alloy. As can be seen, the two curves coincide throughout the elastic region. Although true stress and strain are more accurate, almost all engineering work is based on engineering stress and strain, which is justifiable for two reasons: (a) design using ductile materials is limited to the elastic region where engineering and true values differ little, and (b) the reduction in area of most parts at their service stresses is not known; only the original area is known.

**Figure 48.9** *True and Engineering Stresses and Strains for a Ferrous Alloy*

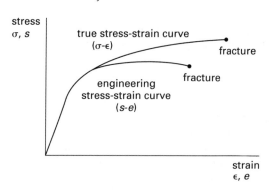

### Example 48.1

The engineering stress in a solid tension member was 47,000 $\mathrm{lbf/in^2}$ at failure. The reduction in area was 80%. What were the true stress and strain at failure?

*Solution*

Since engineering stress, $s$, is $F/A_o$, from Eq. 48.5 the true stress is

$$\sigma = \frac{s}{1-q}$$

$$= \frac{47,000 \, \dfrac{\mathrm{lbf}}{\mathrm{in}^2}}{1-0.80} = 235,000 \, \mathrm{lbf/in^2}$$

From Eq. 48.9, the true strain is

$$\epsilon = \ln\left(\frac{1}{1-0.80}\right) = 1.61 \quad (161\%)$$

## 9. DUCTILITY

A material that deforms and elongates a great deal before failure is said to be a *ductile material*.[4] (Steel, for example, is a ductile material.) The *percent elongation*, short for *percent elongation at failure*, is the total plastic strain at failure (see Fig. 48.10). (Percent elongation does not include the elastic strain, because even at ultimate failure the material snaps back an amount equal to the elastic strain.)

$$\text{percent elongation} = \frac{L_f - L_o}{L_o} \times 100\%$$

$$= e_f \times 100\% \qquad \textit{48.11}$$

The value of the final strain to be used in Eq. 48.11 is found by extending a line from the failure point

[4]The words "brittle" and "ductile" are antonyms.

**Figure 48.10** *Percent Elongation*

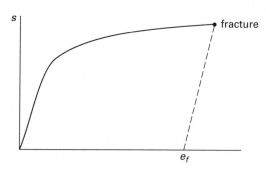

downward to the strain axis, parallel to the linear portion of the curve. This is equivalent to putting the two broken specimen pieces together and measuring the total length.

Highly ductile materials exhibit large percent elongations. However, percent elongation is not the same as *ductility*.

$$\text{ductility} = \frac{\text{ultimate failure strain}}{\text{yielding strain}} \qquad \textit{48.12}$$

The *reduction in area* (at the point of failure), expressed as a percentage or decimal fraction, is a third measure of a material's ductility. The reduction in area due to necking down will be 50% or greater for ductile materials and less than 10% for brittle materials.[5]

$$\text{reduction in area} = \frac{A_o - A_f}{A_o} \times 100\% \qquad \textit{48.13}$$

## 10. STRAIN ENERGY

*Strain energy*, also known as *internal work*, is the energy per unit volume stored in a deformed material. The strain energy is equivalent to the work done by the applied tensile force. Simple work is calculated as the product of a force moving through a distance.

$$\text{work} = \text{force} \times \text{distance} = \int F \, dL \qquad \textit{48.14}$$

$$\text{work per unit volume} = \int \frac{F \, dL}{AL} = \int_0^{\epsilon_{\text{final}}} \sigma \, d\epsilon \qquad \textit{48.15}$$

This work per unit volume corresponds to the area under the true stress-strain curve. Units are $\mathrm{in\text{-}lbf/in^3}$ (i.e., inch-pounds (a unit of energy) per cubic inch (a unit of volume)), usually shortened to $\mathrm{lbf/in^2}$ (MPa). (Equation 48.15 cannot be simplified further because stress is not proportional to strain for the entire curve.)

[5]*Notch-brittle materials* have reductions in area that are moderate (e.g., 25% to 35%) when tested in the usual manner, but close to zero when the test specimen is given a small notch or crack.

## 11. RESILIENCE

A *resilient material* is able to absorb and release *strain energy* without permanent deformation. *Resilience* is measured by the *modulus of resilience*, also known as the *elastic toughness*, which is the strain energy per unit volume required to reach the yield point. This is represented by the area under the stress-strain curve up to the yield point. Since the stress-strain curve is essentially a straight line up to that point, the area is triangular.

$$U_R = \int_0^{\epsilon_y} \sigma \, d\epsilon = E \int_0^{\epsilon_y} \epsilon \, d\epsilon = \frac{E\epsilon_y^2}{2}$$
$$= \frac{S_y \epsilon_y}{2} \qquad \qquad 48.16$$

The modulus of resilience (see Fig. 48.11) varies greatly for steel. It can be more than ten times higher for high-carbon spring steel ($U_R = 320 \text{ lbf/in}^2$, 2.2 MPa) than for low-carbon steel ($U_R = 20 \text{ lbf/in}^2$, 0.14 MPa).

*Figure 48.11* Modulus of Resilience

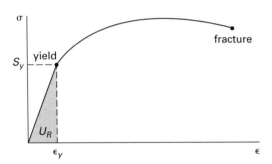

## 12. TOUGHNESS

A *tough material* will be able to withstand occasional high stresses without fracturing. Products subjected to sudden loading, such as chains, crane hooks, railroad couplings, and so on, should be tough. One measure of a material's *toughness* is the *modulus of toughness* (i.e, the strain energy or work per unit volume required to cause fracture). (See Fig. 48.12.) This is the total area under the stress-strain curve, given the symbol $U_T$. Since the area is irregular, the modulus of toughness cannot be exactly calculated by a simple formula. However, the modulus of toughness of ductile materials (with large strains at failure) can be approximately calculated from either Eq. 48.17 or Eq. 48.18.

$$U_T \approx S_u \epsilon_u \qquad \text{[ductile]} \qquad 48.17$$

$$U_T \approx \left( \frac{S_y + S_u}{2} \right) \epsilon_u \qquad \text{[ductile]} \qquad 48.18$$

For brittle materials, the stress-strain curve may be either linear or parabolic. If the curve is parabolic, Eq. 48.19 approximates the modulus of toughness.

$$U_T \approx \tfrac{2}{3} S_u \epsilon_u \qquad \text{[brittle]} \qquad 48.19$$

*Figure 48.12* Modulus of Toughness

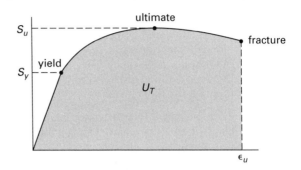

## 13. UNLOADING AND RELOADING

If the load is removed after a specimen is stressed elastically, the material will return to its original state. If the load is removed after a specimen is stressed into the plastic region, the *unloading curve* will follow a sloped path back to zero stress. The slope of the unloading curve will be equal to the original modulus of elasticity, $E$, illustrated by Fig. 48.13.

If this same material is subsequently reloaded, the *reloading curve* will follow the previous unloading curve up to the continuation of the original stress-strain curve. Therefore, the *apparent yield stress* of the reloaded specimen will be higher. This extra strength is the result of the strain hardening that has occurred.[6] Although the material will have a higher strength, its ductility and toughness will have been reduced.

*Figure 48.13* Unloading and Reloading Curves

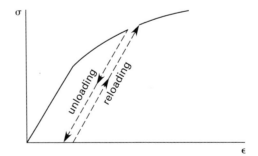

## 14. COMPRESSIVE STRENGTH

*Compressive strength*, $S_{uc}$ (i.e., ultimate strength in compression), is an important property for brittle

---

[6]The additional strength is lost if the material is subsequently annealed.

materials such as concrete and cast iron that are primarily loaded in compression only. ($f'_c$ is commonly used as the symbol for the compressive strength of concrete.) The compressive strengths of these materials are much greater than their tensile strengths, whereas the compressive strengths for ductile materials, such as steel, are the same as their tensile yield strengths.

Within the linear (elastic) region, Hooke's law is valid for compression of both brittle and ductile materials.

The failure mechanism for ductile materials is plastic deformation alone. Such materials do not rupture in compression. Thus, a ductile material can support a load long after the material is distorted beyond a useful shape.

The failure mechanism for brittle materials is shear along an inclined plane. The characteristic plane and hourglass failures for brittle materials are shown in Fig. 48.14.[7] Theoretically, only *cohesion* contributes to compressive strength, and the *angle of rupture* (i.e., the incline angle), $\theta$, should be 45°. In real materials, however, internal friction also contributes strength. The angles of rupture for cast iron, concrete, brick, and so on vary roughly between 50° and 60°. If the *angle of internal friction*, $\phi$, is known for the material, the angle of rupture can be calculated exactly from *Mohr's theory of rupture*.

$$\theta = 45° + \frac{\phi}{2} \qquad 48.20$$

**Figure 48.14** Compressive Failures

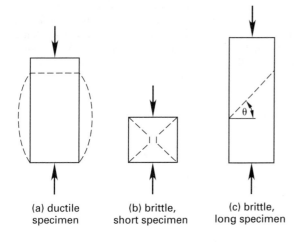

|  |  |  |
|:-:|:-:|:-:|
| (a) ductile specimen | (b) brittle, short specimen | (c) brittle, long specimen |

## 15. TORSION TEST

Figure 48.15 illustrates a simple cube loaded by a shear stress, $\tau$. The volume of the cube does not decrease when loaded, but the shape changes. The *shear strain* is the angle, $\theta$, expressed in radians. The shear strain is proportional to the shear stress, analogous to Hooke's law for tensile loading. $G$ is the *shear modulus*, also known

[7]The *hourglass failure* appears when the material is too short for a complete failure surface to develop.

as the *modulus of shear, modulus of elasticity in shear,* and *modulus of rigidity.*

$$\tau = G\theta \qquad 48.21$$

**Figure 48.15** Cube Loaded in Shear

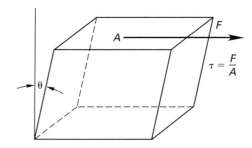

The shear modulus can be calculated from the modulus of elasticity and Poisson's ratio and, therefore, can be derived from the results of a tensile test. (See Table 48.5.)

$$G = \frac{E}{2(1 + \nu)} \qquad 48.22$$

**Table 48.5** Approximate Values of Shear Modulus

| material | lbf/in$^2$ | GPa |
|---|---|---|
| aluminum | $3.8 \times 10^6$ | 26 |
| brass | $5.5 \times 10^6$ | 38 |
| copper | $6.2 \times 10^6$ | 43 |
| cast iron | $8.0 \times 10^6$ | 55 |
| magnesium | $2.4 \times 10^6$ | 17 |
| steel | $11.5 \times 10^6$ | 79 |
| stainless steel | $10.6 \times 10^6$ | 73 |
| titanium | $6.0 \times 10^6$ | 41 |
| glass | $4.2 \times 10^6$ | 29 |

The shear stress can also be calculated from a torsion test, as illustrated by Fig. 48.16. Equation 48.23 relates the angle of twist (in radians) to the shear modulus.

**Figure 48.16** Uniform Bar in Torsion

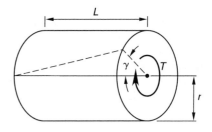

$$\gamma = \frac{TL}{JG} = \frac{\tau L}{rG} \quad \text{[radians]} \qquad 48.23$$

The *shear strength*, $S_s$ or $S_{ys}$, of a material is the maximum shear stress that the material can support without

**Table 48.6** Relationships Between Elastic Constants

| elastic constants | in terms of | | | | |
|---|---|---|---|---|---|
| | $E, \nu$ | $E, G$ | $B, \nu$ | $B, G$ | $E, B$ |
| $E$ | – | – | $3(1-2\nu)B$ | $\dfrac{9BG}{3B+G}$ | – |
| $\nu$ | – | $\dfrac{E}{2G}-1$ | – | $\dfrac{3B-2G}{2(3B+G)}$ | $\dfrac{3B-E}{6B}$ |
| $G$ | $\dfrac{E}{2(1+\nu)}$ | – | $\dfrac{3(1-2\nu)B}{2(1+\nu)}$ | – | $\dfrac{3EB}{9B-E}$ |
| $B$ | $\dfrac{E}{3(1-2\nu)}$ | $\dfrac{GE}{3(3G-E)}$ | – | – | – |

yielding in shear. (The ultimate shear strength, $S_{us}$, is rarely encountered.) For ductile materials, *maximum shear stress theory* predicts the shear strength as one-half of the tensile yield strength. A more accurate relationship is derived from the *distortion energy theory* (also known as *von Mises theory*).

$$S_{ys} = \frac{S_{yt}}{\sqrt{3}} = 0.577 S_{yt} \qquad 48.24$$

## 16. RELATIONSHIP BETWEEN THE ELASTIC CONSTANTS

The elastic constants (modulus of elasticity, shear modulus, bulk modulus, and Poisson's ratio) are related in elastic materials. Table 48.6 lists the common relationships.

## 17. FATIGUE TESTING

A material can fail after repeated stress loadings even if the stress level never exceeds the ultimate strength, a condition known as *fatigue failure*.

The behavior of a material under repeated loadings is evaluated by a fatigue test. A specimen is loaded repeatedly to a specific stress amplitude, $S$, and the number of applications of that stress required to cause failure, $N$, is counted. Rotating beam tests that load the specimen in bending are more common than alternating deflection and push-pull tests but are limited to round specimens (see Fig. 48.17).[8]

This procedure is repeated for different stresses, using eight to fifteen specimens. The results of these tests are graphed, resulting in an *S-N curve* (i.e., stress-number of cycles), also known as a *Wöhler curve*, shown in Fig. 48.18.

---

[8]In the design of ductile steel buildings, the static case is assumed up to 20,000 cycles. However, in critical applications (such as nuclear steam vessels, turbines, and so on) that experience temperature swings, fatigue failure can occur with a smaller number of cycles due to *cyclic strain*, not due to *cyclic stress*.

**Figure 48.17** Rotating Beam Test

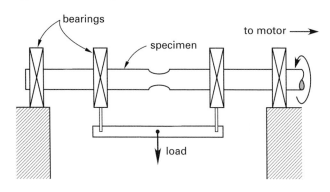

For an alternating stress test, the stress plotted on the *S-N* curve can be the maximum, minimum, or mean value. The choice depends on the method of testing as well as the intended application. The maximum stress should be used in rotating beam tests, since the mean stress is zero. For cyclic, one-dimensional bending, the maximum and mean stresses are commonly used.

For a specific stress level, say $S_p$ in Fig. 48.18, the number of cycles required to cause failure, $N_p$, is the *fatigue life*. $S_p$ is the *fatigue strength* corresponding to $N_p$.

**Figure 48.18** Typical S-N Curve for Steel

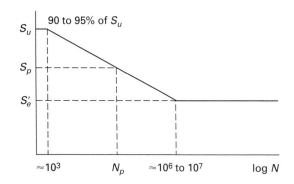

For steel in bending subjected to fewer than approximately $10^3$ loadings, the fatigue strength starts at the ultimate strength and drops to 90% to 95% of the ultimate strength at $10^3$ cycles.[9] (Although *low-cycle fatigue* theory has its own peculiarities, a part experiencing a small number of cycles can usually be designed or analyzed for static loading.) The curve is linear between $10^3$ and $10^6$ cycles if a logarithmic $N$-scale is used. Beyond $10^6$ to $10^7$ cycles, there is no further decrease in strength.

Therefore, below a certain stress level, called the *endurance limit* or *endurance strength*, $S'_e$, the material will withstand an almost infinite number of loadings without experiencing failure.[10] This is characteristic of steel and titanium. Therefore, if a dynamically loaded part is to have an infinite life, the stress must be kept below the endurance limit. The ratio $S'_e/S_u$ is known as the *endurance ratio* or *fatigue ratio*. For carbon steel, the endurance ratio is approximately 0.4 for pearlitic, 0.60 for ferritic, and 0.25 for martensitic microstructures. For martensitic alloy steels, it is approximately 0.35.

For steel whose microstructure is unknown, the endurance strength is given approximately by Eq. 48.25.[11]

$$S'_{e,\text{steel}} \begin{cases} = 0.5S_u & [S_u < 200{,}000 \text{ lbf/in}^2] \\ & [S_u < 1.4 \text{ GPa}] \\ = 100{,}000 \text{ lbf/in}^2 & [S_u > 200{,}000 \text{ lbf/in}^2] \\ (700 \text{ MPa}) & [S_u > 1.4 \text{ GPa}] \end{cases}$$

$$48.25$$

For cast iron, the endurance ratio is lower.

$$S'_{e,\text{cast iron}} = 0.4S_u \qquad \textbf{48.26}$$

Steel and titanium are the most important engineering materials that have well-defined endurance limits. Many nonferrous metals and alloys, such as aluminum, magnesium, and copper alloys, do not have well-defined endurance limits (see Fig. 48.19). The strength continues to decrease with cyclic loading and never levels off. In such cases, the endurance limit is taken as the stress that causes failure at $10^7$ loadings (less typically, at $10^8$ or $5 \times 10^8$ loadings). Alternatively, the endurance strength is approximated by Eq. 48.27.

$$S'_{e,\text{aluminum}} = \begin{cases} 0.3S_u & [\text{cast}] \\ 0.4S_u & [\text{wrought}] \end{cases} \qquad \textbf{48.27}$$

The yield strength is an irrelevant factor in cyclic loading. Fatigue failures are fracture failures; they are not yielding failures. They start with microscopic cracks at the material surface. Some of the cracks are present

initially; others form when repeated cold working reduces the ductility in strain-hardened areas. These cracks grow minutely with each loading. Since cracks start at the location of surface defects, the endurance limit is increased by proper treatment of the surface. Such treatments include polishing, surface hardening, shot peening, and filleting joints.

*Figure 48.19* Typical S-N Curve for Aluminum

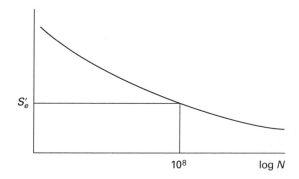

The endurance limit is not a true property of the material since the other significant influences, particularly surface finish, are never eliminated. However, representative values of $S'_e$ obtained from ground and polished specimens provide a baseline to which other factors can be applied to account for the effects of surface finish, temperature, stress concentration, notch sensitivity, size, environment, and desired reliability. These other influences are accounted for by fatigue strength reduction (derating) factors, $k_i$, which are used to calculate a working endurance strength, $S_e$, for the material. (See Fig. 48.20.)

$$S_e = \prod k_i S'_e \qquad \textbf{48.28}$$

Since a rough surface significantly decreases the endurance strength of a specimen, it is not surprising that notches (and other features that produce stress concentration) do so as well. In some cases, the theoretical tensile *stress concentration factor*, $K_t$, due to notches and other features can be determined theoretically or experimentally. The ratio of the fatigue strength of a polished specimen to the fatigue strength of a notched specimen at the same number of cycles is known as the *fatigue notch factor*, $K_f$, also known as the *fatigue stress concentration factor*. The *fatigue notch sensitivity*, $q$, is a measure of the degree of agreement between the stress concentration factor and the fatigue notch factor.

$$q = \frac{K_f - 1}{K_t - 1} \qquad [K_f > 1] \qquad \textbf{48.29}$$

## 18. TESTING OF PLASTICS

With reasonable variations, mechanical properties of plastics are evaluated using the same methods as for

---

[9]Steel in tension has a lower fatigue life at $10^3$ cycles, approximately 72–75% of $S_u$.

[10]Most endurance tests use some form of sinusoidal loading. However, the fatigue and endurance strengths do not depend much on the shape of the loading curve. Only the maximum amplitude of the stress is relevant. Therefore, the endurance limit can be used with other types of loading (sawtooth, square wave, random, etc.).

[11]The coefficient in Eq. 48.25 actually varies between 0.25 and 0.6. However, 0.5 is commonly quoted.

**Figure 48.20** Surface Finish Reduction Factors for Endurance Strength

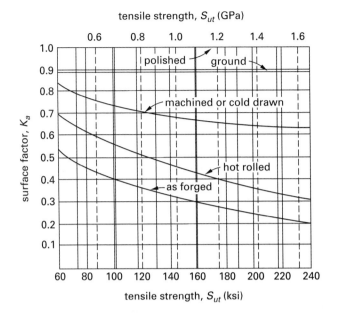

Reprinted with permission from *Mechanical Engineering Design*, 3rd ed. by Joseph Edward Shigley, © 1977, The McGraw-Hill Companies.

and some permanent deformation. The recovery might be complete if the load is removed within 10 hours, but if the loading is longer (e.g., 100 hours), recovery may be only partial. Because of this behavior, plastics are subjected to various other tests.

**Figure 48.21** Typical Tensile Test Performance for Plastics (loaded at 2 in/min)

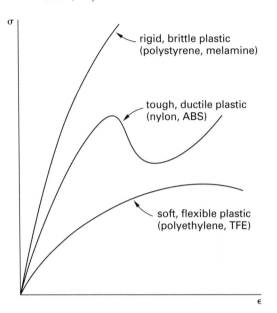

metals.[12] Although temperature is an important factor in the testing of plastics, tests for tensile strength, endurance, hardness, toughness, and creep rate are similar or the same as for metals. Figure 48.21 illustrates typical tensile test results.[13]

Unlike metals, which follow Hooke's law, plastics are non-Hookean. (They may be Hookean for a short-duration loading.) The modulus of elasticity, for example, changes with stress level, temperature, time, and chemical environment. A plastic that appears to be satisfactory under one set of conditions can fail quickly under slightly different conditions. Therefore, properties of plastics determined from testing (and from tables) should be used only to compare similar materials, not to predict long-term behavior. Plastic tests are used to determine material specifications, not performance specifications.

On a short-term basis, plastics behave elastically. They distort when loaded and spring back when unloaded. Under prolonged loading, however, creep (cold flow) becomes significant. When loading is removed, there is some instantaneous recovery, some delayed recovery,

Additional tests used to determine the mechanical properties of plastics include deflection temperature, long-term (e.g., 3000 hours) tensile creep, creep rupture, and *stress-relaxation* (long-duration, constant-strain tensile testing at elevated temperatures).[14] Because some plastics deteriorate when exposed to light, plasma, or chemicals, performance under these conditions can be evaluated, as can be the insulating and dielectric properties.

The *creep modulus* (also known as *apparent modulus*), determined from tensile creep testing, is the instantaneous ratio of stress to creep strain. The creep modulus decreases with time. The *deflection temperature* test indicates the dimensional stability of a plastic at high temperatures. A plastic bar is loaded laterally (as a beam) to a known stress level, and the temperature of the bar is gradually increased. The temperature at which the deflection reaches 0.010 in (0.254 mm) is taken as the *thermal deflection temperature* (TDT). The *Vicat softening point*, primarily used with polyethylenes, is the temperature at which a loaded standard needle penetrates 1 mm when the temperature is uniformly increased at a standard rate.

---

[12]For example, plastic specimens for tensile testing can be produced by injection molding as well as by machining from compression-molded plaques, rather than by machining from bar stock.

[13]Plastics are sensitive to the rate of loading. Figure 48.21 illustrates tensile performance based on a 2 in/min loading rate. However, for a fast loading rate (e.g., 2 in/sec), most plastics would exhibit brittle performance. On the other hand, given a slow loading rate (e.g., 2 in/month), most would behave as a soft and flexible plastic. Therefore, with different rates of loading, all three types of stress-strain performance shown in Fig. 48.21 can be obtained from the same plastic.

---

[14]Plastic pipes have their own special tests (e.g., ASTM D1598, D1785, and D2444).

## 19. NONDESTRUCTIVE TESTING

*Nondestructive testing* (NDT) or *nondestructive evaluation* (NDE) is used when it is impractical or uneconomical to perform destructive sampling on manufactured products and their parts. Typical applications of NDT are inspection of helicopter blades, cast aluminum wheels, and welds in nuclear pressure vessels. Some procedures are particularly useful in providing quality monitoring on a continuous, real-time basis. In addition to visual processes, the main types of nondestructive testing are magnetic particle, eddy current, liquid penetrant, ultrasonic imaging, acoustic emission, and infrared testing, as well as radiography.

The *visual-optical* process differs from normal visual inspection in the use of optical scanning systems, borescopes, magnifiers, and holographic equipment. Flaws are identified as changes in light intensity (reflected, transmitted, or refracted), color changes, polarization changes, or phase changes. This method is limited to the identification of surface flaws or interior flaws in transparent materials.

*Liquid penetrant testing* is based on a fluorescent dye being drawn by capillary action into surface defects. A developer substance is commonly used to aid in visual inspection. This method can be used with any nonporous material, including metals, plastics, and glazed ceramics. It is capable of finding cracks, porosities, pits, seams, and laps.

Liquid penetrant tests are simple, can be used with complex shapes, and can be performed on site. Workpieces must be clean and nonporous. However, only small surface defects are detectable.

*Magnetic particle testing* takes advantage of the attraction of ferromagnetic powders (e.g., the *Magnaflux*[TM] *process*) and fluorescent particles (e.g., the *Magnaglow*[TM] *process*) to leakage flux at surface flaws in magnetic materials. The particles accumulate and become visible at such flaws when an intense magnetic field is set up in a workpiece.

This method can locate most surface flaws (such as cracks, laps, and seams) and, in some special cases, subsurface flaws. The procedure is fast and simple to interpret. However, workpieces must be ferromagnetic and clean. Following the test, demagnetization may be required. A high-current power source is required.

*Eddy current testing* uses alternating current from a test coil to induce eddy currents in electrically conducting, metallic objects. Flaws and other material properties affect the current flow. The change in current is monitored by a detection circuit or on a meter or screen. This method can be used to locate defects of many types, including cracks, voids, inclusions, and weld defects, as well as to find changes in composition, structure, hardness, and porosity. Intimate contact between the material and the test coil is not required. Operation can be continuous, automatic, and monitored electronically. Sensitivity is easily adjusted. Therefore, this

method is ideal for unattended continuous processing. Many variables, however, can affect the current flow, and only electrically conducting materials can be tested with this method.

With *infrared testing*, infrared radiation emitted from objects can be detected and correlated with quality. Any discontinuities that interrupt heat flow, such as flaws, voids, and inclusions, can be detected.

Infrared testing requires access to only one side and is highly sensitive. It is applicable to complex shapes and assemblies of dissimilar components but is relatively slow. The detection can be performed electronically. Results are affected by variations in material size, coatings, and colors, and hot spots can be hidden by cool surface layers.

In *ultrasound imaging testing (ultrasonics)*, mechanical vibrations in the 0.1 MHz to 50 MHz range are induced by pressing a piezoelectric transducer against a workpiece.[15] The transmitted waves are normally reflected back, but the waves are scattered by interior defects. The results are interpreted by reading a screen or meter. The method can be used for metals, plastics, glass, rubber, graphite, and concrete. It is excellent for detecting internal defects such as inclusions, cracks, porosities, laminations, and changes in material structure.

Ultrasound testing is extremely flexible. It can be automated and is very fast. Results can be recorded or interpreted electronically. Penetration through thick steel layers is possible. Direct contact (or immersion in a fluid) is required, but only one surface needs to be accessible. Rough surfaces and complex shapes may cause difficulties, however. A related method, *acoustic emission monitoring*, is used to test pressurized systems.

*Radiography* (i.e., *nuclear sensing*) uses neutron, X-ray, gamma-ray (e.g., Ce-137), and isotope (e.g., Co-60) sources. (When neutrons are used, the method is known as *neutron radiography* or *neutron gaging*). The intensity of emitted radiation is changed when the rays pass through defects, and the intensity changes are monitored on a fluoroscope or recorded on film. This method can be used to detect internal defects, changes in material structure, thickness, and the absence of internal workpieces. It is also used to check liquid levels in filled containers.

Up to 30 in (0.75 m) of steel can be penetrated by X-ray sources. Gamma sources, which are more portable and lower in cost than X-ray sources, can be used with steel up to 10 in (0.25 m).

Radiography requires access to both sides of the workpiece. Radiography involves some health risk, and there may be government standards associated with its use. Electrical power and cooling water may be required in large installations. Shielding and film processing are also

---

[15]Theoretically, any frequency can be used. However, as frequency goes up, the detail available increases, while the penetration decreases. Biomedical applications operate below 8 MHz, and industrial NDT uses 2–10 MHz waves.

**Table 48.7** Hardness Penetration Tests

| test | penetrator | diagram | measured dimension | hardness |
|---|---|---|---|---|
| Brinell | sphere | (a) | diameter, $d$ | $\mathrm{BHN} = \dfrac{2P}{\pi D(D - \sqrt{D^2 - d^2})}$ |
| Rockwell[*] | sphere or penetrator | (b) | depth, $t$ | $R = C_1 - C_2 t$ |
| Vickers | square pyramid | (b) | mean diagonal, $d_1$ | $\mathrm{VHN} = \dfrac{1.854P}{d_1^2}$ |
| Meyer | sphere | (a) | diameter, $d$ | $\mathrm{MHN} = \dfrac{4P}{\pi d^2}$ |
| Meyer-Vickers | square pyramid | (b) | mean diagonal, $d_1$ | $M_V = \dfrac{2P}{d_1^2}$ |
| Knoop | asymmetrical pyramid | (c) | long diagonal, $L$ | $K = \dfrac{14.2P}{L^2}$ |

(a)    (b)    (c)

[*]$C_1$ and $C_2$ are constants that depend on the scale.

required, making this the most expensive form of non-destructive testing.

There are two types of *holographic NDT methods.* *Acoustic holography* is a form of ultrasonic testing that passes an ultrasonic beam through the workpiece (or through a medium such as water surrounding the workpiece) and measures the displacement of the workpiece (or medium). With suitable processing, a three-dimensional hologram is formed that can be visually inspected.

In one form of *optical holography*, a hologram of the unloaded workpiece is imposed on the actual workpiece. If the workpiece is then loaded (stressed), the observed changes (e.g., deflections) from the holographic image will be non-uniform when discontinuities and defects are present.

## 20. HARDNESS TESTING

Hardness tests measure the capacity of a surface to resist deformation. (See Table 48.7.) The main use of hardness testing is to verify heat treatments, an important factor in product service life. Through empirical correlations, it is also possible to predict the ultimate strength and toughness of some materials.

The *Brinell hardness test* is used primarily with iron and steel castings, although it can be used with softer materials. (See Table 48.8.) The *Brinell hardness number*, BHN, is determined by pressing a hardened steel ball into the surface of a specimen. The diameter of the resulting depression is correlated to the hardness. The standard ball is 10 mm in diameter and loads are 500 kg and 3000 kg for soft and hard materials, respectively.

The Brinell hardness number is the load per unit contact area. If a load, $P$ (in kilograms), is applied through a steel ball of diameter, $D$ (in millimeters), and produces a depression of diameter, $d$ (in millimeters), and depth, $t$ (in millimeters), the Brinell hardness number can be calculated from Eq. 48.30.

$$\mathrm{BHN} = \frac{P}{A_{\mathrm{contact}}} = \frac{P}{\pi D t}$$
$$= \frac{2P}{\pi D(D - \sqrt{D^2 - d^2})} \qquad 48.30$$

For heat-treated plain-carbon and medium-alloy steels, the ultimate tensile strength in $\mathrm{lbf/in^2}$ can be approximately calculated from the steel's Brinell hardness number.

$$S_u \approx 500(\mathrm{BHN}) \qquad 48.31$$

The *Rockwell hardness test* is similar to the Brinell test. A steel ball or diamond spheroconical penetrator (known as a *brale indenter*) is pressed into the material. The machine applies an initial load (60, 100, or 150 kgf) that sets the penetrator below surface imperfections.[16] Then a significant load is applied. The Rockwell hardness is determined from the depth of penetration and is read directly from a dial.

**Table 48.8** *Correlations Between Hardness Scales for Steel*

| Brinell number | Vickers number | Rockwell numbers | | scleroscope number |
|---|---|---|---|---|
| | | C | B | |
| 780 | 1150 | 70 | ... | 106 |
| 712 | 960 | 66 | ... | 95 |
| 653 | 820 | 62 | ... | 87 |
| 601 | 717 | 58 | ... | 81 |
| 555 | 633 | 55 | 120 | 75 |
| 514 | 567 | 52 | 119 | 70 |
| 477 | 515 | 49 | 117 | 65 |
| 429 | 454 | 45 | 115 | 59 |
| 401 | 420 | 42 | 113 | 55 |
| 363 | 375 | 38 | 110 | 51 |
| 321 | 327 | 34 | 108 | 45 |
| 293 | 296 | 31 | 106 | 42 |
| 277 | 279 | 29 | 104 | 39 |
| 248 | 248 | 24 | 102 | 36 |
| 235 | 235 | 22 | 99 | 34 |
| 223 | 223 | 20 | 97 | 32 |
| 207 | 207 | 16 | 95 | 30 |
| 197 | 197 | 13 | 93 | 29 |
| 183 | 183 | 9 | 90 | 27 |
| 166 | 166 | 4 | 86 | 25 |
| 153 | 153 | ... | 82 | 23 |
| 140 | 140 | ... | 78 | 21 |
| 131 | 131 | ... | 74 | 20 |
| 121 | 121 | ... | 70 | ... |
| 112 | 112 | ... | 66 | ... |
| 105 | 105 | ... | 62 | ... |
| 99 | 99 | ... | 59 | ... |
| 95 | 95 | ... | 56 | ... |

Although a number of Rockwell scales (A through G) exist, the B and C scales are commonly used for steel. The *Rockwell B scale* is used with a steel ball for mild steel and high-strength aluminum. The *Rockwell C scale* is used with the brale indentor for hard steels having ultimate tensile strengths up to 300 ksi (2 GPa). The *Rockwell A scale* has a wide range and can be used with both soft materials (such as annealed brass) and hard materials (such as cemented carbides).

Other penetration hardness tests include the *Meyer, Vickers, Meyer-Vickers,* and *Knoop* tests, as described in Table 48.7.

---

[16]Other Rockwell tests use 15, 30, and 45 kgf. The use of kgf units is traditional, and even modern test equipment is calibrated in kgf. Multiply kgf by 9.80665 to get newtons.

*Cutting hardness* is a measure of the force per unit area to cut a chip at low speed. (See Fig. 48.22.)

$$\text{cutting hardness} = \frac{F}{bt} \qquad 48.32$$

**Figure 48.22** *Cutting Hardness*

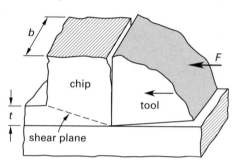

The *scratch hardness test*, also known as the *Mohs test*, compares the hardness of the material to that of minerals. Minerals of increasing hardness are used to scratch the sample. The resulting *Mohs scale* hardness can be used or correlated to other hardness scales, as in Fig. 48.23.

**Figure 48.23** *Mohs Hardness Scale*

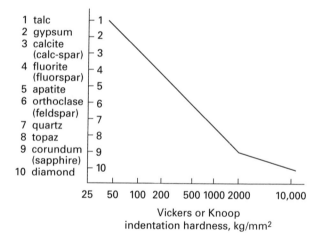

The *file hardness* test is a combination of the cutting and scratch tests. Files of known hardness are drawn across the sample. The file ceases to cut the material when the material and file hardnesses are the same.

All of the preceding hardness tests are *destructive tests* because they mar the material surface. However, *ultrasonic tests* and various *rebound tests* (e.g., the *Shore hardness test* and the *scleroscopic hardness test*) are *nondestructive tests*. In a rebound test, a standard object, usually a diamond-tipped hammer, is dropped from a standard height onto the sample. The height of the rebound is measured and correlated to other hardness scales.

The various hardness tests do not measure identical properties of the material, so correlations between the various scales are not exact. For steel, the Brinell and Vickers hardness numbers are approximately the same below values of 320 Brinell. Also, the Brinell hardness is approximately ten times the Rockwell C hardness ($R_c$) for $R_c > 20$. Table 48.8 is an accepted correlation between several of the scales for steel. The table should not be used for other materials.

## 21. TOUGHNESS TESTING

During World War II, the United States experienced spectacular failures in approximately 25% of its Liberty ships and T-2 tankers. The mild steel plates of these ships were connected by welds that lost their ductility and became brittle in winter temperatures. Some of the ships actually broke into two sections. Such *brittle failures* are most likely to occur when three conditions are met: (a) triaxial stress, (b) low temperature, and (c) rapid loading.

*Toughness* is a measure of the material's ability to yield and absorb highly localized and rapidly applied stresses.

*Notch toughness* is evaluated by measuring the *impact energy* that causes a notched sample to fail.[17]

In the *Charpy test* (see Fig. 48.24), popular in the United States, a standardized beam specimen is given a 45° notch. The specimen is then centered on simple supports with the notch down. A falling pendulum striker hits the center of the specimen. This test is performed several times with different heights and different specimens until a sample fractures.

**Figure 48.24** *Charpy Test*

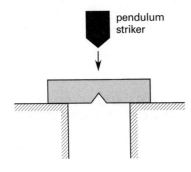

The kinetic energy expended at impact, equal to the initial potential energy less the rebound or follow-through height of the pendulum striker, is calculated from measured heights. It is designated $C_V$ and is expressed in either foot-pounds (ft-lbf) or joules (J).[18]

---

[17]Without a notch, the specimen would experience uniaxial stress (tension and compression) at impact. The notch allows triaxial stresses to develop. Most materials become more brittle under triaxial stresses than under uniaxial stresses.

[18]In Europe, the energy is often expressed per unit cross section of specimen area.

The energy required to cause failure is a measure of toughness.

At 70°F (21°C), the energy required to cause failure ranges from 45 ft-lbf (60 J) for carbon steels to approximately 110 ft-lbf (150 J) for chromium-manganese steels. As temperature is reduced, however, the toughness decreases. In BCC metals, such as steel, at a low enough temperature the toughness decreases sharply. The transition from high-energy ductile failures to low-energy brittle failures begins at the *fracture transition plastic (FTP) temperature.*

Since the transition occurs over a wide temperature range, the *transition temperature* (also known as the *ductile-brittle transition temperature*, DBTT) is taken as the temperature at which an impact of 15 ft-lbf (20.4 J) will cause failure. (15 ft-lbf is used for low-carbon ship steels. Other values may be used with other materials.) This occurs at approximately 30°F (−1°C) for low-carbon steel. Table 48.9 gives ductile transition temperatures for some forms of steel.

**Table 48.9** *Approximate Ductile Transition Temperatures*

| type of steel | transition ductile temperature, °F |
|---|---|
| carbon steel | 30° |
| high-strength, low-alloy steel | 0° to 30° |
| heat-treated, high-strength, carbon steel | −25° |
| heat-treated, construction alloy steel | −40° to −80° |

The appearance of the fractured surface is also used to evaluate the transition temperature. The fracture can be fibrous (from shear fracture) or granular (from cleavage fracture), or a mixture of both. The fracture planes are studied and the percentages of ductile failure plotted against temperature. The temperature at which the failure is 50% fibrous and 50% granular is known as the *fracture appearance transition temperature*, FATT.

Not all materials have a ductile-brittle transition. Aluminum, copper, other FCC metals, and most HCP metals do not lose their toughness abruptly. Figure 48.25 illustrates the failure energy curves for several different types of materials.

**Figure 48.25** *Failure Energy versus Temperature*

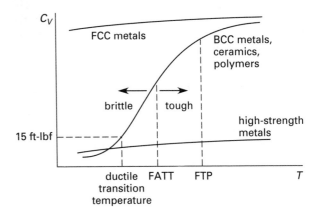

Another toughness test is the *Izod test*. This is illustrated in Fig. 48.26 and is similar to the Charpy test in its use of a notched specimen. The height to which a swinging pendulum follows through after causing the specimen to fail determines the energy of failure.

**Figure 48.26** *Izod Test*

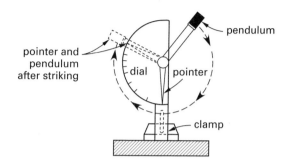

## 22. CREEP TEST

*Creep* or *creep strain* is the continuous yielding of a material under constant stress. For metals, creep is negligible at low temperatures (i.e., less than half of the absolute melting temperature), although the usefulness of nonreinforced plastics as structural materials is seriously limited by creep at room temperature.

During a *creep test*, a low tensile load of constant magnitude is applied to a specimen, and the strain is measured as a function of time. The *creep strength* is the stress that results in a specific creep rate, usually 0.001% or 0.0001% per hour. The *rupture strength*, determined from a *stress-rupture test*, is the stress that results in a failure after a given amount of time, usually 100, 1000, or 10,000 hours.

If strain is plotted as a function of time, three different curvatures will be apparent following the initial elastic extension.[19] (See Fig. 48.27.) During the first stage, the *creep rate* ($d\epsilon/dt$) decreases since strain hardening (dislocation generation and interaction with grain boundaries and other barriers) is occurring at a greater rate than annealing (annihilation of dislocations, climb, cross-slip, and some recrystallization). This is known as *primary creep*.

During the second stage, the creep rate is constant, with strain hardening and annealing occurring at the same rate. This is known as *secondary creep* or *cold flow*. During the third stage, the specimen begins to neck down, and rupture eventually occurs. This region is known as *tertiary creep*.

The secondary creep rate is lower than the primary and tertiary creep rates. The secondary creep rate, represented by the slope (on a log-log scale) of the line during the second stage, is temperature and stress dependent. This slope increases at higher temperatures and stresses.

**Figure 48.27** *Stages of Creep*

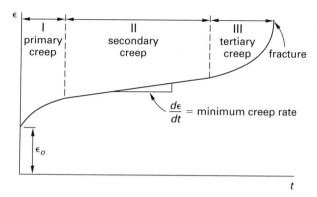

The creep rate curve can be represented by the following empirical equation, known as *Andrade's equation*.

$$\epsilon = \epsilon_o(1 + \beta t^{1/3})e^{kt} \qquad 48.33$$

Dislocation climb (glide and creep) is the primary creep mechanism, although diffusion creep and grain boundary sliding also contribute to creep on a microscopic level. On a larger scale, the mechanisms of creep involve slip, subgrain formation, and grain-boundary sliding. (See Fig. 48.28.)

**Figure 48.28** *Effect of Stress on Creep Rates*

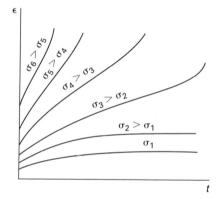

## 23. EFFECTS OF IMPURITIES AND STRAIN ON MECHANICAL PROPERTIES

Anything that restricts the movement of dislocations will increase the strength of metals and reduce ductility. Alloying materials, impurity atoms, imperfections, and other dislocations produce stronger materials. This is illustrated in Fig. 48.29.

Additional dislocations are generated by the plastic deformation (i.e., cold working) of metals, and these dislocations can strain-harden the metal. Figure 48.30 shows the effect of strain-hardening on mechanical properties.

---

[19]In Great Britain, the initial elastic elongation is considered the first stage. Therefore, creep has four stages in British nomenclature.

**Figure 48.29** *Effect of Impurities on Mechanical Properties*

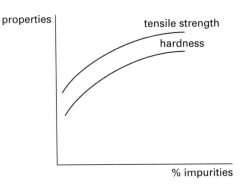

**Figure 48.30** *Effect of Strain-Hardening on Mechanical Properties*

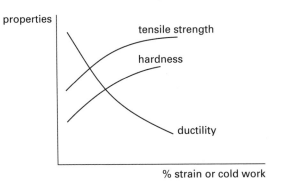

# 49 Thermal Treatment of Metals

## Nomenclature

| | | | |
|---|---|---|---|
| $A$ | atomic fraction | – | – |
| $F$ | degrees of freedom | – | – |
| $G$ | gravimetric fraction | – | – |
| $M$ | martensite transformation temperature | °F | °C |
| $M$ | molecular weight | lbm/lbmol | kg/kmol |
| $N$ | number of elements | – | – |
| $P$ | number of phases | – | – |
| $R_C$ | Rockwell C hardness | – | – |

## Subscripts

| | |
|---|---|
| $f$ | finish |
| $s$ | start |

## 1. SOLUBLE ALLOY EQUILIBRIUM DIAGRAMS

Most engineering materials are not pure elements but are alloys of two or more elements. Alloys of two elements are known as *binary alloys*. Steel, for example, is an alloy of primarily iron and carbon. Usually, one of the elements is present in a much smaller amount, and this element is known as the *alloying ingredient*. The primary ingredient is known as the *host ingredient, base metal*, or *parent ingredient*.

Sometimes, such as with alloys of copper and nickel, the alloying ingredient is 100% soluble in the parent ingredient. Nickel-copper alloy is said to be a *completely miscible alloy* or a *solid-solution alloy*.

The presence of the alloying ingredient changes the thermodynamic properties, notably the freezing (or melting) temperatures of both elements.[1] Usually the freezing temperatures decrease as the percentage of alloying ingredient is increased. Since the freezing points of the two elements are not the same, one of them will start to solidify at a higher temperature than the other. Thus, for any given composition, the alloy might consist of all liquid, all solid, or a combination of solid and liquid, depending on the temperature.

A *phase* of a material at a specific temperature will have a specific composition and crystalline structure and distinct physical, electrical, and thermodynamic properties. (In metallurgy, the word *phase* refers to more than just solid, liquid, and gas phases.)

The regions of an *equilibrium diagram*, also known as a *phase diagram*, illustrate the various alloy phases. The phases are plotted against temperature and composition. (The composition is usually a gravimetric fraction of the alloying ingredient. Only one ingredient's gravimetric fraction needs to be plotted for a binary alloy.) Sometimes, the amount of alloying ingredient is specified in *atomic fraction* or *atomic percent*. The conversions between gravimetric fractions, $G_A$ and $G_B$, and atomic fractions, $A_A$ and $A_B$, depend on the ratio of the molecular weights, $M_A$ and $M_B$.

$$A_A = \frac{M_B G_A}{M_B G_A + M_A G_B} \qquad 49.1$$

$$A_B = \frac{M_A G_B}{M_B G_A + M_A G_B} \qquad 49.2$$

$$G_A = \frac{M_A A_A}{M_A A_A + M_B A_B} \qquad 49.3$$

$$G_B = \frac{M_B A_B}{M_A A_A + M_B A_B} \qquad 49.4$$

The equilibrium conditions do not occur instantaneously, and an equilibrium diagram is applicable only to the case of slow cooling.

Figure 49.1 is an equilibrium diagram for copper-nickel alloy. (Most equilibrium diagrams are much more complex.) The *liquidus line* is the boundary above which no solid can exist. The *solidus line* is the boundary below which no liquid can exist. The area between these two

---

[1]The term *freezing point* or *melting point* is used depending on whether heat is being removed or added, respectively.

lines represents a mixture of solid and liquid phase materials.

**Figure 49.1** Copper-Nickel Phase Diagram

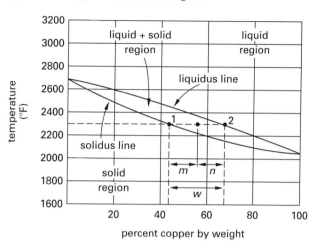

Curve (a) in Fig. 49.2 is a *time-temperature* or *temperature-time curve* for a pure metal cooling from liquid to solid state. At a particular point, the temperature remains constant (i.e., there is a *thermal arrest*). This temperature is the *freezing point* of the liquid, indicated on the graph by a horizontal line known as a *shelf* or *plateau*. The metal continues to lose heat energy—its *heat of fusion*—as the phase change from liquid to solid occurs.

**Figure 49.2** Time-Temperature Curves

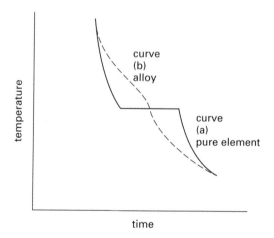

With an alloy of two elements, it is logical to expect two plateaus, since the two elements solidify at different temperatures. However, there is a range of temperatures over which the solidification occurs, and the transformation curve is smooth with an inflection point. This is illustrated by curve (b) in Fig. 49.2.

If the time-temperature curve is plotted for various compositions, the transition temperature will vary with

proportions of the constituents. The locus of the curve's inflection points coincides with the liquidus and solidus lines in the equilibrium diagram (see Fig. 49.3).

**Figure 49.3** Time-Temperature Curves for a Binary Alloy

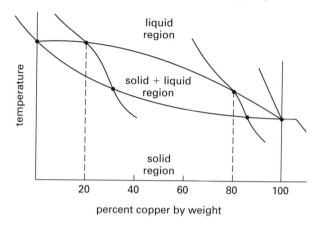

## 2. THE LEVER RULE

Within the liquid-solid region, the percentage of solid and liquid phases is a function of temperature and composition. Near the liquidus line, there is very little solid phase. Near the solidus line, there is very little liquid phase. The *lever rule* is used to find the relative amounts of solid and liquid phase at any composition. These percentages are given in fraction (or percent) by weight.

Figure 49.1 shows an alloy with an average composition of 55% copper at 2300°F. (A horizontal line representing different conditions at a single temperature is known as a *tie line*.) The liquid composition is defined by point 2, and the solid composition is defined by point 1.

The fractions of solid and liquid phases depend on the distances $m$, $n$, and $w$ (equal to $m + n$), which are measured using any convenient scale. (Although the distances can be measured in millimeters or tenths of an inch, it is more convenient to use the percentage alloying ingredient scale. This is illustrated in Ex. 49.1.) Then, the fractions of solid and liquid can be calculated from Eq. 49.5 and Eq. 49.6.

$$\text{fraction solid} = \frac{n}{w} = 1 - \text{fraction liquid} \qquad \textit{49.5}$$

$$\text{fraction liquid} = \frac{m}{w} = 1 - \text{fraction solid} \qquad \textit{49.6}$$

The lever rule and method of determining the composition of the two components are applicable to any solution or mixture, liquid or solid, in which two phases are present.

### Example 49.1

A mixture of 55% copper and 45% nickel exists at 2300°F. What are the fractions of solid and liquid phases and the compositions of each?

*Solution*

Referring back to Fig. 49.1, the solid portion of the mixture will have the composition at point 1 (44% copper), while the liquid will be at composition 2 (68% copper).

The phase fractions are

$$\text{fraction solid} = \frac{68\% - 55\%}{68\% - 44\%} = 0.54 \ (54\%)$$

$$\text{fraction liquid} = 1.00 - 0.54 = 0.46 \ (46\%)$$

## 3. EUTECTIC ALLOY EQUILIBRIUM DIAGRAMS: PARTIAL SOLUBILITY

Just as only a limited amount of salt can be absorbed by water, there are many instances where a limited amount of the alloying ingredient can be absorbed by the solid mixture. The elements of a binary alloy may be completely soluble in the liquid state but only partially soluble in the solid state.

When the alloying ingredient is present in amounts above the maximum solubility percentage, the alloying ingredient precipitates out. In aqueous solutions, the precipitate falls to the bottom of the container. In metallic alloys, the precipitate remains suspended as pure crystals dispersed throughout the primary metal.

Figure 49.4 is typical of an equilibrium diagram for ingredients displaying a limited solubility.

**Figure 49.4** *Equilibrium Diagram of a Limited Solid Solubility Alloy*

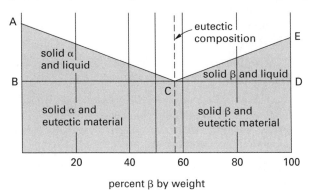

In chemistry, a *mixture* is different from a *solution*. Salt in water forms a solution. Sugar crystals mixed with salt crystals form a mixture. An alloy consisting of a mixture of two solid ingredients with a melting point lower than the melting point of either ingredient is known as a *eutectic alloy*.

In Fig. 49.4, the components $\alpha$ and $\beta$ are perfectly miscible at point C only. This point is known as the *eutectic composition*. The material in the region ABC consists of a mixture of solid component $\alpha$ crystals in a liquid of components $\alpha$ and $\beta$. This liquid is known as

the *eutectic material*, and it will not solidify until the line B–D (the *eutectic line*, *eutectic point*, or *eutectic temperature*) is reached, the lowest point at which the eutectic material can exist in liquid form.[2]

Since the two ingredients do not mix, reducing the temperature below the eutectic line results in crystals (layers or plates) of both pure ingredients forming. This is the microstructure of a solid eutectic alloy: alternating pure crystals of the two ingredients. Since two solid substances are produced from a single liquid substance, the process could be written in chemical reaction format as: liquid $\rightarrow \alpha + \beta$. (Alternatively, upon heating the reaction would be: $\alpha + \beta \rightarrow$ liquid.) For this reason, the phase change is called a *eutectic reaction*.

There are similar reactions involving other phases and states. Table 49.1 and Table 49.2 illustrate these.

**Table 49.1** *Types of Equilibrium Reactions*

| reaction name | type of reaction upon cooling |
|---|---|
| eutectic | liquid $\rightarrow$ solid $\alpha$ + solid $\beta$ |
| peritectic | liquid + solid $\alpha$ $\rightarrow$ solid $\beta$ |
| monotectic | liquid $\alpha$ $\rightarrow$ liquid $\beta$ + solid $\alpha$ |
| eutectoid | solid $\gamma$ $\rightarrow$ solid $\alpha$ + solid $\beta$ |
| peritectoid | solid $\alpha$ + solid $\gamma$ $\rightarrow$ solid $\beta$ |

**Table 49.2** *Typical Appearance of Equilibrium Diagram at Reaction Points*

| reaction name | phase reaction | phase diagram |
|---|---|---|
| eutectic | $L \rightarrow \alpha(s) + \beta(s)$ cooling | |
| peritectic | $\alpha(s) + L \rightarrow \beta(s)$ cooling | |
| eutectoid | $\gamma(s) \rightarrow \alpha(s) + \beta(s)$ cooling | |
| peritectoid | $\alpha(s) + \gamma(s) \rightarrow \beta(s)$ cooling | |

## 4. GIBBS PHASE RULE

The *Gibbs phase rule* defines the relationship between the number of phases and elements in an equilibrium mixture. For such an equilibrium mixture to exist, the alloy must have been slowly cooled, and thermodynamic equilibrium must have been achieved along the way. At equilibrium, and considering both temperature and

---

[2]The term *point* usually can be interpreted as temperature. Thus, the eutectic point really refers to the eutectic temperature.

pressure to be independent variables, the Gibbs phase rule is

$$P + F = N + 2 \qquad \textit{49.7}$$

$P$ is the number of phases existing simultaneously; $F$ is the number of independent variables, known as *degrees of freedom*; and $N$ is the number of elements in the alloy. Composition, temperature, and pressure are examples of degrees of freedom that can be varied.

For example, if water is to be stored in a condition where three phases (solid, liquid, gas) are present simultaneously, then $P = 3$, $N = 1$, and $F = 0$. That is, neither pressure nor temperature can be varied. This state corresponds to the *triple point* of water.

If pressure is constant, then the number of degrees of freedom is reduced by one, and the Gibbs phase rule can be rewritten as

$$P + F = N + 1 \Big|_{\text{constant pressure}} \qquad \textit{49.8}$$

If the Gibbs rule predicts $F = 0$, then an alloy can exist in only one composition.

## 5. ALLOTROPIC CHANGES IN STEEL

*Allotropes* have the same compositions but different atomic structures (microstructures), volumes, electrical resistances, and magnetic properties. In the case of iron, *allotropic changes* are reversible changes that occur at the *critical points* (i.e., *critical temperatures*). (See Table 49.3.)

Iron exists in three allotropic forms: alpha-iron, delta-iron, and gamma-iron. The changes are brought about by varying the temperature of the iron. As shown in Fig. 49.5, heating pure iron from room temperature changes its structure from body-centered cubic (BCC) to face-centered cubic (FCC) and then back to body-centered cubic.

**Figure 49.5** *Allotropic Changes of Iron*

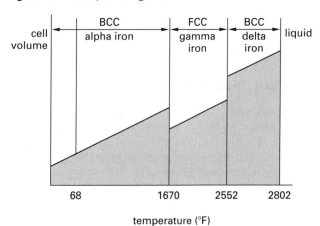

*Alpha-iron*, also known as *ferrite*, is a BCC structure that exists only below the $A_3$ line (defined as follows). The maximum carbon solubility is 0.03%, the lowest of all three allotropic forms. Alpha-iron is stable from $-460°F$ to $1674°F$ ($-273°C$ to $912°C$), soft, and strongly magnetic up to approximately $1414°F$ ($768°C$). (*Beta-iron* is a nonmagnetic form of BCC alpha-iron that exists between $1418°F$ and $1674°F$ ($770°C$ and $912°C$). The distinction between alpha- and beta-iron is not usually made.)

*Gamma-iron* is an FCC arrangement of iron atoms, stable between $1674°F$ and $2541°F$ ($912°C$ and $1394°C$), and nonmagnetic. The maximum carbon solubility for solid iron is 2.11% (also reported as 2.08%, 2.03%, and 1.7%).

*Delta-iron* is a BCC form of iron existing above $2541°F$ ($1394°C$).

**Table 49.3** *Allotropic Points for Pure Iron (upon heating)*

$1674°F$ ($912°C$): alpha (BCC) to gamma (FCC) transition
$2541°F$ ($1394°C$): gamma (FCC) to delta (BCC) transition
$2800°F$ ($1538°C$): delta (BCC) to liquid transition

Since allotropic changes occur at differing temperatures in an iron-carbon mixture dependent on composition, there are *critical lines* but no critical points. Depending on the authority, these critical lines may be labeled $A_c$, $A_r$, or just A.[3] Refer to Fig. 49.6.

- $A_0$: the critical line, about $410°F$ ($210°C$), above which cementite becomes nonmagnetic

- $A_1$: the so-called *lower critical point*, or *eutectoid temperature*, $1333°F$ ($723°C$) and 0.8% carbon, a line above which the austenite-to-ferrite and -cementite transformation occurs

- $A_2$: the critical line, about $1418°F$ ($770°C$), below which the alloy becomes magnetic[4]

- $A_3$: the so-called *upper critical point*, or critical line forming a division between austenite (above) and ferrite (below), with the actual temperature being dependent on composition. (See Fig. 49.6.)

- $A_4$: the critical point, $2541°F$ ($1394°C$), at which the gamma-delta transformation occurs

In some parts of the iron-carbon diagram, two critical lines coincide. For example, the tie line ($1333°F$ or $723°C$) for more than 0.8% carbon is labeled $A_{1,3}$ or $A_{13}$.

---

[3]Labeling the critical lines as $A_{c1}$, $A_{c2}$, and so on is a reference to the French word *chauffage* ("heating"). Such critical temperatures are encountered if iron is slowly heated from room temperature. Critical lines labels of $A_{r1}$, $A_{r2}$, and so on refer to the French word *refroidissement* ("cooling"), as such critical temperatures are observed upon cooling the iron back to room temperature. As the temperatures ($A_{c1}$ and $A_{r1}$, etc.) are approximately the same, the distinction is not always made, as it is not in this book.

[4]The $A_2$ temperature $1418°F$ ($770°C$) is also known as the *Curie point*.

**Figure 49.6** *Iron-Carbon Diagram*

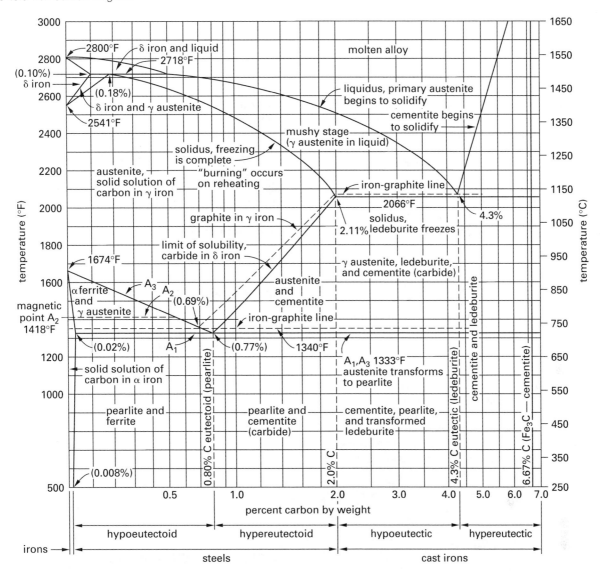

## 6. THE IRON-CARBON DIAGRAM

The *iron-carbon phase diagram* is much more complex than idealized equilibrium diagrams due to the existence of many different phases. Each of these phases has a different microstructure and, therefore, different mechanical properties. By treating the steel in a manner that forces the occurrence of particular phases, steel with desired wear and endurance properties can be produced.

Iron-carbon mixtures are categorized into *steel* (less than 2% carbon) and *cast iron* (more than 2% carbon) according to the amounts of carbon in the mixtures. Iron-carbon alloys are further classified as follows.

- *steel:* iron alloy with less than 2.0% carbon

  *hypoeutectoid steel:* iron alloy with less than 0.8% carbon, consisting of ferrite and pearlite

  *eutectoid steel:* equilibrium iron alloy with 0.8% carbon, consisting of ferrite and pearlite

  *hypereutectoid steel:* iron alloy with 0.8% to 2.0% carbon, consisting of cementite and pearlite

- *cast iron:* iron alloy with more than 2% carbon

  *hypoeutectic cast iron:* iron alloy with 2.0% to 4.3% carbon

  *eutectic cast iron:* iron alloy with 4.3% carbon

  *hypereutectic cast iron:* iron alloy with more than 4.3% carbon

The most important eutectic reaction in the iron-carbon system is the formation of a solid mixture of austenite and cementite at approximately 2065°F (1129°C). *Austenite* is a solid solution of carbon in gamma-iron. It is nonmagnetic, decomposes on slow cooling, and does not normally exist below 1333°F (723°C), though it can be partially preserved by extremely rapid cooling.

*Cementite* ($Fe_3C$), also known as *carbide* or *iron carbide*, has approximately 6.67% carbon. Cementite is the hardest of all forms of iron, has low tensile strength, and is quite brittle. Cementite ceases to be magnetic above the $A_0$ line.

The most important eutectoid reaction in the iron-carbon system is the formation of *pearlite* from the decomposition of austentite at approximately 1333°F (723°C). Pearlite is actually a mixture of two solid components, ferrite and cementite, with the common *lamellar (layered) appearance*. (The name pearlite is derived from similarity in appearance to mother-of-pearl.)

*Ferrite* is essentially pure iron (less than 0.025% carbon) in BCC alpha-iron structure. It is magnetic and has properties complementary to cementite, since it has low hardness, high tensile strength, and high ductility.

### Example 49.2

An iron alloy at 1500°F contains 2.0% carbon by weight. (a) How much carbon is present in each of the phases? (b) What are the percentages of austentite and cementite in the mixture?

*Solution*

(a) Referring to the iron-carbon diagram, Fig. 49.6, compositions of the phases are read from the carbon contents at the intersection of the 1500°F tie line and the phase boundary lines. The austenite will have the composition at point 1 (approximately 1.08% carbon), while the cementite will be at composition 2 (6.67% carbon).

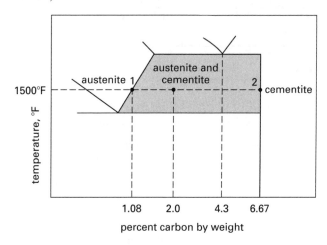

(b) The phase fractions are

$$\text{fraction austenite} = \frac{6.67\% - 2.0\%}{6.67\% - 1.08\%} = 0.835 \ (83.5\%)$$

$$\text{fraction cementite} = \frac{2.0\% - 1.08\%}{6.67\% - 1.08\%} = 0.165 \ (16.5\%)$$

## 7. QUENCHING AND RATES OF COOLING

As steel is heated, the grain sizes remain the same until the $A_1$ line is reached. Between the $A_1$ and $A_2$ lines, the average grain size of austenite in solution decreases.

This characteristic is used in heat treatments wherein steel is heated and then quenched. The *quenching* can be performed with gases (usually air), oil, water, or brine. Agitation or spraying of these fluids during the quenching process increases the severity of the quenching.

The *rate of cooling* determines the hardness and ductility. Rapid cooling in water or brine is necessary to quench low- and medium-carbon steels, since steels with small amounts of pearlite are difficult to harden. Oil is used to quench high-carbon and alloy steel or parts with nonuniform cross sections (to prevent warping). Figure 49.7 illustrates the relative rates of cooling for different quenching media.

**Figure 49.7** *Relative Cooling Rates for Different Quenching Media*

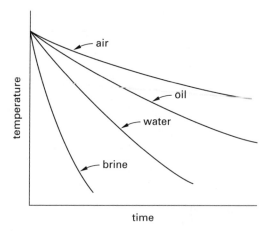

## 8. TTT AND CCT CURVES

*Controlled-cooling-transformation (CCT) curves* and *time-temperature-transformation (TTT) curves* are used to determine how fast an alloy should be cooled to obtain a desired microstructure. Although these curves show different phases, they are not equilibrium diagrams. On the contrary, they show the microstructures that are produced with controlled temperatures or when quenching interrupts the equilibrium process.

TTT curves are determined under ideal, isothermal conditions. For that reason, they are also known as *isothermal transformation diagrams*. CCT curves are experimentally determined under conditions of continuous cooling. Therefore, CCT curves are better suited for designing cooling processes. However, TTT curves are more readily available than CCT curves and are used in lieu of them. Both curves are similar in shape, although the CCT curves are displaced downward and to the right from TTT curves.

Figure 49.8 shows a TTT diagram for a high-carbon (0.80% carbon or more) steel. Curve 1 represents extremely rapid quenching. The transformation begins at 420°F (216°C) and continues for 8–30 seconds, changing all of the austenite to martensite. Such a material is seldom used because martensite has almost

*Figure 49.8* *TTT Diagram for High-Carbon Steel*

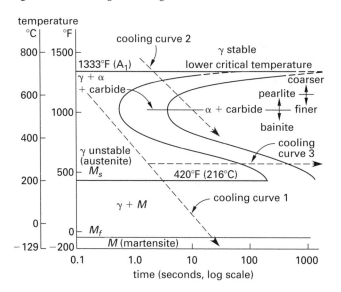

no ductility.

Curve 2 is a slower quench that converts all of the austenite to fine pearlite. This corresponds to a *normalizing process*.

A horizontal line below the critical temperature is a *tempering process*. If the temperature is decreased rapidly along curve 1 to 520°F (270°C) and is then held constant along cooling curve 3, *bainite* is produced. This is the principle of *austempering*. Performing the same procedure at 350°F to 400°F (180°C to 200°C) is *martempering*, which produces *tempered martensite*, a soft and tough steel.

The austenite-martensite transformation is extremely important, and the temperatures at which the transition starts and finishes are sometimes referred to as *critical temperatures*.

- $M_s$: the temperature at which austenite first begins to transform into martensite

- $M_f$: the temperature at which austenite is fully transformed into martensite

## 9. STEEL HARDENING PROCESSES

*Hard steel* resists plastic deformation. Steel is hard if it has a homogeneous, austenitic structure with coarse grains. Some steels (e.g., those that have little carbon) are difficult to harden. The *hardenability* of a steel

specimen can be determined in a standard *Jominy end-quench test*.

The basic hardening processes consist of heating to approximately 100°F (50°C) above the $A_3$ critical line, allowing austenite to form, and then quenching rapidly. Hardened steel consists primarily of martensite or bainite. *Martensite* is a supersaturated solution of carbon in alpha-iron.[5] *Bainite* is not as hard as martensite, but it does have good impact strength and fairly high hardness. Neither martensite nor bainite are equilibrium substances—they are not found on the iron-carbon equilibrium diagram but are formed during the quenching operation.

The maximum hardness obtained depends on the carbon content. An upper limit of $R_C$ 66–67 is reached with approximately 0.5% carbon. No further increase

*Figure 49.9* *Maximum Hardness of Steel*

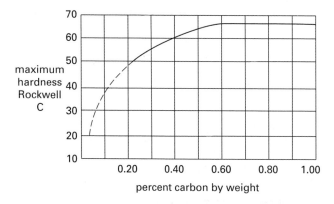

in hardness is achieved by increasing the carbon content. Since hardening is accompanied by a decrease in toughness, it is usually followed by tempering.

There are many steel-hardening processes with special names, listed alphabetically as follows.

- *austempering:* an interrupted quenching process resulting in an austenite-to-bainite transition. Steel is quenched to below approximately 800°F (430°C) but above 400°F (200°C), and is allowed to reach equilibrium. No martensite is formed, and further tempering is not required.

- *austenitizing:* quenching after heating above the $A_3$ line (for steel with up to 0.8% carbon) or above the $A_1$ line (for steel with more than 0.8% carbon).

- *martempering:* an interrupted quenching process resulting in an austenite-to-(tempered) martensite transition. Steel is quenched to below 400°F (200°C) and allowed to reach equilibrium. Further tempering is not required.

---

[5]The carbon in martensite distorts the BCC structure of iron. The distorted BCC lattice is known as a *body-centered tetragonal* (BCT) structure.

## 10. HEAT TREATING CAST IRON

Hardness in low-carbon steels results from the presence of martensite or bainite—supersaturated solutions of carbon in alpha-iron that begin as austenite at higher temperatures and are "frozen" in place by rapid cooling. Cast iron starts as iron carbide (cementite), not austenite, and cast iron is not normally hardened by heating and quenching. Hardness in cast irons is primarily obtained by including alloying ingredients that promote the formation and retention of iron carbide ($Fe_3C$). Although it is hard, iron carbide is also very brittle.

Heating can actually reduce hardness in cast iron. Iron carbide dissociates into iron and graphite at high temperatures. Graphite in flake form (*gray cast iron*) and in spheroidal form (*nodular cast iron*) greatly decreases hardness. The presence of silicon in cast iron greatly affects hardness, since silicon promotes the formation of graphite. Cast irons typically contain $1\frac{1}{2}\%$ or more silicon, and less than 1% is needed to ensure that the carbon in iron carbide dissociates into iron and graphite. Cast iron with no graphite is known as *white cast iron*. Cast iron that has been heated and slowly cooled to permit graphite to form is known as *malleable cast iron*.

Small castings (e.g., with dimensions less than 3 in or 4 in) or white cast iron contain less silicon and can be hardened by rapid quenching. This occurs because rapid cooling prevents the dissociation of iron carbide into iron and graphite. However, the interiors of larger castings cannot be cooled fast enough to prevent the formation of weakening graphite.

## 11. PROPERTIES VERSUS GRAIN SIZE

Many properties are related to grain size, which initially depends on composition but can be changed by heat treatment. Coarse-grained structures have less toughness and ductility but have greater machineability and case hardenability.

As low-carbon steels are heated from room temperature, the grain size remains constant up to the $A_1$ line. Above the $A_1$ line, ferrite and pearlite are transformed into austenite, and the grain size decreases. The grain size is minimum at the upper critical line, $A_3$, and then increases again as the steel is heated above the $A_3$ line.

Aluminum in small quantities is an important alloying ingredient in steel. As a deoxidizer, it raises the temperature at which rapid grain growth takes place. In steels that have been deoxidized with aluminum (e.g., medium-carbon and alloy steels), no grain growth occurs until the *coarsening temperature* is reached, which is well above the critical temperature.

## 12. RECRYSTALLIZATION

*Recrystallization* can be used with all metals to relieve stresses induced during cold working. It involves heating the material in a furnace to a specific temperature (the *recrystallization temperature*) and holding it there for a long time. This induces the formation and growth of strain-free grains within the grains already formed. The resulting microstructure is essentially the microstructure that existed before any cold working but is softer and more ductile than the cold-worked microstructure.

The recrystallization process is more sensitive to temperature than it is to exposure time. Recrystallization will occur naturally over a wide range of temperatures; however, the reaction rates increase at higher temperatures. Table 49.4 lists approximate temperatures that will produce complete recrystallization in one hour.

**Table 49.4** *Approximate Recrystallization Temperatures*

| material | recrystallization temperature | |
|---|---|---|
| | °F | °C |
| copper (99.999% pure) | 250 | 120 |
| (5% zinc) | 600 | 315 |
| (5% aluminum) | 550 | 290 |
| (2% beryllium) | 700 | 370 |
| aluminum (99.999% pure) | 175 | 80 |
| (99.0+% pure) | 550 | 290 |
| (alloys) | 600 | 315 |
| nickel (99.99% pure) | 700 | 370 |
| (99.4% pure) | 1100 | 590 |
| (monel metal) | 1100 | 590 |
| iron (pure) | 750 | 400 |
| (low-carbon steel) | 1000 | 540 |
| magnesium (99.99% pure) | 150 | 65 |
| (alloys) | 450 | 230 |
| zinc | 50 | 10 |
| tin | 25 | −4 |
| lead | 25 | −4 |

## 13. STRESS RELIEF PROCESSES FOR STEEL

The annealing, normalizing, and tempering processes are used to relieve the internal stresses, refine the grain size, and soften the material (to improve machineability). The high temperatures used in these processes allow some of the carbon to migrate out of the martensite, thereby relieving stresses in the crystalline structure. (See Fig. 49.10.)

The basic *full annealing* process involves heating to approximately 100°F (50°C) above the critical $A_3$ point, allowing austenite to form fully, and then cooling slowly in a furnace to produce coarse pearlite. Three separate stages constitute annealing. First, the material is stress-relieved by heating in the *recovery stage*. Next, during the *recrystallization stage*, new crystals form within the existing distorted structure. Finally, during the *grain growth stage*, some of the crystals grow in size (by eliminating smaller grains).

The *partial annealing process*, also known as *process annealing, spheroidize-annealing*, or just *spheroidizing*, also softens the material and relieves stresses, but heating is to below the $A_3$ point. The term *spheroidizing* gets its name from the spherical cementite particles that appear in the ferrite matrix, as opposed to the lamellar structure of cementite and ferrite in pearlite.[6]

**Figure 49.10** *Products of Cooling and Reheating Iron-Carbon Alloys*

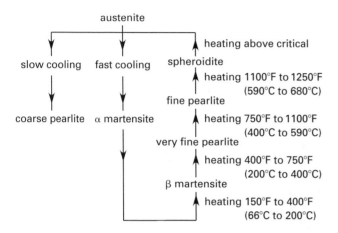

*Normalizing* is similar to annealing but is more rapid than furnace cooling because it uses air cooling. Heating is to approximately 200°F (100°C) higher than the critical point, generally about 1650°F (900°C). The typical air cooling rate is 100 °F/min (50 °C/min). Normalizing produces a harder and stronger steel than full annealing.

*Tempering*, also known as *drawing* or *toughening*, is used with hypoeutectoid steels to change martensite into pearlite. It is used after hardening to produce softer and tougher steel. The steel is heated to below its critical temperature. However, the higher the temperature, the more soft and ductile the steel becomes.

To avoid *blue embrittlement* (named after the resulting blue surface finish) in steel, tempering should not be done between 450°F and 700°F (230°C and 370°C). Within this range, the *notch toughness* of the steel (as determined from an impact test), is lowered considerably. The reason for this effect has been traced to free nitrogen in the steel.

## 14. COLD WORKING VERSUS HOT WORKING

As a material is worked and the dislocations move, plastic strain builds up. If the strain occurs at a high enough temperature (i.e., above the recrystallization temperature), there will be sufficient thermal energy to anneal out the lattice distortions. Forming operations above the recrystallization temperature are known as *hot working*, since annealing occurs simultaneously with the plastic forming. The material remains ductile.

[6]Although the cementite concentrations have changed in shape, the phase is the same: cementite and ferrite.

If the plastic forming takes place at a low temperature (i.e., below the recrystallization temperature), there will be insufficient thermal energy to anneal out the dislocations. The material will become progressively stronger, harder, and more brittle until it eventually fails. This is known as *cold working*.

## 15. HARDENING OF NONALLOTROPIC ALLOYS

The properties of nonferrous substances that do not readily form allotropes cannot be changed by rapid heating followed by controlled cooling. Such substances are known as *nonallotropic alloys* and include aluminum, copper, and magnesium alloys as well as stainless steels containing nickel.

The primary method of hardening nonallotropic alloys is *solution heat treatment*, which consists of two or three steps: precipitation, quenching, and (optionally) artificial aging. Because of these steps, solution heat treating is also known as *precipitation hardening* and *age hardening*.

Precipitation involves the formation of a new crystalline structure through the application of controlled quenching and tempering. Precipitation disperses hard particles throughout the existing more ductile material. These particles disrupt the long dislocation planes of the material, restricting the movement of dislocations and increasing the strength and stiffness of the alloy. The ultimate strength is raised to the rupture strength of either the particles or the surrounding matrix.

Solution heat treatment culminates in rapid quenching. Quenching speeds must be consistent with the size of the object. Massive specimens may require slower processes that use oil or boiling water.

The final step is to hold the material at a specific temperature for a given amount of time. This is known as *aging* or *artificial aging*. Post-treatment cooling for precipitation hardening is relatively unimportant.

It is important not to over-age aluminum. If the precipitation process goes on too long, the precipitates will not be effective in strengthening the material. Precipitation hardening is optimum at the point where the particles are just starting to form.

Table 49.5 lists some of the more common *temper designations* for 2XXX-, 6XXX-, and 7XXX-series aluminum alloys that can be precipitation hardened. For example, 2024-T4 is a widely used alloy having strength and toughness when hardened and aged.

## 16. SURFACE HARDENING

Often, it is desirable to have a hard (wear-resistant) outer surface with a ductile interior. This combination is needed when the product is subjected to fatigue. There are several processes used to *surface harden* (also known as *case harden* and *differential harden*) steel.

**Table 49.5** *Aluminum Tempers*

| temper | description |
|---|---|
| T2 | annealed (castings only) |
| T3 | solution heat-treated, followed by cold working |
| T4 | solution heat-treated, followed by natural aging |
| T5 | artificial aging only |
| T6 | solution heat-treated, followed by artificial aging |
| T7 | solution heat-treated, followed by stabilizing by overaging heat treating |
| T8 | solution heat treated, followed by cold working and subsequent artificial aging |

- *boron diffusion:* exposure to boron (a powerful hardening ingredient) at low temperatures; slow but distortion-free; suitable for high carbon, spring, and tool steels as well as bonded steel carbides and some age-hardenable alloys.

- *carburizing:* heating for up to 24 hours at approximately 1650°F (900°C) in contact with a carbonaceous material (usually carbon monoxide, CO, gas), followed by rapid cooling. Carburizing is used for steels with less than 0.2% carbon. Carburizing is also known as *cementation.*

- *cyaniding:* heating at 1700°F (925°C) in a cyanide-rich atmosphere or immersion in a cyanide salt bath.

- *flame hardening:* supplying flame heat at the surface in quantities and rates higher than can be conducted into the material's interior, followed by drastic spray quenching. Typically used with steels containing more than 0.4% carbon.

- *induction hardening:* using high-frequency electric currents to heat the metal surface, followed by normal quenching. Typically used with steels containing more than 0.4% carbon.

- *nitriding:* heating at 1000°F (540°C) for up to 100 hours (but usually less than 70 hours) in an ammonia atmosphere, followed by slow cooling (no quenching required).

## 17. SHOT-PEENING

*Shot-peening* is the "bombardment" of a metal surface by high-velocity particles (e.g., hard steel shot). As each particle strikes, the target's surface stretches and deforms plastically, creating residual compressive stresses at the surface. The induced compressive stress from shot-peening removes tensile stresses left over from manufacturing operations and offsets the effects of applied tensile operating loads. In gears, the compressive layer improves load-carrying capacity by increasing the bending fatigue strength of the teeth.[7] A 20% improvement in both strength (i.e., endurance limit) and wear is typical for shot-peened parts.

---

[7]Fatigue failure never starts in an area under compressive stress.

# 50 Manufacturing Processes

## Nomenclature

| | | | |
|---|---|---|---|
| $A$ | area | ft$^2$ | m$^2$ |
| $b$ | chip width | ft | m |
| $c$ | specific heat | Btu/lbm-°F | J/kg·°C |
| $d$ | depth of cut | ft | m |
| $D$ | diameter | ft | m |
| $E$ | energy | ft-lbf | J |
| $f$ | feed rate | ft/rev | m/rev |
| $F$ | force | lbf | N |
| $g$ | gravitational acceleration | ft/sec$^2$ | m/s$^2$ |
| $g_c$ | gravitational constant | lbm-ft/lbf-sec$^2$ | n.a. |
| $h$ | height | ft | m |
| $J$ | Joule's constant (778.17) | ft-lbf/Btu | n.a. |
| $J$ | rotational moment of inertia | lbm/ft$^2$ | kg·m$^2$ |
| $L$ | length | ft | m |
| $m$ | mass | lbm | kg |
| $n$ | constant | – | – |
| $n$ | rotational speed | rpm | rpm |
| $N$ | number of items | – | – |
| $p$ | pressure | lbf/ft$^2$ | Pa |
| $P$ | power | ft-lbf/min | W |
| $r$ | chip thickness ratio | – | – |
| $r$ | radius | ft | m |
| $S$ | strength | lbf/ft$^2$ | Pa |
| $t$ | thickness | ft | m |
| $t$ | time | sec | s |
| $T$ | temperature | °F | °C |
| $T$ | time | sec | s |
| $U$ | specific cutting energy | ft-lbf/ft$^3$ | J/m$^3$ |
| v | cutting speed | ft/min | m/s |
| $V$ | volume | ft$^3$ | m$^3$ |
| $Z_w$ | metal removal rate | ft$^3$/min | m$^3$/s |

## Symbols

| | | | |
|---|---|---|---|
| $\alpha$ | true rake angle | deg | deg |
| $\beta$ | rake face resultant angle | deg | deg |
| $\theta$ | clearance angle | deg | deg |
| $\rho$ | density | lbm/ft$^3$ | kg/m$^3$ |
| $\sigma$ | normal stress | lbf/ft$^2$ | Pa |
| $\tau$ | shear stress | lbf/ft$^2$ | Pa |
| $\phi$ | shear angle | deg | deg |
| $\omega$ | wedge angle | deg | deg |
| $\omega$ | rotational speed | rad/sec | rad/s |

## Subscripts

| | |
|---|---|
| $c$ | chip |
| $f$ | final |
| $h$ | horizontal |
| $n$ | normal |
| $o$ | original (undeformed) |
| $p$ | constant pressure |
| $R$ | resultant |
| $s$ | shear |
| $t$ | tangent |
| $u$ | ultimate |
| $v$ | vertical |

## 1. CHIP FORMATION

One of the most common ways that a workpiece can be shaped is by removing material through chip-forming operations such as turning, drilling, multitooth operations (e.g., milling, broaching, sawing, and filing), and specialized processes, such as thread and gear cutting.

The toughness of a workpiece can be determined from the nature of the chips produced. Brittle materials produce discrete fragments, known as *discontinuous chips*, *segmented chips*, or *type-one chips*. Ductile materials form long, helix-coiled string chips, known as *continuous chips* or *type-two chips*.[1] *Chip-breaker* grooves are often ground in the cutting tool face to cause long chips

---

[1]There is also a *type-three chip*, a continuous chip with a *built-up edge* (BUE), which is not discussed here.

to break into shorter, more manageable pieces. Chip formation is optimum when chips are produced in the shapes of sixes and nines.

## 2. CUTTING TOOL SPEEDS AND FORCES

Figure 50.1 shows a chip produced during *orthogonal cutting* (i.e., two-dimensional cutting). Cutting involves both compressive and shear stresses. The chip expands from its undeformed thickness, $t_o$, to $t_c$ (due to the release of compressive stress) as it slides over the cutting tool. The *chip thickness ratio*, $r = t_o/t_c$, is typically around 0.5.

***Figure 50.1*** *Tool-Workpiece-Chip Geometry*

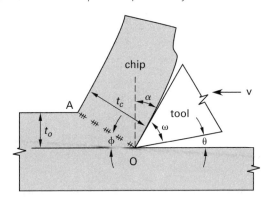

The cutting energy is minimum when the *shear angle*, $\phi$, is approximately given by Eq. 50.1. (The angle $\beta$ is shown in Fig. 50.2. It can be determined from $F_n$ and $F_t$ from toolpost dynamometer data.)

$$\phi_{\text{minimum energy}} = \frac{\pi}{4} + \frac{\alpha}{2} - \frac{\beta}{2} \qquad 50.1$$

The angle at which the tool meets the workpiece is characterized by *true rake angle*, $\alpha$. This angle has a major impact on chip formation, and it depends on the tool shape and set-up orientation of the tool. Small rake angles result in excessive compression, high tool forces, and excessive friction. Chips produced are thick, hot, and highly deformed. Conventional cutting tools are oriented with positive rake angles; sintered carbide and ceramic tools used for cutting steel are frequently designed to be oriented with negative rake angles to provide additional support of the cutting edge.

Figure 50.1 also shows the *clearance angle (relief angle)*, $\theta$, and the *wedge angle*, $\omega$. The sum of the rake, clearance, and wedge angles is $\pi/2$ ($90°$).

$$\alpha + \theta + \omega = \frac{\pi}{2} \qquad 50.2$$

The rake angle, shear angle, and chip thickness ratio are related by Eq. 50.3 and Eq. 50.4.

From Fig. 50.1,

$$r = \frac{t_o}{t_c} = \frac{\sin\phi}{\cos(\phi - \alpha)} \qquad 50.3$$

Rearranging Eq. 50.3 and applying trigonometric identities,

$$\tan\phi = \frac{r\cos\alpha}{1 - r\sin\alpha} \qquad 50.4$$

The relative velocity difference between the tool and the workpiece is the *cutting speed*, v. (See Table 50.1.) The cutting speed can be calculated from the diameter, $D$, of a rotating workpiece and the rotational speed, $n$. If $D$ is in feet and $n$ is in revolutions per minute, the cutting

***Table 50.1*** *Typical Cutting Speeds (ft/min)*

| material | high-speed steel | | carbide | |
| --- | --- | --- | --- | --- |
| | rough | finish | rough | finish |
| cast iron | 50–60 | 80–110 | 120–200 | 350–400 |
| semisteel[*] | 40–50 | 65–90 | 140–160 | 250–300 |
| malleable iron[*] | 80–110 | 110–130 | 250–300 | 300–400 |
| steel casting[*] (0.35C) | 45–60 | 70–90 | 150–180 | 200–250 |
| brass (85-5-5) | 200–300 | 200–300 | 600–1000 | 600–1000 |
| bronze (80-10-10)[*] | 110–150 | 150–180 | 600 | 1000 |
| aluminum | 400 | 700 | 800 | 1000 |
| SAE 1020[*] | 80–100 | 100–120 | 300–400 | 300–400 |
| SAE 1050[*] | 60–80 | 100 | 200 | 200 |
| stainless steel[*] | 100–120 | 100–120 | 240–300 | 240–300 |

(Multiply ft/min by $5.08 \times 10^{-3}$ to obtain m/s.)

[*]Appropriate lubricants are used to achieve listed speeds.

Used with permission from *Manufacturing Processes*, 5th Ed., by Myron L. Begeman and B. H. Amstead, published by John Wiley & Sons, Inc., copyright © 1963.

speed will be in traditional (in the United States) units of feet per minute.[2]

$$v = \pi D n \qquad 50.5$$

The velocity of the chip relative to the tool face is the *chip velocity*, $v_c$. The *shear velocity*, $v_s$, is the velocity of the chip relative to the workpiece.

$$v_c = rv = \frac{v \sin \phi}{\cos(\phi - \alpha)} \qquad 50.6$$

$$v_s = \frac{v \cos \alpha}{\cos(\phi - \alpha)} \qquad 50.7$$

Figure 50.2 illustrates that the resultant force, $F_R$, between the tool and the chip can be resolved into tangential force, $F_t$, and normal force, $F_n$, relative to the rake face of the tool, or alternatively, into a horizontal cutting force, $F_h$, and a vertical thrust force, $F_v$, relative to the cutting surface. The horizontal and vertical forces are commonly measured by a strain-gauge toolpost dynamometer. Equation 50.8 and Eq. 50.9 relate these two sets of forces.

$$F_t = F_h \sin \alpha + F_v \cos \alpha \qquad 50.8$$

$$F_n = F_h \cos \alpha + F_v \sin \alpha \qquad 50.9$$

**Figure 50.2** *Force Components in Orthogonal Cutting*

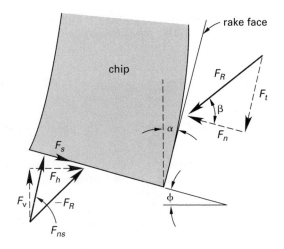

A third set of axes can be chosen relative to the shear plane. The forces parallel, $F_s$, and normal, $F_{ns}$, to the shear plane can be derived from the horizontal and vertical forces. These are also shown in Fig. 50.2.

$$F_s = F_h \cos \phi + F_v \sin \phi \qquad 50.10$$

$$F_{ns} = F_h \sin \phi + F_v \cos \phi \qquad 50.11$$

Shear stress is the primary parameter affecting the cutting energy requirement. The average shear stress, $\tau$, is $F_s$ divided by the area of the shear plane, $A_s$, which depends on the chip width, $b$.

---

[2]The symbol CS is also used for cutting speed.

$$A_s = \frac{b t_o}{\sin \phi} \qquad 50.12$$

$$\tau = \frac{F_s}{A_s} = \frac{F_s \sin \phi}{b t_o} \qquad 50.13$$

The average normal stress is

$$\sigma = \frac{F_{ns}}{A_s} = \frac{F_{ns} \sin \phi}{b t_o} \qquad 50.14$$

The energy required per unit cutting time is the cutting power, $P$.

$$P = F_h v \qquad 50.15$$

If $F_h$ is in pounds, and if v is in feet per minute, the horsepower requirement is

$$\text{cutting horsepower} = \frac{F_h v}{33{,}000} \qquad 50.16$$

The *metal removal rate*, $Z_w$, is

$$Z_w = b t_o v \qquad 50.17$$

The energy expended per unit volume removed, known as the *specific cutting energy*, is

$$U = \frac{P}{Z_w} = \frac{F_h v}{Z_w} = \frac{F_h}{b t_o} \qquad 50.18$$

For a simple lathe (turning) operation on a cylindrical workpiece of diameter $D$, the cutting time to make a cut of (horizontal) length $L$ in a single pass is

$$t_{\min} = \frac{L}{f n_{\text{rpm}}} = \frac{\pi L D}{f v} \qquad 50.19$$

### Example 50.1

A cylindrical cast steel cylinder with a diameter of 6.00 in is faced in a lathe. The procedure removes the outer 0.40 in of the bar. The lathe develops a maximum power of 20 hp. The unit power limit for this material and process is 2.0 hp-min/in$^3$. The cutting pressure is maintained at its maximum limit throughout. What is the approximate minimum time to face the bar?

*Solution*

The bar radius is

$$r = \frac{6.00 \text{ in}}{2} = 3.00 \text{ in}$$

The volume of material to be removed is

$$V = AL = \pi r^2 L = \pi (3.00 \text{ in})^2 (0.4 \text{ in})$$
$$= 11.31 \text{ in}^3$$

If the lathe develops the full 20 hp power throughout the operation, the minimum cutting time is

$$t = V \left( \frac{\text{unit power}}{\text{lathe power}} \right)$$

$$= (11.31 \text{ in}^3) \left( \frac{2.0 \; \frac{\text{hp-min}}{\text{in}^3}}{20 \text{ hp}} \right)$$

$$= 1.13 \text{ min}$$

### Example 50.2

A cylindrical workpiece with a diameter of 4.00 in is turned on a lathe. The cutting speed is 200 ft/min. The depth of cut is 0.20 in. The feed is 0.010 ipr (inches per revolution). How long will it take to make a 14 in cut?

*Solution*

The cutting time is

$$t = \frac{LD}{f\text{v}}$$

$$= \frac{(14 \text{ in})(4.00 \text{ in})}{\left(12 \; \frac{\text{in}}{\text{ft}}\right)\left(0.010 \; \frac{\text{in}}{\text{rev}}\right)\left(200 \; \frac{\text{ft}}{\text{min}}\right)}$$

$$= 2.33 \text{ min}$$

## 3. TOOL MATERIALS

*Carbon tool steel* is plain carbon steel with approximately 0.9% to 1.3% carbon, which has been hardened and tempered. It can be given a good edge but is restricted to use below 400°F to 600°F (200°C to 300°C) to prevent further tempering.

*High-speed steel* (HSS) contains tungsten or chromium and retains its hardness up to approximately 1100°F (600°C), a property known as *red hardness*. The common 18-4-1 formulation contains 18% tungsten, 4% chromium, and 1% vanadium. Other categories include *molybdenum high-speed steels* and *superhigh-speed steels*. Tools made with these steels can be run approximately twice as fast as carbon steel tools.

*Cast nonferrous* cutting tools have similar characteristics to carbides and are used in an as-cast condition. A common composition contains 45% cobalt, 34% chromium, 18% tungsten, and 2% carbon. Cast nonferrous tools are brittle but can be used up to approximately 1700°F (925°C) and operate at speeds twice that of HSS tools.

*Sintered carbides* are produced through powder metallurgy from nonferrous metals (e.g., tungsten carbide and titanium carbide with some cobalt). Carbide tools are commonly of the throw-away type. They are very hard, can be used up to 2200°F (1200°C), and operate at cutting speeds two to five times as fast as HSS tools. However, they are less tough and cannot be used where impact forces are significant.

*Ceramic tools* manufactured from aluminum oxide have the same expected life as carbide tools but can operate at speeds from two to three times higher. They operate below 2000°F (1100°C).

*Diamonds* and diamond dust are used in specific cases, usually in finishing operations.

In addition to speed and temperature considerations, there should be no possibility of welding between the chip and tool material. Diamonds, for example, are soluble in the presence of high-temperature iron. Also, aluminum oxide tools are not satisfactory for machining aluminum.

## 4. TEMPERATURE AND COOLING FLUIDS

Friction is greatly reduced in *free-machining steels* that have had sulfur added as an alloying ingredient. However, only approximately 25% of the heat developed in cutting is due to friction between the tool and the workpiece. The remainder results from compression and shear stresses. Only 20% to 40% of this heat is removed by the tool and workpiece. The remainder must be removed by the chips and cooling fluids.

If all the heat generated goes into the chip (which does not actually occur), the *adiabatic chip temperature* is given by Eq. 50.20.

$$T_c = T_o + \frac{U}{\rho c_p} \qquad \text{[SI]} \qquad 50.20(a)$$

$$T_c = T_o + \frac{U}{\rho c_p J} \qquad \text{[U.S.]} \qquad 50.20(b)$$

Cutting fluids are used to reduce friction, remove heat, remove chips, and protect against corrosion. Gases, such as air, carbon dioxide, and water vapor, can be used, but they do not remove heat well, cannot be reused, and may require an exhaust system. Water is a good heat remover, but it promotes rust. (Addition of *sal soda* to water produces an efficient, inexpensive cutting fluid that does not promote rusting.)

*Straight-cutting oils* (i.e., petroleum-based nonsoluble oils) reduce friction and do not cause rust but are less efficient at heat removal than water. Therefore, emulsions of water and oil or water-miscible fluids (soluble oils) are often used with steel. Kerosene lubricants are commonly used with aluminum.

Chlorinated or sulfurized oils are used to decrease friction.[3] Other additives are used to inhibit rust, clean the

---

[3]Chlorine and sulfur form metallic chlorides and sulfides at cutting temperatures. These compounds have low shear strength, and therefore, friction is reduced. Chlorinated oils work better at low speeds, whereas sulfurized oils work better under severe conditions.

workpiece, soften water, promote film formation, and inhibit bacterial growth.

## 5. TOOL LIFE

Tools wear and fail through abrasion, loss of hardness, and fracture. Three common types of failure are *flank wear*, *crater wear*, and *nose failure*. The life of a tool, $T$ (expressed in minutes), is the length of time it will cut satisfactorily before requiring grinding and depends on the conditions of use. The *tool life equation*, also known as *Taylor's equation*, relates cutting speed, $v$, and tool life, $T$, for a particular combination of tool and workpiece.

$$v T^n = \text{constant} \qquad 50.21$$

The exponent $n$ is an empirical constant that must be determined for each tool-workpiece setup. Typical values are 0.1 for high-speed steel, 0.2 for carbides, and 0.4 for ceramics.

Since the tool feed rate, $f$, and depth of cut, $d$, are also important parameters affecting tool life, Taylor's equation has been expanded into Eq. 50.22. (The depth of cut, $d$, is the same as the chip thickness, $t_o$, shown in Fig. 50.1.)

$$v T^n d^x f^y = \text{constant} \qquad 50.22$$

## 6. ABRASIVES AND GRINDING

*Grinding* (i.e., *abrasive machining*) is used as a finishing operation since very fine and dimensionally accurate surface finishes can be produced. However, grinding is also used for gross material removal. In fact, grinding is the only economical way to cut hardened steel.

Most modern grinding wheels are produced from aluminum oxide. However, grinding wheels can be produced from either *natural abrasives* or *synthetic abrasives*, as described in Table 50.2.

**Table 50.2** Types of Grinding Wheel Abrasives

> *natural abrasives*
>     sandstone
>     solid quartz
>     emery (50% to 60% $Al_2O_3$ plus iron oxide)
>     corundum (75% to 90% $Al_2O_3$ plus iron oxide)
>     garnet
>     diamond
> *synthetic abrasives*
>     silicon carbide, SiC
>     aluminum oxide, $Al_2O_3$
>     boron carbide

Abrasive grit size is measured by the smallest standard-size screen through which the grains will pass. *Coarse grits*, for example, will pass through #6 (i.e., having 6 uniform openings per inch) to #24 screens, inclusive,

but will be retained on any finer screen. Table 50.3 summarizes the abrasive size designations.

**Table 50.3** Abrasive Grit Sizes

| | screen sizes | |
| --- | --- | --- |
| designations | English | metric (mm) |
| coarse | #6–#24 | 4.23–1.06 |
| medium | #30–#60 | 0.847–0.423 |
| fine[*] | #70–#600 | 0.363–0.042 |

(Multiply in by 25.4 to obtain mm.)

[*]Sizes #240 through #600 are also known as *flour grit*.

*Snagging* describes very rough grinding, such as that performed in foundries to remove gates, fins, and risers from castings.

*Honing* is grinding in which very little material, 0.001 in to 0.005 in (0.025 mm to 0.13 mm) is removed. Its purpose is to size the workpiece, to remove tool marks from a prior operation, and to produce very smooth surfaces. Coolants, such as sulfurized mineral-base oils and kerosene, are used to cool the workpiece and to flush away small chips. Because the stones are moved with an oscillatory pattern, honing leaves a characteristic cross-hatch pattern.

*Lapping* is used to produce dimensionally accurate surfaces by removing less than 0.001 in (0.025 mm). Parts are lapped to produce a close fit and to correct minor surface imperfections.

After any cutting or standard grinding operation, the surface of a workpiece will consist of *smear metal* (a fragmented, noncrystalline surface). *Superfinishing* or *ultrafinishing* using light pressure, short but fast oscillations of the stone, and copious amounts of lubricant-coolant, removes the smear metal and leaves a solid crystalline metal surface. The operation is similar to honing, but the stone moves with a different motion. There is essentially no dimensional change in the workpiece.

Other nonprecision methods of abrasion can be used to improve the surface finish and to remove burrs, scale, and oxides. Such methods include buffing, wire brushing, tumbling (i.e., barrel finishing), polishing, and vibratory finishing.

*Centerless grinding* is a method of grinding that does not require clamping, chucking, or holding round workpieces. The workpiece is supported between two abrasive wheels by a work-rest blade. One wheel rotates at the normal speed and does the actual grinding. The other wheel, the *regulating wheel*, is mounted at a slight offset angle and turns more slowly. Its purpose is to rotate and position the workpiece. (See Fig. 50.3.)

**Figure 50.3** *Centerless Grinding*

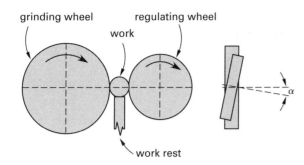

## 7. CHIPLESS (NONTRADITIONAL) MACHINING

*Electrical discharge machining* (EDM), also known as *electrodischarge machining*, *electrospark machining*, and *electronic erosion*, uses high-energy electrical discharges (i.e., sparks) to shape an electrically conducting workpiece. Thousands of controlled sparks are generated per second between a cutting head and the workpiece, while a servomechanism controls the separating gap. EDM requires the cutting to be performed in a dielectric liquid. The final cut surface consists of small craters melted by the arcs.

EDM can be used with all conductive metals, regardless of melting point, toughness, and hardness. Since there is no contact between the tool and the workpiece, delicate and intricate cutting is possible. However, the metal removal rate is low. Also, the tool material is lost much faster than the workpiece material. Wear ratios for the tool and workpiece vary between 20:1 for common brass tools to 4:1 for expensive tool materials.

*Electrochemical machining* (ECM) removes metal by electrolysis in a high-current deplating operation. Current densities of 1500 A/in² to 2000 A/in² (230 A/cm² to 310 A/cm²) are common. A tool electrode (the cathode) with the approximate profile desired to be given to the workpiece (the anode) is brought close to the workpiece. The separation is maintained by a servomechanism. A water-based electrolyte (e.g., sodium chloride solution) is forced between the tool and workpiece. The electrolyte completes the circuit and removes the free ions.

ECM shapes and cuts metal of any hardness or toughness. Relatively high (compared with EDM) metal removal rates are possible. Unlike EDM, the tool is not consumed or changed in shape. The *current efficiency* is defined as the volume of metal removed per unit energy used. Typical units are cubic inches per 1000 ampere-minutes.

*Electrochemical grinding* (ECG), also known as *electrolytic grinding*, a variant of electrochemical machining, is used to shape and sharpen carbide cutting tools. It uses a rotating metal disk electrode with diamond dust (typically) bonded on the surface. Less than 1% of the workpiece material is removed by conventional grinding. The remainder is removed by electrolysis.

*Chemical milling (chem-milling)*, typically used in the manufacture of printed circuit boards, is the selective removal of material not protected by a mask. Some masks are scribed and removed by hand, but most are *photosensitive resists*. When photosensitive resists are used, the workpiece is coated with a light-sensitive emulsion. The emulsion is then exposed through a negative and developed, which removes the unexposed emulsion. Finally, the workpiece is placed in a *reagent* (the *etchant*), which removes only unmasked workpiece metal.

Chemical milling works with almost any metal, such as copper, aluminum, magnesium, and steel. Although the removal rate is low, very large areas can be processed. For highly accurate work, the tendency of the etchant to undercut the mask must be known and compensated for. The *etching radius (etch factor)* is one method of quantifying this tendency.

*Ultrasonic machining* (USM) or *ultrasonic impact machining* works with metallic and nonmetallic materials of any hardness. USM can be used to shape hard and brittle materials such as glass, ceramics, crystals, and gemstones, as well as tool steel and other metals. Ultrasonic energy with a frequency between 15,000 Hz and 30,000 Hz is generated in a *transducer* through magnetostrictive and piezoelectric effects. Wear of the transducer is minimal.

The transducer is separated from the workpiece by a slurry of abrasive particles. The ultrasonic energy generated is used to hurl fine abrasive particles against the workpiece at ultrasonic velocities. The same abrasives used for grinding wheels are used with USM: aluminum oxide, silicon carbide, and boron carbide. Grit sizes of 280 mesh or finer are common.

*Laser machining* is used to cut or burn very small holes in the workpiece with high dimensional accuracy.

## 8. COLD- AND HOT-WORKING OPERATIONS

Whether a workpiece is considered cold worked or hot worked depends on whether the working temperature is below or above the recrystallization temperature, respectively. Table 50.4 categorizes most common forming operations.

## 9. PRESSWORK

*Presswork* is a general term used to denote the blanking, bending and forming, and shearing of thin-gage metals. Presses (also known as *brakes*) are used with dies and punches to form the workpieces. Press forces are very high, and press capacities (known as *tonnage*) are often quoted in tons.

With *progressive dies*, the workpiece advances through a sequence of operations. Each of the press operations is

**Table 50.4** Cold- and Hot-Working Operations

*cold working*
    bending
    coining
    cold forging
    cold rolling
    cutting
    drawing
    drilling
    extruding
    grinding
    hobbing
    peening and burnishing
    riveting and staking
    rolling
    shearing, trimming, blanking, and piercing
    sizing
    spinning
    squeezing (e.g., swaging)
    thread rolling
*hot working*
    bending
    extruding and drawing (bar and wire)
    hot forging
    hot rolling
    piercing
    pipe welding
    spinning and shear forming
    swaging

performed at a *station*. Progressive dies can be of the *strip die* or *transfer die* varieties.

*Shearing operations* (blanking, punching, notching, etc.) cut pieces from flat plates, strips, and coil stock. Since the cutting is a shearing operation, the press force required to blank $N$ items at one time is given by Eq. 50.23.

$$F = NS_{us}Lt \qquad 50.23$$

The distinction between blanking and punching is relative. *Blanking* produces usable pieces (i.e., *blanks*), leaving the source piece behind as scrap. *Punching* is the operation of removing scrap blanks from the workpiece, leaving the source piece as the final product.

Bending and forming operations are often considered in the same category. *Bending dies* are used in press brakes to bend along a straight axis. *Forming dies* bend and form the blank along a curved axis and may incorporate other operations (e.g., notching, piercing, lancing, and cut-offs). There is little or no metal flow in a die-forming operation. The tension and compression on opposite surfaces of the blank are approximately equal.

*Spring-back, bend allowance,* and bending pressure can be calculated for bending and forming operations. However, it is generally necessary to make test runs to determine these values under realistic conditions.

*Drawing* is a cold-forming process that converts a flat blank into a hollow vessel (e.g., beverage cans). Drawing sheet metal blanks results in plastic metal flow along a curved axis. Double-acting presses may be required to accomplish deep draws.

*Coining*, as used in the production of coins, is a severe operation requiring high tonnage, due to the fact that the metal flow is completely confined within the die cavity. Because of this, coining is used mainly to form small parts.

*Embossing* forms shallow raised letters or other designs in relief on the surface of sheet metal blanks. It differs from coining in that the workpiece is not confined.

*Swaging* operations reduce the workpiece area by cold flowing the metal into a die cavity by a high compressive force or impact. It is applicable to small parts requiring close finishes. *Sizing* and *cold heading* of bolts and rivets are related operations.

## 10. FORGING

*Forging* is the repeated hammering of a workpiece to obtain the desired shape. Forging can be a cold-work process but is commonly considered to be a hot-work process when the term forging is used. Hot-work forging is carried out above the recrystallization temperature to produce a strain-free product. Table 50.5 lists the approximate temperatures for hot forging.

**Table 50.5** Approximate Hot-Forging Temperatures

| material | temperature | |
|---|---|---|
| | °F | °C |
| steel | 2000–2300 | 1100–1250 |
| copper alloys | 1400–1700 | 750–925 |
| magnesium alloys | 600 | 300 |
| aluminum alloys | 700–850 | 375–450 |

The oldest form of forging is similar to what is done by blacksmiths. Commercial *hammer forging, smith forging,* or *open die forging* consists of repeatedly hammering the workpiece (known as the *stock*) in a powered forge. Accuracy is low since the shape is not defined by dies, and considerable operator skill is required.

*Drop forging (closed-die forming)* relies on closed-impression dies to produce the desired shape. One-half of the die set is stationary; the other half is attached to the hammer. The metal flows plastically into the die upon impact by the forge hammer. The forging blows are repeated at the rate of several times a minute (for *gravity drop hammers*, also known as *board hammers*) to more than 300 times a minute (for *powered hammers*).[4] Progressive forging operations are used to significantly change the shape of a part over several steps.

---

[4]Steam or compressed air can be used to lift the gravity drop hammer back into place, but the hammer is not powered during its downward travel.

**Material Science**

The total forming energy (work) supplied by a gravity drop hammer is the potential energy of the hammer.

$$E = mgh \qquad \text{[SI]} \quad 50.24(a)$$

$$E = \frac{mgh}{g_c} \qquad \text{[U.S.]} \quad 50.24(b)$$

The total forming energy supplied by a powered drop hammer falling from height $h$ is

$$E = (mg + pA)h \qquad \text{[SI]} \quad 50.25(a)$$

$$E = \left(\frac{mg}{g_c} + pA\right)h \qquad \text{[U.S.]} \quad 50.25(b)$$

The total forming energy supplied by an eccentric-crank press can be calculated from its rotational speeds before and after the impact.

$$\begin{aligned} E &= \tfrac{1}{2}J(\omega_o^2 - \omega_f^2) \\ &= \tfrac{1}{2}J\left(\frac{\pi}{30}\right)^2(n_o^2 - n_f^2) \end{aligned} \qquad \text{[SI]} \quad 50.26(a)$$

$$\begin{aligned} E &= \tfrac{1}{2}\left(\frac{J}{g_c}\right)(\omega_o^2 - \omega_f^2) \\ &= \tfrac{1}{2}\left(\frac{J}{g_c}\right)\left(\frac{\pi}{30}\right)^2(n_o^2 - n_f^2) \end{aligned} \qquad \text{[U.S.]} \quad 50.26(b)$$

In *impactor forging* or *counterblow forging*, the workpiece is held in position while the dies are hammered horizontally into it from both sides. *Upset forming* involves holding and applying pressure to round heated blanks. The part is progressively formed from one end to the other and, characteristic of upset forming, becomes shorter in length but larger in diameter.

With *press forging*, the part is shaped by a slow squeezing action, rather than rapid impacts. This allows more forging energy to be used in shaping rather than in transmission to the machine and foundation. The pressing action can be obtained through screw or hydraulic action.

Following forging, the part will have a thin projection of excess metal known as *flash* at the *parting line*. The flash is trimmed off by *trimmer dies* in a subsequent operation. Also, since the hot-worked part will be covered with scale, the part is cleaned in acid (an operation known as *pickling*). Additional processing may include shotpeening, tumbling, and heat treatment.

## 11. SAND MOLDING

With *sand molding*, a mold is produced by packing sand around a pattern. After the pattern is removed, the remaining cavity has the desired shape. To facilitate the removal of the pattern, all surfaces parallel to the direction of withdrawal are slightly tapered. This taper is called *draft*. After molten metal is poured into the

mold, the sand is removed, exposing the completed cast part.

Various types of foundry sand are used, including green (moist) sand, dry (baked) sand, and carbon dioxide process sands. Carbon dioxide process sands contain approximately 4% silicate of soda ($Na_2SiO_3$), which hardens upon exposure to carbon dioxide. Pure, dry silica sand has no binding capacity and is not suited for molding. Various types of clay can be added to dry sand to improve its bonding characteristics (*cohesiveness*).

Other additives permit gases to escape during casting (*permeability*) and enhance the sand's heat resistance (*refractoriness*). Small amounts of organic matter can be added to the sand to enhance its *collapsibility*. The organic matter burns out when exposed to the hot metal, permitting the mold to be easily removed.

The *gating system* shown in Fig. 50.4 is the set of openings and passages that brings molten metal to the mold cavity in a controlled manner. The metal is poured into a *sprue hole* and enters a vertical passage known as a *downgate*. The metal then passes directly into the distribution passageways, known as *runners*, before entering the cavity. An entrance to the cavity may be constricted to control the rate of fill, and such constrictions are known as *gates*. *Risers* serve as accumulators to feed molten metal into the cavity during initial shrinkage.

**Figure 50.4** *Sand-Casting Gating System*

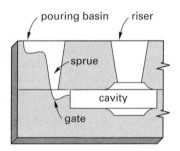

After hardening, the gates and risers can be broken off from iron castings, but a torch or cutting wheel is necessary with steel castings. Raw castings will be covered with sand and scale that must be removed by tumbling or sand blasting.

If a part has a recess or hole, a *core* must be placed in the mold. Cores come in *green sand core* and *dry sand core* varieties. (See Fig. 50.5.) Green sand cores are formed by the pattern itself. Dry sand cores are formed and baked in a separate operation and inserted into the sand mold.

## 12. GRAVITY MOLDING

With *gravity molding (tilt pouring, gravity die casting, or permanent molding)*, molten metal is poured into a

**Figure 50.5** *Green and Dry Sand Cores*

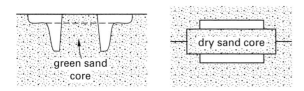

metal or graphite mold. Pressure is not used to fill the mold.[5] The mold may be coated prior to filling to prevent the casting from sticking to the mold's interior. Both ferrous and nonferrous metals (including magnesium, aluminum, and copper alloys) can be gravity molded. This method has the advantage (over sand casting) that a new mold is not required for each casting.

## 13. DIE CASTING

*Die casting (pressure die casting)* is suitable for creating parts of zinc, aluminum, copper, magnesium, and lead/tin alloys. (More than 75% of all die casting uses zinc alloys.) Molten metal is forced under pressure into a permanent metallic mold known as a *die*. Dies that produce one part per injection are known as *single cavity dies*. Dies that produce more than one part per injection are known as *multiple cavity dies*. The metal can be introduced by a plunger or compressed air but never by gravity alone. The casting pressure is maintained until solidification is complete.

Dies for zinc, tin, and lead alloys are usually made from high-carbon and alloy steels, although low-carbon steel can be used for zinc casting dies. Dies for aluminum, magnesium, and copper alloys (which melt at higher temperatures) are made out of heat-resisting alloy steel.

*Hot-chamber die casting* and *cold-chamber die casting* are the two main variations of die casting. The hot-chamber method is limited to alloys (e.g., zinc, tin, and lead) that have melting temperatures below 1000°F (550°C) and that do not attack the injection apparatus. The injection apparatus (the *gooseneck*) is submerged in the molten metal, and low temperatures limit corrosion.

Brass (and other copper alloys), aluminum, and magnesium have high melting temperatures and require higher injection pressures. They also corrode ferrous machine parts and become contaminated by the iron they pick up. Brass and bronze, with their 1600°F to 1900°F (875°C to 1050°C) melting temperatures, particularly attack the steel in die casting machines. These alloys are

usually melted in a separate furnace and ladled into the plunger cavity. This is the principle of the *cold-chamber method.*

After solidification, the sprues, gates, runners, and overflows are cut off in *trimming dies.* Die castings will typically be harder on the outside than on the inside due to the chilling action of the die. Also, gases may have been trapped inside the part, making the interior of the casting porous. Porous castings are brittle and subject to fracture.

## 14. CENTRIFUGAL CASTING

If the mold is rapidly rotated, the molten metal will be forced into the mold by centripetal action while the metal solidifies. This process is known as *centrifugal casting* or *centrifuging.* This method is particularly useful in producing objects with round and symmetrical (e.g., hexagonal) outer surfaces such as gun barrels and brake drums.

## 15. INVESTMENT CASTING

Casting methods that produce a molding cavity from a wax pattern are known as *investment casting, precision casting,* or the *lost-wax process.* These methods are suited to small, complex shapes and casting of precious metals.

A positive image of the part to be cast is created from wax.[6] The image is coated with the *investment material,* which can be finely ground refractory, plaster of paris or another ceramic material, or rubber, which becomes the mold. The mold is heated, and the liquid wax is poured out. The mold is then filled with molten metal. Centrifuging may be used to ensure complete filling. After the metal solidifies, the mold is broken off.

If more than one image is to be cast, a *master pattern* is made out of wood, steel, or plastic. The master pattern is used to make a *master die.* The wax patterns are then made in the master die.

## 16. CONTINUOUS CASTING

Any process in which molten material is continuously poured into a mold is known as *continuous casting.* This method can be used to produce sheets of glass, copper slabs, and brass or bronze bars. Continuous casting of bars is similar to extrusion, except that a cooling apparatus is included as part of the extrusion head.

## 17. PLASTIC MOLDING

Thermosetting compounds are purchased in liquid form, which makes them easy to combine with additives. Thermoplastic materials are commonly purchased in

---

[5]There is limited application of strength-, strain-, and safety-critical components (e.g., the production of railway wheels and some steel ingots) of a process known as *pressure pouring,* in which the molten metal is forced into the mold by air pressure, or drawn into the mold by a vacuum, or both (as in the case of counter-pressure casting). Pressure pouring differs from die casting in that ferrous alloys are used, typically magnesium ferrosilicon-treated ductile iron (MgFeSi iron).

[6]Other variations of this process use plastics with low melting points, lead, and even frozen mercury (which freezes at 40°F (4°C)).

granular form. They are mixed with additives in a *muller* (i.e., a bulk mixer) before transfer to the feed hoppers. Thermoplastic materials can also be molded into small pellets called *preforms* for easier handling in subsequent melting operations. Common additives for plastics are color pigments and tints, stabilizers, plasticizers, fillers, and resins.

The *hot compression molding* process is the oldest plastic forming process but is used extensively only for thermosetting polymers. (Compression molding is similar to the coining process for metals.) A measured amount of plastic material is placed in the open cavity of a heated mold. The mold is closed, and pressure is applied. The plastic flows into the mold, taking on its shape. Compression molding of thermoplastic resins is not very practical.[7]

Some plastic parts can be produced by *cold molding*. In this process, the part is simply cold pressed in a mold and then heated outside the press to fuse the particles.

The main method of forming thermoplastic resins is *injection molding*. The plastic molding compound is gravity fed into a heating chamber, where it is plasticized. Heating and metering is done by the *torpedo*, illustrated in Fig. 50.6. The molten plastic is injected under pressure into a water-cooled mold, where it solidifies almost immediately. (The mold temperature is constant.)

**Figure 50.6** *Injection Molding Equipment*

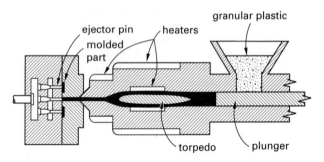

*Transfer molding* involves the heating of thermosetting plastic powder or preforms under pressure outside the mold cavity. The molten plastic is then forced from the transfer chamber through the gate and runner system into the mold cavity. The plastic is cured in the mold by maintaining the pressure and temperature. This method differs from injection molding in that the mold is kept heated and the plastic part is ejected while it is still hot.

*Blow molding* and *vacuum forming* both rely on an air-pressure differential to draw a heated thermoplastic sheet around a pattern or into a mold. The plastic retains the shape of the mold after cooling.

Most thermoplastics can be *extrusion-formed* into shapes (including sheets) of any length.[8] Solid plastic in granulated or powdered form is fed by a screw-feed mechanism into a heating chamber and then extruded through a die. The plastic is cooled by contact with air or water after extrusion. (See Fig. 50.7.)

**Figure 50.7** *Extrusion Process*

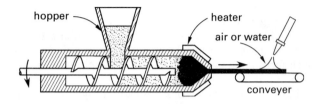

## 18. POWDER METALLURGY

Useful parts can be made by compressing a metal powder into shape and bonding the particles with heat—the principle of *powder metallurgy*. Typical powder metallurgy products include tungsten carbide cutting tools, copper motor brushes, bronze porous bearings, auto connecting rods and transmission parts, iron magnets, and filters.

Compared with machined products, parts made by powder metallurgy are relatively weak and expensive. However, the process is used for parts that cannot be easily produced in other manners. These are porous products, parts with complex shapes, items made from materials that are difficult to machine (e.g., tungsten carbide), and products that require the characteristics of two materials (e.g., copper/graphite electrical contacts).

The two most common types of powders are *iron-based powders* and *copper-based powders*. However, aluminum, nickel, tungsten, and other metals are also used. Bronze powders are regularly used to produce porous bearings; brass and iron are applicable to small machine parts where strength is important.

Metallic powders must be very fine. Most metal powders are produced by *atomizing* (i.e., using a jet of air to break up a fine stream of molten metal). Other methods include reduction of oxides, electrolytic deposition, and precipitation from a liquid or gas.

Pure metal powders are often mixed to improve manufacturing or performance characteristics. Graphite improves lubricating qualities and is added to powders used for bearings and electrical contacts. Cobalt and other metals are added to tungsten carbide to improve bonding.

Powders can be pressed into their final shape with a punch and die. The ejected shape is known as a *briquette* or *green compact*. With *isostatic molding*, hydraulic pressure is applied in all directions to the powder. Other methods of forming the green compact are centrifuging,

---

[7]Unless a mold is cooled before the part is removed, distortion of thermoplastics can result.

[8]Thermosetting plastics harden too quickly to be extruded.

*slip casting* (i.e., slurry casting), extrusion, and rolling. Due to internal friction, the density of a powder metallurgy part will not be consistent throughout but will be higher at the surface.

*Sintering* is heating to 70% to 90% of the melting point of the metal. The temperature is maintained for up to three hours, although the duration is commonly less than an hour. Typical sintering temperatures are 1600°F (875°C) for copper, 2050°F (1120°C) for iron, 2150°F (1175°C) for stainless steel, and 2700°F (1475°C) for tungsten carbide. To prevent the formation of metallic oxides, sintering must be performed in an inert or reducing gas (e.g., nitrogen) atmosphere.

## 19. HIGH ENERGY RATE FORMING

*High energy rate forming* (HERF), also known as *high velocity forming* (HVF), is the name given to several processes that plastically deform metals with blasts of high-pressure shock waves. Although these processes have traditionally been used to form thin metals, they are also applicable to other manufacturing needs such as powder metallurgy, forging, and welding.

The most common HERF method is *explosive forming*, as illustrated in Fig. 50.8. A small amount of low- or high-explosive is detonated. The resulting shock waves travel through the surrounding medium and force the metal into a shape determined by the dies. The medium can be either gas or liquid.

**Figure 50.8** Explosive Forming

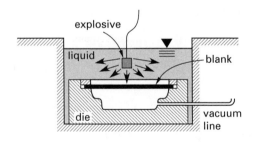

Another process using explosives is *explosive bonding*. A detonation is used to drive two similar metals together. When used with explosives in sheet form, this method has been successful in producing combinations of two metallic sheets (i.e., *cladding*). The resulting bond is almost metallurgically complete.

With *electro-hydraulic forming* (also known as *electro-spark forming*), the forming pressure is obtained from the discharge of massive amounts of stored electricity. The electrical energy is built up in a capacitor bank. The energy used can be changed by adding or removing capacitors from the circuit. The discharge is across a spark gap between two electrodes in a nonconducting medium. Upon discharge, the electrical energy is converted directly into work. The usual medium is liquid.

*Magnetic forming (magnetic pulse forming)* is another example of the direct conversion of electrical energy into work. (See Fig. 50.9.) As with electrospark forming, a large amount of electrical energy is built up in a capacitor bank. A special expendable forming coil is placed around a part to be compressed or within a part to be expanded. (Magnetic forming can also be used for embossing if the workpiece is placed between the forming coil and the embossing die.) When the capacitor bank discharges, the current in the forming coil induces a current and a force in the workpiece. This force stresses the workpiece beyond its elastic limit.

**Figure 50.9** Magnetic Forming

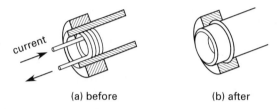

(a) before    (b) after

## 20. GAS WELDING

With *welding*, two metals are fused (i.e., melted) together by localized heat or pressure. This is known as *fusion* or *coalescence*. A welding rod of similar metal can be used to fill large voids between the two pieces but is not always necessary. The main types of welds are the *bead, groove, fillet,* and *plug (spot) welds* shown in Fig. 50.10.

**Figure 50.10** Types of Fusion Welds

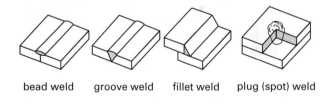

bead weld    groove weld    fillet weld    plug (spot) weld

In *gas welding processes*, a combustible fuel and oxidizing gas are combined in a *torch (blowpipe)*. Although natural gas and hydrogen can be used as fuels, *acetylene gas* ($C_2H_2$) is most common. Oxygen gas is the oxidizer, hence the names *oxyhydrogen welding* and *oxyacetylene welding*. The maximum welding temperature with oxyhydrogen welding is approximately 5100°F (2800°C), and with oxyacetylene welding, it is approximately 6000°F (3300°C).

*MAPP gas (methyacetylene propadiene)* is also extensively used. It is safer to store, is more dense, and provides more energy per unit volume than acetylene.

The proportions of oxygen and acetylene can be adjusted to obtain three different welding conditions: reducing, neutral, and oxidizing flames. Figure 50.11 illustrates the reactions that occur with a *neutral flame*, which is obtained when the oxygen:acetylene

**Figure 50.11** *Neutral Oxyacetylene Flame*

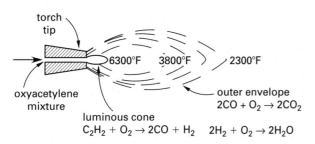

proportions are approximately 1:1 by volume. The inner luminous cone of the flame is distinctly blue in color and is the hottest part of the flame. The outer envelope is only slightly luminous and may be difficult to see. Oxygen for the combustion of the outer envelope comes from the atmosphere.

If the gas proportions are adjusted to an excess of acetylene, the flame is known as a *reducing flame* or *carburizing flame*. This process is used to weld many nonferrous metals (including Monel metal and nickel alloys as well as some alloy steels) and in applying several types of hard surfacing materials.

An *oxidizing flame* requires an excess of oxygen. Oxidizing fusion has some application to brass and bronze but is generally undesirable.[9]

## 21. ARC WELDING

Temperatures of up to 10,000°F (5550°C) can be obtained from *arc welding* using either DC or AC electrical current. The arc is created by first touching the electrode to the workpiece, establishing the current flow, and then moving the electrode slightly away. The current flow is maintained by the arc, and the electrical energy is converted to heat, which melts the metal.

If a *carbon electrode* is used to create an arc, a welding rod must be used to supply the filler material. Alternatively, in *metal electrode welding*, the electrode is itself melted by the arc and becomes the filler material.

Electrodes can be bare, but most have coatings known as *flux* that melt into slag and improve other welding characteristics.[10] The molten slag floats on and covers the molten metal, inhibiting the high temperature formation of oxides that weaken most welds. Welding with coated electrodes is known as *shielded metal arc welding*

---

[9]An oxyacetylene *cutting torch* uses oxygen to cut steel, but the process is different from regular welding. A cutting torch uses a neutral flame to heat the steel. The oxygen used to oxidize (cut) the metal issues from a separate orifice in the torch and does not participate in the combustion of the acetylene.

[10]In addition to providing a protection from oxidation, the next most important function of the coating is to stabilize the arc (i.e., reduce the effect of variations in the separation of electrode and workpiece). Other functions include reducing weld metal splatter, adding alloying ingredients, changing the weld bead shape, improving overhead and vertical weldability, and providing additional filler material.

(SMAW). Most coatings have a significant amount of $SiO_2$ and/or $TiO_2$, plus small amounts of oxides of other metals. Hardened slag is chipped away after the weld has cooled.

In the United States, welding rods are classified according to the tensile strength (in ksi—thousands of $lbf/in^2$) of the deposited material. Thus, an E70 welding rod will produce a bead with a minimum nonstress relieved tensile strength of 70,000 $lbf/in^2$.

With *submerged-arc welding*, the flux is granular and is dispensed from a feed tube ahead of the welding process. The tip of the electrode and the arc are buried in the granular flux. The arc melts the granular flux, forming a coating that protects against oxidation.

Another method of protecting the molten weld metal from oxidation is by shielding with an inert gas. This is done when welding magnesium, aluminum, stainless steels, and some other steels. Argon gas is commonly used, although helium and argon-helium mixtures are also used. (Carbon dioxide gas can be used to shield the weld when working with plain-carbon and low-alloy steels.) This is the principle of *inert gas shielded arc welding*. With *TIG welding (tungsten inert gas)*, the arc issues from an air- or water-cooled, nonconsumable tungsten electrode.

*MIG welding (metal inert gas)* is similar to TIG welding except that a consumable wire is used as the electrode. (See Fig. 50.12.) This method is also known as the *GMAW (gas metal arc welding) process*.

**Figure 50.12** *TIG and MIG Welding Processes*

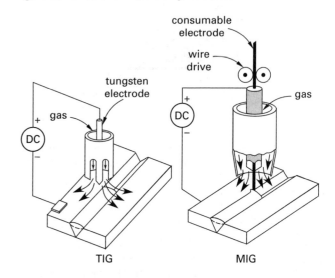

## 22. SOLDERING AND BRAZING

*Soldering* and *brazing* both use a molten dissimilar metal as glue between the two pieces. Generally, soldering uses a lead-tin filler with melting points below 800°F (425°C), whereas brazing uses copper-zinc or silver-based alloys with melting points above 800°F (425°C).

Soldering is not a fusion process, since parts do not melt. Since some alloying occurs, brazing is considered to be a fusion process.

Usually, the solder or brazing material is drawn into the space separating the pieces by capillary action. However, a chemical flux can be used to remove oxides from the surfaces, inhibit additional oxidation, and improve cohesion of the filler material.

## 23. ADHESIVE BONDING

Both thermoplastic and thermosetting polymers are used as structural adhesives. Polymers of both types are sometimes combined to obtain the performance characteristics of both. A low-viscosity primer may be used to prepare the bonding surface for the adhesive. Almost all adhesive bonds are of the *lap-joint variety.*

Structural adhesives can be used to join any similar or dissimilar materials. However, even structural adhesives are limited to low-strength and low-temperature (e.g., less than 500°F or 250°C) applications.[11]

*Thermoplastic adhesives,* such as polyamides, vinyls, and nonvulcanized neoprene rubbers, soften when heated and cannot be used for elevated temperatures. They are generally used for nonstructural applications.

*Thermosetting adhesives* (e.g., epoxies, isocyanate, phenolic rubbers and vinyls, vulcanized rubbers, and neoprene) are used as structural adhesives. They must be used with elevated temperatures and where creep is unacceptable. Heat, pressure, ultraviolet radiation, and/or chemical hardeners must be used to activate the curing process.

The performance of an adhesive is determined by its toughness, tensile strength, peel strength, and temperature resistance. Adhesives with high tensile and shear strengths are usually hard and brittle and have low *peel strengths.* Adhesives with high peel strengths are usually more ductile and have fair tensile and shear strengths.

## 24. MANUFACTURE OF METAL PIPE

Pipes and tubes can be either seamed or seamless. Seamless pipe is made by piercing, whereas seamed pipe is made by forming and butt- or lap-welding the joined edges.

Thin-wall pipe can be formed by drawing a heated flat skelp through a welding bell. Prior to forming, the edges of the skelp are heated by flame or induction to the forging temperature, and joining occurs spontaneously in the welding bell, as shown in Fig. 50.13.

Pipes with larger diameters and wall thicknesses are formed with *roll forming* and *electric butt-welding,* in which the skelp is cold-rolled into circular shape by a series of roller pairs. One of the last rolling operations

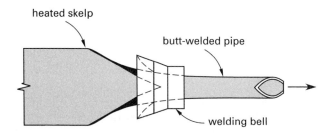

**Figure 50.13** *Butt-Welding in a Welding Bell*

incorporates induction heating to bring the pipe edges up to forging temperature before they are pressed together. The flash is subsequently removed from the inside and outside of the pipe.

The skelp of *lap-welded pipe* passes through rollers that overlap the edges. Unlike roll forming, however, a fixed mandrel is placed inside the pipe. The heated skelp is rolled between the roller and mandrel at high pressure, which welds the heated edges together.

*Seamless pipe* is manufactured by heating a solid round bar known as a *billet* to forging temperature and piercing it with a mandrel. (See Fig. 50.14.) Subsequent operations with rollers strengthen, size, and finish the pipe.

**Figure 50.14** *Production of Seamless Pipe*

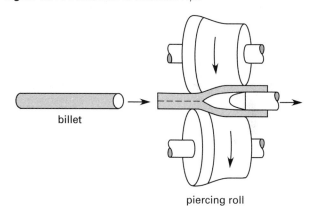

## 25. SURFACE FINISHING AND COATINGS

Finishes protect and improve the appearance of surfaces. Some processes involve material removal, and others involve material addition. Some of the common finishing methods and coating systems are listed as follows.

- *abrasive cleaning:* shooting sand (i.e., *sand blasting*), steel grit, or steel shot against workpieces to remove casting sand, scale, and oxidation.

- *anodizing:* an electroplating-acid bath oxidation process for aluminum and magnesium. The workpiece is the anode in the electrical circuit.

---

[11]Certain *ceramic adhesives* have useful ranges up to 1000°F (550°C).

- *barrel finishing (tumbling):* rotating parts in a barrel filled with an abrasive or non-abrasive medium. Widely used to remove burrs, flash, scale, and oxides.

- *buffing:* a fine finishing operation, similar to polishing, using a very fine polishing compound (e.g., *rouge*).

- *burnishing:* a fine grinding or peening operation designed to leave a characteristic pattern on the surface of the workpiece.

- *calorizing:* the diffusing of aluminum into a steel surface, producing an aluminum oxide that protects the steel from high-temperature corrosion.

- *electroplating:* the electro-deposition of a coating onto the workpiece. Electrical current is used to drive ions in solution to the part. The workpiece is the cathode in the electrical circuit.

- *galvanizing:* a zinc coating applied to low-carbon steel to improve corrosion resistance. The coating can be applied in a hot dip bath, by electroplating, or by dry tumbling (*sheradizing*).

- *hard surfacing:* the creation (by spraying, plating, fusion welding, or heat treatment) of a hard metal surface in a softer product.

- *honing:* a grinding operation using stones moving in a reciprocating pattern. Leaves a characteristic cross-hatch pattern.

- *lapping:* a fine grinding operation used to obtain exact fit and dimensional accuracy.

- *metal spraying:* the spraying of molten metal onto a product. Methods include *metallizing, metal powder spraying,* and *plasma flame spraying.*

- *organic finishes:* the covering of surfaces with an organic film of paint, enamel, or lacquer.

- *painting:* see *organic finishes.*

- *parkerizing:* application of a thin phosphate coating on steel to improve corrosion resistance. This process is known as *bonderizing* when used as a primer for paints.

- *pickling:* a process in which metal is dipped in dilute acid solutions to remove dirt, grease, and oxides.

- *polishing:* abrasion of parts against wheels or belts coated with polishing compounds.

- *sheradizing:* a specific method of zinc galvanizing in which parts are tumbled in zinc dust at high temperatures.

- *superfinishing:* a super-fine grinding operation used to expose nonfragmented, crystalline base metal.

- *tin-plating:* a hot-dip or electroplate application of tin to steel.

# Topic X: Mechanics of Materials

Chapter

# 51 Properties of Areas

## Nomenclature

| | | |
|---|---|---|
| $A$ | area | units$^2$ |
| $b$ | base distance | units |
| $c$ | distance to extreme fiber | units |
| $d$ | separation distance | units |
| $h$ | height distance | units |
| $I$ | moment of inertia | units$^4$ |
| $J$ | polar moment of inertia | units$^4$ |
| $L$ | length | units |
| $P$ | product of inertia | units$^4$ |
| $Q$ | first moment of the area | units$^3$ |
| $r$ | radius | units |
| $r$ | radius of gyration | units |
| $S$ | section modulus | units$^3$ |
| $V$ | volume | units$^3$ |
| $x$ | distance in the $x$-direction | units |
| $y$ | distance in the $y$-direction | units |

## Symbols

| | | |
|---|---|---|
| $\theta$ | angle | degrees |

## Subscripts

| | |
|---|---|
| $c$ | centroidal |
| $o$ | with respect to the origin |

## 1. CENTROID OF AN AREA

The *centroid* of an area is analogous to the center of gravity of a homogeneous body.[1] The centroid is often described as the point at which a thin homogeneous plate would balance. This definition, however, combines the definitions of centroid and center of gravity and

---

[1]The analogy has been simplified. A three-dimensional body also has a centroid. The centroid and center of gravity will coincide when the body is homogeneous.

implies that gravity is required to identify the centroid, which is not true.

The location of the centroid of an area bounded by the $x$- and $y$-axes and the mathematical function $y = f(x)$ can be found by the *integration method* by using Eq. 51.1 through Eq. 51.4. The centroidal location depends only on the geometry of the area and is identified by the coordinates $(x_c, y_c)$. Some references place a bar over the coordinates of the centroid to indicate an average point, such as $(\overline{x}, \overline{y})$.

$$x_c = \frac{\int x\, dA}{A} \qquad 51.1$$

$$y_c = \frac{\int y\, dA}{A} \qquad 51.2$$

$$A = \int f(x)\, dx \qquad 51.3$$

$$dA = f(x)dx = g(y)dy \qquad 51.4$$

The locations of the centroids of *basic shapes*, such as triangles and rectangles, are well known. The most common basic shapes have been included in App. 51.A. There should be no need to derive centroidal locations for these shapes by the integration method.

The centroid of a complex area can be found from Eq. 51.5 and Eq. 51.6 if the area can be divided into the basic shapes in App. 51.A. This process is simplified when all or most of the subareas adjoin the reference axis. Example 51.1 illustrates this method.

$$x_c = \frac{\sum_i A_i x_{ci}}{\sum_i A_i} \qquad 51.5$$

$$y_c = \frac{\sum_i A_i y_{ci}}{\sum_i A_i} \qquad 51.6$$

### Example 51.1

An area is bounded by the $x$- and $y$-axes, the line $x = 2$, and the function $y = e^{2x}$. Find the $x$-component of the centroid.

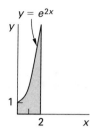

*Solution*

First, use Eq. 51.3 to find the area.

$$A = \int f(x)\,dx = \int_{x=0}^{x=2} e^{2x}\,dx$$

$$= \tfrac{1}{2}e^{2x}\Big|_{0}^{2} = 27.3 - 0.5 = 26.8 \text{ units}^2$$

Since $y$ is a function of $x$, $dA$ must be expressed in terms of $x$. From Eq. 51.4,

$$dA = f(x)\,dx = e^{2x}\,dx$$

Finally, use Eq. 51.1 to find $x_c$.

$$x_c = \frac{\int x\,dA}{A} = \frac{1}{26.8}\int_{x=0}^{x=2} xe^{2x}\,dx$$

$$= \left(\frac{1}{26.8}\right)\tfrac{1}{2}xe^{2x} - \tfrac{1}{4}e^{2x}\Big|_{0}^{2} = 1.54 \text{ units}$$

**Example 51.2**

Find the $y$-coordinate of the centroid of the area shown.

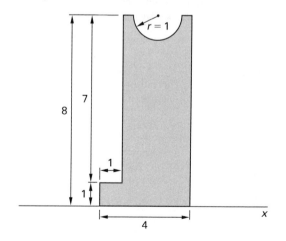

*Solution*

The $x$-axis is the reference axis. The area is divided into basic shapes of a $1 \times 1$ square, a $3 \times 8$ rectangle, and a half-circle of radius 1. (The area could also be divided into $1 \times 4$ and $3 \times 7$ rectangles and the half-circle, but then the $3 \times 7$ rectangle would not adjoin the $x$-axis.)

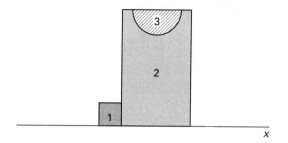

First, calculate the areas of the basic shapes. Notice that the half-circle area is negative since it represents a cut-out.

$$A_1 = (1.0)(1.0) = 1.0 \text{ units}^2$$

$$A_2 = (3.0)(8.0) = 24.0 \text{ units}^2$$

$$A_3 = -\tfrac{1}{2}\pi r^2 = -\tfrac{1}{2}\pi(1.0)^2 = -1.57 \text{ units}^2$$

Next, find the $y$-components of the centroids of the basic shapes. Most are found by inspection, but App. 51.A can be used for the half-circle. Notice that the centroidal location for the half-circle is positive.

$$y_{c1} = 0.5 \text{ units}$$

$$y_{c2} = 4.0 \text{ units}$$

$$y_{c3} = 8.0 - 0.424 = 7.576 \text{ units}$$

Finally, use Eq. 51.6.

$$y_c = \frac{\sum A_i y_{ci}}{\sum A_i}$$

$$= \frac{(1.0)(0.5) + (24.0)(4.0) + (-1.57)(7.576)}{1.0 + 24.0 - 1.57}$$

$$= 3.61 \text{ units}$$

## 2. FIRST MOMENT OF THE AREA

The quantity $\int x\,dA$ is known as the *first moment of the area* or *first area moment* with respect to the $y$-axis. Similarly, $\int y\,dA$ is known as the first moment of the area with respect to the $x$-axis. By rearranging Eq. 51.1 and Eq. 51.2, the first moment of the area can be calculated from the area and centroidal distance.

$$Q_y = \int x\,dA = x_c A \qquad\qquad 51.7$$

$$Q_x = \int y\,dA = y_c A \qquad\qquad 51.8$$

In basic engineering, the two primary applications of the first moment concept are to determine centroidal locations and shear stress distributions. In the latter application, the first moment of the area is known as the *statical moment*.

# 3. CENTROID OF A LINE

The location of the *centroid of a line* is defined by Eq. 51.9 and Eq. 51.10, which are analogous to the equations used for centroids of areas.

$$x_c = \frac{\int x \, dL}{L} \qquad 51.9$$

$$y_c = \frac{\int y \, dL}{L} \qquad 51.10$$

Since equations of lines are typically in the form $y = f(x)$, $dL$ must be expressed in terms of $x$ or $y$.

$$dL = \left( \sqrt{\left(\frac{dy}{dx}\right)^2 + 1} \right) dx \qquad 51.11$$

$$dL = \left( \sqrt{\left(\frac{dx}{dy}\right)^2 + 1} \right) dy \qquad 51.12$$

# 4. THEOREMS OF PAPPUS-GULDINUS

The *Theorems of Pappus-Guldinus* define the surface and volume of revolution (i.e., the surface area and volume generated by revolving a curve around a fixed axis).

- *Theorem I:* The area of a surface of revolution (see Fig. 51.1) is equal to the product of the length of the generating curve and the distance traveled by the centroid of the curve while the surface is being generated.

**Figure 51.1** *Surface of Revolution*

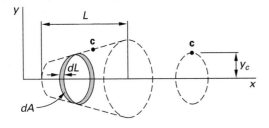

When a part, $dL$, of a line $L$ is revolved about the $x$-axis, a differential ring having surface area $dA$ is generated.

$$dA = 2\pi y \, dL \qquad 51.13$$

$$A = \int dA = 2\pi \int y \, dL = 2\pi y_c L \qquad 51.14$$

- *Theorem II:* The volume of a surface of revolution (see Fig. 51.2) is equal to the generating area times the distance traveled by the centroid of the area in generating the volume.

**Figure 51.2** *Volume of Revolution*

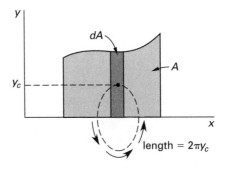

When a differential plane area, $dA$, is revolved about the $x$-axis and does not intersect the $y$-axis, it generates a ring of volume $dV$.

$$dV = \pi y^2 \, dx = \pi y \, dA \qquad 51.15$$

$$V = 2\pi y_c A \qquad 51.16$$

# 5. MOMENT OF INERTIA OF AN AREA

The *moment of inertia*, $I$, of an area is needed in mechanics of materials problems. It is convenient to think of the moment of inertia of a beam's cross-sectional area as a measure of the beam's ability to resist bending. Thus, given equal loads, a beam with a small moment of inertia will bend more than a beam with a large moment of inertia.

Since the moment of inertia represents a resistance to bending, it is always positive. Since a beam can be unsymmetrical (e.g., a rectangular beam) and can be stronger in one direction than another, the moment of inertia depends on orientation. Therefore, a reference axis or direction must be specified.

The moment of inertia taken with respect to one of the axes in the rectangular coordinate system is sometimes referred to as the *rectangular moment of inertia*.

The symbol $I_x$ is used to represent a moment of inertia with respect to the $x$-axis. Similarly, $I_y$ is the moment of inertia with respect to the $y$-axis. $I_x$ and $I_y$ do not normally combine and are not components of some resultant moment of inertia.

Any axis can be chosen as the reference axis, and the value of the moment of inertia will depend on the reference selected. The moment of inertia taken with respect to an axis passing through the area's centroid is known as the *centroidal moment of inertia*, $I_{cx}$ or $I_{cy}$. The centroidal moment of inertia is the smallest possible moment of inertia for the shape.

The *integration method* can be used to calculate the moment of inertia of a function that is bounded by the $x$- and $y$-axes and a curve $y = f(x)$. From Eq. 51.17 and Eq. 51.18, it is apparent why the moment of inertia is also known as the *second moment of the area* or *second area moment*.

$$I_x = \int y^2 \, dA \qquad 51.17$$

$$I_y = \int x^2 \, dA \qquad 51.18$$

$$dA = f(x) \, dx = g(y) \, dy \qquad 51.19$$

The moments of inertia of the *basic shapes* are well known and are listed in App. 51.A.

### Example 51.3

What is the centroidal moment of inertia with respect to the *x*-axis of a rectangle 5.0 units wide and 8.0 units tall?

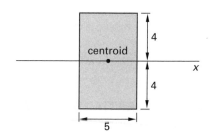

*Solution*

Since the centroidal moment of inertia is needed, the reference line passes through the centroid. From App. 51.A, the centroidal moment of inertia is

$$I_{cx} = \frac{bh^3}{12} = \frac{(5)(8)^3}{12} = 213.3 \text{ units}^4$$

### Example 51.4

What is the moment of inertia with respect to the *y*-axis of the area bounded by the *y*-axis, the line $y = 8.0$, and the parabola $y^2 = 8x$?

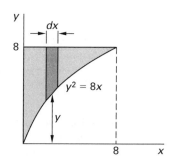

*Solution*

This problem is more complex than it first appears, since the area is above the curve, bounded not by $y = 0$ but by $y = 8$. In particular, $dA$ must be determined correctly.

$$y = \sqrt{8x}$$

Use Eq. 51.4.

$$dA = (8 - f(x)) \, dx = (8 - y) \, dx$$
$$= (8 - \sqrt{8x}) \, dx$$

Equation 51.18 is used to calculate the moment of inertia with respect to the *y*-axis.

$$I_y = \int x^2 \, dA = \int_0^8 x^2 (8 - \sqrt{8x}) \, dx$$
$$= \frac{8}{3} x^3 - \left( \frac{4\sqrt{2}}{7} \right) x^{7/2} \Big|_0^8 = 195.0 \text{ units}^4$$

## 6. PARALLEL AXIS THEOREM

If the moment of inertia is known with respect to one axis, and the moment of inertia with respect to another, the parallel axis can be calculated from the *parallel axis theorem*, also known as the *transfer axis theorem*. In Eq. 51.20, *d* is the distance between the centroidal axis and the second, parallel axis.

$$I_{\text{parallel axis}} = I_c + A d^2 \qquad 51.20$$

The second term in Eq. 51.20 is often much larger than the first term. Areas close to the centroidal axis do not affect the moment of inertia considerably. This principle is exploited by structural steel shapes (see Fig. 51.3) that derive bending resistance from *flanges* located away from the centroidal axis. The *web* does not contribute significantly to the moment of inertia.

**Figure 51.3** *Structural Steel W-Shape*

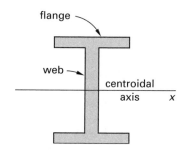

## Example 51.5

Find the moment of inertia about the $x$-axis for the inverted-T area shown.

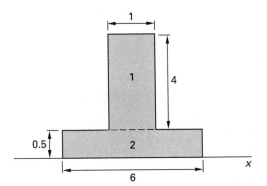

*Solution*

The area is divided into two basic shapes: 1 and 2. From App. 51.A, the moment of inertia of basic shape 2 with respect to the $x$-axis is

$$I_{x2} = \frac{bh^3}{3} = \frac{(6.0)(0.5)^3}{3} = 0.25 \text{ units}^4$$

The moment of inertia of basic shape 1 about its own centroid is

$$I_{cx1} = \frac{bh^3}{12} = \frac{(1)(4)^3}{12} = 5.33 \text{ units}^4$$

The $x$-axis is located 2.5 units from the centroid of basic shape 1. Therefore, from the parallel axis theorem, Eq. 51.20, the moment of inertia of basic shape 1 about the $x$-axis is

$$I_{x1} = 5.33 + (4)(2.5)^2 = 30.33 \text{ units}^4$$

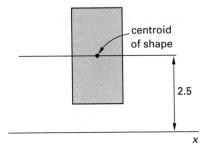

The total moment of inertia of the T-area is

$$I_x = I_{x1} + I_{x2} = 30.33 \text{ units}^4 + 0.25 \text{ units}^4$$
$$= 30.58 \text{ units}^4$$

## Example 51.6

Find the moment of inertia about the horizontal centroidal axis for the inverted-T area shown in Ex. 51.5.

*Solution*

The first step is to find the location of the centroid. The areas and centroidal locations (with respect to the $x$-axis) of the two basic shapes are

$$A_1 = (4.0)(1.0) = 4.0 \text{ units}^2$$

$$A_2 = (0.5)(6.0) = 3.0 \text{ units}^2$$

$$y_{c1} = 2.5 \text{ units}$$

$$y_{c2} = 0.25 \text{ units}$$

From Eq. 51.6, the composite centroid is located at

$$y_c = \frac{A_1 y_{c1} + A_2 y_{c2}}{A_1 + A_2} = \frac{(4.0)(2.5) + (3.0)(0.25)}{4.0 + 3.0}$$
$$= 1.536 \text{ units}$$

The distances between the centroids of the basic shapes and the composite shape are

$$d_1 = 2.5 - 1.536 = 0.964 \text{ units}$$

$$d_2 = 1.536 - 0.25 = 1.286 \text{ units}$$

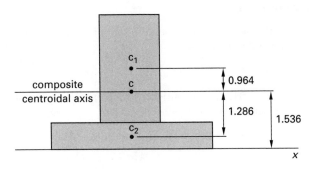

The centroidal moments of inertia of the basic shapes with respect to an axis parallel to the $x$-axis are

$$I_{cx1} = \frac{bh^3}{12} = \frac{(1)(4)^3}{12} = 5.33 \text{ units}^4$$

$$I_{cx2} = \frac{bh^3}{12} = \frac{(6)(0.5)^3}{12} = 0.0625 \text{ units}^4$$

Using Eq. 51.20, the centroidal moment of inertia of the inverted-T area is

$$I_{cx} = I_{cx1} + A_1 d_1^2 + I_{cx2} + A_2 d_2^2$$
$$= 5.33 + (4.0)(0.964)^2 + 0.0625 + (3.0)(1.286)^2$$
$$= 14.07 \text{ units}^4$$

## 7. POLAR MOMENT OF INERTIA

The *polar moment of inertia*, $J$, is required in torsional shear stress calculations.[2] It can be thought of as a measure of an area's resistance to torsion (twisting). The definition of a polar moment of inertia of a two-dimensional area requires three dimensions because the reference axis for a polar moment of inertia of a plane area is perpendicular to the plane area.

The polar moment of inertia can be derived from Eq. 51.21.

$$J = \int (x^2 + y^2)\,dA \qquad 51.21$$

It is often easier to use the *perpendicular axis theorem* to quickly calculate the polar moment of inertia.

- *perpendicular axis theorem:* The polar moment of inertia of a plane area about an axis normal to the plane is equal to the sum of the moments of inertia about any two mutually perpendicular axes lying in the plane and passing through the given axis.

$$J = I_x + I_y \qquad 51.22$$

Since the two perpendicular axes can be chosen arbitrarily, it is most convenient to use the centroidal moments of inertia.

$$J = I_{cx} + I_{cy} \qquad 51.23$$

### Example 51.7

What is the centroidal polar moment of inertia of a circular area of radius $r$?

*Solution*

From App. 51.A, the centroidal moment of inertia of a circle with respect to the $x$-axis is

$$I_{cx} = \frac{\pi r^4}{4}$$

Since the area is symmetrical, $I_{cy}$ and $I_{cx}$ are the same. From Eq. 51.23,

$$J_c = I_{cx} + I_{cy} = \frac{\pi r^4}{4} + \frac{\pi r^4}{4} = \frac{\pi r^4}{2}$$

## 8. RADIUS OF GYRATION

Every nontrivial area has a centroidal moment of inertia. Usually, some portions of the area are close to the centroidal axis and other portions are farther away. The *transverse radius of gyration*, or just *radius of gyration*, $r$, is an imaginary distance from the centroidal axis at which the entire area can be assumed to exist without

[2]The symbols $I_z$ and $I_{xy}$ are also encountered. However, the symbol $J$ is more common.

affecting the moment of inertia. Despite the name "radius," the radius of gyration is not limited to circular shapes or to polar axes. This concept is illustrated in Fig. 51.4.

**Figure 51.4** *Radius of Gyration*

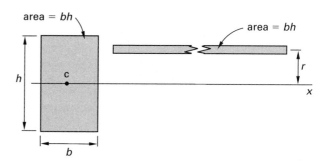

The method of calculating the radius of gyration is based on the parallel axis theorem. If all of the area is located a distance $r$ from the original centroidal axis, there will be no $I_c$ term in Eq. 51.20. Only the $Ad^2$ term will contribute to the moment of inertia.

$$I = r^2 A \qquad 51.24$$

$$r = \sqrt{\frac{I}{A}} \qquad 51.25$$

The concept of *least radius of gyration* comes up frequently in column design problems. (The column will tend to buckle about an axis that produces the smallest radius of gyration.) Usually, finding the least radius of gyration for symmetrical sections means solving Eq. 51.25 twice: once with $I_x$ to find $r_x$ and once with $I_y$ to find $r_y$. The smallest value of $r$ is the least radius of gyration.

The analogous quantity in the polar system is

$$r = \sqrt{\frac{J}{A}} \qquad 51.26$$

Just as the polar moment of inertia, $J$, can be calculated from the two rectangular moments of inertia, the polar radius of gyration can be calculated from the two rectangular radii of gyration.

$$r^2 = r_x^2 + r_y^2 \qquad 51.27$$

### Example 51.8

What is the radius of gyration of the rectangular shape in Ex. 51.3?

*Solution*

The area of the rectangle is

$$A = bh = (5)(8) = 40 \text{ units}^2$$

From Eq. 51.25, the radius of gyration is

$$r_x = \sqrt{\frac{I_x}{A}} = \sqrt{\frac{213.3}{40}} = 2.31 \text{ units}$$

2.31 units is the distance from the centroidal $x$-axis that an infinitely long strip with an area of 40 square units would have to be located in order to have a moment of inertia of 213.3 units[4].

## 9. PRODUCT OF INERTIA

The *product of inertia*, $P_{xy}$, of a two-dimensional area is useful when relating properties of areas evaluated with respect to different axes. It is found by multiplying each differential element of area by its $x$- and $y$-coordinate and then summing over the entire area. (See Fig. 51.5.)

$$P_{xy} = \int xy \, dA \qquad 51.28$$

**Figure 51.5** Calculating the Product of Inertia

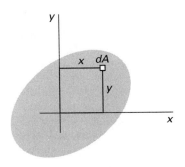

The product of inertia is zero when either axis is an axis of symmetry. Since the axes can be chosen arbitrarily, the area may be in one of the negative quadrants, and the product of inertia may be negative.

The parallel axis theorem for products of inertia is given (see Fig. 51.6) by Eq. 51.29. (Both axes are allowed to move to new positions.) $x'_c$ and $y'_c$ are the coordinates of the centroid in the new coordinate system.

$$P_{x'y'} = P_{c,xy} + x'_c y'_c A \qquad 51.29$$

## 10. SECTION MODULUS

In the analysis of beams, the outer compressive (or tensile) surface is known as the *extreme fiber*. The distance, $c$, from the centroidal axis of the beam cross section to the extreme fiber is the "distance to the extreme fiber." The *section modulus*, $S$, combines the

**Figure 51.6** Parallel Axis Theorem for Products of Inertia

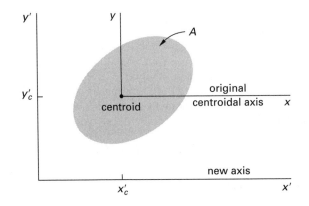

centroidal moment of inertia and the distance to the extreme fiber.

$$S = \frac{I_c}{c} \qquad 51.30$$

## 11. ROTATION OF AXES

Figure 51.7 shows rotation of the $x$-$y$ axes through an angle, $\theta$, into a new set of $u$-$v$ axes, without rotating the area. If the moments and product of inertia of the area are known with respect to the old $x$-$y$ axes, the new properties can be calculated from Eq. 51.31 through Eq. 51.33.

$$\begin{aligned} I_u &= I_x \cos^2\theta - 2P_{xy}\sin\theta\cos\theta + I_y \sin^2\theta \\ &= \tfrac{1}{2}(I_x + I_y) + \tfrac{1}{2}(I_x - I_y)\cos 2\theta \\ &\quad - P_{xy}\sin 2\theta \end{aligned} \qquad 51.31$$

$$\begin{aligned} I_v &= I_x \sin^2\theta + 2P_{xy}\sin\theta\cos\theta + I_y \cos^2\theta \\ &= \tfrac{1}{2}(I_x + I_y) - \tfrac{1}{2}(I_x - I_y)\cos 2\theta \\ &\quad + P_{xy}\sin 2\theta \end{aligned} \qquad 51.32$$

**Figure 51.7** Rotation of Axes

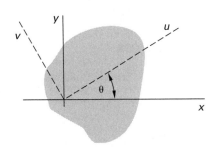

$$P_{uv} = I_x \sin\theta\cos\theta + P_{xy}(\cos^2\theta - \sin^2\theta)$$
$$- I_y \sin\theta\cos\theta$$
$$= \tfrac{1}{2}(I_x - I_y)\sin 2\theta + P_{xy}\cos 2\theta \qquad 51.33$$

Since the polar moment of inertia about a fixed axis perpendicular to any two orthogonal axes in the plane is constant, the polar moment of inertia is unchanged by the rotation.

$$J_{xy} = I_x + I_y = I_u + I_v = J_{uv} \qquad 51.34$$

### Example 51.9

What is the centroidal area moment of inertia of a $6 \times 6$ square that is rotated 45° from its "flat" orientation?

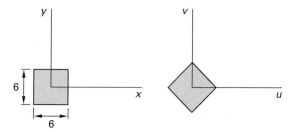

*Solution*

The centroidal moments of inertia with respect to the $x$- and $y$-axes are

$$I_x = I_y = \frac{s^4}{12} = \frac{(6)^4}{12}$$
$$= 108 \text{ units}^4$$

Since the centroidal $x$- and $y$-axes are axes of symmetry, the product of inertia is zero.

Use Eq. 51.31.

$$I_u = I_x \cos^2\theta - 2P_{xy}\sin\theta\cos\theta + I_y\sin^2\theta$$
$$= (108)\cos^2 45° - 0 + (108)\sin^2 45°$$
$$= 108 \text{ units}^4$$

The centroidal moment of inertia of a square is the same regardless of rotation angle.

### 12. PRINCIPAL AXES

Referring to Fig. 51.7, there is one angle, $\theta$, that will maximize the moment of inertia, $I_u$. This angle can be found from calculus by setting $dI_u/d\theta = 0$. The resulting equation defines two angles, one that maximizes $I_u$ and one that minimizes $I_u$.

$$\tan 2\theta = \frac{-2P_{xy}}{I_x - I_y} \qquad 51.35$$

The two angles that satisfy Eq. 51.35 are 90° apart. The set of $u$-$v$ axes defined by Eq. 51.35 are known as *principal axes*. The moments of inertia about the principal axes are defined by Eq. 51.36 and are known as the *principal moments of inertia*.

$$I_{\text{max,min}} = \tfrac{1}{2}(I_x + I_y) \pm \sqrt{\tfrac{1}{4}(I_x - I_y)^2 + P_{xy}^2} \qquad 51.36$$

### 13. MOHR'S CIRCLE

Once $I_x$, $I_y$, and $P_{xy}$ are known, *Mohr's circle* can be drawn to graphically determine the moments of inertia about the principal axes. The procedure for drawing Mohr's circle is given as follows. (See Fig. 51.8 and Fig. 51.9.)

*Figure 51.8* Mohr's Circle

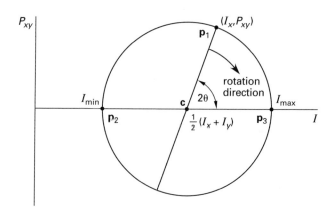

*Figure 51.9* Principal Axes from Mohr's Circle

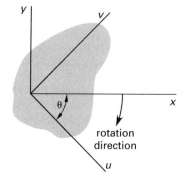

*step 1:* Determine $I_x$, $I_y$, and $P_{xy}$ for the existing set of axes.

*step 2:* Draw a set of $I$-$P_{xy}$ axes.

*step 3:* Plot the center of the circle, point **c**, by calculating distance $c$ along the $I$-axis.

$$c = \tfrac{1}{2}(I_x + I_y) \qquad 51.37$$

*step 4:* Plot the point $\mathbf{p}_1 = (I_x, P_x)$.

*step 5:* Draw a line from point $\mathbf{p}_1$ through center $\mathbf{c}$ and extend it an equal distance below the $I$-axis. This is the diameter of the circle.

*step 6:* Using the center $\mathbf{c}$ and point $\mathbf{p}_1$, draw the circle. An alternate method of constructing the circle is to draw a circle of radius $r$.

$$r = \sqrt{\tfrac{1}{4}(I_x - I_y)^2 + P_{xy}^2} \qquad 51.38$$

*step 7:* Point $\mathbf{p}_2$ defines $I_{\min}$. Point $\mathbf{p}_3$ defines $I_{\max}$.

*step 8:* Determine the angle $\theta$ as half of the angle $2\theta$ on the circle. This angle corresponds to $I_{\max}$. (The axis giving the minimum moment of inertia is perpendicular to the maximum axis.) The sense of this angle and the sense of the rotation are the same. That is, the direction that the diameter would have to be turned in order to coincide with the $I_{\max}$-axis has the same sense as the rotation of the $x$-$y$ axes needed to form the principal $u$-$v$ axes.

# 52 Strength of Materials

## Nomenclature

| | | | |
|---|---|---|---|
| $a$ | width | in | m |
| $A$ | area | in$^2$ | m$^2$ |
| $b$ | width | in | m |
| $c$ | distance to extreme fiber | in | m |
| $C$ | constant | – | – |
| $C$ | couple | in-lbf | N·m |
| $d$ | distance, depth, or diameter | in | m |
| $e$ | eccentricity | in | m |
| $E$ | modulus of elasticity | lbf/in$^2$ | MPa |
| $F$ | force | lbf | N |
| $G$ | shear modulus | lbf/in$^2$ | MPa |
| $h$ | height | in | m |
| $I$ | moment of inertia | in$^4$ | m$^4$ |
| $J$ | polar moment of inertia | in$^4$ | m$^4$ |
| $k$ | spring constant | lbf/in | N/m |
| $K$ | stress concentration factor | – | – |
| $L$ | length | in | m |
| $M$ | moment | in-lbf | N·m |
| $P$ | force | lbf | N |
| $Q$ | statical moment | in$^3$ | m$^3$ |
| $r$ | radius | in | m |
| $R$ | reaction | lbf | N |
| $R$ | rigidity | – | – |
| $S$ | force | lbf | N |
| $S$ | section modulus | in$^3$ | m$^3$ |
| $t$ | thickness | in | m |
| $T$ | temperature | °F | °C |
| $T$ | torque | in-lbf | N·m |
| $u$ | unit force | lbf | N |
| $U$ | energy | in-lbf | N·m |
| $V$ | vertical shear force | lbf | N |
| $V$ | volume | in$^3$ | m$^3$ |
| $w$ | load per unit length | lbf/in | N/m |
| $x$ | location | in | m |
| $y$ | location | in | m |

## Symbols

| | | | |
|---|---|---|---|
| $\alpha$ | coefficient of linear thermal expansion | 1/°F | 1/°C |
| $\beta$ | coefficient of volumetric thermal expansion | 1/°F | 1/°C |
| $\gamma$ | coefficient of area thermal expansion | 1/°F | 1/°C |
| $\delta$ | deformation | in | m |
| $\epsilon$ | strain | – | – |
| $\theta$ | angle | deg | deg |
| $\nu$ | Poisson's ratio | – | – |
| $\rho$ | radius of curvature | in | m |
| $\sigma$ | normal stress | lbf/in$^2$ | MPa |
| $\tau$ | shear stress | lbf/in$^2$ | MPa |
| $\phi$ | angle | rad | rad |

## Subscripts

| | |
|---|---|
| 0 | nominal |
| $a$ | alternating |
| $b$ | bending |
| $c$ | centroidal |
| $i$ | inside |
| $j$ | $j$th member |
| $l$ | left |
| $m$ | mean |
| $o$ | original |
| $r$ | range or right |
| $r$ | right |
| th | thermal |
| $w$ | web |

## 1. BASIC CONCEPTS

*Strength of materials* (known also as *mechanics of materials*) deals with the elastic behavior of loaded engineering materials.[1]

*Stress* is force per unit area, $F/A$. Typical units of stress are $lbf/in^2$, ksi (thousands of pounds per square inch), and MPa. Although there are many names given to stress, there are only two primary types, differing in the orientation of the loaded area. With *normal stress*, $\sigma$, the area is normal to the force carried. With *shear stress*, $\tau$, the area is parallel to the force. (See Fig. 52.1.)

**Figure 52.1** *Normal and Shear Stress*

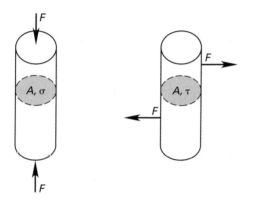

*Strain*, $\epsilon$, is elongation expressed on a fractional or percentage basis. It may be listed as having units of in/in, mm/mm, percent, or no units at all. A strain in one direction will be accompanied by strains in orthogonal directions in accordance with Poisson's ratio. *Dilation* is the sum of the strains in the three coordinate directions.

$$\text{dilation} = \epsilon_x + \epsilon_y + \epsilon_z \qquad 52.1$$

## 2. HOOKE'S LAW

*Hooke's law* is a simple mathematical statement of the relationship between elastic stress and strain: Stress is proportional to strain. (See Fig. 52.2.) For normal stress, the constant of proportionality is the *modulus of elasticity* (*Young's modulus*), $E$.

$$\sigma = E\epsilon \qquad 52.2$$

For shear stress, the constant of proportionality is the *shear modulus*, $G$.

$$\tau = G\phi \qquad 52.3$$

---

[1]Plastic behavior and ultimate strength design are not covered in this chapter.

**Figure 52.2** *Application of Hooke's Law*

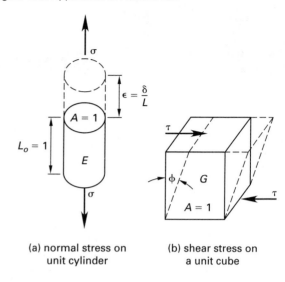

(a) normal stress on unit cylinder

(b) shear stress on a unit cube

## 3. ELASTIC DEFORMATION

Since stress is $F/A$ and strain is $\delta/L_o$, Hooke's law can be rearranged in form to give the elongation of an axially loaded member with a uniform cross section experiencing normal stress. Tension loading is considered positive; compressive loading is negative.

$$\delta = L_o\epsilon = \frac{L_o\sigma}{E} = \frac{L_oF}{EA} \qquad 52.4$$

The actual length of a member under loading is given by Eq. 52.5. The algebraic sign of the deformation must be observed.

$$L = L_o + \delta \qquad 52.5$$

## 4. TOTAL STRAIN ENERGY

The energy stored in a loaded member is equal to the work required to deform the member. Below the proportionality limit, the total *strain energy* for a member loaded in tension or compression is given by Eq. 52.6.

$$U = \tfrac{1}{2}F\delta = \frac{F^2L_o}{2AE} = \frac{\sigma^2 L_o A}{2E} \qquad 52.6$$

## 5. STIFFNESS AND RIGIDITY

*Stiffness* is the amount of force required to cause a unit of deformation (displacement) and is often referred to as a *spring constant*. Typical units are pounds per inch and newtons per meter. The stiffness of a spring or other structure can be calculated from the deformation equation by solving for $F/\delta$. Equation 52.7 is valid for tensile and compressive normal stresses. For torsion and bending, the stiffness equation will depend on how the deflection is calculated.

$$k = \frac{F}{\delta} \quad \text{[general form]} \qquad 52.7(a)$$

$$k = \frac{AE}{L_o} \quad \text{[normal stress form]} \qquad 52.7(b)$$

When more than one spring or resisting member share the load, the relative stiffnesses are known as *rigidities*. Rigidities have no units, and the individual rigidity values have no significance. (See Fig. 52.3.) A ratio of two rigidities, however, indicates how much stiffer one member is compared to another. Equation 52.8 is one method of calculating rigidity in a multi-member structure. (Since rigidities are relative numbers, they can be multiplied by the least common denominator to obtain integer values.)

$$R_j = \frac{k_j}{\sum_i k_i} \qquad 52.8$$

*Rigidity* is proportional to the reciprocal of deflection. (See Table 52.1.) *Flexural rigidity* is the reciprocal of deflection in members that are acted upon by a moment (i.e., are in bending), although that term may also be used to refer to the product, $EI$, of the modulus of elasticity and the moment of inertia.

**Figure 52.3** *Stiffness and Rigidity*

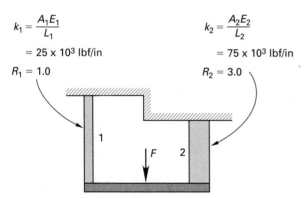

$$k_1 = \frac{A_1 E_1}{L_1}$$
$$= 25 \times 10^3 \text{ lbf/in}$$
$$R_1 = 1.0$$

$$k_2 = \frac{A_2 E_2}{L_2}$$
$$= 75 \times 10^3 \text{ lbf/in}$$
$$R_2 = 3.0$$

## 6. THERMAL DEFORMATION

If the temperature of an object is changed, the object will experience length, area, and volume changes. The magnitude of these changes will depend on the *coefficient of linear expansion*, $\alpha$, which is widely tabulated for solids. (See Table 52.2.) The *coefficient of volumetric expansion*, $\beta$, is encountered less often for solids but is used extensively with liquids and gases.

$$\Delta L = \alpha L_o (T_2 - T_1) \qquad 52.9$$

$$\Delta A = \gamma A_o (T_2 - T_1) \qquad 52.10$$

$$\gamma \approx 2\alpha \qquad 52.11$$

**Table 52.1** *Deflection and Stiffness for Various Systems (due to bending moment alone)*

| system | maximum deflection ($x$) | stiffness ($k$) |
|---|---|---|
| | $\dfrac{Fh}{AE}$ | $\dfrac{AE}{h}$ |
| | $\dfrac{Fh^3}{3EI}$ | $\dfrac{3EI}{h^3}$ |
| | $\dfrac{Fh^3}{12EI}$ | $\dfrac{12EI}{h^3}$ |
| | $\dfrac{wL^4}{8EI}$ | $\dfrac{8EI}{L^3}$ |
| | $\dfrac{Fh^3}{12E(I_1 + I_2)}$ | $\dfrac{12E(I_1 + I_2)}{h^3}$ |
| | $\dfrac{FL^3}{48EI}$ | $\dfrac{48EI}{L^3}$ |
| ($w$ is load per unit length) | $\dfrac{5wL^4}{384EI}$ | $\dfrac{384EI}{5L^3}$ |
| | $\dfrac{FL^3}{192EI}$ | $\dfrac{192EI}{L^3}$ |
| ($w$ is load per unit length) | $\dfrac{wL^4}{384EI}$ | $\dfrac{384EI}{L^3}$ |

*Mechanics of Materials*

***Table 52.2*** *Average Coefficients of Linear Thermal Expansion*
*(multiply all values by $10^{-6}$)*

| substance | 1/°F | 1/°C |
|---|---|---|
| aluminum alloy | 12.8 | 23.0 |
| brass | 10.0 | 18.0 |
| cast iron | 5.6 | 10.1 |
| chromium | 3.8 | 6.8 |
| concrete | 6.7 | 12.0 |
| copper | 8.9 | 16.0 |
| glass (plate) | 4.9 | 8.9 |
| glass (Pyrex$^{TM}$) | 1.8 | 3.2 |
| invar | 0.39 | 0.7 |
| lead | 15.6 | 28.0 |
| magnesium alloy | 14.5 | 26.1 |
| marble | 6.5 | 11.7 |
| platinum | 5.0 | 9.0 |
| quartz, fused | 0.2 | 0.4 |
| steel | 6.5 | 11.7 |
| tin | 14.9 | 26.9 |
| titanium alloy | 4.9 | 8.8 |
| tungsten | 2.4 | 4.4 |
| zinc | 14.6 | 26.3 |

(Multiply 1/°F by 9/5 to obtain 1/°C.)
(Multiply 1/°C by 5/9 to obtain 1/°F.)

$$\Delta V = \beta V_o (T_2 - T_1) \qquad 52.12$$

$$\beta \approx 3\alpha \qquad 52.13$$

It is a common misconception that a hole in a plate will decrease in size when the plate is heated (because the surrounding material "squeezes in" on the hole). However, changes in temperature affect all dimensions the same way. In this case, the circumference of the hole is a linear dimension that follows Eq. 52.9. As the circumference increases, the hole area also increases. (See Fig. 52.4.)

***Figure 52.4*** *Thermal Expansion of an Area*

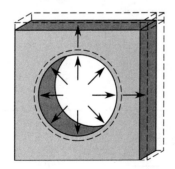

If Eq. 52.9 is rearranged, an expression for the *thermal strain* is obtained.

$$\epsilon_{th} = \frac{\Delta L}{L_o} = \alpha(T_2 - T_1) \qquad 52.14$$

Thermal strain is handled in the same manner as strain due to an applied load. For example, if a bar is heated but is not allowed to expand, the stress can be calculated from the thermal strain and Hooke's law.

$$\sigma_{th} = E\epsilon_{th} \qquad 52.15$$

Low values of the coefficient of expansion, such as with Pyrex$^{TM}$ glassware, result in low thermally induced stresses and high insensitivity to temperature extremes. Differences in the coefficients of expansion of two materials are used in *bimetallic elements*, such as thermostatic springs and strips.

**Example 52.1**

A replacement steel railroad rail ($L = 20.0$ m, $A = 60 \times 10^{-4}$ m$^2$) was installed when its temperature was 5°C. The rail was installed tightly in the line, without an allowance for expansion. If the rail ends are constrained by adjacent rails and if the spikes prevent buckling, what is the compressive force in the rail at 25°C?

*Solution*

From Table 52.2, the coefficient of linear expansion for steel is $11.7 \times 10^{-6}$ 1/°C. From Eq. 52.14, the thermal strain is

$$\epsilon_{th} = \alpha(T_2 - T_1) = \left(11.7 \times 10^{-6} \frac{1}{°C}\right)(25°C - 5°C)$$

$$= 2.34 \times 10^{-4} \text{ m/m}$$

The modulus of elasticity of steel is $20 \times 10^{10}$ N/m$^2$ ($20 \times 10^4$ MPa). The compressive stress is given by Hooke's law. Use Eq. 52.15.

$$\sigma_{th} = E\epsilon_{th} = \left(20 \times 10^{10} \frac{N}{m^2}\right)\left(2.34 \times 10^{-4} \frac{m}{m}\right)$$

$$= 4.68 \times 10^7 \text{ N/m}^2$$

The compressive force is

$$F = \sigma_{th}A = \left(4.68 \times 10^7 \frac{N}{m^2}\right)(60 \times 10^{-4} \text{ m}^2)$$

$$= 281\,000 \text{ N}$$

## 7. STRESS CONCENTRATIONS

A *geometric stress concentration* occurs whenever there is a discontinuity or non-uniformity in an object. Examples of non-uniform shapes are stepped shafts, plates with holes and notches, and shafts with keyways. It is convenient to think of stress as lines of force following streamlines within an object. (See Fig. 52.5.) There will be a stress concentration wherever local geometry forces the streamlines closer together.

**Figure 52.5** *Streamline Analogy to Stress Concentrations*

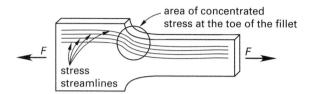

Stress values determined by simplistic $F/A$, $Mc/I$, or $Tr/J$ calculations will be greatly understated. *Stress concentration factors (stress risers)* are correction factors used to account for the non-uniform stress distributions. The symbol $K$ is often used, but this is not universal. The actual stress is determined as the product of the stress concentration factor, $K$, and the *nominal stress*, $\sigma_0$. Values of the stress concentration factor are almost always greater than 1.0 and can run as high as 3.0 and above. The exact value for a given application must be determined from extensive experimentation or from published tabulations of standard configurations.

$$\sigma' = K\sigma_0 \qquad 52.16$$

Stress concentration factors are normally not applied to members with multiple redundancy, for static loading of ductile materials, or where local yielding around the discontinuity reduces the stress. For example, there will be many locations of stress concentration in a lap rivet connection. However, the stresses are kept low by design, and stress concentrations are disregarded.

Stress concentration factors are not applicable to every point on an object; they apply only to the point of maximum stress. For example, with filleted shafts, the maximum stress occurs at the toe of the fillet. Therefore, the stress concentration factor should be applied to the stress calculated from the smaller section's properties. For objects with holes or notches, it is important to know if the nominal stress to which the factor is applied is calculated from an area that includes or excludes the holes or notches.

In addition to geometric stress concentrations, there are also *fatigue stress concentrations*. The *fatigue stress concentration* factor is the ratio of the fatigue strength without a stress concentration to the fatigue stress with a stress concentration. Fatigue stress concentration factors depend on the material, material strength, and geometry of the stress concentration (i.e., radius of the notch). Fatigue stress concentration factors can be less than the geometric factors from which they are computed.

## 8. COMBINED STRESSES (BIAXIAL LOADING)

Loading is rarely confined to a single direction. Many practical cases have different normal and shear stresses on two or more perpendicular planes. Sometimes, one of the stresses may be small enough to be disregarded, reducing the analysis to one dimension. In other cases, however, the shear and normal stresses must be combined to determine the maximum stresses acting on the material.

For any point in a loaded specimen, a plane can be found where the shear stress is zero. The normal stresses associated with this plane are known as the *principal stresses*, which are the maximum and minimum stresses acting at that point in any direction.

For two-dimensional (biaxial) loading (i.e., two normal stresses combined with a shearing stress), the normal and shear stresses on a plane whose normal line is inclined an angle $\theta$ from the horizontal can be found from Eq. 52.17 and Eq. 52.18. Proper sign convention must be adhered to when using the combined stress equations. The positive senses of shear and normal stresses are shown in Fig. 52.6. As is usually the case, tensile normal stresses are positive; compressive normal stresses are negative. In two dimensions, shear stresses are designated as clockwise (positive) or counterclockwise (negative).[2]

$$\sigma_\theta = \tfrac{1}{2}(\sigma_x + \sigma_y) + \tfrac{1}{2}(\sigma_x - \sigma_y)\cos 2\theta + \tau\sin 2\theta$$

$$52.17$$

$$\tau_\theta = -\tfrac{1}{2}(\sigma_x - \sigma_y)\sin 2\theta + \tau\cos 2\theta \qquad 52.18$$

**Figure 52.6** *Sign Convention for Combined Stress*

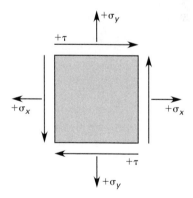

At first glance, the orientation of the shear stresses may seem confusing. However, the arrangement of stresses shown produces equilibrium in the $x$- and $y$-directions without causing rotation. Other than a mirror image or a trivial rotation of the arrangement shown in Fig. 52.6,

---

[2]Some sources refer to the shear stress as $\tau_{xy}$; others use the symbol $\tau_z$. When working in two dimensions only, the subscripts $xy$ and $z$ are unnecessary and confusing conventions.

no other arrangement of shear stresses will produce equilibrium.

The maximum and minimum values (as $\theta$ is varied) of the normal stress, $\sigma_\theta$, are the *principal stresses*, which can be found by differentiating Eq. 52.17 with respect to $\theta$, setting the derivative equal to zero, and substituting $\theta$ back into Eq. 52.17. Equation 52.19 is derived in this manner. A similar procedure is used to derive the *extreme shear stresses* (i.e., maximum and minimum shear stresses) in Eq. 52.20 from Eq. 52.18. (The term *principal stress* implies a normal stress, never a shear stress.)

$$\sigma_1, \sigma_2 = \tfrac{1}{2}(\sigma_x + \sigma_y) \pm \tau_1 \qquad \textit{52.19}$$

$$\tau_1, \tau_2 = \pm \tfrac{1}{2}\sqrt{(\sigma_x - \sigma_y)^2 + (2\tau)^2} \qquad \textit{52.20}$$

The angles of the planes on which the normal stresses are minimum and maximum are given by Eq. 52.21. (See Fig. 52.7.) $\theta$ is measured from the $x$-axis, clockwise if negative and counterclockwise if positive. Equation 52.21 will yield two angles, 90° apart. These angles can be substituted back into Eq. 52.17 and Eq. 52.18 to determine which angle corresponds to the minimum normal stress and which angle corresponds to the maximum normal stress.[3]

$$\theta_{\sigma_1, \sigma_2} = \tfrac{1}{2}\arctan\left(\frac{2\tau}{\sigma_x - \sigma_y}\right) \qquad \textit{52.21}$$

**Figure 52.7** *Stresses on an Inclined Plane*

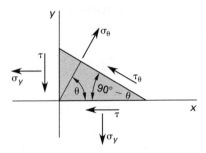

The angles of the planes on which the shear stresses are minimum and maximum are given by Eq. 52.22. These planes will be 90° apart and will be rotated 45° from the planes of principal normal stresses. As with Eq. 52.21, $\theta$ is measured from the $x$-axis, clockwise if negative and counterclockwise if positive. Generally, the sign of a shear stress on an inclined plane will be unimportant.

$$\theta_{\tau_1, \tau_2} = \tfrac{1}{2}\arctan\left(\frac{\sigma_x - \sigma_y}{-2\tau}\right) \qquad \textit{52.22}$$

---

[3]Alternatively, the following procedure can be used to determine the direction of the principal planes. Let $\sigma_x$ be the algebraically larger of the two given normal stresses. The angle between the direction of $\sigma_x$ and the direction of $\sigma_1$, the algebraically larger principal stress, will always be less than 45°.

### Example 52.2

(a) Find the maximum normal and shear stresses on the object shown. (b) Determine the angle of the plane of principal normal stresses.

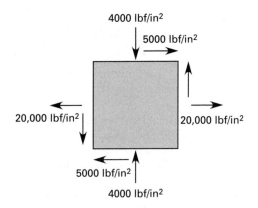

*Solution*

(a) Find the principal shear stresses first. The applied 4000 lbf/in² compressive stress is negative. Equation 52.20 can be used directly.

$$\tau_1 = \tfrac{1}{2}\sqrt{(\sigma_x - \sigma_y)^2 + (2\tau)^2}$$

$$= \tfrac{1}{2}\sqrt{\left(20{,}000\ \frac{\text{lbf}}{\text{in}^2} - \left(-4000\ \frac{\text{lbf}}{\text{in}^2}\right)\right)^2 + \left((2)\left(5000\ \frac{\text{lbf}}{\text{in}^2}\right)\right)^2}$$

$$= 13{,}000\ \text{lbf/in}^2$$

From Eq. 52.19, the maximum normal stress is

$$\sigma_1 = \tfrac{1}{2}(\sigma_x + \sigma_y) + \tau_1$$

$$= \left(\tfrac{1}{2}\right)\left(20{,}000\ \frac{\text{lbf}}{\text{in}^2} + \left(-4000\ \frac{\text{lbf}}{\text{in}^2}\right)\right) + 13{,}000\ \frac{\text{lbf}}{\text{in}^2}$$

$$= 21{,}000\ \text{lbf/in}^2 \quad [\text{tension}]$$

(b) The angle of the principal normal stresses is given by Eq. 52.21.

$$\theta = \tfrac{1}{2}\arctan\left(\frac{2\tau}{\sigma_x - \sigma_y}\right)$$

$$= \tfrac{1}{2}\arctan\left(\frac{(2)\left(5000\ \frac{\text{lbf}}{\text{in}^2}\right)}{20{,}000\ \frac{\text{lbf}}{\text{in}^2} - \left(-4000\ \frac{\text{lbf}}{\text{in}^2}\right)}\right)$$

$$= \left(\tfrac{1}{2}\right)(22.6°, 202.6°)$$

$$= 11.3°, 101.3°$$

It is not obvious which angle produces which normal stress. One of the angles can be substituted back into the general equation (see Eq. 52.17) for $\sigma_\theta$.

$$
\begin{aligned}
\sigma_{11.3°} = {} & \left(\tfrac{1}{2}\right)\left(20{,}000\ \frac{\text{lbf}}{\text{in}^2} + \left(-4000\ \frac{\text{lbf}}{\text{in}^2}\right)\right) \\
& + \left(\tfrac{1}{2}\right)\left(20{,}000\ \frac{\text{lbf}}{\text{in}^2} - \left(-4000\ \frac{\text{lbf}}{\text{in}^2}\right)\right) \\
& \times \cos\left((2)(11.3°)\right) \\
& + \left(5000\ \frac{\text{lbf}}{\text{in}^2}\right)\sin\left((2)(11.3°)\right) \\
= {} & 21{,}000\ \text{lbf/in}^2
\end{aligned}
$$

Thus, the 11.3° angle corresponds to the maximum normal stress of 21,000 lbf/in².

## 9. MOHR'S CIRCLE FOR STRESS

Mohr's circle can be constructed to graphically determine the principal stresses. (See Fig. 52.8.)

**Figure 52.8** Mohr's Circle for Stress

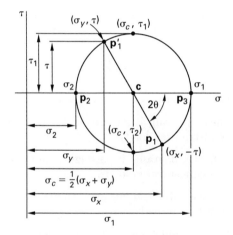

step 1: Determine the applied stresses: $\sigma_x$, $\sigma_y$, and $\tau$. (Tensile normal stresses are positive; compressive normal stresses are negative. Clockwise shear stresses are positive; counterclockwise shear stresses are negative.)

step 2: Draw a set of $\sigma$-$\tau$ axes.

step 3: Plot the center of the circle, point **c**, by calculating $\sigma_c = \tfrac{1}{2}(\sigma_x + \sigma_y)$.

step 4: Plot the point $\mathbf{p}_1 = (\sigma_x, -\tau)$. (Alternatively, plot $\mathbf{p}_1'$ at $(\sigma_y, +\tau)$.)

step 5: Draw a line from point $\mathbf{p}_1$ through center **c** and extend it an equal distance beyond the $\sigma$-axis. This is the diameter of the circle.

step 6: Using the center **c** and point $\mathbf{p}_1$, draw the circle. An alternative method is to draw a circle of radius $r$ about point **c**.

$$
r = \sqrt{\tfrac{1}{4}(\sigma_x - \sigma_y)^2 + \tau^2} \qquad \text{52.23}
$$

step 7: Point $\mathbf{p}_2$ defines the smaller principal stress, $\sigma_2$. Point $\mathbf{p}_3$ defines the larger principal stress, $\sigma_1$.

step 8: Determine the angle $\theta$ as half of the angle $2\theta$ on the circle. This angle corresponds to the larger principal stress, $\sigma_1$. On Mohr's circle, angle $2\theta$ is measured counterclockwise from the $\mathbf{p}_1$-$\mathbf{p}_1'$ line to the horizontal axis.

### Example 52.3

Construct Mohr's circle for Ex. 52.2.

*Solution*

$$
\begin{aligned}
\sigma_c &= \tfrac{1}{2}(\sigma_x + \sigma_y) \\
&= \left(\tfrac{1}{2}\right)\left(20{,}000\ \frac{\text{lbf}}{\text{in}^2} + \left(-4000\ \frac{\text{lbf}}{\text{in}^2}\right)\right) \\
&= 8000\ \text{lbf/in}^2
\end{aligned}
$$

$$
\begin{aligned}
r &= \sqrt{\tfrac{1}{4}(\sigma_x - \sigma_y)^2 + \tau^2} \\
&= \sqrt{\begin{array}{c}\left(\left(\tfrac{1}{4}\right)\left(20{,}000\ \dfrac{\text{lbf}}{\text{in}^2} - \left(-4000\ \dfrac{\text{lbf}}{\text{in}^2}\right)\right)\right)^2 \\[6pt] + \left(5000\ \dfrac{\text{lbf}}{\text{in}^2}\right)^2\end{array}} \\
&= 13{,}000\ \text{lbf/in}^2
\end{aligned}
$$

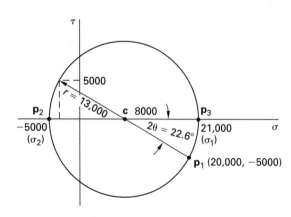

## 10. IMPACT LOADING

If a load is applied to a structure suddenly, the structure's response will be composed of two parts: a transient response (which decays to zero) and a steady-state response. (These two parts are also known as the *dynamic* and *static responses*, respectively.) It is not unusual for the transient loading to be larger than the steady-state response.

Although a *dynamic analysis* of the structure is preferred, the procedure is lengthy and complex. Therefore, arbitrary multiplicative factors may be applied to the steady-state stress to determine the maximum transient. For example, if a load is applied quickly as compared to the natural period of vibration of the structure (e.g., the classic definition of an *impact load*), a dynamic factor of 2.0 might be used. Actual dynamic factors should be determined or validated by testing.

The energy-conservation method (i.e., the work-energy principle) can be used to determine the maximum stress due to a falling mass. The total change in potential energy of the mass from the change in elevation and the deflection $\delta$) is equated to the appropriate expression for total strain energy. (See Sec. 52.4.)

## 11. SHEAR AND MOMENT

*Shear* at a point is the sum of all vertical forces acting on an object. It has units of pounds, kips, tons, newtons, and so on. Shear is not the same as shear stress, since the area of the object is not considered.

A typical application is shear at a point on a beam, $V$, defined as the sum of all vertical forces between the point and one of the ends.[4] The direction (i.e., to the left or right of the point) in which the summation proceeds is not important. Since the values of shear will differ only in sign for summations to the left and right ends, the direction that results in the fewest calculations should be selected.

$$V = \sum_{\substack{\text{point to} \\ \text{one end}}} F_i \qquad 52.24$$

Shear is taken as positive when there is a net upward force to the left of a point and negative when there is a net downward force between the point and the left end.

*Moment* at a point is the total bending moment acting on an object. In the case of a beam, the moment, $M$, will be the algebraic sum of all moments and couples located between the investigation point and one of the beam ends. As with shear, the number of calculations required to calculate the moment can be minimized by careful choice of the beam end.[5]

$$M = \sum_{\substack{\text{point to} \\ \text{one end}}} F_i d_i + \sum_{\substack{\text{point to} \\ \text{one end}}} C_i \qquad 52.25$$

---

[4]The conditions of equilibrium require that the sum of all vertical forces on a beam be zero. However, the *shear* can be nonzero because only a portion of the beam is included in the analysis. Since that portion extends to the beam end in one direction only, shear is sometimes called *resisting shear* or *one-way shear*.

[5]The conditions of equilibrium require that the sum of all moments on a beam be zero. However, the *moment* can be nonzero because only a portion of the beam is included in the analysis. Since that portion extends to the beam end in one direction only, moment is sometimes called *bending moment*, *flexural moment*, *resisting moment*, or *one-way moment*.

Moment is taken as positive when the upper surface of the beam is in compression and the lower surface is in tension. (See Fig. 52.12.) Since the beam ends will usually be higher than the midpoint, it is commonly said that "a positive moment will make the beam smile."

## 12. SHEAR AND BENDING MOMENT DIAGRAMS

The value of the shear and moment, $V$ and $M$, will depend on location along the beam. Both shear and moment can be described mathematically for simple loadings, but the formulas are likely to become discontinuous as the loadings become more complex. It is much more convenient to describe the shear and moment functions graphically. Graphs of shear and moment as functions of position along the beam are known as *shear* and *moment diagrams*. Drawing these diagrams does not require knowing the shape or area of the beam.

The following guidelines and conventions should be observed when constructing a *shear diagram*.

- The shear at any point is equal to the sum of the loads and reactions from the point to the left end.

- The magnitude of the shear at any point is equal to the slope of the moment line at that point.

$$V = \frac{dM}{dx} \qquad 52.26$$

- Loads and reactions acting upward are positive.

- The shear diagram is straight and sloping over uniformly distributed loads.

- The shear diagram is straight and horizontal between concentrated loads.

- The shear is a vertical line and is undefined at points of concentrated loads.

The following guidelines and conventions should be observed when constructing a *bending moment diagram*. By convention, the moment diagram is drawn on the compression side of the beam.

- The moment at any point is equal to the sum of the moments and couples from the point to the left end.[6]

- Clockwise moments about the point are positive.

- The magnitude of the moment at any point is equal to the area under the shear line up to that point. This is equivalent to the integral of the shear function.

$$M = \int V\,dx \qquad 52.27$$

---

[6]If the beam is cantilevered with its built-in end at the left, the fixed-end moment will be unknown. In that case, the moment must be calculated to the right end of the beam.

- The *maximum moment* occurs where the shear is zero.

- The moment diagram is straight and sloping between concentrated loads.

- The moment diagram is curved (parabolic upward) over uniformly distributed loads.

These principles are illustrated in Fig. 52.9.

**Figure 52.9** *Drawing Shear and Moment Diagrams*

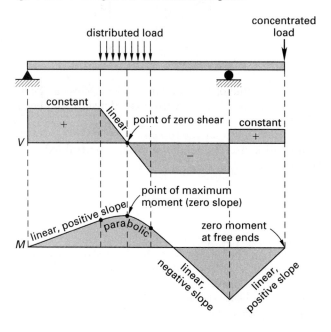

## Example 52.4

Draw the shear and bending moment diagrams for the following beam.

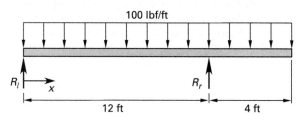

*Solution*

First, determine the reactions. The uniform load of $100x$ can be assumed to be concentrated at $L/2$.

$$R_r = \frac{\left(\frac{1}{2}\right)(16 \text{ ft})(16 \text{ ft})\left(100 \frac{\text{lbf}}{\text{ft}}\right)}{12 \text{ ft}} = 1066.7 \text{ lbf}$$

$$R_l = (16 \text{ ft})\left(100 \frac{\text{lbf}}{\text{ft}}\right) - R_r = 533.3 \text{ lbf}$$

The shear diagram starts at $+533.3$ at the left reaction but decreases linearly at the rate of 100 lbf/ft between

the two reactions. Measuring $x$ from the left end, the shear line goes through zero at

$$x = \frac{533.3 \text{ lbf}}{100 \frac{\text{lbf}}{\text{ft}}} = 5.333 \text{ ft}$$

The shear just to the left of the right reaction is

$$533.3 \text{ lbf} - (12 \text{ ft})\left(100 \frac{\text{lbf}}{\text{ft}}\right) = -666.7 \text{ lbf}$$

The shear just to the right of the right reaction is

$$-666.7 \text{ lbf} + R_r = +400 \text{ lbf}$$

To the right of the right reaction, the shear diagram decreases to zero at the same constant rate: 100 lbf/ft. This is sufficient information to draw the shear diagram.

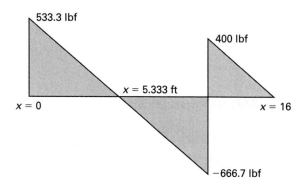

The bending moment at a distance $x$ to the right of the left end has two parts. The left reaction of 533.3 lbf acts with moment arm $x$. The moment between the two reactions is

$$M_x = 533.3x - 100x\left(\frac{x}{2}\right)$$

This equation describes a parabolic section (curved upward) with a peak at $x = 5.333$ ft, where the shear is zero. The maximum moment is

$$M_{x=5.333 \text{ ft}} = (533.3 \text{ lbf})(5.333 \text{ ft})$$
$$- \left(50 \frac{\text{lbf}}{\text{ft}}\right)(5.333 \text{ ft})^2$$
$$= 1422.0 \text{ ft-lbf}$$

The moment at the right reaction (where $x = 12$ ft) is

$$M_{x=12 \text{ ft}} = (533.3 \text{ lbf})(12 \text{ ft}) - \left(50 \frac{\text{lbf}}{\text{ft}}\right)(12 \text{ ft})^2$$
$$= -800 \text{ ft-lbf}$$

The right end is a free end, so the moment is zero. The moment between the right reaction and the right end could be calculated by summing moments to the left end, but it is more convenient to sum moments to the right end. Measuring $x$ from the right end, the moment is derived only from the uniform load.

$$M = 100x\left(\frac{x}{2}\right) = 50x^2$$

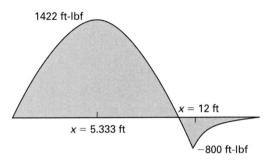

This is sufficient information to draw the moment diagram. Once the maximum moment is located, no attempt is made to determine the exact curvature. The point where $M = 0$ is of limited interest, and no attempt is made to determine the exact location.

Notice that the cross-sectional area of the beam was not needed in this example.

## 13. SHEAR STRESS IN BEAMS

*Shear stress* is generally not the limiting factor in most designs. However, it can control (or be limited by code) in wood, masonry, and concrete beams and in thin tubes.

The average shear stress experienced at a point along the length of a beam depends on the shear, $V$, at that point and the area, $A$, of the beam. The shear can be found from the shear diagram.

$$\tau = \frac{V}{A} \qquad 52.28$$

In most cases, the entire area, $A$, of the beam is used in calculating the average shear stress. However, in flanged beam calculations it is assumed that only the web carries the average shear stress.[7] (See Fig. 52.10.) The flanges are not included in shear stress calculations.

$$\tau = \frac{V}{t_w d} \qquad 52.29$$

Shear stress is also induced in a beam due to flexure (i.e., bending). Figure 52.6 shows that for biaxial loading, identical shear stresses exist simultaneously in all four directions. One set of parallel shears (a couple) counteracts the rotational moment from the other set of parallel shears. The horizontal shear exists even when the loading is vertical (e.g., when a horizontal beam is loaded by

a vertical force). For that reason, the term *horizontal shear* is sometimes used to distinguish it from the applied shear load.

**Figure 52.10** *Web of a Flanged Beam*

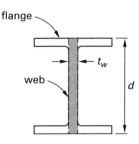

The exact value of the horizontal shear stress is dependent on the location, $y_1$, within the depth of the beam. The shear stress distribution is given by Eq. 52.30. The shear stress is zero at the top and bottom surfaces of the beam and is usually maximum at the neutral axis (i.e., the center).

$$\tau_{y_1} = \frac{QV}{Ib} \qquad 52.30$$

In Eq. 52.30, $V$ is the vertical shear at the point along the length of the beam where the shear stress is wanted. $I$ is the beam's centroidal moment of inertia, and $b$ is the width of the beam at the depth $y_1$ within the beam where the shear stress is wanted. $Q$ is the *statical moment* of the area, as defined by Eq. 52.31.

$$Q = \int_{y_1}^{c} y \, dA \qquad 52.31$$

For rectangular beams, $dA = b \, dy$. Then, the statical moment of the area $A^*$ above layer $y_1$ is equal to the product of the area and the distance from the centroidal axis to the centroid of the area. (See Fig. 52.11.)

$$Q = y^* A^* \qquad 52.32$$

**Figure 52.11** *Shear Stress Distribution Within a Rectangular Beam*

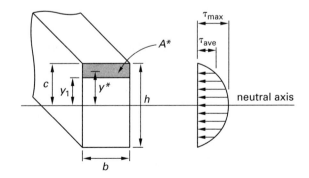

---

[7]This is more than an assumption; it is a fact. There are several reasons the flanges do not contribute to shear resistance, including a non-uniform shear stress distribution in the flanges. This non-uniformity is too complex to be analyzed by elementary methods.

Equation 52.33 calculates the maximum shear stress in a rectangular beam. It is 50% higher than the average shear stress.

$$\tau_{\text{max,rectangular}} = \frac{3V}{2A} = \frac{3V}{2bh} \qquad 52.33$$

For a beam with a circular cross section, the maximum shear stress is

$$\tau_{\text{max,circular}} = \frac{4V}{3A} = \frac{4V}{3\pi r^2} \qquad 52.34$$

For a hollow cylinder used as a beam, the maximum shear stress occurs at the plane of the neutral axis and is

$$\tau_{\text{max,hollow cylinder}} = \frac{2V}{A} \qquad 52.35$$

## 14. BENDING STRESS IN BEAMS

Normal stress occurs in a bending beam, as shown in Fig. 52.12, where the beam is acted upon by a *transverse force*. Although it is a normal stress, the term *bending stress* or *flexural stress* is used to indicate the cause of the stress. The lower surface of the beam experiences tensile stress (which causes lengthening). The upper surface of the beam experiences compressive stress (which causes shortening). There is no normal stress along a horizontal plane passing through the centroid of the cross section, a plane known as the *neutral plane* or the *neutral axis*.

**Figure 52.12** *Normal Stress Due to Bending*

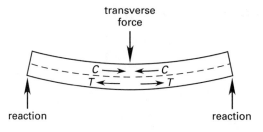

Bending stress varies with location (depth) within the beam. It is zero at the neutral axis and increases linearly with distance from the neutral axis, as predicted by Eq. 52.36. (See Fig. 52.13.)

$$\sigma_b = \frac{-My}{I_c} \qquad 52.36$$

In Eq. 52.36, $M$ is the *bending moment*. $I_c$ is the centroidal moment of inertia of the beam's cross section. The negative sign in Eq. 52.36, required by the convention that compression is negative, is commonly omitted.

Since the maximum stress will govern the design, $y$ can be set equal to $c$ to obtain the *extreme fiber stress*. $c$ is the distance from the neutral axis to the *extreme fiber*

(i.e., the top or bottom surface most distant from the neutral axis).

$$\sigma_{b,\text{max}} = \frac{Mc}{I_c} \qquad 52.37$$

Equation 52.37 shows that the maximum bending stress will occur where the moment along the length of the beam is maximum. The region immediately adjacent to the point of maximum bending moment is called the *dangerous section* of the beam. The dangerous section can be found from a bending moment or shear diagram.

**Figure 52.13** *Bending Stress Distribution in a Beam*

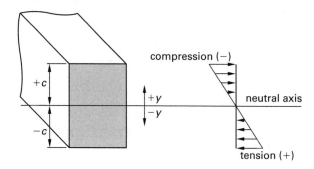

For any given beam cross section, $I_c$ and $c$ are fixed. Therefore, these two terms can be combined into the *section modulus*, $S$.[8]

$$\sigma_{b,\text{max}} = \frac{M}{S} \qquad 52.38$$

$$S = \frac{I_c}{c} \qquad 52.39$$

Since $c = h/2$, the section modulus of a rectangular $b \times h$ section ($I_c = bh^3/12$) is

$$S_{\text{rectangular}} = \frac{bh^2}{6} \qquad 52.40$$

## Example 52.5

The beam in Ex. 52.4 has a 6 in × 8 in cross section. What are the maximum shear and bending stresses in the beam?

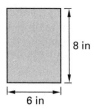

*Solution*

The maximum shear (taken from the shear diagram) is 666.7 lbf. (The negative sign can be disregarded.)

---

[8] The symbol $Z$ is also commonly used for the section modulus.

From Eq. 52.33, the maximum shear stress in a rectangular beam is

$$\tau_{max} = \frac{3V}{2A} = \frac{(3)(666.7 \text{ lbf})}{(2)(6 \text{ in})(8 \text{ in})}$$

$$= 20.8 \text{ lbf/in}^2$$

The centroidal moment of inertia is

$$I_c = \frac{bh^3}{12} = \frac{(6 \text{ in})(8 \text{ in})^3}{12} = 256 \text{ in}^4$$

The maximum bending moment (from the bending moment diagram) is 1422 ft-lbf. From Eq. 52.37, the maximum bending stress is

$$\sigma_{b,max} = \frac{Mc}{I_c} = \frac{(1422 \text{ ft-lbf})\left(12 \frac{\text{in}}{\text{ft}}\right)(4 \text{ in})}{256 \text{ in}^4}$$

$$= 266.6 \text{ lbf/in}^2$$

## 15. STRAIN ENERGY DUE TO BENDING MOMENT

The elastic strain energy due to a bending moment stored in a beam is

$$U = \frac{1}{2EI} \int M^2(x)\, dx \qquad\qquad 52.41$$

The use of Eq. 52.41 is illustrated by Ex. 52.10.

## 16. ECCENTRIC LOADING OF AXIAL MEMBERS

If a load is applied through the centroid of a tension or compression member's cross section, the loading is said to be *axial loading* or *concentric loading. Eccentric loading* occurs when the load is not applied through the centroid.

If an axial member is loaded eccentrically, it will bend and experience bending stress in the same manner as a beam. Since the member experiences both axial stress and bending stress, it is known as a *beam-column*. In Fig. 52.14, $e$ is known as the *eccentricity*.

Both the axial stress and bending stress are normal stresses oriented in the same direction; therefore, simple addition can be used to combine them. Combined stress theory is not applicable. By convention, $F$ is negative if the force compresses the member (as shown in Fig. 52.14).

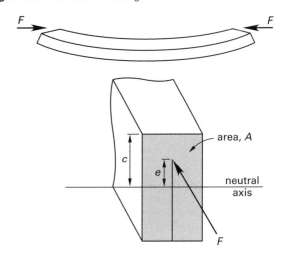

**Figure 52.14** *Eccentric Loading of an Axial Member*

$$\sigma_{max,min} = \frac{F}{A} \pm \frac{Mc}{I_c} \qquad\qquad 52.42$$

$$\sigma_{max,min} = \frac{F}{A} \pm \frac{Fec}{I_c} \qquad\qquad 52.43$$

If a pier or column (primarily designed as a compression member) is loaded with an eccentric compressive load, part of the section can still be placed in tension. (See Fig. 52.15.) Tension will exist when the $Mc/I_c$ term in Eq. 52.42 is larger than the $F/A$ term. It is particularly important to eliminate or severely limit tensile stresses in unreinforced concrete and masonry piers, since these materials cannot support tension.

**Figure 52.15** *Tension in a Pier*

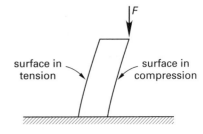

Regardless of the magnitude of the load, there will be no tension as long as the eccentricity is low. In a rectangular member, the load must be kept within a rhombus-shaped area formed from the middle thirds of the centroidal axes. This area is known as the *core, kern,* or *kernel*. Figure 52.16 illustrates the kernel for other cross sections.

**Figure 52.16** *Kerns of Common Cross Sections*

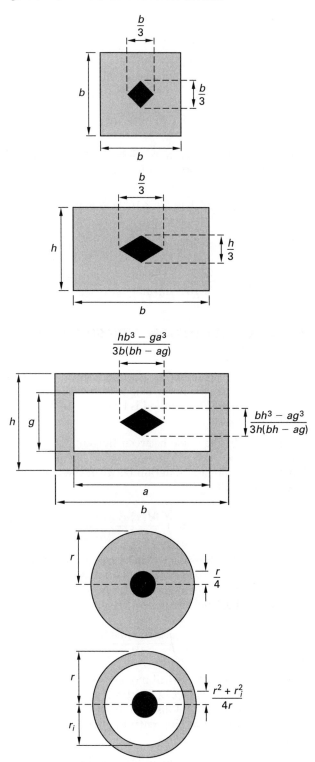

### Example 52.6

A built-in hook with a cross section of 1 in × 1 in carries a load of 500 lbf, but the load is not in line with the centroidal axis of the hook's neck. What are the minimum and maximum stresses in the neck? Is the neck in tension everywhere?

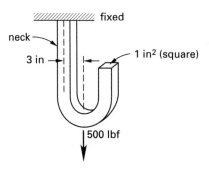

*Solution*

The centroidal moment of inertia of a 1 in × 1 in section is

$$I_c = \frac{bh^3}{12} = \frac{(1 \text{ in})(1 \text{ in})^3}{12} = 0.0833 \text{ in}^4$$

The hook is eccentrically loaded with an eccentricity of 3 in. From Eq. 52.43, the total stress is the sum of the direct axial tension and the bending stress. As the hook bends to reduce the eccentricity, the inner face of the neck will experience a tensile bending stress. The outer face of the neck will experience a compressive bending stress.

$$\begin{aligned}
\sigma_{\text{max,min}} &= \frac{F}{A} \pm \frac{Fec}{I} \\
&= \frac{500 \text{ lbf}}{1 \text{ in}^2} \pm \frac{(500 \text{ lbf})(3 \text{ in})(0.5 \text{ in})}{0.0833 \text{ in}^4} \\
&= 500 \ \frac{\text{lbf}}{\text{in}^2} \pm 9000 \ \frac{\text{lbf}}{\text{in}^2} \\
&= +9500 \text{ lbf/in}^2, \ -8500 \text{ lbf/in}^2
\end{aligned}$$

The 500 lbf/in$^2$ direct stress is tensile, and the inner face experiences a total tensile stress of 9500 lbf/in$^2$. However, the compressive bending stress of 9000 lbf/in$^2$ counteracts the direct tensile stress, resulting in an 8500 lbf/in$^2$ compressive stress at the outer face of the neck.

## 17. BEAM DEFLECTION: DOUBLE INTEGRATION METHOD

The deflection and the slope of a loaded beam are related to the moment and shear by Eq. 52.44 through Eq. 52.48.

$$y = \text{deflection} \qquad \qquad 52.44$$

$$y' = \frac{dy}{dx} = \text{slope} \qquad \qquad 52.45$$

$$y'' = \frac{d^2y}{dx^2} = \frac{M(x)}{EI} \qquad \qquad 52.46$$

$$y''' = \frac{d^3y}{dx^3} = \frac{V(x)}{EI} \qquad \qquad 52.47$$

If the *moment function*, $M(x)$, is known for a section of the beam, the deflection at any point can be found from Eq. 52.48.

$$y = \frac{1}{EI} \int \left( \int M(x)\,dx \right) dx \qquad \textbf{52.48}$$

In order to find the deflection, constants must be introduced during the integration process. For some simple configurations, these constants can be found from Table 52.3.

**Table 52.3** Beam Boundary Conditions

| end condition | $y$ | $y'$ | $y''$ | $V$ | $M$ |
|---|---|---|---|---|---|
| simple support | 0 | | | | 0 |
| built-in support | 0 | 0 | | | |
| free end | | | 0 | 0 | 0 |
| hinge | | | | | 0 |

## Example 52.7

Find the tip deflection of the beam shown. $EI$ is $5 \times 10^{10}$ lbf-in$^2$ everywhere.

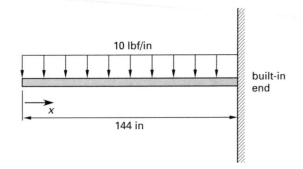

*Solution*

The moment at any point $x$ from the left end of the beam is

$$M(x) = -10x\left(\tfrac{1}{2}x\right) = -5x^2$$

This is negative by the left-hand rule convention. From Eq. 52.46,

$$y'' = \frac{M(x)}{EI}$$

$$EIy'' = M(x) = -5x^2$$

$$EIy' = \int -5x^2\,dx = -\tfrac{5}{3}x^3 + C_1$$

Since $y' = 0$ at a built-in support (Table 52.3) and $x = 144$ in at the built-in support,

$$0 = \left(-\tfrac{5}{3}\right)(144)^3 + C_1$$

$$C_1 = 4.98 \times 10^6$$

$$EIy = \int \left(-\tfrac{5}{3}x^3 + 4.98 \times 10^6\right) dx$$

$$= -\tfrac{5}{12}x^4 + (4.98 \times 10^6)x + C_2$$

Again, $y = 0$ at $x = 144$ in, so $C_2 = -5.38 \times 10^8$ lbf-in$^3$. Therefore, the deflection as a function of $x$ is

$$y = \frac{1}{EI}\left(-\tfrac{5}{12}x^4 + (4.98 \times 10^6)x - 5.38 \times 10^8\right)$$

At the tip $x = 0$, so the deflection is

$$y_{\text{tip}} = \frac{-5.38 \times 10^8 \text{ lbf-in}^3}{5 \times 10^{10} \text{ lbf-in}^2} = -0.0108 \text{ in}$$

## 18. BEAM DEFLECTION: MOMENT AREA METHOD

The moment area method is a semigraphical technique that is applicable whenever slopes of deflection beams are not too great. This method is based on the following two theorems.

- *Theorem I:* The angle between tangents at any two points on the *elastic line* of a beam is equal to the area of the moment diagram between the two points divided by $EI$.

$$\phi = \int \frac{M(x)\,dx}{EI} \qquad \textbf{52.49}$$

- *Theorem II:* One point's deflection away from the tangent of another point is equal to the *statical moment* of the bending moment between those two points divided by $EI$.

$$y = \int \frac{xM(x)\,dx}{EI} \qquad \textbf{52.50}$$

If $EI$ is constant, the statical moment $\int xM(x)\,dx$ can be calculated as the product of the total moment diagram area times the horizontal distance from the point whose deflection is wanted to the centroid of the moment diagram.

If the moment diagram has positive and negative parts (areas above and below the zero line), the statical moment should be taken as the sum of two products, one for each part of the moment diagram.

## Example 52.8

Find the deflection, $y$, and the angle, $\phi$, at the free end of the cantilever beam shown. Neglect the beam weight.

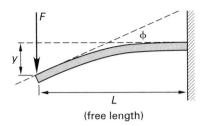

(free length)

*Solution*

The deflection angle, $\phi$, is the angle between the tangents at the free and built-in ends (Theorem I). The moment diagram is

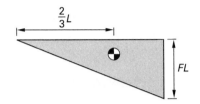

The area of the moment diagram is

$$\tfrac{1}{2}(FL)(L) = \tfrac{1}{2}FL^2$$

From Eq. 52.49,

$$\phi = \frac{FL^2}{2EI}$$

From Eq. 52.50,

$$y = \left(\frac{FL^2}{2EI}\right)\left(\tfrac{2}{3}L\right) = \frac{FL^3}{3EI}$$

### Example 52.9

Find the deflection of the free end of the cantilever beam shown. Neglect the beam weight.

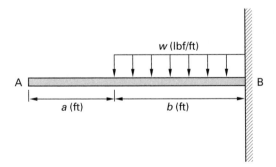

*Solution*

The distance from point A (where the deflection is wanted) to the centroid is $a + 0.75b$. The area of the moment diagram is $wb^3/6$. From Theorem II,

$$y = \left(\frac{wb^3}{6EI}\right)(a + 0.75b)$$

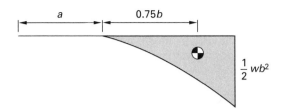

## 19. BEAM DEFLECTION: STRAIN ENERGY METHOD

The deflection at a point of load application can be found by the strain energy method. This method uses the work-energy principle and equates the external work to the total internal strain energy. Since work is a force moving through a distance (which in this case is the deflection), Eq. 52.51 holds true.

$$\tfrac{1}{2}Fy = \sum U \qquad\qquad 52.51$$

### Example 52.10

Find the deflection at the tip of the stepped beam shown.

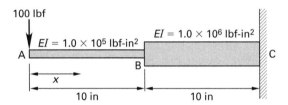

*Solution*

In section AB, $M(x) = 100x$ in-lbf.

From Eq. 52.41,

$$U = \frac{1}{2EI}\int M^2(x)\,dx$$

$$= \frac{1}{(2)(1 \times 10^5 \text{ lbf-in}^2)}\int_0^{10 \text{ in}}(100x)^2 = 16.67 \text{ in-lbf}$$

In section BC, $M = 100x$.

$$U = \frac{1}{(2)(1 \times 10^6 \text{ lbf-in}^2)}\int_{10 \text{ in}}^{20 \text{ in}}(100x)^2 = 11.67 \text{ in-lbf}$$

Equating the internal work ($U$) and the external work,

$$\sum U = W$$

$$16.67 \text{ in-lbf} + 11.67 \text{ in-lbf} = \left(\tfrac{1}{2}\right)(100 \text{ lbf})y$$

$$y = 0.567 \text{ in}$$

## 20. BEAM DEFLECTION: CONJUGATE BEAM METHOD

The *conjugate beam method* changes a deflection problem into one of drawing moment diagrams. The method has the advantage of being able to handle beams of varying cross sections (e.g., stepped beams) and materials. It has the disadvantage of not easily being able to handle beams with two built-in ends. The following steps constitute the conjugate beam method.

*step 1:* Draw the moment diagram for the beam as it is actually loaded.

*step 2:* Construct the $M/EI$ diagram by dividing the value of $M$ at every point along the beam by $EI$ at that point. If the beam is of constant cross section, $EI$ will be constant, and the $M/EI$ diagram will have the same shape as the moment diagram. However, if the beam cross section varies with $x$, $I$ will change. In that case, the $M/EI$ diagram will not look the same as the moment diagram.

*step 3:* Draw a conjugate beam of the same length as the original beam. The material and the cross-sectional area of this conjugate beam are not relevant.

   (a) If the actual beam is simply supported at its ends, the conjugate beam will be simply supported at its ends.

   (b) If the actual beam is simply supported away from its ends, the conjugate beam has hinges at the support points.

   (c) If the actual beam has free ends, the conjugate beam has built-in ends.

   (d) If the actual beam has built-in ends, the conjugate beam has free ends.

*step 4:* Load the conjugate beam with the $M/EI$ diagram. Find the conjugate reactions by methods of statics. Use the superscript * to indicate conjugate parameters.

*step 5:* Find the conjugate moment at the point where the deflection is wanted. The deflection is numerically equal to the moment as calculated from the conjugate beam forces.

### Example 52.11

Find the deflections at the two load points. $EI$ has a constant value of $2.356 \times 10^7$ lbf-in$^2$.

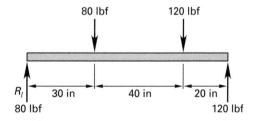

*Solution*

Applying step 1, the moment diagram for the actual beam is

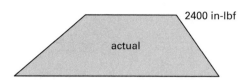

Applying steps 2, 3, and 4, since the beam cross section is constant, the conjugate load has the same shape as the original moment diagram. The peak load on the conjugate beam is

$$\frac{M}{EI} = \frac{2400 \text{ in-lbf}}{2.356 \times 10^7 \text{ lbf-in}^2} = 1.019 \times 10^{-4} \text{ 1/in}$$

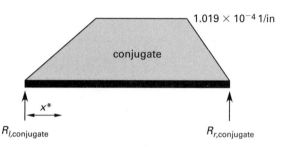

The conjugate reaction, $R_{l,\text{conjugate}}$, is found by the following method. The loading diagram is assumed to be made up of a rectangular load and two negative triangular loads. The area of the rectangular load (which has a centroid at $x_{\text{conjugate}} = 45$ in) is taken as $(90 \text{ in})(1.019 \times 10^{-4} \text{ 1/in}) = 9.171 \times 10^{-3}$.

Similarly, the area of the left triangle (which has a centroid at $x_{\text{conjugate}} = 10$ in) is $(\frac{1}{2})(30 \text{ in})(1.019 \times 10^{-4} \text{ 1/in}) = 1.529 \times 10^{-3}$. The area of the right triangle (which has a centroid at $x_{\text{conjugate}} = 83.33$ in) is taken as $(\frac{1}{2})(20 \text{ in})(1.019 \times 10^{-4} \text{ 1/in}) = 1.019 \times 10^{-3}$.

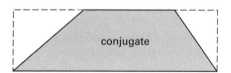

$$\sum M_l^* = (90 \text{ in})R_{r,\text{conjugate}} + (1.019 \times 10^{-3})$$
$$\times (83.3 \text{ in})$$
$$+ (1.529 \times 10^{-3})(10 \text{ in})$$
$$- (9.171 \times 10^{-3})(45 \text{ in})$$
$$= 0$$
$$R_{r,\text{conjugate}} = 3.472 \times 10^{-3}$$

Then,

$$R_{l,\text{conjugate}} = (9.171 - 1.019$$
$$- 1.529 - 3.472) \times 10^{-3}$$
$$= 3.151 \times 10^{-3}$$

In step 5, the conjugate moment at $x_{\text{conjugate}} = 30$ in is the deflection of the actual beam at that point.

$$
\begin{aligned}
M_{\text{conjugate}} &= (3.151 \times 10^{-3})(30 \text{ in}) \\
&\quad + (1.529 \times 10^{-3})(30 \text{ in} - 10 \text{ in}) \\
&\quad - (9.171 \times 10^{-3})\left(\frac{30 \text{ in}}{90 \text{ in}}\right)(15 \text{ in}) \\
&= 7.926 \times 10^{-2} \text{ in}
\end{aligned}
$$

The conjugate moment (the deflection) at the right-most load is

$$
\begin{aligned}
M_{\text{conjugate}} &= (3.472 \times 10^{-3})(20 \text{ in}) \\
&\quad + (1.019 \times 10^{-3})(13.3 \text{ in}) \\
&\quad - (9.171 \times 10^{-3})\left(\frac{20 \text{ in}}{90 \text{ in}}\right)(10 \text{ in}) \\
&= 6.261 \times 10^{-2} \text{ in}
\end{aligned}
$$

## 21. BEAM DEFLECTION: TABLE LOOK-UP METHOD

Appendix 52.A is a compilation of the most commonly used beam deflection formulas. These formulas should never need to be derived and should be used whenever possible. They are particularly useful in calculating deflections due to multiple loads using the principle of superposition.

The actual deflection of very *wide beams* (i.e., those whose widths are larger than 8 or 10 times the thickness) is less than that predicted by the equations in App. 52.A for elastic behavior. (This is particularly true for leaf springs.) The large width prevents lateral expansion and contraction of the beam material, reducing the deflection. For wide beams, the calculated deflection should be reduced by multiplying by $(1 - \nu^2)$.

## 22. BEAM DEFLECTION: SUPERPOSITION

When multiple loads act simultaneously on a beam, all of the loads contribute to deflection. The principle of *superposition* permits the deflections at a point to be calculated as the sum of the deflections from each individual load acting singly.[9] This principle is valid as long as none of the deflections is excessive and all stresses are kept less than the yield point of the beam material.

## 23. INFLECTION POINTS

The *inflection point* (also known as a *point of contraflexure*) on a horizontal beam in elastic bending occurs where the curvature changes from concave up to

concave down, or vice versa. There are three ways of determining the inflection point.

1. If the elastic deflection equation, $y(x)$, is known, the inflection point can be found consistent with normal calculus methods (i.e., by determining the value of $x$ for which $y'(x) = M(x) = 0$).

2. From Eq. 52.46, $y''(x) = M(x)/EI$. $y'$ is also the reciprocal of the *radius of curvature*, $\rho$, of the beam.

$$
y''(x) = \frac{1}{\rho(x)} = \frac{M(x)}{EI} \qquad \textit{52.52}
$$

Since the flexural rigidity, $EI$, is always positive, the radius of curvature, $\rho(x)$, changes sign when the moment equation, $M(x)$, changes sign.

3. If a shear diagram is known, the inflection point can sometimes be found by noting the point at which the positive and negative shear areas on either side of the point balance.

## 24. TRUSS DEFLECTION: STRAIN ENERGY METHOD

The deflection of a truss at the point of a single load application can be found by the *strain energy method* if all member forces are known. This method is illustrated by Ex. 52.12.

**Example 52.12**

Find the vertical deflection of point A under the external load of 707 lbf. $AE = 10^6$ lbf for all members. The internal forces have already been determined.

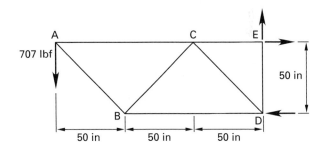

*Solution*

The length of member AB is $\sqrt{(50 \text{ in})^2 + (50 \text{ in})^2} = 70.7$ in. From Eq. 52.6, the internal strain energy in member AB is

$$
U = \frac{F^2 L_o}{2AE} = \frac{(-1000 \text{ lbf})^2(70.7 \text{ in})}{(2)(10^6 \text{ lbf})} = 35.4 \text{ in-lbf}
$$

---

[9]The principle of superposition is not limited to deflections. It can also be used to calculate the shear and moment at a point and to draw the shear and moment diagrams.

Similarly, the energy in all members can be determined.

| member | $L$ (in) | $F$ (lbf) | $U$ (in-lbf) |
|---|---|---|---|
| AB | 70.7 | −1000 | +35.4 |
| BC | 70.7 | +1000 | +35.4 |
| AC | 100 | +707 | +25.0 |
| BD | 100 | −1414 | +100.0 |
| CD | 70.7 | −1000 | +35.4 |
| CE | 50 | +2121 | +112.5 |
| DE | 50 | +707 | +12.5 |
| | | | 356.2 |

The work done by a constant force $F$ moving through a distance $y$ is $Fy$. In this case, the force increases with $y$. The average force is $\frac{1}{2}F$. The external work is $W_{\text{ext}} = \left(\frac{1}{2}\right)(707 \text{ lbf})y$, so

$$\left(\tfrac{1}{2}\right)(707 \text{ lbf})y = 356.2 \text{ in-lbf}$$

$$y = 1 \text{ in}$$

## 25. TRUSS DEFLECTION: VIRTUAL WORK METHOD

The *virtual work method* (also known as the *unit load method*) is an extension of the strain energy method. It can be used to determine the deflection of any point on a truss.

*step 1:* Draw the truss twice.

*step 2:* On the first truss, place all the actual loads.

*step 3:* Find the forces, $S$, due to the actual applied loads in all the members.

*step 4:* On the second truss, place a dummy one-unit load in the direction of the desired displacement.

*step 5:* Find the forces, $u$, due to the one-unit dummy load in all members.

*step 6:* Find the desired displacement from Eq. 52.53. The summation is over all truss members that have nonzero forces in *both* trusses.

$$\delta = \sum \frac{SuL}{AE} \qquad \textit{52.53}$$

### Example 52.13

What is the horizontal deflection of joint F on the truss shown? Use $E = 3 \times 10^7 \text{ lbf/in}^2$. Joint A is restrained horizontally. Member lengths and areas are listed in the accompanying table.

*Solution*

Applying steps 1 and 2, use the truss as drawn.

Applying step 3, the forces in all the truss members are summarized in step 5.

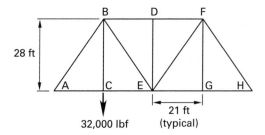

Applying step 4, draw the truss and load it with a unit horizontal force at point F.

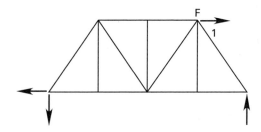

Applying step 5, find the forces, $u$, in all members of the second truss. These are summarized in the following table. Notice the sign convention: + for tension and − for compression.

| member | $S$ (lbf) | $u$ (lbf) | $L$ (ft) | $A$ (in²) | $\frac{SuL}{AE}$ (ft) |
|---|---|---|---|---|---|
| AB | −30,000 | 5/12 | 35 | 17.5 | −8.33 × 10⁻⁴ |
| CB | 32,000 | 0 | 28 | 14 | 0 |
| EB | −10,000 | −5/12 | 35 | 17.5 | 2.75 × 10⁻⁴ |
| ED | 0 | 0 | 28 | 14 | 0 |
| EF | 10,000 | 5/12 | 35 | 17.5 | 2.78 × 10⁻⁴ |
| GF | 0 | 0 | 28 | 14 | 0 |
| HF | −10,000 | −5/12 | 35 | 17.5 | 2.78 × 10⁻⁴ |
| BD | −12,000 | 1/2 | 21 | 10.5 | −4.00 × 10⁻⁴ |
| DF | −12,000 | 1/2 | 21 | 10.5 | −4.00 × 10⁻⁴ |
| AC | 18,000 | 3/4 | 21 | 0.5 | 9.00 × 10⁻⁴ |
| CE | 18,000 | 3/4 | 21 | 10.5 | 9.00 × 10⁻⁴ |
| EG | 6000 | 1/4 | 21 | 10.5 | 1.00 × 10⁻⁴ |
| GH | 6000 | 1/4 | 21 | 10.5 | 1.00 × 10⁻⁴ |
| | | | | | 12.01 × 10⁻⁴ |

Since $12.01 \times 10^{-4}$ is positive, the deflection is in the direction of the dummy unit load. In this case, the deflection is to the right.

## 26. MODES OF BEAM FAILURE

Beams can fail in different ways, including excessive deflection, local buckling, lateral buckling, and rotation.

Excessive deflection occurs when a beam bends more than a permitted amount.[10] The deflection is elastic and no yielding occurs. For this reason, the failure mechanism is sometimes called *elastic failure*. Although the beam does not yield, the excessive deflection may

---

[10]Building codes specify maximum permitted deflections in terms of beam length.

cause cracks in plaster and sheetrock, misalignment of doors and windows, and occupant concern and reduction of confidence in the structure.

*Local buckling* is an overload condition that occurs near large concentrated loads. Such locations include where a column frames into a supporting girder or a reaction point. *Vertical buckling* and *web crippling*, two types of local buckling, can be eliminated by use of *stiffeners*. Such stiffeners can be referred to as *intermediate stiffeners*, *bearing stiffeners*, *web stiffeners*, and *flange stiffeners*, depending on the location and technique of stiffening. (See Fig. 52.17.)

**Figure 52.17** *Local Buckling and Stiffeners*

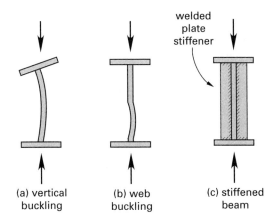

(a) vertical buckling

(b) web buckling

(c) stiffened beam

*Lateral buckling*, such as illustrated in Fig. 52.18, occurs when a long, unsupported member rolls out of its normal plane. To prevent lateral buckling, either the beam's compression flange must be supported continuously or at frequent intervals along its length, or the beam must be restrained against twisting about its longitudinal axis.

**Figure 52.18** *Lateral Buckling and Flange Support*

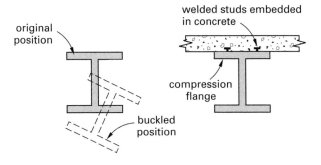

*Rotation* is an inelastic (plastic) failure of the beam. When the bending stress at a point exceeds the strength of the beam material, the material yields. (See Fig. 52.19.) As the beam yields, its slope changes. Since the beam appears to be rotating at a hinge at the yield point, the term *plastic hinge* is used to describe the failure mechanism.

**Figure 52.19** *Beam Failure by Rotation*

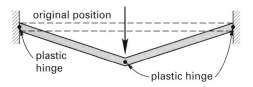

## 27. CURVED BEAMS

Many members (e.g., hooks, chain links, clamps, and machine frames) have curved main axes. The distribution of bending stress in a curved beam is nonlinear. Compared to a straight beam, the stress at the inner radius is higher because the inner radius fibers are shorter. Conversely, the stress at the outer radius is lower because the outer radius fibers are longer. Also, the neutral axis is shifted from the center inward toward the center of curvature.

Since the process of finding the neutral axis and calculating the stress amplification is complex, tables and graphs are used for quick estimates and manual computations. The forms of these computational aids vary, but the straight-beam stress is generally multiplied by factors, $K$, to obtain the stresses at the extreme faces. The factor values depend on the beam cross section and radius of curvature.

$$\sigma_{\text{curved}} = K\sigma_{\text{straight}}$$
$$= \frac{KMc}{I} \qquad \qquad 52.54$$

Table 52.4 is typical of compilations for round and rectangular beams. Factors $K_A$ and $K_B$ are the multipliers for the inner (high stress) and outer (low stress) faces, respectively. The ratio $h/r$ is the fractional distance that the neutral axis shifts inward toward the radius of curvature.

## 28. COMPOSITE STRUCTURES

A *composite structure* is one in which two or more different materials are used. Each material carries part of an applied load. Examples of composite structures include steel-reinforced concrete and steel-plated timber beams.

Most simple composite structures can be analyzed using the *method of consistent deformations*, also known as the *area transformation method*. This method assumes that the strains are the same in both materials at the interface between them. Although the strains are the same, the stresses in the two adjacent materials are not equal, since stresses are proportional to the moduli of elasticity.

The following steps comprise an analysis method based on area transformation.

**Table 52.4** Curved Beam Correction Factors

| solid rectangular section | $r/c$ | $K_A$ | $K_B$ | $h/r$ |
|---|---|---|---|---|
| | 1.2 | 2.89 | 0.57 | 0.305 |
| | 1.4 | 2.13 | 0.63 | 0.204 |
| | 1.6 | 1.79 | 0.67 | 0.149 |
| | 1.8 | 1.63 | 0.70 | 0.112 |
| | 2.0 | 1.52 | 0.73 | 0.090 |
| | 3.0 | 1.30 | 0.81 | 0.041 |
| | 4.0 | 1.20 | 0.85 | 0.021 |
| | 6.0 | 1.12 | 0.90 | 0.0093 |
| | 8.0 | 1.09 | 0.92 | 0.0052 |
| | 10.0 | 1.07 | 0.94 | 0.0033 |

| solid circular section | $r/c$ | $K_A$ | $K_B$ | $h/r$ |
|---|---|---|---|---|
| | 1.2 | 3.41 | 0.54 | 0.224 |
| | 1.4 | 2.40 | 0.60 | 0.151 |
| | 1.6 | 1.96 | 0.65 | 0.108 |
| | 1.8 | 1.75 | 0.68 | 0.084 |
| | 2.0 | 1.62 | 0.71 | 0.069 |
| | 3.0 | 1.33 | 0.79 | 0.03 |
| | 4.0 | 1.23 | 0.84 | 0.016 |
| | 6.0 | 1.14 | 0.89 | 0.007 |
| | 8.0 | 1.10 | 0.91 | 0.0039 |
| | 10.0 | 1.08 | 0.93 | 0.0025 |

*step 1:* Determine the modulus of elasticity for each of the materials used in the structure.

*step 2:* For each of the materials used, calculate the *modular ratio, n*.

$$n = \frac{E}{E_{\text{weakest}}} \qquad 52.55$$

$E_{\text{weakest}}$ is the smallest modulus of elasticity of any of the materials used in the composite structure. For two materials that experience the same strains (i.e., are perfectly bonded), $n$ is also the ratio of stresses.

*step 3:* For all of the materials except the weakest, multiply the actual material stress area by $n$. Consider this expanded (*transformed*) area to have the same composition as the weakest material.

*step 4:* If the structure is a tension or compression member, the distribution or placement of the transformed area is not important. Just assume that the transformed areas carry the axial load. For beams in bending, the transformed area can add to the width of the beam, but it cannot change the depth of the beam or the thickness of the reinforcement.

*step 5:* For compression or tension numbers, calculate the stresses in the weakest and stronger materials.

$$\sigma_{\text{weakest}} = \frac{F}{A_t} \qquad 52.56$$

$$\sigma_{\text{stronger}} = \frac{nF}{A_t} \qquad 52.57$$

*step 6:* For beams in bending, proceed through step 9. Find the centroid of the transformed beam.

*step 7:* Find the centroidal moment of inertia of the transformed beam $I_{c,t}$.

*step 8:* Find $V_{\max}$ and $M_{\max}$ by inspection or from the shear and moment diagrams.

*step 9:* Calculate the stresses in the weakest and stronger materials.

$$\sigma_{\text{weakest}} = \frac{M c_{\text{weakest}}}{I_{c,t}} \qquad 52.58$$

$$\sigma_{\text{stronger}} = \frac{n M c_{\text{stronger}}}{I_{c,t}} \qquad 52.59$$

**Example 52.14**

A short circular steel core is surrounded by a copper tube. The assemblage supports an axial compressive load of 100,000 lbf. The core and tube are well bonded, and the load is applied uniformly. Find the compressive stress in the inner steel core and the outer copper tube.

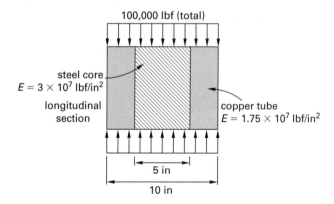

*Solution*

The moduli of elasticity are given in the illustration. From step 2, the modular ratio is

$$n = \frac{E_{\text{steel}}}{E_{\text{copper}}} = \frac{3 \times 10^7 \ \frac{\text{lbf}}{\text{in}^2}}{1.75 \times 10^7 \ \frac{\text{lbf}}{\text{in}^2}} = 1.714$$

The actual cross-sectional area of the steel is

$$A_{\text{steel}} = \frac{\pi}{4} d^2 = \left(\frac{\pi}{4}\right)(5 \text{ in})^2$$

$$= 19.63 \text{ in}^2$$

The actual cross-sectional area of the copper is

$$A_{\text{copper}} = \frac{\pi}{4}(d_o^2 - d_i^2) = \left(\frac{\pi}{4}\right)((10 \text{ in})^2 - (5 \text{ in})^2)$$

$$= 58.90 \text{ in}^2$$

The steel is the stronger material. Its area must be expanded to an equivalent area of copper. From step 3, the total transformed area is

$$A_t = A_{\text{copper}} + nA_{\text{steel}}$$

$$= 58.90 \text{ in}^2 + (1.714)(19.63 \text{ in}^2) = 92.55 \text{ in}^2$$

Since the two pieces are well bonded and the load is applied uniformly, both pieces experience identical strains. From step 5, the compressive stresses are

$$\sigma_{\text{copper}} = \frac{F}{A_t} = \frac{-100,000 \text{ lbf}}{92.55 \text{ in}^2}$$

$$= -1080 \text{ lbf/in}^2 \quad [\text{compression}]$$

$$\sigma_{\text{steel}} = \frac{nF}{A_t} = n\sigma_{\text{copper}} = (1.714)\left(-1080 \ \frac{\text{lbf}}{\text{in}^2}\right)$$

$$= -1851 \text{ lbf/in}^2$$

## Example 52.15

At a particular point along the length of a steel-reinforced wood beam, the moment is 40,000 ft-lbf. Assume the steel reinforcement is lag-bolted to the wood at regular intervals along the beam. What are the maximum bending stresses in the wood and steel?

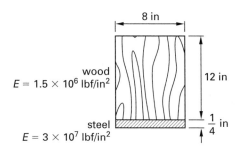

*Solution*

The moduli of elasticity are given in the illustration. From step 2, the modular ratio is

$$n = \frac{E_{\text{steel}}}{E_{\text{wood}}} = \frac{3 \times 10^7 \ \dfrac{\text{lbf}}{\text{in}^2}}{1.5 \times 10^6 \ \dfrac{\text{lbf}}{\text{in}^2}} = 20$$

The actual cross-sectional area of the steel is

$$A_{\text{steel}} = (0.25 \text{ in})(8 \text{ in}) = 2 \text{ in}^2$$

The steel is the stronger material. Its area must be expanded to an equivalent area of wood. Since the depth of the beam and reinforcement cannot be increased (step 4), the width must increase. The width of the transformed steel plate is

$$b' = nb = (20)(8 \text{ in}) = 160 \text{ in}$$

The centroid of the transformed section is 4.45 in from the horizontal axis. The centroidal moment of inertia of the transformed section is $I_{c,t} = 2211.5 \text{ in}^4$. (The calculations for centroidal location and moment of inertia are not presented here.)

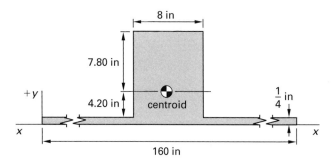

Since the steel plate is bolted to the wood at regular intervals, both pieces experience the same strain. From step 9, the stresses in the wood and steel are

$$\sigma_{\text{max,wood}} = \frac{Mc_{\text{wood}}}{I}$$

$$= \frac{(40,000 \text{ ft-lbf})\left(12 \ \dfrac{\text{in}}{\text{ft}}\right)(7.8 \text{ in})}{2211.5 \text{ in}^4}$$

$$= 1693 \text{ lbf/in}^2$$

$$\sigma_{\text{max,steel}} = \frac{nMc_{\text{steel}}}{I}$$

$$= \frac{(20)(40,000 \text{ ft-lbf})\left(12 \ \dfrac{\text{in}}{\text{ft}}\right)(4.45 \text{ in})}{2211.5 \text{ in}^4}$$

$$= 19,317 \text{ lbf/in}^2$$

# 53 Failure Theories

## Nomenclature

| | | | |
|---|---|---|---|
| $A$ | area | $in^2$ | $m^2$ |
| $C$ | constant | – | – |
| $E$ | modulus of elasticity | $lbf/in^2$ | Pa |
| $F$ | force | lbf | N |
| FS | factor of safety | – | – |
| $K_f$ | stress concentration factor | – | – |
| MR | modulus of resilience | $in\text{-}lbf/in^3$ | $J/m^3$ |
| $n$ | number of cycles | – | – |
| $N$ | endurance life | – | – |
| $S$ | strength | $lbf/in^2$ | Pa |
| $U$ | strain energy | $in\text{-}lbf/in^3$ | $J/m^3$ |

## Symbols

| | | | |
|---|---|---|---|
| $\epsilon$ | strain | in/in | m/m |
| $\nu$ | Poisson's ratio | – | – |
| $\sigma$ | normal stress | $lbf/in^2$ | Pa |
| $\tau$ | shear stress | $lbf/in^2$ | Pa |

## Subscripts

| | |
|---|---|
| $a$ | allowable |
| alt | alternating |
| $c$ | compressive |
| $e$ | endurance |
| eq | equivalent |
| $m$ | mean |
| $r$ | range |
| $s$ | shear |
| $t$ | tensile |
| $u$ | ultimate |
| $y$ | yield |

## 1. STATIC LOADING OF BRITTLE MATERIALS: UNIAXIAL LOADING

Brittle materials such as gray cast iron fail by sudden fracturing, not by yielding. The basic method of designing with brittle materials is to keep the maximum stress below the ultimate strength. Stress concentration factors must be included. The failure criterion is

$$\sigma > S_u \quad \text{[failure criterion]} \qquad 53.1$$

## 2. STATIC LOADING OF BRITTLE MATERIALS: MAXIMUM NORMAL STRESS THEORY

The *maximum normal stress theory* predicts the failure stress reasonably well for brittle materials under static biaxial loading.[1] Failure is assumed to occur either if the largest tensile principal stress, $\sigma_1$, is greater than the ultimate tensile strength, or if the largest compressive principal stress, $\sigma_2$, is greater than the ultimate compressive strength. Brittle materials generally have much higher compressive strengths than tensile strengths, so both tensile and compressive stresses must be checked.

Figure 53.1 illustrates a standard graphical method of describing the safe operating region. Notice that the axes represent the principal stresses, not the stresses in the $x$- and $y$-directions.

Stress concentration factors are applicable to brittle materials under static loading. The *factor of safety*, FS, is the ultimate strength, $S_u$, divided by the actual stress, $\sigma$. (The *margin of safety* is MS = FS − 1.) Where a factor of safety is known in advance, the *allowable stress*, $S_a$, can be calculated by dividing the ultimate strength by it. The allowable operating region can be constructed from the allowable stresses rather than from the ultimate stresses.

$$\text{FS} = \frac{S_u}{\sigma} \qquad 53.2$$

$$S_a = \frac{S_u}{\text{FS}} \qquad 53.3$$

---

[1]As described in subsequent sections, the maximum normal stress theory is limited to cases where both principal stresses have same sign (i.e., both are compressive or both are tensile).

The failure criterion is

$$\sigma_1, \sigma_2 > S_u \quad \text{[failure criterion]} \qquad 53.4$$

**Figure 53.1** *Maximum Normal Stress Theory*

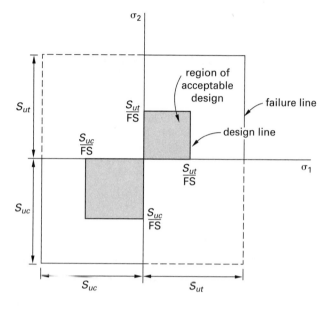

## Example 53.1

A lathe bed is made of gray cast iron. The ultimate strengths for the cast iron are 30,000 lbf/in² (tension) and 110,000 lbf/in² (compression). The bed is subjected to maximum stresses of 17,150 lbf/in² in tension and 42,800 lbf/in² in compression. What are the factors of safety?

*Solution*

The factor of safety in tension is

$$\text{FS} = \frac{S_{ut}}{\sigma_t} = \frac{30{,}000 \ \frac{\text{lbf}}{\text{in}^2}}{17{,}150 \ \frac{\text{lbf}}{\text{in}^2}}$$

$$= 1.75$$

The factor of safety in compression is

$$\text{FS} = \frac{S_{uc}}{\sigma_c} = \frac{110{,}000 \ \frac{\text{lbf}}{\text{in}^2}}{42{,}800 \ \frac{\text{lbf}}{\text{in}^2}}$$

$$= 2.57$$

Tensile failure is the limiting case.

## 3. STATIC LOADING OF BRITTLE MATERIALS: COULOMB-MOHR THEORY

The maximum normal stress theory is somewhat in conflict with experimental evidence. Reliable operation in the second and fourth quadrants (i.e., when the two principal stresses have opposite signs) has not been observed, even though the stresses are less than the ultimate strengths. The *Coulomb-Mohr theory* is a conservative theory that reduces the acceptable operating region in the second and fourth quadrants. As with the maximum normal stress theory, the factor of safety is calculated from Eq. 53.2.

The failure criterion is

$$\frac{\sigma_1}{S_{ut}} + \frac{\sigma_2}{S_{uc}} > 1 \quad \text{[failure criterion]} \qquad 53.5$$

## 4. STATIC LOADING OF BRITTLE MATERIALS: MODIFIED MOHR THEORY

The Coulomb-Mohr theory is considered conservative because failures typically "miss" the diagonal line in Fig. 53.2 by a considerable margin. The *modified Mohr theory* more closely predicts the envelope of observed failures. (See Fig. 53.3.) Though the failure criterion can be based on deriving equations of the straight lines in the second and fourth quadrants, graphical solutions are more common.

**Figure 53.2** *Coulomb-Mohr Theory*

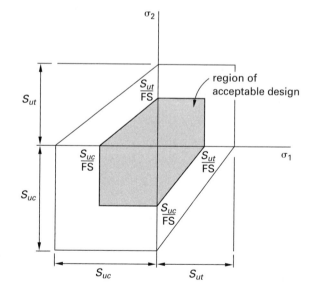

*Figure 53.3* Modified Mohr Theory

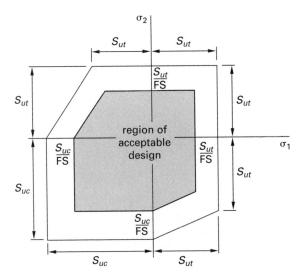

## Example 53.2

A 0.5 in diameter dowel is made from cast iron (ultimate tensile strength of 40,000 lbf/in²; ultimate compressive strength of 135,000 lbf/in²). The dowel supports a compressive load of 15,000 lbf and is subjected simultaneously to an unknown torsional shear stress. Use the modified Mohr theory to determine the torsional shear stress that will cause failure.

*Solution*

The compressive normal stress in the dowel is

$$\sigma_c = \frac{F}{A} = \frac{-15,000 \text{ lbf}}{\left(\dfrac{\pi}{4}\right)(0.5 \text{ in})^2}$$

$$= -76,394 \text{ lbf/in}^2 \quad \text{[negative because compressive]}$$

To use the modified Mohr theory, the principal stresses must be known. However, the shear stress in this problem is the unknown, so the principal stress cannot be calculated directly. Use a trial-and-error approach.

Assume the shear stress is 10,000 lbf/in². With $\tau = 10{,}000$ lbf/in², $\sigma_x = -76{,}400$ lbf/in², and $\sigma_y = 0$, the principal stresses are found (from standard methods) to be 1300 lbf/in² and −77,700 lbf/in². Draw the modified Mohr diagram and plot these principal values. The point is not close to the failure line, so the assumed shear stress of 10,000 lbf/in² is too low.

Repeat the process with shear stresses of 30,000 lbf/in² and 40,000 lbf/in².

| $\tau$ | $\sigma_1$ | $\sigma_2$ |
|--------|-----------|-----------|
| 10,000 | 1300 | −77,700 |
| 30,000 | 10,400 | −86,800 |
| 40,000 | 17,100 | −93,500 |

The third point (corresponding to a shear stress of 40,000 lbf/in²) is essentially on the failure line, so this is the maximum allowable shear stress. (It is a coincidence that this is also the ultimate tensile strength.)

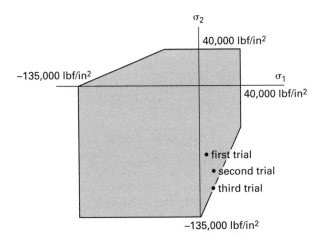

## 5. STATIC LOADING OF DUCTILE MATERIALS: UNIAXIAL LOADING

Ductile materials fail by yielding, not by fracture. The basic method of designing with ductile materials in uniaxial loading is to keep the maximum stress below the yield strength. This is known as the *maximum stress theory*. Stress concentration factors must be included. The failure criterion is

$$\sigma > S_y \quad \text{[failure criterion]} \qquad 53.6$$

The failure criterion for the equivalent *maximum strain theory* is

$$\epsilon > \frac{S_y}{E} \quad \text{[failure criterion]} \qquad 53.7$$

## 6. STATIC LOADING OF DUCTILE MATERIALS: MAXIMUM SHEAR STRESS THEORY

With the conservative *maximum shear stress theory*, shear stress is used to indicate yielding (i.e., failure). Loading is not limited to shear and torsion. Loading can include normal stresses as well as shear stresses. According to the maximum shear stress theory, yielding occurs when the maximum shear stress exceeds the yield strength in shear.[2] It is implicit in this theory that the

---

[2]The application of this theory, as well as those that follow, depends on being able to find the maximum shear stress, $\tau_{max}$. Even in some cases of uniaxial loading, finding the maximum shear stress can be tricky. Two cylindrical rollers or two spheres in contact, for example, represent cases of uniaxial compressive loading, which have well-documented but non-obvious maximum shear stresses.

Mechanics of
Materials

yield strength in shear is half of the tensile yield strength.[3]

$$S_{ys} = \frac{S_{yt}}{2} \qquad 53.8$$

From the combined stress theory, the maximum shear stress, $\tau_{max}$, for *triaxial loading* is the maximum of the three combined shear stresses. (For biaxial loading, only Eq. 53.9 is used.)

$$\tau_{12} = \frac{\sigma_1 - \sigma_2}{2} \qquad 53.9$$

$$\tau_{23} = \frac{\sigma_2 - \sigma_3}{2} \qquad 53.10$$

$$\tau_{13} = \frac{\sigma_1 - \sigma_3}{2} \qquad 53.11$$

$$\tau_{max} = \max(\tau_{12}, \tau_{23}, \tau_{13}) \qquad 53.12$$

The failure criterion is

$$\tau_{max} > S_{ys} = \frac{S_{yt}}{2} \quad \text{[failure criterion]} \qquad 53.13$$

The acceptance region representing combinations of allowable principal stresses for the maximum shear stress theory is shown graphically in Fig. 53.4 for the case of biaxial loading (i.e., $\sigma_3 = 0$). The shape of this failure envelope is similar to that of the Coulomb-Mohr theory for brittle materials, but the limits are based on yield strengths, not on ultimate strengths. Since the

**Figure 53.4** *Maximum Shear Stress Theory*

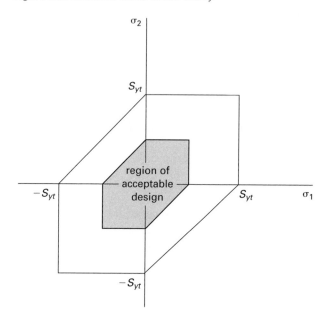

tensile and compressive yield strengths are assumed equal for ductile materials, the failure envelope is symmetrical. In Fig. 53.4, the limits are not divided by 2 as might be suspected from Eq. 53.8 through Eq. 53.13. This is because the failure envelope is used with the principal stress, not with the shear stress.

The factor of safety with the maximum shear stress theory is

$$FS = \frac{S_{ys}}{\tau_{max}} = \frac{S_{yt}}{2\tau_{max}} \qquad 53.14$$

### Example 53.3

Strain gauges attached to a bearing support show the three principal stresses to be 10,600 lbf/in², 2400 lbf/in², and −9200 lbf/in². The support is cast aluminum with a tensile yield strength of 24,000 lbf/in². Use the maximum shear stress theory to determine the factor of safety.

*Solution*

Calculate the combined shear stresses.

$$\tau_{12} = \frac{\sigma_1 - \sigma_2}{2}$$

$$= \frac{10,600 \dfrac{lbf}{in^2} - 2400 \dfrac{lbf}{in^2}}{2}$$

$$= 4100 \ lbf/in^2$$

$$\tau_{23} = \frac{\sigma_2 - \sigma_3}{2}$$

$$= \frac{2400 \dfrac{lbf}{in^2} - \left(-9200 \dfrac{lbf}{in^2}\right)}{2}$$

$$= 5800 \ lbf/in^2$$

$$\tau_{13} = \frac{\sigma_1 - \sigma_3}{2}$$

$$= \frac{10,600 \dfrac{lbf}{in^2} - \left(-9200 \dfrac{lbf}{in^2}\right)}{2}$$

$$= 9900 \ lbf/in^2$$

The maximum shear stress is 9900 lbf/in².

From Eq. 53.14, the factor of safety is

$$FS = \frac{S_{yt}}{2\tau_{max}}$$

$$= \frac{24,000 \dfrac{lbf}{in^2}}{(2)\left(9900 \dfrac{lbf}{in^2}\right)}$$

$$= 1.21$$

---

[3]If the symmetrical shape of the failure envelope is accepted, it is easy to justify the assumption that the yield strength in shear is one-half of the yield strength in tension. For pure shear loading, the two principal stresses will each be equal and opposite (with magnitudes equal to the applied shear stress). Plotting the locus of points with $\sigma_1 = -\sigma_2$, the failure envelope is encountered at $S_{yt}/2$.

## 7. STATIC LOADING OF DUCTILE MATERIALS: STRAIN ENERGY THEORY

The *strain energy theory* calculates the energy per unit volume. This energy is compared to the energy that causes yielding, which is known as the *modulus of resilience*, MR. Since the stress-strain curve is essentially a straight line, as shown by Fig. 53.5, the strain energy (i.e., the area under the curve) for any uniaxial loading less than the yield strength is

$$U = \frac{\sigma\epsilon}{2} \qquad 53.15$$

**Figure 53.5** *Strain Energy as Area Under the Stress-Strain Curve*

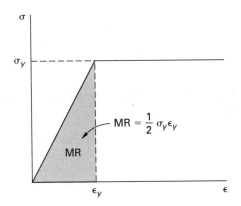

For biaxial loading on the principal planes, the strain energy is calculated from superposition.

$$U = \frac{\sigma_1\epsilon_1 + \sigma_2\epsilon_2}{2} \qquad 53.16$$

Strain is generally not measured directly, but it can be calculated from the stress and material properties.

$$\epsilon_1 = \frac{1}{E}\left(\sigma_1 - \nu\sigma_2\right) \qquad 53.17$$

$$\epsilon_2 = \frac{1}{E}\left(\sigma_2 - \nu\sigma_1\right) \qquad 53.18$$

$$U = \frac{1}{2E}\left(\sigma_1^2 + \sigma_2^2 - 2\nu\sigma_1\sigma_2\right) \qquad 53.19$$

Failure is assumed to occur when the strain energy, $U$, exceeds the modulus of resilience. The modulus of resilience can be determined from a simple tensile test and is calculated as the area under the stress-strain curve up to the yield point.[4]

---

[4]The strain energy theory is seldom used to predict failures. However, it is similar in development to the distortion energy theory, which has supplanted it, so the strain energy theory is presented here as part of the logical progression of failure theories.

At failure in a simple tensile test, $\sigma_1 = S_{yt}$ and $\sigma_2 = 0$. Therefore, from Eq. 53.19, the modulus of resilience is

$$\text{MR} = \frac{1}{2E}\left(S_{yt}^2 + (0)^2 - (2)(0)\right) = \frac{S_{yt}^2}{2E} \qquad 53.20$$

Eliminating the $1/2E$ terms, the failure criterion is

$$\sigma_1^2 + \sigma_2^2 - 2\nu\sigma_1\sigma_2 > S_{yt}^2 \quad \text{[failure criterion]} \qquad 53.21$$

The factor of safety is

$$\text{FS} = \frac{S_{yt}}{\sqrt{\sigma_1^2 + \sigma_2^2 - 2\nu\sigma_1\sigma_2}} \qquad 53.22$$

## 8. STATIC LOADING OF DUCTILE MATERIALS: DISTORTION ENERGY THEORY

The *distortion energy theory* (also known as the *theory of constant energy of distortion, von Mises theory, von Mises-Hencky theory,* and *octahedral shear-stress theory*) is similar in development to the strain energy method but is more strict. It is commonly used to predict tensile and shear failure in steel parts. The *von Mises stress* (also known as the *effective stress*), $\sigma'$, is calculated from the principal stresses. For *biaxial loading*,

$$\sigma' = \sqrt{\sigma_1^2 + \sigma_2^2 - \sigma_1\sigma_2} \qquad 53.23$$

For *triaxial loading* (i.e., where there are three orthogonal normal stresses), the *von Mises stress* is

$$\sigma' = \sqrt{\tfrac{1}{2}\left((\sigma_1 - \sigma_2)^2 + (\sigma_2 - \sigma_3)^2 + (\sigma_3 - \sigma_1)^2\right)} \qquad 53.24$$

The failure criterion is

$$\sigma' > S_{yt} \quad \text{[failure criterion]} \qquad 53.25$$

The factor of safety is

$$\text{FS} = \frac{S_{yt}}{\sigma'} \qquad 53.26$$

If the loading is pure torsion at failure, then $\sigma_1 = -\sigma_2 = \tau_{\max}$, and $\sigma_3 = 0$. If $\tau_{\max}$ is substituted for $\sigma$ in Eq. 53.24 (with $\sigma_3 = 0$), an expression for the yield strength in shear is derived. Equation 53.27 predicts a larger yield strength in shear than does the maximum shear stress theory ($0.5S_{yt}$).

$$S_{ys} = \tau_{\max,\text{failure}} = \frac{S_{yt}}{\sqrt{3}} = 0.577S_{yt} \qquad 53.27$$

**Example 53.4**

The steel used in a shaft has a tensile yield strength of $110{,}000 \text{ lbf/in}^2$ and a Poisson's ratio of 0.3. The shaft is simultaneously acted upon by a longitudinal compressive stress of $60{,}000 \text{ lbf/in}^2$ and by a torsional stress of $40{,}000 \text{ lbf/in}^2$. Compare the factor of safety using (a) maximum shear stress, (b) strain energy, and (c) distortion energy.

*Solution*

From the combined stress theory and using $\sigma_x = -60{,}000 \text{ lbf/in}^2$, the maximum shear stress is

$$\tau_{\max} = \tfrac{1}{2}\sqrt{(\sigma_x - \sigma_y)^2 + (2\tau)^2}$$
$$= \tfrac{1}{2}\sqrt{\left(-60{,}000 \, \frac{\text{lbf}}{\text{in}^2}\right)^2 + \left((2)\left(40{,}000 \, \frac{\text{lbf}}{\text{in}^2}\right)\right)^2}$$
$$= 50{,}000 \text{ lbf/in}^2$$

The principal stresses are

$$\sigma_1, \sigma_2 = \tfrac{1}{2}(\sigma_x + \sigma_y) \pm \tau_{\max}$$
$$= \left(\tfrac{1}{2}\right)\left(-60{,}000 \, \frac{\text{lbf}}{\text{in}^2}\right) \pm 50{,}000 \, \frac{\text{lbf}}{\text{in}^2}$$
$$= 20{,}000 \text{ lbf/in}^2, \, -80{,}000 \text{ lbf/in}^2$$

(a) From Eq. 53.14, the factor of safety for the maximum shear stress theory is

$$\text{FS} = \frac{S_{yt}}{2\tau_{\max}}$$
$$= \frac{110{,}000 \, \dfrac{\text{lbf}}{\text{in}^2}}{(2)\left(50{,}000 \, \dfrac{\text{lbf}}{\text{in}^2}\right)}$$
$$= 1.1$$

(b) From Eq. 53.22, the factor of safety for the strain energy theory is

$$\text{FS} = \frac{S_{yt}}{\sqrt{\sigma_1^2 + \sigma_2^2 - 2\nu\sigma_1\sigma_2}}$$
$$= \frac{110{,}000 \, \dfrac{\text{lbf}}{\text{in}^2}}{\sqrt{\begin{array}{c}\left(-80{,}000 \, \dfrac{\text{lbf}}{\text{in}^2}\right)^2 + \left(20{,}000 \, \dfrac{\text{lbf}}{\text{in}^2}\right)^2 \\ - (2)(0.3)\left(-80{,}000 \, \dfrac{\text{lbf}}{\text{in}^2}\right)\left(20{,}000 \, \dfrac{\text{lbf}}{\text{in}^2}\right)\end{array}}}$$
$$= 1.25$$

(c) From Eq. 53.23, the von Mises stress is

$$\sigma' = \sqrt{\sigma_1^2 + \sigma_2^2 - \sigma_1\sigma_2}$$
$$= \sqrt{\begin{array}{c}\left(-80{,}000 \, \dfrac{\text{lbf}}{\text{in}^2}\right)^2 + \left(20{,}000 \, \dfrac{\text{lbf}}{\text{in}^2}\right)^2 \\ - \left(-80{,}000 \, \dfrac{\text{lbf}}{\text{in}^2}\right)\left(20{,}000 \, \dfrac{\text{lbf}}{\text{in}^2}\right)\end{array}}$$
$$= 91{,}652 \text{ lbf/in}^2$$

From Eq. 53.26, the factor of safety for the distortion energy theory is

$$\text{FS} = \frac{S_{yt}}{\sigma'} = \frac{110{,}000 \, \dfrac{\text{lbf}}{\text{in}^2}}{91{,}652 \, \dfrac{\text{lbf}}{\text{in}^2}}$$
$$= 1.20$$

## 9. ALTERNATING STRESS: SODERBERG LINE

Many parts are subjected to a combination of static and reversed loadings, as illustrated in Fig. 53.6 for sinusoidal loadings. For these parts, failure cannot be determined solely by comparing stresses with the yield strength or endurance limit. The combined effects of the average stress and the amplitude of the reversal must be considered. This is done graphically on a diagram that plots the mean stress versus the alternating stresses.

*Figure 53.6* Sinusoidal Fluctuating Stress

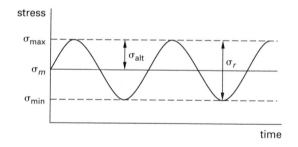

The *mean stress* is

$$\sigma_m = \frac{\sigma_{\max} + \sigma_{\min}}{2} \qquad 53.28$$

The *alternating stress* is half of the *range stress*.

$$\sigma_r = \sigma_{\max} - \sigma_{\min} \qquad 53.29$$
$$\sigma_{\text{alt}} = \tfrac{1}{2}\sigma_r = \tfrac{1}{2}(\sigma_{\max} - \sigma_{\min}) \qquad 53.30$$

A criterion for acceptable design (or for failure) is established by graphically relating the yield strength and the endurance limit. One method of relating this information is a Soderberg line, particularly suited for normal stresses in ductile materials. However, it is the most conservative of the fluctuating stress theories.

Figure 53.7 illustrates how an area of acceptable design is developed by drawing a straight line (the *Soderberg line* or the *failure line*) from the endurance limit, $S_e$, to the yield strength, $S_{yt}$. Both of these values should be divided by a suitable factor of safety to define the allowable design area. If the point $(\sigma_m, \sigma_{alt})$ falls below the allowable stress line, the design is acceptable.

Stress concentration factors, such as $K_f$, are applied to the alternating stress only. This is justified with ductile materials, such as steel, which yield around discontinuities, reducing constant mean stress. Increasing the alternating stress is equivalent to reducing $S_e$ by $K_f$.

Figure 53.7 illustrates how the Soderberg *equivalent stress*, $\sigma_{eq}$, is determined from the operating point. The factor of safety is

$$\text{FS} = \frac{S_e}{\sigma_{eq}} = \frac{S_e}{\sigma_{alt} + \left(\dfrac{S_e}{S_{yt}}\right)\sigma_m} \qquad 53.31$$

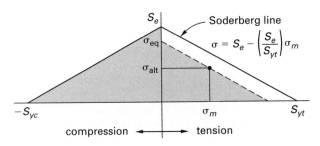

*Figure 53.7* Soderberg Line and Equivalent Stress

## 10. ALTERNATING STRESS: GOODMAN LINE

The Soderberg line is a conservative criterion, and it is not often used. Also, the envelope of failures is not truly linear, but follows more of a parabolic line, named the *Gerber line* or *Gerber parabolic relationship*, extending above the Soderberg line from the endurance limit to the ultimate tensile strength.

The *Goodman line* (also known as the *modified Goodman line*) is less conservative than the Soderberg line and is more easily constructed than the Gerber line.[5] It is applicable for steel, aluminum, titanium, and some

magnesium alloys.[6] Figure 53.8 illustrates the Goodman line, as well as the method for determining the Goodman equivalent stress.[7] The Goodman factor of safety is

$$\text{FS} = \frac{S_e}{\sigma_{eq}} = \frac{S_e}{\sigma_{alt} + \left(\dfrac{S_e}{S_{ut}}\right)\sigma_m} \qquad 53.32$$

As with the Soderberg criterion, only the alternating stress should be increased by $K_f$.

For a ductile material experiencing an alternating shear stress, the yield and endurance strengths in shear can be calculated from distortion energy theory for use in drawing the *Goodman shear line*.[8]

$$S_{ys} = 0.577 S_{yt} \qquad 53.33$$

$$S_{es} = 0.577 S_e \qquad 53.34$$

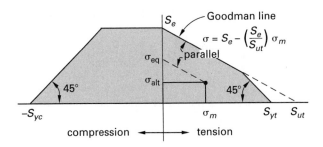

*Figure 53.8* Goodman Line and Equivalent Stress

### Example 53.5

An aircraft bell crank is made from aluminum with the following properties: yield strength, 40,000 $lbf/in^2$; ultimate strength, 45,000 $lbf/in^2$; and endurance limit 13,500 $lbf/in^2$ reduced to 8500 $lbf/in^2$ by various derating factors. The bell crank is subjected to minimum and maximum tensile stresses of 6500 $lbf/in^2$ and 9500 $lbf/in^2$, respectively. The bell crank is to have an indefinite life, and a factor of safety in excess of 3 is required. Is the design acceptable?

---

[5]There is no significant distinction between a *Goodman line* and a *modified Goodman line*. The modifications were made early in the theory's development, and both names are now used to represent a straight line drawn on the fatigue diagram between the endurance and ultimate tensile strengths.

[6]The Goodman line should not be used with gray cast iron and some types of magnesium. Failures for these materials do not follow the theory well. The envelope of failure for these materials is bounded by the parabolic *Smith line*, which is similar to the Gerber line, except that it runs under the straight line connecting the endurance and ultimate strengths. However, brittle materials such as gray cast iron are not usually considered satisfactory for stresses that fluctuate significantly unless very large factors of safety are included.

[7]The Soderberg, Goodman, Gerber, and Smith lines are drawn on *fatigue diagrams*. Strictly speaking, they are Soderberg *lines*, not Soderberg *diagrams*. The distinction is more critical for Goodman lines, because there is a Goodman diagram for fatigue loading, but it is constructed differently. However, the correct names are not consistently applied, and the terms "Soderberg diagram" and "Goodman diagram" are often heard.

[8]Theoretically, the coefficient 0.5 could be used instead of 0.577 if the maximum shear stress theory were to be elected. However, this is probably never done in practice.

*Solution*

The mean and alternating stresses are

$$\sigma_m = \tfrac{1}{2}(\sigma_{\max} + \sigma_{\min})$$
$$= \left(\tfrac{1}{2}\right)\left(9500\ \frac{\text{lbf}}{\text{in}^2} + 6500\ \frac{\text{lbf}}{\text{in}^2}\right)$$
$$= 8000\ \text{lbf/in}^2$$
$$\sigma_a = \tfrac{1}{2}(\sigma_{\max} - \sigma_{\min})$$
$$= \left(\tfrac{1}{2}\right)\left(9500\ \frac{\text{lbf}}{\text{in}^2} - 6500\ \frac{\text{lbf}}{\text{in}^2}\right)$$
$$= 1500\ \text{lbf/in}^2$$

The material properties are divided by the safety factor.

$$\frac{S_{yt}}{3} = \frac{40{,}000\ \dfrac{\text{lbf}}{\text{in}^2}}{3} = 13{,}333\ \text{lbf/in}^2$$

$$\frac{S_{ut}}{3} = \frac{45{,}000\ \dfrac{\text{lbf}}{\text{in}^2}}{3} = 15{,}000\ \text{lbf/in}^2$$

$$\frac{S_e}{3} = \frac{8500\ \dfrac{\text{lbf}}{\text{in}^2}}{3} = 2833\ \text{lbf/in}^2$$

Since the operating point lies above the failure line, the design is not acceptable.

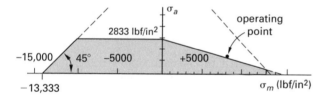

## 11. CUMULATIVE FATIGUE

If a part is subjected to $\sigma_{\max,1}$ for $n_1$ cycles, $\sigma_{\max,2}$ for $n_2$ cycles, and so on, the part will accumulate varying amounts of fatigue damage during each series of cycles. *Miner's rule* (also known as the *Palmgren-Miner cycle ratio summation formula, fatigue interaction formula, and cumulative usage factor rule*) can be used to evaluate cumulative damage.[9] In Eq. 53.35, the $N_i$ are the fatigue lives for the corresponding stress levels.

$$\sum \frac{n_i}{N_i} > C \quad \text{[failure criterion]} \qquad \textbf{53.35}$$

A value of 1.0 is commonly used for $C$. However, the exact value should be determined from experimentation appropriate to the material and application. Values between approximately 0.7 and 2.2 have been reported.

## 12. ALTERNATING COMBINED STRESSES

If there are variations in biaxial or triaxial stresses, a conservative approach is to use a combination of the distortion energy and Goodman line.

*step 1:* Calculate the principal mean stresses from the combined stress theory.

*step 2:* Calculate the principal alternating stresses.

*step 3:* Calculate the mean and alternating von Mises stresses.

$$\sigma'_m = \sqrt{\sigma_{m,1}^2 + \sigma_{m,2}^2 - \sigma_{m,1}\sigma_{m,2}} \qquad \textbf{53.36}$$

$$\sigma'_{\text{alt}} = \sqrt{\sigma_{\text{alt},1}^2 + \sigma_{\text{alt},2}^2 - \sigma_{\text{alt},1}\sigma_{\text{alt},2}} \qquad \textbf{53.37}$$

*step 4:* Plot the mean and alternating von Mises stresses in relationship to a standard Goodman line.

---

[9]Miner's rule does not take into account the increase in the endurance limit that results when virgin material is understressed.

# 54

# Engineering Design

## Nomenclature

| | | | |
|---|---|---|---|
| $A$ | area | ft$^2$ | m$^2$ |
| $b$ | width | ft | m |
| $c$ | distance from neutral axis to extreme fiber | ft | m |
| $C$ | circumference | ft | m |
| $d$ | diameter | ft | m |
| $e$ | eccentricity | ft | m |
| $E$ | energy | ft-lbf | J |
| $E$ | modulus of elasticity | lbf/ft$^2$ | Pa |
| $f$ | coefficient of friction | – | – |
| $F$ | force | lbf | N |
| FS | factor of safety | – | – |
| $g$ | acceleration of gravity | ft/sec$^2$ | m/s$^2$ |
| $g_c$ | gravitational constant | ft-lbm/lbf-sec$^2$ | n.a. |
| $G$ | shear modulus | lbf/ft$^2$ | Pa |
| $h$ | height | ft | m |
| $I$ | interference | ft | ft |
| $I$ | moment of inertia | ft$^4$ | m$^4$ |
| $J$ | polar moment of inertia | ft$^4$ | m$^4$ |
| $k$ | stiffness | lbf/ft | N/m |
| $K$ | end-restraint coefficient | – | – |
| $K$ | stress concentration factor | – | – |

| | | | |
|---|---|---|---|
| $L$ | length | ft | m |
| $m$ | mass | lbm | kg |
| $M$ | moment | ft-lbf | N·m |
| $n$ | modular ratio | – | – |
| $n$ | number of connectors | – | – |
| $n$ | rotational speed | rpm | rpm |
| $N$ | normal force | lbf | N |
| $p$ | perimeter | ft | m |
| $p$ | pressure | lbf/ft$^2$ | Pa |
| $P$ | power | hp | kW |
| $q$ | shear flow | lbf/ft | N/m |
| $r$ | radius | ft | m |
| $r$ | radius of gyration | ft | m |
| $S$ | strength | lbf/ft$^2$ | Pa |
| SR | slenderness ratio | – | – |
| $t$ | thickness | ft | m |
| $T$ | torque | ft-lbf | N·m |
| $U$ | energy | ft-lbf | J |
| $y$ | weld size | ft | m |

## Symbols

| | | | |
|---|---|---|---|
| $\alpha$ | thread half-angle | deg | deg |
| $\delta$ | deflection | in | m |
| $\gamma$ | angle of twist | rad | rad |
| $\epsilon$ | strain | ft/ft | m/m |
| $\theta$ | lead angle | deg | deg |
| $\theta$ | shear strain | rad | rad |
| $\nu$ | Poisson's ratio | – | – |
| $\sigma$ | normal stress | lbf/ft$^2$ | Pa |
| $\tau$ | shear stress | lbf/ft$^2$ | Pa |
| $\phi$ | angle | rad | rad |
| $\phi$ | load factor | – | – |

## Subscripts

| | |
|---|---|
| $a$ | allowable |
| $b$ | bending |
| $c$ | centroidal, circumferential, or collar |
| cr | critical |
| $e$ | Euler or effective |
| eq | equivalent |
| $f$ | flange |
| $h$ | hoop |
| $i$ | inside or initial |
| $l$ | longitudinal |
| $m$ | mean |
| $n$ | normal |
| $o$ | outside |
| $p$ | bearing or potential |
| $r$ | radial or rope |
| sh | sheave |
| $t$ | tension, thread, or transformed |

Mechanics of Materials

$T$  torque
$u$  ultimate
$ut$  ultimate tensile
$v$  vertical
$w$  wire or web
$y$  yield

## 1. ALLOWABLE STRESS DESIGN

Once an actual stress has been determined, it can be compared to the *allowable stress*. In engineering design, the term "allowable" always means that a factor of safety has been applied to the governing material strength.

$$\text{allowable stress} = \frac{\text{material strength}}{\text{factor of safety}} \qquad 54.1$$

For ductile materials, the material strength used is the yield strength. For steel, the factor of safety ranges from 1.5 to 2.5, depending on the type of steel and the application. Higher factors of safety are seldom necessary in normal, noncritical applications, due to steel's predictable and reliable performance.

$$\sigma_a = \frac{S_y}{\text{FS}} \quad [\text{ductile}] \qquad 54.2$$

For brittle materials, the material strength used is the ultimate strength. Since brittle failure is sudden and unpredictable, the factor of safety is high (e.g., in the 6 to 10 range).

$$\sigma_a = \frac{S_u}{\text{FS}} \quad [\text{brittle}] \qquad 54.3$$

If an actual stress is less than the allowable stress, the design is considered acceptable. This is the principle of the *allowable stress design method*, also known as the *working stress design method*.

$$\sigma_{\text{actual}} \leq \sigma_a \qquad 54.4$$

## 2. ULTIMATE STRENGTH DESIGN

The allowable stress method has been replaced in most structural work by the *ultimate strength design method*, also known as the *load factor design method*, *plastic design method*, or just *strength design method*. This design method does not use allowable stresses at all. Rather, the member is designed so that its actual *nominal strength* exceeds the required ultimate strength.[1]

The *ultimate strength* (i.e., the required strength) of a member is calculated from the actual *service loads* and multiplicative factors known as *overload factors* or *load factors*. Usually, a distinction is made between dead loads and live loads.[2] For example, the required ultimate moment-carrying capacity in a concrete beam designed according to ACI 318 would be[3]

$$M_u = 1.2 M_{\text{dead load}} + 1.6 M_{\text{live load}} \qquad 54.5$$

The *nominal strength* (i.e., the actual ultimate strength) of a member is calculated from the dimensions and materials. A *capacity reduction factor*, $\phi$, of 0.70 to 0.90 is included in the calculation to account for typical workmanship and increase required strength. The moment criteria for an acceptable design is

$$M_n \geq \frac{M_u}{\phi} \qquad 54.6$$

## 3. SLENDER COLUMNS

Very short compression members are known as *piers*. Long compression members are known as *columns*. Failure in piers occurs when the applied stress exceeds the yield strength of the material. However, very long columns fail by sideways *buckling* long before the compressive stress reaches the yield strength. Buckling failure is sudden, often without significant initial sideways bending. The load at which a column fails is known as the *critical load* or *Euler load*.

The *Euler load* is the theoretical maximum load that an initially straight column can support without buckling. For columns with frictionless or pinned ends, this load is given by Eq. 54.7. $r$ is the *radius of gyration*.

$$F_e = \frac{\pi^2 EI}{L^2} = \frac{\pi^2 EA}{\left(\dfrac{L}{r}\right)^2} \qquad 54.7$$

The corresponding column stress is given by Eq. 54.8. In order to use Euler's theory, this stress cannot exceed half of the compressive yield strength of the column material.

$$\sigma_e = \frac{F_e}{A} = \frac{\pi^2 E}{\left(\dfrac{L}{r}\right)^2} \qquad 54.8$$

The quantity $L/r$ is known as the *slenderness ratio*. Long columns have high slenderness ratios. The smallest slenderness ratio for which Eq. 54.8 is valid is the *critical slenderness ratio*. Typical critical slenderness ratios range from 80 to 120. The critical slenderness ratio becomes smaller as the compressive yield strength increases.

$L$ is the longest unbraced column length. If a column is braced against buckling at some point between its two

---

[1]It is a characteristic of the ultimate strength design method that the term "strength" actually means load, shear, or moment. Strength seldom, if ever, refers to stress. Thus, the nominal strength of a member might be the load (in pounds, kips, newtons, etc.) or moment (in ft-lbf, ft-kips, or N·m) that the member supports at plastic failure.

[2]*Dead load* is an inert, inactive load, primarily due to the structure's own weight. *Live load* is the weight of all non-permanent objects, including people and furniture, in the structure.
[3]American Concrete Institute publication 318 has been adopted as the source of concrete design rules in the United States.

ends, the column is known as a *braced column*, and $L$ will be less than the full column height. Columns with rectangular cross sections have two radii of gyration, $r_x$ and $r_y$, and therefore, will have two slenderness ratios. The largest slenderness ratio will govern the design.

Columns do not always have frictionless or pinned ends. Often, a column will be fixed ("clamped," "built in," etc.) at its top and base. In such cases, the *effective length*, $L'$, must be used in place of $L$ in Eq. 54.7 and Eq. 54.8.

$$L' = KL \qquad 54.9$$

Equation 54.8 becomes Eq. 54.10.

$$\sigma_e = \frac{F_e}{A} = \frac{\pi^2 E}{\left(\dfrac{L'}{r}\right)^2} \qquad 54.10$$

$K$ is the *end restraint coefficient*, which varies from 0.5 to 2.0 according to Table 54.1. For most real columns, the design values of $K$ should be used since infinite stiffness of the supporting structure is not achievable.

**Table 54.1** *Theoretical End Restraint Coefficients*

| illus. | end conditions | ideal | recommended for design |
|--------|----------------|-------|------------------------|
| (a) | both ends pinned | 1 | 1.0[*] |
| (b) | both ends built in | 0.5 | 0.65[*]–0.90 |
| (c) | one end pinned, one end built in | 0.707 | 0.80[*]–0.90 |
| (d) | one end built in, one end free | 2 | 2.0–2.1[*] |
| (e) | one end built in, one end fixed against rotation but free | 1 | 1.2[*] |
| (f) | one end pinned, one end fixed against rotation but free | 2 | 2.0[*] |

[*]AISC values

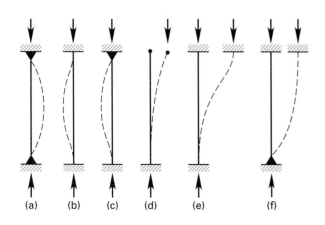

*Euler's curve* for columns, line BCD in Fig. 54.1, is generated by plotting the *Euler stress* (see Eq. 54.8) versus the slenderness ratio. Since the material's compressive yield strength cannot be exceeded, a horizontal line AC is added to limit applications to the region below. Theoretically, members with slenderness ratios less than $(SR)_C$ could be treated as pure compression members. However, this is not done in practice.

Defects in materials, errors in manufacturing, inabilities to achieve theoretical end conditions, and eccentricities frequently combine to cause column failures in the region around point C. Therefore, this region is excluded by designers.

The empirical *Johnson procedure* used to exclude the failure area is to draw a parabolic curve from point A through a tangent point T on the Euler curve at a stress of $^1/_2 S_y$. The corresponding value of the slenderness ratio is

$$(SR)_T = \sqrt{\frac{2\pi^2 E}{S_y}} \qquad 54.11$$

**Figure 54.1** *Euler's Curve*

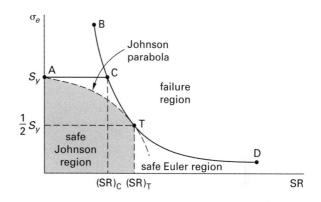

**Example 54.1**

A steel member is used as an 8.5 ft long column. The ends are pinned. What is the maximum allowable compressive stress in order to have a factor of safety of 3.0? Use the following data for the column.

$$E = 2.9 \times 10^7 \text{ lbf/in}^2$$

$$S_{yt} = 36{,}000 \text{ lbf/in}^2$$

$$r = 0.569 \text{ in}$$

*Solution*

First, check the slenderness ratio to see if this is a long column.

$$(\text{SR})_{\text{T}} = \sqrt{\frac{2\pi^2 \left(2.9 \times 10^7 \, \frac{\text{lbf}}{\text{in}^2}\right)}{36{,}000 \, \frac{\text{lbf}}{\text{in}^2}}} = 126.1$$

$$\frac{L}{r} = \frac{(8.5 \text{ ft})\left(12 \, \frac{\text{in}}{\text{ft}}\right)}{0.569 \text{ in}} = 179.3 \quad [>126.1, \text{ so OK}]$$

From Eq. 54.8, the Euler stress is

$$\sigma_e = \frac{\pi^2 E}{\left(\frac{L}{r}\right)^2} = \frac{\pi^2 \left(2.9 \times 10^7 \, \frac{\text{lbf}}{\text{in}^2}\right)}{(179.3)^2}$$

$$= 8903 \text{ lbf/in}^2$$

Since $8903 \text{ lbf/in}^2$ is less than half of the yield strength of $36{,}000 \text{ lbf/in}^2$, the Euler formula is valid. The allowable working stress is

$$\sigma_a = \frac{\sigma_e}{\text{FS}} = \frac{8903 \, \frac{\text{lbf}}{\text{in}^2}}{3}$$

$$= 2968 \text{ lbf/in}^2$$

## 4. INTERMEDIATE COLUMNS

Columns with slenderness ratios less than the critical slenderness ratio, but that are too long to be short piers, are known as *intermediate columns*. The *parabolic formula* (also known as the *J. B. Johnson formula*) is used to describe the parabolic line between points A and T on Fig. 54.1. The critical stress is given by Eq. 54.12, where $a$ and $b$ are curve-fit constants.

$$\sigma_{\text{cr}} = \frac{F_{\text{cr}}}{A} = a - b\left(\frac{KL}{r}\right)^2 \qquad 54.12$$

It is commonly assumed that the stress at point A is $S_y$ and the stress at point T is $S_y/2$. In that case, the parabolic formula becomes

$$\sigma_{\text{cr}} = S_y - \left(\frac{1}{E}\right)\left(\frac{S_y}{2\pi}\right)^2 \left(\frac{KL}{r}\right)^2 \qquad 54.13$$

## 5. ECCENTRICALLY LOADED COLUMNS

Accidental eccentricities are introduced during the course of normal manufacturing, so the load on real columns is rarely axial. The *secant formula* is one of

the methods available for determining the critical column stress and critical load with eccentric loading.[4]

$$\sigma_{\max} = \sigma_{\text{ave}}(1 + \text{amplification factor})$$

$$= \left(\frac{F}{A}\right)\left(1 + \left(\frac{ec}{r^2}\right)\sec\left(\frac{\pi}{2}\sqrt{\frac{F}{F_e}}\right)\right)$$

$$= \left(\frac{F}{A}\right)\left(1 + \left(\frac{ec}{r^2}\right)\sec\left(\frac{L}{2r}\sqrt{\frac{F}{AE}}\right)\right)$$

$$= \left(\frac{F}{A}\right)\left(1 + \left(\frac{ec}{r^2}\right)\sec\phi\right) \qquad 54.14$$

$$\phi = \tfrac{1}{2}\left(\frac{L}{r}\right)\sqrt{\frac{F}{AE}} \qquad 54.15$$

For a given *eccentricity, e,* or eccentricity ratio, $ec/r^2$, and an assumed value of the buckling load, $F$, Eq. 54.15 is solved by trial and error for the slenderness ratio, $L/r$. Equation 54.14 and Eq. 54.15 converge quickly to the known $L/r$ ratio when assumed values of $F$ are substituted. ($L/r$ is smaller when $F$ is larger.)

## 6. THIN-WALLED CYLINDRICAL TANKS

In general, tanks under internal pressure experience circumferential, longitudinal, and radial stresses. If the wall thickness is small, the radial stress component is negligible and can be disregarded. A cylindrical tank is a *thin-walled tank* if its wall thickness-to-internal diameter ratio is less than approximately $0.1$.[5]

$$\frac{t}{d_i} = \frac{t}{2r_i} < 0.1 \quad [\text{thin-walled}] \qquad 54.16$$

The *hoop stress,* $\sigma_h$, also known as *circumferential stress* and *tangential stress*, for a cylindrical thin-walled tank under internal pressure is derived from the free-body diagram of a cylinder half.[6] (See Fig. 54.2.) Since the cylinder is assumed to be thin-walled, it is not important which radius (e.g., inner, mean, or outer) is used in Eq. 54.17. However, the inner radius is used by common convention.

$$\sigma_h = \frac{pr}{t} \qquad 54.17$$

The axial forces on the ends of the cylindrical tank produce a stress, known as the *longitudinal stress* or

---

[4]The design of timber, steel, and reinforced concrete building columns is very code-intensive. None of the theoretical methods presented in this section is acceptable for building design.

[5]There is overlap in the thin-wall/thick-wall criterion. The limiting ratios between thin- and thick-walled cylinders are matters of the accuracy desired. Lamé's solution can always be used for both thick- and thin-walled cylinders.

[6]There is no simple way, including the more exact Lamé solution, of evaluating theoretical stresses in thin-walled cylinders under external pressure, since failure is by collapse, not yielding. However, empirical equations exist for predicting the *collapsing pressure*.

*long stress, $\sigma_l$*, directed along the tank's longitudinal axis.

$$\sigma_l = \frac{pr}{2t} \qquad 54.18$$

**Figure 54.2** *Stresses in a Thin-Walled Tank*

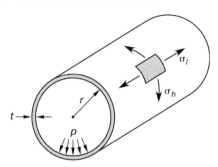

Unless the tank is subject to torsion, there is no shear stress. Accordingly, the hoop and long stresses are the principal stresses. They do not combine into larger stresses. Their combined effect should be evaluated according to the appropriate failure theory.

The increase in length due to pressurization is easily determined from the longitudinal strain.

$$\Delta L = L\epsilon_l$$
$$= L\left(\frac{\sigma_l - \nu\sigma_h}{E}\right) \qquad 54.19$$

The increase in circumference (from which the radial increase can also be determined) due to pressurization is

$$\Delta C = C\epsilon_h$$
$$= \pi d_o\left(\frac{\sigma_h - \nu\sigma_l}{E}\right) \qquad 54.20$$

**Example 54.2**

A thin-walled pressurized tank is supported at both ends, as shown. Points A and B are located midway between the supports and at the upper and lower surfaces, respectively. Evaluate the maximum stresses on the tank.

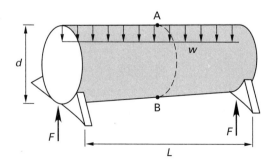

*Solution*

Since the tank is thin-walled, the radial stress is essentially zero. The hoop and longitudinal stresses are given

by Eq. 54.17 and Eq. 54.18, respectively. In addition, point A experiences a bending stress. The bending stress is

$$\sigma_b = \frac{Mc}{I}$$
$$c = \frac{d}{2}$$
$$M = \frac{FL}{2} = \frac{wL^2}{8}$$
$$I = \left(\frac{d}{2}\right)^3 \pi t$$

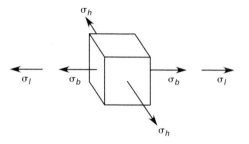

The bending stress is compressive at point A and tensile at point B. At point B, the bending stress has the same sign (tensile) as the longitudinal stress, and these two stresses add to each other. There is no torsional stress, so the resultant normal stresses are the principal stresses.

$$\sigma_1 = \sigma_h$$
$$\sigma_2 = \sigma_l + \sigma_b$$

## 7. THICK-WALLED CYLINDERS

A thick-walled cylinder has a wall thickness-to-radius ratio greater than 0.2 (i.e., a wall thickness-to-diameter ratio greater than 0.1). Figure 54.3 illustrates a thick-walled tank under either internal or external pressure.

In thick-walled tanks, radial stress is significant and cannot be disregarded. In *Lamé's solution*, a thick-walled cylinder is assumed to be made up of thin laminar rings. This method shows that the radial and circumferential stresses vary with location within the tank wall. (The term *circumferential stress* is preferred over *hoop stress* when dealing with thick-walled cylinders.) Compressive stresses are negative.

$$\sigma_c = \frac{r_i^2 p_i - r_o^2 p_o + \dfrac{(p_i - p_o)r_i^2 r_o^2}{r^2}}{r_o^2 - r_i^2} \qquad 54.21$$

$$\sigma_r = \frac{r_i^2 p_i - r_o^2 p_o - \dfrac{(p_i - p_o)r_i^2 r_o^2}{r^2}}{r_o^2 - r_i^2} \qquad 54.22$$

$$\sigma_l = \frac{p_i r_i^2}{r_o^2 - r_i^2} \quad \left[ \begin{array}{c} p_o \text{ does not act} \\ \text{longitudinally on the ends} \end{array} \right] \qquad 54.23$$

At every point in the cylinder, the circumferential, radial, and long stresses are the principal stresses. Unless an external torsional shear stress is added, it is not necessary to use the combined stress equations. Failure theories can be applied directly.

**Figure 54.3** Thick-Walled Cylinder

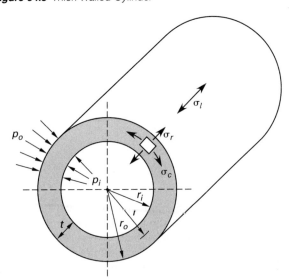

The cases of main interest are those of internal or external pressure only. The stress equations for these cases are summarized in Table 54.2. The maximum shear and normal stresses occur at the inner surface for both internal and external pressure.

**Table 54.2** Stresses in Thick-Walled Cylinders[*]

| stress | external pressure, $p$ | internal pressure, $p$ |
|---|---|---|
| $\sigma_{c,o}$ | $\dfrac{-(r_o^2 + r_i^2)p_o}{r_o^2 - r_i^2}$ | $\dfrac{2r_i^2 p_i}{r_o^2 - r_i^2}$ |
| $\sigma_{r,o}$ | $-p_o$ | $0$ |
| $\sigma_{c,i}$ | $\dfrac{-2r_o^2 p_o}{r_o^2 - r_i^2}$ | $\dfrac{(r_o^2 + r_i^2)p_i}{r_o^2 - r_i^2}$ |
| $\sigma_{r,i}$ | $0$ | $-p_i$ |
| $\tau_{\max}$ | $\frac{1}{2}\sigma_{c,i}$ | $\frac{1}{2}(\sigma_{c,i} + p_i)$ |

[*]Table 54.2 can be used with thin-walled cylinders. However, in most cases it will not be necessary to do so.

The *diametral strain* (which is the same as the *circumferential* and *radial strains*) is given by Eq. 54.24. Radial stresses are always compressive (hence they are negative), and algebraic signs must be observed with Eq. 54.24. Since the circumferential and radial stresses depend on location within the wall thickness, the strain

can be evaluated at inner, outer, and any intermediate locations within the wall.

$$\epsilon = \frac{\Delta d}{d} = \frac{\Delta C}{C} = \frac{\Delta r}{r}$$
$$= \frac{\sigma_c - \nu(\sigma_r + \sigma_l)}{E} \qquad 54.24$$

## 8. THIN-WALLED SPHERICAL TANKS

There is no unique axis in a spherical tank or in the spherical ends of a cylindrical tank. Therefore, the hoop and long stresses are identical.

$$\sigma = \frac{pr}{2t} \qquad 54.25$$

## 9. INTERFERENCE FITS

When assembling two pieces, interference fitting is often more economical than pinning, keying, or splining. The assembly operation can be performed in a hydraulic press, either with both pieces at room temperature or after heating the outer piece and cooling the inner piece. The former case is known as a *press fit* or *interference fit*; the latter as a *shrink fit*.

If two cylinders are pressed together, the pressure acting between them will expand the outer cylinder (placing it into tension) and will compress the inner cylinder. The *interference*, $I$, is the difference in dimensions between the two cylinders. *Diametral interference* and *radial interference* are both used.[7]

$$I_{\text{diametral}} = 2I_{\text{radial}}$$
$$= d_{o,\text{inner}} - d_{i,\text{outer}}$$
$$= |\Delta d_{o,\text{inner}}| + |\Delta d_{i,\text{outer}}| \qquad 54.26$$

If the two cylinders have the same length, the thick-wall cylinder equations can be used. The materials used for the two cylinders do not need to be the same. Since there is no longitudinal stress from an interference fit and since the radial stress is negative, the strain from Eq. 54.24 is

$$\epsilon = \frac{\Delta d}{d} = \frac{\Delta C}{C} = \frac{\Delta r}{r}$$
$$= \frac{\sigma_c - \nu\sigma_r}{E} \qquad 54.27$$

Equation 54.28 applies to the general case where both cylinders are hollow and have different moduli of elasticity and Poisson's ratios. The outer cylinder is

[7]Theoretically, the interference can be given to either the inner or outer cylinder, or it can be shared by both cylinders. However, in the case of a surface-hardened shaft with a standard diameter, all of the interference is usually given to the disk. Otherwise, it may be necessary to machine the shaft and remove some of the hardened surface.

designated as the *hub*; the inner cylinder is designated as the *shaft*. If the shaft is solid, use $r_{i,\text{shaft}} = 0$ in Eq. 54.28.

$$I_{\text{diametral}} = 2I_{\text{radial}}$$

$$= \left(\frac{2pr_{o,\text{shaft}}}{E_{\text{hub}}}\right)\left(\frac{r_{o,\text{hub}}^2 + r_{o,\text{shaft}}^2}{r_{o,\text{hub}}^2 + r_{o,\text{shaft}}^2} + \nu_{\text{hub}}\right)$$

$$+ \left(\frac{2pr_{o,\text{shaft}}}{E_{\text{shaft}}}\right)\left(\frac{r_{o,\text{shaft}}^2 + r_{i,\text{shaft}}^2}{r_{o,\text{shaft}}^2 - r_{i,\text{shaft}}^2} - \nu_{\text{shaft}}\right)$$

$$\textit{54.28}$$

In the special case where the shaft is solid and is made from the same material as the hub, the diametral interference is given by Eq. 54.29.

$$I_{\text{diametral}} = 2I_{\text{diametral}}$$

$$= \left(\frac{4pr_{\text{shaft}}}{E}\right)\left(\frac{1}{1 - \left(\dfrac{r_{\text{shaft}}}{r_{o,\text{hub}}}\right)^2}\right) \qquad \textit{54.29}$$

The maximum assembly force required to overcome friction during a press-fitting operation is given by Eq. 54.30. This relationship is approximate because the coefficient of friction is not known with certainty and the assembly force affects the pressure, $p$, through Poisson's ratio. The coefficient of friction is highly variable. Values in the range of 0.03 to 0.33 have been reported. In the absence of experimental data, it is reasonable to use 0.12 for lightly oiled connections and 0.15 for dry assemblies.

$$F_{\max} = fN = 2\pi f p r_{o,\text{shaft}} L_{\text{interface}} \qquad \textit{54.30}$$

The maximum torque that the press-fitted hub can withstand or transmit is given by Eq. 54.31. This can be greater or less than the shaft's torsional shear capacity. Both values should be calculated.

$$T_{\max} = 2\pi f p r_{o,\text{shaft}}^2 L_{\text{interface}} \qquad \textit{54.31}$$

Most interference fits are designed to keep the contact pressure or the stress below a given value. Designs of interference fits limited by strength generally use the distortion energy failure criterion. That is, the maximum shear stress is compared with the shear strength determined from the failure theory.

### Example 54.3

A steel cylinder has inner and outer diameters of 1.0 in and 2.0 in, respectively. The cylinder is pressurized internally to 10,000 lbf/in². The modulus of elasticity is $2.9 \times 10^7$ lbf/in², and Poisson's ratio is 0.3. What is the radial strain at the inside face?

*Solution*

The longitudinal stress is

$$\sigma_l = \frac{F}{A} = \frac{p_i \pi r_i^2}{\pi(r_o^2 - r_i^2)} = \frac{p_i r_i^2}{r_o^2 - r_i^2}$$

$$= \frac{\left(10,000 \ \dfrac{\text{lbf}}{\text{in}^2}\right)(0.5 \text{ in})^2}{(1.0 \text{ in})^2 - (0.5 \text{ in})^2}$$

$$= 3333 \text{ lbf/in}^2$$

The stresses at the inner face are found from Table 54.2.

$$\sigma_{c,i} = \frac{(r_o^2 + r_i^2)p}{r_o^2 - r_i^2}$$

$$= \frac{\left((1.0 \text{ in})^2 + (0.5 \text{ in})^2\right)\left(10,000 \ \dfrac{\text{lbf}}{\text{in}^2}\right)}{(1.0 \text{ in})^2 - (0.5 \text{ in})^2}$$

$$= 16,667 \text{ lbf/in}^2$$

$$\sigma_{r,i} = -p = -10,000 \text{ lbf/in}^2$$

The circumferential and radial stresses increase the radial strain; the longitudinal stress decreases the radial strain. The radial strain is

$$\frac{\Delta r}{r} = \frac{\sigma_{c,i} - \nu(\sigma_{r,i} + \sigma_l)}{E}$$

$$= \frac{16,667 \ \dfrac{\text{lbf}}{\text{in}^2} - (0.3)\left(-10,000 \ \dfrac{\text{lbf}}{\text{in}^2} + 3333 \ \dfrac{\text{lbf}}{\text{in}^2}\right)}{2.9 \times 10^7 \ \dfrac{\text{lbf}}{\text{in}^2}}$$

$$= 6.44 \times 10^{-4}$$

### Example 54.4

A hollow aluminum cylinder is pressed over a hollow brass cylinder as shown. Both cylinders are 2 in long. The interference is 0.004 in. The average coefficient of friction during assembly is 0.25. (a) What is the maximum shear stress in the brass? (b) What initial disassembly force is required to separate the two cylinders?

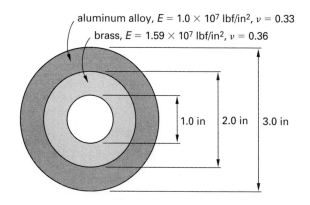

aluminum alloy, $E = 1.0 \times 10^7$ lbf/in², $\nu = 0.33$

brass, $E = 1.59 \times 10^7$ lbf/in², $\nu = 0.36$

1.0 in   2.0 in   3.0 in

*Solution*

(a) Work with the aluminum outer cylinder, which is under internal pressure.

$$\sigma_{c,i} = \frac{(r_o^2 + r_i^2)p}{r_o^2 - r_i^2}$$

$$= \frac{\left((1.5\ \text{in})^2 + (1.0\ \text{in})^2\right)p}{(1.5\ \text{in})^2 - (1.0\ \text{in})^2}$$

$$= 2.6p$$

$$\sigma_{r,i} = -p$$

From Eq. 54.24, the diametral strain is

$$\epsilon = \frac{\sigma_{c,i} - \nu(\sigma_{r,i} + \sigma_l)}{E}$$

$$= \frac{2.6p - (0.33)(-p)}{1.0 \times 10^7\ \dfrac{\text{lbf}}{\text{in}^2}}$$

$$= 2.93 \times 10^{-7}p$$

$$\Delta d = \epsilon d = (2.93 \times 10^{-7}p)(2.0\ \text{in})$$

$$= 5.86 \times 10^{-7}p$$

Now work with the brass inner cylinder, which is under external pressure. Use Table 54.2.

$$\sigma_{c,o} = \frac{-(r_o^2 + r_i^2)p}{r_o^2 - r_i^2}$$

$$= \frac{-\left((1.0\ \text{in})^2 + (0.5\ \text{in})^2\right)p}{(1.0\ \text{in})^2 - (0.5\ \text{in})^2}$$

$$= -1.667p$$

$$\sigma_{r,o} = -p$$

From Eq. 54.24, the diametral strain is

$$\epsilon = \frac{\sigma_{c,o} - \nu(\sigma_{r,o} + \sigma_l)}{E}$$

$$= \frac{-1.667p - (0.36)(-p)}{1.59 \times 10^7\ \dfrac{\text{lbf}}{\text{in}^2}}$$

$$= -0.822 \times 10^{-7}p$$

$$\Delta d = \epsilon d = (-0.822 \times 10^{-7}p)(2.0\ \text{in})$$

$$= -1.644 \times 10^{-7}p$$

The diametral interference is known to be 0.004 in. From Eq. 54.26,

$$I_{\text{diametral}} = |\Delta d_{o,\text{inner}}| + |\Delta d_{i,\text{outer}}|$$

$$0.004\ \text{in} = |5.86 \times 10^{-7}p| + |-1.644 \times 10^{-7}p|$$

$$p = 5330\ \text{lbf/in}^2$$

From Table 54.2, the circumferential stress at the inner face of the brass (under external pressure) is

$$\sigma_{c,i} = \frac{-2r_o^2 p}{r_o^2 - r_i^2}$$

$$= \frac{(-2)(1.0\ \text{in})^2\left(5330\ \dfrac{\text{lbf}}{\text{in}^2}\right)}{(1.0\ \text{in})^2 - (0.5\ \text{in})^2}$$

$$= -14{,}213\ \text{lbf/in}^2$$

Also from Table 54.2, the maximum shear stress is

$$\tau_{\max} = \tfrac{1}{2}\sigma_{c,i} = \left(\tfrac{1}{2}\right)\left(-14{,}213\ \frac{\text{lbf}}{\text{in}^2}\right)$$

$$= -7107\ \text{lbf/in}^2$$

(b) The initial force necessary to disassemble the two cylinders is the same as the maximum assembly force. Use Eq. 54.30.

$$F_{\max} = 2\pi f p r_{\text{shaft}} L_{\text{interface}}$$

$$= (2\pi)(0.25)\left(5330\ \frac{\text{lbf}}{\text{in}^2}\right)(1\ \text{in})(2\ \text{in})$$

$$= 16{,}745\ \text{lbf}$$

## 10. STRESS CONCENTRATIONS FOR PRESS-FITTED SHAFTS IN FLEXURE

When a shaft carrying a press-fitted hub (whose thickness is less than the shaft length) is loaded in flexure, there will be an increase in shaft bending stress in the vicinity of the inner hub edge. The fatigue life of the shaft can be seriously affected by this stress increase. The extent of the increase depends on the magnitude of the bending stress, $\sigma_b$, and the contact pressure, $p$, and can be as high as 2.0 or more.

Some designs attempt to reduce the increase in shaft stress by grooving the disk (to allow the disk to flex). Other designs rely on various treatments to increase the fatigue strength of the shaft. For an unmodified simple press-fit, the multiplicative stress concentration factor (to be applied to the bending stress calculated from $\sigma_b = Mc/I$) is given by Fig. 54.4.

## 11. BOLTS

There are three leading specifications for bolt thread families: ANSI, ISO metric, and DIN metric.[8] ANSI (essentially identical to SAE, ASTM, and ISO-inch standard) is widely used in the United States. DIN (Deutches Institute für Normung) fasteners are widely available and broadly accepted.[9] ISO metric (the

---

[8]Other fastener families include the Italian UNI, Swiss VSM, Japanese JIS, and United Kingdom's BS series.

[9]To add to the confusion, many DIN standards are identical to ISO standards, with only slight differences in the tolerance ranges. However, the standards are not interchangeable in every case.

**Figure 54.4** *Stress Concentration Factors for Press Fits*

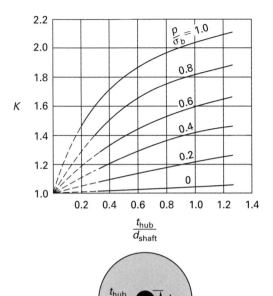

"Fatigue of Shafts at Fitted Members, with a Related Photo-elastic Analysis," reproduced from *Transactions of the ASME*, Vol. 57, © 1935, and Vol. 58, © 1936, with permission of the American Society of Mechanical Engineers.

International Organization for Standardization, which first met in 1961) fasteners are used in large volume by U.S. car manufacturers. The CEN (European Committee for Standardization) standards promulgated by the European Community (EC) have essentially adopted the ISO standards.

An American National (Unified) thread is specified by the sequence of parameters S(×L)-N-F-A-(H-E), where S is the thread outside diameter (nominal size), L is the optional shank length, N is the number of threads per inch, F is the thread pitch family, A is the class (allowance), and H and E are the optional hand and engagement length designations. The letter R can be added to the thread pitch family to indicate that the thread roots are radiused (for better fatigue resistance). For example, a $^3/_8$ × 1-16UNC-2A bolt is $^3/_8$ in in diameter, 1 in in length, and has 16 Unified Coarse threads per inch rolled with a class 2A accuracy.[10] A UNRC bolt would be identical except for radiused roots. Table 54.3 lists some (but not all) values for these parameters.

The *grade* of a bolt indicates the fastener material and is marked on the bolt cap.[11] In this regard, the marking depends on whether an SAE grade or ASTM

**Table 54.3** *Representative American National (United) Bolt Thread Designations*[a]

| | |
|---|---|
| S | Size |
| | 1 through 12 |
| | $^1/_4$ in through $^9/_{16}$ in in $^1/_{16}$ in increments |
| | $^5/_8$ in through $1^1/_2$ in in $^1/_8$ in increments |
| | $1^3/_4$ in through 4 in in $^1/_4$ in increments |
| F | Thread Family |
| | UNC and NC—Unified Coarse[b] |
| | UNF and NF—Unified Fine[b] |
| | UNEF and NEF—Unified Extra Fine[c] |
| | 8N—8 threads per inch |
| | 12UN and 12N—12 threads per inch |
| | 16UN and 16N—16 threads per inch |
| | UN, UNS, and NS—special series |
| A | Allowance (A—external threads, B—internal threads)[d] |
| | 1A and 1B—liberal allowance for each of assembly with dirty or damaged threads |
| | 2A and 2B—normal production allowance (sufficient for plating) |
| | 3A and 3B—close tolerance work with no allowance |
| H | Hand |
| | blank—right-hand thread |
| | LH—left-hand thread |

[a]In addition to fastener thread families, there are other special-use threads such as Acme, stub, square, buttress, and worm series.
[b]Previously known as United States Standard or American Standard.
[c]The UNEF series is the same as the SAE (Society of Automotive Engineers) fine series.
[d]Allowance classes 2 and 3 (without the A and B designation) were used prior to industry transition to the Unified classes.

designation is used. The minimum *proof load* (i.e., the maximum stress the bolt can support without acquiring a permanent set) increases with the grade. (The term *proof strength* is less common. Table 54.4 lists how the caps of the bolts are marked to distinguish among the major grades.[12] If a bolt is manufactured in the United States, its cap must also show the logo or mark of the manufacturer.

A metric thread is specified by an M or MJ and a diameter and a pitch in millimeters, in that order. Thus, M10 × 1.5 is a thread having a nominal major diameter of 10 mm and a pitch of 1.5 mm. The MJ series have rounded root fillets and larger minor diameters.

Head markings on metric bolts indicate their *property class* and correspond to the approximate tensile strength in MPa divided by 100. For example, a bolt marked 8.8 would correspond to a medium carbon, quenched and tempered bolt with an approximate tensile strength of 880 MPa. (The minimum of the tensile strength range for property class 8.8 is 830 MPa.) (See Table 54.5.)

---

[10]Threads are generally rolled, not cut, into a bolt.
[11]The *type* of a structural bolt should not be confused with the *grade* of a structural rivet.

[12]Optional markings can also be used.

*Mechanics of Materials*

**Table 54.4** *Selected Bolt Grades and Designations*
*LC = low-carbon; MC = medium-carbon; Q&T = quenched and tempered; CD = cold-drawn*
*(Subject to change and requires validation for design use.)*
*(Not for use with stainless steel.)*

| standard | head marking[i] | material type | proof load (ksi) | | minimum tensile strength (ksi) | | minimum yield strength (ksi) | |
|---|---|---|---|---|---|---|---|---|
| **SAE grades** | | | | | | | | |
| grade 1 | none | LC or MC | $33^h$ | | 55 | $60^n$ | | $36^n$ |
| grade 2 | none | LC or MC | $55^a$ | $33^b$ | $74^a$ | $60^b$ | $57^a$ | $36^b$ |
| grade 4 | none | CD MC | – | | 115 | | | |
| grade 5 | 3 tics, 360° | Q&T MC | $85^c$ | $74^d$ | $120^c$ | $105^d$ | $92^c$ | $81^d$ |
| grade 5.1 | 3 tics, 180° | | $85^p$ | | $120^p$ | | | |
| grade 5.2 | 3 tics, 120° | Q&T LC martensite | $85^q$ | | $120^q$ | | $92^q$ | |
| grade 7 | 5 tics, 360° | Q&T MC alloy | $105^h$ | | $133^h$ | | $115^h$ | |
| grade 8 | 6 tics, 360° | Q&T MC alloy | $120^h$ | | $150^h$ | | $130^h$ | |
| grade 8.1 | none | | $120^h$ | | $150^h$ | | $130^h$ | |
| grade 8.2 | 6 tics, 180° | Q&T LC martensite | $120^s$ | | $150^s$ | | | |
| **ISO designations** | | | | | | | | |
| class 4.6 | none | LC or MC | | 225 MPa | | 400 MPa | | |
| class 4.8 | | | | 310 MPa | | 420 MPa | | |
| class 5.8 | none | LC or MC | | $380^o$ MPa | | $520^o$ MPa | | |
| class 8.8 | 8.8 or 88 | Q&T MC | $580^j$ | $600^k$ MPa | $800^j$ | $830^k$ MPa | $640^j$ | $660^k$ MPa |
| class 9.8 | 9.8 | | | $650^r$ MPa | | $900^r$ MPa | | |
| class 10.9 | 10.9 or 109 | Q&T alloy steel | | $830^l$ MPa | | $1040^l$ MPa | $940^l$ | MPa |
| class 12.9 | 12.9 | | | $970^m$ MPa | | $1220^m$ MPa | $1100^m$ | MPa |
| **ASTM designations** | | | | | | | | |
| A307 grades A, B | none | LC | – | | 60 | | | |
| A325 type 1 | 3 tics, 360°, A325 | Q&T MC | $85^e$ | $74^f$ | $120^e$ | $105^f$ | $92^e$ | $81^f$ |
| A325 type 2 | 3 tics, 120°, A325 | Q&T LC martensite | $85^e$ | $74^f$ | $120^e$ | $105^f$ | $92^e$ | |
| A325 type 3 | A325 | Q&T weathering steel | $85^e$ | $74^f$ | $120^e$ | $105^f$ | $92^e$ | $81^f$ |
| A354 grade BC | BC | Q&T alloy steel | $105^g$ | | $125^g$ | | $109^g$ | |
| A354 grade BB | BB | Q&T alloy steel | $80^g$ | | $105^g$ | | $83^g$ | |
| A354 grade BD | 6 tics, 360° | Q&T alloy steel | $120^h$ | | $150^h$ | | $130^h$ | |
| A449 | 3 tics, 360° | Q&T MC | $85^c$ | $74^d$ | $120^c$ | $105^d$ | $92^c$ | $81^d$ |
| A490 type 1 | A490 | Q&T alloy steel | $120^t$ | | $150–170^t$ | | $130^t$ | |
| A490 type 3 | A490 | Q&T weathering steel | – | | – | | | |

(Multiply ksi by 6894.8 to obtain kPa.)

$^a 1/4 – 3/4$ in    $^j 5–15$ mm    $^q 1/4 – 1$ in
$^b 3/4 – 1 1/2$ in    $^k 16–72$ mm    $^r 1.6–16$ mm
$^c 1/4 – 1$ in    $^l 5–100$ mm    $^s 1/4 – 1$ in
$^d 1 – 1 1/2$ in    $^m 1.6–100$ mm    $^t 1/2 – 1 1/2$ in
$^e 1/2 – 1$ in    $^n 1/4 – 1 1/4$ in
$^f 1 1/8 – 1 1/2$ in    $^o 5–24$ mm
$^g 1/4 – 2 1/2$ in    $^p$ No. 6 – $3/8$ in
$^h 1/4 – 1 1/2$ in
[i]Tics are spread over the arc indicated.

## 12. RIVET AND BOLT CONNECTIONS

Figure 54.5 illustrates a tension *lap joint* connection using rivet or bolt connectors.[13] Unless the plate material is very thick, the effects of eccentricity are disregarded. A connection of this type can fail in shear, tension, or bearing. A common design procedure is to determine the number of connectors based on shear stress and then to check the bearing and tensile stresses.

*Table 54.5* Dimensions of American Unified Standard Threaded Bolts[*]

| nominal size | threads per inch | major diameter (in) | minor area (in²) | tensile stress area (in²) |
|---|---|---|---|---|
| coarse series | | | | |
| 1/4 | 20 | 0.2500 | 0.0269 | 0.0318 |
| 5/16 | 18 | 0.3125 | 0.0454 | 0.0524 |
| 3/8 | 16 | 0.3750 | 0.0678 | 0.0775 |
| 7/16 | 14 | 0.4375 | 0.0933 | 0.1063 |
| 1/2 | 13 | 0.5000 | 0.1257 | 0.1419 |
| 9/16 | 12 | 0.5625 | 0.162 | 0.182 |
| 5/8 | 11 | 0.6250 | 0.202 | 0.226 |
| 3/4 | 10 | 0.7500 | 0.302 | 0.334 |
| 7/8 | 9 | 0.8750 | 0.419 | 0.462 |
| 1 | 8 | 1.0000 | 0.551 | 0.606 |
| fine series | | | | |
| 1/4 | 28 | 0.2500 | 0.0326 | 0.0364 |
| 5/16 | 24 | 0.3125 | 0.0524 | 0.0580 |
| 3/8 | 24 | 0.3750 | 0.0809 | 0.0878 |
| 7/16 | 20 | 0.4375 | 0.1090 | 0.1187 |
| 1/2 | 20 | 0.5000 | 0.1486 | 0.1599 |
| 9/16 | 18 | 0.5625 | 0.189 | 0.203 |
| 5/8 | 18 | 0.6250 | 0.240 | 0.256 |
| 3/4 | 16 | 0.7500 | 0.351 | 0.373 |
| 7/8 | 14 | 0.8750 | 0.480 | 0.509 |
| 1 | 12 | 1.0000 | 0.625 | 0.663 |

(Multiply in by 25.4 to obtain mm.)
(Multiply in² by 645 to obtain mm².)
[*]Based on ANSI B1.1-1974.

*Figure 54.5* Tension Lap Joint

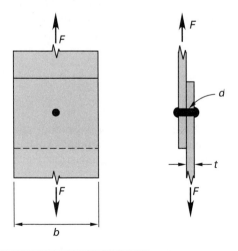

One of the failure modes is shearing of the connectors. In the case of *single shear*, each connector supports its proportionate share of the load. In *double shear*, each connector has two shear planes, and the stress per connector is halved.[14] (See Fig. 54.6.) The shear stress in a cylindrical connector is

$$\tau = \frac{F}{A} = \frac{F}{\frac{\pi}{4} d^2} \qquad 54.32$$

*Figure 54.6* Single and Double Shear

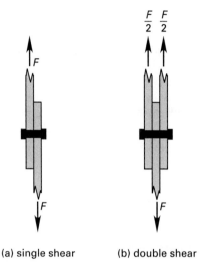

(a) single shear          (b) double shear

The number of required connectors, as determined by shear, is

$$n = \frac{\tau}{\text{allowable shear stress}} \qquad 54.33$$

The plate can fail in tension. If there are $n$ connector holes of diameter $d$ in a line across the width, $b$, of the plate, the cross-sectional area in the plate remaining to resist the tension is

$$A_t = t(b - nd) \qquad 54.34$$

The number of connectors across the plate width must be chosen to keep the tensile stress less than the allowable stress. The maximum tensile stress in the plate will be

$$\sigma_t = \frac{F}{A_t} \qquad 54.35$$

The plate can also fail by *bearing* (i.e., crushing). The number of connectors must be chosen to keep the actual

---

[13]Rivets are no longer used in building construction, but they are still extensively used in manufacturing.

[14]"Double shear" is not the same as "double rivet" or "double butt." *Double shear* means that there are two shear planes in one rivet. *Double rivet* means that there are two rivets along the force path. *Double butt* refers to the use of two backing plates (i.e., "scabs") used on either side to make a tension connection between two plates. Similarly, *single butt* refers to the use of a single backing plate to make a tension connection between two plates.

**Mechanics of Materials**

*bearing stress* below the allowable bearing stress. For one connector, the bearing stress in the plate is

$$\sigma_p = \frac{F}{dt} \qquad 54.36$$

$$n = \frac{\sigma_p}{\sigma_{a,\text{bearing}}} \qquad 54.37$$

The plate can also fail by shear tear-out, as illustrated in Fig. 54.7. The shear stress is

$$\tau = \frac{F}{2A} = \frac{F}{2t\left(L - \dfrac{d}{2}\right)} \qquad 54.38$$

The *joint efficiency* is the ratio of the strength of the joint divided by the strength of a solid (i.e., unpunched or undrilled) plate.

**Figure 54.7** *Shear Tear-Out*

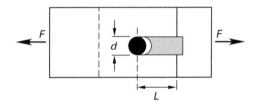

## 13. BOLT PRELOAD

Consider the ungasketed connection shown in Fig. 54.8. The load varies from $F_{\min}$ to $F_{\max}$. If the bolt is initially snug but without initial tension, the force in the bolt also will vary from $F_{\min}$ to $F_{\max}$. If the bolt is tightened so that there is an initial *preload force*, $F_i$, greater than $F_{\max}$ in addition to the applied load, the bolt will be placed in tension and the parts held together will be in compression.[15] When a load is applied, the bolt tension will increase even more, but the compression in the parts will decrease.

The amount of compression in the parts will vary as the applied load varies. Thus, the clamped members will carry some of the applied load, since this varying load has to "uncompress" the clamped part as well as lengthen the bolt. The net result is the reduction of the variation of the force in the bolt. The initial tension produces a larger mean stress, but the overall result is the reduction of the alternating stress. Thus, preloading is an effective method of reducing the alternating stress in bolted tension connections.

It is convenient to define the *spring constant, k*, of the bolt. The *grip, L*, is the thickness of the parts being connected by the bolt (not the bolt length). It is

---

[15]If the initial preload force is less than $F_{\max}$, the bolt may still carry a portion of the applied load. Equation 54.43 can be solved for the value of $F$ that will result in a loss of compression ($F_{\text{parts}} = 0$) and cause the bolt to carry the entire applied load.

**Figure 54.8** *Bolted Tension Joint with Varying Load*

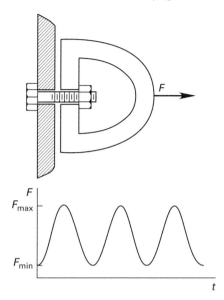

common to use the nominal diameter of the bolt, disregarding the reduction due to threading.

$$k_{\text{bolt}} = \frac{F}{\Delta L} = \frac{A_{\text{bolt}} E_{\text{bolt}}}{L} \qquad 54.39$$

The actual spring constant for a bolted part, $k_{\text{part}}$, is difficult to determine if the clamped area is not small and well defined. The only accurate way to determine the stiffness of a part in a bolted joint is through experimentation. If the clamped parts are flat plates, various theories can be used to calculate the effective load-bearing areas of the flanges, but doing so is a laborious process.

One simple rule of thumb is that the bolt force spreads out to three times the bolt-hole diameter. Of course, the hole diameter needs to be considered (i.e., needs to be subtracted) in calculating the effective force area. If the modulus of elasticity is the same for the bolt and the clamped parts, using this rule of thumb, the larger area results in the parts being eight times stiffer than the bolts.

$$k_{\text{parts}} = \frac{A_{e,\text{parts}} E_{\text{parts}}}{L} \qquad 54.40$$

If the clamped parts have different moduli of elasticity, or if a gasket constitutes one of the layers compressed by the bolt, the composite spring constant can be found from Eq. 54.41.[16]

$$\frac{1}{k_{\text{parts,composite}}} = \frac{1}{k_1} + \frac{1}{k_2} + \frac{1}{k_3} + \cdots \qquad 54.41$$

---

[16]If a soft washer or gasket is used, its spring constant can control Eq. 54.41.

The bolt and the clamped parts all carry parts of the applied load, $F_{applied}$. $F_i$ is the initial preload force.

$$F_{bolt} = F_i + \frac{k_{bolt}F_{applied}}{k_{bolt} + k_{parts}} \qquad 54.42$$

$$F_{parts} = \frac{k_{parts}F_{applied}}{k_{bolt} + k_{parts}} - F_i \qquad 54.43$$

O-ring (metal and elastomeric) seals permit metal-to-metal contact and affect the effective spring constant of the parts very little. However, the seal force tends to separate bolted parts and must be added to the applied force. The seal force can be obtained from the seal deflection and seal stiffness or from manufacturer's literature.

For static loading, recommended amounts of preloading often are specified as a percentage of the *proof strength* (or *proof load*) in psi.[17] For bolts, the proof load is slightly less than the yield strength. Traditionally, preload has been specified conservatively as 75% of proof for reusable connectors and 90% of proof for one-use connectors.[18] Connectors with some ductility can safely be used beyond the yield point, and 100% is now in widespread use.[19] When understood, advantages of preloading to 100% of proof load often outweigh the disadvantages.[20]

If the applied load varies, the forces in the bolt and parts will also vary. In that case, the preload must be determined from an analysis of the Goodman line.

Tightening of a tension bolt will induce a torsional stress in the bolt.[21] Where the bolt is to be locked in place, the torsional stress can be removed without greatly affecting the preload by slightly backing off the bolt. If the bolt is subject to cyclic loading, the bolt will probably slip back by itself, and it is reasonable to neglect the effects of torsion in the bolt altogether. (This is the reason that well-designed connections allow for a loss of 5% to 10% of the initial preload during routine use.)

Stress concentrations at the beginning of the threaded section are significant in cyclic loading.[22] To avoid a

reduction in fatigue life, the alternating stress used in the Goodman line should be multiplied by an appropriate stress concentration factor, $K$. For fasteners with rolled threads, an average factor of 2.2 for SAE grades 0 to 2 (metric grades 3.6 to 5.8) is appropriate. For SAE grades 4 to 8 (metric grades 6.6 to 10.9), an average factor of 3.0 is appropriate. Stress concentration factors for the fillet under the bolt head are different, but lower than these values. Stress concentration factors for cut threads are much higher.

The stress in a bolt depends on its load-carrying area. This area is typically obtained from a table of bolt properties. In practice, except for loading near the bolt's failure load, working stresses are low, and the effects of threads usually are ignored, so the area is based on the major (nominal) diameter.

$$\sigma_{bolt} = \frac{KF}{A} \qquad 54.44$$

## 14. BOLT TORQUE TO OBTAIN PRELOAD

During assembly, the preload tension is not monitored directly. Rather, the torque required to tighten the bolt is used to determine when the proper preload has been reached. Methods of obtaining the required preload include the standard torque wrench, the *run-of-the-nut method* (e.g., turning the bolt some specific angle past snugging torque), *direct-tension indicating* (DTI) washers, and computerized automatic assembly.

The standard manual torque wrench does not provide precise, reliable preloads, since the fraction of the torque going into bolt tension is variable.[23] Torque-, angle-, and time-monitoring equipment, usually part of an automated assembly operation, is essential to obtaining precise preloads on a consistent basis. It automatically applies the snugging torque and specified rotation, then checks the results with torque and rotation sensors. The computer warns of out-of-spec conditions.

The *Maney formula* is a simple relationship between the initial bolt tension, $F_i$, and the installation torque, $T$. The *torque coefficient*, $K_T$ (also known as the *bolt torque factor* and the *nut factor*) used in Eq. 54.46 depends mainly on the coefficient of friction, $f$. The torque coefficient for lubricated bolts generally varies from 0.15 to 0.20, and a value of 0.2 is commonly used.[24] With anti-seize lubrication, it can drop as low as 0.12. (The torque coefficient is not the same as the coefficient of friction.)

$$T = K_T d_{bolt} F_i \qquad 54.45$$

---

[17]This is referred to as a "rule of thumb" specification, because a mathematical analysis is not performed to determine the best preload.
[18]Some U.S. military specifications call for 80% of proof load in tension fasteners and only 30% for shear fasteners. The object of keeping the stresses below yielding is to be able to reuse the bolts.
[19]Even under normal elastic loading of a bolt, local plastic deformation occurs in the bolt-head fillet and thread roots. Since the stress-strain curve is nearly flat at the yield point, a small amount of elongation into the plastic region does not increase the stress or tension in the bolt.
[20]The disadvantages are: (a) Field maintenance probably won't be possible, as manually running up bolts to 100% proof will result in many broken bolts. (b) Bolts should not be reused, as some will have yielded. (c) The highest-strength bolts do not exhibit much plastic elongation and ordinarily should not be run up to 100% proof load.
[21]An argument for the conservative 75% of proof load preload limit is that the residual torsional stress will increase the bolt stress to 90% or higher anyway, and the additional 10% needed to bring the preload up to 100% probably won't improve economic performance much.
[22]Stress concentrations are frequently neglected for static loading.

---

[23]Even with good lubrication, about 50% of the torque goes into overcoming friction between the head and collar/flange, another 40% is lost in thread friction, and only the remaining 10% goes into tensioning the connector.
[24]With a coefficient of friction of 0.15, the torque coefficient is approximately 0.20 for most bolt sizes, regardless of whether the threads are coarse or fine.

$$K_T = \frac{f_c r_c}{d_{\text{bolt}}} + \left(\frac{r_t}{d_{\text{bolt}}}\right)\left(\frac{\tan\theta + f_t \sec\alpha}{1 - f_t \tan\theta \sec\alpha}\right) \qquad 54.46$$

$$\tan\theta = \frac{\text{lead per revolution}}{2\pi r_t} \qquad 54.47$$

$f_c$ is the coefficient of friction at the collar (fastener bearing face). $r_c$ is the mean collar radius (i.e., the effective radius of action of the friction forces on the bearing face). $r_t$ is the effective radius of action of the frictional forces on the thread surfaces. Similarly, $f_t$ is the coefficient of friction between the thread contact surfaces. $\theta$ is the *lead angle*, also known as the *helix thread angle*. $\alpha$ is the *thread half angle* (30° for UNF threads), and $d_{t,m}$ is the mean thread diameter.

## 15. FILLET WELDS

The common *fillet weld* is shown in Fig. 54.9. Such welds are used to connect one plate to another. The applied load, $F$, is assumed to be carried in shear by the *effective weld throat*. The *effective throat size*, $t_e$, is related to the weld size, $y$, by Eq. 54.48.

$$t_e = 0.707y \qquad 54.48$$

**Figure 54.9** *Fillet Lap Weld and Symbol*

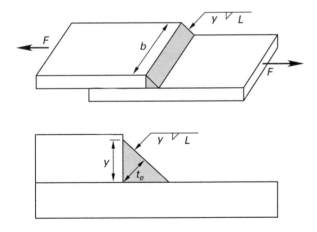

Neglecting any increased stresses due to eccentricity, the shear stress in a fillet lap weld depends on the *effective throat thickness*, $t_e$, and is

$$\tau = \frac{F}{bt_e} \qquad 54.49$$

Weld (filler) metal should have a strength equal to or greater than the base material. Properties of filler metals are readily available from their manufacturers and, for standard rated welding rods, from engineering handbooks.

## 16. CIRCULAR SHAFT DESIGN

Shear stress occurs when a shaft is placed in torsion. The shear stress at the outer surface of a bar of radius $r$, which is torsionally loaded by a torque, $T$, is

$$\tau = G\theta = \frac{Tr}{J} \qquad 54.50$$

The total strain energy due to torsion is

$$U = \frac{T^2 L}{2GJ} \qquad 54.51$$

$J$ is the shaft's polar moment of inertia. For a solid round shaft,

$$J = \frac{\pi r^4}{2} = \frac{\pi d^4}{32} \qquad 54.52$$

For a hollow round shaft,

$$J = \frac{\pi}{2}(r_o^4 - r_i^4) \qquad 54.53$$

If a shaft of length $L$ carries a torque $T$, as in Fig. 54.10, the angle of twist (in radians) will be

$$\gamma = \frac{L\theta}{r} = \frac{TL}{GJ} \qquad 54.54$$

**Figure 54.10** *Torsional Deflection of a Circular Shaft*

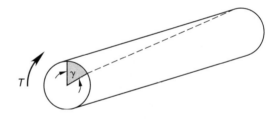

$G$ is the *shear modulus*. For steel, it is approximately $11.5 \times 10^6$ lbf/in² ($8.0 \times 10^4$ MPa). The shear modulus also can be calculated from the modulus of elasticity.

$$G = \frac{E}{2(1 + \nu)} \qquad 54.55$$

The torque, $T$, carried by a shaft spinning at $n$ revolutions per minute is related to the transmitted horsepower.

$$T_{\text{N·m}} = \frac{9549 P_{\text{kW}}}{n_{\text{rpm}}} \qquad \text{[SI]} \quad 54.56(a)$$

$$T_{\text{in-lbf}} = \frac{63{,}025 P_{\text{horsepower}}}{n_{\text{rpm}}} \qquad \text{[U.S.]} \quad 54.56(b)$$

If a statically loaded shaft without axial loading experiences a bending stress, $\sigma_x = Mc/I$ (i.e., is loaded in flexure), in addition to torsional shear stress,

$\tau = Tr/J$, the maximum shear stress from the combined stress theory is

$$\tau_{\max} = \sqrt{\left(\frac{\sigma_x}{2}\right)^2 + \tau^2} \qquad 54.57$$

$$\tau_{\max} = \frac{16}{\pi d^3}\sqrt{M^2 + T^2} \qquad 54.58$$

The equivalent normal stress from the distortion energy theory is

$$\sigma' = \frac{16}{\pi d^3}\sqrt{4M^2 + 3T^2} \qquad 54.59$$

The diameter can be determined by setting the shear and normal stresses equal to the maximum allowable shear (as calculated from the maximum shear stress theory, $S_y/2(\mathrm{FS})$, or from the distortion energy theory, $\sqrt{3}S_y/2(\mathrm{FS})$, and normal stresses, respectively).

Equation 54.57 and Eq. 54.58 should not be used with dynamically loaded shafts (i.e., those that are turning). Fatigue design of shafts should be designed according to a specific code (e.g., ANSI or ASME) or should use a fatigue analysis (e.g., Goodman, Soderberg, or Gerber).

### Example 54.5

The press-fitted, aluminum alloy-brass cylinder described in Ex. 54.4 is used as a shaft. The press fit is adequate to maintain nonslipping contact between the two materials. The shaft carries a steady torque of 24,000 in-lbf. There is no bending stress. What is the maximum torsional shear stress in the (a) aluminum and (b) brass?

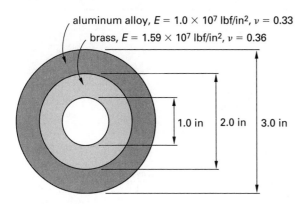

aluminum alloy, $E = 1.0 \times 10^7$ lbf/in², $\nu = 0.33$
brass, $E = 1.59 \times 10^7$ lbf/in², $\nu = 0.36$

*Solution*

The stronger material (as determined from the shear modulus, $G$) should be converted to an equivalent area of the weaker material.

For the aluminum, from Eq. 54.55,

$$G_{\text{aluminum}} = \frac{E}{2(1+\nu)} = \frac{1.0 \times 10^7\, \frac{\text{lbf}}{\text{in}^2}}{(2)(1+0.33)}$$

$$= 3.76 \times 10^6\ \text{lbf/in}^2$$

For the brass, from Eq. 54.55,

$$G_{\text{brass}} = \frac{E}{2(1+\nu)} = \frac{1.59 \times 10^7\, \frac{\text{lbf}}{\text{in}^2}}{(2)(1+0.36)}$$

$$= 5.85 \times 10^6\ \text{lbf/in}^2$$

The brass is the stronger material. The modular shear ratio is

$$n = \frac{G_{\text{brass}}}{G_{\text{aluminum}}} = \frac{5.85 \times 10^6\, \frac{\text{lbf}}{\text{in}^2}}{3.76 \times 10^6\, \frac{\text{lbf}}{\text{in}^2}}$$

$$= 1.56$$

The polar moment of inertia of the aluminum is

$$J_{\text{aluminum}} = \frac{\pi}{2}(r_o^4 - r_i^4)$$

$$= \left(\frac{\pi}{2}\right)\left((1.5\ \text{in})^4 - (1.0\ \text{in})^4\right)$$

$$= 6.38\ \text{in}^4$$

The equivalent polar moment of inertia of the brass is

$$J_{\text{brass}} = n\left(\frac{\pi}{2}\right)(r_o^4 - r_i^4)$$

$$= (1.56)\left(\frac{\pi}{2}\right)\left((1.0\ \text{in})^4 - (0.5\ \text{in})^4\right)$$

$$= 2.30\ \text{in}^4$$

The total equivalent polar moment of inertia is

$$J_{\text{total}} = J_{\text{aluminum}} + J_{\text{brass}}$$

$$= 6.38\ \text{in}^4 + 2.30\ \text{in}^4 = 8.68\ \text{in}^4$$

(a) The maximum torsional shear stress in the aluminum occurs at the outer edge. Use Eq. 54.50.

$$\tau = \frac{Tr}{J} = \frac{(24{,}000\ \text{in-lbf})(1.5\ \text{in})}{8.68\ \text{in}^4}$$

$$= 4147\ \text{lbf/in}^2$$

(b) Using the composite structures analysis methodology, the maximum torsional shear stress in the brass is

$$\tau = \frac{nTr}{J} = \frac{(1.56)(24{,}000\ \text{in-lbf})(1.0\ \text{in})}{8.68\ \text{in}^4}$$

$$= 4313\ \text{lbf/in}^2$$

## 17. TORSION IN THIN-WALLED, NONCIRCULAR SHELLS

Shear stress due to torsion in a thin-walled, noncircular shell (also known as a *closed box*) acts around the perimeter of the shell, as shown in Fig. 54.11. The shear stress, $\tau$, is given by Eq. 54.60. $A$ is the area enclosed by the centerline of the shell.

$$\tau = \frac{T}{2At} \qquad \textit{54.60}$$

**Figure 54.11** Torsion in Thin-Walled Shells

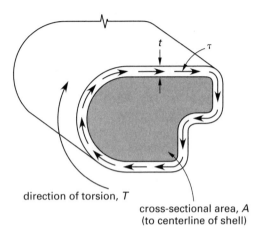

direction of torsion, $T$

cross-sectional area, $A$
(to centerline of shell)

The shear stress at any point is not proportional to the distance from the centroid of the cross section. Rather, the *shear flow, q*, around the shell is constant, regardless of whether the wall thickness is constant or variable.[25] The shear flow is the shear per-unit length of the centerline path.[26] At any point where the shell thickness is $t$,

$$q = \tau t = \frac{T}{2A} \quad \text{[constant]} \qquad \textit{54.61}$$

When the wall thickness, $t$, is constant, the angular twist depends on the perimeter, $p$, of the shell as measured along the centerline of the shell wall.

$$\gamma = \frac{TLp}{4A^2 tG} \qquad \textit{54.62}$$

## 18. TORSION IN SOLID, NONCIRCULAR MEMBERS

When a noncircular solid member is placed in torsion, the shear stress is not proportional to the distance from the centroid of the cross section. The maximum shear usually occurs close to the point on the surface that is nearest the centroid.

Shear stress, $\tau$, and angular deflection, $\gamma$, due to torsion are functions of the cross-sectional shape. They cannot be specified by simple formulas that apply to all sections. Table 54.6 lists the governing equations for

[25]The concept of shear flow can also be applied to a regular beam in bending, although there is little to be gained by doing so. Removing the dimension $b$ in the general beam shear stress equation, $q = VQ/I$.
[26]Shear flow is not analogous to magnetic flux or other similar quantities because the shear flow path does not need to be complete (i.e., does not need to return to its starting point).

several basic cross sections. These formulas have been derived by dividing the member into several concentric thin-walled closed shells and summing the torsional strength provided by each shell.

**Table 54.6** Torsion in Solid, Noncircular Shapes

| cross section | $K$ in formula $\gamma = TL/KG$ | $\tau$(max) |
|---|---|---|
| **ellipse** ($2a$ wide, $2b$ tall) | $\dfrac{\pi a^3 b^3}{a^2 + b^2}$ | $\dfrac{2T}{\pi a b^2}$ (maximum at ends of minor axis) |
| **square** (side $a$) | $0.1406 a^4$ | $\dfrac{T}{0.208 a^3}$ (maximum at midpoint of each side) |
| **rectangle** ($2a$ wide, $2b$ tall) | * | $\dfrac{T(3a + 1.8b)}{8a^2 b^2}$ (maximum at midpoint of each longer side) |
| $*ab^3 \left( \dfrac{16}{3} - \left( \dfrac{3.36b}{a} \right) \left( 1 - \dfrac{b^4}{12a^4} \right) \right)$ | | |
| **equilateral triangle** (side $a$) | $\dfrac{a^4 \sqrt{3}}{80}$ | $\dfrac{20T}{a^3}$ (maximum at midpoint of each side) |
| **slotted tube** (radius $r$, thickness $t$) | $\dfrac{2\pi r t^3}{3}$ | $\dfrac{T(6\pi r + 1.8t)}{4\pi^2 r^2 t^2}$ (maximum along both edges remote from ends) |
| **I-beam** ($b$, $h$, $t_f$, $t_w$) | $\dfrac{2b t_f^3 + h t_w^3}{3}$ | $\dfrac{3T t_f}{2b t_f^3 + h t_w^3}$ $[t_w < t_f]$ |

# 19. SHEAR CENTER FOR BEAMS

A beam with a symmetrical cross section supporting a transverse force that is offset from the longitudinal centroidal axis will be acted upon by a torsional moment, and the beam will tend to "roll" about a longitudinal axis known as the *bending axis* or *torsional axis*. For solid, symmetrical cross sections, this bending axis passes through the centroid of the cross section. However, for an asymmetrical beam (e.g., a channel beam on its side), the bending axis passes through the *shear center* (*torsional center* or *center of twist*), not the centroid. (See Fig. 54.12.) The shear center is a point that does not experience rotation (i.e., is a point about which all other points rotate) when the beam is in torsion.

***Figure 54.12*** *Channel Beam in Pure Bending (shear resultant, V, directed through shear center, O)*

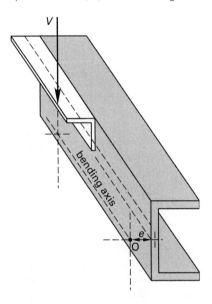

For beams with transverse loading, simple bending without torsion can only occur if the transverse load (*shear resultant* or *shear force of action*) is directed through the shear center. Otherwise, a torsional moment calculated as the product of the shear resultant and the torsional eccentricity will cause the beam to twist. The *torsional eccentricity* is the distance between the line of action of the shear resultant and the shear center.

The location of the shear center for any particular beam geometry is determined by setting the torsional moment equal to the shear resisting moment, as calculated from the total shear flow and the appropriate moment arm. In practice, however, shear centers for common shapes are located in tables similar to Fig. 54.13.

***Figure 54.13*** *Shear Centers of Selected Thin-Walled Open Sections*[*]

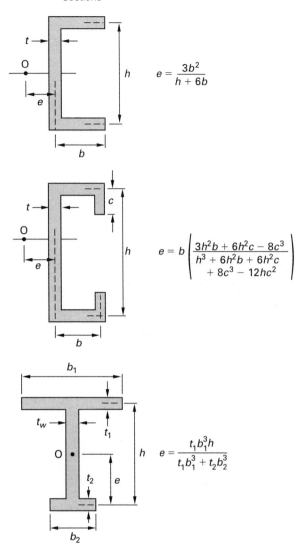

[*]Distance $e$ from shear center, O, measured from the reference point shown. Distance $e$ is not the torsional eccentricity.

# 20. ECCENTRICALLY LOADED BOLTED CONNECTIONS

An eccentrically loaded connection is illustrated in Fig. 54.14. The bracket's natural tendency is to rotate about the centroid of the connector group. The shear stress in the connectors includes both the direct vertical shear and the torsional shear stress. The sum of these shear stresses is limited by the shear strength of the critical connector, which in turn determines the capacity of the connection, as limited by bolt shear strength.[27]

Analysis of an eccentric connection is similar to the analysis of a shaft under torsion. The shaft torque, $T$,

---

[27]This type of analysis is known as an *elastic analysis* of the connection. Although it is traditional, it tends to greatly understate the capacity of the connection.

is analogous to the moment, $Fe$, on the connection. The shaft's radius corresponds to the distance from the centroid of the fastener group to the *critical fastener*. The critical fastener is the one for which the vector sum of the vertical and torsional shear stresses is the greatest.

$$\tau = \frac{Tr}{J} = \frac{Fer}{J} \qquad \qquad 54.63$$

*Figure 54.14* Eccentrically Loaded Connection

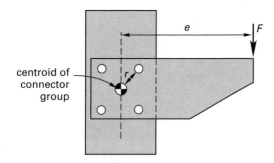

The polar moment of inertia, $J$, is calculated from the parallel axis theorem. Since bolts and rivets have little resistance to twisting in their holes, their individual polar moments of inertia are omitted.[28] Only the $r^2 A$ terms in the parallel axis theorem are used. $r_i$ is the distance from the fastener group centroid to the centroid (i.e., center) of the $i$th fastener, which has an area of $A_i$.

$$J = \sum_i r_i^2 A_i \qquad \qquad 54.64$$

The torsional shear stress is directed perpendicularly to a line between each fastener and the connector group centroid. The direction of the shear stress is the same as the rotation of the connection. (See Fig. 54.15.)

*Figure 54.15* Direction of Torsional Shear Stress

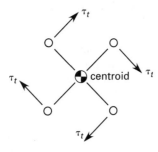

Once the torsional shear stress has been determined in the critical fastener, it is added in a vector sum to the

---

[28]In spot-welded and welded stud connections, the torsional resistance of each connector can be considered.

direct vertical shear stress. The direction of the vertical shear stress is the same as that of the applied force.

$$\tau_v = \frac{F}{nA} \qquad \qquad 54.65$$

Typical connections gain great strength from the frictional slip resistance between the two surfaces. By preloading the connection bolts, the normal force between the plates is greatly increased. The connection strength from friction will rival or exceed the strength from bolt shear in connections that are designed to take advantage of preload.

### Example 54.6

All fasteners used in the bracket shown have a nominal $^1/_2$ in diameter. What is the stress in the most critical fastener?

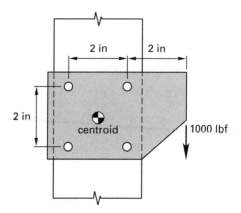

*Solution*

Since the fastener group is symmetrical, the centroid is centered within the four fasteners. This makes the eccentricity of the load equal to 3 in. Each fastener is located a distance $r$ from the centroid, where

$$r = \sqrt{x^2 + y^2} = \sqrt{(1 \text{ in})^2 + (1 \text{ in})^2} = 1.414 \text{ in}$$

The area of each fastener is

$$A_i = \frac{\pi}{4} d^2 = \left(\frac{\pi}{4}\right)(0.5 \text{ in})^2 = 0.1963 \text{ in}^2$$

Using the parallel axis theorem for polar moments of inertia and disregarding the individual torsional resistances of the fasteners,

$$J = \sum_i r_i^2 A_i = (4)(1.414 \text{ in})^2 (0.1963 \text{ in}^2)$$

$$= 1.570 \text{ in}^4$$

The torsional shear stress in each fastener is

$$\tau_t = \frac{Fer}{J} = \frac{(1000 \text{ lbf})(3 \text{ in})(1.414 \text{ in})}{1.570 \text{ in}^4}$$

$$= 2702 \text{ lbf/in}^2$$

Each torsional shear stress can be resolved into a horizontal shear stress, $\tau_{tx}$, and a vertical shear stress, $\tau_{ty}$. Both of these components are equal to

$$\tau_{tx} = \tau_{ty} = \frac{(\sqrt{2})\left(2702 \ \frac{\text{lbf}}{\text{in}^2}\right)}{2}$$
$$= 1911 \ \text{lbf/in}^2$$

The direct vertical shear downward is

$$\tau_v = \frac{F}{nA} = \frac{1000 \ \text{lbf}}{(4)(0.1963 \ \text{in}^2)}$$
$$= 1274 \ \text{lbf/in}^2$$

The two right fasteners have vertical downward components of torsional shear stress. The direct vertical shear is also downward. These downward components add, making both right fasteners critical.

The total stress in each of these fasteners is

$$\tau = \sqrt{\tau_{tx}^2 + (\tau_{ty} + \tau_v)^2}$$
$$= \sqrt{\left(1911 \ \frac{\text{lbf}}{\text{in}^2}\right)^2 + \left(1911 \ \frac{\text{lbf}}{\text{in}^2} + 1274 \ \frac{\text{lbf}}{\text{in}^2}\right)^2}$$
$$= 3714 \ \text{lbf/in}^2$$

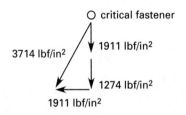

## 21. ECCENTRICALLY LOADED WELDED CONNECTIONS

The traditional elastic analysis of an eccentrically loaded welded connection is virtually the same as for a bolted connection, with the additional complication of having to determine the polar moment of inertia of the welds.[29] This can be done either by taking the welds as lines or by assuming each weld has an arbitrary thickness, $t$. After finding the centroid of the weld group, the rectangular moments of inertia of the individual welds are taken about that centroid using the parallel axis theorem. These rectangular moments of inertia are added to determine the polar moment of inertia. This laborious process can be shortened by use of App. 54.A.

The torsional shear stress, calculated from $Mr/J$ (where $r$ is the distance from the centroid of the weld group to the most distant weld point), is added vectorially to the

[29]Steel building design does not use an elastic analysis to design eccentric brackets, either bolted or welded. The design methodology is highly proceduralized and codified.

direct shear to determine the maximum shear stress at the critical weld point.

## 22. FLAT PLATES

Flat plates under uniform pressure are separated into two edge-support conditions: simply supported and built-in edges.[30] Commonly accepted working equations are summarized in Table 54.7. It is assumed that (a) the plates are of "medium" thickness (meaning that the thickness is equal to or less than one-fourth of the minimum dimension of the plate), (b) the pressure is no more than will produce a maximum deflection of one-half of the thickness, (c) the plates are constructed of isotropic, elastic material, and (d) the stress does not exceed the yield strength.

### Example 54.7

A steel pipe with an inside diameter of 10 in (254 mm) is capped by welding round mild steel plates on its ends. The allowable stress is 11,100 lbf/in² (77 MPa). The internal gage pressure in the pipe is maintained at 500 lbf/in² (3.5 MPa). What plate thickness is required?

*SI Solution*

A fixed edge approximates the welded edges of the plate. From Table 54.7, the maximum bending stress is

$$\sigma_{\text{max}} = \frac{3pr^2}{4t^2}$$

$$t = \sqrt{\frac{3pr^2}{4\sigma_{\text{max}}}} = \sqrt{\frac{(3)(3.5 \ \text{MPa})\left(\dfrac{254 \ \text{mm}}{2}\right)^2}{(4)(77 \ \text{MPa})}}$$
$$= 23.4 \ \text{mm}$$

*Customary U.S. Solution*

A fixed edge approximates the welded edges of the plate. From Table 54.7, the maximum bending stress is

$$\sigma_{\text{max}} = \frac{3pr^2}{4t^2}$$

$$t = \sqrt{\frac{3pr^2}{4\sigma_{\text{max}}}} = \sqrt{\frac{(3)\left(500 \ \dfrac{\text{lbf}}{\text{in}^2}\right)\left(\dfrac{10 \ \text{in}}{2}\right)^2}{(4)\left(11,100 \ \dfrac{\text{lbf}}{\text{in}^2}\right)}}$$
$$= 0.919 \ \text{in}$$

[30]Fixed-edge conditions are theoretical and are seldom achieved in practice. Considering this fact and other simplifying assumptions that are made to justify the use of Table 54.7, a value of $\nu = 0.3$ can be used without loss of generality.

**Table 54.7** Flat Plates Under Uniform Pressure

| shape | edge condition | maximum stress | deflection at center |
|---|---|---|---|
| circular | simply supported | $\dfrac{\frac{3}{8}pr^2(3+\nu)}{t^2}$ (at center) | $\dfrac{\frac{3}{16}pr^4(1-\nu)(5+\nu)}{Et^3}$ |
| | built-in | $\dfrac{\frac{3}{4}pr^2}{t^2}$ (at edge) | $\dfrac{\frac{3}{16}pr^4(1-\nu^2)}{Et^3}$ |
| rectangular | simply supported | $\dfrac{C_1pb^2}{t^2}$ (at center) | $\dfrac{C_2pb^4}{Et^3}$ |
| | built-in | $\dfrac{C_3pb^2}{t^2}$ (at centers of long edges) | $\dfrac{C_4pb^4}{Et^3}$ |

| $\frac{a}{b}$ | 1.0 | 1.2 | 1.4 | 1.6 | 1.8 | 2 | 3 | 4 | 5 | ∞ |
|---|---|---|---|---|---|---|---|---|---|---|
| $C_1$ | 0.287 | 0.376 | 0.453 | 0.517 | 0.569 | 0.610 | 0.713 | 0.741 | 0.748 | 0.750 |
| $C_2$ | 0.044 | 0.062 | 0.077 | 0.091 | 0.102 | 0.111 | 0.134 | 0.140 | 0.142 | 0.142 |
| $C_3$ | 0.308 | 0.383 | 0.436 | 0.487 | 0.497 | 0.500 | 0.500 | 0.500 | 0.500 | 0.500 |
| $C_4$ | 0.0138 | 0.0188 | 0.023 | 0.025 | 0.027 | 0.028 | 0.028 | 0.028 | 0.028 | 0.028 |

## 23. SPRINGS

An *ideal spring* is assumed to be perfectly elastic within its working range. The deflection is assumed to follow *Hooke's law*.[31] The *spring constant, k*, is also known as the *stiffness, spring rate, scale,* and *k-value*.[32]

$$F = k\delta \qquad 54.66$$

$$k = \frac{F_1 - F_2}{\delta_1 - \delta_2} \qquad 54.67$$

A spring stores energy when it is compressed or extended. By the *work-energy principle*, the energy storage is equal to the work required to displace the spring. The potential energy of a spring whose ends have been displaced a total distance $\delta$ is

$$\Delta E_p = \tfrac{1}{2}k\delta^2 \qquad 54.68$$

If a mass, $m$, is dropped from height $h$ onto and captured by a spring, the compression, $\delta$, can be found by

equating the change in potential energy to the energy storage.

$$mg(h+\delta) = \tfrac{1}{2}k\delta^2 \qquad \text{[SI]} \quad 54.69(a)$$

$$m\left(\frac{g}{g_c}\right)(h+\delta) = \tfrac{1}{2}k\delta^2 \qquad \text{[U.S.]} \quad 54.69(b)$$

Within the elastic region, this energy can be recovered by restoring the spring to its original unstressed condition. It is assumed that there is no permanent set, and no energy is lost through external friction or *hysteresis* (internal friction) when the spring returns to its original length.[33]

The entire applied load is felt by each spring in a series of springs linked end-to-end. The *equivalent (composite) spring constant* for springs in series is

$$\frac{1}{k_{\text{eq}}} = \frac{1}{k_1} + \frac{1}{k_2} + \frac{1}{k_3} + \cdots \qquad 54.70$$

Springs in parallel (e.g., concentric springs) share the applied load. The equivalent spring constant for springs in parallel is

$$k_{\text{eq}} = k_1 + k_2 + k_3 + \cdots \qquad 54.71$$

---

[31]A spring can be perfectly elastic even though it does not follow Hooke's law. The deviation from proportionality, if any, occurs at very high loads. The difference in theoretical and actual spring forces is known as the *straight-line error*.

[32]Another unfortunate name for the spring constant, $k$, that is occasionally encountered is the *spring index*. This is not the same as the spring index, $C$, used in helical coil spring design. The units will determine which meaning is intended.

[33]There is essentially no hysteresis in properly formed compression, extension, or open-wound helical torsion springs.

## 24. WIRE ROPE

*Wire rope* is constructed by first winding individual *wires* into *strands* and then winding the strands into rope. (See Fig. 54.16.) Wire rope is specified by its diameter and numbers of strands and wires. The most common *hoisting cable* is 6 × 19, consisting of six strands of 19 wires each, wound around a core. This configuration is sometimes referred to as "standard wire rope." Other common configurations are 6 × 7 (stiff *transmission* or *haulage rope*), 8 × 19 (*extra-flexible hoisting rope*), and the abrasion-resistant 6 × 37. The diameter and area of a wire rope are based on the circle that just encloses the rope.

**Figure 54.16** *Wire Rope Cross Sections*

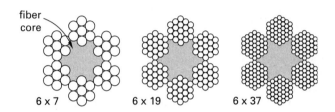

Wire rope can be obtained in a variety of materials and cross sections. In the past, wire ropes were available in iron, cast steel, traction steel (TS), mild plow steel (MPS), and plow steel (PS) grades. Modern wire ropes are generally available only in improved plow steel (IPS) and extra-improved plow steel (EIP) grades.[34] *Monitor* and *blue center steels* are essentially the same as improved plow steel.

While manufacturer's data should be relied on whenever possible, general properties of 6 × 19 wire rope are given in Table 54.9.

In Table 54.9, the ultimate strength, $S_{ut,r}$, is the ultimate tensile load that the rope can carry without breaking. This is different from the ultimate strength, $S_{ut,w}$, of each wire given in Table 54.8. The rope's tensile strength will be only 80% to 95% of the combined tensile

**Table 54.8** *Minimum Strengths of Wire Materials*

| material | ultimate strength, $S_{ut,w}$ (ksi) |
|---|---|
| iron | 65 |
| cast steel | 140 |
| extra-strong cast steel | 160 |
| plow steel | 175–210 |
| improved plow steel | 200–240 |
| extra-improved plow steel | 240–280 |

(Multiply ksi by 6.8947 to obtain MPa.)

[34]The term "plow steel" is somewhat traditional, as hard-drawn AISI 1070 or AISI 1080 might actually be used.

strengths of the individual wires. The modulus of elasticity for steel ropes is more a function of how the rope is constructed than the type of steel used. (See Table 54.10.)

**Table 54.9** *Properties of 6 x 19 Steel Wire Rope (improved plow steel, fiber core)*

| diameter | | mass | | tensile strength[a,b] $S_{ut,r}$ | |
|---|---|---|---|---|---|
| (in) | (mm) | (lbm/ft) | (kg/m) | (tons[c]) | (tonnes) |
| 1/4 | (6.4) | 0.11 | (0.16) | 2.74 | (2.49) |
| 3/8 | (9.5) | 0.24 | (0.35) | 6.10 | (5.53) |
| 1/2 | (13) | 0.42 | (0.63) | 10.7 | (9.71) |
| 5/8 | (16) | 0.66 | (0.98) | 16.7 | (15.1) |
| 7/8 | (22) | 1.29 | (1.92) | 32.2 | (29.2) |
| 1 1/8 | (29) | 2.13 | (3.17) | 52.6 | (47.7) |
| 1 3/8 | (35) | 3.18 | (4.73) | 77.7 | (70.5) |
| 1 5/8 | (42) | 4.44 | (6.61) | 107 | (97.1) |
| 1 7/8 | (48) | 5.91 | (8.80) | 141 | (128) |
| 2 1/8 | (54) | 7.59 | (11.3) | 179 | (162) |
| 2 3/8 | (60) | 9.48 | (14.1) | 222 | (201) |
| 2 5/8 | (67) | 11.6 | (17.3) | 268 | (243) |

(Multiply in by 25.4 to obtain mm.)
(Multiply lbm/ft by 1.488 to obtain kg/m.)
(Multiply tons by 0.9072 to obtain tonnes.)

[a]Add 7 1/2% for wire ropes with steel cores.
[b]Deduct 10% for galvanized wire ropes.
[c]tons of 2000 pounds

**Table 54.10** *Typical Characteristics of Steel Wire Ropes*

| configuration | mass (lbm/ft) | area (in²) | minimum sheave diameter | modulus of elasticity (psi) |
|---|---|---|---|---|
| 6 × 7 | $1.50d_r^2$ | $0.380d_r^2$ | $42–72d_r$ | $14 \times 10^6$ |
| 6 × 19 | $1.60d_r^2$ | $0.404d_r^2$ | $30–45d_r$ | $12 \times 10^6$ |
| 6 × 37 | $1.55d_r^2$ | $0.404d_r^2$ | $18–27d_r$ | $11 \times 10^6$ |
| 8 × 19 | $1.45d_r^2$ | $0.352d_r^2$ | $21–31d_r$ | $10 \times 10^6$ |

(Multiply lbm/ft by 1.49 to obtain kg/m.)
(Multiply in² by 6.45 to obtain cm².)
(Multiply psi by $6.9 \times 10^{-6}$ to obtain GPa.)

The central core can be of natural (e.g., hemp) or synthetic fibers or, for higher-temperature use, steel strands or cable. Core designations are FC for *fiber core*, IWRC for *independent wire rope core*, and WSC for *wire-strand core*. Wire rope is protected against corrosion by lubrication carried in the saturated fiber core. Steel-cored ropes are approximately 7.5% stronger than fiber-cored ropes.

*Structural rope, structural strand,* and *aircraft cabling* are similar in design to wire rope but are intended for permanent installation in bridges and aircraft, respectively. Structural rope and strand are galvanized to prevent corrosion, while aircraft cable is usually manufactured from corrosion-resistant steel.[35,36] Galvanized ropes should not be used for hoisting, as the galvanized coating will be worn off. Structural rope and strand have a nominal tensile strength of 220 ksi (1.5 GPa) and a modulus of elasticity of approximately 20,000 ksi (140 GPa) for diameters between $^3/_8$ and 4 in (0.95 and 10.2 cm).

The most common winding is *regular lay* in which the wires are wound in one direction and the strands are wound around the core in the opposite direction. Regular lay ropes do not readily kink or unwind. Wires and strands in *lang lay* ropes are wound in the same direction, resulting in a wear-resistant rope that is more prone to unwinding. Lang lay ropes should not be used to support loads that are held in free suspension.

In addition to considering the primary tensile dead load, the significant effects of bending and sheave-bearing pressure must be considered when selecting wire rope. Self-weight may also be a factor for long cables. Appropriate dynamic factors should be applied to allow for acceleration, deceleration, and impacts. In general for hoisting and hauling, the working load should not exceed 20% of the breaking strength (i.e., a minimum factor of safety of 5 should be used).[37]

If $d_w$ is the nominal wire strand diameter in inches,[38] $d_r$ is the nominal wire rope diameter in inches, and $d_{\text{sh}}$ is the sheave diameter in inches, the stress from bending around a drum or sheave is given by Eq. 54.72. $E_w$ is the modulus of elasticity of the wire material (approximately $3 \times 10^7$ psi (207 GPa for steel), not the rope's modulus of elasticity, though the latter is widely used in this calculation.[39]

$$\sigma_{\text{bending}} = \frac{d_w E_w}{d_{\text{sh}}} \qquad 54.72$$

To reduce stress and eliminate permanent set in wire ropes, the diameter of the sheave should be kept as large as is practical, ideally 45 to 90 times the rope diameter. Alternatively, the minimum diameter of the sheave or drum may be stated as 400 times the diameter of the individual outer wires in the rope.[40] Table 54.10 lists minimum diameters for specific rope types.

For any allowable bending stress, the allowable load is calculated simply from the aggregate total area of all wire strands.

$$F_a = \sigma_{a,\text{bending}} \times \text{number of strands} \times A_{\text{strand}} \qquad 54.73$$

To prevent wear and fatigue of the sheave or drum, the radial bearing pressure should be kept as low as possible. Actual maximum bearing pressures are highly dependent on the sheave material, type of rope, and application. For 6 × 19 wire ropes, the acceptable bearing pressure can be as low as 500 psi (3.5 MPa) for cast-iron sheaves and as high as 2500 psi (17 MPa) for alloy steel sheaves. The approximate bearing pressure of the wire rope on the sheave or drum depends on the tensile force in the rope and is given by Eq. 54.74.

$$p_{\text{bearing}} = \frac{2F_t}{d_r d_{\text{sh}}} \qquad 54.74$$

Fatigue failure in wire rope can be avoided by keeping the ratio $p_{\text{bearing}}/S_{ut,w}$ below approximately 0.014 for 6 × 19 wire rope.[41] ($S_{ut,w}$ is the ultimate tensile strength of the wire material, not of the rope.)

---

[35]Manufacture of structural rope and strand in the United States are in accordance with ASTM A603 and ASTM A586, respectively.

[36]Galvanizing usually reduces the strength of wire rope by approximately 10%.

[37]Factors of safety are much higher and may be as high as 8 to 12 for elevators and hoists carrying passengers.

[38]For 6 × 19 standard wire rope, the outer wire strands are typically 1/13 to 1/16 of the wire rope diameter. For 6 × 7 haulage rope, the ratio is approximately 1/9.

[39]Although $E$ in Eq. 54.72 is often referred to as "the modulus of elasticity of the wire rope," it is understood that $E$ is actually the modulus of elasticity of the wire rope material.

[40]For elevators and mine hoists, the sheave-to-wire diameter ratio may be as high as 1000.

[41]A maximum ratio of 0.001 is often quoted for wire rope regardless of configuration.

# Topic XI: Dynamics

Dynamics

Chapter

# 55 Properties of Solid Bodies

## Nomenclature

| | | | |
|---|---|---|---|
| $a$ | acceleration | ft/sec$^2$ | m/s$^2$ |
| $d$ | distance | ft | m |
| $g$ | gravitational acceleration | ft/sec$^2$ | m/s$^2$ |
| $g_c$ | gravitational constant | ft-lbm/lbf-sec$^2$ | n.a. |
| $h$ | height | ft | m |
| $I$ | mass moment of inertia | lbm-ft$^2$ | kg·m$^2$ |
| $L$ | length | ft | m |
| $m$ | mass | lbm | kg |
| $r$ | radius | ft | m |
| $r$ | radius of gyration | ft | m |
| $V$ | volume | ft$^3$ | m$^3$ |
| $w$ | weight | lbf | N |

## Symbols

| | | | |
|---|---|---|---|
| $\rho$ | density | lbm/ft$^3$ | kg/m$^3$ |

## Subscripts

| | |
|---|---|
| $c$ | centroidal |
| $i$ | inner |
| $o$ | outer |

## 1. CENTER OF GRAVITY

A solid body will have both a center of gravity and a centroid, but the locations of these two points will not necessarily coincide. The earth's attractive force, called *weight*, can be assumed to act through the *center of gravity* (also known as the *center of mass*). Only when the body is homogeneous will the *centroid of the volume* coincide with the center of gravity.[1]

For simple objects and regular polyhedrons, the location of the center of gravity can be determined by inspection. It will always be located on an axis of symmetry. The location of the center of gravity can also be determined

---

[1]The study of nonhomogeneous bodies is beyond the scope of this book. Homogeneity is assumed for all solid objects.

---

mathematically if the object can be described mathematically.

$$x_c = \frac{\int x \, dm}{m} \qquad 55.1$$

$$y_c = \frac{\int y \, dm}{m} \qquad 55.2$$

$$z_c = \frac{\int z \, dm}{m} \qquad 55.3$$

If the object can be divided into several smaller constituent objects, the location of the composite center of gravity can be calculated from the centers of gravity of each of the constituent objects.

$$x_c = \frac{\sum m_i x_{ci}}{\sum m_i} \qquad 55.4$$

$$y_c = \frac{\sum m_i y_{ci}}{\sum m_i} \qquad 55.5$$

$$z_c = \frac{\sum m_i z_{ci}}{\sum m_i} \qquad 55.6$$

## 2. MASS AND WEIGHT

The *mass*, $m$, of a homogeneous solid object is calculated from its mass density and volume. Mass is independent of the strength of the gravitational field.

$$m = \rho V \qquad 55.7$$

The *weight*, $w$, of an object depends on the strength of the gravitational field, $g$.

$$w = mg \qquad \text{[SI]} \quad 55.8(a)$$

$$w = \frac{mg}{g_c} \qquad \text{[U.S.]} \quad 55.8(b)$$

## 3. INERTIA

Inertia (the *inertial force* or *inertia vector*), $m\mathbf{a}$, is the resistance the object offers to attempt to accelerate it (i.e., change its velocity) in a linear direction. Although the mass, $m$, is a scalar quantity, the acceleration, $\mathbf{a}$, is a vector.

## 4. MASS MOMENT OF INERTIA

The *mass moment of inertia* measures a solid object's resistance to changes in rotational speed about a specific axis. $I_x$, $I_y$, and $I_z$ are the mass moments of inertia with respect to the $x$-, $y$-, and $z$-axes. They are not components of a resultant value.[2]

The *centroidal mass moment of inertia*, $I_c$, is obtained when the origin of the axes coincides with the object's center of gravity. Although it can be found mathematically from Eq. 55.9 through Eq. 55.11, it is easier to use App. 55.A for simple objects.

$$I_x = \int (y^2 + z^2)\, dm \qquad 55.9$$

$$I_y = \int (x^2 + z^2)\, dm \qquad 55.10$$

$$I_z = \int (x^2 + y^2)\, dm \qquad 55.11$$

## 5. PARALLEL AXIS THEOREM

Once the centroidal mass moment of inertia is known, the *parallel axis theorem* is used to find the mass moment of inertia about any parallel axis.

$$I_{\text{any parallel axis}} = I_c + md^2 \qquad 55.12$$

For a composite object, the parallel axis theorem must be applied for each of the constituent objects.

$$I = I_{c,1} + m_1 d_1^2 + I_{c,2} + m_2 d_2^2 + \cdots \qquad 55.13$$

## 6. RADIUS OF GYRATION

The *radius of gyration*, $r$, of a solid object represents the distance from the rotational axis at which the object's entire mass could be located without changing the mass moment of inertia.

$$r = \sqrt{\frac{I}{m}} \qquad 55.14$$

$$I = r^2 m \qquad 55.15$$

## 7. PRINCIPAL AXES

An object's mass moment of inertia depends on the orientation of axes chosen. The *principal axes* are the axes for which the *products of inertia* are zero. Equipment rotating about a principal axis will draw minimum power during speed changes.

Finding the principal axes through calculation is too difficult and time consuming to be used with most rotating equipment. Furthermore, the rotating axis is generally fixed. Therefore, *balancing operations* are used to change the distribution of mass about the rotational axis. A device, such as a rotating shaft, flywheel, or crank, is said to be *statically balanced* if its center of mass lies on the axis of rotation. It is said to be *dynamically balanced* if the center of mass lies on the axis of rotation and the products of inertia are zero.

---

[2]At first, it may be confusing to use the same symbol, $I$, for area and mass moments of inertia. However, the problem types are distinctly dissimilar, and both moments of inertia are seldom used simultaneously.

# 56 Kinematics

## Nomenclature

| | | | |
|---|---|---|---|
| $a$ | acceleration | ft/sec$^2$ | m/s$^2$ |
| $d$ | distance | ft | m |
| $g$ | gravitational acceleration | ft/sec$^2$ | m/s$^2$ |
| $H$ | height | ft | m |
| $l$ | length | ft | m |
| $n$ | rotational speed | rpm | rpm |
| $r$ | radius | ft | m |
| $R$ | earth's radius | ft | m |
| $R$ | range | ft | m |
| $s$ | distance | ft | m |
| $t$ | time | sec | s |
| $T$ | flight time | sec | s |
| $v$ | velocity | ft/sec | m/s |
| $z$ | elevation | ft | m |

## Symbols

| | | | |
|---|---|---|---|
| $\alpha$ | angle | deg | deg |
| $\alpha$ | angular acceleration | rad/sec$^2$ | rad/s$^2$ |
| $\beta$ | angle | deg | deg |
| $\gamma$ | angle | deg | deg |
| $\theta$ | angular position | rad | rad |
| $\phi$ | angle or latitude | deg | deg |
| $\omega$ | angular velocity | rad/sec | rad/s |

## Subscripts

| | |
|---|---|
| 0 | initial |
| $\phi$ | transverse |
| $a$ | acceleration |
| $c$ | Coriolis |
| $H$ | to maximum altitude |
| $n$ | normal |
| $O$ | center |
| $r$ | radial |
| $t$ | tangential |

## 1. INTRODUCTION TO KINEMATICS

*Dynamics* is the study of moving objects. The subject is divided into kinematics and kinetics. *Kinematics* is the study of a body's motion independent of the forces on the body. It is a study of the geometry of motion without consideration of the causes of motion. Kinematics deals only with relationships among position, velocity, acceleration, and time.

## 2. PARTICLES AND RIGID BODIES

Bodies in motion can be considered *particles* if rotation is absent or insignificant. Particles do not possess rotational kinetic energy. All parts of a particle have the same instantaneous displacement, velocity, and acceleration.

A *rigid body* does not deform when loaded and can be considered a combination of two or more particles that remain at a fixed, finite distance from each other. At any given instant, the parts (particles) of a rigid body can have different displacements, velocities, and accelerations.

## 3. COORDINATE SYSTEMS

The position of a particle is specified with reference to a *coordinate system*. The description takes the form of an ordered sequence $(q_1, q_2, q_3, ...)$ of numbers called *coordinates*. A coordinate can represent a position along an axis, as in the rectangular coordinate system, or it can represent an angle, as in the polar, cylindrical, and spherical coordinate systems.

In general, the number of *degrees of freedom* is equal to the number of coordinates required to completely specify the state of an object. If each of the coordinates is independent of the others, the coordinates are known as *holonomic coordinates*.

The state of a particle is completely determined by the particle's location. In three-dimensional space, the locations of particles in a system of $m$ particles must be specified by $3m$ coordinates. However, the number of required coordinates can be reduced in certain cases. The position of each particle constrained to motion on a surface (i.e., on a two-dimensional system) can be specified by only two coordinates. A particle constrained to moving on a curved path requires only one coordinate.[1]

The state of a rigid body is a function of orientation as well as position. Six coordinates are required to specify the state: three for orientation and three for location.

## 4. CONVENTIONS OF REPRESENTATION

Consider the particle shown in Fig. 56.1. Its position (as well as its velocity and acceleration) can be specified in three primary forms: vector form, rectangular coordinate form, and unit vector form.

**Figure 56.1** Position of a Particle

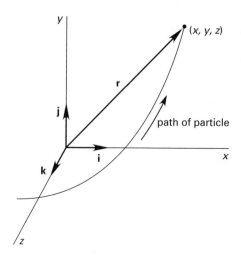

The vector form of the particle's position is $\mathbf{r}$, where the vector $\mathbf{r}$ has both magnitude and direction. The rectangular coordinate form is $(x, y, z)$. The unit vector form is

$$\mathbf{r} = x\mathbf{i} + y\mathbf{j} + z\mathbf{k} \qquad 56.1$$

## 5. LINEAR PARTICLE MOTION

A *linear system* is one in which particles move only in straight lines. (It is also known as a *rectilinear system*.) The relationships among position, velocity, and acceleration for a linear system are given by Eq. 56.2 through Eq. 56.4. When values of $t$ are substituted into these

equations, the position, velocity, and acceleration are known as *instantaneous values*.

$$s(t) = \int v(t)\,dt = \int \left( \int a(t)\,dt \right) dt \qquad 56.2$$

$$v(t) = \frac{ds(t)}{dt} = \int a(t)\,dt \qquad 56.3$$

$$a(t) = \frac{dv(t)}{dt} = \frac{d^2 s(t)}{dt^2} \qquad 56.4$$

The average velocity and acceleration over a period from $t_1$ to $t_2$ are

$$v_{\text{ave}} = \frac{\int_1^2 v(t)\,dt}{t_2 - t_1} = \frac{s_2 - s_1}{t_2 - t_1} \qquad 56.5$$

$$a_{\text{ave}} = \frac{\int_1^2 a(t)\,dt}{t_2 - t_1} = \frac{v_2 - v_1}{t_2 - t_1} \qquad 56.6$$

### Example 56.1

A particle is constrained to move along a straight line. The velocity and location are both zero at $t = 0$. The particle's velocity as a function of time is

$$v(t) = 8t - 6t^2$$

(a) What are the acceleration and position functions?

(b) What is the instantaneous velocity at $t = 5$?

*Solution*

(a) $a(t) = \dfrac{dv(t)}{dt} = \dfrac{d(8t - 6t^2)}{dt}$

$\qquad = 8 - 12t$

$\qquad s(t) = \displaystyle\int v(t)\,dt = \int (8t - 6t^2)\,dt$

$\qquad\qquad = 4t^2 - 2t^3 \quad \text{when } s(t = 0) = 0$

(b) Substituting $t = 5$ into the $v(t)$ function,

$$v(5) = (8)(5) - (6)(5)^2$$

$$= -110 \quad \text{[backward]}$$

## 6. DISTANCE AND SPEED

The terms "displacement" and "distance" have different meanings in kinematics. *Displacement* (or *linear displacement*) is the net change in a particle's position as determined from the position function, $s(t)$. *Distance traveled* is the accumulated length of the path traveled during all direction reversals, and it can be found by adding the path lengths covered during periods in which

---

[1]The curve can be a straight line, as in the case of a mass hanging on a spring and oscillating up and down. In this case, the coordinate will be a linear coordinate.

the velocity sign does not change. Thus, distance is always greater than or equal to displacement.

$$\text{displacement} = s(t_2) - s(t_1) \qquad 56.7$$

Similarly, "velocity" and "speed" have different meanings: *velocity* is a vector, having both magnitude and direction; *speed* is a scalar quantity, equal to the magnitude of velocity. When specifying speed, direction is not considered.

### Example 56.2

What distance is traveled during the period $t = 0$ to $t = 6$ by the particle described in Ex. 56.1?

*Solution*

Start by determining when, if ever, the velocity becomes negative. (This can be done by inspection, graphically, or algebraically.) Solving for the roots of the velocity equation, the velocity changes from positive to negative at

$$t = \tfrac{4}{3}$$

The initial displacement is zero. From the position function, the position at $t = {}^4/_3$ is

$$s\left(\tfrac{4}{3}\right) = (4)\left(\tfrac{4}{3}\right)^2 - (2)\left(\tfrac{4}{3}\right)^3 = 2.37$$

The displacement while the velocity is positive is

$$\Delta s = s\left(\tfrac{4}{3}\right) - s(0) = 2.37 - 0$$
$$= 2.37$$

The position at $t = 6$ is

$$s(6) = (4)(6)^2 - (2)(6)^3 = -288$$

The displacement while the velocity is negative is

$$\Delta s = s(6) - s\left(\tfrac{4}{3}\right) = -288 - 2.37$$
$$= -290.37$$

The total distance traveled is

$$2.37 + 290.37 = 292.74$$

### 7. UNIFORM MOTION

The term *uniform motion* means uniform velocity. The velocity is constant and the acceleration is zero. For a constant velocity system, the position function varies linearly with time. (See Fig. 56.2.)

$$s(t) = s_0 + \mathrm{v}t \qquad 56.8$$
$$\mathrm{v}(t) = \mathrm{v} \qquad 56.9$$
$$a(t) = 0 \qquad 56.10$$

**Figure 56.2** *Constant Velocity*

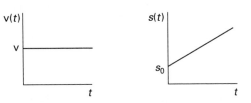

### 8. UNIFORM ACCELERATION

The acceleration is constant in many cases, as shown in Fig. 56.3. (Gravitational acceleration, where $a = g$, is a notable example.) If the acceleration is constant, the $a$ term can be taken out of the integrals in Eq. 56.2 and Eq. 56.3.

$$a(t) = a \qquad 56.11$$
$$\mathrm{v}(t) = a \int dt = \mathrm{v}_0 + at \qquad 56.12$$
$$s(t) = a \int\!\int dt^2 = s_0 + \mathrm{v}_0 t + \tfrac{1}{2}at^2 \qquad 56.13$$

**Figure 56.3** *Uniform Acceleration*

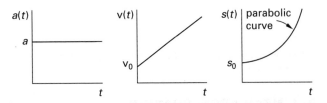

Table 56.1 summarizes the equations required to solve most uniform acceleration problems.

### Example 56.3

A locomotive traveling at 80 km/h locks its wheels and skids 95 m before coming to a complete stop. If the deceleration is constant, how many seconds will it take for the locomotive to come to a standstill?

*Solution*

First, convert the 80 km/h to meters per second.

$$\mathrm{v}_0 = \frac{\left(80\ \dfrac{\text{km}}{\text{h}}\right)\left(1000\ \dfrac{\text{m}}{\text{km}}\right)}{3600\ \dfrac{\text{s}}{\text{h}}} = 22.22\ \text{m/s}$$

In this problem, $\mathrm{v}_0 = 22.2$ m/s, $\mathrm{v} = 0$, and $s = 95$ m are known. $t$ is the unknown. From Table 56.1,

$$t = \frac{2s}{\mathrm{v}_0 + \mathrm{v}} = \frac{(2)(95\ \text{m})}{22.22\ \dfrac{\text{m}}{\text{s}} + 0}$$
$$= 8.55\ \text{s}$$

**Table 56.1** Uniform Acceleration Formulas[*]

| to find | given these | use this equation |
|---|---|---|
| $a$ | $t, v_0, v$ | $a = \dfrac{v - v_0}{t}$ |
| $a$ | $t, v_0, s$ | $a = \dfrac{2s - 2v_0 t}{t^2}$ |
| $a$ | $v_0, v, s$ | $a = \dfrac{v^2 - v_0^2}{2s}$ |
| $s$ | $t, a, v_0$ | $s = v_0 t + \frac{1}{2}at^2$ |
| $s$ | $a, v_0, v$ | $s = \dfrac{v^2 - v_0^2}{2a}$ |
| $s$ | $t, v_0, v$ | $s = \frac{1}{2}t(v_0 + v)$ |
| $t$ | $a, v_0, v$ | $t = \dfrac{v - v_0}{a}$ |
| $t$ | $a, v_0, s$ | $t = \dfrac{\sqrt{v_0^2 + 2as} - v_0}{a}$ |
| $t$ | $v_0, v, s$ | $t = \dfrac{2s}{v_0 + v}$ |
| $v_0$ | $t, a, v$ | $v_0 = v - at$ |
| $v_0$ | $t, a, s$ | $v_0 = \dfrac{s}{t} - \frac{1}{2}at$ |
| $v_0$ | $a, v, s$ | $v_0 = \sqrt{v^2 - 2as}$ |
| $v$ | $t, a, v_0$ | $v = v_0 + at$ |
| $v$ | $a, v_0, s$ | $v = \sqrt{v_0^2 + 2as}$ |

[*]The table can be used for rotational problems by substituting $\alpha$, $\omega$, and $\theta$ for $a$, $v$, and $s$, respectively.

## 9. LINEAR ACCELERATION

*Linear acceleration* means that the acceleration increases uniformly with time. Figure 56.4 shows how the velocity and position vary with time.[2]

**Figure 56.4** Linear Acceleration

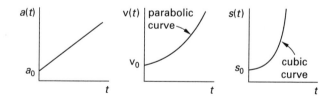

## 10. PROJECTILE MOTION

A *projectile* is placed into motion by an initial impulse. (Kinematics deals only with dynamics during the flight. The force acting on the projectile during the launch phase is covered in kinetics.) Neglecting air drag, once

[2]Because of the successive integrations, if the acceleration function is a polynomial of degree $n$, the velocity function will be a polynomial of degree $n + 1$. Similarly, the position function will be a polynomial of degree $n + 2$.

the projectile is in motion, it is acted upon only by the downward gravitational acceleration (i.e., its own weight). Thus, projectile motion is a special case of motion under constant acceleration.

Consider a general projectile set into motion at an angle of $\phi$ (from the horizontal plane) and initial velocity $v_0$. Its range is $R$, the maximum altitude attained is $H$, and the total flight time is $T$. In the absence of air drag, the following rules apply to the case of a level target.[3]

- The trajectory is parabolic.
- The impact velocity is equal to initial velocity, $v_0$.
- The impact angle is equal to the initial launch angle, $\phi$.
- The range is maximum when $\phi = 45°$.
- The time for the projectile to travel from the launch point to the apex is equal to the time to travel from apex to impact point.
- The time for the projectile to travel from the apex of its flight path to impact is the same time an initially stationary object would take to fall a distance $H$.

Table 56.2 contains the solutions to most common projectile problems. These equations are derived from the laws of uniform acceleration and conservation of energy.

### Example 56.4

A projectile is launched at 600 ft/sec (180 m/s) with a 30° inclination from the horizontal. The launch point is on a plateau 500 ft (150 m) above the plane of impact. Neglecting friction, find the maximum altitude, $H$, above the plane of impact, the total flight time, $T$, and the range, $R$.

*SI Solution*

The maximum altitude above the impact plane includes the height of the plateau and the elevation achieved by the projectile.

$$H = z + \frac{v_0^2 \sin^2 \phi}{2g}$$

$$= 150 \text{ m} + \frac{\left(180 \frac{\text{m}}{\text{s}}\right)^2 (\sin^2 30°)}{(2)\left(9.81 \frac{\text{m}}{\text{s}^2}\right)} = 562.8 \text{ m}$$

[3]The case of projectile motion with air friction cannot be handled in kinematics, since a retarding force acts continuously on the projectile. In kinetics, various assumptions (e.g., friction varies linearly with the velocity or with the square of the velocity) can be made to include the effect of air friction.

The total flight time includes the time to reach the maximum altitude and the time to fall from the maximum altitude to the impact plane below.

$$T = t_H + t_{\text{fall}}$$

$$= \frac{v_0 \sin \phi}{g} + \sqrt{\frac{2H}{g}}$$

$$= \frac{\left(180 \, \frac{m}{s}\right)(\sin 30°)}{9.81 \, \frac{m}{s^2}} + \sqrt{\frac{(2)(562.8 \, m)}{9.81 \, \frac{m}{s^2}}}$$

$$= 19.89 \, s$$

The $x$-component of velocity is

$$v_x = v_0 \cos \phi = \left(180 \, \frac{m}{s}\right)(\cos 30°)$$

$$= 155.9 \, m/s$$

The range is

$$R = v_x T = \left(155.9 \, \frac{m}{s}\right)(19.89 \, s)$$

$$= 3101 \, m$$

*Customary U.S. Solution*

The maximum altitude above the impact plane is given by Table 56.2.

$$H = 500 \, \text{ft} + \frac{\left(600 \, \frac{\text{ft}}{\text{sec}}\right)^2 (\sin^2 30°)}{(2)\left(32.2 \, \frac{\text{ft}}{\text{sec}^2}\right)}$$

$$= 1897.5 \, \text{ft}$$

The total flight time is

$$T = \frac{\left(600 \, \frac{\text{ft}}{\text{sec}}\right)(\sin 30°)}{32.2 \, \frac{\text{ft}}{\text{sec}^2}} + \sqrt{\frac{(2)(1897.5 \, \text{ft})}{32.2 \, \frac{\text{ft}}{\text{sec}^2}}}$$

$$= 20.17 \, \text{sec}$$

The maximum range is

$$R = v_0 T \cos \phi = \left(600 \, \frac{\text{ft}}{\text{sec}}\right)(20.17 \, \text{sec})(\cos 30°)$$

$$= 10{,}481 \, \text{ft} \quad (1.98 \, \text{mi})$$

**Example 56.5**

A bomber flies horizontally at 275 mi/hr at an altitude of 9000 ft. At what viewing angle, $\phi$, from the bomber to the target should the bombs be dropped?

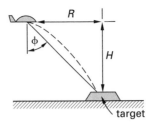

*Solution*

This is a case of horizontal projection. The falling time depends only on the altitude of the bomber. From Table 56.2,

$$T = \sqrt{\frac{2H}{g}} = \sqrt{\frac{(2)(9000 \, \text{ft})}{32.2 \, \frac{\text{ft}}{\text{sec}^2}}}$$

$$= 23.64 \, \text{sec}$$

If air friction is neglected, the bomb has the same horizontal velocity as the bomber. Since the time of flight, $T$, is known, the distance traveled during that time can be calculated.

$$R = v_0 T = \frac{\left(275 \, \frac{\text{mi}}{\text{hr}}\right)\left(5280 \, \frac{\text{ft}}{\text{mi}}\right)(23.64 \, \text{sec})}{3600 \, \frac{\text{sec}}{\text{hr}}}$$

$$= 9535 \, \text{ft}$$

The viewing angle is found from trigonometry.

$$\phi = \arctan\left(\frac{R}{H}\right) = \arctan\left(\frac{9535 \, \text{ft}}{9000 \, \text{ft}}\right)$$

$$= 46.7°$$

## 11. ROTATIONAL PARTICLE MOTION

*Rotational particle motion* (also known as *angular motion* and *circular motion*) is motion of a particle around a circular path. (See Fig. 56.5.) The particle travels through $2\pi$ radians per complete revolution.

**Figure 56.5** *Rotational Particle Motion*

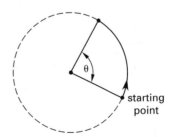

The behavior of a rotating particle is defined by its *angular position*, $\theta$, *angular velocity*, $\omega$, and *angular acceleration*, $\alpha$, functions. These variables are analogous to the $s(t)$, $v(t)$, and $a(t)$ functions for linear systems.

**Table 56.2** *Projectile Motion Equations*
  *($\phi$ may be negative for projection downward)*

|  | level target | target above | target below | horizontal projection |
|---|---|---|---|---|
|  | | | | $v_0 = v_{x'}\ \phi = 0°$ |
| $x(t)$ | $v_0 \cos\phi\, t$ | | | $v_0 t$ |
| $y(t)$ | $v_0 \sin\phi\, t - \frac{1}{2}gt^2$ | | | $H - \frac{1}{2}gt^2$ |
| $v_x(t)^a$ | $v_0 \cos\phi$ | | | $v_0$ |
| $v_y(t)^b$ | $v_0 \sin\phi - gt$ | | | $-gt$ |
| $v(t)^c$ | $\sqrt{v_0^2 - 2gy} = \sqrt{v_0^2 - 2gtv_0\sin\phi + g^2 t^2}$ | | | $\sqrt{v_0^2 + g^2 t^2}$ |
| $v(y)^d$ | $\sqrt{v_0^2 - 2gy}$ | | | $\sqrt{v_0^2 + 2g(H - y)}$ |
| $H$ | $\dfrac{v_0^2 \sin^2\phi}{2g}$ | $\dfrac{v_0^2 \sin^2\phi}{2g}$ | $z + \dfrac{v_0^2 \sin^2\phi}{2g}$ | $\frac{1}{2}g T^2$ |
| $R$ | $\dfrac{v_0^2 \sin 2\phi}{g}$ | $\left(\dfrac{v_0\cos\phi}{g}\right)\left(v_0\sin\phi + \sqrt{v_0^2\sin^2\phi - 2gz}\right)$ | $\left(\dfrac{v_0\cos\phi}{g}\right)\left(v_0\sin\phi + \sqrt{2gz + v_0^2\sin^2\phi}\right)$ | $v_0 T$ |

$v_0 T \cos\phi$

$\dfrac{R}{v_0 \cos\phi}$

| $T$ | $\dfrac{2v_0 \sin\phi}{g}$ | $\dfrac{v_0\sin\phi}{g} + \sqrt{\dfrac{2(H-z)}{g}}$ | $\dfrac{v_0\sin\phi}{g} + \sqrt{\dfrac{2H}{g}}$ | $\sqrt{\dfrac{2H}{g}}$ |
| $t_H$ | $\dfrac{v_0\sin\phi}{g} = \dfrac{T}{2}$ | $\dfrac{v_0\sin\phi}{g}$ | | |

[a] horizontal velocity component
[b] vertical velocity component
[c] resultant velocity as a function of time
[d] resultant velocity as a function of vertical elevation above the launch point

Angular variables can be substituted one-for-one in place of linear variables in most equations.

The relationships among angular position, velocity, and acceleration for a rotational system are given by Eq. 56.14 through Eq. 56.16. When values of $t$ are substituted into these equations, the position, velocity, and acceleration are known as *instantaneous values*.

$$\theta(t) = \int \omega(t)\,dt = \int\int \alpha(t)\,dt^2 \qquad \textbf{56.14}$$

$$\omega(t) = \frac{d\theta(t)}{dt} = \int \alpha(t)\,dt \qquad \textbf{56.15}$$

$$\alpha(t) = \frac{d\omega(t)}{dt} = \frac{d^2\theta(t)}{dt^2} \qquad \textbf{56.16}$$

The average velocity and acceleration are

$$\omega_{\text{ave}} = \frac{\int_1^2 \omega(t)\,dt}{t_2 - t_1} = \frac{\theta_2 - \theta_1}{t_2 - t_1} \qquad \textit{56.17}$$

$$\alpha_{\text{ave}} = \frac{\int_1^2 \alpha(t)\,dt}{t_2 - t_1} = \frac{\omega_2 - \omega_1}{t_2 - t_1} \qquad \textit{56.18}$$

### Example 56.6

A turntable starts from rest and accelerates uniformly at $1.5 \text{ rad/sec}^2$. How many revolutions will it take before a rotational speed of $33^1/_3$ rpm is attained?

*Solution*

First, convert $33^1/_3$ rpm into radians per second. Since there are $2\pi$ radians per complete revolution,

$$\omega = \frac{\left(33\frac{1}{3}\frac{\text{rev}}{\text{min}}\right)\left(2\pi\frac{\text{rad}}{\text{rev}}\right)}{60\frac{\text{sec}}{\text{min}}} = 3.49 \text{ rad/sec}$$

$\alpha$, $\omega_0$, and $\omega$ are known. $\theta$ is unknown. (This is analogous to knowing $a$, $v_0$, and $v$, and not knowing $s$.) From Table 56.1,

$$\theta = \frac{\omega^2 - \omega_0^2}{2\alpha} = \frac{\left(3.49\frac{\text{rad}}{\text{sec}}\right)^2 - \left(0\frac{\text{rad}}{\text{sec}}\right)^2}{(2)\left(1.5\frac{\text{rad}}{\text{sec}^2}\right)}$$

$$= 4.06 \text{ rad}$$

Converting from radians to revolutions,

$$n = \frac{4.06 \text{ rad}}{2\pi\frac{\text{rad}}{\text{rev}}} = 0.646 \text{ rev}$$

### Example 56.7

A flywheel is brought to a standstill from 400 rpm in 8 sec. (a) What was its average angular acceleration in rad/sec during that period? (b) How far (in radians) did the flywheel travel?

*Solution*

(a) The initial rotational speed must be expressed in radians per second. Since there are $2\pi$ radians per revolution,

$$\omega_0 = \frac{\left(400\frac{\text{rev}}{\text{min}}\right)\left(2\pi\frac{\text{rad}}{\text{rev}}\right)}{60\frac{\text{sec}}{\text{min}}}$$

$$= 41.89 \text{ rad/sec}$$

$t$, $\omega_0$, and $\omega$ are known, and $\alpha$ is unknown. These variables are analogous to $t$, $v_0$, $v$, and $a$ in Table 56.1.

$$\alpha = \frac{\omega - \omega_0}{t} = \frac{0 - 41.89\frac{\text{rad}}{\text{sec}}}{8 \text{ sec}}$$

$$= -5.236 \text{ rad/sec}^2$$

(b) $t$, $\omega_0$, and $\omega$ are known, and $\theta$ is unknown. Again, from Table 56.1,

$$\theta = \tfrac{1}{2}t(\omega + \omega_0) = \left(\tfrac{1}{2}\right)(8 \text{ sec})\left(41.89\frac{\text{rad}}{\text{sec}} + 0\right)$$

$$= 167.6 \text{ rad} \quad (26.67 \text{ rev})$$

## 12. RELATIONSHIP BETWEEN LINEAR AND ROTATIONAL VARIABLES

A particle moving in a curvilinear path will also have instantaneous linear velocity and linear acceleration. These linear variables will be directed tangentially to the path and, therefore, are known as *tangential velocity* and *tangential acceleration*, respectively. (See Fig. 56.6.) In general, the linear variables can be obtained by multiplying the rotational variables by the path radius, $r$.

$$v_t = \omega r \qquad \textit{56.19}$$

$$v_{t,x} = v_t \cos\phi = \omega r \cos\phi \qquad \textit{56.20}$$

$$v_{t,y} = v_t \sin\phi = \omega r \sin\phi \qquad \textit{56.21}$$

$$a_t = \frac{dv_t}{dt} = \alpha r \qquad \textit{56.22}$$

*Figure 56.6* Tangential Variables

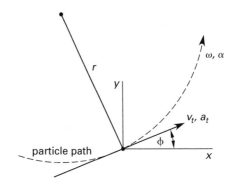

If the path radius is constant, as it would be in rotational motion, the linear distance (i.e., the *arc length*) traveled is

$$s = \theta r \qquad \textit{56.23}$$

**Dynamics**

## 13. NORMAL ACCELERATION

A moving particle will continue tangentially to its path unless constrained otherwise. For example, a rock twirled on a string will move in a circular path only as long as there is tension in the string. When the string is released, the rock will move off tangentially.

The twirled rock is acted upon by the tension in the string. In general, a restraining force will be directed toward the center of rotation. Whenever a mass experiences a force, an acceleration is acting.[4] The acceleration has the same sense as the applied force (i.e., is directed toward the center of rotation). Since the inward acceleration is perpendicular to the tangential velocity and acceleration, it is known as *normal acceleration, $a_n$*.

$$a_n = \frac{\mathrm{v}_t^2}{r} = r\omega^2 = \mathrm{v}_t\omega \qquad 56.24$$

The *resultant acceleration, $a$*, is the vector sum of the tangential and normal accelerations. The magnitude of the resultant acceleration is

$$a = \sqrt{a_t^2 + a_n^2} \qquad 56.25$$

The $x$- and $y$-components of the resultant acceleration are

$$a_x = a_n\sin\phi \pm a_t\cos\phi \qquad 56.26$$

$$a_y = a_n\cos\phi \mp a_t\sin\phi \qquad 56.27$$

The normal and tangential accelerations can be expressed in terms of the $x$- and $y$-components of the resultant acceleration (not shown in Fig. 56.7).

$$a_n = a_x\sin\phi \pm a_y\cos\phi \qquad 56.28$$

$$a_t = a_x\cos\phi \mp a_y\sin\phi \qquad 56.29$$

**Figure 56.7** Normal Acceleration

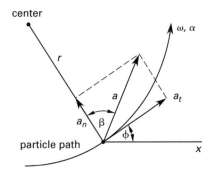

## 14. CORIOLIS ACCELERATION

Consider a particle moving with linear radial velocity $\mathrm{v}_r$ away from the center of a flat disk rotating with

constant velocity $\omega$. Since $\mathrm{v}_t = \omega r$, the particle's tangential velocity will increase as it moves away from the center of rotation. This increase is believed to be produced by the tangential *Coriolis acceleration, $a_c$*. (See Fig. 56.8.)

$$a_c = 2\mathrm{v}_r\omega \qquad 56.30$$

**Figure 56.8** Coriolis Acceleration on a Rotating Disk

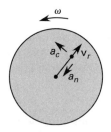

Coriolis acceleration also acts on particles moving on rotating spheres. Consider an aircraft flying with constant air speed v from the equator to the north pole while the earth (a sphere of radius $R$) rotates below it. Three accelerations act on the aircraft: normal, radial, and Coriolis accelerations, shown in Fig. 56.9. The Coriolis acceleration depends on the latitude, $\phi$, because the earth's tangential velocity is less near the poles than at the equator.

$$a_n = r\omega^2 = R\omega^2\cos\phi \qquad 56.31$$

$$a_r = \frac{\mathrm{v}^2}{R} \qquad 56.32$$

$$a_c = 2\omega\mathrm{v}_x = 2\omega\mathrm{v}\sin\phi \qquad 56.33$$

**Figure 56.9** Coriolis Acceleration on a Rotating Sphere

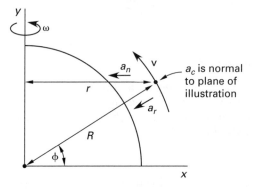

### Example 56.8

A slider moves with a constant velocity of 20 ft/sec along a rod rotating at 5 rad/sec. What is the magnitude of the slider's total acceleration when the slider is 4 ft from the center of rotation?

---

[4]This is a direct result of Newton's second law of motion.

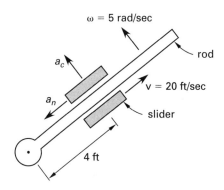

*Solution*

The normal acceleration is given by Eq. 56.31.

$$a_n = r\omega^2 = (4 \text{ ft})\left(5\,\frac{\text{rad}}{\text{sec}}\right)^2$$

$$= 100 \text{ ft/sec}^2$$

The Coriolis acceleration is given by Eq. 56.33.

$$a_c = 2\text{v}\omega = (2)\left(20\,\frac{\text{ft}}{\text{sec}}\right)\left(5\,\frac{\text{rad}}{\text{sec}}\right)$$

$$= 200 \text{ ft/sec}^2$$

The total acceleration is given by Eq. 56.25.

$$a = \sqrt{a_n^2 + a_c^2} = \sqrt{\left(100\,\frac{\text{ft}}{\text{sec}^2}\right)^2 + \left(200\,\frac{\text{ft}}{\text{sec}^2}\right)^2}$$

$$= 223.6 \text{ ft/sec}^2$$

## 15. PARTICLE MOTION IN POLAR COORDINATES

In polar coordinates, the path of a particle is described by a radius vector, **r**, and an angle, $\phi$. Since the velocity of a particle is not usually directed radially out from the center of the coordinate system, it can be divided into two perpendicular components. The terms *normal* and *tangential* are not used with polar coordinates. Rather, the terms *radial* and *transverse* are used. Figure 56.10 illustrates the *radial* and *transverse components* of velocity in a polar coordinate system.

Figure 56.10 also illustrates the unit radial and unit transverse vectors, $\mathbf{e}_r$ and $\mathbf{e}_\phi$, used in the vector forms of the motion equations.

$$\text{position: } \mathbf{r} = r\mathbf{e}_r \qquad 56.34$$

$$\text{velocity: } \mathbf{v} = \text{v}_r\mathbf{e}_r + \text{v}_\phi\mathbf{e}_\phi = \frac{dr}{dt}\mathbf{e}_r + r\frac{d\phi}{dt}\mathbf{e}_\phi \qquad 56.35$$

$$\text{acceleration: } \mathbf{a} = a_r\mathbf{e}_r + a_\phi\mathbf{e}_\phi$$

$$= \left(\frac{d^2 r}{dt^2} - r\left(\frac{d\phi}{dt}\right)^2\right)\mathbf{e}_r$$

$$+ \left(r\frac{d^2\phi}{dt^2} + 2\frac{dr}{dt}\frac{d\phi}{dt}\right)\mathbf{e}_\phi \qquad 56.36$$

**Figure 56.10** *Radial and Transverse Components*

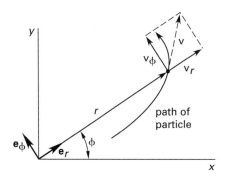

The magnitudes of the radial and transverse components of velocity and acceleration are given by Eq. 56.37 through Eq. 56.40.

$$\text{v}_r = \frac{dr}{dt} \qquad 56.37$$

$$\text{v}_\phi = r\frac{d\phi}{dt} \qquad 56.38$$

$$a_r = \frac{d^2 r}{dt^2} - r\left(\frac{d\phi}{dt}\right)^2 \qquad 56.39$$

$$a_\phi = r\frac{d^2\phi}{dt^2} + 2\frac{dr}{dt}\frac{d\phi}{dt} \qquad 56.40$$

If the radial and transverse components of acceleration and velocity are known, they can be used to calculate the tangential and normal accelerations in a rectangular coordinate system.

$$a_t = \frac{a_r\text{v}_r + a_\phi\text{v}_\phi}{\text{v}_t} \qquad 56.41$$

$$a_n = \frac{a_\phi\text{v}_r - a_r\text{v}_\phi}{\text{v}_t} \qquad 56.42$$

## 16. RELATIVE MOTION

The term *relative motion* is used when motion of a particle is described with respect to something else in motion. The particle's position, velocity, and acceleration may be specified with respect to another moving particle or with respect to a moving frame of reference, known as a *Newtonian* or *inertial frame of reference*.

In Fig. 56.11, two particles, A and B, are moving with different velocities along a straight line. The separation between the two particles at any specific instant is the *relative position*, $s_{B/A}$, of B with respect to A, calculated as the difference between their two *absolute positions*.

$$s_{B/A} = s_B - s_A \qquad 56.43$$

**Figure 56.11** *Relative Positions of Two Particles*

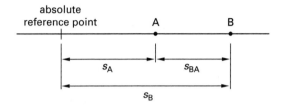

Similarly, the *relative velocity* and *relative acceleration* of B with respect to A are the differences between the two *absolute velocities* and *absolute accelerations*, respectively.

$$v_{B/A} = v_B - v_A \qquad 56.44$$

$$a_{B/A} = a_B - a_A \qquad 56.45$$

Particles A and B are not constrained to move along a straight line. However, the subtraction must be done in vector or graphical form in all but the simplest cases.

$$\mathbf{s}_{B/A} = \mathbf{s}_B - \mathbf{s}_A \qquad 56.46$$

$$\mathbf{v}_{B/A} = \mathbf{v}_B - \mathbf{v}_A \qquad 56.47$$

$$\mathbf{a}_{B/A} = \mathbf{a}_B - \mathbf{a}_A \qquad 56.48$$

Since vector subtraction and addition operations can be performed graphically, many relative motion problems can be solved by a simplified graphical process.

### Example 56.9

A stream flows at 5 km/h. At what upstream angle, $\phi$, should a 10 km/h boat be piloted in order to reach the shore directly opposite the initial point?

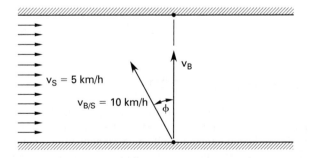

*Solution*

From Eq. 56.47, the absolute velocity of the boat, $v_B$, with respect to the shore is equal to the vector sum of the absolute velocity of the stream, $v_S$, and the relative

velocity of the boat with respect to the stream, $v_{B/S}$. The magnitudes of these two velocities are known.

$$v_B = v_S + v_{B/S}$$

Since vector addition is accomplished graphically by placing the two vectors head to tail, the angle can be determined from trigonometry.

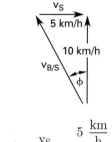

$$\sin \phi = \frac{v_S}{v_{B/S}} = \frac{5 \ \dfrac{km}{h}}{10 \ \dfrac{km}{h}} = 0.5$$

$$\phi = \arcsin 0.5 = 30°$$

### Example 56.10

A stationary member of a marching band tosses a 2.0 ft long balanced baton straight up into the air and then begins walking forward at 4 mi/hr. At a particular moment, the baton is 20 ft in the air and is falling back toward the earth with a velocity of 30 ft/sec. The tip of the baton is rotating at 140 rpm in the orientation shown.

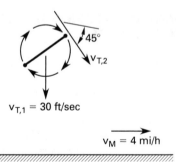

(a) What is the speed of the baton tip with respect to the ground? (b) What is the speed of the baton tip with respect to the band member?

*Solution*

(a) The baton tip has two absolute velocity components. The first, with a magnitude of $v_{T,1} = 30$ ft/sec, is directed vertically downward. The second, with a magnitude of $v_{T,2}$, is directed as shown in the illustration. The baton's radius, $r$, is 1 ft. From Eq. 56.19,

$$v_{T,2} = r\omega = \frac{(1 \text{ ft})\left(140 \frac{\text{rev}}{\text{min}}\right)\left(2\pi \frac{\text{rad}}{\text{rev}}\right)}{60 \frac{\text{sec}}{\text{min}}}$$

$$= 14.7 \text{ ft/sec}$$

The vector sum of these two absolute velocities is the velocity of the tip, $\mathbf{v}_T$, with respect to the earth.

$$\mathbf{v}_T = \mathbf{v}_{T,1} + \mathbf{v}_{T,2}$$

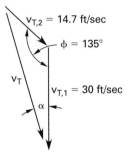

The velocity of the tip is found from the law of cosines.

$$v_T = \sqrt{v_{T,1}^2 + v_{T,2}^2 - 2v_{T,1}v_{T,2}\cos\phi}$$

$$= \sqrt{\begin{array}{l}\left(30 \frac{\text{ft}}{\text{sec}}\right)^2 + \left(14.7 \frac{\text{ft}}{\text{sec}}\right)^2 \\ \quad - (2)\left(30 \frac{\text{ft}}{\text{sec}}\right)\left(14.7 \frac{\text{ft}}{\text{sec}}\right)(\cos 135°)\end{array}}$$

$$= 41.7 \text{ ft/sec}$$

The angle $\alpha$ is found from the law of sines.

$$\frac{\sin\alpha}{14.7 \frac{\text{ft}}{\text{sec}}} = \frac{\sin 135°}{41.7 \frac{\text{ft}}{\text{sec}}}$$

$$\alpha = 14.4°$$

The band member's absolute velocity, $v_M$, is

$$v_M = \frac{\left(4 \frac{\text{mi}}{\text{hr}}\right)\left(5280 \frac{\text{ft}}{\text{mi}}\right)}{3600 \frac{\text{sec}}{\text{hr}}} = 5.87 \text{ ft/sec}$$

(b) From Eq. 56.44, the velocity of the tip with respect to the band member is

$$\mathbf{v}_{T/M} = \mathbf{v}_T - \mathbf{v}_M$$

Subtracting a vector is equivalent to adding its negative. The velocity triangle is as shown. The law of cosines is used again to determine the relative velocity.

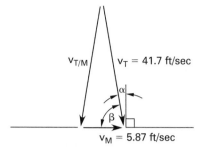

$$\beta = 90° - \alpha = 90° - 14.4°$$

$$= 75.6°$$

$$v_{T/M} = \sqrt{v_T^2 + v_M^2 - 2v_Tv_M\cos\beta}$$

$$= \sqrt{\begin{array}{l}\left(41.7 \frac{\text{ft}}{\text{sec}}\right)^2 + \left(5.87 \frac{\text{ft}}{\text{sec}}\right)^2 \\ \quad - (2)\left(41.7 \frac{\text{ft}}{\text{sec}}\right)\left(5.87 \frac{\text{ft}}{\text{sec}}\right)(\cos 75.6°)\end{array}}$$

$$= 40.6 \text{ ft/sec} \quad (27.7 \text{ mph})$$

## 17. DEPENDENT MOTION

When the position of one particle in a multiple-particle system depends on the position of one or more other particles, the motions are said to be "dependent." A block-and-pulley system with one fixed rope end, as illustrated by Fig. 56.12, is a *dependent system*.

**Figure 56.12** Dependent System

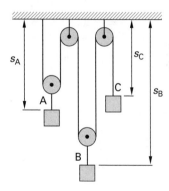

The following statements define the behavior of a dependent block-and-pulley system.

- Since the length of the rope is constant, the sum of the rope segments representing distances between the blocks and pulleys is constant. By convention, the distances are measured from the top of the block to the support point.[5] Since in Fig. 56.12 there are two

---

[5]In measuring distances, the finite diameters of the pulleys and the lengths of rope wrapped around the pulleys are disregarded.

ropes supporting block A, two ropes supporting block B, and one rope supporting block C,

$$2s_A + 2s_B + s_C = \text{constant} \qquad 56.49$$

- Since the position of the $n$th block in an $n$-block system is determined when the remaining $n - 1$ positions are known, the number of *degrees of freedom* is one less than the number of blocks.

- The movement, velocity, and acceleration of a block supported by two ropes are half the same quantities of a block supported by one rope.

- The relative relationships between the blocks' velocities or accelerations are the same as the relationships between the blocks' positions. For Fig. 56.12,

$$2v_A + 2v_B + v_C = 0 \qquad 56.50$$
$$2a_A + 2a_B + a_C = 0 \qquad 56.51$$

## 18. GENERAL PLANE MOTION

Rigid body *plane motion* can be described in two dimensions. Examples include rolling wheels, gear sets, and linkages. Plane motion can be considered as the sum of a translational component and a rotation about a fixed axis, as illustrated by Fig. 56.13.

**Figure 56.13** *Components of Plane Motion*

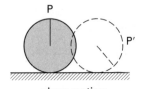

plane motion

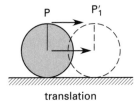 

translation                    rotation

## 19. ROTATION ABOUT A FIXED AXIS

Analysis of the rotational component of a rigid body's plane motion can sometimes be simplified if the location of the body's instantaneous center is known. Using the instantaneous center reduces many relative motion problems to simple geometry. The *instantaneous center* (also known as the *instant center* and IC) is a point at which the body could be fixed (pinned) without changing the instantaneous angular velocities of any point on the body. Thus, with the angular velocities, the body seems to rotate about a fixed instantaneous center.

The instantaneous center is located by finding two points for which the absolute velocity directions are known. Lines drawn perpendicular to these two velocities will intersect at the instantaneous center. (This graphical procedure is slightly different if the two velocities are parallel, as Fig. 56.14 shows. In that case, use is made of the fact that the tangential velocity is proportional to the distance from the instantaneous center.) For a rolling wheel, the instantaneous center is the point of contact with the supporting surface.

**Figure 56.14** *Graphical Method of Finding the Instantaneous Center*

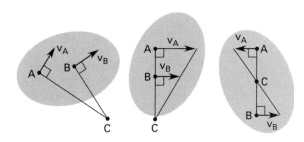

The absolute velocity of any point P on a wheel rolling (see Fig. 56.15) with translational velocity, $v_O$, can be found by geometry. Assume that the wheel is pinned at C and rotates with its actual angular velocity, $\omega = v_O/r$. The direction of the point's velocity will be perpendicular to the line of length $l$ between the instantaneous center and the point.

$$v = l\omega = \frac{lv_O}{r} \qquad 56.52$$

**Figure 56.15** *Instantaneous Center of a Rolling Wheel*

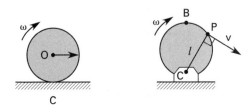

Equation 56.52 is valid only for a velocity referenced to the instantaneous center, point C. Table 56.3 can be used to find the velocities with respect to other points.

**Table 56.3** *Relative Velocities of a Rolling Wheel*

| point | reference point | | |
|---|---|---|---|
| | O | C | B |
| $v_O$ | 0 | $v_O \rightarrow$ | $\leftarrow v_O$ |
| $v_C$ | $\leftarrow v_O$ | 0 | $\leftarrow 2v_O$ |
| $v_B$ | $v_O \rightarrow$ | $2v_O \rightarrow$ | 0 |

## Example 56.11

A truck with 35 in diameter tires travels at a constant 35 mi/hr. What is the absolute velocity of point P on the circumference of the tire?

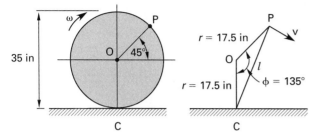

### Solution

The translational velocity of the center of the wheel is

$$v_O = \frac{\left(35\,\frac{mi}{hr}\right)\left(5280\,\frac{ft}{mi}\right)}{3600\,\frac{sec}{hr}} = 51.33 \text{ ft/sec}$$

The wheel radius is

$$r = \frac{35 \text{ in}}{(2)\left(12\,\frac{in}{ft}\right)} = 1.458 \text{ ft}$$

The angular velocity of the wheel is

$$\omega = \frac{v_O}{r} = \frac{51.33\,\frac{ft}{sec}}{1.458 \text{ ft}}$$
$$= 35.21 \text{ rad/sec}$$

The instantaneous center is the contact point, C. The law of cosines is used to find the distance $l$.

$$l^2 = r^2 + r^2 - 2r^2\cos\phi = 2r^2(1 - \cos\phi)$$

$$l = \sqrt{(2)(1.458 \text{ ft})^2(1 - \cos 135°)} = 2.694 \text{ ft}$$

From Eq. 56.52, the absolute velocity of point P is

$$v_P = l\omega = (2.694 \text{ ft})\left(35.21\,\frac{rad}{sec}\right)$$
$$= 94.9 \text{ ft/sec}$$

## 20. INSTANTANEOUS CENTER OF ACCELERATION

The *instantaneous center of acceleration* is used to compute the absolute acceleration of a point as if a body were in pure rotation about that point. It is the same as the instantaneous center of rotation, only for a body starting from rest and accelerating uniformly with angular acceleration, $\alpha$. The absolute acceleration, $a$,

determined from Eq. 56.53 is the same as the *resultant acceleration* in Fig. 56.7.

$$a = l\alpha = \frac{la_O}{r} \qquad 56.53$$

In general, the instantaneous center of acceleration, $C_a$, will be deflected at angle $\beta$ from the absolute acceleration vectors, as shown in Fig. 56.16. The relationship among the angle $\beta$, the instantaneous acceleration, $\alpha$, and the instantaneous velocity, $\omega$, is

$$\tan\beta = \frac{\alpha}{\omega^2} \qquad 56.54$$

**Figure 56.16** *Instantaneous Center of Acceleration*

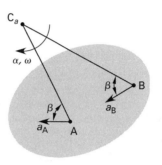

## 21. SLIDER RODS

The absolute velocity of any point, P, on a slider rod assembly can be found from the instantaneous center concept. (See Fig. 56.17.) The instantaneous center, C, is located by extending perpendiculars from the velocity vectors.

**Figure 56.17** *Instantaneous Center of Slider Rod Assembly*

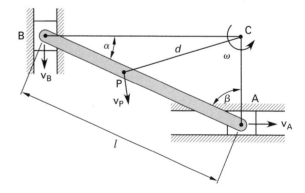

If the velocity with respect to point C of one end of the slider is known, for example $v_A$, then $v_B$ can be found from geometry. Since the slider can be assumed to rotate about point C with angular velocity $\omega$,

$$\omega = \frac{v_A}{AC} = \frac{v_A}{l\cos\beta} = \frac{v_B}{BC} = \frac{v_B}{l\cos\alpha} \qquad 56.55$$

Since $\cos \alpha = \sin \beta$,

$$v_B = v_A \tan \beta \qquad \text{56.56}$$

If the velocity with respect to point C of any other point P is required, it can be found from

$$v_P = d\omega \qquad \text{56.57}$$

## 22. SLIDER-CRANK ASSEMBLIES

Figure 56.18 illustrates a slider-crank assembly for which points A and D are in the same plane and at the same elevation. The instantaneous velocity of any point, P, on the rod can be found if the distance to the instantaneous center is known.

$$v_P = d\omega_1 \qquad \text{56.58}$$

**Figure 56.18** *Slider-Crank Assembly*

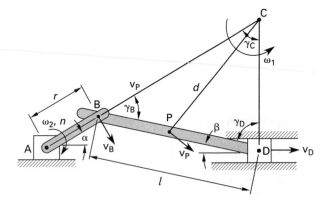

The tangential velocity of point B on the crank with respect to point A is perpendicular to the end of the crank. Slider D moves with a horizontal velocity. The intersection of lines drawn perpendicular to these velocity vectors locates the instantaneous center, point C. At any given instant, the connecting rod seems to rotate about point C with instantaneous angular velocity $\omega_1$.

The velocity of point D, $v_D$, is

$$v_D = (CD)\omega_1 = v_B\left(\frac{CD}{BC}\right) = \omega_2\left(\frac{AB \times CD}{BC}\right) \qquad \text{56.59}$$

Similarly, the velocity of point B, $v_B$, is

$$v_B = (AB)\omega_2 = (BC)\omega_2 = v_D\left(\frac{BC}{CD}\right) \qquad \text{56.60}$$

$$\omega_2 = \frac{2\pi n}{60} \qquad \text{56.61}$$

The following geometric relationships exist between the various angles.

$$\frac{\omega_1}{\omega_2} = \frac{AB}{BC} \qquad \text{56.62}$$

$$\frac{\sin \alpha}{l} = \frac{\sin \beta}{r} \qquad \text{56.63}$$

$$\frac{\sin \gamma_D}{BC} = \frac{\sin \gamma_B}{CD} = \frac{\sin \gamma_C}{l} \qquad \text{56.64}$$

$$\gamma_B = \alpha + \beta \qquad \text{56.65}$$

$$\gamma_D = 90 - \beta \qquad \text{56.66}$$

$$\gamma_C = 90 - \alpha \qquad \text{56.67}$$

# 57 Kinetics

## Nomenclature

| | | | |
|---|---|---|---|
| $a$ | acceleration | ft/sec$^2$ | m/s$^2$ |
| $a$ | coefficient of rolling resistance | ft | m |
| $a$ | semimajor axis length | ft | m |
| $A$ | area | ft$^2$ | m$^2$ |
| $b$ | semiminor axis length | ft | m |
| $C$ | coefficient | – | – |
| $C$ | coefficient of viscous damping (linear) | lbf-sec/ft | N·s/m |
| $C$ | coefficient of viscous damping (quadratic) | lbf-sec$^2$/ft$^2$ | N·s$^2$/m$^2$ |
| $C$ | constant used in space mechanics | 1/ft | 1/m |
| $d$ | diameter | ft | m |

| | | | |
|---|---|---|---|
| $e$ | coefficient of restitution | – | – |
| $e$ | superelevation | ft/ft | m/m |
| $E$ | energy | ft-lbf | J |
| $E$ | modulus of elasticity | lbf/ft$^2$ | Pa |
| $f$ | coefficient of friction | – | – |
| $F$ | force | lbf | N |
| $g$ | acceleration due to gravity | ft/sec$^2$ | m/s$^2$ |
| $g_c$ | gravitational constant | ft-lbm/lbf-sec$^2$ | n.a. |
| $G$ | universal gravitational constant | lbf-ft$^2$/lbm$^2$ | N·m$^2$/kg$^2$ |
| $h$ | angular momentum | ft$^2$-lbm/sec | m$^2$·kg/s |
| $h$ | height | ft | m |
| $I$ | mass moment of inertia | lbm-ft$^2$ | kg·m$^2$ |
| Imp | angular impulse | lbf-ft-sec | N·m·s |
| Imp | linear impulse | lbf-sec | N·s |
| $k$ | spring constant | lbf/ft | N/m |
| $m$ | mass | lbm | kg |
| $\dot{m}$ | mass flow rate | lbm/sec | kg/s |
| $M$ | mass of the earth | lbm | kg |
| $M$ | moment | ft-lbf | N·m |
| $n$ | rotational speed | rev/sec | rev/s |
| $N$ | normal force | lbf | N |
| $p$ | momentum | lbf-sec | N·s |
| $r$ | radius | ft | m |
| $s$ | distance | ft | m |
| $t$ | time | sec | s |
| $T$ | tension | lbf | N |
| $T$ | torque | ft-lbf | N·m |
| v | velocity | ft/sec | m/s |
| $w$ | weight | lbf | – |
| $W$ | work | ft-lbf | J |
| $y$ | superelevation | ft | m |

## Symbols

| | | | |
|---|---|---|---|
| $\alpha$ | angular acceleration | rad/sec$^2$ | rad/s$^2$ |
| $\delta$ | deflection | ft | m |
| $\epsilon$ | eccentricity | – | – |
| $\theta$ | angular position | rad | rad |
| $\sigma$ | stress | lbf/ft$^2$ | Pa |
| $\phi$ | angle | deg or rad | deg or rad |
| $\omega$ | angular velocity | rad/sec | rad/s |

## Subscripts

| | |
|---|---|
| 0 | initial |
| $b$ | braking |
| $c$ | centripetal |
| $C$ | instant center |
| $f$ | final or frictional |
| $k$ | kinetic (dynamic) |

**Dynamics**

| | |
|---|---|
| $n$ | normal |
| $O$ | center or centroidal |
| $p$ | periodic |
| $r$ | rolling |
| $s$ | static |
| $t$ | tangential or terminal |
| $w$ | wedge |

## 1. INTRODUCTION TO KINETICS

*Kinetics* is the study of motion and the forces that cause motion. Kinetics includes an analysis of the relationship between the force and mass for translational motion and between torque and moment of inertia for rotational motion. Newton's laws form the basis of the governing theory in the subject of kinetics.

## 2. RIGID BODY MOTION

The most general type of motion is *rigid body motion*. There are five types.

- *pure translation:* The orientation of the object is unchanged as its position changes. (Motion can be in straight or curved paths.)

- *rotation about a fixed axis:* All particles within the body move in concentric circles around the *axis of rotation*.

- *general plane motion:* The motion can be represented in two dimensions (i.e., in the *plane of motion*).

- *motion about a fixed point:* This describes any three-dimensional motion with one fixed point, such as a spinning top or a truck-mounted crane. The distance from a fixed point to any particle in the body is constant.

- *general motion:* This is any motion not falling into one of the other four categories.

Figure 57.1 illustrates the terms yaw, pitch, and roll as they relate to general motion. *Yaw* is a left or right swinging motion of the leading edge. *Pitch* is an up or down swinging motion of the leading edge. *Roll* is rotation about the leading edge's longitudinal axis.

*Figure 57.1* Yaw, Pitch, and Roll

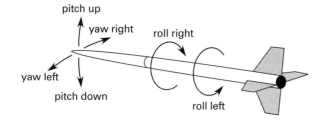

## 3. STABILITY OF EQUILIBRIUM POSITIONS

*Stability* is defined in terms of a body's relationship with an equilibrium position. *Neutral equilibrium* exists if a body, when displaced from its equilibrium position, remains in its displaced state. *Stable equilibrium* exists if the body returns to the original equilibrium position after experiencing a displacement. *Unstable equilibrium* exists if the body moves away from the equilibrium position. These terms are illustrated by Fig. 57.2.

*Figure 57.2* Types of Equilibrium

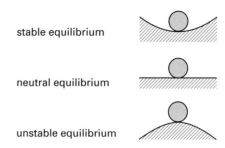

## 4. CONSTANT FORCES

*Force* is a push or a pull that one body exerts on another, including gravitational, electrostatic, magnetic, and contact influences. Forces that do not vary with time are *constant forces*.

Actions of other bodies on a rigid body are known as *external forces*. External forces are responsible for external motion of a body. *Internal forces* hold together parts of a rigid body.

## 5. LINEAR MOMENTUM

The vector *linear momentum* (usually just *momentum*) is defined by Eq. 57.1.[1] It has the same direction as the velocity vector. Momentum has units of force × time (e.g., lbf-sec or N·s).

$$\mathbf{p} = m\mathbf{v} \qquad \text{[SI]} \quad 57.1(a)$$

$$\mathbf{p} = \frac{m\mathbf{v}}{g_c} \qquad \text{[U.S.]} \quad 57.1(b)$$

Momentum is conserved when no external forces act on a particle. If no forces act on the particle, the velocity and direction of the particle are unchanged. The *law of conservation of momentum* states that the linear momentum is unchanged if no unbalanced forces act on the particle. This does not prohibit the mass and velocity from changing, however. Only the product of mass and velocity is constant. Depending on the nature

---

[1]The symbols **P**, **mom**, $mv$, and others are also used for momentum. Some authorities assign no symbol and just use the word momentum.

of the problem, momentum can be conserved in any or all of the three coordinate directions.

$$\sum m_0 \mathbf{v}_0 = \sum m_f \mathbf{v}_f \qquad 57.2$$

## 6. BALLISTIC PENDULUM

Figure 57.3 illustrates a *ballistic pendulum*. A projectile of known mass but unknown velocity is fired into a hanging target (the *pendulum*). The projectile is captured by the pendulum, which moves forward and upward. Kinetic energy is not conserved during impact because some of the projectile's kinetic energy is transformed into heat. However, momentum is conserved during impact, and the movement of the pendulum can be used to calculate the impact velocity of the projectile.[2]

**Figure 57.3** *Ballistic Pendulum*

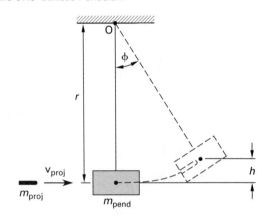

Since no external forces act on the block during impact, the momentum of the system is conserved.

$$\mathbf{p}_{\text{before impact}} = \mathbf{p}_{\text{after impact}} \qquad 57.3$$

$$m_{\text{proj}}\mathbf{v}_{\text{proj}} = (m_{\text{proj}} + m_{\text{pend}})\mathbf{v}_{\text{pend}} \qquad 57.4$$

Although kinetic energy before impact is not conserved, the total remaining energy after impact is conserved. That is, once the projectile has been captured by the pendulum, the kinetic energy of the pendulum-projectile combination is converted totally to potential energy as the pendulum swings upward.

$$\tfrac{1}{2}(m_{\text{proj}} + m_{\text{pend}})\text{v}_{\text{pend}}^2 = (m_{\text{proj}} + m_{\text{pend}})gh \qquad 57.5$$

$$\text{v}_{\text{pend}} = \sqrt{2gh} \qquad 57.6$$

---

[2]In this type of problem, it is important to be specific about when the energy and momentum are evaluated. During impact, kinetic energy is not conserved, but momentum is conserved. After impact, as the pendulum swings, energy is conserved but momentum is not conserved because gravity (an external force) acts on the pendulum during its swing.

The relationship between the rise of the pendulum, $h$, and the swing angle, $\phi$, is

$$h = r(1 - \cos\phi) \qquad 57.7$$

Since the time during which the force acts is not well defined, there is no single equivalent force that can be assumed to initiate the motion. Any force that produces the same impulse over a given contact time will be applicable.

## 7. ANGULAR MOMENTUM

The vector *angular momentum* (also known as *moment of momentum*) taken about a point O is the moment of the linear momentum vector. Angular momentum has units of distance × force × time (e.g., ft-lbf-sec or N·m·s). It has the same direction as the rotation vector and can be determined by use of the right-hand rule. (That is, it acts in a direction perpendicular to the plane containing the position and linear momentum vectors.) (See Fig. 57.4.)

$$\mathbf{h}_O = \mathbf{r} \times m\mathbf{v} \qquad \text{[SI]} \quad 57.8(a)$$

$$\mathbf{h}_O = \frac{\mathbf{r} \times m\mathbf{v}}{g_c} \qquad \text{[U.S.]} \quad 57.8(b)$$

Any of the methods normally used to evaluate cross-products can be used with angular momentum. The scalar form of Eq. 57.8 is

$$\text{h}_O = rm\text{v}\sin\phi \qquad \text{[SI]} \quad 57.9(a)$$

$$\text{h}_O = \frac{rm\text{v}\sin\phi}{g_c} \qquad \text{[U.S.]} \quad 57.9(b)$$

**Figure 57.4** *Angular Momentum*

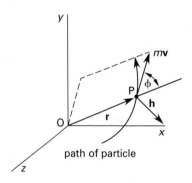

path of particle

For a rigid body rotating about an axis passing through its center of gravity located at point O, the scalar value of angular momentum is given by Eq. 57.10.

$$h_O = I\omega \qquad \text{[SI]} \quad 57.10(a)$$

$$h_O = \frac{I\omega}{g_c} \qquad \text{[U.S.]} \quad 57.10(b)$$

## 8. NEWTON'S FIRST LAW OF MOTION

Much of this chapter is based on Newton's laws of motion. *Newton's first law of motion* can be stated in several forms.

*common form:* A particle will remain in a state of rest or will continue to move with constant velocity unless an unbalanced external force acts on it.

*law of conservation of momentum form:* If the resultant external force acting on a particle is zero, then the linear momentum of the particle is constant.

## 9. NEWTON'S SECOND LAW OF MOTION

*Newton's second law of motion* is stated as follows.

*second law:* The acceleration of a particle is directly proportional to the force acting on it and inversely proportional to the particle mass. The direction of acceleration is the same as the force of direction.

This law can be stated in terms of the force vector required to cause a change in momentum. The resultant force is equal to the rate of change of linear momentum.

$$\mathbf{F} = \frac{d\mathbf{p}}{dt} \qquad 57.11$$

If the mass is constant with respect to time, the scalar form of Eq. 57.11 is[3]

$$F = m\left(\frac{d\mathrm{v}}{dt}\right) = ma \qquad \text{[SI]} \quad 57.12(a)$$

$$F = \left(\frac{m}{g_c}\right)\left(\frac{d\mathrm{v}}{dt}\right) = \frac{ma}{g_c} \qquad \text{[U.S.]} \quad 57.12(b)$$

Equation 57.12 can be written in rectangular coordinates form (i.e., in terms of $x$- and $y$-component forces), in polar coordinates form (i.e., tangential and normal components), and in cylindrical coordinates form (i.e., radial and transverse components).

Although Newton's laws do not specifically deal with rotation, there is an analogous relationship between torque and change in angular momentum. For a rotating body, the torque, $\mathbf{T}$, required to change the angular momentum is

$$\mathbf{T} = \frac{d\mathbf{h}_0}{dt} \qquad 57.13$$

---

[3]Equation 57.12 shows that force is a scalar multiple of acceleration. Any consistent set of units can be used. For example, if both sides are divided by the acceleration of gravity (i.e., so that acceleration in Eq. 57.12 is in gravities), the force will have units of *g-forces* or *gees* (i.e., multiples of the gravitational force).

If the moment of inertia is constant, the scalar form of Eq. 57.13 is

$$T = I\left(\frac{d\omega}{dt}\right) = I\alpha \qquad \text{[SI]} \quad 57.14(a)$$

$$T = \left(\frac{I}{g_c}\right)\left(\frac{d\omega}{dt}\right) = \frac{I\alpha}{g_c} \qquad \text{[U.S.]} \quad 57.14(b)$$

### Example 57.1

The acceleration in $\mathrm{m/s^2}$ of a 40 kg body is specified by the equation

$$a(t) = 8 - 12t$$

What is the instantaneous force acting on the body at $t = 6$ s?

*Solution*

The acceleration is

$$a(6) = 8\ \frac{\mathrm{m}}{\mathrm{s^2}} - \left(12\ \frac{\mathrm{m}}{\mathrm{s^3}}\right)(6\ \mathrm{s}) = -64\ \mathrm{m/s^2}$$

From Newton's second law, the instantaneous force is

$$f = ma = (40\ \mathrm{kg})\left(-64\ \frac{\mathrm{m}}{\mathrm{s^2}}\right)$$
$$= -2560\ \mathrm{N}$$

### Example 57.2

During start-up, a 4.0 ft diameter pulley with centroidal moment of inertia of 1610 lbm-ft$^2$ is subjected to tight-side and loose-side belt tensions of 200 lbf and 100 lbf, respectively. A frictional torque of 15 ft-lbf is acting to resist pulley rotation. (a) What is the angular acceleration? (b) How long will it take the pulley to reach a speed of 120 rpm?

*Solution*

(a) From Eq. 57.24, the net torque is

$$T = rF_{\mathrm{net}} = (2\ \mathrm{ft})(200\ \mathrm{lbf} - 100\ \mathrm{lbf}) - 15\ \mathrm{ft\text{-}lbf}$$
$$= 185\ \mathrm{ft\text{-}lbf}$$

From Eq. 57.14, the angular acceleration is

$$\alpha = \frac{g_c T}{I} = \frac{\left(32.2\ \dfrac{\mathrm{lbm\text{-}ft}}{\mathrm{lbf\text{-}sec^2}}\right)(185\ \mathrm{ft\text{-}lbf})}{1610\ \mathrm{lbm\text{-}ft^2}}$$
$$= 3.7\ \mathrm{rad/sec^2}$$

(b) The rotational speed is

$$\omega = \frac{\left(120 \ \frac{\text{rev}}{\text{min}}\right)\left(2\pi \ \frac{\text{rad}}{\text{rev}}\right)}{60 \ \frac{\text{sec}}{\text{min}}}$$

$$= 12.6 \ \text{rad/sec}$$

This is a case of constant angular acceleration starting from rest.

$$t = \frac{\omega}{\alpha} = \frac{12.6 \ \frac{\text{rad}}{\text{sec}}}{3.7 \ \frac{\text{rad}}{\text{sec}^2}}$$

$$= 3.4 \ \text{sec}$$

## 10. CENTRIPETAL FORCE

Newton's second law says there is a force for every acceleration a body experiences. For a body moving around a curved path, the total acceleration can be separated into tangential and normal components. By Newton's second law, there are corresponding forces in the tangential and normal directions. The force associated with the normal acceleration is known as the *centripetal force*.[4]

$$F_c = ma_n = \frac{mv_t^2}{r} \qquad \text{[SI]} \quad \textbf{57.15(a)}$$

$$F_c = ma_n = \frac{mv_t^2}{g_c r} \qquad \text{[U.S.]} \quad \textbf{57.15(b)}$$

The centripetal force is a real force on the body toward the center of rotation. The so-called *centrifugal force* is an apparent force on the body directed away from the center of rotation. The centripetal and centrifugal forces are equal in magnitude but opposite in sign. (See Fig. 57.5.)

**Figure 57.5** *Centripetal Force*

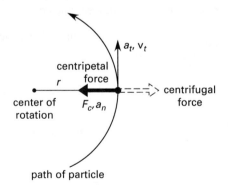

An unbalanced rotating body (vehicle wheel, clutch disk, rotor of an electrical motor, etc.) will experience a dynamic *unbalanced force*. Though the force is essentially centripetal in nature and is given by Eq. 57.15, it is generally difficult to assign a value to the radius. For that reason, the force is often determined directly on the rotating body or from the deflection of its supports.

Since the body is rotating, the force will be experienced in all directions perpendicular to the axis of rotation. If the supports are flexible, the force will cause the body to vibrate, and the frequency of vibration will essentially be the rotational speed. If the supports are rigid, the bearings will carry the unbalanced force and transmit it to other parts of the frame.

### Example 57.3

A 4500 lbm (2000 kg) car travels at 40 mph (65 kph) around a curve with a radius of 200 ft (60 m). What is the centripetal force?

*SI Solution*

The tangential velocity is

$$v_t = \frac{\left(65 \ \frac{\text{km}}{\text{h}}\right)\left(1000 \ \frac{\text{m}}{\text{km}}\right)}{3600 \ \frac{\text{s}}{\text{h}}}$$

$$= 18.06 \ \text{m/s}$$

From Eq. 57.15(a), the centripetal force is

$$F_c = \frac{mv_t^2}{r} = \frac{(2000 \ \text{kg})\left(18.06 \ \frac{\text{m}}{\text{s}}\right)^2}{60 \ \text{m}}$$

$$= 10\,872 \ \text{N}$$

*Customary U.S. Solution*

The tangential velocity is

$$v_t = \frac{\left(40 \ \frac{\text{mi}}{\text{hr}}\right)\left(5280 \ \frac{\text{ft}}{\text{mi}}\right)}{3600 \ \frac{\text{sec}}{\text{hr}}}$$

$$= 58.66 \ \text{ft/sec}$$

From Eq. 57.15(b), the centripetal force is

$$F_c = \frac{mv_t^2}{g_c r} = \frac{(4500 \ \text{lbm})\left(58.66 \ \frac{\text{ft}}{\text{sec}}\right)^2}{\left(32.2 \ \frac{\text{ft-lbm}}{\text{lbf-sec}^2}\right)(200 \ \text{ft})}$$

$$= 2404 \ \text{lbf}$$

---

[4]The term *normal force* is reserved for the plane reaction in friction calculations.

## 11. NEWTON'S THIRD LAW OF MOTION

*Newton's third law of motion* is as follows.

*third law:* For every acting force between two bodies, there is an equal but opposite reacting force on the same line of action.

$$\mathbf{F}_{\text{reacting}} = -\mathbf{F}_{\text{acting}} \qquad 57.16$$

## 12. DYNAMIC EQUILIBRIUM

An accelerating body is not in static equilibrium. Accordingly, the familiar equations of statics ($\Sigma F = 0$ and $\Sigma M = 0$) do not apply. However, if the *inertial force*, $m\mathbf{a}$, is included in the static equilibrium equation, the body is said to be in *dynamic equilibrium*.[5,6] This is known as *D'Alembert's principle*. Since the inertial force acts to oppose changes in motion, it is negative in the summation.

$$\sum \mathbf{F} - m\mathbf{a} = 0 \qquad \text{[SI]} \quad 57.17(a)$$

$$\sum \mathbf{F} - \frac{m\mathbf{a}}{g_c} = 0 \qquad \text{[U.S.]} \quad 57.17(b)$$

It should be clear that D'Alembert's principle is just a different form of Newton's second law, with the $ma$ term transposed to the left-hand side.

The analogous rotational form of the dynamic equilibrium principle is

$$\sum \mathbf{T} - I\alpha = 0 \qquad \text{[SI]} \quad 57.18(a)$$

$$\sum \mathbf{T} - \frac{I\alpha}{g_c} = 0 \qquad \text{[U.S.]} \quad 57.18(b)$$

## 13. FLAT FRICTION

Friction is a force that always resists motion or impending motion. It always acts parallel to the contacting surfaces. The frictional force, $F_f$, exerted on a stationary body is known as *static friction, Coulomb friction,* and *fluid friction.* If the body is moving, the friction is known as *dynamic friction* and is less than the static friction.

The actual magnitude of the frictional force depends on the *normal force, N,* and the *coefficient of friction, f,* between the body and the surface.[7] For a body resting on a horizontal surface, the normal force is the weight of the body. (See Fig. 57.6.)

---

[5]Other names for the inertial force are *inertia vector* (when written as $m\mathbf{a}$), *dynamic reaction,* and *reversed effective force.* The term $\Sigma \mathbf{F}$ is known as the *effective force.*

[6]*Dynamic* and *equilibrium* are contradictory terms. A better term is *simulated equilibrium,* but this form has not caught on.

[7]The symbol $\mu$ is also widely used by engineers to represent the coefficient of friction.

*Figure 57.6* Frictional and Normal Forces

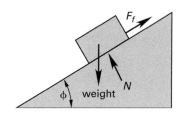

$$N = mg \qquad \text{[SI]} \quad 57.19(a)$$

$$N = \frac{mg}{g_c} \qquad \text{[U.S.]} \quad 57.19(b)$$

If the body rests on an inclined surface, the normal force depends on the incline angle.

$$N = mg \cos \phi \qquad \text{[SI]} \quad 57.20(a)$$

$$N = \frac{mg \cos \phi}{g_c} \qquad \text{[U.S.]} \quad 57.20(b)$$

The maximum static frictional force, $F_f$, is the product of the coefficient of friction, $f$, and the normal force, $N$. (The subscripts $s$ and $k$ are used to distinguish between the static and dynamic (kinetic) coefficients of friction.)

$$F_{f,\text{max}} = f_s N \qquad 57.21$$

The frictional force acts only in response to a disturbing force. If a small disturbing force (i.e., a force less than $F_{f,\text{max}}$) acts on a body, then the frictional force will equal the disturbing force, and the maximum frictional force will not develop. This occurs during the *equilibrium phase.* The *motion impending phase* is when the disturbing force equals the maximum frictional force, $F_{f,\text{max}}$. Once motion begins, however, the coefficient of friction drops slightly, and a lower frictional force opposes movement. These cases are illustrated in Fig. 57.7.

*Figure 57.7* Frictional Force versus Disturbing Force

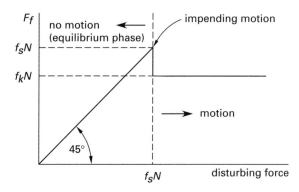

A body on an inclined plane will not begin to slip down the plane until the component of weight parallel to the plane exceeds the frictional force. If the plane's inclination angle can be varied, the body will not slip until the angle reaches a critical angle known as the *angle of repose* or *angle of static friction*, $\phi$. Equation 57.22 relates this angle to the coefficient of static friction.

$$\tan \phi = f_s \qquad 57.22$$

Tabulations of coefficients of friction distinguish between types of surfaces and between static and dynamic cases. They might also list values for dry conditions and oiled conditions. The term *dry* is synonymous with *nonlubricated*. The ambiguous term *wet*, although a natural antonym for *dry*, is sometimes used to mean *oily*. However, it usually means wet with water, as in tires on a wet roadway after a rain. Typical values of the coefficient of friction are given in Table 57.1.[8]

**Table 57.1** *Typical Coefficients of Friction*

| materials | condition | dynamic | static |
|---|---|---|---|
| cast iron on cast iron | dry | 0.15 | 1.00 |
| plastic on steel | dry | 0.35 | 0.45 |
| grooved rubber on pavement | dry | 0.40 | 0.55 |
| bronze on steel | oiled | 0.07 | 0.09 |
| steel on graphite | dry | 0.16 | 0.21 |
| steel on steel | dry | 0.42 | 0.78 |
| steel on steel | oiled | 0.08 | 0.10 |
| steel on asbestos-faced steel | dry | 0.11 | 0.15 |
| steel on asbestos-faced steel | oiled | 0.09 | 0.12 |
| press fits (shaft in hole) | oiled | – | 0.10–0.15 |

A special case of the angle of repose is the *angle of internal friction*, $\phi$, of soil, grain, or other granular material. (See Fig. 57.8.) The angle made by a pile of granular material depends on how much friction there is between the granular particles. Liquids have angles of internal friction of zero, because they do not form piles.

**Figure 57.8** *Angle of Internal Friction of a Pile*

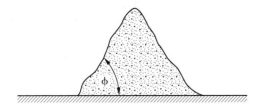

## 14. WEDGES

Wedges are machines that are able to raise heavy loads. The wedge angles are chosen so that friction will keep the wedge in place once it is driven between the load and support. As with any situation where friction is present, the frictional force is parallel to the contacting surfaces. (See Fig. 57.9.)

**Figure 57.9** *Using a Wedge to Raise a Load*

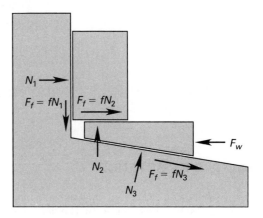

## 15. BELT FRICTION

Friction from a flat belt, rope, or band wrapped around a pulley or sheave is responsible for the transfer of torque. Except at start-up, one side of the belt (the tight side) will have a higher tension than the other (the slack side). The basic relationship between these belt tensions and the coefficient of friction neglects centrifugal effects and is given by Eq. 57.23.[9] (The angle of wrap, $\phi$, must be expressed in radians.) (See Fig. 57.10.)

$$\frac{F_{\max}}{F_{\min}} = e^{f\phi} \qquad 57.23$$

The net transmitted torque is

$$T = (F_{\max} - F_{\min})r \qquad 57.24$$

**Figure 57.10** *Flat Belt Friction*

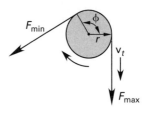

---

[8]Experimental and reported values of the coefficient of friction vary greatly from researcher to researcher and experiment to experiment. The values in Table 57.1 are more for use in solving practice problems than serving as the last word in available data.

[9]This equation does not apply to V-belts. V-belt design and analysis is dependent on the cross-sectional geometry of the belt.

The power transmitted by the belt running at tangential velocity $v_t$ is given by Eq. 57.25.[10]

$$P = (F_{max} - F_{min})v_t \qquad 57.25$$

The change in belt tension caused by centrifugal force should be considered when the velocity or belt mass is very large. Equation 57.26 can be used, where $m$ is the mass per unit length of belt.

$$\frac{F_{max} - mv_t^2}{F_{min} - mv_t^2} = e^{f\phi} \qquad \text{[SI]} \qquad 57.26(a)$$

$$\frac{F_{max} - \dfrac{mv_t^2}{g_c}}{F_{min} - \dfrac{mv_t^2}{g_c}} = e^{f\phi} \qquad \text{[U.S.]} \qquad 57.26(b)$$

## 16. ROLLING RESISTANCE

*Rolling resistance* is a force that opposes motion, but it is not friction. Rather, it is caused by the deformation of the rolling body and the supporting surface. Rolling resistance is characterized by a *coefficient of rolling resistance*, $a$, which has units of length.[11] (See Fig. 57.11.) Since this deformation is very small, the rolling resistance in the direction of motion is

$$F_r = \frac{mga}{r} \qquad \text{[SI]} \qquad 57.27(a)$$

$$F_r = \frac{mga}{rg_c} = \frac{wa}{r} \qquad \text{[U.S.]} \qquad 57.27(b)$$

**Figure 57.11** *Wheel Rolling Resistance*

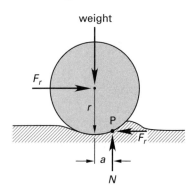

The term *coefficient of rolling friction*, $f_r$, is occasionally encountered, although friction is not the cause of rolling resistance.

$$f_r = \frac{F_r}{w} = \frac{a}{r} \qquad 57.28$$

## 17. ROADWAY BANKING

If a vehicle travels in a circular path with instantaneous radius $r$ and tangential velocity $v_t$, it will experience an apparent centrifugal force. The centrifugal force is resisted by a combination of roadway banking (*superelevation rate*) and *sideways friction*.[12] If the roadway is banked so that friction is not required to resist the centrifugal force, the superelevation angle, $\phi$, can be calculated from Eq. 57.29.[13]

$$\tan\phi = \frac{v_t^2}{gr} \qquad 57.29$$

Equation 57.29 can be solved for the *normal speed* corresponding to the geometry of a curve.

$$v_t = \sqrt{gr\tan\phi} \qquad 57.30$$

When friction is used to counteract some of the centrifugal force, the *side friction factor*, $f$, between the tires and roadway is incorporated into the calculation of the superelevation angle.

$$e = \tan\phi = \frac{v_t^2 - fgr}{gr + fv_t^2} \qquad 57.31$$

If the banking angle, $\phi$, is set to zero, Eq. 57.31 can be used to calculate the maximum velocity of a vehicle making a turn when there is no banking.

For highway design, *superelevation rate*, $e$, is the amount of rise or fall of the cross slope per unit amount of horizontal width (i.e., the tangent of the slope angle above or below horizontal). Customary U.S. units are expressed in feet per foot, such as 0.06 ft/ft, or inch fractions per foot, such as $3/4$ in/ft. SI units are millimeters per meter, such as 60 mm/m. The slope can also be expressed as a percent cross slope, such as 6% cross slope; or as a ratio, 1:17. The *superelevation*, $y$, is the difference in heights of the inside and outside edges of the curve.

When the speed, superelevation, and radius are such that no friction is required to resist sliding, the curve is said to be "balanced." There is no tendency for a vehicle to slide up or down the slope at the *balanced speed*. At any speed other than the balanced speed, some friction is needed to hold the vehicle on the slope. Given a

---

[10]When designing a belt system, the horsepower to be transmitted should be multiplied by a *service factor* to obtain the *design power*. Service factors range from 1.0 to 1.5 and depend on the nature of the power source, the load, and the starting characteristics.

[11]Rolling resistance is traditionally derived by assuming the roller encounters a small step in its path a distance $a$ in front of the center of gravity. The forces acting on the roller are the weight and driving force acting through the centroid and the normal force and rolling resistance acting at the contact point. Equation 57.27 is derived by taking moments about the contact point, P.

[12]The *superelevation rate* is the slope (in ft/ft or m/m) in the transverse direction (i.e., across the roadway).

[13]Generally it is not desirable to rely on roadway banking alone, since a particular superelevation angle would correspond to only a single speed.

friction factor, $f$, the required value of the superelevation rate, $e$, can be calculated. (See Fig. 57.12.)

**Figure 57.12** *Roadway Banking*

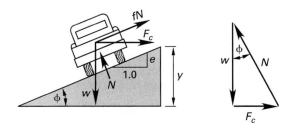

Equation 57.31 can be solved for the curve radius. For small banking angles (i.e., $\phi \leq 8°$), this simplifies to

$$r \approx \frac{v_t^2}{g(e+f)} \qquad \textit{57.32}$$

$$r_m = \frac{v_{km/h}^2}{127(e_{m/m}+f)} \qquad \text{[SI]} \quad \textit{57.33(a)}$$

$$r_{ft} = \frac{v_{mph}^2}{15(e_{ft/ft}+f)} \qquad \text{[U.S.]} \quad \textit{57.33(b)}$$

**Example 57.4**

A 4000 lbm (1800 kg) car travels at 40 mph (65 km/h) around a banked curve with a radius of 500 ft (150 m). What should be the superelevation rate so that the tire friction is not needed to prevent the car from sliding?

*SI Solution*

From Eq. 57.33,

$$e+f = \frac{v^2}{127r}$$

$$e+0 = \frac{\left(65 \frac{km}{hr}\right)^2}{(127)(150 \text{ m})}$$

$$= 0.222 \text{ m/m}$$

*Customary U.S. Solution*

From Eq. 57.33,

$$e+f = \frac{v^2}{15r}$$

$$e+0 = \frac{\left(40 \frac{mi}{hr}\right)^2}{(15)(500 \text{ ft})}$$

$$= 0.213 \text{ ft/ft}$$

**Example 57.5**

A 4000 lbm car travels at 40 mph around a banked curve with a radius of 500 ft. What should be the superelevation angle so that tire friction is not needed to prevent the car from sliding?

*Solution*

The tangential velocity of the car is

$$v_t = \frac{\left(40 \frac{mi}{hr}\right)\left(5280 \frac{ft}{mi}\right)}{3600 \frac{sec}{hr}}$$

$$= 58.66 \text{ ft/sec}$$

From Eq. 57.29,

$$\phi = \arctan\left(\frac{v_t^2}{gr}\right)$$

$$= \arctan\left(\frac{\left(58.66 \frac{ft}{sec}\right)^2}{\left(32.2 \frac{ft}{sec^2}\right)(500 \text{ ft})}\right)$$

$$= 12.06°$$

**Example 57.6**

A vehicle is traveling at 70 mph when it enters a circular curve of a test track. The curve radius is 240 ft. The sideways sliding coefficient of friction between the tires and the roadway is 0.57. (a) At what minimum angle from the horizontal must the curve be banked in order to prevent the vehicle from sliding off the top of the curve? (b) If the roadway is banked at 20° from the horizontal, what is the maximum vehicle speed such that no sliding occurs?

*Solution*

(a) The speed of the vehicle is

$$v_t = \frac{\left(70 \frac{mi}{hr}\right)\left(5280 \frac{ft}{mi}\right)}{3600 \frac{sec}{hr}}$$

$$= 102.7 \text{ ft/sec}$$

Dynamics

**Dynamics**

From Eq. 57.31, the required banking angle is

$$\phi = \arctan\left(\frac{v_t^2 - fgr}{gr + fv_t^2}\right)$$

$$= \arctan\left(\frac{\left(102.7\ \frac{\text{ft}}{\text{sec}}\right)^2 - (0.57)\left(32.2\ \frac{\text{ft}}{\text{sec}^2}\right)(240\ \text{ft})}{\left(32.2\ \frac{\text{ft}}{\text{sec}^2}\right)(240\ \text{ft}) + (0.57)\left(102.7\ \frac{\text{ft}}{\text{sec}}\right)^2}\right)$$

$$= 24.1°$$

(b) Solve Eq. 57.31 for the velocity.

$$v_t = \sqrt{\frac{rg(\tan\phi + f)}{1 - f\tan\phi}}$$

$$= \sqrt{\frac{(240\ \text{ft})\left(32.2\ \frac{\text{ft}}{\text{sec}^2}\right)(\tan 20° + 0.57)}{1 - 0.57\tan 20°}}$$

$$= 95.4\ \text{ft/sec} \quad (65.1\ \text{mi/hr})$$

## 18. MOTION OF RIGID BODIES

When a rigid body experiences pure translation, its position changes without any change in orientation. At any instant, all points on the body have the same displacement, velocity, and acceleration. The behavior of a rigid body in translation is given by Eq. 57.34 and Eq. 57.35. All equations are written for the center of mass. (These equations represent Newton's second law written in component form.)

$$\sum F_x = ma_x \qquad \text{[consistent units]} \qquad 57.34$$

$$\sum F_y = ma_y \qquad \text{[consistent units]} \qquad 57.35$$

When a torque acts on a rigid body, the rotation will be about the center of gravity unless the body is constrained otherwise. In the case of rotation, the torque and angular acceleration are related by Eq. 57.36.

$$T = I\alpha \qquad \text{[SI]} \qquad 57.36(a)$$

$$T = \frac{I\alpha}{g_c} \qquad \text{[U.S.]} \qquad 57.36(b)$$

*Euler's equations of motion* are used to analyze the motion of a rigid body about a fixed point, O. This class of problem is particularly difficult because the mass moments of products of inertia change with time if a fixed set of axes is used. Therefore, it is more convenient to define the *x*-, *y*-, and *z*-axes with respect to the body. Such an action is acceptable because the angular momentum about the origin, $\mathbf{h}_O$, corresponding to a given angular velocity, $\omega$, is independent of the choice of coordinate axes.

An infinite number of axes can be chosen. (A general relationship between moments and angular momentum is given in most dynamics textbooks.) However, if the origin is at the mass center and the *x*-, *y*-, and *z*-axes coincide with the principal axes of inertia of the body (such that the product of inertia is zero), the angular momentum of the body about the origin (i.e., point O at (0,0,0)) is given by the simplified relationship

$$\mathbf{h}_O = I_x\omega_x\mathbf{i} + I_y\omega_y\mathbf{j} + I_z\omega_z\mathbf{k} \qquad 57.37$$

The three scalar Euler equations of motion can be derived from this simplified relationship.

$$\sum M_x = I_x\alpha_x - (I_y - I_z)\omega_y\omega_z \qquad 57.38$$

$$\sum M_y = I_y\alpha_y - (I_z - I_x)\omega_z\omega_x \qquad 57.39$$

$$\sum M_z = I_z\alpha_z - (I_x - I_y)\omega_x\omega_y \qquad 57.40$$

### Example 57.7

A 5000 lbm truck skids with a deceleration of 15 ft/sec². (a) What is the coefficient of sliding friction? (b) What are the frictional forces and normal reactions (per axle) at the tires?

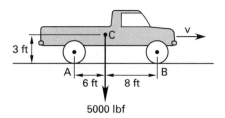

*Solution*

(a) The free-body diagram of the truck in equilibrium with the inertial force is shown.

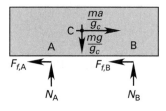

The equation of dynamic equilibrium in the horizontal direction is

$$\sum F_x = \frac{ma}{g_c} - F_{f,\text{A}} - F_{f,\text{B}} = 0$$

$$= \frac{ma}{g_c} - (N_\text{A} + N_\text{B})f = 0$$

$$\frac{(5000\ \text{lbm})\left(15\ \frac{\text{ft}}{\text{sec}^2}\right)}{32.2\ \frac{\text{ft-lbm}}{\text{lbf-sec}^2}} - (5000\ \text{lbf})f = 0$$

The coefficient of friction is

$$f = 0.466$$

(b) The vertical reactions at the tires can be found by taking moments about one of the contact points.

$$\sum M_A: 14 N_B - (6 \text{ ft})(5000 \text{ lbf})$$

$$-(3 \text{ ft})\left(\frac{5000 \text{ lbm}}{32.2 \dfrac{\text{ft-lbm}}{\text{lbf-sec}^2}}\right)\left(15 \dfrac{\text{ft}}{\text{sec}^2}\right) = 0$$

$$N_B = 2642 \text{ lbf}$$

The remaining vertical reaction is found by summing vertical forces.

$$\sum F_y: \qquad N_A + N_B - \frac{mg}{g_c} = 0$$

$$N_A + 2642 \text{ lbf} - 5000 \text{ lbf} = 0$$

$$N_A = 2358 \text{ lbf}$$

The horizontal frictional forces at the front and rear axles are

$$F_{f,A} = (0.466)(2358 \text{ lbf}) = 1099 \text{ lbf}$$

$$F_{f,B} = (0.466)(2642 \text{ lbf}) = 1231 \text{ lbf}$$

## 19. CONSTRAINED MOTION

Figure 57.13 shows a cylinder (or sphere) on an inclined plane. If there is no friction, there will be no torque to start the cylinder rolling. Regardless of the angle, the cylinder will slide down the incline in *unconstrained motion.*

**Figure 57.13** *Unconstrained Motion*

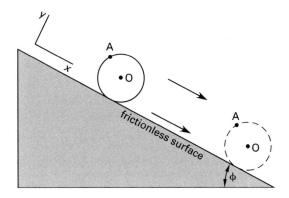

The acceleration sliding down the incline can be calculated by writing Newton's second law for an axis parallel to the plane.[14] Once the acceleration is known, the

velocity can be found from the constant-acceleration equations.

$$ma_{O,x} = mg \sin \phi \qquad 57.41$$

If friction is sufficiently large, or if the inclination is sufficiently small, there will be no slipping. This condition occurs if

$$\phi < \arctan f_s \qquad 57.42$$

The frictional force acting at the cylinder's radius, $r$, supplies a torque that starts and keeps the cylinder rolling. The frictional force is

$$F_f = f N \qquad 57.43$$

$$F_f = fmg \cos \phi \qquad \text{[SI]} \qquad 57.44(a)$$

$$F_f = \frac{fmg \cos \phi}{g_c} \qquad \text{[U.S.]} \qquad 57.44(b)$$

With no slipping, the cylinder has two degrees of freedom (the $x$-directional and angle of rotation), and motion of the center of mass must simultaneously satisfy (i.e., is constrained by) two equations. (This excludes motion perpendicular to the plane.) This is called *constrained motion.* (See Fig. 57.14.)

$$mg \sin \phi - F_f = ma_{O,x} \qquad \text{[consistent units]} \qquad 57.45$$

$$F_f r = I_O \alpha \qquad \text{[consistent units]} \qquad 57.46$$

**Figure 57.14** *Constrained Motion*

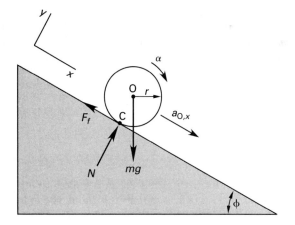

The mass moment of inertia used in calculating angular acceleration can be either the centroidal moment of inertia, $I_O$, or the moment of inertia taken about the contact point, $I_C$, depending on whether torques (moments) are evaluated with respect to point O or point C, respectively.

If moments are evaluated with respect to point O, the coefficient of friction, $f$, must be known, and the centroidal moment of inertia can be used. If moments are

---

[14]Most inclined plane problems are conveniently solved by resolving all forces into components parallel and perpendicular to the plane.

evaluated with respect to the contact point, the frictional and normal forces drop out of the torque summation. The cylinder instantaneously rotates as though it were pinned at point C. (See Fig. 57.15.) If the centroidal moment of inertia, $I_O$, is known, the parallel axis theorem can be used to find the required moment of inertia.

$$I_C = I_O + mr^2 \qquad 57.47$$

**Figure 57.15** *Instantaneous Center of a Constrained Cylinder*

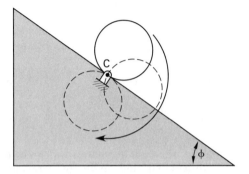

When there is no slipping, the cylinder will roll with constant linear and angular accelerations. The distance traveled by the center of mass can be calculated from the angle of rotation.

$$s_O = r\theta \qquad 57.48$$

If $\phi \geq \arctan f_s$, the cylinder will simultaneously roll and slide down the incline. The analysis is similar to the no-sliding case, except that the coefficient of sliding friction is used. Once sliding has started, the inclination angle can be reduced to $\arctan f_k$, and rolling with sliding will continue.

**Example 57.8**

A 150 kg cylinder with radius 0.3 m is pulled up a plane inclined at 30° as fast as possible without the cylinder slipping. The coefficient of friction is 0.236. There is a groove in the cylinder at radius = 0.2 m. A rope in the groove applies a force of 500 N up the ramp. What is the linear acceleration of the cylinder?

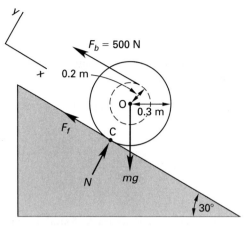

*Solution*

*To solve by summing forces in the x-direction:*

The normal force is

$$N = mg\cos\phi = (150 \text{ kg})\left(9.81 \ \frac{\text{m}}{\text{s}^2}\right)(\cos 30°)$$
$$= 1274.4 \text{ N}$$

The frictional maximum (friction impending) force is

$$F_f = fN = (0.236)(1274.4 \text{ N})$$
$$= 300.8 \text{ N}$$

The summation of forces in the x-direction is

$$ma_{O,x} = mg\sin\phi - F_f - F_b$$

$$a_{O,x} = \frac{(150 \text{ kg})\left(9.81 \ \frac{\text{m}}{\text{s}^2}\right)(\sin 30°) - 300.8 \text{ N} - 500 \text{ N}}{150 \text{ kg}}$$

$$= -0.434 \text{ m/s}^2 \quad \text{[up the incline]}$$

*To solve by taking moment about the contact point:*[15]

From Ftn. 15, the centroidal mass moment of inertia of the cylinder is

$$I_O = \tfrac{1}{2}mr^2 = (0.5)(150 \text{ kg})(0.3 \text{ m})^2$$
$$= 6.75 \text{ kg·m}^2$$

---

[15]This example can also be solved by summing moments about the center. If this is done, the governing equations are

$$I_O = \tfrac{1}{2}mr^2$$
$$M_O = I_O\alpha = F_b r' - F_f r$$
$$\tfrac{1}{2}mr^2\alpha = F_b r' - fmgr\cos\phi$$
$$a_{O,x} = r\alpha$$

The mass moment of inertia with respect to the contact point, C, is given by the parallel axis theorem.

$$I_C = I_O + mr^2 = \tfrac{1}{2}mr^2 + mr^2 = \tfrac{3}{2}mr^2$$
$$= \left(\tfrac{3}{2}\right)(150 \text{ kg})(0.3 \text{ m})^2$$
$$= 20.25 \text{ kg·m}^2$$

The $x$-component of the weight acts through the center of gravity. (This term dropped out when moments were taken with respect to the center of gravity.)

$$(mg)_x = (150 \text{ kg})\left(9.81 \tfrac{\text{m}}{\text{s}^2}\right)(\sin 30°)$$
$$= 735.8 \text{ N}$$

The summation of torques about point C gives the angular acceleration with respect to point C.

$$(735.8 \text{ N})(0.3 \text{ m})$$
$$- (500 \text{ N})(0.3 \text{ m} + 0.2 \text{ m}) = (20.25 \text{ kg·m}^2)\alpha$$
$$\alpha = -1.445 \text{ rad/s}^2$$

The linear acceleration can be calculated from the angular acceleration and the distance between points C and O.

$$a_{O,x} = ra = (0.3 \text{ m})\left(-1.445 \tfrac{\text{rad}}{\text{s}^2}\right)$$
$$= -0.433 \text{ m/s}^2 \quad \text{[up the incline]}$$

## 20. CABLE TENSION FROM AN ACCELERATING SUSPENDED MASS

When a mass hangs motionless from a cable, or when the mass is moving with a uniform velocity, the cable tension will equal the weight of the mass. However, when the mass is accelerating downward, the weight must be reduced by the inertial force. If the mass experiences a downward acceleration equal to the gravitational acceleration, there is no tension in the cable. Thus, the two cases shown in Fig. 57.16 are not the same.

**Figure 57.16** *Cable Tension from a Suspended Mass*

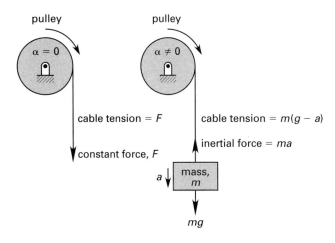

### Example 57.9

A 10.0 lbm (4.6 kg) mass hangs from a rope wrapped around a 2.0 ft (0.6 m) diameter pulley with a centroidal moment of inertia of 70 lbm-ft$^2$ (2.9 kg·m$^2$). (a) What is the angular acceleration of the pulley? (b) What is the linear acceleration of the mass?

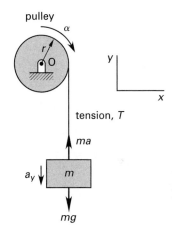

*SI Solution*

(a) The two equations of motion are

$$\sum F_{y,\text{mass}}: \ T + ma - mg = 0$$
$$T - (4.6 \text{ kg})\left(9.81 \tfrac{\text{m}}{\text{s}^2}\right) = -(4.6 \text{ kg})a_y$$

$$\sum M_{O,\text{pulley}}: \ Tr = I\alpha$$
$$T(0.3 \text{ m}) = (2.9 \text{ kg·m}^2)\alpha$$

**Dynamics**

Both $a_y$ and $\alpha$ are unknown but are related by $a_y = r\alpha$. Substituting into the $\Sigma F$ equation and eliminating the tension,

$$\frac{(2.9 \text{ kg·m}^2)\alpha}{0.3 \text{ m}} - 45.1 \text{ N} = -(4.6 \text{ kg})(0.3 \text{ m})\alpha$$

$$9.67\alpha - 45.1 \text{ N} = -1.38\alpha$$

$$\alpha = 4.08 \text{ rad/s}^2$$

(b) The linear acceleration of the mass is

$$a_y = r\alpha = (0.3 \text{ m})\left(4.08 \frac{\text{rad}}{\text{s}^2}\right)$$

$$= 1.22 \text{ m/s}^2$$

*Customary U.S. Solution*

(a) The two equations of motion are

$$\sum F_{y,\text{mass}}: \quad T + \frac{ma_y}{g_c} - \frac{mg}{g_c} = 0$$

$$T - \frac{(10 \text{ lbm})\left(32.2 \frac{\text{ft}}{\text{sec}^2}\right)}{32.2 \frac{\text{ft-lbm}}{\text{lbf-sec}^2}} = -\frac{(10 \text{ lbm})a_y}{32.2 \frac{\text{ft-lbm}}{\text{lbf-sec}^2}}$$

$$\sum M_{O,\text{pulley}}: \quad Tr = I\alpha$$

$$T(1.0 \text{ ft}) = (70 \text{ lbm-ft}^2)\alpha$$

$$a_y = r\alpha$$

(b) Substituting into the $\Sigma F$ equation and eliminating the tension,

$$\frac{(70 \text{ lbm-ft}^2)\alpha}{\left(32.2 \frac{\text{ft-lbm}}{\text{lbf-sec}^2}\right)(1.0 \text{ ft})} - 10 \text{ lbf} = -\frac{(10 \text{ lbm})(1.0 \text{ ft})\alpha}{32.2 \frac{\text{ft-lbm}}{\text{lbf-sec}^2}}$$

$$2.174\alpha - 10 \text{ lbf} = -0.311\alpha$$

$$\alpha = 4.02 \text{ rad/sec}^2$$

$$a_y = r\alpha = (1.0 \text{ ft})\left(4.02 \frac{\text{rad}}{\text{sec}^2}\right)$$

$$= 4.02 \text{ ft/sec}^2$$

**Example 57.10**

A 300 lbm cylinder ($I_O = 710$ lbm-ft$^2$) has a narrow groove cut in it as shown. One end of the cable is wrapped around the cylinder in the groove, while the other end supports a 200 lbm mass. The pulley is massless and frictionless, and there is no slipping. Starting from a standstill, what are the linear accelerations of the 200 lbm mass and the cylinder?

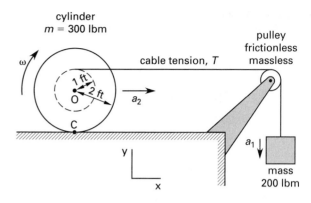

*Solution*

Since there is no slipping, there is friction between the cylinder and the plane. However, the coefficient of friction is not given. Therefore, moments must be taken about the contact point (the instantaneous center). The moment of inertia about the contact point is

$$I_C = I_O + mr^2$$

$$= 710 \text{ lbm-ft}^2 + (300 \text{ lbm})(2 \text{ ft})^2$$

$$= 1910 \text{ lbm-ft}^2$$

The first equation is a summation of forces on the mass.

$$\sum F_y: \quad T + \frac{ma_1}{g_c} - \frac{mg}{g_c} = 0$$

The second equation is a summation of moments about the instantaneous center. The frictional force passes through the instantaneous center and is disregarded.

$$\sum M_C: \quad T(2.0 \text{ ft} + 1.0 \text{ ft}) = \frac{I_C\alpha}{g_c}$$

Since there are three unknowns, a third equation is needed. This is the relationship between the linear and angular accelerations. $a_2$ is the acceleration of point O, located 2 ft from point C. $a_1$ is the acceleration of the cable, whose groove is located 3 ft from point C.

$$\alpha = \frac{a}{r} = \frac{a_2}{2.0 \text{ ft}} = \frac{a_1}{3.0 \text{ ft}}$$

$$a_1 = \tfrac{3}{2}a_2$$

Solving the three equations simultaneously yields

$$\text{cylinder: } a_2 = 10.4 \text{ ft/sec}^2$$

$$\text{mass: } a_1 = 15.6 \text{ ft/sec}^2$$

$$T = 103 \text{ lbf}$$

$$\alpha = 5.2 \text{ rad/sec}^2$$

## 21. IMPULSE

*Impulse*, **Imp**, is a vector quantity equal to the change in momentum.[16] Units of linear impulse are the same as for linear momentum: lbf-sec and N·s. Units of lbf-ft-sec and N·m·s are used for angular impulse. Equation 57.49 and Eq. 57.50 define the scalar magnitudes of *linear impulse* and *angular impulse*. Figure 57.17 illustrates that impulse is represented by the area under the *F-t* (or *T-t*) curve.

$$\text{Imp} = \int_{t_1}^{t_2} F \, dt \quad \text{[linear]} \qquad 57.49$$

$$\text{Imp} = \int_{t_1}^{t_2} T \, dt \quad \text{[angular]} \qquad 57.50$$

**Figure 57.17** *Impulse*

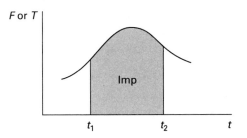

If the applied force or torque is constant, impulse is easily calculated. A large force acting for a very short period of time is known as an *impulsive force*.

$$\text{Imp} = F(t_2 - t_1) \quad \text{[linear]} \qquad 57.51$$

$$\text{Imp} = T(t_2 - t_1) \quad \text{[angular]} \qquad 57.52$$

If the impulse is known, the average force acting over the duration of the impulse is

$$F_{\text{ave}} = \frac{\text{Imp}}{\Delta t} \qquad 57.53$$

## 22. IMPULSE-MOMENTUM PRINCIPLE

The change in momentum is equal to the applied *impulse*. This is known as the *impulse-momentum* principle. For a linear system with constant force and mass, the scalar magnitude form of this principle is

$$\text{Imp} = \Delta p \qquad 57.54$$

$$F(t_2 - t_1) = m(v_2 - v_1) \qquad \text{[SI]} \quad 57.55(a)$$

$$F(t_2 - t_1) = \frac{m(v_2 - v_1)}{g_c} \qquad \text{[U.S.]} \quad 57.55(b)$$

---

[16]Although **Imp** is the most common notation, engineers have no universal symbol for impulse. Some authors use $I$, **I**, and $i$, but these symbols can be mistaken for moment of inertia. Other authors merely use the word *impulse* in their equations.

For an angular system with constant torque and moment of inertia, the analogous equations are

$$T(t_2 - t_1) = I(\omega_2 - \omega_1) \qquad \text{[SI]} \quad 57.56(a)$$

$$T(t_2 - t_1) = \frac{I(\omega_2 - \omega_1)}{g_c} \qquad \text{[U.S.]} \quad 57.56(b)$$

### Example 57.11

A 1.62 oz (0.046 kg) marble attains a velocity of 170 mph (76 m/s) in a hunting slingshot. Contact with the sling is 1/25th of a second. What is the average force on the marble during contact?

*SI Solution*

From Eq. 57.55(a), the average force is

$$F = \frac{m\Delta v}{\Delta t}$$

$$= \frac{(0.046 \text{ kg})\left(76 \frac{m}{s}\right)}{\frac{1}{25} \text{ s}}$$

$$= 87.4 \text{ N}$$

*Customary U.S. Solution*

The mass of the marble is

$$m = \frac{162 \text{ oz}}{16 \frac{\text{oz}}{\text{lbm}}}$$

$$= 0.101 \text{ lbm}$$

The velocity of the marble is

$$v = \frac{\left(170 \frac{\text{mi}}{\text{hr}}\right)\left(5280 \frac{\text{ft}}{\text{mi}}\right)}{3600 \frac{\text{sec}}{\text{hr}}}$$

$$= 249.3 \text{ ft/sec}$$

From Eq. 57.55(b), the average force is

$$F = \frac{m\Delta v}{g_c \Delta t}$$

$$= \frac{(0.101 \text{ lbm})\left(249.3 \frac{\text{ft}}{\text{sec}}\right)}{\left(32.2 \frac{\text{ft-lbm}}{\text{lbf-sec}^2}\right)\left(\frac{1}{25} \text{ sec}\right)}$$

$$= 19.5 \text{ lbf}$$

**Example 57.12**

A 2000 kg cannon fires a 10 kg projectile horizontally at 600 m/s. It takes 0.007 s for the projectile to pass through the barrel and 0.01 s for the cannon to recoil. The cannon has a spring mechanism to absorb the recoil. (a) What is the cannon's initial recoil velocity? (b) What force is exerted on the recoil spring?

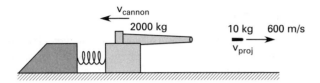

*Solution*

(a) The accelerating force is applied to the projectile quickly, and external forces such as gravity and friction are not significant factors. Therefore, momentum is conserved.

$$\sum p:\ m_{\text{proj}}\Delta v_{\text{proj}} = m_{\text{cannon}}\Delta v_{\text{cannon}}$$

$$(10\ \text{kg})\left(600\ \frac{\text{m}}{\text{s}}\right) = (2000\ \text{kg})(v_{\text{cannon}})$$

$$v_{\text{cannon}} = 3\ \text{m/s}$$

(b) From Eq. 57.55(a), the recoil force is

$$F = \frac{m\Delta v}{\Delta t} = \frac{(2000\ \text{kg})\left(3\ \frac{\text{m}}{\text{s}}\right)}{0.01\ \text{s}}$$

$$= 6 \times 10^5\ \text{N}$$

## 23. IMPULSE-MOMENTUM PRINCIPLE IN OPEN SYSTEMS

The impulse-momentum principle can be used to determine the forces acting on flowing fluids (i.e., in open systems). This is the method used to calculate forces in jet engines and on pipe bends, and forces due to other changes in flow geometry. Equation 57.57 is rearranged in terms of a mass flow rate.

$$F = \frac{m\Delta v}{\Delta t} = \dot m \Delta v \qquad \text{[SI]} \quad 57.57(a)$$

$$F = \frac{m\Delta v}{g_c \Delta t} = \frac{\dot m \Delta v}{g_c} \qquad \text{[U.S.]} \quad 57.57(b)$$

**Example 57.13**

Air enters a jet engine at 1500 ft/sec (450 m/s) and leaves at 3000 ft/sec (900 m/s). The thrust produced is 10,000 lbf (44 500 N). Disregarding the small amount of fuel added during combustion, what is the mass flow rate?

*SI Solution*

From Eq. 57.57(a),

$$\dot m = \frac{F}{\Delta v} = \frac{44\,500\ \text{N}}{900\ \frac{\text{m}}{\text{s}} - 450\ \frac{\text{m}}{\text{s}}}$$

$$= 98.9\ \text{kg/s}$$

*Customary U.S. Solution*

From Eq. 57.57(b),

$$\dot m = \frac{F g_c}{\Delta v} = \frac{(10{,}000\ \text{lbf})\left(32.2\ \frac{\text{ft-lbm}}{\text{lbf-sec}^2}\right)}{3000\ \frac{\text{ft}}{\text{sec}} - 1500\ \frac{\text{ft}}{\text{sec}}}$$

$$= 215\ \text{lbm/sec}$$

**Example 57.14**

20 kg of sand fall continuously each second on a conveyor belt moving horizontally at 0.6 m/s. What power is required to keep the belt moving?

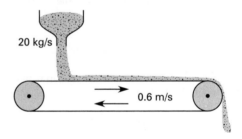

*Solution*

From Eq. 57.57(a), the force on the sand is

$$F = \dot m \Delta v = \left(20\ \frac{\text{kg}}{\text{s}}\right)\left(0.6\ \frac{\text{m}}{\text{s}}\right)$$

$$= 12\ \text{N}$$

The power required is

$$P = F v = (12\ \text{N})\left(0.6\ \frac{\text{m}}{\text{s}}\right)$$

$$= 7.2\ \text{W}$$

*Dynamics* (side tab)

**Example 57.15**

A $6 \times 9$, $^5/_8$ in diameter hoisting cable (area of 0.158 in², modulus of elasticity of $12 \times 10^6$ lbf/in²) carries a 1000 lbm load at its end. The load is being lowered vertically at the rate of 4 ft/sec. When 200 ft of cable have been reeled out, the take-up reel suddenly locks. Neglect the cable mass. What are the (a) cable stretch, (b) maximum dynamic force in the cable, (c) maximum dynamic stress in the cable, and (d) approximate time for the load to come to a stop vertically?

*Solution*

(a) The stiffness of the cable is

$$k = \frac{F}{x} = \frac{AE}{L} = \frac{(0.158\,\text{in}^2)\left(12 \times 10^6\,\dfrac{\text{lbf}}{\text{in}^2}\right)}{(200\,\text{ft})\left(12\,\dfrac{\text{in}}{\text{ft}}\right)}$$

$$= 790\,\text{lbf/in}$$

Neglecting the cable mass, the kinetic energy of the moving load is

$$E_k = \frac{mv^2}{2g_c} = \frac{(1000\,\text{lbm})\left(4\,\dfrac{\text{ft}}{\text{sec}}\right)^2\left(12\,\dfrac{\text{in}}{\text{ft}}\right)}{(2)\left(32.2\,\dfrac{\text{ft-lbm}}{\text{lbf-sec}^2}\right)}$$

$$= 2981\,\text{in-lbf}$$

By the work-energy principle, the decrease in kinetic energy is equal to the work of lengthening the cable (i.e., the energy stored in the spring).

$$\Delta E_k = \tfrac{1}{2}k\delta^2$$

$$2981\,\text{in-lbf} = \left(\tfrac{1}{2}\right)\left(790\,\frac{\text{lbf}}{\text{in}}\right)\delta^2$$

$$\delta = 2.75\,\text{in}$$

(b) The maximum dynamic force in the cable is

$$F = k\delta = \left(790\,\frac{\text{lbf}}{\text{in}}\right)(2.75\,\text{in})$$

$$= 2173\,\text{lbf}$$

(c) The maximum dynamic tensile stress in the cable is

$$\sigma = \frac{F}{A} = \frac{2173\,\text{lbf}}{0.158\,\text{in}^2}$$

$$= 13{,}753\,\text{lbf/in}^2$$

(d) Since the tensile force in the cable increases from zero to the maximum while the load decelerates, the average decelerating force is half of the maximum force. From the impulse momentum principle, Eq. 57.55,

$$F\Delta t = \frac{m\Delta v}{g_c}$$

$$\left(\tfrac{1}{2}\right)(2173\,\text{lbf})\Delta t = \frac{(1000\,\text{lbm})\left(4\,\dfrac{\text{ft}}{\text{sec}}\right)}{32.2\,\dfrac{\text{ft-lbm}}{\text{lbf-sec}^2}}$$

$$\Delta t = 0.114\,\text{sec}$$

## 24. IMPACTS

According to Newton's second law, momentum is conserved unless a body is acted upon by an external force such as gravity or friction from another object. In an *impact* or *collision*, contact is very brief and the effect of external forces is insignificant. Therefore, momentum is conserved, even though energy may be lost through heat generation and deformation of the bodies.

Consider two particles, initially moving with velocities $v_1$ and $v_2$ on a collision path, as shown in Fig. 57.18. The conservation of momentum equation can be used to find the velocities after impact, $v_1'$ and $v_2'$. (Observe algebraic signs with velocities.)

$$m_1 v_1 + m_2 v_2 = m_1 v_1' + m_2 v_2' \qquad 57.58$$

**Figure 57.18** *Direct Central Impact*

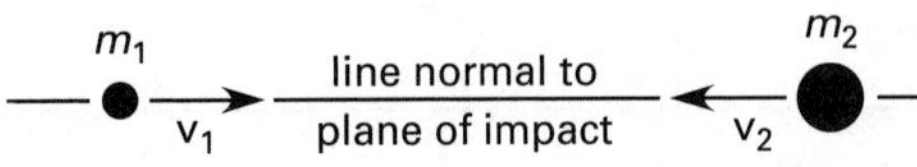

The impact is said to be an *inelastic impact* if kinetic energy is lost. (Other names for an inelastic impact are *plastic impact* and *endoergic impact*.[17]) The impact is said to be *perfectly inelastic* or *perfectly plastic* if the two particles stick together and move on with the same final velocity.[18] The impact is said to be *elastic* only if kinetic energy is conserved.

$$m_1 v_1^2 + m_2 v_2^2 = m_1 v_1'^2 + m_2 v_2'^2 \big|_{\text{elastic impact}} \qquad 57.59$$

## 25. COEFFICIENT OF RESTITUTION

A simple way to determine whether the impact is elastic or inelastic is by calculating the *coefficient of restitution*, $e$. The collision is inelastic if $e < 1.0$, perfectly inelastic if $e = 0$, and elastic if $e = 1.0$. The coefficient of restitution is the ratio of relative velocity differences along a mutual straight line. (When both impact

---

[17]Theoretically, there is also an *exoergic impact* (i.e., one in which kinetic energy is gained during the impact). However, this can occur only in special cases, such as in nuclear reactions.

[18]In traditional textbook problems, clay balls should be considered perfectly inelastic.

velocities are not directed along the same straight line, the coefficient of restitution should be calculated separately for each velocity component.)

$$e = \frac{\text{relative separation velocity}}{\text{relative approach velocity}}$$

$$= \frac{v_1' - v_2'}{v_2 - v_1} \qquad 57.60$$

## 26. REBOUND FROM STATIONARY PLANES

Figure 57.19 illustrates the case of an object rebounding from a massive, stationary plane.[19] This is an impact where $m_2 = \infty$ and $v_2 = 0$. The impact force acts perpendicular to the plane, regardless of whether the impact is elastic or inelastic. Therefore, the $x$-component of velocity is unchanged. Only the $y$-component of velocity is affected, and even then, only if the impact is inelastic.

$$v_x = v_x' \qquad 57.61$$

**Figure 57.19** Rebound from a Stationary Plane

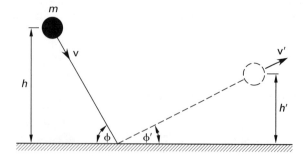

The coefficient of restitution can be used to calculate the *rebound angle, rebound height*, and *rebound velocity*.

$$e = \frac{\sin \phi'}{\sin \phi} = \sqrt{\frac{h'}{h}} = \frac{-v_y'}{v_y} \qquad 57.62$$

### Example 57.16

A golf ball dropped vertically from a height of 8.0 ft (2.4 m) onto a hard surface rebounds to a height of 6.0 ft (1.8 m). What are the (a) impact velocity, (b) rebound velocity, and (c) coefficient of restitution?

*SI Solution*

The impact velocity of the golf ball is found by equating the decrease in potential energy to the increase in kinetic energy.

$$-\tfrac{1}{2}mv^2 = mgh$$

$$v = -\sqrt{2gh} = -\sqrt{(2)\left(9.81 \, \frac{m}{s^2}\right)(2.4 \, m)}$$

$$= -6.86 \text{ m/s} \quad \text{[negative because down]}$$

The rebound velocity can be found from the rebound height.

$$v' = \sqrt{2gh'} = \sqrt{(2)\left(9.81 \, \frac{m}{s^2}\right)(1.8 \, m)}$$

$$= 5.94 \text{ m/s}$$

From Eq. 57.60 with $v_2 = v_2' = 0$ (or Eq. 57.62),

$$e = \frac{5.94 \, \frac{m}{s} - 0}{0 - \left(-6.86 \, \frac{m}{s}\right)}$$

$$= 0.866$$

*Customary U.S. Solution*

The impact velocity is

$$v = -\sqrt{2gh} = -\sqrt{(2)\left(32.2 \, \frac{ft}{sec^2}\right)(8.0 \, ft)}$$

$$= -22.7 \text{ ft/sec} \quad \text{[negative because down]}$$

Similarly, the rebound velocity is

$$v' = \sqrt{(2)\left(32.2 \, \frac{ft}{sec^2}\right)(6.0 \, ft)}$$

$$= 19.7 \text{ ft/sec}$$

From Eq. 57.60 with $v_2 = v_2' = 0$ (or from Eq. 57.62),

$$e = \frac{19.7 \, \frac{ft}{sec} - 0}{0 - \left(-22.7 \, \frac{ft}{sec}\right)}$$

$$= 0.866$$

## 27. COMPLEX IMPACTS

The simplest type of impact is the direct central impact, shown in Fig. 57.18. An impact is said to be a *direct impact* when the velocities of the two bodies are perpendicular to the contacting surfaces. *Central impact* occurs when the force of the impact is along the line of connecting centers of gravity. Round bodies (i.e., spheres) always experience central impact, whether or not the impact is direct.

---

[19]The particle path is shown as a straight line in Fig. 57.19 for convenience. The core will be a straight line only when the particle is dropped straight down. Otherwise, the path will be parabolic.

When the velocities of the bodies are not along the same line, the impact is said to be an *oblique impact*, as illustrated in Fig. 57.20. The coefficient of restitution can be used to find the $x$-components of the resultant velocities. Since impact is central, the $y$-components of velocities will be unaffected by the collision.

$$\text{v}_{1y} = \text{v}'_{1y} \qquad 57.63$$

$$\text{v}_{2y} = \text{v}'_{2y} \qquad 57.64$$

$$e = \frac{\text{v}'_{1x} - \text{v}'_{2x}}{\text{v}_{2x} - \text{v}_{1x}} \qquad 57.65$$

$$m_1\text{v}_{1x} + m_2\text{v}_{2x} = m_1\text{v}'_{1x} + m_2\text{v}'_{2x} \qquad 57.66$$

**Figure 57.20** *Central Oblique Impact*

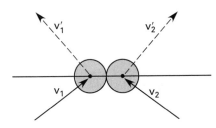

*Eccentric impacts* are neither direct nor central. The coefficient of restitution can be used to calculate the linear velocities immediately after impact along a line normal to the contact surfaces. Since the impact is not central, the bodies will rotate. Other methods must be used to calculate the rate of rotation. (See Fig. 57.21.)

$$e = \frac{\text{v}'_{1n} - \text{v}'_{2n}}{\text{v}_{2n} - \text{v}_{1n}} \qquad 57.67$$

**Figure 57.21** *Eccentric Impact*

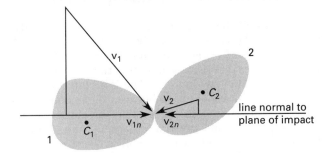

## 28. VELOCITY-DEPENDENT FORCE

A force that is a function of velocity is known as a *velocity-dependent force*. A common example of a velocity-dependent force is the *viscous drag* a particle experiences when falling through a fluid. There are two main cases of viscous drag: linear and quadratic.

A *linear velocity-dependent force* is proportional to the first power of the velocity. A linear relationship is typical of a particle falling slowly through a fluid (i.e., viscous drag in laminar flow). In Eq. 57.68, $C$ is a constant of proportionality known as the *viscous coefficient* or *coefficient of viscous damping*.

$$F_b = C\text{v} \qquad 57.68$$

In the case of a particle falling slowly through a viscous liquid, the differential equation of motion and its solution are derived from Newton's second law.

$$mg - C\text{v} = ma \qquad \text{[SI]} \quad 57.69(a)$$

$$\frac{mg}{g_c} - C\text{v} = \frac{ma}{g_c} \qquad \text{[U.S.]} \quad 57.69(b)$$

$$\text{v}(t) = \text{v}_t(1 - e^{-Ct/m}) \qquad \text{[SI]} \quad 57.70(a)$$

$$\text{v}(t) = \text{v}_t(1 - e^{-Cg_ct/m}) \qquad \text{[U.S.]} \quad 57.70(b)$$

Equation 57.70 shows that the velocity asymptotically approaches a final value known as the *terminal velocity*, $\text{v}_t$. For laminar flow, the terminal velocity is

$$\text{v}_t = \frac{mg}{C} \qquad \text{[SI]} \quad 57.71(a)$$

$$\text{v}_t = \frac{mg}{Cg_c} \qquad \text{[U.S.]} \quad 57.71(b)$$

A *quadratic velocity-dependent force* is proportional to the second power of the velocity. A quadratic relationship is typical of a particle falling quickly through a fluid (i.e., turbulent flow).

$$F_b = C\text{v}^2 \qquad 57.72$$

In the case of a particle falling quickly through a liquid under the influence of gravity, the differential equation of motion is

$$mg - C\text{v}^2 = ma \qquad \text{[SI]} \quad 57.73(a)$$

$$\frac{mg}{g_c} - C\text{v}^2 = \frac{ma}{g_c} \qquad \text{[U.S.]} \quad 57.73(b)$$

For turbulent flow, the terminal velocity is

$$\text{v}_t = \sqrt{\frac{mg}{C}} \qquad \text{[SI]} \quad 57.74(a)$$

$$\text{v}_t = \sqrt{\frac{mg}{Cg_c}} \qquad \text{[U.S.]} \quad 57.74(b)$$

If a skydiver falls far enough, a turbulent terminal velocity of approximately 125 mph (200 kph) will be achieved. With a parachute (but still in turbulent flow), the terminal velocity is reduced to approximately 25 mph (40 kph).

## 29. VARYING MASS

Integral momentum equations must be used when the mass of an object varies with time. Most varying mass problems are complex, but the simplified case of an ideal rocket can be evaluated. This discussion assumes constant gravitational force, constant fuel usage, and constant exhaust velocity. (For brevity of presentation, all of the following equations are presented in consistent form only.)

The forces acting on the rocket are its thrust, $F$, and gravity. Newton's second law is

$$F(t) - F_{\text{gravity}} = \frac{d}{dt}\big(m(t)\mathrm{v}(t)\big) \qquad 57.75$$

If $\dot{m}$ is the constant fuel usage, the thrust and gravitational forces are

$$F(t) = \dot{m}\mathrm{v}_{\text{exhaust,absolute}} = \dot{m}\Big(\mathrm{v}_{\text{exhaust}} - \mathrm{v}(t)\Big) \qquad 57.76$$

$$F_{\text{gravity}} = m(t)g \qquad 57.77$$

The velocity as a function of time is found by solving the following differential equation.

$$\dot{m}\mathrm{v}_{\text{exhaust,absolute}} - m(t)g = \dot{m}\mathrm{v}(t) + m(t)\frac{d\mathrm{v}(t)}{dt} \qquad 57.78$$

$$\mathrm{v}(t) = \mathrm{v}_0 - gt + \mathrm{v}_{\text{exhaust}}\ln\left(\frac{m_0}{m_0 - \dot{m}t}\right) \qquad 57.79$$

The final burnout *velocity*, $\mathrm{v}_f$, depends on the initial mass, $m_0$, and the final mass, $m_f$.

$$\mathrm{v}_f = \mathrm{v}_0 - g\left(\frac{m_0}{\dot{m}}\right)\left(1 - \frac{m_f}{m_0}\right) + \mathrm{v}_{\text{exhaust}}\ln\left(\frac{m_0}{m_f}\right) \qquad 57.80$$

A simple relationship exists for a rocket starting from standstill in a gravity-free environment (i.e., $\mathrm{v}_0 = 0$ and $g = 0$).

$$\frac{m_f}{m_0} = e^{-\mathrm{v}_f/\mathrm{v}_{\text{exhaust}}} \qquad 57.81$$

## 30. CENTRAL FORCE FIELDS

When the force on a particle is always directed toward or away from a fixed point, the particle is moving in a *central force field*. Examples of central force fields are gravitational fields (*inverse-square attractive fields*) and electrostatic fields (*inverse-square repulsive fields*). Particles traveling in inverse-square attractive fields can have circular, elliptical, parabolic, or hyperbolic paths. Particles traveling in inverse-square repulsive fields always travel in hyperbolic paths.

The fixed point, O in Fig. 57.22, is known as the *center of force*. The magnitude of the force depends on the distance between the particle and the center of force.

**Figure 57.22** *Motion in a Central Force Field*

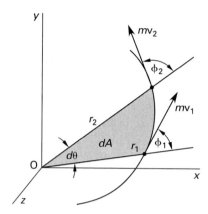

The angular momentum of a particle moving in a central force field is constant.

$$\mathbf{h}_O = \mathbf{r} \times m\mathbf{v} = \text{constant} \qquad 57.82$$

Equation 57.82 can be written in scalar form.

$$r_1 m\mathrm{v}_1 \sin\phi_1 = r_2 m\mathrm{v}_2 \sin\phi_2 \qquad 57.83$$

For a particle moving in a central force field, the *areal velocity* is constant.

$$\text{areal velocity} = \frac{dA}{dt} = \tfrac{1}{2}r^2\frac{d\theta}{dt} = \frac{h_O}{2m} \qquad 57.84$$

## 31. NEWTON'S LAW OF GRAVITATION

For a particle far enough away from a large body, gravity can be considered to be a central force field. *Newton's law of gravitation*, also known as *Newton's law of universal gravitation*, describes the force of attraction between the two masses. The law states that the attractive gravitational force between the two masses is directly proportional to the product of masses, is inversely proportional to the square of the distance between their centers of mass, and is directed along a line passing through the centers of gravity of both masses.

$$F = \frac{Gm_1 m_2}{r^2} \qquad 57.85$$

$G$ is *Newton's gravitational constant* (*Newton's universal constant*). Approximate values of $G$ for the earth are given in Table 57.2 for different sets of units. For an earth-particle combination, the product $Gm_{\text{earth}}$ has the value of $4.39 \times 10^{14}$ lbf-ft$^2$/lbm ($4.00 \times 10^{14}$ N·m$^2$/kg).

**Table 57.2** *Approximate Values of Newton's Gravitational Constant, G*

| | |
|---|---|
| $6.673 \times 10^{-11}$ | $N \cdot m^2/kg^2$ |
| $6.673 \times 10^{-8}$ | $cm^3/g \cdot s^2$ |
| $3.436 \times 30^{-8}$ | $lbf\text{-}ft^2/slug^2$ |
| $3.320 \times 10^{-11}$ | $lbf\text{-}ft^2/lbm^2$ |
| $3.436 \times 10^{-8}$ | $ft^4/lbf\text{-}sec^4$ |

## 32. KEPLER'S LAWS OF PLANETARY MOTION

Kepler's three *laws of planetary motion* are as follows.

- *Law of orbits:* The path of each planet is an ellipse with the sun at one focus.

- *Law of areas:* The radius vector drawn from the sun to a planet sweeps equal areas in equal times. (The areal velocity is constant. This is equivalent to the statement, "The angular velocity is constant.")

$$\frac{dA}{dt} = \text{constant} \qquad 57.86$$

- *Law of periods:* The square of a planet's periodic time is proportional to the cube of the semimajor axis of its orbit.

$$t_p^2 \propto a^3 \qquad 57.87$$

The *periodic time* referenced in Kepler's third law is the time required for a satellite to travel once around the parent body. If the parent body rotates in the same plane and with the same periodic time as the satellite, the satellite will always be above the same point on the parent body. This condition defines a *geostationary orbit*. Referring to Fig. 57.23, the periodic time is

$$t_p = \frac{2\pi ab}{h_0} = \frac{\pi ab}{\dfrac{dA}{dt}} = \frac{\pi ab}{\text{areal velocity}} \qquad 57.88$$

**Figure 57.23** *Planetary Motion*

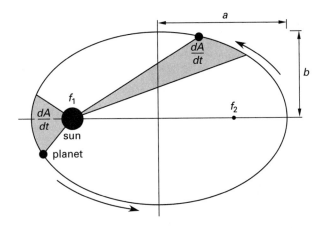

## 33. SPACE MECHANICS

Figure 57.24 illustrates the motion of a satellite that is released in a path parallel to the earth's surface. (The dotted line represents the launch phase and is not relevant to the analysis.)

**Figure 57.24** *Space Mechanics Nomenclature*

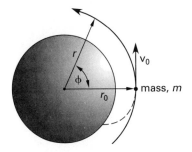

At the instant of release, the magnitude of the angular momentum, **h**, is

$$\mathbf{h} = h_0 = r_0 m v_0 \qquad 57.89$$

The force exerted on the satellite by the earth is given by Newton's law of gravitation. For any angle, $\phi$, swept out by the satellite, the separation distance can be found from Eq. 57.90. $C$ is a constant.

$$\frac{1}{r} = \frac{GM}{h^2} + C\cos\phi \qquad 57.90$$

$$C = \frac{1}{r_0} - \frac{GM}{h^2} \qquad 57.91$$

The *orbit eccentricity*, $\epsilon$, can be calculated and used to determine the type of orbit, as listed in Table 57.3.

$$\epsilon = \frac{Ch^2}{GM} \qquad 57.92$$

**Table 57.3** *Orbit Eccentricities*

| value of $\epsilon$ | type of orbit |
|---|---|
| $> 1.0$ | nonreturning hyperbola |
| $= 1.0$ | nonreturning parabola |
| $< 1.0$ | ellipse |
| $= 0.0$ | circle |

The limiting value of the initial release velocity, $v_{0,max}$, to prevent a nonreturning orbit ($\epsilon = 1.0$) is known as the *escape velocity*. For the earth, the escape velocity is approximately 7.0 mi/sec (25,000 mph, or 11.2 km/s).

$$v_{escape} = v_{0,max} = \sqrt{\frac{2GM}{r_0}} \qquad 57.93$$

The release velocity, $v_{0,circular}$, that results in a circular orbit is

$$v_{0,circular} = \sqrt{\frac{GM}{r_0}} \qquad \textit{57.94}$$

For an elliptical orbit, the minimum separation is known as the *perigee distance*. The maximum separation distance is known as the *apogee*. The terms perigee and apogee are traditionally used for earth satellites. The terms *perihelion* (closest) and *aphelion* (farthest) are used to describe distances between the sun and earth.

**Dynamics**

# 58

# Vibrating Systems

## Nomenclature

| | | | |
|---|---|---|---|
| $a$ | acceleration | ft/sec$^2$ | m/s$^2$ |
| $a$ | plate short side dimension | ft | m |
| $A$ | amplitude | ft | m |
| $A$ | amplitude | rad | rad |
| $A$ | maximum acceleration amplitude | g's | g's |
| AR | amplitude ratio | – | – |
| $b$ | plate long side dimension | ft | m |
| $C$ | coefficient of viscous damping (linear) | lbf-sec/ft | N·s/m |
| $d$ | diameter or distance | ft | m |
| $D$ | displacement | ft | m |
| $D$ | vibration parameter | ft-lbf | N·m |
| $e$ | eccentricity | ft | m |
| $E$ | energy | ft-lbf | J |
| $E$ | modulus of elasticity | lbf/ft$^2$ | Pa |
| $f$ | frequency | Hz | Hz |
| $F$ | force | lbf | N |
| $g$ | gravitational acceleration | ft/sec$^2$ | m/s$^2$ |
| $g_c$ | gravitational constant | ft-lbm/lbf-sec$^2$ | n.a. |
| $G$ | shear modulus | lbf/ft$^2$ | Pa |
| $G$ | shock transmission | g's | g's |
| $h$ | height | ft | m |
| $I$ | area moment of inertia | – | – |
| $I$ | mass moment of inertia | lbm-ft$^2$ | kg·m$^2$ |
| $J$ | polar area moment of inertia | ft$^4$ | m$^4$ |
| $k$ | spring constant | lbf/ft | N/m |
| $k_r$ | torsional spring constant | ft-lbf/rad | N·m/rad |
| $K$ | vibration constant | – | – |
| $L$ | length | ft | m |
| $m$ | mass | lbm | kg |
| $r$ | radius | ft | m |
| $r$ | ratio of forcing to natural frequency | – | – |
| $r$ | root | sec$^{-1}$ | s$^{-1}$ |
| SG | specific gravity | – | – |
| $t$ | thickness of thin plate | ft | m |
| $t$ | time | sec | s |
| $T$ | period | sec | s |
| TR | transmissibility | – | – |
| v | velocity | ft/sec | m/s |
| $w$ | load per unit area | lbf/ft$^2$ | N/m$^2$ |
| $w$ | load per unit length | lbf/ft | N/m |
| $x$ | position | ft | m |

## Symbols

| | | | |
|---|---|---|---|
| $\alpha$ | angular acceleration | rad/sec$^2$ | rad/s$^2$ |
| $\beta$ | magnification factor | – | – |
| $\delta$ | deflection | ft | m |
| $\delta$ | logarithmic decrement | – | – |
| $\zeta$ | damping ratio | – | – |
| $\eta$ | isolation efficiency | – | – |
| $\nu$ | Poisson's ratio | – | – |
| $\rho$ | density | lbm/ft$^3$ | kg/m$^3$ |
| $\phi$ | angle | rad | rad |
| $\omega$ | natural frequency | rad/sec | rad/s |

## Subscripts

| | |
|---|---|
| 0 | initial |
| $b$ | block |
| $c$ | complementary |
| $d$ | damped |
| eq | equivalent |
| $f$ | forced |
| $l$ | liquid |
| $n$ | $n$th mode |
| $p$ | particular |
| $r$ | rotational |
| st | static |

# 1. TYPES OF VIBRATIONS

*Vibration* is an oscillatory motion about an equilibrium point.[1] If the motion is the result of a disturbing force that is applied once and then removed, the motion is known as *natural* (or *free*) *vibration*. If a force of impulse is applied repeatedly to a system, the motion is known as *forced vibration*.

Within both of the categories of natural and forced vibrations are the subcategories of damped and undamped vibrations. If there is no damping (i.e., no friction), a system will experience free vibrations indefinitely. This is known as *free vibration* and *simple harmonic motion*. (See Fig. 58.1.)

**Figure 58.1** *Types of Vibrations*

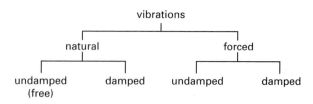

The performance (behavior) of some mechanical systems can be defined by a single variable. Such systems are referred to as *single degree of freedom (SDOF) systems*. For example, the position of a mass hanging from a spring is defined by the one variable $x(t)$.[2] Systems requiring two or more variables to define the positions of all parts are known as *multiple degree of freedom (MDOF) systems*. (See Fig. 58.2.)

**Figure 58.2** *Single and Multiple Degree of Freedom Systems*

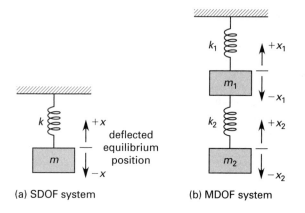

(a) SDOF system          (b) MDOF system

# 2. IDEAL COMPONENTS

When used to describe components in a vibrating system, the adjectives *perfect* and *ideal* generally imply

---

[1]Although this chapter is presented in terms of mechanical vibrations, the concepts are equally applicable to electrical, fluid, and other types of systems.
[2]Although the convention is by no means universal, the variable $x$ is commonly used as the position variable in oscillatory systems, even when the motion is in the vertical ($y$) direction.

*linearity* and the absence of friction and damping. The behavior of a *linear component* can be described by a linear equation. For example, the linear equation $F = kx$ describes a linear spring; however, the quadratic equation $F = Cv^2$ describes a nonlinear dashpot. Similarly, $F = ma$ and $F = Cv$ are linear inertial and viscous forces, respectively.

# 3. STATIC DEFLECTION

An important concept used in calculating the behavior of a vibrating system is the *static deflection*, $\delta_{st}$. This is the deflection of a mechanical system due to gravitational force alone.[3] (The disturbing force is not considered.) In calculating the static deflection, it is extremely important to distinguish between mass and weight. Figure 58.3 illustrates several cases of static deflection.

**Figure 58.3** *Examples of Static Deflection*

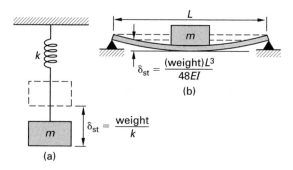

# 4. FREE VIBRATION

The simple mass and ideal spring illustrated in Fig. 58.3 is an example of a system that can experience free vibration. After the mass is displaced and released, it will oscillate up and down. Since there is no friction (i.e., the vibration is undamped), the oscillations will continue forever. (See Fig. 58.4.)

**Figure 58.4** *Free Vibration*

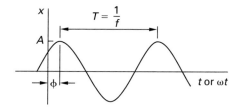

The system is initially at rest. The mass is hanging on the spring, and the *equilibrium position* is at the position of static deflection. The system is disturbed by a downward force (i.e., the mass is pulled downward from its static deflection and released).

After the initial disturbing force is removed, the object will be acted upon by the restoring force $(-kx)$ and the

---

[3]The term *deformation* is used synonymously with *deflection*.

inertial force $(-ma)$. Both of these forces are proportional to the displacement from the equilibrium point, and they are opposite in sign from the displacement. From D'Alembert's principle,

$$\sum F = 0: -ma - kx = 0$$

$$m\frac{d^2x}{dt^2} = -kx \qquad \text{[SI]} \quad 58.1(a)$$

$$\frac{m}{g_c}\frac{d^2x}{dt^2} = -kx \qquad \text{[U.S.]} \quad 58.1(b)$$

The solution to this second-order differential equation is easily derived. $x_0$ and $v_0$ are the initial displacement and initial velocity of the object, respectively.

$$x(t) = x_0 \cos \omega t + \left(\frac{v_0}{\omega}\right)\sin \omega t \qquad 58.2$$

$\omega$ is known as the *angular natural frequency of vibration*. It has units of radians per second. It is not the same as the *linear natural frequency*, $f$, which has units of hertz (formerly known as cycles per second). The *period of oscillation*, $T$, is the reciprocal of the linear frequency.

$$\omega = \sqrt{\frac{k}{m}} \qquad \text{[SI]} \quad 58.3(a)$$

$$\omega = \sqrt{\frac{kg_c}{m}} \qquad \text{[U.S.]} \quad 58.3(b)$$

$$f = \frac{\omega}{2\pi} = \frac{1}{T} \qquad 58.4$$

$$T = \frac{1}{f} = \frac{2\pi}{\omega} \qquad 58.5$$

Equation 58.3 can be written in terms of the weight of the object suspended from the spring or in terms of the static deflection.

$$\text{weight} = mg \qquad \text{[SI]} \quad 58.6(a)$$

$$\text{weight} = \frac{mg}{g_c} \qquad \text{[U.S.]} \quad 58.6(b)$$

$$\omega = \sqrt{\frac{kg}{\text{weight}}} = \sqrt{\frac{g}{\delta_{st}}} \qquad 58.7$$

Equation 58.7 is extremely useful. It is not limited to the simple mass-on-a-spring arrangement that is shown in Fig. 58.3(a). It can be used with a variety of systems, including those involving beams, shafts, and plates. Example 58.2 illustrates how this is done.

Equation 58.8 is an alternate form of the solution to Eq. 58.1. $A$ is the *amplitude* and $\phi$ is the phase angle.

$$x(t) = A\cos(\omega t - \phi) \qquad 58.8$$

$$A = \sqrt{x_0^2 + \left(\frac{v_0}{\omega}\right)^2} \qquad 58.9$$

$$\phi = \arctan\left(\frac{v_0}{\omega x_0}\right) \qquad 58.10$$

The position, velocity, and acceleration are all sinusoidal with time. The maximum values are

$$x_{max} = A \qquad 58.11$$

$$v_{max} = A\omega \qquad 58.12$$

$$a_{max} = A\omega^2 \qquad 58.13$$

## 5. INITIAL CONDITIONS

With natural, undamped vibrations, the initial conditions (i.e., initial position and velocity) do not affect the natural period of oscillation. However, the amplitude of the oscillations will be affected, as indicated by Eq. 58.9.

### Example 58.1

A 120 lbm (54 kg) mass is supported by three springs as shown. The initial displacement is 2.0 in (5.0 cm) downward from the static equilibrium position. No external forces act on the mass after it is released. What are the maximum velocity and acceleration?

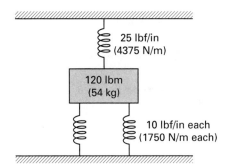

25 lbf/in
(4375 N/m)

120 lbm
(54 kg)

10 lbf/in each
(1750 N/m each)

*SI Solution*

Since the springs are in parallel, they all share the applied load. The equivalent spring constant is

$$k_{eq} = k_1 + k_2 + k_3$$

$$= 4375 \ \frac{N}{m} + 1750 \ \frac{N}{m} + 1750 \ \frac{N}{m} = 7875 \ N/m$$

The static deflection is

$$\delta_{st} = \frac{\text{weight}}{k} = \frac{mg}{k}$$

$$= \frac{(54 \ \text{kg})\left(9.81 \ \frac{m}{s^2}\right)}{7875 \ \frac{N}{m}} = 0.0673 \ m$$

The natural frequency is given by Eq. 58.7. (Compare this to the value calculated from Eq. 58.3.)

$$\omega = \sqrt{\frac{g}{\delta_{st}}} = \sqrt{\frac{9.81 \frac{m}{s^2}}{0.0673 \ m}}$$

$$= 12.07 \ rad/s$$

Since the mass is pulled down and released, the initial conditions are

$$v_0 = 0$$

$$x_0 = 5.0 \ cm \quad (0.05 \ m)$$

From Eq. 58.9, the amplitude of oscillation is $A = 0.05$ m. From Eq. 58.12 and Eq. 58.13, the maximum velocity and acceleration are

$$v_{max} = A\omega = (0.05 \ m)\left(12.07 \ \frac{rad}{s}\right)$$

$$= 0.604 \ m/s$$

$$a_{max} = A\omega^2 = (0.05 \ m)\left(12.07 \ \frac{rad}{s}\right)^2$$

$$= 7.28 \ m/s^2$$

(Radians are dimensionless.)

*Customary U.S. Solution*

The equivalent spring constant is

$$k_{eq} = 25 \ \frac{lbf}{in} + 10 \ \frac{lbf}{in} + 10 \ \frac{lbf}{in} = 45 \ lbf/in$$

Referring to Fig. 58.3 and Eq. 58.6, the static deflection is

$$\delta_{st} = \frac{weight}{k} = \frac{mg}{kg_c}$$

$$= \frac{(120 \ lbm)\left(32.2 \ \frac{ft}{sec^2}\right)}{\left(45 \ \frac{lbf}{in}\right)\left(32.2 \ \frac{ft\text{-}lbm}{lbf\text{-}sec^2}\right)} = 2.67 \ in$$

The natural frequency is given by Eq. 58.7.

$$\omega = \sqrt{\frac{g}{\delta_{st}}} = \sqrt{\frac{\left(32.2 \ \frac{ft}{sec^2}\right)\left(12 \ \frac{in}{ft}\right)}{2.67 \ in}}$$

$$= 12.03 \ rad/sec$$

Since the mass is pulled down and released, the initial conditions are

$$v_0 = 0$$

$$x_0 = \frac{2 \ in}{12 \ \frac{in}{ft}} = 0.167 \ ft$$

From Eq. 58.9, the amplitude of oscillation is $A = 0.167$ ft. From Eq. 58.12 and Eq. 58.13, the maximum velocity and acceleration are

$$v_{max} = A\omega = (0.167 \ ft)\left(12.03 \ \frac{rad}{sec}\right)$$

$$= 2.0 \ ft/sec$$

$$a_{max} = A\omega^2 = (0.167 \ ft)\left(12.03 \ \frac{rad}{sec}\right)^2$$

$$= 24.2 \ ft/sec^2$$

(Radians are dimensionless.)

**Example 58.2**

A diving board is supported by a frictionless pivot at one end and by an unyielding, frictionless fulcrum, as indicated. A diver of mass $m$ stands at the free end and bounces up and down. What is the frequency of oscillation?

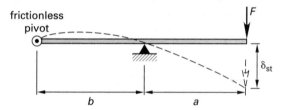

*Solution*

The deflection curve of the beam is shown by the dotted line. The tip force is

$$F = mg \quad [SI]$$

$$F = \frac{mg}{g_c} \quad [U.S.]$$

Use standard beam tables to determine the deflection. If the diver were to stand perfectly still, the static deflection at the tip would be

$$\delta_{st} = \frac{Fa^2(a+b)}{3EI}$$

From Eq. 58.4 and Eq. 58.7, the linear natural frequency is

$$f = \frac{1}{2\pi}\sqrt{\frac{g}{\delta_{st}}} = \frac{1}{2\pi}\sqrt{\frac{3EIg}{Fa^2(a+b)}}$$

## 6. VERTICAL VERSUS HORIZONTAL OSCILLATION

As long as friction is absent, the two cases of oscillation shown in Fig. 58.5 are equivalent (i.e., will have the same frequency and amplitude). Although it may seem that there is an extra gravitational force with vertical motion, the weight of the body is completely canceled by the opposite spring force when the system is in equilibrium. Therefore, vertical oscillations about an equilibrium point are equivalent to horizontal oscillations about the unstressed point.

**Figure 58.5** *Vertical and Horizontal Oscillations*

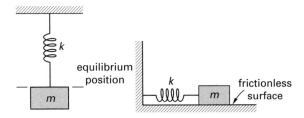

## 7. CONSERVATION OF ENERGY

The conservation of energy in vibrating systems requires the kinetic energy at the static equilibrium position to equal the stored elastic energy at the position of maximum displacement. For the mass-spring system shown in Fig. 58.5, the energy conservation equation is

$$\tfrac{1}{2}kx_{max}^2 = \tfrac{1}{2}mv_{max}^2 \qquad \text{[SI]} \quad 58.14(a)$$

$$\tfrac{1}{2}kx_{max}^2 = \frac{mv_{max}^2}{2g_c} \qquad \text{[U.S.]} \quad 58.14(b)$$

The velocity function is derived by taking the derivative of the position function.

$$x(t) = x_{max}\sin\omega t \qquad\qquad 58.15$$

$$v(t) = \frac{dx(t)}{dt} = \omega x_{max}\cos\omega t \qquad 58.16$$

Equation 58.16 shows that $v_{max} = \omega x_{max}$. Substituting this into Eq. 58.14 derives the natural frequency of vibration.

$$\omega^2 = \frac{k}{m} \qquad\qquad \text{[SI]} \quad 58.17(a)$$

$$\omega^2 = \frac{kg_c}{m} \qquad\qquad \text{[U.S.]} \quad 58.17(b)$$

## 8. FREE ROTATION

The so-called *torsional pendulum* in Fig. 58.6 can be analyzed in a manner similar to the spring-mass combination. Ignoring the mass and moment of inertia of the shaft, the differential equation is

$$-k_r\phi = I\frac{d^2\phi}{dt^2} \qquad\qquad \text{[SI]} \quad 58.18(a)$$

$$-k_r\phi = \frac{I}{g_c}\frac{d^2\phi}{dt^2} \qquad\qquad \text{[U.S.]} \quad 58.18(b)$$

The *torsional spring constant*, $k_r$, used in Eq. 58.18 is

$$k_r = \frac{GJ}{L} = \frac{\pi d^4 G}{32L} \qquad\qquad 58.19$$

If the shaft is connected to the support through a torsional spring, or if the shaft consists of several sections of different diameters (i.e., a *stepped shaft*), the equivalent torsional spring constant can be calculated in the same manner as for springs in series.

$$\frac{1}{k_{eq}} = \frac{1}{k_{r1}} + \frac{1}{k_{r2}} + \cdots \qquad 58.20$$

**Figure 58.6** *Torsional Pendulum*

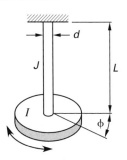

The solution to Eq. 58.18 is directly analogous to the solution for the spring-mass system. Equation 58.21 through Eq. 58.27 summarize the governing equations.

$$\phi(t) = \phi_0\cos\omega t + \left(\frac{\omega_0}{\omega}\right)\sin\omega t = A\cos(\omega t - \phi) \qquad 58.21$$

$$\omega = \sqrt{\frac{k_r}{I}} \qquad\qquad \text{[SI]} \quad 58.22(a)$$

$$\omega = \sqrt{\frac{k_r g_c}{I}} \qquad\qquad \text{[U.S.]} \quad 58.22(b)$$

$$A = \sqrt{\phi_0^2 + \left(\frac{\omega_0}{\omega}\right)^2} \qquad\qquad 58.23$$

$$\phi = \arctan\left(\frac{\omega_0}{\omega\phi_0}\right) \qquad\qquad 58.24$$

$$\phi_{max} = A \qquad\qquad 58.25$$

$$\omega_{max} = A\omega \qquad\qquad 58.26$$

$$\alpha_{max} = A\omega^2 \qquad\qquad 58.27$$

## 9. SUMMARY OF FREE VIBRATION PERFORMANCE EQUATIONS

Most equations of motion can be easily derived for free vibrations without damping. Table 58.1 provides a convenient summary of several common cases.

## 10. RAYLEIGH'S METHOD

Usually, the mass of the spring (beam, bar, shaft, etc.) is disregarded when calculating the frequency or period of vibration of a simple system. This is done to simplify the solution, although the mass of the spring element actually does affect the frequency. The exact solution is generally complex, but *Rayleigh's method* can be used to derive answers that will usually be less than 5% in error.

**Table 58.1** *Performance of Simple Oscillatory Systems (small deflections; consistent units)*[*]

| mechanism | natural frequency ($\omega$) | linear frequency ($f$) | period ($T$) |
|---|---|---|---|
| mass and spring | $\sqrt{\dfrac{k}{m}}$ | $\dfrac{1}{2\pi}\sqrt{\dfrac{k}{m}}$ | $2\pi\sqrt{\dfrac{m}{k}}$ |
| mass on massless beam ($I$ = area moment of inertia of cross section) | $\sqrt{\dfrac{48EI}{mL^3}}$ | $\dfrac{1}{2\pi}\sqrt{\dfrac{48EI}{mL^3}}$ | $2\pi\sqrt{\dfrac{mL^3}{48EI}}$ |
| constrained compound pendulum (massless bar, frictionless pivot) | $\sqrt{\dfrac{mgL + kd^2}{mL^2}}$ | $\dfrac{1}{2\pi}\sqrt{\dfrac{mgL + kd^2}{mL^2}}$ | $2\pi\sqrt{\dfrac{mL^2}{mgL + kd^2}}$ |
| simple pendulum | $\sqrt{\dfrac{g}{L}}$ | $\dfrac{1}{2\pi}\sqrt{\dfrac{g}{L}}$ | $2\pi\sqrt{\dfrac{L}{g}}$ |
| compound pendulum | $\sqrt{\dfrac{mgd}{I_0}}$ | $\dfrac{1}{2\pi}\sqrt{\dfrac{mgd}{I_0}}$ | $2\pi\sqrt{\dfrac{I_0}{mgd}}$ |
| conical pendulum | $\sqrt{\dfrac{g}{h}}$ | $\dfrac{1}{2\pi}\sqrt{\dfrac{g}{h}}$ | $2\pi\sqrt{\dfrac{h}{g}}$ |
| constrained compound pendulum | $\sqrt{\dfrac{kd^2}{mL^2}}$ | $\dfrac{1}{2\pi}\sqrt{\dfrac{kd^2}{mL^2}}$ | $2\pi\sqrt{\dfrac{mL^2}{kd^2}}$ |

**Dynamics**

| mechanism | natural frequency ($\omega$) | linear frequency ($f$) | period ($T$) |
|---|---|---|---|
| two masses and spring | $\sqrt{\dfrac{k(m_1 + m_2)}{m_1 m_2}}$ | $\dfrac{1}{2\pi}\sqrt{\dfrac{k(m_1 + m_2)}{m_1 m_2}}$ | $2\pi\sqrt{\dfrac{m_1 m_2}{k(m_1 + m_2)}}$ |
| torsional mass and spring | $\sqrt{\dfrac{JG}{I_0 L}}$ | $\dfrac{1}{2\pi}\sqrt{\dfrac{JG}{I_0 L}}$ | $2\pi\sqrt{\dfrac{I_0 L}{JG}}$ |
| two torsional masses | $\sqrt{\dfrac{JG(I_1 + I_2)}{I_1 I_2 L}}$ | $\dfrac{1}{2\pi}\sqrt{\dfrac{JG(I_1 + I_2)}{I_1 I_2 L}}$ | $2\pi\sqrt{\dfrac{I_1 I_2 L}{JG(I_1 + I_2)}}$ |
| oscillating pulley | $\sqrt{\dfrac{k}{m_1 + \dfrac{m_2}{2}}}$ | $\dfrac{1}{2\pi}\sqrt{\dfrac{k}{m_1 + \dfrac{m_2}{2}}}$ | $2\pi\sqrt{\dfrac{m_1 + \dfrac{m_2}{2}}{k}}$ |
| floating block | $\sqrt{\dfrac{g(\mathrm{SG}_l)}{L(\mathrm{SG}_b)}}$ | $\dfrac{1}{2\pi}\sqrt{\dfrac{g(\mathrm{SG}_l)}{L(\mathrm{SG}_b)}}$ | $2\pi\sqrt{\dfrac{L(\mathrm{SG}_b)}{g(\mathrm{SG}_l)}}$ |
| U-tube and liquid | $\sqrt{\dfrac{2g}{L}}$ | $\dfrac{1}{2\pi}\sqrt{\dfrac{2g}{L}}$ | $2\pi\sqrt{\dfrac{L}{2g}}$ |
| cantilever spring | $\sqrt{\dfrac{3EI}{mL^3}}$ | $\dfrac{1}{2\pi}\sqrt{\dfrac{3EI}{mL^3}}$ | $2\pi\sqrt{\dfrac{mL^3}{3EI}}$ |

*Replace $m$ with $m/g_c$ for customary U.S. units. Replace $I_o$ with $I_o/g_c$ for customary U.S. units.

Rayleigh's method is to increase the oscillating object's mass by a fraction of the spring mass.

- For spring-mass systems, add $1/3$ of the spring mass to the oscillating object mass.

- For simply supported beams loaded at the center, add $17/35$ of the beam mass to the carried mass.

- For cantilever beams loaded at the free end, add $33/140$ of the beam mass to the carried mass.

- For circular shafts in torsion, add $1/3$ of the shaft mass moment of inertia to the mass moment of inertia of the rotating load.

## 11. TRANSFORMED LINEAR STIFFNESS

Transformers are used in electrical and electronic circuits to convert one voltage to another. The analogous mechanical device is the lever. Figure 58.7 illustrates a simple lever-spring system. The pivot point is frictionless; the springs are ideal; the lever is infinitely stiff and has negligible mass.

**Figure 58.7** *Lever-Coupled Linear System*

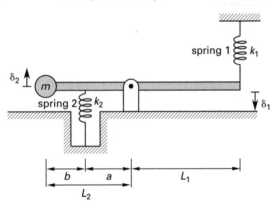

(a) original system

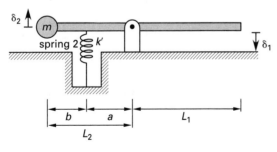

(b) equivalent system

The *lever ratio* is

$$\frac{L_1}{L_2} = \frac{\delta_1}{\delta_2} \qquad 58.28$$

In ideal electrical circuits, the energies per unit time (i.e., the power) transferred across the two transformer

windings are equal. In ideal mechanical systems, energy is similarly conserved. Specifically, the change in a mass' gravitational potential energy is equal to the change in stored spring energy. Referring to Fig. 58.7(a), a unit deflection of the mass $(\delta_2 = 1)$ will result in stored spring energies such that the total energy change is zero. Equation 58.29 disregards internal energy.

$$\frac{\Delta E}{\delta_2} = \tfrac{1}{2}k_2\left(\frac{a}{L_2}\right)^2 + \tfrac{1}{2}k_1\left(\frac{L_1}{L_2}\right)^2 - mg = 0 \qquad \text{[SI]} \quad 58.29(a)$$

$$\frac{\Delta E}{\delta_2} = \tfrac{1}{2}k_2\left(\frac{a}{L_2}\right)^2 + \tfrac{1}{2}k_1\left(\frac{L_1}{L_2}\right)^2 - \frac{mg}{g_c} = 0 \qquad \text{[U.S.]} \quad 58.29(b)$$

An equivalent system is shown in Fig. 58.7(b). All of the spring force is concentrated at the position of spring 2. The angular frequency of the mass is unchanged. The equivalent spring constant, $k'$, referred to spring 2, is

$$k' = k_2 + \left(\frac{L_1}{a}\right)^2 k_1 \qquad 58.30$$

Another transformation (not shown) is not as immediately useful. An equivalent spring constant, referred to spring 1, could be derived. This would have the effect of concentrating all of the spring force at the position of spring 1.

$$k'' = k_1 + \left(\frac{a}{L_1}\right)^2 k_2 \qquad 58.31$$

## 12. TRANSFORMED ANGULAR STIFFNESS

In a meshing gear set, the transmitted force and power are the same for both meshing gears. However, a gear set changes the torque. Therefore, a meshing gear set is a torsional analog to an electrical transformer.

Consider the loaded gear set shown in Fig. 58.8. The masses of shafts 1 and 2 are insignificant compared with the torsional loads, $I_1$ and $I_2$. Also, the tooth stiffness is typically disregarded in analyzing this type of system. Power transfer across the gear set is without loss.

The original system can be converted to the equivalent simple, torsional system shown in Fig. 58.8(b). The equivalent load $I_2'$ of shaft 2 (referred to shaft 1) is

$$I_2' = \left(\frac{\omega_2}{\omega_1}\right)^2 I_2 = \left(\frac{N_1}{N_2}\right)^2 I_2$$

$$= \left(\frac{d_1}{d_2}\right)^2 I_2 \qquad 58.32$$

The equivalent torsional stiffness of shaft 2 is

$$k_2' = \left(\frac{N_1}{N_2}\right)^2 k_2 = \left(\frac{d_1}{d_2}\right)^2 k_2 \qquad 58.33$$

**Figure 58.8** *Gear-Coupled Torsional System*

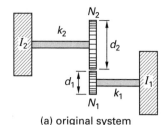

(a) original system

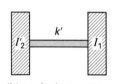

(b) equivalent system

The total equivalent torsional stiffness, $k'$, is

$$\frac{1}{k'} = \frac{1}{k_1} + \frac{1}{k_2'} \qquad 58.34$$

The natural frequency is

$$\omega = \sqrt{k'\left(\frac{I_1 + I_2'}{I_1 I_2'}\right)} \qquad 58.35$$

## 13. DAMPED FREE VIBRATIONS

When friction resists the oscillatory motion, the system is said to be damped. Friction can occur internally, between two surfaces, or due to motion through a liquid. (Air friction can be disregarded in most problems.) The third type of friction is known as *viscous damping*. Figure 58.9 illustrates the *dashpot* symbol used to represent a source of viscous damping.

**Figure 58.9** *Spring-Mass System with Dashpot*

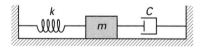

The viscous damping force can be a function of v or $v^2$. If velocity is high through the liquid, the viscous damping force will be a function of $v^2$. Only low-velocity, *linear damping* is covered in this chapter. With linear damping, the damping force is proportional to velocity. The constant of proportionality, $C$, is also known as the *coefficient of viscous damping*.

$$F = Cv = C\frac{dx}{dt} \qquad 58.36$$

The differential equation of motion is

$$m\frac{d^2x}{dt^2} = -kx - C\frac{dx}{dt} \qquad [\text{SI}] \quad 58.37$$

The general solution is given by Eq. 58.38. The constants $A$ and $B$ must be determined from initial conditions.

$$x(t) = Ae^{r_1 t} + Be^{r_2 t} \qquad 58.38$$

The roots of Eq. 58.38 are

$$r_1, r_2 = \frac{-C}{2m} \pm \sqrt{\left(\frac{C}{2m}\right)^2 - \frac{k}{m}} \qquad [\text{SI}] \quad 58.39(a)$$

$$r_1, r_2 = g_c\left(\frac{-C}{2m} \pm \sqrt{\left(\frac{C}{2m}\right)^2 - \frac{k}{mg_c}}\right) \qquad [\text{U.S.}] \quad 58.39(b)$$

The *damping ratio (damping factor)*, $\zeta$, is defined as

$$\zeta = \frac{C}{2m\omega} = \frac{C}{2\sqrt{mk}} = \frac{C}{C_{\text{critical}}} \qquad [\text{SI}] \quad 58.40(a)$$

$$\zeta = \frac{Cg_c}{2m\omega} = \frac{C}{2\sqrt{\dfrac{mk}{g_c}}} = \frac{C}{C_{\text{critical}}} \qquad [\text{U.S.}] \quad 58.40(b)$$

If $\zeta < 1.0$ (i.e., $C < C_{\text{critical}}$), the radical in Eq. 58.39 will be negative, and both roots will be imaginary. The oscillations are said to be *underdamped*. This case is also known as *light damping*. Motion will be oscillatory with diminishing magnitude, as illustrated in Fig. 58.10.

**Figure 58.10** *Underdamped Oscillation*

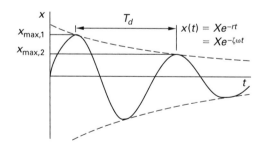

$\omega_d$ is the *damped frequency*. It is not the same as the natural frequency, $\omega$, which is calculated assuming that $C = 0$.

$$\omega_d = \omega\sqrt{1 - \zeta^2} \qquad 58.41$$

The *logarithmic decrement*, $\delta$, is the natural logarithm of the ratio of two successive amplitudes.

$$\delta = \ln \frac{x_n}{x_{n+1}} = \frac{2\pi\zeta}{\sqrt{1-\zeta^2}} \qquad 58.42$$

$$= \zeta\omega T_d$$

If $\zeta > 1.0$ (i.e., $C > C_{\text{critical}}$), the radical is positive, and both roots are real. The motion is said to be *overdamped*. This case is also known as *heavy damping*. There will be a gradual return to the equilibrium, but no oscillation. (See Fig. 58.11.)

**Figure 58.11** *Overdamped Oscillation*

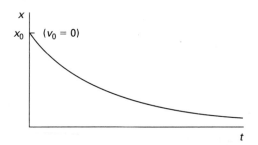

If $\zeta = 1.0$ (i.e., $C = C_{\text{critical}}$), the radical is zero, and the motion is said to be *critically damped*. (See Fig. 58.12.) Such motion is also known as *dead-beat motion*. There is no overshoot, and the return is the fastest of the three types of damped motion. The *critical damping coefficient* is

$$C_{\text{critical}} = 2m\omega = 2\sqrt{km} = \frac{C}{\zeta} \qquad \text{[SI]} \quad 58.43(a)$$

$$C_{\text{critical}} = \frac{2m\omega}{g_c} = 2\sqrt{\frac{km}{g_c}} = \frac{C}{\zeta} \qquad \text{[U.S.]} \quad 58.43(b)$$

**Figure 58.12** *Critically Damped Oscillation*

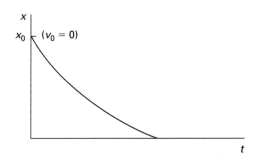

## 14. UNDAMPED FORCED VIBRATIONS

When an external disturbing force, $F(t)$, acts on the system, the system is said to be forced. Although the *forcing function* is usually considered to be periodic, it

need not be (as in the case of impulse, step, and random functions).[4] However, an initial disturbance (i.e., when a mass is displaced and released to oscillate freely) is not an example of forced vibration. (See Fig. 58.13.)

**Figure 58.13** *Forced Vibrations*

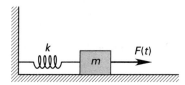

Consider a sinusoidal periodic force with a *forcing frequency* of $\omega_f$ and maximum value of $F_0$.

$$F(t) = F_0 \cos \omega_f t \qquad 58.44$$

The differential equation of motion is

$$m\frac{d^2x}{dt^2} = -kx + F_0 \cos \omega_f t \qquad \text{[SI]} \quad 58.45$$

The solution to Eq. 58.45 consists of the sum of two parts: a complementary solution and a particular solution. The *complementary solution* is obtained by setting $F_0 = 0$ (i.e., solving the homogeneous differential equation). The solution is

$$x_c(t) = A\cos\omega t + B\sin\omega t \qquad 58.46$$

The *particular solution* is found by assuming its form and substituting that function into Eq. 58.47.

$$x_p(t) = D\cos\omega_f t \qquad 58.47$$

$$D = \frac{F_0}{m(\omega^2 - \omega_f^2)} \qquad \text{[SI]} \quad 58.48(a)$$

$$D = \frac{F_0 g_c}{m(\omega^2 - \omega_f^2)} \qquad \text{[U.S.]} \quad 58.48(b)$$

The solution of Eq. 58.45 is

$$x(t) = A\cos\omega t + B\sin\omega t$$
$$+ \left(\frac{F_0}{m(\omega^2 - \omega_f^2)}\right)\cos\omega_f t \qquad \text{[SI]} \quad 58.49(a)$$

$$x(t) = A\cos\omega t + B\sin\omega t$$
$$+ \left(\frac{F_0 g_c}{m(\omega^2 - \omega_f^2)}\right)\cos\omega_f t \qquad \text{[U.S.]} \quad 58.49(b)$$

---

[4]The sinusoidal case is important, since Fourier transforms can be used to model any forcing function in terms of sinusoids.

## 15. MAGNIFICATION FACTOR

The *magnification factor*, $\beta$ (also known as the *amplitude ratio* and *amplification factor*), is defined as the ratio of the steady-state vibration amplitude, $D$, and the *pseudo-static deflection*, $F_0/k$.

$$\beta = \frac{D}{\frac{F_0}{k}} = \left| \frac{1}{1 - \left(\frac{\omega_f}{\omega}\right)^2} \right| \qquad 58.50$$

Figure 58.14 illustrates the magnification factor for various values of $\omega_f/\omega$. *Resonance* occurs when $\omega_f$ equals or nearly equals $\omega$. Oscillations are theoretically infinite.[5] Resonance leads to rapid failure in structures and mechanical equipment.

**Figure 58.14** *Magnification Factor (no damping)*

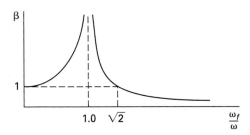

If the ratio $\omega_f/\omega = \sqrt{2}$, the magnification factor is

$$\beta = \left| \frac{1}{1 - (\sqrt{2})^2} \right| = 1 \qquad 58.51$$

Therefore, $\omega_f/\omega$ must be greater than $\sqrt{2}$ for the system to have an oscillation magnitude smaller than the static deflection alone.

When $\omega_f$ is significantly greater than $\omega$, the magnification factor is close to zero, and the system will be nearly stationary.

### Example 58.3

A 250 lbm (113.6 kg) motor turns at a rate of 1000 rpm (16.66 rps). It is mounted on a resilient pad having a stiffness of 3000 lbf/in (525 kN/m). Due to an unbalanced condition, a periodic force of 20 lbf (89 N) is applied in the vertical direction once each revolution. If the motor is constrained to move vertically and damping is negligible, what is the amplitude of vibration?

---

[5]Damping is always present in real systems and keeps the excursions finite.

*SI Solution*

The natural frequency of the system is

$$\omega = \sqrt{\frac{k}{m}} = \sqrt{\frac{\left(525 \ \frac{kN}{m}\right)\left(1000 \ \frac{N}{kN}\right)}{113.6 \ kg}}$$
$$= 67.98 \ rad/s$$

The forcing frequency is

$$\omega_f = \left(16.66 \ \frac{rev}{s}\right)\left(2\pi \ \frac{rad}{rev}\right) = 104.7 \ rad/s$$

The pseudo-static deflection is

$$\frac{F_0}{k} = \frac{89 \ N}{525\,000 \ \frac{N}{m}} = 1.70 \times 10^{-4} \ m$$

The magnification factor is

$$\beta = \left| \frac{1}{1 - \left(\frac{\omega_f}{\omega}\right)^2} \right| = \left| \frac{1}{1 - \left(\frac{104.7 \ \frac{rad}{s}}{67.98 \ \frac{rad}{s}}\right)^2} \right|$$
$$= 0.729$$

The amplitude of oscillation is calculated from Eq. 58.50.

$$D = \beta\left(\frac{F_0}{k}\right) = (0.729)(1.70 \times 10^{-4} \ m)$$
$$= 1.24 \times 10^{-4} \ m$$

*Customary U.S. Solution*

The natural frequency of the system is

$$\omega = \sqrt{\frac{kg_c}{m}} = \sqrt{\frac{\left(3000 \ \frac{lbf}{in}\right)\left(32.2 \ \frac{ft\text{-}lbm}{lbf\text{-}sec^2}\right)\left(12 \ \frac{in}{ft}\right)}{250 \ lbm}}$$
$$= 68.09 \ rad/sec$$

The forcing frequency is

$$\omega_f = \frac{\left(1000 \ \frac{rev}{min}\right)\left(2\pi \ \frac{rad}{rev}\right)}{60 \ \frac{sec}{min}} = 104.7 \ rad/sec$$

The pseudo-static deflection is

$$\frac{F_0}{k} = \frac{20 \ lbf}{3000 \ \frac{lbf}{in}} = 0.00667 \ in$$

Dynamics

The magnification factor is

$$\beta = \left| \frac{1}{1 - \left(\frac{\omega_f}{\omega}\right)^2} \right| = \left| \frac{1}{1 - \left(\frac{104.7 \ \frac{\text{rad}}{\text{sec}}}{68.09 \ \frac{\text{rad}}{\text{sec}}}\right)^2} \right|$$

$$= 0.733$$

The amplitude of oscillation is calculated from Eq. 58.50.

$$D = \beta \left(\frac{F_0}{k}\right) = (0.733)(0.00667 \text{ in})$$

$$= 0.00489 \text{ in}$$

## 16. DAMPED FORCED VIBRATIONS

If a viscous damping force, $Cv = C(dx/dt)$, is added to a sinusoidally forced system, as in Fig. 58.15, the differential equation of motion is

$$m\frac{d^2x}{dt^2} = -kx - C\frac{dx}{dt} + F_0 \cos\omega_f t \quad \text{[SI]} \quad 58.52(a)$$

$$\frac{m}{g_c}\frac{d^2x}{dt^2} = -kx - C\frac{dx}{dt} + F_0 \cos\omega_f t \quad \text{[U.S.]} \quad 58.52(b)$$

**Figure 58.15** Damped Forced Oscillations

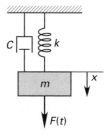

The solution to Eq. 58.52 has several terms. As a result of the damping force, the complementary solution has decaying exponentials. Therefore, the complementary solution is also known as the *transient component* because its contribution to the system performance decreases rapidly. However, the transient terms do contribute to the initial performance. For this reason, initial cycles may experience displacements greater than the steady-state values. The particular solution is known as the *steady-state component*.

Equation 58.53 defines the damped magnification factor, $\beta_d$, for steady-state damped forced vibrations. (See

Fig. 58.16.) The magnification factor for the undamped case (see Eq. 58.50) can be derived by setting $\zeta = 0$.

$$\beta_d = \left| \frac{D}{\frac{F_0}{k}} \right| = \frac{1}{\sqrt{\left(1 - \left(\frac{\omega_f}{\omega}\right)^2\right)^2 + \left(\frac{C\omega_f}{m\omega^2}\right)^2}} \quad \text{[SI only]}$$

$$= \frac{1}{\sqrt{\left(1 - \left(\frac{\omega_f}{\omega}\right)^2\right)^2 + \left(\frac{2C\omega_f}{C_{\text{critical}}\omega}\right)^2}} \quad \text{[U.S. and SI]}$$

$$= \frac{1}{\sqrt{(1 - r^2)^2 + (2\zeta r)^2}} \quad \text{[U.S. and SI]}$$

$$58.53$$

$$r = \frac{\omega_f}{\omega} \qquad 58.54$$

**Figure 58.16** Damped Magnification Factor

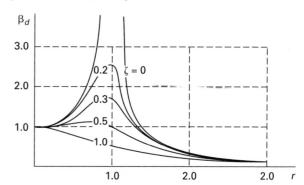

## 17. VIBRATION ISOLATION AND CONTROL

It is often desired to isolate a rotating machine from its surroundings, to limit the vibrations that are transmitted to the supports, and to reduce the amplitude of the machine's vibrations.

The *transmissibility* (i.e., *linear transmissibility*) is the ratio of the transmitted force (i.e., the force transmitted to the supports) to the applied force (i.e., the force from the imbalance). Equation 58.55 is illustrated in Fig. 58.17.

$$\text{TR} = \left| \frac{F_{\text{transmitted}}}{F_{\text{applied}}} \right|$$

$$= \beta_d\sqrt{1 + (2r\zeta)^2} \qquad 58.55$$

$$\beta_d = \frac{1}{\sqrt{(1 - r^2)^2 + (2\zeta r)^2}} \qquad 58.56$$

In some cases, the transmissibility may be reported in units of *decibels*. Unlike linear transmissibility, which is always positive, logarithmic transmissibility can be negative.

$$\text{TR}_{\text{dB}} = 20 \log\left(\text{TR}_{\text{linear}}\right) \qquad 58.57$$

**Figure 58.17** *Linear Transmissibility*

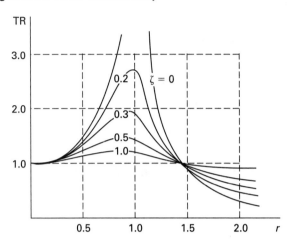

The magnitude of oscillations in vibrating equipment can be reduced and the equipment isolated from the surroundings by mounting on resilient pads or springs. The isolated system must have a natural frequency less than $1/\sqrt{2} = 0.707$ times the disturbing (forcing) frequency. That is, the transmissibility will be reduced below 1.0 only if $\omega_f/\omega > \sqrt{2}$. Otherwise, the attempted isolation will actually increase the transmitted force.

The natural frequency is found from Eq. 58.7, where the static deflection is calculated from the mass and stiffness of the pad or spring.

The amount of isolation is characterized by the *isolation efficiency*, also known as the *percent of isolation* and *degree of isolation*. Suggested isolation efficiencies for satisfactory operation are listed in Table 58.2.

$$\eta = 1 - \left| \frac{1}{\left(\dfrac{f_f}{f}\right)^2 - 1} \right| \qquad 58.58$$

**Table 58.2** *Typical Isolation Efficiencies*

| equipment | isolation efficiency |
|---|---|
| centrifugal compressor | 0.98 |
| reciprocating compressor | |
|   0 to 15 hp | 0.85 |
|   15 to 150 hp | 0.90 |
| centrifugal fans | |
|   800 rpm and higher | 0.90 to 0.95 |
| centrifugal pumps | 0.95 |
| pipe mounts | 0.95 |

Isolation materials and isolator devices have specific deflection characteristics. If the isolation efficiency is known, it can be used to determine the type of isolator or isolation device used based on the static deflection. Table 58.3 lists typical ranges of isolation materials and devices. Table 58.4 lists typical damping factors.

A *tuned system* is one for which the natural frequency of the vibration absorber is equal to the frequency that is to be eliminated (i.e., the forcing frequency). In theory, this is easy to accomplish: the mass and spring constant of the absorber are varied until the desired natural frequency is achieved. This is known as "tuning" the system.

**Table 58.3** *Typical Deflection of Isolation Materials and Devices*

| approximate deflection | | |
|---|---|---|
| in | mm | materials |
| $0-^1/_{16}$ | 0–2 | cork, natural rubber, felt, lead/asbestos, fiberglass |
| $^1/_{16}-^1/_4$ | 2–6 | neoprene pads, neoprene mounts, multiple layers of felt or cork |
| $^1/_4-1^1/_2$ | 6–40 | steel coil springs, multiple layers of natural rubber or neoprene pads |
| $1^1/_2-15$ | 40–380 | steel coil or leaf springs |

(Multiply in by 25.4 to obtain mm.)

**Table 58.4** *Typical Damping Factors of Isolators*

| material | $\zeta$ |
|---|---|
| steel spring | 0.005 |
| natural rubber | 0.05 |
| neoprene | 0.05 |
| cork | 0.06 |
| felt | 0.06 |
| metal mesh | 0.12 |
| air damper | 0.17 |
| friction-damped spring | 0.33 |

**Example 58.4**

A 4000 lbm (1800 kg) machine is rotating at 1000 rpm. The rotating component has an imbalance of 100 lbm (45 kg) acting with an eccentricity of 2 in (50 mm). The machine is already supported by isolation mounts having a combined stiffness of 30,000 lbf/in (5.3 MN/m), but vibration is still excessive. In order to reduce the amplitude of vibration, a viscous damper is connected between the machine and a rigid support. The damping ratio is 0.2. What are the (a) amplitude of oscillation and (b) transmitted force?

*SI Solution*

(a) The static deflection of the mounts is

$$\delta_{st} = \frac{mg}{k}$$

$$= \frac{(1800 \text{ kg})\left(9.81 \frac{\text{m}}{\text{s}^2}\right)}{\left(5.3 \frac{\text{MN}}{\text{m}}\right)\left(10^6 \frac{\text{N}}{\text{MN}}\right)}$$

$$= 0.00333 \text{ m}$$

The undamped natural frequency is

$$f = \frac{1}{2\pi}\sqrt{\frac{g}{\delta_{st}}}$$

$$= \frac{1}{2\pi}\sqrt{\frac{9.81 \frac{\text{m}}{\text{s}^2}}{0.00333 \text{ m}}}$$

$$= 8.638 \text{ Hz}$$

The forcing frequency is

$$f_f = \frac{1000 \frac{\text{rev}}{\text{min}}}{60 \frac{\text{s}}{\text{min}}} = 16.67 \text{ Hz}$$

The angular forcing frequency is

$$\omega_f = 2\pi f_f = 2\pi(16.67 \text{ Hz})$$

$$= 104.7 \text{ rad/sec}$$

The out-of-balance force caused by the rotating eccentric mass is

$$F_f = m\omega^2 r$$

$$= \frac{(45 \text{ kg})\left(104.7 \frac{\text{rad}}{\text{s}}\right)^2 (50 \text{ mm})}{1000 \frac{\text{mm}}{\text{m}}}$$

$$= 24\,665 \text{ N}$$

The ratio of frequencies is

$$r = \frac{\omega_f}{\omega} = \frac{16.67 \text{ Hz}}{8.638 \text{ Hz}}$$

$$= 1.93$$

From Eq. 58.53, the magnification factor is

$$\beta = \left| \frac{1}{\sqrt{(1-r^2)^2 + (2\zeta r)^2}} \right|$$

$$= \frac{1}{\sqrt{\left(1 - (1.93)^2\right)^2 + \left((2)(0.2)(1.93)\right)^2}}$$

$$= 0.353$$

The amplitude of oscillation is

$$A = \beta\left(\frac{F_f}{k}\right) = \frac{(0.353)(24\,665 \text{ N})}{\left(5.3 \frac{\text{MN}}{\text{m}}\right)\left(10^6 \frac{\text{N}}{\text{MN}}\right)}$$

$$= 0.00164 \text{ m}$$

(b) The transmissibility is

$$\text{TR} = \beta\sqrt{1 + (2r\zeta)^2}$$

$$= 0.353\sqrt{1 + \left((2)(1.93)(0.2)\right)^2}$$

$$= 0.446$$

The transmitted force is

$$F = (\text{TR})F_f = (0.446)(24\,665 \text{ N})$$

$$= 11\,000 \text{ N}$$

*Customary U.S. Solution*

(a) The static deflection of the mounts is

$$\delta_{st} = \frac{\text{weight}}{k} = \left(\frac{m}{k}\right)\left(\frac{g}{g_c}\right)$$

$$= \left(\frac{4000 \text{ lbm}}{30{,}000 \frac{\text{lbf}}{\text{in}}}\right)\left(\frac{32.2 \frac{\text{ft}}{\text{sec}^2}}{32.2 \frac{\text{ft-lbm}}{\text{lbf-sec}^2}}\right)$$

$$= 0.1333 \text{ in}$$

The undamped natural frequency is

$$f = \frac{1}{2\pi}\sqrt{\frac{g}{\delta_{st}}}$$

$$= \frac{1}{2\pi}\sqrt{\frac{\left(32.2 \frac{\text{ft}}{\text{sec}^2}\right)\left(12 \frac{\text{in}}{\text{ft}}\right)}{0.1333 \text{ in}}}$$

$$= 8.569 \text{ Hz}$$

The forcing frequency is

$$f_f = \frac{1000 \frac{\text{rev}}{\text{min}}}{60 \frac{\text{s}}{\text{min}}} = 16.67 \text{ Hz}$$

The angular forcing frequency is

$$\omega_f = 2\pi f_f = 2\pi(16.67 \text{ Hz})$$

$$= 104.7 \text{ rad/sec}$$

The out-of-balance force caused by the rotating eccentric mass is

$$F_f = \frac{m\omega_f^2 r}{g_c}$$

$$= \frac{(100 \text{ lbm})\left(104.7 \dfrac{\text{rad}}{\text{sec}}\right)^2 (2 \text{ in})}{\left(32.2 \dfrac{\text{ft-lbm}}{\text{lbf-sec}^2}\right)\left(12 \dfrac{\text{in}}{\text{ft}}\right)}$$

$$= 5674 \text{ lbf}$$

The ratio of frequencies is

$$r = \frac{\omega_f}{\omega} = \frac{16.67 \text{ Hz}}{8.569 \text{ Hz}} = 1.945$$

From Eq. 58.53, the magnification factor is

$$\beta = \left| \frac{1}{\sqrt{(1-r^2)^2 + (2\zeta r)^2}} \right|$$

$$= \frac{1}{\sqrt{\left(1 - (1.945)^2\right)^2 + \left((2)(0.2)(1.945)\right)^2}}$$

$$= 0.346$$

The amplitude of oscillation is

$$A = \beta\left(\frac{F_f}{k}\right) = \frac{(0.346)(5674 \text{ lbf})}{30,000 \dfrac{\text{lbf}}{\text{in}}}$$

$$= 0.0654 \text{ in}$$

(b) The transmissibility is

$$\text{TR} = \beta\sqrt{1 + (2r\zeta)^2}$$

$$= 0.346\sqrt{1 + \left((2)(1.945)(0.2)\right)^2}$$

$$= 0.438$$

The transmitted force is

$$F = (\text{TR})F_f = (0.438)(5674 \text{ lbf})$$

$$= 2485 \text{ lbf}$$

## 18. ISOLATION FROM ACTIVE BASE

In some cases, a machine is to be isolated from an active base. The base (floor, supports, etc.) vibrates, and the magnitude of the vibration seen by the machine is to be limited or reduced. This case is not fundamentally different from the case of a vibrating machine being isolated from a stationary base.

The concept of transmissibility is replaced by the amplitude ratio (magnification factor or amplification factor).

This is the ratio of the transmitted displacement (deflection, excursion, motion, etc.) to the applied displacement. That is, it is the ratio of the maximum mass motion to the maximum base motion. The amplification ratio is numerically identical to the transmissibility calculated in Eq. 58.55.

$$\text{AR} = \frac{\delta_{\text{dynamic}}}{\delta_{\text{static}}} = \frac{\delta_{\text{dynamic}}}{\dfrac{F}{k}} = \text{TR} \qquad 58.59$$

### Example 58.5

The suspension system of a car is modeled as a perfect spring and dashpot connecting a massless wheel and a supported mass, $m$, as shown. The car enters a bumpy area with forward velocity, v, and initial mass position, $y_0$. The profile of the road surface is modeled as a perfect sinusoid with a peak-to-peak distance of $L$. The sinusoid is described mathematically by the equation $y(x) = y_{\text{max}}\sin 2\pi(x - x_0)/L$.

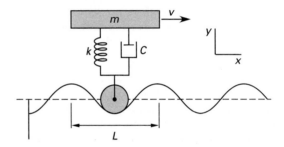

What are the (a) forcing frequency, (b) undamped natural frequency of the system, and (c) horizontal velocity at resonance? (d) For a particular speed, how would the steady-state amplitude of the mass be determined?

*Solution*

(a) Since the horizontal car velocity is v, the frequency of the forcing function will be

$$f_f = \frac{\text{v}}{L}$$

(b) The undamped natural frequency is found from Eq. 58.3. (Use consistent units.)

$$f = \frac{1}{2\pi}\sqrt{\frac{k}{m}}$$

(c) Resonance occurs when the forcing frequency equals the natural frequency. (Use consistent units.)

$$f_f = f$$

$$\frac{\text{v}}{L} = \frac{1}{2\pi}\sqrt{\frac{k}{m}}$$

$$\text{v} = \frac{L}{2\pi}\sqrt{\frac{k}{m}}$$

(d) The damping ratio is

$$\zeta = \frac{C}{C_{\text{critical}}}$$

The amplitude ratio can be found from either Fig. 58.17 or Eq. 58.55. The steady-state amplitude will be $(AR)y_{\max}$.

## 19. VIBRATIONS IN SHAFTS

A shaft's natural frequency of vibration is referred to as the *critical speed*. This is the rotational speed in revolutions per second that just equals the lateral natural frequency of vibration. Therefore, vibration in shafts is basically an extension of lateral vibrations (e.g., whipping "up and down") in beams. Rotation is disregarded, and the shaft is considered only from the standpoint of lateral vibrations.

The shaft will have multiple modes of vibration. General practice is to keep the operating speed well below the first critical speed. For shafts with distributed or multiple loadings, it may be important to know the second critical speed. However, higher critical speeds are usually well out of the range of operation.

For shafts with constant cross-sectional areas and simple loading configurations, the static deflection due to pulleys, gears, and self-weight can be found from beam formulas. Shafts with single antifriction (i.e., ball and roller) bearings at each end can be considered to be simply supported, while shafts with sleeve bearings or two side-by-side antifriction bearings at each shaft end can be considered to have fixed built-in supports.[6]

A shaft carrying no load other than its own weight can be considered as a uniformly loaded beam. The maximum deflection at midspan can be found from beam tables.

Once the deflection is known, Eq. 58.7 can be used to find the critical speed.

The classical analysis of a shaft carrying single or multiple inertial loads (flywheel, pulley, etc.) assumes that the shaft itself is weightless.[7]

Equation 58.60 gives the *critical speed* (i.e., the fundamental frequency of vibration) of a shaft carrying multiple masses in terms of the static deflections at the masses.[8] In theory, all that is necessary is to stop the rotation and measure the deflection (from the horizontal) at each mass. The assumption that the static and rotating deflection curves are identical is not exactly true. However, the speed error is less than approximately 5%, generally on the high side.

$$f = \frac{1}{2\pi}\sqrt{\frac{g\sum m_i\delta_{\text{st},i}}{\sum m_i\delta_{\text{st},i}^2}} \qquad 58.60$$

When there are only two rotating masses on the shaft, the total deflections can be calculated by superposition. However, Eq. 58.60 is laborious to use when there are more than two masses, as the number of deflection calculations is the square of the number of masses. In that case, the *Dunkerley approximation* is used.[9] In Eq. 58.61, the $f_i$ are the natural frequencies of vibration when mass $m_i$ alone is on the shaft. The Dunkerley equation generally underestimates the critical speed.

$$\left(\frac{1}{f}\right)^2 = \sum\left(\frac{1}{f_i}\right)^2 \qquad 58.61$$

If a shaft's critical speed is unsuitable, Eq. 58.62 can be used to calculate the approximate diameter of a shaft that will be acceptable.

$$d_{\text{new}} = d_{\text{old}}\sqrt{\frac{f_{\text{new}}}{f_{\text{old}}}} \qquad 58.62$$

Typical values used in the analysis of steel shafts are: density $(\rho)$, 0.28 lbm/in$^3$ (7750 kg/m$^3$); modulus of elasticity $(E)$, $2.9\times10^7$ lbf/in$^2$ (200 GPa); gravitational constant $(g_c)$, 386.4 in-lbm/lbf-sec$^2$.

### Example 58.6

A 30 lbm (14 kg) flywheel is supported on an overhanging steel shaft as shown. The shaft diameter is 1.0 in (25 mm), and the shaft mass is negligible. What is the critical speed of the shaft?

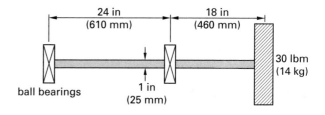

---

[6]Sleeve bearings are assumed to be fixed supports, not because they have the mechanical strength to prevent binding, but because sleeve bearings cannot operate and would not be operating with an angled shaft.

[7]The mass of the shaft can be included with Rayleigh's method and Dunkerley's approximation.

[8]This equation is sometimes known as the *Rayleigh equation* or the *Rayleigh-Ritz equation*, as it is based on the Rayleigh method of equating the maximum kinetic energy to the maximum potential energy.

[9]The spelling "Dunkerly" is also found in the literature.

*SI Solution*

The radius of the shaft is

$$r = \frac{d}{2} = \frac{25 \text{ mm}}{2} = 12.5 \text{ mm}$$

The moment of inertia of the circular cross section is

$$I = \frac{\pi r^4}{4} = \frac{\pi \left( \dfrac{12.5 \text{ mm}}{1000 \ \frac{\text{mm}}{\text{m}}} \right)^4}{4}$$
$$= 1.917 \times 10^{-8} \text{ m}^4$$

The ball bearings prevent vertical but not angular deflection. Consider the shaft to be simply supported. The deflection at the flywheel is

$$\delta_{\text{st}} = \frac{Fa^2(a+b)}{3EI}$$
$$= \frac{mga^2(a+b)}{3EI}$$
$$= \frac{(14 \text{ kg})\left( 9.81 \ \frac{\text{m}}{\text{s}^2} \right)(0.46 \text{ m})^2(0.46 \text{ m} + 0.61 \text{ m})}{(3)(200 \text{ GPa})\left( 10^9 \ \frac{\text{Pa}}{\text{GPa}} \right)(1.917 \times 10^{-8} \text{ m}^4)}$$
$$= 2.703 \times 10^{-3} \text{ m} \quad (2.703 \text{ mm})$$

From Eq. 58.7, the natural frequency is

$$f = \frac{1}{2\pi} \sqrt{\frac{g}{\delta_{\text{st}}}}$$
$$= \frac{1}{2\pi} \sqrt{\frac{9.81 \ \frac{\text{m}}{\text{s}^2}}{2.703 \times 10^{-3} \text{ m}}}$$
$$= 9.59 \text{ Hz}$$

*Customary U.S. Solution*

The radius of the shaft is

$$r = \frac{d}{2} = \frac{1 \text{ in}}{2} = 0.5 \text{ in}$$

The moment of inertia of the circular cross section is

$$I = \frac{\pi r^4}{4} = \frac{\pi (0.5 \text{ in})^4}{4}$$
$$= 0.0491 \text{ in}^4$$

The ball bearings prevent vertical but not angular deflection. Consider the shaft to be simply supported. The deflection at the flywheel is

$$\delta_{\text{st}} = \frac{Fa^2(a+b)}{3EI}$$
$$= \left( \frac{ma^2(a+b)}{3EI} \right) \left( \frac{g}{g_c} \right)$$
$$= \left( \frac{(30 \text{ lbm})(18 \text{ in})^2(18 \text{ in} + 24 \text{ in})}{(3)\left( 2.9 \times 10^7 \ \frac{\text{lbf}}{\text{in}^2} \right)(0.0491 \text{ in}^4)} \right)$$
$$\times \left( \frac{32.2 \ \frac{\text{ft}}{\text{sec}^2}}{32.2 \ \frac{\text{ft-lbm}}{\text{lbf-sec}^2}} \right)$$
$$= 0.0956 \text{ in}$$

From Eq. 58.7, the natural frequency is

$$f = \frac{1}{2\pi} \sqrt{\frac{g}{\delta_{\text{st}}}}$$
$$= \frac{1}{2\pi} \sqrt{\frac{\left( 32.2 \ \frac{\text{ft}}{\text{sec}^2} \right)\left( 12 \ \frac{\text{in}}{\text{ft}} \right)}{0.0956 \text{ in}}}$$
$$= 10.1 \text{ Hz}$$

## 20. SECOND CRITICAL SHAFT SPEED

The second critical speed for a simply supported shaft uniformly loaded along its length is four times the fundamental critical speed.

The second critical speed for a massless shaft carrying two concentrated masses, $m_1$ and $m_2$, can be derived from Eq. 58.63. The positive roots are $1/\omega_1$ and $1/\omega_2$, where $\omega_1$ and $\omega_2$ in this bi-quadratic equation are the first and second natural frequencies (critical speeds), respectively. The $a_{ij}$ influence constants are the deflections at the location of mass $i$ due to a unit force at the location of mass $j$. Only three deflection calculations are needed because $a_{12} = a_{21}$.

$$\frac{1}{\omega^4} - \left( \frac{1}{\omega^2} \right)(a_{11}m_1 + a_{22}m_2)$$
$$+ (a_{11}a_{22} - a_{12}a_{21})m_1 m_2 = 0 \qquad \textit{58.63}$$

## 21. CRITICAL SPEED OF STEPPED SHAFTS

The Dunkerley equation is one of the few simple methods for evaluating the critical speed of a stepped shaft. The mass of each shaft section is assumed to be concentrated at the midpoints of their respective lengths. The "shaft" itself is considered to be massless. This rather crude approximation provides surprisingly good results. As a further simplification, the shaft section with the smallest diameter can be disregarded.

## 22. VIBRATIONS IN THIN PLATES

The natural frequency of thin plates and diaphragms depends on the shape (e.g., circular, square, or rectangular) and the method of mounting (e.g., fixed or free edges). Completely fixed edges are difficult to achieve in practice, so the formulas used to calculate the natural frequencies are derived from a blend of heuristic and theoretical methods.

Equation 58.64 is an approximate equation based on a modification of Eq. 58.7. As an approximation, this equation can be used with any uniformly loaded round, square, rectangular, elliptical, or triangular plates with any edge conditions.[10] The maximum error is small, generally less than 3%. $\delta_{st}$ is the maximum static deflection produced by the self-mass of the plate and any uniformly distributed mass attached to the plate and vibrating with it.

$$f = \frac{1.277}{2\pi} \sqrt{\frac{y}{\delta_{st}}}$$    **58.64**

If certain assumptions are made, the fundamental natural frequency of a plate can be derived. These assumptions are that the plate material is elastic, homogeneous, and of uniform thickness. Also, the plate thickness and deflections are assumed to be small in comparison to the size of the plate. $w$ is the uniform load per unit area, including the self-weight per unit area (calculated as the specific weight times the thickness). In Eq. 58.66, $a$ is the shorter edge distance. Values of $K$ are given in Table 58.5.

$$f = \frac{K}{2\pi} \sqrt{\frac{Dg}{wr^4}}$$    [circular plates]    **58.65**

$$f = \frac{K}{2\pi} \sqrt{\frac{Dg}{wa^4}}$$    [rectangular plates]    **58.66**

$$D = \frac{Et^3}{12(1 - \nu^2)}$$    **58.67**

## 23. BALANCING

To be "balanced," a rotating component must be statically and dynamically in equilibrium. For two masses, $m_i$, rotating at radii, $r_i$, and mounted 180° out of phase, the requirement for *static balance* is given by Eq. 58.68. Static balancing is usually sufficient for rotating disks, thin wheels, and gears.

$$\sum m_i r_i = 0$$    [static balance]    **58.68**

---

[10]The deflection calculated must correspond to the actual edge conditions.

**Table 58.5** *Vibration Coefficients for Plates (fundamental mode)*

| case | | K | | | | | |
|---|---|---|---|---|---|---|---|
| circular plate | | | | | | | |
| fixed edges | | 10.2 | | | | | |
| simply supported edges | | 4.99 | | | | | |
| free edges[*] | | 5.25 | | | | | |
| rectangular plate | | | | | | | |
| $a \times b$ sides | $a/b =$ | 1.0 | 0.8 | 0.6 | 0.4 | 0.2 | 0.0 |
| fixed edges | | 36.0 | 29.9 | 25.9 | 23.6 | 22.6 | 22.4 |
| simply supported edges | | 19.7 | 16.2 | 13.4 | 11.5 | 10.3 | 9.87 |

[*]Supported at inner surface; edges free to flutter.

*Dynamic balance* requires that there also be no unbalanced couples. When all of the masses are in the same plane of rotation, static balance and dynamic balance will be achieved simultaneously. When rotation is not in the same plane, all of the couples must also balance. In general, two balancing masses are required to provide static and dynamic balance where there is an unbalanced couple. (See Fig. 58.18.)

$$\sum m_i r_i x_i = 0$$    [dynamic balance]    **58.69**

**Figure 58.18** *Static and Dynamic Balance*

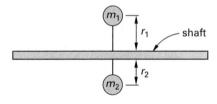

(a) in plane static balance

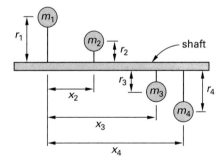

(b) out of plane dynamic balance

## Example 58.7

A 2 ft (610 mm) diameter steel disk is mounted with a 0.33 ft (100 mm) eccentricity on a shaft. The mass of the disk is 570 lbm (260 kg). The assembly is balanced by two counterweights located 1.5 ft (460 mm) and 3.0 ft (910 mm) from the disk, respectively, each mounted at a radius of 1.0 ft (300 mm) from the shaft's centerline. What are the masses of the two counterweights?

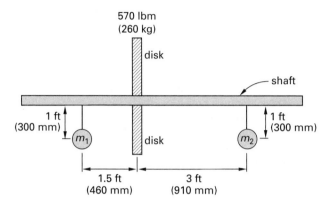

*SI Solution*

Take moments about the position of the left-hand counterweight. The couple produced by the eccentric disk mass must be balanced by the right counterweight. Use Eq. 58.69.

$$m_{\text{disk}} x_{\text{disk}} e = m_{\text{right}} x_{\text{right}} r_{\text{right}}$$

$$\frac{(260 \text{ kg})(460 \text{ mm})}{\times (100 \text{ mm})} = \frac{m_{\text{right}}(460 \text{ mm} + 910 \text{ mm})}{\times (300 \text{ mm})}$$

$$m_{\text{right}} = 29.1 \text{ kg}$$

Use Eq. 58.68 to satisfy the criterion for static balance.

$$m_{\text{left}} r_{\text{left}} + m_{\text{right}} r_{\text{right}} - m_{\text{disk}} e = 0$$

$$m_{\text{left}}(300 \text{ mm}) + (29.1 \text{ kg})(300 \text{ mm})$$

$$- (260 \text{ kg})(100 \text{ mm}) = 0$$

$$m_{\text{left}} = 57.6 \text{ kg}$$

*Customary U.S. Solution*

Take moments about the position of the left-hand counterweight. The couple produced by the eccentric disk mass must be balanced by the right counterweight. Use Eq. 58.69.

$$m_{\text{disk}} x_{\text{disk}} e = m_{\text{right}} x_{\text{right}} r_{\text{right}}$$

$$(570 \text{ lbm})(1.5 \text{ ft})(0.33 \text{ ft}) = m_{\text{right}}(1.5 \text{ ft} + 3 \text{ ft})(1 \text{ ft})$$

$$m_{\text{right}} = 62.7 \text{ lbm}$$

Use Eq. 58.68 to satisfy the criterion for static balance.

$$m_{\text{left}} r_{\text{left}} + m_{\text{right}} r_{\text{right}} - m_{\text{disk}} e = 0$$

$$m_{\text{left}}(1 \text{ ft}) + (62.7 \text{ lbm})(1 \text{ ft})$$

$$- (570 \text{ lbm})(0.33 \text{ ft}) = 0$$

$$m_{\text{left}} = 125.4 \text{ lbm}$$

## 24. MODAL VIBRATIONS

In addition to the fundamental (first) frequency emphasized up to this point, there can be higher-order frequencies (*harmonics* or higher *modes*). Generally, if a system is protected against resonance at its fundamental frequency, it will be protected against resonance at the even higher harmonic frequencies.

When a beam, shaft, or plate vibrates laterally at its fundamental frequency, all of it will be on one side of the equilibrium position at any given moment. This is not true with higher modes. There will be one or more positions (i.e., *nodes*) where the deflection curve will pass through the equilibrium position. The location of the nodes and the *modal shape* are given in handbooks.

The $n$th harmonic frequency for beams and shafts with uniformly distributed masses on simple supports is $n^2$ times the fundamental frequency. If both ends are fixed, then the higher modal frequencies are $\frac{1}{9}(2n+1)^2$ times the fundamental frequency. Cantilever beams with distributed masses are not handled so easily.

Equation 58.70 calculates the $n$th natural frequency for a cantilever beam with uniformly distributed load, $w$, (including its own weight) per unit length over its entire length, $L$. Table 58.6 gives the values of the vibration constant, $K_n$, and the locations of the nodes.

$$f_n = \frac{K_n}{2\pi} \sqrt{\frac{EIg}{wL^4}} \qquad \textit{58.70}$$

***Table 58.6*** *Vibration Constants for Cantilever Beams[*]*

| mode | $K_n$ | position of nodes ($x/L$, from fixed end) | | | | |
|------|-------|------|------|------|------|------|
| 1 | 3.52 | 0.0 | | | | |
| 2 | 22.0 | 0.0 | 0.783 | | | |
| 3 | 61.7 | 0.0 | 0.504 | 0.868 | | |
| 4 | 121 | 0.0 | 0.358 | 0.644 | 0.905 | |
| 5 | 200 | 0.0 | 0.279 | 0.500 | 0.723 | 0.926 |

[*]Any uniform cross section; uniformly distributed load $w$ (including own weight) per unit length; length, $L$.

**Dynamics**

## 25. MULTIPLE DEGREE OF FREEDOM SYSTEMS

A system is designed as a *multiple degree of freedom (MDOF) system* when it takes two or more independent variables to define the position of independent parts of the system.

MDOF systems may oscillate at a single natural frequency (e.g., when all of the masses are moving in the same direction), or different parts of the system may oscillate independently at their own frequencies. The *amplitude ratio* is the ratio of the maximum deflection (excursion, movement, etc.) of one part of the system to the maximum deflection of another.

Closed-form solutions for various simple systems can be calculated, but real-world situations are more difficult. For example, consider a circuit board mounted on a bulkhead (plate or diaphragm). The circuit board has its own vibration characteristics; the bulkhead also has its own. When they are assembled together, a third vibration characteristic is produced. The circuit board and the bulkhead are said to be "coupled" because each affects the other. (See Fig. 58.19.)

**Figure 58.19** *System with Three Degrees of Freedom*

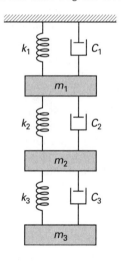

Even when it is difficult to determine the actual performance characteristics of an MDOF system, damage from resonance can still be prevented by proper design. The modes are "uncoupled" as much as possible. The *octave rule* is a simple rule of thumb for uncoupling the modes. The octave rule requires the uncoupled natural frequency of each added component to be at least twice the natural frequency of the element to which it is attached. In other words, the natural frequency should be doubled every time an additional degree of freedom is added to the system.

When space, weight, or other constraints make it impossible to design modal interaction out of the system, testing and other (usually heuristic or graphical) methods must be used to determine the maximum dynamic forces on the assembly.

## 26. SHOCK

Vibration is a steady-state, regular phenomenon. *Shock* is a transient phenomenon. Shock results in a sharp, nearly sudden change in velocity. A *shock pulse (shock impulse)* is a disturbing force characterized by a rise and subsequent decay of acceleration in a very short period of time. A shock pulse is described by its peak amplitude (usually in gravities), its duration (in milliseconds), and an overall shape (triangular, rectangular, half-sine, etc.). Shock pulses, particularly those that are complex, are often depicted graphically as an acceleration-time curve.

The change in velocity, $\Delta v$, can be found as area under an acceleration-time curve or as the area under a force-time curve divided by the mass. For a complex curve, the area can be found by integration or by breaking it into simpler sections. Equations 58.71 through 58.76 give the change in velocity for several regular shocks of maximum acceleration amplitude, $A_0$, and duration, $t_0$.

$$\Delta v = \sqrt{2gh} \quad \text{[inelastic drop from height } h\text{]} \qquad 58.71$$

$$\Delta v = 2\sqrt{2gh} \quad \text{[elastic drop from height } h\text{]} \qquad 58.72$$

$$\Delta v = \frac{2A_0 g t_0}{\pi} \quad \text{[half-sine acceleration]} \qquad 58.73$$

$$\Delta v = A_0 g t_0 \quad \text{[rectangular acceleration]} \qquad 58.74$$

$$\Delta v = \frac{A_0 g t_0}{2} \quad \text{[triangular acceleration]} \qquad 58.75$$

$$\Delta v = \frac{A_0 g t_0}{2} \quad \text{[versed sine acceleration]} \qquad 58.76$$

The *shock transmission (transmitted shock)*, $G$, is an acceleration parameter (with units of gravities) that depends on the natural frequency, $f$, and the change in velocity, $\Delta v$, due to the shock.

$$G = \frac{2\pi f \Delta v}{g} = \frac{\omega \Delta v}{g} \qquad 58.77$$

The *dynamic deflection* of a linear isolator that experiences a shock pulse is

$$\delta = \frac{\Delta v}{2\pi f} = \frac{\Delta v}{\omega} \qquad 58.78$$

Isolating a system against shocks is very different than isolating the system against vibration. Isolators must be capable of absorbing shock energy instantly. The energy may be dissipated in an inelastic isolator (e.g., crush insulation), or the energy may be released at the damped natural frequency of the system later in the cycle.

### Example 58.8

A sensitive piece of electronic equipment is mounted on an isolation system with a natural frequency of 15 Hz. The equipment and mount are subjected to a standard 15 g, 11 msec, half-sine shock test. What are the (a) maximum shock transmission and (b) isolation deflection?

*SI Solution*

(a) From Eq. 58.73, the change in velocity is

$$\Delta v = \frac{2A_0 g t_0}{\pi}$$

$$= \frac{(2)(15 \text{ g})\left(9.81 \, \frac{\text{m}}{\text{s}^2 \cdot \text{g}}\right)(11 \text{ ms})}{\pi\left(1000 \, \frac{\text{ms}}{\text{s}}\right)}$$

$$= 1.03 \text{ m/s}$$

From Eq. 58.77, the shock transmission is

$$G = \frac{2\pi f \Delta v}{g} = \frac{2\pi(15 \text{ Hz})\left(1.03 \, \frac{\text{m}}{\text{s}}\right)}{9.81 \, \frac{\text{m}}{\text{s}^2 \cdot \text{g}}}$$

$$= 9.9 \text{ g's} \quad (9.9 \text{ gravities})$$

(b) Use Eq. 58.78 to find the dynamic linear deflection.

$$\delta = \frac{\Delta v}{2\pi f} = \frac{1.03 \, \frac{\text{m}}{\text{s}}}{2\pi(15 \text{ Hz})}$$

$$= 0.0109 \text{ m}$$

*Customary U.S. Solution*

From Eq. 58.73, the change in velocity is

$$\Delta v = \frac{2A_0 g t_0}{\pi}$$

$$= \frac{(2)(15 \text{ g})\left(32.2 \, \frac{\text{ft}}{\text{sec}^2 \cdot \text{g}}\right)\left(12 \, \frac{\text{in}}{\text{ft}}\right)(11 \text{ msec})}{\pi\left(1000 \, \frac{\text{msec}}{\text{sec}}\right)}$$

$$= 40.59 \text{ in/sec}$$

From Eq. 58.77, the shock transmission is

$$G = \frac{2\pi f \Delta v}{g} = \frac{2\pi(15 \text{ Hz})\left(40.59 \, \frac{\text{in}}{\text{sec}}\right)}{\left(32.2 \, \frac{\text{ft}}{\text{sec}^2 \cdot \text{g}}\right)\left(12 \, \frac{\text{in}}{\text{ft}}\right)}$$

$$= 9.9 \text{ g's} \quad (9.9 \text{ gravities})$$

(b) Use Eq. 58.78 to find the dynamic linear deflection.

$$\delta = \frac{\Delta v}{2\pi f} = \frac{40.59 \, \frac{\text{in}}{\text{sec}}}{2\pi(15 \text{ Hz})}$$

$$= 0.431 \text{ in}$$

## 27. VIBRATION AND SHOCK TESTING

There are two basic types of vibration testing: constant and random-frequency tests. Constant-frequency *sinusoidal tests* (also known as *harmonic tests*) were the earliest types of tests used. They detect, one at a time, the resonant frequencies of an item. Random tests excite all of the resonant frequencies simultaneously and duplicate the buffeting that equipment will typically experience.

Most vibration tests are performed on a *shaker table* (*exciter*). Tables can be electrodynamic, hydraulic, or mechanical. Electrodynamic shakers produce the highest frequencies, but they are limited to shaking smaller pieces of equipment with low forces and displacements. Hydraulic shakers (also known as *hydrashakers*) use hydraulic fluid to drive a piston. Test frequencies are limited to approximately 500 Hz. Forces can be quite large. Mechanical shakers use eccentric cams to produce the traditional test. They cannot produce random vibrations. Mechanical shakers are limited to approximately 55 Hz.

Since there are many types of shock, there are many types of shock tests used. Hammers, spring-loaded rigs, air guns, and drops into sand pits are in use. Shipping containers are often subjected to a crude *drop test*, where the containerized equipment is mounted on a table and then dropped onto a concrete floor. The *swing test* is similar—the test specimen swings into a concrete wall. A common test for aircraft-mounted equipment is a *machine drop* test producing a half-sine shock of 15 g's lasting 11 ms. (The nature of the shock is controlled by the resilience of the contact surface.) Missile components subjected to explosive impulses (e.g., explosive separation of stages) may be tested with a 6 ms, 100 g shock.

## 28. VIBRATION INSTRUMENTATION

There are two main types of *vibration sensors*: proximity and casing devices. *Proximity displacement transducers* (*proximity probes*) do not touch the vibrating item. They sense vibrations by establishing an electric/magnetic field between the probe tip and the object. Changes in the field produced by the minute movement of the conductive surface are detected by the instrument.

*Casing transducers* that touch or are mounted on vibrating equipment come in two varieties: accelerometers and velocity transducers. The output of an *accelerometer* is proportional to acceleration. Accelerometers generally use a piezoelectric crystal with an internally mounted reference mass. The voltage produced by the crystal varies as the attached mass is vibrated. Accelerometers can also be constructed as strain gauges mounted on small flexible members. A *velocity transducer* is an electromagnetic device with a coil and core (or magnet), one of which is stationary and the other that is moving. The voltage produced is proportional to the relative velocity of the core through the coil. The terms *vibration pick-up* and *vibrometer* refer to devices that produce velocity-dependent voltages.

**Dynamics**

# Topic XII: Circuits

Circuits

# 59 Electrostatics and Electromagnetics

## Nomenclature

| | | |
|---|---|---|
| $a$ | acceleration | $m/s^2$ |
| $a$ | radius | m |
| $A$ | area | $m^2$ |
| $b$ | radius | m |
| $B$ | magnetic flux density | T |
| $c$ | speed of light | m/s |
| $C$ | capacitance | F |
| $d$ | distance | m |
| $D$ | displacement | $C/m^2$ |
| $E$ | electric field strength | V/m |
| $F$ | force | N |
| $H$ | magnetic field strength | A/m |
| $I$ | current | A |
| $k$ | dielectric constant | $N \cdot m^2/C^2$ |
| $l$ | length | m |
| $L$ | inductance | H |
| $m$ | mass | kg |
| $m$ | pole strength | Wb |
| $N$ | number of turns | – |
| $P$ | dielectric polarization | $C/m^2$ |
| $Q$ | charge | C |
| $r$ | radius or distance | m |
| $R$ | radius | m |
| $\mathcal{R}$ | reluctance | A/Wb |
| $s$ | distance | m |
| $S$ | Poynting vector | $W/m^2$ |
| $t$ | time | s |
| $T$ | torque | $N \cdot m$ |
| $U$ | energy | J |
| v | velocity | m/s |
| $V$ | voltage[1] | V |
| $W$ | work | J |
| $x$ | position | m |
| $y$ | displacement | m |

### Symbols

| | | |
|---|---|---|
| $\epsilon$ | permittivity | $C^2/N \cdot m^2$ |
| $\theta$ | angle | degrees |
| $\lambda$ | flux density per unit length | C/m |
| $\mu$ | permeability | H/m |
| $\rho$ | flux density per unit volume | $C/m^3$ |
| $\sigma$ | flux density per unit area | $C/m^2$ |
| $\phi$ | magnetic flux | Wb |
| $\Phi$ | electric flux | C |
| $\omega$ | angular velocity | rad/s |

### Subscripts

| | | |
|---|---|---|
| 0 | free space (vacuum) |
| $a$ | accelerating |
| $d$ | displacement |
| $f$ | final |
| $i$ | initial |
| $l$ | per unit length |
| $m$ | magnetic |
| $p$ | pole |
| $r$ | relative or distance $r$ |
| $s$ | per unit area |
| $v$ | per unit volume |

## 1. ELECTROSTATIC CHARGE

The charge on an electron is referred to simply as the *elementary charge* or *fundamental unit of charge*. In the

---
[1]The standard SI symbol for emf is $E$. However, $V$ is used in this chapter to avoid confusion with the electric field strength.

Circuits

old centimeter-gram-second (cgs) system of units, the charge was referred to as one *electrostatic unit* (esu). Charge is measured in the SI system in coulombs (C). (1 C is approximately $6.242 \times 10^{18}$ esu.) Another convenient unit of charge is the faraday, the charge on one mole of electrons and equal to approximately $96\,485$ C.

Static electricity can be created in several common ways. *Vitreous static electricity* (i.e., positive charge) can be produced by rubbing silk on a glass rod. *Resinous static electricity* (i.e., negative charge) is produced by rubbing fur on a rubber, amber, or plastic rod.

## 2. CAPACITORS

A *capacitor* (known as a *condenser* in the distant past) is a component that stores electric charge. It can be constructed as two conducting surfaces separated by an insulator such as oiled paper, mica, or air. A simple type of capacitor (i.e., the *parallel plate capacitor*) is constructed as two parallel plates. If the plates are connected across a voltage potential, charges will build up and create an electric field between the plates.[2] The amount of charge, $Q$, built up is proportional to the applied voltage. The constant of proportionality, $C$, is the *capacitance* in farads (F) and depends on the capacitor construction.[3] Capacitance represents the ability to store charge; the greater the capacitance, the greater the charge stored.

$$Q = CV \qquad 59.1$$

Equation 59.2 gives the capacitance of two parallel plates of equal area $A$ separated by distance $r$. $\epsilon$ is the permittivity of the medium separating the plates (see Sec. 59.6), approximately equal to $8.854 \times 10^{-12}$ F/m (same as $\mathrm{C^2/N{\cdot}m^2}$ and $\mathrm{C^2/J{\cdot}m}$) for a vacuum.

$$C = \frac{\epsilon A}{r} \qquad 59.2$$

The total energy stored in a capacitor is

$$U = \tfrac{1}{2}CV^2 = \tfrac{1}{2}VQ = \frac{Q^2}{2C} \qquad 59.3$$

The energy stored per unit volume of space between two parallel plates is

$$U_v = \frac{U}{\text{volume}} = \frac{CV_{\text{plates}}^2}{2(\text{volume})} = \frac{\epsilon V_{\text{plates}}^2}{2r^2}$$
$$= \frac{\epsilon E^2}{2} \qquad 59.4$$

For capacitors in parallel, the total capacitance is

$$C = C_1 + C_2 + C_3 + \cdots + C_n \qquad 59.5$$

For capacitors in series, the reciprocal of the total capacitance is

$$\frac{1}{C} = \frac{1}{C_1} + \frac{1}{C_2} + \frac{1}{C_3} + \cdots + \frac{1}{C_n} \qquad 59.6$$

## 3. ELECTRIC FLUX

A point charge of $Q$ will generate a radial *electric field*, as illustrated in Fig. 59.1. Imaginary lines of force—*electric flux*, $\Phi$,—are directed from the positive to negative charges and leave or enter at right angles to the surface. (Only a few representative lines are drawn.) The orientations of the field and flux lines always coincide. Flux is numerically equal to the charge.

$$\Phi = Q \qquad 59.7$$

**Figure 59.1** *Flux Lines*

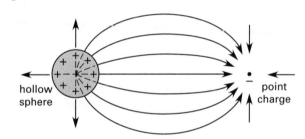

If the charges are the same, the flux outside a hollow charged sphere is the same as the flux from a point charge located at the sphere's geometric center. For the purposes of drawing flux lines and calculating forces, a hollow charged sphere can be replaced by a central point charge of the same charge.

## 4. FLUX DENSITY

The number of flux lines per unit area perpendicular to the flux is the *surface area flux density* or *electric flux density*, $\sigma$, in $\mathrm{C/m^2}$.[4] Flux density is a scalar quantity.[5]

$$\sigma = \frac{\Phi}{A} = \frac{Q}{A} \qquad 59.8$$

Two other quantities also called "flux density" have units of C/m ($\rho_l$ or $\lambda$) and $\mathrm{C/m^3}$ ($\rho$). Care must be exercised to interpret "flux density" correctly.

---

[2]Once the charge has built up on the plates, the battery can be removed.

[3]A farad is a very large capacitance and few capacitors have been constructed with even a 1.0 F capacitance. Most circuit elements will have capacitances in $\mu$F ($10^{-6}$ F) or pF ($10^{-12}$ F).

[4]The symbol $\rho_s$ is also used to represent the surface area (or electric) flux density, $\sigma$.

[5]Flux density is the magnitude of displacement, a vector. (See Sec. 59.8.)

## 5. GAUSS'S LAW

*Gauss's law* says that the electric flux, $\Phi$, passing out of a closed surface (i.e., the Gaussian surface) is equal to the total charge, $Q$, within the surface. This can be expressed by the surface integral Eq. 59.9.

$$\Phi = \oint_S \sigma \, dA = Q \quad \text{[Gauss's law]} \qquad 59.9$$

An application of Gauss's law is calculating the flux density in a spherical surface surrounding a point charge. All of the flux emanating from a point charge passes through a spherical Gaussian surface. The surface area of the sphere depends on the distance, $r$, from the charge.

$$\sigma = \frac{\Phi}{A} = \frac{Q}{4\pi r^2} \quad \text{[spherical]} \qquad 59.10$$

Anticipating Eq. 59.14, which shows that the quotient $Q/A$ is equal to the product of the permittivity, $\epsilon$, and the electric field strength, $E$, Gauss's law can also be written as

$$\Phi = \oint_S \epsilon E \, dA \quad \text{[Gauss's law]} \qquad 59.11$$

## 6. PERMITTIVITY

Electric flux does not pass equally well through all materials. It cannot pass through conductive metals at all, and it is canceled to various degrees by insulating (dielectric) media.[6] A simple interpretation of the *relative permittivity (dielectric constant* or *permittivity ratio)*, $\epsilon_r$, is the ratio of flux in a vacuum to the flux in a medium. Typical values are given in Table 59.1. The *permittivity of free space (of a vacuum)*, $\epsilon_0$, has a value of approximately $8.854 \times 10^{-12}$ $C^2/N \cdot m^2$ (same as F/m and $C^2/J \cdot m$). The formal definition of relative permittivity is based on a ratio of capacitance for a given voltage and separation.

$$\epsilon = \epsilon_r \epsilon_0 \qquad 59.12$$

$$\epsilon_r = \frac{C_{\text{with dielectric}}}{C_{\text{vacuum}}} \qquad 59.13$$

### Example 59.1

Two square ($0.04$ m $\times$ $0.04$ m) parallel plates are separated by a 0.1 cm thick insulator with dielectric constant of 3.4. The plates are connected across 200 V. (a) What is the capacitance? (b) What charge exists on the plates?

---

[6]The polar nature (see Sec. 59.10) of dipoles causes them to form their own electric fields that partially counteract the Coulomb forces between other charged bodies.

*Solution*

(a) From Eq. 59.2,

$$C = \frac{\epsilon A}{r} = \frac{\epsilon_r \epsilon_0 A}{r}$$

$$= \frac{(3.4)\left(8.854 \times 10^{-12} \, \frac{\text{F}}{\text{m}}\right)(0.04 \text{ m})^2}{0.001 \text{ m}}$$

$$= 4.817 \times 10^{-11} \text{ F}$$

(b) From Eq. 59.1,

$$Q = C V_{\text{plates}} = (4.817 \times 10^{-11} \text{ F})(200 \text{ V})$$

$$= 9.63 \times 10^{-9} \text{ C}$$

**Table 59.1** *Typical Relative Permittivities (20°C and 1 atmosphere)*

| material | $\epsilon_r$ | material | $\epsilon_r$ |
|---|---|---|---|
| acetone | 21.3 | mylar | 2.8–3.5 |
| air | 1.00059 | olive oil | 3.11 |
| alcohol | 16–31 | paper | 2.0–2.6 |
| amber | 2.9 | paper (kraft) | 3.5 |
| asbestos paper | 2.7 | paraffin | 1.9–2.5 |
| asphalt | 2.7 | polyethylene | 2.25 |
| bakelite | 3.5–10 | polystyrene | 2.6 |
| benzene | 2.284 | porcelain | 5.7–6.8 |
| carbon dioxide | 1.001 | quartz | 5 |
| carbon tetrachloride | 2.238 | rock | $\approx 5$ |
| castor oil | 4.7 | rubber | 2.3–5.0 |
| diamond | 16.5 | shellac | 2.7–3.7 |
| glass | 5–10 | silicon oil | 2.2–2.7 |
| glycerine | 56.2 | slate | 6.6–7.4 |
| hydrogen | 1.003 | sulfur | 3.6–4.2 |
| lucite | 3.4 | teflon | 2.0–2.2 |
| marble | 8.3 | vacuum | 1.000 |
| methanol | 22 | water | 80.37 |
| mica | 2.5–8 | wood | 2.5–7.7 |
| mineral oil | 2.24 | | |

## 7. ELECTRIC FIELDS

An *electric field*, **E**, with units of N/C (same as V/m) is known as a *vector force field* because a charged object in the field will experience a force in a specific direction. The electric field, **E**, is a vector quantity having both magnitude and direction. The direction is along the flux lines, represented by the unit vector **a**. Equation 59.14 gives the *electric field* in a medium with permittivity $\epsilon$ at a distance $r$ from a point charge.

$$\mathbf{E} = \left(\frac{Q}{4\pi\epsilon r^2}\right)\mathbf{a} \qquad 59.14$$

The energy, $U_v$, per unit volume stored in an electric field is

$$U_v = \tfrac{1}{2}\epsilon E^2 \qquad 59.15$$

Not all electric fields are radial; the electric field depends on the orientation of the surfaces that contain the charge. For example, a *uniform electric field* between two flat plates is the same everywhere between the plates. (See Fig. 59.2.) When there is a potential difference of $V$ volts between two plates separated by a distance $r$, the electric field strength is

$$E_{\text{uniform}} = \frac{V_{\text{plates}}}{r} \qquad 59.16$$

**Figure 59.2** *A Uniform Field*

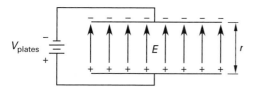

Field strengths for other configurations are given in Table 59.2.

## 8. DISPLACEMENT

*Displacement*, $\mathbf{D}$, is a vector quantity and is defined by Eq. 59.17. It is loosely referred to as the *vector flux density*.

$$\mathbf{D} = \epsilon \mathbf{E} = \sigma \mathbf{a} \qquad 59.17$$

The magnitude of displacement is the flux density. That is, $\sigma = |\mathbf{D}| = D$. (See Sec. 59.4.)

## 9. FORCE ON A CHARGED OBJECT

In general, the force on a charge $Q$ in an electric field is

$$\mathbf{F} = Q\mathbf{E} \qquad 59.18$$

The force experienced by an object with a charge $Q_B$ in an electric field, $\mathbf{E}$, created by charge $Q_A$ is given by Eq. 59.19. Since charges with opposite signs attract, Eq. 59.19 is positive for repulsion and negative for attraction. Although the unit vector $\mathbf{a}$ gives the direction explicitly, the direction of force can usually be found by inspection as the direction the object would move when released. Vector addition (i.e., superposition) can be used with systems of multiple point charges. The constant $k$ in Eq. 59.19 has a value of approximately $8.987 \times 10^9$ N·m²/C². ($\epsilon_r = 1$ for a vacuum.)

$$\mathbf{F}_{A-B} = Q_B \mathbf{E}_A = \left( \frac{k Q_A Q_B}{\epsilon_r r^2} \right) \mathbf{a} \quad \text{[Gaussian]}$$

$$k = \frac{1}{4\pi\epsilon_0} \qquad 59.19$$

Equation 59.19 is the *Gaussian SI form* of *Coulomb's law*. In the analysis of many different field configurations, however, the term $4\pi$ appears. In order to cancel

this term and make subsequent calculations easier, it is incorporated in the *rationalized form* of Coulomb's law.

$$\mathbf{F}_{A-B} = Q_B \mathbf{E}_A = \left( \frac{Q_A Q_B}{4\pi\epsilon_r \epsilon_0 r^2} \right) \mathbf{a} \quad \text{[rationalized]} \qquad 59.20$$

For nonlinear arrangements of point charges, the electrostatic forces must be calculated vectorially or by component addition.

**Table 59.2** *Electric Field and Capacitance for Various Configurations*

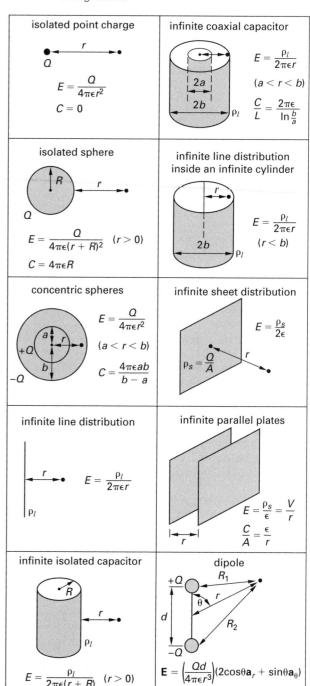

## Example 59.2

Three point charges in a vacuum are arranged in a straight line. Find the force on point charge A.

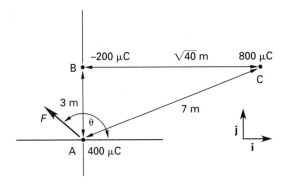

### Solution

Since all three charges are in line, vector analysis is not needed. The force on point charge A is the sum of the individual forces from the other two point charges. Since the sign is negative, the force on A is toward B and C. Use the Gaussian form of Coulomb's law.

$$
\begin{aligned}
F_{A-BC} &= F_{A-B} + F_{A-C} \\
&= \frac{kQ_A Q_B}{r_{A-B}^2} + \frac{kQ_A Q_C}{r_{A-C}^2} \\
&= 8.987 \times 10^9 \; \frac{\text{N·m}^2}{\text{C}^2} \\
&\quad \times \left( \begin{array}{c} \dfrac{(400 \times 10^{-6} \text{ C})(-200 \times 10^{-6})}{(3 \text{ m})^2} \\[2mm] + \dfrac{(400 \times 10^{-6})(800 \times 10^{-6})}{(7 \text{ m})^2} \end{array} \right) \\
&= -79.88 \text{ N} + 58.69 \text{ N} \\
&= -21.19 \text{ N}
\end{aligned}
$$

## Example 59.3

Determine the magnitude and direction of the force on point charge A due to point charges B and C.

### Solution

The charges and distances are the same as in Ex. 59.2.

$$
\begin{aligned}
F_{A-B} &= -80 \text{ N} \quad \text{[attractive]} \\
F_{A-C} &= 58.7 \text{ N} \quad \text{[repulsive]}
\end{aligned}
$$

Determine the unit vectors for these forces. Let point A correspond to the origin. The unit vectors ($\mathbf{a}_{A-B}$ and $\mathbf{a}_{A-C}$) *from* A *to* B and C are determined as follows

$$
\begin{aligned}
\mathbf{a}_{A-B} &= \frac{(x_A - x_B)\mathbf{i} + (y_A - y_B)\mathbf{j}}{\sqrt{(x_A - x_B)^2 + (y_A - y_B)^2}} \\
&= \frac{(0-0)\mathbf{i} + (0-3)\mathbf{j}}{\sqrt{(0-0)^2 + (0-3)^2}} = -\mathbf{j} \\
\mathbf{a}_{A-C} &= \frac{(0-\sqrt{40})\mathbf{i} + (0-3)\mathbf{j}}{\sqrt{(0-\sqrt{40})^2 + (0-3)^2}} \\
&= -0.904\mathbf{i} - 0.429\mathbf{j}
\end{aligned}
$$

The total force is

$$
\begin{aligned}
F_{A-B}\mathbf{a}_{A-B} &+ F_{A-C}\mathbf{a}_{A-C} \\
&= (-80 \text{ N})(-\mathbf{j}) + (58.7 \text{ N})(-0.904\mathbf{i} - 0.429\mathbf{j}) \\
&= -53.06\mathbf{i} + 54.82\mathbf{j} \text{ N}
\end{aligned}
$$

The magnitude of the force is

$$
\mathbf{F}_{A-BC} = \sqrt{(-53.06 \text{ N})^2 + (-54.82 \text{ N})^2} = 76.3 \text{ N}
$$

The direction (counterclockwise angle with respect to the horizontal) is

$$
\theta = 180° - \arctan \frac{54.82 \text{ N}}{53.06 \text{ N}} = 180° - 45.9° = 134.1°
$$

## 10. FORCE ON A DIPOLE

A *dipole* consists of a pair of equal but opposite point charges, $Q$, separated by a small distance, $d$. (See Fig. 59.3.) Substances whose molecules are dipoles are known as *polar substances*. (Water molecules are dipoles. The two hydrogen ions are separated by an angle of 105° with the oxygen molecule at the vertex.) If the molecules of a dielectric are polarized so that they establish an intrinsic electric field, $\mathbf{P}$, of their own (known as the *dielectric polarization*), the total displacement is given by *Maxwell's electric field relation*, Eq. 59.21.

$$
\mathbf{D} = \epsilon\mathbf{E} + \mathbf{P} \qquad \textit{59.21}
$$

The *dipole moment* is

$$
\mathbf{p} = Q\mathbf{d} \qquad \textit{59.22}
$$

The total potential at a distance $r$ (from the dipole's center of symmetry) due to the two charges is[7]

$$
V = \frac{Q(R_2 - R_1)}{4\pi\epsilon R_1 R_2} \approx \frac{Qd\cos\theta}{4\pi\epsilon r^2} \qquad \textit{59.23}
$$

---

[7]When $r \gg d$, the distances from the two dipole charges are essentially the same.

**Figure 59.3** *Dipole*

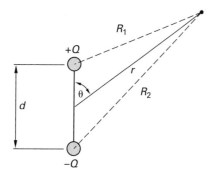

The electric field strength (in cylindrical coordinates) due to the dipole is

$$\mathbf{E} = \left(\frac{Qd}{4\pi\epsilon r^3}\right)(2\cos\theta\mathbf{a}_r + \sin\theta\mathbf{a}_\theta) \qquad 59.24$$

The torque, $T$, experienced by an electric dipole in an electric field (see Fig. 59.4) is

$$T = dF\sin\theta = dEQ\sin\theta \qquad 59.25$$

**Figure 59.4** *Dipole in an Electric Field*

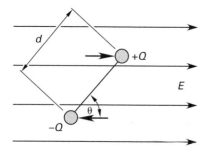

## 11. PARTICLE MOTION PARALLEL TO AN ELECTRIC FIELD

A charged particle (e.g., an electron) moving parallel to the flux lines between two parallel plates separated by distance $r$ will be acted upon by a constant force in the direction of the flux lines. The acceleration will be

$$a = \frac{F}{m} = \frac{EQ}{m} = \frac{V_{\text{plates}}Q}{rm} \qquad 59.26$$

The velocity (parallel to the lines of flux) after accelerating at the rate given by Eq. 59.26 for time $t$ can be found from the uniform acceleration formulas. If the initial velocity (parallel to the lines of flux), $v_i$, and distance traveled, $s$, are known, the velocity is

$$v = \sqrt{2as + v_i^2} = \sqrt{\frac{2EQs}{m} + v_i^2}$$

$$= \sqrt{\frac{2V_{\text{plates}}Qs}{rm} + v_i^2} \qquad 59.27$$

When an electron starts with zero initial velocity at a negative plate in a uniform field and travels to the positive plate, the final velocity depends only on the potential difference across the plates, $V_{\text{plates}}$.

$$v_{\text{m/s}} = 5.931 \times 10^5 \sqrt{V_{\text{plates}}} \qquad \begin{bmatrix} \text{electrons only,} \\ \text{zero initial velocity} \end{bmatrix}$$
$$59.28$$

The velocity of a charged particle moving parallel to an electric field is governed by the conservation of energy law. Specifically, the change in potential energy equals the change in kinetic energy.

$$|\Delta U_{\text{potential}}| = |\Delta U_{\text{kinetic}}| \qquad 59.29$$

$$Q\Delta V = \tfrac{1}{2}m(v_f^2 - v_i^2) \qquad 59.30$$

## 12. PARTICLE MOTION IN A CATHODE RAY TUBE

The deflection, $y$, in the *cathode ray tube* (CRT) shown in Fig. 59.5 depends on the tube dimensions as well as the accelerating and deflecting voltages. In the usual case, electrons boil off from a heated filament. The cloud of electrons surrounding the filament is the *space charge*. Electrons in the cloud start with a zero velocity and are attracted (the *Edison effect*) to a positively charged grid. The electrons follow a parabolic path from the instant they enter the deflection field, $E_d$. When they leave the deflection field, the tangent line (projected back) intersects the $x$-axis at $l_2/2$. Assuming the accelerating voltage, $V_a$, is zero and the electrons enter the deflection field with constant left-to-right velocity, $v_i$, the deflection on a screen distance $l_1$ from the center of the deflection plate is

$$y = l_1 \tan\theta \qquad 59.31$$

$$\tan\theta = \frac{V_d Q l_2}{l_3 m v_i^2} = \frac{E_d Q l_2}{m v_i^2} \qquad 59.32$$

**Figure 59.5** *Deflection in a Cathode Ray Tube*

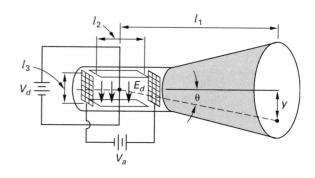

If electrons are accelerated from a standstill by accelerating potential $V_a$ while being deflected vertically, the deflection on the screen is

$$y = l_1 \tan \theta \qquad 59.33$$

$$\tan \theta = \frac{l_2 V_d}{l_3 V_a} \qquad 59.34$$

### 13. WORK IN AN ELECTRIC FIELD

The work, $W$, performed in moving a charge B radially from distance $r_1$ to $r_2$ in a field due to a charge B is given by Eq. 59.35. Work is positive if an external force is required to move the charges (e.g., bringing two repulsive charges together or moving a charge against an electric field). Work is negative if the field does the work (allowing attracting charges to approach each other or allowing repulsive charges to separate).

$$W = -\int_{r_1}^{r_2} \mathbf{F} \cdot d\mathbf{r} = \int_{r_1}^{r_2} \frac{Q_A Q_B}{4\pi \epsilon r^2} dr$$

$$= \left( \frac{Q_A Q_B}{4\pi \epsilon} \right) \left( \frac{1}{r_2} - \frac{1}{r_1} \right) \qquad 59.35$$

In a uniform electric field (as between two charged plates separated by a distance $r$), the work done in moving an object of charge $Q$ a distance $d$ parallel to the field is given by Eq. 59.36. No work is performed in moving the object perpendicular to the field (i.e., across lines of flux or anywhere on an equipotential surface).

$$W = -\mathbf{F} \cdot \mathbf{d} = -EQd = \frac{-V_{\text{plates}} Qd}{r}$$

$$= -Q\Delta V \qquad 59.36$$

#### Example 59.4

How much work is performed in moving an electron ($Q = 1.60 \times 10^{-19}$ C) through a potential difference of 3,000,000 V?

*Solution*

From Eq. 59.36,

$$W = -Q\Delta V = -(1.60 \times 10^{-19} \text{ C})(3 \times 10^6 \text{ V})$$

$$= -4.8 \times 10^{-13} \text{ J}$$

### 14. POTENTIAL ENERGY

The potential energy, $U$, of a system of two similarly charged particles separated by distance $r$ is the work

required to bring one of the charges in from infinity to within a distance $r$ of the other charge. The negative sign in the integral indicates the force opposes the direction of movement.

$$U_{\text{potential}} = W_{\infty-r} = -\int_{\infty}^{r} \mathbf{F} \cdot d\mathbf{r}$$

$$= -\int_{\infty}^{r} \frac{Q_A Q_B}{4\pi \epsilon^2} dr = \frac{Q_A Q_B}{4\pi \epsilon r} \qquad 59.37$$

### 15. ELECTRIC POTENTIAL

Equation 59.38 gives the *electric potential*, $V$, in volts at a point in an electric field, $\mathbf{E}$, where charge A is located. (As written in Eq. 59.38 and Eq. 59.39, the field is produced by $Q_B$. The potential does not depend on $Q_A$, only its location.) Potential is a scalar quantity, so the potential at a point due to several charges is the algebraic sum of the individual potentials. The *potential difference*, $\Delta V$, is the difference in potential between two points. The electric potential difference between two points is one volt if one joule of work is expended in moving one coulomb of charge from one point to another.

No work is performed in moving a charge anywhere on an *equipotential surface* (a surface where all points are at the same potential). Charges on the surface of any conducting surface in an electric field automatically adjust themselves to form an equipotential surface. After this adjustment, there is no net force on any charge on the surface. The potential anywhere inside a hollow charged object is the same as the surface potential.

$$V_A = \frac{U_{\text{potential}}}{Q_A} = \frac{Q_B}{4\pi \epsilon r} \qquad 59.38$$

$$\Delta V_A = \frac{\Delta U_{\text{potential}}}{Q_A} = \frac{W}{Q_A} = \left( \frac{Q_B}{4\pi \epsilon} \right) \left( \frac{1}{r_2} - \frac{1}{r_1} \right) \qquad 59.39$$

The *electric potential gradient* in V/m is the change in potential per unit distance and is identical to the electric field strength in N/C. The negative sign in Eq. 59.40 means the potential decreases as the distance increases.

$$E = \frac{-\Delta V}{\Delta r} \qquad 59.40$$

#### Example 59.5

A +400 $\mu$C charge is located 3.0 m from a +600 $\mu$C charge in a vacuum. (a) What is the system's potential energy? (b) What is the potential of the 400 $\mu$C charge?

*Circuits*

*Solution*

(a) From Eq. 59.39, the potential energy of the system is

$$U_{\text{potential}} = \frac{Q_A Q_B}{4\pi\epsilon r}$$

$$= \frac{(400 \times 10^{-6}\ \text{C})(600 \times 10^{-6}\ \text{C})}{4\pi\left(8.854 \times 10^{-12}\ \dfrac{\text{C}^2}{\text{N·m}^2}\right)(3\ \text{m})}$$

$$= 719.0\ \text{J}$$

(b) The potential is calculated from Eq. 59.38.

$$V = \frac{U_{\text{potential}}}{Q} = \frac{719.0\ \text{J}}{400 \times 10^{-6}\ \text{C}} = 1.80 \times 10^6\ \text{V}$$

## 16. CURRENT

*Current*, $I$, is the movement of charges. By convention, the current moves in a direction opposite to the flow of electrons (i.e., current flows from the positive terminal to the negative terminal). Current is measured in amperes (A) and is the time rate change of charge. That is, the current is equal to the number of coulombs of charge passing a point each second.

$$I = \frac{dQ}{dt} \qquad \text{59.41}$$

## 17. MAGNETIC FLUX DENSITY

The total amount of *magnetic flux* in a magnetic field is $\Phi_m$, measured in webers (Wb). The *magnetic flux density*, $\mathbf{B}$, in teslas (T), equivalent to Wb/m², is one of two measures of the strength of a magnetic field. For this reason, it can be referred to as the *strength of the B-field*. ($\mathbf{B}$ should never be called the magnetic field strength as that name is reserved for $\mathbf{H}$.) For reasons that will become clearer in Sec. 59.25, $\mathbf{B}$ is also known as the *magnetic induction*. The magnetic flux density is found by dividing the magnetic flux by an area perpendicular to it. Magnetic flux density is a vector quantity (analogous in equation form to displacement, $\mathbf{D}$ (see Eq. 59.17) in an electric field).

$$B = |\mathbf{B}| = \frac{\Phi_m}{A} \qquad \text{59.42}$$

$$\mathbf{B} = \left(\frac{\Phi_m}{A}\right)\mathbf{a} \qquad \text{59.43}$$

## 18. GAUSS'S LAW FOR A MAGNETIC FIELD

*Gauss's law* for a magnetic field is

$$\oint \Phi_m = \oint \mathbf{B}\, dA = 0 \qquad \text{59.44}$$

The physical significance of Eq. 59.44 is that magnetic flux lines are continuous and form closed loops only. The net magnetic flux loss or gain is zero. Thus, a magnetic field has no divergence.

$$\nabla \cdot \mathbf{B} = 0 \qquad \text{59.45}$$

## 19. MAGNETIC FIELD STRENGTH

The *magnetic field strength*, $\mathbf{H}$, with units of A/m is derived from the magnetic flux density.

$$\mathbf{H} = \frac{\mathbf{B}}{\mu} = \left(\frac{m}{4\pi\mu r^2}\right)\mathbf{a} \qquad \text{59.46}$$

$\mu$ is the *permeability* of the medium, calculated from the *permeability of free space*, $\mu_0$ (equal to $4\pi \times 10^{-7}$ H/m, same as Wb/A·m, J/A²·m, and N/A²), and the *relative permeability*, $\mu_r$.[8]

$$\mu = \mu_r \mu_0 \qquad \text{59.47}$$

The energy stored per unit volume (in J/m³) in a magnetic field is

$$U_v = \frac{1}{2}\int_V \mathbf{B} \cdot \mathbf{H}\, d\text{volume}$$

$$= \frac{\mu H^2}{2} = \frac{B^2}{2\mu} \qquad \text{59.48}$$

## 20. FORCE BETWEEN TWO MAGNETIC POLES

It is convenient to have a method of calculating the force between two permanent magnet poles that is analogous to Coulomb's law.[9] To do so requires a definition of magnetic *pole strength* (not to be confused with magnetic field strength). The number of *unit-poles*, $m$ (or just *poles*) is numerically equal to the amount of flux, $\Phi_m$, emanating from the end of the magnet. (In Eq. 59.49, $A_p$ is the pole area.)

$$m = \Phi_m = B_p A_p \qquad \text{59.49}$$

Equation 59.50 gives the force between the pole ends of two magnets separated by a distance $r$. The orientation of the magnets is not important. Vector addition and superposition are required if there are more than two poles.

$$\mathbf{F}_{1-2} = m_2 \mathbf{H}_1 = \left(\frac{m_1 m_2}{4\pi\mu r^2}\right)\mathbf{a} \qquad \text{59.50}$$

---

[8]The linear relationship between $B$ and $H$ is, at best, only an approximation. Not only does $\mu_r$ vary with $B$, but $B$ also reaches a maximum when the magnetic material becomes saturated.

[9]It may be convenient and it may work, but there are no isolated magnetic poles.

Circuits

## 21. FORCE ON A PARTICLE MOVING IN A MAGNETIC FIELD

A magnetic field has no effect on a stationary charged particle. The force on a particle of charge $Q$ moving with velocity v through a magnetic field with flux density $B$ is[10]

$$\mathbf{F} = Q\mathbf{v} \times \mathbf{B} \qquad 59.51$$

$$|\mathbf{F}| = Q\mathrm{v}B \qquad 59.52$$

The acceleration is given by Newton's second law, but since Eq. 59.51 is a vector cross-product, the force (and, consequently, the acceleration), will be directed at right angles to both **v** and **B**. The particle will travel in a circular path with radius and angular velocity of

$$r = \frac{m\mathrm{v}}{QB} \qquad 59.53$$

$$\omega = \frac{QB}{m} \qquad 59.54$$

## 22. WORK IN A MAGNETIC FIELD

The *work* required to change the separation distance of two permanent magnet poles is given by Eq. 59.55.

$$W = \int_{r_1}^{r_2} \mathbf{F} \cdot d\mathbf{r} = \left(\frac{-m_A m_B}{4\pi\mu}\right)\left(\frac{1}{r_2} - \frac{1}{r_1}\right) \qquad 59.55$$

## 23. POTENTIAL ENERGY

The total *magnetic potential energy* in a system of two permanent magnet poles separated by a distance $r$ is

$$U_{\text{potential}} = W_{\infty-r} = \int_{\infty}^{r} \mathbf{F} \cdot d\mathbf{r} = \frac{-m_A m_B}{4\pi\mu r} \qquad 59.56$$

## 24. MAGNETIC POTENTIAL

The *magnetic potential*, $U_m$ (with units of amps, A), possessed by a permanent magnet pole of strength $m_A$ is

$$U_m = \frac{U_{\text{potential}}}{m_A} = \frac{-m_B}{4\pi\mu r} \qquad 59.57$$

The *magnetic potential gradient* is

$$H = \frac{-\Delta U_m}{\Delta r} \qquad 59.58$$

## 25. CURRENT-INDUCED MAGNETIC FIELDS

Current flowing in a conductor will induce a magnetic field, **H** (or **B**) around that wire. (See Table 59.3.) The

**Table 59.3** *Magnetic Field and Inductance for Various Configurations*

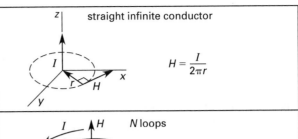

straight infinite conductor

$$H = \frac{I}{2\pi r}$$

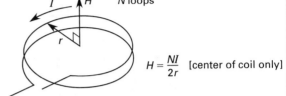

$N$ loops

$$H = \frac{NI}{2r} \quad \text{[center of coil only]}$$

infinite cylindrical coil helix (solenoid)

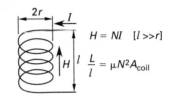

$$H = NI \quad [l \gg r]$$

$$\frac{L}{l} = \mu N^2 A_{\text{coil}}$$

torus (toroidal coil)

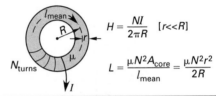

$$H = \frac{NI}{2\pi R} \quad [r \ll R]$$

$$L = \frac{\mu N^2 A_{\text{core}}}{l_{\text{mean}}} = \frac{\mu N^2 r^2}{2R}$$

coaxial cable (high frequencies)

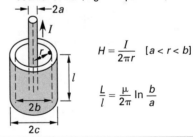

$$H = \frac{I}{2\pi r} \quad [a < r < b]$$

$$\frac{L}{l} = \frac{\mu}{2\pi} \ln \frac{b}{a}$$

parallel transmission lines (high frequencies)

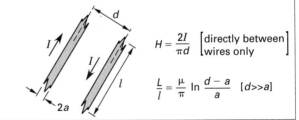

$$H = \frac{2I}{\pi d} \quad \begin{bmatrix} \text{directly between} \\ \text{wires only} \end{bmatrix}$$

$$\frac{L}{l} = \frac{\mu}{\pi} \ln \frac{d-a}{a} \quad [d \gg a]$$

**Circuits**

---

[10]This is a special case of the *Lorentz force equation*. (See Sec. 59.30.)

sense of the magnetic field is given by the right-hand rule.[11] (See Fig. 59.6 and Fig. 59.7.) Assuming constant permeability, the field strength and flux density are proportional to current.

$$\frac{H_1}{H_2} = \frac{B_1}{B_2} = \frac{\Phi_{m,1}}{\Phi_{m,2}} = \frac{I_1}{I_2} \qquad 59.59$$

**Figure 59.6** *Magnetic Field Direction*

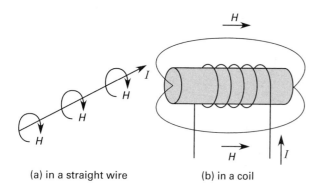

(a) in a straight wire           (b) in a coil

**Figure 59.7** *Right-Hand Rule*

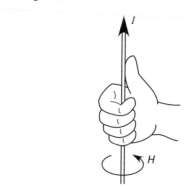

(a) right-hand rule
for straight conductors

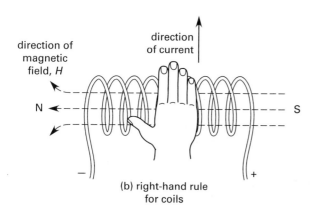

(b) right-hand rule
for coils

The magnetic field a distance $r$ from the center of an infinitely long conductor is given by Eq. 59.60.[12]

$$H = \frac{B}{\mu} = \frac{I}{2\pi r} \qquad 59.60$$

### Example 59.6

A long wire with a radius of 0.003 m carries 4 A of current. Find the magnetic flux density and field strength 0.01 m from the surface of the wire.

*Solution*

From Eq. 59.60, field strength is

$$H = \frac{I}{2\pi r} = \frac{4 \text{ A}}{2\pi(0.003 \text{ m} + 0.01 \text{ m})} = 48.97 \text{ A/m}$$

From Eq. 59.60, the flux density is

$$B = \mu H = \left(4\pi \times 10^{-7} \ \frac{\text{Wb}}{\text{A·m}}\right)\left(48.97 \ \frac{\text{A}}{\text{m}}\right)$$
$$= 6.15 \times 10^{-5} \text{ T}$$

## 26. MAGNETIC FIELD-INDUCED VOLTAGE

Figure 59.8 illustrates a conductor moving orthogonally through a magnetic field, a configuration sometimes called a *linear dynamo*. A voltage known as the *electromotive force* or emf, $V$, proportional to the rate at which the conductors "cut" the flux lines, will be induced in the conductor. The magnitude of this *electromagnetic induction* is given by *Faraday's law*, Eq. 59.61, in which $N$ is the number of conductors and $N\Phi_m$ is the *flux linkage*—the total number of flux lines cut. (The minus sign is specified by *Lenz's law*, which says the induced voltage always acts against the magnetic field. The minus sign is omitted in subsequent equations.)

$$V = -N\frac{d\Phi_m}{dt} \qquad 59.61$$

Since the flux density is $B$, the flux differential can be written in terms of the differential distance traveled.

$$d\Phi_m = B \, dA = Bl \, ds \qquad 59.62$$

The induced voltage can then be written in terms of the velocity of the conductors moving through the field.

$$V = N\frac{d\Phi_m}{dt} = NBl\frac{ds}{dt} = NBl\text{v} \qquad 59.63$$

---

[11]In the case of a straight wire, the thumb indicates the current direction and the fingers curl in the field direction. For a coil, the fingers indicate the current flow and the thumb indicates the field direction.

[12]The same formula can be used for conductors of finite length if *fringing effects* are disregarded.

**Figure 59.8** *Conductor Moving in a Magnetic Field*

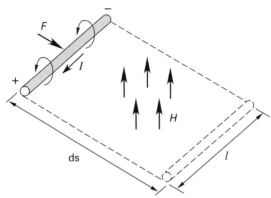

Practical electrical circuits always involve closed loops. A current, $I$, will flow if the conductor is part of a closed loop. The power dissipated in the $N$ conductors is

$$P = IV = NIBlv \qquad 59.64$$

Since power is the time rate of work,

$$P = \frac{dW}{dt} = F\frac{ds}{dt} = Fv \qquad 59.65$$

Setting Eq. 59.64 equal to Eq. 59.65, the force required to push the $N$ conductors through the field is

$$F = NIBl \qquad 59.66$$

Equation 59.66 is applicable only when the conductors cut the magnetic flux at right angles. The more general case, governed by *Ampere's law*, Eq. 59.67, is when the conductor and flux lines meet at an angle $\theta$.

$$F = NIBl\sin\theta \qquad 59.67$$

## 27. FORCE BETWEEN TWO PARALLEL WIRES

The force between two wires carrying currents $I_1$ and $I_2$ is given by Eq. 59.69. This relationship is useful in defining current in terms of force and distance: one ampere of constant current flowing in each of two infinitely long parallel wires one meter apart in a vacuum will produce a force on each wire of $2 \times 10^{-7}$ N per meter of length. From Eq. 59.51,

$$\mathbf{F} = Q\mathbf{v} \times \mathbf{B} = Q\frac{d\mathbf{s}}{dt} \times \mathbf{B} = I\,d\mathbf{s} \times \mathbf{B} \qquad 59.68$$

Then, the force per unit length is found by setting $ds = l = 1$. (**a** is the unit vector from one wire to the other.)

$$\frac{\mathbf{F}}{l} = \mathbf{I} \times \mathbf{B} = \left(\frac{\mu}{2\pi}\right)\left(\frac{I_1 I_2}{r}\right)\mathbf{a} \qquad 59.69$$

Two parallel wires carrying current in the same direction will be attracted to each other. Repulsion will be felt when the currents are in opposite directions.

## 28. INDUCTORS

An *inductor* is basically a coil of wire. When connected across a voltage source, current flows in the coil, establishing a magnetic field that opposes current changes. From Faraday's law, the induced voltage across the ends of the inductor is proportional to the change in flux linkage, which, in turn, is proportional to the current change. The constant of proportionality is the *inductance*, $L$, expressed in units of *henries* (H).

$$V = L\frac{dI(t)}{dt} \qquad 59.70$$

The inductance of an *iron-core inductor* is given by Eq. 59.71, in which $A$ is the cross-sectional area of the core and $l$ is the mean length through the core (corresponding to the mean flux path length).

$$L = \frac{\mu N^2 A}{l} \qquad 59.71$$

The total energy stored in an inductor carrying current $I$ is

$$U = \tfrac{1}{2}LI^2 \qquad 59.72$$

The total inductance of inductors connected in series is

$$L = L_1 + L_2 + L_3 + \cdots + L_n \qquad 59.73$$

The reciprocal of the total inductance of inductors connected in parallel is

$$\frac{1}{L} = \frac{1}{L_1} + \frac{1}{L_2} + \frac{1}{L_3} + \cdots + \frac{1}{L_n} \qquad 59.74$$

## 29. MAGNETIC CIRCUITS

A simple *magnetic circuit* and a simple electrical circuit are analogous. An equation analogous to Ohm's law is written in terms of magnetomotive force and reluctance. The analogy is illustrated in Fig. 59.9.

**Figure 59.9** *Magnetic-Electric Circuit Analogy*

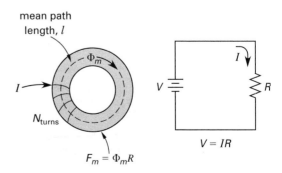

In a magnetic circuit, the flux density is proportional to the *magnetomotive force (mmf)*, $F_m$, with units of amps (A).[13,14]

$$F_m = IN \qquad 59.75$$

*Reluctance*, $\mathcal{R}$, with units of A/Wb (same as the older *rels* unit), is analogous to electrical resistance and depends on the path area, length, and permeability.[15]

$$\mathcal{R} = \frac{l}{\mu A} \qquad 59.76$$

The governing equation (analogous to Ohm's law) for a magnetic circuit is

$$F_m = Hl = \Phi_m \mathcal{R} \qquad 59.77$$

### Example 59.7

A 0.001 m slice is removed from a cast-iron toroidal coil ($\mu_r = 2000$) with a 0.35 m mean toroidal diameter and a 0.07 m core diameter. A steady unknown current flows through 300 turns of wire wrapped around the coil, producing a constant flux density in the core and air gap. What current is required to establish a 10 mWb flux across the air gap?

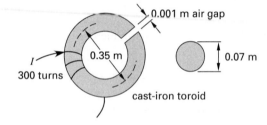

*Solution*

The mean path length through the cast iron is

$$l = \pi(0.35 \text{ m}) - 0.001 \text{ m} = 1.0986 \text{ m}$$

The cross-sectional area of the flux path is

$$A = \frac{\pi}{4} d^2 = \frac{\pi}{4}(0.07 \text{ m})^2 = 0.003849 \text{ m}^2$$

The total reluctance is the sum of the reluctances of the cast iron and air paths.

$$\mathcal{R} = \sum \frac{l}{\mu A} = \frac{1}{\mu_0 A} \sum \frac{l}{\mu_r}$$

$$= \frac{1}{\left(4\pi \times 10^{-7} \dfrac{\text{Wb}}{\text{A·m}}\right)(0.003849 \text{ m}^2)}$$

$$\times \left(\frac{1.0986 \text{ m}}{2000} + \frac{0.001 \text{ m}}{1}\right)$$

$$= 3.203 \times 10^5 \text{ A/Wb}$$

The flux, $\Phi_m$, is given as $10^{-2}$ Wb. From Eq. 59.75 and Eq. 59.77, the current is

$$I = \frac{F_m}{N} = \frac{\Phi_m \mathcal{R}}{N}$$

$$= \frac{(10^{-2} \text{ Wb})\left(3.203 \times 10^5 \dfrac{\text{A}}{\text{Wb}}\right)}{300} = 10.68 \text{ A}$$

## 30. PARTICLES MOVING THROUGH COMBINED MAGNETIC AND ELECTRIC FIELDS

The *Lorentz force equation* gives the force on a particle moving at velocity v when both electric and magnetic fields are present.

$$\mathbf{F} = Q(\mathbf{E} + \mathbf{v} \times \mathbf{B}) \qquad 59.78$$

## 31. POYNTING VECTOR

The *Poynting vector*, **S**, is the instantaneous power density (in W/m²) carried along by an electromagnetic wave. (See Fig. 59.10.) The direction of the Poynting vector is normal to the wave front and in the direction of its motion. In Eq. 59.79, **a** is a unit vector in the direction of the propagation of the wave.

$$\mathbf{S} = \epsilon c E^2 \mathbf{a} = \mathbf{E} \times \mathbf{H} \qquad 59.79$$

**Figure 59.10** *The Poynting Vector*

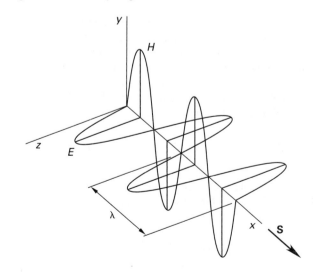

---

[13]The units of Eq. 59.75 appear to be amp-*turns*. However, turns (like *revolutions* and *radians*) are not real units and can be disregarded.

[14]The symbol $V_m$ is also used for magnetomotive force. However, $F_m$ is the standard SI symbol.

[15]In practice, permeability is not constant but depends on the flux density. Therefore, several iterations may be required if either the reluctance or the flux density are unknown.

# 60 Direct-Current Circuits

## Nomenclature

| | | |
|---|---|---|
| $A$ | area | $m^2$ |
| $B$ | magnetic flux density | T |
| $C$ | capacitance | F |
| $d$ | diameter | m |
| $G$ | conductance | S |
| $i, I$ | current | A |
| $l$ | length | m |
| $L$ | inductance | H |
| $N$ | number | – |
| $P$ | power | W |
| $Q$ | charge | C |
| $r$ | radius | m |
| $R$ | resistance | $\Omega$ |
| $t$ | time | s |
| $T$ | temperature | $^\circ$C |
| $T$ | torque | N·m |
| $v, V$ | voltage | V |

## Symbols

| | | |
|---|---|---|
| $\alpha$ | thermal coefficient of resistance | $^\circ$C$^{-1}$ |
| $\rho$ | resistivity | $\Omega$·m |
| $\sigma$ | conductivity | $1/\Omega$·m |
| $\tau$ | time constant | s |

## Subscripts

| | |
|---|---|
| 0 | initial |
| bat | battery |
| $c$ | coil |
| $C$ | capacitor |
| Cu | copper |
| $e$ | equivalent |
| $L$ | inductor |
| $m$ | multiplier |
| $N$ | Norton |
| $R$ | resistor |
| $s$ | shunt or source |
| $t$ | total |
| Th | Thevenin |

## 1. DC VOLTAGE

*Voltage*, V, also known as *electromotive force* (emf), in a *direct-current* (DC) *circuit* may vary in amplitude but not in polarity. In simple problems, its magnitude is also constant (i.e., does not vary with time).[1] Its unit is the *volt* (V), equivalent to combined units of W/A, C/F, J/C, A/S, and Wb/s. Voltage is the same across all points and legs in a parallel circuit.

## 2. RESISTIVITY AND RESISTANCE

*Resistance*, R (measured in ohms, $\Omega$), is the property of a component known as a *resistor* or circuit to impede current flow.[2] A circuit with zero resistance is a *short circuit*, whereas an *open circuit* has infinite resistance. Adjustable resistors are known as *potentiometers* and *rheostats*.

Discrete resistors are usually constructed from carbon compounds, ceramics, oxides, or coiled wire. Resistance depends on the *resistivity*, $\rho$ (in $\Omega$·cm or $\Omega$·in), of the material and the length and cross-sectional area of the resistor. The area, $A$, of circular conductors can be measured in *circular mils*, abbreviated cmils, the area of a one *mil* (0.001 in) diameter circle. In the United

---

[1]The symbol $E$ (electromotive force) is also used to represent voltage.
[2]This is not the same as inductance which is the property of a device to impede a *change* in current flow.

States, 1000 circular mils is abbreviated inconsistently as both MCM and kcmil.

$$R = \frac{\rho l}{A} \qquad 60.1$$

$$A_{\text{cmils}} = \left(\frac{d_{\text{inches}}}{0.001}\right)^2 \qquad 60.2$$

$$A_{\text{in}^2} = 7.854 \times 10^{-7} \times A_{\text{cmils}} \qquad 60.3$$

Resistivity depends on temperature.[3] For most conductors, it increases with temperature, since electron movement through a lattice becomes increasingly difficult at higher temperatures. The variation of resistivity with temperature is specified by the *thermal coefficient of resistance*, $\alpha$, with typical units of $1/°C$.

$$R = R_0(1 + \alpha \Delta T) \qquad 60.4$$

$$\rho = \rho_0(1 + \alpha \Delta T) \qquad 60.5$$

For a combination of resistors in series (i.e., placed end-to-end), the equivalent resistance, $R_e$, is

$$R_e = R_1 + R_2 + R_3 + \cdots \qquad 60.6$$

For a combination of resistors in parallel, the reciprocal of the equivalent resistance is

$$\frac{1}{R_e} = \frac{1}{R_1} + \frac{1}{R_2} + \frac{1}{R_3} + \cdots \qquad 60.7$$

### Example 60.1

What is the resistance of a 2 ft long (61 cm), 0.03 in diameter (0.076 cm) circular wire that has a resistivity of 11 $\Omega$-cmil/ft ($1.83 \times 10^{-6}$ $\Omega$·cm)?

*SI Solution*

From Eq. 60.1,

$$R = \frac{\rho l}{A} = \frac{(1.83 \times 10^{-6} \ \Omega\text{·cm})(61 \ \text{cm})}{\frac{\pi}{4}(0.076 \ \text{cm})^2}$$

$$= 0.0246 \ \Omega$$

*Customary U.S. Solution*

Equation 60.2 gives the cross-sectional area in circular mils.

$$A = \left(\frac{d_{\text{inches}}}{0.001}\right)^2 = \left(\frac{0.03 \ \text{in}}{0.001}\right)^2 = 900 \ \text{cmil}$$

From Eq. 60.1, the resistance is

$$R = \frac{\rho l}{A} = \frac{\left(11 \ \dfrac{\Omega\text{-cmil}}{\text{ft}}\right)(2 \ \text{ft})}{900 \ \text{cmil}} = 0.0244 \ \Omega$$

## 3. CONDUCTIVITY AND CONDUCTANCE

The reciprocals of resistivity and resistance are *conductivity*, $\sigma$, and *conductance*, $G$, respectively. The units of conductance are siemens, S.[4]

$$\sigma = \frac{1}{\rho} \qquad 60.8$$

$$G = \frac{1}{R} \qquad 60.9$$

The *percent conductivity* in Table 60.1 is the ratio of a substance's conductivity to the conductivity of *standard copper*.[5] Alternatively, the conductivity is the ratio of standard copper's resistivity to the substance's resistivity.

$$\% \ \text{conductivity} = \frac{\sigma}{\sigma_{\text{Cu}}} \times 100\% = \frac{\rho_{\text{Cu}}}{\rho} \times 100\% \qquad 60.10$$

**Table 60.1** *Approximate Thermal Coefficients of Resistance and Percent Conductivities*[a]

| material | $\alpha$, $°C^{-1}$ | % IACS conductivity |
|---|---|---|
| aluminum, 99.5% pure | 0.00423 | 63.0 |
| aluminum, 97.5% pure | 0.00435 | 59.8 |
| constantan[b] | 0.00001 | 3.1 |
| copper, IACS (annealed) | 0.00402 | 100.0 |
| copper, pure annealed | 0.00428 | 102.1 |
| copper, hard drawn | 0.00402 | 97.8 |
| gold, 99.9% pure | 0.00377 | 72.6 |
| iron, pure | 0.00625 | 17.5 |
| iron wire, EBB[c] | 0.00463 | 16.2 |
| iron wire, BB[d] | 0.00463 | 13.5 |
| manganin[e] | 0.00000 | 3.41 |
| nickel | 0.00622 | 12.9 |
| platinum, pure | 0.00367 | 14.6 |
| silver, pure annealed | 0.00400 | 108.8 |
| steel wire | 0.00463 | 11.6 |
| tin, pure | 0.00440 | 12.2 |
| zinc, very pure | 0.00406 | 27.7 |

(Multiply $1/°C$ by 0.5556 to obtain $1/°F$.)

[a]temperature between 0°C and 100°C (32°F and 212°F)
[b]58% Cu, 41% Ni, 1% Mn
[c]Extra Best Best grade per ASTM A111 (obsolete)
[d]Best Best grade per ASTM A111 (obsolete)
[e]84% Cu, 4% Ni, 12% Mn

---

[3]A *resistance thermometer* (*resistance temperature detector* or RTD) is a device that uses this characteristic to measure temperature. A *thermistor* is a temperature-sensitive semiconductor.

[4]The siemens is the same as the older unit, the *mho*.
[5]Standard IACS (International Annealed Copper Standard) copper.

The standard IACS resistivity of copper at $20°C$, $\rho_{Cu,20°C}$, is approximately

$$\rho_{Cu,20°C} = 5.8108 \times 10^{-7} \text{ S/m}$$
$$= 1.7241 \times 10^{-6} \text{ Ω·cm}$$
$$= 0.67879 \times 10^{-6} \text{ Ω-in}$$
$$= 10.371 \text{ Ω-cmil/ft} \qquad 60.11$$

For conductances in series, the reciprocal of the *equivalent conductance*, $G_e$, is

$$\frac{1}{G_e} = \frac{1}{G_1} + \frac{1}{G_2} + \frac{1}{G_3} + \cdots \qquad 60.12$$

For conductances in parallel, the equivalent conductance is

$$G_e = G_1 + G_2 + G_3 + \cdots \qquad 60.13$$

## 4. OHM'S LAW

The *voltage drop*, also known as the *IR drop*, across a circuit with resistance $R$ is given by *Ohm's law*.[6]

$$v = iR \qquad 60.14$$

Using Ohm's law implicitly assumes a *linear circuit* (i.e., one consisting of linear elements and linear sources). A *linear element* is a passive element whose performance can be represented by a linear voltage-current relationship. The output of a *linear source* is proportional to the first power of a voltage or current in the circuit.

## 5. ENERGY SOURCES

An *ideal voltage source* supplies power at a constant voltage, regardless of the current drawn. An *ideal current source* is independent of the voltage between its terminals. However, real sources have internal resistances that, at higher currents, decrease the available voltage. Therefore, a *real voltage source* cannot maintain a constant voltage when currents become large. (See Fig. 60.1.) Equation 60.15 gives the *regulation* of a real voltage source.

$$\text{regulation} = \frac{V_{\text{no load}} - V_{\text{full load}}}{V_{\text{full load}}} \times 100\% \qquad 60.15$$

*Independent sources* deliver voltage and current at their rated values regardless of circuit parameters. *Dependent sources* deliver voltage and current at levels determined by a voltage or current elsewhere in the circuit. For example, a voltage source whose output is proportional to the current flowing in a circuit leg is a dependent voltage source.

---

[6]This book uses the convention that uppercase letters represent fixed, maximum, or effective values and lowercase letters represent values that change with time.

**Figure 60.1** *Ideal and Real Energy Sources*

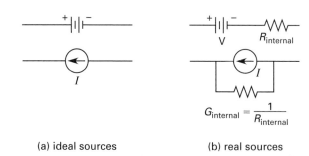

(a) ideal sources　　　　(b) real sources

## 6. VOLTAGE SOURCES IN SERIES AND PARALLEL

Voltage sources connected in series, Fig. 60.2(a), can be reduced to the equivalent circuit shown in Fig. 60.2(b).

$$V_e = \sum V_i \quad \text{[series]} \qquad 60.16$$
$$R_e = \sum R_i \quad \text{[series]} \qquad 60.17$$

**Figure 60.2** *Voltage Sources in Series and Parallel*

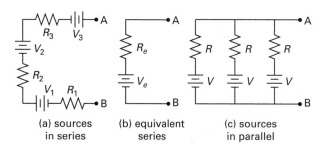

(a) sources in series　(b) equivalent series　(c) sources in parallel

*Millman's theorem* states that $N$ identical voltage sources of voltage $V$ and internal resistance $R$ that are connected in parallel, as in Fig. 60.2(c), can be reduced to the equivalent circuit of Fig. 60.2(b).[7]

$$V_e = V \quad \text{[parallel]} \qquad 60.18$$
$$R_e = \frac{R}{N} \quad \text{[parallel]} \qquad 60.19$$

## 7. SOURCE TRANSFORMATIONS

A voltage source of $V_s$ volts with a series internal resistance of $R_s$ ohms can be replaced by a current source of $I_s$ amperes and internal resistance of (the same) $R_s$ ohms, and vice versa. Therefore, as long as Eq. 60.20 is valid, the two circuits in Fig. 60.3 are equivalent.

$$V_s = I_s R_s \qquad 60.20$$

---

[7]Non-identical sources can be connected in parallel, but the lower-voltage sources may be "charged" by the higher voltage sources. A loop-current analysis (Section 60.23) is needed to determine the currents through the batteries.

Circuits

**Figure 60.3** *Equivalent Sources*

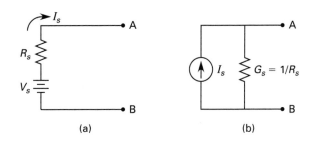

(a)          (b)

## 8. PRACTICAL BATTERIES

Most inexpensive batteries (*dry cells*) for home use operate on the carbon-zinc process. Most automobile batteries operate on the lead-acid process. The *specific energy* of a battery is the energy-to-mass ratio with typical values of 50–75 W·h/kg for lead-acid batteries at room temperature, although modern lithium-ion batteries for hybrid vehicle use have typical values of 100–160 W·h/kg. The *specific peak power* is the power-to-mass ratio, typically in the range of 80–110 W/kg at some standard *degree of discharge*, DoD (usually 50 percent for lead-acid batteries). The specific power for lithium-ion batteries is in the range of 250–350 W/kg.

## 9. THERMOCOUPLES

An electromotive force known as *Seebeck voltage* can be obtained by joining the ends of two dissimilar wires (e.g., iron and constantan) and then keeping the two junctions at different temperatures. Two wires joined in this manner form a *thermocouple*, and the voltage generation is known as the *thermoelectric effect.* (See Fig. 60.4.) If one junction is kept at a fixed reference temperature (say in an ice bath at 0°C) and the other junction is at a temperature *T*, the voltage generated will be approximately

$$V = A + BT + CT^2 \qquad 60.21$$

**Figure 60.4** *A Thermocouple*

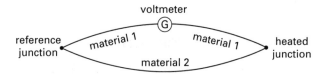

The constants *A*, *B*, and *C* in Eq. 60.21 depend on the particular thermocouple, but once the thermocouple has been calibrated, it can be used to measure temperatures. The voltage increases with temperature up to the *neutral point (temperature)*, after which it decreases back to zero at the *inversion point (temperature)*. Above the inversion temperature, the voltage again increases, but with a reversed polarity. Since the generated voltage varies with temperature, it is convenient to analyze thermocouple data with the aid of published tables. (See Table 60.2.)

**Table 60.2** *Types of Standard Thermocouples*

| type[*] | construction | temperature, °C |
|---|---|---|
| B | Pt-Pt/6%Rh | 0 to +1700 |
| C | tungsten/rhenium | 0 to +2760 |
| E | chromel-constantan | −200 to +870 |
| J | iron-constantan | −200 to +760 |
| K | chromel-alumel | −200 to +1260 |
| M | Ni-Ni/Mo | 0 to +1300 |
| N | nicrosil-Nisil | −200 to +1260 |
| R | Pt-Pt/13%Rh | 0 to +1500 |
| S | Pt-Pt/10%Rh | 0 to +1500 |
| T | copper-constantan | −200 to +400 |

[*]designation given by ISA, ANSI, and ASTM

The output of a thermocouple is in the millivolt range, adequate for measurement but usually inadequate to perform useful work. A *thermopile* is a device constructed as an array of thermocouples (in series or in parallel) to generate a larger voltage or current.

## 10. VOLTAGE AND CURRENT DIVIDERS

Figure 60.5(a) shows a *voltage divider circuit*. The voltage across resistance 2 is

$$v_2 = v\left(\frac{R_2}{R_1 + R_2}\right) \qquad 60.22$$

Figure 60.5(b) shows a *current divider circuit*. The current through resistor 2 is

$$i_2 = i\left(\frac{R_1}{R_1 + R_2}\right) = i\left(\frac{G_2}{G_1 + G_2}\right) \qquad 60.23$$

**Figure 60.5** *Divider Circuits*

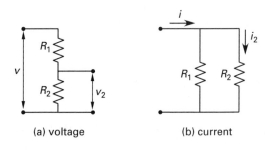

(a) voltage          (b) current

## 11. POWER

The *power* dissipated across two terminals with resistance *R* and voltage drop *v* can be calculated from Eq. 60.24.

$$P = iv = i^2 R = \frac{v^2}{R} = v^2 G \qquad 60.24$$

## 12. DECIBELS

It is customary to express changes in power in *decibels*. Strictly speaking, decibels can be used to express voltage and current ratios in circuits only when the two terminals being measured have the same equivalent resistance. However, it is common practice to use Eq. 60.25 in all cases. When the power ratio is less than 1.0 (i.e., a loss), the power ratio is negative . For cascaded stages in series, the total decibel gain is the sum of the individual stage decibel gains.

$$\text{ratio (in dB)} = 10 \log_{10} \frac{P_2}{P_1}$$

$$\approx 20 \log_{10} \frac{v_2}{v_1} = 20 \log_{10} \frac{i_2}{i_1} \qquad 60.25$$

## 13. MAXIMUM ENERGY TRANSFER

Maximum energy (i.e., maximum "power") transfer from a voltage source occurs when the series source resistance, $R_s$ in Fig. 60.3(a), is reduced to the least possible value (zero if possible) in cases where the load resistance is fixed and the series source resistance can be varied. In cases where the load resistance is variable and the series source resistance is fixed, maximum power transfer occurs when the load resistance equals the series source resistance.

Similarly, maximum energy transfer from a current source occurs when the parallel source resistance, $R_s$ in Fig. 60.3(b), is increased to the highest possible value in cases where the load resistance is fixed. If the load resistance is variable and the source resistance is fixed, then the load resistance must be adjusted to equal the parallel source resistance. The maximum power transfer from a circuit will occur when the load resistance equals the Norton or Thevenin equivalent resistance. (See Sec. 60.21 and Sec. 60.22.)

## 14. KIRCHHOFF'S LAW

*Kirchhoff's current law* (KCL) says that as much current flows out of a *node* (connection) as flows into it.

$$\sum i_{\text{in}} = \sum i_{\text{out}} \qquad 60.26$$

*Kirchhoff's voltage law* (KVL) says that the algebraic sum of voltage drops around any closed path within a circuit is equal to the sum of the applied voltages.

$$\sum V_i = \sum V_j = \sum i R_j \qquad 60.27$$

## 15. SIMPLE SERIES CIRCUITS

In a simple series (*single-loop*) *circuit*, such as shown in Fig. 60.6, the following apply.

- Current is the same through all circuit elements.

$$i = i_{R_1} = i_{R_2} = i_{R_3} \qquad 60.28$$

- The equivalent resistance is the sum of the individual resistances.

$$R_e = R_1 + R_2 + R_3 \qquad 60.29$$

- The equivalent applied voltage is the sum of all voltage sources (polarity considered).

$$V_e = \pm V_1 \pm V_2 \qquad 60.30$$

- The sum of the voltage drops across all components is equal to the equivalent applied voltage (KVL).

$$V_e = I R_e \qquad 60.31$$

**Figure 60.6** *Simple Series Circuit*

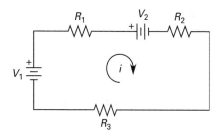

## 16. SIMPLE PARALLEL CIRCUITS

In a simple *parallel circuit* with only one active source, such as shown in Fig. 60.7, the following apply.

- The voltage drop is the same across all legs.

$$V = v_{R_1} = v_{R_2} = v_{R_3}$$
$$= i_1 R_1 = i_2 R_2 = i_3 R_3 \qquad 60.32$$

- The reciprocal of the equivalent resistance is the sum of the reciprocals of the individual resistances.

$$\frac{1}{R_e} = \frac{1}{R_1} + \frac{1}{R_2} + \frac{1}{R_3}$$
$$G_e = G_1 + G_2 + G_3 \qquad 60.33$$

- The total current is the sum of the leg currents (KCL).

$$i = i_1 + i_2 + i_3$$
$$= \frac{V}{R_1} + \frac{V}{R_2} + \frac{V}{R_3} = V(G_1 + G_2 + G_3) \qquad 60.34$$

**Figure 60.7** *Simple Parallel Circuit*

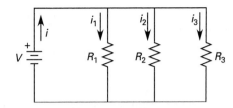

## 17. ANALYSIS OF COMPLEX RESISTIVE NETWORKS

The following procedure should be followed to establish the current and voltage drops in a complex resistive network.

*step 1:* If the circuit is three-dimensional, draw its two-dimensional representation.

*step 2:* Combine series voltage and parallel current sources.

*step 3:* Combine series resistances.

*step 4:* Combine parallel resistances.

*step 5:* Repeat steps 2 through 4 as many times as needed.

*step 6:* If applicable, use the delta-wye transformation. (See Sec. 60.18.)

## 18. DELTA-WYE TRANSFORMATIONS

The equivalent resistances for resistors in *wye* and *delta* configurations (see Fig. 60.8) are

$$R_1 = \frac{R_a R_c}{R_a + R_b + R_c} \qquad 60.35$$

$$R_2 = \frac{R_a R_b}{R_a + R_b + R_c} \qquad 60.36$$

$$R_3 = \frac{R_b R_c}{R_a + R_b + R_c} \qquad 60.37$$

$$R_a = \frac{R_1 R_2 + R_1 R_3 + R_2 R_3}{R_3} \qquad 60.38$$

$$R_b = \frac{R_1 R_2 + R_1 R_3 + R_2 R_3}{R_1} \qquad 60.39$$

$$R_c = \frac{R_1 R_2 + R_1 R_3 + R_2 R_3}{R_2} \qquad 60.40$$

**Figure 60.8** *Wye and Delta Configurations*

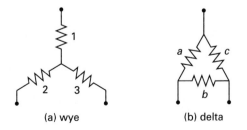

(a) wye          (b) delta

## Example 60.2

Simplify the circuit and determine the total current.

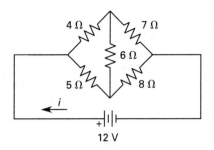

*Solution*

Convert the 4 Ω-5 Ω-6 Ω delta connection to wye form.

$$R_1 = \frac{R_a R_c}{R_a + R_b + R_c} = \frac{(5 \ \Omega)(4 \ \Omega)}{5 \ \Omega + 6 \ \Omega + 4 \ \Omega}$$
$$= 1.33 \ \Omega$$

$$R_2 = \frac{(5 \ \Omega)(6 \ \Omega)}{15 \ \Omega} = 2 \ \Omega$$

$$R_3 = \frac{(6 \ \Omega)(4 \ \Omega)}{15 \ \Omega} = 1.6 \ \Omega$$

The transformed circuit is

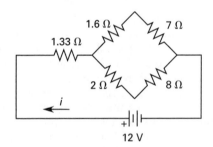

The total equivalent resistance is

$$R_e = 1.33 \ \Omega + \frac{1}{\dfrac{1}{1.6 \ \Omega + 7 \ \Omega} + \dfrac{1}{2 \ \Omega + 8 \ \Omega}} = 5.95 \ \Omega$$

The current is

$$i = \frac{V}{R_e} = \frac{12 \ \text{V}}{5.95 \ \Omega} = 2.02 \ \text{A}$$

## 19. RECIPROCITY THEOREM

The *reciprocity theorem* says that the current present in any branch of a linear, single-voltage source network is interchangeable in location with the voltage source without affecting the current. That is, if a voltage, *v*, in branch 1 causes a current, *i*, to flow in branch 2, then the same voltage placed in branch 2 will cause the same current to flow in branch 1. Alternatively, in a linear resistor network, the ratio of applied voltage to current measured at any point is identical to the ratio obtained if the source and meter locations are exchanged. (See Fig. 60.9.) This ratio is known as the *transfer resistance.*

$$R_{\text{transfer}} = \frac{v_1}{i_1} = \frac{v_2}{i_2} \qquad 60.41$$

**Figure 60.9** *Reciprocal Measurements*

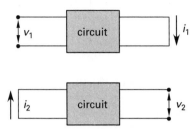

## 20. SUPERPOSITION THEOREM

The *superposition theorem* says that the response of (i.e., voltage across or current through) a linear circuit element fed by two or more independent sources is equal to the response to each source taken individually with all other sources set to zero (i.e., voltage sources shorted and current sources opened).

**Example 60.3**

Determine the current through the center leg.

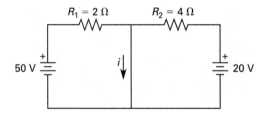

*Solution*

First, work with the left (50 V) battery. Short out the right (20 V) battery.

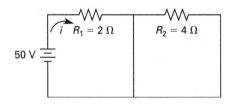

No current will flow through $R_2$. The equivalent resistance and current are

$$R_e = 2 \ \Omega$$
$$i = \frac{V}{R_e} = \frac{50 \ \text{V}}{2 \ \Omega} = 25 \ \text{A}$$

Next, work with the right (20 V) battery. Short out the left (50 V) battery.

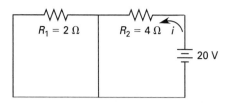

No current will flow through $R_1$. The equivalent resistance and current are

$$R_e = 4 \ \Omega$$
$$i = \frac{V}{R_e} = \frac{20 \ \text{V}}{4} = 5 \ \text{A}$$

Considering both batteries, the total current flowing is 25 A + 5 A = 30 A.

## 21. NORTON'S THEOREM

*Norton's theorem* says that a linear, two-terminal network with dependent or independent sources can be represented by an equivalent circuit consisting of a single current source and resistor in parallel. (See Fig. 60.10.) The *Norton equivalent current*, $i_N$, is the short-circuit current that flows through a shunt across terminals A and B. The *Norton equivalent resistance*, $R_N$, is the resistance across terminals A and B when all independent sources are set to zero (i.e., short-circuiting voltage sources and open-circuiting current sources).[8] The *Norton equivalent voltage*, $v_N$, is measured with terminals open.

$$i_N = \frac{v_N}{R_N} \qquad 60.42$$

**Figure 60.10** *Norton Equivalent Circuit*

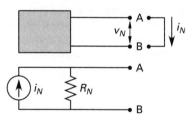

**Example 60.4**

Find the current through resistor $R_3$.

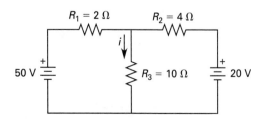

---

[8]This is usually called the *Norton equivalent impedance.*

*Solution*

Remove the load and determine the Norton equivalent resistance.

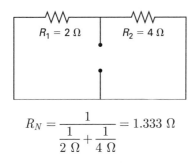

$$R_N = \cfrac{1}{\cfrac{1}{2\ \Omega} + \cfrac{1}{4\ \Omega}} = 1.333\ \Omega$$

Short the terminals and find the Norton equivalent current. The current was found in Ex. 60.3 to be 30 A. The Norton equivalent circuit is

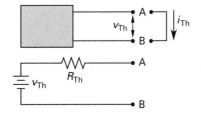

The current through $R_3$ is found from the Norton current and the current division principle (see Sec. 60.10).

$$I_{R_3} = (30\ \text{A})\left(\frac{1.333\ \Omega}{1.333\ \Omega + 10\ \Omega}\right) = 3.53\ \text{A}$$

## 22. THEVENIN'S THEOREM

*Thevenin's theorem* says that a linear, two-terminal network with dependent and independent sources can be represented by a *Thevenin equivalent circuit* consisting of a voltage source in series with a resistor. (See Fig. 60.11.) The *Thevenin equivalent voltage*, $v_{\text{Th}}$ in Eq. 60.43, is the open-circuit voltage across terminals A and B. The *Thevenin equivalent resistance*, $R_{\text{Th}}$, is the resistance across terminals A and B when all independent sources are set to zero (i.e., short-circuiting voltage sources and open-circuiting current sources).[9]

$$v_{\text{Th}} = i_{\text{Th}} R_{\text{Th}} \qquad\qquad 60.43$$

**Figure 60.11** *Thevenin Equivalent Circuit*

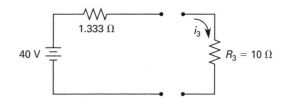

---

[9]This is usually called the *Thevenin equivalent impedance*.

## Example 60.5

Solve Ex. 60.4 using a Thevenin equivalent circuit.

*Solution*

The short-circuit current through the center leg was found in Ex. 60.3 to be 30 A. The equivalent resistance was found in Ex. 60.4 to be 1.333 $\Omega$. Therefore, the Thevenin equivalent voltage is

$$V_{\text{Th}} = i_{\text{Th}} R_{\text{Th}} = (30\ \text{A})(1.333\ \Omega) = 40\ \text{V}$$

The Thevenin equivalent circuit is

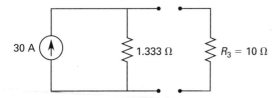

The current $i_3$ is

$$i_3 = \frac{V}{R_e} = \frac{40\ \text{V}}{1.333\ \Omega + 10\ \Omega} = 3.53\ \text{A}$$

## 23. LOOP-CURRENT METHOD

The *loop-current method* (also known as the *mesh current* and *Maxwell loop-current* methods) is a direct extension of Kirchhoff's voltage law and is particularly valuable in determining unknown currents in circuits with several loops and energy sources. It requires writing $n - 1$ simultaneous equations for an $n$-loop system.

*step 1:* Select $n - 1$ loops (i.e., one less than the total number of loops).

*step 2:* Assume current directions for the chosen loops. (Any current whose direction is chosen incorrectly will end up with a negative current in step 4.) Show the direction with an arrow.

*step 3:* Write Kirchhoff's voltage law for each of the $n - 1$ chosen loops. A voltage source is positive when the assumed current direction is from the negative to the positive battery terminal. Voltage ($iR$) drops are always positive.

*step 4:* Solve the $n - 1$ equations (from step 3) for the unknown currents.

## Example 60.6

Find the current in the 0.5 Ω resistor.

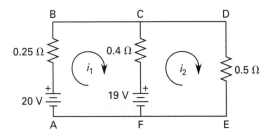

*Solution*

This is a three-loop network—ABCF, FCDE, and ABCDEF. Two simultaneous equations are required. If the first loop is ABCF and the second is FCDE, the assumed currents are $i_1$ and $i_2$. The directions chosen are arbitrary. Kirchhoff's voltage law for loop ABCF is

$$\sum v_i = \sum iR$$
$$20\text{ V} - 19\text{ V} = (0.25\ \Omega)i_1 + (0.4\ \Omega)(i_1 - i_2)$$

The battery polarities determine the signs on the left-hand side of the equation. The polarity of the 19 V battery is opposite to the assumed current flow direction and is negative. For loop FCDE,

$$\sum v_i = \sum iR$$
$$19\text{ V} = (0.4\ \Omega)(i_2 - i_1) + (0.5\ \Omega)i_2$$

Solving these two loop equations results in $i_1 = 20$ A and $i_2 = 30$ A. The current through the 0.5 Ω resistor is 30 A.

## 24. NODE-VOLTAGE METHOD

The *node-voltage method* is an extension of Kirchhoff's current law. While currents can be determined with it, its primary use is in finding voltage potentials at various points (nodes) in the circuit. (A node is a point where three or more wires connect.)

*step 1:* Convert all current sources to voltage sources.

*step 2:* Choose one node as the voltage reference (i.e., 0 V) node. Usually, this will be the circuit ground—a node to which at least one negative battery terminal is connected.

*step 3:* Identify the unknown voltage potentials at all other nodes referred to the reference node.

*step 4:* Write Kirchhoff's current law for all unknown nodes. (This excludes the reference node.)

*step 5:* Write all currents in terms of voltage drops.

*step 6:* Write all voltage drops in terms of the node voltages.

## Example 60.7

Find (a) the voltage potential for node A and (b) the current $i_1$.

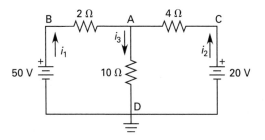

*Solution*

(a) Choose node D as the reference voltage. So, $v_D = 0$. Node A is the only other true node. Kirchhoff's current law for node A is

$$i_1 + i_2 = i_3$$

Write the current equation in terms of the voltage drops and resistances. The directions chosen determine the polarity.

$$\frac{v_{BA}}{2\ \Omega} + \frac{v_{CA}}{4\ \Omega} = \frac{v_{AD}}{10\ \Omega}$$

Write the voltage drops in terms of the node voltages.

$$\frac{v_B - v_A}{2\ \Omega} + \frac{v_C - v_A}{4\ \Omega} = \frac{v_A - v_D}{10\ \Omega}$$

However, node D was chosen as the reference, so $v_D = 0$. Furthermore, $v_B = 50$ V and $v_C = 20$ V. So,

$$\frac{50\text{ V} - v_A}{2\ \Omega} + \frac{20\text{ V} - v_A}{4\ \Omega} = \frac{v_A}{10\ \Omega}$$

Solving, $v_A = 35.3$ V.

(b)
$$i_1 = \frac{v_{BA}}{2\ \Omega} = \frac{v_B - v_A}{2\ \Omega}$$
$$= \frac{50\text{ V} - 35.3\text{ V}}{2\ \Omega}$$
$$= 7.35\text{ A}$$

## 25. D'ARSONVAL METERS

The *d'Arsonval meter* (*galvanometer* or *permanent-magnet moving coil mechanism*) is essentially a current-measuring device. Current flowing through a coil creates a magnetic field that interacts with the field from a permanent magnet to produce a torque to move the indicator needle. The meter can be used to measure current (i.e., an *ammeter*) or voltage (i.e., a *voltmeter*), depending on how an additional resistor is connected to the basic meter coil.

Equation 60.44 gives the torque developed by an $N$-turn coil in a d'Arsonval movement.

$$T_{\text{coil}} = BNi_{\text{coil}}A_{\text{coil}} \qquad 60.44$$

$$A_{\text{coil}} = d_{\text{coil}}l_{\text{coil}} \qquad 60.45$$

The meter can be represented by the simple equivalent circuit shown in Fig. 60.12(a). A current, $i_{\text{max}}$, causes the maximum needle excursion. To measure a greater current, a *shunt resistor*, $R_s$ in Fig. 60.12(b), must be used. If $i_t$ is to cause full-scale movement, a shunt resistance specified by Eq. 60.46 will be required.

$$R_s = \frac{i_{\text{max}}R_c}{i_t - i_{\text{max}}} \qquad 60.46$$

$$I_{\text{actual}} = I_{\text{indicated}} \times \left(\frac{R_s + R_c}{R_s}\right) \qquad 60.47$$

If the meter is used to measure a voltage greater than its maximum rating, $v_{\text{max}}$ (corresponding to $i_{\text{max}}R_c$), then a series *multiplier resistor*, $R_m$ in Fig. 60.12(c), is required.

**Figure 60.12** *Equivalent Meter Circuits*

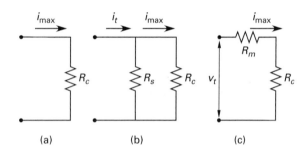

(a)  (b)  (c)

$$R_m = \frac{v_t - i_{\text{max}}R_c}{i_{\text{max}}} \qquad 60.48$$

$$V_{\text{actual}} = V_{\text{indicated}}\left(\frac{R_m + R_c}{R_c}\right) \qquad 60.49$$

## 26. WHEATSTONE BRIDGE

A *Wheatstone bridge* is used to measure an unknown resistance, say $R_1$ in Fig. 60.13. An adjustable resistance is adjusted (i.e., the bridge is "balanced") until no current flows through the *microammeter* (also known as a *null indicator*, hence the name *zero-indicating bridge*). The reciprocity theorem (see Sec. 60.19) can be used if the voltage source and ammeter are reversed. When no current flows through the meter leg, $i_2 = i_4$ and $i_1 = i_3$.

$$v_1 + v_3 = v_2 + v_4 \qquad 60.50$$

$$\frac{R_1}{R_2} = \frac{R_3}{R_4} \qquad 60.51$$

**Figure 60.13** *Wheatstone Bridge*

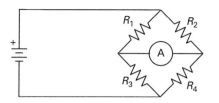

## 27. TRANSIENT ANALYSIS

When a charged capacitor (see Sec. 60.2) is connected across a resistor, the voltage across the capacitor will gradually decrease and approach zero as energy is dissipated in the resistor. Similarly, when an inductor (see Sec. 60.27) through which a steady current is flowing is suddenly connected across a resistor, the current will gradually decrease and approach zero. These gradual decreases are known as *transient behavior*.

Each of these cases assume that any energy sources are disconnected at the time the resistor is connected—hence the name *source-free circuits*. However, only discharging circuits are source-free; charging (energizing) circuits require the instantaneous application of an energy source. Figure 60.14 shows an energizing *series-RL circuit*. Writing Kirchhoff's voltage law around the loop,

$$V = i(t)R + \frac{L\,di(t)}{dt} \qquad 60.52$$

**Figure 60.14** *Energizing Series-RL Circuit*

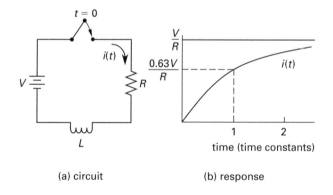

(a) circuit  (b) response

Equation 60.52 is a first-order linear differential equation with the solution of

$$i(t) = \frac{V}{R}\left(1 - e^{-Rt/L}\right) \qquad 60.53$$

The *time constant*, $\tau$, for a series-$LR$ circuit is the time it takes for the current to reach approximately 63.3 percent of its steady-state value, which is $i = V/R$. Substituting $i(\tau) = 0.663\,V/R$ into Eq. 60.53 results in $\tau = L/R$ for a series-$RL$ circuit. (The time constant for a series-$RC$ circuit is $\tau = RC$.) In general, transient variables will have essentially reached their steady-state values after five time constants.

Table 60.3 gives the solutions to simple transient problems. Notice that many response equations can be written in terms of either $t$ or $\tau$. Time (in either seconds or time constants) is assumed to begin when a switch is opened or closed, as appropriate.

**Table 60.3** *Transient Response*

| type of circuit | response |
|---|---|
| series $RC$, charging <br> $\tau = RC$ <br> $e^{-N} = e^{-t/\tau} = e^{-t/RC}$ | $V_{bat} = v_R(t) + v_C(t)$ <br> $i(t) = \dfrac{V_{bat} - V_0}{R} e^{-N}$ <br> $v_R(t) = i(t)R = (V_{bat} - V_0)e^{-N}$ <br> $v_C(t) = V_0 + (V_{bat} - V_0)(1 - e^{-N})$ <br> $Q_C(t) = C(V_0 + (V_{bat} - V_0) $ <br> $(1 - e^{-N}))$ |
| series $RC$, discharging <br> $\tau = RC$ <br> $e^{-N} = e^{-t/\tau} = e^{-t/RC}$ | $0 = v_R(t) + v_C(t)$ <br> $i(t) = \dfrac{V_0}{R} e^{-N}$ <br> $v_R(t) = -V_0 e^{-N}$ <br> $v_C(t) = V_0 e^{-N}$ <br> $Q_C(t) = C V_0 e^{-N}$ |
| series $RL$, charging <br> $\tau = L/R$ <br> $e^{-N} = e^{-t/\tau} = e^{-tR/L}$ | $V_{bat} = v_R(t) + v_L(t)$ <br> $i(t) = I_0 e^{-N} + \dfrac{V_{bat}}{R}(1 - e^{-N})$ <br> $v_R(t) = i(t)R$ <br> $\quad = I_0 R e^{-N} + V_{bat}(1 - e^{-N})$ <br> $v_L(t) = (V_{bat} - I_0 R)e^{-N}$ |
| series $RL$, discharging <br> $\tau = L/R$ <br> $e^{-N} = e^{-t/\tau} = e^{-tR/L}$ | $0 = v_R(t) + v_L(t)$ <br> $i(t) = I_0 e^{-N}$ <br> $v_R(t) = I_0 R e^{-N}$ <br> $v_L(t) = -I_0 R e^{-N}$ |

**Example 60.8**

A series-$RC$ circuit contains a 4 $\Omega$ resistor and an uncharged 300 $\mu$F capacitor. The switch is closed at $t = 0$, connecting the circuit across a 500 V source. What are (a) the time constant, (b) the steady-state current, and (c) the current as a function of time?

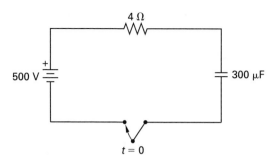

*Solution*

(a) The time constant for a series-$RC$ circuit is

$$\tau = RC = (4\ \Omega)(300 \times 10^{-6}\ \text{F}) = 1.2 \times 10^{-3}\ \text{s}$$

(b) A capacitor acts like an open circuit with a DC voltage source. The steady-state current is zero.

(c) From Table 60.3, the current response is

$$i(t) = \frac{V_{bat} - V_0}{R} e^{-t/\tau} = \frac{500\ \text{V} - 0}{4\ \Omega} e^{-t/1.2 \times 10^{-3}\ \text{s}}$$

$$= 125 e^{-833t}\ \text{A}$$

**Example 60.9**

At $t = 0$, the switch moves from position A to position B. At $t = 6$ s, the switch goes from position B to C. The capacitor is initially uncharged.

(a) What is the voltage on the capacitor just before the switch moves to position C?

(b) What is the current through resistor $R_2$ at $t = 10$ s?

(c) While the switch is in position C, when does the voltage across the capacitor reach 10 volts?

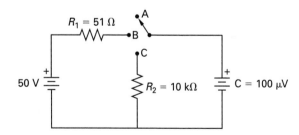

*Solution*

(a) The time constant is

$$\tau = R_1 C = (51 \times 10^3\ \Omega)(100 \times 10^{-6}\ \text{F}) = 5.1\ \text{s}$$

From Table 60.3, the voltage on the capacitor is

$$v_C(t) = V_0 + (V_{bat} - V_0)(1 - e^{-t/\tau})$$

$$= (50\ \text{V})(1 - e^{-6\ \text{s}/5.1\ \text{s}}) = 34.58\ \text{V}$$

Circuits

(b) The duration of discharge is $t' = 10$ s $- 6$ s $= 4$ s. The time constant is

$$\tau = R_2 C = (10 \times 10^3 \ \Omega)(100 \times 10^{-6} \ \text{F}) = 1 \ \text{s}$$

$$v_C(t') = V_0 + (V_{\text{bat}} - V_0)(1 - e^{-t'/\tau})$$

$$= V_0 e^{-t'/\tau} \ \text{when} \ V_{\text{bat}} = 0$$

$$= (34.58 \ \text{V})e^{-4 \ \text{s}/1 \ \text{s}} = 0.633 \ \text{V}$$

$$I = \frac{V}{R_2} = \frac{0.633 \ \text{V}}{10 \times 10^3 \ \Omega} = 6.33 \times 10^{-5} \ \text{A}$$

(c)

$$v_C(t') = V_0 e^{-t'/\tau} \ \text{when} \ V_{\text{bat}} = 0$$

$$10 \ \text{V} = (34.58 \ \text{V})e^{-t'/1}$$

Taking the natural logarithm of both sides,

$$\ln 10 = \ln 34.58 + (-t')$$

$$t' = 1.24 \ \text{s}$$

The true time is

$$t = t' + 6 \ \text{s} = 1.24 \ \text{s} + 6 \ \text{s}$$

$$= 7.24 \ \text{s}$$

Circuits

# 61 Alternating-Current Circuits

## Nomenclature

| | | |
|---|---|---|
| $a$ | inner radius | m |
| $a$ | ratio of transformation | – |
| $b$ | outer radius | m |
| $B$ | magnetic flux density | T |
| $B$ | susceptance | S |
| BW | bandwidth | Hz |
| $c$ | speed of light | m/s |
| $C$ | capacitance | F |
| CF | crest factor | – |
| ERP | effective radiated power | W |
| $f$ | frequency | Hz |
| FF | form factor | – |
| $G$ | conductance | S |
| $h$ | hybrid parameter | various |
| $i, I$ | current[1] | A |
| $k$ | coefficient of coupling | – |
| $k$ | velocity factor | – |
| $L$ | inductance | H |
| $n$ | Steinmetz exponent | – |
| $N$ | number of turns | – |
| $P$ | power or power loss | W |
| $Q$ | quality factor | – |
| $Q$ | reactive power | VAR |
| $R$ | resistance | $\Omega$ |
| $S$ | apparent power | VA |
| SWR | standing wave ratio | – |
| $t$ | time | s |
| $T$ | period | s |
| $U$ | energy | J |
| v | velocity | m/s |
| $v, V$ | voltage[2] | V |
| X | reactance | $\Omega$ |
| $y, Y$ | admittance | S |
| $z, Z$ | impedance | $\Omega$ |

## Symbols

| | | |
|---|---|---|
| $\Gamma$ | reflection coefficient | – |
| $\epsilon$ | permittivity | F/m |
| $\eta$ | efficiency | – |
| $\theta$ | phase angle | rad |
| $\lambda$ | wavelength | m |
| $\mu$ | permeability | H/m |
| $\rho$ | resistivity | $\Omega \cdot$m |
| $\sigma$ | conductivity | $1/\Omega \cdot$m |
| $\tau$ | time constant | s |
| $\phi$ | impedance angle | rad |
| $\phi$ | phase difference angle | rad |
| $\Phi$ | magnetic flux | Wb |
| $\omega$ | angular frequency | rad/s |

## Subscripts

| | | |
|---|---|---|
| 0 | at resonance, characteristic | |
| ave | average | |
| $C$ | capacitor | |
| Cu | copper | |
| $e$ | eddy current or equivalent | |

---

[1]The unsubscripted variables $V$ and $I$ in this chapter represent effective values.
[2]See Ftn. 1.

Circuits

| | |
|---|---|
| eff | effective |
| ep | effective primary |
| ext | external |
| $h$ | hysteresis |
| $i$ | in or imaginary |
| int | internal |
| $l$ | per unit length |
| $L$ | inductor |
| $m$ | maximum |
| $M$ | mutual inductance |
| $o$ | out |
| $p$ | primary |
| $r$ | radiation or real |
| $R$ | resistor |
| $s$ | secondary |
| $t$ | total |
| $w$ | wave |

## 1. AC VOLTAGE

The term *alternating waveform* describes any symmetrical waveform, including square, sawtooth, triangular, and sinusoidal waves, whose polarity varies regularly with time. However, the term "AC" (i.e., *alternating current*) almost always means that the current is produced from the application of a sinusoidal voltage.[3] Sinusoidal variables can be specified without loss of generality as either sines or cosines.[4] If a sine waveform is used, Eq. 61.1 gives the instantaneous voltage as a function of time. $V_m$ is the *maximum value* (also known as the *amplitude*) of the sinusoid. If $V(t)$ is not zero at $t = 0$, a *phase angle*, $\theta$, must be used.[5]

$$v(t) = V_m \sin(\omega t + \theta) \qquad 61.1$$

Figure 61.1 illustrates the *period of the waveform, T*. (Since the horizontal axis corresponds to time and not distance, the waveform does not have a wavelength.)

**Figure 61.1** *Sinusoidal Waveform with Phase Angle*

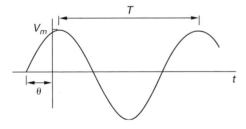

The *frequency, f*, of the sinusoid is the reciprocal of the period in hertz (Hz). *Angular frequency, $\omega$*, in rad/s can

also be used. Table 61.1 describes standard electromagnetic frequency bands.

$$f = \frac{1}{T} = \frac{\omega}{2\pi} \qquad 61.2$$

$$\omega = 2\pi f = \frac{2\pi}{T} \qquad 61.3$$

**Table 61.1** *Standard Electromagnetic Frequency Band Designations*

| designation | meaning | frequency range |
|---|---|---|
| ULF | ultra-low frequency | 0.01 Hz to 10 Hz |
| ELF | extremely low frequency | 10 Hz to 3 kHz |
| VLF | very low frequency | 3 kHz to 30 kHz |
| LF | low frequency | 30 kHz to 300 kHz |
| MF | medium frequency | 300 kHz to 3 MHz |
| HF | high frequency | 3 MHz to 30 MHz |
| VHF | very high frequency | 30 MHz to 300 MHz |
| UHF | ultra-high frequency | 300 MHz to 3 GHz |
| SHF | super high frequency | 3 GHz to 30 GHz |
| EHF | extremely high frequency | 30 GHz to 300 GHz |

The imaginary exponentials $e^{j\theta}$ and $e^{-j\theta}$ can be combined to produce $\sin\theta$ and $\cos\theta$ terms. Therefore, it is not surprising that there are several equivalent methods of indicating a sinusoidal waveform in abbreviated form:

- *trigonometric:* $V_m \sin(\omega t + \theta)$
- *exponential:* $V_m e^{j\theta}$
- *polar* or *phasor:* $V_m \angle \theta$
- *rectangular:* $V_r + jV_i$

## 2. DEFECTS IN AC VOLTAGE

An AC voltage is rarely perfectly consistent in magnitude, phase, and frequency. The following are typical residential AC voltage defects.

- *blackout:* a complete failure that lasts more than one cycle
- *brownout:* a decrease in the steady-state voltage amplitude
- *hunting:* a frequency deviation effect resulting from irregular generation (i.e., uneven speed of the generator's prime mover)
- *line noise:* interference that appears as small magnitude variations "riding" on the regular waveform caused by radio frequency and electromagnetic sources
- *chronic overvoltage:* an increase in the steady-state voltage amplitude that lasts for a long time
- *sag (or dip):* a disturbance (similar to a brownout but of shorter duration) that occurs when the line voltage drops below approximately 80% to 85% of its rated voltage by one or more cycles

---

[3]With few exceptions, only sinusoidal voltages and currents are covered in this chapter.

[4]Since the point at which time begins (i.e., $t = 0$) is irrelevant in steady-state AC circuit problems, it makes no difference whether a sine or cosine waveform is used.

[5]*Phase* is not the same as *phase difference*, the difference in phase between corresponding points on two sinusoids of the same frequency. (See Sec. 61.7.)

- *spike:* a high-voltage (up to 6000 V or more) peak lasting approximately 100 $\mu$s to one-half of a cycle

- *surge:* a disturbance (similar to an overvoltage, but of much shorter duration) that occurs when the line voltage exceeds 110% of its rated voltage for one or more cycles

- *transient:* a high-voltage (up to 20,000 V or more) peak lasting approximately 10 $\mu$s to 100 $\mu$s

## 3. AC VOLTAGE SOURCES

The symbols for an AC voltage source are shown in Fig. 61.2. When two or more AC sources are connected in a circuit, their effects can be determined from the superposition principle.

**Figure 61.2** AC Voltage Sources

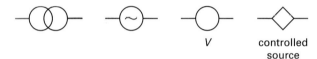

V          controlled
           source

## 4. AVERAGE VALUE

Equation 61.4 calculates the *average value* of any periodic voltage. For waveforms that are symmetrical with respect to the horizontal time axis, however, Eq. 61.4 is zero. Therefore, the average is taken over only half of a cycle and has a value of $2V_m/\pi$ for a sinusoid. (Table 61.2 gives the values for other waveforms.) This is equivalent to taking the average of the *rectified waveform* (i.e., the absolute value of the waveform). A DC current equal to the average value of a rectified AC current has the same electrolytic action (e.g., capacitor charging, plating operations, and ion formation).

$$V_{\text{ave}} = \frac{1}{2\pi}\int_0^{2\pi} v(\theta)\,d\theta = \frac{1}{T}\int_0^{T} v(t)\,dt \qquad 61.4$$

$$V_{\text{ave}} = \frac{1}{\pi}\int_0^{\pi} v(\theta)\,d\theta = \frac{2V_m}{\pi} \quad \text{[rectified sinusoid]} \qquad 61.5$$

The needle movement of a typical DC current meter is proportional to the average current. Unless a sinusoidal waveform is rectified, the reading will be zero.

### Example 61.1

A plating tank with an effective fluid resistance of 100 $\Omega$ is connected to a full-wave diode rectifier. The applied voltage is sinusoidal with a maximum value of 170 V. How much time is required for a total transfer of 0.005 faradays? Assume ideal diodes.

**Table 61.2** *Characteristics of Alternating Waveforms*

| waveform | $\dfrac{V_{\text{ave}}}{V_m}$ | $\dfrac{V_{\text{rms}}}{V_m}$ | FF | CF |
|---|---|---|---|---|
| sinusoid | 0 | $\dfrac{1}{\sqrt{2}}$ | – | $\sqrt{2}$ |
| full-wave rectified sinusoid | $\dfrac{2}{\pi}$ | $\dfrac{1}{\sqrt{2}}$ | $\dfrac{\pi}{2\sqrt{2}}$ | $\sqrt{2}$ |
| half-wave rectified sinusoid | $\dfrac{1}{\pi}$ | $\dfrac{1}{2}$ | $\dfrac{\pi}{2}$ | 2 |
| symmetrical square wave | 0 | 1 | – | 1 |
| unsymmetrical square wave | $\dfrac{t}{T}$ | $\sqrt{\dfrac{t}{T}}$ | $\sqrt{\dfrac{T}{t}}$ | $\sqrt{\dfrac{T}{t}}$ |
| sawtooth and symmetrical triangular | 0 | $\dfrac{1}{\sqrt{3}}$ | – | $\sqrt{3}$ |
| sawtooth and symmetrical triangular | $\dfrac{1}{2}$ | $\dfrac{1}{\sqrt{3}}$ | $\dfrac{2}{\sqrt{3}}$ | $\sqrt{3}$ |

*Solution*

The average value of the rectified sinusoidal voltage is

$$V_{\text{ave}} = \frac{2V_m}{\pi} = \frac{(2)(170\,\text{V})}{\pi} = 108.2\,\text{V}$$

The average current is given by Ohm's law.

$$I_{\text{ave}} = \frac{V_{\text{ave}}}{R} = \frac{108.2\,\text{V}}{100\,\Omega} = 1.082\,\text{A}$$

Since a faraday is equal to 96 485 A·s, the plating time is

$$t = \frac{\text{no. of faradays}}{I_{\text{ave}}} = \frac{(0.005\ \text{faraday})\left(96\,485\ \dfrac{\text{A·s}}{\text{faraday}}\right)}{(1.082\ \text{A})\left(60\ \dfrac{\text{s}}{\text{min}}\right)}$$

$$= 7.43\ \text{min}$$

## 5. EFFECTIVE VALUE

The *effective value* (also known as the *root-mean-square* or rms, *value*) of an alternating waveform is given by Eq. 61.6.[6] A DC current of $I$ produces the same heating effect as an AC current of $I_{\text{eff}}$. (The waveform does not need to be rectified.) For a sinusoidal waveform, $V = V_m/\sqrt{2} \approx 0.707\,V_m$. The scale reading of a typical AC current meter is proportional to the effective current.

$$V = V_{\text{eff}} = \sqrt{\frac{1}{2\pi}\int_0^{2\pi} v^2(\theta)\,d\theta} = \sqrt{\frac{1}{T}\int_0^T v^2(t)\,dt}$$

$$61.6$$

### Example 61.2

A 170 V (maximum value) sinusoidal voltage is connected across a 4 Ω resistor. What power is dissipated in the resistor?

*Solution*

From Table 61.2, the effective voltage is

$$V = \frac{V_m}{\sqrt{2}} = \frac{170\ \text{V}}{\sqrt{2}} = 120.2\ \text{V}$$

The dissipated power is

$$P = \frac{V^2}{R} = \frac{(120.2\ \text{V})^2}{4\ \Omega} = 3612\ \text{W}$$

## 6. FORM AND CREST FACTOR

The *form factor*, FF, is

$$\text{FF} = \frac{V_{\text{eff}}}{V_{\text{ave}}} \qquad\qquad 61.7$$

---

[6]The value of the standard voltage used in the United States, reported as 115–120 V, is an effective value.

The *crest factor*, CF (also known as the *peak factor* and *amplitude factor*), is

$$\text{CF} = \frac{V_m}{V_{\text{eff}}} \qquad\qquad 61.8$$

## 7. PHASE ANGLES

AC circuit elements have the ability to cause a *phase angle difference*, $\phi$, between the applied voltage and resulting current. Ordinarily, the current and voltage sinusoids do not peak at the same time, and the terms *leading* (capacitive) and *lagging* (inductive) are used to describe the relationship between them. In a *leading circuit*, the phase angle difference is positive, and the current reaches its peak before the voltage. (See Fig. 61.3.) In a *lagging circuit*, the phase angle difference is negative, and the current reaches its peak after the voltage.

$$v(t) = V_m \sin(\omega t + \theta) \quad \text{[reference]} \qquad 61.9$$

$$i(t) = I_m \sin(\omega t + \theta + \phi) \quad \text{[leading]} \qquad 61.10$$

$$i(t) = I_m \sin(\omega t + \theta - \phi) \quad \text{[lagging]} \qquad 61.11$$

**Figure 61.3** *Leading Phase Angle Difference*

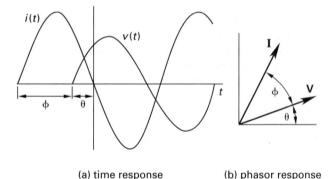

(a) time response      (b) phasor response

## 8. IMPEDANCE

Alternating current circuits can contain three different passive circuit elements—resistors, inductors, and capacitors. (See Table 61.3.) Each AC circuit element is assigned an angle, $\phi$, known as its *impedance angle*, that corresponds to the phase difference angle produced when a sinusoidal voltage is applied across the element alone. The term *impedance*, $Z$ (with units of ohms) describes the effects these elements have on current magnitude and phase.

Impedance is a complex quantity with a magnitude and an associated angle and is usually written in *phasor (polar) form* (e.g., $Z\angle\phi$). However, it can also be written in *rectangular form* as the complex sum of its *resistive* ($R$) and *reactive* ($X$) *components*, both having units of

Circuits

ohms.[7] The resistive and reactive components combine trigonometrically in the *impedance triangle* as shown in Fig. 61.4.

**Table 61.3** *Characteristics of Resistors, Capacitors, and Inductors*[*]

|  | resistor | capacitor | inductor |
|---|---|---|---|
| value | $R\ (\Omega)$ | $C\ (\mathrm{F})$ | $L\ (\mathrm{H})$ |
| reactance, $X$ | $0$ | $\dfrac{-1}{\omega C}$ | $\omega L$ |
| rectangular impedance, $Z$ | $R + j0$ | $0 - \dfrac{j}{\omega C}$ | $0 + j\omega L$ |
| phasor impedance, $Z$ | $R\angle 0°$ | $\dfrac{1}{\omega C}\angle -90°$ | $\omega L \angle 90°$ |
| phase | in-phase | leading | lagging |
| rectangular admittance, $Y$ | $\dfrac{1}{R} + j0$ | $0 + j\omega C$ | $0 - \dfrac{j}{\omega L}$ |
| phasor admittance, $Y$ | $\dfrac{1}{R}\angle 0°$ | $\omega C \angle 90°$ | $\dfrac{1}{\omega L}\angle -90°$ |

[*]As defined by IEEE Std. 280. However, capacitive reactance is generally referred to as a positive number. In vector calculations, its negativity must be considered.

$$\mathbf{Z} \equiv R \pm jX \qquad 61.12$$

$$Z = \sqrt{R^2 + X^2} \qquad 61.13$$

$$R = Z\cos\phi \quad \text{[resistive part]} \qquad 61.14$$

$$X = Z\sin\phi \quad \text{[reactive part]} \qquad 61.15$$

**Figure 61.4** *Leading Impedance Triangle*

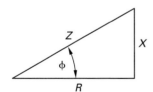

## 9. ADMITTANCE

The reciprocal of complex impedance is the complex quantity *admittance*, $\mathbf{Y}$. Admittance is particularly useful in analyzing parallel circuits, since admittances of parallel circuit elements add together.

$$\mathbf{Y} = \frac{1}{\mathbf{Z}} = \frac{1}{Z}\angle -\phi \qquad 61.16$$

The reciprocal of the resistive part of impedance is *conductance*, $G$. The reciprocal of the reactive part of impedance is *susceptance*, $B$.

$$G = \frac{1}{R} \qquad 61.17$$

$$B = \frac{1}{X} \qquad 61.18$$

By multiplying by a complex conjugate, admittance can be written in terms of resistance and reactance, and vice versa.

$$\mathbf{Y} \equiv G + jB = \frac{R}{R^2 + X^2} - j\frac{X}{R^2 + X^2} \qquad 61.19$$

$$\mathbf{Y} \equiv R + jX = \frac{G}{G^2 + B^2} - j\frac{B}{G^2 + B^2} \qquad 61.20$$

## 10. RESISTORS

An *ideal* (*pure, perfect*, etc.) *resistor* has no inductance or capacitance. Equation 61.21 gives the impedance in phasor form of an ideal resistor. The magnitude of the impedance is the resistance, $R$, and the impedance angle is zero. Therefore, current and voltage are in phase in a purely resistive circuit.

$$\mathbf{Z}_R \equiv R\angle 0° \equiv R + j0 \qquad 61.21$$

## 11. CAPACITORS

An *ideal* (*pure, perfect*, etc.) *capacitor* has no resistance or inductance.[8] Equation 61.22 gives the impedance in phasor form of an ideal capacitor with capacitance $C$. The magnitude of the impedance is the *capacitive reactance*, $X_C$, with units of ohms, and the impedance angle is $-\pi/2$ ($-90°$). Therefore, current leads the voltage by $90°$ in a purely capacitive circuit.[9]

$$\mathbf{Z}_C \equiv X_C\angle -90° \qquad 61.22$$

$$X_C = \frac{1}{\omega C} = \frac{1}{2\pi f C} \qquad 61.23$$

The impedance of an ideal capacitor has a zero resistive part. The rectangular form of a purely capacitive impedance is

$$\mathbf{Z}_C \equiv 0 - jX_C \qquad 61.24$$

---

[7]Since most engineering calculators can convert between rectangular and polar forms, this chapter does not include the operations needed to perform the conversions in the examples.

[8]A *varactor* (*varactor diode, variable voltage capacitor, voltage variable, capacitor,* or *varicap*) is a semiconductor device whose capacitance depends on the applied voltage. It is typically used in turning oscillator circuits.

[9]While the *impedance angle* for a capacitor is negative, the current phase angle difference is positive, hence a leading circuit. This is a direct result of dividing the impedance into the voltage to obtain the current: $\mathbf{I} = \mathbf{V}/\mathbf{Z}$.

## 12. INDUCTORS

An *ideal* (*pure*, *perfect*, etc.) *inductor* has no resistance or capacitance. Equation 61.25 gives the impedance in phasor form of an ideal inductor with inductance $L$. The magnitude of the impedance is the *inductive reactance*, $X_L$, with units of ohms, and the impedance angle is $\pi/2$ (90°). Therefore, current lags the voltage by 90° in a purely inductive circuit.[10]

$$\mathbf{Z}_L \equiv X_L \angle 90° \qquad 61.25$$

$$X_L = \omega L = 2\pi f L \qquad 61.26$$

The impedance of an ideal inductor has a zero resistive part. The rectangular form of a purely inductive impedance is

$$\mathbf{Z}_L \equiv 0 + jX_L \qquad 61.27$$

## 13. COMBINING IMPEDANCES

Impedances in combination are like resistors: impedances in series are added, while the reciprocals of impedances in parallel are added. For series circuits, the resistive and reactive parts of each impedance element are calculated separately and summed. For parallel circuits, the conductance and susceptance of each element are summed. The total impedance is found by a complex addition of the resistive (conductive) and reactive (susceptive) parts. It is convenient to perform the addition in rectangular form.

$$Z_e = \sum Z$$
$$= \sqrt{\left(\sum R\right)^2 + \left(\sum X_L - \sum X_C\right)^2} \quad \text{[series]} \qquad 61.28$$

$$\frac{1}{Z_e} = \sum \frac{1}{Z} = Y_e$$
$$= \sqrt{\left(\sum G\right)^2 + \left(\sum B_C - \sum B_L\right)^2} \quad \text{[parallel]} \qquad 61.29$$

Table 61.4 and Table 61.5 give the impedance, phase angle, and admittance of many common combinations of resistors, inductors, and capacitors.

**Example 61.3**

Determine the impedance and admittance of the following circuits.

(a)

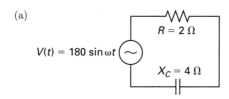

(b)

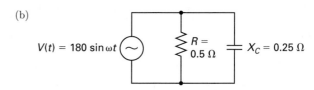

*Solution*

(a) From Eq. 61.28,

$$Z = \sqrt{R^2 + X_C^2} = \sqrt{(2\ \Omega)^2 + (4\ \Omega)^2} = 4.47\ \Omega$$

$$\phi = \arctan \frac{X_C}{R} = \arctan \frac{-4\ \Omega}{2\ \Omega} = -63.4°$$

$$\mathbf{Z} = 4.47\ \Omega \angle -63.4°$$

From Eq. 61.16,

$$\mathbf{Y} = \frac{1}{\mathbf{Z}} = \frac{1}{4.47\ \Omega \angle -63.4°} = 0.224\ \text{S} \angle 63.4°$$

(b) Since this is a parallel circuit, work with the admittances.

$$G = \frac{1}{R} = \frac{1}{0.5\ \Omega} = 2\ \text{S}$$

$$B_C = \frac{1}{X_C} = \frac{1}{0.25\ \Omega} = 4\ \text{S}$$

$$Y = \sqrt{G^2 + B_C^2} = \sqrt{(2\ \text{S})^2 + (4\ \text{S})^2} = 4.47\ \text{S}$$

$$\phi = \arctan \frac{B_C}{G} = \arctan \frac{4\ \text{S}}{2\ \text{S}} = 63.4°$$

$$\mathbf{Y} = 4.47\ \text{S} \angle 63.4°$$

$$\mathbf{Z} = \frac{1}{\mathbf{Y}} = \frac{1}{4.47\ \text{S} \angle 63.4°} = 0.224\ \Omega \angle -63.4°$$

## 14. OHM'S LAW FOR AC CIRCUITS

Ohm's law for AC circuits with linear components is similar to Ohm's law for DC circuits[11,12]

$$\mathbf{V} = \mathbf{IZ} \qquad 61.30$$

$$V \angle \theta_V = I \angle \theta_I \, Z \angle \phi_Z \qquad 61.31$$

It is important to recognize that $V$ and $I$ will either both be maximum values or will both be effective values, and never a combination of the two. If the voltage source is specified by its effective value, then the current calculated from $I = V/Z$ will be an effective value.

---

[10]While the *impedance angle* for an inductor is positive, the current phase angle difference is negative, hence a lagging circuit. This is a direct result of dividing the impedance into the voltage to obtain the current: $\mathbf{I} = \mathbf{V}/\mathbf{Z}$.

[11]This is complex number arithmetic.
[12]Ohm's law can be used with *nonlinear devices* (NLD) within only a limited region in which linear behavior is assumed.

**Table 61.4** *Impedance of Series-Connected Circuit Elements*

| circuit | impedance $Z = R + jX$ (ohms) | magnitude of impedance $\lvert Z \rvert = \sqrt{R^2 + X^2}$ (ohms) | phase angle $\theta = \tan^{-1}\dfrac{X}{R}$ (radians) | admittance $Y = \dfrac{1}{Z}$ (siemens) |
|---|---|---|---|---|
| $R$ | $R$ | $R$ | $0$ | $\dfrac{1}{R}$ |
| $L$ | $j\omega L$ | $\omega L$ | $+\dfrac{\pi}{2}$ | $-j\left(\dfrac{1}{\omega L}\right)$ |
| $C$ | $-j\left(\dfrac{1}{\omega C}\right)$ | $\dfrac{1}{\omega C}$ | $-\dfrac{\pi}{2}$ | $j\omega C$ |
| $R_1 \quad R_2$ | $R_1 + R_2$ | $R_1 + R_2$ | $0$ | $\dfrac{1}{R_1 + R_2}$ |
| $M$ $L_1 \quad L_2$ | $j\omega\left(L_1 + L_2 \pm 2M\right)$ | $\omega\left(L_1 + L_2 \pm 2M\right)$ | $+\dfrac{\pi}{2}$ | $-j\left(\dfrac{1}{\omega\left(L_1 + L_2 \pm 2M\right)}\right)$ |
| $C_1 \quad C_2$ | $-j\left(\dfrac{1}{\omega}\right)\left(\dfrac{C_1 + C_2}{C_1 C_2}\right)$ | $\left(\dfrac{1}{\omega}\right)\left(\dfrac{C_1 + C_2}{C_1 C_2}\right)$ | $-\dfrac{\pi}{2}$ | $j\omega\left(\dfrac{C_1 C_2}{C_1 + C_2}\right)$ |
| $R \quad L$ | $R + j\omega L$ | $\sqrt{R^2 + \omega^2 L^2}$ | $\tan^{-1}\dfrac{\omega L}{R}$ | $\dfrac{R - j\omega L}{R^2 + \omega^2 L^2}$ |
| $R \quad C$ | $R - j\left(\dfrac{1}{\omega C}\right)$ | $\sqrt{\dfrac{\omega^2 C^2 R^2 + 1}{\omega^2 C^2}}$ | $\tan^{-1}\dfrac{1}{\omega RC}$ | $\dfrac{\omega^2 C^2 R + j\omega C}{\omega^2 C^2 R^2 + 1}$ |
| $L \quad C$ | $j\left(\omega L - \dfrac{1}{\omega C}\right)$ | $\omega L - \dfrac{1}{\omega C}$ | $\pm\dfrac{\pi}{2}$ | $-j\left(\dfrac{\omega C}{\omega^2 LC - 1}\right)$ |
| $R \quad L \quad C$ | $R + j\left(\omega L - \dfrac{1}{\omega C}\right)$ | $\sqrt{R^2 + \left(\omega L - \dfrac{1}{\omega C}\right)^2}$ | $\tan^{-1}\dfrac{\omega L - \dfrac{1}{\omega C}}{R}$ | $\dfrac{R - j\left(\omega L - \dfrac{1}{\omega C}\right)}{R^2 + \left(\omega L - \dfrac{1}{\omega C}\right)^2}$ |

**Circuits**

**Table 61.5** *Impedance of Parallel-Connected Circuit Elements*

| circuit | impedance<br>$Z = R + jX$<br>(ohms) | magnitude of impedance<br>$|Z| = \sqrt{R^2 + X^2}$<br>(ohms) |
|---|---|---|
| $R_1$ / $R_2$ | $\dfrac{R_1 R_2}{R_1 + R_2}$ | $\dfrac{R_1 R_2}{R_1 + R_2}$ |
| $L_1$, $M$, $L_2$ | $+j\omega \left( \dfrac{L_1 L_2 - M^2}{L_1 + L_2 \mp 2M} \right)$ | $\omega \left( \dfrac{L_1 L_2 - M^2}{L_1 + L_2 \mp 2M} \right)$ |
| $C_1$ / $C_2$ | $-j \left( \dfrac{1}{\omega(C_1 + C_2)} \right)$ | $\dfrac{1}{\omega(C_1 + C_2)}$ |
| $R$ / $L$ | $\dfrac{\omega^2 L^2 R + j\omega L R^2}{\omega^2 L^2 + R^2}$ | $\dfrac{\omega L R}{\sqrt{\omega^2 L^2 + R^2}}$ |
| $R$ / $C$ | $\dfrac{R - j\omega R^2 C}{1 + \omega^2 R^2 C^2}$ | $\dfrac{R}{\sqrt{1 + \omega^2 R^2 C^2}}$ |
| $L$ / $C$ | $j \left( \dfrac{\omega L}{1 - \omega^2 L C} \right)$ | $\dfrac{\omega L}{1 - \omega^2 L C}$ |
| $R$, $L$, $C$ | $\dfrac{\dfrac{1}{R} - j\left( \omega C - \dfrac{1}{\omega L} \right)}{\left( \dfrac{1}{R} \right)^2 + \left( \omega C - \dfrac{1}{\omega L} \right)^2}$ | $\dfrac{1}{\sqrt{\left( \dfrac{1}{R} \right)^2 + \left( \omega C - \dfrac{1}{\omega L} \right)^2}}$ |
| $R$, $L$ / $C$ | $\dfrac{\dfrac{R}{\omega^2 C^2} - j\left( \dfrac{R^2}{\omega C} + \dfrac{L}{C}\left( \omega L - \dfrac{1}{\omega C} \right) \right)}{R^2 + \left( \omega L - \dfrac{1}{\omega C} \right)^2}$ | $\dfrac{\sqrt{\left( \dfrac{R}{\omega^2 C^2} \right)^2 + \left( \dfrac{R^2}{\omega C} + \left( \dfrac{L}{C} \right)\left( \omega L - \dfrac{1}{\omega C} \right) \right)^2}}{R^2 + \left( \omega L - \dfrac{1}{\omega C} \right)^2}$ |
| $R_1$, $L$ / $R_2$, $C$ | $\dfrac{R_1 R_2 (R_1 + R_2) + \omega^2 L^2 R_2 + \dfrac{R_1}{\omega^2 C^2}}{(R_1 + R_2)^2 + \left( \omega L - \dfrac{1}{\omega C} \right)^2}$<br>$+ j\left( \dfrac{\omega R_2^2 L - \dfrac{R_1^2}{\omega C} - \left( \dfrac{L}{C} \right)\left( \omega L - \dfrac{1}{\omega C} \right)}{(R_1 + R_2)^2 + \left( \omega L - \dfrac{1}{\omega C} \right)^2} \right)$ | $\sqrt{\left( \dfrac{R_1 R_2 (R_1 + R_2) + \omega^2 L^2 R_2 + \dfrac{R_1}{\omega^2 C^2}}{(R_1 + R_2)^2 + \left( \omega L - \dfrac{1}{\omega C} \right)^2} \right)^2 + \left( \dfrac{\omega L R_2^2 - \dfrac{R_1^2}{\omega C} - \dfrac{L}{C}\left( \omega L - \dfrac{1}{\omega C} \right)}{(R_1 + R_2)^2 + \left( \omega L - \dfrac{1}{\omega C} \right)^2} \right)^2}$ |

| circuit | phase angle $\theta = \tan^{-1} \dfrac{X}{R}$ (radians) | admittance $Y = \dfrac{1}{Z}$ (siemens) |
|---|---|---|
| $R_1$ $R_2$ | $0$ | $\dfrac{R_1 + R_2}{R_1 R_2}$ |
| $L_1$ $M$ $L_2$ | $+\dfrac{\pi}{2}$ | $-j\left(\dfrac{1}{\omega}\right)\left(\dfrac{L_1 + L_2 \mp 2M}{L_1 L_2 - M^2}\right)$ |
| $C_1$ $C_2$ | $-\dfrac{\pi}{2}$ | $+j\omega(C_1 + C_2)$ |
| $R$ $L$ | $\tan^{-1}\dfrac{R}{\omega L}$ | $\dfrac{\omega L - jR}{\omega L R}$ |
| $R$ $C$ | $\tan^{-1}(-\omega R C)$ | $\dfrac{1}{R} + j\omega C$ |
| $L$ $C$ | $\pm\dfrac{\pi}{2}$ | $-j\left(\dfrac{1 - \omega^2 LC}{\omega L}\right)$ |
| $R$ $L$ $C$ | $\tan^{-1} R\left(\dfrac{1}{\omega L} - \omega C\right)$ | $\dfrac{1}{R} + j\left(\omega C - \dfrac{1}{\omega L}\right)$ |
| $R$ $L$ $C$ | $\tan^{-1}\left(\dfrac{\dfrac{R^2}{\omega C} + \left(\dfrac{L}{C}\right)\left(\omega L - \dfrac{1}{\omega C}\right)}{\dfrac{R}{\omega^2 C^2}}\right)$ | $\dfrac{R + j\omega\left(R^2 C - L + \omega^2 L^2 C\right)}{R^2 + \omega^2 L^2}$ |
| $R_1$ $L$ $R_2$ $C$ | $\tan^{-1}\left(\dfrac{\omega L R^2 - \dfrac{R_1^2}{\omega C} - \left(\dfrac{L}{C}\right)\left(\omega L - \dfrac{1}{\omega C}\right)}{R_1 R_2(R_1 + R_2) + \omega^2 L^2 R_2 + \dfrac{R_1}{\omega^2 C^2}}\right)$ | $\dfrac{R_1 + \omega^2 R_1 R_2 C^2(R_1 + R_2) + \omega^4 L^2 C^2 R_2}{(R_1^2 + \omega^2 L^2)(\omega^2 R_2^2 C^2 + 1)}$ $+ j\left(\dfrac{\omega\left(R_1^2 C - L + \omega^2 LC(L - R_2^2 C)\right)}{(R_1^2 + \omega^2 L^2)(\omega^2 R_2^2 C^2 + 1)}\right)$ |

**Example 61.4**

What are the currents in the capacitors in Ex. 61.3?

*Solution*

(a) The current is the same in both components.

$$\mathbf{I} = \frac{\mathbf{V}}{\mathbf{Z}} = \frac{180 \text{ V} \angle 0°}{4.47 \text{ } \Omega \angle -63.4°} = 40.3 \text{ A} \angle 63.4°$$

(b) The voltage is the same across both components.

$$\mathbf{I}_C = \frac{\mathbf{V}}{\mathbf{Z}_C} = \frac{180 \text{ V} \angle 0°}{0.25 \text{ } \Omega \angle -90°} = 720 \text{ A} \angle 90°$$

## 15. POWER

The instantaneous power, $P$, dissipated in a purely resistive circuit (see Fig. 61.5) is

$$P_R(t) = i(t)v(t) = (I_m \sin \omega t)(V_m \sin \omega t)$$
$$= I_m V_m \sin^2 \omega t = \tfrac{1}{2} I_m V_m - \tfrac{1}{2} I_m V_m \cos 2\omega t$$
$$61.32$$

The second term in Eq. 61.32 integrates to zero. Therefore, the *average power* dissipated is

$$P_R = \tfrac{1}{2} I_m V_m = \frac{V_m^2}{2R} = IV \qquad 61.33$$

**Figure 61.5** *Power in AC Circuit Elements*

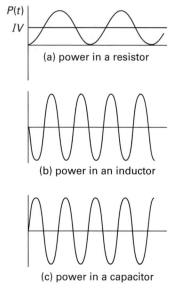

(a) power in a resistor

(b) power in an inductor

(c) power in a capacitor

Since the current leads the voltage in a purely capacitive circuit, it can be represented by a cosine term (since the sine and cosine differ only by a 90° phase). Since $\sin 2\omega t$ averages to zero, no power is dissipated in a capacitor.

However, energy is stored in a capacitor during its charging process.

$$P_C(t) = i(t)v(t) = (I_m \cos \omega t)(V_m \sin \omega t)$$
$$= I_m V_m \sin \omega t \cos \omega t = \tfrac{1}{2} I_m V_m \sin 2\omega t \quad [\text{ave} = 0]$$
$$61.34$$

Similarly, the energy stored in an inductor is

$$P_L(t) = -\tfrac{1}{2} I_m V_m \sin 2\omega t \quad [\text{ave} = 0] \qquad 61.35$$

## 16. ANALYSIS OF AC CIRCUITS

All of the circuit analysis techniques (e.g., Ohm's and Kirchhoff's laws, loop-current and node-voltage methods, superposition, etc.) can be used to evaluate AC circuits as long as complex arithmetic is used.

## 17. SERIES AC CIRCUITS

A series AC circuit consists of circuit elements in series. The current passes simultaneously through all elements, and the voltage drop across each can be found from Ohm's law. The total circuit impedance must be found as the complex sum of individual impedances. It is convenient to convert all known circuit element impedances to rectangular form.

For a *series RL circuit* (no capacitance), the magnitude of the impedance can be found from the impedance triangle. (See Sec. 61.8.) This is a lagging circuit.

$$\mathbf{Z}_{RL} \equiv Z_{RL} \angle \phi_{RL} \equiv R + jX_L$$
$$= Z_{RL} \cos \phi_{RL} + jZ_{RL} \sin \phi_{RL} \qquad 61.36$$

$$Z_{RL} = \sqrt{R^2 + X_L^2} \qquad 61.37$$

$$\tan \phi_{RC} = \frac{X_L}{R} \qquad 61.38$$

For a *series RC circuit* (no inductance), the magnitude of the impedance can also be found from the impedance triangle. This is a leading circuit.

$$\mathbf{Z}_{RC} \equiv Z_{RC} \angle \phi_{RC} \equiv R - jX_C$$
$$= Z_{RC} \cos \phi_{RC} + jZ_{RC} \sin \phi_{RC} \qquad 61.39$$

$$Z_{RC} = \sqrt{R^2 + X_C^2} \qquad 61.40$$

$$\tan \phi_{RC} = \frac{-X_C}{R} \qquad 61.41$$

A *series RLC circuit* consists of inductance, resistance, and capacitance in series and can be either a lagging or leading circuit depending on whether $\omega L$ or $1/\omega C$ is larger, respectively.

$$\mathbf{Z}_{RLC} \equiv Z_{RLC} \angle \phi_{RLC} \equiv R + j(X_L - X_C)$$
$$= Z_{RLC} \cos \phi_{RLC} + jZ_{RLC} \sin \phi_{RLC} \qquad 61.42$$

$$Z_{RLC} = \sqrt{R^2 + (X_L - X_C)^2} \qquad 61.43$$

$$\tan \phi_{RLC} = \frac{X_L - X_C}{R} \qquad 61.44$$

Figure 61.6 shows the impedance triangle for a series *RLC* circuit.

**Figure 61.6** *Impedance Triangle for Series RLC Circuit*

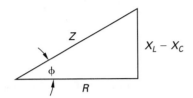

## Example 61.5

A circuit consists of a resistor and a capacitor in series with a sinusoidal voltage source with a 500 V peak voltage. (a) What is the phasor form of the impedance? (b) What is the phasor form of the current?

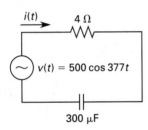

*Solution*

(a) The capacitive reactance is

$$X_C = \frac{1}{\omega C} = \frac{1}{\left(377 \; \frac{\text{rad}}{\text{s}}\right)(300 \times 10^{-6} \; \text{F})} = 8.842 \; \Omega$$

The impedance is

$$Z = \sqrt{R^2 + X_C^2} = \sqrt{(4 \; \Omega)^2 + (8.842 \; \Omega)^2} = 9.70 \; \Omega$$

$$\phi_Z = \arctan \frac{-8.842 \; \Omega}{4 \; \Omega} = -65.7°$$

(b) The peak current is

$$I_m = \frac{V_m}{Z} = \frac{500 \; \text{V}\angle 0°}{9.7 \; \Omega \angle -65.7°} = 51.5 \; \text{A}\angle 65.7°$$

## Example 61.6

A circuit consists of an inductor, resistor, capacitor, and special "black box" impedance element in series with a sinusoidal voltage source.

(a) What is the trigonometric form of the current? (b) Is the current leading or lagging? (c) What is the voltage across the inductor?

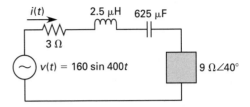

*Solution*

(a) First, determine the rectangular form of the total circuit impedance.

*resistor:* $R = 3 \; \Omega$

*capacitor:* $X_C = \dfrac{1}{\omega C}$

$$= \frac{1}{\left(400 \; \frac{\text{rad}}{\text{s}}\right)(625 \times 10^{-6} \; \text{F})}$$

$$= 4 \; \Omega$$

*inductor:* $X_L = \omega L = \left(400 \; \dfrac{\text{rad}}{\text{s}}\right)(2.5 \times 10^{-3} \; \text{H})$

$$= 1 \; \Omega$$

*box:* $R_{\text{box}} + jX_{\text{box}} = (9 \; \Omega)(\cos 40° + j \sin 40°)$

$$= 6.894 + j5.785 \; \Omega$$

Next, convert the impedance to phasor form.

$$Z = \sqrt{\left(\sum R\right)^2 + \left(\sum X\right)^2}$$

$$= \sqrt{(3 \; \Omega + 6.894 \; \Omega)^2 + (1 \; \Omega + 5.785 \; \Omega - 4 \; \Omega)^2}$$

$$= 10.28 \; \Omega$$

$$\phi = \arctan \frac{2.785 \; \Omega}{9.894 \; \Omega} = 15.72°$$

The voltage source in phasor form is $\mathbf{V} = 160\angle 0°$. (Use maximum values throughout for consistency.) Then, calculate the current from Ohm's law.[13] This current is the same throughout the entire circuit.

$$\mathbf{I} = \frac{\mathbf{V}}{\mathbf{Z}} = \frac{160 \; \text{V}\angle 0°}{10.28 \; \Omega \angle 15.72°} = 15.56 \; \text{A}\angle -15.72°$$

$$\phi = \frac{(-15.72°)(2\pi)}{360°} = -0.274 \; \text{rad}$$

$$I(t) = 15.56 \sin(400t - 0.274 \; \text{rad})$$

(b) Since the current angle is negative, this is a lagging circuit.

---

[13]Notice that the quantity $400t$ is in radians and that 15.75° is in degrees. This mixing of units is permissible.

(c) the voltage across the inductor is

$$\mathbf{V}_L = \mathbf{I}\mathbf{Z}_L = (15.56 \text{ A}\angle{-15.72°})(1 \text{ }\Omega\angle 90°)$$

$$= 15.56 \text{ V}\angle 74.28°$$

$$\phi = \frac{(74.28°)(2\pi)}{360°}$$

$$= 1.296 \text{ rad}$$

$$v_L(t) = 15.56\sin(400t + 1.296 \text{ rad})$$

## 18. PARALLEL CIRCUITS

A parallel circuit (see Fig. 61.7) consists of inductances, resistances, and/or capacitances in parallel. The same voltage appears simultaneously across all elements, and the current through each can be found from Ohm's law. The total circuit impedance must be found as the complex sum of individual admittances. It is convenient to convert all known circuit element impedances to rectangular admittance form.

**Figure 61.7** *Impedance Triangle for Parallel RLC Circuit*

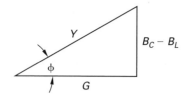

For a *parallel RL circuit* (no capacitance), the magnitude of the impedance can be found from the admittance triangle. This is a lagging circuit.

$$\mathbf{Y}_{RL} \equiv Y_{RL}\angle\phi_{RL} \equiv G - jB_L$$

$$= Y_{RL}\cos\phi_{RL} + jY_{RL}\sin\phi_{RL} \quad 61.45$$

$$Y_{RL} = \sqrt{G^2 + B_L^2} \quad 61.46$$

$$\tan\phi_{RL} = \frac{-B_L}{G} \quad 61.47$$

For a *parallel RC circuit* (no inductance), the magnitude of the admittance can also be found from the admittance triangle. This is a leading circuit.

$$\mathbf{Y}_{RC} \equiv Y_{RC}\angle\phi_{RC} \equiv G + jB_C$$

$$= Y_{RC}\cos\phi_{RC} + jY_{RC}\sin\phi_{RC} \quad 61.48$$

$$Y_{RC} = \sqrt{G^2 + B_C^2} \quad 61.49$$

$$\tan\phi_{RC} = \frac{-B_C}{G} \quad 61.50$$

A *GLC circuit* (i.e., a *parallel RLC circuit*) consists of inductance, resistance, and capacitance in parallel and can be either a lagging or leading circuit.

$$\mathbf{Y}_{GLC} \equiv Y_{GLC}\angle\phi_{GLC} \equiv G + j(B_C - B_L)$$

$$= Y_{GLC}\cos\phi_{GLC} + jY_{GLC}\sin\phi_{GLC} \quad 61.51$$

$$Y_{GLC} = \sqrt{G^2 + (B_C - B_L)^2} \quad 61.52$$

$$\tan\phi_{GLC} = \frac{B_C - B_L}{G} \quad 61.53$$

### Example 61.7

A circuit consists of an inductor, resistor, and capacitor in parallel with a sinusoidal voltage source as shown. The current is $i(t) = 3.2\sin(\omega t + 45°)$.

(a) What is the phasor form of the applied voltage? (b) What is the trigonometric form of the current referenced to a voltage with zero phase angle? (c) Is the current leading or lagging? (d) What is the current in the capacitor measured with respect to a voltage with zero phase angle?

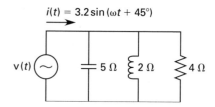

*Solution*

(a) The total admittance of the circuit is the sum of the individual admittances. (It is convenient to work in rectangular form.)

$$\mathbf{Y} = \sum\mathbf{Y}_i = Y_C + Y_L + Y_R$$

$$= (0 + j0.2) + (0 - j0.5) + (0.25 + j0)$$

$$= 0.25 - j0.3$$

Converting to phasor form,

$$\mathbf{Y} = 0.3905\angle{-50.19°}$$

The applied voltage is

$$\mathbf{V} = \mathbf{I}\mathbf{Z} = \frac{\mathbf{I}}{\mathbf{Y}} = \frac{3.2 \text{ A}\angle 45°}{0.3905 \text{ S}\angle{-50.19°}}$$

$$= 8.195 \text{ V}\angle 95.19°$$

(b) If the time scale is changed so that the voltage is zero at $t = 0$, then the phase angle must be subtracted from both the current and voltage expressions.

$$v(t) = 8.195 \sin(\omega t + 95.19° - 95.19°)$$

$$= (8.195 \text{ V}) \sin \omega t$$

$$i(t) = 3.2 \sin(\omega t + 45° - 95.19°)$$

$$= (3.2 \text{ A}) \sin(\omega t - 50.19°)$$

(c) Since the current angle is negative, the circuit is lagging.

(d) The current in the capacitor is

$$\mathbf{I}_C = \frac{\mathbf{V}}{\mathbf{Z}_C} = \frac{8.195 \text{ V} \angle 0°}{5 \text{ } \Omega \angle -90°}$$

$$= 1.639 \text{ A} \angle 90°$$

## 19. RESONANT CIRCUITS

A *resonant circuit* has a zero phase angle difference. This is equivalent to saying the circuit is purely resistive (i.e., the *power factor* described in Sec. 61.24 is equal to 1.0) in its response to an AC voltage. The frequency at which the circuit becomes purely resistive is the *resonant frequency*.

For frequencies below the resonant frequency, a series *RLC* circuit will be capacitive (leading) in nature; above the resonant frequency, the circuit will be inductive (lagging) in nature.

For frequencies below the resonant frequency, a parallel *GLC* circuit will be inductive (lagging) in nature above the resonant frequency, the circuit will be capacitive (leading) in nature.

Circuits can become resonant in two ways. If the frequency of the applied voltage is fixed, the elements must be adjusted so that the capacitive reactance cancels the inductive reactance (i.e., $X_L - X_C = 0$. If the circuit elements are fixed, the frequency must be adjusted.

As Fig. 61.8 and Fig. 61.9 illustrate, a circuit approaches resonant behavior gradually. $\omega_1$ and $\omega_2$ are the *half-power points* (*70 percent points* or *3 dB points*) because at those frequencies, the power dissipated in the resistor is half of the power dissipated at the resonant frequency.

$$Z_{\omega_1} = \sqrt{2}R \qquad 61.54$$

$$I_{\omega_1} = \frac{V}{Z_{\omega_1}} = \frac{V}{\sqrt{2}R} = \frac{I_0}{\sqrt{2}} \qquad 61.55$$

$$P_{\omega_1} = I^2 R = \left(\frac{I_0}{\sqrt{2}}\right)^2 R = \tfrac{1}{2}P_0 \qquad 61.56$$

The frequency difference between the half-power points is the *bandwidth*, BW, a measure of *circuit selectivity*.

The smaller the bandwidth, the more selective the circuit.

$$\text{BW} = f_2 - f_1 \qquad 61.57$$

The *quality factor*, $Q$, for a circuit is a dimensionless ratio that compares the reactive energy stored in an inductor each cycle to the resistive energy dissipated.[14] Fig. 61.8 illustrates the effect the quality factor has on the frequency characteristic.

$$Q = 2\pi \left(\frac{\text{maximum energy stored per cycle}}{\text{energy dissipated per cycle}}\right)$$

$$= \frac{f_0}{(\text{BW})_{\text{Hz}}} = \frac{\omega_0}{(\text{BW})_{\text{rad/s}}} = \frac{f_0}{f_2 - f_1} = \frac{\omega_0}{\omega_2 - \omega_1}$$

$$[\text{parallel or series}]$$
$$61.58$$

Then, the energy stored in the inductor of a series *RLC* circuit each cycle is

$$U = \frac{I_m^2 L}{2} = I^2 L = Q \times \frac{I^2 R}{2\pi f_0} \qquad 61.59$$

The relationships between the half-power points and quality factor are

$$f_1, f_2 = f_0 \left(\sqrt{1 + \frac{1}{4Q^2}} \mp \frac{1}{2Q}\right)$$

$$\approx f_0 \mp \frac{f_0}{2Q} = \frac{\text{BW}}{2} \qquad 61.60$$

Table 61.6 summarizes the most frequently used resonant circuit formulas for series and parallel circuits.

## 20. SERIES RESONANCE

In a resonant series *RLC* circuit,

- impedance is minimum
- impedance equals resistance
- current and voltage are in phase
- current is maximum
- power dissipation is maximum

The total impedance (in rectangular form) of a series *RLC* circuit is $R + j(X_L - X_C)$. At the resonant frequency, $\omega_0 = 2\pi f_0$,

$$X_L = X_C \quad [\text{at resonance}] \qquad 61.61$$

$$\omega_0 L = \frac{1}{\omega_0 C} \qquad 61.62$$

$$\omega_0 = 2\pi f_0 = \frac{1}{\sqrt{LC}} \qquad 61.63$$

---

[14]The name *figure of merit* refers to the quality factor calculated from the inductance and internal resistance of a coil.

Circuits

*Table 61.6* Resonant Circuit Formulas

| unknown quantity | symbol | units | series | parallel |
|---|---|---|---|---|
| resonant frequency | $f_0$ | Hz | $\dfrac{1}{2\pi\sqrt{LC}}$ | |
| | | | $\dfrac{QR}{2\pi L} = \dfrac{1}{2\pi QRC}$ | $\dfrac{R}{2\pi QL} = \dfrac{Q}{2\pi RC}$ |
| | $\omega_0$ | $\dfrac{\text{rad}}{\text{s}}$ | $\dfrac{1}{\sqrt{LC}}$ | |
| | | | $\dfrac{QR}{L} = \dfrac{1}{QRC}$ | $\dfrac{R}{QL} = \dfrac{Q}{RC}$ |
| bandwidth | BW | Hz | $f_2 - f_1 = \dfrac{f_0}{Q}$ | |
| | | | $\dfrac{R}{2\pi L} = \dfrac{1}{2\pi Q^2 RC}$ | $\dfrac{1}{2\pi RC} = \dfrac{R}{2\pi Q^2 L}$ |
| | BW | $\dfrac{\text{rad}}{\text{s}}$ | $\omega_2 - \omega_1 = \dfrac{\omega_0}{Q}$ | |
| | | | $\dfrac{R}{L} = \dfrac{1}{Q^2 RC}$ | $\dfrac{1}{CR} = \dfrac{R}{Q^2 L}$ |
| quality factor | $Q$ | – | $\dfrac{1}{R}\sqrt{\dfrac{L}{C}}$ | $R\sqrt{\dfrac{C}{L}}$ |
| | | | $\dfrac{\omega_0 L}{R} = \dfrac{1}{\omega_0 RC}$ | $\omega_0 RC = \dfrac{R}{\omega_0 L}$ |
| lower half-power point | $f_1$ | Hz | $f_0 - \dfrac{\text{BW}}{2} = f_0\left(1 - \dfrac{1}{2Q}\right)$ | |
| | | | $f_0 - \dfrac{R}{4\pi L}$ | $f_0 - \dfrac{1}{4\pi CR}$ |
| | $\omega_1$ | $\dfrac{\text{rad}}{\text{s}}$ | $\omega_0 - \dfrac{\text{BW}}{2} = \omega_0\left(1 - \dfrac{1}{2Q}\right)$ | |
| | | | $\omega_0 - \dfrac{R}{2L}$ | $\omega_0 - \dfrac{1}{2CR}$ |
| upper half-power point | $f_2$ | Hz | $f_0 + \dfrac{\text{BW}}{2} = f_0\left(1 + \dfrac{1}{2Q}\right)$ | |
| | | | $f_0 + \dfrac{R}{4\pi L}$ | $f_0 + \dfrac{1}{4\pi CR}$ |
| | $\omega_2$ | $\dfrac{\text{rad}}{\text{s}}$ | $\omega_0 + \dfrac{\text{BW}}{2} = \omega_0\left(1 + \dfrac{1}{2Q}\right)$ | |
| | | | $\omega_0 + \dfrac{R}{2L}$ | $\omega_0 + \dfrac{1}{2CR}$ |

Circuits

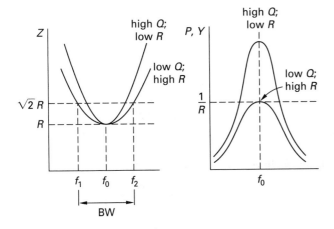

**Figure 61.8** *Series Resonance (band-pass filter)*

The power dissipation in the resistor is

$$P = \tfrac{1}{2}I_m^2 R = \frac{V_m^2}{2R} = I^2 R = \frac{V^2}{R} \qquad \textit{61.64}$$

The quality factor for a series $RLC$ circuit is

$$Q = \frac{X}{R} = \frac{\omega_0 L}{R} = \frac{1}{\omega_0 RC} = \frac{1}{R}\sqrt{\frac{L}{C}}$$
$$= \frac{\omega_0}{(\mathrm{BW})_{\mathrm{rad/s}}} = \frac{f_0}{(\mathrm{BW})_{\mathrm{Hz}}} = G\omega_0 L = \frac{G}{\omega_0 C} \qquad \textit{61.65}$$

## Example 61.8

A series $RLC$ circuit is connected across a sinusoidal voltage with peak of 20 V.

(a) What is the resonant frequency (in rad/s)? (b) What are the half-power points (in rad/s)? (c) What is the peak current at resonance? (d) What is the peak voltage across each component at resonance?

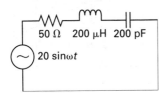

*Solution*

(a) Equation 61.63 gives the resonant frequency.

$$\omega_0 = \frac{1}{\sqrt{LC}} = \frac{1}{\sqrt{(200\times10^{-6}\ \mathrm{H})(200\times10^{-12}\ \mathrm{F})}}$$
$$= 5\times10^6\ \mathrm{rad/s}$$

(b) The half-power points are given by Eq. 61.60.

$$\omega_1,\omega_2 = \omega_0 \mp \frac{\mathrm{BW}}{2} = \omega_0 \mp \frac{\omega_0}{2Q} = \omega_0 \mp \frac{R}{2L}$$
$$= 5\times10^6 \mp \frac{50\ \Omega}{(2)(200\times10^{-6}\ \mathrm{H})}$$
$$= 5.125\times10^6\ \mathrm{rad/s},\ 4.875\times10^6\ \mathrm{rad/s}$$

(c) The total impedance at resonance is just the resistance. The peak resonant current is

$$I_0 = \frac{V_m}{Z_0} = \frac{V_m}{R} = \frac{20\ \mathrm{V}\angle0^\circ}{50\ \Omega} = 0.4\ \mathrm{A}\angle0^\circ$$

(d) The peak voltages across the components are

$$V_R = I_0 R = (0.4\ \mathrm{A}\angle0^\circ)(50\ \Omega) = 20\ \mathrm{V}\angle0^\circ$$
$$V_L = I_0 X_L = I_0 j\omega_0 L$$
$$= j(0.4\ \mathrm{A}\angle0^\circ)\left(5\times10^6\ \frac{\mathrm{rad}}{\mathrm{s}}\right)(200\times10^{-6}\ \mathrm{H})$$
$$= 400\ \mathrm{V}\angle90^\circ$$
$$V_C = I_0 X_C = \frac{I_0}{j\omega_0 C} = \frac{0.4\ \mathrm{A}\angle0^\circ}{j\left(5\times10^6\ \frac{\mathrm{rad}}{\mathrm{s}}\right)(200\times10^{-12}\ \mathrm{F})}$$
$$= 400\ \mathrm{V}\angle-90^\circ$$

## 21. PARALLEL RESONANCE

In a resonant parallel $GLC$ circuit,

- impedance is maximum
- impedance equals resistance
- current and voltage are in phase
- current is minimum
- power dissipation is minimum

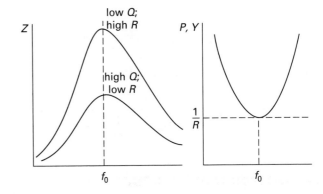

**Figure 61.9** *Parallel Resonance (band-reject filter)*

Circuits

The total admittance (in rectangular form) of a parallel $GLC$ circuit is $G + j(B_C - B_L)$. At resonance,

$$B_L = B_C \quad \text{[at resonance]} \qquad 61.66$$

$$\omega_0 L = \frac{1}{\omega_0 C} \qquad 61.67$$

$$\omega_0 = 2\pi f_0 = \frac{1}{\sqrt{LC}} \qquad 61.68$$

The power dissipation in the resistor is

$$P = \tfrac{1}{2}I_m^2 R = \frac{V_m^2}{2R} = I^2 R = \frac{V^2}{R} \qquad 61.69$$

The quality factor for a parallel $RLC$ circuit is

$$Q = \frac{R}{X} = \omega_0 RC = \frac{R}{\omega_0 L} = R\sqrt{\frac{C}{L}}$$

$$= \frac{\omega_0}{(\text{BW})_{\text{rad/s}}} = \frac{f_0}{(\text{BW})_{\text{Hz}}} = \frac{\omega_0 C}{G} = \frac{1}{G\omega_0 L} \qquad 61.70$$

### Example 61.9

A parallel $RLC$ circuit containing a 10 $\Omega$ resistor has a resonant frequency of 1 MHz and also a bandwidth of 10 kHz. To what should the resistor be changed in order to increase the bandwidth to 20 kHz without changing the resonant frequency?

*Solution*

From Eq. 61.58 and Eq. 61.70,

$$Q_{\text{old}} = \frac{f_0}{\text{BW}} = 2\pi f_0 RC$$

$$C = \frac{1}{2\pi R(\text{BW})} = \frac{1}{(2\pi)(10 \ \Omega)(10 \times 10^3 \ \text{Hz})}$$

$$= \frac{1 \times 10^{-5}}{2\pi} \ \text{F}$$

The new quality factor is

$$Q_{\text{new}} = \frac{f_0}{\text{BW}} = \frac{10^6 \ \text{Hz}}{20 \times 10^3 \ \text{Hz}} = 50$$

From Eq. 61.70, the required resistance is

$$R = \frac{Q}{2\pi f_0 C} = \frac{50}{(2\pi)(10^6 \ \text{Hz})\left(\dfrac{1 \times 10^{-5}}{2\pi} \ \text{F}\right)} = 5 \ \Omega$$

### 22. ANTIRESONANT CIRCUIT

Figure 61.10 illustrates an *antiresonant circuit* (also known as a *tank circuit* and *electrical resonator circuit*). The resistance in the inductive branch may be a discrete element or an internal resistance of a non-ideal inductor. The antiresonant frequency is

$$\omega_0 = 2\pi f_0 = \sqrt{\frac{1}{LC} - \frac{R^2}{L^2}}$$

$$\approx \frac{1}{\sqrt{LC}} \quad [R \ll \omega_0 L] \qquad 61.71$$

$$Q = \frac{1}{\omega_0 CR} = \frac{X_C}{R} = \frac{X_L}{R} \qquad 61.72$$

**Figure 61.10** *Antiresonant (Tank) Circuit*

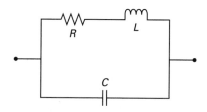

At resonance, the capacitor and inductor trade the same stored energy on alternate half cycles. When the capacitor discharges, the inductor charges, and vice versa. At the antiresonant frequency, the tank circuit presents a high impedance to the primary circuit current, even though the current within the tank is high. Power is dissipated only in the resistance.

The antiresonant circuit is equivalent to a parallel $RLC$ circuit whose resistance is $Q^2 R$.

### Example 61.10

A practical tank circuit (see Fig. 61.10) containing a 10 $\Omega$ resistor has a resonant frequency of 1 MHz and a bandwidth of 10 kHz. What size resistor should be added in parallel with the circuit in order to increase the bandwidth to 20 kHz?

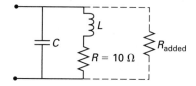

*Solution*

From Eq. 61.58 for the unmodified circuit,

$$Q = \frac{f_0}{\text{BW}} = \frac{10^6 \ \text{Hz}}{10 \times 10^3 \ \text{Hz}} = 100$$

Replace the existing tank circuit with its equivalent parallel $RLC$ resonant circuit. The resistance is

$$R_e = Q^2 R = (100)^2 (10 \ \Omega) = 10^5 \ \Omega$$

From Eq. 61.70, the capacitance in the parallel (and original) circuit is

$$C = \frac{Q}{\omega_0 R_e} = \frac{100}{\omega_0(10^5\ \Omega)} = \frac{10^{-3}}{\omega_0}$$

Calculate the quality factor for a circuit with a 20 kHz bandwidth.

$$Q = \frac{f_0}{\mathrm{BW}} = \frac{10^6\ \mathrm{Hz}}{20 \times 10^3\ \mathrm{Hz}} = 50$$

This bandwidth requires a total parallel resistance of

$$R_t = \frac{Q}{\omega_0 C} = \frac{50}{\omega_0\left(\dfrac{10^{-3}}{\omega_0}\right)} = 5 \times 10^4\ \Omega$$

Since the required additional resistance is in parallel with the equivalent resistance,

$$\frac{1}{R_t} = \frac{1}{R_e} + \frac{1}{R_{\mathrm{added}}}$$

$$\frac{1}{5 \times 10^4\ \Omega} = \frac{1}{10^5\ \Omega} + \frac{1}{R_{\mathrm{added}}}$$

$$R_{\mathrm{added}} = 10^5\ \Omega$$

## 23. FILTERS

Resonant circuits are frequency-selective since they tend to favor (disfavor) particular frequencies. A *filter* (*tuned circuit*) is a fixed-element circuit that favors a particular range of frequencies. *Low-pass filters* (see Fig. 61.11) pass only low frequencies (i.e., 0 Hz up to some cut-off frequency), whereas *high-pass filters* (see Fig. 61.12) pass frequencies higher than some cut-off frequency.[15] *Band-pass filters*, such as a series $RLC$ circuit or those that can be constructed as low-pass and high-pass filters in series, pass a range of frequencies between low and high cut-off frequencies. *Band-reject* (*notch*) filters such as a parallel $RLC$ circuit pass all but a selected range of frequencies.

A typical filter is the series $RC$ circuit shown in Fig. 61.11. If the output terminals are across the capacitor, the result is a *low-pass circuit*. At very low frequencies, the output voltage equals the input voltage since $X_C \gg R$. At very high frequencies, the output voltage is approximately zero. The *cut-off frequency* (*half-power point*), $f_{\mathrm{cut-off}}$, is the frequency at which the output is 0.707 of the input. It is calculated from the *time constant*, $\tau$, of the filter.

$$\mathbf{V}_o(t) = \mathbf{V}_i(t)\left(\frac{-jX_C}{R - jX_C}\right) \qquad 61.73$$

---

[15]A *differentiator* is a high-pass filter with a short time constant with respect to the period of the input voltage. An *integrator* is a low-pass filter with a long time constant with respect to the period of the input voltage.

$$f_{\mathrm{cut-off}} = \frac{1}{2\pi RC} = \frac{X_C}{R} = \frac{1}{2\pi\tau} \qquad 61.74$$

$$\tau = RC \qquad 61.75$$

**Figure 61.11** *Low-Pass RC Filter*

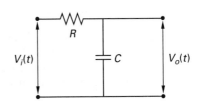

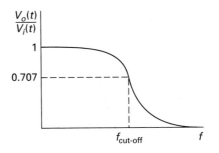

**Figure 61.12** *High-Pass RC Filter*

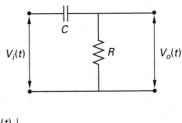

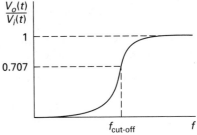

A *step input* voltage to a low-pass filter at $t = 0$ is equivalent to a transient circuit in which the battery is switched into the circuit at $t = 0$. The voltage across the capacitor gradually increases, approaching the step height.

If the output terminals are across the resistor, the result is a *high-pass circuit*. At very low frequencies, the output voltage is zero. At very high frequencies, the output voltage is equal to the input voltage.

$$\mathbf{V}_o(t) = \mathbf{V}_i(t)\left(\frac{R}{R - jX_C}\right) \qquad 61.76$$

$$f_{\text{cut-off}} = \frac{1}{2\pi RC} = \frac{1}{2\pi\tau} \qquad 61.77$$

A *step input* voltage to a high-pass filter at $t = 0$ is equivalent to a transient circuit in which the battery is switched into the circuit at $t = 0$. The voltage across the resistor is initially equal to the step height but gradually decays to zero.

## 24. COMPLEX POWER

The *complex power vector*, $\mathbf{S}$, is the vector sum of the *real (true) power vector*, $\mathbf{P}$, and the imaginary *reactive power vector*, $\mathbf{Q}$. Its magnitude can be calculated as Eq. 61.78 from the information at the voltage source terminals.[16] In general, $\mathbf{S} = \mathbf{I}^*\mathbf{V}$ where $\mathbf{I}^*$ is the complex conjugate of the current (i.e., the current with the phase difference angle reversed).[17]

$$S = IV = \tfrac{1}{2}I_m V_m \qquad 61.78$$

The *apparent power*, $S$, in volt-amps (VA) is the magnitude of the complex power vector. The *average power*, $P$, in watts (W) is the magnitude of the *real power vector*, which coincides with the real part of $\mathbf{S}$. The *reactive power*, $Q$, in volt-amps reactive (VAR) is the magnitude of the reactive power vector, which coincides with the imaginary part of $\mathbf{S}$.

Since the complex conjugate of the current is used in calculating $\mathbf{S}$, the *power angle*, $\phi$, is the same as the overall *impedance angle*. The *power factor*, pf, also known as the *phase factor* (usually given in percent) for a sinusoidal voltage is $\cos\phi$.[18] For a purely resistive load, pf $= 1$; for a purely reactive load, pf $= 0$.

Energy dissipation occurs only in the resistive part of a circuit since inductors and capacitors merely store and release energy. The average power, $P$, is calculated from the currents passing through the resistors or from the apparent power and the power factor.

$$P = \sum I_R V_R = \tfrac{1}{2}\sum I_{R,m} V_{R,m}$$
$$= S\cos\phi \qquad 61.79$$

$$\phi = \phi_Z \qquad 61.80$$

---

[16]Although Eq. 61.78 gives power for both effective and maximum voltages, it is common to use effective values (almost exclusively) in specifying power systems.
[17]This is an arbitrary convention, but it is used to keep the sign of the impedance angle equal to the sign of the power angle.
[18]The cosine is positive for positive and negative angles. Therefore, the descriptions "lagging" (for an inductive circuit) and "leading" (for a capacitive circuit) must be used with the power factor.

Equation 61.81 gives the *reactive power*, $Q$, in volt-amps reactive (VAR) stored in inductors and capacitors each cycle. $\sin\phi$ is the *reactive factor*.

$$Q = -\sum I_X V_X = -\tfrac{1}{2}\sum I_{X,m} V_{X,m}$$
$$= S\sin\phi \qquad 61.81$$

If power is graphed in the complex plane, as shown in Fig. 61.13, the three power vectors close on themselves to make the *power triangle* (which is congruent to the *impedance triangle*).

$$S^2 = P^2 + Q^2 \qquad 61.82$$

**Figure 61.13** *A Lagging Power Triangle*

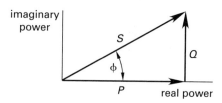

**Example 61.11**

For the circuit shown, find the (a) apparent power, (b) real power, and (c) reactive power, and (d) draw the power triangle.

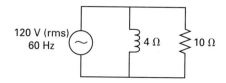

*Solution*

(a) The equivalent impedance is

$$\frac{1}{Z} = \frac{1}{j4\ \Omega} + \frac{1}{10\ \Omega} = -j0.25 + 0.10\ \text{S}$$

$$Z = 3.714\ \Omega$$

$$\phi_Z = \arctan\frac{0.25\ \text{S}}{0.10\ \text{S}} = 68.2°$$

The total current is

$$I = \frac{V}{Z} = \frac{120\ \text{V}\angle 0°}{3.714\ \Omega\angle 68.2°} = 32.31\ \text{A}\angle -68.2°$$

The apparent power is

$$S = I^*V = (32.31\ \text{A}\angle 68.2°)(120\ \text{V}\angle 0°)$$
$$= 3877\ \text{VA}\angle 68.2°$$

(b) The real power is

$$P = \frac{V_R^2}{R} = \frac{(120\ \text{V}\angle 0°)^2}{10\ \Omega} = 1440\ \text{W}$$

Alternatively, the real power can be calculated from Eq. 61.79.

$$P = S\cos\phi = (3877 \text{ VA})\cos 68.2° = 1440 \text{ W}$$

(c) The reactive power is

$$Q = \frac{V_L^2}{X_L} = \frac{(120 \text{ V})^2}{4 \text{ } \Omega} = 3600 \text{ VAR}$$

Alternatively, the reactive power can be calculated from Eq. 61.81.

$$Q = S\sin\phi = (3877 \text{ VA})\sin 68.2° = 3600 \text{ VAR}$$

(d) The real power is represented by the vector (in rectangular form) of $1440 + j0$. The reactive power is represented by the vector $0 + j3600$. The apparent power is represented (in phasor form) by the vector $3877 \text{ VA}\angle 68.2°$. The power triangle is

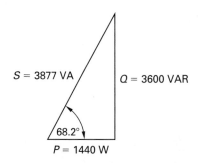

$$S = 3877 \text{ VA}$$
$$Q = 3600 \text{ VAR}$$
$$68.2°$$
$$P = 1440 \text{ W}$$

## 25. MAXIMUM ENERGY AND POWER TRANSFER

Assuming a fixed primary impedance, maximum energy (power) transfer in a complex circuit occurs when the source and load resistances are equal and the reactances are opposite. This is equivalent to having a resonant circuit.

$$R_{\text{load}} = R_{\text{source}} \qquad 61.83$$

$$X_{\text{load}} = -X_{\text{source}} \qquad 61.84$$

## 26. COST OF ELECTRICAL ENERGY

Except for very large users, electrical utilities measure electrical power usage on the basis of real energy. In Eq. 61.85, $C$ is the cost per kW·h used.

$$\text{cost} = C \times P_{\text{kW·h}} \qquad 61.85$$

Although only real power is dissipated or transformed into other forms, the reactive power contributes to total current. Reactive power results from an actual draw of current supplying the magnetization energy in motors and charge on capacitors. Therefore, the distribution system (wire, transformers, etc.) must be sized to carry this current. When real power is measured at the service location, the power factor is routinely monitored in a

larger (geographical) region and its effect built into the charge per kW·h.

## 27. POWER FACTOR CORRECTION

Inasmuch as apparent power is paid for but only real power is dissipated, it may be possible to reduce electrical utility charges if the power angle is reduced without changing the real power. This operation, known as *power factor correction*, is routinely accomplished by changing the circuit reactance in order to reduce the reactive power. The change in reactive power needed to change the power angle from $\phi_1$ to $\phi_2$ is

$$\Delta Q = P(\tan\phi_1 - \tan\phi_2) \qquad 61.86$$

When a circuit is capacitive (i.e., leading), induction motors can be connected across the line to improve the power factor. When a circuit is inductive (i.e., lagging), capacitors can be added across the line. The size (in farads) capacitor required is

$$C = \frac{\Delta Q}{\pi f V_m^2} \qquad [V_m \text{ maximum}] \qquad 61.87$$

$$C = \frac{\Delta Q}{2\pi f V^2} \qquad [V \text{ effective}] \qquad 61.88$$

### Example 61.12

A 60 Hz, 5 hp induction motor draws 53 A (rms) at 117 V (rms) with a 78.5% electrical-to-mechanical energy conversion efficiency. What capacitance should be connected across the line to increase the power factor to 92%?

*Solution*

The apparent power is found from the observed voltage and current.

$$S = IV = (53 \text{ A})(117 \text{ V}) = 6201 \text{ VA}$$

The real power drawn from the line is calculated from the real work done by the motor.

$$P = P_{\text{in}} = \frac{P_{\text{out}}}{\eta} = \frac{(5 \text{ hp})\left(746 \text{ } \frac{\text{W}}{\text{hp}}\right)}{0.785} = 4752 \text{ W}$$

The reactive power and power angle are calculated from the real and apparent powers.

$$Q_1 = \sqrt{S^2 - P^2} = \sqrt{(6201 \text{ VA})^2 - (4752 \text{ W})^2}$$
$$= 3984 \text{ VAR}$$
$$\phi_1 = \arccos\frac{4752 \text{ W}}{6201 \text{ VA}} = 39.97°$$

**Circuits**

The desired power factor angle is

$$\phi_2 = \arccos 0.92 = 23.07°$$

The reactive power after the capacitor is installed is

$$Q_2 = P \tan \phi_2 = (4752 \text{ W})\tan 23.07° = 2024 \text{ VAR}$$

The required capacitance is found from Eq. 61.88.

$$C = \frac{\Delta Q}{2\pi f V^2} = \frac{3984 \text{ VAR} - 2024 \text{ VAR}}{(2\pi)(60 \text{ Hz})(117 \text{ V})^2}$$
$$= 3.8 \times 10^{-4} \text{ F} \quad (380 \text{ } \mu\text{F})$$

## 28. MUTUAL INDUCTANCE

When a coil of wire is energized by a sinusoidal source, the induced magnetic flux is also sinusoidal.

$$\Phi(t) = \Phi_m \sin \omega t \qquad 61.89$$

Faraday's law gives the voltage induced when some of the magnetic flux produced in one coil (having $N_p$ turns) passes through a second coil of wire (having $N_s$ turns).

$$v_s(t) = -N_s \frac{d\Phi}{dt} = -N_s \omega \Phi_m \cos \omega t \qquad 61.90$$

The effective induced voltage is

$$v_s(t) = -N_s \left(\frac{2\pi}{\sqrt{2}}\right) f \Phi_m \cos \omega t$$
$$= -4.44 N_s f \Phi_m \cos \omega t \qquad 61.91$$

The flux that passes through both coils is the *mutual flux*. The *mutual inductance*, $M$ (see Fig. 61.14), is the proportionality constant between the induced voltage and the rate of current change. The *mutual reactance*, $X_M$, is analogous to inductive reactance.

$$v_p(t) = M \frac{di_s(t)}{dt} \qquad 61.92$$

$$v_s(t) = M \frac{di_p(t)}{dt} \qquad 61.93$$

$$X_M = \omega M \qquad 61.94$$

The *coefficient of coupling* (*coupling coefficient*), $k$, is the fraction of total flux that links both coils and varies from near zero (e.g., for radio coils) to near 1.0 (e.g., for iron-core transformers).[19] Depending on the coil winding and current direction, it can be positive or negative. (The coil winding direction may be shown in circuit diagrams by dots placed at one of the coil ends.) In

Eq. 61.95, $L_p$ and $L_s$ are the *self-inductances* of the primary and secondary windings, respectively.

$$k = \frac{M}{\sqrt{L_p L_s}} = \frac{X_M}{\sqrt{X_p X_s}} \qquad 61.95$$

**Figure 61.14** *Positive and Negative Mutual Inductance*

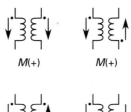

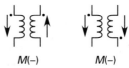

For the flux-linked circuit of Fig. 61.15, the primary and secondary loop equations are

$$V_p = I_p Z_p + I_s j X_m \quad \text{[primary]} \qquad 61.96$$
$$0 = I_s Z_s + I_p j X_m \quad \text{[secondary]} \qquad 61.97$$

**Figure 61.15** *Flux-Linked Circuit with Mutual Inductance*

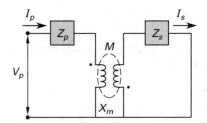

active equivalent circuit

## 29. IDEAL TRANSFORMERS

Transformers are used to change voltages, match impedances, and isolate circuits. They consist of coils of wire wound on a magnetically permeable core. The primary current produces a magnetic flux in the core, which induces a current in the secondary coil. Figure 61.16 shows two common designs—*shell* and *core transformers*.[20]

An *ideal transformer* possesses the following characteristics.

- The coefficient of coupling is 1.0 (i.e., all flux passes through each coil, and there is no leakage flux).

- The coils have no resistance.

- All flux is contained inside a path of constant area.

---

[19]An iron-core transformer is indicated by one or more lines drawn between the coils on circuit diagrams.

[20]An *autotransformer* is a transformer in which part of the winding is common to both primary and secondary circuits.

**Figure 61.16** *Shell and Core Transformers*

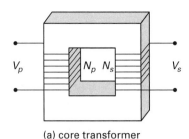

(a) core transformer

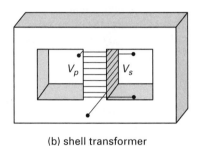

(b) shell transformer

The ratio of the numbers of primary to secondary windings is the *turns ratio (ratio of transformation)*, *a*. If the turns ratio is greater than unity, the transformer decreases voltage and is a *step-down transformer*. If the turns ratio is less than unity, the transformer increases voltage and is a *step-up transformer*.

$$a = \frac{N_p}{N_s} \qquad 61.98$$

In a lossless (i.e., 100% efficient) transformer, the power transfer from the primary side equals the power transfer to the secondary side, so

$$I_p V_p = I_s V_s \quad \text{[lossless]} \qquad 61.99$$

$$a = \frac{N_p}{N_s} = \frac{V_p}{V_s} = \frac{I_s}{I_p} = \sqrt{\frac{Z_p}{Z_s}} \quad \text{[lossless]} \qquad 61.100$$

## 30. IMPEDANCE MATCHING BY TRANSFORMER

A transformer can be used to match impedances for maximum power transfer. Considering Fig. 61.17, the secondary current is

$$I_s = \frac{V_s}{Z_s} \qquad 61.101$$

The impedance "seen" by the source, $Z_{\text{ep}}$, is known as the *effective primary impedance* or *reflected impedance* and is given by Eq. 61.102. This equation can be used to match impedances.

$$Z_{\text{ep}} = \frac{V_p}{I_p} = Z_p + a^2 Z_s \qquad 61.102$$

$$a = \sqrt{\frac{Z_p}{Z_s}} \qquad 61.103$$

Assuming a fixed primary impedance, maximum power consumption in the load occurs when the secondary impedance is the conjugate of the primary impedance. Therefore, for maximum power transfer with a transformer, the matched conditions are

$$R_p = a^2 R_s \qquad 61.104$$

$$X_p = -a^2 X_s \qquad 61.105$$

**Figure 61.17** *Impedance Matching*

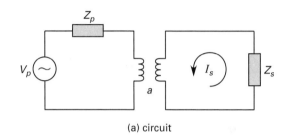

(a) circuit

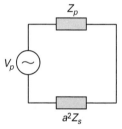

(b) equivalent circuit

### Example 61.13

An 8 Ω speaker is to be impedance-matched to a 10 kΩ amplifier. What should be the turns ratio of the transformer?

*Solution*

From Eq. 61.103,

$$a = \sqrt{\frac{Z_p}{Z_s}} = \sqrt{\frac{10 \times 10^3 \ \Omega}{8 \ \Omega}}$$

$$= 35.4$$

### Example 61.14

An ideal step-down transformer has 200 primary turns and 50 secondary turns in its coils. 440 V are applied across the primary. The secondary load resistance is 5 Ω. (a) What is the secondary current? (b) What is the effective primary resistance?

*Solution*

(a) The turns ratio is

$$a = \frac{N_p}{N_s} = \frac{200}{50} = 4$$

The secondary voltage and current are

$$V_s = \frac{V_p}{a} = \frac{440 \text{ V}}{4} = 110 \text{ V}$$

$$I_s = \frac{V_s}{R_s} = \frac{110 \text{ V}}{5 \text{ }\Omega} = 22 \text{ A}$$

(b) From Eq. 61.104, the resistance seen by the primary source is

$$R = a^2 R_s = (4)^2 (5 \text{ }\Omega) = 80 \text{ }\Omega$$

## 31. REAL TRANSFORMERS

*Real transformers* have winding resistance and flux leakage. (See Fig. 61.18.) *Primary leakage flux* links only the primary coil, while *secondary leakage flux* links only the secondary coil. According to Faraday's law, each leakage flux corresponds to a loss of induced voltage in its corresponding coil. Each induced voltage loss can be considered to have been caused by an equivalent reactance (inductance). The *primary* and *secondary leakage reactances* are identified as

$$X_p = \omega L_p = \frac{V_{p,\text{loss}}}{I_p} = \frac{4.44 f \Phi_{m,\text{loss}} N_p}{I_p} \qquad \textit{61.106}$$

$$X_s = \omega L_s = \frac{V_{s,\text{loss}}}{I_s} \qquad \textit{61.107}$$

Equation 61.108 relates the turns ratio to the leakage inductances.

$$a = \sqrt{\frac{L_p}{L_s}} \qquad \textit{61.108}$$

**Figure 61.18** *Equivalent Real Transformer*

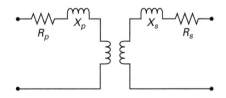

## 32. TRANSFORMER RATING

Transformer *nameplate ratings* normally include the two winding voltages (either of which can be the primary), frequency, and kVA rating. Continuous operation at the rated values will not result in excessive heat build-up. (Apparent power in kVA, not real power, is used in the rating because heating is proportional to the square of the supply current.)

The two main sources of power loss in transformers are the *core losses* (*iron losses*) and *copper losses* (*winding losses*). Core losses consist of *hysteresis loss*, $P_h$ (losses due to cyclic changes in the magnetic state of iron), and *eddy-current losses*, $P_e$ (losses caused by the flow of microscopic currents in the iron). Both losses are constant in magnitude and independent of load (i.e., do not vary between no-load and full-load conditions). In Eq. 61.109, $n$ is the *Steinmetz exponent*, which varies from 1.5 to 2.5 and is typically taken as 1.6. The maximum magnetic flux density, $B_m$, can be calculated from Faraday's law.

$$P_h = k_h f B_m^n \quad \text{[per unit volume of core]} \qquad \textit{61.109}$$

$$P_e = k_e f^2 B_m^2 \quad \text{[per unit volume of core]} \qquad \textit{61.110}$$

*Copper losses* (due to resistance heating in the wire), $P_{\text{Cu}}$, are

$$P_{\text{Cu}} = I^2 R \qquad \textit{61.111}$$

The *transformer efficiency* is the ratio of output power to input power. The maximum efficiency occurs when the copper losses equal the core losses. The *all-day efficiency* is the ratio of energy delivered by the transformer in a 24-hour period to the energy input during the same period of time.

$$\eta = \frac{P_{\text{out}}}{P_{\text{in}}} = \frac{P_{\text{in}} - \sum P_{\text{losses}}}{P_{\text{in}}} \qquad \textit{61.112}$$

The *regulation* of a transformer is the percentage change in output voltage between no-load and full-load conditions.

$$\text{regulation} = \frac{V_{\text{no load}} - V_{\text{full load}}}{V_{\text{full load}}} \times 100\% \qquad \textit{61.113}$$

## 33. TWO-PORT TRANSFORMER MODEL

An iron-core transformer can be represented by the two-port *hybrid parameter model* (see Sec. 61.39) shown in Fig. 61.19.[21] The admittance, $Y$, accounts for power loss. The inductive (negative) susceptance, $B$, accounts for energy storage. Resistance, $R$, includes the effects of both windings. The inductance, $L$, is an equivalent leakage inductance. The governing equations are

$$\mathbf{I}_1 = \mathbf{V}_1 \mathbf{Y} + a\mathbf{I}_2 \qquad \textit{61.114}$$

$$\mathbf{V}_2 = a\mathbf{V}_1 - \mathbf{Z}\mathbf{I}_2 \qquad \textit{61.115}$$

---

[21]This is a simple model of a transformer. Other more complex models predict transformer performance more precisely.

*Circuits*

**Figure 61.19** *Two-Port Transformer Model*

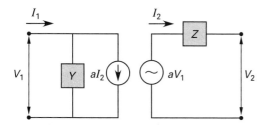

## 34. TRANSFORMER TESTING

Referring to Fig. 61.19, if the secondary is open, $I_2 = 0$ and

$$\mathbf{Y} = G + jB \quad \begin{bmatrix} \text{open} \\ \text{secondary} \end{bmatrix} \qquad 61.116$$

With the transformer loaded, the power lost in the winding resistance and leakage reactance is significant.

$$\mathbf{Z} = R + jX_L = R + j\omega L \qquad 61.117$$

Based on these characteristics, testing of a transformer to determine its rating, efficiency, and equivalent circuit parameters is standardized and divided into two parts: open- and closed-circuit tests. The distinction between primary and secondary may be replaced by high-voltage (HV) and low-voltage (LV), or vice versa, designations.[22]

*Open-circuit test:* The secondary terminals are left open. (Actually, a meter across the terminals measures secondary voltage, $V_{2o}$, during the test, but the meter has such a high resistance that it appears as an open circuit.) The rated voltage, $V_{1o}$, is applied to the primary terminals.[23] The current, $I_{1o}$ (the *exciting current*), in the primary side is measured, while a wattmeter measures the total input power, $P_o$.

$$G = \frac{P_o}{V_{1o}^2} \qquad 61.118$$

$$Y = \frac{I_{1o}}{V_{1o}} \qquad 61.119$$

$$B = -\sqrt{Y^2 - G^2} = \frac{-\sqrt{I_{1o}^2 V_{1o}^2 - P_o^2}}{V_{1o}^2} \qquad 61.120$$

$$a = \frac{V_{1o}}{V_{2o}} \qquad 61.121$$

*Short-circuit test:* The secondary terminals are shorted. (An ammeter connected to the terminals may measure the current, $V_{2s}$, through the secondary side.[24] However, the ammeter has a very low resistance and appears as a

short circuit.) A voltage is applied to the primary windings such that the rated current (i.e., the volt-amp rating divided by the voltage rating) flows. This voltage will be relatively low because the secondary has no significant impedance. (Since the voltage is low, the core losses are also small, and the primary admittance, $\mathbf{Y}$, is omitted from the model.) A voltmeter measures the applied voltage while a wattmeter measures the copper loss only.

$$R = \frac{P_s}{a^2 I_{1s}^2} \qquad 61.122$$

$$Z = \frac{V_{1s}}{aI_{2s}} = \frac{V_{1s}}{a^2 I_{1s}} \qquad 61.123$$

$$X = \sqrt{Z^2 - R^2} \qquad 61.124$$

$$a = \frac{I_{2s}}{I_{1s}} \qquad 61.125$$

## 35. POWER AND COMMUNICATION TRANSMISSION LINES

Transmission lines are used to carry electrical power and communication signals long distances with minimum losses. Common transmission lines include optical fibers, coaxial cables with polyethylene dielectric, two-wire (balanced) lines, and 75 $\Omega$, 150 $\Omega$, or 300 $\Omega$ twin-lead lines.

*Electric power transmission* is accomplished, in most cases, with *overhead transmission lines* carried on structures (poles and towers). These lines carry three-phase voltages of 110 kV or higher, since (consistent with *Joule's law*) higher voltages reduce current and power losses.[25] Most power lines are not insulated. Aluminum is preferred over copper because of its cost and resistance/weight ratio. Aluminum strands wrapped around a steel core, known as *aluminum conductor steel reinforced* (ACSR), provide more tensile strength than plain aluminum wire. Since high-frequency current migrates to the outer layer of wires, known as the *skin effect*, the higher-resistance steel strand does not participate in power transmission. *Strain insulators* separate the transmission line from the supporting structure. The network of transmission power distribution paths is known as the *power grid*. Between structures, lines follow the shape of a catenary cable.

Losses in power transmission are primarily resistive. Due to the skin effect, resistance can be decreased only slightly by using larger-diameter conductors, but cost also increases. An optimum diameter can be approximated from *Kelvin's law*, which balances the cost of

---

[22]The terms *high-tension* and *low-tension* are synonymous with HV and LV, respectively.

[23]The open-circuit test is performed at the rated voltage because the core losses are dependent on the magnetic flux density, $B_m$.

[24]The ammeter is not essential, but the ratio of currents provides a check on the accuracy of the turns ratio calculation.

[25]Power transmission voltages can exceed 800 kV. Voltages less than 1000 V are referred to as *low voltage*; 1000 V–33 kV as *medium voltage, distribution level*; 33–230 kV as *high voltage, sub-transmission and transmission levels*; 230–800 kV as *very high voltage*; and 800 kV and above as *ultra-high voltage*. Lines carrying ultra-high voltages are prone to *corona discharge* losses that can negate the benefits of the higher losses.

**Circuits**

power loss against the debt service (i.e., interest) paid for the line. Line capacitance and inductance affect the *power factor*. The line must be sized to carry the apparent power (VA), but the reactive power (VARS) does not benefit the load. High power factors are maintained by *static VAR compensators* (i.e., fast-switching capacitor banks), *phase-shifting transformers*, and electronic source modification (*flexible AC transmission systems* or FACTS).

Performance of a transmission line is determined by many parameters: characteristic impedance, $Z_0$; resistance per unit length, $R_l$; capacitance per unit length, $C_l$; inductance per unit length, $L_l$; shunt conductance, $G_l$; internal inductance, $L_{int}$; and external inductance, $L_{ext}$; at high and low frequencies. Other parameters include the characteristic impedance, standing wave ratio, and reflection coefficient.[26]

In order to eliminate any power being reflected back to the source, a transmission line should be terminated with a resistance that is equal to the *characteristic impedance*, $Z_0$.

$$Z_0 = \sqrt{\frac{L_l}{C_l}} \qquad 61.126$$

If the transmission line is not properly terminated (with a resistance equal to its characteristic impedance), signals from the generator will be reflected to some degree at the end of the line. The reflected waves combine with incoming waves to form *standing waves*. The *standing wave ratio*, SWR, is the ratio of maximum to minimum voltages (currents) encountered along the line and is a number typically greater than unity. (A perfect match between transmission line and termination results in a SWR of unit. In that case, all of the input power is absorbed by the load.) Voltage and current standing wave ratios, VSWR and ISWR, are equal.

$$\text{SWR} = \text{VSWR} = \text{ISWR} = \max\begin{cases} Z_0/R_l \\ R_l/Z_0 \end{cases} \qquad 61.127$$

The *reflection coefficient*, $\Gamma$, is similar to the standing wave ratio in that it is a ratio of reflected to incident voltage (current). The fraction of the incident power that is reflected back from the load is $\Gamma^2$.

$$\Gamma = \frac{V_{\text{reflected}}}{V_{\text{incident}}} = \frac{I_{\text{reflected}}}{I_{\text{incident}}} = \left|\frac{Z_0 - R_l}{Z_0 + R_l}\right| \qquad 61.128$$

The relationship between the standing wave ratio and the reflection coefficient is

$$\Gamma = \frac{\text{SWR} - 1}{\text{SWR} + 1} \qquad 61.129$$

The *velocity of propagation*, $v_w$, is the velocity at which an electromagnetic wave travels along the transmission line. The units of Eq. 61.130 depend on the units of $L_l$ and $C_l$. The *velocity factor* is the ratio of velocity of propagation to the speed of light in free space.[27]

$$v_w = \frac{1}{\sqrt{L_l C_l}} \qquad 61.130$$

Whereas a DC current density is uniform across the cross-sectional area of a conductor, AC current density is nonuniform. More current is carried near the outer surface than in the interior, and this phenomenon (known as the *skin effect*) becomes more pronounced the higher the frequency. At extremely high frequencies (the exact value of which depends on the conductor size, permeability, and conductivity), almost all current is carried in a very thin layer at the surface, resulting in a correspondingly high resistance.

## 36. ANTENNAS

*Antennas* are characterized by their actual configuration and length, often specified in terms of wavelengths. For example, a *dipole antenna (Hertz antenna)* is a half-wavelength conductor that radiates primarily in two opposite directions. A *Marconi antenna* (typically used in *whip antennas*) is a vertical quarter-wavelength conductor with a grounded base that radiates in an omnidirectional pattern.

*Short antennas* have lengths less than half of a wavelength; *long antennas* are longer. Long antennas are classified as *resonant antennas* (when their lengths are even multiples of half-wavelengths) and *nonresonant antennas* otherwise.

The wavelength and frequency are related by Eq. 61.131. The *velocity factor*, $k$, is the ratio of the speeds of electromagnetic waves in the transmission medium and in a vacuum. For air, $k$ is essentially 1.0.

$$f\lambda = kc_{\text{vacuum}} = c_{\text{medium}} \qquad 61.131$$

An antenna that is tuned to one frequency can also be used at a different (higher) frequency by including traps to shorten the actual antenna length. A *trap* is a frequency-dependent circuit that appears as an open or closed circuit depending on the frequency. (See Fig. 61.20.)

The *gain (merit)* in decibels of an antenna is the ratio of the powers that must be supplied to a standard antenna and the test antennas to produce identical field strengths at a given location.

$$\text{gain} = 10\log_{10}\frac{P_{\text{std}}}{P_{\text{test}}} = 20\log_{10}\frac{V_{\text{std}}}{V_{\text{test}}} \quad [\text{in dB}] \qquad 61.132$$

---

[26]The Smith chart is a convenient method of evaluating these latter three parameters with low-loss lines.

[27]The speed of light in free space is approximately $3 \times 10^8$ m/s ($9.82 \times 10^8$ ft/sec).

**Figure 61.20** *Dipole Antenna with Traps*

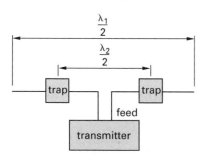

**Figure 61.21** *Beam Width*

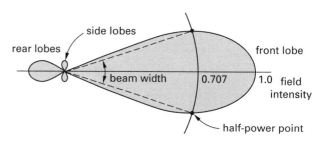

The *effective radiated power*, ERP, is the product of the antenna input power and gain in decibels.

$$\text{gain} = 10 \log_{10} \frac{\text{ERP}}{P_{\text{input}}} \qquad 61.133$$

An *omnidirectional antenna* radiates equally in all directions. A *parasitic array* is a *directional antenna* that is constructed from a dipole antenna and one or more *parasitic elements* (i.e., elements that are not connected to the driven dipole). A *director* is a parasitic element that is shorter than the dipole length, while a *refractor* is longer. When a dipole is placed between a director and a reflector, wave propagation is enhanced in the direction of the director and reduced in the direction of the reflector. The *front-to-back ratio* is the ratio of powers (in decibels) radiated in the favored (forward) and disfavored (backward) directions.

An antenna appears to the transmitter as a circuit element with a complex impedance. At the frequency for which the antenna was tuned, $f_0$, the antenna appears totally resistive (the resistance consisting of radiation and wire resistances) and the current is maximized.[28] The *quality factor*, $Q$, of an antenna depends on the current bandwidth and is

$$Q = \frac{f_0}{\text{BW}} \qquad 61.134$$

The *beam width* (not to be confused with the *bandwidth*) of a directional antenna is the width (in degrees) of the major lobe between the two directions at which the relative radiated power is one-half of its peak value. (See Fig. 61.21.) At these *half-power points*, the field intensity is 0.707 times the peak value, or down 3 dB from the peak value.

The *radiation resistance*, $R_r$, with units of ohms, is an intangible resistance seen by the transmitter. The *total radiated power*, $P_r$, and the radiation resistance are related by the current flow in the antenna feed.

$$P_r = \tfrac{1}{2} I_m^2 R_r = I^2 R_r \qquad 61.135$$

---

[28]*Log-periodic antennas* have very wide bandwidths and are, therefore, referred to as *frequency-independent antennas*.

## 37. IMPEDANCE MODEL OF TWO-PORT NETWORK

The impedance parameters of an *impedance model* (also known as a *z-parameter model* and an *open-circuit*

**Figure 61.22** *Impedance Model Parameters*

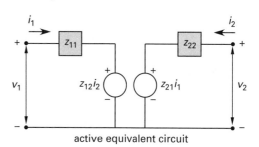

active equivalent circuit

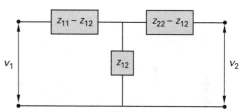

passisve T-model equivalent circuit

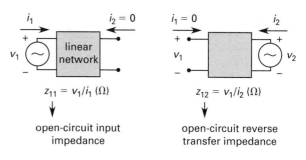

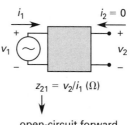

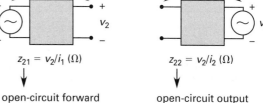

*impedance model*) are defined by Eq. 61.136 and Eq. 61.137. The names and method of determining the z-parameters are shown in Fig. 61.22. The standard and *T-circuit (Y-circuit) equivalent models* based on z-parameters are also shown.

$$v_1 = z_{11}i_1 + z_{12}i_2 \qquad 61.136$$

$$v_2 = z_{21}i_1 + z_{22}i_2 \qquad 61.137$$

## 38. ADMITTANCE MODEL OF TWO-PORT NETWORK

In an *admittance model*, the governing equations are given by Eq. 61.138 and by Eq. 61.139. Figure 61.23

**Figure 61.23** *Admittance Model Parameters*

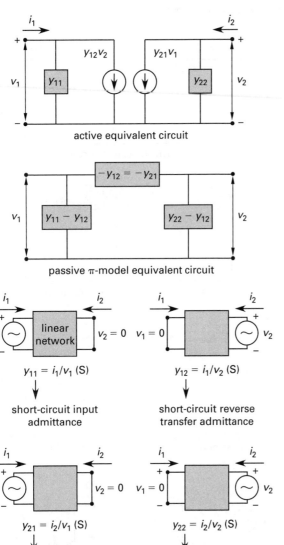

illustrates the standard and *pi-circuit (delta-circuit) equivalent models* based on the *y*-parameters.

$$i_1 = y_{11}v_1 + y_{12}v_2 \qquad 61.138$$

$$i_2 = y_{21}v_1 + y_{22}v_2 \qquad 61.139$$

## 39. HYBRID MODEL OF TWO-PORT NETWORK

Equation 61.140 and Eq. 61.141 describe the performance of a *hybrid model* shown in Fig. 61.24 (also known as an *h-parameter model*), so named because one of the governing equations is based on current and the other on voltage.

$$v_1 = h_{11}i_1 + h_{12}v_2 \qquad 61.140$$

$$i_2 = h_{21}i_1 + h_{22}v_2 \qquad 61.141$$

**Figure 61.24** *Hybrid Circuit Model*

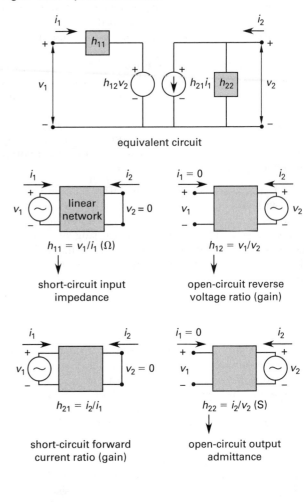

# 62 Three-Phase Electricity

## Nomenclature

| | | |
|---|---|---|
| $a$ | ratio of transformation | – |
| $I$ | current | A |
| $N$ | number of turns | – |
| $P$ | power | W |
| $Q$ | reactive power | VAR |
| $R$ | resistance | $\Omega$ |
| $S$ | apparent power | VA |
| $t$ | time | s |
| $V$ | voltage | V |
| $X$ | reactance | $\Omega$ |
| $Z$ | impedance | $\Omega$ |

## Symbols

| | | |
|---|---|---|
| $\phi$ | impedance angle | rad |

## Subscripts

| | |
|---|---|
| $L$ | inductive |
| $p$ | phase or primary |
| pu | per-unit |
| $s$ | secondary |
| $t$ | total |

## 1. BENEFITS OF THREE-PHASE POWER

Three-phase energy distribution systems use fewer and smaller conductors and, therefore, are more efficient than multiple single-phase systems providing the same power. Three-phase motors provide a uniform torque, not a pulsating torque as do single-phase motors. Three-phase induction motors do not require additional starting windings or associated switches. When rectified, three-phase voltage has a smoother waveform and less ripple to be filtered out.

## 2. GENERATION OF THREE-PHASE POTENTIAL

Figure 62.1(a) shows the symbolic representation of an AC generator that produces three equal sinusoidal voltages. The generated voltage in each coil is known as the *phase voltage*, $V_p$, or *coil voltage*. (Three-phase voltages are almost always stated as effective values.) Due primarily to the location of the windings, the three sinusoids are 120° apart in phase as shown in Fig. 62.1(b). If $\mathbf{V}_a$ is chosen as the reference voltage, then Eq. 62.1 through Eq. 62.3 represent the phasor forms of the three sinusoids. At any moment, the vector sum of these three voltages is zero.

$$\mathbf{V}_a = V_p \angle 0° \qquad 62.1$$

$$\mathbf{V}_b = V_p \angle -120° \qquad 62.2$$

$$\mathbf{V}_c = V_p \angle -240° \qquad 62.3$$

Equation 62.1 through Eq. 62.3 define an *ABC* or *positive sequence*. That is, $\mathbf{V}_a$ reaches its peak before $\mathbf{V}_b$, and $\mathbf{V}_b$ peaks before $\mathbf{V}_c$. With a *CBA* (also written as *ACB*) or *negative sequence*, obtained by rotating the

**Figure 62.1** *Three-Phase Voltage*

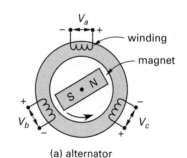

(a) alternator

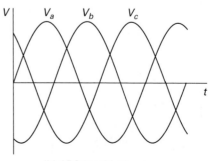

(b) ABC (positive) sequence

*field magnet* in the opposite direction, the order of the delivered sinusoids is reversed (i.e., $\mathbf{V}_c$, $\mathbf{V}_b$, $\mathbf{V}_a$).

Although a six-conductor transmission line could be used to transmit the three voltages, it is more efficient to interconnect the windings. The two methods are commonly referred to as *delta (mesh)* and *wye (star)* connections.

Figure 62.2(a) illustrates delta source connections. The voltage across any two of the lines is known as the *line voltage (system voltage)* and is equal to the phase voltage. Any of the coils can be selected as the reference as long as the sequence is maintained. For a positive (ABC) sequence,

$$\mathbf{V}_{CA} = V_p \angle 0° \qquad 62.4$$

$$\mathbf{V}_{AB} = V_p \angle -120° \qquad 62.5$$

$$\mathbf{V}_{BC} = V_p \angle -240° \qquad 62.6$$

Wye-connected sources are illustrated in Fig. 62.2(b). While the *line-to-neutral voltages* are equal to the phase voltage, the line voltages are greater—$\sqrt{3}$ times the phase voltage. The *ground wire (neutral)* is needed to carry current only if the system is unbalanced.[1] (See Sec. 62.3.) For an ABC sequence, the line voltages are

$$\mathbf{V}_{AB} = \sqrt{3}\, V_p \angle 30° \qquad 62.7$$

$$\mathbf{V}_{BC} = \sqrt{3}\, V_p \angle -90° \qquad 62.8$$

$$\mathbf{V}_{CA} = \sqrt{3}\, V_p \angle -210° \qquad 62.9$$

**Figure 62.2** *Delta and Wye Source Connections*

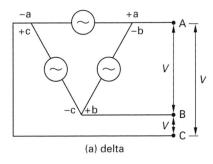

(a) delta

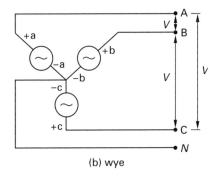

(b) wye

---

[1]The neutral wire is usually kept to provide for minor imbalance.

Although the magnitude of the line voltage depends on whether the generator coils are delta- or wye-connected, each connection results in three equal sinusoidal voltages, each 120 degrees out of phase with one another.

## 3. DISTRIBUTION SYSTEMS

Three-phase power is delivered by three-wire and four-wire systems. A *three-wire system* contains three power conductors. A *four-wire system* consists of three power conductors and a neutral conductor.

Utility power distribution starts with generation. The generator is connected through step-up *subtransmission transformers* that supply *transmission lines*. The actual transmission line voltage depends on the distance between the subtransmission transformers and the user. Distribution *substation transformers* reduce the voltage from the transmission line level to approximately 35 kV. The *primary distribution system* delivers power to *distribution transformers* that further reduce voltage to 120 V to 600 V.

## 4. BALANCED LOADS

Three impedances are required to fully load a three-phase voltage source. The impedances in a three-phase system are *balanced* when they are identical in magnitude and angle. The voltages and line current, and real, apparent (kVA), and reactive powers are all identical in a balanced system. Also, the power factor is the same for each phase. Therefore, balanced systems can be analyzed on a per-phase basis. Such calculations are known as *one-line analyses*.

Figure 62.3 illustrates the vector diagram for a balanced delta three-phase system. The phase voltages, $\mathbf{V}$, are separated by 120° phase angles, as are the phase currents, $\mathbf{I}$. The phase difference angle, $\phi$, between a phase voltage and its respective phase current depends on the phase impedance. With delta-connected resistive loads, the phase and line currents differ in phase by 30°.

**Figure 62.3** *Positive ABC Balanced Delta Load Vector Diagram*

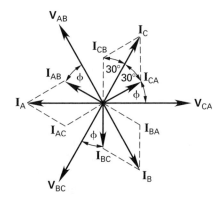

## 5. DELTA-CONNECTED LOADS

Figure 62.4 illustrates delta-connected loads. The line and phase voltages are equal. However, the line and phase currents are different.

**Figure 62.4** *Delta-Connected Loads*

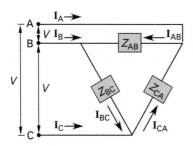

The *phase currents* for a balanced system are calculated from the line voltage (same as the phase voltage). For a positive (ABC) sequence,

$$\mathbf{I}_{AB} = \frac{\mathbf{V}_{AB}}{\mathbf{Z}_{AB}} = \frac{V\angle{-120°}}{Z\angle\phi} = \frac{V}{Z}\angle{-120° - \phi} \quad 62.10$$

$$\mathbf{I}_{BC} = \frac{\mathbf{V}_{BC}}{\mathbf{Z}_{BC}} = \frac{V}{Z}\angle{-240° - \phi} \quad 62.11$$

$$\mathbf{I}_{CA} = \frac{\mathbf{V}_{CA}}{\mathbf{Z}_{CA}} = \frac{V}{Z}\angle{-\phi} \quad 62.12$$

The *line currents* are not the same as the phase currents but are $\sqrt{3}$ times the phase current and displaced $-30°$ in phase from the phase currents.

$$|\mathbf{I}_A| = |\mathbf{I}_{AB} - \mathbf{I}_{CA}| = \sqrt{3}I_{AB} \quad 62.13$$

$$|\mathbf{I}_B| = |\mathbf{I}_{BC} - \mathbf{I}_{AB}| = \sqrt{3}I_{BC} \quad 62.14$$

$$|\mathbf{I}_C| = |\mathbf{I}_{CA} = \mathbf{I}_{BC}| = \sqrt{3}I_{CA} \quad 62.15$$

$$I_A = I_B = I_C \quad \text{[balanced]} \quad 62.16$$

Each impedance in a balanced system dissipates the same real *phase power*, $P_p$. The total power dissipated is three times the phase power. (This is the same as for wye-connected loads.)

$$P_t = 3P_p = 3V_pI_p\cos\phi = \sqrt{3}VI\cos\phi \quad 62.17$$

### Example 62.1

Three identical impedances are connected in delta across a three-phase system with 240 V (rms) line voltages in an ABC sequence. Find (a) phase current $\mathbf{I}_{AB}$, (b) phase real power $P_p$, (c) line current $\mathbf{I}_B$, and (d) total real power.

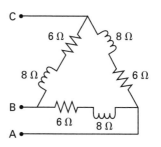

*Solution*

(a) The phase impedance is

$$\mathbf{Z}_p = \sqrt{R^2 + X_L^2}\angle\arctan\left(\frac{X}{R}\right)$$
$$= \sqrt{(6\ \Omega)^2 + (8\ \Omega)^2}\angle\arctan\left(\frac{8\ \Omega}{6\ \Omega}\right)$$
$$= 10\ \Omega\angle53.13°$$

The phase current is

$$\mathbf{I}_{AB} = \frac{\mathbf{V}_{AB}}{\mathbf{Z}_{AB}} = \frac{240\ V\angle{-120°}}{10\ \Omega\angle53.13°} = 24\ A\angle{-173.13°}$$

(b) The phase real power is

$$P_p = I_p^2R = (24\ A)^2(6\ \Omega) = 3456\ W$$

(c) The phase current $I_{BC}$ contributes to the line current $I_B$ as well as $I_{AB}$.

$$\mathbf{I}_{BC} = \frac{\mathbf{V}_{BC}}{\mathbf{Z}_{BC}} = \frac{240\ V\angle{-240°}}{10\ \Omega\angle53.13°}$$
$$= 24\ A\angle{-293.13°}$$
$$\mathbf{I}_B = \mathbf{I}_{BC} - \mathbf{I}_{AB}$$
$$= 24\ A\angle{-293.13°} - 24\ A\angle{-173.13°}$$
$$= 41.57\ A\angle36.87°$$

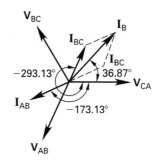

(d) The total real power is three times the phase power.

$$P_t = 3P_p = (3)(3456\ W) = 10{,}368\ W$$

## 6. WYE-CONNECTED LOADS

Figure 62.5 illustrates three equal impedances connected in wye configuration. The line and phase currents are equal. However, the phase voltage is less than the line voltage since two phases are connected across two lines. The line and phase currents are

$$\mathbf{I}_A = \mathbf{I}_{AN} = \frac{\mathbf{V}_{AN}}{\mathbf{Z}_{AN}} = \frac{\mathbf{V}}{\sqrt{3}\mathbf{Z}_{AN}} \qquad 62.18$$

$$\mathbf{I}_B = \mathbf{I}_{BN} = \frac{\mathbf{V}_{BN}}{\mathbf{Z}_{BN}} = \frac{\mathbf{V}}{\sqrt{3}\mathbf{Z}_{BN}} \qquad 62.19$$

$$\mathbf{I}_C = \mathbf{I}_{CN} = \frac{\mathbf{V}_{CN}}{\mathbf{Z}_{CN}} = \frac{\mathbf{V}}{\sqrt{3}\mathbf{Z}_{CN}} \qquad 62.20$$

$$\mathbf{I}_N = 0 \quad \text{[balanced]} \qquad 62.21$$

**Figure 62.5** *Wye-Connected Loads*

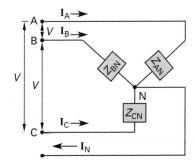

The total power dissipated in a balanced wye-connected system is three times the phase power. (This is the same as for the delta-connection.)

$$P_t = 3P_p = 3V_pI_p\cos\phi = \sqrt{3}\,VI\cos\phi \qquad 62.22$$

### Example 62.2

A three-phase, 480 V, 250 hp motor has an efficiency of 94% and a power factor of 90%. What is the line current?

*Solution*

It does not matter whether the motor's windings are delta- or wye-connected. From Eq. 62.22, the line current is

$$I = \frac{P_t}{\eta\sqrt{3}\,V\cos\phi} = \frac{(250\text{ hp})\left(745.7\,\dfrac{\text{W}}{\text{hp}}\right)}{(0.94)\sqrt{3}(480\text{ V})(0.9)}$$

$$= 265\text{ A}$$

### 7. DELTA-WYE CONVERSIONS

It is occasionally convenient to convert a delta system to a wye system, and vice versa. The equations used to convert between delta and wye resistor networks are applicable if impedances are substituted for resistances.

### 8. PER-UNIT CALCULATIONS

In power systems, the *per-unit system* is regularly used to express voltage, current, impedance, and kVA ratings. The common bases are line voltage (in kV) and apparent power (in kVA). Since currents, voltages, and power may differ between line and phase in three-phase systems, it is important to recognize the difference between line-to-neutral (subscript *ln*) and phase (subscript *p*) values when using Eq. 62.23 through Eq. 62.28 to calculate the bases. For example, the base power is the phase power, one-third of the total power: $S_p = S_t/3$.

$$S_{\text{base}} = S_p \qquad 62.23$$

$$V_{\text{base}} = V_p \qquad 62.24$$

$$I_{\text{base}} = \frac{S_{\text{base}}}{V_{\text{base}}} = \frac{S_p}{V_p} \qquad 62.25$$

$$Z_{\text{base}} = \frac{V_{\text{base}}}{I_{\text{base}}} = \frac{V_p^2}{S_p} \qquad 62.26$$

$$P_{\text{base}} = S_p \qquad 62.27$$

$$Q_{\text{base}} = S_p \qquad 62.28$$

The per-unit values are

$$I_{\text{pu}} = \frac{I_{\text{actual}}}{I_{\text{base}}} \qquad 62.29$$

$$V_{\text{pu}} = \frac{V_{\text{actual}}}{V_{\text{base}}} \qquad 62.30$$

$$Z_{\text{pu}} = \frac{Z_{\text{actual}}}{Z_{\text{base}}} \qquad 62.31$$

$$P_{\text{pu}} = \frac{P_{\text{actual}}}{P_{\text{base}}} \qquad 62.32$$

$$Q_{\text{pu}} = \frac{Q_{\text{actual}}}{Q_{\text{base}}} \qquad 62.33$$

Ohm's law and other circuit analysis methods can be used with the per-unit quantities.

$$V_{\text{pu}} = I_{\text{pu}}Z_{\text{pu}} \qquad 62.34$$

### Example 62.3

A wye-connected three-phase device is rated to draw a total of 300 kVA when connected to a line-to-line voltage of 15 kV. The device's per-unit impedance is $0.1414 + j0.9900$. What is the actual impedance?

*Solution*

The bases are

$$V_{\text{base}} = V_p = \frac{V}{\sqrt{3}} = \frac{15\text{ kV}}{\sqrt{3}} = 8.66\text{ kV}$$

$$S_{\text{base}} = S_p = \frac{300\text{ kVA}}{3} = 100\text{ kVA}$$

Circuits

The base current is

$$I_{\text{base}} = \frac{S_{\text{base}}}{V_{\text{base}}} = \frac{100 \text{ kVA}}{8.66 \text{ kV}} = 11.55 \text{ A}$$

The base impedance is

$$Z_{\text{base}} = \frac{V_{\text{base}}}{I_{\text{base}}} = \frac{8660 \text{ V}}{11.55 \text{ A}} = 750 \text{ } \Omega$$

From Eq. 62.31, the actual impedance is

$$Z_{\text{actual}} = Z_{\text{pu}} Z_{\text{base}} = (0.1414 + j0.9900)(750 \text{ } \Omega)$$
$$= 106.1 + j742.5 \text{ } \Omega$$

## 9. UNBALANCED LOADS

The three-phase loads are unequal in an unbalanced system. A fourth conductor, the *neutral conductor*, is required for the line voltages to be constant. Such a system is known as a *four-wire system*. Without the neutral conductor (i.e., in a *three-wire system*), the common point of the load connections is not at zero potential, and the voltages across the three impedances vary from the line-to-neutral voltage. The voltage at the common point is known as the *displacement neutral voltage*.

Regardless of whether the line voltages are equal, the line currents are not the same nor do they have a 120° phase difference. Unbalanced systems are evaluated by computing the phase currents and then applying Kirchhoff's current law (in vector form) to obtain the line currents. The *neutral current* is

$$\mathbf{I}_N = -(\mathbf{I}_A + \mathbf{I}_B + \mathbf{I}_C) \qquad \textit{62.35}$$

### Example 62.4

Unequal impedances are connected to a 208 V (rms) three-phase system. The sequence is ABC with $\angle V_{\text{CA}} = 0°$. (a) What is the line current $I_A$? (b) What is the total real power dissipated?

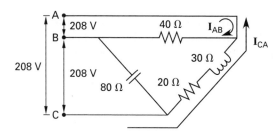

*Solution*

(a) First, calculate the phase current $\mathbf{I}_{AB}$.

$$\mathbf{I}_{AB} = \frac{\mathbf{V}_{AB}}{\mathbf{Z}_{AB}} = \frac{208 \text{ V} \angle -120°}{40 \text{ } \Omega \angle 0°} = 5.20 \text{ A} \angle -120°$$

The phase impedance $\mathbf{Z}_{CA}$ is

$$\mathbf{Z}_{CA} = \sqrt{R^2 + X_L^2} \angle \arctan\left(\frac{X_L}{R}\right)$$
$$= \sqrt{(20 \text{ } \Omega)^2 + (30 \text{ } \Omega)^2} \angle \arctan\left(\frac{30 \text{ } \Omega}{20 \text{ } \Omega}\right)$$
$$= 36.06 \text{ } \Omega \angle 56.31°$$

Next, calculate the phase current $\mathbf{I}_{CA}$.

$$\mathbf{I}_{CA} = \frac{\mathbf{V}_{CA}}{\mathbf{Z}_{CA}} = \frac{208 \text{ V} \angle 0°}{36.06 \text{ } \Omega \angle 56.31°} = 5.77 \text{ A} \angle -56.31°$$

The line current $\mathbf{I}_A$ is

$$\mathbf{I}_A = \mathbf{I}_{AB} - \mathbf{I}_{CA} = \mathbf{I}_{AB} + \mathbf{I}_{AC}$$
$$= 5.20 \text{ A} \angle -120° - 5.77 \text{ A} \angle -56.31°$$
$$= 5.77 \text{ A} \angle -182.95°$$

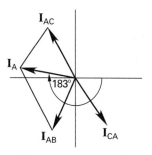

(b) The total real power dissipated is

$$P = \sum I^2 R = (5.2 \text{ A})^2 (40 \text{ } \Omega)$$
$$+ (5.77 \text{ A})^2 (20 \text{ } \Omega)$$
$$= 1747 \text{ W}$$

## 10. THREE-PHASE TRANSFORMERS

*Transformer banks* for three-phase systems have three primary and three secondary windings. Each primary-secondary set is referred to as a *transformer*, as distinguished from the *bank*. Each side can be connected in delta or wye configuration, making a total of the four (i.e., delta-delta, delta-wye, wye-delta, and wye-wye) possible transformer configurations shown in Fig. 62.6. The turns ratio, $a$, is the same for each winding. Each transformer provides one-third of the total kVA rating, regardless of connection configuration. However, the secondary voltages and currents depend on the configuration, as the figure shows.

$$a = \frac{N_p}{N_s} \qquad \textit{62.36}$$

**Figure 62.6** *Three-Phase Transformer Configurations*

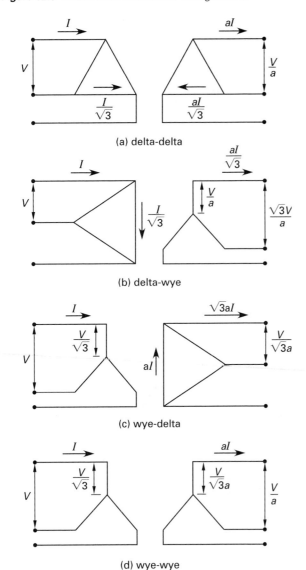

(a) delta-delta

(b) delta-wye

(c) wye-delta

(d) wye-wye

## Example 62.5

A 240 V (rms) three-phase system drawing 1200 kVA is supplied by a 2400 V (primary-side) transformer bank. Each transformer is connected in a wye-delta (primary-secondary) configuration. What are the (a) ratio of transformation, and high-side and low-side, (b) winding voltages, (c) winding currents, and (d) kVA rating?

*Solution*

This is case (c) in Fig. 62.6.

(a) The primary line voltage is 2400 V.

$$240 \text{ V} = \frac{V}{\sqrt{3}a} = \frac{2400 \text{ V}}{\sqrt{3}a}$$

The ratio of transformation is

$$a = \frac{2400 \text{ V}}{(240 \text{ V})\sqrt{3}} = 5.774$$

(b) The winding (phase) voltages are

$$V_{\text{high}} = \frac{V}{\sqrt{3}} = \frac{2400 \text{ V}}{\sqrt{3}} = 1386 \text{ V}$$

$$V_{\text{low}} = 240 \text{ V} \quad \text{[given]}$$

(c) The winding (phase) currents are

$$I_{\text{high}} = \frac{S_p}{V_p} = \frac{\sqrt{3}S_t}{3V}$$

$$= \frac{\sqrt{3}(1200 \times 10^3 \text{ VA})}{(3)(2400 \text{ V})} = 288.7 \text{ A}$$

$$I_{\text{low}} = aI_{\text{high}} = (5.774)(288.7 \text{ A})$$

$$= 1667 \text{ A}$$

(d) The transformer winding kVA rating (per phase) is

$$S_{\text{high}} = I_{\text{high}} V_{\text{high}} = (288.7 \text{ A})(1386 \text{ V})$$

$$= 4 \times 10^5 \text{ VA} \quad (400 \text{ kVA})$$

## 11. TWO-WATTMETER METHOD

Regardless of how the load impedances are connected, two *wattmeters* can be used to determine the total real power in a three-wire system when connected in the *two-wattmeter configuration* shown in Fig. 62.7. Since one or both of the power readings can be negative, the additions in this section must be performed algebraically.

$$P_t = P_1 + P_2 \qquad \textit{62.37}$$

**Figure 62.7** *Connections for the Two-Wattmeter Method*

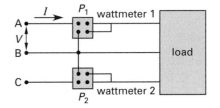

The system does not need to be balanced in order to use Eq. 62.37. However, if it is, Eq. 62.38 can be used. $\phi$ is the power factor angle of one phase of the load.

$$P_t = IV \cos(\phi - 30°) + IV \cos(\phi + 30°) \qquad \textit{62.38}$$

The two-wattmeter readings also determine the reactive and apparent power.

$$Q = \sqrt{3}IV \sin\phi = \sqrt{3}(P_1 - P_2) \qquad \textit{62.39}$$

$$S^2 = P^2 + Q^2 \qquad \textit{62.40}$$

## 12. FAULTS AND FAULT CURRENT

A *fault* is an unwanted connection (i.e., a short circuit) between a line and ground or another line. Although the *fault current* is usually very high before circuit breakers trip, it is not infinite because the transformers and transmission line have finite impedance up to the fault point. If the line impedance is known, the fault current can be found by Ohm's law.[2]

$$V = I_{\text{fault}} Z \qquad\qquad 62.41$$

---

[2]Equation 62.41 ignores the relatively small current flowing in the wire before the fault occurs and the transient (DC) current component. Although the fault current has a transient component, it dies out so quickly as to be insignificant.

# 63 Rotating Electrical Machines

## Nomenclature

| | | |
|---|---|---|
| $a$ | number of parallel armature paths | – |
| $a$ | ratio of transformation | – |
| $A$ | area | $m^2$ |
| $B$ | magnetic flux density | T |
| $B$ | susceptance | S |
| $d$ | diameter | m |
| $E$ | generated emf | V |
| $f$ | electrical frequency | Hz |
| $G$ | conductance | S |
| $I$ | current | A |
| $k$ | constant | various |
| $n$ | rotational speed | rev/min |
| $N$ | number of series armature paths | – |
| $p$ | total number of poles | – |
| $P$ | power | W |
| $q$ | number of loops | – |
| $r$ | radius | m |
| $R$ | resistance | $\Omega$ |
| $s$ | slip | – |
| $S$ | apparent power | VA |
| SR | speed regulation | – |
| $t$ | time | s |
| $T$ | period | s |
| $T$ | torque | N·m |
| $V$ | line voltage | V |
| VR | voltage regulation | – |
| $X$ | reactance | $\Omega$ |
| $Y$ | admittance | S |
| $z$ | total number of conductors | – |
| $Z$ | impedance | $\Omega$ |

## Symbols

| | | |
|---|---|---|
| $\delta$ | torque angle | rad |
| $\eta$ | efficiency | – |
| $\theta$ | phase difference angle | rad |
| $\Phi$ | magnetic flux (per pole) | Wb |
| $\omega$ | electrical frequency | rad/s |
| $\Omega$ | armature rotational speed | rad/s |

## Subscripts

| | |
|---|---|
| 0 | no load |
| $a$ | armature |
| adj | adjusted |
| $b$ | blocked rotor |
| Cu | copper |
| $E$ | emf |
| $f$ | field |
| $m$ | maximum |
| $p$ | per phase |
| $r$ | rotor |
| $s$ | synchronous |
| st | stator |
| $t$ | total |
| $T$ | torque |

## 1. ROTATING MACHINES

Rotating machines are broadly categorized as AC and DC machines. Both categories include machines that use power (i.e., motors) and those that generate power (alternators and generators). Most machines can be constructed in either single-phase or polyphase configurations, although single-phase machines may be outclassed in terms of economics and efficiency.

Circuits

Types of small AC motors include split-phase, repulsion-induction, universal, capacitor, and series motors. Large AC motors are almost always three-phase, but it is necessary to analyze only one phase of the motor. Torque and power are divided evenly among the three phases. Machines can be wye- or delta-wired or both.[1] It is common to refer to line-to-line voltage as the *terminal voltage, V*. Regardless of the wiring, each phase contributes one-third of the total torque, real power, and apparent power.

$$\mathbf{V}_p = \begin{cases} V & \text{[delta-wired]} \\ \dfrac{V}{\sqrt{3}} & \text{[wye-wired]} \end{cases} \qquad 63.1$$

$$T_p = \frac{T_t}{3} \qquad 63.2$$

$$P_p = \frac{P_t}{3} \qquad 63.3$$

$$S_p = \frac{S_t}{3} \qquad 63.4$$

## 2. TORQUE AND POWER

Torque and power are operating parameters. It takes power to turn an alternator or generator. A motor converts electrical power into mechanical power. In the SI system, power is given in kilowatts (kW). One horsepower is equivalent to 745.7 watts. The relationship between torque and power is

$$T_{\text{ft-lbf}} = \frac{5252 \times P_{\text{horsepower}}}{n_{\text{rpm}}} \qquad 63.5$$

$$T_{\text{N·m}} = \frac{1000 \times P_{\text{kW}}}{\Omega} = \frac{9549 \times P_{\text{kW}}}{n_{\text{rpm}}} \qquad 63.6$$

There are many important torque parameters for motors. The *starting torque* (also known as *static torque*, *breakaway torque*, and *locked-rotor torque*) is the turning effort exerted in starting a load from rest. *Pull-up torque* (*acceleration torque*) is the minimum torque developed during the period of acceleration from rest to full speed. *Pull-in torque* (as developed in synchronous motors) is the maximum torque that brings the motor back to synchronous speed. (Nominal pull-in torque is the torque that is developed at 95% of synchronous speed.) The *steady-state torque* must be provided to the load on a continuous basis and establishes the temperature increase that the motor must be able to withstand without deterioration. The *rated torque* is developed at rated speed and rated horsepower. The maximum torque a motor can develop at

its synchronous speed is the *pull-out torque*. *Breakdown torque* is the maximum torque the motor can develop without stalling (i.e., without coming rapidly to a complete stop).

Equation 63.7 is the general torque expression for a rotating machine with $N$ coils of cross-sectional area $A$, each carrying current $I$ through a magnetic field of strength $B$.

$$T = NBAI \cos \omega t \qquad 63.7$$

## 3. MOTOR NAMEPLATES

As specified by the National Electrical Manufacturers Association (NEMA), motor nameplates are required to contain the following information.

- rated voltage or voltages
- rated full-load amps for each voltage
- frequency
- phase
- rated full-load speed
- insulation class and rated ambient temperature
- rated horsepower
- time rating
- locked rotor code
- manufacturer's name and address

In addition to this required information, motor nameplates may also include the frame size, NEMA design letter, service factor, full-load efficiency, power factor, serial numbers, manufacturing codes, and other identifying information.

## 4. SERVICE FACTOR

The horsepower and torque ratings listed on the *nameplate* of a motor can be provided on a continuous basis without overheating. However, a motor can be operated at a slightly higher load without exceeding a safe temperature rise.[2] The ratio of the safe to standard loads is the *service factor*, usually expressed as a decimal. Service factors vary from 1.15 to 1.4, with the lower values going to larger, more efficient motors.

$$\text{service factor} = \frac{\text{safe load}}{\text{nameplate load}} \qquad 63.8$$

---

[1]Wye connections have several benefits. (1) A neutral (ground) wire is intrinsically part of the circuit. (2) Higher-order (harmonic) terms are not shorted out. (3) Starting current is lower. Nevertheless, high horsepower motors are usually run in delta because the full line voltage appears across each coil, resulting in greater power. To avoid large starting currents, the motor can be started in wye and switched over to delta.

[2]The higher temperature has a deteriorating effect on the winding insulation, however. A general rule of thumb is that a motor loses two or three hours of useful life for each hour run at the factored load.

## 5. MOTOR CLASSIFICATIONS

The National Electrical Manufacturers Association (NEMA) has categorized motors in several ways: *speed classification* (constant-, adjustable-, multi-, varying-speed, etc.), *service classification* (general, definite, and special purpose), and *motor class*. Motor class is a primary indicator of the maximum motor operating temperature, which, in turn, depends on the type of insulation used on the conductors. Motor class is actually an *insulation class*, as the type of insulation used on the windings determines the maximum operating temperature. The classes are as follows: Class A, 105°C (221°F); Class B, 130°C (266°F); Class F, 155°C (311°F); and Class H, 180°C (356°F).

## 6. POWER LOSSES

The losses for all rotating machines can be divided into four categories. *Copper losses*, $P_{Cu}$, are real power losses due to wire and winding resistance. In a DC machine, copper losses are due to resistance in the armature and field windings as well as from the brush contact resistance. In an AC machine, copper losses occur in the armature and exciter field windings. There are no brush losses in an induction machine.

$$P_{Cu} = \sum I^2 R \qquad \text{63.9}$$

*Core losses*, including hysteresis and eddy current losses, are constant losses that are independent of the load and, for that reason, are also known as *open-circuit* and *no-load losses*. In DC and synchronous AC machines, core losses occur in the armature iron. In induction machines, core losses occur in the stator iron.

*Mechanical losses* (also known as *rotational losses*) include brush and bearing friction and *windage* (air friction). (Windage is a no-load loss but is not an electrical core loss.) Mechanical losses are determined by measuring the power input at rated speed and no load.

*Stray losses* constitute the fourth category and are due to non-uniform current distribution in the conductors. Stray losses are approximately 1% for DC machines and zero for AC machines.

Real power only is used to compute the *efficiency* of a rotating machine.

$$\eta = \frac{\text{output}}{\text{input}} = \frac{\text{output}}{\text{output} + \text{losses}} = \frac{\text{input} - \text{losses}}{\text{input}} \qquad \text{63.10}$$

## 7. REGULATION

The *voltage regulation*, VR, is

$$VR = \frac{\text{no-load voltage} - \text{full-load voltage}}{\text{full-load voltage}} \times 100\% \qquad \text{63.11}$$

The *speed regulation*, SR, is

$$SR = \frac{\text{no-load speed} - \text{full-load voltage}}{\text{full-load speed}} \times 100\% \qquad \text{63.12}$$

## 8. NO-LOAD CONDITIONS

The meaning of the term *no load* is different for generators and motors. For unloaded shunt-wired alternators and generators (see Sec. 63.26), there is no electrical load connected across the output terminals, so although the field current flows, the line current, $I$, is zero. For unloaded shunt-wired motors, the work performed is zero, but line current is still drawn to keep the motor turning. All of the current is field current, however, and (neglecting mechanical losses) the armature current, $I_a$, is zero.

$$I = 0; I_f \neq 0; \ I_a = I_f \quad \text{[generator; no load]} \qquad \text{63.13}$$

$$I \neq 0; I_f = I; \ I_a = 0 \quad \text{[motor; no load]} \qquad \text{63.14}$$

## 9. PRODUCTION OF AC POTENTIAL

A potential of alternating polarity is produced by an *alternator* (*AC generator*). A permanent magnet or DC electromagnet produces a constant magnetic field. Figure 63.1 illustrates how several loops of wire can be combined into a rotating *induction coil* or *armature* to produce a continuously varying potential in a *dynamo* (*coil dynamo*).

*Figure 63.1* Elementary Two-Pole, Single-Coil Dynamo

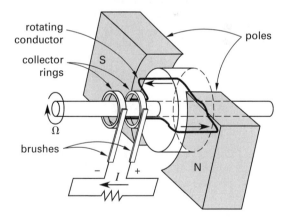

The induced voltage, $E$, is commonly called *electromotive force* (emf). In an elementary alternator, emf is the desired end result and is picked off by stationary brushes making contact with *slip rings* on the rotating shaft.[3] In a motor, emf is also produced but is referred to as *back-emf* (*counter-emf*) since it opposes the input current.

---

[3]Practical alternators, described in Sec. 63.12, use slip rings to feed the field, not to transfer the generated voltage.

Circuits

Assuming the magnetic field flux density under each pole face, $B$, is uniform, the maximum flux linked by a coil with $N$ turns and area $A$ is $NAB$. Since the coil rotates, the flux linkage is a function of the projected coil area. Instantaneous induced voltage is predicted by *Faraday's law*. Note the distinctions in Eq. 63.15 and Eq. 63.16 between the rotational speeds of the armature ($\Omega = 2\pi n/60$) and electrical waveform ($\omega = 2\pi f$), and between the effective (rms) voltage, $V$, and maximum voltage, $V_m$.

$$V(t) = V_m \sin \omega t = \omega NAB \sin \omega t$$

$$= \frac{p}{2} \Omega NAB \sin \left( \frac{p\Omega}{2} t \right)$$

$$= \frac{\pi n p N A B}{60} \sin \left( \frac{p\Omega}{2} t \right) \qquad \text{63.15}$$

$$V = \frac{V_m}{\sqrt{2}} = \frac{\omega NAB}{\sqrt{2}} = \frac{p\Omega NAB}{2\sqrt{2}}$$

$$= \frac{\pi n p N A B}{\sqrt{2}(60)} \quad \text{[effective]} \qquad \text{63.16}$$

Alternators are characterized by the number of magnetic poles, $p$. (Both north and south poles are counted to distinguish the quantity from the number of *pole pairs*.)

The *pole arc* is the distance a pole face spans along the periphery of the armature. While it could be expressed in degrees, it is usually expressed as a linear distance (e.g., inches or millimeters).

The *pole pitch* is the arc distance (along the periphery of the armature) between centers of adjacent poles. While it could be expressed in degrees, calculated as 360° divided by the number of poles, it is usually expressed as a linear distance (e.g., inches or millimeters). An alternative definition of pole pitch is the number of armature conductors or armature slots divided by the number of poles.

The *coil pitch (coil span)* is the number of armature conductors or slots spanned by the coil face. It is common to report coil pitch rounded down to an integer value. In *full-pitch coils*, coil pitch is equal to the pole pitch. In *fractional-pitched coils*, it is less than the pole pitch. Coil pitches measuring as low as eight-tenths of the pole pitch are used without significant reduction in the generated voltage. Fractional-pitched windings are intentionally used to reduce the cost of materials (copper) and to improve commutation.

The *winding pitch* is the distance as measured along the periphery of the armature between where one conductor appears and reappears subsequently (i.e., between two successive turns of a conductor).

A two-pole alternator produces one complete sinusoidal cycle per revolution. An alternator with $p$ poles produces $p/2$ cycles per revolution. Since the armature normally turns at a constant speed known as the *synchronous*

speed, $n_s$, the *electrical frequency*, $f$, of the generated potential is given by Eq. 63.18. The actual rotational speed, $n$, is known as the *mechanical frequency*.

$$n_s = \frac{120f}{p} = \frac{60\Omega}{2\pi} = \frac{60\omega}{\pi p} \quad \begin{bmatrix} \text{synchronous} \\ \text{speed} \end{bmatrix} \qquad \text{63.17}$$

$$f = \frac{1}{T} = \frac{\omega}{2\pi} = \frac{p n_s}{120} \qquad \text{63.18}$$

Since the coils in an alternator have inductance as well as resistance (see Sec. 63.17), the rated capacity of an AC machine is reported as apparent power at some rated voltage and power factor.

## Example 63.1

A four-pole alternator produces a 60 Hz potential. (a) What is the mechanical speed of the armature? (b) What is the angular velocity of the potential? (c) What is the angular velocity of the armature?

*Solution*

(a) From Eq. 63.17, the rotational speed is

$$n = n_s = \frac{120f}{p} = \frac{\left( 120 \, \frac{\text{pole·s}}{\text{min}} \right)(60 \text{ Hz})}{4 \text{ poles}}$$

$$= 1800 \text{ rpm}$$

(c) From Eq. 63.15, the armature's angular velocity is

$$\Omega = \frac{2\omega}{p} = \frac{2\pi n}{60} = \frac{\left( 2\pi \, \frac{\text{rad}}{\text{rev}} \right)\left( 1800 \, \frac{\text{rev}}{\text{min}} \right)}{60 \, \frac{\text{s}}{\text{min}}}$$

$$= 188.5 \text{ rad/s}$$

(b) The angular velocity of the 60 Hz potential is

$$\omega = 2\pi f = \left( 2\pi \, \frac{\text{rad}}{\text{cycle}} \right)(60 \text{ Hz}) = 377 \text{ rad/s}$$

## Example 63.2

The rotor of a single-phase, four-pole alternator rotates at 1200 rpm. The effective diameter and length of the 20 turn (loop) coil are 0.12 m and 0.24 m, respectively. The magnetic flux density is 1.2 T. What is the effective voltage produced?

*Solution*

The coil area is

$$A = dl = (0.12 \text{ m})(0.24 \text{ m}) = 0.0288 \text{ m}^2$$

From Eq. 63.16, the effective voltage is

$$V = \frac{p\pi nNAB}{\sqrt{2}(60)}$$

$$= \frac{(4)\left(1200 \ \frac{\text{rev}}{\text{min}}\right)\left(2\pi \ \frac{\text{rad}}{\text{rev}}\right)(20)(0.0288 \ \text{m}^2)(1.2 \ \text{T})}{2\sqrt{2}\left(60 \ \frac{\text{s}}{\text{min}}\right)}$$

$$= 122.8 \ \text{V}$$

## 10. SINGLE-PHASE AC ALTERNATORS

A single-phase AC alternator has only one set of windings. (Three-phase alternators have three sets of independent windings). However, the alternator can have more than one set of poles. The variable $p$, used in alternator and generator formulas, refers to the total number of poles (always an even number), not the number of pole pairs.

The distinction between coils, turns, and inductors is not always clearly made. A *coil (winding)* consists of one or more *turns (loops)* of wire. Each turn contains two *series conductors (inductors, bars, etc.)*. Each conductor generates an emf. The variable $N$, used in alternator and generator formulas, usually refers to the total number of turns, not the number of conductors.

The sinusoidal voltage developed in a single-phase alternator is

$$V(t) = V_m \sin \omega t \qquad \text{63.19}$$

Care must be taken to distinguish between the armature speed, $n$ (in rpm), the angular armature speed, $\Omega$ (in rad/s), and the linear and angular voltage frequencies, $f$ and $\omega$ (in Hz and rad/s, respectively). Table 63.1 summarizes the most frequently used formulas for a single-phase AC alternator.

## 11. ARMATURE WINDINGS

An armature winding consists of several *coils* of continuous wire formed into $q$ loops. Each loop in an armature contributes two parallel *conductors*, known as *bars*. The number of conductors, $z$, is

$$z = 2q \qquad \text{63.20}$$

Voltage is induced only in conductors and inductors that are parallel to the armature shaft (i.e., that cut magnetic flux lines). The number of *series paths*, $N$, between positive and negative brush sets depends on how the coils are wound and connected to the commutator.

$$N = \frac{z}{a} = \frac{2q}{a} \qquad \text{63.21}$$

The number of parallel armature paths between each pair of brushes, $a$, equals the number of poles, $p$, in a *lap-wound armature*.[4] (The number of poles is also equal to the number of brushes.) The coil connections are made to adjacent commutator segments. This type of winding is commonly used in DC machines, induction motors, and, to a lesser extent, in AC generators.

In a *wave-wound armature* (i.e., *two-circuit*, *zig-zag*, and *series-drum armature*), $a = 2$ and the coil connections are on commutator segments on opposite sides of the armature regardless of the number of poles. Only two brushes are needed, although more can be used. This configuration is commonly used in DC armatures requiring the generation or use of higher voltages than lap windings can tolerate.

In general, a high-current armature is lap-wound to provide a large number of parallel paths, and a low-current armature is wave-wound to provide a small number of parallel paths. For any number of poles, $p$, and armature conductors, $z$, a wave-wound armature will produce a higher terminal voltage than a lap-wound armature because there are more conductors in series. On the other hand, a lap winding has a greater current capacity than a wave winding because it has more parallel paths. In small machines, the capacity of the armature conductors is not critical, and in order to produce reasonable voltages, wave windings are used. In large machines, the large number of armature conductors easily produces suitable voltages, and the current carrying capacity is more critical. Accordingly, lap windings are used in large machines.

Lap and wave windings can be simplex, duplex, triplex, etc. The adjectives simplex and multiplex (duplex, triplex, etc.) refer to the number of plural paths between brushes (winding terminals). A simplex lap-wound armature will have as many parallel paths as the number of poles. A simplex wave-wound armature will have two parallel paths irrespective of the number of poles. An 8-pole machine with simplex windings will either have eight parallel circuits (lap) or two (wave). If the armature must carry a larger current for the same number of poles, a greater number of parallel paths will be needed. Multiplex windings are used for that purpose. Multiplexing a winding has a similar effect to increasing the wire diameter.[5]

For a lap winding, the number of parallel paths will be the number of poles times the plurality (plex or multiplexity). A duplex lap winding will have $2p$ parallel paths; a triplex will have $3p$, and so on. For a wave winding, the number of parallel paths will be two times

---

[4]Auxiliary windings are used in *shaded pole motors; split-phase motors* have two separate windings; *capacitor motors* have capacitors in series with an auxiliary winding.

[5]From a practical standpoint, there is only so much room on the armature (in the armature slots). So, if the plex is increased, the number of parallel paths in series, and consequently the generated voltage, must decrease. A duplex lap winding will have twice the current capacity but will generate half the voltage, simply because the total number of conductors is fixed for a given wire diameter and armature size.

**Table 63.1** *Single-Phase AC Alternator Formulas*

|  | in terms of $n$ | in terms of $\Omega$ | in terms of $f$ | in terms of $\omega$ |
|---|---|---|---|---|
| $n$ (rpm) | — | $\dfrac{30\Omega}{\pi}$ | $\dfrac{120f}{p}$ | $\dfrac{60\omega}{\pi p}$ |
| $\Omega$ (mechanical rad/s) | $\dfrac{\pi n}{30}$ | — | $\dfrac{4\pi f}{p}$ | $\dfrac{2\omega}{p}$ |
| $f$ (Hz) | $\dfrac{pn}{120}$ | $\dfrac{p\Omega}{4\pi}$ | — | $\dfrac{\omega}{2\pi}$ |
| $\omega$ (electrical rad/s) | $\dfrac{\pi pn}{60}$ | $\dfrac{p\Omega}{2}$ | $2\pi f$ | — |
| single-phase $V_m$ (volts) | $\dfrac{\pi pnNAB}{60}$ | $\dfrac{\Omega pNAB}{2}$ | $2\pi fNAB$ | $\omega NAB$ |
| single-phase $V_{\text{eff}}$ (volts) | $\dfrac{\pi pnNAB}{60\sqrt{2}}$ | $\dfrac{\Omega pNAB}{2\sqrt{2}}$ | $\dfrac{2\pi fNAB}{\sqrt{2}}$ | $\dfrac{\omega NAB}{\sqrt{2}}$ |
| single-phase $V_{\text{ave}}$ (volts) | $\dfrac{pnNAB}{30}$ | $\dfrac{\Omega pNAB}{\pi}$ | $4fNAB$ | $\dfrac{2\omega NAB}{\pi}$ |

the multiplexity. The number of poles is not a factor. (See Table 63.2.)

In the armature, armature coils contain all of the loops between their respective poles. Each coil will consist of all of the parallel paths, "coming" and "going," between its two poles. The number of coils is $N/2$, equal to the number of pole pairs.

**Table 63.2** *Multiplex Coil Parameters*

| type | lap winding | wave winding |
|---|---|---|
| simplex | $a = p$ | $a = 2$ |
| duplex | $a = 2p$ | $a = 4$ |
| triplex | $a = 3p$ | $a = 6$ |
| quadraplex | $a = 4p$ | $a = 8$ |

## 12. PRACTICAL ALTERNATORS

There are several reasons why large alternators are not designed with stationary magnetic fields and rotating coils, as shown in Fig. 63.1. (1) The rotating coils must be well insulated to prevent shorting with the high voltages that are induced. (2) Structural bracing is required to counteract the large centrifugal force resulting from rotating many coils of wire. (3) It is difficult to make efficient high-voltage, high-power connections through slip rings.

For these reasons, practical alternators reverse the locations of the field and induction coils so that the low-voltage field revolves and the high-voltage is induced in stationary coils. (See Fig. 63.2.) The magnetic field is produced by *field windings* whose DC *magnetization current* is supplied through brushes and slip rings. Such an armature containing field coils is known as a *rotor*. The stationary induction coils are placed in slots a small distance from the rotor and are known as the *stator*. The closer the stator and rotor are, the smaller the magnetization current required.

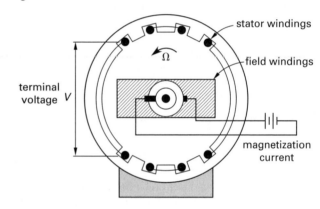

**Figure 63.2** *A Practical Alternator*

With *cylindrical rotors* (*wound rotors*), the field windings are embedded in an externally smooth-surfaced cylindrical rotor. This makes them suitable for high-speed (i.e., greater than 1800 rpm) operation since windage losses are reduced. The poles of *salient-pole rotors*, on the other hand, resemble exposed wound electromagnets. This is adequate for 1800 rpm and below.

## 13. SYNCHRONOUS MOTORS

*Synchronous motors* are essentially dynamo alternators operating in reverse. Alternating current is supplied to the stationary stator windings. DC current is applied to the field windings in the rotor through brushes and slip rings as in the alternator.[6] The field current interacts with the stator field, causing the armature to turn. Since the stator field frequency is fixed, the motor runs only at a single *synchronous speed* (see Eq. 63.17).

---

[6]The magnetization energy can also be generated through induction. In that case, the rotor includes diodes to rectify the induced AC potential.

Important features of synchronous motors follow.

- They turn at constant speeds, regardless of the load.[7] Stalling occurs when a motor's counter-torque is exceeded.

- The power factor can be adjusted manually without losing synchronization by varying the field current. A unity power factor occurs with *normal excitation* current. The power factor is leading (lagging) when the current is more (less) than normal, and this is known as being *over-* (*under-*) *excited*.

- They can draw leading currents and be used for power factor correction, in which case they are known as *synchronous capacitors* (*synchronous condensers*).

Since the starting torque is zero, it is necessary to bring a synchronous motor up to speed by some other means. The most common method is to include auxiliary windings in the pole faces so that the motor can be started as an induction motor. At synchronous speed, the auxiliary windings draw no current. If the motor speed becomes nonsynchronous, the auxiliary windings draw power to resynchronize the rotor.

## 14. OPERATING CHARACTERISTICS OF SYNCHRONOUS MOTORS

Since synchronous motors run at only one speed, curves of horsepower (or other characteristics) versus speed are not used. Instead, motors are categorized on the basis of speed, torque, and horsepower at a specific power factor.

## 15. SYNCHRONOUS MACHINE EQUIVALENT CIRCUIT

Figure 63.3 illustrates a simple equivalent circuit for a synchronous machine. The vector voltage relationship is defined by Eq. 63.22 and Eq. 63.23, which incorporate the equivalent *synchronous inductance*, $X_s$, of each phase. $V_p$ is the phase voltage. The series armature resistance, $R_a$, is small and normally disregarded.

$$\mathbf{E} = \mathbf{V}_p + (R_a + jX_s)\mathbf{I}_a$$
$$\approx \mathbf{V}_p + jX_s\mathbf{I}_a \quad \text{[alternator]} \qquad 63.22$$
$$\mathbf{V}_p = \mathbf{E} + (R_a + jX_s)\mathbf{I}_a$$
$$\approx \mathbf{E} + jX_s\mathbf{I}_a \quad \text{[motor]} \qquad 63.23$$

Equation 63.24 gives the real power generated per phase. For a motor, the power factor is determined by $E$ and $I_f$. For an alternator, the power factor determines $E$ and $I_f$. (Equation 63.24 mixes variables with different units.)

$$P_p = T_p\Omega = S\cos\theta = VI\cos\theta = \frac{VE}{X_s}\sin\delta \qquad 63.24$$

Circuits

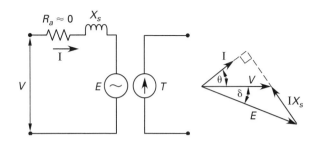

**Figure 63.3** *Synchronous Motor Equivalent Circuit*

Equation 63.25 gives the torque produced per phase. $\Phi_r$ and $\Phi_{st}$ are the internal rotor and stator fluxes, respectively. The *torque angle*, $\delta$ (also known as the *power angle* and *displacement angle*), is the phase difference angle between the applied voltage, $V$, and the generated emf, $E$. It is positive for an alternator and negative for a motor. Torque is maximum when $\delta = 90°$, a condition equivalent to a unity power factor and known as *pull-out torque*. The *pull-out power* is found by setting $\delta = 90°$ in Eq. 63.24.

$$T_p = k_T\Phi_r\Phi_{st}\sin\delta \qquad 63.25$$

The apparent power per phase is

$$S = VI = \frac{P_p}{\cos\theta} \qquad 63.26$$

### Example 63.3

A six-pole motor is connected to a three-phase 240 V (rms) 60 Hz line. Its stator windings are connected in wye configuration. The motor has a synchronous reactance of 3 ohms per phase and is rated as 10 kVA at its synchronous speed and 100% power factor. What are the (a) synchronous speed, (b) phase voltage, (c) line current, (d) voltage drop across the synchronous reactance, (e) generated back emf, and (f) torque angle?

*Solution*

(a) Equation 63.17 gives the synchronous speed.

$$n_s = \frac{120f}{p} = \frac{\left(120\,\dfrac{\text{pole·s}}{\text{min}}\right)(60\text{ Hz})}{6\text{ poles}} = 1200\text{ rpm}$$

(b) Since the windings are in a wye configuration, the phase voltage is

$$V_p = \frac{V}{\sqrt{3}} = \frac{240\text{ V}}{\sqrt{3}} = 138.6\text{ V}$$

(c) The line current is the same as the phase current, which can be calculated from the apparent power. Each phase draws one-third of the apparent power. The real power is

$$P_p = \frac{S}{3}\cos\theta = \frac{(10{,}000 \text{ VA})(1)}{3} = 3333 \text{ W}$$

$$I_p = \frac{P_p}{V_p} = \frac{3333 \text{ W}}{138.6 \text{ V}} = 24.05 \text{ A}$$

(d) The voltage drop across each winding is

$$V_p = I_p X_p = (24.05 \text{ A})(3 \text{ }\Omega) = 72.15 \text{ V}$$

(e) The back emf vector is

$$\mathbf{E} = \mathbf{V}_p - jIX = 138.6 \text{ V} - j72.15 \text{ V}$$
$$= 156.3 \text{ V}\angle -27.50°$$

The back emf is 156.3 V.

(f) The torque angle was found in part (e) to be $\delta = -27.50°$.

## 16. INDUCTION MOTORS

Induction motors are essentially constant-speed devices that receive power through induction—there are no brushes or slip rings. A motor can be considered as a rotating transformer secondary (the *rotor*) with a stationary primary (the *stator*). The stator field rotates at the synchronous speed given in Eq. 63.17. An emf is induced as the stator field moves past the rotor conductors. Since the rotor windings have reactance, the rotor field lags the induced emf.

In order to have a change in flux linkage, the rotor must turn slower than the synchronous speed. The difference in speed is small but essential. *Percent slip*, $s$, typically 2% to 5%, is the percentage difference in speed between the rotor and stator field.[8] Percent slip and *percent synchronism* are complements (i.e., add to 100%). *Slip* in rpm is the difference between actual and synchronous speeds.

$$s = \frac{n_s - n}{n_s} = \frac{\Omega_s - \Omega}{\Omega_s} \qquad 63.27$$

The stator is identical to that of an alternator or synchronous motor. In a *wound rotor*, the rotor is similar to an armature winding in a dynamo.[9] However, there are no wire windings at all in a *squirrel-cage rotor*.[10] The rotor consists of copper or aluminum bars embedded in slots in the cylindrical iron core of the rotor. The ends of the bars are shorted by conductive rings, as illustrated in Fig. 63.4, to form loops. The rotor and its

components, connections, and lines constitute the *rotor circuit* or *secondary circuit*.

**Figure 63.4** *A Squirrel-Cage Rotor Induction Motor*

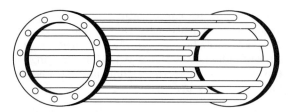

### Example 63.4

An induction motor developing 10 hp is connected to a three-phase 240 V (rms) 60 Hz power line. The stator windings are connected in wye configuration. The synchronous speed is 1800 rpm, but the motor turns at 1738 rpm when loaded. Its energy efficiency is 80%, and the power factor (pf) is 70%. Calculate the (a) slip, (b) number of poles, (c) line current drawn, and (d) phase voltage.

*Solution*

(a) From Eq. 63.27, the slip is

$$s = \frac{n_s - n}{n_s} = \frac{1800 \text{ rpm} - 1738 \text{ rpm}}{1800 \text{ rpm}} = 0.03444$$

(b) The number of poles is calculated from the synchronous speed.

$$p = \frac{120f}{n_s} = \frac{\left(120 \dfrac{\text{pole·s}}{\text{min}}\right)(60 \text{ Hz})}{1800 \dfrac{\text{rev}}{\text{min}}} = 4$$

(c) The total real power input is

$$p = \frac{(10 \text{ hp})\left(745.7 \dfrac{\text{W}}{\text{hp}}\right)}{0.8} = 9321 \text{ W}$$

The power per phase is

$$P_p = \frac{P}{3} = \frac{9321 \text{ W}}{3} = 3107 \text{ W}$$

Since the phase and line currents are identical in wye-connected loads, the line current is

$$I = I_p = \frac{P_p}{V_p(\text{pf})} = \frac{3107 \text{ W}}{\left(\dfrac{240 \text{ V}}{\sqrt{3}}\right)(0.70)} = 32.03 \text{ A}$$

(d) The voltage across each phase is

$$V_p = \frac{V}{\sqrt{3}} = \frac{240 \text{ V}}{\sqrt{3}} = 138.6 \text{ V}$$

---

[8]Slip can be expressed as either a decimal or a fraction (e.g., 0.05 slip or 5% slip).

[9]Windings can be incorporated in the rotor to obtain better speed control or high torque.

[10]If the slip rings of a wound rotor are shorted, the motor behaves as a squirrel-cage motor.

## 17. INDUCTION MOTOR EQUIVALENT CIRCUIT

Figure 63.5(a) illustrates the equivalent circuit for an induction motor.[11] It is very similar to the equivalent circuit for a transformer. $R_1$ and $X_1$ represent the equivalent stator resistance and reactance, respectively. $G_0$ and $B_0$ represent the stator core loss and susceptance, respectively, as determined from no-load testing. The dashed line represents the air gap across which energy is transferred to the rotor. $R_2$ and $X_2$ are the equivalent rotor resistance and reactance, respectively. There is no element to model the rotor core loss, which is negligible. Rotational losses are included in the stator core loss element, $G$. Any equivalent load resistance is included in the rotor resistance. The ratio of transformation, $a$, is taken as 1.0 for a squirrel-cage motor.[12]

Using an adjusted voltage, $V_{\text{adj}}$, simplifies the model, as shown in Fig. 63.5(b). The relationship between the applied terminal voltage, $V_1$, and the adjusted voltage is

$$\mathbf{V}_{\text{adj}} = \mathbf{V}_1 - \mathbf{I}_0(R_1 + jX_1) \qquad 63.28$$

$$V_{\text{adj}} \approx V_1 - I_0\sqrt{R_1^2 + X_1^2} \qquad 63.29$$

**Figure 63.5** Equivalent Circuits of an Induction Motor

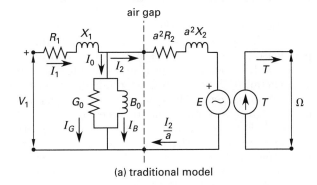

(a) traditional model

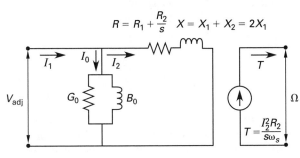

(b) simplified model ($a = 1$)

---

[11]This equivalent circuit cannot be used for a double squirrel-cage motor.

[12]With a phase-wound rotor, the brushes can be lifted from the slip rings. The ratio of transformation, $a$, can be determined as the ratio of applied voltage to voltage across the slip rings. This cannot be done with a squirrel-cage motor, so the equivalent circuit parameters are redefined with the ratio of transformation.

The total series resistance per phase, $R$, is

$$R = R_1 + \frac{R_2}{s} \qquad 63.30$$

Equation 63.31 gives the torque-speed relationship predicted by this model.

$$T_p = \frac{I_2^2 R_2}{s\omega_s} = \frac{V_{\text{adj}}^2 R_2}{s\omega_s\left(\left(R_1 + \dfrac{R_2}{s}\right)^2 + X^2\right)} \qquad 63.31$$

$$T_t = 3T_p \qquad 63.32$$

## 18. OPERATING CHARACTERISTICS OF INDUCTION MOTORS

Under normal operating conditions, slip is small (less than 0.05) and Eq. 63.33 predicts that the torque will be directly proportional to slip, $s$, and inversely proportional to the rotor resistance, $R_2$. At low speeds, the reactive term is larger than the resistive term in Eq. 63.34.

$$T_p \approx \frac{V_{\text{adj}}^2 s}{\omega_s R_2} \quad \text{[high speed, per phase]} \qquad 63.33$$

$$T_p \approx \frac{V_{\text{adj}}^2 R_2}{s\omega_s X^2} \quad \text{[low speed, per phase]} \qquad 63.34$$

The starting torque per phase is proportional to rotor winding resistance and terminal voltage and can be found by setting $s = 1$ in Eq. 63.34.

$$T_{\text{starting}} \approx \frac{V_{\text{adj}}^2 R_2}{\omega_s X^2} \quad \text{[starting, per phase]} \qquad 63.35$$

The maximum torque (known as *breakdown torque*) is independent of the rotor circuit resistance, but the rotor resistance does affect the speed at which the maximum torque occurs. A large rotor circuit resistance only causes the maximum torque to occur at a larger slip. Maximum torque varies directly with the square of the stator voltage.

$$T_{\text{max}} = \frac{V_{\text{adj}}^2}{2\omega_s\left(R_1 + \sqrt{R_1^2 + X^2}\right)} \quad \text{[per phase]} \qquad 63.36$$

$$s_{\text{max }T} = \frac{R_2}{\sqrt{R_1^2 + X^2}} \qquad 63.37$$

Figure 63.6 illustrates typical characteristic curves of an induction motor. Such curves are conventionally provided for the motor running at its optimum efficiency and power factor. Curves can be provided for polyphase or single-phase operation.

**Figure 63.6** *Characteristic Curves for an Induction Motor*

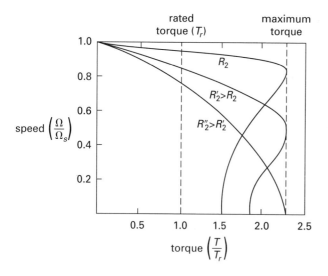

## 19. TESTING INDUCTION MOTORS

Performance of induction motors is evaluated in a manner analogous to transformer testing. The *no-load motor test (running-light test)* corresponds to an open-circuit transformer test. This test determines the values of $B$ and $G$ in the equivalent circuit. The *locked-rotor test* corresponds to a closed circuit transformer test and determines the values of $R$ and $X$.

*No-load test:* The motor is run at the rated voltage without load. The line voltage, $V$, line current, $I_0$, and power per phase, $P_0$, are measured.

$$\text{no-load power factor} = \cos\theta_0 = \frac{P_0}{V_{\text{adj}}I_0} \qquad 63.38$$

Referring to Fig. 63.5, the *magnetization (quadrature) current*, $I_B$, and the in-phase component of the *exciting current*, $I_G$, are

$$I_B = I_0 \sin\theta_0 \qquad 63.39$$

$$I_G = I_0 \cos\theta_0 \qquad 63.40$$

The stator parameters are

$$G_0 = \frac{P_0}{V_{\text{adj}}^2} \qquad 63.41$$

$$B_0 = -\sqrt{Y_0^2 - G_0^2} = -\sqrt{\left(\frac{I_0}{V_0}\right)^2 - G_0^2}$$
$$\approx \frac{I_B}{V_{\text{adj}}} \qquad 63.42$$

*Locked-rotor test:* The rotor is blocked. A (low) voltage is applied (and measured) such that the rated current flows. The phase current, $I_b$, power per phase, $P_b$, and phase voltage, $V_b$, are measured.

The phase current lags the phase voltage by the angle

$$\cos\theta_b = \frac{P_b}{I_b V_b} \qquad 63.43$$

The stator reactance, $X_1$, is usually assumed to be one-half of $X$, the remainder being the rotor reactance. However, since performance depends greatly on $R_2$, the division of $R$ must be less arbitrary. One approach is to measure the DC resistance between the terminals and use this as an approximate value of $2R_1$.[13]

(This assumes wye-connected motor windings.) The remainder of $R$ is given to $R_2$.

$$X = X_1 + X_2 = 2X_1 = \frac{V_b}{I_b \sin\theta_b}$$
$$= \sqrt{Z^2 - R^2} \qquad 63.44$$

$$R = R_1 + R_2 = \frac{P_b}{I_b^2} \qquad 63.45$$

$$Z = \frac{V_b}{I_b} \qquad 63.46$$

## 20. STARTING INDUCTION MOTORS

Induction motors draw their maximum currents when starting (i.e., when slip, $s$, is 1). The starting torque for a polyphase motor varies directly with the square of the stator voltage but also depends on the rotor resistance and reactance. (Starting torque for a single-phase induction motor is zero. Therefore, single-phase induction motors must be brought up to speed by some other means.)[14] For a given stator voltage, there is a particular rotor resistance that maximizes starting torque. Increasing or decreasing resistance from this optimum value decreases the starting torque. Starting current can be calculated from Eq. 63.47 by setting $s = 1$.

$$I_{\text{starting}} = I_1 = I_0 + I_2$$
$$= I_0 + \frac{V_{\text{adj}}}{R_1 + \dfrac{R_2}{s} + jX} \qquad 63.47$$

The *blocked- (locked-) rotor current* is the current drawn by the motor with the rotor held stationary. It is a worst-case starting current since the rotor begins to move immediately upon starting, reducing the current drawn. *Free-rotor starting current* is approximately 75% of the blocked-rotor current.

---

[13]This value is corrected empirically, but the correction is beyond the scope of this book.

[14]Polyphase induction motors will run but will not start on a single phase. There is a danger of current overload if synchronization is lost during single-phase operation. Protective elements (e.g., circuit breakers) are used to limit current.

## 21. SPEED CONTROL FOR INDUCTION MOTORS

The rotational speed of an induction motor depends on the number of poles, line voltage, supply frequency, and rotor circuit resistance. For a given machine, the number of poles cannot be varied without excessive complexity in winding switching and increased manufacturing cost. Since the breakdown torque is proportional to the square of the voltage, reducing the voltage may stall the motor, so voltage speed control is rarely used. Changing the supply frequency is also impractical in most instances. Introducing a resistance in series with the rotor decreases the motor speed but is applicable only to wound-rotor motors.

Speed control is commonly accomplished by introducing a foreign voltage in the secondary (rotor) circuit. If the foreign voltage opposes the voltage induced in the secondary circuit, the motor speed will be reduced, and vice versa.

If two induction motors are available, one can be used to control the other by connecting them in *cascade*. The shafts are rigidly connected, and the rotor and stator windings are interconnected.[15]

## 22. POWER TRANSFER IN INDUCTION MOTORS

Figure 63.7 illustrates the power transfer in an induction motor. Equation 63.48 through Eq. 63.53 are per phase.

$$\text{input power} = V_1 I_1 \cos\theta \qquad 63.48$$

$$\text{stator copper losses} = I_1^2 R_1 \qquad 63.49$$

$$\text{rotor input power} = \frac{I_2^2 R_2}{s} \qquad 63.50$$

$$\text{rotor copper losses} = I_2^2 R_2 \qquad 63.51$$

$$\text{electrical power delivered} = I_2^2 R_2 \left(\frac{1-s}{s}\right) \qquad 63.52$$

$$\text{shaft output power} = T\Omega \qquad 63.53$$

**Figure 63.7** *Induction Motor Power Transfer*

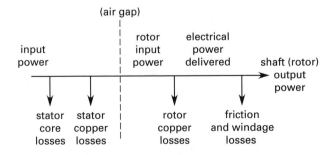

Core losses are constant. (See Sec. 63.6.) Copper losses are proportional to the square of the delivered current.

### Example 63.5

A 15 hp induction motor with six poles operates at 80% efficiency on a three-phase 240 V (rms) 60 Hz line. The following losses are observed for full-load operation.

| | |
|---|---|
| stator copper loss | 540 W |
| friction/windage loss | 975 W |
| core loss | 675 W |

What are the (a) speed and (b) torque when the motor delivers half power?

*Solution*

(a) The full-load output power is

$$P = (15 \text{ hp})\left(745.7 \; \frac{\text{W}}{\text{hp}}\right) = 11{,}186 \text{ W}$$

The input power is

$$P_{\text{in}} = \frac{P_{\text{out}}}{\eta} = \frac{11{,}186 \text{ W}}{0.80} = 13{,}982 \text{ W}$$

The full-load rotor copper loss is

$$\begin{aligned} \text{rotor copper} \atop \text{loss} &= \left( \begin{array}{c} 13{,}982 \text{ W} - 11{,}186 \text{ W} - 540 \text{ W} \\ - \; 975 \text{ W} - 675 \text{ W} \end{array} \right) \\ &= 606 \text{ W} \quad \text{[full power]} \end{aligned}$$

The output power at half-load is

$$P = \tfrac{1}{2}(15 \text{ hp})\left(745.7 \; \frac{\text{W}}{\text{hp}}\right) = 5593 \text{ W}$$

At half-load, the friction and core losses are unchanged. From Eq. 63.51 and Eq. 63.52, the ratio of actual to full-load copper losses is equal to the square of the ratio of actual to full-load output power.

$$\text{rotor copper loss} = \left(\tfrac{1}{2}\right)^2 (606 \text{ W}) = 152 \text{ W}$$

$$\text{stator copper loss} = \left(\tfrac{1}{2}\right)^2 (540 \text{ W}) = 135 \text{ W}$$

The rotor input power is

$$P_{\text{in}} = 5593 \text{ W} + 975 \text{ W} + 152 \text{ W} = 6720 \text{ W}$$

The synchronous speed is

$$n_s = \frac{120f}{p} = \frac{\left(120 \; \dfrac{\text{pole·s}}{\text{min}}\right)(60 \text{ Hz})}{6 \text{ poles}} = 1200 \text{ rpm}$$

The slip at half-power is found from Eq. 63.50.

$$s = \frac{\text{rotor copper loss}}{\text{rotor input power}} = \frac{152 \text{ W}}{6720 \text{ W}} = 0.0226$$

---

[15]The two armatures can also be constructed on a single shaft.

The speed at half-load is

$$n = n_s(1-s) = \left(1200 \ \frac{\text{rev}}{\text{min}}\right)(1-0.0226)$$
$$= 1173 \text{ rpm}$$

(b) The torque is found from Eq. 63.5.

$$T = \frac{5252P}{n} = \frac{(5252)\left(\frac{1}{2}\right)(15 \text{ hp})}{1173 \ \frac{\text{rev}}{\text{min}}} = 33.6 \text{ ft-lbf}$$

## 23. PRODUCTION OF DC POTENTIAL

A *generator* is a device that produces DC potential. The actual voltage induced is sinusoidal (i.e., AC). However, brushes on *split-ring commutators* make the connection to the rotating armature and rectify the AC potential.[16] The more coils there are, the smoother the DC voltage. Two commutator segments are needed for each coil, and each coil produces its own sine wave. Since mechanical commutation is difficult unless the emf is produced in a rotating armature, DC generators are based on the simple design shown in Fig. 63.8 and suffer from the same limitations mentioned in Sec. 63.12.

**Figure 63.8** Commutator Action

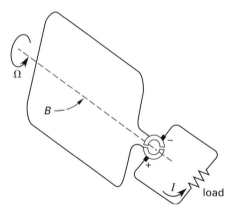

The average emf, $E$, for a DC machine (motor or generator) is given by Eq. 63.54. ($\Phi$ is the flux per pole. Other variables are discussed in Sec. 63.11.) For a generator, the emf is greater than the armature voltage ($E > V_a$). For a motor, the armature voltage is greater than the back emf ($V_a > E$).

$$E = \frac{Np\Phi n}{60} = \frac{zp\Phi n}{60a} = k_E \Phi n \qquad 63.54$$

$$k_E = \frac{zp}{60a} \qquad 63.55$$

[16] *Sparking* is one of the problems associated with commutators and occurs at points of low resistance. Brush resistance is nonlinear and drops as current increases.

## 24. DC GENERATORS

The simple DC generator armature consists of a single coil with several turns (loops) of wire. The two ends of the coil terminate at a *commutator*. The commutator consists of a single ring split into two halves known as *segments*. (This arrangement is shown in Fig. 63.8.) The brushes slide on the commutator and make contact with the adjacent segment every half-rotation of the coil. This produces a rectified (though not constant) potential, as shown in Fig. 63.9.

**Figure 63.9** Rectified DC Voltage Induced in a Single Coil

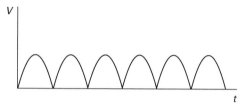

Modern DC generators contain multiple coils connected in series, an arrangement known as *closed coil winding*. (See Fig. 63.10.) The coils are spaced uniformly around the armature core. The single-ring commutator is divided into as many pairs of segments as there are coils. There are only two brushes, however, located on opposite sides of the commutator. Since there are many coils, as the armature rotates, the brushes always make contact with two segments of the commutator that are in nearly the same positions relative to the magnetic field. The average induced emf, $\overline{E}$, for a finite number of coils approaches the voltage induced with an infinite number of coils. $N$ is the total number of series turns in all of the coils.

$$\overline{E} \approx E_\infty = 2\left(\frac{n}{60}\right)NAB$$

**Figure 63.10** Two-Coil, Four Segment Closed-Coil Armature

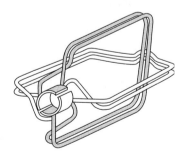

Since the coils are connected in series (in the modern closed-coil winding arrangement), the emf induced is the sum of the emfs induced in the individual coils. The voltage induced in each coil of a DC generator with multiple coils is still sinusoidal, but the terminal output is nearly constant, not a (rectified) sinusoid. The slight variations in the voltage are known as *ripple*. Since the output is nearly constant, the concept of electrical frequency has no meaning, and distinctions between maximum, effective, and average voltages are not made. (See Fig. 63.11.)

**Figure 63.11** *Rectified DC Voltage from a Two-Coil, Four Segment Generator*

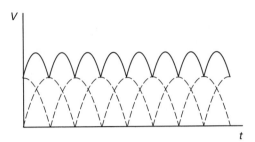

It is also possible to use an *open-coil connection* (see Fig. 63.12), though this is seldom done. Lower voltages are produced, as the average induced voltage is the maximum voltage from a single coil.

**Figure 63.12** *Two-Coil, Four-Segment Open-Coil Armature*

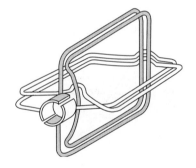

## 25. SERIES-WIRED DC MACHINES

The equivalent circuit of a *series-wired DC machine* is shown in Fig. 63.13. The only components in the circuit are the field and armature resistances in series (from which the name is derived). The *brush resistance* is also considered to be in series but is often included in the armature resistance specification. The governing equations are given by Eq. 63.56 through Eq. 63.59. $I_a$ is positive for a motor and negative for a generator. The magnetic flux varies with the armature current.

$$E = k'_E n\phi = k_E n I_a = V - I_a(R_a + R_f) \qquad 63.56$$

$$V = E + I_a(R_a + R_f) \qquad 63.57$$

The speed, torque, and current are related according to

$$\frac{T_1}{T_2} = \left(\frac{I_{a,1}}{I_{a,2}}\right)^2 \approx \left(\frac{n_2}{n_1}\right)^2 \qquad 63.58$$

The torque is

$$T = k'_T \Phi I_a = k_T I_a^2$$

$$= k_T \left(\frac{V}{k_E n + R_a + R_f}\right)^2 \qquad 63.59$$

**Figure 63.13** *Series-Wired DC Motor Equivalent Circuit*

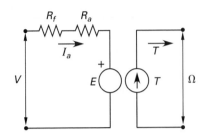

For a motor, a reduction in load (torque) causes a corresponding reduction in armature current. However, since the field and armature currents are identical, the flux is also reduced and the speed increases to maintain Eq. 63.57. Therefore, a series motor is not a constant-speed device. A load should never be completely removed from a running DC motor (as the motor will "run away," potentially damaging itself), and gears (not belts, which can slip) are the preferred method of connecting DC motors to their loads.

The back-emf, $E$, is zero when the motor starts from rest. Therefore, the armature current, $I_a$, must be excessively high in order to keep Eq. 63.57 valid. Thus, reduced voltages are required when starting, and the field resistance, $R_f$, is often a rheostat or switchable resistor bank.

At high speeds, the back-emf, $E$, counteracts the applied voltage, increasing as the rotational speed increases. The *stall speed* is the speed at which Eq. 63.60 becomes valid.

$$E = I_a(R_a + R_f) = \frac{V}{2} \quad \text{[stall]} \qquad 63.60$$

## 26. SHUNT-WIRED DC MACHINES

*Shunt-wired DC machines* have constant (but adjustable) magnetic fields since the field currents are constant. For shunt-wired motors, this results in a relatively constant speed. The magnetic coil is fed from the same line as the armature (as it is in Fig. 63.14) in a *self-excited machine*; in a *separately excited machine*, the field coil is fed from another source. In Eq. 63.61 through Eq. 63.67, $I_a$ is positive for a motor and negative for a generator.

$$E = k_E n \Phi \qquad 63.61$$

$$V = E + I_a R_a = I_f R_f \qquad 63.62$$

$$I = I_a + I_f \quad \text{[motor]} \qquad 63.63$$

$$I = I_a - I_f \quad \text{[generator]} \qquad 63.64$$

*Figure 63.14* Shunt-Wired DC Motor Equivalent Circuit

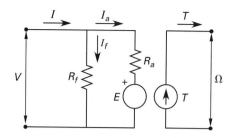

The speed current relationship is

$$n = n_0 - k_n T = \frac{V - I_a R_a}{k_E \Phi} \qquad 63.65$$

The torque is

$$T = k_T \Phi I_a \qquad 63.66$$

The torque and current are proportional.

$$\frac{T_1}{T_2} = \frac{I_{a,1}}{I_{a,2}} \qquad 63.67$$

## 27. COMPOUND DC MACHINES

*Compound DC machines* have both series and shunt windings. Their performance is intermediate between those of shunt and series machines.

### Example 63.6

A two-pole DC generator with a simplex lap-wound armature is turned at 1800 rpm. There are 100 conductors between the brushes. The average magnetic flux density in the air gap between the pole faces and armature is 1.2 T. The pole faces have areas of 0.03 m². What is the no-load terminal voltage?

*Solution*

The flux per pole is

$$\Phi = BA = (1.2\ \text{T})(0.03\ \text{m}^2) = 0.036\ \text{Wb}$$

From Sec. 63.8, the term "no load" for a generator means the line current is zero. From Sec. 63.11, $N = z/a = z/p$ for a simplex lap-wound armature.

$$E = \frac{zp\Phi n}{60a} = \frac{Np\Phi n}{60}$$

$$= \frac{(100)(2)(0.036\ \text{Wb})\left(1800\ \dfrac{\text{rev}}{\text{min}}\right)}{60\ \dfrac{\text{s}}{\text{min}}} = 216\ \text{V}$$

## 28. VOLTAGE-CURRENT CHARACTERISTICS FOR DC GENERATORS

Figure 63.15 illustrates the voltage-current characteristics for a DC generator.

*Figure 63.15* DC Generator Voltage-Current Characteristics

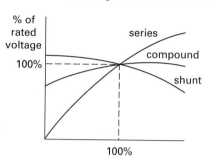

## 29. TORQUE CHARACTERISTICS FOR DC MOTORS

The torque produced by a DC motor is illustrated by Fig. 63.16.

*Figure 63.16* DC Motor Torque Characteristics

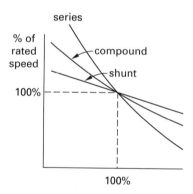

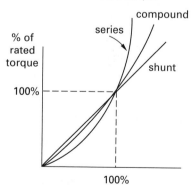

Circuits

## 30. STARTING DC MOTORS

DC motors have very low armature resistances. At rest, there is no back-emf, $E$. If connected across the full line voltage, the high current could damage the motor, hence such motors are almost always started with a resistance in series with the armature winding.[17] The initial resistance is chosen so that the starting current is limited to approximately 150% of the full-load current. As the motor builds up speed, the back-emf opposes the line voltage, reducing the current. The starting resistance can then be gradually reduced and removed.

Since torque depends on the current, starting torque is limited to approximately 225%, 175%, and 150% of the full-load torque for series, compound, and shunt motors, respectively, when the starting current is 150% of full-load current.

## 31. SPEED CONTROL FOR DC MOTORS

The speed of a DC motor can be controlled by changing the armature conditions, field conditions, or both. Changes can be made manually or automatically.

Automatic starters may switch between positions based on time, current, voltage, or magnetic field.

*Armature control* techniques include (1) placing a variable resistance in series or parallel with the armature, and (2) changing the voltage across it. The field voltage is held constant. For a constant torque load, the speed varies in approximate proportion to the voltage impressed on the armature.

*Field control (field weakening)* techniques include (1) changing the resistance of the field winding (series or shunt) and (2) changing the voltage across it. The armature voltage is held constant. Reducing the field current reduces the flux and increases the motor speed.

Electronic control of DC motors is achieved by supplying either the armature or field (or both) through rectifiers (e.g., silicon-controlled rectifiers—SCRs). Other electronic control methods include *pulse-width modulation* (PWM), *current chopping* using a transistor bridge, and digital or *incremental control* using a *shaft encoder*. Through a feedback element or other control mechanism, output is automatically adjusted to maintain the requirements of the machine or process.

---

[17]Small fractional horsepower ($1/4$ hp or smaller) motors are exceptions.

# 64 Electronics and Amplifiers

## Nomenclature

| | | |
|---|---|---|
| $A$ | gain | – |
| $C$ | capacitance | F |
| $g_m$ | transconductance | S |
| $h$ | hybrid parameter | various |
| $i, I$ | current | A |
| $P$ | power | W |
| $q$ | charge on an electron | C |
| $r, R$ | resistance | $\Omega$ |
| $S$ | sensitivity | – |
| $t$ | time | s |
| $T$ | absolute temperature | K |
| $v, V$ | voltage | V |

## Symbols

| | | |
|---|---|---|
| $\alpha$ | CB forward current ratio | – |
| $\alpha$ | temperature coefficient (Zener diode) | %/K |
| $\beta$ | CE forward current ratio | – |
| $\kappa$ | Boltzmann constant | J/K |
| $\eta$ | empirical factor | – |

## Subscripts

| | |
|---|---|
| 0 | saturation |
| $B$ | base |
| BB | base bias |
| $c$ | common collector |
| co | cut-off |
| $C$ | collector |
| $d$ | dynamic |
| $D$ | diode or drain |
| DD | drain bias |
| $e$ | common emitter |
| $E$ | emitter |
| EE | emitter bias |
| $f$ | forward |
| $G$ | gate |
| $i$ | input |
| $I$ | current |
| $L$ | load or inductance |
| $m$ | maximum |
| $o$ | output or open circuit |
| $O$ | thermally generated |
| $P$ | power or pinch off |
| $Q$ | quiescent |
| $r$ | reverse |
| $R$ | resistance |
| $s$ | saturation |
| $S$ | source or saturation |
| $T$ | at temperature $T$ |
| $V$ | voltage |
| VZ | voltage (Zener) |
| $Z$ | impedance |

## 1. CHAPTER CONVENTIONS

In this chapter, uppercase letters represent DC and fixed values (e.g., battery voltages and sinusoidal rms values). Lowercase letters represent AC or instantaneous values. Equivalent parameters (e.g., equivalent resistances) are represented by lowercase letters. The reference to $t$ is omitted for functions of time (e.g., $v(t)$ is represented simply as $v$).

Subscripts on current variables refer to the terminals into which the current flows. Subscripts on voltage variables refer to the terminals across which the voltage appears.

Bias battery naming is slightly arbitrary. For bipolar junction transistor circuits, $V_{BB}$ is base bias battery, $V_{CC}$ is the collector bias battery, and $V_{EE}$ is the emitter bias battery. Other names are also used.

## 2. SEMICONDUCTOR MATERIALS

Silicon (Si) and germanium (Ge), which form the basis for most semiconductors, are slightly more conductive than insulators but not nearly as conductive as metals. Both elements are from Group 4 in the periodic chart and possess four valence electrons. Normally, covalent bonding is used to fill the outer shell of eight and create a crystalline lattice.

The lattice will have many defects—free electrons and missing electrons (*holes*). In an *intrinsic semiconductor*, both electron and hole movement contribute to current (i.e., "carry" charge) when a voltage is applied.[1] The formation of holes and free electrons increases with temperature and is known as *thermal carrier generation*.

The addition of minute amounts (e.g., 10 ppb) of impurities (*dopes*) with either three or five valence electrons results in *doped semiconductors* with increased conductivity. Typical Group 5 dopes are phosphorus (P) and arsenic (As).

When a five-valence dope such as arsenic is used in a lattice needing only four to fill the octet, an extra *negative charge carrier* (i.e., an electron) becomes available for conduction. Arsenic is, therefore, a *donor*, and the resultant semiconductor material is called an *n-type* because electrons are the *majority carriers*. Some holes are present, of course, and these are the *minority carriers*.

Typical Group 3 dopes, aluminum (Al), boron (B), indium (I), and gallium (Ga), have only three valence electrons. A *vacancy* (*hole*) is created when one of these elements is used as a dope. A vacancy is essentially the same as a positive charge of $1.6 \times 10^{-19}$ C and is easily shifted from one atom to another. These dopes are *acceptors*, and semiconductors manufactured with acceptor dopes are called *p-types* because positive holes are the majority carriers. Electrons are the minority carriers.

## 3. HALL VOLTAGE

When a current-carrying conductor is placed in a magnetic field such that the magnetic field is perpendicular to the direction of flow, the flowing electrons will be forced to one side of the conductor. The flowing

vacancies (holes) will be forced to the other side of the conductor. The separation of the two oppositely charged current carriers produces a voltage known as the *Hall voltage* across the width of the conductor in the direction of the magnetic field. The presence of this voltage is empirical evidence for the existence of positive current carriers.

## 4. DEVICE PERFORMANCE CHARACTERISTICS

Most electronic devices can be modeled as two-port devices. There are two variables—current and voltage—for each port. The relationship between these variables depends on the semiconductor device and can be expressed mathematically (as with MOSFETs), modeled in equivalent circuits (as with transistor *h*-parameters),or described graphically (as with BJT characteristic curves).

Semiconductor devices are inherently *nonlinear devices*. While the curves are nonlinear, performance within a limited range can still be assumed to be linear if the variations in incoming *small signals* are much less than the average (steady, DC, etc.) values. If operation is well outside the linear region or if the input signal is large compared to the average value, the device distorts the input signal. This is known as *nonlinear operation*.

Figure 64.1 illustrates a characteristic curve for an ideal transistor operating as a current amplifying device. The voltage-current graph is divided into various regions known by various names, such as the *saturation* (or *on*), *cut-off* (or *off*), *active, breakdown, avalanche,* and *pinch-off* regions. The locations of these regions depend on the type of transistor (e.g., BJT or FET) and its polarity. (See Fig. 64.13 and Fig. 64.20.) Though amplifier operation is normally in the linear active region, operation in other regions is possible in digital and radio frequency (rf) applications.

**Figure 64.1** *Typical Semiconductor Performance*

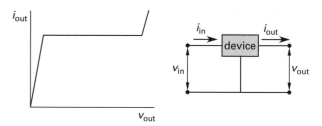

## 5. BIAS

In the terminology of electronics, *bias* is used as both a verb and a noun.

To bias a semiconductor device means to establish its *quiescent operating point* (i.e., its operating point with no input signal). In particular, with linear circuits, to bias the device means to establish the DC voltages and

---

[1]The term "intrinsic" means natural and extremely pure. Intrinsic semiconductors core undoped; the only impurities they have are naturally occurring.

currents that will exist at the device's terminals when the input signal is zero (or nearly so).

With respect to p-n junctions, *forward bias* (also known as the *on condition*) is the application of a positive voltage to the p-type material, or equivalently, the flow of a current from the p-type to the n-type material. In a small semiconductor device, forward bias will result in a current in the milliampere range.

*Reverse bias* (also known as the *off condition*) is the application of a negative voltage to the p-type material, or equivalently, the flow of a current from the n-type to the p-type material. In a small semiconductor device, reverse bias will result in a current in the nanoampere range.

A *self-biasing* circuit employs an input bias voltage that is derived from the amplifier's output circuit. This negative feedback control regulates the output current and voltage against variations in transistor parameters.

## 6. AMPLIFIERS

An *amplifier* produces an output signal from the input signal. The input and output signals can be either voltage or current. The output can be either smaller or larger (the usual case) than the input in magnitude. While most amplifiers merely scale the input voltage or current upward, the amplification process can include a sign change, phase change, or a complete phase shift of 180°.[2] The ratio of the output to the input is known as the *gain* or *amplification factor*, $A$. A *voltage amplification factor*, $A_V$, and *current amplification factor*, $A_I$ or $\beta$, can be calculated for an amplifier.

Figure 64.2 illustrates a simplified current amplifier with *current amplification factor* $\beta$. The additional current leaving the amplifier is provided by the bias battery, $V_2$.

$$i_{\text{out}} = \beta i_{\text{in}} \qquad 64.1$$

A capacitor, $C$, is placed in the output terminal to force all DC current to travel through the *load resistor*, $R_L$. Kirchhoff's voltage law for loop abcd is

$$V_2 = i_{\text{out}} R_L + V_{ac} = \beta i_{\text{in}} R_L + V_{ac} \qquad 64.2$$

If there is no input signal (i.e., $i_{\text{in}} = 0$, then $i_{\text{out}} = 0$ and the entire battery voltage appears across terminals $ac$ ($V_{ac} = V_2$). If the voltage across terminals $ac$ is zero, then the entire battery voltage appears across $R_L$ so that $i_{\text{out}} = V_2/R_L$.

## 7. AMPLIFIER CLASS

The amplifier class depends on how much of the input is translated into the output. (A sinusoidal input signal is assumed.) The output of an amplifier depends on the

**Figure 64.2** General Amplifier

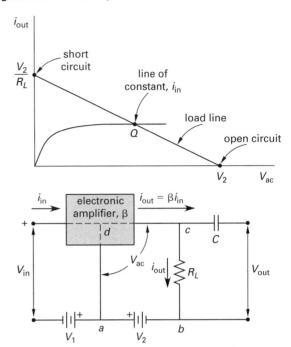

bias setting, which in turn establishes the quiescent point.

A *Class A amplifier* (see Fig. 64.3) has a quiescent point in the center of the active region of the operating characteristics. Class A amplifiers have the greatest linearity and the least distortion. Load current flows throughout the full input signal cycle. Since the load resistance of a properly designed amplifier will equal the Thevenin equivalent source resistance, the maximum power conversion efficiency of an ideal Class A amplifier is 50%.

**Figure 64.3** Class A Amplifier

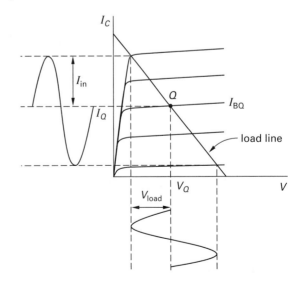

---

[2]An *inverting amplifier* is one for which $v_{\text{out}} = -A_V v_{\text{in}}$. For a sinusoidal input, this is equivalent to a phase shift of 180 degrees (i.e., $v_{\text{out}} = A_V v_{\text{in}} \angle -180°$.

Circuits

For *Class B amplifiers* (see Fig. 64.4), the quiescent point is established at the cut-off point. A load current flows only if the signal drives the amplifier into its active region, and the circuit acts like an amplifying half-wave rectifier. Class B amplifiers are usually combined in pairs, each amplifying the signal in its respective half of the input cycle. This is known as *push-pull operation*. The output waveform will be sinusoidal except for the small amount of crossover distortion that occurs as the signal processing transfers from one amplifier to the other. The maximum power conversion efficiency of an ideal Class B push-pull amplifier is approximately 78%.

**Figure 64.4** *Class B Amplifier*

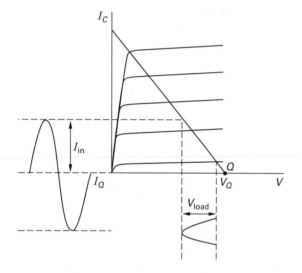

The intermediate *Class AB amplifier* has a quiescent point somewhat above cut-off but where a portion of the input signal still produces no load current. The output current flows for more than half of the input cycle. AB amplifiers are also used in push-pull circuits.

*Class C amplifiers* (see Fig. 64.5) have quiescent points well into the cut-off region. Load current flows during less than one-half of the input cycle. For a purely resistive load, the output would be decidedly nonsinusoidal. However, if the input frequency is constant, as in radio frequency (rf) power circuits, the load can be a parallel LRC tank circuit tuned to be resonant at the signal frequency. The LRC circuit stores electrical energy, converting the output signal to a sinusoid. The power conversion efficiency of an ideal Class C amplifier is 100%.

## 8. LOAD LINE AND QUIESCENT POINT

The $i_{in}$–$v_{in}$ curves illustrate how amplification occurs. The two known points, $(v_{out}, i_{out}) = (V_2, 0)$ and $(v_{out}, i_{out}) = (0, V_2/R_L)$, are plotted on the voltage-current characteristic curve. The straight *load line* is drawn between them. The change in output voltage

**Figure 64.5** *Class C Amplifier*

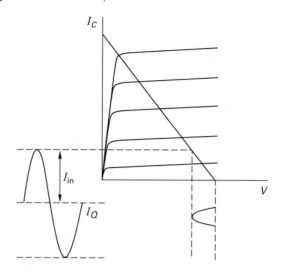

(the horizontal axis) due to a change in input voltage can be determined. Equation 64.3 gives the voltage *gain* (*amplification factor*).[3,4]

$$A_V = \frac{\alpha v_{out}}{\alpha v_{in}} \approx \frac{\Delta v_{out}}{\Delta v_{in}} \qquad 64.3$$

Usually, a nominal current (the *quiescent current*) flows in the *abcd* circuit even when there is no signal. The point on the load line corresponding to this current is the *quiescent point* (*Q-point* or *operating point*). It is common to represent the quiescent parameters with uppercase letters (sometimes with a subscript $Q$) and to write instantaneous values in terms of small changes to the quiescent conditions.

$$v_{in} = V_Q + \Delta v_{in} \qquad 64.4$$

$$v_{out} = V_{out} + \Delta v_{out} \qquad 64.5$$

$$i_{out} = I_{out} + \Delta i_{out} \qquad 64.6$$

Since it is a straight line, the load line can also be drawn if the quiescent point and any other point, usually $(V_2, 0)$, are known.

The ideal voltage amplifier has an infinite *input impedance* (so that all of $v_{in}$ appears across the amplifier and no current or power is drawn from the source) and zero *output impedance* (so that all of the output current flows through the load resistor).

## 9. P-N JUNCTIONS

A *p-n junction* (see Fig. 64.6) consists of a p-type semiconductor (the *anode*) and an n-type semiconductor (the *cathode*) bonded together. Due to the

---

[3]Gain can be increased by increasing the load resistance, but a larger biasing battery, $V_2$, is required. The choice of battery size depends on the amplifier circuit devices, space considerations, and economic constraints.

[4]A *high-gain* amplifier has a gain in the tens or hundreds of thousands.

concentration gradient of donor and acceptor atoms, a *diffusion current*, $I_{\text{diffusion}}$, (also known as *recombination current* and *injection current*) consisting of both holes and electrons flows across the junction. Free holes and electrons combine in the vicinity of the junction, leaving behind immobile, unneutralized ions in the semiconductor.

**Figure 64.6** *Semiconductor p-n Junction*

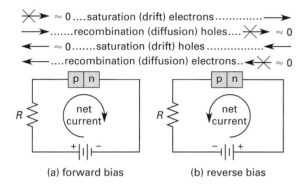

The unneutralized ions establish a potential difference known as the *barrier voltage* and a corresponding electrostatic field that causes an opposing *drift current*, $I_s$ (also known as *saturation current, thermal current,* or *reverse saturation current*).[5] Once equilibrium has been reached, the drift and diffusion currents are equal but in opposite directions. Thus, at equilibrium (and without any applied voltage), the current crossing the junction is zero.

$$I_{\text{junction}} = I_{\text{diffusion}} + I_s = 0 \quad \text{[algebraic sum]} \qquad 64.7$$

The region in the vicinity of the junction is called the *depletion region* since the concentrations of both types of carriers are reduced.

The equilibrium is changed when a battery is connected across the junction. With forward bias (i.e., the positive terminal connected to the p-type material), holes are repelled across the junction into the n-type material, and electrons are repelled across the junction into the p-type material. A forward-bias voltage, $V_{0f}$, of approximately 0.5–0.7 V for silicon and approximately 0.2–0.3 V for germanium is required to overcome the barrier voltage.[6] Once overcome, the junction current increases significantly due to increases in diffusion current. The drift current component depends primarily on temperature and is not affected.

With reverse bias (i.e., the positive battery terminal connected to the n-type material), the battery adds to the barrier voltage and increases the electrostatic field. Diffusion current is reduced to almost zero. Only a small (e.g., $10^{-9}$ A) reverse current flows. This has two components: (1) *drift (reverse saturation) current* composed of thermally generated minority carriers, and (2) *surface leakage current* flowing on the surface of the semiconductor.

If the reverse bias voltage is increased to the reverse breakdown voltage, the reverse current will increase dramatically. The value of the reverse breakdown voltage and the mechanism that causes reverse breakdown depend on the device, doping concentrations, and geometry.

At high doping concentrations, the mechanism for *breakdown* is the *Zener effect*. Zener breakdown occurs at reverse voltages of less than 6 V. The electrostatic field in the semiconductor is so high that electrons are directly excited into conduction. At lower doping concentrations, the breakdown mechanism is the *avalanche effect*, which occurs at reverse voltages greater than 6 V. (See Fig. 64.7.) Carriers passing through the depletion region acquire enough energy from the electric field to eject multiple electrons from the lattice atoms with which they collide.

## 10. DIODE PERFORMANCE CHARACTERISTICS

A *diode* is a device that passes current in one direction only. An *ideal diode*, approximated by a p-n junction, has a zero voltage drop (i.e., has no forward resistance) and acts as a short circuit when forward biased (i.e., when "on"). It has an infinite resistance and acts as an open circuit when reverse biased (i.e., when "off").

Figure 64.7 illustrates a typical real *semiconductor diode* performance characteristic. The *reverse bias voltage* is the voltage below which the current is very small (i.e., less than 1 percent of the maximum rated current). The *peak inverse (reverse) voltage*, PIV or PRV, is the maximum reverse bias the diode can withstand without damage. The forward current is also limited.[7]

**Figure 64.7** *Semiconductor Diode Characteristics and Symbol*

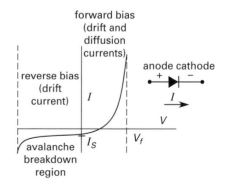

---

[5]The symbols $I_0$ and $I_{\text{co}}$ are also used.

[6]The exact value depends on the manufacturing method and temperature.

[7]Maximum forward current and peak inverse voltage for silicon diode rectifiers are approximately 600 A and 1000 V, respectively.

Equation 64.8 is based on the *Fermi-Dirac probability function* and gives the approximate current when the diode is reverse-biased with voltage $V$ excluding the breakdown region. ($V$ is negative with reverse bias.) $V_T$ is the *volt equivalent* of the temperature. $\eta$ is an empirical constant equal to 1 for germanium and 2 for silicon.

$$I = I_s(e^{qV/\eta\kappa T} - 1) = I_s(e^{V/\eta V_T} - 1) \qquad 64.8$$

$$V_T = \frac{\kappa T}{q} \qquad 64.9$$

Due to the flatness of the curve, any value of $I$ with a small reverse bias (e.g., between 0 V and $-1$ V) can be used as the saturation current. For silicon, $I_s \approx 10^{-9}$ A; for germanium, $I_s \approx 10^{-6}$ A. At room temperature, $T = 293$K. Also, $q = 1.6 \times 10^{-19}$ C and $\kappa = 1.38 \times 10^{-23}$ J/K. Substituting into Eq. 64.8,

$$I \approx I_s(e^{40\ V/\eta} - 1) \qquad 64.10$$

For both silicon and germanium diodes, the saturation current doubles for each 10°C increase in temperature.

$$\frac{I_{s2}}{I_{s1}} = 2^{(T_2 - T_1)/10} \qquad 64.11$$

Figure 64.8 shows that a *real diode* can be modeled in low-speed circuits as an ideal diode, a resistor, and a voltage source, $V_0$. The voltage source in the model accounts for the barrier voltage. It is approximately 0.7 V for silicon and 0.3 V for germanium.

The *dynamic forward resistance*, $r_f$, is the inverse of the slope of the current-voltage characteristic at the operating point in the forward-bias region. Disregarding lead contact resistance (less than 2 $\Omega$), it can be shown that Eq. 64.12 gives the dynamic forward resistance (in ohms) at room temperature (approximately 300K).[8]

$$r_f = \frac{0.026\eta}{I_D} \quad [I_D \text{ in A}] \qquad 64.12$$

There are two ways of determining an unknown forward resistance, $r_f$. One way is by iteration, by first assuming $r_f = 0$, then finding $I_D$ from Eq. 64.13, and finally using Eq. 64.12 to get an approximate $r_f$. The second way is to use the load line (see Fig. 64.9) and find the slope of the diode characteristic at the operating point.

The *dynamic reverse resistance*, $r_r$, is the inverse of the slope at a point in the reverse-bias region.[9] Since the reverse current is very small, $r_r$ is essentially infinite.

A diode's *static forward resistance* is the average resistance, $r_{\text{static}} = V_D/I_D$, at a given operating point. It is not used in the model shown in Fig. 64.8.

---

[8]This is derived from Eq. 64.9.
[9]Specifying the reverse current, $I_{\text{co}}$, is equivalent to specifying the reverse resistance.

**Figure 64.8** *Diode Equivalent Circuit*

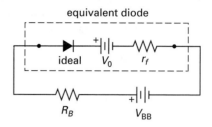

This simple model assumes (1) the reverse-bias current is sufficiently small such that the diode acts as an open circuit in that direction, (2) the reverse bias voltage does not exceed the breakdown voltage, and (3) the switching time is essentially instantaneous.

## 11. DIODE LOAD LINE

Figure 64.9 shows a forward-biased real diode in a simple circuit. $V_{BB}$ is the *bias battery* (hence the subscripts), and $R_B$ is a current-limiting resistor. $r_f$ and $V_0$ are equivalent diode parameters, not discrete components. If $r_f$ is known in the vicinity of the operating point, the diode current, $I_D$, is found from Kirchhoff's voltage law. The diode voltage is found from Ohm's law: $V_D = I_D r_f$.

$$V_{BB} - V_0 = I_D(R_B + r_f) \qquad 64.13$$

**Figure 64.9** *Diode Load Line and Equivalent Circuit*

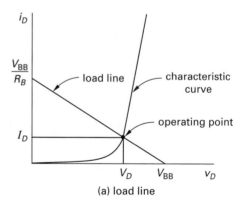

(a) load line

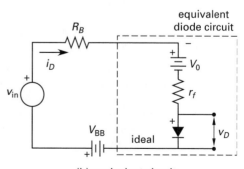

(b) equivalent circuit

The diode current and voltage drop can also be found graphically from the *diode characteristic curve* (see Fig. 64.9) and the *load line*, a straight line representing the locus of points satisfying Eq. 64.13.[10] The load line is defined by two points. If the diode current is zero, all of the battery voltage appears across the diode, hence the point $(V_{BB}, 0)$. If the voltage drop across the diode is zero, all of the voltage appears across the current-limiting resistor, hence the point $(0, V_{BB}/R_B)$. (Since the diode characteristic curve implicitly includes the effects of $V_0$ and $r_f$, these terms should be omitted.) The no-signal *operating point*, also known as the *quiescent point*, is the intersection of the diode characteristic curve and the load line.

The *static load line* is derived assuming there is no signal (i.e., $v_{in} = 0$. With a signal, the *dynamic load line* shifts left or right while keeping the same slope. This is equivalent to solving Eq. 64.13 with an additional voltage source.

$$i_D = I_D + \Delta i_D = \frac{V_{BB} - V_0 + v_{in}}{R_B + r_f} \qquad 64.14$$

## 12. DIODE APPLICATIONS

Diodes are readily integrated into *rectifier, clipping*, and *clamping* circuits. A *clipping circuit* cuts the peaks off of waveforms; a *clamping circuit* shifts the DC (average) component of the signal. Figure 64.10 illustrates the response to a sinusoid with peak voltage $V_m$ for several simple circuits.

## 13. SCHOTTKY, ZENER, AND TUNNEL DIODES

A *Schottky barrier diode* (*hot-carrier diode*) is formed by bonding platinum metal to n-type silicon. The platinum serves as the acceptor material. It has negligible charge storage and is used in high-speed switching applications. *Tunnel diodes* (*Esaki diodes*) exhibit negative resistance when suitably biased. *Zener diodes* have almost constant voltage characteristics in the reverse-bias region (in which they usually operate). (See Fig. 64.11.) This makes them ideal for voltage regulation applications. The *avalanche behavior* is not, in itself, destructive to the diode as long as the power rating is not exceeded. The maximum reverse voltage, the *Zener voltage* ($V_Z$), varies with temperature. The *temperature coefficient* is

$$\alpha_{VZ} = \frac{\Delta V_Z}{V_Z \Delta T} \times 100\% \qquad 64.15$$

---

**Figure 64.10** *Outputs from Simple Diode Circuits*

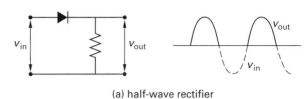

(a) half-wave rectifier

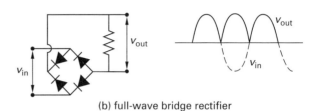

(b) full-wave bridge rectifier

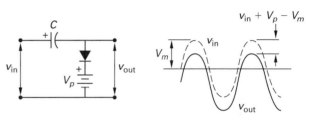

(c) clamping circuit (*C* charges to $V_m$)

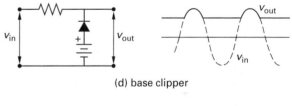

(d) base clipper

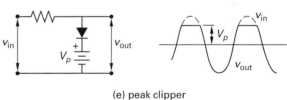

(e) peak clipper

**Figure 64.11** *Special Kinds of Diodes*

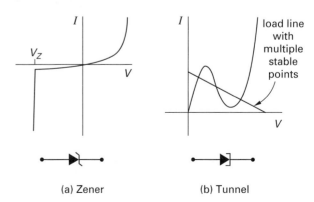

(a) Zener

(b) Tunnel

---

[10]Notice that the horizontal axis voltage is the voltage across the diode (modeled as a resistor).

## 14. PNP AND NPN JUNCTIONS

A *bipolar junction transistor* (BJT) (see Fig. 64.12) consists of a thin n-type semiconductor between two p-typesemiconductors (pnp) or a thin p-type semiconductor between two n-type semiconductors (npn). Regardless of type, the thin center semiconductor is called the *base* (B), and the outer semiconductors are the *emitter* (E) and *collector* (C). An arrow in the transistor symbol always points toward the n-type material. (The symbol "npn" can be thought to mean "not pointing in.")

**Figure 64.12** *Bipolar Junction Transistors*

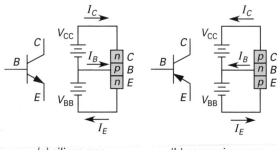

(a) silicon npn          (b) germanium pnp

For silicon (npn) and germanium (pnp) transistors, the emitter-base junction is normally forward-biased. The majority carriers (see Sec. 64.2) originate in the emitter and are injected into the base region. Minority carriers originate in much smaller numbers in the base and enter the emitter. Some holes combine in the base region with electrons and are lost. (Since hole-electron combination in the base is undesirable, the base is kept very thin.) However, many majority carriers make it across the base to the collector.

The collector-base junction is normally reverse-biased, which prohibits majority carrier movement from the collector to the base. However, majority carriers are attracted in the opposite direction—toward the collector—where they combine with minority carriers from the reverse-bias battery $V_{CC}$.

## 15. BJT TRANSISTOR PERFORMANCE CHARACTERISTICS

When the base-emitter junction is forward-biased and the collector-base junction is reverse-biased, the transistor is said to be operating in the *active region*. (See Fig. 64.13.)

When the base-emitter junction is not forward-biased (as when $V_{BB}$ is zero), the base current will be nearly zero, and the transistor acts like a simple switch. This is known as being *off* or *open*, and operating in the *cut-off mode*.[11] Also, the collector and emitter currents are zero

---

[11]Except for digital and switching applications, this condition usually results from improper selection of circuit resistances.

**Figure 64.13** *BJT Operating Regions*

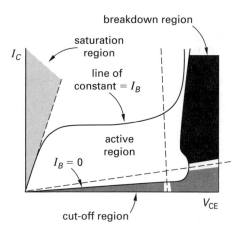

when the transistor operates in the *cut-off region*. However, a very small input voltage (in place of $V_{BB}$) will forward bias the base-emitter junction, which because $I_B$ is so low, instantly forces the transistor into its saturation region. This results in a large collector current.

When the collector-emitter voltage is very low (usually about 0.3 V for silicon and 0.1 V for germanium), the transistor operates in its *saturation region*. Regardless of the collector current, the transistor operates as a closed switch (i.e., a short circuit between the collector and emitter). This is known as being *on* or *closed*.

$$V_{CE} \approx 0 \quad \text{[saturation]} \qquad 64.16$$

Transistors are manufactured from silicon and germanium, although silicon transistors have a higher temperature operating range. While the collector cut-off current is very small at room temperature, it doubles every 10°C (see Eq. 64.11), rendering germanium transistors useless around 100°C. Silicon transistors remain useful up to approximately 200°C. While germanium has a lower collector-emitter saturation voltage and may outperform silicon in high-speed and high-frequency devices, silicon is nevertheless the usual material for most semiconductor devices and integrated circuit systems.

## 16. BJT TRANSISTOR PARAMETERS

Equation 64.17 is Kirchhoff's current law taking the transistor as a node. Usually, the collector current is proportional to, and two or three orders of magnitude larger than, the base current, $I_B$. Thus, a small change in base current of, for example, 1 mA, can produce a change in collector current of, for example, 100 mA. The *current (amplification) ratio*, $\beta_{DC}$, is the ratio of collector-base currents.

$$I_E = I_C + I_B \qquad 64.17$$

$$\beta_{\text{DC}} = \frac{I_C}{I_B} = \frac{\alpha_{\text{DC}}}{1 - \alpha_{\text{DC}}} \qquad 64.18$$

$$\alpha_{\text{DC}} = \frac{I_C}{I_E} = \frac{\beta_{\text{DC}}}{1 + \beta_{\text{DC}}} \qquad 64.19$$

Both $\alpha_{\text{DC}}$ and $\beta_{\text{DC}}$ are for DC signals only. The corresponding values for small signals are designated $\alpha_{\text{AC}}$ and $\beta_{\text{AC}}$, respectively, and are calculated from differentials. (The difference between $\beta_{\text{AC}}$ and $\beta_{\text{DC}}$ is very small, and the two are not usually distinguished.)

$$\beta_{\text{AC}} = \frac{\Delta I_C}{\Delta I_B} = \frac{i_C}{i_B} \qquad 64.20$$

$$\alpha_{\text{AC}} = \frac{\Delta I_C}{\Delta I_E} = \frac{i_C}{i_E} \qquad 64.21$$

Thermal (saturation) current is small but always present and can be included in Eq. 64.17. $I_{\text{CBO}}$ is the thermal current at the collector-base junction.

$$I_C = I_E - I_B \qquad 64.22$$

$$I_C = \alpha I_E - I_{\text{CBO}} \approx \alpha I_E \qquad 64.23$$

Figure 64.14 illustrates a family of curves for various base currents. The DC amplification factor, $\beta_{\text{DC}}$ can be found (for a wide range of $V_{\text{CE}}$ values) by taking a point on any line in the active (horizontal line) region and calculating the ratio of the coordinates, $I_C/I_B$. The small-signal amplification factor, $\beta_{\text{AC}}$, is calculated as the difference in the $I_C$ between two $I_B$ lines divided by the differences in $I_B$.[12]

**Figure 64.14** BJT Characteristic Curves

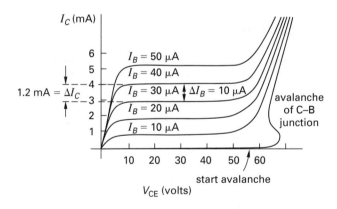

Another important transistor parameter is the *collector-emitter resistance*. The symbols $R_{\text{CE}}$ and $r_{\text{CE}}$ are used to represent this resistance for DC and AC applications, respectively.[13] Since the base is thin, the CE junction is basically a diode. (See Sec. 64.10.) Therefore, $R_{\text{CE}}$ is small and may be disregarded in determining biasing

[12]$\beta_{\text{DC}}$ and $\beta_{\text{AC}}$ are specified by many manufacturers as $h_{\text{FE}}$ and $h_{\text{fe}}$, respectively.
[13]$R_{\text{CE}}$ is specified by many manufacturers as $h_{\text{oe}}$, where $r_{\text{CE}} \approx 1/h_{\text{oe}}$.

requirements. If required, it can be calculated (as $r_f$) from Eq. 64.12 or from the following two equations.

$$R_{\text{CE}} = \frac{V_{\text{CE}}}{I_C} \quad [I_B \text{ constant}] \qquad 64.24$$

$$r_{\text{CE}} = \frac{\Delta V_{\text{CE}}}{\Delta I_C} \quad [I_B \text{ constant}] \qquad 64.25$$

## 17. BJT TRANSISTOR CIRCUIT CONFIGURATIONS

There are six ways a BJT transistor can be connected in a circuit, depending on which leads serve as input and output. Only three configurations have significant practical use, however. (See Fig. 64.15.) The terminal not used for either input or output is referred to as the *common* terminal. For example, in a *common emitter* circuit the base receives the input signal and the output signal is at the collector. For the *common collector* (also known as an *emitter follower*) circuit, the input is to the base and the output is from the emitter. For a *common base* circuit, the input is to the emitter and the output is from the collector.

**Figure 64.15** Transistor Connections (bypass capacitors not shown)

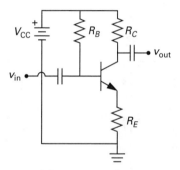

(a) common emitter

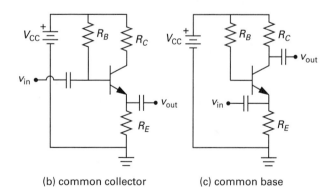

(b) common collector   (c) common base

## 18. BJT BIASING CIRCUITS

To maximize the transistor's operating range when large-swing signals are expected, the quiescent point should be approximately centered in the active region.

Circuits

The purpose of biasing is to establish the base current and, in conjunction with the load line, the quiescent point. Figure 64.16 illustrates several typical biasing methods: fixed bias, fixed bias with feedback, self-bias, voltage-divider bias, multiple-battery bias, and switching circuit (cut-off) bias. All of the methods can be used with all three common-lead configurations and with both npn (shown) and pnp transistors.[14] It is common to omit the bias battery (shown in Fig. 64.16(a) in dashed lines) in transistor circuits.

**Figure 64.16** *DC Biasing Methods*

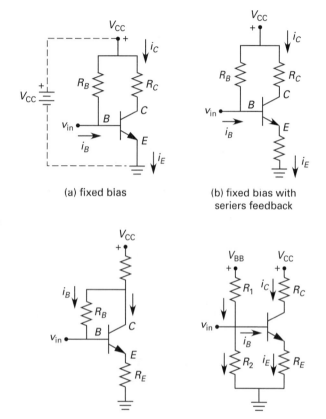

(a) fixed bias

(b) fixed bias with series feedback

(c) self-bias with shunt feedback

(d) voltage-divider biasing

(e) multiple-battery bias

(f) cut-off biasing

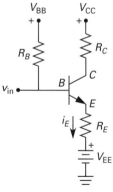

The base current can be found by writing Kirchhoff's voltage law around the input loop, including the bias battery, external resistances, and the $V_{BE}$ barrier voltage (see Sec. 64.9) that opposes the bias battery. Since the base is thin, there is negligible resistance from the base to the emitter, and $V_{BE}$ (being less than 1 V) may be omitted as well. For the case of *fixed bias with feedback* (*fixed bias with emitter resistance*) illustrated in Fig. 64.16(b), the base current is found from

$$V_{CC} = I_B R_B + V_{BE} + I_E R_E \quad [v_{in} = 0] \qquad 64.26$$

$$I_B = \frac{V_{CC} - V_{BE} - I_E R_E}{R_B} \quad [v_{in} = 0]$$

$$= \frac{V_{CC} - V_{BE}}{R_B + \dfrac{\beta}{\alpha} R_E} = \frac{V_{CC} - V_{BE}}{R_B + (1+\beta)R_E} \quad [v_{in} = 0] \qquad 64.27$$

The design of a fixed-bias amplifier with emitter resistance starts by choosing the quiescent collector current, $I_{CQ}$. $R_E$ is selected so that the voltage across $R_E$ is approximately three to five times the intrinsic $V_{BE}$ voltage (i.e., 0.3 V for germanium and 0.7 V for silicon). Once $R_E$ is found, $R_B$ can be calculated from Eq. 64.26.

$$R_E \approx \frac{3 V_{BE}}{I_{CQ}} \qquad 64.28$$

The *bias stability*, with respect to any quantity $M$, is given by Eq. 64.29. Variable $M$ commonly represents temperature ($T$), current amplification ($\beta$), collector-base cut-off ($I_{CBO}$), and base-emitter voltage ($V_{BE}$).

$$S_M = \frac{\Delta I_C / I_{CQ}}{\Delta M / M} \qquad 64.29$$

## 19. BJT LOAD LINE

Figure 64.17 illustrates part of a simple common emitter transistor amplifier circuit. The bias battery and emitter and collector resistances define the *load line*. If the emitter-collector junction could act as a short circuit (i.e., $V_{CE} = 0$), the collector current would be $V_{CC}/(R_C + R_E)$. If the signal is large enough, it can completely oppose the battery-induced current, in which case, the net collector current is zero and the full bias battery voltage appears across the emitter-collector junction. The intersection of the load line and the base current curve defines the *quiescent point*. The load line, base current, and quiescent point are illustrated in Fig. 64.17.

## 20. CASCADED AMPLIFIERS

Several amplifiers arranged so that the output of one is the input to the next are said to be *cascaded*

---

[14]The polarity of the DC supply voltages must be reversed to convert the circuits shown for use with pnp transistors.

**Figure 64.17** *Load Line and Quiescent Point*

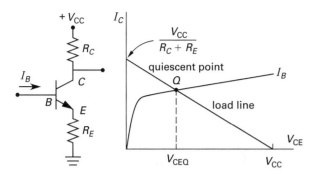

amplifiers.[15] When each *amplifier stage* is properly coupled to the following, the overall gain is

$$A_{\text{total}} = A_{V,1} A_{V,2} A_{V,3} \ldots \qquad 64.30$$

Capacitors are used in amplifier circuits to isolate stages and pass small signals. (This is known as *capacitive coupling*.)[16] A capacitor appears to a steady (DC) voltage as an open circuit. However, it appears to a small (AC) voltage as a short circuit, allowing the input and output signals to pass through, leaving the DC portion behind.

## 21. AMPLIFIER GAIN AND POWER

The *voltage-, current-, resistance-,* and *power-gain* are

$$A_V = \frac{\Delta V_{\text{out}}}{\Delta V_{\text{in}}} = \frac{v_{\text{out}}}{v_{\text{in}}} = \beta A_R \qquad 64.31$$

$$A_I = \frac{\Delta I_{\text{out}}}{\Delta I_{\text{in}}} = \frac{i_{\text{out}}}{i_{\text{in}}} = \beta \qquad 64.32$$

$$A_R = \frac{Z_{\text{out}}}{Z_{\text{in}}} = \frac{A_V}{\beta} \qquad 64.33$$

$$A_P = \frac{P_{\text{out}}}{P_{\text{in}}} = \beta^2 A_R = A_I A_V \qquad 64.34$$

The collector power dissipation, $P_C$, should not exceed the rated value. (This restriction applies to all points on the load line.)

$$P_C = \tfrac{1}{2} I_C V_{\text{CE}} \quad [\text{rms values}] \qquad 64.35$$

## 22. BJT EQUIVALENT *h*-PARAMETERS

A transistor can be modeled by a variety of equivalent two-port circuits. However, the hybrid parameters are used most often due to the ease of making appropriate measurements. Four *h-parameters* are needed for the model.

---

[15]A cascade amplifier should not be confused with a *cascode amplifier* (a high-gain, low-noise amplifier with two transistors directly connected in common emitter and common base configurations).
[16]The term *resistor-capacitor coupling* is also used.

$h_i$ = input impedence with output shorted ($\Omega$)
$h_f$ = forward transfer current ratio with output shorted (dimensionless)
$h_r$ = reverse transfer voltage ratio with input open (dimensionless)
$h_o$ = output admittance with input open (S)

The governing equations for this *small-signal circuit model* are

$$v_i = h_i i_i + h_r v_o \qquad 64.36$$

$$i_o = h_f i_i + h_o v_o \qquad 64.37$$

While a single subscript could be used (as in Eq. 64.36 and Eq. 64.37), a second subscript is needed to indicate the configuration. For example, $h_{\text{ie}}$ is the equivalent input impedance in the common emitter configuration. (See Fig. 64.18.) Transistor manufacturers provide data for the common emitter configuration only. Table 64.1 can be used to derive the *h*-parameters for other configurations.

**Figure 64.18** *Common Emitter Equivalent Hybrid Transistor Circuit*

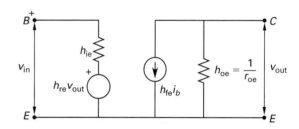

**Table 64.1** *Equivalent Circuit Parameters*

| symbol | common emitter | common collector | common base |
|---|---|---|---|
| $h_{11}, h_{\text{ie}}$ | $h_{\text{ie}}$ | $h_{\text{ic}}$ | $\dfrac{h_{\text{ib}}}{1 + h_{\text{fb}}}$ |
| $h_{12}, h_{\text{re}}$ | $h_{\text{re}}$ | $1 - h_{\text{rc}}$ | $\dfrac{h_{\text{ib}} h_{\text{ob}}}{1 + h_{\text{fb}}} - h_{\text{rb}}$ |
| $h_{21}, h_{\text{fe}}$ | $h_{\text{fe}}$ | $-1 - h_{\text{fc}}$ | $\dfrac{-h_{\text{fb}}}{1 + h_{\text{fb}}}$ |
| $h_{22}, h_{\text{oe}}$ | $h_{\text{oe}}$ | $h_{\text{oc}}$ | $\dfrac{h_{\text{ob}}}{1 + h_{\text{fb}}}$ |
| $h_{11}, h_{\text{ib}}$ | $\dfrac{h_{\text{ie}}}{1 + h_{\text{fe}}}$ | $\dfrac{-h_{\text{ic}}}{h_{\text{fc}}}$ | $h_{\text{ib}}$ |
| $h_{12}, h_{\text{rb}}$ | $\dfrac{h_{\text{ie}} h_{\text{oe}}}{1 + h_{\text{fe}}} - h_{\text{re}}$ | $h_{\text{rc}} - \dfrac{h_{\text{ic}} h_{\text{oc}}}{h_{\text{fc}}} - 1$ | $h_{\text{rb}}$ |
| $h_{21}, h_{\text{fb}}$ | $\dfrac{-h_{\text{fe}}}{1 + h_{\text{fe}}}$ | $\dfrac{-1 + h_{\text{fc}}}{h_{\text{fc}}}$ | $h_{\text{fb}}$ |
| $h_{22}, h_{\text{ob}}$ | $\dfrac{h_{\text{oe}}}{1 + h_{\text{fe}}}$ | $\dfrac{-h_{\text{oc}}}{h_{\text{fc}}}$ | $h_{\text{ob}}$ |
| $h_{11}, h_{\text{ic}}$ | $h_{\text{ie}}$ | $h_{\text{ic}}$ | $\dfrac{h_{\text{ib}}}{1 + h_{\text{fb}}}$ |
| $h_{12}, h_{\text{rc}}$ | $1 - h_{\text{re}}$ | $h_{\text{rc}}$ | $1$ |
| $h_{21}, h_{\text{fc}}$ | $-1 - h_{\text{fe}}$ | $h_{\text{fc}}$ | $\dfrac{-1}{1 + h_{\text{fb}}}$ |
| $h_{22}, h_{\text{oc}}$ | $h_{\text{oe}}$ | $h_{\text{oc}}$ | $\dfrac{h_{\text{ob}}}{1 + h_{\text{fb}}}$ |

**Circuits**

## 23. BJT EQUIVALENT CIRCUITS

Table 64.2 summarizes the governing equations for the equivalent circuits of a bipolar junction transistor.

### Example 64.1

A transistor is used in a common-emitter amplifier circuit as shown. Assume the inductor has infinite impedance and the capacitors have zero impedance to AC signals. The transistor $h$-parameters are $h_{ie} = 750\ \Omega$, $h_{oe} = 9.09 \times 10^{-5}$ S, $h_{fe} = 184$, and $h_{re} = 1.25 \times 10^{-4}$.

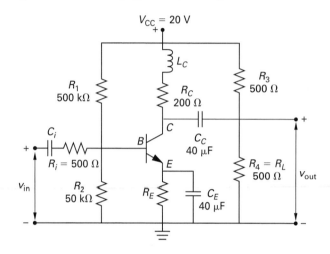

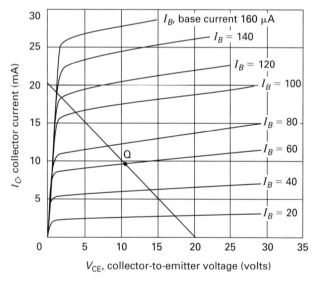

(a) If a quiescent point is wanted at approximately $V_{CE} = 11$ V and $I_C = 10$ mA, what should $R_E$ be?

For parts (b) through (g), assume $R_E = 800\ \Omega$.

(b) Draw the DC load line. (c) If $i_B = 80 \times 10^{-6}$ A, what is $i_C$? (d) What is the AC circuit voltage gain? (e) What is the AC circuit current gain? (f) What is the input impedance? (g) What is the output impedance? (h) Given $v_{in} = 0.25 \sin 400t$ V, what is $v_{out}$? (i) What is the purpose of the inductor, $L_C$?

*Solution*

(a) Write the equation for the voltage drop in the C-E circuit. Disregard the inductor (which passes DC signals). Use $h_{fe}$ for $h_{FE}$ since they are essentially the same and both are large. Kirchhoff's voltage law is

$$\alpha = \frac{\beta}{1+\beta} \approx \frac{h_{fe}}{1+h_{fe}} = \frac{184}{1+184} \approx 1$$

$$V_{CC} = I_C R_C + I_E R_E + V_{CE}$$

$$= I_C R_C + \left(\frac{I_C}{\alpha}\right) R_E + V_{CE}$$

$$\approx I_C (R_C + R_E) + V_{CE}$$

$$R_E = \frac{V_{CC} - V_{CE}}{I_C} - R_C$$

$$= \frac{20\ \text{V} - 11\ \text{V}}{10 \times 10^{-3}\ \text{A}} - 200\ \Omega$$

$$\approx 700\ \Omega$$

(b) If $I_C = 0$, then $V_{CE} = V_{CC}$. This is one point on the load line. If $V_{CE} = 0$, then

$$I_C = \frac{V_{CC}}{R_C + R_E} = \frac{20\ \text{V}}{200\ \Omega + 800\ \Omega} = 0.02\ \text{A}$$

These two points define the DC load line.

(c) From Eq. 64.18,

$$i_C = \beta i_B = h_{fe} i_B = (184)(80 \times 10^{-6}\ \text{A}) = 14.7\ \text{mA}$$

(d) To determine the AC circuit voltage gain, it is necessary to simplify the circuit. The bias battery $V_{CC}$ is shorted out. (This is because the battery merely shifts the signal without affecting the signal swing.) Therefore, many of the resistors connect directly to ground. The inductor has infinite impedance, so $R_C$ is disconnected. Both capacitors act as short circuits, so $R_E$ is bypassed.

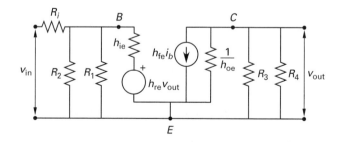

To simplify the circuit further, recognize that $R_1$, $R_2$, and $1/h_{oe}$ are very large and can be treated as infinite impedances.

**Table 64.2** *BJT Equivalent Circuits*

| common connection | equivalent circuit | network equations |
|---|---|---|
| **CE** 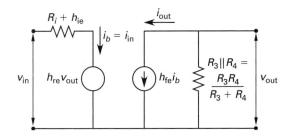 | **CE** | common emitter $$v_{be} = h_{ie}i_b + h_{re}v_{ce}$$ $$i_c = h_{fe}i_b + h_{oe}v_{ce}$$ |
| **CC** | **CC** | common collector $$v_{bc} = h_{ic}i_b + h_{rc}v_{ec}$$ $$i_e = h_{fc}i_b + h_{oc}v_{ec}$$ |
| **CB** | **CB** | common base $$v_{eb} = h_{ib}i_e + h_{rb}v_{cb}$$ $$i_c = h_{fb}i_e + h_{ob}v_{cb}$$ |

Continuing with the simplified analysis,

$$v_{out} = h_{fe}i_b \left( \frac{R_3 R_4}{R_3 + R_4} \right)$$

$$= 184i_b \left( \frac{(500\ \Omega)(500\ \Omega)}{500\ \Omega + 500\ \Omega} \right) = 46,000i_b$$

$$v_{in} = i_b(R_i + h_{ie}) + h_{re}v_{out}$$

$$= i_b(500\ \Omega + 750\ \Omega) + (1.25 \times 10^{-4})(46,000i_b)$$

$$= 1256i_b$$

(To perform an exact analysis, $h_{oe}$ and $h_{re}$ must be considered. Convert $R_i$ to its Norton equivalent resistance and place in parallel with $R_1$, $R_2$, and $h_{ie}$. Use the current-divider concept to calculate $i_b$.)

The voltage gain is

$$A_V = \frac{v_{out}}{v_{in}} \approx \frac{46,000i_b}{1256i_b} = 36.6$$

(e) The current gain is

$$A_I = \frac{i_{out}}{v_{in}} \approx \frac{h_{fe}i_b}{i_b} = 184$$

(f) The input impedance (resistance) is

$$R_{in} = \frac{v_{in}}{i_{in}} \approx \frac{1256i_b}{i_b} = 1256\ \Omega$$

(g) The output impedance is effectively the Thevenin equivalent resistance of the output circuit. The load resistance ($R_4$ in this instance) is removed. The independent source voltage ($v_{in}$) is shorted, which effectively opens the controlled source $h_{fe}i_b$. The remaining resistance between collector and ground is

$$R_{out} = R_3 = 500 \ \Omega$$

(h) The output voltage is

$$v_{out} = A_V v_{in} = (36.6)(0.25 \ \text{V}) \sin 1400t$$
$$= 9.15 \sin 1400t \ \text{V}$$

(i) There are several possible uses for the inductor. It might be included to limit voltage extremes, such as high-voltage spikes, which might damage the transistor when the amplifier is turned on or off. Alternatively, it might prevent AC current from being drawn across $R_C$, and in so doing, it holds $V_C$ at a constant value.

## 24. FIELD-EFFECT TRANSISTORS

*Field-effect transistors* (FETs), also known as *unipolar transistors*, are constructed from an n-type channel surrounded by a p-type gate (or vice versa) as shown in Fig. 64.19. (To distinguish them from the bipolar junction transistor, the leads are renamed *source, gate,* and *drain.*) Except for the JFETs, which have three leads, all FETs have a fourth lead connected to the substrate which may be called the *substrate, body, base,* or *bulk lead.* All FETs can be manufactured as n- and p-channel types.

There are two major categories of FETs—the *JFET* (*junction FET*) and the *MOSFET* (*metallic oxide semiconductor FET*), also known as the *IGFET* (*insulated gate FET*). Several MOSFET varieties exist, including the *vertical MOSFET* (*VMOSFET*). The short channels in a VMOSFET can sustain high currents (in the range of tens of amps or more), making them useful in power amplifier applications. VMOSFET and MOSFET operating characteristics are otherwise similar.

Unlike bipolar junction transistors that draw and are controlled by a base current, field-effect transistors draw negligible gate current.[17] The channel is controlled by the gate *voltage.*

## 25. JFETS

Junction field-effect transistors (JFETs) rely on a reverse-biased p-n junction for control. The mechanism of control is the width of the *depletion region* (which increases with increasing reverse bias). The thickness of the depletion region controls the resistance of the channel between the drain and the source. When the bottom of the depletion region meets the p⁻ substrate, the FET is said to be *pinched off.* (It then behaves similarly to the bipolar junction transistor.)

---

[17]A few nanoamps (insignificant in almost all cases) of gate current is passed due to surface leakage. This parameter is known as the *gate leakage current, $I_{GSS}$.*

***Figure 64.19** Field-Effect Transistors*

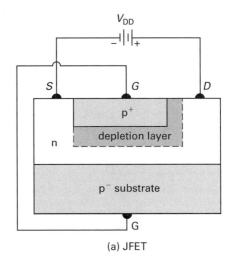

(a) JFET

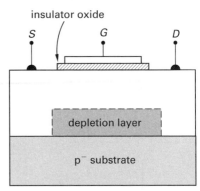

(b) MOSFET

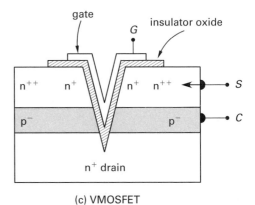

(c) VMOSFET

Figure 64.20 illustrates typical performance characteristics for an *n-channel JFET*. The curves are divided into the *resistive (ohmic) region, pinch-off region,* and *avalanche region* (not shown). For a fixed value of $V_{GS}$, the drain-source voltage separating the resistive and pinch-off regions is the *pinch-off voltage*, $V_P \approx V_{P0} + V_{GS}$. As Fig. 64.20 shows, there is a value for $V_{GS}$ for which no drain current flows. This is also referred to as the pinch-off voltage but is designated $V_{GS(OFF)}$. The drain current corresponding to the horizontal part of a curve (for

**Figure 64.20** *JFET Symbols and Characteristics*

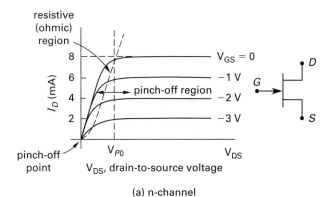

(a) n-channel

(b) p-channel

a given value of $V_{GS}$) is the *saturation current*, represented by $I_{DSS}$.

Some manufacturers do not provide characteristic curves, choosing instead to indicate only $I_{DSS}$ and $V_{GS(OFF)}$. The characteristic curves can be derived, as necessary, from *Shockley's equation*.

$$I_D = I_{DSS}\left(1 - \frac{V_{GS}}{V_P}\right)^2 \qquad 64.38$$

The *transconductance*, $g_m$, is defined for small-signal analysis by Eq. 64.39.

$$g_m = \frac{\Delta I_D}{\Delta V_{GS}} = \frac{i_D}{v_{GS}}$$

$$= \frac{-2I_{DSS}}{V_P}\left(1 - \frac{V_{GS}}{V_P}\right) = g_{mo}\left(1 - \frac{V_{GS}}{V_P}\right)$$

$$\approx \frac{A_V}{R_{out}} \qquad 64.39$$

The drain-source resistance can be obtained from the slope of the $V_{GS}$ characteristic.

$$r_d = r_{DS} = \frac{\Delta V_{DS}}{\Delta I_D} = \frac{v_{DS}}{i_D} \qquad 64.40$$

## 26. JFET PINCH-OFF VOLTAGE

The term "pinch-off voltage" and the symbol $V_P$ are ambiguous, as the actual pinch-off voltage in a circuit depends on the gate-source voltage, $V_{GS}$. When $V_{GS}$ is

zero, the pinch-off voltage is represented unambiguously by $V_{P0}$ (where the zero refers the value of $V_{GS}$). For other values of $V_{GS}$,

$$V_P = V_{P0} + V_{GS}$$

Some manufacturers do not adhere to this convention when reporting the pinch-off voltage for their JFETs. They may give a value for $V_{P0}$ and refer to it as $V_P$. The absence of a value for $V_{GS}$ implies that the values given is actually $V_{P0}$.

## 27. JFET BIASING

JFETs operate with a reverse-biased gate-source junction. The quiescent point is established by choosing $V_{GSQ}$. $I_D$ is then determined by Shockley's equation. (See Eq. 64.38.)

Figure 64.21 shows a JFET self-biasing circuit. Since the gate current is negligible, $I_D = I_S$, and the load line equation is

$$V_{DD} = I_S(R_D + R_S) + V_{DS}$$

$$= I_D(R_D + R_S) + V_{DS} \qquad 64.41$$

At the quiescent point, $V_{in} = 0$. From Kirchhoff's voltage law, around the input loop,

$$V_{GS} = -I_S R_S = -I_D R_S \qquad 64.42$$

**Figure 64.21** *Self-Biasing JFET Circuit*

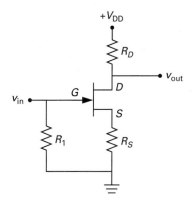

### Example 64.2

A JFET with the characteristics shown operates as a small-signal amplifier. The supply voltage is 24 V. A quiescent bias source current of 5 mA is desired at a bias voltage of $V_{DS} = 15$ V. Design a self-biasing circuit similar to Fig. 64.21.

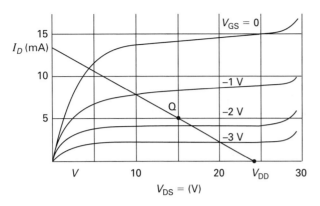

## Solution

Since the gate draws negligible current, $I_D = I_S$. At the quiescent point, $I_{GS} = -1.75$ V. From Eq. 64.42,

$$R_S = \frac{-V_{GS}}{I_D} = \frac{-(-1.75 \text{ V})}{0.005 \text{ A}} = 350 \; \Omega$$

From Eq. 64.41,

$$R_D = \frac{V_{DD} - V_{DS}}{I_S} - R_S = \frac{24 \text{ V} - 15 \text{ V}}{0.005 \text{ A}} - 350 \; \Omega$$
$$= 1450 \; \Omega$$

## 28. MOSFETS

The construction of a MOSFET was illustrated in Figure 64.19. The gate is completely insulated from the channel by a layer of silicon oxide. Unlike JFETs, MOSFETs may operate in the *depletion mode* (normally on current is reduced by the gate voltage) or *enhancement mode* (normally off current is increased by the gate voltage) as shown in Fig. 64.22. Shockley's equation (see Eq. 64.38) is valid, as is Eq. 64.39 for the transconductance. Characteristic curves are similar to those of the JFET.

**Figure 64.22** *MOSFET Symbols*

(a) n-channel (depletion)   (b) n-channel (enhancement)

(c) p-channel (depletion)   (d) p-channel (enhancement)

## 29. MOSFET BIASING

A typical MOSFET biasing circuit is illustrated in Figure 64.23. Since MOSFETs do not have a p-n junction between the gate and channel, biasing can be either forward or reverse. The polarity of the gate-source bias voltage depends on whether the transistor is to operate in the depletion mode or enhancement mode. As no gate current flows, the resistor $R_G$ is used merely to control the input impedance. When $R_G$ is very large, the input impedance is essentially $R_2$.

**Figure 64.23** *Typical MOSFET Biasing Circuit*

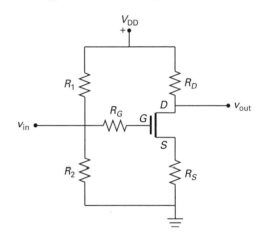

Together, $R_1$ and $R_2$ form a voltage divider. Since the gate current is zero, the voltage divider is unloaded. Therefore, the gate voltage is

$$V_G = V_{DD}\left(\frac{R_2}{R_1 + R_2}\right) = V_{GS} + I_S R_S \qquad 64.43$$

The load line equation is found from Kirchhoff's voltage law and is the same as for the JFET.

$$V_{DD} = I_S(R_D + R_S) + V_{DS} \qquad 64.44$$

## Example 64.3

A MOSFET is used in an amplifier as shown. All capacitors have zero impedance to AC signals. The no-signal drain current $(I_D)$ is 20 mA. The performance of the transistor is defined by

$$I_D = (14 + V_{GS})^2 \quad \text{[in mA]}$$

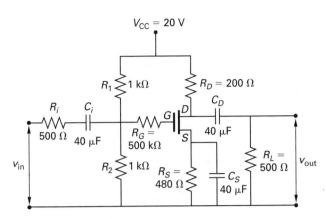

(a) What is the potential ($V_D$) at point D with no signal? (b) What is $V_S$ with no signal? (c) What is $V_{DSQ}$? (d) If $V_G = 0$, then what is $V_{GS}$? (e) What is the input impedance? (f) What is the voltage gain? (g) What is the output impedance?

*Solution*

(a) $\quad V_D = V_{CC} - I_D R_D$

$$= 20\text{ V} - (20 \times 10^{-3}\text{ A})(200\ \Omega) = 16\text{ V}$$

(b) The voltage drop between the source and the ground is through $R_S$. Since the gate draws negligible current, the drain and source currents are the same.

$$V_S = I_S R_S = (20 \times 10^{-3}\text{ A})(480\ \Omega) = 9.6\text{ V}$$

(c) The quiescent voltage drop across the drain-source junction is

$$V_{DSQ} = V_D - V_S = 16\text{ V} - 9.6\text{ V} = 6.4\text{ V}$$

(d) If the gate is at zero potential and the source is at 9.6 V potential, then

$$V_{GS} = V_G - V_S = 0 - 9.6\text{ V} = -9.6\text{ V}$$

(e) To simplify the circuit, short out the bias battery $V_{CC}$. Consider the capacitors as short circuits to AC signals.

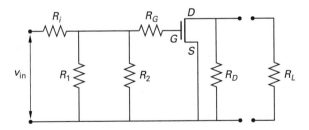

Since $R_G$ is so large, it effectively is an open circuit. The input impedance (resistance) is

$$R_{in} = R_i + R_1 \| R_2 = 500\ \Omega + \frac{(1000\ \Omega)(1000\ \Omega)}{1000\ \Omega + 1000\ \Omega}$$

$$= 1000\ \Omega$$

(f) An FET has properties similar to a vacuum tube. The voltage gain is normally calculated from the transconductance.

$$A_V \approx g_m R_{out}$$

The transconductance is not known, but it can be calculated from Eq. 64.39 and the performance equation.

$$g_m = \frac{dI_D}{dV_{GS}} = \frac{d(14 + V_{GS})^2 \times 10^{-3}\ \frac{\text{A}}{\text{mA}}}{dV_{GS}}$$

$$= (2)(14 + V_{GS}) \times 10^{-3}$$

Since $V_{GS} = -9.6$ V at the quiescent point,

$$g_m = (2)(14 - 9.6\text{ V}) \times 10^{-3} = 8.8 \times 10^{-3}\text{ S}$$

The resistance is a parallel combination of $R_D$, $R_L$, and $r_d$. The drain-source resistance, $r_d$, is normally very large and, since it was not given in this problem, is disregarded. Then,

$$R = R_D \| R_L = \frac{R_D R_L}{R_D + R_L} = \frac{(200\ \Omega)(500\ \Omega)}{200\ \Omega + 500\ \Omega} = 143\ \Omega$$

$$A_V = g_m R = (8.8 \times 10^{-3}\text{ S})(143\ \Omega) = 1.26$$

This is a small gain. If advantage is not being taken of other properties possessed by the circuit, the resistances should be adjusted to increase the voltage gain.

(g) Proceeding as in the solution to Ex. 64.1, part (g), the output impedance is $R_D = 200\ \Omega$.

## 30. POWER SEMICONDUCTORS

While current capacity is limited in most semiconductors, a *power semiconductor* is capable of controlling or passing large amounts of current. A *thyristor* is a device that exhibits positive feedback (*regenerative feedback*). A short pulse is sufficient to turn the thyristor on, and the feedback provides the holding current to keep it on. The internal power dissipated is very small compared to the power the thyristor controls. *Silicon-controlled rectifiers* (SCR's) and *Shockley diodes* are traditional types of thyristors. Modern power semiconductors include the *high-power bipolar junction transistor* (HPBT), power MOSFET, *gate turn-off* thyristor (GTO), and *insulated gate bipolar transistor* (IGBT), sometimes called a *conductivity-modulated field-effect transistor* (COMFET).

Power semiconductors can be categorized as trigger and control devices. *Control class devices* include the BJTs and FETs used in full-range amplifiers. *Trigger class devices*, such as GTOs, start conduction by some trigger input and, then, behave as diodes.

A *TRIAC* (*triode for alternating current*), also known as a bidirectional triode thyristor, is a device that conducts (large amounts of) current in either direction once it has

*Circuits*

been triggered by a small gate voltage of either polarity. Modern implementations are essentially two SCRs in parallel, though reversed in polarity, and with common gates.

## 31. OPERATIONAL AMPLIFIERS

An *operational amplifier* (*op amp*) is a high-gain DC amplifier. The characteristics of an ideal op amp are infinite positive gain, infinite input impedance, zero output impedance, and infinite bandwidth.[18] Since the input impedance is infinite, ideal op amps draw no current. (See Fig. 64.24.) An op amp has two terminals—an *inverting terminal* marked "−" and a *non-inverting terminal* marked "+". (The output of an ideal op amp is zero when the two inputs are at equal potentials.)

$$v_{\text{out}} = A_V(v_{\text{in}}^+ - v_{\text{in}}^-) \qquad 64.45$$

**Figure 64.24** *Ideal Operational Amplifier with Feedback*

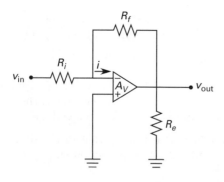

The gain from real op amps is not infinite but is, nevertheless, very large (i.e., $A_V = 10^5$ to $10^8$). Most op amps work in the ± 10–15 V range. Keeping $v_{\text{out}}$ less than (approximately) 10 V when the gain is $10^8$ means that $v_{\text{in}}$ is very small (0.1 $\mu$V). Thus, for the purpose of analyzing a circuit, the input is a *virtual ground*.

Depending on the method of feedback, the op amp can be made to perform a number of different operations, which are illustrated in Fig. 64.25. The gain of an op amp by itself is positive. An op amp with a negative gain is assumed to be connected in such a manner as to achieve negative feedback.[19]

## 32. INPUT IMPEDANCE OF OP AMP CIRCUITS

The input impedance of an op amp circuit is the ratio of the applied voltage to current drawn $(v_{\text{in}}/i_{\text{in}})$. In practical circuits, the input impedance is determined by

---

[18]This means that the gain is constant for all frequencies down to zero Hz.

[19]Op amps can be configured to perform many other functions, including phase shifting, clipping, voltage following, voltage-to-current conversion, current-to-voltage conversion, etc.

assuming that the op amp itself draws no current; any current drawn is by the remainder of the biasing and feedback circuits. Kirchhoff's voltage law is written for the signal-to-ground circuit.

### Example 64.4

The circuit shown uses an ideal op amp and receives a 1 $\mu$V signal. Find the (a) current through the feedback resistor, (b) voltage gain, and (c) current through the load resistor.

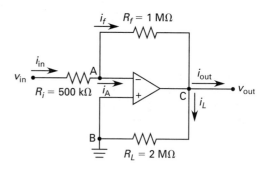

*Solution*

(a) Since the op amp is ideal, $i_A = 0$ and $v_A = v_B = 0$. The current through the input resistor is calculated from the voltage drop across it.

$$v_{\text{in}} - v_A = v_{\text{in}} - 0 = i_{\text{in}} R_i$$

$$i_{\text{in}} = \frac{v_{\text{in}}}{R_i} = \frac{1 \times 10^{-6} \text{ V}}{500 \times 10^3 \text{ } \Omega} = 2 \times 10^{-12} \text{ A}$$

However, $i_A = 0$, so $i_f = i_{\text{in}} = 2 \times 10^{-12}$ A.

(b) Similarly,

$$v_{\text{out}} = v_C = -i_f R_f$$

$$A_V = \frac{v_{\text{out}}}{v_{\text{in}}} = \frac{-i_f R_f}{i_{\text{in}} R_i} = \frac{-R_f}{R_i}$$

$$= \frac{-1 \times 10^6 \text{ } \Omega}{500 \times 10^3 \text{ } \Omega} = -2$$

(c) Since $v_A$ and $v_B$ are both zero, $v_C$ can be calculated in two ways.

$$v_C = i_f R_f = i_L R_L$$

$$i_L = i_f \left( \frac{R_f}{R_L} \right) = \left( 2 \times 10^{-12} \text{ A} \right) \left( \frac{1 \text{ M}\Omega}{2 \text{ M}\Omega} \right)$$

$$= 1 \times 10^{-12} \text{ A}$$

## 33. PHOTOVOLTAIC CELLS

A *photovoltaic cell* (*PV cell* or *solar cell*) develops a voltage from incident light, usually light in the visible region. Traditional PV cells are grown from silicon and (when price is no object) from gallium arsenide (GaAs) in crystalline form. Efficiency, as measured by electrical energy output as a fraction of solar energy input, is approximately 20–30% for single crystals and 15–20%

**Figure 64.25** *Operational Amplifier Circuits*

$$\frac{v_{out}}{v_{in}} = \frac{A}{1 + AH}$$

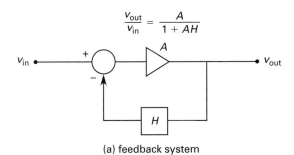

(a) feedback system

$$\frac{v_{out}}{v_{in}} = -\frac{R_f}{R_i}$$

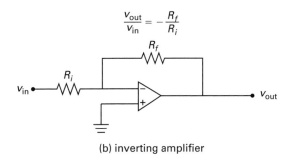

(b) inverting amplifier

$$\frac{v_{out}}{v_{in}} = \frac{R_f + R_i}{R_f}$$

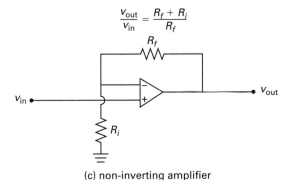

(c) non-inverting amplifier

$$v_{out} = -R_f \left( \frac{v_1}{R_1} + \frac{v_2}{R_2} + \frac{v_3}{R_3} \right)$$

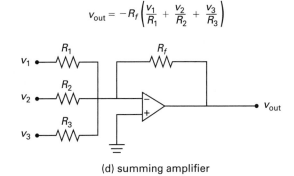

(d) summing amplifier

$$v_{out} = \frac{-1}{RC} \int v_{in} dt$$

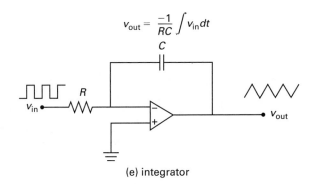

(e) integrator

$$v_{out} = -RC \frac{dv_{in}}{dt}$$

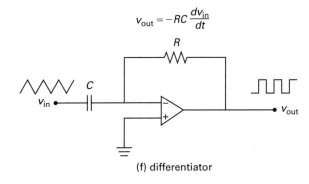

(f) differentiator

$$\frac{v_{out}}{v_{in}} = \frac{-R_f}{R_i(1 + j\omega R_f C)} \text{ [sinusoidal input]}$$

(g) low-pass filter

$$v_{out} = \frac{R_3}{R_2} v_2 - \frac{R_f}{R_1} v_1$$

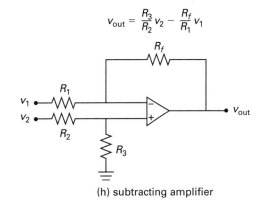

(h) subtracting amplifier

for polycrystalline cells.[20] Second generation thin semi-conductor films, such as copper indium (di)selenide (CIS) and cadmium telluride, are much less inexpensive but have efficiencies in the range of 3–15%.

---

[20]Approximately 1350 W/m² strike the earth's atmosphere, an amount known as the *solar constant*. However, reflection and absorption reduce this amount to approximately 540–950 W/m² at the earth's surface. 1000 W/m² is often used for initial estimates of performance and efficiency.

# 65 Pulse Circuits: Waveform Shaping and Logic

## Nomenclature

| | | |
|---|---|---|
| $E$ | error | V |
| $h$ | hybrid parameter | various |
| $I$ | current | A |
| $m$ | minterm | – |
| $M$ | maxterm | – |
| $n$ | integer | – |
| N | binary number | – |
| $R$ | resistance | $\Omega$ |
| $S$ | binary selection word | – |
| $V$ | voltage | V |

## Subscripts

| | |
|---|---|
| $B$ | base |
| $C$ | collector |
| CC | collector supply |
| DD | drain supply |
| $E$ | emitter |
| $f$ | feedback |
| F | forward (DC component) |
| $i$ | input |
| $O$ | thermally generated |
| ref | reference |
| $S$ | saturation |
| sat | saturation |

## 1. DIGITAL CIRCUITS

Signals in a digital circuit are limited to two voltages only, corresponding to on and off (high and low, etc.) states. The type of logic circuit determines if the on condition is a high or a low voltage and if the off condition is a high or low voltage. (It is traditional to refer to the states as 1 and 0. The actual voltages are known as the *logic levels*.) An *off circuit* has an infinite resistance and passes very little current, known as *the cut-off condition*. An *on circuit* has low resistance and passes a large current, known as the *saturated condition*. Very little current flows below the *threshold voltage*; full current flows above the *saturation voltage*.[1]

One of the logic levels will represent a *true* or 1 condition. One of the voltage levels will represent a *false* or 0 condition. If the 1 condition voltage level is higher than the 0 condition voltage level, then the logic is called *positive logic* (e.g., 1 = 5 V and 0 = 0 V, or 1 = 0 V and 0 = −3.3 V). In positive logic, a 1 is also a high. If the 1 condition voltage level is lower than the 0 condition voltage level, then the logic is called *negative logic* (e.g., 1 = 0 V and 0 = 5 V, or 1 = −3.3 V and 0 = 0 V). In negative logic a 1 is also a low. Positive logic is more common in most applications.

The actual values of the logic levels are forced to different values and might be 0 V and +5 V, −5 V and +5 V, −5 V and +3.5 V, etc. Newer technologies tend to use voltages closer to 0 V, which usually results in less power consumption and less heat dissipation. The logic levels are specified in ranges of voltages that are acceptable for a 1 and for a 0. For example, a device might have a 1 defined as greater than 2 V and a 0 as less than 0.8 V. Any voltage between 0.8 and 2.0 V is an undefined state, meaning the designer of the device gives no assurances about whether the output of the device will be 1 or 0 or some undefined voltage in between. The upper voltage for the 1 and the lower voltage for the 0 have limits to avoid damaging or destroying the device. The circuit designer must keep the voltages at the inputs to the logic devices within the defined voltage ranges at all times.

---

[1]For example, $V_{\text{BET}} = 0.5$ V and $V_{\text{BES}} = 0.7$ V for a silicon BJT transistor. The corresponding values for germanium are 0.1 V and 0.3 V. For MOSFETs, the threshold voltage is the lowest gate-to-source at which the channel becomes conductive.

Circuits

## 2. INTEGRATED CIRCUITS

Integrated circuits combine active and passive elements and their interconnections onto a single chip. In *monolithic construction*, all components are formed on the same silicon substrate using the production processes of thermal oxidation (oxide growth), photo-lithography, etching, ion implantation (masked impurity diffusion and chemical vapor deposition, CVD), thermal redistribution, insulation and epitaxial growth, and metallization (evaporization or sputtering).[2] Resistors, capacitors, diodes, and transistors can be produced on a single substrate by these processes.

The *integration level* is the number of logic gates on a single chip. With *small-scale integration* (SSI), there are fewer than 12 logic gates in a device. With *medium-scale integration* (MSI), there are fewer than 100 logic gates in a device. With *large-scale integration* (LSI), there are more than 1000 logic gates. With *very large-scale integration* (VLSI), the integration level is in the tens and hundreds of thousands. With *ultra large-scale integration* (ULSI), the integration level is greater than one million.

By convention, the term *half-pitch* is used by the semiconductor industry to indicate half of the average minimum required separation (i.e., the *pitch*) between adjacent features. Terms such as "45 nm process," "45 nm geometry," and "45 nm technology" describe the half-pitch.[3] Half-pitch depends on many variables, including the wavelength of light used during manufacturing. Using *immersion lithography*[4] and argon-fluoride lasers producing *deep ultraviolet light* (DUV) with wavelengths between 248 nm and 193 nm, half-pitch values of 45 nm are routine, and sub-20 nm values are possible. Thus, the half-pitch is considerably less than the wavelength of the light used. *Geometry* and *isolation* refer to the actual separation distance between features. In order to prevent bleed-over of current between adjacent features, isolation may need to be greater than twice the half-pitch.

The size of the smallest feature that can be created using photolithography is known as both *minimum feature size* and *critical dimension* (CD). For example, a transistor will be created as different areas on a wafer. Those areas can be no narrower than the CD.

The *fan-out* is the number of compatible logic inputs (or other devices) that can be driven (i.e., connected to the integrated circuit's output leads) without exceeding its output current specifications. The *fan-in* is the number of inputs the gate has.

---

[2]*Epitaxial growth* produces a crystalline substance that continues the atomic structure of the substrate.

[3]In practice, a term such as "45 nm" refers to the capability of the process, not the size of a feature. This has led to half-pitch values being used to refer to feature size, feature separation, and even device (e.g., transistor) size.

[4]In immersion lithography, the air gap between the light source lens and the wafer is replaced with water or another liquid. The liquid's index of refraction helps bend and focus the image.

In *hybrid construction*, two or more basically different types of components are combined in a circuit to perform similar functions. For example, hybrid circuits may integrate circuits and discrete components. The components are usually mounted on a substrate such as a printed circuit board, ceramic wafer, or alumina ($Al_2O_3$) plate.

## 3. LOGIC FAMILIES

The circuit elements and manner in which they are combined determine the integrated circuit family. *Passive logic circuits* have no transistors (i.e., have only resistors and diodes), compared to *active logic circuits*, which have transistors.

Passive logic circuits are implemented as discrete devices in simple circuits involving separate resistors and diodes. Simple logic functions can be implemented with diodes controlling the voltage of the output. The output can be connected such that if any of the diodes are forward biased, the output will be a high for a logical OR. The output can be connected such that if all of the diodes are reverse biased, the output will be a high for a logical AND. Passive logic is not effective for complicated logical functions and is not effective for miniaturization.

Active logic circuits always include transistors. The transistors will either be "off" or "on," in contrast to the amplifier circuits where the transistor is operated somewhere in between. Much more complicated logic functions are possible only with active logic. Active logic is very effective for miniaturization.

One of the earliest logic families was *diode-transistor logic* (DTL). *High-threshold logic* (HTL) is similar to DTL but uses higher voltages and better diodes to produce greater noise immunity. *Variable-threshold logic* (VTL) is a variation of DTL with adjustable noise immunity. *Direct-coupled transistor logic* (DCTL) is an unsophisticated connection of transistors with poor immunity to noise and low logic levels. Since there are no resistors, DCTL has low power dissipation. *Resistor-transistor logic* (RTL) is similar to DCTL, with the addition of resistors to limit current hogging. *Resistor-capacitor transistor logic* (RCTL) is a low-cost compromise between speed and power. *Transistor-transistor logic* (TTL or $T^2L$) is the fastest of all logic families, using transistors operating in the saturated region. The fastest of all logic families using bipolar junction transistors is *emitter-coupled logic* (ECL), also known as *current mode logic* (CML). ECL transistors do not operate in the saturation region. ECL circuits consume more power than TTL.

*Metallic oxide semiconductor* (MOS) circuitry contains MOSFETs. It requires few resistors and has high packing density. *Complementary construction* uses both n-channel and p-channel transistors in the same circuit. NMOS logic uses n-channel MOSFETs, while PMOS uses p-channel MOSFETs. (NMOS is approximately six times slower than PMOS.) *Complementary metallic oxide semiconductor* (CMOS), also known as *complementary transistor logic* (CTL), circuits have high packing density and reliability. They are capable of higher switching speeds than MOS.

Circuitry based on bipolar transistors switches faster, offers greater current drive, and is more suitable for analog applications than CMOS circuitry.[5] On the other hand, CMOS consumes less power, dissipates less heat, and has a higher packing density. Bipolar and CMOS circuitry are combined in *bipolar complementary metallic oxide semiconductor* (biCMOS) integrated circuits.

## 4. RESISTOR-TRANSISTOR LOGIC (RTL)

The basic gate of *resistor-transistor logic* (RTL) is the NOR gate shown in Fig. 65.1(a). Each transistor forms a NOT function; if the input to the base is zero, the transistor output is essentially an open circuit. If the input saturates the base, the transistor output is essentially a short circuit. When combined as in Fig. 65.1(a), or with any number of transistors connected to the output, each transistor can short the output to ground if its base is saturated. Figure 65.1(b) shows the logic symbol for positive logic, and Fig. 65.1(c) is the truth table for the gate. If positive logic is used, L = 0 and H = 1. If negative logic had been used, L = 1 and H = 0. If negative logic had been used, this gate would be a NAND gate.

## 5. DIODE-TRANSISTOR LOGIC (DTL)

The basic gate of *diode-transistor logic* (DTL) is the NAND gate shown in Fig. 65.2(a). Two diodes are shown connected to point B, but more diodes can be connected for increased fan-in. When either of the diodes, $D$, is forward biased (connected to a low voltage) the result is in a low voltage at point B. Diodes $D_1$ and $D_2$ will be reverse biased; the transistor output will be essentially an open circuit, and the output will be high. Only when all the input diodes are reverse biased will $D_1$ and $D_2$ be forward biased, and the base will be driven into saturation. The transistor output will essentially be a short circuit, and the output will be a low voltage. Figure 65.2(b) shows the logic symbol for positive logic, and Fig. 65.2(c) is the truth table for the gate. If negative logic had been used, this gate would be a NOR gate.

## 6. TRANSISTOR-TRANSISTOR LOGIC (TTL OR T²L)

The basic gate of *transistor-transistor logic* (TTL or T²L) is the NAND gate shown in Fig. 65.3(a). The basic TTL NAND gate can be considered a modified DTL NAND gate. In TTL, the input diodes, $D$, and the base diodes, $D_1$ and $D_2$, in Fig. 65.3 are replaced by a multi-input transistor, $Q_1$. A fan-in of two is shown, but a

---

[5]CMOS is equivalent to TTL in speed but slower than ECL.

**Figure 65.1** *Resistor-Transistor Logic*

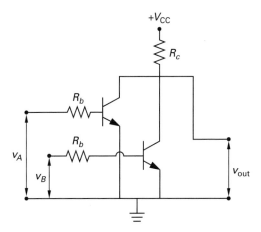

(a) basic gate: NOR

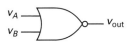

(b) logic symbol

| $v_A$ | $v_B$ | $v_{out}$ |
|---|---|---|
| L | L | H |
| L | H | L |
| H | L | L |
| H | H | L |

(c) voltage truth table

greater fan-in is possible with a transistor with multiple emitters. When at least one input is low voltage, the emitter of $Q_1$ is forward biased. Assuming $Q_2$ is cut off, the current supplied to the collector of $Q_1$ is the reverse saturation current, $I_{EBO}$, of the emitter-base junction of $Q_2$. Since this current is small, $I_{B1} > I_{EBO}/h_{FE}$, and $Q_1$ is in saturation. The voltage at node $B_2$ is then equal to the input voltage, $v_A$ or $v_B$, summing with $V_{CE,sat}$, about 0.4 V to 0.8 V. This voltage is not enough to forward bias $Q_2$ and $Q_3$, since about 0.7 V is required for each. $Q_3$ is cut off, and the output is high ($+V_{CC}$). When all the inputs are high, the emitter-base junction of $Q_1$ is reverse biased. $Q_1$ is cut off, and the base-collector junction is forward biased by approximately $+V_{CC}$. This voltage is high enough, about 2.1 V, to forward bias the base-collector junction of $Q_1$ and the base-emitter junctions of $Q_2$ and $Q_3$. Thus, $Q_3$ is driven into saturation, and the output is low. Figure 65.3(b) shows the logic symbol for positive logic, and Fig. 65.3(c) is the truth table for the gate. If negative logic had been used, this gate would be a NOR gate.

TTL logic is the fastest saturating-logic gate. The fall time of TTL circuits is improved by reducing the collector current in the saturation state. This is accomplished using a *totem-pole configuration*, also called *active*

**Figure 65.2** *Diode-Transistor Logic*

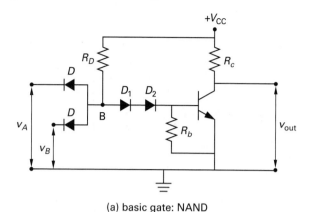

(a) basic gate: NAND

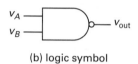

(b) logic symbol

| $v_A$ | $v_B$ | $v_{out}$ |
|---|---|---|
| L | L | H |
| L | H | H |
| H | L | H |
| H | H | L |

(c) voltage truth table

**Figure 65.3** *Transistor-Transistor Logic*

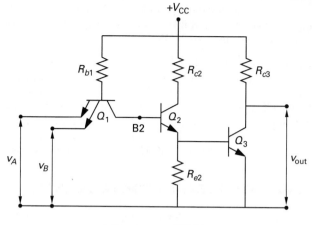

(a) basic gate: NAND

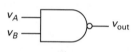

(b) logic symbol

| $v_A$ | $v_B$ | $v_{out}$ |
|---|---|---|
| L | L | H |
| H | L | H |
| L | H | H |
| H | H | L |

(c) voltage truth table

**Figure 65.4** *TTL Totem-Pole Configuration*

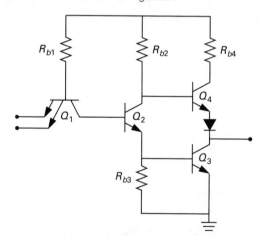

*pull-up*, which uses an additional transistor and diode in series with $R_{c3}$.[6] Fig. 65.4 shows the totem-pole configuration for a TTL gate. Additionally, the collector of $Q_3$ can be left open, in which case it is called an *open-collector TTL gate*. The open-collector TTL gate is used with wired logic or collector logic, since the collector outputs are wired to the next gate's output without intervening circuitry.

## 7. EMITTER-COUPLED LOGIC (ECL)

The basic gate of *emitter-coupled logic* (ECL) is the OR-NOR gate shown in Fig. 65.5(a). The basic building block of the family is the *differential amplifier* ($Q_2$ and $Q_3$ of Fig. 65.5). The transistors do not saturate. Instead, they remain in the active region at the input voltages $v_A$ and $v_B$, operating above and below (high and low) relative to the reference voltage, $V_R$. Since the emitter current through $R_e$ remains approximately constant, when both inputs are low (i.e., below $V_R$) the emitter currents from $Q_1$ and $Q_2$ are minimal, and the emitter current of $Q_3$ is minimal. This causes the voltage output at $v_{OR}$ to be high and the voltage output at $v_{NOR}$ to be low. Since there are two outputs with reverse logic, this gate works with either positive or negative logic as

an OR function. Figure 65.5(b) shows the logic symbol for positive logic, and Fig. 65.5(c) is the truth table for the gate. If negative logic had been used, $v_{NOR}$ would be the OR output, and $v_{OR}$ would be the NOR output.

ECL improves on the propagation delay of TTL by not driving the transistors into saturation. This minimizes the switching time, since charge carriers do not have to be cleared from the base junction. However, the power

---

[6]The configuration is called "totem-pole" because the additional transistor sits on top of $Q_3$ in the diagram.

consumption of ECL is greater than TTL. ECL is also called *current-mode logic* (CML).

**Figure 65.5** *Emitter-Coupled Logic*

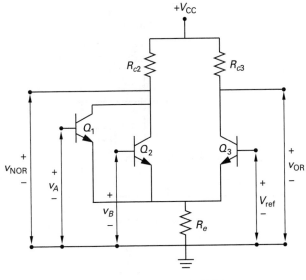

(a) basic gate: OR–NOR

(b) logic symbol

| $v_A$ | $v_B$ | $v_{OR}$ | $v_{NOR}$ |
|---|---|---|---|
| L | L | L | H |
| H | L | H | L |
| L | H | H | L |
| H | H | H | L |

(c) voltage truth table

## 8. MOS LOGIC

The basic gate of MOS logic is the NAND gate shown in Fig. 65.6(a). The basic building block for the family is the *inverter* ($Q_1$ and $Q_2$ in Fig. 65.6(a)). A fan-in of two is shown, but multiple inputs are possible. When either input is logic 0 (low voltage), the associated transistor $Q_2$ or $Q_3$, is off. No current flows, and $V_{out}$ is high. When both of the inputs are logic 1 (high voltage), both transistors are on. Current flows, and the output voltage is low. Figure 65.6(b) shows the logic symbol for positive logic, and Fig. 65.6(c) is the truth table for the gate. If negative logic had been used, this gate would then be a NOR gate.

MOS logic is simple and requires no external resistors or capacitors, making it well-suited for realization in integrated circuit form. The NMOS positive logic form shown in Fig. 65.6(a) draws power in only one state. The PMOS positive logic (NOR) draws power in three

states. In both cases, the power drawn is less than for other logic families, allowing for much higher device densities. NMOS is faster, since electron mobility in n-type material is greater than hole mobility in p-type material. PMOS is less expensive. Technology advances have reduced the propagation delays of MOS gates to values comparable to other logic families.

**Figure 65.6** *MOS Logic*

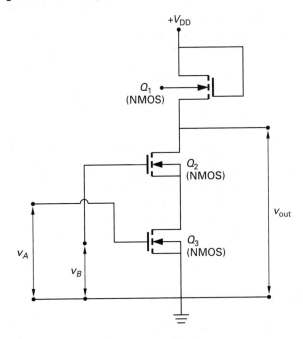

(a) basic gate: NAND

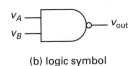

(b) logic symbol

| $v_A$ | $v_B$ | $v_{out}$ |
|---|---|---|
| L | L | H |
| L | H | H |
| H | L | H |
| H | H | L |

(c) voltage truth table

## 9. CMOS LOGIC

The basic gate of CMOS logic may be either the NOR gate or the NAND gate as shown in Fig. 65.7(a) and Fig. 65.7(b). A fan-in of two is shown, but a greater fan-in is possible with more identical transistors. For the NOR gate in Fig. 65.7(a), when both the inputs are low,

Circuits

$Q_1$ and $Q_2$ are on, and $Q_3$ and $Q_4$ are off. The output is high due to the connection to the drain supply via the conducting transistors $Q_1$ and $Q_2$. When either or both of the inputs are low, one or both of the upper transistors ($Q_1$ and/or $Q_2$) are off. When one or both of the lower transistors ($Q_3$ and/or $Q_4$) are on, there is a connection from the output to ground via the conducting channel, and the output is low. For the NAND gate in Fig. 65.7(b), when one or both the inputs are low, one or both upper transistors will be on ($Q_1$ and/or $Q_2$) connecting the output to the drain supply voltage. The output is high. If both inputs are high, the lower series transistors, $Q_3$ and $Q_4$, are on and connect the output to ground. The output is low. Figure 65.7(c) shows the logic symbol for positive logic, and Fig. 65.7(d) is the truth table for the gate. If negative logic had been used, the NOR gate would be a NAND gate, and the NAND gate would be a NOR.

CMOS logic requires a single supply voltage only and is relatively easy to fabricate. The only input current flow is that required to change the gate-channel capacitances and any leakage through the off transistor. Thus, the power consumption is extremely low. Because of the low power (and thereby low current), the fan-out to other CMOS gates is very high. Speeds are comparable to ECL. CMOS has the best overall properties of any logic family.

## 10. BOOLEAN ALGEBRA

The rules of *Boolean algebra* are used to write and simplify expressions of binary variables (i.e., variables constrained to two values). The basic laws governing Boolean variables are listed as follows.

commutative:

$$A + B = B + A \qquad \text{65.1}$$

$$A \cdot B = B \cdot A \qquad \text{65.2}$$

associative:

$$A + (B + C) = (A + B) + C \qquad \text{65.3}$$

$$A \cdot (B \cdot C) = (A \cdot B) \cdot C \qquad \text{65.4}$$

distributive:

$$A \cdot (B + C) = (A \cdot B) + (A \cdot C) \qquad \text{65.5}$$

$$A + (B \cdot C) = (A + B) \cdot (A + C) \qquad \text{65.6}$$

absorptive:

$$A + (A \cdot B) = A \qquad \text{65.7}$$

$$A \cdot (A + B) = A \qquad \text{65.8}$$

*De Morgan's theorems* are

$$\overline{(A + B)} = \overline{A} \cdot \overline{B} \qquad \text{65.9}$$

$$\overline{A \cdot B} = \overline{A} \cdot \overline{B} \qquad \text{65.10}$$

**Figure 65.7** *CMOS Logic*

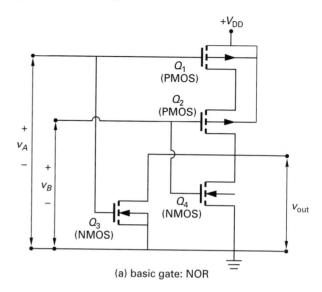

(a) basic gate: NOR

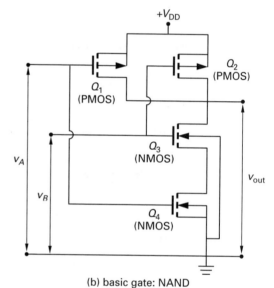

(b) basic gate: NAND

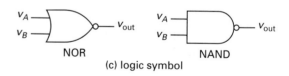

(c) logic symbol

| $v_A$ | $v_B$ | $v_{out}$ |
|---|---|---|
| L | L | H |
| L | H | L |
| H | L | L |
| H | H | L |

NOR

| $v_A$ | $v_B$ | $v_{out}$ |
|---|---|---|
| L | L | H |
| L | H | H |
| H | L | H |
| H | H | L |

NAND

(d) voltage truth tables

The following basic identities are used to simplify Boolean expressions.

$$0 + 0 = 0 \qquad 65.11$$

$$0 + 1 = 1 \qquad 65.12$$

$$1 + 0 = 1 \qquad 65.13$$

$$1 + 1 = 1 \qquad 65.14$$

$$0 \cdot 0 = 0 \qquad 65.15$$

$$0 \cdot 1 = 0 \qquad 65.16$$

$$1 \cdot 0 = 0 \qquad 65.17$$

$$1 \cdot 1 = 1 \qquad 65.18$$

$$A + 0 = A \qquad 65.19$$

$$A + 1 = 1 \qquad 65.20$$

$$A + A = A \qquad 65.21$$

$$A + \overline{A} = 1 \qquad 65.22$$

$$A \cdot 0 = 0 \qquad 65.23$$

$$A \cdot 1 = A \qquad 65.24$$

$$A \cdot A = A \qquad 65.25$$

$$A \cdot \overline{A} = 0 \qquad 65.26$$

$$\overline{0} = 1 \qquad 65.27$$

$$\overline{1} = 0 \qquad 65.28$$

$$\overline{\overline{A}} = A \qquad 65.29$$

### Example 65.1

Simplify and write the truth table for the following network of logic gates.

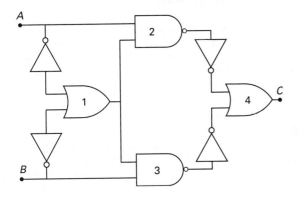

Determine the inputs and output of each gate in turn.

| gate | inputs | output |
|---|---|---|
| 1 | $\overline{A}, \overline{B}$ | $(\overline{A} + \overline{B})$ |
| 2 | $A, (\overline{A} + B)$ | $\overline{A \cdot (\overline{A} + \overline{B})}$ |
| 3 | $B, (\overline{A} + B)$ | $\overline{B \cdot (\overline{A} + \overline{B})}$ |
| 4 | $A \cdot (\overline{A} + \overline{B}), B \cdot (\overline{A} + \overline{B})$ | $A \cdot (\overline{A} + \overline{B}) + B \cdot (\overline{A} + \overline{B})$ |

Simplify the output of gate 4.

$$A \cdot (\overline{A} + \overline{B}) + B \cdot (\overline{A} + \overline{B}) \quad \text{[original]}$$

$$A \cdot \overline{A} + A \cdot \overline{B} + B \cdot \overline{A} + B \cdot \overline{B} \quad \text{[distributive]}$$

$$A \cdot \overline{B} + B \cdot \overline{A} \quad \text{[since } A \cdot \overline{A} = 0\text{]}$$

$$A \oplus B \quad \text{[definition]}$$

The truth table is

| $A$ | $B$ | $C$ |
|---|---|---|
| 0 | 0 | 0 |
| 0 | 1 | 1 |
| 1 | 0 | 1 |
| 1 | 1 | 0 |

## 11. MINTERMS AND MAXTERMS

A Boolean expression can be written as the *sum of products* (SOP) of all the possible combinations of products of the variables that give a true result. Products that give the true results are *minterms*. A Boolean expression can be written as the *product of sums* (POS) of all the possible combinations of sums of the variables that give a false result. These sums that give the false results are *maxterms*. An *implicant* is a Boolean term, either in sum or product form, which contains one or more maxterms or minterms of a Boolean function. A *prime implicant* is an implicant which is not entirely contained within another implicant in the Boolean function. An *essential prime implicant* is a prime implicant which contains a minterm or maxterm not contained in any other prime implicant. A Boolean expression is minimized when all the implicants are both prime and essential.

A Boolean function can be expressed as a sum of the minterms that are true, called a *canonical sum of products* (SOP) form. In SOP, conditions that make the output true are ORed together to define the function. A Boolean function can also be expressed as the product of the maxterms that are false, called a *canonical product of sums* (POS) form. In SOP, conditions that make the function false are inverted (assume the function is true everywhere except where that term makes it false) and then ANDed with other terms. A function in SOP form is often represented as a minterm list, and a function in POS form is often represented as a maxterm list. Table 65.1 shows the minterms and maxterms for a three-variable function.

A sum of minterms (SOP) for a function of $A$, $B$, $C$, and $D$ is

$$F(ABCD) = \sum m(h, i, j, ...)$$
$$= m_h + m_i + m_j + ... \qquad 65.30$$

**Table 65.1** *Minterms and Maxterms*

| decimal row number | binary input combinations $ABC$ | minterm (product term) | maxterm (sum term) |
|---|---|---|---|
| 0 | 000 | $m_0 = \overline{A}\,\overline{B}\,\overline{C}$ | $M_0 = A + B + C$ |
| 1 | 001 | $m_1 = \overline{A}\,\overline{B}C$ | $M_1 = A + B + \overline{C}$ |
| 2 | 010 | $m_2 = \overline{A}B\overline{C}$ | $M_2 = A + \overline{B} + C$ |
| 3 | 011 | $m_3 = \overline{A}BC$ | $M_3 = A + \overline{B} + \overline{C}$ |
| 4 | 100 | $m_4 = A\overline{B}\,\overline{C}$ | $M_4 = \overline{A} + B + C$ |
| 5 | 101 | $m_5 = A\overline{B}C$ | $M_5 = \overline{A} + B + \overline{C}$ |
| 6 | 110 | $m_6 = AB\overline{C}$ | $M_6 = \overline{A} + \overline{B} + C$ |
| 7 | 111 | $m_7 = ABC$ | $M_7 = \overline{A} + \overline{B} + \overline{C}$ |

A product of maxterms (POS) for a function of $A$, $B$, $C$, and $D$ is

$$F(ABCD) = \sum M(h, i, j, \ldots)$$
$$= M_h + M_i + M_j + \ldots \qquad 65.31$$

### Example 65.2

A Boolean function, $F$, is defined by the truth table shown. Find (a) the minterms, (b) the maxterms, (c) the SOP representation, (d) the POS representation, (e) the expressions that are SOP implicants, (f) the expressions that are not implicants, (g) the implicants that are not prime, and (h) the implicants that are both prime and essential.

| $X$ $Y$ $Z$ | $F$ |
|---|---|
| 0 0 0 | 0 |
| 0 0 1 | 1 |
| 0 1 0 | 0 |
| 0 1 1 | 1 |
| 1 0 0 | 1 |
| 1 0 1 | 0 |
| 1 1 0 | 0 |

*Solution*

(a) and (b) The minterms and maxterms can be derived directly from the conditions that make the function true and false, respectively.

| $X$ $Y$ $Z$ | $F$ | minterms | maxterms |
|---|---|---|---|
| 0 0 0 | 0 | | $X + Y + Z$ |
| 0 0 1 | 1 | $\overline{X}\cdot\overline{Y}\cdot Z$ | |
| 0 1 0 | 0 | | $X + \overline{Y} + Z$ |
| 0 1 1 | 1 | $\overline{X}\cdot Y\cdot Z$ | |
| 1 0 0 | 1 | $X\cdot\overline{Y}\cdot\overline{Z}$ | |
| 1 0 1 | 0 | | $\overline{X} + Y + \overline{Z}$ |
| 1 1 0 | 0 | | $\overline{X} + \overline{Y} + Z$ |
| 1 1 1 | 1 | $X\cdot Y\cdot Z$ | |

The function can be represented either as the SOP of the minterms that are true or as the POS of its maxterms that are false, and both expressions are equivalent.

(c) $F = \overline{X}\cdot\overline{Y}\cdot Z + \overline{X}\cdot Y\cdot Z + X\cdot\overline{Y}\cdot\overline{Z} + X\cdot Y\cdot Z$ [minterm]

(d) $F = (X + Y + Z) \cdot (X + \overline{Y} + Z) \cdot (\overline{X} + Y + \overline{Z})$
$\cdot (\overline{X} + \overline{Y} + Z)$ [maxterms]

(e) Using the distributive theorem $A\cdot B + A\cdot\overline{B}$ $= A\cdot(B + \overline{B}) = A\cdot 1 = A$, the first and second terms of the minterm form can be combined into $\overline{X}\cdot\overline{Y}\cdot Z$ $+ \overline{X}\cdot Y\cdot Z = \overline{X}\cdot Z$, and the second and fourth terms can be combined into $\overline{X}\cdot Y\cdot Z + X\cdot Y\cdot Z = Y\cdot Z$. The result is

$$F = \overline{X}\cdot Z + X\cdot\overline{Y}\cdot\overline{Z} + Y\cdot Z$$

All of the following expressions are implicants in the SOP form.

$$\overline{X}\cdot\overline{Y}\cdot Z, \ \overline{X}\cdot Y\cdot Z, \ X\cdot\overline{Y}\cdot\overline{Z}, \ X\cdot Y\cdot Z, \ \overline{X}\cdot Z, \ Y\cdot Z$$

(f) The following are NOT implicants in the SOP form because they are not true for this Boolean function.

$$X\cdot Y\cdot Z, \ \overline{X}\cdot Y\cdot\overline{Z}, \ X\cdot\overline{Y}\cdot Z, \ X\cdot Y\cdot\overline{Z}$$

(g) The following implicants are not prime because they are contained entirely in the implicants.

$$\overline{X}\cdot Z, \ Y\cdot Z, \ \overline{X}\cdot\overline{Y}\cdot Z, \ \overline{X}\cdot Y\cdot Z, \ X\cdot Y\cdot Z$$

(h) The remaining implicants are both prime and essential, so the expression is in its simplest form.

$$F = \overline{X}\cdot Z + X\cdot\overline{Y}\cdot\overline{Z} + Y\cdot Z$$

## 12. KARNAUGH MAP

A *Karnaugh map* (*K-map*) is a graphical representation of a Boolean function. There is a similar representation called a *Veitch diagram.*[7] Each adjacent row and column of the K-map changes by only one variable. The K-map allows visual identification of the essential prime implicants of a function. The minterms or maxterms next to each other in the K-map have both the true and false conditions of one variable. The top row is next to the bottom row, the left column is next to the right column, and the corners are next to the opposing vertical and horizontal corners. Two cells next to each other eliminate one variable. Four cells in a row, column, or square eliminate two variables, and eight cells together eliminate three variables.

Figure 65.8 is a K-map for four variables. The arrangement of the variables is critical to filling out the K-map correctly. The variable A is the *most significant bit*

---

[7]The distinction between the K-map and Veitch diagram is subtle and not important to the method discussed here.

(MSB), and D is the *least significant bit* (LSB). If the minterm $\overline{A}\cdot B\cdot C\cdot D$ is true, then the minterm $0111 = 7$ is true, so a 1 would be put in the cell with $m_7$.

**Figure 65.8** *Karnaugh Map for Four Variables*

| AB\CD | 00 | 01 | 11 | 10 |
|---|---|---|---|---|
| 00 | $m_0$ | $m_1$ | $m_3$ | $m_2$ |
| 01 | $m_4$ | $m_5$ | $m_7$ | $m_6$ |
| 11 | $m_{12}$ | $m_{13}$ | $m_{15}$ | $m_{14}$ |
| 10 | $m_8$ | $m_5$ | $m_{11}$ | $m_{10}$ |

**Example 65.3**

Minimize the following Karnaugh map using SOP and POS. (The following is segment "a" of the seven-segment display, the top bar.) An X in a cell indicates a "don't care."

| AB\CD | 00 | 01 | 11 | 10 |
|---|---|---|---|---|
| 00 | 1 | 0 | 1 | 1 |
| 01 | 0 | 1 | 1 | 1 |
| 11 | X | X | X | X |
| 10 | 1 | 1 | X | X |

*Solution*

A "don't care" is a condition that is never used. As in this case, the seven-segment display only displays numbers between 0 and 9, so 10 through 15 do not occur. A "don't care" can also be a condition where the inputs can occur, but the output is not used, so it does not matter.

The objective in SOP is to cover all the 1's using as few groupings as possible and groupings that are as large as possible. The "don't cares" can be assumed to be 1 or 0 to make groupings, but once the state of the "don't care" is chosen for one grouping, it must remain the same.

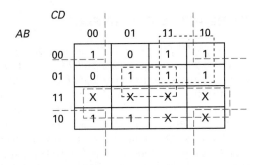

Minimize using SOP.

The bottom two rows can be grouped with the two 1's and all the "don't cares" if the "don't cares" are all assumed to be 1. This grouping has all combinations of

C and D, B is both 0 and 1, and A is always 1, so for this group, the implicant is A.

The two columns on the right can be grouped together. This grouping has all combinations of A and B, D is both 0 and 1, and C is always 1, so the implicant for this group is C.

The four squares in the center can be grouped together. Even though one of the 1's is covered by the implicant C, it is still beneficial to make the grouping bigger by covering that 1 again. This grouping has A both 0 and 1, C is both 0 and 1, C is always 1, and B is always 1, so the implicant for this group is $B \cdot D$.

There is one minterm in the upper left corner that is not part of any of the groupings. This corner can be grouped with the other three corners. This grouping has A both 0 and 1, C is both 0 and 1, C is always 0, and B is always 0, so the implicant for this group is $\overline{B} \cdot \overline{D}$.

The final form for the expression is

$$a = A + C + B\cdot D + \overline{B}\cdot\overline{D}$$

Minimize using POS.

| AB\CD | 00 | 01 | 11 | 10 |
|---|---|---|---|---|
| 00 | 1 | 0 | 1 | 1 |
| 01 | 0 | 1 | 1 | 1 |
| 11 | X | X | X | X |
| 10 | 1 | 1 | X | X |

The 0 in the first row, second column is surrounded by 1's on all sides, so it must be represented as a single maxterm. A is 0, B is 0, C is 0, and D is 1, so the implicant is the inverse, or $A + B + C + \overline{D}$. The 0 in the second row, first column can be grouped with the "don't care" below it. A is both 0 and 1, B is 1, C is 0, and D is 0, so the implicant is the inverse, or $\overline{B} + C + D$.

The final form for the expression is

$$a = (A + B + C + \overline{D})\cdot(\overline{B} + C + D)$$

## 13. LOGIC GATES

A *gate* performs a logical operation on one or more inputs. The inputs (labeled A, B, C, etc.) and output are limited to the values of 0 or 1. A listing of the output values for all possible input values is known as a *truth table*. Table 65.2 combines the symbols, names, and truth tables for the most common gates.

The appearance of the gate symbols for the NAND and NOR functions reflect De Morgan's theorems. In Fig. 65.9, the NAND function is represented both with NOT inputs on an OR gate and as an AND gate with an inverted output. The results are equivalent. In Fig. 65.10, the NOR function is represented both with an

OR gate with an inverted output and as NOT inputs on an AND gate. The results are equivalent. This is traditionally done to produce an inverted output connected to an inverted input on the logic symbol diagrams.

**Table 65.2** Logic Gates

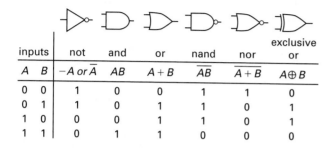

| inputs | | not | and | or | nand | nor | exclusive or |
|---|---|---|---|---|---|---|---|
| $A$ | $B$ | $-A$ or $\overline{A}$ | $AB$ | $A+B$ | $\overline{AB}$ | $\overline{A+B}$ | $A \oplus B$ |
| 0 | 0 | 1 | 0 | 0 | 1 | 1 | 0 |
| 0 | 1 | 1 | 0 | 1 | 1 | 0 | 1 |
| 1 | 0 | 0 | 0 | 1 | 1 | 0 | 1 |
| 1 | 1 | 0 | 1 | 1 | 0 | 0 | 0 |

**Figure 65.9** NAND Gate Alternate Symbols

**Figure 65.10** NOR Gate Alternate Symbols

## Example 65.4

Realize the logic $(A \cdot \overline{C}) + (B \cdot \overline{C})$ using only two-input NAND gates. The positive assertion of $A$ and $B$ and the negative assertion of $C$ are available.

*Solution*

In many practical designs, it is desirable to use logic gates of the same type to realize a logic function because the commercially available packages include multiple logic gates of the same type. In this problem, the output gate must be a NAND, so it is shown as an OR with inverted inputs. The inputs to the output gate must be $\overline{A \cdot \overline{C}}$ and $\overline{B \cdot \overline{C}}$. Each of these inputs can be realized with the AND symbol with an inverted input. The logic symbol realization with two-input NAND gates is

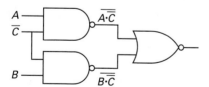

## 14. FLIP-FLOPS

A *flip-flop* is a *bistable multivibrator* circuit that is either on or off. It switches rapidly from one state to the other, and once in a state, remains there; therefore, a flip-flop can be used as a memory device. The outputs of flip-flops are used for a variety of logic functions. Often, the outputs are connected to a series of gates to produce desired logic variables.

There are several types of flip-flops: $D$, $T$, $R$-$S$, and $J$-$K$. A *synchronous flip-flop* changes only when set in synchronization with a clock pulse; an *asynchronous flip-flop* can change any time. The flip-flop has a present state that is either a 1 (true) or 0 (false), and this state is present on the output $Q$ while the complement (or opposite) state will be on the $\overline{Q}$ output. The present state of the flip-flop is represented by $Q$ or $Q_n$ and the next state is represented by $Q^+$ or $Q_{n+1}$ in most references.

When a clock pulse occurs for a synchronous flip-flop or when one or both of the inputs change for an asynchronous flip-flop, the flip-flop will go to its next state, which may be the same as its current state or it may be the opposite. All inputs will be stable during clock transitions with proper design. The synchronous flip-flops will maintain their current state while the next state is determined during either the leading or trailing edge of the clock pulse, whichever the device uses, but not both. The outputs of the synchronous flip-flop will remain constant at all other times regardless of changes in the states of the inputs. Table 65.3 lists the next state transition for each of the flip-flop types to be described in the following sections.

**Table 65.3** Composite Flip-Flop State Transition[*]

| $Q_n$ | $Q_{n+1}$ | $S$ | $R$ | $T$ | $J$ | $K$ | $D$ |
|---|---|---|---|---|---|---|---|
| 0 | 0 | 0 | X | 0 | 0 | X | 0 |
| 0 | 1 | 1 | 0 | 1 | 1 | X | 1 |
| 1 | 0 | 0 | 1 | 1 | X | 1 | 0 |
| 1 | 1 | X | 0 | 0 | X | 0 | 1 |

[*]The X in the table is labeled a "don't care" and can be 0 or 1—it makes no difference to the next state.

## 15. SET-RESET (S-R) FLIP-FLOPS

The *set-reset (S-R) flip-flop* either sets or resets a memory state and is also known as a *set-clear (S-C) flip-flop*. Equation 65.32 gives the Boolean logic for the next state of the $S$-$R$ flip-flop. Figure 65.11 shows the symbol for the $S$-$R$ flip-flop with a clock input. The next state of the $S$-$R$ flip-flop depends both on the current state and the $S$ and $R$ inputs. The $S$ input sets the flip-flop to the 1 state and the $R$ input resets the flip-flop to the 0 state. If both inputs are 0, then the flip-flop keeps the current state. The designer of the input circuit must ensure that no conditions allow both the $S$ and $R$ inputs to be 1 at the same time because this is an indeterminate state. Table 65.4 is the next state table for the *S-R* flip-flop.

$$Q_{n+1} = (S + \overline{R}) \cdot Q_n \qquad 65.32$$

**Figure 65.11** *Clocked S-R Flip-Flop Symbol*

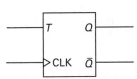

**Table 65.4** *Next-State for S-R Flip-Flop*

| present | | | next |
|---|---|---|---|
| $S$ | $R$ | $Q_n$ | $Q_{n+1}$ |
| 0 | 0 | 1 | 1 (no change) |
| 0 | 0 | 0 | 0 (no change) |
| 0 | 1 | X | 0 |
| 1 | 0 | X | 1 |
| 1 | 1 | invalid | not allowed |

*The X in the table is labeled a "don't care" and can be 0 or 1—it makes no difference to the next state.

## 16. TOGGLE (*T*) FLIP-FLOPS

The *toggle* (*T*) *flip-flop* has only one data input. It is clocked so it will only change states when the clock pulse is present. The next state of the flip-flop will be unchanged if the *T* input is 0 during a clock pulse. The next state of the flip-flop will switch to the opposite state or toggle if the *T* input is 1 during a clock pulse. Equation 65.33 gives the Boolean logic for the next state of the *T* flip-flop. Figure 65.12 shows the symbol for the *T* flip-flop. Table 65.5 is the next state table for the *T* flip-flop. *T* flip-flops are not commercially available and must be built from one of the other types of flip-flops with added logic.

$$Q_{n+1} = T \oplus Q_n = \overline{T} \cdot Q_n + T \cdot \overline{Q_n} \qquad 65.33$$

**Figure 65.12** *T Flip-Flop Symbol*

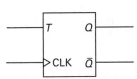

**Table 65.5** *Next-State for T Flip-Flop*

| present | | next |
|---|---|---|
| $T$ | $Q_n$ | $Q_{n+1}$ |
| 0 | 0 | 0 |
| 1 | 0 | 1 |
| 1 | 1 | 0 |
| 0 | 1 | 1 |

*The X in the table is labeled a "don't care" and can be 0 or 1—it makes no difference to the next state.

## 17. *J-K* FLIP-FLOPS

The *J-K* *flip-flop* takes its initials from the inventors. The *J-K* flip-flop is similar to the *S-R* flip-flop, but the state with both inputs 1 is allowed and makes the next state toggle. If the state with both inputs 1 is not used, then the *J-K* flip-flop behaves like a clocked *S-R* flip-flop. The *J* input acts to set the flip-flop, and the *K* input acts to reset the flip-flop. If the *J* and *K* inputs are tied together to provide only one input, then the *J-K* flip-flop behaves like a *T* flip-flop. Equation 65.29 gives the Boolean logic for the next state of the *J-K* flip-flop.

Figure 65.13 shows the symbol for the *J-K* flip-flop. The triangle at the clock indicates the device is *edge triggered*. The small circle at the input of the clock indicates it is *trailing edge triggered* (changes state on a transition from higher to lower voltage). Without the small circle, the flip-flop is *leading edge triggered* (changes state on a transition from lower to higher voltage). Table 65.6 is the next state table for the *J-K* flip-flop. *J-K* flip-flops may also have *S-R* inputs to allow asynchronous selection of the state of the flip-flop.

*Master-slave flip-flops* are not leading edge or trailing edge triggered. The master-slave flip-flop actually has two flip-flops, the master, which is active when the clock is true, and the slave, which is active only when the clock is false. The master will be exposed to the set or reset conditions of the inputs during the time the clock is true so the inputs must remain stable while the clock is true for a predictable result. When the clock is false, the slave will be exposed to the set or reset conditions of the master, but the master conditions are stable since the master is not active, so the slave will be set or reset accordingly.

Ordinary flip-flops are more complicated devices, but they are easier to use since the inputs need only be stable just before and just after the triggering edge of the clock pulse.

$$Q_{n+1} = J \cdot \overline{Q_n} + \overline{K} \cdot Q_n \qquad 65.34$$

**Figure 65.13** *J-K Flip-Flop*

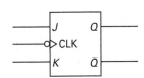

**Table 65.6** *Next-State for J-K Flip-Flop*

| present | | | next |
|---|---|---|---|
| $J$ | $K$ | $Q_n$ | $Q_{n+1}$ |
| 0 | 0 | 0 | 0 (no change) |
| 0 | 0 | 1 | 1 (no change) |
| 0 | 1 | X | 0 (reset) |
| 1 | 0 | X | 1 (set) |
| 1 | 1 | 0 | 1 (toggle) |
| 1 | 1 | 1 | 0 (toggle) |

## 18. DATA (*D*) FLIP-FLOPS

The *delay* or *data* (*D*) *flip-flop* holds one bit of data between clock pulses. The output $Q$ will follow (one period later) the input for the next state regardless of the current state. The $D$ flip-flop is always a clocked device. Equation 65.35 gives the Boolean logic for the next state of the $D$ flip-flop. Figure 65.14 shows the symbol for the $D$ flip-flop. Table 65.7 is the next state table for the $D$ flip-flop.

$$Q_{n+1} = D \qquad 65.35$$

**Figure 65.14** *D Flip-Flop Symbol*

**Table 65.7** *Next-State for D Flip-Flop*

| present | next | |
|---|---|---|
| $D$ | $Q_n$ | $Q_{n+1}$ |
| 0 | 0 | 0 |
| 0 | 1 | 0 |
| 1 | 0 | 1 |
| 1 | 1 | 1 |

## 19. MULTIPLEXERS

A *multiplexer*[8] is any device that takes data from two or more sources and makes it available on a single output. In digital systems, a multiplexer is a device that directs one of the inputs to the output depending on binary coded sector lines. The multiplexer is also called a *data selector* because it selects which data to represent on the output. A commonly used abbreviation for the multiplexer is *MUX*. Figure 65.15(a) is the simplified block diagram for a multiplexer.

Figure 65.15(b) is the switch analogy for the multiplexer. The switch $S$ is positioned based on the value of the binary section word connecting the output $Y$ to the appropriate input. Figure 65.15(c) is the logic diagram for a simple four input multiplexer. If the selector word is $S = 00_2$, then $s_1 = 0$, $s_0 = 0$; the top AND gate will be true if $x_0$ is true and false if $x_0$ is false. All the other gates will be false. If the selector word is $S = 01_2$, then $s_1 = 0$, $s_0 = 1$; the second from the top AND gate will be true if $x_1$ is true and false if $x_1$ is false. All the other gates will be false. Likewise, $S = 10_2$ will select $x_2$, and $S = 11_2$ will select $x_3$.

---

[8]Multiplexer is also spelled *multiplexor.*

**Figure 65.15** *Multiplexer*

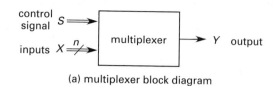

(a) multiplexer block diagram

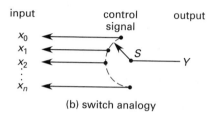

(b) switch analogy

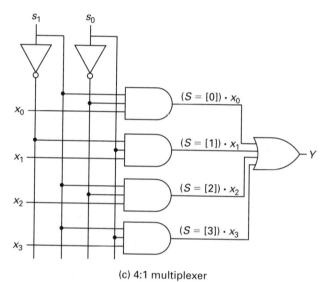

(c) 4:1 multiplexer

## 20. DEMULTIPLEXERS

A *demultiplexer* takes a multiplexed input and provides separate outputs. In digital systems, a demultiplexer is a device that directs the input to one of the outputs depending on binary coded sector lines. The demultiplexer is also called a data selector because it selects the data to represent on each output. Commonly used abbreviations for the multiplexer are *DEMUX* or *DMUX*. Figure 65.16(a) is the simplified block diagram for a demultiplexer.

Figure 65.16(b) is the switch analogy for the demultiplexer. The switch $S$ is positioned based on the value of the binary section word connecting the input $X$ to the appropriate output. Figure 65.16(c) is the logic diagram for a four input demultiplexer. If the selector word is $S = 00_2$, then $s_1 = 0$, $s_0 = 0$; the top AND gate will be true if $X$ is true and false if $X$ is false. All the other gates will be false. If the selector word is $S = 01_2$, then $s_1 = 0$, $s_0 = 1$; the second from the top AND gate will be true if $X$ is true and false if $X$ is false. All the other gates will be false. Likewise, $S = 10_2$ will select the third AND gate, and $S = 11_2$ will select the bottom AND gate. If the input of the demultiplexer is connected to the output of a multiplexer and both have synchronized data selection words, then each output of the demultiplexer will be the same as the corresponding input of the multiplexer when the selection word for that input is present.

**Figure 65.16** Demultiplexer

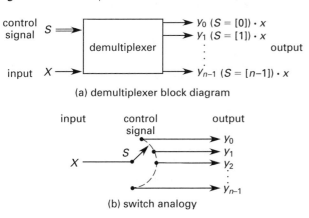

(a) demultiplexer block diagram

(b) switch analogy

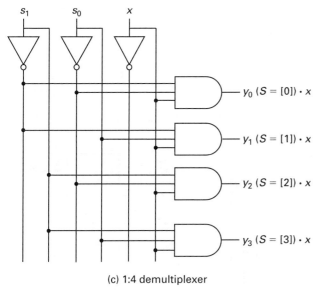

(c) 1:4 demultiplexer

## 21. REGISTERS

Digital systems (including computers) must store data as digital bits. A *register* is a collection of flip-flops for temporary storage of digitally encoded information. The data content of the register is the state of the flip-flops. Registers are organized in digital systems to represent digital words. The register contents can then be used for logic operations, commands, calculations, data input, data output, and so on.

A common type of register is the *shift register*. Figure 65.17 shows a two-bit shift register implemented using two *J-K* flip-flops. If the input on the left is true when the clock pulse occurs, the $J_1$ input will be true, and the $K_1$ input will be false, so $Q_1$ will set. If the input is false, $Q_1$ will reset. Each clock pulse, $Q_2$ will take the state that $Q_1$ had on the previous clock pulse. In this way,

data shifts from $Q_1$ to $Q_2$. If there are more flip-flops, more bits can be shifted. Shift registers can be selectable to shift right or left. One application of shift registers is serial-to-parallel conversion or parallel-to-serial conversion. Data can also be loaded into a register in parallel or retrieved in parallel.

**Figure 65.17** Two-Bit Shift Register

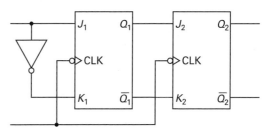

## 22. COUNTERS

*Counters* are sets of flip-flops, normally of the same type and triggered by the same edge of the clock signal. The number of states terminates the count in the sequence. The maximum number of states possible is $2^n$, where $n$ is the number of flip-flops in the circuit. Connections between the flip-flops and possibly logic circuits between them determine the next state of the counter, not external inputs. Typically a counter has no inputs, although the initial state may be set to a desired value before counting starts. In some applications, the clock pulses do not occur at regular intervals but occur only when certain events occur or certain conditions are present. When a counter starts at zero and counts to $2^n - 1$, and then resets to zero, it is called a *binary up counter*, *modulo* ($2^n$) *counter*, or *ripple counter*. A ripple counter is shown in Fig. 65.18.

The ripple counter shown in Fig. 65.18 is constructed of *J-K* flip-flops with the inputs true, so each flip-flop will toggle when clocked. The output of stage 0 is the clock for stage 1, so there is a true-to-false transition at the clock of stage 1 for every other clock pulse to stage 0. So stage 1 will toggle every other clock pulse to stage 0. Likewise, stage 2 will toggle every other clock pulse to stage 1. Figure 65.18(b) shows the timing diagram for the ripple counter.

A ripple counter is *asynchronous* because the stage 1 and 2 flip-flops are not actually triggered by the clock. Since the count ripples from one stage to the next, there is a propagation delay through each flip-flop. This makes it necessary to design the circuits that use the clock to ignore the clock near the clock transitions.

In Fig. 65.18, the flip-flops start with $Q = 0$ and operate on the falling edge of the clock. The initial clock signal (arbitrarily labeled zero) is applied to the state 0 flip-flop, which is triggered on the falling edge. On clock signal 1, the stage 0 flip-flop switches from high to low, that is, $Q = 1$ to $Q = 0$, and the stage 1 flip-flop is triggered from the output. On clock signal 2, the stage 0 flip-flop switches from low to high, and the stage 2 flip-flop remains unchanged. On clock signal 3, the stage 0 flip-flop switches from high to low, as does the stage 1 flip-flop. The stage 2 flip-flop is triggered from the output of the stage 1 flip-flop. The process continues as shown through decimal 7 and then resets.

**Figure 65.18** *Ripple Counter*

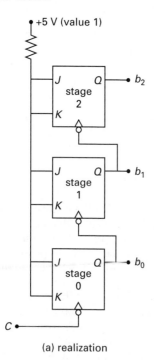

(a) realization

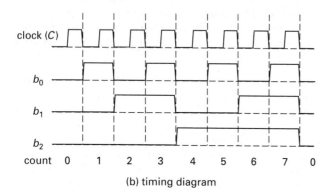

(b) timing diagram

## 23. SYNCHRONOUS BINARY COUNTERS

A binary counter is *synchronous* when all flip-flops transition with the same edge of the same clock signal. Synchronous counters can include any number of flip-flops. The circuit using the clock must ignore the clock very close to the transition edge. For synchronous counters, a *state table*, referred to as an *excitation table*, can be created in which the counter next state is determined by the current state. A synchronous binary counter is shown in Fig. 65.19.

**Figure 65.19** *Synchronous Binary Counter*

| present state $Q_2Q_1Q_0$ | next state $Q_2^+Q_1^+Q_0^+$ | required $T$ inputs $T_2T_1T_0$ |
|---|---|---|
| 000 | 001 | 001 |
| 001 | 010 | 011 |
| 010 | 011 | 001 |
| 011 | 100 | 111 |
| 100 | 101 | 001 |
| 101 | 110 | 011 |
| 110 | 111 | 001 |
| 111 | 000 | 111 |

(a) excitation table

$$T_2 = Q_0 \cdot Q_1$$

$$T_1 = Q_0$$

$$T_0 = 1$$

(b) control equations

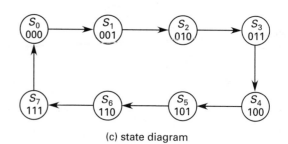

(c) state diagram

(d) realization

The state table is constructed by listing the desired states in the desired sequence as in Fig. 65.19(a). The next state and required inputs are also listed. The logic to generate the required inputs is normally determined using a Karnaugh map or Veitch diagram. For this counter, the logic can be determined by inspection. The resulting control equations are shown in Fig. 65.19(b). To visualize the transitions and aid in the design, state diagrams, as shown in Fig. 65.19(c), are useful tools. The states are indicated by the $S$'s and generally do not have the same subscripts as the counts. The realization of the network using $J$-$K$ flip-flops configured as toggles is shown in Fig. 65.19(d).

A problem with this type of design is that more than one flip-flop can change state at the same clock transition. Consequently, due to differences in transition speeds, a false value can occur during the transition process. This can be avoided by allowing only one flip-flop to change state during each clock period.[9] This is equivalent to ensuring that state assignments are logically adjacent to the present state.[10]

## 24. ARITHMETIC AND LOGIC UNITS

An *arithmetic and logic unit* (ALU) of a digital system (e.g., computer processor) is a combinatorial circuit[11] for performing the basic operations, such as addition, subtraction, and Boolean logic functions (e.g., AND, OR, NOT, XOR). The operation performed by the ALU depends on a command or *instruction* from a controlling circuit. One possible implementation of the ALU is that it performs all the operations in parallel, and a multiplexer selects one of the outputs according to the operation commanded.

One of the most useful functions of the ALU is binary addition. Binary addition is accomplished with the following rules.[12]

- $0 + 0 = $ sum 0 carry 0
- $0 + 1 = $ sum 1 carry 0
- $1 + 0 = $ sum 1 carry 0
- $1 + 1 = $ sum 1 carry 1

Given these rules, adder truth tables are constructed as shown in Fig. 65.20. From the truth tables, the necessary combinatorial logic circuits are realized. For the *half-adder*, the sum is XOR logic, and the carry is AND logic. The *full-adder* adds the bits $a_i$ and $b_i$ and the carry-in bit from the next lower stage of the adder.

To add an $n$-bit number, adders are cascaded to form a *ripple-carry adder*. (See Fig. 65.21.) The term *ovr* at the stage $n-1$ indicates an *overflow condition*, which occurs if the value is outside the range of numbers that can be stored.

---

[9]This is called a *Gray code*.
[10]This is not always possible, but it is a desirable design goal.
[11]A combinatorial circuit is a circuit has $m$ inputs and $n$ outputs.
[12]In this context, the "+" sign means addition, not Boolean OR.

***Figure 65.20*** *Adders*

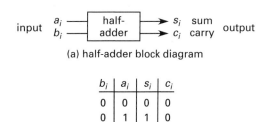

(a) half-adder block diagram

| $b_i$ | $a_i$ | $s_i$ | $c_i$ |
|---|---|---|---|
| 0 | 0 | 0 | 0 |
| 0 | 1 | 1 | 0 |
| 1 | 0 | 1 | 0 |
| 1 | 1 | 0 | 1 |

(b) half-adder truth table

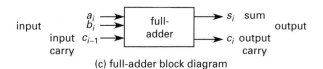

(c) full-adder block diagram

| $b_i$ | $a_i$ | $c_{i-1}$ | $s_i$ | $c_i$ |
|---|---|---|---|---|
| 0 | 0 | 0 | 0 | 0 |
| 0 | 0 | 1 | 1 | 0 |
| 0 | 1 | 0 | 1 | 0 |
| 0 | 1 | 1 | 0 | 1 |
| 1 | 0 | 0 | 1 | 0 |
| 1 | 0 | 1 | 0 | 1 |
| 1 | 1 | 0 | 0 | 1 |
| 1 | 1 | 1 | 1 | 1 |

(d) full-adder truth table

***Figure 65.21*** *Ripple-Carry Adder*

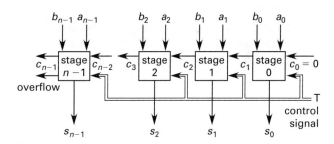

Subtraction is accomplished using one's complement arithmetic. That is, negative numbers are changed to one's complement (i.e., by converting every 1 to 0 and every 0 to 1) before being added.

## 25. DIGITAL-TO-ANALOG CONVERSION

Digital data in words often represent analog voltages. When the analog voltage is needed, conversion is necessary. The conversion is accomplished with a *digital-to-analog* (D/A) *converter*, also abbreviated DAC. Similarly, the conversion from analog voltage to digitally encoded words is called *analog-to-digital* (A/D) *conversion*.

Digital information is considered to be binary and weighted according to its position as a power of two. For example, the value of a positive $n$-bit binary number $N$ is

$$N = b_{n-1}2^{n-1} + b_{n-2}2^{n-2} + \cdots + b_2 2^2 + b_1 2^1 + b_0 2^0$$

$$65.36$$

The binary bits represented by the $b$'s have a value of zero or one. When coded, $N$ is

$$N = b_{n-1}b_{n-2}...b_2 b_1 b_0 \qquad 65.37$$

The digit $b_{n-1}$ is the *most significant bit*, MSB, while $b_0$ is the *least significant bit*, LSB. The maximum value that can be represented by the code of an n-bit word is

$$\text{maximum value} = 2(\text{MSB}) - \text{LSB} = 2n - 1 \qquad 65.38$$

A DAC converts the binary word by effectively multiplying the value by an analog voltage scale factor. In the DAC, the value of each bit controls one switch. If the bit is true, the switch is closed. If the bit is false, the switch is open. A DAC using binary-weighted resistors is shown in Fig. 65.22. The device is an inverting configuration; the output will be the value of the digital word times the negative of the reference voltage. The switches are shown as mechanical switches for simplicity, but in most digital systems, the switches are electronic. One possible configuration of an electronic switch for a DAC is shown in Fig. 65.23.

### Example 65.5

A DAC similar to that shown in Fig. 65.23 is used with a reference voltage of 10.0 V $\pm$ 0.5 V. Each of the resistors has a tolerance of $\pm 1\%$. Assume the op amp is ideal. What is the maximum error if only one bit is true?

*Solution*

The maximum error will occur if bit $b_7$ is true since that is the MSB and results in the largest nominal output voltage. The ideal voltage out would be exactly $-10$ V. The output voltage is

$$V_{\text{out}} = -\frac{R_f}{R_{\text{in}}} V_{\text{in}}$$

If the input voltage and the feedback resistance are at their minimum values, and the input resistance is at its maximum value, the error is

$$E_{\text{max}} = V_{\text{ideal}} - \left(-\frac{R_{f,\text{min}}}{R_{\text{in,max}}}\right) V_{\text{in,min}}$$

$$= -10\,\text{V} + \left(\frac{10\,\text{k}\Omega \times 0.99}{10\,\text{k}\Omega \times 1.01}\right)(10.0\,\text{V} - 0.5\,\text{V})$$

$$= -0.69\,\text{V}$$

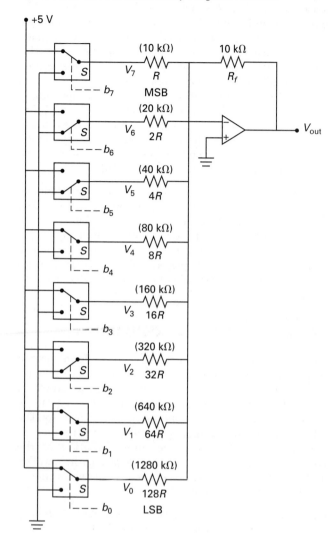

**Figure 65.22** *D/A Converter: Binary Weighted Resistors*

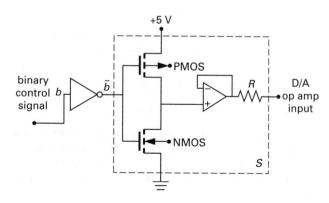

**Figure 65.23** *Digitally Controlled Switch*

If the input voltage and the feedback resistance are at their maximum values and the input resistance is at its minimum value, the error is

$$
\begin{aligned}
E_{\text{max}} &= V_{\text{ideal}} - \left( -\frac{R_{f,\text{max}}}{R_{\text{in,min}}} \right) V_{\text{in,max}} \\
&= -10\,\text{V} + \left( \frac{10\,\text{k}\Omega \times 1.01}{10\,\text{k}\Omega \times 0.99} \right) (10.0\,\text{V} + 0.5\,\text{V}) \\
&= 0.71\,\text{V}
\end{aligned}
$$

## 26. LADDER-TYPE D/A CONVERSION

A *ladder-type D/A converter* is based on a cell concept. Each succeeding cell has twice the weight of the previous one, hence the term "ladder."[13] The design of the first cell, which corresponds to the LSB, is shown in Fig. 65.24(a). A typical cell is shown in Fig. 65.24(b). Either the reference voltage, $V_{\text{ref}}$, or 0 V is applied, depending on the switch position controlled by the bit value.

**Figure 65.24** *Cells of a Ladder-Type D/A Converter*

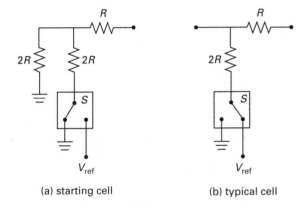

(a) starting cell          (b) typical cell

The Thevenin equivalent circuits of the starting cell with the output taken as $V_0$, the succeeding cell with the output taken as $V_1$, and the combination of the two are shown in Fig. 65.25.

A ladder-type DAC with seven cells has an output of

$$
\begin{aligned}
V_{\text{out}} &= \frac{V_7}{2} + \frac{V_6}{4} + \frac{V_5}{8} + \frac{V_4}{16} + \frac{V_3}{32} \\
&\quad + \frac{V_2}{64} + \frac{V_1}{128} + \frac{V_0}{256}
\end{aligned} \qquad 65.39
$$

The voltages in Eq. 65.39 are either $V_{\text{ref}}$ or 0 V, depending on the value of the bits. The converter is completed by connection to the noninverting terminal of the operational amplifier, which provides buffering, as shown in Fig. 65.26.

---

[13]The ladder-type DAC uses twice the number of resistors for a given number of bits as the binary weighted DAC, but the resistors used have only two values, $R$ and $2R$.

**Figure 65.25** *D/A Converter Cells' Thevenin Equivalents*

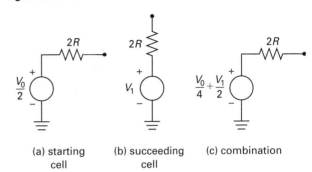

(a) starting      (b) succeeding      (c) combination
cell                 cell

When any switch in Fig. 65.26 is closed, there is a transient. A propagation delay occurs due to capacitance in the circuit, and the delay is greater the further down the ladder. That is, the LSB, $S_0$ has a greater delay in causing the output to change than the MSB, $S_7$. These transients and delays are avoided by the inverting-type DAC shown in Fig. 65.27.

Each of the switches will be connected to zero potential regardless of the position because the output of the operational amplifier will produce a current to keep the minus input at zero potential. Therefore, the current through each switch will always be the same. The two $2R$ resistors at $S_0$ divide half the current coming from the left. The equivalent resistance of the resistors to the right of $S_1$ is $2R$, so half the current coming from the left will go through $S_1$. Likewise, each switch will have twice the current of the switch to the right. Therefore, the current into the inverting terminal is

$$
I = \frac{V_{\text{ref}}}{2R} \left( \begin{aligned} & b_7 + \tfrac{1}{2}b_6 + \tfrac{1}{4}b_5 + \tfrac{1}{8}b_4 + \tfrac{1}{16}b_3 \\ & + \tfrac{1}{32}b_2 + \tfrac{1}{64}b_1 + \tfrac{1}{128}b_0 \end{aligned} \right) \qquad 65.40
$$

Assuming an ideal operational amplifier, the current through the feedback resistor is equal and opposite to the current from the ladder circuit, so the output voltage is

$$
\begin{aligned}
V_{\text{out}} &= -2RI \\
&= -V_{\text{ref}} \left( \begin{aligned} & b_7 + \tfrac{1}{2}b_6 + \tfrac{1}{4}b_5 + \tfrac{1}{8}b_4 + \tfrac{1}{16}b_3 \\ & + \tfrac{1}{32}b_2 + \tfrac{1}{64}b_1 + \tfrac{1}{128}b_0 \end{aligned} \right)
\end{aligned} \qquad 65.41
$$

The maximum output occurs when all of the binary input values are true. In this condition, the magnitude of the maximum output for an inverting ladder-type converter of $n$-bit size is

$$
|V_{\text{out,max}}| = V_{\text{ref}} \left( 2 - \frac{1}{2^{n-1}} \right) \qquad 65.42
$$

## 27. MULTIPLYING-TYPE D/A CONVERTER

If the reference voltage of the noninverting D/A converter in Fig. 65.26 is replaced by an analog voltage signal, the output becomes a product of the analog signal and the binary input controlling the electronic switches. Such a device is called *a multiplying-type D/A converter*. This arrangement is also called a *programmable attenuator* because the output voltage is a fraction of the analog input, with the attenuation controlled by the digital system or computer supplying the binary signal.

The inverting ladder-type D/A converter of Fig. 65.27 can also be used as a programmable attenuator by changing the feedback resistance from $2R$ to $R$. The binary weighted D/A converter of Fig. 65.22 can also be used as a programmable attenuator by changing the feedback resistance from 10 k$\Omega$ to 5 k$\Omega$. In both cases, the incoming voltage, $V_{\text{ref}}$ or +5 V, is replaced by the analog input voltage to be attenuated.

**Figure 65.26** *Ladder-Type D/A Converter*

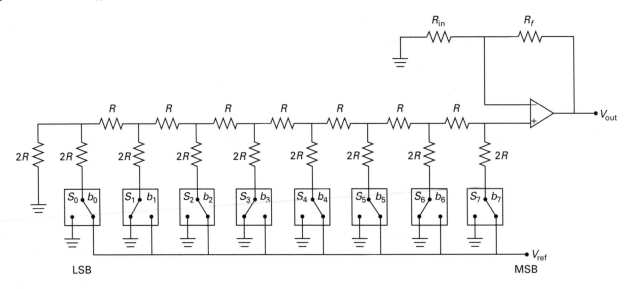

**Figure 65.27** *Inverting-Type D/A Converter*

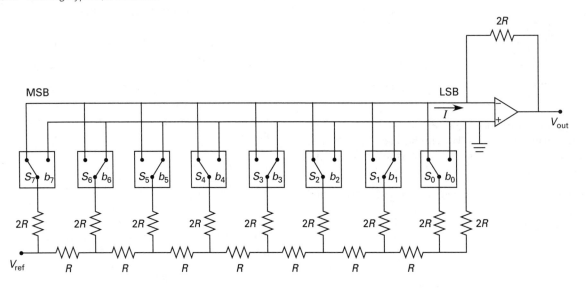

# Topic XIII: Physics

Physics

# 66 Light, Illumination, and Optics

## Nomenclature

| | | | |
|---|---|---|---|
| $A$ | area | $ft^2$ | $m^2$ |
| $c$ | speed of light | ft/sec | m/s |
| $d$ | slit spacing | ft | m |
| $E$ | illuminance | $lm/ft^2$ | $lm/m^2$ |
| $f$ | focal length | ft | m |
| $f$ | frequency | Hz | Hz |
| $i$ | distance to image | ft | m |
| $I$ | luminous intensity | lm/sr | lm/sr |
| $m$ | interference order | – | – |
| $m$ | magnification | – | – |
| $M$ | luminous emittance | $lm/ft^2$ | $lm/m^2$ |
| $n$ | absolute index of refraction | – | – |
| $o$ | distance to object | ft | m |
| $P$ | power | diopter | diopter |
| $Q$ | quantity of light | W | W |
| $r$ | radius | ft | m |
| $s$ | separation distance | ft | m |
| $t$ | film thickness | ft | m |
| $t$ | time | sec | s |
| v | velocity | ft/sec | m/s |
| $x$ | horizontal distance | ft | m |
| $y$ | vertical distance | ft | m |

## Symbols

| | | | |
|---|---|---|---|
| $\alpha$ | angle | deg | rad |
| $\eta$ | efficiency (luminous) | lm/W | lm/W |
| $\theta$ | angle | deg | rad |
| $\lambda$ | wavelength | ft | m |
| $\Phi$ | luminous flux | lm | lm |
| $\omega$ | solid angle | sr | sr |

## Subscripts

| | |
|---|---|
| $c$ | critical |
| $i$ | incident |
| $l$ | luminous |
| $r$ | reflected |
| $t$ | total |
| $v$ | visible |

## 1. VISIBLE LIGHT

*Visible light* consists of electromagnetic radiation with wavelengths approximately between $3.8 \times 10^{-7}$ m and $7.5 \times 10^{-7}$ m. It is bounded on either side of the electromagnetic spectrum by infrared and ultraviolet radiation. For the average observer, the wavelength of light at the center of the *visible spectrum* is $5.550 \times 10^{-7}$ m.

The visual sensation at the lower and upper limits of the visible spectrum are intense violet and red, respectively. These and intermediate color sensations are given in Table 66.1.

**Table 66.1** *Color Versus Wavelength*

| color | wavelength (m) |
|---|---|
| violet | $< 4.5 \times 10^{-7}$ |
| blue | 4.5 to $5.0 \times 10^{-7}$ |
| green | 5.0 to $5.7 \times 10^{-7}$ |
| yellow | 5.7 to $5.9 \times 10^{-7}$ |
| orange | 5.9 to $6.1 \times 10^{-7}$ |
| red | $> 6.1 \times 10^{-7}$ |

(Multiply m by $10^{-9}$ to obtain nm.)

The principal natural light source is the photosphere of the sun, a plasma envelope with a temperature of approximately 6300K. Artificial light sources include mercury arcs (5000K), carbon arcs (4000K), incandescent (tungsten filament) lamps (600K), and fluorescent lamps (300K).

## 2. ELECTROMAGNETIC RADIATION

According to *Maxwell's electromagnetic theory*, accelerating electric charges generate and radiate energy in the form of *electromagnetic* (e-m) *waves*. This e-m energy travels through a vacuum at the speed of light. Figure 66.1 illustrates the *wavelength*, $\lambda$, and the electrical field strength, $E$, as a function of time (or distance) from the source. Wavelength should be expressed in meters but is often expressed in *microns* ($10^{-6}$ m), millimicrons and nanometers (both $10^{-9}$ m), and *angstroms* ($10^{-10}$ m).

The distance, $s$, traveled through a medium in time $t$ by an electromagnetic wave of frequency $f$ and wavelength $\lambda$ is $s = f\lambda t$. The distance traveled in one second is found by setting $t = 1$ and corresponds to the speed of light, $c$. Equation 66.1 is valid for all types of electromagnetic radiation.

$$c = \lambda f \qquad\qquad 66.1$$

When an electromagnetic wave (see Fig. 66.1) passes into a new medium (e.g., air, glass), the frequency remains constant. The speed and wavelength both change. In materials opaque to electromagnetic transmission, the speed and wavelength are both zero.

**Figure 66.1** *An Electromagnetic Wave*

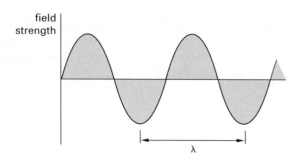

Table 66.2 lists the approximate ranges of frequencies for common types of electromagnetic radiation. Frequencies up to approximately $10^{12}$ Hz are categorized as *radio waves*. (Use Eq. 66.1 to calculate the corresponding wavelengths.)

### Example 66.1

What is the length of a quarter-wave antenna that operates at 100 MHz?

**Table 66.2** *The Electromagnetic Spectrum*

| approximate frequency range (Hz) | name or use |
|---|---|
| $10^4$ | lowest communication frequency |
| $5.5 \times 10^5$–$1.6 \times 10^6$ | AM broadcast band |
| $2 \times 10^6$–$5 \times 10^7$ | shortwave radio band |
| $5.4 \times 10^7$–$8.9 \times 10^8$ | television band |
| $8.8 \times 10^7$–$1.1 \times 10^8$ | FM broadcast band |
| $10^9$–$10^{11}$ | radar and microwaves |
| $10^{12}$ | highest experimental frequency |
| $10^{11}$–$3.8 \times 10^{14}$ | infrared radiation |
| $3.8 \times 10^{14}$–$7.5 \times 10^{14}$ | human visible spectrum |
| $7.5 \times 10^{14}$–$3 \times 10^{17}$ | ultraviolet radiation |
| $3 \times 10^{17}$–$3 \times 10^{19}$ | X-rays |
| above $3 \times 10^{19}$ | gamma rays |

*Solution*

Use Eq. 66.1.

$$\text{length} = \frac{\lambda}{4} = \frac{c}{4f} = \frac{3 \times 10^8 \ \frac{\text{m}}{\text{s}}}{(4)(10^8 \ \text{Hz})} = 0.75 \text{ m}$$

## 3. SPEED OF LIGHT

Electromagnetic radiation of all types propagates in a vacuum at the *speed of light*, $c = 2.9979 \times 10^8$ m/s, usually just rounded to $3 \times 10^8$ m/s (186,000 mi/sec). This velocity is independent of both the observer's reference frame and the velocity of the observer.[1] While the product of terms in Eq. 66.1 is always constant, the frequency and observed wavelength of the electromagnetic radiation depend on the relative motion between source and observer. If the energy is emitted at frequency $f$ and the relative separation velocity is $v$, the observed frequency and wavelength will be

$$\frac{f'}{f} = \frac{\lambda}{\lambda'} = \frac{1 - \dfrac{v}{c}}{\sqrt{1 - \left(\dfrac{v}{c}\right)^2}} \qquad\qquad 66.2$$

$$\frac{v}{c} = \frac{f^2 - f'^2}{f^2 + f'^2} = \frac{\lambda'^2 - \lambda^2}{\lambda'^2 + \lambda^2} \qquad\qquad 66.3$$

For example, when a source rapidly recedes from an observer, the observed frequency will be less than the emitted frequency. Accordingly, the observed wavelength will increase and the observed light (visible radiation) will exhibit a *red shift* (that is, will appear more

---

[1]This effect is a direct consequence of Einstein's special theory of relativity.

Physics

red). Such a change in frequency is known as a *Doppler shift*.[2]

## Example 66.2

An astronomer tracking a comet observes wavelengths that are 0.3% higher than they are known to be. What is the comet's velocity with respect to the earth?

*Solution*

The astronomer observes $\lambda' = 1.003\lambda$. Solving Eq. 66.2 for v,

$$\frac{1}{1.003} = \frac{1 - \dfrac{v}{3 \times 10^8 \frac{m}{s}}}{\sqrt{1 - \left(\dfrac{v}{3 \times 10^8 \frac{m}{s}}\right)^2}}$$

This example has two solutions: $9.04 \times 10^5$ m/s and $2.98 \times 10^8$ m/s. Since the second value is essentially the speed of light, it is discarded. Since the velocity is positive, the comet is receding. This is consistent with the fact that wavelengths are shifted higher when objects are separating.

## 4. SOURCE ENERGY

The amount of visible light emitted from a source is the *luminous flux*, $\Phi$, with units of lumens (lm).[3] Not all of the energy emitted will be in the visible region, though. Therefore, the source power is not a good indicator of the lighting effect. The ratio of luminous flux to total radiant energy (also known as the *quantity of light*), $Q$, is the *luminous efficacy* or *luminous efficiency*, $\eta_l$. Table 66.3 gives typical values.

$$\eta_l = \frac{\Phi}{Q} \qquad \qquad 66.4$$

The *luminous emittance* (also known as *brightness*), $M$, of a source is the luminous flux per unit area of the source. (Compare this with *illuminance* covered in Sec. 66.6, which is the flux per unit of receiving area.) In the

**Table 66.3** Typical Luminous Efficiencies

| source | | $\eta_l$ (lm/W) |
|---|---|---|
| | candle | 0.1 |
| 25 W | tungsten lamp | 10 |
| 100 W | tungsten lamp | 16 |
| 1000 W | tungsten lamp | 22 |
| 40 W | fluorescent lamp | 80 |
| 400 W | mercury fluorescent lamp | 58 |
| 1000 W | carbon arc | 60 |
| 100 W | mercury arc | 35 |
| 1000 W | mercury arc | 65 |
| 400 W | metal halide lamp | 85 |
| 1000 W | metal halide lamp | 100 |
| 400 W | high-pressure sodium lamp | 125 |
| 1000 W | high-pressure sodium lamp | 130 |
| 180 W | low-pressure sodium lamp | 180 |

United States, the unit of brightness is the $lm/ft^2$ (foot-lambert).[4]

$$M_v = \frac{\Phi}{A_{source}} \qquad \qquad 66.5$$

## Example 66.3

A 100 W lamp has a luminous flux of 4400 lumens. What is the lamp's luminous efficiency?

*Solution*

From Eq. 66.4,

$$\eta_l = \frac{\Phi}{Q} = \frac{4400 \text{ lm}}{100 \text{ W}} = 44 \text{ lm/W}$$

## Example 66.4

A tungsten filament emitting 1600 lm has a surface area of 0.35 $in^2$ ($2.26 \times 10^{-4}$ $m^2$). What is its brightness?

*SI Solution*

$$M_v = \frac{\Phi}{A_{surface}} = \frac{1600 \text{ lm}}{2.26 \times 10^{-4} \text{ m}^2} = 7.08 \times 10^6 \text{ lm/m}^2$$

*Customary U.S. Solution*

$$M_v = \frac{\Phi}{A_{surface}} = \frac{1600 \text{ lm}}{\dfrac{0.35 \text{ in}^2}{144 \frac{\text{in}^2}{\text{ft}^2}}} = 6.58 \times 10^5 \text{ lm/ft}^2$$

---

[2]There is another way to obtain the appearance of a red shift, known as the *Wolf effect*, but it requires coherent light such as that produced from lasers. If light from two coherent sources (or two split beams from one coherent source) are recombined in just the right way, some of the frequency components will cancel and others will be reinforced, thereby shifting the spectrum toward the red or blue. Natural light sources are not coherent, so cosmological red shifts are assumed to be caused by relative separation.

[3]A *lumen* is approximately equal to $1.47 \times 10^{-3}$ W of visible light with a wavelength of $5.55 \times 10^{-7}$ W.

[4]The term *lambert* in the U.S. unit is a source of confusion. The old lambert unit is the same as $lm/cm^2$. A foot-lambert is the same as $lm/ft^2$.

## 5. LUMINOUS INTENSITY

The *luminous intensity*, $I$, of a source is a measure of flux per unit solid angle. Its units are lumens per steradian (lm/sr).[5] Equation 66.6 relates flux from an omnidirectional point source to intensity measured on a spherical receptor of radius $r$.

$$I_v = \frac{\Phi}{\omega} = \frac{\Phi r^2}{A} = \frac{\Phi_t r^2}{4\pi r^2} = \frac{\Phi_t}{4\pi} \qquad 66.6$$

### Example 66.5

Two lumens pass through a circular hole (diameter = 0.5 m) in a screen located 6.0 m from an omnidirectional source. (a) What is the luminous intensity of the source? (b) What luminous flux is emitted by the source?

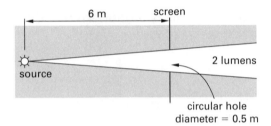

*Solution*

(a) The solid angle subtended by the hole is

$$\omega = \frac{A}{r^2} = \frac{\frac{\pi}{4}(0.5 \text{ m})^2}{(6 \text{ m})^2} = 0.005454 \text{ sr}$$

From Eq. 66.6, the intensity is

$$I_v = \frac{\Phi}{\omega} = \frac{2 \text{ lm}}{0.005454 \text{ sr}} = 366.7 \text{ lm/sr}$$

(b) From Eq. 66.6, the luminous flux is

$$\Phi_t = 4\pi I = (4\pi \text{ sr})\left(366.7 \frac{\text{lm}}{\text{sr}}\right)$$
$$= 4608 \text{ lm}$$

## 6. ILLUMINANCE

The *illuminance* (*illumination*), $E_v$, is a measure of received light, measured as luminous flux per incident area. Typical units are lux (lm/m²) and lm/ft² (foot-candles or foot-lamberts). (See Table 66.4.) The name *irradiance* is used when the units are W/m².

$$E_v = \frac{\Phi}{A_{\text{receptor}}} \qquad 66.7$$

---

[5]Intensity in candle power (*candles*) is numerically equal to intensity in lumens per steradian. However, intensity is not a measure of power, and the term *candlepower* should not be used.

**Table 66.4** *Recommended Illuminance*

| location | lm/ft² | lux |
|---|---|---|
| roadway | 1–2 | 10–20 |
| living room | 5–15 | 50–150 |
| library (reading) | 30–70 | 300–700 |
| evening sports | 30–100 | 300–1000 |
| factory (assembly) | 100–200 | 1000–2000 |
| office | 100–200 | 1000–2000 |
| factory (fine assembly) | 500–1000 | 5000–10 000 |
| hospital operating room | 2000–2500 | 20 000–25 000 |

(Multiply lux by 0.0929 to obtain lm/ft².)

For an omnidirectional source and a spherical receptor of radius $r$,

$$E_v = \frac{\Phi_t}{4\pi r^2} \qquad 66.8$$

Equation 66.8 shows that illumination follows the *distance squared law*.

$$E_{v,1} r_1^2 = E_{v,2} r_2^2 \qquad 66.9$$

### Example 66.6

A lamp radiating hemispherically and rated at 2000 lm is positioned 20 ft (6 m) above the ground. What is the illumination on a walkway directly below the lamp?

*SI Solution*

$$E_v = \frac{\Phi}{A} = \frac{\Phi}{\frac{1}{2}A_{\text{sphere}}} = \frac{2000 \text{ lm}}{\left(\frac{1}{2}\right)(4\pi)(6 \text{ m})^2} = 8.84 \text{ lux}$$

*Customary U.S. Solution*

$$E_v = \frac{\Phi}{A} = \frac{2000 \text{ lm}}{\left(\frac{1}{2}\right)(4\pi)(20 \text{ ft})^2} = 0.796 \text{ lm/ft}^2 \text{ (ft-c)}$$

## 7. POLYCHROMATIC LIGHT

Light with only one color (i.e., one wavelength) is *monochromatic light*. *Polychromatic light* (literally "many colors") contains light of different wavelengths. *White light* contains wavelengths of all or nearly all of the colors. Since light of different wavelengths is refracted at slightly different angles, polychromatic light can be split into rays of individual colors by a prism. (See Sec. 66.16.)

## 8. COHERENT LIGHT

Normally, light sources produce waves with random phasing. However, all of the waves in the *coherent light* produced by a *laser* (an acronym for light amplification

Physics

by stimulated emission of radiation) are in phase.[6] A *resonant cavity* is the most common method for generating coherent radiation. This is a transparent cylinder with mirrored interior ends. (One of the ends is semi-transparent to allow the escape of the laser beam.) The tube can be solid (e.g., a ruby laser) or hollow and filled with a gas (e.g., a helium-neon laser). Solid lasers commonly operate in high-power *strobe (burst) mode* (i.e., in discrete flashes), while gas lasers operate at lower power in *continuous mode*.

If the tube length is an integral number of wavelengths, any monochromatic light introduced by a gas discharge (within the tube) or flash coil (surrounding the tube) will be reflected back and forth by the mirrors at the tube ends. Standing waves will be produced and the resultant radiation will be coherent.

Ruby or gas atoms in the tube are normally in their lowest (ground) energy state. An atom at a higher energy state will emit energy when it drops back to the ground state. The frequency of the emitted energy is predicted by the quantum equation.

Laser performance depends on stimulating high energy atoms to release their energy. When a photon hits a high energy atom, it causes the atom to drop to a lower energy state. This energy release, in the form of another photon, is in phase with the incident photon and is known as *stimulated emission*, a form of resonance. (Of course, incoherent spontaneous decay from high to lower energy states also occurs. For this reason, it is essential to have a high density of photons.)

For light to be emitted, the number of atoms at the higher energy state must exceed the number at the lower state. The normal proportion of high- and low-energy atoms (i.e., a *normal population*) must be reversed, a process known as *population inversion*. High energy atoms are produced by supplying external energy to the laser.

A *hologram* is a three-dimensional image of an object illuminated by a broad band of coherent light, such that some of the light is viewed (or exposes a piece of film) directly while another portion is seen (or recorded) as reflections of the object. This is illustrated in Fig. 66.2.

## 9. INTERACTION OF LIGHT WITH MATTER

Light travels through a vacuum as an electromagnetic wave. When the light makes contact with matter, some of the wave energy is absorbed by the matter, causing electrons to jump into higher energy states. (A polished metal surface will absorb only about 10% of the incident energy, reflecting the remaining 90% away.) Some of this absorbed energy is re-emitted when the electrons drop back to a lower energy level. Generally, the re-emitted light will not be at its original wavelength.

---

[6]When the coherent waves are beyond the visible region, the term *maser* (microwave amplification by stimulated emission of radiation) may be used. Such a device is also known as a *paramagnetic amplifier*.

**Figure 66.2** *A Hologram*

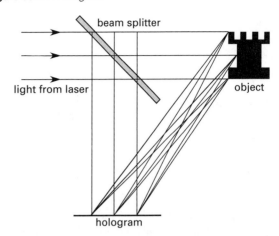

If the reflecting surface is smooth, the *reflection angle* for most of the light will be the same as the *incident angle*, and the light is said to be *regularly reflected* (i.e., the case of *specular reflection*). If the surface is rough, however, the light will be scattered and reflected randomly (the case of *diffuse reflection*). (See Fig. 66.3.)

**Figure 66.3** *Specular and Diffuse Reflection*

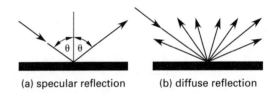

(a) specular reflection    (b) diffuse reflection

The energy that is absorbed is said to be *refracted*. In the case of an *opaque material*, the refracted energy is absorbed within a very thin layer and converted to heat. However, the light is able to pass through a *transparent material* without being absorbed. Light is partially absorbed in a *translucent material*.

## 10. INDEX OF REFRACTION OF A SUBSTANCE

While the speed of light in a vacuum is constant, it is different in transparent media. The absolute *index of refraction (refractive index)*, $n$, is the ratio of the speed of light in a vacuum (essentially the same as in air) to the speed of light in a particular medium.[7] It is not strictly constant but varies 1% to 2% over the visible light spectrum. (See Table 66.5.) This variation is disregarded, however, in simple studies.

$$n = \frac{c_{\text{vacuum}}}{c_{\text{medium}}} = \frac{3 \times 10^8 \, \frac{\text{m}}{\text{s}}}{c_{\text{medium,m/s}}} \qquad 66.10$$

---

[7]The *relative index of refraction*, the ratio of the speeds of light in two different media, is also encountered.

Physics

**Table 66.5** *Approximate Absolute Indices of Refraction (at λ = 5.893 × 10⁻⁷ m)*

| medium | $n$ |
|---|---|
| air (20°C, 1 atm) | 1.00029 |
| benzene | 1.50 |
| borosilicate crown glass | 1.5243 |
| diamond | 2.417 |
| flint glass, dense | 1.6555 |
| hydrogen (20°C, 1 atm) | 1.00013 |
| ice | 1.31 |
| quartz, fused | 1.46 |
| salt | 1.53 |
| water (20°C) | 1.3330 |

Most transparent substances, including glass, water, air, and polymethyl metacrylate (Lucite™), are *isotropic*; that is, light travels at the same speed in all directions. However, some transparent crystals are *anisotropic*.

**Example 66.7**

At a particular frequency corresponding to red light, the refractive index of water is approximately 1.3300. What is the speed of this light in water?

*Solution*

From Eq. 66.10,

$$c_{\text{water}} = \frac{c_{\text{vacuum}}}{n} = \frac{3 \times 10^8 \; \frac{\text{m}}{\text{s}}}{1.3300} = 2.26 \times 10^8 \text{ m/s}$$

## 11. POLARIZED LIGHT

Light can be considered to be a transverse wave, and almost all sources produce light with waves in randomly oriented planes. *Polarized light* has waves in only one plane and can be produced in several ways. These include scattering light from small particles, sending light through a filter that passes waves in only one plane, and reflecting light from a glass surface at a special incident angle.[8] (See Fig. 66.4.) The incident angle required to obtain polarized light by reflection is known as *Brewster's angle* and is given by Eq. 66.11.

$$\tan \theta = \frac{n_2}{n_1} \qquad 66.11$$

Polarized light is obtained from a Polaroid™ film. Such a filter consists of a transparent sheet with many tiny crystals of a quinine compound all oriented in the same direction.

The crystals have the property of transmitting only light whose plane of vibration coincides with the orientation of the crystals.

Due to the spiral nature of its structure, an *optically active* material has the ability to change the plane of

[8]Some minerals (e.g., tourmaline) are natural polarizing filters.

polarization. Quartz, some sugars, and many organic and inorganic solutions have this property. The amount of rotation depends on the sample length (depth) and is given in deg/mm for solids and deg/dm for liquids. This property can be used to determine the sugar concentration of commercial syrups.

**Figure 66.4** *Polarized Light*

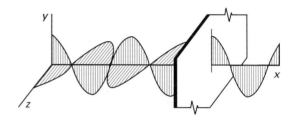

## 12. INTERFERENCE

*Interference* (also known as *cancellation*) occurs when two waves combine so that one subtracts from the other. It is also possible for two waves to add to each other, which is known as *reinforcement* or *superposition*.

Most light sources emit light waves with random phase relationships. However, a source of parallel rays can be obtained by allowing light to pass through a narrow slit. According to *Huygens' principle* (also known as *Huygens' construction*), each point in a wave front can be considered as a source of waves. Therefore, the slit will act as a light source, as shown in Fig. 66.5. The circular lines represent the wave maxima (crests) and the spaces represent the wave minima (troughs).

Interference can be obtained by allowing the parallel rays to pass through two additional slits, which themselves act as two sources. However, these secondary sources are in phase because the light was derived from a single primary slit.

**Figure 66.5** *Huygens' Principle*

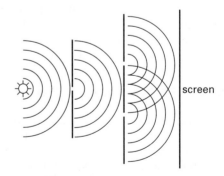

## 13. INTERFERENCE FROM SLITS

The visual effect on a distant screen of combining two in-phase sources will be regions of darkness and light. (The arrangement shown in Fig. 66.6 depicts *Young's experiment*.) A bright spot or band means that two wave crests or two wave troughs coincide (i.e., are in phase at that point) and reinforce each other. If a trough coincides with a crest, the two sources are 180 degrees out of phase, and the result is a dark spot or band.

**Figure 66.6** *Reinforcement and Interference*

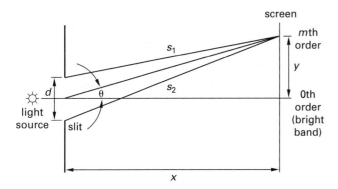

For reinforcement (i.e., a bright band) to occur, the difference in path lengths must be a whole number, $m$, of wavelengths. The number $m$ is known as the *order of interference*. The central bright band is of the zeroth order, and there are two of each $m$th order image, one on each side of the center. When $x \gg d$, small angle approximations are valid and $y/x = (s_2 - s_1)/d$. For the $m$th reinforcement (light band, maximum, etc.),

$$s_1 - s_2 = m\lambda \quad \text{[in phase]} \qquad 66.12$$

$$y = \frac{m\lambda x}{d} \quad \text{[in phase]} \qquad 66.13$$

Between the bright bands are dark bands of cancellation. For cancellation, the difference in path lengths is an odd number of half wavelengths. For the $m$th cancellation (dark band, minimum, etc.),

$$s_1 - s_2 = \frac{(2m + 1)\lambda}{2} \quad \text{[out of phase]} \qquad 66.14$$

$$y = \frac{(2m + 1)\lambda x}{2d} \quad \text{[out of phase]} \qquad 66.15$$

**Example 66.8**

A screen is placed 2.4 m from a sodium gas discharge tube emitting two beams of in-phase light ($\lambda = 5.893 \times 10^{-7}$ m) separated by 0.0005 m. What is the distance from the central image to the first reinforcement?

*Solution*

From Eq. 66.13 with $m = 1$,

$$y = \frac{m\lambda x}{d} = \frac{(1)(5.893 \times 10^{-7} \text{ m})(2.4 \text{ m})}{0.0005 \text{ m}}$$

$$= 0.00282 \text{ m}$$

## 14. INTERFERENCE FROM THIN FILMS

Interference occurs when light passes through thin films (walls of soap bubbles, layers of oil on water, etc.)[9] If the incident beam is white light (i.e., is composed of light of all colors), each wavelength will produce its own interference pattern, resulting in a rainbow effect.

Figure 66.7 shows a solid reflective surface covered by a transparent film of thickness $t$. Ray D is composed of a partial reflection of ray B and the remainder of ray A. Because of the 180 degree phase reversal a point G, cancellation along ray D requires that the path FGH be an integral number of wavelengths. However, the path length is also approximately equal to twice the film thickness. Equation 66.16 and Eq. 66.18 define the relationships for the $m$th order reinforcement and cancellation, respectively. It is essential that the wavelength in the film, $\lambda_{\text{film}}$, be used, not the free space wavelength.

$$\left(m + \tfrac{1}{2}\right)\lambda_{\text{film}} = 2t \quad \text{[in phase]} \qquad 66.16$$

$$m\lambda_{\text{film}} = 2t \quad \text{[out of phase]} \qquad 66.17$$

$$\lambda_{\text{film}} = \frac{\lambda_{\text{vacuum}}}{n_{\text{film}}} \qquad 66.18$$

**Figure 66.7** *Interference from a Thin Film*

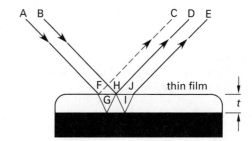

## 15. DIFFRACTION

*Diffracted light* is light whose path has been changed by passing around corners or through narrow slits. As a consequence of *Huygens' principle*, each edge of a single diffraction slit (shown in Fig. 66.8) acts as a source of light waves and is capable of producing interference. For diffraction of monochromatic light by a single slit, the $m$th order reinforcements and cancellations are the same as predicted by Eq. 66.12 through Eq. 66.15.

The same equations can also be used for a *diffraction grating*, a transparent sheet containing numerous slits (scratches) spaced a distance $d$ (known as the *grating space* or *grating constant*) apart and typically found in a *diffraction grating spectrometer*.[10,11] The angle at which the spectrometer must be turned to view an $m$th order

---

[9]A circular pattern of dark bands known as *Newton's rings* is created by interference in a film of thin air trapped between a flat reflecting surface and a plano-convex lens placed flat side up.
[10]The number of slits per centimeter ranges from 400 to 6000.
[11]While the number of slits (one, two, or hundreds) does not affect the position of the image, it does affect the brightness of the image.

**Figure 66.8** *Diffraction Around a Corner*

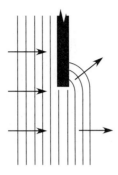

image can be calculated from $\sin\theta = y/s \approx y/x$. The maximum number of orders of interference produced can be determined by setting $\theta$ equal to $90°$.

$$\sin\theta = \frac{m\lambda}{d} \quad \text{[in phase]} \qquad 66.19$$

$$\sin\theta = \frac{(2m+1)\lambda}{2d} \quad \text{[out of phase]} \qquad 66.20$$

## 16. REFRACTION

*Refraction* is the bending of light as it passes from one transparent medium into another. *Snell's law*, Eq. 66.21, relates the incident and refracted angles and predicts that the light will bend *toward the normal* when it enters an optically denser material. For a vacuum (and, for practical purposes, air), $n_1 = 1$.

$$n_{\text{relative}} = \frac{n_2}{n_1} = \frac{\sin\theta_1}{\sin\theta_2} \qquad 66.21$$

If a light beam passes through a transparent medium with parallel surfaces, the emergent beam will be parallel to the incident beam, as illustrated in Fig. 66.9(a). However, the refraction due to a prism with apex angle $\alpha$, shown in Fig. 66.9(b), is more complex. Equation 66.22 predicts the minimum *angle of refraction (angle of deviation)* from the original path.[12]

$$n_{\text{relative}} = \frac{\sin\frac{1}{2}(\alpha + \theta)}{\sin\frac{1}{2}\alpha} \qquad 66.22$$

Due to refraction, submerged (underwater) objects appear closer to the surface than they actually are. Equation 66.23 gives the apparent depth.

$$\text{apparent depth} = \frac{\text{actual depth}}{n_{\text{relative}}} \qquad 66.23$$

---

[12]Different wavelengths (colors) will be reflected at slightly different angles. Equation 66.22 gives the minimum deviation angle.

**Figure 66.9** *Refraction of Light*

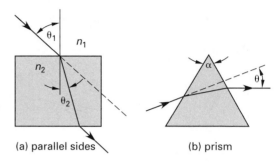

(a) parallel sides          (b) prism

## 17. LENSES

Lenses use refraction to form images of objects. Lenses are classified as converging or diverging depending on what they do to parallel light rays. A *converging lens* is thicker at its center, and parallel light rays will converge to the *focus*, F. Regardless of the direction of incoming rays, the image will always appear on the *focal plane*. A converging lens is classified as *bi-convex* (if both surfaces curve outward as shown in Fig. 66.10), *plano-convex* (if it has one flat surface), or *convexo-concave* (if one surface curves inward and the other curves outward). A *diverging lens* is thicker at its edges, and parallel rays will appear to originate from the *virtual focus*.

**Figure 66.10** *Focal Plane of a Converging Lens*

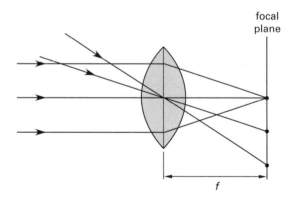

A *thin lens* has a thickness that is small compared to its diameter. Because they are easiest to produce, most thin lenses are *spherical lenses*, although the two surfaces do not necessarily have the same radii of curvature. Only thin lenses are covered in this chapter.

The *principal axis* is a line that passes through the centers of curvature, points $C_1$ and $C_2$, and the foci, points F and F'. The foci are equidistant from the *optical center*, the midpoint of the thickness along the principal axis, and distance $f$ is the *focal length*.[13] All light rays that pass through the optical center, regardless of the direction from which they come, pass through

---

[13]If the lens is assumed to be thin, little error will be introduced by measuring distances from the outside surface.

without deviation. Figure 66.11 illustrates common lens nomenclature.

**Figure 66.11** *Lens Nomenclature*

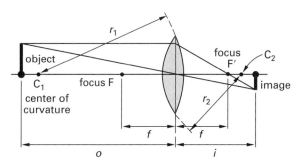

## 18. SPHERICAL AND CHROMATIC ABERRATION

The surfaces of most conventional lenses are spherical, primarily because it is prohibitively time-consuming and expensive to make lenses with curvatures that change from center to edge. However, such a shape introduces *spherical aberration*: light rays passing through the edges of the lens converge before those that go through the center, blurring the image.

Because different wavelengths are refracted at slightly different angles, a conventional lens will act like a prism, and beams of different colors will focus at different points. White light contains different colors, so the image from a common lens will have a slight rainbow effect, known as *chromatic aberration*, around the edges. Special combinations of lens materials can be formed into *compound lenses*, however, to create an *achromatic lens.*[14]

A third problem associated with lenses is *coma*, which gives points of light—light bulbs, for example—comet-like tails.

*Gradient-index glass*—glass whose density (and hence index of refraction) varies with the location within the lens—is difficult to manufacture but is capable of solving these problems. Spherical aberration can be cured by making the edges of a lens less dense than the center. Chromatic aberration can be cured by varying the density from front to back.[15]

## 19. POWER OF A LENS

The *power* of a lens is a function of its curvature and is calculated as the reciprocal of the focal length, *f*. By convention, the focal length is positive for a converging lens and negative for a diverging lens. Power is

measured in *diopters*, the reciprocal of the focal length in meters.[16]

$$P = \frac{1}{f_{\text{in meters}}} \qquad 66.24$$

## 20. IMAGE TYPE AND POSITION

Images of objects formed by converging rays are real, and images formed by diverging rays are virtual. Light actually passes through a *real image* but does not originate from a *virtual image*. (The location of a virtual image can be found by tracing diverging rays back to a single point.) An *erect image* will be oriented the same as the object (e.g., both right side up), while an *inverted image* will be reversed. For any single lens, a real image will be inverted and a virtual image will be erect. These relationships are summarized in Table 66.6.

## 21. LENS MAKER'S EQUATION

The *lens maker's equation*, Eq. 66.25, relates the focal length, radii of curvature, and refractive index of the lens material. By convention, the focal length, *f*, is positive for a converging lens and negative for a diverging lens. The radius of curvature is positive for a convex surface, negative for a concave surface, and infinite for a plane surface.

$$\frac{1}{f} = (n-1)\left(\frac{1}{r_2} + \frac{1}{r_2}\right) \qquad 66.25$$

## 22. LENS EQUATION

The *lens equation*, Eq. 66.26, determines the image position (i.e., the distance from the optical center to the image).[17] Distances *o* (to the object) and *i* (to the image) are positive on opposite sides of the lens. By convention, the focal length, *f*, is positive for a converging lens and negative for a diverging lens. (Also, see Table 66.6.)

$$\frac{1}{o} + \frac{1}{i} = \frac{1}{f} \qquad 66.26$$

## 23. MAGNIFICATION

Equation 66.27 gives the *magnification*, *m*, of a lens. A negative magnification means the image is inverted. For a bi-convex lens with object and image on opposite sides, both *o* and *i* will be positive (since *o* and *i* are

---

[14]High quality photographic lenses may incorporate as many as six different lenses. A zoom lens may have up to 12 or 14 separate lenses, yet the image may still be less than perfect.

[15]Correction of all the aberrations simultaneously is too complex for even a gradient-index lens, however.

[16]The power of the human eye is approximately +60 diopters. Eyeglasses used to correct small vision errors have powers in the range of −5 to +5 diopters.

[17]Another method, the *method of curvatures*, can also determine the image position but is not covered in this book.

**Table 66.6** *Images from Spherical Lenses*[*]

| type of lens | object position | image position | image type | image size | image orientation |
| --- | --- | --- | --- | --- | --- |
| converging | $0 < o < f$ | $-\infty < i < 0$ | virtual | larger | erect |
| | $o = f$ | $i = \infty$ | – | – | – |
| | $f < o < 2f$ | $2f < i < \infty$ | real | larger | inverted |
| | $o = 2f$ | $i = 2f$ | real | same | inverted |
| | $2f < o < \infty$ | $f < i < 2f$ | real | smaller | inverted |
| | $o = \infty$ | $i = f$ | real | zero | – |
| diverging | $0 < o < \infty$ | $-f < i < 0$ | virtual | smaller | erect |
| | $o = \infty$ | $i = -f$ | virtual | zero | – |

[*]Distances $o$ and $i$ are positive in opposite directions.

positive on opposite sides) and the magnification will be negative.

$$m = -\frac{i}{o} \qquad 66.27$$

### Example 66.9

A bi-convex lens is manufactured from glass with an index of refraction of 1.52. The radii of curvature for the two surfaces are 0.14 m and 0.20 m. A 2 cm high object is placed 0.28 m from the lens. What are the (a) focal length, (b) position of the image, and (c) magnification?

*Solution*

(a) The focal length is calculated from the lens maker's equation, Eq. 66.25.

$$\frac{1}{f} = (n-1)\left(\frac{1}{r_1} + \frac{1}{r_2}\right)$$
$$= (1.52 - 1)\left(\frac{1}{0.14\text{ m}} + \frac{1}{0.20\text{ m}}\right) = 6.314 \text{ 1/m}$$
$$f = \frac{1}{6.314\,\frac{1}{\text{m}}} = 0.158 \text{ m}$$

(b) The image position is calculated from the lens equation, Eq. 66.26.

$$\frac{1}{i} = \frac{1}{f} - \frac{1}{o} = 6.314\,\frac{1}{\text{m}} - \frac{1}{0.28\text{ m}} = 2.743 \text{ 1/m}$$
$$i = \frac{1}{2.743\,\frac{1}{\text{m}}} = 0.365 \text{ m}$$

(c) The magnification is calculated from Eq. 66.27.

$$m = -\frac{i}{o} = -\frac{0.365\text{ m}}{0.28\text{ m}} = -1.30$$

Since $|m| > 1$, the image is larger than the object. Since $m < 0$, the image is inverted.

### 24. LENSES IN SERIES

The performance of most *compound lenses* and two or more lenses in series is difficult to evaluate. One exception is the case where two lenses are placed in contact with coinciding principal axes. In that case, Eq. 66.28 predicts the focal length of the lens combination.

$$\frac{1}{f} = \frac{1}{f_1} + \frac{1}{f_2} \qquad 66.28$$

When the two lenses are separated by a distance $s$, the image from the first lens should be found (independently of the second lens) and used as the object for the second lens. The magnification will be the product of the individual magnifications. Equation 66.29 calculates the focal length of a combination of separated lenses.

$$m = m_1 m_2 \qquad 66.29$$
$$\frac{1}{f} = \frac{1}{f_1} + \frac{1}{f_2} - \frac{s}{f_1 f_2} \qquad 66.30$$

### 25. REFLECTION

Reflected light leaves a reflecting surface at the same angle (as measured from a normal line) it approaches. (See Fig. 66.12.) While reflections are normally associated with smooth opaque surfaces, light can also be totally reflected from a transparent surface if the incident angle is sufficiently large. Equation 66.31 gives the *critical incident angle* at which an optically transparent surface becomes totally reflecting. Total reflecting prisms are used in place of silvered mirrors when precise reflection is required. Total internal reflection is also the principle by which *optical fibers (light pipes)* transmit light.

$$\sin \theta_c = \frac{1}{n} \qquad 66.31$$

### 26. PLANE MIRRORS

The method of determining the image position of an object in a plane mirror is illustrated in Fig. 66.13. The image appears to be at point I but is a *virtual image* because no rays of light actually pass through or originate from that point. Furthermore, no image would be projected onto a screen or photographic plate placed at point I. Conversely, a *real image* would project onto a

Physics

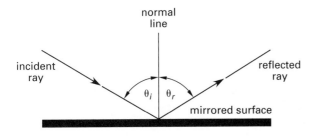

**Figure 66.12** *Reflection from a Surface*

screen. Distances $s$ and $s'$ are equal, but the image is reversed from right to left.

**Figure 66.13** *Plane Mirror Image*

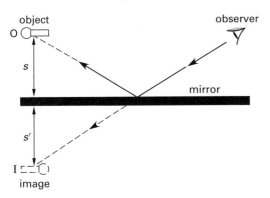

## 27. REFLECTION FROM A ROTATING MIRROR

When a flat mirror receiving a beam of light from a stationary source is rotated through an angle $\alpha$, the reflected beam will rotate $2\alpha$ from its original position. (See Fig. 66.14.)

**Figure 66.14** *Image from a Rotating Mirror*

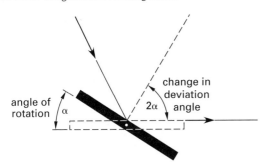

## 28. SPHERICAL MIRRORS

A *spherical mirror* can be either concave (when the inside is reflecting), or convex (when the outside is reflecting), as shown in Fig. 66.15. The *principal axis* is the line through the *center of curvature*, point C, and the *vertex*, point V. The *focal length*, $f$, is half of the radius of curvature. ($f = r/2$ is the full statement of the

*mirror maker's equation.*) The *mirror power* is $1/f$ (in diopters) as with lenses.

**Figure 66.15** *Spherical Mirrors*

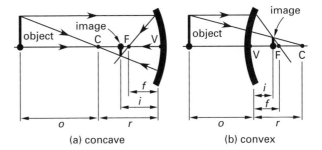

(a) concave        (b) convex

Spherical mirrors are subject to *spherical aberration.* (See Sec. 66.18.) If all incident rays are parallel, a *parabolic mirror* can be used in place of a spherical mirror to eliminate this problem.

## 29. IMAGES FROM A SPHERICAL MIRROR

Unlike a plane mirror, the image from a spherical mirror will not normally be the same size as the object. Table 66.7 lists the types, sizes, and positions of images formed from spherical mirrors. The focal length, $f$, is negative for a convex mirror. Images from convex mirrors are always reduced, erect, virtual images.

## 30. MIRROR EQUATION

The *mirror equation*, Eq. 66.32, relates the focal length to the object and image distances. This is the same as Eq. 66.26 for lenses, but the sign convention is different. The distances are positive on one side of the mirror and negative on the other, the convention corresponding to the sign of the focal length. The focal length is positive for a concave mirror and negative for a convex mirror. Therefore, distances to the left of the concave mirror in Fig. 66.15(a) are positive (because $f$ is positive and point F is located to the left of the mirror) and distances to the right are negative. For the convex mirror shown in Fig. 66.15(b), distances to the right are negative (because $f$ is negative and point F is located to the right of the mirror) and distances to the left are positive.

$$\frac{1}{o} + \frac{1}{i} = \frac{1}{f} = \frac{2}{r} \qquad 66.32$$

## 31. MIRROR MAGNIFICATION

Equation 66.33 calculates the magnification, $m$, produced by a mirror. An erect image will have a positive

**Physics**

**Table 66.7** *Images from Spherical Mirrors*
*(f is negative for a convex mirror.)*

| type of mirror | object position | image position | image type | image size | image orientation |
|---|---|---|---|---|---|
| concave | $o = 0$ | $i = 0$ | – | same | erect |
| | $f > o > 0$ | $0 > i > -\infty$ | virtual | larger | erect |
| | $o = f$ | $i = -\infty$ | – | – | – |
| | $r > o > f$ | $\infty > i > r$ | real | larger | inverted |
| | $o = r$ | $i = r$ | real | same | inverted |
| | $\infty > o > r$ | $r > i > f$ | real | smaller | inverted |
| | $o = \infty$ | $i = f$ | real | zero | – |
| convex | $o = 0$ | $i = 0$ | virtual | same | erect |
| | $\infty > o > 0$ | $0 > i > f$ | virtual | smaller | erect |
| | $o = \infty$ | $i = f$ | virtual | zero | – |

magnification; an inverted image will have a negative magnification.

$$m = -\frac{i}{o} = \frac{f}{f - o} = \frac{f - i}{f} \qquad 66.33$$

## Example 66.10

An object is placed 0.50 m from a convex mirror with a focal length of 0.2 m. What are the (a) image distance and (b) magnification?

*Solution*

(a) The image distance is calculated from Eq. 66.32. It is negative, indicating that the image appears to originate behind the mirror.

$$\frac{1}{i} = \frac{1}{f} - \frac{1}{o} = \frac{1}{-0.2 \text{ m}} - \frac{1}{0.5 \text{ m}}$$

$$= -7.0 \text{ m}^{-1}$$

$$i = -0.143 \text{ m}$$

(b) The magnification is calculated from Eq. 66.33. It is positive, indicating an erect image.

$$m = -\frac{i}{o} = -\frac{-0.143 \text{ m}}{0.50 \text{ m}}$$

$$= 0.286$$

# 67 Waves and Sound

## Nomenclature

| | | | |
|---|---|---|---|
| $a$ | speed of sound | ft/sec | m/s |
| $E$ | modulus of elasticity | $lbf/ft^2$ | Pa |
| $f$ | frequency | Hz | Hz |
| $F$ | wire tension | lbf | N |
| $g_c$ | gravitational constant | ft-lbm/lbf-sec$^2$ | n.a. |
| $I$ | sound intensity | dB | dB |
| $k$ | ratio of specific heats | – | – |
| $L$ | length | ft | m |
| $m_l$ | mass per unit length | lbm/ft | kg/m |
| MW | molecular weight | lbm/lbmole | kg/kmol |
| $R$ | specific gas constant | ft-lbf/lbm-°R | kJ/kg·K |
| $R^*$ | universal gas constant | ft-lbf/lbmole-°R | kJ/kmol·K |
| $T$ | period | sec | s |
| $T$ | temperature | °R | K |
| v | velocity | ft/sec | m/s |

## Symbols

| | | | |
|---|---|---|---|
| $\lambda$ | wavelength | ft | m |
| $\rho$ | density | $lbm/ft^3$ | $kg/m^3$ |

## Subscripts

| | |
|---|---|
| $l$ | per unit length |
| $n$ | $n$th harmonic |
| $o$ | reference or observer |
| $s$ | source |

## 1. PHYSIOLOGICAL BASIS OF SOUND

Longitudinal sound waves in the 20 Hz to 20,000 Hz range can be heard by most people. The average human ear is most sensitive to frequencies around 3000 Hz. Even in this range, however, the intensity must be great enough for a sound to be detected. *Intensity* is the amount of energy passing through a unit area each second. For example, the *threshold of audibility* at 3000 Hz is approximately $10^{-12}$ W/m$^2$. If the sound is too intense (i.e., the intensity is above the *threshold of pain*, appproximately 10 W/m$^2$), the sound will be painful. The thresholds of audibility and pain are both somewhat frequency-dependent.

*Loudness* is a qualitative sensation that can be quantified approximately by comparing the sound intensity to a reference intensity (usually the assumed threshold of audibility, $10^{-12}$ W/m$^2$). Loudness is perceived by most individuals approximately logarithmically and is calculated in that manner. (See Table 67.1.)

$$\text{loudness} = 10 \log\left(\frac{I}{I_o}\right)$$
$$= 10 \log\left(\frac{I}{10^{-12} \dfrac{\text{W}}{\text{m}^2}}\right) \qquad 67.1$$

**Table 67.1** *Approximate Loudness of Various Sources*

| source | loudness (dB) | perception |
|---|---|---|
| jet engine, thunder | 120 | painful |
| jackhammer | 110 | deafening |
| sheet metal shop | 90 | very loud |
| street noise | 70 | loud |
| office noise, normal speech | 50 | moderate |
| quiet conversation | 30 | quiet |
| whisper | 20 | faint |
| anechoic room | 10 | very faint |
| none | 0 | silence |

## 2. LONGITUDINAL SOUND WAVES

Sound, like light and electromagnetic radiation, is a wave phenomenon. However, sound has two distinct differences from electromagnetic waves. First, sound is transmitted as *longitudinal waves*, also known as *compression waves*. Such waves are alternating compressions and expansions of the medium through which they travel. In comparison, electromagnetic waves are *transverse waves*. Second, longitudinal waves require a medium through which to travel—they cannot travel through a vacuum, as can electromagnetic waves.

Physics

For ease of visualization, longitudinal waves may be represented by transverse waves. The *wavelength* is the distance between successive compressions (or expansions), as shown in Fig. 67.1.

**Figure 67.1** *Longitudinal Waves*

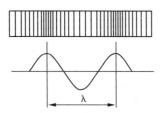

The usual relationships among wavelength, propagation velocity, period, and frequency are valid.

$$a = f\lambda \qquad \textit{67.2}$$

$$T = \frac{1}{f} \qquad \textit{67.3}$$

## 3. PROPAGATION VELOCITY OF LONGITUDINAL SOUND WAVES

The *propagation velocity* of longitudinal waves, commonly known as the *speed of sound* and *sonic velocity*, depends on the compressibility of the supporting medium. For solids, the compressibility is accounted for by the elastic modulus, $E$. For liquids, it is customary to refer to the elastic modulus as the *bulk modulus*.

$$a = \sqrt{\frac{E}{\rho}} \qquad [\text{SI}] \quad \textit{67.4(a)}$$

$$a = \sqrt{\frac{Eg_c}{\rho}} \qquad [\text{U.S}] \quad \textit{67.4(b)}$$

For ideal gases, the propagation velocity is calculated from Eq. 67.5.

$$a = \sqrt{kRT} = \sqrt{\frac{kR^* T}{\text{MW}}} \qquad [\text{SI}] \quad \textit{67.5(a)}$$

$$a = \sqrt{kg_c RT} = \sqrt{\frac{kg_c R^* T}{\text{MW}}} \qquad [\text{U.S.}] \quad \textit{67.5(b)}$$

Table 67.2 lists approximate values for the speed of sound in various media.

## 4. INTERVALS AND OCTAVES

An *interval* is a ratio (2:3, 3:5, etc.) of the frequencies of two pure tones. An *octave interval* is an integer ratio (2:1, 3:1, etc.) or a whole fraction ratio (1:2, 1:3, etc.). The basis of most music is our perception that combinations of pure tones with intervals in ratios of small whole numbers are pleasant. Pleasant combinations are known

**Table 67.2** *Approximate Speeds of Sound (at one atmospheric pressure).*

| material | speed of sound | |
|---|---|---|
| | m/s | ft/sec |
| air | 330 at 0°C | 1130 at 70°F |
| aluminum | 4990 | 16,400 |
| carbon dioxide | 260 at 0°C | 870 at 70°F |
| hydrogen | 1260 at 0°C | 3310 at 70°F |
| steel | 5150 | 16,900 |
| water | 1490 at 20°C | 4880 at 70°F |

(Multiply ft/sec by 0.3048 to obtain m/s.)

as *consonant intervals*; unpleasant combinations are known as *discordant* or *dissonant intervals*.[1] The principal intervals are listed in Table 67.3.

**Table 67.3** *Principal Musical Intervals*

| interval | name |
|---|---|
| 1:2 | octave |
| 2:3 | fifth |
| 3:4 | fourth |
| 4:5 | major third |
| 5:6 | minor third |
| 3:5 | major sixth |
| 5:8 | minor sixth |
| 7:8 | (dissonant) |

## 5. MUSICAL SCALES

A *musical scale* is a series of tones ("do-re-mi," etc.), known as *notes*, that sound good when used in combination. The first note in the scale is known as the *root note*. The pitch of musical tones is standardized so that performances will be independent of instrument tuning.[2] (In music, the word *pitch* is used in place of *frequency*.) For example, the pitch of *middle C* is (approximately) 260 Hz; A above middle C has a pitch of 440 Hz.

Ideally, all intervals would be exactly consonant, as it is with the *diatonic scale*, which has a root note of C. However, the intervals between successive pairs of diatonic tones are not constant, making changes from the key of C difficult. With the *equally tempered scale*, the frequencies are changed slightly to equally space the notes, and the intervals between pairs are no longer exactly consonant. However, the dissonance is not readily apparent.

## 6. HARMONICS AND OVERTONES

Musical sounds are seldom pure tones; they are combinations of several *partial tones*—a fundamental tone and its octave harmonics (overtones). An *overtone* is any tone produced by a sound source in addition to the fundamental tone. The overtones of stringed and

---

[1]Whether an interval is consonant or dissonant is a matter of custom and education.
[2]The difference between the *American concert pitch* (A = 440 Hz) and the *international pitch* (A = 435 Hz) is small.

Physics

wind instruments are octaves of the fundamental tone. *Ohm's acoustic law* states that the quality of a complex musical sound depends on the number of relative strengths of the various partials and not on the differences in phase. *Quality*, also known as *timbre*, is a qualitative term.

## 7. TAUT WIRES

Figure 67.2 illustrates several vibrational *modes* (ways of vibrating) for a taut wire. The points where the wire crosses over its position from a previous cycle are known as *nodes*. *Antinodes* are located halfway between nodes.

**Figure 67.2** *Harmonics (Overtones) in a Taut Wire*

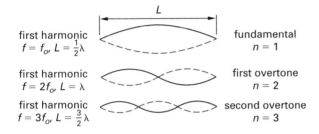

The *fundamental* (*first harmonic* or *first mode*) *wavelength* generated by a taut wire of length $L$ is $\lambda_1 = 2L$. The wavelength of the $n$th harmonic ($n$th mode or $(n-1)$th overtone) is

$$\lambda_n = \frac{2L}{n} \qquad 67.6$$

The fundamental and harmonic frequencies produced by a taut wire are given by Eq. 67.7. As in Eq. 67.6, $n = 1$ for the fundamental frequency.

$$f = \frac{n \mathrm{v}_{\text{transverse}}}{2L} \qquad 67.7$$

The velocity of a transverse wave along a taut wire depends on the wire tension, $F$, and the mass per unit length, $m_l$.

$$\mathrm{v}_{\text{transverse}} = \sqrt{\frac{F}{m_l}} \qquad \text{[SI]} \quad 67.8(a)$$

$$\mathrm{v}_{\text{transverse}} = \sqrt{\frac{F g_c}{m_l}} \qquad \text{[U.S.]} \quad 67.8(b)$$

## 8. RESONANCE AND BEATS

*Resonance* (also known as *sympathetic vibration*) occurs when a body (e.g., a wire or gas column) capable of producing sound receives impulses at one of its fundamental or harmonic frequencies.

If two sources produce sound with slightly different frequencies, the combined sound will have a beating

effect. The effect is due to successive reinforcement and interference of the waves from both sources. Audible *beats* occur only when the two frequencies are very close. The beat frequency is the difference of the two frequencies.

$$f_{\text{beats}} = f_1 - f_2 \qquad 67.9$$

## 9. RESONANCE IN GAS COLUMNS

In resonant gas columns, a node will always form at a closed end, and an antinode will form at an open end. The gas in a hollow cylinder closed at one end and open at the other will be resonant if its length is any quarter multiple of the wavelength. Thus, for resonance at the fundamental frequency, the length of the cylinder must be $\lambda/4$. In general, the wavelength and length of cylinder (closed at one end) are related to the harmonic number.

$$\lambda = \frac{4L}{2n-1} \qquad 67.10$$

If the cylinder is open at both ends, the length must be $\lambda/2$ for resonance to occur. The cylinder length, wavelength, and harmonic number are related by Eq. 67.11. (Longitudinal sound waves are represented as transverse waves in Fig. 67.3 for convenience.)

$$\lambda = \frac{2L}{n} \qquad 67.11$$

**Figure 67.3** *Resonance in Gas Columns*

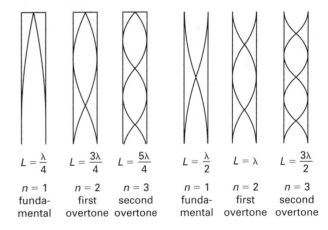

## 10. DOPPLER EFFECT

If the distance between a sound source and a listener is changing, the frequency heard will differ from the frequency emitted. If the separation distance is decreasing, the frequency will be shifted higher; if the separation distance is increasing, the frequency will be shifted lower. This shifting is known as the *Doppler effect*. The ratio of observed frequency ($f'$) to emitted frequency ($f$) depends on the local speed of sound ($a$) and the absolute velocities of the source and observer ($v_s$ and $v_o$). In Eq. 67.12, $v_s$ is positive if the source moves away from the observer and is negative otherwise; $v_o$ is positive if the observer moves toward the source and negative otherwise.

$$\frac{f'}{f} = \frac{a + v_o}{a + v_s} \qquad 67.12$$

## Example 67.1

On a particular day, the local speed of sound is 1130 ft/sec (344 m/s). What frequency is heard by a stationary pedestrian when a 1200 Hz siren on an emergency vehicle moves away at 45 mph (72 km/h)?

*SI Solution*

The observer's velocity is zero. The source's velocity is positive because the vehicle is moving away. From Eq. 67.12, the frequency heard is

$$f' = f\left(\frac{a + v_o}{a + v_s}\right)$$

$$= (1200 \text{ Hz})\left(\frac{344 \frac{m}{s} + 0}{344 \frac{m}{s} + \frac{\left(72 \frac{km}{h}\right)\left(1000 \frac{m}{km}\right)}{3600 \frac{s}{h}}}\right)$$

$$= 1134 \text{ Hz}$$

*Customary U.S. Solution*

The source velocity is positive and must be converted from mi/hr to ft/sec.

$$v_s = \frac{\left(45 \frac{mi}{hr}\right)\left(5280 \frac{ft}{mi}\right)}{3600 \frac{sec}{hr}} = 66 \text{ ft/sec}$$

From Eq. 67.12,

$$f' = f\left(\frac{a + v_o}{a + v_s}\right) = (1200 \text{ Hz})\left(\frac{1130 \frac{ft}{sec} + 0}{1130 \frac{ft}{sec} + 66 \frac{ft}{sec}}\right)$$

$$= 1134 \text{ Hz}$$

# 68 Instrumentation and Measurements

## Nomenclature

| | | | |
|---|---|---|---|
| $A$ | area | $in^2$ | $m^2$ |
| $b$ | base length | in | m |
| BC | bridge constant | – | – |
| $c$ | distance from neutral axis | in | m |
| $C$ | concentration | various | various |
| $d$ | diameter | in | m |
| $E$ | modulus of elasticity | $lbf/in^2$ | Pa |
| $F$ | force | lbf | N |
| $G$ | shear modulus | $lbf/in^2$ | Pa |
| GF | gage factor | – | – |
| $h$ | height | in | m |
| $I$ | current | A | A |
| $I$ | moment of inertia | $in^4$ | $m^4$ |
| $J$ | polar moment of inertia | $in^4$ | $m^4$ |
| $k$ | constant | various | various |
| $k$ | deflection constant | lbf/in | N/m |
| $K$ | factor | – | – |
| $L$ | shaft length | in | m |
| $M$ | moment | in-lbf | N·m |
| $n$ | number | – | – |
| $n$ | rotational speed | rpm | rpm |
| $p$ | pressure | $lbf/in^2$ | Pa |
| $P$ | permeability | various | various |
| $P$ | power | hp | kW |
| $Q$ | statical moment | $in^3$ | $m^3$ |
| $r$ | radius | in | m |
| $R$ | resistance | $\Omega$ | $\Omega$ |
| $t$ | thickness | in | m |
| $T$ | temperature | °R | K |
| $T$ | torque | in-lbf | N·m |
| $V$ | voltage | V | V |
| VR | voltage ratio | – | – |
| $y$ | deflection | in | m |

### Symbols

| | | | |
|---|---|---|---|
| $\alpha$ | temperature coefficient | 1/°R | 1/K |
| $\beta$ | constant | °R | K |
| $\beta$ | temperature coefficient | $1/°R^2$ | $1/K^2$ |
| $\gamma$ | shear strain | – | – |
| $\epsilon$ | strain | – | – |
| $\eta$ | efficiency | – | – |
| $\theta$ | angle of twist | deg | deg |
| $\nu$ | Poisson's ratio | – | – |
| $\rho$ | resistivity | $\Omega$-in | $\Omega$·cm |
| $\sigma$ | stress | $lbf/in^2$ | Pa |
| $\tau$ | shear stress | $lbf/in^2$ | Pa |
| $\phi$ | angle of twist | rad | rad |

### Subscripts

| | |
|---|---|
| $b$ | battery |
| $g$ | gage |
| $o$ | original |
| $r$ | ratio |
| ref | reference |
| $t$ | transverse or total |
| $T$ | at temperature $T$ |
| $x$ | in $x$-direction |
| $y$ | in $y$-direction |

## 1. ACCURACY

A measurement is said to be *accurate* if it is substantially unaffected by (i.e., is insensitive to) all variation outside of the measurer's control.

For example, suppose a rifle is aimed at a point on a distant target and several shots are fired. The target point represents the "true value" of a measurement—the value that should be obtained. The impact points represent the measured values—what is obtained. The distance from the centroid of the points of impact to the

Physics

target point is a measure of the alignment accuracy between the barrel and the sights. This difference between the true and measured values is known as the measurement *bias*.

## 2. PRECISION

*Precision* is not synonymous with *accuracy*. Precision is concerned with the repeatability of the measured results. If a measurement is repeated with identical results, the experiment is said to be precise. The average distance of each impact from the centroid of the impact group is a measure of precision. Thus, it is possible to take highly precise measurements and still be inaccurate (i.e., have a large bias).

Most measurement techniques (e.g., taking multiple measurements and refining the measurement methods or procedures) that are intended to improve accuracy actually increase the precision.

Sometimes the term *reliability* is used with regard to the precision of a measurement. A *reliable measurement* is the same as a *precise estimate*.

## 3. STABILITY

*Stability* and *insensitivity* are synonymous terms. (Conversely, *instability* and *sensitivity* are synonymous.) A stable measurement is insensitive to minor changes in the measurement process.

### Example 68.1

At 65°F (18°C), the centroid of an impact group on a rifle target is 2.1 in (5.3 cm) from the sight-in point. At 80°F (27°C), the distance is 2.3 in (5.8 cm). What is the sensitivity to temperature?

*SI Solution*

$$\text{sensitivity to temperature} = \frac{\Delta \text{ measurement}}{\Delta \text{ temperature}}$$
$$= \frac{5.8 \text{ cm} - 5.3 \text{ cm}}{27°\text{C} - 18°\text{C}}$$
$$= 0.0556 \text{ cm/}°\text{C}$$

*Customary U.S. Solution*

$$\text{sensitivity to temperature} = \frac{\Delta \text{ measurement}}{\Delta \text{ temperature}}$$
$$= \frac{2.3 \text{ in} - 2.1 \text{ in}}{80°\text{F} - 65°\text{F}}$$
$$= 0.0133 \text{ in/}°\text{F}$$

## 4. CALIBRATION

*Calibration* is used to determine or verify the scale of the measurement device. In order to calibrate a measurement device, one or more known values of the quantity to be measured (temperature, force, torque, etc.) are applied to the device and the behavior of the device is noted. (If the measurement device is linear, it may be adequate to use just a single calibration value. This is known as *single-point calibration*.)

Once a measurement device has been calibrated, the calibration signal should be reapplied as often as necessary to prove the reliability of the measurements. In some electronic measurement equipment, the calibration signal is applied continuously.

## 5. ERROR TYPES

Measurement errors can be categorized as *systematic (fixed) errors*, *random (accidental) errors*, *illegitimate errors*, and *chaotic errors*.

*Systematic errors*, such as improper calibration, use of the wrong scale, and incorrect (though consistent) technique, are essentially constant or similar in nature over time. *Loading errors* are systematic errors and occur when the act of measuring alters the true value.[1] Some *human errors*, if present in each repetition of the measurement, are also systematic. Systematic errors can be reduced or eliminated by refinement of the experimental method.

*Random errors* are caused by random and irregular influences generally outside the control of the measurer. Such errors are introduced by fluctuations in the environment, changes in the experimental method, and variations in materials and equipment operation. Since the occurrence of these errors is irregular, their effects can be reduced or eliminated by multiple repetitions of the experiment.

There is no reason to expect or tolerate *illegitimate errors* (e.g., errors in computations and other blunders). These are essentially mistakes that can be avoided through proper care and attention.

*Chaotic errors*, such as resonance, vibration, or experimental "noise," essentially mask or entirely invalidate the experimental results. Unlike the random errors previously mentioned, chaotic disturbances are sufficiently large to reduce the experimental results to meaninglessness.[2] Chaotic errors must be eliminated.

---

[1]For example, inserting an air probe into a duct will change the flow pattern and velocity around the probe.

[2]Much has been written in recent years about *chaos theory*. This theory holds that, for many processes, the ending state is dependent on imperceptible differences in the starting state. Future weather conditions and the landing orientation of a finely balanced spinning top are often used as examples of states that are greatly affected by their starting conditions.

## 6. ERROR MAGNITUDES

If a single measurement is taken of some quantity whose true value is known, the *error* is simply the difference between the true and measured values. However, the true value is never known in an experiment, and measurements are usually taken several times, not just once. Therefore, many conventions exist for estimating the unknown error.

When most experimental quantities are measured, the measurements tend to cluster around some "average value." The measurements will be distributed according to some distribution, such as linear, normal, Poisson, and so on. The measurements can be graphed in a *histogram* and the distribution inferred. Usually the data will be normally distributed.[3]

Certain error terms used with normally distributed data have been standardized. These are listed in Table 68.1.

**Table 68.1** *Normal Distribution Error Terms*

| term | number of standard deviations | percent certainty | approximate odds of being incorrect |
|------|------|------|------|
| probable error | 0.6745 | 50 | 1 in 2 |
| mean deviation | 0.6745 | 50 | 1 in 2 |
| standard deviation | 1.000 | 68.3 | 1 in 3 |
| one-sigma error | 1.000 | 68.3 | 1 in 3 |
| 90% error | 1.6449 | 90 | 1 in 10 |
| two-sigma error | 2.000 | 95 | 1 in 20 |
| three-sigma error | 3.000 | 99.7 | 1 in 370 |
| maximum error[*] | 3.29 | 99.9+ | 1 in 1000 |

[*]The true maximum error is theoretically infinite.

## 7. POTENTIOMETERS

A *potentiometer (potentiometer transducer, variable resistor)* is a resistor with a sliding third contact. It converts linear or rotary motion into a variable resistance (voltage).[4] It consists of a resistance oriented in a linear or angular manner and a variable-position contact point known as the *tap*. A voltage is applied across the entire resistance, causing current to flow through the resistance. The voltage at the tap will vary with tap position. (See Fig. 68.1.)

---

[3]The results of all numerical experiments are not automatically normally distributed. The throw of a die (one of two dice) is linearly distributed. Emissive power of a heated radiator is skewed with respect to wavelength. However, the means of sets of experimental data generally will be normally distributed, even if the raw measurements are not.

[4]There is a voltage-balancing device that shares the name *potentiometer (potentiometer circuit)*. An unknown voltage source can be measured by adjusting a calibrated voltage until a null reading is obtained on a voltage meter. The applications are sufficiently different that no confusion occurs when the "pot is adjusted."

**Figure 68.1** *Potentiometer Circuit Diagram*

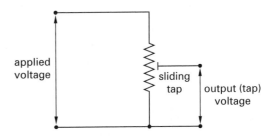

## 8. TRANSDUCERS

Physical quantities are often measured with transducers (*detector-transducers*). A *transducer* converts one variable into another. For example, a Bourdon tube pressure gauge converts pressure to angular displacement; a strain gauge converts stress to resistance change. Transducers are primarily mechanical in nature (e.g., pitot tube, spring devices, Bourdon tube pressure gauge) or electrical in nature (e.g., thermocouple, strain gauge, moving-core transformer).

## 9. SENSORS

While the term "transducer" is commonly used for devices that respond to mechanical input (force, pressure, torque, etc.), the term *sensor* is commonly applied to devices that respond to varying chemical conditions.[5] For example, an electrochemical sensor might respond to a specific gas, compound, or ion (known as a *target substance* or *species*). Two types of electrochemical sensors are in use today: potentiometric and amperometric.

*Potentiometric sensors* generate a measurable voltage at their terminals. In electrochemical sensors taking advantage of half-cell reactions at electrodes, the generated voltage is proportional to the absolute temperature, $T$, and is inversely proportional to the number of electrons, $n$, taking part in the chemical reaction at the half-cell. In Eq. 68.1, $p_1$ is the partial pressure of the target substance at the measurement electrode; $p_2$ is the partial pressure of the target substance at the reference electrode.

$$V \propto \left( \frac{T_{\text{absolute}}}{n} \right) \ln \frac{p_1}{p_2} \qquad 68.1$$

*Amperometric sensors* (also known as *voltammetric sensors*) generate a measurable current at their terminals. In the conventional electrochemical sensors known as *diffusion-controlled cells*, a high-conductivity acid or alkaline liquid electrolyte is used with a gas-permeable membrane that transmits ions from the outside to the inside of the sensor. A reference voltage is applied to two terminals within the electrolyte, and the current generated at a (third) sensing electrode is measured.

---

[5]The categorization is common but not universal. The terms "transducer," "sensor," and "pickup" are often used loosely.

Physics

The maximum current generated is known as the *limiting current*. Current is proportional to the concentration ($C$) of the target substance, the permeability ($P$), the exposed sensor (membrane) area ($A$), and the number of electrons transferred per molecule detected ($n$). The current is inversely proportional to the membrane thickness ($t$).

$$I \propto \frac{nPCA}{t} \qquad \qquad 68.2$$

## 10. VARIABLE-INDUCTANCE TRANSDUCERS

*Inductive transducers* contain a wire coil and a moving permeable *core*.[6] As the core moves, the flux linkage through the coil changes. The change in inductance affects the overall impedance of the detector circuit.

The *differential transformer* or *linear variable differential transformer* (LVDT) is an important type of *variable-inductance transducer*. (See Fig. 68.2.) It converts linear motion into a change in voltage. The transformer is supplied with a low AC voltage. When the core is centered between the two secondary windings, the LVDT is said to be in its *null position*.

**Figure 68.2** *Linear Variable Differential Transformer Schematic and Performance Characteristic*

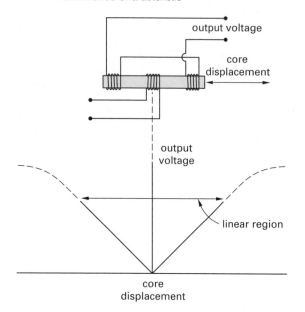

Movement of the core changes the magnetic flux linkage between the primary and secondary windings. Over a reasonable displacement range, the output voltage is proportional to the displacement of the core from the null position, hence the "linear" designation. The voltage changes phase (by 180°) as the core passes through the null position.

---

[6]The term *core* is used even if the cross sections of the coil and core are not circular.

Sensitivity of a LVDT is measured in mV/in (mV/cm) of core movement. The sensitivity and output voltage depend on the frequency of the applied voltage (i.e., the *carrier frequency*) and are directly proportional to the magnitude of the applied voltage.

## 11. VARIABLE-RELUCTANCE TRANSDUCERS

A *variable-reluctance transducer (pickup)* is essentially a permanent magnet and a coil in the vicinity of the process being monitored.[7] There are no moving parts in this type of transducer. However, some of the magnet's magnetic flux passes through the surroundings, and the presence or absence of the process changes the coil voltage. Two typical applications of variable-reluctance pickups are measuring liquid levels and determining the rotational speed of a gear.

## 12. VARIABLE-CAPACITANCE TRANSDUCERS

In *variable-capacitance transducers*, the capacitance of a device can be modified by changing the plate separation, plate area, or dielectric constant of the medium separating the plates.

## 13. OTHER ELECTRICAL TRANSDUCERS

The *piezoelectric effect* is the name given to the generation of an electrical voltage when placed under stress.[8] *Piezoelectric transducers* generate a small voltage when stressed (strained). Since voltage is developed during the application of changing strain but not while strain is constant, piezoelectric transducers are limited to dynamic applications. Piezoelectric transducers may suffer from low voltage output, instability, and limited ranges in operating temperature and humidity.

The *photoelectric effect* is the generation of an electrical voltage when a material is exposed to light.[9] Devices using this effect are known as *photocells, photovoltaic cells, photosensors,* or *light-sensitive detectors*, depending on the applications. The sensitivity need not be to light in the visible spectrum. Photoelectric detectors can be made that respond to infrared and ultraviolet radiation. The magnitude of the voltage (or of the current in an attached circuit) will depend on the amount of illumination. If the cell

---

[7]*Reluctance* depends on the area, length, and permeability of the medium through which the magnetic flux passes.

[8]Quartz, table sugar, potassium sodium tartarate (Rochelle salt), and barium titanate are examples of piezoelectric materials. Quartz is commonly used to provide a stable frequency in electronic oscillators. Barium titanate is used in some ultrasonic cleaners and sonar-like equipment.

[9]While the *photogenerative (photovoltaic)* definition is the most common definition, the term "photoelectric" can also be used with *photoconductive devices* (those whose resistance changes with light) and *photoemissive devices* (those that emit light when a voltage is applied).

isreverse-biased by an external battery, its operation is similar to a constant-current source.[10] (See Fig. 68.3.)

**Figure 68.3** *Photovoltaic Device Characteristic Curves*

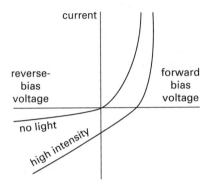

## 14. PHOTOSENSITIVE CONDUCTORS

*Cadmium sulfide* and *cadmium selenide* are two compounds that decrease in resistance when exposed to light. Cadmium sulfide is most sensitive to light in the 5000 Å to 6000 Å (0.5 $\mu$m to 0.6 $\mu$m) range, while cadmium selenide shows peak sensitivities in the 7000 Å to 8000 Å range (0.7 $\mu$m to 0.8 $\mu$m). These compounds are used in *photosensitive conductors*. Due to hysteresis, photosensitive conductors do not react instantaneously to changes in intensity.[11] High-speed operation requires high light intensities and careful design.

## 15. RESISTANCE TEMPERATURE DETECTORS

*Resistance temperature detectors* (RTDs), also known as *resistance thermometers*, make use of changes in their resistance to determine changes in temperature. A fine wire is wrapped around a form and protected with glass or a ceramic coating. Nickel and copper are commonly used for industrial RTDs. 100 $\Omega$ and 1000 $\Omega$ platinum is used when precision resistance thermometry is required. RTDs are connected through resistance bridges (see Sec. 68.19) to compensate for lead resistance.

Resistance in most conductors increases with temperature. The resistance at a given temperature can be calculated from the *coefficients of thermal resistance*, $\alpha$ and $\beta$.[12] The variation of resistance with temperature

---

[10]A semiconductor device is *reverse-biased* when a negative battery terminal is connected to a p-type semiconductor material in the device, or when a positive battery terminal is connected to an n-type semiconductor material.

[11]*Hysteresis* is the tendency for the transducer to continue to respond (i.e., indicate) when the load is removed. Alternatively, hysteresis is the difference in transducer outputs when a specific load is approached from above and from below. Hysteresis is usually expressed in percent of the full-load reading during any single calibration cycle.

[12]Higher-order terms (third, fourth, etc.) are used when extreme accuracy is required.

is nonlinear, though $\beta$ is small and is often insignificant over short temperature ranges. Therefore, a linear relationship is often assumed and only $\alpha$ is used. In Eq. 68.3, $R_{\text{ref}}$ is the resistance at the reference temperature, $T_{\text{ref}}$, usually 100 $\Omega$ at 32°F (0°C).

$$R_T \approx R_{\text{ref}} \left(1 + \alpha \Delta T + \beta \Delta T^2\right) \qquad 68.3$$

$$\Delta T = T - T_{\text{ref}} \qquad 68.4$$

In commercial RTDs, $\alpha$ is referred to by the literal term *alpha-value*. There are two applicable alpha values for platinum, depending on the purity. Commercial platinum RTDs produced in the United States generally have alpha values of 0.00391 1/°C, while RTDs produced in Europe and other countries generally have alpha values of 0.00385 1/°C.

## 16. THERMISTORS

*Thermistors* are temperature-sensitive semiconductors constructed from oxides of manganese, nickel, and cobalt, and from sulfides of iron, aluminum, and copper. Thermistor materials are encapsulated in glass or ceramic materials to prevent penetration of moisture. Unlike RTDs, the resistance of thermistors decreases as the temperature increases.

Thermistor temperature-resistance characteristics are exponential. Depending on the brand, material, and construction, $\beta$ typically varies between 3400K and 3900K.

$$R = R_o e^k \qquad 68.5$$

$$k = \beta \left(\frac{1}{T} - \frac{1}{T_o}\right) \quad [T \text{ in K}] \qquad 68.6$$

Thermistors can be connected to measurement circuits with copper wire and soldered connections. Compensation of lead wire effects is not required because resistance of thermistors is very large, far greater than the resistance of the leads. Since the negative temperature characteristic makes it difficult to design customized detection circuits, some thermistor and instrumentation standardization has occurred. The most common thermistors have resistances of 2252 $\Omega$ at 77°F (25°C), and most instrumentation is compatible with them. Other standardized resistances are 3000 $\Omega$, 5000 $\Omega$, 10,000 $\Omega$, and 30,000 $\Omega$ at 77°F (25°C). (See Table 68.2.)

Thermistors typically are less precise and more unpredictable than metallic resistors. Since resistance varies exponentially, most thermistors are suitable for use only up to approximately 550°F (290°C).

**Physics**

**Table 68.2** *Typical Resistivities and Coefficients of Thermal Resistance*[a]

| conductor | resistivity[b] ($\Omega$·cm) | $\alpha^{c,d}$ (1/°C) |
|---|---|---|
| alumel[e] | $28.1 \times 10^{-6}$ | 0.0024 @ 212°F (100°C) |
| aluminum | $2.82 \times 10^{-6}$ | 0.0039 @ 68°F (20°C) |
| | | 0.0040 @ 70°F (21°C) |
| brass | $7 \times 10^{-6}$ | 0.002 @ 68°F (20°C) |
| constantan[f,g] | $49 \times 10^{-6}$ | 0.00001 @ 68°F (20°C) |
| chromel[h] | $0.706 \times 10^{-6}$ | 0.00032 @ 68°F (20°C) |
| copper, annealed | $1.724 \times 10^{-6}$ | 0.0043 @ 32°F (0°C) |
| | | 0.0039 @ 70°F (21°C) |
| | | 0.0037 @ 100°F (38°C) |
| | | 0.0031 @ 200°F (93°C) |
| gold | $2.44 \times 10^{-6}$ | 0.0034 @ 68°F (20°C) |
| iron (99.98% pure) | $10 \times 10^{-6}$ | 0.005 @ 68°F (20°C) |
| isoelastic[i] | $112 \times 10^{-6}$ | 0.00047 |
| lead | $22 \times 10^{-6}$ | 0.0039 |
| magnesium | $4.6 \times 10^{-6}$ | 0.004 @ 68°F (20°C) |
| manganin[j] | $44 \times 10^{-6}$ | 0.0000 @ 68°F (20°C) |
| monel[k] | $42 \times 10^{-6}$ | 0.002 @ 68°F (20°C) |
| nichrome[l] | $100 \times 10^{-6}$ | 0.0004 @ 68°F (20°C) |
| nickel | $7.8 \times 10^{-6}$ | 0.006 @ 68°F (20°C) |
| platinum | $10 \times 10^{-6}$ | 0.0039 @ 32°F (0°C) |
| | | 0.0036 @ 70°F (21°C) |
| platinum-iridium[m] | $24 \times 10^{-6}$ | 0.0013 |
| platinum-rhodium[n] | $18 \times 10^{-6}$ | 0.0017 @ 212°F (100°C) |
| silver | $1.59 \times 10^{-6}$ | 0.004 @ 68°F (20°C) |
| tin | $11.5 \times 10^{-6}$ | 0.0042 @ 68°F (20°C) |
| tungsten (drawn) | $5.8 \times 10^{-6}$ | 0.0045 @ 70°F (21°C) |

[a]Compiled from various sources. Data is not to be taken too literally, as values depend on composition and cold working.
[b]At 20°C (68°F)
[c]Values vary with temperature. Common values given when no temperature is specified.
[d]Multiply 1/°C by 5/9 to obtain 1/°F. Multiply ppm/°F by $1.8 \times 10^{-6}$ to obtain 1/°C.
[e]Trade name for 94% Ni, 2.5% Mn, 2% Al, 1% Si, 0.5% Fe (TM of Hoskins Manufacturing Co.)
[f]60% Cu, 40% Ni, also known by trade names *Advance, Eureka*, and *Ideal*.
[g]Constantan is also the name given to the composition 55% Cu and 45% Ni, an alloy with slightly different properties.
[h]Trade name for 90% Ni, 10% Cr (TM of Hoskins Manufacturing Co.)
[i]36% Ni, 8% Cr, 0.5% Mo, remainder Fe
[j]9% to 18% Mn, 11% to 4% Ni, remainder Cu
[k]33% Cu, 67% Ni
[l]75% Ni, 12% Fe, 11% Cr, 2% Mn
[m]95% Pt, 5% Ir
[n]90% Pt, 10% Rh

## 17. THERMOCOUPLES

A *thermocouple* consists of two wires of dissimilar metals joined at both ends.[13] One set of ends, typically called a *junction*, is kept at a known *reference temperature* while the other junction is exposed to the unknown temperature.[14] (See Fig. 68.4.) In a laboratory, the reference junction is often maintained at the *ice point*, 32°F (0°C), in an ice/water bath for convenience in later analysis. In commercial applications, the reference temperature can be any value, with appropriate compensation being made.

**Figure 68.4** *Thermocouple Circuits*

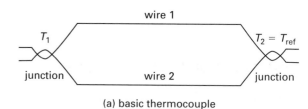

(a) basic thermocouple

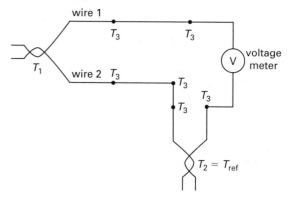

(b) thermocouple in measurement circuit

Thermocouple materials, standard ANSI designations, and approximate useful temperature ranges are given in Table 68.3.[15] The "usable" temperature range can be much larger than the useful range. The most significant factors limiting the useful temperature range, sometimes referred to as the *error limits range*, are linearity, the rate at which the material will erode due to oxidation at higher temperatures, irreversible magnetic effects above magnetic critical points, and longer stabilization periods at higher temperatures.

[13]The joint may be made by simply twisting the ends together. However, to achieve a higher mechanical strength and a better electrical connection, the ends should be soldered, brazed, or welded.
[14]The *ice point* is the temperature at which liquid water and ice are in equilibrium. Other standardized temperature references are: *oxygen point*, −297.346°F (90.19K); *steam point*, 212.0°F (373.16K); *sulfur point*, 832.28°F (717.76K); *silver point*, 1761.4°F (1233.96K); and *gold point*, 1945.4°F (1336.16K).
[15]It is not uncommon to list the two thermocouple materials with "vs." (as in *versus*). For example, a copper/constantan thermocouple might be designated as "copper vs. constantan."

**Table 68.3** *Typical Temperature Ranges of Thermocouple Materials[a]*

| materials | ANSI designation | useful range, °F (°C) |
|---|---|---|
| copper-constantan | T | −300 to 700 (−180 to 370) |
| chromel-constantan | E | 32 to 1600 (0 to 870) |
| iron-constantan[b] | J | 32 to 1400 (0 to 760) |
| chromel-alumel | K | 32 to 2300 (0 to 1260) |
| platinum-<br>10% rhodium | S | 32 to 2700 (0 to 1480) |
| platinum-<br>13% rhodium | R | 32 to 2700 (0 to 1480) |
| Pt-6% Rh-<br>Pt-30% Rh | B | 1600 to 3100 (870 to 1700) |
| tungsten-<br>Tu-25% rhenium | – | to 4200[c] (2320) |
| Tu-5% rhenium-<br>Tu-26% rhenium | – | to 4200[c] (2320) |
| Tu-3% rhenium-<br>Tu-25% rhenium | – | to 4200[c] (2320) |
| iridium-rhodium | – | to 3500[c] (1930) |
| nichrome-<br>constantan | – | to 1600[c] (870) |
| nichrome-alumel | – | to 2200[c] (1200) |

[a]Actual values will depend on wire gauge, atmosphere (oxidizing or reducing), use (continuous or intermittent), and manufacturer.
[b]Nonoxidizing atmospheres only.
[c]Approximate usable temperature range. Error limit range is less.

A voltage is generated when the temperatures of the two junctions are different. This phenomenon is known as the *Seebeck effect*.[16] Referring to the polarities of the voltage generated, one metal is known as the *positive element* while the other is the *negative element*. The generated voltage is small, and thermocouples are calibrated in $\mu V/°F$ or $\mu V/°C$. An amplifier may be required to provide usable signal levels, although thermocouples can be connected in series (a *thermopile*) to increase the value.[17] The accuracy (referred to as the *calibration*) of thermocouples is approximately $^1/_2\%$ to $^3/_4\%$, though manufacturers produce thermocouples with various guaranteed accuracies.

The voltage generated by a thermocouple is given by Eq. 68.7. Since the *thermoelectric constant*, $k_T$, varies with temperature, thermocouple problems are solved with published tables of total generated voltage versus temperature. (See App. 68.A.)

$$V = k_T(T - T_{\text{ref}}) \qquad 68.7$$

---

[16]The inverse of the Seebeck effect, that current flowing through a junction of dissimilar metals will cause either heating or cooling, is the *Peltier effect*, though the term is generally used in regard to cooling applications. An extension of the Peltier effect, known as the *Thompson effect*, is that heat will be carried along the conductor. Both the Peltier and Thompson effects occur simultaneously with the Seebeck effect. However, the Peltier and Thompson effects are so minuscule that they can be disregarded.
[17]There is no special name for a combination of thermocouples connected in parallel.

Generation of thermocouple voltage in a measurement circuit is governed by three laws. The *law of homogeneous circuits* states that the temperature distribution along one or both of the thermocouple leads is irrelevant. Only the junction temperatures contribute to the generated voltage.

The *law of intermediate metals* states that an intermediate length of wire placed within one leg or at the junction of the thermocouple circuit will not affect the voltage generated as long as the two new junctions are at the same temperature. This law permits the use of a measuring device, soldered connections, and extension leads.

The *law of intermediate temperatures* states that if a thermocouple generates voltage $V_1$ when its junctions are at $T_1$ and $T_2$, and it generates voltage $V_2$ when its junctions are at $T_2$ and $T_3$, then it will generate voltage $V_1 + V_2$ when its junctions are at $T_1$ and $T_3$.

### Example 68.2

A type-K (chromel-alumel) thermocouple produces a voltage of 10.79 mV. The "cold" junction is kept at 32°F (0°C) by an ice bath. What is the temperature of the hot junction?

*Solution*

Since the cold junction temperature corresponds to the reference temperature, the hot junction temperature is read directly from App. 68.A as 510°F.

### Example 68.3

A type-K (chromel-alumel) thermocouple produces a voltage of 10.87 mV. The "cold" junction is at 70°F. What is the temperature of the hot junction?

*Solution*

Use the law of intermediate temperatures. From App. 68.A, the thermoelectric constant for 70°F is 0.84 mV. If the cold junction had been at 32°C, the generated voltage would have been higher. The corrected reading is

$$10.87 \text{ mV} + 0.84 \text{ mV} = 11.71 \text{ mV}$$

The temperature corresponding to this voltage is 550°F.

### Example 68.4

A type-T (copper-constantan) thermocouple is connected directly to a voltage meter. The temperature of the meter's screw-terminals is measured by a nearby thermometer as 70°F. The thermocouple generates 5.262 mV. What is the temperature of its hot junction?

*Solution*

There are two connections at the meter. However, both connections are at the same temperature, so the meter can be considered to be a length of different wire, and the law of intermediate metals applies. Since the meter connections are not at 32°F, the law of intermediate temperatures applies. The corrected voltage is

$$5.262 \text{ mV} + 0.832 \text{ mV} = 6.094 \text{ mV}$$

From App. 68.A, 6.094 mV corresponds to 280°F.

## 18. STRAIN GAUGES

A *bonded strain gauge* is a metallic resistance device that is cemented to the surface of the unstressed member.[18] The gauge consists of a metallic conductor (known as the *grid*) on a backing (known as the *substrate*).[19] The substrate and grid experience the same strain as the surface of the member. The resistance of the gauge changes as the member is stressed due to changes in conductor cross section and intrinsic changes in resistivity with strain. Temperature effects must be compensated for by the circuitry or by using a second unstrained gauge as part of the bridge measurement system. (See Sec. 68.19.)

When simultaneous strain measurements in two or more directions are needed, it is convenient to use a commercial *rosette strain gauge*. A rosette consists of two or more *grids* properly oriented for application as a single unit. (See Fig. 68.5.)

**Figure 68.5** *Strain Gauge*

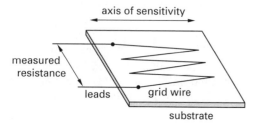

(a) folded-wire strain gauge

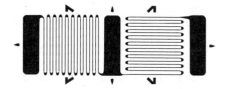

(b) commercial two-element rosette

---

[18]A *bonded strain gauge* is constructed by bonding the conductor to the surface of the member. An *unbonded strain gauge* is constructed by wrapping the conductor tightly around the member or between two points on the member.

  Strain gauges on rotating shafts are usually connected through *slip rings* to the measurement circuitry.

[19]The grids of strain gauges were originally of the folded-wire variety. For example, nichrome wire with a total resistance under 1000 Ω was commonly used. Modern strain gauges are generally of the foil type manufactured by printed circuit techniques. Semiconductor gauges are also used when extreme sensitivity (i.e., gage factors in excess of 100) is required. However, semiconductor gauges are extremely temperature-sensitive.

The *gage factor (strain sensitivity factor)*, GF, is the ratio of the fractional change in resistance to the fractional change in length (strain) along the detecting axis of the gauge. The higher the gage factor, the greater the sensitivity of the gauge. The gage factor is a function of the gauge material. It can be calculated from the grid material's properties and configuration. From a practical standpoint, however, the gage factor and gage resistance are provided by the gauge manufacturer. Only the change in resistance is measured. (See Table 68.4.)

$$
\begin{aligned}
\text{GF} &= 1 + 2\nu + \frac{\dfrac{\Delta \rho}{\rho_o}}{\dfrac{\Delta L}{L_o}} \\[2mm]
&= \frac{\dfrac{\Delta R_g}{R_g}}{\dfrac{\Delta L}{L_o}} = \frac{\dfrac{\Delta R_g}{R_g}}{\epsilon}
\end{aligned}
\qquad \textit{68.8}
$$

**Table 68.4** *Approximate Gage Factors* [a]

| material | GF |
|---|---|
| constantan | 2.0 |
| iron, soft | 4.2 |
| isoelastic | 3.5 |
| manganin | 0.47 |
| monel | 1.9 |
| nichrome | 2.0 |
| nickel | $-12$[b] |
| platinum | 4.8 |
| platinum-iridium | 5.1 |

[a]Other properties of strain gauge materials are listed in Table 68.2.
[b]Value depends on amount of preprocessing and cold working.

Constantan and isoelastic wires along with metal foil with gage factors of approximately 2 and initial resistances of less than 1000 Ω (typically 120 Ω, 350 Ω, 600 Ω, and 700 Ω) are commonly used. In practice, the gage factor and initial gage resistance, $R_g$, are specified by the manufacturer of the gauge. Once the strain sensitivity factor is known, the strain, $\epsilon$, can be determined from the change in resistance. Strain is often reported in units of $\mu\text{in/in}$ ($\mu\text{m/m}$) and is given the name *microstrain*.

$$\epsilon = \frac{\Delta R_g}{(\text{GF})R_g} \qquad \textit{68.9}$$

Theoretically, a strain gauge should not respond to strain in its transverse direction. However, the turn-around end-loops are also made of strain-sensitive material, and the end-loop material contributes to a nonzero sensitivity to strain in the transverse direction. Equation 68.10 defines the *transverse sensitivity factor*, $K_t$, which is of academic interest in most problems. The transverse sensitivity factor is seldom greater than 2%.

$$K_t = \frac{(\text{GF})_{\text{transverse}}}{(\text{GF})_{\text{longitudinal}}} \qquad 68.10$$

## Example 68.5

A strain gauge with a nominal resistance of 120 $\Omega$ and gage factor of 2.0 is used to measure a strain of 1 $\mu$in/in. What is the change in resistance?

*Solution*

From Eq. 68.9,

$$\Delta R_g = (\text{GF})R_g\epsilon = (2.0)(120 \ \Omega)\left(1 \times 10^{-6} \ \frac{\text{in}}{\text{in}}\right)$$
$$= 2.4 \times 10^{-4} \ \Omega$$

## 19. WHEATSTONE BRIDGES

The *Wheatstone bridge*, shown in Fig. 68.6, is one type of *resistance bridge*.[20] The bridge can be used to determine the unknown resistance of a resistance transducer (e.g., thermistor or resistance-type strain gauge), say $R_1$ in Fig. 68.6. The potentiometer is adjusted (i.e., the bridge is "balanced") until no current flows through the meter or until there is no voltage across the meter (hence the name *null indicator*).[21,22] When the bridge is balanced and no current flows through the meter leg, Eq. 68.11 through Eq. 68.14 are applicable.

$$I_2 = I_4 \quad \text{[balanced]} \qquad 68.11$$
$$I_1 = I_3 \quad \text{[balanced]} \qquad 68.12$$
$$V_1 + V_3 = V_2 + V_4 \quad \text{[balanced]} \qquad 68.13$$
$$\frac{R_1}{R_2} = \frac{R_3}{R_4} \quad \text{[balanced]} \qquad 68.14$$

**Figure 68.6** *Wheatstone Bridge*

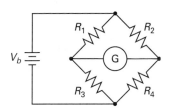

---

[20]Other types of resistance bridges are the *differential series balance bridge, shunt balance bridge,* and *differential shunt balance bridge.* These differ in the manner in which the adjustable resistor is incorporated into the circuit.
[21]This gives rise to the alternate names of *zero-indicating bridge* and *null-indicating bridge.*
[22]The unknown resistance can also be determined from the amount of voltage unbalance shown by the meter reading, in which case, the bridge is known as a *deflection bridge* rather than a null-indicating bridge. Deflection bridges are described in Sec. 68.20.

Since any one of the four resistances can be the unknown, up to three of the remaining resistances can be fixed or adjustable, and the battery and meter can be connected to either of two diagonal corners, it is sometimes confusing to apply Eq. 68.14 literally. However, the following bridge law statement can be used to help formulate the proper relationship: *When a series Wheatstone bridge is null-balanced, the ratio of resistance of any two adjacent arms equals the ratio of resistance of the remaining two arms, taken in the same sense.* In this statement, "taken in the same sense" means that both ratios must be formed reading either left to right, right to left, top to bottom, or bottom to top.

## 20. STRAIN GAUGE DETECTION CIRCUITS

The resistance of a strain gauge can be measured by placing the gauge in either a ballast circuit or bridge circuit. A *ballast circuit* consists of a voltage source ($V_b$) of less than 10 V (typical), a current-limiting ballast resistance ($R_b$), and the strain gauge of known resistance ($R_g$) in series. (See Fig. 68.7.) This is essentially a voltage-divider circuit. The change in voltage ($\Delta V_g$) across the strain gauge is measured. The strain ($\epsilon$) can be determined from Eq. 68.15.

$$\Delta V_g = \frac{(\text{GF})\epsilon V_b R_b R_g}{(R_b + R_g)^2} \qquad 68.15$$

**Figure 68.7** *Ballast Circuit*

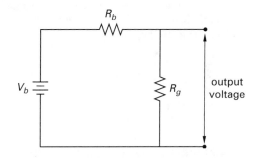

Ballast circuits do not provide temperature compensation, nor is their sensitivity adequate for measuring static strain. Ballast circuits, where used, are often limited to measurement of transient strains. A bridge detection circuit overcomes these limitations.

Figure 68.8 illustrates how a strain gauge can be used with a resistance bridge. Gauge 1 measures the strain, while *dummy gauge* 2 provides temperature compensation.[23] The meter voltage is a function of the input (battery) voltage and the resistors. (As with bridge circuits, the input voltage is typically less than 10 V.) The variable resistance is used for balancing the bridge

---

[23]This is a "quarter-bridge" or "1/4-bridge" configuration, as described in Sec. 68.23. The strain gauge used for temperature compensation is not active.

prior to the strain. When the bridge is balanced, $V_{\text{meter}}$ is zero.

**Figure 68.8** *Strain Gauge in Resistance Bridge*

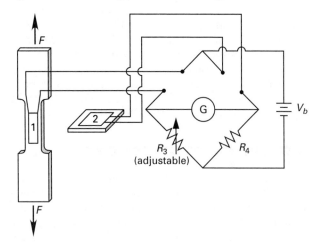

When the gauge is strained, the bridge becomes unbalanced. Assuming the bridge is initially balanced, the voltage at the meter (known as the *voltage deflection* from the null condition) will be[24]

$$V_{\text{meter}} = V_b \left( \frac{R_1}{R_1 + R_3} - \frac{R_2}{R_2 + R_4} \right) \quad [\text{$\frac{1}{4}$-bridge}]$$

$$68.16$$

For a single strain gauge in a resistance bridge and neglecting lead resistance, the voltage deflection is related to the strain by Eq. 68.17.

$$V_{\text{meter}} = \frac{(\text{GF})\epsilon V_b}{4 + 2(\text{GF})\epsilon}$$

$$\approx \tfrac{1}{4}(\text{GF})\epsilon V_b \quad [\text{$\frac{1}{4}$-bridge}] \qquad 68.17$$

## 21. STRAIN GAUGE IN UNBALANCED RESISTANCE BRIDGE

A resistance bridge does not need to be balanced prior to use as long as an accurate digital voltmeter is used in the detection circuit. The voltage ratio difference, $\Delta\text{VR}$, is defined as the fractional change in the output voltage from the unstrained to the strained condition.

$$\Delta\text{VR} = \left( \frac{V_{\text{meter}}}{V_b} \right)_{\text{strained}} - \left( \frac{V_{\text{meter}}}{V_b} \right)_{\text{unstrained}} \qquad 68.18$$

If the only resistance change between the strained and unstrained conditions is in the strain gauge and lead resistance is disregarded, the fractional change in gage

---

[24]Equation 68.16 applies to the unstrained condition as well. However, if the gauge is unstrained and the bridge is balanced, the resistance term in parentheses is zero.

resistance for a single strain gauge in a resistance bridge is

$$\frac{\Delta R_g}{R_g} = \frac{-4\Delta(\text{VR})}{1 + 2\Delta(\text{VR})} \quad [\text{$\frac{1}{4}$-bridge}] \qquad 68.19$$

Since the fractional change in gage resistance also occurs in the definition of the gage factor (see Eq. 68.9), the strain is

$$\epsilon = \frac{-4\Delta\text{VR}}{(\text{GF})(1 + 2\Delta\text{VR})} \quad [\text{$\frac{1}{4}$-bridge}] \qquad 68.20$$

## 22. BRIDGE CONSTANT

The voltage deflection can be doubled (or quadrupled) by using two (or four) strain gauges in the bridge circuit. The larger voltage deflection is more easily detected, resulting in more accurate measurements.

Use of multiple strain gauges is generally limited to configurations where symmetrical strain is available on the member. For example, a beam in bending experiences the same strain on the top and bottom faces. Therefore, if the temperature-compensation strain gauge shown in Fig. 68.8 is bonded to the bottom of the beam, the resistance change would double.

The *bridge constant* (BC) is the ratio of the actual voltage deflection to the voltage deflection from a single gage. Depending on the number and orientation of the gages used, bridge constants of 1.0, 1.3, 2.0, 2.6, and 4.0 may be encountered (for materials with a Poisson's ratio of 0.3).

Figure 68.9 illustrates how (up to) four strain gauges can be connected in a Wheatstone bridge circuit. The total strain indicated will be the algebraic sum of the four strains detected. For example, if all four strains are equal in magnitude, $\epsilon_1$ and $\epsilon_4$ are tensile (i.e., positive), and $\epsilon_2$ and $\epsilon_3$ are compressive (i.e., negative), then the bridge constant would be 4.

$$\epsilon_t = \epsilon_1 - \epsilon_2 - \epsilon_3 + \epsilon_4 \qquad 68.21$$

**Figure 68.9** *Wheatstone Bridge Strain Gauge Circuit**

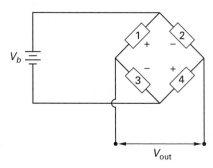

---

*The 45° orientations shown are figurative. Actual gauge orientation can be in any direction.

## 23. STRESS MEASUREMENTS IN KNOWN DIRECTIONS

Strain gauges are the most frequently used method of determining the stress in a member. Stress can be calculated from strain, or the measurement circuitry can be calibrated to give the stress directly.

For stress in only one direction (i.e., the *uniaxial stress* case), such as a simple bar in tension, only one strain gage is required. The stress can be calculated from *Hooke's law*.

$$\sigma = E\epsilon \qquad 68.22$$

When a surface, such as that of a pressure vessel, experiences simultaneous stresses in two directions (the *biaxial stress* case), the strain in one direction affects the strain in the other direction.[25] Therefore, two strain gauges are required, even if the stress in only one direction is needed. The strains actually measured by the gauges are known as the *net strains*.

$$\epsilon_x = \frac{\sigma_x - \nu\sigma_y}{E} \qquad 68.23$$

$$\epsilon_y = \frac{\sigma_y - \nu\sigma_x}{E} \qquad 68.24$$

The stresses are determined by solving Eq. 68.23 and Eq. 68.24 simultaneously.

$$\sigma_x = \frac{E(\epsilon_x + \nu\epsilon_y)}{1 - \nu^2} \qquad 68.25$$

$$\sigma_y = \frac{E(\epsilon_y + \nu\epsilon_x)}{1 - \nu^2} \qquad 68.26$$

Figure 68.9 shows how four strain gauges can be interconnected in a bridge circuit. Figure 68.10 shows how (up to) four strain gauges would be physically oriented on a test specimen to measure different types of stress. Not all four gauges are needed in all cases. If four gauges are used, the arrangement is said to be a *full bridge*. If only one or two gauges are used, the terms *quarter-bridge* ($^1/_4$-bridge) and *half-bridge* ($^1/_2$-bridge), respectively, apply.

In the case of up to four gauges applied to detect bending strain (see Fig. 68.10(a)), the bridge constant (BC) can be 1.0 (one gauge in position 1), 2.0 (two gauges in positions 1 and 2), or 4.0 (all four gauges). The relationships between the stress, strain, and applied force are

$$\sigma = E\epsilon = \frac{E\epsilon_t}{BC} \qquad 68.27$$

$$\sigma = \frac{Mc}{I} = \frac{Mh}{2I} \qquad 68.28$$

$$I = \frac{bh^3}{12} \quad \text{[rectangular section]} \qquad 68.29$$

For axial strain (see Fig. 68.10(b)) and a material with a Poisson's ratio of 0.3, the bridge constant can be 1.0 (one gauge in position 1), 1.3 (two gauges in positions 1 and 2), 2.0 (two gauges in positions 1 and 3), or 2.6 (all four gauges).

$$\sigma = E\epsilon = \frac{E\epsilon_t}{BC} \qquad 68.30$$

$$\sigma = \frac{F}{A} \qquad 68.31$$

$$A = bh \quad \text{[rectangular section]} \qquad 68.32$$

**Figure 68.10** *Orientation of Strain Gauges*

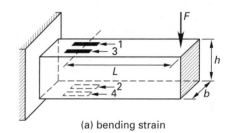

(a) bending strain

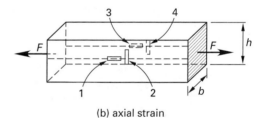

(b) axial strain

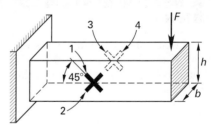

(c) shear strain

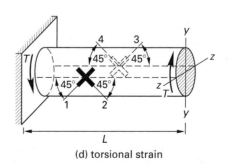

(d) torsional strain

---

[25]Thin-wall pressure vessel theory shows that the hoop stress is twice the longitudinal stress. However, the ratio of circumferential to longitudinal strains is closer to 4:1 than to 2:1.

For shear strain (see Fig. 68.10(c)) and a material with a Poisson's ratio of 0.3, the bridge constant can be 2.0 (two gauges in positions 1 and 2) or 4.0 (all four gauges). The shear strain is twice the axial strain at 45°.

$$\tau = G\gamma = 2G\epsilon = \frac{2G\epsilon_t}{\text{BC}} \qquad 68.33$$

$$\tau_{\max} = \frac{FQ_{\max}}{bI} \qquad 68.34$$

$$\gamma = 2\epsilon \quad [\text{at } 45°] \qquad 68.35$$

$$Q_{\max} = \frac{bh^2}{8} \quad [\text{rectangular section}] \qquad 68.36$$

$$G = \frac{E}{2(1+\nu)} \qquad 68.37$$

For torsional strain (see Fig. 68.10(d)) and a material with a Poisson's ratio of 0.3, the bridge constant can be 2.0 (two gauges in positions 1 and 2) or 4.0 (all four gauges). The shear strain is twice the axial strain at 45°.

$$\tau = G\gamma = 2G\epsilon = \frac{2G\epsilon_t}{\text{BC}} \qquad 68.38$$

$$\tau = \frac{Tr}{J} = \frac{Td}{2J} \qquad 68.39$$

$$\gamma = 2\epsilon \quad [\text{at } 45°] \qquad 68.40$$

$$J = \frac{\pi d^4}{32} \quad [\text{solid circular}] \qquad 68.41$$

$$\phi = \frac{TL}{JG} \qquad 68.42$$

$$G = \frac{E}{2(1+\nu)} \qquad 68.43$$

## 24. STRESS MEASUREMENTS IN UNKNOWN DIRECTIONS

In order to calculate the maximum stresses (i.e., the principal stresses) on the surface shown in Sec. 68.23, the gauges would be oriented in the known directions of the principal stresses.

In most cases, however, the directions of the principal stresses are not known. Therefore, rosettes of at least three gauges are used to obtain information in a third direction. Rosettes of three gauges (*rectangular* and *equiangular (delta) rosettes*) are used for this purpose. *T-delta rosettes* include a fourth strain gauge to refine and validate the results of the three primary gauges. Table 68.5 can be used for calculating principal stresses.

## 25. LOAD CELLS

*Load cells* are used to measure forces. A load cell is a transducer that converts a tensile or compressive force into an electrical signal. Though the details of the load cell will vary with the application, the basic elements are (a) a member that is strained by the force and (b) a strain detection system (e.g., strain gauge). The force is

calculated from the observed deflection, $y$. In Eq. 68.44, the spring constant, $k$, is known as the load cell's *deflection constant*.

$$F = ky \qquad 68.44$$

Because of their low cost and simple construction, *bending beam load cells* are the most common variety of load cells. Two strain gauges, one on the top and the other mounted on the bottom of a cantilever bar, are used. *Shear beam load cells* (which detect force by measuring the shear stress) can be used where the shear does not vary considerably with location, as in the web of an I-beam cross section.[26] The common S-shaped load cell constructed from a machined steel block can be instrumented as either a bending beam or shear beam load cell.

Load cell applications are categorized into grades or accuracy classes, with class III (500 to 10,000 scale divisions) being the most common. Commercial load cells meet standardized limits on errors due to temperature, nonlinearity, and hysteresis. The *temperature effect on output* (TEO) is typically stated in percentage change per 100°F (55.5°C) change in temperature.

Nonlinearity errors are reduced in proportion to the load cell's derating (i.e., using the load cell to measure forces less than its rated force). For example, a 2:1 derating will reduce the nonlinearity errors by 50%. Hysteresis is not normally reduced by derating.

The overall error of force measurement can be reduced by a factor of $1/\sqrt{n}$ (where $n$ is the number of load cells that share the load equally) by using more than one load cell. Conversely, the applied force can vary by $\sqrt{n}$ times the known accuracy of a single load cell without decreasing the error.

## 26. DYNAMOMETERS

Torque from large motors and engines is measured by a *dynamometer*. *Absorption dynamometers* (e.g., the simple *friction brake, Prony brake, water brake,* and *fan brake*) dissipate energy as the torque is measured. Opposing torque in pumps and compressors must be supplied by a *driving dynamometer*, which has its own power input. *Transmission dynamometers* (e.g., *torque meters, torsion dynamometers*) use strain gauges to sense torque. They do not absorb or provide energy.

Using a brake dynamometer involves measuring a force, a moment arm, and the angular speed of rotation. The

---

[26]While shear in a rectangular beam varies parabolically with distance from the neutral axis, shear in the web of an I-beam is essentially constant at $F/A$. The flanges carry very little of the shear load.
Other advantages of the shear beam load cell include protection from the load and environment, high side load rejection, lower creep, faster RTZ (return to zero) after load removal, and higher tolerance of vibration, dynamic forces, and noise.

**Table 68.5** *Stress-Strain Relationships for Strain Gauge Rosettes*[a]

| | rectangular | equiangular (delta) | T-delta |
|---|---|---|---|
| type of rosette | | | |
| principal strains, $\epsilon_p$, $\epsilon_q$ | $\dfrac{1}{2}\left( \dfrac{\epsilon_a + \epsilon_c}{\pm \sqrt{2(\epsilon_a - \epsilon_b)^2 + 2(\epsilon_b - \epsilon_c)^2}} \right)$ | $\dfrac{1}{3}\left( \dfrac{\epsilon_a + \epsilon_b + \epsilon_c}{\pm \sqrt{2(\epsilon_a - \epsilon_b)^2 + 2(\epsilon_b - \epsilon_c)^2 + 2(\epsilon_c - \epsilon_a)^2}} \right)$ | $\dfrac{1}{2}\left( \dfrac{\epsilon_a + \epsilon_d}{\pm \sqrt{(\epsilon_a - \epsilon_d)^2 + \left(\frac{4}{3}\right)(\epsilon_b - \epsilon_c)^2}} \right)$ |
| principal stresses, $\sigma_1$, $\sigma_2$ | $\dfrac{E}{2}\left( \dfrac{\epsilon_a + \epsilon_c}{1 - \nu} \pm \dfrac{1}{1 + \nu} \atop \times \sqrt{2(\epsilon_a - \epsilon_b)^2 + 2(\epsilon_b - \epsilon_c)^2} \right)$ | $\dfrac{E}{3}\left( \dfrac{\epsilon_a + \epsilon_b + \epsilon_c}{1 - \nu} \pm \dfrac{1}{1 + \nu} \atop \times \sqrt{2(\epsilon_a - \epsilon_b)^2 + 2(\epsilon_b - \epsilon_c)^2 + 2(\epsilon_c - \epsilon_a)^2} \right)$ | $\dfrac{E}{2}\left( \dfrac{\epsilon_a + \epsilon_d}{1 - \nu} \pm \dfrac{1}{1 + \nu} \atop \times \sqrt{(\epsilon_a - \epsilon_d)^2 + \left(\frac{4}{3}\right)(\epsilon_b - \epsilon_c)^2} \right)$ |
| maximum shear, $\tau_{max}$ | $\dfrac{E}{2(1 + \nu)}$ $\times \sqrt{2(\epsilon_a - \epsilon_b)^2 + 2(\epsilon_b - \epsilon_c)^2}$ | $\dfrac{E}{3(1 + \nu)}$ $\times \sqrt{2(\epsilon_a - \epsilon_b)^2 + 2(\epsilon_b - \epsilon_c)^2 + 2(\epsilon_c - \epsilon_a)^2}$ | $\dfrac{E}{2(1 + \nu)}$ $\times \sqrt{(\epsilon_a - \epsilon_d)^2 + \left(\frac{4}{3}\right)(\epsilon_b - \epsilon_c)^2}$ |
| $\tan 2\theta$[b] | $\dfrac{2\epsilon_b - \epsilon_a - \epsilon_c}{\epsilon_a - \epsilon_c}$ | $\dfrac{\sqrt{3}(\epsilon_c - \epsilon_b)}{2\epsilon_a - \epsilon_b - \epsilon_c}$ | $\dfrac{2}{\sqrt{3}}\left( \dfrac{\epsilon_c - \epsilon_b}{\epsilon_a - \epsilon_d} \right)$ |
| $0 < \theta < +90°$ | $\epsilon_b > \dfrac{\epsilon_a + \epsilon_c}{2}$ | $\epsilon_c > \epsilon_b$ | $\epsilon_c > \epsilon_b$ |

[a]$\theta$ is measured in the counterclockwise direction from the $a$-axis of the rosette to the axis of the algebraically larger stress.
[b]$\theta$ is the angle from $a$-axis to axis of maximum normal stress.

familiar torque-power-speed relationships are used with absorption dynamometers.

$$T = Fr \qquad 68.45$$

$$P_{\text{ft-lbf/min}} = 2\pi T_{\text{ft-lbf}} n_{\text{rpm}} \qquad 68.46$$

$$P_{\text{kW}} = \dfrac{T_{\text{N·m}} n_{\text{rpm}}}{9549} \qquad [\text{SI}] \qquad 68.47(a)$$

$$P_{\text{hp}} = \dfrac{2\pi F_{\text{lbf}} r_{\text{ft}} n_{\text{rpm}}}{33{,}000}$$

$$= \dfrac{2\pi T_{\text{ft-lbf}} n_{\text{rpm}}}{33{,}000} \qquad [\text{U.S}] \qquad 68.47(b)$$

Some brakes and dynamometers are constructed with a "standard" brake arm whose length is 5.252 ft. In that case, the horsepower calculation conveniently reduces to

$$P_{\text{hp}} = \dfrac{F_{\text{lbf}} n_{\text{rpm}}}{1000} \qquad [\text{"standard arm" brake}] \qquad 68.48$$

If an absorption dynamometer uses a DC generator to dissipate energy, the generated voltage ($V$ in volts) and line current ($I$ in amps) are used to determine the power. Equation 68.47 and Eq. 68.48 are used to determine the torque.

$$P_{\text{hp}} = \dfrac{I V}{\eta\left(1000 \, \dfrac{\text{W}}{\text{kW}}\right)\left(0.7457 \, \dfrac{\text{W}}{\text{hp}}\right)} \qquad [\text{absorption}]$$
$$68.49$$

For a driving dynamometer using a DC motor,

$$P_{\text{hp}} = \dfrac{I V \eta}{\left(1000 \, \dfrac{\text{W}}{\text{kW}}\right)\left(0.7457 \, \dfrac{\text{W}}{\text{hp}}\right)} \qquad [\text{driving}] \qquad 68.50$$

Torque can be measured directly by a *torque meter* mounted to the power shaft. Either the angle of twist ($\phi$) or the shear strain ($\tau/G$) is measured. The torque in a solid shaft of diameter $d$ and length $L$ is

$$T = \dfrac{J G \phi}{L} = \left(\dfrac{\pi}{32}\right) d^4 \left(\dfrac{G \phi}{L}\right)$$

$$= \left(\dfrac{\pi}{16}\right) d^3 \tau \qquad [\text{solid round}] \qquad 68.51$$

## 27. INDICATOR DIAGRAMS

*Indicator diagrams* are plots of pressure versus volume and are encountered in the testing of reciprocating engines. In the past, indicator diagrams were actually drawn on an *indicator card* wrapped around a drum through a mechanical linkage of arms and springs. This method is mechanically complex and is not suitable for rotational speeds above 2000 rpm. Modern records of pressure and volume are produced by signals from electronic transducers recorded in real time by computers.

Analysis of indicator diagrams produced by mechanical devices requires knowing the spring constant, also known as the *spring scale*. The *mean effective pressure* (MEP) is calculated by dividing the area of the diagram by the width of the plot and then multiplying by the spring constant.

Physics

# Topic XIV: Systems Analysis

Systems
Analysis

# 69 Modeling of Engineering Systems

## Nomenclature

| | | | |
|---|---|---|---|
| $a$ | acceleration | ft/sec$^2$ | m/s$^2$ |
| $a$ | ratio of transformation | – | – |
| $A$ | area | ft$^2$ | m$^2$ |
| $C$ | damping coefficient | lbf-sec/ft | N·s/m |
| $C$ | capacitance | F | F |
| $C$ | capacitance (fluid) | ft$^5$/lbf | m$^5$/N |
| $C_r$ | rotational damping coefficient | ft-lbf-sec/rad | N·m·s/rad |
| $f$ | forcing function | various | various |
| $F$ | force | lbf | N |
| $g_c$ | gravitational constant | ft-lbm/lbf-sec$^2$ | n.a. |
| $h$ | depth | ft | m |
| $I$ | current | A | A |
| $I$ | inertance (fluid) | lbf/ft$^4$ | N/m$^4$ |
| $I$ | mass moment of inertia | lbm-ft$^2$ | kg·m$^2$ |
| $k$ | spring constant | lbf/ft | N/m |
| $k_r$ | rotational spring constant | ft-lbf/rad | N·m/rad |
| $l$ | length | ft | m |
| $L$ | inductance | H | H |
| $m$ | mass | lbm | kg |
| $n$ | number of teeth | – | – |
| $p$ | pressure | lbf/ft$^2$ | N/m$^2$ |
| $Q$ | flow rate | ft$^3$/sec | m$^3$/s |
| $r$ | response function | various | various |
| $R$ | resistance | $\Omega$ | $\Omega$ |
| $t$ | time | sec | s |
| $T$ | torque | ft-lbf | N·m |
| v | velocity | ft/sec | m/s |
| $V$ | voltage | V | V |
| $x$ | position | ft | m |

## Symbols

| | | | |
|---|---|---|---|
| $\alpha$ | angular acceleration | rad/sec$^2$ | rad/s$^2$ |
| $\gamma$ | specific weight | lbf/ft$^3$ | N/m$^3$ |
| $\theta$ | angular position | rad | rad |
| $\omega$ | angular velocity | rad/sec | rad/s |

## Subscripts

| | |
|---|---|
| $C$ | dashpot |
| $f$ | fluid |
| $k$ | spring |
| $L$ | liquid or load |
| $m$ | mass or motor |
| $r$ | rotational |

## 1. INTRODUCTION TO ENGINEERING SYSTEMS MODELING

The ultimate benefit derived from a model is the ability to predict the behavior (known as the *response*) of a real-world system. Some models (scale models, working models, mock-ups, etc.) are physical, but others (such as the ones in this chapter) are mathematical. The goal of modeling is to develop a differential equation or other mathematical function, known as the *response function*, that predicts how the system will behave (i.e., what position, acceleration, or velocity it will have). The response function is commonly transformed into the Laplace *s*-variable domain.

It is rarely possible to write the response function by observation, and several methods of developing the response function are available. This chapter takes the *two-port black box* (see Fig. 69.1) approach—"two-port" because there are two pairs of connections to the model, and "black box" because the inner workings of the model are irrelevant once the response function is known.[1]

**Figure 69.1** *Two-Port Black Box Model*

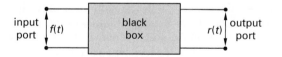

Many types of real-world systems (e.g., long-term weather prediction) are too complex to be modeled mathematically. Others lend themselves to special types of modeling theory. This chapter is limited to systems that can be modeled by *idealized (linear) elements* (components, devices, etc.).[2] (See Sec. 69.2.) In some cases, *nonlinear elements* can be considered linear in

---

[1]It is important to recognize that the mathematical response function is the model. Drawing system diagrams and taking other steps to derive the response function are merely developmental aids.

[2]Almost any linear flow process can be modeled in this manner. Fluid flow and heat transfer systems are other common applications.

limited operation ranges. Elements such as mechanical springs and electrical resistors that absorb and dissipate energy are known as *passive elements*. Energy sources are *active elements*.

The response function for a model is derived from a *system equation*, which is usually a differential equation. The *order of the system (model)* is the highest order derivative in the system equation.

It is not necessary to work separate problems to find position, velocity, and acceleration response functions for a system. If the position function $x(t)$ is known, it can be differentiated to give $v(t)$ and $a(t)$. If any one of the three response functions is known, the other two can be derived.

A *single degree of freedom* (SDOF) *system* can be completely defined by one response variable. A single mass on a spring, a swinging pendulum, and a rotating pulley are examples of SDOF systems. In each case, one variable ($x$ or $\theta$) defines the position of the major system element.

Systems in which multiple components have their own values of the dependent variable are known as *multiple degree of freedom* (MDOF) *systems*. The degree of freedom of the system is the number of unrelated dependent variables needed to specify the behavior of all major system elements.

## 2. ELEMENTS

Each physical device in the system is an *element*. All of the ideal elements are considered to be two-port devices. Springs and dashpots have two ends, each of which can have a different value of the response variable. For example, the velocity of both ends of a shock absorber need not be the same. Even though masses do not have ends in the traditional sense, they are considered to be two-port devices as well.

A dependent variable that describes the performance of an element is a *response variable* (also known as a *state variable*). Each element in the model will have its own response variable, such as position, velocity, and acceleration in mechanical systems and voltage in electrical systems. Time is the independent variable.

## 3. ENERGY SOURCES AND FORCING FUNCTIONS

Some systems start with and gradually use up stored energy; other systems receive energy on a one-time basis. Still others receive energy on a continuous basis. The equation describing the amount of energy introduced as a function of time is the *forcing function*. Figure 69.2 illustrates common forcing function profiles. Although a wide variety of forcing functions is possible, engineering systems easily become too complicated for manual solutions unless limited to simple types.

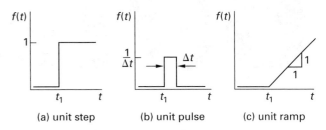

**Figure 69.2** *Common Forcing Function Profiles*

An energy source need not be an actual physical component such as a wound spring, battery, or fuel cell. Anything that produces motion in the system, including potential energy or an applied force, can be an energy source.

In modeling, the source or method of energy generation may not be known. For example, a velocity source may produce a specific velocity regardless of the system mass, or a current source may produce a specific current regardless of circuit elements. How this is accomplished need not be explicitly known.

The *homogeneous forcing function* is the zero function (i.e., no energy at all).[3] The homogeneous case does not preclude an initial disturbance or a previous amount of potential energy. However, the forcing function is homogeneous if it ceases to act as soon as the system begins to move. For example, a spring-mass system that is displaced, released, and allowed to oscillate freely experiences a zero force after the release.

A *unit step* has zero magnitude up to a particular instant ($t_1$) and a magnitude of 1 thereafter. It can be multiplied by a scalar if the magnitude of the actual forcing function has any other value.

$$f(t) = \begin{cases} 0 & t < t_1 \\ 1 & t \geq t_1 \end{cases} \qquad 69.1$$

Hitting a bell with a hammer is an example of an impulse. A *unit impulse* (also known as a *pulse*) is a limited-duration force whose total impulse (that is, $f\Delta t$) is 1. The unit pulse can be multiplied by a scalar if the actual pulse has an impulse different than 1. Usually the time, $\Delta t$, is extremely short.

$$f(t) = \begin{cases} 0 & t < t_1 \\ \dfrac{1}{\Delta t} & t_1 \leq t < t_1 + \Delta t \\ 0 & t \geq t + \Delta t \end{cases} \qquad 69.2$$

Some forces increase with time. The slope of a *unit ramp* forcing function is 1. If a forcing function changes at any other rate, the unit ramp can be multiplied by a scalar.

---

[3]This is consistent with homogeneous differential equations defined in Sec. 69.1.

The unit step, ramp, impulse, parabola, and so on, are known as *singularity functions* because of the singularity that exists at the point of discontinuity in each function. The effects of integrating and differentiating these functions are listed in Table 69.1.

**Table 69.1** *Operations on Forcing Functions*

| function | function when differentiated | function when integrated |
|---|---|---|
| unit impulse | – | unit step |
| unit step | unit impulse | unit ramp |
| unit ramp | unit step | unit parabola |
| unit parabola | unit ramp | (third degree) |
| unit exponential | unit exponential | unit exponential |
| unit sinusoid | unit sinusoid | unit sinusoid |

The most common forcing functions used in the analysis of engineering systems are sinusoids. When combined with Fourier series analysis, sinusoids can be used to approximate all other forcing functions.

## 4. THROUGH- AND ACROSS-VARIABLES

*Through-variables* have different values at the two ends of an element; *across-variables* have the same value. For example, force is a through-variable since it is passed through objects. Consider a mass hanging on a spring. The gravitational force on the mass is passed through the spring to the support.

The velocity of a mass, however, is measured with respect to an inertial (stationary) frame of reference. Since the "ground" is connected to a mass (see Rule 1, Sec. 69.5), velocity is the across-variable.

## 5. SYSTEM DIAGRAMS

The system elements are interconnected to produce a system diagram with the following procedure. The diagram is then used to derive the system equation. System diagrams for mechanical systems do not always resemble the topology of the systems they represent. Example 69.1(a) illustrates a system of two elements (a mass and dashpot) connected in series that actually has a parallel system diagram.

*step 1:* Decide on a dependent response variable. Position ($x$ or $\theta$) is preferred in mechanical systems, but velocity (v or $\omega$) and acceleration ($a$ or $\alpha$) can be used if desired.

*step 2:* Identify all parts of the system that have different values of the response variable. It is not necessary to know the actual values, only to recognize where the variable changes. One of the values will always be zero (corresponding to zero velocity) and is known as the *ground level*.

*step 3:* Start the system diagram by drawing a horizontal line for each different value of the dependent variable identified in step 2.

*step 4:* Insert and connect the passive elements (masses, springs, etc.) to the appropriate horizontal lines.

  *Rule 1:* One end of a mass always connects to the lowest (ground) level.

*step 5:* Insert and connect the active (energy) sources to the appropriate horizontal lines.

  *Rule 2:* One end of an energy source always connects to the lowest (ground) level.

A *line diagram* is a variation of the system diagram in which the levels associated with different values of the response variable (the horizontal lines from step 3) are replaced with nodes, and the element symbols are replaced with their values.

## 6. SYSTEM EQUATIONS

The *system equation* is derived from a system or line diagram. The system equation is a differential equation containing the dependent variable, but it does not explicitly give the response variable. Traditional methods and Laplace transforms can be used to derive the response variable.

The principle used to obtain a system equation from a system diagram is a conservation law analogous to Kirchhoff's current law that says "...what goes in must come out...."[4] For mechanical systems, force is the conserved quantity. Specifically, the total force supplied by the source equals the forces leaving through all parallel branches (known as *legs*) in the system diagram. Use of the following two rules is illustrated in Ex. 69.1.

  *Rule 3:* The force passing through a leg consisting of elements in series can be determined from the conditions across any of the elements in that leg.

  *Corollary to Rule 3:* The force passing through a leg consisting of elements in series can be determined from the conditions across the first element in that leg.

  *Rule 4:* When writing a difference in response variable values (e.g., $v_2-v_1$ or $x_2-x_1$), the first subscript is the same as the node number for which the equation is being written.

## 7. MECHANICAL SYSTEMS

A mechanical system consists of interconnected lumped masses, linear springs, linear dashpots, and energy sources. The response variable for a mechanical system

---

[4]Electrical current is a through-variable. Kirchhoff's current law applies to any through-variable.

is usually the position of one of the masses, although velocity and acceleration can also be chosen.

A *lumped mass* is a rigid body that acts like a particle. All parts of the mass experience identical displacements, velocities, and accelerations. An *ideal lumped mass* is one for which Newtonian (i.e., nonrelativistic) physics applies. The governing equation for a lumped mass is Newton's second law.[5]

$$F = ma = mx'' \quad \text{[ideal lumped mass]} \quad \text{[SI]} \quad \textbf{69.3(a)}$$

$$F = \frac{ma}{g_c} = \frac{mx''}{g_c} \quad \text{[ideal lumped mass]} \quad \text{[U.S.]} \quad \textbf{69.3(b)}$$

An ideal spring is massless, has a constant stiffness ($k$), and is immune to set, creep, and fatigue. Hooke's law predicts its performance but is rewritten to explicitly include the positions of both spring ends. (If both ends are displaced the same distance, there will be no change in extension, compression, or force.)

$$F = k(x_2 - x_1) \quad \text{[ideal spring]} \quad \textbf{69.4}$$

A *dashpot (damper)*, of which an automobile shock absorber is an example, is a device that opposes motion and dissipates energy. An ideal linear dashpot is massless and applies a force opposing motion that is proportional to velocity. Since both ends of the dashpot move, the governing equation is

$$F = C(v_2 - v_1)$$
$$= C(x_2' - x_1') \quad \text{[ideal dashpot]} \quad \textbf{69.5}$$

### Example 69.1

For the three systems shown, draw the system diagram, draw the line diagram, and write the system equation. Use consistent units.

(a) A mass is connected through a damper to a solid wall as shown. The mass slides without friction on its support and is acted upon by a force.

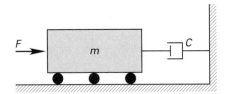

(b) A force is applied through a damper to a mass. The mass rolls on frictionless bearings.

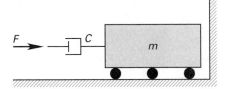

(c) A shock absorber with an integral coil spring is acted upon by a force applied to one end.

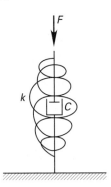

*Solution*

(a) Choose velocity as the response variable. All parts of the mass move with the same velocity: Call it $v_1$. The plunger also moves with velocity $v_1$ since it is attached to the mass. The body of the damper is attached to the stationary wall. Call this velocity $v = 0$.

Two horizontal lines are drawn: the top line for $v_1$ and the bottom line for $v = 0$. One end of the damper travels at $v_1$; the other end travels at $v = 0$. Therefore, connect the dashpot to these lines. The mass moves at $v_1$, so one of its lines connects to $v_1$. By Rule 1 (Section 69.5), the other end connects to $v = 0$. The force contacts the mass, so one end of the force connects to $v_1$. By Rule 2, the other end connects to $v = 0$.

The system and line diagrams are

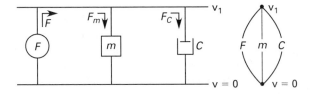

The force leaving the "source" splits—some of it going through the mass and some of it going through the dashpot. The conservation law is written to conserve the force in the $v_1$ line.

$$F = F_m + F_C$$

Expanding Rule 4 (see Sec. 69.6) with Eq. 69.3 and Eq. 69.5,

$$F = ma_1 + C(v_1 - 0)$$

---

[5]It is understood that force, $F$, and acceleration, $a$, are functions of time.

The differential equation is

$$F = mx_1'' + Cx_1'$$

(b) There are three velocities in this example: the velocity of the plunger (call this $v_1$), the velocity of the dashpot body and mass (call this $v_2$), and $v = 0$. The system line diagrams are

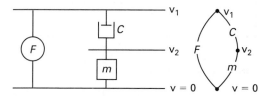

The system equation is written to conserve force in the $v_1$ line. The force passing through the dashpot is the same force experienced by the mass and is not additive. (This is the essence of Rule 3.) Rule 4 is used to write the dashpot force as $C(v_1 - v_2)$.

$$F = F_C$$
$$F = C(v_1 - v_2) = C(x_1' - x_2')$$

Since the same force is experienced by both the dashpot and mass, the system equation could be written in terms of the governing equation of the mass. Whether or not this is a better choice will depend on the initial conditions and other information that is available.

$$F = F_m = ma_2 = mx_2''$$

(c) There are only two velocities here. The system and line diagrams are

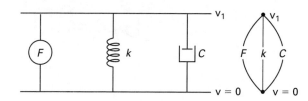

Force in the $v_1$ line is conserved. The system equation is

$$F = F_C + F_k$$
$$= Cv_1 + kx_1 = Cx_1' + kx_1$$

## 8. MECHANICAL ENERGY TRANSFORMATIONS

Levers transform one force into another at the expense of the distance the ends travel. The ratio of transformation depends on the lengths of the lever on both sides of the fulcrum. Whether the ratio is less than or greater than one is a matter of preference as long as the transformed force and velocity are correct. The ratio will be negative because the direction of the force and velocity is reversed by a lever. The transformation is represented in system diagrams by the symbol for an electrical transformer.

## Example 69.2

What are the system equations for the system shown?

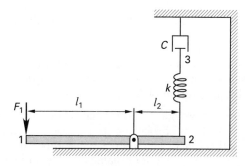

*Solution*

The lever transforms the force and displacement at point 1 into force and displacement at point 2. The system diagram is

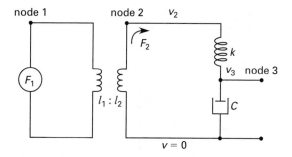

One of the system equations is based on node 2.

$$F_2 = k(x_2 - x_3)$$

The following additional equations are based on the lever transformation ratio. The minus signs indicate that the displacement and force at opposite ends of the lever are in opposite directions.

$$x_2 = \left(-\frac{l_2}{l_1}\right)x_1$$

$$F_2 = \left(-\frac{l_2}{l_1}\right)F_1$$

## 9. ROTATIONAL SYSTEMS

*Rotational systems* are directly analogous to mechanical systems. Flywheels have rotational mass moments of inertia ($I$), torsion springs have torsional stiffness ($k_r$), and fluid couplings have rotational viscous damping ($C_r$). The governing equations for these elements are

$$T = I\alpha \quad \text{[ideal flywheel]} \qquad \text{[SI]} \quad 69.6(a)$$

$$T = \frac{I_\alpha}{g_c} \quad \text{[ideal flywheel]} \qquad \text{[U.S.]} \quad 69.6(b)$$

$$T = k_r(\theta_2 - \theta_1) \quad \text{[ideal torsion spring]} \qquad 69.7$$

$$T = C_r(\omega_2 - \omega_1) \quad \text{[ideal fluid coupling]} \qquad 69.8$$

Angular velocity ($\omega$ in rad/s) is usually chosen as the response variable, although angular position ($\theta$ in radians) or acceleration ($\alpha$ in rad/s$^2$) can be used if desired. As with the translational mechanical systems, any one of these three response variables can be used to write the system equation, after which the others can be found by integration or differentiation.

Energy can be provided to rotational systems by constant-torque or constant-velocity sources. Gear sets transform torque and velocity. The ratio of transformation, $a$, is the ratio of numbers of teeth or diameters. The ratio is negative because each set of gears reverses the direction of rotation. Whether or not the ratio $n_1/n_2$ or $n_2/n_1$ is used is not important as long as the transformed dependent variable is correct. When a gear set increases the rotational speed, the torque decreases, and vice versa. (See Fig. 69.3.)

$$a = -\frac{T_1}{T_2} = -\frac{\omega_2}{\omega_1} = \frac{n_1}{n_2} \qquad 69.9$$

**Figure 69.3** Rotational Transformer

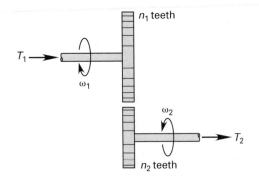

**Example 69.3**

Two flywheels are connected by a flexible shaft. The second flywheel is acted upon by a linear viscous force. Draw the system diagram and write the system equation.

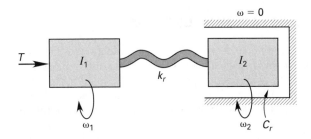

*Solution*

Choose angular velocity as the response variable. There are three different angular velocities—two for the flywheels and one for the stationary reference. The system diagram is

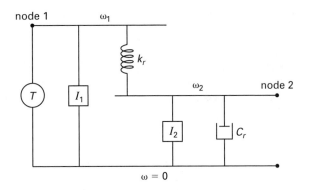

The system equations are

$$\begin{aligned} node\ 1: \quad T &= I_1\alpha_1 + k_r(\theta_1 - \theta_2) \\ &= I_1\theta_1'' + k_r(\theta_1 - \theta_2) \\ node\ 2: \quad 0 &= C_r\omega_2 + I_2\alpha_2 + k_r(\theta_2 - \theta_1) \\ &= C_r\theta_2' + I_2\theta_2'' + k_r(\theta_2 - \theta_1) \end{aligned}$$

**Example 69.4**

A motor drives a flywheel through a set of reduction gears. The flywheel is connected to the driven gear by a flexible shaft. All other gears and shafts have infinite stiffness. What is the system equation?

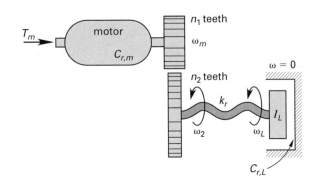

*Solution*

Angular velocity, $\omega$, is chosen as the response variable. The gear set is represented by the symbol for an electrical transformer. The ratio of transformation is less than 1. This means that the motor's torque will be decreased while the rotational speed is increased. The system equations are

$$\begin{aligned} node\ m: \quad & T_m = I_m\alpha_m + C_{r,m}\omega_m + aT_2 \\ node\ 2: \quad & T_2 = k_r(\theta_2 - \theta_L) \\ node\ L: \quad & 0 = C_{r,L}\omega_L + I_L\alpha_L + k_r(\theta_L - \theta_2) \end{aligned}$$

Since $\omega_m$, $\omega_2$, $\omega_L$, and $T_2$ are all unknown, a fourth equation is needed.

$$a\omega_m = -\omega_2$$

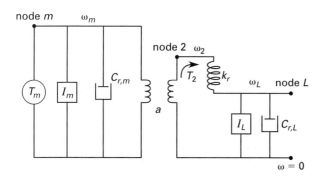

## 10. FLUID SYSTEMS

Figure 69.4 illustrates an open fluid *reservoir* with a constant vertical cross section. The flow rate out of the reservoir can be calculated from the cross-sectional area and the rate of change in the surface elevation. In Eq. 69.10, $C_f$ is known as the *fluid capacitance*. In system diagrams, open reservoirs always connect to the lowest (ground) level.

$$Q = A\left(\frac{dh}{dt}\right) = \frac{A\left(\frac{dp}{dt}\right)}{\gamma} = C_f\left(\frac{dp}{dt}\right) \quad\quad 69.10$$

$$C_f = \frac{A}{\gamma} \quad\quad 69.11$$

**Figure 69.4** *Fluid Reservoir*

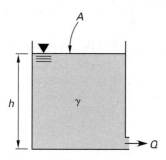

The Darcy equation indicates that *flow resistance* is proportional to the square of the flow quantity. As a simplification over a narrow range of flows, the flow resistance in simple systems analysis problems is assumed to be proportional to the flow quantity. $R_f$ is the *fluid resistance coefficient* and is often found by experimentation.

$$p_2 - p_1 = R_f Q \quad\quad 69.12$$

Newton's second law (or the impulse-momentum principle) is the basis for defining the *fluid inertance* (*fluid inductance*), $I$, which accounts for the inertia of the fluid flow.

$$F = m\frac{d\mathrm{v}}{dt} \quad\quad 69.13$$

$$A(p_2 - p_1) = \gamma A l \frac{\dfrac{dQ}{dt}}{A} \quad\quad 69.14$$

$$p_2 - p_1 = \frac{\gamma l}{A}\frac{dQ}{dt} = I\frac{dQ}{dt} \quad\quad 69.15$$

$$I = \frac{\gamma l}{A} \quad\quad 69.16$$

Equation 69.15 is integrated to obtain an expression for the flow quantity.

$$Q = \frac{1}{I}\int (p_2 - p_1)\, dt \quad\quad 69.17$$

### Example 69.5

A pump is used to keep liquid flowing through a filter pack. Pipe friction and potential head are insignificant. What is the system equation?

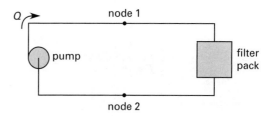

*Solution*

Pressure varies along the flow path and is the response variable. There are two different pressures: before and after the filter pack. This is analogous to a series electrical circuit of a battery and resistor. The system diagram is

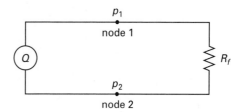

The system equation is

$$Q = \frac{1}{R_f}(p_2 - p_1)$$

## Example 69.6

A reservoir discharges through a long pipe and is not refilled. What is the system equation that defines the pressure at the tank bottom as a function of time?

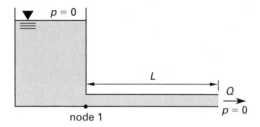

*Solution*

Although there is flow, there is no external energy source (i.e., there is no pump) in this system. This is analogous to an electrical capacitor discharging through a resistor. The system diagram is

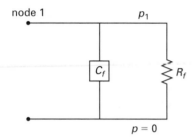

The system equation is

$$0 = \frac{p_1}{R_f} + C_f \frac{dp_1}{dt}$$

## Example 69.7

A pump transfers liquid from one reservoir to another. Pipe friction is insignificant. What are the system equations that define the pressures at the bottoms of the reservoirs?

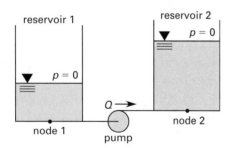

*Solution*

Both open reservoirs connect to the lowest level. The system diagram is

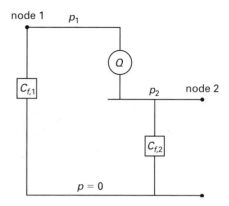

The system equations are

$$node\ 1: \quad Q = C_{f,1} \frac{dp_1}{dt}$$

$$node\ 2: \quad Q = C_{f,2} \frac{dp_2}{dt}$$

## 11. ELECTRICAL SYSTEMS

Resistors, capacitors, and inductors are passive elements in an electrical circuit. Voltage (an across-variable) is chosen as a convenient dependent variable, and the governing equations are written in terms of the conserved quantity, current (a through-variable).

$$I = \frac{V_1 - V_2}{R} \quad \text{[ideal resistor]} \qquad 69.18$$

$$I = C \frac{d(V_1 - V_2)}{dt} \quad \text{[ideal capacitor]} \qquad 69.19$$

$$I = \frac{1}{L} \int (V_1 - V_2)\, dt \quad \text{[ideal inductor]} \qquad 69.20$$

The system diagram is the same as the electrical circuit. It is not necessary to draw a different diagram.

## Example 69.8

What are the system equations for the electrical circuits shown?

(a)

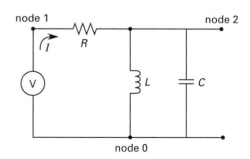

(b)

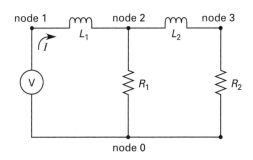

node 1    $L_1$    node 2    $L_2$    node 3

$V$    $R_1$    $R_2$

node 0

*Solution*

(a) Nodes 1 and 2 are at different potentials (voltages).

*node 1:* $I = \dfrac{V_1 - V_2}{R}$

*node 2:* $0 = C\dfrac{dV_2}{dt} + \dfrac{1}{L}\int V_2 \, dt + \dfrac{V_2 - V_1}{R}$

(b) Nodes 1, 2, and 3 have different potentials.

*node 1:* $I = \dfrac{1}{L_1}\int (V_1 - V_2)\, dt$

*node 2:*

$0 = \dfrac{V_2}{R_1} + \dfrac{1}{L_2}\int (V_2 - V_3)\, dt + \dfrac{1}{L_1}\int (V_2 - V_1)\, dt$

*node 3:* $0 = \dfrac{1}{L_2}\int (V_3 - V_2)\, dt + \dfrac{V_3}{R_2}$

**Systems Analysis**

# 70 Analysis of Engineering Systems

## Nomenclature

| | |
|---|---|
| $A$ | steady-state response |
| $B$ | damping coefficient |
| BW | bandwidth |
| $C$ | capacitance or constant |
| $e(t)$ | error |
| $E(s)$ | error, $\mathcal{L}(e(t))$ |
| $f(t)$ | forcing function |
| $F(s)$ | forcing function, $\mathcal{L}(f(t))$ |
| $G(s)$ | forward transfer function |
| $h$ | step height |
| $H(s)$ | reverse transfer function |
| $i(t)$ | current |
| $I(s)$ | current, $\mathcal{L}(i(t))$ |
| $j$ | $\sqrt{-1}$ |
| $k$ | spring stiffness |
| $K$ | gain |
| $L$ | inductance |
| $L$ | length of line |
| $m$ | mass |
| $M$ | fraction overshoot or magnitude |
| $n$ | degrees of freedom |
| $n$ | order of the system |
| $n$ | system type |
| $N$ | Nyquist's number |
| $p(t)$ | arbitrary function |
| $P(s)$ | arbitrary function, $\mathcal{L}(p(t))$ |
| $P$ | number of poles |
| $Q$ | quality factor |
| $r$ | real value (root) |
| $r(t)$ | time response |
| $R(s)$ | response function, $\mathcal{L}(r(t))$ |
| $S$ | sensitivity |
| $t$ | time |
| $T(t)$ | transfer function |
| $T(s)$ | transfer function, $\mathcal{L}(T(t))$ |
| $u$ | input variable |
| $v$ | velocity |
| $V$ | voltage |
| $x$ | position |
| $x$ | state variable |
| $y$ | output variable |
| $Z$ | number of zeros |

## Symbols

| | |
|---|---|
| $\alpha$ | pole angle |
| $\beta$ | zero angle |
| $\delta$ | a small number |
| $\epsilon$ | a small number |
| $\zeta$ | damping ratio |
| $\tau$ | time constant |
| $\omega$ | natural frequency |

## Subscripts

| | |
|---|---|
| $d$ | damped |
| $f$ | forced or feedback |
| $i$ | in |
| $n$ | natural |
| $o$ | out |
| $p$ | peak or pole |
| $r$ | rise |
| $s$ | settling |
| $t$ | total |
| $z$ | zero |

## 1. TYPES OF RESPONSE

*Natural response* (also known as *initial condition response*, *homogeneous response*, and *unforced response*) is how a system behaves when energy is applied and is subsequently removed. The system is left alone and is allowed to do what it would naturally, without the application of further disturbing forces. In the absence of friction, natural response is characterized by a sinusoidal response function. With friction, the response function will contain exponentially decaying sinusoids.

*Forced response* is the behavior of a system that is acted on by a force that is applied periodically. Forced response in the absence of friction is characterized by sinusoidal terms having the same frequency as the forcing function.

Natural and forced responses are present simultaneously in forced systems. The sum of the two responses is the *total response*.[1] This is the reason that differential equations are solved by adding a particular solution to the homogeneous solution. The homogeneous solution corresponds to the natural response; the particular solution corresponds to the forced response.

$$\frac{\text{total}}{\text{response}} = \frac{\text{natural}}{\text{response}} + \frac{\text{forced}}{\text{response}} \qquad 70.1$$

Since the influence of decaying functions disappears after a few cycles, natural response is sometimes referred to as *transient response*. Once the transient response effects have died out, the total response will consist entirely of forced terms. This is the *steady-state response*.

## 2. GRAPHICAL SOLUTION

Graphical solutions are available in limited cases, particularly those with homogeneous, step, and sinusoidal inputs. When the system equation is a homogeneous second-order linear differential equation with constant coefficients in the form of Eq. 70.2, the natural time response can be determined from Fig. 70.1. $\omega$ is the natural frequency, and $\zeta$ is the damping ratio.

$$x'' + 2\zeta\omega x' + \omega^2 x = 0 \qquad 70.2$$

**Figure 70.1** *Natural Response*

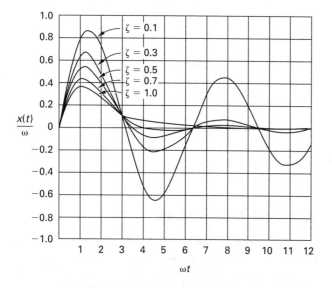

When the system equation is a second-order linear differential equation with constant coefficients, and the forcing function is a step of height $h$ (as in Eq. 70.3), the time response can be determined from Fig. 70.2.

$$x'' + 2\zeta\omega x' + \omega^2 x = \omega^2 h \qquad 70.3$$

**Figure 70.2** *Response to a Unit Step*

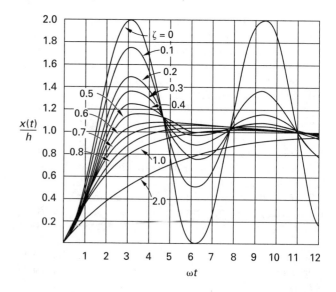

Figure 70.2 illustrates that a system responding to a step will eventually settle to the steady-state position of the step, but that when damping is low ($\zeta < 1$), there will be *overshoot*. Figure 70.3 illustrates this and other parameters of second-order response to a step input. The settling time, $t_s$, depends on the *tolerance* (i.e., the separation of the actual and steady-state responses). The *time delay*, $t_d$, in Fig. 70.3 is the time to reach 50% of the steady-state value. The *time constant*, $\tau$, is the time to reach approximately 63% of the steady-state value.

$$\omega_d = \text{damped frequency} = \omega\sqrt{1-\zeta^2} \qquad 70.4$$

$$t_r = \text{rise time} = \frac{\pi - \arccos\zeta}{\omega_d} \qquad 70.5$$

$$t_p = \text{peak time} = \frac{\pi}{\omega_d} \qquad 70.6$$

$$M_p = \text{peak gain} \quad [\text{fraction overshoot}]$$

$$= \exp\left(\frac{-\pi\zeta}{\sqrt{1-\zeta^2}}\right) \qquad 70.7$$

$$t_s = \text{settling time} = \frac{3.91}{\zeta\omega} \quad [\text{2\% criterion}] \qquad 70.8$$

$$t_s = \frac{3.00}{\zeta\omega} \quad [\text{5\% criterion}] \qquad 70.9$$

$$\tau = \text{time constant} = \frac{1}{\zeta\omega} \qquad 70.10$$

---

[1]This response is a function of time, and, therefore, is referred to as *time response* to distinguish it from *frequency response*. (See Sec. 70.13.)

**Figure 70.3** *Second-Order Step Time Response Parameters*

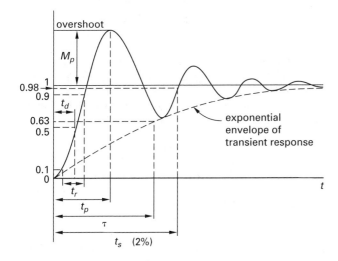

## 3. CLASSICAL SOLUTION METHOD

A system model can be thought of as a block where the *input signal* is the forcing function, $f(t)$, and the *output signal* is the response function, $r(t)$. The standard analytical method of determining the response of a system from its system equation uses Laplace transforms. Accordingly, the classical solution approach does not determine the output signal or response function directly. Rather, it derives the *transfer function*, $T(s)$.[2] The transfer function is also known as the *rational function*.

$$T(s) = \mathcal{L}\left(\frac{r(t)}{f(t)}\right) \qquad 70.11$$

Transfer functions in the *s*-domain are Laplace transformations of the corresponding time-domain functions. Transforming $r(t)/f(t)$ into $T(s)$ is more than a simple change of variables. The *s* symbol can be thought of as a derivative operator; similarly, the integration operator is represented as $1/s$. By convention, Laplace transforms are represented by uppercase letters, while operand functions are represented by lowercase letters.

### Example 70.1

A system equation in differential equation form has been derived for the output force from a mechanical network. Convert the system equation to the *s*-domain.

$$f(t) = k\int (v_1 - v_2)\,dt + m\,\frac{d(v_2 - v_1)}{dt}$$

---

[2]Strictly speaking, $r(t)/f(t)$ is the *transfer function* and $\mathcal{L}(r(t)/f(t))$ is the *transform of the transfer function*. However, the distinction is seldom made.

*Solution*

Replace all derivative operators by $s$; replace all integration operators by $1/s$. Replace time velocity, $v(t)$, with *s*-domain velocity, $v(s)$.

$$F(s) = \frac{kv_1}{s} - \frac{kv_2}{s} + smv_2 - smv_1$$

### Example 70.2

Determine the transfer function for the mechanical system shown and draw its black box representation.

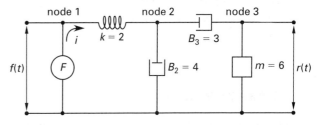

*Solution*

First, write the system equations. There are two loop "currents," so two simultaneous equations will be needed. One system equation can be written for each of the three nodes, so there will be one redundant system equation. (At this point, it is not obvious which of the two system equations can be used to derive the transfer function.)

*node 1:*
$$f(t) = k\int (v_1 - v_2)\,dt = 2\int (v_1 - v_2)\,dt$$

*node 2:*
$$0 = k\int (v_2 - v_1)\,dt + B_2 v_2 + B_3(v_2 - v_3)$$
$$= 2\int (v_2 - v_1)\,dt + 4v_2 + 3(v_2 - v_3)$$

*node 3:*
$$0 = m\frac{dv_3}{dt} + B_3(v_3 - v_2)$$
$$= 6\frac{dv_3}{dt} + 3(v_3 - v_2)$$

Next, convert the system equations to the *s*-domain by substituting $s$ for derivative operations and $1/s$ for integration operations.

*node 1:*
$$F(s) = \frac{2(v_1 - v_2)}{s}$$

*node 2:*
$$0 = \frac{2(v_2 - v_1)}{s} + 4v_2 + 3(v_2 - v_3)$$

*node 3:*
$$0 = 6sv_3 + 3(v_3 - v_2)$$

The transfer function is the ratio of the output to input velocities, $v_3/v_1$. It does not depend on $v_2$. The equation for node 1 cannot be used unless $i(t)$ is known, which it (generally) is not. $v_2$ is eliminated from the equations for nodes 2 and 3. From the second and third nodes,

$$v_2 = \frac{2v_1 + 3sv_3}{2 + 7s}$$
$$= v_3(1 + 2s)$$

The transfer function, $T(s)$, is found by equating these two expressions and solving for $v_3/v_1$.

$$T(s) = \frac{v_3}{v_1} = \frac{1}{7s^2 + 4s + 1}$$

The black box representation of this system is

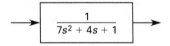

## 4. FEEDBACK THEORY

The output signal is returned as input in a feedback loop (feedback system). As illustrated in Fig. 70.4, a basic feedback system consists of two black box units (a *dynamic unit* and a *feedback unit*), a *pick-off point* (*take-off point*), and a *summing point* (*comparator* or *summer*). The summing point is assumed to perform positive addition unless a minus sign is present. The incoming signal, $V_i$, is combined with the feedback signal, $V_f$, to give the *error* (*error signal*), $E$. Whether addition or subtraction is used in Eq. 70.12 depends on whether the summing point is additive (i.e., a positive feedback system) or subtractive (i.e., a negative feedback system), respectively. $E(s)$ is the *error transfer function* (*error gain*).

$$E(s) = \mathcal{L}(e(t)) = V_i(s) \pm V_f(s)$$
$$= V_i(s) \pm H(s)V_o(s) \qquad 70.12$$

**Figure 70.4** *Feedback System*

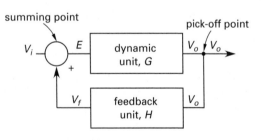

The ratio $E(s)/V_i(s)$ is the *error ratio (actuating signal ratio)*.

$$\frac{E(s)}{V_i(s)} = \frac{1}{1 + G(s)H(s)} \quad \text{[negative feedback]} \qquad 70.13$$

$$\frac{E(s)}{V_i(s)} = \frac{1}{1 - G(s)H(s)} \quad \text{[positive feedback]} \qquad 70.14$$

Since the dynamic and feedback units are black boxes, each has an associated transfer function. The transfer function of the dynamic unit is known as the *forward transfer function (direct transfer function)*, $G(s)$. In most feedback systems—amplifier circuits in particular—the magnitude of the forward transfer function is known as the *forward gain* or *direct gain*. $G(s)$ can be a scalar if the dynamic unit merely scales the error. However, $G(s)$ is normally a complex operator that changes both the magnitude and the phase of the error.

$$V_o(s) = G(s)E(s) \qquad 70.15$$

The pick-off point transmits the output signal, $V_o$, from the dynamic unit back to the feedback element. The output of the dynamic unit is not reduced by the pick-off point. The transfer function of the feedback unit is the *reverse transfer function (feedback transfer function, feedback gain, etc.)*, $H(s)$, which can be a simple magnitude-changing scalar or a phase-shifting function.

$$V_f(s) = H(s)V_o(s) \qquad 70.16$$

The ratio $V_f(s)/V_i(s)$ is the *feedback ratio (primary feedback ratio)*.

$$\frac{V_f(s)}{V_i(s)} = \frac{G(s)H(s)}{1 + G(s)H(s)} \quad \text{[negative feedback]} \qquad 70.17$$

$$\frac{V_f(s)}{V_i(s)} = \frac{G(s)H(s)}{1 - G(s)H(s)} \quad \text{[positive feedback]} \qquad 70.18$$

The *loop transfer function (loop gain, open-loop gain, or open-loop transfer function)* is the gain after going around the loop one time, $\pm G(s)H(s)$.

The *overall transfer function (closed-loop transfer function, control ratio, system function, closed-loop gain, etc.)*, $G_{\text{loop}}(s)$, is the overall transfer function of the feedback system. The quantity $1 + G(s)H(s) = 0$ is the *characteristic equation*. The *order of the system* is the largest exponent of $s$ in the characteristic equation. (This corresponds to the highest-order derivative in the system equation.)

$$G_{\text{loop}}(s) = \frac{V_o(s)}{V_i(s)} = \frac{G(s)}{1 + G(s)H(s)} \quad \text{[negative feedback]}$$
$$70.19$$

$$G_{\text{loop}}(s) = \frac{G(s)}{1 - G(s)H(s)} \quad \text{[positive feedback]} \qquad 70.20$$

With positive feedback and $G(s)H(s)$ less than 1.0, $G_{\text{loop}}$ will be larger than $G(s)$. This increase in gain is a characteristic of positive feedback systems. As $G(s)H(s)$ approaches 1.0, the closed-loop transfer function increases without bound, usually an undesirable effect.

In a negative feedback system, the denominator of Eq. 70.19 will be greater than 1.0. Although the closed-loop transfer function will be less than $G(s)$, there may be other desirable effects. Generally, a system with negative feedback will be less sensitive to variations in temperature, circuit component values, input signal frequency, and signal noise. Other benefits include distortion reduction, increased stability, and impedance matching.[3]

### Example 70.3

A high-gain, non-inverting operational amplifier has a gain of $10^6$. Feedback is provided by a resistor voltage divider. What is the closed-loop gain?

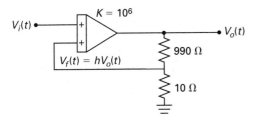

*Solution*

The fraction of the output signal appearing at the summing point depends on the resistances in the divider circuit.

$$h = \frac{10 \ \Omega}{10 \ \Omega + 990 \ \Omega} = 0.01$$

Since the feedback path only scales the feedback signal, the feedback will be positive. From Eq. 70.19, the closed-loop gain is

$$K_{\text{loop}} = \frac{K}{1 - Kh} = \frac{10^6}{1 - (10^6)(0.01)} \approx -100$$

## 5. SENSITIVITY

For large loop gains ($G(s)H(s) \gg 1$) in negative feedback systems, the overall gain will be approximately $1/H(s)$. Thus, the forward gain will not be a factor, and by choice of $H(s)$, the output can be made insensitive to variations in $G(s)$.

In general, the *sensitivity* of any variable, $A$, with respect to changes in another parameter, $B$, is

$$S_B^A = \frac{d \ln A}{d \ln B} = \frac{\frac{dA}{A}}{\frac{dB}{B}} \approx \left(\frac{\Delta A}{\Delta B}\right)\left(\frac{B}{A}\right) \qquad \textit{70.21}$$

---

[3]For circuits to be directly connected in series without affecting their performance, all input impedances must be infinite, and all output impedances must be zero.

The sensitivity of the loop transfer function with respect to the forward transfer function is

$$S_{G(s)}^{G_{\text{loop}}(s)} = \left(\frac{\Delta G_{\text{loop}}(s)}{\Delta G(s)}\right)\left(\frac{G(s)}{G_{\text{loop}}(s)}\right)$$

$$= \frac{1}{1 + G(s)H(s)} \qquad \text{[negative feedback]} \qquad \textit{70.22}$$

$$S_{G(s)}^{G_{\text{loop}}(s)} = \frac{1}{1 - G(s)H(s)} \qquad \text{[positive feedback]} \qquad \textit{70.23}$$

### Example 70.4

A closed-loop gain of $-100$ is required from a circuit, and the output signal must not vary by more than $\pm 1\%$. An amplifier is available, but its output varies by $\pm 20\%$. How can this amplifier be used?

*Solution*

A closed-loop sensitivity of 0.01 is required. This means

$$\frac{\Delta V_o}{V_o} = \frac{\Delta G_{\text{loop}}}{G_{\text{loop}}}\frac{V_i}{V_i} = \frac{\Delta G_{\text{loop}}}{G_{\text{loop}}} = 0.01$$

Similarly, the existing amplifier has a sensitivity of 20%. This means

$$\frac{\Delta G}{G} = 0.20$$

The ratio of these variations corresponds to the definition of sensitivity (Equation 70.21). For positive feedback,

$$\frac{\Delta G_{\text{loop}}}{\Delta G}\frac{G}{G_{\text{loop}}} = \frac{0.01}{0.20} = \frac{1}{1 - GH}$$

Solving, $GH = -19$.

Solving for $G$ from Eq. 70.20,

$$G_{\text{loop}} = -100 = \frac{G}{1 - GH} = \frac{G}{1 - (-19)}$$

Solving, $G = -2000$. Finally, solve for $H$.

$$H = \frac{GH}{G} = \frac{-19}{-2000} = 0.0095$$

## 6. BLOCK DIAGRAM ALGEBRA

The functions represented by several interconnected black boxes (*cascaded blocks*) can be simplified into a single block operation. Some of the most important simplification rules of block diagram algebra are shown in Fig. 70.5. Case 3 represents the standard feedback model.

**Figure 70.5** *Rules of Simplifying Block Diagrams*

| case | original structure | equivalent structure |
|------|--------------------|-----------------------|

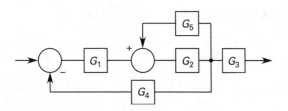

*(Figure 70.5 table of block diagram simplification rules, cases 1 through 8)*

### Example 70.5

A complex block system is constructed from five blocks and two summing points. What is the overall transfer function?

*Solution*

Use case 5 to move the second summing point back to the first summing point.

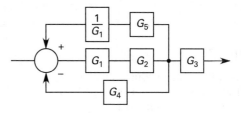

Use case 1 to combine boxes in series.

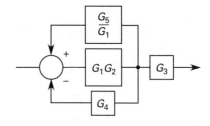

Use case 2 to combine the two feedback loops.

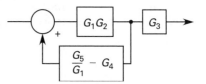

Use case 8 to move the pick-off point outside the $G_3$ box.

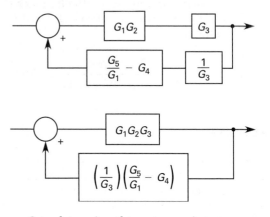

Use case 3 to determine the system gain.

$$G_{\text{loop}} = \cfrac{G_1 G_2 G_3}{1 - G_1 G_2 G_3 \left(\dfrac{1}{G_3}\right)\left(\dfrac{G_5}{G_1} - G_4\right)}$$

$$= \frac{G_1 G_2 G_3}{1 - G_2 G_5 + G_1 G_2 G_4}$$

## 7. PREDICTING SYSTEM TIME RESPONSE

The transfer function is derived without knowledge of the input and is insufficient to predict the time response of the system. The system time response will depend on the form of the input function. Since the transfer function is expressed in the $s$-domain, the forcing and response functions must be also.

$$R(s) = T(s)F(s) \qquad 70.24$$

The time-based response function, $r(t)$, is found by performing the inverse Laplace transform.

$$r(t) = \mathcal{L}^{-1}(R(s)) \qquad 70.25$$

### Example 70.6

A mechanical system is acted upon by a constant force of eight units starting at $t = 0$. Determine the time-based response function, $r(t)$, if the transfer function is

$$T(s) = \frac{6}{(s+2)(s+4)}$$

*Solution*

The forcing function is a step of height 8 at $t = 0$. The Laplace transform of a unit step is $1/s$. Therefore, $F(s) = 8/s$.

From Eq. 70.24,

$$R(s) = T(s)F(s) = \left(\frac{6}{(s+2)(s+4)}\right)\left(\frac{8}{s}\right)$$

$$= \frac{48}{s(s+2)(s+4)}$$

The response function is found from Eq. 70.25 and a table of Laplace transforms. (A product of linear terms in the denominator of $R(s)$ is equivalent to a sum of terms in $r(t)$.)

$$r(t)\mathcal{L}^{-1} = \frac{48}{s(s+2)(s+4)}$$

$$= 6 - 12e^{-2t} + 6e^{-4t}$$

The last two terms are decaying exponentials, which represent the transient natural response. The first term does not vary with time; it is the steady-state response.

## 8. PREDICTING TIME RESPONSE FROM A RELATED RESPONSE

In some cases, it may be possible to use a known response to one input to determine the response to another input. For example, the impulse function is the derivative of the step function, so the response to an impulse is the derivative of the response to a step function.

## 9. INITIAL AND FINAL VALUES

The initial and final (steady-state) values of any function, $P(s)$, can be found from the *initial* and *final value*

*theorems*, respectively, providing the limits exist. Equation 70.26 and Eq. 70.27 are particularly valuable in determining the steady-state response (substitute $R(s)$ for $P(s)$) and the steady-state error (substitute $E(s)$ for $P(s)$).

$$\lim_{t \to 0^+} p(t) = \lim_{s \to \infty}(sP(s)) \quad \text{[initial value]} \qquad 70.26$$

$$\lim_{t \to \infty} p(t) = \lim_{s \to 0}(sP(s)) \quad \text{[final value]} \qquad 70.27$$

### Example 70.7

Determine the final value of the response function $r(t)$ if

$$R(s) = \frac{1}{s(s+1)}$$

*Solution*

From Eq. 70.27, $R(s)$ is multiplied by $s$ and the limit taken as $s$ tends to zero.

$$R(\infty) = \lim_{s \to 0}\left(\frac{s}{s(s+1)}\right) = \lim_{s \to 0}\left(\frac{1}{s+1}\right) = 1$$

## 10. SPECIAL CASES OF STEADY-STATE RESPONSE

In addition to determining the steady-state response from the final value theorem (see Sec. 70.9), the steady-state response to a specific input can be easily derived from the transfer function, $T(s)$, in a few specialized cases. For example, the steady-state response function for a system acted upon by an impulse is simply the transfer function. That is, a pulse has no long-term effect on a system.

$$R(\infty) = T(s) \quad \text{[pulse input]} \qquad 70.28$$

The steady-state response for a *step input* (often referred to as a *DC input*) is obtained by substituting 0 for $s$ everywhere in the transfer function. (If the step has magnitude $h$, the steady-state response is multiplied by $h$.)

$$R(\infty) = T(0) \quad \text{[unit step input]} \qquad 70.29$$

The steady-state response for a sinusoidal input is obtained by substituting $j\omega_f$ for $s$ everywhere in the transfer function, $T(s)$. The output will have the same frequency as the input. It is particularly convenient to perform sinusoidal calculations using phasor notation (as illustrated in Ex. 70.9).

$$R(\infty) = T(j\omega_f) \qquad 70.30$$

### Example 70.8

What is the steady-state response of the system in Ex. 70.6 when acted upon by a step of height 8?

*Solution*

Substitute 0 for $s$ in $T(s)$ and multiply by 8.

$$R(\infty) = (8)(T(0)) = (8)\left(\frac{6}{(0+2)(0+4)}\right) = 6$$

### Example 70.9

Determine the steady-state response when a sinusoidal forcing function of $4\sin(2t+45°)$ is applied to a system whose transfer function is

$$T(s) = \frac{-1}{7s^2 + 7s + 1}$$

*Solution*

The angular frequency of the forcing function is $\omega_f = 2$ rad/s. Substitute $j2$ for $s$ in $T(s)$, and simplify the expression by recognizing that $j^2 = -1$.

$$T(j2) = \frac{-1}{7(j2)^2 + 7(j2) + 1}$$

$$= \frac{-1}{-28 + 14j + 1} = \frac{-1}{-27 + 14j}$$

Next, convert $T(j2)$ to phasor (polar) form. The magnitude and angle of the denominator are

$$\text{magnitude} = \sqrt{(14)^2 + (-27)^2} = 30.41$$

$$\text{angle} = 180° - \arctan\frac{14}{27} = 152.6°$$

However, this is the negative reciprocal of $T(j2)$

$$T(j2) = \frac{-1}{30.41\angle 152.6°} = -0.0329\angle -152.6°$$

The forcing function expressed in phasor form is $4\angle 45°$. From Eq. 70.11, the steady-state response is

$$v(t) = T(t)f(t) = (-0.0329\angle -152.6°)(4\angle 45°)$$

$$= -0.1316\angle -107.6°$$

### 11. POLES AND ZEROS

A *pole* is a value of $s$ that makes a function, $P(s)$, infinite. Specifically, a pole makes the denominator of $P(s)$ zero.[4] A *zero* of the function makes the numerator of $P(s)$ (and hence $P(s)$ itself) zero. Poles and zeros need not be real or unique; they can be imaginary and repeated within a function.

A *pole-zero diagram* is a plot of poles and zeros in the *s-plane*—a rectangular coordinate system with real and imaginary axes. Zeros are represented by $\bigcirc$s; poles are

---

[4]Pole values are the system *eigenvalues*.

represented as $\times$s. Poles off the real axis always occur in conjugate pairs known as *pole pairs*.

Sometimes it is necessary to derive the function $P(s)$ from its pole-zero diagram. This will be only partially successful since repeating identical poles and zeros are not usually indicated on the diagram. Also, scale factors (scalar constants) are not shown.

### Example 70.10

Draw the pole-zero diagram for the following transfer function.

$$T(s) = \frac{5(s+3)}{(s+2)(s^2+2s+2)}$$

*Solution*

The numerator is zero when $s = -3$. This is the only zero of the transfer function.

The denominator is zero when $s = -2$ and $s = -1 \pm j$. These three values are the poles of the transfer function.

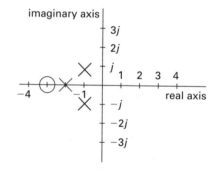

### Example 70.11

A pole-zero diagram for a transfer function $T(s)$ has a single pole at $s = -2$ and a single zero at $s = -7$. What is the corresponding function?

*Solution*

$$T(s) = \frac{K(s+7)}{s+2}$$

The scale factor $K$ must be determined by some other means.

### 12. PREDICTING SYSTEM TIME RESPONSE FROM RESPONSE POLE-ZERO DIAGRAMS

A response pole-zero diagram (see Fig. 70.6) based on $R(s)$ can be used to predict how the system responds to a specific input. (Note that this pole-zero diagram must be based on the product $T(s)F(s)$ since that is how $R(s)$ is calculated. Plotting the product $T(s)F(s)$ is equivalent to plotting $T(s)$ and $F(s)$ separately on the same diagram.)

The system will experience an *exponential decay* when a single pole falls on the real axis. A pole with a value of $-r$, corresponding to the linear term $(s + r)$, will decay at the rate of $e^{-rt}$. The quantity $1/r$ is the decay *time constant*, the time for the response to achieve approximately 63% of its steady-state value. Thus, the farther left the point is located from the vertical imaginary axis, the faster the motion will die out.

*Undamped sinusoidal oscillation* will occur if a pole pair falls on the imaginary axis. A conjugate pole pair with the value of $\pm j\omega$ indicates oscillation with a natural frequency of $\omega$ rad/s.

Pole pairs to the left of the imaginary axis represent *decaying sinusoidal* response. The closer the poles are to the real (horizontal) axis, the slower will be the oscillations. The closer the poles are to the imaginary (vertical) axis, the slower will be the decay. The *natural frequency*, $\omega$, of undamped oscillation can be determined from a *conjugate pole pair* having values of $r \pm \omega_f$.

$$\omega = \sqrt{r^2 + \omega_f^2} \qquad 70.31$$

**Figure 70.6** *Types of Response Determined by Pole Location*

$e^{-rt} \sin \omega_2 t$ term
at $-r + j\omega_2$

$\sin \omega_1 t$ term
at $+j\omega_1$

$e^{-rt}$ term
at $-r$

$\sin \omega_1 t$ term
at $-j\omega_1$

$e^{-rt} \sin \omega_2 t$ term
at $-r - j\omega_2$

The magnitude and phase shift can be determined for any input frequency from the pole-zero diagram (see Fig. 70.7) with the following procedure: Locate the angular frequency, $\omega_f$, on the imaginary axis. Draw a line from each pole (i.e., a pole-line) and from each zero (i.e., a zero-line) of $T(s)$ to this point. The angle of each of these lines is the angle between it and the horizontal real axis. The overall magnitude is the product of the lengths of the zero-lines divided by the product of the lengths of the pole-lines. (The scale factor must also be included because it is not shown on the pole-zero diagram.) The phase is the sum of the pole-angles less the sum of the zero-angles.

$$|R| = \frac{K \prod_z |L_z|}{\prod_p |L_p|} \qquad 70.32$$

$$\angle R = \sum_p \alpha - \sum_z \beta \qquad 70.33$$

**Figure 70.7** *Calculating Magnitude and Phase from a Pole-Zero Diagram*

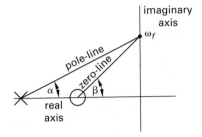

### Example 70.12

What is the response of a system with the response pole-zero diagram representing $R(s) = T(s)F(s)$ shown?

*Solution*

The poles are at $r = -\frac{1}{2}$ and $r = -4$. The response is

$$r(t) = C_1 e^{-\frac{1}{2}t} + C_2 e^{-4t}$$

Constants $C_1$ and $C_2$ must be found from other data.

### Example 70.13

What is the system response if $T(s) = (s + 2)/(s + 3)$ and the input is a unit step?

*Solution*

The transform of a unit step is $1/s$. The response is

$$R(s) = T(s)F(s) = \frac{s + 2}{s(s + 3)}$$

The response pole-zero diagram is

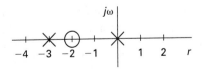

The pole at $r = 0$ contributes the exponential $C_1 e^{-0t}$ (or just $C_1$) to the total response. The pole at $r = -3$ contributes the term $C_2 e^{-3t}$. The total response is

$$r(t) = C_1 + C_2 e^{-3t}$$

Constants $C_1$ and $C_2$ must be found from other data.

## 13. FREQUENCY RESPONSE

The gain and phase angle frequency response of a system will change as the forcing frequency is varied. (The dependence of $r(t)$ on $\omega_f$ was illustrated in Ex. 70.9.) The *frequency response* is the variation in these parameters, always with a sinusoidal input. *Gain* and *phase characteristics* are plots of the steady-state gain and phase angle responses with a sinusoidal input versus frequency. While a linear frequency scale can be used, frequency response is almost always presented with a logarithmic frequency scale.

The steady-state gain response is expressed in decibels, while the steady-state phase angle response is expressed in degrees. The gain is calculated from Eq. 70.34, where $|T(s)|$ is the absolute value of the steady-state response.

$$\text{gain} = 20\log|T(j\omega)| \quad [\text{in dB}] \qquad 70.34$$

A doubling of $|T(j\omega)|$ is referred to as an *octave* and corresponds to a 6.02 dB increase. A tenfold increase in $|T(j\omega)|$ is a *decade* and corresponds to a 20 dB increase.

$$\text{number of octaves} = \frac{\text{gain}_2 - \text{gain}_1}{6.02} \quad [\text{in dB}]$$
$$= 3.32 \times \text{number of decades} \qquad 70.35$$

$$\text{number of decades} = \frac{\text{gain}_2 - \text{gain}_1}{20} \quad [\text{in dB}]$$
$$= 0.301 \times \text{number of octaves} \qquad 70.36$$

## 14. GAIN CHARACTERISTIC

The *gain characteristic* (*M*-curve for magnitude) is a plot of the gain as $\omega_f$ is varied. It is possible to make a rough sketch of the gain characteristic by calculating the gain at a few points (pole frequencies, $\omega = 0$, $\omega = \infty$, etc.). The curve will usually be asymptotic to several lines. The frequencies at which these asymptotes intersect are *corner frequencies*. The peak gain, $M_p$, occurs at the natural (resonant) frequency of the system.[5] Large peak gains indicate lowered stability and large overshoots. The *gain crossover point*, if any, is the frequency at which $\log(\text{gain}) = 0$.

The *half-power points (cut-off frequencies)* are the frequencies for which the gain is 0.707 (i.e., $\sqrt{2}/2$) times the peak value. (This is equivalent to saying the gain is 3 dB less than the peak gain.) The *cut-off rate* is the slope of the gain characteristic in dB/octave at a half-power point. The frequency difference between the half-power points is the *bandwidth*, BW. (See Fig. 70.8.) The *closed-loop bandwidth* is the frequency range

over which the closed-loop gain falls 3 dB below its value at $\omega = 0$. (The term "bandwidth" often means closed-loop bandwidth.) The *quality factor*, $Q$, is

$$Q = \frac{\omega_n}{\text{BW}} \qquad 70.37$$

**Figure 70.8** Bandwidth

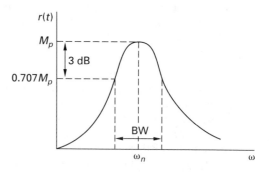

Since a low or negative gain (compared to higher parts of the curve) effectively represents attenuation, the gain characteristic can be used to distinguish between low- and high-pass filters. A low-pass filter will have a large gain at low frequencies and a small gain at high frequencies. Conversely, a high-pass filter will have a high gain at high frequencies and a low gain at low frequencies.

It may be possible to determine certain parameters (e.g., the natural frequency and bandwidth) from the transfer function directly. For example, when the denominator of $T(s)$ is a single linear term of the form $s + r$, the bandwidth will be equal to $r$. Thus, the bandwidth and time constant are reciprocals.

Another important case is when $T(s)$ has the form of Eq. 70.38. (Compare the form of $T(s)$ here to its form in Eq. 70.11 in Sec. 70.3.) The coefficient of the $s^2$ term must be 1, and $\omega_n$ must be much larger than BW so that the pole is close to the imaginary axis. The zero defined by constants $a$ and $b$ in the numerator is not significant.

$$T(s) = \frac{as + b}{s^2 + (\text{BW})s + \omega_n^2} \qquad 70.38$$

### Example 70.14

Determine the maximum gain, bandwidth, upper half-power frequency, and half-power gain of a system whose transfer function is

$$T(s) = \frac{1}{s + 5}$$

*Solution*

The steady-state response to sinusoidal input is determined by substituting $j\omega$ for $s$ in $T(s)$. When $\omega = 0$, $T(s)$ will have a value of $1/5 = 0.2$, and $T(s)$ decreases

---

[5]The gain characteristic peaks when the forcing frequency equals the natural frequency. It is also said that this peak corresponds to the resonant frequency. Strictly speaking, this is true, although the gain may not actually be resonant (i.e., be infinite).

thereafter. The maximum gain is 0.2. The bandwidth is 5 rad/s. Since the lower half-power frequency is implicitly 0 rad/s, the upper half-power frequency is 5 rad/s, at which point the gain will be $(0.707)(0.2) = 0.141$.

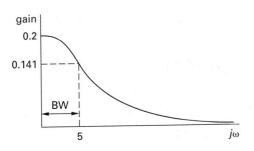

**Example 70.15**

Predict the natural frequency and bandwidth for the following transfer function.

$$T(s) = \frac{s + 19}{s^2 + 7s + 1000}$$

*Solution*

The form of this equation is the same as Eq. 70.38. The bandwidth is BW = 7, and the natural frequency is $\sqrt{1000} = 31.62$ rad/s.

## 15. PHASE CHARACTERISTIC

The phase angle response will also change as the forcing frequency is varied. The *phase characteristic* ($\alpha$ *curve*) is a plot of the phase angle as $\omega_f$ is varied.

## 16. STABILITY

A stable system will remain at rest unless disturbed by external influence, and it will return to a rest position once the disturbance is removed. A pole with a value of $-r$ on the real axis corresponds to an exponential response of $e^{-rt}$. Since $e^{-rt}$ is a decaying signal, the system is stable. Similarly, a pole of $+r$ on the real axis corresponds to an exponential response of $e^{rt}$. Since $e^{rt}$ increases without limit, the system is unstable.

Since any pole to the right of the imaginary axis corresponds to a positive exponential, a *stable system* will have poles only in the left half of the $s$-plane. If there is an isolated pole on the imaginary axis, the response is stable. However, a conjugate pole pair on the imaginary axis corresponds to a sinusoid that does not decay with time. Such a system is considered to be unstable.

*Passive systems* (i.e., the homogeneous case) are not acted upon by a forcing function and are always stable. In the absence of an energy source, exponential growth cannot occur. *Active systems* contain one or more energy sources and may be stable or unstable.

There are several *frequency response (domain) analysis* techniques for determining the stability of a system,

including Bode plot, root-locus diagram, Routh stability criterion, Hurwitz test, and Nichols chart. The term *frequency response* almost always means the steady-state response to a sinusoidal input.

The value of the denominator of $T(s)$ is the primary factor affecting stability. When the denominator approaches zero, the system increases without bound. In the typical feedback loop, the denominator is $1 \pm GH$, which can be zero only if $|GH| = 1$. It is logical, then, that most of the methods for investigating stability (e.g., Bode plots, root-locus, Nyquist analysis, and the Nichols chart) investigate the value of the open-loop transfer function, $GH$. Since $\log(1) = 0$, the requirement for stability is that $\log(GH)$ must not equal 0 dB.

A negative feedback system will also become unstable if it changes to a positive feedback system, which can occur when the feedback signal is changed in phase more than 180°. Therefore, another requirement for stability is that the phase angle change must not exceed 180°.

## 17. BODE PLOTS

*Bode plots* are gain and phase characteristics for the open-loop $G(s)H(s)$ transfer function used to determine the *relative stability* of a system. (See Fig. 70.9.) The gain characteristic is a plot of $20 \log(|G(s)H(s)|)$ versus $\omega$ for a sinusoidal input. (It is important to recognize that the Bode plots, though similar in appearance to the gain and phase frequency response charts, are used to evaluate stability and do not describe the closed-loop system response.)

*Figure 70.9 Gain and Phase Margin Bode Plots*

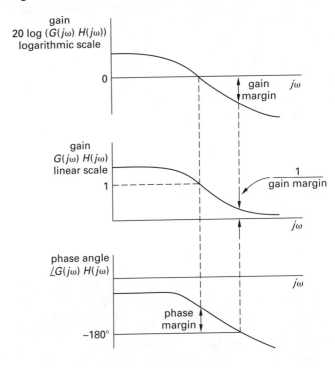

The *gain margin* is the number of decibels that the open-loop transfer function, $G(s)H(s)$, is below 0 dB at the *phase crossover frequency* (i.e., where the phase angle is $-180°$). (If the gain happens to be plotted on a linear scale, the gain margin is the reciprocal of the gain at the phase crossover point.) The gain margin must be positive for a stable system, and the larger it is, the more stable the system will be.

The *phase margin* is the number of degrees the phase angle is above $-180°$ at the *gain crossover point* (i.e., where the logarithmic gain is 0 dB or the actual gain is 1).

In most cases, large positive gain and phase margins will ensure a stable system. However, the margins could have been measured at other than the crossover frequencies. Therefore, a Nyquist stability plot is needed to verify the absolute stability of a system.

## 18. ROOT-LOCUS DIAGRAMS

A *root-locus diagram* is a pole-zero diagram showing how the poles of $G(s)H(s)$ move when one of the system parameters (e.g., the gain factor) in the transfer function is varied. The diagram gets its name from the need to find the roots of the denominator (i.e., the poles). The locus of points defined by the various poles is a line or curve that can be used to predict *points of instability* or other critical operating points. A point of instability is reached when the line crosses the imaginary axis into the right-hand side of the pole-zero diagram.

A root-locus curve may not be contiguous, and multiple curves will exist for different sets of roots. Sometimes the curve splits into two branches. In other cases, the curve leaves the real axis at *breakaway points* and continues on with constant or varying slopes approaching asymptotes. One branch of the curve will start at each open-loop pole and end at an open-loop zero.

### Example 70.16

Draw the root-locus diagram for a feedback system with open-loop transfer function $G(s)H(s)$. $K$ is a scalar constant that can be varied.

$$G(s)H(s) = \frac{Ks(s+1)(s+2)}{s(s+2) + K(s+1)}$$

*Solution*

The poles are the zeros of the denominator.

$$s_1, s_2 = -\tfrac{1}{2}(2+K) \pm \sqrt{1 + \tfrac{1}{4}K^2}$$

Since the second term can be either added or subtracted, there are two roots for each value of $K$. Allowing $K$ to vary from zero to infinity produces a root-locus diagram with two distinct branches. The first branch extends from the pole at the origin to the zero at $s = -1$. The second branch extends from the pole at $s = -2$ to $-\infty$.

All poles and zeros are not shown. Since neither branch crosses into the right half, the system is stable for all values of $K$.

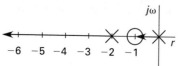

## 19. HURWITZ TEST

A stable system has poles only in the left half of the $s$-plane. These poles correspond to roots of the *characteristic equation* (see Sec. 70.4). The characteristic equation can be expanded into a polynomial of the form

$$a_0 s^n + a_1 s^{n-1} + \cdots + a_{n-1}s + a_n = 0 \qquad 70.39$$

The *Hurwitz stability criterion* requires that all coefficients be present and have the same sign (which is equivalent to requiring all coefficients to be positive). If the coefficients differ in sign, the system is unstable. If the coefficients are all alike in sign, the system may or may not be stable. The Routh criterion (test) should be used in that case.

## 20. ROUTH CRITERION

The *Routh criterion*, like the Hurwitz test, uses the coefficients of the polynomial characteristic equation. A table (the *Routh table*) of these coefficients is formed. The Routh-Hurwitz criterion states that the number of sign changes in the first column of the table equals the number of positive (unstable) roots. Therefore, the system will be stable if all entries in the first column have the same sign.

The table is organized in the following manner.

$$
\begin{array}{cccc}
a_0 & a_2 & a_4 & a_6 \cdots \\
a_1 & a_3 & a_5 & a_7 \cdots \\
b_1 & b_2 & b_3 & b_4 \cdots \\
c_1 & c_2 & c_3 & c_4 \cdots \\
\vdots & \vdots & \vdots & \vdots
\end{array}
$$

The remaining coefficients are calculated in the following pattern until all values are zero.

$$b_1 = \frac{a_1 a_2 - a_0 a_3}{a_1} \qquad 70.40$$

$$b_2 = \frac{a_1 a_4 - a_0 a_5}{a_1} \qquad 70.41$$

$$b_3 = \frac{a_1 a_6 - a_0 a_7}{a_1} \qquad 70.42$$

$$c_1 = \frac{b_1 a_3 - a_1 b_2}{b_1} \qquad 70.43$$

$$c_2 = \frac{b_1 a_5 - a_1 b_3}{b_1} \qquad 70.44$$

Systems Analysis

Special methods are used if there is a zero in the first column but nowhere else in that row. One of the methods is to substitute a small number, represented by $\epsilon$ or $\delta$, for the zero and calculate the remaining coefficients as usual.

**Example 70.17**

Evaluate the stability of a system that has a characteristic equation of

$$s^3 + 5s^2 + 6s + C = 0$$

*Solution*

All of the polynomial terms are present (Hurwitz criterion), so the Routh table is

$$
\begin{array}{c|cc}
s^3 & 1 & 6 \\
s^2 & 5 & C \\
s^1 & \dfrac{30 - C}{5} & 0 \\
s^0 & C & 0
\end{array}
$$

To be stable, all of the entries in the first column must be positive, which requires $0 < C < 30$.

## 21. NYQUIST ANALYSIS

The Nyquist analysis is a particularly useful graphical method when time delays are present in a system or when frequency response data are available. *Nyquist's stability criterion* is $N = P - Z$, where $P$ is the number of poles in the right half of the $s$-plane, $Z$ is the number of zeros in the right half of the $s$-plane, and $N$ is the number of encirclements (revolutions) of $1 + G(s)H(s)$ around the critical point. $N$ may be positive, negative, or zero. Essentially, if $P$ is zero, then $N$ must be zero. Otherwise, $N$ and $P$ must be equal for the system to be stable, which is another way of saying the number of zeros in the right half of the pole-zero diagram must be zero.

## 22. APPLICATION TO CONTROL SYSTEMS

A control system monitors a process and makes adjustments to maintain performance within certain acceptable limits. Feedback is implicitly a part of all control systems.[6] The *controller (control element)* is the part of the control system that establishes the acceptable limits of performance, usually by setting its own reference inputs. The controller transfer function for a

proportional controller is a constant: $G_1(s) = K$.[7] The *plant (controlled system)* is the part of the system that responds to the controller. Both of these are in the forward loop. The input signal, $R(s)$, in Fig. 70.10 is known in a control system as the *command* or *reference value*. Figure 70.10 is known as a *control logic diagram* or *control logic block diagram*.

**Figure 70.10** *Typical Feedback Control System*

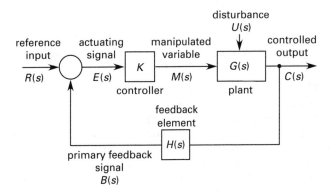

A *servomechanism* is a special type of control system in which the controlled variable is mechanical position, velocity, or acceleration. In many servomechanisms, $H(s) = 1$ (i.e., unity feedback), and it is desired to keep the output equal to the reference input (i.e., maintain a zero error function). If the input, $R(s)$, is constant, the descriptive terms *regulator* and *regulating system* are used.

## 23. APPLICATION TO TACHOMETER CONTROL

*Tachometers* are used to measure rotational speeds. Each rotation of the tachometer shaft produces one or more inductive or photoelectric pulses. The rotational speed is determined by counting and scaling the number of pulses per period.

To maintain a particular rotational speed, the output of the tachometer is fed back to a control circuit. The control logic block diagram of a *constant-speed control system* is shown in Fig. 70.11. Control circuits may contain various electrical and electronic devices.[8]

The desired speed is usually set by adjusting a voltage level. This is illustrated as an input potentiometer, which feeds the desired speed into a *summing point*.

---

[6]Not all controlled systems are feedback systems. The positions of many precision devices (e.g., print heads in dot-matrix printers, or cutting heads on some numerically controlled machines) are controlled by precision *stepper motors*. However, unless the device has feedback (e.g., a position sensor), it will have no way of knowing if it gets out of control.

[7]Sometimes the notation $K_n$ is used for $K$, where $n$ is the type of the system. In a *type 0 system*, a constant error signal results in a constant value of the output signal. In a *type 1 system*, a constant error signal results in a constant rate of change of the output signal. In a *type 2 system*, a constant error signal produces a constant second derivative of the output variable.

[8]The electronic circuit may contain low-pass filters, differential amplifiers, booster amplifiers, an adjustable voltage source, and so on. The actual integration of these devices generally requires specific product knowledge.

**Figure 70.11** *Block Diagram for Constant-Speed Motor Control*

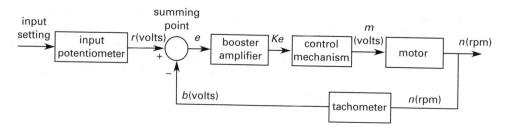

The tachometer feeds the actual speed into the summing point as well, which actually computes the difference (known as the *error, e*) between the two signals.

The error is amplified from $e$ to $Ke$, and the amplified signal is fed into an appropriate *control mechanism*. There are many methods, both analog and digital, by which the signal can be translated by the control mechanism into a change in motor input. Figure 70.11 only requires that the transfer function (i.e., from $Ke$ to $m$) be known. The changed motor speed, $n$, is converted to a tachometer signal, $b$, which is fed back to the summing point to complete the loop.

## 24. STATE MODEL REPRESENTATION

While the classical methods of designing and analyzing control systems are adequate for most situations, state model representations are preferred for more complex cases, particularly those with multiple inputs and outputs or when behavior is nonlinear or varies with time.

The state variables completely define the dynamic state (position, voltage, pressure, etc.), $x_i(t)$, of the system at time $t$. (In simple problems, the number of state variables corresponds to the number of *degrees of freedom*, $n$, of the system.) The $n$ state variables are written in matrix form as a state vector, $\mathbf{X}$.

$$\mathbf{X} = \begin{pmatrix} x_1 \\ x_2 \\ x_3 \\ \vdots \\ x_n \end{pmatrix} \qquad 70.45$$

It is a characteristic of state models that the state vector is acted upon by a first-degree derivative operator, $d/dt$, to produce a differential term, $\mathbf{X}'$, of order 1.

$$\mathbf{X}' = \frac{d\mathbf{X}}{dt} \qquad 70.46$$

Equation 70.47 and Eq. 70.48 show the general form of a state model representation: $\mathbf{U}$ is an $r$-dimensional (i.e., an $r \times 1$ matrix) *control vector*; $\mathbf{Y}$ is an $m$-dimensional (i.e., an $m \times 1$ matrix) *output vector*; $\mathbf{A}$ is an $n \times n$ *system matrix*; $\mathbf{B}$ is an $n \times r$ *control matrix*; and $\mathbf{C}$ is an $m \times n$ *output matrix*. The actual unknowns are the $x_i$ variables. The $y_i$ variables, which may not be needed in all problems, are only linear combinations of the $x_i$ variables. (For example, the $x$ variables might represent spring end positions; the $y$ variables might represent stresses in the spring. Then, $y = k\Delta x$.) Equation 70.47 is the *state equation*, and Eq. 70.48 is the *response equation*.

$$\mathbf{X}' = \mathbf{AX} + \mathbf{BU} \qquad 70.47$$

$$\mathbf{Y} = \mathbf{CX} \qquad 70.48$$

A conventional block diagram can be modified to show the multiplicity of signals in a state model, as shown in Fig. 70.12.[9] The actual physical system does not need to be a feedback system. The form of Eq. 70.47 and Eq. 70.48 is the sole reason that a feedback diagram is appropriate.

**Figure 70.12** *State Variable Diagram*

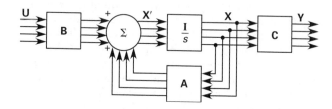

A state variable model permits only first-degree derivatives, so additional $x_i$ state variables are used for higher-order terms (e.g., acceleration).

*System controllability* exists if all of the system states can be controlled by the inputs, $\mathbf{U}$. In state model language, system controllability means that an arbitrary initial state can be steered to an arbitrary target state in a finite amount of time. *System observability* exists if the initial system states can be predicted from knowing the inputs, $\mathbf{U}$, and observing the outputs, $\mathbf{Y}$.[10]

---

[9]The block $\mathbf{I}/s$ is a diagonal identity matrix with elements of $1/s$. This effectively is an integration operator.

[10]*Kalman's theorem*, based on matrix rank, is used to determine system controllability and observability.

## Example 70.18

Write the state variable formulation of a mechanical system's transfer function $T(s)$.

$$T(s) = \frac{7s^2 + 3s + 1}{s^3 + 4s^2 + 6s + 2}$$

*Solution*

Recognize the transfer function as the quotient of two terms, and multiply $T(s)$ by the dimensionless quantity $x/x$.

$$T(s) = \frac{Y(s)}{U(s)} = \left(\frac{7s^2 + 3s + 1}{s^3 + 4s^2 + 6s + 2}\right)\left(\frac{x}{x}\right)$$

$$Y(s) = 7s^2x + 3sx + x$$

$$U(s) = s^3x + 4s^2x + 6sx + 2x$$

$Y(s)$ and $U(s)$ represent the following differential equations.

$$y(t) = 7x''(t) + 3x'(t) + x(t)$$

$$u(t) = x'''(t) + 4x''(t) + 6x'(t) + 2x(t)$$

Make the following substitutions. ($x_4(t)$ is not needed because one level of differentiation is built into the state model.)

$$x_1(t) = x(t)$$

$$x_2(t) = x'(t) = x_1'(t)$$

$$x_3(t) = x''(t) = x_2'(t)$$

Write the first derivative variables in terms of the $x_i(t)$ to get the **A** matrix entries.

$$x_1' = 0x_1(t) + 1x_2(t) + 0x_3(t) + 0$$

$$x_2' = 0x_1(t) + 0x_2(t) + 1x_3(t) + 0$$

$$x_3' = -2x_1(t) - 6x_2(t) - 4x_3(t) + u(t)$$

Determine the coefficients of the **C** matrix by rewriting $y(t)$ in the same variable order.

$$y(t) = 1x_1(t) + 3x_2(t) + 7x_3(t)$$

From Eq. 70.47 and Eq. 70.48, the state variable representation of $T(s)$ is

$$\begin{bmatrix} x_1' \\ x_2' \\ x_3' \end{bmatrix} = \begin{bmatrix} 0 & 1 & 0 \\ 0 & 0 & 1 \\ -2 & -6 & -4 \end{bmatrix} \begin{bmatrix} x_1 \\ x_2 \\ x_3 \end{bmatrix} + \begin{bmatrix} 0 \\ 0 \\ 1 \end{bmatrix} [r(t)]$$

$$[y(t)] = [1 \; 3 \; 7] \begin{bmatrix} x_1 \\ x_2 \\ x_3 \end{bmatrix}$$

# 71 Analog Modeling and Simulation

## 1. ANALOG MODELS

In contrast to the usual case of using equations to predict the behavior of model physical systems, it is possible to build electrical circuits that simulate the solutions of equations, particularly ordinary linear differential equations with constant coefficients.[1] These circuits are *analog models*. Voltages in such circuits correspond (are analogous) to unknown variables in the equations being investigated.

Rather than taking a breadboard, or experimental, approach in building such circuits, analog circuitry made for simulation has traditionally been implemented in an *analog computer*—a collection of components (op amps, potentiometers, oscilloscopes, etc.) in a common housing. The components are interconnected by programmable switch matrices. Although a few analog computers remain in use for teaching purposes in linear circuits, controls, and mechatronics labs, most modern simulation is performed digitally in Matlab, Simulink, and VisSim.

## 2. STRUCTURAL ELEMENTS

When voltage represents a variable, most of the standard mathematical operations (e.g., addition, multiplication, differentiation) with that variable can be performed electrically by passing the voltage through different components.[2] For example, the change in voltage when a potentiometer is placed in the circuit is analogous to scalar multiplication. Table 71.1 lists the primary structural elements of an analog model and the operations they perform. High-gain direct current operational amplifiers are used to implement most of these elements.

---

[1]The mathematical equations, in turn, are generally used to model mechanical, fluid, thermal, and electrical systems.
[2]While possible, differentiation is seldom simulated due to the tendency differentiators have to develop electrical noise.

*Table 71.1* Structural Elements of an Analog Model

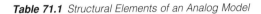

The external input voltage to an analog circuit model corresponds to the forcing function of the nonhomogeneous differential equation being modeled. For example, a sinusoidal voltage will correspond to a sinusoidal forcing function, and so on. Such inputs are typically generated externally or as outputs from other analog model circuits.

Initial conditions are obtained by setting dedicated potentiometers on their corresponding integrators.

### Example 71.1

An ideal operational amplifier is used as an inverting integrator. Its input represents a variable $x(t)$ whose initial value is $x(0) = 5$. The variable is acted upon by a step input of height 3. (a) How would this be modeled? (b) What would be the output of the model at $t = 4$?

*Solution*

(a) The symbolic representation of a model for this situation is

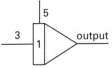

(b) Integrating a constant 3 results in $3t$. Adding in the initial value, the function is $x(t) = 3t + 5$. At $t = 4$, the function value is

$$x(4) = (3)(4) + 5 = 17$$

The inverted output is $-17$.

## 3. INDIRECT PROGRAMMING

The act of arranging the elements into a simulation model is known as *indirect programming* or *mechanization*. (The procedure given here assumes that there are no differentiating operational amplifiers in the mechanization of the differential equation.)

*step 1:* Normalize the differential equation so that the coefficient of the highest-order term is 1.0.

*step 2:* Solve for the negative of the highest-order term. (This accounts for the inversion performed by the final summing amplifier.)

*step 3:* Assume that the highest-order derivative exists. Route it through successive integrators to obtain lower-order raw derivatives.

*step 4:* Run raw derivatives through scaling potentiometers to obtain all required scaled derivatives.

*step 5:* Feed the forcing function and all scaled derivatives into the final summer.

*step 6:* Since the highest-order derivative is equal to the sum of all other terms (step 2), complete the circuit by connecting the output of the final summer to the point where the highest order derivative was assumed to exist in step 3.

*step 7:* Set the initial conditions on the potentiometers corresponding to the integrators used.[3] Set the scaling potentiometers to the constant coefficients of their respective terms.

### Example 71.2

Construct the analog model circuit that solves the differential equation.

$$2x'' + 0.32x' + 1.28x = 0.6$$

*Solution*

*step 1:* $x'' + 0.16x' + 0.64x = 0.3$

*step 2:* $-x'' = 0.16x' + 0.64x - 0.3$

*step 3:*

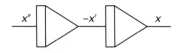

---

[3]Depending on the design of the analog circuit, the sign of the initial conditions may need to be reversed since integrators change the sign of their signals.

*step 4:*

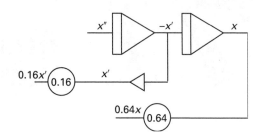

*step 5:*

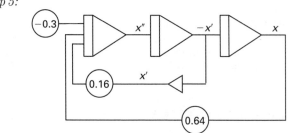

## 4. TIME SCALING

The *real-time* response of some models will be too fast or too slow to be conveniently observed. Therefore, a *time-scaling factor*, $h$, is used so that the *machine time* (i.e., *simulated time*) is $T = ht$. If $h > 1$, the analog model runs slower than real time; if $h < 1$, the model runs faster. *Time scaling* is based on the fact that

$$\frac{df(t)}{dt} = h \frac{df\left(\frac{T}{h}\right)}{dT} \qquad 71.1$$

Substituting $T/h$ for $t$ everywhere (which does not change the equation being solved) is accomplished by reducing the gain of each integrator by $h$ and by changing the time scale in the supplied forcing function from $t$ to $T/h$. (Time scaling does not change magnitude scaling. See Sec. 71.5.)

For each closed loop in the model, the product of all scaling coefficients (known as the *loop gain*) is changed by the $n$th power of the time-scale factor, where $n$ is the number of integrators in the loop.

### Example 71.3

Construct the analog model circuit that simulates the following differential equation three times slower than its real time performance. Neglect initial conditions.

$$-x'' = 0.8x' + 0.3x - \sin(t)$$

*Solution*

The time-scaled differential equation is

$$-(3)^2 x'' = (3)(0.8)x' + 0.3x - \sin\left(\frac{T}{3}\right)$$

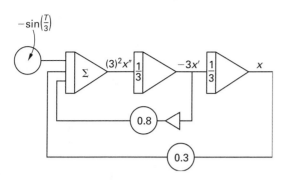

## 5. MAGNITUDE SCALING

The operational amplifiers used in the structural elements are generally limited to a voltage range of approximately $\pm 0.1$ volts to $\pm 10$ volts. Noise is a problem below the lower limit, and amplification becomes distorted and nonlinear above the upper limit. Also, the potentiometers are limited to scalar multiplication between 0 and 1.0. These limitations may require *magnitude scaling* of the differential equation. (Magnitude scaling does not change time scaling.)

In order to properly scale inputs and outputs, the maximum expected value of each variable, derivative, and forcing function must be known. In the case of a variable expected to achieve a value of $x_{max}$ that passes through an amplifier limited to 10 volts, the scaling factor, $k$, is chosen so that $k \leq 10/x_{max}$. (The scaling factors themselves are limited to maximum values determined by the circuitry.) Each structural element used will have its own scaling factor. Magnitude scaling is based on the fact that

$$\int \frac{df(t)}{dt} = \frac{1}{k} \int \frac{dkf(t)}{dt} \qquad 71.2$$

For each closed loop in the model, the *loop gain* must remain the same as in an unscaled implementation.

### Example 71.4

Construct an analog model circuit that simulates the following differential equation. The expected maximum values are: $x'$, 400; $x$, 500; $f(t)$, 1000. The maximum value of any simulation variable or scaling factor is 10. Neglect time scaling.

$$20x' + 300x = 800e^{-0.6t}$$

$$x(0) = 50$$

*Solution*

First, divide through by 20 and isolate the derivative term.

$$-x' = 15x - 40e^{-0.6t}$$

The scale factors are not unique.

$$k_1 = \frac{10}{400} = 0.025$$

$$k_0 = \frac{10}{500} = 0.02$$

$$k_f = \frac{10}{1000} = 0.01$$

The scaled initial condition is

$$k_0 x_0 = (0.02)(50) = 1.0$$

The simulation equation is written by first multiplying all terms by 0.025. Then, each term is multiplied *and* divided by its respective scaling factor.

$$-0.025x' = \frac{(0.025)(15)(0.02x)}{0.02} - \frac{(0.025)(40)(0.01)e^{-0.6t}}{0.01}$$

This is simplified somewhat, leaving the scaling factors and implementing all values greater than ten as products of lesser numbers. (The input function is not part of a loop and can be completely implemented.)

$$-0.025x' = (1.875)(10)(0.02x) - e^{-0.6}$$

The magnitude-scaled analog model circuit includes scalar multiplication *between* stages. This can be accomplished in several ways, but inasmuch as the differentiators are actually variable-gain amplifiers, they are used. The setting corresponds to the ratio of the output to input scale coefficients (in this case, $0.02/0.025 = 0.8$).

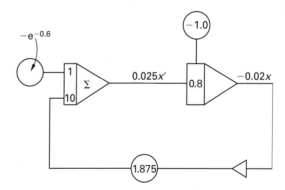

## 6. SIMULTANEOUS DIFFERENTIAL EQUATIONS

Since a differential equation is solved with a loop of operational amplifiers, simultaneous differential equations can be handled by interconnecting loops for each equation. If time and magnitude scaling are used, all loops must be scaled by the same factors.

**Example 71.5**

Construct the analog model circuit that simultaneously solves the two differential equations

$$-x_1' = 0.8x_1 - 0.7x_2 - 0.3$$
$$-x_2' = x_2 - 0.4x_1$$

*Solution*

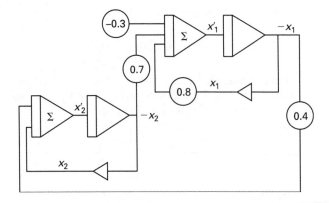

# Topic XV: Computer Programming

Computer
Programming

# 72 Computer Hardware

## 1. EVOLUTION OF COMPUTER HARDWARE

The term *hardware* encompasses the equipment and devices that perform data preparation, input, computation, control, primary and secondary storage, and output functions, but it does not include the programs, routines, and applications (i.e., computer *software*) that control the computer.[1]

*Digital computers* are generally acknowledged to have gone through six major evolutionary stages.[2,3]

- *zeroth generation:* electromechanical calculators

- *first generation:* vacuum tube computers

- *second generation:* transistor computers

- *third generation:* integrated circuit computers

- *fourth generation:* VLSI (very large-scale integration) and ULSI (ultra large-scale integration) computers

The term *fifth generation* refers to the efforts (largely on the part of Japan's Ministry of International Trade and Industry) throughout the 1980s to produce massively parallel computers with artificial intelligence that ran on top of extensive specialized databases and that were easier to program and use. These efforts were commercially unsuccessful.

---

[1]The term "software program" is redundant.

[2]The categorization depends on who is counting and what characteristics are considered evolutionary. Some writers omit electromechanical calculators from the evolution.

[3]Before the days of calculators and computers, some companies were large enough to have employees who did nothing but crank through calculations for engineers and designers. These people were called the company's "computers."

The *sixth generation* of computers draws its utility from generalization, its speed from parallel processing, circuitry minimization, reduced instruction set computers (RISC), its user-friendliness from window-style graphical user interfaces (GUI) and "What You See is What You Get" (WYSIWYG) applications, and its database from network technology (local area network (LAN), wide area network (WAN), and internet), all in stark contrast to fifth generation designs. The application of current technology to social networking devices (e.g., cell phones and online gaming equipment) and applications (e.g., texting, Facebook®, Twitter™) epitomizes the sixth generation.

Gordon Moore, founder of Intel, predicted in 1965 that the number of transistors on a chip would double every 12 months. In 1975, he amended the prediction to 24 months, and the amended statement is known as *Moore's Law*. Although the prediction essentially held true for decades, the doubling rate has slowed due to economic pressures, decline in demand, and technological limitations (e.g., feature size and heat dissipation).[4]

The goal of integrating a natural-language human-machine interface with speech recognition and response, cognition, learning, problem solving abilities, image processing, and visual perception remains elusive, as do locomotion and other robotics features. Future generations of computers will have the benefits of research into biological processing at the molecular level, nanotechnology, optical and laser switching, and energy and particle transitions operating at the atomic and quantum levels.

## 2. COMPUTER CATEGORIZATION

*Supercomputers* are extremely powerful computers, usually constructed with massively parallel architecture for specific functions (e.g., engineering design, number crunching, and analysis of strategies).

*Mainframe computers* are general-purpose computers used in large, centralized data processing complexes, supporting many programs running simultaneously, with access from remote "dumb" terminals and personal computers running terminal applications.

---

[4]Feature size appears to be limited by lithography methods, heat dissipation, and quantum tunneling (leakage) of up to 10-20 nm.

*Minicomputers* are smaller computers, usually dedicated to business at a single site, though otherwise similar to mainframes.

*Microcomputers*, commonly referred to as *personal computers* (PCs), are single-user computers.

Due to increasing microprocessor capabilities (e.g., *dual-core* and *quad-core* central processing units (CPU's) that combine two and four, respectively, or more CPU's together onto a single chip), and high-throughput LAN and WAN connections to *servers* and the internet, the distinction between these categories based on computing power is indistinct and unimportant.

## 3. COMPUTER ARCHITECTURE

All digital computers, from giant supercomputers to the smallest microcomputers, contain three main components—a central processing unit (CPU), main memory, and external (peripheral) devices. Figure 72.1 illustrates a typical integration of these functions.

**Figure 72.1** *Simplified Computer Architecture*

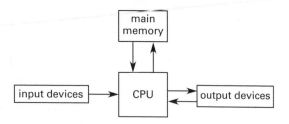

## 4. MICROPROCESSORS

A *microprocessor* is a *central processing unit* (CPU) on a single chip. (See Fig. 72.2.) With *large-scale integration* (LSI), most microprocessors are contained on one chip, although other chips in the set can be used for memory and input/output control. The most prevalent microprocessors are produced by Intel (Pentium family) and Motorola (680X0 family), although comparable and work-alike microprocessors are produced by others (e.g., Advanced Micro Devices (AMD), Cyrix, and Sun).

The basic microprocessor CPU consists of an arithmetic and logic unit, several accumulators, one or more registers, stacks, and a control unit. The *control unit* fetches and decodes the incoming instructions and generates the signals necessary for the arithmetic and logic unit to perform the intended function. The *arithmetic and logic unit* (ALU) executes commands and manipulates data.

*Accumulators* hold data and instructions for further manipulation in the ALU. Registers are used for temporary storage of instructions or data. The *program counter* (PC) is a special register that always points to (contains) the address of the next instruction to be executed. Another special register is the *instruction register* (IR), which holds the current instruction during its execution. *Stacks* provide temporary data storage in sequential order—usually on a *last-in, first-out* (LIFO)

**Figure 72.2** *Microprocessor Architecture*

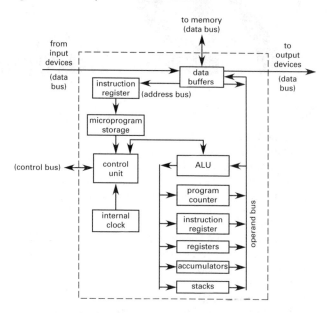

basis. Because their operation is analogous to spring-loaded tray holders in cafeterias, the name *pushdown stack* is also used.

Microprocessors communicate with support chips and peripherals through connections in a *bus* or *channel*, which is logically subdivided into three different functions.[5] The *address bus* directs memory and input/output device transfers. The *data bus* carries the actual data and is the busiest bus. The *control bus* communicates control and status information. The number of lines in the address bus determines the amount of random access memory that can be directly addressed. When there are $n$ address lines in the bus, $2^n$ words of memory can be addressed.

Since the speed at which data can be transferred within the microprocessor is an order of magnitude greater than the speed of external memory, a hierarchy of *cache memories* within the chip is used to accumulate and hold data temporarily. A cache is a relatively small memory stack, typically 32 bytes, situated electronically between the processing core and the main memory. Cache memory gains its advantage from *spatial locality* (i.e., speed due to the nearness of bytes) and *temporal locality* (i.e., speed due to recentness of access).

All microprocessors use a *clock* to control instruction and data movements. The core *clock rate* is specified in microprocessor cycles per second (e.g., 3 GHz). Ideally, the clock rate is the number of instructions the microprocessor can execute per second. However, one or more cycles may be required for complex instructions (*macrocommands*). For example, executing a complex instruction may require one or more cycles each to fetch the instruction from memory, decode the instruction to

---

[5]The term *bus* refers to the physical path (i.e., wires or circuit board traces) along which the signal travels. The term *channel* refers to the logical path.

see what to do, execute the instruction, and store (write) the result.[6] Since operations on floating point numbers are macrocommands, the speed of a microprocessor can also be specified in *flops*, the number of floating point arithmetic operations it can perform per second. Similar units of processing speed are *mips* (millions of instructions per second).

*Pipelining (pipeline architecture)* is the term that describes sending multiple instructions through a microprocessor in parallel and having the processor execute overlapping instructions simultaneously. During execution, instructions pass through sequential stages that process parts of an instruction in succession while other stages are processing parts of other instructions. Intermediate results are stored in multiple registers. All processing stages are active simultaneously, though they are not working on the same instruction at the same time. The result is that completed instructions emerge from the processing pipeline in quick succession.

*Superscaled architecture* describes microprocessors that have multiple pipelines. Superscaled computers can execute more than one instruction per clock cycle. Instructions are assigned to executive units (the pipelines) according to instruction type or other rationale.

Most microprocessors are rich in complex executable commands (i.e., the *command set*) and are known as *complex instruction-set computing* (CISC) microprocessors. In order to increase the operating speed, however, *reduced instruction-set computing* (RISC) microprocessors are limited to performing simple, standardized format instructions but are otherwise fully featured. Unlike CISC units, RISC microprocessors require four separate instructions for the common fetch-decode-execute-write sequence.

Microprocessors can be designed to operate on 4-bit, 8-bit, 16-bit, 32-bit, and 64-bit words, although microprocessors with smaller than 16-bit words are now used primarily only in process control applications.

Some microprocessors can emulate the operation of other microprocessors. *Emulation mode* is also referred to as *virtual mode*.

Power consumption and heat dissipation currently limit microprocessor speeds and component densities. Heating effects are particularly severe in chips with multiple cores and in overclocked chips (i.e., chips that are made to run at speeds greater than those intended by the manufacturer or integrator). Microprocessors must be cooled in order to keep them from burning themselves out. Heat effects are managed by judicious power regulation within the computer, capacitive electrical storage, limiting clock speeds, use of heat sinks, cooling fans, and liquid cooling.

## 5. CONTROL OF COMPUTER OPERATION

The user interface and basic operation of a computer are controlled by the *operating system* (OS), also known as the *monitor program*. The operating system is a program that controls the computer at its most basic level and provides the environment for application programs. The operating system manages the memory, schedules processing operations, accesses peripheral devices, communicates with the user/operator, and (in multi-tasking environments) resolves conflicting requirements for resources.

Parts of the operating system are obtained from read-only memory, and parts are subsequently loaded from peripherals. The most basic communications between the microprocessor are controlled by the *basic input/output system* (BIOS), usually stored in battery-maintained, programmable read-only memory. The BIOS provides just enough utility for the computer to begin operating. Under control of the BIOS, more detailed operating system instructions and communications protocols are loaded from the external peripherals.

In early computers, a small part of executable code that was used to initiate data transfers and logical operations when the computer was first started was known as a *bootstrap loader*. Although modern start-up operations are more sophisticated, the phrase "booting the computer" is still used today.

During program operation, peripherals and other parts of the computer signal the operating system through interrupts. An *interrupt (interrupt request,* IRQ) is an asynchronous hardware or software signal that stops the execution of the current instruction (or, in some cases, the current program) and transfers control to another memory location, subroutine, or program. Software interrupts signal error conditions such as division by zero, overflow and underflow, and syntax errors. Hardware interrupts signal accessibility status and data transfer from/to peripherals and components. The operating system intercepts, decodes, and acts on these interrupts.

## 6. COMPUTER MEMORY

Computer memory consists of many equally sized storage locations, each of which has an associated address. The contents of a storage location may change, but the address does not.

The total number of storage locations in a computer can be measured in various ways. A *bit* (binary digit) is the smallest changeable data unit. Bits can only have values of 1 or 0. Bits are combined into *nibbles* (4 bits), *bytes* (8 bits, the smallest number of bits that can represent

---

[6]In some cases, other chips (e.g., *memory management units*) can perform some of these tasks.

one alphanumeric character), *half-words* (8 and 16 bits), *words* (8, 16, 32, and 64 bits) and *doublewords* (16, 32, 64, and 128 bits).[7]

The number of memory storage locations is always a multiple of 2. The abbreviations K, M, and G are used to designate the quantities $2^{10}$ (1024), $2^{20}$ (1,048,576), and $2^{30}$ (1,073,741,824), respectively. For example, a 6G memory would contain $6 \times 2^{30}$ bytes. (K and M do not mean one thousand and one million exactly.)

Most of the memory locations are used for applications and data. However, portions of the memory may be used for video memory, I/O cache memory, the BIOS, and other purposes. *Video memory* (known as VRAM) contains the contents displayed on the screen. Since the screen is refreshed many times per second, the screen information must be repeatedly read from video memory. *Cache memory* holds the most recently read and frequently read data in memory, making subsequent retrieval of that data much faster than reading from a disk drive, or even from main memory.[8] *OS memory* contains the BIOS that is read in when the computer is first started. *Scratchpad memory* is high-speed cache memory used to store a small amount of data temporarily so that the data can be retrieved quickly.

Modern memory hardware is semiconductor based.[9] Memory is designated as RAM (random access memory), ROM (read-only memory), PROM (programmable read-only memory), and EPROM (erasable programmable read-only memory). While data in RAM is easily changed, data in ROM cannot be altered. PROMs are initially blank but once filled, they cannot be changed. EPROMs are initially blank but can be filled, erased, and refilled repeatedly.[10] The term *firmware* is used to describe programs stored in ROMs and EPROMs.

The contents of a *volatile memory* are lost when the power is turned off. RAM is usually volatile, while ROM, PROM, and EPROM are *non-volatile*. With *static memory*, data does not need to be refreshed and remains as long as the power stays on. With *dynamic memory*, data must be continually refreshed. Static and dynamic RAM (i.e., DRAM) are both volatile.

*Virtual memory (storage)* (VS) is a technique by which programs and data larger than main memory can be accessed by the computer. (Virtual memory is not synonymous with *virtual machine*.) In virtual memory systems, some of the disk space is used as an extension of the semiconductor memory. A large application or program is divided into modules of equal size called *pages*. Each page is switched into (and out of) RAM from (and back to) disk storage as needed, a process known as *paging*. This interchange is largely transparent to the user. Of course, access to data stored on a disk drive is much slower than semiconductor memory access. *Thrashing* is a deadlock situation that occurs when a program references a different page for almost every instruction, and there is not even enough real memory to hold most of the virtual memory.

Most memory locations are filled and managed by the CPU. However, *direct memory access* (DMA) is a powerful I/O technique that allows peripherals (e.g., graphics, network, sound cards, and disk drives) to transfer data directly into and out of memory without affecting the CPU. Although special DMA hardware is required, DMA does not require explicit program instructions, making data transfer faster.

## 7. PARITY AND ECC MEMORY

*Parity* is a technique used to ensure that the bits within a memory byte are correct. For every eight data bits, there is a ninth bit—the parity bit—that serves as a *check bit*. The nine bits together constitute a *frame*. In *odd-parity recording*, the parity bit will be set so there is an odd number of one-bits. In *even-parity recording*, the parity bit will be set so that there is an even number of one-bits. When the data are read, the nine bits are checked to ensure valid data.[11]

Parity random-access memory (RAM) has been all but superseded by memory chips with error-correcting code (ECC). ECC RAM chips use additional circuitry, not parity bits, to detect and correct memory errors. This feature makes ECC memory more expensive and slower than both parity and nonparity memory, and accordingly, ECC memory is only used on uptime-sensitive applications.

Because modern memory is extremely reliable and less costly than ECC memory, personal computers commonly use neither parity nor ECC memory.

## 8. INPUT/OUTPUT DEVICES

Devices that feed data to, or receive data from, the computer are known as *input/output* (I/O) *devices* or *peripherals*. Monitors, keyboards, game controllers, light pens, digitizers, printers and plotters, scanners,

**Computer Programming**

---

[7]The distinction between doublewords, words, and half-words depends on the computer. Sixteen bits would be a word in a 16-bit computer but would be a half-word in a 32-bit computer. Furthermore, *double-precision* (*double-length*) *words* double the number of bytes normally used. The abbreviations KB (*kilobytes*) and KW (*kilowords*) used by some manufacturers do not help much to clarify the ambiguity.

[8]A high-speed mainframe computer may require 200–500 nanoseconds to access main memory but only 20–50 nanoseconds to access cache memory.

[9]The term "core," derived from the ferrite cores used in early computers, is seldom used today.

[10]Most EPROMs can be erased by exposing them to ultraviolet light with a wavelength of 253.7 nm.

---

[11]This does not detect two of the bits in the frame being incorrect, however.

and tape and disk drives are common peripherals.[12] Point-of-sale (POS) devices, bar code readers, and magnetic ink character recognition (MICR) and optical character recognition (OCR) readers are less common devices.

Peripherals are connected to their computer through multi-line cables. With a *parallel interface* (used in a *parallel device*), there are as many separate lines in the cable as there are bits (typically seven, eight, or nine) in the code representing a character. An additional line is used as the *strobe signal* to carry a timing signal. With a *serial interface* (used in a *serial device*), all bits pass one at a time along a single line in the cable. The *transmission speed (baud rate)* in bps is the number of bits that pass through the data line per second.[13]

Peripherals such as scanners and printers typically do not have large memories. They only need memories large enough to store the information before the data are displayed or printed. The small memories are known as *buffers*. The peripheral can send the status (i.e., full, empty, off-line, etc.) of its buffer to the computer in several different ways. This is known as *flow control* or *handshaking*.

If the computer and peripheral are configured so that each can send and receive data, the peripheral can send a single character (e.g., the XOFF *character* for transmission off) to the computer when its buffer is full. Similarly, a different character (e.g., the XON *character* for transmission on) can be sent when the peripheral is ready for more data. The computer must monitor the incoming data line for these characters. This is known as *software flow control* or *software handshaking*.

If there are enough separate lines between the computer and the peripheral, one or more of them can be used for *hardware handshaking*.[14] In this method of flow control, the peripheral keeps the voltage on one of the lines high (or low) when it is able to accept more data. The computer monitors the voltage on this line.

In order for computers and peripherals to communicate, they must use common hardware connectors, pin-outs, and communication protocols. For example, early parallel ports and devices, primarily printers, used the *Centronics standard*. Faster parallel devices communicated using the *small computer systems interface* (SCSI

interface). Serial ports and devices that communicate serially use the Electronics Industry Association Recommended Standard 232 (RS-232). Modern personal computers use *universal serial bus* (USB) connections for virtually all external connections (e.g., keyboards, monitors, mice, and drives). The Institute of Electrical and Electronics Engineers (IEEE) 1394 serial communications standard (also known as Apple's FireWire, Sony's i.Link, and Texas Instrument's Lynx) is used to connect external devices that require the highest data transfer rates. Transmission/transfer speeds are given in bits per second, abbreviated bps and bits/s. Maximum transfer rates are 12 Mbps for USB 1.0, 480 Mbps for USB 2.0, 5 Gbps for USB 3.0 ("SuperSpeed USB"), and 400 Mbps for the original IEEE 1394 S400 standard, now increased to 3.2 Gbps for S3200. Speeds for Ethernet connected and optically connected devices can be much higher, up to and beyond 100 Gbps.

Most peripheral devices are connected to the computer by a dedicated channel (cable). However, a pair of *multiplexers* (*statistical multiplexers* or *concentrators*) can be used to carry data for several peripherals along a single cable which is known as the *composite link*. (See Fig. 72.3.) There are two methods of achieving multiplexed transmission: *frequency division multiplexing* (FDM) and *time division multiplexing* (TDM). With FDM, the available transmission band is divided into narrower bands, each used for a separate channel. In TDM, the connecting channel is operated at a much higher clock rate (proportional to the capacity of the multiplexer), and each peripheral shares equally in the available cycles.

***Figure 72.3*** *Multiplexed Peripherals*

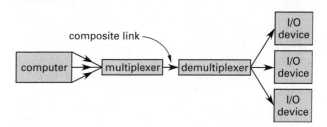

## 9. RANDOM STORAGE DEVICES

*Random access (direct access) storage devices*, also known as *mass storage devices*, include magnetic and optical disk drive units. They are random access because individual records can be accessed without having to read through the entire file.

Magnetic disk drives (*hard drives*) are composed of several *platters*, each with one or more read/write heads. The platters typically turn at 4500–7200 rpm, although 10,000 and 15,000 rpm drives are used in server ("enterprise") installations. Data on a surface are organized into tracks, sectors, and cylinders. (See Fig. 72.4.) *Tracks* are the concentric storage areas. *Sectors* are pie-shaped subdivisions of each track. A *cylinder*

---

[12]A *monitor* consists only of a viewing screen. Modern monitors are of the liquid crystal display (LCD) and plasma flat-screen varieties. Monitors generally do not include keyboards. Early monitors were of *cathode ray tube* (CRT) construction, and "CRT" is still a common synonym for "monitor." Early remote access work stations combined CRTs and keyboards into a single unit known as a *terminal*. Today, the name "digital video terminal" (DVT) generally encompasses all of these and is taken to mean a viewing screen and keyboard together.

[13]The name *baud rate* is derived from the use of the *Baudot code*. One baud is one modulation per second. If there is a one-to-one correspondence between modulations and bits, one baud unit is the same as one bit per second (bps). In general, the unit bps should always be used.

[14]RS-232 devices usually require only two or three lines—data in, data out, and ground. Most serial computer cables contain more lines than this, and one of the extra lines can be used for DTR (*data terminal ready*) or CTS (*clear to send*) handshaking.

**Figure 72.4** *Tracks, Sectors, and Cylinders*

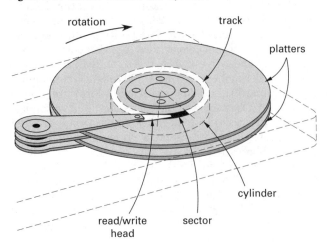

consists of the same numbered track on all drive platters. Some drives and *disk packs* are removable, but most hard drives are fixed (i.e., non-removable).

Depending on the media, *optical disk drives* (e.g. CD-ROM and DVD) can be *read only* (R/O) or *read/write* (R/W) in nature. WORM drives (write once, read many) can be written by the user, while others can only be read.[15] Some media are writable only once; other media are rewritable.

In addition to *storage capacity* (usually specified in gigabytes—GB) there are several parameters that describe the performance of a disk drive. *Areal density* is a measure of the number of data bits stored per square inch of disk surface. It is calculated by multiplying the number of bits per track by the number of tracks per (radial) inch. The *average seek time* is the average time it takes to move a head from one location to a new location. The *track-to-track seek time* is the time required to move a head from one track to an adjacent track. *Latency* or *rotational delay* is the time it takes for a disk to spin a particular sector under the head for reading. On the average, latency is one-half of the time to spin a full revolution. The *average access time* is the time to move to a new sector and read the data. Access time is the sum of average latency and average seek time.

Early drives used Seagate Technology-506 (ST-506) and enhanced small disk interface (ESDI) interfaces. For many years thereafter, small computer systems interface (SCSI), and integrated (or intelligent) drive electronics/advanced technology attachment (IDE/ATA) interfaces were used for connections and communications between drives and microprocessors. IDE hardware controllers are limited to two channels, each controlling two devices. Each channel is limited in speed to the slowest device on that channel. IDE speeds are designated in Mbps (e.g., ATA/33 transfers data at 33 Mbps). SCSI speeds are considerably faster, so SCSI is used for most server and redundant array of independent disks

---

[15]The WORM acronym is also interpreted as write once-read mostly.

(RAID) installations. However, serial ATA (SATA) drives have replaced IDE/ATA drives in most modern personal computers, primarily for reasons of performance. External drives using USB and IEEE 1394 connections are common, as are USB *flash drives* (e.g., *thumb drives, portable drives, stick drives*).

RAID technology combines two or three drives into a single unit to achieve greater reliability. The computer sees a RAID drive as a single disk, even though the data are being separated and/or duplicated on different drives. *Striping* refers to splitting data bits among multiple drives. *Mirroring* refers to duplicating data on all drives so that any single redundant drive can fail without data loss. RAID 0 involves striping without parity memory, resulting in faster speed but risk of loss of all data upon the failure of any one drive. RAID 1 involves mirroring. RAID 2 uses striping and *Hamming error correcting code*. RAID 3, 4, 5, and 6 involve various combinations of striping with single parity and dual parity memory. RAID 10 (RAID 1+0) involves striping with mirroring.

Low capacity *floppy disks* (*diskettes*) are random access storage devices. Magnetic diskettes typically store less than one million characters, although higher capacity removable diskettes have been produced. CD-ROM, DVD, and USB flash drives have made floppy disk drives and diskettes obsolete.

## 10. SEQUENTIAL STORAGE DEVICES

Tape units are *sequential access devices* because a computer cannot access information stored at the end of a tape without first reading or passing by the information stored at the front of the tape. Some tapes use *indexed sequential formats* in which a directory of files on the tape is placed at the start of the tape. The tape can be rapidly wound to (near) the start of the target file without having to read everything in between.

The obsolete nine-track tape drives that were once common in "data centers" have been replaced by drives based on the quarter-inch cartridge (QIC) and digital audio tape (DAT) format. These drives are used for backing up hard disks while operating in *streaming mode*. A streaming tape reads and writes while running continuously.

## 11. PROCESSING TERMINOLOGY

*In real time (interactive) mode processing*, jobs (also referred to as processes, programs, and operations) are received by a computer sequentially as they are submitted. In *batch mode processing*, jobs are grouped into efficient batches and held until a convenient or standard time.

*Multiprocessing (multitasking, multiprogramming)* occurs when a computer runs multiple applications simultaneously. The computer user's "current" application runs in the foreground window while other programs run in background windows, hence the terms *foreground processing* and *background processing.* (Background applications are often not visible at all.) When different users are sharing time (*time sharing*) on a single computer, each enjoys a *virtual machine* and is unaware of the other users. Multiprocessing is usually accomplished by segmenting the main computer memory and/or dividing the microprocessor cycles among the applications.

## 12. TELECOMMUNICATION

Access to and between remote computers can occur using wired or wireless equipment. Wired connections use phone lines (*plain old phone*, POP), high speed digital subscriber lines (DSL), Trunk-carrier 1 and Trunk-carrier 3 (T1 and T3) phone, coaxial (coax) video cable, and various forms of Ethernet LAN/WAN wiring technology. Wireless communications can use infrared, radio frequency, microwave, and other optical signals. In order to use phone and video coax wiring designed to transmit analog signals, the computer's digital signals must be converted to analog signals using a *modem* (*modulator/demodulator*).

With *simplex communication*, transmission is only in one direction. With *half-duplex communication*, data can be transmitted in both directions, but only in one direction at a time. With *duplex* or *full duplex communication*, data can be transmitted in both directions simultaneously. 33,600 bps is the maximum practical transmission speed over voice-grade phone lines without data compression. Much higher rates, however, are possible over dedicated data lines and wide-band lines.

In *asynchronous* or *start-stop transmission*, each character is preceded and followed by special signals (i.e., *start* and *stop bits*). Thus, every 8-bit character is actually transmitted as 10 bits, and the character transmission rate is one-tenth of the transmission speed in bits per second.[16] With asynchronous transmission, it is possible to distinguish the beginning and end of each character from the bit stream itself.

*Synchronous equipment* transmits a block of data continuously without pause and requires a built-in clock to maintain synchronization. Synchronous transmission is preceded and interwoven with special clock-synchronizing characters, and the separation of a bit stream into individual characters is done by counting bits from the start of the previous character. Since start and stop bits are not used, synchronous communication is approximately 20% faster than asynchronous communication.

There are three classes of communication lines—narrowband, voice-grade, and wide-band—depending on the bandwidth (i.e., range of frequencies) available for signaling.

*Narrow-band* may only support a single channel of communication, as the bandwidth is too narrow for modulation. *Wide-band channels* support the highest transfer rates, since the bandwidth can be divided into individual channels. *Voice-grade lines*, supporting frequencies between 300 Hz and 3000–3300 Hz, are midrange in bandwidth.

Errors in transmission can easily occur over voice grade lines at the rate of 1 in 10,000. In general, methods of ensuring the accuracy of transmitted and received data are known as *communications protocols* and *transmission standards*. A simple way of checking the transmission is to have the receiver send each block of data back to the sender. This process is known as *loop checking* or *echo checking*. If the characters in a block do not match, they are re-sent. While accurate, this method requires sending each block of data twice.

Another method of checking the accuracy of transmitted data is for both the receiver and sender to calculate a *check digit*, *checksum*, or *block check character* derived from each block of *characters* sent. (A common transmission block size is 128 characters.) With CRC (*cycling redundancy checking*), the block check character is the remainder after dividing all the serialized bits in a transmission block by a predetermined binary number, a process known as *hashing*. Then, the block check character is sent and compared after each block of data.

## 13. LOCAL AND WIDE AREA NETWORKS

A local group of independent computers sharing peripherals (e.g., an internet connection) is known as a *local area network*. LANs connected by registered jack-45 (RJ45) Cat 5 ethernet cabling typically communicate at speeds between 200 kbps and 100 Mbps, with 10 Mbps a typical speed for legacy equipment. *Wireless connections* based on 802.11g and 802.11n standards communicate at up to 54 Mbps and 300 Mbps (net rates), respectively, although local transmission path blockages and electromagnetic interference can degrade these speeds.

Connections between LANs, distant computers, and the internet are made through a *wide area network*. While LAN equipment generally belongs to the user, WANs are usually owned by *internet service providers* (ISPs). A LAN connects through a *router* (part of the LAN) to a *hub* (part of the WAN). Communication speeds vary depending on the modem, router, and hub. Most residential users connecting through routers to DSL and cable modems are limited to a range of 1–10 Mbps, while voice grade modems with compression are limited to 56 kbps or slower.

Various communication protocols can be used in LANs and WANs, although the *transmission control protocol/ internet protocol* (TCP/IP) is dominant.

Computer
Programming

---

[16]For historical reasons, a second stop bit is used when data are sent at 10 characters per second. This is referred to as 110 bps.

# 73 Data Structures and Program Design

## 1. CHARACTER CODING

*Alphanumeric data* refers to characters that can be displayed or printed, including numerals and symbols (\$, %, &, etc.) but excluding *control characters* (tab, carriage return, form feed, etc.). Since computers can handle binary numbers only, all symbolic data must be represented by binary codes. *Coding* refers to the manner in which alphanumeric data and control characters are represented by sequences of bits.

In the past, the standard method for coding data on 80-column, 12-row punch cards was the *Hollerith code*. The *Baudot code* is a five-bit code that was used in Telex and teletypewriter (TWX and TTY) communications. By shifting to an alternate character set (numerals versus letters), it has a maximum of 64 ($2 \times 2^5$) characters. In some early computers, characters were represented by six-bit combinations known as *binary coded decimal* (BCD). The 64 ($2^6$) different combinations, however, proved insufficient to represent all necessary characters.

The *Extended Binary Coded Decimal Interchange Code* (EBCDIC) was used in early IBM mainframe computers.[1] It uses eight bits (a byte) for each character, allowing a maximum of 256 ($2^8$) different characters.

---

[1]EBCDIC is pronounced eb'-sih-dik.

Since strings of bits are difficult to read, the *packed decimal* format is used to simplify working with EBCDIC data. Each byte is converted into two strings of four bits each. The two strings are then converted to hexadecimal format. Since $(1111)_2 = (15)_{10} = (F)_{16}$, the largest possible EBCDIC character is coded FF in packed decimal.

Except when EBCDIC is needed for backward compatibility, almost all modern computing and teleprocessing equipment and many internet applications use the *American Standard Code for Information Interchange*, (ASCII), a seven-bit code permitting 128 ($2^7$) different combinations. Of these combinations, only 94 ASCII character types are printable (i.e., upper and lower case letters, numerals, punctuation marks, typographical and mathematical symbols, and the space). The remaining combinations include many obsolete *control characters*, such as decimal code 7 (BEL) that rings the bell on a teletype. The high-order (8th) bit in an 8-bit byte (octet) is not standardized. In early applications, it was used for parity checking. Others toggled it to code characters for non-English languages or to create proprietary sets of graphical characters. *8-bit clean* applications, such as *Simple Mail Transfer Protocol* (SMTP), set the 8th bit to zero.

English language internet text can be coded in ASCII, but *Unicode* (*Universal Character Set*, UCS) provides more flexibility and is dominant. There are more than 100,000 Unicode characters, representing all of the world's languages, defined in the standard ISO 8859. The symbols are divided into 17 sets known as *planes*, each with 65,536 ($2^{16}$) characters. The first plane, Plane 0, containing characters for most languages, is known as the *Basic Multilingual Plane* (BMP). A character is defined by its code (known as a *code point*), but the character's representation is a function of the user's application (e.g., web browser and the user's choice of fonts).

Unicode is implemented by encoding characters in multiples of 8-bit bytes. The Unicode Transformation Format-8 (UTF-8) is a variable-width encoding format that can be implemented with one to four bytes. If UTF-8 one-byte encoding is used, the result is essentially the English language ASCII character set. Since UTF-8 is backward compatible with ASCII, it is often implemented in web and email applications. A code point in the BMP is referenced by the prefix "U+" followed by a four-character hexadecimal number. For example,

Computer Programming

"U+263A" represents a smiley face on a white background ("Have a nice day!").

To encode foreign languages in UTF-8, additional bytes are required. UTF-16 is used to encode the entire ISO 8859 character set. In UTF-16, bytes are combined into 16-bit *surrogate pairs* (*code units*). UTF-16 is widely implemented in modern applications and operating systems.

### Example 73.1

The number $(7)_{10}$ is represented as 11110111 in EBCDIC. What is the packed decimal representation?

*Solution*

The first four bits are 1111, which is $(15)_{10}$ or $(F)_{16}$. The last four bits are 0111, which is $(7)_{10}$ or $(7)_{16}$. The packed decimal representation is F7.

## 2. PROGRAM DESIGN

A *program* is a sequence of computer instructions that performs some function. The program is designed to implement an *algorithm*, which is a procedure consisting of a finite set of well-defined steps. Each step in the algorithm usually is implemented by one or more instructions (e.g., READ, GOTO, OPEN, etc.) entered by the programmer. These original "human-readable" instructions are known as *source code statements*.

Except in rare cases, a computer will not understand source code statements. Therefore, the source code is translated into machine-readable *object code* and absolute memory locations by an assembler or computer program in order to produce an executable program.

If the executable program is loaded from CD-ROM, disk, or the web, it is referred to as *software*. If the program is loaded from ROM or EPROM, it is referred to as *firmware*. The computer mechanism itself is known as the *hardware*.

## 3. FLOWCHARTING SYMBOLS

A *flowchart* is a step-by-step drawing representing a specific procedure or algorithm. Figure 73.1 illustrates the most common flowcharting symbols. The terminal symbol begins and ends a flowchart. The input/output symbol defines an I/O operation, including those to and from the keyboard, printer, memory, and disk. The processing symbol and predefined process symbol refer to calculations and data manipulation. The decision symbol indicates a point where a decision must be made or two items are compared. The connector symbol indicates that the flowchart continues elsewhere. The off-page symbol indicates that the flowchart continues on the following page. Comments can be added in an annotation symbol.

**Figure 73.1** *Flowcharting Symbols*

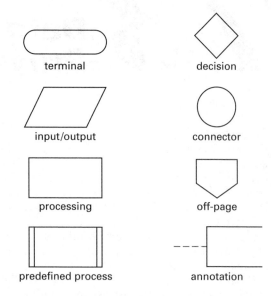

## 4. LOW-LEVEL LANGUAGES

Programs are written in specific languages, of which there are two general types: low-level and high-level. *Low-level languages* include machine language and assembly language.

*Machine language (machine code) instructions* are intrinsically compatible with and understood by the computer's CPU. They are the CPU's native language. An instruction normally consists of two parts: the operation to be performed (*op-code*) and the operand expressed as a storage location. Each instruction must ultimately be expressed as a series of bits, a form known as *intrinsic machine code*. However, octal and hexadecimal coding and representation are more convenient. In either case, coding a machine language program is tedious and seldom done by hand. (See Table 73.1.)

**Table 73.1** *Comparison of Typical ADD Commands*

| language | instruction |
|---|---|
| intrinsic machine code | 1111 0001 |
| machine language | 1A |
| assembly language | AR |
| high-level language | + |

*Assembly language* is more sophisticated (i.e., is more symbolic) than machine language. Mnemonic codes are used to specify the operations. The operands are referred to by variable names rather than the addresses. Blocks of code that are to be repeated verbatim at multiple locations in the program are known as *macros (macro instructions)*. Macros are written only once and are referred to by a symbolic name in the source code.

Assembly language code is translated into machine language by an *assembler* (*macro-assembler* if macros are supported). After assembly, portions of other programs or function libraries may be combined by a *linker*. In order to run, the program must be placed in the computer's memory by a *loader*. Assembly language programs are preferred for highly efficient, fast, and small programs. However, the coding inconvenience outweighs this advantage for most applications, particularly when *optimizing compilers* are used with high-level languages.

## 5. HIGH-LEVEL LANGUAGES

*High-level languages* are easier to use than low-level languages because the instructions resemble English. High-level statements are translated into machine language by either an interpreter or a compiler. A *compiler* performs the checking and conversion functions on all instructions only when the compiler is invoked. A true stand-alone executable program is created. An *interpreter*, however, checks the instructions and converts them line by line into machine code during execution but produces no stand-alone program capable of being used without the interpreter.[2,3]

Many high-level languages have been developed, though not all have been commercial successes. The internet has necessitated the development of many new languages that facilitate the control of appearance. Some languages are general, and others are tailored to specific purposes. Some languages are *domain specific*, such as *Mathematica* and those used with computer operating systems and proprietary databases, and some are specific to hardware. Some languages, such as Tcl and Ajax, are essentially *toolkits*, collections of subroutines, and are therefore higher level than others. Some dinosaurs, by virtue of their installed base in government and industry, are "here forever," despite the advantages of newer languages. Many obsolete languages have fervent supporters. It is impossible to list them all, but Table 73.2 lists some of the most interesting high-level languages.

## 6. HORIZONTAL AND VERTICAL MARKETS

Some programs (e.g., word processing, spreadsheets, drawing, and presentation programs) are written for *horizontal markets* (i.e., general consumers). Contact management software (CMS), customer relationship management (CRM), and human resources information systems (HRIS) are examples of horizontal business markets. Other programs are written for specific types of business, known as *vertical markets*. For example, because most automobile parts distributors operate similarly, a vertical market program can be designed to help distributors manage their businesses. And, while there are innumerable business categories, some of these fit into general vertical market categories: artificial

---

[2]Some interpreters check syntax as each statement is entered by the programmer.
[3]Some languages and implementations of other languages blur the distinction between interpreters and compilers. Terms such as *pseudo-compiler* and *incremental compiler* are used in these cases.

*Table 73.2* Important High-Level Languages

| | |
|---|---|
| business languages | Ada, Algol (ALGOL)*, COBOL, RPG |
| general purpose | BASIC and Visual BASIC, C, C++, C#, Eiffel, Haskell, Jade, Java, Modula*, Modula 2*, Pascal*, Turbo Pascal*, Perl, PL/I* (PL/1)*, Ruby, Tcl |
| scientific | APL*, Forth (FORTH)*, Fortran (FORTRAN) |
| functional | ML, Scala |
| special purpose | Awk*, GPSS, LISP, Common LISP, LOGO, Prolog*, SIMSCRIPT, Simula*, Smalltalk, SNOBOL* |
| web scripting | Ajax, ASP*, CGI, Java script, PHP, Python |
| web representation | HTML, XHTML, MathML, XML |
| database processing | Access, Delphi, FoxPro, Oracle, SQL, MYSQL |
| operating system | JCL*, CPM*, Mac OS, Solaris, Unix, Linux, Windows |

*essentially obsolete

intelligence (AI), computer-aided design (CAD), computer-aided engineering (CAE), computer-aided manufacturing (CAM), computer-aided instruction (CAI), database and database management system (DB and DBMS), executive information system (EIS), electronic medical records (EMR), healthcare information system (HIS), management information system (MIS), and point of sale (POS).

## 7. RELATIVE COMPUTATIONAL SPEED

Certain languages are more efficient (i.e., execute faster) than others. While it is impossible to be specific and exceptions abound, assembly language programs are fastest, followed in order of decreasing speed by compiled, pseudo-compiled, and interpreted programs.

Similarly, certain program structures are more efficient than others. For example, when performing a repetitive operation, the most efficient structure will be a single equation, followed in order of decreasing speed by a stand-alone loop and a loop within a subroutine. Incrementing the loop variables and managing the exit and entry points are known as *overhead*. Overhead takes time during execution.

Programs written in high-level languages can enjoy the efficiencies associated with low-level languages when they are processed by an *optimizing compiler*. The portion of a program that is optimized is known as the program *scope*. The scope can vary from one line to an entire program. Optimizing is generally limited to one dependent variable such as execution speed or memory usage.

**Computer Programming**

Program optimization can be performed automatically or by clever coding by experienced programmers. Some of the methods used to improve program efficiency include programming for the most common cases, storing results for reuse, eliminating intermediate results (combining instructions), reducing branching, and using adjacent memory and stack locations. Many methods are used to optimize loops, memory usage, and recursive operations. *Strength reduction* is a term used to describe replacing a complex set of operations with a simpler set. Examples of strength reduction include left-shifting bits instead of multiplying by two, and replacing division with multiplication by a reciprocal.

## 8. STRUCTURE, DATA TYPING, AND PORTABILITY

A language is said to be *structured* if subroutines and other procedures each have one specific entry point and one specific return point.[4] A language has *strong data types* if integer and real numbers cannot be combined in arithmetic statements.

A *static programming language* is one in which constraints on variables and expressions are checked at the time of compiling. C++, Java, Fortran, and Pascal use *static typing*. A *dynamic language* is one in which the variables are checked at execution. Javascript, Perl, Python, and Ruby use *dynamic typing*.

Most computer languages, including Fortran and BASIC, are well suited for *imperative programming* (also called *declarative programming*). Imperative programming is concerned with changes of state. Programs are constructed from routines that return different values depending on the calling parameters. The programming is "imperative" because the routines always execute the same arithmetic, looping, and branching instructions in the same sequence, regardless of the calling parameters. The same routine can have different outcomes depending on the calling parameters. Imperative programs are said to have *side effects*, which include changing the values of variables, displaying a character, writing to a file, and calling other functions.

In contrast to imperative languages, *(multi) functional languages* such as Standard ML, Scheme, Scala, LISP, and APL, database query languages such as SQL, and domain-specific languages such as Mathematica, eschew or prohibit side effects. Pure functions that are called by the user do not depend on histories of other values, what has appeared on the screen, what has gone on before the call, or the length of time required for the function to complete (no *temporality*). A *pure function* has *residential transparency* (i.e., the function can always be replaced by its value).

A *portable language* can be implemented on different machines. Most portable languages are either

sufficiently rigidly defined (as in the cases of ADA and C) to eliminate variants and extensions, or (as in the case of Java and Pascal) are compiled into an intermediate, machine-independent form. This so-called *pseudocode* (*p-code*) is neither source nor object code. It is the assembly language of a hypothetical, generic CPU. The language is said to have been "ported to a new machine" when an interpreter is written that converts p-code to the appropriate machine code and supplies specific drivers for input, output, printers, and disk use.[5]

## 9. STRUCTURED PROGRAMMING

*Structured programming* (also known as *top-down programming*, *procedure-oriented programming*, and *GOTO-less programming*) divides a procedure or algorithm into parts known as subprograms, subroutines, modules, blocks, or procedures.[6] Internal subprograms are written by the programmer; external subprograms are supplied in a library by another source. Ideally, the mainline program will consist entirely of a series of calls (references) to these subprograms. Liberal use is made of FOR/NEXT, DO/WHILE, and DO/UNTIL commands. Labels and GOTO commands are avoided as much as possible.

Very efficient programs can be constructed in languages that support *recursive calls* (i.e., permit a subprogram to call itself). Some languages (e.g., Pascal and PL/I) permit recursion; others, such as Fortran, do not. Multifunctional languages permit *recursion* and implement it transparently to the user.

Variables whose values are accessible strictly within the subprogram are *local variables*. *Global variables* can be referred to by the main program and all other subprograms.

## 10. MODULAR AND OBJECT-ORIENTED PROGRAMMING

*Modular programming* builds efficient larger applications from smaller parts known as *modules*. Ideally, modules deal with unique functions and concerns that are not dealt with by other modules. For example, a module that looks up airline flight numbers could be shared by a reservation-booking module and seat-assigning module. It would not be efficient for the booking and seating modules to each have their own flight number search routine.

Modular programming is intrinsic to *object-oriented programming* (OOP). Java, Python, C++, some variants of Pascal, Visual Basic .NET, and Ruby are *object-oriented programming languages* (OOPL). OOP defines objects with predefined characteristics and behaviors

---

[4]Contrast this with BASIC, which permits (1) a GOSUB to a specific subroutine with a return from anywhere within the subroutine and (2) unlimited GOTO statements to anywhere in the main program.

[5]Some companies have produced Pascal engines that run p-code directly.
[6]The format and readability of the source code—improved by indenting nested structures, for example—are generally present but do not define structured programming.

Computer Programming

and also defines all the interactions with, functions of, and operations on the objects. Objects are essentially data structures (i.e., databases) and their associated functionalities. OOP exhibits characteristics such as class, inheritance, data hiding, data abstraction, polymorphism, encapsulation, interface, and package.

For example, consider a program called "Business Trip Reservation System" created in an OOP environment. "Aircraft" would be an *object class* that has *object variables* of aircraft name, passenger capacity, speed, and travel cost per distance. A particular instance (component or record) would have an object variable name value of "Learjet," an associated passenger capacity of 12, a speed of 300 knots/hr, and an operating cost of $100 per mile traveled. Other related object classes in the program might also be established for "Customer" and "Reservation." One of the advantages of OOP is modularity. That is, an existing module for one class can be used without reprogramming for another class. For example, "Short-Haul Aircraft" would be a *subclass* of the "Aircraft" class and would have the same characteristics and functionality as the "Aircraft" class by virtue of *inheritance*. In a *package* or *namespace*, the related classes of "Aircraft," "Customer," and "Reservation" would be integrated in such ways as to prevent duplication of object variable names and to produce good code organization.

As a result of *data abstraction*, an end user could specify the value of a key to be used to retrieve "Learjet" characteristics, but the end user would have no knowledge of the look-up method (e.g., hashing, binary tree, linear search). As a result of *encapsulation*, an end user would not see or have access to the actual data structure (i.e., internal representation) being accessed. Only the object itself is allowed to modify component variable values, and only by using the functionality built into the object. As a result of *data hiding*, if "Learjet" had other characteristics (e.g., fuel capacity) in the database, an object-oriented program working with a subclass of "Short-Haul Aircraft" would not be able to access, change, or corrupt those characteristics in the database, since that data would be invisible to it.

Objects also define and contain the *activities* (i.e., interactions with and operations to be performed on the data types). Thus, when designing objects it is necessary to anticipate all ways in which the objects will be used. For example, it might be possible to view characteristics, compare characteristics, reserve, pay for, and cancel a reservation for a Learjet. How the object's functionality is presented to the user is determined by the object's *interface*. As a result of *polymorphism*, a common form of functionality may be used with different objects (e.g., highlighting and clicking on a specific aircraft, a reservation, or a customer will all result in information being displayed, even though the information values and their formats differ.)

## 11. INTERACTIVE SYSTEMS, EXPERT SYSTEMS, AND ARTIFICIAL INTELLIGENCE

An *expert system* is software that may, through its interactions with the user, partially or completely replace a human *subject matter expert*. An expert system will be limited to a specific type of situation, usually a problem requiring a specific professional or technical *body of knowledge* (e.g., medical knowledge). The user interacts with an expert system through a serious of sequential queries and responses, though not necessarily text based as has characterized many expert system interfaces. *Fuzzy logic* may be used to approximate human reasoning when the user-provided inputs are incomplete, imprecise, or contradictory. An expert system may or may not be able to learn and modify its database by collecting information about a user's specific situation. Essentially, an expert system simulates a human expert's problem-solving skills and information recall, with the goal that the system's consistency and enhanced knowledge base will compensate for the inevitable lack of intuition and illogical insights that define human brilliance.

Expert systems represent one of three major areas in *artificial intelligence* (AI): interactive systems, expert systems, and robotics. One of the earliest and best known instances of an interactive expert system was ELIZA, a "therapeutic" program that responded to key words within the user's responses. The goal of building a thinking machine that includes sensing, cognition, reasoning, planning, knowledge recall, learning, communication, physical object relocation, and self-locomotion is elusive. Researchers have, thus far, had to limit their work to smaller, specialized subsets of the entire goal.

## 12. FIELDS, RECORDS, AND FILE TYPES

A collection of *fields* is known as a *record*. For example, name, age, and address might be fields in a personnel record. Groups of records are stored in a *file* (*table* or *database*).

A *sequential file* structure contains consecutive records and must be read starting at the beginning. An *indexed sequential file* is one for which a separate index file (see Sec. 73.13) is maintained to help locate records.

With a *random (direct access) file structure*, any record can be accessed without starting at the beginning of the file.

## 13. FILE INDEXING

It is usually inefficient to maintain the records of an entire file in order. (A good example is a mailing list with thousands of names. It is more efficient to keep the names in the order of entry than to sort the list each time names are added or deleted.) *Indexing* is a means of

**Computer Programming**

specifying the order of the records without actually changing the order of those records.

An *index* (*key* or *keyword*) *file* is analogous to the index at the end of a book. It is an ordered list of items with references ("pointers") to the complete records. One field in the data record is selected as the *key field* (index).[7] The sorted keys are usually kept in a file separate from the data file. One of the standard search techniques is used to find a specific key.

## 14. SORTING

*Sorting routines* place data in ascending or descending numerical or alphabetical order.

With the method of *successive minima*, a list is searched sequentially until the smallest element is found and brought to the top of the list. That element is then ignored, and the remaining elements are searched for the smallest element, which, when found, is placed after previous minimum, and so on. A total of $n(n-1)/2$ comparisons will be required.[8] (See Fig. 73.2.)

**Figure 73.2** Key and Data Files

key file

| key | record |
|--------|--------|
| ADAMS | 3 |
| JONES | 2 |
| SMITH | 1 |
| THOMAS | 4 |

data file

| record | last name | first name | age |
|--------|-----------|------------|-----|
| 1 | SMITH | JOHN | 27 |
| 2 | JONES | WANDA | 39 |
| 3 | ADAMS | HENRY | 58 |
| 4 | THOMAS | SUSAN | 18 |

In a *bubble sort*, each element in the list is compared with the element immediately following it. If the first element is larger, the positions of the two elements are reversed (swapped). In effect, the smaller element "bubbles" to the top of the list. The comparisons continue to be made until the bottom of the list is reached. If no swaps are made in a pass, the list is sorted. A total of

approximately $n^2/2$ comparisons are needed, on the average, to sort a list in this manner.[9]

In an *insertion sort*, the elements are ordered by rewriting them in the proper sequence. After the proper position of an element is found, all elements below that position are bumped down one place in the sequence. The resulting vacancy is filled by the inserted element. At worst, approximately $n^2/2$ comparisons will be required. On the average, there will be approximately $n^2/4$ comparisons.

Disregarding the number of swaps, the number of comparisons required by the successive minima, bubble, and insertion sorts is on the order of $n^2$. When $n$ is large, these methods are too slow. The *quicksort* is more complex but reduces the average number of comparisons (with random data) to approximately $n \times \log n/\log 2$, generally considered as being on the order of $n \times \log n$.[10] The maximum number of comparisons for a heap sort is $n \times \log n/\log 2$, but it is likely that even fewer comparisons will be needed.

## 15. SEARCHING

If a group of records (i.e., a list) is randomly organized, a particular element in the list can be found only by a *linear search* (*sequential search*). At best, only one comparison and, at worst, $n$ comparisons will be required to find something (an event known as a *hit*) in a list of $n$ elements. The average is $n/2$ comparisons, described as being "on the order of $n$."

If the records are in ascending or descending order, a *binary search* will be more efficient.[11] The search begins by looking at the middle element in the list. If the middle element is the sought-for element, the search is over. If not, half of the list can be disregarded in further searching since elements in that portion will be either too large or too small. The middle element in the remaining part of the list is investigated, and the procedure continues until a hit occurs or the list is exhausted. The number of required comparisons in a list of $n$ elements will be $\log n/\log 2$ (i.e., on the order of $\log n$).

## 16. HASHING

An index file is not needed if the record number (i.e., the storage location for a read or write operation) can be calculated directly from the key, a technique known as *hashing*.[12] The procedure by which a numeric or non-numeric key (e.g., a last name) is converted into a

---

[7]More than one field can be indexed. However, each field will require its own index file.

[8]When $n$ is large, $n^2/2$ is sometimes given as the number of comparisons.

[9]This is about the same as for the successive minima approach. However, swapping occurs more frequently in the bubble sort, slowing it down.

[10]However, the quicksort falters (in speed) when the elements are in near-perfect order.

[11]A binary search is unrelated to a binary tree. A binary tree structure (see Sec. 73.17) greatly reduces search time but does not use a sorted list.

[12]Of course, finding a record in this manner requires it to have been written in a location determined by the same hashing routine.

Computer Programming

record number is called the *hashing function* or *hashing algorithm*. Most hashing algorithms use a *remaindering modulus*—the remainder after dividing the key by the number of records, *n*, in the list. Excellent results are obtained if *n* is a prime number; poor results occur if *n* is a power of 2.

Not all hashed record numbers will be correct. A *collision* occurs when an attempt is made to use a record number that is already in use. Chaining, linear probing, and double hashing are techniques used to resolve such collisions.

## 17. DATABASE STRUCTURES

Databases can be implemented as indexed files, linked lists, and tree structures; in all three cases, the records are written and remain in the order of entry.

An indexed file such as that shown in Fig. 73.2 keeps the data in one file and maintains separate index files (usually in sorted order) for each key field. The index file must be recreated each time records are added to the file. A *flat file* has only one key field by which records can be located. Searching techniques (see Sec. 73.15) are used to locate a particular record. In a *linked list (threaded list)*, each record has an associated *pointer* (usually a record number or memory address) to the next record in key sequence. Only two pointers are changed when a record is added or deleted. Generally, a linear search following the links through the records is used. (See Fig. 73.3.)

**Figure 73.3** *Database Structures*

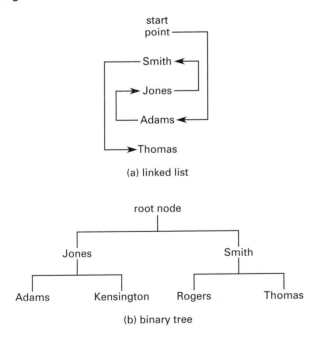

(a) linked list

(b) binary tree

Pointers are also used in *tree structures*. Each record has one or more pointers to other records that meet certain criteria. In a *binary tree structure*, each record has two

pointers—usually to records that are lower and higher, respectively, in key sequence. In general, records in a tree structure are referred to as *nodes*. The first record in the file is called the *root node*. A particular node will have one node above it (the *parent* or *ancestor*) and one or more nodes below it (the *daughters* or *offspring*). Records are found in a tree by starting at the root node and moving sequentially according to the tree structure. The number of comparisons required to find a particular element is $1 + (\log n / \log 2)$, which is on the order of $\log n$.

## 18. HIERARCHICAL, RELATIONAL, AND OBJECT-ORIENTED DATABASES

Databases can be organized hierarchically or relationally, with the distinction being in the ways data are organized and retrieved. The earliest databases were hierarchical in nature. A *hierarchical database* relates information in an organized, structured format. (See Fig. 73.4.) A logical tree structure and the terms "parent" and "daughter" (or "parent" and "child") are often appropriate. For example, an insurance company may have two databases, one for insured primary adults and another for insured children who are covered under their parents' plans. A parent record may have many daughter records, but a daughter record may have only a single parent record, hence the appearance of branching tree structure. This is known as a *one-to-many* and *N-mapping* relationship. In a pure hierarchical management system, daughter records are accessible only through their parent records. The hierarchy of records and sub-records gives hierarchical databases their name. (Parent and daughter databases contain sets of records with fields. It is the hierarchical nature of the records, not the table-like record-field arrangements of the files, that characterizes hierarchical database management schemes.)

**Figure 73.4** *A Hierarchical Personnel File*

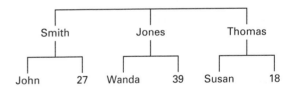

Hierarchical databases are the fastest databases, but they are specialized and limited to the applications for which they are specifically built. Modern databases (e.g., Oracle, SQL, MySQL, DB2) are relational in nature. In a *relational database*, the end user is unaware of the actual structure, interacting with the database through *queries* that are processed by the database management system transparently. For the purpose of visualization, information is characterized as being stored in a *table*—the equivalent of a matrix. Each row in the table is known as a *tuple*. Columns contain *attribute* information. (See Fig. 73.5.)

**Figure 73.5** A Relational Personnel File

| rec. no. | last | first | age | |
|----------|--------|-------|-----|---|
| 1 | Smith | John | 27 | |
| 2 | Jones | Wanda | 39 | |
| 3 | Thomas | Susan | 18 | |

Relational databases can support queries about any type of information contained within them, as essentially all of the information is accessible without "going through" any other piece of information. Any element of information can be accessed directly by referring to the field name and field value. Virtually all relational databases are accessed and manipulated entirely through the *structured query language* (SQL).

An *object-oriented database* contains objects as well as information. For example, video clips, PowerPoint presentations, and image collections are objects that can be included with traditional data objects such as name, address, and date of birth. Object-oriented databases are usually relational in nature. They exhibit characteristics typical of object-oriented programming such as classes and inheritance.

Computer
Programming

# 74 Fortran

## 1. INTRODUCTION TO FORTRAN

Fortran is a high-level language originally designed for scientific calculations. Its name is derived from the phrase *formula translation*. Versions of the language exist for personal computers through the largest super-computers. ANSI (American National Standards Institute) Fortran, also known as *Standard Fortran, Full Fortran*, etc., is the basis for most versions. Fortrans that are identical in operation to ANSI Fortran are known as *full implementations*. Fortrans that accept a subset of ANSI Fortran commands are known as *partial implementations*. Fortrans that add their own commands not in ANSI Fortran are known as *extended implementations*. Almost all versions implement Fortran as a *compiled language*.

This chapter is intended as a brief documentation of the Fortran language, not as a lesson in programming. However, due to the many versions of Fortran, some of the instructions listed in this chapter may not be compatible with all compilers.

## 2. STRUCTURAL ELEMENTS

Symbols in Fortran statements are limited to uppercase letters, the digits 0 through 9, the blank, and the following characters.

```
+  =  -  *  /  (  )  ,  .  $
```

Fortran compilers pack the characters; therefore, blanks can be inserted anywhere in most statements. For example, the following two statements compile identically.

```
IF  (AGE  .  LT  .  YEARS)  GO  TO  10
IF(AGE.LT.YEARS)GOTO10
```

Line length in free-form implementations is essentially unlimited. In fixed-form implementations, statements are prepared in 80-position (80-column) lines.[1] Table 74.1 describes how the various positions are allocated. While statements can be optionally numbered, they are executed sequentially regardless of the order of statement numbers.

**Table 74.1** Column Positions in Fixed-Form Fortran Statements

| position | use |
|---|---|
| 1 | The letter "C" in position 1 indicates a comment. |
| 2–5 | The *statement number*, if used, can run from 1 through 9999. |
| 6 | Any character except zero indicates a continuation from the previous line. |
| 7–72 | The Fortran statement appears here. |
| 73–80 | The last eight positions are available for any use and are ignored by the compiler.* |

*These positions can be used to number the final debugged program sequentially or identify the program name (acronym) or programmer.

## 3. REPRESENTATION OF CONSTANTS

Numerical data can be of either real or integer types. *Integers* have no decimal points. *Real numbers*, also known as *floating point numbers*, are distinguished from integers by a decimal point and may contain fractional parts. *Single-* and *double-precision* floating point numbers are indicated by the letters E and D, respectively. Constants in Fortran are positive unless preceded by a minus sign. Commas are not allowed in any number. (See Table 74.2.)

Constants will also be invalid if they are defined too large or too small. The maximum values that constants can assume vary greatly and are functions of the computer word size. As a consequence of the binary arithmetic used in integer operations, however, a computer with a standard word size of $n$ bits will usually support single precision integers in the range of $-2^{n-1}$ to

---

[1]This line length is an inheritance from the punch card era.

Computer Programming

**Table 74.2** *Correct and Incorrect Fortran Constants*

| type | correct representation | incorrect representation | |
|---|---|---|---|
| integer | 90000 | 90,000 | (comma) |
| integer | 14 | 14. | (decimal point) |
| real | 2. | 2 | (missing decimal) |
| real | 90000. | 90,000. | (comma) |
| real | 7.0 E–4 | 7 E–4 | (missing decimal)* |
| real | 7.1*10**2.3 | 7.1E2.3 | (non-integer exponent) |
| real | 1000. | E3 | (exponent alone) |

*May be permitted by some Fortran implementations.

$2^{n-1} - 1$. The implementation of *double precision*, which doubles the number of bytes used to store constants, varies.[2]

## 4. INTEGER AND REAL VARIABLES[3]

Variable names are formed from up to six alphanumeric characters. The first character must be a letter. Variable names starting with the letters I, J, K, L, M, and N are assumed by the compiler to be integer variables. This is known as *implicit typing*. All variables not implicitly or explicitly integer are real.

An implicit type can be overridden by declaring a variable name in an *explicit typing* statement (e.g., INTEGER or REAL) at the beginning of the program. (Variables following the implicit type conventions do not need to be declared.) The order of the declarations is not important. For example, the statements INTEGER TIME and REAL INSTANT would establish TIME and INSTANT as integer and real variables, respectively. (See Table 74.3.)

**Table 74.3** *Correct and Incorrect Fortran Variables*

| type | correct representation | incorrect representation | |
|---|---|---|---|
| integer | JIM | SLIM | (real) |
| integer | M200 | 200M | (begins with a number) |
| integer | KK | K&K | (& not allowed) |
| integer | IYOU | I$YOU | ($ not allowed)* |
| integer | IBGONE | IBEGONE | (7 characters) |
| real | SLIP | 1SLIP | (begins with a number) |
| real | RS | R–S | (– not allowed) |
| real | TIMEB | TIMEBASE | (8 characters) |

*May be permitted by some Fortran implementations.

---

[2]In some implementations, *half-precision* numbers are available. The INTEGER*2 statement allocates two bytes instead of the normal four used for integer storage. Similarly, the REAL*4 statement allocates four bytes instead of the normal eight used for real numbers.
[3]Other Fortran variable types include characters, complex numbers, and logical variables defined by the statements CHARACTER, COMPLEX, and LOGICAL, respectively. LOGICAL variables can assume only the values of true and false.

## 5. ARRAY VARIABLES

*Arrays* (also known as matrices) are defined by a DIMENSION statement placed at the beginning of a program. The maximum number of dimensions depends on the Fortran implementation and computer. Unless explicitly typed, the variable type for array entries will correspond to the name of the array according to the rules prescribed in Sec. 74.4. For example, the following statements establish a 1-row×5-column real array called SAMPLE and a 2-row×7-column integer array called INCOME.

DIMENSION SAMPLE(5)
DIMENSION INCOME(2,7)

Elements of arrays are addressed by placing the identifying subscripts in parentheses (e.g., SAMPLE(2) and INCOME(1,5)). For a two-dimensional array, the first number defines the row and the second number defines the column. Zero and negative numbers are not permitted as subscripts. Subscripts can also be integer variables (e.g., SAMPLE(K)) as long as the variable is within the range of the array size.

## 6. INITIAL VALUES

Variables and arrays, once defined and declared, are not automatically initialized. The DATA statement is used if it is necessary to assign an initial value. For example, in the following statements, variables X, Y, and Z will be initialized to 1.0. All elements in the array ONEDIM will be equal to 0.0. The array TWODIM will be set to

$$\begin{pmatrix} 1.0 & 2.0 & 3.0 \\ 4.0 & 5.0 & 6.0 \end{pmatrix}$$

DIMENSION ONEDIM(5)
DIMENSION TWODIM(2,3)
DATA X,Y,Z/3*1.0/(ONEDIM(I),I = 1,5)/5*0.0/
1((TWODIM(I,J),J = 1,3),I = 1,2)/1.,2.,3.,4.,5.,6./

After being initialized with a DATA statement, variables can have their values changed by arithmetic operations.

## 7. ARITHMETIC OPERATIONS

The standard arithmetic operations listed in Table 74.4 are supported in Fortran.

The equals sign replaces one quantity with another but does not imply equality. For example, the statement $Z = Z + 1$ is algebraically incorrect but is a valid Fortran statement. (See Table 74.5.)

Each operation must be explicitly and unambiguously stated; two operations in a row are also unacceptable. Table 74.6 illustrates several common errors in writing Fortran statements.

**Table 74.4** *Fortran Arithmetic Operations*

| symbol | meaning |
|--------|---------|
| = | replacement (assignment) |
| + | addition |
| − | subtraction |
| * | multiplication |
| / | division |
| ** | exponentiation |
| ( ) | preferred operation |

**Table 74.5** *Correct and Incorrect Fortran Assignments*

| correct representation | incorrect representation |
|------------------------|--------------------------|
| A = A + B | A + B = A |
| B = A*A | A**2 = A*A |
| A = 3; B= 3 | A = B = 3 |

**Table 74.6** *Correct and Incorrect Fortran Operations*

| correct representation | incorrect representation |
|------------------------|--------------------------|
| A*B | AB |
| A*B | AXB |
| A + (−B) | A + −B |
| A*(−B) | A*−B |
| A/(−2.0) | A/−2.0 |
| 1.7**2.7 | −1.7**2.7 |

All variables and constants can be raised to real exponents (e.g., 2.0**3.1). However, only positive real numbers can be raised to real exponents (e.g., 2.0**3.1 but not −2.0**3.1). This is a consequence of using logarithms to perform the exponentiation, since logarithms of negative numbers do not exist. However, a number raised to an integer exponent is calculated by successive multiplications (rather than with logarithms), so this limitation does not exist for integer exponents (i.e., −2.0**3 is permitted).

Most, but not all, Fortran implementations permit *mixed-mode arithmetic* (i.e., the mixing of real and integer variables in a single arithmetic statement).[4] Integer variable values are first converted to real values by adding a zero fractional part. Real arithmetic is then used to evaluate the expression.

When mixed-mode arithmetic is permitted, care must be taken to observe the conversion of real data to integer mode. Within mixed-mode expressions, intermediate results will be real numbers. Division between two integers, however, results in truncation. Also, real numbers are truncated to their whole number portions upon assignment to an integer variable.

## 8. HIERARCHY OF OPERATIONS

Operations in an arithmetic statement are performed in the order of exponentiation first, multiplication and division second, and addition and subtraction third. In the event there are two consecutive operations with the same hierarchy (e.g., a multiplication followed by a division), the operations are performed in the order encountered, normally left to right (except for exponentiation, which is right to left).[5] Parentheses can modify this order; operations within parentheses are always evaluated before operations outside. If nested parentheses are present in an expression, the expression is evaluated outward starting from the innermost pair.

### Example 74.1

Evaluate J in the following expression. Mixed-mode arithmetic is permitted, and expressions are scanned left to right.

$$J = (6.0 + 3.0)^*3.0/6.0 + 5.0 - 6.0^{**}2.0$$

*Solution*

The expression within the parentheses is evaluated first.

$$9.0^*3.0/6.0 + 5.0 - 6.0^{**}2.0$$

The exponentiation is performed next.

$$9.0^*3.0/6.0 + 5.0 - 36.0$$

The multiplication and division are performed next.

$$4.5 + 5.0 - 36.0$$

The addition and subtraction are performed last.

$$-26.5$$

However, J is an integer variable, and the assignment of J = −26.5 results in truncating the fractional part. Ultimately, J has the value of −26.

## 9. NON-EXECUTABLE STATEMENTS

Several Fortran commands are used to establish and initialize data locations and define and control the environment in which the program runs, but they are not part of the program logic. Most of the *non-executable instructions* shown in Table 74.7 must be placed at the beginning of a program, before the executable instructions.

---

[4]Raising a real number to an integer exponent is not mixed-mode arithmetic.

[5]In most implementations, a statement will be scanned from left to right. Once a left-to-right scan is complete, some implementations then scan from right to left; others return to the equals sign and start a second left-to-right scan. Parentheses should be used to define the intended order of operations.

*Computer Programming*

**Table 74.7** *Non-executable Statements*

| | |
|---|---|
| BLOCK DATA | initializes variables in COMMON |
| COMMON | establishes memory locations accessible to two or more programs and subprograms |
| DATA | initializes variables |
| DIMENSION | establishes arrays |
| END | defines the end of a program or subprogram |
| EQUIVALENCE | specifies multiple names for a single variable |
| EXTERNAL | passes the name of one subprogram to another subprogram |
| FUNCTION | establishes the name of a subprogram |
| INTEGER | defines variables as integers |
| PROGRAM | identifies the main program |
| REAL | defines variables as real numbers |
| SUBROUTINE | establishes the name of a subroutine |

## 10. LOGICAL STATEMENTS

A *logical statement* is a relational expression using one of the operators in Table 74.8. The use of these operators is illustrated in Ex. 74.2.

**Table 74.8** *Fortran Logical Operators*

| | |
|---|---|
| .AND. | and |
| .EQ. | equal |
| .GE. | greater than or equal to |
| .GT. | greater than |
| .LE. | less than or equal to |
| .LT. | less than |
| .NE. | not equal to |
| .NOT. | not |
| .OR. | or |

## 11. CONTROL STATEMENTS

The STOP statement indicates the logical end of the program (i.e., completion of the algorithm). The format is

$$s \quad \text{STOP}$$

Depending on the branching of the algorithm, a program may have more than one STOP statement. When a STOP statement is reached, program execution stops and control is turned over to the operating system.[6] Normally, the program cannot be continued once a STOP has been executed. The statement number, s, is optional, but if present, the value is made known to the user.

The CALL statement is used to transfer execution to a subroutine. CALL EXIT will terminate execution and turn control over to the operating system. (In this regard, CALL EXIT and STOP have identical functions.)

[6]Unless it is an END statement, the line following a STOP must be numbered. Otherwise, it will be impossible to get to that statement.

The RETURN statement ends execution of a subroutine and passes control back to the main program.

The CONTINUE statement does nothing. It can be used with a statement number as the last line of a DO loop.

The END statement is required as the last physical statement in a source program and in any subprogram. A program cannot be compiled without an END statement, as it tells the compiler that there are no more lines in the program. END statements cannot be numbered.

The explicit GO TO s statement transfers control to statement s (s is known as a *label*.) The computed GO TO has the form GO TO $(s_1, s_2, \ldots, s_n)$ [expression]. The value of [expression] (i.e., $1, 2, \ldots, n$) determines which label is to be used.

The arithmetic IF statement is written

$$\text{IF [expression] } s_1, s_2, s_3$$

[expression] is any numerical variable or set of arithmetic operations. The $s_i$ are statement numbers (*labels*). Transfer is to statement $s_1$, $s_2$, or $s_3$ according to whether [expression] is negative, zero, or positive, respectively.

The logical IF statement has the form

$$\text{IF [logical expression] [statement]}$$

The [statement] can be any executable statement except DO and IF. If [logical expression] is true, [statement] will execute; otherwise, the next instruction will be executed. Some Fortran implementations have an IF... THEN... ELSE statement.

The DO statement (covered in Sec. 74.12) is also a control statement.

### Example 74.2

Determine the meaning of the following two statements:

(a) IF (A.GT.25.6) A = 27.0

(b) IF (Z.EQ.(T − 4.0).OR.Z.EQ.0.) GO TO 17

*Solution*

(a) If A is greater than 25.6, then set A equal to 27.0.

(b) If Z is equal to (T − 4.0) or if Z is equal to zero, then go to statement 17.

## 12. PROGRAM LOOPS

A *loop* consists of one or more instructions intended to be executed one or more times. A *nested loop* is a loop within a larger loop, as illustrated in Fig. 74.1.[7] Loops can be constructed from IF and GO TO statements;

[7]Although not required by instruction syntax, nested DO loops should be indented to enhance program readability.

Computer Programming

however, the DO statement is a convenient method of defining a loop.

**Figure 74.1** *Nested DO Loops*

```
┌──── DO 100 I = 1, 10
│  ┌──── DO 90 J = 2, 6, 2
│  │  ┌──── DO 80 K = 1, 10, 3
│  │  └── 80 CONTINUE
│  └── 90 CONTINUE
└── 100  CONTINUE
```

The general form of the DO statement is

$$DO\ s\ i = j,k,l$$

where  s   is a statement number
       i   is the integer loop variable (the *counter*)
       j   is the initial value assigned to i
       k   is the inclusive upper bound (the *test*) of i,
          which must exceed j
       l   is the *increment* for i, with a default of 1

Statement s must be an executable statement, including the dummy CONTINUE statement, but it cannot be a control statement. The DO statement causes the statements immediately following it through statement s to be executed until i equals or exceeds k. When i equals or exceeds k, the statement following s is executed. However, the loop can also be exited before i reaches k.[8]

## 13. INPUT/OUTPUT STATEMENTS

The READ, PRINT, WRITE, and FORMAT statements are Fortran's main input/output (I/O) commands. The structures of the READ and WRITE statements are

$$READ\ \ \ (u_1, s)\ \ [list]$$
$$WRITE\ (u_2, s)\ \ [list]$$

$u_1$ is a *unit number* (e.g., 5) that designates the file or input device (e.g., disk or keyboard). This number varies with installation. $u_2$ is a unit number that designates the file or output device (e.g., a printer or monitor). s is the statement number corresponding to a FORMAT statement, and [list] is a list of variables separated by commas whose values are being read or written. For example,

$$READ\ (5, 100)\ X, Y, Z$$
$$100\ FORMAT\ (F5.0)$$

The [list] can also include an implicit DO loop, particularly valuable in loading initial values of array variables. The following example reads seven values, the first six into the array WOW and the last into TOP.

$$READ\ (5, 85)\ (WOW(J), J = 1, 6), TOP$$
$$85\ \ \ \ FORMAT\ (7F5.0)$$

---

[8]While control can be transferred out of an executing DO loop, control should never be transferred into a DO loop, other than to the DO statement.

The purpose of the FORMAT statement is to define the location, size, and type of data being read. The FORMAT statement has the form of

$$s\ \ \ \ \ FORMAT\ [field\ list]$$

s is the statement number (i.e., the *line label*). [field list] contains format codes set apart by commas defining the fields. [field list] can be longer than [list] in the corresponding READ or WRITE statement.

The format code for integers is nIw where w is the number of character positions and n is an optional repeat counter indicating the number of consecutive variables having the same format.

The format codes for real values are nFw.d (fixed) and nEw.d (exponential). w is the number of character positions allocated including the space required for the decimal point, sign (if negative), and (in scientific notation) the exponent E or D. d is the number of spaces to the right of the decimal point to be printed. When data are read in, decimal points in any position take precedence over the value of d. Values are truncated after d decimal positions. Rounding is not automatic.

The F format will print up to (w−1) digits, sign, and blanks. The E format will print a total of (w−2) digits, sign, and blanks and give the data in a standard scientific notation with an exponent.

Table 74.9 lists these and other codes that can be used in a FORMAT statement.

**Table 74.9** *Fortran FORMAT Codes*

| code | use |
|------|-----|
| A | text string |
| D | real number, double precision |
| E | real number in scientific notation |
| F | real number |
| H | alphanumeric character (obsolete) |
| I | integer |
| L | logical variable |
| P | decimal point modifier |
| T | position (tab column) number |
| X | blank (horizontal space) |
| Z | hexadecimal number |
| / | skipped line |
| ' ' | literal data |

The WRITE command provides the flexibility to send alphanumeric characters and other content to a file, printer, or monitor. The monitor is the default output device in most Fortran implementations, and results can be displayed on the default output device by simply using the PRINT command. The format of the PRINT command used to display a list of variables is

$$PRINT\ *,\ [list]$$

Computer Programming

Following this command, variables in a list will be displayed in the sequence of the list. The asterisk (*) indicates that the variables will be displayed in accordance with their types; this is sometimes called *free-format display*. If more control over the display format is needed, a WRITE/FORMAT combination should be used.

The asterisk may also be used with READ and WRITE statements. In the following statements, the first asterisk means the input comes from the keyboard in a READ statement and goes to the monitor in a WRITE statement. The second asterisk means the computer decides how the variables will be displayed based on their types.

```
READ (*,*) [list]
WRITE (*,*) [list]
```

## 14. PRINTER CONTROL

The usual output device in early implementations was a line printer with 133 logical print positions (columns). The first print position was used for *carriage control*, leaving 132 actual physical print positions.[9] Although line length is no longer an issue, Fortran remains line-oriented. The first print position is still reserved. The character in the first output position will control the printer advance according to the rules in Table 74.10. Carriage control is usually accomplished by the use of literal data within a FORMAT statement.

*Table 74.10* Fortran Printer Control Characters

| character | function |
| --- | --- |
| blank | advance one line |
| 0 | advance two lines |
| 1 | form feed (skip to line one on the next page) |
| + | carriage return (overprint—do not advance) |

### Example 74.3

What is printed by the following program segment?

```
        K = 193
        WRITE (6, 100) K
        WRITE (6, 101) K
100     FORMAT(' ', I3)
101     FORMAT(I3)
```

*Solution*

The number 193 would be printed on the next line of the current page. The number 93 would be printed on the first line of the following page.

---

[9]*Carriage control* is an inheritance from the early use of typewriter-style printers as operator keyboards. High-speed printers no longer have carriages.

## 15. STANDARD LIBRARY FUNCTIONS

Fortran supplies many standard *library functions* (*supplied functions, intrinsic functions,* or *canned functions*) for mathematical operations. The single precision library functions listed in Table 74.11 are available in most Fortran implementations. Most functions are accessed by placing the argument in parentheses after the function name (e.g., SIN(THETA)). In most cases, placing the letter D before the function name will cause the calculation to be performed in double precision (e.g., DATAN is the arctangent in double precision). Arguments for trigonometric functions must be expressed in radians.

*Table 74.11* Representative Fortran Library Functions

| function | use |
| --- | --- |
| ABS | real absolute value |
| ALOG | natural logarithm |
| ALOG10 | common logarithm |
| AMOD | real remaindering modulus |
| ARCOS | arccosine |
| ARSIN | arcsine |
| ATAN | arctangent |
| CONJG | complex conjugate |
| COS | cosine |
| COSH | hyperbolic cosine |
| EXP | $e^x$ |
| FIX | convert real to integer |
| FLOAT | convert integer to real |
| IABS | integer absolute value |
| INT | convert real to integer |
| MAX | maximum value |
| MIN | minimum value |
| MOD | integer remaindering modulus |
| SIN | sine |
| SINH | hyperbolic sine |
| SQRT | square root |
| TAN | tangent |
| TANH | hyperbolic tangent |

## 16. STATEMENT FUNCTIONS

In Fortran, the term *subprogram* refers to an instruction or group of instructions that is used repeatedly. The term encompasses user-defined functions, function subprograms, and subroutines.

A user-defined *statement function* processes one or more input numbers (*arguments* or *parameters*) and returns a single numerical value. It must be defined prior to use, and the definition must fit on one statement. For example, the function PROD calculates the product of two numbers in the following code segment.

```
        main program
        PROD(X1,X2) = X1*X2
        HEIGHT = 2.5
        WIDTH = 7.5
        AREA = PROD(HEIGHT,WIDTH)
        STOP
        END
```

## 17. FUNCTION SUBPROGRAMS

A user-defined *function subprogram* (*external function*) processes one or more input numbers (*arguments* or *parameters*) and returns a single numerical value. It is not limited to a single line but is created with the FUNCTION statement according to the following rules.

- In the main program, the function is defined as a variable.

- When used in the main program, the function is followed by its actual arguments in parentheses.

- Within the function itself, the function name is type declared and defined by the FUNCTION statement.

- The arguments need not have the same names in the main program and function subprogram, but they must agree in number, sequence, type, and precision (length).

- Both the function subprogram and the main program have END statements.

- Only the function subprogram ends with a RETURN statement.

- The function may, but need not, have a STOP statement.

For example, the product of two numbers is calculated by the function PROD in the following code segment.

```
main program
REAL PROD
HEIGHT = 2.5
WIDTH = 7.5
AREA = PROD(HEIGHT,WIDTH)
STOP
END

function subprogram
REAL FUNCTION PROD(X1,X2)
PROD = X1*X2
RETURN
END
```

## 18. SUBROUTINES

A *subroutine* is a user-defined subprogram. It is not limited to mathematical calculations, nor is it limited to returning to the statement immediately following the calling statement. Thus, a subroutine is more versatile than a function or function subprogram. Subroutines are governed by the following rules.

- The subroutine has no type and does not assume any value.

- The subroutine is activated by a CALL statement.

- Both the subroutine and the main program have END statements.

- Only the subroutine has a RETURN statement.

- The subroutine may, but need not, have a STOP statement.

- The arguments need not have the same names in the main program and subroutine.

- The arguments must agree in number, sequence, type, and precision (length).

The following code segment illustrates a subroutine GETNUM that obtains two numbers for subsequent calculations.

```
main program
CALL GETNUM(HEIGHT,WIDTH)
AREA = HEIGHT*WIDTH
STOP
END

subroutine
SUBROUTINE GETNUM(X1,X2)
READ(5,100)X1,X2
100   FORMAT(2F5.1)
RETURN
END
```

## 19. COMMON VARIABLES

Variables in functions and subroutines (and in other programs) are independent of the main program. They may have the same names as variables in the main program but their values can be different. Values of variables can be shared, however, if the *regular* COMMON statement is used to establish a link between the variables. The COMMON statement has the format

COMMON [variable list]

The main program and the subroutines must all have COMMON statements. The COMMON statement merely assigns memory locations to be shared by the main program and all of its subroutines. While it is convenient to do so, it is not necessary to use the same variable names. The order of the common variables establishes their positions in memory.

The *named* COMMON statement shares values with only some of the subroutines. While there can be only one regular COMMON statement per program, there can be many named COMMON statements. The format of the named COMMON statement is

COMMON/[name]/[variable list]

**Computer Programming**

# 75 Digital Logic

## Contents

## Nomenclature

| | |
|---|---|
| $m$ | minterm |
| $M$ | maxterm |
| $n$ | number of bits |
| $N$ | total number of possible values |

## 1. DIGITAL INFORMATION REPRESENTATION

Computers process information in digital form. The simplest representation of digital information is that of a single binary variable called a *scalar*. The notation for a scalar is $x_i$. A *binary scalar* takes on a value of zero or one, and represents one bit of information. The smallest information storage unit is termed a *cell*.

At the electronic device level, a zero or one is defined in terms of a voltage output specific to the logic family chosen. For transistor-transistor-logic (TTL), the range of values of the zero and one, called the *state* of the cell, is shown in Fig. 75.1. In TTL logic, zero is represented by the voltage range 0–0.8 V, while one is represented by the voltage range 2.0–5.0 V.

**Figure 75.1** *Definition of Binary States for TTL*

A grouping of $n$ binary variables is called an *n-tuple* and is represented by $[x_1, x_2, x_2, ..., x_n]$. The value of $n$ is called the *word length*. The cells can be listed and numbered left to right or right to left. Left to right is the normal notation. An exception occurs when representing a binary number, in which case the cells are listed right to left, from zero to $n-1$, corresponding to $2^0, 2^1, ..., 2^{n-1}$. Because each binary variable in the $n$-tuple takes on only one of two values, the total number of values that can be represented, $N$, is

$$N = 2^n \qquad 75.1$$

An ordered collection of cells is called a *register*. The information within each register is encoded to represent numerical, logical, or character information. The variables within the register are called *state variables* because they define the value (state) of the register (vector). Particular values within a register are called *states of the register*. For example, a 2-tuple register has four states: [0,0], [0,1], [1,0], and [1,1].

### Example 75.1

A certain computer uses a 32-bit word length. What is the maximum number of possible state variables?

*Solution*

The maximum number of state variables is given by Equation 75.1.

$$N = 2^n = 2^{32} = 4{,}294{,}967{,}296$$

## 2. ELECTRONIC LOGIC DEVICE LEVELS AND LIMITS

Logic one is considered to indicate a true condition. Logic zero indicates a false condition. Logic one is normally associated with the higher voltage in an electronic circuit, and logic zero is associated with the lower voltage. This type of association is called *positive logic*. When the low-voltage condition represents the true statement, the logic is called *negative logic*. In order to avoid confusion, truth tables are sometimes given in terms of high (H) and low (L) voltages, rather than one and zero.

An alternative is to use *assertion levels*. When the assertion (true) condition is the high-voltage level, the term *active high* is used. This corresponds to positive logic. When the assertion (true) condition is the low-voltage level, the term *active low* is used.

In practice, the type of device and its limits need to be considered. Depending on the circuit, the high- and low-voltage levels will be represented by a range of values. Also, the fan-in and fan-out capabilities limit the number of variable inputs and the number of logic gates that can be attached to the output. Additionally, the propagation delays must be accounted for to ensure that levels of logic have time to respond, providing the proper output before the next set of variables arrives. The propagation delays, therefore, determine the operating speed of the logic network. Timing diagrams are used to map these delays and ensure proper operation.

## 3. FUNDAMENTAL LOGIC OPERATIONS

Information to be processed is combined with, and compared to, other information. The logic that does this in such a manner that the output depends only on the value of the inputs is called *combinational logic*.[1] *Logic variables* and *logic constants* have one of two possible values, zero or one. By convention, zero is considered "false," and one is considered "true."

Logic functions are commonly represented in *truth tables*. Because an *n*-tuple can take on $2^n$ values, a truth table has $2^n$ rows. Each of these rows has an associated output that can be either zero or one. Combining these two facts gives the number of scalar functions that can be generated from *n* logic variables. A logic operator involving two operands is called a *binary operator*. For a two-variable input, that is, a 2-tuple, there are 16 possible functions.

$$\text{no. of scalar functions} = (2)^{2^n} \qquad 75.2$$

Not all the scalar functions are useful. In fact, only six are of primary interest. Further, all logic functions can be represented in terms of only two of these, NAND and NOR. These often-used functions are called *logic operators*. A logic operator involving a single variable or operand is called a *unary operation*. A unary function is shown in Table 75.1.

**Table 75.1** *Unary Function*

| $x$ | $f_1(x) = 0$ | $f_2(x) = x$ | $f_3(x) = \overline{x}$ | $f_4(x) = 1$ |
|---|---|---|---|---|
| 0 | 0 | 0 | 1 | 1 |
| 1 | 0 | 1 | 0 | 1 |

The function $f_2(x)$ is called the *identity function*. The function $f_3(x)$ is called the NOT, or negation, function. The symbol and truth table for the NOT function are shown in Fig. 75.2.

Six logic operators (or gates) are used so often that they have been given their own symbols: AND, OR, XOR, NAND, NOR, and XNOR (or coincidence). The X indicates the word "exclusive," while N indicates the word

---

[1]When the output has a memory, that is, when the output is used as feedback to the input, the logic is called *sequential*.

**Figure 75.2** *NOT Logic*

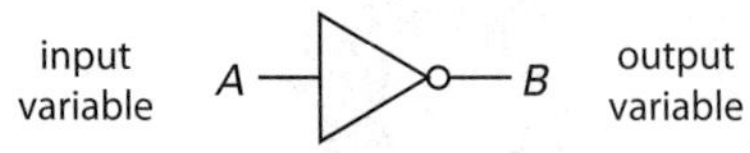

(a) symbol

| $A$ | $B$ |
|---|---|
| 0 | 1 |
| 1 | 0 |

(b) truth table

$$B = \overline{A}$$

(c) equation

"not." The logic operator AND is indicated by a dot "·" or by "∧". The dot is commonly not used when all the variables are represented by single letters. (For example, $A \cdot B$ can be written as $AB$.) The logic operator OR is indicated by "+" or "∨". The small circle on the end of any gate symbol, or on the input or output, indicates the NOT function. The standard symbol for the NOT operation is $\overline{x}$. Alternate symbols for the NOT operation include $-x$ or $x'$. The NOT operator applied over an expression has the effect of enclosing the expression in parentheses. For example,

$$\overline{x_1 x_2} = \overline{(x_1 x_2)}$$

A circle around the AND or OR symbol indicates "exclusive" operation, $\otimes$ or $\oplus$. The fundamental logic operations and their properties are shown in Table 75.2.[2]

When combinations of logic operations are to be analyzed, the following *order of precedence* is applicable. If in doubt when writing an expression, use parentheses to establish the proper order of precedence.

*step 1:* Evaluate those items within parentheses first.

*step 2:* Evaluate logic expressions from left to right, applying all instances of the NOT operations first, the AND operations second, and the OR operations last.

Two functions, $f$ and $g$, are *logically equivalent* when they are defined for the same arguments and

$$f(x_n, ..., x_1) = g(x_n, ..., x_1) \qquad 75.3$$

Equation 75.3 must be valid for all combinations on the *n*-tuple. Logical equivalence is easily verified by using a truth table for both functions and ensuring that the results are identical. Additionally, the rules of *switching*

---

[2]Argument inputs and results are commonly shown using capital letters. They can be scalars or vectors. Scalars have only two values. Vectors represent information and may have numerous values. The properties and laws applicable to logic variables are also applicable to logic vectors. Capital letters are sometimes reserved for logic vectors in educational texts. The symbols of *A*, *B*, *C*, or *X*, *Y*, *Z* are used for inputs and outputs with no special significance attached.

**Table 75.2** *Logic Operators*

| operator | symbol | truth table | equation |
|---|---|---|---|

AND
$A$, $B$ → $C$

| A | B | C |
|---|---|---|
| 0 | 0 | 0 |
| 0 | 1 | 0 |
| 1 | 0 | 0 |
| 1 | 1 | 1 |

$A \cdot B = C$

OR
$A$, $B$ → $C$

| A | B | C |
|---|---|---|
| 0 | 0 | 0 |
| 0 | 1 | 1 |
| 1 | 0 | 1 |
| 1 | 1 | 1 |

$A + B = C$

XOR
$A$, $B$ → $C$

| A | B | C |
|---|---|---|
| 0 | 0 | 0 |
| 0 | 1 | 1 |
| 1 | 0 | 1 |
| 1 | 1 | 0 |

$A \oplus B = C$

NAND
$A$, $B$ → $C$

| A | B | C |
|---|---|---|
| 0 | 0 | 1 |
| 0 | 1 | 1 |
| 1 | 0 | 1 |
| 1 | 1 | 0 |

$\overline{A \cdot B} = C$

NOR
$A$, $B$ → $C$

| A | B | C |
|---|---|---|
| 0 | 0 | 1 |
| 0 | 1 | 0 |
| 1 | 0 | 0 |
| 1 | 1 | 0 |

$\overline{A + B} = C$

XNOR or coincidence
$A$, $B$ → $C$
$A$, $B$ → $C$

| A | B | C |
|---|---|---|
| 0 | 0 | 1 |
| 0 | 1 | 0 |
| 1 | 0 | 0 |
| 1 | 1 | 1 |

$A \odot B = C$
or
$A \otimes B = C$

*algebra*, usually referred to as *Boolean algebra*, can be used to transform one of the functions into the other, which proves equivalence.

## Example 75.2

If $A = 1$ and $B = 0$, what is the output of the following expression?

$$\overline{\left(\overline{(AB)}\ \overline{(AB)}\right)}$$

*Solution*

Applying the rules of precedence, the expression is reduced as follows.

$$
\begin{aligned}
\overline{\left(\overline{(AB)}\ \overline{(AB)}\right)} &= \overline{\left(\overline{(1 \cdot 0)}\ \overline{(1 \cdot 0)}\right)} \\
&= \overline{\left(\overline{(0)}\ \overline{(0)}\right)} \\
&= \overline{(1 \cdot 1)} \\
&= \overline{1} \\
&= 0
\end{aligned}
$$

## 4. LOGIC GATES

A *gate* performs a logical operation on one or more inputs. The inputs, labeled $A$, $B$, $C$, etc., and output are limited to the values of 0 or 1. A listing of the output value for all possible input values is known as a *truth table*. Table 75.3 combines the symbols, names, and truth tables for the most common gates.

**Table 75.3** *Logic Gates*

| inputs | | NOT | AND | OR | NAND | NOR | exclusive OR |
|---|---|---|---|---|---|---|---|
| A | B | $-A$ or $\overline{A}$ | $AB$ | $A + B$ | $\overline{AB}$ | $\overline{A + B}$ | $A \oplus B$ |
| 0 | 0 | 1 | 0 | 0 | 1 | 1 | 0 |
| 0 | 1 | 1 | 0 | 1 | 1 | 0 | 1 |
| 1 | 0 | 0 | 0 | 1 | 1 | 0 | 1 |
| 1 | 1 | 0 | 1 | 1 | 0 | 0 | 0 |

## 5. BOOLEAN ALGEBRA

The rules of Boolean algebra are used to write and simplify expressions of binary variables (i.e., variables constrained to two values). The basic laws governing Boolean variables are listed here.

*commutative:*

$$A + B = B + A \qquad 75.4$$
$$A \cdot B = B \cdot A \qquad 75.5$$

*associative:*

$$A + (B + C) = (A + B) + C \qquad 75.6$$
$$A \cdot (B \cdot C) = (A \cdot B) \cdot C \qquad 75.7$$

*distributive:*

$$A \cdot (B + C) = (A \cdot B) + (A \cdot C) \qquad 75.8$$
$$A + (B \cdot C) = (A + B) \cdot (A + C) \qquad 75.9$$

*absorptive:*

$$A + (A \cdot B) = A \qquad 75.10$$
$$A \cdot (A + B) = A \qquad 75.11$$

**Computer Programming**

*De Morgan's theorems* are

$$\overline{(A + B)} = \overline{A} \cdot \overline{B} \qquad \text{75.12}$$

$$\overline{A \cdot B} = \overline{A} + \overline{B} \qquad \text{75.13}$$

The following basic identities are used to simplify Boolean expressions.

$$0 + 0 = 0 \qquad \text{75.14}$$

$$0 + 1 = 1 \qquad \text{75.15}$$

$$1 + 0 = 1 \qquad \text{75.16}$$

$$1 + 1 = 1 \qquad \text{75.17}$$

$$0 \cdot 0 = 0 \qquad \text{75.18}$$

$$0 \cdot 1 = 0 \qquad \text{75.19}$$

$$1 \cdot 0 = 0 \qquad \text{75.20}$$

$$1 \cdot 1 = 1 \qquad \text{75.21}$$

$$A + 0 = A \qquad \text{75.22}$$

$$A + 1 = 1 \qquad \text{75.23}$$

$$A + A = A \qquad \text{75.24}$$

$$A + \overline{A} = 1 \qquad \text{75.25}$$

$$A \cdot 0 = 0 \qquad \text{75.26}$$

$$A \cdot 1 = A \qquad \text{75.27}$$

$$A \cdot A = A \qquad \text{75.28}$$

$$A \cdot \overline{A} = 0 \qquad \text{75.29}$$

$$-0 = 1 \qquad \text{75.30}$$

$$-1 = 0 \qquad \text{75.31}$$

$$-\overline{A} = A \qquad \text{75.32}$$

**Example 75.3**

Simplify and write the truth tables for the following network of logic gates.

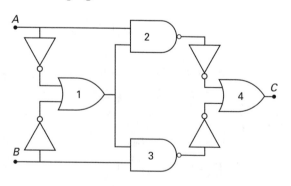

*Solution*

Determine the inputs and output of each gate in turn.

| gate | inputs | output |
|------|--------|--------|
| 1 | $\overline{A}, \overline{B}$ | $(\overline{A} + \overline{B})$ |
| 2 | $A, (\overline{A} + \overline{B})$ | $\overline{A \cdot (\overline{A} + \overline{B})}$ |
| 3 | $B, (\overline{A} + \overline{B})$ | $\overline{B \cdot (\overline{A} + \overline{B})}$ |
| 4 | $A \cdot (\overline{A} + \overline{B}), B \cdot (\overline{A} + \overline{B})$ | $A \cdot (\overline{A} + \overline{B}) + B \cdot (\overline{A} + \overline{B})$ |

Simplify the output of gate 4.

| | |
|---|---|
| $A \cdot (\overline{A} + \overline{B}) + B \cdot (\overline{A} + \overline{B})$ | [original] |
| $A \cdot \overline{A} + A \cdot \overline{B} + B \cdot \overline{A} + B \cdot \overline{B}$ | [distributive] |
| $A \cdot \overline{B} + B \cdot \overline{A}$ | [since $A \cdot \overline{A} = 0$] |
| $A \oplus B$ | [definition] |

The truth table is

| $A$ | $B$ | $C$ |
|-----|-----|-----|
| 0 | 0 | 0 |
| 0 | 1 | 1 |
| 1 | 0 | 1 |
| 1 | 1 | 0 |

## 6. SWITCHING ALGEBRA

*Switching algebra* is a particular form of Boolean algebra, though the distinction is seldom made. Three basic postulates define switching algebra.

- *postulate 1*: A Boolean variable has two possible values, zero and one, and these values are exclusive. That is,

$$\text{if } x = 0 \text{ then } x \neq 1, \text{and}$$

$$\text{if } x = 1 \text{ then } x \neq 0$$

- *postulate 2*: The NOT operation is defined as

$$\overline{0} = 1 \text{ and } \overline{1} = 0$$

- *postulate 3*: The logic operations AND and OR are defined as in Table 75.2.

Using these postulates, the special properties, laws, and theorems of switching algebra are developed. The properties, laws, and theorems of switching algebra are summarized in Table 75.4.

*De Morgan's theorem*, stated in Table 75.4, has many applications and is used when constructing logic networks from NAND and NOR logic. NAND and NOR are *universal operations*, sometimes called *complete sets*. The term "complete set" indicates that any logic expression can be represented solely by NAND logic, or solely by NOR logic. Because any logic expression is a combination of NOT, AND, and OR operations, proving that NAND logic can represent any of these three proves universality. This proof is shown in Eq. 75.33 through

**Table 75.4** Basic Properties of Switching Algebra (Boolean algebra)

| name | property | dual |
|---|---|---|
| special properties: 0 | $0 + A = A$ | $0 \cdot A = 0$ |
| special properties: 1 | $1 + A = 1$ | $1 \cdot A = A$ |
| idempotence law | $A + A = A$ | $A \cdot A = A$ |
| complementation law | $A + \overline{A} = 1$ | $A \cdot \overline{A} = 0$ |
| involution | $(\overline{\overline{A}}) = A$ | |
| commutative law | $A + B = B + A$ | $A \cdot B = B \cdot A$ |
| associative law | $A + (B + C) = (A + B) + C$ | $A \cdot (B \cdot C) = (A \cdot B) \cdot C$ |
| distributive law | $A \cdot (B + C) = (A \cdot B) + (A \cdot C)$ | $A + (B \cdot C) = (A + B)(A + C)$ |
| absorption law | $A + (A \cdot B) = A$ | $A \cdot (A + B) = A$ |
| | $A + (\overline{A} \cdot B) = A + B$ | $A \cdot (\overline{A} + B) = A \cdot B$ |
| De Morgan's theorem | $\overline{(A + B)} = \overline{A} \cdot \overline{B}$ | $\overline{(A \cdot B)} = \overline{A} + \overline{B}$ |

Eq. 75.35 for the NOT, AND, and OR operations, respectively.

$$\overline{(A \cdot A)} = \overline{A} + \overline{A} = \overline{A} \qquad 75.33$$

$$\overline{\overline{(A \cdot B)} \cdot \overline{(A \cdot B)}} = (A \cdot B) + (A \cdot B)$$
$$= A \cdot B \qquad 75.34$$

$$\overline{\overline{(A \cdot A)} \cdot \overline{(B \cdot B)}} = (A \cdot A) + (B \cdot B)$$
$$= A + B \qquad 75.35$$

Using the dual of Eq. 75.33 through Eq. 75.35 proves the universality of the NOR operation. The dual of each of the items in Table 75.2 is obtained using the *principle of duality*. The dual of an expression is obtained by applying the following steps, without changing the variables involved.

*step 1:* If present, change zeros to ones and ones to zeros.

*step 2:* Change the original OR operations to AND.

*step 3:* Change the original AND operations to OR.

## 7. MINTERMS AND MAXTERMS

A *product term* is a function defined by a logical AND set of terms with either a variable, $x_i$, or its negation, $\overline{x_i}$. A product term where all the variables appear, but only once, is called a *minterm*. The name occurs because the minterms take on a value of one for only one of the $2^n$ combinations of an $n$-tuple, that is, for the input variables, and a value of zero for all other combinations. Expressed in terms of true/false notation, the minterms give the rows of a truth table in which the function is true. Minterms for a three-variable function are shown in Table 75.5.

A *sum term* is a function defined by a logical OR set of terms with either a variable $x_i$ or its negation $\overline{x_i}$. A sum term where all the variables appear, but only once, is called a *maxterm*. The name occurs because the maxterms take on a value of zero for only one of the $2^n$ combinations of an $n$-tuple, that is, for the input variables, and a value of one for all other combinations.

**Table 75.5** Minterms and Maxterms

| decimal row number | binary input combinations $ABC$ | minterm (product term) | maxterm (sum term) |
|---|---|---|---|
| 0 | 000 | $m_0 = \overline{A}\,\overline{B}\,\overline{C}$ | $M_0 = A + B + C$ |
| 1 | 001 | $m_1 = \overline{A}\,\overline{B}C$ | $M_1 = A + B + \overline{C}$ |
| 2 | 010 | $m_2 = \overline{A}B\overline{C}$ | $M_2 = A + \overline{B} + C$ |
| 3 | 011 | $m_3 = \overline{A}BC$ | $M_3 = A + \overline{B} + \overline{C}$ |
| 4 | 100 | $m_4 = A\overline{B}\,\overline{C}$ | $M_4 = \overline{A} + B + C$ |
| 5 | 101 | $m_5 = A\overline{B}C$ | $M_5 = \overline{A} + B + \overline{C}$ |
| 6 | 110 | $m_6 = AB\overline{C}$ | $M_6 = \overline{A} + \overline{B} + C$ |
| 7 | 111 | $m_7 = ABC$ | $M_7 = \overline{A} + \overline{B} + \overline{C}$ |

Expressed in terms of true/false notation, the maxterms give the rows of a truth table in which the function is false. Maxterms for a three-variable function are shown in Table 75.5.

### Example 75.4

Consider the truth table for the two-variable function $F(A, B)$ shown. List the minterms.

| $A$ | $B$ | $F(A, B)$ |
|---|---|---|
| 0 | 0 | 0 |
| 0 | 1 | 1 |
| 1 | 0 | 1 |
| 1 | 1 | 0 |

*Solution*

The minterms occur where the value of the function is true (that is, one). Thus, the minterms are

$$m_1 = \overline{A}B$$
$$m_2 = A\overline{B}$$

### Example 75.5

For the function in Ex. 75.4, list the maxterms.

*Solution*

The maxterms occur where the value of the function is false (that is, zero). Thus, the maxterms are

$$M_0 = A + B$$

$$M_3 = \overline{A} + \overline{B}$$

## 8. CANONICAL REPRESENTATION OF LOGIC FUNCTIONS

Any arbitrary function described by a truth table can be represented using minterms and maxterms. For example, the *minterm form* of the three-variable function in Table 75.5 is

$$F(A, B, C) = \sum_{i=0}^{7} m_i \qquad \textit{75.36}$$

When Eq. 75.4 is expanded, it is referred to as the *canonical sum-of-product form* (SOP) of the function. For example,

$$F(A, B, C) = m_0 + m_1 + m_2 + m_3 + m_4 + m_5$$
$$+ m_6 + m_7$$
$$= \overline{A}\,\overline{B}\,\overline{C} + \overline{A}\,\overline{B}\,C + \overline{A}\,B\,\overline{C} + \overline{A}\,B\,C$$
$$+ A\overline{B}\,\overline{C} + A\overline{B}\,C + A\,B\,\overline{C} + A\,B\,C$$
$$\textit{75.37}$$

The terms are called *minterms* because they have a value of one for only one of the $2^i$ possible values of Eq. 75.4. (These are the terms of a function that are true; that is, logic one.)

The *maxterm form* of the three-variable function in Table 75.5 is

$$F(A, B, C) = \prod_{i=0}^{7} M_i \qquad \textit{75.38}$$

When Eq. 75.6 is expanded, it is referred to as the *canonical product-of-sum form* (POS) of the function. For example,

$$F(A, B, C) = M_0 M_1 M_2 M_3 M_4 M_5 M_6 M_7$$
$$= (A + B + C)(A + B + \overline{C})$$
$$\times (A + \overline{B} + C)(A + \overline{B} + \overline{C})$$
$$\times (\overline{A} + B + C)(\overline{A} + B + \overline{C})$$
$$\times (\overline{A} + \overline{B} + C)(\overline{A} + \overline{B} + \overline{C}) \qquad \textit{75.39}$$

The terms are called *maxterms* because they have a value of zero for only one of the $2^i$ possible values of Eq. 75.6. (These are the terms of a function that are false; that is, logic zero.)

The term "canonical" in this context means a simple or significant form of an equation or function. Canonical

forms contain all the input variables in each term. Thus, SOP and POS forms may not be the simplest expressions with regard to realizing these functions in terms of logic gates and may require minimization.

### Example 75.6

Write the function in Ex. 75.4 in SOP form.

*Solution*

The minterms were determined in Ex. 75.4. The SOP form, using the format of Eq. 75.5 is

$$F(A, B) = \overline{A}B + A\overline{B}$$

### Example 75.7

Write the function of Ex. 75.4 in POS form.

*Solution*

The maxterms were determined in Ex. 75.5. The POS form, using the format of Eq. 75.7 is

$$F(A, B) = (A + B) + (\overline{A} + \overline{B})$$

### Example 75.8

Consider the truth table shown. What are the (a) SOP and (b) POS forms of the function?

| $X$ | $Y$ | $Z$ | $F(X, Y, Z)$ |
|-----|-----|-----|--------------|
| 0 | 0 | 0 | 0 |
| 0 | 0 | 1 | 1 |
| 0 | 1 | 0 | 1 |
| 0 | 1 | 1 | 0 |
| 1 | 0 | 0 | 1 |
| 1 | 0 | 1 | 0 |
| 1 | 1 | 0 | 0 |
| 1 | 1 | 1 | 1 |

*Solution*

(a) Minterms (logic value one) exist in rows 1, 2, 4, and 7. The SOP form is

$$F(X, Y, Z) = m_1 + m_2 + m_4 + m_7$$
$$= \overline{X}\,\overline{Y}Z + \overline{X}\,Y\overline{Z} + X\overline{Y}\,\overline{Z} + X\,Y\,Z$$

(b) Maxterms (logic value zero) exist in rows 0, 3, 5, and 6. The POS form is

$$F(X, Y, Z) = M_0 M_3 M_5 M_6$$
$$= (X + Y + Z)(X + \overline{Y} + \overline{Z})$$
$$\times (\overline{X} + Y + \overline{Z})(\overline{X} + \overline{Y} + Z)$$

## 9. CANONICAL REALIZATION OF LOGIC FUNCTIONS: SOP

*Realization* is the process of creating a logic diagram from the truth table for a given function. Any scalar function can be represented in either canonical sum-of-product or product-of-sum form.[3] Thus, any function can be realized by a two-level logic circuit. The connection points are shown with a dot. Undotted line crossings indicate nonconnecting crossings. It is normally assumed that the input variables and their complements are available.

The SOP form of the function in Ex. 75.8 indicates a two-level AND-OR realization, which is shown in Fig. 75.3.

**Figure 75.3** SOP Realization

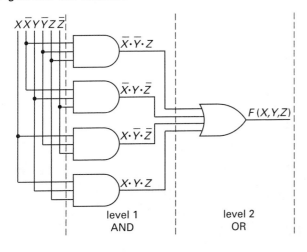

In Fig. 75.3, the AND gates are in logic level one. The OR gate is in logic level two.

The same realization can occur using only NAND gates. Consider the equivalence shown in Fig. 75.4.

The NAND gate can be represented in two ways: the standard way or as an OR gate with the inputs inverted. Using this equivalence, the SOP form of the function in Ex. 75.8 can be realized with NAND gates only, as shown in Fig. 75.5.

The realization in Fig. 75.5 is AND-OR. This can be seen by considering the inversions represented by the bubbles (small circles) on the output of the level-one logic to cancel with the inversions on the input to the level-two logic.[4]

**Figure 75.4** NAND Equivalence

| X | Y | Z | $\overline{X \cdot Y \cdot Z}$ | $\overline{X} + \overline{Y} + \overline{Z}$ |
|---|---|---|---|---|
| 0 | 0 | 0 | 1 | 1 |
| 0 | 0 | 1 | 1 | 1 |
| 0 | 1 | 0 | 1 | 1 |
| 0 | 1 | 1 | 1 | 1 |
| 1 | 0 | 0 | 1 | 1 |
| 1 | 0 | 1 | 1 | 1 |
| 1 | 1 | 0 | 1 | 1 |
| 1 | 1 | 1 | 0 | 0 |

(a) truth table

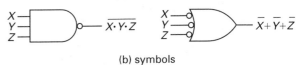

(b) symbols

**Figure 75.5** SOP NAND-NAND Realization

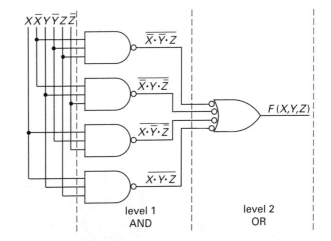

## 10. CANONICAL REALIZATION OF LOGIC FUNCTIONS: POS

The POS form of the function in Ex. 75.8 indicates a two-level OR-AND realization, which is shown in Fig. 75.6.

In Fig. 75.6, the OR gates are in logic level one. The AND gate is in logic level two.

The same realization can occur using only NOR gates. Consider the equivalence shown in Fig. 75.7.

The NOR gate can be represented in two ways: the standard way or as an AND gate with the inputs inverted. Using this equivalence, the POS form of the function in Ex. 75.8 can be realized with NOR gates only, as shown in Fig. 75.8.

The realization in Fig. 75.8 is OR-AND. This can be understood by considering the inversions represented by the bubbles on the output of the level-one logic to cancel with the inversions on the input to the level-two logic.[5]

---

[3]Canonical forms of logic realization are useful in the design of read-only memory (ROM) and programmable logic array (PLA) circuits.
[4]Figure 75.5 is not a minimum realization because the function is the XOR of X, Y, and Z and could be realized with two two-input XOR gates.

[5]Figure 75.8 is also not a minimum realization because the function is the XOR of X, Y, and Z and could be realized with two two-input XOR gates.

**Figure 75.6** POS Realization

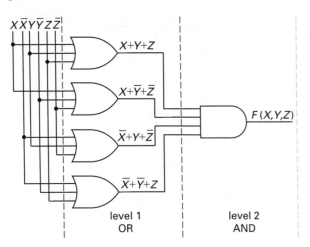

level 1
OR

level 2
AND

**Figure 75.8** POS NOR-NOR Realization

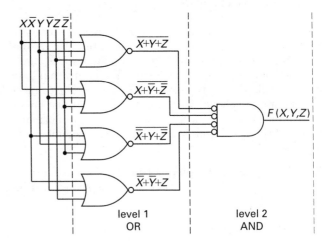

level 1
OR

level 2
AND

**Figure 75.7** NOR Equivalence

| X | Y | Z | $\overline{X+Y+Z}$ | $\overline{X}\cdot\overline{Y}\cdot\overline{Z}$ |
|---|---|---|---|---|
| 0 | 0 | 0 | 1 | 1 |
| 0 | 0 | 1 | 0 | 0 |
| 0 | 1 | 0 | 0 | 0 |
| 0 | 1 | 1 | 0 | 0 |
| 1 | 0 | 0 | 0 | 0 |
| 1 | 0 | 1 | 0 | 0 |
| 1 | 1 | 0 | 0 | 0 |
| 1 | 1 | 1 | 0 | 0 |

(a) truth table

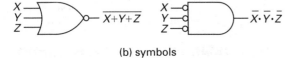

(b) symbols

Computer
Programming

# Topic XVI: Atomic Theory

Chapter

# 76 Atomic Theory and Nuclear Engineering

## Nomenclature

| | | |
|---|---|---|
| $a$ | acceleration | $m/s^2$ |
| $A$ | activity | Bq |
| $A$ | area | $m^2$ |
| $A$ | atomic weight | kg/kmol |
| $B$ | magnetic flux density | T |
| $c$ | speed of light | m/s |
| $D$ | absorbed dose | Gy |
| $e$ | charge on electron | C |
| $E$ | electric field | V |
| $E$ | energy | J |
| $f$ | thermal utilization | – |
| $f$ | switching frequency | Hz |
| $\bar{f}$ | wave number | 1/m |
| $F$ | force | N |
| $h$ | Planck's constant | J·s |
| $\hbar$ | reduced Planck's constant $(h/2\pi)$ | J·s |
| $H$ | equivalent dose | rem |
| $H$ | magnetic field | A/m |
| $J$ | beam intensity | $1/m^2 \cdot s$ |
| $k$ | constant in electrostatic equation | $J \cdot m/C^2$ |
| $k$ | production rate | 1/s |
| $k$ | relativistic factor | – |
| $k_\infty$ | infinite-$k$ | – |
| $l$ | azimuthal quantum number | – |
| $L$ | length | m |
| $m_{1/2}$ | half-value mass | $kg/m^2$ |
| $m$ | mass | kg |
| $m_l$ | magnetic quantum number | – |
| $m_s$ | electron spin quantum number | – |
| $n$ | number of moles | – |
| $n$ | principal quantum number | – |
| $N$ | neutron number | – |
| $N$ | number of atoms | – |
| $N$ | number of atoms per unit volume | $1/m^3$ |
| $N_A$ | Avogadro's number | 1/mol |
| $p$ | linear momentum | $kg \cdot m/s$ |
| $p$ | resonance escape probability | – |
| $q$ | charge | C |
| $Q$ | binding energy | J |
| $r$ | radius | m |
| $R_\infty$ | Rydberg constant | 1/m |
| $t$ | time | s |
| $t_{1/2}$ | half-life | s |
| $T$ | temperature | K |
| $v$ | velocity | m/s |
| $W_r$ | radiation weighting factor | rem/rd (Sv/Gy) |

**Atomic Theory**

| | | |
|---|---|---|
| $x$ | distance | m |
| $x_{1/2}$ | half-value thickness | m |
| $Z$ | atomic number | – |

## Symbols

| | | |
|---|---|---|
| $\Delta$ | mass defect | kg |
| $\epsilon$ | error or uncertainty | various |
| $\epsilon$ | fast fission factor | – |
| $\epsilon$ | permittivity | $C^2/N{\cdot}m^2$ |
| $\eta$ | multiplication constant | – |
| $\lambda$ | decay constant | 1/s |
| $\lambda$ | wavelength | m |
| $\mu$ | magnetic moment | J/T |
| $\mu_l$ | linear attenuation coefficient | 1/m |
| $\mu_m$ | mass attenuation coefficient | $m^2/kg$ |
| $\nu$ | frequency | Hz |
| $\rho$ | density | $kg/m^3$ |
| $\sigma$ | cross section | $m^2$ |
| $\Sigma$ | macroscopic cross section | 1/m |
| $\tau$ | average life | s |
| $\phi$ | work function | J |
| $\Psi$ | wave function | m |
| $\omega$ | angular velocity | rad/s |

## Subscripts

| | |
|---|---|
| 0 | original |
| $\alpha$ | alpha decay |
| $a$ | absorption |
| $A$ | Avogadro |
| $B$ | Bohr |
| $c$ | capture |
| $es$ | elastic scattering |
| $E$ | energy |
| $f$ | fission |
| $is$ | inelastic scattering |
| $l$ | magnetic |
| $p$ | momentum or proton decay |
| $s$ | scattering or spin |
| $t$ | time, at time $t$, or total |
| $v$ | at velocity v |
| $x$ | position |
| $w$ | wave |

## 1. MASS, CHARGE, AND ENERGY UNITS

Calculations involving atomic and nuclear reactions use the SI system. Therefore, the kilogram is the unit of mass. The *unified atomic mass unit* (amu), also known simply as a *mass unit* (u), is approximately equal to $1.66054 \times 10^{-27}$ kg.[1]

The standard SI unit of charge is the coulomb (C). However, it is convenient also to define the *electronic charge unit*, equal to $1.60218 \times 10^{-10}$ C, the charge on an electron.[2]

The standard SI energy unit is the joule (J). However, since charged particles are accelerated in an electric field, it is convenient to define the *electron volt* (eV) equal to the energy required to take one electronic charge across a potential of one volt.

Since an electron volt is small, units of MeV and GeV are used frequently.[3] Other units of energy occasionally encountered are listed in Table 76.1.

**Table 76.1** *Units of Energy*

| name | number equivalent to one eV |
|---|---|
| atomic unit[*] | 0.03675 |
| BeV | $10^{-9}$ |
| cal (IT) | $3.827 \times 10^{-20}$ |
| erg | $1.60218 \times 10^{-12}$ |
| GeV | $10^{-9}$ |
| joule | $1.60218 \times 10^{-19}$ |
| kcal (mean) | $3.827 \times 10^{-23}$ |
| MeV | $10^{-6}$ |

[*]This is not the same as the *atomic mass unit*. It is also known as *Hartree energy*.

## 2. MASS AND ENERGY EQUIVALENCE

*Einstein's equation* specifies the equivalence of mass and energy.[4] Energy, $E$, is in joules when mass, $m$, is in kilograms. The speed of light in a vacuum, $c$, is approximately $2.9979 \times 10^8$ m/s.

$$E = mc^2 \qquad 76.1$$

Equation 76.1 says that one kilogram is equivalent to $8.98755 \times 10^{16}$ J. Since one atomic mass unit is equal to $1.66054 \times 10^{-27}$ kg and one electron volt is equal to $1.60218 \times 10^{-19}$ J, one atomic mass unit corresponds to approximately 931.494 MeV. Table 76.2 lists other common conversion factors.

## 3. MAXWELL'S ELECTROMAGNETIC THEORY

In 1864, James Clerk Maxwell was the first to recognize the structure of electromagnet waves, predicting in his *electromagnetic theory of light* that such waves (including light) exist simultaneously as high-frequency electrical and magnetic waves of the same magnitude. The oscillations of charged particles (as in a high-frequency oscillator circuit) produce electric and magnetic fields. The electric field is always at right angles to the magnetic field, and both are at right angles to the direction of propagation. This is illustrated in Fig. 76.1.

---

[1]Based on an atomic mass of carbon-12 equal to 12 adopted in 1961. The older oxygen-16 scale is obsolete. The atomic mass unit (in grams) is the reciprocal of Avogadro's number.

[2]One of the earliest measurements of the electron charge came from Millikan's oil drop experiment.

[3]The once-common abbreviation BeV stands for "billion electron volts." However, the prefix "B" is not a standard SI prefix, and GeV is the proper SI symbol for $10^9$ electron volts.

[4]It is important to recognize the difference between *equivalence* and *duality*. Einstein's theory says matter can be converted into energy and vice versa. It does not say that matter exists simultaneously as energy.

**Table 76.2** Mass and Energy Conversion Factors*
(multiply ↓ to obtain →)

|      | kg                      | amu                     | J                       | eV                      |
|------|-------------------------|-------------------------|-------------------------|-------------------------|
| kg   | 1                       | $6.0221 \times 10^{26}$ | $8.98755 \times 10^{16}$ | $5.6096 \times 10^{35}$ |
| amu  | $1.66054 \times 10^{-27}$ | 1                     | $1.4924 \times 10^{-10}$ | $9.3149 \times 10^{8}$  |
| J    | $1.1127 \times 10^{-17}$  | $6.7005 \times 10^{9}$  | 1                       | $6.2415 \times 10^{18}$ |
| eV   | $1.7827 \times 10^{-36}$  | $1.0735 \times 10^{-9}$ | $1.60218 \times 10^{-19}$ | 1                     |

\* based on $c = 2.997925 \times 10^8$ m/s

1 eV = $1.60218 \times 10^{-19}$ J

1 amu = $1.66054 \times 10^{-27}$ kg

1 kg = $8.98755 \times 10^{16}$ J

**Figure 76.1** Propagation of an Electromagnetic Wave

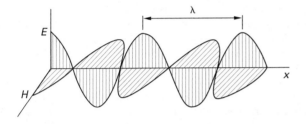

Electromagnetic radiation with frequencies around $10^{19}$ Hz and wavelengths comparable to the interatomic spacing in solids are classified as *X-rays*. (X-rays are produced when one of the outer electrons in a heavy atom jumps down to fill an empty inner electron position.) Electromagnetic radiation of even higher frequency is known as *gamma radiation*. The highest-energy gamma rays are observed as *cosmic rays*.

Since the speed of light, $c$, in $\nu = c/\lambda$ is a constant, it can be omitted (as being understood), resulting in a definition of the *wave number, $\bar{f}$*.

$$\bar{f} = \frac{1}{\lambda}$$ 76.2

## 4. ENERGY EMISSION FROM ACCELERATING CHARGED PARTICLES

As a consequence of the Maxwell laws, an accelerating charged particle radiates energy. According to this theory, an electron or other charged particle following a curved path radiates energy because it is accelerating toward the instantaneous center of rotation. As a consequence of this radiation, the orbit will decay, with the particle gradually spiraling inward.

## 5. ENERGY QUANTA

Most observed properties of light and other radiant energy are consistent with a wave nature. In interactions with matter, however, electromagnetic radiation behaves as though it consists of discrete pieces of energy known as *quanta, Planck quanta, bundles, packets, corpuscles,* and *photons*. *Quantum theory* explains the interaction phenomena observed between matter and these waves that were not explained by Maxwell's theory.

The amount of energy in a quantum is not constant but depends on the wave frequency.[5] In Eq. 76.3, $h$ is *Planck's constant*, which is approximately equal to $6.626069 \times 10^{-34}$ J·s, and $\nu$ is the frequency of the electromagnetic radiation.[6,7]

$$E = h\nu = \frac{hc}{\lambda}$$ 76.3

Quantum theory is useful in explaining the energy that leaves a system when the system changes its atomic state. In that case, the quantum energy is equal to the change in the system energy.

$$E_{\text{quantum}} = E_{2,\text{system}} - E_{1,\text{system}} = h\nu$$ 76.4

It is not difficult to calculate the wavelength of a photon with energy $E$. If $E$ and $\lambda$ are in electron volts and meters, respectively, the product $hc$ has a value of approximately $1.24 \times 10^{-6}$ m·eV. (Relativistic energies must be used for high-speed particles.)

$$\lambda = \frac{hc}{E}$$ 76.5

## 6. PHOTOELECTRIC EFFECT

The *photoelectric effect* is the phenomenon whereby a short-wavelength photon (e.g., gamma radiation, X-ray, etc.) hitting an atom on the surface of a substance causes an electron to be ejected.[8] An electrical current can be generated in this manner. Some solids (e.g., potassium and cesium) are photoelectric in the visible region. But, for a specific substance, the photon wavelength must be less than a critical value (in the ultraviolet region for most metals). The corresponding frequency is known as the *threshold frequency*.

Albert Einstein reasoned that an amount of energy—known as the *work function, $\phi$*—is needed to remove the electron from the surface, with the remainder of the incident photon's energy contributing to the kinetic

[5]A wave concept (i.e., frequency) is used here to define the radiant energy as a discrete unit. Such inconsistencies support the *dual nature* or *duality* of energy as being both a wave and a particle.
[6]The traditional nuclear symbol for frequency, $\nu$, is the same as frequency, $f$.
[7]The symbol $\hbar$, equal to $h/2\pi$, is also known as Planck's constant.
[8]This is different from the *photovoltaic effect*, which is the generation of a voltage potential between a thin film of one substance (e.g., selenium) deposited on the surface of another substance (e.g., iron).

Atomic Theory

energy of the electron. If the photon energy is less than the work function, an electron will not be ejected.

$$h\nu = \phi + \tfrac{1}{2}mv^2 \qquad 76.6$$

## 7. ANGULAR MOMENTUM QUANTA

Atomic theory recognizes that angular momentum, $mvr$, is quantized with a fundamental unit of $h/2\pi$.[9] The number of momentum quanta an electron can have around a nucleus is $nh/2\pi$. The number $n$ is known as the *principal quantum number*.

$$mvr = \frac{nh}{2\pi} = n\hbar \qquad 76.7$$

The neutron, proton, and electron each have, by definition, angular momentum of $h/4\pi$, a quantity referred to as "spin of one-half."

## 8. RUTHERFORD'S EXPERIMENT

In 1911, Ernest Rutherford used collimated alpha particles from radium to bombard gold foil. Most of the incident beam passed through the foil; however, a small number of particles was deflected, some through every large angles approaching 180 degrees. Rutherford concluded that atoms are massive positive charges concentrated in a *nucleus* surrounded by mostly empty space.

By neglecting the interaction of the alpha particles and electrons, and by assuming that the nucleus remained stationary, Rutherford calculated the electrostatic radius of the atom to be approximately $10^{-14}$ m. He reasoned that the kinetic energy of an alpha particle that was reflected 180 degrees would have to be exactly balanced by the *Coulomb (electrostatic) repulsion* from the target nucleus. In Eq. 76.8, $r$ is the *effective electrostatic radius*, and $2e$ and $Ze$ are the charges on the alpha particle and gold nucleus, respectively.

$$\tfrac{1}{2}mv^2 = \frac{(2e)(Ze)}{4\pi\epsilon r} \qquad 76.8$$

## 9. ATOMIC PARTICLES

Electrons, protons, and neutrons are the primary atomic particles. (The negative electron is also known as a *negatron*.) The proton mass is approximately 1836 times the electron mass. Other simple particles important in atomic theory are the *deuteron* (nucleus of deuterium) and *alpha particle* (nucleus of helium). Table 76.3 lists the approximate mass and charge of these and other particles.

---

[9]The symbol **h** is sometimes used for the vector angular momentum. However, in this chapter, $h$ is Planck's constant, and no symbol is used for angular momentum.

*Table 76.3* Properties of Atomic Particles (based on C-12 = 12)

| particle | rest mass (kg) | rest mass (u) | charge |
|---|---|---|---|
| alpha, $\alpha$ | $6.644656 \times 10^{-27}$ | 4.001506 | +2 |
| beta, $\beta$, e | $9.109382 \times 10^{-31}$ | 0.0005485979 | −1 |
| deuteron, d | $3.34358 \times 10^{-27}$ | 2.01410 | +1 |
| electron, e | $9.109382 \times 10^{-31}$ | 0.0005485979 | −1 |
| gamma ray, $\gamma$ | 0 | 0 | 0 |
| neutrino, $\nu$ | 0 | 0 | 0 |
| neutron, n | $1.674927 \times 10^{-27}$ | 1.0086649 | 0 |
| positron, +e | $9.109382 \times 10^{-31}$ | 0.0005485979 | +1 |
| proton, p | $1.672621 \times 10^{-27}$ | 1.00727647 | +1 |

(Multiply charge given by $1.60218 \times 10^{-19}$ to obtain coulombs.)
(Multiply mass units (u) by $1.66054 \times 10^{-27}$ to obtain kilograms.)

## 10. NUCLEONS AND ISOTOPES

The particles (i.e., protons and neutrons) in the nucleus of an atom are known as *nucleons*. The number of nucleons is the *mass number*, essentially the same as the *atomic weight*. The number of each type of nucleon can be determined from the *atomic number*, $Z$, and atomic weight, $A$. The number of protons in the nucleus is equal to the atomic number. The number of neutrons (the *neutron number*) is equal to the difference $A - Z$. In a neutral (un-ionized) atom, the number of electrons is equal to the number of protons, but the mass contributed by the electrons is very small.

There are several different shorthand ways of designating the nuclear structure, including E-A, $_ZE^A$, $_Z^AE$, and $_Z^AE_N$ where E is the element symbol, $A$ is the atomic weight, $Z$ is the atomic number, and $N$ is the number of neutrons. Generally, specifying the number of neutrons is superfluous, since $A = N + Z$.

Most elements exist in different forms known as *isotopes*, differing in the number of neutrons each has. Isotopes have the same atomic numbers but different atomic weights (e.g., $_6^{14}C$ and $_6^{12}C$). For example, hydrogen has three isotopes: regular hydrogen (H-1), deuterium (H-2), and tritium (H-3). Typically, two or more of the isotopes exist in nature simultaneously and will be intermixed in a naturally occurring sample. The atomic weight of each an element may be reported for a specific isotope or as an average weighted by the *relative abundance* (probability of occurrence) of each isotope. In chemical reactions, different isotopes typically react at different rates, but not in different manners. In nuclear reactions, however, different isotopes behave in decidedly different manners.

### Example 76.1

How many protons, neutrons, and electrons are in the carbon isotope $_6^{14}C$?

*Solution*

The atomic number is 6, so there are six protons and six electrons. The six protons contribute 6 to the atomic weight, and the difference is made up of $14 - 6 = 8$ neutrons.

## 11. NUMBER OF NUCLEI PER UNIT VOLUME

Equation 76.9 calculates the *molecular density*, the number of nuclei (atoms) per unit volume. $N_A$ is Avogadro's number, which is approximately $6.0221 \times 10^{23}$ atoms/mol.

$$N = \frac{\rho N_A}{A} \qquad\qquad 76.9$$

## 12. STRONG AND WEAK INTERACTIONS

The nucleus consists of protons and neutrons. On the basis of electrostatics solely, the protons should repel one another and the nucleus should disintegrate. However, there is a stronger attractive force between the nucleons that is roughly 100 times stronger than the electrostatic repulsion. This *strong interaction*, also called the *strong nuclear force*, is charge-independent (i.e., is the same for proton-proton, neutron-neutron, and proton-neutron pairs). The strong force acts only at short ranges (approximately $10^{-14}$ m between closest nucleons), is noncentral (i.e., does not follow the *inverse square law*), and exhibits *saturation* (i.e., once a stable structure is achieved, additional nucleons are not strongly held).

The *weak interaction* (*weak nuclear force, Fermi interaction, beta interaction*) is a force that is several orders of magnitude weaker than electrostatic interaction and affects all particles equally, regardless of mass. Its effect is to transform all particles into electrons and neutrinos through $\beta$ decay processes. The *law of conservation of heavy particles* prohibits such decay in protons and neutrons.

## 13. NUCLEON BINDING ENERGY

As stated in Sec. 76.12, nucleons are held together by the *strong nuclear force*. The *binding energy*, $Q$, can be found as the energy equivalent (see Sec. 76.2) of the nucleus *mass defect*, $\Delta$—the difference between the total nucleus mass and the mass of the nucleons. (The units of $Q$ and $\Delta$ are somewhat interchangeable, being kg, amu, J, and MeV as is convenient). The *binding energy per nucleon*, approximately 8 MeV for all but the lighter elements, is obtained by dividing this energy by the number of nucleons.

$$Q = \Delta \times c^2 \qquad\qquad 76.10$$

## Example 76.2

What is the binding energy per nucleon in a deuteron (deuterium nucleus) in (a) joules and (b) electron volts?

*Solution*

A deuterium nucleus consists of a proton and a neutron. (The electron is not a nucleon). The mass defect is the difference between the total nucleus mass and the sum of the proton and neutron masses found in Table 76.3. The negative sign indicates that energy would have to be added to break the deuterium nucleus into a proton and a neutron.

$$\begin{aligned}
\Delta &= m_d - (m_p + m_n) \\
&= \left(3.34358 - (1.672621 + 1.674927)\right) \times 10^{-27} \\
&= -0.003968 \times 10^{-27} \text{ kg}
\end{aligned}$$

(a) There are two nucleons. Therefore, the binding energy per nucleon is

$$\begin{aligned}
Q &= \tfrac{1}{2}\Delta \times c^2 \\
&= (0.5)\left(0.003968 \times 10^{-27} \text{ kg}\right)\left(3 \times 10^8 \ \tfrac{\text{m}}{\text{s}}\right)^2 \\
&= 1.7856 \times 10^{-13} \text{ J}
\end{aligned}$$

(b) Converting to electron volts,

$$Q = \frac{1.7856 \times 10^{-13} \text{ J}}{1.60218 \times 10^{-19} \ \frac{\text{J}}{\text{eV}}} = 1.11 \times 10^6 \text{ eV}$$

## 14. STABILITY OF NUCLEI

Some nuclei are unstable and will disintegrate spontaneously by a process known as *radioactivity*. The instability is due to too many or too few neutrons in the nucleus. While the neutrons have zero electrostatic effect, they contribute to the strong nuclear force needed to balance proton repulsion.

If a nucleus has too many neutrons, one neutron may spontaneously transform into a proton. An electron is also released to retain charge neutrality. (In addition, gamma radiation may be emitted to decrease the binding energy.) This electron emission is known as $-\beta$ *decay*.

$$ {}_{0}^{1}\text{n} \rightarrow {}_{1}^{1}\text{p} + {}_{-1}^{0}\text{e} \qquad\qquad 76.11$$

If the nucleus has too few neutrons, a proton transforms into a neutron with a positron emission. This is known as $+\beta$ *decay*.

$$ {}_{1}^{1}\text{p} \rightarrow {}_{0}^{1}\text{n} + {}_{+1}^{0}\text{e} \qquad\qquad 76.12$$

**Atomic Theory**

$\alpha$ *decay* decreases the number of both protons and neutrons by two and may also result in a stable nucleus.

$$2{}_0^1n + 2{}_1^1p \rightarrow {}_2^4He \qquad 76.13$$

Figure 76.2 is a general plot of stable isotopes. (A complete diagram showing all known stable and unstable isotopes is called a *chart of the nuclides*.) Nuclei above the curved line will have too many protons to be stable and must move downward (by electron capture, positron emission, alpha emission, or spontaneous fission) to reach the stable zone.[10] (Moving to the right is possible but would amount to a neutron capture, which is unlikely since neutrons are neutral.) Nuclei below the line will have too many neutrons and must move upward (by electron emission) or to the left to reach the stable zone.

**Figure 76.2** *Stable Isotopes*

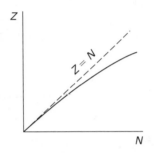

Figure 76.3 summarizes the various types of reactions required to move in a particular direction on a chart of the nuclides.

**Figure 76.3** *Relative Location versus Decay Process*

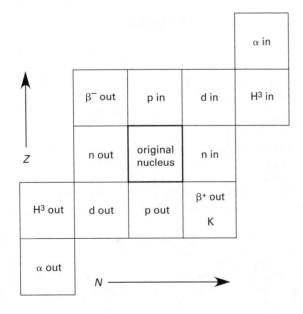

[10]Emission of alpha particles and spontaneous fission are important only for heavy elements.

## 15. RADIOACTIVITY

The disintegration of radioactive isotopes is described by a negative exponential law. The rate of radioactive decay is specified by a *decay constant*, $\lambda$ (not to be confused with wavelength, which uses the same symbol), that is essentially independent of the local environment (i.e., pressure, temperature, etc.). The decay constant can be used to calculate the *half-life*, $t_{1/2}$, the time required for half of the original atoms to decay. (See Table 76.4) Half-life is a function of the decay constant. (The half-life should not be confused with the *average life (mean life expectancy)*, $\tau$.)

$$t_{1/2} = \frac{-\ln(0.5)}{\lambda} \approx \frac{0.6931}{\lambda} \qquad 76.14$$

$$\tau = \frac{1}{\lambda} \qquad 76.15$$

**Table 76.4** *Approximate Half-Lives of Selected Isotopes*

| isotope | $t_{1/2}$ (years) | $\lambda$ (years$^{-1}$) |
|---------|-------------------|--------------------------|
| C-14    | 5730              | $1.21 \times 10^{-4}$    |
| Co-60   | 5.27              | 0.131                    |
| H-3     | 12.3              | $5.62 \times 10^{-2}$    |
| Pu-239  | $2.41 \times 10^4$ | $2.87 \times 10^{-5}$   |
| Ra-226  | 1602              | $4.33 \times 10^{-4}$    |
| Th-232  | $1.40 \times 10^{10}$ | $4.93 \times 10^{-11}$ |
| U-234   | $2.46 \times 10^5$ | $2.82 \times 10^{-6}$   |
| U-235   | $7.04 \times 10^8$ | $9.85 \times 10^{-10}$  |
| U-238   | $4.46 \times 10^9$ | $1.55 \times 10^{-10}$  |

The decay constant can be calculated if the mass (activity, number of atoms, etc.) is known at two times.

$$\lambda = \frac{-\ln\left(\dfrac{m_t}{m_0}\right)}{t - t_0} \qquad 76.16$$

The number of spontaneous disintegrations per second (dis sec$^{-1}$, dis/s, dps, or just 1/s) is the *activity*, $A$, typically measured in units of *becquerels* (Bq) or *curies* (Ci). (1.0 Ci is equal to $3.700 \times 10^{10}$ Bq.) The fraction of atoms present at time $t$ is calculated from Eq. 76.18. Time, $t$, is normally in seconds, minutes, or years. However, it can also be given in half-lives if the decay constant has units of 1/half-life.

$$A_t = \lambda N_t = A_0 e^{-\lambda t} \qquad 76.17$$

$$\frac{A_t}{A_0} = \frac{m_t}{m_0} = \frac{N_t}{N_0} = e^{-\lambda t} \qquad 76.18$$

Atomic Theory

The length of time between two activities, masses, etc., is

$$t_2 - t_1 = \frac{-\ln\left(\dfrac{m_2}{m_1}\right)}{\lambda} \qquad \textit{76.19}$$

### Example 76.3

What will be the activity of 60 kg of Pu-239 (half-life of 24,100 years) after 2000 years?

*Solution*

From Eq. 76.14, the decay constant is

$$\lambda = \frac{0.6931}{t_{1/2}} = \frac{0.6931}{24{,}100 \text{ yr}}$$

$$= 2.876 \times 10^{-5} \text{ yr}^{-1}$$

The mass, number of moles, and number of atoms of Pu-239 after 2000 years will be

$$m_{2000} = m_0 e^{-\lambda t} = (60 \text{ kg}) e^{-(2.876 \times 10^{-5} \text{ 1/yr})(2000 \text{ yr})}$$

$$= 56.65 \text{ kg}$$

$$n_{2000} = \frac{m}{A} = \frac{56.65 \text{ kg}}{239 \dfrac{\text{kg}}{\text{kmol}}} = 0.2370 \text{ kmol}$$

$$N_{2000} = n \times N_A = (0.2370 \text{ kmol})\left(1000 \frac{\text{mol}}{\text{kmol}}\right)$$

$$\times \left(6.0221 \times 10^{23} \frac{\text{atoms}}{\text{mol}}\right)$$

$$= 1.427 \times 10^{26} \text{ atoms}$$

From Eq. 76.17, the activity after 2000 years will be

$$A_{2000} = \lambda N_{2000}$$

$$= \frac{\left(2.876 \times 10^{-5} \dfrac{1}{\text{yr}}\right)(1.427 \times 10^{26} \text{ atoms})}{\left(365 \dfrac{\text{d}}{\text{yr}}\right)\left(24 \dfrac{\text{h}}{\text{d}}\right)\left(3600 \dfrac{\text{s}}{\text{h}}\right)}$$

$$= 1.302 \times 10^{14} \text{ Bq}$$

## 16. RADIOCARBON DATING

Neutrons from cosmic rays combine with atmospheric nitrogen to form radioactive carbon, C-14, which has a half-life of 5730 years and a corresponding decay constant of $1.21 \times 10^{-4}$ 1/yr.[11]

$$^{14}_{7}\text{N} + ^{1}_{0}\text{n} \rightarrow ^{14}_{6}\text{C} + x^{1}_{1}\text{H} \qquad \textit{76.20}$$

Some of the resulting radioactive carbon eventually forms radioactive carbon dioxide and becomes part of

---

[11]The half-life of C-14 is 5730 ± 40 years, so different values can be reported in the literature.

the food chain. The concentration of radioactive carbon in living plants and animals is stable at approximately 1 part in $7.75 \times 10^{11}$ ($1.29 \times 10^{-12}$ parts) with an equilibrium activity of 0.255 Bq (i.e., 15 disintegrations per minute) per gram of carbon, but it decreases upon death since the decaying radioactive carbon is not replaced by breathing. The approximate length of time since death (i.e., "age") is found from Eq. 76.18 and the current concentration of activity of C-14.

Tritium is also produced by cosmic rays in the atmosphere. Since its half-life is approximately 12.3 years, it can be used to date more contemporary objects (groundwater, wine, etc.).

### Example 76.4

A 0.470 kg sample of charcoal recovered from the excavation of an ancient campfire has an activity of $0.5 \times 10^{-3}$ $\mu$Ci. What is its approximate age?

*Solution*

The current activity is

$$A_t = \frac{\left(0.5 \times 10^{-3} \ \mu\text{Ci}\right)\left(3.700 \times 10^{10} \ \dfrac{\text{Bq}}{\text{Ci}}\right)}{10^6 \ \dfrac{\mu\text{Ci}}{\text{Ci}}} = 18.5 \text{ Bq}$$

Since the equilibrium activity is 0.255 Bq per gram of carbon, Eq. 76.18 is

$$\frac{A_t}{A_0} = e^{-\lambda t}$$

$$\frac{18.5 \text{ Bq}}{(470 \text{ g})\left(0.255 \ \dfrac{\text{Bq}}{\text{g}}\right)} = e^{-(1.209 \times 10^{-4} \ 1/\text{yr})t}$$

$$t = \frac{-\ln(0.1544)}{1.209 \times 10^{-4} \ \dfrac{1}{\text{yr}}} = 15{,}455 \text{ 1/yr}$$

## 17. DECAY WITH IRRADIATION

If a radioactive isotope is *irradiated* (bombarded with the proper particles), it can be replenished. Since it will be disintegrating while it is being irradiated, there will be simultaneous decay and replenishment. An important special case is that in which a radioactive isotope is produced at some constant rate, $k$. With sufficiently prolonged irradiation, the disintegration rate approaches the production rate, $k$, a situation known as *saturation*. The differential equation and solution are

$$\frac{dN}{dt} = k - \lambda N \qquad \textit{76.21}$$

$$A_t = k(1 - e^{-\lambda t}) \qquad \textit{76.22}$$

## 18. EMISSION SPECTRA

It is well known that visible (polychromatic) light can be split into color bands by a prism. The colors obtained depend on the makeup of the incident light beam, since only the colors originally present can be separated out. Light from an electric gas discharge or arc can be split into different wavelengths also, with a diffraction grating commonly used to produce solid bright *spectral lines (spectra)*, not color bands.[12]

The separation and placement of these spectral lines is unique for each element, with the placement of each spectral line corresponding to a different discharge frequency. Some of the discharge frequencies correspond to invisible electromagnetic radiation, hence each element has infrared and ultraviolet as well as visible spectra.

*Emission spectra* are commonly created, measured, and photographed in a *spectrometer (spectroscope)*. The term *emission* indicates that the spectra are caused by the release of certain frequencies of energy (supplied by the electrical input to the spectrometer) when an electron moves to a lower orbit. *Absorption spectra* are similar, consisting of dark bands from energy of particular frequencies being absorbed when an electron moves to a higher orbit. (See Fig. 76.4.)

**Figure 76.4** *Balmer Hydrogen Spectra*

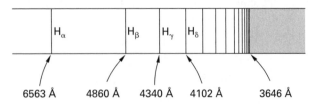

The *wave number*, $\overline{f}$ (see Sec. 76.3), of the visible spectra produced by a hydrogen gas discharge was shown by early investigators to be described by Eq. 76.23. $R_\infty$ is the empirical *Rydberg constant* for hydrogen and has a value of approximately $1.09737 \times 10^7$ m$^{-1}$. While Eq. 76.23 describes the hydrogen spectra, it was originally derived as a mathematical exercise, not from atomic structure theory.

$$\overline{f} = R_\infty \left( \frac{1}{(2)^2} - \frac{1}{n^2} \right) \qquad 76.23$$

## 19. BOHR HYDROGEN ATOM

In 1913, Niels Bohr considered the hydrogen atom to be a positively charged nucleus with a negatively charged orbiting electron and, in so doing, was able to calculate an electron velocity and radius that compared favorably

with experimental measurements. For stability, the outward centrifugal force has to equal the inward Coulomb attraction. (In Eq. 76.24, $k$ incorporates all of the constants in the formula for electrostatic attraction.) The more energy the electron had, the farther from the nucleus it would orbit.

$$\frac{m\mathrm{v}^2}{r} = \frac{ke^2}{r^2} \qquad 76.24$$

To support the theory and to eliminate certain complications, Bohr formulated four postulates.

*first postulate:* Electron orbits are discrete and nonradiating (see Sec. 76.4), and an electron may not remain between these orbits.

*second postulate:* The energy change experienced by an electron changing from one orbit to another is quantized.[13] (See Sec. 76.5.)

*third postulate:* Classical mechanics does not hold when the electron is between orbits.

*fourth postulate:* Angular momentum (see Sec. 76.7) is quantized.

## 20. ELECTRON BINDING ENERGY

To explain atomic spectra, Bohr reasoned that each bright band in the hydrogen spectrum is caused by energy emission when an electron drops from orbit $n_1$ to orbit $n_2$. Each time an electron makes a specific transition, it releases the same amount of energy and, from Eq. 76.26, the band has the same wavelength and frequency.

The energy needed to increase an electron's orbit is the *excitation energy*. The amount of energy needed to completely remove an electron from its nucleus depends on the orbit it is originally in. The farther the electron is from its nucleus, the less energy is required. The energy to remove the electron is the *binding (ionization) energy*.

The *binding energy* holding the electron to the nucleus (approximately 13.6 eV for hydrogen in its ground state) is the difference between the kinetic energy of the electron and the electrostatic potential energy of the electron-nucleus combination.[14] The attraction term is larger than the kinetic term; otherwise, the electron would escape.

$$E_{\text{binding}} = \tfrac{1}{2}m\mathrm{v}^2 - \left| \frac{ke^2}{r} \right| \qquad 76.25$$

Bohr derived an expression for the change in the atom's energy as a function of the change in its principal

---

[12]Spectra are generated by incandescent gases but not by incandescent solids. A tungsten filament will have no identifiable spectra, only a continuous spectrum.

[13]Free electrons, however, can have any reasonable amount of energy. Only when an electron is captured by a nucleus does its energy become quantized. The difference between its original energy and the quantized energy, which can be any amount, is released. This release shows up as a fuzzy area to the right of the spectral record. (See Fig. 76.4.)
[14]It is trivial to derive Eq. 76.25 from Eq. 76.24.

quantum number, $n$ (i.e., specifically when an electron jumped from the $n_1$ to the $n_2$ orbit). This energy change is related to the *wave number*, $\bar{f}$, by Eq. 76.26. Equation 76.27 is known as *Bohr's equation*, and the value of the constant $R_\infty$ compares favorably with the Rydberg constant.[15] (See Sec. 76.18.)

$$\bar{f} = R_\infty \left( \frac{1}{n_2^2} - \frac{1}{n_1^2} \right) \qquad 76.26$$

$$R_\infty = \frac{2\pi^2 m e^4 k^2}{ch^3} \qquad 76.27$$

The *ground state* (i.e., the electron orbit closest to the nucleus) corresponds to $n = 1$ in Eq. 76.26. An electron in any other state constitutes an *excited state*.

The Bohr model is able to explain the ultraviolet (*Lyman series*) and infrared (*Paschen, Brackett,* and *Pfund series*) spectra. In addition to explaining the spectra, it derives the Rydberg constant, calculates a satisfactory value for the hydrogen atom radius, and accounts quantitatively for ionization and excitation energies.

However, Bohr's model of the hydrogen atom, despite its correlation with experimental observations on hydrogen, is not sufficiently complex to explain the structure of other atoms, or for that matter, even other hydrogen details (i.e., *spectral line intensities* and *fine structure*).[16] Certain *Bohr model extensions* were more successful by postulating additional quantum numbers, elliptical orbits, use of relativistic mass, and arbitrary rules. However, like initial calculations of the Rydberg number, these extensions were postulated without underlying theory. They predicted what happened but did not explain why.

### 21. BOHR MAGNETON

An electron traveling in a circular orbit corresponding to principal quantum number $n$ around a nucleus will have a *magnetic moment*, $\mu$, given by Eq. 76.28. The magnetic moment is directed along the same line but reversed in direction from the angular momentum vector. Substituting the known values of $e$, $h$, and $m$ for a hydrogen atom's electron in the ground state ($n = 1$)

gives the value of $\mu = 9.274 \times 10^{-24}$ J·T, a quantity known as the *Bohr magneton*, $\mu_B$.

$$\mu = \frac{ehn}{4\pi m} \qquad 76.28$$

The magnetic moment is useful in explaining diamagnetism, but the magnetic moment of the spinning electron is needed to explain paramagnetism and ferromagnetism. The magnetic moment due to the spin of an electron is arbitrarily defined as 1 Bohr magneton as well.

### 22. WAVE NATURE OF AN ELECTRON

Louis de Broglie suggested a dual particle/wave character for the electron in 1924. It was particularly useful to have the electron be a standing wave, since this character eliminated the need to define orbits arbitrarily as nonradiating. (See Sec. 76.4 and Sec. 76.19.) The *de Broglie wavelength* of an electron follows from the requirement that the electron wave be a standing wave with average radius of $r_B$ (*Bohr's radius*) around the nucleus.[17] Since there must be an integral number of wavelengths,

$$2\pi r_B = n\lambda \qquad 76.29$$

Equation 76.30 calculates the de Broglie *wave velocity*. Since the linear velocity, $\mathrm{v}$, of a particle is always less than the speed of light, the wave velocity will always be greater than the speed of light.

$$\mathrm{v}_w = \nu\lambda = \frac{c^2}{\mathrm{v}} \qquad 76.30$$

Multiplying by $m/m$,

$$\mathrm{v}_w = \frac{mc^2}{m\mathrm{v}} = \frac{E}{p} = \frac{h\nu}{p} \qquad 76.31$$

Therefore, the de Broglie wavelength is

$$\lambda = \frac{h}{p} \qquad 76.32$$

### 23. HEISENBERG'S UNCERTAINTY PRINCIPLE

Werner Heisenberg stated that measuring the position or energy of a particle disturbs the particle. His *uncertainty principle* says the product of the error in the determination of the position, $x$, of any particle and its linear momentum, $p_x = m\mathrm{v}_x$, is on the order of Planck's constant, $h$ ($6.626068 \times 10^{-34}$ J·s). The more accurate the former, the less accurate will be the latter. A similar statement was made for the uncertainty in determining

---

[15]Actually, while Bohr's Rydberg constant closely matches empirical data, it contains a small error introduced by his assumption that the orbit is circular, which is equivalent to saying the nucleus has infinite mass. This error can be eliminated by using a *reduced electron mass* in Eq. 76.27.

$$m_{\text{reduced}} = \frac{m_{\text{nucleus}} m_{\text{electron}}}{m_{\text{nucleus}} + m_{\text{electron}}}$$

[16]Detailed analyses show that spectral lines are not simple but possess *fine structure* (i.e., they consist of a number of closely spaced lines that blend together).

[17]The standard nuclear symbol for the *Bohr radius* is $a_0$.

the time, $t$, at which an event occurs and the change in total energy, $E_2 - E_1$, for that event.

$$\epsilon_x \epsilon_p \approx h \qquad 76.33$$

$$\epsilon_t \epsilon_{\Delta_E} \approx h \qquad 76.34$$

### Example 76.5

The velocity of an electron is measured as 297 m/s with an accuracy of one part in 10,000. What is the maximum theoretical accuracy with which the electron's position can be located?

*Solution*

The uncertainty in velocity is $297/10,000 = 0.0297$ m/s. The uncertainty in momentum is

$$\epsilon_p = m\epsilon_v = \left(9.1094 \times 10^{-31} \text{ kg}\right)\left(0.0297 \, \frac{\text{m}}{\text{s}}\right)$$

$$= 2.7055 \times 10^{-32} \text{ kg·m/s}$$

The approximate uncertainty in position is found from Eq. 76.33.

$$\epsilon_x \approx \frac{h}{\epsilon_p} = \frac{6.626 \times 10^{-34} \text{ J·s}}{2.7055 \times 10^{-32} \, \frac{\text{kg·m}}{\text{s}}} = 2.45 \times 10^{-2} \text{ m}$$

## 24. WAVE MECHANICS

*Wave mechanics* (*quantum mechanics*), developed by E. Schrödinger and W. Heisenberg somewhat independently, overcomes all of the deficiencies of the Bohr atom. Each *wave function*, $\Psi$, mathematically describes an *orbital*, a region around the nucleus where an electron at a certain energy level is most likely to be found. The square of the wave function, $\Psi^2$, is a probability density function describing the probability per unit volume of finding the electron in a given region of space.

Equation 76.35 is the classical form of the wave equation (the *Schrödinger equation*). $\Psi$ is the amplitude of the wave (with wavelength $\lambda$ and frequency $\nu$) at point $x$ and time $t$.

$$\Psi = A \exp\left[2\pi i\left(\frac{x}{\lambda} - \nu t\right)\right] \qquad 76.35$$

## 25. WAVE MECHANICS QUANTUM NUMBERS

The theory of wave mechanics, which does not depend on circular orbits, derives the four quantum numbers listed in Table 76.5 rather than postulating them as the Bohr extensions did.[18] These four quantum numbers are sufficient to entirely explain the periodic chart, element structures, and all known spectral characteristics.

[18]The development of the wave theory and derivation of these quantum numbers is fascinating but beyond the scope of this book.

**Table 76.5** *Wave Mechanics Quantum Numbers*

| symbol | name | allowed values |
|---|---|---|
| $n$ | principal quantum number | $1, 2, 3, \ldots$ |
| $l$ | azimuthal quantum number | $0, 1, 2, \ldots (n-1)$ |
| $m_l$ | magnetic quantum number | $-l, \ldots 0, \ldots + l$ |
| $m_s$ | electron spin quantum number | $-\frac{1}{2}, +\frac{1}{2}$ |

The *principal (total) quantum number* defines the total energy of the electron. All electrons with the same principal quantum number, $n$, are in the same *shell*. Shells are named with the letters K, L, M, N, O, P, and Q as shown in Table 76.6. For the K shell, $n = 1$, so $l = m_l = 0$ and $m_s = \pm 1/2$. Thus, there is a maximum of two electrons in the K shell.

**Table 76.6** *Number of Shell Electrons*

| $n$ | shell name | maximum number of electrons $(2n^2)$ |
|---|---|---|
| 1 | K | 2 |
| 2 | L | 8 |
| 3 | M | 18 |
| 4 | N | 32 |
| 5 | O | 50 |
| 6 | P | 72 |
| 7 | Q | 98 |

For the L shell, $n = 2$, and there are two subshells corresponding to $l$ of 0 and 1. The s subshell contains two electrons corresponding to $m_l = 0$ and $m_s = \pm 1/2$. The p subshell contains six electrons corresponding to $m_l = -1$, 0, and +1 with two $m_s$ values for each. Thus, there is a total of eight L shell electrons.

The *azimuthal quantum number*, $l$, defines the magnitude of the electron's angular momentum vector. All electrons with the same azimuthal quantum number are in the same *subshell*.[19] Subshells are named s, p, d, f, g, h, and i as shown in Table 76.7. The letters s, p, d, and f correspond to the spectral line types of *sharp, principal, diffuse,* and *fundamental.*

**Table 76.7** *Number of Subshell Electrons*

| $l$ | subshell name | maximum number of electrons $(2(2l+1))$ |
|---|---|---|
| 0 | s | 2 |
| 1 | p | 6 |
| 2 | d | 10 |
| 3 | f | 14 |
| 4 | g | 18 |
| 5 | h | 22 |
| 6 | i | 26 |

[19]The azimuthal quantum number is also known as the *orbital quantum number* and *orbital angular momentum quantum number*, but this name is misleading since $l$ determines the subshell, not the orbital.

Atomic Theory

The *magnetic quantum number*, $m_l$, defines the direction of the electron's angular momentum vector. All electrons with the same magnetic quantum number are in the same orbital. The p subshell is divided into three equal-energy *orbitals*—$m_l = -1$ ($p_y$), $m_l = 0$ ($p_z$), and $m_l = +1$ ($p_x$).

The *spin quantum number*, $m_s$, defines the spin angular momentum vector. Each orbital contains two electrons corresponding to the values $m_s = \pm^1/_2$. There are $n^2$ orbitals for each principal quantum number, $n$. Thus, there are $2n^2$ electrons per principal quantum number, which is consistent with Table 76.6.

Quantum theory successfully deals with spectral intensities and fine structure. Spectral intensity is explained by the probabilistic nature of a transition from one energy level to another. The more probable a transition, the greater the intensity of its line. Fine structure is explained as transitions from different subshells with the same principal quantum number.

## 26. PAULI EXCLUSION PRINCIPLE

According to the *Pauli exclusion principle*, no two electrons can have the same set of four quantum numbers. Thus, the maximum number of electrons with any given principal quantum number, $n$, is $2n^2$. The maximum number of electrons with the same magnetic quantum number, $m_l$, is $2(2l + 1)$. The maximum number of electrons in any one orbital is 2.

## 27. ELECTRONIC CONFIGURATION

The *electronic configuration* of an atom can be written in shorthand notation as a series of modular designations of the form $n$(subshell name)$^N$ where $n$ is the principal quantum number and $N$ is the total number of electrons in the subshell whose name is listed. (When $N = 1$, the superscript is normally omitted.) Thus, helium (which has a full K shell) would be designated as $1s^2$. Three electrons in a p subshell would be written as $2p^3$. The electronic configurations for the elements are given in Fig. 76.5.

The periodic table of the elements lists the number of electrons in the shells for each element, but it does not distinguish between the subshells. Figure 76.5 provides a convenient means of writing the electronic configuration of any element. The horizontal rows, identified as $n = 1$, 2, 3, etc., represent the principal quantum number, n. The vertical columns, numbered $A = 2, 4, 12$, etc., represent the atomic number, $A$. Within the table, the letters (s, p, d, etc.) are the subshells. The superscripts are the maximum number of electrons each subshell can hold.

To use the table, start at the upper left-hand corner and move down the columns from left to right, adding the superscripts until the atomic number is reached. For example, the electronic structure of chromium, Cr ($A = 24$), is $1s^2 2s^2 2p^6 3s^2 3p^6 4s^2 3d^4$.

**Figure 76.5** *Electronic Configuration of the Elements*

| n | | | | | | | | | | | | n |
|---|---|---|---|---|---|---|---|---|---|---|---|---|
| 1 | $s^2$ | | | | | | | | | | | 1 |
| 2 | | $s^2$ | $p^6$ | | | | | | | | | 2 |
| 3 | | | $s^2$ | $p^6$ | $d^{10}$ | | | | | | | 3 |
| 4 | | | | $s^2$ | $p^6$ | $d^{10}$ | $f^{14}$ | | | | | 4 |
| 5 | | | | | $s^2$ | $p^6$ | $d^{10}$ | $f^{14}$ | $g^{18}$ | | | 5 |
| 6 | | | | | | $s^2$ | $p^6$ | $d^{10}$ | $f^{14}$ | $g^{18}$ | $h^{22}$ | 6 |
| 7 | | | | | | | $s^2$ | $p^6$ | $d^{10}$ | $f^{14}$ | $g^{18}$ | 7 |
| 8 | | | | | | | | $s^2$ | $p^6$ | $d^{10}$ | $f^{14}$ | 8 |
| 9 | | | | | | | | | $s^2$ | $p^6$ | $d^{10}$ | 9 |
| 10 | | | | | | | | | | $s^2$ | $p^6$ | 10 |
| 11 | | | | | | | | | | | $s^2$ | 11 |
| A | 2 | 4 | 12 | 20 | 38 | 56 | 88 | 120 | 170 | 220 | 292 | |

All 10 electrons in the 3d subshell were not used because the accumulation stopped with 24 electrons. (Notice that the 4s subshell filled before the 3d subshell began filling.)

The last subshell is full in inert elements. Inert elements are indicated by shaded boxes. For example, argon, Ar ($A = 18$) has an electronic structure of $1s^2 2s^2 2p^6 3s^2 3p^6$.

Electrons do not fill up the subshells in increasing numerical order as the atomic number increases. Rather, they fill the subgroups that will result in the most stable atomic configuration (i.e., the lowest energy). In order of increasing energy, the subshells are: 1s, 2s, 2p, 3s, 3p, 4s, 3d, 4p, 5s, 4d, 5p, 6s, 4f, 5d, 6p, 7s, 5f, 6d, etc. The most stable configurations are the *noble (inert) gases*, whose shells are completely filled.

*Hund's rule* states that, for a set of equal-energy orbitals, each orbital is occupied by one electron before any orbital has two electrons. Therefore, the first electrons to occupy orbitals within a subshell have parallel spins. For example, the configuration $p_y^1 p_z^1$ occurs before $p_y^2$.

## 28. NEUTRINOS

A *neutrino*, symbol $\nu$, is an elementary particle that has no charge, has near-zero rest mass, and travels at the speed of light. Its interaction with other particles is so weak as to be almost unobservable. Neutrinos are typically generated during $\beta$ decay, in which case they are called *electron neutrinos*, $\nu_e$. As with electrons, protons, and neutrons, the neutrino has a spin of $^1/_2$. (See Sec. 76.7.)

**Atomic Theory**

## 29. ANTIMATTER

The quantum theory predicts the existence of antiparticles, particles with exactly the same masses as the positive particles but opposite charges. The first antimatter particle discovered was the *positron* (positive electron), $e^+$.[20,21] Antiparticles can survive by moving rapidly, but once they come to rest, they are quickly annihilated and converted entirely and directly to energy. For example, an electron and positron are annihilated in the reaction

$$e^+ + e^- \rightarrow 2\gamma \qquad\qquad 76.36$$

Since the atomic weight of both the electron and positron is 0.0005486 amu (corresponding to approximately 0.5 MeV), as a consequence of the total conversion of matter to energy, each $\gamma$ ray will also have an *annihilation energy* of 0.5 MeV.

## 30. FAMILIES OF PARTICLES

A *fundamental (elementary) particle* cannot be split into smaller particles. The electron, proton, and neutron were originally thought to be fundamental, but now hundreds of particles have been documented. Particles are categorized into families according to their properties (mass, charge, spin angular momentum, lifetime, parity, hypercharge, isotopic spin, strangeness, charm, etc.). Figure 76.6 broadly classifies the particles into types. *Baryons* are heavy particles that include the proton and neutron; *mesons* have rest masses ranging down to about 0.1 amu; *leptons* include the electron, *mu* particle, and neutrino. All mesons and baryons, in turn, can be constructed from *quarks*.

**Figure 76.6** *Types of Particles*

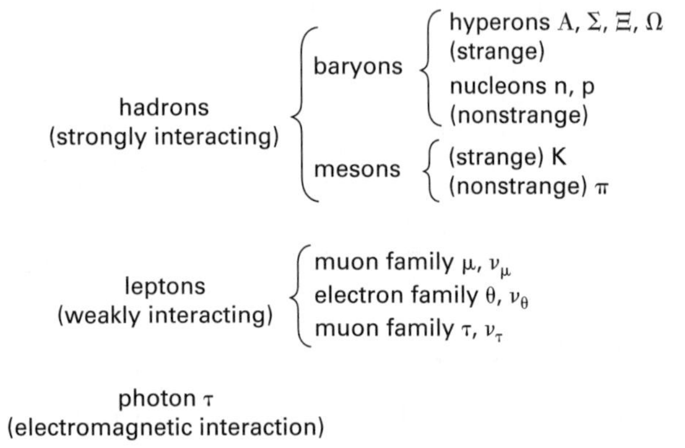

hadrons
(strongly interacting)
- baryons
  - hyperons $\Lambda, \Sigma, \Xi, \Omega$ (strange)
  - nucleons n, p (nonstrange)
- mesons
  - (strange) K
  - (nonstrange) $\pi$

leptons
(weakly interacting)
- muon family $\mu, \nu_\mu$
- electron family $\theta, \nu_\theta$
- muon family $\tau, \nu_\tau$

photon $\tau$
(electromagnetic interaction)

Leptons and quarks are considered by most nuclear investigators to be the true elementary particles, of which there are twelve. All other known particles can be constructed from them. The twelve elementary particles are known as *fermions* and are divided into three families (not to be confused with the types shown in Figure 76.6) known as *flavors* in nuclear studies. The first family consists of the electron, the electron's neutrino, and the *up and down quarks*. This family makes up ordinary matter. For example, a proton consists of two up and one down quarks. A neutron consists of two down and one up quarks.

The second and third families are observed only fleetingly in high-energy collisions. The second family consists of the muon, its neutrino, and the *strange* and *charmed quarks*. (Particles containing a strange quark decay more slowly than originally expected and were considered to be "strange." Other particles slow to decay were said to "lead a charmed life.") The third family consists of the massive *tau (tauon)* particle and its neutrino and the even more massive *top* (also known as *truth*) and *bottom* (also known as *beauty*) *quarks*.

Each of the twelve elementary particles has an appropriate anti-particle. When a quark and its antiquark combine, their flavors cancel. For example, in the *psi/J* particle, consisting of a charmed quark and its antiparticle, all charm is canceled, a situation known as *hidden charm*. Particles with only one charmed quark are said to have *naked charm*. Analogous combinations of other quarks result in *hidden beauty, naked bottoms*, etc.

## 31. NUCLEAR TRANSFORMATIONS

A *nuclear transformation (reaction)* involves a change in atomic structure of one or more particles. Nuclear reaction equations are similar to chemical reaction equations. Each particle is assigned a symbol with subscripts and superscripts as described in Sec. 76.10. In nuclear reactions, both the superscripts and subscripts must balance. If the products have more mass than the reactants (or vice versa), the excess must correspond to the kinetic energy of the reactants (products). In the following example, note the shorthand methods of indicating the reacting and product particles.

$$^{14}_{7}N + ^{1}_{1}H \rightarrow ^{11}_{6}C + ^{4}_{2}He$$

$$^{14}_{7}N + ^{1}_{1}H \xrightarrow{\alpha} ^{11}_{6}C$$

$$N^{14}(p, \alpha)C^{11}$$

**Example 76.6**

What is emitted in the following nuclear reaction?

$$^{13}_{7}N \rightarrow ^{12}_{6}C + ?$$

---

[20]Positrons are easily produced when high-energy photons strike a nucleus and are converted to equal numbers of electrons and positrons, a process called *pair production*. This is the source of positrons in colliding beam ring colliders. (See Sec. 76.49.)

$$2\gamma \rightarrow e^+ + e^-$$

[21]The symbol $(e^+)$ for the positron is an exception, as the standard symbol for an antiparticle is a bar over the symbol.

Atomic Theory

*Solution*

Since the subscripts and superscripts must balance, the required particle is a proton, as it has a charge of $7 - 6 = +1$ and mass of $13 - 12 = +1$.

## 32. INTERACTIONS OF NUCLEAR PARTICLES WITH MATTER

Whether or not a moving particle interacts with a nearby atom depends on the effective size of the target atom. The effective size (for the purpose of interaction analyses) of a target atom is its *cross section*, $\sigma$, usually measured in units of *barns*. (One barn is equal to $10^{-28}$ m$^2$.) Furthermore, the quantity $N\sigma$ occurs so often in nuclear calculations that it is given the special symbol $\Sigma$ and the name *macroscopic cross section*.

Most particles can interact with atoms in two different ways—they can be absorbed or scattered. The *total cross section* is the sum of the *absorption* and *scattering cross sections*.

$$\sigma_t = \sigma_a + \sigma_s \qquad 76.37$$

It is useful to think of cross sections as interaction probabilities per unit path length. Then, the probability of a particle being absorbed by a target atom is $p\{\text{absorption}\} = \sigma_a/\sigma_t$. The probability of any interaction at all when a particle travels a distance $x$ through a substance depends on the number of atoms per unit volume, $N$. (See Sec. 76.11.)

$$p\{\text{interaction}\} = \sigma_t N x = \Sigma_t x \qquad 76.38$$

If a particle beam has an incident density, $J$, and area, $A$, the average number of interactions with the target per second is

$$\text{interaction rate} = JAN\sigma_t x = JA\Sigma_t x \qquad 76.39$$

### Example 76.7

A carbon-12 graphite target block is 0.05 cm thick and 0.5 cm$^2$ in area. It is struck with a 0.1 cm$^2$ neutron beam with an intensity of $5 \times 10^8$ neutrons/cm$^2$·s. The graphite has a density of 1.6 g/cm$^3$ and a total cross section of 2.6 barns. What are the (a) probability that there will be an interaction and (b) interaction rate?

*Solution*

(a) From Eq. 76.9, the number of atoms per unit volume is

$$N = \frac{\rho N_A}{A} = \frac{\left(1.6 \ \frac{\text{g}}{\text{cm}^3}\right)\left(6.0221 \times 10^{23} \ \frac{\text{atoms}}{\text{mol}}\right)}{12 \ \frac{\text{g}}{\text{mol}}}$$

$$= 8.029 \times 10^{22} \ \text{atoms/cm}^3$$

The probability of an interaction is calculated from Eq. 76.38.

$$p\{\text{interaction}\} = \sigma_t N x$$

$$= (2.6 \ \text{b})\left(10^{-28} \ \frac{\text{m}^2}{\text{b}}\right)\left(100 \ \frac{\text{cm}}{\text{m}}\right)^2$$

$$\times \left(8.029 \times 10^{22} \ \frac{\text{atoms}}{\text{cm}^3}\right)(0.05 \ \text{cm})$$

$$= 0.01044 \quad (\text{about } 1\%)$$

(b) The interaction rate is calculated from Eq. 76.39.

$$\text{interaction rate} = JA(\sigma_t N x)$$

$$= \left(5 \times 10^8 \ \frac{1}{\text{cm}^2 \cdot \text{s}}\right)(0.1 \ \text{cm}^2)(0.01044)$$

$$= 5.22 \times 10^5 \ \text{interactions/s}$$

## 33. NEUTRON INTERACTIONS

Moving neutrons can be absorbed or scattered in several different ways.

*scattering:* Inelastic scattering (subscript *is*) occurs when the nucleus is struck by a neutron, gains energy, then decays by emitting a gamma ray. This reaction is written as $(n, n')$ since the reflected neutron has a different energy than the incident neutron. With *elastic scattering* (subscript *es*), also known as the *Compton effect* and *Compton scattering*, the target's electron increases in energy at the expense of the incident particle's kinetic energy. This reaction is written as $(n, n)$.

*absorption:* This includes *radiative capture* (subscript *c* or $\gamma$) and is written as $(n, \gamma)$, which occurs when the neutron is captured by a nucleus that then emits gamma radiation. *Fission* (subscript *f*) occurs when the nucleus splits into two smaller fragments. *Proton decay* $(m, p)$ and $\alpha$ *decay*, subscripts *p* and $\alpha$, respectively, occur when a neutron is absorbed and a proton or alpha particle, respectively, is emitted.

The total cross section equation can be written as

$$\sigma_t = \sigma_{is} + \sigma_{es} + \sigma_c + \sigma_f + \sigma_p + \sigma_\alpha \qquad 76.40$$

It is customary to describe neutrons according to their kinetic energies as in Table 76.8.

*Table 76.8* Neutron Energy Categories

| energy (eV) | description |
| --- | --- |
| 0.001 | cold |
| 0.25 | thermal |
| 1 | slow (resonant) |
| 100 | slow |
| $10^4$ | intermediate |
| $10^6$ | fast |
| $10^8$ | ultrafast |
| $10^{10}$ | relativistic |

**Atomic Theory**

Most nuclear reactors use *thermal neutrons*. This means that the neutrons have experienced a sufficient number of collisions to bring them down to the same energy level as that of the surrounding atoms, usually assumed to be 20°C. The $(n, p)$, $(n, \alpha)$, and $(n, n')$ reactions are rare at the thermal energy level, so Eq. 76.40 becomes

$$\sigma_t = \sigma_{es} + \sigma_c + \sigma_f \qquad 76.41$$

All of the neutrons do not travel at the same speed, even in a thermal population. The velocities are distributed according to a Maxwellian population, similar to the distribution used in kinetic gas theory. Equation 76.42 gives the *most probable velocity*.

$$v_{\text{probable}} \approx 128\sqrt{T} \qquad 76.42$$

### Example 76.8

What is the most probable kinetic energy (in joules) of a neutron in a 50°C environment?

*Solution*

From Eq. 76.42, the most probable velocity is proportional to the square root of the absolute temperature.

$$v_{\text{probable}} \approx 128\sqrt{T} = 128\sqrt{50°C + 273°}$$
$$= 2300 \text{ m/s}$$

The kinetic energy is

$$E_{\text{kinetic}} = \tfrac{1}{2}mv^2 = (0.5)\left(1.675 \times 10^{-27} \text{ kg}\right)\left(2300 \ \frac{\text{m}}{\text{s}}\right)^2$$
$$= 4.43 \times 10^{-21} \text{ J}$$

## 34. ATTENUATION OF RADIATION

*Attenuation* is a reduction in the intensity of radiation accomplished with shielding. The shield thickness required to attenuate the radiation to a particular level depends on the particle, particle energy, and shielding material. Particles of *alpha radiation* are easily stopped within a few millimeters because their double charges generate path ionization and are susceptible to electrostatic interaction. *Beta radiation* consists of singly charged particles that penetrate to intermediate distances. *Gamma radiation* has no charge and produces no ionization. Therefore, it is difficult to attenuate and poses the major health threat.

*Fast neutrons* (see Table 76.8) can be scattered but not absorbed until slowed to thermal velocities. (Fortunately, significant sources of fast neutrons do not exist on earth.) The ability of a substance to scatter neutrons is determined by its *scattering cross section*, $\sigma_s$. Hydrogen, heavy water ($D_2O$), and iron have high scattering cross sections.

The ability of a substance to absorb neutrons depends on its *absorption cross section*, $\sigma_a$. Cadmium ($^{113}_{48}Cd$),

boron ($^{10}_{5}B$), and lithium have extraordinarily high values, and the first two are used as *emergency (scram) rods*. Because of their low costs, carbon (graphite) rods and water are used for normal nuclear fission *moderators*.[22]

## 35. ATTENUATION OF GAMMA RADIATION

Gamma radiation is attenuated by the photoelectric effect (see Sec. 76.6), pair production (see Sec. 76.29), and Compton scattering (see Sec. 76.33). For gamma radiation, then, the total cross section is

$$\sigma_t = \sigma_{\text{Compton}} + \sigma_{\text{photoelectric}} + \sigma_{\text{pair}} \qquad 76.43$$

Traditionally, the macroscopic cross section (see Sec. 76.32) has been called the *attenuation coefficient* or *linear absorption coefficient*, $\mu_l$, with units of $1/m$. The *mass attenuation coefficient (mass absorption coefficient)*, $\mu_m$, with units of $m^2/kg$ incorporates the density of the material. Both of these parameters depend on the energy of the incident radiation.

$$\mu_l = \sigma_t N = \Sigma_t \qquad 76.44$$

$$\mu_m = \frac{\mu_l}{\rho} \qquad 76.45$$

The attenuated intensity of a monoenergetic, narrow-beam gamma source passing through a shield of thickness $x$ is

$$I = I_0 e^{-\mu_l x} = I_0 e^{-\mu_m \rho x} \qquad 76.46$$

The *half-value thickness* and *half-value mass* are the shield thickness and mass that reduce the radiation to half of its original value.

$$x_{1/2} = \frac{-\ln(0.5)}{\mu_l} \approx \frac{0.6931}{\mu_l} \qquad 76.47$$

$$m_{1/2} = \frac{-\ln(0.5)}{\mu_m} \approx \frac{0.6931}{\mu_m} \qquad 76.48$$

The thickness and mass of shielding required for any other attenuation is

$$x = \frac{-\ln(\text{remaining fraction})}{\mu_l} \qquad 76.49$$

$$m = \frac{-\ln(\text{remaining fraction})}{\mu_m} \qquad 76.50$$

There are two reasons that Eq. 76.47 through Eq. 76.50 cannot be used to determine shield thicknesses for living organisms. First, a radiation source is rarely a collimated beam, and second, the attenuated beam is often not lost, just scattered into a wider beam. These difficulties are handled by introducing a linear *build-up*

---

[22]Substances such as boron and cadmium have such a high neutron affinity that they are referred to as *poisons*.

*factor*, $B$, into Eq. 76.46. The build-up factor depends on the shielding material and the particle energy.

## 36. RADIATION EXPOSURE

*Exposure* is a measure of gamma radiation at the surface of an object and can be determined by the number of ionized air molecules in C/kg (coulombs per kilogram of air) at that point. The *roentgen* (R) exposure unit is equal to $2.58 \times 10^{-4}$ C/kg. The *exposure rate*, in R/s, is the number of ions generated per unit time.

Exposure is a measure of ionization surrounding a person, but biological damage is dependent on the amount of energy actually absorbed. The *absorbed dose*, $D$, typically given in *grays* (Gy), equal to 1.0 J/kg, or *rads* (rd), equal to 0.01 J/kg, is a measure of energy absorbed per unit tissue mass.

Some types of radiation are more damaging than others. To account for this, the *radiation weighting factor*, $W_r$, (see Table 76.9) previously known as the *relative biological effectiveness* (RBE) or *quality factor*, in rems/rd, is used to calculate the *equivalent dose*, $H$, from the absorbed dose, $D$. The *rem* (radiation effective man) is the unit of equivalent dose. (The *sievert* unit, abbreviated Sv, is equal to 100 rems.) The largest equivalent dose permitted by law during a period of time is known as the *maximum permissible dose*, MPD. (The dose from cosmic radiation, natural radioactivity in the earth, and other natural sources amounts to approximately 0.1 rems per year.)

$$H = W_r D \qquad 76.51$$

**Table 76.9** *Approximate Radiation Weighting Factors*

| type of radiation | $W_r$ (rems/rd) |
|---|---|
| X-rays | 1 |
| gamma radiation | 1 |
| $\beta$ particles, $> 0.03$ MeV | 1 |
| $\beta$ particles, $< 0.03$ MeV | 2 |
| thermal neutrons | 5 |
| fast neutrons | 10 |
| protons | 10 |
| $\alpha$ particles | 20 |

## Example 76.9

A 1 MeV gamma radiation source deposits energy in biological tissue at the rate of $10^{-6}$ J/kg·s. What is the equivalent absorbed dose rate?

*Solution*

From Table 76.9, $W_r = 1$ rem/rd. The equivalent dose rate is

$$\frac{dH}{dt} = W_r \frac{dD}{dt} = \frac{\left(1\ \frac{\text{rem}}{\text{rd}}\right)\left(10^{-6}\ \frac{\text{J}}{\text{kg·s}}\right)}{0.01\ \frac{\text{J}}{\text{kg·rd}}}$$

$$= 1 \times 10^{-4}\ \text{rem/s}$$

## 37. FISSION

*Fission* occurs when a nucleus splits into smaller fragments, but it can only occur in heavy nuclei such as uranium, plutonium, and thorium.[23] Fission can be induced by an impacting neutron (as normally happens in a fission reactor), and two or three additional neutrons are ejected from the fissioning nucleus.[24] The *multiplication constant*, *(reproduction factor)* $\eta$, is the number of neutrons emitted per neutron absorbed. Once thermalized, these neutrons may themselves cause other nuclei to fission, the principle of a *chain reaction*. The smaller fragments of the original nucleus continue to decay several more times. Although the exact fission fragments vary considerably, a typical fission reaction involving U-235 is

$$^{235}_{92}\text{U} + ^{1}_{0}\text{n} \rightarrow ^{140}_{54}\text{Xe} + ^{94}_{38}\text{Sr} + 2^{1}_{0}\text{n} + \gamma + 200\ \text{MeV}$$

The energy release in fission reactions involving uranium, plutonium, and thorium is essentially the same —around 200 MeV per fission. This energy is divided into kinetic energy of the fragments (84%), further decay of fission fragments (11%), kinetic energy of ejected neutrons ($2^1/_2$%), and prompt gamma rays ($2^1/_2$%). All but the neutrino energy (approximately 6% of the total) are useful.

Thermal neutrons will fission fuel having odd atomic masses (e.g., U-235 and Pu-239) because they have high fission cross sections, $\sigma_f$. These isotopes are *fissionable* and *fissile*. However, fast neutrons are required to fission stable *fissionable* and *non-fissile* isotopes such as U-238 and Pu-240. U-235 occurs naturally, but only to the extent of 1 part in 140. Ordinary uranium is almost all U-238. An *enriched fuel* is uranium (in the form of uranium dioxide, $UO_2$) to which additional U-235 (about 3% by weight) has been added.

---

[23]Bombardment with neutrons is not the only method of causing fission. *Photofission* occurs when fissionable substances are exposed to high-energy gamma radiation and X-rays.
[24]Because the fission fragments also decay and emit delayed neutrons, the average number of neutrons obtained in uranium U-235 fission is approximately 2.43. In plutonium fission, the value is approximately 2.95. In natural uranium, the value is approximately 1.3.

## 38. NUCLEAR REACTOR DYNAMICS

Enriched nuclear fuels are normally a mixture of U-238 and a small amount of U-235. However, only fast neutrons (most of which are moderated to thermal velocity in a nuclear reactor) will fission the U-238. The ratio of the total number of neutrons to the neutrons produced by U-235 reactions alone is the *fast fission factor*, $\epsilon$. Not all neutrons cause fission; some neutrons experience non-fission absorptions.

A minority fraction $(1-p)$ of the moderated neutrons experiences *radiative capture*. This transforms U-238 into U-239 and eventually into Pu-239 but does not contribute to fission. $p$ is the *resonance escape probability*. Of the remaining neutrons, a small fraction $(1-f)$ is absorbed by the moderator and structure. $f$ is the *thermal utilization*, the neutron fraction actually entering the fuel.

The *four-factor formula* determines whether or not fission can sustain itself in a reactor. The formula calculates *infinite-k*, which must be greater than 1.0 for a chain reaction.[25]

$$k_\infty = \eta \epsilon p f \qquad 76.52$$

## 39. BREEDER REACTORS

In a *breeder reactor*, fissionable material is produced by placing a blanket, of a fertile isotope (e.g., U-238 and Th-232) around a concentrated fissionable core. Neutrons are captured by the blanket, and the fertile isotope becomes fissionable according to the following reactions. The *doubling time* is the time (usually expressed in days) to double the mass of fissionable plutonium.

$$^{238}_{92}\text{U} + {}_0 n^1_{\text{fast}} \rightarrow {}^{239}_{92}\text{U} \xrightarrow[\beta-]{(23\ \text{min})} {}^{239}_{93}\text{Np} \xrightarrow[\beta-]{(2.3\ \text{days})} {}^{239}_{94}\text{Pu}$$

$$^{232}_{90}\text{Th} + {}_0 n^1_{\text{fast}} \xrightarrow[\beta-]{(23\ \text{min})} {}^{233}_{90}\text{Th} \rightarrow {}^{233}_{91}\text{Pa} \xrightarrow[\beta-]{(27\ \text{days})} {}^{233}_{92}\text{U}$$

## 40. FUSION

If two or more light atoms have sufficient energy (available only at high temperatures or velocities), they may fuse together to form a heavier nucleus. During fusion (also known as a *thermonuclear reaction*), mass is lost and converted to energy according to Einstein's law, Equation 76.1. This energy appears as kinetic energy of the fusion products and, theoretically, can be converted to useful heat.

---

[25]Normally, $k$ is kept just slightly (i.e., a few percent) above 1.0 by absorbing the excess neutrons in a moderator. In the absence of such a moderator, the reactor will *run away*. Fission bombs (atomic bombs) are built from U-235 or Pu-239 metal without a moderator, and virtually all of the nuclear fuel is consumed instantaneously.

Cold ($2 \times 10^6$ K) stars produce energy by the *proton-proton cycle* which consists of three related fusion reactions.

$$^1_1\text{H} + {}^1_1\text{H} \rightarrow {}^2_1\text{H} + e^+ + 0.4\ \text{MeV}$$

$$^1_1\text{H} + {}^2_1\text{H} \rightarrow {}^3_2\text{He} + 5.5\ \text{MeV}$$

$$^3_2\text{He} + {}^3_2\text{He} \rightarrow {}^4_2\text{He} + 2{}^1_1\text{H} + 12.9\ \text{MeV}$$

Since the first two reactions must each occur twice for each ${}^4_2\text{He}$ produced, the total energy is even more than is shown. The energy released per helium atom produced (less neutrino energy) is approximately 26.2 MeV.

Hotter stars fuse from the *carbon cycle*, producing approximately 24.7 MeV per helium atom. This reaction starts with ${}^{12}_6\text{C}$ and hydrogen, forming helium in a series of steps. The ${}^{12}_6\text{C}$ is regenerated in the last step and is not used up.

The most promising commercial cycles are simpler. Two possibilities are listed here. Both reactions have approximately the same probability of occurring.

$$^2_1\text{H} + {}^2_1\text{H} \rightarrow {}^3_1\text{H} + {}^1_1\text{H} + 4.0\ \text{MeV}$$

$$^2_1\text{H} + {}^2_1\text{H} \rightarrow {}^3_2\text{He} + {}^1_0\text{n} + 3.3\ \text{MeV}$$

At high enough ($10^4$ K and above) temperatures, all atoms break down into *plasma*, a mixture of nuclei and free electrons. Containment of a plasma is essential since plasma loses its thermal energy upon contact with a container wall, but containment is difficult at fusion temperatures ($10^8$ K, although ignition may occur in the $10^6$ K region). One form of containment is the *magnetic confinement bottle*, which is a confined space created by magnetic fields. The doughnut-shaped *tokamak* invented in the Soviet Union forms the basis for most magnetic confinement research. The force exerted on a plasma particle of charge $q$ moving at velocity $\mathbf{v}$ in a magnetic field $\mathbf{B}$ is

$$\mathbf{F} = q\mathbf{v} \times \mathbf{B} \quad \text{[vector cross-product]} \qquad 76.53$$

*Inertial confinement*, another technique, uses powerful laser beams to confine and ignite small pellets of hydrogen.

## 41. SPECIAL RELATIVITY

Einstein's principle of *special relativity* (usually just referred to as *relativity*) is simply that velocity is relative. The two basic postulates are

(1) The fundamental laws of physics are the same in all *inertial* (i.e., non-accelerating) *frames of reference*. (That is, when you perform an experiment of any kind

Atomic Theory

while moving at a constant speed, you will get the same results as when you are "stationary.")

(2) The observed speed of light is independent of the observer's motion.[26] No signal or energy can be transmitted with a speed greater than the speed of light.

The practical applications of special relativity are mass increase, length contraction, time dilation, and relativistic velocity sums.[27] At low velocities, these effects are negligible. For example, an object must be moving at 30 percent of the speed of light to experience a 5 percent increase in mass. Table 76.10 summarizes the relativistic effect.

**Table 76.10** *Approximate Relativistic Effect*

| fraction of speed of light | mass increase |
|---|---|
| 0.10 | 0.5% |
| 0.20 | 2.0% |
| 0.30 | 4.8% |
| 0.40 | 9.1% |
| 0.50 | 15% |
| 0.60 | 25% |
| 0.70 | 40% |
| 0.80 | 67% |
| 0.90 | 130% |
| 1.00 | $\infty$ |

## 42. MASS INCREASE

The mass of an object (as measured from a stationary reference) increases as the object's velocity approaches the speed of light. (See Table 76.10.) For an object with rest mass $m_0$, the *relativistic mass*, $m_v$, is

$$m_v = km_0 \qquad 76.54$$

$$k = \frac{1}{\sqrt{1 - \left(\frac{v}{c}\right)^2}} \qquad 76.55$$

The total relativistic energy of a particle is the sum of its rest mass energy and its kinetic energy. (See Table 76.11.)

$$E_{total} = E_{rest\ mass} + E_{kinetic} \qquad 76.56$$

$$E_{rest\ mass} = m_0 c^2 \qquad 76.57$$

$$E_{kinetic} = (m_v - m_0)c^2 = m_0(k-1)c^2 \qquad 76.58$$

Thus, the total relativistic energy is

$$E_{total} = m_v c^2 = km_0 c^2 \qquad 76.59$$

---

[26]This is also the logical conclusion of the Michelson-Morley experiment.
[27]The theory also explains magnetic force as a relativistic correction to Coulomb's law.

**Table 76.11** *Approximate Velocities of Particles with Various Energies*

| kinetic energy | velocity (fraction of $c$) | | |
|---|---|---|---|
| | electron | proton | deuteron |
| 0.1 MeV | 0.553 | 0.0146 | 0.0103 |
| 1 MeV | 0.941 | 0.0462 | 0.0326 |
| 10 MeV | 0.9988 | 0.145 | 0.103 |
| 100 MeV | 0.999987 | 0.429 | 0.316 |
| 1 GeV | 0.99999987 | 0.876 | 0.760 |
| 10 GeV | 0.9999999987 | 0.996 | 0.988 |
| 100 GeV | 0.99999999999987 | 0.999957 | 0.99983 |

## 43. LENGTH CONTRACTION

The length of an object (as measured from a stationary reference) decreases as the velocity increases. ($k$ is defined in Eq. 76.55.)

$$L_v = \frac{L_0}{k} \qquad 76.60$$

### Example 76.10

A 100 m long spacecraft travels at $0.7c$ relative to a 10 000 km diameter planet. (a) What is the apparent length of the spacecraft as viewed from the spacecraft itself? (b) What is the apparent length of the spacecraft as seen from the planet? (c) What is the apparent diameter of the planet as seen from the spacecraft?

*Solution*

(a) Since the observer moves with the spacecraft, the relative velocity is zero. Thus, $k = 1$, and the spacecraft appears to be 100 m long.

(b) From Eq. 76.55 at v = $0.7c$,

$$k = \frac{1}{\sqrt{1 - (0.7)^2}} = 1.40$$

The apparent length of the spacecraft is

$$L_v = \frac{100\ m}{1.4} = 71.4\ m$$

(c) Since velocity is relative, the length contraction will be the same to either observer. The apparent diameter of the planet will be 10 000 km/1.40 = 7140 km.

### Example 76.11

An electron moves with linear velocity of $2 \times 10^7$ m/s. What are its (a) wave velocity, (b) total relativistic energy, (c) linear momentum, and (d) de Broglie wavelength?

*Solution*

(a) The wave velocity is calculated from Eq. 76.30.

$$v_w = \frac{c^2}{v} = \frac{\left(3 \times 10^8 \; \frac{m}{s}\right)^2}{2 \times 10^7 \; \frac{m}{s}} = 4.5 \times 10^9 \; m/s$$

(b) The total relativistic energy is

$$k = \frac{1}{\sqrt{1 - \left(\dfrac{2 \times 10^7 \; \frac{m}{s}}{3 \times 10^8 \; \frac{m}{s}}\right)^2}} = 1.0022$$

$$E_{total} = km_0 c^2$$

$$= \frac{(1.0022)\left(9.1094 \times 10^{-31} \; kg\right)\left(3.0 \times 10^8 \; \frac{m}{s}\right)^2}{1.60218 \times 10^{-19} \; \frac{J}{eV}}$$

$$= 5.128 \times 10^5 \; eV$$

(c) The linear momentum is given by Eq. 76.31.

$$p = \frac{E_{total}}{v_w} = \frac{5.128 \times 10^5 \; eV}{4.5 \times 10^9 \; \frac{m}{s}} = 1.140 \times 10^{-4} \; eV \cdot s/m$$

(d) The de Broglie wavelength is calculated from Equation 76.32.

$$\lambda = \frac{h}{p} = \frac{6.626 \times 10^{-34} \; J \cdot s}{\left(1.140 \times 10^{-4} \; \frac{eV \cdot s}{m}\right)\left(1.60218 \times 10^{-19} \; \frac{J}{eV}\right)}$$

$$= 3.648 \times 10^{-11} \; m$$

## 44. TIME DILATION

A duration of time $t_0$ (as measured in a moving frame of reference) will appear to be longer when measured from a stationary reference.

$$t_0 = kt_v \qquad 76.61$$

One of the effects of time dilation is to affect the observed frequency of waves. If $\nu_0$ is the frequency as measured by an observer moving with the source at velocity v, the frequency observed by a stationary observer will be given by the relativistic *Doppler effect* equation.

$$\nu_v = k\nu_0\left(1 - \frac{v}{c}\right) \qquad 76.62$$

**Example 76.12**

A beam of pions attains a velocity of $0.6c$. The pions have a stationary half-life of $1.8 \times 10^{-8}$ s. How far will the beam travel before half of the pions have decayed?

*Solution*

From Eq. 76.55 or Table 76.10 at $v = 0.6c$, $k = 1.25$. From the standpoint of a moving pion, the half-life is unchanged. From the standpoint of the stationary observer, the half-life is increased by 1.25. The distance traveled is

$$x = vt = (0.6)\left(3.0 \times 10^8 \; \frac{m}{s}\right)(1.25)(1.8 \times 10^{-8} \; s)$$

$$= 4.05 \; m$$

## 45. RELATIVISTIC VELOCITY

The *relativistic velocity* as a function of acceleration and time is

$$v = \frac{at}{\sqrt{1 + \left(\dfrac{at}{c}\right)^2}} \qquad 76.63$$

If objects A and B are approaching (or separating from) each other with velocities $v_A$ and $v_B$ (measured with respect to some stationary point), the approach (separation) velocity measured from either of the moving frames of reference is the *relativistic velocity sum* (Observe consistent sign conventions when using Eq. 76.64.)

$$v = \frac{v_A - v_B}{1 - \dfrac{v_A v_B}{c^2}} \qquad 76.64$$

**Example 76.13**

A spacecraft travels at velocity $0.4c$ toward a star emitting photons (velocity $= 1.0c$). What is the photon velocity as measured by the spacecraft?

*Solution*

This is a relativistic velocity sum. From Eq. 76.64,

$$v = \frac{0.4c - (-1.0c)}{1 - \dfrac{(0.4c)(-1.0c)}{c^2}} = 1.0c$$

Thus, the speed of light is constant regardless of the observer's velocity.

## 46. GENERAL RELATIVITY

Einstein's *general relativity* theory is actually a modern, relativistic theory of gravitation. It addresses the requirement that changes in gravity cannot be transmitted instantaneously, as nothing can travel faster than the speed of light. According to general relativity, being in a gravitational field is equivalent to being accelerated. Einstein's *theory of equivalence* describes a gravitational field as a distortion of space around it. All objects, including photons of light, are affected by this

Atomic Theory

*curvature of space.* The classical verification of the theory is the refraction of light seen around the edge of a star during an eclipse.

## 47. CIRCULAR ACCELERATION OF CHARGED PARTICLES

Charged particles move in circular paths when passing through a magnetic field. Equation 76.65 calculates the path radius.[28]

$$r = \frac{m\mathrm{v}}{Bq} \qquad 76.65$$

The linear and angular velocities are found by solving Eq. 76.65 for v.

$$\mathrm{v} = \frac{Bqr}{m} \qquad 76.66$$

$$\omega = \frac{\mathrm{v}}{r} = \frac{Bq}{m} \qquad 76.67$$

The kinetic energy of a particle is

$$E_{\text{kinetic}} = \tfrac{1}{2}m\mathrm{v}^2 = \frac{(Bqr)^2}{2m} = \frac{m(r\omega)^2}{2} \qquad 76.68$$

In a *cyclotron (magnetic resonance accelerator)*, a large electromagnet supplies the magnetic field needed to keep the particles traveling in circles inside two semicircular flat cavities called *dees* (because of their shape). The particles are accelerated by an electric field whose polarity is reversed every half revolution. (Two polarity reversals constitute one cycle.) The *switching frequency*, *f*, is the reciprocal of the time per revolution, *t*. Equation 76.69 shows that the switching frequency is independent of the path radius. Therefore, for any given values of $B$, $m$, and $q$, the same switching frequency will work for slow particles just starting their outward spiral and fast particles entering the target zone.

$$f = \frac{\mathrm{v}}{2\pi r} = \frac{Bq}{2\pi m} \qquad 76.69$$

Imparted energies in early conventional cyclotrons were limited to approximately 10 MeV for protons, 20 MeV for deuterons, and 40 MeV for $\alpha$ particles. Velocities higher than approximately $0.2c$ result in significant mass changes (which affect *f*) and spoil the timing of regular cyclotrons. Varying the timing and increasing the magnetic field strength as the particles spiral outward extend the useful range to 500 MeV and above.

*Synchrotrons* achieve energies in the GeV range by taking advantage of the fact that particle velocities, no matter how high their energies, cannot exceed the speed of light. Therefore, the switching frequency is fixed at approximately $c/2\pi r$. Auxiliary means are used to obtain particles at near-*c* velocities for use in the synchrotron. Energy is added to the particle beam each revolution but the velocity does not increase appreciably.

Since a charged particle traveling in a synchrotron is constantly accelerating toward the center of rotation, it will radiate electromagnetic energy known as *synchrotron radiation* (see Sec. 76.4). The particle will have reached its maximum energy when the synchrotron radiation loss per revolution equals the energy input per cycle.

### Example 76.14

A cyclotron switches at $1.3 \times 10^7$ Hz and has a dee radius of 0.5 m. What total energy (in eV) can a deuteron achieve? Disregard the relativistic mass increase.

*Solution*

Equation 76.69 calculates the required magnetic flux density.

$$B = \frac{2\pi mf}{q} = \frac{(2\pi)(3.34358 \times 10^{-27}\ \text{kg})(1.3 \times 10^7\ \text{Hz})}{1.60218 \times 10^{-19}\ \text{C}}$$
$$= 1.705\ \text{T}$$

Equation 76.68 calculates the imparted energy.

$$E_{\text{kinetic}} = \frac{(Bqr)^2}{2m}$$
$$= \frac{(1.705\ \text{T})^2(1.60218 \times 10^{-19}\ \text{C})^2(0.5\ \text{m})^2}{(2)(3.34358 \times 10^{-27}\ \text{kg})\left(1.60218 \times 10^{-19}\ \frac{\text{J}}{\text{eV}}\right)}$$
$$= 1.74 \times 10^7\ \text{eV}$$

## 48. LINEAR ACCELERATION OF CHARGED PARTICLES

Charged particles move in a straight line in *linear accelerators*, eliminating synchrotron radiation loss. The equations governing relativistic mass increase can be used directly.

### Example 76.15

A linear accelerator delivers 50 MeV electrons. What velocity does this correspond to?

*Solution*

First, convert 50 MeV to joules.

$$E = (50 \times 10^6\ \text{eV})(1.60218 \times 10^{-19}\ \text{C})$$
$$= 8.011 \times 10^{-12}\ \text{J}$$

---

[28]Relativistic masses should be used if known. If not known (because the velocity is not known, for example), an iterative solution will be required.

**Atomic Theory**

Next, use Eq. 76.59 to calculate $k$.

$$k = \frac{E_{\text{total}}}{m_0 c^2} = \frac{8.011 \times 10^{-12} \text{ J}}{\left(9.1094 \times 10^{-31} \text{ kg}\right)\left(3.0 \times 10^8 \frac{\text{m}}{\text{s}}\right)^2}$$

$$= 97.71$$

Finally, determine v from Eq. 76.55.

$$\text{v} = c\sqrt{1 - \left(\frac{1}{k}\right)^2}$$

$$= c\sqrt{1 - \left(\frac{1}{97.71}\right)^2} = 0.9999476c$$

## 49. COLLIDING BEAM ACCELERATORS

It is common in high-energy physics to bombard a stationary target with high-velocity particles obtained in a synchrotron or linear accelerator. In keeping with the conservation of momentum, the target fragments and reaction products recoil forward, transforming most of the incident beam's energy into kinetic energy of the reaction products.

This problem is eliminated in *colliding beam storage ring devices* in which two beams of particles moving with equal speeds but in different directions collide. Colliding devices can be single pass (the *single-pass collider*) or particles can make many cycles before colliding.[29] In that manner, two beams of 30 GeV each (releasing a total of 60 GeV of energy) can produce results equivalent to a 2000 GeV beam striking a stationary target. Even higher energies are realized when positive matter and antimatter (e.g., electrons and positrons) are collided.

## 50. PARTICLE DETECTION DEVICES

The collisions of high-energy particles created in particle accelerators are monitored by a variety of detection devices. Essentially all such devices rely on detecting the ionized molecules produced when a moving charged particle passes through a particular medium, hence the name *ionization detectors*. *Cloud chambers* (also known as *expansion chambers* or *fog chambers*—using air supersaturated with water vapor), *bubble chambers* (using a liquid such as propane or hydrogen), and *wire chambers* (using criss-crossing wires in a gas-filled environment) have been commonly used in the past. Modern *surface barrier detectors* are solid-state ionization detectors.

In early cloud chambers, photographic images of the ionization trails were made. Photographic images can also be created directly by having the particles pass through a *photographic emulsion*.

---

[29]The CERN (French abbreviation for European Laboratory for Particle Physics) large electron-positron collider and the cancelled U.S. Superconducting Supercollider (SSC) in Texas are examples of this type.

# Topic XVII: Engineering Management

# 77 Economic and Financial Analysis

## Nomenclature

| | | |
|---|---|---|
| $A$ | annual amount | $ |
| $B$ | present worth of all benefits | $ |
| $BV_j$ | book value at end of the $j$th year | $ |
| $C$ | cost or present worth of all costs | $ |
| $d$ | declining balance depreciation rate | decimal |
| $D$ | demand | various |
| $D$ | depreciation | $ |
| DR | present worth of after-tax depreciation recovery | $ |
| $e$ | constant inflation rate | decimal |
| $\mathcal{E}$ | expected value | various |
| $E_0$ | initial amount of an exponentially growing cash flow | $ |
| EAA | equivalent annual amount | $ |
| EUAC | equivalent uniform annual cost | $ |
| $f$ | federal income tax rate | decimal |
| $F$ | forecasted quantity | various |
| $F$ | future worth | $ |
| $g$ | exponential growth rate | decimal |
| $G$ | uniform gradient amount | $ |
| $i$ | effective interest rate | decimal |
| $i\%$ | effective interest rate | % |
| $i'$ | effective interest rate corrected for inflation | decimal |

| | | |
|---|---|---|
| $k$ | number of compounding periods per year | – |
| $m$ | an integer | – |
| $n$ | number of compounding periods or years in life of asset | – |
| $p$ | probability | decimal |
| $P$ | present worth | $ |
| $r$ | nominal rate per year (rate per annum) | decimal per unit time |
| ROI | return on investment | $ |
| ROR | rate of return | decimal per unit time |
| $s$ | state income tax rate | decimal |
| $S_n$ | expected salvage value in year $n$ | $ |
| $t$ | composite tax rate | decimal |
| $t$ | time | years (typical) |
| $T$ | a quantity equal to $\frac{1}{2}n(n+1)$ | – |
| TC | tax credit | $ |
| $z$ | a quantity equal to $\frac{1+i}{1-d}$ | decimal |

**Symbols**

| | | |
|---|---|---|
| $\alpha$ | smoothing coefficient for forecasts | – |
| $\phi$ | effective rate per period $(r/k)$ | decimal |

**Subscripts**

| | |
|---|---|
| 0 | initial |
| $j$ | at time $j$ |
| $n$ | at time $n$ |
| $t$ | at time $t$ |

## 1. IRRELEVANT CHARACTERISTICS

In its simplest form, an *engineering economic analysis* is a study of the desirability of making an investment.[1] The decision-making principles in this chapter can be applied by individuals as well as by companies. The nature of the spending opportunity or industry is not important. Farming equipment, personal investments, and multimillion dollar factory improvements can all be evaluated using the same principles.

Similarly, the applicable principles are insensitive to the monetary units. Although *dollars* are used in this chapter, it is equally convenient to use pounds, yen, or euros.

Finally, this chapter may give the impression that investment alternatives must be evaluated on a year-by-year basis. Actually, the *effective period* can be defined as a day, month, century, or any other convenient period of time.

## 2. MULTIPLICITY OF SOLUTION METHODS

Most economic conclusions can be reached in more than one manner. There are usually several different analyses

that will eventually result in identical answers.[2] Other than the pursuit of elegant solutions in a timely manner, there is no reason to favor one procedural method over another.[3]

## 3. PRECISION AND SIGNIFICANT DIGITS

The full potential of electronic calculators will never be realized in engineering economic analyses. Considering that calculations are based on estimates of far-future cash flows and that unrealistic assumptions (no inflation, identical cost structures of replacement assets, etc.) are routinely made, it makes little sense to carry cents along in calculations.

The calculations in this chapter have been designed to illustrate and review the principles presented. Because of this, greater precision than is normally necessary in everyday problems may be used. Though used, such precision is not warranted.

Unless there is some compelling reason to strive for greater precision, the following rules are presented for use in reporting final answers to engineering economic analysis problems.

- Omit fractional parts of the dollar (i.e., cents).

- Report and record a number to a maximum of four significant digits unless the first digit of that number is 1, in which case, a maximum of five significant digits should be written. For example,

| | | |
|---|---|---|
| $49 | not | $49.43 |
| $93,450 | not | $93,453 |
| $1,289,700 | not | $1,289,673 |

## 4. NONQUANTIFIABLE FACTORS

An engineering economic analysis is a quantitative analysis. Some factors cannot be introduced as numbers into the calculations. Such factors are known as *nonquantitative factors*, *judgment factors*, and *irreducible factors*. Typical nonquantifiable factors are

- preferences

- political ramifications

- urgency

- goodwill

- prestige

- utility

- corporate strategy

---

[1]This subject is also known as *engineering economics* and *engineering economy*. There is very little, if any, true economics in this subject.

[2]Because of round-off errors, particularly when factors are taken from tables, these different calculations will produce slightly different numerical results (e.g., $49.49 versus $49.50). However, this type of divergence is well known and accepted in engineering economic analysis.

[3]This does not imply that approximate methods, simplifications, and rules of thumb are acceptable.

- environmental effects

- health and safety rules

- reliability

- political risks

Since these factors are not included in the calculations, the policy is to disregard the issues entirely. Of course, the factors should be discussed in a final report. The factors are particularly useful in breaking ties between competing alternatives that are economically equivalent.

## 5. YEAR-END AND OTHER CONVENTIONS

Except in short-term transactions, it is simpler to assume that all receipts and disbursements (cash flows) take place at the end of the year in which they occur.[4] This is known as the *year-end convention*. The exceptions to the year-end convention are initial project cost (purchase cost), trade-in allowance, and other cash flows that are associated with the inception of the project at $t = 0$.

On the surface, such a convention appears grossly inappropriate since repair expenses, interest payments, corporate taxes, and so on seldom coincide with the end of a year. However, the convention greatly simplifies engineering economic analysis problems, and it is justifiable on the basis that the increased precision associated with a more rigorous analysis is not warranted (due to the numerous other simplifying assumptions and estimates initially made in the problem).

There are various established procedures, known as *rules* or *conventions*, imposed by the Internal Revenue Service on U.S. taxpayers. An example is the *half-year rule*, which permits only half of the first-year depreciation to be taken in the first year of an asset's life when certain methods of depreciation are used. These rules are subject to constantly changing legislation and are not covered in this book. The implementation of such rules is outside the scope of engineering practice and is best left to accounting professionals.

## 6. CASH FLOW DIAGRAMS

Although they are not always necessary in simple problems (and they are often unwieldy in very complex problems), *cash flow diagrams* can be drawn to help visualize and simplify problems having diverse receipts and disbursements.

The following conventions are used to standardize cash flow diagrams.

- The horizontal (time) axis is marked off in equal increments, one per period, up to the duration (or *horizon*) of the project.

- Two or more transfers in the same period are placed end to end, and these may be combined.

- Expenses incurred before $t = 0$ are called *sunk costs*. Sunk costs are not relevant to the problem unless they have tax consequences in an after-tax analysis.

- *Receipts* are represented by arrows directed upward. *Disbursements* are represented by arrows directed downward. The arrow length is proportional to the magnitude of the cash flow.

### Example 77.1

A mechanical device will cost $20,000 when purchased. Maintenance will cost $1000 each year. The device will generate revenues of $5000 each year for five years, after which the salvage value is expected to be $7000. Draw and simplify the cash flow diagram.

*Solution*

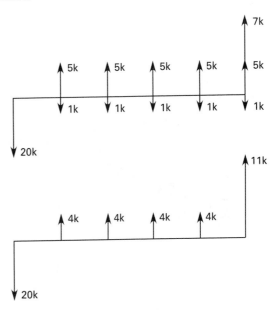

## 7. TYPES OF CASH FLOWS

To evaluate a real-world project, it is necessary to present the project's cash flows in terms of standard cash flows that can be handled by engineering economic analysis techniques. The standard cash flows are single payment cash flow, uniform series cash flow, gradient series cash flow, and the infrequently encountered exponential gradient series cash flow. (See Fig. 77.1.)

A *single payment cash flow* can occur at the beginning of the time line (designated as $t = 0$), at the end of the time line (designated as $t = n$), or at any time in between.

---

[4]A *short-term transaction* typically has a lifetime of five years or less and has payments or compounding that are more frequent than once per year.

The *uniform series cash flow* consists of a series of equal transactions starting at $t = 1$ and ending at $t = n$. The symbol $A$ is typically given to the magnitude of each individual cash flow.[5]

The *gradient series cash flow* starts with a cash flow (typically given the symbol $G$) at $t = 2$ and increases by $G$ each year until $t = n$, at which time the final cash flow is $(n - 1)G$.

An *exponential gradient series cash flow* is based on a phantom value (typically given the symbol $E_0$) at $t = 0$ and grows or decays exponentially according to the following relationship.[6]

$$\text{amount at time } t = E_t = E_0(1 + g)^t$$

$$[t = 1, 2, 3, \ldots, n] \qquad 77.1$$

**Figure 77.1** *Standard Cash Flows*

(a) single payment

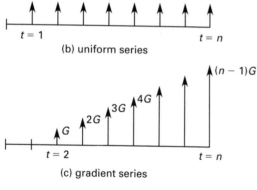

(b) uniform series

(c) gradient series

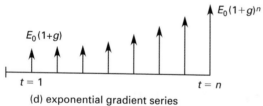

(d) exponential gradient series

In Eq. 77.1, $g$ is the *exponential growth rate*, which can be either positive or negative. Exponential gradient cash flows are rarely seen in economic justification projects assigned to engineers.[7]

## 8. TYPICAL PROBLEM TYPES

There is a wide variety of problem types that, collectively, are considered to be engineering economic analysis problems.

By far, the majority of engineering economic analysis problems are *alternative comparisons*. In these problems, two or more mutually exclusive investments compete for limited funds. A variation of this is a *replacement/retirement analysis*, which is repeated each year to determine if an existing asset should be replaced. Finding the percentage return on an investment is a *rate of return problem*, one of the alternative comparison solution methods.

Investigating interest and principal amounts in loan payments is a *loan repayment problem*. An *economic life analysis* will determine when an asset should be retired. In addition, there are miscellaneous problems involving economic order quantity, learning curves, break-even points, product costs, and so on.

## 9. IMPLICIT ASSUMPTIONS

Several assumptions are implicitly made when solving engineering economic analysis problems. Some of these assumptions are made with the knowledge that they are or will be poor approximations of what really will happen. The assumptions are made, regardless, for the benefit of obtaining a solution.

The most common assumptions are the following.

- The year-end convention is applicable.

- There is no inflation now, nor will there be any during the lifetime of the project.

- Unless otherwise specifically called for, a before-tax analysis is needed.

- The effective interest rate used in the problem will be constant during the lifetime of the project.

- Nonquantifiable factors can be disregarded.

- Funds invested in a project are available and are not urgently needed elsewhere.

- Excess funds continue to earn interest at the effective rate used in the analysis.

This last assumption, like most of the assumptions listed, is almost never specifically mentioned in the body of a solution. However, it is a key assumption when

---

[5]Notice that the cash flows do not begin at $t = 0$. This is an important concept with all of the series cash flows. This convention has been established to accommodate the timing of annual maintenance (and similar) cash flows for which the year-end convention is applicable.

[6]Notice the convention for an exponential cash flow series: The first cash flow, $E_0$, is at $t = 1$, as in the uniform annual series. However, the first cash flow is $E_0(1 + g)$. The cash flow of $E_0$ at $t = 0$ is absent (i.e., is a *phantom cash flow*).

[7]For one of the few discussions on exponential cash flow, see *Capital Budgeting*, Robert V. Oakford, The Ronald Press Company, New York, 1970.

comparing two alternatives that have different initial costs.

For example, suppose two investments, one costing $10,000 and the other costing $8000, are to be compared at 10%. It is obvious that $10,000 in funds is available, otherwise the costlier investment would not be under consideration. If the smaller investment is chosen, what is done with the remaining $2000? The last assumption yields the answer: the $2000 is "put to work" in some investment earning (in this case) 10%.

## 10. EQUIVALENCE

Industrial decision makers using engineering economic analysis are concerned with the magnitude and timing of a project's cash flow as well as with the total profitability of that project. In this situation, a method is required to compare projects involving receipts and disbursements occurring at different times.

By way of illustration, consider $100 placed in a bank account that pays 5% effective annual interest at the end of each year. After the first year, the account will have grown to $105. After the second year, the account will have grown to $110.25.

Assume that you will have no need for money during the next two years, and any money received will immediately go into your 5% bank account. Then, which of the following options would be more desirable?

option A: $100 now

option B: $105 to be delivered in one year

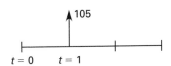

option C: $110.25 to be delivered in two years

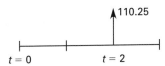

As illustrated, none of the options is superior under the assumptions given. If the first option is chosen, you will immediately place $100 into a 5% account, and in two years the account will have grown to $110.25. In fact, the account will contain $110.25 at the end of two years regardless of the option chosen. Therefore, these alternatives are said to be *equivalent*.

Equivalence may or may not be the case, depending on the interest rate. Thus, an alternative that is acceptable to one decision maker may be unacceptable to another. The interest rate that is used in actual calculations is known as the *effective interest rate*.[8] If compounding is once a year, it is known as the *effective annual interest rate*. However, effective quarterly, monthly, daily, and so on, interest rates are also used.

The fact that $100 today grows to $105 in one year (at 5% annual interest) is an example of what is known as the *time value of money* principle. This principle simply articulates what is obvious: Funds placed in a secure investment will increase to an equivalent future amount. The procedure for determining the present investment from the equivalent future amount is known as *discounting*.

## 11. SINGLE-PAYMENT EQUIVALENCE

The equivalence of any present amount, $P$, at $t = 0$, to any future amount, $F$, at $t = n$, is called the *future worth* and can be calculated from Eq. 77.2.

$$F = P(1 + i)^n \qquad 77.2$$

The factor $(1 + i)^n$ is known as the *single payment compound amount factor* and has been tabulated in App. 77.B for various combinations of $i$ and $n$.

Similarly, the equivalence of any future amount to any present amount is called the *present worth* and can be calculated from Eq. 77.3.

$$P = F(1 + i)^{-n} = \frac{F}{(1 + i)^n} \qquad 77.3$$

The factor $(1 + i)^{-n}$ is known as the *single payment present worth factor*.[9]

The interest rate used in Eq. 77.2 and Eq. 77.3 must be the effective rate per period. Also, the basis of the rate (annually, monthly, etc.) must agree with the type of period used to count $n$. Thus, it would be incorrect to use an effective annual interest rate if $n$ was the number of compounding periods in months.

### Example 77.2

How much should you put into a 10% (effective annual rate) savings account in order to have $10,000 in five years?

---

[8]The adjective *effective* distinguishes this interest rate from other interest rates (e.g., nominal interest rates) that are not meant to be used directly in calculating equivalent amounts.

[9]The *present worth* is also called the *present value* and *net present value*. These terms are used interchangeably and no significance should be attached to the terms *value, worth,* and *net*.

*Solution*

This problem could also be stated: What is the equivalent present worth of $10,000 five years from now if money is worth 10% per year?

$$P = F(1+i)^{-n} = (\$10{,}000)(1+0.10)^{-5}$$
$$= \$6209$$

The factor 0.6209 would usually be obtained from the tables.

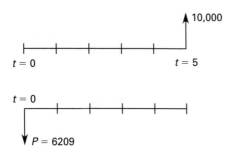

## 12. STANDARD CASH FLOW FACTORS AND SYMBOLS

Equation 77.2 and Eq. 77.3 may give the impression that solving engineering economic analysis problems involves a lot of calculator use, and, in particular, a lot of exponentiation. Such calculations may be necessary from time to time, but most problems are simplified by the use of tabulated values of the factors.

Rather than actually writing the formula for the compound amount factor (which converts a present amount to a future amount), it is common convention to substitute the standard functional notation of $(F/P, i\%, n)$. Thus, the future value in $n$ periods of a present amount would be symbolically written as

$$F = P(F/P, i\%, n) \qquad 77.4$$

Similarly, the present worth factor has a functional notation of $(P/F, i\%, n)$. The present worth of a future amount $n$ periods hence would be symbolically written as

$$P = F(P/F, i\%, n) \qquad 77.5$$

Values of these *cash flow (discounting) factors* are tabulated in App. 77.B. There is often initial confusion about whether the $(F/P)$ or $(P/F)$ column should be used in a particular problem. There are several ways of remembering what the functional notations mean.

One method of remembering which factor should be used is to think of the factors as conditional probabilities. The conditional probability of event **A** given that event **B** has occurred is written as $p\{\mathbf{A}|\mathbf{B}\}$, where the given event comes after the vertical bar. In the standard notational form of discounting factors, the given amount is similarly placed after the slash. What you

want comes before the slash. $(F/P)$ would be a factor to find $F$ given $P$.

Another method of remembering the notation is to interpret the factors algebraically. Thus, the $(F/P)$ factor could be thought of as the fraction $F/P$. Algebraically, Eq. 77.4 would be

$$F = P\left(\frac{F}{P}\right) \qquad 77.6$$

This algebraic approach is actually more than an interpretation. The numerical values of the discounting factors are consistent with this algebraic manipulation. Thus, the $(F/A)$ factor could be calculated as $(F/P) \times (P/A)$. This consistent relationship can be used to calculate other factors that might be occasionally needed, such as $(F/G)$ or $(G/P)$. For instance, the annual cash flow that would be equivalent to a uniform gradient may be found from

$$A = G(P/G, i\%, n)(A/P, i\%, n) \qquad 77.7$$

Formulas for the compounding and discounting factors are contained in Table 77.1. Normally, it will not be necessary to calculate factors from the formulas. Appendix 77.B is adequate for solving most problems.

### Example 77.3

What factor will convert a gradient cash flow ending at $t = 8$ to a future value at $t = 8$? (That is, what is the $(F/G, i\%, 8)$ factor?) The effective annual interest rate is 10%.

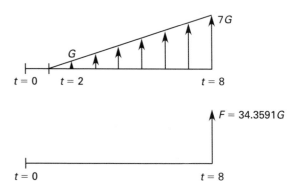

*Solution*

*method 1:*

From Table 77.1, the $(F/G, 10\%, 8)$ factor is

$$(F/G, 10\%, 8)$$
$$= \frac{(1+i)^n - 1}{i^2} - \frac{n}{i}$$
$$= \frac{(1+0.10)^8 - 1}{(0.10)^2} - \frac{8}{0.10}$$
$$= 34.3589$$

**Table 77.1** *Discount Factors for Discrete Compounding*

| factor name | converts | symbol | formula |
|---|---|---|---|
| single payment compound amount | $P$ to $F$ | $(F/P, i\%, n)$ | $(1 + i)^n$ |
| single payment present worth | $F$ to $P$ | $(P/F, i\%, n)$ | $(1 + i)^{-n}$ |
| uniform series sinking fund | $F$ to $A$ | $(A/F, i\%, n)$ | $\dfrac{i}{(1+i)^n - 1}$ |
| capital recovery | $P$ to $A$ | $(A/P, i\%, n)$ | $\dfrac{i(1+i)^n}{(1+i)^n - 1}$ |
| uniform series compound amount | $A$ to $F$ | $(F/A, i\%, n)$ | $\dfrac{(1+i)^n - 1}{i}$ |
| uniform series present worth | $A$ to $P$ | $(P/A, i\%, n)$ | $\dfrac{(1+i)^n - 1}{i(1+i)^n}$ |
| uniform gradient present worth | $G$ to $P$ | $(P/G, i\%, n)$ | $\dfrac{(1+i)^n - 1}{i^2(1+i)^n} - \dfrac{n}{i(1+i)^n}$ |
| uniform gradient future worth | $G$ to $F$ | $(F/G, i\%, n)$ | $\dfrac{(1+i)^n - 1}{i^2} - \dfrac{n}{i}$ |
| uniform gradient uniform series | $G$ to $A$ | $(A/G, i\%, n)$ | $\dfrac{1}{i} - \dfrac{n}{(1+i)^n - 1}$ |

*method 2:*

The tabulated values of $(P/G)$ and $(F/P)$ in App. 77.B can be used to calculate the factor.

$$(F/G, 10\%, 8)$$
$$= (P/G, 10\%, 8)(F/P, 10\%, 8)$$
$$= (16.0287)(2.1436)$$
$$= 34.3591$$

The $(F/G)$ factor could also have been calculated as the product of the $(A/G)$ and $(F/A)$ factors.

## 13. CALCULATING UNIFORM SERIES EQUIVALENCE

A cash flow that repeats each year for $n$ years without change in amount is known as an *annual amount* and is given the symbol $A$. As an example, a piece of equipment may require annual maintenance, and the maintenance cost will be an annual amount. Although the equivalent value for each of the $n$ annual amounts could be calculated and then summed, it is more expedient to use one of the uniform series factors. For example, it is possible to convert from an annual amount to a future amount by use of the $(F/A)$ factor.

$$F = A(F/A, i\%, n) \qquad 77.8$$

A *sinking fund* is a fund or account into which annual deposits of $A$ are made in order to accumulate $F$ at $t = n$

in the future. Since the annual deposit is calculated as $A = F(A/F, i\%, n)$, the $(A/F)$ factor is known as the *sinking fund factor*. An *annuity* is a series of equal payments $(A)$ made over a period of time.[10] Usually, it is necessary to "buy into" an investment (a bond, an insurance policy, etc.) in order to ensure the annuity. In the simplest case of an annuity that starts at the end of the first year and continues for $n$ years, the purchase price $(P)$ is

$$P = A(P/A, i\%, n) \qquad 77.9$$

The present worth of an *infinite (perpetual) series* of annual amounts is known as a *capitalized cost*. There is no $(P/A, i\%, \infty)$ factor in the tables, but the capitalized cost can be calculated simply as

$$P = \frac{A}{i} \qquad [i \text{ in decimal form}] \qquad 77.10$$

Alternatives with different lives will generally be compared by way of *equivalent uniform annual cost* (EUAC). An EUAC is the annual amount that is equivalent to all of the cash flows in the alternative. The EUAC differs in sign from all of the other cash flows. Costs and expenses expressed as EUACs, which would normally be considered negative, are actually positive. The term *cost* in the designation EUAC serves to make clear the meaning of a positive number.

---

[10]An annuity may also consist of a lump sum payment made at some future time. However, this interpretation is not considered in this chapter.

### Example 77.4

Maintenance costs for a machine are $250 each year. What is the present worth of these maintenance costs over a 12-year period if the interest rate is 8%?

*Solution*

$$P = A(P/A, 8\%, 12) = (-\$250)(7.5361)$$
$$= -\$1884$$

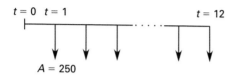

### 14. FINDING PAST VALUES

From time to time, it will be necessary to determine an amount in the past equivalent to some current (or future) amount. For example, you might have to calculate the original investment made 15 years ago given a current annuity payment.

Such problems are solved by placing the $t = 0$ point at the time of the original investment, and then calculating the past amount as a $P$ value. For example, the original investment, $P$, can be extracted from the annuity, $A$, by using the standard cash flow factors.

$$P = A(P/A, i\%, n) \qquad 77.11$$

The choice of $t = 0$ is flexible. As a general rule, the $t = 0$ point should be selected for convenience in solving a problem.

### Example 77.5

You currently pay $250 per month to lease your office phone equipment. You have three years (36 months) left on the five-year (60-month) lease. What would have been an equivalent purchase price two years ago? The effective interest rate per month is 1%.

*Solution*

The solution of this example is not affected by the fact that investigation is being performed in the middle of the horizon. This is a simple calculation of present worth.

$$P = A(P/A, 1\%, 60)$$
$$= (-\$250)(44.9550) = -\$11,239$$

### 15. TIMES TO DOUBLE AND TRIPLE AN INVESTMENT

If an investment doubles in value (in $n$ compounding periods and with $i\%$ effective interest), the ratio of current value to past investment will be 2.

$$F/P = (1 + i)^n = 2 \qquad 77.12$$

Similarly, the ratio of current value to past investment will be 3 if an investment triples in value. This can be written as

$$F/P = (1 + i)^n = 3 \qquad 77.13$$

It is a simple matter to extract the number of periods, $n$, from Eq. 77.12 and Eq. 77.13 to determine the *doubling time* and *tripling time*, respectively. (See Table 77.2.) For example, the doubling time is

$$n = \frac{\log 2}{\log(1 + i)} \qquad 77.14$$

When a quick estimate of the doubling time is needed, the *rule of 72* can be used. The doubling time is approximately $72/i$.

The tripling time is

$$n = \frac{\log 3}{\log(1 + i)} \qquad 77.15$$

**Table 77.2** *Doubling and Tripling Times for Various Interest Rates*

| interest rate (%) | doubling time (periods) | tripling time (periods) |
|---|---|---|
| 1 | 69.7 | 110.4 |
| 2 | 35.0 | 55.5 |
| 3 | 23.4 | 37.2 |
| 4 | 17.7 | 28.0 |
| 5 | 14.2 | 22.5 |
| 6 | 11.9 | 18.9 |
| 7 | 10.2 | 16.2 |
| 8 | 9.01 | 14.3 |
| 9 | 8.04 | 12.7 |
| 10 | 7.27 | 11.5 |
| 11 | 6.64 | 10.5 |
| 12 | 6.12 | 9.69 |
| 13 | 5.67 | 8.99 |
| 14 | 5.29 | 8.38 |
| 15 | 4.96 | 7.86 |
| 16 | 4.67 | 7.40 |
| 17 | 4.41 | 7.00 |
| 18 | 4.19 | 6.64 |
| 19 | 3.98 | 6.32 |
| 20 | 3.80 | 6.03 |

## 16. VARIED AND NONSTANDARD CASH FLOWS

### Gradient Cash Flow

A common situation involves a uniformly increasing cash flow. If the cash flow has the proper form, its present worth can be determined by using the *uniform gradient factor*, $(P/G, i\%, n)$. The uniform gradient factor finds the present worth of a uniformly increasing cash flow that starts in year two (not in year one).

**Figure 77.2** *Positive and Negative Gradient Cash Flows*

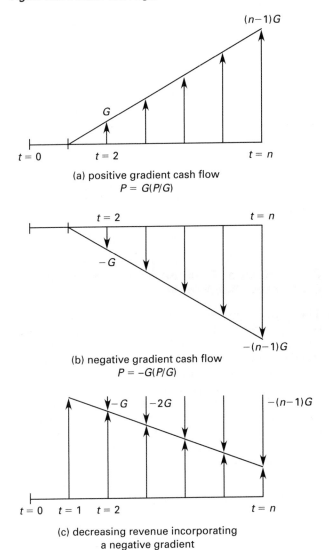

(a) positive gradient cash flow
$P = G(P/G)$

(b) negative gradient cash flow
$P = -G(P/G)$

(c) decreasing revenue incorporating a negative gradient
$P = A(P/A) - G(P/G)$

There are three common difficulties associated with the form of the uniform gradient. The first difficulty is that the initial cash flow occurs at $t = 2$. This convention recognizes that annual costs, if they increase uniformly, begin with some value at $t = 1$ (due to the year-end convention) but do not begin to increase until $t = 2$. The tabulated values of $(P/G)$ have been calculated to find the present worth of only the increasing part of the

annual expense. The present worth of the base expense incurred at $t = 1$ must be found separately with the $(P/A)$ factor.

The second difficulty is that, even though the factor $(P/G, i\%, n)$ is used, there are only $n - 1$ actual cash flows. It is clear that $n$ must be interpreted as the *period number* in which the last gradient cash flow occurs, not the number of gradient cash flows.

Finally, the sign convention used with gradient cash flows may seem confusing. If an expense increases each year (as in Ex. 77.6), the gradient will be negative, since it is an expense. If a revenue increases each year, the gradient will be positive. (See Fig. 77.2.) In most cases, the sign of the gradient depends on whether the cash flow is an expense or a revenue.[11]

### Example 77.6

Maintenance on an old machine is $100 this year but is expected to increase by $25 each year thereafter. What is the present worth of five years of the costs of maintenance? Use an interest rate of 10%.

*Solution*

In this problem, the cash flow must be broken down into parts. (Notice that the five-year gradient factor is used even though there are only four nonzero gradient cash flows.)

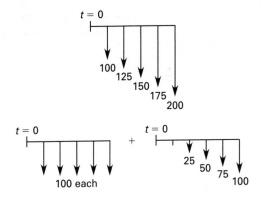

$$P = A(P/A, 10\%, 5) + G(P/G, 10\%, 5)$$
$$= (-\$100)(3.7908) - (\$25)(6.8618)$$
$$= -\$551$$

### Stepped Cash Flows

*Stepped cash flows* are easily handled by the technique of *superposition of cash flows*. This technique is illustrated by Ex. 77.7.

---

[11]This is not a universal rule. It is possible to have a uniformly decreasing revenue as in Fig. 77.2(c). In this case, the gradient would be negative.

## Example 77.7

An investment costing $1000 returns $100 for the first five years and $200 for the following five years. How would the present worth of this investment be calculated?

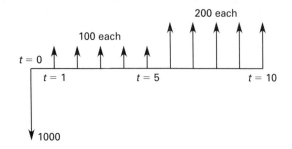

*Solution*

Using the principle of superposition, the revenue cash flow can be thought of as $200 each year from $t = 1$ to $t = 10$, with a negative revenue of $100 from $t = 1$ to $t = 5$. Superimposed, these two cash flows make up the actual performance cash flow.

$$P = -\$1000 + (\$200)(P/A, i\%, 10) - (\$100)(P/A, i\%, 5)$$

### Missing and Extra Parts of Standard Cash Flows

A missing or extra part of a standard cash flow can also be handled by superposition. For example, suppose an annual expense is incurred each year for ten years, except in the ninth year. (The cash flow is illustrated in Fig. 77.3.) The present worth could be calculated as a subtractive process.

$$P = A(P/A, i\%, 10) - A(P/F, i\%, 9) \qquad 77.16$$

**Figure 77.3** *Cash Flow with a Missing Part*

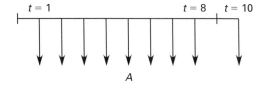

Alternatively, the present worth could be calculated as an additive process.

$$P = A(P/A, i\%, 8) + A(P/F, i\%, 10) \qquad 77.17$$

### Delayed and Premature Cash Flows

There are cases when a cash flow matches a standard cash flow exactly, except that the cash flow is delayed or starts sooner than it should. Often, such cash flows can be handled with superposition. At other times, it may be more convenient to shift the time axis. This shift is

known as the *projection method*. Example 77.8 illustrates the projection method.

## Example 77.8

An expense of $75 is incurred starting at $t = 3$ and continues until $t = 9$. There are no expenses or receipts until $t = 3$. Use the projection method to determine the present worth of this stream of expenses.

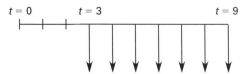

*Solution*

First, determine a cash flow at $t = 2$ that is equivalent to the entire expense stream. If $t = 0$ was where $t = 2$ actually is, the present worth of the expense stream would be

$$P' = (-\$75)(P/A, i\%, 7)$$

$P'$ is a cash flow at $t = 2$. It is now simple to find the present worth (at $t = 0$) of this future amount.

$$P = P'(P/F, i\%, 2) = (-\$75)(P/A, i\%, 7)(P/F, i\%, 2)$$

### Cash Flows at Beginnings of Years: The Christmas Club Problem

This type of problem is characterized by a stream of equal payments (or expenses) starting at $t = 0$ and ending at $t = n - 1$. (See Fig. 77.4.) It differs from the standard annual cash flow in the existence of a cash flow at $t = 0$ and the absence of a cash flow at $t = n$. This problem gets its name from the service provided by some savings institutions whereby money is automatically deposited each week or month (starting immediately, when the savings plan is opened) in order to accumulate money to purchase Christmas presents at the end of the year.

**Figure 77.4** *Cash Flows at Beginnings of Years*

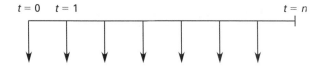

It may seem that the present worth of the savings stream can be determined by directly applying the $(P/A)$ factor. However, this is not the case, since the Christmas Club cash flow and the standard annual cash flow differ. The Christmas Club problem is easily handled by superposition, as illustrated by Ex. 77.9.

## Example 77.9

How much can you expect to accumulate by $t = 10$ for a child's college education if you deposit $300 at the beginning of each year for a total of ten payments?

*Solution*

Notice that the first payment is made at $t = 0$ and that there is no payment at $t = 10$. The future worth of the first payment is calculated with the $(F/P)$ factor. The absence of the payment at $t = 10$ is handled by superposition. Notice that this "correction" is not multiplied by a factor.

$$F = (\$300)(F/P, i\%, 10) + (\$300)(F/A, i\%, 10) - \$300$$
$$= (\$300)(F/A, i\%, 11) - \$300$$

## 17. THE MEANING OF PRESENT WORTH AND *i*

If $100 is invested in a 5% bank account (using annual compounding), you can remove $105 one year from now; if this investment is made, you will receive a *return on investment* (ROI) of $5. The cash flow diagram (see Figure 77.5) and the present worth of the two transactions are

$$P = -\$100 + (\$105)(P/F, 5\%, 1)$$
$$= -\$100 + (\$105)(0.9524)$$
$$= 0$$

**Figure 77.5** *Cash Flow Diagram*

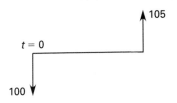

Notice that the present worth is zero even though you will receive a $5 return on your investment.

However, if you are offered $120 for the use of $100 over a one-year period, the cash flow diagram (see Fig. 77.6) and present worth (at 5%) would be

$$P = -\$100 + (\$120)(P/F, 5\%, 1)$$
$$= -\$100 + (\$120)(0.9524) = \$14.29$$

Therefore, the present worth of an alternative is seen to be equal to the equivalent value at $t = 0$ of the increase in return above that which you would be able to earn in an investment offering $i\%$ per period. In the previous case, $14.29 is the present worth of ($20 − $5), the difference in the two ROIs.

**Figure 77.6** *Cash Flow Diagram*

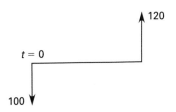

The present worth is also the amount that you would have to be given to dissuade you from making an investment, since placing the initial investment amount along with the present worth into a bank account earning $i\%$ will yield the same eventual return on investment. Relating this to the previous paragraphs, you could be dissuaded from investing $100 in an alternative that would return $120 in one year by a $t = 0$ payment of $14.29. Clearly, ($100 + $14.29) invested at $t = 0$ will also yield $120 in one year at 5%.

Income-producing alternatives with negative present worths are undesirable, and alternatives with positive present worths are desirable because they increase the average earning power of invested capital. (In some cases, such as municipal and public works projects, the present worths of all alternatives are negative, in which case, the least negative alternative is best.)

The selection of the interest rate is difficult in engineering economics problems. Usually, it is taken as the average rate of return that an individual or business organization has realized in past investments. Alternatively, the interest rate may be associated with a particular level of risk. Usually, $i$ for individuals is the interest rate that can be earned in relatively *risk-free investments*.

## 18. SIMPLE AND COMPOUND INTEREST

If $100 is invested at 5%, it will grow to $105 in one year. During the second year, 5% interest continues to be accrued, but on $105, not on $100. This is the principle of *compound interest*: The interest accrues interest.[12]

If only the original principal accrues interest, the interest is said to be *simple interest*. Simple interest is rarely encountered in engineering economic analyses, but the concept may be incorporated into short-term transactions.

## 19. EXTRACTING THE INTEREST RATE: RATE OF RETURN

An intuitive definition of the *rate of return* (ROR) is the effective annual interest rate at which an investment accrues income. That is, the rate of return of a project

---

[12]This assumes, of course, that the interest remains in the account. If the interest is removed and spent, only the remaining funds accumulate interest.

is the interest rate that would yield identical profits if all money were invested at that rate. Although this definition is correct, it does not provide a method of determining the rate of return.

It was previously seen that the present worth of a $100 investment invested at 5% is zero when $i = 5\%$ is used to determine equivalence. Thus, a working definition of rate of return would be the effective annual interest rate that makes the present worth of the investment zero. Alternatively, rate of return could be defined as the effective annual interest rate that will discount all cash flows to a total present worth equal to the required initial investment.

It is tempting, but impractical, to determine a rate of return analytically. It is simply too difficult to extract the interest rate from the equivalence equation. For example, consider a $100 investment that pays back $75 at the end of each of the first two years. The present worth equivalence equation (set equal to zero in order to determine the rate of return) is

$$P = 0 = -\$100 + (\$75)(1+i)^{-1} + (\$75)(1+i)^{-2}$$

*77.18*

Solving Eq. 77.18 requires finding the roots of a quadratic equation. In general, for an investment or project spanning $n$ years, the roots of an $n$th-order polynomial would have to be found. It should be obvious that an analytical solution would be essentially impossible for more complex cash flows. (The rate of return in this example is 31.87%.)

If the rate of return is needed, it can be found from a trial-and-error solution. To find the rate of return of an investment, proceed as follows.

*step 1:* Set up the problem as if to calculate the present worth.

*step 2:* Arbitrarily select a reasonable value for $i$. Calculate the present worth.

*step 3:* Choose another value of $i$ (not too close to the original value), and again solve for the present worth.

*step 4:* Interpolate or extrapolate the value of $i$ that gives a zero present worth.

*step 5:* For increased accuracy, repeat steps 2 and 3 with two more values that straddle the value found in step 4.

A common, although incorrect, method of calculating the rate of return involves dividing the annual receipts or returns by the initial investment. (See Sec. 77.81.) However, this technique ignores such items as salvage, depreciation, taxes, and the time value of money. This technique also is inadequate when the annual returns vary.

It is possible that more than one interest rate will satisfy the zero present worth criteria. This confusing situation

occurs whenever there is more than one change in sign in the investment's cash flow.[13] Table 77.3 indicates the numbers of possible interest rates as a function of the number of sign reversals in the investment's cash flow.

**Table 77.3** *Multiplicity of Rates of Return*

| number of sign reversals | number of distinct rates of return |
|---|---|
| 0 | 0 |
| 1 | 0 or 1 |
| 2 | 0, 1, or 2 |
| 3 | 0, 1, 2, or 3 |
| 4 | 0, 1, 2, 3, or 4 |
| $m$ | $0, 1, 2, 3, \ldots, m-1, m$ |

Difficulties associated with interpreting the meaning of multiple rates of return can be handled with the concepts of external investment and external rate of return. An *external investment* is an investment that is distinct from the investment being evaluated (which becomes known as the internal investment). The *external rate of return*, which is the rate of return earned by the external investment, does not need to be the same as the rate earned by the internal investment.

Generally, the multiple rates of return indicate that the analysis must proceed as though money will be invested outside of the project. The mechanics of how this is done are not covered here.

### Example 77.10

What is the rate of return on invested capital if $1000 is invested now with $500 being returned in year 4 and $1000 being returned in year 8?

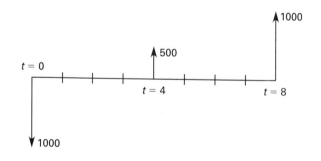

*Solution*

First, set up the problem as a present worth calculation.

---

[13]There will always be at least one change of sign in the cash flow of a legitimate investment. (This excludes municipal and other tax-supported functions.) At $t = 0$, an investment is made (a negative cash flow). Hopefully, the investment will begin to return money (a positive cash flow) at $t = 1$ or shortly thereafter. Although it is possible to conceive of an investment in which all of the cash flows were negative, such an investment would probably be classified as a *hobby*.

Try $i = 5\%$.

$$P = -\$1000 + (\$500)(P/F, 5\%, 4)$$
$$+ (\$1000)(P/F, 5\%, 8)$$
$$= -\$1000 + (\$500)(0.8227) + (\$1000)(0.6768)$$
$$= \$88$$

Next, try a larger value of $i$ to reduce the present worth. If $i = 10\%$,

$$P = -\$1000 + (\$500)(P/F, 10\%, 4)$$
$$+ (\$1000)(P/F, 10\%, 8)$$
$$= -\$1000 + (\$500)(0.6830) + (\$1000)(0.4665)$$
$$= -\$192$$

Using simple interpolation, the rate of return is

$$\text{ROR} = 5\% + \left(\frac{\$88}{\$88 + \$192}\right)(10\% - 5\%)$$
$$= 6.57\%$$

A second iteration between 6% and 7% yields 6.39%.

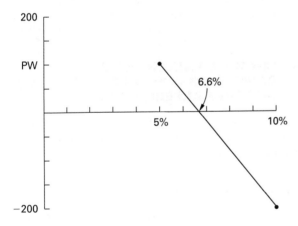

**Example 77.11**

A biomedical company is developing a new drug. A venture capital firm gives the company $25 million initially and $55 million more at the end of the first year. The drug patent will be sold at the end of year 5 to the highest bidder, and the biomedical company will receive $80 million. (The venture capital firm will receive everything in excess of $80 million.) The firm invests unused money in short-term commercial paper earning 10% effective interest per year through its bank. In the meantime, the biomedical company incurs development expenses of $50 million annually for the first three years. The drug is to be evaluated by a government agency and there will be neither expenses nor revenues during the fourth year. What is the biomedical company's rate of return on this investment?

*Solution*

Normally, the rate of return is determined by setting up a present worth problem and varying the interest rate until the present worth is zero. Writing the cash flows, though, shows that there are two reversals of sign: one at $t = 2$ (positive to negative) and the other at $t = 5$ (negative to positive). Therefore, there could be two interest rates that produce a zero present worth. (In fact, there actually are two interest rates: 10.7% and 41.4%.)

| time | cash flow (millions) |
|------|----------------------|
| 0 | +25 |
| 1 | +55 − 15 = +5 |
| 2 | −50 |
| 3 | −50 |
| 4 | 0 |
| 5 | +80 |

However, this problem can be reduced to one with only one sign reversal in the cash flow series. The initial $25 million is invested in commercial paper (an *external investment* having nothing to do with the drug development process) during the first year at 10%. The accumulation of interest and principal after one year is

$$(25)(1 + 0.10) = 27.5$$

This 27.5 is combined with the 5 (the money remaining after all expenses are paid at $t = 1$) and invested externally, again at 10%. The accumulation of interest and principal after one year (i.e., at $t = 2$) is

$$(27.5 + 5)(1 + 0.10) = 35.75$$

This 35.75 is combined with the development cost paid at $t = 2$.

The cash flow for the development project (the internal investment) is

| time | cash flow (millions) |
|------|----------------------|
| 0 | 0 |
| 1 | 0 |
| 2 | 35.75 − 50 = − 14.25 |
| 3 | − 50 |
| 4 | 0 |
| 5 | +80 |

Now, there is only one sign reversal in the cash flow series. The *internal rate of return* on this development project is found by the traditional method to be 10.3%. Notice that this is different from the rate the company can earn from investing externally in commercial paper.

## 20. RATE OF RETURN VERSUS RETURN ON INVESTMENT

*Rate of return* (ROR) is an effective annual interest rate, typically stated in percent per year. *Return on*

investment (ROI) is a dollar amount. Thus, *rate of return* and *return on investment* are not synonymous.

Return on investment can be calculated in two different ways. The accounting method is to subtract the total of all investment costs from the total of all net profits (i.e., revenues less expenses). The time value of money is not considered.

In engineering economic analysis, the return on investment is calculated from equivalent values. Specifically, the present worth (at $t = 0$) of all investment costs is subtracted from the future worth (at $t = n$) of all net profits.

When there are only two cash flows, a single investment amount and a single payback, the two definitions of return on investment yield the same numerical value. When there are more than two cash flows, the returns on investment will be different depending on which definition is used.

## 21. MINIMUM ATTRACTIVE RATE OF RETURN

A company may not know what effective interest rate, $i$, to use in engineering economic analysis. In such a case, the company can establish a minimum level of economic performance that it would like to realize on all investments. This criterion is known as the *minimum attractive rate of return* (MARR). Unlike the effective interest rate, $i$, the minimum attractive rate of return is not used in numerical calculations.[14] It is used only in comparisons with the rate of return.

Once a rate of return for an investment is known, it can be compared to the minimum attractive rate of return. To be a viable alternative, the rate of return must be greater than the minimum attractive rate of return.

The advantage of using comparisons to the minimum attractive rate of return is that an effective interest rate, $i$, never needs to be known. The minimum attractive rate of return becomes the correct interest rate for use in present worth and equivalent uniform annual cost calculations.

## 22. TYPICAL ALTERNATIVE-COMPARISON PROBLEM FORMATS

With the exception of some investment and rate of return problems, the typical problem involving engineering economics will have the following characteristics.

- An interest rate will be given.

- Two or more alternatives will be competing for funding.

- Each alternative will have its own cash flows.

- It is necessary to select the best alternative.

## 23. DURATIONS OF INVESTMENTS

Because they are handled differently, short-term investments and short-lived assets need to be distinguished from investments and assets that constitute an infinitely lived project. Short-term investments are easily identified: a drill press that is needed for three years or a temporary factory building that is being constructed to last five years.

Investments with perpetual cash flows are also (usually) easily identified: maintenance on a large flood control dam and revenues from a long-span toll bridge. Furthermore, some items with finite lives can expect renewal on a repeated basis.[15] For example, a major freeway with a pavement life of 20 years is unlikely to be abandoned; it will be resurfaced or replaced every 20 years.

Actually, if an investment's finite lifespan is long enough, it can be considered an infinite investment because money 50 or more years from now has little impact on current decisions. The $(P/F, 10\%, 50)$ factor, for example, is 0.0085. Thus, one dollar at $t = 50$ has an equivalent present worth of less than one penny. Since these far-future cash flows are eclipsed by present cash flows, long-term investments can be considered finite or infinite without significant impact on the calculations.

## 24. CHOICE OF ALTERNATIVES: COMPARING ONE ALTERNATIVE WITH ANOTHER ALTERNATIVE

Several methods exist for selecting a superior alternative from among a group of proposals. Each method has its own merits and applications.

### Present Worth Method

When two or more alternatives are capable of performing the same functions, the superior alternative will have the largest present worth. The *present worth method* is restricted to evaluating alternatives that are mutually exclusive and that have the same lives. This method is suitable for ranking the desirability of alternatives.

### Example 77.12

Investment A costs $10,000 today and pays back $11,500 two years from now. Investment B costs $8000 today and pays back $4500 each year for two years. If an interest rate of 5% is used, which alternative is superior?

---

[14]Not everyone adheres to this rule. Some people use "minimum attractive rate of return" and "effective interest rate" interchangeably.

[15]The term *renewal* can be interpreted to mean replacement or repair.

*Solution*

$$P(A) = -\$10,000 + (\$11,500)(P/F, 5\%, 2)$$
$$= -\$10,000 + (\$11,500)(0.9070)$$
$$= \$431$$
$$P(B) = -\$8000 + (\$4500)(P/A, 5\%, 2)$$
$$= -\$8000 + (\$4500)(1.8594)$$
$$= \$367$$

Alternative A is superior and should be chosen.

## Capitalized Cost Method

The present worth of a project with an infinite life is known as the *capitalized cost* or *life cycle cost*. Capitalized cost is the amount of money at $t = 0$ needed to perpetually support the project on the earned interest only. Capitalized cost is a positive number when expenses exceed income.

In comparing two alternatives, each of which is infinitely lived, the superior alternative will have the lowest capitalized cost.

Normally, it would be difficult to work with an infinite stream of cash flows since most economics tables do not list factors for periods in excess of 100 years. However, the $(A/P)$ discounting factor approaches the interest rate as $n$ becomes large. Since the $(P/A)$ and $(A/P)$ factors are reciprocals of each other, it is possible to divide an infinite series of equal cash flows by the interest rate in order to calculate the present worth of the infinite series. This is the basis of Eq. 77.19.

$$\text{capitalized cost} = \text{initial cost} + \frac{\text{annual costs}}{i} \qquad 77.19$$

Equation 77.19 can be used when the annual costs are equal in every year. If the operating and maintenance costs occur irregularly instead of annually, or if the costs vary from year to year, it will be necessary to somehow determine a cash flow of equal annual amounts (EAA) that is equivalent to the stream of original costs.

The equal annual amount may be calculated in the usual manner by first finding the present worth of all the actual costs and then multiplying the present worth by the interest rate (the $(A/P)$ factor for an infinite series). However, it is not even necessary to convert the present worth to an equal annual amount since Eq. 77.20 will convert the equal amount back to the present worth.

$$\text{capitalized cost} = \text{initial cost} + \frac{\text{EAA}}{i}$$
$$= \text{initial cost} + \frac{\text{present worth}}{\text{of all expenses}} \qquad 77.20$$

## Example 77.13

What is the capitalized cost of a public works project that will cost \$25,000,000 now and will require \$2,000,000 in maintenance annually? The effective annual interest rate is 12%.

*Solution*

Worked in millions of dollars, from Eq. 77.19, the capitalized cost is

$$\text{capitalized cost} = 25 + (2)(P/A, 12\%, \infty)$$
$$= 25 + \frac{2}{0.12} = 41.67$$

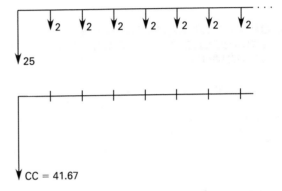

## Annual Cost Method

Alternatives that accomplish the same purpose but that have unequal lives must be compared by the *annual cost method*.[16] The annual cost method assumes that each alternative will be replaced by an identical twin at the end of its useful life (infinite renewal). This method, which may also be used to rank alternatives according to their desirability, is also called the *annual return method* or *capital recovery method*.

Restrictions are that the alternatives must be mutually exclusive and repeatedly renewed up to the duration of the longest-lived alternative. The calculated annual cost is known as the *equivalent uniform annual cost* (EUAC) or just *equivalent annual cost*. Cost is a positive number when expenses exceed income.

## Example 77.14

Which of the following alternatives is superior over a 30-year period if the interest rate is 7%?

|  | alternative A | alternative B |
|---|---|---|
| type | brick | wood |
| life | 30 years | 10 years |
| initial cost | \$1800 | \$450 |
| maintenance | \$5/year | \$20/year |

---

[16]Of course, the annual cost method can be used to determine the superiority of assets with identical lives as well.

*Solution*

$$EUAC(A) = (\$1800)(A/P, 7\%, 30) + \$5$$
$$= (\$1800)(0.0806) + \$5$$
$$= \$150$$
$$EUAC(B) = (\$450)(A/P, 7\%, 10) + \$20$$
$$= (\$450)(0.1424) + \$20$$
$$= \$84$$

Alternative B is superior since its annual cost of operation is the lowest. It is assumed that three wood facilities, each with a life of 10 years and a cost of $450, will be built to span the 30-year period.

## 25. CHOICE OF ALTERNATIVES: COMPARING AN ALTERNATIVE WITH A STANDARD

With specific economic performance criteria, it is possible to qualify an investment as acceptable or unacceptable without having to compare it with another investment. Two such performance criteria are the benefit-cost ratio and the minimum attractive rate of return.

### Benefit-Cost Ratio Method

The *benefit-cost ratio* method is often used in municipal project evaluations where benefits and costs accrue to different segments of the community. With this method, the present worth of all benefits (irrespective of the beneficiaries) is divided by the present worth of all costs. The project is considered acceptable if the ratio equals or exceeds 1.0, that is, if $B/C \geq 1.0$.

When the benefit-cost ratio method is used, disbursements by the initiators or sponsors are *costs*. Disbursements by the users of the project are known as *disbenefits*. It is often difficult to determine whether a cash flow is a cost or a disbenefit (whether to place it in the numerator or denominator of the benefit-cost ratio calculation).

Regardless of where the cash flow is placed, an acceptable project will always have a benefit-cost ratio greater than or equal to 1.0, although the actual numerical result will depend on the placement. For this reason, the benefit-cost ratio method should not be used to rank competing projects.

The benefit-cost ratio method of comparing alternatives is used extensively in transportation engineering where the ratio is often (but not necessarily) written in terms of annual benefits and annual costs instead of present worths. Another characteristic of highway benefit-cost ratios is that the route (road, highway, etc.) is usually already in place and that various alternative upgrades are being considered. There will be existing benefits and costs associated with the current route. Therefore, the

*change* (usually an increase) in benefits and costs is used to calculate the benefit-cost ratio.[17]

$$B/C = \frac{\Delta_{\text{benefits}}^{\text{user}}}{\Delta_{\text{cost}}^{\text{investment}} + \Delta\text{maintenance} - \Delta_{\text{value}}^{\text{residual}}}$$

$$77.21$$

Notice that the change in *residual value (terminal value)* appears in the denominator as a negative item. An increase in the residual value would decrease the denominator.

### Example 77.15

By building a bridge over a ravine, a state department of transportation can shorten the time it takes to drive through a mountainous area. Estimates of costs and benefits (due to decreased travel time, fewer accidents, reduced gas usage, etc.) have been prepared. Should the bridge be built? Use the benefit-cost ratio method of comparison.

| | millions |
|---|---|
| initial cost | 40 |
| capitalized cost of perpetual annual maintenance | 12 |
| capitalized value of annual user benefits | 49 |
| residual value | 0 |

*Solution*

If Eq. 77.21 is used, the benefit-cost ratio is

$$B/C = \frac{49}{40 + 12 + 0} = 0.942$$

Since the benefit-cost ratio is less than 1.00, the bridge should not be built.

If the maintenance costs are placed in the numerator (per Ftn. 17), the benefit-cost ratio value will be different, but the conclusion will not change.

$$B/C_{\text{alternate method}} = \frac{49 - 12}{40} = 0.925$$

### Rate of Return Method

The minimum attractive rate of return (MARR) has already been introduced as a standard of performance against which an investment's actual *rate of return* (ROR) is compared. If the rate of return is equal to or exceeds the minimum attractive rate of return, the

---

[17]This discussion of highway benefit-cost ratios is not meant to imply that everyone agrees with Eq. 77.21. In *Economic Analysis for Highways* (International Textbook Company, Scranton, PA, 1969), author Robley Winfrey took a strong stand on one aspect of the benefits versus disbenefits issue: highway maintenance. According to Winfrey, regular highway maintenance costs should be placed in the numerator as a subtraction from the user benefits. Some have called this mandate the *Winfrey method*.

investment is qualified. This is the basis for the *rate of return method* of alternative selection.

Finding the rate of return can be a long, iterative process. Usually, the actual numerical value of rate of return is not needed; it is sufficient to know whether or not the rate of return exceeds the minimum attractive rate of return. This *comparative analysis* can be accomplished without calculating the rate of return simply by finding the present worth of the investment using the minimum attractive rate of return as the effective interest rate (i.e., $i = $ MARR). If the present worth is zero or positive, the investment is qualified. If the present worth is negative, the rate of return is less than the minimum attractive rate of return.

## 26. RANKING MUTUALLY EXCLUSIVE MULTIPLE PROJECTS

Ranking of multiple investment alternatives is required when there is sufficient funding for more than one investment. Since the best investments should be selected first, it is necessary to place all investments into an ordered list.

Ranking is relatively easy if the present worths, future worths, capitalized costs, or equivalent uniform annual costs have been calculated for all the investments. The highest ranked investment will be the one with the largest present or future worth, or the smallest capitalized or annual cost. Present worth, future worth, capitalized cost, and equivalent uniform annual cost can all be used to rank multiple investment alternatives.

However, neither rates of return nor benefit-cost ratios should be used to rank multiple investment alternatives. Specifically, if two alternatives both have rates of return exceeding the minimum acceptable rate of return, it is not sufficient to select the alternative with the highest rate of return.

An *incremental analysis*, also known as a *rate of return on added investment study*, should be performed if rate of return is used to select between investments. An incremental analysis starts by ranking the alternatives in order of increasing initial investment. Then, the cash flows for the investment with the lower initial cost are subtracted from the cash flows for the higher-priced alternative on a year-by-year basis. This produces, in effect, a third alternative representing the costs and benefits of the added investment. The added expense of the higher-priced investment is not warranted unless the rate of return of this third alternative exceeds the minimum attractive rate of return as well. The choice criterion is to select the alternative with the higher initial investment if the incremental rate of return exceeds the minimum attractive rate of return.

An incremental analysis is also required if ranking is to be done by the benefit-cost ratio method. The incremental analysis is accomplished by calculating the ratio of differences in benefits to differences in costs for each possible pair of alternatives. If the ratio exceeds 1.0,

alternative 2 is superior to alternative 1. Otherwise, alternative 1 is superior.[18]

$$\frac{B_2 - B_1}{C_2 - C_1} \geq 1 \quad \text{[alternative 2 superior]} \qquad 77.22$$

## 27. ALTERNATIVES WITH DIFFERENT LIVES

Comparison of two alternatives is relatively simple when both alternatives have the same life. For example, a problem might be stated: "Which would you rather have: car A with a life of three years, or car B with a life of five years?"

However, care must be taken to understand what is going on when the two alternatives have different lives. If car A has a life of three years and car B has a life of five years, what happens at $t = 3$ if the five-year car is chosen? If a car is needed for five years, what happens at $t = 3$ if the three-year car is chosen?

In this type of situation, it is necessary to distinguish between the length of the need (the *analysis horizon*) and the lives of the alternatives or assets intended to meet that need. The lives do not have to be the same as the horizon.

### Finite Horizon with Incomplete Asset Lives

If an asset with a five-year life is chosen for a three-year need, the disposition of the asset at $t = 3$ must be known in order to evaluate the alternative. If the asset is sold at $t = 3$, the salvage value is entered into the analysis (at $t = 3$) and the alternative is evaluated as a three-year investment. The fact that the asset is sold when it has some useful life remaining does not affect the analysis horizon.

Similarly, if a three-year asset is chosen for a five-year need, something about how the need is satisfied during the last two years must be known. Perhaps a rental asset will be used. Or, perhaps the function will be "farmed out" to an outside firm. In any case, the costs of satisfying the need during the last two years enter the analysis, and the alternative is evaluated as a five-year investment.

If both alternatives are "converted" to the same life, any of the alternative selection criteria (present worth method, annual cost method, etc.) can be used to determine which alternative is superior.

### Finite Horizon with Integer Multiple Asset Lives

It is common to have a long-term horizon (need) that must be met with short-lived assets. In special instances, the horizon will be an integer number of asset lives. For example, a company may be making a 12-year

---

[18]It goes without saying that the benefit-cost ratios for all investment alternatives by themselves must also be equal to or greater than 1.0.

transportation plan and may be evaluating two cars: one with a three-year life, and another with a four-year life.

In this example, four of the first car or three of the second car are needed to reach the end of the 12-year horizon.

If the horizon is an integer number of asset lives, any of the alternative selection criteria can be used to determine which is superior. If the present worth method is used, all alternatives must be evaluated over the entire horizon. (In this example, the present worth of 12 years of car purchases and use must be determined for both alternatives.)

If the equivalent uniform annual cost method is used, it may be possible to base the calculation of annual cost on one lifespan of each alternative only. It may not be necessary to incorporate all of the cash flows into the analysis. (In the running example, the annual cost over three years would be determined for the first car; the annual cost over four years would be determined for the second car.) This simplification is justified if the subsequent asset replacements (renewals) have the same cost and cash flow structure as the original asset. This assumption is typically made implicitly when the annual cost method of comparison is used.

### Infinite Horizon

If the need horizon is infinite, it is not necessary to impose the restriction that asset lives of alternatives be integer multiples of the horizon. The superior alternative will be replaced (renewed) whenever it is necessary to do so, forever.

Infinite horizon problems are almost always solved with either the annual cost or capitalized cost method. It is common to (implicitly) assume that the cost and cash flow structure of the asset replacements (renewals) are the same as the original asset.

### 28. OPPORTUNITY COSTS

An *opportunity cost* is an imaginary cost representing what will not be received if a particular strategy is rejected. It is what you will lose if you do or do not do something. As an example, consider a growing company with an existing operational computer system. If the company trades in its existing computer as part of an upgrade plan, it will receive a *trade-in allowance*. (In other problems, a *salvage value* may be involved.)

If one of the alternatives being evaluated is not to upgrade the computer system at all, the trade-in allowance (or, salvage value in other problems) will not be realized. The amount of the trade-in allowance is an opportunity cost that must be included in the problem analysis.

Similarly, if one of the alternatives being evaluated is to wait one year before upgrading the computer, the *difference in trade-in allowances* is an opportunity cost that must be included in the problem analysis.

### 29. REPLACEMENT STUDIES

An investigation into the retirement of an existing process or piece of equipment is known as a *replacement study*. Replacement studies are similar in most respects to other alternative comparison problems: An interest rate is given, two alternatives exist, and one of the previously mentioned methods of comparing alternatives is used to choose the superior alternative. Usually, the annual cost method is used on a year-by-year basis.

In replacement studies, the existing process or piece of equipment is known as the *defender*. The new process or piece of equipment being considered for purchase is known as the *challenger*.

### 30. TREATMENT OF SALVAGE VALUE IN REPLACEMENT STUDIES

Since most defenders still have some market value when they are retired, the problem of what to do with the salvage arises. It seems logical to use the salvage value of the defender to reduce the initial purchase cost of the challenger. This is consistent with what would actually happen if the defender were to be retired.

By convention, however, the defender's salvage value is subtracted from the defender's present value. This does not seem logical, but it is done to keep all costs and benefits related to the defender with the defender. In this case, the salvage value is treated as an opportunity cost that would be incurred if the defender is not retired.

If the defender and the challenger have the same lives and a present worth study is used to choose the superior alternative, the placement of the salvage value will have no effect on the net difference between present worths for the challenger and defender. Although the values of the two present worths will be different depending on the placement, the difference in present worths will be the same.

If the defender and the challenger have different lives, an annual cost comparison must be made. Since the salvage value would be "spread over" a different number of years depending on its placement, it is important to abide by the conventions listed in this section.

There are a number of ways to handle salvage value in retirement studies. The best way is to calculate the cost of keeping the defender one more year. In addition to the usual operating and maintenance costs, that cost includes an opportunity interest cost incurred by not selling the defender, and also a drop in the salvage value

Engineering Management

if the defender is kept for one additional year. Specifically,

$$\text{EUAC (defender)} = \text{next year's maintenance costs}$$
$$+ \, i(\text{current salvage value})$$
$$+ \, \text{current salvage}$$
$$- \, \text{next year's salvage}$$

$$77.23$$

It is important in retirement studies not to double count the salvage value. That is, it would be incorrect to add the salvage value to the defender and at the same time subtract it from the challenger.

Equation 77.23 contains the difference in salvage value between two consecutive years. This calculation shows that the defender/challenger decision must be made on a year-by-year basis. One application of Eq. 77.23 will not usually answer the question of whether the defender should remain in service indefinitely. The calculation must be repeatedly made as long as there is a drop in salvage value from one year to the next.

## 31. ECONOMIC LIFE: RETIREMENT AT MINIMUM COST

As an asset grows older, its operating and maintenance costs typically increase. Eventually, the cost to keep the asset in operation becomes prohibitive, and the asset is retired or replaced. However, it is not always obvious when an asset should be retired or replaced.

As the asset's maintenance cost is increasing each year, the amortized cost of its initial purchase is decreasing. It is the sum of these two costs that should be evaluated to determine the point at which the asset should be retired or replaced. Since an asset's initial purchase price is likely to be high, the amortized cost will be the controlling factor in those years when the maintenance costs are low. Therefore, the EUAC of the asset will decrease in the initial part of its life. (See Fig. 77.7.)

**Figure 77.7** *EUAC Versus Age at Retirement*

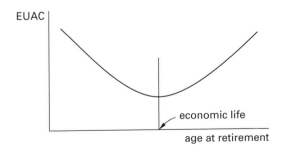

However, as the asset grows older, the change in its amortized cost decreases while maintenance cost increases. Eventually, the sum of the two costs reaches a minimum and then starts to increase. The age of the asset at the minimum cost point is known as the *economic life* of the asset. The economic life generally is less than the length of need and the technological lifetime of the asset.

The determination of an asset's economic life is illustrated by Ex. 77.16.

### Example 77.16

Buses in a municipal transit system have the characteristics listed. In order to minimize its annual operating expenses, when should the city replace its buses if money can be borrowed at 8%?

initial cost of bus: $120,000

| year | maintenance cost | salvage value |
|------|------------------|---------------|
| 1 | 35,000 | 60,000 |
| 2 | 38,000 | 55,000 |
| 3 | 43,000 | 45,000 |
| 4 | 50,000 | 25,000 |
| 5 | 65,000 | 15,000 |

*Solution*

The annual maintenance is different each year. Each maintenance cost must be spread over the life of the bus. This is done by first finding the present worth and then amortizing the maintenance costs. If a bus is kept for one year and then sold, the annual cost will be

$$\text{EUAC}(1) = (\$120{,}000)(A/P, 8\%, 1)$$
$$+ (\$35{,}000)(A/F, 8\%, 1)$$
$$- (\$60{,}000)(A/F, 8\%, 1)$$
$$= (\$120{,}000)(1.0800) + (\$35{,}000)(1.000)$$
$$- (\$60{,}000)(1.000)$$
$$= \$104{,}600$$

If a bus is kept for two years and then sold, the annual cost will be

$$\text{EUAC}(2) = (\$120{,}000 + (\$35{,}000)(P/F, 8\%, 1))$$
$$\times (A/P, 8\%, 2)$$
$$+ (\$38{,}000 - \$55{,}000)(A/F, 8\%, 2)$$
$$= (\$120{,}000 + (\$35{,}000)(0.9259))(0.5608)$$
$$+ (\$38{,}000 - \$55{,}000)(0.4808)$$
$$= \$77{,}296$$

If a bus is kept for three years and then sold, the annual cost will be

$$\begin{aligned}
\text{EUAC}(3) &= (\$120{,}000 + (35{,}000)(P/F, 8\%, 1) \\
&\quad + (\$38{,}000)(P/F, 8\%, 2))(A/P, 8\%, 3) \\
&\quad + (\$43{,}000 - \$45{,}000)(A/F, 8\%, 3) \\
&= (\$120{,}000 + (\$35{,}000)(0.9259) \\
&\quad + (\$38{,}000)(0.8573))(0.3880) \\
&\quad - (\$2000)(0.3080) \\
&= \$71{,}158
\end{aligned}$$

This process is continued until the annual cost begins to increase. In this example, EUAC(4) is $71,700. Therefore, the buses should be retired after three years.

## 32. LIFE-CYCLE COST

The *life-cycle cost* of an alternative is the equivalent value (at $t = 0$) of the alternative's cash flow over the alternative's lifespan. Since the present worth is evaluated using an effective interest rate of $i$ (which would be the interest rate used for all engineering economic analyses), the life-cycle cost is the same as the alternative's present worth. If the alternative has an infinite horizon, the life-cycle cost and capitalized cost will be identical.

## 33. CAPITALIZED ASSETS VERSUS EXPENSES

High expenses reduce profit, which in turn reduces income tax. It seems logical to label each and every expenditure, even an asset purchase, as an expense. As an alternative to this *expensing the asset*, it may be decided to capitalize the asset. *Capitalizing the asset* means that the cost of the asset is divided into equal or unequal parts, and only one of these parts is taken as an expense each year. Expensing is clearly the more desirable alternative, since the after-tax profit is increased early in the asset's life.

There are long-standing accounting conventions as to what can be expensed and what must be capitalized.[19] Some companies capitalize everything—regardless of cost —with expected lifetimes greater than one year. Most companies, however, expense items whose purchase costs are below a cut-off value. A cut-off value in the range of $250 to $500, depending on the size of the company, is chosen as the maximum purchase cost of an expensed asset. Assets costing more than this are capitalized.

It is not necessary for a large corporation to keep track of every lamp, desk, and chair for which the purchase price is greater than the cut-off value. Such assets, all of which

have the same lives and have been purchased in the same year, can be placed into groups or *asset classes*. A group cost, equal to the sum total of the purchase costs of all items in the group, is capitalized as though the group was an identifiable and distinct asset itself.

## 34. PURPOSE OF DEPRECIATION

*Depreciation* is an *artificial expense* that spreads the purchase price of an asset or other property over a number of years.[20] Depreciating an asset is an example of capitalization, as previously defined. The inclusion of depreciation in engineering economic analysis problems will increase the after-tax present worth (profitability) of an asset. The larger the depreciation, the greater will be the profitability. Therefore, individuals and companies eligible to utilize depreciation want to maximize and accelerate the depreciation available to them.

Although the entire property purchase price is eventually recognized as an expense, the net recovery from the expense stream never equals the original cost of the asset. That is, depreciation cannot realistically be thought of as a fund (an annuity or sinking fund) that accumulates capital to purchase a replacement at the end of the asset's life. The primary reason for this is that the depreciation expense is reduced significantly by the impact of income taxes, as will be seen in later sections.

## 35. DEPRECIATION BASIS OF AN ASSET

The *depreciation basis* of an asset is the part of the asset's purchase price that is spread over the *depreciation period*, also known as the *service life*.[21] Usually, the depreciation basis and the purchase price are not the same.

A common depreciation basis is the difference between the purchase price and the expected salvage value at the end of the depreciation period. That is,

$$\text{depreciation basis} = C - S_n \qquad 77.24$$

There are several methods of calculating the year-by-year depreciation of an asset. Equation 77.24 is not universally compatible with all depreciation methods. Some methods do not consider the salvage value. This

---

[19]For example, purchased vehicles must be capitalized; payments for leased vehicles can be expensed. Repainting a building with paint that will last five years is an expense, but the replacement cost of a leaking roof must be capitalized.

[20]In the United States, the tax regulations of Internal Revenue Service (IRS) allow depreciation on almost all forms of *business property* except land. The following types of property are distinguished: *real* (e.g., buildings used for business), *residential* (e.g., buildings used as rental property), and *personal* (e.g., equipment used for business). Personal property does *not* include items for personal use (such as a personal residence), despite its name. *Tangible personal property* is distinguished from *intangible property* (goodwill, copyrights, patents, trademarks, franchises, agreements not to compete, etc.).

[21]The *depreciation period* is selected to be as short as possible within recognized limits. This depreciation will not normally coincide with the *economic life* or *useful life* of an asset. For example, a car may be capitalized over a depreciation period of three years. It may become uneconomical to maintain and use at the end of an economic life of nine years. However, the car may be capable of operation over a useful life of 25 years.

is known as an *unadjusted basis*. When the depreciation method is known, the depreciation basis can be rigorously defined.[22]

## 36. DEPRECIATION METHODS

Generally, tax regulations do not allow the cost of an asset to be treated as a deductible expense in the year of purchase. Rather, portions of the depreciation basis must be allocated to each of the $n$ years of the asset's depreciation period. The amount that is allocated each year is called the *depreciation*.

Various methods exist for calculating an asset's depreciation each year.[23] Although the depreciation calculations may be considered independently (for the purpose of determining book value or as an academic exercise), it is important to recognize that depreciation has no effect on engineering economic analyses unless income taxes are also considered.

### Straight Line Method

With the *straight line method*, depreciation is the same each year. The depreciation basis $(C - S_n)$ is allocated uniformly to all of the $n$ years in the depreciation period. Each year, the depreciation will be

$$D = \frac{C - S_n}{n} \qquad 77.25$$

### Constant Percentage Method

The *constant percentage method*[24] is similar to the straight line method in that the depreciation is the same each year. If the fraction of the basis used as depreciation is $1/n$, there is no difference between the constant percentage and straight line methods. The two methods differ only in what information is available. (With the straight line method, the life is known. With the constant percentage method, the depreciation fraction is known.)

Each year, the depreciation will be

$$D = (\text{depreciation fraction})(\text{depreciation basis})$$
$$= (\text{depreciation fraction})(C - S_n) \qquad 77.26$$

### Sum-of-the-Years' Digits Method

In *sum-of-the-years' digits* (SOYD) depreciation, the digits from 1 to $n$ inclusive are summed. The total, $T$, can also be calculated from

$$T = \tfrac{1}{2}n(n+1) \qquad 77.27$$

The depreciation in year $j$ can be found from Eq. 77.28. Notice that the depreciation in year $j$, $D_j$, decreases by a constant amount each year.

$$D_j = \frac{(C - S_n)(n - j + 1)}{T} \qquad 77.28$$

### Double Declining Balance Method[25]

*Double declining balance*[26] (DDB) depreciation is independent of salvage value. Furthermore, the book value never stops decreasing, although the depreciation decreases in magnitude. Usually, any book value in excess of the salvage value is written off in the last year of the asset's depreciation period. Unlike any of the other depreciation methods, double declining balance depends on accumulated depreciation.

$$D_{\text{first year}} = \frac{2C}{n} \qquad 77.29$$

$$D_j = \frac{2\left(C - \displaystyle\sum_{m=1}^{j-1} D_m\right)}{n} \qquad 77.30$$

Calculating the depreciation in the middle of an asset's life appears particularly difficult with double declining balance, since all previous years' depreciation amounts seem to be required. It appears that the depreciation in the sixth year, for example, cannot be calculated unless the values of depreciation for the first five years are calculated. However, this is not true.

Depreciation in the middle of an asset's life can be found from the following equations. ($d$ is known as the *depreciation rate*.)

$$d = \frac{2}{n} \qquad 77.31$$

$$D_j = dC(1 - d)^{j-1} \qquad 77.32$$

### Statutory Depreciation Systems

In the United States, property placed into service in 1981 and thereafter must use the *Accelerated Cost Recovery System* (ACRS), and after 1986, the *Modified*

---

[22]For example, with the Accelerated Cost Recovery System (ACRS) the *depreciation basis* is the total purchase cost, regardless of the expected salvage value. With declining balance methods, the depreciation basis is the purchase cost less any previously taken depreciation.
[23]This discussion gives the impression that any form of depreciation may be chosen regardless of the nature and circumstances of the purchase. In reality, the IRS tax regulations place restrictions on the higher-rate (accelerated) methods, such as declining balance and sum-of-the-years' digits methods. Furthermore, the *Economic Recovery Act of 1981* and the *Tax Reform Act of 1986* substantially changed the laws relating to personal and corporate income taxes.
[24]The *constant percentage method* should not be confused with the declining balance method, which used to be known as the *fixed percentage on diminishing balance method*.

[25]In the past, the *declining balance method* has also been known as the *fixed percentage of book value* and *fixed percentage on diminishing balance method*.
[26]Double declining balance depreciation is a particular form of *declining balance depreciation*, as defined by the IRS tax regulations. Declining balance depreciation includes 125% declining balance and 150% declining balance depreciations that can be calculated by substituting 1.25 and 1.50, respectively, for the 2 in Eq. 77.29.

*Accelerated Cost Recovery System* (MACRS) or other statutory method. Other methods (straight line, declining balance, etc.) cannot be used except in special cases.

Property placed into service in 1980 or before must continue to be depreciated according to the method originally chosen (e.g., straight line, declining balance, or sum-of-the-years' digits). ACRS and MACRS cannot be used.

Under ACRS and MACRS, the cost recovery amount in the $j$th year of an asset's cost recovery period is calculated by multiplying the initial cost by a factor.

$$D_j = C \times \text{factor} \qquad 77.33$$

The initial cost used is not reduced by the asset's salvage value for ACRS and MACRS calculations. The factor used depends on the asset's cost recovery period. (See Table 77.4.) Such factors are subject to continuing legislation changes. Current tax publications should be consulted before using this method.

**Table 77.4** *Representative MACRS Depreciation Factors**

| | depreciation rate for recovery period ($n$) | | | |
|---|---|---|---|---|
| year ($j$) | 3 years | 5 years | 7 years | 10 years |
| 1 | 33.33% | 20.00% | 14.29% | 10.00% |
| 2 | 44.45% | 32.00% | 24.49% | 18.00% |
| 3 | 14.81% | 19.20% | 17.49% | 14.40% |
| 4 | 7.41% | 11.52% | 12.49% | 11.52% |
| 5 | | 11.52% | 8.93% | 9.22% |
| 6 | | 5.76% | 8.92% | 7.37% |
| 7 | | | 8.93% | 6.55% |
| 8 | | | 4.46% | 6.55% |
| 9 | | | | 6.56% |
| 10 | | | | 6.55% |
| 11 | | | | 3.28% |

*Values are for the "half-year" convention. This table gives typical values only. Since these factors are subject to continuing revision, they should not be used without consulting an accounting professional.

### Production or Service Output Method

If an asset has been purchased for a specific task and that task is associated with a specific lifetime amount of output or production, the depreciation may be calculated by the fraction of total production produced during the year. Under the *units of production* method, the depreciation is not expected to be the same each year.

$$D_j = (C - S_n)\left(\frac{\text{actual output in year } j}{\text{estimated lifetime output}}\right) \qquad 77.34$$

### Sinking Fund Method

The *sinking fund method* is seldom used in industry because the initial depreciation is low. The formula for sinking fund depreciation (which increases each year) is

$$D_j = (C - S_n)(A/F, i\%, n)(F/P, i\%, j - 1) \qquad 77.35$$

## Example 77.17

An asset is purchased for $9000. Its estimated economic life is ten years, after which it will be sold for $200. Find the depreciation in the first three years using straight line, double declining balance, and sum-of-the-years' digits depreciation methods.

*Solution*

$$\text{SL: } D = \frac{\$9000 - \$200}{10} = \$880 \text{ each year}$$

$$\text{DDB: } D_1 = \frac{(2)(\$9000)}{10} = \$1800 \text{ in year 1}$$

$$D_2 = \frac{(2)(\$9000 - \$1800)}{10} = \$1440 \text{ in year 2}$$

$$D_3 = \frac{(2)(\$9000 - \$3240)}{10} = \$1152 \text{ in year 3}$$

$$\text{SOYD: } T = \left(\tfrac{1}{2}\right)(10)(11) = 55$$

$$D_1 = \left(\tfrac{10}{55}\right)(\$9000 - \$200) = \$1600 \text{ in year 1}$$

$$D_2 = \left(\tfrac{9}{55}\right)(\$8800) = \$1440 \text{ in year 2}$$

$$D_3 = \left(\tfrac{8}{55}\right)(\$8800) = \$1280 \text{ in year 3}$$

## 37. ACCELERATED DEPRECIATION METHODS

An *accelerated depreciation method* is one that calculates a depreciation amount greater than a straight line amount. Double declining balance and sum-of-the-years' digits methods are accelerated methods. The ACRS and MACRS methods are explicitly accelerated methods. Straight line and sinking fund methods are not accelerated methods.

Use of an accelerated depreciation method may result in unexpected tax consequences when the depreciated asset or property is disposed of. Professional tax advice should be obtained in this area.

## 38. BOOK VALUE

The difference between original purchase price and accumulated depreciation is known as *book value*.[27] At the end of each year, the book value (which is initially equal to the purchase price) is reduced by the depreciation in that year.

It is important to properly synchronize depreciation calculations. It is difficult to answer the question, "What is the book value in the fifth year?" unless the timing of

---

[27]The balance sheet of a corporation usually has two asset accounts: the *equipment account* and the *accumulated depreciation account*. There is no book value account on this financial statement, other than the implicit value obtained from subtracting the accumulated depreciation account from the equipment account. The book values of various assets, as well as their original purchase cost, date of purchase, salvage value, and so on, and accumulated depreciation appear on detail sheets or other peripheral records for each asset.

the book value change is mutually agreed upon. It is better to be specific about an inquiry by identifying when the book value change occurs. For example, the following question is unambiguous: "What is the book value at the end of year 5, after subtracting depreciation in the fifth year?" or "What is the book value after five years?"

Unfortunately, this type of care is seldom taken in book value inquiries, and it is up to the respondent to exercise reasonable care in distinguishing between beginning-of-year book value and end-of-year book value. To be consistent, the book value equations in this chapter have been written in such a way that the year subscript $(j)$ has the same meaning in book value and depreciation calculations. That is, $BV_5$ means the book value at the end of the fifth year, after five years of depreciation, including $D_5$, have been subtracted from the original purchase price.

There can be a great difference between the book value of an asset and the *market value* of that asset. There is no legal requirement for the two values to coincide, and no intent for book value to be a reasonable measure of market value.[28] Therefore, it is apparent that book value is merely an accounting convention with little practical use. Even when a depreciated asset is disposed of, the book value is used to determine the consequences of disposal, not the price the asset should bring at sale.

The calculation of book value is relatively easy, even for the case of the declining balance depreciation method.

For the straight line depreciation method, the book value at the end of the $j$th year, after the $j$th depreciation deduction has been made, is

$$BV_j = C - \frac{j(C - S_n)}{n} = C - jD \qquad 77.36$$

For the sum-of-the-years' digits method, the book value is

$$BV_j = (C - S_n)\left(1 - \frac{j(2n + 1 - j)}{n(n + 1)}\right) + S_n \qquad 77.37$$

For the declining balance method, including double declining balance (see Ftn. 26), the book value is

$$BV_j = C(1 - d)^j \qquad 77.38$$

For the sinking fund method, the book value is calculated directly as

$$BV_j = C - (C - S_n)(A/F, i\%, n)(F/A, i\%, j) \qquad 77.39$$

Of course, the book value at the end of year $j$ can always be calculated for any method by successive subtractions

(i.e., subtraction of the accumulated depreciation), as Eq. 77.40 illustrates.

$$BV_j = C - \sum_{m=1}^{j} D_m \qquad 77.40$$

Figure 77.8 illustrates the book value of a hypothetical asset depreciated using several depreciation methods. Notice that the double declining balance method initially produces the fastest write-off, while the sinking fund method produces the slowest write-off. Note also that the book value does not automatically equal the salvage value at the end of an asset's depreciation period with the double declining balance method.[29]

**Figure 77.8** *Book Value with Different Depreciation Methods*

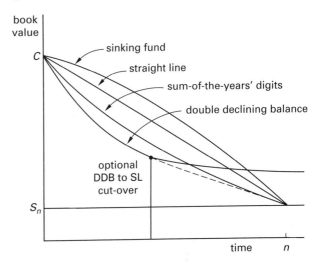

**Example 77.18**

For the asset described in Ex. 77.17, calculate the book value at the end of the first three years if sum-of-the-years' digits depreciation is used. The book value at the beginning of year 1 is $9000.

*Solution*

From Eq. 77.40,

$$BV_1 = \$9000 - \$1600 = \$7400$$
$$BV_2 = \$7400 - \$1440 = \$5960$$
$$BV_3 = \$5960 - \$1280 = \$4680$$

---

[28]Common examples of assets with great divergences of book and market values are buildings (rental houses, apartment complexes, factories, etc.) and company luxury automobiles (Porsches, Mercedes, etc.) during periods of inflation. Book values decrease, but actual values increase.

[29]This means that the straight line method of depreciation may result in a lower book value at some point in the depreciation period than if double declining balance is used. A *cut-over* from double declining balance to straight line may be permitted in certain cases. Finding the *cut-over point*, however, is usually done by comparing book values determined by both methods. The analytical method is complicated.

## 39. AMORTIZATION

Amortization and depreciation are similar in that they both divide up the cost basis or value of an asset. In fact, in certain cases, the term "amortization" may be used in place of the term "depreciation." However, depreciation is a specific form of amortization.

Amortization spreads the cost basis or value of an asset over some base. The base can be time, units of production, number of customers, and so on. The asset can be tangible (e.g., a delivery truck or building) or intangible (e.g., goodwill or a patent).

If the asset is tangible, if the base is time, and if the length of time is consistent with accounting standards and taxation guidelines, then the term "depreciation" is appropriate. However, if the asset is intangible, if the base is some other variable, or if some length of time other than the customary period is used, then the term "amortization" is more appropriate.[30]

### Example 77.19

A company purchases complete and exclusive patent rights to an invention for $1,200,000. It is estimated that once commercially produced, the invention will have a specific but limited market of 1200 units. For the purpose of allocating the patent right cost to production cost, what is the amortization rate in dollars per unit?

*Solution*

The patent should be amortized at the rate of

$$\frac{\$1,200,000}{1200 \text{ units}} = \$1000 \text{ per unit}$$

## 40. DEPLETION

*Depletion* is another artificial deductible operating expense, designed to compensate mining organizations for decreasing mineral reserves. Since original and remaining quantities of minerals are seldom known accurately, the *depletion allowance* is calculated as a fixed percentage of the organization's gross income. These percentages are usually in the 10% to 20% range and apply to such mineral deposits as oil, natural gas, coal, uranium, and most metal ores.

## 41. BASIC INCOME TAX CONSIDERATIONS

The issue of income taxes is often overlooked in academic engineering economic analysis exercises. Such a position is justifiable when an organization (e.g., a non-profit school, a church, or the government) pays no

income taxes. However, if an individual or organization is subject to income taxes, the income taxes must be included in an economic analysis of investment alternatives.

Assume that an organization pays a fraction $f$ of its profits to the federal government as income taxes. If the organization also pays a fraction $s$ of its profits as state income taxes and if state taxes paid are recognized by the federal government as tax-deductible expenses, then the composite tax rate is

$$t = s + f - sf \qquad 77.41$$

The basic principles used to incorporate taxation into engineering economic analyses are the following.

- Initial purchase expenditures are unaffected by income taxes.

- Salvage revenues are unaffected by income taxes.

- Deductible expenses, such as operating costs, maintenance costs, and interest payments, are reduced by the fraction $t$ (i.e., multiplied by the quantity $(1-t)$).

- Revenues are reduced by the fraction $t$ (i.e., multiplied by the quantity $(1-t)$).

- Since tax regulations allow the depreciation in any year to be handled as if it were an actual operating expense, and since operating expenses are deductible from the income base prior to taxation, the after-tax profits will be increased. If $D$ is the depreciation, the net result to the after-tax cash flow will be the addition of $tD$. Depreciation is multiplied by $t$ and added to the appropriate year's cash flow, increasing that year's present worth.

For simplicity, most engineering economics practice problems involving income taxes specify a single income tax rate. In practice, however, federal and most state tax rates depend on the income level. Each range of incomes and its associated tax rate are known as *income bracket* and *tax bracket*, respectively. For example, the state income tax rate might be 4% for incomes up to and including $30,000, and 5% for incomes above $30,000. The income tax for a taxpaying entity with an income of $50,000 would have to be calculated in two parts.

$$\text{tax} = (0.04)(\$30,000) + (0.05)(\$50,000 - \$30,000)$$
$$= \$2200$$

Income taxes and depreciation have no bearing on municipal or governmental projects since municipalities, states, and the U.S. government pay no taxes.

### Example 77.20

A corporation that pays 53% of its profit in income taxes invests $10,000 in an asset that will produce a $3000 annual revenue for eight years. If the annual expenses are $700, salvage after eight years is $500,

---

[30]From time to time, the U.S. Congress has allowed certain types of facilities (e.g., emergency, grain storage, and pollution control) to be written off more rapidly than would otherwise be permitted in order to encourage investment in such facilities. The term "amortization" has been used with such write-off periods.

and 9% interest is used, what is the after-tax present worth? Disregard depreciation.

*Solution*

$$P = -\$10,000 + (\$3000)(P/A, 9\%, 8)(1 - 0.53)$$
$$- (\$700)(P/A, 9\%, 8)(1 - 0.53)$$
$$+ (\$500)(P/F, 9\%, 8)$$
$$= -\$10,000 + (\$3000)(5.5348)(0.47)$$
$$- (\$700)(5.5348)(0.47) + (\$500)(0.5019)$$
$$= -\$3766$$

## 42. TAXATION AT THE TIMES OF ASSET PURCHASE AND SALE

There are numerous rules and conventions that governmental tax codes and the accounting profession impose on organizations. Engineers are not normally expected to be aware of most of the rules and conventions, but occasionally it may be necessary to incorporate their effects into an engineering economic analysis.

### Tax Credit

A *tax credit* (also known as an *investment tax credit* or *investment credit*) is a one-time credit against income taxes.[31] Therefore, it is added to the after-tax present worth as a last step in an engineering economic analysis. Such tax credits may be allowed by the government from time to time for equipment purchases, employment of various classes of workers, rehabilitation of historic landmarks, and so on.

A tax credit is usually calculated as a fraction of the initial purchase price or cost of an asset or activity.

$$\text{TC} = \text{fraction} \times \text{initial cost} \qquad 77.42$$

When the tax credit is applicable, the fraction used is subject to legislation. A professional tax expert or accountant should be consulted prior to applying the tax credit concept to engineering economic analysis problems.

Since the investment tax credit reduces the buyer's tax liability, a tax credit should be included only in after-tax engineering economic analyses. The credit is assumed to be received at the end of the year.

### Gain or Loss on the Sale of a Depreciated Asset

If an asset that has been depreciated over a number of prior years is sold for more than its current book value, the difference between the book value and selling price is taxable income in the year of the sale. Alternatively, if the asset is sold for less than its current book value, the difference between the selling price and book value is an expense in the year of the sale.

### Example 77.21

One year, a company makes a $5000 investment in a historic building. The investment is not depreciable, but it does qualify for a one-time 20% tax credit. In that same year, revenue is $45,000 and expenses (exclusive of the $5000 investment) are $25,000. The company pays a total of 53% in income taxes. What is the after-tax present worth of this year's activities if the company's interest rate for investment is 10%?

*Solution*

The tax credit is

$$\text{TC} = (0.20)(\$5000) = \$1000$$

This tax credit is assumed to be received at the end of the year. The after-tax present worth is

$$P = -\$5000 + (\$45,000 - \$25,000)(1 - 0.53)$$
$$\times (P/F, 10\%, 1) + (\$1000)(P/F, 10\%, 1)$$
$$= -\$5000 + (\$20,000)(0.47)(0.9091)$$
$$+ (\$1000)(0.9091)$$
$$= \$4455$$

## 43. DEPRECIATION RECOVERY

The economic effect of depreciation is to reduce the income tax in year $j$ by $tD_j$. The present worth of the asset is also affected: The present worth is increased by $tD_j(P/F, i\%, j)$. The after-tax present worth of all depreciation effects over the depreciation period of the asset is called the *depreciation recovery* (DR).[32]

$$\text{DR} = t \sum_{j=1}^{n} D_j(P/F, i\%, j) \qquad 77.43$$

There are multiple ways depreciation can be calculated, as summarized in Table 77.5. *Straight line depreciation recovery* from an asset is easily calculated, since the depreciation is the same each year. Assuming the asset has a constant depreciation of $D$ and depreciation period of $n$ years, the depreciation recovery is

$$\text{DR} = tD(P/A, i\%, n) \qquad 77.44$$

$$D = \frac{C - S_n}{n} \qquad 77.45$$

---

[31]Strictly, *tax credit* is the more general term, and applies to a credit for doing anything creditable. An *investment tax credit* requires an investment in something (usually real property or equipment).

[32]Since the depreciation benefit is reduced by taxation, depreciation cannot be thought of as an annuity to fund a replacement asset.

**Table 77.5** Depreciation Calculation Summary

| method | depreciation basis | depreciation in year $j$ ($D_j$) | book value after $j$th depreciation ($BV_j$) | after-tax depreciation recovery (DR) | supplementary formulas |
|---|---|---|---|---|---|
| straight line (SL) | $C - S_n$ | $\dfrac{C - S_n}{n}$ (constant) | $C - jD$ | $tD(P/A, i\%, n)$ | |
| constant percentage | $C - S_n$ | fraction $\times (C - S_n)$ (constant) | $C - jD$ | $tD(P/A, i\%, n)$ | |
| sum-of-the-years' digits (SOYD) | $C - S_n$ | $\dfrac{(C - S_n) \times (n - j + 1)}{T}$ | $(C - S_n) \times \left(1 - \dfrac{j(2n + 1 - j)}{n(n+1)}\right) + S_n$ | $\dfrac{t(C - S_n)}{T} \times (n(P/A, i\%, n) - (P/G, i\%, n))$ | $T = \frac{1}{2}n(n+1)$ |
| double declining balance (DDB) | $C$ | $dC(1-d)^{j-1}$ | $C(1-d)^j$ | $tC\left(\dfrac{d}{1-d}\right) \times (P/EG, z-1, n)$ | $d = \dfrac{2}{n}$; $z = \dfrac{1+i}{1-d}$ $(P/EG, z-1, n) = \dfrac{z^n - 1}{z^n(z-1)}$ |
| sinking fund (SF) | $C - S_n$ | $(C - S_n) \times (A/F, i\%, n) \times (F/P, i\%, j-1)$ | $C - (C - S_n) \times (A/F, i\%, n) \times (F/A, i\%, j)$ | $\dfrac{t(C - S_n)(A/F, i\%, n)}{1+i}$ | |
| accelerated cost recovery system (ACRS/MACRS) | $C$ | $C \times$ factor | $C - \sum\limits_{m=1}^{j} D_m$ | $t \sum\limits_{j=1}^{n} D_j(P/F, i\%, j)$ | |
| units of production or service output | $C - S_n$ | $(C - S_n) \times \left(\dfrac{\text{actual output in year } j}{\text{lifetime output}}\right)$ | $C - \sum\limits_{m=1}^{j} D_m$ | $t \sum\limits_{j=1}^{n} D_j(P/F, i\%, j)$ | |

*Sum-of-the-years' digits depreciation recovery* is also relatively easily calculated, since the depreciation decreases uniformly each year.

$$\text{DR} = \left(\frac{t(C - S_n)}{T}\right)(n(P/A, i\%, n) - (P/G, i\%, n)) \qquad 77.46$$

Finding *declining balance depreciation recovery* is more involved. There are three difficulties. The first (the apparent need to calculate all previous depreciations in order to determine the subsequent depreciation) has already been addressed by Eq. 77.32.

The second difficulty is that there is no way to ensure (that is, to force) the book value to be $S_n$ at $t = n$. Therefore, it is common to write off the remaining book value (down to $S_n$) at $t = n$ in one lump sum. This assumes $BV_n \geq S_n$.

The third difficulty is that of finding the present worth of an *exponentially decreasing cash flow*. Although the proof is omitted here, such exponential cash flows can be handled with the *exponential gradient factor*, $(P/EG)$.[33]

$$(P/EG, z-1, n) = \frac{z^n - 1}{z^n(z-1)} \qquad 77.47$$

$$z = \frac{1+i}{1-d} \qquad 77.48$$

---

[33]The $(P/A)$ columns in App. 77.B can be used for $(P/EG)$ as long as the interest rate is assumed to be $z - 1$.

Then, as long as $BV_n > S_n$, the declining balance depreciation recovery is

$$DR = tC\left(\frac{d}{1-d}\right)(P/EG, z-1, n) \qquad 77.49$$

## Example 77.22

For the asset described in Ex. 77.17, calculate the after-tax depreciation recovery with straight line and sum-of-the-years' digits depreciation methods. Use 6% interest with 48% income taxes.

*Solution*

SL: $\quad DR = (0.48)(\$880)(P/A, 6\%, 10)$

$\qquad = (0.48)(\$880)(7.3601)$

$\qquad = \$3109$

SOYD: The depreciation series can be thought of as a constant \$1600 term with a negative \$160 gradient.

$$DR = (0.48)(\$1600)(P/A, 6\%, 10)$$
$$\qquad - (0.48)(\$160)(P/G, 6\%, 10)$$
$$= (0.48)(\$1600)(7.3601)$$
$$\qquad - (0.48)(\$160)(29.6023)$$
$$= \$3379$$

Notice that the ten-year $(P/G)$ factor is used even though there are only nine years in which the gradient reduces the initial \$1600 amount.

## Example 77.23

What is the after-tax present worth of the asset described in Ex. 77.20 if straight line, sum-of-the-years' digits, and double declining balance depreciation methods are used?

*Solution*

Using SL, the depreciation recovery is

$$DR = (0.53)\left(\frac{\$10,000 - \$500}{8}\right)(P/A, 9\%, 8)$$

$$= (0.53)\left(\frac{\$9500}{8}\right)(5.5348)$$

$$= \$3483$$

Using SOYD, the depreciation recovery is calculated as follows.

$$T = \left(\tfrac{1}{2}\right)(8)(9) = 36$$

$$\text{depreciation base} = \$10,000 - \$500 = \$9500$$

$$D_1 = \left(\tfrac{8}{36}\right)(\$9500) = \$2111$$

$$G = \left(\tfrac{1}{36}\right)(\$9500) = \$264$$

$$DR = (0.53)\Big(($2111)(P/A, 9\%, 8)$$
$$\qquad - ($264)(P/G, 9\%, 8)\Big)$$

$$= (0.53)\Big(($2111)(5.5348)$$
$$\qquad - ($264)(16.8877)\Big)$$

$$= \$3830$$

Using DDB, the depreciation recovery is calculated as follows.[34]

$$d = \frac{2}{8} = 0.25$$

$$z = \frac{1 + 0.09}{1 - 0.25} = 1.4533$$

$$(P/EG, z-1, n) = \frac{(1.4533)^8 - 1}{(1.4533)^8(0.4533)} = 2.095$$

From Eq. 77.49,

$$DR = (0.53)\left(\frac{(0.25)(\$10,000)}{0.75}\right)(2.095)$$

$$= \$3701$$

The after-tax present worth, neglecting depreciation, was previously found to be $-\$3766$.

The after-tax present worths, including depreciation recovery, are

SL: $\quad P = -\$3766 + \$3483 = \quad -\$283$

SOYD: $P = -\$3766 + \$3830 = \quad \$64$

DDB: $\quad P = -\$3766 + \$3701 = \quad -\$65$

## 44. OTHER INTEREST RATES

The *effective interest rate per period*, $i$ (also called *yield* by banks), is the only interest rate that should be used in equivalence equations. The interest rates at the top of the factor tables in App. 77.B are implicitly all effective interest rates. Usually, the period will be one year, hence

---

[34]This method should start by checking that the book value at the end of the depreciation period is greater than the salvage value. In this example, such is the case. However, the step is not shown.

the name *effective annual interest rate*. However, there are other interest rates in use as well.

The term *nominal interest rate, r (rate per annum)*, is encountered when compounding is more than once per year. The nominal rate does not include the effect of compounding and is not the same as the effective rate. And, since the effective interest rate can be calculated from the nominal rate only if the number of compounding periods per year is known, nominal rates cannot be compared unless the method of compounding is specified. The only practical use for a nominal rate per year is for calculating the effective rate per period.

## 45. RATE AND PERIOD CHANGES

If there are $k$ compounding periods during the year (two for semiannual compounding, four for quarterly compounding, twelve for monthly compounding, etc.) and the nominal rate is $r$, the *effective rate per compounding period* is

$$\phi = \frac{r}{k} \qquad 77.50$$

The effective annual rate, $i$, can be calculated from the effective rate per period, $\phi$, by using Eq. 77.51.

$$i = (1 + \phi)^k - 1$$
$$= \left(1 + \frac{r}{k}\right)^k - 1 \qquad 77.51$$

Sometimes, only the effective rate per period (e.g., per month) is known. However, that will be a simple problem since compounding for $n$ periods at an effective rate per period is not affected by the definition or length of the period.

The following rules may be used to determine which interest rate is given.

- Unless specifically qualified, the interest rate given is an annual rate.

- If the compounding is annual, the rate given is the effective rate. If compounding is other than annual, the rate given is the nominal rate.

The effective annual interest rate determined on a *daily compounding basis* will not be significantly different than if *continuous compounding* is assumed.[35] In the case of continuous (or daily) compounding, the discounting factors can be calculated directly from the nominal interest rate and number of years, without having to find the effective interest rate per period. Table 77.6 can be used to determine the discount factors for continuous compounding.

---

[35]The number of *banking days in a year* (250, 360, etc.) must be specifically known.

**Table 77.6** *Discount Factors for Continuous Compounding (n is the number of years)*

| symbol | formula |
|--------|---------|
| $(F/P, r\%, n)$ | $e^{rn}$ |
| $(P/F, r\%, n)$ | $e^{-rn}$ |
| $(A/F, r\%, n)$ | $\dfrac{e^r - 1}{e^{rn} - 1}$ |
| $(F/A, r\%, n)$ | $\dfrac{e^{rn} - 1}{e^r - 1}$ |
| $(A/P, r\%, n)$ | $\dfrac{e^r - 1}{1 - e^{-rn}}$ |
| $(P/A, r\%, n)$ | $\dfrac{1 - e^{-rn}}{e^r - 1}$ |

### Example 77.24

A savings and loan offers a nominal rate of 5.25% compounded daily over 365 days in a year. What is the effective annual rate?

*Solution*

*method 1:* Use Eq. 77.51.

$$r = 0.0525, \quad k = 365$$
$$i = \left(1 + \frac{0.0525}{365}\right)^{365} - 1 = 0.0539$$

*method 2:* Assume daily compounding is the same as continuous compounding.

$$i = (F/P, r\%, 1) - 1$$
$$= e^{0.0525} - 1 = 0.0539$$

### Example 77.25

A real estate investment trust pays $7,000,000 for a 100-unit apartment complex. The trust expects to sell the complex in ten years for $15,000,000. In the meantime, it expects to receive an average rent of $900 per month from each apartment. Operating expenses are expected to be $200 per month per occupied apartment. A 95% occupancy rate is predicted. In similar investments, the trust has realized a 15% effective annual return on its investment. Compare to those past investments the expected present worth of this investment when calculated assuming (a) annual compounding (i.e., the year-end convention), and (b) monthly compounding. Disregard taxes, depreciation, and all other factors.

*Solution*

(a) The net annual income will be

$$(0.95)(100 \text{ units})\left(\frac{\$900}{\text{unit-mo}} - \frac{\$200}{\text{unit-mo}}\right)\left(12 \frac{\text{mo}}{\text{yr}}\right)$$
$$= \$798,000/\text{yr}$$

The present worth of ten years of operation is

$$P = -\$7,000,000 + (\$798,000)(P/A, 15\%, 10)$$
$$+ (\$15,000,000)(P/F, 15\%, 10)$$
$$= -\$7,000,000 + (\$798,000)(5.0188)$$
$$+ (\$15,000,000)(0.2472)$$
$$= \$713,000$$

(b) The net monthly income is

$$(0.95)(100 \text{ units})\left( \frac{\$900}{\text{unit-mo}} - \frac{\$200}{\text{unit-mo}} \right)$$
$$= \$66,500/\text{mo}$$

Equation 77.51 is used to calculate the effective monthly rate, $\phi$, from the effective annual rate, $i = 15\%$, and the number of compounding periods per year, $k = 12$.

$$\phi = (1 + i)^{1/k} - 1$$
$$= (1 + 0.15)^{1/12} - 1 = 0.011715 \ (1.1715\%)$$

The number of compounding periods in ten years is

$$n = (10 \text{ yr})\left( 12 \ \frac{\text{mo}}{\text{yr}} \right) = 120 \text{ mo}$$

The present worth of 120 months of operation is

$$P = -\$7,000,000 + (\$66,500)(P/A, 1.1715\%, 120)$$
$$+ (\$15,000,000)(P/F, 1.1715\%, 120)$$

Since table values for 1.1715% discounting factors are not available, the factors are calculated from Table 77.1.

$$(P/A, 1.1715\%, 120) = \frac{(1 + i)^n - 1}{i(1 + i)^n}$$
$$= \frac{(1 + 0.011715)^{120} - 1}{(0.011715)(1 + 0.011715)^{120}}$$
$$= 64.261$$

$$(P/F, 1.1715\%, 120) = (1 + i)^{-n} = (1 + 0.011715)^{-120}$$
$$= 0.2472$$

The present worth over 120 monthly compounding periods is

$$P = -\$7,000,000 + (\$66,500)(64.261)$$
$$+ (\$15,000,000)(0.2472)$$
$$= \$981,357$$

## 46. BONDS

A *bond* is a method of long-term financing commonly used by governments, states, municipalities, and very large corporations.[36] The bond represents a contract to pay the bondholder specific amounts of money at specific times. The holder purchases the bond in exchange for specific payments of interest and principal. Typical municipal bonds call for quarterly or semiannual interest payments and a payment of the *face value of the bond* on the *date of maturity* (end of the bond period).[37] Due to the practice of discounting in the bond market, a bond's face value and its purchase price generally will not coincide.

In the past, a bondholder had to submit a coupon or ticket in order to receive an interim interest payment. This has given rise to the term *coupon rate*, which is the nominal annual interest rate on which the interest payments are made. Coupon books are seldom used with modern bonds, but the term survives. The coupon rate determines the magnitude of the semiannual (or otherwise) interest payments during the life of the bond. The bondholder's own effective interest rate should be used for economic decisions about the bond.

Actual *bond yield* is the bondholder's actual rate of return of the bond, considering the purchase price, interest payments, and face value payment (or, value realized if the bond is sold before it matures). By convention, bond yield is calculated as a nominal rate (rate per annum), not an effective rate per year. The bond yield should be determined by finding the effective rate of return per payment period (e.g., per semiannual interest payment) as a conventional rate of return problem. Then, the nominal rate can be found by multiplying the effective rate per period by the number of payments per year, as in Eq. 77.51.

### Example 77.26

What is the maximum amount an investor should pay for a 25-year bond with a $20,000 face value and 8% coupon rate (interest only paid semiannually)? The bond will be kept to maturity. The investor's effective annual interest rate for economic decisions is 10%.

*Solution*

For this problem, take the compounding period to be six months. Then, there are 50 compounding periods. Since 8% is a nominal rate, the effective bond rate per period is calculated from Eq. 77.50 as

$$\phi_{\text{bond}} = \frac{r}{k} = \frac{8\%}{2} = 4\%$$

---

[36]In the past, 30-year bonds were typical. Shorter term 10-year, 15-year, 20-year, and 25-year bonds are also commonly issued.
[37]A *fully amortized bond* pays back interest and principal throughout the life of the bond. There is no balloon payment.

The bond payment received semiannually is

$$(0.04)(\$20,000) = \$800$$

10% is the investor's effective rate per year, so Eq. 77.51 is again used to calculate the effective analysis rate per period.

$$0.10 = (1 + \phi)^2 - 1$$

$$\phi = 0.04881 \quad (4.88\%)$$

The maximum amount that the investor should be willing to pay is the present worth of the investment.

$$P = (\$800)(P/A, 4.88\%, 50)$$
$$+ (\$20,000)(P/F, 4.88\%, 50)$$

Table 77.1 can be used to calculate the following factors.

$$(P/A, 4.88\%, 50) = \frac{(1 + 0.0488)^{50} - 1}{(0.0488)(1.0488)^{50}} = 18.600$$

$$(P/F, 4.88\%, 50) = \frac{1}{(1 + 0.0488)^{50}} = 0.09233$$

Then, the present worth is

$$P = (\$800)(18.600) + (\$20,000)(0.09233)$$
$$= \$16,727$$

## 47. PROBABILISTIC PROBLEMS

If an alternative's cash flows are specified by an implicit or explicit probability distribution rather than being known exactly, the problem is *probabilistic*.

Probabilistic problems typically possess the following characteristics.

- There is a chance of loss that must be minimized (or, rarely, a chance of gain that must be maximized) by selection of one of the alternatives.

- There are multiple alternatives. Each alternative offers a different degree of protection from the loss. Usually, the alternatives with the greatest protection will be the most expensive.

- The magnitude of loss or gain is independent of the alternative selected.

Probabilistic problems are typically solved using annual costs and expected values. An *expected value* is similar to an *average value* since it is calculated as the mean of the given probability distribution. If cost 1 has a probability of occurrence, $p_1$, cost 2 has a probability of occurrence, $p_2$, and so on, the expected value is

$$\mathcal{E}\{\text{cost}\} = p_1(\text{cost } 1) + p_2(\text{cost } 2) + \cdots \quad \textit{77.52}$$

### Example 77.27

Flood damage in any year is given according to the following table. What is the present worth of flood damage for a ten-year period? Use 6% as the effective annual interest rate.

| damage | probability |
|---|---|
| 0 | 0.75 |
| $10,000 | 0.20 |
| $20,000 | 0.04 |
| $30,000 | 0.01 |

*Solution*

The expected value of flood damage in any given year is

$$\mathcal{E}\{\text{damage}\} = (0)(0.75) + (\$10,000)(0.20)$$
$$+ (\$20,000)(0.04) + (\$30,000)(0.01)$$
$$= \$3100$$

The present worth of ten years of expected flood damage is

$$\text{present worth} = (\$3100)(P/A, 6\%, 10)$$
$$= (\$3100)(7.3601)$$
$$= \$22,816$$

### Example 77.28

A dam is being considered on a river that periodically overflows and causes $600,000 damage. The damage is essentially the same each time the river causes flooding. The project horizon is 40 years. A 10% interest rate is being used.

Three different designs are available, each with different costs and storage capacities.

| design alternative | cost | maximum capacity |
|---|---|---|
| A | $500,000 | 1 unit |
| B | $625,000 | 1.5 units |
| C | $900,000 | 2.0 units |

The National Weather Service has provided a statistical analysis of annual rainfall runoff from the watershed draining into the river.

| units annual rainfall | probability |
|---|---|
| 0 | 0.10 |
| 0.1 to 0.5 | 0.60 |
| 0.6 to 1.0 | 0.15 |
| 1.1 to 1.5 | 0.10 |
| 1.6 to 2.0 | 0.04 |
| 2.1 or more | 0.01 |

Which design alternative would you choose assuming the dam is essentially empty at the start of each rainfall season?

*Solution*

The sum of the construction cost and the expected damage should be minimized. If alternative A is chosen, it will have a capacity of 1 unit. Its capacity will be exceeded (causing $600,000 damage) when the annual rainfall exceeds 1 unit. Therefore, the expected value of the annual cost of alternative A is

$$
\begin{aligned}
\mathcal{E}\{\text{EUAC(A)}\} \\
&= (\$500{,}000)(A/P, 10\%, 40) \\
&\quad + (\$600{,}000)(0.10 + 0.40 + 0.01) \\
&= (\$500{,}000)(0.1023) + (\$600{,}000)(0.15) \\
&= \$141{,}150
\end{aligned}
$$

Similarly,

$$
\begin{aligned}
\mathcal{E}\{\text{EUAC(B)}\} &= (\$625{,}000)(A/P, 10\%, 40) \\
&\quad + (\$600{,}000)(0.04 + 0.01) \\
&= (\$625{,}000)(0.1023) + (\$600{,}000)(0.05) \\
&= \$93{,}938
\end{aligned}
$$

$$
\begin{aligned}
\mathcal{E}\{\text{EUAC(C)}\} &= (\$900{,}000)(A/P, 10\%, 40) \\
&\quad + (\$600{,}000)(0.01) \\
&= (\$900{,}000)(0.1023) + (\$600{,}000)(0.01) \\
&= \$98{,}070
\end{aligned}
$$

Alternative B should be chosen.

## 48. FIXED AND VARIABLE COSTS

The distinction between fixed and variable costs depends on how these costs vary when an independent variable changes. For example, factory or machine production is frequently the independent variable. However, it could just as easily be vehicle miles driven, hours of operation, or quantity (mass, volume, etc.).

If a cost is a function of the independent variable, the cost is said to be a *variable cost*. (See Table 77.7.) The change in cost per unit variable change (i.e., what is usually called the *slope*) is known as the *incremental cost*. Material and labor costs are examples of variable costs. They increase in proportion to the number of product units manufactured.

*Table 77.7* Summary of Fixed and Variable Costs

*fixed costs*
  rent
  property taxes
  interest on loans
  insurance
  janitorial service expense
  tooling expense
  setup, cleanup, and tear-down expenses
  depreciation expense
  marketing and selling costs
  cost of utilities
  general burden and overhead expense
*variable costs*
  direct material costs
  direct labor costs
  cost of miscellaneous supplies
  payroll benefit costs
  income taxes
  supervision costs

If a cost is not a function of the independent variable, the cost is said to be a *fixed cost*. Rent and lease payments are typical fixed costs. These costs will be incurred regardless of production levels.

Some costs have both fixed and variable components, as Fig. 77.9 illustrates. The fixed portion can be determined by calculating the cost at zero production.

An additional category of cost is the *semivariable cost*. This type of cost increases stepwise. Semivariable cost structures are typical of situations where *excess capacity* exists. For example, supervisory cost is a stepwise function of the number of production shifts. Also, labor cost for truck drivers is a stepwise function of weight (volume) transported. As long as a truck has room left (i.e., excess capacity), no additional driver is needed. As soon as the truck is filled, labor cost will increase.

*Figure 77.9* Fixed and Variable Costs

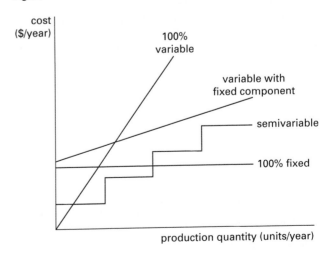

cost ($/year)

100% variable

variable with fixed component

semivariable

100% fixed

production quantity (units/year)

## 49. ACCOUNTING COSTS AND EXPENSE TERMS

The accounting profession has developed special terms for certain groups of costs. When annual costs are incurred due to the functioning of a piece of equipment, they are known as *operating and maintenance (O&M) costs*. The annual costs associated with operating a business (other than the costs directly attributable to production) are known as *general, selling, and administrative (GS&A) expenses*.

*Direct labor costs* are costs incurred in the factory, such as assembly, machining, and painting labor costs. *Direct material costs* are the costs of all materials that go into production.[38] Typically, both direct labor and direct material costs are given on a per-unit or per-item basis. The sum of the direct labor and direct material costs is known as the *prime cost*.

There are certain additional expenses incurred in the factory, such as the costs of factory supervision, stock-picking, quality control, factory utilities, and miscellaneous supplies (cleaning fluids, assembly lubricants, routing tags, etc.) that are not incorporated into the final product. Such costs are known as *indirect manufacturing expenses* (IME) or *indirect material and labor costs*.[39] The sum of the per-unit indirect manufacturing expense and prime cost is known as the *factory cost*.

*Research and development* (R&D) *costs* and *administrative expenses* are added to the factory cost to give the *manufacturing cost* of the product.

Additional costs are incurred in marketing the product. Such costs are known as *selling expenses* or *marketing expenses*. The sum of the selling expenses and manufacturing cost is the *total cost* of the product.

Figure 77.10 illustrates these terms.[40] Table 77.8 lists typical classifications of expenses.

The distinctions among the various forms of cost (particularly with overhead costs) are not standardized. Each company must develop a classification system to deal with the various cost factors in a consistent manner. There are also other terms in use (e.g., *raw materials, operating supplies, general plant overhead*), but these terms must be interpreted within the framework of each company's classification system. Table 77.8 is typical of such classification systems.

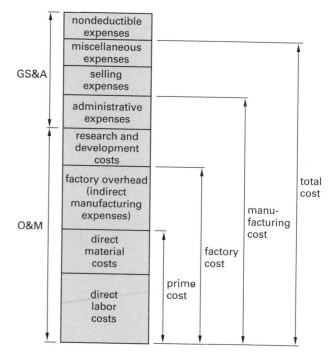

**Figure 77.10** *Costs and Expenses Combined*

## 50. ACCOUNTING PRINCIPLES

### Basic Bookkeeping

An accounting or *bookkeeping system* is used to record historical financial transactions. The resultant records are used for product costing, satisfaction of statutory requirements, reporting of profit for income tax purposes, and general company management.

Bookkeeping consists of two main steps: recording the transactions, followed by categorization of the transactions.[41] The transactions (receipts and disbursements) are recorded in a *journal (book of original entry)* to complete the first step. Such a journal is organized in a simple chronological and sequential manner. The transactions are then categorized (into interest income, advertising expense, etc.) and posted (i.e., entered or written) into the appropriate *ledger account*.[42]

The ledger accounts together constitute the *general ledger* or *ledger*. All ledger accounts can be classified into one of three types: *asset accounts, liability accounts,* and *owners' equity accounts*. Strictly speaking, income and expense accounts, kept in a separate journal, are included within the classification of owners' equity accounts.

---

[38]There may be problems with pricing the material when it is purchased from an outside vendor and the stock on hand derives from several shipments purchased at different prices.
[39]The *indirect material and labor costs* usually exclude costs incurred in the office area.
[40]Notice that *total cost* does not include income taxes.

[41]These two steps are not to be confused with the *double-entry bookkeeping method*.
[42]The two-step process is more typical of a *manual bookkeeping system* than a computerized *general ledger system*. However, even most computerized systems produce reports in journal entry order, as well as account summaries.

**Table 77.8** *Typical Classification of Expenses*

*direct labor expenses*
  machining and forming
  assembly
  finishing
  inspection
  testing
*direct material expenses*
  items purchased from other vendors
  manufactured assemblies
*factory overhead expenses (indirect manufacturing expenses)*
  supervision
  benefits
    pension
    medical insurance
    vacations
  wages overhead
    unemployment compensation taxes
    social security taxes
    disability taxes
  stock-picking
  quality control and inspection
  expediting
  rework
  maintenance
  miscellaneous supplies
    routing tags
    assembly lubricants
    cleaning fluids
    wiping cloths
    janitorial supplies
  packaging (materials and labor)
  factory utilities
  laboratory
  depreciation on factory equipment
*research and development expenses*
  engineering (labor)
  patents
  testing
  prototypes (material and labor)
  drafting
  O&M of R&D facility
*administrative expenses*
  corporate officers
  accounting
  secretarial/clerical/reception
  security (protection)
  medical (nurse)
  employment (personnel)
  reproduction
  data processing
  production control
  depreciation on nonfactory equipment
  office supplies
  office utilities
  O&M of offices
*selling expenses*
  marketing (labor)
  advertising
  transportation (if not paid by customer)
  outside sales force (labor and expenses)
  demonstration units
  commissions
  technical service and support
  order processing
  branch office expenses
*miscellaneous expenses*
  insurance
  property taxes
  interest on loans
*nondeductible expenses*
  federal income taxes
  fines and penalties

Together, the journal and ledger are known simply as "the books" of the company, regardless of whether bound volumes of pages are actually involved.

## Balancing the Books

In a business environment, *balancing the books* means more than reconciling the checkbook and bank statements. All accounting entries must be posted in such a way as to maintain the equality of the *basic accounting equation*,

$$\text{assets} = \text{liability} + \text{owner's equity} \qquad 77.53$$

In a *double-entry bookkeeping system*, the equality is maintained within the ledger system by entering each transaction into two balancing ledger accounts. For example, paying a utility bill would decrease the cash account (an asset account) and decrease the utility expense account (a liability account) by the same amount.

Transactions are either *debits* or *credits*, depending on their sign. Increases in asset accounts are debits; decreases are credits. For liability and equity accounts, the opposite is true: Increases are credits, and decreases are debits.[43]

## Cash and Accrual Systems[44]

The simplest form of bookkeeping is based on the *cash system*. The only transactions that are entered into the journal are those that represent cash receipts and disbursements. In effect, a checkbook register or bank deposit book could serve as the journal.

During a given period (e.g., month or quarter), expense liabilities may be incurred even though the payments for those expenses have not been made. For example, an invoice (bill) may have been received but not paid. Under the *accrual system*, the obligation is posted into the appropriate expense account before it is paid.[45] Analogous to expenses, under the accrual system, income will be claimed before payment is received. Specifically, a sales transaction can be recorded as income when the customer's order is received, when the

---

[43]There is a difference in sign between asset and liability accounts. Thus, an increase in an expense account is actually a decrease. The accounting profession, apparently, is comfortable with the common confusion that exists between debits and credits.

[44]There is also a distinction made between cash flows that are known and those that are expected. It is a *standard accounting principle* to record losses in full, at the time they are recognized, even before their occurrence. In the construction industry, for example, losses are recognized in full and projected to the end of a project as soon as they are foreseeable. Profits, on the other hand, are recognized only as they are realized (typically, as a percentage of project completion). The difference between cash and accrual systems is a matter of *bookkeeping*. The difference between loss and profit recognition is a matter of *accounting convention*. Engineers seldom need to be concerned with the accounting tradition.

[45]The expense for an item or service might be accrued even *before* the invoice is received. It might be recorded when the purchase order for the item or service is generated, or when the item or service is received.

outgoing invoice is generated, or when the merchandise is shipped.

## Financial Statements

Each period, two types of corporate financial statements are typically generated: the *balance sheet* and *profit and loss (P&L) statements*.[46] The profit and loss statement, also known as a *statement of income and retained earnings*, is a summary of sources of *income* or *revenue* (interest, sales, fees charged, etc.) and *expenses* (utilities, advertising, repairs, etc.) for the period. The expenses are subtracted from the revenues to give a *net income* (generally, before taxes).[47] Figure 77.11 illustrates a simple profit and loss statement.

**Figure 77.11** Simplified Profit and Loss Statement

| *revenue* | | |
|---|---|---|
| interest | 2000 | |
| sales | 237,000 | |
| returns | (23,000) | |
| net revenue | | 216,000 |
| *expenses* | | |
| salaries | 149,000 | |
| utilities | 6000 | |
| advertising | 28,000 | |
| insurance | 4000 | |
| supplies | 1000 | |
| net expenses | | 188,000 |
| *period net income* | 28,000 | |
| *beginning retained earnings* | 63,000 | |
| *net year-to-date earnings* | | 91,000 |

The *balance sheet* presents the *basic accounting equation* in tabular form. The balance sheet lists the major categories of assets and outstanding liabilities. The difference between asset values and liabilities is the *equity*, as defined in Eq. 77.53. This equity represents what would be left over after satisfying all debts by liquidating the company.

There are several terms that appear regularly on balance sheets.

- *current assets:* cash and other assets that can be converted quickly into cash, such as accounts receivable, notes receivable, and merchandise (inventory). Also known as *liquid assets.*

- *fixed assets:* relatively permanent assets used in the operation of the business and relatively difficult to convert into cash. Examples are land, buildings, and equipment. Also known as *nonliquid assets.*

[46]Other types of financial statements (*statements of changes in financial position, cost of sales statements*, inventory and asset reports, etc.) also will be generated, depending on the needs of the company.
[47]Financial statements also can be prepared with percentages (of total assets and net revenue) instead of dollars, in which case they are known as *common size financial statements.*

- *current liabilities:* liabilities due within a short period of time (e.g., within one year) and typically paid out of current assets. Examples are accounts payable, notes payable, and other accrued liabilities.

- *long-term liabilities:* obligations that are not totally payable within a short period of time (e.g., within one year).

Figure 77.12 is a simplified balance sheet.

**Figure 77.12** Simplified Balance Sheet

| ASSETS | | |
|---|---|---|
| *current assets* | | |
| cash | 14,000 | |
| accounts receivable | 36,000 | |
| notes receivable | 20,000 | |
| inventory | 89,000 | |
| prepaid expenses | 3000 | |
| total current assets | | 162,000 |
| *plant, property, and equipment* | | |
| land and buildings | 217,000 | |
| motor vehicles | 31,000 | |
| equipment | 94,000 | |
| accumulated depreciation | (52,000) | |
| total fixed assets | | 290,000 |
| **total assets** | | 452,000 |
| LIABILITIES AND OWNERS' EQUITY | | |
| *current liabilities* | | |
| accounts payable | 66,000 | |
| accrued income taxes | 17,000 | |
| accrued expenses | 8000 | |
| total current liabilities | | 91,000 |
| *long-term debt* | | |
| notes payable | 117,000 | |
| mortgage | 23,000 | |
| total long-term debt | | 140,000 |
| *owners' and stockholders' equity* | | |
| stock | 130,000 | |
| retained earnings | 91,000 | |
| total owners' equity | | 221,000 |
| **total liabilities and owners' equity** | | 452,000 |

## Analysis of Financial Statements

Financial statements are evaluated by management, lenders, stockholders, potential investors, and many other groups for the purpose of determining the *health of the company*. The health can be measured in terms of *liquidity* (ability to convert assets to cash quickly), *solvency* (ability to meet debts as they become due), and *relative risk* (of which one measure is *leverage*—the portion of total capital contributed by owners).

The analysis of financial statements involves several common ratios, usually expressed as percentages. The following are some frequently encountered ratios.

- *current ratio:* an index of short-term paying ability.

$$\text{current ratio} = \frac{\text{current assets}}{\text{current liabilities}} \qquad 77.54$$

- *quick (or acid-test) ratio:* a more stringent measure of short-term debt-paying ability. The *quick assets* are defined to be current assets minus inventories and prepaid expenses.

$$\text{quick ratio} = \frac{\text{quick assets}}{\text{current liabilities}} \qquad 77.55$$

- *receivable turnover:* a measure of the average speed with which accounts receivable are collected.

$$\text{receivable turnover} = \frac{\text{net credit sales}}{\text{average net receivables}} \qquad 77.56$$

- *average age of receivables:* number of days, on the average, in which receivables are collected.

$$\text{average age of receivables} = \frac{365}{\text{receivable turnover}} \qquad 77.57$$

- *inventory turnover:* a measure of the speed with which inventory is sold, on the average.

$$\text{inventory turnover} = \frac{\text{cost of goods sold}}{\text{average cost of inventory on hand}} \qquad 77.58$$

- *days supply of inventory on hand:* number of days, on the average, that the current inventory would last.

$$\text{days supply of inventory on hand} = \frac{365}{\text{inventory turnover}} \qquad 77.59$$

- *book value per share of common stock:* number of dollars represented by the balance sheet owners' equity for each share of common stock outstanding.

$$\begin{aligned} &\text{book value per share of common stock} \\ &= \frac{\text{common shareholders' equity}}{\text{number of outstanding shares}} \qquad 77.60 \end{aligned}$$

- *gross margin:* gross profit as a percentage of sales. (Gross profit is sales less cost of goods sold.)

$$\text{gross margin} = \frac{\text{gross profit}}{\text{net sales}} \qquad 77.61$$

- *profit margin ratio:* percentage of each dollar of sales that is net income.

$$\text{profit margin} = \frac{\text{net income before taxes}}{\text{net sales}} \qquad 77.62$$

- *return on investment ratio:* shows the percent return on owners' investment.

$$\text{return on investment} = \frac{\text{net income}}{\text{owners' equity}} \qquad 77.63$$

- *price-earnings ratio:* indication of relationship between earnings and market price per share of common stock, useful in comparisons between alternative investments.

$$\text{price-earnings} = \frac{\text{market price per share}}{\text{earnings per share}} \qquad 77.64$$

## 51. COST ACCOUNTING

*Cost accounting* is the system that determines the cost of manufactured products. Cost accounting is called *job cost accounting* if costs are accumulated by part number or contract. It is called *process cost accounting* if costs are accumulated by departments or manufacturing processes.

Cost accounting is dependent on historical and recorded data. The unit product cost is determined from actual expenses and numbers of units produced. Allowances (i.e., budgets) for future costs are based on these historical figures. Any deviation from historical figures is called a *variance*. Where adequate records are available, variances can be divided into *labor variance* and *material variance*.

When determining a unit product cost, the direct material and direct labor costs are generally clear-cut and easily determined. Furthermore, these costs are 100% variable costs. However, the indirect cost per unit of product is not as easily determined. Indirect costs (*burden*, *overhead*, etc.) can be fixed or semivariable costs. The amount of indirect cost allocated to a unit will depend on the unknown future overhead expense as well as the unknown future production (*vehicle size*).

A typical method of allocating indirect costs to a product is as follows.

*step 1:* Estimate the total expected indirect (and overhead) costs for the upcoming year.

*step 2:* Determine the most appropriate vehicle (basis) for allocating the overhead to production. Usually, this vehicle is either the number of units expected to be produced or the number of direct hours expected to be worked in the upcoming year.

*step 3:* Estimate the quantity or size of the overhead vehicle.

*step 4:* Divide expected overhead costs by the expected overhead vehicle to obtain the unit overhead.

*step 5:* Regardless of the true size of the overhead vehicle during the upcoming year, one unit of overhead cost is allocated per unit of overhead vehicle.

Once the prime cost has been determined and the indirect cost calculated based on projections, the two are combined into a *standard factory cost* or *standard cost*, which remains in effect until the next budgeting period (usually a year).

During the subsequent manufacturing year, the standard cost of a product is not generally changed merely because it is found that an error in projected indirect costs or production quantity (vehicle size) has been made. The allocation of indirect costs to a product is assumed to be independent of errors in forecasts. Rather, the difference between the expected and actual expenses, known as the *burden (overhead) variance*, experienced during the year is posted to one or more *variance accounts*.

Burden (overhead) variance is caused by errors in forecasting both the actual indirect expense for the upcoming year and the overhead vehicle size. In the former case, the variance is called *burden budget variance*; in the latter, it is called *burden capacity variance*.

### Example 77.29

A company expects to produce 8000 items in the coming year. The current material cost is $4.54 each. Sixteen minutes of direct labor are required per unit. Workers are paid $7.50 per hour. 2133 direct labor hours are forecasted for the product. Miscellaneous overhead costs are estimated at $45,000.

Find the per-unit (a) expected direct material cost, (b) direct labor cost, (c) prime cost, (d) burden as a function of production and direct labor, and (e) total cost.

*Solution*

(a) The direct material cost was given as $4.54.

(b) The direct labor cost is

$$\left(\frac{16 \text{ min}}{60 \frac{\text{min}}{\text{hr}}}\right)\left(\frac{\$7.50}{\text{hr}}\right) = \$2.00$$

(c) The prime cost is

$$\$4.54 + \$2.00 = \$6.54$$

(d) If the burden vehicle is production, the burden rate is $45,000/8000 = $5.63 per item.

If the burden vehicle is direct labor hours, the burden rate is $45,000/2133 = $21.10 per hour.

(e) If the burden vehicle is production, the total cost is

$$\$4.54 + \$2.00 + \$5.63 = \$12.17$$

If the burden vehicle is direct labor hours, the total cost is

$$\$4.54 + \$2.00 + \left(\frac{16 \text{ min}}{60 \frac{\text{min}}{\text{hr}}}\right)\left(\frac{\$21.10}{\text{hr}}\right) = \$12.17$$

### Example 77.30

The actual performance of the company in Ex. 77.29 is given by the following figures.

actual production: 7560

actual overhead costs: $47,000

What are the burden budget variance and the burden capacity variance?

*Solution*

The burden capacity variance is

$$\$45,000 - (7560)(\$5.63) = \$2437$$

The burden budget variance is

$$\$47,000 - \$45,000 = \$2000$$

The overall burden variance is

$$\$47,000 - (7560)(\$5.63) = \$4437$$

The sum of the burden capacity and burden budget variances should equal the overall burden variance.

$$\$2437 + \$2000 = \$4437$$

### 52. COST OF GOODS SOLD

*Cost of goods sold* (COGS) is an accounting term that represents an inventory account adjustment.[48] Cost of goods sold is the difference between the starting and ending inventory valuations. That is,

$$\text{COGS} = \text{starting inventory valuation}$$
$$- \text{ending inventory valuation} \qquad 77.65$$

Cost of goods sold is subtracted from *gross profit* to determine the *net profit* of a company. Despite the fact that cost of goods sold can be a significant element in the profit equation, the inventory adjustment may not be made each accounting period (e.g., each month) due to the difficulty in obtaining an accurate inventory valuation.

With a *perpetual inventory system*, a company automatically maintains up-to-date inventory records, either through an efficient stocking and stock-releasing system

---

[48]The cost of goods sold inventory adjustment is posted to the COGS *expense account.*

or through a *point of sale* (POS) *system* integrated with the inventory records. If a company only counts its inventory (i.e., takes a *physical inventory*) at regular intervals (e.g., once a year), it is said to be operating on a *periodic inventory system*.

Inventory accounting is a source of many difficulties. The inventory value is calculated by multiplying the quantity on hand by the standard cost. In the case of completed items actually assembled or manufactured at the company, this standard cost usually is the manufacturing cost, although factory cost also can be used. In the case of purchased items, the standard cost will be the cost per item charged by the supplying vendor. In some cases, delivery and transportation costs will be included in this standard cost.

It is not unusual for the elements in an item's inventory to come from more than one vendor, or from one vendor in more than one order. Inventory valuation is more difficult if the price paid is different for these different purchases. There are four methods of determining the cost of elements in inventory. Any of these methods can be used (if applicable), but the method must be used consistently from year to year. The four methods are as follows.

- *specific identification method:* Each element can be uniquely associated with a cost. Inventory elements with serial numbers fit into this costing scheme. Stock, production, and sales records must include the serial number.

- *average cost method:* The standard cost of an item is the average of (recent or all) purchase costs for that item.

- *first-in, first-out* (FIFO) *method:* This method keeps track of how many of each item were purchased each time and the number remaining out of each purchase, as well as the price paid at each purchase. The inventory system assumes that the oldest elements are issued first.[49] Inventory value is a weighted average dependent on the number of elements from each purchase remaining. Items issued no longer contribute to the inventory value.

- *last-in, first-out* (LIFO) *method:* This method keeps track of how many of each item were purchased each time and the number remaining out of each purchase, as well as the price paid at each purchase.[50] The inventory value is a weighted average dependent on the number of elements from each purchase remaining. Items issued no longer contribute to the inventory value.

---

[49]If all elements in an item's inventory are identical, and if all shipments of that item are agglomerated, there will be no way to guarantee that the oldest element in inventory is issued first. But, unless *spoilage* is a problem, it really does not matter.

[50]See previous footnote.

## 53. BREAK-EVEN ANALYSIS

*Special Nomenclature*

| | |
|---|---|
| $a$ | *incremental cost* to produce one additional item (also called *marginal cost* or *differential cost*) |
| $C$ | total cost |
| $f$ | fixed cost that does not vary with production |
| $p$ | *incremental value* (price) |
| $Q$ | quantity sold |
| $Q^*$ | quantity at break-even point |
| $R$ | total revenue |

*Break-even analysis* is a method of determining when the value of one alternative becomes equal to the value of another. A common application is that of determining when costs exactly equal revenue. If the manufactured quantity is less than the break-even quantity, a loss is incurred. If the manufactured quantity is greater than the break-even quantity, a profit is made. (See Fig. 77.13.)

Assuming no change in the inventory, the *break-even point* can be found by setting costs equal to revenue $(C = R)$.

$$C = f + aQ \qquad 77.66$$

$$R = pQ \qquad 77.67$$

$$Q^* = \frac{f}{p - a} \qquad 77.68$$

**Figure 77.13** *Break-Even Quality*

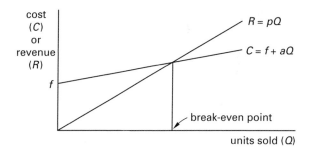

An alternative form of the break-even problem is to find the number of units per period for which two alternatives have the same total costs. Fixed costs are to be spread over a period longer than one year using the equivalent uniform annual cost (EUAC) concept. One of the alternatives will have a lower cost if production is less than the break-even point. The other will have a lower cost for production greater than the break-even point.

### Example 77.31

Two plans are available for a company to obtain automobiles for its sales representatives. How many miles must the cars be driven each year for the two plans to have the same costs? Use an interest rate of 10%. (Use the year-end convention for all costs.)

*plan 1:* Lease the cars and pay $0.15 per mile.

*plan 2:* Purchase the cars for $5000. Each car has an economic life of three years, after which it can be sold for $1200. Gas and oil cost $0.04 per mile. Insurance is $500 per year.

*Solution*

Let $x$ be the number of miles driven per year. Then, the EUAC for both alternatives is

$$\text{EUAC(A)} = 0.15x$$
$$\begin{aligned}\text{EUAC(B)} &= 0.04x + \$500 + (\$5000)(A/P, 10\%, 3) \\ &\quad - (\$1200)(A/F, 10\%, 3) \\ &= 0.04x + \$500 + (\$5000)(0.4021) \\ &\quad - (\$1200)(0.3021) \\ &= 0.04x + 2148\end{aligned}$$

Setting EUAC(A) and EUAC(B) equal and solving for $x$ yields 19,527 miles per year as the break-even point.

## 54. PAY-BACK PERIOD

The *pay-back period* is defined as the length of time, usually in years, for the cumulative net annual profit to equal the initial investment. It is tempting to introduce equivalence into pay-back period calculations, but by convention, this is generally not done.[51]

$$\text{pay-back period} = \frac{\text{initial investment}}{\text{net annual profit}} \qquad 77.69$$

**Example 77.32**

A ski resort installs two new ski lifts at a total cost of $1,800,000. The resort expects the annual gross revenue to increase by $500,000 while it incurs an annual expense of $50,000 for lift operation and maintenance. What is the pay-back period?

*Solution*

From Eq. 77.69,

$$\text{pay-back period} = \frac{\$1,800,000}{\$500,000 - \$50,000} = 4 \text{ years}$$

---

[51]Equivalence (i.e., interest and compounding) generally is not considered when calculating the "pay-back period." However, if it is desirable to include equivalence, then the term *pay-back period* should not be used. Other terms, such as *cost recovery period* or *life of an equivalent investment*, should be used. Unfortunately, this convention is not always followed in practice.

## 55. MANAGEMENT GOALS

Depending on many factors (market position, age of the company, age of the industry, perceived marketing and sales windows, etc.), a company may select one of many production and marketing strategic goals. Three such strategic goals are

- maximization of product demand
- minimization of cost
- maximization of profit

Such goals require knowledge of how the dependent variable (e.g., demand quantity or quantity sold) varies as a function of the independent variable (e.g., price). Unfortunately, these three goals are not usually satisfied simultaneously. For example, minimization of product cost may require a large production run to realize economies of scale, while the actual demand is too small to take advantage of such economies of scale.

If sufficient data are available to plot the independent and dependent variables, it may be possible to optimize the dependent variable graphically. (See Fig. 77.14.) Of course, if the relationship between independent and dependent variables is known algebraically, the dependent variable can be optimized by taking derivatives or by use of other numerical methods.

***Figure 77.14*** *Graphs of Management Goal Functions*

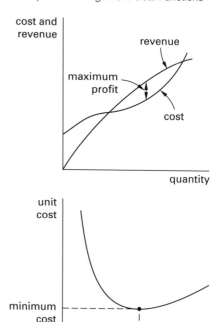

## 56. INFLATION

It is important to perform economic studies in terms of *constant value dollars*. One method of converting all cash flows to constant value dollars is to divide the flows by some annual *economic indicator* or price index.

If indicators are not available, cash flows can be adjusted by assuming that inflation is constant at a decimal rate ($e$) per year. Then, all cash flows can be converted to $t = 0$ dollars by dividing by $(1 + e)^n$, where $n$ is the year of the cash flow.

An alternative is to replace the effective annual interest rate ($i$) with a value corrected for inflation. This corrected value ($i'$) is

$$i' = i + e + ie \qquad 77.70$$

This method has the advantage of simplifying the calculations. However, precalculated factors are not available for the non-integer values of $i'$. Therefore, Table 77.1 must be used to calculate the factors.

### Example 77.33

What is the uninflated present worth of a $2000 future value in two years if the average inflation rate is 6% and $i$ is 10%?

*Solution*

$$P = \frac{F}{(1 + i)^n (1 + e)^n}$$

$$= \frac{\$2000}{(1 + 0.10)^2 (1 + 0.06)^2} = \$1471$$

### Example 77.34

Repeat Ex. 77.33 using Eq. 77.70.

*Solution*

$$i' = i + e + ie$$

$$= 0.10 + 0.06 + (0.10)(0.06) = 0.166$$

$$P = \frac{\$2000}{(1 + 0.166)^2} = \$1471$$

## 57. CONSUMER LOANS

*Special Nomenclature*

| | |
|---|---|
| $BAL_j$ | balance after the $j$th payment |
| $j$ | payment or period number |
| LV | principal total value loaned (cost minus down payment) |
| $N$ | total number of payments to pay off the loan |
| $PI_j$ | $j$th interest payment |
| $PP_j$ | $j$th principal payment |
| $PT_j$ | $j$th total payment |
| $\phi$ | effective rate per period ($r/k$) |

Many different arrangements can be made between a borrower and a lender. With the advent of creative financing concepts, it often seems that there are as many variations of loans as there are loans made. Nevertheless,

there are several traditional types of transactions. Real estate or investment texts, or a financial consultant, should be consulted for more complex problems.

### Simple Interest

Interest due does not compound with a *simple interest loan*. The interest due is merely proportional to the length of time that the principal is outstanding. Because of this, simple interest loans are seldom made for long periods (e.g., more than one year). (For loans less than one year, it is commonly assumed that a year consists of 12 months of 30 days each.)

### Example 77.35

A $12,000 simple interest loan is taken out at 16% per annum interest rate. The loan matures in two years with no intermediate payments. How much will be due at the end of the second year?

*Solution*

The interest each year is

$$PI = (0.16)(\$12,000) = \$1920$$

The total amount due in two years is

$$PT = \$12,000 + (2)(\$1920) = \$15,840$$

### Example 77.36

$4000 is borrowed for 75 days at 16% per annum simple interest. How much will be due at the end of 75 days?

*Solution*

$$\text{amount due} = \$4000 + (0.16)\left(\frac{75 \text{ days}}{360 \frac{\text{days}}{\text{bank yr}}}\right)(\$4000)$$

$$= \$4133$$

### Loans with Constant Amount Paid Toward Principal

With this loan type, the payment is not the same each period. The amount paid toward the principal is constant, but the interest varies from period to period. (See Fig. 77.15.) The equations that govern this type of loan are

$$BAL_j = LV - (j)(PP) \qquad 77.71$$

$$PI_j = \phi(BAL)_{j-1} \qquad 77.72$$

$$PT_j = PP + PI_j \qquad 77.73$$

$$PP = \frac{LV}{N} \qquad 77.74$$

$$N = \frac{LV}{PP} \qquad 77.75$$

$$LV = (PP + PI_1)(P/A, \phi, N)$$
$$- PI_N(P/G, \phi, N) \qquad 77.76$$

$$1 = \left(\frac{1}{N} + \phi\right)(P/A, \phi, N)$$
$$- \left(\frac{\phi}{N}\right)(P/G, \phi, N) \qquad 77.77$$

**Figure 77.15** *Loan with Constant Amount Paid Toward Principal*

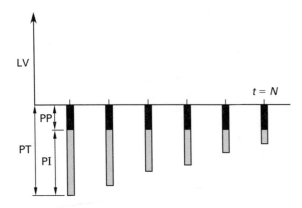

## Example 77.37

A $12,000 six-year loan is taken from a bank that charges 15% effective annual interest. Payments toward the principal are uniform, and repayments are made at the end of each year. Tabulate the interest, total payments, and the balance remaining after each payment is made.

*Solution*

The amount of each principal payment is

$$PP = \frac{\$12,000}{6} = \$2000$$

At the end of the first year (before the first payment is made), the principal balance is $12,000 (i.e., $BAL_0 = \$12,000$). From Eq. 77.72, the interest payment is

$$PI_1 = (0.15)(\$12,000) = \$1800$$

The total first payment is

$$PT_1 = PP + PI = \$2000 + \$1800$$
$$= \$3800$$

The following table is similarly constructed.

| $j$ | $BAL_j$ | $PP_j$ | $PI_j$ | $PT_j$ |
|---|---|---|---|---|
| | (in dollars) | | | |
| 0 | 12,000 | – | – | – |
| 1 | 10,000 | 2000 | 1800 | 3800 |
| 2 | 8000 | 2000 | 1500 | 3500 |
| 3 | 6000 | 2000 | 1200 | 3200 |
| 4 | 4000 | 2000 | 900 | 2900 |
| 5 | 2000 | 2000 | 600 | 2600 |
| 6 | 0 | 2000 | 300 | 2300 |

## Direct Reduction Loans

This is the typical "interest paid on unpaid balance" loan. The amount of the periodic payment is constant, but the amounts paid toward the principal and interest both vary. (See Fig. 77.16.)

$$BAL_{j-1} = PT\left(\frac{1 - (1 + \phi)^{j-1-N}}{\phi}\right) \qquad 77.78$$

$$PI_j = \phi(BAL)_{j-1} \qquad 77.79$$

$$PP_j = PT - PI_j \qquad 77.80$$

$$BAL_j = BAL_{j-1} - PP_j \qquad 77.81$$

$$N = \frac{-\ln\left(1 - \frac{\phi(LV)}{PT}\right)}{\ln(1 + \phi)} \qquad 77.82$$

Equation 77.82 calculates the number of payments necessary to pay off a loan. This equation can be solved with effort for the total periodic payment (PT) or the initial value of the loan (LV). It is easier, however, to use the $(A/P, i\%, n)$ factor to find the payment and loan value.

$$PT = LV(A/P, \phi\%, N) \qquad 77.83$$

If the loan is repaid in yearly installments, then $i$ is the effective annual rate. If the loan is paid off monthly, then $i$ should be replaced by the effective rate per month ($\phi$ from Eq. 77.51). For monthly payments, $N$ is the number of months in the loan period.

**Figure 77.16** *Direct Reduction Loan*

## Example 77.38

A $45,000 loan is financed at 9.25% per annum. The monthly payment is $385. What are the amounts paid toward interest and principal in the 14th period? What is the remaining principal balance after the 14th payment has been made?

*Solution*

The effective rate per month is

$$\phi = \frac{r}{k} = \frac{0.0925}{12}$$
$$= 0.0077083 \ldots \quad [\text{say } 0.007708]$$

$$N = \frac{-\ln\left(1 - \frac{(0.007708)(45,000)}{385}\right)}{\ln(1 + 0.007708)} = 301$$

$$\text{BAL}_{13} = (\$385)\left(\frac{1 - (1 + 0.007708)^{14-1-301}}{0.007708}\right)$$
$$= \$44,476.39$$

$$\text{PI}_{14} = (0.007708)(\$44,476.39) = \$342.82$$

$$\text{PP}_{14} = \$385 - \$342.82 = \$42.18$$

$$\text{BAL}_{14} = \$44,476.39 - \$42.18 = \$44,434.21$$

## Direct Reduction Loans with Balloon Payments

This type of loan has a constant periodic payment, but the duration of the loan is insufficient to completely pay back the principal (i.e, the loan is not fully amortized). Therefore, all remaining unpaid principal must be paid back in a lump sum when the loan matures. This large payment is known as a *balloon payment*.[52] (See Fig. 77.17.)

Equation 77.78 through Eq. 77.82 also can be used with this type of loan. The remaining balance after the last payment is the balloon payment. This balloon payment must be repaid along with the last regular payment calculated.

## 58. FORECASTING

There are many types of forecasting models, although most are variations of the basic types.[53] All models produce a *forecast* ($F_{t+1}$) of some quantity (*demand* is used in this section) in the next period based on actual measurements ($D_j$) in current and prior periods. All of the models also try to provide *smoothing* (or *damping*) of extreme data points.

---

[52]The term *balloon payment* may include the final interest payment as well. Generally, the problem statement will indicate whether the balloon payment is inclusive or exclusive of the regular payment made at the end of the loan period.

[53]For example, forecasting models that take into consideration steady (linear), cyclical, annual, and seasonal trends are typically variations of the exponentially weighted model. A truly different forecasting tool, however, is *Monte Carlo simulation*.

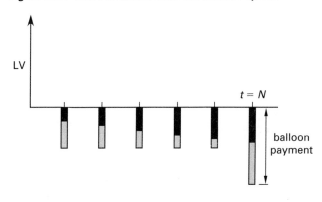

**Figure 77.17** *Direct Reduction Loan with Balloon Payment*

## Forecasts by Moving Averages

The method of *moving average forecasting* weights all previous demand data points equally and provides some smoothing of extreme data points. The amount of smoothing increases as the number of data points, $n$, increases.

$$F_{t+1} = \frac{1}{n} \sum_{m=t+1-n}^{t} D_m \qquad 77.84$$

## Forecasts by Exponentially Weighted Averages

With *exponentially weighted forecasts*, the more current (most recent) data points receive more weight. This method uses a *weighting factor* ($\alpha$), also known as a *smoothing coefficient*, which typically varies between 0.01 and 0.30. An initial forecast is needed to start the method. Forecasts immediately following are sensitive to the accuracy of this first forecast. It is common to choose $F_0 = D_1$ to get started.

$$F_{t+1} = \alpha D_t + (1 - \alpha)F_t \qquad 77.85$$

## 59. LEARNING CURVES

*Special Nomenclature*

| | |
|---|---|
| $b$ | learning curve constant |
| $n$ | total number of items produced |
| $R$ | decimal learning curve rate ($2^{-b}$) |
| $T_1$ | time or cost for the first item |
| $T_n$ | time or cost for the $n$th item |

The more products that are made, the more efficient the operation becomes due to experience gained. Therefore, direct labor costs decrease.[54] Usually, a *learning curve* is specified by the decrease in cost each time the cumulative quantity produced doubles. If there is a 20% decrease per doubling, the curve is said to be an 80% learning curve (i.e., the *learning curve rate*, $R$, is 80%).

---

[54]Remember that learning curve reductions apply only to direct labor costs. They are not applied to indirect labor or direct material costs.

Then, the time to produce the $n$th item is

$$T_n = T_1 n^{-b} \qquad 77.86$$

The total time to produce units from quantity $n_1$ to $n_2$ inclusive is approximately given by Eq. 77.87. $T_1$ is a constant, the time for item 1, and does not correspond to $n$ unless $n_1 = 1$.

$$\int_{n_1}^{n_2} T_n \, dn \approx \left(\frac{T_1}{1-b}\right)\left(\left(n_2 + \tfrac{1}{2}\right)^{1-b} - \left(n_1 - \tfrac{1}{2}\right)^{1-b}\right)$$
$$77.87$$

The *average time per unit* over the production from $n_1$ to $n_2$ is the above total time from Eq. 77.87 divided by the quantity produced, $(n_2 - n_1 + 1)$.

$$T_{\text{ave}} = \frac{\displaystyle\int_{n_1}^{n_2} T_n \, dn}{n_2 - n_1 + 1} \qquad 77.88$$

Table 77.9 lists representative values of the *learning curve constant* ($b$). For learning curve rates not listed in the table, Eq. 77.89 can be used to find $b$.

$$b = \frac{-\log_{10} R}{\log_{10}(2)} = \frac{-\log_{10} R}{0.301} \qquad 77.89$$

**Table 77.9** Learning Curve Constants

| learning curve rate ($R$) | $b$ |
|---|---|
| 0.70 (70%) | 0.515 |
| 0.75 (75%) | 0.415 |
| 0.80 (80%) | 0.322 |
| 0.85 (85%) | 0.234 |
| 0.90 (90%) | 0.152 |
| 0.95 (95%) | 0.074 |

### Example 77.39

A 70% learning curve is used with an item whose first production time is 1.47 hr. (a) How long will it take to produce the 11th item? (b) How long will it take to produce the 11th through 27th items?

*Solution*

(a) From Eq. 77.86,

$$T_{11} = (1.47 \text{ hr})(11)^{-0.515} = 0.428 \text{ hr}$$

(b) The time to produce the 11th item through 27th item is given by Eq. 77.87.

$$T \approx \left(\frac{1.47 \text{ hr}}{1 - 0.515}\right)\left((27.5)^{1-0.515} - (10.5)^{1-0.515}\right)$$
$$= 5.643 \text{ hr}$$

## 60. ECONOMIC ORDER QUANTITY

*Special Nomenclature*

| | |
|---|---|
| $a$ | constant depletion rate (items/unit time) |
| $h$ | inventory storage cost ($/item-unit time) |
| $H$ | total inventory storage cost between orders ($) |
| $K$ | fixed cost of placing an order ($) |
| $Q$ | order quantity (original quantity on hand) |
| $t^*$ | time at depletion |

The *economic order quantity* (EOQ) is the order quantity that minimizes the inventory costs per unit time. Although there are many different EOQ models, the simplest is based on the following assumptions.

- Reordering is instantaneous. The time between order placement and receipt is zero. (See Fig. 77.18.)

- Shortages are not allowed.

- Demand for the inventory item is deterministic (i.e., is not a random variable).

- Demand is constant with respect to time.

- An order is placed when the inventory is zero.

**Figure 77.18** Inventory with Instantaneous Reorder

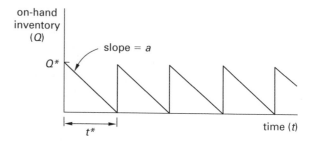

If the original quantity on hand is $Q$, the stock will be depleted at

$$t^* = \frac{Q}{a} \qquad 77.90$$

The total inventory storage cost between $t_0$ and $t^*$ is

$$H = \tfrac{1}{2}Qht^* = \frac{Q^2 h}{2a} \qquad 77.91$$

The total inventory and ordering cost per unit time is

$$C_t = \frac{aK}{Q} + \frac{hQ}{2} \qquad 77.92$$

$C_t$ can be minimized with respect to $Q$. The economic order quantity and time between orders are

$$Q^* = \sqrt{\frac{2aK}{h}} \qquad 77.93$$

$$t^* = \frac{Q^*}{a} \qquad 77.94$$

## 61. SENSITIVITY ANALYSIS

Data analysis and forecasts in economic studies require estimates of costs that will occur in the future. There are always uncertainties about these costs. However, these uncertainties are insufficient reason not to make the best possible estimates of the costs. Nevertheless, a decision between alternatives often can be made more confidently if it is known whether or not the conclusion is sensitive to moderate changes in data forecasts. Sensitivity analysis provides this extra dimension to an economic analysis.

The sensitivity of a decision is determined by inserting a range of estimates for critical cash flows and other parameters. If radical changes can be made to a cash flow without changing the decision, the decision is said to be *insensitive* to uncertainties regarding that cash flow. However, if a small change in the estimate of a cash flow will alter the decision, that decision is said to be very *sensitive* to changes in the estimate. If the decision is sensitive only for a limited range of cash flow values, the term *variable sensitivity* is used. Figure 77.19 illustrates these terms.

**Figure 77.19** *Types of Sensitivity*

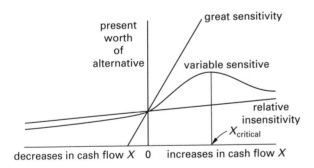

An established semantic tradition distinguishes between risk analysis and uncertainty analysis. *Risk analysis* addresses variables that have a known or estimated probability distribution. In this regard, statistics and probability theory can be used to determine the probability of a cash flow varying between given limits. On the other hand, *uncertainty analysis* is concerned with situations in which there is not enough information to determine the probability or frequency distribution for the variables involved.

As a first step, sensitivity analysis should be applied one at a time to the dominant factors. Dominant cost factors are those that have the most significant impact on the present value of the alternative.[55] If warranted, additional investigation can be used to determine the sensitivity to several cash flows varying simultaneously. Significant judgment is needed, however, to successfully determine the proper combinations of cash flows to vary. It is common to plot the dependency of the present value on the cash flow being varied in a two-dimensional graph. Simple linear interpolation is used (within reason) to determine the critical value of the cash flow being varied.

## 62. VALUE ENGINEERING

The *value* of an investment is defined as the ratio of its return (performance or utility) to its cost (effort or investment). The basic object of *value engineering* (VE, also referred to as *value analysis*) is to obtain the maximum per-unit value.[56]

Value engineering concepts often are used to reduce the cost of mass-produced manufactured products. This is done by eliminating unnecessary, redundant, or superfluous features, by redesigning the product for a less expensive manufacturing method, and by including features for easier assembly without sacrificing utility and function.[57] However, the concepts are equally applicable to one-time investments, such as buildings, chemical processing plants, and space vehicles. In particular, value engineering has become an important element in all federally funded work.[58]

Typical examples of large-scale value engineering work are using stock-sized bearings and motors (instead of custom manufactured units), replacing rectangular concrete columns with round columns (which are easier to form), and substituting custom buildings with prefabricated structures.

Value engineering is usually a team effort. And, while the original designers may be on the team, usually outside consultants are utilized. The cost of value engineering is usually returned many times over through reduced construction and life-cycle costs.

[55]In particular, engineering economic analysis problems are sensitive to the choice of effective interest rate ($i$) and to accuracy in cash flows at or near the beginning of the horizon. The problems will be less sensitive to accuracy in far-future cash flows, such as salvage value and subsequent generation replacement costs.

[56]Value analysis, the methodology that has become today's value engineering, was developed in the early 1950s by Lawrence D. Miles, an analyst at General Electric.

[57]Some people say that value engineering is "the act of going over the plans and taking out everything that is interesting."

[58]U.S. Government Office of Management and Budget Circular A-131 outlines value engineering for federally funded construction projects.

# 78

# Management Science

## Nomenclature

| | | |
|---|---|---|
| $c$ | cycle time | various |
| $c$ | number of failures (defects) | – |
| $c_2$ | ratio of $\overline{\sigma}/\sigma'$ | – |
| $C$ | confidence level | – |
| $C$ | cost | various |
| $d_2$ | ratio of $\overline{R}/\sigma'$ | – |
| $D$ | duration | various |
| $k$ | number | – |
| $L$ | expected line length | – |
| $L_q$ | expected queue length | – |
| $M$ | number of consecutive samples | – |
| MTBF | mean time between failures | various |
| MTBFO | mean time before failure outage | various |
| MTTF | mean time to failure | various |
| MTTR | mean time to repair | various |
| $n$ | actual number of observations | – |
| $n$ | number | – |
| $N$ | sample size | – |
| $N$ | theoretical number of observations | – |
| $p$ | observed fraction ($n/N$) | – |
| $p$ | true fraction defective | – |
| $p\{x\}$ | probability of event $x$ | – |
| $P$ | precision | – |
| $Q$ | quantity | – |
| $R$ | range | various |

| | | |
|---|---|---|
| $R$ | reliability | – |
| RB | relative bonus | – |
| RE | relative earnings | – |
| $s$ | number of servers in the system | – |
| $s$ | sample standard deviation | various |
| $t$ | time | various |
| $t_C$ | Student's $t$-distribution factor | – |
| $W$ | waiting time in the system | various |
| $W_q$ | waiting time in the queue | various |
| $x$ | general variable | – |
| $z$ | standard normal variable | – |
| $Z$ | objective function | – |

## Symbols

| | | |
|---|---|---|
| $\alpha$ | producer's risk | – |
| $\beta$ | consumer's risk | – |
| $\epsilon$ | absolute error | various |
| $\lambda$ | mean arrival rate | 1/time |
| $\lambda$ | mean failure rate | 1/time |
| $\lambda$ | mean of the Poisson distribution | – |
| $\mu$ | distribution mean | various |
| $\mu$ | mean service (repair) rate | 1/time |
| $\nu$ | number of degrees of freedom | – |
| $\rho$ | utilization factor, $\lambda/s\mu$ | – |
| $\sigma$ | standard deviation (of the sample) | various |
| $\sigma'$ | standard deviation of the population | various |

## Subscripts

| | | |
|---|---|---|
| ave | average |
| $C$ | at confidence level $C$ |
| $i$ | period $i$ |
| max | maximum |
| min | minimum |
| $p$ | fraction |
| $q$ | queue |
| std | standard |
| $t$ | total, or at time $t$ |

## 1. INTRODUCTION

*Management science*, also known as *quantitative business analysis, operations research,* and *management systems modeling,* is used to develop mathematical models of real-world situations. This chapter presents various quantitative business analysis techniques used to model and analyze manufacturing and industrial environments. Accordingly, this chapter is more concerned with solutions to problems than with explaining why the problems need to be solved or with listing advantages and disadvantages of solutions. Though they may seem to be obscure, all of the techniques presented in this chapter are commonly taught in operations research (OR), industrial engineering (IE), and MBA curricula.[1]

A *deterministic model* is a mathematical model that is built around a set of fixed rules such that any given input always results in a specific output. If an input can produce a variety of outputs determined by rules of probability, the model is known as a *probabilistic* or *stochastic model.*

A common aspect of most management science techniques is the goal of arriving at an optimum solution (regardless of whether the goal is actually realized in practice). The process of optimizing is unique to each type of problem. Calculus is not generally used in optimizing.[2] Optimizing real-world problems always requires a computer, though optimization by hand is possible with simple problems.[3]

Some management science methods attempt to optimize a specific mathematical function known as the *objective function, Z.* If $Z$ is a profit function, it is optimized by maximization; if $Z$ is a cost or time function, it is optimized by minimization.[4] Some management science techniques can maximize only, so in cases requiring minimization, the negative of the objective function is maximized.

Objective functions are restricted from increasing without being bound by *constraints* placed on one or more of the function's variables. These constraints are typically mathematical representations of how resources are limited or combined. Non-negativity constraints are common in mathematical programming problems.

If the objective function and its constraints are linear combinations of the independent variables, the model is said to be a *linear model.* Otherwise, the model is nonlinear.

Not all manufacturing management problems need to be solved by complex or obscure procedures. Some problems (e.g., facilities layout) do not have a general solution procedure and must be solved by exhaustive enumeration. Many problems, such as Ex. 78.1 and

Ex. 78.2, can be solved simply by using common sense and logical thinking to minimize the total cost.

### Example 78.1

A particular part is used by a company at a uniform rate of 120 units per month. The part is obtained from the supplier at a cost of $20 per unit. The company's cost of stocking the product is $0.06 per unit per month. The prorated cost of placing an order and putting shipments into inventory is $0.07 per unit over all normal order quantities. The company's effective monthly interest rate on borrowed money is 0.5% per month. The inventory is initially full. Orders arrive instantaneously when needed, and shortages do not occur. What is the optimum stocking quantity of the product?

*Solution*

This is essentially an economic order quantity (EOQ) problem. However, the ordering cost is expressed per unit ordered, and therefore, the total ordering cost is initially unknown. An iterative approach is necessary.

The interest expense on money tied up in a unit of inventory is

$$\left(\frac{0.005}{\text{month}}\right)\left(\frac{\$20}{\text{unit}}\right) = \frac{\$0.10}{\text{unit-month}}$$

Let $Q$ be the quantity ordered. The total cost per month is

$$C_t = \text{cost of ordering} + \text{cost of stocking}$$
$$= \left(\frac{\text{no. of orders}}{\text{month}}\right)\left(\frac{\text{cost}}{\text{order}}\right)$$
$$+ (\text{average monthly inventory})$$
$$\times (\text{stocking} + \text{interest costs})$$

Initially assume that the order quantity, $Q$, is 100. Then, the cost of placing an order will be

$$\left(\frac{100 \text{ units}}{\text{order}}\right)\left(\frac{\$0.07}{\text{unit}}\right) = \$7.00/\text{order}$$

The inventory drops linearly from $Q$ to zero over time, so the average inventory at any moment is $Q/2$.

$$C_t = \left(\frac{\dfrac{120}{\text{month}}}{Q}\right)(\$7.00) + \left(\frac{Q}{2}\right)(\$0.06 + \$0.10)$$
$$= \frac{\$840}{Q} + \$0.08Q$$

---

[1]*Operations research* developed as a field of its own during World War II when optimizing modeling techniques were used to determine the best way for a submarine to patrol a specific region.

[2]One obvious exception is how the economic order quantity is calculated. (See Ex. 78.1.) The EOQ formula is derived by taking the derivative of the total cost function.

[3]Some management science techniques, though interesting, are too obtuse, time consuming, or complex for solving by hand. Subjects that have been omitted from or given only a mere mention in this book include nonlinear programming, dynamic programming, and integer programming. Furthermore, most management science subjects have many complicated variations that are omitted from this chapter. Simple forecasting and the economic order quantity (EOQ) model are also traditional management science subjects.

[4]There is an important difference between *cost* and *price*. Both represent an amount paid, but the distinction depends on who makes the payment and when the payment is made. To one party, the cost of materials incorporated into a manufactured item is the price paid by that party for those materials. That is, there is no difference. However, the cost to one party to acquire or produce an item is much lower than the price at which the item is later sold to a second party.

This is minimized by setting the derivative of the cost function equal to zero.

$$\frac{dC_t}{dQ} = \frac{-\$840}{Q^2} + 0.08 = 0$$

$$Q = 102.5 \text{ units}$$

This is close to the initial estimate of $Q$. (Further iterations refine the value to exactly 105 units.)

### Example 78.2

Three different semi-automatic machines are being evaluated as a replacement to a completely manual operation. Each machine is mutually exclusive, and only one machine can be selected. Each machine has a different level of automation, cost of operation, and fraction of generated defects. Which machine should be selected in each production quantity range?

| machine | set-up cost | per unit material cost | per unit labor cost | fraction defective |
|---------|-------------|------------------------|---------------------|--------------------|
| A | $200 | $0.47 | $0.56 | 0.06 |
| B | $700 | $0.52 | $0.35 | 0.03 |
| C | $1200 | $0.54 | $0.27 | 0.02 |

*Solution*

When a process has a *scrap rate* greater than zero or a *yield* less than 100% (i.e., a fraction of the items produced are defective), the total cost per saleable item produced should be minimized.

$$\frac{\text{cost}}{\text{saleable item}} = \frac{\dfrac{\text{cost}}{\text{aggregate item}}}{1 - \text{fraction defective}}$$

Solving this example requires determining each machine's cost per saleable item as a function of production quantity. The setup cost is a *fixed cost*, allocated over all units produced. The material and labor costs are *variable costs*. The fraction defective is a scale factor that increases the cost of all saleable items produced. Let $x$ represent the total number of all items (saleable and nonsaleable) produced. The unit cost per saleable item is

$$C = \frac{\text{material cost} + \text{labor cost} + \dfrac{\text{set-up cost}}{x}}{1 - \text{fraction defective}}$$

$$C_A = \frac{\$0.47 + \$0.56 + \dfrac{\$200}{x}}{1 - 0.06}$$

$$= \$1.10 + \frac{\$212.77}{x}$$

$$C_B = \frac{\$0.52 + \$0.35 + \dfrac{\$700}{x}}{1 - 0.03}$$

$$= \$0.90 + \frac{\$721.65}{x}$$

$$C_C = \frac{\$0.54 + \$0.27 + \dfrac{\$1200}{x}}{1 - 0.02}$$

$$= \$0.83 + \frac{\$1224.49}{x}$$

The three cost equations are not linear functions, and straight lines cannot be drawn between two points on the curves. Graphical or algebraic operations show that $C_A$ is minimum for $0 < x < 2544$, $C_B$ is minimum for $2544 < x < 7183$, and $C_C$ is minimum for $x > 7183$.

## 2. CRITICAL PATH TECHNIQUES

### Definitions

*activity:* any subdivision of a project whose execution requires time and other resources.

*critical path:* a path connecting all activities that have minimum or zero slack times. The critical path is the longest path through the network.

*duration:* the time required to perform an activity. All durations are *normal durations* unless otherwise referred to as *crash durations*.

*event:* the beginning or completion of an activity.

*event time:* actual time at which an event occurs.

*float:* same as slack time.

*slack time:* the minimum time that an activity can be delayed without causing the project to fall behind schedule. Slack time is always minimum or zero along the critical path.

### Introduction

*Critical path techniques* are used to graphically represent the multiple relationships between stages in a complicated project. The graphical network shows the *precedence relationships* between the various activities. The graphical network can be used to control and monitor the progress, cost, and resources of a project. A critical path technique will also identify the most critical activities in the project.

Critical path techniques use *directed graphs* to represent a project. These graphs are made up of *arcs* (arrows) and *nodes* (junctions). The placement of the arcs and nodes completely specifies the precedences of the project. Durations and precedences are usually given in a *precedence table (precedence matrix)*.

## Activity-on-Node Networks: The Critical Path Method

The *critical path method*, CPM, is a deterministic method applicable when all activity durations are known in advance. CPM is usually represented as an *activity-on-node* model since arcs are used to specify precedence and the nodes represent the activities. Two *dummy nodes* taking zero time may be used to specify the start and finish of the project. Other starting and finishing events are not represented on the graph, other than as the heads and tails of the arcs.

## Solving a CPM Problem

The solution to a critical path method problem reveals the earliest and latest times that an activity can be started and finished. It also identifies the *critical path* and generates the *slack times* for each activity.

The following procedure may be used to solve a CPM problem. To facilitate the solution, each node should be replaced by a square that has been quartered. The compartments have the meanings indicated by the key.

| ES | EF |
|----|----|
| LS | LF |

key
**ES** Earliest Start
**EF** Earliest Finish

**LS** Latest Start
**LF** Latest Finish

*step 1:* Place the project start time or date in the **ES** and **EF** positions of the start activity. The start time is zero for relative calculations.

*step 2:* Consider any unmarked activity, all of whose predecessors have been marked in the **EF** and **ES** positions. (Go to step 4 if there are no such activities.) Mark in its **ES** position the largest number marked in the **EF** position of those predecessors.

*step 3:* Add the activity time to the **ES** time and write this in the **EF** box. Go to step 2.

*step 4:* Place the value of the latest finish date in the **LS** and **LF** boxes of the finish mode.

*step 5:* Consider unmarked predecessors whose successors have all been marked. Their **LF** is the smallest **LS** of the successors. Go to step 7 if there are no unmarked predecessors.

*step 6:* The **LS** for the new node is **LF** minus its activity time. Go to step 5.

*step 7:* The slack time for each node is **LS** − **ES** or **LF** − **EF**.

*step 8:* The critical path encompasses nodes for which the slack time equals **LS** − **ES** from the start node. There may be more than one critical path.

## Example 78.3

Using the precedence table given, construct the precedence matrix and draw an activity-on-node network.

| activity | duration (days) | predecessors |
|----------|-----------------|--------------|
| A, start | 0 | – |
| B | 7 | A |
| C | 6 | A |
| D | 3 | B |
| E | 9 | B, C |
| F | 1 | D, E |
| G | 4 | C |
| H, finish | 0 | F, G |

*Solution*

The precedence matrix is

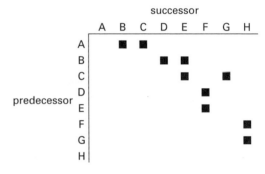

The activity-on-node network is

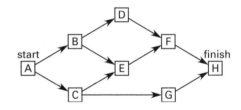

## Example 78.4

Complete the network for the previous example and find the critical path. Assume the desired completion date is in 19 days.

*Solution*

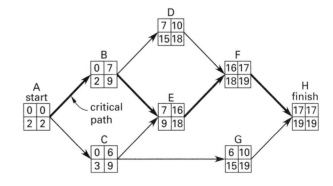

## Activity-on-Branch Networks

In an *activity-on-branch* network, the arcs represent the activities, which are labeled with letters of the alphabet, and the nodes represent events, which are numbered. The activity durations may appear in parentheses near the activity letter.

The activity-on-branch method is complicated by the frequent requirement for *dummy activities* and *dummy nodes* to maintain precedence. Consider the following part of a precedence table:

| activity | predecessors |
|----------|--------------|
| L | – |
| M | – |
| N | L, M |
| P | M |

Note that activity P depends on the completion of only activity M. Figure 78.1(a) is an activity-on-branch representation of this precedence. However, N depends on the completion of both L and M. It would be incorrect to draw the network as Fig. 78.1(b) since the activity N appears twice. To represent the project, the dummy activity X must be used, as shown in Fig. 78.1(c).

**Figure 78.1** *Activity-on-Branch Networks*

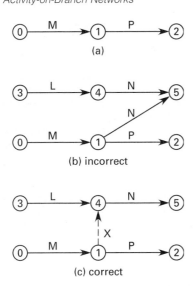

If two activities have the same starting and ending events, a dummy node is required to give one activity a uniquely identifiable completion event. This is illustrated in Fig. 78.2(b).

The solution method for an activity-on-branch problem is essentially the same as for the activity-on-node problem, requiring forward and reverse passes to determine earliest and latest dates.

**Figure 78.2** *Use of a Dummy Node*

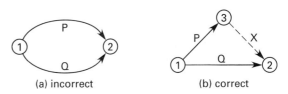

### Example 78.5

Represent the project in Ex. 78.3 as an activity-on-branch network.

*Solution*

| event | event description |
|-------|-------------------|
| 0 | start project |
| 1 | finish B, start D |
| 2 | finish C, start G |
| 3 | finish B and C, start E |
| 4 | finish D and E, start F |
| 5 | finish F and G |

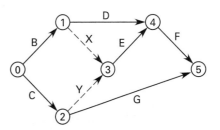

## Stochastic Critical Path Models

*Stochastic models* differ from deterministic models only in the way in which the activity durations are found. Whereas durations are known explicitly for the deterministic model, the time for a stochastic activity is distributed as a random variable.

This stochastic nature complicates the problem greatly since the actual distribution is often unknown. Such problems are solved as a deterministic model using the mean of an assumed duration distribution as the activity duration.

The most common stochastic critical path model is PERT, which stands for *program evaluation and review technique*. In PERT, all duration variables are assumed to come from a *beta distribution*, with mean and standard deviation given by Eq. 78.1 and Eq. 78.2, respectively.

$$\mu = \tfrac{1}{6}\left(t_{\min} + 4t_{\text{most likely}} + t_{\max}\right) \qquad 78.1$$

$$\sigma = \tfrac{1}{6}\left(t_{\max} - t_{\min}\right) \qquad 78.2$$

The project *completion time* for large projects is assumed to be normally distributed with mean equal to the critical path length and with overall variance equal to the sum of the variances along the critical path.

The probability that a project duration will exceed some length, $D$, can be found from Eq. 78.3. $z$ is the standard normal variable.

$$p\{\text{duration} > D\} = p\{t > z\} \qquad \textit{78.3}$$

$$z = \frac{D - \mu_{\text{critical path}}}{\sigma_{\text{critical path}}} \qquad \textit{78.4}$$

### Example 78.6

The mean times and variances for activities along a PERT critical path are given. What is the probability that the project's completion time will be (a) less than 14 days, (b) more than 14 days, (c) more than 23 days, and (d) between 14 and 23 days?

| activity mean time (days) | activity standard deviation (days) |
|---|---|
| 9 | 1.3 |
| 4 | 0.5 |
| 7 | 2.6 |

*Solution*

The most likely completion time is the sum of the mean activity times.

$$\mu_{\text{critical path}} = 9 \text{ days} + 4 \text{ days} + 7 \text{ days}$$
$$= 20 \text{ days}$$

The variance of the project's completion times is the sum of the variances along the critical path. Variance, $\sigma^2$, is the square of the standard deviation, $\sigma$.

$$(\sigma_{\text{critical path}})^2 = (1.3 \text{ days})^2 + (0.5 \text{ days})^2$$
$$+ (2.6 \text{ days})^2$$
$$= (8.7 \text{ days})^2$$
$$\sigma_{\text{critical path}} = \sqrt{(8.7 \text{ days})^2}$$
$$= 2.95 \text{ days} \quad [\text{use 3 days}]$$

The standard normal variable corresponding to 14 days is

$$z = \frac{D - \mu_{\text{critical path}}}{\sigma_{\text{critical path}}}$$
$$= \frac{14 \text{ days} - 20 \text{ days}}{3 \text{ days}} = -2.0$$

The area under the standard normal curve is 0.4772 for $0.0 < z < -2.0$. Since the normal curve is symmetrical, the negative sign is irrelevant in determining the area.

The standard normal variable corresponding to 23 days is

$$z = \frac{D - \mu_{\text{critical path}}}{\sigma_{\text{critical path}}}$$
$$= \frac{23 \text{ days} - 20 \text{ days}}{3 \text{ days}} = 1.0$$

The area under the standard normal curve is 0.3413 for $0.0 < z < 1.0$.

(a) $p\{\text{duration} < 14\} = p\{z < -2.0\} = 0.5 - 0.4772$
$$= 0.0228 \quad (2.28\%)$$

(b) $p\{\text{duration} > 14\} = p\{z > -2.0\} = 0.4772 + 0.5$
$$= 0.9772 \quad (97.72\%)$$

(c) $p\{\text{duration} > 23\} = p\{z > 1.0\} = 0.5 - 0.3413$
$$= 0.1587 \quad (15.87\%)$$

(d) $p\{14 < \text{duration} < 23\} = p\{-2.0 < z < 1.0\}$
$$= 0.4772 + 0.3413$$
$$= 0.8185 \quad (81.85\%)$$

## 3. QUEUING THEORY

### Introduction

A *queue* is a waiting line. *Queuing theory* can predict the length of a waiting line, the average time a customer can expect to spend in the queue, and the probability that $n$ customers will be in the queue.

Different queuing models have been developed to account for differences in the distribution of arrivals, distribution of service times, number of servers, size of the calling population, and order in which the customers are served.

Most of the models are too complicated to be presented here. However, two models are given after a brief listing of the general relationships. The relationships are for *steady-state operation*. This means that the service facility has been open and in operation long enough for the queue to stabilize.

### General Relationships

$$L = \lambda W \qquad \textit{78.5}$$

$$L_q = \lambda W_q \qquad \textit{78.6}$$

$$W = W_q + \frac{1}{\mu} \qquad \textit{78.7}$$

$$\lambda < \mu s \qquad \textit{78.8}$$

$$\text{average service time} = 1/\mu \qquad \textit{78.9}$$

$$\text{average time between arrivals} = 1/\lambda \qquad \textit{78.10}$$

## The M/M/1 System

The following points are assumed in the $M/M/1$ model.[5]

- There is only one server $(s = 1)$.

- The calling population is infinite.

- The service times are exponentially distributed with mean service rate $\mu$. That is, the probability of a customer's remaining service time exceeding $h$ (after already spending time with the server) is

$$p\{t > h\} = e^{-\mu h} \qquad 78.11$$

Notice that Eq. 78.11 is independent of the time already spent with the server.

- The arrival rate is distributed as Poisson with mean $\lambda$. The probability of $x$ customers arriving in the next period (the time period used to express $\mu$ and $\lambda$) is

$$p\{x\} = \frac{e^{-\lambda}\lambda^x}{x!} \qquad 78.12$$

The following relationships are valid for the $M/M/1$ system.

$$p\{0\} = 1 - \rho = \frac{\mu - \lambda}{\mu} \qquad 78.13$$

$$p\{n\} = p\{0\}\rho^n \qquad 78.14$$

$$W = \frac{1}{\mu - \lambda} = W_q + \frac{1}{\mu} = \frac{L}{\lambda} \qquad 78.15$$

$$W_q = \frac{\rho}{\mu - \lambda} = \frac{L_q}{\lambda} \qquad 78.16$$

$$L = \frac{\lambda}{\mu - \lambda} = L_q + \rho \qquad 78.17$$

$$L_q = \frac{\rho\lambda}{\mu - \lambda} \qquad 78.18$$

### Example 78.7

Given an $M/M/1$ system with an average service rate of $\mu = 20$ customers per hr and an average arrival rate of $\lambda = 12$ per hr, find the steady-state values of $W$, $W_q$, $L$, and $L_q$. What is the probability that there will be five customers in the system?

*Solution*

$$\rho = \frac{\lambda}{\mu} = \frac{12}{20} = 0.6$$

$$W = \frac{1}{\mu - \lambda} = \frac{1}{20 - 12} = 0.125 \text{ hr}$$

$$W_q = \frac{\rho}{\mu - \lambda} = \frac{0.6}{20 - 12} = 0.075 \text{ hr}$$

$$L = \frac{\lambda}{\mu - \lambda} = \frac{12}{20 - 12} = 1.5 \text{ customers}$$

$$L_q = \frac{\rho\lambda}{\mu - \lambda} = \frac{(0.6)(12)}{20 - 12} = 0.9 \text{ customers}$$

$$p\{0\} = 1 - \rho = 1 - 0.6 = 0.4$$

$$p\{5\} = p\{0\}\rho^n = (0.4)(0.6)^5 = 0.031$$

## The M/M/s System

The same assumptions are used for the $M/M/s$ system as were used for the $M/M/1$ system, except that there are $s$ servers instead of only one. Each server has a mean service rate $\mu$. All servers draw from a single line so that the first person in line goes to the first available server. Each server does not have its own line. However, if customers are allowed to change the lines they are in so that they may go to any available server, this model may also be used to predict the performance of a multiple server system where each server has its own line.

$$\rho = \frac{\lambda}{s\mu} \qquad 78.19$$

$$W = W_q + \frac{1}{\mu} \qquad 78.20$$

$$W_q = \frac{L_q}{\lambda} \qquad 78.21$$

$$L_q = \frac{p\{0\}\left(\dfrac{\lambda}{\mu}\right)^s \rho}{s!(1 - \rho)^2} \qquad 78.22$$

$$L = L_q + \frac{\lambda}{\mu} \qquad 78.23$$

$$p\{0\} = \frac{1}{\dfrac{\left(\dfrac{\lambda}{\mu}\right)^s}{s!(1 - \rho)} + \displaystyle\sum_{j=0}^{s-1} \dfrac{\left(\dfrac{\lambda}{\mu}\right)^j}{j!}} \qquad 78.24$$

$$p\{n\} = \frac{p\{0\}\left(\dfrac{\lambda}{\mu}\right)^n}{n!} \quad [n \le s] \qquad 78.25$$

$$p\{n\} = \frac{p\{0\}\left(\dfrac{\lambda}{\mu}\right)^n}{s! s^{n-s}} \quad [n > s] \qquad 78.26$$

Figure 78.3 is a graphical solution to the $M/M/s$ multiple server model.

---

[5]In the nomenclature of queuing theory, an exponentially distributed interarrival or service time is given the symbol $M$ (for Markovian). Other options include $D$ (deterministic, constant, regular, etc.), $E$ or $K$ (for Erlangian or gamma distribution), and $G$ (general distribution).

**Figure 78.3** *Mean Number in System for M/M/s System*

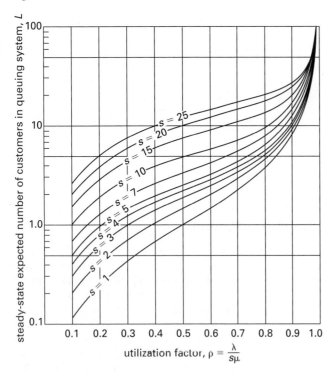

Reprinted from *Operations Research*, 6th Ed., by Frederick S. Hillier and Gerald J. Lieberman, Holden-Day, Inc., with permission from the McGraw-Hill Companies, © 1974.

### Example 78.8

A company has several identical machines operating in parallel. The average breakdown rate is 0.7 machines per week. There is one repair station for the entire company. It takes a maintenance worker one entire week to repair a machine, although the time is reduced in proportion to the number of maintenance workers assigned to the repair. Each maintenance worker is paid $400 per week. Machine downtime is valued at $800 per week. Other costs (additional tools, etc.) are to be disregarded. What is the optimum number of maintenance workers at the repair station?

*Solution*

Assume the number of breakdowns per week can be represented by a Poisson distribution. Then, this example can be solved with queuing theory. Using two or more maintenance workers only decreases the repair time, so this is a single-server model, even with multiple maintenance workers.

The average number of machines breaking down each week is the mean arrival rate, $\lambda = 0.7$ per week. With one worker, the repair rate, $\mu$, is 1.0 per week.

The average time a machine is out of service (waiting for its turn to be repaired and during the repair) is $W$, the "time in the system."

$$W = \frac{1}{\mu - \lambda} = \frac{1}{1 - 0.7}$$
$$= 3.33 \text{ weeks}$$

With one maintenance worker, the average downtime cost in a week is

$$\left(\frac{\text{downtime cost}}{\text{machine-week}}\right)(\text{no. of machines})(\text{no. of weeks})$$
$$= (\text{downtime cost})\lambda W$$
$$= \left(\frac{\$800}{\text{machine-week}}\right)(0.7 \text{ machines})(3.33 \text{ weeks})$$
$$= \$1865$$

However, the product of $\lambda$ and $W$ is the same as the average number of machines in the system, $L$. The average number of machines in the system, $L$ (i.e., being repaired or waiting for repair), is

$$L_1 = \frac{\lambda}{\mu - \lambda} = \frac{0.7}{1 - 0.7}$$
$$= 2.33$$

The total average weekly cost with one worker is the sum of the costs of the worker and the downtime.

$$C_{t,1} = (1 \text{ worker})\left(\frac{\$400}{\text{week}}\right) + (2.33)(\$800) = \$2264$$

With two workers, the values are

$$\mu = (2 \text{ workers})\left(\frac{1}{\text{worker-week}}\right) = 2/\text{week}$$

$$L_2 = \frac{0.7}{2 - 0.7} = 0.538$$

$$C_{t,2} = (2)\left(\frac{\$400}{\text{week}}\right) + (0.538)(\$800) = \$1230$$

With three workers, the values are

$$\mu = (3 \text{ workers})\left(\frac{1}{\text{worker-week}}\right) = 3/\text{week}$$

$$L_3 = \frac{0.7}{3 - 0.7} = 0.304$$

$$C_{t,3} = (3)\left(\frac{\$400}{\text{week}}\right) + (0.304)(\$800) = \$1443$$

Adding workers will increase the costs above $C_3$. Two maintenance workers should staff the maintenance station as this number minimizes the weekly cost.

## Example 78.9

Twenty identical machines are in operation. The hourly reliability for any one machine is 90%. (That is, the probability of a machine breaking down in any given hour is 10%.) The cost of downtime is $5 per hour. Each broken machine requires one technician for repair, and the average repair time is one hour. If all technicians are busy, broken machines wait idle. Each technician costs $2.5 per hour. How many separate technicians should be used?

*Solution*

This is a multiple-server model. It is assumed that the $M/M/s$ assumptions are satisfied. The mean arrival rate, $\lambda$, is $(0.10)(20) = 2$ per hour. The repair rate, $\mu$, is 1 per hour. Clearly, one technician cannot handle the workload, nor can two technicians. The average number of machines in the system, $L$ (i.e., being repaired or waiting for repair), is calculated for two, three, four, and five servers. The total cost per hour is calculated as

$$C = 2.5s + 5L$$

| $s$ | $p_0$ | $L_q$ | $L$ | cost per hour |
|---|---|---|---|---|
| 2 | – | – | – | infinite |
| 3 | 0.11 | 0.91 | 2.91 | 22.0 |
| 4 | 0.13 | 0.17 | 2.17 | 20.9 |
| 5 | 0.13 | 0.04 | 2.04 | 22.7 |

The minimum hourly cost is achieved with four technicians.

## 4. RELIABILITY

A *fault* in a machine or other system is a known cause of breakdown. An *error* is an undesired state within the machine that might lead to improper operation. A *failure* occurs when the machine fails to operate as expected or intended. A *fault-tolerant system* contains provisions to avoid failures after faults occur.

In the most common case, units fail permanently and are neither repaired nor replaced. Reliability of a single item (machine, unit, piece of equipment, etc.) is characterized by its *mean time to failure*, MTTF. The term *mean time between forced outages*, MTBFO, is used with redundant systems in place of MTTF. *Coverage* is the probability of the system reconfiguring itself when a fault occurs. *Redundancy* is the primary tool used to increase reliability and coverage. Systems with two units in parallel are known as *duplex systems*. Systems with three units are known as *triple modular redundancy*, TMR, systems.[6]

The exponential distribution is most frequently used in reliability calculations.[7] The *failure rate*, $\lambda$, is the expected number of failures per unit time. The *reliability* is

$$R_t = e^{-\lambda t} \qquad 78.27$$

The MTTF is found by integrating the reliability function. Therefore, the MTTF is the reciprocal of the failure (arrival) rate.

$$\text{MTTF} = \int_0^\infty R_t \, dt = \int_0^\infty e^{-\lambda t} \, dt$$
$$= \frac{1}{\lambda} \qquad 78.28$$

The probability of exactly $c$ failures in time $t$ is given by the Poisson distribution. A cumulative probability chart can be used to find the probability of $c$ or fewer failures in time $t$.

$$p\{c\} = \frac{\left(\dfrac{t}{\text{MTTF}}\right)^c e^{-t/\text{MTTF}}}{c!} \qquad 78.29$$

For a 1-out-of-$n$ *fully redundant system* (a parallel system with $n$ identical redundant units, only one of which needs to be operational for the system to operate), the reliability is

$$R_{\text{1-out-of-}n \text{ system}} = 1 - (1 - R_t)^n \qquad 78.30$$

For a 1-out-of-2 redundant system (also known as a *2-1-0 system*),

$$R_{\text{1-out-of-2}} = 2e^{-\lambda t} - e^{-2\lambda t} \qquad 78.31$$

$$\text{MTTF}_{\text{1-out-of-2}} = \int_0^\infty R_t \, dt$$
$$= \int_0^\infty 2e^{-\lambda t} - e^{-2\lambda t}$$
$$= \frac{1.5}{\lambda} \qquad 78.32$$

For a 1-out-of-3 fully redundant system (also known as a *3-2-1-0 system*),

$$R_{\text{1-out-of-3}} = 3e^{-\lambda t} - 3e^{-2\lambda t} + e^{-3\lambda t} \qquad 78.33$$

$$\text{MTTF}_{\text{1-out-of-3}} = \frac{11}{6\lambda} \qquad 78.34$$

In some cases, failed units are repaired online. The average repair time is the *mean time to repair*, MTTR, and is the reciprocal of the repair rate, $\mu$. The *mean time to failure*, MTTF, is the average time a unit operates before failing. The *mean time between failures*, MTBF,

---

[6]With TMR systems, only one unit is required for successful operation. When one unit fails, the system becomes a duplex system until the failed unit is repaired. With logic, software, electronic, and computer systems, failure can be determined by comparing the output of each of the three units. In effect, the two good units "vote" to determine which unit is faulty and should be shut down.

[7]The three-parameter *Weibull distribution* is more descriptive, flexible, and powerful than the negative exponential distribution. It has gained acceptance primarily in the aerospace industry because of its ability to model the failure distribution more exactly. Its complexity, however, makes application to noncritical applications cumbersome.

is the length of time between when the original and repaired units start.

$$MTBF = MTTF + MTTR \qquad 78.35$$

$$MTTR = \frac{1}{\mu} \qquad 78.36$$

*Availability* is the probability that a system will be operating at any given time. The system *uptime* is calculated by multiplying the availability by the theoretically maximum number of operational hours (e.g., 8760 hours per year).[8] *Unavailability* and *downtime* are similarly calculated.

$$availability = \frac{MTTF}{MTBF} \qquad 78.37$$

$$unavailability = 1 - availability \qquad 78.38$$

$$uptime = (availability)\left(8760 \; \frac{hr}{yr}\right) \qquad 78.39$$

$$downtime = (unavailability)\left(8760 \; \frac{hr}{yr}\right) \qquad 78.40$$

The MTBF values for fully redundant systems can be calculated from a Markov model of the system. (See Sec. 78.24.)

$$MTBF_{1\text{-out-of-2}} = \frac{\mu}{2\lambda^2} \qquad 78.41$$

$$MTBF_{1\text{-out-of-3}} = \frac{\mu^2}{3\lambda^3} \qquad 78.42$$

Repairing a machine as soon as it breaks down is always the preferred course of action. In some cases, however, a faulty machine can be repaired only at regular intervals. This is particularly true for unattended equipment that is inspected only at periodic intervals, often called the *proof test interval*, PTI. A complete failure will occur if the system's redundancy is not adequate to sustain multiple faults during the PTI.

It is not unexpected that the system's MTBF is a function of the PTI. A small decrease in the PTI can greatly increase the MTBF. The MTBFs, availabilities, and *average downtime*, ADT, for periodically repaired equipment operating with full redundancy is

$$ADT_{1\text{-out-of-2}} = \frac{PTI}{2} \qquad 78.43$$

$$MTBF_{1\text{-out-of-2}} = \frac{1}{\lambda^2(PTI)} \qquad 78.44$$

$$availability_{1\text{-out-of-2}} = \frac{1 - \lambda^2(PTI)^2}{3} \qquad 78.45$$

$$ADT_{1\text{-out-of-3}} = \frac{PTI}{\sqrt{3}} \qquad 78.46$$

$$MTBF_{1\text{-out-of-3}} = \frac{1}{\lambda^3(PTI)^2} \qquad 78.47$$

$$availability_{1\text{-out-of-3}} = \frac{1 - \lambda^3(PTI)^3}{4} \qquad 78.48$$

The *mean down time*, MDT, for a *k*-out-of-*n* system with periodic repair is

$$MDT = \frac{PTI}{n - k + 2} \qquad 78.49$$

### Example 78.10

One hundred items are tested to failure. Two failed at $t = 1$, 5 at $t = 2$, 7 at $t = 3$, 20 at $t = 4$, 35 at $t = 5$, and 31 at $t = 6$. Find the probability of failure in any period, the conditional probability of failure, and the mean time to failure.

*Solution*

| elapsed time $t$ | failures $F(t)$ | survivors $S(t)$ | probability of failure $0.01F(t)$ | conditional probability of failure $F(t)/S(t-1)$ |
|---|---|---|---|---|
| 0 | 0 | 100 | 0 | 0 |
| 1 | 2 | 98 | 0.02 | 0.02 |
| 2 | 5 | 93 | 0.05 | 0.051 |
| 3 | 7 | 86 | 0.07 | 0.075 |
| 4 | 20 | 66 | 0.20 | 0.233 |
| 5 | 35 | 31 | 0.35 | 0.530 |
| 6 | 31 | 0 | 0.31 | 1.00 |

The mean time to failure is

$$MTTF = \frac{\begin{array}{c}(2)(1) + (5)(2) + (7)(3) \\ + (20)(4) + (35)(5) + (31)(6)\end{array}}{100}$$
$$= 4.74$$

### Example 78.11

The overall reliability of a fully redundant system must be at least 0.99 over a year's time. The system will be designed to operate as long as any one of multiple identical, parallel units is operational. The MTTF of each unit is 0.8 years. What level of redundancy is required?

*Solution*

For each unit,

$$\lambda = \frac{1}{MTTF} = \frac{1}{0.8 \; yr} = 1.25 \; 1/yr$$

$$R_{1 \; yr} = e^{-\lambda t} = e^{(-1.25 \; 1/yr)(1 \; yr)}$$

$$= 0.2865$$

---

[8]If the machine does not operate 24 hours per day or 365 days per year, the number of hours will be accordingly reduced.

From Eq. 78.30,

$$R_{\text{1-out-of-}n \text{ system}} = 1 - (1 - R_t)^n$$
$$0.99 = 1 - (1 - 0.2865)^n$$
$$n = 13.6 \quad [\text{use } 14]$$

## 5. PREVENTATIVE MAINTENANCE

The value of *preventative maintenance*, PM, to prevent breakdowns is undisputed.[9] However, it is not as easy to decide on the frequency and timing of PM, the size of maintenance facilities and number of staff, location and centralization issues, and the quantity of spares to be carried. Quantitative business analysis techniques can be used to formulate some of these PM policies.[10]

The general goal in optimizing PM policies is to minimize the total cost of operation, taking into consideration the costs of preventative maintenance, downtime, and repair. Sometimes the costs are fixed, as when specific penalties must be paid when output is not achieved. At other times, the costs are related to hourly rates and the duration of downtime. The time to failure of a machine and the times for both repair and preventative maintenance are generally not fixed, and they are not always normally distributed either. However, unless simulation is used, it is almost always necessary to work with the average times (e.g., *mean time to failure*, MTTF).

The following guidelines should be considered when establishing PM policies, particularly for single machines. When there are several identical machines operating in parallel, the problem more closely resembles a waiting-line (queuing) problem. Breakdowns are comparable to arrivals in the line, and repair stations (repair crews) are the stations. The optimum solution takes into consideration the costs of idle maintenance crews.

- PM is more applicable when the time-to-breakdown distribution has low variability because the time before a breakdown can be more accurately predicted.

- PM is only useful when its cost is less than the cost of the breakdown. In the absence of cost information, PM is useful when the average PM time is less than the average repair time.

- PM is more applicable when there is little or no inventory of the item produced by the broken machine.

---

[9]Maintenance to correct disrepair is known as *remedial maintenance*.
[10]Queuing theories are covered in Sec. 78.3. Replacement policies are covered in Sec. 78.6. Even when there are no specific quantitative techniques for solving a particular problem type, simulation can be used to determine the outcome of most preventative maintenance policies.

## 6. REPLACEMENT

### Introduction

*Replacement* and *renewal models* determine the most economical time to replace existing equipment. Replacement processes fall into two categories, depending on the life pattern of the equipment, which either deteriorates gradually (becomes obsolete or less efficient) or fails suddenly.

In the case of gradual deterioration, the solution consists of balancing the cost of new equipment against the cost of maintenance or decreased efficiency of the old equipment. Several models are available for cases with specialized assumptions, but no general solution methods exist.

In the case of sudden failure (e.g., light bulbs), the solution method consists of finding a replacement frequency that minimizes the costs of the required new items, the labor for replacement, and the expected cost of failure. The solution is made difficult by the probabilistic nature of the life spans.

### Deterioration Models

The replacement decision criterion with deterioration models is the present worth of all future costs associated with each policy. Solution is by trial and error, calculating the present worth of each policy and incrementing the replacement period by one time period for each iteration.

### Example 78.12

Item A is currently in use. Its maintenance cost is $400 this year and is increasing each year by $30. Item A can be replaced by item B at a current cost of $3500. However, the purchase cost of B is increasing by $50 each year. Item B has no maintenance costs. Disregarding income taxes, find the optimum replacement year. Use 10% as the interest rate.

*Solution*

Calculate the present worth, $P$, of the various policies.

*policy 1:* Replacement at $t = 5$ (starting the 6th year)

$$P(A) = (-\$400)(P/A, 10\%, 5) - (\$30)(P/G, 10\%, 5)$$
$$= (-\$400)(3.7908) - (\$30)(6.8618)$$
$$= -\$1722$$
$$P(B) = -(\$3500 + (5)(\$50))(P/F, 10\%, 5)$$
$$= -(\$3500 + (5)(\$50))(0.6209)$$
$$= -\$2328$$

*policy 2:* Replacement at $t = 6$

$$P(A) = (-\$400)(P/A, 10\%, 6) - (\$30)(P/G, 10\%, 6)$$
$$= (-\$400)(4.353) - (\$30)(9.6842)$$
$$= -\$2032$$

$$P(B) = -\big(\$3500 + (6)(\$50)\big)(P/F, 10\%, 6)$$
$$= -\big(\$3500 + (6)(\$50)\big)(0.5645)$$
$$= -\$2145$$

*policy 3:* Replacement at $t = 7$

$$P(A) = (-\$400)(P/A, 10\%, 7) - (\$30)(P/G, 10\%, 7)$$
$$= (-\$400)(4.8684) - (\$30)(12.7631)$$
$$= -\$2330$$

$$P(B) = -\big(\$3500 + (7)(\$50)\big)(P/F, 10\%, 7)$$
$$= -\big(\$3500 + (7)(\$50)\big)(0.5132)$$
$$= -\$1976$$

The present worth of A drops below the present worth of B sometime between $t = 6$ and $t = 7$. Replacement should take place at that time.

## Sudden Failure Models

The time between installation and failure is not constant for members in the general equipment population. Therefore, in order to solve a sudden failure model, it is necessary to have the distribution of individual item lives (*mortality curve*). The conditional probability of failure in a small time interval, say from $t$ to $t + \delta t$, is calculated from the mortality curve. This probability is *conditional* since it is conditioned on nonfailure up to time $t$.

The conditional probability of failure may decrease with time (as with *infant mortality*), remain constant (as with an exponential reliability distribution and failure from random causes), or increase with time (as with items that deteriorate with use). If the conditional probability of failure decreases or remains constant over time, operating items should never be replaced prior to failure.

It is usually assumed that all failures occur at the end of a period. The problem is to find the period that minimizes the total cost. The expression for the number of units failing in time $t$ is

$$F(t) = n\left( p\{t\} + \sum_{i=1}^{t-1} p\{i\} p\{t - i\} \right.$$
$$\left. + \sum_{j=2}^{t-1}\left( \sum_{i=1}^{j-1} p\{i\} p\{j - 1\} p\{t - j\} + ... \right) \right) \qquad 78.50$$

The term $np\{t\}$ gives the number of failures in time $t$ from the original group.

The term $n\sum p\{i\} p\{t - i\}$ gives the number of failures in time $t$ from the set of items that replaced the original items.

The third probability term times $n$ gives the number of failures in time $t$ from the set of items that replaced the first replacement set.

It can be shown that $F(t)$ with replacement will converge to a steady state limiting rate of

$$\overline{F(t)} = \frac{n}{\text{MTBF}} \qquad 78.51$$

The optimum policy is to replace all items in the group, including items just recently installed, when the total cost per period is minimized. That is, try to find $T$ such that $K(T)/T$ is minimized.

$$K(T) = nC_1 + C_2 \sum_{t=0}^{T-1} F(t) \qquad 78.52$$

Discounting (i.e., taking into consideration the time value of money) is usually not included in the total cost formula since the time periods are considered short. If the equipment has an unusually long life, discounting would be required.

There are some cases where group replacement is always less expensive than replacing the failures as they occur. Group replacement will be the most economical policy if Eq. 78.53 is valid.

$$C_2\left(\overline{F(t)}\right) > \left. \frac{K(T)}{T} \right|_{\min} \qquad 78.53$$

If the opposite inequality holds, group replacement may still be the optimum policy. Further analysis is then required.

## 7. LINEAR PROGRAMMING

### Introduction

*Mathematical programming* is a modeling procedure applicable to problems for which the objective and resource limitations can be described mathematically.

If the objective function and all resource constraints are linear (polynomials of degree 1 only), the procedure is known as *linear programming*.

If the variables can take on only integer values, a procedure known as *integer programming* is required. If the objective or constraint functions include higher-order polynomials or other functions, then nonlinear optimization techniques such as *gradient search* or *dynamic programming* must be used.

## Formulation of a Linear Programming Problem

All linear programming problems have a common format. Each has an *objective function* that is to be optimized. Usually the objective function is to be maximized, as in the case of a profit function. If the objective is to minimize some function, such as cost, the problem may be turned into a maximization problem by maximizing the negative of the original function.

Each linear programming problem also has a set of functions called *constraints* that limits the independent variables. All variables may take on noninteger values. In some problems, *non-negativity constraints* limit variables to positive values.

## Solving Linear Programming Problems

Linear programming problems are generally solved by computer, although some simple problems may be solved by hand with a procedure known as the *simplex method*. Specialized methods providing easy manual solutions are available for certain classes of problems, primarily the *transportation* and *assignment problems*.

The *simplex method* is an iterative procedure for solving linear programming problems with any number of variables. It works with a matrix (called a *tableau*) of values formed from the resource limitations. Since the method is iterative and systematic, it is easily computerized. While it is possible to solve even multidimensional linear programming problems manually using the simplex method, the computational effort is burdensome, the time requirement is high, and the possibility for accidental errors is high.[11] Therefore, using the simplex method for more than familiarization with the theory is not recommended.

If a linear programming problem can be formulated in terms of only two variables, $x_1$ and $x_2$, it can be solved graphically by the following procedure.

*step 1:* Graph all of the constraints and determine the *feasible region*. Usually this will result in a *convex polygon*.

*step 2:* Evaluate the objective function, $Z$, at each corner of the polygon.

*step 3:* The values of $x_1$ and $x_2$ that optimize $Z$ are the optimal solution.

Once a solution is found, it is possible to determine the effect on the objective function of changing one of the program parameters. This is known as *sensitivity analysis* and is very important in instances where the accuracy of collected data is unknown.

### Example 78.13

A cattle rancher uses a mixture of three types of cattle food and wants to minimize the cost of feeding his cattle. The costs per pound of the three food types are

| food type | cost per pound |
|-----------|----------------|
| 1 | 1.5 |
| 2 | 2.5 |
| 3 | 3.5 |

The rancher is also concerned with meeting published nutritional information on minimum daily requirements (MDR) given in milligrams per animal. The composition of each food type is known and the contributions for each vitamin in milligrams per pound are

| vitamin | MDR (mg) | food type 1 | 2 | 3 |
|---------|----------|-------------|---|---|
| A | 100 | 1 | 7 | 13 |
| B | 200 | 3 | 9 | 15 |
| C | 300 | 5 | 11 | 17 |

It is also physically impossible for an animal to eat more than the following amounts per day.

| food type | maximum feeding (pounds) |
|-----------|--------------------------|
| 1 | 50 |
| 2 | 40 |
| 3 | 30 |

Formulate this problem as a linear programming model to determine how much of each food the rancher should use.

*Solution*

Let $x_i$ be the number of pounds of food $i$ purchased per animal. Then, the objective function to be minimized is the total cost per animal.

$$Z = 1.5x_1 + 2.5x_2 + 3.5x_3$$

The constraints on this problem are[12]

$$
\begin{aligned}
x_1 + 7x_2 + 13x_3 &\geq 100 \\
3x_1 + 9x_2 + 15x_3 &\geq 200 \\
5x_1 + 11x_2 + 17x_3 &\geq 300 \\
x_1 &\leq 50 \\
x_2 &\leq 40 \\
x_3 &\leq 30 \\
x_1 &\geq 0 \\
x_2 &\geq 0 \\
x_3 &\geq 0
\end{aligned}
$$

---

[11]Also, the learning time is high.

[12]Depending on the reason for the maximum feeding limits, it may be more appropriate to replace the three single constraints with a single *interaction equation*: $x_1/50 + x_2/40 + x_3/30 \leq 1$.

**Example 78.14**

Solve the following linear programming problem.

$$\max Z = 2x_1 + x_2$$

Constraints:

| (a) | $x_1 + 4x_2$ | $\leq$ | 24 |
| (b) | $x_1 + 2x_2$ | $\leq$ | 14 |
| (c) | $2x_1 - x_2$ | $\leq$ | 8 |
| (d) | $x_1 - x_2$ | $\leq$ | 3 |
| (e) | $x_1$ | $\geq$ | 0 |
| (f) | $x_2$ | $\geq$ | 0 |

*Solution*

Since there are only two variables, this problem can be solved graphically.

The region enclosed by the constraints is shown.

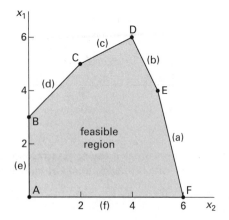

The coordinates and $Z$-values for each corner are

| corner | coordinates $(x_2, x_1)$ | $Z$ |
|--------|--------------------------|-----|
| A | (0, 0) | 0 |
| B | (0, 3) | 6 |
| C | (2, 5) | 12 |
| D | (4, 6) | 16 |
| E | (5, 4) | 13 |
| F | (6, 0) | 6 |

$Z$ is maximized when $x_1 = 6$ and $x_2 = 4$.

## 8. PLANT LOCATION

The reasons for locating a single new facility at a particular location are as varied as the industries represented. Some reasons are quantitative; others are not. Some reasons look to immediate gains; others look to the future. However, the traditional management science goal in locating a single plant is to minimize the sum of all the quantifiable costs affected by a choice of location. These costs include, among other things, facilities acquisition, labor, taxes, utilities, shipping, and transportation.

The traditional "*plant location problem*" is much more narrowly defined. In fact, the name for this problem type is a misnomer, with the name "multiplant" *fixed-charge problem* being more appropriate. The plant location problem determines where to locate one or more new plants within the framework of existing *plants (sources)* and existing demand locations or *distribution centers (destinations, demands, sinks,* etc.). All locations produce the same product; all distribution centers accept the same product.

Typically, production and distribution expenses are the major cost elements. All other costs, including initial fixed costs, are omitted unless they can be represented as per-unit costs associated with each unique combination of sources and destinations. The object is to minimize the total cost of production and distribution by specifying which new plant(s) should open and how many units each plant should ship to each distribution center.

Constraints ensure that the total demands are satisfied and that the total production capacities are not exceeded. In real-world problems, individual shipments may also be limited, as when the path between one plant and its destination has a limited capacity. These limitations are known as *arc capacities*.

As defined, the plant location problem is a specific case of linear programming known as the *transportation problem*. The transportation problem is concerned with allocating shipments from sources to destinations so that the total transportation cost is minimized. Specific algorithms, manual and computerized, exist for solving this problem.

**Example 78.15**

Existing plants A and B have insufficient combined capacity to meet product demand. Plants C and D are being evaluated as possible new locations. Either or both can be selected. Either or both of the existing two plants can be closed. The capacities of each plant location are: A, 8; B, 6; C, 5; D, 7.

Warehouses 1, 2, and 3 have demands of 9, 4, and 5, respectively.

The per-unit cost, $C_{ij}$, associated with each combination of plant ($i$) and warehouse ($j$) are given in the following table.

| warehouse | plant location A | B | C | D |
|-----------|------------------|-----|-----|-----|
| 1 | 17 | 15 | 19 | 12 |
| 2 | 22 | 24 | 10 | 8 |
| 3 | 39 | 32 | 45 | 16 |

Formulate this problem as a linear programming program.

*Solution*

This solution is the classical linear programming formulation of the *transportation problem*, not a fixed-charge problem. Let the variable $Q_{ij}$ represent the quantity shipped from plant location $i$ to warehouse $j$. Then, the objective function is

$$\text{minimize } Z = \sum_{i=A}^{D} \sum_{j=1}^{3} C_{ij} Q_{ij}$$

The production capacity constraints are

$$Q_{A1} + Q_{A2} + Q_{A3} \leq 8$$
$$Q_{B1} + Q_{B2} + Q_{B3} \leq 6$$
$$Q_{C1} + Q_{C2} + Q_{C3} \leq 5$$
$$Q_{D1} + Q_{D2} + Q_{D3} \leq 7$$

The demand constraints are

$$Q_{A1} + Q_{B1} + Q_{C1} \geq 9$$
$$Q_{A2} + Q_{B2} + Q_{C2} \geq 4$$
$$Q_{A3} + Q_{B3} + Q_{C3} \geq 5$$

The non-negativity constraints are

$$\text{all } Q_{ij} \geq 0$$

## 9. FACILITIES LAYOUT

*Facilities layout (plant layout)* problems are numerous in variety and complexity. Laying out facilities involves locating departments and/or operations with respect to one another. In traditional *process layout*, machines with the same function are grouped together. In *product layout* (product-oriented layout), the layout depends on the sequencing of production operations. If the same equipment is used at two different times, it is duplicated in a product layout.

Some computerized methods exist for exhaustively evaluating alternatives. Manual layout techniques are even more limited. Often, paper-cutting is combined with intuition to come up with a layout. Departments are sized to a particular scale and are cut out of paper. The pieces of paper are slid around until a layout "works."

Except for the artificial case of a small number of equally sized, equally shaped departments or operations whose locations are limited to a rectangular grid, it is unlikely that all possible layouts will be considered.[13] The "optimum" layout may actually be merely the best that could be found given the amount of time available.

An alternate manual method is to construct a graph whose "vertices" (nodes) are the departments or operations. The "edges" (line segments) are drawn between two vertices if adjacent associated departments are desired. The edges may be weighted to indicate the level of traffic between the departments. The goal is to rearrange the vertices so that no edges cross. If this can be done, then the layout can be planar. If the departments are somewhat flexible in terms of size and shape, it is possible to have the desired adjacencies.

Certain simplifying assumptions are usually made with both computerized and manual methods. For example, all layouts may be required to be two dimensional. Departments may be assumed to be square or rectangular. When the locations of specific pieces of equipment within the department are unknown, it is assumed that all movement into and out of the department originates and terminates at the centroid of the departmental area. Also, only highly repetitive movements between departments are considered. Once-in-a-while travel is excluded from the analysis.

Almost all facility layout procedures—manual and computerized, exact, trial-and-error, and heuristic— attempt to minimize the transportation cost, sometimes referred to as *movement*.[14] In simple cases, this may mean minimizing the product of trips between departments and the distances between their centroids. In more complex cases, the product of trips and distances may also be multiplied by volumes, weights, and labor rates.

Nonquantitative factors also need to be considered. Sometimes, as when equipment, records, or personnel are shared, it is absolutely necessary that departments be located next to each other. In other cases, as when safety is compromised, it may be absolutely essential to separate departments. In most cases, the *nearness priorities* characterize the adjacency requirements between being absolutely necessary and being absolutely undesirable.[15] The ways that nonquantitative factors are presented and incorporated into the solution vary from case to case.

## 10. WORK MEASUREMENT

*Work measurement* determines how long it takes to complete an operation. The *standard time* (also known as the *product standard* or *production standard*) for an operation is the time required by a qualified and trained worker working at a normal pace. Standard times can be determined in a variety of ways, including by statistical

---

[13]The number of layout variations, including mirror images, with $n$ equally sized square departments is $n!$.

[14]A *trial-and-error method* depends on insight, intuition, and ingenuity to come up with a solution. A *heuristic method* follows a procedure and/or uses rules of thumb to derive an answer. Neither is an optimizing technique.

[15]The *Muther nearness priorities* (developed by Richard Muther in the 1950s) are (1) absolutely necessary, (2) very important, (3) important, (4) ok (ordinary importance), (5) unimportant, and (6) undesirable.

analysis of actual measurements (*time studies*), by analysis of historical data, or by analysis of the physical motions associated with the process (i.e., *predetermined systems*). *Learning curves* can be incorporated where appropriate.

Standard times are used to develop production schedules (*factory loading*), estimate costs and output, determine worker effectiveness, and establish the basis for incentive labor plans. Standard times are combined with actual labor rates to develop *standard costs* of products.

Time standards are usually not developed directly for large operations (e.g., building a television). For various reasons (including reduction of variance), standards are developed for each significant *element* (also known as a *task*) of the operation (e.g., tightening a cover screw or soldering a resistor to the printed circuit board). Such task data can then be used to determine standard times for other products with identical tasks.

In determining the tasks requiring time standards, it is normal to separate fixed and variable tasks. For example, setting up the workplace at the beginning of each shift is a fixed task that would be studied separately. Applying adhesive strips to the product on an assembly line is a variable task. It is also common to separate handling and material transport tasks from processing tasks.

Actual time measurements during normal operation can be taken by hand with a stopwatch or derived from the analysis of video recordings.[16] Operations can be continuously observed, or measurements can be taken on a random basis.[17] To establish a truly representative average, measurements should be taken from as many individuals as possible.

The raw average time for an operation is not used as a production standard for two reasons. First, different people work at different levels of effort, and workers do not always work at a normal pace when being observed. This is taken into consideration by the observer who applies a subjective *performance factor (performance rating)* to the raw data.[18] Expressing the performance factor as a decimal, the *normal time* is

$$t_{\text{normal}} = t_{\text{ave}}(\text{performance factor}) \qquad 78.54$$

Second, *time allowances* must be added for personal time, normal delays, defective material, and fatigue.[19] Such allowances, usually expressed as percentages, depend greatly on the nature of the operation. Allowances for hard manual labor are much larger than allowances for sitting workers. Expressing the performance factor and allowance as decimals, the standard time is[20]

$$t_{\text{std}} = \frac{t_{\text{ave}}(\text{performance factor})}{1 - \text{allowance}}$$

$$= \frac{t_{\text{normal}}}{1 - \text{allowance}} \qquad 78.55$$

The variability of time study data is commonly expressed in terms of a *coefficient of variation*. This is simply the percentage variation calculated from Eq. 78.56. Notice that the sample standard deviation, $s_t$, of the time measurements is used.

$$\text{coefficient of variation} = \left(\frac{s_t}{t_{\text{ave}}}\right) \times 100\% \qquad 78.56$$

$$s_t = \sqrt{\frac{\sum t^2 - \frac{\left(\sum t\right)^2}{N}}{N - 1}} \qquad 78.57$$

The number of observations is determined from the desired precision of the final answer and the observed variability of the times measured. The more variable the times and the greater the required *precision* ($\pm P$), the more observations will be required to achieve a desired *confidence level* ($C$).[21,22] The most common confidence level is 95%, although 90% and 99% are also used.[23] A 95% confidence level (i.e., a 0.95 *confidence coefficient*) means that the probability is 95% that the desired precision will be met.

Equation 78.58 can be used for estimating the sample size required. If $N$ is already greater than $n$, no further sampling is required.

$$N = \left(\frac{t_C s_t}{P t_{\text{ave}}}\right)^2 \qquad 78.58$$

---

[16]The stopwatch is used in two different ways. In *snapback recording*, the stopwatch is zeroed out after each measurement. In *continuous recording*, the elapsed times are recorded. The interval times are determined later by subtraction.

[17]Though the sampling is random, the random schedule is established in advance with the aid of random numbers. The investigator follows a predetermined observation schedule that is unknown to the worker being observed.

[18]In theory, either the standard time or the performance rating must be known. However, the investigator traditionally observes the time and rates the performance simultaneously.

[19]A minimum of 5% should be added for personal time. Allowances for normal delays are determined by observation. Allowances for fatigue are highly subjective and depend on the nature of the operation.

[20]It is also common to simply add the allowance percentage, though the two methods are not equivalent. The method used is usually a matter of company policy.

$$t_{\text{std}} = t_{\text{normal}}(1 + \text{allowance})$$

[21]The *confidence interval* and *confidence level* are different. The confidence interval is twice the precision, expressed in absolute terms. If a $\pm 2\%$ precision is required for a 1.0 minute operation, the confidence interval will be 0.04 minute.

[22]The precision is sometimes called the *required accuracy*, even though a required accuracy of $\pm 5\%$ is obviously inappropriate. "Required maximum inaccuracy" is more descriptive of the actual intent of the term.

[23]The actual value used will be a matter of company policy.

The actual maximum inaccuracy (error), $\epsilon$, is calculated from Eq. 78.59, which uses a two-tail distribution factor, $t_C$, from Table 78.1 or App. 78.A.[24]

$$\epsilon = Pt_{\text{ave}} = t_C s_t \qquad 78.59$$

$$P = \frac{\epsilon}{t_{\text{ave}}} = \frac{t_C s_t}{t_{\text{ave}}} \qquad 78.60$$

**Table 78.1** *Two-Tail Distribution Factors*[*]

| confidence level, $C$ | $z_C$ or $t_C(\infty)$ |
|---|---|
| 90% | 1.645 |
| 95% | 1.960 |
| 97.5% | 2.240 |
| 98% | 2.326 |
| 99% | 2.476 |
| 99.9% | 3.270 |

[*]Standard normal distribution is equivalent to a Student's $t$-distribution with infinite degrees of freedom.

Time standards can also be developed from systems of *standard data*, also known as *universal standard data* and *synthetic standard data*.[25] Standard data systems are particularly valuable when there is no operation to observe, as when the product is still at the blueprint stage. An analyst (who is familiar with the steps and equipment necessary to complete an operation) subdivides the operation into tasks and the tasks into even finer micromotions.

For example, the task of moving a small object from a conveyor belt to a workstation might be subdivided into the micromotions of eye focus, reach, pickup and grasp, move, position, and release. Times for these micromotions are taken from extensive tables of standard data. Such tables may list time in minutes or in TMUs (*time measurement units*) equal to 0.0006 minutes.

### Example 78.16

A highly experienced worker was repeatedly observed in a time study. The average time taken by the worker to complete a task was 0.80 min. The observing analyst gave the worker a performance rating of 120%. The standard allowance is 10%. What is the standard time for the task?

---

[24]When the standard deviation of the underlying population is not known and must be estimated from the sample, as is usually the case, *Student's t-distribution* should be used in place of the normal distribution. This level of sophistication is particularly necessary when the number of observations is small. Values of $t_C$ are determined from App. 78.A. When the *actual* error is being calculated from $n$ actual observations, $t_C$ should be determined with $n-1$ degrees of freedom. When the required number of observations, $N$, is being calculated, and the degrees of freedom is unknown (i.e., $N-1$ is unknown) or expected to be large, the degrees of freedom is assumed to be infinite. Values from Student's $t$-distribution for infinite degrees of freedom are identical to $z$-values from the normal table. (See Table 78.1.)

[25]*Methods-Time-Measurement* (MTM), *Work Factor*, and MODAPTS are commercial systems of predetermined standard data systems.

*Solution*

From Eq. 78.55, the standard time is

$$t_{\text{std}} = \frac{t_{\text{ave}}(\text{performance factor})}{1 - \text{allowance}}$$

$$= \frac{(0.80 \text{ min})(1.2)}{1 - 0.10} = 1.07 \text{ min}$$

### Example 78.17

The estimate of an operation's time is to be obtained from 10 measurements. (a) Given a required confidence level of 90%, what is the precision of the estimate? (b) Given a required precision of 5%, what is the confidence level? (c) How many additional observations are required to achieve the 5% precision with a 90% confidence level?

times: $0.32, 0.35, 0.34, 0.36, 0.38, 0.40, 0.40,$
$$0.36, 0.31, 0.38$$

*Solution*

$$\sum t = 0.32 + 0.35 + 0.34 + 0.36 + 0.38 + 0.40$$
$$+ 0.40 + 0.36 + 0.31 + 0.38$$
$$= 3.6$$
$$\sum t^2 = (0.32)^2 + (0.35)^2 + (0.34)^2 + (0.36)^2$$
$$+ (0.38)^2 + (0.40)^2 + (0.40)^2 + (0.36)^2$$
$$+ (0.31)^2 + (0.38)^2$$
$$= 1.3046$$

(a) The average time and sample standard deviation are

$$t_{\text{ave}} = \frac{\sum t}{N} = \frac{3.60}{10} = 0.36$$

From Eq. 78.57, the sample standard deviation is

$$s_t = \sqrt{\frac{\sum t^2 - \dfrac{\left(\sum t\right)^2}{N}}{N-1}}$$

$$= \sqrt{\frac{1.3046 - \dfrac{(3.6)^2}{10}}{10-1}}$$

$$= 0.03091$$

Since the number of samples is less than 20 (or 25 or 30), the normal curve should not be used. From App. 78.A, the $t_C$ factor for 90% two-tailed confidence limits with

$n - 1 = 10 - 1 = 9$ degrees of freedom is 1.83. The maximum error (with 90% confidence) is

$$\epsilon = t_C s_t = (1.83)(0.03091)$$
$$= 0.0566$$

The precision is

$$P = \frac{\epsilon}{t_{\text{ave}}} = \frac{0.0566}{0.36}$$
$$= 0.157 \quad (\pm 15.7\%)$$

(b) With a precision of $\pm P$, the maximum error is

$$\epsilon = P t_{\text{ave}} = (0.05)(0.36) = 0.018$$

The value of $t_C$ is

$$t_C = \frac{\epsilon}{s_t} = \frac{0.018}{0.03091}$$
$$= 0.582$$

From App. 78.A with $n - 1 = 9$ degrees of freedom, the value of 0.582 can be interpolated between the 40% and 50% (two-tail) columns. The confidence level is between 40% and 50%.

(c) From Eq. 78.58, the total number of observations required is

$$N = \left(\frac{t_C s_t}{P t_{\text{ave}}}\right)^2$$
$$= \left(\frac{(1.83)(0.03091)}{(0.05)(0.36)}\right)^2$$
$$= 9.9 \quad [\text{use 10 observations}]$$

In this case, no additional observations are required.

## 11. WORK SAMPLING

*Work sampling* is a method of achieving the results of a stopwatch study without using a stopwatch.[26] It directly determines the fraction of time spent in a particular operation. Work sampling can be used to indirectly establish time standards and to rate performance.

The basic procedure is to observe a worker at random times over a specific time period (e.g., 8 or 40 hours) and to simply note whether or not the worker was performing a given task. The fraction of observations is identical to the fraction of time the worker spends on the task.

---

[26]Work sampling is particularly useful in determining the fraction of time that a worker or group of workers is idle.

This fraction is combined with the time period to determine the average time for the task.

$$t_{\text{std}} = \frac{\begin{array}{c}(\text{fixed time period})(\text{fraction observed}) \\ \times (\text{performance factor})\end{array}}{(1 - \text{allowance})(\text{no. of pieces produced})} \quad \text{78.61}$$

The accuracy of the average depends on the total number of observations, $N$. It is a fundamental principle in work sampling studies that the total required number of observations is proportional to the amount of time spent on the given task. Since the time is initially unknown, an estimate is made and the number of observations is iteratively refined as observations are taken. The observed fraction of observations of the task is

$$\overline{p} = \frac{n}{N} = \frac{\text{no. of observations of task}}{\text{total no. of observations}} \quad \text{78.62}$$

The sample standard deviation of the observed fraction is the standard deviation of the binomial distribution.

$$s_p = \sqrt{\frac{\overline{p}(1 - \overline{p})}{N}} \quad \text{78.63}$$

It is important not to confuse the maximum fractional error, commonly referred to as the *precision*, with the maximum absolute error. Since $\overline{p}$ varies from 0 to 1, the fractional and absolute errors are both less than 1.00. The maximum actual inaccuracy is

$$\epsilon = P\overline{p} = t_C s_p \quad \text{78.64}$$

The total number of observations required to achieve the desired precision and confidence is

$$N = \left(\frac{t_C \overline{p}(1 - \overline{p})}{\epsilon}\right)^2$$
$$= t_C^2 \left(\frac{1 - \overline{p}}{P^2 \overline{p}}\right) \quad \text{78.65}$$

### Example 78.18

Work sampling is to be used to determine the fraction of time that a worker is idle. The initial estimate of idleness is 12%. The required precision is $\pm 10\%$. The required confidence level is 95%. (a) Determine the number of observations that should be initially scheduled. (b) Assuming the fraction of time idle is 12%, with the given confidence level, what is the range of values of the fraction of idleness?

*Solution*

(a) The initial estimate of the fraction is $\overline{p} = 0.12$, and the required precision, $P$, is 0.10. Assuming a large number of samples, the two-tail normal curve parameter for a 95% confidence level is found in Table 78.1 to be 1.960.

From Eq. 78.65, the number of observations is

$$N = \frac{t_C^2(1 - \bar{p})}{P^2 \bar{p}}$$

$$= \frac{(1.960)^2(1.00 - 0.12)}{(0.10)^2(0.12)}$$

$$= 2817$$

(b) With a precision (i.e., fractional error) of 10%, the absolute error is

$$\epsilon = P\bar{p} = (0.10)(0.12) = 0.012$$

The range of values will be

$$\bar{p} \pm \epsilon = 0.12 \pm 0.012$$

$$= 0.132, \ 0.108$$

## 12. ASSEMBLY LINE BALANCING

*Line balancing* determines which tasks will be performed progressively at multiple assembly stations. Some tasks must precede others; some tasks can be performed at any point in the assembly; some tasks (e.g., installation of fasteners and final tightening) can be split between stations. Line balancing can also determine how many stations are needed and which tasks will be performed in parallel to increase throughput.

The cumulative durations of all the tasks at a particular station cannot exceed the cycle time, which is the reciprocal of the *production rate*. Generally, the *cycle time* is known.[27]

$$\text{cycle time} = \frac{1}{\text{production rate}} \qquad 78.66$$

Line balancing establishes the capacity of the line. Work cannot pass down the line any faster than it can pass through the *bottleneck* of the line (i.e., the station with the longest set of tasks). Therefore, the longest set of tasks establishes the minimum cycle time. The *target cycle time* is the ideal case and can be calculated if the number of stations is known.

$$\text{target cycle time} = \frac{\sum(\text{all task times})}{\text{no. of stations}} \qquad 78.67$$

As a management science problem, the simultaneous goals are to maximize the throughput of the line, minimize the cumulative idle time of all the stations, minimize the total labor cost, and minimize the initial investment in station equipment. In most initial studies,

only the goal of minimizing idle time is considered. The total idle time is

$$\text{idle time} = (\text{no. of stations})(\text{cycle time})$$
$$- \sum(\text{all task times}) \qquad 78.68$$

The percentage *delay* (also known as *balance delay*) is defined by Eq. 78.69. The best arrangement will have the lowest balance delay. However, an optimal solution does not require a zero balance delay. In fact, a zero balance delay may not be possible, as when the sum of task times is not an integer multiple of the cycle time. Optimality is still assured if the number of stations is the smallest integer greater than the sum of task times divided by the cycle time.

$$\text{balance delay} = \frac{\text{idle time} \times 100\%}{(\text{no. of stations})(\text{cycle time})} \qquad 78.69$$

Trial-and-error methods of balancing lines by hand are of the shuffle-and-reshuffle variety. Heuristic methods also require considerable personal ingenuity. The number of possibilities is often very large, and therefore, modern linear and dynamic programming models, as well as techniques based on exhaustive enumeration, are solved with computers. Regardless of the method used, solutions are often not unique.

The following three different heuristic methods are used to balance lines.

- The task with the largest time (that will fit in the station's available time) is assigned to that station.

- The task with the most successors (that will fit in the station's available time) is assigned to that station.

- The task with the greatest sum of task times of its successor tasks (that will fit in the station's available time) is assigned to that station.

*Standard times* (see Sec. 78.10) to perform the various station tasks are used in the balancing process. However, except for robot operators and machine-controlled lines, task times are actually random. Sometimes tasks take longer; sometimes they take less time. This implies that almost every line will be unavoidably unbalanced. If the work is rigidly paced to the cycle time (as it would be if the work were permanently attached to the conveyance system), some work pieces might be left unfinished if the previous part required more time than usual. This scenario introduces what is probably one of the most important requirements for maximizing the line throughput: station inventory.

Line throughput will be maximized if each station has a backlog of unfinished work.[28] In this case, the conveyance system is used merely to bring work to stations rather than to pace the stations. If a station is busy when a new piece of work arrives, the station operator

---

[27]The problem is much more difficult when the cycle time is not known. Not only do all arrangements of tasks need to be evaluated, but also the arrangements need to be evaluated over all reasonable cycle times.

[28]Measurable increases in line output have been reported by selectively unbalancing the line and ensuring that each station has a backlog of work.

merely places that piece into his or her inventory of unfinished work. If all tasks are finished early, before a new piece of work arrives, the operator begins on a piece from the inventory. The station is never idle.

### Example 78.19

(a) Design a heuristically balanced line for a cycle time of 10. (b) Calculate the idle time. (c) Calculate the balance delay. (d) What would be the effect of decreasing the cycle time to 9?

| task | predecessors | time |
|------|-------------|------|
| start | none | 0 |
| A | start | 7 |
| B | start | 4 |
| C | B | 5 |
| D | A, C | 2 |
| E | C | 8 |
| F | D, E | 4 |
| finish | F | 0 |

*Solution*

(a) First, draw a simple precedence diagram to visualize the precedences.

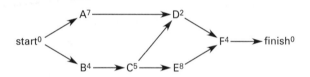

*Station 1:* The task with the largest time without predecessors will be assigned. (The "start" task, used merely for consistency in drawing directed graphs, is disregarded.) Task A is the largest and is assigned to station 1. That takes up 7 of the cycle time, which leaves $10 - 7 = 3$ still available. B is too long, and C's predecessor is incomplete. D's predecessor, C, is incomplete. No tasks can be added to fill the remaining time.

*Station 2:* Task A is taken; tasks B and D are the only possibilities. Task B, with the larger time of 4, is assigned to station 2. Continuing down the list, tasks C and D can be performed. Task C has a larger task time than D. Assigning task C to station 2 increases the station time to a total of 9. No further tasks can be added.

*Station 3:* Tasks A, B, and C have been taken. Tasks D and E are available and both are assigned to station 3.

*Station 4:* Tasks A, B, C, D, and E have been taken. Task F is assigned to station 4.

The heuristic solution is

| Station 1: | A | (busy 7; idle 3) |
| Station 2: | B, C | (busy 9; idle 1) |
| Station 3: | D, E | (busy 10; idle 0) |
| Station 4: | F | (busy 4; idle 6) |

(b) The sum of task times is

$$7 + 4 + 5 + 2 + 8 + 4 = 30$$

The idle time is

$$\text{idle time} = (\text{no. of stations})(\text{cycle time})$$
$$- \sum(\text{all task times})$$
$$= (4 \text{ stations})(10) - 30 = 10$$

(c) The balance delay is

$$\text{balance delay} = \frac{\text{idle time} \times 100\%}{(\text{no. of stations})(\text{cycle time})}$$
$$= \frac{(10)(100\%)}{(4)(10)} = 25\%$$

(d) Since station 3 has no idle time, decreasing the cycle time would require rebalancing the line. Task D could be shifted to station 4, leaving station 3 only with task E. The assignments to stations 1 and 2 would not change. The production rate would be increased, and the percentage balance delay would be decreased.

$$\text{balance delay} = \frac{(36 - 30)(100\%)}{(4)(9)} = 16.7\%$$

## 13. WAGE INCENTIVE PLANS

The goal of *wage incentive plans* is to motivate workers to produce more by allowing them to earn more.[29] *Individual incentive plans* are tied to each worker's output; *group incentive plans* are tied to the output of a department or other group. Individual plans can usually be modified for use as group plans.

The simplest wage incentive plan is *piecework* or *piecerate*, in which workers are paid in proportion to the number of pieces completed. Piecework payment can be very effective, but it is difficult to administer under minimum-wage laws.

*Standard-hour plans* are similar to piecework, but they guarantee workers a *base wage* regardless of output. Above the *performance standard* (i.e., the output corresponding to the standard payment), workers are paid in proportion to the output. The difference between the base and actual wages is the *bonus. Productivity* (also

---

[29]As a very general rule, the output before wage incentives are implemented will be 60% to 90% of the output based on standard times. With wage incentives, productivity will increase to 130% or more. Although workers may not work significantly faster with wage incentives, they reduce their idle time.

called *efficiency*) is the percentage of the standard output achieved. Output may be expressed in units or standard hours.

The bonus can be 100% of the increase in output above the standard or it can be less (though usually no less than 50% of the increase).[30] Plans with a 100% bonus are known as *one-for-one plans* or *100% premium plans*, while plans with bonuses less than 100% are known as *gain-sharing plans*. The percentage of the earned bonus kept by the worker is known as *labor's participation* in the plan.

Effective plans allow workers to earn bonuses of at least 25% to 30% of their base wages (i.e., the *relative earnings* are 125% to 130%). However, depending on the standard, some workers may be unable to consistently achieve even the standard output. Rather than encouraging higher output, such plans might discourage workers, resulting in even lower output than without the incentive program. For this reason, many incentive programs begin paying a bonus at output less than 100% of standard. Such *reduced-standard plans* can be one-for-one or gain sharing.

Wage incentive plans can be identified in one of three ways: by name, by specifying the participation and productivity at which the bonus begins, or by specifying the standard bonus. The *standard bonus* is the bonus earned at a productivity of 100%. Obviously, the standard bonus is zero for any 100% standard plan. For any linear plan, the standard bonus depends on the *bonus factor* defined by Eq. 78.70.

$$\text{standard bonus} = (\text{bonus factor})(\text{participation})$$

$$[\text{linear plans}]$$
$$78.70$$

The *relative earnings*, RE, are

$$RE = 1 - \text{participation}$$
$$+ (\text{productivity})(\text{participation})(1 + \text{bonus factor})$$
$$78.71$$

The *relative bonus*, RB, is

$$RB = RE - 1 \qquad 78.72$$

Figure 78.4 illustrates four types of *linear incentive plans*. *Curvilinear plans*, also known as *self-regulating plans*, are also used, particularly when there is the possibility of runaway production or cheating. (See Fig. 78.5.) Payment under curvilinear plans asymptotically approaches a maximum "ceiling" wage. The maximum relative earnings is one plus the fractional participation. The nature of the curve is specified by giving the productivity level at which the bonus begins and at least one other point on the curve.

[30]The lower bonus may be justified on the basis that the remainder is needed to pay for the administrative costs of running the incentive program.

**Figure 78.4** *Four Types of Linear Incentive Plans*

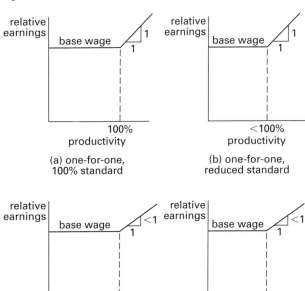

(a) one-for-one, 100% standard

(b) one-for-one, reduced standard

(c) gain sharing, 100% standard

(d) gain sharing, reduced standard

**Figure 78.5** *Curvilinear Plan*

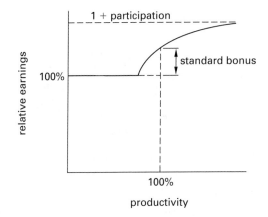

For curvilinear plans, the standard bonus and relative earnings are

$$\text{standard bonus} = \frac{(\text{bonus factor})(\text{participation})}{1 + \text{bonus factor}}$$

$$[\text{curvilinear plans}]$$
$$78.73$$

$$RE = 1 + \frac{(\text{participation})\Big(\text{productivity} + (\text{productivity})(\text{bonus factor}) - 1\Big)}{\text{productivity} + (\text{productivity})(\text{bonus factor})}$$

$$[\text{curvilinear plans}]$$
$$78.74$$

For both linear and curvilinear reduced-standard plans, the productivity at which the bonus begins can be found from Eq. 78.75.

$$\begin{array}{c} \text{minimum bonus} \\ \text{productivity} \end{array} = \frac{1}{1 + \text{bonus factor}}$$

$$\begin{bmatrix} \text{linear and} \\ \text{curvilinear plans} \end{bmatrix} \qquad 78.75$$

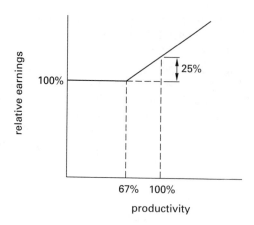

### Example 78.20

A linear wage incentive program provides for 50% participation and a 25% standard bonus. A worker produces 1900 units in 8 hr. The standard time for each unit is 0.004 hr. What are (a) the bonus factor, (b) the worker's relative earnings, and (c) the worker's relative bonus?

*Solution*

(a) From Eq. 78.70, the bonus factor is

$$\text{bonus factor} = \frac{\text{standard bonus}}{\text{participation}} = \frac{0.25}{0.50}$$

$$= 0.50 \quad (50\%)$$

(b) The worker's productivity is

$$\text{productivity} = \frac{\text{standard time}}{\text{actual time}}$$

$$= \frac{(1900 \text{ units})\left(0.004 \dfrac{\text{hr}}{\text{unit}}\right)}{8 \text{ hr}}$$

$$= 0.95 \quad (95\%)$$

From Eq. 78.71, the relative earnings are

$$\text{RE} = 1 - \text{participation}$$
$$+ (\text{productivity})(\text{participation})(1 + \text{bonus factor})$$
$$= 1 - 0.50 + (0.95)(0.50)(1 + 0.50)$$
$$= 1.2125 \quad (121.25\%)$$

(c) From Eq. 78.72, relative bonus is

$$\text{RB} = \text{RE} - 1 = 1.2125 - 1$$
$$= 0.2125 \quad (21.25\%)$$

### Example 78.21

A curvilinear wage incentive program provides for 75% participation and a 37.5% standard bonus. A worker produces 1900 units in 8 hr. The standard time for each unit is 0.004 hr. What are the worker's relative earnings?

*Solution*

As in Ex. 78.20, the productivity is 95%. The bonus factor is calculated from Eq. 78.73.

$$\text{standard bonus} = \frac{(\text{bonus factor})(\text{participation})}{1 + \text{bonus factor}}$$

$$0.375 = \frac{(\text{bonus factor})(0.75)}{1 + \text{bonus factor}}$$

$$\text{bonus factor} = 1.00$$

From Eq. 78.74, the relative earnings are

$$\text{RE} = 1 + \frac{(\text{participation})\left(\begin{array}{c} \text{productivity} \\ + (\text{productivity})(\text{bonus factor}) - 1 \end{array}\right)}{\text{productivity} + (\text{productivity})(\text{bonus factor})}$$

$$= 1 + (0.75)\left(\frac{0.95 + (0.95)(1.0) - 1.0}{0.95 + (0.95)(1.00)}\right)$$

$$= 1.355 \quad (135.5\%)$$

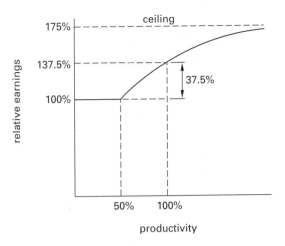

## 14. QUALITY CONTROL CHARTS

*Statistical quality control*, SQC, also known as *statistical process control*, SPC, uses several techniques to ensure that a minimum quality level is consistently obtained from production processes. Typical SQC tasks include routine monitoring of process output, sampling incoming raw materials, and testing finished work.

Monitoring process output and charting the results are often the most visible aspects of SQC. Small samples of work are tested at regular or random intervals, and the results are shown graphically.[31] The graphs are known as *control charts* or *Shewhart control charts* because they show, in addition to the measured values, the *control limits* (i.e., the limits of acceptable values).[32] (See Fig. 78.6.)

***Figure 78.6*** *Interpretation of SPC Charts*

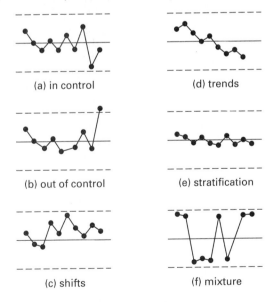

Control charts can be prepared for the average value of some process variable (the *x-bar chart*), for the range or other measures of dispersion (*R-chart, s-chart*, or *σ-chart*), and for the fraction defective (*p-chart*), the number of defects per unit (*c-chart*), or any combination thereof. Charts that require a measurement of a process variable are known as *variable charts*. The *p*-chart and *c*-chart are examples of *attribute charts*, where only the condition of an item needs to be determined.

It is a basic assumption that random effects are present in every process. Variation within certain limits is inevitable, and if the magnitudes of the limits or the variation are unacceptable, they must be reduced by changes in manufacturing or product design. If the magnitudes are acceptable, then corrections are required only when the results exceed the magnitude expected on the basis of random effects.

For a process in control, the variable is assumed to be normally distributed with population mean $\mu$ and population standard deviation $\sigma'$. (See Eq. 78.76.) When the process changes by more than $\pm 2\sigma'$ (the *warning limits*), a change might be occurring, and the process should be watched. When the process changes by more than $\pm 3\sigma'$ (the *action limits*), the process is considered to have gone out of control.[33] In that case, the process is evaluated and, if necessary, changed.[34]

The *average run length*, ARL, is the average number of samples from an in-control process that will be tested before a point outside the control limits will be encountered.

$$\text{ARL} = \frac{1}{1 - C} \qquad 78.76$$

A process is "in control" as long as the process variable is within the control limits and there are no suspicious events. A *suspicious event* (also known as a *shift* in central tendency) indicates that the process average has shifted. A suspicious event occurs when more than seven consecutive points are on one side of the average, even though the points are within the control limits. Other suspicious events occur when 10 out of 11, 12 out of 14, 14 out of 17, and 16 out of 20 points are on the same side of the average. These criteria are known as *zone tests*.

*Stratification* occurs when almost all of the points on the chart are near the centerline. This is usually too good to be true, as samples from normal populations do not react in this manner. Usually, stratification indicates that the control limits have been incorrectly calculated.

A *mixture pattern* of points is essentially the opposite of stratification. A mixture pattern occurs when all of the points are near the two control limits. Normal populations do not behave in this manner either, but bimodal

---

[31]The graphs are often conspicuously posted at the entrances of departments.
[32]Control limits have nothing to do with *specification limits*. Specification limits determine if the product is acceptable to the customer. Control limits determine if the process is statistically in control.

[33]With $3\sigma$ control limits, no more than 0.27% (i.e., 27 times out of 1000) of the points are expected to be outside of the control limits.
[34]Although three standard deviations is the usual value, any multiple of standard deviations can be chosen for the action limits.

distributions do. The usual interpretation is that the product is coming from two different statistical populations. Differences in equipment, raw materials, adjustment, or operators can explain the bimodality.

In a typical control chart application, a sample of $N$ objects is taken and the quality parameter, $x$, is measured for each of the $N$ items. The average of the sample, $\overline{x}$, is calculated. This process is repeated $M$ times (e.g., over $M$ days, etc.).[35] The grand average, $\overline{\overline{x}}$, and (in theory) the standard deviation, $\sigma_{\overline{x}}$, of the process averages are calculated from the $M$ values of $\overline{x}$.[36]

The "*x-bar chart*" ($\overline{x}$ chart) for process averages is drawn with a process average of $\overline{\overline{x}}$ and horizontal control limits of $\pm 3\sigma_{\overline{x}}$. As long as the process is in control, the sample averages will be within the *upper control limit*, UCL, and *lower control limit*, LCL, defined by Eq. 78.77.

$$\text{LCL}_{x\text{-bar}}, \ \text{UCL}_{x\text{-bar}} = \overline{\overline{x}} \pm 3\sigma_{\overline{x}} \qquad \textit{78.77}$$

If the process variable (e.g., diameter) cannot be negative but the lower control limit in Eq. 78.77 is negative, that limit is set to zero. If one or more of the $M$ averages is found to be outside of the control limits, those values are excluded and the procedure is repeated to find the control chart values for a process in control.

The unbiased estimators of the population average, $\mu$, and standard deviation, $\sigma'$, of the $N \times M$ individual values (not the $M$ averages) are $\overline{\overline{x}}$ and $\sqrt{N}\sigma_{\overline{x}}$. Therefore, the upper and lower control limits can also be written as Eq. 78.78, if these population parameters are known.

$$\text{LCL}_{x\text{-bar}}, \ \text{UCL}_{x\text{-bar}} = \overline{\overline{x}} \pm \frac{3\sigma'}{\sqrt{N}} \qquad \textit{78.78}$$

For various reasons, including timely response and job enrichment, the steps and calculations necessary to maintain x-bar (and other) charts are often performed by workers who are not trained in higher mathematics or statistical analysis. Therefore, a simplified method based on the *range* of values (i.e., the maximum measurement less the minimum measurement) can be used to calculate the standard deviation.[37]

$$R = \text{maximum sample value} - \text{minimum sample value}$$
$$\textit{78.79}$$

The range is calculated for each of the $M$ samples. Then, the average range, $\overline{R}$, is calculated and converted to the

population standard deviation, $\sigma'$. A similar calculation can be done if the individual standard deviation, $\sigma$, is calculated for each of the $M$ samples. The factors $c_2$ and $d_2$ used in Eq. 78.80 depend on the sample size, $N$. (See Table 78.2.)

$$\sigma' = \frac{\overline{R}}{d_2} = \frac{\overline{\sigma}}{c_2} \qquad \textit{78.80}$$

The lower and upper control limits for the x-bar chart can be written as

$$\text{LCL}_{x\text{-bar}}, \ \text{UCL}_{x\text{-bar}} = \overline{\overline{x}} \pm \frac{3\overline{R}}{d_2\sqrt{N}} \qquad \textit{78.81}$$

***Table 78.2*** $d_2$ and $c_2$ Factors for Estimating $\sigma'$ from $\overline{R}$ and $\overline{\sigma}$[*]

| $N$ | $d_2$ | $c_2$ |
|---|---|---|
| 2 | 1.128 | 0.5642 |
| 3 | 1.693 | 0.7236 |
| 4 | 2.059 | 0.7979 |
| 5 | 2.326 | 0.8407 |
| 6 | 2.534 | 0.8686 |
| 7 | 2.704 | 0.8882 |
| 8 | 2.847 | 0.9027 |
| 9 | 2.970 | 0.9139 |
| 10 | 3.078 | 0.9227 |
| 11 | 3.173 | 0.9300 |
| 12 | 3.258 | 0.9359 |
| 13 | 3.336 | 0.9410 |
| 14 | 3.407 | 0.9453 |
| 15 | 3.472 | 0.9490 |
| 16 | 3.532 | 0.9523 |
| 17 | 3.588 | 0.9551 |
| 18 | 3.640 | 0.9576 |
| 19 | 3.689 | 0.9599 |
| 20 | 3.735 | 0.9619 |
| 21 | 3.778 | 0.9638 |
| 22 | 3.819 | 0.9655 |
| 23 | 3.858 | 0.9670 |
| 24 | 3.895 | 0.9684 |
| 25 | 3.931 | 0.9696 |
| 30 | 4.086 | 0.9748 |
| 35 | 4.213 | 0.9784 |
| 40 | 4.322 | 0.9811 |
| 45 | 4.415 | 0.9832 |
| 50 | 4.498 | 0.9849 |
| 55 | 4.572 | 0.9863 |
| 60 | 4.639 | 0.9874 |
| 65 | 4.699 | 0.9884 |
| 70 | 4.755 | 0.9892 |
| 75 | 4.806 | 0.9900 |
| 80 | 4.854 | 0.9906 |
| 85 | 4.898 | 0.9912 |
| 90 | 4.939 | 0.9916 |
| 95 | 4.978 | 0.9921 |
| 100 | 5.015 | 0.9925 |

[*]Sampling from a normally distributed population assumed.

---

[35]For the assumption of a normal distribution to be valid, $M$ must be greater than approximately 20.

[36]Though $s^2$ is an unbiased estimator of the population variance, $\sigma'^2$, $s$ is a biased estimator of $\sigma'$ and is not used. However, if $s$ is calculated for each sample, it can be used if corrected according to

$$\sigma' = \frac{1}{c_2}\sqrt{\left(\frac{N-1}{N}\right)}\overline{s}$$

[37]Most quality monitoring today, even on the "factory floor," uses the computer. The R-chart is still an option on most quality software, but there is little to justify its use. The s-chart and σ-chart are more powerful and detect changes in variability faster.

The variability of the process can also be tracked by charting the range values in an *R-chart*.[38] The average range is assumed to be zero, and Eq. 78.82 and Eq. 78.83 give the upper and lower control limits, respectively. The values depend on the confidence level (i.e., the percentage of points that are expected to fall outside the limits when the process is in control). Since control limits on the *R*-chart are not symmetrical, values of $D_{\rm LCL}$ and $D_{\rm UCL}$ are tabulated in Table 78.3.[39]

$$\mathrm{UCL}_p = D_{\rm UCL}\overline{R} \qquad 78.82$$

$$\mathrm{LCL}_p = D_{\rm LCL}\overline{R} \qquad 78.83$$

**Table 78.3** Control Limit D-Factors for R-Charts
(for percentage outside limits)

| | $D_{\rm LCL}$ | | | $D_{\rm UCL}$ | | |
|---|---|---|---|---|---|---|
| N | 0.1% | 0.5% | 2.5% | 2.5% | 0.5% | 0.1% |
| 2 | 0.00 | 0.01 | 0.04 | 3.17 | 3.97 | 4.65 |
| 3 | 0.06 | 0.13 | 0.30 | 3.68 | 4.42 | 5.06 |
| 4 | 0.20 | 0.34 | 0.59 | 3.98 | 4.69 | 5.31 |
| 5 | 0.37 | 0.55 | 0.85 | 4.20 | 4.89 | 5.48 |
| 6 | 0.53 | 0.75 | 1.07 | 4.36 | 5.03 | 5.62 |
| 7 | 0.69 | 0.92 | 1.25 | 4.49 | 5.15 | 5.73 |
| 8 | 0.83 | 1.08 | 1.41 | 4.60 | 5.25 | 5.82 |
| 9 | 0.97 | 1.21 | 1.55 | 4.70 | 5.34 | 5.90 |
| 10 | 1.08 | 1.33 | 1.67 | 4.78 | 5.42 | 5.97 |

The *p-chart* is particularly useful in monitoring the fraction of nonquantitative rejects (i.e., *nonconformances*—the quantity of units that don't work or "just aren't right" for any reason). The *np-chart* monitors the number of rejects in a standard sample size. The variable $p$ is the fraction (or percentage) defective in each sample of $N$ items. $\overline{p}$ is the average fraction defective taken over $M$ samples, which is equal to the total number of defects found in $M$ samples divided by the total number of samples, $N \times M$.

The average and control limits both depend on $\overline{p}$, which is determined historically. The lower and upper control limits for the $p$-chart are given by Eq. 78.84. $N_i$ is the number of samples in period $i$, usually a constant. The second term in Eq. 78.84 is three times the standard deviation of a binomial distribution.[40]

$$\mathrm{LCL}_p,\ \mathrm{UCL}_p = \overline{p} \pm 3\sqrt{\frac{\overline{p}(1-\overline{p})}{N_i}} \qquad 78.84$$

The *c-chart* tracks the number of defects (i.e., *nonconformities*) per sample of constant size. (The sample size may be one complete assembly or single unit.) The defects may be of any variety, anywhere in the assembly. Over a period of $M$ samples, the average number of defects, $\overline{c}$, is determined. The probability of any number of defects per sample is assumed to be a Poisson distribution, and since the mean and variance of a Poisson distribution are the same, the upper and lower control limits are given by Eq. 78.85.[41]

$$\mathrm{LCL}_c,\ \mathrm{UCL}_c = \overline{c} \pm 3\sqrt{\overline{c}} \qquad 78.85$$

For most other charts, the average and limits can be allowed to change and shift over time. However, for an in-control process monitored by a *c*-chart, the average and limits are fixed. If quality deterioration occurs and $\overline{c}$ increases, changing the average and limits would imply that the process is still in control when it is not.

There is an important difference between nonconformances and nonconformities. A *nonconformance* is a *defective*—an item that does not meet specification. A *nonconformity* is a *defect*. For example, an acceptance specification may say that a painted object may not have more than three paint defects. An object with two bubbles in its paint will have two nonconformities but will not be a nonconformance.

There are three common variations with time of control charts: (1) When the sample size changes from time to time, the horizontal control limits will move inward or outward with the changes. (2) The average may show a steady, shifting trend represented by an inclined average line, as when there is uniform tool wear. The parallel control limits will be similarly inclined. (3) In continuous processes, such as liquid product manufacturing or refining, an argument can be made for using a moving average (i.e., recalculating the average over the most recent $M$ samples). The smoothing effect of a moving average accurately represents the effect of product blending.

### Example 78.22

An *x*-bar chart with $\pm 2.5\sigma$ limits is maintained for a process. If the process is in control, how many samples would you expect to be taken before a point outside the control limits is seen?

*Solution*

From a standard normal table, the fractional confidence level is the fraction inside $z = \pm 2.5\sigma$.

$$C = (2)(0.4938) = 0.9876$$

---

[38]$\sigma$ charts are also used. For more information on $\sigma$ charts, refer to a textbook on statistical quality control.
[39]Though the factors are not symmetrical, the probabilities are. For example, with the 95% factors, 2.5% will be outside of each control limit. However, asymmetrical probabilities could be chosen if desired.
[40]The *p*-chart and *np*-chart use the normal approximation to the binomial distribution. For this approximation to be reliable, the sample size should be greater than approximately 25, $\overline{np}$ should be greater than or equal to approximately 4, and $np$ should be greater than or equal to approximately 1.

[41]Unlike the normal distribution, the Poisson distribution is not symmetrical. Since the control limits in the *c*-chart are symmetrical, they do not represent equal probabilities of exceedance.

From Eq. 78.76, the average run length is

$$\text{ARL} = \frac{1}{1-C} = \frac{1}{1-0.9876}$$
$$= 80.65$$

## 15. QUALITY ACCEPTANCE SAMPLING

*Acceptance sampling* is the testing of samples taken from a *lot* (batch or process) in order to determine if the entire lot should be accepted or rejected. Acceptance sampling is appropriate when testing is destructive or when 100% testing would be too expensive. To design an *acceptance plan* (also known as a *Dodge-Romig plan*), the number in the sample and the *acceptance number* (i.e., the maximum allowable number of defects in the sample) must be specified.

In *single acceptance plans*, a sample of size $n$ out of a total lot size of $N$ is tested. If the number of defects is equal to or less than $c$, the lot is accepted. The plan can be described graphically by an *operating characteristic (OC) curve* (see Fig. 78.7), which plots the *probability of acceptance* (also known as the *producer's acceptance risk*) versus the *lot quality* (i.e., the true fraction or percentage defective).[42] Points on the OC curve are determined from the binomial or, more preferably, from the Poisson approximation to the binomial. In practice, however, acceptance plans are generally designed by referring to tables of predetermined plans.

**Figure 78.7** *Acceptance Plan Operation Characteristic Curve*

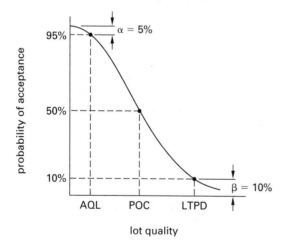

The *producer's risk*, $\alpha$, also known as the *supplier's risk*, is the probability of a *Type I error* (i.e., rejecting a good lot—that is, of finding trouble when none exists). The *consumer's risk* is the probability of a *Type II error* (i.e., accepting a bad lot—that is, of not finding trouble when it exists). Once a sampling plan has been determined

(i.e., the OC curve is established) and the actual incoming fraction defective is known, Eq. 78.86 can be applied.

$$\alpha + \beta = 100\% \qquad \textit{78.86}$$

The only way to decrease the consumer's risk without increasing the producer's risk is to increase the sample size (i.e., to change the sampling plan). Sampling plans based on large samples have steeper OC curves. (The ideal curve, corresponding to 100% sampling, is a vertical line at the lot quality.) Plans with smaller samples have more gradually inclined curves. When the sample history shows unsatisfactory quality, a *tightened plan* with more samples can be implemented. When the part or supplier has a history of extremely high quality, a *reduced plan* with fewer samples can be used.

Three points, usually those corresponding to 95%, 50%, and 10% probabilities, are given special consideration on the OC curve. The lot quality corresponding to a 95% probability of acceptance (i.e., a 95% consumer's risk) is known as the *acceptable quality level*, AQL.[43] The AQL is sometimes written as the "$p_{95\%}$ quality" or the "$p_{0.95}$ point." The AQL is the percentage defective that is "satisfactory" (i.e., a lot with percentage defective equal to $p_{95\%}$ or less is a good lot). The probability will be $1 - \alpha = 5\%$ that the lot will be rejected if the fraction defective is $p_{95\%}$ or less.

The lot quality corresponding to a 50% probability of acceptance (i.e., producer's and consumer's risks both equal to 50%) is known as the "$p_{50\%}$ quality" or "$p_{0.50}$ point," sometimes referred to as the "*point of control*" (POC) or "*indifference quality level.*"

The lot quality corresponding to a 10% probability of acceptance (i.e., a 10% consumer's risk) is known as the *lot tolerance percent (fraction) defective*, LTPD, also known as the "$p_{10\%}$ quality" or, less frequently, the *rejectable quality level*, RQL.

It is generally easier to design a sampling plan that has specific values of AQL and LTPD than to find the AQL and LTPD for a known plan. There are few published tables for the latter case. The traditional method is to plot the OC curve and read the values of fraction defective corresponding to probabilities of acceptance of 95% and 10%. (See Ex. 78.24.)

A Poisson distribution is used to describe the probability of any item being defective. The average number of defects in a sample is

$$\lambda = (\text{fraction defective})(\text{sample size}) \qquad \textit{78.87}$$

$$p\{m \text{ defects}\} = \frac{\lambda^m e^{-\lambda}}{m!} \qquad \textit{78.88}$$

---

[42]Operating characteristic curves are actually a series of discontinuous points since lot items are finite and discrete. However, they are never drawn in that manner.

[43]Any probability could be used. However, the 95% consumer's risk (5% producer's risk) is traditional.

The probability that a lot will be accepted is

$$p\{\text{acceptance}\} = \sum_{m=0}^{c} p\{m \text{ defects}\} \qquad 78.89$$

In practice, most acceptance plans are "designed" by using published tables or, at the very least, by using *cumulative summation ("cumsum") tables* for a particular distribution. However, if the $p_{50\%}$ quality is known, a single-sampling plan can be easily designed. Once the number of defects, $c$, has been (arbitrarily) chosen, Eq. 78.90 is used to calculate the sample size. There will be a family of plans (with different values of $c$) that satisfies Eq. 78.90. Plans with large values of $N$ and $c$ are better than plans with small values, as the slope of the OC curve is steeper near the $p_{50\%}$ point. In Eq. 78.90, $p_{50\%}$ is expressed in decimal form.

$$N = \frac{c + 0.67}{p_{50\%}} \qquad 78.90$$

If all items in rejected lots are subsequently screened (i.e., 100% tested) with only good items being passed, the screened lots will contain no defectives.[44] When the various lots are combined, some lots will have been acceptance-sample passed and some will have been 100% screened. The *average outgoing quality limit*, AOQL, is the maximum fraction (percentage) defective of accepted items in the combined lots. In the long run, regardless of incoming quality, item quality will be better than or equal to the AOQL.[45]

The AOQL cannot be read directly from the OC chart. Given a specific OC curve, the AOQL is the maximum value of the product of the lot quality and the probability of acceptance (i.e., the abscissa and ordinate of points on the OC curve). The AOQL can be rapidly found from the curve by trial and error.

With a *double-acceptance plan*, a sample of $n_1$ is taken out of the original lot size of $N$. The lot is accepted if the number of defects is $c_1$ or fewer and rejected if greater than $c_2$. Otherwise, a second sample of $n_2$ is taken. The lot is accepted if the total number of defectives from both samples is $c_2$ or fewer.

---

[44]Theoretically, each defective item should be replaced with a good item. However, if the lot size is very large, it doesn't really make any difference if a defective item is merely discarded.
[45]Exceptions are possible in the short run.

### Example 78.23

A sample of 100 items is taken from a large lot known to have a 3% defect rate. The acceptance number is 2. What is the probability of acceptance?

*Solution*

The lot will be accepted if 0, 1, or 2 defects are found. The Poisson probability distributed is assumed. The average number of defects in the sample is

$$\lambda = (\text{fraction defective})(\text{sample size})$$
$$= (0.03)(100)$$
$$= 3$$
$$p\{m \text{ defects}\} = \frac{\lambda^m e^{-\lambda}}{m!}$$

$$p\{\text{acceptance}\} = p\{0 \text{ defects}\} + p\{1 \text{ defect}\}$$
$$+ p\{2 \text{ defects}\}$$
$$= \frac{3^0 e^{-3}}{0!} + \frac{3^1 e^{-3}}{1!} + \frac{3^2 e^{-3}}{2!}$$
$$= 0.0498 + 0.1494 + 0.2240$$
$$= 0.4232 \ (42.32\%)$$

(This is the same value that is obtained from a cumulative probability chart with $\lambda = 3$ and $c = 2$.)

### Example 78.24

A sampling plan is designed to accept the entire lot if there are 2 or fewer defective in a sample of 100 items. The actual fraction defective of the lot is unknown. (a) Draw the operating characteristic curve. (b) What is the approximate AQL for a producer's risk of 5%? (c) What is the approximate LTPD for a consumer's risk of 10? (d) What is the approximate AOQL?

*Solution*

(a) Points on the curve can be generated by repeating Ex. 78.23 for different values of the fraction defective, $p$.

| fraction defective, $p$ | $Np$ | $p$ (acceptance) |
|---|---|---|
| 0.00 | 0.00 | 1.000 |
| 0.01 | 1.00 | 0.920 |
| 0.02 | 2.00 | 0.677 |
| 0.03 | 3.00 | 0.423 |
| 0.04 | 4.00 | 0.238 |
| 0.05 | 5.00 | 0.125 |
| 0.06 | 6.00 | 0.062 |

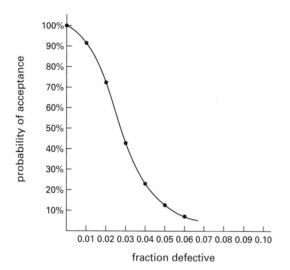

(b) The producer's risk is 5%, so the probability of acceptance is 95%. The AQL is between 0 and 1%. (The actual value is 0.82%.)

(c) The probability of acceptance is the same as the consumer's risk: 10%. The LTPD is between 5% and 6%. (The actual value is 5.3%.)

(d) The AOQL is the maximum product of the fraction defective and the probability of acceptance. This is approximately $(0.02)(0.677) = 0.01354$ (1.35%).

## 16. RISK ASSESSMENT

*Risk assessment* is an analytical process that determines the probability that some mishap will occur. It is applied primarily to human health risks and safety hazards or accidents from industrial practices.

*Health risk assessment* is based on toxicological studies. Data are usually gathered from exposure of laboratory animals to chemicals or processes, although human exposure data can be used if available.[46] Tests are run to determine if a chemical is a *carcinogen* (i.e., causes cancer), a *mutagen* (i.e., causes mutations), a *teratogen* (i.e., causes birth defects), a *nephrotoxin* (i.e., harms the kidneys), a *neurotoxin* (i.e., damages the brain or nervous system), or a *genotoxin* (i.e., harms the genes).

*Safety risk assessment* of potentially unsafe equipment or practices is traditionally called *hazard analysis*. Attention has traditionally focused on processes and equipment. However, hazard analysis is also applicable to material storage, shipping, transportation, and waste disposal. An analysis of each point in the process is performed using a fault-tree or event-tree.

*Fault-tree analysis*, FTA, is a deductive logic modeling method. Unwanted events (i.e., failures or accidents) are

first assumed. Then, the conditions are identified that could bring about the events. All possible contributors to an unwanted event are considered. The event is shown graphically at the top of a network. It is linked through various event statements, logic gates, and probabilities to more basic fault events located laterally and below, producing a graphical tree having branches of sequences and system conditions.

*Failure modes and effect analysis*, FMEA, is essentially the reverse of fault-tree analysis. It starts with the components of the system and, by focusing on the weaknesses and failure susceptibilities, evaluates how the components can contribute to the unwanted event. The basis for FMEA is, essentially, a parts list showing how each assembly is broken down into subassemblies and basic components. The appearance of an FMEA analysis is tabular, with the columns being used for failure modes, causes, symptoms, redundancies, consequences of failure, frequencies, probabilities, and so on.

*Life-cycle analysis* is used to prevent failures of items that have limited lives or that accumulate damage. Though the probability of failure may ideally be low, the history of the equipment may also be unknown. Life cycle analysis commonly focuses on time, history, and condition to determine the *degree of damage* (i.e., any reduction in useful life) present in a unit. One approach is to use reporting systems (i.e., operating logs and historical data). Another approach uses ongoing *material conditions monitoring*, MCM, and testing of the actual equipment.

## 17. WORK METHODS

The goal of *work methods (methods engineering)* is the reduction of fabrication and assembly time, worker effort, and manufacturing cost. This is accomplished in a variety of ways, including initial *design for manufacture and assembly* (DFMA) selection of methods to be used, human factors engineering, work measurement, plant layout, assembly line balancing, and administration of the manufacturing process. Work methods, which is more worker- and workplace-oriented, goes beyond traditional *value engineering*, which is product-design oriented.

## 18. BEHAVIORAL SCIENCE

*Behavioral science* deals with the psychological aspects of jobs to increase productivity. The subject draws from at least three major disciplines: psychology of the individual, sociology of the group, and anthropology of the culture. It is an outgrowth of the "human relations" thinking of the 1930s, whose goal was to increase the happiness of each worker.[47] Current behavioral science,

---

[46]The best-known exposure test is the *Ames test*, whereby microbes, cells, or test animals are exposed to a chemical in order to determine its carcinogenicity.

[47]"A happy employee is a productive employee" was the thinking.

however, seeks to minimize the tensions that limit productivity.[48] The following material briefly summarizes several popular behavioral science theories.

Improvements in worker satisfaction and decreases in worker dissatisfaction are often accomplished by modifying the structure of the job.[49] *Job flexibility* is a technique giving a worker the ability to move from job task to job task. *Job enlargement* is a technique extending the number of tasks a worker performs within a job.[50] *Horizontal job enlargement* adds new production activities to a job. *Vertical job enlargement* adds planning, inspection, and other nonproduction tasks to the job. *Job enrichment* is a subjective *result* felt by the worker when some technique (perhaps flexibility or enlargement) is used.

Dr. Abraham Maslow's *need hierarchy theory* (see Table 78.4) holds that certain needs become dominant only when lesser needs are satisfied. Although some needs can be sublimated and others overlapped, the theory requires the low-level needs to be satisfied before the higher-level needs can be realized.[51] The need hierarchy theory explains why money is a poor motivator of an affluent individual.

*Table 78.4* Maslow's Need Hierarchy (in order of lower to higher needs)

- *physiological needs:* air, food, water

- *safety needs:* protection against danger, threat, deprivation, arbitrary decisions; need for security in a dependent relationship

- *social needs:* belonging, association, acceptance, giving and receiving love, friendship

- *ego needs:* self-respect, confidence, achievement, self-image, group image, reputation, status, recognition, appreciation

- *self-fulfillment needs:* realizing self-potential, self-development, creativity

The *theory of influence* attempts to explain the effectiveness of supervisors. The most effective supervisors are those who help their workers benefit. For example, supervisors who are close (socially) to their workers and side with them in disputes are effective only if they have enough influence to help the workers. Consequently, knowledge and training are useless unless supervisors have the power to implement what they have learned.

The theory of influence includes five main tenets: (1) Employees think well of supervisors who help them reach their goals and meet their needs. (2) An influential supervisor will be able to help workers. (3) An influential supervisor who is also a disciplinarian will cause dissatisfaction. (4) A supervisor with no influence will not be able to affect worker satisfaction in any way. (5) Increases in supervisor influence are necessary to increase worker satisfaction.

Frederick Herzberg's *motivation-maintenance theory* explains worker satisfaction in terms of satisfiers and dissatisfiers. The *dissatisfiers* (also called *maintenance/ motivation factors*) do not motivate employees; they can only dissatisfy them. Dissatisfiers include salary, fringe benefits, company policy, administration, supervision, working conditions, and interpersonal relations. *Satisfiers* (also known as *motivators*) determine job satisfaction. Common satisfiers are achievement, recognition, the type and nature of the work, responsibility, and advancement. Dissatisfiers must be eliminated before the satisfiers can work.

Two ways of thinking, named "Theory X" and "Theory Y," have been proposed to describe the extremes of management style. The largely pessimistic *Theory X* is based on the assumption that workers inherently dislike and avoid working. Therefore, workers must be coerced into working by threats of punishment. Rewards are not sufficient.

Theory X assumes that the average employee wants to be directed, avoids responsibility, and seeks the security of an employee-employer relationship.[52] Theory X is pessimistic about the effectiveness of employers to satisfy or motivate their employees. According to the theory, by satisfying the physiological and safety (lower level) needs, employees shift the emphasis to higher needs, which cannot be satisfied. Employees, unable to derive satisfaction from their work, behave according to Theory X.

*Theory Y* assumes that workers find expenditure of effort to be natural and not inherently distasteful. It assumes that the average worker learns to accept and enjoy responsibility. It assumes that creativity is widely distributed among employees and that the potentials of average employees are only partially realized. Theory Y places the blame for worker laziness, indifference, and lack of cooperation in the lap of management.

---

[48]There are many who believe that motivational programs are not "honest," since management tries to convince employees to do what management wants. Wage incentive programs encourage employee dishonesty and errors by emphasizing quantity, not quality. Theory X management, with its implied punitive action if goals are not achieved, has never been effective.

[49]There is little or no evidence that workers want a social aspect to their jobs, job enlargement, or more autonomy. Though the relationship between satisfaction, absenteeism, and turnover has been established, the relationship between satisfaction and productivity is questionable.

[50]There are advantages to keeping a job small in scope. Learning time is low, mental effort is reduced, the pay rate can be lower for untrained workers, and supervision is reduced. However, such simple jobs also result in high turnover, increased absenteeism, and low pride in the job (and, subsequently, low quality). Job enlargement generally increases pride and quality, reduces inspection and material handling, and decreases turnover and absenteeism. However, training time is greater, tooling costs are higher, and record-keeping is more complex.

[51]The ego and self-fulfillment needs are rarely satisfied.

[52]Theory X is supported by much evidence. Workers exist in a continuum of wants, needs, and desires. Many of the needs are satisfied only off the job. Therefore, work is considered by the worker as a punishment or a price paid for off-the-job satisfaction.

A subsequent *Theory Z* was developed from observations, in the 1970s and 1980s, of Japanese companies that were more productive, profitable, stable, and quality-conscious than their United States counterparts.[53] Theory Z describes the traditional Japanese working environment where an employee stays with a company for his or her entire career, identifies with the company, and is cared for by the company.[54] In a Theory Z environment, employees desire and strive to build happy and intimate working relationships with everyone —co-workers, subordinates, and superiors. Theory Z employees need and want to be supported by the company. They value a working environment in which such things as personal family, personal cultures, personal and corporate traditions, and social institutions are as important as the work and indistinguishable from it. Theory Z employees have a highly developed sense of order, discipline, strong company loyalty, trust in the company, a moral obligation to work hard, and a cohesive bonding with their fellow employees. In exchange for management that supports and looks out for their well-being, Theory Z employees attend diligently and competently to their responsibilities.

Theory Z environments typically exhibit participatory management, with non-management employees being involved in consensus decision making. Implicit in a Theory Z environment are the high confidence management has in its employees and in the employees' underlying competence in decision making. In order for employees to make good decisions in a wide variety of situations, they must be (or become) generalists in their knowledge of the company. Such an environment tends to rotate non-management employees through positions, and with this lateral movement, vertical movement (promotion) tends to be slower. When an employee finally receives a management assignment, the employee will be familiar with all aspects of the company and will be experienced in maintaining a Theory Z environment.

Theories X, Y, and Z explain what employees want and why they do what they do. However, these explanations are not as much motivational theories as they are descriptive behavioral theories. Accordingly, the theories may be used to shape company policies, but motivation is assumed to be internal to the employees.

The philosophy of a *zero-defect program* is to expect perfect work from everybody. This counteracts the thinking of workers who have been conditioned to believe that everybody is imperfect and that errors are natural.[55] In a true zero-defect program, standards of performance are set for each worker. Workers are periodically checked against these performance requirements, and recognition is given when the goals are met.

Zero-defect programs develop constant, conscious desires and effort to do the job correctly the first time. This is accomplished by emphasizing what employees have for their own: pride and desire. Employees are continually reminded that their jobs are important, that the product is important, and that management thinks their efforts are important. The challenge of perfection is presented, and the importance of perfection is explained. Management sets an example by expecting zero defects from itself.

In *quality circle programs*, also known as *quality control circles* (QCC) and *total quality management (TQM) programs*, worker groups actively participate in measuring and improving departmental operations. Regular (weekly or monthly) meetings are held, with attendance usually being voluntary. Workers who attend are encouraged to suggest topics and ideas for improvement. Topics include effective ways to improve quality, safety, satisfaction, and productivity. Because the lowest-level worker is involved in goal-establishment, quality circles are known as a *bottom-up approach* to goal setting.[56]

*Management by objectives*, MBO, is a *top-down approach* to goal setting. In theory, the senior company officer establishes realistic, attainable goals for company performance and for his or her immediate managers (who accept the goals). These managers establish realistic goals for lower-level supervisors (who accept their goals). The supervisors, in turn, work out obtainable goals for their subordinates, and so on.

## 19. HUMAN FACTORS ENGINEERING

Unlike behavioral science (see Sec. 78.18) that deals with the psychological effects of work on the individual, *human factors engineering*, also known as *human engineering* and *ergonomics*, is concerned with the physical effects of work. Human factors attempts to increase or optimize the efficiency and reliability of the *machine-worker system* (also known as the *man-machine system*). The "optimization" process used is not mathematical but is based on thoughtful design technique.[57]

The three most common goals of human factors engineering are protecting the operator, minimizing the required conscious thought and effort, and assuring that

---

[53]Theory Z was popularized in the 1980s by Dr. William Ouchi.

[54]In the wake of global market competition, global economics and recession, the ever-increasing pace of development, and employees' own desire for increased financial wealth, status, and possessions, the traditional family-oriented company environment has been severely tested even in Japan. Many Japanese companies have been unable to provide employment security to their employees, and those that do increasingly need to transfer, reassign, or geographically relocate their employees. Such disruptions have challenged the trust and loyalty Japanese employees have traditionally felt toward their employers.

[55]Indeed, perfection is the standard for some professions (e.g., doctors, lawyers, and engineers).

[56]Subordinate workers do not dictate to supervisors in *bottom-up* methods. Workers develop recommendations that are passed up, but the decisions are made by supervisors.

[57]It is sometimes difficult to decide when improving the system involves human factors engineering and when it does not. Human factors engineering encompasses the subtopics of safety, industrial hygiene, design of training, and management and supervision, as well as system design.

the machine-worker system will operate successfully. These goals are usually accomplished by foolproofing the *machine-worker interface* (i.e., the controls and physical devices used) and designing the work environment for long-term health and safety.

Typical foolproofing techniques include locating controls in visible and accessible locations, varying the control handle and knob shapes for different functions, sizing and locating levers and wheels so that adequate operating force can be applied, mounting switches with uniform on-off directions, and arranging gauges so that their dials point in a uniform direction during proper operation.

*Anthropometric data* may be used to correctly size the equipment and work environment. Such items as range of motion, sitting height, leg room, and speed of motion are extensively tabulated by age, gender, nationality, and population percentile. Since it is impossible to design the machine-worker interface for every individual, the reach (as one example) required to operate a control should be no greater than the shortest reach of all workers expected to operate it. Designs frequently intentionally exclude the upper and lower 5% of the population sizes.

*Static anthropometry* deals with human dimensions such as height, length of forearms, foot size, and so on. *Dynamic anthropometry* deals with limits of motion, such as how far a person can reach in front of them. Both types of data are used in workplace design. Table 78.5 lists typical dimensions developed from anthropometric data for video display terminals, VDTs.[58]

**Table 78.5** *Recommended Dimensions for VDT Workstations*

|  | inches | centimeters |
|---|---|---|
| workstation surface |  |  |
| minimum height | 29 | 74 |
| width | 48–64 | 122–163 |
| depth | 32 | 81 |
| legroom |  |  |
| minimum height | 26 | 66 |
| minimum width | 26 | 66 |
| eye-to-screen distance | 20–30 | 51–76 |
| minimum height of home-row |  |  |
| key from floor | 29 | 74 |
| keyboard thickness | 1–2 | 2.5–5 |

(Multiply in by 2.54 to obtain cm.)

Attention is also given to the work environment, including temperature, humidity, illumination, and noise level. The "comfort range" for temperature and humidity depend on the nature of the work, dress, and the ventilation. Special consideration is required when the working temperature is not in the range of 65°F to 80°F (26.4°C to 38.4°C). Ear protection should be provided,

used, and required when the noise level is above an action level of 85 dB.[59] The level of illumination provided depends on the nature of the work. Typically, storage areas require 5 fc to 20 fc (54 lux to 229 lux), office work requires 100 fc to 200 fc (1100 lux to 2200 lux), and fine assembly requires 500 to 1000 fc (5400 lux to 1100 lux).[60]

## 20. MACHINE SAFEGUARDING

Most machines must be safeguarded to prevent injury to workers. For example, for a machine that is belt-driven, the belt must be enclosed by a safeguard (i.e., a *guard* or *shield*) to prevent a worker's clothing, hair, body, tools, or workpieces from being drawn into the drive mechanism. In general, safeguards should (1) prevent contact, (2) be secure, (3) protect from entering objects, (4) create no new hazards, (5) create no interference with work, and (6) allow safe repair and maintenance.

A worker should not be able to easily remove or tamper with a safeguard. A safeguard that can be made ineffective accidentally or in order to speed up operation is little better than no safeguard. Safeguards and safety devices should be made of durable materials that will withstand the conditions of normal use. They must be firmly secured to the machine.

The safeguard should ensure that no objects can fall into moving parts. A small tool that is dropped into a rotating machine can become a dangerous projectile.

A safeguard defeats its own purpose if it creates a hazard of its own. Safeguards should not contain jagged holes or unfinished, sharp edges. Edges should be rolled or bolted in such a way that sharp edges are eliminated.

A safeguard that impedes a worker from performing the job quickly and comfortably might be overridden or disregarded. Proper safeguarding should enhance efficiency as it relieves worker apprehension about injury.

It should be possible to lubricate, maintain, and repair equipment without removing any safeguards. Locating an oil reservoir outside the guard with a supply line leading to the lubrication point will eliminate the need for a worker or maintenance worker to enter hazardous areas.

There are many ways to safeguard machinery. The type of operation, nature of the raw material, method of handling, physical layout of the work area (see Fig. 78.8), and the production requirements or limitations all have to be taken into consideration. In general, power transmission apparatus is best protected by fixed guards that enclose the danger area. For hazards at the point of operation, where moving equipment performs

---

[58]Local laws, regulations, and other standards may apply.

[59]The U.S. Occupational Safety and Health Act (OSHA) of 1970 established 90 dB as the maximum noise level for an 8 hour exposure. Higher levels, up to a maximum of 115 dB, are permitted for shorter intervals, as short as 15 minutes.

[60]The *foot-candle* is equal to a lumen per square foot. The unit of illumination in SI units is the *lux*, equal to a lumen per square meter.

**Figure 78.8** *Typical Design of Computer Workstations*

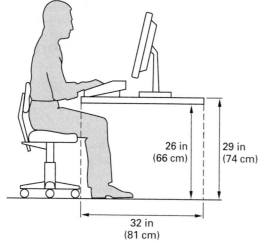

(a) side view

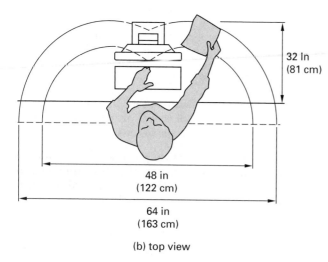

(b) top view

work on stock, Table 78.6 lists options for safeguarding. Miscellaneous methods, such as signs and awareness barriers, should also be considered.

## 21. SIMULATION

### Introduction

*Simulation* is a technique of performing sampling experiments on a model of the system. Simulation is appropriate when experimenting with the real system would be unsafe, inconvenient, expensive, or time-consuming, or when analytical techniques are not available.

### Time Control

Time control can either be *fixed-interval incrementing* (also known as *uniform time flow*) or next-event incrementing (also known as *variable time flow*). With fixed-interval incrementing, the model's master clock is

**Table 78.6** *Machine Safeguarding Options*

- *guard types*
  fixed
  interlocked
  adjustable
  self-adjusting
- *presence-sensing devices*
  electromechanical (pressure-sensitive body bar,
    floor mat, etc.)
  foot switch
  photoelectric (optical)
  infrared
  capacitive (radio-frequency)
- *actuation control*
  proper presence
  proper sequence
  two-switch (two-hand) actuation
  remote actuation
- *feeding and ejection options*
  automatic
  semiautomatic
  remote (robotic)

advanced by one unit and the system model is updated by determining what has happened during the elapsed time.

With next-event incrementing, the master clock is incremented by a variable amount of time. In this case, the computer actually proceeds by keeping track of when the simulated events occur and jumping ahead to the next of these events.

### Generating Random Numbers

A truly random process cannot be repeated. However, since all *random number* sequences in a simulation test can be duplicated, such sequences are called *pseudo-random number* sequences. A mathematical procedure for creating this sequence is known as a *random number generator*. This generator should have the following properties.

- The numbers generated should come from a uniform distribution.
- The generation should be fast.
- The sequence should not repeat.
- The sequence should not deteriorate to a single value.

Historically, the *midsquare, midproduct*, and *Fibonacci methods* have been used to generate random numbers, but these do not always yield satisfactory results. The *linear congruential method* is frequently used in modern simulation programs.

A number $D$ is said to be congruent with $N$ with modulus $M$ if $(D - N)/M$ is an integer.

The *mixed congruential method* calculates the next random variate, $r_{i+1}$, from the current variate, $r_i$, by using Eq. 78.91.

$$r_{i+1} = (ar_i + c)(\textbf{mod } T) \qquad 78.91$$

**mod** is the remaindering modulus function. This means that $(ar_i + c)$ is to be divided by $T$ and $r_{i+1}$ set equal to the remainder. The starting value, $r_0$, is known as the *seed* and must be supplied by the user. If $c = 0$, the method is known as a *multiplicative congruential method*. If $a = 1$, it is an *additive congruential method*.

The constants, $a$, $c$, and $T$ are usually chosen according to established rules to ensure the desired properties of the generator.

The *Mersenne twister algorithm* generates random number sequences that are higher in quality than linear congruential methods.

## Variance Reduction Techniques

Methods of increasing the accuracy of the sample estimation without increasing computer time are called *variance reduction techniques*. Two such techniques, *stratified sampling* and *complementary random numbers*, are complicated. They are used when extreme accuracy is required.

## 22. MONTE CARLO SIMULATION

### Introduction

*Monte Carlo simulation* evaluates the interactions of random variables whose distributions are known but whose combined effects are too complex to be specified mathematically. *Crude Monte Carlo* simulation is essentially random sampling from distributions to obtain values of each interacting variable. The following steps are similar in all Monte Carlo problems.

*step 1:* Establish the probability distribution for each variable in the study. This does not need to be a mathematical formula—a histogram is sufficient.

*step 2:* Form the cumulative distribution function for each variable.

*step 3:* Multiply the cumulative variable by 100 to obtain a cumulative axis that runs from 0 to 100 instead of 0 to 1.

*step 4:* Generate random numbers between 0 and 100. Either a random number table or a pseudo-random number generator may be used.

*step 5:* For each random number generated, locate the corresponding variable value. If the original distribution is continuous, take the midpoint of the range as the variable value.

A typical application of the Monte Carlo technique is finding the average line size, average waiting time, or percent idleness of servers in queuing problems.

### Example 78.25

A small bank has only one teller. The distribution of service times is

| time | probability |
|------|-------------|
| $1/4$ minute | 0.55 |
| $1/2$ minute | 0.40 |
| 1 minute | 0.05 |

Use the crude Monte Carlo method to simulate the operation of the teller's waiting line. What is the average time the teller spends with a customer?

*Solution*

*step 1:* Given.

*step 2:* The cumulative and converted distributions are

| time (minutes) | cumulative probability | (100) (cumulative probability) | range |
|----------------|------------------------|-------------------------------|-------|
| $1/4$ | 0.55 | 55 | 1–55 |
| $1/2$ | 0.95 | 95 | 56–95 |
| 1 | 1.00 | 100 | 96–100 |

*step 3:* See step 2.

*step 4:* Select ten random numbers. For this problem, use the sequence 01, 90, 25, 29, 09, 37, 67, 07, 15, 38.

*step 5:* The first random number is 01. This corresponds to a service time of $1/4$ minute. The random number 90 corresponds to a service time of $1/2$ minute, etc. The times of the first 10 customers are found similarly: $1/4$, $1/2$, $1/4$, $1/4$, $1/4$, $1/4$, $1/2$, $1/4$, $1/4$, $1/4$.

The average service time of the first ten customers is

$$\bar{t} = \left(\tfrac{1}{10}\right)\left(\tfrac{1}{4} + \tfrac{1}{2} + \tfrac{1}{4} + \tfrac{1}{4} + \tfrac{1}{4} + \tfrac{1}{4} + \tfrac{1}{2} + \tfrac{1}{4} + \tfrac{1}{4} + \tfrac{1}{4}\right)$$
$$= 0.3 \text{ minutes}$$

(The actual average service time could have been found as the expected value. However, more iterations would be required to obtain this value by the simulation method.)

$$\bar{t} = (0.55)(0.25 \text{ min}) + (0.4)(0.5) + (0.05)(1)$$
$$= 0.388 \text{ min}$$

### Variance Reduction

In Ex. 78.25, many of the random numbers were located nearer to 0 than to 100. Since the number formed by subtracting a random number from 100 is also a random number, it can be used to improve the accuracy of the estimation. This is known as the *complementary Monte Carlo technique* or *complementary random number technique.*

### Example 78.26

Repeat Ex. 78.25 using the complementary Monte Carlo technique.

*Solution*

The ten complementary random numbers are 99, 10, 75, 71, 91, 63, 33, 93, 85, and 62. The ten new service times are: 1, $\frac{1}{4}$, $\frac{1}{2}$, $\frac{1}{2}$, $\frac{1}{2}$, $\frac{1}{2}$, $\frac{1}{4}$, $\frac{1}{2}$, $\frac{1}{2}$, and $\frac{1}{2}$ minutes.

Combining these ten new service times with the original ten times gives an average of 0.4 minutes, which is closer to the true mean of 0.388.

## 23. DECISION THEORY

### Decision Trees

A *tree diagram* is a graphical method of enumerating all of the possible outcomes of a sequence of actions. The total number of outcomes is given by the *fundamental principle of counting*:

$$N = \prod n_i \qquad 78.92$$

In Eq. 78.92, $n_i$ is the number of elements in the $i$th set. The value calculated assumes that no outcomes are restricted and that all outcomes are distinctly different.

If a tree is used to model a decision process, it is called a *decision tree*. Usually decision trees have probabilities, losses, and rewards associated with the various possible outcomes. The tree is divided into segments called *generations* or *stages*. At the end of any given stage, the current status is described by the value of the *state variable*.

### Example 78.27

Find the possible values of product $P = A \times B$, where $A$ can take on the values (7, 8) and $B$ can take on the values (1, 3, 5). There are no restricted combinations.

*Solution*

There are two elements in the first set and three in the second. From Eq. 78.92, the number of possible outcomes (combinations) is

$$N = \prod n_i = (2)(3) = 6$$

The tree diagram is

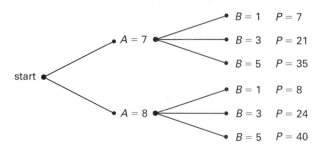

Therefore, the set of products is $P = (7, 8, 21, 24, 35, 40)$.

### Decision Theory Problem Formulation

When decisions have to be made, there is usually a finite number of alternatives to be selected from. The set of alternatives is known as the *action space*. Each alternative is associated with a *benefit* or *efficiency*. Usually these efficiencies are expressed in monetary terms. If the benefit is negative, the term *loss* is used in place of efficiency. In general, it is necessary to choose from a set of alternatives, $A$.

$$A: (a_1, a_2, a_3, ..., a_n)$$

The state of nature set is $\Theta$.

$$\Theta: (\theta_1, \theta_2, \theta_3, ..., \theta_m)$$

Decision theory commonly deals in *loss functions*, as opposed to *efficiency functions*. The difference is in sign only. The loss incurred depends only on the alternative chosen and the state of nature. The loss function is given by $L(a_i, \theta_j)$. If the loss function depends on a random variable, the expected loss should be used.

### Decision-Making Criteria

The *minimax criterion* selects the alternative that minimizes the maximum loss. The criterion is very conservative and is seldom used since it assumes that nature is a conscious, malevolent opponent that inflicts maximum damage on the decision maker.

As an alternative to the minimax criterion, *Bayes' principle* requires knowledge of the probability distribution of the possible states of nature. Specifically, let $p\{\theta_j\}$ be the prior probability of the $j$th state. Then, the expected loss associated with the $i$th alternative is

$$E\{L(a_i)\} = p\{\theta_1\}L(a_i, \theta_1) + p\{\theta_2\}L(a_i, \theta_2)$$
$$+ \cdots + p\{\theta_m\}L(a_i, \theta_m) \qquad 78.93$$

Bayes' principle chooses the alternative that minimizes the expected loss.

Engineering Management

## 24. MARKOV PROCESSES

### Introduction

A *Markov process* is a random process in which the state variable may have one of a limited number of values at any given time. Since the process is random, it is not possible to predict with certainty which value the random variable will take in the next time increment. One way to describe the future or history of the process is the sequence of *state variables* ($T_0$, $T_1$, $T_2$, ..., $T_n$).

All possible values of the state variable comprise the *state space*. Changes of state are called *transitions*. If the state actually changes, it is a *real transition*; otherwise, it is a *virtual transition*.

At any given time, the probability of moving to one of the other states depends on the current state and the current time. Thus, the $T_j$ are not always identically distributed nor are they independent.

The process is known as a *Markov process* if the current value of the state variable, $T_i$, depends only on the previous value, $T_{i-1}$, and affects only the upcoming value, $T_{i+1}$. Such a linking of the values is sometimes called a *Markov chain*.

### Transition and State Probabilities

The usual method of recording the transition probabilities is in a *transition matrix*, $\mathbf{P}$. For a process with only three states,

$$\mathbf{P} = \begin{bmatrix} p_{11} & p_{12} & p_{13} \\ p_{21} & p_{22} & p_{23} \\ p_{31} & p_{32} & p_{33} \end{bmatrix}$$

In general, $p_{jk}$ is the probability of making a move to state $k$ given that the variable is currently in state $j$. Each row must sum to one, but the columns may not. Because the move from $j$ to $k$ is made in one step or jump, the above matrix is also called the *one-step transition matrix*, $\mathbf{P}^{(1)}$. If the matrix is the same from time increment to time increment, it is said to be a *stationary matrix*.

A *two-step transition matrix*, $\mathbf{P}^{(2)}$, contains values of $p_{jk}$ that are probabilities of moving from state $j$ to $k$ in exactly two steps. The matrix $\mathbf{P}^{(2)}$ can be found by squaring $\mathbf{P}$ using matrix multiplication. The *n-step transition matrix* is found by raising the one-step matrix to the $n$th power. This is a specific case of the *Chapman-Kolmogorov equation*.

$$\mathbf{P}^{(n)} = (\mathbf{P})^n \qquad 78.94$$

If the value of $T_0$ is known, the probability of moving from state $j$ to $k$ in $n$ steps can be found directly from the $n$-step transition matrix. However, often $T_0$ is unknown—given by its own probability distribution $\mathbf{P}_i^{(0)}$ where $\mathbf{P}_i^{(0)}$ is the probability of the state variable

having the value $i$ at time 0. Then, the probability of being in state $k$ in $n$ steps regardless of the initial state is given by Eq. 78.95.

$$p_k^{(n)} = p_1^{(0)} p_{1k}^{(n)} + p_2^{(0)} p_{2k}^{(n)} + p_3^{(0)} p_{3k}^{(n)} + ... \qquad 78.95$$

Notice that the state probabilities are the probabilities of *being* in a state after some number of steps, whereas the transition probabilities are the probabilities of moving *between* two states.

### Steady-State Probabilities

The transition probabilities stabilize as the number of steps increases. That is, the starting position has increasingly less influence on the transition probabilities as $n$ increases. The *steady-state transition probabilities* ($\pi_1, \pi_2, \pi_3, \ldots, \pi_n$) can be found by solving the simultaneous equations specified by matrix equation Eq. 78.96 and the *normalizing equation*, Eq. 78.97.

$$[\pi_1 \ \pi_2 \ \pi_3] = [\pi_1 \ \pi_2 \ \pi_3]\mathbf{P} \qquad 78.96$$

$$\pi_1 + \pi_2 + \pi_3 = 1 \qquad 78.97$$

The reciprocals of the steady-state probabilities are the *average times between reassignments* to the $i$th state.

### Special Cases

If the transition from state $j$ to $k$ is possible, we say that state $k$ is *reachable* from state $j$. (See Fig. 78.9.) If $j$ is also reachable from $k$, we say that the two states *communicate*. If all states in the process belong to a single communicating class, the process is said to be *irreducible*. Otherwise, it is *reducible*. A state is an *absorbing state* if a transition to another state is impossible, as in $p_{jj} = 1$.

**Figure 78.9** *Types of Markov States*

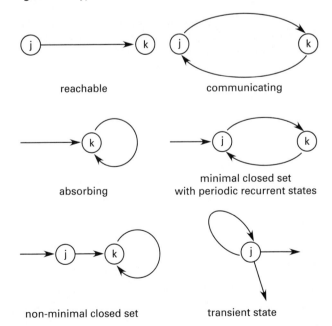

reachable

communicating

absorbing

minimal closed set with periodic recurrent states

non-minimal closed set

transient state

A *closed set of states* is a set such that no state outside the set is reachable from a state within the set. If there are no absorbing states in the closed set, it is called a *minimal closed set*. States in a minimal closed set are called *recurrent states*. States outside of a closed set are called *transient states*.

If a recurrent state must be reached with a regular frequency due to the structure of the process, it is said to be *periodic*; otherwise, it is *aperiodic*.

A state that is not transient, not periodic, and has a finite mean recurrent time is called an *ergodic state*.

It is possible that the simultaneous equation method for finding steady-state probabilities will fail since these probabilities may not exist. However, no difficulties will arise if all states are ergodic.

### Example 78.28

A parrot in an oasis has three trees from which to choose. On any given day, the parrot can be in any tree, even the previous day's tree, but it may be in only one tree per day.

Once the parrot gets to tree 1, it will not move directly to tree 3. Moving to tree 2 or staying in tree 1 are equally likely. If the parrot gets to tree 2, it will not move directly to tree 1 but may stay in tree 2 or move to tree 3 with equal likelihood. If the parrot gets to tree 3, it will not stay in tree 3 but will move to tree 1 three times as often as to tree 2.

The one-step transition matrix is

$$\mathbf{P} = \begin{bmatrix} 0.50 & 0.50 & 0.00 \\ 0.00 & 0.50 & 0.50 \\ 0.75 & 0.25 & 0.00 \end{bmatrix}$$

(a) Given that the parrot is in tree 1, what is the probability that it will be in tree 2 after two moves? (b) Find the steady-state transition probabilities.

*Solution*

(a) The state variable $T_j$ is a random variable representing the tree the parrot is in at time $j$. One way to describe the future or history of the parrot's wanderings is by the sequence $(T_0, T_1, T_2, \ldots, T_n)$.

*method 1:* Enumeration

There are three paths the parrot can take to get from tree 1 to tree 2 in two days.

$$
\begin{aligned}
T_0&=1, T_1=1, T_2=2; \text{ probability} = (0.5)(0.5) &= 0.25 \\
T_0&=1, T_1=2, T_2=2; \text{ probability} = (0.5)(0.5) &= 0.25 \\
T_0&=1, T_1=3, T_2=2; \text{ probability} = (0)(0.25) &\underline{= 0.00} \\
& & p_{12}^{(2)} = 0.50
\end{aligned}
$$

*method 2:* Using $\mathbf{P}^{(2)}$

Squaring the one-step transition matrix gives

$$\mathbf{P}^{(2)} = (\mathbf{P})^2 = \begin{bmatrix} 0.250 & 0.500 & 0.250 \\ 0.375 & 0.375 & 0.250 \\ 0.375 & 0.500 & 0.125 \end{bmatrix}$$

The two-step transition probability of $p_{12}^{(2)}$ is 0.500.

(b) Solve the following four equations simultaneously.

$$
\begin{aligned}
\pi_1 &= 0.5\pi_1 && + 0.75\pi_3 \\
\pi_2 &= 0.5\pi_1 & + 0.5\pi_2 & + 0.25\pi_3 \\
\pi_3 &= & 0.5\pi_2 & \\
1 &= \pi_1 & + \pi_2 & + \pi_3
\end{aligned}
$$

The solution is

$$
\begin{aligned}
\pi_1 &= 0.333\ldots \\
\pi_2 &= 0.444\ldots \\
\pi_3 &= 0.222\ldots
\end{aligned}
$$

# 79 Construction and Jobsite Safety

## 1. INTRODUCTION

Common sense can go a long way in preventing many jobsite accidents. However, scheduling, economics, lack of concern, carelessness, and laxity are often more likely to guide actions than common sense. For that reason, the "science" of accident prevention is highly regulated. In the United States, workers' safety is regulated by the federal Occupational Safety and Health Act (OSHA). State divisions (e.g., Cal-OSHA) are charged with enforcing federal and state safety regulations. Surface and underground mines, which share many hazards with the construction industry and which have others specific to them, are regulated by the federal Mine Safety and Health Act (MSHA). Other countries may have more restrictive standards.

All of the federal OSHA regulations are published in the Congressional Federal Register (CFR). The two main categories of standards are the "1910 standards," which apply to general industry, and the "1926 standards," which apply to the construction industry.

This chapter covers only a minuscule portion of the standards and safety issues facing engineers and construction workers. Only the briefest descriptions of these complex issues and the regulations that govern them are provided.

## 2. TRENCHING AND EXCAVATION

Soils are classified as stable rock and types A, B, or C, with type C being the most unstable. Type A soils are cohesive soils with unconfined compressive strengths of 1.5 tons per square foot (tsf) (144 kPa) or greater. Type B soils are cohesive soils with compressive strengths of 0.5 to 1.5 tsf (48 to 144 kPa) and some granular soils. Type C soils are cohesive soils with compressive strengths less than 0.5 tsf (48 to 144 kPa), and some

granular soils. Soils classified as types A and B may need to be reclassified as type C following rain and flooding [OSHA 1926.652, App. A to Subpart P].

Except for excavations entirely in stable rock, excavations deeper than 5 ft (1.5 m) in all types of earth must be protected from cave-in and collapse [OSHA 1926.652]. Excavations less than 5 ft (1.5 m) deep are usually exempt but may also need to be protected when inspection indicates that hazardous ground movement is possible.

Timber and aluminum shoring (hydraulic, pneumatic, and screw-jacked) and trench shields that meet the requirements of OSHA 1926.652 App. C to Subpart P may be used in excavations up to 20 ft (6 m) deep.

Sloping and benching the trench walls may be substituted for shoring. For long-term use, the maximum allowable slope (H:V) for type A soils is $3/4$:1 (53° from the horizontal). For type B soils, it is 1:1 (45°). For type C soils, it is $1^1/2$:1 (34°). These slopes must be reduced 50% if the soil shows signs of distress. Sloped walls in excavations deeper than 20 ft (6 m) must be designed by a professional engineer. Greater slopes are permitted for short-term usage in excavations less than 12 ft (3.67 m) deep.

In trenches 4 ft (1.2 m) deep or more, ladders, stairways, or ramps are required with a maximum lateral spacing of 25 ft (7.5 m) [OSHA 1926.651(b)(2)].

Spoils and other equipment that could fall into a trench or an excavation must be kept at least 2 ft (0.6 m) from the edge of a trench unless secured in some other fashion [OSHA 1926.651(l)].

## 3. CHEMICAL HAZARDS

OSHA's Hazard Communication Standard [OSHA 1910.1200] requires that the dangers of all chemicals purchased, used, or manufactured be known to employees. The hazards are communicated in a variety of ways, including labeling containers, training employees, and providing ready access to *material safety data sheets* (MSDSs).

OSHA has suggested a nonmandatory standard form for the MSDS, but other forms are acceptable as long as they contain the same (or more) information. The information contained on an MSDS consists of following categories.

*Chemical identity:* the identity of the substance as it appears on the label.

*Section I: Manufacturer's name and contact information:* manufacturer's name, address, telephone number, and emergency phone number; date the MSDS was prepared; and an optional signature of the preparer.

*Section II: Hazardous ingredients/identity information:* list of the hazardous components by chemical identity and other common names; OSHA *permissible exposure limit* (PEL), ACGIH *threshold level value* (TLV), and other recommended exposure limits; percentage listings of the hazardous components is optional.

*Section III: Physical/chemical characteristics:* boiling point, vapor pressure, vapor density, specific gravity, melting point, evaporation rate, solubility in water, physical appearance, and odor.

*Section IV: Fire and explosion hazard data:* flash point (and method used to determine it), flammability limits, extinguishing media, special fire-fighting procedures, and unusual fire and explosion hazards.

*Section V: Reactivity data:* stability, conditions to avoid, incompatibility (materials to avoid), hazardous decomposition or by-products, and hazardous polymerization (and conditions to avoid).

*Section VI: Health hazard data:* routes of entry into the body (inhalation, skin, ingestion), health hazards (acute = immediate, or chronic = builds up over time), carcinogenicity (NTP, IARC monographs, OSHA regulated), signs and symptoms of exposure, medical conditions generally aggravated by exposure, and emergency and first-aid procedures.

*Section VII: Precautions for safe handling and use:* steps to be taken in case the material is released or spilled, waste disposal method, precautions to be taken in handling or storage, and other precautions.

*Section VIII: Control measures:* respiratory protection (type to be specified), ventilation (local, mechanical exhaust, special, or other), protective gloves, eye protection, other protective clothing or equipment, and work/hygienic practices.

*Section IX: Special precautions and comments:* safe storage and handling, types of labels or markings for containers, and Department of Transportation (DOT) policies for handling the material.

## 4. CONFINED SPACES AND HAZARDOUS ATMOSPHERES

Employees entering confined spaces (e.g., excavations, sewers, tanks) must be properly trained, supervised, and equipped. Atmospheres in confined spaces must be monitored for oxygen content and other harmful contaminants. Oxygen content in confined spaces must be maintained at 19.5% or higher unless a breathing apparatus is provided [OSHA 1910.146]. Employees entering deep confined excavations must wear harnesses with lifelines [OSHA 1926.651].

## 5. POWER LINE HAZARDS

Employees operating cranes or other overhead material-handling equipment (e.g., concrete boom trucks, backhoe arms, and raised dumptruck boxes) must be aware of the possibility of inadvertent power line contact. Prior to operation, the site must be thoroughly inspected for the danger of power line contact. OSHA provides specific minimum requirements for safe operating distances. For example, for lines of 50 kV or less, all parts of the equipment must be kept at least 10 ft (3 m) from the power line [OSHA 1926.550(a)(15)]. A good rule of thumb for voltages greater than 50 kV is a clearance of 35 ft (10.5 m). However, the exact OSHA clearance requirement can be calculated from Eq. 79.1.

$$\text{line clearance} = 3 \text{ m} + (10.2 \text{ mm})(V_{kV} - 50 \text{ kV})$$
$$[\text{SI}] \quad 79.1(a)$$

$$\text{line clearance} = 10 \text{ ft} + (0.4 \text{ in})(V_{kV} - 50 \text{ kV})$$
$$[\text{U.S.}] \quad 79.1(b)$$

## 6. FALL AND IMPACT PROTECTION

Fall protection can take the form of barricades, walkways, bridges (with guardrails), nets, and fall-arrest systems. Personal fall-arrest systems include lifelines, lanyards, and deceleration devices. Such equipment is attached to an anchorage at one end and to the body-belt or body hardness at the other. All equipment is to be properly used, certified, and maintained. Employees are to be properly trained in the equipment's use and operation.

Employees must be protected from impalement hazards from exposed rebar [OSHA 1926.701(b)]. A widely used method for covering rebar ends has been plastic *mushroom caps*, often orange or yellow in color. However, OSHA no longer considers plastic caps adequate for anything more than scratch protection. Commercially available steel-reinforced caps and wooden troughs capable of withstanding a 250 lbm/10 ft drop test without breakthrough can still be used.

*Head protection,* usually in the form of a helmet (*hardhat*), is part of a worker's *personal protective equipment* (PPE). Head protection is required where there is a danger of head injuries from impact, flying or falling objects, electrical shock, or burns [OSHA 1910.132(a) and (c)]. Head protection should be nonconductive when there is electrical or thermal danger [OSHA 1910.335(a)(1)(iv)].

## 7. NOISE

OSHA sets maximum limits on daily sound exposure. The "all-day" 8 hr noise level limit is 90 dBA. This is higher than the maximum level permitted in most other countries (e.g., noise control threshold of 85 dBA in the United Kingdom, Germany, and Japan). In the United States, employees may not be exposed to steady sound levels above 115 dBA, regardless of the duration. Impact sound levels are limited to 140 dBA.

Hearing protection, educational programs, periodic examinations, and other actions are required for workers whose 8 hr exposure is more than 85 dBA or whose noise dose exceeds 50% of the *action levels*. (See Table 79.1.)

**Table 79.1** *Typical Permissible Noise Exposure Levels*[*]

| sound level (dBA) | exposure (hr/day) |
|---|---|
| 90 | 8 |
| 92 | 6 |
| 95 | 4 |
| 97 | 3 |
| 100 | 2 |
| 102 | $1\frac{1}{2}$ |
| 105 | 1 |
| 110 | $\frac{1}{2}$ |
| 115 | $\frac{1}{4}$ or less |

[*]without hearing protection

Source: OSHA, Sec. 1910.95, Table G-16

## 8. SCAFFOLDS

Scaffolds are any temporary elevated platform (supported or suspended) and its supporting structure (including points of anchorage) used for supporting employees, materials, or both. Construction and use of scaffolds are regulated in detail by OSHA Std. 1926.451. A few of the regulations are summarized in the following paragraphs.

Each employee who performs work on a scaffold must be trained by a person qualified to recognize the hazards associated with the type of scaffold used and to understand the procedures to control or minimize those hazards. The training must include such topics as the nature of any electrical hazards, fall hazards, falling object hazards, the maintenance and disassembly of the fall protection systems, the use of the scaffold, handling of materials, the capacity, and the maximum intended load.

Fall protection (guardrail systems and personal fall arrest systems) must be provided for each employee on a scaffold more than 10 ft (3.1 m) above a lower level.

Each scaffold and scaffold component must have the capacity to support its own weight and at least 4 times the maximum intended load applied or transmitted to it. Suspension ropes and connecting hardware must support 6 times the intended load. Scaffolds and scaffold components may not be loaded in excess of their maximum intended loads or rated capacities, whichever is less.

The scaffold platform must be planked or decked as fully as possible. It cannot deflect more than $^1/_{60}$ of the span when loaded.

The work area for each scaffold platform and walkway must be at least 18 in (46 cm) wide. When the work area must be less than 18 in (46 cm) wide, guardrails and/or personal fall arrest systems must still be used.

Access must be provided when the scaffold platforms are more than 2 ft (0.6 m) above or below a point of access. Direct access is acceptable when the scaffold is not more than 14 in (36 cm) horizontally and not more than 24 in (61 cm) vertically from the other surfaces. Cross braces cannot be used as a means of access.

A competent person with the authority to require prompt corrective action is required to inspect the scaffold, scaffold components, and ropes on suspended scaffolds before each work shift and after any occurrence which could affect the structural integrity.

# 80

# Engineering Law

## 1. FORMS OF COMPANY OWNERSHIP[1]

There are three basic forms of company ownership in the United States: (a) sole proprietorship, (b) partnership, and (c) corporation.[2] Each of these forms of ownership has advantages and disadvantages.

## 2. SOLE PROPRIETORSHIPS

A *sole proprietorship (single proprietorship)* is the easiest form of ownership to establish. Other than the necessary licenses, filings, and notices (which apply to all forms of ownership), no legal formalities are required to start business operations. A sole proprietor (the owner) has virtually total control of the business and makes all supervisory and management decisions.

Legally, there is no distinction between the sole proprietor and the sole proprietorship (the business). This is the greatest disadvantage of this form of business. The owner is solely responsible for the operation of the business, even if the owner hires others for assistance. The owner assumes personal, legal, and financial liability for all acts and debts of the company. If the company debts remain unpaid, or in the event there is a legal judgment against the company, the owner's personal assets (home, car, savings, etc.) can be seized or attached.

Another disadvantage of the sole proprietorship is the lack of significant organizational structure. In times of business crisis or trouble, there may be no one to share the responsibility or to help make decisions. When the owner is sick or dies, there may be no way to continue the business.

There is also no distinction between the incomes of the business and the owner. Therefore, the business income is taxed at the owner's income tax rate. Depending on the owner's financial position, the success of the business, and the tax structure, this can be an advantage or a disadvantage.[3]

## 3. PARTNERSHIPS

A *partnership* (also known as a *general partnership*) is ownership by two or more persons known as *general partners*. Legally, this form is very similar to a sole proprietorship, and the two forms of business have many of the same advantages and disadvantages. For example, with the exception of an optional *partnership agreement*, there are a minimum of formalities to setting up business. The partners make all business and management decisions themselves according to an agreed-upon process. The business income is split among the partners and taxed at the partners' individual tax rates.[4] Continuity of the business is still a problem since most partnerships are automatically dissolved upon the withdrawal or death of one of the partners.[5]

One advantage of a partnership over a sole proprietorship is the increase in available funding. Not only do more partners bring in more start-up capital, but the

---

[1]This chapter is not intended to be a substitute for professional advice. Law is not always black and white. For every rule there are exceptions. For every legal principle, there are variations. For every type of injury, there are numerous legal precedents. This chapter covers the superficial basics of a small subset of U.S. law affecting engineers.

[2]The discussion of forms of company ownership in Sec. 80.2, Sec. 80.3, and Sec. 80.4 applies equally to service-oriented companies (e.g., consulting engineering firms) and product-oriented companies.

[3]To use a simplistic example, if the corporate tax rates are higher than the individual tax rates, it would be *financially* better to be a sole proprietor because the company income would be taxed at a lower rate.

[4]The percentage split is specified in the partnership agreement.

[5]Some or all of the remaining partners may want to form a new partnership, but this is not always possible.

resource pool may make business credit easier to obtain. Also, the partners bring a diversity of skills and talents.

Unless the partnership agreement states otherwise, each partner can individually obligate (i.e., *bind*) the partnership without the consent of the other partners. Similarly, each partner has personal responsibility and liability for the acts and debts of the partnership company, just as sole proprietors do. In fact, each partner assumes the *sole* responsibility, not just a proportionate share. If one or more partners are unable to pay, the remaining partners shoulder the entire debt. The possibility of one partner having to pay for the actions of another partner must be considered when choosing this form of business ownership.

A *limited partnership* differs from a general partnership in that one or more of the partners are silent. The *limited partners* make a financial contribution to the business and receive a share of the profit but do not participate in the management and cannot bind the partnership. While *general partners* have unlimited personal liabilities, limited partners are generally liable only to the extent of their investment.[6] A written partnership agreement is required, and the agreement must be filed with the proper authorities.

## 4. CORPORATIONS

A corporation is a legal entity (i.e., a legal person) distinct from the founders and owners. The separation of ownership and management makes the corporation a fundamentally different business form than a sole proprietorship or partnership, with very different advantages and disadvantages.

A corporation becomes legally distinct from its founders upon formation and proper registration. Ownership of the corporation is through shares of stock, distributed to the founders and investors according to some agreed-upon investment and distribution rule. Thus, the founders and investors become the stockholders (i.e., owners) of the corporation. A *closely held* (*private*) corporation is one in which all stock is owned by a family or small group of co-investors. A *public corporation* is one whose stock is available for the public-at-large to purchase.

There is no mandatory connection between ownership and management functions. The decision-making power is vested in the executive officers and a *board of directors* that governs by majority vote. The stockholders elect the board of directors which, in turn, hires the executive officers, management, and other employees. Employees of the corporation may or may not be stockholders.

Disadvantages (at least for a person or persons who could form a partnership or sole proprietorship) include the higher corporate tax rate, difficulty and complexity

of formation (some states require a minimum number of persons on the board of directors), and additional legal and accounting paperwork.

However, since a corporation is distinctly separate from its founders and investors, those individuals are not liable for the acts and debts of the corporation. Debts are paid from the corporate assets. Income to the corporation is not taxable income to the owners. (Only the salaries, if any, paid to the employees by the corporation are taxable to the employees.) Even if the corporation were to go bankrupt, the assets of the owners would not ordinarily be subject to seizure or attachment.

A corporation offers the best guarantee of continuity of operation in the event of the death, incapacity, or retirement of the founders since, as a legal entity, it is distinct from the founders and owners.

## 5. LIMITED LIABILITY ENTITIES

A variety of other legal entities have been established that blur the lines between the three traditional forms of business (i.e., proprietorship, partnership, and corporation). The *limited liability partnership*, LLP, extends a measure of corporate-like protection to professionals while permitting partnership-like personal participation in management decisions. Since LLPs are formed and operated under state laws, actual details vary from state to state. However, most LLPs allow the members to participate in management decisions, while not being responsible for the misdeeds of other partners. As in a corporation, the debts of the LLP do not become debts of the members.[7] The *double taxation* characteristic of traditional corporations is avoided, as profits to the LLP flow through to the members.

For engineers and architects (as well as doctors, lawyers, and accountants), the *professional corporation*, PC, offers protection from the actions (e.g., malpractice) of other professionals within a shared environment, such as a design firm. While a PC does not shield the individual from responsibility for personal negligence or malpractice, it does permit the professional to be associated with a larger entity, such as a partnership of other PCs, without accepting responsibility for the actions of the other members. In that sense, the protection is similar to that of an LLP. Unlike a traditional corporation, a PC may have a board of directors consisting of only a single individual, the professional.

The *limited liability company*[8], LLC, also combines advantages from partnerships and corporations. In an LLC, the members are shielded from debts of the LLC

---

[6]That is, if the partnership fails or is liquidated to pay debts, the limited partners lose no more than their initial investments.

[7]Depending on the state, the shield may be complete or limited. It is common that the protection only applies to negligence-related claims, as opposed to intentional tort claims, contract-related obligations, and day-to-day operating expenses such as rent, utilities, and employees.

[8]LLC does not mean *limited liability corporation*. LLCs are not corporations.

while enjoying the pass-through of all profits.[9] Like a partnership or shareholder, a member's obligation is limited to the *membership interest* in (i.e., contribution to) the LLC. LLCs are directed and controlled by one or more managers who may also be members. A variation of the LLC specifically for design, medical, and other professionals is the *professional limited liability company*, PLLC.

The traditional corporation, as described in Sec. 80.4, is referred to as a *Subchapter C corporation* or "*C corp.*"[10] A variant is the *S corporation* ("*S corp*") which combines characteristics of the C corporation with pass-through for taxation. S corporations can be limited or treated differently than C corporations by state and federal law.

## 6. PIERCING THE CORPORATE VEIL

An individual operating as a corporation, LLP, LLC, or PC entity may lose all protection if his or her actions are fraudulent, or if the court decides the business is an "alter ego" of the individual. Basically, this requires the business to be run as a business. Business and personal assets cannot be intermingled, and business decisions must be made and documented in a business-like manner. If operated fraudulently or loosely, a court may assign liability directly to an individual, an action known as *piercing the corporate veil*.

## 7. AGENCY

In some contracts, decision-making authority and right of action are transferred from one party (the owner, or *principal*) who would normally have that authority to another person (the *agent*). For example, in construction contracts, the engineer may be the agent of the owner for certain transactions. Agents are limited in what they can do by the scope of the agency agreement. Within that scope, however, an agent acts on behalf of the principal, and the principal is liable for the acts of the agent and is bound by contracts made in the principal's name by the agent.

Agents are required to execute their work with care, skill, and diligence. Specifically, agents have *fiduciary responsibility* toward their principal, meaning that agent must be honest and loyal. Agents are liable for damages resulting from a lack of diligence, loyalty, and/or honesty. If the agents misrepresented their skills when obtaining the agency, they can be liable for breach of contract or fraud.

## 8. GENERAL CONTRACTS

A *contract* is a legally binding agreement or promise to exchange goods or services.[11] A written contract is merely a documentation of the agreement. Some agreements must be in writing, but most agreements for engineering services can be verbal, particularly if the parties to the agreement know each other well.[12] Written contract documents do not need to contain intimidating legal language, but all agreements must satisfy three basic requirements to be enforceable (binding).

- There must be a clear, specific, and definite *offer* with no room for ambiguity or misunderstanding.

- There must be some form of conditional future *consideration* (i.e., payment).[13]

- There must be an *acceptance* of the offer.

There are other conditions that the agreement must meet to be enforceable. These conditions are not normally part of the explicit agreement but represent the conditions under which the agreement was made.

- The agreement must be *voluntary* for all parties.

- All parties must have *legal capacity* (i.e., be mentally competent, of legal age, and uninfluenced by drugs).

- The purpose of the agreement must be *legal*.

For small projects, a simple *letter of agreement* on one party's stationery may suffice. For larger, complex projects, a more formal document may be required. Some clients prefer to use a *purchase order*, which can function as a contract if all basic requirements are met.

Regardless of the format of the written document— letter of agreement, purchase order, or standard form —a contract should include the following features.[14]

- introduction, preamble, or preface indicating the purpose of the contract

- name, address, and business forms of both contracting parties

- signature date of the agreement

- effective date of the agreement (if different from the signature date)

---

[11]Not all agreements are legally binding (i.e., enforceable). Two parties may agree on something, but unless the agreement meets all of the requirements and conditions of a contract, the parties cannot hold each other to the agreement.

[12]All states have a *statute of frauds* that, among other things, specifies what types of contracts must be in writing to be enforceable. These include contracts for the sale of land, contracts requiring more than one year for performance, contracts for the sale of goods over $500 in value, contracts to satisfy the debts of another, and marriage contracts. Contracts to provide engineering services do not fall under the statute of frauds.

[13]Actions taken or payments made prior to the agreement are irrelevant. Also, it does not matter to the courts whether the exchange is based on equal value or not.

[14]*Construction contracts* are unique unto themselves. Items that might also be included as part of the *contract documents* are the agreement form, the general conditions, drawings, specifications, and addenda.

---

[9]LLCs enjoy *check the box taxation*, which means they can elect to be taxed as sole proprietorships, partnerships, or corporations.

[10]The reference is to subchapter C of the Internal Revenue Code.

- duties and obligations of both parties
- deadlines and required service dates
- fee amount
- fee schedule and payment terms
- agreement expiration date
- standard boilerplate clauses
- signatures of parties or their agents
- declaration of authority of the signatories to bind the contracting parties
- supporting documents

## 9. STANDARD BOILERPLATE CLAUSES

It is common for full-length contract documents to include important *boilerplate clauses*. These clauses have specific wordings that should not normally be changed, hence the name "boilerplate." Some of the most common boilerplate clauses are paraphrased here.

- Delays and inadequate performance due to war, strikes, and acts of God and nature are forgiven (*force majeure*).

- The contract document is the complete agreement, superseding all previous verbal and written agreements.

- The contract can be modified or canceled only in writing.

- Parts of the contract that are determined to be void or unenforceable will not affect the enforceability of the remainder of the contract (*severability*). Alternatively, parts of the contract that are determined to be void or unenforceable will be rewritten to accomplish their intended purpose without affecting the remainder of the contract.

- None (or one, or both) of the parties can (or cannot) assign its (or their) rights and responsibilities under the contract (*assignment*).

- All notices provided for in the agreement must be in writing and sent to the address in the agreement.

- Time is of the essence.[15]

- The subject headings of the agreement paragraphs are for convenience only and do not control the meaning of the paragraphs.

- The laws of the state in which the contract is signed must be used to interpret and govern the contract.

- Disagreements shall be arbitrated according to the rules of the American Arbitration Association.

- Any lawsuits related to the contract must be filed in the county and state in which the contract is signed.

- Obligations under the agreement are unique, and in the event of a breach, the defaulting party waives the defense that the loss can be adequately compensated by monetary damages (*specific performance*).

- In the event of a lawsuit, the prevailing party is entitled to an award of reasonable attorneys' and court fees.[16]

- Consequential damages are not recoverable in a lawsuit.

## 10. SUBCONTRACTS

When a party to a contract engages a third party to perform the work in the original contract, the contract with the third party is known as a *subcontract*. Whether or not responsibilities can be subcontracted under the original contract depends on the content of the *assignment clause* in the original contract.

## 11. PARTIES TO A CONSTRUCTION CONTRACT

A specific set of terms has developed for referring to parties in consulting and construction contracts. The *owner* of a construction project is the person, partnership, or corporation that actually owns the land, assumes the financial risk, and ends up with the completed project. The *developer* contracts with the architect and/or engineer for the design and with the contractors for the construction of the project. In some cases, the owner and developer are the same, in which case the term *owner-developer* can be used.

The *architect* designs the project according to established codes and guidelines but leaves most stress and capacity calculations to the *engineer*.[17] Depending on the construction contract, the engineer may work for the architect, or vice versa, or both may work for the developer.

Once there are approved plans, the developer hires *contractors* to do the construction. Usually, the entire construction project is awarded to a *general contractor*. Due to the nature of the construction industry, separate *subcontracts* are used for different tasks (electrical, plumbing, mechanical, framing, fire sprinkler installation, finishing, etc.). The general contractor who hires all of these different *subcontractors* is known as the *prime contractor* (or *prime*). (The subcontractors can also work directly for the owner-developer, although this is less common.) The prime contractor is responsible for all of the acts of the subcontractors and is liable for any damage suffered by the owner-developer due to those acts.

---

[15]Without this clause in writing, damages for delay cannot be claimed.

[16]Without this clause in writing, attorneys' fees and court costs are rarely recoverable.

[17]On simple small projects, such as wood-framed residential units, the design may be developed by a *building designer*. The legal capacities of building designers vary from state to state.

Construction is managed by an agent of the owner-developer known as the *construction manager*, who may be the engineer, the architect, or someone else.

## 12. STANDARD CONTRACTS FOR DESIGN PROFESSIONALS

Several professional organizations have produced standard agreement forms and other standard documents for design professionals.[18] Among other standard forms, notices, and agreements, the following standard contracts are available.[19]

- standard contract between engineer and client

- standard contract between engineer and architect

- standard contract between engineer and contractor

- standard contract between owner and construction manager

Besides completeness, the major advantage of a standard contract is that the meanings of the clauses are well established, not only among the design professionals and their clients but also in the courts. The clauses in these contracts have already been litigated many times. Where a clause has been found to be unclear or ambiguous, it has been rewritten to accomplish its intended purpose.

## 13. CONSULTING FEE STRUCTURE

Compensation for consulting engineering services can incorporate one or more of the following concepts.

- *lump-sum fee:* This is a predetermined fee agreed upon by client and engineer. This payment can be used for small projects where the scope of work is clearly defined.

- *cost plus fixed fee:* All costs (labor, material, travel, etc.) incurred by the engineer are paid by the client. The client also pays a predetermined fee as profit. This method has an advantage when the scope of services cannot be determined accurately in advance. Detailed records must be kept by the engineer in order to allocate costs among different clients.

- *per diem fee:* The engineer is paid a specific sum for each day spent on the job. Usually, certain direct expenses (e.g., travel and reproduction) are billed in addition to the per diem rate.

- *salary plus:* The client pays for the employees on an engineer's payroll (the salary) plus an additional percentage to cover indirect overhead and profit plus certain direct expenses.

- *retainer:* This is a minimum amount paid by the client, usually in total and in advance, for a normal amount of work expected during an agreed-upon period. None of the retainer is returned, regardless of how little work the engineer performs. The engineer can be paid for additional work beyond what is normal, however. Some direct costs, such as travel and reproduction expenses, may be billed directly to the client.

- *percentage of construction cost:* This method, which is widely used in construction design contracts, pays the architect and/or the engineer a percentage of the final total cost of the project. Costs of land, financing, and legal fees are generally not included in the construction cost, and other costs (plan revisions, project management labor, value engineering, etc.) are billed separately.

## 14. MECHANIC'S LIENS

For various reasons, providers and material, labor, and design services to construction sites may not be promptly paid or even paid at all. Such providers have, of course, the right to file a lawsuit demanding payment, but due to the nature of the construction industry, such relief may be insufficient or untimely. Therefore, such providers have the right to file a *mechanic's lien* (also known as a *construction lien, materialman's lien, supplier's lien,* or *laborer's lien*) against the property. Although there are strict requirements for deadlines, filing, and notices, the procedure for obtaining (and removing) such a lien is simple. The lien establishes the supplier's security interest in the property. Although the details depend on the state, essentially the property owner is prevented from transferring title (i.e., selling) the property until the lien has been removed by the supplier. The act of filing a lawsuit to obtain payment is known as "perfecting the lien." Liens are perfected by forcing a judicial foreclosure sale. The court orders the property sold, and the proceeds are used to pay off any lienholders.

---

[18]There are two main sources of standardized construction and design agreements: EJCDC and AIA. Consensus documents, known as *ConsensusDOCS*, for every conceivable situation have been developed by the *Engineers Joint Contracts Documents Committee*, EJCDC. EJCDC includes the American Society of Civil Engineers (ASCE), the American Council of Engineering Companies (ACEC), National Society of Professional Engineers' (NSPE's) Professional Engineers in Private Practice Division, Associated General Contractors of America (AGC), and more than fifteen other participating professional engineering design, construction, owner, legal, and risk management organizations including the Associated Builders and Contractors, American Subcontractors Association, Construction Users Roundtable, National Roofing Contractors Association, Mechanical Contractors Association of America, and National Plumbing, Heating-Cooling Contractors Association. The American Institute of Architects, AIA, has developed its own standardized agreements in a less-collaborative manner. Though popular with architects, AIA provisions are considered less favorable to engineers, contractors, and subcontractors who believe the AIA documents assign too much authority to architects, too much risk and liability to contractors, and too little flexibility in how construction disputes are addressed and resolved.

[19]The Construction Specifications Institute (CSI) has produced standard specifications for materials. The standards have been organized according to a UNIFORMAT structure consistent with ASTM Standard E1557.

## 15. DISCHARGE OF A CONTRACT

A contract is normally discharged when all parties have satisfied their obligations. However, a contract can also be terminated for the following reasons:

- mutual agreement of all parties to the contract

- impossibility of performance (e.g., death of a party to the contract)

- illegality of the contract

- material breach by one or more parties to the contract

- fraud on the part of one or more parties

- failure (i.e., loss or destruction) of consideration (e.g., the burning of a building one party expected to own or occupy upon satisfaction of the obligations)

Some contracts may be dissolved by actions of the court (e.g., bankruptcy), passage of new laws and public acts, or a declaration of war.

*Extreme difficulty* (including economic hardship) in satisfying the contract does not discharge it, even if it becomes more costly or less profitable than originally anticipated.

## 16. TORTS

A *tort* is a civil wrong committed by one person causing damage to another person or person's property, emotional well-being, or reputation.[20] It is a breach of the rights of an individual to be secure in person or property. In order to correct the wrong, a civil lawsuit (*tort action* or *civil complaint*) is brought by the alleged injured party (the *plaintiff*) against the *defendant*. To be a valid *tort action* (i.e., lawsuit), there must have been injury (i.e., damage). Generally, there will be no contract between the two parties, so the tort action cannot claim a breach of contract.[21]

Tort law is concerned with compensation for the injury, not punishment. Therefore, tort awards usually consist of general, compensatory, and special damages and rarely include punitive and exemplary damages. (See Sec. 80.20 for definitions of these damages.)

## 17. BREACH OF CONTRACT, NEGLIGENCE, MISREPRESENTATION, AND FRAUD

A *breach of contract* occurs when one of the parties fails to satisfy all of its obligations under a contract. The breach can be *willful* (as in a contractor walking off a construction job) or *unintentional* (as in providing less than adequate quality work or materials). A *material breach* is defined as nonperformance that results in the injured party receiving something substantially less than or different from what the contract intended.

Normally, the only redress that an *injured party* has through the courts in the event of a breach of contract is to force the breaching party to provide *specific performance*—that is, to satisfy all remaining contract provisions and to pay for any damage caused. Normally, *punitive damages* (to punish the breaching party) are unavailable.

*Negligence* is an action, willful or unwillful, taken without proper care or consideration for safety, resulting in damages to property or injury to persons. "Proper care" is a subjective term, but in general it is the diligence that would be exercised by a reasonably prudent person.[22] Damages sustained by a negligent act are recoverable in a tort action. (See Sec. 80.16.) If the plaintiff is partially at fault (as in the case of *comparative negligence*), the defendant will be liable only for the portion of the damage caused by the defendant.

Punitive damages are available, however, if the breaching party was fraudulent in obtaining the contract. In addition, the injured party has the right to void (nullify) the contract entirely. A *fraudulent act* is basically a special case of misrepresentation (i.e., an intentionally false statement known to be false at the time it is made). Misrepresentation that does not result in a contract is a tort. When a contract is involved, misrepresentation can be a breach of that contract (i.e., *fraud*).

Unfortunately, it is extremely difficult to prove *compensatory fraud* (i.e., fraud for which damages are available). Proving fraud requires showing *beyond a reasonable doubt* (a) a reckless or intentional misstatement of a material fact, (b) an intention to deceive, (c) it resulted in misleading the innocent party to contract, and (d) it was to the innocent party's detriment.

For example, if an engineer claims to have experience in designing steel buildings but actually has none, the court might consider the misrepresentation a fraudulent action. If, however, the engineer has some experience, but an insufficient amount to do an adequate job, the engineer probably will not be considered to have acted fraudulently.

---

[20]The difference between a *civil tort (lawsuit)* and a *criminal lawsuit* is the alleged injured party. A *crime* is a wrong against society. A criminal lawsuit is brought by the state against a defendant.

[21]It is possible for an injury to be both a breach of contract and a tort. Suppose an owner has an agreement with a contractor to construct a building, and the contract requires the contractor to comply with all state and federal safety regulations. If the owner is subsequently injured on a stairway because there was no guardrail, the injury could be recoverable both as a tort and as a breach of contract. If a third party unrelated to the contract was injured, however, that party could recover only through a tort action.

---

[22]Negligence of a design professional (e.g., an engineer or architect) is the absence of a *standard of care* (i.e., customary and normal care and attention) that would have been provided by other engineers. It is highly subjective.

## 18. STRICT LIABILITY IN TORT

*Strict liability in tort* means that the injured party wins if the injury can be proven. It is not necessary to prove negligence, breach of explicit or implicit warranty, or the existence of a contract (*privity of contract*). Strict liability in tort is most commonly encountered in product liability cases. A defect in a product, regardless of how the defect got there, is sufficient to create strict liability in tort.

Case law surrounding defective products has developed and refined the following requirements for winning a strict liability in tort case. The following points must be proved:

- The product was defective in manufacture, design, labeling, and so on.

- The product was defective when used.

- The defect rendered the product unreasonably dangerous.

- The defect caused the injury.

- The specific use of the product that caused the damage was reasonably foreseeable.

## 19. MANUFACTURING AND DESIGN LIABILITY

Case law makes a distinction between *design professionals* (architects, structural engineers, building designers, etc.) and manufacturers of consumer products. Design professionals are generally consultants whose primary product is a design service sold to sophisticated clients. Consumer product manufacturers produce specific product lines sold through wholesalers and retailers to the unsophisticated public.

The law treats design professionals favorably. Such professionals are expected to meet a *standard of care* and skill that can be measured by comparison with the conduct of other professionals. However, professionals are not expected to be infallible. In the absence of a contract provision to the contrary, design professionals are not held to be guarantors of their work in the strict sense of legal liability. Damages incurred due to design errors are recoverable through tort actions, but proving a breach of contract requires showing negligence (i.e., not meeting the standard of care).

On the other hand, the law is much stricter with consumer product manufacturers, and perfection is (essentially) expected of them. They are held to the standard of strict liability in tort without regard to negligence. A manufacturer is held liable for all phases of the design and manufacturing of a product being marketed to the public.[23]

Prior to 1916, the court's position toward product defects was exemplified by the expression *caveat emptor* ("let the buyer beware").[24] Subsequent court rulings have clarified that "... a manufacturer is strictly liable in tort when an article [it] places on the market, knowing that it will be used without inspection, proves to have a defect that causes injury to a human being."[25]

Although all defectively designed products can be traced back to a design engineer or team, only the manufacturing company is usually held liable for injury caused by the product. This is more a matter of economics than justice. The company has liability insurance; the product design engineer (who is merely an employee of the company) probably does not. Unless the product design or manufacturing process is intentionally defective, or unless the defect is known in advance and covered up, the product design engineer will rarely be punished by the courts.[26]

## 20. DAMAGES

An injured party can sue for *damages* as well as for specific performance. Damages are the award made by the court for losses incurred by the injured party.

- *General* or *compensatory damages* are awarded to make up for the injury that was sustained.

- *Special damages* are awarded for the direct financial loss due to the breach of contract.

- *Nominal damages* are awarded when responsibility has been established but the injury is so slight as to be inconsequential.

- *Liquidated damages* are amounts that are specified in the contract document itself for nonperformance.

- *Punitive* or *exemplary damages* are awarded, usually in tort and fraud cases, to punish and make an example of the defendant (i.e., to deter others from doing the same thing).

- *Consequential damages* provide compensation for indirect losses incurred by the injured party but not directly related to the contract.

---

[23]The reason for this is that the public is not considered to be as sophisticated as a client who contracts with a design professional for building plans.

[24]1916, *McPherson vs. Buick*. McPherson bought a Buick from a car dealer. The car had a defective wooden steering wheel, and there was evidence that reasonable inspection would have uncovered the defect. The steering wheel injured McPherson, who then sued Buick. Buick defended itself under the ancient *prerequisite of privity* (i.e., the requirement of a face-to-face contractual relationship in order for liability to exist), since the dealer, not Buick, had sold the car to McPherson, and no contract between Buick and McPherson existed. The judge disagreed, thus establishing the concept of *third party liability* (i.e., manufacturers are responsible to consumers even though consumers do not buy directly from manufacturers).

[25]1963, *Greenman vs. Yuba Power Products*. Greenman purchased and was injured by an electric power tool.

[26]The engineer can expect to be discharged from the company. However, for strategic reasons, this discharge probably will not occur until after the company loses the case.

## 21. INSURANCE

Most design firms and many independent design professionals carry *errors and omissions insurance* to protect them from claims due to their mistakes. Such policies are costly, and for that reason, some professionals choose to "go bare."[27] Policies protect against inadvertent mistakes only, not against willful, knowing, or conscious efforts to defraud or deceive.

---

[27]Going bare appears foolish at first glance, but there is a perverted logic behind the strategy. One-person consulting firms (and perhaps, firms that are not profitable) are "judgment-proof." Without insurance or other assets, these firms would be unable to pay any large judgments against them. When damage victims (and their lawyers) find this out in advance, they know that judgments will be uncollectable. So, often the lawsuit never makes its way to trial.

# 81 Engineering Ethics

## 1. CREEDS, CODES, CANONS, STATUTES, AND RULES

It is generally conceded that an individual acting on his or her own cannot be counted on to always act in a proper and moral manner. Creeds, statutes, rules, and codes all attempt to complete the guidance needed for an engineer to do "...the correct thing."

A *creed* is a statement or oath, often religious in nature, taken or assented to by an individual in ceremonies. For example, the *Engineers' Creed* adopted by the National Society of Professional Engineers in 1954 is[1]

> As a professional engineer, I dedicate my professional knowledge and skill to the advancement and betterment of human welfare.
>
> I pledge...
>
> ... to give the utmost of performance;
>
> ... to participate in none but honest enterprise;
>
> ... to live and work according to the laws of man and the highest standards of professional conduct;
>
> ... to place service before profit, the honor and standing of the profession before personal advantage, and the public welfare above all other considerations.
>
> In humility and with need for Divine Guidance, I make this pledge.

A *code* is a system of nonstatutory, nonmandatory canons of personal conduct. A *canon* is a fundamental belief that usually encompasses several rules. For example, the code of ethics of the American Society of Civil Engineers (ASCE) contains seven canons.

[1]The *Faith of an Engineer* adopted by the Accreditation Board for Engineering and Technology (ABET), formerly the Engineer's Council for Professional Development (ECPD), is a similar but more detailed creed.

1. Engineers shall hold paramount the safety, health, and welfare of the public and shall strive to comply with the principles of sustainable development in the performance of their professional duties.

2. Engineers shall perform services only in areas of their competence.

3. Engineers shall issue public statements only in an objective and truthful manner.

4. Engineers shall act in professional matters for each employer or client as faithful agents or trustees and shall avoid conflicts of interest.

5. Engineers shall build their professional reputation on the merit of their service and shall not compete unfairly with others.

6. Engineers shall act in such a manner as to uphold and enhance the honor, integrity, and dignity of the engineering profession, and shall act with zero tolerance for bribery, fraud, and corruption.

7. Engineers shall continue their professional development throughout their careers and shall provide opportunities for the professional development of those engineers under their supervision.

A *rule* is a guide (principle, standard, or norm) for conduct and action in a certain situation. A *statutory rule* is enacted by the legislative branch of a state or federal government and carries the weight of law. Some U.S. engineering registration boards have statutory *rules of professional conduct*.

## 2. PURPOSE OF A CODE OF ETHICS

Many different sets of *codes of ethics* (*canons of ethics, rules of professional conduct*, etc.) have been produced by various engineering societies, registration boards, and other organizations.[2] The purpose of these ethical guidelines is to guide the conduct and decision making of engineers. Most codes are primarily educational. Nevertheless, from time to time they have been used

[2]All of the major engineering technical and professional societies in the United States (ASCE, IEEE, ASME, AIChE, NSPE, etc.) and throughout the world have adopted codes of ethics. Most U.S. societies have endorsed the *Code of Ethics of Engineers* developed by the Accreditation Board for Engineering and Technology (ABET). The National Council of Examiners for Engineering and Surveying (NCEES) has developed its *Model Rules of Professional Conduct* as a guide for state registration boards in developing guidelines for the professional engineers in those states.

by the societies and regulatory agencies as the basis for disciplinary actions.

Fundamental to ethical codes is the requirement that engineers render faithful, honest, professional service. In providing such service, engineers must represent the interests of their employers or clients and, at the same time, protect public health, safety, and welfare.

There is an important distinction between what is legal and what is ethical. Many legal actions can be violations of codes of ethical or professional behavior.[3] For example, an engineer's contract with a client may give the engineer the right to assign the engineer's responsibilities, but doing so without informing the client would be unethical.

Ethical guidelines can be categorized on the basis of who is affected by the engineer's actions—the client, vendors and suppliers, other engineers, or the public at large.[4]

## 3. ETHICAL PRIORITIES

There are frequently conflicting demands on engineers. While it is impossible to use a single decision-making process to solve every ethical dilemma, it is clear that ethical considerations will force engineers to subjugate their own self-interests. Specifically, the ethics of engineers dealing with others need to be considered in the following order from highest to lowest priority.

- society and the public
- the law
- the engineering profession
- the engineer's client
- the engineer's firm
- other involved engineers
- the engineer personally

## 4. DEALING WITH CLIENTS AND EMPLOYERS

The most common ethical guidelines affecting engineers' interactions with their employer (the *client*) can be summarized as follows.[5]

---

[3]Whether the guidelines emphasize ethical behavior or professional conduct is a matter of wording. The intention is the same: to provide guidelines that transcend the requirements of the law.

[4]Some authorities also include ethical guidelines for dealing with the employees of an engineer. However, these guidelines are no different for an engineering employer than they are for a supermarket, automobile assembly line, or airline employer. Ethics is not a unique issue when it comes to employees.

[5]These general guidelines contain references to contractors, plans, specifications, and contract documents. This language is common, though not unique, to the situation of an engineer supplying design services to an owner-developer or architect. However, most of the ethical guidelines core general enough to apply to engineers in industry as well.

- Engineers should not accept assignments for which they do not have the skill, knowledge, or time.

- Engineers must recognize their own limitations. They should use associates and other experts when the design requirements exceed their abilities.

- The client's interests must be protected. The extent of this protection exceeds normal business relationships and transcends the legal requirements of the engineer-client contract.

- Engineers must not be bound by what the client wants in instances where such desires would be unsuccessful, dishonest, unethical, unhealthy, or unsafe.

- Confidential client information remains the property of the client and must be kept confidential.

- Engineers must avoid conflicts of interest and should inform the client of any business connections or interests that might influence their judgment. Engineers should also avoid the *appearance* of a conflict of interest when such an appearance would be detrimental to the profession, their client, or themselves.

- The engineers' sole source of income for a particular project should be the fee paid by their client. Engineers should not accept compensation in any form from more than one party for the same services.

- If the client rejects the engineer's recommendations, the engineer should fully explain the consequences to the client.

- Engineers must freely and openly admit to the client any errors made.

All courts of law have required an engineer to perform in a manner consistent with normal professional standards. This is not the same as saying an engineer's work must be error-free. If an engineer completes a design, has the design and calculations checked by another competent engineer, and an error is subsequently shown to have been made, the engineer may be held responsible, but the engineer will probably not be considered negligent.

## 5. DEALING WITH SUPPLIERS

Engineers routinely deal with manufacturers, contractors, and vendors (*suppliers*). In this regard, engineers have great responsibility and influence. Such a relationship requires that engineers deal justly with both clients and suppliers.

An engineer will often have an interest in maintaining good relationships with suppliers since this often leads to future work. Nevertheless, relationships with suppliers must remain highly ethical. Suppliers should not be encouraged to feel that they have any special favors coming to them because of a long-standing relationship with the engineer.

The ethical responsibilities relating to suppliers are listed as follows.

- The engineer must not accept or solicit gifts or other valuable considerations from a supplier during, prior to, or after any job. An engineer should not accept discounts, allowances, commissions, or any other indirect compensation from suppliers, contractors, or other engineers in connection with any work or recommendations.

- The engineer must enforce the plans and specifications (i.e., the *contract documents*) but must also interpret the contract documents fairly.

- Plans and specifications developed by the engineer on behalf of the client must be complete, definite, and specific.

- Suppliers should not be required to spend time or furnish materials that are not called for in the plans and contract documents.

- The engineer should not unduly delay the performance of suppliers.

## 6. DEALING WITH OTHER ENGINEERS

Engineers should try to protect the engineering profession as a whole, to strengthen it, and to enhance its public stature. The following ethical guidelines apply.

- An engineer should not attempt to maliciously injure the professional reputation, business practice, or employment position of another engineer. However, if there is proof that another engineer has acted unethically or illegally, the engineer should advise the proper authority.

- An engineer should not review someone else's work while the other engineer is still employed unless the other engineer is made aware of the review.

- An engineer should not try to replace another engineer once the other engineer has received employment.

- An engineer should not use the advantages of a salaried position to compete unfairly (i.e., moonlight) with other engineers who have to charge more for the same consulting services.

- Subject to legal and proprietary restraints, an engineer should freely report, publish, and distribute information that would be useful to other engineers.

## 7. DEALING WITH (AND AFFECTING) THE PUBLIC

In regard to the social consequences of engineering, the relationship between an engineer and the public is essentially straightforward. Responsibilities to the public demand that the engineer place service to humankind above personal gain. Furthermore, proper ethical behavior requires that an engineer avoid association with projects that are contrary to public health and welfare or that are of questionable legal character.

- Engineers must consider the safety, health, and welfare of the public in all work performed.

- Engineers must uphold the honor and dignity of their profession by refraining from self-laudatory advertising, by explaining (when required) their work to the public, and by expressing opinions only in areas of knowledge.

- When engineers issue a public statement, they must clearly indicate if the statement is being made on anyone's behalf (i.e., if anyone is benefitting from their position).

- Engineers must keep their skills at a state-of-the-art level.

- Engineers should develop public knowledge and appreciation of the engineering profession and its achievements.

- Engineers must notify the proper authorities when decisions adversely affecting public safety and welfare are made.[6]

## 8. COMPETITIVE BIDDING

The ethical guidelines for dealing with other engineers presented here and in more detailed codes of ethics no longer include a prohibition on *competitive bidding*. Until 1971, most codes of ethics for engineers considered competitive bidding detrimental to public welfare, since cost cutting normally results in a lower quality design.

However, in a 1971 case against the National Society of Professional Engineers that went all the way to the U.S. Supreme Court, the prohibition against competitive bidding was determined to be a violation of the Sherman Antitrust Act (i.e., it was an unreasonable restraint of trade).

The opinion of the Supreme Court does not *require* competitive bidding—it merely forbids a prohibition against competitive bidding in NSPE's code of ethics. The following points must be considered.

- Engineers and design firms may individually continue to refuse to bid competitively on engineering services.

- Clients are not required to seek competitive bids for design services.

- Federal, state, and local statutes governing the procedures for procuring engineering design services, even those statutes that prohibit competitive bidding, are not affected.

---

[6]This practice has come to be known as *whistle-blowing*.

- Any prohibitions against competitive bidding in individual state engineering registration laws remain unaffected.

- Engineers and their societies may actively and aggressively lobby for legislation that would prohibit competitive bidding for design services by public agencies.

# Topic XVIII: Engineering Licensure

Engineering
Licensure

# 82 The FE Exam

## 1. ENGINEERING REGISTRATION

*Engineering registration* (also known as *engineering licensing*) in the United States is an examination process by which a state's board of engineering licensing (i.e., registration board) determines and certifies that you have achieved a minimum level of competence.[1] This process protects the public by preventing unqualified individuals from offering engineering services.

Most engineers do not need to be registered.[2] In particular, most engineers who work for companies that design and manufacture products are exempt from the licensing requirement. This is known as the *industrial exemption*.[3] Nevertheless, there are many good reasons for registering. For example, you cannot offer consulting engineering design services in any state unless you are registered in that state. Even within a product-oriented corporation, you may find that employment, advancement, or managerial positions are limited to registered engineers.

Once you have met the registration requirements, you will be allowed to use the titles Professional Engineer (PE), Registered Engineer (RE), and Consulting Engineer (CE).

Although the registration process is similar in all 50 states, each state has its own registration law. Unless you offer consulting engineering services in more than one state, however, you will not need to register in other states.

## The U.S. Registration Procedure

The registration procedure is similar in most states. You will take two eight-hour written examinations. The first is the *Fundamentals of Engineering Examination*, also known as the *Engineer-In-Training Examination* and the *Intern Engineer Exam*. The initials FE, EIT, and IE are also used.[4] This examination covers basic subjects from all of the mathematics, physics, chemistry, and engineering classes you took during your first four university years.

In rare cases, you may be allowed to skip this first examination. However, the actual details of registration qualifications, experience requirements, minimum education levels, fees, and examination schedules vary from state to state. Contact your state's registration board for additional information. At **www.ppi2pass.com/stateboards**, you'll find contact information (websites, telephone numbers, email addresses, etc.) for all U.S. state and territorial boards of registration.

The second eight-hour examination is the *Principles and Practice of Engineering Exam*. The initials PE are also used. This examination covers subjects only from your areas of specialty.

## National Council of Examiners for Engineering and Surveying

The National Council of Examiners for Engineering and Surveying (NCEES) in Clemson, South Carolina, produces, distributes, and scores the national FE and PE examinations.[5] The individual states purchase the examinations from NCEES and administer them themselves in a uniform, controlled environment dictated by NCEES.

## Reciprocity Among States

All states use the NCEES examinations. If you take and pass the FE or PE examination in one state, your certificate will be honored by all of the other states. Although there may be other special requirements imposed by a state, it will not be necessary to retake

---

[1] Licensing of engineers is not unique to the United States. However, the practice of requiring a degreed engineer to take an examination is not common in other countries. Licensing in many countries requires a degree and may also require experience, references, and demonstrated knowledge of ethics and law, but no technical examination.

[2] Less than one-third of the degreed engineers in the United States are licensed.

[3] Only one or two states have abolished the industrial exemption. There has always been a lot of "talk" among engineers about abolishing it, but there has been little success in actually trying to do so. One of the reasons is that manufacturers' lobbies are very strong.

---

[4] The terms *engineering intern* (EI) and *intern engineer* (IE) have also been used in the past to designate the status of an engineer who has passed the first exam. These uses are rarer but may still be encountered in some states.

[5] National Council of Examiners for Engineering and Surveying, P.O. Box 1686, Clemson, SC 29633, (803) 654-6824.

the FE and PE examinations.[6] The issuance of an engineering license based on another state's license is known as *reciprocity* or *comity*.

With minor exceptions, having a license from one state will not permit you to practice engineering in another state. You must have a professional engineering license from each state in which you work. For most engineers, this is not a problem, but for some, it is. Luckily, it is not too difficult to get a license from every state you work in once you have a license from one state.

The simultaneous administration of identical examinations in all states has led to the term *uniform examination*. However, each state is still free to choose its own minimum passing score and to add special questions and requirements to the examination process. Therefore, the use of a uniform examination has not, by itself, ensured reciprocity among states.

## 2. THE FE EXAMINATION

### Applying for the Examination

While NCEES has standardized the FE exam and exam environment in all states, the application process may vary from state to state. For many years, each state had its own application form, which was obtained directly from the state's licensing board or from a designated agent, such as the examinee's university if the examinee was still a student. Each state was (and still is) able to establish the exam and administration fees for its examinees and set requirements of age, experience, and education that differ from those of other states. Many did and continue to do just that. Traditionally, when taking the FE exam, all of the examinee's contact and communications throughout the application, acceptance, and notification stages would be directly with the state licensing board.

At least 39 states, particularly those with the fewest number of annual applicants, now choose to allow NCEES to administer a partial (and in some cases unilateral) online application process. Thus, the examinee's state will determine his or her eligibility to take the exam, while NCEES will coordinate the examinee's presence. A list of participating states is included on the NCEES website, www.ncees.org.

For some of these participating states, the examinee must complete an online pre-approval application on the NCEES website and pay the appropriate fee to NCEES. Upon notification that his or her education and other qualifications meet minimum requirements, in order to reserve a seat in the examination room, the examinee is required to establish an online account with *NCEES Exam Administration Services* (formerly known

as ELSES), submit the required information, pay a separate examination fee to NCEES, and wait for NCEES to send an email or postal Exam Authorization for Admittance notice. This email or notice is generally sent 3–4 weeks prior to the examination date.

In other states participating in the NCEES Exam Administration Services program, the process for taking the FE exam consists of applying directly to the examinee's state in the traditional fashion—either online at the state's website or with a paper form, paying the appropriate application fee directly to the state, and waiting for the state to confirm receipt of a complete application. At that point, the examinee may establish an online account with NCEES Exam Administration Services, as previously described.

The Exam Authorization for Admittance notice will include information about the date, time, and location of the exam, as well as information about what may be brought into the exam room, how examinees are expected to behave, and what they can expect to experience.

Approximately 10–12 weeks after the exam, NCEES will release the results of the examinee's examination to the state, which will notify him or her directly. This notification may include information about additional state-specific steps in the process.

Keep a copy of your examination application, and send the original application by certified mail, requesting a receipt of delivery. Keep your proof of mailing and delivery with your copy of the application.

All states make special arrangements for persons who are physically challenged or who have religious or other special requirements. Be sure to communicate your situation to your state or to NCEES well in advance of your examination date.

### Examination Dates

The national FE and PE examinations are administered twice a year (usually in mid-April and late October), on the same weekends in all states. Check for a current exam schedule at **www.ppi2pass.com/fefaq**.

### FE Examination Format

The NCEES Fundamentals of Engineering examination has the following format and characteristics.

- There are two four-hour sessions separated by a one-hour lunch.

- Examination questions are distributed in a bound examination booklet. A different examination booklet is used for each of these two sessions.

- The morning session (also known as the *A.M. session*) has 120 multiple-choice questions, each with four possible answers lettered (A) to (D). Each problem in the morning session is worth one point. The total score possible in the morning is 120 points.

---

[6]For example, California requires all civil engineering applicants to pass special examinations in seismic design and surveying in addition to their regular eight-hour PE exams. Licensed engineers from other states only have to pass these two special exams. They do not need to retake the PE exam.

Guessing is valid; no points are subtracted for incorrect answers.

There are questions on the morning session examination from most of the undergraduate engineering degree program subjects. Questions from the same subject are all grouped together, and the subjects are labeled. The percentages of questions on each subject in the morning session are given in Table 82.1.

**Table 82.1** *Morning FE Exam Subjects*

| subject | percentage of total questions (%) |
|---|---|
| chemistry | 9 |
| computers | 7 |
| electricity and magnetism | 9 |
| engineering economics | 8 |
| engineering probability and statistics | 7 |
| engineering mechanics (statics and dynamics) | 10 |
| ethics and business practices | 7 |
| fluid mechanics | 7 |
| materials properties | 7 |
| mathematics | 15 |
| strength of materials | 7 |
| thermodynamics | 7 |

- There are seven different versions of the afternoon session (also known as the *P.M. session*). Six of the versions correspond to a specific engineering discipline (chemical, civil, electrical, environmental, industrial, or mechanical engineering). The seventh version of the afternoon examination is a general examination suitable for anyone, but in particular, for engineers whose specialties are not one of the other six disciplines. NCEES calls the first six *discipline-specific exams* and the seventh version the "Other Disciplines" exam.

Each version of the afternoon session consists of 60 questions worth two points each, for a total of 120 points. All questions are mandatory. Each question consists of a problem statement followed by multiple-choice questions. Four answer choices lettered (A) through (D) are given, from which you must choose the best answer. Questions in each subject may be grouped into related problem sets containing between two and ten questions each.

Questions on the afternoon examination are intended to cover concepts learned in the last two years of a four-year degree program. Unlike morning questions, these questions may deal with more than one basic concept per question.

The percentages of questions on each subject in the "other disciplines" afternoon session examination are given in Table 82.2.

**Table 82.2** *Afternoon FE Exam Subjects ("Other Disciplines" Exam)*

| subject | percentage of total questions (%) |
|---|---|
| advanced engineering mathematics | 10 |
| engineering probability and statistics | 9 |
| biology | 5 |
| engineering economics | 10 |
| application of engineering mechanics | 13 |
| engineering materials | 11 |
| fluids | 15 |
| electricity and magnetism | 12 |
| thermodynamics and heat transfer | 15 |

The percentages of questions on each subject in the discipline-specific afternoon session examination are listed in Table 82.3. The discipline-specific afternoon examinations cover substantially different bodies of knowledge than the morning examination. Formulas and tables of data needed to solve questions in these examinations will be included in either the NCEFS Handbook or in the body of the question statement itself.

- The scores from the morning and afternoon sessions are added together to determine your total score. No points are subtracted for guessing or incorrect answers. Both sessions are given equal weight. It is not necessary to achieve any minimum score on either the morning or afternoon sessions.

- All grading is done by computer optical sensing. Responses must be recorded on special answer sheets with the mechanical pencils provided to examinees by NCEES.

## Use of SI Units on the FE Exam

Metric, or SI, units are used in virtually all subjects, except some civil engineering and surveying subjects that typically use only customary U.S. (i.e., English) units. Some of the remaining questions may be presented on the exam in both metric and English units. These questions are actually stated twice, once in metric units and once in English units. Dual dimensioning is not used. However, these dual-statement questions are rapidly being phased out.

It is the goal of NCEES to use SI units that are consistent with ANSI/IEEE standard 268-1992 (the American Standard for Metric Practice). Non-SI metric units might still be used when commonly used as such or where needed for consistency with tabulated data (e.g., use of bars in pressure measurement).

## Grading and Scoring the FE Exam

The FE exam is not graded on a curve, and there is no guarantee that a certain percent of examinees will pass. Rather, NCEES uses a modification of the *Angoff procedure* to determine the suggested passing score (the cutoff point or *cut score*).

**Table 82.3** *Afternoon FE Exam Subjects (Discipline-Specific Exams)*

**Chemical Engineering**

| subject | percentage of total questions (%) |
|---|---|
| chemistry | 10 |
| material/energy balances | 15 |
| chemical engineering thermodynamics | 10 |
| fluid dynamics | 10 |
| heat transfer | 10 |
| mass transfer | 10 |
| chemical reaction engineering | 10 |
| process design and economic optimization | 10 |
| computer usage in chemical engineering | 5 |
| process control | 5 |
| safety, health, and environmental | 5 |

**Civil Engineering**

| subject | percentage of total questions (%) |
|---|---|
| surveying | 11 |
| hydraulics and hydrologic systems | 12 |
| soil mechanics and foundations | 15 |
| environmental engineering | 12 |
| transportation | 12 |
| structural analysis | 10 |
| structural design | 10 |
| construction management | 10 |
| materials | 8 |

**Electrical Engineering**

| subject | percentage of total questions (%) |
|---|---|
| circuits | 16 |
| power | 13 |
| electromagnetics | 7 |
| control systems | 10 |
| communications | 9 |
| signal processing | 8 |
| electronics | 15 |
| digital systems | 12 |
| computer systems | 10 |

**Environmental Engineering**

| subject | percentage of total questions (%) |
|---|---|
| water resources | 25 |
| water and wastewater engineering | 30 |
| air quality engineering | 15 |
| solid and hazardous waste engineering | 15 |
| environmental science and management | 15 |

**Industrial Engineering**

| subject | percentage of total questions (%) |
|---|---|
| engineering economics | 15 |
| probability and statistics | 15 |
| modeling and computation | 12 |
| industrial management | 10 |
| manufacturing and production systems | 13 |
| facilities and logistics | 12 |
| human factors, productivity, ergonomics and work design | 12 |
| quality | 11 |

**Mechanical Engineering**

| subject | percentage of total questions (%) |
|---|---|
| mechanical design and analysis | 15 |
| kinematics, dynamics, and vibrations | 15 |
| materials and processing | 10 |
| measurements, instrumentation, and controls | 10 |
| thermodynamics and energy conservation processes | 15 |
| fluid mechanics and fluid machinery | 15 |
| heat transfer | 10 |
| refrigeration and HVAC | 10 |

With this method, a group of engineering professors and other experts estimate the fraction of minimally qualified engineers that will be able to answer each question correctly. The summation of the estimated fractions for all test questions becomes the passing score. The passing score in recent years has been somewhat less than 50% (i.e., a raw score of approximately 110 points out of 240). Because the law in most states requires engineers to achieve a score of 70% to become licensed, you may be reported as having achieved a score of 70% if your raw score is greater than the passing score established by NCEES, regardless of the raw percentage. The actual score may be slightly more or slightly less than 110 as determined from the performance of all examinees on the equating subtest.

Approximately 20% of each FE exam consists of questions repeated from previous examinations—this is the *equating subtest*. Since the performance of previous examinees on the equating subtest is known, comparisons can be made between the two examinations and examinee populations. These comparisons are used to adjust the passing score.

The individual states are free to adopt their own passing scores, but all adopt NCEES's suggested passing score because the states believe this cutoff score can be defended if challenged.

You will generally receive the results of your examination from your state board (not NCEES) by mail. Allow at least three months for notification. Candidates will receive a pass or fail notice only, and not a numerical score. A diagnostic report is provided to those who fail.

See **www.ppi2pass.com/fepassrates** for recent and historic FE pass rates.

## Permitted Reference Material

Since October 1993, the FE examination has been what NCEES calls a "limited-reference" exam. This means that no books or references other than those supplied by NCEES may be used. Therefore, the FE examination is really an "NCEES-publication only" exam. NCEES provides its own handbook for use during the examination. No books from other publishers may be used.

## What Does *Most Nearly* Really Mean?

One of the more disquieting aspects of these questions is that the available answer choices are seldom exact. Answer choices generally have only two or three significant digits. Exam questions instruct you to complete the sentence, "The value is most nearly ...", or ask "Which answer choice is closest to the correct value?" A lot of self-confidence is required to move on to the next question if you don't find an exact match for the answer you calculated, or if you have to split the difference because no available answer choice is close.

As NCEES describes it: "Many of the questions on NCEES exams require calculations to arrive at a numerical answer. Depending on the method of calculation used, it is very possible that examinees working correctly will arrive at a range of answers. The phrase 'most nearly' is used to accommodate all these answers that have been derived correctly but which may be slightly different from the correct answer choice given on the exam. You should use good engineering judgment when selecting your choice of answer. For example, if the question asks you to calculate an electrical current or determine the load on a beam, you should literally select the answer option that is most nearly what you calculated, regardless of whether it is more or less than your calculated value. However, if the question asks you to select a fuse or circuit breaker to protect against a calculated current or to size a beam to carry a load, you should select an answer option that will safely carry the current or load. Typically, this requires selecting a value that is closest to but larger than the current or load."

The difference is significant. Suppose you were asked to calculate "most nearly" the volumetric pure water flow required to dilute a contaminated stream to an acceptable concentration. Suppose, also, that you calculated 823 gpm. If the answer choices were (A) 600 gpm, (B) 800 gpm, (C) 1000 gpm, and (D) 1200 gpm, you would go with answer choice (B), because it is most nearly what you calculated. If, however, you were asked to select a pump or pipe with the same rated capacities, you would have to go with choice (C). Got it? If not, stop reading until you understand the distinction.

## 3. CALCULATORS

To prevent unauthorized transcription and distribution of the examination questions, calculators with communicating and text editing capabilities have been banned by NCEES. Calculators permitted by the NCEES are listed at **www.ppi2pass.com/calculators**. You cannot share calculators with other examinees.

It is essential that a calculator used for engineering examinations have the following functions.

- trigonometric functions
- inverse trigonometric functions
- hyperbolic functions
- pi
- square root and $x^2$
- common and natural logarithms
- $y^x$ and $e^x$

For maximum speed, your calculator should also have or be programmed for the following functions.

- extracting roots of quadratic and higher-order equations
- converting between polar (phasor) and rectangular vectors
- finding standard deviations and variances
- calculating determinants of $3 \times 3$ matrices
- linear regression
- economic analysis and other financial functions

## 4. STRATEGIES FOR PASSING THE FE EXAM

The most successful strategy to pass the FE exam is to prepare in all of the examination subjects. Do not limit the number of subjects you study, hoping to find enough questions in your particular areas of knowledge to pass.

Fast recall and stamina are essential to doing well. You must be able to quickly recall solution procedures, formulas, and important data. You will not have time during the exam to derive solutions methods—you must know them instinctively. This ability must be maintained for eight hours. The best way to develop fast recall and stamina is to work numerous practice problems. Be sure to gain familiarity with the NCEES Handbook by using it as your only reference for most of the practice problems you work.

In order to get exposure to all examination subjects, it is imperative that you develop and adhere to a review schedule. If you are not taking a classroom review course (where the order of your preparation is determined by the lectures), prepare your own review schedule. For example, plan on covering this book at the rate of one chapter every few days in order to finish before the examination date.

There are also physical demands on your body during the examination. It is very difficult to remain tense, alert, and attentive for eight hours or more. Unfortunately, the more time you study, the less time you have to maintain your physical condition. Thus, most examinees arrive at the examination site in peak mental condition but in deteriorated physical condition. While preparing for the FE exam is not the only good reason for embarking on a physical conditioning program, it can serve as a good incentive to get in shape.

It will be helpful to make a few simple decisions prior to starting your review. You should be aware of the different options available to you. For example, you should decide early on to

- use SI units in your preparation

- perform electrical calculations with effective (rms) or maximum values

- take calculations out to a maximum of four significant digits

- prepare in all examination subjects, not just your specialty areas

At the beginning of your review program, you should locate a spare calculator. It is not necessary to buy a spare if you can arrange to borrow one from a friend or the office. However, if possible, your primary and spare calculators should be identical. If your spare calculator is not identical to the primary calculator, spend a few minutes familiarizing yourself with its functions.

### A Few Days Before the Exam

There are a few things you should do a week or so before the examination date. Visit the exam site in order to find the building, parking areas, examination room, and rest rooms. You should also make arrangements for child care and transportation. Since the examination does not always start or end at the designated times, make sure your child care and transportation arrangements can tolerate a later-than-usual completion.

Second in importance to your scholastic preparation is the preparation of your two examination kits. The first kit consists of a bag or box containing items to bring with you into the examination room. NCEES provides mechanical pencils for use in the exam. It is not necessary (nor is it permitted) for you to bring your own pencils or erasers.

```
[ ]    letter admitting you to the examination
[ ]    photographic identification
[ ]    main calculator
[ ]    spare calculator
[ ]    extra calculator batteries
[ ]    unobtrusive snacks
[ ]    travel pack of tissues
[ ]    headache remedy
[ ]    $2.00 in change
[ ]    light, comfortable sweater
[ ]    loose shoes or slippers
[ ]    cushion for your chair
[ ]    small hand towel
[ ]    earplugs
```

```
[ ]    wristwatch
[ ]    wire coat hanger
[ ]    extra set of car keys
```

The second kit consists of the following items and should be left in a separate bag or box in your car in case they are needed.

```
[ ]    copy of your application
[ ]    proof of delivery
[ ]    this book
[ ]    other references
[ ]    regular dictionary
[ ]    scientific dictionary
[ ]    course notes in three-ring binders
[ ]    cardboard box (use as a bookcase)
[ ]    instruction booklets for all your calculators
[ ]    light lunch
[ ]    beverages in thermos and cans
[ ]    sunglasses
[ ]    extra pair of prescription glasses
[ ]    raincoat, boots, gloves, hat, and umbrella
[ ]    street map of the examination site
[ ]    note to the parking patrol for your windshield
[ ]    battery powered desk lamp
```

### The Day Before the Exam

Take off from work the day before the examination to relax. Do not cram the last night. A good prior night's sleep is the best way to start the examination. If you live far from the examination site, consider getting a hotel room in which to spend the night.

Make sure your exam kits are packed and ready to go.

### The Day of the Exam

You should arrive at least 30 minutes before the examination starts. This will allow time for finding a convenient parking place, bringing your materials to the examination room, and making room and seating changes. Be prepared, though, to find that the examination room is not open or ready at the designated time.

Once the examination has started, observe the following suggestions.

- Set your wristwatch alarm for five minutes before the end of each four-hour session and use that remaining time to guess at all of the remaining unsolved problems. Do not work up until the very end. You will be successful with about 25% of your guesses, and these points will more than make up for the few points you might earn by working during the last five minutes.

- Do not spend more than two minutes per morning question. (The average time available per problem is two minutes.) If you have not finished a question in that time, make a note of it and continue on.

Engineering Licensure

- Do not spend time trying to ask your proctors technical questions. Even if they are knowledgeable in engineering, they will not be permitted to answer your questions.

- Make a quick mental note about any problems for which you cannot find a correct response or for which you believe there are two correct answers. Errors in the exam are rare, but they do occur. Being able to point out an error later might give you the margin you need to pass. Since such problems are almost always discovered during the scoring process and discounted from the examination, it is not necessary to tell your proctor, but be sure to mark the one best answer before moving on.

- Make sure all of your responses on the answer sheet are dark and completely fill the bubbles.

Engineering
Licensure

# Appendices
# Table of Contents

## APPENDIX 2.A
Conversion Factors

| multiply | by | to obtain | multiply | by | to obtain |
|---|---|---|---|---|---|
| acres | 0.40468 | hectares | feet | 30.48 | centimeters |
| | 43,560.0 | square feet | | 0.3048 | meters |
| | $1.5625 \times 10^{-3}$ | square miles | | $1.645 \times 10^{-4}$ | miles (nautical) |
| ampere-hours | 3600.0 | coulombs | | $1.894 \times 10^{-4}$ | miles (statute) |
| angstrom units | $3.937 \times 10^{-9}$ | inches | feet/min | 0.5080 | centimeters/sec |
| | $1 \times 10^{-4}$ | microns | feet/sec | 0.592 | knots |
| astronomical units | $1.496 \times 10^{8}$ | kilometers | | 0.6818 | miles/hr |
| atmospheres | 76.0 | centimeters of | foot-pounds | $1.285 \times 10^{-3}$ | Btu |
| | | mercury | | $5.051 \times 10^{-7}$ | horsepower-hours |
| atomic mass unit | $9.3149 \times 10^{8}$ | electron-volts | | $3.766 \times 10^{-7}$ | kilowatt-hours |
| | $1.4924 \times 10^{-10}$ | joules | foot-pound/sec | 4.6272 | Btu/hr |
| | $1.6605 \times 10^{-27}$ | kilograms | | $1.818 \times 10^{-3}$ | horsepower |
| BeV (also GeV) | $1 \times 10^{9}$ | electron-volts | | $1.356 \times 10^{-3}$ | kilowatts |
| Btu | $3.93 \times 10^{-4}$ | horsepower-hours | furlongs | 660.0 | feet |
| | 778.2 | foot-pounds | | 0.125 | miles (statute) |
| | $2.931 \times 10^{-4}$ | kilowatt hours | gallons | 0.1337 | cubic feet |
| | $1.0 \times 10^{-5}$ | therms | | 3.785 | liters |
| Btu/hr | 0.2161 | foot-pounds/sec | gallons $H_2O$ | 8.3453 | pounds $H_2O$ |
| | $3.929 \times 10^{-4}$ | horsepower | gallons/min | 8.0208 | cubic feet/hr |
| | 0.293 | watts | | 0.002228 | cubic feet/sec |
| bushels | 2150.4 | cubic inches | GeV (also BeV) | $1 \times 10^{9}$ | electron-volts |
| calories, gram (mean) | $3.9683 \times 10^{-3}$ | Btu (mean) | grams | $1 \times 10^{-3}$ | kilograms |
| centares | 1.0 | square meters | | $3.527 \times 10^{-2}$ | ounces (avoirdupois) |
| centimeters | $1 \times 10^{-5}$ | kilometers | | $3.215 \times 10^{-2}$ | ounces (troy) |
| | $1 \times 10^{-2}$ | meters | | $2.205 \times 10^{-3}$ | pounds |
| | 10.0 | millimeters | hectares | 2.471 | acres |
| | $3.281 \times 10^{-2}$ | feet | | $1.076 \times 10^{5}$ | square feet |
| | 0.3937 | inches | horsepower | 2545.0 | Btu/hr |
| chains | 792.0 | inches | | 42.44 | Btu/min |
| coulombs | $1.036 \times 10^{-5}$ | faradays | | 550 | foot-pounds/sec |
| cubic centimeters | 0.06102 | cubic inches | | 0.7457 | kilowatts |
| | $2.113 \times 10^{-3}$ | pints (U.S. liquid) | | 745.7 | watts |
| cubic feet | 0.02832 | cubic meters | horsepower-hours | 2545.0 | Btu |
| | 7.4805 | gallons | | $1.976 \times 10^{-6}$ | foot-pounds |
| cubic feet/min | 62.43 | pounds $H_2O$/min | | 0.7457 | kilowatt-hours |
| cubic feet/sec | 448.831 | gallons/min | hours | $4.167 \times 10^{-2}$ | days |
| | 0.64632 | millions of gallons | | $5.952 \times 10^{-3}$ | weeks |
| | | per day | inches | 2.540 | centimeters |
| cubits | 18.0 | inches | | $1.578 \times 10^{-5}$ | miles |
| days | 86,400.0 | seconds | inches, $H_2O$ | 5.199 | pounds force/ft$^2$ |
| degrees (angle) | $1.745 \times 10^{-2}$ | radians | | 0.0361 | psi |
| degrees/sec | 0.1667 | revolutions/min | | 0.0735 | inches, mercury |
| dynes | $1 \times 10^{-5}$ | newtons | inches, mercury | 70.7 | pounds force/ft$^2$ |
| electron-volts | $1.074 \times 10^{-9}$ | atomic mass units | | 0.491 | pounds force/in$^2$ |
| | $1 \times 10^{-9}$ | BeV (also GeV) | | 13.60 | inches, $H_2O$ |
| | $1.60218 \times 10^{-19}$ | joules | joules | $6.705 \times 10^{9}$ | atomic mass units |
| | $1.78266 \times 10^{-36}$ | kilograms | | $9.478 \times 10^{-4}$ | Btu |
| | $1 \times 10^{-6}$ | MeV | | $1 \times 10^{7}$ | ergs |
| faradays/sec | 96,485 | amperes (absolute) | | $6.2415 \times 10^{18}$ | electron-volts |
| fathoms | 6.0 | feet | | $1.1127 \times 10^{-17}$ | kilograms |

*(continued)*

**APPENDIX 2.A** *(continued)*
Conversion Factors

| multiply | by | to obtain | multiply | by | to obtain |
|---|---|---|---|---|---|
| kilograms | $6.0221 \times 10^{26}$ | atomic mass units | pascal-sec | 1000 | centipoise |
| | $5.6096 \times 10^{35}$ | electron-volts | | 10 | poise |
| | $8.9875 \times 10^{16}$ | joules | | 0.02089 | pound force-sec/ft$^2$ |
| | 2.205 | pounds | | 0.6720 | pound mass/ft-sec |
| kilometers | 3281.0 | feet | | 0.02089 | slug/ft-sec |
| | 1000.0 | meters | pints (liquid) | 473.2 | cubic centimeters |
| | 0.6214 | miles | | 28.87 | cubic inches |
| kilometers/hr | 0.5396 | knots | | 0.125 | gallons |
| kilowatts | 3412.9 | Btu/hr | | 0.5 | quarts (liquid) |
| | 737.6 | foot-pounds/sec | poise | 0.002089 | pound-sec/ft$^2$ |
| | 1.341 | horsepower | pounds | 0.4536 | kilograms |
| kilowatt-hours | 3413.0 | Btu | | 16.0 | ounces |
| knots | 6076.0 | feet/hr | | 14.5833 | ounces (troy) |
| | 1.0 | nautical miles/hr | | 1.21528 | pounds (troy) |
| | 1.151 | statute miles/hr | pounds/ft$^2$ | 0.006944 | pounds/in$^2$ |
| light years | $5.9 \times 10^{12}$ | miles | pounds/in$^2$ | 2.308 | feet, H$_2$O |
| links (surveyor) | 7.92 | inches | | 27.7 | inches, H$_2$O |
| liters | 1000.0 | cubic centimeters | | 2.037 | inches, mercury |
| | 61.02 | cubic inches | | 144 | pounds/ft$^2$ |
| | 0.2642 | gallons (U.S. liquid) | quarts (dry) | 67.20 | cubic inches |
| | 1000.0 | milliliters | quarts (liquid) | 57.75 | cubic inches |
| | 2.113 | pints | | 0.25 | gallons |
| MeV | $1 \times 10^6$ | electron-volts | | 0.9463 | liters |
| meters | 100.0 | centimeters | radians | 57.30 | degrees |
| | 3.281 | feet | | 3438.0 | minutes |
| | $1 \times 10^{-3}$ | kilometers | revolutions | 360.0 | degrees |
| | $5.396 \times 10^{-4}$ | miles (nautical) | revolutions/min | 6.0 | degrees/sec |
| | $6.214 \times 10^{-4}$ | miles (statute) | rods | 16.5 | feet |
| | 1000.0 | millimeters | | 5.029 | meters |
| microns | $1 \times 10^{-6}$ | meters | rods (surveyor) | 5.5 | yards |
| miles (nautical) | 6076 | feet | seconds | $1.667 \times 10^{-2}$ | minutes |
| | 1.853 | kilometers | square meters/sec | $1 \times 10^6$ | centistokes |
| | 1.1516 | miles (statute) | | 10.76 | square feet/sec |
| miles (statute) | 5280.0 | feet | | $1 \times 10^4$ | stokes |
| | 1.609 | kilometers | slugs | 32.174 | pounds mass |
| | 0.8684 | miles (nautical) | stokes | 0.0010764 | square feet/sec |
| miles/hr | 88.0 | feet/min | tons (long) | 1016.0 | kilograms |
| milligrams/liter | 1.0 | parts/million | | 2240.0 | pounds |
| milliliters | $1 \times 10^{-3}$ | liters | | 1.120 | tons (short) |
| millimeters | $3.937 \times 10^{-2}$ | inches | tons (short) | 907.1848 | kilograms |
| newtons | $1 \times 10^5$ | dynes | | 2000.0 | pounds |
| ohms (international) | 1.0005 | ohms (absolute) | | 0.89287 | tons (long) |
| ounces | 28.349527 | grams | volts (absolute) | $3.336 \times 10^{-3}$ | statvolts |
| | $6.25 \times 10^{-2}$ | pounds | watts | 3.4129 | Btu/hr |
| ounces (troy) | 1.09714 | ounces (avoirdupois) | | $1.341 \times 10^{-3}$ | horsepower |
| parsecs | $3.086 \times 10^{13}$ | kilometers | yards | 0.9144 | meters |
| | $1.9 \times 10^{13}$ | miles | | $4.934 \times 10^{-4}$ | miles (nautical) |
| | | | | $5.682 \times 10^{-4}$ | miles (statute) |

## APPENDIX 2.B
Common SI Unit Conversion Factors

| multiply | by | to obtain |
|---|---|---|
| **AREA** | | |
| circular mil | 506.7 | square micrometer |
| square feet | 0.0929 | square meter |
| square kilometer | 0.3861 | square mile |
| square meter | 10.764 | square feet |
| | 1.196 | square yard |
| square micrometer | 0.001974 | circular mil |
| square mile | 2.590 | square kilometer |
| square yard | 0.8361 | square meter |
| **ENERGY** | | |
| Btu (international) | 1.0551 | kilojoule |
| erg | 0.1 | microjoule |
| foot-pound | 1.3558 | joule |
| horsepower-hour | 2.6485 | megajoule |
| joule | 0.7376 | foot-pound |
| | 0.10197 | meter-kilogram force |
| kilogram-calorie (international) | 4.1868 | kilojoule |
| kilojoule | 0.9478 | Btu |
| | 0.2388 | kilogram-calorie |
| kilowatt-hour | 3.6 | megajoule |
| megajoule | 0.3725 | horsepower-hour |
| | 0.2778 | kilowatt-hour |
| | 0.009478 | therm |
| meter-kilogram force | 9.8067 | joule |
| microjoule | 10.0 | erg |
| therm | 105.506 | megajoule |
| **FORCE** | | |
| dyne | 10.0 | micronewton |
| kilogram force | 9.8067 | newton |
| kip | 4448.2 | newton |
| micronewton | 0.1 | dyne |
| newton | 0.10197 | kilogram force |
| | 0.0002248 | kip |
| | 3.597 | ounce force |
| | 0.2248 | pound force |
| ounce force | 0.2780 | newton |
| pound force | 4.4482 | newton |
| **HEAT** | | |
| $Btu/ft^2$-hr | 3.1546 | $watt/m^2$ |
| $Btu/ft^2$-hr-$°F$ | 5.6783 | $watt/m^2·°C$ |
| $Btu/ft^3$ | 0.0373 | $megajoule/m^3$ |
| $Btu/ft^3$-$°F$ | 0.06707 | $megajoule/m^3·°C$ |
| Btu/hr | 0.2931 | watt |
| Btu/lbm | 2326 | joule/kg |
| Btu/lbm-$°F$ | 4186.8 | joule/kg·$°C$ |
| Btu-inch/$ft^2$-hr-$°F$ | 0.1442 | watt/meter·$°C$ |
| joule/kg | 0.000430 | Btu/lbm |
| joule/kg·$°C$ | 0.0002388 | Btu/lbm-$°F$ |
| $megajoule/m^3$ | 26.839 | $Btu/ft^3$ |
| $megajoule/m^3·°C$ | 14.911 | $Btu/ft^3$-$°F$ |
| watt | 3.4129 | Btu/hr |

*(continued)*

**APPENDIX 2.B** *(continued)*
Common SI Unit Conversion Factors

| multiply | by | to obtain |
|---|---|---|
| **HEAT** *(continued)* | | |
| watt/m·°C | 6.933 | Btu-inch/ft$^2$-hr-°F |
| watt/m$^2$ | 0.3170 | Btu/ft$^2$-hr |
| watt/m$^2$·°C | 0.1761 | Btu/ft$^2$-hr |
| **LENGTH** | | |
| Angstrom | 0.1 | nanometer |
| foot | 0.3048 | meter |
| inch | 25.4 | millimeter |
| kilometer | 0.6214 | mile |
| | 0.540 | mile (international nautical) |
| meter | 3.2808 | foot |
| | 1.0936 | yard |
| micrometer | 1.0 | micron |
| micron | 1.0 | micrometer |
| mil | 0.0254 | millimeter |
| mile | 1.6093 | kilometer |
| mile (international nautical) | 1.852 | kilometer |
| millimeter | 0.0394 | inch |
| | 39.370 | mil |
| nanometer | 10.0 | Angstrom |
| yard | 0.9144 | meter |
| **MASS** | | |
| grain | 64.799 | milligram |
| gram | 0.0353 | ounce (avoirdupois) |
| | 0.03215 | ounce (troy) |
| kilogram | 2.2046 | pounds-mass |
| | 0.068522 | slug |
| | 0.0009842 | ton (long—2240 lbm) |
| | 0.001102 | ton (short—2000 lbm) |
| milligram | 0.0154 | grain |
| ounce (avoirdupois) | 28.350 | gram |
| ounce (troy) | 31.1035 | gram |
| pounds-mass | 0.4536 | kilogram |
| slug | 14.5939 | kilogram |
| ton (long—2240 lbm) | 1016.047 | kilogram |
| ton (short—2000 lbm) | 907.185 | kilogram |
| **PRESSURE** | | |
| bar | 100.0 | kilopascal |
| inch, H$_2$O (20°C) | 0.2486 | kilopascal |
| inch, Hg (20°C) | 3.3741 | kilopascal |
| kilogram force/cm$^2$ | 98.067 | kilopascal |
| kilopascal | 0.01 | bar |
| | 4.0219 | inch, H$_2$O (20°C) |
| | 0.2964 | inch, Hg (20°C) |
| | 0.0102 | kilogram force/cm$^2$ |
| | 7.528 | millimeter Hg (20°C) |
| | 0.1450 | psi |
| | 0.009869 | standard atmosphere (760 torr) |
| | 7.5006 | torr |
| millimeter Hg (20°C) | 0.13284 | kilopascal |
| psi | 6.8948 | kilopascal |

*(continued)*

**APPENDIX 2.B** *(continued)*
Common SI Unit Conversion Factors

| multiply | by | to obtain |
|---|---|---|
| **PRESSURE** *(continued)* | | |
| standard atmosphere (760 torr) | 101.325 | kilopascal |
| torr | 0.13332 | kilopascal |
| **POWER** | | |
| Btu (international)/hr | 0.2931 | watt |
| foot-pound/sec | 1.3558 | watt |
| horsepower | 0.7457 | kilowatt |
| kilowatt | 1.341 | horsepower |
| | 0.2843 | tons of refrigeration |
| meter-kilogram force/second | 9.8067 | watt |
| tons of refrigeration | 3.517 | kilowatt |
| watt | 3.4122 | Btu (international)/hr |
| | 0.7376 | foot-pound/sec |
| | 0.10197 | meter-kilogram force/second |
| **TEMPERATURE** | | |
| Celsius | $\frac{9}{5}\,^\circ C + 32^\circ$ | Fahrenheit |
| Fahrenheit | $\frac{5}{9}(^\circ F - 32^\circ)$ | Celsius |
| Kelvin | $\frac{9}{5}$ | Rankine |
| Rankine | $\frac{5}{9}$ | Kelvin |
| **TORQUE** | | |
| gram force centimeter | 0.098067 | millinewton·meter |
| kilogram force meter | 9.8067 | newton-meter |
| millinewton | 10.197 | gram force centimeter |
| newton-meter | 0.10197 | kilogram force meter |
| | 0.7376 | foot pound |
| | 8.8495 | inch pound |
| foot-pound | 1.3558 | newton-meter |
| inch-pound | 0.1130 | newton-meter |
| **VELOCITY** | | |
| feet/sec | 0.3048 | meters/second |
| kilometers/hr | 0.6214 | miles/hr |
| meters/second | 3.2808 | feet/sec |
| | 2.2369 | miles/hr |
| miles/hr | 1.60934 | kilometers/hr |
| | 0.44704 | meters/second |
| **VISCOSITY** | | |
| centipoise | 0.001 | pascal-seconds |
| centistoke | $1 \times 10^{-6}$ | meter$^2$/second |
| pascal-second | 1000 | centipoise |
| meter$^2$/second | $1 \times 10^{6}$ | centistoke |
| **VOLUME (capacity)** | | |
| cubic centimeter | 0.06102 | cubic inch |
| cubic foot | 28.3168 | liter |
| cubic inch | 16.3871 | cubic centimeter |
| cubic meter | 1.308 | cubic yard |
| cubic yard | 0.7646 | cubic meter |
| gallon (U.S.) | 3.785 | liter |

*(continued)*

**APPENDIX 2.B** *(continued)*
Common SI Unit Conversion Factors

| multiply | by | to obtain |
|---|---|---|
| **VOLUME (capacity)** *(continued)* | | |
| liter | 0.2642 | gallon (U.S.) |
| | 2.113 | pint (U.S. fluid) |
| | 1.0567 | quart (U.S. fluid) |
| | 0.03531 | cubic foot |
| milliliter | 0.0338 | ounce (U.S. fluid) |
| ounce (U.S. fluid) | 29.574 | milliliter |
| pint (U.S. fluid) | 0.4732 | liter |
| quart (U.S. fluid) | 0.9464 | liter |
| **VOLUME FLOW (gas–air)** | | |
| cubic meter/sec | 2119 | standard cubic foot/min |
| liter/sec | 2.119 | standard cubic foot/min |
| microliter/sec | 0.000127 | standard cubic foot/hr |
| milliliter/sec | 0.002119 | standard cubic foot/min |
| | 0.127133 | standard cubic foot/hr |
| standard cubic foot/min | 0.0004719 | cubic meter/sec |
| | 0.4719 | liter/sec |
| | 471.947 | milliliter/sec |
| standard cubic foot/hr | 7866 | microliter/sec |
| | 7.8658 | milliliter/sec |
| **VOLUME FLOW (liquid)** | | |
| gallon/hr (U.S.) | 0.001052 | liter/sec |
| gallon/min (U.S.) | 0.06309 | liter/sec |
| liter/sec | 951.02 | gallon/hr (U.S.) |
| | 15.850 | gallon/min (U.S.) |

## APPENDIX 9.A
Mensuration of Two-Dimensional Areas

### Nomenclature
$A$    total surface area
$b$    base
$c$    chord length
$d$    distance
$h$    height
$L$    length
$p$    perimeter
$r$    radius
$s$    side (edge) length, arc length
$\theta$    vertex angle, in radians
$\phi$    central angle, in radians

### Circular Sector

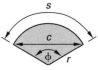

$$A = \tfrac{1}{2}\phi r^2 = \tfrac{1}{2}sr$$

$$\phi = \frac{s}{r}$$

$$s = r\phi$$

$$c = 2r\sin\left(\frac{\phi}{2}\right)$$

### Triangle

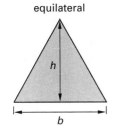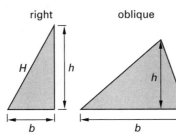

equilateral      right      oblique

$$A = \tfrac{1}{2}bh = \frac{\sqrt{3}}{4}b^2 \qquad A = \tfrac{1}{2}bh \qquad A = \tfrac{1}{2}bh$$

$$h = \frac{\sqrt{3}}{2}b \qquad\qquad H^2 = b^2 + h^2$$

### Parabola

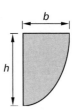

$$A = \tfrac{2}{3}bh$$

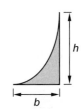

$$A = \tfrac{1}{3}bh$$

### Circle

$$p = 2\pi r$$

$$A = \pi r^2 = \frac{p^2}{4\pi}$$

### Circular Segment

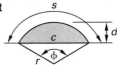

$$A = \tfrac{1}{2}r^2(\phi - \sin\phi)$$

$$\phi = \frac{s}{r} = 2\left(\arccos\frac{r-d}{r}\right)$$

$$c = 2r\sin\left(\frac{\phi}{2}\right)$$

### Ellipse

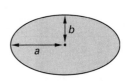

$$A = \pi ab$$

$$p \approx 2\pi\sqrt{\tfrac{1}{2}(a^2 + b^2)} \qquad \begin{bmatrix} \text{Euler's} \\ \text{upper bound} \end{bmatrix}$$

*(continued)*

**APPENDIX 9.A** *(continued)*
Mensuration of Two-Dimensional Areas

**Trapezoid**

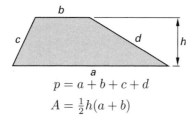

$$p = a + b + c + d$$
$$A = \tfrac{1}{2}h(a + b)$$

The trapezoid is isosceles if $c = d$.

**Parallelogram**

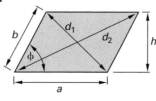

$$p = 2(a + b)$$
$$d_1 = \sqrt{a^2 + b^2 - 2ab\cos\phi}$$
$$d_2 = \sqrt{a^2 + b^2 + 2ab\cos\phi}$$
$$d_1^2 + d_2^2 = 2(a^2 + b^2)$$
$$A = ah = ab\sin\phi$$

If $a = b$, the parallelogram is a rhombus.

**Regular Polygon (*n* equal sides)**

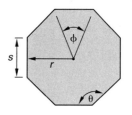

$$\phi = \frac{2\pi}{n}$$
$$\theta = \frac{\pi(n-2)}{n} = \pi - \phi$$
$$p = ns$$
$$s = 2r\tan\left(\frac{\theta}{2}\right)$$
$$A = \tfrac{1}{2}nsr$$

| regular polygons | | area $(A)$ when diameter of inscribed circle $= 1$ | area $(A)$ when side $= 1$ | radius $(r)$ of circumscribed circle when side $= 1$ | length $(L)$ of side when radius $(r)$ of circumscribed circle $= 1$ | length $(L)$ of side when perpendicular to center $= 1$ | perpendicular $(p)$ to center when side $= 1$ |
|---|---|---|---|---|---|---|---|
| sides | name | | | | | | |
| 3 | triangle | 1.299 | 0.433 | 0.577 | 1.732 | 3.464 | 0.289 |
| 4 | square | 1.000 | 1.000 | 0.707 | 1.414 | 2.000 | 0.500 |
| 5 | pentagon | 0.908 | 1.720 | 0.851 | 1.176 | 1.453 | 0.688 |
| 6 | hexagon | 0.866 | 2.598 | 1.000 | 1.000 | 1.155 | 0.866 |
| 7 | heptagon | 0.843 | 3.634 | 1.152 | 0.868 | 0.963 | 1.038 |
| 8 | octagon | 0.828 | 4.828 | 1.307 | 0.765 | 0.828 | 1.207 |
| 9 | nonagon | 0.819 | 6.182 | 1.462 | 0.684 | 0.728 | 1.374 |
| 10 | decagon | 0.812 | 7.694 | 1.618 | 0.618 | 0.650 | 1.539 |
| 11 | undecagon | 0.807 | 9.366 | 1.775 | 0.563 | 0.587 | 1.703 |
| 12 | dodecagon | 0.804 | 11.196 | 1.932 | 0.518 | 0.536 | 1.866 |

**APPENDIX 9.B**
Mensuration of Three-Dimensional Volumes

### Nomenclature
$A$    area
$b$    base
$h$    height
$r$    radius
$R$    radius
$s$    side (edge) length
$V$    volume

### Sphere

$$V = \frac{4\pi r^3}{3}$$
$$A = 4\pi r^2$$

### Right Circular Cone

$$V = \frac{\pi r^2 h}{3}$$
$$A = \pi r \sqrt{r^2 + h^2}$$

(does not include base area)

### Right Circular Cylinder

$$V = \pi r^2 h$$
$$A = 2\pi rh$$

(does not include end area)

### Spherical Segment (Spherical Cap)

Surface area of a spherical segment of radius $r$ cut out by an angle $\theta_0$ rotated from the center about a radius, $r$, is

$$A = 2\pi r^2 \left(1 - \cos\theta_0\right)$$
$$\omega = \frac{A}{r^2} = 2\pi \left(1 - \cos\theta_0\right)$$

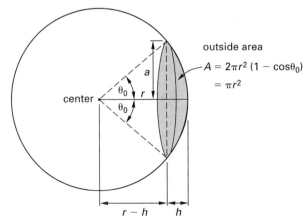

$$V_{\text{cap}} = \frac{\pi}{6}h(3a^2 + h^2)$$
$$= \frac{\pi}{3}h^2(3r - h)$$
$$a = \sqrt{h(2r - h)}$$

### Paraboloid of Revolution

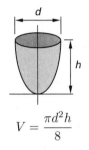

$$V = \frac{\pi d^2 h}{8}$$

### Torus

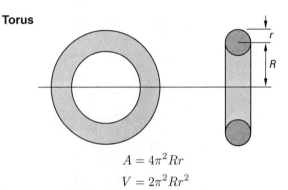

$$A = 4\pi^2 Rr$$
$$V = 2\pi^2 Rr^2$$

### Regular Polyhedra (identical faces)

| name | number of faces | form of faces | total surface area | volume |
|------|------|------|------|------|
| tetrahedron | 4 | equilateral triangle | $1.7321s^2$ | $0.1179s^3$ |
| cube | 6 | square | $6.0000s^2$ | $1.0000s^3$ |
| octahedron | 8 | equilateral triangle | $3.4641s^2$ | $0.4714s^3$ |
| dodecahedron | 12 | regular pentagon | $20.6457s^2$ | $7.6631s^3$ |
| isosahedron | 20 | equilateral triangle | $8.6603s^2$ | $2.1817s^3$ |

The radius of a sphere inscribed within a regular polyhedron is

$$r = \frac{3V_{\text{polyhedron}}}{A_{\text{polyhedron}}}$$

## APPENDIX 11.A
Abbreviated Table of Indefinite Integrals
(In each case, add a constant of integration.
All angles are measured in radians.)

### GENERAL FORMULAS

1.  $\int dx = x$
2.  $\int c\, dx = c \int dx$
3.  $\int (dx + dy) = \int dx + \int dy$
4.  $\int u\, dv = uv - \int v\, du$ (integration by parts)

### ALGEBRAIC FORMS

5.  $\int x^n dx = \dfrac{x^{n+1}}{n+1} \quad [n \neq -1]$

6.  $\int x^{-1} dx = \int \dfrac{dx}{x} = \ln|x|$

7.  $\int (ax+b)^n dx = \dfrac{(ax+b)^{n+1}}{a(n+1)} \quad [n \neq -1]$

8.  $\int \dfrac{dx}{ax+b} = \dfrac{1}{a} \ln(ax+b)$

9.  $\int \dfrac{x\, dx}{ax+b} = \dfrac{1}{a^2} \left( ax+b - b\ln(ax+b) \right)$

10. $\int \dfrac{x\, dx}{(ax+b)^2} = \dfrac{1}{a^2} \left( \dfrac{b}{ax+b} + \ln(ax+b) \right)$

11. $\int \dfrac{dx}{x(ax+b)} = \dfrac{1}{b} \ln \left( \dfrac{x}{ax+b} \right)$

12. $\int \dfrac{dx}{x(ax+b)^2} = \dfrac{1}{b(ax+b)} + \dfrac{1}{b^2} \ln \left( \dfrac{x}{ax+b} \right)$

13. $\int \dfrac{dx}{x^2+a^2} = \dfrac{1}{a} \tan^{-1} \left( \dfrac{x}{a} \right)$

14. $\int \dfrac{dx}{a^2-x^2} = \dfrac{1}{a} \tanh^{-1} \left( \dfrac{x}{a} \right)$

15. $\int \dfrac{x\, dx}{ax^2+b} = \dfrac{1}{2a} \ln(ax^2+b)$

16. $\int \dfrac{dx}{x(ax^n+b)} = \dfrac{1}{bn} \ln \left( \dfrac{x^n}{ax^n+b} \right)$

17. $\int \dfrac{dx}{ax^2+bx+c} = \dfrac{1}{\sqrt{b^2-4ac}} \ln \left( \dfrac{2ax+b-\sqrt{b^2-4ac}}{2ax+b+\sqrt{b^2-4ac}} \right) \quad [b^2 > 4ac]$

18. $\int \dfrac{dx}{ax^2+bx+c} = \dfrac{2}{\sqrt{4ac-b^2}} \tan^{-1} \left( \dfrac{2ax+b}{\sqrt{4ac-b^2}} \right) \quad [b^2 > 4ac]$

19. $\int \sqrt{a^2-x^2}\, dx = \dfrac{x}{2} \sqrt{a^2-x^2} + \dfrac{a^2}{2} \sin^{-1} \left( \dfrac{x}{a} \right)$

20. $\int x\sqrt{a^2-x^2}\, dx = -\tfrac{1}{3}(a^2-x^2)^{3/2}$

21. $\int \dfrac{dx}{\sqrt{a^2-x^2}} = \sin^{-1} \left( \dfrac{x}{a} \right)$

22. $\int \dfrac{x\, dx}{\sqrt{a^2-x^2}} = -\sqrt{a^2-x^2}$

## APPENDIX 12.A
Laplace Transforms

| $f(t)$ | $\mathcal{L}(f(t))$ | $f(t)$ | $\mathcal{L}(f(t))$ |
|---|---|---|---|
| $\delta(t)$  [unit impulse at $t = 0$] | $1$ | $e^{at}$ | $\dfrac{1}{s-a}$ |
| $\delta(t-c)$  [unit impulse at $t = c$] | $e^{-cs}$ | $e^{at}\sin bt$ | $\dfrac{b}{(s-a)^2+b^2}$ |
| $1$ or $u_0$  [unit step at $t = 0$] | $\dfrac{1}{s}$ | $e^{at}\cos bt$ | $\dfrac{s-a}{(s-a)^2+b^2}$ |
| $u_c$  [unit step at $t = c$] | $\dfrac{e^{-cs}}{s}$ | $e^{at}t^n$  ($n$ is a positive integer) | $\dfrac{n!}{(s-a)^{n+1}}$ |
| $t$  [unit ramp at $t = 0$] | $\dfrac{1}{s^2}$ | $1-e^{-at}$ | $\dfrac{a}{s(s+a)}$ |
| $\dfrac{t^{n-1}}{(n-1)!}$ | $\dfrac{1}{s^n}$ | $e^{-at}+at-1$ | $\dfrac{a^2}{s^2(s+a)}$ |
| $\sin at$ | $\dfrac{a}{s^2+a^2}$ | $\dfrac{e^{-at}-e^{-bt}}{b-a}$ | $\dfrac{1}{(s+a)(s+b)}$ |
| $at-\sin at$ | $\dfrac{a^3}{s^2(s^2+a^2)}$ | $\dfrac{(c-a)e^{-at}-(c-b)e^{-bt}}{b-a}$ | $\dfrac{s+c}{(s+a)(s+b)}$ |
| $\sinh at$ | $\dfrac{a}{s^2-a^2}$ | $\dfrac{1}{ab}+\dfrac{be^{-at}-ae^{-bt}}{ab(a-b)}$ | $\dfrac{1}{s(s+a)(s+b)}$ |
| $t\sin at$ | $\dfrac{2as}{(s^2+a^2)^2}$ | $t\sinh at$ | $\dfrac{2as}{(s^2-a^2)^2}$ |
| $\cos at$ | $\dfrac{s}{s^2+a^2}$ | $t\cosh at$ | $\dfrac{s^2+a^2}{(s^2-a^2)^2}$ |
| $1-\cos at$ | $\dfrac{a^2}{s(s^2+a^2)}$ | rectangular pulse, magnitude $M$, duration $a$ | $\left(\dfrac{M}{s}\right)(1-e^{-as})$ |
| $\cosh at$ | $\dfrac{s}{s^2-a^2}$ | triangular pulse, magnitude $M$, duration $2a$ | $\left(\dfrac{M}{as^2}\right)(1-e^{-as})^2$ |
| $t\cos at$ | $\dfrac{s^2-a^2}{(s^2+a^2)^2}$ | sawtooth pulse, magnitude $M$, duration $a$ | $\left(\dfrac{M}{as^2}\right)(1-(as+1)e^{-as})$ |
| $t^n$  ($n$ is a positive integer) | $\dfrac{n!}{s^{n+1}}$ | sinusoidal pulse, magnitude $M$, duration $\pi/a$ | $\left(\dfrac{Ma}{s^2+a^2}\right)(1+e^{-\pi s/a})$ |

**Appendices**

**APPENDIX 13.A**
Areas Under the Standard Normal Curve
(0 to $z$)

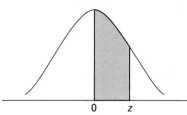

| $z$ | 0 | 1 | 2 | 3 | 4 | 5 | 6 | 7 | 8 | 9 |
|---|---|---|---|---|---|---|---|---|---|---|
| 0.0 | 0.0000 | 0.0040 | 0.0080 | 0.0120 | 0.0160 | 0.0199 | 0.0239 | 0.0279 | 0.0319 | 0.0359 |
| 0.1 | 0.0398 | 0.0438 | 0.0478 | 0.0517 | 0.0557 | 0.0596 | 0.0636 | 0.0675 | 0.0714 | 0.0754 |
| 0.2 | 0.0793 | 0.0832 | 0.0871 | 0.0910 | 0.0948 | 0.0987 | 0.1026 | 0.1064 | 0.1103 | 0.1141 |
| 0.3 | 0.1179 | 0.1217 | 0.1255 | 0.1293 | 0.1331 | 0.1368 | 0.1406 | 0.1443 | 0.1480 | 0.1517 |
| 0.4 | 0.1554 | 0.1591 | 0.1628 | 0.1664 | 0.1700 | 0.1736 | 0.1772 | 0.1808 | 0.1844 | 0.1879 |
| 0.5 | 0.1915 | 0.1950 | 0.1985 | 0.2019 | 0.2054 | 0.2088 | 0.2123 | 0.2157 | 0.2190 | 0.2224 |
| 0.6 | 0.2258 | 0.2291 | 0.2324 | 0.2357 | 0.2389 | 0.2422 | 0.2454 | 0.2486 | 0.2518 | 0.2549 |
| 0.7 | 0.2580 | 0.2612 | 0.2642 | 0.2673 | 0.2704 | 0.2734 | 0.2764 | 0.2794 | 0.2823 | 0.2852 |
| 0.8 | 0.2881 | 0.2910 | 0.2939 | 0.2967 | 0.2996 | 0.3023 | 0.3051 | 0.3078 | 0.3106 | 0.3133 |
| 0.9 | 0.3159 | 0.3186 | 0.3212 | 0.3238 | 0.3264 | 0.3289 | 0.3315 | 0.3340 | 0.3365 | 0.3389 |
| 1.0 | 0.3413 | 0.3438 | 0.3461 | 0.3485 | 0.3508 | 0.3531 | 0.3554 | 0.3577 | 0.3599 | 0.3621 |
| 1.1 | 0.3643 | 0.3665 | 0.3686 | 0.3708 | 0.3729 | 0.3749 | 0.3770 | 0.3790 | 0.3810 | 0.3830 |
| 1.2 | 0.3849 | 0.3869 | 0.3888 | 0.3907 | 0.3925 | 0.3944 | 0.3962 | 0.3980 | 0.3997 | 0.4015 |
| 1.3 | 0.4032 | 0.4049 | 0.4066 | 0.4082 | 0.4099 | 0.4115 | 0.4131 | 0.4147 | 0.4162 | 0.4177 |
| 1.4 | 0.4192 | 0.4207 | 0.4222 | 0.4236 | 0.4251 | 0.4265 | 0.4279 | 0.4292 | 0.4306 | 0.4319 |
| 1.5 | 0.4332 | 0.4345 | 0.4357 | 0.4370 | 0.4382 | 0.4394 | 0.4406 | 0.4418 | 0.4429 | 0.4441 |
| 1.6 | 0.4452 | 0.4463 | 0.4474 | 0.4484 | 0.4495 | 0.4505 | 0.4515 | 0.4525 | 0.4535 | 0.4545 |
| 1.7 | 0.4554 | 0.4564 | 0.4573 | 0.4582 | 0.4591 | 0.4599 | 0.4608 | 0.4616 | 0.4625 | 0.4633 |
| 1.8 | 0.4641 | 0.4649 | 0.4656 | 0.4664 | 0.4671 | 0.4678 | 0.4686 | 0.4693 | 0.4699 | 0.4706 |
| 1.9 | 0.4713 | 0.4719 | 0.4726 | 0.4732 | 0.4738 | 0.4744 | 0.4750 | 0.4756 | 0.4761 | 0.4767 |
| 2.0 | 0.4772 | 0.4778 | 0.4783 | 0.4788 | 0.4793 | 0.4798 | 0.4803 | 0.4808 | 0.4812 | 0.4817 |
| 2.1 | 0.4821 | 0.4826 | 0.4830 | 0.4834 | 0.4838 | 0.4842 | 0.4846 | 0.4850 | 0.4854 | 0.4857 |
| 2.2 | 0.4861 | 0.4864 | 0.4868 | 0.4871 | 0.4875 | 0.4878 | 0.4881 | 0.4884 | 0.4887 | 0.4890 |
| 2.3 | 0.4893 | 0.4896 | 0.4898 | 0.4901 | 0.4904 | 0.4906 | 0.4909 | 0.4911 | 0.4913 | 0.4916 |
| 2.4 | 0.4918 | 0.4920 | 0.4922 | 0.4925 | 0.4927 | 0.4929 | 0.4931 | 0.4932 | 0.4934 | 0.4936 |
| 2.5 | 0.4938 | 0.4940 | 0.4941 | 0.4943 | 0.4945 | 0.4946 | 0.4948 | 0.4949 | 0.4951 | 0.4952 |
| 2.6 | 0.4953 | 0.4955 | 0.4956 | 0.4957 | 0.4959 | 0.4960 | 0.4961 | 0.4962 | 0.4963 | 0.4964 |
| 2.7 | 0.4965 | 0.4966 | 0.4967 | 0.4968 | 0.4969 | 0.4970 | 0.4971 | 0.4972 | 0.4973 | 0.4974 |
| 2.8 | 0.4974 | 0.4975 | 0.4976 | 0.4977 | 0.4977 | 0.4978 | 0.4979 | 0.4979 | 0.4980 | 0.4981 |
| 2.9 | 0.4981 | 0.4982 | 0.4982 | 0.4983 | 0.4984 | 0.4984 | 0.4985 | 0.4985 | 0.4986 | 0.4986 |
| 3.0 | 0.4987 | 0.4987 | 0.4987 | 0.4988 | 0.4988 | 0.4989 | 0.4989 | 0.4989 | 0.4990 | 0.4990 |
| 3.1 | 0.4990 | 0.4991 | 0.4991 | 0.4991 | 0.4992 | 0.4992 | 0.4992 | 0.4992 | 0.4993 | 0.4993 |
| 3.2 | 0.4993 | 0.4993 | 0.4994 | 0.4994 | 0.4994 | 0.4994 | 0.4994 | 0.4995 | 0.4995 | 0.4995 |
| 3.3 | 0.4995 | 0.4995 | 0.4996 | 0.4996 | 0.4996 | 0.4996 | 0.4996 | 0.4996 | 0.4996 | 0.4997 |
| 3.4 | 0.4997 | 0.4997 | 0.4997 | 0.4997 | 0.4997 | 0.4997 | 0.4997 | 0.4997 | 0.4997 | 0.4998 |
| 3.5 | 0.4998 | 0.4998 | 0.4998 | 0.4998 | 0.4998 | 0.4998 | 0.4998 | 0.4998 | 0.4998 | 0.4998 |
| 3.6 | 0.4998 | 0.4998 | 0.4999 | 0.4999 | 0.4999 | 0.4999 | 0.4999 | 0.4999 | 0.4999 | 0.4999 |
| 3.7 | 0.4999 | 0.4999 | 0.4999 | 0.4999 | 0.4999 | 0.4999 | 0.4999 | 0.4999 | 0.4999 | 0.4999 |
| 3.8 | 0.4999 | 0.4999 | 0.4999 | 0.4999 | 0.4999 | 0.4999 | 0.4999 | 0.4999 | 0.4999 | 0.4999 |
| 3.9 | 0.5000 | 0.5000 | 0.5000 | 0.5000 | 0.5000 | 0.5000 | 0.5000 | 0.5000 | 0.5000 | 0.5000 |

## APPENDIX 16.A
### Properties of Water at Atmospheric Pressure
### (customary U.S. units)

| temperature ($°F$) | density ($lbm/ft^3$) | absolute viscosity ($lbf\text{-}sec/ft^2$) | kinematic viscosity ($ft^2/sec$) | surface tension ($lbf/ft$) | vapor pressure head[a,b,c] ($ft$) | bulk modulus ($lbf/in^2$) |
|---|---|---|---|---|---|---|
| 32 | 62.42 | $3.746 \times 10^{-5}$ | $1.931 \times 10^{-5}$ | $0.518 \times 10^{-2}$ | 0.20 | $293 \times 10^3$ |
| 40 | 62.43 | $3.229 \times 10^{-5}$ | $1.664 \times 10^{-5}$ | $0.514 \times 10^{-2}$ | 0.28 | $294 \times 10^3$ |
| 50 | 62.41 | $2.735 \times 10^{-5}$ | $1.410 \times 10^{-5}$ | $0.509 \times 10^{-2}$ | 0.41 | $305 \times 10^3$ |
| 60 | 62.37 | $2.359 \times 10^{-5}$ | $1.217 \times 10^{-5}$ | $0.504 \times 10^{-2}$ | 0.59 | $311 \times 10^3$ |
| 70 | 62.30 | $2.050 \times 10^{-5}$ | $1.059 \times 10^{-5}$ | $0.500 \times 10^{-2}$ | 0.84 | $320 \times 10^3$ |
| 80 | 62.22 | $1.799 \times 10^{-5}$ | $0.930 \times 10^{-5}$ | $0.492 \times 10^{-2}$ | 1.17 | $322 \times 10^3$ |
| 90 | 62.11 | $1.595 \times 10^{-5}$ | $0.826 \times 10^{-5}$ | $0.486 \times 10^{-2}$ | 1.62 | $323 \times 10^3$ |
| 100 | 62.00 | $1.424 \times 10^{-5}$ | $0.739 \times 10^{-5}$ | $0.480 \times 10^{-2}$ | 2.21 | $327 \times 10^3$ |
| 110 | 61.86 | $1.284 \times 10^{-5}$ | $0.667 \times 10^{-5}$ | $0.473 \times 10^{-2}$ | 2.97 | $331 \times 10^3$ |
| 120 | 61.71 | $1.168 \times 10^{-5}$ | $0.609 \times 10^{-5}$ | $0.465 \times 10^{-2}$ | 3.96 | $333 \times 10^3$ |
| 130 | 61.55 | $1.069 \times 10^{-5}$ | $0.558 \times 10^{-5}$ | $0.460 \times 10^{-2}$ | 5.21 | $334 \times 10^3$ |
| 140 | 61.38 | $0.981 \times 10^{-5}$ | $0.514 \times 10^{-5}$ | $0.454 \times 10^{-2}$ | 6.78 | $330 \times 10^3$ |
| 150 | 61.20 | $0.905 \times 10^{-5}$ | $0.476 \times 10^{-5}$ | $0.447 \times 10^{-2}$ | 8.76 | $328 \times 10^3$ |
| 160 | 61.00 | $0.838 \times 10^{-5}$ | $0.442 \times 10^{-5}$ | $0.441 \times 10^{-2}$ | 11.21 | $326 \times 10^3$ |
| 170 | 60.80 | $0.780 \times 10^{-5}$ | $0.413 \times 10^{-5}$ | $0.433 \times 10^{-2}$ | 14.20 | $322 \times 10^3$ |
| 180 | 60.58 | $0.726 \times 10^{-5}$ | $0.385 \times 10^{-5}$ | $0.426 \times 10^{-2}$ | 17.87 | $313 \times 10^3$ |
| 190 | 60.36 | $0.678 \times 10^{-5}$ | $0.362 \times 10^{-5}$ | $0.419 \times 10^{-2}$ | 22.29 | $313 \times 10^3$ |
| 200 | 60.12 | $0.637 \times 10^{-5}$ | $0.341 \times 10^{-5}$ | $0.412 \times 10^{-2}$ | 27.61 | $308 \times 10^3$ |
| 212 | 59.83 | $0.593 \times 10^{-5}$ | $0.319 \times 10^{-5}$ | $0.404 \times 10^{-2}$ | 35.38 | $300 \times 10^3$ |

[a]Based on actual densities, not on standard "cold, clear water."
[b]Can also be calculated from steam tables as $(p_{saturation})(144 \text{ in}^2/\text{ft}^2)(v)(g/g_c)$.
[c]Multiply the vapor pressure head by the density and divide by 144 $\text{in}^2/\text{ft}^2$ to obtain psi.

**Appendices**

**APPENDIX 16.B**
Properties of Water at Atmospheric Pressure
(SI units)

| temperature (°C) | density (kg/m³) | absolute viscosity (Pa·s) | kinematic viscosity (m²/s) | vapor pressure (kPa) | bulk modulus (kPa) |
|---|---|---|---|---|---|
| 0 | 999.87 | $1.7921 \times 10^{-3}$ | $1.792 \times 10^{-6}$ | 0.611 | $204 \times 10^4$ |
| 4 | 1000.00 | $1.5674 \times 10^{-3}$ | $1.567 \times 10^{-6}$ | 0.813 | $206 \times 10^4$ |
| 10 | 999.73 | $1.3077 \times 10^{-3}$ | $1.371 \times 10^{-6}$ | 1.228 | $211 \times 10^4$ |
| 20 | 998.23 | $1.0050 \times 10^{-3}$ | $1.007 \times 10^{-6}$ | 2.338 | $220 \times 10^4$ |
| 25 | 997.08 | $0.8937 \times 10^{-3}$ | $8.963 \times 10^{-7}$ | 3.168 | $221 \times 10^4$ |
| 30 | 995.68 | $0.8007 \times 10^{-3}$ | $8.042 \times 10^{-7}$ | 4.242 | $223 \times 10^4$ |
| 40 | 992.25 | $0.6560 \times 10^{-3}$ | $6.611 \times 10^{-7}$ | 7.375 | $227 \times 10^4$ |
| 50 | 988.07 | $0.5494 \times 10^{-3}$ | $5.560 \times 10^{-7}$ | 12.333 | $230 \times 10^4$ |
| 60 | 983.24 | $0.4688 \times 10^{-3}$ | $4.768 \times 10^{-7}$ | 19.92 | $228 \times 10^4$ |
| 70 | 977.81 | $0.4061 \times 10^{-3}$ | $4.153 \times 10^{-7}$ | 31.16 | $225 \times 10^4$ |
| 80 | 971.83 | $0.3565 \times 10^{-3}$ | $3.668 \times 10^{-7}$ | 47.34 | $221 \times 10^4$ |
| 90 | 965.34 | $0.3165 \times 10^{-3}$ | $3.279 \times 10^{-7}$ | 70.10 | $216 \times 10^4$ |
| 100 | 958.38 | $0.2838 \times 10^{-3}$ | $2.961 \times 10^{-7}$ | 101.325 | $207 \times 10^4$ |

Compiled from various sources.

**Appendices**

### APPENDIX 16.C
Viscosity of Water in Other Units
(customary U.S. units)

| temperature | absolute viscosity | kinematic viscosity | |
|---|---|---|---|
| (°F) | (cP) | (cSt) | (SSU) |
| 32 | 1.79 | 1.79 | 33.0 |
| 50 | 1.31 | 1.31 | 31.6 |
| 60 | 1.12 | 1.12 | 31.2 |
| 70 | 0.98 | 0.98 | 30.9 |
| 80 | 0.86 | 0.86 | 30.6 |
| 85 | 0.81 | 0.81 | 30.4 |
| 100 | 0.68 | 0.69 | 30.2 |
| 120 | 0.56 | 0.57 | 30.0 |
| 140 | 0.47 | 0.48 | 29.7 |
| 160 | 0.40 | 0.41 | 29.6 |
| 180 | 0.35 | 0.36 | 29.5 |
| 212 | 0.28 | 0.29 | 29.3 |

Reprinted with permission from *Hydraulic Handbook*, 10th ed., by Colt Industries, Fairbanks Morse Pump Division, Kansas City, Kansas, 1977.

**Appendices**

**APPENDIX 16.D**
Properties of Air at Atmospheric Pressure
(customary U.S. units)

| temperature ($^\circ$F) | density (lbm/ft$^3$) | absolute viscosity (lbf-sec/ft$^2$) | kinematic viscosity (ft$^2$/sec) |
|---|---|---|---|
| 0 | 0.0862 | $3.28 \times 10^{-7}$ | $12.6 \times 10^{-5}$ |
| 20 | 0.0827 | $3.50 \times 10^{-7}$ | $13.6 \times 10^{-5}$ |
| 40 | 0.0794 | $3.62 \times 10^{-7}$ | $14.6 \times 10^{-5}$ |
| 60 | 0.0763 | $3.74 \times 10^{-7}$ | $15.8 \times 10^{-5}$ |
| 68 | 0.0752 | $3.78 \times 10^{-7}$ | $16.0 \times 10^{-5}$ |
| 80 | 0.0735 | $3.85 \times 10^{-7}$ | $16.9 \times 10^{-5}$ |
| 100 | 0.0709 | $3.96 \times 10^{-7}$ | $18.0 \times 10^{-5}$ |
| 120 | 0.0684 | $4.07 \times 10^{-7}$ | $18.9 \times 10^{-5}$ |
| 250 | 0.0559 | $4.74 \times 10^{-7}$ | $27.3 \times 10^{-5}$ |

## APPENDIX 16.E
Properties of Air at Atmospheric Pressure
(SI units)

| temperature (°C) | density (kg/m$^3$) | absolute viscosity (Pa·s) | kinematic viscosity (m$^2$/s) |
|---|---|---|---|
| 0 | 1.293 | $1.709 \times 10^{-5}$ | $1.322 \times 10^{-5}$ |
| 20 | 1.20 | $1.80 \ \times 10^{-5}$ | $1.51 \ \times 10^{-5}$ |
| 50 | 1.093 | $1.951 \times 10^{-5}$ | $1.785 \times 10^{-5}$ |
| 100 | 0.946 | $2.175 \times 10^{-5}$ | $2.474 \times 10^{-5}$ |
| 150 | 0.834 | $2.385 \times 10^{-5}$ | $3.077 \times 10^{-5}$ |
| 200 | 0.746 | $2.582 \times 10^{-5}$ | $3.724 \times 10^{-5}$ |
| 250 | 0.675 | $2.770 \times 10^{-5}$ | $4.416 \times 10^{-5}$ |
| 300 | 0.616 | $2.946 \times 10^{-5}$ | $5.145 \times 10^{-5}$ |
| 350 | 0.567 | $3.113 \times 10^{-5}$ | $5.907 \times 10^{-5}$ |
| 400 | 0.525 | $3.277 \times 10^{-5}$ | $6.721 \times 10^{-5}$ |
| 450 | 0.488 | $3.433 \times 10^{-5}$ | $7.750 \times 10^{-5}$ |
| 500 | 0.457 | $3.583 \times 10^{-5}$ | $8.436 \times 10^{-5}$ |

Appendices

### APPENDIX 18.A
Area, Wetted Perimeter, and Hydraulic Radius of Partially Filled Circular Pipes

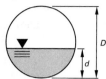

| $\dfrac{d}{D}$ | $\dfrac{\text{area}}{D^2}$ | $\dfrac{\text{wetted perimeter}}{D}$ | $\dfrac{r_h}{D}$ | $\dfrac{d}{D}$ | $\dfrac{\text{area}}{D^2}$ | $\dfrac{\text{wetted perimeter}}{D}$ | $\dfrac{r_h}{D}$ |
|---|---|---|---|---|---|---|---|
| 0.01 | 0.0013 | 0.2003 | 0.0066 | 0.51 | 0.4027 | 1.5908 | 0.2531 |
| 0.02 | 0.0037 | 0.2838 | 0.0132 | 0.52 | 0.4127 | 1.6108 | 0.2561 |
| 0.03 | 0.0069 | 0.3482 | 0.0197 | 0.53 | 0.4227 | 1.6308 | 0.2591 |
| 0.04 | 0.0105 | 0.4027 | 0.0262 | 0.54 | 0.4327 | 1.6509 | 0.2620 |
| 0.05 | 0.0147 | 0.4510 | 0.0326 | 0.55 | 0.4426 | 1.6710 | 0.2649 |
| 0.06 | 0.0192 | 0.4949 | 0.0389 | 0.56 | 0.4526 | 1.6911 | 0.2676 |
| 0.07 | 0.0242 | 0.5355 | 0.0451 | 0.57 | 0.4625 | 1.7113 | 0.2703 |
| 0.08 | 0.0294 | 0.5735 | 0.0513 | 0.58 | 0.4723 | 1.7315 | 0.2728 |
| 0.09 | 0.0350 | 0.6094 | 0.0574 | 0.59 | 0.4822 | 1.7518 | 0.2753 |
| 0.10 | 0.0409 | 0.6435 | 0.0635 | 0.60 | 0.4920 | 1.7722 | 0.2776 |
| 0.11 | 0.0470 | 0.6761 | 0.0695 | 0.61 | 0.5018 | 1.7926 | 0.2797 |
| 0.12 | 0.0534 | 0.7075 | 0.0754 | 0.62 | 0.5115 | 1.8132 | 0.2818 |
| 0.13 | 0.0600 | 0.7377 | 0.0813 | 0.63 | 0.5212 | 1.8338 | 0.2839 |
| 0.14 | 0.0688 | 0.7670 | 0.0871 | 0.64 | 0.5308 | 1.8546 | 0.2860 |
| 0.15 | 0.0739 | 0.7954 | 0.0929 | 0.65 | 0.5404 | 1.8755 | 0.2881 |
| 0.16 | 0.0811 | 0.8230 | 0.0986 | 0.66 | 0.5499 | 1.8965 | 0.2899 |
| 0.17 | 0.0885 | 0.8500 | 0.1042 | 0.67 | 0.5594 | 1.9177 | 0.2917 |
| 0.18 | 0.0961 | 0.8763 | 0.1097 | 0.68 | 0.5687 | 1.9391 | 0.2935 |
| 0.19 | 0.1039 | 0.9020 | 0.1152 | 0.69 | 0.5780 | 1.9606 | 0.2950 |
| 0.20 | 0.1118 | 0.9273 | 0.1206 | 0.70 | 0.5872 | 1.9823 | 0.2962 |
| 0.21 | 0.1199 | 0.9521 | 0.1259 | 0.71 | 0.5964 | 2.0042 | 0.2973 |
| 0.22 | 0.1281 | 0.9764 | 0.1312 | 0.72 | 0.6054 | 2.0264 | 0.2984 |
| 0.23 | 0.1365 | 1.0003 | 0.1364 | 0.73 | 0.6143 | 2.0488 | 0.2995 |
| 0.24 | 0.1449 | 1.0239 | 0.1416 | 0.74 | 0.6231 | 2.0714 | 0.3006 |
| 0.25 | 0.1535 | 1.0472 | 0.1466 | 0.75 | 0.6318 | 2.0944 | 0.3017 |
| 0.26 | 0.1623 | 1.0701 | 0.1516 | 0.76 | 0.6404 | 2.1176 | 0.3025 |
| 0.27 | 0.1711 | 1.0928 | 0.1566 | 0.77 | 0.6489 | 2.1412 | 0.3032 |
| 0.28 | 0.1800 | 1.1152 | 0.1614 | 0.78 | 0.6573 | 2.1652 | 0.3037 |
| 0.29 | 0.1890 | 1.1373 | 0.1662 | 0.79 | 0.6655 | 2.1895 | 0.3040 |
| 0.30 | 0.1982 | 1.1593 | 0.1709 | 0.80 | 0.6736 | 2.2143 | 0.3042 |
| 0.31 | 0.2074 | 1.1810 | 0.1755 | 0.81 | 0.6815 | 2.2395 | 0.3044 |
| 0.32 | 0.2167 | 1.2025 | 0.1801 | 0.82 | 0.6893 | 2.2653 | 0.3043 |
| 0.33 | 0.2260 | 1.2239 | 0.1848 | 0.83 | 0.6969 | 2.2916 | 0.3041 |
| 0.34 | 0.2355 | 1.2451 | 0.1891 | 0.84 | 0.7043 | 2.3186 | 0.3038 |
| 0.35 | 0.2450 | 1.2661 | 0.1935 | 0.85 | 0.7115 | 2.3462 | 0.3033 |
| 0.36 | 0.2546 | 1.2870 | 0.1978 | 0.86 | 0.7186 | 2.3746 | 0.3026 |
| 0.37 | 0.2642 | 1.3078 | 0.2020 | 0.87 | 0.7254 | 2.4038 | 0.3017 |
| 0.38 | 0.2739 | 1.3284 | 0.2061 | 0.88 | 0.7320 | 2.4341 | 0.3008 |
| 0.39 | 0.2836 | 1.3490 | 0.2102 | 0.89 | 0.7384 | 2.4655 | 0.2995 |
| 0.40 | 0.2934 | 1.3694 | 0.2142 | 0.90 | 0.7445 | 2.4981 | 0.2980 |
| 0.41 | 0.3032 | 1.3898 | 0.2181 | 0.91 | 0.7504 | 2.5322 | 0.2963 |
| 0.42 | 0.3130 | 1.4101 | 0.2220 | 0.92 | 0.7560 | 2.5681 | 0.2944 |
| 0.43 | 0.3229 | 1.4303 | 0.2257 | 0.93 | 0.7612 | 2.6061 | 0.2922 |
| 0.44 | 0.3328 | 1.4505 | 0.2294 | 0.94 | 0.7662 | 2.6467 | 0.2896 |
| 0.45 | 0.3428 | 1.4706 | 0.2331 | 0.95 | 0.7707 | 2.6906 | 0.2864 |
| 0.46 | 0.3527 | 1.4907 | 0.2366 | 0.96 | 0.7749 | 2.7389 | 0.2830 |
| 0.47 | 0.3627 | 1.5108 | 0.2400 | 0.97 | 0.7785 | 2.7934 | 0.2787 |
| 0.48 | 0.3727 | 1.5308 | 0.2434 | 0.98 | 0.7816 | 2.8578 | 0.2735 |
| 0.49 | 0.3827 | 1.5508 | 0.2467 | 0.99 | 0.7841 | 2.9412 | 0.2665 |
| 0.50 | 0.3927 | 1.5708 | 0.2500 | 1.00 | 0.7854 | 3.1416 | 0.2500 |

**APPENDIX 18.B**
Dimensions of Welded and Seamless Steel Pipe[a,b]
(selected sizes)[c]
(customary U.S. units)

| nominal diameter (in) | schedule | outside diameter (in) | wall thickness (in) | internal diameter (in) | internal area (in²) | internal diameter (ft) | internal area (ft²) |
|---|---|---|---|---|---|---|---|
| $\frac{1}{8}$ | 40 (S) | 0.405 | 0.068 | 0.269 | 0.0568 | 0.0224 | 0.00039 |
|  | 80 (X) |  | 0.095 | 0.215 | 0.0363 | 0.0179 | 0.00025 |
| $\frac{1}{4}$ | 40 (S) | 0.540 | 0.088 | 0.364 | 0.1041 | 0.0303 | 0.00072 |
|  | 80 (X) |  | 0.119 | 0.302 | 0.0716 | 0.0252 | 0.00050 |
| $\frac{3}{8}$ | 40 (S) | 0.675 | 0.091 | 0.493 | 0.1909 | 0.0411 | 0.00133 |
|  | 80 (X) |  | 0.126 | 0.423 | 0.1405 | 0.0353 | 0.00098 |
| $\frac{1}{2}$ | 40 (S) | 0.840 | 0.109 | 0.622 | 0.3039 | 0.0518 | 0.00211 |
|  | 80 (X) |  | 0.147 | 0.546 | 0.2341 | 0.0455 | 0.00163 |
|  | 160 |  | 0.187 | 0.466 | 0.1706 | 0.0388 | 0.00118 |
|  | (XX) |  | 0.294 | 0.252 | 0.0499 | 0.0210 | 0.00035 |
| $\frac{3}{4}$ | 40 (S) | 1.050 | 0.113 | 0.824 | 0.5333 | 0.0687 | 0.00370 |
|  | 80 (X) |  | 0.154 | 0.742 | 0.4324 | 0.0618 | 0.00300 |
|  | 160 |  | 0.218 | 0.614 | 0.2961 | 0.0512 | 0.00206 |
|  | (XX) |  | 0.308 | 0.434 | 0.1479 | 0.0362 | 0.00103 |
| 1 | 40 (S) | 1.315 | 0.133 | 1.049 | 0.8643 | 0.0874 | 0.00600 |
|  | 80 (X) |  | 0.179 | 0.957 | 0.7193 | 0.0798 | 0.00500 |
|  | 160 |  | 0.250 | 0.815 | 0.5217 | 0.0679 | 0.00362 |
|  | (XX) |  | 0.358 | 0.599 | 0.2818 | 0.0499 | 0.00196 |
| $1\frac{1}{4}$ | 40 (S) | 1.660 | 0.140 | 1.380 | 1.496 | 0.1150 | 0.01039 |
|  | 80 (X) |  | 0.191 | 1.278 | 1.283 | 0.1065 | 0.00890 |
|  | 160 |  | 0.250 | 1.160 | 1.057 | 0.0967 | 0.00734 |
|  | (XX) |  | 0.382 | 0.896 | 0.6305 | 0.0747 | 0.00438 |
| $1\frac{1}{2}$ | 40 (S) | 1.900 | 0.145 | 1.610 | 2.036 | 0.1342 | 0.01414 |
|  | 80 (X) |  | 0.200 | 1.500 | 1.767 | 0.1250 | 0.01227 |
|  | 160 |  | 0.281 | 1.338 | 1.406 | 0.1115 | 0.00976 |
|  | (XX) |  | 0.400 | 1.100 | 0.9503 | 0.0917 | 0.00660 |
| 2 | 40 (S) | 2.375 | 0.154 | 2.067 | 3.356 | 0.1723 | 0.02330 |
|  | 80 (X) |  | 0.218 | 1.939 | 2.953 | 0.1616 | 0.02051 |
|  | 160 |  | 0.343 | 1.689 | 2.240 | 0.1408 | 0.01556 |
|  | (XX) |  | 0.436 | 1.503 | 1.774 | 0.1253 | 0.01232 |
| $2\frac{1}{2}$ | 40 (S) | 2.875 | 0.203 | 2.469 | 4.788 | 0.2058 | 0.03325 |
|  | 80 (X) |  | 0.276 | 2.323 | 4.238 | 0.1936 | 0.02943 |
|  | 160 |  | 0.375 | 2.125 | 3.547 | 0.1771 | 0.02463 |
|  | (XX) |  | 0.552 | 1.771 | 2.464 | 0.1476 | 0.01711 |
| 3 | 40 (S) | 3.500 | 0.216 | 3.068 | 7.393 | 0.2557 | 0.05134 |
|  | 80 (X) |  | 0.300 | 2.900 | 6.605 | 0.2417 | 0.04587 |
|  | 160 |  | 0.437 | 2.626 | 5.416 | 0.2188 | 0.03761 |
|  | (XX) |  | 0.600 | 2.300 | 4.155 | 0.1917 | 0.02885 |
| $3\frac{1}{2}$ | 40 (S) | 4.000 | 0.226 | 3.548 | 9.887 | 0.2957 | 0.06866 |
|  | 80 (X) |  | 0.318 | 3.364 | 8.888 | 0.2803 | 0.06172 |
|  | (XX) |  | 0.636 | 2.728 | 5.845 | 0.2273 | 0.04059 |

*(continued)*

**APPENDIX 18.B** *(continued)*
Dimensions of Welded and Seamless Steel Pipe[a,b]
(selected sizes)[c]
(customary U.S. units)

| nominal diameter (in) | schedule | outside diameter (in) | wall thickness (in) | internal diameter (in) | internal area (in²) | internal diameter (ft) | internal area (ft²) |
|---|---|---|---|---|---|---|---|
| 4 | 40 (S) | 4.500 | 0.237 | 4.026 | 12.73 | 0.3355 | 0.08841 |
|  | 80 (X) |  | 0.337 | 3.826 | 11.50 | 0.3188 | 0.07984 |
|  | 120 |  | 0.437 | 3.626 | 10.33 | 0.3022 | 0.07171 |
|  | 160 |  | 0.531 | 3.438 | 9.283 | 0.2865 | 0.06447 |
|  | (XX) |  | 0.674 | 3.152 | 7.803 | 0.2627 | 0.05419 |
| 5 | 40 (S) | 5.563 | 0.258 | 5.047 | 20.01 | 0.4206 | 0.1389 |
|  | 80 (X) |  | 0.375 | 4.813 | 18.19 | 0.4011 | 0.1263 |
|  | 120 |  | 0.500 | 4.563 | 16.35 | 0.3803 | 0.1136 |
|  | 160 |  | 0.625 | 4.313 | 14.61 | 0.3594 | 0.1015 |
|  | (XX) |  | 0.750 | 4.063 | 12.97 | 0.3386 | 0.09004 |
| 6 | 40 (S) | 6.625 | 0.280 | 6.065 | 28.89 | 0.5054 | 0.2006 |
|  | 80 (X) |  | 0.432 | 5.761 | 26.07 | 0.4801 | 0.1810 |
|  | 120 |  | 0.562 | 5.501 | 23.77 | 0.4584 | 0.1650 |
|  | 160 |  | 0.718 | 5.189 | 21.15 | 0.4324 | 0.1469 |
|  | (XX) |  | 0.864 | 4.897 | 18.83 | 0.4081 | 0.1308 |
| 8 | 20 | 8.625 | 0.250 | 8.125 | 51.85 | 0.6771 | 0.3601 |
|  | 30 |  | 0.277 | 8.071 | 51.16 | 0.6726 | 0.3553 |
|  | 40 (S) |  | 0.322 | 7.981 | 50.03 | 0.6651 | 0.3474 |
|  | 60 |  | 0.406 | 7.813 | 47.94 | 0.6511 | 0.3329 |
|  | 80 (X) |  | 0.500 | 7.625 | 45.66 | 0.6354 | 0.3171 |
|  | 100 |  | 0.593 | 7.439 | 43.46 | 0.6199 | 0.3018 |
|  | 120 |  | 0.718 | 7.189 | 40.59 | 0.5990 | 0.2819 |
|  | 140 |  | 0.812 | 7.001 | 38.50 | 0.5834 | 0.2673 |
|  | (XX) |  | 0.875 | 6.875 | 37.12 | 0.5729 | 0.2578 |
|  | 160 |  | 0.906 | 6.813 | 36.46 | 0.5678 | 0.2532 |
| 10 | 20 | 10.75 | 0.250 | 10.250 | 82.52 | 0.85417 | 0.5730 |
|  | 30 |  | 0.307 | 10.136 | 80.69 | 0.84467 | 0.5604 |
|  | 40 (S) |  | 0.365 | 10.020 | 78.85 | 0.83500 | 0.5476 |
|  | 60 (X) |  | 0.500 | 9.750 | 74.66 | 0.8125 | 0.5185 |
|  | 80 |  | 0.593 | 9.564 | 71.84 | 0.7970 | 0.4989 |
|  | 100 |  | 0.718 | 9.314 | 68.13 | 0.7762 | 0.4732 |
|  | 120 |  | 0.843 | 9.064 | 64.53 | 0.7553 | 0.4481 |
|  | 140 (XX) |  | 1.000 | 8.750 | 60.13 | 0.7292 | 0.4176 |
|  | 160 |  | 1.125 | 8.500 | 56.75 | 0.7083 | 0.3941 |
| 12 | 20 | 12.75 | 0.250 | 12.250 | 117.86 | 1.0208 | 0.8185 |
|  | 30 |  | 0.330 | 12.090 | 114.80 | 1.0075 | 0.7972 |
|  | (S) |  | 0.375 | 12.000 | 113.10 | 1.0000 | 0.7854 |
|  | 40 |  | 0.406 | 11.938 | 111.93 | 0.99483 | 0.7773 |
|  | (X) |  | 0.500 | 11.750 | 108.43 | 0.97917 | 0.7530 |
|  | 60 |  | 0.562 | 11.626 | 106.16 | 0.96883 | 0.7372 |
|  | 80 |  | 0.687 | 11.376 | 101.64 | 0.94800 | 0.7058 |
|  | 100 |  | 0.843 | 11.064 | 96.14 | 0.92200 | 0.6677 |
|  | 120 (XX) |  | 1.000 | 10.750 | 90.76 | 0.89583 | 0.6303 |
|  | 140 |  | 1.125 | 10.500 | 86.59 | 0.87500 | 0.6013 |
|  | 160 |  | 1.312 | 10.126 | 80.53 | 0.84383 | 0.5592 |

*(continued)*

**APPENDIX 18.B** (continued)
Dimensions of Welded and Seamless Steel Pipe[a,b]
(selected sizes)[c]
(customary U.S. units)

| nominal diameter (in) | schedule | outside diameter (in) | wall thickness (in) | internal diameter (in) | internal area (in²) | internal diameter (ft) | internal area (ft²) |
|---|---|---|---|---|---|---|---|
| 14 O.D. | 10 | 14.00 | 0.250 | 13.500 | 143.14 | 1.1250 | 0.9940 |
| | 20 | | 0.312 | 13.376 | 140.52 | 1.1147 | 0.9758 |
| | 30 (S) | | 0.375 | 13.250 | 137.89 | 1.1042 | 0.9575 |
| | 40 | | 0.437 | 13.126 | 135.32 | 1.0938 | 0.9397 |
| | (X) | | 0.500 | 13.000 | 132.67 | 1.0833 | 0.9213 |
| | 60 | | 0.593 | 12.814 | 128.96 | 1.0679 | 0.8956 |
| | 80 | | 0.750 | 12.500 | 122.72 | 1.0417 | 0.8522 |
| | 100 | | 0.937 | 12.126 | 115.48 | 1.0105 | 0.8020 |
| | 120 | | 1.093 | 11.814 | 109.62 | 0.98450 | 0.7612 |
| | 140 | | 1.250 | 11.500 | 103.87 | 0.95833 | 0.7213 |
| | 160 | | 1.406 | 11.188 | 98.31 | 0.93233 | 0.6827 |
| 16 O.D. | 10 | 16.00 | 0.250 | 15.500 | 188.69 | 1.2917 | 1.3104 |
| | 20 | | 0.312 | 15.376 | 185.69 | 1.2813 | 1.2895 |
| | 30 (S) | | 0.375 | 15.250 | 182.65 | 1.2708 | 1.2684 |
| | 40 (X) | | 0.500 | 15.000 | 176.72 | 1.2500 | 1.2272 |
| | 60 | | 0.656 | 14.688 | 169.44 | 1.2240 | 1.1767 |
| | 80 | | 0.843 | 14.314 | 160.92 | 1.1928 | 1.1175 |
| | 100 | | 1.031 | 13.938 | 152.58 | 1.1615 | 1.0596 |
| | 120 | | 1.218 | 13.564 | 144.50 | 1.1303 | 1.0035 |
| | 140 | | 1.437 | 13.126 | 135.32 | 1.0938 | 0.9397 |
| | 160 | | 1.593 | 12.814 | 128.96 | 1.0678 | 0.8956 |
| 18 O.D. | 10 | 18.00 | 0.250 | 17.500 | 240.53 | 1.4583 | 1.6703 |
| | 20 | | 0.312 | 17.376 | 237.13 | 1.4480 | 1.6467 |
| | (S) | | 0.375 | 17.250 | 233.71 | 1.4375 | 1.6230 |
| | 30 | | 0.437 | 17.126 | 230.36 | 1.4272 | 1.5997 |
| | (X) | | 0.500 | 17.000 | 226.98 | 1.4167 | 1.5762 |
| | 40 | | 0.562 | 16.876 | 223.68 | 1.4063 | 1.5533 |
| | 60 | | 0.750 | 16.500 | 213.83 | 1.3750 | 1.4849 |
| | 80 | | 0.937 | 16.126 | 204.24 | 1.3438 | 1.4183 |
| | 100 | | 1.156 | 15.688 | 193.30 | 1.3073 | 1.3423 |
| | 120 | | 1.375 | 15.250 | 182.65 | 1.2708 | 1.2684 |
| | 140 | | 1.562 | 14.876 | 173.81 | 1.2397 | 1.2070 |
| | 160 | | 1.781 | 14.438 | 163.72 | 1.2032 | 1.1370 |
| 20 O.D. | 10 | 20.00 | 0.250 | 19.500 | 298.65 | 1.6250 | 2.0739 |
| | 20 (S) | | 0.375 | 19.250 | 291.04 | 1.6042 | 2.0211 |
| | 30 (X) | | 0.500 | 19.000 | 283.53 | 1.5833 | 1.9689 |
| | 40 | | 0.593 | 18.814 | 278.00 | 1.5678 | 1.9306 |
| | 60 | | 0.812 | 18.376 | 265.21 | 1.5313 | 1.8417 |
| | 80 | | 1.031 | 17.938 | 252.72 | 1.4948 | 1.7550 |
| | 100 | | 1.281 | 17.438 | 238.83 | 1.4532 | 1.6585 |
| | 120 | | 1.500 | 17.000 | 226.98 | 1.4167 | 1.5762 |
| | 140 | | 1.750 | 16.500 | 213.83 | 1.3750 | 1.4849 |
| | 160 | | 1.968 | 16.064 | 202.67 | 1.3387 | 1.4075 |

(continued)

**APPENDIX 18.B** *(continued)*
Dimensions of Welded and Seamless Steel Pipe[a,b]
(selected sizes)[c]
(customary U.S. units)

| nominal diameter (in) | schedule | outside diameter (in) | wall thickness (in) | internal diameter (in) | internal area (in$^2$) | internal diameter (ft) | internal area (ft$^2$) |
|---|---|---|---|---|---|---|---|
| 24 O.D. | 10 | 24.00 | 0.250 | 23.500 | 433.74 | 1.9583 | 3.0121 |
| | 20 (S) | | 0.375 | 23.250 | 424.56 | 1.9375 | 2.9483 |
| | (X) | | 0.500 | 23.000 | 415.48 | 1.9167 | 2.8852 |
| | 30 | | 0.562 | 22.876 | 411.01 | 1.9063 | 2.8542 |
| | 40 | | 0.687 | 22.626 | 402.07 | 1.8855 | 2.7922 |
| | 60 | | 0.968 | 22.060 | 382.20 | 1.8383 | 2.6542 |
| | 80 | | 1.218 | 21.564 | 365.21 | 1.7970 | 2.5362 |
| | 100 | | 1.531 | 20.938 | 344.32 | 1.7448 | 2.3911 |
| | 120 | | 1.812 | 20.376 | 326.92 | 1.6980 | 2.2645 |
| | 140 | | 2.062 | 19.876 | 310.28 | 1.6563 | 2.1547 |
| | 160 | | 2.343 | 19.310 | 292.87 | 1.6092 | 2.0337 |
| 30 O.D. | 10 | 30.00 | 0.312 | 29.376 | 677.76 | 2.4480 | 4.7067 |
| | (S) | | 0.375 | 29.250 | 671.62 | 2.4375 | 4.6640 |
| | 20 (X) | | 0.500 | 29.000 | 660.52 | 2.4167 | 4.5869 |
| | 30 | | 0.625 | 28.750 | 649.18 | 2.3958 | 4.5082 |

(Multiply in by 25.4 to obtain mm.)
(Multiply in$^2$ by 645 to obtain mm$^2$.)
[a]Designations are per ANSI B36.10.
[b]The "S" wall thickness was formerly designated as "standard weight." Standard weight and schedule-40 are the same for all diameters through 10 in. For diameters between 12 in and 24 in, standard weight pipe has a wall thickness of 0.375 in. The "X" wall thickness was formerly designated as "extra strong." Extra strong weight and schedule-80 are the same for all diameters through 8 in. For diameters between 10 in and 24 in, extra strong weight pipe has a wall thickness of 0.50 in. The "XX" wall thickness was formerly designed as "double extra strong." Double extra strong weight pipe does not have a corresponding schedule number.
[c]Pipe sizes and weights in most common usage are listed. Other weights and sizes exist.

**APPENDIX 18.C**
Dimensions of Welded and Seamless Steel Pipe
Schedules 40 and 80
(SI units)

| nominal pipe size (in) | schedule number | wall thickness (mm) | inside diameter (mm) | inside cross-sectional area (m²) |
|---|---|---|---|---|
| $\frac{1}{8}$ | 40 | 1.73 | 6.83 | $0.3664 \times 10^{-4}$ |
|  | 80 | 2.41 | 5.46 | $0.2341 \times 10^{-4}$ |
| $\frac{1}{4}$ | 40 | 2.24 | 9.25 | $0.6720 \times 10^{-4}$ |
|  | 80 | 3.02 | 7.67 | $0.4620 \times 10^{-4}$ |
| $\frac{3}{8}$ | 40 | 2.31 | 12.52 | $1.231 \times 10^{-4}$ |
|  | 80 | 3.20 | 10.74 | $0.9059 \times 10^{-4}$ |
| $\frac{1}{2}$ | 40 | 2.77 | 15.80 | $1.961 \times 10^{-4}$ |
|  | 80 | 3.73 | 13.87 | $1.511 \times 10^{-4}$ |
| $\frac{3}{4}$ | 40 | 2.87 | 20.93 | $3.441 \times 10^{-4}$ |
|  | 80 | 3.91 | 18.85 | $2.791 \times 10^{-4}$ |
| 1 | 40 | 3.38 | 26.64 | $5.574 \times 10^{-4}$ |
|  | 80 | 4.45 | 24.31 | $4.641 \times 10^{-4}$ |
| $1\frac{1}{4}$ | 40 | 3.56 | 35.05 | $9.648 \times 10^{-4}$ |
|  | 80 | 4.85 | 32.46 | $8.275 \times 10^{-4}$ |
| $1\frac{1}{2}$ | 40 | 3.68 | 40.89 | $13.13 \times 10^{-4}$ |
|  | 80 | 5.08 | 38.10 | $11.40 \times 10^{-4}$ |
| 2 | 40 | 3.91 | 52.50 | $21.65 \times 10^{-4}$ |
|  | 80 | 5.54 | 49.25 | $19.05 \times 10^{-4}$ |
| $2\frac{1}{2}$ | 40 | 5.16 | 62.71 | $30.89 \times 10^{-4}$ |
|  | 80 | 7.01 | 59.00 | $27.30 \times 10^{-4}$ |
| 3 | 40 | 5.49 | 77.92 | $47.69 \times 10^{-4}$ |
|  | 80 | 7.62 | 73.66 | $42.61 \times 10^{-4}$ |
| $3\frac{1}{2}$ | 40 | 5.74 | 90.12 | $63.79 \times 10^{-4}$ |
|  | 80 | 8.08 | 85.45 | $57.35 \times 10^{-4}$ |
| 4 | 40 | 6.02 | 102.3 | $82.19 \times 10^{-4}$ |
|  | 80 | 8.56 | 97.18 | $74.17 \times 10^{-4}$ |
| 5 | 40 | 6.55 | 128.2 | $129.1 \times 10^{-4}$ |
|  | 80 | 9.53 | 122.3 | $117.5 \times 10^{-4}$ |
| 6 | 40 | 7.11 | 154.1 | $186.5 \times 10^{-4}$ |
|  | 80 | 10.97 | 146.3 | $168.1 \times 10^{-4}$ |
| 8 | 40 | 8.18 | 202.7 | $322.7 \times 10^{-4}$ |
|  | 80 | 12.70 | 193.7 | $294.7 \times 10^{-4}$ |

*(continued)*

**APPENDIX 18.C** *(continued)*
Dimensions of Welded and Seamless Steel Pipe
Schedules 40 and 80
(SI units)

| nominal pipe size (in) | schedule number | wall thickness (mm) | inside diameter (mm) | inside cross-sectional area (m²) |
|---|---|---|---|---|
| 10 | 40 | 9.27 | 254.5 | $508.6 \times 10^{-4}$ |
|    | 80 | 15.06 | 242.9 | $463.4 \times 10^{-4}$ |
| 12 | 40 | 10.31 | 303.2 | $721.9 \times 10^{-4}$ |
|    | 80 | 17.45 | 289.0 | $655.6 \times 10^{-4}$ |
| 14 | 40 | 11.10 | 333.4 | $872.8 \times 10^{-4}$ |
|    | 80 | 19.05 | 317.5 | $791.5 \times 10^{-4}$ |
| 16 | 40 | 12.70 | 381.0 | $1140 \times 10^{-4}$ |
|    | 80 | 21.41 | 363.6 | $1038 \times 10^{-4}$ |
| 18 | 40 | 14.05 | 428.7 | $1443 \times 10^{-4}$ |
|    | 80 | 23.80 | 409.6 | $1317 \times 10^{-4}$ |
| 20 | 40 | 15.06 | 477.9 | $1793 \times 10^{-4}$ |
|    | 80 | 26.19 | 455.6 | $1630 \times 10^{-4}$ |
| 24 | 40 | 17.45 | 574.7 | $2593 \times 10^{-4}$ |
|    | 80 | 30.94 | 547.7 | $2356 \times 10^{-4}$ |

**APPENDIX 18.D**
Dimensions of Small Diameter PVC Pipe[a,b]
(customary U.S. units)

| nominal size (in) | schedule | wall thickness[c] (in) | outside diameter (in) | inside diameter (in) |
|---|---|---|---|---|
| $\frac{1}{4}$ | 40 | 0.088 | 0.540 | 0.364 |
| | 80 | 0.119 | 0.540 | 0.302 |
| $\frac{1}{2}$ | 40 | 0.109 | 0.840 | 0.622 |
| | 80 | 0.147 | 0.840 | 0.546 |
| $\frac{3}{4}$ | 40 | 0.113 | 1.050 | 0.824 |
| | 80 | 0.154 | 1.050 | 0.742 |
| 1 | 40 | 0.133 | 1.315 | 1.049 |
| | 80 | 0.179 | 1.315 | 0.957 |
| $1\frac{1}{4}$ | 40 | 0.140 | 1.660 | 1.380 |
| | 80 | 0.191 | 1.660 | 1.278 |
| $1\frac{1}{2}$ | 40 | 0.145 | 1.900 | 1.610 |
| | 80 | 0.200 | 1.900 | 1.500 |
| 2 | 40 | 0.154 | 2.375 | 2.067 |
| | 80 | 0.218 | 2.375 | 1.939 |
| 3 | 40 | 0.216 | 3.500 | 3.068 |
| | 80 | 0.300 | 3.500 | 2.900 |
| 4 | 40 | 0.237 | 4.500 | 4.026 |
| | 80 | 0.337 | 4.500 | 3.826 |

(Multiply in by 25.4 to obtain mm.)
(Multiply in$^2$ by 645 to obtain mm$^2$.)
[a]abstracted from ASTM Specification D1785-85
[b]Two strengths of PVC are in use. A maximum (bursting) stress of 6400 psig (44 MPa) applies to PVC types 1120, 1220, and 4116. A maximum (bursting) stress of 5000 psig (35 MPa) applies to PVC types 2112, 2116, and 2120.
[c]minimum wall thickness, with tolerances of −0%, +10%

**APPENDIX 18.E**
Dimensions of Large Diameter PVC Sewer and Water Pipe
(customary U.S. units)

| nominal size (in) | designation | minimum wall thickness (in) | outside diameter (in) |
|---|---|---|---|
| ASTM 2729 | | | |
| 3 | | 0.070 | 3.250 |
| 4 | | 0.075 | 4.215 |
| 6 | | 0.100 | 6.275 |
| | | | |
| ASTM D3034 | | | |
| 4 | regular | 0.120 | 4.215 |
| | HW | 0.162 | 4.215 |
| 6 | regular | 0.180 | 6.275 |
| | HW | 0.241 | 6.275 |
| 8 | regular | 0.240 | 8.400 |
| | HW | 0.323 | 8.400 |
| 10 | regular | 0.300 | 10.50 |
| | HW | 0.404 | 10.50 |
| 12 | regular | 0.360 | 12.50 |
| | HW | 0.481 | 12.50 |
| 15 | regular | 0.437 | 15.30 |
| | HW | 0.588 | 15.30 |
| 18 | regular | | 18.701 |
| | HW | | 18.701 |
| 21 | regular | | 22.047 |
| | HW | | 22.047 |
| 24 | regular | | 24.803 |
| | HW | | 24.803 |
| 27 | regular | | 27.95 |
| | HW | | 27.95 |
| | | | |
| ASTM F679 | | | |
| 18 | regular | 0.536 | 18.701 |
| | HW | 0.719 | 18.801 |
| | | | |
| ASTM AWWA C900 | | | |
| 4 | CL150 | 0.267 | 4.800 |
| 6 | CL150 | 0.383 | 6.900 |
| 8 | CL150 | 0.503 | 9.060 |
| 10 | CL150 | 0.617 | 11.100 |
| 12 | CL150 | 0.733 | 13.200 |

(Multiply in by 25.4 to obtain mm.)

**APPENDIX 18.F**
Dimensions and Weights of Concrete Sewer Pipe
(customary U.S. units)

| 3000 psi | | | 3500 psi | | | 4000 psi | | |
|---|---|---|---|---|---|---|---|---|
| internal diameter (in) | minimum shell thickness (in) | mass per foot* (lbm/ft) | internal diameter (in) | minimum shell thickness (in) | mass per foot* (lbm/ft) | internal diameter (in) | minimum shell thickness (in) | mass per foot* (lbm/ft) |
| 12 | 2 | 93 | 12 | $1\frac{3}{4}$ | 79 | 12 | | |
| 15 | $2\frac{1}{4}$ | 127 | 15 | 2 | 111 | 15 | | |
| 18 | $2\frac{1}{2}$ | 168 | 18 | | | 18 | 2 | 131 |
| 21 | $2\frac{3}{4}$ | 214 | 21 | | | 21 | $2\frac{1}{4}$ | 171 |
| 24 | 3 | 264 | 24 | $2\frac{5}{8}$ | 229 | 24 | $2\frac{1}{2}$ | 217 |
| 27 | 3 | 295 | 27 | $2\frac{3}{4}$ | 268 | 27 | $2\frac{5}{8}$ | 255 |
| 30 | $3\frac{1}{2}$ | 384 | 30 | 3 | 324 | 30 | $2\frac{3}{4}$ | 295 |
| 33 | $3\frac{3}{4}$ | 451 | 33 | $3\frac{1}{4}$ | 396 | 33 | $2\frac{3}{4}$ | 322 |
| 36 | 4 | 524 | 36 | $3\frac{3}{8}$ | 435 | 36 | 3 | 383 |
| 42 | $4\frac{1}{2}$ | 686 | 42 | $3\frac{3}{4}$ | 561 | 42 | $3\frac{3}{8}$ | 500 |
| 48 | 5 | 867 | 48 | $4\frac{1}{4}$ | 727 | 48 | $3\frac{3}{4}$ | 635 |
| 54 | $5\frac{1}{2}$ | 1068 | 54 | $4\frac{5}{8}$ | 887 | 54 | $4\frac{1}{4}$ | 810 |
| 60 | 6 | 1295 | 60 | 5 | 1064 | 60 | $4\frac{1}{2}$ | 950 |
| 66 | $6\frac{1}{2}$ | 1542 | 66 | $5\frac{3}{8}$ | 1256 | 66 | $4\frac{3}{4}$ | 1100 |
| 72 | 7 | 1811 | 72 | $5\frac{3}{4}$ | 1463 | 72 | 5 | 1260 |
| 78 | $7\frac{1}{2}$ | 2100 | | | | | | |
| 84 | 8 | 2409 | | | | | | |
| 90 | 8 | 2565 | | | | | | |
| 96 | $8\frac{1}{2}$ | 2906 | | | | | | |
| 108 | 9 | 3446 | | | | | | |

(Multiply in by 25.4 to obtain mm.)
(Multiply psi by 6.89 to obtain kPa.)
(Multiply lbm/ft by 1.488 to obtain kg/m.)
*Masses given are for reinforced tongue-and-groove pipe. Reinforced bell-and-spigot pipe is heavier. Based on 150 lbm/ft$^3$ concrete.

**APPENDIX 18.G**
Dimensions of Cast Iron Pipe
(customary U.S. units)
(all dimensions in inches)

| nominal diameter | class A 100 ft head 43 psig outside diameter | class A 100 ft head 43 psig inside diameter | class B 200 ft head 86 psig outside diameter | class B 200 ft head 86 psig inside diameter | class C 300 ft head 130 psig outside diameter | class C 300 ft head 130 psig inside diameter | class D 400 ft head 173 psig outside diameter | class D 400 ft head 173 psig inside diameter |
|---|---|---|---|---|---|---|---|---|
| 3 | 3.80 | 3.02 | 3.96 | 3.12 | 3.96 | 3.06 | 3.96 | 3.00 |
| 4 | 4.80 | 3.96 | 5.00 | 4.10 | 5.00 | 4.04 | 5.00 | 3.96 |
| 6 | 6.90 | 6.02 | 7.10 | 6.14 | 7.10 | 6.08 | 7.10 | 6.00 |
| 8 | 9.05 | 8.13 | 9.05 | 8.03 | 9.30 | 8.18 | 9.30 | 8.10 |
| 10 | 11.10 | 10.10 | 11.10 | 9.96 | 11.40 | 10.16 | 11.40 | 10.04 |
| 12 | 13.20 | 12.12 | 13.20 | 11.96 | 13.50 | 12.14 | 13.50 | 12.00 |
| 14 | 15.30 | 14.16 | 15.30 | 13.98 | 15.65 | 14.17 | 15.65 | 14.01 |
| 16 | 17.40 | 16.20 | 17.40 | 16.00 | 17.80 | 16.20 | 17.80 | 16.02 |
| 18 | 19.50 | 18.22 | 19.50 | 18.00 | 19.92 | 18.18 | 19.92 | 18.00 |
| 20 | 21.60 | 20.26 | 21.60 | 20.00 | 22.06 | 20.22 | 22.06 | 20.00 |
| 24 | 25.80 | 24.28 | 25.80 | 24.02 | 26.32 | 24.22 | 26.32 | 24.00 |
| 30 | 31.74 | 29.98 | 32.00 | 29.94 | 32.40 | 30.00 | 32.74 | 30.00 |
| 36 | 37.96 | 35.98 | 38.30 | 36.00 | 38.70 | 39.98 | 39.16 | 36.00 |
| 42 | 44.20 | 42.00 | 44.50 | 41.94 | 45.10 | 42.02 | 45.58 | 42.02 |
| 48 | 50.50 | 47.98 | 50.80 | 47.96 | 51.40 | 47.98 | 51.98 | 48.06 |
| 54 | 56.66 | 53.96 | 57.10 | 54.00 | 57.80 | 54.00 | 58.40 | 53.94 |
| 60 | 62.80 | 60.02 | 63.40 | 60.06 | 64.20 | 60.20 | 64.82 | 60.06 |
| 72 | 75.34 | 72.10 | 76.00 | 72.10 | 76.88 | 72.10 | | |
| 84 | 87.54 | 84.10 | 88.54 | 84.10 | | | | |

| nominal diameter | class E 500 ft head 217 psig outside diameter | class E 500 ft head 217 psig inside diameter | class F 600 ft head 260 psig outside diameter | class F 600 ft head 260 psig inside diameter | class G 700 ft head 304 psig outside diameter | class G 700 ft head 304 psig inside diameter | class H 800 ft head 347 psig outside diameter | class H 800 ft head 347 psig inside diameter |
|---|---|---|---|---|---|---|---|---|
| 6 | 7.22 | 6.06 | 7.22 | 6.00 | 7.38 | 6.08 | 7.38 | 6.00 |
| 8 | 9.42 | 8.10 | 9.42 | 8.00 | 9.60 | 8.10 | 9.60 | 8.00 |
| 10 | 11.60 | 10.12 | 11.60 | 10.00 | 11.84 | 10.12 | 11.84 | 10.00 |
| 12 | 13.78 | 12.14 | 13.78 | 12.00 | 14.08 | 12.14 | 14.08 | 12.00 |
| 14 | 15.98 | 14.18 | 15.98 | 14.00 | 16.32 | 14.18 | 16.32 | 14.00 |
| 16 | 18.16 | 16.20 | 18.16 | 16.00 | 18.54 | 16.18 | 18.54 | 16.00 |
| 18 | 20.34 | 18.20 | 20.34 | 18.00 | 20.78 | 18.22 | 20.78 | 18.00 |
| 20 | 22.54 | 20.24 | 22.54 | 20.00 | 23.02 | 20.24 | 23.02 | 20.00 |
| 24 | 26.90 | 24.28 | 26.90 | 24.00 | 27.76 | 24.26 | 27.76 | 24.00 |
| 30 | 33.10 | 30.00 | 33.46 | 30.00 | | | | |
| 36 | 39.60 | 36.00 | 40.04 | 36.00 | | | | |

(Multiply in by 25.4 to obtain mm.)

**APPENDIX 18.H**
Standard ANSI Piping Symbols

| | flanged | screwed | bell and spigot | welded | soldered |
|---|---|---|---|---|---|
| joint | | | | | |
| elbow—90° | | | | | |
| elbow—45° | | | | | |
| elbow—turned up | | | | | |
| elbow—turned down | | | | | |
| elbow—long radius | | | | | |
| reducing elbow | | | | | |
| tee | | | | | |
| tee—outlet up | | | | | |
| tee—outlet down | | | | | |
| side outlet tee—outlet up | | | | | |
| cross | | | | | |
| reducer—concentric | | | | | |
| reducer—eccentric | | | | | |
| lateral | | | | | |
| gate valve | | | | | |
| globe valve | | | | | |
| check valve | | | | | |
| stop cock | | | | | |
| safety valve | | | | | |
| expansion joint | | | | | |
| union | | | | | |
| sleeve | | | | | |
| bushing | | | | | |

## APPENDIX 19.A
Specific Roughness and Hazen-Williams Constants for Various Water Pipe Materials[a]

| type of pipe or surface | $\epsilon$ (ft) range | $\epsilon$ (ft) design | $C$ range | $C$ clean | $C$ design[b] |
|---|---|---|---|---|---|
| steel | | | | | |
| welded and seamless | 0.0001–0.0003 | 0.0002 | 150–80 | 140 | 100 |
| interior riveted, no projecting rivets | | | | 139 | 100 |
| projecting girth rivets | | | | 130 | 100 |
| projecting girth and horizontal rivets | | | | 115 | 100 |
| vitrified, spiral-riveted, flow with lap | | | | 110 | 100 |
| vitrified, spiral-riveted, flow against lap | | | | 100 | 90 |
| corrugated | | | | 60 | 60 |
| | | | | | |
| mineral | | | | | |
| concrete | 0.001–0.01 | 0.004 | 152–85 | 120 | 100 |
| cement-asbestos | | | 160–140 | 150 | 140 |
| vitrified clays | | | | | 110 |
| brick sewer | | | | | 100 |
| | | | | | |
| iron | | | | | |
| cast, plain | 0.0004–0.002 | 0.0008 | 150–80 | 130 | 100 |
| cast, tar (asphalt) coated | 0.0002–0.0006 | 0.0004 | 145–50 | 130 | 100 |
| cast, cement lined | 0.000008 | 0.000008 | | 150 | 140 |
| cast, bituminous lined | 0.000008 | 0.000008 | 160–130 | 148 | 140 |
| cast, centrifugally spun | 0.00001 | 0.00001 | | | |
| galvanized, plain | 0.0002–0.0008 | 0.0005 | | | |
| wrought, plain | 0.0001–0.0003 | 0.0002 | 150–80 | 130 | 100 |
| | | | | | |
| miscellaneous | | | | | |
| copper and brass | 0.000005 | 0.000005 | 150–120 | 140 | 130 |
| wood stave | 0.0006–0.003 | 0.002 | 145–110 | 120 | 110 |
| transite | 0.000008 | 0.000008 | | | |
| lead, tin, glass | | 0.000005 | 150–120 | 140 | 130 |
| plastic (PVC and ABS) | | 0.000005 | 150–120 | 150–140 | 130 |
| fiberglass | 0.000017 | 0.000017 | 160–150 | 150 | 150 |

(Multiply ft by 0.3 to obtain m.)

[a]$C$ values for sludge are 20% to 40% less than the corresponding water pipe values.

[b]The following guidelines are provided for selecting Hazen-Williams coefficients for cast-iron pipes of different ages. Values for welded steel pipe are similar to those of cast-iron pipe 5 years older. New pipe, all sizes: $C = 130$. 5 year-old pipe: $C = 120$ ($d < 24$ in); $C = 115$ ($d \geq 24$ in). 10 year-old pipe: $C = 105$ ($d = 4$ in); $C = 110$ ($d = 12$ in); $C = 85$ ($d \geq 30$ in). 40 year-old pipe: $C = 65$ ($d = 4$ in); $C = 80$ ($d = 16$ in).

**APPENDIX 19.B**
Darcy Friction Factors (turbulent flow)

relative roughness ($\epsilon/D$)

| Reynolds no. | 0.00000 | 0.000001 | 0.0000015 | 0.00001 | 0.00002 | 0.00004 | 0.00005 | 0.00006 | 0.00008 |
|---|---|---|---|---|---|---|---|---|---|
| $2 \times 10^3$ | 0.0495 | 0.0495 | 0.0495 | 0.0495 | 0.0495 | 0.0495 | 0.0495 | 0.0495 | 0.0495 |
| $2.5 \times 10^3$ | 0.0461 | 0.0461 | 0.0461 | 0.0461 | 0.0461 | 0.0461 | 0.0461 | 0.0461 | 0.0461 |
| $3 \times 10^3$ | 0.0435 | 0.0435 | 0.0435 | 0.0435 | 0.0435 | 0.0436 | 0.0436 | 0.0436 | 0.0436 |
| $4 \times 10^3$ | 0.0399 | 0.0399 | 0.0399 | 0.0399 | 0.0399 | 0.0399 | 0.0400 | 0.0400 | 0.0400 |
| $5 \times 10^3$ | 0.0374 | 0.0374 | 0.0374 | 0.0374 | 0.0374 | 0.0374 | 0.0374 | 0.0375 | 0.0375 |
| $6 \times 10^3$ | 0.0355 | 0.0355 | 0.0355 | 0.0355 | 0.0355 | 0.0356 | 0.0356 | 0.0356 | 0.0356 |
| $7 \times 10^3$ | 0.0340 | 0.0340 | 0.0340 | 0.0340 | 0.0340 | 0.0341 | 0.0341 | 0.0341 | 0.0341 |
| $8 \times 10^3$ | 0.0328 | 0.0328 | 0.0328 | 0.0328 | 0.0328 | 0.0328 | 0.0329 | 0.0329 | 0.0329 |
| $9 \times 10^3$ | 0.0318 | 0.0318 | 0.0318 | 0.0318 | 0.0318 | 0.0318 | 0.0318 | 0.0319 | 0.0319 |
| $1 \times 10^4$ | 0.0309 | 0.0309 | 0.0309 | 0.0309 | 0.0309 | 0.0309 | 0.0310 | 0.0310 | 0.0310 |
| $1.5 \times 10^4$ | 0.0278 | 0.0278 | 0.0278 | 0.0278 | 0.0278 | 0.0279 | 0.0279 | 0.0279 | 0.0280 |
| $2 \times 10^4$ | 0.0259 | 0.0259 | 0.0259 | 0.0259 | 0.0259 | 0.0260 | 0.0260 | 0.0260 | 0.0261 |
| $2.5 \times 10^4$ | 0.0245 | 0.0245 | 0.0245 | 0.0245 | 0.0246 | 0.0246 | 0.0246 | 0.0247 | 0.0247 |
| $3 \times 10^4$ | 0.0235 | 0.0235 | 0.0235 | 0.0235 | 0.0235 | 0.0236 | 0.0236 | 0.0236 | 0.0237 |
| $4 \times 10^4$ | 0.0220 | 0.0220 | 0.0220 | 0.0220 | 0.0220 | 0.0221 | 0.0221 | 0.0222 | 0.0222 |
| $5 \times 10^4$ | 0.0209 | 0.0209 | 0.0209 | 0.0209 | 0.0210 | 0.0210 | 0.0211 | 0.0211 | 0.0212 |
| $6 \times 10^4$ | 0.0201 | 0.0201 | 0.0201 | 0.0201 | 0.0201 | 0.0202 | 0.0203 | 0.0203 | 0.0204 |
| $7 \times 10^4$ | 0.0194 | 0.0194 | 0.0194 | 0.0194 | 0.0195 | 0.0196 | 0.0196 | 0.0197 | 0.0197 |
| $8 \times 10^4$ | 0.0189 | 0.0189 | 0.0189 | 0.0189 | 0.0190 | 0.0190 | 0.0191 | 0.0191 | 0.0192 |
| $9 \times 10^4$ | 0.0184 | 0.0184 | 0.0184 | 0.0184 | 0.0185 | 0.0186 | 0.0186 | 0.0187 | 0.0188 |
| $1 \times 10^5$ | 0.0180 | 0.0180 | 0.0180 | 0.0180 | 0.0181 | 0.0182 | 0.0183 | 0.0183 | 0.0184 |
| $1.5 \times 10^5$ | 0.0166 | 0.0166 | 0.0166 | 0.0166 | 0.0167 | 0.0168 | 0.0169 | 0.0170 | 0.0171 |
| $2 \times 10^5$ | 0.0156 | 0.0156 | 0.0156 | 0.0157 | 0.0158 | 0.0160 | 0.0160 | 0.0161 | 0.0163 |
| $2.5 \times 10^5$ | 0.0150 | 0.0150 | 0.0150 | 0.0151 | 0.0152 | 0.0153 | 0.0154 | 0.0155 | 0.0157 |
| $3 \times 10^5$ | 0.0145 | 0.0145 | 0.0145 | 0.0146 | 0.0147 | 0.0149 | 0.0150 | 0.0151 | 0.0153 |
| $4 \times 10^5$ | 0.0137 | 0.0137 | 0.0137 | 0.0138 | 0.0140 | 0.0142 | 0.0143 | 0.0144 | 0.0146 |
| $5 \times 10^5$ | 0.0132 | 0.0132 | 0.0132 | 0.0133 | 0.0134 | 0.0137 | 0.0138 | 0.0140 | 0.0142 |
| $6 \times 10^5$ | 0.0127 | 0.0128 | 0.0128 | 0.0129 | 0.0131 | 0.0133 | 0.0135 | 0.0136 | 0.0139 |
| $7 \times 10^5$ | 0.0124 | 0.0124 | 0.0124 | 0.0126 | 0.0127 | 0.0131 | 0.0132 | 0.0134 | 0.0136 |
| $8 \times 10^5$ | 0.0121 | 0.0121 | 0.0121 | 0.0123 | 0.0125 | 0.0128 | 0.0130 | 0.0131 | 0.0134 |
| $9 \times 10^5$ | 0.0119 | 0.0119 | 0.0119 | 0.0121 | 0.0123 | 0.0126 | 0.0128 | 0.0130 | 0.0133 |
| $1 \times 10^6$ | 0.0116 | 0.0117 | 0.0117 | 0.0119 | 0.0121 | 0.0125 | 0.0126 | 0.0128 | 0.0131 |
| $1.5 \times 10^6$ | 0.0109 | 0.0109 | 0.0109 | 0.0112 | 0.0114 | 0.0119 | 0.0121 | 0.0123 | 0.0127 |
| $2 \times 10^6$ | 0.0104 | 0.0104 | 0.0104 | 0.0107 | 0.0110 | 0.0116 | 0.0118 | 0.0120 | 0.0124 |
| $2.5 \times 10^6$ | 0.0100 | 0.0100 | 0.0101 | 0.0104 | 0.0108 | 0.0113 | 0.0116 | 0.0118 | 0.0123 |
| $3 \times 10^6$ | 0.0097 | 0.0098 | 0.0098 | 0.0102 | 0.0105 | 0.0112 | 0.0115 | 0.0117 | 0.0122 |
| $4 \times 10^6$ | 0.0093 | 0.0094 | 0.0094 | 0.0098 | 0.0103 | 0.0110 | 0.0113 | 0.0115 | 0.0120 |
| $5 \times 10^6$ | 0.0090 | 0.0091 | 0.0091 | 0.0096 | 0.0101 | 0.0108 | 0.0111 | 0.0114 | 0.0119 |
| $6 \times 10^6$ | 0.0087 | 0.0088 | 0.0089 | 0.0094 | 0.0099 | 0.0107 | 0.0110 | 0.0113 | 0.0118 |
| $7 \times 10^6$ | 0.0085 | 0.0086 | 0.0087 | 0.0093 | 0.0098 | 0.0106 | 0.0110 | 0.0113 | 0.0118 |
| $8 \times 10^6$ | 0.0084 | 0.0085 | 0.0085 | 0.0092 | 0.0097 | 0.0106 | 0.0109 | 0.0112 | 0.0118 |
| $9 \times 10^6$ | 0.0082 | 0.0083 | 0.0084 | 0.0091 | 0.0097 | 0.0105 | 0.0109 | 0.0112 | 0.0117 |
| $1 \times 10^7$ | 0.0081 | 0.0082 | 0.0083 | 0.0090 | 0.0096 | 0.0105 | 0.0109 | 0.0112 | 0.0117 |
| $1.5 \times 10^7$ | 0.0076 | 0.0078 | 0.0079 | 0.0087 | 0.0094 | 0.0104 | 0.0108 | 0.0111 | 0.0116 |
| $2 \times 10^7$ | 0.0073 | 0.0075 | 0.0076 | 0.0086 | 0.0093 | 0.0103 | 0.0107 | 0.0110 | 0.0116 |
| $2.5 \times 10^7$ | 0.0071 | 0.0073 | 0.0074 | 0.0085 | 0.0093 | 0.0103 | 0.0107 | 0.0110 | 0.0116 |
| $3 \times 10^7$ | 0.0069 | 0.0072 | 0.0073 | 0.0084 | 0.0092 | 0.0103 | 0.0107 | 0.0110 | 0.0116 |
| $4 \times 10^7$ | 0.0067 | 0.0070 | 0.0071 | 0.0084 | 0.0092 | 0.0102 | 0.0106 | 0.0110 | 0.0115 |
| $5 \times 10^7$ | 0.0065 | 0.0068 | 0.0070 | 0.0083 | 0.0092 | 0.0102 | 0.0106 | 0.0110 | 0.0115 |

**APPENDIX 19.B** *(continued)*
Darcy Friction Factors (turbulent flow)

relative roughness ($\epsilon/D$)

| Reynolds no. | 0.0001 | 0.00015 | 0.00020 | 0.00025 | 0.00030 | 0.00035 | 0.0004 | 0.0006 | 0.0008 |
|---|---|---|---|---|---|---|---|---|---|
| $2 \times 10^3$ | 0.0495 | 0.0496 | 0.0496 | 0.0496 | 0.0497 | 0.0497 | 0.0498 | 0.0499 | 0.0501 |
| $2.5 \times 10^3$ | 0.0461 | 0.0462 | 0.0462 | 0.0463 | 0.0463 | 0.0463 | 0.0464 | 0.0466 | 0.0467 |
| $3 \times 10^3$ | 0.0436 | 0.0437 | 0.0437 | 0.0437 | 0.0438 | 0.0438 | 0.0439 | 0.0441 | 0.0442 |
| $4 \times 10^3$ | 0.0400 | 0.0401 | 0.0401 | 0.0402 | 0.0402 | 0.0403 | 0.0403 | 0.0405 | 0.0407 |
| $5 \times 10^3$ | 0.0375 | 0.0376 | 0.0376 | 0.0377 | 0.0377 | 0.0378 | 0.0378 | 0.0381 | 0.0383 |
| $6 \times 10^3$ | 0.0356 | 0.0357 | 0.0357 | 0.0358 | 0.0359 | 0.0359 | 0.0360 | 0.0362 | 0.0365 |
| $7 \times 10^3$ | 0.0341 | 0.0342 | 0.0343 | 0.0343 | 0.0344 | 0.0345 | 0.0345 | 0.0348 | 0.0350 |
| $8 \times 10^3$ | 0.0329 | 0.0330 | 0.0331 | 0.0331 | 0.0332 | 0.0333 | 0.0333 | 0.0336 | 0.0339 |
| $9 \times 10^3$ | 0.0319 | 0.0320 | 0.0321 | 0.0321 | 0.0322 | 0.0323 | 0.0323 | 0.0326 | 0.0329 |
| $1 \times 10^4$ | 0.0310 | 0.0311 | 0.0312 | 0.0313 | 0.0313 | 0.0314 | 0.0315 | 0.0318 | 0.0321 |
| $1.5 \times 10^4$ | 0.0280 | 0.0281 | 0.0282 | 0.0283 | 0.0284 | 0.0285 | 0.0285 | 0.0289 | 0.0293 |
| $2 \times 10^4$ | 0.0261 | 0.0262 | 0.0263 | 0.0264 | 0.0265 | 0.0266 | 0.0267 | 0.0272 | 0.0276 |
| $2.5 \times 10^4$ | 0.0248 | 0.0249 | 0.0250 | 0.0251 | 0.0252 | 0.0254 | 0.0255 | 0.0259 | 0.0264 |
| $3 \times 10^4$ | 0.0238 | 0.0239 | 0.0240 | 0.0241 | 0.0243 | 0.0244 | 0.0245 | 0.0250 | 0.0255 |
| $4 \times 10^4$ | 0.0223 | 0.0224 | 0.0226 | 0.0227 | 0.0229 | 0.0230 | 0.0232 | 0.0237 | 0.0243 |
| $5 \times 10^4$ | 0.0212 | 0.0214 | 0.0216 | 0.0218 | 0.0219 | 0.0221 | 0.0223 | 0.0229 | 0.0235 |
| $6 \times 10^4$ | 0.0205 | 0.0207 | 0.0208 | 0.0210 | 0.0212 | 0.0214 | 0.0216 | 0.0222 | 0.0229 |
| $7 \times 10^4$ | 0.0198 | 0.0200 | 0.0202 | 0.0204 | 0.0206 | 0.0208 | 0.0210 | 0.0217 | 0.0224 |
| $8 \times 10^4$ | 0.0193 | 0.0195 | 0.0198 | 0.0200 | 0.0202 | 0.0204 | 0.0206 | 0.0213 | 0.0220 |
| $9 \times 10^4$ | 0.0189 | 0.0191 | 0.0194 | 0.0196 | 0.0198 | 0.0200 | 0.0202 | 0.0210 | 0.0217 |
| $1 \times 10^5$ | 0.0185 | 0.0188 | 0.0190 | 0.0192 | 0.0195 | 0.0197 | 0.0199 | 0.0207 | 0.0215 |
| $1.5 \times 10^5$ | 0.0172 | 0.0175 | 0.0178 | 0.0181 | 0.0184 | 0.0186 | 0.0189 | 0.0198 | 0.0207 |
| $2 \times 10^5$ | 0.0164 | 0.0168 | 0.0171 | 0.0174 | 0.0177 | 0.0180 | 0.0183 | 0.0193 | 0.0202 |
| $2.5 \times 10^5$ | 0.0158 | 0.0162 | 0.0166 | 0.0170 | 0.0173 | 0.0176 | 0.0179 | 0.0190 | 0.0199 |
| $3 \times 10^5$ | 0.0154 | 0.0159 | 0.0163 | 0.0166 | 0.0170 | 0.0173 | 0.0176 | 0.0188 | 0.0197 |
| $4 \times 10^5$ | 0.0148 | 0.0153 | 0.0158 | 0.0162 | 0.0166 | 0.0169 | 0.0172 | 0.0184 | 0.0195 |
| $5 \times 10^5$ | 0.0144 | 0.0150 | 0.0154 | 0.0159 | 0.0163 | 0.0167 | 0.0170 | 0.0183 | 0.0193 |
| $6 \times 10^5$ | 0.0141 | 0.0147 | 0.0152 | 0.0157 | 0.0161 | 0.0165 | 0.0168 | 0.0181 | 0.0192 |
| $7 \times 10^5$ | 0.0139 | 0.0145 | 0.0150 | 0.0155 | 0.0159 | 0.0163 | 0.0167 | 0.0180 | 0.0191 |
| $8 \times 10^5$ | 0.0137 | 0.0143 | 0.0149 | 0.0154 | 0.0158 | 0.0162 | 0.0166 | 0.0180 | 0.0191 |
| $9 \times 10^5$ | 0.0136 | 0.0142 | 0.0148 | 0.0153 | 0.0157 | 0.0162 | 0.0165 | 0.0179 | 0.0190 |
| $1 \times 10^6$ | 0.0134 | 0.0141 | 0.0147 | 0.0152 | 0.0157 | 0.0161 | 0.0165 | 0.0178 | 0.0190 |
| $1.5 \times 10^6$ | 0.0130 | 0.0138 | 0.0144 | 0.0149 | 0.0154 | 0.0159 | 0.0163 | 0.0177 | 0.0189 |
| $2 \times 10^6$ | 0.0128 | 0.0136 | 0.0142 | 0.0148 | 0.0153 | 0.0158 | 0.0162 | 0.0176 | 0.0188 |
| $2.5 \times 10^6$ | 0.0127 | 0.0135 | 0.0141 | 0.0147 | 0.0152 | 0.0157 | 0.0161 | 0.0176 | 0.0188 |
| $3 \times 10^6$ | 0.0126 | 0.0134 | 0.0141 | 0.0147 | 0.0152 | 0.0157 | 0.0161 | 0.0176 | 0.0187 |
| $4 \times 10^6$ | 0.0124 | 0.0133 | 0.0140 | 0.0146 | 0.0151 | 0.0156 | 0.0161 | 0.0175 | 0.0187 |
| $5 \times 10^6$ | 0.0123 | 0.0132 | 0.0139 | 0.0146 | 0.0151 | 0.0156 | 0.0160 | 0.0175 | 0.0187 |
| $6 \times 10^6$ | 0.0123 | 0.0132 | 0.0139 | 0.0145 | 0.0151 | 0.0156 | 0.0160 | 0.0175 | 0.0187 |
| $7 \times 10^6$ | 0.0122 | 0.0132 | 0.0139 | 0.0145 | 0.0151 | 0.0155 | 0.0160 | 0.0175 | 0.0187 |
| $8 \times 10^6$ | 0.0122 | 0.0131 | 0.0139 | 0.0145 | 0.0150 | 0.0155 | 0.0160 | 0.0175 | 0.0187 |
| $9 \times 10^6$ | 0.0122 | 0.0131 | 0.0139 | 0.0145 | 0.0150 | 0.0155 | 0.0160 | 0.0175 | 0.0187 |
| $1 \times 10^7$ | 0.0122 | 0.0131 | 0.0138 | 0.0145 | 0.0150 | 0.0155 | 0.0160 | 0.0175 | 0.0186 |
| $1.5 \times 10^7$ | 0.0121 | 0.0131 | 0.0138 | 0.0144 | 0.0150 | 0.0155 | 0.0159 | 0.0174 | 0.0186 |
| $2 \times 10^7$ | 0.0121 | 0.0130 | 0.0138 | 0.0144 | 0.0150 | 0.0155 | 0.0159 | 0.0174 | 0.0186 |
| $2.5 \times 10^7$ | 0.0121 | 0.0130 | 0.0138 | 0.0144 | 0.0150 | 0.0155 | 0.0159 | 0.0174 | 0.0186 |
| $3 \times 10^7$ | 0.0120 | 0.0130 | 0.0138 | 0.0144 | 0.0150 | 0.0155 | 0.0159 | 0.0174 | 0.0186 |
| $4 \times 10^7$ | 0.0120 | 0.0130 | 0.0138 | 0.0144 | 0.0150 | 0.0155 | 0.0159 | 0.0174 | 0.0186 |
| $5 \times 10^7$ | 0.0120 | 0.0130 | 0.0138 | 0.0144 | 0.0150 | 0.0155 | 0.0159 | 0.0174 | 0.0186 |

**APPENDIX 19.B** *(continued)*
Darcy Friction Factors (turbulent flow)

relative roughness ($\epsilon/D$)

| Reynolds no. | 0.001 | 0.0015 | 0.002 | 0.0025 | 0.003 | 0.0035 | 0.004 | 0.006 | 0.008 |
|---|---|---|---|---|---|---|---|---|---|
| $2 \times 10^3$ | 0.0502 | 0.0506 | 0.0510 | 0.0513 | 0.0517 | 0.0521 | 0.0525 | 0.0539 | 0.0554 |
| $2.5 \times 10^3$ | 0.0469 | 0.0473 | 0.0477 | 0.0481 | 0.0485 | 0.0489 | 0.0493 | 0.0509 | 0.0524 |
| $3 \times 10^3$ | 0.0444 | 0.0449 | 0.0453 | 0.0457 | 0.0462 | 0.0466 | 0.0470 | 0.0487 | 0.0503 |
| $4 \times 10^3$ | 0.0409 | 0.0414 | 0.0419 | 0.0424 | 0.0429 | 0.0433 | 0.0438 | 0.0456 | 0.0474 |
| $5 \times 10^3$ | 0.0385 | 0.0390 | 0.0396 | 0.0401 | 0.0406 | 0.0411 | 0.0416 | 0.0436 | 0.0455 |
| $6 \times 10^3$ | 0.0367 | 0.0373 | 0.0378 | 0.0384 | 0.0390 | 0.0395 | 0.0400 | 0.0421 | 0.0441 |
| $7 \times 10^3$ | 0.0353 | 0.0359 | 0.0365 | 0.0371 | 0.0377 | 0.0383 | 0.0388 | 0.0410 | 0.0430 |
| $8 \times 10^3$ | 0.0341 | 0.0348 | 0.0354 | 0.0361 | 0.0367 | 0.0373 | 0.0379 | 0.0401 | 0.0422 |
| $9 \times 10^3$ | 0.0332 | 0.0339 | 0.0345 | 0.0352 | 0.0358 | 0.0365 | 0.0371 | 0.0394 | 0.0416 |
| $1 \times 10^4$ | 0.0324 | 0.0331 | 0.0338 | 0.0345 | 0.0351 | 0.0358 | 0.0364 | 0.0388 | 0.0410 |
| $1.5 \times 10^4$ | 0.0296 | 0.0305 | 0.0313 | 0.0320 | 0.0328 | 0.0335 | 0.0342 | 0.0369 | 0.0393 |
| $2 \times 10^4$ | 0.0279 | 0.0289 | 0.0298 | 0.0306 | 0.0315 | 0.0323 | 0.0330 | 0.0358 | 0.0384 |
| $2.5 \times 10^4$ | 0.0268 | 0.0278 | 0.0288 | 0.0297 | 0.0306 | 0.0314 | 0.0322 | 0.0352 | 0.0378 |
| $3 \times 10^4$ | 0.0260 | 0.0271 | 0.0281 | 0.0291 | 0.0300 | 0.0308 | 0.0317 | 0.0347 | 0.0374 |
| $4 \times 10^4$ | 0.0248 | 0.0260 | 0.0271 | 0.0282 | 0.0291 | 0.0301 | 0.0309 | 0.0341 | 0.0369 |
| $5 \times 10^4$ | 0.0240 | 0.0253 | 0.0265 | 0.0276 | 0.0286 | 0.0296 | 0.0305 | 0.0337 | 0.0365 |
| $6 \times 10^4$ | 0.0235 | 0.0248 | 0.0261 | 0.0272 | 0.0283 | 0.0292 | 0.0302 | 0.0335 | 0.0363 |
| $7 \times 10^4$ | 0.0230 | 0.0245 | 0.0257 | 0.0269 | 0.0280 | 0.0290 | 0.0299 | 0.0333 | 0.0362 |
| $8 \times 10^4$ | 0.0227 | 0.0242 | 0.0255 | 0.0267 | 0.0278 | 0.0288 | 0.0298 | 0.0331 | 0.0361 |
| $9 \times 10^4$ | 0.0224 | 0.0239 | 0.0253 | 0.0265 | 0.0276 | 0.0286 | 0.0296 | 0.0330 | 0.0360 |
| $1 \times 10^5$ | 0.0222 | 0.0237 | 0.0251 | 0.0263 | 0.0275 | 0.0285 | 0.0295 | 0.0329 | 0.0359 |
| $1.5 \times 10^5$ | 0.0214 | 0.0231 | 0.0246 | 0.0259 | 0.0271 | 0.0281 | 0.0292 | 0.0327 | 0.0357 |
| $2 \times 10^5$ | 0.0210 | 0.0228 | 0.0243 | 0.0256 | 0.0268 | 0.0279 | 0.0290 | 0.0325 | 0.0355 |
| $2.5 \times 10^5$ | 0.0208 | 0.0226 | 0.0241 | 0.0255 | 0.0267 | 0.0278 | 0.0289 | 0.0325 | 0.0355 |
| $3 \times 10^5$ | 0.0206 | 0.0225 | 0.0240 | 0.0254 | 0.0266 | 0.0277 | 0.0288 | 0.0324 | 0.0354 |
| $4 \times 10^5$ | 0.0204 | 0.0223 | 0.0239 | 0.0253 | 0.0265 | 0.0276 | 0.0287 | 0.0323 | 0.0354 |
| $5 \times 10^5$ | 0.0202 | 0.0222 | 0.0238 | 0.0252 | 0.0264 | 0.0276 | 0.0286 | 0.0323 | 0.0353 |
| $6 \times 10^5$ | 0.0201 | 0.0221 | 0.0237 | 0.0251 | 0.0264 | 0.0275 | 0.0286 | 0.0323 | 0.0353 |
| $7 \times 10^5$ | 0.0201 | 0.0221 | 0.0237 | 0.0251 | 0.0264 | 0.0275 | 0.0286 | 0.0322 | 0.0353 |
| $8 \times 10^5$ | 0.0200 | 0.0220 | 0.0237 | 0.0251 | 0.0263 | 0.0275 | 0.0286 | 0.0322 | 0.0353 |
| $9 \times 10^5$ | 0.0200 | 0.0220 | 0.0236 | 0.0251 | 0.0263 | 0.0275 | 0.0285 | 0.0322 | 0.0353 |
| $1 \times 10^6$ | 0.0199 | 0.0220 | 0.0236 | 0.0250 | 0.0263 | 0.0275 | 0.0285 | 0.0322 | 0.0353 |
| $1.5 \times 10^6$ | 0.0198 | 0.0219 | 0.0235 | 0.0250 | 0.0263 | 0.0274 | 0.0285 | 0.0322 | 0.0352 |
| $2 \times 10^6$ | 0.0198 | 0.0218 | 0.0235 | 0.0250 | 0.0262 | 0.0274 | 0.0285 | 0.0322 | 0.0352 |
| $2.5 \times 10^6$ | 0.0198 | 0.0218 | 0.0235 | 0.0249 | 0.0262 | 0.0274 | 0.0285 | 0.0322 | 0.0352 |
| $3 \times 10^6$ | 0.0197 | 0.0218 | 0.0235 | 0.0249 | 0.0262 | 0.0274 | 0.0285 | 0.0321 | 0.0352 |
| $4 \times 10^6$ | 0.0197 | 0.0218 | 0.0235 | 0.0249 | 0.0262 | 0.0274 | 0.0284 | 0.0321 | 0.0352 |
| $5 \times 10^6$ | 0.0197 | 0.0218 | 0.0235 | 0.0249 | 0.0262 | 0.0274 | 0.0284 | 0.0321 | 0.0352 |
| $6 \times 10^6$ | 0.0197 | 0.0218 | 0.0235 | 0.0249 | 0.0262 | 0.0274 | 0.0284 | 0.0321 | 0.0352 |
| $7 \times 10^6$ | 0.0197 | 0.0218 | 0.0234 | 0.0249 | 0.0262 | 0.0274 | 0.0284 | 0.0321 | 0.0352 |
| $8 \times 10^6$ | 0.0197 | 0.0218 | 0.0234 | 0.0249 | 0.0262 | 0.0274 | 0.0284 | 0.0321 | 0.0352 |
| $9 \times 10^6$ | 0.0197 | 0.0218 | 0.0234 | 0.0249 | 0.0262 | 0.0274 | 0.0284 | 0.0321 | 0.0352 |
| $1 \times 10^7$ | 0.0197 | 0.0218 | 0.0234 | 0.0249 | 0.0262 | 0.0273 | 0.0284 | 0.0321 | 0.0352 |
| $1.5 \times 10^7$ | 0.0197 | 0.0217 | 0.0234 | 0.0249 | 0.0262 | 0.0273 | 0.0284 | 0.0321 | 0.0352 |
| $2 \times 10^7$ | 0.0197 | 0.0217 | 0.0234 | 0.0249 | 0.0262 | 0.0273 | 0.0284 | 0.0321 | 0.0352 |
| $2.5 \times 10^7$ | 0.0196 | 0.0217 | 0.0234 | 0.0249 | 0.0262 | 0.0273 | 0.0284 | 0.0321 | 0.0352 |
| $3 \times 10^7$ | 0.0196 | 0.0217 | 0.0234 | 0.0249 | 0.0262 | 0.0273 | 0.0284 | 0.0321 | 0.0352 |
| $4 \times 10^7$ | 0.0196 | 0.0217 | 0.0234 | 0.0249 | 0.0262 | 0.0273 | 0.0284 | 0.0321 | 0.0352 |
| $5 \times 10^7$ | 0.0196 | 0.0217 | 0.0234 | 0.0249 | 0.0262 | 0.0273 | 0.0284 | 0.0321 | 0.0352 |

Appendices

**APPENDIX 19.B** *(continued)*
Darcy Friction Factors (turbulent flow)

| Reynolds no. | relative roughness ($\epsilon/D$) | | | | | | | | |
| --- | --- | --- | --- | --- | --- | --- | --- | --- | --- |
| | 0.01 | 0.015 | 0.02 | 0.025 | 0.03 | 0.035 | 0.04 | 0.045 | 0.05 |
| $2 \times 10^3$ | 0.0568 | 0.0602 | 0.0635 | 0.0668 | 0.0699 | 0.0730 | 0.0760 | 0.0790 | 0.0819 |
| $2.5 \times 10^3$ | 0.0539 | 0.0576 | 0.0610 | 0.0644 | 0.0677 | 0.0709 | 0.0740 | 0.0770 | 0.0800 |
| $3 \times 10^3$ | 0.0519 | 0.0557 | 0.0593 | 0.0628 | 0.0661 | 0.0694 | 0.0725 | 0.0756 | 0.0787 |
| $4 \times 10^3$ | 0.0491 | 0.0531 | 0.0570 | 0.0606 | 0.0641 | 0.0674 | 0.0707 | 0.0739 | 0.0770 |
| $5 \times 10^3$ | 0.0473 | 0.0515 | 0.0555 | 0.0592 | 0.0628 | 0.0662 | 0.0696 | 0.0728 | 0.0759 |
| $6 \times 10^3$ | 0.0460 | 0.0504 | 0.0544 | 0.0583 | 0.0619 | 0.0654 | 0.0688 | 0.0721 | 0.0752 |
| $7 \times 10^3$ | 0.0450 | 0.0495 | 0.0537 | 0.0576 | 0.0613 | 0.0648 | 0.0682 | 0.0715 | 0.0747 |
| $8 \times 10^3$ | 0.0442 | 0.0489 | 0.0531 | 0.0571 | 0.0608 | 0.0644 | 0.0678 | 0.0711 | 0.0743 |
| $9 \times 10^3$ | 0.0436 | 0.0484 | 0.0526 | 0.0566 | 0.0604 | 0.0640 | 0.0675 | 0.0708 | 0.0740 |
| $1 \times 10^4$ | 0.0431 | 0.0479 | 0.0523 | 0.0563 | 0.0601 | 0.0637 | 0.0672 | 0.0705 | 0.0738 |
| $1.5 \times 10^4$ | 0.0415 | 0.0466 | 0.0511 | 0.0553 | 0.0592 | 0.0628 | 0.0664 | 0.0698 | 0.0731 |
| $2 \times 10^4$ | 0.0407 | 0.0459 | 0.0505 | 0.0547 | 0.0587 | 0.0624 | 0.0660 | 0.0694 | 0.0727 |
| $2.5 \times 10^4$ | 0.0402 | 0.0455 | 0.0502 | 0.0544 | 0.0584 | 0.0621 | 0.0657 | 0.0691 | 0.0725 |
| $3 \times 10^4$ | 0.0398 | 0.0452 | 0.0499 | 0.0542 | 0.0582 | 0.0619 | 0.0655 | 0.0690 | 0.0723 |
| $4 \times 10^4$ | 0.0394 | 0.0448 | 0.0496 | 0.0539 | 0.0579 | 0.0617 | 0.0653 | 0.0688 | 0.0721 |
| $5 \times 10^4$ | 0.0391 | 0.0446 | 0.0494 | 0.0538 | 0.0578 | 0.0616 | 0.0652 | 0.0687 | 0.0720 |
| $6 \times 10^4$ | 0.0389 | 0.0445 | 0.0493 | 0.0536 | 0.0577 | 0.0615 | 0.0651 | 0.0686 | 0.0719 |
| $7 \times 10^4$ | 0.0388 | 0.0443 | 0.0492 | 0.0536 | 0.0576 | 0.0614 | 0.0650 | 0.0685 | 0.0719 |
| $8 \times 10^4$ | 0.0387 | 0.0443 | 0.0491 | 0.0535 | 0.0576 | 0.0614 | 0.0650 | 0.0685 | 0.0718 |
| $9 \times 10^4$ | 0.0386 | 0.0442 | 0.0491 | 0.0535 | 0.0575 | 0.0613 | 0.0650 | 0.0684 | 0.0718 |
| $1 \times 10^5$ | 0.0385 | 0.0442 | 0.0490 | 0.0534 | 0.0575 | 0.0613 | 0.0649 | 0.0684 | 0.0718 |
| $1.5 \times 10^5$ | 0.0383 | 0.0440 | 0.0489 | 0.0533 | 0.0574 | 0.0612 | 0.0648 | 0.0683 | 0.0717 |
| $2 \times 10^5$ | 0.0382 | 0.0439 | 0.0488 | 0.0532 | 0.0573 | 0.0612 | 0.0648 | 0.0683 | 0.0717 |
| $2.5 \times 10^5$ | 0.0381 | 0.0439 | 0.0488 | 0.0532 | 0.0573 | 0.0611 | 0.0648 | 0.0683 | 0.0716 |
| $3 \times 10^5$ | 0.0381 | 0.0438 | 0.0488 | 0.0532 | 0.0573 | 0.0611 | 0.0648 | 0.0683 | 0.0716 |
| $4 \times 10^5$ | 0.0381 | 0.0438 | 0.0487 | 0.0532 | 0.0573 | 0.0611 | 0.0647 | 0.0682 | 0.0716 |
| $5 \times 10^5$ | 0.0380 | 0.0438 | 0.0487 | 0.0531 | 0.0572 | 0.0611 | 0.0647 | 0.0682 | 0.0716 |
| $6 \times 10^5$ | 0.0380 | 0.0438 | 0.0487 | 0.0531 | 0.0572 | 0.0611 | 0.0647 | 0.0682 | 0.0716 |
| $7 \times 10^5$ | 0.0380 | 0.0438 | 0.0487 | 0.0531 | 0.0572 | 0.0611 | 0.0647 | 0.0682 | 0.0716 |
| $8 \times 10^5$ | 0.0380 | 0.0437 | 0.0487 | 0.0531 | 0.0572 | 0.0611 | 0.0647 | 0.0682 | 0.0716 |
| $9 \times 10^5$ | 0.0380 | 0.0437 | 0.0487 | 0.0531 | 0.0572 | 0.0610 | 0.0647 | 0.0682 | 0.0716 |
| $1 \times 10^6$ | 0.0380 | 0.0437 | 0.0487 | 0.0531 | 0.0572 | 0.0610 | 0.0647 | 0.0682 | 0.0716 |
| $1.5 \times 10^6$ | 0.0379 | 0.0437 | 0.0487 | 0.0531 | 0.0572 | 0.0610 | 0.0647 | 0.0682 | 0.0716 |
| $2 \times 10^6$ | 0.0379 | 0.0437 | 0.0487 | 0.0531 | 0.0572 | 0.0610 | 0.0647 | 0.0682 | 0.0716 |
| $2.5 \times 10^6$ | 0.0379 | 0.0437 | 0.0487 | 0.0531 | 0.0572 | 0.0610 | 0.0647 | 0.0682 | 0.0716 |
| $3 \times 10^6$ | 0.0379 | 0.0437 | 0.0487 | 0.0531 | 0.0572 | 0.0610 | 0.0647 | 0.0682 | 0.0716 |
| $4 \times 10^6$ | 0.0379 | 0.0437 | 0.0486 | 0.0531 | 0.0572 | 0.0610 | 0.0647 | 0.0682 | 0.0716 |
| $5 \times 10^6$ | 0.0379 | 0.0437 | 0.0486 | 0.0531 | 0.0572 | 0.0610 | 0.0647 | 0.0682 | 0.0716 |
| $6 \times 10^6$ | 0.0379 | 0.0437 | 0.0486 | 0.0531 | 0.0572 | 0.0610 | 0.0647 | 0.0682 | 0.0716 |
| $7 \times 10^6$ | 0.0379 | 0.0437 | 0.0486 | 0.0531 | 0.0572 | 0.0610 | 0.0647 | 0.0682 | 0.0716 |
| $8 \times 10^6$ | 0.0379 | 0.0437 | 0.0486 | 0.0531 | 0.0572 | 0.0610 | 0.0647 | 0.0682 | 0.0716 |
| $9 \times 10^6$ | 0.0379 | 0.0437 | 0.0486 | 0.0531 | 0.0572 | 0.0610 | 0.0647 | 0.0682 | 0.0716 |
| $1 \times 10^7$ | 0.0379 | 0.0437 | 0.0486 | 0.0531 | 0.0572 | 0.0610 | 0.0647 | 0.0682 | 0.0716 |
| $1.5 \times 10^7$ | 0.0379 | 0.0437 | 0.0486 | 0.0531 | 0.0572 | 0.0610 | 0.0647 | 0.0682 | 0.0716 |
| $2 \times 10^7$ | 0.0379 | 0.0437 | 0.0486 | 0.0531 | 0.0572 | 0.0610 | 0.0647 | 0.0682 | 0.0716 |
| $2.5 \times 10^7$ | 0.0379 | 0.0437 | 0.0486 | 0.0531 | 0.0572 | 0.0610 | 0.0647 | 0.0682 | 0.0716 |
| $3 \times 10^7$ | 0.0379 | 0.0437 | 0.0486 | 0.0531 | 0.0572 | 0.0610 | 0.0647 | 0.0682 | 0.0716 |
| $4 \times 10^7$ | 0.0379 | 0.0437 | 0.0486 | 0.0531 | 0.0572 | 0.0610 | 0.0647 | 0.0682 | 0.0716 |
| $5 \times 10^7$ | 0.0379 | 0.0437 | 0.0486 | 0.0531 | 0.0572 | 0.0610 | 0.0647 | 0.0682 | 0.0716 |

# APPENDIX 19.C

Water Pressure Drop in Schedule-40 Steel Pipe

| | pressure drop per 1000 ft of schedule-40 steel pipe. in lbf/in$^2$ | | | | | | | | | | | | | | | | | |
|---|---|---|---|---|---|---|---|---|---|---|---|---|---|---|---|---|---|---|
| discharge (gal/min) | velocity (ft/sec) | pressure drop | velocity (ft/sec) | pressure drop | velocity (ft/sec) | pressure drop | velocity (ft/sec) | pressure drop | velocity (ft/sec) | pressure drop | velocity (ft/sec) | pressure drop | velocity (ft/sec) | pressure drop | velocity (ft/sec) | pressure drop | velocity (ft/sec) | pressure drop |
| | 1 in | | | | | | | | | | | | | | | | | |
| 1 | 0.37 | 0.49 | 1¼ in | | | | | | | | | | | | | | | |
| 2 | 0.74 | 1.70 | 0.43 | 0.45 | 1½ in | | | | | | | | | | | | | |
| 3 | 1.12 | 3.53 | 0.64 | 0.94 | 0.47 | 0.44 | | | | | | | | | | | | |
| 4 | 1.49 | 5.94 | 0.86 | 1.55 | 0.63 | 0.74 | | | | | | | | | | | | |
| 5 | 1.86 | 9.02 | 1.07 | 2.36 | 0.79 | 1.12 | 2 in | | | | | | | | | | | |
| 6 | 2.24 | 12.25 | 1.28 | 3.30 | 0.95 | 1.53 | 0.57 | 0.46 | | | | | | | | | | |
| 8 | 2.98 | 21.1 | 1.72 | 5.52 | 1.26 | 2.63 | 0.76 | 0.75 | 2½ in | | | | | | | | | |
| 10 | 3.72 | 30.8 | 2.14 | 8.34 | 1.57 | 3.86 | 0.96 | 1.14 | 0.67 | 0.48 | | | | | | | | |
| 15 | 5.60 | 64.6 | 3.21 | 17.6 | 2.36 | 8.13 | 1.43 | 2.33 | 1.00 | 0.99 | 3 in | | | | | | | |
| 20 | 7.44 | 110.5 | 4.29 | 29.1 | 3.15 | 13.5 | 1.91 | 3.86 | 1.34 | 1.64 | 0.87 | 0.59 | 3½ in | | | | | |
| 25 | | | 5.36 | 43.7 | 3.94 | 20.2 | 2.39 | 5.81 | 1.68 | 2.48 | 1.08 | 0.67 | 0.81 | 0.42 | | | | |
| 30 | | | 6.43 | 62.9 | 4.72 | 29.1 | 2.87 | 8.04 | 2.01 | 3.43 | 1.30 | 1.21 | 0.97 | 0.60 | 4 in | | | |
| 35 | | | 7.51 | 82.5 | 5.51 | 38.2 | 3.35 | 10.95 | 2.35 | 4.49 | 1.52 | 1.58 | 1.14 | 0.79 | 0.88 | 0.42 | | |
| 40 | | | | | 6.30 | 47.8 | 3.82 | 13.7 | 2.68 | 5.88 | 1.74 | 2.06 | 1.30 | 1.00 | 1.01 | 0.53 | | |
| 45 | | | | | 7.08 | 60.6 | 4.30 | 17.4 | 3.00 | 7.14 | 1.95 | 2.51 | 1.46 | 1.21 | 1.13 | 0.67 | | |
| 50 | | | | | 7.87 | 74.7 | 4.78 | 20.6 | 3.35 | 8.82 | 2.17 | 3.10 | 1.62 | 1.44 | 1.26 | 0.80 | | |
| 60 | | | | | | | 5.74 | 29.6 | 4.02 | 12.2 | 2.60 | 4.29 | 1.95 | 2.07 | 1.51 | 1.10 | 5 in | |
| 70 | | | | | | | 6.69 | 38.6 | 4.69 | 15.3 | 3.04 | 5.84 | 2.27 | 2.71 | 1.76 | 1.50 | 1.12 | 0.48 |
| 80 | | | | | | | 7.65 | 50.3 | 5.37 | 21.7 | 3.48 | 7.62 | 2.59 | 3.53 | 2.01 | 1.87 | 1.28 | 0.63 |
| 90 | 6 in | | | | | | 8.60 | 63.6 | 6.04 | 26.1 | 3.91 | 9.22 | 2.92 | 4.46 | 2.26 | 2.37 | 1.44 | 0.80 |
| 100 | 1.11 | 0.39 | | | | | 9.56 | 75.1 | 6.71 | 32.3 | 4.34 | 11.4 | 3.24 | 5.27 | 2.52 | 2.81 | 1.60 | 0.95 |
| 125 | 1.39 | 0.56 | | | | | | | 8.38 | 48.2 | 5.42 | 17.1 | 4.05 | 7.86 | 3.15 | 4.38 | 2.00 | 1.48 |
| 150 | 1.67 | 0.78 | | | | | | | 10.06 | 60.4 | 6.51 | 23.5 | 4.86 | 11.3 | 3.78 | 6.02 | 2.41 | 2.04 |
| 175 | 1.94 | 1.06 | | | | | | | 11.73 | 90.0 | 7.59 | 32.0 | 5.67 | 14.7 | 4.41 | 8.20 | 2.81 | 2.78 |
| 200 | 2.22 | 1.32 | 8 in | | | | | | | | 8.68 | 39.7 | 6.48 | 19.2 | 5.04 | 10.2 | 3.21 | 3.46 |
| 225 | 2.50 | 1.66 | 1.44 | 0.44 | | | | | | | 9.77 | 50.2 | 7.29 | 23.1 | 5.67 | 12.9 | 3.61 | 4.37 |
| 250 | 2.78 | 2.05 | 1.60 | 0.55 | | | | | | | 10.85 | 61.9 | 8.10 | 28.5 | 6.30 | 15.9 | 4.01 | 5.14 |
| 275 | 3.06 | 2.36 | 1.76 | 0.63 | | | | | | | 11.94 | 75.0 | 8.91 | 34.4 | 6.93 | 18.3 | 4.41 | 6.22 |
| 300 | 3.33 | 2.80 | 1.92 | 0.75 | | | | | | | 13.02 | 84.7 | 9.72 | 40.9 | 7.56 | 21.8 | 4.81 | 7.41 |
| 325 | 3.61 | 3.29 | 2.08 | 0.88 | | | | | | | | | 10.53 | 45.5 | 8.18 | 25.5 | 5.21 | 8.25 |
| 350 | 3.89 | 3.62 | 2.24 | 0.97 | | | | | | | | | 11.35 | 52.7 | 8.82 | 29.7 | 5.61 | 9.57 |
| 375 | 4.16 | 4.16 | 2.40 | 1.11 | | | | | | | | | 12.17 | 60.7 | 9.45 | 32.3 | 6.01 | 11.0 |
| 400 | 4.44 | 4.72 | 2.56 | 1.27 | | | | | | | | | 12.97 | 68.9 | 10.08 | 36.7 | 6.41 | 12.5 |
| 425 | 4.72 | 5.34 | 2.72 | 1.43 | | | | | | | | | 13.78 | 77.8 | 10.70 | 41.5 | 6.82 | 14.1 |
| 450 | 5.00 | 5.96 | 2.88 | 1.60 | 10 in | | | | | | | | 14.59 | 87.3 | 11.33 | 46.5 | 7.22 | 15.0 |
| 475 | 5.27 | 6.66 | 3.04 | 1.69 | 1.93 | 0.30 | | | | | | | | | 11.96 | 51.7 | 7.62 | 16.7 |
| 500 | 5.55 | 7.39 | 3.20 | 1.87 | 2.04 | 0.63 | | | | | | | | | 12.59 | 57.3 | 8.02 | 18.5 |
| 550 | 6.11 | 8.94 | 3.53 | 2.26 | 2.24 | 0.70 | | | | | | | | | 13.84 | 69.3 | 8.82 | 22.4 |
| 600 | 6.66 | 10.6 | 3.85 | 2.70 | 2.44 | 0.86 | | | | | | | | | 15.10 | 82.5 | 9.62 | 26.7 |
| 650 | 7.21 | 11.8 | 4.17 | 3.16 | 2.65 | 1.01 | 12 in | | | | | | | | | | 10.42 | 31.3 |
| 700 | 7.77 | 13.7 | 4.49 | 3.69 | 2.85 | 1.18 | 2.01 | 0.48 | | | | | | | | | 11.22 | 36.3 |
| 750 | 8.32 | 15.7 | 4.81 | 4.21 | 3.05 | 1.35 | 2.15 | 0.55 | | | | | | | | | 12.02 | 41.6 |
| 800 | 8.88 | 17.8 | 5.13 | 4.79 | 3.26 | 1.54 | 2.29 | 0.62 | 14 in | | | | | | | | 12.82 | 44.7 |
| 850 | 9.44 | 20.2 | 5.45 | 5.11 | 3.46 | 1.74 | 2.44 | 0.70 | 2.02 | 0.43 | | | | | | | 13.62 | 50.5 |
| 900 | 10.00 | 22.6 | 5.77 | 5.73 | 3.66 | 1.94 | 2.58 | 0.79 | 2.14 | 0.48 | | | | | | | 14.42 | 56.6 |
| 950 | 10.55 | 23.7 | 6.09 | 6.38 | 3.87 | 2.23 | 2.72 | 0.88 | 2.25 | 0.53 | | | | | | | 15.22 | 63.1 |
| 1000 | 11.10 | 26.3 | 6.41 | 7.08 | 4.07 | 2.40 | 2.87 | 0.98 | 2.38 | 0.59 | | | | | | | 16.02 | 70.0 |
| 1100 | 12.22 | 31.8 | 7.05 | 8.56 | 4.48 | 2.74 | 3.16 | 1.18 | 2.61 | 0.68 | 16 in | | | | | | 17.63 | 84.6 |
| 1200 | 13.32 | 37.8 | 7.69 | 10.2 | 4.88 | 3.27 | 3.45 | 1.40 | 2.85 | 0.81 | 2.18 | 0.40 | | | | | | |
| 1300 | 14.43 | 44.4 | 8.33 | 11.3 | 5.29 | 3.86 | 3.73 | 1.56 | 3.09 | 0.95 | 2.36 | 0.47 | | | | | | |
| 1400 | 15.54 | 51.5 | 8.97 | 13.0 | 5.70 | 4.44 | 4.02 | 1.80 | 3.32 | 1.10 | 2.54 | 0.54 | | | | | | |
| 1500 | 16.65 | 55.5 | 9.62 | 15.0 | 6.10 | 5.11 | 4.30 | 2.07 | 3.55 | 1.19 | 2.73 | 0.62 | | | | | | |
| 1600 | 17.76 | 63.1 | 10.26 | 17.0 | 6.51 | 5.46 | 4.59 | 2.36 | 3.80 | 1.35 | 2.91 | 0.71 | 18 in | | | | | |
| 1800 | 19.98 | 79.8 | 11.54 | 21.6 | 7.32 | 6.91 | 5.16 | 2.98 | 4.27 | 1.71 | 3.27 | 0.85 | 2.58 | 0.48 | | | | |
| 2000 | 22.20 | 98.5 | 12.83 | 25.0 | 8.13 | 8.54 | 5.73 | 3.47 | 4.74 | 2.11 | 3.63 | 1.05 | 2.88 | 0.56 | | | | |
| 2500 | | | 16.03 | 39.0 | 10.18 | 12.5 | 7.17 | 5.41 | 5.92 | 3.09 | 4.54 | 1.63 | 3.59 | 0.88 | 20 in | | | |
| 3000 | | | 19.24 | 52.4 | 12.21 | 18.0 | 8.60 | 7.31 | 7.12 | 4.45 | 5.45 | 2.21 | 4.31 | 1.27 | 3.45 | 0.73 | | |
| 3500 | | | 22.43 | 71.4 | 14.25 | 22.9 | 10.03 | 9.95 | 8.32 | 6.18 | 6.35 | 3.00 | 5.03 | 1.52 | 4.03 | 0.94 | 24 in | |
| 4000 | | | 25.65 | 93.3 | 16.28 | 29.9 | 11.48 | 13.0 | 9.49 | 7.92 | 7.25 | 3.92 | 5.74 | 2.12 | 4.61 | 1.22 | 3.19 | 0.51 |
| 4500 | | | | | 18.31 | 37.8 | 12.90 | 15.4 | 10.67 | 9.36 | 8.17 | 4.97 | 6.47 | 2.50 | 5.19 | 1.55 | 3.59 | 0.60 |
| 5000 | | | | | 20.35 | 46.7 | 14.34 | 18.9 | 11.84 | 11.6 | 9.08 | 5.72 | 7.17 | 3.08 | 5.76 | 1.78 | 3.99 | 0.74 |
| 6000 | | | | | 24.42 | 67.2 | 17.21 | 27.3 | 14.32 | 15.4 | 10.88 | 8.24 | 8.62 | 4.45 | 6.92 | 2.57 | 4.80 | 1.00 |
| 7000 | | | | | 28.50 | 85.1 | 20.08 | 37.2 | 16.60 | 21.0 | 12.69 | 12.2 | 10.04 | 6.06 | 8.06 | 3.50 | 5.68 | 1.36 |
| 8000 | | | | | | | 22.95 | 45.1 | 18.98 | 27.4 | 14.52 | 13.6 | 11.48 | 7.34 | 9.23 | 4.57 | 6.38 | 1.78 |
| 9000 | | | | | | | 25.80 | 57.0 | 21.35 | 34.7 | 16.32 | 17.2 | 12.92 | 9.20 | 10.37 | 5.36 | 7.19 | 2.25 |
| 10,000 | | | | | | | 28.63 | 70.4 | 23.75 | 42.9 | 18.16 | 21.2 | 14.37 | 11.5 | 11.53 | 6.63 | 7.96 | 2.78 |
| 12,000 | | | | | | | 34.38 | 93.6 | 28.50 | 61.8 | 21.80 | 30.9 | 17.23 | 16.5 | 13.83 | 9.54 | 9.57 | 3.71 |
| 14,000 | | | | | | | | | 33.20 | 84.0 | 25.42 | 41.6 | 20.10 | 20.7 | 16.14 | 12.0 | 11.18 | 5.05 |
| 16,000 | | | | | | | | | | | 29.05 | 54.4 | 22.96 | 27.1 | 18.43 | 15.7 | 12.77 | 6.60 |

(Multiply gal/min by 0.0631 to obtain L/s.) (Multiply in by 25.4 to obtain mm.)
(Multiply ft/sec by 0.3 to obtain m/s.) (Multiply lbf/in$^2$-1000 ft by 2.3 to obtain kPa/100 m.)

Reproduced with permission from *Design of Fluid Systems Hook-Ups*, published by Spirax Sarco, Inc., © 1997.

Appendices

### APPENDIX 19.D
Water Flow in Steel Pipe

Assumptions:  new, schedule-40 pipe
$\epsilon = 0.00015$ ft
water temperature = 60°F
water viscosity = $1.217 \times 10^{-5}$ ft$^2$/sec

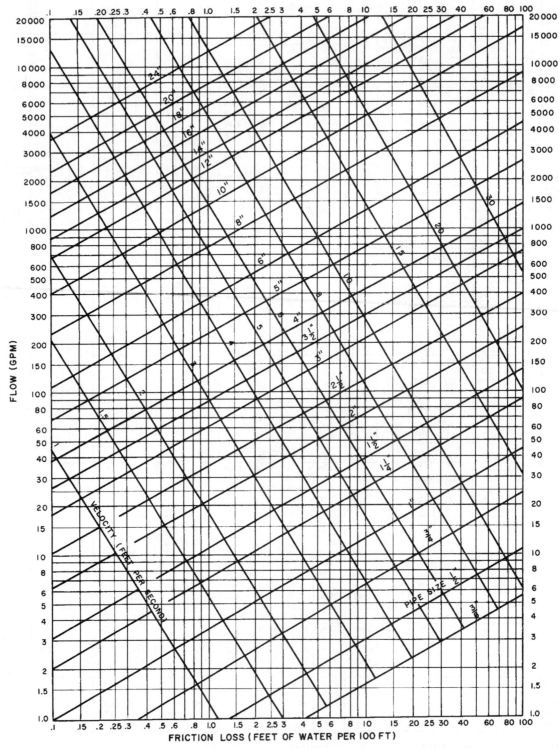

Reprinted with permission from *Handbook of Air Conditioning System Design*,
Carrier Air Conditioning Company, copyright © 1965 McGraw-Hill Book Company, Inc.

## APPENDIX 19.E
### Equivalent Length of Straight Pipe for Various (Generic) Fittings
### (in feet, turbulent flow only, for any fluid)

| fittings | | | ¼ | ⅜ | ½ | ¾ | 1 | 1¼ | 1½ | 2 | 2½ | 3 | 4 | 5 | 6 | 8 | 10 | 12 | 14 | 16 | 18 | 20 | 24 |
|---|---|---|---|---|---|---|---|---|---|---|---|---|---|---|---|---|---|---|---|---|---|---|---|
| regular 90° ell | screwed | steel | 2.3 | 3.1 | 3.6 | 4.4 | 5.2 | 6.6 | 7.4 | 8.5 | 9.3 | 11.0 | 13.0 | | | | | | | | | | |
| | | cast iron | | | | | | | | | | 9.0 | 11.0 | | | | | | | | | | |
| | flanged | steel | | | 0.92 | 1.2 | 1.6 | 2.1 | 2.4 | 3.1 | 3.6 | 4.4 | 5.9 | 7.3 | 8.9 | 12.0 | 14.0 | 17.0 | 18.0 | 21.0 | 23.0 | 25.0 | 30.0 |
| | | cast iron | | | | | | | | | | 3.6 | 4.8 | | 7.2 | 9.8 | 12.0 | 15.0 | 17.0 | 19.0 | 22.0 | 24.0 | 28.0 |
| long radius 90° ell | screwed | steel | 1.5 | 2.0 | 2.2 | 2.3 | 2.7 | 3.2 | 3.4 | 3.6 | 3.6 | 4.0 | 4.6 | | | | | | | | | | |
| | | cast iron | | | | | | | | | | 3.3 | 3.7 | | | | | | | | | | |
| | flanged | steel | | | 1.1 | 1.3 | 1.6 | 2.0 | 2.3 | 2.7 | 2.9 | 3.4 | 4.2 | 5.0 | 5.7 | 7.0 | 8.0 | 9.0 | 9.4 | 10.0 | 11.0 | 12.0 | 14.0 |
| | | cast iron | | | | | | | | | | 2.8 | 3.4 | | 4.7 | 5.7 | 6.8 | 7.8 | 8.6 | 9.6 | 11.0 | 11.0 | 13.0 |
| regular 45° ell | screwed | steel | 0.34 | 0.52 | 0.71 | 0.92 | 1.3 | 1.7 | 2.1 | 2.7 | 3.2 | 4.0 | 5.5 | | | | | | | | | | |
| | | cast iron | | | | | | | | | | 3.3 | 4.5 | | | | | | | | | | |
| | flanged | steel | | | 0.45 | 0.59 | 0.81 | 1.1 | 1.3 | 1.7 | 2.0 | 2.6 | 3.5 | 4.5 | 5.6 | 7.7 | 9.0 | 11.0 | 13.0 | 15.0 | 16.0 | 18.0 | 22.0 |
| | | cast iron | | | | | | | | | | 2.1 | 2.9 | | 4.5 | 6.3 | 8.1 | 9.7 | 12.0 | 13.0 | 15.0 | 17.0 | 20.0 |
| tee-line flow | screwed | steel | 0.79 | 1.2 | 1.7 | 2.4 | 3.2 | 4.6 | 5.6 | 7.7 | 9.3 | 12.0 | 17.0 | | | | | | | | | | |
| | | cast iron | | | | | | | | | | 9.9 | 14.0 | | | | | | | | | | |
| | flanged | steel | | | 0.69 | 0.82 | 1.0 | 1.3 | 1.5 | 1.8 | 1.9 | 2.2 | 2.8 | 3.3 | 3.8 | 4.7 | 5.2 | 6.0 | 6.4 | 7.2 | 7.6 | 8.2 | 9.6 |
| | | cast iron | | | | | | | | | | 1.9 | 2.2 | | 3.1 | 3.9 | 4.6 | 5.2 | 5.9 | 6.5 | 7.2 | 7.7 | 8.8 |
| tee-branch flow | screwed | steel | 2.4 | 3.5 | 4.2 | 5.3 | 6.6 | 8.7 | 9.9 | 12.0 | 13.0 | 17.0 | 21.0 | | | | | | | | | | |
| | | cast iron | | | | | | | | | | 14.0 | 17.0 | | | | | | | | | | |
| | flanged | steel | | | 2.0 | 2.6 | 3.3 | 4.4 | 5.2 | 6.6 | 7.5 | 9.4 | 12.0 | 15.0 | 18.0 | 24.0 | 30.0 | 34.0 | 37.0 | 43.0 | 47.0 | 52.0 | 62.0 |
| | | cast iron | | | | | | | | | | 7.7 | 10.0 | | 15.0 | 20.0 | 25.0 | 30.0 | 35.0 | 39.0 | 44.0 | 49.0 | 57.0 |
| 180° return bend | screwed | steel | 2.3 | 3.1 | 3.6 | 4.4 | 5.2 | 6.6 | 7.4 | 8.5 | 9.3 | 11.0 | 13.0 | | | | | | | | | | |
| | | cast iron | | | | | | | | | | 9.0 | 11.0 | | | | | | | | | | |
| | regular flanged | steel | | | 0.92 | 1.2 | 1.6 | 2.1 | 2.4 | 3.1 | 3.6 | 4.4 | 5.9 | 7.3 | 8.9 | 12.0 | 14.0 | 17.0 | 18.0 | 21.0 | 23.0 | 25.0 | 30.0 |
| | | cast iron | | | | | | | | | | 3.6 | 4.8 | | 7.2 | 9.8 | 12.0 | 15.0 | 17.0 | 19.0 | 22.0 | 24.0 | 28.0 |
| | long radius flanged | steel | | | 1.1 | 1.3 | 1.6 | 2.0 | 2.3 | 2.7 | 2.9 | 3.4 | 4.2 | 5.0 | 5.7 | 7.0 | 8.0 | 9.0 | 9.4 | 10.0 | 11.0 | 12.0 | 14.0 |
| | | cast iron | | | | | | | | | | 2.8 | 3.4 | | 4.7 | 5.7 | 6.8 | 7.8 | 8.6 | 9.6 | 11.0 | 11.0 | 13.0 |
| globe valve | screwed | steel | 21.0 | 22.0 | 22.0 | 24.0 | 29.0 | 37.0 | 42.0 | 54.0 | 62.0 | 79.0 | 110.0 | | | | | | | | | | |
| | | cast iron | | | | | | | | | | 65.0 | 86.0 | | | | | | | | | | |
| | flanged | steel | | | 38.0 | 40.0 | 45.0 | 54.0 | 59.0 | 70.0 | 77.0 | 94.0 | 120.0 | 150.0 | 190.0 | 260.0 | 310.0 | 390.0 | | | | | |
| | | cast iron | | | | | | | | | | 77.0 | 99.0 | | 150.0 | 210.0 | 270.0 | 330.0 | | | | | |
| gate valve | screwed | steel | 0.32 | 0.45 | 0.56 | 0.67 | 0.84 | 1.1 | 1.2 | 1.5 | 1.7 | 1.9 | 2.5 | | | | | | | | | | |
| | | cast iron | | | | | | | | | | 1.6 | 2.0 | | | | | | | | | | |
| | flanged | steel | | | | | | | | 2.6 | 2.7 | 2.8 | 2.9 | 3.1 | 3.2 | 3.2 | 3.2 | 3.2 | 3.2 | 3.2 | 3.2 | 3.2 | 3.2 |
| | | cast iron | | | | | | | | | | 2.3 | 2.4 | | 2.6 | 2.7 | 2.8 | 2.9 | 2.9 | 3.0 | 3.0 | 3.0 | 3.0 |
| angle valve | screwed | steel | 12.8 | 15.0 | 15.0 | 15.0 | 17.0 | 18.0 | 18.0 | 18.0 | 18.0 | 18.0 | 18.0 | | | | | | | | | | |
| | | cast iron | | | | | | | | | | 15.0 | 15.0 | | | | | | | | | | |
| | flanged | steel | | | 15.0 | 15.0 | 17.0 | 18.0 | 18.0 | 21.0 | 22.0 | 28.0 | 38.0 | 50.0 | 63.0 | 90.0 | 120.0 | 140.0 | 160.0 | 190.0 | 210.0 | 240.0 | 300.0 |
| | | cast iron | | | | | | | | | | 23.0 | 31.0 | | 52.0 | 74.0 | 98.0 | 120.0 | 150.0 | 170.0 | 200.0 | 230.0 | 280.0 |
| swing check valve | screwed | steel | 7.2 | 7.3 | 8.0 | 8.8 | 11.0 | 13.0 | 15.0 | 19.0 | 22.0 | 27.0 | 38.0 | | | | | | | | | | |
| | | cast iron | | | | | | | | | | 22.0 | 31.0 | | | | | | | | | | |
| | flanged | steel | | | 3.8 | 5.3 | 7.2 | 10.0 | 12.0 | 17.0 | 21.0 | 27.0 | 38.0 | 50.0 | 63.0 | 90.0 | 120.0 | 140.0 | | | | | |
| | | cast iron | | | | | | | | | | 22.0 | 31.0 | | 52.0 | 74.0 | 98.0 | 120.0 | | | | | |
| coupling or union | screwed | steel | 0.14 | 0.18 | 0.21 | 0.24 | 0.29 | 0.36 | 0.39 | 0.45 | 0.47 | 0.53 | 0.65 | | | | | | | | | | |
| | | cast iron | | | | | | | | | | 0.44 | 0.52 | | | | | | | | | | |
| inlet | bell mouth inlet | steel | 0.04 | 0.07 | 0.10 | 0.13 | 0.18 | 0.26 | 0.31 | 0.43 | 0.52 | 0.67 | 0.95 | 1.3 | 1.6 | 2.3 | 2.9 | 3.5 | 4.0 | 4.7 | 5.3 | 6.1 | 7.6 |
| | | cast iron | | | | | | | | | | 0.55 | 0.77 | | 1.3 | 1.9 | 2.4 | 3.0 | 3.6 | 4.3 | 5.0 | 5.7 | 7.0 |
| | square mouth inlet | steel | 0.44 | 0.68 | 0.96 | 1.3 | 1.8 | 2.6 | 3.1 | 4.3 | 5.2 | 6.7 | 9.5 | 13.0 | 16.0 | 23.0 | 29.0 | 35.0 | 40.0 | 47.0 | 53.0 | 61.0 | 76.0 |
| | | cast iron | | | | | | | | | | 5.5 | 7.7 | | 13.0 | 19.0 | 24.0 | 30.0 | 36.0 | 43.0 | 50.0 | 57.0 | 70.0 |
| | re-entrant pipe | steel | 0.88 | 1.4 | 1.9 | 2.6 | 3.6 | 5.1 | 6.2 | 8.5 | 10.0 | 13.0 | 19.0 | 25.0 | 32.0 | 45.0 | 58.0 | 70.0 | 80.0 | 95.0 | 110.0 | 120.0 | 150.0 |
| | | cast iron | | | | | | | | | | 11.0 | 15.0 | | 26.0 | 37.0 | 49.0 | 61.0 | 73.0 | 86.0 | 100.0 | 110.0 | 140.0 |

(Multiply in by 25.4 to obtain mm. Multiply ft by 0.3 to obtain m.)

## APPENDIX 19.F
### Hazen-Williams Nomograph ($C = 100$)

Quantity (i.e., flow rate) and velocity are proportional to the $C$-value. For values of $C$ other than 100, the quantity and velocity must be converted according to $\dot{V}_{\text{actual}} = \dot{V}_{\text{chart}} C_{\text{actual}}/100$. When quantity is the unknown, use the chart with known values of diameter, slope, or velocity to find $\dot{V}_{\text{chart}}$, and then convert to $\dot{V}_{\text{actual}}$. When velocity is the unknown, use the chart with the known values of diameter, slope, or quantity to find $\dot{V}_{\text{chart}}$, then convert to $\dot{V}_{\text{actual}}$. If $\dot{V}_{\text{actual}}$ is known, it must be converted to $\dot{V}_{\text{chart}}$ before this nomograph can be used. In that case, the diameter, loss, and quantity are as read from this chart.

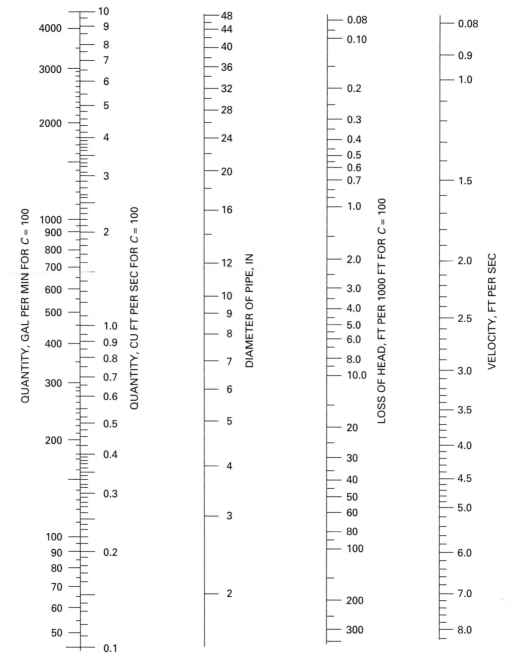

(Multiply gal/min by 0.0631 to obtain L/s.)
(Multiply ft³/sec by 28.3 to obtain L/s.)
(Multiply in by 25.4 to obtain mm.)
(Multiply ft/1000 ft by 0.1 to obtain m/100 m.)
(Multiply ft/sec by 0.3 to obtain m/s.)

## APPENDIX 21.A
Manning's Roughness Coefficient[a,b]
*(design use)*

| channel material | $n$ |
|---|---|
| plastic (PVC and ABS) | 0.009 |
| clean, uncoated cast iron | 0.013–0.015 |
| clean, coated cast iron | 0.012–0.014 |
| dirty, tuberculated cast iron | 0.015–0.035 |
| riveted steel | 0.015–0.017 |
| lock-bar and welded steel pipe | 0.012–0.013 |
| galvanized iron | 0.015–0.017 |
| brass and glass | 0.009–0.013 |
| wood stave | |
|    small diameter | 0.011–0.012 |
|    large diameter | 0.012–0.013 |
| concrete | |
|    average value used | 0.013 |
|    typical commercial, ball and spigot | |
|       rubber gasketed end connections | |
|         – full (pressurized and wet) | 0.010 |
|         – partially full | 0.0085 |
|    with rough joints | 0.016–0.017 |
|    dry mix, rough forms | 0.015–0.016 |
|    wet mix, steel forms | 0.012–0.014 |
|    very smooth, finished | 0.011–0.012 |
| vitrified sewer | 0.013–0.015 |
| common-clay drainage tile | 0.012–0.014 |
| asbestos | 0.011 |
| planed timber (flume) | 0.012 (0.010–0.014) |
| canvas | 0.012 |
| unplaned timber (flume) | 0.013 (0.011–0.015) |
| brick | 0.016 |
| rubble masonry | 0.017 |
| smooth earth | 0.018 |
| firm gravel | 0.023 |
| corrugated metal pipe (CMP) | 0.022–0.033 (See Sec. 18.19.) |
| natural channels, good condition | 0.025 |
| rip rap | 0.035 |
| natural channels with stones and weeds | 0.035 |
| very poor natural channels | 0.060 |

[a]compiled from various sources
[b]Values outside these ranges have been observed, but these values are typical.

**Appendices**

**APPENDIX 21.B**
Manning Equation Nomograph
$\left(\text{solves } v = \dfrac{1.486}{n} R^{2/3} \sqrt{S}\right)$

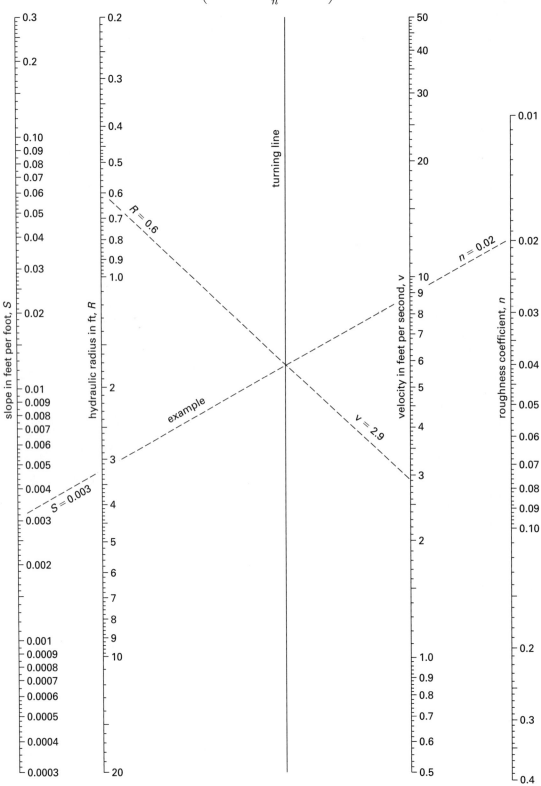

## APPENDIX 21.C
Circular Channel Ratios[*]

Experiments have shown that $n$ varies slightly with depth. This figure gives velocity and flow rate ratios for varying $n$ (solid line) and constant $n$ (broken line) assumptions.

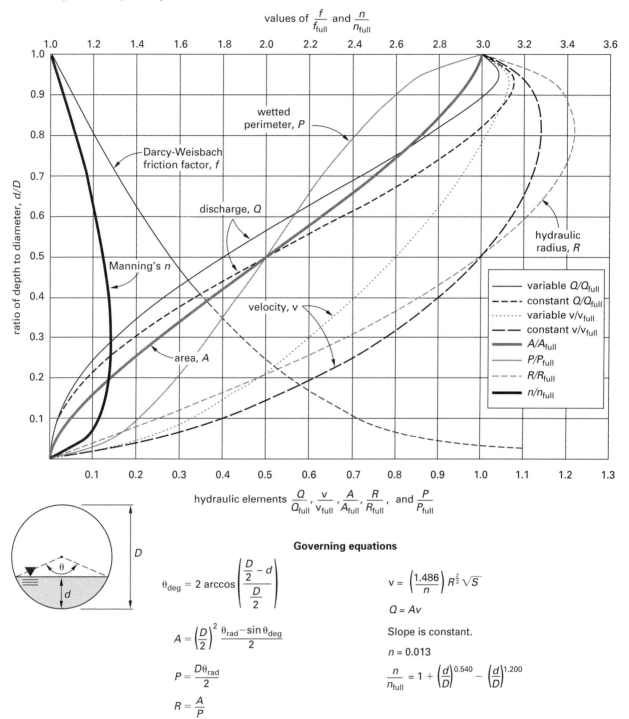

**Governing equations**

$$\theta_{deg} = 2\arccos\left(\frac{\frac{D}{2} - d}{\frac{D}{2}}\right)$$

$$v = \left(\frac{1.486}{n}\right) R^{\frac{2}{3}}\sqrt{S}$$

$$A = \left(\frac{D}{2}\right)^2 \frac{\theta_{rad} - \sin\theta_{deg}}{2}$$

$$Q = Av$$

Slope is constant.

$$n = 0.013$$

$$P = \frac{D\theta_{rad}}{2}$$

$$\frac{n}{n_{full}} = 1 + \left(\frac{d}{D}\right)^{0.540} - \left(\frac{d}{D}\right)^{1.200}$$

$$R = \frac{A}{P}$$

[*]For $n = 0.013$.

Adapted from *Design and Construction of Sanitary and Storm Sewers*, p. 87, ASCE, 1969, as originally presented in "Design of Sewers to Facilitate Flow", Camp, T. R., *Sewage Works Journal*, 18, 3 (1946).

## APPENDIX 21.D
Critical Depths in Circular Channels

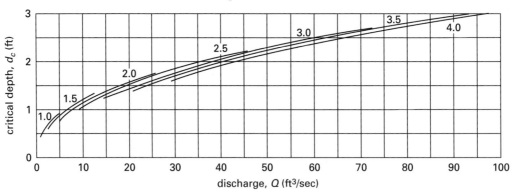

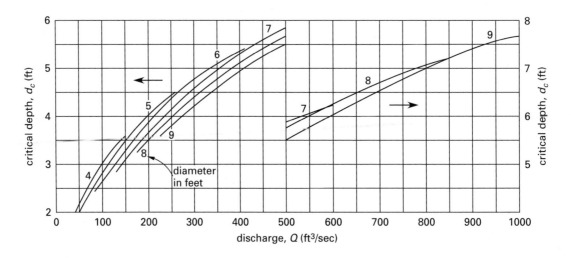

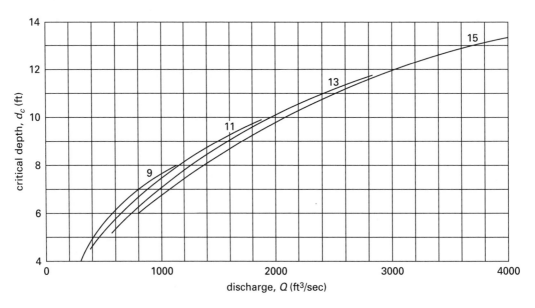

## APPENDIX 21.E
Conveyance Factor, $K$
Symmetrical Rectangular,[a] Trapezoidal, and V-Notch[b] Open Channels
(use for determining $Q$ or $b$ when $d$ is known)
(customary U.S. units[c,d])

$$K \text{ in } Q = K\left(\frac{1}{n}\right)d^{8/3}\sqrt{S}$$

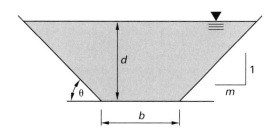

| | | | | | $m$ and $\theta$ | | | | | |
|---|---|---|---|---|---|---|---|---|---|---|
| | 0.0 | 0.25 | 0.5 | 0.75 | 1.0 | 1.5 | 2.0 | 2.5 | 3.0 | 4.0 |
| $x = d/b$ | 90° | 76.0° | 63.4° | 53.1° | 45.0° | 33.7° | 26.6° | 21.8° | 18.4° | 14.0° |
| 0.01 | 146.7 | 147.2 | 147.6 | 148.0 | 148.3 | 148.8 | 149.2 | 149.5 | 149.9 | 150.5 |
| 0.02 | 72.4 | 72.9 | 73.4 | 73.7 | 74.0 | 74.5 | 74.9 | 75.3 | 75.6 | 76.3 |
| 0.03 | 47.6 | 48.2 | 48.6 | 49.0 | 49.8 | 49.8 | 50.2 | 50.6 | 50.9 | 51.6 |
| 0.04 | 35.3 | 35.8 | 36.3 | 36.6 | 36.9 | 37.4 | 37.8 | 38.2 | 38.6 | 39.3 |
| 0.05 | 27.9 | 28.4 | 28.9 | 29.2 | 29.5 | 30.0 | 30.5 | 30.9 | 31.2 | 32.0 |
| 0.06 | 23.0 | 23.5 | 23.9 | 24.3 | 24.6 | 25.1 | 25.5 | 26.0 | 26.3 | 27.1 |
| 0.07 | 19.5 | 20.0 | 20.4 | 20.8 | 21.1 | 21.6 | 22.0 | 22.4 | 22.8 | 23.6 |
| 0.08 | 16.8 | 17.3 | 17.8 | 18.1 | 18.4 | 18.9 | 19.4 | 19.8 | 20.2 | 21.0 |
| 0.09 | 14.8 | 15.3 | 15.7 | 16.1 | 16.4 | 16.9 | 17.4 | 17.8 | 18.2 | 19.0 |
| 0.10 | 13.2 | 13.7 | 14.1 | 14.4 | 14.8 | 15.3 | 15.7 | 16.2 | 16.6 | 17.4 |
| 0.11 | 11.83 | 12.33 | 12.76 | 13.11 | 13.42 | 13.9 | 14.4 | 14.9 | 15.3 | 16.1 |
| 0.12 | 10.73 | 11.23 | 11.65 | 12.00 | 12.31 | 12.8 | 13.3 | 13.8 | 14.2 | 15.0 |
| 0.13 | 9.80 | 10.29 | 10.71 | 11.06 | 11.37 | 11.9 | 12.4 | 12.8 | 13.3 | 14.1 |
| 0.14 | 9.00 | 9.49 | 9.91 | 10.26 | 10.57 | 11.1 | 11.6 | 12.0 | 12.5 | 13.4 |
| 0.15 | 8.32 | 8.80 | 9.22 | 9.67 | 9.88 | 10.4 | 10.9 | 11.4 | 11.8 | 12.7 |
| 0.16 | 7.72 | 8.20 | 8.61 | 8.96 | 9.27 | 9.81 | 10.29 | 10.75 | 11.2 | 12.1 |
| 0.17 | 7.19 | 7.67 | 8.08 | 8.43 | 8.74 | 9.28 | 9.77 | 10.23 | 10.68 | 11.6 |
| 0.18 | 6.73 | 7.20 | 7.61 | 7.96 | 8.27 | 8.81 | 9.30 | 9.76 | 10.21 | 11.1 |
| 0.19 | 6.31 | 6.78 | 7.19 | 7.54 | 7.85 | 8.39 | 8.88 | 9.34 | 9.80 | 10.7 |
| 0.20 | 5.94 | 6.40 | 6.81 | 7.16 | 7.47 | 8.01 | 8.50 | 8.97 | 9.43 | 10.3 |
| 0.22 | 5.30 | 5.76 | 6.16 | 6.51 | 6.82 | 7.36 | 7.86 | 8.33 | 8.79 | 9.70 |
| 0.24 | 4.77 | 5.22 | 5.62 | 5.96 | 6.27 | 6.82 | 7.32 | 7.79 | 8.26 | 9.18 |
| 0.26 | 4.32 | 4.77 | 5.16 | 5.51 | 5.82 | 6.37 | 6.87 | 7.35 | 7.81 | 8.74 |
| 0.28 | 3.95 | 4.38 | 4.77 | 5.12 | 5.48 | 5.98 | 6.48 | 6.96 | 7.43 | 8.36 |
| 0.30 | 3.62 | 4.05 | 4.44 | 4.78 | 5.09 | 5.64 | 6.15 | 6.63 | 7.10 | 8.04 |
| 0.32 | 3.34 | 3.77 | 4.15 | 4.49 | 4.80 | 5.35 | 5.86 | 6.34 | 6.82 | 7.75 |
| 0.34 | 3.09 | 3.51 | 3.89 | 4.23 | 4.54 | 5.10 | 5.60 | 6.09 | 6.56 | 7.50 |
| 0.36 | 2.88 | 3.29 | 3.67 | 4.01 | 4.31 | 4.87 | 5.38 | 5.86 | 6.34 | 7.28 |
| 0.38 | 2.68 | 3.09 | 3.47 | 3.81 | 4.11 | 4.67 | 5.17 | 5.66 | 6.14 | 7.09 |
| 0.40 | 2.51 | 2.92 | 3.29 | 3.62 | 3.93 | 4.48 | 4.99 | 5.48 | 5.96 | 6.91 |
| 0.42 | 2.36 | 2.76 | 3.13 | 3.46 | 3.77 | 4.32 | 4.83 | 5.32 | 5.80 | 6.75 |
| 0.44 | 2.22 | 2.61 | 2.98 | 3.31 | 3.62 | 4.17 | 4.68 | 5.17 | 5.66 | 6.60 |
| 0.46 | 2.09 | 2.48 | 2.85 | 3.18 | 3.48 | 4.04 | 4.55 | 5.04 | 5.52 | 6.47 |
| 0.48 | 1.98 | 2.36 | 2.72 | 3.06 | 3.36 | 3.91 | 4.43 | 4.92 | 5.40 | 6.35 |
| 0.50 | 1.87 | 2.26 | 2.61 | 2.94 | 3.25 | 3.80 | 4.31 | 4.81 | 5.29 | 6.24 |
| 0.55 | 1.65 | 2.02 | 2.37 | 2.70 | 3.00 | 3.55 | 4.07 | 4.56 | 5.05 | 6.00 |
| 0.60 | 1.46 | 1.83 | 2.17 | 2.50 | 2.80 | 3.35 | 3.86 | 4.36 | 4.84 | 5.80 |

*(continued)*

**APPENDIX 21.E** *(continued)*
Conveyance Factor, $K$
Symmetrical Rectangular,[a] Trapezoidal, and V-Notch[b] Open Channels
(use for determining $Q$ or $b$ when $d$ is known)
(customary U.S. units[c,d])

$$K \text{ in } Q = K\left(\frac{1}{n}\right) d^{8/3}\sqrt{S}$$

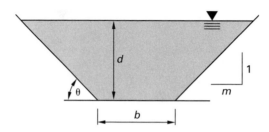

| | $m$ and $\theta$ | | | | | | | | | |
|---|---|---|---|---|---|---|---|---|---|---|
| | 0.0 | 0.25 | 0.5 | 0.75 | 1.0 | 1.5 | 2.0 | 2.5 | 3.0 | 4.0 |
| $x = d/b$ | 90° | 76.0° | 63.4° | 53.1° | 45.0° | 33.7° | 26.6° | 21.8° | 18.4° | 14.0° |
| 0.70 | 1.18 | 1.53 | 1.87 | 2.19 | 2.48 | 3.03 | 3.55 | 4.04 | 4.53 | 5.49 |
| 0.80 | 0.982 | 1.31 | 1.64 | 1.95 | 2.25 | 2.80 | 3.31 | 3.81 | 4.30 | 5.26 |
| 0.90 | 0.831 | 1.15 | 1.47 | 1.78 | 2.07 | 2.62 | 3.13 | 3.63 | 4.12 | 5.08 |
| 1.00 | 0.714 | 1.02 | 1.33 | 1.64 | 1.93 | 2.47 | 2.99 | 3.48 | 3.97 | 4.93 |
| 1.20 | 0.548 | 0.836 | 1.14 | 1.43 | 1.72 | 2.26 | 2.77 | 3.27 | 3.76 | 4.72 |
| 1.40 | 0.436 | 0.708 | 0.998 | 1.29 | 1.57 | 2.11 | 2.62 | 3.12 | 3.60 | 4.57 |
| 1.60 | 0.357 | 0.616 | 0.897 | 1.18 | 1.46 | 2.00 | 2.51 | 3.00 | 3.49 | 4.45 |
| 1.80 | 0.298 | 0.546 | 0.820 | 1.10 | 1.38 | 1.91 | 2.42 | 2.91 | 3.40 | 4.36 |
| 2.00 | 0.254 | 0.491 | 0.760 | 1.04 | 1.31 | 1.84 | 2.35 | 2.84 | 3.33 | 4.29 |
| 2.25 | 0.212 | 0.439 | 0.700 | 0.973 | 1.24 | 1.77 | 2.28 | 2.77 | 3.26 | 4.22 |
| $\infty$ | 0.00 | 0.091 | 0.274 | 0.499 | 0.743 | 1.24 | 1.74 | 2.23 | 2.71 | 3.67 |

[a]For rectangular channels, use the 0.0 (90°, vertical sides) column.
[b]For V-notch triangular channels, use the $d/b = \infty$ row.
[c]$Q$ = flow rate, ft³/sec; $d$ = depth of flow, ft; $b$ = bottom width of channel, ft; $S$ = geometric slope, ft/ft; $n$ = Manning's roughness constant.
[d]For SI units (i.e., $Q$ in m³/s and $d$ and $b$ in m), divide each table value by 1.486.

## APPENDIX 21.F
Conveyance Factor, $K'$
Symmetrical Rectangular,[a] Trapezoidal Open Channels
(use for determining $Q$ or $d$ when $b$ is known)
(customary U.S. units[b,c])

$$K' \text{ in } Q = K'\left(\frac{1}{n}\right)b^{8/3}\sqrt{S}$$

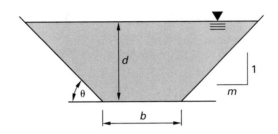

| | | | | | $m$ and $\theta$ | | | | | |
|---|---|---|---|---|---|---|---|---|---|---|
| | 0.0 | 0.25 | 0.5 | 0.75 | 1.0 | 1.5 | 2.0 | 2.5 | 3.0 | 4.0 |
| $x = d/b$ | 90° | 76.0° | 63.4° | 53.1° | 45.0° | 33.7° | 26.6° | 21.8° | 18.4° | 14.0° |
| 0.01 | 0.00068 | 0.00068 | 0.00069 | 0.00069 | 0.00069 | 0.00069 | 0.00069 | 0.00069 | 0.00070 | 0.00070 |
| 0.02 | 0.00213 | 0.00215 | 0.00216 | 0.00217 | 0.00218 | 0.00220 | 0.00221 | 0.00222 | 0.00223 | 0.00225 |
| 0.03 | 0.00414 | 0.00419 | 0.00423 | 0.00426 | 0.00428 | 0.00433 | 0.00436 | 0.00439 | 0.00443 | 0.00449 |
| 0.04 | 0.00660 | 0.00670 | 0.00679 | 0.00685 | 0.00691 | 0.00700 | 0.00708 | 0.00716 | 0.00723 | 0.00736 |
| 0.05 | 0.00946 | 0.00964 | 0.00979 | 0.00991 | 0.01002 | 0.01019 | 0.01033 | 0.01047 | 0.01060 | 0.01086 |
| 0.06 | 0.0127 | 0.0130 | 0.0132 | 0.0134 | 0.0136 | 0.0138 | 0.0141 | 0.0148 | 0.0145 | 0.0150 |
| 0.07 | 0.0162 | 0.0166 | 0.0170 | 0.0173 | 0.0175 | 0.0180 | 0.0183 | 0.0187 | 0.0190 | 0.0197 |
| 0.08 | 0.0200 | 0.0206 | 0.0211 | 0.0215 | 0.0219 | 0.0225 | 0.0231 | 0.0236 | 0.0240 | 0.0250 |
| 0.09 | 0.0241 | 0.0249 | 0.0256 | 0.0262 | 0.0267 | 0.0275 | 0.0282 | 0.0289 | 0.0296 | 0.0310 |
| 0.10 | 0.0284 | 0.0294 | 0.0304 | 0.0311 | 0.0318 | 0.0329 | 0.0339 | 0.0348 | 0.0358 | 0.0376 |
| 0.11 | 0.0329 | 0.0343 | 0.0354 | 0.0364 | 0.0373 | 0.0387 | 0.0400 | 0.0413 | 0.0424 | 0.0448 |
| 0.12 | 0.0376 | 0.0393 | 0.0408 | 0.0420 | 0.0431 | 0.0450 | 0.0466 | 0.0482 | 0.0497 | 0.0527 |
| 0.13 | 0.0425 | 0.0446 | 0.0464 | 0.0480 | 0.0493 | 0.0516 | 0.0537 | 0.0556 | 0.0575 | 0.0613 |
| 0.14 | 0.0476 | 0.0502 | 0.0524 | 0.0542 | 0.0559 | 0.0587 | 0.0612 | 0.0636 | 0.0659 | 0.0706 |
| 0.15 | 0.0528 | 0.0559 | 0.0585 | 0.0608 | 0.0627 | 0.0662 | 0.0692 | 0.0721 | 0.0749 | 0.0805 |
| 0.16 | 0.0582 | 0.0619 | 0.0650 | 0.0676 | 0.0700 | 0.0740 | 0.0777 | 0.0811 | 0.0845 | 0.0912 |
| 0.17 | 0.0638 | 0.0680 | 0.0716 | 0.0748 | 0.0775 | 0.0823 | 0.0866 | 0.0907 | 0.0947 | 0.1026 |
| 0.18 | 0.0695 | 0.0744 | 0.0786 | 0.0822 | 0.0854 | 0.0910 | 0.0960 | 0.1008 | 0.1055 | 0.1148 |
| 0.19 | 0.0753 | 0.0809 | 0.0857 | 0.0899 | 0.0936 | 0.1001 | 0.1059 | 0.1115 | 0.1169 | 0.1277 |
| 0.20 | 0.0812 | 0.0876 | 0.0931 | 0.0979 | 0.1021 | 0.1096 | 0.1163 | 0.1227 | 0.1290 | 0.1414 |
| 0.22 | 0.0934 | 0.1015 | 0.109 | 0.115 | 0.120 | 0.130 | 0.139 | 0.147 | 0.155 | 0.171 |
| 0.24 | 0.1061 | 0.1161 | 0.125 | 0.133 | 0.140 | 0.152 | 0.163 | 0.173 | 0.184 | 0.204 |
| 0.26 | 0.119 | 0.131 | 0.142 | 0.152 | 0.160 | 0.175 | 0.189 | 0.202 | 0.215 | 0.241 |
| 0.28 | 0.132 | 0.147 | 0.160 | 0.172 | 0.182 | 0.201 | 0.217 | 0.234 | 0.249 | 0.281 |
| 0.30 | 0.146 | 0.163 | 0.179 | 0.193 | 0.205 | 0.228 | 0.248 | 0.267 | 0.287 | 0.324 |
| 0.32 | 0.160 | 0.180 | 0.199 | 0.215 | 0.230 | 0.256 | 0.281 | 0.304 | 0.327 | 0.371 |
| 0.34 | 0.174 | 0.198 | 0.219 | 0.238 | 0.256 | 0.287 | 0.316 | 0.343 | 0.370 | 0.423 |
| 0.36 | 0.189 | 0.216 | 0.241 | 0.263 | 0.283 | 0.319 | 0.353 | 0.385 | 0.416 | 0.478 |
| 0.38 | 0.203 | 0.234 | 0.263 | 0.288 | 0.312 | 0.353 | 0.392 | 0.429 | 0.465 | 0.537 |
| 0.40 | 0.218 | 0.253 | 0.286 | 0.315 | 0.341 | 0.389 | 0.434 | 0.476 | 0.518 | 0.600 |
| 0.42 | 0.233 | 0.273 | 0.309 | 0.342 | 0.373 | 0.427 | 0.478 | 0.526 | 0.574 | 0.668 |
| 0.44 | 0.248 | 0.293 | 0.334 | 0.371 | 0.405 | 0.467 | 0.525 | 0.580 | 0.633 | 0.740 |
| 0.46 | 0.264 | 0.313 | 0.359 | 0.401 | 0.439 | 0.509 | 0.574 | 0.636 | 0.696 | 0.816 |
| 0.48 | 0.279 | 0.334 | 0.385 | 0.432 | 0.474 | 0.553 | 0.625 | 0.695 | 0.763 | 0.897 |
| 0.50 | 0.295 | 0.355 | 0.412 | 0.463 | 0.511 | 0.598 | 0.679 | 0.757 | 0.833 | 0.983 |
| 0.55 | 0.335 | 0.410 | 0.482 | 0.548 | 0.609 | 0.722 | 0.826 | 0.926 | 1.025 | 1.22 |
| 0.60 | 0.375 | 0.468 | 0.557 | 0.640 | 0.717 | 0.858 | 0.990 | 1.117 | 1.24 | 1.49 |

*(continued)*

**APPENDIX 21.F** *(continued)*
Conveyance Factor, $K'$
Symmetrical Rectangular,[a] Trapezoidal Open Channels
(use for determining $Q$ or $d$ when $b$ is known)
(customary U.S. units[b,c])

$$K' \text{ in } Q = K'\left(\frac{1}{n}\right)b^{8/3}\sqrt{S}$$

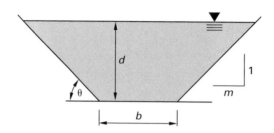

| | $m$ and $\theta$ | | | | | | | | | |
|---|---|---|---|---|---|---|---|---|---|---|
| | 0.0 | 0.25 | 0.5 | 0.75 | 1.0 | 1.5 | 2.0 | 2.5 | 3.0 | 4.0 |
| $x = d/b$ | 90° | 76.0° | 63.4° | 53.1° | 45.0° | 33.7° | 26.6° | 21.8° | 18.4° | 14.0° |
| 0.70 | 0.457 | 0.592 | 0.722 | 0.844 | 0.959 | 1.17 | 1.37 | 1.56 | 1.75 | 2.12 |
| 0.80 | 0.542 | 0.725 | 0.906 | 1.078 | 1.24 | 1.54 | 1.83 | 2.10 | 2.37 | 2.90 |
| 0.90 | 0.628 | 0.869 | 1.11 | 1.34 | 1.56 | 1.98 | 2.36 | 2.74 | 3.11 | 3.83 |
| 1.00 | 0.714 | 1.022 | 1.33 | 1.64 | 1.93 | 2.47 | 2.99 | 3.48 | 3.97 | 4.93 |
| 1.20 | 0.891 | 1.36 | 1.85 | 2.33 | 2.79 | 3.67 | 4.51 | 5.32 | 6.11 | 7.67 |
| 1.40 | 1.07 | 1.74 | 2.45 | 3.16 | 3.85 | 5.17 | 6.42 | 7.64 | 8.84 | 11.2 |
| 1.60 | 1.25 | 2.16 | 3.14 | 4.14 | 5.12 | 6.99 | 8.78 | 10.52 | 12.2 | 15.6 |
| 1.80 | 1.43 | 2.62 | 3.93 | 5.28 | 6.60 | 9.15 | 11.6 | 14.0 | 16.3 | 20.9 |
| 2.00 | 1.61 | 3.12 | 4.82 | 6.58 | 8.32 | 11.7 | 14.9 | 18.1 | 21.2 | 27.3 |
| 2.25 | 1.84 | 3.81 | 6.09 | 8.46 | 10.8 | 15.4 | 19.8 | 24.1 | 28.4 | 36.7 |

[a]For rectangular channels, use the 0.0 (90°, vertical sides) column.
[b]$Q$ = flow rate, ft³/sec; $d$ = depth of flow, ft; $b$ = bottom width of channel, ft; $S$ = geometric slope, ft/ft; $n$ = Manning's roughness constant.
[c]For SI units (i.e., $Q$ in m³/s, and $d$ and $b$ in m), divide each table value by 1.486.

## APPENDIX 22.A
Properties of Saturated Steam by Temperature
(customary U.S. units)

| temp. (°F) | absolute pressure (lbf/in²) | specific volume (ft³/lbm) | | internal energy (Btu/lbm) | | enthalpy (Btu/lbm) | | | entropy (Btu/lbm-°R) | | temp. (°F) |
|---|---|---|---|---|---|---|---|---|---|---|---|
| | | sat. liquid $(v_f)$ | sat. vapor $(v_g)$ | sat. liquid $(u_f)$ | sat. vapor $(u_g)$ | sat. liquid $(h_f)$ | evap. $(h_{fg})$ | sat. vapor $(h_g)$ | sat. liquid $(s_f)$ | sat. vapor $(s_g)$ | |
| 32 | 0.0886 | 0.01602 | 3302 | −0.02 | 1021.0 | −0.02 | 1075.2 | 1075.2 | −0.0004 | 2.1868 | 32 |
| 34 | 0.0961 | 0.01602 | 3059 | 2.00 | 1021.7 | 2.00 | 1074.1 | 1076.1 | 0.00405 | 2.1797 | 34 |
| 36 | 0.1040 | 0.01602 | 2836 | 4.01 | 1022.3 | 4.01 | 1072.9 | 1076.9 | 0.00812 | 2.1727 | 36 |
| 38 | 0.1126 | 0.01602 | 2632 | 6.02 | 1023.0 | 6.02 | 1071.8 | 1077.8 | 0.01217 | 2.1658 | 38 |
| 40 | 0.1217 | 0.01602 | 2443 | 8.03 | 1023.7 | 8.03 | 1070.7 | 1078.7 | 0.01620 | 2.1589 | 40 |
| 42 | 0.1316 | 0.01602 | 2270 | 10.04 | 1024.3 | 10.04 | 1069.5 | 1079.6 | 0.02022 | 2.1522 | 42 |
| 44 | 0.1421 | 0.01602 | 2111 | 12.05 | 1025.0 | 12.05 | 1068.4 | 1080.4 | 0.02421 | 2.1454 | 44 |
| 46 | 0.1533 | 0.01602 | 1964 | 14.06 | 1025.6 | 14.06 | 1067.3 | 1081.3 | 0.02819 | 2.1388 | 46 |
| 48 | 0.1653 | 0.01602 | 1828 | 16.06 | 1026.3 | 16.06 | 1066.1 | 1082.2 | 0.03215 | 2.1322 | 48 |
| 50 | 0.1781 | 0.01602 | 1703 | 18.07 | 1026.9 | 18.07 | 1065.0 | 1083.1 | 0.03609 | 2.1257 | 50 |
| 52 | 0.1918 | 0.01603 | 1587 | 20.07 | 1027.6 | 20.07 | 1063.8 | 1083.9 | 0.04001 | 2.1192 | 52 |
| 54 | 0.2065 | 0.01603 | 1481 | 22.07 | 1028.2 | 22.07 | 1062.7 | 1084.8 | 0.04392 | 2.1128 | 54 |
| 56 | 0.2221 | 0.01603 | 1382 | 24.07 | 1028.9 | 24.08 | 1061.6 | 1085.7 | 0.04781 | 2.1065 | 56 |
| 58 | 0.2387 | 0.01603 | 1291 | 26.08 | 1029.6 | 26.08 | 1060.5 | 1086.6 | 0.05168 | 2.1002 | 58 |
| 60 | 0.2564 | 0.01604 | 1206 | 28.08 | 1030.2 | 28.08 | 1059.3 | 1087.4 | 0.05554 | 2.0940 | 60 |
| 62 | 0.2752 | 0.01604 | 1128 | 30.08 | 1030.9 | 30.08 | 1058.2 | 1088.3 | 0.05938 | 2.0879 | 62 |
| 64 | 0.2953 | 0.01604 | 1055 | 32.08 | 1031.5 | 32.08 | 1057.1 | 1089.2 | 0.06321 | 2.0818 | 64 |
| 66 | 0.3166 | 0.01604 | 987.7 | 34.08 | 1032.2 | 34.08 | 1055.9 | 1090.0 | 0.06702 | 2.0758 | 66 |
| 68 | 0.3393 | 0.01605 | 925.2 | 36.08 | 1032.8 | 36.08 | 1054.8 | 1090.9 | 0.07081 | 2.0698 | 68 |
| 70 | 0.3634 | 0.01605 | 867.1 | 38.07 | 1033.5 | 38.08 | 1053.7 | 1091.8 | 0.07459 | 2.0639 | 70 |
| 72 | 0.3889 | 0.01606 | 813.2 | 40.07 | 1034.1 | 40.07 | 1052.5 | 1092.6 | 0.07836 | 2.0581 | 72 |
| 74 | 0.4160 | 0.01606 | 763.0 | 42.07 | 1034.8 | 42.07 | 1051.4 | 1093.5 | 0.08211 | 2.0523 | 74 |
| 76 | 0.4448 | 0.01606 | 716.3 | 44.07 | 1035.4 | 44.07 | 1050.3 | 1094.4 | 0.08585 | 2.0466 | 76 |
| 78 | 0.4752 | 0.01607 | 672.9 | 46.07 | 1036.1 | 46.07 | 1049.1 | 1095.2 | 0.08957 | 2.0409 | 78 |
| 80 | 0.5075 | 0.01607 | 632.4 | 48.06 | 1036.7 | 48.07 | 1048.0 | 1096.1 | 0.09328 | 2.0353 | 80 |
| 82 | 0.5416 | 0.01608 | 594.7 | 50.06 | 1037.4 | 50.06 | 1046.9 | 1097.0 | 0.09697 | 2.0297 | 82 |
| 84 | 0.5778 | 0.01608 | 559.5 | 52.06 | 1038.0 | 52.06 | 1045.7 | 1097.8 | 0.1007 | 2.0242 | 84 |
| 86 | 0.6160 | 0.01609 | 526.7 | 54.05 | 1038.7 | 54.06 | 1044.6 | 1098.7 | 0.1043 | 2.0187 | 86 |
| 88 | 0.6564 | 0.01610 | 496.0 | 56.05 | 1039.3 | 56.05 | 1043.4 | 1099.5 | 0.1080 | 2.0133 | 88 |
| 90 | 0.6990 | 0.01610 | 467.4 | 58.05 | 1039.9 | 58.05 | 1042.4 | 1100.4 | 0.1116 | 2.0079 | 90 |
| 92 | 0.7441 | 0.01611 | 440.7 | 60.04 | 1040.6 | 60.05 | 1041.3 | 1101.3 | 0.1152 | 2.0026 | 92 |
| 94 | 0.7917 | 0.01611 | 415.6 | 62.04 | 1041.2 | 62.04 | 1040.1 | 1102.1 | 0.1189 | 1.9974 | 94 |
| 96 | 0.8418 | 0.01612 | 392.3 | 64.04 | 1041.9 | 64.04 | 1039.0 | 1103.0 | 0.1225 | 1.9922 | 96 |
| 98 | 0.8947 | 0.01613 | 370.3 | 66.03 | 1042.5 | 66.04 | 1037.8 | 1103.8 | 0.1260 | 1.9870 | 98 |
| 100 | 0.9505 | 0.01613 | 349.8 | 68.03 | 1043.2 | 68.03 | 1036.7 | 1104.7 | 0.1296 | 1.9819 | 100 |
| 105 | 1.103 | 0.01615 | 304.0 | 73.02 | 1044.8 | 73.03 | 1033.8 | 1106.8 | 0.1385 | 1.9693 | 105 |
| 110 | 1.277 | 0.01617 | 265.0 | 78.01 | 1046.4 | 78.02 | 1031.0 | 1109.0 | 0.1473 | 1.9570 | 110 |
| 115 | 1.473 | 0.01619 | 231.6 | 83.01 | 1048.0 | 83.01 | 1028.1 | 1111.1 | 0.1560 | 1.9450 | 115 |
| 120 | 1.695 | 0.01621 | 203.0 | 88.00 | 1049.6 | 88.00 | 1025.2 | 1113.2 | 0.1647 | 1.9333 | 120 |
| 125 | 1.945 | 0.01623 | 178.3 | 92.99 | 1051.1 | 93.00 | 1022.3 | 1115.3 | 0.1732 | 1.9218 | 125 |

*(continued)*

**APPENDIX 22.A** *(continued)*
Properties of Saturated Steam by Temperature
(customary U.S. units)

| temp. (°F) | absolute pressure (lbf/in²) | specific volume (ft³/lbm) | | internal energy (Btu/lbm) | | enthalpy (Btu/lbm) | | | entropy (Btu/lbm-°R) | | temp. (°F) |
|---|---|---|---|---|---|---|---|---|---|---|---|
| | | sat. liquid ($v_f$) | sat. vapor ($v_g$) | sat. liquid ($u_f$) | sat. vapor ($u_g$) | sat. liquid ($h_f$) | evap. ($h_{fg}$) | sat. vapor ($h_g$) | sat. liquid ($s_f$) | sat. vapor ($s_g$) | |
| 130 | 2.226 | 0.01625 | 157.1 | 97.99 | 1052.7 | 97.99 | 1019.4 | 1117.4 | 0.1818 | 1.9106 | 130 |
| 135 | 2.541 | 0.01627 | 138.7 | 102.98 | 1054.3 | 102.99 | 1016.5 | 1119.5 | 0.1902 | 1.8996 | 135 |
| 140 | 2.893 | 0.01629 | 122.8 | 107.98 | 1055.8 | 107.99 | 1013.6 | 1121.6 | 0.1986 | 1.8888 | 140 |
| 145 | 3.286 | 0.01632 | 109.0 | 112.98 | 1057.4 | 112.99 | 1010.7 | 1123.7 | 0.2069 | 1.8783 | 145 |
| 150 | 3.723 | 0.01634 | 96.93 | 117.98 | 1059.0 | 117.99 | 1007.7 | 1125.7 | 0.2151 | 1.8680 | 150 |
| 155 | 4.209 | 0.01637 | 86.40 | 122.98 | 1060.5 | 122.99 | 1004.8 | 1127.8 | 0.2233 | 1.8580 | 155 |
| 160 | 4.747 | 0.01639 | 77.18 | 127.98 | 1062.0 | 128.00 | 1001.8 | 1129.8 | 0.2314 | 1.8481 | 160 |
| 165 | 5.343 | 0.01642 | 69.09 | 132.99 | 1063.6 | 133.01 | 998.9 | 1131.9 | 0.2394 | 1.8384 | 165 |
| 170 | 6.000 | 0.01645 | 61.98 | 138.00 | 1065.1 | 138.01 | 995.9 | 1133.9 | 0.2474 | 1.8290 | 170 |
| 175 | 6.724 | 0.01648 | 55.71 | 143.01 | 1066.6 | 143.03 | 992.9 | 1135.9 | 0.2553 | 1.8197 | 175 |
| 180 | 7.520 | 0.01651 | 50.17 | 148.02 | 1068.1 | 148.04 | 989.9 | 1137.9 | 0.2632 | 1.8106 | 180 |
| 185 | 8.393 | 0.01654 | 45.27 | 153.03 | 1069.6 | 153.06 | 986.8 | 1139.9 | 0.2710 | 1.8017 | 185 |
| 190 | 9.350 | 0.01657 | 40.92 | 158.05 | 1071.0 | 158.08 | 983.7 | 1141.8 | 0.2788 | 1.7930 | 190 |
| 195 | 10.396 | 0.01660 | 37.05 | 163.07 | 1072.5 | 163.10 | 980.7 | 1143.8 | 0.2865 | 1.7844 | 195 |
| 200 | 11.538 | 0.01663 | 33.61 | 168.09 | 1074.0 | 168.13 | 977.6 | 1145.7 | 0.2941 | 1.7760 | 200 |
| 205 | 12.782 | 0.01667 | 30.54 | 173.12 | 1075.4 | 173.16 | 974.4 | 1147.6 | 0.3017 | 1.7678 | 205 |
| 210 | 14.14 | 0.01670 | 27.79 | 178.2 | 1076.8 | 178.2 | 971.3 | 1149.5 | 0.3092 | 1.7597 | 210 |
| 212 | 14.71 | 0.01672 | 26.78 | 180.2 | 1077.4 | 180.2 | 970.1 | 1150.3 | 0.3122 | 1.7565 | 212 |
| 220 | 17.20 | 0.01677 | 23.13 | 188.2 | 1079.6 | 188.3 | 965.0 | 1153.3 | 0.3242 | 1.7440 | 220 |
| 230 | 20.80 | 0.01685 | 19.37 | 198.3 | 1082.4 | 198.4 | 958.5 | 1156.9 | 0.3389 | 1.7288 | 230 |
| 240 | 24.99 | 0.01692 | 16.31 | 208.4 | 1085.1 | 208.5 | 952.0 | 1160.5 | 0.3534 | 1.7141 | 240 |
| 250 | 29.84 | 0.01700 | 13.82 | 218.5 | 1087.7 | 218.6 | 945.4 | 1164.0 | 0.3678 | 1.7000 | 250 |
| 260 | 35.45 | 0.01708 | 11.76 | 228.7 | 1090.3 | 228.8 | 938.6 | 1167.4 | 0.3820 | 1.6863 | 260 |
| 270 | 41.88 | 0.01717 | 10.06 | 238.9 | 1092.8 | 239.0 | 931.7 | 1170.7 | 0.3960 | 1.6730 | 270 |
| 280 | 49.22 | 0.01726 | 8.64 | 249.0 | 1095.2 | 249.2 | 924.7 | 1173.9 | 0.4099 | 1.6601 | 280 |
| 290 | 57.57 | 0.01735 | 7.46 | 259.3 | 1097.5 | 259.5 | 917.6 | 1177.0 | 0.4236 | 1.6476 | 290 |
| 300 | 67.03 | 0.01745 | 6.466 | 269.5 | 1099.8 | 269.7 | 910.3 | 1180.0 | 0.4372 | 1.6354 | 300 |
| 310 | 77.69 | 0.01755 | 5.626 | 279.8 | 1101.9 | 280.1 | 902.8 | 1182.8 | 0.4507 | 1.6236 | 310 |
| 320 | 89.67 | 0.01765 | 4.914 | 290.1 | 1103.9 | 290.4 | 895.1 | 1185.5 | 0.4640 | 1.6120 | 320 |
| 330 | 103.07 | 0.01776 | 4.308 | 300.5 | 1105.9 | 300.8 | 887.2 | 1188.0 | 0.4772 | 1.6007 | 330 |
| 340 | 118.02 | 0.01787 | 3.788 | 310.9 | 1107.7 | 311.2 | 879.3 | 1190.5 | 0.4903 | 1.5897 | 340 |
| 350 | 134.63 | 0.01799 | 3.343 | 321.3 | 1109.4 | 321.7 | 871.0 | 1192.7 | 0.5032 | 1.5789 | 350 |
| 360 | 153.03 | 0.01811 | 2.958 | 331.8 | 1111.0 | 332.3 | 862.5 | 1194.8 | 0.5161 | 1.5684 | 360 |
| 370 | 173.36 | 0.01823 | 2.625 | 342.3 | 1112.5 | 342.9 | 853.8 | 1196.7 | 0.5289 | 1.5580 | 370 |
| 380 | 195.74 | 0.01836 | 2.336 | 352.9 | 1113.9 | 353.5 | 845.0 | 1198.5 | 0.5415 | 1.5478 | 380 |
| 390 | 220.3 | 0.01850 | 2.084 | 363.5 | 1115.1 | 364.3 | 835.8 | 1200.1 | 0.5541 | 1.5378 | 390 |
| 400 | 247.3 | 0.01864 | 1.864 | 374.2 | 1116.2 | 375.1 | 826.4 | 1201.4 | 0.5667 | 1.5279 | 400 |
| 410 | 276.7 | 0.01879 | 1.671 | 385.0 | 1117.1 | 385.9 | 816.7 | 1202.6 | 0.5791 | 1.5182 | 410 |
| 420 | 308.8 | 0.01894 | 1.501 | 395.8 | 1117.8 | 396.9 | 806.8 | 1203.6 | 0.5915 | 1.5085 | 420 |
| 430 | 343.6 | 0.01910 | 1.351 | 406.7 | 1118.4 | 407.9 | 796.4 | 1204.3 | 0.6038 | 1.4990 | 430 |

*(continued)*

**APPENDIX 22.A** *(continued)*
Properties of Saturated Steam by Temperature
(customary U.S. units)

| temp. (°F) | absolute pressure (lbf/in²) | specific volume (ft³/lbm) | | internal energy (Btu/lbm) | | enthalpy (Btu/lbm) | | | entropy (Btu/lbm-°R) | | temp. (°F) |
|---|---|---|---|---|---|---|---|---|---|---|---|
| | | sat. liquid ($v_f$) | sat. vapor ($v_g$) | sat. liquid ($u_f$) | sat. vapor ($u_g$) | sat. liquid ($h_f$) | evap. ($h_{fg}$) | sat. vapor ($h_g$) | sat. liquid ($s_f$) | sat. vapor ($s_g$) | |
| 440 | 381.5 | 0.01926 | 1.218 | 417.6 | 1118.9 | 419.0 | 785.8 | 1204.8 | 0.6161 | 1.4895 | 440 |
| 450 | 422.5 | 0.01944 | 1.0999 | 428.7 | 1119.1 | 430.2 | 774.9 | 1205.1 | 0.6283 | 1.4802 | 450 |
| 460 | 466.8 | 0.01962 | 0.9951 | 439.8 | 1119.2 | 441.5 | 763.6 | 1205.1 | 0.6405 | 1.4708 | 460 |
| 470 | 514.5 | 0.01981 | 0.9015 | 451.1 | 1119.0 | 452.9 | 752.0 | 1204.9 | 0.6527 | 1.4615 | 470 |
| 480 | 566.0 | 0.02001 | 0.8179 | 462.4 | 1118.7 | 464.5 | 739.8 | 1204.3 | 0.6648 | 1.4522 | 480 |
| 490 | 621.2 | 0.02022 | 0.7429 | 473.8 | 1118.1 | 476.1 | 727.4 | 1203.5 | 0.6769 | 1.4428 | 490 |
| 500 | 680.6 | 0.02044 | 0.6756 | 485.4 | 1117.2 | 487.9 | 714.4 | 1202.3 | 0.6891 | 1.4335 | 500 |
| 510 | 744.1 | 0.02068 | 0.6149 | 497.1 | 1116.2 | 499.9 | 700.9 | 1200.8 | 0.7012 | 1.4240 | 510 |
| 520 | 812.1 | 0.02092 | 0.5601 | 508.9 | 1114.8 | 512.0 | 686.9 | 1198.9 | 0.7134 | 1.4146 | 520 |
| 530 | 884.7 | 0.02119 | 0.5105 | 520.8 | 1113.1 | 524.3 | 672.4 | 1196.7 | 0.7256 | 1.4050 | 530 |
| 540 | 962.2 | 0.02146 | 0.4655 | 533.0 | 1111.1 | 536.8 | 657.2 | 1194.0 | 0.7378 | 1.3952 | 540 |
| 550 | 1044.8 | 0.02176 | 0.4247 | 545.3 | 1108.8 | 549.5 | 641.4 | 1190.9 | 0.7501 | 1.3853 | 550 |
| 560 | 1132.7 | 0.02208 | 0.3874 | 557.8 | 1106.0 | 562.4 | 624.8 | 1187.2 | 0.7625 | 1.3753 | 560 |
| 570 | 1226.2 | 0.02242 | 0.3534 | 570.5 | 1102.8 | 575.6 | 607.4 | 1183.0 | 0.7750 | 1.3649 | 570 |
| 580 | 1325.5 | 0.02279 | 0.3223 | 583.5 | 1099.2 | 589.1 | 589.3 | 1178.3 | 0.7876 | 1.3543 | 580 |
| 590 | 1430.8 | 0.02319 | 0.2937 | 596.7 | 1095.0 | 602.8 | 570.0 | 1172.8 | 0.8004 | 1.3434 | 590 |
| 600 | 1542.5 | 0.02363 | 0.2674 | 610.3 | 1090.3 | 617.0 | 549.6 | 1166.6 | 0.8133 | 1.3320 | 600 |
| 610 | 1660.9 | 0.02411 | 0.2431 | 624.2 | 1084.8 | 631.6 | 528.0 | 1159.6 | 0.8266 | 1.3201 | 610 |
| 620 | 1786.2 | 0.02465 | 0.2206 | 638.5 | 1078.6 | 646.7 | 504.8 | 1151.5 | 0.8401 | 1.3077 | 620 |
| 630 | 1918.9 | 0.02525 | 0.1997 | 653.4 | 1071.5 | 662.4 | 480.1 | 1142.4 | 0.8539 | 1.2945 | 630 |
| 640 | 2059.2 | 0.02593 | 0.1802 | 668.9 | 1063.2 | 678.7 | 453.1 | 1131.8 | 0.8683 | 1.2803 | 640 |
| 650 | 2207.8 | 0.02672 | 0.1618 | 685.1 | 1053.6 | 696.0 | 423.7 | 1119.7 | 0.8833 | 1.2651 | 650 |
| 660 | 2364.9 | 0.02766 | 0.1444 | 702.4 | 1042.2 | 714.5 | 390.8 | 1105.3 | 0.8991 | 1.2482 | 660 |
| 670 | 2531.2 | 0.02883 | 0.1277 | 721.1 | 1028.4 | 734.6 | 353.6 | 1088.2 | 0.9163 | 1.2292 | 670 |
| 680 | 2707.3 | 0.03036 | 0.1113 | 742.2 | 1011.1 | 757.4 | 309.4 | 1066.8 | 0.9355 | 1.2070 | 680 |
| 690 | 2894 | 0.03258 | 0.0946 | 767.1 | 987.7 | 784.5 | 253.8 | 1038.3 | 0.9582 | 1.1790 | 690 |
| 700 | 3093 | 0.03665 | 0.0748 | 801.5 | 948.2 | 822.5 | 168.5 | 991.0 | 0.9900 | 1.1353 | 700 |
| 705.103 | 3200.11 | 0.04975 | 0.04975 | 866.6 | 866.6 | 896.1 | 0 | 896.1 | 1.0526 | 1.0526 | 705.103 |

Values in this table were calculated from *NIST Standard Reference Database 10*, "NIST/ASME Steam Properties," Ver. 2.11, National Institute of Standards and Technology, U.S. Department of Commerce, Gaithersburg, MD, 1997, which has been licensed to PPI.

**APPENDIX 22.B**
Properties of Saturated Steam by Pressure
(customary U.S. units)

| absolute press (lbf/in$^2$) | temp. (°F) | specific volume (ft$^3$/lbm) | | internal energy (Btu/lbm) | | enthalpy (Btu/lbm) | | | entropy (Btu/lbm-°R) | | | absolute press (lbf/in$^2$) |
|---|---|---|---|---|---|---|---|---|---|---|---|---|
| | | sat. liquid $v_f$ | sat. vapor $v_g$ | sat. liquid $u_f$ | sat. vapor $u_g$ | sat. liquid $h_f$ | evap. $h_{fg}$ | sat. vapor $h_g$ | sat. liquid $s_f$ | evap. $s_{fg}$ | sat. vapor $s_g$ | |
| 0.4 | 72.83 | 0.01606 | 791.8 | 40.91 | 1034.4 | 40.91 | 1052.1 | 1093.0 | 0.0799 | 1.9758 | 2.0557 | 0.4 |
| 0.6 | 85.18 | 0.01609 | 539.9 | 53.23 | 1038.4 | 53.24 | 1045.1 | 1098.3 | 0.1028 | 1.9182 | 2.0210 | 0.6 |
| 0.8 | 94.34 | 0.01611 | 411.6 | 62.38 | 1041.3 | 62.38 | 1039.9 | 1102.3 | 0.1195 | 1.8770 | 1.9965 | 0.8 |
| 1.0 | 101.69 | 0.01614 | 333.5 | 69.72 | 1043.7 | 69.72 | 1035.7 | 1105.4 | 0.1326 | 1.8450 | 1.9776 | 1.0 |
| 1.2 | 107.87 | 0.01616 | 280.9 | 75.88 | 1045.7 | 75.89 | 1032.2 | 1108.1 | 0.1435 | 1.8187 | 1.9622 | 1.2 |
| 1.5 | 115.64 | 0.01619 | 227.7 | 83.64 | 1048.2 | 83.65 | 1027.8 | 1111.4 | 0.1571 | 1.7864 | 1.9435 | 1.5 |
| 2.0 | 126.03 | 0.01623 | 173.7 | 94.02 | 1051.5 | 94.02 | 1021.8 | 1115.8 | 0.1750 | 1.7445 | 1.9195 | 2.0 |
| 3.0 | 141.42 | 0.01630 | 118.7 | 109.4 | 1056.3 | 109.4 | 1012.8 | 1122.2 | 0.2009 | 1.6849 | 1.8858 | 3.0 |
| 4.0 | 152.91 | 0.01636 | 90.63 | 120.9 | 1059.9 | 120.9 | 1006.0 | 1126.9 | 0.2199 | 1.6423 | 1.8621 | 4.0 |
| 5.0 | 162.18 | 0.01641 | 73.52 | 130.2 | 1062.7 | 130.2 | 1000.5 | 1130.7 | 0.2349 | 1.6089 | 1.8438 | 5.0 |
| 6.0 | 170.00 | 0.01645 | 61.98 | 138.0 | 1065.1 | 138.0 | 995.9 | 1133.9 | 0.2474 | 1.5816 | 1.8290 | 6.0 |
| 7.0 | 176.79 | 0.01649 | 53.65 | 144.8 | 1067.1 | 144.8 | 991.8 | 1136.6 | 0.2581 | 1.5583 | 1.8164 | 7.0 |
| 8.0 | 182.81 | 0.01652 | 47.34 | 150.8 | 1068.9 | 150.9 | 988.1 | 1139.0 | 0.2676 | 1.5380 | 1.8056 | 8.0 |
| 9.0 | 188.22 | 0.01656 | 42.40 | 156.3 | 1070.5 | 156.3 | 984.8 | 1141.1 | 0.2760 | 1.5201 | 1.7961 | 9.0 |
| 10 | 193.16 | 0.01659 | 38.42 | 161.2 | 1072.0 | 161.3 | 981.9 | 1143.1 | 0.2836 | 1.5040 | 1.7876 | 10 |
| 14.696 | 211.95 | 0.01671 | 26.80 | 180.1 | 1077.4 | 180.2 | 970.1 | 1150.3 | 0.3122 | 1.4445 | 1.7566 | 14.696 |
| 15 | 212.99 | 0.01672 | 26.29 | 181.2 | 1077.7 | 181.2 | 969.5 | 1150.7 | 0.3137 | 1.4412 | 1.7549 | 15 |
| 20 | 227.92 | 0.01683 | 20.09 | 196.2 | 1081.8 | 196.3 | 959.9 | 1156.2 | 0.3358 | 1.3961 | 1.7319 | 20 |
| 25 | 240.03 | 0.01692 | 16.31 | 208.4 | 1085.1 | 208.5 | 952.0 | 1160.5 | 0.3535 | 1.3606 | 1.7141 | 25 |
| 30 | 250.30 | 0.01700 | 13.75 | 218.8 | 1087.8 | 218.9 | 945.2 | 1164.1 | 0.3682 | 1.3313 | 1.6995 | 30 |
| 35 | 259.25 | 0.01708 | 11.90 | 227.9 | 1090.1 | 228.0 | 939.2 | 1167.2 | 0.3809 | 1.3064 | 1.6873 | 35 |
| 40 | 267.22 | 0.01715 | 10.50 | 236.0 | 1092.1 | 236.1 | 933.7 | 1169.8 | 0.3921 | 1.2845 | 1.6766 | 40 |
| 45 | 274.41 | 0.01721 | 9.402 | 243.3 | 1093.9 | 243.5 | 928.7 | 1172.2 | 0.4022 | 1.2650 | 1.6672 | 45 |
| 50 | 280.99 | 0.01727 | 8.517 | 250.1 | 1095.4 | 250.2 | 924.0 | 1174.2 | 0.4113 | 1.2475 | 1.6588 | 50 |
| 55 | 287.05 | 0.01732 | 7.788 | 256.2 | 1096.8 | 256.4 | 919.7 | 1176.1 | 0.4196 | 1.2316 | 1.6512 | 55 |
| 60 | 292.68 | 0.01738 | 7.176 | 262.0 | 1098.1 | 262.2 | 915.6 | 1177.8 | 0.4273 | 1.2170 | 1.6443 | 60 |
| 65 | 297.95 | 0.01743 | 6.656 | 267.4 | 1099.3 | 267.6 | 911.8 | 1179.4 | 0.4344 | 1.2035 | 1.6379 | 65 |
| 70 | 302.91 | 0.01748 | 6.207 | 272.5 | 1100.4 | 272.7 | 908.1 | 1180.8 | 0.4411 | 1.1908 | 1.6319 | 70 |
| 75 | 307.58 | 0.01752 | 5.816 | 277.3 | 1101.4 | 277.6 | 904.6 | 1182.1 | 0.4474 | 1.1790 | 1.6264 | 75 |
| 80 | 312.02 | 0.01757 | 5.473 | 281.9 | 1102.3 | 282.1 | 901.2 | 1183.3 | 0.4534 | 1.1679 | 1.6212 | 80 |
| 85 | 316.24 | 0.01761 | 5.169 | 286.2 | 1103.2 | 286.5 | 898.0 | 1184.5 | 0.4590 | 1.1573 | 1.6163 | 85 |
| 90 | 320.26 | 0.01765 | 4.897 | 290.4 | 1104.0 | 290.7 | 894.9 | 1185.6 | 0.4643 | 1.1474 | 1.6117 | 90 |
| 95 | 324.11 | 0.01770 | 4.653 | 294.4 | 1104.8 | 294.7 | 891.9 | 1186.6 | 0.4694 | 1.1379 | 1.6073 | 95 |
| 100 | 327.81 | 0.01774 | 4.433 | 298.2 | 1105.5 | 298.5 | 889.0 | 1187.5 | 0.4743 | 1.1289 | 1.6032 | 100 |
| 110 | 334.77 | 0.01781 | 4.050 | 305.4 | 1106.8 | 305.8 | 883.4 | 1189.2 | 0.4834 | 1.1120 | 1.5954 | 110 |
| 120 | 341.25 | 0.01789 | 3.729 | 312.2 | 1107.9 | 312.6 | 878.2 | 1190.7 | 0.4919 | 1.0965 | 1.5884 | 120 |
| 130 | 347.32 | 0.01796 | 3.456 | 318.5 | 1109.0 | 318.9 | 873.2 | 1192.1 | 0.4998 | 1.0821 | 1.5818 | 130 |
| 140 | 353.03 | 0.01802 | 3.220 | 324.5 | 1109.9 | 324.9 | 868.5 | 1193.4 | 0.5071 | 1.0686 | 1.5757 | 140 |
| 150 | 358.42 | 0.01809 | 3.015 | 330.1 | 1110.8 | 330.6 | 863.9 | 1194.5 | 0.5141 | 1.0559 | 1.5700 | 150 |
| 160 | 363.54 | 0.01815 | 2.835 | 335.5 | 1111.6 | 336.0 | 859.5 | 1195.5 | 0.5206 | 1.0441 | 1.5647 | 160 |
| 170 | 368.41 | 0.01821 | 2.675 | 340.6 | 1112.3 | 341.2 | 855.2 | 1196.4 | 0.5268 | 1.0328 | 1.5596 | 170 |
| 180 | 373.07 | 0.01827 | 2.532 | 345.5 | 1112.9 | 346.1 | 851.2 | 1197.3 | 0.5328 | 1.0222 | 1.5549 | 180 |
| 190 | 377.52 | 0.01833 | 2.404 | 350.3 | 1113.5 | 350.9 | 847.2 | 1198.1 | 0.5384 | 1.0119 | 1.5503 | 190 |
| 200 | 381.80 | 0.01839 | 2.288 | 354.8 | 1114.1 | 355.5 | 843.3 | 1198.8 | 0.5438 | 1.0022 | 1.5460 | 200 |
| 250 | 400.97 | 0.01865 | 1.844 | 375.2 | 1116.2 | 376.1 | 825.5 | 1201.6 | 0.5679 | 0.9591 | 1.5270 | 250 |

*(continued)*

**APPENDIX 22.B** *(continued)*
Properties of Saturated Steam by Pressure
(customary U.S. units)

| absolute press ($lbf/in^2$) | temp. (°F) | specific volume ($ft^3$/lbm) | | internal energy (Btu/lbm) | | enthalpy (Btu/lbm) | | | entropy (Btu/lbm-°R) | | | absolute press ($lbf/in^2$) |
|---|---|---|---|---|---|---|---|---|---|---|---|---|
| | | sat. liquid $v_f$ | sat. vapor $v_g$ | sat. liquid $u_f$ | sat. vapor $u_g$ | sat. liquid $h_f$ | evap. $h_{fg}$ | sat. vapor $h_g$ | sat. liquid $s_f$ | evap. $s_{fg}$ | sat. vapor $s_g$ | |
| 300 | 417.35 | 0.01890 | 1.544 | 392.9 | 1117.7 | 394.0 | 809.4 | 1203.3 | 0.5882 | 0.9229 | 1.5111 | 300 |
| 350 | 431.74 | 0.01913 | 1.326 | 408.6 | 1118.5 | 409.8 | 794.6 | 1204.4 | 0.6059 | 0.8915 | 1.4974 | 350 |
| 400 | 444.62 | 0.01934 | 1.162 | 422.7 | 1119.0 | 424.2 | 780.9 | 1205.0 | 0.6217 | 0.8635 | 1.4852 | 400 |
| 450 | 456.31 | 0.01955 | 1.032 | 435.7 | 1119.2 | 437.3 | 767.8 | 1205.1 | 0.6360 | 0.8382 | 1.4742 | 450 |
| 500 | 467.04 | 0.01975 | 0.928 | 447.7 | 1119.1 | 449.5 | 755.5 | 1205.0 | 0.6491 | 0.8152 | 1.4642 | 500 |
| 550 | 476.98 | 0.01995 | 0.842 | 458.9 | 1118.8 | 461.0 | 743.5 | 1204.5 | 0.6611 | 0.7939 | 1.4550 | 550 |
| 600 | 486.24 | 0.02014 | 0.770 | 469.5 | 1118.3 | 471.7 | 732.1 | 1203.8 | 0.6724 | 0.7739 | 1.4463 | 600 |
| 700 | 503.13 | 0.02051 | 0.656 | 489.0 | 1116.9 | 491.7 | 710.2 | 1201.9 | 0.6929 | 0.7376 | 1.4305 | 700 |
| 800 | 518.27 | 0.02088 | 0.569 | 506.8 | 1115.0 | 509.9 | 689.4 | 1199.3 | 0.7113 | 0.7050 | 1.4162 | 800 |
| 900 | 532.02 | 0.02124 | 0.501 | 523.3 | 1112.7 | 526.8 | 669.4 | 1196.2 | 0.7280 | 0.6750 | 1.4030 | 900 |
| 1000 | 544.65 | 0.02160 | 0.446 | 538.7 | 1110.1 | 542.7 | 649.9 | 1192.6 | 0.7435 | 0.6472 | 1.3907 | 1000 |
| 1100 | 556.35 | 0.02196 | 0.401 | 553.2 | 1107.1 | 557.7 | 631.0 | 1188.6 | 0.7580 | 0.6211 | 1.3790 | 1100 |
| 1200 | 567.26 | 0.02233 | 0.362 | 567.0 | 1103.8 | 571.9 | 612.3 | 1184.2 | 0.7715 | 0.5963 | 1.3678 | 1200 |
| 1300 | 577.49 | 0.02270 | 0.330 | 580.2 | 1100.2 | 585.6 | 593.9 | 1179.5 | 0.7844 | 0.5726 | 1.3570 | 1300 |
| 1400 | 587.14 | 0.02307 | 0.302 | 592.9 | 1096.3 | 598.9 | 575.5 | 1174.4 | 0.7967 | 0.5498 | 1.3465 | 1400 |
| 1500 | 596.26 | 0.02346 | 0.277 | 605.2 | 1092.1 | 611.7 | 557.3 | 1169.0 | 0.8085 | 0.5278 | 1.3363 | 1500 |
| 1600 | 604.93 | 0.02386 | 0.255 | 617.1 | 1087.7 | 624.1 | 539.1 | 1163.2 | 0.8198 | 0.5064 | 1.3262 | 1600 |
| 1700 | 613.18 | 0.02428 | 0.236 | 628.7 | 1082.9 | 636.3 | 520.8 | 1157.1 | 0.8308 | 0.4854 | 1.3162 | 1700 |
| 1800 | 621.07 | 0.02471 | 0.218 | 640.1 | 1077.9 | 648.3 | 502.3 | 1150.6 | 0.8415 | 0.4648 | 1.3063 | 1800 |
| 1900 | 628.61 | 0.02516 | 0.203 | 651.3 | 1072.5 | 660.1 | 483.6 | 1143.7 | 0.8520 | 0.4443 | 1.2963 | 1900 |
| 2000 | 635.85 | 0.02564 | 0.188 | 662.4 | 1066.8 | 671.8 | 464.6 | 1136.4 | 0.8623 | 0.4240 | 1.2863 | 2000 |
| 2250 | 652.74 | 0.02696 | 0.157 | 689.7 | 1050.6 | 701.0 | 415.1 | 1116.0 | 0.8875 | 0.3731 | 1.2606 | 2250 |
| 2500 | 668.17 | 0.02859 | 0.131 | 717.6 | 1031.1 | 730.8 | 360.8 | 1091.6 | 0.9130 | 0.3199 | 1.2329 | 2500 |
| 2750 | 682.34 | 0.03080 | 0.108 | 747.6 | 1006.3 | 763.2 | 297.8 | 1061.0 | 0.9404 | 0.2607 | 1.2011 | 2750 |
| 3000 | 695.41 | 0.03434 | 0.085 | 783.4 | 969.9 | 802.5 | 214.4 | 1016.9 | 0.9733 | 0.1856 | 1.1589 | 3000 |
| 3200.11 | 705.1028 | 0.04975 | 0.04975 | 866.6 | 866.6 | 896.1 | 0 | 896.1 | 1.0526 | 0 | 1.0526 | 3200.11 |

Values in this table were calculated from *NIST Standard Reference Database 10*, "NIST/ASME Steam Properties," Ver. 2.11, National Institute of Standards and Technology, U.S. Department of Commerce, Gaithersburg, MD, 1997, which has been licensed to PPI.

## APPENDIX 22.C
Properties of Superheated Steam
(customary U.S. units)
specific volume ($v$) in ft³/lbm; enthalpy ($h$) in Btu/lbm; entropy ($s$) in Btu/lbm-°R

| absolute pressure (psia) (sat. temp., °F) | | temperature (°F) | | | | | | | | |
|---|---|---|---|---|---|---|---|---|---|---|
| | | 200 | 300 | 400 | 500 | 600 | 700 | 800 | 900 | 1000 |
| 1.0 | $v$ | 392.5 | 452.3 | 511.9 | 571.5 | 631.1 | 690.7 | 750.3 | 809.9 | 869.5 |
| (101.69) | $h$ | 1150.1 | 1195.7 | 1241.8 | 1288.6 | 1336.2 | 1384.6 | 1433.9 | 1484.1 | 1535.1 |
| | $s$ | 2.0510 | 2.1153 | 2.1723 | 2.2238 | 2.2710 | 2.3146 | 2.3554 | 2.3937 | 2.4299 |
| 5.0 | $v$ | 78.15 | 90.25 | 102.25 | 114.21 | 126.15 | 138.09 | 150.02 | 161.94 | 173.86 |
| (162.18) | $h$ | 1148.5 | 1194.8 | 1241.3 | 1288.2 | 1335.9 | 1384.4 | 1433.7 | 1483.9 | 1535.0 |
| | $s$ | 1.8716 | 1.9370 | 1.9944 | 2.0461 | 2.0934 | 2.1371 | 2.1779 | 2.2162 | 2.2525 |
| 10.0 | $v$ | 38.85 | 44.99 | 51.04 | 57.04 | 63.03 | 69.01 | 74.98 | 80.95 | 86.91 |
| (193.16) | $h$ | 1146.4 | 1193.8 | 1240.6 | 1287.8 | 1335.6 | 1384.2 | 1433.5 | 1483.8 | 1534.9 |
| | $s$ | 1.7926 | 1.8595 | 1.9174 | 1.9693 | 2.0167 | 2.0605 | 2.1014 | 2.1397 | 2.1760 |
| 14.696 | $v$ | ....... | 30.53 | 34.67 | 38.77 | 42.86 | 46.93 | 51.00 | 55.07 | 59.13 |
| (211.95) | $h$ | ....... | 1192.7 | 1240.0 | 1287.4 | 1335.3 | 1383.9 | 1433.3 | 1483.6 | 1534.8 |
| | $s$ | ....... | 1.8160 | 1.8744 | 1.9266 | 1.9741 | 2.0179 | 2.0588 | 2.0972 | 2.1335 |
| 20.0 | $v$ | ....... | 22.36 | 25.43 | 28.46 | 31.47 | 34.47 | 37.46 | 40.45 | 43.44 |
| (227.92) | $h$ | ....... | 1191.6 | 1239.3 | 1286.9 | 1334.9 | 1383.6 | 1433.1 | 1483.4 | 1534.6 |
| | $s$ | ....... | 1.7808 | 1.8398 | 1.8922 | 1.9399 | 1.9838 | 2.0247 | 2.0631 | 2.0994 |
| 60.0 | $v$ | ....... | 7.260 | 8.355 | 9.400 | 10.426 | 11.440 | 12.448 | 13.453 | 14.454 |
| (292.68) | $h$ | ....... | 1181.9 | 1233.7 | 1283.1 | 1332.2 | 1381.6 | 1431.5 | 1482.1 | 1533.5 |
| | $s$ | ....... | 1.6496 | 1.7138 | 1.7682 | 1.8168 | 1.8613 | 1.9026 | 1.9413 | 1.9778 |
| 100.0 | $v$ | ....... | ....... | 4.936 | 5.588 | 6.217 | 6.834 | 7.446 | 8.053 | 8.658 |
| (327.81) | $h$ | ....... | ....... | 1227.7 | 1279.3 | 1329.4 | 1379.5 | 1429.8 | 1480.7 | 1532.4 |
| | $s$ | ....... | ....... | 1.6521 | 1.7089 | 1.7586 | 1.8037 | 1.8453 | 1.8842 | 1.9209 |
| 150.0 | $v$ | ....... | ....... | 3.222 | 3.680 | 4.112 | 4.531 | 4.944 | 5.353 | 5.759 |
| (358.42) | $h$ | ....... | ....... | 1219.7 | 1274.3 | 1325.9 | 1376.8 | 1427.7 | 1479.0 | 1531.0 |
| | $s$ | ....... | ....... | 1.6001 | 1.6602 | 1.7114 | 1.7573 | 1.7994 | 1.8386 | 1.8755 |
| 200.0 | $v$ | ....... | ....... | 2.362 | 2.725 | 3.059 | 3.380 | 3.693 | 4.003 | 4.310 |
| (381.80) | $h$ | ....... | ....... | 1210.9 | 1269.0 | 1322.3 | 1374.1 | 1425.6 | 1477.3 | 1529.6 |
| | $s$ | ....... | ....... | 1.5602 | 1.6243 | 1.6771 | 1.7238 | 1.7664 | 1.8060 | 1.8430 |
| 250.0 | $v$ | ....... | ....... | ....... | 2.151 | 2.426 | 2.688 | 2.943 | 3.193 | 3.440 |
| (400.97) | $h$ | ....... | ....... | ....... | 1263.6 | 1318.6 | 1371.3 | 1423.5 | 1475.6 | 1528.1 |
| | $s$ | ....... | ....... | ....... | 1.5953 | 1.6499 | 1.6975 | 1.7406 | 1.7804 | 1.8177 |
| 300.0 | $v$ | ....... | ....... | ....... | 1.767 | 2.005 | 2.227 | 2.442 | 2.653 | 2.861 |
| (417.35) | $h$ | ....... | ....... | ....... | 1257.9 | 1314.8 | 1368.6 | 1421.3 | 1473.9 | 1526.7 |
| | $s$ | ....... | ....... | ....... | 1.5706 | 1.6271 | 1.6756 | 1.7192 | 1.7594 | 1.7969 |
| 400.0 | $v$ | ....... | ....... | ....... | 1.285 | 1.477 | 1.651 | 1.817 | 1.978 | 2.136 |
| (444.62) | $h$ | ....... | ....... | ....... | 1245.6 | 1306.9 | 1362.9 | 1416.9 | 1470.4 | 1523.9 |
| | $s$ | ....... | ....... | ....... | 1.5288 | 1.5897 | 1.6402 | 1.6849 | 1.7257 | 1.7637 |

**APPENDIX 22.C** *(continued)*
Properties of Superheated Steam
(customary U.S. units)
specific volume ($v$) in ft$^3$/lbm; enthalpy ($h$) in Btu/lbm; entropy ($s$) in Btu/lbm-°R

| absolute pressure (psia) (sat. temp., °F) | | temperature (°F) | | | | | | | | | |
|---|---|---|---|---|---|---|---|---|---|---|---|
| | | 500 | 600 | 700 | 800 | 900 | 1000 | 1100 | 1200 | 1400 | 1600 |
| 450.0 | $v$ | 1.123 | 1.300 | 1.458 | 1.608 | 1.753 | 1.894 | 2.034 | 2.172 | 2.445 | |
| (456.31) | $h$ | 1238.9 | 1302.8 | 1360.0 | 1414.7 | 1468.6 | 1522.4 | 1576.5 | 1631.0 | 1742.0 | |
| | $s$ | 1.5103 | 1.5737 | 1.6253 | 1.6706 | 1.7118 | 1.7499 | 1.7858 | 1.8196 | 1.8828 | |
| 500.0 | $v$ | 0.993 | 1.159 | 1.304 | 1.441 | 1.573 | 1.701 | 1.827 | 1.952 | 2.199 | |
| (467.04) | $h$ | 1231.9 | 1298.6 | 1357.0 | 1412.5 | 1466.9 | 1521.0 | 1575.3 | 1630.0 | 1741.2 | |
| | $s$ | 1.4928 | 1.5591 | 1.6118 | 1.6576 | 1.6992 | 1.7376 | 1.7736 | 1.8076 | 1.8708 | |
| 600.0 | $v$ | 0.795 | 0.946 | 1.073 | 1.190 | 1.302 | 1.411 | 1.518 | 1.623 | 1.830 | |
| (486.24) | $h$ | 1216.5 | 1289.9 | 1351.0 | 1408.0 | 1463.3 | 1518.1 | 1572.8 | 1627.9 | 1739.7 | |
| | $s$ | 1.4596 | 1.5326 | 1.5877 | 1.6348 | 1.6771 | 1.7160 | 1.7523 | 1.7865 | 1.8501 | |
| 700.0 | $v$ | ....... | 0.793 | 0.908 | 1.011 | 1.109 | 1.204 | 1.296 | 1.387 | 1.566 | |
| (503.13) | $h$ | ....... | 1280.7 | 1344.8 | 1403.4 | 1459.7 | 1515.1 | 1570.4 | 1625.9 | 1738.2 | |
| | $s$ | ....... | 1.5087 | 1.5666 | 1.6151 | 1.6581 | 1.6975 | 1.7341 | 1.7686 | 1.8324 | |
| 800.0 | $v$ | ....... | 0.678 | 0.783 | 0.877 | 0.964 | 1.048 | 1.130 | 1.211 | 1.368 | |
| (518.27) | $h$ | ....... | 1270.9 | 1338.4 | 1398.7 | 1456.0 | 1512.2 | 1568.0 | 1623.8 | 1736.7 | |
| | $s$ | ....... | 1.4867 | 1.5476 | 1.5975 | 1.6414 | 1.6812 | 1.7182 | 1.7529 | 1.8171 | |
| 900.0 | $v$ | ....... | 0.588 | 0.686 | 0.772 | 0.852 | 0.928 | 1.001 | 1.073 | 1.214 | |
| (532.02) | $h$ | ....... | 1260.4 | 1331.8 | 1393.9 | 1452.3 | 1509.2 | 1565.5 | 1621.8 | 1735.2 | |
| | $s$ | ....... | 1.4658 | 1.5302 | 1.5816 | 1.6263 | 1.6667 | 1.7040 | 1.7389 | 1.8034 | |
| 1000 | $v$ | ....... | 0.514 | 0.608 | 0.688 | 0.761 | 0.831 | 0.898 | 0.963 | 1.091 | 1.216 |
| (544.65) | $h$ | ....... | 1249.3 | 1325.0 | 1389.0 | 1448.6 | 1506.2 | 1563.0 | 1619.7 | 1733.6 | 1849.6 |
| | $s$ | ....... | 1.4457 | 1.5140 | 1.5671 | 1.6126 | 1.6535 | 1.6912 | 1.7264 | 1.7912 | 1.8504 |
| 1200 | $v$ | ....... | 0.402 | 0.491 | 0.562 | 0.626 | 0.686 | 0.743 | 0.798 | 0.906 | 1.012 |
| (567.26) | $h$ | ....... | 1224.2 | 1310.6 | 1379.0 | 1441.0 | 1500.1 | 1558.1 | 1615.5 | 1730.6 | 1847.3 |
| | $s$ | ....... | 1.4061 | 1.4842 | 1.5408 | 1.5882 | 1.6302 | 1.6686 | 1.7043 | 1.7698 | 1.8294 |
| 1400 | $v$ | ....... | 0.318 | 0.406 | 0.472 | 0.529 | 0.582 | 0.632 | 0.681 | 0.774 | 0.866 |
| (587.14) | $h$ | ....... | 1193.8 | 1295.1 | 1368.5 | 1433.1 | 1494.0 | 1553.0 | 1611.3 | 1727.5 | 1845.0 |
| | $s$ | ....... | 1.3649 | 1.4566 | 1.5174 | 1.5668 | 1.6100 | 1.6491 | 1.6853 | 1.7515 | 1.8115 |
| 1600 | $v$ | ....... | ....... | 0.342 | 0.404 | 0.456 | 0.504 | 0.549 | 0.592 | 0.676 | 0.756 |
| (604.93) | $h$ | ....... | ....... | 1278.3 | 1357.6 | 1425.1 | 1487.7 | 1547.9 | 1607.1 | 1724.5 | 1842.7 |
| | $s$ | ....... | ....... | 1.4302 | 1.4959 | 1.5475 | 1.5920 | 1.6319 | 1.6686 | 1.7354 | 1.7958 |
| 1800 | $v$ | ....... | ....... | 0.291 | 0.350 | 0.399 | 0.443 | 0.484 | 0.524 | 0.599 | 0.671 |
| (621.07) | $h$ | ....... | ....... | 1260.0 | 1346.2 | 1416.9 | 1481.3 | 1542.8 | 1602.8 | 1721.4 | 1840.3 |
| | $s$ | ....... | ....... | 1.4044 | 1.4759 | 1.5299 | 1.5756 | 1.6164 | 1.6537 | 1.7211 | 1.7819 |
| 2000 | $v$ | ....... | ....... | 0.249 | 0.308 | 0.354 | 0.395 | 0.433 | 0.469 | 0.537 | 0.603 |
| (635.85) | $h$ | ....... | ....... | 1239.7 | 1334.3 | 1408.5 | 1474.8 | 1537.6 | 1598.5 | 1718.3 | 1838.0 |
| | $s$ | ....... | ....... | 1.3783 | 1.4568 | 1.5135 | 1.5606 | 1.6022 | 1.6400 | 1.7082 | 1.7693 |

Values in this table were calculated from *NIST Standard Reference Database 10*, "NIST/ASME Steam Properties," Ver. 2.11, National Institute of Standards and Technology, U.S. Department of Commerce, Gaithersburg, MD, 1997, which has been licensed to PPI.

### APPENDIX 22.D
Properties of Compressed Water
(customary U.S. units)

| T (°F) | p (psia) | ρ (lbm/ft³) | v (ft³/lbm) | x | h (Btu/lbm) | s (Btu/lbm-°R) | u (Btu/lbm) |
|---|---|---|---|---|---|---|---|
| 32 | 200 | 62.46 | 0.01601 | subcooled | 0.5852 | −0.00001534 | −0.007334 |
| 100 | 200 | 62.03 | 0.01612 | subcooled | 68.56 | 0.1295 | 67.96 |
| 200 | 200 | 60.16 | 0.01662 | subcooled | 168.6 | 0.2939 | 167.9 |
| 300 | 200 | 57.34 | 0.01744 | subcooled | 270 | 0.437 | 269.3 |
| 381.8 | 200 | 54.39 | 0.01839 | 0 | 355.5 | 0.5438 | 354.8 |
| 381.8 | 200 | 0.437 | 2.288 | 1 | 1199 | 1.546 | 1114 |
| | | | | | | | |
| 32 | 400 | 62.5 | 0.016 | subcooled | 1.187 | 0.00000448 | 0.003032 |
| 100 | 400 | 62.07 | 0.01611 | subcooled | 69.09 | 0.1294 | 67.89 |
| 200 | 400 | 60.2 | 0.01661 | subcooled | 169 | 0.2936 | 167.8 |
| 300 | 400 | 57.39 | 0.01742 | subcooled | 270.3 | 0.4366 | 269.1 |
| 400 | 400 | 53.7 | 0.01862 | subcooled | 375.2 | 0.5662 | 373.8 |
| 444.6 | 400 | 51.7 | 0.01934 | 0 | 424.2 | 0.6217 | 422.7 |
| 444.6 | 400 | 0.8609 | 1.162 | 1 | 1205 | 1.485 | 1119 |
| | | | | | | | |
| 32 | 600 | 62.55 | 0.01599 | subcooled | 1.788 | 0.00002258 | 0.01296 |
| 100 | 600 | 62.1 | 0.0161 | subcooled | 69.61 | 0.1292 | 67.83 |
| 200 | 600 | 60.24 | 0.0166 | subcooled | 169.5 | 0.2934 | 167.6 |
| 300 | 600 | 57.44 | 0.01741 | subcooled | 270.7 | 0.4362 | 268.8 |
| 400 | 600 | 53.77 | 0.0186 | subcooled | 375.4 | 0.5657 | 373.4 |
| 486.2 | 600 | 49.65 | 0.02014 | 0 | 471.7 | 0.6724 | 469.5 |
| 486.2 | 600 | 1.298 | 0.7702 | 1 | 1204 | 1.446 | 1118 |
| | | | | | | | |
| 32 | 800 | 62.59 | 0.01598 | subcooled | 2.388 | 0.00003896 | 0.02244 |
| 100 | 800 | 62.14 | 0.01609 | subcooled | 70.14 | 0.1291 | 67.76 |
| 200 | 800 | 60.28 | 0.01659 | subcooled | 169.9 | 0.2931 | 167.5 |
| 300 | 800 | 57.49 | 0.01739 | subcooled | 271.1 | 0.4359 | 268.5 |
| 400 | 800 | 53.84 | 0.01857 | subcooled | 375.7 | 0.5652 | 372.9 |
| 500 | 800 | 48.99 | 0.02041 | subcooled | 487.9 | 0.6885 | 484.9 |
| 518.3 | 800 | 47.9 | 0.02088 | 0 | 509.9 | 0.7112 | 506.8 |
| 518.3 | 800 | 1.757 | 0.5692 | 1 | 1199 | 1.416 | 1115 |
| | | | | | | | |
| 32 | 1000 | 62.63 | 0.01597 | subcooled | 2.986 | 0.00005365 | 0.0315 |
| 100 | 1000 | 62.18 | 0.01608 | subcooled | 70.67 | 0.129 | 67.69 |
| 200 | 1000 | 60.31 | 0.01658 | subcooled | 170.4 | 0.2929 | 167.3 |
| 300 | 1000 | 57.54 | 0.01738 | subcooled | 271.5 | 0.4355 | 268.2 |
| 400 | 1000 | 53.9 | 0.01855 | subcooled | 375.9 | 0.5646 | 372.5 |
| 500 | 1000 | 49.1 | 0.02037 | subcooled | 487.8 | 0.6876 | 484 |
| 544.6 | 1000 | 46.3 | 0.0216 | 0 | 542.7 | 0.7435 | 538.7 |
| 544.6 | 1000 | 2.242 | 0.446 | 1 | 1193 | 1.391 | 1110 |
| | | | | | | | |
| 32 | 1500 | 62.74 | 0.01594 | subcooled | 4.477 | 0.00008309 | 0.05229 |
| 100 | 1500 | 62.27 | 0.01606 | subcooled | 71.98 | 0.1287 | 67.53 |
| 200 | 1500 | 60.41 | 0.01655 | subcooled | 171.5 | 0.2923 | 166.9 |
| 300 | 1500 | 57.65 | 0.01734 | subcooled | 272.4 | 0.4346 | 267.6 |
| 400 | 1500 | 54.06 | 0.0185 | subcooled | 376.5 | 0.5633 | 371.4 |
| 500 | 1500 | 49.36 | 0.02026 | subcooled | 487.6 | 0.6855 | 482 |
| 596.3 | 1500 | 42.62 | 0.02346 | 0 | 611.7 | 0.8085 | 605.2 |
| 596.3 | 1500 | 3.61 | 0.277 | 1 | 1169 | 1.336 | 1092 |

*(continued)*

**APPENDIX 22.D** *(continued)*
Properties of Compressed Water
(customary U.S. units)

| $T$ (°F) | $p$ (psia) | $\rho$ (lbm/ft³) | $v$ (ft³/lbm) | $x$ | $h$ (Btu/lbm) | $s$ (Btu/lbm-°R) | $u$ (Btu/lbm) |
|------|------|------|------|------|------|------|------|
| 32 | 2000 | 62.85 | 0.01591 | subcooled | 5.959 | 0.0001023 | 0.0705 |
| 100 | 2000 | 62.36 | 0.01603 | subcooled | 73.3 | 0.1284 | 67.36 |
| 200 | 2000 | 60.51 | 0.01653 | subcooled | 172.7 | 0.2917 | 166.5 |
| 300 | 2000 | 57.77 | 0.01731 | subcooled | 273.3 | 0.4338 | 266.9 |
| 400 | 2000 | 54.22 | 0.01844 | subcooled | 377.1 | 0.5621 | 370.3 |
| 500 | 2000 | 49.62 | 0.02015 | subcooled | 487.5 | 0.6835 | 480.1 |
| 600 | 2000 | 42.89 | 0.02332 | subcooled | 614.4 | 0.809 | 605.8 |
| 635.8 | 2000 | 39.01 | 0.02564 | 0 | 671.8 | 0.8623 | 662.3 |
| 635.8 | 2000 | 5.316 | 0.1881 | 1 | 1136 | 1.286 | 1067 |
| | | | | | | | |
| 32 | 3000 | 63.06 | 0.01586 | subcooled | 8.903 | 0.0001113 | 0.09948 |
| 100 | 3000 | 62.55 | 0.01599 | subcooled | 75.91 | 0.1278 | 67.04 |
| 200 | 3000 | 60.7 | 0.01648 | subcooled | 174.9 | 0.2905 | 165.8 |
| 300 | 3000 | 58 | 0.01724 | subcooled | 275.2 | 0.4321 | 265.7 |
| 400 | 3000 | 54.53 | 0.01834 | subcooled | 378.4 | 0.5596 | 368.2 |
| 500 | 3000 | 50.1 | 0.01996 | subcooled | 487.5 | 0.6796 | 476.4 |
| 600 | 3000 | 43.94 | 0.02276 | subcooled | 610.1 | 0.8009 | 597.4 |
| 695.4 | 3000 | 29.12 | 0.03434 | 0 | 802.5 | 0.9733 | 783.4 |
| 695.4 | 3000 | 11.81 | 0.08466 | 1 | 1017 | 1.159 | 969.9 |
| | | | | | | | |
| 32 | 4000 | 63.26 | 0.01581 | subcooled | 11.82 | 0.00008277 | 0.119 |
| 100 | 4000 | 62.73 | 0.01594 | subcooled | 78.52 | 0.1271 | 66.72 |
| 200 | 4000 | 60.88 | 0.01642 | subcooled | 177.2 | 0.2894 | 165.1 |
| 300 | 4000 | 58.22 | 0.01718 | subcooled | 277.1 | 0.4304 | 264.4 |
| 400 | 4000 | 54.83 | 0.01824 | subcooled | 379.7 | 0.5572 | 366.2 |
| 500 | 4000 | 50.55 | 0.01978 | subcooled | 487.7 | 0.676 | 473.1 |
| 600 | 4000 | 44.81 | 0.02231 | subcooled | 607 | 0.794 | 590.5 |
| 700 | 4000 | 34.83 | 0.02871 | subcooled | 763.6 | 0.9347 | 742.3 |

Values in this table were calculated from *NIST Standard Reference Database 10*, "NIST/ASME Steam Properties," Ver. 2.11, National Institute of Standards and Technology, U.S. Department of Commerce, Gaithersburg, MD, 1997, which has been licensed to PPI.

### APPENDIX 22.E
Enthalpy-Entropy (Mollier) Diagram for Steam
(customary U.S. units)

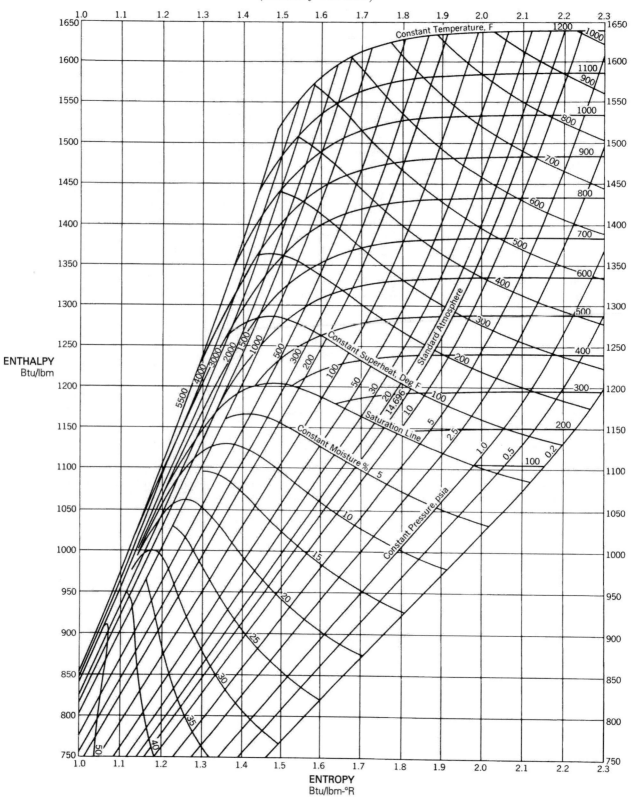

ENTHALPY
Btu/lbm

ENTROPY
Btu/lbm-°R

Used with permission from *Steam: Its generation and use*, 41st ed., edited by S. C. Schultz and J. B. Kitto, copyright © 2005, by The Babcock & Wilcox Company.

**APPENDIX 22.F**
Properties of Low-Pressure Air
(customary U.S. units)

$T$ in °R; $h$ and $u$ in Btu/lbm; $\phi$ in Btu/lbm-°R

| $T$ | $h$ | $p_r$ | $u$ | $v_r$ | $\phi$ |
|---|---|---|---|---|---|
| 360 | 85.97 | 0.3363 | 61.29 | 396.6 | 0.50369 |
| 380 | 90.75 | 0.4061 | 64.70 | 346.6 | 0.51663 |
| 400 | 95.53 | 0.4858 | 68.11 | 305.0 | 0.52890 |
| 420 | 100.32 | 0.5760 | 71.52 | 270.1 | 0.54058 |
| 440 | 105.11 | 0.6776 | 74.93 | 240.6 | 0.55172 |
| 460 | 109.90 | 0.7913 | 78.36 | 215.33 | 0.56235 |
| 480 | 114.69 | 0.9182 | 81.77 | 193.65 | 0.57255 |
| 500 | 119.48 | 1.0590 | 85.20 | 174.90 | 0.58233 |
| 520 | 124.27 | 1.2147 | 88.62 | 158.58 | 0.59172 |
| 537 | 128.34 | 1.3593 | 91.53 | 146.34 | 0.59945 |
| 540 | 129.06 | 1.3860 | 92.04 | 144.32 | 0.60078 |
| 560 | 133.86 | 1.5742 | 95.47 | 131.78 | 0.60950 |
| 580 | 138.66 | 1.7800 | 98.90 | 120.70 | 0.61793 |
| 600 | 143.47 | 2.005 | 102.34 | 110.88 | 0.62607 |
| 620 | 148.28 | 2.249 | 105.78 | 102.12 | 0.63395 |
| 640 | 153.09 | 2.514 | 109.21 | 94.30 | 0.64159 |
| 660 | 157.92 | 2.801 | 112.67 | 87.27 | 0.64902 |
| 680 | 162.74 | 3.111 | 116.12 | 80.96 | 0.65621 |
| 700 | 167.56 | 3.446 | 119.58 | 75.25 | 0.66321 |
| 720 | 172.39 | 3.806 | 123.04 | 70.07 | 0.67002 |
| 740 | 177.23 | 4.193 | 126.51 | 65.38 | 0.67665 |
| 760 | 182.08 | 4.607 | 129.99 | 61.10 | 0.68312 |
| 780 | 186.94 | 5.051 | 133.47 | 57.20 | 0.68942 |
| 800 | 191.81 | 5.526 | 136.97 | 53.63 | 0.69558 |
| 820 | 196.69 | 6.033 | 140.47 | 50.35 | 0.70160 |
| 840 | 201.56 | 6.573 | 143.98 | 47.34 | 0.70747 |
| 860 | 206.46 | 7.149 | 147.50 | 44.57 | 0.71323 |
| 880 | 211.35 | 7.761 | 151.02 | 42.01 | 0.71886 |
| 900 | 216.26 | 8.411 | 154.57 | 39.64 | 0.72438 |
| 920 | 221.18 | 9.102 | 158.12 | 37.44 | 0.72979 |
| 940 | 226.11 | 9.834 | 161.68 | 35.41 | 0.73509 |
| 960 | 231.06 | 10.61 | 165.26 | 33.52 | 0.74030 |
| 980 | 236.02 | 11.43 | 168.83 | 31.76 | 0.74540 |
| 1000 | 240.98 | 12.30 | 172.43 | 30.12 | 0.75042 |
| 1040 | 250.95 | 14.18 | 179.66 | 27.17 | 0.76019 |
| 1080 | 260.97 | 16.28 | 186.93 | 24.58 | 0.76964 |
| 1120 | 271.03 | 18.60 | 194.25 | 22.30 | 0.77880 |
| 1160 | 281.14 | 21.18 | 201.63 | 20.29 | 0.78767 |
| 1200 | 291.30 | 24.01 | 209.05 | 18.51 | 0.79628 |
| 1240 | 301.52 | 27.13 | 216.53 | 16.93 | 0.80466 |
| 1280 | 311.79 | 30.55 | 224.05 | 15.52 | 0.81280 |
| 1320 | 322.11 | 34.31 | 231.63 | 14.25 | 0.82075 |
| 1360 | 332.48 | 38.41 | 239.25 | 13.12 | 0.82848 |
| 1400 | 342.90 | 42.88 | 246.93 | 12.10 | 0.83604 |
| 1440 | 353.37 | 47.75 | 254.66 | 11.17 | 0.84341 |

*(continued)*

**APPENDIX 22.F** *(continued)*
Properties of Low-Pressure Air
(customary U.S. units)

$T$ in °R; $h$ and $u$ in Btu/lbm; $\phi$ in Btu/lbm-°R

| $T$ | $h$ | $p_r$ | $u$ | $v_r$ | $\phi$ |
|------|--------|--------|--------|--------|---------|
| 1480 | 363.89 | 53.04 | 262.44 | 10.34 | 0.85062 |
| 1520 | 374.47 | 58.78 | 270.26 | 9.578 | 0.85767 |
| 1560 | 385.08 | 65.00 | 278.13 | 8.890 | 0.86456 |
| 1600 | 395.74 | 71.73 | 286.06 | 8.263 | 0.87130 |
| 1650 | 409.13 | 80.89 | 296.03 | 7.556 | 0.87954 |
| 1700 | 422.59 | 90.95 | 306.06 | 6.924 | 0.88758 |
| 1750 | 436.12 | 101.98 | 316.16 | 6.357 | 0.89542 |
| 1800 | 449.71 | 114.0 | 326.32 | 5.847 | 0.90308 |
| 1850 | 463.37 | 127.2 | 336.55 | 5.388 | 0.91056 |
| 1900 | 477.09 | 141.5 | 346.85 | 4.974 | 0.91788 |
| 1950 | 490.88 | 157.1 | 357.20 | 4.598 | 0.92504 |
| 2000 | 504.71 | 174.0 | 367.61 | 4.258 | 0.93205 |
| 2050 | 518.61 | 192.3 | 378.08 | 3.949 | 0.93891 |
| 2100 | 532.55 | 212.1 | 388.60 | 3.667 | 0.94564 |
| 2150 | 546.54 | 233.5 | 399.17 | 3.410 | 0.95222 |
| 2200 | 560.59 | 256.6 | 409.78 | 3.176 | 0.95868 |
| 2250 | 574.69 | 281.4 | 420.46 | 2.961 | 0.96501 |
| 2300 | 588.82 | 308.1 | 431.16 | 2.765 | 0.97123 |
| 2350 | 603.00 | 336.8 | 441.91 | 2.585 | 0.97732 |
| 2400 | 617.22 | 367.6 | 452.70 | 2.419 | 0.98331 |
| 2450 | 631.48 | 400.5 | 463.54 | 2.266 | 0.98919 |
| 2500 | 645.78 | 435.7 | 474.40 | 2.125 | 0.99497 |
| 2550 | 660.12 | 473.3 | 485.31 | 1.996 | 1.00064 |
| 2600 | 674.49 | 513.5 | 496.26 | 1.876 | 1.00623 |
| 2650 | 688.90 | 556.3 | 507.25 | 1.765 | 1.01172 |
| 2700 | 703.35 | 601.9 | 518.26 | 1.662 | 1.01712 |
| 2750 | 717.83 | 650.4 | 529.31 | 1.566 | 1.02244 |
| 2800 | 732.33 | 702.0 | 540.40 | 1.478 | 1.02767 |
| 2850 | 746.88 | 756.7 | 551.52 | 1.395 | 1.03282 |
| 2900 | 761.45 | 814.8 | 562.66 | 1.318 | 1.03788 |
| 2950 | 776.05 | 876.4 | 573.84 | 1.247 | 1.04288 |
| 3000 | 790.68 | 941.4 | 585.04 | 1.180 | 1.04779 |
| 3050 | 805.34 | 1011 | 596.28 | 1.118 | 1.05264 |
| 3100 | 820.03 | 1083 | 607.53 | 1.060 | 1.05741 |
| 3150 | 834.75 | 1161 | 618.82 | 1.006 | 1.06212 |
| 3200 | 849.48 | 1242 | 630.12 | 0.9546 | 1.06676 |
| 3250 | 864.24 | 1328 | 641.46 | 0.9069 | 1.07134 |
| 3300 | 879.02 | 1418 | 652.81 | 0.8621 | 1.07585 |
| 3350 | 893.83 | 1513 | 664.20 | 0.8202 | 1.08031 |
| 3400 | 908.66 | 1613 | 675.60 | 0.7807 | 1.08470 |
| 3450 | 923.52 | 1719 | 687.04 | 0.7436 | 1.08904 |
| 3500 | 938.40 | 1829 | 698.48 | 0.7087 | 1.09332 |
| 3550 | 953.30 | 1946 | 709.95 | 0.6759 | 1.09755 |
| 3600 | 968.21 | 2068 | 721.44 | 0.6449 | 1.10172 |
| 3650 | 983.15 | 2196 | 732.95 | 0.6157 | 1.10584 |

*(continued)*

**APPENDIX 22.F** *(continued)*
Properties of Low-Pressure Air
(customary U.S. units)

$T$ in °R; $h$ and $u$ in Btu/lbm; $\phi$ in Btu/lbm-°R

| $T$ | $h$ | $p_r$ | $u$ | $v_r$ | $\phi$ |
|------|--------|--------|---------|--------|---------|
| 3700 | 998.11 | 2330 | 744.48 | 0.5882 | 1.10991 |
| 3750 | 1013.1 | 2471 | 756.04 | 0.5621 | 1.11393 |
| 3800 | 1028.1 | 2618 | 767.60 | 0.5376 | 1.11791 |
| 3850 | 1043.1 | 2773 | 779.19 | 0.5143 | 1.12183 |
| 3900 | 1058.1 | 2934 | 790.80 | 0.4923 | 1.12571 |
| | | | | | |
| 3950 | 1073.2 | 3103 | 802.43 | 0.4715 | 1.12955 |
| 4000 | 1088.3 | 3280 | 814.06 | 0.4518 | 1.13334 |
| 4050 | 1103.4 | 3464 | 825.72 | 0.4331 | 1.13709 |
| 4100 | 1118.5 | 3656 | 837.40 | 0.4154 | 1.14079 |
| 4150 | 1133.6 | 3858 | 849.09 | 0.3985 | 1.14446 |
| | | | | | |
| 4200 | 1148.7 | 4067 | 860.81 | 0.3826 | 1.14809 |
| 4300 | 1179.0 | 4513 | 884.28 | 0.3529 | 1.15522 |
| 4400 | 1209.4 | 4997 | 907.81 | 0.3262 | 1.16221 |
| 4500 | 1239.9 | 5521 | 931.39 | 0.3019 | 1.16905 |
| 4600 | 1270.4 | 6089 | 955.04 | 0.2799 | 1.17575 |
| | | | | | |
| 4700 | 1300.9 | 6701 | 978.73 | 0.2598 | 1.18232 |
| 4800 | 1331.5 | 7362 | 1002.5 | 0.2415 | 1.18876 |
| 4900 | 1362.2 | 8073 | 1026.3 | 0.2248 | 1.19508 |
| 5000 | 1392.9 | 8837 | 1050.1 | 0.2096 | 1.20129 |
| 5100 | 1423.6 | 9658 | 1074.0 | 0.1956 | 1.20738 |
| | | | | | |
| 5200 | 1454.4 | 10539 | 1098.0 | 0.1828 | 1.21336 |
| 5300 | 1485.3 | 11481 | 1122.0 | 0.1710 | 1.21923 |

*Gas Tables: Thermodynamic Properties of Air, Products of Combustion and Component Gases, Compressible Flow Functions*, 2nd Edition, Joseph H. Keenan, Jing Chao, and Joseph Kaye, copyright © 1980. Reproduced with permission of John Wiley & Sons, Inc.

### APPENDIX 22.G
Properties of Saturated Refrigerant-12 (R-12) by Temperature
(customary U.S. units)

| temp. (°F) | pressure (psia) | specific volume (ft³/lbm) | | enthalpy (Btu/lbm) | | | entropy (Btu/lbm-°R) | | | temp. (°F) |
| | | sat. liquid | sat. vapor | sat. liquid | evap. | sat. vapor | sat. liquid | evap. | sat. vapor | |
| --- | --- | --- | --- | --- | --- | --- | --- | --- | --- | --- |
| $T$ | $p$ | $v_f$ | $v_g$ | $h_f$ | $h_{fg}$ | $h_g$ | $s_f$ | $s_{fg}$ | $s_g$ | $T$ |
| −60 | 5.37 | 0.01036 | 6.516 | −4.20 | 75.33 | 71.13 | −0.0102 | 0.1681 | 0.1783 | −60 |
| −50 | 7.13 | 0.01047 | 5.012 | −2.11 | 74.42 | 72.31 | −0.0050 | 0.1717 | 0.1767 | −50 |
| −40 | 9.32 | 0.0106 | 3.911 | 0.00 | 73.50 | 73.50 | 0.00000 | 0.17517 | 0.17517 | −40 |
| −30 | 12.02 | 0.0107 | 3.088 | 2.03 | 72.67 | 74.70 | 0.00471 | 0.16916 | 0.17387 | −30 |
| −20 | 15.28 | 0.0108 | 2.474 | 4.07 | 71.80 | 75.87 | 0.00940 | 0.16335 | 0.17275 | −20 |
| −10 | 19.20 | 0.0109 | 2.003 | 6.14 | 70.91 | 77.05 | 0.01403 | 0.15772 | 0.17175 | −10 |
| 0 | 23.87 | 0.0110 | 1.637 | 8.25 | 69.96 | 78.21 | 0.01869 | 0.15222 | 0.17091 | 0 |
| 5 | 26.51 | 0.0111 | 1.485 | 9.32 | 69.47 | 78.79 | 0.02097 | 1.14955 | 0.17052 | 5 |
| 10 | 29.35 | 0.0112 | 1.351 | 10.39 | 68.97 | 79.36 | 0.02328 | 0.14687 | 0.17015 | 10 |
| 20 | 35.75 | 0.0113 | 1.121 | 12.55 | 67.94 | 80.49 | 0.02783 | 0.14166 | 0.16949 | 20 |
| 30 | 43.16 | 0.0115 | 0.939 | 14.76 | 66.85 | 81.61 | 0.03233 | 0.13654 | 0.16887 | 30 |
| 40 | 51.68 | 0.0116 | 0.792 | 17.00 | 65.71 | 82.71 | 0.03680 | 0.13153 | 0.16833 | 40 |
| 50 | 61.39 | 0.0118 | 0.673 | 19.27 | 64.51 | 83.78 | 0.04126 | 0.12659 | 0.16785 | 50 |
| 60 | 72.41 | 0.0119 | 0.575 | 21.57 | 63.25 | 84.82 | 0.04568 | 0.12173 | 0.16741 | 60 |
| 70 | 84.82 | 0.0121 | 0.493 | 23.90 | 61.92 | 85.82 | 0.05009 | 0.11692 | 0.16701 | 70 |
| 80 | 98.76 | 0.0123 | 0.425 | 26.28 | 60.52 | 86.80 | 0.05446 | 0.11215 | 0.16662 | 80 |
| 86 | 107.9 | 0.0124 | 0.389 | 27.72 | 59.65 | 87.37 | 0.05708 | 0.10932 | 0.16640 | 86 |
| 90 | 114.3 | 0.0125 | 0.368 | 28.70 | 59.04 | 87.74 | 0.05882 | 0.10742 | 0.16624 | 90 |
| 100 | 131.6 | 0.0127 | 0.319 | 31.16 | 57.46 | 88.62 | 0.06316 | 0.10268 | 0.16584 | 100 |
| 110 | 150.7 | 0.0129 | 0.277 | 33.65 | 55.78 | 89.43 | 0.06749 | 0.09793 | 0.16542 | 110 |
| 120 | 171.8 | 0.0132 | 0.240 | 36.16 | 53.99 | 90.15 | 0.07180 | 0.09315 | 0.16495 | 120 |
| 233 | 596.9 | 0.02870 | 0.02870 | 78.86 | 0 | 78.86 | 0.1359 | 0 | 0.1359 | 233 |

Reproduced with permission from the DuPont Company.

**APPENDIX 22.H**
Properties of Saturated Refrigerant-12 (R-12) by Pressure
(customary U.S. units)

| pressure (psia) | temp. (°F) | specific volume (ft³/lbm) | | enthalpy (Btu/lbm) | | | entropy (Btu/lbm-°R) | | | pressure (psia) |
| | | sat. liquid | sat. vapor | sat. liquid | evap. | sat. vapor | sat. liquid | evap. | sat. vapor | |
|---|---|---|---|---|---|---|---|---|---|---|
| $p$ | $T$ | $v_f$ | $vg$ | $h_f$ | $h_{fg}$ | $h_g$ | $s_f$ | $s_{fg}$ | $s_g$ | $p$ |
| 5 | −62.5 | 0.01034 | 6.953 | −4.73 | 75.56 | 70.83 | −0.0115 | 0.1943 | 0.1788 | 5 |
| 10 | −37.3 | 0.0106 | 3.662 | 0.54 | 73.28 | 73.82 | 0.00127 | 0.17360 | 0.17487 | 10 |
| 15 | −20.8 | 0.0108 | 2.518 | 3.91 | 71.87 | 75.78 | 0.00902 | 0.16381 | 0.17283 | 15 |
| 20 | −8.2 | 0.0109 | 1.925 | 6.53 | 70.74 | 77.27 | 0.01488 | 0.15672 | 0.17160 | 20 |
| 30 | 11.1 | 0.0112 | 1.324 | 10.62 | 68.86 | 79.48 | 0.02410 | 0.14597 | 0.17007 | 30 |
| 40 | 25.9 | 0.0114 | 1.009 | 13.86 | 67.30 | 81.16 | 0.03049 | 0.13865 | 0.16914 | 40 |
| 50 | 38.3 | 0.0116 | 0.817 | 16.58 | 65.94 | 82.52 | 0.03597 | 0.13244 | 0.16841 | 50 |
| 60 | 48.7 | 0.0117 | 0.688 | 18.96 | 64.69 | 83.65 | 0.04065 | 0.12726 | 0.16791 | 60 |
| 80 | 66.3 | 0.0120 | 0.521 | 23.01 | 62.44 | 85.45 | 0.04844 | 0.11872 | 0.16716 | 80 |
| 100 | 80.9 | 0.0123 | 0.419 | 26.49 | 60.40 | 86.89 | 0.05483 | 0.11176 | 0.16659 | 100 |
| 120 | 93.4 | 0.0126 | 0.419 | 29.53 | 58.52 | 88.05 | 0.06030 | 0.10580 | 0.16610 | 120 |
| 140 | 104.5 | 0.0128 | 0.298 | 32.28 | 56.71 | 88.99 | 0.06513 | 0.10053 | 0.16566 | 140 |
| 160 | 114.5 | 0.0130 | 0.260 | 34.78 | 54.99 | 89.77 | 0.06958 | 0.09564 | 0.16522 | 160 |
| 180 | 123.7 | 0.0133 | 0.228 | 37.07 | 53.31 | 90.38 | 0.07337 | 0.09139 | 0.16476 | 180 |
| 200 | 132.1 | 0.0135 | 0.202 | 39.21 | 51.65 | 90.86 | 0.07694 | 0.08730 | 0.16424 | 200 |
| 220 | 139.9 | 0.0138 | 0.181 | 41.22 | 50.28 | 91.50 | 0.08021 | 0.08354 | 0.16375 | 220 |
| 596 | 233.6 | 0.02870 | 0.02870 | 78.86 | 0 | 78.86 | 0.1359 | 0 | 0.1359 | 596 |

Reproduced with permission from the DuPont Company.

## APPENDIX 22.I
### Properties of Superheated Refrigerant-12 (R-12)
(customary U.S. units)
specific volume ($v$) in ft$^3$/lbm; enthalpy ($h$) in Btu/lbm; entropy ($s$) in Btu/lbm-°R

| pressure (psia) (sat. temp.) | | temperature (°F) | | | | | | | | | | | |
|---|---|---|---|---|---|---|---|---|---|---|---|---|---|
| | | −40 | −20 | 0 | 20 | 40 | 60 | 80 | 100 | 150 | 200 | 250 | 300 |
| 5 | $v$ | 7.363 | 7.726 | 8.088 | 8.450 | 8.812 | 9.173 | 9.533 | 9.893 | 10.79 | 11.69 | ...... | ...... |
| | $h$ | 73.72 | 76.36 | 79.05 | 81.78 | 84.56 | 87.41 | 90.30 | 93.25 | 100.84 | 108.75 | ...... | ...... |
| (−62.5) | $s$ | 0.1859 | 0.1920 | 0.1979 | 0.2038 | 0.2095 | 0.2150 | 0.2205 | 0.2258 | 0.2388 | 0.2518 | ...... | ...... |
| 10 | $v$ | ...... | 3.821 | 4.006 | 4.189 | 4.371 | 4.556 | 4.740 | 4.923 | 5.379 | 5.831 | 6.281 | ...... |
| | $h$ | ...... | 76.11 | 78.81 | 81.56 | 84.35 | 87.19 | 90.11 | 93.05 | 100.66 | 108.63 | 116.88 | ...... |
| (−37.3) | $s$ | ...... | 0.1801 | 0.1861 | 0.1919 | 0.1977 | 0.2033 | 0.2087 | 0.2141 | 0.2271 | 0.2396 | 0.2517 | ...... |
| 15 | $v$ | ...... | 2.521 | 2.646 | 2.771 | 2.895 | 3.019 | 3.143 | 3.266 | 3.571 | 3.877 | 4.191 | ...... |
| | $h$ | ...... | 75.89 | 78.59 | 81.37 | 84.18 | 87.03 | 89.94 | 92.91 | 100.53 | 108.49 | 116.78 | ...... |
| (−20.8) | $s$ | ...... | 0.17307 | 0.17913 | 0.18499 | 0.19074 | 0.19635 | 0.20185 | 0.20723 | 0.22028 | 0.23282 | 0.24491 | ...... |
| 20 | $v$ | ...... | ...... | 1.965 | 2.060 | 2.155 | 2.250 | 2.343 | 2.437 | 2.669 | 2.901 | 3.130 | ...... |
| | $h$ | ...... | ...... | 78.39 | 81.14 | 83.97 | 86.85 | 89.78 | 92.75 | 100.40 | 108.38 | 116.67 | ...... |
| (−8.2) | $s$ | ...... | ...... | 0.17407 | 0.17996 | 0.18573 | 0.19138 | 0.19688 | 0.20229 | 0.21537 | 0.22794 | 0.24005 | ...... |
| 25 | $v$ | ...... | ...... | ...... | 1.712 | 1.793 | 1.873 | 1.952 | 2.031 | 2.227 | 2.422 | 2.615 | ...... |
| | $h$ | ...... | ...... | ...... | 80.95 | 83.78 | 86.67 | 89.61 | 92.56 | 100.26 | 108.26 | 116.56 | ...... |
| (2.2) | $s$ | ...... | ...... | ...... | 0.17637 | 0.18216 | 0.18783 | 0.19336 | 0.19748 | 0.21190 | 0.22450 | 0.23665 | ...... |
| 30 | $v$ | ...... | ...... | ...... | 1.364 | 1.430 | 1.495 | 1.560 | 1.624 | 1.784 | 1.943 | 2.099 | ...... |
| | $h$ | ...... | ...... | ...... | 80.75 | 83.59 | 86.49 | 89.43 | 92.42 | 100.12 | 108.13 | 116.45 | ...... |
| (11.1) | $s$ | ...... | ...... | ...... | 0.17278 | 0.17859 | 0.18429 | 0.18983 | 0.19527 | 0.20843 | 0.22105 | 0.23325 | ...... |
| 35 | $v$ | ...... | ...... | ...... | 1.109 | 1.237 | 1.295 | 1.352 | 1.409 | 1.550 | 1.689 | 1.827 | ...... |
| | $h$ | ...... | ...... | ...... | 80.49 | 83.40 | 86.30 | 89.26 | 92.26 | 99.98 | 108.01 | 116.33 | ...... |
| (18.9) | $s$ | ...... | ...... | ...... | 0.16963 | 0.17591 | 0.18162 | 0.18719 | 0.19266 | 0.20584 | 0.21849 | 0.23069 | ...... |
| 40 | $v$ | ...... | ...... | ...... | ...... | 1.044 | 1.095 | 1.144 | 1.194 | 1.315 | 1.435 | 1.554 | ...... |
| | $h$ | ...... | ...... | ...... | ...... | 83.20 | 86.11 | 89.09 | 92.09 | 99.83 | 107.88 | 116.21 | ...... |
| (25.9) | $s$ | ...... | ...... | ...... | ...... | 0.17322 | 0.17896 | 0.18455 | 0.19004 | 0.20325 | 0.21592 | 0.22813 | ...... |
| 50 | $v$ | ...... | ...... | ...... | ...... | 0.821 | 0.863 | 0.904 | 0.944 | 1.044 | 1.142 | 1.239 | 1.332 |
| | $h$ | ...... | ...... | ...... | ...... | 82.76 | 85.72 | 88.72 | 91.75 | 99.54 | 107.62 | 116.00 | 124.69 |
| (38.3) | $s$ | ...... | ...... | ...... | ...... | 0.16895 | 0.17475 | 0.19040 | 0.18591 | 0.19923 | 0.21196 | 0.22419 | 0.23600 |
| 60 | $v$ | ...... | ...... | ...... | ...... | ...... | 0.708 | 0.743 | 0.778 | 0.863 | 0.946 | 1.028 | 1.108 |
| | $h$ | ...... | ...... | ...... | ...... | ...... | 85.33 | 88.35 | 91.41 | 99.24 | 107.36 | 115.54 | 124.29 |
| (48.7) | $s$ | ...... | ...... | ...... | ...... | ...... | 0.17120 | 0.17689 | 0.18246 | 0.19585 | 0.20865 | 0.22094 | 0.23280 |
| 70 | $v$ | ...... | ...... | ...... | ...... | ...... | 0.553 | 0.642 | 0.673 | 0.750 | 0.824 | 0.896 | 0.967 |
| | $h$ | ...... | ...... | ...... | ...... | ...... | 84.94 | 87.96 | 91.05 | 98.94 | 107.10 | 115.54 | 124.29 |
| (57.9) | $s$ | ...... | ...... | ...... | ...... | ...... | 0.16765 | 0.17399 | 0.17961 | 0.19310 | 0.20597 | 0.21830 | 0.23020 |
| 80 | $v$ | ...... | ...... | ...... | ...... | ...... | ...... | 0.540 | 0.568 | 0.636 | 0.701 | 0.764 | 0.826 |
| | $h$ | ...... | ...... | ...... | ...... | ...... | ...... | 87.56 | 90.68 | 98.64 | 106.84 | 115.30 | 124.08 |
| (66.3) | $s$ | ...... | ...... | ...... | ...... | ...... | ...... | 0.17108 | 0.17675 | 0.19035 | 0.20328 | 0.21566 | 0.22760 |

*(continued)*

**APPENDIX 22.I** *(continued)*
Properties of Superheated Refrigerant-12 (R-12)
(customary U.S. units)
specific volume ($v$) in ft³/lbm; enthalpy ($h$) in Btu/lbm; entropy ($s$) in Btu/lbm-°R

| pressure (psia) (sat. temp.) | | temperature (°F) | | | | | | | | | | | |
|---|---|---|---|---|---|---|---|---|---|---|---|---|---|
| | | −40 | −20 | 0 | 20 | 40 | 60 | 80 | 100 | 150 | 200 | 250 | 300 |
| 90 (73.6) | $v$ | ...... | ...... | ...... | ...... | ...... | ...... | ...... | 0.505 | 0.568 | 0.627 | 0.685 | 0.742 |
| | $h$ | ...... | ...... | ...... | ...... | ...... | ...... | ...... | 90.31 | 98.32 | 106.56 | 115.07 | 123.88 |
| | $s$ | ...... | ...... | ...... | ...... | ...... | ...... | ...... | 0.17443 | 0.18813 | 0.20111 | 0.21356 | 0.22554 |
| 100 (80.9) | $v$ | ...... | ...... | ...... | ...... | ...... | ...... | ...... | 0.442 | 0.499 | 0.553 | 0.606 | 0.657 |
| | $h$ | ...... | ...... | ...... | ...... | ...... | ...... | ...... | 89.93 | 97.99 | 106.29 | 114.84 | 123.67 |
| | $s$ | ...... | ...... | ...... | ...... | ...... | ...... | ...... | 0.17210 | 0.18590 | 0.19894 | 0.21145 | 0.22347 |
| 120 (93.4) | $v$ | ...... | ...... | ...... | ...... | ...... | ...... | ...... | 0.357 | 0.407 | 0.454 | 0.500 | 0.543 |
| | $h$ | ...... | ...... | ...... | ...... | ...... | ...... | ...... | 89.13 | 97.30 | 105.70 | 114.35 | 123.25 |
| | $s$ | ...... | ...... | ...... | ...... | ...... | ...... | ...... | 0.16803 | 0.18207 | 0.19529 | 0.20792 | 0.22000 |
| 140 (104.5) | $v$ | ...... | ...... | ...... | ...... | ...... | ...... | ...... | ...... | 0.341 | 0.383 | 0.423 | 0.462 |
| | $h$ | ...... | ...... | ...... | ...... | ...... | ...... | ...... | ...... | 96.65 | 105.14 | 113.85 | 122.85 |
| | $s$ | ...... | ...... | ...... | ...... | ...... | ...... | ...... | ...... | 0.17868 | 0.19205 | 0.20479 | 0.21701 |
| 160 (114.5) | $v$ | ...... | ...... | ...... | ...... | ...... | ...... | ...... | ...... | 0.318 | 0.335 | 0.372 | 0.408 |
| | $h$ | ...... | ...... | ...... | ...... | ...... | ...... | ...... | ...... | 95.82 | 104.50 | 113.33 | 122.39 |
| | $s$ | ...... | ...... | ...... | ...... | ...... | ...... | ...... | ...... | 0.17561 | 0.18927 | 0.20213 | 0.21444 |
| 180 (123.7) | $v$ | ...... | ...... | ...... | ...... | ...... | ...... | ...... | ...... | 0.294 | 0.287 | 0.321 | 0.353 |
| | $h$ | ...... | ...... | ...... | ...... | ...... | ...... | ...... | ...... | 94.99 | 103.85 | 112.81 | 121.92 |
| | $s$ | ...... | ...... | ...... | ...... | ...... | ...... | ...... | ...... | 0.17254 | 0.18648 | 0.19947 | 0.21187 |
| 200 (132.1) | $v$ | ...... | ...... | ...... | ...... | ...... | ...... | ...... | ...... | 0.241 | 0.255 | 0.288 | 0.317 |
| | $h$ | ...... | ...... | ...... | ...... | ...... | ...... | ...... | ...... | 94.16 | 103.12 | 112.20 | 121.42 |
| | $s$ | ...... | ...... | ...... | ...... | ...... | ...... | ...... | ...... | 0.16970 | 0.18395 | 0.19717 | 0.20970 |
| 220 (139.9) | $v$ | ...... | ...... | ...... | ...... | ...... | ...... | ...... | ...... | 0.188 | 0.232 | 0.254 | 0.282 |
| | $h$ | ...... | ...... | ...... | ...... | ...... | ...... | ...... | ...... | 93.32 | 102.39 | 111.59 | 120.91 |
| | $s$ | ...... | ...... | ...... | ...... | ...... | ...... | ...... | ...... | 0.16685 | 0.18142 | 0.19387 | 0.20753 |

Reproduced with permission from the DuPont Company.

**APPENDIX 22.J**
Pressure-Enthalpy Diagram for Refrigerant-12 (R-12)
(customary U.S. units)

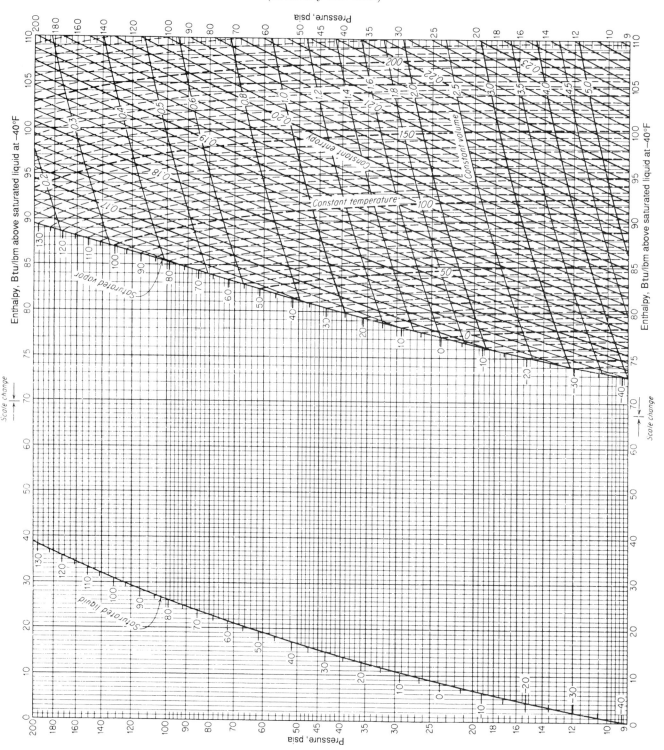

Reproduced with permission from the DuPont Company.

## APPENDIX 22.K
Properties of Saturated Refrigerant-22 (R-22) by Temperature
(customary U.S. units)

| temp. (°F) | saturation pressure (psia) | saturation pressure (psig) | volume (ft³/lbm) $v_g$ | density (lbm/ft³) $\rho_g$ | enthalpy (Btu/lbm) $h_f$ | enthalpy (Btu/lbm) $h_g$ | entropy (Btu/lbm-°R) $s_f$ | entropy (Btu/lbm-°R) $s_g$ |
|---|---|---|---|---|---|---|---|---|
| −130 | 0.68858 | 28.519[a] | 59.170 | 96.313 | −24.388 | 89.888 | −0.065456 | 0.28118 |
| −120 | 1.0725 | 27.738[a] | 39.078 | 95.416 | −21.538 | 91.049 | −0.056942 | 0.27452 |
| −110 | 1.6199 | 26.623[a] | 26.578 | 94.509 | −18.738 | 92.211 | −0.048818 | 0.26848 |
| −100 | 2.3802 | 25.075[a] | 18.558 | 93.590 | −15.980 | 93.371 | −0.041046 | 0.26298 |
| −90 | 3.4111 | 22.976[a] | 13.268 | 92.660 | −13.259 | 94.527 | −0.033590 | 0.25798 |
| −80 | 4.7793 | 20.191[a] | 9.6902 | 91.717 | −10.570 | 95.676 | −0.026418 | 0.25342 |
| −75 | 5.6131 | 18.493[a] | 8.3419 | 91.241 | −9.2346 | 96.247 | −0.022929 | 0.25128 |
| −70 | 6.5603 | 16.564[a] | 7.2139 | 90.761 | −7.9050 | 96.815 | −0.019500 | 0.24924 |
| −65 | 7.6317 | 14.383[a] | 6.2655 | 90.278 | −6.5802 | 97.380 | −0.016128 | 0.24728 |
| −60 | 8.8386 | 11.926[a] | 5.4641 | 89.791 | −5.2593 | 97.942 | −0.012808 | 0.24541 |
| −50 | 11.707 | 6.0851[a] | 4.2039 | 88.807 | −2.6263 | 99.055 | −0.006316 | 0.24189 |
| −48 | 12.361 | 4.7548[a] | 3.9962 | 88.608 | −2.1007 | 99.275 | −0.005039 | 0.24122 |
| −46 | 13.042 | 3.3666[a] | 3.8007 | 88.408 | −1.5753 | 99.495 | −0.003770 | 0.24056 |
| −44 | 13.754 | 1.9186[a] | 3.6168 | 88.208 | −1.0501 | 99.714 | −0.002507 | 0.23991 |
| −42 | 14.495 | 0.4090[a] | 3.4437 | 88.007 | −0.5250 | 99.932 | −0.001250 | 0.23927 |
| −41.47 | 14.696 | 0.0 | 3.3997 | 87.954 | −0.3865 | 99.990 | −0.000920 | 0.23910 |
| −40 | 15.268 | 0.5717 | 3.2805 | 87.806 | 0.0 | 100.15 | 0.0 | 0.23864 |
| −38 | 16.072 | 1.3763 | 3.1267 | 87.604 | 0.5250 | 100.37 | 0.001244 | 0.23802 |
| −36 | 16.910 | 2.2138 | 2.9816 | 87.401 | 1.0500 | 100.58 | 0.002482 | 0.23741 |
| −34 | 17.781 | 3.0852 | 2.8446 | 87.197 | 1.5751 | 100.80 | 0.003714 | 0.23681 |
| −32 | 18.687 | 3.9914 | 2.7152 | 86.993 | 2.1003 | 101.01 | 0.004940 | 0.23622 |
| −30 | 19.629 | 4.9333 | 2.5930 | 86.788 | 2.6257 | 101.22 | 0.006161 | 0.23564 |
| −28 | 20.608 | 5.9119 | 2.4774 | 86.582 | 3.1512 | 101.44 | 0.007377 | 0.23506 |
| −26 | 21.624 | 6.9283 | 2.3680 | 86.375 | 3.6771 | 101.65 | 0.008587 | 0.23450 |
| −24 | 22.679 | 7.9832 | 2.2645 | 86.168 | 4.2032 | 101.86 | 0.009792 | 0.23394 |
| −22 | 23.774 | 9.0778 | 2.1664 | 85.960 | 4.7297 | 102.07 | 0.010993 | 0.23340 |
| −20 | 24.909 | 10.213 | 2.0735 | 85.751 | 5.2566 | 102.28 | 0.012188 | 0.23285 |
| −18 | 26.086 | 11.390 | 1.9854 | 85.542 | 5.7840 | 102.48 | 0.013379 | 0.23232 |
| −16 | 27.306 | 12.610 | 1.9018 | 85.331 | 6.3119 | 102.69 | 0.014566 | 0.23180 |
| −14 | 28.569 | 13.873 | 1.8225 | 85.120 | 6.8403 | 102.90 | 0.015748 | 0.23128 |
| −12 | 29.877 | 15.181 | 1.7472 | 84.908 | 7.3693 | 103.10 | 0.016926 | 0.23077 |
| −10 | 31.231 | 16.535 | 1.6757 | 84.695 | 7.8989 | 103.30 | 0.018100 | 0.23027 |
| −8 | 32.632 | 17.936 | 1.6077 | 84.481 | 8.4292 | 103.51 | 0.019270 | 0.22977 |
| −6 | 34.081 | 19.385 | 1.5430 | 84.266 | 8.9603 | 103.71 | 0.020436 | 0.22928 |
| −4 | 35.579 | 20.883 | 1.4815 | 84.051 | 9.4921 | 103.91 | 0.021598 | 0.22880 |
| −2 | 37.127 | 22.431 | 1.4230 | 83.834 | 10.025 | 104.10 | 0.022757 | 0.22832 |
| 0 | 38.726 | 24.030 | 1.3672 | 83.617 | 10.558 | 104.30 | 0.023912 | 0.022785 |
| 2 | 40.378 | 25.682 | 1.3141 | 83.399 | 11.093 | 104.50 | 0.025064 | 0.22738 |
| 4 | 42.083 | 27.387 | 1.2635 | 83.179 | 11.628 | 104.69 | 0.026213 | 0.22693 |
| 6 | 43.843 | 29.147 | 1.2152 | 82.959 | 12.164 | 104.89 | 0.027359 | 0.22647 |
| 8 | 45.658 | 30.962 | 1.1692 | 82.738 | 12.702 | 105.08 | 0.028502 | 0.22602 |

*(continued)*

**APPENDIX 22.K** (continued)
Properties of Saturated Refrigerant-22 (R-22) by Temperature
(customary U.S. units)

| temp. (°F) | saturation pressure | | volume ($\text{ft}^3$/lbm) | density (lbm/$\text{ft}^3$) | enthalpy (Btu/lbm) | | entropy (Btu/lbm-°R) | |
|---|---|---|---|---|---|---|---|---|
| | (psia) | (psig) | $v_g$ | $\rho_g$ | $h_f$ | $h_g$ | $s_f$ | $s_g$ |
| 10 | 47.530 | 32.834 | 1.1253 | 82.516 | 13.240 | 105.27 | 0.029642 | 0.22558 |
| 12 | 49.461 | 34.765 | 1.0833 | 82.292 | 13.779 | 105.46 | 0.030779 | 0.22515 |
| 14 | 51.450 | 36.754 | 1.0433 | 82.068 | 14.320 | 105.64 | 0.031913 | 0.22471 |
| 16 | 53.501 | 38.805 | 1.0050 | 81.843 | 14.862 | 105.83 | 0.033045 | 0.22429 |
| 18 | 55.612 | 40.916 | 0.96841 | 81.616 | 15.405 | 106.02 | 0.034175 | 0.22387 |
| 20 | 57.786 | 43.090 | 0.93343 | 81.389 | 15.950 | 106.20 | 0.035302 | 0.22345 |
| 25 | 63.505 | 48.809 | 0.85246 | 80.815 | 17.317 | 106.65 | 0.038110 | 0.22243 |
| 30 | 69.641 | 54.945 | 0.77984 | 80.234 | 18.693 | 107.09 | 0.040905 | 0.22143 |
| 35 | 76.215 | 61.519 | 0.71454 | 79.645 | 20.078 | 107.52 | 0.043689 | 0.22046 |
| 40 | 83.246 | 68.550 | 0.65571 | 79.049 | 21.474 | 107.94 | 0.046464 | 0.21951 |
| 45 | 90.754 | 76.058 | 0.60258 | 78.443 | 22.880 | 108.35 | 0.049229 | 0.21858 |
| 50 | 98.758 | 84.062 | 0.55451 | 77.829 | 24.298 | 108.74 | 0.051987 | 0.21767 |
| 55 | 107.28 | 92.583 | 0.51093 | 77.206 | 25.728 | 109.12 | 0.054739 | 0.21677 |
| 60 | 116.34 | 101.64 | 0.47134 | 76.572 | 27.170 | 109.49 | 0.057486 | 0.21589 |
| 65 | 125.95 | 111.26 | 0.43531 | 75.928 | 28.626 | 109.84 | 0.060228 | 0.21502 |
| 70 | 136.15 | 121.45 | 0.40245 | 75.273 | 30.095 | 110.18 | 0.062968 | 0.21416 |
| 80 | 158.36 | 143.66 | 0.34497 | 73.926 | 33.077 | 110.80 | 0.068441 | 0.21246 |
| 90 | 183.14 | 168.44 | 0.29668 | 72.525 | 36.121 | 111.35 | 0.073911 | 0.21077 |
| 100 | 210.67 | 195.97 | 0.25582 | 71.061 | 39.233 | 111.81 | 0.079400 | 0.20907 |
| 110 | 241.13 | 226.44 | 0.22102 | 69.524 | 42.422 | 112.17 | 0.084906 | 0.20734 |
| 120 | 274.73 | 260.03 | 0.19118 | 67.901 | 45.694 | 112.42 | 0.090444 | 0.20554 |
| 130 | 311.66 | 296.96 | 0.16542 | 66.174 | 49.064 | 112.52 | 0.096033 | 0.20365 |
| 140 | 352.14 | 337.45 | 0.14300 | 64.319 | 52.550 | 112.47 | 0.10170 | 0.20161 |
| 150 | 396.42 | 381.72 | 0.12334 | 62.301 | 56.177 | 112.20 | 0.10749 | 0.19938 |
| 160 | 444.75 | 430.06 | 0.10590 | 60.068 | 59.989 | 111.67 | 0.11345 | 0.19684 |
| 170 | 497.46 | 482.76 | 0.090228 | 57.532 | 64.055 | 110.76 | 0.11970 | 0.19386 |
| 180 | 554.89 | 540.19 | 0.075819 | 54.533 | 68.504 | 109.30 | 0.12640 | 0.19018 |
| 190 | 617.52 | 602.82 | 0.061991 | 50.703 | 73.617 | 106.88 | 0.13399 | 0.18518 |
| 200 | 686.02 | 671.32 | 0.046923 | 44.671 | 80.406 | 101.99 | 0.14394 | 0.17666 |
| 205.07[b] | 723.4 | 708.7 | 0.03123 | 32.03 | 91.58 | 91.58 | 0.1605 | 0.1605 |

[a]in Hg vacuum
[b]critical point

Reprinted with permission from *1989 ASHRAE Handbook—Fundamentals*, American Society of Heating, Refrigerating, and Air-Conditioning Engineers, Inc. (ASHRAE), www.ashrae.org, Alanta, GA, © 1989.

**APPENDIX 22.L**
Pressure-Enthalpy Diagram for Refrigerant-22 (R-22)
(customary U.S. units)

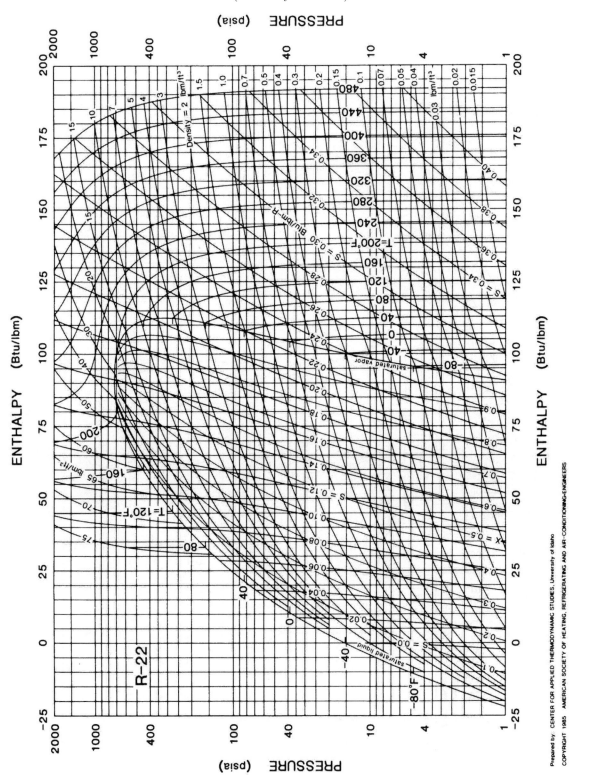

### APPENDIX 22.M
Pressure-Enthalpy Diagram for Refrigerant HFC-134a
(customary U.S. units)

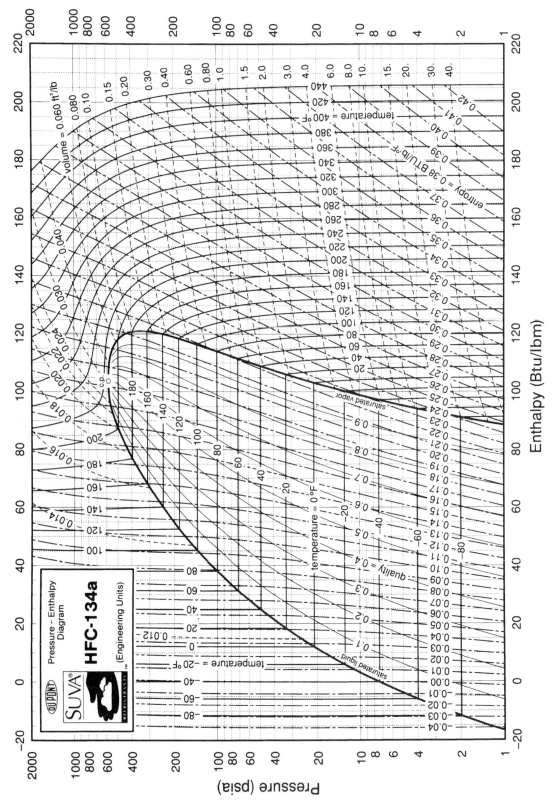

Reproduced with permission from the DuPont Company.

## APPENDIX 22.N
### Properties of Saturated Steam by Temperature
(SI units)

| temp. (°C) | absolute pressure (bars) | specific volume (cm³/g) | | internal energy (kJ/kg) | | enthalpy (kJ/kg) | | | entropy (kJ/kg·K) | | temp. (°C) |
|---|---|---|---|---|---|---|---|---|---|---|---|
| | | sat. liquid $v_f$ | sat. vapor $v_g$ | sat. liquid $u_f$ | sat. vapor $u_g$ | sat. liquid $h_f$ | evap. $h_{fg}$ | sat. vapor $h_g$ | sat. liquid $s_f$ | sat. vapor $s_g$ | |
| 0.01 | 0.006117 | 1.0002 | 205991 | 0.00 | 2374.9 | 0.0006 | 2500.9 | 2500.9 | 0.0000 | 9.1555 | 0.01 |
| 4 | 0.00814 | 1.0001 | 157116 | 16.81 | 2380.4 | 16.81 | 2491.4 | 2508.2 | 0.0611 | 9.0505 | 4 |
| 5 | 0.00873 | 1.0001 | 147011 | 21.02 | 2381.8 | 21.02 | 2489.1 | 2510.1 | 0.0763 | 9.0248 | 5 |
| 6 | 0.00935 | 1.0001 | 137633 | 25.22 | 2383.2 | 25.22 | 2486.7 | 2511.9 | 0.0913 | 8.9993 | 6 |
| 8 | 0.01073 | 1.0002 | 120829 | 33.63 | 2385.9 | 33.63 | 2482.0 | 2515.6 | 0.1213 | 8.9491 | 8 |
| 10 | 0.01228 | 1.0003 | 106303 | 42.02 | 2388.6 | 42.02 | 2477.2 | 2519.2 | 0.1511 | 8.8998 | 10 |
| 11 | 0.01313 | 1.0004 | 99787 | 46.22 | 2390.0 | 46.22 | 2474.8 | 2521.0 | 0.1659 | 8.8754 | 11 |
| 12 | 0.01403 | 1.0005 | 93719 | 50.41 | 2391.4 | 50.41 | 2472.5 | 2522.9 | 0.1806 | 8.8513 | 12 |
| 13 | 0.01498 | 1.0007 | 88064 | 54.60 | 2392.8 | 54.60 | 2470.1 | 2524.7 | 0.1953 | 8.8274 | 13 |
| 14 | 0.01599 | 1.0008 | 82793 | 58.79 | 2394.1 | 58.79 | 2467.7 | 2526.5 | 0.2099 | 8.8037 | 14 |
| 15 | 0.01706 | 1.0009 | 77875 | 62.98 | 2395.5 | 62.98 | 2465.3 | 2528.3 | 0.2245 | 8.7803 | 15 |
| 16 | 0.01819 | 1.0011 | 73286 | 67.17 | 2396.9 | 67.17 | 2463.0 | 2530.2 | 0.2390 | 8.7570 | 16 |
| 17 | 0.01938 | 1.0013 | 69001 | 71.36 | 2398.2 | 71.36 | 2460.6 | 2532.0 | 0.2534 | 8.7339 | 17 |
| 18 | 0.02065 | 1.0014 | 64998 | 75.54 | 2399.6 | 75.54 | 2458.3 | 2533.8 | 0.2678 | 8.7111 | 18 |
| 19 | 0.02198 | 1.0016 | 61256 | 79.73 | 2401.0 | 79.73 | 2455.9 | 2535.6 | 0.2822 | 8.6884 | 19 |
| 20 | 0.02339 | 1.0018 | 57757 | 83.91 | 2402.3 | 83.91 | 2453.5 | 2537.4 | 0.2965 | 8.6660 | 20 |
| 21 | 0.02488 | 1.0021 | 54483 | 88.10 | 2403.7 | 88.10 | 2451.2 | 2539.3 | 0.3107 | 8.6437 | 21 |
| 22 | 0.02645 | 1.0023 | 51418 | 92.28 | 2405.0 | 92.28 | 2448.8 | 2541.1 | 0.3249 | 8.6217 | 22 |
| 23 | 0.02811 | 1.0025 | 48548 | 96.46 | 2406.4 | 96.47 | 2446.4 | 2542.9 | 0.3391 | 8.5998 | 23 |
| 24 | 0.02986 | 1.0028 | 45858 | 100.64 | 2407.8 | 100.65 | 2444.1 | 2544.7 | 0.3532 | 8.5781 | 24 |
| 25 | 0.03170 | 1.0030 | 43337 | 104.83 | 2409.1 | 104.83 | 2441.7 | 2546.5 | 0.3672 | 8.5566 | 25 |
| 26 | 0.03364 | 1.0033 | 40973 | 109.01 | 2410.5 | 109.01 | 2439.3 | 2548.3 | 0.3812 | 8.5353 | 26 |
| 27 | 0.03568 | 1.0035 | 38754 | 113.19 | 2411.8 | 113.19 | 2436.9 | 2550.1 | 0.3952 | 8.5142 | 27 |
| 28 | 0.03783 | 1.0038 | 36672 | 117.37 | 2413.2 | 117.37 | 2434.5 | 2551.9 | 0.4091 | 8.4933 | 28 |
| 29 | 0.04009 | 1.0041 | 34716 | 121.55 | 2414.6 | 121.55 | 2432.2 | 2553.7 | 0.4229 | 8.4725 | 29 |
| 30 | 0.04247 | 1.0044 | 32878 | 125.73 | 2415.9 | 125.73 | 2429.8 | 2555.5 | 0.4368 | 8.4520 | 30 |
| 31 | 0.04497 | 1.0047 | 31151 | 129.91 | 2417.3 | 129.91 | 2427.4 | 2557.3 | 0.4505 | 8.4316 | 31 |
| 32 | 0.04760 | 1.0050 | 29526 | 134.09 | 2418.6 | 134.09 | 2425.1 | 2559.2 | 0.4642 | 8.4113 | 32 |
| 33 | 0.05035 | 1.0054 | 27998 | 138.27 | 2420.0 | 138.27 | 2422.7 | 2561.0 | 0.4779 | 8.3913 | 33 |
| 34 | 0.05325 | 1.0057 | 26560 | 142.45 | 2421.3 | 142.45 | 2420.4 | 2562.8 | 0.4916 | 8.3714 | 34 |
| 35 | 0.05629 | 1.0060 | 25205 | 146.63 | 2422.7 | 146.63 | 2417.9 | 2564.5 | 0.5051 | 8.3517 | 35 |
| 36 | 0.05948 | 1.0064 | 23929 | 150.81 | 2424.0 | 150.81 | 2415.5 | 2566.3 | 0.5187 | 8.3321 | 36 |
| 38 | 0.06633 | 1.0071 | 21593 | 159.17 | 2426.7 | 159.17 | 2410.7 | 2569.9 | 0.5456 | 8.2935 | 38 |
| 40 | 0.07385 | 1.0079 | 19515 | 167.53 | 2429.4 | 167.53 | 2406.0 | 2573.5 | 0.5724 | 8.2555 | 40 |
| 45 | 0.09595 | 1.0099 | 15252 | 188.43 | 2436.1 | 188.43 | 2394.0 | 2582.4 | 0.6386 | 8.1633 | 45 |
| 50 | 0.1235 | 1.0121 | 12027 | 209.33 | 2442.7 | 209.34 | 2382.0 | 2591.3 | 0.7038 | 8.0748 | 50 |
| 55 | 0.1576 | 1.0146 | 9564 | 230.24 | 2449.3 | 230.26 | 2369.8 | 2600.1 | 0.7680 | 7.9898 | 55 |
| 60 | 0.1995 | 1.0171 | 7667 | 251.16 | 2455.9 | 251.18 | 2357.6 | 2608.8 | 0.8313 | 7.9081 | 60 |
| 65 | 0.2504 | 1.0199 | 6194 | 272.09 | 2462.4 | 272.12 | 2345.4 | 2617.5 | 0.8937 | 7.8296 | 65 |
| 70 | 0.3120 | 1.0228 | 5040 | 293.03 | 2468.9 | 293.07 | 2333.0 | 2626.1 | 0.9551 | 7.7540 | 70 |

*(continued)*

### APPENDIX 22.N *(continued)*
Properties of Saturated Steam by Temperature
(SI units)

| temp. (°C) | absolute pressure (bars) | specific volume (cm³/g) | | internal energy (kJ/kg) | | enthalpy (kJ/kg) | | | entropy (kJ/kg·K) | | temp. (°C) |
|---|---|---|---|---|---|---|---|---|---|---|---|
| | | sat. liquid $v_f$ | sat. vapor $v_g$ | sat. liquid $u_f$ | sat. vapor $u_g$ | sat. liquid $h_f$ | evap. $h_{fg}$ | sat. vapor $h_g$ | sat. liquid $s_f$ | sat. vapor $s_g$ | |
| 75 | 0.3860 | 1.0258 | 4129 | 313.99 | 2475.2 | 314.03 | 2320.6 | 2634.6 | 1.0158 | 7.6812 | 75 |
| 80 | 0.4741 | 1.0291 | 3405 | 334.96 | 2481.6 | 335.01 | 2308.0 | 2643.0 | 1.0756 | 7.6111 | 80 |
| 85 | 0.5787 | 1.0324 | 2826 | 355.95 | 2487.8 | 356.01 | 2295.3 | 2651.3 | 1.1346 | 7.5434 | 85 |
| 90 | 0.7018 | 1.0360 | 2359 | 376.97 | 2494.0 | 377.04 | 2282.5 | 2659.5 | 1.1929 | 7.4781 | 90 |
| 95 | 0.8461 | 1.0396 | 1981 | 398.00 | 2500.0 | 398.09 | 2269.5 | 2667.6 | 1.2504 | 7.4151 | 95 |
| 100 | 1.014 | 1.0435 | 1672 | 419.06 | 2506.0 | 419.17 | 2256.4 | 2675.6 | 1.3072 | 7.3541 | 100 |
| 105 | 1.209 | 1.0474 | 1418 | 440.15 | 2511.9 | 440.27 | 2243.1 | 2683.4 | 1.3633 | 7.2952 | 105 |
| 110 | 1.434 | 1.0516 | 1209 | 461.26 | 2517.7 | 461.42 | 2229.6 | 2691.1 | 1.4188 | 7.2381 | 110 |
| 115 | 1.692 | 1.0559 | 1036 | 482.42 | 2523.3 | 482.59 | 2216.0 | 2698.6 | 1.4737 | 7.1828 | 115 |
| 120 | 1.987 | 1.0603 | 891.2 | 503.60 | 2528.9 | 503.81 | 2202.1 | 2705.9 | 1.5279 | 7.1291 | 120 |
| 125 | 2.322 | 1.0649 | 770.0 | 524.83 | 2534.3 | 525.07 | 2188.0 | 2713.1 | 1.5816 | 7.0770 | 125 |
| 130 | 2.703 | 1.0697 | 668.0 | 546.10 | 2539.5 | 546.38 | 2173.7 | 2720.1 | 1.6346 | 7.0264 | 130 |
| 135 | 3.132 | 1.0746 | 581.7 | 567.41 | 2544.7 | 567.74 | 2159.1 | 2726.9 | 1.6872 | 6.9772 | 135 |
| 140 | 3.615 | 1.0798 | 508.5 | 588.77 | 2549.6 | 589.16 | 2144.3 | 2733.4 | 1.7392 | 6.9293 | 140 |
| 145 | 4.157 | 1.0850 | 446.0 | 610.19 | 2554.4 | 610.64 | 2129.2 | 2739.8 | 1.7907 | 6.8826 | 145 |
| 150 | 4.762 | 1.0905 | 392.5 | 631.66 | 2559.1 | 632.18 | 2113.8 | 2745.9 | 1.8418 | 6.8371 | 150 |
| 155 | 5.435 | 1.0962 | 346.5 | 653.19 | 2563.5 | 653.79 | 2098.0 | 2751.8 | 1.8924 | 6.7926 | 155 |
| 160 | 6.182 | 1.1020 | 306.8 | 674.79 | 2567.8 | 675.47 | 2082.0 | 2757.4 | 1.9426 | 6.7491 | 160 |
| 165 | 7.009 | 1.1080 | 272.4 | 696.46 | 2571.9 | 697.24 | 2065.6 | 2762.8 | 1.9923 | 6.7066 | 165 |
| 170 | 7.922 | 1.1143 | 242.6 | 718.20 | 2575.7 | 719.08 | 2048.8 | 2767.9 | 2.0417 | 6.6650 | 170 |
| 175 | 8.926 | 1.1207 | 216.6 | 740.02 | 2579.4 | 741.02 | 2031.7 | 2772.7 | 2.0906 | 6.6241 | 175 |
| 180 | 10.03 | 1.1274 | 193.8 | 761.92 | 2582.8 | 763.05 | 2014.2 | 2777.2 | 2.1392 | 6.5840 | 180 |
| 185 | 11.23 | 1.1343 | 173.9 | 783.91 | 2586.0 | 785.19 | 1996.2 | 2781.4 | 2.1875 | 6.5447 | 185 |
| 190 | 12.55 | 1.1415 | 156.4 | 806.00 | 2589.0 | 807.43 | 1977.9 | 2785.3 | 2.2355 | 6.5059 | 190 |
| 195 | 13.99 | 1.1489 | 140.9 | 828.18 | 2591.7 | 829.79 | 1959.0 | 2788.8 | 2.2832 | 6.4678 | 195 |
| 200 | 15.55 | 1.1565 | 127.2 | 850.47 | 2594.2 | 852.27 | 1939.7 | 2792.0 | 2.3305 | 6.4302 | 200 |
| 205 | 17.24 | 1.1645 | 115.1 | 872.87 | 2596.4 | 874.88 | 1920.0 | 2794.8 | 2.3777 | 6.3930 | 205 |
| 210 | 19.08 | 1.1727 | 104.3 | 895.39 | 2598.3 | 897.63 | 1899.6 | 2797.3 | 2.4245 | 6.3563 | 210 |
| 215 | 21.06 | 1.1813 | 94.68 | 918.04 | 2599.9 | 920.53 | 1878.8 | 2799.3 | 2.4712 | 6.3200 | 215 |
| 220 | 23.20 | 1.1902 | 86.09 | 940.82 | 2601.3 | 943.58 | 1857.4 | 2801.0 | 2.5177 | 6.2840 | 220 |
| 225 | 25.50 | 1.1994 | 78.40 | 963.74 | 2602.2 | 966.80 | 1835.4 | 2802.2 | 2.5640 | 6.2483 | 225 |
| 230 | 27.97 | 1.2090 | 71.50 | 986.81 | 2602.9 | 990.19 | 1812.7 | 2802.9 | 2.6101 | 6.2128 | 230 |
| 235 | 30.63 | 1.2190 | 65.30 | 1010.0 | 2603.2 | 1013.8 | 1789.4 | 2803.2 | 2.6561 | 6.1775 | 235 |
| 240 | 33.47 | 1.2295 | 59.71 | 1033.4 | 2603.1 | 1037.6 | 1765.4 | 2803.0 | 2.7020 | 6.1423 | 240 |
| 245 | 36.51 | 1.2403 | 54.65 | 1057.0 | 2602.7 | 1061.6 | 1740.7 | 2802.2 | 2.7478 | 6.1072 | 245 |
| 250 | 39.76 | 1.2517 | 50.08 | 1080.8 | 2601.8 | 1085.8 | 1715.2 | 2800.9 | 2.7935 | 6.0721 | 250 |
| 255 | 43.23 | 1.2636 | 45.94 | 1104.8 | 2600.5 | 1110.2 | 1688.8 | 2799.1 | 2.8392 | 6.0369 | 255 |
| 260 | 46.92 | 1.2761 | 42.17 | 1129.0 | 2598.7 | 1135.0 | 1661.6 | 2796.6 | 2.8849 | 6.0016 | 260 |
| 265 | 50.85 | 1.2892 | 38.75 | 1153.4 | 2596.5 | 1160.0 | 1633.5 | 2793.5 | 2.9307 | 5.9661 | 265 |
| 270 | 55.03 | 1.3030 | 35.62 | 1178.1 | 2593.7 | 1185.3 | 1604.4 | 2789.7 | 2.9765 | 5.9304 | 270 |

*(continued)*

## APPENDIX 22.N *(continued)*
### Properties of Saturated Steam by Temperature
### (SI units)

| temp. (°C) | absolute pressure (bars) | specific volume (cm³/g) | | internal energy (kJ/kg) | | enthalpy (kJ/kg) | | | entropy (kJ/kg·K) | | temp. (°C) |
|---|---|---|---|---|---|---|---|---|---|---|---|
| | | sat. liquid $v_f$ | sat. vapor $v_g$ | sat. liquid $u_f$ | sat. vapor $u_g$ | sat. liquid $h_f$ | evap. $h_{fg}$ | sat. vapor $h_g$ | sat. liquid $s_f$ | sat. vapor $s_g$ | |
| 275 | 59.46 | 1.3175 | 32.77 | 1203.1 | 2590.3 | 1210.9 | 1574.3 | 2785.2 | 3.0224 | 5.8944 | 275 |
| 280 | 64.17 | 1.3328 | 30.15 | 1228.3 | 2586.4 | 1236.9 | 1543.0 | 2779.9 | 3.0685 | 5.8579 | 280 |
| 285 | 69.15 | 1.3491 | 27.76 | 1253.9 | 2581.8 | 1263.3 | 1510.5 | 2773.7 | 3.1147 | 5.8209 | 285 |
| 290 | 74.42 | 1.3663 | 25.56 | 1279.9 | 2576.5 | 1290.0 | 1476.7 | 2766.7 | 3.1612 | 5.7834 | 290 |
| 295 | 79.99 | 1.3846 | 23.53 | 1306.2 | 2570.5 | 1317.3 | 1441.4 | 2758.7 | 3.2080 | 5.7451 | 295 |
| 300 | 85.88 | 1.4042 | 21.66 | 1332.9 | 2563.6 | 1345.0 | 1404.6 | 2749.6 | 3.2552 | 5.7059 | 300 |
| 305 | 92.09 | 1.4252 | 19.93 | 1360.2 | 2555.9 | 1373.3 | 1366.1 | 2739.4 | 3.3028 | 5.6657 | 305 |
| 310 | 98.65 | 1.4479 | 18.34 | 1387.9 | 2547.1 | 1402.2 | 1325.7 | 2728.0 | 3.3510 | 5.6244 | 310 |
| 315 | 105.6 | 1.4724 | 16.85 | 1416.3 | 2537.2 | 1431.8 | 1283.2 | 2715.1 | 3.3998 | 5.5816 | 315 |
| 320 | 112.8 | 1.4990 | 15.47 | 1445.3 | 2526.0 | 1462.2 | 1238.4 | 2700.6 | 3.4494 | 5.5372 | 320 |
| 325 | 120.5 | 1.5283 | 14.18 | 1475.1 | 2513.4 | 1493.5 | 1190.8 | 2684.3 | 3.5000 | 5.4908 | 325 |
| 330 | 128.6 | 1.5606 | 12.98 | 1505.8 | 2499.2 | 1525.9 | 1140.2 | 2666.0 | 3.5518 | 5.4422 | 330 |
| 335 | 137.1 | 1.5967 | 11.85 | 1537.6 | 2483.0 | 1559.5 | 1085.9 | 2645.4 | 3.6050 | 5.3906 | 335 |
| 340 | 146.0 | 1.6376 | 10.78 | 1570.6 | 2464.4 | 1594.5 | 1027.3 | 2621.9 | 3.6601 | 5.3356 | 340 |
| 345 | 155.4 | 1.6846 | 9.769 | 1605.3 | 2443.1 | 1631.5 | 963.4 | 2594.9 | 3.7176 | 5.2762 | 345 |
| 350 | 165.3 | 1.7400 | 8.802 | 1642.1 | 2418.1 | 1670.9 | 892.8 | 2563.6 | 3.7784 | 5.2110 | 350 |
| 355 | 175.7 | 1.8079 | 7.868 | 1682.0 | 2388.4 | 1713.7 | 812.9 | 2526.7 | 3.8439 | 5.1380 | 355 |
| 360 | 186.7 | 1.8954 | 6.949 | 1726.3 | 2351.8 | 1761.7 | 719.8 | 2481.5 | 3.9167 | 5.0536 | 360 |
| 365 | 198.2 | 2.0172 | 6.012 | 1777.8 | 2303.8 | 1817.8 | 605.2 | 2423.0 | 4.0014 | 4.9497 | 365 |
| 370 | 210.4 | 2.2152 | 4.954 | 1844.1 | 2230.3 | 1890.7 | 443.8 | 2334.5 | 4.1112 | 4.8012 | 370 |
| 374 | 220.64000 | 3.1056 | 3.1056 | 2015.70 | 2015.70 | 2084.30 | 0 | 2084.3 | 4.4070 | 4.4070 | 373.9 |

(Multiply MPa by 10 to obtain bars.)

Values in this table were calculated from *NIST Standard Reference Database 10*, "NIST/ASME Steam Properties," Ver. 2.11, National Institute of Standards and Technology, U.S. Department of Commerce, Gaithersburg, MD, 1997, which has been licensed to PPI.

## APPENDIX 22.O
Properties of Saturated Steam by Pressure
(SI units)

| absolute pressure (bars) | temp. (°C) | specific volume (cm³/g) | | internal energy (kJ/kg) | | enthalpy (kJ/kg) | | | entropy (kJ/kg·K) | | absolute pressure (bars) |
|---|---|---|---|---|---|---|---|---|---|---|---|
| | | sat. liquid $v_f$ | sat. vapor $v_g$ | sat. liquid $u_f$ | sat. vapor $u_g$ | sat. liquid $h_f$ | evap. $h_{fg}$ | sat. vapor $h_g$ | sat. liquid $s_f$ | sat. vapor $s_g$ | |
| 0.04 | 28.96 | 1.0041 | 34791 | 121.38 | 2414.5 | 121.39 | 2432.3 | 2553.7 | 0.4224 | 8.4734 | 0.04 |
| 0.06 | 36.16 | 1.0065 | 23733 | 151.47 | 2424.2 | 151.48 | 2415.2 | 2566.6 | 0.5208 | 8.3290 | 0.06 |
| 0.08 | 41.51 | 1.0085 | 18099 | 173.83 | 2431.4 | 173.84 | 2402.4 | 2576.2 | 0.5925 | 8.2273 | 0.08 |
| 0.10 | 45.81 | 1.0103 | 14670 | 191.80 | 2437.2 | 191.81 | 2392.1 | 2583.9 | 0.6492 | 8.1488 | 0.10 |
| 0.20 | 60.06 | 1.0172 | 7648 | 251.40 | 2456.0 | 251.42 | 2357.5 | 2608.9 | 0.8320 | 7.9072 | 0.20 |
| 0.30 | 69.10 | 1.0222 | 5228 | 289.24 | 2467.7 | 289.27 | 2335.3 | 2624.5 | 0.9441 | 7.7675 | 0.30 |
| 0.40 | 75.86 | 1.0264 | 3993 | 317.58 | 2476.3 | 317.62 | 2318.4 | 2636.1 | 1.0261 | 7.6690 | 0.40 |
| 0.50 | 81.32 | 1.0299 | 3240 | 340.49 | 2483.2 | 340.54 | 2304.7 | 2645.2 | 1.0912 | 7.5930 | 0.50 |
| 0.60 | 85.93 | 1.0331 | 2732 | 359.84 | 2489.0 | 359.91 | 2292.9 | 2652.9 | 1.1454 | 7.5311 | 0.60 |
| 0.70 | 89.93 | 1.0359 | 2365 | 376.68 | 2493.9 | 376.75 | 2282.7 | 2659.4 | 1.1921 | 7.4790 | 0.70 |
| 0.80 | 93.49 | 1.0385 | 2087 | 391.63 | 2498.2 | 391.71 | 2273.5 | 2665.2 | 1.2330 | 7.4339 | 0.80 |
| 0.90 | 96.69 | 1.0409 | 1869 | 405.10 | 2502.1 | 405.20 | 2265.1 | 2670.3 | 1.2696 | 7.3943 | 0.90 |
| 1.00 | 99.61 | 1.0432 | 1694 | 417.40 | 2505.6 | 417.50 | 2257.4 | 2674.9 | 1.3028 | 7.3588 | 1.00 |
| 1.01325 | 99.97 | 1.0434 | 1673 | 418.95 | 2506.0 | 419.06 | 2256.5 | 2675.5 | 1.3069 | 7.3544 | 1.01325 |
| 1.50 | 111.3 | 1.0527 | 1159 | 466.97 | 2519.2 | 467.13 | 2226.0 | 2693.1 | 1.4337 | 7.2230 | 1.50 |
| 2.00 | 120.2 | 1.0605 | 885.7 | 504.49 | 2529.1 | 504.70 | 2201.5 | 2706.2 | 1.5302 | 7.1269 | 2.00 |
| 2.50 | 127.4 | 1.0672 | 718.7 | 535.08 | 2536.8 | 535.35 | 2181.1 | 2716.5 | 1.6072 | 7.0524 | 2.50 |
| 3.00 | 133.5 | 1.0732 | 605.8 | 561.10 | 2543.2 | 561.43 | 2163.5 | 2724.9 | 1.6717 | 6.9916 | 3.00 |
| 3.50 | 138.9 | 1.0786 | 524.2 | 583.88 | 2548.5 | 584.26 | 2147.7 | 2732.0 | 1.7274 | 6.9401 | 3.50 |
| 4.00 | 143.6 | 1.0836 | 462.4 | 604.22 | 2553.1 | 604.65 | 2133.4 | 2738.1 | 1.7765 | 6.8955 | 4.00 |
| 4.50 | 147.9 | 1.0882 | 413.9 | 622.65 | 2557.1 | 623.14 | 2120.2 | 2743.4 | 1.8205 | 6.8560 | 4.50 |
| 5.00 | 151.8 | 1.0926 | 374.8 | 639.54 | 2560.7 | 640.09 | 2108.0 | 2748.1 | 1.8604 | 6.8207 | 5.00 |
| 6.00 | 158.8 | 1.1006 | 315.6 | 669.72 | 2566.8 | 670.38 | 2085.8 | 2756.1 | 1.9308 | 6.7592 | 6.00 |
| 7.00 | 164.9 | 1.1080 | 272.8 | 696.23 | 2571.8 | 697.00 | 2065.8 | 2762.8 | 1.9918 | 6.7071 | 7.00 |
| 8.00 | 170.4 | 1.1148 | 240.3 | 719.97 | 2576.0 | 720.86 | 2047.4 | 2768.3 | 2.0457 | 6.6616 | 8.00 |
| 9.00 | 175.4 | 1.1212 | 214.9 | 741.55 | 2579.6 | 742.56 | 2030.5 | 2773.0 | 2.0940 | 6.6213 | 9.00 |
| 10.0 | 179.9 | 1.1272 | 194.4 | 761.39 | 2582.7 | 762.52 | 2014.6 | 2777.1 | 2.1381 | 6.5850 | 10.0 |
| 15.0 | 198.3 | 1.1539 | 131.7 | 842.83 | 2593.4 | 844.56 | 1946.4 | 2791.0 | 2.3143 | 6.4430 | 15.0 |
| 20.0 | 212.4 | 1.1767 | 99.59 | 906.14 | 2599.1 | 908.50 | 1889.8 | 2798.3 | 2.4468 | 6.3390 | 20.0 |
| 25.0 | 224.0 | 1.1974 | 79.95 | 958.91 | 2602.1 | 961.91 | 1840.0 | 2801.9 | 2.5543 | 6.2558 | 25.0 |
| 30.0 | 233.9 | 1.2167 | 66.66 | 1004.7 | 2603.2 | 1008.3 | 1794.9 | 2803.2 | 2.6455 | 6.1856 | 30.0 |
| 35.0 | 242.6 | 1.2350 | 57.06 | 1045.5 | 2602.9 | 1049.8 | 1752.8 | 2802.6 | 2.7254 | 6.1243 | 35.0 |
| 40.0 | 250.4 | 1.2526 | 49.78 | 1082.5 | 2601.7 | 1087.5 | 1713.3 | 2800.8 | 2.7968 | 6.0696 | 40.0 |
| 45.0 | 257.4 | 1.2696 | 44.06 | 1116.5 | 2599.7 | 1122.3 | 1675.7 | 2797.9 | 2.8615 | 6.0197 | 45.0 |
| 50.0 | 263.9 | 1.2864 | 39.45 | 1148.2 | 2597.0 | 1154.6 | 1639.6 | 2794.2 | 2.9210 | 5.9737 | 50.0 |
| 55.0 | 270.0 | 1.3029 | 35.64 | 1177.9 | 2593.7 | 1185.1 | 1604.6 | 2789.7 | 2.9762 | 5.9307 | 55.0 |
| 60.0 | 275.6 | 1.3193 | 32.45 | 1206.0 | 2589.9 | 1213.9 | 1570.7 | 2784.6 | 3.0278 | 5.8901 | 60.0 |
| 65.0 | 280.9 | 1.3356 | 29.73 | 1232.7 | 2585.7 | 1241.4 | 1537.5 | 2778.9 | 3.0764 | 5.8516 | 65.0 |
| 70.0 | 285.8 | 1.3519 | 27.38 | 1258.2 | 2581.0 | 1267.7 | 1504.9 | 2772.6 | 3.1224 | 5.8148 | 70.0 |
| 75.0 | 290.5 | 1.3682 | 25.33 | 1282.7 | 2575.9 | 1292.9 | 1473.0 | 2765.9 | 3.1662 | 5.7793 | 75.0 |

*(continued)*

**APPENDIX 22.O** *(continued)*
Properties of Saturated Steam by Pressure
(SI units)

| absolute pressure (bars) | temp. (°C) | specific volume (cm³/g) | | internal energy (kJ/kg) | | enthalpy (kJ/kg) | | | entropy (kJ/kg·K) | | absolute pressure (bars) |
|---|---|---|---|---|---|---|---|---|---|---|---|
| | | sat. liquid $v_f$ | sat. vapor $v_g$ | sat. liquid $u_f$ | sat. vapor $u_g$ | sat. liquid $h_f$ | evap. $h_{fg}$ | sat. vapor $h_g$ | sat. liquid $s_f$ | sat. vapor $s_g$ | |
| 80.0 | 295.0 | 1.3847 | 23.53 | 1306.2 | 2570.5 | 1317.3 | 1441.4 | 2758.7 | 3.2081 | 5.7450 | 80.0 |
| 85.0 | 299.3 | 1.4013 | 21.92 | 1329.0 | 2564.7 | 1340.9 | 1410.1 | 2751.0 | 3.2483 | 5.7117 | 85.0 |
| 90.0 | 303.3 | 1.4181 | 20.49 | 1351.1 | 2558.5 | 1363.9 | 1379.0 | 2742.9 | 3.2870 | 5.6791 | 90.0 |
| 95.0 | 307.2 | 1.4352 | 19.20 | 1372.6 | 2552.0 | 1386.2 | 1348.2 | 2734.4 | 3.3244 | 5.6473 | 95.0 |
| 100 | 311.0 | 1.4526 | 18.03 | 1393.5 | 2545.2 | 1408.1 | 1317.4 | 2725.5 | 3.3606 | 5.6160 | 100 |
| 100 | 311.0 | 1.4526 | 18.03 | 1393.5 | 2545.2 | 1408.1 | 1317.4 | 2725.5 | 3.3606 | 5.6160 | 100 |
| 110 | 318.1 | 1.4885 | 15.99 | 1434.1 | 2530.5 | 1450.4 | 1255.9 | 2706.3 | 3.4303 | 5.5545 | 110 |
| 120 | 324.7 | 1.5263 | 14.26 | 1473.1 | 2514.3 | 1491.5 | 1193.9 | 2685.4 | 3.4967 | 5.4939 | 120 |
| 130 | 330.9 | 1.5665 | 12.78 | 1511.1 | 2496.5 | 1531.5 | 1131.2 | 2662.7 | 3.5608 | 5.4336 | 130 |
| 140 | 336.7 | 1.6097 | 11.49 | 1548.4 | 2477.1 | 1571.0 | 1066.9 | 2637.9 | 3.6232 | 5.3727 | 140 |
| 150 | 342.2 | 1.6570 | 10.34 | 1585.3 | 2455.6 | 1610.2 | 1000.5 | 2610.7 | 3.6846 | 5.3106 | 150 |
| 160 | 347.4 | 1.7094 | 9.309 | 1622.3 | 2431.8 | 1649.7 | 931.1 | 2580.8 | 3.7457 | 5.2463 | 160 |
| 170 | 352.3 | 1.7693 | 8.371 | 1659.9 | 2405.2 | 1690.0 | 857.5 | 2547.5 | 3.8077 | 5.1787 | 170 |
| 180 | 357.0 | 1.8398 | 7.502 | 1699.0 | 2374.8 | 1732.1 | 777.7 | 2509.8 | 3.8718 | 5.1061 | 180 |
| 190 | 361.5 | 1.9268 | 6.677 | 1740.5 | 2339.1 | 1777.2 | 688.8 | 2466.0 | 3.9401 | 5.0256 | 190 |
| 200 | 365.7 | 2.0400 | 5.865 | 1786.4 | 2295.0 | 1827.2 | 585.1 | 2412.3 | 4.0156 | 4.9314 | 200 |
| 210 | 369.8 | 2.2055 | 4.996 | 1841.2 | 2233.7 | 1887.6 | 451.0 | 2338.6 | 4.1064 | 4.8079 | 210 |
| 220.64 | 373.95 | 3.1056 | 3.1056 | 2015.7 | 2015.7 | 2084.3 | 0 | 2084.3 | 4.4070 | 4.4070 | 220.64 |

(Multiply MPa by 10 to obtain bars.)

Values in this table were calculated from *NIST Standard Reference Database 10*, "NIST/ASME Steam Properties," Ver. 2.11, National Institute of Standards and Technology, U.S. Department of Commerce, Gaithersburg, MD, 1997, which has been licensed to PPI.

## APPENDIX 22.P
Properties of Superheated Steam
(SI units)
specific volume ($v$) in m³/kg; enthalpy ($h$) in kJ/kg; entropy ($s$) in kJ/kg·K

| absolute pressure (kPa) (sat. temp. °C) | | temperature °C | | | | | | | |
|---|---|---|---|---|---|---|---|---|---|
| | | 100 | 150 | 200 | 250 | 300 | 360 | 420 | 500 |
| 10 (45.81) | $v$ | 17.196 | 19.513 | 21.826 | 24.136 | 26.446 | 29.216 | 31.986 | 35.680 |
| | $h$ | 2687.5 | 2783.0 | 2879.6 | 2977.4 | 3076.7 | 3197.9 | 3321.4 | 3489.7 |
| | $s$ | 8.4489 | 8.6892 | 8.9049 | 9.1015 | 9.2827 | 9.4837 | 9.6700 | 9.8998 |
| 50 (81.32) | $v$ | 3.419 | 3.890 | 4.356 | 4.821 | 5.284 | 5.839 | 6.394 | 7.134 |
| | $h$ | 2682.4 | 2780.2 | 2877.8 | 2976.1 | 3075.8 | 3197.2 | 3320.8 | 3489.3 |
| | $s$ | 7.6953 | 7.9413 | 8.1592 | 8.3568 | 8.5386 | 8.7401 | 8.9266 | 9.1566 |
| 75 (91.76) | $v$ | 2.270 | 2.588 | 2.900 | 3.211 | 3.521 | 3.891 | 4.262 | 4.755 |
| | $h$ | 2679.2 | 2778.4 | 2876.6 | 2975.3 | 3075.1 | 3196.7 | 3320.4 | 3489.0 |
| | $s$ | 7.5011 | 7.7509 | 7.9702 | 8.1685 | 8.3507 | 8.5524 | 8.7391 | 8.9692 |
| 100 (99.61) | $v$ | 1.6959 | 1.9367 | 2.172 | 2.406 | 2.639 | 2.917 | 3.195 | 3.566 |
| | $h$ | 2675.8 | 2776.6 | 2875.5 | 2974.5 | 3074.5 | 3196.3 | 3320.1 | 3488.7 |
| | $s$ | 7.3610 | 7.6148 | 7.8356 | 8.0346 | 8.2172 | 8.4191 | 8.6059 | 8.8361 |
| 150 (111.35) | $v$ | ...... | 1.2855 | 1.4445 | 1.6013 | 1.7571 | 1.9433 | 2.129 | 2.376 |
| | $h$ | ...... | 2772.9 | 2873.1 | 2972.9 | 3073.3 | 3195.3 | 3319.4 | 3488.2 |
| | $s$ | ...... | 7.4208 | 7.6447 | 7.8451 | 8.0284 | 8.2309 | 8.4180 | 8.6485 |
| 400 (143.61) | $v$ | ...... | 0.4709 | 0.5343 | 0.5952 | 0.6549 | 0.7257 | 0.7961 | 0.8894 |
| | $h$ | ...... | 2752.8 | 2860.9 | 2964.5 | 3067.1 | 3190.7 | 3315.8 | 3485.5 |
| | $s$ | ...... | 6.9306 | 7.1723 | 7.3804 | 7.5677 | 7.7728 | 7.9615 | 8.1933 |
| 700 (164.95) | $v$ | ...... | ...... | 0.3000 | 0.3364 | 0.3714 | 0.4126 | 0.4533 | 0.5070 |
| | $h$ | ...... | ...... | 2845.3 | 2954.0 | 3059.4 | 3185.1 | 3311.5 | 3482.3 |
| | $s$ | ...... | ...... | 6.8884 | 7.1070 | 7.2995 | 7.5080 | 7.6986 | 7.9319 |
| 1000 (179.88) | $v$ | ...... | ...... | 0.2060 | 0.2328 | 0.2580 | 0.2874 | 0.3162 | 0.3541 |
| | $h$ | ...... | ...... | 2828.3 | 2943.1 | 3051.6 | 3179.4 | 3307.1 | 3479.1 |
| | $s$ | ...... | ...... | 6.6955 | 6.9265 | 7.1246 | 7.3367 | 7.5294 | 7.7641 |
| 1500 (198.29) | $v$ | ...... | ...... | 0.13245 | 0.15201 | 0.16971 | 0.18990 | 0.2095 | 0.2352 |
| | $h$ | ...... | ...... | 2796.0 | 2923.9 | 3038.2 | 3169.8 | 3299.8 | 3473.7 |
| | $s$ | ...... | ...... | 6.4536 | 6.7111 | 6.9198 | 7.1382 | 7.3343 | 7.5718 |
| 2000 (212.38) | $v$ | ...... | ...... | ...... | 0.11150 | 0.12551 | 0.14115 | 0.15617 | 0.17568 |
| | $h$ | ...... | ...... | ...... | 2903.2 | 3024.2 | 3159.9 | 3292.3 | 3468.2 |
| | $s$ | ...... | ...... | ...... | 6.5475 | 6.7684 | 6.9937 | 7.1935 | 7.4337 |
| 2500 (223.95) | $v$ | ...... | ...... | ...... | 0.08705 | 0.09894 | 0.11188 | 0.12416 | 0.13999 |
| | $h$ | ...... | ...... | ...... | 2880.9 | 3009.6 | 3149.8 | 3284.8 | 3462.7 |
| | $s$ | ...... | ...... | ...... | 6.4107 | 6.6459 | 6.8788 | 7.0824 | 7.3254 |

*(continued)*

**APPENDIX 22.P** *(continued)*
Properties of Superheated Steam
(SI units)
specific volume ($v$) in m³/kg; enthalpy ($h$) in kJ/kg; entropy ($s$) in kJ/kg·K

| absolute pressure (kPa) (sat. temp. °C) | | temperature °C | | | | | | | |
|---|---|---|---|---|---|---|---|---|---|
| | | 100 | 150 | 200 | 250 | 300 | 360 | 420 | 500 |
| 3000 | $v$ | ...... | ...... | ...... | 0.07063 | 0.08118 | 0.09236 | 0.10281 | 0.11620 |
| (233.85) | $h$ | ...... | ...... | ...... | 2856.5 | 2994.3 | 3139.5 | 3277.1 | 3457.2 |
| | $s$ | ...... | ...... | ...... | 6.2893 | 6.5412 | 6.7823 | 6.9900 | 7.2359 |

(Multiply kPa by 0.1 to obtain bars.)

Values in this table were calculated from *NIST Standard Reference Database 10*, "NIST/ASME Steam Properties," Ver. 2.11, National Institute of Standards and Technology, U.S. Department of Commerce, Gaithersburg, MD, 1997, which has been licensed to PPI.

**Appendices**

## APPENDIX 22.Q
Properties of Compressed Water
(SI units)

| $T$ (°C) | $p$ (bars) | $\rho$ (kg/cm³) | $v$ (cm³/kg) | $x$ | $h$ (kJ/kg) | $s$ (kJ/kg·K) | $u$ (kJ/kg) |
|---|---|---|---|---|---|---|---|
| 0 | 25 | 0.0010011 | 998.94 | subcooled | 2.5 | 0.00000380 | 0.0026204 |
| 25 | 25 | 0.00099813 | 1001.9 | subcooled | 107.14 | 0.36658 | 104.63 |
| 50 | 25 | 0.00098908 | 1011 | subcooled | 211.49 | 0.70266 | 208.96 |
| 75 | 25 | 0.00097591 | 1024.7 | subcooled | 316.02 | 1.0142 | 313.45 |
| 100 | 25 | 0.00095947 | 1042.2 | subcooled | 420.97 | 1.3053 | 418.36 |
| 125 | 25 | 0.00094018 | 1063.6 | subcooled | 526.64 | 1.5794 | 523.98 |
| 150 | 25 | 0.00091815 | 1089.1 | subcooled | 633.43 | 1.8395 | 630.71 |
| 175 | 25 | 0.00089332 | 1119.4 | subcooled | 741.87 | 2.0885 | 739.07 |
| 200 | 25 | 0.00086538 | 1155.6 | subcooled | 852.65 | 2.329 | 849.76 |
| 223.95 | 25 | 0.00083512 | 1197.4 | 0 | 961.91 | 2.5543 | 958.91 |
| 223.95 | 25 | 0.000012508 | 79949 | 1 | 2801.9 | 6.2558 | 2602.1 |
| | | | | | | | |
| 0 | 50 | 0.0010023 | 997.68 | subcooled | 5.0325 | 0.0001383 | 0.044068 |
| 25 | 50 | 0.00099925 | 1000.8 | subcooled | 109.45 | 0.36592 | 104.44 |
| 50 | 50 | 0.00099016 | 1009.9 | subcooled | 213.64 | 0.7015 | 208.59 |
| 75 | 50 | 0.00097701 | 1023.5 | subcooled | 318.03 | 1.0127 | 312.92 |
| 100 | 50 | 0.00096063 | 1041 | subcooled | 422.85 | 1.3034 | 417.64 |
| 125 | 50 | 0.00094144 | 1062.2 | subcooled | 528.37 | 1.5771 | 523.06 |
| 150 | 50 | 0.00091956 | 1087.5 | subcooled | 634.98 | 1.8368 | 629.55 |
| 175 | 50 | 0.00089493 | 1117.4 | subcooled | 743.19 | 2.0852 | 737.61 |
| 200 | 50 | 0.00086726 | 1153.1 | subcooled | 853.68 | 2.3251 | 847.91 |
| 225 | 50 | 0.00083599 | 1196.2 | subcooled | 967.38 | 2.5592 | 961.4 |
| 250 | 50 | 0.00080009 | 1249.9 | subcooled | 1085.7 | 2.791 | 1079.5 |
| 263.94 | 50 | 0.00077737 | 1286.4 | 0 | 1154.6 | 2.921 | 1148.2 |
| 263.94 | 50 | 0.000025351 | 39446 | 1 | 2794.2 | 5.9737 | 2597 |
| | | | | | | | |
| 0 | 75 | 0.0010036 | 996.44 | subcooled | 7.5555 | 0.0002494 | 0.082204 |
| 25 | 75 | 0.0010004 | 999.64 | subcooled | 111.75 | 0.36526 | 104.25 |
| 50 | 75 | 0.00099124 | 1008.8 | subcooled | 215.79 | 0.70035 | 208.22 |
| 75 | 75 | 0.0009781 | 1022.4 | subcooled | 320.05 | 1.0111 | 312.38 |
| 100 | 75 | 0.00096179 | 1039.7 | subcooled | 424.73 | 1.3015 | 416.93 |
| 125 | 75 | 0.00094269 | 1060.8 | subcooled | 530.11 | 1.5748 | 522.15 |
| 150 | 75 | 0.00092095 | 1085.8 | subcooled | 636.54 | 1.8341 | 628.4 |
| 175 | 75 | 0.00089651 | 1115.4 | subcooled | 744.54 | 2.082 | 736.17 |
| 200 | 75 | 0.00086911 | 1150.6 | subcooled | 854.73 | 2.3212 | 846.1 |
| 225 | 75 | 0.00083824 | 1193 | subcooled | 968.01 | 2.5545 | 959.06 |
| 250 | 75 | 0.00080293 | 1245.4 | subcooled | 1085.7 | 2.7851 | 1076.4 |
| 275 | 75 | 0.0007614 | 1313.4 | subcooled | 1210.2 | 3.0174 | 1200.4 |
| 290.54 | 75 | 0.00073088 | 1368.2 | 0 | 1292.9 | 3.1662 | 1282.7 |
| 290.54 | 75 | 0.000039479 | 25330 | 1 | 2765.9 | 5.7793 | 2575.9 |
| | | | | | | | |
| 0 | 100 | 0.0010048 | 995.2 | subcooled | 10.069 | 0.00033757 | 0.1171 |
| 25 | 100 | 0.0010015 | 998.54 | subcooled | 114.05 | 0.3646 | 104.06 |
| 50 | 100 | 0.00099231 | 1007.8 | subcooled | 217.94 | 0.6992 | 207.86 |
| 75 | 100 | 0.00097919 | 1021.3 | subcooled | 322.07 | 1.0096 | 311.85 |
| 100 | 100 | 0.00096293 | 1038.5 | subcooled | 426.62 | 1.2996 | 416.23 |
| 125 | 100 | 0.00094393 | 1059.4 | subcooled | 531.84 | 1.5725 | 521.25 |
| 150 | 100 | 0.00092232 | 1084.2 | subcooled | 638.11 | 1.8313 | 627.27 |
| 175 | 100 | 0.00089807 | 1113.5 | subcooled | 745.89 | 2.0788 | 734.75 |
| 200 | 100 | 0.00087094 | 1148.2 | subcooled | 855.8 | 2.3174 | 844.31 |
| 225 | 100 | 0.00084044 | 1189.9 | subcooled | 968.68 | 2.5499 | 956.78 |
| 250 | 100 | 0.0008057 | 1241.2 | subcooled | 1085.8 | 2.7792 | 1073.4 |
| 275 | 100 | 0.00076513 | 1307 | subcooled | 1209.3 | 3.0097 | 1196.2 |
| 300 | 100 | 0.00071529 | 1398 | subcooled | 1343.3 | 3.2488 | 1329.4 |

*(continued)*

**APPENDIX 22.Q** *(continued)*
Properties of Compressed Water
(SI units)

| $T$ (°C) | $p$ (bars) | $\rho$ (kg/cm$^3$) | $\nu$ (cm$^3$/kg) | $x$ | $h$ (kJ/kg) | $s$ (kJ/kg·K) | $\mu$ (kJ/kg) |
|------|------|------|------|------|------|------|------|
| 0 | 150 | 0.0010073 | 992.76 | subcooled | 15.069 | 0.00044686 | 0.17746 |
| 25 | 150 | 0.0010037 | 996.35 | subcooled | 118.63 | 0.36325 | 103.69 |
| 50 | 150 | 0.00099443 | 1005.6 | subcooled | 222.23 | 0.6969 | 207.15 |
| 75 | 150 | 0.00098135 | 1019 | subcooled | 326.1 | 1.0065 | 310.81 |
| 100 | 150 | 0.0009652 | 1036.1 | subcooled | 430.39 | 1.2958 | 414.85 |
| 125 | 150 | 0.00094638 | 1056.7 | subcooled | 535.33 | 1.568 | 519.48 |
| 150 | 150 | 0.00092503 | 1081 | subcooled | 641.27 | 1.826 | 625.05 |
| 175 | 150 | 0.00090114 | 1109.7 | subcooled | 748.63 | 2.0725 | 731.98 |
| 200 | 150 | 0.0008745 | 1143.5 | subcooled | 857.99 | 2.31 | 840.84 |
| 225 | 150 | 0.00084471 | 1183.8 | subcooled | 970.12 | 2.5409 | 952.36 |
| 250 | 150 | 0.00081103 | 1233 | subcooled | 1086.1 | 2.768 | 1067.6 |
| 275 | 150 | 0.00077216 | 1295.1 | subcooled | 1207.8 | 2.9951 | 1188.3 |
| 300 | 150 | 0.00072555 | 1378.3 | subcooled | 1338.3 | 3.2279 | 1317.6 |
| 0 | 200 | 0.0010097 | 990.36 | subcooled | 20.033 | 0.00046962 | 0.22569 |
| 25 | 200 | 0.0010058 | 994.19 | subcooled | 123.2 | 0.36187 | 103.32 |
| 50 | 200 | 0.00099653 | 1003.5 | subcooled | 226.51 | 0.69461 | 206.44 |
| 75 | 200 | 0.00098348 | 1016.8 | subcooled | 330.13 | 1.0035 | 309.79 |
| 100 | 200 | 0.00096744 | 1033.7 | subcooled | 434.17 | 1.292 | 413.5 |
| 125 | 200 | 0.00094879 | 1054 | subcooled | 538.84 | 1.5635 | 517.76 |
| 150 | 200 | 0.00092769 | 1077.9 | subcooled | 644.45 | 1.8208 | 622.89 |
| 175 | 200 | 0.00090414 | 1106 | subcooled | 751.42 | 2.0664 | 729.3 |
| 200 | 200 | 0.00087797 | 1139 | subcooled | 860.27 | 2.3027 | 837.49 |
| 225 | 200 | 0.00084882 | 1178.1 | subcooled | 971.69 | 2.5322 | 948.13 |
| 250 | 200 | 0.00081609 | 1225.4 | subcooled | 1086.7 | 2.7573 | 1062.2 |
| 275 | 200 | 0.00077871 | 1284.2 | subcooled | 1206.7 | 2.9814 | 1181 |
| 300 | 200 | 0.00073471 | 1361.1 | subcooled | 1334.4 | 3.2091 | 1307.1 |
| 0 | 250 | 0.0010122 | 988 | subcooled | 24.962 | 0.00040919 | 0.26234 |
| 25 | 250 | 0.001008 | 992.07 | subcooled | 127.75 | 0.36047 | 102.95 |
| 50 | 250 | 0.00099861 | 1001.4 | subcooled | 230.79 | 0.69233 | 205.75 |
| 75 | 250 | 0.00098559 | 1014.6 | subcooled | 334.16 | 1.0004 | 308.79 |
| 100 | 250 | 0.00096965 | 1031.3 | subcooled | 437.95 | 1.2883 | 412.17 |
| 125 | 250 | 0.00095116 | 1051.3 | subcooled | 542.35 | 1.5591 | 516.07 |
| 150 | 250 | 0.0009303 | 1074.9 | subcooled | 647.66 | 1.8156 | 620.78 |
| 175 | 250 | 0.00090707 | 1102.5 | subcooled | 754.25 | 2.0604 | 726.69 |
| 200 | 250 | 0.00088133 | 1134.6 | subcooled | 862.61 | 2.2956 | 834.24 |
| 225 | 250 | 0.00085279 | 1172.6 | subcooled | 973.38 | 2.5237 | 944.06 |
| 250 | 250 | 0.00082092 | 1218.1 | subcooled | 1087.4 | 2.7471 | 1057 |
| 275 | 250 | 0.00078485 | 1274.1 | subcooled | 1206 | 2.9685 | 1174.2 |
| 300 | 250 | 0.00074302 | 1345.9 | subcooled | 1331.3 | 3.1919 | 1297.6 |
| 0 | 300 | 0.0010145 | 985.67 | subcooled | 29.858 | 0.00026879 | 0.28791 |
| 25 | 300 | 0.0010101 | 989.98 | subcooled | 132.28 | 0.35905 | 102.58 |
| 50 | 300 | 0.0010007 | 999.33 | subcooled | 235.05 | 0.69005 | 205.07 |
| 75 | 300 | 0.00098767 | 1012.5 | subcooled | 338.19 | 0.99746 | 307.81 |
| 100 | 300 | 0.00097182 | 1029 | subcooled | 441.74 | 1.2847 | 410.87 |
| 125 | 300 | 0.0009535 | 1048.8 | subcooled | 545.88 | 1.5548 | 514.42 |
| 150 | 300 | 0.00093286 | 1072 | subcooled | 650.89 | 1.8106 | 618.73 |
| 175 | 300 | 0.00090994 | 1099 | subcooled | 757.11 | 2.0545 | 724.15 |
| 200 | 300 | 0.00088462 | 1130.4 | subcooled | 865.02 | 2.2888 | 831.1 |
| 225 | 300 | 0.00085664 | 1167.4 | subcooled | 975.17 | 2.5156 | 940.15 |
| 250 | 300 | 0.00082556 | 1211.3 | subcooled | 1088.4 | 2.7373 | 1052 |
| 275 | 300 | 0.00079064 | 1264.8 | subcooled | 1205.7 | 2.9563 | 1167.7 |
| 300 | 300 | 0.00075066 | 1332.2 | subcooled | 1328.9 | 3.176 | 1288.9 |

(Multiply MPa by 10 to obtain bars.)

Values in this table were calculated from *NIST Standard Reference Database 10*, "NIST/ASME Steam Properties," Ver. 2.11, National Institute of Standards and Technology, U.S. Department of Commerce, Gaithersburg, MD, 1997, which has been licensed to PPI.

## APPENDIX 22.R
### Enthalpy-Entropy (Mollier) Diagram for Steam
### (SI units)

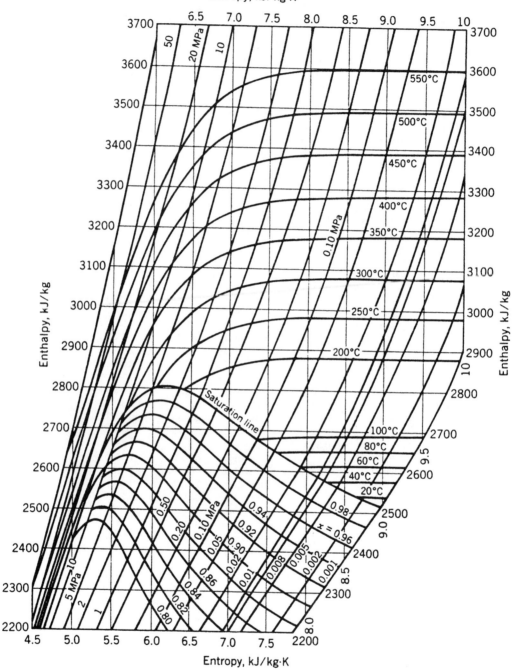

(Multiply MPa by 10 to obtain bars.)

## APPENDIX 22.S
Properties of Low-Pressure Air
(SI units)

$T$ in $K$; $h$ and $u$ in kJ/kg; $\phi$ in kJ/kg·K

| $T$ | $h$ | $p_r$ | $u$ | $v_r$ | $\phi$ |
|-----|-----|-------|-----|-------|--------|
| 200 | 199.97 | 0.3363 | 142.56 | 1707 | 1.29559 |
| 210 | 209.97 | 0.3987 | 149.69 | 1512 | 1.34444 |
| 220 | 219.97 | 0.4690 | 156.82 | 1346 | 1.39105 |
| 230 | 230.02 | 0.5477 | 164.00 | 1205 | 1.43557 |
| 240 | 240.02 | 0.6355 | 171.13 | 1084 | 1.47824 |
| 250 | 250.05 | 0.7329 | 178.28 | 979 | 1.51917 |
| 260 | 260.09 | 0.8405 | 185.45 | 887.8 | 1.55848 |
| 270 | 270.11 | 0.9590 | 192.60 | 808.0 | 1.59634 |
| 280 | 280.13 | 1.0889 | 199.75 | 738.0 | 1.63279 |
| 285 | 285.14 | 1.1584 | 203.33 | 706.1 | 1.65055 |
| 290 | 290.16 | 1.2311 | 206.91 | 676.1 | 1.66802 |
| 295 | 295.17 | 1.3068 | 210.49 | 647.9 | 1.68515 |
| 300 | 300.19 | 1.3860 | 214.07 | 621.2 | 1.70203 |
| 305 | 305.22 | 1.4686 | 217.67 | 596.0 | 1.71865 |
| 310 | 310.24 | 1.5546 | 221.25 | 572.3 | 1.73498 |
| 315 | 315.27 | 1.6442 | 224.85 | 549.8 | 1.75106 |
| 320 | 320.29 | 1.7375 | 228.42 | 528.6 | 1.76690 |
| 325 | 325.31 | 1.8345 | 232.02 | 508.4 | 1.78249 |
| 330 | 330.34 | 1.9352 | 235.61 | 489.4 | 1.79783 |
| 340 | 340.42 | 2.149 | 242.82 | 454.1 | 1.82790 |
| 350 | 350.49 | 2.379 | 250.02 | 422.2 | 1.85708 |
| 360 | 360.58 | 2.626 | 257.24 | 393.4 | 1.88543 |
| 370 | 370.67 | 2.892 | 264.46 | 367.2 | 1.91313 |
| 380 | 380.77 | 3.176 | 271.69 | 343.4 | 1.94001 |
| 390 | 390.88 | 3.481 | 278.93 | 321.5 | 1.96633 |
| 400 | 400.98 | 3.806 | 286.16 | 301.6 | 1.99194 |
| 410 | 411.12 | 4.153 | 293.43 | 283.3 | 2.01699 |
| 420 | 421.26 | 4.522 | 300.69 | 266.6 | 2.04142 |
| 430 | 431.43 | 4.915 | 307.99 | 251.1 | 2.06533 |
| 440 | 441.61 | 5.332 | 315.30 | 236.8 | 2.08870 |
| 450 | 451.80 | 5.775 | 322.62 | 223.6 | 2.11161 |
| 460 | 462.02 | 6.245 | 329.97 | 211.4 | 2.13407 |
| 470 | 472.24 | 6.742 | 337.32 | 200.1 | 2.15604 |
| 480 | 482.49 | 7.268 | 344.70 | 189.5 | 2.17760 |
| 490 | 492.74 | 7.824 | 352.08 | 179.7 | 2.19876 |
| 500 | 503.02 | 8.411 | 359.49 | 170.6 | 2.21952 |
| 510 | 513.32 | 9.031 | 366.92 | 162.1 | 2.23993 |
| 520 | 523.63 | 9.684 | 374.36 | 154.1 | 2.25997 |
| 530 | 533.98 | 10.37 | 381.84 | 146.7 | 2.27967 |
| 540 | 544.35 | 11.10 | 389.34 | 139.7 | 2.29906 |
| 550 | 554.74 | 11.86 | 396.86 | 133.1 | 2.31809 |
| 560 | 565.17 | 12.66 | 404.42 | 127.0 | 2.33685 |
| 570 | 575.59 | 13.50 | 411.97 | 121.2 | 2.35531 |
| 580 | 586.04 | 14.38 | 419.55 | 115.7 | 2.37348 |
| 590 | 596.52 | 15.31 | 427.15 | 110.6 | 2.39140 |

*(continued)*

**APPENDIX 22.S** *(continued)*
Properties of Low-Pressure Air
(SI units)

*T* in *K*; *h* and *u* in kJ/kg; $\phi$ in kJ/kg·K

| T | h | $p_r$ | u | $v_r$ | $\phi$ |
|---|---|---|---|---|---|
| 600 | 607.02 | 16.28 | 434.78 | 105.8 | 2.40902 |
| 610 | 617.53 | 17.30 | 442.42 | 101.2 | 2.42644 |
| 620 | 628.07 | 18.36 | 450.09 | 96.92 | 2.44356 |
| 630 | 638.63 | 19.84 | 457.78 | 92.84 | 2.46048 |
| 640 | 649.22 | 20.64 | 465.50 | 88.99 | 2.47716 |
| 650 | 659.84 | 21.86 | 473.25 | 85.34 | 2.49364 |
| 660 | 670.47 | 23.13 | 481.01 | 81.89 | 2.50985 |
| 670 | 681.14 | 24.46 | 488.81 | 78.61 | 2.52589 |
| 680 | 691.82 | 25.85 | 496.62 | 75.50 | 2.54175 |
| 690 | 702.52 | 27.29 | 504.45 | 72.56 | 2.55731 |
| 700 | 713.27 | 28.80 | 512.33 | 69.76 | 2.57277 |
| 710 | 724.04 | 30.38 | 520.23 | 67.07 | 2.58810 |
| 720 | 734.82 | 32.02 | 528.14 | 64.53 | 2.60319 |
| 730 | 745.62 | 33.72 | 536.07 | 62.13 | 2.61803 |
| 740 | 756.44 | 35.50 | 544.02 | 59.82 | 2.63280 |
| 750 | 767.29 | 37.35 | 551.99 | 57.63 | 2.64737 |
| 760 | 778.18 | 39.27 | 560.01 | 55.54 | 2.66176 |
| 770 | 789.11 | 41.31 | 568.07 | 53.39 | 2.67595 |
| 780 | 800.03 | 43.35 | 576.12 | 51.64 | 2.69013 |
| 790 | 810.99 | 45.55 | 584.21 | 49.86 | 2.70400 |
| 800 | 821.95 | 47.75 | 592.30 | 48.08 | 2.71787 |
| 820 | 843.98 | 52.59 | 608.59 | 44.84 | 2.74504 |
| 840 | 866.08 | 57.60 | 624.95 | 41.85 | 2.77170 |
| 860 | 888.27 | 63.09 | 641.40 | 39.12 | 2.79783 |
| 880 | 910.56 | 68.98 | 657.95 | 36.61 | 2.82344 |
| 900 | 932.93 | 75.29 | 674.58 | 34.31 | 2.84856 |
| 920 | 955.38 | 82.05 | 691.28 | 32.18 | 2.87324 |
| 940 | 977.92 | 89.28 | 708.08 | 30.22 | 2.89748 |
| 960 | 1000.55 | 97.00 | 725.02 | 28.40 | 2.92128 |
| 980 | 1023.25 | 105.2 | 741.98 | 26.73 | 2.94468 |
| 1000 | 1046.04 | 114.0 | 758.94 | 25.17 | 2.96770 |
| 1020 | 1068.89 | 123.4 | 776.10 | 23.72 | 2.99034 |
| 1040 | 1091.85 | 133.3 | 793.36 | 22.39 | 3.01260 |
| 1060 | 1114.86 | 143.9 | 810.62 | 21.14 | 3.03449 |
| 1080 | 1137.89 | 155.2 | 827.88 | 19.98 | 3.05608 |
| 1100 | 1161.07 | 167.1 | 845.33 | 18.896 | 3.07732 |
| 1120 | 1184.28 | 179.7 | 862.79 | 17.886 | 3.09825 |
| 1140 | 1207.57 | 193.1 | 880.35 | 16.946 | 3.11883 |
| 1160 | 1230.92 | 207.2 | 897.91 | 16.064 | 3.13916 |
| 1180 | 1254.34 | 222.2 | 915.57 | 15.241 | 3.15916 |
| 1200 | 1277.79 | 238.0 | 933.33 | 14.470 | 3.17888 |
| 1220 | 1301.31 | 254.7 | 951.09 | 13.747 | 3.19834 |
| 1240 | 1324.93 | 272.3 | 968.95 | 13.069 | 3.21751 |
| 1260 | 1348.55 | 290.8 | 986.90 | 12.435 | 3.23638 |
| 1280 | 1372.24 | 310.4 | 1004.76 | 11.835 | 3.25510 |

*(continued)*

**APPENDIX 22.S** *(continued)*
Properties of Low-Pressure Air
(SI units)

$T$ in $K$; $h$ and $u$ in kJ/kg; $\phi$ in kJ/kg·K

| $T$ | $h$ | $p_r$ | $u$ | $v_r$ | $\phi$ |
|------|---------|-------|---------|--------|---------|
| 1300 | 1395.97 | 330.9 | 1022.82 | 11.275 | 3.27345 |
| 1320 | 1419.76 | 352.5 | 1040.88 | 10.747 | 3.29160 |
| 1340 | 1443.60 | 375.3 | 1058.94 | 10.247 | 3.30959 |
| 1360 | 1467.49 | 399.1 | 1077.10 | 9.780  | 3.32724 |
| 1380 | 1491.44 | 424.2 | 1095.26 | 9.337  | 3.34474 |
| 1400 | 1515.42 | 450.5 | 1113.52 | 8.919  | 3.36200 |
| 1420 | 1539.44 | 478.0 | 1131.77 | 8.526  | 3.37901 |
| 1440 | 1563.51 | 506.9 | 1150.13 | 8.153  | 3.39586 |
| 1460 | 1587.63 | 537.1 | 1168.49 | 7.801  | 3.41247 |
| 1480 | 1611.79 | 568.8 | 1186.95 | 7.468  | 3.42892 |
| 1500 | 1635.97 | 601.9 | 1205.41 | 7.152  | 3.44516 |
| 1520 | 1660.23 | 636.5 | 1223.87 | 6.854  | 3.46120 |
| 1540 | 1684.51 | 672.8 | 1242.43 | 6.569  | 3.47712 |
| 1560 | 1708.82 | 710.5 | 1260.99 | 6.301  | 3.49276 |
| 1580 | 1733.17 | 750.0 | 1279.65 | 6.046  | 3.50829 |
| 1600 | 1757.57 | 791.2 | 1298.30 | 5.804  | 3.52364 |
| 1620 | 1782.00 | 834.1 | 1316.96 | 5.574  | 3.53879 |
| 1640 | 1806.46 | 878.9 | 1335.72 | 5.355  | 3.55381 |
| 1660 | 1830.96 | 925.6 | 1354.48 | 5.147  | 3.56867 |
| 1680 | 1855.50 | 974.2 | 1373.24 | 4.949  | 3.58335 |
| 1700 | 1880.1  | 1025  | 1392.7  | 4.761  | 3.5979  |
| 1750 | 1941.6  | 1161  | 1439.8  | 4.328  | 3.6336  |
| 1800 | 2003.3  | 1310  | 1487.2  | 3.944  | 3.6684  |
| 1850 | 2065.3  | 1475  | 1534.9  | 3.601  | 3.7023  |
| 1900 | 2127.4  | 1655  | 1582.6  | 3.295  | 3.7354  |
| 1950 | 2189.7  | 1852  | 1630.6  | 3.022  | 3.7677  |
| 2000 | 2252.1  | 2068  | 1678.7  | 2.776  | 3.7994  |
| 2050 | 2314.6  | 2303  | 1726.8  | 2.555  | 3.8303  |
| 2100 | 2377.4  | 2559  | 1775.3  | 2.356  | 3.8605  |
| 2150 | 2440.3  | 2837  | 1823.8  | 2.175  | 3.8901  |
| 2200 | 2503.2  | 3138  | 1872.4  | 2.012  | 3.9191  |
| 2250 | 2566.4  | 3464  | 1921.3  | 1.864  | 3.9474  |

*Gas Tables: International Version—Thermodynamic Properties of Air, Products of Combustion and Component Gases, Compressible Flow Functions,* Second Edition, Joseph H. Keenan, Jing Chao, and Joseph Kaye, copyright © 1983. Reproduced with permission of John Wiley & Sons, Inc.

## APPENDIX 22.T
Properties of Saturated Refrigerant-12 (R-12) by Temperature
(SI units)

| temp. (°C) | pressure (MPa) | volume vapor (m³/kg) | density liquid (kg/m³) | enthalpy liquid (kJ/kg) | enthalpy vapor (kJ/kg) | entropy liquid (kJ/kg·K) | entropy vapor (kJ/kg·K) |
|---|---|---|---|---|---|---|---|
| −100 | 0.001174 | 10.122 | 1678.0 | 112.69 | 306.46 | 0.60441 | 1.7235 |
| −95 | 0.001851 | 6.6005 | 1665.0 | 116.92 | 308.67 | 0.62845 | 1.7048 |
| −90 | 0.002836 | 4.4264 | 1651.9 | 121.14 | 310.90 | 0.65182 | 1.6879 |
| −85 | 0.004230 | 3.0449 | 1638.7 | 125.36 | 313.16 | 0.67457 | 1.6727 |
| −80 | 0.006160 | 2.1438 | 1625.5 | 129.59 | 315.44 | 0.69673 | 1.6589 |
| −75 | 0.008774 | 1.5416 | 1612.1 | 133.81 | 317.74 | 0.71835 | 1.6465 |
| −70 | 0.012246 | 1.1301 | 1598.7 | 138.06 | 320.05 | 0.73948 | 1.6353 |
| −65 | 0.016776 | 0.84332 | 1585.2 | 142.32 | 322.38 | 0.76015 | 1.6252 |
| −60 | 0.022591 | 0.63956 | 1571.5 | 146.58 | 324.71 | 0.78040 | 1.6161 |
| −55 | 0.029944 | 0.49230 | 1557.8 | 150.87 | 327.05 | 0.80025 | 1.6079 |
| −50 | 0.039115 | 0.38415 | 1543.9 | 155.18 | 329.40 | 0.81974 | 1.6005 |
| −45 | 0.050408 | 0.30355 | 1529.9 | 159.51 | 331.74 | 0.83890 | 1.5938 |
| −40 | 0.064152 | 0.24264 | 1515.7 | 163.86 | 334.09 | 0.85775 | 1.5879 |
| −35 | 0.080701 | 0.19603 | 1501.4 | 168.25 | 336.43 | 0.87632 | 1.5825 |
| −30 | 0.10043 | 0.15993 | 1486.9 | 172.67 | 338.76 | 0.89462 | 1.5777 |
| −29.79 | 0.101325 | 0.15861 | 1486.3 | 172.85 | 338.86 | 0.89538 | 1.5775 |
| −25 | 0.12373 | 0.13166 | 1472.3 | 177.12 | 341.08 | 0.91269 | 1.5734 |
| −20 | 0.15101 | 0.10929 | 1457.4 | 181.61 | 343.39 | 0.93053 | 1.5696 |
| −15 | 0.18272 | 0.09142 | 1442.4 | 186.14 | 345.69 | 0.94817 | 1.5662 |
| −10 | 0.21928 | 0.07702 | 1427.1 | 190.72 | 347.96 | 0.96561 | 1.5632 |
| −5 | 0.26117 | 0.06531 | 1411.5 | 195.33 | 350.22 | 0.98289 | 1.5605 |
| 0 | 0.30885 | 0.05571 | 1395.6 | 200.00 | 352.44 | 1.0000 | 1.5581 |
| 2 | 0.32966 | 0.05236 | 1389.2 | 201.88 | 353.32 | 1.0068 | 1.5572 |
| 4 | 0.35150 | 0.04925 | 1382.7 | 203.77 | 354.20 | 1.0136 | 1.5564 |
| 6 | 0.37441 | 0.04637 | 1376.2 | 205.66 | 355.07 | 1.0203 | 1.5556 |
| 8 | 0.39842 | 0.04369 | 1369.6 | 207.57 | 355.93 | 1.0271 | 1.5548 |
| 10 | 0.42356 | 0.04119 | 1363.0 | 209.48 | 356.79 | 1.0338 | 1.5541 |
| 12 | 0.44986 | 0.03887 | 1356.2 | 211.40 | 357.65 | 1.0405 | 1.5533 |
| 14 | 0.47737 | 0.03670 | 1349.5 | 213.33 | 358.49 | 1.0472 | 1.5527 |
| 16 | 0.50610 | 0.03468 | 1342.6 | 215.27 | 359.33 | 1.0538 | 1.5520 |
| 18 | 0.53610 | 0.03279 | 1335.7 | 217.22 | 360.16 | 1.0605 | 1.5514 |
| 20 | 0.56740 | 0.03102 | 1328.7 | 219.18 | 360.98 | 1.0671 | 1.5508 |
| 22 | 0.60003 | 0.02937 | 1321.6 | 221.14 | 361.80 | 1.0737 | 1.5502 |
| 24 | 0.63403 | 0.02782 | 1314.5 | 223.12 | 362.60 | 1.0803 | 1.5497 |
| 26 | 0.66943 | 0.02637 | 1307.3 | 225.11 | 363.40 | 1.0868 | 1.5491 |
| 28 | 0.70626 | 0.02500 | 1299.9 | 227.10 | 364.19 | 1.0934 | 1.5486 |
| 30 | 0.74457 | 0.02372 | 1292.5 | 229.11 | 364.96 | 1.0999 | 1.5481 |
| 32 | 0.78439 | 0.02252 | 1285.0 | 231.12 | 365.73 | 1.1064 | 1.5476 |
| 34 | 0.82574 | 0.02138 | 1277.4 | 233.15 | 366.48 | 1.1130 | 1.5471 |
| 36 | 0.86868 | 0.02032 | 1269.7 | 235.18 | 367.22 | 1.1195 | 1.5466 |
| 38 | 0.91324 | 0.01931 | 1261.9 | 237.23 | 367.95 | 1.1259 | 1.5461 |

*(continued)*

**APPENDIX 22.T** *(continued)*
Properties of Saturated Refrigerant-12 (R-12) by Temperature
(SI units)

| temp. (°C) | pressure (MPa) | volume vapor (m³/kg) | density liquid (kg/m³) | enthalpy liquid (kJ/kg) | vapor (kJ/kg) | entropy liquid (kJ/kg·K) | vapor (kJ/kg·K) |
|---|---|---|---|---|---|---|---|
| 40 | 0.95944 | 0.01836 | 1253.9 | 239.29 | 368.67 | 1.1324 | 1.5456 |
| 42 | 1.0073 | 0.01746 | 1245.9 | 241.36 | 369.37 | 1.1389 | 1.5451 |
| 44 | 1.0570 | 0.01662 | 1237.7 | 243.44 | 370.06 | 1.1453 | 1.5446 |
| 46 | 1.1084 | 0.01581 | 1299.3 | 245.54 | 370.73 | 1.1518 | 1.5441 |
| 48 | 0.1616 | 0.01506 | 1220.9 | 247.64 | 371.38 | 1.1582 | 1.5435 |
| 50 | 1.2167 | 0.01434 | 1212.2 | 249.76 | 372.02 | 1.1647 | 1.5430 |
| 52 | 1.2736 | 0.01366 | 1203.5 | 251.90 | 372.64 | 1.1711 | 1.5425 |
| 54 | 1.3325 | 0.01301 | 1194.5 | 254.04 | 373.24 | 1.1776 | 1.5419 |
| 56 | 1.3934 | 0.01239 | 1185.4 | 256.21 | 373.82 | 1.1840 | 1.5413 |
| 58 | 1.4562 | 0.01181 | 1176.1 | 258.38 | 374.38 | 1.1904 | 1.5407 |
| 60 | 1.5212 | 0.01126 | 1166.6 | 260.58 | 374.91 | 1.1969 | 1.5401 |
| 62 | 1.5883 | 0.01073 | 1156.9 | 262.79 | 375.42 | 1.2033 | 1.5394 |
| 64 | 1.6575 | 0.01023 | 1146.9 | 265.02 | 375.90 | 1.2098 | 1.5387 |
| 66 | 1.7289 | 0.009746 | 1136.7 | 267.27 | 376.36 | 1.2162 | 1.5379 |
| 68 | 1.8026 | 0.009289 | 1126.3 | 269.54 | 376.78 | 1.2227 | 1.5371 |
| 70 | 1.8786 | 0.008852 | 1115.6 | 271.83 | 377.17 | 1.2292 | 1.5362 |
| 72 | 1.9570 | 0.008434 | 1104.6 | 274.15 | 377.53 | 1.2357 | 1.5353 |
| 74 | 2.0378 | 0.008034 | 1093.3 | 276.49 | 377.85 | 1.2423 | 1.5343 |
| 76 | 2.1210 | 0.007651 | 1081.6 | 278.86 | 378.13 | 1.2489 | 1.5332 |
| 78 | 2.2069 | 0.007283 | 1069.6 | 281.27 | 378.36 | 1.2555 | 1.5320 |
| 80 | 2.2953 | 0.006931 | 1057.2 | 283.70 | 378.54 | 1.2622 | 1.5308 |
| 85 | 2.5282 | 0.006106 | 1024.1 | 289.97 | 378.75 | 1.2792 | 1.5271 |
| 90 | 2.7790 | 0.005350 | 987.60 | 296.56 | 378.52 | 1.2968 | 1.5225 |
| 95 | 3.0490 | 0.004648 | 946.44 | 303.58 | 377.67 | 1.3152 | 1.5165 |
| 100 | 3.3399 | 0.003980 | 898.55 | 311.26 | 375.88 | 1.3351 | 1.5083 |
| 105 | 3.6538 | 0.003311 | 839.10 | 320.08 | 372.41 | 1.3576 | 1.4960 |
| 110 | 3.9943 | 0.002517 | 746.58 | 331.91 | 364.02 | 1.3875 | 1.4713 |
| *111.80 | 4.125 | 0.00179 | 558 | 348.4 | 348.4 | 1.430 | 1.430 |

*critical point

Reprinted with permission from *1993 ASHRAE Handbook–Fundamentals, SI Edition*, American Society of Heating, Refrigerating, and Air-Conditioning Engineers, Inc. (ASHRAE), www.ashrae.org, Atlanta, GA, ©1993.

## APPENDIX 22.U
Pressure-Enthalpy Diagram for Refrigerant-12 (R-12)
(SI units)

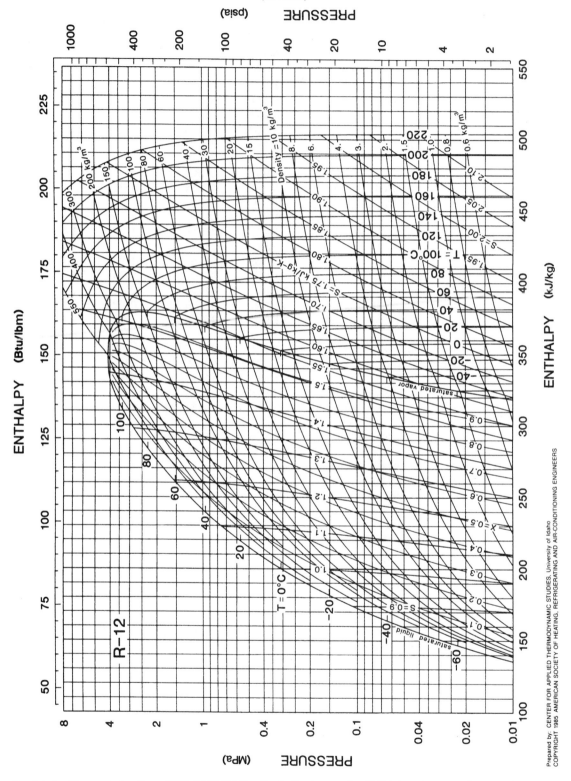

Reprinted with permission from *1993 ASHRAE Handbook–Fundamentals, SI Edition*, American Society of Heating, Refrigerating, and Air-Conditioning Engineers, Inc. (ASHRAE), www.ashrae.org, Atlanta, GA © 1993.

**APPENDIX 22.V**
Properties of Saturated
Refrigerant-22 (R-22) by Temperature
(SI units)

| temp. (°C) | pressure (MPa) | volume vapor (m³/kg) | density liquid (kg/m³) | enthalpy liquid (kJ/kg) | enthalpy vapor (kJ/kg) | entropy liquid (kJ/kg·K) | entropy vapor (kJ/kg·K) |
|---|---|---|---|---|---|---|---|
| −90 | 0.004748 | 3.6939 | 1542.8 | 98.575 | 364.20 | 0.55032 | 2.006 |
| −85 | 0.007084 | 2.5394 | 1529.9 | 104.54 | 366.63 | 0.58246 | 1.9754 |
| −80 | 0.010308 | 1.7883 | 1516.8 | 110.42 | 369.06 | 0.61326 | 1.9524 |
| −75 | 0.014662 | 1.2870 | 1503.6 | 116.21 | 371.49 | 0.64284 | 1.9312 |
| −70 | 0.020424 | 0.94477 | 1490.3 | 121.92 | 373.91 | 0.67132 | 1.9117 |
| −65 | 0.027914 | 0.70609 | 1476.7 | 127.58 | 376.32 | 0.69879 | 1.8938 |
| −60 | 0.037491 | 0.53641 | 1463.1 | 133.18 | 378.72 | 0.72535 | 1.8773 |
| −55 | 0.049556 | 0.41362 | 1449.2 | 138.74 | 381.10 | 0.75109 | 1.8621 |
| −50 | 0.064549 | 0.32330 | 1435.2 | 144.27 | 383.45 | 0.77610 | 1.8479 |
| −45 | 0.082947 | 0.25586 | 1421.0 | 149.77 | 385.77 | 0.80044 | 1.8348 |
| −40.82 | 0.101325 | 0.21223 | 1408.9 | 154.37 | 387.69 | 0.82034 | 1.8246 |
| −40 | 0.10527 | 0.20480 | 1406.5 | 155.26 | 388.06 | 0.82419 | 1.8227 |
| −38 | 0.11542 | 0.18790 | 1400.7 | 157.46 | 388.96 | 0.83354 | 1.8180 |
| −36 | 0.12632 | 0.17268 | 1394.8 | 159.66 | 389.86 | 0.84281 | 1.8135 |
| −34 | 0.13801 | 0.15894 | 1388.9 | 161.86 | 390.75 | 0.85200 | 1.8091 |
| −32 | 0.15053 | 0.14651 | 1382.9 | 164.06 | 391.64 | 0.86113 | 1.8049 |
| −30 | 0.16391 | 0.13524 | 1376.9 | 166.26 | 392.52 | 0.87018 | 1.8007 |
| −28 | 0.17821 | 0.12502 | 1370.9 | 168.46 | 393.39 | 0.87917 | 1.7967 |
| −26 | 0.19346 | 0.11573 | 1364.8 | 170.67 | 394.25 | 0.88810 | 1.7927 |
| −24 | 0.20969 | 0.10726 | 1358.7 | 172.89 | 395.10 | 0.89697 | 1.7889 |
| −22 | 0.22696 | 0.09954 | 1352.6 | 175.10 | 395.95 | 0.90579 | 1.7851 |
| −20 | 0.24531 | 0.09249 | 1346.4 | 177.33 | 396.79 | 0.91455 | 1.7815 |
| −18 | 0.26477 | 0.08603 | 1340.1 | 179.56 | 397.62 | 0.92327 | 1.7779 |
| −16 | 0.28540 | 0.08012 | 1333.8 | 181.79 | 398.43 | 0.93194 | 1.7744 |
| −14 | 0.30724 | 0.07470 | 1327.5 | 184.04 | 399.24 | 0.94057 | 1.7710 |
| −12 | 0.33034 | 0.06971 | 1321.1 | 186.29 | 400.04 | 0.94916 | 1.7677 |
| −10 | 0.35474 | 0.06513 | 1314.6 | 188.55 | 400.83 | 0.95771 | 1.7644 |
| −8 | 0.38049 | 0.06090 | 1308.1 | 190.82 | 401.61 | 0.96623 | 1.7612 |
| −6 | 0.40763 | 0.05701 | 1301.5 | 193.10 | 402.37 | 0.97471 | 1.7581 |
| −4 | 0.43622 | 0.05341 | 1294.9 | 195.39 | 403.12 | 0.98317 | 1.7550 |
| −2 | 0.46630 | 0.05008 | 1288.2 | 197.69 | 403.87 | 0.99160 | 1.7520 |
| 0 | 0.49792 | 0.04700 | 1281.5 | 200.00 | 404.59 | 1.0000 | 1.7490 |
| 2 | 0.53113 | 0.04415 | 1274.7 | 202.32 | 405.31 | 1.0084 | 1.7461 |
| 4 | 0.56599 | 0.04150 | 1267.8 | 204.66 | 406.01 | 1.0167 | 1.7432 |
| 6 | 0.60254 | 0.03904 | 1260.8 | 207.01 | 406.70 | 1.0251 | 1.7404 |
| 8 | 0.64083 | 0.03675 | 1253.8 | 209.37 | 407.37 | 1.0334 | 1.7376 |
| 10 | 0.68091 | 0.03462 | 1246.7 | 211.74 | 408.03 | 1.0417 | 1.7349 |
| 12 | 0.72285 | 0.03263 | 1239.5 | 214.13 | 408.67 | 1.0500 | 1.7322 |
| 14 | 0.76668 | 0.03078 | 1232.3 | 216.54 | 409.29 | 1.0583 | 1.7295 |
| 16 | 0.81246 | 0.02905 | 1224.9 | 218.96 | 409.90 | 1.0665 | 1.7269 |
| 18 | 0.86025 | 0.02743 | 1217.5 | 221.40 | 410.49 | 1.0748 | 1.7243 |

*(continued)*

**APPENDIX 22.V** *(continued)*
Properties of Saturated
Refrigerant-22 (R-22) by Temperature
(SI units)

| temp. (°C) | pressure (MPa) | volume vapor (m³/kg) | density liquid (kg/m³) | enthalpy liquid (kJ/kg) | enthalpy vapor (kJ/kg) | entropy liquid (kJ/kg·K) | entropy vapor (kJ/kg·K) |
|---|---|---|---|---|---|---|---|
| 20 | 0.91009 | 0.02592 | 1210.0 | 223.85 | 411.06 | 1.0831 | 1.7217 |
| 22 | 0.96205 | 0.02451 | 1202.4 | 226.32 | 411.61 | 1.0913 | 1.7191 |
| 24 | 1.0162 | 0.02318 | 1194.6 | 228.80 | 412.14 | 1.0996 | 1.7165 |
| 26 | 1.0725 | 0.02193 | 1186.8 | 231.31 | 412.65 | 1.1078 | 1.7140 |
| 28 | 1.1312 | 0.02076 | 1178.9 | 233.83 | 413.13 | 1.1160 | 1.7114 |
| 30 | 1.1921 | 0.01967 | 1170.8 | 236.38 | 413.60 | 1.1243 | 1.7089 |
| 32 | 1.2555 | 0.01863 | 1162.6 | 238.94 | 414.03 | 1.1325 | 1.7063 |
| 34 | 1.3213 | 0.01766 | 1154.3 | 241.52 | 414.45 | 1.1408 | 1.7038 |
| 36 | 1.3896 | 0.01674 | 1145.9 | 244.13 | 414.83 | 1.1490 | 1.7012 |
| 38 | 1.4605 | 0.01588 | 1137.3 | 246.75 | 415.19 | 1.1573 | 1.6987 |
| 40 | 1.5340 | 0.01506 | 1128.6 | 249.40 | 415.52 | 1.1656 | 1.6961 |
| 42 | 1.6102 | 0.01429 | 1119.7 | 252.07 | 415.82 | 1.1739 | 1.6934 |
| 44 | 1.6892 | 0.01356 | 1110.6 | 254.77 | 416.08 | 1.1822 | 1.6908 |
| 46 | 1.7710 | 0.01287 | 1101.4 | 257.49 | 416.31 | 1.1905 | 1.6881 |
| 48 | 1.8556 | 0.01221 | 1091.9 | 260.24 | 416.50 | 1.1989 | 1.6854 |
| 50 | 1.9432 | 0.01159 | 1082.3 | 263.02 | 416.65 | 1.2072 | 1.6826 |
| 52 | 2.0339 | 0.01101 | 1072.4 | 265.83 | 416.75 | 1.2156 | 1.6798 |
| 54 | 2.1276 | 0.01045 | 1062.3 | 268.67 | 416.81 | 1.2241 | 1.6769 |
| 56 | 2.2244 | 0.009915 | 1051.9 | 271.55 | 416.83 | 1.2326 | 1.6739 |
| 58 | 2.3245 | 0.009409 | 1041.3 | 274.46 | 416.79 | 1.2411 | 1.6709 |
| 60 | 2.4279 | 0.008927 | 1030.3 | 277.41 | 416.69 | 1.2497 | 1.6677 |
| 65 | 2.7015 | 0.007816 | 1001.3 | 284.98 | 416.16 | 1.2714 | 1.6594 |
| 70 | 2.9975 | 0.006819 | 969.68 | 292.88 | 416.14 | 1.2937 | 1.6500 |
| 75 | 3.3175 | 0.005917 | 934.38 | 301.22 | 413.46 | 1.3169 | 1.6393 |
| 80 | 3.6633 | 0.005086 | 893.89 | 310.18 | 410.88 | 1.3414 | 1.6265 |
| 85 | 4.0370 | 0.004301 | 845.17 | 320.13 | 406.90 | 1.3681 | 1.6104 |
| 90 | 4.4413 | 0.003517 | 780.60 | 331.96 | 400.28 | 1.3996 | 1.5877 |
| 95 | 4.8808 | 0.002547 | 660.94 | 350.67 | 384.95 | 1.4490 | 1.5421 |
| *96.15 | 4.988 | 0.00195 | 513 | 368.1 | 368.1 | 1.496 | 1.496 |

*critical point

Reprinted with permission from *1993 ASHRAE Handbook–Fundamentals, SI Edition*, American Society of Heating, Refrigerating, and Air-Conditioning Engineers, Inc. (ASHRAE), www.ashrae.org, Atlanta, GA, ©1993.

**APPENDIX 22.W**
Pressure-Enthalpy Diagram for Refrigerant-22 (R-22)
(SI units)

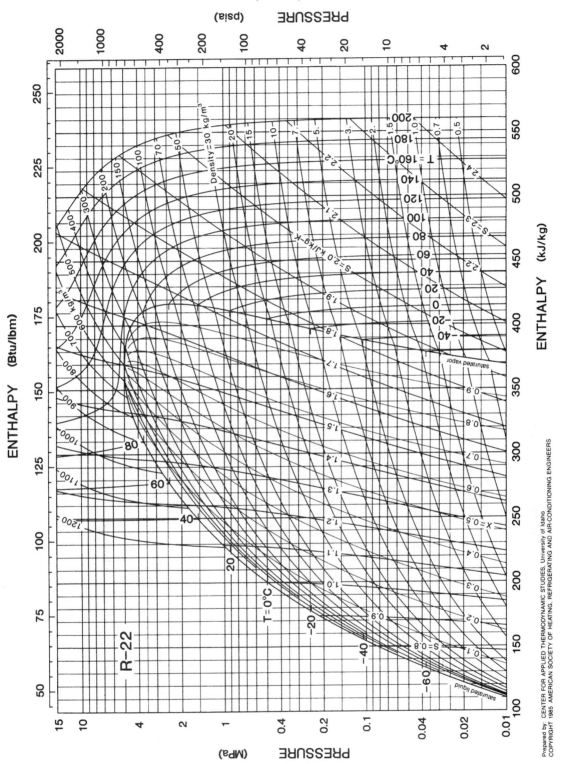

## APPENDIX 22.X
Pressure-Enthalpy Diagram for Refrigerant HFC-134a
(SI units)

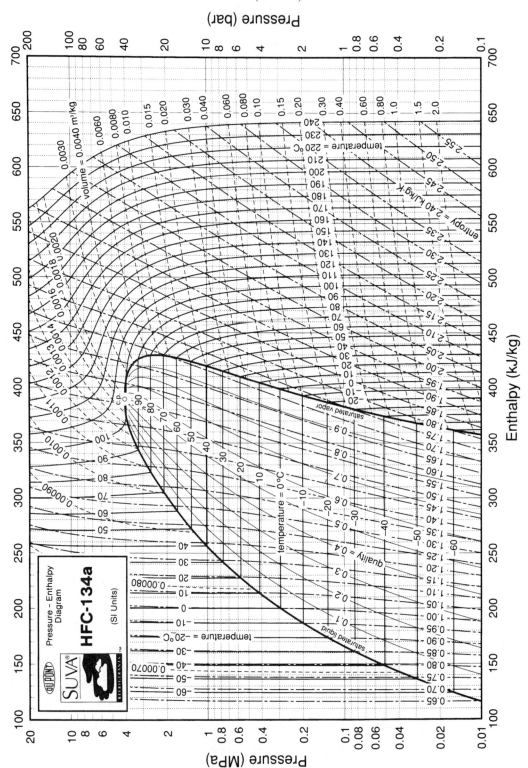

Reproduced with permission from the DuPont Company.

**APPENDIX 22.Y**
Physical Properties of Selected Solids
(customary U.S. units)

| material | $\rho$ (lbm/ft$^3$) (68°F) (20°C) | $c_p$ (Btu/lbm-°F) (68°F) (20°C) | $\alpha$ (ft$^2$/hr) (68°F) (20°C) | $k$ (Btu/hr-ft-°F) (68°F) (20°C) | $k$ (212°F) (100°C) | $k$ (572°F) (300°C) |
|---|---|---|---|---|---|---|
| *metals* | | | | | | |
| aluminum | 168.6 | 0.224 | 3.55 | 132 | 132 | 133 |
| copper | 555 | 0.092 | 3.98 | 223 | 219 | 213 |
| gold | 1206 | 0.031 | 4.52 | 169 | 170 | 172 |
| iron | 492 | 0.122 | 0.83 | 42.3 | 39.0 | 31.6 |
| lead | 708 | 0.030 | 0.80 | 20.3 | 19.3 | 17.2 |
| magnesium | 109 | 0.248 | 3.68 | 99.5 | 96.8 | 91.4 |
| nickel | 556 | 0.111 | 0.87 | 53.7 | 47.7 | 36.9 |
| platinum | 1340 | 0.032 | 0.09 | 40.5 | 41.9 | 43.5 |
| silver | 656 | 0.057 | 6.42 | 240 | 237 | 209 |
| tin | 450 | 0.051 | 1.57 | 36 | 36 | |
| tungsten | 1206 | 0.032 | 2.44 | 94 | 87 | 77 |
| uranium $\alpha$ | 1167 | 0.027 | 0.53 | 16.9 | 17.2 | 19.6 |
| zinc | 446 | 0.094 | 1.55 | 65 | 63 | 58 |
| | | | | | | |
| *alloys* | | | | | | |
| aluminum 2024 | 173 | 0.23 | 1.76 | 70.2 | | |
| brass (70% Cu, 30% Zn) | 532 | 0.091 | 1.27 | 61.8 | 73.9 | 85.3 |
| constantan (60% Cu, 40% Ni) | 557 | 0.098 | 0.24 | 13.1 | 15.4 | |
| iron, cast | 455 | 0.100 | 0.65 | 29.6 | 26.8 | |
| nichrome V | 530 | 0.106 | 0.12 | 7.06 | 7.99 | 9.94 |
| stainless steel | 488 | 0.110 | 0.17 | 9.4 | 10.0 | 13 |
| steel, mild (1% C) | 488 | 0.113 | 0.45 | 24.8 | 24.8 | 22.9 |
| | | | | | | |
| *nonmetals* | | | | | | |
| asbestos | 36 | 0.25 | | 0.092 | 0.11 | 0.125 |
| brick (fire clay) | 144 | 0.22 | | | 0.65 | |
| brick (masonry) | 106 | 0.20 | | 0.38 | | |
| brick (chrome) | 188 | 0.20 | | | 0.67 | |
| concrete | 150 | 0.21 | | 0.70 | | |
| corkboard | 10 | 0.4 | | 0.025 | | |
| diatomaceous earth, powdered | 14 | 0.2 | | 0.03 | | |
| glass, window | 170 | 0.2 | | 0.45 | | |
| glass, Pyrex™ | 140 | 0.2 | | 0.63 | 0.67 | 0.84 |
| Kaolin firebrick | 19 | | | | | 0.052 |
| 85% magnesia | 17 | | | 0.038 | 0.041 | |
| sandy loam, 4% H$_2$O | 104 | ~0.4 | | 0.54 | | |
| sandy loam, 10% H$_2$O | 121 | | | 1.08 | | |
| rock wool | ~10 | 0.2 | | 0.023 | 0.033 | |
| wood, oak, $\perp$ to grain | 51 | 0.57 | | 0.12 | | |
| wood, oak, —— to grain | 51 | 0.57 | | 0.23 | | |

(Multiply lbm/ft$^3$ by 16.018 to obtain kg/m$^3$.)
(Multiply Btu/lbm-°F by 4186.8 to obtain J/kg·K.)
(Multiply ft$^2$/hr by 0.0929 to obtain m$^2$/h.)
(Multiply Btu/hr-ft-°F by 1.7307 to obtain W/m·K.)

*Engineering Heat Transfer*, James R. Welty, copyright © 1974. Reproduced with permission of John Wiley & Sons, Inc.

### APPENDIX 22.Z
Generalized Compressibility Charts

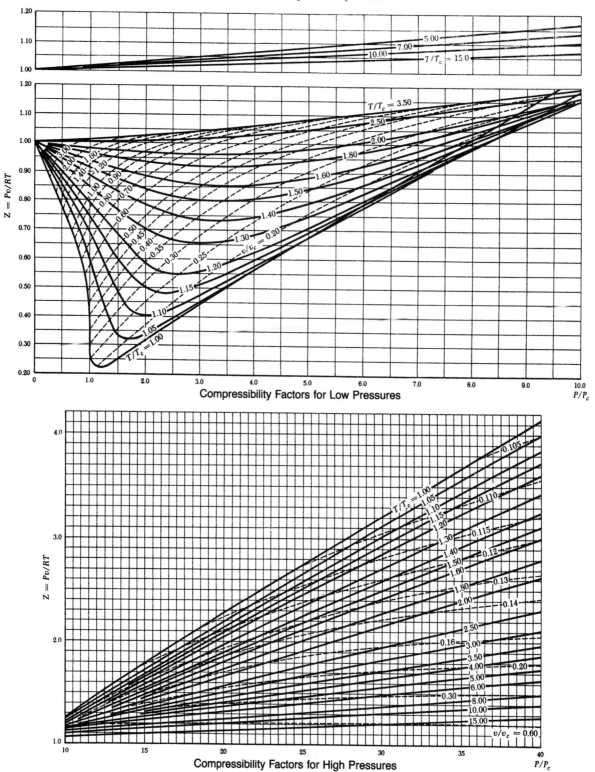

Reproduced with permission from "Generalized PVT Properties of Gases," *ASME Transactions*, Vol. 76, pp. 1057-1066, Edward F. Obert and L.C. Nelson, American Society of Mechanical Engineers, 1954.

# APPENDIX 24.A
ASHRAE Psychrometric Chart No. 1, Normal Temperature—
Sea Level (32–120°F) (customary U.S. units)

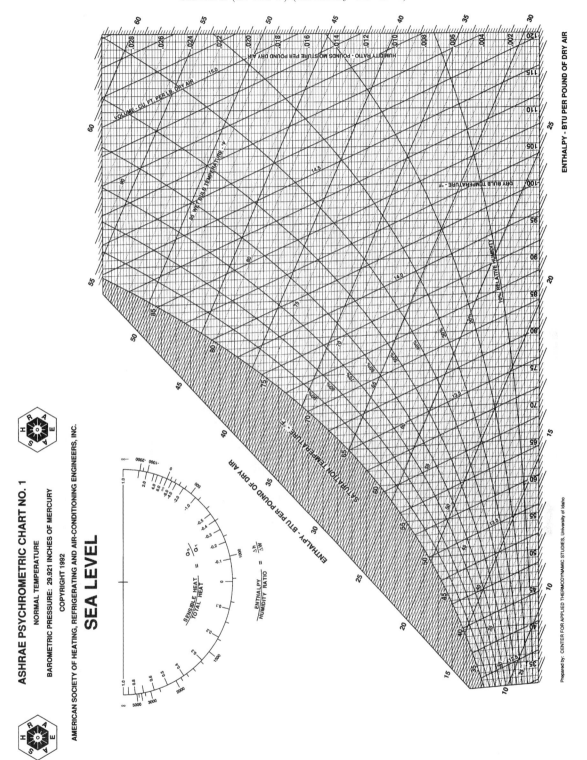

## APPENDIX 24.B
ASHRAE Psychrometric Chart No. 1, Normal Temperature—
Sea Level (0–50°C) (SI units)

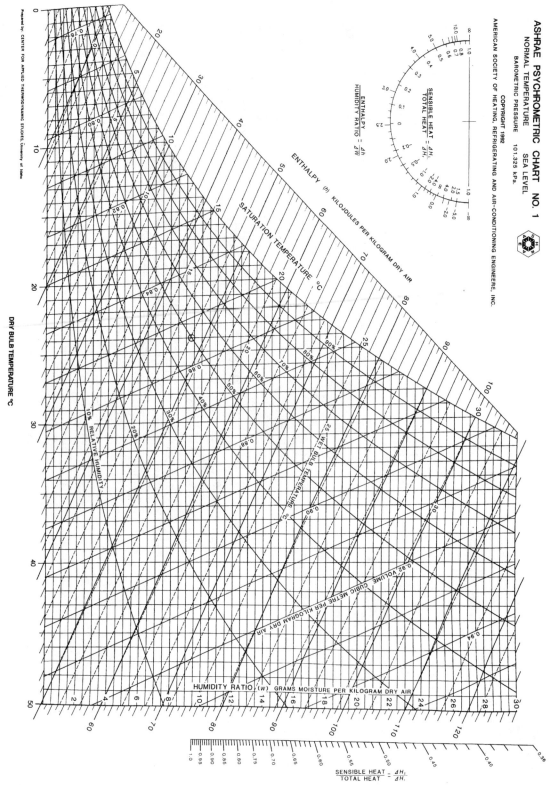

## APPENDIX 24.C
ASHRAE Psychrometric Chart No. 2, Low Temperature—
Sea Level (−40–50°F) (customary U.S. units)

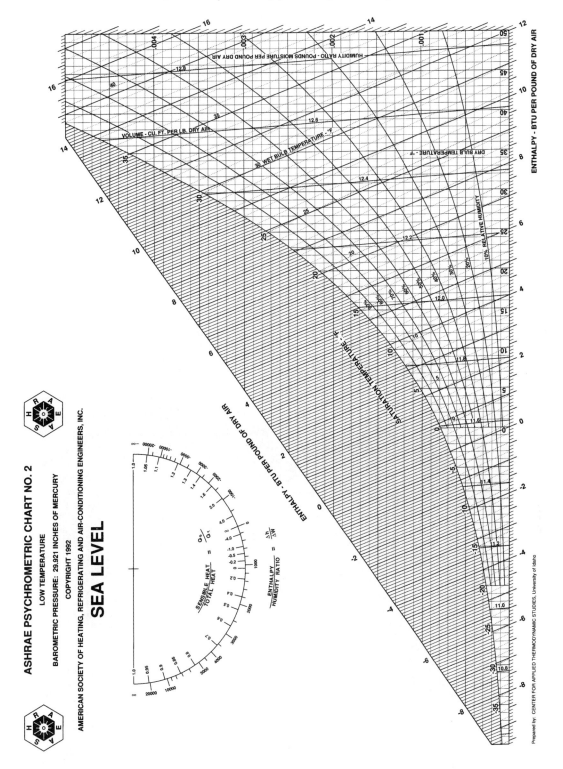

Copyright © 1992 by the American Society of Heating, Refrigerating and Air-Conditioning Engineers, Inc. Used by permission.

### APPENDIX 24.D
ASHRAE Psychrometric Chart No. 3, High Temperature—
Sea Level (60–250°F) (customary U.S. units)

**ASHRAE PSYCHROMETRIC CHART NO. 3**

HIGH TEMPERATURE

BAROMETRIC PRESSURE: 29.921 INCHES OF MERCURY

COPYRIGHT 1992

AMERICAN SOCIETY OF HEATING, REFRIGERATING AND AIR-CONDITIONING ENGINEERS, INC.

## SEA LEVEL

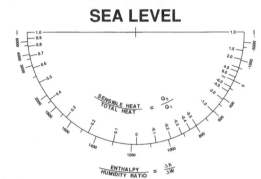

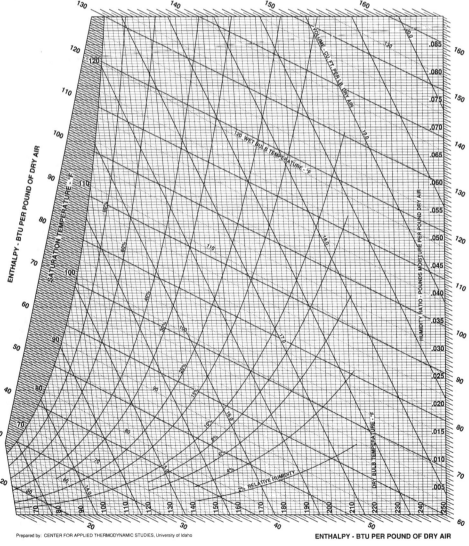

Prepared by: CENTER FOR APPLIED THERMODYNAMIC STUDIES, University of Idaho

**ENTHALPY - BTU PER POUND OF DRY AIR**

## APPENDIX 25.A
Isentropic Flow Factors
$(k = 1.4)$

M = Mach number, $p$ = pressure, $p_0$ = total pressure, $\rho$ = density, $\rho_0$ = total density, $T$ = temperature, $T_0$ = total temperature, $A$ = area at a point, $A^*$ = throat area for M = 1, v = velocity, and $a^*$ = speed of sound at throat.

| M | $p/p_0$ | $\rho/\rho_0$ | $T/T_0$ | $A/A^*$ | $v/a^*$ | M | $p/p_0$ | $\rho/\rho_0$ | $T/T_0$ | $A/A^*$ | $v/a^*$ |
|---|---|---|---|---|---|---|---|---|---|---|---|
| 0.00 | 1.0000 | 1.0000 | 1.0000 | — | 0.0000 | 0.51 | 0.8374 | 0.8809 | 0.9506 | 1.3212 | 0.5447 |
| 0.01 | 0.9999 | 1.0000 | 1.0000 | 57.8737 | 0.0110 | 0.52 | 0.8317 | 0.8766 | 0.9487 | 1.3034 | 0.5548 |
| 0.02 | 0.9997 | 0.9998 | 0.9999 | 28.9420 | 0.0219 | 0.53 | 0.8259 | 0.8723 | 0.9468 | 1.2865 | 0.5649 |
| 0.03 | 0.9994 | 0.9996 | 0.9998 | 19.3005 | 0.0329 | 0.54 | 0.8201 | 0.8679 | 0.9449 | 1.2703 | 0.5750 |
| 0.04 | 0.9989 | 0.9992 | 0.9997 | 14.4815 | 0.0438 | 0.55 | 0.8142 | 0.8634 | 0.9430 | 1.2549 | 0.5851 |
| 0.05 | 0.9983 | 0.9988 | 0.9995 | 11.5914 | 0.0548 | 0.56 | 0.8082 | 0.8589 | 0.9410 | 1.2403 | 0.5951 |
| 0.06 | 0.9975 | 0.9982 | 0.9993 | 9.6659 | 0.0657 | 0.57 | 0.8022 | 0.8544 | 0.9390 | 1.2263 | 0.6051 |
| 0.07 | 0.9966 | 0.9976 | 0.9990 | 8.2915 | 0.0766 | 0.58 | 0.7962 | 0.8498 | 0.9370 | 1.2130 | 0.6150 |
| 0.08 | 0.9955 | 0.9968 | 0.9987 | 7.2616 | 0.0876 | 0.59 | 0.7901 | 0.8451 | 0.9349 | 1.2003 | 0.6249 |
| 0.09 | 0.9944 | 0.9960 | 0.9984 | 6.4613 | 0.0985 | 0.60 | 0.7840 | 0.8405 | 0.9328 | 1.1882 | 0.6348 |
| 0.10 | 0.9930 | 0.9950 | 0.9980 | 5.8218 | 0.1094 | 0.61 | 0.7778 | 0.8357 | 0.9307 | 1.1767 | 0.6447 |
| 0.11 | 0.9916 | 0.9940 | 0.9976 | 5.2992 | 0.1204 | 0.62 | 0.7716 | 0.8310 | 0.9286 | 1.1656 | 0.6545 |
| 0.12 | 0.9900 | 0.9928 | 0.9971 | 4.8643 | 0.1313 | 0.63 | 0.7654 | 0.8262 | 0.9265 | 1.1551 | 0.6643 |
| 0.13 | 0.9883 | 0.9916 | 0.9966 | 4.4969 | 0.1422 | 0.64 | 0.7591 | 0.8213 | 0.9243 | 1.1451 | 0.6740 |
| 0.14 | 0.9864 | 0.9903 | 0.9961 | 4.1824 | 0.1531 | 0.65 | 0.7528 | 0.8164 | 0.9221 | 1.1356 | 0.6837 |
| 0.15 | 0.9844 | 0.9888 | 0.9955 | 3.9103 | 0.1639 | 0.66 | 0.7465 | 0.8115 | 0.9199 | 1.1265 | 0.6934 |
| 0.16 | 0.9823 | 0.9873 | 0.9949 | 3.6727 | 0.1748 | 0.67 | 0.7401 | 0.8066 | 0.9176 | 1.1179 | 0.7031 |
| 0.17 | 0.9800 | 0.9857 | 0.9943 | 3.4635 | 0.1857 | 0.68 | 0.7338 | 0.8016 | 0.9153 | 1.1097 | 0.7127 |
| 0.18 | 0.9776 | 0.9840 | 0.9936 | 3.2779 | 0.1965 | 0.69 | 0.7274 | 0.7966 | 0.9131 | 1.1018 | 0.7223 |
| 0.19 | 0.9751 | 0.9822 | 0.9928 | 3.1123 | 0.2074 | 0.70 | 0.7209 | 0.7916 | 0.9107 | 1.0944 | 0.7318 |
| 0.20 | 0.9725 | 0.9803 | 0.9921 | 2.9635 | 0.2182 | 0.71 | 0.7145 | 0.7865 | 0.9084 | 1.0873 | 0.7413 |
| 0.21 | 0.9697 | 0.9783 | 0.9913 | 2.8293 | 0.2290 | 0.72 | 0.7080 | 0.7814 | 0.9061 | 1.0806 | 0.7508 |
| 0.22 | 0.9668 | 0.9762 | 0.9904 | 2.7076 | 0.2398 | 0.73 | 0.7016 | 0.7763 | 0.9037 | 1.0742 | 0.7602 |
| 0.23 | 0.9638 | 0.9740 | 0.9895 | 2.5968 | 0.2506 | 0.74 | 0.6951 | 0.7712 | 0.9013 | 1.0681 | 0.7696 |
| 0.24 | 0.9607 | 0.9718 | 0.9886 | 2.4956 | 0.2614 | 0.75 | 0.6886 | 0.7660 | 0.8989 | 1.0624 | 0.7789 |
| 0.25 | 0.9575 | 0.9694 | 0.9877 | 2.4027 | 0.2722 | 0.76 | 0.6821 | 0.7609 | 0.8964 | 1.0570 | 0.7883 |
| 0.26 | 0.9541 | 0.9670 | 0.9867 | 2.3173 | 0.2829 | 0.77 | 0.6756 | 0.7557 | 0.8940 | 1.0519 | 0.7975 |
| 0.27 | 0.9506 | 0.9645 | 0.9856 | 2.2385 | 0.2936 | 0.78 | 0.6691 | 0.7505 | 0.8915 | 1.0471 | 0.8068 |
| 0.28 | 0.9470 | 0.9619 | 0.9846 | 2.1656 | 0.3043 | 0.79 | 0.6625 | 0.7452 | 0.8890 | 1.0425 | 0.8160 |
| 0.29 | 0.9433 | 0.9592 | 0.9835 | 2.0979 | 0.3150 | 0.80 | 0.6560 | 0.7400 | 0.8865 | 1.0382 | 0.8251 |
| 0.30 | 0.9395 | 0.9564 | 0.9823 | 2.0351 | 0.3257 | 0.81 | 0.6495 | 0.7347 | 0.8840 | 1.0342 | 0.8343 |
| 0.31 | 0.9355 | 0.9535 | 0.9811 | 1.9765 | 0.3364 | 0.82 | 0.6430 | 0.7295 | 0.8815 | 1.0305 | 0.8433 |
| 0.32 | 0.9315 | 0.9506 | 0.9799 | 1.9218 | 0.3470 | 0.83 | 0.6365 | 0.7242 | 0.8789 | 1.0270 | 0.8524 |
| 0.33 | 0.9274 | 0.9476 | 0.9787 | 1.8707 | 0.3576 | 0.84 | 0.6300 | 0.7189 | 0.8763 | 1.0237 | 0.8614 |
| 0.34 | 0.9231 | 0.9445 | 0.9774 | 1.8229 | 0.3682 | 0.85 | 0.6235 | 0.7136 | 0.8737 | 1.0207 | 0.8704 |
| 0.35 | 0.9188 | 0.9413 | 0.9761 | 1.7780 | 0.3788 | 0.86 | 0.6170 | 0.7083 | 0.8711 | 1.0179 | 0.8793 |
| 0.36 | 0.9143 | 0.9380 | 0.9747 | 1.7358 | 0.3893 | 0.87 | 0.6106 | 0.7030 | 0.8685 | 1.0153 | 0.8882 |
| 0.37 | 0.9098 | 0.9347 | 0.9734 | 1.6961 | 0.3999 | 0.88 | 0.6041 | 0.6977 | 0.8659 | 1.0129 | 0.8970 |
| 0.38 | 0.9052 | 0.9313 | 0.9719 | 1.6587 | 0.4104 | 0.89 | 0.5977 | 0.6924 | 0.8632 | 1.0108 | 0.9058 |
| 0.39 | 0.9004 | 0.9278 | 0.9705 | 1.6234 | 0.4209 | 0.90 | 0.5913 | 0.6870 | 0.8606 | 1.0089 | 0.9146 |
| 0.40 | 0.8956 | 0.9243 | 0.9690 | 1.5901 | 0.4313 | 0.91 | 0.5849 | 0.6817 | 0.8579 | 1.0071 | 0.9233 |
| 0.41 | 0.8907 | 0.9207 | 0.9675 | 1.5587 | 0.4418 | 0.92 | 0.5785 | 0.6764 | 0.8552 | 1.0056 | 0.9320 |
| 0.42 | 0.8857 | 0.9170 | 0.9659 | 1.5289 | 0.4522 | 0.93 | 0.5721 | 0.6711 | 0.8525 | 1.0043 | 0.9406 |
| 0.43 | 0.8807 | 0.9132 | 0.9643 | 1.5007 | 0.4626 | 0.94 | 0.5658 | 0.6658 | 0.8498 | 1.0031 | 0.9493 |
| 0.44 | 0.8755 | 0.9094 | 0.9627 | 1.4740 | 0.4729 | 0.95 | 0.5595 | 0.6604 | 0.8471 | 1.0021 | 0.9578 |
| 0.45 | 0.8703 | 0.9055 | 0.9611 | 1.4487 | 0.4833 | 0.96 | 0.5532 | 0.6551 | 0.8444 | 1.0014 | 0.9663 |
| 0.46 | 0.8650 | 0.9016 | 0.9594 | 1.4246 | 0.4936 | 0.97 | 0.5469 | 0.6498 | 0.8416 | 1.0008 | 0.9748 |
| 0.47 | 0.8596 | 0.8976 | 0.9577 | 1.4018 | 0.5038 | 0.98 | 0.5407 | 0.6445 | 0.8389 | 1.0003 | 0.9832 |
| 0.48 | 0.8541 | 0.8935 | 0.9560 | 1.3801 | 0.5141 | 0.99 | 0.5345 | 0.6392 | 0.8361 | 1.0001 | 0.9916 |
| 0.49 | 0.8486 | 0.8894 | 0.9542 | 1.3595 | 0.5243 | 1.00 | 0.5283 | 0.6339 | 0.8333 | 1.0000 | 1.0000 |
| 0.50 | 0.8430 | 0.8852 | 0.9524 | 1.3398 | 0.5345 | | | | | | |

# APPENDIX 25.B

## Isentropic Flow and Normal Shock Parameters ($k = 1.4$)

($x$ refers to upstream conditions. $y$ refers to downstream conditions. For example, $p_x/p_{0,y}$ is the ratio of static pressure before the shock wave to total pressure after the shock wave.)

| M | $p/p_0$ | $\rho/\rho_0$ | $T/T_0$ | $A/A^*$ | $v/a^*$ | $M_y$ | $p_y/p_x$ | $\rho_y/\rho_x$ | $T_y/T_x$ | $p_{0,y}/p_{0,x}$ | $p_x/p_{0,y}$ |
|---|---|---|---|---|---|---|---|---|---|---|---|
| 1.00 | 0.5283 | 0.6339 | 0.8333 | 1.0000 | 1.0000 | 1.0000 | $0.1000\times10^{1}$ | $0.1000\times10^{1}$ | $0.1000\times10^{1}$ | 1.0000 | 0.5283 |
| 1.10 | 0.4684 | 0.5817 | 0.8052 | 1.0079 | 1.0812 | 0.9118 | $0.1245\times10^{1}$ | $0.1169\times10^{1}$ | $0.1065\times10^{1}$ | 0.9989 | 0.4689 |
| 1.20 | 0.4124 | 0.5311 | 0.7764 | 1.0304 | 1.1583 | 0.8422 | $0.1513\times10^{1}$ | $0.1342\times10^{1}$ | $0.1128\times10^{1}$ | 0.9928 | 0.4154 |
| 1.30 | 0.3609 | 0.4829 | 0.7474 | 1.0663 | 1.2311 | 0.7860 | $0.1805\times10^{1}$ | $0.1516\times10^{1}$ | $0.1191\times10^{1}$ | 0.9794 | 0.3685 |
| 1.40 | 0.3142 | 0.4374 | 0.7184 | 1.1149 | 1.2999 | 0.7397 | $0.2120\times10^{1}$ | $0.1690\times10^{1}$ | $0.1255\times10^{1}$ | 0.9582 | 0.3280 |
| 1.50 | 0.2724 | 0.3950 | 0.6897 | 1.1762 | 1.3646 | 0.7011 | $0.2458\times10^{1}$ | $0.1862\times10^{1}$ | $0.1320\times10^{1}$ | 0.9298 | 0.2930 |
| 1.60 | 0.2353 | 0.3557 | 0.6614 | 1.2502 | 1.4254 | 0.6684 | $0.2820\times10^{1}$ | $0.2032\times10^{1}$ | $0.1388\times10^{1}$ | 0.8952 | 0.2628 |
| 1.70 | 0.2026 | 0.3197 | 0.6337 | 1.3376 | 1.4825 | 0.6405 | $0.3205\times10^{1}$ | $0.2198\times10^{1}$ | $0.1458\times10^{1}$ | 0.8557 | 0.2368 |
| 1.80 | 0.1740 | 0.2868 | 0.6068 | 1.4390 | 1.5360 | 0.6165 | $0.3613\times10^{1}$ | $0.2359\times10^{1}$ | $0.1532\times10^{1}$ | 0.8127 | 0.2142 |
| 1.90 | 0.1492 | 0.2570 | 0.5807 | 1.5553 | 1.5861 | 0.5956 | $0.4045\times10^{1}$ | $0.2516\times10^{1}$ | $0.1608\times10^{1}$ | 0.7674 | 0.1945 |
| 2.00 | 0.1278 | 0.2300 | 0.5556 | 1.6875 | 1.6330 | 0.5774 | $0.4500\times10^{1}$ | $0.2667\times10^{1}$ | $0.1687\times10^{1}$ | 0.7209 | 0.1773 |
| 2.10 | 0.1094 | 0.2058 | 0.5313 | 1.8369 | 1.6769 | 0.5613 | $0.4978\times10^{1}$ | $0.2812\times10^{1}$ | $0.1770\times10^{1}$ | 0.6742 | 0.1622 |
| 2.20 | $0.9352\times10^{-1}$ | 0.1841 | 0.5081 | 2.0050 | 1.7179 | 0.5471 | $0.5480\times10^{1}$ | $0.2951\times10^{1}$ | $0.1857\times10^{1}$ | 0.6281 | 0.1489 |
| 2.30 | $0.7997\times10^{-1}$ | 0.1646 | 0.4859 | 2.1931 | 1.7563 | 0.5344 | $0.6005\times10^{1}$ | $0.3085\times10^{1}$ | $0.1947\times10^{1}$ | 0.5833 | 0.1371 |
| 2.40 | $0.6840\times10^{-1}$ | 0.1472 | 0.4647 | 2.4031 | 1.7922 | 0.5231 | $0.6553\times10^{1}$ | $0.3212\times10^{1}$ | $0.2040\times10^{1}$ | 0.5401 | 0.1266 |
| 2.50 | $0.5853\times10^{-1}$ | 0.1317 | 0.4444 | 2.6367 | 1.8257 | 0.5130 | $0.7125\times10^{1}$ | $0.3333\times10^{1}$ | $0.2137\times10^{1}$ | 0.4990 | 0.1173 |
| 2.60 | $0.5012\times10^{-1}$ | 0.1179 | 0.4252 | 2.8960 | 1.8571 | 0.5039 | $0.7720\times10^{1}$ | $0.3449\times10^{1}$ | $0.2238\times10^{1}$ | 0.4601 | 0.1089 |
| 2.70 | $0.4295\times10^{-1}$ | 0.1056 | 0.4068 | 3.1830 | 1.8865 | 0.4956 | $0.8338\times10^{1}$ | $0.3559\times10^{1}$ | $0.2343\times10^{1}$ | 0.4236 | 0.1014 |
| 2.80 | $0.3685\times10^{-1}$ | $0.9463\times10^{-1}$ | 0.3894 | 3.5001 | 1.9140 | 0.4882 | $0.8980\times10^{1}$ | $0.3664\times10^{1}$ | $0.2451\times10^{1}$ | 0.3895 | $0.9461\times10^{-1}$ |
| 2.90 | $0.3165\times10^{-1}$ | $0.8489\times10^{-1}$ | 0.3729 | 3.8498 | 1.9398 | 0.4814 | $0.9645\times10^{1}$ | $0.3763\times10^{1}$ | $0.2563\times10^{1}$ | 0.3577 | $0.8848\times10^{-1}$ |
| 3.00 | $0.2722\times10^{-1}$ | $0.7623\times10^{-1}$ | 0.3571 | 4.2346 | 1.9640 | 0.4752 | $0.1033\times10^{2}$ | $0.3857\times10^{1}$ | $0.2679\times10^{1}$ | 0.3283 | $0.8291\times10^{-1}$ |
| 3.10 | $0.2345\times10^{-1}$ | $0.6852\times10^{-1}$ | 0.3422 | 4.6573 | 1.9866 | 0.4695 | $0.1104\times10^{2}$ | $0.3947\times10^{1}$ | $0.2799\times10^{1}$ | 0.3012 | $0.7785\times10^{-1}$ |
| 3.20 | $0.2023\times10^{-1}$ | $0.6165\times10^{-1}$ | 0.3281 | 5.1210 | 2.0079 | 0.4643 | $0.1178\times10^{2}$ | $0.4031\times10^{1}$ | $0.2922\times10^{1}$ | 0.2762 | $0.7323\times10^{-1}$ |
| 3.30 | $0.1748\times10^{-1}$ | $0.5554\times10^{-1}$ | 0.3147 | 5.6286 | 2.0278 | 0.4596 | $0.1254\times10^{2}$ | $0.4112\times10^{1}$ | $0.3049\times10^{1}$ | 0.2533 | $0.6900\times10^{-1}$ |
| 3.40 | $0.1512\times10^{-1}$ | $0.5009\times10^{-1}$ | 0.3019 | 6.1837 | 2.0466 | 0.4552 | $0.1332\times10^{2}$ | $0.4188\times10^{1}$ | $0.3180\times10^{1}$ | 0.2322 | $0.6513\times10^{-1}$ |
| 3.50 | $0.1311\times10^{-1}$ | $0.4523\times10^{-1}$ | 0.2899 | 6.7896 | 2.0642 | 0.4512 | $0.1412\times10^{2}$ | $0.4261\times10^{1}$ | $0.3315\times10^{1}$ | 0.2129 | $0.6157\times10^{-1}$ |
| 3.60 | $0.1138\times10^{-1}$ | $0.4089\times10^{-1}$ | 0.2784 | 7.4501 | 2.0808 | 0.4474 | $0.1495\times10^{2}$ | $0.4330\times10^{1}$ | $0.3454\times10^{1}$ | 0.1953 | $0.5829\times10^{-1}$ |
| 3.70 | $0.9903\times10^{-2}$ | $0.3702\times10^{-1}$ | 0.2675 | 8.1691 | 2.0964 | 0.4439 | $0.1580\times10^{2}$ | $0.4395\times10^{1}$ | $0.3596\times10^{1}$ | 0.1792 | $0.5526\times10^{-1}$ |
| 3.80 | $0.8629\times10^{-2}$ | $0.3355\times10^{-1}$ | 0.2572 | 8.9506 | 2.1111 | 0.4407 | $0.1668\times10^{2}$ | $0.4457\times10^{1}$ | $0.3743\times10^{1}$ | 0.1645 | $0.5247\times10^{-1}$ |
| 3.90 | $0.7532\times10^{-2}$ | $0.3044\times10^{-1}$ | 0.2474 | 9.7990 | 2.1250 | 0.4377 | $0.1758\times10^{2}$ | $0.4516\times10^{1}$ | $0.3893\times10^{1}$ | 0.1510 | $0.4987\times10^{-1}$ |
| 4.00 | $0.6586\times10^{-2}$ | $0.2766\times10^{-1}$ | 0.2381 | 10.7187 | 2.1381 | 0.4350 | $0.1850\times10^{2}$ | $0.4571\times10^{1}$ | $0.4047\times10^{1}$ | 0.1388 | $0.4747\times10^{-1}$ |
| 4.10 | $0.5769\times10^{-2}$ | $0.2516\times10^{-1}$ | 0.2293 | 11.7147 | 2.1505 | 0.4324 | $0.1944\times10^{2}$ | $0.4624\times10^{1}$ | $0.4205\times10^{1}$ | 0.1276 | $0.4523\times10^{-1}$ |
| 4.20 | $0.5062\times10^{-2}$ | $0.2292\times10^{-1}$ | 0.2208 | 12.7916 | 2.1622 | 0.4299 | $0.2041\times10^{2}$ | $0.4675\times10^{1}$ | $0.4367\times10^{1}$ | 0.1173 | $0.4314\times10^{-1}$ |
| 4.30 | $0.4449\times10^{-2}$ | $0.2090\times10^{-1}$ | 0.2129 | 13.9549 | 2.1732 | 0.4277 | $0.2140\times10^{2}$ | $0.4723\times10^{1}$ | $0.4532\times10^{1}$ | 0.1080 | $0.4120\times10^{-1}$ |
| 4.40 | $0.3918\times10^{-2}$ | $0.1909\times10^{-1}$ | 0.2053 | 15.2099 | 2.1837 | 0.4255 | $0.2242\times10^{2}$ | $0.4768\times10^{1}$ | $0.4702\times10^{1}$ | $0.9948\times10^{-1}$ | $0.3938\times10^{-1}$ |
| 4.50 | $0.3455\times10^{-2}$ | $0.1745\times10^{-1}$ | 0.1980 | 16.5622 | 2.1936 | 0.4236 | $0.2346\times10^{2}$ | $0.4812\times10^{1}$ | $0.4875\times10^{1}$ | $0.9170\times10^{-1}$ | $0.3768\times10^{-1}$ |
| 4.60 | $0.3053\times10^{-2}$ | $0.1597\times10^{-1}$ | 0.1911 | 18.0178 | 2.2030 | 0.4217 | $0.2452\times10^{2}$ | $0.4853\times10^{1}$ | $0.5052\times10^{1}$ | $0.8459\times10^{-1}$ | $0.3609\times10^{-1}$ |
| 4.70 | $0.2701\times10^{-2}$ | $0.1464\times10^{-1}$ | 0.1846 | 19.5828 | 2.2119 | 0.4199 | $0.2560\times10^{2}$ | $0.4893\times10^{1}$ | $0.5233\times10^{1}$ | $0.7809\times10^{-1}$ | $0.3459\times10^{-1}$ |
| 4.80 | $0.2394\times10^{-2}$ | $0.1343\times10^{-1}$ | 0.1783 | 21.2637 | 2.2204 | 0.4183 | $0.2671\times10^{2}$ | $0.4930\times10^{1}$ | $0.5418\times10^{1}$ | $0.7214\times10^{-1}$ | $0.3319\times10^{-1}$ |
| 4.90 | $0.2126\times10^{-2}$ | $0.1233\times10^{-1}$ | 0.1724 | 23.0671 | 2.2284 | 0.4167 | $0.2784\times10^{2}$ | $0.4966\times10^{1}$ | $0.5607\times10^{1}$ | $0.6670\times10^{-1}$ | $0.3187\times10^{-1}$ |
| 5.00 | $0.1890\times10^{-2}$ | $0.1134\times10^{-1}$ | 0.1667 | 25.0000 | 2.2361 | 0.4152 | $0.2900\times10^{2}$ | $0.5000\times10^{1}$ | $0.5800\times10^{1}$ | $0.6172\times10^{-1}$ | $0.3062\times10^{-1}$ |
| 5.10 | $0.1683\times10^{-2}$ | $0.1044\times10^{-1}$ | 0.1612 | 27.0696 | 2.2433 | 0.4138 | $0.3018\times10^{2}$ | $0.5033\times10^{1}$ | $0.5997\times10^{1}$ | $0.5715\times10^{-1}$ | $0.2945\times10^{-1}$ |
| 5.20 | $0.1501\times10^{-2}$ | $0.9620\times10^{-2}$ | 0.1561 | 29.2833 | 2.2503 | 0.4125 | $0.3138\times10^{2}$ | $0.5064\times10^{1}$ | $0.6197\times10^{1}$ | $0.5297\times10^{-1}$ | $0.2834\times10^{-1}$ |
| 5.30 | $0.1341\times10^{-2}$ | $0.8875\times10^{-2}$ | 0.1511 | 31.6491 | 2.2569 | 0.4113 | $0.3260\times10^{2}$ | $0.5093\times10^{1}$ | $0.6401\times10^{1}$ | $0.4913\times10^{-1}$ | $0.2730\times10^{-1}$ |
| 5.40 | $0.1200\times10^{-2}$ | $0.8197\times10^{-2}$ | 0.1464 | 34.1748 | 2.2631 | 0.4101 | $0.3385\times10^{2}$ | $0.5122\times10^{1}$ | $0.6610\times10^{1}$ | $0.4560\times10^{-1}$ | $0.2631\times10^{-1}$ |
| 5.50 | $0.1075\times10^{-2}$ | $0.7578\times10^{-2}$ | 0.1418 | 36.8690 | 2.2691 | 0.4090 | $0.3512\times10^{2}$ | $0.5149\times10^{1}$ | $0.6822\times10^{1}$ | $0.4236\times10^{-1}$ | $0.2537\times10^{-1}$ |
| 5.60 | $0.9643\times10^{-3}$ | $0.7012\times10^{-2}$ | 0.1375 | 39.7402 | 2.2748 | 0.4079 | $0.3642\times10^{2}$ | $0.5175\times10^{1}$ | $0.7038\times10^{1}$ | $0.3938\times10^{-1}$ | $0.2448\times10^{-1}$ |
| 5.70 | $0.8663\times10^{-3}$ | $0.6496\times10^{-2}$ | 0.1334 | 42.7974 | 2.2803 | 0.4069 | $0.3774\times10^{2}$ | $0.5200\times10^{1}$ | $0.7258\times10^{1}$ | $0.3664\times10^{-1}$ | $0.2364\times10^{-1}$ |
| 5.80 | $0.7794\times10^{-3}$ | $0.6023\times10^{-2}$ | 0.1294 | 46.0500 | 2.2855 | 0.4059 | $0.3908\times10^{2}$ | $0.5224\times10^{1}$ | $0.7481\times10^{1}$ | $0.3412\times10^{-1}$ | $0.2284\times10^{-1}$ |
| 5.90 | $0.7021\times10^{-3}$ | $0.5590\times10^{-2}$ | 0.1256 | 49.5075 | 2.2905 | 0.4050 | $0.4044\times10^{2}$ | $0.5246\times10^{1}$ | $0.7709\times10^{1}$ | $0.3179\times10^{-1}$ | $0.2208\times10^{-1}$ |
| 6.00 | $0.6334\times10^{-3}$ | $0.5194\times10^{-2}$ | 0.1220 | 53.1798 | 2.2953 | 0.4042 | $0.4183\times10^{2}$ | $0.5268\times10^{1}$ | $0.7941\times10^{1}$ | $0.2965\times10^{-1}$ | $0.2136\times10^{-1}$ |
| 6.10 | $0.5721\times10^{-3}$ | $0.4829\times10^{-2}$ | 0.1185 | 57.0772 | 2.2998 | 0.4033 | $0.4324\times10^{2}$ | $0.5289\times10^{1}$ | $0.8176\times10^{1}$ | $0.2767\times10^{-1}$ | $0.2067\times10^{-1}$ |
| 6.20 | $0.5173\times10^{-3}$ | $0.4495\times10^{-2}$ | 0.1151 | 61.2102 | 2.3042 | 0.4025 | $0.4468\times10^{2}$ | $0.5309\times10^{1}$ | $0.8415\times10^{1}$ | $0.2584\times10^{-1}$ | $0.2002\times10^{-1}$ |
| 6.30 | $0.4684\times10^{-3}$ | $0.4187\times10^{-2}$ | 0.1119 | 65.5899 | 2.3084 | 0.4018 | $0.4614\times10^{2}$ | $0.5329\times10^{1}$ | $0.8658\times10^{1}$ | $0.2416\times10^{-1}$ | $0.1939\times10^{-1}$ |
| 6.40 | $0.4247\times10^{-3}$ | $0.3904\times10^{-2}$ | 0.1088 | 70.2274 | 2.3124 | 0.4011 | $0.4762\times10^{2}$ | $0.5347\times10^{1}$ | $0.8905\times10^{1}$ | $0.2259\times10^{-1}$ | $0.1880\times10^{-1}$ |
| 6.50 | $0.3855\times10^{-3}$ | $0.3643\times10^{-2}$ | 0.1058 | 75.1343 | 2.3163 | 0.4004 | $0.4912\times10^{2}$ | $0.5365\times10^{1}$ | $0.9156\times10^{1}$ | $0.2115\times10^{-1}$ | $0.1823\times10^{-1}$ |
| 6.60 | $0.3503\times10^{-3}$ | $0.3402\times10^{-2}$ | 0.1030 | 80.3227 | 2.3200 | 0.3997 | $0.5065\times10^{2}$ | $0.5382\times10^{1}$ | $0.9411\times10^{1}$ | $0.1981\times10^{-1}$ | $0.1768\times10^{-1}$ |
| 6.70 | $0.3187\times10^{-3}$ | $0.3180\times10^{-2}$ | 0.1002 | 85.8049 | 2.3235 | 0.3991 | $0.5220\times10^{2}$ | $0.5399\times10^{1}$ | $0.9670\times10^{1}$ | $0.1857\times10^{-1}$ | $0.1716\times10^{-1}$ |
| 6.80 | $0.2902\times10^{-3}$ | $0.2974\times10^{-2}$ | $0.9758\times10^{-1}$ | 91.5935 | 2.3269 | 0.3985 | $0.5378\times10^{2}$ | $0.5415\times10^{1}$ | $0.9933\times10^{1}$ | $0.1741\times10^{-1}$ | $0.1667\times10^{-1}$ |
| 6.90 | $0.2646\times10^{-3}$ | $0.2785\times10^{-2}$ | $0.9504\times10^{-1}$ | 97.9017 | 2.3302 | 0.3979 | $0.5538\times10^{2}$ | $0.5430\times10^{1}$ | $0.1020\times10^{2}$ | $0.1634\times10^{-1}$ | $0.1619\times10^{-1}$ |
| 7.00 | $0.2416\times10^{-3}$ | $0.2609\times10^{-2}$ | $0.9259\times10^{-1}$ | 104.1429 | 2.3333 | 0.3974 | $0.5700\times10^{2}$ | $0.5444\times10^{1}$ | $0.1047\times10^{2}$ | $0.1535\times10^{-1}$ | $0.1573\times10^{-1}$ |
| 7.10 | $0.2207\times10^{-3}$ | $0.2446\times10^{-2}$ | $0.9024\times10^{-1}$ | 110.9309 | 2.3364 | 0.3968 | $0.5864\times10^{2}$ | $0.5459\times10^{1}$ | $0.1074\times10^{2}$ | $0.1443\times10^{-1}$ | $0.1530\times10^{-1}$ |
| 7.20 | $0.2019\times10^{-3}$ | $0.2295\times10^{-2}$ | $0.8797\times10^{-1}$ | 118.0799 | 2.3393 | 0.3963 | $0.6031\times10^{2}$ | $0.5472\times10^{1}$ | $0.1102\times10^{2}$ | $0.1357\times10^{-1}$ | $0.1488\times10^{-1}$ |
| 7.30 | $0.1848\times10^{-3}$ | $0.2155\times10^{-2}$ | $0.8578\times10^{-1}$ | 125.6046 | 2.3421 | 0.3958 | $0.6200\times10^{2}$ | $0.5485\times10^{1}$ | $0.1130\times10^{2}$ | $0.1277\times10^{-1}$ | $0.1448\times10^{-1}$ |
| 7.40 | $0.1694\times10^{-3}$ | $0.2025\times10^{-2}$ | $0.8367\times10^{-1}$ | 133.5200 | 2.3448 | 0.3954 | $0.6372\times10^{2}$ | $0.5498\times10^{1}$ | $0.1159\times10^{2}$ | $0.1202\times10^{-1}$ | $0.1409\times10^{-1}$ |
| 7.50 | $0.1554\times10^{-3}$ | $0.1904\times10^{-2}$ | $0.8163\times10^{-1}$ | 141.8415 | 2.3474 | 0.3949 | $0.6546\times10^{2}$ | $0.5510\times10^{1}$ | $0.1188\times10^{2}$ | $0.1133\times10^{-1}$ | $0.1372\times10^{-1}$ |
| 7.60 | $0.1427\times10^{-3}$ | $0.1792\times10^{-2}$ | $0.7967\times10^{-1}$ | 150.5849 | 2.3499 | 0.3945 | $0.6722\times10^{2}$ | $0.5522\times10^{1}$ | $0.1217\times10^{2}$ | $0.1068\times10^{-1}$ | $0.1336\times10^{-1}$ |
| 7.70 | $0.1312\times10^{-3}$ | $0.1687\times10^{-2}$ | $0.7777\times10^{-1}$ | 159.7665 | 2.3523 | 0.3941 | $0.6900\times10^{2}$ | $0.5533\times10^{1}$ | $0.1247\times10^{2}$ | $0.1008\times10^{-1}$ | $0.1302\times10^{-1}$ |
| 7.80 | $0.1207\times10^{-3}$ | $0.1589\times10^{-2}$ | $0.7594\times10^{-1}$ | 169.4030 | 2.3546 | 0.3937 | $0.7081\times10^{2}$ | $0.5544\times10^{1}$ | $0.1277\times10^{2}$ | $0.9510\times10^{-2}$ | $0.1269\times10^{-1}$ |
| 7.90 | $0.1111\times10^{-3}$ | $0.1498\times10^{-2}$ | $0.7417\times10^{-1}$ | 179.5114 | 2.3569 | 0.3933 | $0.7264\times10^{2}$ | $0.5555\times10^{1}$ | $0.1308\times10^{2}$ | $0.8982\times10^{-2}$ | $0.1237\times10^{-1}$ |
| 8.00 | $0.1024\times10^{-3}$ | $0.1414\times10^{-2}$ | $0.7246\times10^{-1}$ | 190.1094 | 2.3591 | 0.3929 | $0.7450\times10^{2}$ | $0.5565\times10^{1}$ | $0.1339\times10^{2}$ | $0.8488\times10^{-2}$ | $0.1207\times10^{-1}$ |
| 8.10 | $0.9449\times10^{-4}$ | $0.1334\times10^{-2}$ | $0.7081\times10^{-1}$ | 201.2148 | 2.3612 | 0.3925 | $0.7638\times10^{2}$ | $0.5575\times10^{1}$ | $0.1370\times10^{2}$ | $0.8025\times10^{-2}$ | $0.1177\times10^{-1}$ |
| 8.20 | $0.8723\times10^{-4}$ | $0.1260\times10^{-2}$ | $0.6921\times10^{-1}$ | 212.8461 | 2.3632 | 0.3922 | $0.7828\times10^{2}$ | $0.5585\times10^{1}$ | $0.1402\times10^{2}$ | $0.7592\times10^{-2}$ | $0.1149\times10^{-1}$ |
| 8.30 | $0.8060\times10^{-4}$ | $0.1191\times10^{-2}$ | $0.6767\times10^{-1}$ | 225.0221 | 2.3652 | 0.3918 | $0.8020\times10^{2}$ | $0.5594\times10^{1}$ | $0.1434\times10^{2}$ | $0.7187\times10^{-2}$ | $0.1122\times10^{-1}$ |
| 8.40 | $0.7454\times10^{-4}$ | $0.1126\times10^{-2}$ | $0.6617\times10^{-1}$ | 237.7622 | 2.3671 | 0.3915 | $0.8215\times10^{2}$ | $0.5603\times10^{1}$ | $0.1466\times10^{2}$ | $0.6806\times10^{-2}$ | $0.1095\times10^{-1}$ |
| 8.50 | $0.6898\times10^{-4}$ | $0.1066\times10^{-2}$ | $0.6472\times10^{-1}$ | 251.0862 | 2.3689 | 0.3912 | $0.8412\times10^{2}$ | $0.5612\times10^{1}$ | $0.1499\times10^{2}$ | $0.6449\times10^{-2}$ | $0.1070\times10^{-1}$ |

**APPENDIX 25.C**
Fanno Flow Factors
($k = 1.4$)

| M | $p/p^*$ | $a/a^* = \rho^*/\rho$ | $T/T^*$ | $p_0/p_0^*$ | $4fL/D$ |
|---|---------|----------------------|---------|-------------|---------|
| 0.00 | $\infty$ | 0.0000 | 1.200 | $\infty$ | $\infty$ |
| 0.05 | 21.903 | 0.0547 | 1.199 | 11.592 | 280.02 |
| 0.10 | 10.944 | 0.1094 | 1.197 | 5.822 | 66.922 |
| 0.12 | 9.116 | 0.131 | 1.1965 | 4.864 | 45.408 |
| 0.14 | 7.809 | 0.153 | 1.195 | 4.182 | 32.511 |
| 0.16 | 6.829 | 0.175 | 1.194 | 3.673 | 24.198 |
| 0.18 | 6.066 | 0.196 | 1.192 | 3.278 | 18.543 |
| 0.20 | 5.455 | 0.218 | 1.1905 | 2.963 | 14.533 |
| 0.25 | 4.355 | 0.272 | 1.185 | 2.403 | 8.483 |
| 0.30 | 3.619 | 0.3257 | 1.178 | 2.035 | 5.299 |
| 0.35 | 3.092 | 0.379 | 1.171 | 1.778 | 3.453 |
| 0.40 | 2.696 | 0.431 | 1.163 | 1.590 | 2.308 |
| 0.45 | 2.386 | 0.483 | 1.153 | 1.448 | 1.566 |
| 0.50 | 2.138 | 0.534 | 1.143 | 1.340 | 1.069 |
| 0.52 | 2.052 | 0.555 | 1.138 | 1.303 | 0.917 |
| 0.54 | 1.972 | 0.575 | 1.134 | 1.270 | 0.787 |
| 0.56 | 1.897 | 0.595 | 1.129 | 1.240 | 0.673 |
| 0.58 | 1.828 | 0.615 | 1.124 | 1.213 | 0.576 |
| 0.60 | 1.763 | 0.635 | 1.119 | 1.188 | 0.491 |
| 0.65 | 1.618 | 0.684 | 1.106 | 1.135 | 0.325 |
| 0.70 | 1.493 | 0.732 | 1.093 | 1.094 | 0.208 |
| 0.75 | 1.385 | 0.779 | 1.078 | 1.062 | 0.127 |
| 0.80 | 1.289 | 0.825 | 1.064 | 1.038 | 0.072 |
| 0.85 | 1.205 | 0.870 | 1.048 | 1.020 | 0.0363 |
| 0.90 | 1.129 | 0.914 | 1.0327 | 1.009 | 0.0145 |
| 0.95 | 1.061 | 0.958 | 1.0165 | 1.002 | 0.0033 |
| 1.00 | 1.000 | 1.000 | 1.000 | 1.000 | 0.000 |
| 1.20 | 0.804 | 1.158 | 0.932 | 1.030 | 0.0336 |
| 1.50 | 0.606 | 1.365 | 0.827 | 1.176 | 0.136 |
| 1.60 | 0.557 | 1.425 | 0.794 | 1.250 | 0.172 |
| 1.70 | 0.513 | 1.483 | 0.760 | 1.338 | 0.208 |
| 1.80 | 0.474 | 1.536 | 0.728 | 1.439 | 0.242 |
| 1.90 | 0.439 | 1.586 | 0.697 | 1.555 | 0.274 |
| 2.00 | 0.408 | 1.633 | 0.667 | 1.687 | 0.305 |
| 2.50 | 0.292 | 1.826 | 0.533 | 2.637 | 0.432 |
| 3.00 | 0.218 | 1.964 | 0.428 | 4.235 | 0.522 |
| 3.50 | 0.1685 | 2.064 | 0.348 | 6.789 | 0.586 |
| 4.00 | 0.134 | 2.138 | 0.286 | 10.719 | 0.633 |
| 4.50 | 0.108 | 2.194 | 0.237 | 16.562 | 0.667 |
| 5.00 | 0.0894 | 2.236 | 0.200 | 25.000 | 0.694 |

## APPENDIX 25.D
Rayleigh Flow Factors
$(k = 1.4)$

| M | $p/p^*$ | $p_0/p_0^*$ | $T/T^*$ | $T_0/T_0^*$ | $a/a^* = \rho^*/\rho$ |
|------|--------|--------|--------|--------|--------|
| 0.00 | 2.400 | 1.268 | 0.000 | 0.000 | 0.000 |
| 0.05 | 2.392 | 1.266 | 0.0143 | 0.0119 | 0.00598 |
| 0.10 | 2.367 | 1.259 | 0.056 | 0.0468 | 0.0237 |
| 0.12 | 2.353 | 1.255 | 0.079 | 0.0667 | 0.0339 |
| 0.14 | 2.336 | 1.251 | 0.107 | 0.089 | 0.0458 |
| 0.16 | 2.317 | 1.246 | 0.137 | 0.115 | 0.0593 |
| 0.18 | 2.296 | 1.241 | 0.1708 | 0.143 | 0.0744 |
| 0.20 | 2.273 | 1.235 | 0.2066 | 0.1735 | 0.091 |
| 0.25 | 2.207 | 1.218 | 0.304 | 0.257 | 0.138 |
| 0.30 | 2.131 | 1.198 | 0.409 | 0.3468 | 0.192 |
| 0.35 | 2.048 | 1.178 | 0.514 | 0.439 | 0.251 |
| 0.40 | 1.961 | 1.157 | 0.615 | 0.529 | 0.314 |
| 0.45 | 1.870 | 1.135 | 0.708 | 0.614 | 0.378 |
| 0.50 | 1.778 | 1.114 | 0.790 | 0.691 | 0.444 |
| 0.52 | 1.741 | 1.106 | 0.819 | 0.720 | 0.470 |
| 0.54 | 1.704 | 1.098 | 0.847 | 0.747 | 0.497 |
| 0.56 | 1.668 | 1.090 | 0.872 | 0.772 | 0.523 |
| 0.58 | 1.632 | 1.083 | 0.896 | 0.796 | 0.549 |
| 0.60 | 1.596 | 1.075 | 0.917 | 0.819 | 0.574 |
| 0.65 | 1.508 | 1.058 | 0.961 | 0.868 | 0.637 |
| 0.70 | 1.424 | 1.043 | 0.993 | 0.908 | 0.697 |
| 0.75 | 1.343 | 1.030 | 1.014 | 0.940 | 0.755 |
| 0.80 | 1.266 | 1.019 | 1.025 | 0.964 | 0.810 |
| 0.85 | 1.193 | 1.011 | 1.028 | 0.981 | 0.862 |
| 0.90 | 1.125 | 1.005 | 1.0245 | 0.992 | 0.911 |
| 0.95 | 1.060 | 1.001 | 1.0146 | 0.998 | 0.957 |
| 1.00 | 1.000 | 1.000 | 1.000 | 1.000 | 1.000 |
| 1.20 | 0.796 | 1.0194 | 0.912 | 0.978 | 1.146 |
| 1.50 | 0.578 | 1.122 | 0.753 | 0.909 | 1.301 |
| 1.60 | 0.523 | 1.176 | 0.702 | 0.884 | 1.340 |
| 1.70 | 0.475 | 1.240 | 0.654 | 0.859 | 1.375 |
| 1.80 | 0.433 | 1.316 | 0.609 | 0.836 | 1.405 |
| 1.90 | 0.396 | 1.403 | 0.567 | 0.814 | 1.431 |
| 2.00 | 0.363 | 1.503 | 0.529 | 0.794 | 1.455 |
| 2.50 | 0.246 | 2.222 | 0.378 | 0.710 | 1.538 |
| 3.00 | 0.176 | 3.424 | 0.280 | 0.654 | 1.588 |
| 3.50 | 0.132 | 5.328 | 0.214 | 0.616 | 1.619 |
| 4.00 | 0.1025 | 8.227 | 0.168 | 0.589 | 1.641 |
| 4.50 | 0.0818 | 12.502 | 0.135 | 0.569 | 1.656 |
| 5.00 | 0.0667 | 18.634 | 0.111 | 0.555 | 1.667 |

## APPENDIX 25.E
International Standard Atmosphere

| customary U.S. units | | | | SI units | | |
|---|---|---|---|---|---|---|
| altitude (ft) | temperature (°R) | pressure (psia) | | altitude (m) | temperature (K) | pressure (bar) |
| 0 | 518.7 | 14.696 | | 0 | 288.15 | 1.01325 |
| 1000 | 515.1 | 14.175 | | | | |
| 2000 | 511.6 | 13.664 | | 500 | 284.9 | 0.9546 |
| 3000 | 508.0 | 13.168 | | 1000 | 281.7 | 0.8988 |
| 4000 | 504.4 | 12.692 | | 1500 | 278.4 | 0.8456 |
| | | | | 2000 | 275.2 | 0.7950 |
| 5000 | 500.9 | 12.225 | | 2500 | 271.9 | 0.7469 |
| 6000 | 497.3 | 11.778 | | | | |
| 7000 | 493.7 | 11.341 | | 3000 | 268.7 | 0.7012 |
| 8000 | 490.2 | 10.914 | | 3500 | 265.4 | 0.6578 |
| 9000 | 486.6 | 10.501 | | 4000 | 262.2 | 0.6166 |
| | | | | 4500 | 258.9 | 0.5775 |
| 10,000 | 483.0 | 10.108 | | 5000 | 255.7 | 0.5405 |
| 11,000 | 479.5 | 9.720 | | | | |
| 12,000 | 475.9 | 9.347 | | 5500 | 252.4 | 0.5054 |
| 13,000 | 472.3 | 8.983 | | 6000 | 249.2 | 0.4722 |
| 14,000 | 468.8 | 8.630 | | 6500 | 245.9 | 0.4408 |
| | | | | 7000 | 242.7 | 0.4111 |
| 15,000 | 465.2 | 8.291 | | 7500 | 239.5 | 0.3830 |
| 16,000 | 461.6 | 7.962 | | | | |
| 17,000 | 458.1 | 7.642 | | 8000 | 236.2 | 0.3565 |
| 18,000 | 454.5 | 7.338 | | 8500 | 233.0 | 0.3315 |
| 19,000 | 450.9 | 7.038 | | 9000 | 229.7 | 0.3080 |
| | | | | 9500 | 226.5 | 0.2858 |
| 20,000 | 447.4 | 6.753 | | 10 000 | 223.3 | 0.2650 |
| 21,000 | 443.8 | 6.473 | | | | |
| 22,000 | 440.2 | 6.203 | | 10 500 | 220.0 | 0.2454 |
| 23,000 | 436.7 | 5.943 | | 11 000 | 216.8 | 0.2270 |
| 24,000 | 433.1 | 5.693 | | 11 500 | 216.7 | 0.2098 |
| | | | | 12 000 | 216.7 | 0.1940 |
| 25,000 | 429.5 | 5.452 | | 12 500 | 216.7 | 0.1793 |
| 26,000 | 426.0 | 5.216 | | | | |
| 27,000 | 422.4 | 4.990 | | 13 000 | 216.7 | 0.1658 |
| 28,000 | 418.8 | 4.774 | | 13 500 | 216.7 | 0.1533 |
| 29,000 | 415.3 | 4.563 | | 14 000 | 216.7 | 0.1417 |
| | | | | 14 500 | 216.7 | 0.1310 |
| 30,000 | 411.7 | 4.362 | | 15 000 | 216.7 | 0.1211 |
| 31,000 | 408.1 | 4.165 | | | | |
| 32,000 | 404.6 | 3.978 | | 15 500 | 216.7 | 0.1120 |
| 33,000 | 401.0 | 3.797 | | 16 000 | 216.7 | 0.1035 |
| 34,000 | 397.5 | 3.625 | | 16 500 | 216.7 | 0.09572 |
| | | | | 17 000 | 216.7 | 0.08850 |
| 35,000 | 393.9 | 3.458 | | 17 500 | 216.7 | 0.08182 |
| 36,000 | 392.7 | 3.296 | | | | |
| 37,000 | 392.7 | 3.143 | | 18 000 | 216.7 | 0.07565 |
| 38,000 | 392.7 | 2.996 | | 18 500 | 216.7 | 0.06995 |
| 39,000 | 392.7 | 2.854 | | 19 000 | 216.7 | 0.06467 |
| | | | | 19 500 | 216.7 | 0.05980 |
| 40,000 | 392.7 | 2.721 | | 20 000 | 216.7 | 0.05529 |
| 41,000 | 392.7 | 2.593 | | | | |
| 42,000 | 392.7 | 2.475 | | 22 000 | 218.6 | 0.04047 |
| 43,000 | 392.7 | 2.358 | | 24 000 | 220.6 | 0.02972 |
| 44,000 | 392.7 | 2.250 | | 26 000 | 222.5 | 0.02188 |
| | | | | 28 000 | 224.5 | 0.01616 |
| 45,000 | 392.7 | 2.141 | | 30 000 | 226.5 | 0.01197 |
| 46,000 | 392.7 | 2.043 | | | | |
| 47,000 | 392.7 | 1.950 | | 32 000 | 228.5 | 0.00889 |
| 48,000 | 392.7 | 1.857 | | | | |
| 49,000 | 392.7 | 1.768 | | | | |
| 50,000 | 392.7 | 1.690 | | | | |
| 51,000 | 392.7 | 1.611 | | | | |
| 52,000 | 392.7 | 1.532 | | | | |
| 53,000 | 392.7 | 1.464 | | | | |
| 54,000 | 392.7 | 1.395 | | | | |
| 55,000 | 392.7 | 1.331 | | | | |
| 56,000 | 392.7 | 1.267 | | | | |
| 57,000 | 392.7 | 1.208 | | | | |
| 58,000 | 392.7 | 1.154 | | | | |
| 59,000 | 392.7 | 1.100 | | | | |
| 60,000 | 392.7 | 1.046 | | | | |
| 61,000 | 392.7 | 0.997 | | | | |
| 62,000 | 392.7 | 0.953 | | | | |
| 63,000 | 392.7 | 0.909 | | | | |
| 64,000 | 392.7 | 0.864 | | | | |
| 65,000 | 392.7 | 0.825 | | | | |

(U.S. units) troposphere, tropopause, stratosphere (to approximately 160,000 ft)

(SI units) troposphere, tropopause, stratosphere (to approximately 50 000 m)

**APPENDIX 33.A**
Atomic Numbers and Weights of the Elements
(referred to Carbon-12)

| name | symbol | atomic number | atomic weight | name | symbol | atomic number | atomic weight |
|---|---|---|---|---|---|---|---|
| actinium | Ac | 89 | – | mercury | Hg | 80 | 200.59 |
| aluminum | Al | 13 | 26.9815 | molybdenum | Mo | 42 | 95.94 |
| americium | Am | 95 | – | neodymium | Nd | 60 | 144.24 |
| antimony | Sb | 51 | 121.75 | neon | Ne | 10 | 20.183 |
| argon | Ar | 18 | 39.948 | neptunium | Np | 93 | 237.048 |
| arsenic | As | 33 | 74.9216 | nickel | Ni | 28 | 58.71 |
| astatine | At | 85 | – | niobium | Nb | 41 | 92.906 |
| barium | Ba | 56 | 137.34 | nitrogen | N | 7 | 14.0067 |
| berkelium | Bk | 97 | – | nobelium | No | 102 | – |
| beryllium | Be | 4 | 9.0122 | osmium | Os | 76 | 190.2 |
| bismuth | Bi | 83 | 208.980 | oxygen | O | 8 | 15.9994 |
| boron | B | 5 | 10.811 | palladium | Pd | 46 | 106.4 |
| bromine | Br | 35 | 79.904 | phosphorus | P | 15 | 30.9738 |
| cadmium | Cd | 48 | 112.40 | platinum | Pt | 78 | 195.09 |
| calcium | Ca | 20 | 40.08 | plutonium | Pu | 94 | – |
| californium | Cf | 98 | – | polonium | Po | 84 | – |
| carbon | C | 6 | 12.01115 | potassium | K | 19 | 39.102 |
| cerium | Ce | 58 | 140.12 | praseodymium | Pr | 59 | 140.907 |
| cesium | Cs | 55 | 132.905 | promethium | Pm | 61 | – |
| chlorine | Cl | 17 | 35.453 | protactinium | Pa | 91 | 231.036 |
| chromium | Cr | 24 | 51.996 | radium | Ra | 88 | – |
| cobalt | Co | 27 | 58.9332 | radon | Rn | 86 | 226.025 |
| copper | Cu | 29 | 63.546 | rhenium | Re | 75 | 186.2 |
| curium | Cm | 96 | – | rhodium | Rh | 45 | 102.905 |
| dysprosium | Dy | 66 | 162.50 | rubidium | Rb | 37 | 85.47 |
| einsteinium | Es | 99 | – | ruthenium | Ru | 44 | 101.07 |
| erbium | Er | 68 | 167.26 | samarium | Sm | 62 | 150.35 |
| europium | Eu | 63 | 151.96 | scandium | Sc | 21 | 44.956 |
| fermium | Fm | 100 | – | selenium | Se | 34 | 78.96 |
| fluorine | F | 9 | 18.9984 | silicon | Si | 14 | 28.086 |
| francium | Fr | 87 | – | silver | Ag | 47 | 107.868 |
| gadolinium | Gd | 64 | 157.25 | sodium | Na | 11 | 22.9898 |
| gallium | Ga | 31 | 69.72 | strontium | Sr | 38 | 87.62 |
| germanium | Ge | 32 | 72.59 | sulfur | S | 16 | 32.064 |
| gold | Au | 79 | 196.967 | tantalum | Ta | 73 | 180.948 |
| hafnium | Hf | 72 | 178.49 | technetium | Tc | 43 | – |
| helium | He | 2 | 4.0026 | tellurium | Te | 52 | 127.60 |
| holmium | Ho | 67 | 164.930 | terbium | Tb | 65 | 158.924 |
| hydrogen | H | 1 | 1.00797 | thallium | Tl | 81 | 204.37 |
| indium | In | 49 | 114.82 | thorium | Th | 90 | 232.038 |
| iodine | I | 53 | 126.9044 | thulium | Tm | 69 | 168.934 |
| iridium | Ir | 77 | 192.2 | tin | Sn | 50 | 118.69 |
| iron | Fe | 26 | 55.847 | titanium | Ti | 22 | 47.90 |
| krypton | Kr | 36 | 83.80 | tungsten | W | 74 | 183.85 |
| lanthanum | La | 57 | 138.91 | uranium | U | 92 | 238.03 |
| lead | Pb | 82 | 207.19 | vanadium | V | 23 | 50.942 |
| lithium | Li | 3 | 6.939 | xenon | Xe | 54 | 131.30 |
| lutetium | Lu | 71 | 174.97 | ytterbium | Yb | 70 | 173.04 |
| magnesium | Mg | 12 | 24.312 | yttrium | Y | 39 | 88.905 |
| manganese | Mn | 25 | 54.9380 | zinc | Zn | 30 | 65.37 |
| mendelevium | Md | 101 | – | zirconium | Zr | 40 | 91.22 |

## APPENDIX 33.B
### Water Chemistry CaCO$_3$ Equivalents

| cations | formula | ionic weight | equivalent weight | factor |
|---|---|---|---|---|
| aluminum | $Al^{+3}$ | 27.0 | 9.0 | 5.56 |
| ammonium | $NH_4^+$ | 18.0 | 18.0 | 2.78 |
| calcium | $Ca^{+2}$ | 40.1 | 20.0 | 2.50 |
| cupric copper | $Cu^{+2}$ | 63.6 | 31.8 | 1.57 |
| cuprous copper | $Cu^{+3}$ | 63.6 | 21.2 | 2.36 |
| ferric iron | $Fe^{+3}$ | 55.8 | 18.6 | 2.69 |
| ferrous iron | $Fe^{+2}$ | 55.8 | 27.9 | 1.79 |
| hydrogen | $H^+$ | 1.0 | 1.0 | 50.00 |
| manganese | $Mn^{+2}$ | 54.9 | 27.5 | 1.82 |
| magnesium | $Mg^{+2}$ | 24.3 | 12.2 | 4.10 |
| potassium | $K^+$ | 39.1 | 39.1 | 1.28 |
| sodium | $Na^+$ | 23.0 | 23.0 | 2.18 |

| anions | formula | ionic weight | equivalent weight | factor |
|---|---|---|---|---|
| bicarbonate | $HCO_3^-$ | 61.0 | 61.0 | 0.82 |
| carbonate | $CO_3^{-2}$ | 60.0 | 30.0 | 1.67 |
| chloride | $Cl^-$ | 35.5 | 35.5 | 1.41 |
| fluoride | $F^-$ | 19.0 | 19.0 | 2.66 |
| hydroxide | $OH^-$ | 17.0 | 17.0 | 2.94 |
| nitrate | $NO_3^-$ | 62.0 | 62.0 | 0.81 |
| phosphate (tribasic) | $PO_4^{-3}$ | 95.0 | 31.7 | 1.58 |
| phosphate (dibasic) | $HPO_4^{-2}$ | 96.0 | 48.0 | 1.04 |
| phosphate (monobasic) | $H_2PO_4^-$ | 97.0 | 97.0 | 0.52 |
| sulfate | $SO_4^{-2}$ | 96.1 | 48.0 | 1.04 |
| sulfite | $SO_3^{-2}$ | 80.1 | 40.0 | 1.25 |

| compounds | formula | molecular weight | equivalent weight | factor |
|---|---|---|---|---|
| aluminum hydroxide | $Al(OH)_3$ | 78.0 | 26.0 | 1.92 |
| aluminum sulfate | $Al_2(SO_4)_3$ | 342.1 | 57.0 | 0.88 |
| aluminum sulfate | $Al_2(SO_4)_3 \cdot 18H_2O$ | 666.1 | 111.0 | 0.45 |
| alumina | $Al_2O_3$ | 102.0 | 17.0 | 2.94 |
| sodium aluminate | $Na_2Al_2O_4$ | 164.0 | 27.3 | 1.83 |
| calcium bicarbonate | $Ca(HCO_3)_2$ | 162.1 | 81.1 | 0.62 |
| calcium carbonate | $CaCO_3$ | 100.1 | 50.1 | 1.00 |
| calcium chloride | $CaCl_2$ | 111.0 | 55.5 | 0.90 |
| calcium hydroxide (pure) | $Ca(OH)_2$ | 74.1 | 37.1 | 1.35 |
| calcium hydroxide (90%) | $Ca(OH)_2$ | – | 41.1 | 1.22 |
| calcium oxide (lime) | $CaO$ | 56.1 | 28.0 | 1.79 |
| calcium sulfate (anhydrous) | $CaSO_4$ | 136.2 | 68.1 | 0.74 |
| calcium sulfate (gypsum) | $CaSO_4 \cdot 2H_2O$ | 172.2 | 86.1 | 0.58 |
| calcium phosphate | $Ca_3(PO_4)_2$ | 310.3 | 51.7 | 0.97 |
| disodium phosphate | $Na_2HPO_4 \cdot 12H_2O$ | 358.2 | 119.4 | 0.42 |
| disodium phosphate (anhydrous) | $Na_2HPO_4$ | 142.0 | 47.3 | 1.06 |
| ferric oxide | $Fe_2O_3$ | 159.6 | 26.6 | 1.88 |
| iron oxide (magnetic) | $Fe_3O_4$ | 321.4 | – | – |
| ferrous sulfate (copperas) | $FeSO_4 \cdot 7H_2O$ | 278.0 | 139.0 | 0.36 |
| magnesium oxide | $MgO$ | 40.3 | 20.2 | 2.48 |

*(continued)*

**APPENDIX 33.B** *(continued)*
Water Chemistry CaCO₃ Equivalents

| compounds | formula | molecular weight | equivalent weight | factor |
|---|---|---|---|---|
| magnesium bicarbonate | $Mg(HCO_3)_2$ | 146.3 | 73.2 | 0.68 |
| magnesium carbonate | $MgCO_3$ | 84.3 | 42.2 | 1.19 |
| magnesium chloride | $MgCl_2$ | 95.2 | 47.6 | 1.05 |
| magnesium hydroxide | $Mg(OH)_2$ | 58.3 | 29.2 | 1.71 |
| magnesium phosphate | $Mg_3(PO_4)_2$ | 263.0 | 43.8 | 1.14 |
| magnesium sulfate | $MgSO_4$ | 120.4 | 60.2 | 0.83 |
| monosodium phosphate | $NaH_2PO_4 \cdot H_2O$ | 138.1 | 46.0 | 1.09 |
| monosodium phosphate (anhydrous) | $NaH_2PO_4$ | 120.1 | 40.0 | 1.25 |
| metaphosphate | $NaPO_3$ | 102.0 | 34.0 | 1.47 |
| silica | $SiO_2$ | 60.1 | 30.0 | 1.67 |
| sodium bicarbonate | $NaHCO_3$ | 84.0 | 84.0 | 0.60 |
| sodium carbonate | $Na_2CO_3$ | 106.0 | 53.0 | 0.94 |
| sodium chloride | $NaCl$ | 58.5 | 58.5 | 0.85 |
| sodium hydroxide | $NaOH$ | 40.0 | 40.0 | 1.25 |
| sodium nitrate | $NaNO_3$ | 85.0 | 85.0 | 0.59 |
| sodium sulfate | $Na_2SO_4$ | 142.0 | 71.0 | 0.70 |
| sodium sulfite | $Na_2SO_3$ | 126.1 | 63.0 | 0.79 |
| tetrasodium EDTA | $(CH_2)_2N_2(CH_2COONa)_4$ | 380.2 | 95.1 | 0.53 |
| trisodium phosphate | $Na_3PO_4 \cdot 12H_2O$ | 380.2 | 126.7 | 0.40 |
| trisodium phosphate (anhydrous) | $Na_3PO_4$ | 164.0 | 54.7 | 0.91 |
| trisodium NTA | $(CH_2)_3N(COONa)_3$ | 257.1 | 85.7 | 0.58 |

| gases | formula | molecular weight | equivalent weight | factor |
|---|---|---|---|---|
| ammonia | $NH_3$ | 17 | 17 | 2.94 |
| carbon dioxide | $CO_2$ | 44 | 22 | 2.27 |
| hydrogen | $H_2$ | 2 | 1 | 50.00 |
| hydrogen sulfide | $H_2S$ | 34 | 17 | 2.94 |
| oxygen | $O_2$ | 32 | 8 | 6.25 |

| acids | formula | molecular weight | equivalent weight | factor |
|---|---|---|---|---|
| carbonic | $H_2CO_3$ | 62.0 | 31.0 | 1.61 |
| hydrochloric | $HCl$ | 36.5 | 36.5 | 1.37 |
| phosphoric | $H_3PO_4$ | 98.0 | 32.7 | 1.53 |
| sulfuric | $H_2SO_4$ | 98.1 | 49.1 | 1.02 |

(Multiply the concentration (in mg/L) of the substance by the corresponding factors to obtain the equivalent concentration in mg/L as CaCO₃. For example, 70 mg/L of $Mg^{++}$ would be $(70 \text{ mg/L})(4.1) = 287$ mg/L as CaCO₃.)

## APPENDIX 33.C
### Saturation Values of Dissolved Oxygen in Water[*]

| temperature (°C) | chloride concentration in water (mg/L) | | | difference per 100 mg chloride | vapor pressure (mm Hg) |
|---|---|---|---|---|---|
| | 0 | 5000 | 10,000 | | |
| | dissolved oxygen (mg/L) | | | | |
| 0 | 14.62 | 13.79 | 12.97 | 0.017 | 5 |
| 1 | 14.23 | 13.41 | 12.61 | 0.016 | 5 |
| 2 | 13.84 | 13.05 | 12.28 | 0.015 | 5 |
| 3 | 13.48 | 12.72 | 11.98 | 0.015 | 6 |
| 4 | 13.13 | 12.41 | 11.69 | 0.014 | 6 |
| 5 | 12.80 | 12.09 | 11.39 | 0.014 | 7 |
| 6 | 12.48 | 11.79 | 11.12 | 0.014 | 7 |
| 7 | 12.17 | 11.51 | 10.85 | 0.013 | 8 |
| 8 | 11.87 | 11.24 | 10.61 | 0.013 | 8 |
| 9 | 11.59 | 10.97 | 10.36 | 0.012 | 9 |
| 10 | 11.33 | 10.73 | 10.13 | 0.012 | 9 |
| 11 | 11.08 | 10.49 | 9.92 | 0.011 | 10 |
| 12 | 10.83 | 10.28 | 9.72 | 0.011 | 11 |
| 13 | 10.60 | 10.05 | 9.52 | 0.011 | 11 |
| 14 | 10.37 | 9.85 | 9.32 | 0.010 | 12 |
| 15 | 10.15 | 9.65 | 9.14 | 0.010 | 13 |
| 16 | 9.95 | 9.46 | 8.96 | 0.010 | 14 |
| 17 | 9.74 | 9.26 | 8.78 | 0.010 | 15 |
| 18 | 9.54 | 9.07 | 8.62 | 0.009 | 16 |
| 19 | 9.35 | 8.89 | 8.45 | 0.009 | 17 |
| 20 | 9.17 | 8.73 | 8.30 | 0.009 | 18 |
| 21 | 8.99 | 8.57 | 8.14 | 0.009 | 19 |
| 22 | 8.83 | 8.42 | 7.99 | 0.008 | 20 |
| 23 | 8.68 | 8.27 | 7.85 | 0.008 | 21 |
| 24 | 8.53 | 8.12 | 7.71 | 0.008 | 22 |
| 25 | 8.38 | 7.96 | 7.56 | 0.008 | 24 |
| 26 | 8.22 | 7.81 | 7.42 | 0.008 | 25 |
| 27 | 8.07 | 7.67 | 7.28 | 0.008 | 27 |
| 28 | 7.92 | 7.53 | 7.14 | 0.008 | 28 |
| 29 | 7.77 | 7.39 | 7.00 | 0.008 | 30 |
| 30 | 7.63 | 7.25 | 6.86 | 0.008 | 32 |

[*]For saturation at barometric pressures other than 760 mm Hg (29.92 in Hg), $C'_s$ is related to the corresponding tabulated values, $C_s$, by the following equation.

$$C'_s = C_s \left( \frac{P - p}{760 - p} \right)$$

$C'_s$ = solubility at barometric pressure $P$ and given temperature, mg/L
$C_s$ = saturation at given temperature from table, mg/L
$P$ = barometric pressure, mm Hg
$p$ = pressure of saturated water vapor at temperature of the water selected from table, mm Hg

**APPENDIX 33.D**
Names and Formulas of Important Chemicals

| common name | chemical name | chemical formula |
|---|---|---|
| acetone | acetone | $(CH_3)_2CO$ |
| acetylene | acetylene | $C_2H_2$ |
| ammonia | ammonia | $NH_3$ |
| ammonium | ammonium hydroxide | $NH_4OH$ |
| aniline | aniline | $C_6H_5NH_2$ |
| bauxite | hydrated aluminum oxide | $Al_2O_3 \cdot 2H_2O$ |
| bleach | calcium hypochlorite | $CaCl(OCl)$ |
| borax | sodium tetraborate | $Na_2B_4O_7 \cdot 10H_2O$ |
| carbide | calcium carbide | $CaC_2$ |
| carbolic acid | phenol | $C_6H_5OH$ |
| carbon dioxide | carbon dioxide | $CO_2$ |
| carborundum | silicon carbide | $SiC$ |
| caustic potash | potassium hydroxide | $KOH$ |
| caustic soda/lye | sodium hydroxide | $NaOH$ |
| chalk | calcium carbonate | $CaCO_3$ |
| cinnabar | mercuric sulfide | $HgS$ |
| ether | diethyl ether | $(C_2H_5)_2O$ |
| Glauber's salt | decahydrated sodium sulfate | $Na_2SO_4 \cdot 10H_2O$ |
| glycerine | glycerine | $C_3H_5(OH)_3$ |
| grain alcohol | ethanol | $C_2H_5OH$ |
| graphite | crystaline carbon | $C$ |
| gypsum | calcium sulfate | $CaSO_4 \cdot 2H_2O$ |
| halite | sodium chloride | $NaCl$ |
| iron chloride | ferrous chloride | $FeCl_2 \cdot 4H_2O$ |
| laughing gas | nitrous oxide | $N_2O$ |
| limestone | calcium carbonate | $CaCO_3$ |
| magnesia | magnesium oxide | $MgO$ |
| marsh gas | methane | $CH_4$ |
| muriate of potash | potassium chloride | $KCl$ |
| muriatic acid | hydrochloric acid | $HCl$ |
| niter | sodium nitrate | $NaNO_3$ |
| niter cake | sodium bisulfate | $NaHSO_4$ |
| oleum | fuming sulfuric acid | $SO_3$ in $H_2SO_4$ |
| potash | potassium carbonate | $K_2CO_3$ |
| prussic acid | hydrogen cyanide | $HCN$ |
| pyrites | ferrous sulfide | $FeS$ |
| pyrolusite | manganese dioxide | $MnO_2$ |
| quicklime | calcium oxide | $CaO$ |
| sal soda | decahydrated sodium carbonate | $NaCO_3 \cdot 10H_2O$ |
| salammoniac | ammonium chloride | $NH_4Cl$ |
| sand or silica | silicon dioxide | $SiO_2$ |
| salt cake | sodium sulfate (crude) | $Na_2SO_4$ |
| slaked lime | calcium hydroxide | $Ca(OH)_2$ |
| soda ash | sodium carbonate | $Na_2CO_3$ |
| soot | amorphous carbon | $C$ |
| stannous chloride | stannous chloride | $SnCl_2 \cdot 2H_2O$ |
| superphosphate | monohydrated primary calcium phosphate | $Ca(H_2PO_4)_2 \cdot H_2O$ |
| table salt | sodium chloride | $NaCl$ |
| table sugar | sucrose | $C_{12}H_{22}O_{11}$ |
| trilene | trichloroethylene | $C_2HCl_3$ |
| urea | urea | $CO(NH_2)_2$ |
| washing soda | decahydrated sodium carbonate | $Na_2CO_3 \cdot 10H_2O$ |
| wood alcohol | methanol | $CH_3OH$ |
| zinc blende | zinc sulfide | $ZnS$ |

**APPENDIX 33.E**
Dissociation Constants of Acids at 25°C

| acid | | $K_a$ |
|---|---|---|
| acetic | $K_1$ | $1.8 \times 10^{-5}$ |
| arsenic | $K_1$ | $5.6 \times 10^{-3}$ |
| | $K_2$ | $1.2 \times 10^{-7}$ |
| | $K_3$ | $3 \times 10^{-12}$ |
| arsenious | $K_1$ | $1.4 \times 10^{-9}$ |
| benzoic | $K_1$ | $6.3 \times 10^{-5}$ |
| boric | $K_1$ | $5.9 \times 10^{-10}$ |
| carbonic | $K_1{}^*$ | $4.5 \times 10^{-7}$ |
| | $K_2$ | $5.6 \times 10^{-11}$ |
| chloroacetic | $K_1$ | $1.4 \times 10^{-3}$ |
| chromic | $K_2$ | $3 \times 10^{-7}$ |
| citric | $K_1$ | $7.4 \times 10^{-4}$ |
| | $K_2$ | $1.7 \times 10^{-5}$ |
| | $K_3$ | $3.9 \times 10^{-7}$ |
| ethylenedinitrilotetracetic | $K_1$ | $1 \times 10^{-2}$ |
| | $K_2$ | $2.1 \times 10^{-3}$ |
| | $K_3$ | $6.9 \times 10^{-7}$ |
| | $K_4$ | $7.4 \times 10^{-11}$ |
| formic | $K_1$ | $1.8 \times 10^{-4}$ |
| hydrocyanic | $K_1$ | $5 \times 10^{-10}$ |
| hydrofluoric | $K_1$ | $6 \times 10^{-4}$ |
| hydrogen sulfide | $K_1$ | $1.0 \times 10^{-8}$ |
| | $K_2$ | $1.2 \times 10^{-14}$ |
| hypochlorous | $K_1$ | $2.8 \times 10^{-8}$ |
| iodic | $K_1$ | $1.8 \times 10^{-1}$ |
| nitrous | $K_1$ | $5 \times 10^{-4}$ |
| oxalic | $K_1$ | $5.4 \times 10^{-2}$ |
| | $K_2$ | $5.1 \times 10^{-5}$ |
| phenol | $K_1$ | $1.1 \times 10^{-10}$ |
| o-phosphoric (ortho) | $K_1$ | $7.1 \times 10^{-3}$ |
| | $K_2$ | $6.3 \times 10^{-8}$ |
| | $K_3$ | $4.4 \times 10^{-13}$ |
| o-phthalic (ortho) | $K_1$ | $1.1 \times 10^{-3}$ |
| | $K_2$ | $3.9 \times 10^{-6}$ |
| salicylic | $K_1$ | $1.0 \times 10^{-3}$ |
| | $K_2$ | $4 \times 10^{-14}$ |
| sulfamic | $K_1$ | $1.0 \times 10^{-1}$ |
| sulfuric | $K_1$ | $1.1 \times 10^{-2}$ |
| sulfurous | $K_1$ | $1.7 \times 10^{-2}$ |
| | $K_2$ | $6.3 \times 10^{-8}$ |
| tartaric | $K_1$ | $9.2 \times 10^{-4}$ |
| | $K_2$ | $4.3 \times 10^{-5}$ |
| thiocyanic | $K_1$ | $1.4 \times 10^{-1}$ |

*apparent constant based on $C_{H_2CO_3} = [CO_2] + [H_2CO_3]$

Appendices

## APPENDIX 33.F
### Dissociation Constants of Bases at 25°C

| base | | $K_b$ |
|------|------|------|
| 2-amino-2-(hydroxymethyl)-1,3-propanediol | $K_1$ | $1.2 \times 10^{-6}$ |
| ammonia | $K_1$ | $1.8 \times 10^{-5}$ |
| aniline | $K_1$ | $4.2 \times 10^{-10}$ |
| diethylamine | $K_1$ | $1.3 \times 10^{-3}$ |
| hexamethylenetetramine | $K_1$ | $1 \times 10^{-9}$ |
| hydrazine | $K_1$ | $9.8 \times 10^{-7}$ |
| hydroxylamine | $K_1$ | $9.6 \times 10^{-9}$ |
| lead hydroxide | $K_1$ | $1.2 \times 10^{-4}$ |
| piperidine | $K_1$ | $1.3 \times 10^{-3}$ |
| pyridine | $K_1$ | $1.5 \times 10^{-9}$ |
| silver hydroxide | $K_1$ | $6.0 \times 10^{-5}$ |

Appendices

**APPENDIX 33.G**
Approximate Solubility Products at Room Temperature

| compound | $K_{sp}$ |
|---|---|
| AgCl | $1.56 \times 10^{-10}$ |
| AgBr | $3.3 \times 10^{-13}$ |
| AgI | $8.5 \times 10^{-17}$ |
| $Ag_3AsO_4$ | $1 \times 10^{-23}$ |
| $Ag_2CrO_4$ | $1.1 \times 10^{-12}$ |
| $Ag_3PO_4$ | $1.6 \times 10^{-18}$ |
| $BaCO_3$ | $8.1 \times 10^{-9}$ |
| $BaC_2O_4$ | $1 \times 10^{-7}$ |
| $BaCrO_4$ | $2 \times 10^{-10}$ |
| $BaSO_4$ | $1.1 \times 10^{-10}$ |
| $CaCO_3$ | $8.7 \times 10^{-9}$ |
| $CaSO_4$ | $6.1 \times 10^{-5}$ |
| $Ca_3(PO_4)_2$ | $1 \times 10^{-25}$ |
| $CaC_2O_4$ | $2.6 \times 10^{-9}$ |
| $Hg_2Cl_2$ | $1.1 \times 10^{-18}$ |
| $Hg_2Br_2$ | $5.2 \times 10^{-23}$ |
| $Hg_2I_2$ | $4.5 \times 10^{-29}$ |
| $MgCO_3$ | $2.6 \times 10^{-5}$ |
| $MgNH_4PO_4$ | $2.5 \times 10^{-13}$ |
| $PbCl_2$ | $2.4 \times 10^{-4}$ |
| $PbBr_2$ | $6.3 \times 10^{-6}$ |
| $PbI_2$ | $8.7 \times 10^{-9}$ |
| $PbCO_3$ | $1.5 \times 10^{-13}$ |
| $PbCrO_4$ | $1.8 \times 10^{-14}$ |
| $PbSO_4$ | $1.8 \times 10^{-8}$ |
| $Pb_3(PO_4)_2$ | $3 \times 10^{-44}$ |
| $SrCO_3$ | $9 \times 10^{-10}$ |
| $SrC_2O_4$ | $5.6 \times 10^{-8}$ |
| $SrCrO_4$ | $3.6 \times 10^{-5}$ |
| $SrSO_4$ | $2.8 \times 10^{-7}$ |
| $Al(OH)_3$ | $1.9 \times 10^{-33}$ |
| $Cd(OH)_2$ | $1.2 \times 10^{-14}$ |
| $Co(OH)_2$ | $2 \times 10^{-16}$ |
| $Cr(OH)_3$ | $6.7 \times 10^{-31}$ |
| $Fe(OH)_3$ | $4 \times 10^{-38}$ |
| $Mg(OH)_2$ | $5.5 \times 10^{-12}$ |
| $Mn(OH)_2$ | $4.5 \times 10^{-14}$ |
| $Ni(OH)_2$ | $1.6 \times 10^{-14}$ |
| $Pb(OH)_2$ | $2.8 \times 10^{-16}$ |
| $Sn(OH)_2$ | $5 \times 10^{-26}$ |
| $Zn(OH)_2$ | $4.5 \times 10^{-17}$ |
| $Ag_2S$ | $1 \times 10^{-51}$ |
| $Bi_2S_3$ | $1.6 \times 10^{-72}$ |
| CdS | $3.6 \times 10^{-29}$ |
| CoS | $2 \times 10^{-27}$ |
| CuS | $4 \times 10^{-38}$ |
| FeS | $3.7 \times 10^{-19}$ |
| HgS | $3 \times 10^{-54}$ |
| MnS | $1.4 \times 10^{-15}$ |
| NiS | $1 \times 10^{-26}$ |
| PbS | $3.4 \times 10^{-28}$ |
| SnS | $8 \times 10^{-29}$ |
| ZnS | $4.5 \times 10^{-24}$ |
| $H_2S$ | $1.1 \times 10^{-23}$ |

## APPENDIX 35.A
Heats of Combustion for Common Compounds[a]

| substance | formula | molecular weight | specific volume (ft³/lbm) | heat of combustion Btu/ft³ gross (high) | Btu/ft³ net (low) | Btu/lbm gross (high) | Btu/lbm net (low) |
|---|---|---|---|---|---|---|---|
| carbon | C | 12.01 | | | | 14,093 | 14,093 |
| carbon dioxide | $CO_2$ | 44.01 | 8.548 | | | | |
| carbon monoxide | CO | 28.01 | 13.506 | 322 | 322 | 4347 | 4347 |
| hydrogen | $H_2$ | 2.016 | 187.723 | 325 | 275 | 60,958 | 51,623 |
| nitrogen | $N_2$ | 28.016 | 13.443 | | | | |
| oxygen | $O_2$ | 32.000 | 11.819 | | | | |
| paraffin series (alkanes) | | | | | | | |
| methane | $CH_4$ | 16.041 | 23.565 | 1013 | 913 | 23,879 | 21,520 |
| ethane | $C_2H_6$ | 30.067 | 12.455 | 1792 | 1641 | 22,320 | 20,432 |
| propane | $C_3H_8$ | 44.092 | 8.365 | 2590 | 2385 | 21,661 | 19,944 |
| n-butane | $C_4H_{10}$ | 58.118 | 6.321 | 3370 | 3113 | 21,308 | 19,680 |
| isobutane | $C_4H_{10}$ | 58.118 | 6.321 | 3363 | 3105 | 21,257 | 19,629 |
| n-pentane | $C_5H_{12}$ | 72.144 | 5.252 | 4016 | 3709 | 21,091 | 19,517 |
| isopentane | $C_5H_{12}$ | 72.144 | 5.252 | 4008 | 3716 | 21,052 | 19,478 |
| neopentane | $C_5H_{12}$ | 72.144 | 5.252 | 3993 | 3693 | 20,970 | 19,396 |
| n-hexane | $C_6H_{14}$ | 86.169 | 4.398 | 4762 | 4412 | 20,940 | 19,403 |
| olefin series (alkenes and alkynes) | | | | | | | |
| ethylene | $C_2H_4$ | 28.051 | 13.412 | 1614 | 1513 | 21,644 | 20,295 |
| propylene | $C_3H_6$ | 42.077 | 9.007 | 2336 | 2186 | 21,041 | 19,691 |
| n-butene | $C_4H_8$ | 56.102 | 6.756 | 3084 | 2885 | 20,840 | 19,496 |
| isobutene | $C_4H_8$ | 56.102 | 6.756 | 3068 | 2869 | 20,730 | 19,382 |
| n-pentene | $C_5H_{10}$ | 70.128 | 5.400 | 3836 | 3586 | 20,712 | 19,363 |
| aromatic series | | | | | | | |
| benzene | $C_6H_6$ | 78.107 | 4.852 | 3751 | 3601 | 18,210 | 17,480 |
| toluene | $C_7H_8$ | 92.132 | 4.113 | 4484 | 4284 | 18,440 | 17,620 |
| xylene | $C_8H_{10}$ | 106.158 | 3.567 | 5230 | 4980 | 18,650 | 17,760 |
| miscellaneous gases | | | | | | | |
| acetylene | $C_2H_2$ | 26.036 | 14.344 | 1499 | 1448 | 21,500 | 20,776 |
| air | | 28.9 | 13.063 | | | | |
| ammonia | $NH_3$ | 17.031 | 21.914 | 441 | 365 | 9668 | 8001 |
| digester gas[b] | – | 25.8 | 18.3 | 658 | 593 | 15,521 | 13,988 |
| ethyl alcohol | $C_2H_5OH$ | 46.067 | 8.221 | 1600 | 1451 | 13,161 | 11,929 |
| hydrogen sulfide | $H_2S$ | 34.076 | 10.979 | 647 | 596 | 7100 | 6545 |
| iso-octane | $C_8H_{18}$ | 114.2 | 0.0232[c] | | 98.9[c] | 20,590 | 19,160 |
| methyl alcohol | $CH_3OH$ | 32.041 | 11.820 | 868 | 768 | 10,259 | 9078 |
| naphthalene | $C_{10}H_8$ | 128.162 | 2.955 | 5854 | 5654 | 17,298 | 16,708 |
| sulfur | S | 32.06 | | | | 3983 | 3983 |
| sulfur dioxide | $SO_2$ | 64.06 | 5.770 | | | | |
| water vapor | $H_2O$ | 18.016 | 21.017 | | | | |

(Multiply Btu/lbm by 2.326 to obtain kJ/kg.)
(Multiply Btu/ft³ by 37.25 to obtain kJ/m³.)
[a]Gas volumes listed are at 60°F (16°C) and 1 atm.
[b]Digester gas from wastewater treatment plants is approximately 65% methane and 35% carbon dioxide by volume. Use composite properties of these two gases.
[c]liquid form

Appendices

**APPENDIX 40.A**
Representative Thermal Conductivity[a,b]
(at 32°F (0°C) unless specified otherwise)

| material | $\dfrac{\text{Btu}}{\text{hr-ft}^2\text{-}°\text{F}}$ | $\dfrac{\text{W}}{\text{m·K}}$ |
|---|---|---|
| air | 0.014 | 0.024 |
| aluminum | 117 | 202 |
| asbestos | 0.087 | 0.15 |
| brass | 56 | 97 |
| brick, fire clay (400°F) | 0.58 | 1.0 |
| concrete | 0.5 | 0.9 |
| copper | 224 | 388 |
| cork | 0.025 | 0.043 |
| fiberglass | 0.03 | 0.05 |
| glass | 0.63 | 1.1 |
| glass, Pyrex™ | 0.68 | 1.2 |
| gold (68°F) | 169 | 292 |
| hydrogen (100°F) | 0.11 | 0.19 |
| ice | 1.3 | 2.2 |
| iron, cast (4% C, 68°F) | 30 | 52 |
| iron, pure | 36 | 62 |
| lead (70°F) | 20 | 35 |
| mercury | 4.83 | 8.36 |
| nickel | 34.4 | 59.5 |
| oxygen | 0.016 | 0.028 |
| rubber, soft | 0.10 | 0.17 |
| silver | 242 | 419 |
| steel (1% C)[c] | 26–37 | 45–64 |
| tungsten | 92 | 160 |
| water | 0.32 | 0.55 |
| zinc | 65 | 110 |

(Multiply Btu-ft/hr-ft²-°F by 12 to get Btu-in/hr-ft²-°F.)
(Multiply Btu-ft/hr-ft²-°F by 1.7307 to get W/m·K.)
(Multiply Btu-ft/hr-ft²-°F by 4.1365 × 10⁻³ to get cal·cm/s·cm²·°C.)
[a]Values of thermal conductivity are typically accurate to only ±20%, although in some cases the error may be as small as ±10%.
[b]Values are compiled from a variety of sources.
[c]depends greatly on alloy and carbon content, as well as temperature

**APPENDIX 40.B**
Properties of Metals and Alloys[a,b]

| metal | thermal conductivity, $k$ $\left(\dfrac{Btu}{hr\text{-}ft^2\text{-}°F}\right)$ | | | | $c_p$ $\left(\dfrac{Btu}{lbm\text{-}°F}\right)$ | $\rho$ $\left(\dfrac{lbm}{ft^3}\right)$ |
|---|---|---|---|---|---|---|
| | 32°F (0°C) | 212°F (100°C) | 572°F (300°C) | 932°F (500°C) | 32°F (0°C) | 32°F (0°C) |
| aluminum alloy | 92 | 104 | – | – | – | – |
| aluminum, pure | 117 | 119 | 133 | 155 | 0.208 | 169 |
| brass (70% Cu, 30% Zn) | 58 | 60 | 66 | – | 0.092 | 532 |
| bronze (75% Cu, 25% Sn) | 15 | – | – | – | 0.082 | 540 |
| copper, pure | 224 | 218 | 212 | 207 | 0.091 | 558 |
| iron, cast, alloy | 30 | 28.3 | 27.0 | – | 0.10 | 455 |
| iron, cast, plain | 33 | 31.8 | 27.7 | 24.8 | 0.11 | 474 |
| iron, pure | 35.8 | 36.6 | – | – | 0.104 | 491 |
| lead | 20.1 | 19 | 18 | – | 0.030 | 705 |
| magnesium | 91 | 92 | – | – | 0.232 | 109 |
| nickel/chrome | 7.5 | 9.2 | – | – | – | – |
| silver | 242 | 238 | – | – | 0.056 | 655 |
| steel, carbon (1% C) | 26.5 | 26.0 | 25.0 | 22.0 | 0.11 | 490 |
| steel, stainless | 8.0 | 9.3 | 11.0 | 12.8 | 0.11 | 488 |
| tin | 36 | 34 | – | – | 0.054 | 456 |
| zinc | 65 | 64 | 59 | – | 0.091 | 446 |

(Multiply Btu-ft/hr-ft$^2$-°F by 12 to get Btu-in/hr-ft$^2$-°F.)
(Multiply Btu-ft/hr-ft$^2$-°F by 1.7307 to get W/m·K.)
(Multiply Btu-ft/hr-ft$^2$-°F by $4.1365 \times 10^{-3}$ to get cal·cm/s·cm$^2$·°C.)
(Multiply Btu/lbm-°F by 4186.8 to obtain J/kg·K.)
(Multiply lbm/ft$^3$ by 16.0185 to obtain kg/m$^3$.)
[a]Values of thermal conductivity are typically accurate to only ±20%, although in some cases the error may be as small as ±10%.
[b]Values are compiled from a variety of sources.

Appendices

### APPENDIX 40.C
Properties of Nonmetals[a,b]

| material | average temperature (°F) | $k$ $\left(\dfrac{\text{Btu}}{\text{hr-ft}^2\text{-}°\text{F}}\right)$ | $c_p$ $\left(\dfrac{\text{Btu}}{\text{lbm-}°\text{F}}\right)$ | $\rho$ $\left(\dfrac{\text{lbm}}{\text{ft}^3}\right)$ |
|---|---|---|---|---|
| asbestos | 32 | 0.087 | 0.25 | 36 |
|  | 392 | 0.12 |  |  |
| brick, building | 70 | 0.38 | 0.20 | 106 |
| brick, fire-clay | 392 | 0.58 | 0.20 | 144 |
|  | 1832 | 0.95 |  |  |
| brick, Kaolin |  |  |  |  |
|   insulating | 932 | 0.15 |  | 27 |
|  | 2102 | 0.26 |  |  |
|   firebrick | 392 | 0.05 |  | 19 |
|  | 1400 | 0.11 |  |  |
| concrete, stone | 70 | 0.54 | 0.20 | 144 |
|   with 10% moisture | 70 | 0.70 |  | 140 |
| diatomaceous earth |  |  |  |  |
|   powdered | 100 | 0.030 | 0.21 | 14 |
|  | 300 | 0.036 |  |  |
|  | 600 | 0.046 |  |  |
| glass, window | 70 | 0.45 | 0.2 | 170 |
| glass wool (fine) | 20 | 0.022 |  |  |
|  | 100 | 0.031 |  | 1.5 |
|  | 200 | 0.043 |  |  |
| glass wool (packed) | 20 | 0.016 |  |  |
|  | 100 | 0.022 |  | 6.0 |
|  | 200 | 0.029 |  |  |
| ice | 32 | 1.28 | 0.46 | 57 |
| magnesia, 85%, molded pipe | 32 | 0.032 |  | 17 |
|   covering ($T < 600°\text{F}$) | 200 | 0.037 |  |  |
| molded pipe covering, | 400 | 0.051 |  | 26 |
|   diatomaceous silica | 1600 | 0.088 |  |  |
| sand, dry | 68 | 0.20 |  | 95 |
| sand, with 10% water | 68 | 0.60 |  | 100 |
| soil, dry | 70 | 0.20 | 0.44 |  |
| soil, wet | 70 | 1.5 |  |  |
| wood, oak |  |  |  |  |
|   perpendicular to grain | 70 | 0.12 | 0.57 | 51 |
|   parallel to grain | 70 | 0.20 | 0.57 |  |
| wood, pine |  |  |  |  |
|   perpendicular to grain | 70 | 0.06 | 0.67 | 31 |
|   parallel to grain | 70 | 0.14 | 0.67 |  |

(Multiply Btu-ft/hr-ft$^2$-°F by 12 to get Btu-in/hr-ft$^2$-°F.)
(Multiply Btu-ft/hr-ft$^2$-°F by 1.7307 to get W/m·K.)
(Multiply Btu-ft/hr-ft$^2$-°F by 4.1365 × 10$^{-3}$ to get cal·cm/s·cm$^2$·°C.)
(Multiply Btu/lbm-°F by 4186.8 to obtain J/kg·K.)
(Multiply lbm/ft$^3$ by 16.0185 to obtain kg/m$^3$.)
[a]Values of thermal conductivity are typically accurate to only ±20%, although in some cases the error may be as small as ±10%.
[b]Values are compiled from a variety of sources.

**APPENDIX 41.A**
Properties of Saturated Water
(customary U.S. units)

| $T$ (°F) | $\rho$ (lbm/ft$^3$) | $c_p$ (Btu/lbm-°F) | $\mu$ (lbm/ft-sec) | $\nu$ (ft$^2$/sec) | $k$ (Btu/hr-ft-°F) | Pr | $\beta$ (1/°F) | $\dfrac{g\beta\rho^2}{\mu^2}$ (1/ft$^3$-F) |
|---|---|---|---|---|---|---|---|---|
| 32 | 62.4 | 1.01 | $1.20 \times 10^{-3}$ | $1.93 \times 10^{-5}$ | 0.319 | 13.7 | $-0.37 \times 10^{-4}$ | |
| 40 | 62.4 | 1.00 | $1.04 \times 10^{-3}$ | $1.67 \times 10^{-5}$ | 0.325 | 11.6 | $0.20 \times 10^{-4}$ | |
| 50 | 62.4 | 1.00 | $0.88 \times 10^{-3}$ | $1.40 \times 10^{-5}$ | 0.332 | 9.55 | $0.49 \times 10^{-4}$ | $2.3 \times 10^6$ |
| 60 | 62.3 | 0.999 | $0.76 \times 10^{-3}$ | $1.22 \times 10^{-5}$ | 0.340 | 8.03 | $0.85 \times 10^{-4}$ | $8.0 \times 10^6$ |
| 70 | 62.3 | 0.998 | $0.658 \times 10^{-3}$ | $1.06 \times 10^{-5}$ | 0.347 | 6.82 | $1.2 \times 10^{-4}$ | $18.4 \times 10^6$ |
| 80 | 62.2 | 0.998 | $0.578 \times 10^{-3}$ | $0.93 \times 10^{-5}$ | 0.353 | 5.89 | $1.5 \times 10^{-4}$ | $34.6 \times 10^6$ |
| 90 | 62.1 | 0.997 | $0.514 \times 10^{-3}$ | $0.825 \times 10^{-5}$ | 0.359 | 5.13 | $1.8 \times 10^{-4}$ | $56.0 \times 10^6$ |
| 100 | 62.0 | 0.998 | $0.458 \times 10^{-3}$ | $0.740 \times 10^{-5}$ | 0.364 | 4.52 | $2.0 \times 10^{-4}$ | $85.0 \times 10^6$ |
| 150 | 61.2 | 1.00 | $0.292 \times 10^{-3}$ | $0.477 \times 10^{-5}$ | 0.384 | 2.74 | $3.1 \times 10^{-4}$ | $118 \times 10^6$ |
| 200 | 60.1 | 1.00 | $0.205 \times 10^{-3}$ | $0.341 \times 10^{-5}$ | 0.394 | 1.88 | $4.0 \times 10^{-4}$ | $440 \times 10^6$ |
| 250 | 58.8 | 1.01 | $0.158 \times 10^{-3}$ | $0.269 \times 10^{-5}$ | 0.396 | 1.45 | $4.8 \times 10^{-4}$ | $1.11 \times 10^9$ |
| 300 | 57.3 | 1.03 | $0.126 \times 10^{-3}$ | $0.220 \times 10^{-5}$ | 0.395 | 1.18 | $6.0 \times 10^{-4}$ | $2.14 \times 10^9$ |
| 350 | 55.6 | 1.05 | $0.105 \times 10^{-3}$ | $0.189 \times 10^{-5}$ | 0.391 | 1.02 | $6.9 \times 10^{-4}$ | $4.00 \times 10^9$ |
| 400 | 53.6 | 1.08 | $0.091 \times 10^{-3}$ | $0.170 \times 10^{-5}$ | 0.381 | 0.927 | $8.0 \times 10^{-4}$ | $6.24 \times 10^9$ |
| 450 | 51.6 | 1.12 | $0.080 \times 10^{-3}$ | $0.155 \times 10^{-5}$ | 0.367 | 0.876 | $9.0 \times 10^{-4}$ | $8.95 \times 10^9$ |
| 500 | 49.0 | 1.19 | $0.071 \times 10^{-3}$ | $0.145 \times 10^{-5}$ | 0.349 | 0.87 | $10.0 \times 10^{-4}$ | $12.1 \times 10^9$ |
| 550 | 45.9 | 1.31 | $0.064 \times 10^{-3}$ | $0.139 \times 10^{-5}$ | 0.325 | 0.93 | $11.0 \times 10^{-4}$ | $15.3 \times 10^9$ |
| 600 | 42.4 | 1.51 | $0.058 \times 10^{-3}$ | $0.137 \times 10^{-5}$ | 0.292 | 1.09 | $12.0 \times 10^{-4}$ | $17.8 \times 10^9$ |
| | | | | | | | | $20.6 \times 10^9$ |

(Multiply Btu/lbm°F by 4187 to obtain J/kg·K.)
(Multiply lbm/sec-ft by 3600 to obtain lbm/hr-ft.)
(Multiply lbm/sec-ft by 1.488 to obtain kg/s·m.)
(Multiply ft$^2$/sec by 0.0929 to obtain m$^2$/s.)
(Multiply Btu/hr-ft-°F by 1.730 to obtain W/m·K.)
(Multiply 1/°F by 5/9 to obtain 1/K.)
(Multiply 1/ft$^3$-°F by 19.611 to obtain 1/m$^3$·K.)

Source: *Handbook of Thermodynamic Tables and Charts* by K. Raznjevic, copyright © 1976.

## APPENDIX 41.B
Properties of Saturated Water
(SI units)

| $T$ (°C) | $T$ (K) | $\rho$ (kg/m$^3$) | $c_p$ (kJ/kg·K) | $\mu$ (kg/m·s) | $k$ (W/m·K) | Pr | $\beta$ (1/K) | $\dfrac{g\beta\rho^2}{\mu^2}$ (1/K·m$^3$) |
|---|---|---|---|---|---|---|---|---|
| 0 | 273.2 | 999.6 | 4.229 | $1.786 \times 10^{-3}$ | 0.5694 | 13.3 | $-0.630 \times 10^{-4}$ | |
| 15.6 | 288.8 | 998.0 | 4.187 | $1.131 \times 10^{-3}$ | 0.5884 | 8.07 | $1.44 \times 10^{-4}$ | $10.93 \times 10^8$ |
| 26.7 | 299.9 | 996.4 | 4.183 | $0.860 \times 10^{-3}$ | 0.6109 | 5.89 | $2.34 \times 10^{-4}$ | $30.70 \times 10^8$ |
| 37.8 | 311.0 | 994.7 | 4.183 | $0.682 \times 10^{-3}$ | 0.6283 | 4.51 | $3.24 \times 10^{-4}$ | $68.0 \times 10^8$ |
| 65.6 | 338.8 | 981.9 | 4.187 | $0.432 \times 10^{-3}$ | 0.6629 | 2.72 | $5.04 \times 10^{-4}$ | $256.2 \times 10^8$ |
| 93.3 | 366.5 | 962.7 | 4.229 | $0.3066 \times 10^{-3}$ | 0.6802 | 1.91 | $6.66 \times 10^{-4}$ | $642 \times 10^8$ |
| 121.1 | 394.3 | 943.5 | 4.271 | $0.2381 \times 10^{-3}$ | 0.6836 | 1.49 | $8.46 \times 10^{-4}$ | $1300 \times 10^8$ |
| 148.9 | 422.1 | 917.9 | 4.312 | $0.1935 \times 10^{-3}$ | 0.6836 | 1.22 | $10.08 \times 10^{-4}$ | $2231 \times 10^8$ |
| 204.4 | 477.6 | 858.6 | 4.522 | $0.1384 \times 10^{-3}$ | 0.6611 | 0.950 | $14.04 \times 10^{-4}$ | $5308 \times 10^8$ |
| 260.0 | 533.2 | 784.9 | 4.982 | $0.1042 \times 10^{-3}$ | 0.6040 | 0.859 | $19.8 \times 10^{-4}$ | $11\,030 \times 10^8$ |
| 315.6 | 588.8 | 679.4 | 6.322 | $0.0862 \times 10^{-3}$ | 0.5071 | 1.07 | $31.5 \times 10^{-4}$ | $19\,260 \times 10^8$ |

(Multiply lbm/ft$^3$ by 16.0185 to obtain kg/m$^3$.)
(Multiply Btu/lbm-°F by 4187 to obtain J/kg·K.)
(Multiply lbm/sec-ft by 3600 to obtain lbm/hr-ft.)
(Multiply lbm/sec-ft by 1.488 to obtain kg/s·m.)
(Multiply Btu/hr-ft-°F by 1.730 to obtain W/m·K.)
(Multiply 1/°F by 5/9 to obtain 1/K.)
(Multiply 1/ft$^3$-°F by 19.611 to obtain 1/m$^3$·K.)
(Multiply ft$^2$/sec ty 0.0929 to obtain m$^2$/s.)

Geankoplis, Christie John, *Transport Processes and Separation Process Principles*, 4th ed., copyright © 2003. Reprinted by permission of Pearson Education, Inc., Upper Saddle River, NJ.

**APPENDIX 41.C**
Properties of Atmospheric Air[*]
(customary U.S. units)

| $T$ ($°F$) | $\rho$ (lbm/ft³) | $c_p$ (Btu/lbm-°F) | $\mu$ (lbm/ft-sec) | $\nu$ (ft²/sec) | $k$ (Btu/hr-ft-°F) | Pr | $\beta$ (1/°F) | $\frac{g\beta\rho^2}{\mu^2}$ (1/ft³-F) |
|---|---|---|---|---|---|---|---|---|
| 0 | 0.086 | 0.239 | $1.110 \times 10^{-5}$ | $0.130 \times 10^{-3}$ | 0.0133 | 0.73 | $2.18 \times 10^{-3}$ | $4.2 \times 10^6$ |
| 32 | 0.081 | 0.240 | $1.165 \times 10^{-5}$ | $0.145 \times 10^{-3}$ | 0.0140 | 0.72 | $2.03 \times 10^{-3}$ | $3.16 \times 10^6$ |
| 100 | 0.071 | 0.240 | $1.285 \times 10^{-5}$ | $0.180 \times 10^{-3}$ | 0.0154 | 0.72 | $1.79 \times 10^{-3}$ | $1.76 \times 10^6$ |
| 200 | 0.060 | 0.241 | $1.440 \times 10^{-5}$ | $0.239 \times 10^{-3}$ | 0.0174 | 0.72 | $1.52 \times 10^{-3}$ | $0.850 \times 10^6$ |
| 300 | 0.052 | 0.243 | $1.610 \times 10^{-5}$ | $0.306 \times 10^{-3}$ | 0.0193 | 0.71 | $1.32 \times 10^{-3}$ | $0.444 \times 10^6$ |
| 400 | 0.046 | 0.245 | $1.75 \times 10^{-5}$ | $0.378 \times 10^{-3}$ | 0.0212 | 0.689 | $1.16 \times 10^{-3}$ | $0.258 \times 10^6$ |
| 500 | 0.0412 | 0.247 | $1.890 \times 10^{-5}$ | $0.455 \times 10^{-3}$ | 0.0231 | 0.683 | $1.04 \times 10^{-3}$ | $0.159 \times 10^6$ |
| 600 | 0.0373 | 0.250 | $2.000 \times 10^{-5}$ | $0.540 \times 10^{-3}$ | 0.0250 | 0.685 | $0.943 \times 10^{-3}$ | $0.106 \times 10^6$ |
| 700 | 0.0341 | 0.253 | $2.14 \times 10^{-5}$ | $0.625 \times 10^{-3}$ | 0.0268 | 0.690 | $0.862 \times 10^{-3}$ | $70.4 \times 10^3$ |
| 800 | 0.0314 | 0.256 | $2.25 \times 10^{-5}$ | $0.717 \times 10^{-3}$ | 0.0286 | 0.697 | $0.794 \times 10^{-3}$ | $49.8 \times 10^3$ |
| 900 | 0.0291 | 0.259 | $2.36 \times 10^{-5}$ | $0.815 \times 10^{-3}$ | 0.0303 | 0.705 | $0.735 \times 10^{-3}$ | $36.0 \times 10^3$ |
| 1000 | 0.0271 | 0.262 | $2.47 \times 10^{-5}$ | $0.917 \times 10^{-3}$ | 0.0319 | 0.713 | $0.685 \times 10^{-3}$ | $26.5 \times 10^3$ |
| 1500 | 0.0202 | 0.276 | $3.00 \times 10^{-5}$ | $1.47 \times 10^{-3}$ | 0.0400 | 0.739 | $0.510 \times 10^{-3}$ | $7.45 \times 10^3$ |
| 2000 | 0.0161 | 0.286 | $3.54 \times 10^{-5}$ | $2.14 \times 10^{-3}$ | 0.0471 | 0.753 | $0.406 \times 10^{-3}$ | $2.84 \times 10^3$ |
| 2500 | 0.0133 | 0.292 | $3.69 \times 10^{-5}$ | $2.80 \times 10^{-3}$ | 0.051 | 0.763 | $0.338 \times 10^{-3}$ | $1.41 \times 10^3$ |
| 3000 | 0.0114 | 0.297 | $3.85 \times 10^{-5}$ | $3.39 \times 10^{-3}$ | 0.054 | 0.765 | $0.289 \times 10^{-3}$ | $0.815 \times 10^3$ |

(Multiply lbm/ft³ by 16.0185 to obtain kg/m³.)
(Multiply Btu/lbm-°F by 4187 to obtain J/kg·K.)
(Multiply lbm/sec-ft by 3600 to obtain lbm/hr-ft.)
(Multiply lbm/sec-ft by 1.488 to obtain kg/s·m.)
(Multiply Btu/hr-ft-°F by 1.730 to obtain W/m·K.)
(Multiply 1/°F by 5/9 to obtain 1/K.)
(Multiply 1/ft³-°F by 19.611 to obtain 1/m³·K.)
(Multiply ft²/sec ty 0.0929 to obtain m²/s.)

[*] $\mu$, $k$, $c_p$, and Pr do not greatly depend on pressure and may be used over a wide range of pressures.

Source: *Handbook of Thermodynamic Tables and Charts* by K. Raznjevic, copyright © 1976.

### APPENDIX 41.D
Properties of Atmospheric Air[*]
(SI units)

| $T$ (°C) | $T$ (K) | $\rho$ (kg/m³) | $c_p$ (kJ/kg·K) | $\mu$ (kg/m·s) | $k$ (W/m·K) | Pr | $\beta$ (1/K) | $\dfrac{g\beta\rho^2}{\mu^2}$ (1/K·m³) |
|---|---|---|---|---|---|---|---|---|
| −17.8 | 255.4 | 1.379 | 1.0048 | $1.62 \times 10^{-5}$ | 0.02250 | 0.720 | $3.92 \times 10^{-3}$ | $2.79 \times 10^{8}$ |
| 0 | 273.2 | 1.293 | 1.0048 | $1.72 \times 10^{-5}$ | 0.02423 | 0.715 | $3.65 \times 10^{-3}$ | $2.04 \times 10^{8}$ |
| 10.0 | 283.2 | 1.246 | 1.0048 | $1.78 \times 10^{-5}$ | 0.02492 | 0.713 | $3.53 \times 10^{-3}$ | $1.72 \times 10^{8}$ |
| 37.8 | 311.0 | 1.137 | 1.0048 | $1.90 \times 10^{-5}$ | 0.02700 | 0.705 | $3.22 \times 10^{-3}$ | $1.12 \times 10^{8}$ |
| 65.6 | 338.8 | 1.043 | 1.0090 | $2.03 \times 10^{-5}$ | 0.02925 | 0.702 | $2.95 \times 10^{-3}$ | $0.775 \times 10^{8}$ |
| 93.3 | 366.5 | 0.964 | 1.0090 | $2.15 \times 10^{-5}$ | 0.03115 | 0.694 | $2.74 \times 10^{-3}$ | $0.534 \times 10^{8}$ |
| 121.1 | 394.3 | 0.895 | 1.0132 | $2.27 \times 10^{-5}$ | 0.03323 | 0.692 | $2.54 \times 10^{-3}$ | $0.386 \times 10^{8}$ |
| 148.9 | 422.1 | 0.838 | 1.0174 | $2.37 \times 10^{-5}$ | 0.03531 | 0.689 | $2.38 \times 10^{-3}$ | $0.289 \times 10^{8}$ |
| 176.7 | 449.9 | 0.785 | 1.0216 | $2.50 \times 10^{-5}$ | 0.03721 | 0.687 | $2.21 \times 10^{-3}$ | $0.214 \times 10^{8}$ |
| 204.4 | 477.6 | 0.740 | 1.0258 | $2.60 \times 10^{-5}$ | 0.03894 | 0.686 | $2.09 \times 10^{-3}$ | $0.168 \times 10^{8}$ |
| 232.2 | 505.4 | 0.700 | 1.0300 | $2.71 \times 10^{-5}$ | 0.04084 | 0.684 | $1.98 \times 10^{-3}$ | $0.130 \times 10^{8}$ |
| 260.0 | 533.2 | 0.662 | 1.0341 | $2.80 \times 10^{-5}$ | 0.04258 | 0.680 | $1.87 \times 10^{-3}$ | $0.104 \times 10^{8}$ |

(Multiply lbm/ft³ by 16.0185 to obtain kg/m³.)
(Multiply Btu/lbm-°F by 4187 to obtain J/kg·K.)
(Multiply lbm/sec-ft by 3600 to obtain lbm/hr-ft.)
(Multiply lbm/sec-ft by 1.488 to obtain kg/s·m.)
(Multiply ft²/sec ty 0.0929 to obtain m²/s.)
(Multiply Btu/hr-ft-°F by 1.730 to obtain W/m·K.)
(Multiply 1/°F by 5/9 to obtain 1/K.)
(Multiply 1/ft³-°F by 19.611 to obtain 1/m³·K.)
[*]$\mu$, $k$, $c_p$, and Pr do not greatly depend on pressure and may be used over a wide range of pressures.

Geankoplis, Christie John, *Transport Processes and Separation Process Principles*, 4th ed., copyright © 2003. Reprinted by permission of Pearson Education, Inc., Upper Saddle River, NJ.

**Appendices**

**APPENDIX 41.E**
Properties of Steam at One Atmosphere[*]
(customary U.S. units)

| $T$ (°F) | $\rho$ (lbm/ft$^3$) | $c_p$ (Btu/lbm-°F) | $\mu$ (lbm/ft-sec) | $\nu$ (ft$^2$/sec) | $k$ (Btu/hr-ft-°F) | Pr | $\beta$ (1/°F) | $\dfrac{g\beta\rho^2}{\mu^2}$ (1/ft$^3$-F) |
|---|---|---|---|---|---|---|---|---|
| 212 | 0.0372 | 0.451 | $0.870 \times 10^{-5}$ | $0.234 \times 10^{-3}$ | 0.0145 | 0.96 | $1.49 \times 10^{-3}$ | $0.877 \times 10^{6}$ |
| 300 | 0.0328 | 0.456 | $1.00 \times 10^{-5}$ | $0.303 \times 10^{-3}$ | 0.0171 | 0.95 | $1.32 \times 10^{-3}$ | $0.459 \times 10^{6}$ |
| 400 | 0.0288 | 0.462 | $1.13 \times 10^{-5}$ | $0.395 \times 10^{-3}$ | 0.0200 | 0.94 | $1.16 \times 10^{-3}$ | $0.243 \times 10^{6}$ |
| 500 | 0.0258 | 0.470 | $1.265 \times 10^{-5}$ | $0.490 \times 10^{-3}$ | 0.0228 | 0.94 | $1.04 \times 10^{-3}$ | $0.139 \times 10^{6}$ |
| 600 | 0.0233 | 0.477 | $1.420 \times 10^{-5}$ | $0.610 \times 10^{-3}$ | 0.0257 | 0.94 | $0.943 \times 10^{-3}$ | $82 \times 10^{3}$ |
| 700 | 0.0213 | 0.485 | $1.555 \times 10^{-5}$ | $0.725 \times 10^{-3}$ | 0.0288 | 0.93 | $0.862 \times 10^{-3}$ | $52.1 \times 10^{3}$ |
| 800 | 0.0196 | 0.494 | $1.70 \times 10^{-5}$ | $0.855 \times 10^{-3}$ | 0.0321 | 0.92 | $0.794 \times 10^{-3}$ | $34.0 \times 10^{3}$ |
| 900 | 0.0181 | 0.50 | $1.810 \times 10^{-5}$ | $0.987 \times 10^{-3}$ | 0.0355 | 0.91 | $0.735 \times 10^{-3}$ | $23.6 \times 10^{3}$ |
| 1000 | 0.0169 | 0.51 | $1.920 \times 10^{-5}$ | $1.13 \times 10^{-3}$ | 0.0388 | 0.91 | $0.685 \times 10^{-3}$ | $17.1 \times 10^{3}$ |
| 1200 | 0.0149 | 0.53 | $2.14 \times 10^{-5}$ | $1.44 \times 10^{-3}$ | 0.0457 | 0.88 | $0.603 \times 10^{-3}$ | $9.4 \times 10^{3}$ |
| 1400 | 0.0133 | 0.55 | $2.36 \times 10^{-5}$ | $1.78 \times 10^{-3}$ | 0.053 | 0.87 | $0.537 \times 10^{-3}$ | $5.49 \times 10^{3}$ |
| 1600 | 0.0120 | 0.56 | $2.58 \times 10^{-5}$ | $2.14 \times 10^{-3}$ | 0.061 | 0.87 | $0.485 \times 10^{-3}$ | $3.38 \times 10^{3}$ |
| 1800 | 0.0109 | 0.58 | $2.81 \times 10^{-5}$ | $2.58 \times 10^{-3}$ | 0.068 | 0.87 | $0.442 \times 10^{-3}$ | $2.14 \times 10^{3}$ |
| 2000 | 0.0100 | 0.60 | $3.03 \times 10^{-5}$ | $3.03 \times 10^{-3}$ | 0.076 | 0.86 | $0.406 \times 10^{-3}$ | $1.43 \times 10^{3}$ |
| 2500 | 0.0083 | 0.64 | $3.58 \times 10^{-5}$ | $4.30 \times 10^{-3}$ | 0.096 | 0.86 | $0.338 \times 10^{-3}$ | $0.603 \times 10^{3}$ |
| 3000 | 0.0071 | 0.67 | $4.00 \times 10^{-5}$ | $5.75 \times 10^{-3}$ | 0.114 | 0.86 | $0.289 \times 10^{-3}$ | $0.293 \times 10^{3}$ |

(Multiply lbm/ft$^3$ by 16.0185 to obtain kg/m$^3$.)
(Multiply Btu/lbm-°F by 4187 to obtain J/kg·K.)
(Multiply lbm/sec-ft by 3600 to obtain lbm/hr-ft.)
(Multiply lbm/sec-ft by 1.488 to obtain kg/s·m.)
(Multiply ft$^2$/sec ty 0.0929 to obtain m$^2$/s.)
(Multiply Btu/hr-ft-°F by 1.730 to obtain W/m·K.)
(Multiply 1/°F by 5/9 to obtain 1/K.)
(Multiply 1/ft$^3$-°F by 19.611 to obtain 1/m$^3$·K.)
[*] $\mu$, $k$, $c_p$, and Pr do not greatly depend on pressure and may be used over a wide range of pressures.

Reprinted with permission from E. R. G. Eckert and R. M. Drake, *Analysis of Heat and Mass Transfer*, copyright © 1972.

Appendices

**APPENDIX 41.F**
Properties of Steam at One Atmosphere*
(SI units)

| $T$ (°C) | $T$ (K) | $\rho$ (kg/m$^3$) | $c_p$ (kJ/kg·K) | $\mu$ (kg/m·s) | $k$ (W/m·K) | Pr | $\beta$ (1/K) | $\dfrac{g\beta\rho^2}{\mu^2}$ (1/K·m$^3$) |
|---|---|---|---|---|---|---|---|---|
| 100.0 | 373.2 | 0.596 | 1.888 | $1.295 \times 10^{-5}$ | 0.02510 | 0.96 | $2.68 \times 10^{-3}$ | $0.557 \times 10^8$ |
| 148.9 | 422.1 | 0.525 | 1.909 | $1.488 \times 10^{-5}$ | 0.02960 | 0.95 | $2.38 \times 10^{-3}$ | $0.292 \times 10^8$ |
| 204.4 | 477.6 | 0.461 | 1.934 | $1.682 \times 10^{-5}$ | 0.03462 | 0.94 | $2.09 \times 10^{-3}$ | $0.154 \times 10^8$ |
| 260.0 | 533.2 | 0.413 | 1.968 | $1.883 \times 10^{-5}$ | 0.03946 | 0.94 | $1.87 \times 10^{-3}$ | $0.0883 \times 10^8$ |
| 315.6 | 588.8 | 0.373 | 1.997 | $2.113 \times 10^{-5}$ | 0.04448 | 0.94 | $1.70 \times 10^{-3}$ | $52.1 \times 10^5$ |
| 371.1 | 644.3 | 0.341 | 2.030 | $2.314 \times 10^{-5}$ | 0.04985 | 0.93 | $1.55 \times 10^{-3}$ | $33.1 \times 10^5$ |
| 426.7 | 699.9 | 0.314 | 2.068 | $2.529 \times 10^{-5}$ | 0.05556 | 0.92 | $1.43 \times 10^{-3}$ | $21.6 \times 10^5$ |

(Multiply lbm/ft$^3$ by 16.0185 to obtain kg/m$^3$.)
(Multiply Btu/lbm-°F by 4187 to obtain J/kg·K.)
(Multiply lbm/sec-ft by 3600 to obtain lbm/hr-ft.)
(Multiply lbm/sec-ft by 1.488 to obtain kg/s·m.)
(Multiply ft$^2$/sec ty 0.0929 to obtain m$^2$/s.)
(Multiply Btu/hr-ft-°F by 1.730 to obtain W/m·K.)
(Multiply 1/°F by 5/9 to obtain 1/K.)
(Multiply 1/ft$^3$-°F by 19.611 to obtain 1/m$^3$·K.)
*$\mu$, $k$, $c_p$, and Pr do not greatly depend on pressure and may be used over a wide range of pressures.

Geankoplis, Christie John, *Transport Processes and Separation Process Principles*, 4th ed., copyright © 2003. Reprinted by permission of Pearson Education, Inc., Upper Saddle River, NJ.

**Appendices**

## APPENDIX 42.A
Correction Factor, $F_c$, for the Logarithmic Mean Temperature Difference
(one shell pass, even number of tube passes)

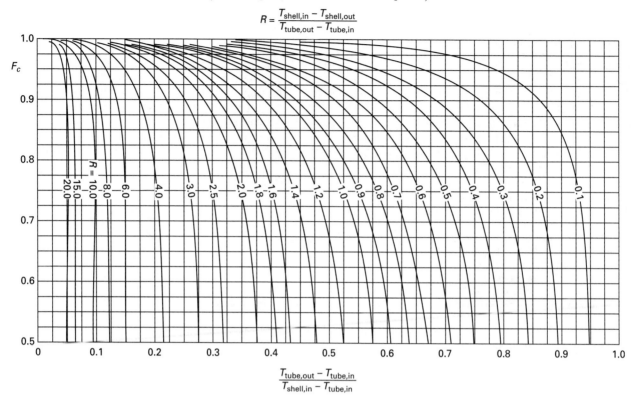

$$R = \frac{T_{shell,in} - T_{shell,out}}{T_{tube,out} - T_{tube,in}}$$

$$\frac{T_{tube,out} - T_{tube,in}}{T_{shell,in} - T_{tube,in}}$$

© 1999 by Tubular Exchanger Manufacturers Association.

**APPENDIX 42.B**
Correction Factor, $F_c$, for the Logarithmic Mean Temperature Difference
(two shell passes, multiple of four tube passes)

$$R = \frac{T_{shell,in} - T_{shell,out}}{T_{tube,out} - T_{tube,in}}$$

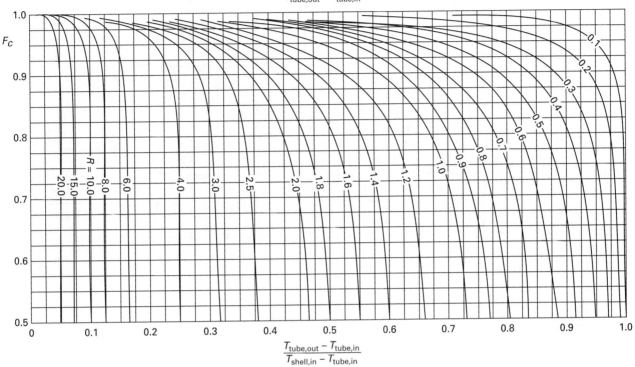

$$\frac{T_{tube,out} - T_{tube,in}}{T_{shell,in} - T_{tube,in}}$$

## APPENDIX 42.C
Correction Factor, $F_c$, for the Logarithmic Mean Temperature Difference
(one shell pass, three tube passes, two counter and one cocurrent)

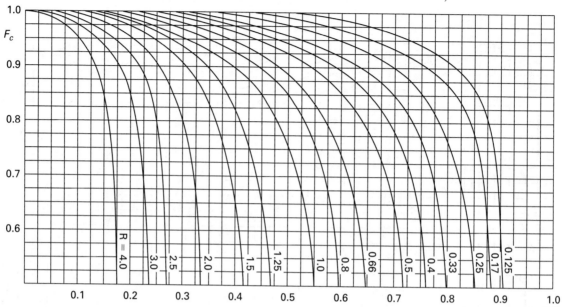

© 1999 by Tubular Exchanger Manufacturers Association.

## APPENDIX 42.D
Correction Factor, $F_c$, for the Logarithmic Mean Temperature Difference
(crossflow shell, one tube pass)

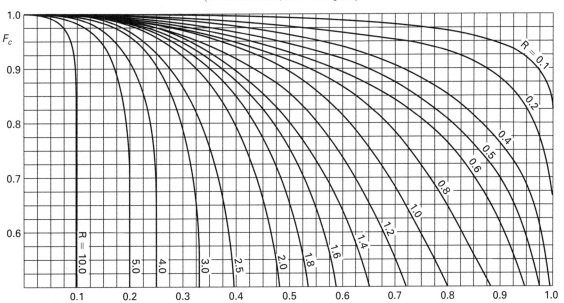

© 1999 by Tubular Exchanger Manufacturers Association.

**APPENDIX 42.E**
Characteristics of Birmingham Wire Gage (BWG) Size Tubing

| tube outside diameter (in) | BWG (gage) | tube inside diameter (in) | thickness (in) | internal area (in²) | external surface per ft length (ft²/ft) | internal surface per ft length (ft²/ft) |
|---|---|---|---|---|---|---|
| 1/4 | 22 | 0.194 | 0.028 | 0.0296 | 0.0654 | 0.0508 |
| 1/4 | 24 | 0.206 | 0.022 | 0.0333 | 0.0654 | 0.0539 |
| 1/4 | 26 | 0.214 | 0.018 | 0.0360 | 0.0654 | 0.0560 |
| 1/4 | 27 | 0.218 | 0.016 | 0.0373 | 0.0654 | 0.0571 |
| | | | | | | |
| 3/8 | 18 | 0.277 | 0.049 | 0.0603 | 0.0935 | 0.0725 |
| 3/8 | 20 | 0.305 | 0.035 | 0.0731 | 0.0935 | 0.0798 |
| 3/8 | 22 | 0.319 | 0.028 | 0.0799 | 0.0935 | 0.0835 |
| 3/8 | 24 | 0.331 | 0.022 | 0.0860 | 0.0935 | 0.0867 |
| | | | | | | |
| 1/2 | 16 | 0.370 | 0.065 | 0.1075 | 0.1309 | 0.0969 |
| 1/2 | 18 | 0.402 | 0.049 | 0.1269 | 0.1309 | 0.1052 |
| 1/2 | 20 | 0.430 | 0.035 | 0.1452 | 0.1309 | 0.1126 |
| 1/2 | 22 | 0.444 | 0.028 | 0.1548 | 0.1309 | 0.1162 |
| | | | | | | |
| 5/8 | 12 | 0.407 | 0.109 | 0.1301 | 0.1636 | 0.1066 |
| 5/8 | 13 | 0.435 | 0.095 | 0.1486 | 0.1636 | 0.1139 |
| 5/8 | 14 | 0.459 | 0.083 | 0.1655 | 0.1636 | 0.1202 |
| 5/8 | 15 | 0.481 | 0.072 | 0.1817 | 0.1636 | 0.1259 |
| 5/8 | 16 | 0.495 | 0.065 | 0.1924 | 0.1636 | 0.1296 |
| 5/8 | 17 | 0.509 | 0.058 | 0.2035 | 0.1636 | 0.1333 |
| 5/8 | 18 | 0.527 | 0.049 | 0.2181 | 0.1636 | 0.1380 |
| 5/8 | 19 | 0.541 | 0.042 | 0.2299 | 0.1636 | 0.1416 |
| 5/8 | 20 | 0.555 | 0.035 | 0.2419 | 0.1636 | 0.1453 |
| | | | | | | |
| 3/4 | 10 | 0.482 | 0.134 | 0.1825 | 0.1963 | 0.1262 |
| 3/4 | 11 | 0.510 | 0.120 | 0.2043 | 0.1963 | 0.1335 |
| 3/4 | 12 | 0.532 | 0.109 | 0.2223 | 0.1963 | 0.1393 |
| 3/4 | 13 | 0.560 | 0.095 | 0.2463 | 0.1963 | 0.1466 |
| 3/4 | 14 | 0.584 | 0.083 | 0.2679 | 0.1963 | 0.1529 |
| 3/4 | 15 | 0.606 | 0.072 | 0.2884 | 0.1963 | 0.1587 |
| 3/4 | 16 | 0.620 | 0.065 | 0.3019 | 0.1963 | 0.1623 |
| 3/4 | 17 | 0.634 | 0.058 | 0.3157 | 0.1963 | 0.1660 |
| 3/4 | 18 | 0.652 | 0.049 | 0.3339 | 0.1963 | 0.1707 |
| 3/4 | 20 | 0.680 | 0.035 | 0.3632 | 0.1963 | 0.1780 |
| | | | | | | |
| 1 | 8 | 0.670 | 0.165 | 0.3526 | 0.2618 | 0.1754 |
| 1 | 10 | 0.732 | 0.134 | 0.4208 | 0.2618 | 0.1916 |
| 1 | 11 | 0.760 | 0.120 | 0.4536 | 0.2618 | 0.1990 |
| 1 | 12 | 0.782 | 0.109 | 0.4803 | 0.2618 | 0.2047 |
| 1 | 13 | 0.810 | 0.095 | 0.5153 | 0.2618 | 0.2121 |
| 1 | 14 | 0.834 | 0.083 | 0.5463 | 0.2618 | 0.2183 |
| 1 | 15 | 0.856 | 0.072 | 0.5755 | 0.2618 | 0.2241 |
| 1 | 16 | 0.870 | 0.065 | 0.5945 | 0.2618 | 0.2278 |
| 1 | 18 | 0.902 | 0.049 | 0.6390 | 0.2618 | 0.2361 |
| 1 | 20 | 0.930 | 0.035 | 0.6793 | 0.2618 | 0.2435 |

*(continued)*

**APPENDIX 42.E** *(continued)*
Characteristics of Birmingham Wire Gage (BWG) Size Tubing

| tube outside diameter (in) | BWG (gage) | tube inside diameter (in) | thickness (in) | internal area (in$^2$) | external surface per ft length (ft$^2$/ft) | internal surface per ft length (ft$^2$/ft) |
|---|---|---|---|---|---|---|
| 1$^1$/$_4$ | 7 | 0.890 | 0.180 | 0.6221 | 0.3272 | 0.2330 |
| 1$^1$/$_4$ | 8 | 0.920 | 0.165 | 0.6648 | 0.3272 | 0.2409 |
| 1$^1$/$_4$ | 10 | 0.982 | 0.134 | 0.7574 | 0.3272 | 0.2571 |
| 1$^1$/$_4$ | 11 | 1.010 | 0.120 | 0.8012 | 0.3272 | 0.2644 |
| 1$^1$/$_4$ | 12 | 1.062 | 0.109 | 0.8365 | 0.3272 | 0.2702 |
| 1$^1$/$_4$ | 13 | 1.060 | 0.095 | 0.8825 | 0.3272 | 0.2775 |
| 1$^1$/$_4$ | 14 | 1.084 | 0.083 | 0.9229 | 0.3272 | 0.2838 |
| 1$^1$/$_4$ | 16 | 1.120 | 0.065 | 0.9852 | 0.3272 | 0.2932 |
| 1$^1$/$_4$ | 18 | 1.152 | 0.049 | 1.0432 | 0.3272 | 0.3016 |
| 1$^1$/$_4$ | 20 | 1.180 | 0.035 | 1.0936 | 0.3272 | 0.3089 |
| | | | | | | |
| 1$^1$/$_2$ | 10 | 1.232 | 0.134 | 1.1921 | 0.3927 | 0.3225 |
| 1$^1$/$_2$ | 12 | 1.282 | 0.109 | 1.2908 | 0.3927 | 0.3356 |
| 1$^1$/$_2$ | 14 | 1.334 | 0.083 | 1.3977 | 0.3927 | 0.3492 |
| 1$^1$/$_2$ | 16 | 1.370 | 0.065 | 1.4741 | 0.3927 | 0.3587 |
| | | | | | | |
| 2 | 11 | 1.760 | 0.120 | 2.4328 | 0.5236 | 0.4608 |
| 2 | 12 | 1.782 | 0.109 | 2.4941 | 0.5236 | 0.4665 |
| 2 | 13 | 1.810 | 0.095 | 2.5730 | 0.5236 | 0.4739 |
| 2 | 14 | 1.834 | 0.083 | 2.6417 | 0.5236 | 0.4801 |

Appendices

## APPENDIX 43.A
### Emissivities of Various Surfaces[*]

| material | wavelength 9.3 $\mu$m<br>average temperature 100°F<br>38°C | 5.4 $\mu$m<br>500°F<br>260°C | 3.6 $\mu$m<br>1000°F<br>538°C | 1.8 $\mu$m<br>2500°F<br>1371°C | 0.6 $\mu$m<br>solar |
|---|---|---|---|---|---|
| aluminum, polished | 0.04 | 0.05 | 0.08 | 0.19 | |
| aluminum, oxidized | 0.11 | 0.12 | 0.18 | | |
| aluminum (2024T), weathered | 0.4 | 0.32 | 0.27 | | |
| aluminum, surface roofing | 0.22 | | | | |
| aluminum, anodized | 0.94 | 0.42 | 0.60 | 0.34 | |
| brass, polished | 0.10 | 0.10 | | | |
| brass, oxidized | 0.61 | | | | |
| brick, red | 0.93 | | | | 0.7 |
| brick, fire clay | 0.9 | | 0.7 | 0.75 | |
| brick, silica | 0.9 | | 0.75 | 0.84 | |
| brick, magnesite refractory | 0.9 | | | 0.4 | |
| chromium, polished | 0.08 | 0.17 | 0.26 | 0.40 | 0.49 |
| copper, polished | 0.04 | 0.05 | 0.18 | 0.17 | |
| copper, oxidized | 0.87 | 0.83 | 0.77 | | |
| enamel, white | 0.9 | | | | |
| glass | 0.9 | | | | |
| ice (at 32°F) | 0.97 | | | | |
| iron, polished | 0.06 | 0.08 | 0.13 | 0.25 | 0.45 |
| iron, cast, oxidized | 0.63 | 0.66 | 0.76 | | |
| iron, galvanized, new | 0.23 | | | 0.42 | 0.66 |
| iron, galvanized, dirty | 0.28 | | | 0.90 | 0.89 |
| iron, oxide | 0.96 | | 0.85 | | 0.74 |
| iron, molten | | | | 0.3–0.4 | |
| magnesium | 0.07 | 0.13 | 0.18 | 0.24 | 0.30 |
| paper, white | 0.95 | | | 0.25 | 0.28 |
| paint, aluminized lacquer | 0.65 | 0.65 | | | |
| paint, lacquer, black, flat | 0.96 | 0.98 | | | |
| paint, lacquer, white | 0.9 | | | | |
| paint, lampback | 0.96 | 0.97 | | 0.97 | 0.97 |
| paint, white (ZnO) | 0.95 | | 0.91 | | 0.18 |
| paint, enamel, white | 0.9 | | | | |
| stainless steel, 18-8, polished | 0.15 | 0.18 | 0.22 | | |
| stainless steel, 18-8, weathered | 0.85 | 0.85 | 0.85 | | |
| steel tube, oxidized | | 0.8 | | | |
| steel plate, rough | 0.94 | 0.97 | 0.98 | | |
| tungsten filament | 0.03 | | | 0.18 | 0.35 (6000°F) |
| water | 0.99 | | | | |
| wood | 0.93 | | | | |
| zinc, polished | 0.02 | 0.03 | 0.04 | 0.06 | 0.46 |
| zinc, galvanized sheet | 0.25 | | | | |

[*]compiled from various sources

**APPENDIX 48.A**
Typical Properties of Structural Steel,
Aluminum, and Magnesium
(all values in ksi)

structural steel

| designation | application | $S_u$ | $S_y$ | approximate $S_e$ |
|---|---|---|---|---|
| A36 | shapes | 58–80 | 36 | 29–40 |
| | plates | 58–80 | 36 | 29–40 |
| A53 | pipe | 60 | 35 | 30 |
| A242 | shapes | 70 | 50 | 35 |
| | plates to $\frac{3}{4}$ in | 70 | 50 | 35 |
| A440 | shapes | 70 | 50 | 35 |
| | plates to $\frac{3}{4}$ in | 70 | 50 | 35 |
| A441 | shapes | 70 | 50 | 35 |
| | plates to $\frac{3}{4}$ in | 70 | 50 | 35 |
| A500 | tubes | 45 | 33 | 22 |
| A501 | tubes | 58 | 36 | 29 |
| A514 | plates to $\frac{3}{4}$ in | 115–135 | 100 | 55 |
| A529 | shapes | 60–85 | 42 | 30–42 |
| | plates to $\frac{1}{2}$ in | 60–85 | 42 | 30–42 |
| A570 | sheet/strip | 55 | 40 | 27 |
| A572 | shapes | 65 | 50 | 30 |
| | plates | 60 | 42 | 30 |
| A588 | shapes | 70 | 50 | 35 |
| | plates to 4 in | 70 | 50 | 35 |
| A606 | hot-rolled sheet | 70 | 50 | 35 |
| | cold-rolled sheet | 65 | 45 | 32 |
| A607 | sheet | 60 | 45 | 30 |
| A618 | shapes | 70 | 50 | 35 |
| | tubes | 70 | 50 | 35 |
| A913 | shapes | 65 | 50 | 30 |
| A992 | shapes | 65 | 50 | – |

structural aluminum

| designation | application | $S_u$ | $S_y$ | approximate $S_e$ (at $10^8$ cyc.) |
|---|---|---|---|---|
| 2014-T6 | shapes/bars | 63 | 55 | 19 |
| 6061-T6 | all | 42 | 35 | 14.5 |

structural magnesium

| designation | application | $S_u$ | $S_y$ | approximate $S_e$ (at $10^7$ cyc.) |
|---|---|---|---|---|
| AZ31 | shapes | 38 | 29 | 19 |
| AZ61 | shapes | 45 | 33 | 19 |
| AZ80 | shapes | 55 | 40 | – |

(Multiply ksi by 6.895 to obtain MPa.)

**APPENDIX 48.B**
Typical Mechanical Properties of Representative Metals (room temperature)

The following mechanical properties are not guaranteed since they are averages for various sizes, product forms, and methods of manufacture. Thus, this data is not for design use, but is intended only as a basis for comparing alloys and tempers.

| material designation, composition, typical use, and source if applicable | condition, heat treatment | $S_{ut}$ (ksi) | $S_{yt}$ (ksi) |
|---|---|---|---|
| IRON BASED | | | |
| Armco ingot iron, for fresh and saltwater piping | normalized | 44 | 24 |
| AISI 1020, plain carbon steel, for general machine parts and screws and carburized parts | hot rolled<br>cold worked | 65<br>78 | 43<br>66 |
| AISI 1030, plain carbon steel, for gears, shafts, levers, seamless tubing, and carburized parts | cold drawn | 87 | 74 |
| AISI 1040, plain carbon steel, for high-strength parts, shafts, gears, studs, connecting rods, axles, and crane hooks | hot rolled<br>cold worked<br>hardened | 91<br>100<br>113 | 58<br>88<br>86 |
| AISI 1095, plain carbon steel, for handtools, music wire springs, leaf springs, knives, saws, and agricultural tools such as plows and disks | annealed<br>hot rolled<br>hardened | 100<br>142<br>180 | 53<br>84<br>118 |
| AISI 1330, manganese steel, for axles and drive shafts | annealed<br>cold drawn<br>hardened | 97<br>113<br>122 | 83<br>93<br>100 |
| AISI 4130, chromium-molybdenum steel, for high-strength aircraft structures | annealed<br>hardened | 81<br>161 | 52<br>137 |
| AISI 4340, nickel-chromium-molybdenum steel, for large-scale, heavy-duty, high-strength structures | annealed<br>as rolled<br>hardened | 119<br>192<br>220 | 99<br>147<br>200 |
| AISI 2315, nickel steel, for carburized parts | as rolled<br>cold drawn | 85<br>95 | 56<br>75 |
| AISI 2330, nickel steel | as rolled<br>cold drawn<br>annealed<br>normalized | 98<br>110<br>80<br>95 | 65<br>90<br>50<br>61 |
| AISI 3115, nickel-chromium steel for carburized parts | cold drawn<br>as rolled<br>annealed | 95<br>75<br>71 | 70<br>60<br>62 |
| STAINLESS STEELS | | | |
| AISI 302, stainless steel, most widely used, same as 18-8 | annealed<br>cold drawn | 90<br>105 | 35<br>60 |
| AISI 303, austenitic stainless steel, good machineability | annealed<br>cold worked | 90<br>110 | 35<br>75 |
| AISI 304, austenitic stainless steel, good machineability and weldability | annealed<br>cold worked | 85<br>110 | 30<br>75 |
| AISI 309, stainless steel, good weldability, high strength at high temperatures, used in furnaces and ovens | annealed<br>cold drawn | 90<br>110 | 35<br>65 |
| AISI 316, stainless steel, excellent corrosion resistance | annealed<br>cold drawn | 85<br>105 | 35<br>60 |
| AISI 410, magnetic, martensitic, can be quenched and tempered to give varying strength | annealed<br>cold drawn<br>oil quenched and drawn | 60<br>180<br><br>110 | 32<br>150<br><br>91 |
| AISI 430, magnetic, ferritic, used for auto and architectural trim and for equipment in food and chemical industries | annealed<br>cold drawn | 60<br>100 | 35 |
| AISI 502, magnetic, ferritic, low cost, widely used in oil refineries | annealed | 60 | 25 |

*(continued)*

**APPENDIX 48.B** *(continued)*
Typical Mechanical Properties of Representative Metals (room temperature)

| material designation, composition, typical use, and source if applicable | condition, heat treatment | $S_{ut}$ (ksi) | $S_{yt}$ (ksi) |
|---|---|---|---|
| **ALUMINUM BASED** | | | |
| 2011, for screw machine parts, excellent machineability, but not weldable, and corrosion sensitive | T3 | 55 | 43 |
| | T8 | 59 | 45 |
| 2014, for aircraft structures, weldable | T3 | 63 | 40 |
| | T4, T451 | 61 | 37 |
| | T6, T651 | 68 | 60 |
| 2017, for screw machine parts | T4, T451 | 62 | 40 |
| 2018, for engine cylinders, heads, and pistons | T61 | 61 | 46 |
| 2024, for truck wheels, screw machine parts, and aircraft structures | T3 | 65 | 45 |
| | T4, T351 | 64 | 42 |
| | T361 | 72 | 57 |
| 2025, for forgings | T6 | 58 | 37 |
| 2117, for rivets | T4 | 43 | 24 |
| 2219, high-temperature applications (up to 600°F), excellent weldability and machineabilty | T31, T351 | 52 | 36 |
| | T37 | 57 | 46 |
| | T42 | 52 | 27 |
| 3003, for pressure vessels and storage tanks, poor machine-ability but good weldability, excellent corrosion resistance | 0 | 16 | 6 |
| | H12 | 19 | 18 |
| | H14 | 22 | 21 |
| | H16 | 26 | 25 |
| 3004, same characteristics as 3003 | 0 | 26 | 10 |
| | H32 | 31 | 25 |
| | H34 | 35 | 29 |
| | H36 | 38 | 33 |
| 4032, pistons | T6 | 55 | 46 |
| 5083, unfired pressure vessels, cryogenics, towers, and drilling rigs | 0 | 42 | 21 |
| | H116, H117, H321 | 46 | 33 |
| 5154, saltwater services, welded structures, and storage tanks | 0 | 35 | 17 |
| | H32 | 39 | 30 |
| | H34 | 42 | 33 |
| 5454, same characteristics as 5154 | 0 | 36 | 17 |
| | H32 | 40 | 30 |
| | H34 | 44 | 35 |
| 5456, same characteristics as 5154 | 0 | 45 | 23 |
| | H111 | 47 | 33 |
| | H321, H116, H117 | 51 | 37 |
| 6061, corrosion resistant and good weldability, used in railroad cars | T4 | 33 | 19 |
| | T6 | 42 | 37 |
| 7178, Alclad, corrosion-resistant | 0 | 33 | 15 |
| | T6 | 88 | 78 |

| CAST IRON (note redefinition of columns) | | $S_{ut}$ (ksi) | $S_{us}$ (ksi) | $S_{uc}$ (ksi) |
|---|---|---|---|---|
| gray cast iron | class 20 | 30 | 32.5 | 30 |
| | class 25 | 25 | 34 | 100 |
| | class 30 | 30 | 41 | 110 |
| | class 35 | 35 | 49 | 125 |
| | class 40 | 40 | 52 | 135 |
| | class 50 | 50 | 64 | 160 |
| | class 60 | 60 | 60 | 150 |

*(continued)*

**APPENDIX 48.B** (continued)

Typical Mechanical Properties of Representative Metals (room temperature)

| material designation, composition, typical use, and source if applicable | condition, heat treatment | $S_{ut}$ (ksi) | $S_{yt}$ (ksi) |
|---|---|---|---|
| **COPPER BASED** | | | |
| copper, commercial purity | annealed (furnace cool from 400°C) | 32 | 10 |
| | cold drawn | 45 | 40 |
| cartridge brass: 70% Cu, 30% Zn | cold rolled (annealed 400°C, furnace cool) | 76 | 63 |
| copper-beryllium (1.9% Be, 0.25% Co) | annealed, wqf 1450°F | 70 | |
| | cold rolled | 200 | |
| | hardened after annealing | 200 | 150 |
| phosphor-bronze, for springs | wire, 0.025 in and under | 145 | |
| | 0.025 in to 0.0625 in | 135 | |
| | 0.125 in to 0.250 in | 125 | |
| monel metal | cold-drawn bars, annealed | 70 | 30 |
| red brass | sheet and strip    half-hard | 51 | |
| | hard | 63 | |
| | spring | 78 | |
| yellow brass | sheet and strip    half-hard | 55 | |
| | hard | 68 | |
| | spring | 86 | |
| **NICKEL BASED** | | | |
| pure nickel, magnetic, high corrosion resistance | annealed (ht 1400°F, acrt) | 46 | 8.5 |
| | annealed at 2050°F | 125 | 75 |
| Inconel X, type 550, excellent high temperature properties | annealed and age hardened | 175 | 110 |
| K-monel, excellent high temperature properties and corrosion resistance | annealed (wqf 1600°F) | 100 | 45 |
| | age hardened spring stock | 185 | 160 |
| Invar, 36% Ni, 64% Fe, low coefficient of expansion (1.2 × 10⁻⁶ 1/°C, 0-200°C) | annealed (wqf 800°C) | 71 | 40 |
| **REFRACTORY METALS** (properties at room temperature) | | | |
| molybdenum | as rolled | 100 | 75 |
| tantalum | annealed at 1050°C in vacuum | 60 | 45 |
| | as rolled | 110 | 100 |
| titanium, commercial purity | annealed at 1200°F | 95 | 80 |
| titanium, 6% Al, 4% V | annealed at 1400°F, acrt | 135 | 130 |
| | heat treated (wqf 1750°F, ht 1000°F, acrt) | 170 | 150 |
| titanium, 4% Al, 4% Mn OR 5% Al, 2.75% Cr, 1.25% Fe OR 5% Al, 1.5% Fe, 1.4% Cr, 1.2% Mo | wqf 1450°F, ht 900°F, acrt | 185 | 170 |
| tungsten, commercial purity | hard wire | 600 | 540 |

| MAGNESIUM | | $S_{ut}$ (ksi) | $S_{yt}$ (ksi) | $S_{us}$ (ksi) |
|---|---|---|---|---|
| AZ92, for sand and permanent-mold casting | as cast | 24 | 14 | |
| | solution treated | 39 | 14 | |
| | aged | 39 | 21 | |
| AZ91, for die casting | as cast | 33 | 21 | |
| AZ31X (sheet) | annealed | 35 | 20 | |
| | hard | 40 | 31 | |
| AZ80X, for structural shapes | extruded | 48 | 32 | |
| | extruded and aged | 52 | 37 | |
| ZK60A, for structural shapes | extruded | 49 | 38 | |
| | extruded and aged | 51 | 42 | |
| AZ31B (sheet and plate), for structural shapes in use below 300°F | temper 0 | 32 | 15 | 17 |
| | temper 1124 | 34 | 18 | 18 |
| | temper 1126 | 35 | 21 | 18 |
| | temper F | 32 | 16 | 17 |

Abbreviations:
   wqf: water-quench from
   acrt: air-cooled to room temperature
   ht: heated to

(Multiply ksi by 6.895 to obtain MPa.)

## APPENDIX 51.A
Centroids and Area Moments of Inertia for Basic Shapes

| shape | | centroidal location | | area, $A$ | area moment of inertia (rectangular and polar), $I$, $J$ | radius of gyration, $r$ |
|---|---|---|---|---|---|---|
| | | $x_c$ | $y_c$ | | | |
| rectangle | | $\dfrac{b}{2}$ | $\dfrac{h}{2}$ | $bh$ | $I_x = \dfrac{bh^3}{3}$ $I_{c,x} = \dfrac{bh^3}{12}$ $J_c = \left(\dfrac{1}{12}\right)bh(b^2 + h^2)*$ | $r_x = \dfrac{h}{\sqrt{3}}$ $r_{c,x} = \dfrac{h}{2\sqrt{3}}$ |
| triangular area | | | $\dfrac{h}{3}$ | $\dfrac{bh}{2}$ | $I_x = \dfrac{bh^3}{12}$ $I_{c,x} = \dfrac{bh^3}{36}$ | $r_x = \dfrac{h}{\sqrt{6}}$ $r_{c,x} = \dfrac{h}{3\sqrt{2}}$ |
| trapezoid | | | $h\left(\dfrac{b + 2t}{3b+3t}\right)$ | $\dfrac{(b+t)h}{2}$ | $I_x = \dfrac{(b + 3t)h^3}{12}$ $I_{c,x} = \dfrac{(b^2 + 4bt + t^2)h^3}{36(b + t)}$ | $r_x = \left(\dfrac{h}{\sqrt{6}}\right)\sqrt{\dfrac{b + 3t}{b + t}}$ $r_{c,x} = \dfrac{h\sqrt{2(b^2 + 4bt + t^2)}}{6(b + t)}$ |
| circle | | $0$ | $0$ | $\pi r^2$ | $I_x = I_y = \dfrac{\pi r^4}{4}$ $J_c = \dfrac{\pi r^4}{2}$ | $r_x = \dfrac{r}{2}$ |
| quarter-circular area | | $\dfrac{4r}{3\pi}$ | $\dfrac{4r}{3\pi}$ | $\dfrac{\pi r^2}{4}$ | $I_x = I_y = \dfrac{\pi r^4}{16}$ $J_o = \dfrac{\pi r^4}{8}$ | |
| semicircular area | | $0$ | $\dfrac{4r}{3\pi}$ | $\dfrac{\pi r^2}{2}$ | $I_x = I_y = \dfrac{\pi r^4}{8}$ $I_{c,x} = 0.1098r^4$ $J_o = \dfrac{\pi r^4}{4}$ $J_c = 0.5025r^4$ | $r_x = \dfrac{r}{2}$ $r_{c,x} = 0.264r$ |
| quarter-elliptical area | | $\dfrac{4a}{3\pi}$ | $\dfrac{4b}{3\pi}$ | $\dfrac{\pi ab}{4}$ | $I_x = \dfrac{\pi ab^3}{8}$ $I_y = \dfrac{\pi a^3 b}{8}$ | |
| semielliptical area | | $0$ | $\dfrac{4b}{3\pi}$ | $\dfrac{\pi ab}{2}$ | $J_o = \dfrac{\pi ab(a^2 + b^2)}{8}$ | |
| semiparabolic area | | $\dfrac{3a}{8}$ | $\dfrac{3h}{5}$ | $\dfrac{2ah}{3}$ | | |
| parabolic area | | $0$ | $\dfrac{3h}{5}$ | $\dfrac{4ah}{3}$ | $I_x = \dfrac{4ah^3}{7}$ $I_y = \dfrac{4ha^3}{15}$ $I_{c,x} = \dfrac{16ah^3}{175}$ | $r_x = h\sqrt{\dfrac{3}{7}}$ $r_y = \dfrac{a}{\sqrt{5}}$ |
| parabolic spandrel | $y = kx^2$ | $\dfrac{3a}{4}$ | $\dfrac{3h}{10}$ | $\dfrac{ah}{3}$ | $I_x = \dfrac{ah^3}{21}$ $I_y = \dfrac{3ha^3}{15}$ | |
| general spandrel | $y = kx^n$ | $\left(\dfrac{n + 1}{n + 2}\right)a$ | $\left(\dfrac{n + 1}{4n + 2}\right)h$ | $\dfrac{ah}{n + 1}$ | *Theoretical definition based on $J = I_x + I_y$. However, in torsion, not all parts of the shape are effective. Effective values will be lower. $$J = C\left(\dfrac{b^2 + h^2}{b^3 h^3}\right)$$ | |
| circular sector [$\alpha$ in radians] | | $\dfrac{2r \sin\alpha}{3\alpha}$ | $0$ | $\alpha r^2$ | | |

| $b/h$ | $C$ |
|---|---|
| 1 | 3.56 |
| 2 | 3.50 |
| 4 | 3.34 |
| 8 | 3.21 |

## APPENDIX 52.A
Elastic Beam Deflection Equations
($w$ is the load per unit length.)
($y$ is positive downward.)

**Case 1: Cantilever with End Load**

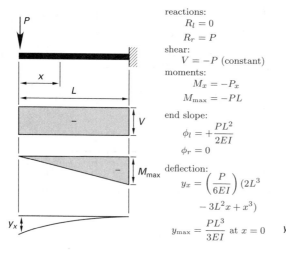

reactions:
$$R_l = 0$$
$$R_r = P$$
shear:
$$V = -P \text{ (constant)}$$
moments:
$$M_x = -Px$$
$$M_{max} = -PL$$
end slope:
$$\phi_l = +\frac{PL^2}{2EI}$$
$$\phi_r = 0$$
deflection:
$$y_x = \left(\frac{P}{6EI}\right)(2L^3 - 3L^2 x + x^3)$$
$$y_{max} = \frac{PL^3}{3EI} \text{ at } x = 0$$

**Case 2: Cantilever with Uniform Load**

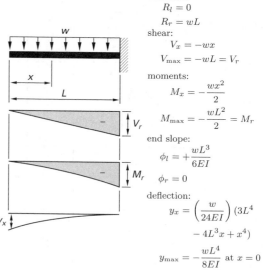

reactions:
$$R_l = 0$$
$$R_r = wL$$
shear:
$$V_x = -wx$$
$$V_{max} = -wL = V_r$$
moments:
$$M_x = -\frac{wx^2}{2}$$
$$M_{max} = -\frac{wL^2}{2} = M_r$$
end slope:
$$\phi_l = +\frac{wL^3}{6EI}$$
$$\phi_r = 0$$
deflection:
$$y_x = \left(\frac{w}{24EI}\right)(3L^4 - 4L^3 x + x^4)$$
$$y_{max} = -\frac{wL^4}{8EI} \text{ at } x = 0$$

**Case 3: Cantilever with Triangular Load**

reactions:
$$R_l = 0$$
$$R_r = \frac{wL}{2}$$
shear:
$$V_x = -\frac{wx^2}{2L}$$
$$V_{max} = \frac{wL}{2} \text{ at } x = L$$
moments:
$$M_x = -\frac{wx^3}{6L}$$
$$M_{max} = \frac{wL^2}{6} \text{ at } x = L$$
end slope:
$$\phi_l = +\frac{wL^3}{24EI}$$
$$\phi_r = 0$$
deflection:
$$y_x = \left(\frac{w}{120EIL}\right)(4L^5 - 5L^4 x + x^5)$$
$$y_{max} = \frac{wL^4}{30EI} \text{ at } x = 0$$

**Case 4: Propped Cantilever with Uniform Load**

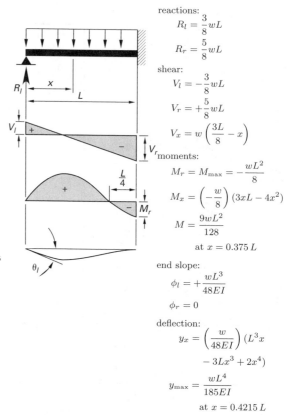

reactions:
$$R_l = \frac{3}{8}wL$$
$$R_r = \frac{5}{8}wL$$
shear:
$$V_l = -\frac{3}{8}wL$$
$$V_r = +\frac{5}{8}wL$$
$$V_x = w\left(\frac{3L}{8} - x\right)$$
moments:
$$M_r = M_{max} = -\frac{wL^2}{8}$$
$$M_x = \left(-\frac{w}{8}\right)(3xL - 4x^2)$$
$$M = \frac{9wL^2}{128} \text{ at } x = 0.375\,L$$
end slope:
$$\phi_l = +\frac{wL^3}{48EI}$$
$$\phi_r = 0$$
deflection:
$$y_x = \left(\frac{w}{48EI}\right)(L^3 x - 3Lx^3 + 2x^4)$$
$$y_{max} = \frac{wL^4}{185EI} \text{ at } x = 0.4215\,L$$

*(continued)*

**APPENDIX 52.A** *(continued)*
Elastic Beam Deflection Equations
($w$ is the load per unit length.)
($y$ is positive downward.)

**Case 5: Cantilever with End Moment**

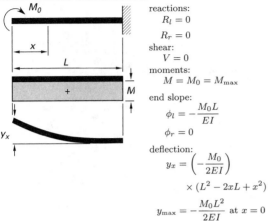

reactions:
$$R_l = 0$$
$$R_r = 0$$
shear:
$$V = 0$$
moments:
$$M = M_0 = M_{max}$$
end slope:
$$\phi_l = -\frac{M_0 L}{EI}$$
$$\phi_r = 0$$
deflection:
$$y_x = \left(-\frac{M_0}{2EI}\right)$$
$$\times (L^2 - 2xL + x^2)$$
$$y_{max} = -\frac{M_0 L^2}{2EI} \text{ at } x = 0$$

**Case 6: Simple Beam with Center Load**

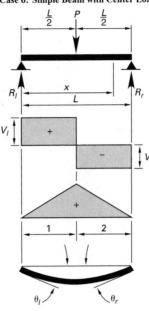

reactions:
$$R_l = R_r = \frac{P}{2}$$
shear:
$$V_l = \frac{P}{2}$$
$$V_r = -\frac{P}{2}$$
moments:
$$M_{x1} = \frac{Px}{2}$$
$$M_{x2} = \left(\frac{P}{2}\right)(L - x)$$
$$M_{max} = \frac{PL}{4}$$
end slope:
$$\phi_l = -\frac{PL^2}{16EI}$$
$$\phi_r = +\frac{PL^2}{16EI}$$
deflection:
$$y_{x1} \left(x < \frac{L}{2}\right)$$
$$= \left(\frac{P}{48EI}\right)(3xL^2 - 4x^3)$$
$$y_{max} = \frac{PL^3}{48EI} \text{ at } x = \frac{L}{2}$$

**Case 7: Simple Beam with Intermediate Load**

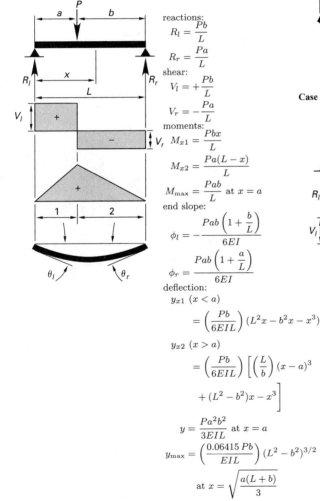

reactions:
$$R_l = \frac{Pb}{L}$$
$$R_r = \frac{Pa}{L}$$
shear:
$$V_l = +\frac{Pb}{L}$$
$$V_r = -\frac{Pa}{L}$$
moments:
$$M_{x1} = \frac{Pbx}{L}$$
$$M_{x2} = \frac{Pa(L-x)}{L}$$
$$M_{max} = \frac{Pab}{L} \text{ at } x = a$$
end slope:
$$\phi_l = -\frac{Pab\left(1 + \dfrac{b}{L}\right)}{6EI}$$
$$\phi_r = \frac{Pab\left(1 + \dfrac{a}{L}\right)}{6EI}$$
deflection:
$$y_{x1} \ (x < a)$$
$$= \left(\frac{Pb}{6EIL}\right)(L^2 x - b^2 x - x^3)$$
$$y_{x2} \ (x > a)$$
$$= \left(\frac{Pb}{6EIL}\right)\left[\left(\frac{L}{b}\right)(x-a)^3\right.$$
$$\left. + (L^2 - b^2)x - x^3\right]$$
$$y = \frac{Pa^2 b^2}{3EIL} \text{ at } x = a$$
$$y_{max} = \left(\frac{0.06415 \, Pb}{EIL}\right)(L^2 - b^2)^{3/2}$$
$$\text{at } x = \sqrt{\frac{a(L+b)}{3}}$$

**Case 8: Simple Beam with Two Loads**

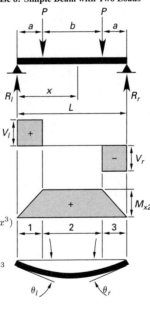

reactions:
$$R_l = R_r = P$$
shear:
$$V_l = +P$$
$$V_r = -P$$
moments:
$$M_{x1} = Px$$
$$M_{x2} = Pa$$
$$M_{x3} = P(L - x)$$
end slope:
$$\phi_l = -\frac{Pa(a+b)}{2EI}$$
$$\phi_r = +\frac{Pa(a+b)}{2EI}$$
deflection:
$$y_{x1} \ (x < a)$$
$$= \left(\frac{P}{6EI}\right)(3Lax$$
$$- 3a^2 x - x^3)$$
$$y_{x2} \ (a < x < a + b)$$
$$= \left(\frac{P}{6EI}\right)(3Lax$$
$$- 3ax^2 - a^3)$$
$$y_{max} = \left(\frac{P}{24EI}\right)(3L^2 a - 4a^3)$$
$$\text{at } x = \frac{L}{2}$$

*(continued)*

**APPENDIX 52.A** *(continued)*
Elastic Beam Deflection Equations
(*w* is the load per unit length.)
(*y* is positive downward.)

**Case 9: Simple Beam with Uniform Load**

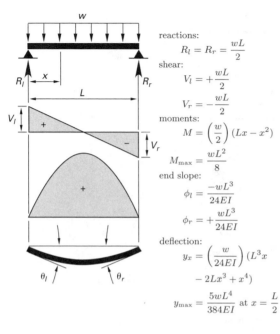

reactions:
$$R_l = R_r = \frac{wL}{2}$$

shear:
$$V_l = +\frac{wL}{2}$$
$$V_r = -\frac{wL}{2}$$

moments:
$$M = \left(\frac{w}{2}\right)(Lx - x^2)$$
$$M_{max} = \frac{wL^2}{8}$$

end slope:
$$\phi_l = \frac{-wL^3}{24EI}$$
$$\phi_r = +\frac{wL^3}{24EI}$$

deflection:
$$y_x = \left(\frac{w}{24EI}\right)(L^3 x - 2Lx^3 + x^4)$$
$$y_{max} = \frac{5wL^4}{384EI} \text{ at } x = \frac{L}{2}$$

**Case 10: Simple Beam with Triangular Load**
(*w* is the maximum loading per unit length at the right end, not the total load, $W = \frac{1}{2}Lw$.)

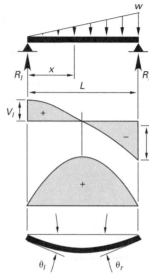

reactions:
$$R_l = \frac{wL}{6}$$
$$R_r = \frac{wL}{3}$$

shear:
$$V_l = +\frac{wL}{6}$$
$$V_r = -\frac{wL}{3}$$
$$V_x = \left(\frac{wL}{6}\right)\left(1 - 3\left(\frac{x}{L}\right)^2\right)$$

moments:
$$M_x = \left(\frac{w}{6}\right)\left(Lx - \frac{x^3}{L}\right)$$
$$M_{max} = \frac{wL^2}{9\sqrt{3}} = 0.0642wL^2$$
$$\text{at } x = 0.577L$$

end slope:
$$\phi_l = \frac{-7wL^3}{360EI}$$
$$\phi_r = +\frac{wL^3}{45EI}$$

deflection:
$$y_x = \left(\frac{w}{360EI}\right)\left(7L^3 x - 10Lx^3 + \frac{3x^5}{L}\right)$$
$$y_{max} = (0.00652)\left(\frac{wL^4}{EI}\right)$$
$$\text{at } x = 0.519L$$

*(continued)*

## APPENDIX 52.A *(continued)*
### Elastic Beam Deflection Equations
(*w* is the load per unit length.)
(*y* is positive downward.)

**Case 11: Simple Beam with Overhung Load**

reactions:

$$R_l = \left(\frac{P}{b}\right)(b+a)$$

$$R_r = \frac{-Pa}{b}$$

shear:

$$V_l = -P$$

$$V_r = \frac{Pa}{b}$$

moments:

$$M_a = Px_a$$

$$M_b = \left(\frac{Pa}{b}\right)(b - x_b)$$

$$M_{\max} = Fa \text{ at } x_a = a$$

deflections:

$$y_a = \left(\frac{F}{3EI}\right)$$
$$\times \left((a^2 + ab)(a - x_a) + \left(\frac{x_a}{2}\right)(x_a^2 - a^2)\right)$$

$$y_b = \left(\frac{Fax_b}{6EI}\right)\left(3x_b - \left(\frac{x_b^2}{b}\right) - 2b\right)$$

$$y_{\text{tip}} = \left(\frac{Fa^2}{3EI}\right)(a + b) \quad [\text{max down}]$$

$$y_{\max} = (0.06415)\left(\frac{Fab^2}{EI}\right) \text{ at } x_b$$
$$= 0.4226b \quad [\text{max up}]$$

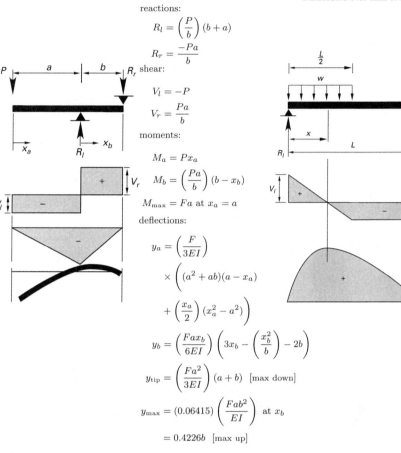

**Case 12: Simple Beam with Uniform Load Distributed over Half of Beam**

reactions:

$$R_l = \frac{3wL}{8}$$

$$R_r = \frac{wL}{8}$$

shear:

$$V_l = +\frac{3wL}{8}$$

$$V_r = -\frac{wL}{8}$$

$$V_x = w\left(\frac{3L}{8} - x\right) \quad \left[x < \frac{L}{2}\right]$$

moments:

$$M_x = \left(\frac{w}{8}\right)\left(3Lx - 4x^2\right) \quad \left[x < \frac{L}{2}\right]$$

$$M_x = \left(\frac{wL^2}{8}\right)\left(1 - \frac{x}{L}\right) \quad \left[x > \frac{L}{2}\right]$$

$$M_{\max} = \frac{9wL^2}{128} \quad \left[x = \frac{3L}{8}\right]$$

deflections:

$$y_x = \left(\frac{wx}{384EI}\right)(9L^3 - 24Lx^2 + 16x^3)$$
$$\left[x < \frac{L}{2}\right]$$

$$y_x = \left(\frac{wL(L-x)}{384EI}\right)(16xL - 8x^2 - L^2)$$
$$\left[x > \frac{L}{2}\right]$$

**APPENDIX 54.A**
Properties of Weld Groups
(treated as lines)

| weld configuration | centroid location | section modulus $S = I_{c,x}/\overline{y}$ | popular moment of inertia $J = I_{c,x} + I_{c,y}$ |
|---|---|---|---|
| 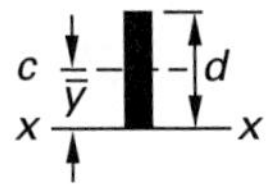 | $\overline{y} = \dfrac{d}{2}$ | $\dfrac{d^2}{6}$ | $\dfrac{d^3}{12}$ |
| 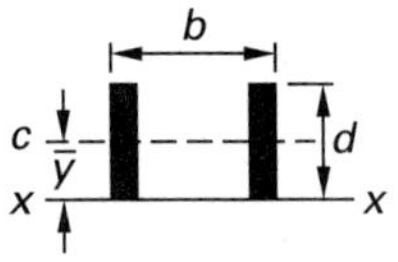 | $\overline{y} = \dfrac{d}{2}$ | $\dfrac{d^2}{3}$ | $\dfrac{d(3b^2 + d^2)}{6}$ |
| 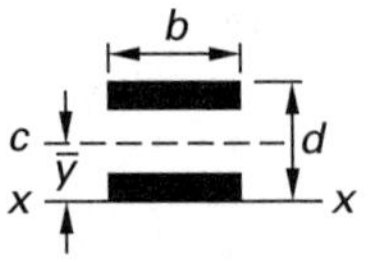 | $\overline{y} = \dfrac{d}{2}$ | $bd$ | $\dfrac{b(3d^2 + b^2)}{6}$ |
| 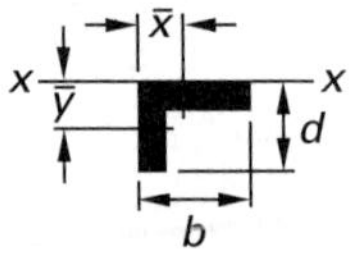 | $\overline{y} = \dfrac{d^2}{2(b+d)}$ $\overline{x} = \dfrac{b^2}{2(b+d)}$ | $\dfrac{4bd + d^2}{6}$ | $\dfrac{(b+d)^4 - 6b^2d^2}{12(b+d)}$ |
| 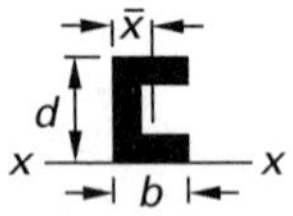 | $\overline{x} = \dfrac{b^2}{2b+d}$ | $bd + \dfrac{d^2}{6}$ | $\dfrac{8b^3 + 6bd^2 + d^3}{12} - \dfrac{b^4}{2b+d}$ |
| 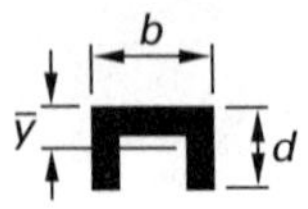 | $\overline{y} = \dfrac{d^2}{b+2d}$ | $\dfrac{2bd + d^2}{3}$ | $\dfrac{b^3 + 6b^2d + 8d^3}{12} - \dfrac{d^4}{2d+b}$ |
| 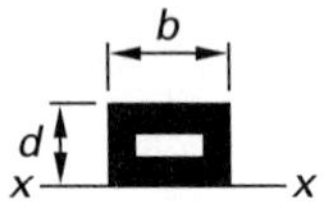 | $\overline{y} = \dfrac{d}{2}$ | $bd + \dfrac{d^2}{3}$ | $\dfrac{(b+d)^3}{6}$ |
| 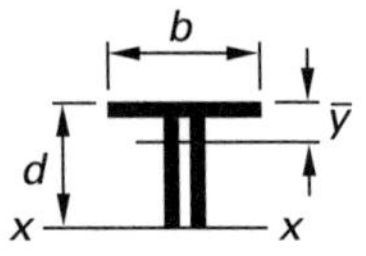 | $\overline{y} = \dfrac{d^2}{b+2d}$ | $\dfrac{2bd + d^2}{3}$ | $\dfrac{b^3 + 8d^3}{12} - \dfrac{d^4}{b+2d}$ |
| 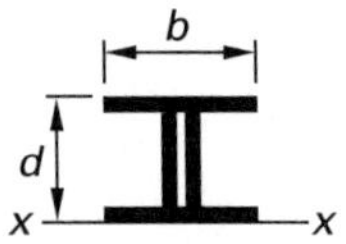 | $\overline{y} = \dfrac{d}{2}$ | $bd + \dfrac{d^2}{3}$ | $\dfrac{b^3 + 3bd^2 + d^3}{6}$ |
| 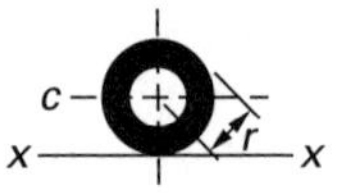 | $\overline{y} = r$ | $\pi r^2$ | $2\pi r^3$ |

## APPENDIX 55.A
Mass Moments of Inertia
(centroids at points labeled $C$)

| | | |
|---|---|---|
| slender rod | | $I_y = I_z = \dfrac{mL^2}{12}$ <br><br> $I_{y'} = I_{z'} = \dfrac{mL^2}{3}$ |
| solid circular cylinder, radius $r$ | | $I_x = \dfrac{mr^2}{2}$ <br><br> $I_y = I_z = \dfrac{m(3r^2 + L^2)}{12}$ |
| hollow circular cylinder, inner radius $r_i$, outer radius $r_o$ | | $I_x = \dfrac{m\left(r_o^2 + r_i^2\right)}{2}$ <br> $\quad = \left(\dfrac{\pi \rho L}{2}\right)\left(r_o^4 - r_i^4\right)$ <br><br> $I_y = I_z = \dfrac{\pi \rho L}{12}\left(3(r_2^4 - r_1^4)\right.$ <br> $\left. \qquad + L^2(r_2^2 - r_1^2)\right)$ |
| thin disk, radius $r$ | | $I_x = \dfrac{mr^2}{2}$ <br><br> $I_y = I_z = \dfrac{mr^2}{4}$ |
| solid circular cone, base radius $r$ | | $I_x = \dfrac{3mr^2}{10}$ <br><br> $I_y = I_z = \left(\dfrac{3m}{5}\right)\left(\dfrac{r^2}{4} + h^2\right)$ |
| thin rectangular plate | | $I_x = \dfrac{m(b^2 + c^2)}{12}$ <br> $I_y = \dfrac{mc^2}{12}$ <br> $I_z = \dfrac{mb^2}{12}$ |
| rectangular parallelepiped | | $I_x = \dfrac{m(b^2 + c^2)}{12}$ <br> $I_y = \dfrac{m(c^2 + a^2)}{12}$ <br> $I_z = \dfrac{m(a^2 + b^2)}{12}$ <br> $I_{x'} = \dfrac{m(4b^2 + c^2)}{12}$ |
| sphere, radius $r$ | | $I_x = I_y = I_z = \dfrac{2mr^2}{5}$ |

**Appendices**

# APPENDIX 68.A
Thermoelectric Constants for Thermocouples
(mV, reference 32°F (0°C))

### (a) chromel-alumel (type K)

| °F | 0 | 10 | 20 | 30 | 40 | 50 | 60 | 70 | 80 | 90 |
|---|---|---|---|---|---|---|---|---|---|---|
| −300 | −5.51 | −5.60 | | | millivolts | | | | | |
| −200 | −4.29 | −4.44 | −4.58 | −4.71 | −4.84 | −4.96 | −5.08 | −5.20 | −5.30 | −5.41 |
| −100 | −2.65 | −2.83 | −3.01 | −3.19 | −3.36 | −3.52 | −3.69 | −3.84 | −4.00 | −4.15 |
| −0 | −0.68 | −0.89 | −1.10 | −1.30 | −1.50 | −1.70 | −1.90 | −2.09 | −2.28 | −2.47 |
| +0 | −0.68 | −0.49 | −0.26 | −0.04 | 0.18 | 0.40 | 0.62 | 0.84 | 1.06 | 1.29 |
| 100 | 1.52 | 1.74 | 1.97 | 2.20 | 2.43 | 2.66 | 2.89 | 3.12 | 3.36 | 3.59 |
| 200 | 3.82 | 4.05 | 4.28 | 4.51 | 4.74 | 4.97 | 5.20 | 5.42 | 5.65 | 5.87 |
| 300 | 6.09 | 6.31 | 6.53 | 6.76 | 6.98 | 7.20 | 7.42 | 7.64 | 7.87 | 8.09 |
| 400 | 8.31 | 8.54 | 8.76 | 8.98 | 9.21 | 9.43 | 9.66 | 9.88 | 10.11 | 10.34 |
| 500 | 10.57 | 10.79 | 11.02 | 11.25 | 11.48 | 11.71 | 11.94 | 12.17 | 12.40 | 12.63 |
| 600 | 12.86 | 13.09 | 13.32 | 13.55 | 13.78 | 14.02 | 14.25 | 14.48 | 14.71 | 14.95 |
| 700 | 15.18 | 15.41 | 15.65 | 15.88 | 16.12 | 16.35 | 16.59 | 16.82 | 17.06 | 17.29 |
| 800 | 17.53 | 17.76 | 18.00 | 18.23 | 18.47 | 18.70 | 18.94 | 19.18 | 19.41 | 19.65 |
| 900 | 19.89 | 20.13 | 20.36 | 20.60 | 20.84 | 21.07 | 21.31 | 21.54 | 21.78 | 22.02 |
| 1000 | 22.26 | 22.49 | 22.73 | 22.97 | 23.20 | 23.44 | 23.68 | 23.91 | 24.15 | 24.39 |
| 1100 | 24.63 | 24.86 | 25.10 | 25.34 | 25.57 | 25.81 | 26.05 | 26.28 | 26.52 | 26.75 |
| 1200 | 26.98 | 27.22 | 27.45 | 27.69 | 27.92 | 28.15 | 28.39 | 28.62 | 28.86 | 29.09 |
| 1300 | 29.32 | 29.56 | 29.79 | 30.02 | 30.25 | 30.49 | 30.72 | 30.95 | 31.18 | 31.42 |
| 1400 | 31.65 | 31.88 | 32.11 | 32.34 | 32.57 | 32.80 | 33.02 | 33.25 | 33.48 | 33.71 |
| 1500 | 33.93 | 34.16 | 34.39 | 34.62 | 34.84 | 35.07 | 35.29 | 35.52 | 35.75 | 35.97 |
| 1600 | 36.19 | 36.42 | 36.64 | 36.87 | 37.09 | 37.31 | 37.54 | 37.76 | 37.98 | 38.20 |
| 1700 | 38.43 | 38.65 | 38.87 | 39.09 | 39.31 | 39.53 | 39.75 | 39.96 | 40.18 | 40.40 |
| 1800 | 40.62 | 40.84 | 41.05 | 41.27 | 41.49 | 41.70 | 41.92 | 42.14 | 42.35 | 42.57 |
| 1900 | 42.78 | 42.99 | 43.21 | 43.42 | 43.63 | 43.85 | 44.06 | 44.27 | 44.49 | 44.70 |
| 2000 | 44.91 | 45.12 | 45.33 | 45.54 | 45.75 | 45.96 | 46.17 | 46.38 | 46.58 | 46.79 |

### (b) iron-constantan (type J)

| °F | 0 | 10 | 20 | 30 | 40 | 50 | 60 | 70 | 80 | 90 |
|---|---|---|---|---|---|---|---|---|---|---|
| −300 | −7.52 | −7.66 | | | millivolts | | | | | |
| −200 | −5.76 | −5.96 | −6.16 | −6.35 | −6.53 | −6.71 | −6.89 | −7.06 | −7.22 | −7.38 |
| −100 | −3.49 | −3.73 | −3.97 | −4.21 | −4.44 | −4.68 | −4.90 | −5.12 | −5.34 | −5.55 |
| −0 | −0.89 | −1.16 | −1.43 | −1.70 | −1.96 | −2.22 | −2.48 | −2.74 | −2.99 | −3.24 |
| +0 | −0.89 | −0.61 | −0.34 | −0.06 | 0.22 | 0.50 | 0.79 | 1.07 | 1.36 | 1.65 |
| 100 | 1.94 | 2.23 | 2.52 | 2.82 | 3.11 | 3.41 | 3.71 | 4.01 | 4.31 | 4.61 |
| 200 | 4.91 | 5.21 | 5.51 | 5.81 | 6.11 | 6.42 | 6.72 | 7.03 | 7.33 | 7.64 |
| 300 | 7.94 | 8.25 | 8.56 | 8.87 | 9.17 | 9.48 | 9.79 | 10.10 | 10.41 | 10.72 |
| 400 | 11.03 | 11.34 | 11.65 | 11.96 | 12.26 | 12.57 | 12.88 | 13.19 | 13.50 | 13.81 |
| 500 | 14.12 | 14.42 | 14.73 | 15.04 | 15.34 | 15.65 | 15.96 | 16.26 | 16.57 | 16.88 |
| 600 | 17.18 | 17.49 | 17.80 | 18.11 | 18.41 | 18.72 | 19.03 | 19.34 | 19.64 | 19.95 |
| 700 | 20.26 | 20.56 | 20.87 | 21.18 | 21.48 | 21.79 | 22.10 | 22.40 | 22.71 | 23.01 |
| 800 | 23.32 | 23.63 | 23.93 | 24.24 | 24.55 | 24.85 | 25.16 | 25.47 | 25.78 | 26.09 |
| 900 | 26.40 | 26.70 | 27.02 | 27.33 | 27.64 | 27.95 | 28.26 | 28.58 | 28.89 | 29.21 |
| 1000 | 29.52 | 29.84 | 30.16 | 30.48 | 30.80 | 31.12 | 31.44 | 31.76 | 32.08 | 32.40 |
| 1100 | 32.72 | 33.05 | 33.37 | 33.70 | 34.03 | 34.36 | 34.68 | 35.01 | 35.35 | 35.68 |
| 1200 | 36.01 | 36.35 | 36.69 | 37.02 | 37.36 | 37.71 | 38.05 | 38.39 | 38.74 | 39.08 |
| 1300 | 39.43 | 39.78 | 40.13 | 40.48 | 40.83 | 41.19 | 41.54 | 41.90 | 42.25 | 42.61 |

### (c) copper-constantan (type T)

| °F | 0 | 10 | 20 | 30 | 40 | 50 | 60 | 70 | 80 | 90 |
|---|---|---|---|---|---|---|---|---|---|---|
| −300 | −5.284 | −5.379 | | | millivolts | | | | | |
| −200 | −4.111 | −4.246 | −4.377 | −4.504 | −4.627 | −4.747 | −4.863 | −4.974 | −5.081 | −5.185 |
| −100 | −2.559 | −2.730 | −2.897 | −3.062 | −3.223 | −3.380 | −3.533 | −3.684 | −3.829 | −3.972 |
| −0 | −0.670 | −0.872 | −1.072 | −1.270 | −1.463 | −1.654 | −1.842 | −2.026 | −2.207 | −2.385 |
| +0 | −0.670 | −0.463 | −0.254 | −0.042 | 0.171 | 0.389 | 0.609 | 0.832 | 1.057 | 1.286 |
| 100 | 1.517 | 1.751 | 1.987 | 2.226 | 2.467 | 2.711 | 2.958 | 3.207 | 3.458 | 3.712 |
| 200 | 3.967 | 4.225 | 4.486 | 4.749 | 5.014 | 5.280 | 5.550 | 5.821 | 6.094 | 6.370 |
| 300 | 6.647 | 6.926 | 7.208 | 7.491 | 7.776 | 8.064 | 8.352 | 8.642 | 8.935 | 9.229 |
| 400 | 9.525 | 9.823 | 10.123 | 10.423 | 10.726 | 11.030 | 11.336 | 11.643 | 11.953 | 12.263 |
| 500 | 12.575 | 12.888 | 13.203 | 13.520 | 13.838 | 14.157 | 14.477 | 14.799 | 15.122 | 15.447 |

Based on National Bureau of Standards Circular No. 561, April 1955.

## APPENDIX 77.A
Standard Cash Flow Factors

| multiply | by | to obtain |
|---|---|---|

$$P = F(1+i)^{-n}$$
$$(P/F, i\%, n)$$

$F$         $P$

$$F = P(1+i)^n$$
$$(F/P, i\%, n)$$

$P$         $F$

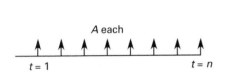

$$P = A\left(\frac{(1+i)^n - 1}{i(1+i)^n}\right)$$
$$(P/A, i\%, n)$$

$A$         $P$

$$A = P\left(\frac{i(1+i)^n}{(1+i)^n - 1}\right)$$
$$(A/P, i\%, n)$$

$P$         $A$

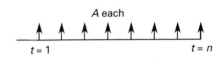

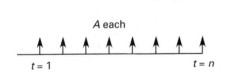

$$F = A\left(\frac{(1+i)^n - 1}{i}\right)$$
$$(F/A, i\%, n)$$

$A$         $F$

$$A = F\left(\frac{i}{(1+i)^n - 1}\right)$$
$$(A/F, i\%, n)$$

$F$         $A$

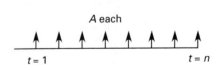

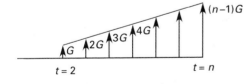

$$P = G\left(\frac{(1+i)^n - 1}{i^2(1+i)^n} - \frac{n}{i(1+i)^n}\right)$$
$$(P/G, i\%, n)$$

$G$         $P$

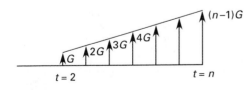

$$A = G\left(\frac{1}{i} - \frac{n}{(1+i)^n - 1}\right)$$
$$(A/G, i\%, n)$$

$G$         $A$

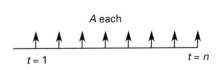

## APPENDIX 77.B
### Cash Flow Equivalent Factors

$$i = 0.50\%$$

| n | P/F | P/A | P/G | F/P | F/A | A/P | A/F | A/G | n |
|---|-----|-----|-----|-----|-----|-----|-----|-----|---|
| 1 | 0.9950 | 0.9950 | 0.0000 | 1.0050 | 1.0000 | 1.0050 | 1.0000 | 0.0000 | 1 |
| 2 | 0.9901 | 0.9851 | 0.9901 | 1.0100 | 2.0050 | 0.5038 | 0.4988 | 0.4988 | 2 |
| 3 | 0.9851 | 2.9702 | 2.9604 | 1.0151 | 3.0150 | 0.3367 | 0.3317 | 0.9967 | 3 |
| 4 | 0.9802 | 3.9505 | 5.9011 | 1.0202 | 4.0301 | 0.2531 | 0.2481 | 1.4938 | 4 |
| 5 | 0.9754 | 4.9259 | 9.8026 | 1.0253 | 5.0503 | 0.2030 | 0.1980 | 1.9900 | 5 |
| 6 | 0.9705 | 5.8964 | 14.6552 | 1.0304 | 6.0755 | 0.1696 | 0.1646 | 2.4855 | 6 |
| 7 | 0.9657 | 6.8621 | 20.4493 | 1.0355 | 7.1059 | 0.1457 | 0.1407 | 2.9801 | 7 |
| 8 | 0.9609 | 7.8230 | 27.1755 | 1.0407 | 8.1414 | 0.1278 | 0.1228 | 3.4738 | 8 |
| 9 | 0.9561 | 8.7791 | 34.8244 | 1.0459 | 9.1821 | 0.1139 | 0.1089 | 3.9668 | 9 |
| 10 | 0.9513 | 9.7304 | 43.3865 | 1.0511 | 10.2280 | 0.1028 | 0.0978 | 4.4589 | 10 |
| 11 | 0.9466 | 10.6770 | 52.8526 | 1.0564 | 11.2792 | 0.0937 | 0.0887 | 4.9501 | 11 |
| 12 | 0.9419 | 11.6189 | 63.2136 | 1.0617 | 12.3356 | 0.0861 | 0.0811 | 5.4406 | 12 |
| 13 | 0.9372 | 12.5562 | 74.4602 | 1.0670 | 13.3972 | 0.0796 | 0.0746 | 5.9302 | 13 |
| 14 | 0.9326 | 13.4887 | 86.5835 | 1.0723 | 14.4642 | 0.0741 | 0.0691 | 6.4190 | 14 |
| 15 | 0.9279 | 14.4166 | 99.5743 | 1.0777 | 15.5365 | 0.0694 | 0.0644 | 6.9069 | 15 |
| 16 | 0.9233 | 15.3399 | 113.4238 | 1.0831 | 16.6142 | 0.0652 | 0.0602 | 7.3940 | 16 |
| 17 | 0.9187 | 16.2586 | 128.1231 | 1.0885 | 17.6973 | 0.0615 | 0.0565 | 7.8803 | 17 |
| 18 | 0.9141 | 17.1728 | 143.6634 | 1.0939 | 18.7858 | 0.0582 | 0.0532 | 8.3658 | 18 |
| 19 | 0.9096 | 18.0824 | 160.0360 | 1.0994 | 19.8797 | 0.0553 | 0.0503 | 8.8504 | 19 |
| 20 | 0.9051 | 18.9874 | 177.2322 | 1.1049 | 20.9791 | 0.0527 | 0.0477 | 9.3342 | 20 |
| 21 | 0.9006 | 19.8880 | 195.2434 | 1.1104 | 22.0840 | 0.0503 | 0.0453 | 9.8172 | 21 |
| 22 | 0.8961 | 20.7841 | 214.0611 | 1.1160 | 23.1944 | 0.0481 | 0.0431 | 10.2993 | 22 |
| 23 | 0.8916 | 21.6757 | 233.6768 | 1.1216 | 24.3104 | 0.0461 | 0.0411 | 10.7806 | 23 |
| 24 | 0.8872 | 22.5629 | 254.0820 | 1.1272 | 25.4320 | 0.0443 | 0.0393 | 11.2611 | 24 |
| 25 | 0.8828 | 23.4456 | 275.2686 | 1.1328 | 26.5591 | 0.0427 | 0.0377 | 11.7407 | 25 |
| 26 | 0.8784 | 24.3240 | 297.2281 | 1.1385 | 27.6919 | 0.0411 | 0.0361 | 12.2195 | 26 |
| 27 | 0.8740 | 25.1980 | 319.9523 | 1.1442 | 28.8304 | 0.0397 | 0.0347 | 12.6975 | 27 |
| 28 | 0.8697 | 26.0677 | 343.4332 | 1.1499 | 29.9745 | 0.0384 | 0.0334 | 13.1747 | 28 |
| 29 | 0.8653 | 26.9330 | 367.6625 | 1.1556 | 31.1244 | 0.0371 | 0.0321 | 13.6510 | 29 |
| 30 | 0.8610 | 27.7941 | 392.6324 | 1.1614 | 32.2800 | 0.0360 | 0.0310 | 14.1265 | 30 |
| 31 | 0.8567 | 28.6508 | 418.3348 | 1.1672 | 33.4414 | 0.0349 | 0.0299 | 14.6012 | 31 |
| 32 | 0.8525 | 29.5033 | 444.7618 | 1.1730 | 34.6086 | 0.0339 | 0.0289 | 15.0750 | 32 |
| 33 | 0.8482 | 30.3515 | 471.9055 | 1.1789 | 35.7817 | 0.0329 | 0.0279 | 15.5480 | 33 |
| 34 | 0.8440 | 31.1955 | 499.7583 | 1.1848 | 36.9606 | 0.0321 | 0.0271 | 16.0202 | 34 |
| 35 | 0.8398 | 32.0354 | 528.3123 | 1.1907 | 38.1454 | 0.0312 | 0.0262 | 16.4915 | 35 |
| 36 | 0.8356 | 32.8710 | 557.5598 | 1.1967 | 39.3361 | 0.0304 | 0.0254 | 16.9621 | 36 |
| 37 | 0.8315 | 33.7025 | 587.4934 | 1.2027 | 40.5328 | 0.0297 | 0.0247 | 17.4317 | 37 |
| 38 | 0.8274 | 34.5299 | 618.1054 | 1.2087 | 41.7354 | 0.0290 | 0.0240 | 17.9006 | 38 |
| 39 | 0.8232 | 35.3531 | 649.3883 | 1.2147 | 42.9441 | 0.0283 | 0.0233 | 18.3686 | 39 |
| 40 | 0.8191 | 36.1722 | 681.3347 | 1.2208 | 44.1588 | 0.0276 | 0.0226 | 18.8359 | 40 |
| 41 | 0.8151 | 36.9873 | 713.9372 | 1.2269 | 45.3796 | 0.0270 | 0.0220 | 19.3022 | 41 |
| 42 | 0.8110 | 37.7983 | 747.1886 | 1.2330 | 46.6065 | 0.0265 | 0.0215 | 19.7678 | 42 |
| 43 | 0.8070 | 38.6053 | 781.0815 | 1.2392 | 47.8396 | 0.0259 | 0.0209 | 20.2325 | 43 |
| 44 | 0.8030 | 39.4082 | 815.6087 | 1.2454 | 49.0788 | 0.0254 | 0.0204 | 20.6964 | 44 |
| 45 | 0.7990 | 40.2072 | 850.7631 | 1.2516 | 50.3242 | 0.0249 | 0.0199 | 21.1595 | 45 |
| 46 | 0.7950 | 41.0022 | 886.5376 | 1.2579 | 51.5758 | 0.0244 | 0.0194 | 21.6217 | 46 |
| 47 | 0.7910 | 41.7932 | 922.9252 | 1.2642 | 52.8337 | 0.0239 | 0.0189 | 22.0831 | 47 |
| 48 | 0.7871 | 42.5803 | 959.9188 | 1.2705 | 54.0978 | 0.0235 | 0.0185 | 22.5437 | 48 |
| 49 | 0.7832 | 43.3635 | 997.5116 | 1.2768 | 55.3683 | 0.0231 | 0.0181 | 23.0035 | 49 |
| 50 | 0.7793 | 44.1428 | 1035.6966 | 1.2832 | 56.6452 | 0.0227 | 0.0177 | 23.4624 | 50 |
| 51 | 0.7754 | 44.9182 | 1074.4670 | 1.2896 | 57.9284 | 0.0223 | 0.0173 | 23.9205 | 51 |
| 52 | 0.7716 | 45.6897 | 1113.8162 | 1.2961 | 59.2180 | 0.0219 | 0.0169 | 24.3778 | 52 |
| 53 | 0.7677 | 46.4575 | 1153.7372 | 1.3026 | 60.5141 | 0.0215 | 0.0165 | 24.8343 | 53 |
| 54 | 0.7639 | 47.2214 | 1194.2236 | 1.3091 | 61.8167 | 0.0212 | 0.0162 | 25.2899 | 54 |
| 55 | 0.7601 | 47.9814 | 1235.2686 | 1.3156 | 63.1258 | 0.0208 | 0.0158 | 25.7447 | 55 |
| 60 | 0.7414 | 51.7256 | 1448.6458 | 1.3489 | 69.7700 | 0.0193 | 0.0143 | 28.0064 | 60 |
| 65 | 0.7231 | 55.3775 | 1675.0272 | 1.3829 | 76.5821 | 0.0181 | 0.0131 | 30.2475 | 65 |
| 70 | 0.7053 | 58.9394 | 1913.6427 | 1.4178 | 83.5661 | 0.0170 | 0.0120 | 32.4680 | 70 |
| 75 | 0.6879 | 62.4136 | 2163.7525 | 1.4536 | 90.7265 | 0.0160 | 0.0110 | 34.6679 | 75 |
| 80 | 0.6710 | 65.8023 | 2424.6455 | 1.4903 | 98.0677 | 0.0152 | 0.0102 | 36.8474 | 80 |
| 85 | 0.6545 | 69.1075 | 2695.6389 | 1.5280 | 105.5943 | 0.0145 | 0.0095 | 39.0065 | 85 |
| 90 | 0.6383 | 72.3313 | 2976.0769 | 1.5666 | 113.3109 | 0.0138 | 0.0088 | 41.1451 | 90 |
| 95 | 0.6226 | 75.4757 | 3265.3298 | 1.6061 | 121.2224 | 0.0132 | 0.0082 | 43.2633 | 95 |
| 100 | 0.6073 | 78.5426 | 3562.7934 | 1.6467 | 129.3337 | 0.0127 | 0.0077 | 45.3613 | 100 |

**APPENDIX 77.B** (*continued*)
Cash Flow Equivalent Factors

$i = 0.75\%$

| n | P/F | P/A | P/G | F/P | F/A | A/P | A/F | A/G | n |
|---|-----|-----|-----|-----|-----|-----|-----|-----|---|
| 1 | 0.9926 | 0.9926 | 0.0000 | 1.0075 | 1.0000 | 1.0075 | 1.0000 | 0.0000 | 1 |
| 2 | 0.9852 | 1.9777 | 0.9852 | 1.0151 | 2.0075 | 0.5056 | 0.4981 | 0.4981 | 2 |
| 3 | 0.9778 | 2.9556 | 2.9408 | 1.0227 | 3.0226 | 0.3383 | 0.3308 | 0.9950 | 3 |
| 4 | 0.9706 | 3.9261 | 5.8525 | 1.0303 | 4.0452 | 0.2547 | 0.2472 | 1.4907 | 4 |
| 5 | 0.9633 | 4.8894 | 9.7058 | 1.0381 | 5.0756 | 0.2045 | 0.1970 | 1.9851 | 5 |
| 6 | 0.9562 | 5.8456 | 14.4866 | 1.0459 | 6.1136 | 0.1711 | 0.1636 | 2.4782 | 6 |
| 7 | 0.9490 | 6.7946 | 20.1808 | 1.0537 | 7.1595 | 0.1472 | 0.1397 | 2.9701 | 7 |
| 8 | 0.9420 | 7.7366 | 26.7747 | 1.0616 | 8.2132 | 0.1293 | 0.1218 | 3.4608 | 8 |
| 9 | 0.9350 | 8.6716 | 34.2544 | 1.0696 | 9.2748 | 0.1153 | 0.1078 | 3.9502 | 9 |
| 10 | 0.9280 | 9.5996 | 42.6064 | 1.0776 | 10.3443 | 0.1042 | 0.0967 | 4.4384 | 10 |
| 11 | 0.9211 | 10.5207 | 51.8174 | 1.0857 | 11.4219 | 0.0951 | 0.0876 | 4.9253 | 11 |
| 12 | 0.9142 | 11.4349 | 61.8740 | 1.0938 | 12.5076 | 0.0875 | 0.0800 | 5.4110 | 12 |
| 13 | 0.9074 | 12.3423 | 72.7632 | 1.1020 | 13.6014 | 0.0810 | 0.0735 | 5.8954 | 13 |
| 14 | 0.9007 | 13.2430 | 84.4720 | 1.1103 | 14.7034 | 0.0755 | 0.0680 | 6.3786 | 14 |
| 15 | 0.8940 | 14.1370 | 96.9876 | 1.1186 | 15.8137 | 0.0707 | 0.0632 | 6.8606 | 15 |
| 16 | 0.8873 | 15.0243 | 110.2973 | 1.1270 | 16.9323 | 0.0666 | 0.0591 | 7.3413 | 16 |
| 17 | 0.8807 | 15.9050 | 124.3887 | 1.1354 | 18.0593 | 0.0629 | 0.0554 | 7.8207 | 17 |
| 18 | 0.8742 | 16.7792 | 139.2494 | 1.1440 | 19.1947 | 0.0596 | 0.0521 | 8.2989 | 18 |
| 19 | 0.8676 | 17.6468 | 154.8671 | 1.1525 | 20.3387 | 0.0567 | 0.0492 | 8.7759 | 19 |
| 20 | 0.8612 | 18.5080 | 171.2297 | 1.1612 | 21.4912 | 0.0540 | 0.0465 | 9.2516 | 20 |
| 21 | 0.8548 | 19.3628 | 188.3253 | 1.1699 | 22.6524 | 0.0516 | 0.0441 | 9.7261 | 21 |
| 22 | 0.8484 | 20.2112 | 206.1420 | 1.1787 | 23.8223 | 0.0495 | 0.0420 | 10.1994 | 22 |
| 23 | 0.8421 | 21.0533 | 224.6682 | 1.1875 | 25.0010 | 0.0475 | 0.0400 | 10.6714 | 23 |
| 24 | 0.8358 | 21.8891 | 243.8923 | 1.1964 | 26.1885 | 0.0457 | 0.0382 | 11.1422 | 24 |
| 25 | 0.8296 | 22.7188 | 263.8029 | 1.2054 | 27.3849 | 0.0440 | 0.0365 | 11.6117 | 25 |
| 26 | 0.8234 | 23.5422 | 284.3888 | 1.2144 | 28.5903 | 0.0425 | 0.0350 | 12.0800 | 26 |
| 27 | 0.8173 | 24.3595 | 305.6387 | 1.2235 | 29.8047 | 0.0411 | 0.0336 | 12.5470 | 27 |
| 28 | 0.8112 | 25.1707 | 327.5416 | 1.2327 | 31.0282 | 0.0397 | 0.0322 | 13.0128 | 28 |
| 29 | 0.8052 | 25.9759 | 350.0867 | 1.2420 | 32.2609 | 0.0385 | 0.0310 | 13.4774 | 29 |
| 30 | 0.7992 | 26.7751 | 373.2631 | 1.2513 | 33.5029 | 0.0373 | 0.0298 | 13.9407 | 30 |
| 31 | 0.7932 | 27.5683 | 397.0602 | 1.2607 | 34.7542 | 0.0363 | 0.0288 | 14.4028 | 31 |
| 32 | 0.7873 | 28.3557 | 421.4675 | 1.2701 | 36.0148 | 0.0353 | 0.0278 | 14.8636 | 32 |
| 33 | 0.7815 | 29.1371 | 446.4746 | 1.2796 | 37.2849 | 0.0343 | 0.0268 | 15.3232 | 33 |
| 34 | 0.7757 | 29.9128 | 472.0712 | 1.2892 | 38.5646 | 0.0334 | 0.0259 | 15.7816 | 34 |
| 35 | 0.7699 | 30.6827 | 498.2471 | 1.2989 | 39.8538 | 0.0326 | 0.0251 | 16.2387 | 35 |
| 36 | 0.7641 | 31.4468 | 524.9924 | 1.3086 | 41.1527 | 0.0318 | 0.0243 | 16.6946 | 36 |
| 37 | 0.7585 | 32.2053 | 552.2969 | 1.3185 | 42.4614 | 0.0311 | 0.0236 | 17.1493 | 37 |
| 38 | 0.7528 | 32.9581 | 580.1511 | 1.3283 | 43.7798 | 0.0303 | 0.0228 | 17.6027 | 38 |
| 39 | 0.7472 | 33.7053 | 608.5451 | 1.3383 | 45.1082 | 0.0297 | 0.0222 | 18.0549 | 39 |
| 40 | 0.7416 | 34.4469 | 637.4693 | 1.3483 | 46.4465 | 0.0290 | 0.0215 | 18.5058 | 40 |
| 41 | 0.7361 | 35.1831 | 666.9144 | 1.3585 | 47.7948 | 0.0284 | 0.0209 | 18.9556 | 41 |
| 42 | 0.7306 | 35.9137 | 696.8709 | 1.3686 | 49.1533 | 0.0278 | 0.0203 | 19.4040 | 42 |
| 43 | 0.7252 | 36.6389 | 727.3297 | 1.3789 | 50.5219 | 0.0273 | 0.0198 | 19.8513 | 43 |
| 44 | 0.7198 | 37.3587 | 758.2815 | 1.3893 | 51.9009 | 0.0268 | 0.0193 | 20.2973 | 44 |
| 45 | 0.7145 | 38.0732 | 789.7173 | 1.3997 | 53.2901 | 0.0263 | 0.0188 | 20.7421 | 45 |
| 46 | 0.7091 | 38.7823 | 821.6283 | 1.4102 | 54.6898 | 0.0258 | 0.0183 | 21.1856 | 46 |
| 47 | 0.7039 | 39.4862 | 854.0056 | 1.4207 | 56.1000 | 0.0253 | 0.0178 | 21.6280 | 47 |
| 48 | 0.6986 | 40.1848 | 886.8404 | 1.4314 | 57.5207 | 0.0249 | 0.0174 | 22.0691 | 48 |
| 49 | 0.6934 | 40.8782 | 920.1243 | 1.4421 | 58.9521 | 0.0245 | 0.0170 | 22.5089 | 49 |
| 50 | 0.6883 | 41.5664 | 953.8486 | 1.4530 | 60.3943 | 0.0241 | 0.0166 | 22.9476 | 50 |
| 51 | 0.6831 | 42.2496 | 988.0050 | 1.4639 | 61.8472 | 0.0237 | 0.0162 | 23.3850 | 51 |
| 52 | 0.6780 | 42.9276 | 1022.5852 | 1.4748 | 63.3111 | 0.0233 | 0.0158 | 23.8211 | 52 |
| 53 | 0.6730 | 43.6006 | 1057.5810 | 1.4859 | 64.7859 | 0.0229 | 0.0154 | 24.2561 | 53 |
| 54 | 0.6680 | 44.2686 | 1092.9842 | 1.4970 | 66.2718 | 0.0226 | 0.0151 | 24.6898 | 54 |
| 55 | 0.6630 | 44.9316 | 1128.7869 | 1.5083 | 67.7688 | 0.0223 | 0.0148 | 25.1223 | 55 |
| 60 | 0.6387 | 48.1734 | 1313.5189 | 1.5657 | 75.4241 | 0.0208 | 0.0133 | 27.2665 | 60 |
| 65 | 0.6153 | 51.2963 | 1507.0910 | 1.6253 | 83.3709 | 0.0195 | 0.0120 | 29.3801 | 65 |
| 70 | 0.5927 | 54.3046 | 1708.6065 | 1.6872 | 91.6201 | 0.0184 | 0.0109 | 31.4634 | 70 |
| 75 | 0.5710 | 57.2027 | 1917.2225 | 1.7514 | 100.1833 | 0.0175 | 0.0100 | 33.5163 | 75 |
| 80 | 0.5500 | 59.9944 | 2132.1472 | 1.8180 | 109.0725 | 0.0167 | 0.0092 | 35.5391 | 80 |
| 85 | 0.5299 | 62.6838 | 2352.6375 | 1.8873 | 118.3001 | 0.0160 | 0.0085 | 37.5318 | 85 |
| 90 | 0.5104 | 65.2746 | 2577.9961 | 1.9591 | 127.8790 | 0.0153 | 0.0078 | 39.4946 | 90 |
| 95 | 0.4917 | 67.7704 | 2807.5694 | 2.0337 | 137.8225 | 0.0148 | 0.0073 | 41.4277 | 95 |
| 100 | 0.4737 | 70.1746 | 3040.7453 | 2.1111 | 148.1445 | 0.0143 | 0.0068 | 43.3311 | 100 |

Appendices

**APPENDIX 77.B** *(continued)*
Cash Flow Equivalent Factors

$i = 1.00\%$

| n | P/F | P/A | P/G | F/P | F/A | A/P | A/F | A/G | n |
|---|-----|-----|-----|-----|-----|-----|-----|-----|---|
| 1 | 0.9901 | 0.9901 | 0.0000 | 1.0100 | 1.0000 | 1.0100 | 1.0000 | 0.0000 | 1 |
| 2 | 0.9803 | 1.9704 | 0.9803 | 1.0201 | 2.0100 | 0.5075 | 0.4975 | 0.4975 | 2 |
| 3 | 0.9706 | 2.9410 | 2.9215 | 1.0303 | 3.0301 | 0.3400 | 0.3300 | 0.9934 | 3 |
| 4 | 0.9610 | 3.9020 | 5.8044 | 1.0406 | 4.0604 | 0.2563 | 0.2463 | 1.4876 | 4 |
| 5 | 0.9515 | 4.8534 | 9.6103 | 1.0510 | 5.1010 | 0.2060 | 0.1960 | 1.9801 | 5 |
| 6 | 0.9420 | 5.7955 | 14.3205 | 1.0615 | 6.1520 | 0.1725 | 0.1625 | 2.4710 | 6 |
| 7 | 0.9327 | 6.7282 | 19.9168 | 1.0721 | 7.2135 | 0.1486 | 0.1386 | 2.9602 | 7 |
| 8 | 0.9235 | 7.6517 | 26.3812 | 1.0829 | 8.2857 | 0.1307 | 0.1207 | 3.4478 | 8 |
| 9 | 0.9143 | 8.5660 | 33.6959 | 1.0937 | 9.3685 | 0.1167 | 0.1067 | 3.9337 | 9 |
| 10 | 0.9053 | 9.4713 | 41.8435 | 1.1046 | 10.4622 | 0.1056 | 0.0956 | 4.4179 | 10 |
| 11 | 0.8963 | 10.3676 | 50.8067 | 1.1157 | 11.5668 | 0.0965 | 0.0865 | 4.9005 | 11 |
| 12 | 0.8874 | 11.2551 | 60.5687 | 1.1268 | 12.6825 | 0.0888 | 0.0788 | 5.3815 | 12 |
| 13 | 0.8787 | 12.1337 | 71.1126 | 1.1381 | 13.8093 | 0.0824 | 0.0724 | 5.8607 | 13 |
| 14 | 0.8700 | 13.0037 | 82.4221 | 1.1495 | 14.9474 | 0.0769 | 0.0669 | 6.3384 | 14 |
| 15 | 0.8613 | 13.8651 | 94.4810 | 1.1610 | 16.0969 | 0.0721 | 0.0621 | 6.8143 | 15 |
| 16 | 0.8528 | 14.7179 | 107.2734 | 1.1726 | 17.2579 | 0.0679 | 0.0579 | 7.2886 | 16 |
| 17 | 0.8444 | 15.5623 | 120.7834 | 1.1843 | 18.4304 | 0.0643 | 0.0543 | 7.7613 | 17 |
| 18 | 0.8360 | 16.3983 | 134.9957 | 1.1961 | 19.6147 | 0.0610 | 0.0510 | 8.2323 | 18 |
| 19 | 0.8277 | 17.2260 | 149.8950 | 1.2081 | 20.8109 | 0.0581 | 0.0481 | 8.7017 | 19 |
| 20 | 0.8195 | 18.0456 | 165.4664 | 1.2202 | 22.0190 | 0.0554 | 0.0454 | 9.1694 | 20 |
| 21 | 0.8114 | 18.8570 | 181.6950 | 1.2324 | 23.2392 | 0.0530 | 0.0430 | 9.6354 | 21 |
| 22 | 0.8034 | 19.6604 | 198.5663 | 1.2447 | 24.4716 | 0.0509 | 0.0409 | 10.0998 | 22 |
| 23 | 0.7954 | 20.4558 | 216.0660 | 1.2572 | 25.7163 | 0.0489 | 0.0389 | 10.5626 | 23 |
| 24 | 0.7876 | 21.2434 | 234.1800 | 1.2697 | 26.9735 | 0.0471 | 0.0371 | 11.0237 | 24 |
| 25 | 0.7798 | 22.0232 | 252.8945 | 1.2824 | 28.2432 | 0.0454 | 0.0354 | 11.4831 | 25 |
| 26 | 0.7720 | 22.7952 | 272.1957 | 1.2953 | 29.5256 | 0.0439 | 0.0339 | 11.9409 | 26 |
| 27 | 0.7644 | 23.5596 | 292.0702 | 1.3082 | 30.8209 | 0.0424 | 0.0324 | 12.3971 | 27 |
| 28 | 0.7568 | 24.3164 | 312.5047 | 1.3213 | 32.1291 | 0.0411 | 0.0311 | 12.8516 | 28 |
| 29 | 0.7493 | 25.0658 | 333.4863 | 1.3345 | 33.4504 | 0.0399 | 0.0299 | 13.3044 | 29 |
| 30 | 0.7419 | 25.8077 | 355.0021 | 1.3478 | 34.7849 | 0.0387 | 0.0287 | 13.7557 | 30 |
| 31 | 0.7346 | 26.5423 | 377.0394 | 1.3613 | 36.1327 | 0.0377 | 0.0277 | 14.2052 | 31 |
| 32 | 0.7273 | 27.2696 | 399.5858 | 1.3749 | 37.4941 | 0.0367 | 0.0267 | 14.6532 | 32 |
| 33 | 0.7201 | 27.9897 | 422.6291 | 1.3887 | 38.8690 | 0.0357 | 0.0257 | 15.0995 | 33 |
| 34 | 0.7130 | 28.7027 | 446.1572 | 1.4026 | 40.2577 | 0.0348 | 0.0248 | 15.5441 | 34 |
| 35 | 0.7059 | 29.4086 | 470.1583 | 1.4166 | 41.6603 | 0.0340 | 0.0240 | 15.9871 | 35 |
| 36 | 0.6989 | 30.1075 | 494.6207 | 1.4308 | 43.0769 | 0.0332 | 0.0232 | 16.4285 | 36 |
| 37 | 0.6920 | 30.7995 | 519.5329 | 1.4451 | 44.5076 | 0.0325 | 0.0225 | 16.8682 | 37 |
| 38 | 0.6852 | 31.4847 | 544.8835 | 1.4595 | 45.9527 | 0.0318 | 0.0218 | 17.3063 | 38 |
| 39 | 0.6784 | 32.1630 | 570.6616 | 1.4741 | 47.4123 | 0.0311 | 0.0211 | 17.7428 | 39 |
| 40 | 0.6717 | 32.8347 | 596.8561 | 1.4889 | 48.8864 | 0.0305 | 0.0205 | 18.1776 | 40 |
| 41 | 0.6650 | 33.4997 | 623.4562 | 1.5038 | 50.3752 | 0.0299 | 0.0199 | 18.6108 | 41 |
| 42 | 0.6584 | 34.1581 | 650.4514 | 1.5188 | 51.8790 | 0.0293 | 0.0193 | 19.0424 | 42 |
| 43 | 0.6519 | 34.8100 | 677.8312 | 1.5340 | 53.3978 | 0.0287 | 0.0187 | 19.4723 | 43 |
| 44 | 0.6454 | 35.4555 | 705.5853 | 1.5493 | 54.9318 | 0.0282 | 0.0182 | 19.9006 | 44 |
| 45 | 0.6391 | 36.0945 | 733.7037 | 1.5648 | 56.4811 | 0.0277 | 0.0177 | 20.3273 | 45 |
| 46 | 0.6327 | 36.7272 | 762.1765 | 1.5805 | 58.0459 | 0.0272 | 0.0172 | 20.7524 | 46 |
| 47 | 0.6265 | 37.3537 | 790.9938 | 1.5963 | 59.6263 | 0.0268 | 0.0168 | 21.1758 | 47 |
| 48 | 0.6203 | 37.9740 | 820.1460 | 1.6122 | 61.2226 | 0.0263 | 0.0163 | 21.5976 | 48 |
| 49 | 0.6141 | 38.5881 | 849.6237 | 1.6283 | 62.8348 | 0.0259 | 0.0159 | 22.0178 | 49 |
| 50 | 0.6080 | 39.1961 | 879.4176 | 1.6446 | 64.4632 | 0.0255 | 0.0155 | 22.4363 | 50 |
| 51 | 0.6020 | 39.7981 | 909.5186 | 1.6611 | 66.1078 | 0.0251 | 0.0151 | 22.8533 | 51 |
| 52 | 0.5961 | 40.3942 | 939.9175 | 1.6777 | 67.7689 | 0.0248 | 0.0148 | 23.2686 | 52 |
| 53 | 0.5902 | 40.9844 | 970.6057 | 1.6945 | 69.4466 | 0.0244 | 0.0144 | 23.6823 | 53 |
| 54 | 0.5843 | 41.5687 | 1001.5743 | 1.7114 | 71.1410 | 0.0241 | 0.0141 | 24.0945 | 54 |
| 55 | 0.5785 | 42.1472 | 1032.8148 | 1.7285 | 72.8525 | 0.0237 | 0.0137 | 24.5049 | 55 |
| 60 | 0.5504 | 44.9550 | 1192.8061 | 1.8167 | 81.6697 | 0.0222 | 0.0122 | 26.5333 | 60 |
| 65 | 0.5237 | 47.6266 | 1358.3903 | 1.9094 | 90.9366 | 0.0210 | 0.0110 | 28.5217 | 65 |
| 70 | 0.4983 | 50.1685 | 1528.6474 | 2.0068 | 100.6763 | 0.0199 | 0.0099 | 30.4703 | 70 |
| 75 | 0.4741 | 52.5871 | 1702.7340 | 2.1091 | 110.9128 | 0.0190 | 0.0090 | 32.3793 | 75 |
| 80 | 0.4511 | 54.8882 | 1879.8771 | 2.2167 | 121.6715 | 0.0182 | 0.0082 | 34.2492 | 80 |
| 85 | 0.4292 | 57.0777 | 2059.3701 | 2.3298 | 132.9790 | 0.0175 | 0.0075 | 36.0801 | 85 |
| 90 | 0.4084 | 59.1609 | 2240.5675 | 2.4486 | 144.8633 | 0.0169 | 0.0069 | 37.8724 | 90 |
| 95 | 0.3886 | 61.1430 | 2422.8811 | 2.5735 | 157.3538 | 0.0164 | 0.0064 | 39.6265 | 95 |
| 100 | 0.3697 | 63.0289 | 2605.7758 | 2.7048 | 170.4814 | 0.0159 | 0.0059 | 41.3426 | 100 |

## APPENDIX 77.B *(continued)*
### Cash Flow Equivalent Factors

$$i = 1.50\%$$

| n | P/F | P/A | P/G | F/P | F/A | A/P | A/F | A/G | n |
|---|-----|-----|-----|-----|-----|-----|-----|-----|---|
| 1 | 0.9852 | 0.9852 | 0.0000 | 1.0150 | 1.0000 | 1.0150 | 1.0000 | 0.0000 | 1 |
| 2 | 0.9707 | 1.9559 | 0.9707 | 1.0302 | 2.0150 | 0.5113 | 0.4963 | 0.4963 | 2 |
| 3 | 0.9563 | 2.9122 | 2.8833 | 1.0457 | 3.0452 | 0.3434 | 0.3284 | 0.9901 | 3 |
| 4 | 0.9422 | 3.8544 | 5.7098 | 1.0614 | 4.0909 | 0.2594 | 0.2444 | 1.4814 | 4 |
| 5 | 0.9283 | 4.7826 | 9.4229 | 1.0773 | 5.1523 | 0.2091 | 0.1941 | 1.9702 | 5 |
| 6 | 0.9145 | 5.6972 | 13.9956 | 1.0934 | 6.2296 | 0.1755 | 0.1605 | 2.4566 | 6 |
| 7 | 0.9010 | 6.5982 | 19.4018 | 1.1098 | 7.3230 | 0.1516 | 0.1366 | 2.9405 | 7 |
| 8 | 0.8877 | 7.4859 | 25.6157 | 1.1265 | 8.4328 | 0.1336 | 0.1186 | 3.4219 | 8 |
| 9 | 0.8746 | 8.3605 | 32.6125 | 1.1434 | 9.5593 | 0.1196 | 0.1046 | 3.9008 | 9 |
| 10 | 0.8617 | 9.2222 | 40.3675 | 1.1605 | 10.7027 | 0.1084 | 0.0934 | 4.3772 | 10 |
| 11 | 0.8489 | 10.0711 | 48.8568 | 1.1779 | 11.8633 | 0.0993 | 0.0843 | 4.8512 | 11 |
| 12 | 0.8364 | 10.9075 | 58.0571 | 1.1956 | 13.0412 | 0.0917 | 0.0767 | 5.3227 | 12 |
| 13 | 0.8240 | 11.7315 | 67.9454 | 1.2136 | 14.2368 | 0.0852 | 0.0702 | 5.7917 | 13 |
| 14 | 0.8118 | 12.5434 | 78.4994 | 1.2318 | 15.4504 | 0.0797 | 0.0647 | 6.2582 | 14 |
| 15 | 0.7999 | 13.3432 | 89.6974 | 1.2502 | 16.6821 | 0.0749 | 0.0599 | 6.7223 | 15 |
| 16 | 0.7880 | 14.1313 | 101.5178 | 1.2690 | 17.9324 | 0.0708 | 0.0558 | 7.1839 | 16 |
| 17 | 0.7764 | 14.9076 | 113.9400 | 1.2880 | 19.2014 | 0.0671 | 0.0521 | 7.6431 | 17 |
| 18 | 0.7649 | 15.6726 | 126.9435 | 1.3073 | 20.4894 | 0.0638 | 0.0488 | 8.0997 | 18 |
| 19 | 0.7536 | 16.4262 | 140.5084 | 1.3270 | 21.7967 | 0.0609 | 0.0459 | 8.5539 | 19 |
| 20 | 0.7425 | 17.1686 | 154.6154 | 1.3469 | 23.1237 | 0.0582 | 0.0432 | 9.0057 | 20 |
| 21 | 0.7315 | 17.9001 | 169.2453 | 1.3671 | 24.4705 | 0.0559 | 0.0409 | 9.4550 | 21 |
| 22 | 0.7207 | 18.6208 | 184.3798 | 1.3876 | 25.8376 | 0.0537 | 0.0387 | 9.9018 | 22 |
| 23 | 0.7100 | 19.3309 | 200.0006 | 1.4084 | 27.2251 | 0.0517 | 0.0367 | 10.3462 | 23 |
| 24 | 0.6995 | 20.0304 | 216.0901 | 1.4295 | 28.6335 | 0.0499 | 0.0349 | 10.7881 | 24 |
| 25 | 0.6892 | 20.7196 | 232.6310 | 1.4509 | 30.0630 | 0.0483 | 0.0333 | 11.2276 | 25 |
| 26 | 0.6790 | 21.3986 | 249.6065 | 1.4727 | 31.5140 | 0.0467 | 0.0317 | 11.6646 | 26 |
| 27 | 0.6690 | 22.0676 | 267.0002 | 1.4948 | 32.9867 | 0.0453 | 0.0303 | 12.0992 | 27 |
| 28 | 0.6591 | 22.7267 | 284.7958 | 1.5172 | 34.4815 | 0.0440 | 0.0290 | 12.5313 | 28 |
| 29 | 0.6494 | 23.3761 | 302.9779 | 1.5400 | 35.9987 | 0.0428 | 0.0278 | 12.9610 | 29 |
| 30 | 0.6398 | 24.0158 | 321.5310 | 1.5631 | 37.5387 | 0.0416 | 0.0266 | 13.3883 | 30 |
| 31 | 0.6303 | 24.6461 | 340.4402 | 1.5865 | 39.1018 | 0.0406 | 0.0256 | 13.8131 | 31 |
| 32 | 0.6210 | 25.2671 | 359.6910 | 1.6103 | 40.6883 | 0.0396 | 0.0246 | 14.2355 | 32 |
| 33 | 0.6118 | 25.8790 | 379.2691 | 1.6345 | 42.2986 | 0.0386 | 0.0236 | 14.6555 | 33 |
| 34 | 0.6028 | 26.4817 | 399.1607 | 1.6590 | 43.9331 | 0.0378 | 0.0228 | 15.0731 | 34 |
| 35 | 0.5939 | 27.0756 | 419.3521 | 1.6839 | 45.5921 | 0.0369 | 0.0219 | 15.4882 | 35 |
| 36 | 0.5851 | 27.6607 | 439.8303 | 1.7091 | 47.2760 | 0.0362 | 0.0212 | 15.9009 | 36 |
| 37 | 0.5764 | 28.2371 | 460.5822 | 1.7348 | 48.9851 | 0.0354 | 0.0204 | 16.3112 | 37 |
| 38 | 0.5679 | 28.8051 | 481.5954 | 1.7608 | 50.7199 | 0.0347 | 0.0197 | 16.7191 | 38 |
| 39 | 0.5595 | 29.3646 | 502.8576 | 1.7872 | 52.4807 | 0.0341 | 0.0191 | 17.1246 | 39 |
| 40 | 0.5513 | 29.9158 | 524.3568 | 1.8140 | 54.2679 | 0.0334 | 0.0184 | 17.5277 | 40 |
| 41 | 0.5431 | 30.4590 | 546.0814 | 1.8412 | 56.0819 | 0.0328 | 0.0178 | 17.9284 | 41 |
| 42 | 0.5351 | 30.9941 | 568.0201 | 1.8688 | 57.9231 | 0.0323 | 0.0173 | 18.3267 | 42 |
| 43 | 0.5272 | 31.5212 | 590.1617 | 1.8969 | 59.7920 | 0.0317 | 0.0167 | 18.7227 | 43 |
| 44 | 0.5194 | 32.0406 | 612.4955 | 1.9253 | 61.6889 | 0.0312 | 0.0162 | 19.1162 | 44 |
| 45 | 0.5117 | 32.5523 | 635.0110 | 1.9542 | 63.6142 | 0.0307 | 0.0157 | 19.5074 | 45 |
| 46 | 0.5042 | 33.0565 | 657.6979 | 1.9835 | 65.5684 | 0.0303 | 0.0153 | 19.8962 | 46 |
| 47 | 0.4967 | 33.5532 | 680.5462 | 2.0133 | 67.5519 | 0.0298 | 0.0148 | 20.2826 | 47 |
| 48 | 0.4894 | 34.0426 | 703.5462 | 2.0435 | 69.5652 | 0.0294 | 0.0144 | 20.6667 | 48 |
| 49 | 0.4821 | 34.5247 | 726.6884 | 2.0741 | 71.6087 | 0.0290 | 0.0140 | 21.0484 | 49 |
| 50 | 0.4750 | 34.9997 | 749.9636 | 2.1052 | 73.6828 | 0.0286 | 0.0136 | 21.4277 | 50 |
| 51 | 0.4680 | 35.4677 | 773.3629 | 2.1368 | 75.7881 | 0.0282 | 0.0132 | 21.8047 | 51 |
| 52 | 0.4611 | 35.9287 | 796.8774 | 2.1689 | 77.9249 | 0.0278 | 0.0128 | 22.1794 | 52 |
| 53 | 0.4543 | 36.3830 | 820.4986 | 2.2014 | 80.0938 | 0.0275 | 0.0125 | 22.5517 | 53 |
| 54 | 0.4475 | 36.8305 | 844.2184 | 2.2344 | 82.2952 | 0.0272 | 0.0122 | 22.9217 | 54 |
| 55 | 0.4409 | 37.2715 | 868.0285 | 2.2679 | 84.5296 | 0.0268 | 0.0118 | 23.2894 | 55 |
| 60 | 0.4093 | 39.3803 | 988.1674 | 2.4432 | 96.2147 | 0.0254 | 0.0104 | 25.0930 | 60 |
| 65 | 0.3799 | 41.3378 | 1109.4752 | 2.6320 | 108.8028 | 0.0242 | 0.0092 | 26.8393 | 65 |
| 70 | 0.3527 | 43.1549 | 1231.1658 | 2.8355 | 122.3638 | 0.0232 | 0.0082 | 28.5290 | 70 |
| 75 | 0.3274 | 44.8416 | 1352.5600 | 3.0546 | 136.9728 | 0.0223 | 0.0073 | 30.1631 | 75 |
| 80 | 0.3039 | 46.4073 | 1473.0741 | 3.2907 | 152.7109 | 0.0215 | 0.0065 | 31.7423 | 80 |
| 85 | 0.2821 | 47.8607 | 1592.2095 | 3.5450 | 169.6652 | 0.0209 | 0.0059 | 33.2676 | 85 |
| 90 | 0.2619 | 49.2099 | 1709.5439 | 3.8189 | 187.9299 | 0.0203 | 0.0053 | 34.7399 | 90 |
| 95 | 0.2431 | 50.4622 | 1824.7224 | 4.1141 | 207.6061 | 0.0198 | 0.0048 | 36.1602 | 95 |
| 100 | 0.2256 | 51.6247 | 1937.4506 | 4.4320 | 228.8030 | 0.0194 | 0.0044 | 37.5295 | 100 |

## APPENDIX 77.B *(continued)*
### Cash Flow Equivalent Factors

$$i = 2.00\%$$

| n | P/F | P/A | P/G | F/P | F/A | A/P | A/F | A/G | n |
|---|-----|-----|-----|-----|-----|-----|-----|-----|---|
| 1 | 0.9804 | 0.9804 | 0.0000 | 1.0200 | 1.0000 | 1.0200 | 1.0000 | 0.0000 | 1 |
| 2 | 0.9612 | 1.9416 | 0.9612 | 1.0404 | 2.0200 | 0.5150 | 0.4950 | 0.4950 | 2 |
| 3 | 0.9423 | 2.8839 | 2.8458 | 1.0612 | 3.0604 | 0.3468 | 0.3268 | 0.9868 | 3 |
| 4 | 0.9238 | 3.8077 | 5.6173 | 1.0824 | 4.1216 | 0.2626 | 0.2426 | 1.4752 | 4 |
| 5 | 0.9057 | 4.7135 | 9.2403 | 1.1041 | 5.2040 | 0.2122 | 0.1922 | 1.9604 | 5 |
| 6 | 0.8880 | 5.6014 | 13.6801 | 1.1262 | 6.3081 | 0.1785 | 0.1585 | 2.4423 | 6 |
| 7 | 0.8706 | 6.4720 | 18.9035 | 1.1487 | 7.4343 | 0.1545 | 0.1345 | 2.9208 | 7 |
| 8 | 0.8535 | 7.3255 | 24.8779 | 1.1717 | 8.5830 | 0.1365 | 0.1165 | 3.3961 | 8 |
| 9 | 0.8368 | 8.1622 | 31.5720 | 1.1951 | 9.7546 | 0.1225 | 0.1025 | 3.8681 | 9 |
| 10 | 0.8203 | 8.9826 | 38.9551 | 1.2190 | 10.9497 | 0.1113 | 0.0913 | 4.3367 | 10 |
| 11 | 0.8043 | 9.7868 | 46.9977 | 1.2434 | 12.1687 | 0.1022 | 0.0822 | 4.8021 | 11 |
| 12 | 0.7885 | 10.5753 | 55.6712 | 1.2682 | 13.4121 | 0.0946 | 0.0746 | 5.2642 | 12 |
| 13 | 0.7730 | 11.3484 | 64.9475 | 1.2936 | 14.6803 | 0.0881 | 0.0681 | 5.7231 | 13 |
| 14 | 0.7579 | 12.1062 | 74.7999 | 1.3195 | 15.9739 | 0.0826 | 0.0626 | 6.1786 | 14 |
| 15 | 0.7430 | 12.8493 | 85.2021 | 1.3459 | 17.2934 | 0.0778 | 0.0578 | 6.6309 | 15 |
| 16 | 0.7284 | 13.5777 | 96.1288 | 1.3728 | 18.6393 | 0.0737 | 0.0537 | 7.0799 | 16 |
| 17 | 0.7142 | 14.2919 | 107.5554 | 1.4002 | 20.0121 | 0.0700 | 0.0500 | 7.5256 | 17 |
| 18 | 0.7002 | 14.9920 | 119.4581 | 1.4282 | 21.4123 | 0.0667 | 0.0467 | 7.9681 | 18 |
| 19 | 0.6864 | 15.6785 | 131.8139 | 1.4568 | 22.8406 | 0.0638 | 0.0438 | 8.4073 | 19 |
| 20 | 0.6730 | 16.3514 | 144.6003 | 1.4859 | 24.2974 | 0.0612 | 0.0412 | 8.8433 | 20 |
| 21 | 0.6598 | 17.0112 | 157.7959 | 1.5157 | 25.7833 | 0.0588 | 0.0388 | 9.2760 | 21 |
| 22 | 0.6468 | 17.6580 | 171.3795 | 1.5460 | 27.2990 | 0.0566 | 0.0366 | 9.7055 | 22 |
| 23 | 0.6342 | 18.2922 | 185.3309 | 1.5769 | 28.8450 | 0.0547 | 0.0347 | 10.1317 | 23 |
| 24 | 0.6217 | 18.9139 | 199.6305 | 1.6084 | 30.4219 | 0.0529 | 0.0329 | 10.5547 | 24 |
| 25 | 0.6095 | 19.5235 | 214.2592 | 1.6406 | 32.0303 | 0.0512 | 0.0312 | 10.9745 | 25 |
| 26 | 0.5976 | 20.1210 | 229.1987 | 1.6734 | 33.6709 | 0.0497 | 0.0297 | 11.3910 | 26 |
| 27 | 0.5859 | 20.7069 | 244.4311 | 1.7069 | 35.3443 | 0.0483 | 0.0283 | 11.8043 | 27 |
| 28 | 0.5744 | 21.2813 | 259.9392 | 1.7410 | 37.0512 | 0.0470 | 0.0270 | 12.2145 | 28 |
| 29 | 0.5631 | 21.8444 | 275.7064 | 1.7758 | 38.7922 | 0.0458 | 0.0258 | 12.6214 | 29 |
| 30 | 0.5521 | 22.3965 | 291.7164 | 1.8114 | 40.5681 | 0.0446 | 0.0246 | 13.0251 | 30 |
| 31 | 0.5412 | 22.9377 | 307.9538 | 1.8476 | 42.3794 | 0.0436 | 0.0236 | 13.4257 | 31 |
| 32 | 0.5306 | 23.4683 | 324.4035 | 1.8845 | 44.2270 | 0.0426 | 0.0226 | 13.8230 | 32 |
| 33 | 0.5202 | 23.9886 | 341.0508 | 1.9222 | 46.1116 | 0.0417 | 0.0217 | 14.2172 | 33 |
| 34 | 0.5100 | 24.4986 | 357.8817 | 1.9607 | 48.0338 | 0.0408 | 0.0208 | 14.6083 | 34 |
| 35 | 0.5000 | 24.9986 | 374.8826 | 1.9999 | 49.9945 | 0.0400 | 0.0200 | 14.9961 | 35 |
| 36 | 0.4902 | 25.4888 | 392.0405 | 2.0399 | 51.9944 | 0.0392 | 0.0192 | 15.3809 | 36 |
| 37 | 0.4806 | 25.9695 | 409.3424 | 2.0807 | 54.0343 | 0.0385 | 0.0185 | 15.7625 | 37 |
| 38 | 0.4712 | 26.4406 | 426.7764 | 2.1223 | 56.1149 | 0.0378 | 0.0178 | 16.1409 | 38 |
| 39 | 0.4619 | 26.9026 | 444.3304 | 2.1647 | 58.2372 | 0.0372 | 0.0172 | 16.5163 | 39 |
| 40 | 0.4529 | 27.3555 | 461.9931 | 2.2080 | 60.4020 | 0.0366 | 0.0166 | 16.8885 | 40 |
| 41 | 0.4440 | 27.7995 | 479.7535 | 2.2522 | 62.6100 | 0.0360 | 0.0160 | 17.2576 | 41 |
| 42 | 0.4353 | 28.2348 | 497.6010 | 2.2972 | 64.8622 | 0.0354 | 0.0154 | 17.6237 | 42 |
| 43 | 0.4268 | 28.6616 | 515.5253 | 2.3432 | 67.1595 | 0.0349 | 0.0149 | 17.9866 | 43 |
| 44 | 0.4184 | 29.0800 | 533.5165 | 2.3901 | 69.5027 | 0.0344 | 0.0144 | 18.3465 | 44 |
| 45 | 0.4102 | 29.4902 | 551.5652 | 2.4379 | 71.8927 | 0.0339 | 0.0139 | 18.7034 | 45 |
| 46 | 0.4022 | 29.8923 | 569.6621 | 2.4866 | 74.3306 | 0.0335 | 0.0135 | 19.0571 | 46 |
| 47 | 0.3943 | 30.2866 | 587.7985 | 2.5363 | 76.8172 | 0.0330 | 0.0130 | 19.4079 | 47 |
| 48 | 0.3865 | 30.6731 | 605.9657 | 2.5871 | 79.3535 | 0.0326 | 0.0126 | 19.7556 | 48 |
| 49 | 0.3790 | 31.0521 | 624.1557 | 2.6388 | 81.9406 | 0.0322 | 0.0122 | 20.1003 | 49 |
| 50 | 0.3715 | 31.4236 | 642.3606 | 2.6916 | 84.5794 | 0.0318 | 0.0118 | 20.4420 | 50 |
| 51 | 0.3642 | 31.7878 | 660.5727 | 2.7454 | 87.2710 | 0.0315 | 0.0115 | 20.7807 | 51 |
| 52 | 0.3571 | 32.1449 | 678.7849 | 2.8003 | 90.0164 | 0.0311 | 0.0111 | 21.1164 | 52 |
| 53 | 0.3501 | 32.4950 | 696.9900 | 2.8563 | 92.8167 | 0.0308 | 0.0108 | 21.4491 | 53 |
| 54 | 0.3432 | 32.8383 | 715.1815 | 2.9135 | 95.6731 | 0.0305 | 0.0105 | 21.7789 | 54 |
| 55 | 0.3365 | 33.1748 | 733.3527 | 2.9717 | 98.5865 | 0.0301 | 0.0101 | 22.1057 | 55 |
| 60 | 0.3048 | 34.7609 | 823.6975 | 3.2810 | 114.0515 | 0.0288 | 0.0088 | 23.6961 | 60 |
| 65 | 0.2761 | 36.1975 | 912.7085 | 3.6225 | 131.1262 | 0.0276 | 0.0076 | 25.2147 | 65 |
| 70 | 0.2500 | 37.4986 | 999.8343 | 3.9996 | 149.9779 | 0.0267 | 0.0067 | 26.6632 | 70 |
| 75 | 0.2265 | 38.6771 | 1084.6393 | 4.4158 | 170.7918 | 0.0259 | 0.0059 | 28.0434 | 75 |
| 80 | 0.2051 | 39.7445 | 1166.7868 | 4.8754 | 193.7720 | 0.0252 | 0.0052 | 29.3572 | 80 |
| 85 | 0.1858 | 40.7113 | 1246.0241 | 5.3829 | 219.1439 | 0.0246 | 0.0046 | 30.6064 | 85 |
| 90 | 0.1683 | 41.5869 | 1322.1701 | 5.9431 | 247.1567 | 0.0240 | 0.0040 | 31.7929 | 90 |
| 95 | 0.1524 | 42.3800 | 1395.1033 | 6.5617 | 278.0850 | 0.0236 | 0.0036 | 32.9189 | 95 |
| 100 | 0.1380 | 43.0984 | 1464.7527 | 7.2446 | 312.2323 | 0.0232 | 0.0032 | 33.9863 | 100 |

**APPENDIX 77.B** *(continued)*
Cash Flow Equivalent Factors

$i = 3.00\%$

| n | P/F | P/A | P/G | F/P | F/A | A/P | A/F | A/G | n |
|---|-----|-----|-----|-----|-----|-----|-----|-----|---|
| 1 | 0.9709 | 0.9709 | 0.0000 | 1.0300 | 1.0000 | 1.0300 | 1.0000 | 0.0000 | 1 |
| 2 | 0.9426 | 1.9135 | 0.9426 | 1.0609 | 2.0300 | 0.5226 | 0.4926 | 0.4926 | 2 |
| 3 | 0.9151 | 2.8286 | 2.7729 | 1.0927 | 3.0909 | 0.3535 | 0.3235 | 0.9803 | 3 |
| 4 | 0.8885 | 3.7171 | 5.4383 | 1.1255 | 4.1836 | 0.2690 | 0.2390 | 1.4631 | 4 |
| 5 | 0.8626 | 4.5797 | 8.8888 | 1.1593 | 5.3091 | 0.2184 | 0.1884 | 1.9409 | 5 |
| 6 | 0.8375 | 5.4172 | 13.0762 | 1.1941 | 6.4684 | 0.1846 | 0.1546 | 2.4138 | 6 |
| 7 | 0.8131 | 6.2303 | 17.9547 | 1.2299 | 7.6625 | 0.1605 | 0.1305 | 2.8819 | 7 |
| 8 | 0.7894 | 7.0197 | 23.4806 | 1.2668 | 8.8923 | 0.1425 | 0.1125 | 3.3450 | 8 |
| 9 | 0.7664 | 7.7861 | 29.6119 | 1.3048 | 10.1591 | 0.1284 | 0.0984 | 3.8032 | 9 |
| 10 | 0.7441 | 8.5302 | 36.3088 | 1.3439 | 11.4639 | 0.1172 | 0.0872 | 4.2565 | 10 |
| 11 | 0.7224 | 9.2526 | 43.5330 | 1.3842 | 12.8078 | 0.1081 | 0.0781 | 4.7049 | 11 |
| 12 | 0.7014 | 9.9540 | 51.2482 | 1.4258 | 14.1920 | 0.1005 | 0.0705 | 5.1485 | 12 |
| 13 | 0.6810 | 10.6350 | 59.4196 | 1.4685 | 15.6178 | 0.0940 | 0.0640 | 5.5872 | 13 |
| 14 | 0.6611 | 11.2961 | 68.0141 | 1.5126 | 17.0863 | 0.0885 | 0.0585 | 6.0210 | 14 |
| 15 | 0.6419 | 11.9379 | 77.0002 | 1.5580 | 18.5989 | 0.0838 | 0.0538 | 6.4500 | 15 |
| 16 | 0.6232 | 12.5611 | 86.3477 | 1.6047 | 20.1569 | 0.0796 | 0.0496 | 6.8742 | 16 |
| 17 | 0.6050 | 13.1661 | 96.0280 | 1.6528 | 21.7616 | 0.0760 | 0.0460 | 7.2936 | 17 |
| 18 | 0.5874 | 13.7535 | 106.0137 | 1.7024 | 23.4144 | 0.0727 | 0.0427 | 7.7081 | 18 |
| 19 | 0.5703 | 14.3238 | 116.2788 | 1.7535 | 25.1169 | 0.0698 | 0.0398 | 8.1179 | 19 |
| 20 | 0.5537 | 14.8775 | 126.7987 | 1.8061 | 26.8704 | 0.0672 | 0.0372 | 8.5229 | 20 |
| 21 | 0.5375 | 15.4150 | 137.5496 | 1.8603 | 28.6765 | 0.0649 | 0.0349 | 8.9231 | 21 |
| 22 | 0.5219 | 15.9369 | 148.5094 | 1.9161 | 30.5368 | 0.0627 | 0.0327 | 9.3186 | 22 |
| 23 | 0.5067 | 16.4436 | 159.6566 | 1.9736 | 32.4529 | 0.0608 | 0.0308 | 9.7093 | 23 |
| 24 | 0.4919 | 16.9355 | 170.9711 | 2.0328 | 34.4265 | 0.0590 | 0.0290 | 10.0954 | 24 |
| 25 | 0.4776 | 17.4131 | 182.4336 | 2.0938 | 36.4593 | 0.0574 | 0.0274 | 10.4768 | 25 |
| 26 | 0.4637 | 17.8768 | 194.0260 | 2.1566 | 38.5530 | 0.0559 | 0.0259 | 10.8535 | 26 |
| 27 | 0.4502 | 18.3270 | 205.7309 | 2.2213 | 40.7096 | 0.0546 | 0.0246 | 11.2255 | 27 |
| 28 | 0.4371 | 18.7641 | 217.5320 | 2.2879 | 42.9309 | 0.0533 | 0.0233 | 11.5930 | 28 |
| 29 | 0.4243 | 19.1885 | 229.4137 | 2.3566 | 45.2189 | 0.0521 | 0.0221 | 11.9558 | 29 |
| 30 | 0.4120 | 19.6004 | 241.3613 | 2.4273 | 47.5754 | 0.0510 | 0.0210 | 12.3141 | 30 |
| 31 | 0.4000 | 20.0004 | 253.3609 | 2.5001 | 50.0027 | 0.0500 | 0.0200 | 12.6678 | 31 |
| 32 | 0.3883 | 20.3888 | 265.3993 | 2.5751 | 52.5028 | 0.0490 | 0.0190 | 13.0169 | 32 |
| 33 | 0.3770 | 20.7658 | 277.4642 | 2.6523 | 55.0778 | 0.0482 | 0.0182 | 13.3616 | 33 |
| 34 | 0.3660 | 21.1318 | 289.5437 | 2.7319 | 57.7302 | 0.0473 | 0.0173 | 13.7018 | 34 |
| 35 | 0.3554 | 21.4872 | 301.6267 | 2.8139 | 60.4621 | 0.0465 | 0.0165 | 14.0375 | 35 |
| 36 | 0.3450 | 21.8323 | 313.7028 | 2.8983 | 63.2759 | 0.0458 | 0.0158 | 14.3688 | 36 |
| 37 | 0.3350 | 22.1672 | 325.7622 | 2.9852 | 66.1742 | 0.0451 | 0.0151 | 14.6957 | 37 |
| 38 | 0.3252 | 22.4925 | 337.7956 | 3.0748 | 69.1594 | 0.0445 | 0.0145 | 15.0182 | 38 |
| 39 | 0.3158 | 22.8082 | 349.7942 | 3.1670 | 72.2342 | 0.0438 | 0.0138 | 15.3363 | 39 |
| 40 | 0.3066 | 23.1148 | 361.7499 | 3.2620 | 75.4013 | 0.0433 | 0.0133 | 15.6502 | 40 |
| 41 | 0.2976 | 23.4124 | 373.6551 | 3.3599 | 78.6633 | 0.0427 | 0.0127 | 15.9597 | 41 |
| 42 | 0.2890 | 23.7014 | 385.5024 | 3.4607 | 82.0232 | 0.0422 | 0.0122 | 16.2650 | 42 |
| 43 | 0.2805 | 23.9819 | 397.2852 | 3.5645 | 85.4839 | 0.0417 | 0.0117 | 16.5660 | 43 |
| 44 | 0.2724 | 24.2543 | 408.9972 | 3.6715 | 89.0484 | 0.0412 | 0.0112 | 16.8629 | 44 |
| 45 | 0.2644 | 24.5187 | 420.6325 | 3.7816 | 92.7199 | 0.0408 | 0.0108 | 17.1556 | 45 |
| 46 | 0.2567 | 24.7754 | 432.1856 | 3.8950 | 96.5015 | 0.0404 | 0.0104 | 17.4441 | 46 |
| 47 | 0.2493 | 25.0247 | 443.6515 | 4.0119 | 100.3965 | 0.0400 | 0.0100 | 17.7285 | 47 |
| 48 | 0.2420 | 25.2667 | 455.0255 | 4.1323 | 104.4084 | 0.0396 | 0.0096 | 18.0089 | 48 |
| 49 | 0.2350 | 25.5017 | 466.3031 | 4.2562 | 108.5406 | 0.0392 | 0.0092 | 18.2852 | 49 |
| 50 | 0.2281 | 25.7298 | 477.4803 | 4.3839 | 112.7969 | 0.0389 | 0.0089 | 18.5575 | 50 |
| 51 | 0.2215 | 25.9512 | 488.5535 | 4.5154 | 117.1808 | 0.0385 | 0.0085 | 18.8258 | 51 |
| 52 | 0.2150 | 26.1662 | 499.5191 | 4.6509 | 121.6962 | 0.0382 | 0.0082 | 19.0902 | 52 |
| 53 | 0.2088 | 26.3750 | 510.3742 | 4.7904 | 126.3471 | 0.0379 | 0.0079 | 19.3507 | 53 |
| 54 | 0.2027 | 26.5777 | 521.1157 | 4.9341 | 131.1375 | 0.0376 | 0.0076 | 19.6073 | 54 |
| 55 | 0.1968 | 26.7744 | 531.7411 | 5.0821 | 136.0716 | 0.0373 | 0.0073 | 19.8600 | 55 |
| 60 | 0.1697 | 27.6756 | 583.0526 | 5.8916 | 163.0534 | 0.0361 | 0.0061 | 21.0674 | 60 |
| 65 | 0.1464 | 28.4529 | 631.2010 | 6.8300 | 194.3328 | 0.0351 | 0.0051 | 22.1841 | 65 |
| 70 | 0.1263 | 29.1234 | 676.0869 | 7.9178 | 230.5941 | 0.0343 | 0.0043 | 23.2145 | 70 |
| 75 | 0.1089 | 29.7018 | 717.6978 | 9.1789 | 272.6309 | 0.0337 | 0.0037 | 24.1634 | 75 |
| 80 | 0.0940 | 30.2008 | 756.0865 | 10.6409 | 321.3630 | 0.0331 | 0.0031 | 25.0353 | 80 |
| 85 | 0.0811 | 30.6312 | 791.3529 | 12.3357 | 377.8570 | 0.0326 | 0.0026 | 25.8349 | 85 |
| 90 | 0.0699 | 31.0024 | 823.6302 | 14.3005 | 443.3489 | 0.0323 | 0.0023 | 26.5667 | 90 |
| 95 | 0.0603 | 31.3227 | 853.0742 | 16.5782 | 519.2720 | 0.0319 | 0.0019 | 27.2351 | 95 |
| 100 | 0.0520 | 31.5989 | 879.8540 | 19.2186 | 607.2877 | 0.0316 | 0.0016 | 27.8444 | 100 |

Appendices

**APPENDIX 77.B** *(continued)*
Cash Flow Equivalent Factors

$$i = 4.00\%$$

| n | P/F | P/A | P/G | F/P | F/A | A/P | A/F | A/G | n |
|---|---|---|---|---|---|---|---|---|---|
| 1 | 0.9615 | 0.9615 | 0.0000 | 1.0400 | 1.0000 | 1.0400 | 1.0000 | 0.0000 | 1 |
| 2 | 0.9246 | 1.8861 | 0.9246 | 1.0816 | 2.0400 | 0.5302 | 0.4902 | 0.4902 | 2 |
| 3 | 0.8890 | 2.7751 | 2.7025 | 1.1249 | 3.1216 | 0.3603 | 0.3203 | 0.9739 | 3 |
| 4 | 0.8548 | 3.6299 | 5.2670 | 1.1699 | 4.2465 | 0.2755 | 0.2355 | 1.4510 | 4 |
| 5 | 0.8219 | 4.4518 | 8.5547 | 1.2167 | 5.4163 | 0.2246 | 0.1846 | 1.9216 | 5 |
| 6 | 0.7903 | 5.2421 | 12.5062 | 1.2653 | 6.6330 | 0.1908 | 0.1508 | 2.3857 | 6 |
| 7 | 0.7599 | 6.0021 | 17.0657 | 1.3159 | 7.8983 | 0.1666 | 0.1266 | 2.8433 | 7 |
| 8 | 0.7307 | 6.7327 | 22.1806 | 1.3686 | 9.2142 | 0.1485 | 0.1085 | 3.2944 | 8 |
| 9 | 0.7026 | 7.4353 | 27.8013 | 1.4233 | 10.5828 | 0.1345 | 0.0945 | 3.7391 | 9 |
| 10 | 0.6756 | 8.1109 | 33.8814 | 1.4802 | 12.0061 | 0.1233 | 0.0833 | 4.1773 | 10 |
| 11 | 0.6496 | 8.7605 | 40.3772 | 1.5395 | 13.4864 | 0.1141 | 0.0741 | 4.6090 | 11 |
| 12 | 0.6246 | 9.3851 | 47.2477 | 1.6010 | 15.0258 | 0.1066 | 0.0666 | 5.0343 | 12 |
| 13 | 0.6006 | 9.9856 | 54.4546 | 1.6651 | 16.6268 | 0.1001 | 0.0601 | 5.4533 | 13 |
| 14 | 0.5775 | 10.5631 | 61.9618 | 1.7317 | 18.2919 | 0.0947 | 0.0547 | 5.8659 | 14 |
| 15 | 0.5553 | 11.1184 | 69.7355 | 1.8009 | 20.0236 | 0.0899 | 0.0499 | 6.2721 | 15 |
| 16 | 0.5339 | 11.6523 | 77.7441 | 1.8730 | 21.8245 | 0.0858 | 0.0458 | 6.6720 | 16 |
| 17 | 0.5134 | 12.1657 | 85.9581 | 1.9479 | 23.6975 | 0.0822 | 0.0422 | 7.0656 | 17 |
| 18 | 0.4936 | 12.6593 | 94.3498 | 2.0258 | 25.6454 | 0.0790 | 0.0390 | 7.4530 | 18 |
| 19 | 0.4746 | 13.1339 | 102.8933 | 2.1068 | 27.6712 | 0.0761 | 0.0361 | 7.8342 | 19 |
| 20 | 0.4564 | 13.5903 | 111.5647 | 2.1911 | 29.7781 | 0.0736 | 0.0336 | 8.2091 | 20 |
| 21 | 0.4388 | 14.0292 | 120.3414 | 2.2788 | 31.9692 | 0.0713 | 0.0313 | 8.5779 | 21 |
| 22 | 0.4220 | 14.4511 | 129.2024 | 2.3699 | 34.2480 | 0.0692 | 0.0292 | 8.9407 | 22 |
| 23 | 0.4057 | 14.8568 | 138.1284 | 2.4647 | 36.6179 | 0.0673 | 0.0273 | 9.2973 | 23 |
| 24 | 0.3901 | 15.2470 | 147.1012 | 2.5633 | 39.0826 | 0.0656 | 0.0256 | 9.6479 | 24 |
| 25 | 0.3751 | 15.6221 | 156.1040 | 2.6658 | 41.6459 | 0.0640 | 0.0240 | 9.9925 | 25 |
| 26 | 0.3607 | 15.9828 | 165.1212 | 2.7725 | 44.3117 | 0.0626 | 0.0226 | 10.3312 | 26 |
| 27 | 0.3468 | 16.3296 | 174.1385 | 2.8834 | 47.0842 | 0.0612 | 0.0212 | 10.6640 | 27 |
| 28 | 0.3335 | 16.6631 | 183.1424 | 2.9987 | 49.9676 | 0.0600 | 0.0200 | 10.9909 | 28 |
| 29 | 0.3207 | 16.9837 | 192.1206 | 3.1187 | 52.9663 | 0.0589 | 0.0189 | 11.3120 | 29 |
| 30 | 0.3083 | 17.2920 | 201.0618 | 3.2434 | 56.0849 | 0.0578 | 0.0178 | 11.6274 | 30 |
| 31 | 0.2965 | 17.5885 | 209.9556 | 3.3731 | 59.3283 | 0.0569 | 0.0169 | 11.9371 | 31 |
| 32 | 0.2851 | 17.8736 | 218.7924 | 3.5081 | 62.7015 | 0.0559 | 0.0159 | 12.2411 | 32 |
| 33 | 0.2741 | 18.1476 | 227.5634 | 3.6484 | 66.2095 | 0.0551 | 0.0151 | 12.5396 | 33 |
| 34 | 0.2636 | 18.4112 | 236.2607 | 3.7943 | 69.8579 | 0.0543 | 0.0143 | 12.8324 | 34 |
| 35 | 0.2534 | 18.6646 | 244.8768 | 3.9461 | 73.6522 | 0.0536 | 0.0136 | 13.1198 | 35 |
| 36 | 0.2437 | 18.9083 | 253.4052 | 4.1039 | 77.5983 | 0.0529 | 0.0129 | 13.4018 | 36 |
| 37 | 0.2343 | 19.1426 | 261.8399 | 4.2681 | 81.7022 | 0.0522 | 0.0122 | 13.6784 | 37 |
| 38 | 0.2253 | 19.3679 | 270.1754 | 4.4388 | 85.9703 | 0.0516 | 0.0116 | 13.9497 | 38 |
| 39 | 0.2166 | 19.5845 | 278.4070 | 4.6164 | 90.4091 | 0.0511 | 0.0111 | 14.2157 | 39 |
| 40 | 0.2083 | 19.7928 | 286.5303 | 4.8010 | 95.0255 | 0.0505 | 0.0105 | 14.4765 | 40 |
| 41 | 0.2003 | 19.9931 | 294.5414 | 4.9931 | 99.8265 | 0.0500 | 0.0100 | 14.7322 | 41 |
| 42 | 0.1926 | 20.1856 | 302.4370 | 5.1928 | 104.8196 | 0.0495 | 0.0095 | 14.9828 | 42 |
| 43 | 0.1852 | 20.3708 | 310.2141 | 5.4005 | 110.0124 | 0.0491 | 0.0091 | 15.2284 | 43 |
| 44 | 0.1780 | 20.5488 | 317.8700 | 5.6165 | 115.4129 | 0.0487 | 0.0087 | 15.4690 | 44 |
| 45 | 0.1712 | 20.7200 | 325.4028 | 5.8412 | 121.0294 | 0.0483 | 0.0083 | 15.7047 | 45 |
| 46 | 0.1646 | 20.8847 | 332.8104 | 6.0748 | 126.8706 | 0.0479 | 0.0079 | 15.9356 | 46 |
| 47 | 0.1583 | 21.0429 | 340.0914 | 6.3178 | 132.9454 | 0.0475 | 0.0075 | 16.1618 | 47 |
| 48 | 0.1522 | 21.1951 | 347.2446 | 6.5705 | 139.2632 | 0.0472 | 0.0072 | 16.3832 | 48 |
| 49 | 0.1463 | 21.3415 | 354.2689 | 6.8333 | 145.8337 | 0.0469 | 0.0069 | 16.6000 | 49 |
| 50 | 0.1407 | 21.4822 | 361.1638 | 7.1067 | 152.6671 | 0.0466 | 0.0066 | 16.8122 | 50 |
| 51 | 0.1353 | 21.6175 | 367.9289 | 7.3910 | 159.7738 | 0.0463 | 0.0063 | 17.0200 | 51 |
| 52 | 0.1301 | 21.7476 | 374.5638 | 7.6866 | 167.1647 | 0.0460 | 0.0060 | 17.2232 | 52 |
| 53 | 0.1251 | 21.8727 | 381.0686 | 7.9941 | 174.8513 | 0.0457 | 0.0057 | 17.4221 | 53 |
| 54 | 0.1203 | 21.9930 | 387.4436 | 8.3138 | 182.8454 | 0.0455 | 0.0055 | 17.6167 | 54 |
| 55 | 0.1157 | 22.1086 | 393.6890 | 8.6464 | 191.1592 | 0.0452 | 0.0052 | 17.8070 | 55 |
| 60 | 0.0951 | 22.6235 | 422.9966 | 10.5196 | 237.9907 | 0.0442 | 0.0042 | 18.6972 | 60 |
| 65 | 0.0781 | 23.0467 | 449.2014 | 12.7987 | 294.9684 | 0.0434 | 0.0034 | 19.4909 | 65 |
| 70 | 0.0642 | 23.3945 | 472.4789 | 15.5716 | 364.2905 | 0.0427 | 0.0027 | 20.1961 | 70 |
| 75 | 0.0528 | 23.6804 | 493.0408 | 18.9453 | 448.6314 | 0.0422 | 0.0022 | 20.8206 | 75 |
| 80 | 0.0434 | 23.9154 | 511.1161 | 23.0498 | 551.2450 | 0.0418 | 0.0018 | 21.3718 | 80 |
| 85 | 0.0357 | 24.1085 | 526.9384 | 28.0436 | 676.0901 | 0.0415 | 0.0015 | 21.8569 | 85 |
| 90 | 0.0293 | 24.2673 | 540.7369 | 34.1193 | 827.9833 | 0.0412 | 0.0012 | 22.2826 | 90 |
| 95 | 0.0241 | 24.3978 | 552.7307 | 41.5114 | 1012.7846 | 0.0410 | 0.0010 | 22.6550 | 95 |
| 100 | 0.0198 | 24.5050 | 563.1249 | 50.5049 | 1237.6237 | 0.0408 | 0.0008 | 22.9800 | 100 |

**APPENDIX 77.B** *(continued)*
Cash Flow Equivalent Factors

$i = 5.00\%$

| n | P/F | P/A | P/G | F/P | F/A | A/P | A/F | A/G | n |
|---|-----|-----|-----|-----|-----|-----|-----|-----|---|
| 1 | 0.9524 | 0.9524 | 0.0000 | 1.0500 | 1.0000 | 1.0500 | 1.0000 | 0.0000 | 1 |
| 2 | 0.9070 | 1.8594 | 0.9070 | 1.1025 | 2.0500 | 0.5378 | 0.4878 | 0.4878 | 2 |
| 3 | 0.8638 | 2.7232 | 2.6347 | 1.1576 | 3.1525 | 0.3672 | 0.3172 | 0.9675 | 3 |
| 4 | 0.8227 | 3.5460 | 5.1028 | 1.2155 | 4.3101 | 0.2820 | 0.2320 | 1.4391 | 4 |
| 5 | 0.7835 | 4.3295 | 8.2369 | 1.2763 | 5.5256 | 0.2310 | 0.1810 | 1.9025 | 5 |
| 6 | 0.7462 | 5.0757 | 11.9680 | 1.3401 | 6.8019 | 0.1970 | 0.1470 | 2.3579 | 6 |
| 7 | 0.7107 | 5.7864 | 16.2321 | 1.4071 | 8.1420 | 0.1728 | 0.1228 | 2.8052 | 7 |
| 8 | 0.6768 | 6.4632 | 20.9700 | 1.4775 | 9.5491 | 0.1547 | 0.1047 | 3.2445 | 8 |
| 9 | 0.6446 | 7.1078 | 26.1268 | 1.5513 | 11.0266 | 0.1407 | 0.0907 | 3.6758 | 9 |
| 10 | 0.6139 | 7.7217 | 31.6520 | 1.6289 | 12.5779 | 0.1295 | 0.0795 | 4.0991 | 10 |
| 11 | 0.5847 | 8.3064 | 37.4988 | 1.7103 | 14.2068 | 0.1204 | 0.0704 | 4.5144 | 11 |
| 12 | 0.5568 | 8.8633 | 43.6241 | 1.7959 | 15.9171 | 0.1128 | 0.0628 | 4.9219 | 12 |
| 13 | 0.5303 | 9.3936 | 49.9879 | 1.8856 | 17.7130 | 0.1065 | 0.0565 | 5.3215 | 13 |
| 14 | 0.5051 | 9.8986 | 56.5538 | 1.9799 | 19.5986 | 0.1010 | 0.0510 | 5.7133 | 14 |
| 15 | 0.4810 | 10.3797 | 63.2880 | 2.0789 | 21.5786 | 0.0963 | 0.0463 | 6.0973 | 15 |
| 16 | 0.4581 | 10.8378 | 70.1597 | 2.1829 | 23.6575 | 0.0923 | 0.0423 | 6.4736 | 16 |
| 17 | 0.4363 | 11.2741 | 77.1405 | 2.2920 | 25.8404 | 0.0887 | 0.0387 | 6.8423 | 17 |
| 18 | 0.4155 | 11.6896 | 84.2043 | 2.4066 | 28.1324 | 0.0855 | 0.0355 | 7.2034 | 18 |
| 19 | 0.3957 | 12.0853 | 91.3275 | 2.5270 | 30.5390 | 0.0827 | 0.0327 | 7.5569 | 19 |
| 20 | 0.3769 | 12.4622 | 98.4884 | 2.6533 | 33.0660 | 0.0802 | 0.0302 | 7.9030 | 20 |
| 21 | 0.3589 | 12.8212 | 105.6673 | 2.7860 | 35.7193 | 0.0780 | 0.0280 | 8.2416 | 21 |
| 22 | 0.3418 | 13.1630 | 112.8461 | 2.9253 | 38.5052 | 0.0760 | 0.0260 | 8.5730 | 22 |
| 23 | 0.3256 | 13.4886 | 120.0087 | 3.0715 | 41.4305 | 0.0741 | 0.0241 | 8.8971 | 23 |
| 24 | 0.3101 | 13.7986 | 127.1402 | 3.2251 | 44.5020 | 0.0725 | 0.0225 | 9.2140 | 24 |
| 25 | 0.2953 | 14.0939 | 134.2275 | 3.3864 | 47.7271 | 0.0710 | 0.0210 | 9.5238 | 25 |
| 26 | 0.2812 | 14.3752 | 141.2585 | 3.5557 | 51.1135 | 0.0696 | 0.0196 | 9.8266 | 26 |
| 27 | 0.2678 | 14.6430 | 148.2226 | 3.7335 | 54.6691 | 0.0683 | 0.0183 | 10.1224 | 27 |
| 28 | 0.2551 | 14.8981 | 155.1101 | 3.9201 | 58.4026 | 0.0671 | 0.0171 | 10.4114 | 28 |
| 29 | 0.2429 | 15.1411 | 161.9126 | 4.1161 | 62.3227 | 0.0660 | 0.0160 | 10.6936 | 29 |
| 30 | 0.2314 | 15.3725 | 168.6226 | 4.3219 | 66.4388 | 0.0651 | 0.0151 | 10.9691 | 30 |
| 31 | 0.2204 | 15.5928 | 175.2333 | 4.5380 | 70.7608 | 0.0641 | 0.0141 | 11.2381 | 31 |
| 32 | 0.2099 | 15.8027 | 181.7392 | 4.7649 | 75.2988 | 0.0633 | 0.0133 | 11.5005 | 32 |
| 33 | 0.1999 | 16.0025 | 188.1351 | 5.0032 | 80.0638 | 0.0625 | 0.0125 | 11.7566 | 33 |
| 34 | 0.1904 | 16.1929 | 194.4168 | 5.2533 | 85.0670 | 0.0618 | 0.0118 | 12.0063 | 34 |
| 35 | 0.1813 | 16.3742 | 200.5807 | 5.5160 | 90.3203 | 0.0611 | 0.0111 | 12.2498 | 35 |
| 36 | 0.1727 | 16.5469 | 206.6237 | 5.7918 | 95.8363 | 0.0604 | 0.0104 | 12.4872 | 36 |
| 37 | 0.1644 | 16.7113 | 212.5434 | 6.0814 | 101.6281 | 0.0598 | 0.0098 | 12.7186 | 37 |
| 38 | 0.1566 | 16.8679 | 218.3378 | 6.3855 | 107.7095 | 0.0593 | 0.0093 | 12.9440 | 38 |
| 39 | 0.1491 | 17.0170 | 224.0054 | 6.7048 | 114.0950 | 0.0588 | 0.0088 | 13.1636 | 39 |
| 40 | 0.1420 | 17.1591 | 229.5452 | 7.0400 | 120.7998 | 0.0583 | 0.0083 | 13.3775 | 40 |
| 41 | 0.1353 | 17.2944 | 234.9564 | 7.3920 | 127.8398 | 0.0578 | 0.0078 | 13.5857 | 41 |
| 42 | 0.1288 | 17.4232 | 240.2389 | 7.7616 | 135.2318 | 0.0574 | 0.0074 | 13.7884 | 42 |
| 43 | 0.1227 | 17.5459 | 245.3925 | 8.1497 | 142.9933 | 0.0570 | 0.0070 | 13.9857 | 43 |
| 44 | 0.1169 | 17.6628 | 250.4175 | 8.5572 | 151.1430 | 0.0566 | 0.0066 | 14.1777 | 44 |
| 45 | 0.1113 | 17.7741 | 255.3145 | 8.9850 | 159.7002 | 0.0563 | 0.0063 | 14.3644 | 45 |
| 46 | 0.1060 | 17.8801 | 260.0844 | 9.4343 | 168.6852 | 0.0559 | 0.0059 | 14.5461 | 46 |
| 47 | 0.1009 | 17.9810 | 264.7281 | 9.9060 | 178.1194 | 0.0556 | 0.0056 | 14.7226 | 47 |
| 48 | 0.0961 | 18.0772 | 269.2467 | 10.4013 | 188.0254 | 0.0553 | 0.0053 | 14.8943 | 48 |
| 49 | 0.0916 | 18.1687 | 273.6418 | 10.9213 | 198.4267 | 0.0550 | 0.0050 | 15.0611 | 49 |
| 50 | 0.0872 | 18.2559 | 277.9148 | 11.4674 | 209.3480 | 0.0548 | 0.0048 | 15.2233 | 50 |
| 51 | 0.0831 | 18.3390 | 282.0673 | 12.0408 | 220.8154 | 0.0545 | 0.0045 | 15.3808 | 51 |
| 52 | 0.0791 | 18.4181 | 286.1013 | 12.6428 | 232.8562 | 0.0543 | 0.0043 | 15.5337 | 52 |
| 53 | 0.0753 | 18.4934 | 290.0184 | 13.2749 | 245.4990 | 0.0541 | 0.0041 | 15.6823 | 53 |
| 54 | 0.0717 | 18.5651 | 293.8208 | 13.9387 | 258.7739 | 0.0539 | 0.0039 | 15.8265 | 54 |
| 55 | 0.0683 | 18.6335 | 297.5104 | 14.6356 | 272.7126 | 0.0537 | 0.0037 | 15.9664 | 55 |
| 60 | 0.0535 | 18.9293 | 314.3432 | 18.6792 | 353.5837 | 0.0528 | 0.0028 | 16.6062 | 60 |
| 65 | 0.0419 | 19.1611 | 328.6910 | 23.8399 | 456.7980 | 0.0522 | 0.0022 | 17.1541 | 65 |
| 70 | 0.0329 | 19.3427 | 340.8409 | 30.4264 | 588.5285 | 0.0517 | 0.0017 | 17.6212 | 70 |
| 75 | 0.0258 | 19.4850 | 351.0721 | 38.8327 | 756.6537 | 0.0513 | 0.0013 | 18.0176 | 75 |
| 80 | 0.0202 | 19.5965 | 359.6460 | 49.5614 | 971.2288 | 0.0510 | 0.0010 | 18.3526 | 80 |
| 85 | 0.0158 | 19.6838 | 366.8007 | 63.2544 | 1245.0871 | 0.0508 | 0.0008 | 18.6346 | 85 |
| 90 | 0.0124 | 19.7523 | 372.7488 | 80.7304 | 1597.6073 | 0.0506 | 0.0006 | 18.8712 | 90 |
| 95 | 0.0097 | 19.8059 | 377.6774 | 103.0347 | 2040.6935 | 0.0505 | 0.0005 | 19.0689 | 95 |
| 100 | 0.0076 | 19.8479 | 381.7492 | 131.5013 | 2610.0252 | 0.0504 | 0.0004 | 19.2337 | 100 |

Appendices

**APPENDIX 77.B** *(continued)*
Cash Flow Equivalent Factors

$i = 6.00\%$

| n | P/F | P/A | P/G | F/P | F/A | A/P | A/F | A/G | n |
|---|-----|-----|-----|-----|-----|-----|-----|-----|---|
| 1 | 0.9434 | 0.9434 | 0.0000 | 1.0600 | 1.0000 | 1.0600 | 1.0000 | 0.0000 | 1 |
| 2 | 0.8900 | 1.8334 | 0.8900 | 1.1236 | 2.0600 | 0.5454 | 0.4854 | 0.4854 | 2 |
| 3 | 0.8396 | 2.6730 | 2.5692 | 1.1910 | 3.1836 | 0.3741 | 0.3141 | 0.9612 | 3 |
| 4 | 0.7921 | 3.4651 | 4.9455 | 1.2625 | 4.3746 | 0.2886 | 0.2286 | 1.4272 | 4 |
| 5 | 0.7473 | 4.2124 | 7.9345 | 1.3382 | 5.6371 | 0.2374 | 0.1774 | 1.8836 | 5 |
| 6 | 0.7050 | 4.9173 | 11.4594 | 1.4185 | 6.9753 | 0.2034 | 0.1434 | 2.3304 | 6 |
| 7 | 0.6651 | 5.5824 | 15.4497 | 1.5036 | 8.3938 | 0.1791 | 0.1191 | 2.7676 | 7 |
| 8 | 0.6274 | 6.2098 | 19.8416 | 1.5938 | 9.8975 | 0.1610 | 0.1010 | 3.1952 | 8 |
| 9 | 0.5919 | 6.8017 | 24.5768 | 1.6895 | 11.4913 | 0.1470 | 0.0870 | 3.6133 | 9 |
| 10 | 0.5584 | 7.3601 | 29.6023 | 1.7908 | 13.1808 | 0.1359 | 0.0759 | 4.0220 | 10 |
| 11 | 0.5268 | 7.8869 | 34.8702 | 1.8983 | 14.9716 | 0.1268 | 0.0668 | 4.4213 | 11 |
| 12 | 0.4970 | 8.3838 | 40.3369 | 2.0122 | 16.8699 | 0.1193 | 0.0593 | 4.8113 | 12 |
| 13 | 0.4688 | 8.8527 | 45.9629 | 2.1329 | 18.8821 | 0.1130 | 0.0530 | 5.1920 | 13 |
| 14 | 0.4423 | 9.2950 | 51.7128 | 2.2609 | 21.0151 | 0.1076 | 0.0476 | 5.5635 | 14 |
| 15 | 0.4173 | 9.7122 | 57.5546 | 2.3966 | 23.2760 | 0.1030 | 0.0430 | 5.9260 | 15 |
| 16 | 0.3936 | 10.1059 | 63.4592 | 2.5404 | 25.6725 | 0.0990 | 0.0390 | 6.2794 | 16 |
| 17 | 0.3714 | 10.4773 | 69.4011 | 2.6928 | 28.2129 | 0.0954 | 0.0354 | 6.6240 | 17 |
| 18 | 0.3503 | 10.8276 | 75.3569 | 2.8543 | 30.9057 | 0.0924 | 0.0324 | 6.9597 | 18 |
| 19 | 0.3305 | 11.1581 | 81.3062 | 3.0256 | 33.7600 | 0.0896 | 0.0296 | 7.2867 | 19 |
| 20 | 0.3118 | 11.4699 | 87.2304 | 3.2071 | 36.7856 | 0.0872 | 0.0272 | 7.6051 | 20 |
| 21 | 0.2942 | 11.7641 | 93.1136 | 3.3996 | 39.9927 | 0.0850 | 0.0250 | 7.9151 | 21 |
| 22 | 0.2775 | 12.0416 | 98.9412 | 3.6035 | 43.3923 | 0.0830 | 0.0230 | 8.2166 | 22 |
| 23 | 0.2618 | 12.3034 | 104.7007 | 3.8197 | 46.9958 | 0.0813 | 0.0213 | 8.5099 | 23 |
| 24 | 0.2470 | 12.5504 | 110.3812 | 4.0489 | 50.8156 | 0.0797 | 0.0197 | 8.7951 | 24 |
| 25 | 0.2330 | 12.7834 | 115.9732 | 4.2919 | 54.8645 | 0.0782 | 0.0182 | 9.0722 | 25 |
| 26 | 0.2198 | 13.0032 | 121.4684 | 4.5494 | 59.1564 | 0.0769 | 0.0169 | 9.3414 | 26 |
| 27 | 0.2074 | 13.2105 | 126.8600 | 4.8223 | 63.7058 | 0.0757 | 0.0157 | 9.6029 | 27 |
| 28 | 0.1956 | 13.4062 | 132.1420 | 5.1117 | 68.5281 | 0.0746 | 0.0146 | 9.8568 | 28 |
| 29 | 0.1846 | 13.5907 | 137.3096 | 5.4184 | 73.6398 | 0.0736 | 0.0136 | 10.1032 | 29 |
| 30 | 0.1741 | 13.7648 | 142.3588 | 5.7435 | 79.0582 | 0.0726 | 0.0126 | 10.3422 | 30 |
| 31 | 0.1643 | 13.9291 | 147.2864 | 6.0881 | 84.8017 | 0.0718 | 0.0118 | 10.5740 | 31 |
| 32 | 0.1550 | 14.0840 | 152.0901 | 6.4534 | 90.8898 | 0.0710 | 0.0110 | 10.7988 | 32 |
| 33 | 0.1462 | 14.2302 | 156.7681 | 6.8406 | 97.3432 | 0.0703 | 0.0103 | 11.0166 | 33 |
| 34 | 0.1379 | 14.3681 | 161.3192 | 7.2510 | 104.1838 | 0.0696 | 0.0096 | 11.2276 | 34 |
| 35 | 0.1301 | 14.4982 | 165.7427 | 7.6861 | 111.4348 | 0.0690 | 0.0090 | 11.4319 | 35 |
| 36 | 0.1227 | 14.6210 | 170.0387 | 8.1473 | 119.1209 | 0.0684 | 0.0084 | 11.6298 | 36 |
| 37 | 0.1158 | 14.7368 | 174.2072 | 8.6361 | 127.2681 | 0.0679 | 0.0079 | 11.8213 | 37 |
| 38 | 0.1092 | 14.8460 | 178.2490 | 9.1543 | 135.9042 | 0.0674 | 0.0074 | 12.0065 | 38 |
| 39 | 0.1031 | 14.9491 | 182.1652 | 9.7035 | 145.0585 | 0.0669 | 0.0069 | 12.1857 | 39 |
| 40 | 0.0972 | 15.0463 | 185.9568 | 10.2857 | 154.7620 | 0.0665 | 0.0065 | 12.3590 | 40 |
| 41 | 0.0917 | 15.1380 | 189.6256 | 10.9029 | 165.0477 | 0.0661 | 0.0061 | 12.5264 | 41 |
| 42 | 0.0865 | 15.2245 | 193.1732 | 11.5570 | 175.9505 | 0.0657 | 0.0057 | 12.6883 | 42 |
| 43 | 0.0816 | 15.3062 | 196.6017 | 12.2505 | 187.5076 | 0.0653 | 0.0053 | 12.8446 | 43 |
| 44 | 0.0770 | 15.3832 | 199.9130 | 12.9855 | 199.7580 | 0.0650 | 0.0050 | 12.9956 | 44 |
| 45 | 0.0727 | 15.4558 | 203.1096 | 13.7646 | 212.7435 | 0.0647 | 0.0047 | 13.1413 | 45 |
| 46 | 0.0685 | 15.5244 | 206.1938 | 14.5905 | 226.5081 | 0.0644 | 0.0044 | 13.2819 | 46 |
| 47 | 0.0647 | 15.5890 | 209.1681 | 15.4659 | 241.0986 | 0.0641 | 0.0041 | 13.4177 | 47 |
| 48 | 0.0610 | 15.6500 | 212.0351 | 16.3939 | 256.5645 | 0.0639 | 0.0039 | 13.5485 | 48 |
| 49 | 0.0575 | 15.7076 | 214.7972 | 17.3775 | 272.9584 | 0.0637 | 0.0037 | 13.6748 | 49 |
| 50 | 0.0543 | 15.7619 | 217.4574 | 18.4202 | 290.3359 | 0.0634 | 0.0034 | 13.7964 | 50 |
| 51 | 0.0512 | 15.8131 | 220.0181 | 19.5254 | 308.7561 | 0.0632 | 0.0032 | 13.9137 | 51 |
| 52 | 0.0483 | 15.8614 | 222.4823 | 20.6969 | 328.2814 | 0.0630 | 0.0030 | 14.0267 | 52 |
| 53 | 0.0456 | 15.9070 | 224.8525 | 21.9387 | 348.9783 | 0.0629 | 0.0029 | 14.1355 | 53 |
| 54 | 0.0430 | 15.9500 | 227.1316 | 23.2550 | 370.9170 | 0.0627 | 0.0027 | 14.2402 | 54 |
| 55 | 0.0406 | 15.9905 | 229.3222 | 24.6503 | 394.1720 | 0.0625 | 0.0025 | 14.3411 | 55 |
| 60 | 0.0303 | 16.1614 | 239.0428 | 32.9877 | 533.1282 | 0.0619 | 0.0019 | 14.7909 | 60 |
| 65 | 0.0227 | 16.2891 | 246.9450 | 44.1450 | 719.0829 | 0.0614 | 0.0014 | 15.1601 | 65 |
| 70 | 0.0169 | 16.3845 | 253.3271 | 59.0759 | 967.9322 | 0.0610 | 0.0010 | 15.4613 | 70 |
| 75 | 0.0126 | 16.4558 | 258.4527 | 79.0569 | 1300.9487 | 0.0608 | 0.0008 | 15.7058 | 75 |
| 80 | 0.0095 | 16.5091 | 262.5493 | 105.7960 | 1746.5999 | 0.0606 | 0.0006 | 15.9033 | 80 |
| 85 | 0.0071 | 16.5489 | 265.8096 | 141.5789 | 2342.9817 | 0.0604 | 0.0004 | 16.0620 | 85 |
| 90 | 0.0053 | 16.5787 | 268.3946 | 189.4645 | 3141.0752 | 0.0603 | 0.0003 | 16.1891 | 90 |
| 95 | 0.0039 | 16.6009 | 270.4375 | 253.5463 | 4209.1042 | 0.0602 | 0.0002 | 16.2905 | 95 |
| 100 | 0.0029 | 16.6175 | 272.0471 | 339.3021 | 5638.3681 | 0.0602 | 0.0002 | 16.3711 | 100 |

Appendices

**APPENDIX 77.B** *(continued)*
Cash Flow Equivalent Factors

$i = 7.00\%$

| n | P/F | P/A | P/G | F/P | F/A | A/P | A/F | A/G | n |
|---|-----|-----|-----|-----|-----|-----|-----|-----|---|
| 1 | 0.9346 | 0.9346 | 0.0000 | 1.0700 | 1.0000 | 1.0700 | 1.0000 | 0.0000 | 1 |
| 2 | 0.8734 | 1.8080 | 0.8734 | 1.1449 | 2.0700 | 0.5531 | 0.4831 | 0.4831 | 2 |
| 3 | 0.8163 | 2.6243 | 2.5060 | 1.2250 | 3.2149 | 0.3811 | 0.3111 | 0.9549 | 3 |
| 4 | 0.7629 | 3.3872 | 4.7947 | 1.3108 | 4.4399 | 0.2952 | 0.2252 | 1.4155 | 4 |
| 5 | 0.7130 | 4.1002 | 7.6467 | 1.4026 | 5.7507 | 0.2439 | 0.1739 | 1.8650 | 5 |
| 6 | 0.6663 | 4.7665 | 10.9784 | 1.5007 | 7.1533 | 0.2098 | 0.1398 | 2.3032 | 6 |
| 7 | 0.6227 | 5.3893 | 14.7149 | 1.6058 | 8.6540 | 0.1856 | 0.1156 | 2.7304 | 7 |
| 8 | 0.5820 | 5.9713 | 18.7889 | 1.7182 | 10.2598 | 0.1675 | 0.0975 | 3.1465 | 8 |
| 9 | 0.5439 | 6.5152 | 23.1404 | 1.8385 | 11.9780 | 0.1535 | 0.0835 | 3.5517 | 9 |
| 10 | 0.5083 | 7.0236 | 27.7156 | 1.9672 | 13.8164 | 0.1424 | 0.0724 | 3.9461 | 10 |
| 11 | 0.4751 | 7.4987 | 32.4665 | 2.1049 | 15.7836 | 0.1334 | 0.0634 | 4.3296 | 11 |
| 12 | 0.4440 | 7.9427 | 37.3506 | 2.2522 | 17.8885 | 0.1259 | 0.0559 | 4.7025 | 12 |
| 13 | 0.4150 | 8.3577 | 42.3302 | 2.4098 | 20.1406 | 0.1197 | 0.0497 | 5.0648 | 13 |
| 14 | 0.3878 | 8.7455 | 47.3718 | 2.5785 | 22.5505 | 0.1143 | 0.0443 | 5.4167 | 14 |
| 15 | 0.3624 | 9.1079 | 52.4461 | 2.7590 | 25.1290 | 0.1098 | 0.0398 | 5.7583 | 15 |
| 16 | 0.3387 | 9.4466 | 57.5271 | 2.9522 | 27.8881 | 0.1059 | 0.0359 | 6.0897 | 16 |
| 17 | 0.3166 | 9.7632 | 62.5923 | 3.1588 | 30.8402 | 0.1024 | 0.0324 | 6.4110 | 17 |
| 18 | 0.2959 | 10.0591 | 67.6219 | 3.3799 | 33.9990 | 0.0994 | 0.0294 | 6.7225 | 18 |
| 19 | 0.2765 | 10.3356 | 72.5991 | 3.6165 | 37.3790 | 0.0968 | 0.0268 | 7.0242 | 19 |
| 20 | 0.2584 | 10.5940 | 77.5091 | 3.8697 | 40.9955 | 0.0944 | 0.0244 | 7.3163 | 20 |
| 21 | 0.2415 | 10.8355 | 82.3393 | 4.1406 | 44.8652 | 0.0923 | 0.0223 | 7.5990 | 21 |
| 22 | 0.2257 | 11.0612 | 87.0793 | 4.4304 | 49.0057 | 0.0904 | 0.0204 | 7.8725 | 22 |
| 23 | 0.2109 | 11.2722 | 91.7201 | 4.7405 | 53.4361 | 0.0887 | 0.0187 | 8.1369 | 23 |
| 24 | 0.1971 | 11.4693 | 96.2545 | 5.0724 | 58.1767 | 0.0872 | 0.0172 | 8.3923 | 24 |
| 25 | 0.1842 | 11.6536 | 100.6765 | 5.4274 | 63.2490 | 0.0858 | 0.0158 | 8.6391 | 25 |
| 26 | 0.1722 | 11.8258 | 104.9814 | 5.8074 | 68.6765 | 0.0846 | 0.0146 | 8.8773 | 26 |
| 27 | 0.1609 | 11.9867 | 109.1656 | 6.2139 | 74.4838 | 0.0834 | 0.0134 | 9.1072 | 27 |
| 28 | 0.1504 | 12.1371 | 113.2264 | 6.6488 | 80.6977 | 0.0824 | 0.0124 | 9.3289 | 28 |
| 29 | 0.1406 | 12.2777 | 117.1622 | 7.1143 | 87.3465 | 0.0814 | 0.0114 | 9.5427 | 29 |
| 30 | 0.1314 | 12.4090 | 120.9718 | 7.6123 | 94.4608 | 0.0806 | 0.0106 | 9.7487 | 30 |
| 31 | 0.1228 | 12.5318 | 124.6550 | 8.1451 | 102.0730 | 0.0798 | 0.0098 | 9.9471 | 31 |
| 32 | 0.1147 | 12.6466 | 128.2120 | 8.7153 | 110.2182 | 0.0791 | 0.0091 | 10.1381 | 32 |
| 33 | 0.1072 | 12.7538 | 131.6435 | 9.3253 | 118.9334 | 0.0784 | 0.0084 | 10.3219 | 33 |
| 34 | 0.1002 | 12.8540 | 134.9507 | 9.9781 | 128.2588 | 0.0778 | 0.0078 | 10.4987 | 34 |
| 35 | 0.0937 | 12.9477 | 138.1353 | 10.6766 | 138.2369 | 0.0772 | 0.0072 | 10.6687 | 35 |
| 36 | 0.0875 | 13.0352 | 141.1990 | 11.4239 | 148.9135 | 0.0767 | 0.0067 | 10.8321 | 36 |
| 37 | 0.0818 | 13.1170 | 144.1441 | 12.2236 | 160.3374 | 0.0762 | 0.0062 | 10.9891 | 37 |
| 38 | 0.0765 | 13.1935 | 146.9730 | 13.0793 | 172.5610 | 0.0758 | 0.0058 | 11.1398 | 38 |
| 39 | 0.0715 | 13.2649 | 149.6883 | 13.9948 | 185.6403 | 0.0754 | 0.0054 | 11.2845 | 39 |
| 40 | 0.0668 | 13.3317 | 152.2928 | 14.9745 | 199.6351 | 0.0750 | 0.0050 | 11.4233 | 40 |
| 41 | 0.0624 | 13.3941 | 154.7892 | 16.0227 | 214.6096 | 0.0747 | 0.0047 | 11.5565 | 41 |
| 42 | 0.0583 | 13.4524 | 157.1807 | 17.1443 | 230.6322 | 0.0743 | 0.0043 | 11.6842 | 42 |
| 43 | 0.0545 | 13.5070 | 159.4702 | 18.3444 | 247.7765 | 0.0740 | 0.0040 | 11.8065 | 43 |
| 44 | 0.0509 | 13.5579 | 161.6609 | 19.6285 | 266.1209 | 0.0738 | 0.0038 | 11.9237 | 44 |
| 45 | 0.0476 | 13.6055 | 163.7559 | 21.0025 | 285.7493 | 0.0735 | 0.0035 | 12.0360 | 45 |
| 46 | 0.0445 | 13.6500 | 165.7584 | 22.4726 | 306.7518 | 0.0733 | 0.0033 | 12.1435 | 46 |
| 47 | 0.0416 | 13.6916 | 167.6714 | 24.0457 | 329.2244 | 0.0730 | 0.0030 | 12.2463 | 47 |
| 48 | 0.0389 | 13.7305 | 169.4981 | 25.7289 | 353.2701 | 0.0728 | 0.0028 | 12.3447 | 48 |
| 49 | 0.0363 | 13.7668 | 171.2417 | 27.5299 | 378.9990 | 0.0726 | 0.0026 | 12.4387 | 49 |
| 50 | 0.0339 | 13.8007 | 172.9051 | 29.4570 | 406.5289 | 0.0725 | 0.0025 | 12.5287 | 50 |
| 51 | 0.0317 | 13.8325 | 174.4915 | 31.5190 | 435.9860 | 0.0723 | 0.0023 | 12.6146 | 51 |
| 52 | 0.0297 | 13.8621 | 176.0037 | 33.7253 | 467.5050 | 0.0721 | 0.0021 | 12.6967 | 52 |
| 53 | 0.0277 | 13.8898 | 177.4447 | 36.0861 | 501.2303 | 0.0720 | 0.0020 | 12.7751 | 53 |
| 54 | 0.0259 | 13.9157 | 178.8173 | 38.6122 | 537.3164 | 0.0719 | 0.0019 | 12.8500 | 54 |
| 55 | 0.0242 | 13.9399 | 180.1243 | 41.3150 | 575.9286 | 0.0717 | 0.0017 | 12.9215 | 55 |
| 60 | 0.0173 | 14.0392 | 185.7677 | 57.9464 | 813.5204 | 0.0712 | 0.0012 | 13.2321 | 60 |
| 65 | 0.0123 | 14.1099 | 190.1452 | 81.2729 | 1146.7552 | 0.0709 | 0.0009 | 13.4760 | 65 |
| 70 | 0.0088 | 14.1604 | 193.5185 | 113.9894 | 1614.1342 | 0.0706 | 0.0006 | 13.6662 | 70 |
| 75 | 0.0063 | 14.1964 | 196.1035 | 159.8760 | 2269.6574 | 0.0704 | 0.0004 | 13.8136 | 75 |
| 80 | 0.0045 | 14.2220 | 198.0748 | 224.2344 | 3189.0627 | 0.0703 | 0.0003 | 13.9273 | 80 |
| 85 | 0.0032 | 14.2403 | 199.5717 | 314.5003 | 4478.5761 | 0.0702 | 0.0002 | 14.0146 | 85 |
| 90 | 0.0023 | 14.2533 | 200.7042 | 441.1030 | 6287.1854 | 0.0702 | 0.0002 | 14.0812 | 90 |
| 95 | 0.0016 | 14.2626 | 201.5581 | 618.6697 | 8823.8535 | 0.0701 | 0.0001 | 14.1319 | 95 |
| 100 | 0.0012 | 14.2693 | 202.2001 | 867.7163 | 12381.6618 | 0.0701 | 0.0001 | 14.1703 | 100 |

Appendices

## APPENDIX 77.B (continued)
### Cash Flow Equivalent Factors

$$i = 8.00\%$$

| n | P/F | P/A | P/G | F/P | F/A | A/P | A/F | A/G | n |
|---|-----|-----|-----|-----|-----|-----|-----|-----|---|
| 1 | 0.9259 | 0.9259 | 0.0000 | 1.0800 | 1.0000 | 1.0800 | 1.0000 | 0.0000 | 1 |
| 2 | 0.8573 | 1.7833 | 0.8573 | 1.1664 | 2.0800 | 0.5608 | 0.4808 | 0.4808 | 2 |
| 3 | 0.7938 | 2.5771 | 2.4450 | 1.2597 | 3.2464 | 0.3880 | 0.3080 | 0.9487 | 3 |
| 4 | 0.7350 | 3.3121 | 4.6501 | 1.3605 | 4.5061 | 0.3019 | 0.2219 | 1.4040 | 4 |
| 5 | 0.6806 | 3.9927 | 7.3724 | 1.4693 | 5.8666 | 0.2505 | 0.1705 | 1.8465 | 5 |
| 6 | 0.6302 | 4.6229 | 10.5233 | 1.5869 | 7.3359 | 0.2163 | 0.1363 | 2.2763 | 6 |
| 7 | 0.5835 | 5.2064 | 14.0242 | 1.7138 | 8.9228 | 0.1921 | 0.1121 | 2.6937 | 7 |
| 8 | 0.5403 | 5.7466 | 17.8061 | 1.8509 | 10.6366 | 0.1740 | 0.0940 | 3.0985 | 8 |
| 9 | 0.5002 | 6.2469 | 21.8081 | 1.9990 | 12.4876 | 0.1601 | 0.0801 | 3.4910 | 9 |
| 10 | 0.4632 | 6.7101 | 25.9768 | 2.1589 | 14.4866 | 0.1490 | 0.0690 | 3.8713 | 10 |
| 11 | 0.4289 | 7.1390 | 30.2657 | 2.3316 | 16.6455 | 0.1401 | 0.0601 | 4.2395 | 11 |
| 12 | 0.3971 | 7.5361 | 34.6339 | 2.5182 | 18.9771 | 0.1327 | 0.0527 | 4.5957 | 12 |
| 13 | 0.3677 | 7.9038 | 39.0463 | 2.7196 | 21.4953 | 0.1265 | 0.0465 | 4.9402 | 13 |
| 14 | 0.3405 | 8.2442 | 43.4723 | 2.9372 | 24.2149 | 0.1213 | 0.0413 | 5.2731 | 14 |
| 15 | 0.3152 | 8.5595 | 47.8857 | 3.1722 | 27.1521 | 0.1168 | 0.0368 | 5.5945 | 15 |
| 16 | 0.2919 | 8.8514 | 52.2640 | 3.4259 | 30.3243 | 0.1130 | 0.0330 | 5.9046 | 16 |
| 17 | 0.2703 | 9.1216 | 56.5883 | 3.7000 | 33.7502 | 0.1096 | 0.0296 | 6.2037 | 17 |
| 18 | 0.2502 | 9.3719 | 60.8426 | 3.9960 | 37.4502 | 0.1067 | 0.0267 | 6.4920 | 18 |
| 19 | 0.2317 | 9.6036 | 65.0134 | 4.3157 | 41.4463 | 0.1041 | 0.0241 | 6.7697 | 19 |
| 20 | 0.2145 | 9.8181 | 69.0898 | 4.6610 | 45.7620 | 0.1019 | 0.0219 | 7.0369 | 20 |
| 21 | 0.1987 | 10.0168 | 73.0629 | 5.0338 | 50.4229 | 0.0998 | 0.0198 | 7.2940 | 21 |
| 22 | 0.1839 | 10.2007 | 76.9257 | 5.4365 | 55.4568 | 0.0980 | 0.0180 | 7.5412 | 22 |
| 23 | 0.1703 | 10.3711 | 80.6726 | 5.8715 | 60.8933 | 0.0964 | 0.0164 | 7.7786 | 23 |
| 24 | 0.1577 | 10.5288 | 84.2997 | 6.3412 | 66.7648 | 0.0950 | 0.0150 | 8.0066 | 24 |
| 25 | 0.1460 | 10.6748 | 87.8041 | 6.8485 | 73.1059 | 0.0937 | 0.0137 | 8.2254 | 25 |
| 26 | 0.1352 | 10.8100 | 91.1842 | 7.3964 | 79.9544 | 0.0925 | 0.0125 | 8.4352 | 26 |
| 27 | 0.1252 | 10.9352 | 94.4390 | 7.9881 | 87.3508 | 0.0914 | 0.0114 | 8.6363 | 27 |
| 28 | 0.1159 | 11.0511 | 97.5687 | 8.6271 | 95.3388 | 0.0905 | 0.0105 | 8.8289 | 28 |
| 29 | 0.1073 | 11.1584 | 100.5738 | 9.3173 | 103.9659 | 0.0896 | 0.0096 | 9.0133 | 29 |
| 30 | 0.0994 | 11.2578 | 103.4558 | 10.0627 | 113.2832 | 0.0888 | 0.0088 | 9.1897 | 30 |
| 31 | 0.0920 | 11.3498 | 106.2163 | 10.8677 | 123.3459 | 0.0881 | 0.0081 | 9.3584 | 31 |
| 32 | 0.0852 | 11.4350 | 108.8575 | 11.7371 | 134.2135 | 0.0875 | 0.0075 | 9.5197 | 32 |
| 33 | 0.0789 | 11.5139 | 111.3819 | 12.6760 | 145.9506 | 0.0869 | 0.0069 | 9.6737 | 33 |
| 34 | 0.0730 | 11.5869 | 113.7924 | 13.6901 | 158.6267 | 0.0863 | 0.0063 | 9.8208 | 34 |
| 35 | 0.0676 | 11.6546 | 116.0920 | 14.7853 | 172.3168 | 0.0858 | 0.0058 | 9.9611 | 35 |
| 36 | 0.0626 | 11.7172 | 118.2839 | 15.9682 | 187.1021 | 0.0853 | 0.0053 | 10.0949 | 36 |
| 37 | 0.0580 | 11.7752 | 120.3713 | 17.2456 | 203.0703 | 0.0849 | 0.0049 | 10.2225 | 37 |
| 38 | 0.0537 | 11.8289 | 122.3579 | 18.6253 | 220.3159 | 0.0845 | 0.0045 | 10.3440 | 38 |
| 39 | 0.0497 | 11.8786 | 124.2470 | 20.1153 | 238.9412 | 0.0842 | 0.0042 | 10.4597 | 39 |
| 40 | 0.0460 | 11.9246 | 126.0422 | 21.7245 | 259.0565 | 0.0839 | 0.0039 | 10.5699 | 40 |
| 41 | 0.0426 | 11.9672 | 127.7470 | 23.4625 | 280.7810 | 0.0836 | 0.0036 | 10.6747 | 41 |
| 42 | 0.0395 | 12.0067 | 129.3651 | 25.3395 | 304.2435 | 0.0833 | 0.0033 | 10.7744 | 42 |
| 43 | 0.0365 | 12.0432 | 130.8998 | 27.3666 | 329.5830 | 0.0830 | 0.0030 | 10.8692 | 43 |
| 44 | 0.0338 | 12.0771 | 132.3547 | 29.5560 | 356.9496 | 0.0828 | 0.0028 | 10.9592 | 44 |
| 45 | 0.0313 | 12.1084 | 133.7331 | 31.9204 | 386.5056 | 0.0826 | 0.0026 | 11.0447 | 45 |
| 46 | 0.0290 | 12.1374 | 135.0384 | 34.4741 | 418.4261 | 0.0824 | 0.0024 | 11.1258 | 46 |
| 47 | 0.0269 | 12.1643 | 136.2739 | 37.2320 | 452.9002 | 0.0822 | 0.0022 | 11.2028 | 47 |
| 48 | 0.0249 | 12.1891 | 137.4428 | 40.2106 | 490.1322 | 0.0820 | 0.0020 | 11.2758 | 48 |
| 49 | 0.0230 | 12.2122 | 138.5480 | 43.4274 | 530.3427 | 0.0819 | 0.0019 | 11.3451 | 49 |
| 50 | 0.0213 | 12.2335 | 139.5928 | 46.9016 | 573.7702 | 0.0817 | 0.0017 | 11.4107 | 50 |
| 51 | 0.0197 | 12.2532 | 140.5799 | 50.6537 | 620.6718 | 0.0816 | 0.0016 | 11.4729 | 51 |
| 52 | 0.0183 | 12.2715 | 141.5121 | 54.7060 | 671.3255 | 0.0815 | 0.0015 | 11.5318 | 52 |
| 53 | 0.0169 | 12.2884 | 142.3923 | 59.0825 | 726.0316 | 0.0814 | 0.0014 | 11.5875 | 53 |
| 54 | 0.0157 | 12.3041 | 143.2229 | 63.8091 | 785.1141 | 0.0813 | 0.0013 | 11.6403 | 54 |
| 55 | 0.0145 | 12.3186 | 144.0065 | 68.9139 | 848.9232 | 0.0812 | 0.0012 | 11.6902 | 55 |
| 60 | 0.0099 | 12.3766 | 147.3000 | 101.2571 | 1253.2133 | 0.0808 | 0.0008 | 11.9015 | 60 |
| 65 | 0.0067 | 12.4160 | 149.7387 | 148.7798 | 1847.2481 | 0.0805 | 0.0005 | 12.0602 | 65 |
| 70 | 0.0046 | 12.4428 | 151.5326 | 218.6064 | 2720.0801 | 0.0804 | 0.0004 | 12.1783 | 70 |
| 75 | 0.0031 | 12.4611 | 152.8448 | 321.2045 | 4002.5566 | 0.0802 | 0.0002 | 12.2658 | 75 |
| 80 | 0.0021 | 12.4735 | 153.8001 | 471.9548 | 5886.9354 | 0.0802 | 0.0002 | 12.3301 | 80 |
| 85 | 0.0014 | 12.4820 | 154.4925 | 693.4565 | 8655.7061 | 0.0801 | 0.0001 | 12.3772 | 85 |
| 90 | 0.0010 | 12.4877 | 154.9925 | 1018.9151 | 12723.9386 | 0.0801 | 0.0001 | 12.4116 | 90 |
| 95 | 0.0007 | 12.4917 | 155.3524 | 1497.1205 | 18701.5069 | 0.0801 | 0.0001 | 12.4365 | 95 |
| 100 | 0.0005 | 12.4943 | 155.6107 | 2199.7613 | 27484.5157 | 0.0800 | 0.0000 | 12.4545 | 100 |

**APPENDIX 77.B** *(continued)*
Cash Flow Equivalent Factors

$i = 9.00\%$

| n | P/F | P/A | P/G | F/P | F/A | A/P | A/F | A/G | n |
|---|-----|-----|-----|-----|-----|-----|-----|-----|---|
| 1 | 0.9174 | 0.9174 | 0.0000 | 1.0900 | 1.0000 | 1.0900 | 1.0000 | 0.0000 | 1 |
| 2 | 0.8417 | 1.7591 | 0.8417 | 1.1881 | 2.0900 | 0.5685 | 0.4785 | 0.4785 | 2 |
| 3 | 0.7722 | 2.5313 | 2.3860 | 1.2950 | 3.2781 | 0.3951 | 0.3051 | 0.9426 | 3 |
| 4 | 0.7084 | 3.2397 | 4.5113 | 1.4116 | 4.5731 | 0.3087 | 0.2187 | 1.3925 | 4 |
| 5 | 0.6499 | 3.8897 | 7.1110 | 1.5386 | 5.9847 | 0.2571 | 0.1671 | 1.8282 | 5 |
| 6 | 0.5963 | 4.4859 | 10.0924 | 1.6771 | 7.5233 | 0.2229 | 0.1329 | 2.2498 | 6 |
| 7 | 0.5470 | 5.0330 | 13.3746 | 1.8280 | 9.2004 | 0.1987 | 0.1087 | 2.6574 | 7 |
| 8 | 0.5019 | 5.5348 | 16.8877 | 1.9926 | 11.0285 | 0.1807 | 0.0907 | 3.0512 | 8 |
| 9 | 0.4604 | 5.9952 | 20.5711 | 2.1719 | 13.0210 | 0.1668 | 0.0768 | 3.4312 | 9 |
| 10 | 0.4224 | 6.4177 | 24.3728 | 2.3674 | 15.1929 | 0.1558 | 0.0658 | 3.7978 | 10 |
| 11 | 0.3875 | 6.8052 | 28.2481 | 2.5804 | 17.5603 | 0.1469 | 0.0569 | 4.1510 | 11 |
| 12 | 0.3555 | 7.1607 | 32.1590 | 2.8127 | 20.1407 | 0.1397 | 0.0497 | 4.4910 | 12 |
| 13 | 0.3262 | 7.4869 | 36.0731 | 3.0658 | 22.9534 | 0.1336 | 0.0436 | 4.8182 | 13 |
| 14 | 0.2992 | 7.7862 | 39.9633 | 3.3417 | 26.0192 | 0.1284 | 0.0384 | 5.1326 | 14 |
| 15 | 0.2745 | 8.0607 | 43.8069 | 3.6425 | 29.3609 | 0.1241 | 0.0341 | 5.4346 | 15 |
| 16 | 0.2519 | 8.3126 | 47.5849 | 3.9703 | 33.0034 | 0.1203 | 0.0303 | 5.7245 | 16 |
| 17 | 0.2311 | 8.5436 | 51.2821 | 4.3276 | 36.9737 | 0.1170 | 0.0270 | 6.0024 | 17 |
| 18 | 0.2120 | 8.7556 | 54.8860 | 4.7171 | 41.3013 | 0.1142 | 0.0242 | 6.2687 | 18 |
| 19 | 0.1945 | 8.9501 | 58.3868 | 5.1417 | 46.0185 | 0.1117 | 0.0217 | 6.5236 | 19 |
| 20 | 0.1784 | 9.1285 | 61.7770 | 5.6044 | 51.1601 | 0.1095 | 0.0195 | 6.7674 | 20 |
| 21 | 0.1637 | 9.2922 | 65.0509 | 6.1088 | 56.7645 | 0.1076 | 0.0176 | 7.0006 | 21 |
| 22 | 0.1502 | 9.4424 | 68.2048 | 6.6586 | 62.8733 | 0.1059 | 0.0159 | 7.2232 | 22 |
| 23 | 0.1378 | 9.5802 | 71.2359 | 7.2579 | 69.5319 | 0.1044 | 0.0144 | 7.4357 | 23 |
| 24 | 0.1264 | 9.7066 | 74.1433 | 7.9111 | 76.7898 | 0.1030 | 0.0130 | 7.6384 | 24 |
| 25 | 0.1160 | 9.8226 | 76.9265 | 8.6231 | 84.7009 | 0.1018 | 0.0118 | 7.8316 | 25 |
| 26 | 0.1064 | 9.9290 | 79.5863 | 9.3992 | 93.3240 | 0.1007 | 0.0107 | 8.0156 | 26 |
| 27 | 0.0976 | 10.0266 | 82.1241 | 10.2451 | 102.7231 | 0.0997 | 0.0097 | 8.1906 | 27 |
| 28 | 0.0895 | 10.1161 | 84.5419 | 11.1671 | 112.9682 | 0.0989 | 0.0089 | 8.3571 | 28 |
| 29 | 0.0822 | 10.1983 | 86.8422 | 12.1722 | 124.1354 | 0.0981 | 0.0081 | 8.5154 | 29 |
| 30 | 0.0754 | 10.2737 | 89.0280 | 13.2677 | 136.3076 | 0.0973 | 0.0073 | 8.6657 | 30 |
| 31 | 0.0691 | 10.3428 | 91.1024 | 14.4618 | 149.5752 | 0.0967 | 0.0067 | 8.8083 | 31 |
| 32 | 0.0634 | 10.4062 | 93.0690 | 15.7633 | 164.0370 | 0.0961 | 0.0061 | 8.9436 | 32 |
| 33 | 0.0582 | 10.4644 | 94.9314 | 17.1820 | 179.8003 | 0.0956 | 0.0056 | 9.0718 | 33 |
| 34 | 0.0534 | 10.5178 | 96.6935 | 18.7284 | 196.9823 | 0.0951 | 0.0051 | 9.1933 | 34 |
| 35 | 0.0490 | 10.5668 | 98.3590 | 20.4140 | 215.7108 | 0.0946 | 0.0046 | 9.3083 | 35 |
| 36 | 0.0449 | 10.6118 | 99.9319 | 22.2512 | 236.1247 | 0.0942 | 0.0042 | 9.4171 | 36 |
| 37 | 0.0412 | 10.6530 | 101.4162 | 24.2538 | 258.3759 | 0.0939 | 0.0039 | 9.5200 | 37 |
| 38 | 0.0378 | 10.6908 | 102.8158 | 26.4367 | 282.6298 | 0.0935 | 0.0035 | 9.6172 | 38 |
| 39 | 0.0347 | 10.7255 | 104.1345 | 28.8160 | 309.0665 | 0.0932 | 0.0032 | 9.7090 | 39 |
| 40 | 0.0318 | 10.7574 | 105.3762 | 31.4094 | 337.8824 | 0.0930 | 0.0030 | 9.7957 | 40 |
| 41 | 0.0292 | 10.7866 | 106.5445 | 34.2363 | 369.2919 | 0.0927 | 0.0027 | 9.8775 | 41 |
| 42 | 0.0268 | 10.8134 | 107.6432 | 37.3175 | 403.5281 | 0.0925 | 0.0025 | 9.9546 | 42 |
| 43 | 0.0246 | 10.8380 | 108.6758 | 40.6761 | 440.8457 | 0.0923 | 0.0023 | 10.0273 | 43 |
| 44 | 0.0226 | 10.8605 | 109.6456 | 44.3370 | 481.5218 | 0.0921 | 0.0021 | 10.0958 | 44 |
| 45 | 0.0207 | 10.8812 | 110.5561 | 48.3273 | 525.8587 | 0.0919 | 0.0019 | 10.1603 | 45 |
| 46 | 0.0190 | 10.9002 | 111.4103 | 52.6767 | 574.1860 | 0.0917 | 0.0017 | 10.2210 | 46 |
| 47 | 0.0174 | 10.9176 | 112.2115 | 57.4176 | 626.8628 | 0.0916 | 0.0016 | 10.2780 | 47 |
| 48 | 0.0160 | 10.9336 | 112.9625 | 62.5852 | 684.2804 | 0.0915 | 0.0015 | 10.3317 | 48 |
| 49 | 0.0147 | 10.9482 | 113.6661 | 68.2179 | 746.8656 | 0.0913 | 0.0013 | 10.3821 | 49 |
| 50 | 0.0134 | 10.9617 | 114.3251 | 74.3575 | 815.0836 | 0.0912 | 0.0012 | 10.4295 | 50 |
| 51 | 0.0123 | 10.9740 | 114.9420 | 81.0497 | 889.4411 | 0.0911 | 0.0011 | 10.4740 | 51 |
| 52 | 0.0113 | 10.9853 | 115.5193 | 88.3442 | 970.4908 | 0.0910 | 0.0010 | 10.5158 | 52 |
| 53 | 0.0104 | 10.9957 | 116.0593 | 96.2951 | 1058.8349 | 0.0909 | 0.0009 | 10.5549 | 53 |
| 54 | 0.0095 | 11.0053 | 116.5642 | 104.9617 | 1155.1301 | 0.0909 | 0.0009 | 10.5917 | 54 |
| 55 | 0.0087 | 11.0140 | 117.0362 | 114.4083 | 1260.0918 | 0.0908 | 0.0008 | 10.6261 | 55 |
| 60 | 0.0057 | 11.0480 | 118.9683 | 176.0313 | 1944.7921 | 0.0905 | 0.0005 | 10.7683 | 60 |
| 65 | 0.0037 | 11.0701 | 120.3344 | 270.8460 | 2998.2885 | 0.0903 | 0.0003 | 10.8702 | 65 |
| 70 | 0.0024 | 11.0844 | 121.2942 | 416.7301 | 4619.2232 | 0.0902 | 0.0002 | 10.9427 | 70 |
| 75 | 0.0016 | 11.0938 | 121.9646 | 641.1909 | 7113.2321 | 0.0901 | 0.0001 | 10.9940 | 75 |
| 80 | 0.0010 | 11.0998 | 122.4306 | 986.5517 | 10950.5741 | 0.0901 | 0.0001 | 11.0299 | 80 |
| 85 | 0.0007 | 11.1038 | 122.7533 | 1517.9320 | 16854.8003 | 0.0901 | 0.0001 | 11.0551 | 85 |
| 90 | 0.0004 | 11.1064 | 122.9758 | 2335.5266 | 25939.1842 | 0.0900 | 0.0000 | 11.0726 | 90 |
| 95 | 0.0003 | 11.1080 | 123.1287 | 3593.4971 | 39916.6350 | 0.0900 | 0.0000 | 11.0847 | 95 |
| 100 | 0.0002 | 11.1091 | 123.2335 | 5529.0408 | 61422.6755 | 0.0900 | 0.0000 | 11.0930 | 100 |

**APPENDIX 77.B** (continued)
Cash Flow Equivalent Factors

$i = 10.00\%$

| n | P/F | P/A | P/G | F/P | F/A | A/P | A/F | A/G | n |
|---|---|---|---|---|---|---|---|---|---|
| 1 | 0.9091 | 0.9091 | 0.0000 | 1.1000 | 1.0000 | 1.1000 | 1.0000 | 0.0000 | 1 |
| 2 | 0.8264 | 1.7355 | 0.8264 | 1.2100 | 2.1000 | 0.5762 | 0.4762 | 0.4762 | 2 |
| 3 | 0.7513 | 2.4869 | 2.3291 | 1.3310 | 3.3100 | 0.4021 | 0.3021 | 0.9366 | 3 |
| 4 | 0.6830 | 3.1699 | 4.3781 | 1.4641 | 4.6410 | 0.3155 | 0.2155 | 1.3812 | 4 |
| 5 | 0.6209 | 3.7908 | 6.8618 | 1.6105 | 6.1051 | 0.2638 | 0.1638 | 1.8101 | 5 |
| 6 | 0.5645 | 4.3553 | 9.6842 | 1.7716 | 7.7156 | 0.2296 | 0.1296 | 2.2236 | 6 |
| 7 | 0.5132 | 4.8684 | 12.7631 | 1.9487 | 9.4872 | 0.2054 | 0.1054 | 2.6216 | 7 |
| 8 | 0.4665 | 5.3349 | 16.0287 | 2.1436 | 11.4359 | 0.1874 | 0.0874 | 3.0045 | 8 |
| 9 | 0.4241 | 5.7590 | 19.4215 | 2.3579 | 13.5795 | 0.1736 | 0.0736 | 3.3724 | 9 |
| 10 | 0.3855 | 6.1446 | 22.8913 | 2.5937 | 15.9374 | 0.1627 | 0.0627 | 3.7255 | 10 |
| 11 | 0.3505 | 6.4951 | 26.3963 | 2.8531 | 18.5312 | 0.1540 | 0.0540 | 4.0641 | 11 |
| 12 | 0.3186 | 6.8137 | 29.9012 | 3.1384 | 21.3843 | 0.1468 | 0.0468 | 4.3884 | 12 |
| 13 | 0.2897 | 7.1034 | 33.3772 | 3.4523 | 24.5227 | 0.1408 | 0.0408 | 4.6988 | 13 |
| 14 | 0.2633 | 7.3667 | 36.8005 | 3.7975 | 27.9750 | 0.1357 | 0.0357 | 4.9955 | 14 |
| 15 | 0.2394 | 7.6061 | 40.1520 | 4.1772 | 31.7725 | 0.1315 | 0.0315 | 5.2789 | 15 |
| 16 | 0.2176 | 7.8237 | 43.4164 | 4.5950 | 35.9497 | 0.1278 | 0.0278 | 5.5493 | 16 |
| 17 | 0.1978 | 8.0216 | 46.5819 | 5.0545 | 40.5447 | 0.1247 | 0.0247 | 5.8071 | 17 |
| 18 | 0.1799 | 8.2014 | 49.6395 | 5.5599 | 45.5992 | 0.1219 | 0.0219 | 6.0526 | 18 |
| 19 | 0.1635 | 8.3649 | 52.5827 | 6.1159 | 51.1591 | 0.1195 | 0.0195 | 6.2861 | 19 |
| 20 | 0.1486 | 8.5136 | 55.4069 | 6.7275 | 57.2750 | 0.1175 | 0.0175 | 6.5081 | 20 |
| 21 | 0.1351 | 8.6487 | 58.1095 | 7.4002 | 64.0025 | 0.1156 | 0.0156 | 6.7189 | 21 |
| 22 | 0.1228 | 8.7715 | 60.6893 | 8.1403 | 71.4027 | 0.1140 | 0.0140 | 6.9189 | 22 |
| 23 | 0.1117 | 8.8832 | 63.1462 | 8.9543 | 79.5430 | 0.1126 | 0.0126 | 7.1085 | 23 |
| 24 | 0.1015 | 8.9847 | 65.4813 | 9.8497 | 88.4973 | 0.1113 | 0.0113 | 7.2881 | 24 |
| 25 | 0.0923 | 9.0770 | 67.6964 | 10.8347 | 98.3471 | 0.1102 | 0.0102 | 7.4580 | 25 |
| 26 | 0.0839 | 9.1609 | 69.7940 | 11.9182 | 109.1818 | 0.1092 | 0.0092 | 7.6186 | 26 |
| 27 | 0.0763 | 9.2372 | 71.7773 | 13.1100 | 121.0999 | 0.1083 | 0.0083 | 7.7704 | 27 |
| 28 | 0.0693 | 9.3066 | 73.6495 | 14.4210 | 134.2099 | 0.1075 | 0.0075 | 7.9137 | 28 |
| 29 | 0.0630 | 9.3696 | 75.4146 | 15.8631 | 148.6309 | 0.1067 | 0.0067 | 8.0489 | 29 |
| 30 | 0.0573 | 9.4269 | 77.0766 | 17.4494 | 164.4940 | 0.1061 | 0.0061 | 8.1762 | 30 |
| 31 | 0.0521 | 9.4790 | 78.6395 | 19.1943 | 181.9434 | 0.1055 | 0.0055 | 8.2962 | 31 |
| 32 | 0.0474 | 9.5264 | 80.1078 | 21.1138 | 201.1378 | 0.1050 | 0.0050 | 8.4091 | 32 |
| 33 | 0.0431 | 9.5694 | 81.4856 | 23.2252 | 222.2515 | 0.1045 | 0.0045 | 8.5152 | 33 |
| 34 | 0.0391 | 9.6086 | 82.7773 | 25.5477 | 245.4767 | 0.1041 | 0.0041 | 8.6149 | 34 |
| 35 | 0.0356 | 9.6442 | 83.9872 | 28.1024 | 271.0244 | 0.1037 | 0.0037 | 8.7086 | 35 |
| 36 | 0.0323 | 9.6765 | 85.1194 | 30.9127 | 299.1268 | 0.1033 | 0.0033 | 8.7965 | 36 |
| 37 | 0.0294 | 9.7059 | 86.1781 | 34.0039 | 330.0395 | 0.1030 | 0.0030 | 8.8789 | 37 |
| 38 | 0.0267 | 9.7327 | 87.1673 | 37.4043 | 364.0434 | 0.1027 | 0.0027 | 8.9562 | 38 |
| 39 | 0.0243 | 9.7570 | 88.0908 | 41.1448 | 401.4478 | 0.0125 | 0.0025 | 9.0285 | 39 |
| 40 | 0.0221 | 9.7791 | 88.9525 | 45.2593 | 442.5926 | 0.1023 | 0.0023 | 9.0962 | 40 |
| 41 | 0.0201 | 9.7991 | 89.7560 | 49.7852 | 487.8518 | 0.1020 | 0.0020 | 9.1596 | 41 |
| 42 | 0.0183 | 9.8174 | 90.5047 | 54.7637 | 537.6370 | 0.1019 | 0.0019 | 9.2188 | 42 |
| 43 | 0.0166 | 9.8340 | 91.2019 | 60.2401 | 592.4007 | 0.1017 | 0.0017 | 9.2741 | 43 |
| 44 | 0.0151 | 9.8491 | 91.8508 | 66.2641 | 652.6408 | 0.1015 | 0.0015 | 9.3258 | 44 |
| 45 | 0.0137 | 9.8628 | 92.4544 | 72.8905 | 718.9048 | 0.1014 | 0.0014 | 9.3740 | 45 |
| 46 | 0.0125 | 9.8753 | 93.0157 | 80.1795 | 791.7953 | 0.1013 | 0.0013 | 9.4190 | 46 |
| 47 | 0.0113 | 9.8866 | 93.5372 | 88.1975 | 871.9749 | 0.1011 | 0.0011 | 9.4610 | 47 |
| 48 | 0.0103 | 9.8969 | 94.0217 | 97.0172 | 960.1723 | 0.1010 | 0.0010 | 9.5001 | 48 |
| 49 | 0.0094 | 9.9063 | 94.4715 | 106.7190 | 1057.1896 | 0.1009 | 0.0009 | 9.5365 | 49 |
| 50 | 0.0085 | 9.9148 | 94.8889 | 117.3909 | 1163.9085 | 0.1009 | 0.0009 | 9.5704 | 50 |
| 51 | 0.0077 | 9.9226 | 95.2761 | 129.1299 | 1281.2994 | 0.1008 | 0.0008 | 9.6020 | 51 |
| 52 | 0.0070 | 9.9296 | 95.6351 | 142.0429 | 1410.4293 | 0.1007 | 0.0007 | 9.6313 | 52 |
| 53 | 0.0064 | 9.9360 | 95.9679 | 156.2472 | 1552.4723 | 0.1006 | 0.0006 | 9.6586 | 53 |
| 54 | 0.0058 | 9.9418 | 96.2763 | 171.8719 | 1708.7195 | 0.1006 | 0.0006 | 9.6840 | 54 |
| 55 | 0.0053 | 9.9471 | 96.5619 | 189.0591 | 1880.5914 | 0.1005 | 0.0005 | 9.7075 | 55 |
| 60 | 0.0033 | 9.9672 | 97.7010 | 304.4816 | 3034.8164 | 0.1003 | 0.0003 | 9.8023 | 60 |
| 65 | 0.0020 | 9.9796 | 98.4705 | 490.3707 | 4893.7073 | 0.1002 | 0.0002 | 9.8672 | 65 |
| 70 | 0.0013 | 9.9873 | 98.9870 | 789.7470 | 7887.4696 | 0.1001 | 0.0001 | 9.9113 | 70 |
| 75 | 0.0008 | 9.9921 | 99.3317 | 1271.8954 | 12708.9537 | 0.1001 | 0.0001 | 9.9410 | 75 |
| 80 | 0.0005 | 9.9951 | 99.5606 | 2048.4002 | 20474.0021 | 0.1000 | 0.0000 | 9.9609 | 80 |
| 85 | 0.0003 | 9.9970 | 99.7120 | 3298.9690 | 32979.6903 | 0.1000 | 0.0000 | 9.9742 | 85 |
| 90 | 0.0002 | 9.9981 | 99.8118 | 5313.0226 | 53120.2261 | 0.1000 | 0.0000 | 9.9831 | 90 |
| 95 | 0.0001 | 9.9988 | 99.8773 | 8556.6760 | 85556.7605 | 0.1000 | 0.0000 | 9.9889 | 95 |
| 100 | 0.0001 | 9.9993 | 99.9202 | 13780.6123 | 137796.1234 | 0.1000 | 0.0000 | 9.9927 | 100 |

## APPENDIX 77.B (continued)
Cash Flow Equivalent Factors

$i = 12.00\%$

| $n$ | $P/F$ | $P/A$ | $P/G$ | $F/P$ | $F/A$ | $A/P$ | $A/F$ | $A/G$ | $n$ |
|---|---|---|---|---|---|---|---|---|---|
| 1 | 0.8929 | 0.8929 | 0.0000 | 1.1200 | 1.0000 | 1.1200 | 1.0000 | 0.0000 | 1 |
| 2 | 0.7972 | 1.6901 | 0.7972 | 1.2544 | 2.1200 | 0.5917 | 0.4717 | 0.4717 | 2 |
| 3 | 0.7118 | 2.4018 | 2.2208 | 1.4049 | 3.3744 | 0.4163 | 0.2963 | 0.9246 | 3 |
| 4 | 0.6355 | 3.0373 | 4.1273 | 1.5735 | 4.7793 | 0.3292 | 0.2092 | 1.3589 | 4 |
| 5 | 0.5674 | 3.6048 | 6.3970 | 1.7623 | 6.3528 | 0.2774 | 0.1574 | 1.7746 | 5 |
| 6 | 0.5066 | 4.1114 | 8.9302 | 1.9738 | 8.1152 | 0.2432 | 0.1232 | 2.1720 | 6 |
| 7 | 0.4523 | 4.5638 | 11.6443 | 2.2107 | 10.0890 | 0.2191 | 0.0991 | 2.5515 | 7 |
| 8 | 0.4039 | 4.9676 | 14.4714 | 2.4760 | 12.2997 | 0.2013 | 0.0813 | 2.9131 | 8 |
| 9 | 0.3606 | 5.3282 | 17.3563 | 2.7731 | 14.7757 | 0.1877 | 0.0677 | 3.2574 | 9 |
| 10 | 0.3220 | 5.6502 | 20.2541 | 3.1058 | 17.5487 | 0.1770 | 0.0570 | 3.5847 | 10 |
| 11 | 0.2875 | 5.9377 | 23.1288 | 3.4785 | 20.6546 | 0.1684 | 0.0484 | 3.8953 | 11 |
| 12 | 0.2567 | 6.1944 | 25.9523 | 3.8960 | 24.1331 | 0.1614 | 0.0414 | 4.1897 | 12 |
| 13 | 0.2292 | 6.4235 | 28.7024 | 4.3635 | 28.0291 | 0.1557 | 0.0357 | 4.4683 | 13 |
| 14 | 0.2046 | 6.6282 | 31.3624 | 4.8871 | 32.3926 | 0.1509 | 0.0309 | 4.7317 | 14 |
| 15 | 0.1827 | 6.8109 | 33.9202 | 5.4736 | 37.2797 | 0.1468 | 0.0268 | 4.9803 | 15 |
| 16 | 0.1631 | 6.9740 | 36.3670 | 6.1304 | 42.7533 | 0.1434 | 0.0234 | 5.2147 | 16 |
| 17 | 0.1456 | 7.1196 | 38.6973 | 6.8660 | 48.8837 | 0.1405 | 0.0205 | 5.4353 | 17 |
| 18 | 0.1300 | 7.2497 | 40.9080 | 7.6900 | 55.7497 | 0.1379 | 0.0179 | 5.6427 | 18 |
| 19 | 0.1161 | 7.3658 | 42.9979 | 8.6128 | 63.4397 | 0.1358 | 0.0158 | 6.8375 | 19 |
| 20 | 0.1037 | 7.4694 | 44.9676 | 9.6463 | 72.0524 | 0.1339 | 0.0139 | 6.0202 | 20 |
| 21 | 0.0926 | 7.5620 | 46.8188 | 10.8038 | 81.6987 | 0.1322 | 0.0122 | 6.1913 | 21 |
| 22 | 0.0826 | 7.6446 | 48.5543 | 12.1003 | 92.5026 | 0.1308 | 0.0108 | 6.3514 | 22 |
| 23 | 0.0738 | 7.7184 | 50.1776 | 13.5523 | 104.6029 | 0.1296 | 0.0096 | 6.5010 | 23 |
| 24 | 0.0659 | 7.7843 | 51.6929 | 15.1786 | 118.1552 | 0.1285 | 0.0085 | 6.6406 | 24 |
| 25 | 0.0588 | 7.8431 | 53.1046 | 17.0001 | 133.3339 | 0.1275 | 0.0075 | 6.7708 | 25 |
| 26 | 0.0525 | 7.8957 | 54.4177 | 19.0401 | 150.3339 | 0.1267 | 0.0067 | 6.8921 | 26 |
| 27 | 0.0469 | 7.9426 | 55.6369 | 21.3249 | 169.3740 | 0.1259 | 0.0059 | 7.0049 | 27 |
| 28 | 0.0419 | 7.9844 | 56.7674 | 23.8839 | 190.6989 | 0.1252 | 0.0052 | 7.1098 | 28 |
| 29 | 0.0374 | 8.0218 | 57.8141 | 26.7499 | 214.5828 | 0.1247 | 0.0047 | 7.2071 | 29 |
| 30 | 0.0334 | 8.0552 | 58.7821 | 29.9599 | 241.3327 | 0.1241 | 0.0041 | 7.2974 | 30 |
| 31 | 0.0298 | 8.0850 | 59.6761 | 33.5551 | 271.2926 | 0.1237 | 0.0037 | 7.3811 | 31 |
| 32 | 0.0266 | 8.1116 | 60.5010 | 37.5817 | 304.8477 | 0.1233 | 0.0033 | 7.4586 | 32 |
| 33 | 0.0238 | 8.1354 | 61.2612 | 42.0915 | 342.4294 | 0.1229 | 0.0029 | 7.5302 | 33 |
| 34 | 0.0212 | 8.1566 | 61.9612 | 47.1425 | 384.5210 | 0.1226 | 0.0026 | 7.5965 | 34 |
| 35 | 0.0189 | 8.1755 | 62.6052 | 52.7996 | 431.6635 | 0.1223 | 0.0023 | 7.6577 | 35 |
| 36 | 0.0169 | 8.1924 | 63.1970 | 59.1356 | 484.4631 | 0.1221 | 0.0021 | 7.7141 | 36 |
| 37 | 0.0151 | 8.2075 | 63.7406 | 66.2318 | 543.5987 | 0.1218 | 0.0018 | 7.7661 | 37 |
| 38 | 0.0135 | 8.2210 | 64.2394 | 74.1797 | 609.8305 | 0.1216 | 0.0016 | 7.8141 | 38 |
| 39 | 0.0120 | 8.2330 | 64.6967 | 83.0812 | 684.0102 | 0.1215 | 0.0015 | 7.8582 | 39 |
| 40 | 0.0107 | 8.2438 | 65.1159 | 93.0510 | 767.0914 | 0.1213 | 0.0013 | 7.8988 | 40 |
| 41 | 0.0096 | 8.2534 | 65.4997 | 104.2171 | 860.1424 | 0.1212 | 0.0012 | 7.9361 | 41 |
| 42 | 0.0086 | 8.2619 | 65.8509 | 116.7231 | 964.3595 | 0.1210 | 0.0010 | 7.9704 | 42 |
| 43 | 0.0076 | 8.2696 | 66.1722 | 130.7299 | 1081.0826 | 0.1209 | 0.0009 | 8.0019 | 43 |
| 44 | 0.0068 | 8.2764 | 66.4659 | 146.4175 | 1211.8125 | 0.1208 | 0.0008 | 8.0308 | 44 |
| 45 | 0.0061 | 8.2825 | 66.7342 | 163.9876 | 1358.2300 | 0.1207 | 0.0007 | 8.0572 | 45 |
| 46 | 0.0054 | 8.2880 | 66.9792 | 183.6661 | 1522.2176 | 0.1207 | 0.0007 | 8.0815 | 46 |
| 47 | 0.0049 | 8.2928 | 67.2028 | 205.7061 | 1705.8838 | 0.1206 | 0.0006 | 8.1037 | 47 |
| 48 | 0.0043 | 8.2972 | 67.4068 | 230.3908 | 1911.5898 | 0.1205 | 0.0005 | 8.1241 | 48 |
| 49 | 0.0039 | 8.3010 | 67.5929 | 258.0377 | 2141.9806 | 0.1205 | 0.0005 | 8.1427 | 49 |
| 50 | 0.0035 | 8.3045 | 67.7624 | 289.0022 | 2400.0182 | 0.1204 | 0.0004 | 8.1597 | 50 |
| 51 | 0.0031 | 8.3076 | 67.9169 | 323.6825 | 2689.0204 | 0.1204 | 0.0004 | 8.1753 | 51 |
| 52 | 0.0028 | 8.3103 | 68.0576 | 362.5243 | 3012.7029 | 0.1203 | 0.0003 | 8.1895 | 52 |
| 53 | 0.0025 | 8.3128 | 68.1856 | 406.0273 | 3375.2272 | 0.1203 | 0.0003 | 8.2025 | 53 |
| 54 | 0.0022 | 8.3150 | 68.3022 | 454.7505 | 3781.2545 | 0.1203 | 0.0003 | 8.2143 | 54 |
| 55 | 0.0020 | 8.3170 | 68.4082 | 509.3206 | 4236.0050 | 0.1202 | 0.0002 | 8.2251 | 55 |
| 60 | 0.0011 | 8.3240 | 68.8100 | 897.5969 | 7471.6411 | 0.1201 | 0.0001 | 8.2664 | 60 |
| 65 | 0.0006 | 8.3281 | 69.0581 | 1581.8725 | 13173.9374 | 0.1201 | 0.0001 | 8.2922 | 65 |
| 70 | 0.0004 | 8.3303 | 69.2103 | 2787.7998 | 23223.3319 | 0.1200 | 0.0000 | 8.3082 | 70 |
| 75 | 0.0002 | 8.3316 | 69.3031 | 4913.0558 | 40933.7987 | 0.1200 | 0.0000 | 8.3181 | 75 |
| 80 | 0.0001 | 8.3324 | 69.3594 | 8658.4831 | 72145.6925 | 0.1200 | 0.0000 | 8.3241 | 80 |
| 85 | 0.0001 | 8.3328 | 69.3935 | 15259.2057 | 127151.7140 | 0.1200 | 0.0000 | 8.3278 | 85 |
| 90 | 0.0000 | 8.3330 | 69.4140 | 26891.9342 | 224091.1185 | 0.1200 | 0.0000 | 8.3300 | 90 |
| 95 | 0.0000 | 8.3332 | 69.4263 | 47392.7766 | 394931.4719 | 0.1200 | 0.0000 | 8.3313 | 95 |
| 100 | 0.0000 | 8.3332 | 69.4336 | 83522.2657 | 696010.5477 | 0.1200 | 0.0000 | 8.3321 | 100 |

Appendices

## APPENDIX 77.B *(continued)*
### Cash Flow Equivalent Factors

$i = 15.00\%$

| n | P/F | P/A | P/G | F/P | F/A | A/P | A/F | A/G | n |
|---|-----|-----|-----|-----|-----|-----|-----|-----|---|
| 1 | 0.8696 | 0.8696 | 0.0000 | 1.1500 | 1.0000 | 1.1500 | 1.0000 | 0.0000 | 1 |
| 2 | 0.7561 | 1.6257 | 0.7561 | 1.3225 | 2.1500 | 0.6151 | 0.4651 | 0.4651 | 2 |
| 3 | 0.6575 | 2.2832 | 2.0712 | 1.5209 | 3.4725 | 0.4380 | 0.2880 | 0.9071 | 3 |
| 4 | 0.5718 | 2.8550 | 3.7864 | 1.7490 | 4.9934 | 0.3503 | 0.2003 | 1.3263 | 4 |
| 5 | 0.4972 | 3.3522 | 5.7751 | 2.0114 | 6.7424 | 0.2983 | 0.1483 | 1.7228 | 5 |
| 6 | 0.4323 | 3.7845 | 7.9368 | 2.3131 | 8.7537 | 0.2642 | 0.1142 | 2.0972 | 6 |
| 7 | 0.3759 | 4.1604 | 10.1924 | 2.6600 | 11.0668 | 0.2404 | 0.0904 | 2.4498 | 7 |
| 8 | 0.3269 | 4.4873 | 12.4807 | 3.0590 | 13.7268 | 0.2229 | 0.0729 | 2.7813 | 8 |
| 9 | 0.2843 | 4.7716 | 14.7548 | 3.5179 | 16.7858 | 0.2096 | 0.0596 | 3.0922 | 9 |
| 10 | 0.2472 | 5.0188 | 16.9795 | 4.0456 | 20.3037 | 0.1993 | 0.0493 | 3.3832 | 10 |
| 11 | 0.2149 | 5.2337 | 19.1289 | 4.6524 | 24.3493 | 0.1911 | 0.0411 | 3.6549 | 11 |
| 12 | 0.1869 | 5.4206 | 21.1849 | 5.3503 | 29.0017 | 0.1845 | 0.0345 | 3.9082 | 12 |
| 13 | 0.1625 | 5.5831 | 23.1352 | 6.1528 | 34.3519 | 0.1791 | 0.0291 | 4.1438 | 13 |
| 14 | 0.1413 | 5.7245 | 24.9725 | 7.0757 | 40.5047 | 0.1747 | 0.0247 | 4.3624 | 14 |
| 15 | 0.1229 | 5.8474 | 26.9630 | 8.1371 | 47.5804 | 0.1710 | 0.0210 | 4.5650 | 15 |
| 16 | 0.1069 | 5.9542 | 28.2960 | 9.3576 | 55.7175 | 0.1679 | 0.0179 | 4.7522 | 16 |
| 17 | 0.0929 | 6.0472 | 29.7828 | 10.7613 | 65.0751 | 0.1654 | 0.0154 | 4.9251 | 17 |
| 18 | 0.0808 | 6.1280 | 31.1565 | 12.3755 | 75.8364 | 0.1632 | 0.0132 | 5.0843 | 18 |
| 19 | 0.0703 | 6.1982 | 32.4213 | 14.2318 | 88.2118 | 0.1613 | 0.0113 | 5.2307 | 19 |
| 20 | 0.0611 | 6.2593 | 33.5822 | 16.3665 | 102.4436 | 0.1598 | 0.0098 | 5.3651 | 20 |
| 21 | 0.0531 | 6.3125 | 34.6448 | 18.8215 | 118.8101 | 0.1584 | 0.0084 | 5.4883 | 21 |
| 22 | 0.0462 | 6.3587 | 35.6150 | 21.6447 | 137.6316 | 0.1573 | 0.0073 | 5.6010 | 22 |
| 23 | 0.0402 | 6.3988 | 36.4988 | 24.8915 | 159.2764 | 0.1563 | 0.0063 | 5.7040 | 23 |
| 24 | 0.0349 | 6.4338 | 37.3023 | 28.6252 | 184.1678 | 0.1554 | 0.0054 | 5.7979 | 24 |
| 25 | 0.0304 | 6.4641 | 38.0314 | 32.9190 | 212.7930 | 0.1547 | 0.0047 | 5.8834 | 25 |
| 26 | 0.0264 | 6.4906 | 38.6918 | 37.8568 | 245.7120 | 0.1541 | 0.0041 | 5.9612 | 26 |
| 27 | 0.0230 | 6.5135 | 39.2890 | 43.5353 | 283.5688 | 0.1535 | 0.0035 | 6.0319 | 27 |
| 28 | 0.0200 | 6.5335 | 39.8283 | 50.0656 | 327.1041 | 0.1531 | 0.0031 | 6.0960 | 28 |
| 29 | 0.0174 | 6.5509 | 40.3146 | 57.5755 | 377.1697 | 0.1527 | 0.0027 | 6.1541 | 29 |
| 30 | 0.0151 | 6.5660 | 40.7526 | 66.2118 | 434.7451 | 0.1523 | 0.0023 | 6.2066 | 30 |
| 31 | 0.0131 | 6.5791 | 41.1466 | 76.1435 | 500.9569 | 0.1520 | 0.0020 | 6.2541 | 31 |
| 32 | 0.0114 | 6.5905 | 41.5006 | 87.5651 | 577.1005 | 0.1517 | 0.0017 | 6.2970 | 32 |
| 33 | 0.0099 | 6.6005 | 41.8184 | 100.6998 | 664.6655 | 0.1515 | 0.0015 | 6.3357 | 33 |
| 34 | 0.0086 | 6.6091 | 42.1033 | 115.8048 | 765.3654 | 0.1513 | 0.0013 | 6.3705 | 34 |
| 35 | 0.0075 | 6.6166 | 42.3586 | 133.1755 | 881.1702 | 0.1511 | 0.0011 | 6.4019 | 35 |
| 36 | 0.0065 | 6.6231 | 42.5872 | 153.1519 | 1014.3457 | 0.1510 | 0.0010 | 6.4301 | 36 |
| 37 | 0.0057 | 6.6288 | 42.7916 | 176.1246 | 1167.4975 | 0.1509 | 0.0009 | 6.4554 | 37 |
| 38 | 0.0049 | 6.6338 | 42.9743 | 202.5433 | 1343.6222 | 0.1507 | 0.0007 | 6.4781 | 38 |
| 39 | 0.0043 | 6.6380 | 43.1374 | 232.9248 | 1546.1655 | 0.1506 | 0.0006 | 6.4985 | 39 |
| 40 | 0.0037 | 6.6418 | 43.2830 | 267.8635 | 1779.0903 | 0.1506 | 0.0006 | 6.5168 | 40 |
| 41 | 0.0032 | 6.6450 | 43.4128 | 308.0431 | 2046.9539 | 0.1505 | 0.0005 | 6.5331 | 41 |
| 42 | 0.0028 | 6.6478 | 43.5286 | 354.2495 | 2354.9969 | 0.1504 | 0.0004 | 6.5478 | 42 |
| 43 | 0.0025 | 6.6503 | 43.6317 | 407.3870 | 2709.2465 | 0.1504 | 0.0004 | 6.5609 | 43 |
| 44 | 0.0021 | 6.6524 | 43.7235 | 468.4950 | 3116.6334 | 0.1503 | 0.0003 | 6.5725 | 44 |
| 45 | 0.0019 | 6.6543 | 43.8051 | 538.7693 | 3585.1285 | 0.1503 | 0.0003 | 6.5830 | 45 |
| 46 | 0.0016 | 6.6559 | 43.8778 | 619.5847 | 4123.8977 | 0.1502 | 0.0002 | 6.5923 | 46 |
| 47 | 0.0014 | 6.6573 | 43.9423 | 712.5224 | 4743.4824 | 0.1502 | 0.0002 | 6.6006 | 47 |
| 48 | 0.0012 | 6.6585 | 43.9997 | 819.4007 | 5456.0047 | 0.1502 | 0.0002 | 6.6080 | 48 |
| 49 | 0.0011 | 6.6596 | 44.0506 | 942.3108 | 6275.4055 | 0.1502 | 0.0002 | 6.6146 | 49 |
| 50 | 0.0009 | 6.6605 | 44.0958 | 1083.6574 | 7217.7163 | 0.1501 | 0.0001 | 6.6205 | 50 |
| 51 | 0.0008 | 6.6613 | 44.1360 | 1246.2061 | 8301.3737 | 0.1501 | 0.0001 | 6.6257 | 51 |
| 52 | 0.0007 | 6.6620 | 44.1715 | 1433.1370 | 9547.5798 | 0.1501 | 0.0001 | 6.6304 | 52 |
| 53 | 0.0006 | 6.6626 | 44.2031 | 1648.1075 | 10980.7167 | 0.1501 | 0.0001 | 6.6345 | 53 |
| 54 | 0.0005 | 6.6631 | 44.2311 | 1895.3236 | 12628.8243 | 0.1501 | 0.0001 | 6.6382 | 54 |
| 55 | 0.0005 | 6.6636 | 44.2558 | 2179.6222 | 14524.1479 | 0.1501 | 0.0001 | 6.6414 | 55 |
| 60 | 0.0002 | 6.6651 | 44.3431 | 4383.9987 | 29219.9916 | 0.1500 | 0.0000 | 6.6530 | 60 |
| 65 | 0.0001 | 6.6659 | 44.3903 | 8817.7874 | 58778.5826 | 0.1500 | 0.0000 | 6.6593 | 65 |
| 70 | 0.0001 | 6.6663 | 44.4156 | 17735.7200 | 118231.4669 | 0.1500 | 0.0000 | 6.6627 | 70 |
| 75 | 0.0000 | 6.6665 | 44.4292 | 35672.8680 | 237812.4532 | 0.1500 | 0.0000 | 6.6646 | 75 |
| 80 | 0.0000 | 6.6666 | 44.4364 | 71750.8794 | 478332.5293 | 0.1500 | 0.0000 | 6.6656 | 80 |
| 85 | 0.0000 | 6.6666 | 44.4402 | 144316.6470 | 962104.3133 | 0.1500 | 0.0000 | 6.6661 | 85 |
| 90 | 0.0000 | 6.6666 | 44.4422 | 290272.3252 | 1935142.1680 | 0.1500 | 0.0000 | 6.6664 | 90 |
| 95 | 0.0000 | 6.6667 | 44.4433 | 583841.3276 | 3892268.8509 | 0.1500 | 0.0000 | 6.6665 | 95 |
| 100 | 0.0000 | 6.6667 | 44.4438 | 1174313.4507 | 7828749.6713 | 0.1500 | 0.0000 | 6.6666 | 100 |

**APPENDIX 77.B** (continued)
Cash Flow Equivalent Factors

$$i = 20.00\%$$

| $n$ | $P/F$ | $P/A$ | $P/G$ | $F/P$ | $F/A$ | $A/P$ | $A/F$ | $A/G$ | $n$ |
|---|---|---|---|---|---|---|---|---|---|
| 1 | 0.8333 | 0.8333 | 0.0000 | 1.2000 | 1.0000 | 1.2000 | 1.0000 | 0.0000 | 1 |
| 2 | 0.6944 | 1.5278 | 0.6944 | 1.4400 | 2.2000 | 0.6545 | 0.4545 | 0.4545 | 2 |
| 3 | 0.5787 | 2.1065 | 1.8519 | 1.7280 | 3.6400 | 0.4747 | 0.2747 | 0.8791 | 3 |
| 4 | 0.4823 | 2.5887 | 3.2986 | 2.0736 | 5.3680 | 0.3863 | 0.1863 | 1.2742 | 4 |
| 5 | 0.4019 | 2.9906 | 4.9061 | 2.4883 | 7.4416 | 0.3344 | 0.1344 | 1.6405 | 5 |
| 6 | 0.3349 | 3.3255 | 6.5806 | 2.9860 | 9.9299 | 0.3007 | 0.1007 | 1.9788 | 6 |
| 7 | 0.2791 | 3.6046 | 8.2551 | 3.5832 | 12.9159 | 0.2774 | 0.0774 | 2.2902 | 7 |
| 8 | 0.2326 | 3.8372 | 9.8831 | 4.2998 | 16.4991 | 0.2606 | 0.0606 | 2.5756 | 8 |
| 9 | 0.1938 | 4.0310 | 11.4335 | 5.1598 | 20.7989 | 0.2481 | 0.0481 | 2.8364 | 9 |
| 10 | 0.1615 | 4.1925 | 12.8871 | 6.1917 | 25.9587 | 0.2385 | 0.0385 | 3.0739 | 10 |
| 11 | 0.1346 | 4.3271 | 14.2330 | 7.4301 | 32.1504 | 0.2311 | 0.0311 | 3.2893 | 11 |
| 12 | 0.1122 | 4.4392 | 15.4667 | 8.9161 | 39.5805 | 0.2253 | 0.0253 | 3.4841 | 12 |
| 13 | 0.0935 | 4.5327 | 16.5883 | 10.6993 | 48.4966 | 0.2206 | 0.0206 | 3.6597 | 13 |
| 14 | 0.0779 | 4.6106 | 17.6008 | 12.8392 | 59.1959 | 0.2169 | 0.0169 | 3.8175 | 14 |
| 15 | 0.0649 | 4.6755 | 18.5095 | 15.4070 | 72.0351 | 0.2139 | 0.0139 | 3.9588 | 15 |
| 16 | 0.0541 | 4.7296 | 19.3208 | 18.4884 | 87.4421 | 0.2114 | 0.0114 | 4.0851 | 16 |
| 17 | 0.0451 | 4.7746 | 20.0419 | 22.1861 | 105.9306 | 0.2094 | 0.0094 | 4.1976 | 17 |
| 18 | 0.0376 | 4.8122 | 20.6805 | 26.6233 | 128.1167 | 0.2078 | 0.0078 | 4.2975 | 18 |
| 19 | 0.0313 | 4.8435 | 21.2439 | 31.9480 | 154.7400 | 0.2065 | 0.0065 | 4.3861 | 19 |
| 20 | 0.0261 | 4.8696 | 21.7395 | 38.3376 | 186.6880 | 0.2054 | 0.0054 | 4.4643 | 20 |
| 21 | 0.0217 | 4.8913 | 22.1742 | 46.0051 | 225.0256 | 0.2044 | 0.0044 | 4.5334 | 21 |
| 22 | 0.0181 | 4.9094 | 22.5546 | 55.2061 | 271.0307 | 0.2037 | 0.0037 | 4.5941 | 22 |
| 23 | 0.0151 | 4.9245 | 22.8867 | 66.2474 | 326.2369 | 0.2031 | 0.0031 | 4.6475 | 23 |
| 24 | 0.0126 | 4.9371 | 23.1760 | 79.4968 | 392.4842 | 0.2025 | 0.0025 | 4.6943 | 24 |
| 25 | 0.0105 | 4.9476 | 23.4276 | 95.3962 | 471.9811 | 0.2021 | 0.0021 | 4.7352 | 25 |
| 26 | 0.0087 | 4.9563 | 23.6460 | 114.4755 | 567.3773 | 0.2018 | 0.0018 | 4.7709 | 26 |
| 27 | 0.0073 | 4.9636 | 23.8353 | 137.3706 | 681.8528 | 0.2015 | 0.0015 | 4.8020 | 27 |
| 28 | 0.0061 | 4.9697 | 23.9991 | 164.8447 | 819.2233 | 0.2012 | 0.0012 | 4.8291 | 28 |
| 29 | 0.0051 | 4.9747 | 24.1406 | 197.8136 | 984.0680 | 0.2010 | 0.0010 | 4.8527 | 29 |
| 30 | 0.0042 | 4.9789 | 24.2628 | 237.3763 | 1181.8816 | 0.2008 | 0.0008 | 4.8731 | 30 |
| 31 | 0.0035 | 4.9824 | 24.3681 | 284.8516 | 1419.2579 | 0.2007 | 0.0007 | 4.8908 | 31 |
| 32 | 0.0029 | 4.9854 | 24.4588 | 341.8219 | 1704.1095 | 0.2006 | 0.0006 | 4.9061 | 32 |
| 33 | 0.0024 | 4.9878 | 24.5368 | 410.1863 | 2045.9314 | 0.2005 | 0.0005 | 4.9194 | 33 |
| 34 | 0.0020 | 4.9898 | 24.6038 | 492.2235 | 2456.1176 | 0.2004 | 0.0004 | 4.9308 | 34 |
| 35 | 0.0017 | 4.9915 | 24.6614 | 590.6682 | 2948.3411 | 0.2003 | 0.0003 | 4.9406 | 35 |
| 36 | 0.0014 | 4.9929 | 24.7108 | 708.8019 | 3539.0094 | 0.2003 | 0.0003 | 4.9491 | 36 |
| 37 | 0.0012 | 4.9941 | 24.7531 | 850.5622 | 4247.8112 | 0.2002 | 0.0002 | 4.9564 | 37 |
| 38 | 0.0010 | 4.9951 | 24.7894 | 1020.6747 | 5098.3735 | 0.2002 | 0.0002 | 4.9627 | 38 |
| 39 | 0.0008 | 4.9959 | 24.8204 | 1224.8096 | 6119.0482 | 0.2002 | 0.0002 | 4.9681 | 39 |
| 40 | 0.0007 | 4.9966 | 24.8469 | 1469.7716 | 7343.8578 | 0.2001 | 0.0001 | 4.9728 | 40 |
| 41 | 0.0006 | 4.9972 | 24.8696 | 1763.7259 | 8813.6294 | 0.2001 | 0.0001 | 4.9767 | 41 |
| 42 | 0.0005 | 4.9976 | 24.8890 | 2116.4711 | 10577.3553 | 0.2001 | 0.0001 | 4.9801 | 42 |
| 43 | 0.0004 | 4.9980 | 24.9055 | 2539.7653 | 12693.8263 | 0.2001 | 0.0001 | 4.9831 | 43 |
| 44 | 0.0003 | 4.9984 | 24.9196 | 3047.7183 | 15233.5916 | 0.2001 | 0.0001 | 4.9856 | 44 |
| 45 | 0.0003 | 4.9986 | 24.9316 | 3657.2620 | 18281.3099 | 0.2001 | 0.0001 | 4.9877 | 45 |
| 46 | 0.0002 | 4.9989 | 24.9419 | 4388.7144 | 21938.5719 | 0.2000 | 0.0000 | 4.9895 | 46 |
| 47 | 0.0002 | 4.9991 | 24.9506 | 5266.4573 | 26327.2863 | 0.2000 | 0.0000 | 4.9911 | 47 |
| 48 | 0.0002 | 4.9992 | 24.9581 | 6319.7487 | 31593.7436 | 0.2000 | 0.0000 | 4.9924 | 48 |
| 49 | 0.0001 | 4.9993 | 24.9644 | 7583.6985 | 37913.4923 | 0.2000 | 0.0000 | 4.9935 | 49 |
| 50 | 0.0001 | 4.9995 | 24.9698 | 9100.4382 | 45497.1908 | 0.2000 | 0.0000 | 4.9945 | 50 |
| 51 | 0.0001 | 4.9995 | 24.9744 | 10920.5258 | 54597.6289 | 0.2000 | 0.0000 | 4.9953 | 51 |
| 52 | 0.0001 | 4.9996 | 24.9783 | 13104.6309 | 65518.1547 | 0.2000 | 0.0000 | 4.9960 | 52 |
| 53 | 0.0001 | 4.9997 | 24.9816 | 15725.5571 | 78622.7856 | 0.2000 | 0.0000 | 4.9966 | 53 |
| 54 | 0.0001 | 4.9997 | 24.9844 | 18870.6685 | 94348.3427 | 0.2000 | 0.0000 | 4.9971 | 54 |
| 55 | 0.0000 | 4.9998 | 24.9868 | 22644.8023 | 113219.0113 | 0.2000 | 0.0000 | 4.9976 | 55 |
| 60 | 0.0000 | 4.9999 | 24.9942 | 56347.5144 | 281732.5718 | 0.2000 | 0.0000 | 4.9989 | 60 |
| 65 | 0.0000 | 5.0000 | 24.9975 | 140210.6469 | 701048.2346 | 0.2000 | 0.0000 | 4.9995 | 65 |
| 70 | 0.0000 | 5.0000 | 24.9989 | 348888.9569 | 1744439.7847 | 0.2000 | 0.0000 | 4.9998 | 70 |
| 75 | 0.0000 | 5.0000 | 24.9995 | 868147.3693 | 4340731.8466 | 0.2000 | 0.0000 | 4.9999 | 75 |

## APPENDIX 77.B *(continued)*
### Cash Flow Equivalent Factors

$$i = 25.00\%$$

| n | P/F | P/A | P/G | F/P | F/A | A/P | A/F | A/G | n |
|---|-----|-----|-----|-----|-----|-----|-----|-----|---|
| 1 | 0.8000 | 0.8000 | 0.0000 | 1.2500 | 1.0000 | 1.2500 | 1.0000 | 0.0000 | 1 |
| 2 | 0.6400 | 1.4400 | 0.6400 | 1.5625 | 2.2500 | 0.6944 | 0.0444 | 0.4444 | 2 |
| 3 | 0.5120 | 1.9520 | 1.6640 | 1.9531 | 3.8125 | 0.5123 | 0.2623 | 0.8525 | 3 |
| 4 | 0.4096 | 2.3616 | 2.8928 | 2.4414 | 5.7656 | 0.4234 | 0.1734 | 1.2249 | 4 |
| 5 | 0.3277 | 2.6893 | 4.2035 | 3.0518 | 8.2070 | 0.3718 | 0.1218 | 1.5631 | 5 |
| 6 | 0.2621 | 2.9514 | 5.5142 | 3.8147 | 11.2588 | 0.3383 | 0.0888 | 1.8683 | 6 |
| 7 | 0.2097 | 3.1611 | 6.7725 | 4.7684 | 15.0735 | 0.3163 | 0.0663 | 2.1424 | 7 |
| 8 | 0.1678 | 3.3289 | 7.9469 | 5.9605 | 19.8419 | 0.3004 | 0.0504 | 2.3872 | 8 |
| 9 | 0.1342 | 3.4631 | 9.0207 | 7.4506 | 25.8023 | 0.2888 | 0.0388 | 2.6048 | 9 |
| 10 | 0.1074 | 3.5705 | 9.9870 | 9.3132 | 33.2529 | 0.2801 | 0.0301 | 2.7971 | 10 |
| 11 | 0.0859 | 3.6564 | 10.8460 | 11.6415 | 42.5661 | 0.2735 | 0.0235 | 2.9663 | 11 |
| 12 | 0.0687 | 3.7251 | 11.6020 | 14.5519 | 54.2077 | 0.2684 | 0.0184 | 3.1145 | 12 |
| 13 | 0.0550 | 3.7801 | 12.2617 | 18.1899 | 68.7596 | 0.2645 | 0.0145 | 3.2437 | 13 |
| 14 | 0.0440 | 3.8241 | 12.8334 | 22.7374 | 86.9495 | 0.2615 | 0.0115 | 3.3559 | 14 |
| 15 | 0.0352 | 3.8593 | 13.3260 | 28.4217 | 109.6868 | 0.2591 | 0.0091 | 3.4530 | 15 |
| 16 | 0.0281 | 3.8874 | 13.7482 | 35.5271 | 138.1085 | 0.2572 | 0.0072 | 3.5366 | 16 |
| 17 | 0.0225 | 3.9099 | 14.1085 | 44.4089 | 173.6357 | 0.2558 | 0.0058 | 3.6084 | 17 |
| 18 | 0.0180 | 3.9279 | 14.4147 | 55.5112 | 218.0446 | 0.2546 | 0.0046 | 3.6698 | 18 |
| 19 | 0.0144 | 3.9424 | 14.6741 | 69.3889 | 273.5558 | 0.2537 | 0.0037 | 3.7222 | 19 |
| 20 | 0.0115 | 3.9539 | 14.8932 | 86.7362 | 342.9447 | 0.2529 | 0.0029 | 3.7667 | 20 |
| 21 | 0.0092 | 3.9631 | 15.0777 | 108.4202 | 429.6809 | 0.2523 | 0.0023 | 3.8045 | 21 |
| 22 | 0.0074 | 3.9705 | 15.2326 | 135.5253 | 538.1011 | 0.2519 | 0.0019 | 3.8365 | 22 |
| 23 | 0.0059 | 3.9764 | 15.3625 | 169.4066 | 673.6264 | 0.2515 | 0.0015 | 3.8634 | 23 |
| 24 | 0.0047 | 3.9811 | 15.4711 | 211.7582 | 843.0329 | 0.2512 | 0.0012 | 3.8861 | 24 |
| 25 | 0.0038 | 3.9849 | 15.5618 | 264.6978 | 1054.7912 | 0.2509 | 0.0009 | 3.9052 | 25 |
| 26 | 0.0030 | 3.9879 | 15.6373 | 330.8722 | 1319.4890 | 0.2508 | 0.0008 | 3.9212 | 26 |
| 27 | 0.0024 | 3.9903 | 15.7002 | 413.5903 | 1650.3612 | 0.2506 | 0.0006 | 3.9346 | 27 |
| 28 | 0.0019 | 3.9923 | 15.7524 | 516.9879 | 2063.9515 | 0.2505 | 0.0005 | 3.9457 | 28 |
| 29 | 0.0015 | 3.9938 | 15.7957 | 646.2349 | 2580.9394 | 0.2504 | 0.0004 | 3.9551 | 29 |
| 30 | 0.0012 | 3.9950 | 15.8316 | 807.7936 | 3227.1743 | 0.2503 | 0.0003 | 3.9628 | 30 |
| 31 | 0.0010 | 3.9960 | 15.8614 | 1009.7420 | 4034.9678 | 0.2502 | 0.0002 | 3.9693 | 31 |
| 32 | 0.0008 | 3.9968 | 15.8859 | 1262.1774 | 5044.7098 | 0.2502 | 0.0002 | 3.9746 | 32 |
| 33 | 0.0006 | 3.9975 | 15.9062 | 1577.7218 | 6306.8872 | 0.2502 | 0.0002 | 3.9791 | 33 |
| 34 | 0.0005 | 3.9980 | 15.9229 | 1972.1523 | 7884.6091 | 0.2501 | 0.0001 | 3.9828 | 34 |
| 35 | 0.0004 | 3.9984 | 15.9367 | 2465.1903 | 9856.7613 | 0.2501 | 0.0001 | 3.9858 | 35 |
| 36 | 0.0003 | 3.9987 | 15.9481 | 3081.4879 | 12321.9516 | 0.2501 | 0.0001 | 3.9883 | 36 |
| 37 | 0.0003 | 3.9990 | 15.9574 | 3851.8599 | 15403.4396 | 0.2501 | 0.0001 | 3.9904 | 37 |
| 38 | 0.0002 | 3.9992 | 15.9651 | 4814.8249 | 19255.2994 | 0.2501 | 0.0001 | 3.9921 | 38 |
| 39 | 0.0002 | 3.9993 | 15.9714 | 6018.5311 | 24070.1243 | 0.2500 | 0.0000 | 3.9935 | 39 |
| 40 | 0.0001 | 3.9995 | 15.9766 | 7523.1638 | 30088.6554 | 0.2500 | 0.0000 | 3.9947 | 40 |
| 41 | 0.0001 | 3.9996 | 15.9809 | 9403.9548 | 37611.8192 | 0.2500 | 0.0000 | 3.9956 | 41 |
| 42 | 0.0001 | 3.9997 | 15.9843 | 11754.9435 | 47015.7740 | 0.2500 | 0.0000 | 3.9964 | 42 |
| 43 | 0.0001 | 3.9997 | 15.9872 | 14693.6794 | 58770.7175 | 0.2500 | 0.0000 | 3.9971 | 43 |
| 44 | 0.0001 | 3.9998 | 15.9895 | 18367.0992 | 73464.3969 | 0.2500 | 0.0000 | 3.9976 | 44 |
| 45 | 0.0000 | 3.9998 | 15.9915 | 22958.8740 | 91831.4962 | 0.2500 | 0.0000 | 3.9980 | 45 |
| 46 | 0.0000 | 3.9999 | 15.9930 | 28698.5925 | 114790.3702 | 0.2500 | 0.0000 | 3.9984 | 46 |
| 47 | 0.0000 | 3.9999 | 15.9943 | 35873.2407 | 143488.9627 | 0.2500 | 0.0000 | 3.9987 | 47 |
| 48 | 0.0000 | 3.9999 | 15.9954 | 44841.5509 | 179362.2034 | 0.2500 | 0.0000 | 3.9989 | 48 |
| 49 | 0.0000 | 3.9999 | 15.9962 | 56051.9386 | 224203.7543 | 0.2500 | 0.0000 | 3.9991 | 49 |
| 50 | 0.0000 | 3.9999 | 15.9969 | 70064.9232 | 280255.6929 | 0.2500 | 0.0000 | 3.9993 | 50 |
| 51 | 0.0000 | 4.0000 | 15.9975 | 87581.1540 | 350320.6161 | 0.2500 | 0.0000 | 3.9994 | 51 |
| 52 | 0.0000 | 4.0000 | 15.9980 | 109476.4425 | 437901.7701 | 0.2500 | 0.0000 | 3.9995 | 52 |
| 53 | 0.0000 | 4.0000 | 15.9983 | 136845.5532 | 547378.2126 | 0.2500 | 0.0000 | 3.9996 | 53 |
| 54 | 0.0000 | 4.0000 | 15.9986 | 171056.9414 | 684223.7658 | 0.2500 | 0.0000 | 3.9997 | 54 |
| 55 | 0.0000 | 4.0000 | 15.9989 | 213821.1768 | 855280.7072 | 0.2500 | 0.0000 | 3.9997 | 55 |
| 60 | 0.0000 | 4.0000 | 15.9996 | 652530.4468 | 2610117.7872 | 0.2500 | 0.0000 | 3.9999 | 60 |

**APPENDIX 77.B** *(continued)*
Cash Flow Equivalent Factors

$i = 30.00\%$

| $n$ | $P/F$ | $P/A$ | $P/G$ | $F/P$ | $F/A$ | $A/P$ | $A/F$ | $A/G$ | $n$ |
|---|---|---|---|---|---|---|---|---|---|
| 1 | 0.7692 | 0.7692 | 0.0000 | 1.3000 | 1.0000 | 1.3000 | 1.0000 | 0.000 | 1 |
| 2 | 0.5917 | 1.3609 | 0.5917 | 1.6900 | 2.3000 | 0.7348 | 0.4348 | 0.434 | 2 |
| 3 | 0.4552 | 1.8161 | 1.5020 | 2.1970 | 3.9900 | 0.5506 | 0.2506 | 0.827 | 3 |
| 4 | 0.3501 | 2.1662 | 2.5524 | 2.8561 | 6.1870 | 0.4616 | 0.1616 | 1.178 | 4 |
| 5 | 0.2693 | 2.4356 | 3.6297 | 3.7129 | 9.0431 | 0.4106 | 0.1106 | 1.490 | 5 |
| 6 | 0.2072 | 2.6427 | 4.6656 | 4.8268 | 12.7560 | 0.3784 | 0.0784 | 1.765 | 6 |
| 7 | 0.1594 | 2.8021 | 5.6218 | 6.2749 | 17.5828 | 0.3569 | 0.0569 | 2.006 | 7 |
| 8 | 0.1226 | 2.9247 | 6.4800 | 8.1573 | 23.8577 | 0.3419 | 0.0419 | 2.216 | 8 |
| 9 | 0.0943 | 3.0190 | 7.2343 | 10.6045 | 32.0150 | 0.3312 | 0.0312 | 2.396 | 9 |
| 10 | 0.0725 | 3.0915 | 7.8872 | 13.7858 | 42.6195 | 0.3235 | 0.0235 | 2.551 | 10 |
| 11 | 0.0558 | 3.1473 | 8.4452 | 17.9216 | 56.4053 | 0.3177 | 0.0177 | 2.683 | 11 |
| 12 | 0.0429 | 3.1903 | 8.9173 | 23.2981 | 74.3270 | 0.3135 | 0.0135 | 2.795 | 12 |
| 13 | 0.0330 | 3.2233 | 9.3135 | 30.2875 | 97.6250 | 0.3102 | 0.0102 | 2.889 | 13 |
| 14 | 0.0254 | 3.2487 | 9.6437 | 39.3738 | 127.9125 | 0.3078 | 0.0078 | 2.968 | 14 |
| 15 | 0.0195 | 3.2682 | 9.9172 | 51.1859 | 167.2863 | 0.3060 | 0.0060 | 3.034 | 15 |
| 16 | 0.0150 | 3.2832 | 10.1426 | 66.5417 | 218.4722 | 0.3046 | 0.0046 | 3.089 | 16 |
| 17 | 0.0116 | 3.2948 | 10.3276 | 86.5042 | 285.0139 | 0.3035 | 0.0035 | 3.134 | 17 |
| 18 | 0.0089 | 3.3037 | 10.4788 | 112.4554 | 371.5180 | 0.3027 | 0.0027 | 3.171 | 18 |
| 19 | 0.0068 | 3.3105 | 10.6019 | 146.1920 | 483.9734 | 0.3021 | 0.0021 | 3.202 | 19 |
| 20 | 0.0053 | 3.3158 | 10.7019 | 190.0496 | 630.1655 | 0.3016 | 0.0016 | 3.227 | 20 |
| 21 | 0.0040 | 3.3198 | 10.7828 | 247.0645 | 820.2151 | 0.3012 | 0.0012 | 3.248 | 21 |
| 22 | 0.0031 | 3.3230 | 10.8482 | 321.1839 | 1067.2796 | 0.3009 | 0.0009 | 3.264 | 22 |
| 23 | 0.0024 | 3.3254 | 10.9009 | 417.5391 | 1388.4635 | 0.3007 | 0.0007 | 3.278 | 23 |
| 24 | 0.0018 | 3.3272 | 10.9433 | 542.8008 | 1806.0026 | 0.3006 | 0.0006 | 3.289 | 24 |
| 25 | 0.0014 | 3.3286 | 10.9773 | 705.6410 | 2348.8033 | 0.3004 | 0.0004 | 3.297 | 25 |
| 26 | 0.0011 | 3.3297 | 11.0045 | 917.3333 | 3054.4443 | 0.3003 | 0.0003 | 3.305 | 26 |
| 27 | 0.0008 | 3.3305 | 11.0263 | 1192.5333 | 3971.7776 | 0.3003 | 0.0003 | 3.310 | 27 |
| 28 | 0.0006 | 3.3312 | 11.0437 | 1550.2933 | 5164.3109 | 0.3002 | 0.0002 | 3.315 | 28 |
| 29 | 0.0005 | 3.3317 | 11.0576 | 2015.3813 | 6714.6042 | 0.3001 | 0.0001 | 3.318 | 29 |
| 30 | 0.0004 | 3.3321 | 11.0687 | 2619.9956 | 8729.9855 | 0.3001 | 0.0001 | 3.321 | 30 |
| 31 | 0.0003 | 3.3324 | 11.0775 | 3405.9943 | 11349.9811 | 0.3001 | 0.0001 | 3.324 | 31 |
| 32 | 0.0002 | 3.3326 | 11.0845 | 4427.7926 | 14755.9755 | 0.3001 | 0.0001 | 3.326 | 32 |
| 33 | 0.0002 | 3.3328 | 11.0901 | 5756.1304 | 19183.7681 | 0.3001 | 0.0001 | 3.327 | 33 |
| 34 | 0.0001 | 3.3329 | 11.0945 | 7482.9696 | 24939.8985 | 0.3000 | 0.0000 | 3.328 | 34 |
| 35 | 0.0001 | 3.3330 | 11.0980 | 9727.8604 | 32422.8681 | 0.3000 | 0.0000 | 3.329 | 35 |
| 36 | 0.0001 | 3.3331 | 11.1007 | 12646.2186 | 42150.7285 | 0.3000 | 0.0000 | 3.330 | 36 |
| 37 | 0.0001 | 3.3331 | 11.1029 | 16440.0841 | 54796.9471 | 0.3000 | 0.0000 | 3.331 | 37 |
| 38 | 0.0000 | 3.3332 | 11.1047 | 21372.1094 | 71237.0312 | 0.3000 | 0.0000 | 3.331 | 38 |
| 39 | 0.0000 | 3.3332 | 11.1060 | 27783.7422 | 92609.1405 | 0.3000 | 0.0000 | 3.331 | 39 |
| 40 | 0.0000 | 3.3332 | 11.1071 | 36118.8648 | 120392.8827 | 0.3000 | 0.0000 | 3.332 | 40 |
| 41 | 0.0000 | 3.3333 | 11.1080 | 46954.5243 | 156511.7475 | 0.3000 | 0.0000 | 3.332 | 41 |
| 42 | 0.0000 | 3.3333 | 11.1086 | 61040.8815 | 203466.2718 | 0.3000 | 0.0000 | 3.332 | 42 |
| 43 | 0.0000 | 3.3333 | 11.1092 | 79353.1460 | 264507.1533 | 0.3000 | 0.0000 | 3.332 | 43 |
| 44 | 0.0000 | 3.3333 | 11.1096 | 103159.0898 | 343860.2993 | 0.3000 | 0.0000 | 3.332 | 44 |
| 45 | 0.0000 | 3.3333 | 11.1099 | 134106.8167 | 447019.3890 | 0.3000 | 0.0000 | 3.333 | 45 |
| 46 | 0.0000 | 3.3333 | 11.1102 | 174338.8617 | 581126.2058 | 0.3000 | 0.0000 | 3.333 | 46 |
| 47 | 0.0000 | 3.3333 | 11.1104 | 226640.5202 | 755465.0675 | 0.3000 | 0.0000 | 3.333 | 47 |
| 48 | 0.0000 | 3.3333 | 11.1105 | 294632.6763 | 982105.5877 | 0.3000 | 0.0000 | 3.333 | 48 |
| 49 | 0.0000 | 3.3333 | 11.1107 | 383022.4792 | 1276738.2640 | 0.3000 | 0.0000 | 3.333 | 49 |
| 50 | 0.0000 | 3.3333 | 11.1108 | 497929.2230 | 1659760.7433 | 0.3000 | 0.0000 | 3.333 | 50 |

Appendices

**APPENDIX 77.B** *(continued)*
Cash Flow Equivalent Factors

$i = 40.00\%$

| $n$ | $P/F$ | $P/A$ | $P/G$ | $F/P$ | $F/A$ | $A/P$ | $A/F$ | $A/G$ | $n$ |
|---|---|---|---|---|---|---|---|---|---|
| 1 | 0.7143 | 0.7143 | 0.0000 | 1.4000 | 1.0000 | 1.4000 | 1.0000 | 0.000 | 1 |
| 2 | 0.5102 | 1.2245 | 0.5102 | 1.9600 | 2.4000 | 0.8167 | 0.4167 | 0.416 | 2 |
| 3 | 0.3644 | 1.5889 | 1.2391 | 2.7440 | 4.3600 | 0.6294 | 0.2294 | 0.779 | 3 |
| 4 | 0.2603 | 1.8492 | 2.0200 | 3.8416 | 7.1040 | 0.5408 | 0.1408 | 1.092 | 4 |
| 5 | 0.1859 | 2.0352 | 2.7637 | 5.3782 | 10.9456 | 0.4914 | 0.0914 | 1.358 | 5 |
| 6 | 0.1328 | 2.1680 | 3.4278 | 7.5295 | 16.3238 | 0.4613 | 0.0613 | 1.581 | 6 |
| 7 | 0.0949 | 2.2628 | 3.9970 | 10.5414 | 23.8534 | 0.4419 | 0.0419 | 1.766 | 7 |
| 8 | 0.0678 | 2.3306 | 4.4713 | 14.7579 | 34.3947 | 0.4291 | 0.0291 | 1.918 | 8 |
| 9 | 0.0484 | 2.3790 | 4.8585 | 20.6610 | 49.1526 | 0.4203 | 0.0203 | 2.042 | 9 |
| 10 | 0.0346 | 2.4136 | 5.1696 | 28.9255 | 69.8137 | 0.4143 | 0.0143 | 2.141 | 10 |
| 11 | 0.0247 | 2.4383 | 5.4166 | 40.4957 | 98.7391 | 0.4101 | 0.0101 | 2.221 | 11 |
| 12 | 0.0176 | 2.4559 | 5.6106 | 56.6939 | 139.2348 | 0.4072 | 0.0072 | 2.284 | 12 |
| 13 | 0.0126 | 2.4685 | 5.7618 | 79.3715 | 195.9287 | 0.4051 | 0.0051 | 2.334 | 13 |
| 14 | 0.0090 | 2.4775 | 5.8788 | 111.1201 | 275.3002 | 0.4036 | 0.0036 | 2.372 | 14 |
| 15 | 0.0064 | 2.4839 | 5.9688 | 155.5681 | 386.4202 | 0.4026 | 0.0026 | 2.403 | 15 |
| 16 | 0.0046 | 2.4885 | 6.0376 | 217.7953 | 541.9883 | 0.4018 | 0.0018 | 2.426 | 16 |
| 17 | 0.0033 | 2.4918 | 6.0901 | 304.9135 | 759.7837 | 0.4013 | 0.0013 | 2.444 | 17 |
| 18 | 0.0023 | 2.4941 | 6.1299 | 426.8789 | 1064.6971 | 0.4009 | 0.0009 | 2.457 | 18 |
| 19 | 0.0017 | 2.4958 | 6.1601 | 597.6304 | 1491.5760 | 0.4007 | 0.0007 | 2.468 | 19 |
| 20 | 0.0012 | 2.4970 | 6.1828 | 836.6826 | 2089.2064 | 0.4005 | 0.0005 | 2.476 | 20 |
| 21 | 0.0009 | 2.4979 | 6.1998 | 1171.3556 | 2925.8889 | 0.4003 | 0.0003 | 2.482 | 21 |
| 22 | 0.0006 | 2.4985 | 6.2127 | 1639.8978 | 4097.2445 | 0.4002 | 0.0002 | 2.486 | 22 |
| 23 | 0.0004 | 2.4989 | 6.2222 | 2295.8569 | 5737.1423 | 0.4002 | 0.0002 | 2.490 | 23 |
| 24 | 0.0003 | 2.4992 | 6.2294 | 3214.1997 | 8032.9993 | 0.4001 | 0.0001 | 2.492 | 24 |
| 25 | 0.0002 | 2.4994 | 6.2347 | 4499.8796 | 11247.1990 | 0.4001 | 0.0001 | 2.494 | 25 |
| 26 | 0.0002 | 2.4996 | 6.2387 | 6299.8314 | 15747.0785 | 0.4001 | 0.0001 | 2.495 | 26 |
| 27 | 0.0001 | 2.4997 | 6.2416 | 8819.7640 | 22046.9099 | 0.4000 | 0.0000 | 2.496 | 27 |
| 28 | 0.0001 | 2.4998 | 6.2438 | 12347.6696 | 30866.6739 | 0.4000 | 0.0000 | 2.497 | 28 |
| 29 | 0.0001 | 2.4999 | 6.2454 | 17286.7374 | 43214.3435 | 0.4000 | 0.0000 | 2.498 | 29 |
| 30 | 0.0000 | 2.4999 | 6.2466 | 24201.4324 | 60501.0809 | 0.4000 | 0.0000 | 2.498 | 30 |
| 31 | 0.0000 | 2.4999 | 6.2475 | 33882.0053 | 84702.5132 | 0.4000 | 0.0000 | 2.499 | 31 |
| 32 | 0.0000 | 2.4999 | 6.2482 | 47434.8074 | 118584.5185 | 0.4000 | 0.0000 | 2.499 | 32 |
| 33 | 0.0000 | 2.5000 | 6.2487 | 66408.7304 | 166019.3260 | 0.4000 | 0.0000 | 2.499 | 33 |
| 34 | 0.0000 | 2.5000 | 6.2490 | 92972.2225 | 232428.0563 | 0.4000 | 0.0000 | 2.499 | 34 |
| 35 | 0.0000 | 2.5000 | 6.2493 | 130161.1116 | 325400.2789 | 0.4000 | 0.0000 | 2.499 | 35 |
| 36 | 0.0000 | 2.5000 | 6.2495 | 182225.5562 | 455561.3904 | 0.4000 | 0.0000 | 2.499 | 36 |
| 37 | 0.0000 | 2.5000 | 6.2496 | 255115.7786 | 637786.9466 | 0.4000 | 0.0000 | 2.499 | 37 |
| 38 | 0.0000 | 2.5000 | 6.2497 | 357162.0901 | 892902.7252 | 0.4000 | 0.0000 | 2.499 | 38 |
| 39 | 0.0000 | 2.5000 | 6.2498 | 500026.9261 | 1250064.8153 | 0.4000 | 0.0000 | 2.499 | 39 |
| 40 | 0.0000 | 2.5000 | 6.2498 | 700037.6966 | 1750091.7415 | 0.4000 | 0.0000 | 2.499 | 40 |
| 41 | 0.0000 | 2.5000 | 6.2499 | 980052.7752 | 2450129.4381 | 0.4000 | 0.0000 | 2.500 | 41 |
| 42 | 0.0000 | 2.5000 | 6.2499 | 1372073.8853 | 3430182.2133 | 0.4000 | 0.0000 | 2.500 | 42 |
| 43 | 0.0000 | 2.5000 | 6.2499 | 1920903.4394 | 4802256.0986 | 0.4000 | 0.0000 | 2.500 | 43 |
| 44 | 0.0000 | 2.5000 | 6.2500 | 2689264.8152 | 6723159.5381 | 0.4000 | 0.0000 | 2.500 | 44 |
| 45 | 0.0000 | 2.5000 | 6.2500 | 3764970.7413 | 9412424.3533 | 0.4000 | 0.0000 | 2.500 | 45 |

Appendices

## APPENDIX 77.C
Consumer Price Index
All Urban Consumers–(CPI-U)
U.S. City Averages, All Items
(1982-84 = 100)

| Year | CPI | Year | CPI | Year | CPI |
|------|------|------|------|------|-------|
| 1913 | 9.9  | 1945 | 18.0 | 1977 | 60.6  |
| 1914 | 10.0 | 1946 | 19.5 | 1978 | 65.2  |
| 1915 | 10.1 | 1947 | 22.3 | 1979 | 72.6  |
| 1916 | 10.9 | 1948 | 24.1 | 1980 | 82.4  |
| 1917 | 12.8 | 1949 | 23.8 | 1981 | 90.9  |
| 1918 | 15.1 | 1950 | 24.1 | 1982 | 96.5  |
| 1919 | 17.3 | 1951 | 26.0 | 1983 | 99.6  |
| 1920 | 20.0 | 1952 | 26.5 | 1984 | 103.9 |
| 1921 | 17.9 | 1953 | 26.7 | 1985 | 107.6 |
| 1922 | 16.8 | 1954 | 26.9 | 1986 | 109.6 |
| 1923 | 17.1 | 1955 | 26.8 | 1987 | 113.6 |
| 1924 | 17.1 | 1956 | 27.2 | 1988 | 118.3 |
| 1925 | 17.5 | 1957 | 28.1 | 1989 | 124.0 |
| 1926 | 17.7 | 1958 | 28.9 | 1990 | 130.7 |
| 1927 | 17.4 | 1959 | 29.1 | 1991 | 136.2 |
| 1928 | 17.1 | 1960 | 29.6 | 1992 | 140.3 |
| 1929 | 17.1 | 1961 | 29.9 | 1993 | 144.5 |
| 1930 | 16.7 | 1962 | 30.2 | 1994 | 148.2 |
| 1931 | 15.2 | 1963 | 30.6 | 1995 | 152.4 |
| 1932 | 13.7 | 1964 | 31.0 | 1996 | 156.9 |
| 1933 | 13.0 | 1965 | 31.5 | 1997 | 160.5 |
| 1934 | 13.4 | 1966 | 32.4 | 1998 | 163.0 |
| 1935 | 13.7 | 1967 | 33.4 | 1999 | 166.6 |
| 1936 | 13.9 | 1968 | 34.8 | 2000 | 172.2 |
| 1937 | 14.4 | 1969 | 36.7 | 2001 | 177.1 |
| 1938 | 14.1 | 1970 | 38.8 | 2002 | 179.9 |
| 1939 | 13.9 | 1971 | 40.5 | 2003 | 184.0 |
| 1940 | 14.0 | 1972 | 41.8 | 2004 | 188.9 |
| 1941 | 14.7 | 1973 | 44.4 | 2005 | 195.3 |
| 1942 | 16.3 | 1974 | 49.3 | 2006 | 201.6 |
| 1943 | 17.3 | 1975 | 53.8 | 2007 | 207.3 |
| 1944 | 17.6 | 1976 | 56.9 | 2008 | 215.3 |
|      |      |      |      | 2009 | 214.5 |

Source: U.S. Department of Labor, Bureau of Labor Statistics, Washington, DC 20212.

**Appendices**

## APPENDIX 78.A
Percentile Values for Student's $t$-Distribution
($\nu$ degrees of freedom; confidence level $C$; shaded area $= p$)

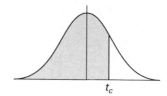

$t_c$

| | two-tail | | | | | | | | | |
|---|---|---|---|---|---|---|---|---|---|---|
| | $t_{99\%}$ | $t_{98\%}$ | $t_{95\%}$ | $t_{90\%}$ | $t_{80\%}$ | $t_{60\%}$ | $t_{50\%}$ | $t_{40\%}$ | $t_{20\%}$ | $t_{10\%}$ |
| | one-tail | | | | | | | | | |
| $\nu$ | $t_{99.5\%}$ | $t_{99\%}$ | $t_{97.5\%}$ | $t_{95\%}$ | $t_{90\%}$ | $t_{80\%}$ | $t_{75\%}$ | $t_{70\%}$ | $t_{60\%}$ | $t_{55\%}$ |
| 1 | 63.66 | 31.82 | 12.71 | 6.31 | 3.08 | 1.376 | 1.000 | 0.727 | 0.325 | 0.158 |
| 2 | 9.92 | 6.96 | 4.30 | 2.92 | 1.89 | 1.061 | 0.816 | 0.617 | 0.289 | 0.142 |
| 3 | 5.84 | 4.54 | 3.18 | 2.35 | 1.64 | 0.978 | 0.765 | 0.584 | 0.277 | 0.137 |
| 4 | 4.60 | 3.75 | 2.78 | 2.13 | 1.53 | 0.941 | 0.741 | 0.569 | 0.271 | 0.134 |
| 5 | 4.03 | 3.36 | 2.57 | 2.02 | 1.48 | 0.920 | 0.727 | 0.559 | 0.267 | 0.132 |
| 6 | 3.71 | 3.14 | 2.45 | 1.94 | 1.44 | 0.906 | 0.718 | 0.553 | 0.265 | 0.131 |
| 7 | 3.50 | 3.00 | 2.36 | 1.90 | 1.42 | 0.896 | 0.711 | 0.549 | 0.263 | 0.130 |
| 8 | 3.36 | 2.90 | 2.31 | 1.86 | 1.40 | 0.889 | 0.706 | 0.546 | 0.262 | 0.130 |
| 9 | 3.25 | 2.82 | 2.26 | 1.83 | 1.38 | 0.883 | 0.703 | 0.543 | 0.261 | 0.129 |
| 10 | 3.17 | 2.76 | 2.23 | 1.81 | 1.37 | 0.879 | 0.700 | 0.542 | 0.260 | 0.129 |
| 11 | 3.11 | 2.72 | 2.20 | 1.80 | 1.36 | 0.876 | 0.697 | 0.540 | 0.260 | 0.129 |
| 12 | 3.06 | 2.68 | 2.18 | 1.78 | 1.36 | 0.873 | 0.695 | 0.539 | 0.259 | 0.128 |
| 13 | 3.01 | 2.65 | 2.16 | 1.77 | 1.35 | 0.870 | 0.694 | 0.538 | 0.259 | 0.128 |
| 14 | 2.98 | 2.62 | 2.14 | 1.76 | 1.34 | 0.868 | 0.692 | 0.537 | 0.258 | 0.128 |
| 15 | 2.95 | 2.60 | 2.13 | 1.75 | 1.34 | 0.866 | 0.691 | 0.536 | 0.258 | 0.128 |
| 16 | 2.92 | 2.58 | 2.12 | 1.75 | 1.34 | 0.865 | 0.690 | 0.535 | 0.258 | 0.128 |
| 17 | 2.90 | 2.57 | 2.11 | 1.74 | 1.33 | 0.863 | 0.689 | 0.534 | 0.257 | 0.128 |
| 18 | 2.88 | 2.55 | 2.10 | 1.73 | 1.33 | 0.862 | 0.688 | 0.534 | 0.257 | 0.127 |
| 19 | 2.86 | 2.54 | 2.09 | 1.73 | 1.33 | 0.861 | 0.688 | 0.533 | 0.257 | 0.127 |
| 20 | 2.84 | 2.53 | 2.09 | 1.72 | 1.32 | 0.860 | 0.687 | 0.533 | 0.257 | 0.127 |
| 21 | 2.83 | 2.52 | 2.08 | 1.72 | 1.32 | 0.859 | 0.686 | 0.532 | 0.257 | 0.127 |
| 22 | 2.82 | 2.51 | 2.07 | 1.72 | 1.32 | 0.858 | 0.686 | 0.532 | 0.256 | 0.127 |
| 23 | 2.81 | 2.50 | 2.07 | 1.71 | 1.32 | 0.858 | 0.685 | 0.532 | 0.256 | 0.127 |
| 24 | 2.80 | 2.49 | 2.06 | 1.71 | 1.32 | 0.857 | 0.685 | 0.531 | 0.256 | 0.127 |
| 25 | 2.79 | 2.48 | 2.06 | 1.71 | 1.32 | 0.856 | 0.684 | 0.531 | 0.256 | 0.127 |
| 26 | 2.78 | 2.48 | 2.06 | 1.71 | 1.32 | 0.856 | 0.684 | 0.531 | 0.256 | 0.127 |
| 27 | 2.77 | 2.47 | 2.05 | 1.70 | 1.31 | 0.855 | 0.684 | 0.531 | 0.256 | 0.127 |
| 28 | 2.76 | 2.47 | 2.05 | 1.70 | 1.31 | 0.855 | 0.683 | 0.530 | 0.256 | 0.127 |
| 29 | 2.76 | 2.46 | 2.04 | 1.70 | 1.31 | 0.854 | 0.683 | 0.530 | 0.256 | 0.127 |
| 30 | 2.75 | 2.46 | 2.04 | 1.70 | 1.31 | 0.854 | 0.683 | 0.530 | 0.256 | 0.127 |
| 40 | 2.70 | 2.42 | 2.02 | 1.68 | 1.30 | 0.851 | 0.681 | 0.529 | 0.255 | 0.126 |
| 60 | 2.66 | 2.39 | 2.00 | 1.67 | 1.30 | 0.848 | 0.679 | 0.527 | 0.254 | 0.126 |
| 120 | 2.62 | 2.36 | 1.98 | 1.66 | 1.29 | 0.845 | 0.677 | 0.526 | 0.254 | 0.126 |
| $\infty$ | 2.58 | 2.33 | 1.96 | 1.645 | 1.28 | 0.842 | 0.674 | 0.524 | 0.253 | 0.126 |

## APPENDIX 78.B
Single Sampling Plan Table for Various Producer's and Consumer's Risks
(values of $np$)

Choose the column for the appropriate consumer's risk, $\beta$. If the producer's risk, $\alpha$, is known, calculate $\beta$ as $1 - \alpha$. Choose the row corresponding to the number of defects, $c$. Read the table value. Calculate the sample size, $N$, as the table value divided by the percent of defectives, $p$, in the lot.

| | | | | $\beta$ | | | |
|---|---|---|---|---|---|---|---|
| $c$ | 99% | 95% | 90% | 50% | 10% | 5% | 1% |
| 0 | 0.010 | 0.051 | 0.105 | 0.693 | 2.303 | 2.996 | 4.605 |
| 1 | 0.149 | 0.355 | 0.532 | 1.678 | 3.890 | 4.744 | 6.638 |
| 2 | 0.436 | 0.818 | 1.102 | 2.674 | 5.322 | 6.296 | 8.406 |
| 3 | 0.823 | 1.366 | 1.745 | 3.672 | 6.681 | 7.754 | 10.045 |
| 4 | 1.279 | 1.970 | 2.433 | 4.671 | 7.994 | 9.154 | 11.605 |
| 5 | 1.785 | 2.613 | 3.152 | 5.670 | 9.275 | 10.513 | 13.108 |
| 6 | 2.330 | 3.286 | 3.895 | 6.670 | 10.532 | 11.842 | 14.571 |
| 7 | 2.906 | 3.981 | 4.656 | 7.669 | 11.771 | 13.148 | 16.000 |
| 8 | 3.507 | 4.695 | 5.432 | 8.669 | 12.995 | 14.434 | 17.403 |
| 9 | 4.130 | 5.426 | 6.221 | 9.669 | 14.206 | 15.705 | 18.783 |
| 10 | 4.771 | 6.169 | 7.021 | 10.668 | 15.407 | 16.962 | 20.145 |
| 11 | 5.428 | 6.924 | 7.829 | 11.668 | 16.598 | 18.208 | 21.490 |
| 12 | 6.099 | 7.690 | 8.646 | 12.668 | 17.782 | 19.442 | 22.821 |
| 13 | 6.782 | 8.464 | 9.470 | 13.668 | 18.958 | 20.668 | 24.139 |
| 14 | 7.477 | 9.246 | 10.300 | 14.668 | 20.128 | 21.886 | 25.446 |
| 15 | 8.181 | 10.035 | 11.135 | 15.668 | 21.292 | 23.098 | 26.743 |
| 16 | 8.895 | 10.831 | 11.976 | 16.668 | 22.452 | 24.302 | 28.031 |
| 17 | 9.616 | 11.633 | 12.822 | 17.668 | 23.606 | 25.500 | 29.310 |
| 18 | 10.346 | 12.442 | 13.672 | 18.668 | 24.756 | 26.692 | 30.581 |
| 19 | 11.082 | 13.254 | 14.525 | 19.668 | 25.902 | 27.879 | 31.845 |
| 20 | 11.825 | 14.072 | 15.383 | 20.668 | 27.045 | 29.062 | 33.103 |
| 21 | 12.574 | 14.894 | 16.244 | 21.668 | 28.184 | 30.241 | 34.355 |
| 22 | 13.329 | 15.719 | 17.108 | 22.668 | 29.320 | 31.416 | 35.601 |
| 23 | 14.088 | 16.548 | 17.975 | 23.668 | 30.453 | 32.586 | 36.841 |
| 24 | 14.853 | 17.382 | 18.844 | 24.668 | 31.584 | 33.752 | 38.077 |
| 25 | 15.623 | 18.218 | 19.717 | 25.667 | 32.711 | 34.916 | 39.308 |
| 30 | 19.532 | 22.444 | 24.113 | 30.667 | 38.315 | 40.690 | 45.401 |
| 35 | 23.525 | 26.731 | 28.556 | 35.667 | 43.872 | 46.404 | 51.409 |
| 40 | 27.587 | 31.066 | 33.038 | 40.667 | 49.390 | 52.069 | 57.347 |
| 45 | 31.704 | 35.441 | 37.550 | 45.667 | 54.878 | 57.695 | 63.231 |
| 50 | 35.867 | 39.849 | 42.089 | 50.667 | 60.339 | 63.287 | 69.066 |

# Index

INDEX - B

INDEX - c

INDEX - D

INDEX - D

INDEX - E

INDEX - E

Newton's interpolating, 15-3
standard form, 5-3
Polyprotic acid, 33-26
Polysaccharide, 34-2 (tbl)
Polystyrene resin zeolite, 33-35
Polytropic
closed system, 23-9
compression, 17-13
efficiency, 31-5
equation of state, 23-2
exponent, 17-13, 23-2
exponent, compression, 31-3
head, 31-5
process, 23-2
specific heat, 23-2
steady-flow, 23-9
Polyvinylchloride (PVC) pipe, 18-11, A-25, A-26
Pond
facultative, 39-3
stabilization, 39-3
Poppet valve, 28-2 (ftn), 28-3 (ftn)
Population, 13-3
inversion, 66-5
Portable language, 73-4
Portland cement, 46-21
concrete, 46-21
types, 46-21
Position
absolute, 56-10
angular, 56-5
equilibrium, 47-2, 58-2
relative, 56-10
vector, 44-2
Positional numbering systems, 14-1
Positive
displacement pump, 20-2
edge dislocation, 47-9
feedback, 70-4
logic, 65-1, 75-1
sequence, 62-1
Positron, 76-12
Post, end, 44-12
Potency factor, carcinogen, 37-5, 37-6
Potential
cell, 33-28, 33-30, 33-31
difference, 59-7
electric, 59-7
energy, 4-3, 18-2
energy, electric, 59-7
energy, magnetic field, 59-9
gradient, 59-7
gradient, magnetic, 59-9
half-cell, 33-29
magnetic, 59-9
oxidation, 33-29
reduction, 33-29
standard oxidation, 33-29
stream, 19-3
Potentiometer, 60-1, 68-3
Potentiometric sensor, 68-3
Poundal, 2-4
Pound, mole, 33-6
Pour point, 35-7
Powder
metallic, 50-10
metallurgy, 50-10
Power, 4-6
angle, 61-18, 63-7
-boosting, 28-15
brake, 28-3
coefficient, 30-5
complex, 61-18
conversions, 4-6
curve, 48-5
cutting, 50-3
cycle, 23-11
cycle, nuclear, 29-1
cycle, solar, 30-3
cycle, vapor, 27-1
design, 57-7 (ftn), 57-8 (ftn)
effective radiated, 61-24, 61-25
electrical, 60-4, 61-10

emissive, 43-1
engine, 28-3
factor, 61-13, 61-18, 61-24
factor correction, 61-19
friction, 20-9, 28-3
gain, 64-11
grid, 61-23
hydraulic, 20-8
indicated, 28-3
induction motor, 63-11
law, 1/7, 18-8
law, viscosity, 19-11
level sound, 38-7
line, 61-23
line, safety, 79-2
loss, motor, 63-3
of a lens, 66-9
of a mirror, 66-11
phase, 62-3
plant, electrical rating, 26-1
plant, thermal rating, 26-1
pump, 20-2, 20-7
semiconductor, 64-17
series, 10-8
specific, battery, 30-10
stroke, 28-2
tidal, 30-6
-to-heat ratio, 28-22, 28-23
-tower system, 30-3
transmission, electric, 61-23
transmission, voltage, 61-23 (ftn)
triangle, 61-18
turbine, 19-35, 20-22
water, 20-8
wave, 30-6
wind, 30-4
Powerhouse, 20-21
Poynting vector, 59-12
Pozzolanic additive, 46-23
Prandtl
boundary layer theory, 19-43
number, 3-2, 3-4 (tbl), 41-2, 42-2
Precedence
order of, 75-2
relationship, 78-3
table, 78-3
Prechamber, 28-3
Precipitation, 33-16
-hardened stainless steel, 46-8 (ftn)
hardening, 46-8 (ftn), 46-11 (ftn), 49-9
softening, 33-35
Precision, 13-13, 68-2, 78-16, 78-18
casting, 50-9
Precombustion chamber, 28-3
Precursor, disinfection by product, 33-36
Predetermined system, 78-16
Prefix, SI, 2-7
Preform, 50-10
Preheater
air, 26-15
convection, 26-15
regenerative, 26-15
Preload
bolt, 54-12
force, 54-12
Prerequisite of privity, 80-7 (ftn)
Present worth, 77-5, 77-11
method, 77-14
Press, 50-6
capacity, 50-6
fit, 54-6
forging, 50-8
hydraulic, 17-14
Pressure, 16-2, 22-5
absolute, 16-3
amplifier, 31-2
at a depth, 17-6 (ftn)
atmospheric, 16-2
average, 17-7
back, 25-13, 26-12
barometric, 16-3
booster, 31-2
brake mean effective, 28-3, 28-8

center of, 17-4, 17-9, 17-10, 44-5
closed system, constant, 23-7
collapsing, 54-4 (ftn)
conduit, 18-6
constant, 23-2
cut-in, 31-2
cut-out, 31-2
cylinder, peak, 28-9 (ftn)
Dalton's law of partial, 22-18
design, 25-14
die casting, 50-9
differential, 16-3
drag, 19-40
-drop endpoint, 33-36
drop, friction, 19-4 (ftn)
drop, steel pipe, A-35
drop, water, steel pipe, A-35
energy, 18-2
-enthalpy diagram, 22-12
external, 17-14
fit, stress concentration factor, 54-9
flow, 18-6
fluid, 16-2
from gas, 17-12
gage, 17-2
gauge, 16-2
gauge, Bourdon, 17-2
head, 20-6 (ftn)
hydrostatic, 17-4
indicated mean effective, 28-3, 28-8
intensifier, 31-2
isentropic, 25-14
level, sound, 38-7
mean effective, 28-9
measurement, 17-1
measuring devices, 17-2
multiple liquid, 17-12
of a gas, partial, 22-18
on a dam, 17-11
on curved surface, 17-10
on plane surface, 17-6, 17-7, 17-8
osmotic, 16-10, 33-19
partial, 22-18
peak cylinder, 28-9 (ftn)
peak firing, 28-9 (ftn)
pouring, 50-9 (ftn)
ratio, 22-11, 28-12, 31-1
ratio, critical, 25-4
ratio, design, 25-5
ratio, isentropic, 28-4
-relief valve, 19-37
root-mean-square, 38-7
saturation, 16-9
sound, 38-6, 38-7
surface, 17-6, 17-7, 17-8, 17-10
total, 18-3, 22-18
transverse, 19-27
unit, 16-2 (fig)
vapor, 16-9, 24-2, 38-2, 38-3
Pressure-enthalpy diagram, 22-12
HFC-134a, A-68, A-88
R-12, A-64, A-67, A-84
R-22, A-87
Pressurized
fluidized bed combustion, 35-5
liquid, 17-14
liquid, temperature increase, 26-7
slagging combustor, 28-16 (ftn)
tank, 19-17
water reactor, 29-1
Presswork, 50-6
Preventative maintenance, 78-11
Price, 78-2 (ftn)
-earnings ratio, 77-35
index, consumer, A-157
Primary
air, 26-3
creep, 48-16
dimension, 2-7
distribution system, 62-2
feedback ratio, 70-4
leakage, 61-22

INDEX - R

INDEX - s

INDEX - T

INDEX - V

# CONVERSION FACTORS

(Atmospheres are standard; calories are gram-calories; gallons are U.S. liquid; miles are statute; pounds-mass are avoirdupois.)

| multiply | by | to obtain |
|---|---|---|
| acre | 43,560 | $ft^2$ |
| angstrom | $1 \times 10^{-10}$ | m |
| atm | 1.01325 | bar |
| atm | 76.0 | cm Hg |
| atm | 33.90 | ft water |
| atm | 29.92 | in Hg |
| atm | 14.696 | $lbf/in^2$ |
| atm | 101.3 | kPa |
| atm | $1.013 \times 10^5$ | Pa |
| bar | 0.9869 | atm |
| bar | $10^5$ | Pa |
| Btu | 778.17 | ft-lbf |
| Btu | 1055 | J |
| Btu | $2.928 \times 10^{-4}$ | kW-hr |
| Btu | $10^{-5}$ | therm |
| Btu/hr | 0.216 | ft-lbf/sec |
| Btu/hr | $3.929 \times 10^{-4}$ | hp |
| Btu/hr | 0.2931 | W |
| Btu/lbm | 2.326 | kJ/kg |
| Btu/lbm-°R | 4.1868 | kJ/kg·K |
| cal | $3.968 \times 10^{-3}$ | Btu |
| cal | 4.1868 | J |
| cm | 0.03281 | ft |
| cm | 0.3937 | in |
| eV | $1.602 \times 10^{-19}$ | J |
| ft | 0.3048 | m |
| $ft^2$ | $2.2957 \times 10^{-5}$ | acre |
| $ft^3$ | 7.481 | gal |
| ft-lbf | $1.285 \times 10^{-3}$ | Btu |
| ft-lbf | 1.35582 | J |
| ft-lbf | $3.766 \times 10^{-7}$ | kW-hr |
| ft-lbf | 1.3558 | N·m |
| gal | 0.13368 | $ft^3$ |
| gal | 3.785 | L |
| gal | $3.7854 \times 10^{-3}$ | $m^3$ |
| gal/min | 0.002228 | $ft^3$/sec |
| $g/cm^3$ | 1000 | $kg/m^3$ |
| $g/cm^3$ | 62.428 | $lbm/ft^3$ |
| hp | 2545 | Btu/hr |
| hp | 33,000 | ft-lbf/min |
| hp | 550 | ft-lbf/sec |
| hp | 0.7457 | kW |
| hp-hr | 2545 | Btu |
| in | 2.54 | cm |
| J | $9.478 \times 10^{-4}$ | Btu |
| J | $6.2415 \times 10^{18}$ | eV |
| J | 0.73756 | ft-lbf |
| J | 1.0 | N·m |
| J/s | 1.0 | W |

| multiply | by | to obtain |
|---|---|---|
| kg | 2.20462 | lbm |
| kip | 1000 | lbf |
| kip | 4448 | N |
| kJ | 0.9478 | Btu |
| kJ | 737.56 | ft-lbf |
| kJ/kg | 0.42992 | Btu/lbm |
| kJ/kg·K | 0.23885 | Btu/lbm-°R |
| km | 3280.8 | ft |
| km | 0.6214 | mi |
| km/hr | 0.6214 | mi/hr |
| kPa | $9.8693 \times 10^{-3}$ | atm |
| kPa | 0.01 | bar |
| kPa | 0.14504 | $lbf/in^2$ |
| ksi | 6894.8 | kPa |
| kW | 3413 | Btu/hr |
| kW | 0.9478 | Btu/sec |
| kW | 737.6 | ft-lbf/sec |
| kW | 1.341 | hp |
| kW-hr | 3413 | Btu |
| kW-hr | $3.6 \times 10^6$ | J |
| L | 0.03531 | $ft^3$ |
| L | 61.02 | $in^3$ |
| L | 0.2642 | gal |
| L | 0.001 | $m^3$ |
| L/s | 2.119 | $ft^3$/min |
| L/s | 15.85 | gal/min |
| lbf | 4.4482 | N |
| $lbf/in^2$ | 0.06805 | atm |
| $lbf/in^2$ | 2.307 | ft water |
| $lbf/in^2$ | 2.036 | in Hg |
| $lbf/in^2$ | 6894.8 | Pa |
| lbm | 0.4536 | kg |
| $lbm/ft^3$ | 0.016018 | $g/cm^3$ |
| $lbm/ft^3$ | 16.018 | $kg/m^3$ |
| m | 3.28083 | ft |
| m/sec | 196.8 | ft/min |
| mi | 5280 | ft |
| mi | 1.6093 | km |
| micron | $1 \times 10^{-6}$ | m |
| N | 0.22481 | lbf |
| N·m | 0.7376 | ft-lbf |
| N·m | 1.0 | J |
| Pa | $1.4504 \times 10^{-4}$ | $lbf/in^2$ |
| therm | $10^5$ | Btu |
| W | 3.4121 | Btu/hr |
| W | 0.7376 | ft-lbf/sec |
| W | $1.341 \times 10^{-3}$ | hp |
| W | 1.0 | J/s |